Amphibians and Reptiles of Wisconsin

Publication of this book was made possible, in part, through support from the Sadie Nolan Amphibian and Reptile Education and Conservation Memorial Fund of the Wisconsin Department of Natural Resources, Milwaukee Public Museum, University of Wisconsin Zoological Museum, Stantec, College of Letters & Sciences at the University of Wisconsin–Whitewater, The Nature Conservancy, Natural Resources Foundation, Penny and Gary Shackelford, Cardno, Lake Koshkonong Wetlands Association, and Robert Henderson.

Amphibians and Reptiles of Wisconsin

Edited by Joshua M. Kapfer and Donald J. Brown

Illustrations by Erik R. Wild

THE UNIVERSITY OF WISCONSIN PRESS

The University of Wisconsin Press
728 State Street, Suite 443
Madison, Wisconsin 53706
uwpress.wisc.edu

Gray's Inn House, 127 Clerkenwell Road
London EC1R 5DB, United Kingdom
eurospanbookstore.com

Printed in China

This book may be available in a digital edition.

Library of Congress Cataloging-in-Publication Data
Names: Kapfer, Joshua M., editor. | Brown, Donald J., 1981– editor. | Wild, Erik Russell, illustrator.
Title: Amphibians and reptiles of Wisconsin / edited by Joshua M. Kapfer and Donald J. Brown ; illustrations by Erik R. Wild.
Description: Madison, Wisconsin : The University of Wisconsin Press, [2022] | Includes bibliographical references and index.
Identifiers: LCCN 2021041581 | ISBN 9780299335205 (hardcover)
Subjects: LCSH: Reptiles—Wisconsin. | Amphibians—Wisconsin.
Classification: LCC QL653.W6 A474 2022 | DDC 597.909775—dc23
LC record available at https://lccn.loc.gov/2021041581

To my understanding and loving wife, Jennifer, without whose support I could not have completed this (or any) project.

To my wonderful kiddo, Violet, who I hope will grow up to enjoy wildlife as much as her daddy.

To my wonderful parents, Max and Judy, who nurtured my interest in wildlife and supported me in my decided career path. I certainly could not have become a professional wildlife biologist without them. My dad, who passed away in January 2019, was very excited about this project, and it saddens me greatly that he did not get to see the finished product. It is also noteworthy that Dad grew up afraid of snakes but hid his phobia from me when I was young (eventually even overcoming it). I'm very thankful for this act of selflessness on his part. Who knows where I'd be today if he had forced his fears onto me as a kid?

To my caring family, Amanda, Bjorn, Viggo, Solveig, Sandy, Sarah, Brian, Oliver, Parker, Scarlett, Chris, and Laura.

To my professional mentors, James Coggins, Mark Sandheinrich, and the late Daniel Sutherland, whose advice and instruction got me to where I am.

To Robert (Bob) Hay, who is largely responsible for teaching me how to conduct fieldwork on Wisconsin's herpetofauna.

To my late pooch, Sampson, whom I didn't think I would miss so much until he was gone.

. . . and to all of those who appreciate wild things and the wild places where they live . . .

—**JOSHUA M. KAPFER**

To the late James R. Dixon, who taught me how to catch turtles and whose passion for amphibian and reptile research and conservation never faded. You are missed by many.

To Michael R. J. Forstner, who first provided me with the opportunity to study amphibians and reptiles and gave me the freedom necessary to develop into an independent scientist.

To Thomas K. Pauley, who in recent years has welcomed and facilitated my desire to study woodland salamanders. Your mentorship, knowledge, kindness, enthusiasm, and contributions to the natural history of amphibians and reptiles in West Virginia continue to inspire me.

To my life partner, Kaycie Stushek, who supports all of my passions and reminds me to live in the present.

—**DONALD J. BROWN**

Contents

Foreword

Michael J. Lannoo

IT BEGAN WITH PHYSICS ENVY—when interactions such as the relationship between energy, mass, and the speed of light began to be described with simple equations such as $E = mc^2$. Before that, the study of natural history and the nascent field of ecology were descriptive disciplines just starting to become experimental. In the United States, ecology began in the nineteenth century in the Midwest, with Stephen Forbes in Illinois. Ecology then spread to include two great botany schools—Charles Bessey's at the University of Nebraska and John Merle Coulter's at the University of Chicago—and continued with W. C. Allee's great animal ecology school, also at Chicago, and Edward Birge's and Chancy Juday's school of limnology at the University of Wisconsin. Once ecology got started, the newly formed midwestern land-grant universities, like apostles, spread the gospel. During this time, ecology was best described as 'organisms having relationships.' This perspective soon faded, however. In 1939 Frederic Clements and Victor Shelford published their classic, *Bio-ecology*. In reviewing this book, G. Evelyn Hutchinson criticized Clements and Shelford's failure to address biogeochemical and metabolic (nutrient flow) processes. After Hutchison's review, with a clear eye toward clean mathematics, the overriding ecological paradigm became 'relationships having organisms' (so-called modern ecology). Despite George E. P. Box's quip about ecological models ("all models are lies; some of them are useful"), mathematical models still dominate ecology (Box and Draper 1987). Ríos-Saldaña et al. (2018) discovered that in the past four decades, fieldwork-based ecological studies have decreased by 20%, while over the same timespan modeling studies increased by 600%.

Even though it became a sidelight, fieldwork never quite disappeared from ecology. Ecology always drew from people interested in the natural world, people who enjoyed being outdoors and wanted to know what they saw when they were tromping around. To meet this demand, field guides were created. To stay small, most were local, usually delimited by state boundaries. These early works were black-and-white affairs with simple line drawings and stippled maps to indicate range. In herpetology, Phil Smith's (1961) Illinois book and Sherman Minton's (1972) Indiana book are fine examples of this first generation of field guide / reference book. Over time and with better publishing technologies, these books became gaudier and more comprehensive—compendiums. You could take them in the field, but they were so big and beautiful that you generally left them back in the truck for reference afterward. The first and among the all-time best of these second-generation field guide / reference books was Dick Vogt's *Natural History of Amphibians and Reptiles in Wisconsin* (Vogt 1981).

Vogt's book came as no surprise. Wisconsin has always been a leader in the disciplines that rely on data collected by field biologists. In 1933, with the publication of *Game Management*, Aldo Leopold founded the discipline of wildlife biology. Two years later, on 17 June 1934, Leopold spoke at the University of Wisconsin's Arboretum dedication (Court 2012). Leopold's idea for the arboretum was "to re-construct . . . a sample of original Wisconsin—a sample of what Dane County looked like when our ancestors arrived here during the 1840s." It would not be done for "amusement" but instead for research. Leopold envisioned an oak savanna growing in a field of prairie grasses and forbs to "serve as a bench mark, a starting point, in the long and laborious job of building a permanent and mutually beneficial relationship between civilized men and a civilized landscape." With this prairie reconstruction, Leopold and his colleagues Norman Fassett and Ted Sperry began the field of restoration ecology. As if all this wasn't enough, Leopold also provided the philosophical underpinning for conservation biology with his Land Ethic: "A thing is right when it tends to preserve the integrity, stability, and beauty of the biotic community. It is wrong when it tends otherwise."

This book by Kapfer and Brown fulfills three desperate needs. First and most pragmatically, it updates our factual, field-based knowledge of Wisconsin's amphibians and reptiles—in the more than 40 years since Vogt's book, over 700 new county records have been published. Second, it continues the old Wisconsin tradition of Leopold and his colleagues by turning the ecological focus back on the animals that live the ecology we study—on the frogs and turtles and snakes and the salamanders and lizards that fascinated us as kids and drew our attention to the natural world. Third, the information in this book blunts the juggernaut of mathematical models laying claim to best representing our natural world. Nature alone creates the biochemical cascade of delight that shivers through the brain of a kid holding a treefrog tadpole or a milksnake for the first time. By creating a one-stop source for almost everything we know about these amphibians and reptiles, Kapfer and Brown open a window to a world otherwise unknown to most modern Badgers. And to generalize, Kapfer and Brown's effort reminds us that we do not own this world, that instead we share it with the rest of creation. Such perspective is humbling and gives us insight into our better, more inclusive selves.

Acknowledgments

A WORK OF THIS TYPE is not possible without the efforts of many, many individuals. From assistance with museum specimens to lending an ear and providing advice, numerous professionals deserve acknowledgment for their direct and indirect contributions.

Julia Colby (Milwaukee Public Museum), Laura Monahan (University of Wisconsin–Madison Zoological Museum), and Emily Halverson (University of Wisconsin–Madison Zoological Museum) facilitated our access to their specimen collections, helped us locate specimens, or determined the ultimate fate of specimens once housed at their institutions. These individuals were incredibly helpful in this regard and filled a nearly criminal number of requests about specimens housed at their respective institutions. Many additional individuals provided a substantial amount of assistance in gathering information on specimens in their respective collections, including Alan Resetar and Rachel Grill (Field Museum of Natural History, Chicago) and Dawn Roberts, Allison Sacerdote-Velat, and John Archer (Chicago Academy of Sciences Peggy Notebaert Nature Museum). Dawn Roberts and Allison Sacerdote-Velat, in particular, helped confirm the identity of many Wisconsin specimens in their institution's collection for this work. Chris Phillips, Christine Mayer, and Dan Wylie (Illinois Natural History Survey) also helped in this regard on countless occasions. Both Chris Phillips and Alan Resetar, in particular, answered many questions about specimens and the distribution of herpetofauna in Illinois. Numerous other individuals responded to many requests and helped confirm the identification of specimens, including Peter Zani and students (University of Wisconsin–Stevens Point); Steve Rogers (Carnegie Museum of Natural History); James Poindexter, Kenneth Tighe, and Jeremy Jacobs (Smithsonian Institution US National Museum); David Kizirian and Abigail Cramer (American Museum of Natural History); Ken Kozak and Benjamin Lowe (University of Minnesota James Ford Bell Museum); Thomas Erdman (University of Wisconsin–Green Bay Richter Museum of Natural History); Jeffrey Stafford (Henry Vilas Zoo); David Mifsud (Michigan Herpetological Atlas); and James Harding (Michigan State University).

Many additional individuals also deserve mention for helping us compile information on specimens or accessing collections: Susan Lewis and Roberto Brenes (Carroll University); Martin St. Maurice and James Courtwright (Marquette University); Nyree Zerega, John Mordacq, Gary Galbreath, and Teresa Horton (Northwestern University); Christine Chandler (Putnum Museum); Meredith Mahoney (Illinois State Museum); Rich King (Northern Illinois University); Fred Janzen (Iowa State University); Stephen Mullin (Eastern Illinois University); Lyle Shannon (University of Minnesota–Duluth);

John Peterson (University of Wisconsin–Platteville); Michael Weil and Christina Chan-Weiher (University of Wisconsin–Eau Claire); Jeff Boundy (Louisiana State University Museum of Natural Sciences); Laura Abraczinskas (Michigan State University Museum); Scott Williams (Burpee Museum); Greg Schneider (University of Michigan Museum of Zoology); Ken Krysko (University of Florida Museum of Natural History); Tony Gamble (University of Minnesota James Ford Bell Museum [currently at Marquette University]); Thomas Labedz (University of Nebraska State Museum); Erick Rickart (Natural History Museum of Utah). We also frequently perused online specimen databases for species distribution data. These included HerpNet (now VertNet), the Smithsonian National Museum of Natural History, and the American Museum of Natural History. We asked many other individuals from institutions throughout the Midwest to check their collections for specimens from Wisconsin, and most responded, even if they had no Wisconsin specimens. We greatly appreciate their time.

Steven Kuehn of the Illinois State Archaeological Survey (University of Illinois at Urbana-Champaign) graciously reviewed the literature and data he possessed on the prehistoric distribution of herpetofauna in Wisconsin. Gayle J. Martinson of the Wisconsin Historical Society did a tremendous job helping us find information on historical figures important to Wisconsin herpetology, and we cannot thank her enough for her time. Claudia Jacobson, registrar at the Milwaukee Public Museum, and Susan Borkin answered many questions and helped find information on past curators, while Sara Podejko of Milwaukee Public Museum assisted with photo acquisition. Thomas Meyer, Kathryn Kahler, and William Cosh (Wisconsin Department of Natural Resources) assisted with acquisition of historic photos and information. The interlibrary loan staff at University of Wisconsin–Whitewater's Anderson Library filled countless requests for literature.

Numerous individuals provided reviews for early drafts of chapters or species accounts in this book, in addition to the accounts that many of them authored, including Andrew Badje (Wisconsin Department of Natural Resources), Lena Carlson (Wisconsin Department of Natural Resources), Tony Gamble (Marquette University), Eric Gustafson (US Forest Service Northern Research Station), Robert Hay (Turtles for Tomorrow), Robert Henderson (Milwaukee Public Museum), Peter Jacobs (University of Wisconsin–Whitewater), Michael Lannoo (Indiana University School of Medicine), Sue Lannoo (Burr Oak Biological), Jeffrey Lorch (US Geological Survey), John Moriarty (Three Rivers Park District), Emily Moriarty Lemmon (Florida State University), Max Nickerson (University of Florida Museum of Natural History), Rori Paloski (Wisconsin Department of Natural Resources), Gregor Schuurman (National Park Service), Richard Staffen (Wisconsin Department of Natural Resources), and Erik Wild (University of Wisconsin–Stevens Point). Furthermore, Robert Henderson (Milwaukee Public Museum) and Michael Lannoo (Indiana University School of Medicine) gave extensive and much-appreciated sage advice on a variety of subjects.

Numerous people graciously donated the use of their photographs for this book, and we are exceedingly grateful (all photographers are identified in the figure captions). One of our initial goals was to only use amphibian and reptile photographs from Wisconsin. This turned out to be much harder than we imagined, and although we did not fully meet this goal, the kind contributions of many photographers brought us very close.

Many individuals provided valuable field assistance in various forms during collection of data for this book. Tim Muehlfeld was a frequent field companion of Joshua Kapfer, and Tim's undying enthusiasm always energized him during countless trips to find county records, gather data, or collect photographs. Many undergraduate students assisted Joshua Kapfer either directly or indirectly over the years. These efforts included conducting fieldwork that led to observations or data for species accounts, searching existing collections for specimens, and measuring museum specimens. From University of Wisconsin–Whitewater: Cody Hallcox, Katie Hausmann, Chris Heneghan, Andrew Ibach, Jackie Kincanon, Jarod Lorenz, Lisa Mitchem, Joseph Mozuch, Philip Pieper, Kari Rebman, Nick Rudolph, Karl Rutzen, Sarah Wyrick, and Zach Zimmerman. From University of Wisconsin–Stevens Point: Alex Thomas. From University of Wisconsin–Waukesha: Matt Schneider. From University of Wisconsin–Madison: Gabriella Struger. Andrew Ibach, in particular, should be given special recognition for measuring several hundred museum specimens for this book.

It is critically important that we acknowledge those who came before us and made substantial contributions to the present-day understanding of Wisconsin's herpetofauna. These individuals are reviewed in an introductory chapter of this book. In some way, all of those who work with Wisconsin's amphibians and reptiles today have stood on the shoulders of these individuals.

Finally, most important to the success of this book was the effort expended by its many contributors. There is no single individual in this fine state who has expertise in all or even most of its native herpetofaunal species. The quality of the species accounts herein stems directly from the fact that the contributing authors were able to draw upon their personal experiences, knowledge, and enthusiasm. Without this group effort and a willingness by each contributing individual to impart their knowledge while summarizing the pertinent literature, this book would not have happened. It is because of their efforts that we have produced a product that we are confident will serve as a lasting contribution to the ecology and conservation of Wisconsin's herpetofauna.

Contributors

Dr. Noah J. Anderson
Assistant Professor, Biology Department
Winona State University,
Winona, Minnesota

Andrew F. Badje
Conservation Biologist, Bureau of
Natural Heritage Conservation
Wisconsin Department of Natural
Resources, Madison, Wisconsin

Craig S. Berg
Aquarium and Reptile Curator (retired),
Milwaukee County Zoo,
Milwaukee, Wisconsin

Tara L. E. Bergeson
Conservation Biologist, Bureau of
Natural Heritage Conservation
Wisconsin Department of Natural
Resources, Madison, Wisconsin

Danielle R. Bradke
Ph.D. Student, Warnell School of
Forestry and Natural Resources
University of Georgia, Athens, Georgia

Dr. Robert D. Brodman
Professor Emeritus of Biology and Environmental Science, Saint
Joseph's College, Rensselaer, Indiana

Dr. Donald J. Brown
Research Assistant Professor of Wildlife Resources, West Virginia
University, School of Natural Resources,
Morgantown, West Virginia
US Forest Service, Northern Research
Station, Parsons, West Virginia

Jeff Canfield
Chaska, Minnesota

Dr. Rebecca A. Christoffel
Sole Proprietor, Christoffel
Conservation, Madison, Wisconsin
President, Snake Conservation
Society, Ames, Iowa
Codirector, Turtles for Tomorrow,
Madison, Wisconsin

Dr. Deahn M. Donner
Project Leader and Research Landscape Ecologist, Institute for
Applied Ecosystem Studies, US
Forest Service, Northern Research
Station, Rhinelander, Wisconsin

Samantha Foster
Milton, Wisconsin

Dr. Tony Gamble
Assistant Professor, Department
of Biological Sciences
Marquette University,
Milwaukee, Wisconsin
Research Associate, Vertebrate Zoology
Milwaukee Public Museum,
Milwaukee, Wisconsin
Research Associate, Bell Museum
of Natural History
University of Minnesota,
Saint Paul, Minnesota

James H. Harding
Instructor and Outreach Specialist,
Department of Integrative Biology
Michigan State University,
East Lansing, Michigan

Billie C. Harrison
Technician, Milwaukee Public
Museum, Milwaukee, Wisconsin
Graduate Student, Conservation
Biology Department
University of Wisconsin–
Stout, Stout, Wisconsin

Robert W. Hay
Director, Turtles for Tomorrow
Cold-Blooded Specialist (retired)
Wisconsin Department of Natural
Resources, Madison, Wisconsin

Dr. Eric T. Hileman
Assistant Research Professor, Department
of Wildlife, Fisheries and Aquaculture
Mississippi State University,
Starkville, Mississippi

Dr. Collin P. Jaeger
Instructor, Biology Department
McHenry County College,
Crystal Lake, Illinois

Tom R. Johnson
State Herpetologist (retired),
Missouri Department of Conservation,
Jefferson City, Missouri

Dr. Emmet J. Judziewicz
Emeritus Professor of Biology,
Department of Biology
University of Wisconsin–Stevens
Point, Stevens Point, Wisconsin

Heather M. Kaarakka
Conservation Biologist, Bureau of
Natural Heritage Conservation
Wisconsin Department of Natural
Resources, Madison, Wisconsin

Dr. Joshua M. Kapfer
Professor, Department of
Biological Sciences
University of Wisconsin–Whitewater,
Whitewater, Wisconsin

Dr. Daniel E. Keyler, FAACT
Department of Medicine (retired),
Hennepin County Medical Center,
Minneapolis, Minnesota
Adjunct Professor, Department of
Experimental & Clinical Pharmacology
University of Minnesota,
Minneapolis, Minnesota

Dr. Richard B. King
Distinguished Research Professor,
Department of Biological Sciences
and Institute for the Study of the
Environment, Sustainability and Energy
Northern Illinois University,
DeKalb, Illinois

Richard S. King
De Soto, Wisconsin

Dr. Jaimie L. Klemish
Postdoctoral Fellow, Department
of Biological Sciences
University of Lethbridge,
Lethbridge, Alberta, Canada

Dr. Michael J. Lannoo
Professor, Department of
Anatomy and Cell Biology
Indiana University School of
Medicine, Indianapolis, Indiana

Carly N. Lapin
Regional Ecologist, Bureau of
Natural Heritage Conservation
Wisconsin Department of Natural
Resources, Rhinelander, Wisconsin

Dr. Jeffrey M. Lorch
Madison, Wisconsin

Elizabeth Mack McKeag
North Idaho Field Representative,
Idaho Wildlife Federation
Grangeville, Idaho

Gabriel J. Miller
Environmental Program Manager,
Department of Land & Environment
Prairie Island Indian Community,
Welch, Minnesota

Michael J. Mossman
Ecologist (retired), Bureau
of Science Services
Wisconsin Department of Natural
Resources, Madison, Wisconsin

Dr. Patrick C. Owen
Associate Professor, Biology Department
University of Cincinnati, Blue
Ash College, Blue Ash, Ohio

R. Neal Paisley
Wildlife Biologist (retired),
Sandhill Wildlife Area
Wisconsin Department of Natural
Resources, Babcock, Wisconsin

Rori A. Paloski
Conservation Biologist, Bureau of
Natural Heritage Conservation
Wisconsin Department of Natural
Resources, Madison, Wisconsin

Dr. John D. Peterson
Associate Professor, Biology Department
University of Wisconsin–Platteville,
Platteville, Wisconsin

Dr. Steven J. Price
*Associate Professor of Stream and
Riparian Ecology,* Department of
Forestry and Natural Resources
University of Kentucky,
Lexington, Kentucky

Corey Raimond
Virocqua, Wisconsin

Dr. Christine A. Ribic
Unit Leader, US Geological
Survey, Wisconsin Cooperative
Wildlife Research Unit
Department of Forest and Wildlife
Ecology, University of Wisconsin–
Madison, Madison, Wisconsin

Dr. Tracy A. G. Rittenhouse
Associate Professor of Wildlife Ecology,
Center for Wildlife and Fisheries
Conservation Center, Department of
Natural Resources & the Environment
University of Connecticut,
Storrs, Connecticut

Dr. Allison B. Sacerdote-Velat
Curator of Herpetology, Chicago
Academy of Sciences, Chicago, Illinois

Richard A. Sajdak
Aquarium and Reptile Curator (retired),
Milwaukee County Zoo,
Milwaukee, Wisconsin

Dr. Gregor W. Schuurman
Ecologist, Natural Resource
Stewardship and Science
US National Park Service,
Fort Collins, Colorado

Spencer R. Siddons
Graduate Assistant, Department
of Biological Sciences
Purdue University, West
Lafayette, Indiana

Paul M. Skawinski
Outreach Specialist and Coordinator,
Wisconsin Citizen Lake Monitoring
Network, University of Wisconsin–
Extensions Program
University of Wisconsin–Stevens
Point, Stevens Point, Wisconsin

Richard A. Staffen
Zoologist/Conservation Biologist, Bureau
of Natural Heritage Conservation
Wisconsin Department of Natural
Resources, Madison, Wisconsin

Dr. Rochelle M. Stiles
Conservation Special Projects Supervisor,
San Francisco Zoological Society,
San Francisco, California

Jonathan W. Swan
Livingston Manor, New York

Phoebe Vanselow
Gustavus, Alaska

Sara E. Viernum
Wildlife Biologist, Stantec,
Portland, Oregon

Dr. Richard C. Vogt
*Pesquisador III, Curator of the
Collection of Reptiles and Amphibians,*
Instituto Nacional de Pesquisas da
Amazônia, Manaus, Amazonas, Brazil

Dreux J. Watermolen
Chief, Analysis Services
Wisconsin Department of Natural
Resources, Madison, Wisconsin

Dr. Erik R. Wild
Honorary Associate,
Department of Biology
University of Wisconsin–Stevens
Point, Stevens Point, Wisconsin

Acronyms and Abbreviations

SEVERAL AGENCIES, ORGANIZATIONS, UNIVERSITIES, AND LOCATIONS are referenced extensively across the chapters and species accounts (e.g., Wisconsin Department of Natural Resources). To reduce the amount of text devoted to these references, we refer to them by an acronym (Table 1).

Table 1. Acronyms for agencies, organizations, universities, and locations.

Name	Acronym
Bay Beach Wildlife Sanctuary	BBWS
Brigham Young University, Monte L. Bean Life Science Museum	BYU
Carnegie Museum of Natural History	CM
Convention on International Trade in Endangered Species	CITES
Field Museum of Natural History	FMNH
Illinois Natural History Survey	INHS
Intergovernmental Panel on Climate Change	IPCC
International Union for Conservation of Nature	IUCN
Michigan State University Museum	MSUM
Milwaukee Public Museum	MPM
National Land Cover Dataset	NLCD
National Museum of Natural History, Smithsonian Institution	USNM
National Park Service	NPS
Natural Resources Conservation Service	NRCS
The Nature Conservancy	TNC
United States Department of Agriculture	USDA
United States Fish and Wildlife Service	USFWS
United States Forest Service	USFS
United States Geological Survey	USGS
University of Michigan Museum of Zoology	UMMZ
University of Wisconsin	UW
University of Wisconsin Zoology Museum	UWZM
Wisconsin Department of Natural Resources	WDNR

Similarly, we use abbreviations for common units of measurement (Table 2) and acronyms for commonly measured morphological features of amphibians and reptiles (Table 3).

Table 2. Abbreviations for units of measurement.

Type of Unit	Unit	Abbreviation
Length (metric)	Millimeter	mm
Length (metric)	Centimeter	cm
Length (metric)	Meter	m
Length (metric)	Kilometer	km
Length (metric)	Nanometer	nm
Length (English)	Inch	in
Length (English)	Foot	ft
Length (English)	Mile	mi
Area (metric)	Hectare	ha
Area (English)	Acre	ac
Area (home range)	Minimum convex polygon	MCP
Volume (metric)	Milliliter	mL
Volume (metric)	Liter	L
Volume (English)	Ounce	oz
Volume (English)	Gallon	gal
Weight (metric)	Microgram	µg
Weight (metric)	Gram	g
Weight (metric)	Kilogram	kg
Weight (English)	Ounce	oz
Weight (English)	Pound	lb
Temperature (metric)	Celsius	°C
Temperature (English)	Fahrenheit	°F
Time (metric)	Kilometers per hour	kph
Time (English)	Miles per hour	mph
Time	Second	sec
Time	Minute	min
Time	Hour	hr
Time	Week	wk
Time	Month	mo
Time	Year	yr
Time	Millions of years	my
Time	Million years ago	mya
Time	Years before present	ybp
Statistic	Confidence interval	CI
Statistic	Mean	$\bar{x}$
Statistic	Standard deviation	SD
Statistic	Standard error	SE

Table 3. Acronyms for commonly measured morphological features.

Name	Acronym
Carapace length	CL
Carapace width	CW
Midline carapace length	MCL
Plastron length	PL
Snout-to-rump length	SRL
Snout-to-urostyle length	SUL
Snout-to-vent length	SVL
Total length[a]	TL

[a] 'Tail length' is spelled out in accounts.

Museums are referred to by the acronym provided in Sabaj (2016). Additional abbreviations used in the book include ca. (circa, meaning approximately), vs. (versus), *n* (sample size), *N* (population size or population census).

Amphibians and Reptiles of Wisconsin

Introduction

Joshua M. Kapfer and Donald J. Brown

Evolution has not resulted in an even distribution of biodiversity across the planet. The state of Wisconsin, with its 18 species of amphibians and 36 species of reptiles, is not considered a center of herpetofaunal diversity in the United States. In fact, among states, Wisconsin ranks 38th and 32nd in species richness for amphibians and reptiles, respectively. Nonetheless, amphibians and reptiles are important components of Wisconsin's natural communities. Furthermore, the continental distributions of numerous species are influenced by landscape elements found within the state's boundaries.

Unlike many states, Wisconsin is largely defined not by arbitrary political boundaries but rather by geologically and biologically important features that give it scientific significance beyond its borders. To the north and east, Wisconsin is mostly bounded by Lakes Superior and Michigan, respectively, two of the largest freshwater lakes in the world. To the west, Wisconsin is bordered by two important North American river systems, the St. Croix and the upper reaches of the Mississippi. Within these natural boundaries, there exist geological features and climatic variability that have resulted in numerous available terrestrial and aquatic habitats for herpetofauna (see Amphibians and Reptiles in Wisconsin Plant Communities chapter).

Wisconsin's most recent glacial activity has led to a complex and variable topography that influences the geographic distributions of its species. The cooler and wetter northern region of the state contains a high concentration of freshwater lakes, while the warmer and drier southern region contains the Kettle Moraine in the east and the Driftless Area in the west, the latter being a nonglaciated region that likely served as an amphibian and reptile refugium during the Pleistocene (see Past, Present, and Potential Future Landscapes of Wisconsin chapter). As a result of the state's geological history, its present landscape, and its current climatic conditions, Wisconsin represents the continental range limits for many species in North America. For example, the northernmost populations of Timber Rattlesnakes (*Crotalus horridus*), Slender Glass Lizards (*Ophisaurus attenuatus*), Ouachita Map Turtles (*Graptemys ouachitensis*), Six-lined Racerunners (*Aspidoscelis sexlineata*), and Western Wormsnakes (*Carphophis vermis*) are in Wisconsin, as are the easternmost populations of Prairie Skinks (*Plestiodon septentrionalis*) and the westernmost populations of Butler's Gartersnakes (*Thamnophis butleri*). Furthermore, some of these populations are disjunct and far removed from the rest of their more contiguous distributions. So although the number of amphibian and reptile species in Wisconsin may not be overwhelming, the nature of their occurrence in the state has broad significance to scientists, conservation biologists, natural resource managers, naturalists, and the general populace of the Upper Midwest.

All of this lays the groundwork for fascinating discussions of how herpetofauna exist and persist in Wisconsin, which should instill a continued desire for knowledge among its citizens. Unfortunately, it has been ca. 40 years since the last comprehensive treatment of the state's amphibians and reptiles was published. Richard Vogt's *Natural History of Amphibians and Reptiles in Wisconsin* (Vogt 1981) has remained a respected and endearing treatise on the topic, despite being long out of print. Undoubtedly, its popularity stems, in part, from the captivating subject matter. Beyond this, however, is the book's combination of scientific rigor and appeal to a broad audience, including scientists, wildlife managers, naturalists, schoolchildren, and the general public. Although other publications with a focus on Wisconsin's amphibians and reptiles have been released since Vogt's, they are also dated and either were targeted primarily to the general public (e.g., the WDNR series *Snakes of Wisconsin* [Christoffel et al. 2000], *Amphibians of Wisconsin* [Christoffel et al. 2009], and *Turtles and Lizards of Wisconsin* [Christoffel et al. 2002]) or addressed only a limited aspect of Wisconsin's herpetofaunal ecology (e.g., *Geographic Distributions of the Amphibians and Reptiles of Wisconsin* [Casper 1996]). As such, there remains no single, current source that summarizes the existing scientific knowledge of the state's amphibians and reptiles.

These past publications notwithstanding, the rapid increase in information that characterizes contemporary science and the increase in conservation concerns for Wisconsin's natural legacy call for a new comprehensive synthesis of the state's amphibians and reptiles. It was for these reasons that we recruited numerous experts in a coordinated effort to produce *Amphibians and Reptiles of Wisconsin*. Our intention was to generate a single book that could serve as the definitive source of scientific information regarding herpetofauna in the state but in a way that also appealed to the general public.

FULFILLING A SCIENTIFIC NEED

Fulfilling a scientific need is one of the primary goals of this book. The past few decades have seen a substantial increase in scientific knowledge regarding the distribution and ecology of Wisconsin's amphibians and reptiles. For example, over 700 county records have been published since the *Natural History of Amphibians and Reptiles in Wisconsin* (Vogt 1981) appeared. New natural history information has accumulated, and many in-depth studies on species biology and conservation have been published (see Figure 33 in Introduction to Conservation and Management of Wisconsin's Amphibians and Reptiles chapter). Advances in molecular and systematic biology have resulted in numerous phylogeographic, biogeographic, and taxonomic studies that have shed light on Wisconsin species, resulting in, among other advances, changes in the scientific names of nearly a third. Furthermore, fieldwork continues to make new discoveries, and a new species was even added to the state's list of reptiles in 2012.

Although our knowledge has increased, much of this valuable information was published in small notes and governmental reports, or it exists in the minds and field notes of experienced field biologists. We believed it was time to see all of this valuable information brought together to facilitate its accessibility and actualize its value. As such, this book's editors and contributors put substantial time and effort into compiling and synthesizing important unpublished data, as well as hard-to-find reports and publications. This information, coupled with a synthesis of the peer-reviewed scientific literature, will provide a strong foundation for future scientific endeavors and conservation initiatives focused on amphibians and reptiles in Wisconsin.

PROMOTING WISCONSIN'S NATURAL LEGACY

Promoting Wisconsin's natural legacy is another primary goal of this book. Wisconsin has a populace that deeply appreciates the natural legacy of their state, founded in the likes of Aldo Leopold, John Muir, and Gaylord Nelson, among others. Furthermore, numerous Wisconsin Indigenous cultures maintain deep traditions and associations with nature. Natural history information is not only the basis upon which all further scientific understanding emanates but also a primary source of curiosity about, appreciation for, and fascination with the natural world to which we belong. As access to information expands, so does public knowledge, which stimulates the desire to learn. This interest is enhanced when conservation issues gain public attention, such as the global decline of amphibians and reptiles (e.g., the near extirpation of the Blanchard's Cricket Frog [*Acris blanchardi*] from Wisconsin), the spread of pathogenic fungi among amphibians and snakes, the potential development of iron-ore mines in pristine areas (such as in the Crandon and Penokee Range regions), and the political, economic, and conservation conflicts that directly or indirectly result from human population growth.

Although basic introductions to Wisconsin's amphibians and reptiles are available, even the most novice individual who is motivated to learn would quickly exhaust the information these works provide. It was, therefore, clear to the editors of this book that an updated source was needed, one that continued Vogt's (1981) initiative to not only inform professional scientists but also remind the general public about Wisconsin's amphibians and reptiles in a way that encourages further learning and enhances appreciation for the state's natural legacy.

OUR OBJECTIVE AND APPROACH

Our objective was to produce a single, comprehensive volume that consolidated the current state of scientific knowledge of Wisconsin's amphibians and reptiles and to present it in a manner suitable for as broad an audience as possible. Toward this end, we included several introductory chapters that address how some general topics relate to amphibians and reptiles in Wisconsin. "Systematics and Classification" describes the placement of amphibians and reptiles in the tree of life and the evolutionary relationships of amphibians and reptiles to one another. Next, "Some Prominent Figures in Wisconsin Herpetology" highlights prominent scientists who have documented, described, and studied the natural history of amphibians and reptiles in the state. "Past, Present, and Potential Future Landscapes of Wisconsin, with Emphasis on Amphibian and Reptile Distributions" and "Amphibians and Reptiles in Wisconsin Plant Communities" together describe the ecological, geological, and floral characteristics of the state as they relate to these animals. Finally, "Introduction to Conservation and Management of Wisconsin's Amphibians and Reptiles" addresses the responsibility we have as stewards of our natural legacy and strategies to promote the persistence of these species.

The individual species accounts serve as the core component of the book. The accounts were written by expert professionals and include in-depth syntheses of the ecology of each species, with a focus on research from Wisconsin and the Upper Midwest. Associated with the species accounts for Wisconsin's two rattlesnakes, the chapter "Snakebite Envenomation in Wisconsin and the Upper Mississippi River Valley," written by venomous snake toxicologist Daniel Keyler, discusses history, biology, and treatment of venomous snake bites. Following the species accounts, the chapter "Hypothetical and Enigmatic Species of Wisconsin" discusses species that were historically reported as part of the state's herpetofauna but for which there is not strong evidence of established populations. This chapter also includes a list of species that are currently undocumented but could occur, given their known presence in adjacent states.

SUMMARY

We believe this book will serve as an important contribution to the collection of natural history books focused on the state of Wisconsin. The scope of coverage exceeds all previous treatments of amphibians and reptiles in the state, and thus we believe it will be an invaluable source of information for a wide variety of individuals who work with or are interested in amphibians and reptiles. We anticipate it will allow practicing herpetologists—young and old, professional and novice—the opportunity to expand their knowledge and appreciation of Wisconsin's amphibians and reptiles. In addition, we hope this book will foster an enjoyment for learning about and enhance an appreciation of Wisconsin's natural legacy. Ultimately, we hope this book will bring out the herpetologist in everyone.

Systematics and Classification

Erik R. Wild and Tony Gamble

> *The affinities of all the beings of the same class have sometimes been represented by a great tree. I believe this simile largely speaks the truth. The green and budding twigs may represent existing species; and those produced during each former year may represent the long succession of extinct species.*
>
> —CHARLES DARWIN, *On the Origin of Species* (1859)

Every species of amphibian and reptile that occurs in Wisconsin has an account in this book. But what is a species? And how are they related to one another? The branch of biology that describes species and their relationships is called systematics. Most systematists, the scientists who study systematics, consider a species an evolutionary lineage of ancestor-descendant related organisms (e.g., Frost and Hillis 1990, Mayden 1997, de Queiroz 1998, Wiley and Lieberman 2011). These lineages have originated through speciation from a common ancestor and are related to one another. This is analogous to a human genealogy or family tree. These relationships among species are what Darwin (1859) referred to as "a great tree," where "the green and budding twigs" represent existing species connected through the various twigs, branches, and trunks, symbolizing common ancestry. Herein we provide accounts for the few "green and budding twigs" that denote two of the major branches on the tree of life, amphibians and reptiles, that have persisted in Wisconsin to this day (Figure 1).

Scientists typically refer to species by their scientific names, which are composed of two words: the first is the genus (generic name), which describes a group of closely related species; the second is the species name (specific epithet), which distinguishes the individual species within a genus. These names are often derived from Latin or Greek words or are Latinized words from other languages. The generic name is always capitalized, and the species name is always lowercase, and both names are always italicized or underlined in print. A scientific name can describe the organism's traits or its location of occurrence, or it can honor a person or persons. The scientific name for humans, for example, is *Homo sapiens*:

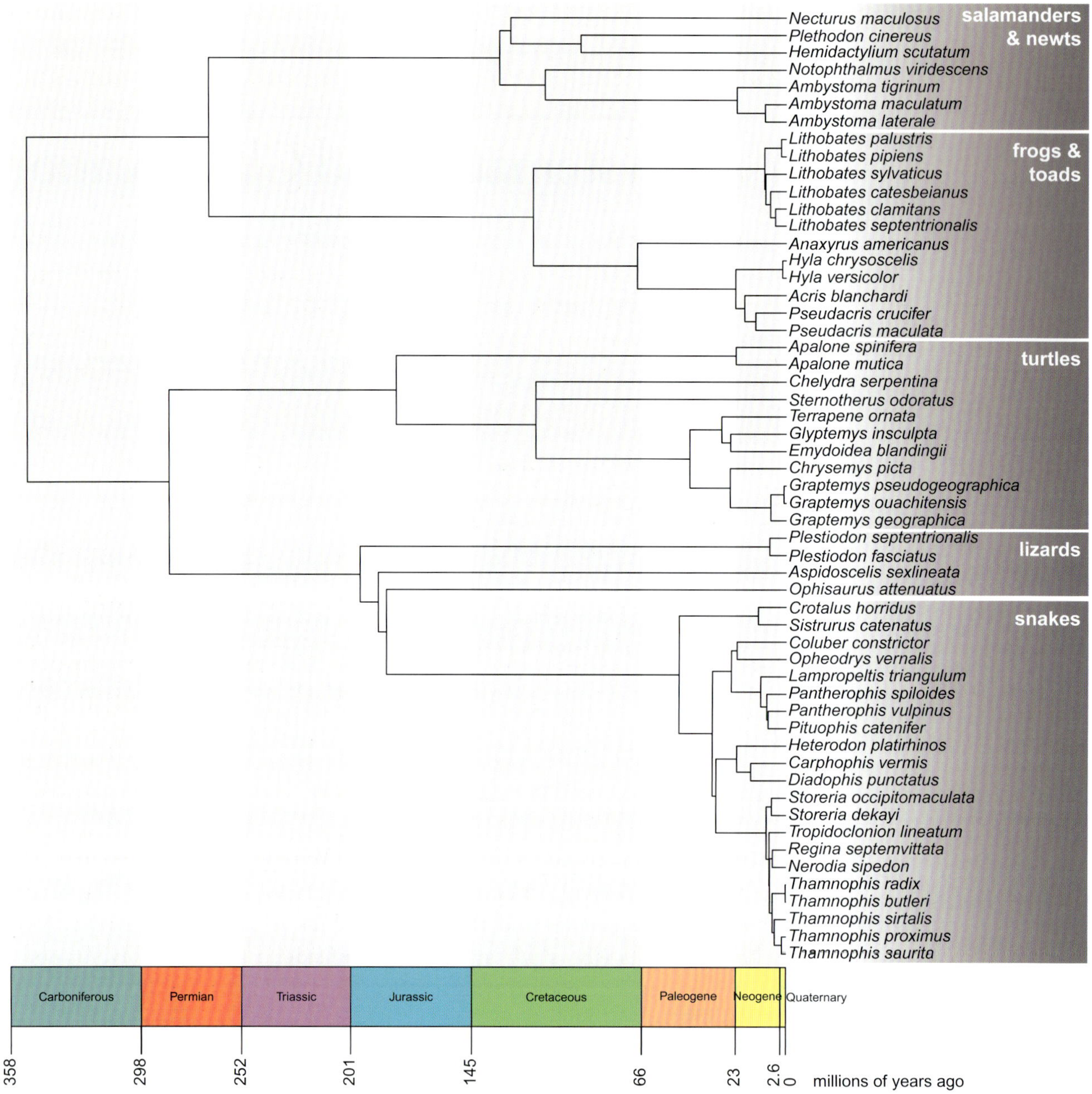

Figure 1. A phylogenetic tree showing the evolutionary relationships among all amphibian and reptile species that occur in Wisconsin. Branch lengths represent approximate divergence times among species with a geologic time scale at the bottom of the figure. This tree is a composite derived from previously published, time-calibrated phylogenetic analyses using molecular genetic data (Jaffe et al. 2011, Pyron and Wiens 2013, Zheng and Wiens 2016).

Homo is the generic name, and *sapiens* is the species name. Scientists will often abbreviate the generic name after its first appearance when more than one species is referred to in the same paragraph, so *Homo sapiens* is abbreviated as *H. sapiens*. We follow this convention throughout the book.

Species represent hypotheses that a particular lineage is distinct from other lineages. These hypotheses are originally put forth in a species description, a published work of largely descriptive data that link a particular specimen (called the holotype) with a particular scientific name. Any individual organism that originates from the same lineage as the holotype belongs to that species. Most species descriptions will also include a list of characteristics that are unique to that species and can differentiate it from other, related species. We have included many of these diagnostic traits in the species accounts to help the reader properly identify each of Wisconsin's amphibians and reptiles and distinguish them from their close relatives.

There exist 8,384 described amphibian species (https://AmphibiaWeb.org) and 11,570 reptile species (excluding birds; http://www.reptile-database.org) as of October 2021. These numbers are increasing every year as systematists describe newly discovered species and use new types of data and methods to better distinguish existing species. To facilitate communication and research, biologists have developed various systems of classification beyond just species. These systems include categories such as genus, family, order, class, phylum, and kingdom. Contemporary biological classification is based on Darwin's (1859) conclusion that names should reflect the evolutionary relationships (phylogeny). Thus, names are applied to groups of all and only those species that have descended from a common ancestor (called monophyletic groups or clades) or, to follow Darwin's tree of life simile, a complete branch of the tree of life. The field of systematics has been revolutionized in recent years through a combination of advances in the philosophy of systematic biology, especially regarding classification (i.e., the development of a monophyletic classification), and the abundance of new data and analytical methods, mostly in the form of molecular genetic data such as DNA sequences and the development of fast, powerful computers to analyze all these new data. These advances have resulted in numerous changes to species names, including nearly a third of Wisconsin's amphibian and reptile species. While such changes require learning new names for familiar creatures and may lead to complaints among naturalists and outdoor enthusiasts, it is important to remember that these changes in taxonomy represent advances in our knowledge about how species are related to each other. So before grumbling about the latest taxonomic change to your favorite species, try to remember that such changes mean we now know a little bit more about the evolutionary history of that species and its relatives, which is a good thing.

A classification based on phylogeny provides many benefits. Among these is the interpretation of species characteristics and natural history from an evolutionary perspective. Species possess a mosaic of characters inherited from both distant and more recent common ancestors. Of particular interest are those traits that evolved in the common ancestor of a particular clade, because these can be associated with the clade's diversification. These evolutionary novelties are termed 'synapomorphies' and usually have persisted in most of the descendants of that clade, although sometimes they can be modified or lost. For example, having four limbs is a synapomorphy for the tetrapods, which contains amphibians, reptiles (including birds), and mammals (Figure 2). However, there are several lineages within the clade Tetrapoda that do not have limbs, such as snakes. This does not mean snakes are not a member of that clade or that they are not tetrapods. It does, however, call attention to the evolutionary distinctiveness of snakes having lost limbs.

Another consequence of a phylogeny-based classification is that some traditional groups will be redefined as we better understand evolutionary history. An excellent example of this is the clade Reptilia, the reptiles. Traditionally, reptiles consisted of lizards, snakes, turtles, and crocodilians. However, multiple lines of evidence, including genetic data, morphology, and fossils, indicate that birds are the closest living relatives to crocodiles and alligators (see Figure 2). Thus, the inclusion of birds as reptiles is required to keep reptiles a clade.

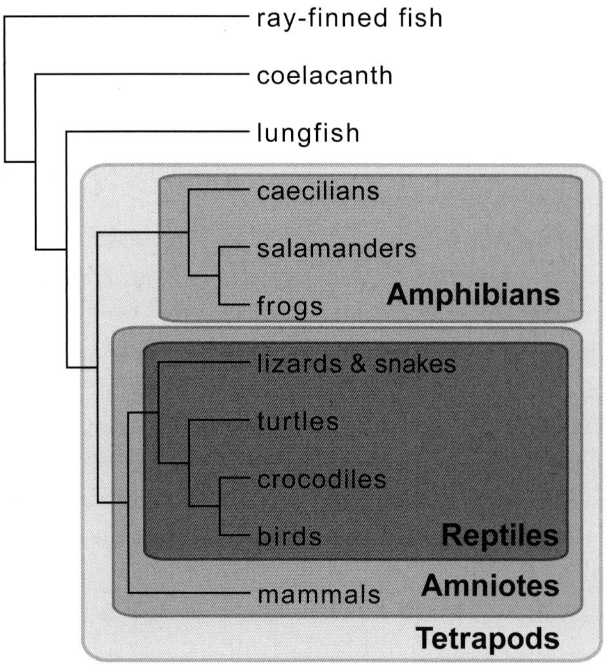

Figure 2. Phylogeny showing the relationships among the major clades of tetrapods (Tetrapoda) and their closest living relatives. Note that many limbless organisms (caecilians, snakes, many lizards, amphisbaenians) have descended from the common ancestor of Tetrapoda.

Both amphibians and reptiles belong to the clade Vertebrata, the vertebrates (having, among other synapomorphies, a vertebral column), and, with the exclusion of fishes and their allies, the clade of their most recent common ancestor, Tetrapoda (having, among other synapomorphies, four pentadactyl [five digits] limbs). However, that is the end of their similarities, as amphibians and amniotes (tetrapods having an amniotic sac surrounding their developing egg [i.e., reptiles and mammals]) diverged. Information on species-level systematics for amphibians and reptiles (exclusive of birds) found in Wisconsin is included in the individual species accounts.

Some Prominent Figures in Wisconsin Herpetology

Joshua M. Kapfer

Throughout history, several individuals have played key roles in advancing our understanding of amphibian and reptile biology, ecology, and distribution in Wisconsin. The goal of this chapter is to chronologically review those individuals who have had substantial impacts on herpetology in the state and to acknowledge the importance of their contributions. Lengthy reviews are restricted to those who have conducted important previous work, published substantially, collected many specimens, or contributed significantly to the knowledge and/or conservation of the state's herpetofauna. Others who have made important but perhaps less extensive contributions to the understanding of herpetology in Wisconsin are discussed more briefly.

EARLY TIMES

It is noteworthy that evidence from the pre-Columbian era suggests that the Native peoples of Wisconsin both made use of and perhaps revered local herpetofauna. Archaeological investigations have found the remains of numerous species both consumed and occasionally worn as ornaments by Indigenous peoples during prehistoric times (e.g., Parmalee 1959, 1960, 1963; Theler 2000). Furthermore, a number of ancient burial mounds throughout the state are shaped to resemble amphibians or reptiles (e.g., Butterfield 1880). In later eras (the 1600s and 1700s), Wisconsin's amphibians and reptiles were sporadically and fleetingly mentioned by various authors (Hennepin 1698, and Carver 1781, reviewed by Vogt 1981). Even through the first half of the 1800s, only a handful of sources mention local herpetofauna, but these typically relate to very specific locations and often focus on the danger of rattlesnakes (e.g., Canfield 1861, Butterfield 1880, Western Historical Company 1882). The Wisconsin naturalist Increase A. Lapham appears to have made the first rigorous attempt to review the state's amphibians and reptiles by publishing a checklist of species in his famous *Fauna and Flora of Wisconsin* (Lapham 1853). This sparse list included 15 reptiles and 5 amphibians. Accompanying each species on Lapham's list was limited information: the scientific name, the common name, synonyms, and the taxonomic authority. Lapham also reported locations where he (or a trusted associate) had observed the species in question, which were often a single county. Clearly, this information was useful, particularly since no attempt to review all of the state's species had been made prior to Lapham (1853). More importantly, this simple checklist laid fertile ground for the substantial work on the state's herpetofauna that followed.

Philo Romayne Hoy (1816–1892)

Philo Romayne Hoy was one of Wisconsin's greatest early natural historians and likely the first to produce any substantial publication on the state's amphibians and reptiles. Hoy was trained as a medical doctor and had a successful practice in Ohio when he moved his family to Racine, Wisconsin, in 1846 (ABPC 1877). The natural history of Wisconsin clearly became one of Hoy's passions, and he was knowledgeable on a diversity of topics that ranged from geology to zoology. During this time, Hoy contributed to the understanding of Wisconsin's natural history in various ways. He was a state fish commissioner for a time (ABPC 1877) and one of several individuals who developed the Wisconsin Academy of Sciences, Arts and Letters, of which he was president for three years (Anonymous1892). He purportedly collected thousands of specimens, including shells, fossils, insects, and many vertebrates (ABPC 1877, Schorger 1944). In addition, Hoy was apparently well-respected by many zoologists of his time. Spencer F. Baird (a zoologist and an eventual curator at the Smithsonian National Museum of Natural History) and Jared Kirtland (a prominent physician and natural historian of Ohio; Kirtland's Snake [*Clonophis kirtlandii*] is named for him) requested Hoy's company as they traveled across southern Wisconsin in 1853 collecting specimens (Schorger 1944, Deiss 1985). There are also several species named for Hoy, such as the Pygmy Shrew (*Sorex hoyi*) and the Bloater (*Coregonus hoyi*; Schorger 1944, Greenburg 2008). The general admiration expressed for Hoy by professionals and community members is impressive. In fact, a memorial published in the *Racine Journal* (2 May 1922) stated that his "value to the state and nation will live for centuries," while Schorger (1944) professed that with his death "passed Wisconsin's greatest pioneer zoologist."

Although Hoy is often heralded for his ornithological and ichthyological contributions, he collected many herpetological specimens. He then used this information as the basis for the creation of the most complete published review of the state's herpetofauna at that time (Hoy 1883). His manuscript included not only an expanded list of native species (more than 30 reptiles and more than 20 amphibians) but also simple descriptions of state distributions and associated habitat (e.g., the entry for the Eastern Hog-nosed Snake [*Heterodon platirhinos*] includes the description "Dry, sandy locations. Not uncommon in the southwestern counties"). Today Hoy's observations and collections provide an important historical record of southern Wisconsin's natural communities from a time when extensive areas of undisturbed wild places still existed in the state. According to Schorger's (1944) description, Hoy was active when the "strip of timber along the shore of Lake Michigan was virtually intact, while west of it was the high, rolling prairie." The forest Schorger referenced was part of the mesic hardwood forest community that historically stretched across the eastern United States, hugged the southern tip of Lake Michigan, and entered Wisconsin along the lake's western edge (Curtis 1959, Dyer 2006). The convergence of these habitat communities in southeastern Wisconsin, coupled with the lower levels of anthropogenic disturbance at that time, possibly contributed to Hoy's observation or collection of numerous species that today are considered unheard-of in the state. These species include several birds that are currently very rare, extinct, or extirpated from Wisconsin, such as the Passenger Pigeon (*Ectopistes migratorius*), Whooping Crane (*Grus americana*), Wood Stork (*Mycteria americana*), and Royal Tern (*Sterna maxima*; Schorger 1944, William Mueller, Western Great Lakes Bird and Bat Observatory, personal communication). Likewise, Hoy collected or reported several amphibians and reptiles that have not been documented in Wisconsin since (although subsequent authors have challenged the validity of these reports; see Hypothetical and Enigmatic Species of Wisconsin chapter). Regardless, Hoy's records likely fueled the imagination of many Wisconsin naturalists since his time, inspiring them to wonder what previously existed or perhaps awaited rediscovery within the state's boundaries. Furthermore, his manuscript (Hoy 1883) remained the only sizeable work on the state's herpetofauna until Pope and Dickinson (1928) was published many years later. (Note that Higley [1889] mostly reiterated Hoy's earlier manuscript and added little but taxonomic keys.)

Thomas Edmund Burt Pope (1879–1958)

Thomas Edmund Burt Pope was a curator of lower zoology at the Milwaukee Public Museum (MPM) from 1913 to 1939. Given that the museum opened in 1882 (Lurie 1983), he was likely the first curator to put substantial effort into the amphibians and reptiles of Wisconsin (Figure 3). Prior to Pope, only Sigmund Graenicher (Ph.D., M.D.) appears to have directly worked with the museum's herpetological collections. Graenicher was a physician who first took the title of honorary curator of ichthyology and herpetology but was later hired as a curator of vertebrate zoology in 1909 (Claudia Jacobson, MPM, personal communication). Yet Graenicher ultimately became acting curator of invertebrates, and most information suggests he was primarily interested in entomology (he was even a charter member of the Entomological Society of America; Cockerell 1937). Although Graenicher collected herpetological specimens in Wisconsin (some of which still exist at the museum), his only known contribution to literature in this area was a short review of lizard reports (Graenicher 1911).

Pope attended Brown University for graduate school and received his master's degree in 1902 (Anonymous 1958). After graduation, he was employed for 10 years as a scientific assistant with the US Bureau of Fisheries, which eventually became part of the US Fish and Wildlife Service (USFWS) (NOAA 2015), and was director of the US Fisheries Laboratory in Woods Hole, Massachusetts (Anonymous 1958). As a result, Pope's early work focused on a variety of aquatic taxa, but particularly fishes and mollusks, including research on effective practices for oyster culture (Moore and Pope 1910). In 1913 he was hired as an associate lecturer at the MPM and was named curator four years later (Anonymous

Figure 3. Thomas Edmund Burt Pope (ca. 1939) shortly before he retired as curator of lower zoology at the MPM (MPM Photo 422423b, used with permission from MPM).

1958). Although as a museum employee he continued to work with fishes (e.g., Hubbs and Pope 1937), he also turned his attention toward Wisconsin's amphibians and reptiles. During his 26-year career at the museum, Pope published several manuscripts on the state's herpetofauna, including a large review of native species that he coauthored with a young William Dickinson (see William Edmund Dickinson entry; Pope and Dickinson 1928). Pope and Dickinson (1928) was the first manuscript that attempted to cover all of Wisconsin's species in considerable detail. It included taxonomic keys, natural history information, photographs, and drawings (by Dickinson). It also covered state distributions in greater detail and reviewed which counties had representative specimens (although it did not always report specimen numbers). Later, Pope published additional species records and distributional updates, including several interesting reviews of purported Western Diamondback Rattlesnake (*Crotalus atrox*) observations in the state (Pope 1928, 1930, 1931; see also Hypothetical and Enigmatic Species of Wisconsin chapter). Pope remained curator at the MPM until he retired in 1939, after which he was granted emeritus status for several years before eventually moving to California, where he died in 1958 (Dickinson 1958).

Unfortunately, one could argue that Pope's herpetological publications have limited value today. The large manuscript he coauthored with Dickinson, for example, included numerous unlikely species that had not been confirmed to occur in the state, as well as incorrect range maps. Furthermore, only a handful of actual specimen numbers accompanied their species accounts, and those reported are mostly assigned to photographic negatives of live specimens pictured in the manuscript rather than county record vouchers. As Vogt (1981) pointed out, "Many of the specimens listed by both T. E. B. Pope and Dickinson with cited museum numbers are no longer in the MPM or are not the number listed. Because so many of the catalogued specimens used by them were incorrectly identified, I have ignored all of their published distribution records unless I have reexamined the specimens." Yet Pope and Dickinson were generally explicit about the rationale and evidence (or lack thereof) they used for the inclusion of the more enigmatic species in their work, which frequently stemmed from reports in prior publications. Furthermore, Pope's manuscripts provide a useful historical record of several currently rare species, such as the Timber Rattlesnake (*Crotalus horridus*; Pope 1923). Pope and Dickinson (1928) was also the most thorough published attempt to review Wisconsin species up to that time. Pope further added numerous specimens to the herpetological collection at the MPM, some of which still exist, so the influence of his work on the subject of herpetology in Wisconsin should not be undersold.

William Edmund Dickinson (1902–1974)

William E. Dickinson was the curator of lower zoology at the MPM from 1940 to 1969, following the retirement of Thomas E. B. Pope (Anonymous 1970; Figure 4). Dickinson was born in 1902 in Florence, Wisconsin, and moved to Milwaukee as a boy (Anonymous 1974). His

Figure 4. William Edmund Dickinson (ca. 1941) early in his career as curator of lower zoology at the MPM (MPM photo 425159a, used with permission from MPM).

Figure 5. A party of Wisconsin rattlesnake "hunters" on a collecting trip for the MPM in 1923 (MPM Photo 46304, used with permission from MPM). Included in this photograph is Thomas E. B. Pope (*second from left*), who was curator of lower zoology at the time, and a young William E. Dickinson (*third from left*), who would later move into the curator position in 1940. According to the description of this photo in Pope (1923), the box nearest Pope (acting as a table for the party's lunch) contained Timber Rattlesnakes collected in Sauk County for a museum display meant to depict a "typical" rattlesnake den in Wisconsin. The remaining party members are identified as (*to the left of Pope*) E. C. Keitel (a hunter and trapper from Merrimac), (*to the right of Dickinson*) Mr. Gastrow, E. R. Tyrrell (an artist who created sketches of the landscape for the proposed museum diorama), and E. D. Ochsner (a taxidermist from Prairie Du Sac). The diorama was completed and displayed in an older building that no longer houses the museum (MPM 1925).

father left when he was only a few years old, and he was raised by a single mother. Dickinson became a 'helper' at the MPM in 1923 (Anonymous 1970). Early in his career he worked with Thomas E. B. Pope on many projects, which included trips to collect Wisconsin rattlesnakes (Figure 5; Pope 1923) and the publication of an early review of Wisconsin's herpetofauna (Pope and Dickinson 1928).

During his time as a research assistant, Dickinson was granted leave to work on a bachelor's degree at UW–Madison, returning to the MPM to continue his duties during school vacations (Susan Borkin and Claudia Jacobson, MPM, personal communication). He graduated in 1932 after completion of his bachelor's thesis, entitled "The Mosquitoes of Wisconsin" (Poser 1932, Egstad 1933), which was later published by the museum (Dickinson 1944). He continued to work for the MPM after graduation but resigned in 1936 to become director of the newly organized Kenosha Public Museum (S. Borkin and C. Jacobson, personal communication). While in Kenosha, Dickinson reportedly "begged, borrowed, and hinted broadly for cases, exhibits, and help in the fledgling natural history museum" (Anonymous 1970) and was important to the museum's early success.

Dickinson came back to the MPM as the curator of lower zoology in 1940 shortly after Pope retired. Upon his return, Dickinson continued with much of the work that he and Pope began previously. In 1946, while a full-time curator, he received a master's degree from Marquette University (Milwaukee, Wisconsin), with his thesis titled "The Lizards and Snakes of Wisconsin" (Dickinson 1946, Michelle Sweetser, Marquette University archivist, personal communication). Over the next 23 years, Dickinson made numerous contributions to the museum's exhibits and collections. In addition to amphibians and reptiles, Dickinson was deeply interested in fishes and invertebrates, publishing works on a variety of taxa during his nearly three-decade tenure with the museum (Dickinson 1944, 1960; MPM 2015). Dickinson's influence at the MPM can still be felt through the continued presence of the Living Oceans exhibit on the third floor, which he helped organize and construct. He was also a member of the Wisconsin Academy of Arts, Sciences and Letters and of the Midwest Museum Conference (Anonymous 1974). He retired in 1969 but maintained ties with the museum as curator emeritus (Anonymous 1974).

In addition to Pope and Dickinson (1928), Dickinson's most significant herpetological contributions were handbooks for species identification (e.g., Dickinson 1949, 1965, the former of which appears to be a reworked version of his master's thesis [Dickinson 1946]). These handbooks included taxonomic keys, color illustrations (which were Dickinson's own), and range maps. Today, his handbooks, much like the works of T. E. B. Pope, are sometimes criticized, because current knowledge indicates they contain many inaccuracies, and he rarely reported specimen numbers to support his county records. Yet Dickinson's contribution to herpetology in Wisconsin should not be forgotten. For decades prior to his time, few others had placed any substantial effort into investigations of the state's herpetofauna. Much like Pope, he is also laudable as a well-rounded natural historian, with interests that spanned many taxa. His career further encapsulated a number of interesting professional stories. The following gem was uncovered in the pages of the *Oshkosh Northwestern* newspaper (published on 8 November 1934): "BITTEN BY BLACK DIAMOND RATTLESNAKE AT MUSEUM. Milwaukee—Opening a box at the Milwaukee public museum yesterday, William E. Dickinson, assistant curator of zoology, was bitten on the end of the thumb by a six-foot diamond back rattlesnake which had been shipped from Texas. Dickinson quickly applied a tourniquet, cut the wound open with his knife and then sucked out the venom. He later was taken to the Deaconess hospital and given a serum treatment. The snake had been sent to Prof. Carl Hauser of the Lutheran high school who had planned to exhibit it alive to his zoology students. On receipt of the box, Hauser took it to Dickinson and asked him to put the snake in a cage."

George J. Knudsen (1922–2004)

Unfortunately, few younger herpetological enthusiasts in Wisconsin today will likely recognize the name George Knudsen. Yet Knudsen was a prominent figure in Wisconsin herpetology throughout the mid- and later 1900s. His work heavily influenced public perception and understanding of the native herpetofauna through the 1950s and 1960s while he was a Wisconsin Department of Natural Resources (WDNR) employee, but it also helped create the foundation upon which the department's current conservation and management strategies are built.

Knudsen was born in Easton, Wisconsin, in 1922 and grew up during the Depression era (Ley 2004). As a child he took on various odd jobs to help generate income for his family, such as guiding hikes and boat tours of the rock formations in the Wisconsin Dells area (Lange 2004, Ley 2004). This time along the Wisconsin River as a child fueled an interest in the natural world that would follow him throughout his life. According to Knudsen, "I was on that river, or in that river, half the time . . . fishing or swimming, and collecting turtles and snakes and all kinds of critters because I just liked to have them around" (Lange 2004).

As he grew older and his interests intensified, Knudsen made regular visits to the library and taught himself to identify species of interest and learn about their natural history (Ley 2004). Knudsen's mother died when he was only ten, and his father was a "businessman" (Ley 2004). Yet although his father was not instrumental in teaching Knudsen about the outdoors, he was supportive of his son's interests (Ley 2004). In the 1930s Knudsen's father took a job as a director of the local Resettlement Administration (Lange 2004). This was one of the New Deal programs implemented by the federal government at that time, and its primary focus was the relocation of farmers trying to make a living on the poor soil in central Wisconsin to other areas (Lange 2004, Ley 2004). As the result of his job, Knudsen's father interacted with biologists and foresters, including the prominent Wisconsin conservationist Aldo Leopold, who was an advisor for the relocation program (Lange 2004). Leopold occasionally visited the Knudsen family and during these visits noted George's insect collection, which eventually totaled some 7,200 specimens (Lange 2004). This must have impressed Leopold, who encouraged Knudsen to attend UW–Madison for a college education (Lange 2004, Ley 2004). Knudsen followed this advice and enrolled at the university in 1940 to study economic entomology before taking a several-year military hiatus (Ley 2004). Knudson returned to the university, supported by the G.I. Bill in 1945, and shifted his focus toward wildlife ecology and field zoology (Ley 2004).

During his college career, Knudsen took courses with the likes of Aldo Leopold, Robert A. McCabe, and John T. Curtis (Lange 2004). He also spent much time in the outdoors, taking detailed field notes on local amphibians, reptiles, and insects, all of which honed his skills as a naturalist, even prior to completion of his degree in 1949 (Ley 2004). In 1950 he was hired as a wildlife biologist by the Wisconsin Conservation Department (now the WDNR; Ley 2004). More than a decade of his early career involved work with mammals (Figure 6), which generated numerous important technical reports (e.g., Knudsen 1962, 1963; Knudsen and Hale 1970). Roughly 12 years after he began with the department, he transitioned into the position of chief state naturalist (Knudsen

Figure 6. George J. Knudsen (August 1958) monitoring an American Black Bear (*Ursus americanus*) as it recovers from anesthetization after receiving an ear tag in Vilas County (photo courtesy of WDNR).

1962; Ley 2004), which he held for over 20 years until he retired in 1983 (Wisconsin Civil Service Information Form Series 2198).

Less of a preservationist and more of an active conservationist, Knudsen was not particularly interested in experimentation or publishing scientific literature but rather careful observation, fieldwork, and applied management (Ley 2004). This is likely the reason for his lack of authored scientific journal articles and explains why he is not so well-known among present-day herpetologists. Regardless, he was considered the expert on the natural history of Wisconsin's herpetofauna for many years. He recorded observations, took photographs, and collected numerous specimens within the state, many of which are still housed at the University of Wisconsin Zoological Museum (UWZM). In addition, he wrote over 40 popular articles on Wisconsin's herpetofauna that were published in the *Wisconsin Conservation Bulletin* (now *Wisconsin Natural Resources Magazine*) throughout the 1950s. Although he had many professional achievements, Knudsen was heavily involved in the early Scientific Areas Preservation Council (now the Natural Areas Preservation Council; Cliff Germain, WDNR [retired], personal communication). Today, the Natural Areas Preservation Council assists in the decision-making process for acquisition, protection, and management of state natural areas, which can include important habitat for amphibians and reptiles. He further used his knowledge and expertise to make recommendations on the addition of species to the state's endangered species list (Ley 2004). In fact, many species currently listed as Threatened or Endangered in the state received their status during Knudsen's tenure at the WDNR. Knudsen also indirectly influenced the first book on Wisconsin's herpetofauna, written by Richard Vogt (Vogt 1981). The following quote from Vogt's book illustrates this: "[Knudsen] planned to write [a book on amphibians and reptiles of Wisconsin] himself. He had written articles and made numerous films and slide shows for the Wisconsin DNR. When I finally received financial backing to finish researching and writing this manuscript [Vogt 1981], George bigheartedly . . . offered me all his range maps, notes, and specimens that he had accumulated over the last 25 years." Although Vogt clearly added a substantial amount of his own data to the final product, George Knudsen's influence is worthy of recognition.

PRESENT DAY

Richard Carl Vogt (1949–2021)

Richard Carl Vogt is well-known among herpetologists and Wisconsin naturalists for publishing many peer-reviewed manuscripts on local herpetofauna, including his book *Natural History of Amphibians and Reptiles in Wisconsin* (Vogt 1981; Figure 7). Vogt was born in the town of Blooming Grove, Wisconsin, east of Madison, in 1949 and exhibited a general interest in all wildlife at an early age (Figure 8; Richard Vogt, Instituto Nacional de Pesquisas da Amazônia, personal communication). After reading *Thrills of a*

Figure 7. Richard Carl Vogt (ca. 1977) conducting horizontal starch gel electrophoresis in the lab. This work was conducted as part of his doctoral research on *Graptemys* species systematics at UW–Madison (photo by L. Mattick).

Above, top: Figure 8. Richard Carl Vogt (ca. 1967) capturing a Massasauga at a site in Juneau County (photo by R. Durnford).

Above, bottom: Figure 9. An early Painted Turtle (*Chrysemys picta*) specimen collected by Richard Carl Vogt in 1971 (MPM 11708). His field notes stated: "On 2 July 1971 I collected in the East Branch of the Milwaukee River, from 0800–1300h, capturing *C. p. marginata*, south of the mouth of a lake, *Acris* and *Rana clamitans* were calling, collected *Rana pipiens* and saw a few *Rana catesbieana*" (R. Vogt, personal communication).

Naturalist's Quest, a book by the famous herpetologist Raymond L. Ditmars (1939), Vogt's interests focused more squarely on amphibians and reptiles (Vogt 1981). In 1967 Vogt began work toward a zoology degree from UW–Madison. It was during his undergraduate career that he met his long-term mentor, Dr. William G. Reeder, who was the director of the university's zoology museum (R. Vogt, personal communication).

By 1971 Vogt had completed his bachelor's degree and started work on a master's degree at UW–Madison with Reeder as his mentor. That year, Vogt also met Dr. Max A. Nickerson, then the newly appointed curator of vertebrate zoology at the MPM, during the Society for the Study of Amphibians and Reptiles' annual meeting in New Mexico (Vogt 1981). Nickerson and Vogt developed a working relationship where Vogt received funding from MPM to finance two years of Wisconsin herpetological inventory work, with collected specimens to be deposited at UWZM and MPM (Figure 9; Vogt 1981). These surveys not only bolstered the collections of amphibians and reptiles from Wisconsin for both institutions but also provided Vogt with ample opportunity to collect natural history information that would end up in his book on Wisconsin herpetofauna. Vogt completed these surveys and his master's degree by 1974 and soon transitioned into a doctoral research project at UW–Madison (again under Reeder's direction) focused on map turtle systematics and ecology (R. Vogt, personal communication). This work allowed Vogt to make further observations and collect more data that would be relevant to his book.

Shortly after completing data collection and analyses for his dissertation (Vogt 1978), Vogt took a postdoctoral position at the Carnegie Museum of Natural History in Pittsburgh, Pennsylvania, where he continued to work on map turtles under C. J. "Jack" McCoy until 1980 (R. Vogt, personal communication), but he also undertook a substantial effort to complete his book on Wisconsin herpetofauna. Ultimately, his book was published in 1981 by the MPM Press after he had completed his postdoctoral work and already moved to Mexico to start his new position as the curator of the Herpetology Collection at the Estación de Biología, Tropical Los Tuxtlas, Instituto de Biología, Universidad Nacional Autónoma de México, San Andrés Tuxtla in Veracruz, which he held from 1981 to 2000. He later took the position of research professor and curator of reptiles and amphibians at the Instituto Nacional de Pesquisas da Amazônia, Manaus, Amazonas, Brazil, which he held until his death in January 2021. In addition to his many works focused on Wisconsin species, Vogt published a large number of manuscripts on the herpetofauna of Mexico, Central America, and South America, including an authoritative book on the turtles of Mexico (Legler and Vogt 2013).

Robert Wayne Hay

Robert Wayne Hay was the cold-blooded species manager with what is now the Bureau of Natural Heritage Conservation of the WDNR from 1990 to 2008 (Figure 10; Robert Hay, Turtles for Tomorrow, personal communication). Hay was born in Madison, Wisconsin (1950), and received a bachelor's degree in integrated natural resource management from UW–Stevens Point in 1973. After finishing college, Hay undertook successful careers in restaurant management, production control, and construction but returned to the wildlife profession as a WDNR employee nearly 20 years after completing his degree.

Hay started at the WDNR when wildlife managers were still often relicts of the 'hook and bullet' era. Given that most Wisconsin amphibians and reptiles are nongame species, this led to challenges in garnering internal support for his early conservation efforts. One of Hay's most substantial impacts on herpetology in Wisconsin is likely the regulations he created while at the WDNR. These include regulations related to 'take' (i.e., mortality and collection) of native herpetofaunal species that took effect in 2000 (R. Hay, personal communication). Prior to Hay, no limits existed on the removal of most native species from the wild. Although the WDNR Fisheries Management division had developed some regulations for turtle harvest prior to Hay, he revised these to better protect turtles in 1997 (R. Hay, personal communication). It was also during Hay's time that a number of species were pushed forward to receive greater legal protection in Wisconsin, such as the Timber Rattlesnake.

Figure 10. Robert Wayne Hay (Turtles for Tomorrow; *left*), Armund Bartz (WDNR; *center*), and Craig Berg (Milwaukee County Zoo, retired; *right*) process Timber Rattlesnakes for radio telemetry research in La Crosse County (2000; photo by R. Sajdak).

Hay undertook a variety of important projects that helped develop better conservation strategies for Wisconsin's amphibians and reptiles, including early work on turtle headstarting and experimentation with translocation as a management tool. These are notable achievements, considering that active research was not always a priority of wildlife managers and regulators in Wisconsin during the early 1990s. Hay further made great efforts to work in conjunction with other professionals both within and outside of the WDNR. He educated and trained countless WDNR, USFWS, and Natural Resources Conservation Service employees, as well as members of private conservation organizations and citizens, on amphibian and reptile conservation in Wisconsin. He is a coauthor on several peer-reviewed publications and a very popular WDNR booklet series on Wisconsin's amphibians and reptiles (i.e., Christoffel et al. 2002, 2008, 2009). Of equal significance is that Hay worked closely and easily with members of the public to successfully enact numerous conservation and management strategies (Joshua Kapfer, UW–Whitewater, personal observation). After retirement, Hay has continued to conduct active conservation work through his nonprofit organization Turtles for Tomorrow (R. Hay, personal communication).

Gary Casper

Gary Casper was the senior collections manager of vertebrate zoology at the MPM from 1984 to 2005. Casper, originally from Minnesota, completed his Ph.D. at UW–Milwaukee in 2003 under the direction of Peter Dunn (Casper 1996, 2003). Not long after starting at MPM, Casper initiated a project through the museum that attempted to better delineate the distributions of Wisconsin's amphibians and reptiles. This effort, named the Wisconsin Herpetological Atlas Project, was partially supported by both the WDNR and the MPM (Casper 1996) and built upon the existing work of Vogt (1981) to document species distributions in the state. It also used observations, specimens, and photographs submitted by Wisconsin's citizens and other professionals, as well as those generated through Casper's own field surveys. Casper published numerous county records in the 1990s that resulted from atlas reports, which created a clearer picture of the distribution of amphibians and reptiles in Wisconsin. He was also responsible for instituting the photographic voucher system at MPM over 30 years ago, a method of cataloging species observations that was rare among museums at the time but is currently common practice (Max Nickerson, University of Florida, Florida Museum of Natural History, personal communication). In addition to his work with the herpetological atlas, Casper was a coinvestigator on projects focused on the conservation and biogeography of several species in Wisconsin, with particular focus on herpetological island biogeography (Hecnar et al. 2002) and biogeography of gartersnakes (*Thamnophis* spp.) in the Upper Midwest (Placyk et al. 2007).

OTHER PAST AND RECENT WISCONSIN HERPETOLOGISTS OF IMPORTANCE

Many other past professionals have conducted work or directed graduate student research on native amphibians and reptiles in Wisconsin (Natural History Box: Wisconsin's Lost Herpetologist?). There is not enough space to list them all here, but their activities should not be slighted. Early curators at the UWZM and the Wisconsin Geological and Natural History Survey organized statewide herpetofaunal surveys and collections from 1917 to 1920 (Conger 1921, UWZM 2015). Hundreds of specimens were taken as the result of this effort, and several key contributors included A. J. Poole, A. I. Ortenburger, H. H. Sheldon, F. G. Hall, and H. H. T. Jackson (who eventually published an early book on Wisconsin's mammals; Jackson 1961). These collections subsequently stimulated the efforts of two UW–Madison graduate students, who worked under the direction of George Wagner (Vogt 1981), to outline the distribution of all amphibian and reptile species in the state (Conger 1921, Rose 1922). Although the distribution maps were largely incomplete and neither

NATURAL HISTORY BOX: Wisconsin's Lost Herpetologist?

A name rarely mentioned in conjunction with herpetology in Wisconsin is Franklin J. W. Schmidt. Schmidt was one of Aldo Leopold's first graduate students at UW–Madison, and he is best remembered today for his work on Greater Prairie Chickens (*Tympanuchus cupido*) in central Wisconsin (Leopold 1936, Meine 2010). Yet Schmidt was a broadly knowledgeable naturalist and excellent field biologist who could apparently disappear into the wild on his own to collect data for weeks at a time, only to emerge later unscathed (Meine 2010). Unfortunately, Schmidt died in his thirties while attempting to save his mother from a fire that started at her home in Stanley, Wisconsin. Also unfortunate was that Schmidt's death occurred before he finished numerous manuscripts, and the fire claimed much of his unpublished data and notes (Leopold 1936).

Often forgotten about Schmidt is that prior to his death, he authored or coauthored two manuscripts on Wisconsin's herpetofauna (Schmidt 1926, Bishop and Schmidt 1931). The latter publication, in particular, was important for its time as an early attempt to taxonomically organize painted turtles (*Chrysemys* spp.) in the midwestern and eastern United States as a single species with various subspecies, as opposed to a multispecies genus, as it was frequently considered up to that point (Bishop and Schmidt 1931). Luckily, not all of Schmidt's notes were lost, and those that remained were eventually archived at the UW–Madison library. These salvaged documents include notes and data Schmidt had compiled on numerous herpetological species during his fieldwork (Leopold 1887–1948), many of which related to specimens he deposited at the Field Museum of Natural History (Cook County, Illinois). Schmidt's field notes were detailed; for example, he recorded body dimensions, scale and femoral pore counts, and stomach contents of numerous lizards he encountered in central Wisconsin (Leopold 1887–1948). His archived notes also include his handwritten thoughts on the recently published manuscript by Pope and Dickinson (1928).

Although the archived notes only cover his review of the Anuran section in Pope and Dickinson (1928), these notes indicate that Schmidt questioned numerous aspects of what they reported. For example, he disagreed with their suggestion that the distribution of Blanchard's Cricket Frog (*Acris blanchardi*) was statewide. He wrote, "It is found along the Mississippi River as far north as Dunn County [a county where the species is currently not otherwise documented], but I found no traces of this frog in Eau Claire, Chippewa, Burnette, or Clark County. In eastern Wisconsin, I found *Acris* in Waupaca County, but it was restricted to a few shallow kettle holes grown up w/ bull-rushes and not in the rivers and lakes as it is at Madison. The Field Museum has specimens from Waupaca County, which I collected during the summer of 1929." George Wagner (past zoology professor at UW–Madison) had noticed Schmidt's herpetological prowess. For example, in a 1950 letter to a "Mr. Nero," he noted that Schmidt was "rapidly developing into an excellent herpetologist" (copies of correspondence provided by Laura Monahan, collections manager, UWZM). Aldo Leopold was particularly impressed with Schmidt (Meine 2010) and wrote in his obituary that it would be "no small task for those who survive him to live even partially up to his mark" (Leopold 1936). One cannot help but wonder what contributions Franklin Schmidt would have made to herpetology in Wisconsin had he lived longer. It is also noteworthy that Schmidt was survived by his brother, Karl P. Schmidt, who went on to be a prominent herpetologist and curator at the Field Museum of Natural History.

Joshua M. Kapfer, UW–Whitewater

NATURAL HISTORY BOX: Prominent MPM Curators with Substantial Contributions in Herpetology outside of Wisconsin

Max A. Nickerson was curator of vertebrate zoology at the MPM from 1971 to 1990 and is still an active and prominent herpetologist. He received his Ph.D. in zoology from Arizona State University in 1968. Nickerson is currently curator of herpetology at the Florida Museum of Natural History (among many other titles) and has authored and coauthored countless publications in the field of herpetology. Although few of his hundreds of publications focus on Wisconsin species, his impact on herpetology in Wisconsin is important nonetheless. By the time Nickerson arrived at MPM, the specimen collections had received little attention for several years. Therefore, he quickly began the substantial task of increasing the collection's functionality and relevance. This included correcting dated taxonomy, addressing numerous specimens that had been misidentified, creating better access to the collection, and working toward expanding the collection's holdings, efforts that Nickerson worked at for the duration of his tenure at MPM. His labors and achievements while at the MPM were, in part, responsible for the size and stature of the museum's vertebrate collection. Even now (decades after his parting), the specimens at MPM likely remain the most important collection of Wisconsin amphibians and reptiles that exists and is a collection he is largely responsible for facilitating. Nickerson's abetment included the provision of important support for Richard Vogt to conduct surveys that ultimately led to many specimens in MPM's holdings, while also generating data for Vogt's book on Wisconsin herpetofauna (Vogt 1981). Nickerson further secured funds from various sources that helped Vogt publish his book through MPM. He was the founder and first president of the (now defunct) Wisconsin Herpetological Society in 1972 (Max Nickerson, University of Florida, Florida Museum of Natural History, personal communication).

Another prominent past MPM curator with an impressive body of herpetological work (but little focus on Wisconsin) is Robert W. Henderson. Henderson is currently an emeritus curator of vertebrate zoology at the MPM and has worked with the herpetology collections at the museum since 1973. Henderson attended the University of Kansas, completing his master's degree under the great snake ecologist Henry S. Fitch in 1974, and received an honorary doctoral degree from UW–Milwaukee in 2020. Most of Henderson's career has focused on West Indian herpetofauna (e.g., Henderson and Powell 2009), and he is the world expert on Neotropical Treeboas (*Corallus*; e.g., Henderson 2015). At the time of this writing, Henderson had by far more peer-reviewed publications to his name than any active herpetologist living in Wisconsin, including ca. 260 scientific articles, 15 books, and 14 book chapters (Robert Henderson, MPM, personal communication).

Joshua M. Kapfer, UW–Whitewater

thesis was ultimately published in a peer-reviewed outlet, these manuscripts helped pave the way for the work that Pope and Dickinson (1928) published shortly thereafter.

Additional noteworthy professionals include Warren Porter and Scott Craven (UW–Madison), John Minnich and James Coggins (UW–Milwaukee), Jeffery Briggs (UW–Whitewater), and Ruth L. Hine (WDNR; in 1949 she was the first woman to receive a Ph.D. in zoology from UW–Madison and was also responsible for what is possibly the most thorough investigation of Northern Leopard Frog biology in the Upper Midwest to date; Hine et al. 1981). Philip Cochran, who earned a Ph.D. from UW–Madison (1984) and was on the faculty at St. Norbert College in De Pere, Wisconsin, from 1984 to 2000, published a substantial number of papers on Wisconsin's herpetofauna and added many county records to the state (Watermolen 2015a; see the Appendix). Tom Johnson, retired state herpetologist for the Missouri Department of Conservation, collected a great number of specimens while an undergraduate at UW–Stevens Point; they are currently housed in the university's large herpetological collection. Richard Sajdak and Craig Berg (both past cold-blooded curators at the Milwaukee County Zoo) were involved in important research with Robert Hay on Timber Rattlesnakes in Wisconsin during the early 1990s (see Figure 10). Sajdak further conducted an extensive master's research project on Mudpuppies (*Necturus maculosus*) at UW–Milwaukee under the direction of James Coggins (Sajdak 1982), was responsible for discovering a new reptile species in the state (see Western Wormsnake, *Carphophis vermis* species account), and has numerous publications to his name, including a book on arboreal snake natural history (Sajdak 2010).

The contributors to this book (several of which are discussed in this chapter) all have some level of expertise on the species found in the state. Some of these individuals have conducted and published research on herpetofauna in Wisconsin or adjacent states, while others have worked toward the conservation and management of these species in the state. Although it is not possible to provide a detailed review of their professional contributions, they are no less important. Herpetological research is currently conducted at many academic (e.g., UW–Green Bay, UW–Madison, UW–Milwaukee, UW–Platteville, UW–Stevens Point, UW–Whitewater), state (e.g., WDNR), and federal (e.g., the US Geological Survey [USGS] and the US Fish and Wildlife Service [USFS]) institutions. The ongoing work at these and other institutions in the state is important and with proper support will continue to contribute to our understanding of amphibians and reptiles in Wisconsin.

In addition to the individuals discussed in this chapter and the contributors to this book, there are a number of still-active professionals in the state who have worked on amphibians and reptiles in Wisconsin or made contributions to the more general field of herpetology (Natural History Box: Prominent MPM Curators with Substantial Contributions in Herpetology Outside of Wisconsin). These include but are not limited to William Karasov (UW–Madison), Walter Sadinski (USGS, La Crosse, Wisconsin), Gerlinde Höbel (UW–Milwaukee), Peter Zani (UW–Stevens Point), Greg Mayer (UW–Parkside), Ellen Censky (Milwaukee Public Museum), and Robert Howe (UW–Green Bay).

Past, Present, and Potential Future Landscapes of Wisconsin, with Emphasis on Amphibian and Reptile Distributions

Donald J. Brown and Christine A. Ribic

The geological history of Wisconsin, particularly glacial and permafrost cycles (Clayton et al. 2001, Syverson and Colgan 2004), shaped the current topography of the state. Bedrock and soil types, rivers and natural lakes, hills and valleys are all products of past environments (e.g., Rawling et al. 2008, Lytwyn 2010, Jacobs et al. 2011, Schaetzl and Attig 2013). These geological legacies, along with climatic gradients, strongly influence the modern plant and animal diversity in Wisconsin (Albert 1995, Schulte et al. 2002, Cutright et al. 2006). Furthermore, plant communities, which are a major component of amphibian and reptile habitats, are constantly changing through natural processes. Plant communities change in the short term due to natural disturbances, weather variability, and biotic interactions (e.g., competition, predation, and facilitation). In the long term, plant communities respond to direct and indirect impacts of climate change, resulting in distributional shifts (Stohlgren et al. 2005, Yurkonis and Meiners 2006, Brooker et al. 2008, Iverson et al. 2008, Crookston et al. 2010). Humans have also strongly affected Wisconsin's vegetative communities over the past two centuries as a result of land-use change (e.g., conversion of savannas and prairies to farmland) and management actions (e.g., wildfire suppression; Radeloff et al. 1999, Grossmann and Mladenoff 2008, Rhemtulla et al. 2009). These anthropogenic landscape changes have affected local abundance and occupancy of amphibians and reptiles, in some cases resulting in steep declines (e.g., Blanchard's Cricket Frog [*Acris blanchardi*]; Hay 1998, Gray and Brown 2005).

Like all organisms, amphibians and reptiles have abiotic and biotic requirements, tolerance levels, and preferences. These factors, coupled with biological interactions and anthropogenic effects, bound the ecological niche of each species. The niche of a species can be defined as "the environmental conditions that allow a species to satisfy its minimum requirements so that birth rate of a local population is equal to or greater than its death rate along with the set of per capita impacts of that species on these environmental conditions" (Chase and Leibold 2003). Short-term, small-scale habitat changes can temporarily impact local abundance and occupancy, sometimes resulting in local extirpation, while long-term climate and landscape-scale habitat changes can impact the geographic distribution of a species (Thomas 2010, Lawing and Polly 2011). Furthermore, because each species occupies a unique ecological niche, it is not unusual for habitat change to positively impact some amphibian and reptile species while negatively impacting others (Russell et al. 1999, Adams et al. 2003, Brown et al. 2012b). Understanding the relationships between habitat condition and occurrence/abundance of amphibian and reptile species helps us to effectively manage current populations, as well as understand past and project future geographic distributions.

In this chapter we discuss past, present, and potential future landscapes of Wisconsin. We broadly discuss amphibian and reptile assemblages within these landscapes, but with the exception of potential climate change impacts, we intentionally leave the detailed species-level information to individual species accounts. Our goal is to provide a framework to understand past, present, and potential future amphibian and reptile distributions in Wisconsin.

PAST LANDSCAPES

Although many glacial-interglacial cycles have occurred in the Midwest over the last 2.5 million years, our knowledge of amphibian and reptile distributions is generally limited to the period following the Wisconsin glaciation, which occurred between ca. 75,000 and 9,500 ybp (Syverson and Colgan 2004). During this period, environmental conditions throughout most of the state were likely unsuitable for amphibians and reptiles (Holman 1992a). However, a nonglaciated refugium for more cold-tolerant species existed in southwestern Wisconsin. This refugium covered ca. 25% of the land area in the state and is known today as the Driftless Area (see Figure 12; Syverson and Colgan 2004). Species such as the Common Gartersnake (*Thamnophis sirtalis*), Prairie Skink (*Plestiodon septentrionalis*), and American Toad (*Anaxyrus americanus*) might have survived the last glacial period in this refugium (Holman 1992a, Fuerst and Austin 2004, Placyk et al. 2007), and evidence suggests that the Wood Frog (*Lithobates sylvaticus*) almost certainly used the area as a glacial refuge (Lee-Yaw et al. 2008).

Following the glacial retreat, boreal forest, dominated by spruce (*Picea* spp.), covered most of Wisconsin (Webb 1987, Panyushkina and Leavitt 2013). By 7,000 to 5,000 ybp, mesic hardwood forests were prominent in eastern and southern portions of the state (Webb 1987, Long et al. 2011). Southwestern Wisconsin transitioned to primarily oak (*Quercus* spp.) savanna and prairie ca. 5,500 ybp and then to more closed-canopy oak and birch (*Betula* spp.) forests beginning ca. 3,500 ybp (Baker et al. 1992, Wright 1992, Williams et al. 2009a). Northern Wisconsin at this time shifted to a pine-dominated (*Pinus* spp.) forest due to climate warming (Yu 2000). Birch (*Betula* spp.) and Eastern Hemlock (*Tsuga canadensis*) also became well established in northern Wisconsin ca. 3,500 ybp (Webb 1974). Today, northern Wisconsin is a transition zone between southern boreal forest to the north and northern hardwood forest to the south (Pastor and Mladenoff 1992).

The majority of amphibian and reptile species currently in Wisconsin likely colonized the state within the last 10,000 years, and fossil evidence indicates that the modern amphibian and reptile community in the central Great Lakes region was mostly present by 6,000–4,000 ybp (Holman 1992a). West and Dallman (1980) reported fossil remains of amphibians and reptiles in the Driftless Area that ranged from 17,000 to 2,400 ybp. Cleland (1966) analyzed fossils from a bog in Sauk County (southwestern Wisconsin in the Driftless Area refugium) and found amphibians from the lowest level (ca. 9,500 ybp) and reptiles from 4,500 to 3,000 ybp. Theler (2000) studied fossil remains from Native American archaeological sites in western Wisconsin and identified 19 amphibian and reptile species, including 4 species of turtles dated between 11,000 and 2,000 ybp and an additional 8 amphibian and reptile species dated between 2,000 and 800 ybp. Holman (2001) identified fossils from Iowa County that date from ca. 17,000 ybp, which included the American Toad, Northern Leopard Frog (*Lithobates pipiens*), Common Gartersnake, Eastern Foxsnake (*Pantherophis vulpinus*), and Eastern Milksnake (*Lampropeltis triangulum*).

Most species colonized Wisconsin from the south (Smith 1957, Holman 1992a, Church et al. 2003, Starkey et al. 2003, Moriarty Lemmon et al. 2007a), but recent genetic studies have shown that there are exceptions to this generalization. For example, Wood Turtles (*Glyptemys insculpta*) likely colonized from the east, moving across the Upper Peninsula of Michigan (Amato et al. 2008). Similarly, it appears that Spring Peepers (*Pseudacris crucifer*) dispersed into the Upper Midwest from eastern North America. The species moved north of the Great Lakes and then south into Minnesota and northwestern Wisconsin; meanwhile, a separate clade colonized from the south (Austin et al. 2002). These examples illustrate the often complex biogeographic histories of Wisconsin's amphibian and reptile species.

RECENT HISTORICAL AND PRESENT LANDSCAPE

The modern distribution of amphibian and reptile species in Wisconsin is influenced by both historical and current habitat conditions. In this section, we first summarize major recent changes to the Wisconsin landscape by comparing pre-European settlement land cover classes (mid-1800s) to recent land cover classes (2006). We then discuss the current status of amphibians and reptiles in the state, focusing primarily on species that have noticeably declined since European settlement. We follow this discussion with a summary of the present landscape of Wisconsin based on natural ecological units, including information on amphibian and reptile species richness, and briefly discuss assemblage differences among the ecological units. We conclude this section by projecting statewide patterns in amphibian and reptile species richness based on county-level data, which provide finer-scale information than ecological units.

Recent Changes to the Wisconsin Landscape

Several studies have attempted to create detailed reconstructions of the Wisconsin landscape prior to major European settlement in the mid- to late 1800s using information from Public Land Surveys conducted by the US General Land Office between 1832 and 1866 (e.g., Schulte et al. 2002, Bolliger and Mladenoff 2005, Liu et al. 2011). Here, we compare the pre-European settlement and current landscape based on six general land use categories: developed land, forest, grassland/shrubby, open water, wetland, and working land (i.e., farmland and pastures). We used Finley (1976) as a basis for categorizing and quantifying the historic landscape, which was based on the General Land Office records. We used the 2006 National Land Cover Dataset (NLCD; Fry et al. 2011) for categorizing and quantifying the current landscape. Based on these sources, there was a large reduction in forest cover over the last ca. 150 years (Figure 11; Table 4), from ca. 74% in the mid-1800s to ca. 36% today. This change was primarily due to conversion of forest to farmland in southern and central Wisconsin in the 1800s (Maizel et al. 1998), much of which was oak savanna (Curtis 1959). Grassland/shrubby habitat was reduced from ca. 8% to ca. 3%. The amount of wetland habitat in the state has remained roughly the same (14%–15%), but note that Dahl (1990) estimated that 27.3% of Wisconsin was wetland in the 1780s. The two categories absent prior to European settlement, working land and developed land, currently account for ca. 37% and ca. 7% of the landscape, respectively.

The structure and composition of Wisconsin's forests have also changed since European settlement, primarily due to changes in land use (e.g., logging and farming) and land management (e.g., fire suppression; Rhemtulla et al. 2007). After the mid-1800s, hemlock and White Pine (*Pinus strobus*) cover in Wisconsin decreased from ca. 22% to 1%, while the average diameter of all tree species in the state also declined (Rhemtulla et al. 2009). In addition, much of the oak savanna in southern Wisconsin that was not embedded in working lands has transitioned to mesic hardwood forest due to succession (Rhemtulla et al. 2009). White and Mladenoff (1994) studied forest change in the Border Lakes region of northern Wisconsin (Vilas County) since the 1860s and concluded that forest cover has changed from primarily old-growth hemlock/hardwoods to second-growth hardwoods/conifers. The northwestern pine barrens ecoregion has shifted from a historically open

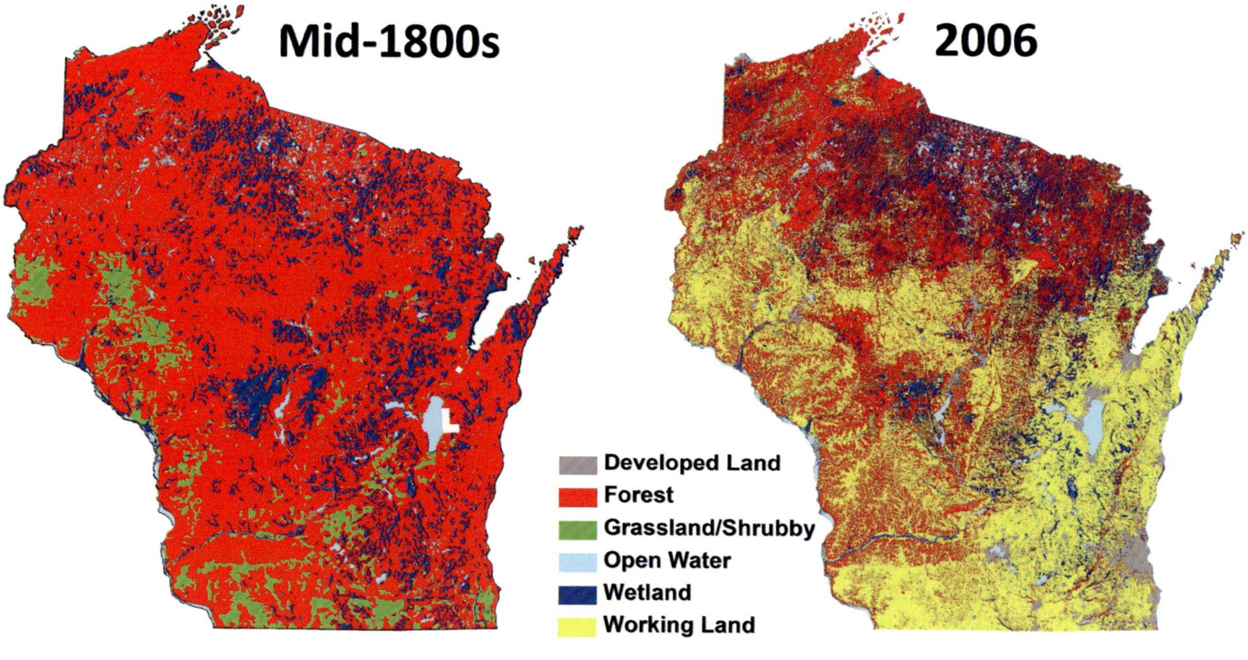

Figure 11. Comparison of recent historical and current land cover classes in Wisconsin. The historical land cover map was derived from Finley (1976), and the current land cover map was derived from the 2006 National Land Cover Dataset (NLCD; Fry et al. 2011). The comparison indicates a dramatic reduction in forest cover over the last ca. 150 years, primarily replaced by working land (farmland and pastures). Much of the forested land lost in southern and southwestern Wisconsin was oak savanna (Curtis 1959).

Table 4. Current land cover in Wisconsin based on the 2006 National Land Cover Dataset (NLCD) and pre-European settlement land cover based on Finley (1976). Included are the percentages of Wisconsin for each of the 15 NLCD classes, followed by a summary of the current and recent historic landscape using broader land cover groups. Numbers next to each NLCD class denote their broader land cover group.

Land Class	**Current Land Cover (%)**	**Historic Land Cover (%)**
Cultivated crops (1)	26.2	
Hay/pasture (1)	10.7	
Deciduous forest (2)	30.0	
Evergreen forest (2)	2.3	
Mixed forest (2)	3.6	
Emergent herbaceous wetlands (3)	2.9	
Woody wetlands (3)	10.8	
Open water (4)	3.8	
Barren land (5)	0.1	
Shrub/scrub (5)	1.3	
Herbaceous (5)	1.3	
Developed, open space (6)	4.1	
Developed, low intensity (6)	2.2	
Developed, medium intensity (6)	0.6	
Developed, high intensity (6)	0.2	
Working land (1)	36.8	0.0
Forest (2)	35.8	73.9
Wetland (3)	13.7	14.6
Open water (4)	3.8	3.3
Grassland/shrubby (5)	2.7	7.9
Developed land (6)	7.2	0.0

pine-dominated landscape to a more closed-canopy pine-oak landscape since the mid-1800s due to anthropogenic fire suppression (Radeloff et al. 1998, Radeloff et al. 1999).

Amphibians and Reptiles in the Current Wisconsin Landscape

Little information exists on the distributions of amphibians and reptiles in nineteenth-century Wisconsin. This was a key period during which the landscape was transformed by European settlers (Rhemtulla et al. 2007), and the absence of firsthand reports makes it difficult to quantify the immediate impacts of settlement on amphibians and reptiles in the state. Casper (2008c) reviewed survey records and land use changes and concluded that species richness and abundance of amphibians and reptiles have declined substantially in Milwaukee County since it was settled. Furthermore, several recent studies concluded that occupancy and relative abundance of anurans in Wisconsin were negatively associated with urbanization (Knutson et al. 1999, Woodford and Meyer 2003, Price et al. 2004). Hecnar (2004) hypothesized that 50%–90% of amphibian populations have been lost across the

Great Lakes basin since settlement, primarily due to wetland loss. While land conversion to agriculture has had a negative impact on many species, some anurans appear to be thriving in the agriculture-dominated areas of Wisconsin, potentially due to the presence of residual forest patches (Knutson et al. 1999, Paloski 2008).

Broad-scale population declines have been noted for several amphibian and reptile species in Wisconsin over the last half century. Probably the most notable of these was the decline of Blanchard's Cricket Frogs between the 1950s and 1970s (Hay 1998). Blanchard's Cricket Frogs were historically abundant throughout southern Wisconsin (Vogt 1981) but have been extirpated from a large portion of their historical distribution within the state (see Blanchard's Cricket Frog, *Acris blanchardi* species account), with concurrent declines observed throughout the Upper Midwest (Gray 2000a, 2000b, Gray and Brown 2005). Declines of several other anurans were also observed during this time period, including Northern Leopard Frogs, American Bullfrogs (*Lithobates catesbeianus*), and Pickerel Frogs (*Lithobates palustris*; Hine et al. 1975, 1981). These declines were the major impetus driving the initiation of the Wisconsin Frog and Toad Survey program in 1981 (Mossman et al. 1998).

The proposed explanations for these anuran declines vary. Northern Leopard Frog declines have been linked to redleg syndrome (a bacterial disease; Densmore and Green 2007) and atrazine (a widely used herbicide), which impacts development and survivorship (Hine et al. 1981, Casper 1998). Northern Leopard Frogs appear to be particularly susceptible to the effects of environmental contamination and degradation, which negatively impact their development and survival (see Northern Leopard Frog, *Lithobates pipiens* species account). The factor(s) that caused Blanchard's Cricket Frog declines are not fully understood (Lehtinen and Skinner 2006). Leading hypotheses include habitat degradation and the negative impacts of prolonged droughts on reproduction and winter survival (Jung 1993, Lannoo 1998, Russell et al. 2002, Gray and Brown 2005, Irwin 2005).

Notable declines have also occurred for several reptile species in the state. Venomous snakes (i.e., Timber Rattlesnake [*Crotalus horridus*] and Massasauga [*Sistrurus catenatus*]) were once common throughout the southern and western portions of the state (Pope 1923, Pope and Dickinson 1928, Schorger 1967–68) but have experienced steep declines over the last two centuries. For example, the Massasauga was historically considered abundant in southern Wisconsin (Pope 1926, Schorger 1967–68) but has declined to the point that reintroduction is now a practiced management tool (King et al. 2004). The Timber Rattlesnake was also historically reported in large numbers but has substantially declined across much of its known range over the last century (Oldfield and Keyler 1989). The reasons for these declines include habitat loss (particularly associated with suitable overwintering locations) and extensive removal by humans (Vogt 1981, Oldfield and Keyler 1989). A Timber Rattlesnake bounty existed in Wisconsin until 1975 and was nearly reinstated in 1983 (Casper 1983), while a Massasauga bounty existed in Wisconsin between 1952 and 1972 (Vogt 1981). These factors, coupled with low recruitment rates for these species, have hindered the ability of populations to recover in Wisconsin, despite legal protection.

The conversion of oak savannas and prairies to agricultural lands in southern and western Wisconsin has affected several species (Brewster 1984). Two species in particular, the Ornate Box Turtle (*Terrapene ornata*) and the Slender Glass Lizard (*Ophisaurus attenuatus*), have declined to the point of being considered Endangered in the state (WDNR 2015c). Adler (1968) reviewed archaeological evidence to conclude that the current distribution of Ornate Box Turtles in Wisconsin (and Illinois) is a small fraction of its former range. Doroff and Keith (1990) found that individual Ornate Box Turtles avoided cropland and were highly selective for remnant prairie with deep sandy soil, while Moore (1993) speculated that these turtles will likely be extirpated from Wisconsin without proactive conservation measures. Little research has been conducted on the Slender Glass Lizard in the region (Pleyte 1975), probably due to its long-standing rarity and secretive nature (Anton 1999). Yet it appears this lizard declined rapidly following European settlement and associated land use changes (Graenicher 1911, Wagner 1922). Similar to the Ornate Box Turtle, the Slender Glass Lizard requires open prairie and savanna with sandy soils (Breitenbach 2011), a habitat type that is currently rare in Wisconsin (Figure 11; Table 4).

While it is clear that many amphibian and reptile species have been affected by the landscape changes following European settlement, this problem is not unique to Wisconsin. Amphibians and reptiles are declining worldwide (Gibbons et al. 2000, Reading et al. 2010, Wake 2012a). On a positive note, there have been no verified species extirpations in the state since European settlement (see Hypothetical and Enigmatic Species of Wisconsin chapter for possible exceptions). However, populations can sometimes persist for decades or even centuries under suboptimal conditions, slowly trending toward extinction (Blaustein et al. 1994, Daszak et al. 2005, Adams et al. 2013). Continuous monitoring programs such as the Wisconsin Frog and Toad Survey (Mossman et al. 1998) are important because they provide the long-term data necessary to detect long-term population declines that would not be apparent from short-term monitoring data. The added effects of climate change will make monitoring efforts even more important in the future.

Current Natural Ecological Units

To describe the current natural ecological units of Wisconsin, we used the US Forest Service's Ecological Subregions classification system (Cleland et al. 2007). While there are other classification systems developed specifically for Wisconsin (e.g., Curtis 1959, Albert 1995), we used this classification system because it is widely employed and encompasses the entire conterminous United States. The Ecological Subregions classification system is hierarchical, with province boundaries primarily defined based on climate gradients, nested section boundaries defined based on physical geography and geology, and further nested subsection boundaries defined based on potential vegetation, soil, and hydrography (Cleland et al. 2007). The two provinces in Wisconsin are split along what is known as the ecological 'tension zone' (Anderson 2005). This zone transects several states in the Great Lakes region and occurs where two primary climatic conditions converge in Wisconsin: the cooler, moister conditions from the north, and the warmer, drier conditions from the south. The tension zone also contains an intermingling of the vegetative communities and disturbance histories associated with both northern and southern environments (Curtis 1959).

Below we provide brief descriptions of the climate, geology, and vegetation that are characteristic of the current ecological provinces ($N = 2$) and sections ($N = 11$) of Wisconsin (Figure 12). Additional information on the pre-European settlement vegetation structure specific to each section can be found in Schulte et al. (2002). The province and section descriptions were taken directly from McNab et al. (2007), with page numbers included in parentheses following description quotes. For amphibian and reptile species richness data, we used county records through December 2018 (including vouchered records and additional nonvouchered records shown in species accounts).

For each ecological unit, amphibian and reptile species richness values were obtained by overlaying county boundaries on the ecological units and determining the total number

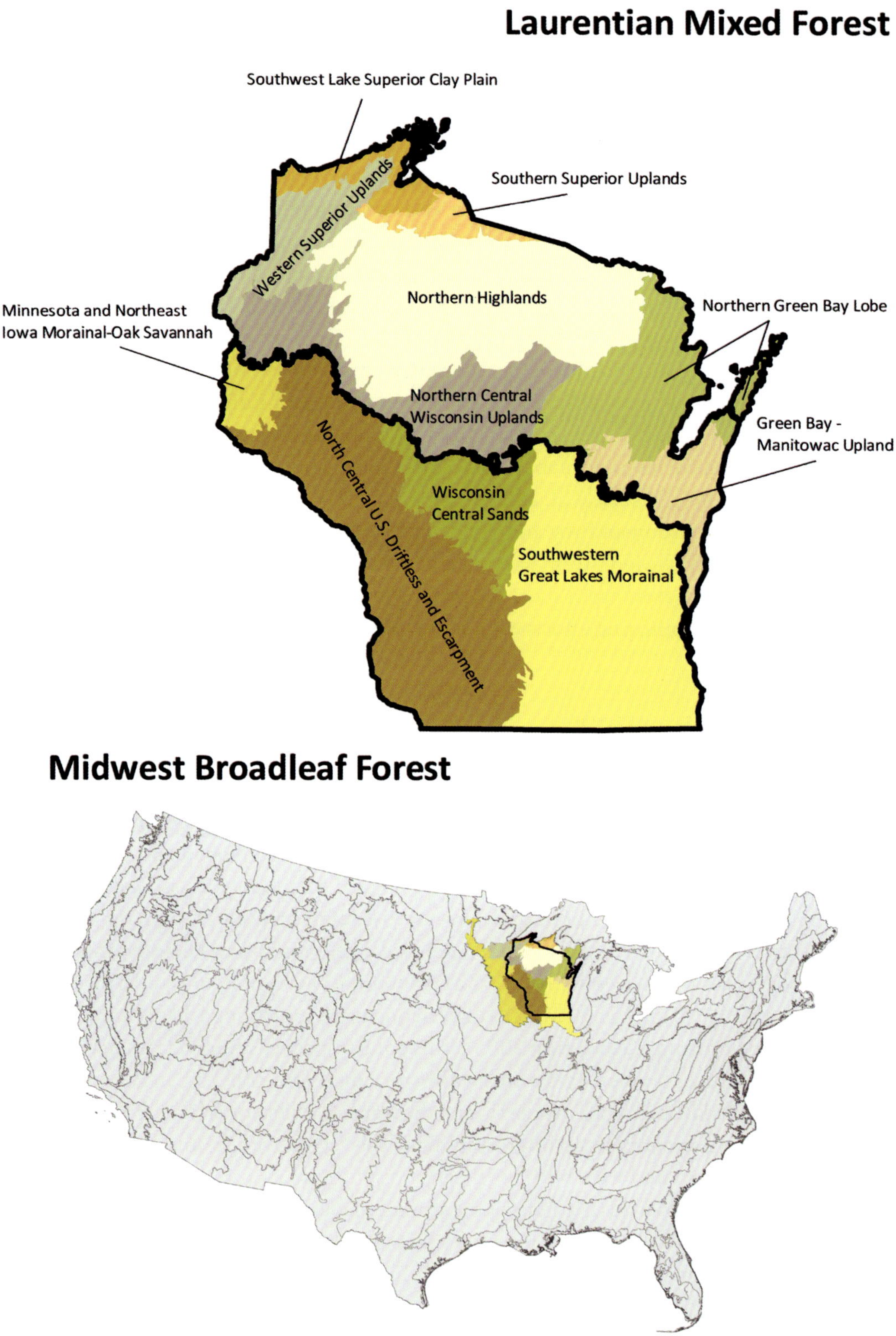

Figure 12. Natural ecological units of Wisconsin, including ecological provinces ($N = 2$) and sections ($N = 11$), based on the US Forest Service's Ecological Subregions classification system. The two provinces (i.e., Laurentian Mixed Forest and Midwest Broadleaf Forest) are separated along the 'tension zone' that crosses central Wisconsin, with a cooler/moister climate to the north and a warmer/drier climate to the south. The continental map depicts the ecological section boundaries in the US.

of documented native amphibians and reptiles in all counties at least partly included in a unit. Note that records for counties bisected by adjacent ecological units were included in species richness tallies for both units, and thus species richness estimates for some ecological units may be biased high. Because of the large variation in size of ecological units in Wisconsin (i.e., sections range from 250,747 to 3,003,647 ha [619,609 to 7,422,173 ac]), we calculated total species richness for amphibians and reptiles per unit area, standardized to a 100,000 ha area (i.e., number of species / 100,000 ha [247,104 ac]). We also calculated species richness by order: Anura (frogs and toads), Caudata (salamanders), Squamata (lizards and snakes), and Testudines (turtles). The nonstandardized species counts are provided in parentheses.

Laurentian Mixed Forest Province

Description: "The province is characterized by a continental climatic regime with maritime influence along the Great Lakes. Winters moderately long with continual ground snow cover; summers warm. Most precipitation occurs during summer. Low-relief, hilly landscapes are a product of past glaciation. Vegetation consists of forests that are a transition between boreal and broadleaf deciduous" (p. 5).

Area: 7,488,543 ha (18,504,593 ac)

Total amphibian species: 0.25 (19); Anura: 0.16 (12); Caudata: 0.09 (7)

Total reptile species: 0.38 (29); Squamata: 0.27 (20); Testudines: 0.12 (9)

Green Bay–Manitowoc Upland Section

Description: "The influence of Lake Michigan results in milder temperatures and more precipitation than sections farther inland. Terrain of this unit ranges from nearly level areas of lacustrine deposits to rolling hills of ground moraines. Bedrock is mostly shales and dolomite that has weathered to form a distinctive and extensive escarpment of cliffs. Soils include clays, loams, sands, and gravels, and many are calcareous in mineral content. Forest vegetation was largely aspen-birch and white-red-jack pine cover types" (pp. 7–8).

Area: 710,191 ha (1,754,920 ac)

Total amphibian species: 2.68 (19); Anura: 1.69 (12); Caudata: 0.99 (7)

Total reptile species: 3.10 (22); Squamata: 1.97 (14); Testudines: 1.13 (8)

North-Central Wisconsin Uplands

Description: "The topography of this section is largely a glacial till plain with portions of moraines. Bedrock formations are sandstones with some gneiss. Soils are variable, ranging from sandy to silt loams that are poorly to well drained. Historic vegetation was forests consisting of maple-beech-birch, aspen-birch, and oak-hickory cover types" (p. 6).

Area: 1,469,923 ha (3,632,259 ac)

Total amphibian species: 1.30 (19); Anura: 0.82 (12); Caudata: 0.48 (7)

Total reptile species: 1.63 (24); Squamata: 1.09 (16); Testudines: 0.54 (8)

Northern Green Bay Lobe

Description: "The landscape is variable, consisting mainly of ground moraines, with lesser areas of lacustrine plains, escarpment, sand dunes, and beach ridges; glacial outwash sand plains with rock outcrops are present along the western edge of this section. Rock formations of sandstones, limestones, and dolomite (bedrock outcrops of basalt or granites along western edge) are covered by diverse soils of lacustrine sands, loamy and poorly drained sands, with well-drained to excessively drained sands to the west. Historic vegetation was mostly forests of aspen-birch, maple-beech-birch, and spruce-fir cover types" (p. 7).

Area: 1,246,550 ha (3,080,292 ac)

Total amphibian species: 1.52 (19); Anura: 0.96 (12); Caudata: 0.56 (7)

Total reptile species: 1.76 (22); Squamata: 1.04 (13); Testudines: 0.72 (9)

Northern Highlands

Description: "This section is a dissected glacial pitted outwash plain, end and ground moraines, and smaller areas of hilly terrain; kettle lakes and depressions are common in the northern part. Bedrock is mostly sandstones and shales with small areas of gneiss and metavolcanics. Soils formed in till deposits and range from sandy loam and silts to poorly or well-drained sands. Vegetation on much of the section was a combination of aspen-birch, maple-beech-birch, white-red-jack pine, and spruce-fir cover types" (p. 7).

Area: 2,745,508 ha (6,784,298 ac)

Total amphibian species: 0.65 (18); Anura: 0.40 (11); Caudata: 0.25 (7)

Total reptile species: 0.84 (23); Squamata: 0.55 (15); Testudines: 0.29 (8)

Southern Superior Uplands Section

Description: "This section has glacial landscapes of level to gently rolling lowlands and flat lacustrine plains with equal areas of hillier uplands and escarpments. Bedrock, which is complex, includes felsic and mafic igneous formations; metamorphics of slate, schist, marble, and amphibolites; and some areas of sandstones. Rock outcrops are common in upland areas. Vegetation is forests of maple-beech-birch and aspen-birch cover types" (p. 5).

Area: 250,747 ha (619,609 ac)

Total amphibian species: 5.98 (15); Anura: 3.59 (9); Caudata: 2.39 (6)

Total reptile species: 4.79 (12); Squamata: 3.19 (8); Testudines: 1.60 (4)

Southwest Lake Superior Clay Plain

Description: "Lake Superior affects the climate of this section resulting in greater precipitation and milder temperature regimes compared to areas farther inland. Landscape is variously dissected, nearly level lake plain of lacustrine clays. Bedrock is mostly sandstones and shale that is covered by water-reworked moraine that formed mostly well-drained soils; some basalts occur in the north part of this unit. Historic vegetation was a forest dominated by conifers [and] aspen-birch cover types" (p. 7).

Area: 366,555 ha (905,777 ac)

Total amphibian species: 4.64 (17); Anura: 3.00 (11); Caudata: 1.64 (6)

Total reptile species: 3.82 (14); Squamata: 2.46 (9); Testudines: 1.36 (5)

Western Superior Uplands

Description: "This section has relatively uniform, undulating, poorly drained, level to rolling landscape of glacial drift plains consisting of ground and end moraines, and local drumlins; large areas of deep sands occur along the eastern edge of the section. Loam soils formed in thick till covering most of bedrock, which consists of sandstones and shales. Forest vegetation consists of aspen-birch, maple-beech-birch, and spruce-fir cover types" (p. 6).

Area: 699,003 ha (1,727,274 ac)

Total amphibian species: 2.57 (18); Anura: 1.57 (11); Caudata: 1.00 (7)

Total reptile species: 2.72 (19); Squamata: 1.72 (12); Testudines: 1.00 (7)

Midwest Broadleaf Forest Province

Description: "The climate in this province is continental with warm to hot summers. Frequent growing season water deficits. Flat to hilly terrain with features associated with former glaciation. Vegetation consists of cold-deciduous, hardwood-dominated forests with a high proportion of species able to tolerate mild, brief, periodic drought during the late summer" (p. 10).

Area: 7,041,446 ha (17,399,792 ac)

Total amphibian species: 0.27 (19); Anura: 0.17 (12); Caudata: 0.10 (7)

Total reptile species: 0.52 (36); Squamata: 0.36 (25); Testudines: 0.16 (11)

Minnesota and Northeast Iowa Morainal and Oak Savanna

Description: "The landscape is level plains and low, rolling hills resulting from glaciation. Soils formed in glacial drift and thin, discontinuous layer of loess. Bedrock is mixture of granites, greenstones, and metasediments with equal area dolomite and sandstone. Existing cover type is primarily agricultural; historic vegetation was a mosaic of oak-hickory, prairie, maple-beech-birch, and elm-ash-cottonwood cover types" (p. 11).

Area: 282,129 ha (697,156 ac)

Total amphibian species: 6.38 (18); Anura: 3.90 (11); Caudata: 2.48 (7)

Total reptile species: 7.44 (21); Squamata: 4.96 (14); Testudines: 2.48 (7)

North-Central US Driftless and Escarpment

Description: "Parts of this section receive above average snowfall resulting from its location bordering two Great Lakes. Section is a maturely dissected unglaciated upland plateau with steep-sided bedrock ridges and mounds. Loess covers much of bedrock, which consists of sandstones, dolomite, and shales; soils are silt loams and sandy loams. Mounds and ridges are capped by resistant dolomite. Current cover types are mostly urban and agricultural; historical vegetation consisted of oak-hickory and prairie cover types; the elm-ash-cottonwood cover type was along rivers" (p. 11).

Area: 3,003,647 ha (7,422,173 ac)

Total amphibian species: 0.63 (19); Anura: 0.40 (12); Caudata: 0.23 (7)

Total reptile species: 1.20 (36); Squamata: 0.83 (25); Testudines: 0.37 (11)

Southwestern Great Lakes Morainal

Description: "This section has relatively uniform, undulating landscape of glacial drift plains consisting of ground and end moraines, and local drumlins; large areas of deep sands occur along eastern edge of the section. Loam soils formed in thick till covering most of bedrock, which consists of sandstones and shales. Current vegetation is agricultural and urban cover types; historic vegetation was oak-hickory, maple-beech-birch, and prairie cover types" (p. 11).

Area: 2,877,570 ha (7,110,630 ac)

Total amphibian species: 0.62 (18); Anura: 0.38 (11); Caudata: 0.24 (7)

Total reptile species: 1.22 (35); Squamata: 0.83 (24); Testudines: 0.38 (11)

Wisconsin Central Sands

Description: "This section is a flat to slightly rolling lake (Glacial Lake Wisconsin) plain with scattered, exposed sandstone buttes. Underlying geology is mostly erosion-resistant sandstone with localized areas of gneiss and granitic rocks. Soils formed in glacial outwash sands and are mostly excessively drained, but small areas of poorly drained swamps also occur. Historic vegetation was a mixture of aspen-birch, oak-hickory, and white-red-jack pine cover types" (pp. 11–12).

Area: 878,101 ha (2,169,835 ac)

Total amphibian species: 2.16 (19); Anura: 1.37 (12); Caudata: 0.80 (7)

Total reptile species: 3.76 (33); Squamata: 2.51 (22); Testudines: 1.25 (11)

Amphibian and reptile assemblages

All of the amphibian and reptile species assemblages present in the Laurentian Mixed Forest Province are also present in the Midwest Broadleaf Forest Province (Table 5). The Midwest Broadleaf Forest Province contains an additional seven reptiles not found in the Laurentian Mixed Forest Province, including five species in the order Squamata (Gray Ratsnake [*Pantherophis spiloides*], Lined Snake [*Tropidoclonion lineatum*], Slender Glass Lizard, Timber Rattlesnake, Western Wormsnake [*Carphophis vermis*]) and two species in the order Testudines (Ornate Box Turtle [*Terrapene ornata*], Smooth Softshell [*Apalone mutica*]).

Table 5. Amphibian and reptile species found in ecological provinces ($N=2$) and sections ($N=11$) of Wisconsin, based on the US Forest Service's Ecological Subregions classification system.

	Laurentian Mixed Forest							Midwest Broadleaf Forest			
	Green Bay—Manitowoc Upland	North Central Wisconsin Uplands	Northern Green Bay Lobe	Northern Highlands	Southern Superior Uplands	Southwest Lake Superior Clay Plain	Western Superior Uplands	Minnesota and Northeast Iowa Morainal-Oak Savanna	North Central U.S. Driftless and Escarpment	Southwestern Great Lakes Morainal	Wisconsin Central Sands
Amphibians											
Anura											
American Bullfrog, *Lithobates catesbeianus*	X	X	X	X	X	X	X	X	X	X	X
American Toad, *Anaxyrus americanus*	X	X	X	X	X	X	X	X	X	X	X
Blanchard's Cricket Frog, *Acris blanchardi*	X	X	X	—	—	—	—	—	X	X	X
Boreal Chorus Frog, *Pseudacris maculata*	X	X	X	X	X	X	X	X	X	X	X
Cope's Treefrog, *Hyla chrysoscelis*	X	X	X	X	—	X	X	X	X	X	X
Gray Treefrog, *Hyla versicolor*	X	X	X	X	X	X	X	X	X	X	X
Green Frog, *Lithobates clamitans*	X	X	X	X	X	X	X	X	X	X	X
Mink Frog, *Lithobates septentrionalis*	X	X	X	X	X	X	X	X	X	—	X
Northern Leopard Frog, *Lithobates pipiens*	X	X	X	X	X	X	X	X	X	X	X
Pickerel Frog, *Lithobates palustris*	X	X	X	X	—	X	X	X	X	X	X
Spring Peeper, *Pseudacris crucifer*	X	X	X	X	X	X	X	X	X	X	X
Wood Frog, *Lithobates sylvaticus*	X	X	X	X	X	X	X	X	X	X	X
Caudata											
Blue-spotted Salamander, *Ambystoma laterale*	X	X	X	X	X	X	X	X	X	X	X
Eastern Newt, *Notophthalmus viridescens*	X	X	X	X	X	X	X	X	X	X	X
Eastern Red-backed Salamander, *Plethodon cinereus*	X	X	X	X	X	X	X	X	X	X	X
Eastern Tiger Salamander, *Ambystoma tigrinum*	X	X	X	X	—	—	X	X	X	X	X

Continued

Table 5. (*Continued*)

	Laurentian Mixed Forest							Midwest Broadleaf Forest			
	Green Bay—Manitowoc Upland	North Central Wisconsin Uplands	Northern Green Bay Lobe	Northern Highlands	Southern Superior Uplands	Southwest Lake Superior Clay Plain	Western Superior Uplands	Minnesota and Northeast Iowa Morainal-Oak Savanna	North Central U.S. Driftless and Escarpment	Southwestern Great Lakes Morainal	Wisconsin Central Sands
Four-toed Salamander, *Hemidactylium scutatum*	X	X	X	X	X	X	X	X	X	X	X
Mudpuppy, *Necturus maculosus*	X	X	X	X	X	X	X	X	X	X	X
Spotted Salamander, *Ambystoma maculatum*	X	X	X	X	X	X	X	X	X	X	X
Total amphibian species	**19**	**19**	**19**	**18**	**15**	**17**	**18**	**18**	**19**	**18**	**19**
Reptiles											
Squamata											
Gophersnake, *Pituophis catenifer*	—	X	—	—	—	—	X	X	X	X	X
Butler's Gartersnake, *Thamnophis butleri*	X	—	—	—	—	—	—	—	X	X	X
Common Five-lined Skink, *Plestiodon fasciatus*	X	X	X	X	—	—	X	X	X	X	X
Common Gartersnake, *Thamnophis sirtalis*	X	X	X	X	X	X	X	X	X	X	X
Dekay's Brownsnake, *Storeria dekayi*	X	X	X	X	—	X	X	X	X	X	X
Eastern Foxsnake, *Pantherophis vulpinus*	X	X	X	X	X	X	X	X	X	X	X
Eastern Hog-nosed Snake, *Heterodon platirhinos*	X	X	X	X	X	X	X	X	X	X	X
Eastern Ribbonsnake, *Thamnophis saurita*	X	—	X	X	—	—	—	—	X	X	X
Gray Ratsnake, *Pantherophis spiloides*	—	—	—	—	—	—	—	—	X	X	X
Lined Snake, *Tropidoclonion lineatum*	—	—	—	—	—	—	—	—	X	X	—
Massasauga, *Sistrurus catenatus*	—	X	X	X	—	—	—	—	X	X	X
Eastern Milksnake, *Lampropeltis triangulum*	X	X	X	X	—	—	X	X	X	X	X
North American Racer, *Coluber constrictor*	—	X	—	X	—	—	—	X	X	X	X

Common Watersnake, *Nerodia sipedon*	X	X	X	X	X	X	X	X	X	X	X
Plains Gartersnake, *Thamnophis radix*	X	—	—	—	—	—	—	X	X	X	X
Prairie Skink, *Plestiodon septentrionalis*	—	X	X	X	X	X	X	X	X	X	X
Queensnake, *Regina septemvittata*	X	—	—	—	—	—	—	—	X	X	—
Red-bellied Snake, *Storeria occipitomaculata*	X	X	X	X	X	X	X	X	X	X	X
Ring-necked Snake, *Diadophis punctatus*	X	X	X	X	X	X	X	—	X	X	X
Six-lined Racerunner, *Aspidoscelis sexlineata*	—	X	—	—	—	—	—	X	X	X	X
Slender Glass Lizard, *Ophisaurus attenuatus*	—	—	—	—	—	—	—	—	X	X	X
Smooth Greensnake, *Opheodrys vernalis*	X	X	X	X	X	X	X	—	X	X	X
Timber Rattlesnake, *Crotalus horridus*	—	—	—	—	—	—	—	X	X	X	X
Western Ribbonsnake, *Thamnophis proximus*	—	X	—	X	—	—	—	—	X	X	X
Western Wormsnake, *Carphophis vermis*	—	—	—	—	—	—	—	—	X	—	—
Testudines											
Blanding's Turtle, *Emydoidea blandingii*	X	X	X	X	—	X	X	X	X	X	X
Eastern Musk Turtle, *Sternotherus odoratus*	X	—	X	X	—	—	—	—	X	X	X
False Map Turtle, *Graptemys pseudogeographica*	X	X	X	X	—	—	X	X	X	X	X
Northern Map Turtle, *Graptemys geographica*	X	X	X	X	—	—	X	X	X	X	X
Ornate Box Turtle, *Terrapene ornata*	—	—	—	—	—	—	—	—	X	X	X
Ouachita Map Turtle, *Graptemys ouachitensis*	—	X	X	—	—	—	—	—	X	X	X
Painted Turtle, *Chrysemys picta*	X	X	X	X	X	X	X	X	X	X	X
Smooth Softshell, *Apalone mutica*	—	—	—	—	—	—	—	—	X	X	X
Snapping Turtle, *Chelydra serpentina*	X	X	X	X	X	X	X	X	X	X	X
Spiny Softshell, *Apalone spinifera*	X	X	X	X	X	X	X	X	X	X	X
Wood Turtle, *Glyptemys insculpta*	X	X	X	X	X	X	X	X	X	X	X
Total reptile species	**22**	**24**	**22**	**23**	**12**	**14**	**19**	**21**	**36**	**35**	**33**

Statewide Patterns in Amphibian and Reptile Species Richness

To visualize statewide patterns in species richness, we used county-level species richness data. We created a net of regularly spaced points (10 × 10 km [6.2 × 6.2 mi]) across the state and computed the total number of documented native species at each point per class and order based on county records (including vouchered records and additional nonvouchered records shown in species accounts). We modeled statewide species richness patterns using kriging. This approach uses semivariograms to estimate spatial autocorrelation relationships for point data and creates prediction surface maps based on the modeled relationships (Brunsdon and Comber 2015). We used a Gaussian semivariance model, with a search radius of 50 km (31 mi) and an output cell size of 10 km^2 (6.2 mi^2) to project broad species richness patterns for each class and order. We performed these analyses using ArcGIS 10.3 (Environmental Systems Research Institute, Redlands, California).

Projected amphibian species richness was highest in the west-central region and lowest in the east-central and southwestern regions (Figure 13). Anura and Caudata showed different species richness patterns. Anuran richness was generally highest in the western and southern portions of the state, whereas salamander richness was generally highest in the northern and eastern portions of the state (Figure 13). Projected reptile species richness was highest in the western and south-central regions (Figure 14). Squamata and Testudines showed similar species richness patterns. Richness was highest in the south-central portion of the state and generally lowest in the northern and eastern portions of the state (Figure 14).

THE POTENTIAL FUTURE FOR WISCONSIN'S LANDSCAPES

It is widely recognized that rapid climate changes are occurring globally, and measurable changes will continue to occur for the foreseeable future (IPCC 2013). Understanding climate change at the state level is the focus of the Wisconsin Initiative on Climate Change Impacts (WICCI). Specifically, WICCI was formed in 2007 to promote information exchange, proactive planning, and adaptation to climate change in the state (https://wicci.wisc.edu/). In this section, we summarize observed recent climate changes and future climate projections for Wisconsin, potential concomitant changes to aquatic and terrestrial habitats, and how these changes could affect Wisconsin's amphibian and reptiles.

Between 1950 and 2006, the annual average temperature in Wisconsin increased by ca. 0.6°C (1.1°F), with the highest average warming in northwestern Wisconsin (+1.4°C [2.5°F]; WICCI 2011). Temperature is projected to continue increasing throughout the state, with annual average temperature increases of 2.2°C–5°C (4°F–9°F) and average winter temperature increases of 2.8°C–6.1°C (5°F–11°F) by around 2050, which will likely result in earlier annual first-bloom dates and longer growing seasons for native plants (Bradley et al. 1999, Zhao and Schwartz 2003, WICCI 2011). Between 1950 and 2006, annual average precipitation in Wisconsin increased by ca. 10% (7.9 cm [3.1 in]; WICCI 2011). Increases were primarily in southern and western portions of the state and during the autumn; annual precipitation decreased in the north (WICCI 2011). Annual average precipitation is also projected to increase in Wisconsin, primarily during the spring and autumn, but models vary substantially in projected precipitation levels (WICCI 2011). Schuster et al. (2012) projected that extreme rainfall events will increase in the state, particularly in the Northeast.

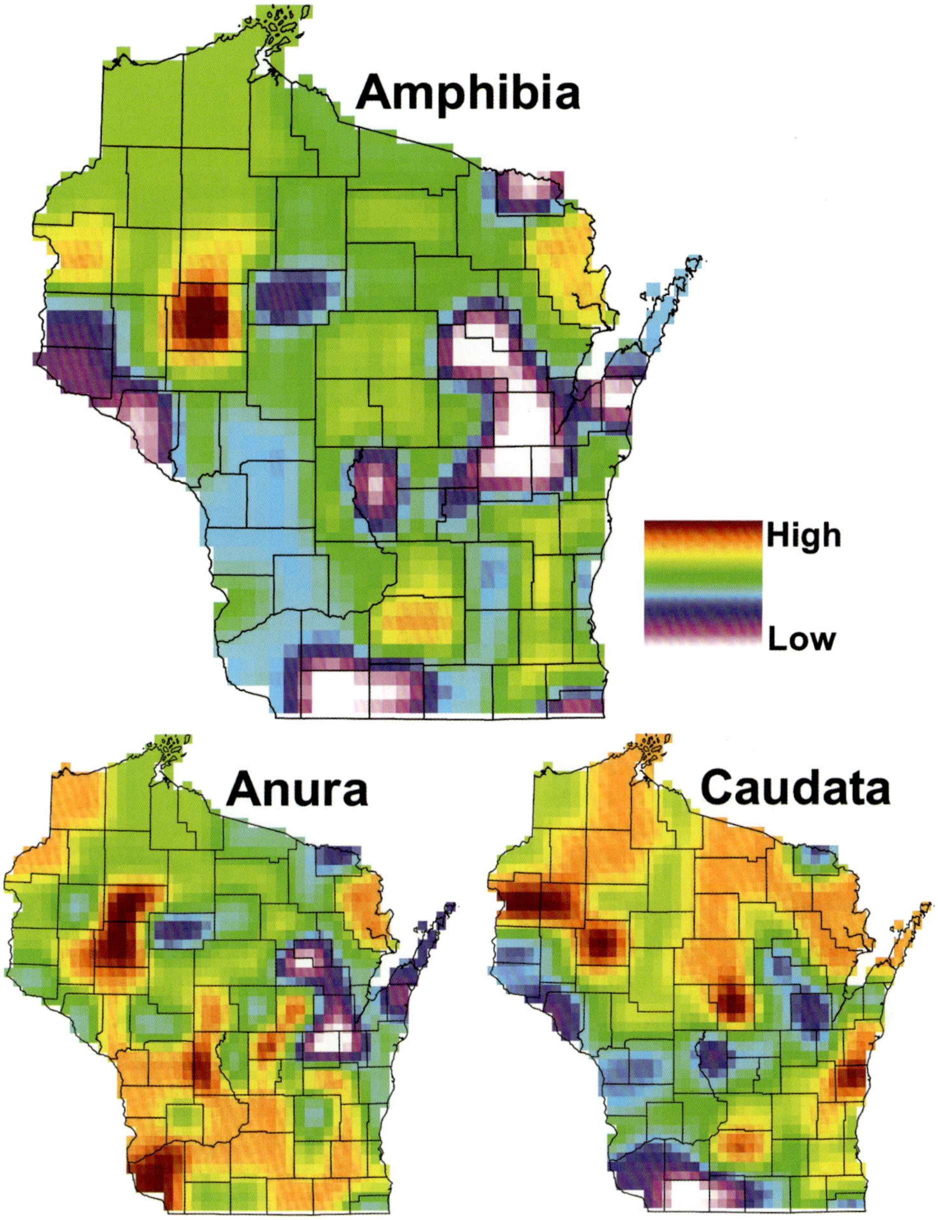

Figure 13. Projected species richness patterns for the class Amphibia and the orders Anura (frogs and toads) and Caudata (salamanders) in Wisconsin. For each map, we used county-level occurrence data and kriging to project statewide richness patterns (see main text for details). We emphasize that counties represented the finest level of data resolution for the analyses; thus, the prediction surface maps are useful only to infer broad spatial patterns.

Climate change may affect the availability and dynamics of the aquatic habitats on which most of the state's amphibians and aquatic/semiaquatic reptiles rely. Magnuson et al. (1997) completed an extensive modeling exercise of potential aquatic changes in the Great Lakes region and concluded that many lakes that traditionally experience complete ice cover during the winter could be ice-free more often. They also surmised that the duration of iced-over conditions on many lakes will be reduced. Magnuson et al. (1997) further

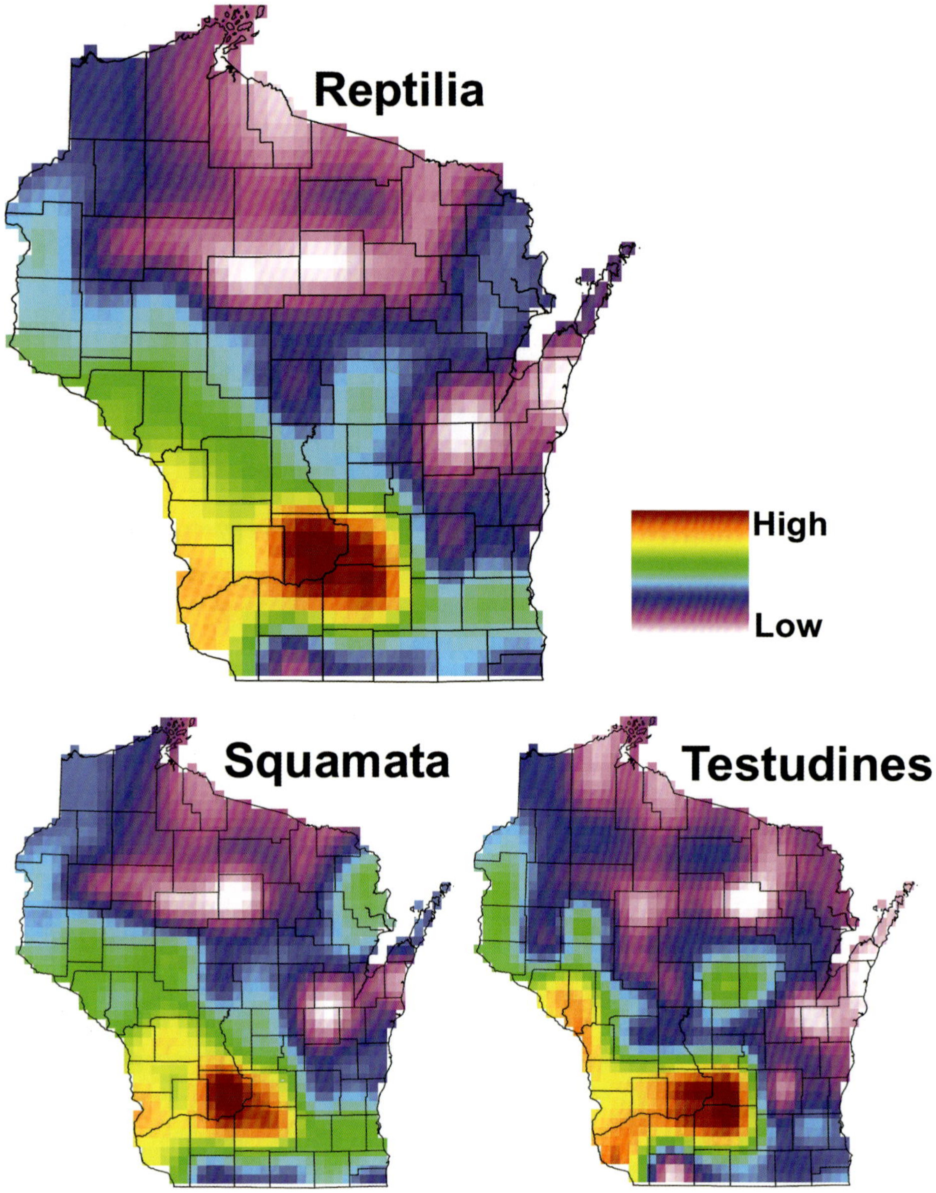

Figure 14. Projected species richness patterns for the class Reptilia and the orders Squamata (lizards and snakes) and Testudines (turtles) in Wisconsin. For each map, we used county-level occurrence data and kriging to project statewide richness patterns (see main text for details). We emphasize that counties represented the finest level of data resolution for the analyses; thus, the prediction surface maps are useful only to infer broad spatial patterns.

estimated that spatial and temporal wetland coverage will experience future declines due to increased summer evaporation. Anticipated hydrologic changes to ephemeral ponds include shifts to wetter, deeper types, increased flood duration in low-lying areas, and higher initial water levels. Such shifts could also result in these systems being connected to other, more permanent waterbodies (WICCI 2011).

As annual temperature and precipitation continue to increase in northern Wisconsin, the climate will likely become suitable for more cold-intolerant and mesic tree species, such as Black Oak (*Quercus velutina*; Iverson et al. 2008, 2013). Based on the Climate Change Tree Atlas (Swanston et al. 2011), habitat suitability in northern Wisconsin will likely decrease substantially for 12 tree species: Balsam Fir (*Abies balsamea*), Bigtooth Aspen (*Populus grandidentata*), Black Ash (*Fraxinus nigra*), Black Spruce (*Picea mariana*), Eastern Hemlock (*Tsuga canadensis*), Northern White-cedar (*Thuja occidentalis*), Paper Birch (*Betula papyrifera*), Quaking Aspen (*Populus tremuloides*), Sugar Maple (*Acer saccharum*), Tamarack (*Larix laricina*), White Spruce (*Picea glauca*), and Yellow Birch (*Betula alleghaniensis*). In contrast, 17 species are projected to have large increases in habitat suitability, all of which are deciduous trees, with the exception of Eastern Red Cedar (*Juniperus virginiana*; Swanston et al. 2011). Forest change modeling that included climate change, logging and wind disturbances, and forest fragmentation impacts on seed dispersal in northern Wisconsin concluded that five species could severely decline (Balsam Fir, White Spruce, Jack Pine [*Pinus banksiana*], Red Pine [*Pinus resinosa*], and Paper Birch), with a likely decline of five other species (Northern White-cedar, Quaking Aspen, Yellow Birch, Red Maple [*Acer rubrum*], and Red Oak [*Quercus rubra*]; Scheller and Mladenoff 2005, Swanston et al. 2011). Future vegetation projections by Williams et al. (2012a) indicated a decline in evergreen tree cover and an increase in deciduous tree cover in northern Wisconsin. Thus, current forest change models are in agreement that the prevalence of northern and boreal tree species in northern Wisconsin will decline in the future, replaced by southern and more mesic hardwoods. Vegetation projections in southern and western Wisconsin indicated that forests could be replaced by grassland prairie due to increased soil desiccation (Notaro et al. 2012).

Potential Climate Change Impacts on Amphibians and Reptiles

The wildlife subcommittee of WICCI assessed the vulnerability of Wisconsin's amphibians and reptiles to climate change as part of a project to prioritize actions for Wisconsin's wildlife (LeDee and Ribic 2015). Based on an extensive literature review, all vertebrates were scored for potential sensitivity to climate change using seven traits: specialized habitat and/or microhabitat requirements, narrow environmental thresholds likely to be exceeded, dependence on an abiotic cue to prompt physiological or behavioral response, dependence on interspecific interaction likely to be altered by changing climate, poor dispersal ability, susceptibility to infectious agents, and behavioral traits unsuited for future climate conditions (maladaptive behavior; LeDee and Ribic 2015). At the class level, amphibians and reptiles showed similar overall vulnerability to climate change based on the seven traits ($\chi^2_6 = 7.2$, $P = 0.27$); this was also found when species were categorized by order ($\chi^2_{18} = 23.7$, $P = 0.14$). Amphibians and reptiles were sensitive to climate change via specialized habitat/microhabitat requirements (94.7% of amphibians and 51.4% of reptiles), dependence on an abiotic cue (63.1% of amphibians and 40% of reptiles), and susceptibility to infectious agents (57.8% of amphibians and 17.1% of reptiles; Table 6). In terms of individual species, amphibians had more climate-sensitive traits than reptiles ($\chi^2_5 = 19.4$, $P < 0.001$). All amphibians were sensitive for either three (31.6%) or four (15.8% of amphibian species) traits, while only 8.6% of reptiles were sensitive for three traits and none for four. For reptiles, squamates were not sensitive (29.2%) or were sensitive for only one trait (41.1% of species), while 73% of Testudines were sensitive for one or two traits (Table 6). In terms of priority actions, the WICCI wildlife subcommittee recommended more research for reptiles, with targeted action or monitoring recommended for amphibians (LeDee and

Table 6. Species-specific results from an assessment of Wisconsin amphibian and reptile vulnerability to climate change based on species traits derived from the scientific literature, which were collated into seven trait groups. A summary of these results can be found in LeDee and Ribic (2015).

	Specialized Habitat and/or Microhabitat Requirements	**Narrow Environmental Thresholds Likely to Be Exceeded**	**Dependence on Environmental Cue**	**Dependence on Interspecific Interaction Likely to Be Altered**	**Poor Dispersal Ability**	**Disease/Parasitism**	**Maladaptive Behavior**	***Total***
Amphibians								
Anura								
American Bullfrog, *Lithobates catesbeianus*	1	0	1	0	0	0	0	2
American Toad, *Anaxyrus americanus*	1	0	1	0	0	1	0	3
Blanchard's Cricket Frog, *Acris blanchardi*	1	0	1	0	0	0	0	2
Boreal Chorus Frog, *Pseudacris maculata*	1	1	1	0	0	0	0	3
Cope's Treefrog, *Hyla chrysoscelis*	1	0	0	0	0	1	0	2
Gray Treefrog, *Hyla versicolor*	1	0	0	0	0	1	0	2
Green Frog, *Lithobates clamitans*	1	0	0	0	0	1	0	2
Mink Frog, *Lithobates septentrionalis*	1	1	1	0	0	1	0	4
Northern Leopard Frog, *Lithobates pipiens*	0	1	0	0	0	1	0	2
Pickerel Frog, *Lithobates palustris*	1	0	1	0	0	1	0	3
Spring Peeper, *Pseudacris crucifer*	1	0	1	0	0	1	0	3
Wood Frog, *Lithobates sylvaticus*	1	1	1	0	0	1	0	4
Caudata								
Blue-spotted Salamander, *Ambystoma laterale*	1	0	1	0	0	0	0	2
Eastern Newt, *Notophthalmus viridescens*	1	0	0	0	0	0	0	1
Eastern Red-backed Salamander, *Plethodon cinereus*	1	1	1	0	0	0	0	3
Eastern Tiger Salamander, *Ambystoma tigrinum*	1	0	1	0	0	1	0	3
Four-toed Salamander, *Hemidactylium scutatum*	1	0	0	0	1	0	0	2

Mudpuppy, *Necturus maculosus*	1	0	0	0	0	1	0	2
Spotted Salamander, *Ambystoma maculatum*	1	1	1	1	0	0	0	4
Total amphibian species	**18**	**6**	**12**	**1**	**1**	**11**	**0**	**49**
Reptiles								
Squamata								
Gophersnake, *Pituophis catenifer*	1	0	0	0	0	0	1	2
Butler's Gartersnake, *Thamnophis butleri*	0	0	0	1	0	0	0	1
Common Five-lined Skink, *Plestiodon fasciatus*	0	0	1	0	0	0	0	1
Common Gartersnake, *Thamnophis sirtalis*	0	0	0	0	0	1	0	1
Dekay's Brownsnake, *Storeria dekayi*	0	0	0	0	0	0	0	0
Eastern Foxsnake, *Pantherophis vulpinus*	0	0	0	0	0	0	0	0
Eastern Hog-nosed Snake, *Heterodon platirhinos*	1	0	0	1	0	0	0	2
Eastern Ribbonsnake, *Thamnophis saurita*	1	0	0	0	0	0	0	1
Gray Ratsnake, *Pantherophis spiloides*	0	0	1	0	0	0	0	1
Massasauga, *Sistrurus catenatus*	0	1	1	1	1	1	0	5
Eastern Milksnake, *Lampropeltis triangulum*	0	0	0	0	0	1	0	1
North American Racer, *Coluber constrictor*	0	0	0	0	0	0	0	0
Common Watersnake, *Nerodia sipedon*	1	0	1	0	0	0	0	2
Plains Gartersnake, *Thamnophis radix*	0	0	0	0	0	0	0	0
Prairie Skink, *Plestiodon septentrionalis*	1	0	0	0	0	0	0	1
Queensnake, *Regina septemvittata*	1	0	0	1	0	0	0	2
Red-bellied Snake, *Storeria occipitomaculata*	1	0	1	0	0	0	0	2
Ring-necked Snake, *Diadophis punctatus*	0	0	1	0	0	0	0	1
Six-lined Racerunner, *Aspidoscelis sexlineata*	1	0	1	0	0	0	0	2
Slender Glass Lizard, *Ophisaurus attenuatus*	0	0	0	0	0	0	0	0
Smooth Greensnake, *Opheodrys vernalis*	0	0	0	0	0	0	0	0
Timber Rattlesnake, *Crotalus horridus*	0	0	0	0	0	0	0	0
Western Ribbonsnake, *Thamnophis proximus*	0	0	0	1	0	0	0	1
Western Wormsnake, *Carphophis vermis*	0	0	0	1	0	0	0	1

Continued

Table 6. (*Continued*)

	Specialized Habitat and/or Microhabitat Requirements	**Narrow Environmental Thresholds Likely to Be Exceeded**	**Dependence on Environmental Cue**	**Dependence on Interspecific Interaction Likely to Be Altered**	**Poor Dispersal Ability**	**Disease/Parasitism**	**Maladaptive Behavior**	***Total***
Testudines								
Blanding's Turtle, *Emydoidea blandingii*	1	1	1	0	0	0	0	3
Eastern Musk Turtle, *Sternotherus odoratus*	1	0	0	0	0	1	0	2
False Map Turtle, *Graptemys pseudogeographica*	0	0	1	0	0	0	0	1
Northern Map Turtle, *Graptemys geographica*	1	0	1	0	0	0	0	2
Ornate Box Turtle, *Terrapene ornata*	1	0	0	0	0	0	0	1
Ouachita Map Turtle, *Graptemys ouachitensis*	1	0	1	0	0	0	0	2
Painted Turtle, *Chrysemys picta*	1	1	1	0	0	0	0	3
Smooth Softshell, *Apalone mutica*	1	0	1	0	0	0	0	2
Snapping Turtle, *Chelydra serpentina*	1	0	1	0	0	1	0	3
Spiny Softshell, *Apalone spinifera*	1	0	0	0	0	1	0	2
Wood Turtle, *Glyptemys insculpta*	1	0	0	0	0	0	0	1
Total reptile species	**18**	**3**	**14**	**6**	**1**	**6**	**1**	**49**

Ribic 2015). These recommendations were likely due to the lack of information about reptiles; there were fewer papers addressing climate-sensitive traits found per reptile species ($\bar{x}=24$ papers/species, SD = 37, $n=35$) compared to amphibian species ($\bar{x}=70$ papers/species, SD = 73, $n=19$).

The WICCI wildlife subcommittee also developed an overview of climate change and Wisconsin's wildlife (LeDee et al. 2013); the report used the Eastern Red-backed Salamander (*Plethodon cinereus*) and Wood Frog as examples of how different aspects of climate change might affect species distributions. LeDee et al. (2013) concluded that the Eastern Red-backed Salamander could decline in Wisconsin due to stress associated with the projected increase in summer temperature and decrease in summer precipitation. Furthermore, the Wood Frog was projected to decline in northern Wisconsin due to potential increases in frequency and severity of extreme droughts (LeDee et al. 2013).

For reptiles, Pomara et al. (2014) assessed vulnerability of Massasaugas to climate change across the species' distribution (including Wisconsin) and found that they are vulnerable to winter climate extremes (i.e., drought and flood events, which are projected to increase in the Upper Midwest) because of narrow thermal and hydration requirements in winter hibernacula. King and Niiro (2013) investigated potential distributional shifts for reptiles of conservation concern in the Great Lakes region, including seven species found in Wisconsin (Butler's Gartersnake [*Thamnophis butleri*], foxsnakes [*Pantherophis* spp.], Massasauga [*Sistrurus catenatus*], Queensnake [*Regina septemvittata*], Blanding's Turtle [*Emydoidea blandingii*], Northern Map Turtle [*Graptemys geographica*], and Wood Turtle [*Glyptemys insculpta*]). King and Niiro (2013) found that annual mean temperature was a key predictor for all of the Wisconsin species and that the optimal temperature relationship was unimodal (i.e., temperature had an optimal value, and habitat suitability decreased when conditions were higher or lower than optimal). Specifically in Wisconsin, habitat suitability for the Queen Snake and the Northern Map Turtle could increase, but habitat suitability for the Butler's Gartersnake, foxsnakes, Massasauga, Blanding's Turtle, and Wood Turtle could decrease. In addition, King and Niiro (2013) reported that Wisconsin could become more suitable for colonization by the Spotted Turtle (*Clemmys guttata*; see Hypothetical and Enigmatic Species of Wisconsin chapter for additional information on the Spotted Turtle).

In addition to potential distributional shifts, climate change will likely affect many biological processes, individual behaviors, and biotic interactions for amphibians and reptiles (see Corn [2005] and Li et al. [2013] for amphibian literature syntheses). Because amphibians and reptiles are ectothermic vertebrates, their biological and behavioral processes are strongly affected by temperature. Several studies have found changes in anuran breeding phenology in response to warming climate (Gibbs and Breisch 2001, Beebee 2002, Todd et al. 2011), and Byer et al. (2020) found that inter-year variation in median nesting date for Blanding's Turtles in Wisconsin was linked to heating degree-days in March. Donner et al. (2015) found that Blue-spotted Salamanders (*Ambystoma laterale*) and Spotted Salamanders (*A. maculatum*) moved into northern Wisconsin breeding ponds 12 and 10 days earlier, respectively, when mean April temperatures were warmer. Leach (2012) studied correlations between weather variables and Wood Frog breeding in northern Minnesota and found that average March temperature had the strongest influence on predicting when males begin to call. Thus, she predicted that Wood Frogs will spawn earlier as temperatures increase in the region. Similarly, Blaustein et al. (2001) noted a significant negative relationship between temperature and breeding for Spring Peepers (*Pseudacris crucifer*) in Michigan, indicating they might breed earlier as temperatures increase. Many aquatic and semiaquatic turtles in Wisconsin have temperature-dependent sex determination (Bull and Vogt 1979, 1981), and climate warming could potentially create female-biased populations (Ihlow et al. 2012). The warming of Wisconsin's climate could also lead to reductions in average amphibian and reptile body size (adhering to Bergmann's Rule). Daufresne et al. (2009) obtained evidence of this impact for fishes in the Baltic Sea and proposed that it be considered a "universal ecological response to global warming." In accordance, Caruso et al. (2014) reported an 8% reduction in average *Plethodon* salamander body size in the eastern United States over a 55-year period, with low-altitude sites with increasing temperature and decreasing precipitation trends showing the greatest body size reductions.

In conclusion, climate change will likely result in novel biological communities, with associated shifts in species assemblages and trophic (i.e., food chain) interactions (Lawler et al. 2010, 2013, Lurgi et al. 2012, LeDee et al. 2013). Amphibians and reptiles are highly influential in top-down and bottom-up ecosystem processes, serving as major predators (e.g., American Bullfrogs [*Lithobates catesbeianus*], Snapping Turtles [*Chelydra serpentina*]), major biomass components, and nutrient links between aquatic and terrestrial systems. Thus, the direct and indirect impacts of climate change on amphibian and reptile communities will have far-reaching effects on community composition and interactions at all trophic levels. We expect there will be a proliferation of future studies focused on how climate change affects amphibians and reptiles in the Midwest. These studies will be important for guiding proactive conservation and management of amphibians, reptiles, and their habitats in Wisconsin.

Amphibians and Reptiles in Wisconsin Plant Communities

The primary purpose of this chapter is to briefly describe the major Wisconsin plant communities available for use by resident amphibians and reptiles (Figure 15). We do not intend to provide a comprehensive list of all possible habitat categories or of all the plant species that do or could potentially occur within them. Rather, we attempt to identify those plant species that are typical of, highly associated with, or restricted to each community. We also do not elaborate on the habitat use of herpetofaunal species within these communities, as this information is covered in the species accounts.

TERRESTRIAL PLANT COMMUNITIES

Emmet J. Judziewicz and Erik R. Wild

The description of the terrestrial plant communities provided here draws freely on those presented in Black and Judziewicz (2009), Judziewicz et al. (2014), and WDNR (2018e). For further information on other plants and plant communities, we recommend the following books and resources: Curtis (1971), Cochrane and Iltis (2000), Hoffman (2002), Cochrane et al. (2006), and Waller and Rooney (2008).

Prairies

Prairies occupied ca. 5%–7% of Wisconsin's pre-settlement landscape (Curtis 1971; presently much less than 1%), and, along with members of the sunflower family (Asteraceae), grasses dominated prairie landscapes. The following is a very brief overview of Wisconsin prairies; for a more complete discussion, see Cochrane and Iltis (2000) and Cochrane et al. (2006). Generally, however, prairies are open habitats dominated by shorter herbaceous plant species and thus possess a more limited abundance of large woody plants compared to forests or woodlands. The natural disturbance regime in prairies, which restricts the ingrowth of large woody plants, featured frequent fires, both natural (due to lightning strike) and human-caused (Native Americans burning to clear land for hunting and other reasons),

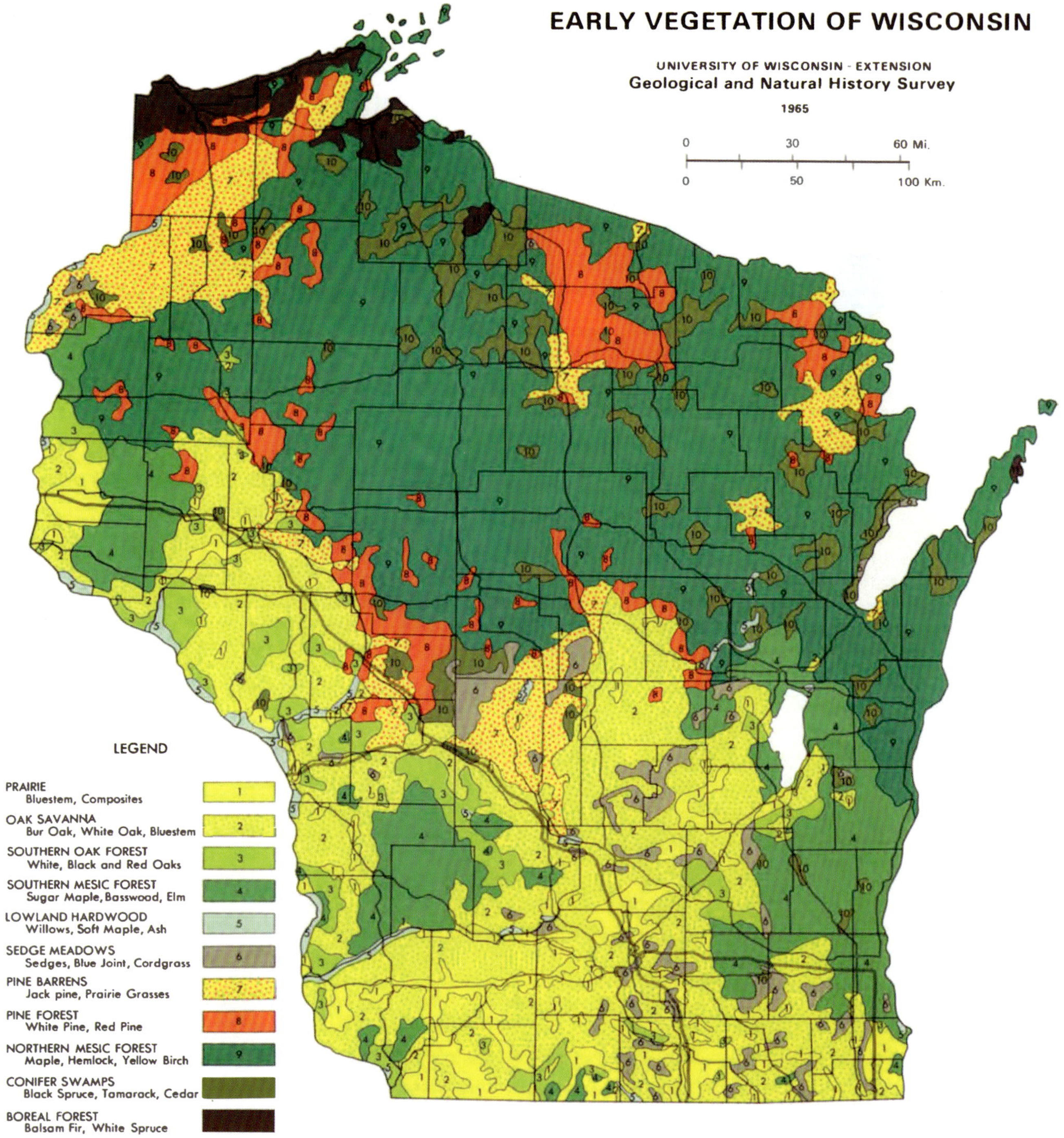

Figure 15. A map depicting the early distribution of vegetation in Wisconsin (map provided courtesy of Wisconsin Geological and Natural History Survey).

Figure 16. Dry prairies. (a) Grant County (photo by E. Judziewicz). (b) Crawford County (photo by J. Kapfer).

as well as grazing by large herbivorous mammals, such as the now extirpated North American Bison (*Bison bison*; Jackson 1961, Curtis 1971, Long 2008). After settlement, prairies were degraded in several ways. Settlers first turned cows loose in increasingly large densities, which resulted in disturbance rates beyond natural levels and led to a reduction in plant diversity. The prairie sod was then plowed and destroyed to plant crops. If the area was poorly drained, it was ditched and tiled, which altered hydrology and caused the decline of many plant species. Finally, fire suppression in many prairies, especially those too steep to graze, led to invasions of woody vegetation, particularly various shrubs and Eastern Red Cedar (*Juniperus virginiana*; Curtis 1971).

Dry prairies

In pre-settlement times, dry prairies covered ca. 2% of southern and western Wisconsin, most frequently on steep south- or west-facing sandstone and dolostone bluffs above major rivers such as the Wisconsin, Mississippi, Chippewa, and St. Croix, but also as 'sand prairies' on flat alluvial terraces along these same rivers. Gravelly dry hill prairies are also found in the Southern Kettle Moraine State Forest in Waukesha and Walworth Counties. Yet since settlement, dry prairies have fared better than either mesic or wet prairies (Cochrane and Iltis 2000). Common grasses are Little Bluestem (*Schizachyrium scoparium*), Big Bluestem (*Andropogon gerardii*), Side-oats Grama Grass (*Bouteloua curtipendula*), and Needle Grass (*Hesperostipa spartea*). Common wildflowers in this community include asters (*Symphyotrichum* spp.), goldenrods (*Solidago* spp.), Flowering Spurge (*Euphorbia corollata*), Common Spiderwort (*Tradescantia ohioensis*), bush-clovers (*Lespedeza* spp.), Pasque Flower (*Anemone patens*), Bird's-foot Violet (*Viola pedata*), Prairie Coreopsis (*Coreopsis palmata*), prairie-clovers (*Dalea* spp.), Lead-plant (*Amorpha canescens*), and blazing-stars (*Liatris* spp.).

Sand barrens

Sand barrens are extreme variants of dry prairies. They are typically blowouts (treeless sandy areas containing little to no established vegetation) and are most commonly found on the sandy terraces along the Mississippi and lower Wisconsin Rivers but sometimes also on abandoned farms in central Wisconsin, the latter due to historic disturbance of vegetation on sandy soils that has not reestablished. With fire suppression and failed agriculture on these poor soils, sand barrens, never very large, are fast disappearing because of succession to woody plants (Curtis 1971). Where still found, these communities have a suite of interesting grasses. Common woody associates are Jack Pine (*Pinus banksiana*), False

Heather (*Hudsonia tomentosa*), Plains Prickly-pear Cactus (*Opuntia macrorhiza*), and Common Juniper (*Juniperus communis*).

Jack pine barrens

Another plant community with strong affinities to dry prairies is jack pine barrens. This savanna community is adapted to frequent fires and is characterized by an open to dense forest of scrubby Jack Pine, often with admixtures of Hill's Oak (*Quercus ellipsoidalis*). The community covered 6%–7% of Wisconsin's presettlement landscape (presently ca. 1%–2%), mostly on sandy soils in the central (Jackson, Juneau, and Wood Counties), far northwestern (Douglas, Bayfield, Burnett, and Washburn Counties), far north-central (Lincoln, Oneida, and Vilas Counties), and far northeastern (Marinette County) parts of the state. Jack pine barrens may also be found on sand terraces along the lower Wisconsin River, such as in the Lone Rock area. Many dry prairie grasses and wildflowers are present, especially in central and far northwestern Wisconsin barrens.

Dry cliffs

This subset of dry prairies features steep dry vertical bedrock exposures that occur on many different rock types, including sandstone, dolostone (dolomite), quartzite, granite, and basalt. Scattered pines, oaks, or shrubs often occur on outcrops, but the canopy is open or nearly so. The most characteristic plants are Polypody Fern (*Polypodium vulgare*), Columbine (*Aquilegia canadensis*), Harebell (*Campanula rotundifolia*), Pale Corydalis (*Capnoides sempervirens*), juniper (*Juniperus* spp.), juneberry (*Amelanchier* spp.), and Bush-honeysuckle (*Diervilla lonicera*).

Right, top: Figure 17. Sand barrens. (a) With clumps of False Heather (*Hudsonia tomentosa*); turtle nesting habitat (Grant County; photo by J. Kapfer). (b) Sauk County (photo by E. Judziewicz).

Right, bottom: Figure 18. Jack pine barrens. (a) Moquah Barrens (Bayfield County; photo by E. Judziewicz). (b) Jackson County (photo by E. Judziewicz).

Mesic (moist) prairies

Mesic prairies covered ca. 2%–3% of southern Wisconsin in pre-settlement times, mostly on flat to gently rolling, rich-soiled, and well-drained sites that were quickly grazed, then plowed for agricultural uses. As a result, this is one of the rarest natural communities in Wisconsin, with perhaps only a couple hundred remnant hectares still in existence, often in long narrow strips along railroads (Cochrane and Iltis 2000, Cochrane et al. 2006). This is the classic "tallgrass prairie" and is the Wisconsin natural community most dominated by grasses; it shares many wildflower associates with those listed under dry prairies and wet prairies. Common grasses are Big Bluestem, Switch Grass (*Panicum virgatum*), and Indian Grass (*Sorghastrum nutans*). Forb diversity is great, with many giant composite herbs such as Wild Quinine (*Parthenium integrifolium*), Prairie Dock (*Silphium terebinthinaceum*), and Compass Plant (*Silphium laciniatum*).

Wet prairies

Wet prairies covered from 1% to 2% of southern Wisconsin in pre-settlement times on rich but poorly drained soils that were ultimately drained with ditches and converted to agriculture. Like mesic prairies, present-day remnants are small, rare, and scattered (Cochrane and Iltis 2000, Cochrane et al. 2006). Wet prairies intergrade with sedge meadows and calcareous fens; if sedges are present, the dominant species are likely to be Tussock Sedge (*Carex stricta*), Common Lake Sedge (*C. lacustris*), Broad-leaved Woolly Sedge (*C. pellita*), and Hairy-fruit Lake Sedge (*C. trichocarpa*). Common grasses are Big Bluestem, Cord Grass (*Spartina pectinata*), and Canada Bluejoint (*Calamagrostis canadensis*). Common wildflowers

Left, top: Figure 19. Dry cliffs. (a) Sandstone (Clark County; photo by E. Judziewicz). (b) Dolostone cliff topped by Eastern Red Cedar (Pepin County; photo by E. Epstein).

Left, bottom: Figure 20. Mesic and wet prairies. (a) Mesic prairie (Rock County; photo by J. Kapfer). (b) Wet prairie (Rock County; photo by J. Kapfer).

include Mountain-mint (*Pycnanthemum virginianum*), Prairie Phlox (*Phlox pilosa*), Black-eyed Susan (*Rudbeckia hirta*), Culver's-root (*Veronicastrum virginicum*), Shooting-star (*Primula meadia*), Prairie Blazing-star (*Liatris pycnostachya*), and golden alexanders (*Zizia* spp.), as well as various asters (*Symphyotrichum* spp.) and goldenrods (*Solidago* spp.).

Forests

Closed-canopy forests covered about 62% of Wisconsin's pre-settlement landscape (Curtis 1971). Interestingly, the current area of coverage is approximately 46%; deforestation has been more or less balanced by fire suppression and abandonment of previously cleared agricultural lands, which has led to succession in prairies (Kassulke and Esposito 2005; Perry et al. 2009). However, forest composition has changed greatly. Previous old-growth stands dominated by Sugar Maple (*Acer saccharum*) and Eastern Hemlock (*Tsuga canadensis*) in northern Wisconsin have now transitioned to a younger forest (maintained by periodic logging) dominated by birch (*Betula* spp.), aspen (*Populus* spp.), and Red Maple (*Acer rubrum*). These forests have few invasive plant threats, although fire suppression has limited pine (*Pinus* spp.) regeneration. In southern Wisconsin, oaks (*Quercus* spp.) have remained the dominant trees, but they are threatened by agricultural expansion, cattle grazing, fire suppression, residential development, and competition with large invasive shrubs such as Common Buckthorn (*Rhamnus cathartica*), Glossy Buckthorn (*Frangula alnus*), and honeysuckles (*Lonicera* spp.).

Southern mesic (moist) forests

Southern mesic forests are upland deciduous tree communities growing on rich well-drained soils on sites not susceptible to fire. In pre-settlement times, this community type covered ca. 10% of Wisconsin (presently ca. 5%). It is especially common in areas of dolostone bedrock, such as in the Lake Michigan counties, parts of the Driftless Area of the southwest, and far western Wisconsin near the confluence of the Mississippi and St. Croix Rivers. The dominant tree species are Sugar Maple, with codominants Basswood (*Tilia americana*), Red Oak (*Quercus rubra*), Hackberry (*Celtis occidentalis*), Bitternut Hickory (*Carya cordiformis*), Black Walnut (*Juglans nigra*), White Ash (*Fraxinus americana*), and, near Lake Michigan, American Beech (*Fagus grandifolia*). These forests typically have

Figure 21. Southern mesic forests. (a) Black River State Forest (Jackson County; photo by E. Epstein). (b) Mesic floodplain terrace and ephemeral pond, Black River State Forest (Jackson County; photo by E. Judziewicz).

fine displays of spring ephemeral wildflowers such as Spring-beauty (*Claytonia virginica*), trout-lilies (*Erythronium* spp.), trilliums (*Trillium* spp.), violets (*Viola* spp.), Bloodroot (*Sanguinaria canadensis*), Mayapple (*Podophyllum peltatum*), and Dutchman's-breeches (*Dicentra cucullaria*); Garlic Mustard (*Alliaria petiolata*) is a serious invasive threat.

Southern dry forests

Southern dry forests are deciduous communities growing on dry loamy to sandy upland soils; these sites are subject to occasional fires on a time scale of years or decades. In pre-settlement times, this community type covered ca. 4% of southern Wisconsin and intergraded with oak savannas and openings. Deciduous trees such as oaks are dominant and include White Oak (*Quercus alba*), Black Oak (*Q. velutina*), Red Oak, and Bur Oak (*Q. macrocarpa*); other common trees are Black Cherry (*Prunus serotina*), Shagbark Hickory (*Carya ovata*), and Red Maple. Common herbaceous understory species are Penn Sedge (*Carex pensylvanica*), Wild Geranium (*Geranium maculatum*), White Snakeroot (*Ageratina altissima*), Hog-peanut (*Amphicarpaea bracteata*), and tick-trefoils (*Desmodium* spp.). Spring ephemeral displays are absent from this community, and most herbs are 'summer greens.'

Oak savannas

As defined by Curtis (1971), oak savannas are communities in which there is less than 50% tree canopy; they comprised ca. 16% of Wisconsin in pre-settlement times (presently less than 0.1%, according to Cochrane and Iltis [2000]). Historically, they occurred on wet-mesic to dry sites with a history of fire. The few extant remnants are mostly on drier sites, with the mesic and wet-mesic openings almost

Left, top: Figure 22. Oak savannas. (a) Waukesha County (photo by E. Epstein). (b) Hugh Iltis Prairie (Waushara County; photo by E. Judziewicz).

Left, bottom: Figure 23. Oak savannas and southern dry forests. (a) Oak opening/savanna (Dunn County; photo by E. Epstein). (b) Southern dry (oak) forest. Fort McCoy (Monroe County; photo by J. M. Kapfer).

totally destroyed by conversion to agricultural or residential uses and by the encroachment of other woody plants due to fire suppression. Bur, White, and Black Oaks are dominant in mature stands as large, open-grown trees with distinctive limb architecture. Shagbark Hickory is sometimes present. American Hazelnut (*Corylus americana*) is a common shrub, and while the herb layer is similar to those found in oak forests and prairies, with many of the same grasses and forbs present, there are some plants that reach their optimal abundance in the 'openings.' The understory was open and grassy under a natural fire regime but with fire suppression became choked with Smooth Sumac (*Rhus glabra*), Gray Dogwood (*Cornus racemosa*), various cherries and plums (*Prunus* spp.), and brambles (*Rubus* spp.), as well as exotics such as honeysuckles and Common Buckthorn. These oak communities grade into dry and moist prairies.

Floodplain forests

These are hardwood forests that occur along rivers that flood periodically. They are most prominent along the larger rivers in southern Wisconsin but can be found essentially throughout the state. This community probably occupied ca. 1% of Wisconsin in pre-settlement times. Tree canopy dominants are Silver Maple (*Acer saccharinum*), River Birch (*Betula nigra*), Green Ash (*Fraxinus pennsylvanica*), Swamp White Oak (*Quercus bicolor*), Cottonwood (*Populus deltoides*), and Box-elder (*Acer negundo*). Common understory plants are Wood-nettle (*Laportea canadensis*) and Ostrich Fern (*Matteuccia struthiopteris*). Climbing woody vines such as Virginia creepers (*Parthenocissus* spp.), River Grape (*Vitis riparia*), and Eastern Poison-ivy (*Toxicodendron radicans*) are often common.

Figure 24. Floodplain forests. (a) Wolf River (Waupaca County; photo by E. Epstein). (b) Wisconsin River (Grant County; photo by J. Kapfer).

Northern mesic (moist) forests

Northern mesic forests dominated most of the northern half of Wisconsin and covered ca. 35% of the state in pre-settlement times (presently ca. 20%). They were mixed forests, with both deciduous and coniferous trees codominant. Sugar Maple and Eastern Hemlock were the dominant trees in pre-settlement times, covering fully one-third of the state. Yellow Birch (*Betula alleghaniensis*), White Pine (*Pinus strobus*), Basswood, and White Ash were also important forest trees. American Beech was an important tree species in the cooler Lake Michigan counties and extending inland to Forest, Langlade, Shawano, and Waupaca Counties. Common herbs, with species diversity ranging from poor in pure Eastern Hemlock to rich in pure maple (*Acer* spp.) stands, include clubmosses (*Lycopodium* spp.), wood ferns (*Dryopteris* spp.), Lady Fern (*Athyrium filix-femina*), Large-leaved Aster (*Eurybia macrophylla*), Wild Sarsaparilla (*Aralia nudicaulis*), Canada Mayflower (*Maianthemum canadense*), and Bluebead Lily (*Clintonia borealis*), as well as many of the spring ephemerals listed under southern mesic forests.

Northern dry forests

These are the pine forests currently found in northern Wisconsin that cover ca. 1%–2% of the state. However, in pre-settlement times this community type actually covered ca. 6%–7% of the state, mostly on nutrient-poor sites with excessively drained sandy or rocky soils. Relict stands of this forest type extend south to the Iowa and Illinois borders and beyond, often on moist north-facing hillsides or in river gorges. Dominant trees include White Pine, Red Pine (*Pinus resinosa*), Jack Pine, and Hill's Oak; on the dry end this community type intergraded with pine barrens. Large acreages of northern dry forests were cut and burned during the catastrophic logging of the late nineteenth and early twentieth centuries. Much of this land was subsequently colonized by White Birch (*Betula papyrifera*), Red Oak, Red Maple, and Quaking and Bigtooth Aspen (*Populus tremuloides* and *P. grandidentata*, respectively). Vast acreages of open 'barrens' were also planted to pine or naturally succeeded to densely stocked 'dry' forests. Common herbs include Bracken Fern (*Pteridium aquilinium*),

Left, top: Figure 25. Ephemeral ponds. (a) Ephemeral pond in southern mesic forest (Ozaukee County; photo by E. Judziewicz). (b) Ephemeral wooded wetland (Rock County; photo by J. Kapfer).

Left, bottom: Figure 26. Northern mesic and northern dry forests. (a) Northern mesic forest, hemlock/hardwoods, Flambeau River State Forest (Sawyer County; photo by E. Judziewicz). (b) Northern dry forest (natural Red Pine) (Forest County; photo by E. Judziewicz).

Wood Anemone (*Anemone quinquefolia*), Partridgeberry (*Mitchella repens*), Large-leaved Aster, Wintergreen (*Gaultheria procumbens*), and Shinleaf (*Pyrola elliptica*).

Northern conifer swamps

Northern conifer swamps are forested wetlands (swamps) in northern Wisconsin that are dominated by conifers such as Northern White Cedar (*Thuja occidentalis*), Balsam Fir (*Abies balsamea*), Black Spruce (*Picea mariana*), and Tamarack (*Larix laricina*); Tag Alder (*Alnus incana*) is often a common understory shrub. This community covered ca. 6%–7% of Wisconsin's pre-settlement landscape (presently ca. 2%–3%), mostly in the northern part of the state but extending south to the bed of Glacial Lake Wisconsin in Jackson County and as small relicts elsewhere throughout the south.

Figure 27. Northern conifer swamps. (a) Northern conifer swamp (Black Spruce and Tamarack), Flambeau River State Forest (Sawyer County; photo by E. Judziewicz). (b) Northern conifer swamp (Northern White Cedar; Door County; photo by E. Judziewicz).

Meadows, Bogs, and Ephemeral Ponds

Although these could be classified as wetlands (see Aquatic Plant Communities section), we follow Curtis (1971) in treating meadows and bogs as on the terrestrial side of the gradient of conditions resulting from the relationship between soil surface and the surface of the water table. These communities are submerged for considerable periods in the spring but are immersed for most of the growing season (Curtis 1971).

Sedge meadows

Sedge meadows covered ca. 3% of Wisconsin in pre-settlement times. However, this natural community grades into wet prairie in the southern part of Wisconsin (where it was very common) and into open bog in the northern part of the state. Changes in hydrology caused by draining and ditching and invasion by exotic plants have greatly reduced the extent and health of sedge meadows in Wisconsin, especially in the south. Reed Canary Grass (*Phalaris arundinacea*) may be dominant in grazed and/or ditched stands. The dominant plants in this open wetland community are Tussock Sedge, other species of sedges such as Long-bracted Tussock Sedge (*Carex aquatilis*), Common Lake Sedge, Few-seeded Hop Sedge (*C. oligosperma*), Broad-leaved Woolly Sedge, and Common Yellow Lake Sedge (*C. utriculata*), as well as Black Bulrush (*Scirpus atrovirens*), Wool-grass (*S. cyperinus*), and Canada Bluejoint. Cattails (*Typha* spp.) are also common, as may be Sensitive Fern (*Onoclea sensibilis*) and Marsh Fern (*Thelypteris palustris*), along with many wildflowers such as Spotted Joe-pye-weed (*Eutrochium maculatum*), Boneset (*Eupatorium perfoliatum*), Blue

Figure 28. Sedge meadows and open bogs. (a) Sedge meadow, Black River State Forest (Jackson County; photo by E. Epstein). (b) Open bog (Marathon County; photo by E. Judziewicz).

Vervain (*Verbena hastata*), Grass-leaved Goldentop (*Euthamia graminifolia*), Marsh Bellflower (*Campanula aparinoides*), Swamp Milkweed (*Asclepias incarnata*), Orange Jewelweed, and species of beggar-ticks (*Bidens* spp.), smartweeds, and water docks (*Rumex* spp.).

Open bogs

These cold, poorly drained peatlands developed in areas where the only source of nutrients was rainwater (Curtis 1971). Bog understories are composed mostly of *Sphagnum* mosses (*Sphagnum* spp., often in deep layers that accumulate to form thick 'bouncy' mats over open water), sedges, cottongrasses (*Eriophorum* spp., not a true grass), and 'ericaceous' shrubs such as Leatherleaf (*Chamaedaphne calyculata*), Labrador-tea (*Rhododendron groenlandicum*), Bog Laurel (*Kalmia polifolia*), Bog Rosemary (*Andromeda glaucophylla*), and blueberries and branberries (*Vaccinium* spp.). This type of bog grades into muskegs, semitreed bogs with scattered small individuals of Black Spruce and Tamarack, along with the shrubs Tag Alder and Bog Birch (*Betula pumila*). Orchids and insectivorous plants such as sundews (*Drosera* spp.) and Pitcher Plants (*Sarracenia purpurea*) are characteristic of this habitat.

Ephemeral ponds

Because they are dry for a portion of the growing season, we review ephemeral ponds in the Terrestrial Plant Communities section. These small (typically <1 ha [2.5 ac]) ponded depressions with poor drainage and water-retaining soils (usually embedded in forests) hold water for a length of time following spring snowmelt but typically dry out by midsummer (Epstein et al. 2002). Common aquatic plants of these habitats include Canada Bluejoint, manna grasses (*Glyceria* spp.), smartweeds (*Persicaria* spp.), Orange Jewelweed (*Impatiens capensis*), and sedges (*Carex* spp.). Although ephemeral ponds have a statewide aggregate area of only perhaps several thousand hectares, they provide critical breeding habitat for many species of amphibians, as well as for their invertebrate food. Being free of fish, ephemeral ponds provide habitats that lack a major predator of amphibian eggs and larvae.

Anthropogenic Areas

Anthropogenic areas are often open disturbed areas that result from and are maintained by human activity. These include pastures, meadows, old fields, active agricultural fields, tree plantations, lawns, greenspaces (e.g., city parks), railroad rights-of-way, roadsides, ditches (agricultural and roadside), and stormwater retention ponds in urban, suburban, or industrial landscapes. They are often dominated by plant species considered invasive or ornamental or those that tolerate high levels of disturbance, such as Kentucky Blue Grass (*Poa pratensis*), Common Dandelion (*Taraxacum officinale*), Smooth Brome Grass (*Bromus inermis*), Garlic Mustard, Common Buckthorn, goldenrods, Raspberry (*Rubus idaeus*), Black Locust (*Robinia pseudoacacia*), pine plantations, Reed Canary Grass (*Phalaris arundinacea*), Cattail (*Typha* spp.), and Common Reed (*Phragmites australis*). Anthropogenic areas can attract a variety of generalist amphibians and reptiles, although such habitats can be ecological traps (Battin 2004).

Figure 29. Anthropogenic areas. (a) Old field (Rock County; photo by J. Kapfer). (b) Roadside ditch (Vilas County; photo by R. Freckmann).

AQUATIC PLANT COMMUNITIES

Joshua M. Kapfer and Paul M. Skawinski

The aquatic environments and associated plant communities of Wisconsin are highly variable. Nonetheless, they can be categorized, generally, by whether they are lentic (i.e., still water) or lotic (i.e., flowing water) systems. Beyond this simple distinction, the assortment of waterbodies and aquatic plant community types that exist in the state is diverse and defined by many additional factors, such as water depth, color, temperature, and productivity. It is outside the scope and intention of this book to carefully define the differences among the vast array of aquatic systems in the state, so we provide only a cursory overview here. Note that in keeping with Curtis (1971), sedge meadows and bogs are treated previously in the Terrestrial Plant Communities section; furthermore, because ephemeral wetlands are dry for a portion of the year, they are also reviewed in the previous section.

Lentic Aquatic Communities

Plant communities in lentic aquatic systems are often influenced not only by overall waterbody size and water depth but also by a region's geological history, soil type, nutrient levels, and the likelihood of groundwater exposure at the soil surface (Curtis 1971, Richardson and Brinson 2000). Lentic aquatic systems are highly variable, running the gamut from small wetlands to large lakes (Reid 1961, Wetzel 2001). Generally, however, wetlands are a transitional area between upland and aquatic plant communities (Curtis 1971). Typically,

Figure 30. Lentic aquatic communities. (a) Large open wetland with emergent vegetation (Rock County; photo by J. Kapfer). (b) Lake (Washburn County; photo by E. Wild). (c) Lake (Dane County; photo by J. Kapfer).

they contain shallow standing water or damp saturated soils, with substantial quantities of emergent or submersed aquatic vegetation (Curtis 1971, Kalff 2002). Smaller lentic communities, such as wetlands, can be ephemeral and are even sometimes associated with the peripheries of larger lentic communities, such as lakes, or lotic aquatic systems (i.e., see previous Ephemeral Ponds entry; Kalff 2002). Because wetlands often have shallow standing water, light easily penetrates to the bottom of the water column. This warms the water and provides energy for photosynthetic organisms, such as plants, that act as structure for other species to use as cover and also convert solar energy into the consumable chemical energy necessary to support aquatic food webs (Wetzel 2001). These shallower, warmer aquatic systems tend to produce greater amounts of plant and animal biomass than do deeper water systems and are dominated by emergent plant species that prefer these conditions (Kalf 2002; Figure 30). Typical emergent plants are often robust macrophytes that exist in pure single-species stands or in diverse stands of species such as bur-reeds (*Sparganium* spp.), bulrushes (*Schoenoplectus* spp.), cattails (*Typha* spp.), Pickerelweed (*Pontederia cordata*), water plantains (*Alisma* spp.), Annual Wild Rice (*Zizania aquatica*), Common Spikerush (*Eleocharis palustris*), Tussock Sedge (*Carex stricta*), Lake Sedge (*C. lacustris*), Three-way Sedge (*Dulichium arundinaceum*), Rattlesnake Manna Grass (*Glyceria canadensis*), Water Lilies (especially *Nymphaea odorata* and *Nuphar variegata*), and exotic and native ecotypes of Common Reed (*Phragmites australis*). Small floating macrophytes such as Small Duckweed (*Lemna minor*), Large Duckweed (*Spirodela polyrrhiza*), Slender Riccia (*Riccia fluitans*), and Purple-fringed Riccia (*Ricciocarpos natans*) often occur between the emergent macrophytes. Emergent vegetation in shallow wetlands is frequently extensive enough to span the total wetland area, thereby limiting the amount of open water (Reid 1961). However, larger, deeper wetlands often possess greater areas of open water, which makes differentiating large wetlands from other lentic systems (e.g., lakes) a challenge (Reid 1961).

As water depth increases in lentic systems, the plant species composition also changes. This can lead to a greater presence of plants that are rooted in the sediments but possess floating leaves, such as water lilies (*Nymphaea* and *Nuphar* spp.), Watershield (*Brasenia schreberi*), Amphibious Smartweed (*Persicaria amphibia*), and Floating-leaf Pondweed (*Potamogeton natans*). Eventually, a depth is attained that prohibits the growth of emergent and floating plants (Curtis 1971). The maximum depth where floating-leaved species persist is typically 2 m (6.5 ft) or less, depending on water clarity. Beyond this depth, submersed aquatic plant communities are more typical (Curtis 1971) and are often dominated by pondweeds (*Potamogeton* spp.), watermilfoils (*Myriophyllum* spp.), naiads (*Najas* spp.), and macroalgae of the family Characeae, especially of the genera *Chara* and *Nitella*. In acidic, nutrient-poor systems, bladderworts (*Utricularia* spp.) are able to employ carnivory, as well as photosynthesis, to become more dominant plants.

Lotic Aquatic Communities

Lotic aquatic systems contain flowing water. This flowing water can originate from a groundwater source, or water that is held in underground rock pores and crevices that breaches the soil surface at specific points on the landscape. Lotic systems can also be fed by surface water, or water that originates from precipitation but drains off adjacent terrestrial landscapes to collect in valleys or channels of various sizes (USGS 2016). These channels dictate water movement as water flows from higher to lower altitude (due to gravity), with the speed of flow related to channel slope, width, and substrates (USGS 2016). Given this, lotic waterbodies are typically characterized by the area of a landscape that drains into them or feeds them (i.e., surface and groundwater sheds), their width and depth, and their rate of water discharge (Allan and Castillo 2007). Lotic systems can be further defined by their 'order,' which is a classification system that relates, in part, to increasing stream width and depth (Strahler 1952, Allan and Castillo 2007). Smaller headwater streams are lower-order lotic systems whose origination is at a point farthest from where that lotic system terminates. (Note: an example of a termination point could be the confluence where one stream or river flows into and joins with another.) Low-order streams are typically narrow and shallow but have high-velocity flow rates (Allan and Castillo 2007; Figure 31) that can be cold if fed by groundwater (Edington 1965). When two or more low-order streams converge, or confluence, the stream that results from this convergence could be larger and, therefore, of a higher order (Strahler 1952). Eventually, after numerous convergences, lotic systems achieve very high order rankings, such as those associated with large rivers that possess wider channels and deeper water (Strahler 1952; Figure 31).

Higher-order lotic systems tend to be more biologically productive than lower-order systems. They are wider, so they receive more direct solar radiation for heat and also support greater rates of photosynthesis than the low-order streams that are easily shaded by shoreline vegetation (Allan and Castillo 2007). Their large volume also results in slower flow rates and longer water retention times, so they are often warm, with an abundance of photosynthetic organisms (Allan and Castillo 2007). In addition, nutrient input from precipitation that runs off adjacent upland soils is deposited in lotic systems and accumulates as water moves downstream from lower to higher order (Allan and Castillo 2007). For these reasons, higher-order lotic systems are typically warm, nutrient-rich, and productive, which provides the energy to support diverse aquatic communities that can differ substantially from low-order streams (Allan and Castillo 2007). Stream order can, therefore, be related to the 'river continuum concept,' which describes the changes in biotic communities that occur due to changes in the physical characteristics of lotic environments as they transition

Figure 31. Lotic aquatic communities. (a) Low-order lotic system, Paradise Springs (Waukesha County; photo by J. Kapfer). (b) Low-order lotic system (Polk County; photo by E. Wild). (c) Mid-order lotic system, Totagatic River (Washburn County; photo by E. Wild). (d) High-order lotic system, Mississippi River (Crawford County; photo by J. Kapfer).

Figure 32. River backwaters. Wisconsin River (Sauk County; photo by J. Kapfer).

from lower- to higher-order systems (Vannote et al. 1980). It is noteworthy that peripheral wetlands frequently form along the course of lotic environments in areas where river velocity drastically decreases and standing water persists (such as where a bend occurs in the river), which encourages the accumulation of nutrients. These low-current regions promote the presence of plant and animal species that also associate with lentic environments, and such locations are sometimes referred to as 'backwaters' or 'sloughs' (Figure 32; see also previous Floodplain Forests entry). Backwater areas, with their shallower water, warm temperatures and accumulated nutrients, can support an abundance of submersed aquatic vegetation, often including dense beds of Common Waterweed (*Elodea canadensis*), Coontail (*Ceratophyllum demersum*), Yellow Water Crowfoot (*Ranunculus flabellaris*), muskgrasses (*Chara* spp.), stoneworts (*Nitella* spp.), Water Stargrass (*Heteranthera dubia*), small duckweeds (*Lemna* spp.), Large Duckweed (*Spirodela polyrrhiza*), watermeal (*Wolffia* spp.), Slender Riccia (*Riccia fluitans*), and Purple-fringed Riccia (*Ricciocarpos natans*).

Low-order and high-order lotic systems can be vastly different in their plant community composition, each comprised of the species that can typically withstand the associated physical characteristics. For example, Watercress (*Nasturtium officinale*), Water Speedwell (*Veronica anagallis-aquatica*), White Water Crowfoot (*Ranunculus aquatilis*), and Cut-leaf Water Parsnip (*Berula erecta*) are frequently found in association with high-velocity, shallow, clear, and cold water. In contrast, Common Waterweed (*Elodea canadensis*), Sago Pondweed (*Stuckenia pectinata*), Water Celery (*Vallisneria americana*), Long-leaf Pondweed (*Potamogeton nodosus*), bur-reeds (*Sparganium* spp.), and many lentic species dominate the larger, warmer rivers and streams with higher nutrient loads and lower flow rates.

Introduction to Conservation and Management of Wisconsin's Amphibians and Reptiles

Joshua M. Kapfer

The field of conservation biology is inexorably linked to Wisconsin. This is because, throughout history, several Wisconsin residents have helped shape modern conservation philosophies and practices. For example, the famous Wisconsinites Increase Lapham and John Muir were among the earliest environmentalists in America to vigorously argue for the protection of natural resources as far back as the mid-1800s (Quaife 1921, Fox 1981). Robert M. La Follette, a legendary Wisconsin politician and symbol of national progressive reform during the late 1800s and early 1900s, made early attempts to preserve state forests through legislation. Later, the celebrated conservationist and UW–Madison professor Aldo Leopold published his book *Game Management* (Leopold 1933), essentially creating the field of wildlife management. Leopold also famously developed important ideas about ecosystem-level protection and the role of humans as land stewards, which helped lay the foundations upon which many modern concepts in conservation biology were built (Meine 2010). In 1970 Gaylord Nelson, a senator from Wisconsin, founded national Earth Day, which became a driving force in the modern environmental movement. With respect to amphibian and reptile conservation, Wisconsin was the first state to implement an annual statewide program to monitor the calls of breeding frogs and toads (i.e., 1981; Natural History Box: Notes on the Origin of the Wisconsin Frog and Toad Survey), which has served as a model for many other state amphibian-monitoring initiatives (e.g., Minnesota [Moriarty 1998] and Missouri [Johnson 1998]).

Despite Wisconsin's strong history of leadership in conserving natural resources, its resident amphibians and reptiles currently face numerous threats to their well-being. In fact, the need for conservation of herpetofauna from the local to the global scale is perhaps greater now than ever in the past. Experts frequently place amphibians and reptiles at the forefront of the modern global biodiversity crisis (reviewed by Gardner et al. 2007). To find support for this notion, one need only peruse a smattering of studies that outline the dramatic losses experienced by amphibian and reptile populations worldwide (e.g., see reviews by Gibbons et al. 2000, Stuart et al. 2004, Wake and Vredenburg 2008, Kumar et al. 2014). This pattern

NATURAL HISTORY BOX: Notes on the Origin of the Wisconsin Frog and Toad Survey

Ruth Hine and I began this survey in 1981—the first of its kind anywhere, at least in North America. Ruth was a wildlife ecologist and then editor in the WDNR's Research Bureau, and she cared deeply about Wisconsin amphibians back when they received little attention. At the time, she was just finishing the editing of Dick Vogt's landmark *Natural History of Amphibians and Reptiles in Wisconsin*. She made that book happen! The Wisconsin Frog and Toad Survey was her brainchild: to experiment with a roadside survey modeled after the successful and well-established federal Breeding Bird Survey. At that time, we were concerned with possible declines of several species, such as Blanchard's Cricket Frog and Leopard Frog, which had suffered some major die-offs, as well as being harvested regularly for biological supply houses. There were other species that we just didn't know much about as well, for example, Pickerel Frog, which had a limited distribution, and Bullfrog, which seemed often to appear or disappear unpredictably from specific sites. (Lakeside residents really noticed when they did!) Ruth enlisted the help of Professor Ray Anderson and Debra Jansen, a graduate student at UW–Stevens Point, and they developed the system we have since come to rely on. It is based on 3 survey periods to cover the phenology of all 12 species and a "Call Index" value to account for the fact that you can't count frogs calling the way you can with birds. They also produced an instructional recording on cassette tape, which many hundreds of interested frog watchers have since heard, with Ray's distinctive voice describing each call.

We enlisted reliable field observers from within and outside the WDNR, many of whom had helped in other volunteer wildlife surveys. (We now call them 'citizen scientists,' but not back then.) Together we took three years to experiment with the technique, such as varying the number of roadside stops and time duration. Some observers set up repeated nightly surveys at wetlands near their homes to better document call phenology and effects of weather and water temperatures in different types of wetlands around the state. The field observers also helped us understand how to encourage and accommodate volunteers. Then, in 1984 the survey was off and running.

We spent a lot of time in the early years sending letters and making phone calls to our current and potential volunteers, way before the advent of email—everything from mundane reminders to everyone to carefully document their routes and station locations (no GPS!) and to ensure correct ID to eliciting feedback and just thanking everyone. We learned a lot from one another. Not only did we build the nation's longest-running auditory anuran monitoring program, but we made great progress on one of the most important goals of the project from the outset: to get people out at night, patiently and quietly listening to frogs, engaging in the realm of nocturnal wildlife, and building deeper connections with these animals and the wetlands they rely on us to protect.

Michael J. Mossman, WDNR (retired)

of global loss is mirrored at the local scale as well. For example, >50% of Wisconsin's herpetofaunal species (7 amphibians and 24 reptiles) are listed as Special Concern, Threatened, or Endangered by the WDNR (Tables 7 and 8), and several have recently experienced substantial population declines (e.g., Blanchard's Cricket Frog [*Acris blanchardi*]).

The conservation of Wisconsin's amphibians and reptiles is beset with challenges. These challenges could, in part, be blamed on the state's position in the Upper Midwest. On the one hand, the state's geographic placement relative to various ecological units in North America contributes to a large diversity of available habitats for herpetofauna (see review in Past, Present, and Potential Future Landscapes of Wisconsin and Amphibians and Reptiles in Wisconsin Plant Communities chapters). On the other hand, Wisconsin is situated at the periphery of many species' North American distributions, which can influence their in-state populations and natural history patterns. For example, Bahn et al. (2006) found evidence that populations are often smaller when they occur along the peripheries rather than at the core of a species' distribution. Swihart et al. (2006) further found that the probability of site occupancy decreased for some species as one moved toward the peripheries of their geographic distribution, a concept that the authors concluded was applicable to several Wisconsin species (e.g., Wood Frog [*Lithobates sylvaticus*] and Blanding's Turtle [*Emydoidea blandingii*]). It is not surprising, then, that Swihart et al. (2003) revealed that species at the peripheries of their geographic distributions were more susceptible to the effects of anthropogenic habitat alteration. The peripheral range position of many herpetofaunal populations in Wisconsin, coupled with past anthropogenic disturbance, has likely contributed to the currently patchy distributions for numerous species in the state (Natural History Box: Historical Anecdotes Describing Human Interactions with Reptiles in Early Wisconsin; see Past, Present, and Potential Future Landscapes of Wisconsin chapter for landscape changes following European settlement).

The development of strategies to conserve Wisconsin herpetofauna is also hampered by limited information. Historically, the conservation of amphibians and reptiles in Wisconsin has not garnered the same attention as that of their vertebrate counterparts (fishes, birds, and mammals). This is, in part, because they are often collectively considered nongame species and so have not received great interest or support from the general public and regulatory agencies in the past. Compounding the problem is the fact that these organisms are frequently treated with disdain or feared by humans rather than embraced (Gibbons 1988, Knight 2008, reviewed by Ceriaco 2012). These points could explain why an active rattlesnake bounty existed in the state for many years (Natural History Box: The Wisconsin Rattlesnake Bounty as an Example of Wildlife Persecution) and why complete regulations for the wild collection of herpetofauna in Wisconsin were not implemented until 2000 (Robert Hay, Turtles for Tomorrow, personal communication). Wisconsin is surely not the only state in the United States that has been sluggish to enact protection for its native herpetofauna. Yet the state's slowly changing attitudes and practices have likely contributed to the relative scarcity of past research on the ecology, conservation, and management of resident herpetofauna. This scarcity, in turn, has hindered progress on the development of science-based strategies to protect these species. For example, it was not until the 1970s that the state's rattlesnake bounty was discontinued and the first herpetofaunal species were listed as Threatened or Endangered by the state (WDNR 2014a). Since this time, however, the amount of research and number of scientific publications focused on herpetofauna in Wisconsin has increased (Figure 33). Such shifts are promising and have helped preserve state populations of these often forgotten or maligned species. Yet much more work is necessary.

Table 7. Conservation status for amphibians and reptiles in Wisconsin. The Species of Greatest Conservation Need list was taken from the 2015 Wisconsin Wildlife Action Plan. State statuses and ranks were taken from the July 2016 Wisconsin Natural Heritage Working List (WDNR 2016a) and the 2015 Wisconsin Wildlife Action Plan. Federal status indicates listing under the Endangered Species Act (USFWS 2018). Global status refers to the IUCN Red List (IUCN 2020), and global rank refers to the NatureServe (2020) status.

Scientific Name	Common Name	Species of Greatest Conservation Need	Status Status	State Rank	Federal Status	Global Status/Rank
Anurans						
Acris blanchardi	Blanchard's Cricket Frog	Yes	Endangered	S1		Least Concern / G5
Anaxyrus americanus	American Toad			S5		Least Concern / G5
Hyla chrysoscelis	Cope's Gray Treefrog			S5		Least Concern / G5
Hyla versicolor	Gray Treefrog			S5		Least Concern / G5
Lithobates catesbeianus	American Bullfrog		Special Concern (H)	S3/S4		Least Concern / G5
Lithobates clamitans	Green Frog			S5		Least Concern / G5
Lithobates palustris	Pickerel Frog	Yes	Special Concern (H)	S3(?)		Least Concern / G5
Lithobates pipiens	Northern Leopard Frog			S4(?)		Least Concern / G5
Lithobates septentrionalis	Mink Frog	Yes	Special Concern (H)	S3		Least Concern / G5
Lithobates sylvaticus	Wood Frog			S4		Least Concern / G5
Pseudacris crucifer	Spring Peeper			S5		Least Concern / G5
Pseudacris maculata	Boreal Chorus Frog			S5		Least Concern / G5
Salamanders						
Ambystoma laterale	Blue-spotted Salamander			S4/S5		Least Concern / G5
Ambystoma maculatum	Spotted Salamander			S4		Least Concern / G5
Ambystoma tigrinum	Eastern Tiger Salamander			S4		Least Concern / G5
Hemidactylium scutatum	Four-toed Salamander	Yes	Special Concern (H)	S3(?)		Least Concern / G5
Necturus maculosus	Mudpuppy		Special Concern (H)	S3/S4		Least Concern / G5

Continued

Table 7. (*Continued*)

Scientific Name	Common Name	Species of Greatest Conservation Need	Status Status	State Rank	Federal Status	Global Status/Rank
Notophthalmus viridescens	Eastern Newt			S4		Least Concern / G5
Plethodon cinereus	Eastern Red-backed Salamander			S4		Least Concern / G5
Lizards						
Aspidoscelis sexlineata	Six-lined Racerunner	Yes	Special Concern (H)	S2/S3		Least Concern / G5
Ophisaurus attenuatus	Slender Glass Lizard	Yes	Endangered	S1		Least Concern / G5
Plestiodon fasciatus	Common Five-lined Skink		Special Concern (H)	S3/S4		Least Concern / G5
Plestiodon septentrionalis	Prairie Skink	Yes	Special Concern (H)	S3		Least Concern / G5
Snakes						
Carphophis vermis	Western Wormsnake	Yes	Special Concern (H)	S1		Least Concern / G5
Coluber constrictor	North American Racer	Yes	Special Concern (P)	S2		Least Concern / G5
Crotalus horridus	Timber Rattlesnake	Yes	Special Concern (P)	S2/S3		Least Concern / G4
Diadophis punctatus arnyi	Prairie Ring-necked Snake	Yes	Special Concern (H)	S2/S3		Least Concern / G5
Diadophis punctatus edwardsii	Northern Ring-necked Snake		Special Concern (H)	S3		Least Concern / G5
Heterodon platirhinos	Eastern Hog-nosed Snake		Special Concern (H)	S3/S4		Least Concern / G5
Lampropeltis triangulum	Eastern Milksnake			S4		Least Concern / G5
Nerodia sipedon	Common Watersnake			S4		Least Concern / G5
Opheodrys vernalis	Smooth Greensnake			S4		Least Concern / G5
Pantherophis spiloides	Gray Ratsnake	Yes	Special Concern (P)	S3		Least Concern / G4
Pantherophis vulpinus	Eastern Foxsnake			S4/S5		Least Concern / G5
Pituophis catenifer	Gophersnake	Yes	Special Concern (P)	S2/S3		Least Concern / G5
Regina septemvittata	Queensnake	Yes	Endangered	S1		Least Concern / G5

Continued

Table 7. (*Continued*)

Scientific Name	Common Name	Species of Greatest Conservation Need	Status Status	State Rank	Federal Status	Global Status/Rank
Sistrurus catenatus	Massasauga	Yes	Endangered	S1	Threatened	Least Concern / G3
Storeria dekayi	Dekay's Brownsnake			S5		Least Concern / G5
Storeria occipitomaculata	Red-bellied Snake			S4/S5		Least Concern / G5
Thamnophis butleri	Butler's Gartersnake	Yes	Special Concern (H)	S3/S4		Least Concern / G4
Thamnophis proximus	Western Ribbonsnake	Yes	Endangered	S1		Least Concern / G5
Thamnophis radix	Plains Gartersnake	Yes	Special Concern (H)	S2(?)		Least Concern / G5
Thamnophis saurita	Eastern Ribbonsnake	Yes	Endangered	S1		Least Concern / G5
Thamnophis sirtalis	Common Gartersnake			S4/S5		Least Concern / G5
Tropidoclonion lineatum	Lined Snake	Yes	Special Concern (H)	S1/S2		Least Concern / G5
Turtles						
Apalone mutica[a]	Smooth Softshell	Yes	Special Concern (H)	S3		Least Concern / G5
Apalone spinifera[a]	Spiny Softshell			S4/S5		Least Concern / G5
Chelydra serpentina[a]	Snapping Turtle			S4/S5		Least Concern / G5
Chrysemys picta	Painted Turtle			S5		Least Concern / G5
Emydoidea blandingii	Blanding's Turtle	Yes	Special Concern (P)	S3/S4	Under Review	Endangered / G4
Glyptemys insculpta	Wood Turtle	Yes	Threatened	S3	Under Review	Endangered / G3
Graptemys geographica	Northern Map Turtle			S4/S5		Least Concern / G5
Graptemys ouachitensis	Ouachita Map Turtle			S4		Least Concern / G5
Graptemys pseudogeographica	False Map Turtle		Special Concern (H)	S3(?)		Least Concern / G5
Sternotherus odoratus	Eastern Musk Turtle			S4		Least Concern / G5
Terrapene ornata[b]	Ornate Box Turtle	Yes	Endangered	S1		Near Threatened / G5

[a] CITES Appendix III protection.

[b] CITES Appendix II protection.

Table 8. Definitions for the status of each species given in Table 7. State Status definitions were taken from the Wisconsin Endangered and Threatened Species Laws and List (Endangered and Threatened; WDNR 2014a), and Wisconsin Natural Heritage Working List (Special Concern; WDNR 2016a). State Rank and Global Rank definitions were taken from the Wisconsin Natural Heritage Working List (WDNR 2016a). Occurrences in the definitions refer to recorded observations (e.g., inventories, reports, published studies) that have passed through a quality assurance process.

Column	Status	Definition
State status	Endangered	"Any species whose continued existence as a viable component of this state's wild animals or wild plants is determined by the Department to be in jeopardy on the basis of scientific evidence."
State status	Threatened	"Any species which appears likely, within the foreseeable future, on the basis of scientific evidence to become endangered."
State status	Special Concern	"Those species about which some problem of abundance or distribution is suspected but not yet proven."
State status	Special Concern (H)	"Take regulated by establishment of open closed seasons."
State status	Special Concern (P)	"Protected wild animal."
State rank	S1	"Critically imperiled in Wisconsin because of extreme rarity (5 or fewer occurrences or very few remaining individuals or acres) or because of some factor(s) making it especially vulnerable to extirpation from the state."
State rank	S2	"Imperiled in Wisconsin because of rarity (6 to 20 occurrences or few remaining individuals or acres) or because of some factor(s) making it very vulnerable to extirpation from the state."
State rank	S3	"Rare or uncommon in Wisconsin (21 to 100 occurrences)."
State rank	S4	"Apparently secure in Wisconsin, with many occurrences."
State rank	S5	"Demonstrably secure in Wisconsin and essentially ineradicable under present conditions."
Global rank	G1	"Critically imperiled globally because of extreme rarity (5 or fewer occurrences or very few remaining individuals or acres) or because of some factor(s) making it especially vulnerable to extinction."
Global rank	G2	"Imperiled globally because of rarity (6 to 20 occurrences or few remaining individuals or acres) or because of some factor(s) making it very vulnerable to extinction throughout its range."
Global rank	G3	"Either very rare and local throughout its range or found locally (even abundantly at some of its locations) in a restricted range (e.g., a single state or physiographic region), or because of other factors making it vulnerable to extinction throughout its range; in terms of occurrences, in the range of 21 to 100."
Global rank	G4	"Apparently globally secure, though it may be quite rare in parts of its range, especially at the periphery."
Global rank	G5	"Demonstrably secure globally, though it may be quite rare in parts of its range, especially at the periphery."

NATURAL HISTORY BOX: Historical Anecdotes Describing Human Interactions with Reptiles in Early Wisconsin

Amphibians and reptiles in Wisconsin have experienced negative anthropogenic impacts since the 1800s. For example, a published account on life in Walworth County during the 1830s and 1840s stated that "the Massasauga, a torpid reptile of the rattlesnake species, was quite numerous on the wet lands. Its bite was very venomous, but, owing to its warning rattle, which was promptly and loudly given, few serious results occurred from encounters with them. They were soon thinned out, and ceased to be a source of anxiety, but while plenty, as they were for the first two or three years, they were certainly annoying, if not dangerous" (Western Historical Company 1882). Published correspondence from one Edmond Rendtorff about a trip to Sauk County in 1840 revealed similar treatment of what is likely the Timber Rattlesnake (*Crotalus horridus*): "At one time, leaving the settlement, we started for Honey Creek valley, in which at that time, not one living soul could be found, but a plenty of rattlesnakes. We killed many of them; also lots of pigeons and prairie chickens, which made us fine soups" (Canfield 1861). Accounts from Dane County in 1878 further reveal early practices for harvesting turtles that could have had population-level impacts, depending on how frequently they were employed. "We had also turtle soup suppers, the turtles caught by cutting holes through the ice on what was called 'Mud Lake,' brought to us by Abel Rasdall. Mr. Peck sent some of them to a Mineral Point hotel-keeper, who informed us he netted $50 on the sale of the soup. The turtles were frozen solid, and rattled together like stones. They were put in the cellar to thaw, before we could dress them, and, going down a few days after, I found they had thawed and were crawling around on the bottom of the cellar" (Western Historical Company 1880). Unfortunately, these anecdotes do not provide data on historical abundances of the species in question and cannot be used as a true population baseline to estimate the impact that humans have had on state populations since the early 1800s. Still, these examples and others like them provide evidence that Wisconsin's amphibians and reptiles were subjected to early persecution by European settlers. In some instances, these impacts were likely substantial and contributed to the current need for conservation initiatives and strategies.

Joshua M. Kapfer, UW–Whitewater

Given these points, there is both a precedent set and a need for a book that summarizes appropriate conservation information that relates to the state's herpetofauna. The goals of this chapter are to compile current relevant material that not only reviews general conservation threats but also summarizes possible management strategies to offset these threats. Many of the conservation and management concepts reviewed here apply to an array of amphibian and reptile species. For this reason, any general threats and management options that apply to many species are covered in this chapter, and readers are referred to it, where applicable, by the text of individual species accounts to reduce redundancy. In some cases, the subjects covered below have been relatively well-studied (e.g., amphibian ecotoxicology), while other subjects are poorly understood (e.g., the influence of social media on collecting species from the wild). I did not attempt to provide a detailed discussion of every possible threat and management strategy. Rather, my purpose was to synthesize the main threats and management considerations relevant to Wisconsin. This chapter should

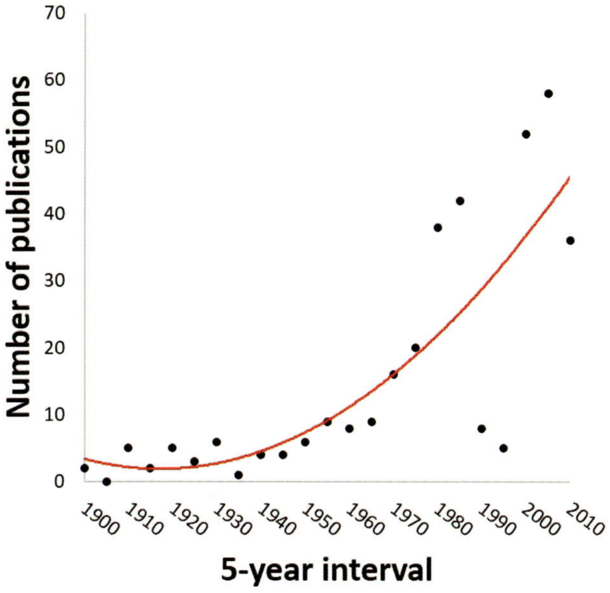

Figure 33. Estimated number of scientific publications involving amphibians and/or reptiles in Wisconsin between 1900 and 2015 ($N = 339$). For this visualization of scientific output over time, we included articles published in peer-reviewed journals, including field research studies, notes (primarily natural history observations), laboratory studies (primarily investigations of pesticide impacts), and studies that incorporated data from multiple states (primarily genetics and taxonomy research). We excluded county record publications, books and book chapters, and publications in non-peer-reviewed outlets (including publications where peer review was questionable). Number of publications is grouped into five-year intervals.

be viewed as a starting point from which interested parties can gain an understanding of herpetofaunal conservation and management in Wisconsin; readers are strongly encouraged to further investigate the greater body of literature on particular topics of interest.

CONSERVATION THREATS

Extinction, extirpation, and the loss of biodiversity due to human activities are concepts clearly linked to conservation. To minimize probability of extinction and enact conservation, it is important to recognize the mechanisms by which extinction occurs. To this end, Diamond (1989) broadly outlined the primary causes of modern extinctions: habitat destruction/fragmentation, 'overkill' (i.e., mortality due to persecution or overharvesting), introduced species, and 'secondary extinctions' (e.g., the extinction of a particular organism because another species it depends on is eliminated). This "evil quartet" (Diamond's term) is applicable across a wide array of situations and geographic locations, so it is relevant to amphibians and reptiles in Wisconsin. The majority of conservation threats outlined below can be directly or indirectly linked to one of Diamond's categories; thus, their potential to negatively influence amphibian and reptile populations in the state should not be underestimated.

Habitat Loss, Fragmentation, and Degradation

Habitat loss, fragmentation, and degradation are firmly identified as primary causes for the decline of many taxa in the United States (reviewed by Wilcove et al. 1998 and Fischer and Lindenmayer 2007). Some authors have even suggested that, among living organisms, amphibians and reptiles are particularly susceptible to the negative impacts of habitat loss (e.g., White et al. 1997, reviewed by Cushman 2006 and Gardner et al. 2007). Yet the full effects of such phenomena are difficult to understand, because there is great variation in how species respond to loss and degradation of habitat (e.g., Hager 1998). Difficulty also arises because there is historical disparity in the global effort put forth to study effects of habitat loss on various groups. A review of existing research suggests that the impacts of habitat loss on amphibians and reptiles have garnered much less attention from researchers than for birds or mammals (Gardner et al. 2007). Furthermore, between amphibians and reptiles, the former has received greater scrutiny, with snakes being particularly overlooked (Gardner et al. 2007). Such large knowledge gaps make design of effective conservation strategies especially challenging.

Despite a lack of detailed information on how the loss of habitat influences Wisconsin herpetofauna, it is still possible to work toward species conservation in the state through an understanding of existing evidence. As such, a reasonable starting point is first to recognize that substantial habitat destruction occurs primarily due to the activities of people (Gallant et al. 2007). The conversion of natural landscapes to those dominated by agriculture, for example, has occurred at a considerable rate in Wisconsin since the time of European settlement (see review in Past, Present, and Potential Future Landscapes of Wisconsin chapter). Similarly, habitat fragmentation (i.e., the division of large intact tracts of suitable habitat into smaller parcels) has been extensive in Wisconsin and also negatively impacts the stability of resident herpetofaunal populations in numerous ways. Fragmentation of habitat could restrict a species' access to habitat types required to meet their life history needs. For example, many Wisconsin species require the presence of both aquatic and terrestrial habitats for survival and reproduction. Disturbances to only one of these or fragmentation that restricts movement between them can have population-level impacts that eventually result in extirpation. Past studies have implicated this phenomenon, known as habitat split, as a major factor behind amphibian declines (Becker et al. 2007). The fragmentation of habitat can further fracture the associated resident populations into small isolated groups of individuals, creating what is known as a metapopulation. If movement among these fractured subpopulations is constrained, it also restricts the exchange of individuals and genetic material. Thus, the smaller subgroups may lack both the stable demographics required for population sustainability and the genetic diversity necessary to be viable (reviewed by Forman et al. 2003). Roads are commonly identified as the cause of anthropogenic habitat fragmentation and habitat split (Saunders et al. 2002), but patterns of anthropogenic land use, whether due to agriculture or to residential development, can also fragment habitat (see relevant subsections related to roads and habitat conversion below).

Because habitat loss and alteration are serious threats to amphibians and reptiles, it follows that these phenomena have recently received some attention by researchers. There is also a current push to better understand landscape-level threats to habitat (rather than site-specific threats and effects) in an effort to realize how shifting land use mosaics affect biodiversity (e.g., Knutson et al. 1999, Mazerolle and Villard 1999, Bennett et al. 2006, Steen et al. 2012a). Regardless, further attention to the influence of habitat change across landscapes and its mitigation is likely critical to amphibian and reptile conservation in Wisconsin. Here is a review of general concepts associated with habitat loss and fragmentation, as well as information on specific habitat types that are relevant to Wisconsin.

Loss of habitat due to agriculture and urban, suburban, and exurban sprawl

Habitat loss due to expansion of agriculture and urban, suburban, and exurban development (USED) is a major concern for amphibian and reptile conservation. For decades,

Figure 34. Two examples of agricultural land (foreground of both pictures) immediately adjacent to snake habitat (background of both pictures; southwestern Wisconsin; photos by J. Kapfer).

past researchers have suspected that habitat loss due to land conversion has negative impacts on herpetofauna (e.g., Minton 1968), a notion that more recent investigations have confirmed (e.g., Gardner et al. 2007, Ribeiro et al. 2009, see multiple chapters in Mitchell et al. [2008]). At a global scale, the current impacts that expanding agricultural landscapes have on biodiversity are well documented (reviewed by Tscharntke et al. 2005), but understanding the full effects of land conversion remains a challenge. For example, USED can cause direct mortality through loss of natural habitat. Yet subtle effects of USED can also occur due to increased prevalence of habitat edges and associated impacts (reviewed by Hansen et al. 2005), high levels of environmental contaminants (e.g., road deicing salts and fertilizers/herbicides applied to ornamental plants; Karraker 2008), large impacts of artificial light on nocturnal activity patterns (Perry et al. 2008), heavy infestations with exotic species (reviewed by Hansen et al. 2005), and high potential for direct contact with humans or domestic animals (reviewed by Adams and Lindsey 2010). It is these subtle effects that can be especially difficult to fully comprehend.

Agricultural and developed lands currently encompass nearly half (44%) of Wisconsin's land cover, which is a dramatic increase since the time of European settlement (reviewed in the Past, Present, and Potential Future Landscapes of Wisconsin chapter). Of these, cultivated lands and pasture are particularly pervasive, comprising an estimated 36.8% of the state's land cover (Figure 34). Rigorous investigations of the interplay between agricultural land and Wisconsin's herpetofauna are sparse and often focused on single species or small spatial scales. For example, Kapfer et al. (2008e) reported that Gophersnakes (*Pituophis catenifer*) avoided agricultural land, but the spatial scale of this work was relatively small. Although rare, landscape-level analyses in upper midwestern states, including Wisconsin, have revealed a negative relationship between amphibian communities and USED (Knutson et al. 1999, Lehtinen et al. 1999). Interestingly, Knutson et al. (1999) found that amphibians in Wisconsin were positively associated with agricultural land that still possessed pockets of forest, indicating that appropriately managed parcels within cultivated properties could provide suitable habitat for herpetofauna.

Developed land alone currently covers 7.2% of Wisconsin's landscape (see Past, Present, and Potential Future Landscapes of Wisconsin chapter). Although all forms of USED likely have negative impacts, the area of exurban development (e.g., residential development in rural settings) in the United States has increased at a greater rate than has urban or agriculture land since 1974 (reviewed by Hansen et al. 2005). LaGro (1996) revealed that exurban development is especially prevalent in Wisconsin and can be most problematic when it borders public land (e.g., state forests, state parks, state natural areas, etc.), a situation desired by many homeowners for aesthetic and recreational purposes. Unfortunately, some public lands represent the only remaining natural habitat for wildlife, so exurban development adjacent to public land increases the likelihood of conflict between humans and wildlife (reviewed by Hansen et al. 2005). Despite these concerns, the influence of exurban development on amphibians and reptiles in Wisconsin is poorly studied and warrants further attention.

Loss of prairies, grasslands, and savannas due to succession

Samson and Knopf (1994) stated that "the largest vegetative province in North America is the native prairie," but this province has also experienced dramatic losses in Wisconsin since European settlement (e.g., Samson and Knopf [1994] reported a decline of 99.9% in-state). The most important cause of prairie, grassland, and savanna declines in the Upper Midwest is conversion of these habitats to agricultural and developed land. Yet the anthropogenic factors that impact prairies, grasslands, and savannas are varied. For example, their existence depends on natural disturbance, which likely took various forms prior to European colonization of North America. These included drought, episodic fire due to lightning strikes, and grazing from herbivores that are now extinct or greatly reduced in abundance (reviewed by Howe 1994). When natural disturbance does not occur in prairies, grasslands, and savannas, the result is unabated expansion of woody vegetation that replaces typical herbaceous prairie species (a process called 'succession'; Anderson 1983). The rate of succession in these open habitats increased dramatically after Europeans settled in North America due to the suppression of natural disturbances (Briggs et al. 2002) and the spread of woody plant seeds by livestock (Brown and Archer 1999).

In addition to the elimination of typical prairie plant species, succession can alter normal habitat function and reduce small-scale structural heterogeneity that could be important for cover and thermoregulation (Lett and Knapp 2005, Brudvig and Asbjornsen 2009). Prairies impacted by succession also have long recovery times, even after woody vegetation is removed. For example, Lett and Knapp (2005) found that recovery of native grassland plants at a site in Kansas had not occurred by two years post-removal of the shrub *Cornus drummondii*, a plant species that has relatives in Wisconsin (e.g., *C. stolonifera*). Given this information, it is not difficult to envision that succession can have long-term negative impacts on both prairie habitats and native herpetofaunal species that occupy them. In fact, Fitch (1999) observed that the progression of succession he documented for over 50 years at a site in Kansas was concomitant with a decline in prairie reptile species such as Gophersnakes and North American Racers (*Coluber constrictor*). In Wisconsin, numerous reptile species use prairies, grasslands, and savannas extensively, including several species that are very rare (e.g., Ornate Box Turtles [*Terrapene ornata*] and Slender Glass Lizards [*Ophisaurus attenuatus*]), which highlights the serious threat that succession poses to Wisconsin's prairie herpetofauna.

Anthropogenic disturbances in forests and woodlands

Forests and woodlands are characterized by the structure and composition of resident vegetation. These parameters are based on forest age and developmental stage, which are dictated by the intensity and frequency of disturbance (reviewed by MacNeil et al. 2013).

Although natural disturbances that fell trees (e.g., fire, wind, disease) are important in forest dynamics, humans also cause unnatural disturbances (e.g., conversion of forests to agriculture and urban lands, timber harvest), which have been extensive since the time of European settlement (Fischer et al. 2013). Generally, more amphibian than reptile species associate with forests and woodlands in Wisconsin, although exceptions exist (e.g., Wood Turtles [*Glyptemys insculpta*]). It is not surprising then that amphibian species richness is positively correlated with forest maturity at large spatial scales (e.g., Jacobs and Houlahan 2011). Mature forests are, unfortunately, far less abundant in Wisconsin now than historically (reviewed by Past, Present, and Potential Future Landscapes of Wisconsin chapter). For example, Gonzalez-Abraham et al. (2007) estimated that a 60% decline had occurred in deciduous forests associated with housing clusters in northern Wisconsin from 1937 to 1999. Furthermore, disturbance that reduces overstory tree canopy cover leads to increased light penetration and temperature, with associated declines in moisture levels that can negatively impact amphibians (reviewed by Semlitsch et al. 2009, Blomquist and Hunter 2010).

The impacts of forest and woodland loss to land conversion should be considered in concert with loss due to commercial timber harvest. The impacts of timber harvest on amphibians and reptiles have received much attention across the US and Canada, although limited work has occurred in Wisconsin. Current evidence suggests that the relationship between habitat structure that remains post-harvest and amphibian populations is an important conservation issue that can be affected by the timber harvest strategy employed. Timber harvest is carried out via several potential practices (see Management section, Silvicultural strategies subsection), which can have different impacts on amphibian populations (reviewed by deMaynadier and Hunter 1995). Despite this, most studies support that clear-cutting and promotion of even-aged tree stands generally have negative impacts on amphibian populations (for further description of clear-cutting and even-aged, see Management section, Silvicultural strategies subsection). DeMaynadier and Hunter (1995) reported that amphibian abundance was 3.5-fold less in clear-cut stands vs. control stands (reported as median difference). Semlitsch et al. (2009) further found that amphibians responded negatively to areas harvested with clear-cut techniques compared to those where trees were selectively harvested. Petranka et al. (1993) reported that clear-cutting in the US national forests of western North Carolina resulted in a loss of ca. 14 million salamanders annually and that recovery of salamander populations could take 50–70 years.

However, documented responses of amphibian communities to clear-cut harvest techniques are occasionally counterintuitive. For example, species that typically associate with open-canopy wetlands actually exhibit greater calling rates and reproductive output in clear-cut vs. undisturbed forested locations (Semlitsch et al. 2009). Some species also do not avoid clear-cuts when migrating to breeding ponds but do avoid them when returning from breeding ponds (Todd et al. 2009). Furthermore, salamanders might be more likely to avoid clear-cut areas than frogs (e.g., Todd et al. 2009). The negative influence of habitat edges along the boundaries of clear-cut areas is another important consideration (see Management section, Silvicultural strategies subsection). These points highlight the complexity of this issue and the need for further research in Wisconsin.

Both large- and small-scale habitat features are relevant to amphibians and reptiles in woodland habitats. Small-scale features, such as the availability of downed woody debris for cover in forests, for example, can influence occupancy by numerous species, particularly salamanders. An assessment of the woody debris available in harvested vs. old-growth forests (including areas within Wisconsin) found that the volume of downed woody debris was greater in old-growth forests than in woodland stands subjected to various timber harvest practices (Goodburn and Lorimer 1998). This difference was particularly pronounced when only downed wood >40 cm (16 in) in diameter, likely from larger trees that are typically associated with older-growth forests, was analyzed. Furthermore, a chronological sequence of available fallen dead wood occurs as a forest develops, with a decline in downed wood volume from ca. 10 to 70 years after forest establishment (when mortality in dominant larger tree species is infrequent), followed by increases between 80 and 200 years post-establishment as dominant trees begin to die (Spetich et al. 1999). These findings highlight that even if disturbance ceases, the time required for forests to accumulate appreciable downed woody debris for amphibians is substantial.

Wetland loss and modification

Wetlands play a critical role in the ecology of many amphibian and reptile species, most notably related to amphibian reproduction (reviewed by Gibbons et al. 2000 and Hecnar 2004). Unfortunately, wetland loss, fragmentation, and degradation have occurred at astounding rates since European settlement. For example, Dahl (1990) estimated that 46% of Wisconsin's wetlands were lost from the 1780s to the 1980s, with the quality of much remaining wetland habitat experiencing further degradation. Such extensive habitat loss and degradation likely has large impacts on many amphibian and reptile species that associate with wetlands, even generalist species that are considered resilient to human disturbance (Rustigian et al. 2003). Along these lines, Brazner et al. (2007) concluded that Spring Peepers (*Pseudacris crucifer*) are an indicator species for ecological stress in Great Lakes coastal wetlands, which is noteworthy, considering Mossman et al. (1998) estimated that this species experienced a 26% decline in Wisconsin over 50 years.

There are numerous specific conservation threats applicable to herpetofauna in Wisconsin wetlands that are worth consideration. For example, the quality of what little wetland habitat remains can be degraded by invasion of exotic species or alteration of wetland hydrology that can allow troublesome species to become established (see Exotic and Native Pest Species section). Environmental contaminants can also be a concern for wetland inhabitants in Wisconsin, particularly in wetlands that experience runoff from adjacent agriculture or urban land parcels (see Pollution section). Amphibians and reptiles appear to benefit from landscapes that possess many wetlands of variable size, depth, and hydroperiod and that are connected through intact adjacent uplands (see review in Management section; see also Lentic Aquatic Communities section). The concept of 'habitat split' (Becker et al. 2007) is also important to consider here, particularly when that split restricts the movement of amphibians between upland habitat (where part of the year is spent) and wetlands (where reproduction occurs).

Degradation of lakes and lakeshore development

The impacts of human recreation and shoreline development are likely primary concerns for herpetofauna that use Wisconsin lakes. For example, the propellers of motorized water craft can injure or kill turtles (Bulté et al. 2010), while heavy human activity can modify normal turtle behaviors (e.g., Pitt and Nickerson 2012). Analyses of residential development patterns in northern Wisconsin found that disturbance is greatest in regions where lakes and streams are more common (Radeloff et al. 2001), with development be-

Figure 35. Developed lakeshores that could have negative impacts on amphibians and reptiles. (a) High densities of residential buildings (Jefferson County; photo by R. Paloski). (b) Mowed grass up to the water's edge (Portage County; photo by R. Paloski).

ing particularly dense within 100 m (328 ft) of lakes (Gonzalez-Abraham et al. 2007). As the development of lakeside habitat progresses, additional habitat modifications that negatively impact wildlife become more prevalent. For example, residential developments on lakeshores sometimes deploy riprap or large rocks placed along shorelines to stabilize banks and control floodwaters (which could also restrict movement of herpetofauna, see Control of soil and bank erosion subsection) and mow lawns to the water's edge (which reduces cover for herpetofauna; Figure 35). It is, therefore, not surprising that lakeshore development in northern Wisconsin negatively affects amphibian populations. Woodford and Meyer (2003) reported that relative abundance of adult Green Frogs (*Lithobates clamitans*) decreased as the area of suitable lakeshore habitat decreased due to development. However, the interplay between lakeshore development and amphibians in Wisconsin can be complex. Paloski (2008) found no relationship between lakeshore development and amphibian diversity at study sites in Portage County, as well as a positive relationship between agricultural land and amphibian diversity. Yet in that study amphibians were more often observed in long stretches of shoreline that were not fragmented by development, and the single lake completely surrounded by urban development in the study contained only one species (American Toad [*Anaxyrus americanus*]; Paloski 2008). Thus, scale likely plays an important role in assessing the impacts of lakeshore development.

Alteration of streams and rivers

Many amphibians and reptiles in Wisconsin use streams and rivers for periods of time to meet a number of their ecological needs. There are also species that associate almost exclusively with rivers or streams for the majority of their lives (e.g., map turtles [*Graptemys* spp.], Smooth Softshells [*Apalone mutica*], and Queensnakes [*Regina septemvittata*]). Unfortunately, the suitability of river and stream habitats for amphibians and reptiles is affected by a diverse array of factors, so comprehension of all associated threats can be challenging. Relevant threats include not only invasive species and environmental contaminants but also factors that influence habitat structure, adjacent landscape use, and suitable connectivity (further reviewed in Management section). For example, disturbance to riverbank or streambank vegetation at specific locations can affect bank stabilization (Figure 36) or alter the extent of woody debris in waterways that influence water flow and create structure for herpetofauna (Allan and Castillo 2007). At larger spatial scales, assessment of entire drainage basins (i.e., all land that drains water into a stream or river) allows for inclusion of disturbance to habitats adjacent to the main channel, including backwaters and uplands. For example, Sterrett et al. (2010) found that the composition of turtle communities changed as landscape-level disturbance increased along riparian corridors (higher species evenness associated with undisturbed land cover). Drainage basin investigations also allow for the consideration of fragmentation that separates rivers from nearby uplands and influences species that spend time in both (e.g., nesting turtles). Yet assessing effects at large scales can make identifying the source of impacts at isolated points difficult (reviewed by Allan 2004).

Figure 36. An example of a bank along the Black River experiencing heavy erosion. This type of erosion can occur due to removal of vegetation associated with manicured lawns in residential developments (La Crosse County; photo by R. Staffen).

Many negative effects that main channel and floodplain habitats experience stem from attempts by humans to control variation and natural disturbances in streams and rivers (Ward et al. 1999). For example, humans often artificially stabilize banks and straighten riparian channels to accommodate watercraft, which prevents rivers from meandering and reduces heterogeneity in floodplain habitats (Ward et al. 1999). VanDeWalle (1993) reported that riprap to control erosion and flooding could inhibit migration of nesting turtles along rivers. Another important example is the use of dams to regulate water flow and depth in streams and rivers. Dams are often implemented to reduce the potential for flood-related damage to human infrastructure or to control water levels for ship traffic and commerce (Ward et al. 1999, Alexander et al. 2012). Unfortunately, dams can act as barriers to movement and migration of species, thus fragmenting populations, particularly for fishes and turtles. Dams also cause upstream water levels to rise, so that larger areas of the associated landscape are permanently inundated, while drastically altering downstream habitats that experience reduced water levels (reviewed by Rood et al. 2005). Several studies support that these types of disturbance can impact herpetofauna. Tucker et al. (2008) captured fewer turtles in both the impoundments behind dams and the tailwaters downstream compared to adjacent backwater habitat along the upper Mississippi River System. Furthermore, the natural variation in water levels (that dams can otherwise restrict) results in spring floods that create important floodplain wetlands for breeding amphibians and low water levels in summer that expose sandy shores for nesting turtles. Along these lines, VanDeWalle (1993) suggested that artificially high water levels due to dikes in upper midwestern rivers altered water flow and current, which then redistributed sand from banks used by nesting turtles and deposited it elsewhere. Lenhart et al. (2013) also found that impoundments along the upper Mississippi River reduced available habitat for nesting turtles as exposed sandbars once suitable for nesting became inundated. Recreational use of sandbars for camping and anchoring watercraft could also negatively impact nesting turtles. Heavy motorboat traffic and camping associated with sandbars could discourage turtles from laying eggs in locations they would otherwise select, while recreationalists could also inadvertently crush turtle nests.

Figure 37. A female Blanding's Turtle (*Emydoidea blandingii*) killed along a road by motor vehicles prior to laying eggs (Iowa County; photo by C. Raimond).

Habitat fragmentation and population impacts due to roads

Amphibians and reptiles are particularly susceptible to the impacts of roads, a threat that can be applicable to all of the habitat types outlined previously (Rytwinski and Fahrig 2012; Figure 37). The most obvious impact of roads is direct mortality, which reduces abundance and potentially alters population structure (e.g., Steen and Gibbs 2004, Patrick and Gibbs 2010) and can be further influenced by road density or traffic intensity (e.g., Fahrig et al. 1995). Although population-level effects of roads are easy to comprehend, other impacts from roads are more subtle than outright mortality. Roads fragment habitats and bisect migration corridors, which can restrict movement (Patrick and Gibbs 2010) and alter population genetic structure (Jackson and Fahrig 2011). Such fragmentation can also negatively impact community characteristics, such as species richness, because some species are more vulnerable to road mortality than others (reviewed by Forman and Alexander 1998, Lehtinen et al. 1999, Andrews et al. 2008, Beebee 2013, Marsh et al. 2017). Furthermore, a species' vagility could influence its probability of road mortality, with more mobile species being more likely to encounter roads and vehicles (Carr and Fahrig 2001). Behavioral responses to and likelihood of associated mortality from road traffic can also vary by species (e.g., some species tend to remain motionless on roads rather than fleeing in response to approaching vehicles; Andrews and Gibbons 2005, Mazerolle et al. 2005). Due to a lack of local research, further work on potential influences that roads have on amphibians and reptiles in Wisconsin is warranted.

Exotic and Native Pest Species

Diamond (1989) identified introduced species as part of his proposed "evil quartet" responsible for modern extinctions. Richardson et al. (2000) defined an 'introduced' species as one transported across a geographical barrier by humans to a location outside of its native range. Once outside of their native range, introduced species are also referred to as 'exotic' (as a poignant example, Lannoo [1997] stated that "European Americans are the ultimate exotics"). Richardson et al. (2000) further classified introduced species as 'invasive' if they produced reproductive offspring in these new areas. However, not all problematic species are exotic. In fact, Richardson et al. (2000) indicated that, whether native or exotic, a species could be considered a 'pest' if it caused environmental or economic harm. Human disturbance can actually result in native species having negative and unnatural impacts on other native species. As an example, native species relocated to a habitat in which they did not originate but that is still within their normal geographic distribution can become problematic. There are also native species that become problematic because their populations grow to unusually large sizes as the result of human activities (see Exotic and native pest animals subsection). Several investigations of the impacts that problematic species (invasive or native) have on amphibians and reptiles exist (e.g., reviewed by Gibbons et al. 2000), although relatively little work has been conducted in Wisconsin.

Exotic invasive plants

The invasion of exotic plants across the US has become a large, seemingly insurmountable, and costly problem (Pimentel et al. 2005). Numerous exotic plant species native to Europe and Asia are currently prevalent in Wisconsin and likely cause extensive ecological damage. Examples include Reed Canary Grass (*Phalaris arundinacea*) and *Phragmites* spp. in wetlands, Eurasian Water-Milfoil (*Myriophyllum spicatum*) in lakes and rivers, European (Common) Buckthorn (*Rhamnus cathartica*) and Garlic Mustard (*Alliaria petiolata*) in woodlands, Japanese Honeysuckle (*Lonicera japonica*) in floodplains and woodlands, and Spotted Knapweed (*Centaurea maculosa*) in prairies. Exotic plant species can cause declines in the richness and abundance of native plant species (Tyser and Key 1988), which modifies the composition and structure of habitat for amphibians and reptiles. For example, Garlic Mustard is a common woodland exotic in Wisconsin that reduces native plant species richness and modifies habitat structure by influencing the establishment of native tree seedlings (Stinson et al. 2007). Similarly, European Buckthorn is an exotic woody species that can dominate forests in southern Wisconsin, altering woodland structure and plant species composition (Mascaro and Schnitzer 2007; Figure 38a).

The impacts of introduced exotic plants can be further exacerbated by human activities. For example, Parker et al. (1993) found that exotic plants dominated anthropogenically disturbed survey locations in southeastern Wisconsin, which could suggest that undisturbed communities are less susceptible to invasion. The authors also found that the difference in soil characteristics between disturbed and undisturbed locations was a particularly important underlying factor in the presence of exotic plants. In contrast, habitats that experience frequent 'natural' disturbance are apparently more resilient to invasion by exotic plants. Predick and Turner (2008) found that exotic shrubs (*Lonicera* spp. and *Rhamnus* spp.) were less resilient than native plants to disturbance in the form of floods in lowland forests along the Wisconsin River. Invasive plants are also more prevalent in small isolated patches of habitat and along edges (such as those associated with roads) than in large contiguous patches of established native vegetation (e.g., Predick and Turner 2008).

Figure 38. Problematic invasive plant species in Wisconsin. (a) A dense stand of European (Common) Buckthorn (*Rhamnus cathartica*) at the edge of a field in southern Wisconsin (Rock County; photo by J. Kapfer). (b) A wetland dominated by Reed Canary Grass (*Phalaris arundinacea*; Jefferson County; photo by J. Kapfer).

The effects that exotic plants have on amphibians and reptiles are somewhat poorly understood, and data specifically related to impacts on upper midwestern herpetofauna are extremely sparse. Mifsud and Thomas (2013) reported that amphibian species richness was negatively associated with the presence of Garlic Mustard and European Buckthorn along an urbanized river in Michigan. Work in Illinois found that European Buckthorn also releases a secondary metabolite that causes malformation and mortality in developing amphibians (Sacerdote and King 2014). Research on the interaction between an invasive wetland plant (Reed Canary Grass) and gartersnakes (*Thamnophis* spp.) in Wisconsin determined that rarer gartersnake species used dense stands of Reed Canary Grass less often than more common gartersnake species (Kapfer et al. 2013b; Figure 38b). On the other hand, Rittenhouse (2001) reported little evidence that Reed Canary Grass influenced habitat quality for anuran larvae collected in Wisconsin, although she suggested that excess decomposition of dead plant biomass in dense stands of Reed Canary Grass could cause nutrient spikes that may reduce survival of anuran larvae. Along these lines, controlled experiments revealed that the quality of detritus produced by decaying invasive wetland plants (based on carbon: nitrogen ratios) influenced the survival to metamorphosis of anuran larvae, with invasive plants contributing to lower rates of metamorphosis in several species native to Wisconsin (Maerz et al. 2010). Yet a recent study in Oregon found evidence that Reed Canary Grass provided certain benefits to both adult and larval Pacific Chorus Frogs (*Pseudacris regilla*; Holzer and Lawler 2015). Thus, the importance of species-specific responses must be considered when assessing impacts of exotic invasive plants on amphibians and reptiles. Furthermore, the breadth of threats posed by invasive plants is difficult to understand, in part because their impacts can be indirect. Several authors suggest that exotic plants create ecological or evolutionary 'traps' for native species (e.g., an exotic plant that attracts native wildlife as potential habitat but is unsuitable and ultimately results in mortality or reduced fitness; Battin 2004, Schlaepfer et al. 2005b). Along these lines, research in Ontario found that an exotic invasive wetland plant (Common Reed [*Phragmites australis*]) that also occurs in Wisconsin did not cause direct turtle mortality, but it reduced the availability of suitable nesting habitat for aquatic turtles (Bolton and Brooks 2010). Furthermore, the spread of *Phragmites* at this site was so rapid that between the time of turtle egg deposition (which occurred prior to plant expansion) and hatchling emergence, either nests were shaded or plant root systems had grown through them.

Exotic and native pest animals

Several animal species deserve mention here, although their potential impacts on amphibians and reptiles in Wisconsin are largely unknown. These include not only invertebrates such as Zebra Mussels (*Dreissena polymorpha*), Spiny Waterfleas (*Bythotrephes longimanus*), and exotic earthworms but also various vertebrates that range from fishes to mammals. American Bullfrogs (*Lithobates catesbeianus*), for example, are a highly invasive species in parts of North America, particularly in western states where they are exotic (reviewed by Funk et al. 2011). When invasive, American Bullfrogs have negative impacts on local species, including other amphibians (e.g., Stiles et al. 2016). Although American Bullfrogs are naturally found in Wisconsin, Casper (1996) suggested that introductions have commonly occurred due to recreational fishing in the state (i.e., frogs used as bait that were discarded and became established). These introduced populations could stem from nonnative stock and have numerous effects on local species, from the intermixing of genes with native

American Bullfrog populations to predation of other native amphibians. Yet the origin of state populations and possible impacts of introductions in Wisconsin are virtually unknown. Similarly, feral and free-roaming cats (*Felis catus*) have drastic negative impacts on wildlife globally (reviewed by Loss et al. 2013). They are estimated to kill hundreds of millions of amphibians and reptiles in the US alone (Loss et al. 2013), a phenomenon that certainly occurs in Wisconsin but is also currently unquantified. Other exotic vertebrates, such as Pond Sliders (*Trachemys scripta*) and Feral Swine (*Sus scrofa*), are also sporadically observed in the state (Taylor 1999, Brown et al. 2012c, WDNR 2018a). However, confirmation of viable Wisconsin populations of both species and assessment of their potential impacts to native herpetofauna do not exist (see further review of Pond Sliders in Hypothetical and Enigmatic Species of Wisconsin chapter).

Past studies indicate a negative correlation between midwestern amphibians and the presence of fishes, whether piscivorous, planktivorous, or herbivorous (Knutzon et al. 2004, Herwig et al. 2013). A number of exotic fish species that are established in Wisconsin, including Bighead Carp (*Hypophthalmichthys nobilis*) and Silver Carp (*Hypophthalmichthys molitrix*) could have substantial impacts on resident herpetofauna. Yet regardless of whether fishes are exotic or native, amphibians typically do not fare well in habitats where fishes are established (Stiles et al. 2016). Lannoo (1998) stated that "predatory interactions between midwestern fishes and tadpoles will always favor fishes" but also noted that large salamander larvae may be more resistant to fish predation, depending on the size of the interacting individuals. Yet under natural circumstances, fishes do not occur in ephemeral wetlands frequented by amphibians, because these habitats often experience conditions that fish species cannot tolerate, such as hypoxia (insufficient oxygen levels) or desiccation during certain times of year (reviewed by Lannoo 1998). However, anthropogenic alterations in hydrology that favor the permanency of wetlands can make otherwise fishless wetlands conducive to colonization by fishes (reviewed by Lannoo 1998) and American Bullfrogs (reviewed by Brown et al. 2012b). Anthropogenic alterations to hydrology occur for a variety of purposes, such as hydropower dams, aquaculture, land development, and agriculture. Such alterations are well-documented in Wisconsin and other midwestern states.

Several additional native vertebrate species (particularly mammalian) are of concern because their populations have grown unnaturally large in areas with high human densities. Populations of midsized mammalian predators, such as Raccoons (*Procyon lotor*), are substantially larger in urban vs. rural locations because they are effective at exploiting resources in anthropogenically altered landscapes (reviewed by Prange and Gehrt 2004). High rates of turtle nest predation by Raccoons and other mammals is well-documented (e.g., Marchand and Litvaitis 2004a; Figure 39) and could be an important stressor on urban turtle populations. Yet nest predation in less disturbed habitats also occurs, which is documented to alter the overall size and structure of the resident turtle populations (Browne and Hecnar 2007). These negative impacts, coupled with the

Figure 39. A common sight in southern Wisconsin: a clutch of freshly laid turtle eggs raided by a predator, likely a midsized mammalian carnivore such as a Raccoon (*Procyon lotor*; Rock County; photo by J. Kapfer).

fact that urban landscapes are prevalent (particularly in southern Wisconsin), support that human-subsidized populations of mammals such as Raccoons is a conservation concern that should not be overlooked.

Pollution

Pollution occurs when unwanted constituents, called contaminants, enter the environment and cause an undesirable change (Botkin and Keller 2003). These contaminants are accidentally or deliberately introduced and can damage living organisms such as amphibians and reptiles. Contaminants that negatively impact biota are often referred to as toxicants or xenobiotics. Toxicants can take a variety of forms, some of which are obvious (e.g., excessive pesticides or herbicides introduced into aquatic ecosystems). On the other hand, some compounds that we normally do not associate with toxicosis can become toxic at unusually high levels. Nutrients such as phosphorus and nitrogen are essential to living organisms and required in moderate amounts. Yet if large doses of nutrients enter the environment (perhaps through runoff of fertilizers from agricultural fields), they can reach toxic levels and have negative impacts on living organisms, which is well documented in amphibians (e.g., Houlahan and Findlay 2003, Knutson et al. 2004).

The list of compounds that can be potentially toxic to amphibians and reptiles is long (reviewed by Sparling et al. 2010). Amphibians, in particular, are susceptible to many environmental contaminants (reviewed by Bridges and Semlitsch 2005), in part because their skin is highly permeable (Quaranta et al. 2009), which facilitates uptake of possible toxicants. Still, assessing the risk that contaminants present to amphibians and reptiles in Wisconsin is difficult, in part because the past and present state of environmental pollution in the Upper Midwest is complex (reviewed by Wiener and Sandheinrich 2010). Furthermore, while government regulations have led to a reduction in environmental pollution from some contaminants (such as heavy metals), other emerging contaminants are poorly regulated, and their impacts are not fully understood (such as endocrine disruptors; Wiener and Sandheinrich 2010). A holistic understanding of how contaminants influence all herpetofauna in Wisconsin is further complicated because, globally, the toxicology of amphibians has received more attention than that of reptiles (but see example reviews on reptiles by Campbell and Campbell 2000, 2001, Gardner and Oberdorster 2005).

Sublethal effects of contaminants on amphibians and reptiles

Understanding the effects of contaminants is especially challenging because these impacts can run the gamut from overt to understated. As a hypothetical example, it is easy to grasp the cause and effect of an accidental pesticide spill that seeps into a nearby pond and leads to mass mortality of frogs. Yet the impacts of environmental contaminants can be much more subtle. For instance, it is more difficult to recognize the effect of a contaminant when it does not cause mortality but instead causes a hormonal imbalance that alters behavior, growth, or development (reviewed by Blaustein et al. 2011). These subtle effects, referred to as sublethal effects, can be just as insidious as the more overt impacts but could go unnoticed. For example, contaminants accumulated in the tissues and eggs of Snapping Turtles (*Chelydra serpentina*) from Canada were not substantial enough to cause mortality but were high enough to potentially depress proper function of the endocrine or other organ systems (de Solla et al. 2007, Letcher et al. 2015). Additionally, when researchers do not address sublethal effects, the results of studies focused on contaminants can appear contradictory. Allran and Karasov (2001) exposed wild amphibian embryos and adults to the herbicide atrazine and found it did not cause mortality or impact swimming speed. Yet later research found that while exposure to low concentrations of atrazine may not kill frogs, it can produce hermaphroditic and chemically castrated individuals (Hayes et al. 2002a, reviewed by Hayes 2004). Atrazine has been used for over 50 years in Wisconsin (WDATCP 2015). Although its use is regulated (enforced standard for levels of atrazine in Wisconsin groundwater is 3.0 parts per billion at the time of this writing), it persists in the environment at levels that can have sublethal effects on amphibians (WDATCP 2015). A study of 2,177 rural wells across southern and western Wisconsin found that 27% contained atrazine levels of 0.1 parts per billion or greater, and some contained levels that exceeded 3.0 parts per billion (Brady et al. 1995). Unfortunately, past studies reported that atrazine concentrations as low as 0.1 to 2.5 parts per billion can cause hermaphroditism or other adverse impacts to the reproductive success of frogs (Hayes et al. 2002b; Hayes et al. 2010). Comprehension of the threat atrazine presents can further be challenging because the likelihood of amphibian exposure to this toxicant can vary based on landscape disturbance and climate. For example, Sadinski et al. (2014) reported that amphibians in the Midwest (including Wisconsin) were at greater risk of exposure to atrazine in wetlands near areas of high corn production. They also speculated that differences in weather patterns and even farming practices (e.g., use of annual crop rotation) could lead to variation in the likelihood of exposure.

Research on the impacts of environmental pollution on amphibians from Wisconsin

Although numerous studies have investigated the influence of contaminants on upper midwestern herpetofauna (e.g., Ling et al. 1986, Helgen et al. 1998, Diana et al. 2000, Knutson et al. 2004), research focused on Wisconsin, particularly concerning reptiles, is sparse. Howe et al. (1998) found that when mixed, the pesticides atrazine and alachlor appeared to cause synergistic toxic impacts to amphibians from western Wisconsin. The previously mentioned laboratory tests on the effects of atrazine by Allran and Karasov (2001) also included amphibians collected in Wisconsin. Jofré and Karasov (1999) tested the effects of ammonia on the survival, malformation rate, and development of anuran embryos and larvae collected in Minnesota, the Dakotas, and Wisconsin. They observed negative impacts at exposure concentrations of ≥0.6 mg/L (0.00008 oz/gal; Green Frogs) and 1.5 mg/L (0.0002 oz/gal; Northern Leopard Frogs [*Lithobates pipiens*]). Interestingly, American Toads were unaffected, which supports the hypothesis that the effects of toxicants vary among amphibian species. The authors determined that some natural Wisconsin waterways can have ammonia concentrations ≥0.04 mg/L (0.000005 oz/gal). Karasov et al. (2005) reported that frog egg hatchability and frog species richness were negatively correlated with contamination levels at experimental locations in Brown and Winnebago Counties. In addition, a series of laboratory studies of contaminant impacts on Northern Leopard Frog larvae (using eggs collected in Wisconsin) showed that environmentally relevant concentrations of lead (Chen et al. 2006), copper (Chen et al. 2007), arsenic (Chen et al. 2009), cadmium (Gross et al. 2007, 2009), polychlorinated biphenyls (Huang et al. 2007, Jofré and Karasov 2008), and polybrominated diphenyl ether (Cary Coyle and Karasov 2010) had sublethal impacts on the species. Exposing larvae to lead (0.1 mg/L

[0.00001 oz/gal]) resulted in spinal deformities and delayed time to metamorphosis (Chen et al. 2006). Similarly, laboratory studies on Wood Frog larvae (using eggs collected in Wisconsin) found sublethal impacts from exposure to the chemical thiosemicarbazide (Riley and Weil 1986). The effects of water acidity on amphibian communities in northern Wisconsin were investigated by several authors. For example, Kutka and Bachmann (1990) suggested that acid deposition in breeding ponds from purported acid rain and snowmelt could be of concern to ambystomatid salamanders in larger ponds that receive substantial input from rainwater. Furthermore, Kutka (1994) conducted laboratory experiments with wild collected salamander larvae from Wisconsin and reported that individuals exposed to lower pH experienced reduced survival or higher activity levels.

Amphibian malformations

It is intuitive to associate the phenomenon of amphibian malformations with environmental pollution (reviewed by Souder 1999, Lannoo 2008). In the past, the phenomenon of malformed amphibians was popularized through the media and received much attention, with particular focus on the Upper Midwest (reviewed by Souder 1999). Although malformations occur naturally (at a rate of ca. 5%), starting in the mid-1990s, a surprising number of amphibian populations were identified globally with malformation rates that were well beyond normal levels (reviewed by Blaustein and Johnson 2003). Malformations are usually associated with frogs and can range from missing eyes to many extra limbs (reviewed by Lannoo 2008). Malformations have been identified as the cause of frog population declines in several situations (Hoppe 2005), but often the public's greatest concerns about amphibian malformations center on what they represent for human health. (If water causes frog malformations, is it dangerous to humans?) A substantial amount of research on the causes of amphibian malformations was conducted (mostly outside of Wisconsin) through the late 1990s and early 2000s (reviewed by Lannoo 2008). These studies proposed various potential causes for malformations, such as environmental contaminants, parasites, and UV-B radiation exposure, all of which can disrupt normal amphibian development (e.g., Blaustein et al. 2003, Lannoo et al. 2003, Sparling et al. 2010). Yet as Lannoo (2008) pointed out, many of these causes could be connected to a single factor: runoff of tainted water from adjacent uplands into wetlands. Pollution in runoff can lead to wetland contamination that causes malformations. Furthermore, excess nutrients in runoff, such as nitrogen, can cause nutrient spikes that support higher populations of hosts (e.g., aquatic snails) for parasites that cause amphibian malformations. Thus, controlling the runoff of polluted water into aquatic systems could address the root of the malformation problem in many situations (Lannoo 2008).

Figure 40. A malformed Blanchard's Cricket Frog (*Acris blanchardi*) with an abnormal fleshy extension growing off the rump (Iowa County; photo by R. Paloski).

A review of data compiled by the WDNR revealed reports of malformed amphibians from 47 of Wisconsin's 72 counties (R. Paloski, WDNR, unpublished data; Figure 40). A cursory review by DuBois (1996) suggested that malformation reports were more frequent in Wisconsin by the late 1990s, with only two reports made to the WDNR prior to 1992. Data on malformation rates at Wisconsin sites are rare, although Green et al. (2002) reported a 2% malformation rate among recently metamorphosed Green Frogs analyzed from an undisclosed Wisconsin location. Furthermore, Hartson et al. (2011) reported anuran malformations at an agriculture pond in west-central Wisconsin. Yet the overall number of sites with identified malformations in Wisconsin is generally low, especially compared to other nearby states such as Minnesota (reviewed by Lannoo 2008). Furthermore, relatively few malformed individuals are found in populations from Wisconsin sites where abnormalities are documented (i.e., there is a low incidence of malformations at sites where they occur; R. Paloski, unpublished data). One exception is a Trempealeau County site discussed by Lannoo et al. (2003). This location produced a relatively high number of unique malformations, which prompted the authors to suspect that the cause of these malformations was not typical (but also not identified).

Disease

Disease is a natural occurrence. Most disease hosts have evolved under pressure from commonly encountered pathogens; thus, there is not typically a large negative impact on populations as a result of infection. However, 'emerging infectious diseases,' or diseases that have recently appeared within a population of hosts or that have experienced increases in incidence or geographic range, are of greater concern. When emerging infectious diseases attack naive host immune systems the effects are more serious, potentially leading to population collapse or extinction of hosts (Briggs et al. 2005). Emerging infectious diseases that impact amphibians and reptiles have become a large conservation concern in recent years, and several are currently known to affect populations in Wisconsin.

Chytrid fungus

Batrachochytrium dendrobatidis (Bd) and, to a lesser extent, *B. salamandrivorans* (Bsal; Martel et al. 2013) are fungal pathogens that have caused global amphibian declines and extinctions (Skerratt et al. 2007). They affect the epidermis and disrupt the ability of infected individuals to achieve osmotic balance, which results in mortality (reviewed by Ortiz-Santaliestra et al. 2013). The fungi spread via small structures called zoospores, which can be transmitted by direct contact between amphibians, as well as by contact with infected water or soil (Fisher et al. 2009). Zoospores can also be accidentally transported by humans on clothing and equipment (Johnson and Speare 2005). There is large variation in the factors that apparently contribute to Bd prevalence, and recent work suggests that susceptibility varies by species (e.g., Talley et al. 2015). For example, the observed prevalence of Bd is often lower in leopard frogs (*Lithobates* spp.) than other species sampled (e.g., Sadinski et al. 2010, Klemish et al. 2012, Ortiz-Santaliestra et al. 2013), potentially because leopard frogs produce skin secretions that provide some level of resistance (Rollins-Smith and Conlon 2005). Susceptibility can also vary based on amphibian developmental stage. Ortiz-Santaliestra et al. (2013) reported that mortality from Bd infection was higher in recently metamorphosed American Toads (70%) than juveniles exposed at four weeks post-metamorphosis. The prevalence of Bd infection also appears to differ due to variation in habitat and environmental factors. For example, amphibians that spend substantial time in upland habitats can be particularly prone to Bd infection (e.g., Ouellet et al. 2005).

Infection rates also appear to be high in wetter or colder months, when the immune function of ectothermic organisms is likely depressed (Ouellet et al. 2005, Lannoo et al. 2011). Recent models predict that the incidence and intensity of Bd infection will shift north as climate change progresses and increases the availability of suitable environments for the fungus, which suggests that temperate populations could experience greater impacts from this disease in the future (Xie et al. 2016).

Several studies have investigated the prevalence of Bd in wild anurans sampled from midwestern locations such as Wisconsin (Loda and Otis 2009, Rodriguez et al. 2009, Sadinski et al. 2010, Klemish et al. 2012, Standish et al. 2018). Klemish et al. (2012) found Green Frogs infected with Bd at 28% of 50 sample sites across Wisconsin. (Note: in 14% of sites, Bd presence/absence could not be determined.) Sadinski et al. (2010) detected Bd at 45% of 75 sites sampled across several midwestern states, including Wisconsin. Bd was also detected in a 2014 sample of dead Northern Leopard Frogs collected from Trempealeau County (Wildlife Health Information Sharing Partnership 2017). More recently, Standish et al. (2018) tested for the presence of Bd and Bsal in 13 amphibian species, including various life stages, across several counties in Wisconsin. While they detected no evidence of Bsal in sampled amphibians, Bd prevalence was high in some species (e.g., it exceeded 70% in sampled Mudpuppies [*Necturus maculosus*]). It is noteworthy, however, that the authors did not report observing cases of chytridiomycosis (the disease caused by Bd infection) in animals from infected populations. A study in Illinois concluded that drought limited Bd prevalence and infection intensity (Beyer et al. 2015). The authors further predicted a lower prevalence of Bd in wetlands with more open canopies due to higher sun penetration (heat is not conducive to fungal growth). However, the results of the study were complex and not consistent within or among species.

Ranaviruses

Pathogens of the genus *Ranavirus* affect cold-blooded vertebrates (i.e., amphibians, reptiles, and fishes), and several cases of mass amphibian mortality are attributable to them (reviewed by Gray et al. 2009). Ranaviruses are spread through many routes, including contact with contaminated water, soil, or infected individuals, and cause gross lesions, hemorrhaging, and chronic cell death. This can result in mortality within days or weeks of infection (reviewed by Gray et al. 2009). Furthermore, Brenes et al. (2014) confirmed that transmission can occur across vertebrate taxa, such as amphibians, reptiles, and fishes. In that study, amphibians that received a ranavirus from contact with infected reptiles and fishes typically experienced mortality, while infected fishes and reptiles did not. This suggests that the latter two groups can act as disease reservoirs in aquatic systems. It is difficult to control ranaviruses, as they can be spread among waterbodies by humans (on boots or equipment) or from stocking fish, and they can persist for weeks outside of hosts within aquatic environments (reviewed by Gray et al. 2009). Records of mass mortality due to a ranavirus exist from Wisconsin, and it was determined as the cause of death in two cases from Vilas County, including ca. 800 frogs in 2009 (Green Frog, Northern Leopard Frog, American Bullfrog) and ca. 300 frogs in 2014 (American Bullfrogs), with an additional report of 119 frogs from Dane County in 2016 (Wood Frogs; Wildlife Health Information Sharing Partnership 2017). A ranavirus was also suspected as the cause of death in 18 Green Frogs analyzed in 1997 from Columbia County (Wildlife Health Information Sharing Partnership 2017). Sekowska et al. (2014) reported an infected American Bullfrog sampled from Grant County. The most extensive sampling for ranaviruses in Wisconsin to date was conducted by Stadish et al. (2018). Their investigation of 13 amphibian species, spanning numerous counties, detected ranaviruses in 6 species, with high estimates of prevalence in some, causing further concern for the threat this disease poses to resident amphibians.

Figure 41. A Timber Rattlesnake (*Crotalus horridus*) with snake fungal disease (SFD). Note the thickened scabby scales around the mouth (*left*) and on the ventral scales (*right*; Grant County; photo by C. Henaghan).

Snake fungal disease

Snake fungal disease (SFD) is an infectious disease caused by the fungus *Ophidiomyces ophiodiicola*. The disease has been confirmed in wild populations of snakes in the eastern and midwestern US (Allender et al. 2015, Lorch et al. 2015, Baker et al. 2018). Clinical symptoms of infected snakes can be subtle, including swollen and slightly discolored scales; symptoms in severe cases include lesions and severe dermatitis that results in scabs, crusty scales, and skin nodules (Sleeman 2013, Lorch et al. 2015). Symptoms of infection are often most apparent on the head, with severe infections causing nasal passages to swell shut or swollen oral tissues that affect the snake's ability to close its mouth, but lesions can occur at many locations on the body (Allender et al. 2015, Lorch et al. 2015).

SFD has been confirmed in snake populations at various Wisconsin locations (Lorch et al. 2016; Figure 41). The fungus seems to be most common in snakes that use communal hibernacula and has been detected in Wisconsin on Timber Rattlesnakes (*Crotalus horridus*), Eastern Milksnakes (*Lampropeltis triangulum*), and Eastern Foxsnakes (*Pantherophis vulpinus*; Lorch et al. 2016). As climate change results in warmer winter conditions in Wisconsin snake hibernacula, potentially leading to more desirable conditions for fungal growth, snakes could experience heavier infection with SFD than is currently known (Allender et al. 2015). There are limited recommendations available for how to offset the impacts or threats associated with this disease. Baker et al. (2016) suggested that its potential for spread could be reduced if researchers and habitat managers disinfected equipment and footwear when moving between sites. The authors also suggested that the possible impacts of SFD be considered during the creation of snake conservation strategies.

Removal from the Wild

Diamond (1989) believed that excessive harvest of wild organisms was also one of several primary causes for global extinctions. Not surprisingly, the removal of amphibians and reptiles from the wild for food, skins, pets, traditional medicines, and research is a global conservation concern. An estimated 14.7 million wild-caught amphibians and 26 million wild-caught reptiles were imported or exported in the US alone from 1998 to 2002. (Note: these figures include only whole animals, not body parts or products; reviewed by Schlaepfer et al. 2005a.) In addition to affecting wild population stability, trade in herpetofauna is implicated in the spread of infectious wildlife diseases (e.g., chytrid fungus) and invasive species (reviewed by Fisher and Garner 2007). Quantifying the impacts of the global trade of herpetofauna presents many challenges because not all countries regulate such trade, and some (including the US) do not always record species-level data for organisms that are internationally transported (Schlaepfer et al. 2005a).

At a more local scale, mention of 'game species' probably does not trigger images of amphibians and reptiles for the average Wisconsin citizen. Indeed, few amphibians and reptiles are subject to the intentional, large-scale harvest experienced by traditional game species such as White-Tailed Deer (*Odocoileus virginianus*), waterfowl, and game fish. Despite this, intentional and accidental harvest of herpetofauna in Wisconsin poses a potentially serious threat to numerous species. Unfortunately, it can be difficult for conservation biologists to understand and offset the potential impacts because they are largely unquantified in Wisconsin. These realities highlight the importance of WDNR regulations on the harvest and collection of amphibians and reptiles in Wisconsin (see Conservation Status and Regulations section). Limited resources and funding exist to support ongoing effective monitoring and enforcement of these regulations in the state. Therefore, diligence on the part of conservation-minded citizens and ethical behavior on the part of amphibian and reptile enthusiasts in Wisconsin are critical to the preservation of many species.

Legal collection that leads to overharvesting

Obvious examples of intentional harvest that can impact native herpetofauna include commercial collection of wild amphibians (for food or by biological supply companies) and the accidental or intentional removal of turtles by citizens and commercial harvesters. In the Upper Midwest, for example, Lannoo et al. (1994) reviewed historical accounts to report that some 20 million Leopard Frogs (*Lithobates pipiens*) were harvested annually from a single county in Iowa during the early 1900s. They reported that such removals were a likely cause for the nearly three orders of magnitude decline in Leopard Frog populations they estimated in the same county ca. 100 years later (Lannoo et al. 1994).

In Wisconsin, turtles provide a prime example of legally harvested species whose populations could be impacted by commercial collection, given the results of research from other states (Moll and Moll 2000, Ceballos and Fitzgerald 2004, Barko and Briggler 2006, Zimmer-Shaffer et al. 2014). Due to the typically low survival of nests and hatchlings, coupled with slow development, several studies have found that turtle population stability is particularly susceptible to removal of adults and subadults (Congdon et al. 1993, 1994, Galbraith et al. 1997, Heppell 1998, Zhour and Jiang 2008). Unfortunately, adults are the most desirable age class for harvest because turtles are often sold by weight in food markets (Brown et al. 2011a, Zimmer-Shaffer et al. 2014). In Wisconsin alone, an estimated 112,236 kg (247,438 lb) of turtles were harvested from the Mississippi River between 1953 and 2013 (WDNR, unpublished data). In 1996 Wisconsin's Natural Resources Board videotaped anonymous testimony from a commercial fisherman working on the Mississippi River (R. Hay, personal communication). This individual recalled loading train cars full of turtles every afternoon as a boy during summers in the 1920s and 1930s for shipment to markets in the eastern US. The fisherman further stated that turtle captures in later years (1980s–1990s) were extremely low compared to captures in his early years of fishing. Whether turtles are the intended target of commercial harvest or not, they appear susceptible to mortality when captured as bycatch, or accidentally, during commercial fishing operations. Barko et al. (2004) found that 29%–36% of softshells (*Apalone* spp.) died when accidentally captured by passive fishing techniques in the Mississippi River.

Illegal or unethical collection

It is probable that populations of several amphibian and reptile species in Wisconsin suffer from illegal or unethical wild collection (primarily for pets). The full impact of illegal or unethical collection is, obviously, difficult to quantify because these activities are not frequently reported. However, anecdotal accounts suggest that illegal and unethical collection has likely led to substantial population declines in several species in Wisconsin since the 1970s (e.g., Ornate Box Turtles; Vogt 1981, R. Hay, personal communication). Unfortunately, because these species are also affected by other threats, such as habitat destruction, it is difficult to clearly link wild collection to observed declines. Along these lines, field surveys that employ unethical methods to find herpetofauna can degrade habitat quality, even when catch and release is the goal. For example, compared to undamaged locations, Goode et al. (2004, 2005) found fewer reptiles in rock outcrops where rocks were roughly overturned without replacement or existing cracks were pried open with pry bars by enthusiasts searching for reptiles. It is possible that the impacts of the illegal/unethical pet trade in Wisconsin can be reduced if local pet store employees, owners, and individuals interested in herpetoculture are particularly familiar with and abide by the Wisconsin regulations for collection and sale of native species (see Conservation Status and Regulations section).

Persecution

Amphibians and reptiles, particularly snakes, are often viewed as undesirable by humans (Knight 2008, Ceriaco 2012). This negative sentiment is so strong that herpetofauna observed along roads are sometimes intentionally targeted by passing motorists (Langley et al. 1989, Ashley et al. 2007). Historically, Wisconsin's citizens have persecuted amphibians and reptiles, as is exemplified by past treatment of the state's rattlesnake species (e.g., Camp 1970). Persecution still surely occurs today but, similar to illegal harvesting, is difficult to quantify due to a lack of reporting. Possible exceptions involve (limited) data generated from Wisconsin's past bounty on rattlesnakes. Until the early 1970s, several counties paid bounties for dead Timber Rattlesnakes (*Crotalus horridus*) and Massasaugas (*Sistrurus catenatus*). Admittedly, animals subject to bounties probably experience heavier persecution than those simply targeted out of fear or dislike because monetary gain is the driving motivator. Regardless, the data available from several counties indicate that thousands of rattlesnakes were killed for the collection of bounties (Natural History Box: The Wisconsin Rattlesnake Bounty as an Example of Wildlife Persecution), which likely contributed to the currently rare status of both species in Wisconsin.

Impacts of social media

The importance of withholding specific locality data to protect amphibian and reptile populations from exploitation is not a new concept (Casper 1982). Yet conservation biologists must now also consider the additional threats posed by social media in this regard. Many internet forums that are organized and used by herpetological enthusiasts currently exist, some of which even focus primarily on reporting field observation of amphibians and reptiles in the Midwest (J. Kapfer, UW–Whitewater, personal observation). While these forums are generally overseen and perused by ethical individuals with a strong interest in

NATURAL HISTORY BOX: The Wisconsin Rattlesnake Bounty as an Example of Wildlife Persecution

Wisconsin's rattlesnake bounty, although currently inactive, led to the mortality of thousands of snakes from the time it began through its termination in the early 1970s. It is easy to imagine that a government-funded program, one that doled out monetary rewards to Wisconsin citizens who killed rattlesnakes, played an important role in the present-day scarcity of Timber Rattlesnakes and Massasaugas in the state. Yet quantifying the full impact of the bounty is difficult, because determining precisely how it operated and pinpointing the exact date on which it was enacted pose challenges. Much of the existing information on the rattlesnake bounty is from copies of old correspondence and intradepartmental reports obtained from WDNR employees (Rori Paloski, WDNR, personal communication). For example, a letter sent from Otis S. Bersing to J. R. Smith on 29 March 1957 indicates that although "the state pays bounties on predatory animals, the county board may authorize payment of an additional bounty for the killing of any animal found to be destructive," and species often considered "destructive" included rattlesnakes. Based on this, the decision to pay rattlesnake bounties was apparently left up to individual counties. As such, the year that bounties were initiated likely varied across the state but could have started in some locations by the early 1950s. An unpublished intradepartmental WDNR report written by Richard P. Thiel (WDNR [retired]) in the early 1990s used Wood County as an example and quotes the following resolution: "Pursuant to provisions of Sect. 29.61 of 1949 Wisconsin Statutes that Wood Co. pay bounty of 50 cents for each rattlesnake killed on or after Jan. 1 1952."

Although some information on the bounty's impact could be gleaned from counties where the date of instigation is known, understanding the bounty's effects also requires numbers. Unfortunately, either county governments did not carefully record the number of rattlesnakes removed for the bounty or those records were lost. As such, estimates on the number of rattlesnakes bountied could be forced to consider anecdotal information. Although anecdotal information is less desirable than hard data, it can provide nuggets of insight, especially if carefully recorded numbers do not exist. An excellent example of this comes from Messeling (1953). Mr. Messeling, a Crawford County resident and rattlesnake hunter in the early to mid-1900s, summarized his recollections of killing Timber Rattlesnakes for the Wisconsin bounty. He described removing up to "3,000 rattlesnakes a year, but most years . . . only 1,000," and as many as 51 individuals from under a single rock. Certainly, the accuracy of Mr. Messeling's numbers can be questioned, but his account strongly suggested that a staggering number of snakes were possibly killed by bounty hunters. In addition to these anecdotal reports, some early attempts were made to actually quantify how many rattlesnakes were killed for the bounty. Unfortunately, a number of the reports associated with these attempts are difficult to interpret (e.g., they report the number of bounties paid but include ambiguous time periods). Yet useful information on the extensive mortality can be assembled from the counties that apparently kept better records on rattlesnake bounties paid. According to WDNR files, the late Dr. Ruth Hine was involved in efforts to end the rattlesnake bounty in the early 1970s. During her time as the chairman of the WDNR's Endangered Species Committee, Hine compiled notes and correspondence on rattlesnake bounties to assess the program's impact. Her important efforts uncovered actual numbers from several counties. For example, her notes indicate that bounties were paid on ca. 1,372 rattlesnakes in Buffalo County from 1967 to 1973. Her notes also include a letter from Richard Vogt (see Some Prominent Figures in Wisconsin Herpetology chapter) on 28 February 1978, who reported that from 1968 to 1972 Crawford County records revealed that 12,444 rattlesnake bounties were paid (although Vogt questioned the validity of this number, indicating possible fraud on the part of the rattlesnake hunters). A WDNR intradepartmental memorandum from Richard Johnson to Paul Brandt on 29 March 1974 reported that from 1964 to 1973 a total of 1,017 and 371 rattlesnake bounties were paid in Grant and Richland Counties, respectively. Although record keepers were not required to distinguish between Timber Rattlesnakes and Massasaugas, the county in which bounties were paid provides some clues. As Thiel reviewed in his previously mentioned report, bounty records from Juneau County (where Timber Rattlesnakes are not documented) are likely all Massasaugas. These bounties, which were paid for snakes collected in only a four-township area, totaled 4,286 from 1952 to 1972 (R. Thiel, unpublished data). A comparison of the bounties paid over time gives some insight into the severe decline that this species experienced there. For example, from 1952 to 1960 the annual mean number of rattlesnake bounties paid in Juneau County was 396. However, from 1961 to 1972 this mean number dropped to only 60 per year, with the last several years producing very few bountied snakes (R. Thiel, unpublished data).

In the early 1970s, the rattlesnake bounty was terminated by the WDNR, although later attempts were made to reinstate the bounty without success. (A renewed effort in the early 1980s was defeated by a distressingly narrow margin [Casper 1983].) Yet cessation of the bounty, unfortunately, led to conflict between conservation biologists and Wisconsin citizens. A letter from a Crawford County resident to the WDNR on 21 March 1978 stated of the bounty's termination: "Counties should be allowed to handle questions of local significance without interference from power-hungry Departments and Organizations not affected by the consequences. . . . Concern for HUMAN welfare should pre-empt concern for welfare of venomous snakes." A follow–up letter by this same Crawford County resident to Ruth Hine on 10 April 1978 stated that "you are wasting your sympathy—the rattler would not return your sentiments" and that the "DNR and Legislators protecting rattlers are causing us unnecessary worry, thus depriving us of our Constitutional rights to freedom from fear and harassment." It was not only average citizens who vehemently opposed the bounty's termination. A letter from the Crawford County Board chairman to the WDNR on 23 March 1978 stated that the Timber Rattlesnake was a threat to human life and produced a "psychological effect" on Wisconsin citizens. The chairman further requested that the WDNR "remove all Timber Rattlesnakes from Crawford County to be placed in a reserve of your choice or any zoos for public display." Fortunately, however, the bounty was not reinstated, or both rattlesnake species, which are still quite rare, could be nothing more than memories of a past Wisconsin fauna. It is also noteworthy that, despite the end of the rattlesnake bounty and the protections afforded these snakes, venomous snakebite in the state remains exceedingly rare (see Snakebite Envenomation in Wisconsin and the Northern Upper Mississippi River Valley chapter).

Joshua M. Kapfer, UW–Whitewater, and Robert W. Hay, Turtles for Tomorrow

amphibian and reptile conservation, there is the potential for exploitation to occur. The level of threat posed by social media is relatively unknown, with no robust data available on its impacts and no mitigation measures considered in Wisconsin to date. As a result, the regulation of illicit activity on these forums often falls on forum administrators. In an effort to promote principled behavior, it could be helpful to encourage forums and citizen-based herpetological societies to adopt a code of ethics that is clearly identified on their affiliated website and/or in their newsletters (a tactic that many such entities have already adopted). These societal codes should include guidelines for withholding the locality of species observed in the field and appropriate field etiquette. We also encourage these entities to take a strong stance against wild collection of herpetofauna and perhaps create 'penalties' for those who do not follow societal guidelines (e.g., banning transgressors from participating in certain events or from posting on forums for a specific period of time). Appropriate behavior by societal members and participants could further be encouraged through rewards, such as an internal certification process that requires oversight and vetting by societal administrators (e.g., 'Certified Midwestern Herpetological Enthusiast').

Climate Change

See Past, Present, and Potential Future Landscapes of Wisconsin chapter.

MANAGEMENT

Natural resource management can cover a variety of topics: (1) active management of populations through reintroductions, augmentation (e.g., captive breeding and release programs), resource supplementation (e.g., supplemental feeding), or regulation of harvest; (2) attempts to offset known threats or sources of mortality (e.g., mitigation of road mortality); and (3) attempts to protect, restore, and manage existing habitat so that it naturally provides adequate resources to support viable populations and communities. Of these topics, the first two are often centered on focal species or focal groups, while the last lends itself to the broader concept of ecosystem management.

Management will more likely succeed if a well-defined preemptive plan with clearly outlined strategies and goals for success is in place. Perhaps, then, an important first step toward successful management is the identification of a project's intended scale for the methods it employs and its projected outcomes. Plans designed at a local scale for specific organisms might not also be as effective when applied to large landscapes and vice versa. Despite this, there is the potential for overlap, where effective management strategies can have positive impacts at multiple spatial and ecological scales. For example, at the broader landscape scale, management that promotes habitat heterogeneity to re-create how landscapes existed historically is likely optimal for supporting diverse wildlife communities (reviewed by Tews et al. 2004). These broad management plans, if they effectively re-create heterogeneous habitats with diverse species assemblages, can also positively influence individual sites and specific organisms that occur within the landscape. Conversely, promotion of site-specific heterogeneity in habitat structure and vegetative species composition at local sites can also help improve the overall quality of communities at broader scales if implemented appropriately (reviewed by Fuhlendorf and Engle 2001).

Given the pervasiveness of habitat loss across large geographic areas, broad-scale management of landscapes should be the ultimate goal of any strategy. Management that is too site-specific or taxa-specific might not as effectively contribute to these broader, more meaningful outcomes. Unfortunately, the scope of management goals is sometimes constrained depending on the resources available to implement management initiatives. Yet even if management projects are applied at a small scale they can still be initially designed with both local and landscape-level benefits in mind. If so, then it is possible to manage for specific taxa at certain sites in a way that also benefits the entire biological community at the landscape level. Such a management philosophy also aligns with the great Wisconsin conservationist Aldo Leopold's concept of "thinking like a mountain." In other words, regardless of specific management agendas, natural resource managers and conservation biologists can think broadly and consider many taxa, their needs, and possible ecological interactions when managing habitats or populations. Below we provide an overview of habitat and population management strategies for amphibians and reptiles from a broader, more community-focused perspective. Specific management actions related to individual species are covered in the species accounts.

Habitat Management

It is cogent to first mention that the most important conservation measure for many species is the protection of existing natural habitat from destruction and degradation. As stated by Fischer and Lindenmayer (2007), extinctions will more likely occur in "landscapes with low native vegetation cover, low landscape connectivity, degraded native vegetation and intensive land use in modified areas." Given the rate of habitat loss since European settlement, there will soon be little left to manage if the intact habitat that remains is not spared from development, conversion to agriculture/pasture, and resource extraction.

Prior to discussion of habitat management, it is prudent to define *habitat*. Here, I follow the definition of Morrison et al. (2006): "an area with a combination of resources and environmental conditions that promote occupancy of individuals of a given species (or population) and allows those individuals to survive and reproduce." This definition provides a clear (albeit general) framework upon which citizens and professional biologists can build conservation initiatives and strategies. It also dictates that habitat management considers all developmental and annual lifecycle stages (e.g., overwintering vs. foraging vs. breeding). Note that suitable habitat and natural vegetation communities are not always the same thing, although vegetation is often a critical part of habitat suitability (reviewed by Fischer and Lindenmayer 2007). As such, vegetation management is a major component of habitat management. However, managers must consider all of the resources and conditions necessary for wildlife species to exist and reproduce, which can extend beyond their associated vegetation communities.

The concept of habitat management can include the reclamation, restoration, or maintenance of habitat for Wisconsin's amphibians and reptiles. Management that upholds the previously stated definition by Morrison et al. (2006) is particularly important given the battery of anthropogenic influences facing most natural resources (see Conservation Threats section). Yet there is much to learn about effective habitat management for these organisms, despite the potential importance. As Shoemaker et al. (2009) stated, "Habitat manipulation is often uncritically embraced as a practical management 'fix.'" Relatively little rigorous published research exists that has investigated effective habitat management strategies for amphibians and reptiles in Wisconsin (or globally, for that matter; as an example, see review of the sparse literature available on snakes in Shoemaker et al. [2009]). This is, in part, due to a lack of attention given to the impact of management practices on local herpetofauna. For example, an early investigation of the impacts that prescribed fire, herbicide application, and mowing had on grassland wildlife during land management in central Wisconsin investigated birds, mammals, and invertebrates but did not consider amphibians and reptiles (Halvorsen and Anderson 1980).

A primary goal of this section is to provide guidance on habitat management for amphibians and reptiles based on scientific research and species ecology where possible. Although specific information and hard data are sometimes lacking, I hope the summaries in this chapter (and the subsequent species accounts) will provide general guidance or a springboard from which interested practitioners can find additional resources. The information below is organized to first cover broad habitat management concepts, regardless of habitat type. The subsequent information is then separated first by general habitat class (e.g., Prairies, Grasslands, and Savannas) and then by topics relevant to that class (e.g., Prescribed Fire). It is important that readers of this section recognize that many of the strategies outlined below are *recommendations*. These are backed with references to rigorous research where possible, but they are sometimes based on anecdotes or logical conclusions drawn from knowledge on the species and habitats in question. I do not suggest that these recommendations be treated as dogma. It is sensible that any management recommendation be further investigated, supported, expanded upon, or invalidated to refine best management practices for Wisconsin's amphibians and reptiles. *We also strongly encourage readers to become familiar with laws, regulations, permits, and ordinances that are relevant at the city, county, state, or federal level before implementing any management strategy outlined below. We further recommend that readers consult WDNR staff to ensure the legality of their plans, particularly when rare species are involved.*

Management Concepts That Are Generally Applicable across Taxa and Habitats

Reference conditions and benchmarks

The goal of habitat management and restoration projects is often to return vegetation communities on a land parcel to some previous state (e.g., prior to disturbance, prior to European settlement). However, reverting a property to some approximation of its historical situation requires knowledge of the original condition. Whether the focus is terrestrial or aquatic habitat, estimation of a site's reference conditions can be used to create benchmarks for restoration project success but can be challenging to accurately determine (reviewed by Stoddard et al. 2006). Some studies have estimated reference conditions through assessment of existing adjacent parcels that contain habitat in a presumably undisturbed state (e.g., Prober et al. 2002). Unfortunately, examples of parcels with undisturbed habitat are typically difficult to locate. Use of historical aerial photographs and original surveyor documents, if available, can provide information on the past condition of a particular property. This method has been employed to create goals for oak savanna restoration at sites in Wisconsin, with some apparent success (Brock and Brock 2004). Yet this strategy might not always provide a picture of conditions prior to European settlement. As such, acquisition of fossilized pollen data (if feasible) from a given area could offer guidance on prehistoric conditions (reviewed by Humphries and Bourgeron 2001, Hupy and Yansa 2009). It is noteworthy, however, that some findings from Wisconsin suggest that the frequency of fossilized tree pollen in samples may not accurately reflect the frequency of those trees on the prehistoric landscape (Schwartz 1989). As a result, use of fossilized pollen samples to guide habitat restoration efforts may only lead to an approximation of pre-European conditions. For further review on the appropriate application of paleoecological data in ecological restoration, see Froyd and Willis (2008).

Maintenance of landscape connectivity

Hilty et al. (2006) defined habitat connectivity as the "extent to which a species or population can move among landscape elements in a mosaic of habitat types." The value of highly connected landscapes has received some attention, particularly as it pertains to the general concepts of wildlife conservation and management (Foreman 2004). Research specific to amphibians and reptiles in the Upper Midwest also supports that the presence of diverse landscapes with many well-connected habitat types is beneficial. For example, Attum et al. (2008) reported that rare wetland reptiles, including Blanding's Turtles (*Emydoidea blandingii*), profit from maintenance of landscapes with numerous connected wetlands and accessible adjacent uplands. Lehtinen et al. (1999) found that amphibian species richness was higher in Minnesota wetlands that were better connected and less isolated. Other research supports that amphibians, which typically avoid disturbed areas with no cover and dry substrates (Mazerolle and Descrochers 2005), must be able to freely move between uplands and adjacent wetlands for reproduction and to facilitate migration between subpopulations (e.g., Gibbs 1998a, Werner et al. 2007, reviewed by Cushman 2006). In accordance, Trenham et al. (2003) indicated that amphibian populations in Wisconsin benefited from the presence of many wetlands connected by undisturbed uplands that support their naturally mobile and dynamic populations (i.e., high turnover due to frequent localized extinctions and recolonizations). Promotion of many diverse and highly connected wetlands is also more likely to meet the assorted ecological needs of the various amphibian and reptile species that could occupy a given location and not only the rare species that are often used as focal taxa for initiation of management endeavors (Snodgrass et al. 2000).

High connectivity across a landscape requires that linkages be made between important landscape elements. Such linkages are achieved with, for example, corridors that facilitate movement between otherwise isolated fragments of suitable habitat (reviewed by Rosenberg et al. 1997, Hilty et al. 2006; see subsequent descriptions of corridor types). Unfortunately, appropriate levels of landscape connectivity for herpetofauna are difficult to determine, let alone achieve, in part because the ecological needs of the different species at a particular location vary. As an example, Price et al. (2004) found not only that amphibian presence in Great Lakes coastal wetlands was heavily influenced by landscape-level habitat composition but also that the extent of this influence varied by species. They reported that Gray Treefrogs (*Hyla versicolor*) and Spring Peepers (*Pseudacris crucifer*) both exhibited strong associations with intact woodland habitat variables within 0.5 km (0.3 mi) of wetlands. On the other hand, the presence of some species was better predicted by other habitat variables that occurred as distant as 1 km (0.6 mi) from wetlands (Price 2004). The influence of a species' life history strategy is also important. For example, sensitivity to habitat loss and to the isolation of wetlands among amphibians that migrate to breeding locations were related to a combination of their longevity, fecundity, and how intact upland migration pathways remained (Harper et al. 2008). Species with shorter life spans and high fecundity (such as the Wood Frog) were more sensitive than species with longer life spans and low fecundity (such as the Spotted Salamander [*Ambystoma maculatum*]).

Although the effectiveness of various corridor types is debatable and depends on context, these types can range from strips of fallow or native vegetation that allow passage through otherwise unsuitable disturbed landscapes to anthropogenic structures that connect habitat on either side of a barrier, such as highway underpasses (reviewed by Hilty et al. 2006). Burbrink et al. (1998) stated that the design of appropriate corridors for amphibians and reptiles required managers to consider the specific ecological needs of the various species on a given site. Maisonneuve and Rioux (2001) found evidence that small mammals and herpetofauna could benefit from the presence of corridors with structurally diverse vegetation along riparian zones. It is important to note that there is also a positive relationship between the size of a suitable habitat area and herpetofaunal species richness in the Great Lakes region (Hecnar et al. 2002). Thus, strategies for effective landscape connectivity should also strive to connect several core habitats that are large in size in an effort to promote the greatest number of species across the landscape. Finally, effectively connecting suitable habitat with corridors could require a significant amount of land, including that of both public and private ownership, which may necessitate the combined efforts of professional biologists and concerned citizens.

The importance of habitat heterogeneity and complex habitat structure

Habitat heterogeneity is described by Fuhlendorf and Engle (2001) as "variability in vegetation stature, composition, density, and biomass . . . [which is] the precursor to biological diversity at most levels of ecological organization and should serve as the foundation for conservation and ecosystem management." The 'habitat heterogeneity hypothesis' dictates that habitats with high structural diversity provide more opportunities for organisms to exploit resources, which should positively influence species richness (MacArthur and Wilson 1967). In short, promotion of numerous broad habitat types with high structural variation should in turn accommodate high species richness and diversity. This concept applies to both terrestrial (e.g., Price et al. 2010, Steen et al. 2012a) and aquatic (e.g., Smiley et al. 2009) habitats but should also be considered at multiple spatial scales.

Several studies support that amphibians and reptiles can benefit from habitats with structural complexities at small spatial scales (e.g., the presence of downed woody debris, leaf litter, rocks, brush piles, woody snags in rivers all influence structural complexity; e.g., Shoemaker et al. 2009, Blomquist and Hunter 2010, Griesbaum 2012). Specific examples can be found in the individual habitat sections discussed subsequently. It is noteworthy that ground debris and underwater snags can be created as a by-product of combating succession. For example, as woody vegetation is removed during prairie management, a portion of the resultant debris can be left behind to create structure, which could be especially useful in areas where other vegetative cover is sparse or aquatic woody debris used for thermoregulation by turtles is limited (Shew et al. 2012). It is noteworthy, however, that Sacerdote and King (2014) found compounds that leach from European Buckthorn (an invasive shrub that is often removed during management) can have negative impacts on amphibians (Sacerdote and King 2014). Woody debris produced through European Buckthorn removal should only be used to create aquatic habitat structure after careful consideration.

Amphibians and reptiles also respond positively to habitat variability at larger spatial scales. For example, Steen et al. (2012a) suggested that variability in habitat at the landscape level was important to support diverse snake communities. Knutson et al. (1999) detected positive associations between amphibian abundance and landscapes with variable wetland characteristics in Wisconsin, while Trenham et al. (2003) similarly found that

Wisconsin amphibians benefit from diverse landscapes (see Maintenance of landscape connectivity subsection). Again, however, understanding the overall role of habitat complexity is complicated by the fact that species vary in what habitat factors are important to them. Some amphibians, for example, respond positively to the presence of forest patches on the landscape, while others respond more to wetland proximity and wetland permanency (Kolozvary and Swihart 1999). Implementing strategies that accommodate many species and promote diverse communities is, therefore, a challenge. Although it is probably not feasible to develop management strategies that maintain all possible habitat conditions, management to generally promote heterogeneous landscapes could act as an umbrella that encapsulates a plethora of species and habitats. Furthermore, enacting strategies that attain management goals while limiting the possible negative impacts that management practices could have on resident species is obviously important (reviewed by Zavaleta et al. 2001).

There are several specific habitat management techniques that, when applied too frequently or aggressively, could result in a loss of habitat complexity, such as techniques implemented to control vegetative succession (e.g., prescribed fire, herbicides, mowing, and mechanical removal of vegetation by hand or heavy machinery; Dix and Butler 1954, Russell et al. 1999, Kost and de Steven 2000). At sites where prescribed fire is applied over too large an area annually or removal of downed woody debris is a common practice, important structural variability and cover for amphibians and reptiles are lost (see Prescribed fire subsection). As such, it is reasonable to reserve frequent implementation of very intense strategies for especially problematic sites (e.g., with heavy infestations of invasive plants or where extensive unregulated succession has occurred). Impacts to herpetofauna could also be reduced if management strategies implement intense techniques in a patchwork or mosaic pattern (e.g., manage different subplots within a given site in alternate years). Similar reduction in impacts could occur if intense management activities are scheduled to avoid the active seasons of local herpetofauna (see Timing subsection). It is noteworthy that conducting both pre- and post-management surveys can be an informative strategy for assessing impacts of employed techniques.

Timing

Regardless of the habitat type where they are employed, some management techniques can result in incidental mortality of amphibians and reptiles. Examples include unintentional strikes with heavy machinery and entrapment in prescribed fires. Given that any level of mortality could have profoundly negative impacts on populations of rare herpetofauna, management strategies should be scheduled for times when there is a lower likelihood that they will kill amphibians and reptiles. For example, avoidance of periods when individuals might migrate through a management area (e.g., mass amphibian migration to breeding ponds in spring) or periods when most amphibians and reptiles are generally active (i.e., spring through early autumn). Rules of thumb to avoid primary active seasons for many species could include scheduling management activities from October 20 to April 5 in southern Wisconsin and October 15 to April 15 in central and northern Wisconsin. However, annual variation will exist, depending on weather (e.g., unseasonably warm conditions) and habitat type (e.g., management near communal snake hibernacula must consider that snakes could be active beyond the general time period mentioned previously). It is advisable to consult with herpetological experts or WDNR staff (the latter especially for Endangered, Threatened, or Special Concern species) to develop appropriate management schedules.

Unfortunately, the periods when there is highest certainty that amphibians and reptiles will be inactive are not always the optimal periods to conduct all habitat management activities. Of course, restricting the use of some management techniques to certain time periods might not greatly limit their effectiveness. Removal of woody vegetation by hand or with machinery, for example, can be productive regardless of the season implemented. On the other hand, the efficacy of strategies, such as the application of prescribed fire, might be hampered when they are only conducted during specific seasons (see Prescribed fire subsection). A reasonable practice could be to schedule the use of these techniques based on monitoring the activity patterns of local populations and refrain from use of some when surveys confirm herpetofauna are active at a management location. This is obviously not fool-proof and could require substantial effort to enact, depending on size of the proposed impact area. If habitat management must occur at a time when individuals could be active or at sites where rare species exist, one possibility is, again, to conduct management in a mosaic pattern that impacts some areas and not others, for example, installing fire breaks to ensure that a mosaic of burned and unburned patches remains after application of prescribed fire. This ostensibly leaves refuges that amphibians and reptiles at the site can use as cover from management activities. It also creates smaller management parcels that could be more reasonable to survey for active herpetofauna immediately prior to activities.

Prairies, Grasslands, and Savannas

Management of prairies, grasslands, and savannas for amphibians and reptiles must consider two primary and sometimes competing factors: combating encroachment of woody vegetation and invasive plant species vs. overintense management that creates a structurally homogenized landscape. Obviously, it is important to limit the growth of woody vegetation and invasive plants if the goal is to maintain the structure of open habitats, and maintenance of open conditions can clearly benefit herpetofauna that use these habitat types. There are several common methods employed to achieve these goals: vegetation removal by hand or with machinery, application of herbicides, and prescribed fire. Yet these must be administered appropriately to avoid directly killing local herpetofauna and avoid creation of conditions they find unfavorable.

Figure 42. A prescribed fire employed to manage a prairie on state land (Iowa County; photo by R. Staffen).

Prescribed fire

Prescribed fire is an effective and commonly employed technique used to manage open habitats (reviewed by Wright 1974; Figure 42). Not only does prescribed fire inhibit vegetative succession in open habitats, it can also control invasive plants (reviewed by DiTomaso et al. 2006). Furthermore, some native plants, particularly prairie species, require fire to complete their life cycles. The general outcome of prescribed fire application can vary based on several factors, such as timing, burn area, and burn frequency. Howe (1994) indicated that autumn and spring burns better control woody species and promote forbs, while annual spring burns promote late warm-season grasses. Longer intervals between burn periods promote forbs, while summer burns could promote species normally suppressed by large C4 grasses. Emery and Gross (2005) further found that annual application of prescribed fire in the summer reduced abundances of the invasive Spotted Knapweed significantly more than in spring or autumn. Howe (1994) suggested that variability in seasonal application of prescribed fire could promote greater plant diversity. Unfortunately, these 'optimal' seasons for vegetation management do not always coincide with those for other organisms (see Timing subsection).

Prescribed fire can have both positive and negative impacts on amphibians and reptiles, so it must be carefully planned and applied. In favor of prescribed fire, Russell et al. (1999) reviewed the results of various studies and concluded that this technique is "an appropriate management tool that can be used with other treatments to benefit herpetofauna." Harner and Geluso (2011) provided evidence that species richness and relative abundance of prairie herpetofauna in Nebraska increased in areas that experienced prescribed fire compared to 'rested pastures,' which were neither grazed nor burned. Wilgers and Horne (2006) reported that various amphibian and reptile species in Kansas benefited from different fire regimes although their sample sizes were very small. Regardless, some species were more prevalent in long-term unburned areas (e.g., Ring-necked Snake [*Diadophis punctatus*]; $n=73$), while others were more prevalent in areas that received annual burn treatments (e.g., Great Plains Skink [*Plestiodon obsoletus*]; $n=16$). The data available on how fire affects amphibians that occupy open habitats are more sparse (reviewed by Pilliod et al. 2003), but large variability likely exists based on species-specific ecologies and life history strategies.

The negative impacts of prescribed fire can include fire-related mortality (e.g., Erwin and Stasiak 1979). In Wisconsin, numerous rare reptile species are found in prairies and savannas that are managed through prescribed fire. It is important to remember that even minimal mortality due to fire can have major population effects for rare species that develop slowly and have low reproductive potential (e.g., see Ornate Box Turtle, *Terrapene ornata* species account). In addition to direct mortality, the overly intense application of fire reduces variability in habitat composition and structure (Figure 43). Griesbaum (2012) reported that the depth of vegetative litter (or the ground debris accumulated due to the vegetative growth of previous seasons) was positively correlated with grassland snake diversity at sites in Illinois. Price et al. (2010) found that Australian reptiles of tropical savanna landscapes responded negatively to more barren, homogenized habitats dominated by bare ground or only grass. These studies exemplify the potential danger of excessive application of prescribed fire if it leads to insufficient cover and reduced habitat variability. Accordingly, Price et al. (2010) recommended implementation of prescribed fire in a way that promotes high habitat variability at the landscape level. This included a mosaic-style fire pattern, where patches of habitat are burned periodically while others are left unburned to promote high habitat heterogeneity and species richness.

Figure 43. An example of a recently burned prairie in Wisconsin. When large parcels of prairie are subjected to prescribed burning, the result can be a lack of refugia that leaves many small vertebrates vulnerable (Sauk County; photo by J. Kapfer).

The temporal dynamics of prescribed fire application can influence responses of amphibian and reptile populations. In addition to implementing prescribed fire during the inactive season (see Timing subsection), the frequency of fire application across years is an important consideration. A multiyear study by Griesbaum (2012) investigated grassland snake response to various habitat factors (such as succession) in Illinois. He recommended that prescribed fires occur on a rotation of two to four years, with the four-year schedule applied to sites where higher-quality grassland currently existed in order to keep important ground litter intact. Additional work in Australia supports that reptile abundances are influenced by the amount of time that has passed after application of a prescribed burn (Smith et al. 2013). Some species reached high abundances quickly after a prescribed burn was conducted, but it took other species many years to achieve high post-fire abundances.

Some research supports that prescribed fire can be applied infrequently and still yield desirable management outcomes. Peterson and Reich (2001) suggested that as few as three prescribed fires per decade could prevent formation of sapling layers and canopy ingrowth in Upper Midwest oak savannas. Careful monitoring of local sites by managers can guide appropriate timing of application, which can optimize management practices and direct future decisions. Note that the WDNR's current grassland and savanna management protocols provide important information on scheduling prescribed fire in areas with rare species (WDNR 2017a). In general, the recommendation to design a variable prescribed fire schedule so that a "homogeneous fire regime" is avoided may be an appropriate rule of thumb (Smith et al. 2013).

Control of vegetation by hand removal, machinery, or herbicide application

Encroachment of woody vegetation in prairies and savannas can also be offset by removal of brush with machinery, by hand, or with the application of herbicides. Unfortunately, the costs or benefits of these techniques to herpetofauna in Wisconsin have not received in-depth study. Appropriate scheduling of these activities is clearly a key consideration (see Timing subsection). Furthermore, these methods can be employed alone or in concert with prescribed fire, but the outcomes related to amphibians and reptiles can be highly variable. For example, studies on savannas in Oklahoma found that lizards were most abundant in untreated areas, amphibians were most abundant in both untreated areas and those treated only with herbicide, while snakes were most abundant in areas treated with both fire and herbicide (Jones et al. 2000). Work in Michigan savannas found that lower-intensity strategies (i.e., burning only rather than in concert with other methods) were effective at restoring savanna, but the positive impacts manifested more slowly than with high-intensity strategies (i.e., burning plus mechanical removal of woody species; Lettow et al. 2014). The authors suggested that decisions to apply low-intensity vs. high-intensity management strategies should consider budget constraints and the current state of site restoration. Lower-intensity strategies may be less likely to result in incidental mortality of resident herpetofauna.

Heavy machinery is potentially useful in some management endeavors, particularly where trees or shrubs must be cut and removed from a site or dense grasses must be mowed. It is likely critical to employ caution when managing habitat with heavy machinery so that accidental mortality of cryptic amphibians and reptiles does not occur. Unfortunately, research on the most effective strategies to implement heavy machinery that also consider the safety of herpetofauna is extremely limited. There is evidence to support that motor-driven lawn mowers with blades set at a height of 20 cm (8 in) reduce management-related mortality of Massasaugas (Durbian et al. 2008). Flail-type mowers are also sometimes recommended because they create little suction that could extract small vertebrates from hiding places in vegetation compared to conventional mowers (Durbian et al. 2008). As with prescribed fire, mortality could also be reduced by mowing in a pattern that leaves refugia or escape corridors for fleeing herpetofauna. The daily and seasonal timing of when heavy machinery is employed can be important (see Timing subsection). Mower-related mortalities may decline if mowing occurs during the warmest parts of the day in summer as opposed to during cooler summer periods (ectothermic vertebrates likely seek refuge from high temperatures by retreating to burrows or moving under cover, but are often out thermoregulating on mild days).

Control of invasive plants in Wisconsin prairies and savannas

The previously reviewed techniques could be used to control the spread of invasive plants. Unfortunately, complete eradication of invasive species, regardless of habitat, is extremely difficult to achieve (e.g., reviewed by Zavaleta et al. 2001). Even if eradication of invasive species were possible, subsequent recovery of native species could take decades (reviewed by Kats and Ferrer 2003). Many competing forces must be factored in when managing habitats to control exotic or problematic plants such as Spotted Knapweed. In addition to prescribed fire, herbicide is frequently employed to achieve management goals. Recent research suggests that herbicide application should be implemented cautiously, given the long-term negative effects on native plants (Rinella et al. 2009). Furthermore, the potential toxicological effects that herbicides have on wildlife cannot be ignored (see Pollution subsection).

Due to the aggressive nature of many invasive species, reliance on a single strategy for control could be insufficient. For example, MacDonald et al. (2013) reported that persistent hand pulling of plants after an infested area has been mowed or after a combination of mowing plus herbicide application was an effective (but labor-intensive) means to control Spotted Knapweed. These strategies will likely be most effective for amphibians and reptiles if they are conducted with appropriate consideration of activity patterns and management intensity.

Special consideration for communal snake hibernacula on bluffside prairies

An important and somewhat unique situation that site managers should be aware of involves habitat associated with communal snake hibernacula. The most obvious examples occur along bluffsides in the driftless area of southwestern Wisconsin, where many rare snake species congregate for overwintering, such as Timber Rattlesnakes, Gophersnakes, Gray Ratsnakes (*Pantherophis spiloides*), and North American Racers. In such locations, maintenance of open tree canopy in the immediate area is preferable, primarily because open bluffsides can quickly become overgrown with woody vegetation such as Eastern Red Cedar (*Juniperus virginiana*) when management is not enacted to control it (Figure 44). Hibernacula that have become overgrown may not experience appropriate sunlight and heat penetration into associated soil, which is necessary to promote a normal schedule of snake egress from winter retreats.

Figure 44. A communal snake hibernaculum site that has become overgrown with woody vegetation (Sauk County; photo by J. Kapfer).

Unfortunately, research on the benefits of woody vegetation removal to snake hibernation in the Midwest is limited. Work in Australia found that in the spring following removal of overhead canopy trees, snake occupancy of retreats on-site increased, and higher temperatures were recorded within retreat locations (Webb et al. 2005). The removal of trees and shrubs on bluffsides is challenging because they can be difficult to access, and operation of equipment such as chainsaws on bluff slopes can be dangerous. The timing of activities is also important, as snakes emerging from hibernation may be sluggish and also congregate in areas that need management. Management activities that accidentally harm snakes could impact a large proportion of populations if many individuals are congregated near hibernacula. Although removal of living woody vegetation near hibernacula is useful, it is likely beneficial to leave some downed woody debris scattered on open bluffsides and promote structural variability in herbaceous vegetation to provide sluggish snakes cover from predators. Knoot and Best (2011) even recommend creation of 50 m (165 ft) buffers around hibernacula and promotion of intact vegetative corridors (30 m [98 ft] in width) that connect hibernacula to active season habitat used by snakes in Iowa.

Woodlands and Forests

There are numerous management considerations for forests and woodlands, with timber harvest practices among the most important. Silviculture is defined as "the art and science of controlling the establishment, growth, composition, health and quality of forests and woodlands to meet the diverse needs and values of landowners and society on a sustainable basis" (USFS 2014). Silviculture programs can greatly influence the suitability of forest habitat for amphibians and reptiles, and it is important to consider this when designing timber harvest and management strategies.

Silvicultural strategies

Long-term effects of timber harvest on amphibians can differ based on harvest strategy (deMaynadier and Hunter 1995), and negative impacts can be mitigated if appropriate microhabitat structure remains intact as part of harvest and subsequent management. Various common harvest practices exist, such as even-aged and uneven-aged tree stand management. Even-aged management during timber harvest produces woodlands comprised of trees that are roughly the same age and size because they have grown from a plot that experienced past removal of all trees at the same time (i.e., past clear-cutting; USFS 2015). On the other hand, selective removal of individual trees or removal of small groups of trees in a harvested area results in uneven-aged stands, which provide greater habitat variability for resident amphibians and reptiles. Numerous studies support that clear-cut harvest practices have negative impacts on amphibians, and these methods should be avoided (see review of impacts in Anthropogenic disturbances in forests and woodlands subsection). Thompson et al. (2003) also predicted that intensive forest management (IFM) practices, as opposed to natural forest regeneration after harvest, would negatively impact salamanders. Some research further supports that changes in tree species composition within forests can negatively affect salamanders (Belasen et al. 2013). Thus, silvicultural strategies should consider original forest tree composition when developing plans so that tree communities are not greatly altered after regeneration from timber harvest and any associated impacts on amphibians are reduced.

The size of the area impacted by tree harvest plays an important role in amphibian conservation in the Upper Midwest. Regosin et al. (2005) recommended maintenance of woodlands within a 200 m (656.2 ft) buffer around breeding wetlands for "pool-breeding, forest-dwelling amphibians." Research in Georgia found that small clear-cut areas (<4 ha [10 ac]) did not act as a barrier for migrating amphibians (Todd et al. 2009). However, research in the Upper Midwest is lacking. In addition to the size of a clear-cut area, management should consider maintenance of downed woody debris in woodland habitats. Ground debris creates critical cover for many amphibians, although the importance varies by species (Semlitsch et al. 2009). It is also imperative to recognize that the effects of timber harvest are not restricted to the immediate area where trees are removed. Removal of trees can result in higher penetration of solar radiation at both the removal site but also near the adjacent edges of harvest zones relative to interior forest, often resulting in increased temperature and reduced moisture across an expanded area. The work of deMaynadier and Hunter (1998) supports that these woodland 'edge effects' can penetrate from 25 to 35 m (82 to 115 ft) into interior woodlands adjacent to disturbance. Given these points, rules of thumb for tree harvest that minimizes negative impacts on amphibians and reptiles could include avoiding even-aged clear-cut strategies, minimizing the area of disturbance, limiting the edge effects related to disturbance, retaining downed woody debris, and considering a site's historic plant species composition.

Control of invasive plant species in Wisconsin woodlands and forests

Even the partial removal of invasive plants from a woodland location can result in a substantial increase in native plant diversity and equitability of native species in forest understories (Stinson et al. 2007). Control strategies could include a multifaceted approach that involves a combination of mechanical removal (e.g., brush-cutting) that is followed by application of herbicide. For example, Reinartz (1997) reported that winter mechanical removal of the invasive shrub Glossy Buckthorn (*Rhamnus frangula*) in Wisconsin was only effective if followed by immediate application of glyphosate herbicide to the stump, which resulted in 92%–100% mortality. Application of glyphosate herbicide also reduced Garlic Mustard in woodlands and did not negatively impact native plant species, according to Carlson and Gorchov (2004). However, positive results from herbicide application were not recognizable until at least two years following application, which supports the importance of long-term plans for invasive plant control (Carlson and Gorchov 2004).

Yet as with open habitats, management of woodlands must reconcile the need to enact management with the potential for incidental mortality of herpetofauna with chosen management strategies. For example, although glyphosate herbicides can be effective in control of invasive plants, they (or the compounds they are mixed with for invasive plant removal) can negatively impact amphibian larvae and juveniles (Howe et al. 2004, Relyea 2005b). Furthermore, some amphibians avoid areas treated with glyphosate-based herbicides applied at concentrations as low as 10% (Gertzog et al. 2011). Thus, care must be used when applying herbicides, and more research on the impacts to amphibians and reptiles is warranted. Specific strategies also exist for the most effective disposal of certain woodland invasive species after removal. For example, Solis (1998) found that hand-removed Garlic Mustard must be placed in bags and destroyed if collected when in flower to prevent post-mortem seed dispersal.

Aquatic Habitats

In this section I review some management considerations broadly applicable to all aquatic habitats and then provide information regarding specific habitat types. Note that the control of pollution is a general management consideration that is critical for all aquatic habitats (see Pollution subsection). Yet the complicated nature of managing aquatic toxicants is a subject too large for summary here and readers are encouraged to investigate other sources for more information (e.g., Sparling et al. 2010).

Protection of adjacent upland habitats and buffers around aquatic habitats

Many amphibians and reptiles frequently use both aquatic habitats and adjacent uplands, often alternating between them for various purposes (reviewed by Gibbons 2003, Semlitsch and Bodie 2003). For example, most amphibians in Wisconsin breed in wetlands but overwinter in adjacent uplands or move between wetlands and uplands while foraging during the active season. Landscape-level analyses also support that diverse wetlands and adjacent uplands across large areas are important to Wisconsin amphibians (Trenham et al. 2003, Price et al. 2004; see Maintenance of landscape connectivity subsection). In addition, freshwater turtle species spend most of their time in aquatic habitats but lay eggs on land. Some turtles, like the Blanding's Turtle (*Emydoidea blandingii*), frequently cross upland habitat while migrating between wetlands throughout the active season and prosper in diverse landscapes with many wetlands connected through natural upland corridors (Attum et al. 2008).

Because both aquatic and adjacent terrestrial habitats are critical to many amphibian and reptile species, conservation strategies should seek not only to protect and manage both habitats but also to maintain connections between them. Even if both habitats are protected, but isolated so that individuals cannot move freely between the aquatic and terrestrial sphere, species could still experience population declines (albeit more slowly than if these habitats were completely destroyed). Although protection of terrestrial buffers around wetlands is a sound practice, buffer width becomes a critical factor. Wetland regulations typically include protections for ≤30 m (98 ft) of adjacent upland habitat (Gamble et al. 2006). Protected buffers around aquatic systems must be wide enough to provide suitable protection. Yet determining acceptable buffer widths can be challenging for various reasons. First, adequate buffer widths vary by species, and knowing these widths often requires a thorough understanding of an organism's ecology (which might not be possible, given available information). According to Harper (2008) the ≤30 m (98 ft) previously mentioned would not be sufficient to protect many pool-breeding amphibians. Semlitsch and Bodie (2003) reviewed numerous studies in an effort to create quantified recommendations for the width of buffers around aquatic habitats that would protect a wide array of species. They determined that buffer distances necessary to protect core upland habitats near aquatic systems (wetland or riparian) were 159–290 m (522–951 ft) for amphibians and 127–289 m (417–948 ft) for reptiles.

Control of soil and bank erosion

Bank erosion is a natural and important process in some aquatic systems, such as rivers (Florsheim et al. 2008). However, anthropogenic modifications to rivers and streams, such as channelization, have resulted in unnaturally high erosion rates in some areas that can lead to degradation of aquatic habitats (Kondolf 1997, Shields et al. 1998). A common tactic to slow erosion is deployment of erosion control mesh, which is laid flat on top of recently disturbed ground to stabilize soil, seed banks, and vegetation. Erosion control mesh products vary dramatically in mesh materials, mesh construction, mesh aperture size, and typical implementation (reviewed by Kapfer and Paloski 2011). Some types of erosion control mesh can entangle amphibians and reptiles, causing mortality, although certain types appear to pose greater risks than others. For example, a review of erosion control products suggests that mesh possessing woven synthetic fibers with bonded intersections are particularly prone to entangling snakes (Kapfer and Paloski 2011). Yet rigorous experimental work is needed to determine the safest products in the various categories of use in order to provide sound guidance to manufacturers and users of these products.

Another technique that is frequently used to stabilize aquatic shorelines is deployment of large rocks (i.e., riprap) or concrete and wood retaining walls that protect banks against waves or water flow. Unfortunately, these can restrict movement of some species that need free access to both terrestrial and aquatic habitats, including nesting turtles, hatchling turtles returning to the water, breeding amphibians, and newly metamorphosed amphibians that leave the water. I was unable to find any published research on the best application of riprap and retaining walls to avoid impacting amphibians and reptiles. Restricting construction of retaining walls in areas where turtles or amphibians frequently move between the aquatic-terrestrial interface is probably advisable (Figure 45). Presumably, use of smaller-circumference rocks as riprap along shorelines, or perhaps as fill added to spaces between larger rocks, would increase the ease with which migrating turtles could negotiate them. Maintenance of bankside access points for herpetofauna along shorelines that are otherwise riprapped or have retaining walls could be a solution. In theory, these access points should contain intact vegetation along gently sloping stretches of shoreline that permit easy movement for nesting or migration. These might also be most effective if used in conjunction with barriers that direct moving herpetofauna toward such access points. Still, research that investigates the best management practices for these strategies is warranted.

Figure 45. Example of a retaining wall along a developed lakeshore that would likely restrict the movement of amphibians and nesting turtles (Fond du Lac County; photo by R. Paloski).

Road underpasses

Although use of underpasses to reduce road mortality could be applicable to many habitat types, it is particularly relevant to aquatic environments bisected by roads. Fragmentation of aquatic habitats and mortality caused by roads could be offset through the use of underpasses or culverts that allow amphibians and reptiles safe passage (also referred to as 'eco-passages'). Evidence supports that some amphibian and reptile species will use underpasses if available, although the extent to which underpasses mitigate road mortality varies (Jackson and Tyning 1989, Beebee 2013, Hamer et al. 2014, Colley et al. 2017, Dillon et al. 2020). A review by Clevenger and Huijser (2011) provided recommendations for tunnel dimensions and characteristics suitable for amphibians and reptiles. They recommended tunnels that range from 1 m (3.3 ft) in width or diameter (depending on shape), up to ca. 2.4 m (8 ft), depending on the overall distance a proposed tunnel would span below a given road (i.e., longer underpasses should be larger in diameter). Some sources further suggest that underpasses be lined with soil from near the tunnel location to increase use by herpetofauna (reviewed by Clevenger and Huijser 2011). Despite this, Patrick et al. (2010) found that some amphibians show no preference for the size, length, or substrate type associated with underpasses. Clevenger and Huijser (2011) further reported that tunnel interiors should receive a modicum of natural light and airflow but also remain moist, which can be facilitated if underpass tunnels possess grates near the road surface. Note, however, that grates could allow road contaminants, such as road de-icing salt residue, to accumulate in tunnels. Thus, it could be important to design underpasses with slopes that allow rain and snowmelt to rinse such residues from tunnel interiors.

Patrick et al. (2010) determined that selection of optimal underpass locations along roadways could be challenging, given that some species show little specificity as to where they cross a road if it bisects a swath of habitat. Yet amphibians and reptiles can be successfully funneled to underpasses or culverts via fences or walls placed along roadways (Jackson and Tyning 1989, Markle et al. 2017; Figure 46), and underpasses are likely more effective when organisms cannot bypass them (Dillon et al. 2020). Jackson (1996) found that salamanders contacting fences at distances of up to 30 m (98.4 ft) from tunnel

Figure 46. Fencing used along a roadside near a waterbody to exclude turtles from the road (Waupaca County; photos by A. Badje). (a) Linear fencing. (b) Bend in the fence (sometimes called a 'turnaround') to direct turtles back toward the water.

entrances still navigated toward underpasses and successfully used them to cross roads. Patrick et al. (2010) suggested that if underpasses for amphibians also allow passage of flowing water from streams or rivers, high water flow rates could impact the likelihood of safe passage through tunnels. In these instances, above-water structure (higher banks of soil, rock, or vegetation) could be maintained within tunnels to allow movement options for semi-aquatic and terrestrial organisms (Figure 47).

Figure 47. An example of a stream crossing with an underpass that includes structure to facilitate terrestrial passage by wildlife (Milwaukee County; photo by K. Leithoff).

Lentic Aquatic Communities

Wetlands

As with all habitats, management of wetlands can be considered at multiple spatial scales. Because amphibians and reptiles collectively show wide variation in habitat preferences, management plans that promote landscapes with a diverse array of wetland types are optimal for maintaining high species richness (e.g., Knutson et al. 1999, Houlahan and Findlay 2003, Jacobs and Houlahan 2011). Yet there are a variety of small-scale wetland characteristics that habitat managers can focus on to potentially promote amphibian and reptile occupancy. For example, maintenance of logs and woody snags in wetlands is probably beneficial, as these can function as underwater refugia, locations for basking turtles, and perches for calling anurans. In general, promotion of variable water depths and convoluted shorelines within a waterbody increases habitat complexity, which could promote herpetofaunal diversity. It is of further importance to consider the size and hydrologic cycle maintained in managed wetlands. For example, Houlahan and Findlay (2003) found a positive correlation between amphibian species richness and wetland size in Canada. Research in Maine determined that ephemeral breeding ponds that retained water for extended periods of time (i.e., 16–17 weeks post-breeding) supported large populations of Wood Frogs and Spotted Salamanders (*Ambystoma maculatum*; Baldwin et al. 2006b). Note, however, that there is an upper limit to how long ephemeral wetlands must hold water to benefit specific amphibian species (see Control of problematic species in wetlands subsection).

Unfortunately, wetland management strategies that benefit some wildlife species may actually be detrimental to amphibians and reptiles. For example, wetland strategies that promote waterfowl habitat sometimes involve the intentional reduction of water levels during the winter. This can expose hibernating turtles and amphibians that overwinter underwater, which can result in high mortality. Hall and Cuthbert (2000) studied the impacts of a late autumn water drawdown in a Minnesota lake on turtles. This drawdown was initiated in late October and exposed 75% of the lakebed within 48 hours, resulting in substantial mortality of both Blanding's Turtles (*Emydoidea blandingii*) and Painted Turtles (*Chrysemys picta*). The authors recommended avoidance of autumn and winter drawdowns and that drawdowns be conducted gradually to allow turtles an opportunity to respond to declining water levels. They also suggested that creation of side channels connected to wetlands being drawn down could provide aquatic refugia for turtles to escape receding water levels, so long as side channels remain inundated.

Control of problematic species in wetlands

As with other habitats, control of invasive vegetation is an important but challenging management goal for Wisconsin wetlands. For example, due to its prevalence and large reproductive potential, the invasive Reed Canary Grass poses substantial management challenges. Thick stands of Reed Canary Grass produce many seeds (e.g., 160 kg/ha [180 lb/ac] produced in a single growing season). These seeds can accumulate in wetland soils over time and facilitate rapid reinvasion even after initial management removes standing vegetation (reviewed by Annen et al. 2005). Annen et al. (2005) found that use of the herbicide sethoxydim, which is selective to grasses, can be effective in reduction of both aboveground biomass and seed banks in Reed Canary Grass. Broad-scale herbicides might also be considered at sites with high abundances of native grasses that need preservation because such herbicides will not target native grasses as heavily as more specific herbicides. Although the effect of sethoxydim on herpetofauna is largely unknown, broad-scale herbicides composed of glyphosate formulations with surfactants can negatively affect amphibians (Howe et al. 2004). Other possible management techniques to control invasive wetland plants exist that range from prescribed fire to tilling with heavy machinery. These methods all have advantages and drawbacks, including potential negative impacts on amphibians and reptiles, which were generally covered previously (see Prescribed fire and Control of vegetation by hand removal, machinery, or herbicide application subsections). See further management considerations in Wisconsin Reed Canary Grass Management Working Group (2009).

Unfortunately, problematic species that wetland habitat managers must consider include more than vegetation. Given the negative relationship between the presence of fish and amphibians (Smith et al. 1999, Knutson et al. 2004, Herwig et al. 2013), efforts to ensure that normally fishless wetlands remain so are critical to the persistence of many amphibian species. This involves maintenance of appropriate natural hydrology so that ephemeral wetlands are not altered to permanently hold water. Furthermore, artificial water features (e.g., fountains) and aerators add oxygen to wetlands that would otherwise become anoxic. Because use of aerators promotes fish survival, they should not be implemented in normally fishless wetlands if amphibian conservation is a management goal. Removal of fish introduced to wetlands is possible via appropriate application of commercial piscicides (such as Rotenone). Some evidence supports that Rotenone is less toxic to amphibians than fishes, indicating the former are capable of recovering from exposure (Mullin et al. 2004, Walston and Mullin 2007b).

Lakes

Control of lakeshore development to maintain large stretches of intact habitat and shoreline vegetation likely benefits amphibians and reptiles (e.g., Paloski 2008). Woodford and Meyer (2003) reported that regulations for lakeshore development during their study in Wisconsin, which restricted house densities to 3.3 per 100 m (328 ft) of shoreline, were likely not adequate to protect sensitive amphibians. They further estimated that if more lakes were developed using this standard, it would result in <50% of suitable habitat for Green Frogs, increasing the risk of local extinctions. Unfortunately, the maximum safe density of residential structures along lakeshores (i.e., the density that limits negative impacts) has not been quantified for most herpetofaunal species in Wisconsin.

Figure 48. Example of a shoreline where natural vegetation was allowed to remain near the water interface. Maintaining vegetation in this fashion benefits amphibians and reptiles that use lakes and adjacent uplands (Portage County; photo by R. Paloski).

As previously reviewed, it is important to promote ease of movement between lakes and adjacent uplands along developed shorelines (see Protection of adjacent upland habitats and buffers around aquatic habitats subsection). In addition, shoreline habitat management should include the maintenance of natural emergent vegetation along the water-land interface to act as cover for snakes, turtles, and amphibians (Figure 48). Similarly, preservation of wide vegetated littoral zones adjacent to shorelines near residences could provide amphibians with refugia from predatory fishes in lakes. In some recreational zones, aquatic vegetation and underwater woody snags are actively removed because swimmers find them undesirable or they could damage watercraft. Yet such underwater structure can act as important cover and foraging locations for herpetofaunal species that use lakes, including turtles, snakes, and frogs. Restricting the number of locations where these removals occur to areas of low herpetofaunal use in lakes is likely important to avoid negative impacts, but further research is needed in Wisconsin.

Lotic Aquatic Communities

As with other communities, maintenance of structural complexity and habitat variability in flowing water systems generally benefits amphibians and reptiles. This can be related to management for native vegetation associated with lotic communities, but also maintenance of natural debris (e.g., woody snags) and typical physical characteristics of benthic and shoreline regions. Research in the upper Mississippi River System supported that backwater habitats and sloughs were important to turtles because, compared to main channels, they possessed high structural complexity (e.g., contained areas with shallower water, low flow rates, softer bottoms, and ample vegetation or woody snags; Donner-Wright et al. 1999, Anderson et al. 2002, Tucker et al. 2008). Other studies further support that structural complexity due to the presence of variable basking substrates, such as fallen logs and deadwood, is an important predictor of turtle species richness (e.g., VanDeWalle 1993, VanDeWalle and Christiansen 1996, Lindeman 1999b, Anderson et al. 2002).

Unaltered riparian floodplains experience natural disturbances, which then result in variable habitat structure and habitat complexity. These natural disturbance regimes occur, in part, because uncontrolled riparian systems experience periodic floods, while the scouring action of unregulated flowing water leads to riverbeds and streambeds that laterally meander over time (reviewed by Ward et al. 1999, Allan and Castillo 2007). Such disturbances create variation in the characteristics of lotic systems, which can then support a broad array of biota (reviewed by Ward et al. 1999, Rood et al. 2005), including amphibians (Hoover and Killgore 2002), snakes with strong affinities for aquatic habitats (e.g., Roth and Greene 2006), and turtles (e.g., Donner-Wright et al. 1999, Anderson et al. 2002). Given these points, avoidance of unnatural riparian channelization, coupled with maintenance of variable flow rates and changing water levels that mimic historical patterns of periodic disturbance, is likely important to herpetofaunal conservation in Wisconsin. It could likewise be beneficial to aquatic herpetofauna if this variation conformed to natural seasonal patterns, for example, a seasonal reduction in water levels that exposes sandy banks or sandbars during the nesting period for turtles (reviewed by Bodie 2001). As previously discussed, the avoidance of practices that lower water levels during the winter, which could expose and kill overwintering turtles and amphibians, should also be considered (reviewed by Bodie 2001; see Wetlands subsection).

An important conservation and management strategy could be the promotion of large-scale watershed-level protection programs designed to reduce erosion and runoff while protecting habitats in all Wisconsin streams and rivers. In recent decades, the US Army Corps of Engineers, the USFWS, and cooperating state departments of natural resources have planned and implemented extensive habitat restoration programs that provide foraging, nesting, and nursery habitats for a wide variety of fish and wildlife species in the Upper Mississippi River National Wildlife and Fish Refuge (e.g., Pool 8, south of La Crosse, Wisconsin). The refuge encompasses the entire length of the Mississippi River in Wisconsin and beyond. These programs have contributed greatly to reversing the loss of habitats that resulted from the installation of dams.

Anthropogenic Landscapes

Agricultural land

Although the pervasiveness of agriculture in Wisconsin has had profound negative impacts on amphibians and reptiles, its extensive coverage in the state necessitates that habitat managers consider this land cover type in conservation plans. Some landscape features in agricultural settings are used by amphibians and reptiles. These include natural wetlands and constructed farm ponds that might provide refugia in an otherwise disturbed rural countryside (Knutson et al. 1999, Knutson et al. 2004, Weyrauch and Grubb 2004). Wetlands in agricultural settings are probably most valuable if access to undisturbed natural habitat is limited (particularly breeding habitat for amphibians), hydroperiods and nutrient levels (e.g., nitrogen and ammonia) are appropriate, fish are absent, and access by cattle is restricted (Knutson et al. 2004, Weyrauch and Grubb 2004). Along these lines, pockets of woodlot and grassland in agricultural landscapes are likely valuable to amphibians and reptiles if they are of adequate size (Weyrauch and Grubb 2004). Enrolling and maintaining land in the Farm Bill conservation programs can benefit a variety of wildlife, including amphibians and reptiles, so long as appropriate use of heavy machinery and timing of management actions are considered (R. Hay, personal communication). Management plans should further investigate strategies to create corridors through agricultural properties to adjacent natural lands, which could be facilitated by preserving pockets or strips of intact woodland and grassland habitats on otherwise disturbed rural parcels.

The influence of grazing livestock is surely important to consider when developing management strategies for agricultural lands. Grazed landscapes can act as acceptable habitat for herpetofauna, particularly if grazing is not intense, structural heterogeneity of vegetation is maintained, and variation in the distribution of ground cover is promoted (e.g., native vegetation, rocks, and fallen woody debris; Brown et al. 2011b). However, Homyack and Guiliano (2002) assessed variation in species richness and biomass of amphibians and reptiles at stream crossings in agricultural settings where cattle did and did not have access. They found that "birds, small mammals, and fish[es] appear to benefit relatively quickly from streambank fencing [to exclude cattle], whereas most reptiles and amphibians need >4 years to recover." Knutson et al. (2004) also determined that cattle use of agricultural ponds negatively impacted amphibians. Despite this, other research has found that cattle ponds provided suitable amphibian breeding habitat (Bull et al. 2001, Burton et al. 2009), highlighting the need for additional work on this subject.

Urban habitat and green space

Although it can be difficult to imagine urban landscapes as habitat for amphibians and reptiles, some herpetofaunal species are frequently found in urbanized environments (e.g., Spinks et al. 2003, Brown et al. 2011a, Lambert and Giller 2015). Yet it is important to consider that urban habitats, even those that house high abundances of certain species, could act as ecological traps (Battin 2004). Furthermore, high species richness or population densities in urban locations could simply occur because individuals are forced to congregate in the only available (albeit subpar) undeveloped parcels in an urban or suburban landscape. It is also important to recognize that even if the use of urban habitats does not result in high adult mortality, urban populations might not be viable if all aspects of life history are not supported. For example, although turtles may survive in urban ponds, a lack of adjacent terrestrial habitat for nesting will lead to limited population recruitment that could eventually result in population declines. The influence of potentially high nest predator densities near urban ponds is also difficult to offset and could have substantial impacts on turtle populations (Lambert et al. 2013). Likewise, many amphibian species also use urban ponds for reproduction (e.g., Boone et al. 2008, Brand and Snodgrass 2010), but survival of eggs, larvae, and recently metamorphosed individuals in these ponds may not be sufficient to support populations long term.

Given the pervasiveness of urban development in the Midwest, strategic plans for residential or commercial land use and development should consider ways to integrate wildlife habitat so that connectivity is maintained among urban green spaces and adjacent natural areas across landscapes. For example, Rudd et al. (2002) found that connectivity could be achieved across urban green spaces in Vancouver, but this required access to public land (e.g., public parks), land associated with infrastructure (e.g., transmission line rights-of-way), and private land (e.g., backyard gardens). Educating landowners on the value of wildlife and assisting with design of backyard areas to support native plant and animal species are critical for creating greenspace networks in highly urbanized environments (Rudd et al. 2002). It is also critical to educate and work in partnership with citizens, appointed or elected government committee members, and politicians at numerous levels (e.g., city, county, and state) to design appropriate wildlife habitat and create plans for green space connectivity across urban areas.

Constructed or restored wetlands

Creation and restoration of wetlands are potential strategies used to mitigate natural wetland loss due to land use conversion in the United States (Kaza and BenDor 2013). Some research supports that enacting such strategies can also be an effective component of amphibian and reptile conservation initiatives. For example, work in New York revealed that Blanding's Turtles frequently used constructed wetlands during spring and early summer (Hartwig and Kiviat 2007). Constructed or restored wetlands can also be quickly and effectively colonized by amphibians (Calhoun et al. 2014, Stiles et al. 2016). Work in Wisconsin has even found that amphibian abundances in restored wetlands did not significantly differ between assessments conducted recently after restoration and over a decade after restoration (Manitowoc County; Nedland et al. 2007). However, as a conservation tool, construction of wetlands has limitations, and it cannot be viewed as a fail-safe against species extinction or extirpation. Natural wetland characteristics are difficult to re-create in artificial waterbodies. Furthermore, wetlands constructed in open habitats are unlikely to support woodland amphibians if the terrestrial landscape surrounding the wetlands is not suitable habitat (reviewed by Porej et al. 2004). Mature woodlands require many decades to establish, so their re-creation on a disturbed landscape is only feasible in the long term. In addition, although amphibians will occupy constructed wetlands, their simple presence is not a clear indication of amphibian population health or wetland health (Calhoun et al. 2014). Of critical importance is that natural resource managers, politicians, and the general public recognize that created wetlands will rarely be the equivalent of the natural wetlands they are meant to replace. Therefore, construction of wetlands should only be viewed as a mitigation technique if protection and management of existing natural wetlands are not possible (Calhoun et al. 2014).

Calhoun et al. (2014) outlined many specific considerations that practitioners should make when attempting to create wetlands (specifically, ephemeral pools in woodlands). These include maintenance of adequate wetland size and hydroperiod as well as promotion of appropriate vegetation, soil hydrology, tree canopy closure, and bank slope. Kiviat et al. (2000) recommended that wetlands constructed to mitigate for habitat losses be of larger area than that of the original (presumably destroyed) wetland and also be located within or immediately adjacent to habitat complexes where target species have been confirmed. The importance of hydroperiod, particularly to restrict colonization by fish or American Bullfrogs, has been reiterated by several studies (reviewed by Brown et al. 2012b, Calhoun et al. 2014, Stiles et al. 2016). However, permanent and semipermanent wetlands can benefit species whose larvae overwinter (reviewed by Brown et al. 2012b). Creating a heterogeneous habitat with high structural variability is also an important consideration. For example, creation of basking structures in artificial wetlands, such as large-diameter logs placed perpendicular to the shoreline so they emerge above the water's surface, may provide important thermoregulatory sites for snakes, turtles, and amphibians. Given that some turtle species are shy and quickly abandon basking sites at the approach of humans, Lambert et al. (2013) suggested creation of secluded basking locations or erection of visual barriers so that turtles cannot see people to promote normal turtle basking behavior.

It is also essential to consider protection and management of the terrestrial habitat surrounding constructed wetlands (reviewed by Brown et al. 2012b). As with natural wetlands, buffers of intact vegetation around constructed or restored wetlands should probably be wide (from 205 to 268 m [673 to 879 ft]; Semlitsch and Bodie 2003). Without proper vegetation management, restored wetlands can become dominated by invasive plant species (e.g., Mulhouse and Galatowitsch 2003, Nedland et al. 2007), potentially limiting their ability to function as a natural habitat for many native herpetofauna. Calhoun et al. (2014) identified the importance of the landscape context in which created wetlands occur and highlighted the benefit of numerous pools with varying hydroperiods in close proximity (<500 m [1,640 ft]) so they are accessible by amphibians. The landscape composition adjacent to constructed wetlands is also important for amphibians. For example, Simon et al. (2009) found that the proportions of forest and impervious surface adjacent to constructed, urban wetlands had positive and negative influences on amphibian species richness, respectively. Plans to select locations for artificial wetlands must strongly consider the habitats and disturbance regime across large areas. Dodson (2008) assessed the relationship among numerous environmental variables, land cover, and biodiversity for 23 artificial ponds spanning a 52 km (32 mi) watershed in southern Wisconsin. This research found a negative correlation between amphibian presence and percent land adjacent to wetlands covered by mowed lawns.

Construction or restoration of turtle nesting habitats

Construction of nesting habitat or management of vegetation at existing nesting sites could be a viable conservation strategy for turtle species whose nesting habitats have been altered or lost (Marchand and Litvaitis 2004a, R. Hay, personal communication). Although these techniques have shown some promise (e.g., Kiviat et al. 2000), they necessitate knowledge of local turtle nesting behavior and require substantial effort to implement and maintain. For example, selection of new nesting locations requires the identification of appropriate soils, including those deep and loose enough to promote nest building and with appropriate moisture levels to facilitate embryo development. Removal of encroaching woody vegetation to ensure proper light penetration for egg incubation is a key component of creating suitable nesting locations but requires a commitment to long-term management of vegetation. Marchand and Litvaitis (2004a) suggested that artificial nesting sites should not be placed in close proximity to road crossings and "other potentially hazardous sites." They further recommended that artificial nesting locations be large enough so that numerous female turtles are not tempted to nest in a congregated fashion, where negative impacts to many tightly clustered nests could destroy a large number of eggs and have drastic population level effects.

POPULATION MANAGEMENT

In this section I review several broad concepts and strategies relevant to the active management of herpetofaunal populations in Wisconsin. The considerations summarized below are related to a range of topics, from best practices for harvest to strategies for enhancement of existing populations. In many cases, though, data to support management suggestions are scant due to limited research on these subjects in Wisconsin. Yet the topics discussed subsequently, in conjunction with the state's existing regulations (see Conservation Status and Regulations section), provide a starting point for those interested in learning more.

Considerations for Best Harvest Practices

Despite a general paucity of quantitative information on what constitutes the sustainable removal of amphibians and reptiles from wild populations, some literature exists on best practices for harvest of turtles. Current models indicate that, generally, even low amounts of annual harvest can cause population declines. For example, Zimmer-Shaffer et al. (2014) estimated that harvest rates of ≤2.3% for adults and juveniles were required to "maintain stationary populations" of Snapping Turtles and Spiny Softshells. In addition to declines in abundance, harvest can unnaturally skew body size distributions in exploited populations if individuals of certain sizes are more desirable for harvest (Gamble and Simmons 2004). However, both loss of population members and alterations to average body size distributions could be offset through restrictions placed on maximum size limits for legally harvested individuals (e.g., Gamble and Simmons 2004, Colteaux and Johnson 2017). Unfortunately, comparable data to guide sustainable strategies do not currently exist for all potentially harvested turtle species or other taxonomic groups historically subjected to harvest (e.g., frogs). The development of plans that support population stability in the face of harvest likely also require multiyear investigations, which should be followed by long-term (more than ten years) population monitoring to determine strategy success.

Reduction of Bycatch and Accidental Capture

Barko et al. (2004) suggested that a relationship existed between habitat, temperature, water oxygen levels, and accidental capture (and mortality) of turtles by commercial fishing operations. Therefore, monitoring of habitat and environmental conditions to plan appropriate harvest schedules could offset some accidental turtle mortality during commercial fishing operations. For example, they reported that no turtle mortality occurred when water temperatures were <20°C (68°F). They also recommended that oxygen levels be tested at sampling locations to determine if they are sufficient for survival of turtles accidentally captured when nets are deployed. Barko and Briggler (2006) and Barko et al. (2004) further suggested that a reduction in incidental mortality due to commercial fishing could occur through implementation of turtle exclusion devices on fishing nets. In addition, efforts to determine how long fishing nets can be deployed before turtle mortality occurs could reduce the potential for incidental deaths.

Population Augmentation, Reintroduction, and Translocations

Population augmentation (via release of individuals from captive breeding or rearing programs), translocation (i.e., movement of individuals from one location to another for repatriation or population augmentation), and reintroductions are intriguing but still largely unproven conservation techniques that remain the subject of much debate (see reviews by Burke 1991, Seigel and Dodd 2002, Dodd 2005, Germano and Bishop 2009, Zippel et al. 2011). Best practices for implementation and the effectiveness of population augmentations and translocations are unknown for many species in Wisconsin. Yet, given the information outlined below, it would seem unwise to either fully endorse or outright denounce these strategies without further investigations specific to Wisconsin. Furthermore, regardless of these strategies' conceptual effectiveness, Dodd (2005) stated, "If the real causes of a species' decline and disappearance are not addressed, [these techniques] become halfway solutions that will not result in conservation." As an example, efforts to augment snake populations affected by road mortality could be futile if the root problem (i.e., road mortality) is not corrected. Although a full review of these techniques is not possible here, we encourage readers to use the information summarized and assess existing literature to learn more.

Population augmentation, such as juvenile headstarting programs, has been employed in conservation initiatives for numerous species for years. The most well known examples involve marine turtles (e.g., Heppell et al. 1996), although headstarting has also been employed for freshwater turtle, snake, and amphibian conservation (including some efforts in Wisconsin). Typically, headstarting is a strategy meant to reduce or offset high mortality in young age classes. Often, this includes protection of nests from predators or removal of eggs from wild nests for incubation in the relative safety of captivity so that hatchlings/juveniles can be released at a later date. Despite its theoretical effectiveness, use of this technique has received criticism (reviewed by Spencer et al. 2017). For example, success of headstarting is not often rigorously assessed after implementation, due, in part, to the fact that assessing the contribution of released headstarted individuals to populations requires long-term monitoring. Furthermore, some research suggests that losses of adults have the strongest demographic influence on populations of reptiles such as turtles (e.g., Heppell et al. 1996, Zimmer-Shaffer et al. 2014), which indicates that preservation of juvenile age classes could have a minimal positive impact. Yet recent investigations into the effectiveness of headstarting or similar techniques have found that they can be effective for frogs (e.g., Canessa et al. 2016) and turtles (Spencer et al. 2017). Spencer et al. (2017) further determined that headstarting was the only strategy that "eliminated all risk of extinction" for the Eastern Long-necked Turtle (*Chelodina longicollis*) in Australia. Likewise, some turtle headstarting initiatives in Wisconsin anecdotally appear to have achieved at least partial success. For example, an intense 12-year headstarting program focused on Ornate Box Turtles in Sauk County resulted in headstarted juveniles that survived to adulthood, engaged in breeding, and became gravid (J. Kapfer, personal observation). Furthermore, a study on Blanding's Turtles in Rock County determined that headstarted juveniles experienced high survival immediately after release (Kastle et al. 2021). While these findings are encouraging, the long-term survival rate of released headstarted juveniles, the level of reproductive success that previously headstarted individuals experience as adults, and the survival rates of the offspring they produce remains elusive. More research into the effectiveness of headstarting as a conservation tool in Wisconsin is warranted.

Like other augmentation methods, the effectiveness of translocating individuals from one location to another for conservation purposes is still largely unknown. Although some studies have found translocations to be at least somewhat successful (e.g., Attum et al. 2013, Roe et al. 2015, King et al. 2004), others have not (e.g., Muths et al. 2001). See also examples of translocation attempts in Wisconsin that have had highly variable levels of success in the Ornate Box Turtle species account. In addition, Cook (2008) attempted an extensive repatriation effort over 20 years on a 10,522 ha (26,000 ac) National Park Service property in the New York metropolitan area. During this time, he attempted 40 translocations spread across 20 herpetofaunal species with only marginal success (19 of 40 translocations considered successful, 7 considered probably successful by the author). Not surprisingly, success was greater in more generalist species, particularly those that are often tolerant of urbanized landscapes. The author further suggested that greater translocation success could occur if the relocation site includes wetlands with diverse characteristics (e.g., variation in depth and hydroperiod). He also indicated that success could be more likely if metapopulation dynamics are considered, screening for disease is possible, release of sufficient numbers of animals for population viability is possible, and implementation of rigorous follow-up monitoring occurs. It is important to factor in the prevailing environmental conditions at locations from which translocated individuals are collected and compare these to conditions where they will be released. Wisconsin often experiences harsh winters, so translocated individuals should likewise have come from a location with harsh winters. There must also be adequate resources at the release site to meet the ecological needs of translocated individuals. While the presence of sufficient food is an obvious requirement, many less obvious resources must also be present, such as adequate overwintering locations. It is, therefore, important to understand the historic range and specific habitat/environmental requirements of the species in question prior to implementing a translocation. Translocations of individuals to outside of their historic range or to unsuitable habitat are more likely to fail. In general, much more information is needed before this strategy can be definitively endorsed as a sound conservation measure in Wisconsin, and translocations should not be attempted without careful consideration and discussion.

CONSERVATION STATUS AND REGULATIONS

This section provides information on the conservation status for each of Wisconsin's amphibians and reptiles. Table 7 shows the conservation status of each species at the state, federal, and global scales. Table 8 provides the definitions for the statuses used in Table 7. This section also briefly reviews some current regulatory considerations for amphibians and reptiles in Wisconsin. Monitoring the status (i.e., abundance and trend) of amphibian and reptile species provides a scientific basis for implementing management and recovery actions and setting regulations for 'take' (Wisconsin defines 'take' as "shooting, shooting at, pursuing, hunting, catching or killing any wild animal" [Wisconsin Administrative Rules Chapter NR 27]). Many current regulations on the take, harvest, or possession of Wisconsin's amphibians and reptiles (including those referenced below) can be found in Wisconsin Administrative Rules and Codes related to the Department of Natural Resources (NR). The NR chapters most directly applicable to the state's amphibians and reptiles are Chapters NR 16 (Captive Wildlife), NR 19 (Miscellaneous Fur, Fish, Game and Outdoor Recreation), and NR 27 (Endangered and Threatened Species). These chapters are available where Wisconsin administrative rules and codes are collected for public viewing, including online repositories.

For common species in the state, collection is currently regulated by open/closed seasons and possession limits and in some cases requires a license, such as a Small Game License. A summary of license information and general regulations for possession and collection of amphibians and reptiles in Wisconsin can be found online through WDNR web resources. For Endangered and Threatened species, collection is illegal without one of the following permits, authorizations, or other approvals. For unintentional take related to on-the-ground projects, such as maintenance activities, construction, or development projects, individuals or entities may apply for an Incidental Take Permit/Authorization through the WDNR. These permits are not guaranteed and are granted after review and approval by the WDNR. Species-specific regulatory guidance can be accessed online through WDNR web resources. It is also noteworthy that to best protect rare species from insidious activities, WDNR data on specific locations of occurrence for rare species (including Timber Rattlesnake hibernacula) are currently exempt from the state's Open Records Law. In addition to restrictions on possession and harvest, it is also currently illegal to sell any wild-collected live specimens of reptile species that are native to Wisconsin, regardless of their origin (i.e., whether collected in-state or out-of-state or bred in captivity). The in-state sale of all turtles collected in Wisconsin, other than those collected on Minnesota and Iowa boundary waters (the Mississippi and St. Croix Rivers), is also prohibited.

Those interested in conducting research on rare species must obtain an Endangered and Threatened Species Permit through the WDNR. Researchers who wish to work with native common herpetofauna must still obtain a Scientific Collector's Permit or Research License through the WDNR. Typically, researchers will also need (Institutional) Animal Care and Use Committee approval through their agency or university.

ADDITIONAL RESOURCES

This chapter provided a broad overview of conservation threats and management strategies for amphibians and reptiles, with emphasis on habitats found in Wisconsin. It acts as a foundation for management considerations but did not provide exhaustive coverage of specific management actions and scenarios. Thus, interested individuals should review additional literature when developing management plans. Dozens of books and thousands of peer-reviewed scientific papers exist that can assist with management of amphibians and reptiles. An excellent general reference manual for management of amphibian and reptile habitats in the Midwest was created by the Partners in Amphibian and Reptile Conservation (Kingsbury and Gibson 2012).

Identification Keys for Amphibians and Reptiles in Wisconsin

AMPHIBIA—ANURA

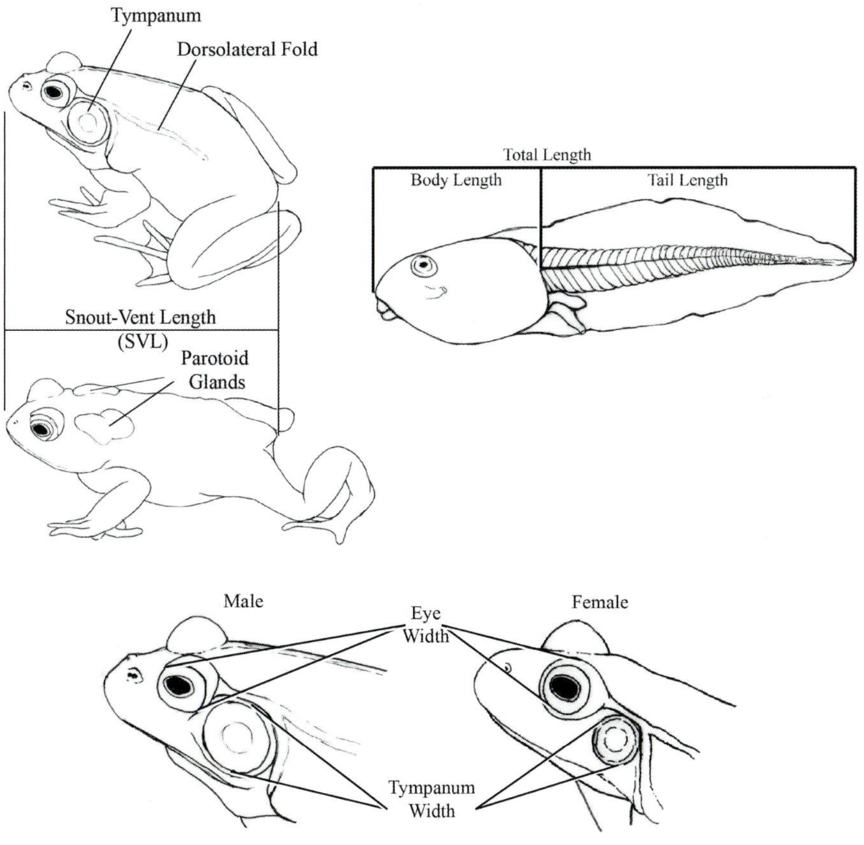

Figure 49. Some structures found among members of Anura that are useful for species identification. The length measurement most frequently used is snout-to-vent length (length from the tip of the snout to the cloacal opening [i.e., vent]). Some researchers use snout-to-urostyle length (SUL), where the measurement is made to the posterior tip of the hipbone (urostyle). For most uses, the difference between these measurements is negligible.

Key to Wisconsin's Frogs and Toads

1A) Skin dry and with parotoid glands and numerous large warts; hop rather than jump or climb; keratinized metatarsal spade on heel of each hind foot.................................... **American Toad (*Anaxyrus americanus*)**

1B) Skin slimy and without parotoid glands or large warts; no keratinized metatarsal spade on heal of each hind foot............................**2**

2A) Terminal toe pads present as in *i* and *ii* (Hylidae).............................**3**

2B) Terminal toe pads absent as in *iii* (Ranidae)...**6**

3A) Toe pads distinctly wider than toes (*i*) **4**

3B) Toe pads scarcely wider than toes (*ii*) **5**

4A) Hidden surface of thighs bright yellow reticulated with brown or black .... **Gray Treefrog (*Hyla versicolor*) or Cope's Gray Treefrog (*Hyla chrysoscelis*)**

4B) Hidden surface of thighs without bright yellow; dorsum with a distinct brown usually X-shaped mark on light brown or tan ground color................. **Spring Peeper (*Pseudacris crucifer*)**

5A) Toes of hind limbs unwebbed or webbing restricted to fringe on each side; dorsum smooth, with three continuous or broken, brown or green stripes .... **Boreal Chorus Frog (*Pseudacris maculata*)**

5B) Toes of hind limbs fully webbed; dorsum with irregularly spaced bumps or warts and usually an interorbital triangle............**Blanchard's Cricket Frog (*Acris blanchardi*)**

6A) Dorsolateral folds absent.....................**7**

6B) Dorsolateral folds present....................**8**

7A) Supratympanic fold prominent.. **American Bullfrog (*Lithobates catesbeianus*)**

7B) Supratympanic fold absent or not prominent; webbing on toes of hind feet extends to last joint of fourth toe and to the tip of the fifth toe; musky odor; upper jaw unmarked green .... **Mink Frog (*Lithobates septentrionalis*)**

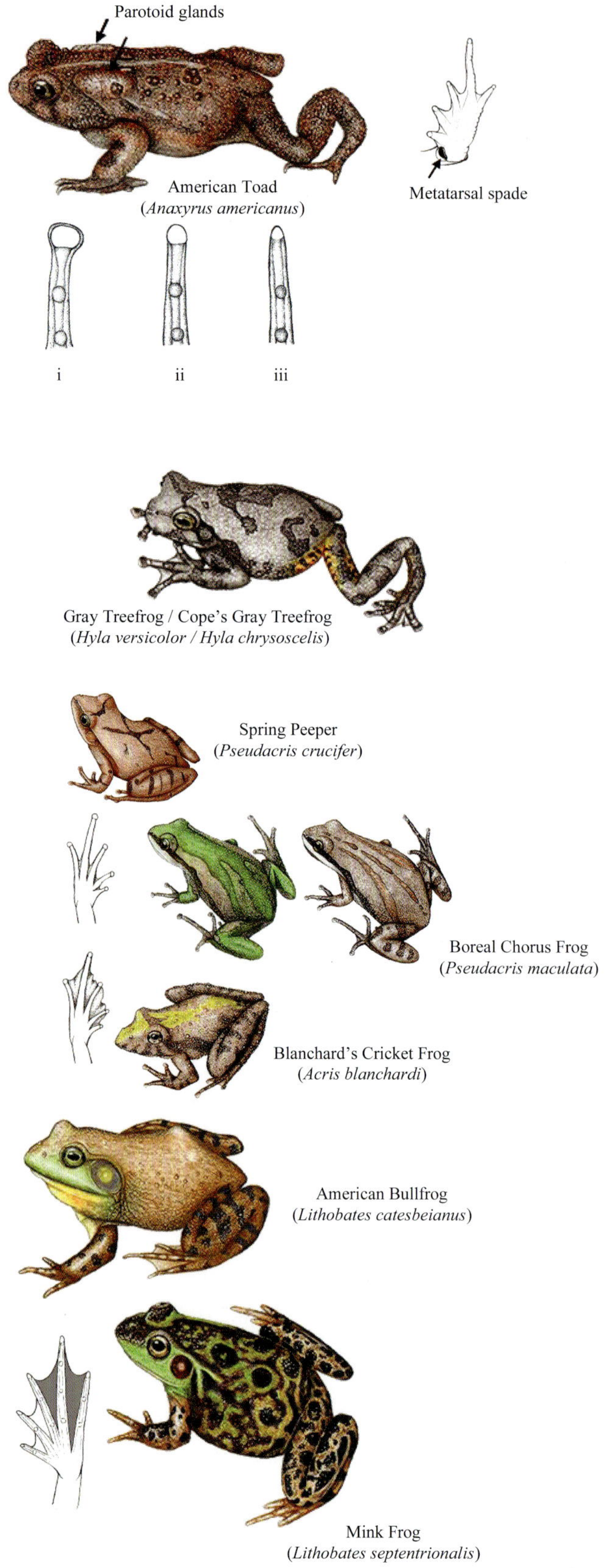

Figure 50. Key adapted from Vogt (1981), with contributions from Powell et al. (1998) and Christoffel et al. (2009).

8A) Dorsolateral folds extending to mid-length of dorsum; webbing on toes of hind feet extends just beyond the second joint of the fourth toe and does not reach the tip of fifth toe; no musky odor; upper jaw mottled with gray or brown . . . . **Green Frog (*Lithobates clamitans*)**

8B) Dorsolateral folds extending entire length of dorsum **9**

9A) Dark mask passing through eye and tympanum; skin between dorsolateral folds smooth; dorsum brown, tan, or reddish brown with no large dark spots . . . . **Wood Frog *Lithobates sylvaticus*)**

9B) Dark mask absent **10**

10A) Dorsum brown or brown with black speckles and without large dark spots . . . . [**Burnsi morph**] **Northern Leopard Frog (*Lithobates pipiens*)**

10B) Dorsum with large dark spots **11**

11A) Dorsal spots usually irregular in size and distribution; thighs and groin white or yellow green, never golden . . . . . . **Northern Leopard Frog (*Lithobates pipiens*)**

11B) Two rows of large square or rectangular paired spots present between dorsolateral folds and two rows of spots lateral to the folds; thighs and groin golden yellow. . . . . **Pickerel Frog (*Lithobates palustris*)**

Green Frog
(*Lithobates clamitans*)

Wood Frog
(*Lithobates sylvaticus*)

Northern Leopard Frog Burnsi morph
(*Lithobates pipiens*)

Northern Leopard Frog
(*Lithobates pipiens*)

Pickerel Frog
(*Lithobates palustris*)

Figure 50. *Continued*

Key to Wisconsin Frog and Toad Larvae (Tadpoles)

1A) Gills internal; body abruptly wider than tail; forelimbs not visible externally **Frogs (ANURA)** **2**

1B) Gills external; body not abruptly wider than tail; forelimbs visible externally. . . . . **Salamanders (CAUDATA)**

2A) Eyes lateral or dorsolateral. **3**

2bB) Eyes dorsal **6**

Figure 51. Key adapted from Watermolen and Gilbertson (1996).

3A) Spiracle tube long and free from body wall; tip of tail usually black .... **Blanchard's Cricket Frog (*Acris blanchardi*)**

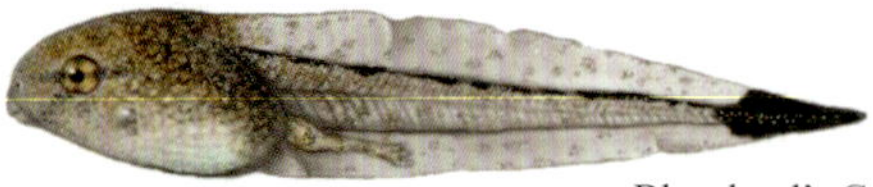

Blanchard's Cricket Frog
(*Acris blanchardi*)

3B) Spiracle tube short and attached to body wall; tip of tail not black **4**

4A) Belly bronze-colored (most conspicuous in sunlight) .... **Boreal Chorus Frog (*Pseudacris maculata*)**

Boreal Chorus Frog
(*Pseudacris maculata*)

4B) Belly white or light-colored **5**

5A) Fins clear with or without black blotches, if present, a clear area near tail musculature; chest and throat lightly colored; belly iridescent.................... **Spring Peeper (*Pseudacris crucifer*)**

Spring Peeper
(*Pseudacris crucifer*)

5B) Fins reddish and heavily mottled with black; belly white or very white cream; throat seldom pigmented.................. **Gray Treefrog (*Hyla versicolor*) or Cope's Gray Treefrog (*Hyla chrysoscelis*)**

Gray/Cope's Gray Treefrog
(*Hyla versicolor/chrysoscelis*)

6A) Vent medial................ **American Toad (*Anaxyrus americanus*)**

American Toad
(*Anaxyrus americanus*)

6B) Vent dextral **7**

7A) White line from nostril to snout.................................. **Northern Leopard Frog (*Lithobates pipiens*)**

Northern Leopard Frog
(*Lithobates pipiens*)

7B) No white line from nostril to snout......... **8**

8A) Tail without spots or mottling, or with only faint small marks........... **Wood Frog (*Lithobates sylvaticus*)**

Wood Frog
(*Lithobates sylvaticus*)

8B) Tail with spots or mottled **9**

9A) Tail with pinkish buff spots....................... **Mink Frog (*Lithobates septentrionalis*)**

Mink Frog
(*Lithobates septentrionalis*)

9B) Tail with dark spots and without pinkish buff spots................................ **10**

10A) Intestinal coil not visible; tail with fine yellow Spots.... **Green Frog (*Lithobates clamitans*)**

Green Frog
(*Lithobates clamitans*)

10B) Intestinal coil visible or slightly visible; tail without fine yellow spots **11**

Figure 51. *Continued*

11A) Body and fins heavily marked, often appearing purple; generally smaller tadpoles (up to 8 cm [3.1 in] TL)............ ....**Pickerel Frog (*Lithobates palustris*)**

11B) Body and fins with distinct black dots, with more dots on the dorsal fin than on the ventral fin; generally larger tadpoles (up to 14 cm [5.5 in] TL).............**American Bullfrog (*Lithobates catesbeianus*)**

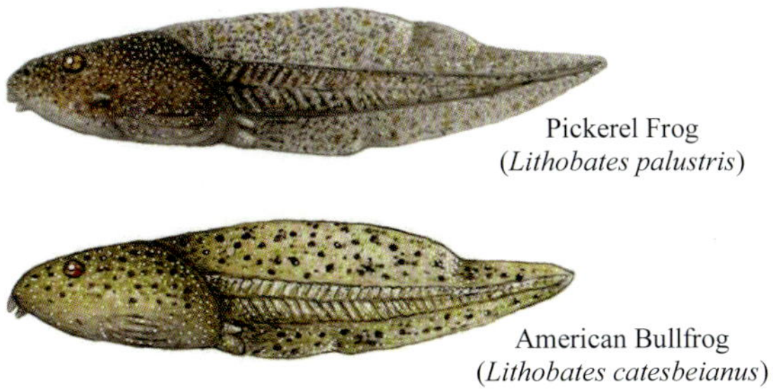

Figure 51. *Continued*

AMPHIBIA—CAUDATA

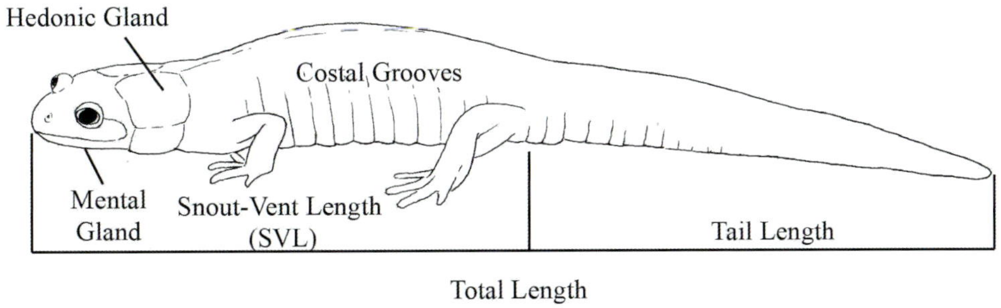

Figure 52. Some structures found among members of Caudata that are useful for species identification.

Key to Wisconsin's Salamanders

1A) Tail musculature with dorsal and ventral fins **2**

1B) Tail musculature without dorsal and ventral fins **3**

2A) External gills present, lower jaw reduced, eyes small, four toes on each hind foot............. .... **Mudpuppy (*Necturus maculosus*)**

2B) External gills absent, lower jaw not reduced, eyes not small, five toes on each hind foot...... .... **Eastern Newt aquatic (*Notophthalmus viridescens*)**

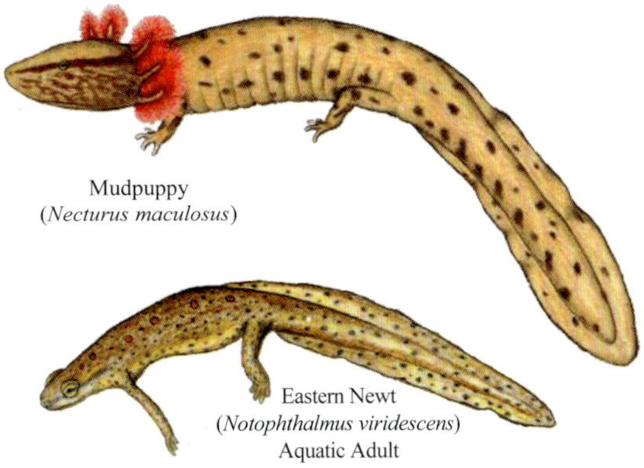

Figure 53. Key adapted from Vogt (1981).

3A) Skin granular, bright orange to brick red to reddish brown. Dorsolateral series of longitudinally aligned red spots bordered in black...... **Eastern Newt eft/terrestrial adult (*Notophthalmus viridescens*)**

3B) Skin not granular or colored bright orange. Dorsolateral series of longitudinally aligned red spots bordered in black absent............**4**

4A) Body long and thin. Size small. Nasolabial grooves present (Plethodontidae) **5**

4B) Body stout and robust. Medium size. Nasolabial grooves absent (Ambystomatidae) **6**

5A) Tail distinctly colored differently than body with a constriction at its base. Venter milky-white with black flecks. Four toes on each hind foot...................**Four-toed Salamander (*Hemidactylium scutatum*)**

5B) Tail not distinctly colored differently than body and with no constriction at its base. Venter light gray anteriorly to dark gray or black posteriorly. Five toes on each hind foot.............................. **Eastern Red-backed Salamander (*Plethodon cinereus*)**

6A) Body color black with light blue or bluish-white markings...........**Blue-spotted Salamander (*Ambystoma laterale*)**

6B) Body color black without light blue or bluish-white markings.............................**7**

7A) Body and tail with dorsolaterally symmetrical series of yellow to orange spots, and gray chin .... **Spotted Salamander (*Ambystoma maculatum*)**

7B) Body and tail with irregular yellow blotches in no distinguishable ordered pattern, and yellow chin............... **Eastern Tiger Salamander (*Ambystoma tigrinum*)**

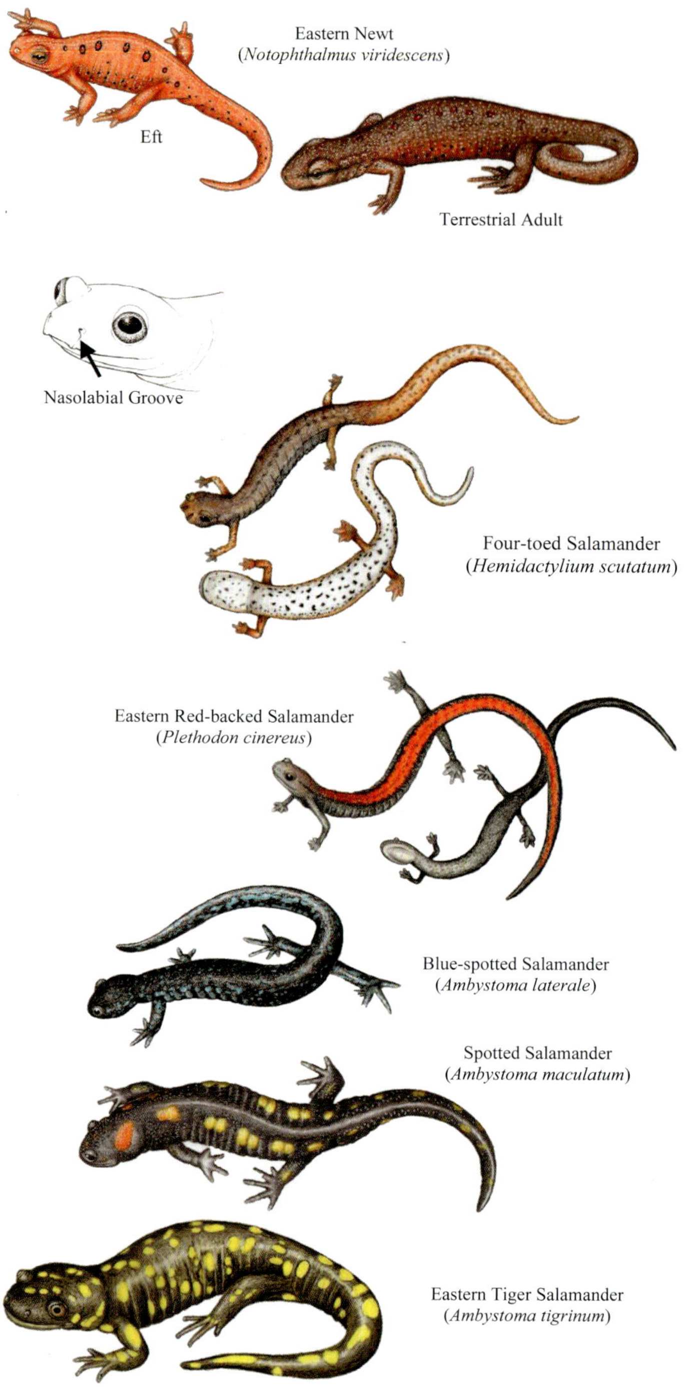

Figure 53. *Continued*

Key to Wisconsin's Salamander Larvae

1A) Gills external; body not abruptly wider than tail; forelimbs visible externally **Salamanders (CAUDATA)** **2**

1B) Gills internal; body abruptly wider than tail; forelimbs not visible externally.. **Frogs (ANURA)**

2A) Dorsal fin extending only to base of tail, near a point above the vent; gills short .... **Mudpuppy (*Necturus maculosus*)**

Mudpuppy
(*Necturus maculosus*)

2B) Dorsal fin extending onto body, nearly to the head; gills long **3**

3A) Ground color gray or yellowish brown, with two rows of dorsal light yellow spots; prominent dark stripe running through eye.................................... **Eastern Newt (*Notophthalmus viridescens*)**

Eastern Newt
(*Notophthalmus viridescens*)

3B) Ground color gray or greenish brown, without two dorsal rows of yellow spots; no or less-prominent dark stripe running through eye... **4**

4A) Four toes on hind feet; light median dorsal band with irregular dark border; total length generally less than 2 cm (0.8 in) .... **Four-toed Salamander (*Hemidactylium scutatum*)**

Four-toed Salamander
(*Hemidactylium scutatum*)

4B) Five toes on hind feet; no light median dorsal band with irregular dark border; total length generally longer than 2 cm (0.8 in)........... **5**

5A) Hatchlings without balancers or rudimentary forelimbs; toes of older larvae flat, broad at base, and pointed at tip; total length often longer than 7 cm (2.8 in) .... **Eastern Tiger Salamander (*Ambystoma tigrinum*)**

Eastern Tiger Salamander
(*Ambystoma tigrinum*)

5B) Hatchlings with balancers and rudimentary forelimbs; toes of older larvae not broad at base, rounded at tip; total length generally 4.5–7.0 cm (1.8–2.8 in)................................. **6**

Figure 54. Key adapted from Watermolen and Gilbertson (1996).

6A) Hatchlings (TL 0.8–1.2 cm [0.3–0.5 in]): dark brown with faint yellowish transverse bars; dorsal and caudal fins almost transparent; older larvae (TL > 3.0 cm [1.2 in]): with yellow midline blotches and yellow lateral stripe .... **Blue-spotted Salamander (*Ambystoma laterale*)**

Blue-spotted Salamander
(*Ambystoma laterale*)

6B) Hatchlings (TL 1.2–1.4 cm [0.5–0.6 in]): yellowish green or brown with dark vermiculation and mottling on body; dorsal and caudal fins darkly mottled; older larvae (TL > 5.0 cm [2.0 in]): with a row of small yellowish spots and light lateral stripe .... **Spotted Salamander (*Ambystoma maculatum*)**

Spotted Salamander
(*Ambystoma maculatum*)

Figure 54. *Continued*

AMPHIBIA—EGGS

Key to Wisconsin's Amphibian Eggs

1A) Eggs deposited out of water (under rocks or logs, or in moss)..**2**

1B) Eggs deposited in water..**3**

2A) Clutches with 3–14 white, spherical eggs; 0.4–0.6 cm (0.16–0.24 in) in diameter; often suspended by a single pedicle; often under rotting pine logs; female generally in attendance of egg clutch...**Eastern Red-backed Salamander (*Plethodon cinereus***

2B) Clutches with 15–25 (possibly as many as 64 or combined with other clutches) cream-colored eggs; 0.4–0.5 cm (0.16–0.20 in) in diameter; suspended in moss (most often *Sphagnum*) 2.0–12.0 cm (0.79–4.72 in) above water; often in communal nests guarded by a female ....**Four-toed Salamander (*Hemidactylium scutatum*)**

3A) Eggs deposited in water singly..**4**

3B) Eggs deposited in water in clumps or clusters **6**

4A) Eggs deposited singly (but in groups of 50–100); 0.5–0.6 cm (0.2 in) in diameter; usually attached beneath objects at 1–3 m (3.3–9.8 ft) depth; yolk not pigmented..........................**Mudpuppy (*Necturus maculosus*)**

4B) Eggs deposited singly; yolk pigmented..**5**

5A) Eggs 0.15–0.25 cm (0.06–0.10 in) in diameter (0.50–0.60 cm [0.20 –0.24 in] in diameter with gelatinous envelope); sometimes in small clusters of up to 4 eggs sometimes in clusters of 7–40); attached to vegetation, rocks, or debris on the pond bottom (see also **11A**) .... **Blue-spotted Salamander (*Ambystoma laterale*)**

Figure 55. Key adapted from Watermolen (1995).

5B) Eggs 0.15 cm (0.06 in) in diameter, with yolk approximately 0.10 cm (0.04 in) in diameter; attached to vegetation .... **Spring Peeper (*Pseudacris crucifer*)**

5C) Eggs 0.20 cm (0.08 in) in diameter; thick jelly-like substance coating each egg (0.65–0.80 cm [0.26–0.31 in] diameter with gelatinous envelope); elliptical in shape; eggs colored brown dorsally, yellow ventrally; often wrapped in aquatic vegetation (e.g., *Chara*)........................ **Eastern Newt (*Notophthalmus viridescens*)**

6A) Egg mass forming a layer on water surface **7**

6B) Egg mass submerged.. **8**

7A) 1000 to 5000 (as many as 20000) black and white eggs; eggs 0.12–0.18 cm (0.05–0.07 in) in diameter; deposited in a large film (up to 1 × 1 m [3.3 × 3.3 ft]) in several layers; in permanent water bodies only.. **American Bullfrog (*Lithobates catesbeianus*)**

7B) 1000 to 4000 (as many as 5000) black and white eggs; eggs 0.10–0.15 cm (0.04–0.06 in) in diameter; deposited in a thin film (up to 30 × 30 cm [11.8 × 11.8 in]); often attached to floating or emergent vegetation... **Green Frog (*Lithobates clamitans*)**

7C) Loose clusters of 10–40 eggs (sometimes laid singly); eggs 0.10–0.12 cm (0.04–0.05 in) in diameter; sometimes attached to vegetation .... **Gray Treefrog (*Hyla versicolor*) or Cope's Gray Treefrog (*Hyla chrysoscelis*)**

7D) 10–15 eggs (sometimes laid singly); eggs 0.20–0.30 cm (0.08–0.12 in) in diameter; attached to vegetation, often in flowing water .... **Blanchard's Cricket Frog (*Acris blanchardi*)**

8A) Eggs in a double string; black; sometimes threaded among emergent vegetation, but never attached.. **American Toad (*Anaxyrus americanus*)**

8B) Eggs in an oval mass resembling a cluster of grapes **9**

9A) Eggs in a single gelatinous envelope.. **10**

9B) Eggs with two distinct gelatinous envelopes... **11**

9C) Eggs with three distinct gelatinous envelopes; 15–110 eggs; eggs approximately 0.30 cm (0.12 in) in diameter (1.00–1.20 cm [0.39–0.47 in] in diameter with gelatinous envelopes); attached to vegetation near the pond bottom .... **Eastern Tiger Salamander (*Ambystoma tigrinum*)**

10A) Light brown eggs; 0.20 cm (0.08 in) in diameter (0.30–0.50 cm [0.12–0.20 in] in diameter with gelatinous envelope); deposited in a loose globular mass (9.0–11.0 cm [3.5–4.3 in] in diameter); attached to vegetation near the surface.. **Pickerel Frog (*Lithobates palustris*)**

10B) 10 to 150 eggs, 0.05–0.13 cm (0.02–0.05 in) in diameter; deposited in a loose, irregularly-shaped cluster (up to 2.5 cm [1.0 in] in diameter); usually attached to vegetation or debris in shallow water, 7.0–20.0 cm (2.8–7.9 in) below the surface....................................... **Boreal Chorus Frog (*Pseudacris maculata*)**

11A) 7–40 eggs (sometimes in small clusters of up to 4 eggs); 0.15–0.25 cm (0.06–0.10 in) in diameter (0.50–0.60 cm [0.20–0.24 in] in diameter gelatinous envelope); attached to vegetation, rocks, or debris on the pond bottom (see also **5A**)..................................... **Blue-spotted Salamander (*Ambystoma laterale*)**

Figure 55. *Continued*

11B) 50–250 eggs; 0.16–0.25 cm (0.06–0.10 in) in diameter (0.60–0.70 cm [0.24–0.28 in] in diameter with gelatinous envelopes); deposited in an oval mass (6.0–10.0 cm [2.4–3.9 in] in diameter); eggs black, but jelly coating is milky white and usually tinted green (due to symbiotic algae in membranes); usually attached to vegetation near the surface; egg mases rather stiff and will hold their shape when lifted from the water . . . .**Spotted Salamander (*Ambystoma maculatum*)**

11C) 500–800 (up to 3000) darkly pigmented (brown to black) eggs; 0.15–0.20 cm (0.06–0.08 in) in diameter; in globular masses 5.0–10.0 cm (2.4–3.9 in) in diameter, attached to submerged branches or plant stems, usually near the surface; often multiple clusters in the same vicinity (communal egg laying); egg masses rather flimsy and do not hold their shape when lifted from the water; the outer jelly capsule may have a bluish tint and often becomes semi-opaque and overgrown with algae prior to hatching. . . . . **Wood Frog (*Lithobates sylvaticus*)**

11D) 2000–4000 brown-black eggs; 0.20 cm (0.08 in) in diameter; in a loose globular mass attached to submerged vegetation, usually 1 m (3.3 ft) or greater in depth. . . . . **Mink Frog (*Lithobates septentrionalis*)**

11E) 1000–6000 black eggs; 0.10–0.18 cm (0.04–0.07 in) in diameter (0.50 cm [0.20 in] in diameter with gelatinous envelope); deposited in a tight, oval mass (10.0–16.0 cm [3.9–6.3 in] in diameter); often attached to vegetation . . . . **Northern Leopard Frog (*Lithobates pipiens*)**

Figure 55. *Continued*

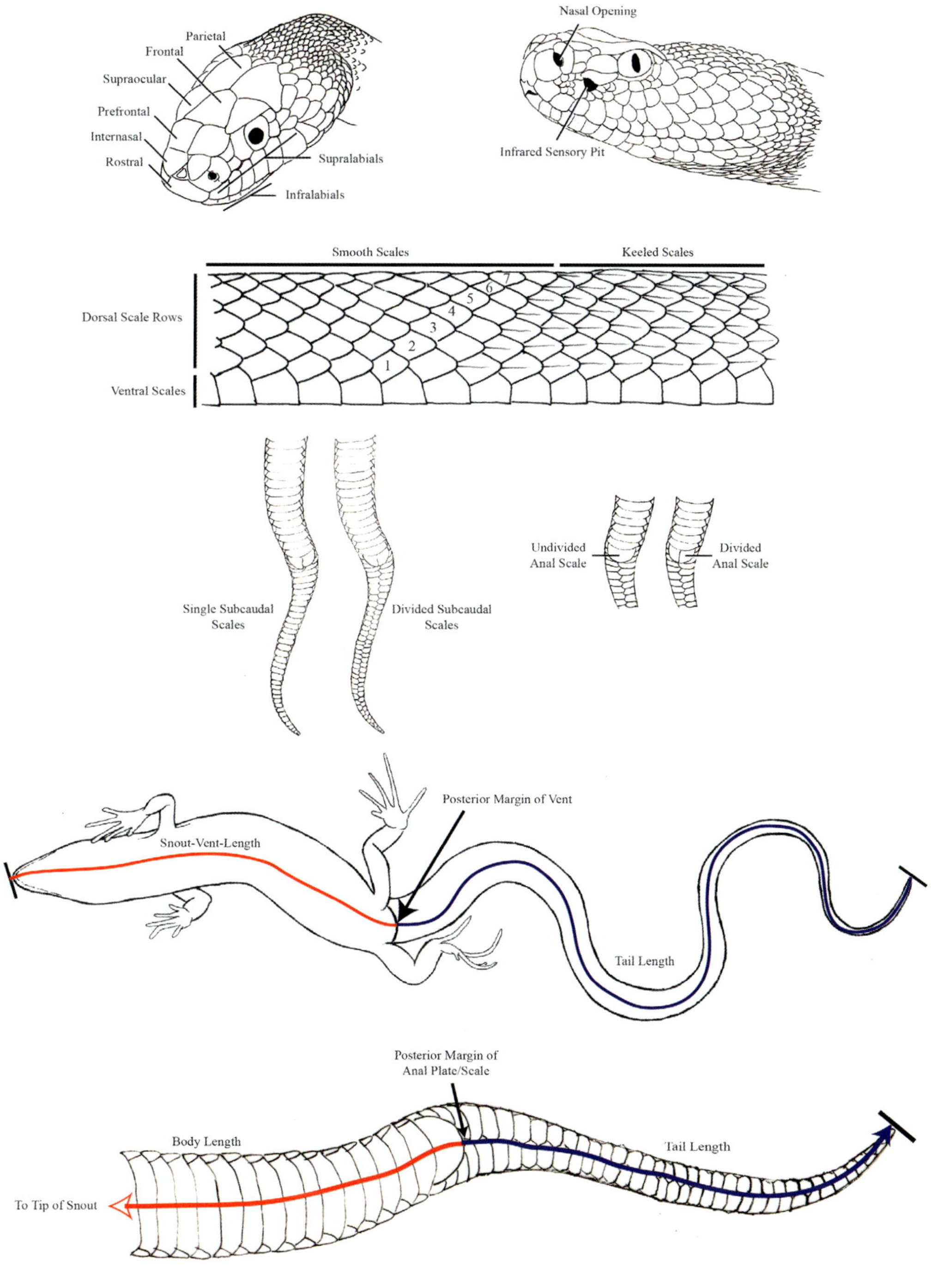

Figure 56. Some structures found among members of Squamata that are useful for species identification.

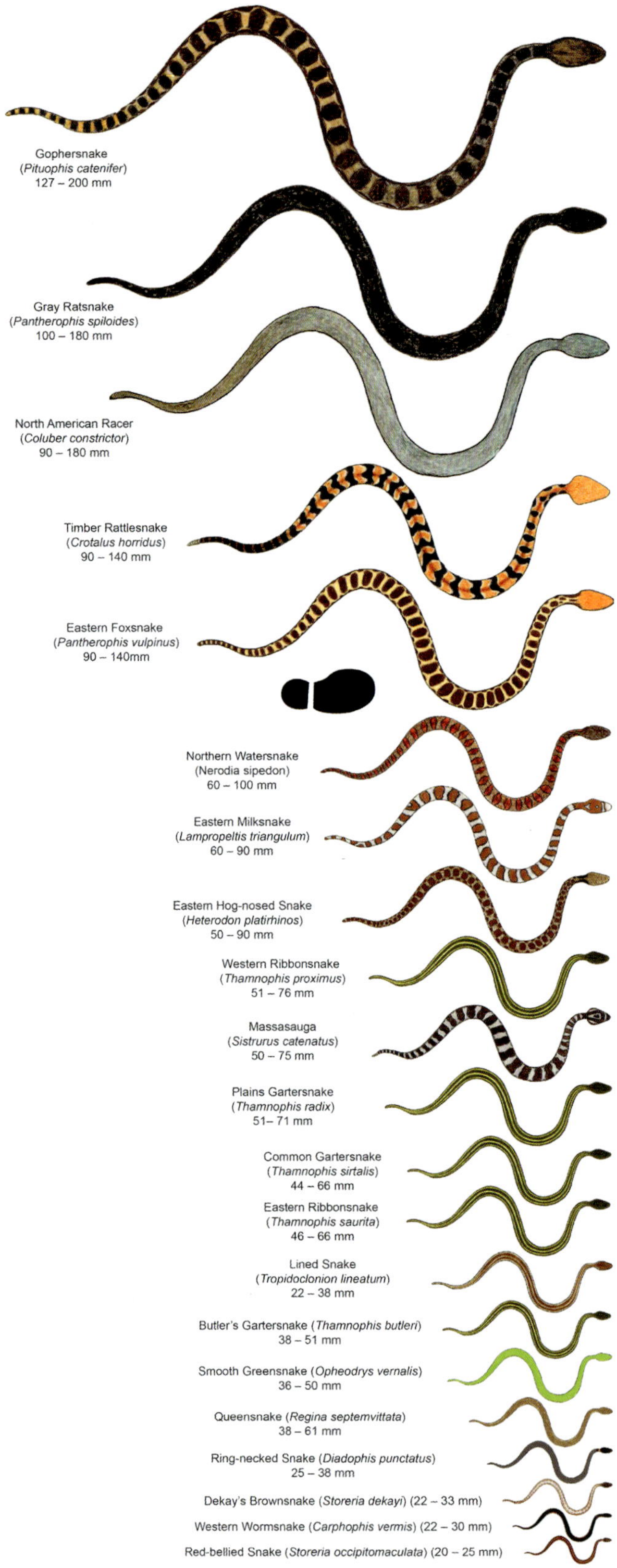

Figure 57. General adult size differences among snake species in Wisconsin based on typical sizes reported in Vogt (1981).

Key to Wisconsin's Snakes and Lizards

1A) Legs present (most Lizards) **2**

1B) Legs absent **4**

2A) Dorsal scales granular; large rectangular ventral scales; tail scales keeled . . . . **Six-lined Racerunner (*Aspidocelis sexlineata*)**

2B) Dorsal scales smooth, large and overlapping; ventral scales not plate-like; tail scales smooth **3**

3A) Two subequal postlabial scales present; dorsolateral light stripe if present no more pronounced than middorsal light stripe . . . . **Common Five-lined Skink (*Plestiodon fasciatus*)**

3B) One postlabial scale present; dorsolateral light stripe pronounced . . . . **Prairie Skink (*Plestiodon septentrionalis*)**

4A) Lateral groove on each side of body; ear openings present; movable eyelids . . . . **Slender Glass Lizard (*Ophisaurus attenuatus*)**

4B) No lateral groove on each side of body; no ear openings; no movable eyelids (snakes) **5**

5A) Rattle on end of tail; pit between eye and nostril on each side of head; most subcaudal scales not divided; pupil vertical **6**

5B) No rattle on end of tail; no pit between eye and nostril; all subcaudal scales divided; pupil round **7**

6A) Top of head with large plate-like scales; body gray with dark brown blotches; tail banded . . . . **Massasauga (*Sistrurus catenatus*)**

6B) Top of head with small asymmetrical scales; body usually yellow or brown with dark jagged bands; tail solid black . . . . **Timber Rattlesnake (*Crotalus horridus*)**

7A) Some or all dorsal scales keeled **8**

7B) All scales smooth **21**

8A) Anal plate divided **9**

8B) Anal plate entire (not divided) **15**

Figure 58. Key adapted from Vogt (1981).

9A) Rostral scale upturned and keeled .... **Eastern Hog-nosed Snake (*Heterodon platirhinos*)**

9B) Rostral scale not upturned or keeled **10**

Eastern Hog-nosed Snake (*Heterodon platirhinos*)

10A) Loreal scale present **11**

10B) Loreal scale absent **14**

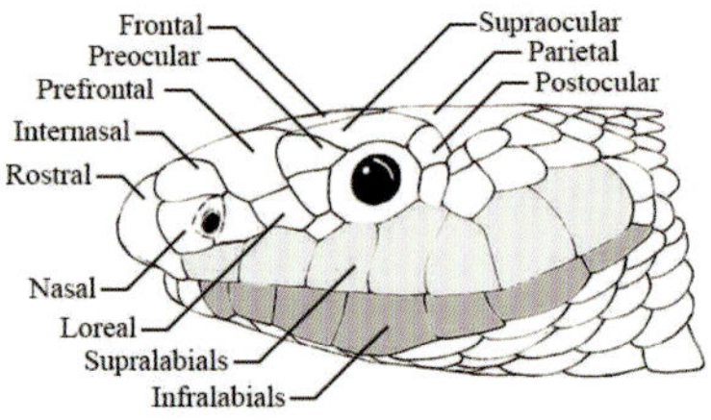

11A) Lateral scales smooth; central scales weekly keeled; 2 postoculars; 25–30 scale rows **12**

11B) All scales heavily keeled; 2 or 3 postoculars; fewer than 25 scale rows **13**

12A) Less than 220 ventrals; head brown or reddish brown; dorsal ground color tan or light brown with dark brown blotches; venter yellow with dark blotches .... **Eastern Foxsnake (*Pantherophis vulpinus*)**

Eastern Foxsnake (*Pantherophis vulpinus*)

12B) More than 220 ventrals; head black; dorsal ground color gray with black blotches; white throat; venter not yellow .... **Gray Ratsnake (*Pantherophis spiloides*)**

Gray Ratsnake (*Pantherophis spiloides*)

13A) Scale rows 19 at midbody, 17 posteriorly; light stripes on rows 1 and 2; two midventral brown stripes or solid brown ventrally .... **Queensnake (*Regina septemvittata*)**

Queensnake (*Regina septemvittata*)

13B) Scale rows 21–25; no light lateral stripes; 30 dorsal bands or spots; venter with red half-moons on light ground color .... **Common Watersnake (*Nerodia sipedon*)**

Common Watersnake (*Nerodia sipedon*)

14A) Scale rows 15; usually 6 supralabials; venter orange or red in life .... **Red-bellied Snake (*Storeria occipitomaculata*)**

14B) Scale rows 17; usually 7 supralabials; venter white or pinkish .... **Dekay's Brownsnake (*Storeria dekayi*)**

Red-bellied Snake (*Storeria occipitomaculata*)

Dekay's Brownsnake (*Storeria dekayi*)

15A) Scale rows less than 27; 2 prefrontals; rostral scute not pronounced or cornified **16**

15B) Scale rows 27 or more at midbody; 3 or more prefrontals; pronounced cornified rostral scute .... **Gophersnake (*Pituophis catenifer*)**

Gophersnake (*Pituophis catenifer*)

Figure 58. *Continued*

16A) Eight or more infralabial scales; venter immaculate **17**

16B) Fewer than eight infralabials; double row of black spots on venter . . . . **Lined Snake (*Tropidoclonion lineatum*)**

Lined Snake (*Tropicoclonion lineatum*)

17A) Light lateral stripe typically on scale rows 2 and 3, but sometimes part of 1 and 4 . . . .**Common Gartersnake (*Thamnophis sirtalis*)**

Common Gartersnake (*Thamnophis sirtalis*)

17B) Light lateral stripe involving fourth scale row **18**

18A) Supralabials white unmarked with black; preorbital white; tail length 27% or more of total length **19**

18B) Supralabials yellow and often marked with black or brown; preorbital not white; tail length less than 27% of total length **20**

19A) Usually 7 supralabials; parietal spots indistinct or absent, never fused; dark ventrolateral stripe covers 1/6 to 1/4 of ventral scales . . . .**Eastern Ribbonsnake (*Thamnophis saurita*)**

Eastern Ribbonsnake (*Thamnophis saurita*)

19B) Usually 8 supralabials; parietal spots may or may not be fused; dark ventrolateral stripe does not, or barely, extends onto ventral scales. . . . . **Western Ribbonsnake (*Thamnnophis proximus*)**

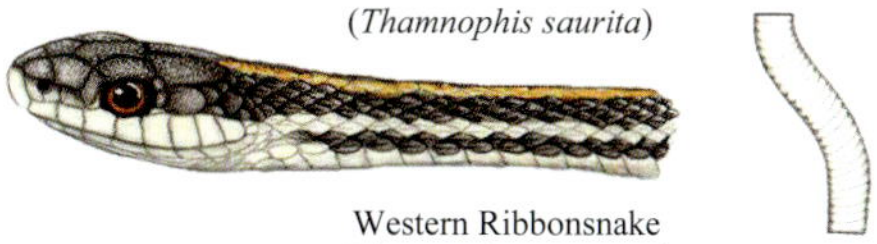

Western Ribbonsnake (*Thamnophis proximus*)

20A) Lateral stripe on second, third, and fourth scale rows; no dark vertical bars on upper labials; 19 scale rows or less . . . .**Butler's Gartersnake (*Thamnophis butleri*)**

Butler's Gartersnake (*Thamnophis butleri*)

20B) Lateral stripes on third and fourth scale rows only; labial sutures margined with black bars; 21 or fewer scale rows . . . **Plains Gartersnake (*Thamnophis radix*)**

Plains Gartersnake (*Thamnophis radix*)

21A) Anal plate entire (not divided). . . . .**Eastern Milksnake (*Lampropeltis triangulum*)**

Eastern Milksnake (*Lampropeltis triangulum*)

21B) Anal plate divided. **22**

Figure 58. *Continued*

22A) Yellow or orange neck ring just behind head. . . . . **Ring-necked Snake** **(*Diadophis punctatus*)**

22B) No neck ring. **23**

23A) Dorsum with dark blotches fading posteriorly. . . . . **juvenile North American Racer (*Coluber constrictor*)**

23B) Uniform dorsal coloration. **24**

24A) Preocular scale absent; dorsal coloration dark gray, ventral surface bright pink. . . . . **Western Wormsnake** **(*Carphophis vermis*)**

24B) Preocular scale present. **25**

25A) Dorsal color blue, gray, or light brown, never green; 2 or 3 anterior temporals on each side of head . . . .**adult North American Racer** **(*Coluber constrictor*)**

25B) Dorsal color green, buff, or bronze; changes rapidly to blue after death; 1 anterior temporal on each side of head . . . . **Smooth Greensnake** **(*Opheodrys vernalis*)**

Figure 58. *Continued*

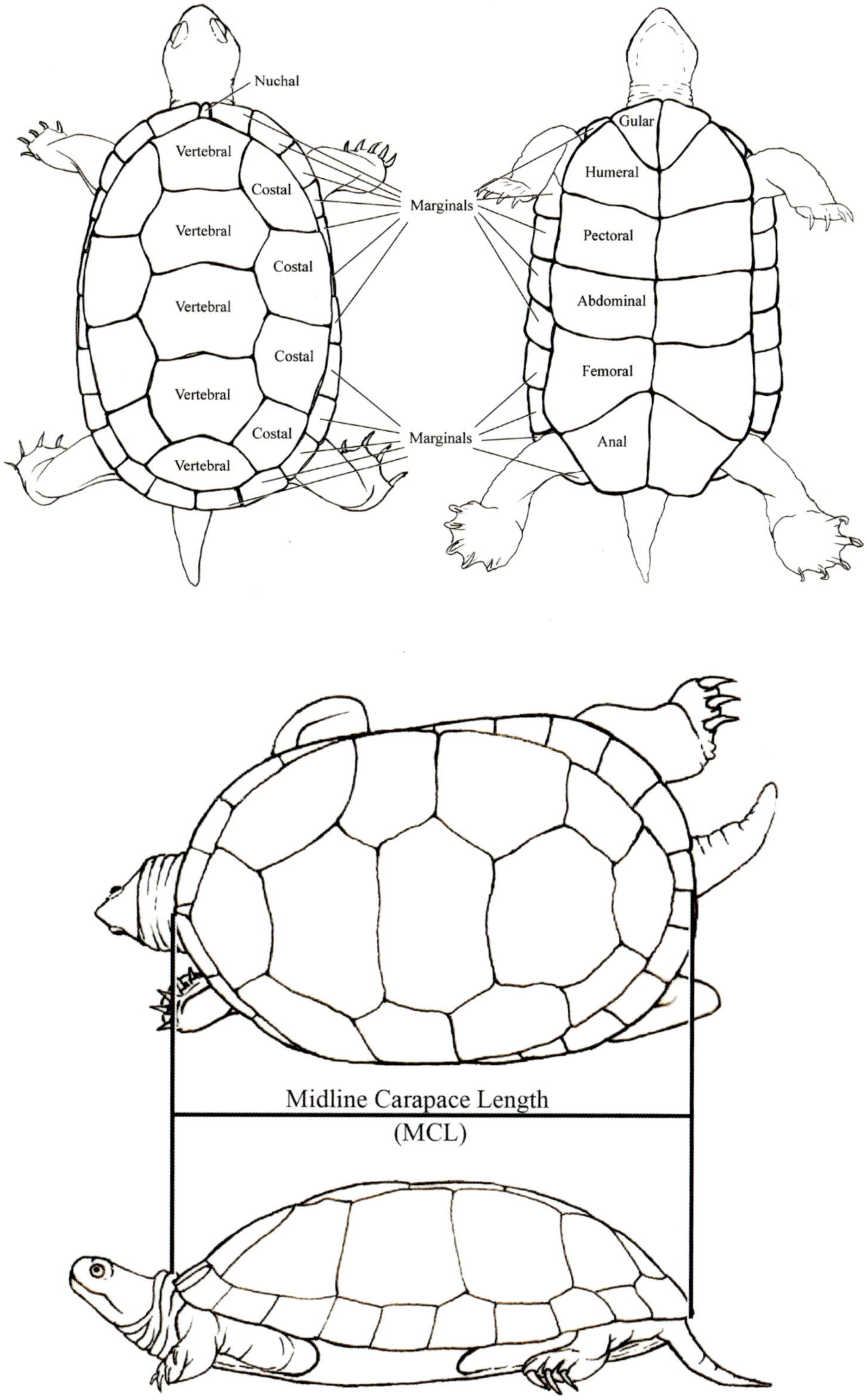

Figure 59. Some structures found among members of Testudines that are useful for species identification.

Key to Wisconsin's Turtles

1A) Carapace covered with keratinized scutes; front feet with 4 or 5 claws. **3**

1B) Carapace lacking keratinized scutes and covered with leathery skin; front feet with 3 claws **2**

2A) Anterior edge of carapace with spiny protuberances; carapace feels like sandpaper; two lateral yellow green stripes on each side of head; nasal septum with horizontal protuberances . . . . **Spiny Softshell (*Apalone spinifera*)**

2B) Anterior edge of carapace without spiny protuberances; carapace smooth; one peach-colored head stripe on each side of head; nasal septum without horizontal protuberances . . . . **Smooth Softshell (*Apalone mutica*)**

3A) Plastron with 11 or fewer scutes **4**

3B) Plastron with 12 scutes **5**

4A) Plastron with 11 scutes; posterior margin of carapace not serrated . . . . **Eastern Musk Turtle (*Sternotherus odoratus*)**

4B) Plastron with 11 or fewer scutes; posterior margin of carapace serrated . . . . **Snapping Turtle (*Chelydra serpentina*)**

5A) Carapace highly domed; plastron with well-developed hinge allowing entire plastron to be closed tightly against the carapace; jaws beaked; toes not webbed . . . . **Ornate Box Turtle (*Terrapene ornata*)**

5B) Carapace not highly domed; greatest height less than 43% of length; plastron with bridge; rear portion of plastron not able to close against carapace; jaws not beaked; toes webbed. **6**

6A) Carapace rounded and blue-black with yellow specks; underside of chin and neck immaculate yellow; front half of plastron hinged . . . . **Blanding's Turtle (*Emydoidea blandingii*)**

6B) Carapace depressed or angular in cross section and not blue-black with yellow specks; underside of chin and neck not immaculate yellow; front half of plastron not hinged **7**

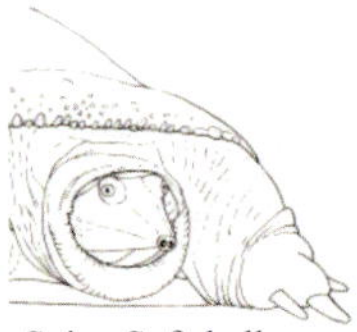
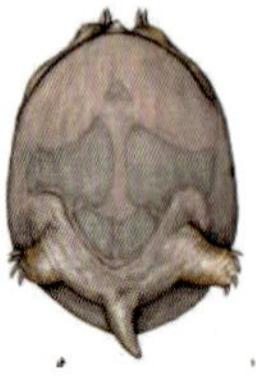

Spiny Softshell
(*Apalone spinifera*)

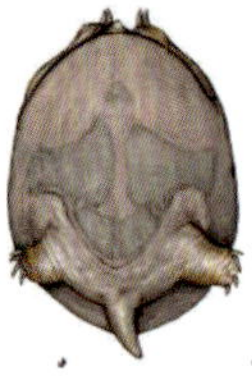

Smooth Softshell
(*Apalone mutica*)

Eastern Musk Turtle
(*Sternotherus odoratus*)

Snapping Turtle
(*Chelydra serpentina*)

Ornate Box Turtle
(*Terrapene ornata*)

Blanding's Turtle
(*Emydoidea blandingii*)

Figure 60. Key adapted from Vogt (1981).

7A) Carapace with strong elevated concentric grooves on each scute; no keel present; plastron yellow with large black blotches on the lateral edge of each scute. . . . . **Wood Turtle (*Glyptemys insculpta*)**

Wood Turtle
(*Glyptemys insculpta*)

7B) Carapace without strong elevated concentric grooves on each scute; dorsal keel may or may not be present; plastron not yellow with large black blotches on the edges of each scute **8**

8A) Upper jaw notched at apex with tooth-like structures on each side; marginal scutes usually marked with red. . . . . **Painted Turtle (*Chrysemys picta*)**

Painted Turtle
(*Chrysemys picta*)

8B) Upper jaw notched at apex without tooth-like structures on each side; marginal scutes marked with yellow, orange, or green. **9**

9A) Carapace smooth with low keel and no definite knobs on vertebral scutes; yellow green or green dorsal spot located posterior to each eye and encircled by 2 or more narrow green or yellow lines; plastron of adult females immaculate yellow . . . . **Northern Map Turtle (*Graptemys geographica*)**

Northern Map Turtle
(*Graptemys geographica*)

9b) Carapace smooth with definite keel and knobs on second, third, and sometimes fourth vertebrals; marking behind eye orange, yellow, or yellow green, never green; plastron of adult females yellow mottled with brown **10**

10A) Large yellow spots, one beneath each eye, one beneath this on lower jaw, and one at symphysis of lower jaw; spot under eye may be fused with wide postorbital spot to form a crescent behind the eye; plastral pattern of young and males primarily follows sutures and covers less than 60% of surface . . . . **Ouachita Map Turtle (*Graptemys ouachitensis*)**

Ouachita Map Turtle
(*Graptemys ouachitensis*)

10B) Yellow spots, if present, are small dots; postorbital spot narrow; plastral pattern of young and males covers 75 to 100% of surface . . . .**False Map Turtle (*Graptemys pseudogeographica*)**

False Map turtle
(*Graptemys pseudogeographica*)

Figure 60. *Continued*

Introduction to Species Accounts

Joshua M. Kapfer and Donald J. Brown

The two primary goals of this book are to fulfill a scientific need and to promote Wisconsin's natural legacy. The species accounts represent the core component of the book and contribute to both goals by providing a thorough review of the biology and ecology of each species while attempting to make the information accessible to a broad audience.

ACCESSIBILITY

To facilitate accessibility, standard English names are used throughout the species accounts. Although common names are arbitrarily assigned to scientifically named species (see Systematics and Classification chapter) and are notorious for varying among languages, cultures, and geographic locations, we believe their use enhances readability for the general public and biologists who are not herpetologists. We provide the scientific name for each species the first time it is mentioned within each chapter or species account. For both common and scientific names, we followed the eighth edition of the *Scientific and Standard English Names of Amphibians and Reptiles of North America North of Mexico* (Crother et al. 2017). However, where relevant we mention other common names regularly used in Wisconsin for these species.

We incorporated several other features designed to enhance the accessibility of this book. The beginning of each species account includes a quick summary in a section titled At a Glance. This section provides a brief overview of the species similar to what one finds in a typical field guide. Many species accounts also include Natural History Boxes. These boxes contain experiences, anecdotes, bits of knowledge, or personal points of view from some of the experts who contributed to this book and additional experts from the Upper Midwest. The boxes occasionally include excerpts from agency reports written by past experts. (Note that because Natural History Boxes are sometimes written from a particular author's point of view or based on their opinions, the content does not always reflect the opinions of the book editors or other account authors. Please direct questions about specific Natural History Boxes to their respective authors.) The boxes are sometimes written in the first person, with the goals of entertaining and educating readers with a variety of backgrounds. In addition, we provide measurement data in both metric (which is standard in scientific writing) and English units. Finally, to facilitate the reader's understanding of scientific verbiage we include a glossary of terms.

CONTENT

Our objective was to provide a robust synthesis of each species' ecology. We emphasized the inclusion of scientific information based on Wisconsin literature, observations, and specimens, but we also consulted the broader body of scientific literature. Specifically, when an aspect of a species' biology was not well-researched in Wisconsin or when we felt that further information was beneficial, we broadened the geographic scope to the Upper Midwest (with a focus on the surrounding states of Minnesota, Michigan, Illinois, and Iowa) and range-wide as needed. In cases of species with large geographic distributions and large bodies of literature, we attempted to focus primarily on literature from studies performed at similar latitudes, under the assumption that the climates and conditions would most likely be comparable to those of Wisconsin. When relevant, we sought to provide the general location where a study or observation occurred, including the county (when in Wisconsin) or the state/province.

The reader of this book will notice that it differs from many state and regional books on amphibians and reptiles in that the accounts are more in-depth and the information is presented in a scientific format in which sources are cited. We felt it was important to support scientific claims or statements with the original source of information. Regional books of this type can fall victim to making statements that are not well-supported by data or citations and thus can perpetuate incorrect or highly situational information. Sources supporting the claims and statements presented in this book are most often the primary literature or rigorous reviews of the literature, but they also occasionally include other types of publications, websites, personal observations, unpublished personal data, and data associated with museum specimens. For published resources, we attempted to avoid the use of 'hidden citations.' In other words, we tried to avoid citing a source that, in turn, cites another source that did not originally reference any literature to back the claim. We also focused on consulting original research rather than simply repeating information from secondary sources. However, we often consulted secondary sources for descriptive information, qualitative observations, personal observations, or unpublished data and as a source to direct the reader to larger literature syntheses.

A substantial benefit of producing an edited volume with numerous contributing authors is that the species accounts are written by individuals or teams with particularly strong personal experience, interest, and regional expertise with the species. Approximately fifty-experts, including university professors, agency personnel, wildlife managers, zoo personnel, and ambitious graduate students, have combined efforts to produce the introductory chapters and species accounts included in this volume. Authors often shared novel personal data, natural history information, experiences, observations, and communications that otherwise may not have reached a wide audience. While literature syntheses are certainly valuable, we believe the inclusion of unpublished data and personal observations enhances the contribution of this book to the discipline of herpetology. Of course, anecdotal information must be integrated with some restraint. As such, our criteria for the use of personal observations and data were that they (1) focused on unique observations or phenomena;

(2) provided data collected on size, weight, or other physical characteristics; (3) described aspects of a species' biology, natural history, ecology, or conservation with local relevance; and (4) supplemented citations of peer-reviewed literature when all published information on the topic was from outside the Upper Midwest.

ORGANIZATION

The species accounts are arranged alphabetically based on taxonomy. Specifically, accounts are ordered in a hierarchical structure by order, family, genus, and specific epithet. For Squamate reptiles (Order Squamata), we also organized accounts first by lizards and then by snakes, with snake accounts further organized by family and subfamily. A brief introduction is provided at the beginning of each order (i.e., Anura, Caudata, Squamata, and Testudines). To maintain consistency with features that have traditionally been used for identification of the amphibian and reptile species in Wisconsin, the keys presented here were developed using information from previous Wisconsin amphibian and reptile keys, including Vogt (1981), Watermolen and Gilbertson (1996), Powell et al. (1998), Parmelee et al. (2002), and Christoffel et al. (2002, 2008, 2009). Accompanying the keys are detailed drawings of the primary features used in identification.

Species accounts are organized in a consistent manner, although they vary in length due to differences in current knowledge and amount of research (especially in Wisconsin). Accounts begin with Wisconsin county-level and North American distribution maps (see Range Maps section for details), followed by a cover photo displaying a typical individual of the species. Next, the At a Glance section provides a synopsis of each species that is particularly directed to the nonscientific audience. A brief Systematics section follows that describes the species' phylogenetic and taxonomic information, including name changes when applicable. Next, the Description section addresses the physical appearance and sizes of all life stages. This section also includes a Diagnosis subsection, which lists morphological features useful for positive identification, based largely on the species keys, and lists useful characteristics to distinguish superficially similar species. The Distribution and Habitat section describes the range-wide and local distribution of the species. Some accounts also review problematic or enigmatic specimens and reports at locations currently considered unusual. We limited maps of species occurrence to the county level for this book because of the sensitive nature of locational data for many of Wisconsin's amphibians and reptiles. This section further addresses the species' habitat associations, requirements, and preferences, including information on specific habitats used to meet particular ecological needs, for example, frequently associated ground cover and general information on breeding habitat (if unique), nesting habitat, and typical habitat used for overwintering (where applicable).

Subsequent sections address additional aspects of the biology and ecology of each species. They summarize what is known and, often, what information is lacking with respect to Reproduction and Development, Activity (i.e., temporal and spatial activity patterns), and Prey and Predators. Each account ends with a Conservation section. This section is designed to build upon the general concepts laid out in the Introduction to Conservation and Management of Wisconsin's Amphibians and Reptiles chapter, with a focus on the conservation status of the species, abundance and demographic information, threats of particular importance, and potential management considerations.

COLLECTION AND OCCURRENCE DATA

Scientific collections served as important sources of several types of data used in this book, including records of unique reports or observations, vouchers for distribution records, and size data for specimens from Wisconsin. In total, 122 institutions were either queried through online databases or contacted directly about their collections. Of these institutions, 51 had specimens collected in Wisconsin, although many of these collections were extremely limited. These data were compiled, and numerous queries were made to curators or collection managers to confirm the presence and/or identification of specimens, especially in cases when the experience of the reported specimen collector was unknown and we could not view the specimen ourselves. We attempted to list all collections and collection staff who provided data in the acknowledgments (see specific review of how institutional collections were used to delineate species distributions in the Range Maps section).

We supplemented size data from personal field observations and published data with measurements made on specimens from Wisconsin in the collections of the University of Wisconsin–Madison Zoology Museum (UWZM) and the Milwaukee Public Museum (MPM) when sufficient material was available. We measured snout-to-vent length (SVL) for frogs and toads; SVL and tail length for salamanders, lizards, and snakes; and midline carapace length (MCL) for turtles (see Identification Keys for Amphibians and Reptiles in Wisconsin chapter for diagrams of these measurements). Measurements were taken with calipers when possible. For long or contorted preserved specimens, particularly snakes, a flexible tape measure was conformed to the length of body or tail to determine length. We used a stiff tape measure to measure MCL in turtles that exceeded caliper length. Sex was determined by probing in snakes (which we found to be effective, even in preserved specimens) and via external characteristics in other taxa, where possible. For example, we used tail length, the distance between the vent and the base of the tail, as well as the position of the vent in relation to the back edge of the carapace to determine turtle sex. In anurans, the presence of enlarged thumb pads and/or darkened vocal sacs was used to determine sex (if possible) in specimens large enough to be mature. We were not able to confidently determine sex in salamander or lizard specimens via external characteristics, with a few exceptions (e.g., Mudpuppies [*Necturus maculosus*], which have obvious genital papillae). We admit that data generated from museum specimens suffer from the inherent prejudices that stem from collection biases and preservation methods and thus may differ from data collected from live specimens. The institutional acronyms used when referring to individual specimens follow Sabaj (2016).

Another valuable source of data is the Wisconsin Department of Natural Resources (WDNR). Unpublished WDNR data were compiled from various monitoring projects, including the results of the Wisconsin Frog and Toad Call Survey, the Turtle Road Crossing Reporting Initiative, and the Gophersnake Sighting Initiative, as well as the Natural Heritage Inventory (NHI) observation database. These data were incorporated, where possible and applicable, in the species accounts. Given the sensitive nature of species tracked by the NHI database, editor access to these data involved a vetting process and special permission from the WDNR. Furthermore, the reproduction of NHI information is restricted, so we did not report location information below the county level for these data.

ANURAN TREND DATA

The Wisconsin Frog and Toad Survey represents the longest-running US state citizen science program for monitoring trends in occupancy and relative abundance of anurans (Weir and Mossman 2005). The program was initiated in 1981 and achieved statewide coverage in 1984 (Mossman et al. 1998), and the number of completed routes increased from ca. 60 to 130 between 1984 and 2016 (Figure 61). According to Mossman et al. (1998), "Each route comprises ten listening stations selected subjectively by a volunteer observer to be within hearing distance of wetlands that represent a range of local anuran breeding habitats, such as ephemeral ponds, lakes, marshes and wooded swamps. . . . Routes [a]re run after sunset under favorable conditions. . . . At each station the observer listen[s] for five minutes (up to ten minutes if necessary due to noise interference) and record[s] one of the following call index values for each species heard: 1 = individuals can be counted, there is space between calls; 2 = calls of individuals are distinguishable, but some calls overlap; 3 = full chorus, calls are constant, continuous, and overlapping." See Mossman et al. (1998) for additional information on the program and monitoring protocols.

For each anuran species, we included a figure showing the statewide trend in occurrence (%) and abundance rank (1–3) at survey sites across the state between 1984 and 2016. We emphasize that this is a citizen science data set; thus, the presence of false positive and negative detection data is likely. In addition, survey routes, observers, and efforts have changed over time, which could influence perceived trends. However, this also represents a large and long-term data set and thus provides useful insights into occupancy and abundance patterns of Wisconsin's anurans.

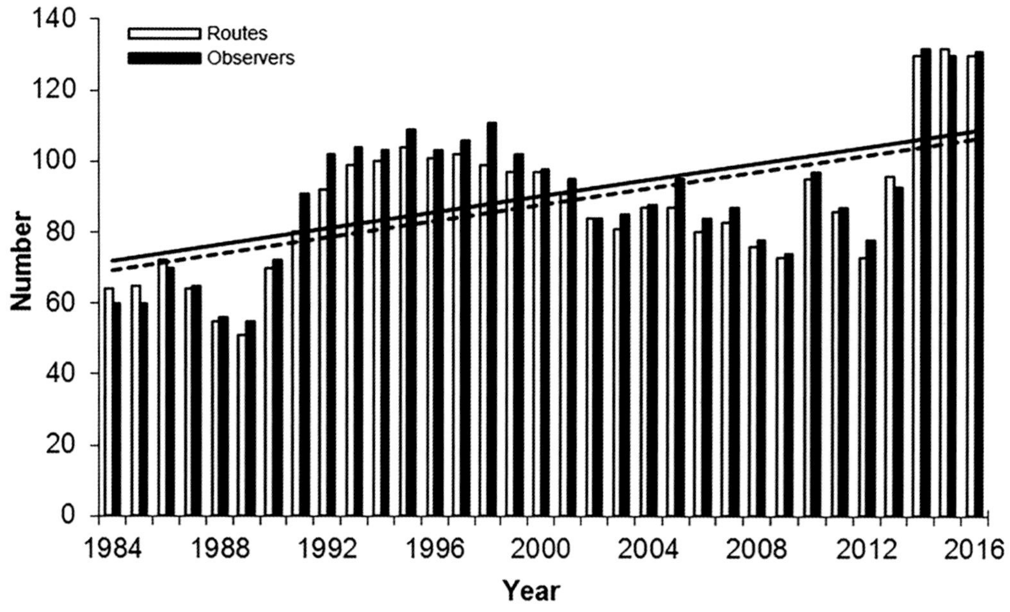

Figure 61. Trend in number of survey routes and lead observers for the Wisconsin Frog and Toad Survey between 1984 and 2016. Note that many secondary observers assist with these monitoring activities annually and are not reported here.

RANGE MAPS

North American range maps were generously provided by Travis Taggart (director, Center for North American Herpetology) and used with permission for this book. These maps are designed to show the general boundaries of where a species occurs in North America and thus should not be taken to indicate that a species is distributed everywhere within that range. Many species have strict habitat requirements and occur in disjunct populations within their range. Nonetheless, these continental distribution maps are useful for giving the reader a broad view of where each species does and does not occur in North America.

We developed county-level range maps to show the distribution of each species and provide a reference for identification of new county records in the state of Wisconsin (based on information available through March 2022). For most species, county-level range maps are based on published county records and institutional voucher numbers for each record (if available) are listed in the Appendix. Exceptions to this rule were typically the result of unusual circumstances, such as misidentification of originally reported vouchers or when published records did not reference voucher material. We used Vogt (1981) as the county record baseline and added subsequently published county records, as well as records from publications prior to Vogt (1981), if these reports included voucher numbers. We note that a county of occurrence listed by Vogt (1981) did not always mean that it was the first source to publish a report of the species in that county.

The counties in most distribution maps are shaded based on two categories: 'Vouchered' (black) and 'Voucher Needed' (gray). 'Vouchered' indicates a published source exists that typically references supporting voucher material. It further indicates we confirmed the presence and accurate species identification of said material. Most voucher material was part of catalogued collections at professional institutions, such as museums or universities. Vouchers typically consisted of preserved whole specimens, but other types of acceptable material (e.g., audio and video recordings and photographs) were referenced by some sources. 'Voucher Needed' indicates that (1) the species was documented in the county by a reputable published source, often Vogt (1981), but catalogued voucher material was not reported (see subsequent discussion on finding replacement material); (2) voucher material once existed but now appears to be lost; (3) the voucher material referenced was not suitable for definitive independent identification; or (4) the information associated with the reported voucher material brought the material's validity into question. Situations where we questioned the validity of a potential voucher are described in the species accounts.

We made extensive attempts to find material for published claims of county occurrence that either lacked vouchers (e.g., records in Vogt [1981]) or referenced unsuitable material. To achieve this goal, we queried numerous online and institutional databases and contacted local institutions to find suitable vouchers. If potential vouchers were housed in local collections, we visited these to confirm their presence and species identity. If we were not able to view the specimens ourselves, we contacted curators or collection managers to assist with the confirmation of a specimen's presence and identification (either through their own review of the material or through photographs of material they sent to us). We attempted to use specimens at local institutions for replacement vouchers when possible (e.g., MPM, UWMZ). Finally, several book contributors and additional community scientists, particularly Andrew Badje (WDNR), put substantial time and energy into field observations to document new county records and obtain vouchers for counties with previous records that were missing vouchers. Our specimen investigations and new field observations resulted in hundreds of new or better supported existing county records for the book.

Although our understanding of the county-level distribution of Wisconsin's amphibians and reptiles is stronger now than it has ever been, there are certainly many county records left to obtain. Species-specific needs for additional distributional information and hypotheses that a species occurs in undocumented or unconfirmed counties are provided in each species account. We hope the publication of this book leads to a rapid increase in the published county records for Wisconsin species. This is particularly important from a conservation standpoint; if a species is not documented in an area, then the species may not be considered during land use or habitat management planning.

CONSERVATION STATUS

We provide current conservation statuses of each species. We include species statuses at several levels and from several sources: global status per the IUCN and NatureServe, federal status per the US Endangered Species Act, and state status per the WDNR. The conservation status of all species can be found in Table 7 in the Introduction to Conservation and Management of Wisconsin's Amphibians and Reptiles chapter. Additional conservation statuses and legal protections relevant to individual species are provided in the individual species accounts (e.g., legal protection through the Convention on International Trade in Endangered Species [CITES]).

Anura
Frogs and Toads

Frogs and toads are members of the class Amphibia and order Anura. All members of the order Anura lack a tail in adults and have hindlimbs that are longer than their forelimbs (see Cannatella [2008] for additional monophyletic characters). The earliest fossil species discovered that contains characteristics representative of modern anurans existed ca. 240 mya, and the oldest modern anuran existed ca. 200 mya (Rocek 2000). There are currently 7,426 recognized anuran species in the world as of October 2021 (https://AmphibiaWeb.org), which are placed in 55 families and 450 genera; the US has 111 species. Three families, five genera, and twelve species occur in Wisconsin, including Bufonidae (*Anaxyrus,* one species), Hylidae (*Acris, Hyla, Pseudacris,* five species), and Ranidae (*Lithobates,* six species).

Figure 62. Wood Frog (*Lithobates sylvaticus*) (illustration by E. Wild).

American Toad

Anaxyrus americanus (Holbrook 1836)

Donald J. Brown

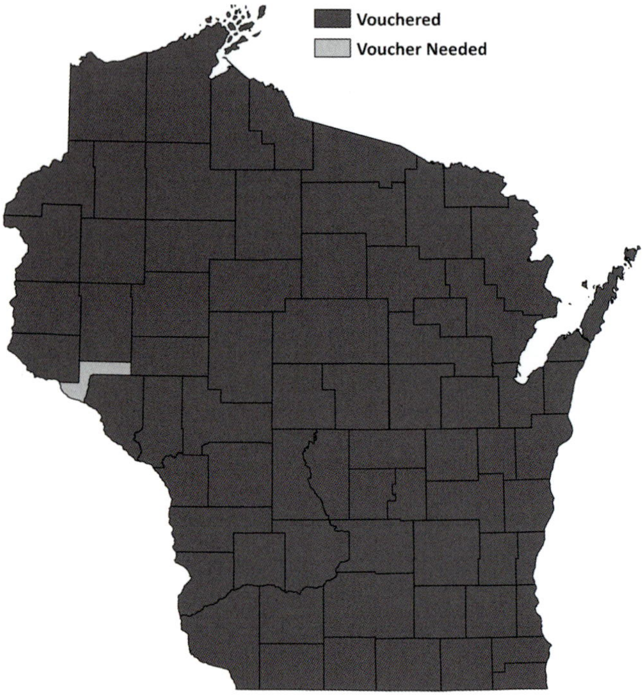

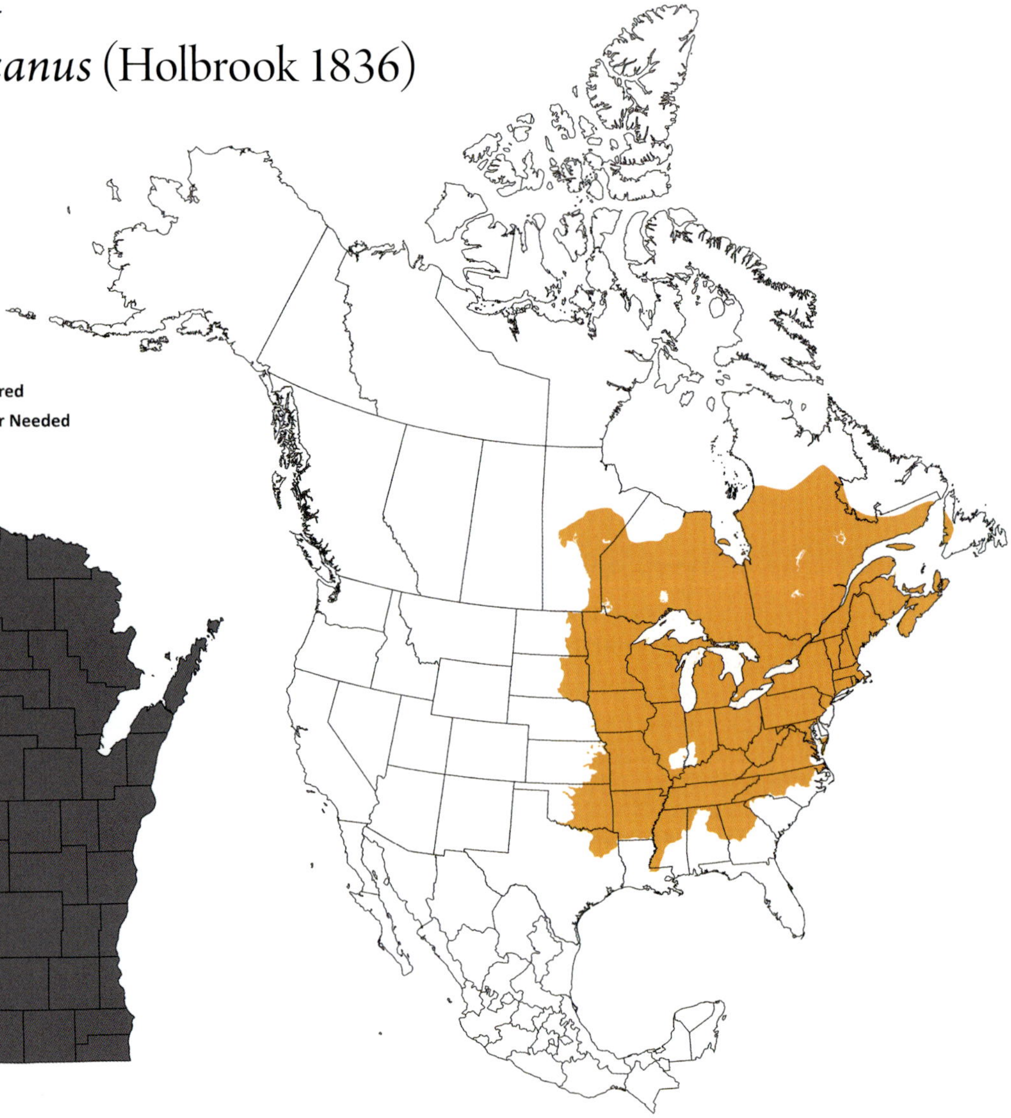

Subadult American Toad (Oneida County; photo by D. Brown).

AT A GLANCE

American Toads are the only toad species found in Wisconsin, and they are distributed throughout the state. They are a fairly large anuran, with a stout body, short limbs, and 'warty' skin that is mostly mottled browns and tans but sometimes reddish. A large gland, called a parotoid gland, is found behind each eye. Larvae are aquatic and often found in natural and man-made wetlands and lakes. Juveniles and adults are terrestrial and can be found in a wide range of forested and nonforested habitats. Adults emerge from hibernation in April, and males typically begin calling to attract females by late April or early May. Mating typically occurs between late April and mid-June, and the male advertisement call, a long-duration high-pitched trill, is arguably the most attractive of Wisconsin's anurans. This species is an explosive breeder. Most reproduction occurs during large breeding congregations on only a few nights per year. Larvae develop rapidly and typically complete metamorphosis by early July. Adults primarily feed on invertebrates, and larvae scrape algae and other microorganisms from substrates. In September and October, individuals select their terrestrial overwintering locations and burrow into the soil for the winter. American Toads are currently considered a common species by the WDNR.

SYSTEMATICS

True toads (family Bufonidae) in North America were historically included in the genus *Bufo*. In an attempt to create monophyletic groups, Frost et al. (2006) proposed placing strictly North American toads in the genus *Anaxyrus*, which is currently accepted (Crother 2017). However, the systematics and classification of North American Bufonidae is an active area of research and contention (Frost et al. 2009, Pauly et al. 2009, Van Bocxlaer et al. 2010), and future classification changes are possible. In addition, hybridization events resulting in viable offspring are common within the genus *Anaxyrus* (reviewed by Dodd 2013). There are currently 19 recognized species of *Anaxyrus* (Crother 2017), with *A. americanus* (American Toad) occurring in Wisconsin. There are currently two recognized subspecies of *A. americanus* (Crother 2017), with *A. a. americanus* (Eastern American Toad) being the subspecies found in Wisconsin. The other subspecies, *A. a. charlesmithi* (Dwarf American Toad), reaches its northern distributional limit in central Illinois (Conant and Collins 1998).

DESCRIPTION

Adults: Dorsal base coloration can vary widely, including shades of gray, green, brown, red, and black (reviewed by Dodd 2013). The dorsum has dark spots, with typically one or two 'warts' (i.e., skin glands) per spot, and usually has a light stripe along the vertebral column (Vogt 1981, reviewed by Dodd 2013). Ventral coloration is white or cream-colored, often with heavy gray, black, or brown mottling (Figure 63). The head contains prominent L-shaped cranial crests, and a large elliptical parotoid gland is present behind each eye (Figure 64). The limbs are short, and the toes on the hind feet are incompletely webbed. A keratinized spade is present on the heel of each hind foot (see Identification Keys for Amphibians and Reptiles in Wisconsin). This is a relatively large anuran. Male SVL typically ranges from 4.5 to 8.5 cm (1.8 to 3.3 in), and female SVL typically ranges from 7.5 to 9.0 cm (3.0 to 3.5 in; Vogt 1981). The mean SVL of adult specimens from Wisconsin housed in the MPM and UWZM collections ranged from 7.5 cm (3.0 in) in females ($n = 74$) to 6.5 cm (2.5 in) in males ($n = 119$). Note that minimum adult size used as threshold for

Figure 63. Adult male American Toad showing its mottled belly and colored throat patch (Jefferson County; photo by J. Kapfer).

Figure 64. Adult American Toad displaying morphological features useful for identification, including cranial crests, parotoid glands, dark dorsal spots, and 'warty' skin (Iron County; photo by S. Viernum).

Figure 65. Metamorph American Toad. Note the relatively smooth skin and lack of cranial crests and parotoid glands but the presence of 'warts' (Dane County; photo by J. Kapfer).

inclusion in mean calculations was based on the work of Acker et al. (1986) from Illinois (see Reproduction and Development section).

Preadult stages: Eggs are bicolored, black to dark brown above and white to cream-colored below (reviewed by Dodd 2013). Eggs are deposited in two long strings that can exceed 60 m (197 ft) in length (Green 2005, reviewed by Dodd 2013). Individual egg diameter ranges from 0.1 to 0.2 cm (0.04 to 0.08 in; Green 2005). Larvae are dark brown to black, with fine silver or gold spots (particularly on the ventral side), and have transparent fins (Vogt 1981, reviewed by Dodd 2013). At hatching, larvae are 0.7–1.2 cm SVL (0.28–0.47 in; Wright and Wright 1949). Size at metamorphosis can vary widely depending on density of individuals and aquatic habitat characteristics (see Reproduction and Development section) but typically does not exceed 2.5 cm (0.98 in; Watermolen and Gilbertson 1996). Vogt (1981) stated that toadlets were 1.0–1.5 cm SVL (0.39–0.59 in) at metamorphosis, with "light yellow or bright red warts." American Toads weigh ca. 0.05 g (0.002 oz) at metamorphosis (Pough and Kamel 1984). Metamorphs and young juveniles lack prominent head characteristics and have comparatively smoother skin (Figure 65).

Variation: In addition to sexual size dimorphism (where females are larger), males differ from females during the breeding season in having dark vocal sacs and nuptial pads on the thumb and forefinger, which aid in grasping females during amplexus. Individuals can also slightly alter their dorsal coloration to more closely match the underlying substrate coloration (Heinen 1994b). A wild albino American Toad was documented in Minnesota in 1992 (Moriarty and Hall 2014), and a leucistic individual was found in North Carolina in 2004 (Brannon 2006). A giant form of the species, where female SVL may exceed 13.0 cm (5.1 in), is restricted to Wisconsin's Lake Michigan islands (Natural History Box: Gigantic Toads). Additional information on morphological variation can be found in Dodd (2013).

Diagnosis: Adult American Toads are characterized by a short and stocky body, dry skin with large warts, presence of parotoid glands, and presence of a keratinized spade on the heel of each hind foot. Larval American Toads are characterized by the presence of dorsally located eyes, a medially located vent, TL <2.5 cm (0.04 in), black coloration (with brown spots in older individuals), and a low rounded tail fin without pigmentation (Watermolen and Gilbertson 1996). American Toads represent the only toad species in Wisconsin, and adults and subadults can easily be distinguished from Wisconsin's frogs by the presence of parotoid glands, cranial crests, and a keratinized spade on the heel.

American Toad eggs can be distinguished from other amphibians in Wisconsin by their encasement

NATURAL HISTORY BOX: Gigantic Toads

The giant form of the American Toad was first documented on a few of Wisconsin's Lake Michigan islands in 1976 by Charles Alan Long (UW–Stevens Point) and his sons (Long and Long 1976). To date, studies of these gigantic morphs have been restricted to females. In 1999 Long obtained additional gigantic toad specimens and recorded more observations to compare with mainland toads from Wisconsin. Of ten females collected, mean SVL was 11.00 cm (4.3 in) and ranged from 7.48 to 12.70 cm (2.9 to 5.0 in). When compared to mainland toads in central Wisconsin, island females were 20%–30% larger and displayed little to no overlap in six morphological characteristics (Long 1982). Elsewhere in their distribution, Conant and Collins (1998) reported much smaller American Toad sizes (range = 5.1–9.0 cm SVL [2.0–3.5 in], maximum SVL = 11.1 cm [4.4 in]). Previous, though unverified, measurements from these islands suggest they could even exceed 13.0 cm SVL (5.1 in) in Wisconsin. Long also noted that the largest females walked rather than hopped. Possibly, this behavioral trait originated on these islands as an adaptive measure to minimize vulnerability to aerial predation, as land-based predation is thought to be insignificant. Long also documented delayed initiation of breeding activity on these islands compared to the mainland, which is likely due to differences in temporal climate patterns caused by the presence of Lake Michigan. Long observed tadpoles in September and noted that a few 'behemoths' were carrying larger egg masses (constituting more eggs) in early to mid-September.

Andrew F. Badje, WDNR (written using personal data and in direct communication with Charles A. Long, UW–Stevens Point, retired)

in long gelatinous strands, while most species lay eggs in globular masses or clusters. American Toad larvae can be distinguished from those of other anurans in Wisconsin by their dorsally located eyes coupled with their black body and tail fin, which possesses a rounded tip and has no pigmentation (Watermolen and Gilbertson 1996). Recently metamorphosed toadlets lack prominent parotoid glands and cranial crests but can be distinguished from Wisconsin's frogs by the presence of 'warts' (see Figure 65).

DISTRIBUTION AND HABITAT

Global distribution: American Toads are present throughout the eastern half of the US and Canada. They range from southern Labrador west to central Manitoba, south to northeastern Texas, and east to the East Coast, loosely following the southern boundary of the Piedmont plateau region (Dodd 2013). Their southernmost extent is located in central Louisiana (Green 2005). In the Appalachian region, the altitudinal distribution limit appears to be <1,800 m (5,905 ft) (Green and Pauley 1987, Dodd 2004).

Distribution in Wisconsin: Published records exist for every county in Wisconsin, although a voucher was not found for Pepin (see the Appendix). American Toads have also been documented on Lake Superior islands off of Ashland and Bayfield Counties (i.e., the Apostle Islands; Hecnar et al. 2002) and Lake Michigan islands off of Door County (see Natural History Box: Gigantic Toads).

Localities of possible occurrence: American Toads have been documented in every county in Wisconsin.

Habitat: Adult American Toads are habitat generalists. Vogt (1981) stated that in Wisconsin they are "commonly found in prairies, marshes, oak savannas, semiopen coniferous and deciduous forests, gardens, and agricultural areas." Individuals have also been documented in the northwestern pine barrens of Wisconsin (Evrard 2000, Evrard and Hoffman 2000), as well as the Weaver Dunes, a sand dune prairie region in southeastern Minnesota (Wabasha County; Cochran 1986b). The tolerance of this species to a wide variety of terrestrial habitats could be related to its effectiveness at absorbing water from the soil, meaning it is less restricted by dry habitats than many other Wisconsin amphibians (Walker and Whitford 1970). In northern Minnesota, Ewert (1969) found that adults used field, forest, and bog habitat patches for foraging after the breeding season, with 132, 97, and 88 total days spent in the habitat types, respectively (n = nine males and six females observed, with 9–45 observation days each). Of all 768 locations recorded between 1 June and 15 July 1966, 36%, 35%, and 29% were in forest, field, and bog habitat types, respectively. However, the results of several studies indicate that dense forest cover could be suboptimal habitat for adults (Guerry and Hunter 2002, Perkins and Hunter 2006, Eigenbrod et al. 2011). Despite their generalist tendencies, American Toads appear to exhibit preferences for certain terrestrial habitat conditions. A mesocosm study found that metamorphs preferred deciduous leaf litter to coniferous leaf litter and coniferous leaf litter to bare soil (Smith and Schulte 2008). Within their chosen habitats, American Toads frequently aestivate within shallow self-dug burrows and crayfish burrows and beneath logs or other ground debris (Moosman and Moosman 2005).

American Toads use seasonal, semipermanent, and permanent wetlands as breeding habitat (reviewed by Dodd 2013). For example, Babbitt (2005) surveyed 103 wetlands encompassing a wide range of sizes and hydroperiods in New Hampshire and documented American Toad larvae in 0.2–2 ha (0.5–5 ac) wetlands, with intermediate (i.e., wetlands held water >4 months and <12 months) and long (i.e., wetlands did not dry) hydroperiods. Backwaters of rivers and streams also provide suitable breeding habitat (Zadnick et al. 2009). American Toads will breed in bogs in Wisconsin and Minnesota (Andrews 1915, Bellis 1959, Ewert 1969). However, a study in northern Minnesota found that American Toad eggs (n = 58 egg masses and 3,038 eggs) and larvae (n = 60) exposed to highly acidic bog water had 0% survival (pH = 4.2, 96-hr exposure; Karns 1992). Tattersall and Wright (1996) found that acidic water disrupts the nitrogen balance in American Toad tadpoles. Adults occasionally deposit eggs in highly ephemeral pools or puddles, despite the low probability of water retention to metamorphosis (Cochran 1982a; Natural History Box: Lending a Helping Hand). In Michigan, Skelly et al. (1999) found that American Toads preferred open-canopy to closed-canopy ponds for breeding. Smith and Rettig (1998) studied habitat associations based on 20 egg strands at a breeding site in Indiana and found that egg strands were more likely to be present in open-canopy areas, but vegetation within the pond was not associated with egg strand location.

After metamorphosis occurs, juveniles remain near their natal waterbody while their aerobic capacity and endurance increase through growth (Taigen and Pough 1981, Pough and Kamel 1984, Taigen and Pough 1985). American Toads hibernate terrestrially in open woodlands and fields (Ewert 1969). In selected overwintering habitats, American Toads use their hind legs to burrow into the soil and retreat below the frost line (Miller 1909b). Ewert (1969) reported that American Toads used a wide range of soil types for hibernacula in northern Minnesota, ranging from coarse sand to a surface layer containing "several inches of nearly pure dense clay."

REPRODUCTION AND DEVELOPMENT

In the Upper Midwest, it appears that males migrate to breeding wetlands prior to females (Ewert 1969, Vogt 1981). Vogt (1981) stated that females arrived at breeding sites one to two weeks after males in Wisconsin. In northern Minnesota, males spent from 0 to 15 days ($\bar{x}$ = 5 days, n = 13) at breeding sites prior to the initiation of calling activity (Ewert 1969). Field experiments in Ontario showed that first-year breeding males were able to locate breeding sites using environmental cues and that individuals did not rely on vocal or olfactory cues for orientation (Oldham 1966). However, translocated individuals required vocal cues to orient them to breeding sites. Although adults will use multiple breeding sites within and between years (e.g., Ewert 1969), some studies have indicated that breeding site fidelity is high for this species. For example, of 264 adults monitored for

NATURAL HISTORY BOX: Lending a Helping Hand

American Toads often breed in semipermanent and permanent wetlands. However, it is not unusual for individuals to deposit eggs in ephemeral waterbodies that will not last long enough for tadpoles to reach metamorphosis. In May 2011 I haphazardly found a swarm of tiny black toad tadpoles packed in a small, muddy pool (ca. 1.2×2.4 m [4×8 ft] wide and 5.1 cm [2 in] deep) on top of a gravel driveway. After locating a suitable larval habitat nearby (a nearby shallow oxbow pond next to a stream), I transferred as many of them as I could (ca. 250 tadpoles) out of the puddle and into the oxbow pond, hoping a few of them would make it to adulthood. On another occasion in 2014, a friend discovered that a puddle in the grass of his lawn had filled overnight with recently hatched American Toad tadpoles. Making it his mission to save these tadpoles from imminent desiccation, he began feeding the puddle with groundwater from the hose in his yard. As the days grew hotter, his efforts became futile as the puddle continued to dwindle in size and depth. In response, he gathered up as many of the tadpoles as he could and released them into a local wetland where they might survive to metamorphosis. These stories illustrate a common situation we face in urban and suburban environments; we encounter scenarios where animals that use disturbed landscapes face a higher risk of mortality, and we have the choice to intervene or let nature take its course. Sometimes offering a bit of friendly assistance can go a long way in helping our wildlife neighbors survive in the environments modified by humans.

Andrew F. Badje, WDNR

two consecutive years in Ontario, 85% bred at the same site both years (mean distance to an alternate breeding site = 100 m [327 ft]; Oldham 1966).

In Wisconsin, Vogt (1981) stated that males initiated calling activity in late April or early May. Christoffel et al. (2009) listed the breeding window in Wisconsin as late April to mid-July. Anderson et al. (1981) recorded the calling period for American Toads in Portage County as May and June. Briggs and Young (1976) noted that American Toads had reproduced and ceased calling by late May in Bayfield County, which is in the most northern region of the state. Similarly, Ewert (1969) documented calling activity at a wetland complex in northern Minnesota from 1966 to 1968 and found that calling activity occurred as early as 15 May and as late as 4 June.

The male advertisement call of the American Toad is a long-duration (≤30 sec) high-pitched trill. Males usually begin calling in the evening near sunset, and calling activity may persist into the daylight hours during peak breeding periods (Dirig 1978, Sullivan 1992, Oseen and Wassersug 2002). Males typically call from land near the water's edge (while facing the waterbody), in shallow water near the water's edge, or on structures (e.g., tree branches) in the water (Figure 66; Oldham 1966). However, Ewert (1969) found that in a shallow lake (maximum depth = 6.7 m [22 ft]) in northern Minnesota, more than 50% of males observed were in open water up to 30.5 m (100 ft) from the shore. Swanson et al. (2007) found that females were attracted to artificial choruses (i.e., playback calls) at distances of up to 40 m (131 ft) in Minnesota.

Although males call to attract females over a prolonged period, American Toads are explosive breeders. As such, reproductive activity occurs within large congregations of individuals over only a few nights each year (Howard 1988). Vogt (1981) noted that for American Toads in Wisconsin, "there is usually a period of intensive breeding activity from May 3 to 16 culminating in a 2- to 3-day orgy." For explosive breeders, mating activity is highly dependent on weather and environmental conditions, such as water temperature (e.g., Oldham 1969, Christein and Taylor 1978, Oseen and Wassersug 2002), and thus annual temporal and

Figure 66. Adult male American Toads calling to attract females for mating. (a) Male calling in shallow water (Wisconsin; photo by N. Melhorn).
(b) Male calling from a branch in shallow water (Wisconsin; photo by N. Melhorn).

spatial variation in timing of the explosive events can be large. A female selects a male for mating based on the characteristics of his call and then navigates to the chosen male and nudges him (Howard 1988). The male then mounts the female's back and engages in amplexus (Figure 67). The grasp is typically axillary (i.e., behind the armpits), but lumbar grasps (i.e., just anterior to the hips) also occur, particularly if the male is substantially smaller than the female (Wright 1914, Aronson 1944, Licht 1976).

Male reproductive success is partially limited by male-male interference competition. For example, larger males can displace smaller males that are already in amplexus with females (Kruse 1982). Males can also intercept and grasp females for reproduction before they reach rival males chosen by the females

Figure 67. American Toads in amplexus (Jackson County; photo by A. Badje).

based on call characteristics (Forester and Thompson 1998). Forester and Thompson (1998) found that males who intercepted females were significantly smaller than males that were found either swimming or calling. In addition, Gatz (1981a) determined that larger males arrived at breeding ponds earlier in the season, spent more nights engaged in calling activity, and had higher mating success than smaller males in Ohio. This is likely related to smaller males having greater energy constraints than larger males (Finkler et al. 2014, but see Wells and Taigen 1984). During explosive breeding events, it is not uncommon for multiple males to mount a single female, creating what are known as 'toad balls.' In Wisconsin, Vogt (1981) documented up to five males in amplexus with a single female. This behavior can result in the males accidentally drowning the female (Wright 1914, Dirig 1978). In addition, males often mount other males, either accidentally or to dislodge them from females. Males that are mounted by rival males emit a release call (a 'chirp') and vibrate to signal release (Schmidt 1990). In addition, Dodd (2013) stated that when males attempt to dislodge another male already clasped to a female, both the female and clasping male will discourage their attempt by "vigorously kicking them with their back legs." If disturbed by humans, the female will swim away (either vertically or horizontally), usually with the male maintaining amplexus (Donald Brown, West Virginia University / USFS, personal observation).

Amplexus can last from a few minutes to over a day before oviposition occurs (Aronson 1944, Wells 1977b). Wells (1977b) considered 12–24 hours to be the normal time spent in amplexus for American Toads, but to my knowledge this topic has not been well-studied. Egg strands are laid on the bottom of waterbodies (10–30 cm [3.9–11.8 in] deep; Dodd 2013) or attached to aquatic vegetation or debris (Miller 1909a). Reported clutch sizes have ranged from 2,000 to 20,000 eggs (reviewed by Green 2005). Kruse (1981) counted eggs laid by 19 females in Illinois and found that clutch size ranged from 2,792 to 13,452 ($\bar{x}$ = 5,673). Females mate only once per year (Ewert 1969, Howard 1988). Ewert (1969) documented within-year male breeding behavior in northern Minnesota and found that six of seven observed individuals engaged in calling activity after reproducing, three of which were observed in amplexus with a second mate. Howard (1988b) also documented males mating up to two times per year in Indiana. In a laboratory study of wild males collected in Illinois, individuals were capable of successfully fertilizing egg strands up to five times in a single breeding season (Kruse and Mounce 1982). In addition, the number of fertilization events did not affect fertility, and two males fertilized one clutch per day for four and five consecutive days, respectively.

Eggs typically develop for two to seven days before larvae hatch (reviewed by Dodd 2013). Smith and Rettig (1998) found the mean time to hatching was 5.9 days (range = 4–8 days, n = 17 egg strands) at a pond in Indiana. In Wisconsin, larvae typically metamorphose and leave the water in late June to early July after a ca. seven-to-eight-week growth period (Knudsen 1957a, Vogt 1981). However, time to metamorphose, as well as larval survival, growth rate, and size at metamorphosis, can vary widely depending on habitat conditions, competition, and predation (e.g., Wilbur 1987, Skelly and Werner 1990, Pearman 1993, Sams and Boone 2010, Reynolds et al. 2011, Rittenhouse 2011). A mesocosm study found that competition with Northern Leopard Frogs (*Lithobates pipiens*) reduced survival to and size at metamorphosis (Purrenhage and Boone 2009). Werner and Glennemeier (1999) found that tadpole growth and survivorship were higher in open-canopy ponds. Another study in Michigan found that tadpoles preferred water temperatures >26°C (79°F) and occupied deeper areas of ponds at night and shallower areas of ponds during the day to maintain higher ambient temperatures (Beiswenger 1977). The tendency of American Toad larvae to form large aggregations in shallow areas during the day appears to be driven largely by individual-level responses to heat and light (Beiswenger 1977).

Recently metamorphosed individuals grow rapidly and disperse from the vicinity of natal wetlands at a weight of ca. 0.5 g (0.02 oz; Taigen and Pough 1981). See Hamilton (1934) for variation in within-year growth rates of juveniles. Males typically become sexually mature in two to three years and females in three to four years (Hamilton 1934, Kalb and Zug 1990). However, Acker et al. (1986) estimated that 3 of 37 females collected from a breeding site in Illinois were two years old. In addition, the smallest mature males and females collected were ca. 5.9 cm SVL (2.3 in) and 6.6 cm SVL (2.6 in), respectively. At a breeding site in Indiana, sexually mature males were two to six years old, ranged from 4.9 to 7.0 cm SVL (1.9 to 2.8 in), and weighed from 14 to 48 g (0.03 to 0.11 oz, n = 400; Howard and Young 1998). Morris and Meyer (1980) reported that a captive male originally collected from the wild in Illinois lived to be 5.5 years old, and according to Dickerson (1906), "Authentic record gives the story of one that lived to be thirty-six years of age and then was killed by accident." Acker et al. (1986) estimated that the oldest males and females collected at a breeding site in Illinois were four and five years old, respectively.

ACTIVITY

Ewert (1969) found that adults emerged from hibernation between 10 April and 8 May in northern Minnesota (n = 6). Allen (1868) also observed individuals on the surface in early April in Massachusetts. In northern Minnesota, individuals remained near their overwintering sites for 1–30 days ($\bar{x}$ = 13 days, n = 12) prior to migrating to breeding sites (Ewert 1969). Breeding migrations and dispersal of juveniles after metamorphosis likely represent the greatest movements that American Toads undertake during the year. In New York, Maynard (1934) documented American Toads migrating along stream corridors up to 4 km (2.5 mi) to breeding ponds. In Maryland, straight-line post-breeding dispersal distance of females ranged from 246 to 1,015 m (807 to 3,330 ft,

$\bar{x}$ = 686 m [2,251 ft], n = 16; Forester et al. 2006). For the 16 females tracked, maximum distance moved per day during the study period ranged from 189 to 610 m (621 to 2,001 ft). In Minnesota, Ewert (1969) documented post-breeding travel distances of 3.7–48.8 m per day (12–160 ft, $\bar{x}$ = 26.1 m [85.5 ft], n = 22), with total dispersal distances ranging from 52 to 1,006 m (170 to 3,300 ft, n = 22). Ewert (1969) noted that dispersal movement was "curtailed during periods of dry days with dry nights and resumed only on rainy or warm, humid evenings." In addition, he documented juvenile American Toads ≥610 m (2,000 ft) from the nearest breeding site. Ewert (1969) also found that 13 of 18 individuals dispersed to and from breeding sites in the same general direction (angle between paths for the 13 individuals = 1°–28°, $\bar{x}$ = 10.3°).

Movement and foraging activity occur before, during, and after the breeding season (Ewert 1969), although movement is minimal once individuals have migrated back to upland habitat following the breeding season. In northern Minnesota, mean movement/day for 18 active but nondispersing individuals ranged from 0.6 to 13.1 m (2 to 43 ft; Ewert 1969). In another study in northern Minnesota during the summer months, minimum distance moved per day ranged from 0.2 to 18.3 m (0.6 to 60 ft, n = 118 total recaptures of 61 marked individuals; Bellis 1959). However, long-distance movement events can occur during migrations to and from breeding sites. Ewert (1969) estimated linear home range sizes for 16 males by documenting locations through the full annual cycle and found that they ranged from 61 to 536 m (200 to 1,760 ft, $\bar{x}$ = 311 m [1,020 ft]).

Adult activity is primarily nocturnal and crepuscular (i.e., peak activity periods occur around dusk and dawn [Higginbotham 1939, FitzGerald and Bider 1974b]). Nightly movements are apparently influenced by moon phase, with the greatest activity during new moons and lowest activity during full moons (FitzGerald and Bider 1974a). According to Dodd (2013), activity is positively correlated with temperature in the early part of the active season, but temperature becomes a less important predictor during the summer months when temperatures are higher. In northern Minnesota, movement activity in autumn was typically diurnal due to subfreezing night temperatures (Ewert 1969). Precipitation can also stimulate activity, particularly breeding activity (Fitch 1956, reviewed by Dodd 2013). In addition, post-metamorphic movement activity has been found to be positively correlated with temperature and precipitation (Gravel et al. 2012).

American Toads aestivate in burrows throughout their annual active period to avoid unfavorable conditions at the ground surface or to conceal themselves from predators. Ewert (1969) observed eight aestivation events between 24 July and 16 August 1967 in northern Minnesota. Five of the individuals buried themselves 5.1–7.6 cm (2–3 in) below the soil, and three individuals buried themselves 12.7–27.9 cm (5–11 in) below the soil. Two of the individuals transitioned from aestivation to hibernation without additional emergences, and he speculated that this is common for American Toads.

Ewert (1969) documented overwintering behavior for 23 individuals in northern Minnesota (17 of which survived the winter). The majority of these entered their hibernacula in September, with 15 October being the latest date of observed surface activity. American Toads overwinter terrestrially, but laboratory studies conducted on individuals collected from Portage County concluded that they are not freeze-tolerant, nor do they have the ability to survive freezing temperatures by supercooling (Holzwart and Hall 1984). Work by Ewert (1969) supported that they survived the winter by burrowing into the soil below the frost line (but staying above the groundwater line), with maximum observed soil depths ranging from 12.7 to 58.4 cm (5 to 23 in). Miller (1909b) found that American Toads in outdoor enclosure experiments burrowed up to 62.2 cm (24.5 in) below the soil surface and also observed that individuals not buried below the frost line did not survive the winter. Ewert (1969) documented spring emergence locations for six of the individuals he monitored in northern Minnesota and found that three of them did not emerge from the spot where they burrowed into the soil for hibernation.

PREY AND PREDATORS

Adult American Toads consume various types of invertebrates, with insects often comprising the majority of the diet. Cook et al. (2002) calculated the mean stomach contents of 405 American Toads collected from throughout their distribution and found that ants (Formicidae) made up the bulk of their spring (February–June) and summer (July–September) diets (17.90% and 23.24% of prey volume, respectively), but Carabidae (order Coleoptera) were also frequently consumed in spring (14.96%) and summer (20.36%). Similarly, in Jack Pine (*Pinus banksiana*) forests in Ontario, ants were found in 88.9% of stomachs analyzed ($n = 36$) and accounted for 37.9% of documented prey items (Bellocq et al. 2000). Other Hymenoptera, Collembola, and Coleoptera species were also frequently consumed. Miller (1909b) estimated that an average-sized adult (36.6 g [1.3 oz]) consumed a mean of 1.12 g (0.04 oz) of invertebrates per day between May and August.

Larvae consume primarily algae and other microorganisms by grazing on substrates with scraping mouthparts or by filter-feeding (Test and McCann 1976, Holomuzki 1997). Larvae also reportedly scavenge on dead animals encountered in wetlands (Gibson and Mitchell 2006), including larval conspecifics (Heinen and Abdella 2005), and even sometimes cannibalize eggs (Hamel 2009). Hamilton (1930) examined the stomach contents of 400 recently metamorphosed individuals (TL 0.8–1.2 cm [0.31–0.47 in]), presumably collected in the eastern US. Primary documented food items (% by volume) included Diptera (primarily larvae, 22%), mites (15.5%), ants (12.8%), beetles (including larvae, 11.8%), thrips (10.1%), and Collembola (6.2%).

A wide variety of vertebrates prey on adult American Toads. Documented predators include many snakes species, such as hog-nosed snakes (*Heterodon* spp.), watersnakes (*Nerodia* spp.), and gartersnakes (*Thamnophis* spp.; Burt 1935, Fitch 1965, Heinen 1993, Heinen 1994a, Forester and Thompson 1998, DeVito 2003). However, American Toads may be toxic to foxsnakes (*Pantherophis* spp.; Karstad 1967). Major mammalian predators include Raccoons (*Procyon lotor*) and skunks (Miller 1909b, Oldham 1966, Schaaf and Garton 1970, Groves 1980). American Toads are also commonly consumed by predatory birds (reviewed by Ross 1989a). They were the most frequently observed prey item at a Broad-winged Hawk (*Buteo platypterus*) nest in Lincoln County (i.e., 14.9% of 107 documented prey items; Rosenfield et al. 1984). Mueller (1980) observed Mallards (*Anas platyrhynchos*) consuming adult American Toads that were diurnally active in the water. Although adults are most vulnerable to predation when they are on the ground surface, Ewert (1969) stated that 2 of 28 individuals tracked through the hibernation period in northern Minnesota died from predation. American Toad eggs are vulnerable to water mold (family Saprolegniaceae; Gomez-Mestre et al. 2006). A wide variety of invertebrates and vertebrates prey on American Toad larvae, such as dragonfly (*Anax* spp.) naiads, predaceous diving beetles (*Dytiscus* spp.), Giant Water Bugs (*Lethocerus americanus*), and Wood Frog (*Lithobates sylvaticus*) larvae (Kruse 1983, Petranka et al. 1994, reviewed by Dodd 2013). In Minnesota, Cochran and Cochran (2007) documented 13 newly metamorphosed American Toads in the stomach of a Brown Trout (*Salmo trutta*).

Predator defense mechanisms employed by various life stages include unpalatability and alterations in behavior. Adults respond to predators primarily by fleeing (if they are in or near water), crouching so that only their toxic dorsal regions are exposed, or inflating themselves to appear larger and increase predator handling time (reviewed by Dodd 2013). Some studies have found that larvae at earlier developmental stages are more palatable than newly hatched and metamorphic-stage larvae (Brodie et al. 1978, Brodie and Formanowicz 1987). The dorsal skin glands of adults contain noxious chemicals, which are particularly concentrated in the parotoid glands, although mammalian predators often avoid these chemicals by consuming adults ventrally, leaving the head and back unconsumed (Schaaf and Garton 1970). Larvae also modify their behavior in response to the presence of predators. In a laboratory experiment using larval American Toads collected from the wild in Michigan, larvae reduced their activity levels in the presence of dragonfly (*Anax junius*) naiads (Anholt et al. 1996). Holomuzki (1995) found that larvae also decreased activity levels in the presence of fish chemical cues. Petranka (1989b) determined that larvae responded to the chemical cues produced by injured conspecifics fleeing the area. Other laboratory studies found that aggregative behavior of larvae was driven by the presence of predators (Gallie et al. 2001, DeVito 2003).

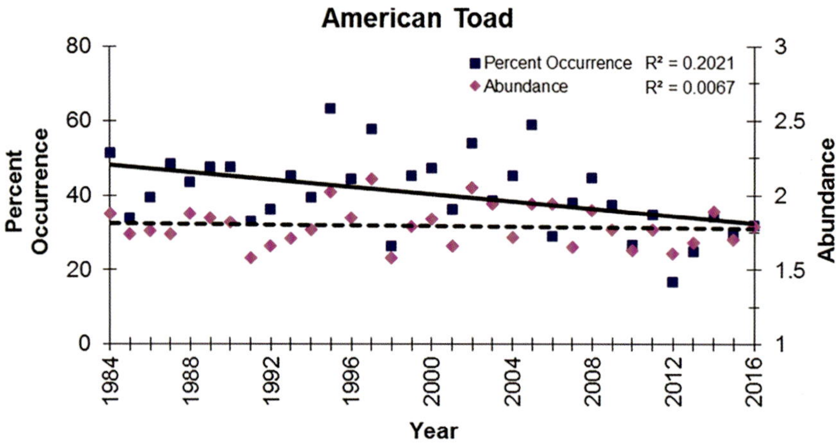

Figure 68. Long-term trend in occurrence (%) and abundance (rank) of American Toads across Wisconsin Frog and Toad Survey statewide monitoring sites. See Introduction to Species Accounts chapter for details on this monitoring program (figure courtesy of WDNR).

Watermolen (2014) compiled the literature on documented parasite infections for the American Toad in Wisconsin, which included 1 fungus (chytrid fungus [*Batrachochytrium dendrobatidis*]; Ortiz-Santaliestra et al. 2013), 1 leech (*Desserobdella picta*; Bolek and Janovy 2005), 14 nematodes, 10 trematodes, and 2 cestodes. Blais (2016) documented the American Medicinal Leech (*Macrobdella decora*) on four adult American Toads at a created wetland in Connecticut. Of the helminth parasites, nematodes are reportedly the most prevalent in American Toads from Wisconsin (Bolek and Coggins 2000, 2003, Yoder and Coggins 2007). At a breeding site in Minnesota where the trematode *Ribeiroia ondatrae* was abundant, 60% of observed metamorph American Toads ($n = 618$) displayed at least one malformation (Johnson and Hartson 2009). In mesocosm studies, survival to metamorphosis decreased when tadpoles were exposed to trematodes (Raffel et al. 2010a, Rohr et al. 2010).

CONSERVATION

Status: Globally, the American Toad is listed as a Least Concern species by the IUCN. It is considered a common species by the WDNR. Based on data collected through the Wisconsin Frog and Toad Survey (1984–95), the status of American Toads in Wisconsin appeared to be stable in 1995, with a significant regional increase observed over the survey period in the Driftless Area (Mossman et al. 1998). However, Wisconsin Frog and Toad Survey data collected over a longer period (1984–2016) indicates that breeding site occupancy may be decreasing over time (Figure 68).

Populations: Population sizes of explosive breeding anurans with short life spans, such as American Toads, can naturally fluctuate widely over time (Blaustein et al. 1994, Pechmann and Wilbur 1994). For example, the number of males detected was >200 in 1982 and <10 in 1984 at a pond in Indiana (Howard 1988b). Such high turnover (or perceived turnover due to lack of breeding activity) can result in dynamic patterns of extirpation and colonization of local breeding sites (e.g., Petranka and Holbrook 2006). Trenham et al. (2003) analyzed American Toad call survey data collected through the Wisconsin Frog and Toad Survey from 1981 to 1998 ($n = 1{,}112$ sites where the species was detected). They found that at 99% of survey sites, American Toads were not detected in at least one of the three survey years. Although false absences (i.e., species were present but not detected) likely accounted for some of the turnover, the authors concluded that "observer error alone [was] an unlikely explanation for the observed detection frequencies."

Reported sex ratios for American Toads are nearly always male-biased. For example, Oldham (1966) reported that only 15% of the 5,937 adults captured at breeding sites in Ontario over a four-year period were female. Mean number of males and females detected annually at a pond in Indiana was 41.3 and 19.3, respectively, over a three-year period (Howard 1988b). However, because reported sex ratios are typically based on captures at breeding sites, there are a number of potential sampling biases associated with the estimates (e.g., males are present at breeding sites for more nights than females, males can mate multiple times per year). However, Christein and Taylor (1978) used pitfall traps to sample American Toads migrating to breeding ponds in Ohio, and results were still highly male-biased (7.17M:1F, *n* = 711). The adult portion of American Toad populations may be inherently male-biased because males typically mature one year earlier than females (see Reproduction and Development section), and thus probability of survival to maturity is likely lower for females. Field sampling coupled with simulation modeling indicated that this was the case for a closely related species, the Houston Toad (*Anaxyrus houstonensis*; Swannack and Forstner 2007).

Threats: Due to their biphasic life cycles, amphibians in Wisconsin are vulnerable to anthropogenic impacts to both aquatic and terrestrial systems (see Introduction to Conservation and Management of Wisconsin's Amphibians and Reptiles chapter). However, unlike many of Wisconsin's amphibians and reptiles, American Toads have persisted through and in some cases benefited from the major anthropogenic landscape changes that have occurred in the region since European settlement. For example, Knutson et al. (2000) found that relative abundance of American Toads was positively associated with agricultural and urban edge habitat in Wisconsin and Iowa. Paloski (2006) did not find a significant relationship between lake shoreline development or surrounding land use and detection/nondetection of American Toads based on call surveys conducted at 14 lakes in Portage County. Burton et al. (2009) found that cattle use of ponds appeared to benefit American Toads in Tennessee. In contrast, Price et al. (2004) found that for populations inhabiting Great Lakes coastal wetlands (i.e., Lake Michigan and Lake Huron), landscape-scale development (and forest cover) was negatively associated with site occupancy by calling males. In Ontario, occupancy of wetlands for breeding in an agricultural landscape was positively associated with amount of forest cover and crop diversity in the surrounding landscape (Collins and Fahrig 2017).

Although they appear somewhat resilient to anthropogenic disturbance, some impacts associated with agricultural or urban/suburban landscapes are subtle and difficult to detect. Many studies have found that agricultural pesticides and herbicides, such as atrazine and carbaryl, can negatively impact larval American Toads (e.g., Boone and James 2003, Storrs and Kiesecker 2004, Distel and Boone 2009). In addition, Storrs Méndez et al. (2009) showed that American Toads are capable of terrestrial uptake of atrazine. In Ontario, agricultural activity was negatively associated with hatching success and positively associated with larval deformities (Bishop et al. 1999). However, a laboratory study using American Toad eggs collected from Dane County found that exposure to atrazine at concentrations typically encountered in the wild did not affect egg hatching success, larval mortality, or larval swimming speed (Allran and Karasov 2001). Using individuals collected from La Crosse County, Howe et al. (1998) found the median lethal concentration (LC50) for atrazine differed between early-stage (66.4 mg/L [0.008 oz/gal]) and late-stage (15.8 mg/L [0.002 oz/gal]) larvae. Ammonium nitrate fertilizer may also negatively affect larval survivorship (Hecnar 1995). However, a laboratory study using American Toad eggs collected from Washburn County found that exposure to a concentration gradient of ammonium did not influence egg hatching success (Jofré and Karasov 1999). A thorough review of documented chemical impacts on American Toad larvae can be found in Dodd (2013).

Exotic invasive plant species have also been found to negatively impact larval American Toads. In New York, replacing Cattail (*Typha latifolia*) with exotic invasive Eurasian Purple Loosestrife (*Lythrum salicaria*) resulted in slower development rates and higher mortality rates for American Toad larvae, possibly due to toxicity of higher tannin concentrations or alteration of the algal community (Brown et al. 2006). In a complementary laboratory study, it was found that the natural compounds released by degrading Eurasian Purple Loosestrife leaves negatively impacted growth and survival (Maerz et al. 2005b). American Toad survival to metamorphosis and mass at metamorphosis were also reduced in experimental mesocosms that contained nonnative compared to native wetland vegetation (Maerz et al. 2010).

Many studies have found that larval habitat quality for American Toads is influenced by the presence of predators and competitors. For example, Babbitt et al. (2003) found that relative abundance of American Toad larvae was negatively associated with the presence of fish in New Hampshire ($n = 42$ wetlands sampled). American Toads avoided breeding in ponds with predatory Wood Frog larvae in North Carolina (Petranka and Holbrook 2006). American Toads are also sensitive to acidic environments (see Habitat subsection). In Ontario, larval mortality was 85%–100% in ponds with pH <4.5, whereas mortality was <25% in ponds with pH >5 (Freda and McDonald 1993). Gibbs et al. (2005) also found that landscape-scale detection of populations in New York was positively correlated with soil pH.

Desiccation is a major source of mortality for juveniles, particularly in open habitats (Rittenhouse et al. 2008). Thus, juveniles appear to benefit from forest cover over grassland when dispersing from their natal ponds (Rothermel and Semlitsch 2002). In addition, like many amphibians, dispersing juvenile and migrating adult American Toads are highly vulnerable to road mortality (Ashley and Robinson 1996, Mazerolle 2004, Mazerolle et al. 2005). Diseases such as redleg disease (e.g., Dusi 1949), ranaviruses (e.g., Driskell et al. 2009), and chytridiomycosis (e.g., Gahl et al. 2012, Wise et al. 2014) could also be detrimental to local populations. For example, Ortiz-Santaliestra et al. (2013) showed that American Toads exposed to *Batrachochytrium dendrobatidis* (Bd) at the beginning of the larval development period exhibited reduced mass and survivorship compared to late-stage exposed and nonexposed individuals. Standish et al. (2018) detected Bd on 30.0% of adult ($n = 10$), 35.3% of juvenile ($n = 17$), and 14.3% of larval ($n = 7$) American Toads sampled in western Wisconsin. In addition, 10.0% of adults, 5.9% of juveniles, and 28.6% of larvae tested positive for a ranavirus. See the chapter Introduction to Conservation and Management of Wisconsin's Amphibians and Reptiles for additional information on threats.

Management: In general, management activities that reduce understory vegetation density, such as mechanical removal and prescribed fire, appear to benefit this species (Kirkland et al. 1996, Greenberg and Waldrop 2008, Matthews et al. 2010). However, post-management availability of cover objects, such as logs, is important (Pitt et al. 2013). Preservation or restoration of breeding wetlands to minimize presence of chemicals and exotic invasive vegetation would benefit this species (see Threats subsection). Efforts to reduce road mortality, such as roadside fencing near breeding wetlands, would also likely benefit this and many other species (Ashley and Robinson 1996, Aresco 2005). See the chapter Introduction to Conservation and Management of Wisconsin's Amphibians and Reptiles for additional information on amphibian management.

Anthropogenically modified or created aquatic environments, such as mitigation wetlands and water impoundments, are commonly used by American Toads (Lehtinen and Galatowitsch 2001, Nedland et al. 2007, reviewed by Brown et al. 2012b). Kline (1998) documented adult or larval American Toads at 14 of 21 created or restored wetlands surveyed in Ozaukee County. Kent (1983) documented American Toads calling in a sewage treatment bog in Bayfield County. Brand and Snodgrass (2010) found that American Toads exclusively used stormwater ponds or artificial wetlands for breeding, despite the availability of natural wetlands, in both suburban and forested watersheds in Maryland. Golf course ponds provided suitable larval habitat in Missouri (Boone et al. 2008). Shulse et al. (2012) determined that production of metamorphs in created wetlands was higher when they contained shallow slopes.

SPECIAL REMARKS

Across the American Toad's distribution, hybridization occurs between American Toads and a variety of other *Anaxyrus* species (reviewed by Dodd 2013). The resulting offspring typically display intermediate morphological and call characteristics (Blair 1963, Green and Pustowka 1997). American Toads are not known to climb trees. However, Cochran (2008b) reported an observation of an American Toad sitting on a rotting tree stump 1 m (3.3 ft) above the ground in a Chippewa River floodplain (Buffalo County). He speculated that the individual either climbed the tree stump or swam to the stump during the high-water period a few months prior and remained there.

Blanchard's Cricket Frog

Acris blanchardi (Harper 1947)

Rori A. Paloski, Andrew F. Badje, and Tara L. E. Bergeson

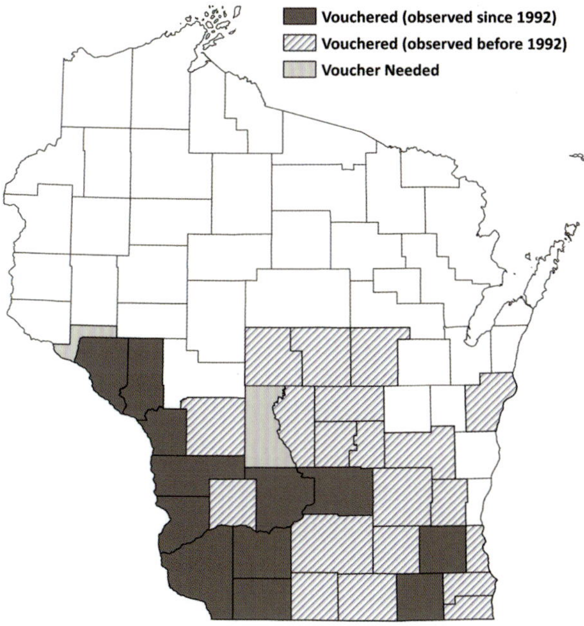

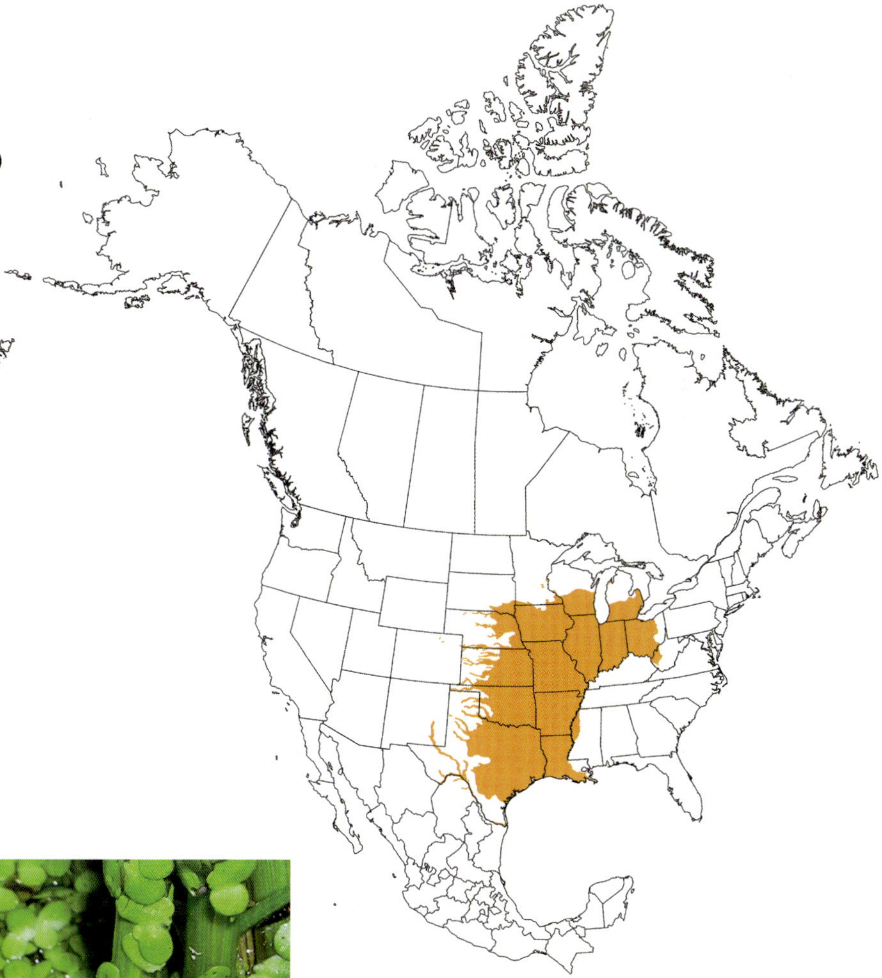

Calling male Blanchard's Cricket Frog (Lafayette County; photo by R. Paloski).

AT A GLANCE

Blanchard's Cricket Frogs are one of Wisconsin's smallest frog species. Their skin appears rough, and they can range in color from brown or tan to green. Some individuals also possess a triangular-shaped marking on top of the head. They were historically very common in the southern region of the state but have experienced dramatic population declines and extirpations since the 1960s. Blanchard's Cricket Frogs are currently most abundant in western Wisconsin (particularly the southwestern counties) but are also found in a handful of southeastern locations. In Wisconsin, Blanchard's Cricket Frogs are typically found near open permanent waterbodies, including slow-moving rivers or streams (and associated floodplain wetlands), where they also breed. They emerge from hibernation in March or April and breed from late spring to early summer. Male advertisement calls are reminiscent of steel marbles clicking together. Larvae develop rapidly after hatching, and metamorphosis can occur from late July through early September. Blanchard's Cricket Frogs consume an array of small invertebrates and are taken by many predators, from snakes and larger frogs to various wading predatory birds (such as herons) and mammals (such as Raccoons). They are incredible jumpers and can leap over 1 m (3.3 ft) to escape predators, which, given their small size is equivalent to a human jumping almost 59 m (194 ft). Despite being adept at escape, they experience high mortality rates and rarely survive longer than 12 months in the wild. By late October, Blanchard's Cricket Frogs typically congregate near or have entered hibernacula, which are cracks, crevices, or small burrows along the shores of aquatic habitats. They cannot tolerate freezing for extended periods, so their hibernacula must possess specific environmental conditions that promote winter survival. Although the factors behind their recent population declines are difficult to identify, likely culprits include habitat loss and changing climate. In Wisconsin, Blanchard's Cricket Frogs are currently listed as an Endangered species and a Species of Greatest Conservation Need by the WDNR.

SYSTEMATICS

The systematics and associated nomenclature for this species have caused confusion in the past. As stated by Dodd (2013), because of this confusion, it can be difficult to determine whether older primary literature sources are discussing *Acris crepitans* (Eastern Cricket Frog) or *A. blanchardi* (Blanchard's Cricket Frog). *A. blanchardi* was originally described by Harper (1947) as *A. gryllus blanchardi* and reclassified as *A. crepitans blanchardi* by Conant (1958) due to suspected differences within the genus *Acris*. McCallum and Trauth (2006) examined the morphology of the subspecies *A. c. blanchardi* and *A. c. crepitans* and did not find substantial differences between the two, recommending elimination of *A. c. blanchardi*. Later, Gamble et al. (2008) analyzed nuclear and mitochondrial DNA to recommend elevation of *A. c. blanchardi* to full species status as *A. blanchardi* for midwestern populations and retention of *A. crepitans* for populations in eastern and south-central portions of the US. Crother (2017) also recognized *A. blanchardi* as a species separate from *A. crepitans* with no subspecies. Various common names have been applied to this species over time, which adds further confusion, including Northern Cricket Frog, Eastern Cricket Frog, Western Cricket Frog, Swamp Cricket Frog, and simply Cricket Frog. Currently, Wisconsin populations are considered *A. blanchardi* (Blanchard's Cricket Frog; Crother 2017). *Acris blanchardi* is a member of the family Hylidae, like Spring Peepers (*Pseudacris crucifer*) and Cope's Gray Treefrogs (*Hyla chrysoscelis*), among others. No additional species in the genus *Acris* occur in Wisconsin.

DESCRIPTION

Adults: Blanchard's Cricket Frogs vary widely in background color, which can be brown, tan, olive, green, or gray; this can also change based on surroundings or environmental conditions (Figures 69, 70, and 71; Knudsen 1958g, Burkett 1984, reviewed by Gray and Brown 2005). The dorsum possesses many small, irregular-sized nodules that range in color from black or other earth tones to bright green (Figure 70; Dodd 2013, Andrew Badje, WDNR,

Above, top: Figure 69. Adult Blanchard's Cricket Frog (Lafayette County; photo by R. Paloski).

Above, bottom: Figure 70. Adult Blanchard's Cricket Frog (Iowa County; photo by R. Paloski).

Figure 71. Adult Blanchard's Cricket Frog (Grant County; photo by T. Brandt).

personal observation). Blanchard's Cricket Frogs often have a vertebral stripe that is bright green, tan, dark brown, gray, or red (Figure 71; see Variation subsection). They also have a dark, ragged stripe on the inside of each thigh (Conant and Collins 1998) and occasionally feature uniform dark bars or variable blotch patterns on the upper and lower hind legs (A. Badje, personal observation). A triangular-shaped interorbital mark is sometimes visible on top of the head between the eyes, while light vertical barring is typically present on the upper lip (Vogt 1981). The granular ventral surface is white, cream, or grayish white with additional brown or black mottling on the throat (Ballinger et al. 2010). Blanchard's Cricket Frogs have long hind legs with fully webbed digits on the hind feet (Ballinger et al. 2010) but not the front feet (Dodd 2013). Compared to several other hylid frogs in Wisconsin, Blanchard's Cricket Frogs' toe pads are quite small and only slightly wider than the digit tips, but these pads still offer some ability to climb vegetation (Rori Paloski, WDNR, personal observation, Vogt 1981). Blanchard's Cricket Frogs are one of Wisconsin's smallest frogs (SVL = 1.95–3.21 cm [0.77–1.26 in]; weight = 0.40–2.90 g [0.01–0.10 oz]; Badje et al. 2021). Mean SVL of adult Wisconsin museum specimens housed in the MPM and UWZM collections was 2.5 cm (1.0 in, $n = 145$; note that only specimens with SVL of ≥2.3 cm [0.9 in] were included). The largest individual among the Wisconsin specimens measured was 3.2 cm SVL (1.25 in; UWZM 24016, collected from Richland County in 1973).

Preadult stages: Blanchard's Cricket Frog eggs are bicolored, being black or brown on one half and tan or white on the other (Knudsen 1958g, Parmelee et al.

Figure 72. Late-stage Blanchard's Cricket Frog larva (Lafayette County; photo by R. Paloski).

Figure 73. Juvenile Blanchard's Cricket Frog (Lafayette County; photo by R. Paloski).

2002, Dodd 2013). Egg diameter, including the vitellus and the inner and outer gelatinous envelopes, ranges from 0.1 to 0.3 cm (0.04 to 0.12 in; Knudsen 1958g, Watermolen 1995). The dorsal color of larvae varies from olive to brown and is mottled throughout with small black dots (Figure 72). Larvae sometimes have a black-tipped tail and a dark narrow band that lines the upper half of the tail musculature (Vogt 1981, Parmelee et al. 2002). The venter of larvae ranges from white to an iridescent pale yellow (Vogt 1981). Larvae have dorsolaterally positioned eyes (Vogt 1981) and a long spiracle tube that is often free from the body wall. Larvae also have two upper and two lower labial tooth rows with papillae encircling the entire mouth (Breckenridge 1944, Vogt 1981). Recently hatched larvae are generally 1 cm (0.4 in) in TL but can reach a TL of 3.5–4.4 cm (1.4–1.7 in) prior to transformation (Breckenridge 1944, Vogt 1981). Juveniles resemble smaller versions of adults but with generally brighter colors and an unmarked throat (Vogt 1981). Recently transformed juveniles measure 0.9–1.2 cm TL (0.35–0.47 in; Figure 73; Breckenridge 1944). Older juvenile and subadult specimens from Wisconsin in the MPM and UWZM collections ranged from 1.2 to 2.2 cm SVL (0.5–0.9 in, $\bar{x}=2.0$ cm [0.8 in], $n=201$).

Variation: Wisconsin individuals do not always possess a triangular interorbital mark on top of the head or a middorsal stripe (Dernehl 1902, R. Paloski, personal observation). When present, the middorsal stripe can range substantially in color from beige or brown to green and even red (Vogt 1981). Although individuals with red dorsal stripes are not frequently reported in Wisconsin today, Dernehl (1902) found that 12% of specimens collected from a Dane County location in October 1901 had stripes with "much" red. Middorsal stripe color can vary geographically and temporally, while multiple color variants can occur in the same population (Nevo 1973, Gray 1983).

Variation in appearance by sex occurs but is most obvious during the breeding season. During this period, males display a single brightly colored vocal sac that can be beige, dusky yellow, green, or mottled gray (see Blanchard's Cricket Frog cover photo). In the Great Lakes region, this feature first becomes prominent from April to early May and can be seen in males by their first October (Gray 1983). Prior to the breeding season, developing eggs can often be seen through the abdominal skin near the groin of gravid females (Gray et al. 2005). Samples collected at the onset of the breeding season in Wisconsin indicated that males were slightly smaller ($\bar{x}$ SVL = 2.44 cm [0.96 in], $n=107$) than females ($\bar{x}$SVL = 2.55 cm [1.00 in], $n=53$). Males also weighed less ($\bar{x}=1.40$ g [0.05 oz], $n=106$) than females ($\bar{x}=1.80$ g [0.06 oz], $n=52$; Badje et al. 2021). Larvae show subtle variation in color, and some individuals possess a black-tipped tail, which is apparently exhibited in the presence of certain predators (see Prey and Predators section).

Diagnosis: Adult and post-metamorphic juvenile Blanchard's Cricket Frogs are characterized by a somewhat slender body, a lack of parotoid glands, and no large warts on the skin (but smaller nodules are present). They possess small toe pads on the terminal end of their digits, which are nearly the same width as the digits. While the digits on the hind feet are fully webbed, the digits of the front feet are not, and there can be a triangular marking in the interorbital region of the head. The larvae are characterized by internal gills and no visible limbs, eyes in a lateral or dorsolateral position on the head, a long spiracle tube that is free from the body wall, and a tail tip that is often black.

As adults, Spring Peepers (*Pseudacris crucifer*) and Boreal Chorus Frogs (*Pseudacris maculata*) are easiest to confuse with Blanchard's Cricket Frogs due to similarities in size. However, both of the former species have smooth skin. Furthermore, Spring Peepers have a characteristic X-shaped marking on the dorsum, and Boreal Chorus Frogs generally feature dark longitudinal lines; both species also lack the interorbital triangular marking that occurs in some Blanchard's Cricket Frogs. Adult Gray Treefrogs (*Hyla versicolor*) and Cope's Gray Treefrogs somewhat resemble Blanchard's Cricket Frogs but are larger in size and possess proportionally larger toe pads. Newly transformed American Toads (*Anaxyrus americanus*) are often confused with Blanchard's Cricket Frogs because of their warty appearance, similar size, and comparable coloration. Yet Blanchard's Cricket Frogs have longer legs in proportion to their body and can leap large distances, whereas toads have shorter legs and tend to hop in short bounds. American Toads also have larger nodules, including large parotoid glands behind the eyes (although parotoid glands are not prominent in recently metamorphic individuals). Juvenile Green Frogs (*Lithobates clamitans*) could be misidentified as Blanchard's Cricket Frogs due to similar size, dorsal coloration, and long legs; however, the tympanum of juvenile Green Frogs is more pronounced and much larger. Green Frogs also have prominent dorsolateral folds along their dorsum, which Blanchard's Cricket Frogs lack. It is noteworthy that the male advertisement call of Blanchard's Cricket Frogs is similar to the call of the Virginia Rail (*Rallus limicola*), King Rail (*Rallus elegans*), and Mink Frog (*Lithobates septentrionalis*); however, the clicking sound of Blanchard's Cricket Frogs is more metallic in quality and usually starts slowly and increases in rapidity rather than remaining at a constant speed (Vogt 1981, Mossman et al. 1998, R. Paloski, personal observation).

Blanchard's Cricket Frog eggs are difficult to differentiate from the eggs of other anurans in the wild. Their eggs are laid in smaller clusters (not long strings, large masses, or sheets), which differentiates them from the eggs of anurans in the genera *Anaxyrus* and *Lithobates*. Furthermore, the diameter of individual Blanchard's Cricket Frog eggs is wider (0.2–0.3 cm [0.08–0.12 in]; Watermolen 1995) than in the eggs of Cope's Gray Treefrogs, Gray Treefrogs, Spring Peepers, and Boreal Chorus Frogs (reviewed by Vogt 1981, Watermolen 1995, Dodd 2013). As with eggs, anuran larvae are hard to distinguish from each other in the wild due to the potential for overlap in size, shape, and general coloration. Yet the eyes of Blanchard's Cricket Frog larvae are positioned more laterally than dorsally, unlike larvae in the genera *Anaxyrus* and *Lithobates*. They further possess a long spiracle tube that is free from the body wall, a characteristic that is not found in *Pseudacris* and *Hyla* larvae.

DISTRIBUTION AND HABITAT

Global distribution: Blanchard's Cricket Frogs are one of the widest-ranging species in the genus *Acris*. The Ohio River and southern portion of the Mississippi River form the eastern boundary of the frog's continental range, although a few isolated populations exist east of these landmarks in West Virginia and Mississippi. From western Ohio, the frog's range extends north into southern Michigan to the border of southwest Ontario, west across southern Wisconsin and portions of Minnesota to South Dakota and Nebraska, and south to Texas (extending along the Pecos River into New Mexico) and northeastern Mexico (Gray et al. 2005, Gamble et al. 2008). A few scattered populations were documented in Colorado, but their current status is apparently questionable.

Distribution in Wisconsin: Based on published county records, Blanchard's Cricket Frogs are documented throughout much of southern Wisconsin, although no voucher specimens were found for Juneau and Pepin Counties (see the Appendix). Furthermore, the only available voucher specimen uncovered for Wood County (UWZM 4051) is a juvenile that is challenging to independently identify. It is noteworthy that Hay (1998) summarized anecdotal records from Juneau and Wood Counties, which further supports their presence in both. Of the counties with published records, recent (i.e., post-1990) confirmation of occurrence exists for Buffalo, Trempealeau, La Crosse, Crawford, Vernon, Grant, Sauk, Columbia, Iowa, Lafayette, Walworth, and Waukesha.

Localities of possible occurrence: Blanchard's Cricket Frogs could have inhabited every county in the southern half of Wisconsin through the early to mid-1900s; however, recent range contractions have made it difficult to document their complete historical distribution. Although these frogs are often associated with southern regions in the state, there are indications that they could have existed farther north. For example, Franklin J. W. Schmidt apparently found them throughout western Wisconsin as far north as Dunn County in the early 1900s (see Some Prominent Figures in Wisconsin Herpetology chapter; Natural History Box: Wisconsin's Lost Herpetologist?). A single Blanchard's Cricket Frog specimen was also reportedly collected near Meridean (Dunn County) sometime between 1917 and 1920 and then deposited at UWZM (Conger 1921, Pope and Dickinson 1928). This specimen, however, is currently missing from the collection, making it impossible to verify its identity (Joshua Kapfer, UW–Whitewater, and Laura Monahan, UWZM, personal communication). Given its historical distribution along the Mississippi River in both Wisconsin and Minnesota, it is also highly plausible that Blanchard's Cricket Frogs once existed in Pierce and possibly St. Croix Counties. Edgren (1944) even stated that he "observed" them still farther north in Ashland County, but no vouchers exist.

The distribution of historical records throughout much of eastern and central Wisconsin further suggests that Blanchard's Cricket Frogs could have occurred in Outagamie, Brown, Kewaunee, Winnebago, Calumet, Sheboygan, and Ozaukee Counties. Unsubstantiated anecdotal records exist from Brown, Calumet, Ozaukee, and Door Counties (WDNR 2016b). Furthermore, Hay (1998) summarized reports from along the border of Waupaca and Outagamie Counties, as well as a cluster of occurrences at the intersection of Calumet, Brown, and Manitowoc Counties, although neither voucher specimens nor audio recordings are known from these specific reports. Suzuki (1951) mentioned a specimen collected from Little Elkhart Lake in Sheboygan County that was deposited at Carroll College (Suzuki 1951), but this specimen was uncataloged and could not be found (J. Kapfer, personal communication).

Habitat: Blanchard's Cricket Frogs occupy many permanent to semipermanent aquatic habitats (Burkett 1984, R. Paloski, personal observation). These include streams, rivers, backwater sloughs, lakes, ephemeral wetlands, wetland scrapes, sedge meadows, drainage ditches, and springs (Knudsen 1958g, Burkett 1984; see Amphibians and Reptiles in Wisconsin Plant Communities chapter). They are frequently associated with habitats that possess varying degrees of tree canopy cover throughout much of their North American range but in Wisconsin are most often found in open-canopy and semi-open-canopy habitats (Burkett 1984, R. Paloski, personal observation). Both Vogt (1981) and Knudsen (1958g) identified the importance of open wet mud flats and open streambanks for this species in Wisconsin (Figure 74); however, numerous sites with apparently sizeable populations in the state lack exposed mud flats and streambanks (R. Paloski, personal observation). Blanchard's Cricket Frogs are prone to desiccation (Ralin and Rogers 1972), so they do not often stray far from water and are strongly associated with wet or damp substrates during most of the active season (e.g., Smith et al. 2003a, Burdick and Swanson 2010). Despite this, they are rarely observed in deeper water or fully submerged but instead occupy the damp zone between standing water and adjacent uplands or rest on floating vegetation near the water's edge (reviewed by Gray et al. 2005).

Blanchard's Cricket Frogs typically breed in wetlands adjacent to flowing water or along lake margins, but they also breed in isolated ponds and

Figure 74. Blanchard's Cricket Frog stream habitat. Note short vegetation and exposed muddy bank (Iowa County; photo by R. Paloski).

slow-moving or stagnant regions of streams and rivers (Dodd 2013, Badje et al. 2021). Breeding locations often possess emergent, submersed, and floating aquatic vegetation near gently sloping, sandy, or muddy banks (reviewed by WDNR 2010, McCallum and Trauth 2021). Larvae are aquatic and reportedly spend most of their time in shallow warm water (Johnson 1991a), where they use aquatic vegetation as cover from predators (Gray et al. 2005). Newly metamorphosed juveniles utilize similar habitat as adults (reviewed by Gray et al. 2005).

Some reports from Wisconsin support that Blanchard's Cricket Frogs overwinter in crevices along south-facing muddy streambanks (Badje et al. 2016). Elsewhere in the Midwest, they overwinter near permanent water sources in crevices, crayfish burrows, and gravel banks but also along shorelines below masses of dead leaves, among cattails and rushes, and under logs (Pope 1944, Walker 1946, Neill 1948, Gray 1971, Swanson and Burdick 2010). Blanchard's Cricket Frogs in Wisconsin likely utilize most, if not all, of these overwintering microhabitats. Selected hibernacula must provide effective thermal buffering for this cold-intolerant frog (see Activity section).

REPRODUCTION AND DEVELOPMENT

Reproduction occurs in spring, and breeding behavior is likely triggered by increasing water temperatures (Minton 1972). In Wisconsin, Blanchard's Cricket Frogs breed once per year from mid-May through mid-July (Vogt 1981), but males sometimes call from April (A. Badje, personal observation) to early August (R. Paloski, personal observation). Males primarily call in the evenings, but it is not uncommon to hear them sporadically during the day (Vogt 1981, R. Paloski, personal observation). The advertisement call of male Blanchard's Cricket Frogs resembles the sound made by clicking steel marbles together, often starting at a slower rate and increasing in speed (Vogt 1981). Males mostly call while close to the shoreline but are occasionally heard farther from the water's edge on floating vegetation (Gray et al. 2005). Males can be highly competitive for access to females at this time and engage in many interesting behaviors during the breeding season. For example, calling males can respond to each other with either direct confrontation, cessation of calling, or apparent tolerance (Burmeister et al. 1999). Males of all sizes can also act as 'satellites' to gain reproductive opportunities. These satellite males do not call; instead, they remain within 50 cm (19.7 in) of actively calling males and attempt to intercept and breed with females attracted to the calling individual (Perrill and Maiger 1988). Further details on the behavioral and competitive interactions of calling males were reviewed by Dodd (2013). Amplexus and external fertilization of eggs typically occur in shallow water near the shoreline among emergent vegetation or on top of floating vegetation (Figure 75; Dodd 2013, R. Paloski, personal observation).

In Wisconsin, female Blanchard's Cricket Frogs lay approximately 200 eggs, divided into small groups of 10–15 (Vogt 1981), although Parmelee et al. (2002) suggested that they often deposit single eggs. Eggs are frequently attached to submerged vegetation but are also found floating on the water's surface and dispersed along the bottoms of waterbodies (Gray et al. 2005, Livezey 1950). In Iowa, clusters of 5–15 eggs, typically attached to vegetation, were located 0.5–2.0 cm (0.2–0.8 in) below the water's surface (Van Gorp 2001). Eggs hatch within several days and metamorphose approximately five to ten weeks thereafter (Burkett 1969, Trauth et al. 2004). In Wisconsin, recently metamorphosed individuals have been observed as early as late July (R. Paloski, personal

Figure 75. Blanchard's Cricket Frogs in amplexus (Lafayette County; photo by R. Paloski).

observation). In Indiana, recently metamorphosed individuals were most commonly encountered in July and August (Minton 1972), while in Ohio and Illinois, they were observed from late July to early September (Walker 1946, Gray 1983).

Growth rates in juvenile Blanchard's Cricket Frogs are approximately 0.01 cm/day (0.004 in/day; Bayless 1969), with some males reaching sexual maturity by the end of their first summer (McCallum et al. 2011). Blanchard's Cricket Frogs in the Upper Midwest attain their maximum adult sizes from May through July following their first winter (Johnson and Christiansen 1976). The mean SVL of breeding adults in an Ohio population was 2.4 cm (0.9 in, SD = 0.17 cm [0.07 in]) for males and 2.6 cm (1.0 in, SD = 0.19 cm [0.07 in]) for females (Lehtinen and MacDonald 2011). Blanchard's Cricket Frogs may live up to ca. four years in captivity (Zippel and Tabaka 2008), but wild individuals experience high mortality, and few survive past their first year (see Populations subsection). Due to their short life span, most Blanchard's Cricket Frogs only reproduce once in their lifetime, which is uncommon among anurans (Burkett 1984, McCallum et al. 2011).

ACTIVITY

In the Midwest, Blanchard's Cricket Frogs are active from March or April through October or November (Johnson and Christiansen 1976, Gray 1983, McCallum and Trauth 2004, Badje et al. 2016). Exact dates of emergence from hibernacula can vary by year and geographic location but are apparently dependent on snow cover, soil temperature, and external ambient air temperatures that near 10°C (50°F; Brenner 1969). In some locations within their continental distribution, they are reportedly active at ambient temperatures of up to 38°C (100.4°F; reviewed by Dodd 2013, Donald Brown, West Virginia University / USFS, personal observation in central Texas). They are also active under cool conditions in Wisconsin (air temperatures as low as 4.5°C [40.1°F]; Badje et al. 2016).

There is limited information available on this species' movements in the Upper Midwest, although available data support that movement patterns are similar across midwestern populations. For example, Gray (1983) found the mean distance moved between captures of marked individuals ranged from 20.7 to 36.1 m (67.9 to 118.4 ft) in central Illinois, while Burkett (1984) reported 19.5–25.0 m (64.0–82.0 ft) in Kansas, with a maximum recorded distance of 213.4 m (700.1 ft). Such findings support that Blanchard's Cricket Frogs are relatively stationary and show an affinity for particular locations. However, past researchers have observed that the frogs can be encouraged to move greater distances by precipitation, humidity, temperature, or the onset of the breeding season (Pyburn 1958, Burkett 1984).

The largest reported movements are typically associated with spring and autumn migration between overwintering and breeding sites. For example, research in southwestern Wisconsin recorded a spring migration distance of 0.66 km (0.41 mi) by a single male Blanchard's Cricket Frog ($\bar{x} = 0.19$ km [0.12 mi] for $n = 37$ dispersing individuals; Badje et al. 2021; Natural History Box: Spring Migrations). Similarly, Gray (1983) documented autumn migration distances in central Illinois of up to 1.3 km (0.8 mi). It is noteworthy that not all populations undertake these seasonal migrations, especially when adequate breeding sites are immediately adjacent to hibernacula or near locations used during the remainder of the active season. Variation in male/female migration patterns has been observed. In 2011 male migration from overwintering to breeding sites in southwestern Wisconsin occurred from mid-April

NATURAL HISTORY BOX: Spring Migrations

In the spring of 2011, Tyler Brandt and I documented Blanchard's Cricket Frog migrations from known overwintering grounds to breeding sites for the WDNR. To ensure the collection of precise individual movements and the identification of subsequently recaptured marked individuals, a preapproved toe-clipping procedure in combination with photographic mark-recapture was utilized. In early April we caught 83 individuals at two locations, photographed their dorsum and two laterals, recorded a GPS point, and clipped one toe digit on either the front or hind foot. After each successive recapture of a marked individual, we photographed the individual and recorded a GPS point. We then searched our cricket frog photo library for an individual with the same toe clip and with matching dorsal and lateral bumps, marks, and color patterns. Once all of these features found their match, we became confident that a specific individual had been recaptured. Over the following months, we recorded shoreline-dependent movements to breeding wetlands beginning in mid-April and lasting into June. Our photographic mark-recapture findings were later spatially mapped, and individual movements (i.e., straight-line distances) were calculated for 37 of the 83 recaptured Blanchard's Cricket Frogs (Site A: $\bar{x}$ movement distance = 0.10 km [0.06 mi], $n = 24$; Site B: $\bar{x}$ movement distance = 0.36 km [0.22 mi], $n = 13$; Badje et al. 2021). Our dispersal distance and seasonal movement findings have led to enhancements in the way Wisconsin's only endangered amphibian is managed and conserved.

Andrew F. Badje, WDNR

to June, whereas female migration began one to four weeks afterward (Badje et al. 2021).

The latest recorded dates of aboveground activity in the Upper Midwest ranged from late October in Iowa to late November in southwestern Wisconsin and central Illinois (Johnson and Christiansen 1976, Gray 1983, Badje et al. 2016). Prior to overwintering, it is common to find large congregations of Blanchard's Cricket Frogs on riverine mud flats and sloping banks near hibernation sites (McCallum and Trauth 2004, Badje et al. 2016). The use of communal hibernacula has been documented in Wisconsin and Arkansas (McCallum and Trauth 2003b, Badje et al. 2016; Natural History Box: Overwintering Ecology).

Blanchard's Cricket Frogs are intolerant of anoxic and hypoxic conditions while submerged, so they cannot effectively overwinter underwater (Gray 1971, Irwin et al. 1999, Swanson and Burdick 2010). As such, they overwinter in damp terrestrial habitats and can physiologically produce glucose and glycerol, which act as cryoprotectants for intercellular tissues (Irwin et al. 1999, Swanson and Burdick 2010). Despite this, they are not highly freeze-tolerant, and they are incapable of supercooling (Gray 1971, Irwin et al. 1999, Swanson and Burdick 2010). Their ability to produce cryoprotectants only offers limited protection, and most cricket frogs cannot survive 24–48-hour durations of mild subzero temperatures between −2.5°C and −0.8°C (27.5°F and 30.6°F) aboveground (Irwin et al. 1999, Swanson and Burdick 2010). Therefore, to survive winter, Blanchard's Cricket Frogs must select overwintering microhabitats (e.g., crayfish burrows or cracks and crevices) where they will not be submerged but also will not freeze for extended periods (e.g., −0.8°C to −0.5°C [27.5°F to 31.1°F] or higher; Irwin et al. 1999, Swanson and Burdick 2010). Gray (1971) documented high survival (85%; $n = 20$) of Blanchard's Cricket Frogs given the opportunity to overwinter in terrestrial cracks in the mud (when also faced with the choice of overwintering aboveground or in water) at an ambient temperature of −2°C (28.4°F) for two weeks (see Natural History Box: Overwintering Ecology). In the central area of their range (i.e., Illinois), Blanchard's Cricket Frogs apparently have considerable overwintering success (Gray

NATURAL HISTORY BOX: Overwintering Ecology

As part of a WDNR study in 2010, Tyler Brandt and I recorded autumn behavior and overwintering habitat for the Blanchard's Cricket Frog in southwestern Wisconsin. Over the course of our study, we documented 20 active overwintering crevices ($\bar{x}$ width = 1.5 cm [0.6 in], $n = 19$, $\bar{x}$ length = 68.1 cm [26.8 in], $n = 7$; Badje et al. 2016) on sloughed mud benches along riverbanks. We found hibernacula by walking the bank edges of rivers (at times, stepping on loose tufts of soil and grass that dropped us feet first into the cold river), which eventually led to our observations of Blanchard's Cricket Frogs exiting nearby cracks in the mud (see Figure 76). Our boot vibrations, penetrating the soil from a few meters away, seemed to signal this exiting response (see also Gray 1971). It seemed obvious to me that the frogs would be safer if they stayed in their cozy crevices; however, the frogs' flight response, presumably from the threat of predation, must have intuitively told them to flee to nearby water. Relying on such instinctive nature, we were able to find that a minimum of 65% of active hibernacula served as communal overwintering sites. One particular 'packed house' contained at least ten Blanchard's Cricket Frogs. As temperatures progressively became colder throughout the autumn, Blanchard's Cricket Frog response time also became slower, which ultimately led to fewer frogs exiting hibernacula. After 19 November, we no longer saw our anuran friends for the year, suggesting they were in their hibernacula physiologically preparing for snow and ice in the upcoming months.

Figure 76. Blanchard's Cricket Frog hibernacula entrance (Iowa County; photo by T. Brandt).

Andrew F. Badje, WDNR

1983), whereas the opposite appears to be the case in South Dakota due to limitations in the availability of suitable hibernacula at the northern periphery of their range (Swanson and Burdick 2010).

PREY AND PREDATORS

Blanchard's Cricket Frogs are opportunistic feeders that consume a variety of invertebrates (Johnson and Christiansen 1976, Vogt 1981, Oldham and Campbell 1990). Primary prey in Wisconsin reportedly include terrestrial arthropods, particularly small insects (Vogt 1981). In other portions of the species' North American range, prey include flies (Diptera), beetles (Coleoptera), true bugs (Hemiptera), butterflies and moths (Lepidoptera), and spiders (Arachnida; Hartman 1906, Jameson 1947, Johnson and Christiansen 1976, Labanick 1976, Burkett 1984). Other occasional

NATURAL HISTORY BOX: Long Jump Champions

Since Blanchard's Cricket Frogs are lower in the food chain, evading predators is a high priority for individuals. Throughout my efforts to document the natural history of this frog in Wisconsin, I've tested their jumping abilities and been witness to many evasion attempts. The majority of my casual observations occurred with this shoreline-dependent species initiating a large leap of ca. 0.9 m (3 ft) into or toward the water. If a frog does not reach water after its first attempt, its subsequent leaps seem much shorter and erratic in nature. Elsewhere, Blanchard's Cricket Frogs have been documented jumping up to 0.9–1.2 m (3–4 ft) in a single bound (Vogt 1981, as reported in Harding 1997). To accomplish the same feat, a 1.8 m (6 ft) human would need to stand and jump a distance of ca. 58.5 m (192 ft)! I've even seen these frogs covering vertical distances of up to 2.1 m (7 ft), partially resembling skydivers, as they leap from the top of banks into the river. Once they reach the water, their normal response is to casually float downstream and then glide over the water, back to the same shore. If predators continue to pursue them downstream (in this case, an intrigued biologist), they will either continue this trend or submerge themselves in soft muck or under leaves in the water. Since Blanchard's Cricket Frogs are not adapted to staying submerged for extended periods of time, I've found they normally resurface after ca. 0.5–1.0 min in nearby emergent or floating vegetation. Here they hide from predators, if the threat continues. Even though this seems to be a simple evasion tactic to us, the massive leaps of this little frog, more often than not, can get it out of drastic circumstances.

Andrew F. Badje, WDNR

prey items of note include plants, inorganic materials, millipedes, small mollusks, and crayfish (Hartman 1906, Jameson 1947, Johnson and Christiansen 1976, Labanick 1976). Johnson and Christiansen (1976) estimated that Blanchard's Cricket Frogs in Iowa consumed 4,800 food items per year. Although the diet of larval Blanchard's Cricket Frogs has not been investigated, they are likely omnivorous, similar to the larvae of other Wisconsin anurans.

Documented predators of adults include American Bullfrogs (*Lithobates catesbeianus*), Green Frogs (*L. clamitans*), Largemouth Bass (*Micropterus salmoides*), turtles (Testudines), Common Gartersnakes (*Thamnophis sirtalis*), Western Ribbonsnakes (*T. proximus*), Common Watersnakes (*Nerodia sipedon*), American Bitterns (*Botaurus lentiginosus*), Great Blue Herons (*Ardea herodias*), Green Herons (*Butorides virescens*), Raccoons (*Procyon lotor*), and Six-spotted Fishing Spiders (*Dolomedes triton*; Carpenter 1952a, Fitch 1958a, Lewis 1962, Fitch 1965, Langlois 1964, Wendelken 1978, Burkett 1984, Perrill and Magier 1988, Johnson 2000b, Gray et al. 2005, Rickard 2006a, Edmond and Bowe 2012). In addition, Virginia Opossums (*Didelphis virginiana*), Killdeer (*Charadrius vociferus*), Common Grackles (*Quiscalus quiscula*), Red-winged Blackbirds (*Agelaius phoeniceus*), Black-crowned Night Herons (*Nycticorax nycticorax*), and Striped Skunks (*Mephitis mephitis*) have been suggested as possible predators (Burkett 1984, Oldham and Campbell 1990). Wolf Spiders (*Tigrosa helluo*) and Six-spotted Fishing Spiders have been observed preying on newly metamorphosed Blanchard's Cricket Frogs (Burkett 1984, Blackburn et al. 2002, Youngquist and Sitvarin 2015), while eggs are consumed by a variety of fishes, and larvae are preyed upon by fishes and dragonfly naiads (Caldwell 1982, Carfango et al. 2011, Dodd 2013).

In addition to relying on their cryptic coloration and access to cover for predator avoidance (Bragg 1943, Pyburn 1958, Dickson 2002), Blanchard's Cricket Frogs are incredible jumpers. They can leap over 1 m (3.3 ft; Vogt 1981; Natural History Box: Long Jump Champions). Although larvae are not unpalatable to most predators (reviewed by Dodd 2013), the black-tipped tails that develop in some populations may be a predator-avoidance response to certain types of local predators. For example, when Blanchard's Cricket Frog larvae co-occur with fish, larvae often lack black-tipped tails (Caldwell 1982). On the contrary, at sites where predatory dragonfly naiads are present rather than fish, Blanchard's Cricket Frog larvae with black-tipped tails are more common. This dark highlight appears to divert dragonfly attention toward the tail tip, the loss of which is less critical to survival (Caldwell 1982, Carfango et al. 2011).

CONSERVATION

Status: Globally, Blanchard's Cricket Frog is considered a Least Concern species by the IUCN. While these frogs are not federally protected in the US, they are considered Endangered and possibly extirpated in Canada (Britton 2001). Blanchard's Cricket Frog is listed as Endangered and considered a Species of Greatest Conservation Need by the WDNR. Current data on long-term trends, including those from the Wisconsin Frog and Toad Survey, suggest that the few remaining state populations have maintained some stability since the 1990s (Figure 77).

Populations: Although robust population data from the Midwest are limited, some sources support that population sizes can be large (or were historically large). For example, a 'minimum living population' of up to 1,260 individuals was recorded at a site in Kansas, with larger population estimates occurring in late summer when juveniles metamorphosed and entered the population with existing adults (Burkett 1984). Despite this, Blanchard's Cricket Frogs have experienced dramatic declines across many northern locations in North America, including Wisconsin (Natural History Box: The Decline of Blanchard's Cricket Frogs in Wisconsin). In Wisconsin, linear shoreline densities of up to 0.6/m (0.2/ft) were estimated at a 0.2 ha (0.4 ac) pond in early September (R. Paloski, unpublished data). In Kansas, the shoreline density of Blanchard's Cricket Frogs reached three to five frogs per meter (0.9 to 1.5 per foot) in late summer (Burkett 1984). Population sex ratios reportedly vary by age class and/or time of year when the estimates are made (Dodd 2013). For example, Badje et al. (2021) documented the gradual changes in the relative abundance and sex ratios of breeding-age individuals (e.g., 15M:1F [20 May 2011], 58M:10F [27 May 2011], and 87M:46F [1 June 2011]) at the onset of the breeding season in a southwestern Wisconsin wetland. In Kansas, Burkett (1984) reported a sex ratio of 1M:4F in newly metamorphosed frogs (i.e., late summer and early autumn) but found that most frogs that were more than three months old were males. Gorman (1986), on the other hand, reported a ca. 1M:1F adult sex ratio in Kansas populations throughout May and June. In Ohio, Dickson (2002) recorded 98 males and 56 females, with an estimated sex ratio of 1.75M:1F, although no time of year was provided.

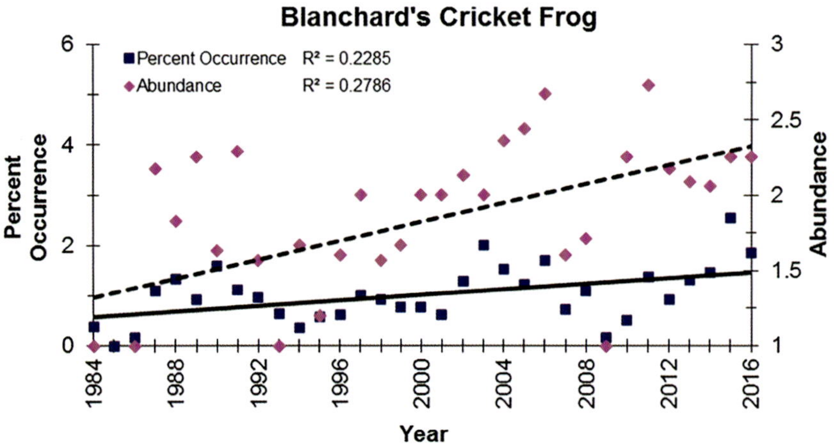

Figure 77. Long-term trend in occurrence (%) and abundance (rank) of Blanchard's Cricket Frogs across Wisconsin Frog and Toad Survey statewide monitoring sites. See Introduction to Species Accounts chapter for details on this monitoring program (figure courtesy of WDNR).

NATURAL HISTORY BOX: The Decline of Blanchard's Cricket Frogs in Wisconsin

Since the 1960s, Blanchard's Cricket Frogs have declined along the northern and western boundaries of their continental range. They are currently considered extirpated from Ontario, West Virginia, Colorado, and Arizona (Frost 1983, Oldham 1992, Hammerson and Livo 1999, COSEWIC 2001, Dickson 2002) and have disappeared from much of their former range in Minnesota (Moriarty 1998), Illinois (Ludwig et al. 1992, Mierzwa 1998, Greenwell et al. 1996), Michigan (Lehtinen 2002), Iowa (Lannoo et al. 1994, Lannoo 1998, Hemesath 1998, LeClere 2013), Ohio (Lehtinen and Skinner 2006), Indiana (Minton et al. 1982, Brodman and Kilmurry 1998, Minton 1998), South Dakota (Burdick and Swanson 2010), and Nebraska (McLeod 2005).

Similar losses have been documented in Wisconsin despite the fact that this species was once considered very abundant. For example, it was historically reported as one of the most common anuran species in southern portions of the state (Pope and Dickinson 1928, Knudsen 1958g). Cahn (1929) even stated it was "the most abundant of the small frogs of the region" in his report of herpetological sightings from Waukesha County in the 1920s. More recently, however, Blanchard's Cricket Frog populations in the state have declined dramatically (Vogt 1981, Hay 1998, Mossman et al. 1998). Mossman et al. (1998) conducted numerous fruitless surveys for this species in its historical range along the Lower Wisconsin River, only detecting the frogs on 3 of ca. 70 calling routes surveyed. Similarly, Jung (1993) detected them at <50% of 40 sites where they historically occurred in Wisconsin. Furthermore, a large number of old specimens exist in the MPM and UWZM collections that were taken from counties where this species has not recently been documented, including Dane ($N=44$ specimens), Rock ($N=56$), and Milwaukee ($N=44$). The large majority of these specimens were collected in 1950 or earlier, with only two taken after 1970 (J. Kapfer, unpublished data). These data collectively support that Blanchard's Cricket Frogs were once more common in the state than today but declined sharply in numbers during the latter half of the 1900s. The specific explanations for these declines remain elusive, although general loss of suitable habitat is likely a key driver.

Yet it is not all bad news for this species in Wisconsin. Since 1990, several populations of Blanchard's Cricket Frogs have been reconfirmed in the western, southwestern, and south-central counties of Buffalo, Columbia, Crawford, Iowa, Grant, La Crosse, Lafayette, Sauk, Trempealeau, and Vernon (Jung 1993, Christoffel and Hay 1995a, Bergeson and Paloski 2004, 2005, Bergeson et al. 2013, Andrew Badje, WDNR, personal communication), with the most numerous observations reported from Grant, Iowa, and Lafayette. As of the mid-2000s, recurrent breeding populations have also been reconfirmed or discovered in Walworth and Waukesha counties (A. Badje, personal communication). Further survey efforts could lead to additional discoveries or confirmations and provide data important to the conservation of this species.

Joshua M. Kapfer, UW–Whitewater

Most individuals survive only one year in the wild but perhaps as little as four months (Johnson and Christiansen 1976, Gray 1983, Lehtinen and MacDonald 2011). In addition, McCallum et al. (2011) estimated that as little as 4.7% of a population might reach its second winter. Several sources further report that populations fluctuate seasonally. In Illinois, the heaviest declines were observed in spring and autumn (Gray 1983), while adult declines in Ohio were most often observed in mid-summer (Lehtinen and MacDonald 2011). In Kansas, Burkett (1984) reported significant population declines in September and October, with as few as 5% of frogs surviving from metamorphosis through their first overwintering period ($n = 3{,}500$). As expected, Blanchard's Cricket Frog populations are highly dynamic, experiencing complete turnover in as few as 16 months, and are sometimes dominated by a single age class (Johnson and Christiansen 1976, Burkett 1984, Lehtinen and MacDonald 2011).

Threats: No single cause can currently explain the statewide decline of this species, but many general threats that face nearly all amphibians in Wisconsin are likely applicable (see Introduction to Conservation and Management of Wisconsin's Amphibians and Reptiles chapter). Specific factors that probably affect local Blanchard's Cricket Frog populations include climate change, extreme weather events, pesticides, diseases, and parasites (or perhaps a combination of several, as suggested by Lannoo 1998). Habitat loss and alteration are also particularly important conservation concerns that reportedly affect this species (Oldham and Campbell 1990, Beasley et al. 2005, Dodd 2013). Despite this, disturbed agricultural lands, including farm ponds and scrapes, as well as riparian corridors that bisect pastures and fields, sometimes yield Blanchard's Cricket Frogs in southwestern Wisconsin (WDNR 2010). Furthermore, Mossman et al. (1998) reported Blanchard's Cricket Frogs from 16 sites in southwestern Wisconsin, 10 of which occurred in "pastured or recently pastured stream bottoms," while the remaining 6 sites were associated with a state park. Hay (1998) also pointed out that, due to this species' frequent association with open muddy flats or banks (see Habitat subsection), the ingrowth of more dense vegetation in these habitats could negatively affect Blanchard's Cricket Frogs. Aquatic habitats and riparian zones that experience low to moderate levels of grazing may possess exposed damp muddy banks, which this species utilizes (A. Badje, personal observation). Despite this, it would be unwise to claim that agricultural activities, such as grazing cattle, and unnatural disturbance benefit Blanchard's Cricket Frogs, as observations indicate that intensive agriculture and its associated direct impacts can negatively affect abundance (A. Badje, personal observation). Although these numbers are not statistically significant, Jung (1993) reported that only 6 of 19 Blanchard's Cricket Frog sites surveyed in Wisconsin were adjacent to agricultural fields (i.e., corn).

Environmental changes, in combination with intrinsic population characteristics, could also negatively affect Blanchard's Cricket Frogs. This species cannot tolerate prolonged exposure to freezing temperatures (see Activity section) or drought (Burkett 1984) while overwintering. They also have naturally short life spans and potentially unstable population structures (see Reproduction and Development section and Populations subsection). These characteristics could make them particularly susceptible to the natural environmental fluctuations that occur in the Upper Midwest (Lannoo 1998). Their narrow tolerances for moisture and temperature levels during hibernation mean that the availability of suitable overwintering locations likely influences this species' distribution and persistence in northern regions (see Natural History Box: Overwintering Ecology). Furthermore, the higher frequency of extreme weather predicted to occur with the progression of climate change could have severe negative impacts on Wisconsin populations, as suggested by Hay (1998). This is corroborated by McCallum (2010), who projected that climate change could result in a significant negative impact to Blanchard's Cricket Frog populations in Arkansas by 2100.

A wide variety of pesticides and other chemical contaminants negatively impact Blanchard's Cricket Frogs (e.g., Flickinger et al. 1980, reviewed by Oldham and Campbell 1990, Reeder et al. 2005, Russell et al. 2002, Ade et al. 2010, reviewed by Dodd 2013, Krynak et al. 2017). In fact, Reeder et al. (2005) reported a correspondence between this species' apparent decline and the period of highest DDT and PCB use in North America (late 1940s through late 1950s). Thought to be related to amphibian declines, the increase in malformation rates among amphibians, including Blanchard's Cricket Frogs, has also been the subject of numerous studies (reviewed by Lannoo 2008). Reports of malformed Blanchard's Cricket Frogs have occurred in Wisconsin (R. Paloski, personal observation), Illinois (Gray 2000a, 2000b), South Dakota (McCallum and Trauth 2004), and Arkansas (McCallum and Trauth 2003a).

Various infectious diseases that impact amphibians worldwide could pose serious threats to Blanchard's Cricket Frogs in Wisconsin (see Introduction to Conservation and Management of Wisconsin's Amphibians and Reptiles chapter). Chytridiomycosis, an infectious disease caused by the fungus *Batrachochytrium dendrobatidis* (Bd), has been detected in Blanchard's Cricket Frogs from Illinois, Indiana, Iowa, Kentucky, Michigan, Ohio, Oklahoma, and Missouri (Steiner and Lehtinen 2008, Zippel and Tabaka 2008, Crawford et al. 2017) but has not yet been documented in Wisconsin (Sadinski et al. 2010). Sekowska et al. (2014) reported that 1 of 50 adult Blanchard's Cricket Frogs sampled from Illinois tested positive for a ranavirus, another infectious agent impacting amphibians globally. Of further importance is that environmental contaminants, including aquatic herbicides, can alter this species' normal skin microbial flora, which provides some natural resistance to infection, potentially leading to greater susceptibility to ingress of pathogens (Krynak et al. 2017). Furthermore, Dodd (2013) reviewed a variety of parasites documented to impact Blanchard's Cricket Frogs, although investigations specific to Wisconsin are unknown.

Interactions with other frog species are also a potential conservation concern that has received little attention. For example, Youngquist et al. (2015) noted that in mesocosm experiments, the survival of Blanchard's Cricket Frog larvae was reduced when they were reared with Green Frog larvae. The authors also found that the presence of Cope's Gray Treefrog larvae (*Hyla chrysoscelis*) increased the time it took for Blanchard's Cricket Frogs to reach metamorphosis. Interestingly, Hay (1998) recounted examples in Wisconsin of ponds that, when originally investigated, harbored many Blanchard's Cricket Frogs that were gradually replaced by Green Frogs (although Blanchard's Cricket Frogs had returned ca. ten years later; R. Paloski, personal observation). The full impact of these interspecies dynamics is unexplored in Wisconsin but worthy of additional investigation.

Management: Numerous management considerations, including appropriate management of preferred habitats, control of environmental pollutants and problematic species, and mitigation of potential impacts from disease, are potentially relevant to Blanchard's Cricket Frogs (see Introduction to Conservation and Management of Wisconsin's Amphibians and Reptiles chapter). Monitoring known state populations and surveying to rediscover populations in historical areas where they are currently considered extirpated are critical for the conservation of this species. Due to its endangered status in Wisconsin, management should be enacted in consultation with the WDNR's Endangered Resources Review Program and consider the protection, creation, or restoration of suitable breeding, nonbreeding, and overwintering habitat (see Habitat subsection). In particular, promotion and protection of exposed banks that could house hibernacula within aquatic habitats where this species occurs are worth consideration, so long as soil erosion is controlled. Biologs and other environmentally friendly erosion stabilization products are strongly recommended over conventional riprap when conducting restoration or erosion control projects, as alterations to shorelines may significantly impact overwintering Blanchard's Cricket Frogs (McCallum 2003b).

It is likely beneficial to maintain adjacent upland vegetation that is short in stature, such as grasses (Table 9), and an open or semi-open tree canopy. Furthermore, past research supports that Blanchard's Cricket Frogs may benefit from natural vegetated buffers that can act as movement corridors through disturbed areas (e.g., Burbrink et al. 1998). The USDA's Conservation Reserve Program (CRP), which encourages farmers to convert agricultural land into uncultivated vegetated buffers on their property, could provide suitable movement pathways if they exist adjacent to wetlands or waterbodies (Youngquist and Boone 2014). It is also likely important to control encroachment of woody vegetation along shorelines

Table 9. Example of a short grass prairie seed mix used in Wisconsin during habitat management for Blanchard's Cricket Frogs.

Species	**Quantity, kg/ha (lb/ac)**
Cover crop of oats	87.16 (16.00)
Creeping Red Fescue	27.24 (5.00)
Sideoats Gramma	5.45 (1.00)
Bergamot	0.38 (0.07)
Black-eyed Susan	0.34 (0.06)
Purple Prairie Clover	0.34 (0.06)

at sites used by this species via mechanical removal of the woody vegetation or prescribed fire. Prescribed burns within Blanchard's Cricket Frog habitat should typically be conducted during the species' inactive period to reduce the likelihood of management-related mortality. Due to the sensitivity of Blanchard's Cricket Frogs to chemical contaminants, use of pesticides to control vegetation should be limited.

Past reports indicate that Blanchard's Cricket Frogs will use restored or constructed wetlands (reviewed by Dodd 2013). Yet because they breed later in the active season than some anurans in Wisconsin, wetlands constructed for Blanchard's Cricket Frogs should be deep enough to ensure that summer droughts do not occur before larvae have metamorphosed. Constructed or reclaimed wetlands and waterbodies should also contain emergent and/or submersed aquatic vegetation and short shoreline vegetation (LeClere 2013, McCallum and Trauth 2021), and they should be free of fishes. Translocation of Blanchard's Cricket Frogs to existing or constructed sites could be a viable management strategy. However, there is not concurrence on the effectiveness (or success) of past projects that have employed this tactic for Blanchard's Cricket Frogs (Rickard 2006b, Sonntag 2006). More research is necessary to validate this strategy before it can be implemented broadly.

The Gray Treefrog Complex in Wisconsin

Erik R. Wild and Spencer R. Siddons

There are two species of gray treefrog in Wisconsin: the Gray Treefrog (*Hyla versicolor*), first described by LeConte (1825), and the Cope's Gray Treefrog (*H. chrysoscelis*), described by Cope (1880). They are cryptic species; that is, they are identical in appearance but differ genetically. Like most living organisms, the Cope's Gray Treefrog (*H. chrysoscelis*) is diploid and possesses two sets of chromosomes ($2n = 24$ total sets of DNA-containing structures within each cell nucleus). The Gray Treefrog (*H. versicolor*), however, is tetraploid, possessing four sets of chromosomes ($4n = 48$ total sets of DNA-containing structures). The diploid-tetraploid nature of this complex was not recognized until Wasserman (1970) reported that the Gray Treefrog was a tetraploid, and Bogart and Wasserman (1972) later confirmed that the Cope's Gray Treefrog was a diploid. More recently, Holloway et al. (2006) found that tetraploidy in Gray Treefrogs originated through three events of allopolyploidy between extant diploid Cope's Gray Treefrogs and two other apparently extinct diploid lineages of treefrogs. (Extant populations of these diploids may exist in the northern regions of the tetraploids' range.) These tetraploid lineages then merged through interbreeding to result in a single species, *H. versicolor*.

Nevertheless, *Hyla versicolor* and *H. chrysoscelis* are so similar in general appearance that they can be difficult to distinguish and correctly identify. Even the originally listed holotype for *H. chrysoscelis* was actually a *H. versicolor* specimen, although this mistake was not realized until subsequent histological examination confirmed the misclassification (Fitzgerald et al. 1981). As expected, such challenges have prompted attempts to identify obvious characteristics that facilitate accurate identification (Table 10). Yet studies like that of Jaslow and Vogt (1977) highlight the difficulty in using morphological differences to distinguish these treefrogs due to extensive intraspecific character variation and the ability of these frogs to alter their coloration. The most reliable methods to differentiate them are to examine their karyotypes to determine ploidy (Olson 1984), compare nucleolus organizer regions (Bachmann and Bogart 1975, Cash and Bogart 1978), contrast erythrocyte size and number (Ralin 1977, Matson 1990a), or conduct other histological investigations (Fitzgerald et al. 1981). Also, Green (1980) used scanning electron microscopy to determine that toe pad cell size is significantly larger in Gray Treefrogs than Cope's

Table 10. Comparisons of distinguishing characteristics to identify Cope's Gray Treefrogs and Gray Treefrogs in field and laboratory settings.

Characteristic	Cope's Gray Treefrog	Gray Treefrog	References
Latin name	*Hyla chrysoscelis*	*Hyla versicolor*	Crother 2017
Chromosome number	Diploid (2*n*)	Tetraploid (4*n*)	Wasserman 1970, Bogart and Wasserman 1972
Breeding call	Fast, sharp	Slow, melodic	Jaslow and Vogt 1977, Gerhardt 2005, Holloway 2006
Adult body length	Short	Long	Jaslow and Vogt 1977, Vogt 1981
Red blood cell size	Small	Large	Ralin 1977, Matson 1990b
Nucleoli	Small, less abundant	Large, more abundant	Cash and Bogart 1978
Toe-pad cell size	Small	Large	Green 1980
Habitat association	Grassland, savannas	Forest, mesic hardwoods	Knutson 2000, 2004; Semlitsch 2008
Urbanization tolerance	Moderate	Low to moderate	Harding 1997, Paloski 2008, Knutson et al. 2000
Wisconsin distribution	Sporadic, with larger gaps in north-central region and along southwestern and eastern borders	Statewide with sporadic gaps and fewer confirmed records in southeastern region	See the Appendix
Wisconsin status	Common, possibly declining	Common, stable	Mossman et al. 1998, Christoffel et al. 2009

Gray Treefrogs. While these methods are useful, they require considerable expertise and resources. They also do not allow immediate identification in the wild, so they are rarely used for distributional records.

The most commonly employed method for distinguishing these treefrogs in the wild is through differences in the characteristics of male advertisement calls. Gray Treefrogs emit a melodic trill that is birdlike with a slower pulse rate, whereas Cope's Gray Treefrogs produce a sharp insect-like trill that has a faster pulse rate (see Vogt [1981] for sonograms of advertisement calls). Thus, when calling in sympatry, these frogs are easily differentiated by anyone with an experienced ear. Yet when calling allopatrically, they can still be difficult to identify when it is not possible to directly compare the call of one putative species to another. The use of male advertisement calls for identification is also muddied by the facts that call pulse rate can be influenced by temperature and call characteristics can vary across the continental distributions of each species (Gerhardt 1974). To further complicate matters, hybridization between these two treefrogs reportedly occurs, and hybrid males exhibit similar, but unique, call characteristics (Gerhardt et al. 1994).

Not surprisingly, the difficulties in differentiating these treefrogs in Wisconsin have confounded attempts to delineate their state distributions (Figure 78). Jaslow and Vogt (1977) made a concerted effort to rectify this by determining characteristics that could facilitate state distributional reviews for each species. They found that male Gray Treefrog SVL was

Figure 78. Examples of potential variation in adult color pattern exhibited by Cope's Gray Treefrogs and Gray Treefrogs (illustrations by E. Wild).

significantly larger than that of male Cope's Gray Treefrog, and the pulse rate of each species' advertisement call did not overlap in Wisconsin across a range of temperatures. Their results support that male advertisement calls represent the most straightforward and practical method for accurately documenting the state range of each species with field-collected data. Despite this, many reported distributional records for both species in the state are still plagued by insufficient voucher material. For example, published reports of one or the other species exist from almost every county in Wisconsin (see the Appendix). Yet many of these records are not supported with voucher material that permits independent verification of species by subsequent researchers, instead relying on whole fluid specimens and/or photographs. These alone are insufficient unless accompanied by recordings or sonograms of advertisement calls or by information on karyotypes, histology, or preserved tissues suitable for genetic analyses.

Given these points, the distributional maps for each species in this work were difficult to accurately create because we could not confirm specimen identity for many published records with available material. As a result, maps of published county records for Cope's Gray Treefrogs and Gray Treefrogs depict three categories: those with records that are confirmed via suitable museum vouchers, those that are missing vouchers, and those where we could not independently confirm species identification from existing voucher material. As the reader will no doubt determine, further work is necessary to accurately document the distribution of these two species in Wisconsin.

Cope's Gray Treefrog

Hyla chrysoscelis (Cope 1880)

Spencer R. Siddons

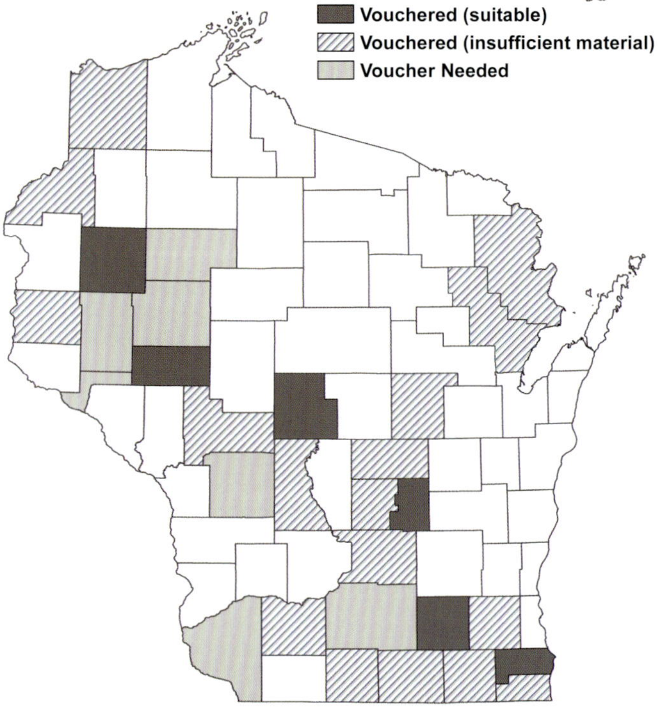

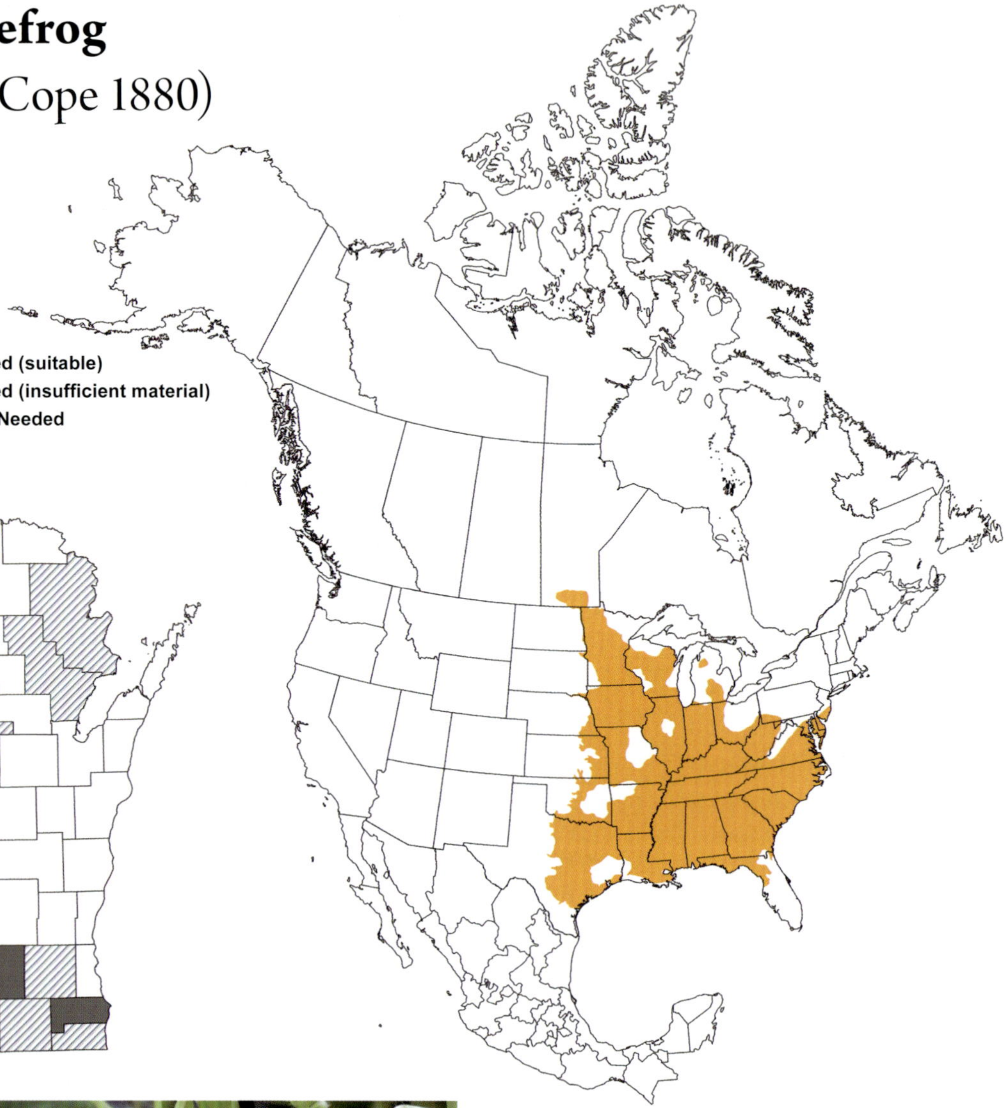

Adult Cope's Gray Treefrog with greenish tinge to dorsal coloration (Jefferson County; photo by J. Kapfer). *Note:* Species identification has not been confirmed, but the individual was observed in a location where only Cope's Gray Treefrogs were heard calling over a five-year period.

AT A GLANCE

Cope's Gray Treefrogs are medium-sized frogs with short bodies, wide heads, and large toe pads that help them adhere to surfaces and aid in climbing. The coloration on their backs ranges from light gray, brown, or tan to greenish gray overlaid with large irregularly shaped darker blotches, while their bellies are white or cream. They often possess darker bands on their outer hind legs and bright yellow streaks covered with dark speckles along their inner thighs. Recently metamorphosed juveniles usually have a solid bright green dorsal surface and white ventral surface. Cope's Gray Treefrogs are difficult to distinguish from Gray Treefrogs but can be identified by differences in the male advertisement calls. (Cope's Gray Treefrogs emit a short sharp trill that resembles a buzzing insect, while Gray Treefrogs emit a more melodic birdlike trill.) They are sporadically distributed across Wisconsin, but reports are sparse in the north-central, southwestern, and eastern border counties. Cope's Gray Treefrogs are often associated with open habitats or lightly wooded areas adjacent to open habitats that include oak savannas and pine barrens. In these habitats, they are frequently found resting in trees or shrubs. Males begin calling to attract females in early May soon after emergence from hibernation and continue calling through June. Preferred breeding sites include open fishless waterbodies, such as sedge meadows and wetlands. Larvae develop rapidly, and metamorphosis typically occurs from June through August in Wisconsin. Adults and juveniles eat a variety of invertebrates, while larvae generally scrape algae and other microorganisms from the surfaces of aquatic vegetation or substrates. Documented predators of adults include numerous reptiles, birds, and mammals, while larvae also fall victim to aquatic invertebrates, fishes, and salamander larvae. By autumn, Cope's Gray Treefrogs have moved upland to their terrestrial overwintering locations. Here, they hibernate under moist forest debris and produce metabolic compounds that protect their cells from damage due to freezing. Cope's Gray Treefrogs are currently considered common by the WDNR but are likely impacted by habitat loss and fragmentation.

SYSTEMATICS

Hyla chrysoscelis (Cope's Gray Treefrog) is positioned in the monophyletic genus *Hyla*, which includes many North American and Eurasian treefrog species (family Hylidae; Faivovich et al. 2005, Crother 2017). Recently, Duellman et al. (2016) proposed placement of North American and East Asian hylids into the resurrected genus *Dryophytes* (Fitzinger 1843), but this taxonomic revision is not currently accepted (Crother 2017). In the US, there are ten recognized species of *Hyla* (Crother 2017), but only *H. chrysoscelis* and *H. versicolor* occur in Wisconsin. There are no recognized subspecies of *H. chrysoscelis* (Crother 2017). Cope's Gray Treefrogs successfully hybridize with other species in the genus *Hyla*, including *H. versicolor* (Gerhardt et al. 1994) and species that do not occur in the state (e.g., Pine Woods Treefrog [*H. femoralis*] and Bird-voiced Treefrog [*H. avivoca*]; Mecham 1960, Pyburn 1960, Jensen et al. 2008). Among Wisconsin frog species, those in the genera *Acris* and *Pseudacris* occur in the same family as *H. chrysoscelis* (Hylidae).

DESCRIPTION

Adults: The dorsal ground color ranges from light gray or light brown to green but can change slightly based on temperature and surrounding habitat (Vogt 1981; Cope's Gray Treefrog cover photo and Figure 79). Large dark blotches that are irregularly shaped and outlined with thin dark lines also occur on the dorsum, while the venter is white. A light-colored patch, or lacrimal spot, is sometimes present immediately below each eye (Vogt 1981). A darker band occasionally extends back from the posterior edge of the eye, surrounding the tympanum, onto the side of the body. Darker bands can also cross the outer surface of the legs, while the inner surfaces of the thighs are bright yellow and covered with darker mottling (Vogt 1981, Dodd 2013; Figure 80). Cope's Gray Treefrogs possess large toe pads that are wider than the tips of their digits (Figure 80). Webbing between the toes is minimal on the front feet and extends only to approximately the first toe joint of the hind feet (Figure 80). They are medium-sized, relative to other anurans in Wisconsin, and reported mean adult male

Above, top: Figure 79. Adult Cope's Gray Treefrog with brown dorsal coloration (Jefferson County; photo by J. Kapfer). *Note*: Species identification was not confirmed, but the individual was observed in a location where only Cope's Gray Treefrogs were heard calling over a five-year period.

Above, bottom: Figure 80. Ventral view of an adult Cope's Gray Treefrog clinging to a window, highlighting its yellow inner thighs, enlarged toe pads, and minimal webbing between toes (Jefferson County; photo by J. Kapfer). *Note:* Species identification confirmed via male advertisement call.

SVL is 3.5 cm (1.4 in; Jaslow and Vogt 1977). Published information on adult weight in Wisconsin is rare. Weight of adult females in a sample from Tennessee was ca. 11.2 g (0.4 oz), while mean male weight was ca. 7.6 g (0.27 oz; Ritke et al. 1990).

Figure 81. Juvenile Cope's Gray Treefrog with bright green dorsal surface (Jefferson County; photo by J. Kapfer). *Note:* Species identification confirmed via male advertisement call.

Preadult stages: Eggs are round and dark gray to black in color (Watermolen 1995). Individual eggs range in diameter from 0.11 to 0.12 cm (0.04 to 0.05 in), with two clear outer gelatinous layers, so that total diameter ranges from 0.4 to 0.6 cm (0.16 to 0.24 in). Upon hatching, larvae are 0.6–0.8 cm TL (0.24–0.31 in; Minton 2001). Similar to other hylid species, their eyes are laterally rather than dorsally positioned on the head (Vogt 1981). Their bodies are dark green to light yellow-green with gold flecks and a white belly (Vogt 1981, Altig and McDiarmid 2015). Their tail fins are deep and fan-shaped, with a distinctly reddish to dark yellow pigment and black mottling toward the tip (Vogt 1981, Parmelee et al. 2002, Altig and McDiarmid 2015). The dorsal surface of recently metamorphosed individuals is typically bright green with no (or few) darker blotches, although the lateral surfaces and toes are often a mottled gray, while the ventral surface is mostly white (Vogt 1981; Figure 81). The mean SVL reported for recently metamorphosed young in Wisconsin was 1.4 cm (0.55 in; Vogt 1981). Published data on the weight of young individuals from Wisconsin are sparse, but mean weight in samples from Tennessee were 0.33 g (0.012 oz; Ritke et al. 1990) and 0.36 g (0.013 oz; weight of individuals from a wild-caught laboratory control group; Parris and Baud 2004). Mean weight at metamorphosis in samples from across several populations in Missouri was 0.67 g (0.02 oz; Ptacek 1996).

Variation: Adult dorsal coloration can vary (see Adults subsection). Adults are also sexually dimorphic, as females are larger than males. A review of several studies showed that mean SVL of adult females was ca. 5.4 cm (2.1 in), whereas mean adult male SVL ranged from 2.7 to 4.2 cm (1.1 to 1.7 in; Jaslow and Vogt 1977, Ritke et al. 1990, Ritke and Semlitsch 1991). Males can be further distinguished by their dark gray vocal sac, while female throats are white (Dodd 2013). Differences in other morphological characteristics have also been reported by geographic location in Wisconsin. For example, Menasco (1975) found evidence that Cope's Gray Treefrogs in Wisconsin had shorter femur and tibia-fibula lengths if they were collected north of the state's ecological tension zone (Curtis 1959). No strong explanation for these findings currently exists, and more extensive morphological sampling is needed for confirmation.

Diagnosis: Cope's Gray Treefrogs lack parotoid glands, large warty tubercules, and keratinized metatarsal spades on the heel of each foot. They possess terminal toe pads that are distinctly wider than the tips of their digits and a yellow coloration on the concealed inner surface of their thighs. Cope's Gray Treefrogs are morphologically similar to Gray Treefrogs, and in the wild they can only be reliably differentiated by characteristics of male advertisement calls. The call of male Cope's Gray Treefrogs is a sharp quick trill that is less melodic than the slower trill of male Gray Treefrogs (Vogt 1981; see Reproduction and Development section). Pulse rates of the advertisement calls for these species are completely nonoverlapping, with mean Cope's Gray Treefrog pulse rates of 35.7 (SD = 1.4) pulses/sec at 20°C (68°F) in populations with the slowest rate and Gray Treefrog pulse rates of 22.9 (SD = 1.5) pulses/sec at 20°C (68°F) in populations with the fastest rate (Gerhardt 2005, Holloway et al. 2006). Adult Cope's Gray Treefrogs are also typically smaller ($\bar{x}$ SVL = 3.5 cm [1.4 in]) than Gray Treefrogs ($\bar{x}$ SVL = 4.3 cm [1.7 in]; Jaslow and Vogt 1977, Vogt and Hine 1977, Vogt 1981). In the absence of audio data, these two species can be reliably differentiated through histological investigations or genetic analyses that compare chromosome numbers (Wasserman 1970, Bogart and Wasserman 1972, Fitzgerald et al. 1981, Holloway et al. 2006; see The Gray Treefrog Complex in Wisconsin account). The adults of three common hylid species in Wisconsin, Blanchard's Cricket Frog (*Acris blanchardi*), Boreal Chorus Frog (*Pseudacris maculata*), and Spring Peeper (*P. crucifer*), exhibit some resemblance to the Cope's Gray Treefrog. However, the former three are much smaller as adults, lack a white patch under their eyes, and do not possess toe pads that are strikingly wider than the tips of their digits.

The eggs and larvae of Cope's Gray Treefrogs and Gray Treefrogs are also similar in appearance. In fact, Vogt (1981) could not determine a reliable way to distinguish the larvae of the two treefrog species in the wild. Treefrog eggs can be distinguished from eggs of anurans in the genera *Anaxyrus* and *Lithobates* because they are laid in small clusters and not in long strings, large globular masses, or sheets at the water's surface. They can also be distinguished from eggs of anurans in the genus *Pseudacris* because individual treefrog eggs are surrounded by two gelatinous envelopes rather than one. Treefrog eggs are reportedly narrower in diameter (0.1–0.12 cm [0.04–0.05 in]) than Blanchard's Cricket Frog eggs (0.2–0.3 cm [0.08–0.12 in]; Watermolen 1995). Several characteristics can be used to generally distinguish treefrog larvae from the larvae of other anurans. For example, hylid larvae have a short spiracle tube that is attached to the body wall and a white or light-colored ventral surface. They have tail fins that are deep and extend far up onto the head, while their eyes are laterally positioned, unlike the larvae of anuran species in the genera *Lithobates* and *Anaxyrus*. They also have a more heavily mottled tail, which is often tinged red or orange, unlike the species in the genera *Acris* and *Pseudacris*.

DISTRIBUTION AND HABITAT

Global distribution: Cope's Gray Treefrogs are found throughout much of the southeastern US (although absent from most of Florida), west to central Texas, and north to Virginia, West Virginia, and central New Jersey. Their distribution expands north into the

Midwest along a band that covers portions of Oklahoma, Kansas, and Nebraska, much of Iowa, Indiana, Illinois, Minnesota, and into Manitoba, but also east across a large portion of Wisconsin; an isolated pocket occurs in the northern Lower Peninsula of Michigan.

Distribution in Wisconsin: Published county records are somewhat sporadically distributed across the state, with larger gaps in occurrence across portions of north-central Wisconsin, as well as along parts of the state's southwestern and eastern borders. Of existing published county records, not all are currently based on voucher material that permits independent confirmation of species (e.g., audio files or sonograms of male advertisement calls; see the Appendix). Voucher material for the Waukesha County record also reportedly included a dissected specimen and prepared blood slides (MPM 29589; Bolek 1997b), although the vouchered blood slides are currently missing (Julia Colby, MPM, personal communication). Future surveys should prioritize the collection of suitable voucher material in counties where Cope's Gray Treefrogs are reported but where supporting evidence is lacking or lost.

Localities of possible occurrence: No confirmed Cope's Gray Treefrog records are known from several regions in Wisconsin (see Distribution in Wisconsin subsection). Surveys of suitable habitat in these regions could reveal the existence of undocumented populations. In particular, Polk, Sauk, and Lafayette Counties seem probable locations due to their close proximity to counties of documented presence. The WDNR Wisconsin Frog and Toad Survey has also generated unconfirmed reports of calling male Cope's Gray Treefrogs in many counties that currently lack published records. These include counties in the western region (Buffalo, Sauk, and Lafayette), central region (Adams), and northern region (Bayfield, Vilas, Oneida, Price, and Marathon) of the state (WIATRI 2016). A number of whole fluid specimens also exist in the MPM collection that were reportedly taken from Bayfield County in the 1970s (e.g., MPM 8401–8416), although species identification of these cannot be confirmed without histological investigations or genetic analyses. Cope's Gray Treefrogs have questionable presence on one of the Apostle Islands (Sand Island; Casper 2008c) and could be present on additional islands if suitable habitat exists. Cochran et al. (1987) reported Gray Treefrogs in Brown County based on the specimen UWZM 22612. However, museum records currently indicate that this specimen was actually identified as a Cope's Gray Treefrog (Emily Halverson, UWZM, personal communication), which is otherwise not currently confirmed in Brown County. Regardless of whether the specimen in question is accurately identified, further surveys are warranted to definitively document Cope's Gray Treefrogs in Brown County.

Habitat: There is limited information on Cope's Gray Treefrog habitat preferences outside of the breeding season in Wisconsin. Some sources report that this species associates with open grassy habitats in the state, including wet prairies (see Amphibians and Reptiles in Wisconsin Plant Communities chapter). Furthermore, Vogt (1981) indicated that individuals often occurred in "oak savannas, dry and dry-mesic northern hardwoods and lowland forests" but that "they do not penetrate the heavily forested regions of north central Wisconsin." Casper (1996) also stated that Cope's Gray Treefrogs were frequently encountered in some "former savanna" regions of southeastern Wisconsin.

Although they use open habitats, adults are frequently found in trees and shrubs, where they seek refuge in tree holes, cracks, and crevices (Ritke and Babb 1991, reviewed by Dodd 2013). Their association with arboreal refugia is so strong that Pittman et al. (2008) reported that they were typically encountered within 2 m (6.6 ft) of trees. Ritke and Babb (1991) further observed that male Cope's Gray Treefrogs from Tennessee commonly occupied tree knotholes (*Sassafras albudum*, *Ulmus* spp.) at substantial heights of 1.9, 2.5, and 5.1 m (6, 8, and 17 ft). Cope's Gray Treefrogs are able to access these high arboreal locations with the aid of specialized skin cells on their toe pads that create tension for clinging to vertical surfaces (Green and Carson 1988, Hanna and Barnes 1991, Federle et al. 2006). When toe pads are pressed against a surface, the soft hexagonal-shaped skin cells spread apart to allow viscous mucous to fill the spaces between, which moistens the cells but also promotes adherence. This adherence allows treefrogs to ascend trees, buildings, windows, and other structures in their surroundings (Green 1981, Hanna and Barnes 1991, Federle et al. 2006; see Figure 80).

During the breeding season, Cope's Gray Treefrogs seek permanent or semipermanent open grassy wetlands, such as sedge meadows, for reproduction (Knutson et al. 2000, 2004; see Reproduction and Development section). Evrard (2000) frequently captured them near open wetlands in pine barren habitats of northwestern Wisconsin during the breeding season. Trenham et al. (2003) also determined that calling male Cope's Gray Treefrogs were positively associated with shrub wetlands or habitats with more herbaceous vegetation in Wisconsin but negatively associated with forested wetlands. Knutson et al. (2000) reported that the presence of calling males was further related to the area of available permanent water in wetlands. Despite this, several studies show that selection of breeding habitat and reproductive output can be altered by the presence of predatory fishes (which generally require more permanent waterbodies; Resetarits and Wilbur 1989, Binckley and Resetarits 2003; see Prey and Predators section). These findings suggest that Cope's Gray Treefrogs prefer breeding sites with specific characteristics: permanent water bodies that either lack fish or include small areas for breeding anurans that fish cannot access. Ritke et al. (1992) determined that adults exhibit high fidelity for the same breeding wetlands over time.

The habitat needs of larvae and juveniles in Wisconsin have received limited attention. Generally speaking, larvae develop in the same wetland habitats that adults select for breeding (reviewed by Altig and McDiarmid 2015). Recently metamorphosed juveniles could use similar habitat to that reportedly used by juvenile Gray Treefrogs in the state (see Gray Treefrog, *Hyla versicolor* species account). Fitch (1958a) observed that juveniles in Kansas stayed near their natal wetlands after metamorphosis and rested on broad-leafed plants.

Cope's Gray Treefrogs overwinter on land in lightly wooded habitats buried under soil, leaf litter, and other ground debris or within fallen logs (Ritke and Babb 1991). Burkholder (1998) observed two individuals overwintering in leaf litter at depths of 2.5 and 5.0 cm (0.98 and 1.97 in). He further found several buried at distances of 0.3–1 m (0.98–3 ft) from the base of Sugar Maples (*Acer saccharum*; air temperature at the time of discovery was 4°C [39°F], and the soil was not frozen). Ritke and Babb (1991) documented adult males hibernating beneath 8 cm (3.15 in) of moist sod. Juveniles could overwinter in similar habitats as adults, although limited information exists. Fitch (1958a) observed juveniles on the large flat leaves of plants located within natal ponds even as air temperatures dropped in autumn, which suggested they overwinter near wetlands. Although the microhabitats that Cope's Gray Treefrogs select for overwintering provide some refuge from harsh winter conditions, these frogs are still exposed to subzero temperatures. As a result, winter survival requires the mobilization and accumulation of cryoprotectant metabolites throughout their bodies (Storey and Storey 1986, Irwin and Lee 2003; see Activity section).

REPRODUCTION AND DEVELOPMENT

In Wisconsin, males typically begin calling to attract females in May when environmental temperature exceeds 15°C (59°F; Vogt 1981, Dodd 2013), although they sometimes call earlier. This species has a prolonged breeding season, with males calling as early as April and sometimes continuing to call well into June (Jaslow and Vogt 1977, Vogt 1981, Mossman and Hine 1984, Mossman et al. 1998).

The advertisement call of male Cope's Gray Treefrogs resonates from a single vocal sac located in the throat region (Figure 82). Their call is a short, sharp, and high-pitched trill, with a reported mean duration of 0.46 sec in Michigan (Bogart and Jaslow 1979) and 0.84 sec in Mississippi (Blair 1958). Jaslow and Vogt (1977) determined that the pulse rates of these trills in Wisconsin males ranged from 41.4 to 71.4 pulses/sec at 18°C–26°C (64°F–79°F). Brown and Brown (1972) reported a mean of 51 pulses/sec at 23°C (73°F) in Illinois. Temperature and pulse rates are positively correlated, which may explain the geographic variation observed in call characteristics (Jaslow and Vogt 1977).

In early spring, males most frequently vocalize at night but also call intermittently during daytime and at dusk as they descend to breeding ponds. Once the breeding season is under way, calling begins around sunset and continues into the night, with less frequent

Figure 82. A calling male Cope's Gray Treefrog. Note the single enlarged vocal sac (Jefferson County; photo by S. Viernum). *Note*: Species identification confirmed via male advertisement call.

calls during the daylight, often after rain (Ritke et al. 1990, Ritke and Semlitsch 1991). Although Cope's Gray Treefrogs breed in wetlands, males will sometimes vocalize near upland overwintering locations in early spring before they migrate to breeding sites (see Activity section). Once at breeding sites, males call from various perches along the water's edge, on top of floating algae mats, or while clinging to vegetation above the water (Vogt 1981, Godwin and Roble 1983). The height of chosen calling perches in Wisconsin varies but is typically <3 m (9.8 ft) above the water (Vogt 1981), yet it can range from 0 to 6 m (0 to 20 ft, $\bar{x} = 1.05$ m [3.45 ft]) high in Kansas (Godwin and Roble 1983). Males will move between several different calling perches per night (Ritke et al. 1990) but tend to exhibit a strong fidelity for certain locations. In Tennessee, Ritke and Semlitsch (1991) recaptured males at the same calling perches 29.6% of the time.

Males that vocalize to attract females are territorial and defend their calling perches against intruding males (Fellers 1979a). Acts of physical aggression between competing males usually last less than a minute and can include head-butting, kicking, shoving, or jumping on the intruder (Ritke and Semlitsch 1991). Despite this, male-male competition does not always include direct physical interactions. For example, silent males will sometimes sit near actively calling males and wait for them to successfully attract a female. Once the resident calling male engages in amplexus with a female and abandons his perch, the nearby silent male will take this presumably optimal calling location (Vogt 1981). These individuals, sometimes referred to as satellite males, typically maintain a crouched posture >10 cm (3.94 in) from a calling male (Roble 1985). In Kansas, <2% of dominant males had a satellite male nearby, but at least one satellite male was found in 20.7% of all choruses of calling males investigated (Roble 1985). Selection of calling locations can be further influenced by interactions with other species, such as the closely related Gray Treefrog. Ptacek (1992) found that when males of both treefrog species called in the same wetland, Cope's Gray Treefrogs selected calling perches at heights of ≤50 cm (19.7 in), but they selected perches above this height in the absence of Gray Treefrogs.

Females arrive at breeding ponds throughout the night, beginning within an hour of when males start to call (Ritke et al. 1990). Male call characteristics are then assessed by females to select individuals for mating (Morris and Yoon 1989, Bee 2008). Females prefer males with higher call rates, lower call frequencies (i.e., lower frequency may indicate a larger male with higher potential fitness), and calls of average or longer duration (Morris and Yoon 1989, Bee 2008). Females are reportedly better able to discriminate call quality among various males in a chorus when calls are spatially separated from sources of background noise (Bee 2007b, 2008a). When males are selected, females initiate amplexus by crawling on the male or nudging him in the side or rear (Ritke et al. 1990). The male then mounts the female in the axillary position (i.e., using his forearms to grasp the female around the armpits; Figure 83).

Amplexus typically lasts three to five hours, during which the pair moves into the water, and oviposition occurs (Ritke et al. 1990). During oviposition, females release eggs into the water that males fertilize externally, a process that can last from ≤10–60 minutes, depending on clutch size (Ritke et al. 1990). The eggs remain near the water's surface, either attached to vegetation or floating in loose clusters (Vogt 1981, Ritke

Figure 83. A pair of Cope's Gray Treefrogs in amplexus during the breeding season (Jefferson County; photo by S. Viernum). *Note:* Species identification confirmed via male advertisement call.

et al. 1990, Binckley and Resetarits 2003). Clutch sizes range from roughly 900 to 2,000 eggs divided into numerous loose clusters of 10–90 eggs (Vogt 1981, Ritke et al. 1990). Work in the southern US supports that females oviposit 1.24 times on average per breeding season (Ritke et al. 1990) but sometimes oviposit up to five clutches per year (Ritke and Lessman 1994). Females that mate early in the breeding season have a greater chance of producing multiple clutches in a season (Godwin and Roble 1983, Ritke et al. 1990). Ritke and Lessman (1994) suggested that the variability in the number of times females oviposit and the prolonged breeding season of this species may be plastic responses to unpredictable environmental conditions. This possible plasticity allows females the ability to reproduce during more favorable environmental conditions, a strategy observed in other temperate-zone anurans (Kyriakopoulou-Sklavounou and Loumbourdis 1990).

Eggs typically develop for three to seven days before hatching (Ritke et al. 1990). Larvae grow rapidly after they hatch, increasing in weight by 0.03 g/day (0.001 oz/day), based on measurements in Missouri (Ptacek 1996). Metamorphosis is completed in four to six weeks (Dodd 2013, Pintar and Resetarits 2017a), which suggests that juveniles should start emerging from wetlands during June in Wisconsin. However, larval developmental rates can be affected by competition with other anuran larvae. For example, Wilbur and Alford (1985) found that Cope's Gray Treefrog larvae took longer to reach metamorphosis and metamorphosed at a lower mass when sympatric with larvae of other anuran species due to competition for food. Larval development can also be influenced by the presence of predators. In wetlands that support Eastern Newts (*Notophthalmus viridescens*), sympatric Cope's Gray Treefrog larvae attained higher body mass at metamorphosis and experienced quicker growth rates, likely because larger larvae are more resilient to predation attempts (Semlitsch 1990).

After metamorphosis, juvenile growth rate is more likely a function of high ratios of protein to carbohydrates in food than the amount of food consumed (Steinwascher and Travis 1983). Juvenile growth rates also reportedly vary by sex. Ritke and Semlitsch (1991) found that in two years of study, mean female SVL increased by 0.28 cm (0.11 in) and 0.2 cm (0.67 in), while mean male SVL increased by 0.19 cm (0.08 in) and 0.27 cm (0.9 in), respectively. Individuals typically reach sexual maturity at two years of age (Wright 1932, Duellman and Trueb 1994). The minimum SVL of calling and presumably mature adult males assessed in Tennessee was 3.9 cm (1.54 in), while mean gravid female SVL was 4.5 cm (1.77 in; Ritke et al. 1990). Robust investigations of longevity in the wild do not exist. In captivity, members of the Gray Treefrog Complex have reportedly lived up to 7.8 years (Snider and Bowler 1992).

ACTIVITY

Cope's Gray Treefrogs typically emerge from hibernation when the soil and leaf litter at their chosen winter refuge thaw in the spring (mid-April to early May; Vogt 1981, Costanzo et al. 1992). After emergence, adults briefly remain in upland habitats to forage and then migrate to breeding wetlands (Pittman et al. 2008). In general, treefrog activity levels are related to individual body temperature, which is affected by environmental temperatures, which vary by time of day and season. During the active season, Cope's Gray Treefrogs are mostly nocturnal, so they remain in refugia such as tree knotholes during the day but forage and mate on warm rainy nights

(Ritke et al. 1990, Ritke and Babb 1991). Cope's Gray Treefrogs from Tennessee emerged from their refugia in tree holes at dusk when air temperature was still ≥15.8°C (60°F) but remained near these refugia as daytime air temperatures exceeded 25°C (77°F; Ritke and Babb 1991). Active summer body temperatures recorded from Kansas individuals were typically between 25.7°C and 33.7°C (78°F and 93°F), with a minimum of 19.7°C (68°F; Fitch 1956). On the other hand, cloacal temperatures of breeding males in Missouri were often lower than summer body temperatures recorded in the study by Fitch (1956), ranging from 18°C to 21°C (64°F to 70°F) in May and 21°C to 24°C (70°F to 75°F) in June (Ptacek 1992).

Greatest movements are probably associated with annual migrations to breeding wetlands in spring and to overwintering locations in autumn. Males can call at distances of up to 25 m (82 ft) from breeding ponds and at heights of 20 m (66 ft) in the treetops before arriving to a breeding locality (Godwin and Roble 1983, Ritke et al. 1990). Similarly, during a six-year period at a site in southern Wisconsin, male advertisement calls during early spring were detected in trees located >150 m (492 ft) from the nearest potential breeding site (J. Kapfer, personal communication). Both examples suggest that males migrated at least these distances for reproduction. Although some adults undertake spring migrations to breed, this behavior is not observed in all populations, particularly in areas where breeding sites are close to suitable terrestrial locations. For example, after the breeding season, some adult Cope's Gray Treefrogs spend periods of time in both breeding and terrestrial habitats, moving only short distances back and forth between them (Dodd 2013). Furthermore, extensive movements in terrestrial foraging habitats after the breeding period are rare. Ritke and Babb (1991) reported that Cope's Gray Treefrogs generally remained within 2 m (6.6 ft) of daytime refugia during nocturnal foraging forays outside of the breeding season.

Cope's Gray Treefrogs are positively phototactic, using both moonlight and sunlight to feed and move within their home range (Jaeger and Hailman 1973). They have true color vision and are sensitive to blue-spectrum light, which might assist in orientation and movement to areas of high illumination, such as open-canopy wetlands in forested areas (Hailman and Jaeger 1974). Information on juvenile movement is limited. It is often assumed that dispersal behavior post metamorphosis is similar between Cope's Gray Treefrog and Gray Treefrog juveniles (see Gray Treefrog, *Hyla versicolor* species account). Observations by Fitch (1958a) suggest that recently metamorphosed individuals do not disperse quickly but remain near natal ponds for several weeks.

As autumn approaches, Cope's Gray Treefrogs move toward terrestrial overwintering locations. In Wisconsin, these movements could begin as early as June (Vogt 1981) and may be stimulated by lower air temperature or the need to forage prior to winter. In Tennessee, Cope's Gray Treefrogs concentrated their daily activities in August on foraging to accumulate size and amass energy reserves for winter (Ritke et al. 1990). Larger individuals potentially experience greater winter survival, as they have lower metabolic rates and will not use energy stores as quickly as smaller frogs (Zeuthen 1953, Maiorana 1976). Some work suggests that Cope's Gray Treefrogs do not travel far from their nonbreeding summer habitat to their overwintering sites. Pittman et al. (2008) found that Cope's Gray Treefrogs in North Carolina occupied the same refugia immediately before and after winter (but not during).

Cope's Gray Treefrogs spend the winter in terrestrial hibernacula that are partially insulated but that still sometimes expose frogs to subfreezing temperatures. To tolerate freezing, individuals synthesize two main cryoprotectants, glycerol and glucose, which depress the freezing point and reduce how much ice can form, which prevents damage from extracellular ice crystals (Costanzo et al. 1992, Irwin and Lee 2003, Zimmerman et al. 2007). Response to freezing appears to vary based on minimum temperature experienced while overwintering. For example, Irwin and Lee (2003) reported that Cope's Gray Treefrogs exposed to freezing temperatures (−2.5°C [27.5°F]) with ice accumulated in over 50% of their body water regained their righting response and limb functionality by 48 hours post-thaw. All individuals survived freezing at −2.9°C for 53.5 hours and regained normal posture 72 hours after thawing (Costanzo et al. 1992). However, those exposed to −5.5°C and −6.5°C

(22.1°F and 20.3°F) took ca. 110 hours to regain righting reflex after thawing (Irwin and Lee 2003).

PREY AND PREDATORS

Adult Cope's Gray Treefrogs are 'sit-and-wait' ambush predators that remain in or near refugia until prey is within reach (Dickerson 1906, Ritke and Babb 1991). They feed on a wide variety of insects, a large portion being beetles (order Coleoptera) and ants (order Hymenoptera), but also caterpillars (order Lepidoptera), roaches (*Parcoblatta* spp.), and crickets (*Ceuthophilus* spp.; Dickerson 1906, Ritke and Babb 1991). Bolek and Coggins (1998) analyzed the stomach contents of 29 adults during the breeding season in Wisconsin and found that 43% of their diet consisted of beetles, 24% was moth and butterfly larvae, 11% was crickets/grasshoppers (order Orthoptera), and 22% of the contents were unidentified insects. Larvae primarily consume algae, fungi, detritus, and bacteria scraped from the surfaces of plants or submerged substrates in breeding ponds (Hoff et al. 1999, Dodd 2013). In addition to scraping surfaces to feed, larvae can consume food from the water column via buccal pumping and filtering the water through their gills, which trap tiny food particles (Kenny 1969, Wassersug and Hoff 1979, Hoff et al. 1999).

Adults are consumed by many types of vertebrates, including gartersnakes (*Thamnophis* spp.), watersnakes (*Nerodia* spp.), turtles (e.g., Snapping Turtle [*Chelydra serpentina*]), birds, large frogs (e.g., American Bullfrog [*Lithobates catesbeianus*]), and various mammals (Jackson 1961, Pentecost and Vogt 1976, reviewed by Dodd 2013). Bolek and Janovy (2005) also observed a leech (*Desserobdella picta*) feeding on an adult during spring migration to a breeding site in Waukesha County. Eggs and larvae are consumed by invertebrates (such as predaceous diving beetles [*Ilybius* spp.]), fishes, and salamanders (larvae and adults; Grubb 1972, Resetarits and Wilbur 1989, Burkett 1991, Resetarits 1998). Eastern Tiger Salamander (*Ambystoma tigrinum*) larvae are particularly voracious predators of anuran larvae (Burkett 1991). The presence of predators can also influence larval morphology and developmental rates (see below and Reproduction and Development section).

Adult Cope's Gray Treefrogs often avoid threats by remaining motionless so that the camouflage provided by skin color and pattern conceals them (Marchisin and Anderson 1978). If disturbed, they will attempt to climb or jump away from the perceived threat but will also inflate their body to appear larger (Marchisin and Anderson 1978). Some authors suggest that jumping away reveals the bright yellow color on the inner thighs, which can startle predators and provide more time for escape (Cook 1984, Holman 2012). Outside of the breeding season, Cope's Gray Treefrogs reportedly remain near selected refugia, which provide cover for quick evasion of predators (Ritke and Babb 1991). If caught, adults emit skin secretions that are potentially noxious (Dodd 2013). In fact, Vogt (1981) reported that the secretions "are irritating to the mucous membranes of the human eyes and nose" and that handling Cope's Gray Treefrogs led to sneezing, watery eyes, and a runny nose.

Larvae have evolved several strategies for predator avoidance. For example, in the presence of predatory dragonfly naiad (order Odonata), Cope's Gray Treefrog larvae can increase tail depth and enhance the tail's red coloration and black mottling (McCollum and Buskirk 1996). These observable changes are likely to attract predators away from the body and toward their less vital tail fins. Although these changes increase chances of survival when predators are present, in the absence of predators, larvae with altered tail morphology experience higher mortality. This suggests that there is an energy and fitness cost associated with alteration of tail morphology (McCollum and Buskirk 1996). Larvae can further change their behavior when they share aquatic systems with certain predators. Kats et al. (1988) found that chemical cues produced by predatory Green Sunfish (*Lepomis cyanellus*) triggered larvae to spend more time hidden in aquatic refugia.

A summary of Cope's Gray Treefrog parasites in Wisconsin was compiled by Watermolen (2014). This summary includes one leech, two protists, four trematodes, three cestodes, and two nematodes (Bolek and Coggins 1998, Bolek and Janovy 2005). Bolek and Coggins (1998) further found that in southeastern Wisconsin, 78% of individuals examined from

April through June 1996 and 1997 were infected with one or more endoparasites.

CONSERVATION

Status: Globally, Cope's Gray Treefrog is listed as a Least Concern species by the IUCN. It is considered common in Wisconsin by the WDNR. Christoffel et al. (2009) reported that Cope's Gray Treefrogs were "locally common but declining" in the state. However, long-term monitoring data based on male advertisement calls from the Wisconsin Frog and Toad Survey supports that existing populations of this species are stable at the state level (Figure 84).

Populations: Population studies that include estimates of abundance, density, and survival in Wisconsin could not be found. Most information on populations is derived from surveys of calling males in the spring. Trenham et al. (2003) analyzed Cope's Gray Treefrog call data collected through the WDNR's Wisconsin Frog and Toad Survey from 1981 to 1998. They considered Cope's Gray Treefrogs widespread throughout the state, documenting their presence at 34% of all survey sites (i.e., 456 of 1,350 sites surveyed). However, their presence was documented in only one to three out of ten survey years at 46% of the sites where the species was detected. This 18-year dataset also indicated that the presence or absence of Cope's Gray Treefrog populations at a given location was dynamic, with gaps in detection occurring every 3.8 years at 63 sites where this species was confirmed. Although some variation in a population's persistence may be due to reported false absences (e.g., calling males were present but simply not heard by surveyors), the authors suggested that surveyor error was not a likely explanation for the detection frequencies they observed.

Reported sex ratios are often male-biased. For example, in Kansas, sex ratios exceeded 2M:1F on 60 nights through the breeding season (May–July; Godwin and Roble 1983). Across two breeding seasons at a site in Tennessee, Ritke and Semlitsch (1991) reported finding 68 males and 14 females and 203 males and 152 females, respectively. Male-biased groups at breeding sites may reflect the fact that males who spend longer times calling (and thus must remain at breeding sites longer) have higher mating success (Ritke and Semlitsch 1991).

Threats: Cope's Gray Treefrogs are likely impacted by many of the general conservation concerns that affect numerous amphibians in Wisconsin, such as habitat disturbance and pollution (see Introduction to Conservation and Management of Wisconsin's Amphibians and Reptiles chapter). For example, studies support that Cope's Gray Treefrogs are sensitive to urbanization (Pillsbury and Miller 2008, Simon et al. 2009), although there is conflicting information on the impacts of human habitat disturbance. In fact, it could be interpreted that this species is rather tolerant of anthropogenic landscapes, as the frogs are found in residential neighborhoods, in commercial areas, and along highways (Vogt 1981, Harding 1997, Mossman et al. 1998, Simon et al. 2009; Natural History Box: Your Friendly Neighborhood

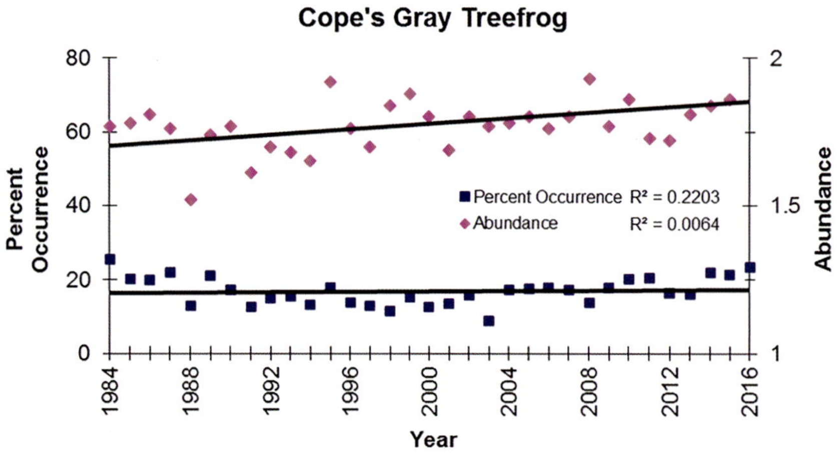

Figure 84. Long-term trend in occurrence (%) and abundance (rank) of Cope's Gray Treefrogs across Wisconsin Frog and Toad Survey statewide monitoring sites. See Introduction to Species Accounts chapter for details on this monitoring program (figure courtesy of WDNR).

NATURAL HISTORY BOX: Your Friendly Neighborhood Treefrog

The ability of Cope's Gray Treefrogs and Gray Treefrogs to occupy human-modified landscapes is a fascinating and puzzling phenomenon. Growing up in the northern suburbs of Minneapolis, Minnesota, I consistently observed treefrogs clinging to the side of my family's home on warm summer nights. I recall our excitement as we would swing open the blinds to our front window and see the belly of a plump treefrog flat against the glass as its fingers appeared to hold it in midair. Returning to my parents' home as an adult, I still find treefrogs hanging on the windows and seeking refuge in the plants and landscaping throughout my mother's garden. In my most surprising treefrog encounter, I tracked down a male calling from the top of a partially filled steeplechase pit at the UW–Stevens Point running track. And now with such rapid communication at our fingertips, I regularly receive pictures of these charming frogs clinging to the homes of family and friends. Human-induced habitat modification is the greatest threat to these treefrogs, but they appear tolerant of many landscape changes. Buildings and homes act as artificial refugia, while house lights and street lamps attract their invertebrate prey items. Small suburban gardens like my mother's appear to provide suitable upland habitat for foraging and overwintering. However, logistical complications impede the ability of scientists to measure treefrog populations in suburbs and cities, making it unclear what environmental factors (natural or artificial) influence persistence of populations throughout developed landscapes. Consequently, scientists are looking for help from 'community science' programs, which provide invaluable information to our understanding of population estimates of suburban-dwelling wildlife and which landscape features are playing a role. Data gathered from community scientists can guide both conservation management and ongoing human development to ensure a more effective coexistence.

Spencer R. Siddons, Purdue University

Treefrog). Paloski (2006) even documented a positive correlation between treefrog presence in Wisconsin and percent of agricultural land use within a 250 m (820 ft) buffer around breeding ponds.

Despite these findings, the negative impacts of human-altered landscapes can be subtle and indirect. For example, higher-quality habitats in disturbed areas are often fragmented by roads or other features typical of degraded landscapes (e.g., agricultural fields). These fragmenting characteristics may not cause immediate mortality, but they can limit the connectivity necessary for treefrogs to meet various life history requirements (e.g., connectivity between uplands used for overwintering and wetlands used for breeding). Along these lines, Pillsbury and Miller (2008) found that Cope's Gray Treefrogs were more frequent in landscapes that possessed unfragmented areas of herbaceous vegetation and were negatively associated with increasing road densities. Furthermore, use of wetlands by Cope's Gray Treefrogs in urban landscapes reportedly declined as the area of impervious surface cover (e.g., concrete or asphalt) increased within 1,500 m (4,921 ft) of wetlands (Simon et al. 2009). Impervious surface coverage in these degraded landscapes can also lead to excessive runoff of harmful contaminants into nearby breeding wetlands. For example, runoff of deicing salts from adjacent roads can increase water conductivity (Simon et al. 2009). High conductivity adversely affected treefrog presence at breeding sites and reportedly resulted in up to 100% mortality of eggs laid in experimental mesocosms (Dobbs et al. 2012). In addition, contamination of breeding wetlands with insecticides is typically more pervasive in degraded vs. undisturbed landscapes.

Exposure to insecticides can significantly reduce the mass of larval Cope's Gray Treefrogs at metamorphosis and lead to slower growth after metamorphosis (Rumschlag and Boone 2015).

Several other conservation concerns likely threaten Cope's Gray Treefrogs in Wisconsin, but these threats have not been thoroughly investigated in the state. Research supports that Cope's Gray Treefrogs are negatively impacted by the presence of fish that prey on amphibian eggs or larvae and avoid breeding in wetlands where fishes occur (Grubb 1972, Binckley and Resetarits 2003). Several investigations have assessed the effects of important disease-causing agents on Cope's Gray Treefrogs, such as the chytrid fungus *Batrachochytrium dendrobatidis* (Bd) and ranaviruses (see Introduction to Conservation and Management of Wisconsin's Amphibians and Reptiles chapter). Although limited, these studies found that the prevalence of Bd infection in sampled treefrogs was relatively low (e.g., Standish et al. 2018, Rothermal et al. 2008) or nonexistent (Sadinski et al. 2010). Others have documented that treefrogs experimentally exposed to Bd experienced very limited disease-related impacts on development or survival (Rumschlag and Boone 2015). The effects of ranaviruses on Cope's Gray Treefrogs reportedly vary. Hoverman et al. (2010) determined that mean mortality rate was 66% for Cope's Gray Treefrog larvae experimentally infected with a ranavirus. Another study found that the prevalence of Gray Treefrog Complex larvae infected with a ranavirus across three ponds in Tennessee ranged from 6% to 43% (O'Bryan et al. 2012). Yet Hoverman et al. (2012) found 0% infection prevalence ($n = 21$ frogs tested) of ranavirus in Cope's Gray Treefrogs sampled in Tennessee, despite infections detected in other amphibian species at the same sites.

Management: Well-vetted management strategies specifically focused on Cope's Gray Treefrogs are limited. Presumably, many general management considerations relevant to anurans in Wisconsin are applicable to Cope's Gray Treefrogs, including control of pollution and protection of suitable habitats (see Introduction to Conservation and Management of Wisconsin's Amphibians and Reptiles chapter). Management activities that focus on protection or restoration of wetlands and maintenance of habitat connectivity that allows unimpeded movement between suitable wetlands and uplands could benefit this species. Landscapes could be managed to preserve numerous wetlands, with undisturbed habitat connections between them, which can facilitate breeding migrations and dispersal of recently metamorphosed juveniles (Pittman et al. 2008).

Management of habitat and hydrology in breeding wetlands must balance this species' association with permanent waterbodies for reproduction (e.g., Knutson et al. 2000) against control of predatory fishes that can persist in permanent water (Grubb 1972). Furthermore, construction of new wetlands is sometimes identified as a possible strategy to offset wetland loss and could benefit Cope's Gray Treefrogs. When compared to older established wetlands, new wetlands often have higher dissolved oxygen, lower conductivity, higher-quality food (phytoplankton, periphyton), and fewer predators and competitors (Wilbur and Collins 1973, Heyer et al. 1975, Wilbur and Alford 1985, Simon et al. 2009), which are all characteristics that could benefit Cope's Gray Treefrogs. This species reportedly bred more often in artificial pools that consistently received new water inputs compared to pools that retained the same water over the course of a breeding season (Pintar and Resetarits 2017b). Furthermore, larvae that developed in recently filled mesocosms experienced greater survival and growth, but they also metamorphosed faster (Wilbur and Alford 1985). Despite this, the true effectiveness of wetland creation as a conservation strategy remains unresolved (see Introduction to Conservation and Management of Wisconsin's Amphibians and Reptiles chapter).

SPECIAL REMARKS

Hybridization between Cope's Gray Treefrogs and Gray Treefrogs is rare (Gerhardt et al. 1994), and the resultant offspring typically experience high mortality (Johnson 1959, 1963, Ralin 1976). Hybridization with other hylid species in Wisconsin, such as Spring Peepers and Boreal Chorus Frogs, is also possible, but reports suggest that offspring of such hybridizations do not develop past larval stages (Moore 1955).

Gray Treefrog

Hyla versicolor (LeConte 1825)

Spencer R. Siddons

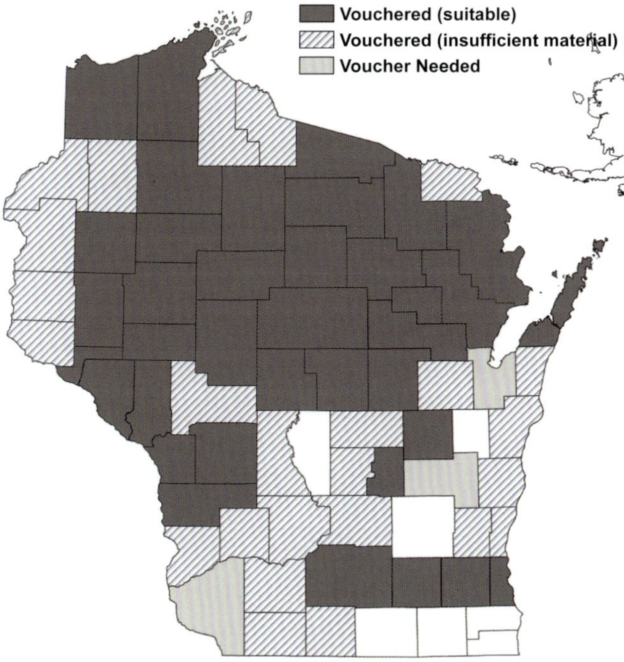

Adult Gray Treefrog (Sauk County; photo by C. Raimond). *Note:* Species identification confirmed via male advertisement call.

AT A GLANCE

Gray Treefrogs are medium-sized, with plump bodies, wide heads, and large toe pads that allow them to climb surfaces (even smooth surfaces such as glass). Their dorsal color ranges from light gray to green, with large darker blotches outlined by thin black lines, and their bellies are generally white or cream. There are darker bands on the outside surfaces of their legs but bright yellow streaks, dappled with dark speckles, along their inner thighs. Juveniles that have recently metamorphosed are usually solid green but with white bellies. Gray Treefrogs are reported throughout much of Wisconsin. Adults and juveniles are terrestrial outside of the breeding season and often associated with woodlands or forested habitat. They occupy dry to mesic hardwood forests and floodplain forests throughout Wisconsin and are mostly arboreal. Gray Treefrogs mate from early May through June, soon after emergence from hibernation, in open fishless wetlands adjacent to wooded habitats. They are very difficult to distinguish from the closely related Cope's Gray Treefrog in Wisconsin. The male advertisement call (which is the easiest way to differentiate the two species) consists of a melodic slow trill that is birdlike in quality, compared to the faster and more insect-like trill of the Cope's Gray Treefrog. Larvae remain in breeding wetlands until metamorphosis, which occurs from June through August in Wisconsin. Adults consume invertebrates, especially insects, while larvae mostly feed on algae and other microorganisms scraped from aquatic surfaces. Adults are preyed upon by numerous vertebrates, including snakes, birds, and mammals; predators of larvae also include large aquatic insects and fishes. In October, Gray Treefrogs move back to their overwintering locations. They hibernate terrestrially below moist ground debris. During winter, they tolerate subzero temperatures via the production of metabolic compounds that protect their body cells against damage from freezing. Gray Treefrogs are currently considered common by the WDNR but are negatively impacted by habitat loss and pollution.

SYSTEMATICS

Like many treefrogs (family Hylidae) in North America, the Gray Treefrog has remained in the monophyletic genus *Hyla* for several decades (Faivovich et al. 2005, Crother 2017). Duellman et al. (2016) suggested that North American and East Asian hylids be placed into the resurrected genus *Dryophytes* (Fitzinger 1843), but this taxonomic revision is not currently accepted (Crother 2017). There are ten species of *Hyla* documented in the US (Crother 2017), but only *H. chrysoscelis* (Cope's Gray Treefrog) and *H. versicolor* occur in Wisconsin. No recognized subspecies of *H. versicolor* currently exist (Crother 2017). Hybridization with *H. chrysoscelis* has been documented (Gerhardt et al. 1994), as well as with other *Hyla* species not found in Wisconsin (e.g., Pine Woods Treefrog [*H. femoralis*] and Bird-voiced Treefrog [*H. avivoca*]; Mecham 1960, Pyburn 1960, Jensen et al. 2008). Both *H. versicolor* and *H. chrysoscelis* occur in the same family as resident Wisconsin species in the genera *Acris* and *Pseudacris* (Hylidae).

DESCRIPTION

Adults: Gray Treefrogs have wide heads and somewhat bumpy dorsal surfaces (Vogt 1981). Dorsal ground color varies from light gray or brown to green, often overlaid with darker blotches outlined in thin black lines (Vogt 1981; Figures 85 and 86). A white or light-colored patch (i.e., lacrimal spot) often occurs immediately below each eye (Figure 85), while the ventral surface is cream or white. The outer surfaces of the legs usually have wide darker cross bands, while the inner thighs possess yellow streaks covered in darker speckles (Vogt 1981; Figure 87). There are large toe pads at the end of each digit and limited webbing between toes, although webbing is more obvious on the hind feet (Vogt 1981). Gray Treefrogs are a medium-sized anuran in Wisconsin but the largest hylid species in the state (Vogt 1981). Adult male SVL ranges from ca. 4.0 to 5.0 cm (1.6 to 2.0 in; Vogt 1981). Jaslow and Vogt (1977) reported that the mean SVL of calling males from Wisconsin was 4.3 cm (1.7 in). The SVL of adult females in Ohio ranged from 4.9 to 5.3 cm (1.9 to 2.1 in; Gatz 1981b).

Above, top: Figure 85. Adult Gray Treefrog with brown and gray dorsal coloration (Oneida County; photo by D. Brown). *Note:* Species identification not confirmed, but the individual was observed in a location with suitable habitat and where only Gray Treefrogs are documented.

Above, bottom: Figure 86. Adult Gray Treefrog with green dorsal coloration (Forest County; photo by D. Donner). *Note:* Species identification not confirmed, but the individual was observed in a location with suitable habitat and where only Gray Treefrogs are documented.

Figure 87. Yellow markings on the inner thighs of Gray Treefrogs (Forest County; photo by D. Brown). *Note:* Species identification not confirmed, but the individual was observed in a location with suitable habitat and where only Gray Treefrogs are documented.

Preadult stages: Individual eggs are darkly pigmented and range in size from 0.11 to 0.12 cm (0.04 to 0.05 in) while possessing two clear gelatinous layers that are 0.4–0.6 cm (0.16–0.24 in) in overall diameter (Watermolen 1995). At hatching, larvae are 0.6–0.8 cm TL (0.24–0.31 in; Minton 2001) and can reach TL of 3.2–3.8 cm (1.3–1.5 in; Parmelee et al. 2002). Like many hylids, larvae possess laterally (rather than dorsally) positioned eyes and have mostly dark green to light yellow-green bodies with gold flecks and white bellies (Vogt 1981). Larvae also have deep fan-shaped tail fins, which are often pigmented with red, orange, or yellow and covered in dark mottling (Vogt 1981). Recently metamorphosed juveniles are solid green dorsally and white ventrally (Vogt 1981), with an SVL that ranges from 1.36 to 2.0 cm (0.54 to 0.79 in) and a mean weight of 0.39 g (0.02 oz; Wright 1914, Grant and Licht 1995). In Wisconsin, reported range of juvenile SVLs from samples collected in Juneau County were 1.4–2.5 cm (0.55–0.98 in; Roble 1979).

Variation: The dorsal color of adults varies from gray or brown to green (see Adults subsection). There is also some variation in appearance by sex, as males typically have gray vocal sacs, whereas female throats are white (Wright 1932, Dodd 2013). Thomas and Höbel (2014) found that in Wisconsin the lacrimal spot (the white patch below each eye) was significantly larger in females than males. Adult females are also reportedly larger in body size than males (Vogt 1981, Matson 1990a).

Diagnosis: Gray Treefrogs do not possess parotoid glands, warty tubercules, or keratinized metatarsal spades on their hind feet. Their terminal toe pads are wider than the digit tips, and their inner thighs have a yellow coloration. Gray Treefrogs cannot be reliably distinguished from Cope's Gray Treefrogs using only external morphological features. The most reliable technique for differentiating them in the wild is by comparing male advertisement calls. Gray Treefrog calls have a slower and more melodic trill compared to the faster and sharper trill of Cope's Gray Treefrogs (Vogt 1981; see Reproduction and Development section). Furthermore, the mean pulse rates of male advertisement calls do not overlap between these two species in Wisconsin, with Gray Treefrog pulse rates being slower than those of Cope's Gray Treefrogs (see Cope's Gray Treefrog, *Hyla chrysoscelis* species account). Adult Gray Treefrogs are also typically larger in SVL (4.3 cm [1.69 in]) than adult Cope's Gray Treefrogs (3.5 cm [1.4 in]) in Wisconsin (Jaslow and Vogt 1977, Vogt and Hine 1977, Vogt 1981). In the absence of auditory data from calling males, definitive identification requires comparison of ploidy or histological investigations. Adult Gray Treefrogs could further be confused with other members of the family Hylidae in Wisconsin, such as Blanchard's Cricket Frogs (*Acris blanchardi*), Boreal Chorus Frogs (*Pseudacris maculata*), and Spring Peepers (*P. crucifer*). However, these other hylids are much smaller as adults (SVL and mass) than Gray Treefrogs, lack a white patch under their eyes, and do not have enlarged toe pads (Vogt 1981).

Treefrog eggs are laid in small clusters and not in long strings, large globular masses, or sheets at the water's surface, so they can be distinguished from eggs of anurans in the genera *Anaxyrus* and *Lithobates*. Individual eggs are surrounded by two gelatinous envelopes, which is unlike eggs of anurans in the genus *Pseudacris*. The diameter of treefrog eggs, excluding the outer envelope, is narrower (0.1–0.12 cm [0.04–0.05 in]) than that of Blanchard's Cricket Frogs (0.2–0.3 cm [0.08–0.12 in]; Watermolen 1995). Much like with adults, the eggs and larvae of Gray Treefrogs are difficult to distinguish from Cope's Gray Treefrogs, and reliable characteristics for differentiating these in the wild are not known (Vogt 1981). The larvae of both hylid species in Wisconsin have internal gills and lack forelimbs. They have laterally positioned eyes, a whitish belly, and a tail fin that is red or orange but mottled with black, white, or light cream. Treefrog larvae generally have more laterally positioned eyes and deeper tail fins than larvae of anurans in the genera *Anaxyrus* and *Lithobates*. They can also be differentiated from larvae of anurans in the genera *Acris* and *Pseudacris*, which do not possess extensive dark mottling and red to yellow pigmentation on their tail fins.

DISTRIBUTION AND HABITAT

Global distribution: Gray Treefrogs are found throughout much of the northeastern US and parts of southern Canada. Their northern range extends from western New Brunswick and eastern Maine west across southern Quebec and Ontario into Manitoba and a small region of eastern Saskatchewan. They occur throughout the northeastern US, south through central Virginia and extreme northern North Carolina, and into the Midwest through the northern two-thirds of Ohio, Indiana, and Illinois, where their range turns south along a band that covers portions of Missouri, northern Arkansas, eastern Oklahoma, and eastern Texas. They are largely absent from the southeastern US, although small pockets of occurrence exist in Arkansas, Tennessee, and Louisiana.

Distribution in Wisconsin: Published Gray Treefrog county records exist across much of Wisconsin, although a handful of these records lack any type of voucher material (see the Appendix). Furthermore, not all published county records are accompanied by voucher material that permits independent confirmation of species identity (e.g., recordings of male advertisement calls, sonogram files; see the Appendix). There is reportedly definitive documentation of Gray Treefrogs on five islands of the Apostle Islands National Lakeshore (Long, Outer, Rocky, Sand, and Stockton, Ashland and Bayfield Counties) and possibly islands off of Door County in Green Bay (Long and Long 1976, Casper 2008c). It is noteworthy that the current voucher specimen reported for Brown County (UWZM 22612; Cochran et al. 1987) was identified as a Cope's Gray Treefrog, according to museum records (Emily Halverson, UWZM, personal communication), and this discrepancy supports that a new voucher is required for the county. As with Cope's Gray Treefrogs, additional sampling and collection of appropriate voucher material, even in already documented counties that lack suitable museum vouchers, are necessary to conclusively delineate the range of this species in Wisconsin.

Localities of possible occurrence: Gray Treefrogs could occur in several undocumented counties that are surrounded by those with published records, such as Adams and Calumet. Anecdotal reports collected during the WDNR's Wisconsin Frog and Toad Survey since 2005 have identified calling males in all Wisconsin counties that currently lack published records (WIATRI 2016). As such, confirmed records of Gray Treefrogs could eventually be published for all counties in the state with additional survey effort and suitable documentation.

Habitat: There is mostly anecdotal information available on habitats used by Gray Treefrogs outside of the breeding season in Wisconsin. In general, this species associates with forested habitats, such as northern and southern hardwood forests (dry and mesic), but also floodplain forests (Vogt 1981; see Amphibians and Reptiles in Wisconsin Plant Communities chapter). Research in the Upper Midwest found a positive relationship between the occurrence of Gray Treefrogs and diverse landscapes with mosaics of deciduous, evergreen, and mixed forests but also wetland-forest and shrub habitats (Knutson et al. 2000, 2004). Although they frequently use forests, Gray Treefrogs sometimes prefer those that are lightly wooded. Eigenbrod et al. (2011) reported that Gray Treefrogs most often use landscapes with 20%–30% forest cover and suggested that their abundances could actually decrease as forest cover nears 100%. The integration of suitable wetlands into wooded landscapes is also important to this species. In Iowa, for example, Gray Treefrogs were associated with well-established clustered wetlands surrounded by woody vegetation (Pillsbury and Miller 2008). Similarly, adult Gray Treefrog populations in New Hampshire were most dense when breeding wetlands were surrounded by 61%–80% forest cover within a 1,500 m (4,921 ft) radius (Herrmann et al. 2005).

When in wooded uplands, Gray Treefrogs spend substantial time exploiting arboreal microhabitats, aided by adhesive toe pads, which facilitate climbing (see description of treefrog toe pads in the Cope's Gray Treefrog, *Hyla chrysoscelis* account). For example, Vogt (1981) reported that they forage at heights of up to 10 m (33 ft) in trees and shrubs in Wisconsin, while Laughlin et al. (2017) documented them in trees at heights of up to 20.5 m (67.3 ft) in the northern part of the state. Yet the selection of arboreal microhabitats and refugia appears to be influenced by environmental temperature and rainfall or humidity. Johnson et al. (2008) determined that radio-tracked individuals occupied lower refugia more often during dry hot conditions and likely moved higher into tree canopies under moist or humid conditions.

Although the nonbreeding season is spent in wooded uplands, Gray Treefrogs breed in permanent, semipermanent, or ephemeral wetlands, highlighting that they require access to suitable terrestrial and aquatic habitats (Johnson 2005). In fact, Trenham et al. (2003) determined that calling males in Wisconsin were positively associated with more open "scrub or shrub wetland patches" and negatively associated with "forested wetland patches." Despite the fact that Gray Treefrogs are often found in forested landscapes, breeding wetlands typically possess little tree cover. Kutka (1990) determined that the percent canopy cover of the most commonly occupied breeding ponds in northern Wisconsin averaged 29.5%. Furthermore, Gray Treefrog populations in Michigan were documented at 25 out of 26 fully or partially open-canopy ponds surveyed but only occurred at 1 out of 11 ponds with closed tree canopies (Skelly et al. 1999). They appear to tolerate breeding habitats with moderately acidic water. For example, in northern Minnesota, breeding adult Gray Treefrogs were found in fen habitats where water pH ranged from 4.4 to 5.0 but completely avoided slightly more acidic bog habitats (e.g., pH 4.0–4.1; Karns 1992).

Larval Gray Treefrogs occupy the same aquatic habitats that adults use for breeding (Skelly et al. 1999, Altig and McDiarmid 2015). Several studies report that larvae further select specific microhabitats within breeding wetlands, but this can be affected by the presence of predators (see also Reproduction and Development and Prey and Predators sections).

After metamorphosis, juveniles move into adjacent upland habitats. Roble (1979) determined that juveniles from Wisconsin remain on herbaceous vegetation near breeding wetlands (e.g., *Carex* spp.) at heights of 20–50 cm (7.9–19.7 in) until metamorphosis is complete and then disperse into the surrounding landscape days to a week later.

Gray Treefrogs overwinter terrestrially in the woodland habitats that are occupied during the nonbreeding season (Johnson et al. 2005, 2007). These upland overwintering sites appear to be randomly distributed around breeding ponds at distances of up to 340 m (1,115.5 ft) away (Johnson et al. 2007a). Johnson et al. (2008) reported that radio-tracked individuals from Missouri overwintered in "the soil beneath several inches of leaf litter." Because they overwinter in terrestrial habitats, winter survival requires physiological adaptations to tolerate freezing conditions (see Activity section). Information on the habitat that juveniles use for hibernation is sparse. According to Fitch (1958a), recently metamorphosed individuals sometimes remain near or within natal ponds as temperatures become cold in autumn, which suggests they overwinter near wetland peripheries.

REPRODUCTION AND DEVELOPMENT

In Wisconsin, Gray Treefrogs typically mate from May through June, although calling is driven by environmental temperature, so males sometimes call from April to early July (Briggs and Young 1976, Jaslow and Vogt 1977, Roble 1979, Engel 1980, Vogt 1981, Mossman et al. 1998). Similarly, adults in a Rhode Island study arrived at breeding ponds from early April through late June, although arrival by late May was more typical (Paton et al. 2000, Paton and Crouch 2002). Christoffel et al. (2009) mentioned that late-season calls are often heard on warm, humid, or rainy nights in Wisconsin but indicated that such "rain calls" did not lead to actual reproduction. Males typically initiate calling at dusk and continue for up to four hours per night (Dodd 2013). Calling activity recorded in Ontario across three breeding seasons occurred at air temperatures as low as 8°C (46°F) and as high as nearly 30°C (86°F; Bertram and Berrill 1997), which is presumably similar to temperature thresholds for calling activity in Wisconsin.

Males emit advertisement calls with the aid of a single vocal sac in the throat region. Advertisement calls are melodic and almost birdlike in quality. Calls are a series of repeated trills that are slow (compared to male Cope's Gray Treefrogs) with pulse rates that range from 14.8 to 29.0 pulses/sec at 18°C–26°C (64°F–79°F; Vogt 1981) and typically last one to three seconds (Anderson et al. 1981). Several studies have reported that call pulse rates increase with rising air temperatures (Blair 1958, Brown and Brown 1972, Gayou 1984). Gayou (1984) also found that intercall intervals (i.e., time between calls) shortened from 7.4 sec at 13.6°C (34°F) to 1.3 sec at 25°C (77°F), with the mean decrease in intercall interval at 0.30 sec per 1°C (34°F) increase. The number of calling males per breeding pond can vary widely. In Indiana, nightly choruses of zero to eight males were documented over a 30-day period (Brodman and Kilmurry 1998). A three-year study in Ontario recorded 1–42 calling males per pond in a single night, with yearly means of 10, 12, and 22 calling males per pond per night (Bertram and Berrill 1997). Males can mate multiple times in a single breeding season. For example, males in a Maine population mated zero to four times per year (Sullivan and Hinshaw 1992).

Males call from perches on vegetation at heights up to 1 m (3.3 ft) above the water or ground (Gerhardt 1978, Vogt 1981). Research from Wisconsin found that males selected calling locations based primarily on environmental temperature, even over the presence of potential competitors and predators (Höbel and Barta 2014). Because environmental temperatures influence both body temperature and male call characteristics, the authors suggested that the selected calling perches were associated with optimal environmental temperature and, thus, enhanced success in mate attraction (Höbel and Barta 2014). Calling males are territorial and will aggressively defend their perches against intruders (Dodd 2013). When a subordinate male moves near a dominant calling male, both will emit back-and-forth encounter calls for less than two minutes (Overton 1914). If the subordinate male does not leave, bouts of head-butting, kicking, shoving, or jumping on the intruder ensue for 30–90 sec until one individual yields (Dodd 2013). Subordinate males (sometimes called 'satellite males') will often remain silent in a low posture near calling males and intercept approaching females that are attracted to the call of a dominant male (Sullivan and Hinshaw 1992). This behavior reduces the potential for direct competition among males while providing reproductive opportunities for subordinate males (Roble 1985). Although environmental conditions are apparently the most important factor influencing the selection of calling perches, the potential for intra- or interspecific male competition also plays a role. For example, when both Gray Treefrogs and Cope's Gray Treefrogs are calling at the same wetland, male Gray Treefrogs typically call at heights of ≥50 cm (19.7 in; Ptacek 1992). However, when Cope's Gray Treefrogs are absent, Gray Treefrogs will most commonly call from perches at heights of ≤50 cm (19.7 in; Ptacek 1992). Larvae of other anuran species also appear to influence where Gray Treefrogs breed. In Missouri, Gray Treefrog reproduction was rare at breeding sites where Eastern Cricket Frogs (*Acris crepitans*), Northern Leopard Frogs (*Lithobates pipiens*), and American Bullfrogs (*L. catesbeianus*) were also present, potentially due to overlap in larval diet (Seale 1980).

Females select males by assessing call characteristics, and prefer males with calls that are longer in duration but emitted at quicker rates (Sullivan and Hinshaw 1992, Gerhardt et al. 1996, 2000, Schwartz et al. 2001). Males with preferred advertisement calls produce offspring that are larger and experience higher survival, which could explain why females select them for breeding (Welch et al. 1998). However, background noise, such as the advertisement calls of conspecific males and even other frog species, potentially interferes with female reception and assessment of calls (Schwartz et al. 2001). In addition to call characteristics, females are attracted to males that have selected easily accessible calling perches, such as those with little vegetation surrounding the perch site (Fellers 1979b). Females reportedly consider male perches that are easily accessible to be of higher quality, as they provide more space for amplexus and more clearly transmit male calls (Littlejohn 1958).

The female initiates contact once a male is selected for reproduction, which prompts him to grasp her from behind around the armpits to assume the mating position known as amplexus (Fellers 1979a, Gatz 1981b). Amplexus typically lasts for three to five hours before amplectant pairs move into the water, where females release eggs, which males fertilize externally (Fellers 1979b). Annually, individual females can produce 1,000 to >2,000 eggs, which are separated among numerous loose clusters of 10–50 eggs that float at the water's surface or are attached to aquatic vegetation (e.g., Wright 1914, Vogt 1981, Johnson and Semlitsch 2003). Females are able to reproduce multiple times per breeding season (Ritke et al. 1990, Johnson and Semlitsch 2003).

Eggs can hatch in as few as three to six days but can take up to two to three weeks, depending on environmental temperature (Vogt 1981, Gosner and Black 1957b). Larval habitat structure, available feeding substrates, competition with other anuran larvae, and the presence of predators can influence developmental rates (reviewed by Dodd 2013; see Prey and Predators section). Under natural conditions, complete metamorphosis of larvae can occur by 40–60 days after hatching (Gosner and Black 1957b). In Wisconsin, recently metamorphosed individuals disperse from wetlands into surrounding uplands from mid-July to mid-August (Roble 1979), although emergence and dispersal can occur later in other states (Paton and Crouch 2002). Prior to entering hibernation in their first year, young Gray Treefrogs can attain an SUL of ca. 2.0–3.0 cm (0.79–1.18 in) and reportedly reach full size and sexual maturity by two years of age (Wright 1932, Babbitt 1937). Wright (1932) reported that the SUL of second-year individuals ranged from 3.0 to 4.1 cm (1.18 to 1.61 in), but after two years the SUL ranged from 4.1 to 5.1 cm (1.61 to 2.01 in). Information on longevity in the wild is sparse, but a captive individual (identified as a member of the Gray Treefrog Complex) reportedly survived for 7.8 years (Snider and Bowler 1992).

ACTIVITY

Gray Treefrogs emerge from hibernation by mid-April to early May. Anecdotal reports suggest that emergence is sometimes correlated with increasing air temperatures and warm spring rains (Christoffel et al. 2009). After emergence, Gray Treefrogs briefly forage in surrounding terrestrial habitats to build energy reserves and then migrate to breeding sites (Oldfield and Moriarity 1994). Gray Treefrog activity levels throughout spring and summer are also positively influenced by humidity (Johnson 2005) and environmental temperature. In New Jersey, for example, Gray Treefrogs were typically active when air temperatures were between 15°C and 34°C (59°F and 93°F) but had a maximum temperature threshold of 38.7°C (101.7°F; John-Alder et al. 1988).

Greatest movements are associated with adult breeding migrations and juvenile dispersal from wetlands after metamorphosis. Adults in Missouri undertook yearly migrations of up to 332 m (1,089 ft) between overwintering upland habitat and wetland habitat used for breeding and on average traveled a maximum distance of 172 m (557 ft) from the wetland edge after breeding (Johnson et al. 2007a). Differences in migration distances and spatial distributions can occur between sexes. In Missouri, females were spatially distributed around breeding ponds at similar distances during both the breeding and nonbreeding seasons, but males were closer to the pond during the breeding and nonbreeding seasons (Johnson et al. 2007a). Outside the breeding season, Gray Treefrogs move sparingly. Johnson et al. (2008) reported that the mean area in which radio-tracked individuals traveled was 530 m^2 (1,739 ft^2). This study also found that "daily activity appeared to be limited to several adjacent trees" in close proximity to overwintering locations and reported that individual mean distance per move was only 6 m (20 ft). Gray Treefrogs apparently exhibit a strong fidelity for specific refuges, which Johnson (2005) found that they returned to for up to six consecutive seasons. Jaeger and Hailman (1973) suggested that Gray Treefrogs are photopositive, using both sunlight and moonlight to forage and move about their habitat. They also appear to orient their movements toward open areas above ponds (i.e., areas of high illumination) because they are sensitive to blue-spectrum light (Hailman and Jaeger 1974).

Larval and juvenile movement patterns are not well-studied in Wisconsin, but research from elsewhere supports that larval activity is influenced by the presence of predatory Eastern Tiger Salamanders (*Ambystoma tigrinum*; Skelly 1992). Recently metamorphosed juveniles are primarily diurnal and active from July to September (Roble 1979). These juveniles typically disperse from wetlands within a week of metamorphosis in Wisconsin, moving a minimum distance of 0–7.6 m (0–25 ft) per day ($\bar{x}$ = 1.6 m [5.2 ft]; Roble 1979), but many do not travel far from wetlands. In Missouri, over 80% of juveniles were found within 35 m (115 ft) of breeding ponds across three years of sampling, and juvenile abundance decreased with increasing distance from breeding ponds (Johnson et al. 2007a). Interestingly, Roble (1979) suggested that juvenile Gray Treefrogs do not ascend into trees during their first year and remain on the forest floor through winter.

In more northern locales, Gray Treefrogs can be active until October (reviewed by Dodd 2013). Because Gray Treefrogs hibernate terrestrially in Wisconsin, they are exposed to freezing conditions and survive by synthesizing cryoprotectant metabolites, which decrease the freezing point of body fluids, thus protecting against cellular damage inflicted by ice crystals (Storey 1997, Layne and Lee 1989, Irwin and Lee 2003). The most abundant cryoprotectant produced by adults is glycerol, while juveniles produce large amounts of both glycerol and glucose (Schmid 1982, Storey and Storey 1986). Glucose accumulates as temperatures cool in preparation for cellular freezing, while the majority of glycerol is produced immediately after internal ice develops in the body (Layne and Lee 1989, Zimmerman 2007). Although overwintering treefrogs are sometimes exposed to freezing temperatures, leaf litter and snow cover can insulate chosen hibernacula so that conditions inside are occasionally warmer than the air aboveground. As such, hibernating treefrogs probably experience a series of freeze-thaw cycles throughout the winter. During the winter in Minnesota, the lowest soil surface temperature associated with overwintering frogs was −7.2°C (19°F; Schmid 1982). During laboratory experiments, Gray Treefrogs survived freezing at −2.5°C (27.5°F) for 12 hours and 36 hours, becoming active after thawing for 2 hours and 4–5 hours, respectively (Storey and Storey 1987). Layne and Lee (1989) determined that Gray Treefrogs survived freezing at −3°C (27°F) and withstood up to 50% of their body water being frozen. Individuals frozen to lower temperatures experienced 100% survival at −5.5°C (22.1°F, $n = 5$) but 29% survival at −6.5°C (20.3°F, $n = 2$ of 7), while only a single individual frozen at −7.5°C (18.5°F) did not survive (Irwin and Lee 2003). Gray Treefrogs frozen at −5.5°C (22.1°F) and −6.5°C (20.3°F) required ca. 110 hours to regain righting reflex (Irwin and Lee 2003).

PREY AND PREDATORS

Adult Gray Treefrogs forage at night and often consume invertebrates that occupy arboreal locations (Mahan and Johnson 2007). Vogt (1981) found that the stomach contents of 87 calling males collected in Wisconsin contained mostly beetles (particularly the family Elateridae) and various caterpillars, as well as numerous ants, spiders, and harvestmen. Although consumption of vertebrates is likely rare, Puckette (1962) documented ophiophagy in wild-caught Gray Treefrogs from Illinois that were reared in captivity. An adult Gray Treefrog measuring ca. 5.0 cm (2.0 in) in SVL reportedly ingested a 32.6 cm (12.8 in) TL Ring-necked Snake (*Diadophis punctatus*) and a deceased Common Gartersnake (*Thamnophis sirtalis*) neonate that was 15.2 cm TL (6.0 in; Puckette 1962). Stomach content mass did not differ between the males and females from a study population in Missouri, but stomach content mass for both sexes increased with distance from breeding ponds (Mahan and Johnson 2007). The authors suggested that adults do not forage extensively at breeding ponds because better feeding opportunities exist farther upland, which underscores the importance of adjacent terrestrial habitat to this species.

Unlike the mostly insectivorous adults, larvae feed on algae, fungi, detritus, and bacteria that are scraped away from plant material or submersed substrates in breeding ponds (Hoff et al. 1999, Dodd 2013). Larvae also reportedly filter particulate matter out of the water column (Hoff et al. 1999). After metamorphosis, juveniles transition to an insectivorous diet, although their primary prey differs from adults. Vogt (1981) determined that the stomach contents of 93 juveniles from Wisconsin contained a large volume of beetle larvae (family Carabidae), march flies (family Bibionidae), fly larvae, harvestmen, and leafhoppers.

Adults and metamorphosed juveniles are consumed by various herpetofaunal species, including gartersnakes (*Thamnophis* spp.), Snapping Turtles (*Chelydra serpentina*), watersnakes (*Nerodia* spp.), and American Bullfrogs (*Lithobates catesbeianus*; Dodd 2013). Numerous mammals and birds also prey upon Gray Treefrogs, such as Raccoons (*Procyon lotor*), Striped Skunks (*Mephitis mephitis*), herons (family Ardeidae), hawks (*Buteo* spp.), and owls (*Otus* spp.; Jackson 1961, Hamerstrom 1983, Cline 2005, Dodd 2013). Known predators of eggs and larvae include Mudminnows (*Umbra limi*), several species of sunfish (e.g., *Lepomis gibbosus, L. macrochirus, L. cyanellus, Pomoxis nigromaculatus*), Eastern Newts (*Notophthalmus viridescens*), Eastern Tiger Salamander larvae (*Ambystoma tigrinum*), dragonfly naiads (*Anax* spp.), and water bugs (*Belostoma* spp.; Morin 1987, Hecnar and M'Closkey 1997b, Smith et al. 1999, Relyea 2001a). Predators can impact larvae beyond causing direct mortality, and several studies have documented these sublethal effects. For example, the presence of Eastern Newts in wetlands can not only induce larvae to develop faster and complete metamorphosis earlier but also reduce larval survivorship (Morin 1987). Furthermore, Relyea (2001a) found that Gray Treefrog larvae experienced increased rates of growth and development when predators were present, such as Mudminnows, dragonfly naiads, Eastern Tiger Salamander larvae, and water bugs. These enhanced developmental rates apparently impact the maximum size larvae attain. Skelly (1992) determined that Gray Treefrog larvae that developed in the presence of larval Eastern Tiger Salamanders were 54% lighter than larvae that did not share habitat with this predator. On the other hand, Morin (1987) reported that larvae were heavier at metamorphosis when they shared habitat with predatory Eastern Newts. The increased mass could enhance their chance of survival, as larger individuals are less likely to fall victim to predators.

To avoid detection by perceived threats, adult Gray Treefrogs typically remain motionless and can slightly modify their skin color to blend in with their surroundings (Roble 1979). They will also jump or climb vegetation to escape potential danger. Jumping reveals the bright yellow streaks on the inner thighs, which could startle or confuse predators and provide an opportunity for escape (Dodd 2013). Adults can apparently release noxious skin secretions (Dodd 2013), which make them unpalatable to predators such as Northern Short-tailed Shrews (*Blarina brevicauda*; Brodie and Formanowicz 1981a). In addition, Gray Treefrogs confronted by predators can exhibit defensive postures that involve flexing their head and orientating their body to extend their hind legs so that the bright yellow patches on their thighs are revealed (Brodie Formanowicz 1981). Field observations of death-feigning were documented in Michigan (Banta and Carl 1967). Gray Treefrogs can also emit a single explosive barking "release call" when bitten or attacked (Brodie and Formanowicz 1981a). This startling call could surprise would-be predators, allowing the frog an opportunity to escape.

There are several adaptations that larvae use to avoid predators. These include morphological changes when predators are present, such as development of shallower bodies, shorter bodies, longer tails, and deeper tails. Larval tail fins can also develop a bright reddish-orange color in response to predator presence, which could attract predators toward the tail fin and away from the more vital body (McCollum 1994, Smith and Van Buskirk 1995, Hecnar and M'Closkey 1997b). In addition to alterations in morphology, larval Gray Treefrogs can avoid predators by modifications to their behavior. For example, Gray Treefrog larvae avoided areas in experimental tanks where predators congregated and reduced overall activity levels in the presence of larval Eastern Tiger Salamanders, Mudminnows, and dragonfly naiads (Skelly 1992, Relyea 2001a).

Information on Gray Treefrog parasites in Wisconsin is sparse (Watermolen 2014), although rangewide reports are summarized by Dodd (2013). Parasites can have substantial (and surprising) impacts on the ecology of Gray Treefrogs. An experimental study found that reproductive success and selection of oviposition sites were influenced by the presence of American Ribbed Fluke Snails (*Pseudosuccinea columella*) that acted as intermediate hosts for trematodes that parasitize Gray Treefrogs (Kiesecker and Skelly 2000). Other studies, however, support that interactions between Gray Treefrogs and their parasites are complicated and sometimes counterintuitive. Hausfater et al. (1990) found no relationship between infection load of nematodes (*Dorylaimus, Physaloptera, Cosmocercela,* and *Polystoma* spp.) and trematodes (Digeneans) with duration or energetic cost of male advertisement calls. In fact, males in amplexus had a higher infection prevalence than unmated males (Hausfater et al. 1990). Wuerthner et al. (2017) determined that Gray Treefrog larvae infected first with trematode parasites and then exposed to a ranavirus carried a reduced viral load and experienced greater survival than larvae not first infected with trematodes.

CONSERVATION

Status: Globally, the Gray Treefrog is listed as a Least Concern species by the IUCN. It is considered common in Wisconsin by the WDNR. Several authors have reported that Gray Treefrogs were common and that their populations were stable or increasing (Mossman et al. 1998, Christoffel et al. 2009). Long-term call survey data from the Wisconsin Frog and Toad Survey suggest that Gray Treefrogs are stable or slightly increasing at the state level (Figure 88).

Populations: Data on populations, including estimates of abundance and density, are apparently sparse. Most available population information is inferred from data collected during surveys that record the intensity or prevalence of male advertisement calls. Trenham et al. (2003) analyzed call survey data of Gray Treefrogs collected through the WDNR Wisconsin Frog and Toad Survey from 1981 to 1998. They documented Gray Treefrog presence at 76% of all survey sites ($n = 1{,}032$ out of 1,350 total sites), which supported the conclusion that this species was widespread throughout the state. Investigation of their 18-year data set revealed that Gray Treefrog presence was more consistent than all other anuran species detected, with gaps in detection occurring on average only every 1.5 years at 83 of the sites where this species was confirmed. In addition, Gray Treefrogs were detected in at least seven of ten years at nearly 70% of sites where their presence was confirmed. When Gray Treefrogs were heard calling, their mean annual abundance rank typically ranged from 2 to 2.5 (out of 3; see Figure 88). Active breeding sites are dominated by male treefrogs, likely because males increase their chance of mating by remaining at wetlands for longer durations than females (Sullivan and Hinshaw 1992). As expected,

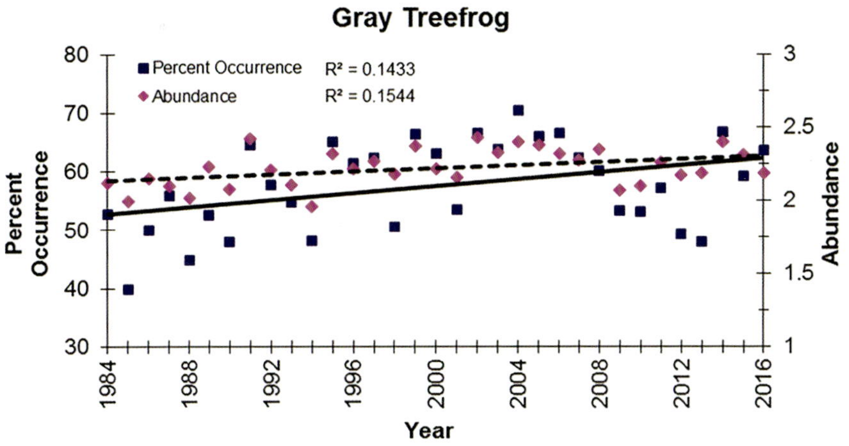

Figure 88. Long-term trend in occurrence (%) and abundance (rank) of Gray Treefrogs across Wisconsin Frog and Toad Survey statewide monitoring sites. See Introduction to Species Accounts chapter for details on this monitoring program (figure courtesy of WDNR).

a study site in Ontario surveyed for three breeding seasons found that sex ratios were heavily male-biased in each year (7.7M:1F, 7.5M:1F, and 3.4M:1F; Bertram and Berrill 1997). Population data on larvae are also uncommon in the literature, although Seale (1980) estimated that the number of larvae in a permanent Missouri pond in June was 3,540.

Threats: Gray Treefrogs are impacted by numerous threats that affect many amphibian species in Wisconsin (see Introduction to Conservation and Management of Wisconsin's Amphibians and Reptiles chapter). For example, they appear to have a low tolerance for urban development (Lehtinen et al. 1999, Pillsbury and Miller 2008). As a result, they are often absent from highly urbanized areas in Wisconsin (Mossman and Hine 1984, Mossman et al. 1998) and are negatively associated with urban edges and other anthropogenic features (e.g., suburban residences, roadways, and commercial and industrial areas; Knutson et al. 2000). There are many potential reasons for this negative interaction. Excessive runoff from impervious surfaces in human-altered landscapes contributes pollutants to potential breeding sites and reduces the likelihood of Gray Treefrog occurrence (Simon et al. 2009).

Gray Treefrogs are also sensitive to habitat fragmentation, which is more prevalent in urban landscapes. Gagne and Fahrig (2007) reported that significantly fewer Gray Treefrogs occurred in urban landscapes that were farther from the nearest forest edge (146.52 m [480.71 ft]) and wetland edge (879.07 m [2,884.09 ft]) than in less disturbed landscapes (9.87 m [32.38 ft] and 467.45 m [1,861.71 ft], respectively). Habitat fragmentation is frequently attributed to roads, and thus it is not surprising that Gray Treefrogs are negatively correlated with increasing road densities. Gray Treefrog abundances in Ontario decreased as road density increased within 750 m (2,460 ft) of breeding ponds (Dam 2005). Similarly, urbanized wetlands in Pennsylvania that were surrounded by greater road densities and limited forest or wetland land cover had proportionally fewer Gray Treefrogs than less disturbed wetlands (Rubbo and Kiesecker 2005). Despite these findings, Gray Treefrogs are still sometimes found in urban or suburban locations. Simon et al. (2009) encountered Gray Treefrogs in residential neighborhoods and commercial areas and along busy highways. They have even been documented using a residential birdhouse in North Carolina (Murphy 1968). Understanding the impacts associated with Gray Treefrog use of anthropogenic habitats is complicated, because disturbed locations occasionally have desirable features that attract frogs but do not promote survival and reproduction (i.e., these places act as ecological traps; Battin 2004).

Agricultural disturbance could negatively impact Gray Treefrogs. Trenham et al. (2003) reported a negative relationship between calling males and land identified as "cultivated or pasture" in Wisconsin. Wetlands in agricultural landscapes often possess higher levels of environmental contaminants that can have negative effects on anurans (e.g., Koprivnikar et al. 2006b). Gray Treefrogs in ponds where cattle had access in Minnesota had lower reproductive success than those surrounded by ungrazed/fallow agricultural land or crops (Knutson et al. 2004). The wetlands bordered by grazed habitat or where cattle had direct access experienced increased (potentially lethal) levels of nutrients, depleted oxygen concentrations, degraded habitat for egg and larvae development, and direct mortality due to trampling. Despite this, available information on how agriculture impacts Gray Treefrogs in Wisconsin is somewhat conflicting. Some studies, for example, support the finding that this treefrog not only occurs in agricultural landscapes (Knutson et al. 2000) but also is positively correlated with percent agricultural land use within 250 m (820 ft) of breeding ponds (Paloski 2006).

Several studies report that environmental contaminants such as pesticides and herbicides associated with urban or agricultural landscapes can negatively impact larval Gray Treefrogs, but reported effects vary widely (Semlitsch et al. 2000, Relyea and Mills 2001, Boone and Bridges-Britton 2006, Gaietto et al. 2014). The commonly used pesticide ingredient carbaryl can be lethal to larval Gray Treefrogs across a range of concentrations (e.g., 0.09–2.5 mg/L [0.000006–0.0003 oz/gal]) after exposure of days to weeks (Zaga et al. 1998, Relyea and Mills 2001) and can lead to reduced SVL of recently metamorphosed individuals (Hanlon and Parris 2014). Exposure to metals and road salts from stormwater runoff can accelerate metamorphosis in larvae, increase size at metamorphosis, and reduce embryo survival (Gosner 1960, Brand et al. 2010). Aquatic habitats can be further degraded by various factors that are also detrimental or lethal to Gray Treefrog larvae, including Glyphosate (Relyea 2005a, 2005b, Relyea et al. 2005), low pH (≤4.3; Gosner and Black 1957b), and high levels of ultraviolet radiation (UVB; Gosner 1960, Crump et al. 1999, Licht 2003). On the other hand, Rowe et al. (2011) reported that some contaminants, such as selenium and vanadium, can accumulate in breeding ponds but have limited impacts on larvae at low concentrations. A thorough review of documented chemical impacts on Gray Treefrog larvae can be found in Dodd (2013).

Various pathogens pose significant threats to global amphibian populations and could also impact Gray Treefrogs in Wisconsin (see Introduction to Conservation and Management of Wisconsin's Amphibians and Reptiles chapter). Yet a review of several studies focused on the effects that pathogens have on this species provides an unclear picture. For example, although ranaviruses were detected in larvae identified as members of the *Hyla chrysoscelis / H. versicolor* complex from Tennessee (infection prevalence of 6%–43%; O'Bryan et al. 2012), another investigation in New York reported that the prevalence of ranavirus infection of treefrogs was 0% (Brunner et al. 2011). Additional studies have found that although Gray Treefrogs are susceptible to the chytrid fungus *Batrachochytrium dendrobatidis* (Bd), the prevalence of Bd infection in treefrogs is low (Ouellet et al. 2005, Longcore et al. 2007, Sadinski et al. 2010, Villamizar-Gomez et al. 2016, Standish et al. 2018). Yet laboratory investigations by Searle et al. (2011) indicate that infection prevalence can vary by treefrog size, with lower infection levels in smaller individuals. Given these results and the fact that many field studies have analyzed only a handful of individual treefrogs, additional investigations on the impacts of Bd are warranted.

Of further concern is that any individual threat to Gray Treefrogs, when coupled with other stressors, can have synergistic effects that might be difficult to predict. Many studies have demonstrated that the combined influence of various stressors can have higher impacts on Gray Treefrog survival than individual stressors alone. In Ontario, agricultural activity surrounding wetlands was positively associated with higher trematode infection prevalence in larval treefrogs, indicating that the effects of one stressor exacerbated the effects of another (Koprivnikar et al. 2006b). Furthermore, impacts to amphibian survival and development can be more severe when individuals are exposed to several stressors acting synergistically, such as occurs when larvae infected with trematodes are also exposed to pesticides (Kiesecker 2002, Johnson and Chase 2004, Koprivnikar et al. 2006b). Relyea and Mills (2001) found that mortality rates of Gray Treefrog larvae exposed to the pesticide carbaryl alone were 10%–60%. Yet this same study reported that when larvae were exposed to additional stressors (i.e., predators and a sublethal dose of carbaryl for 10–16 days), mortality rates jumped to 60%–98%. Similarly, Pauley et al. (2015) reported that survival declined when larvae were exposed to both predatory dragonfly naiads (*Anax junius, Pachydiplax longipennis*) and insecticides, compared to the insecticides alone. Synergistic impacts leading to high mortality have also been found for larvae exposed to both carbaryl and ultraviolet light (UVB [285–320 nm]) when neither stressor alone caused mortality (Zaga et al. 1998). Alternatively, larvae experienced higher survival when exposed to both the pesticide Roundup and Bd than when exposed to Bd alone (Hanlon and Parris 2014). The authors speculated that Roundup depressed Bd growth, which then mitigated the fungus's negative effect on larvae.

Management: In general, management considerations related to protecting and maintaining suitable wetlands and adjacent woodlands are likely relevant to Gray Treefrogs (see Introduction to Conservation and Management of Wisconsin's Amphibians and Reptiles chapter). In addition, promotion of connectivity among suitable habitats to create a matrix of accessible woodlands and wetlands could reduce potential population losses due to stochastic events or human disturbances (Johnson 2005). Viable corridors that sustain movement can also enhance genetic mixing, as well as reduce mortality during migration events. Protection of appropriately sized habitat buffers around breeding wetlands rather than maintenance of only narrow corridors between them could also be an effective management strategy. For example, Johnson and Semlitsch (2003) suggested that a minimum of 60 m (197 ft) of core terrestrial habitat surrounding a breeding pond would be sufficient for reproductive movements during a breeding season. Yet research shows that Gray Treefrogs will occupy terrestrial habitat adjacent to breeding ponds out to 340 m (1,116 ft) during the nonbreeding season (Johnson et al. 2007a). An insufficient amount of terrestrial habitat could disproportionately affect females over males, as females travel farther from ponds and spend less time at breeding wetlands than males (Johnson et al. 2007a).

Due to their sensitivity to environmental pollutants, control of pesticide and herbicide contamination in breeding wetlands is likely an important consideration. Regulations that reduce the risk of contamination and promote maintenance of vegetative buffers around wetlands that can restrict movement of pollutants into the water could benefit Gray Treefrogs. Furthermore, control of livestock activity within or near seasonal wetlands can reduce possible associated impacts. Because fish can consume all Gray Treefrog life stages, removal of stocked fish and prevention of stocking in breeding ponds could help conserve Gray Treefrog populations (Hecnar and M'Closkey 1997b, Smith et al. 1999). Similarly, management strategies that maintain the ephemeral nature of fishless wetlands and avoid alterations that enable permanently standing water (and, indirectly, fish survival) should also be considered.

Several additional management strategies exist that are potentially useful but have not been fully vetted for Gray Treefrogs. Examples include wetland creation, wetland restoration, and translocations of individuals to offset population declines or establish new populations (see Introduction to Conservation and Management of Wisconsin's Amphibians and Reptiles chapter). In Manitowoc County, for example, Gray Treefrogs used restored wetlands immediately after these habitats were restored but also for up to 12 years post-restoration (Nedland et al. 2007). Cook (2008) further reported successful translocation of Gray Treefrog larvae that purportedly resulted in established populations at sites in New York. Yet the appropriateness and success of translocation as a conservation strategy could be dependent on context and various extrinsic factors. As such, its implementation should only occur after careful consideration and with input from experts and regulatory agencies.

SPECIAL REMARKS

Although Gray Treefrogs can hybridize with Cope's Gray Treefrogs (Gerhardt et al. 1994), offspring that result from hybridization between these two species reportedly do not survive (Johnson 1959, 1963, Ralin 1976).

Spring Peeper

Pseudacris crucifer (Wied-Neuwied 1838)

Andrew F. Badje and John D. Peterson

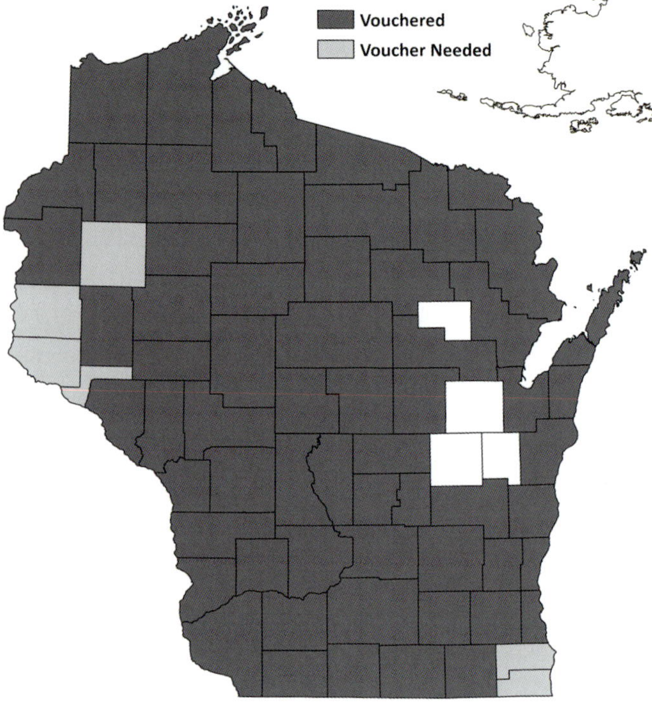

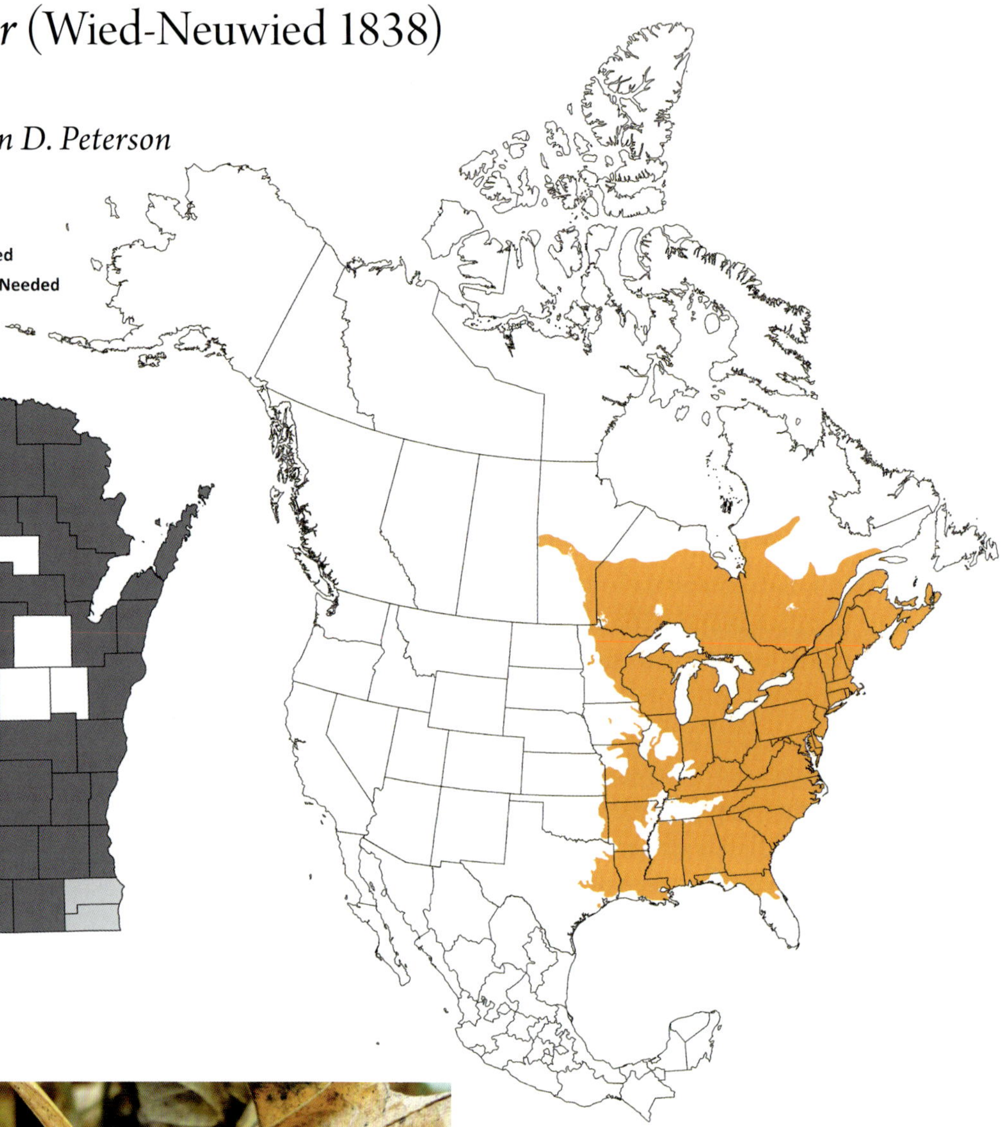

Adult Spring Peeper (Grant County; photo by C. Raimond).

AT A GLANCE

Spring Peepers are small frogs that range in color from gray to light tan. They typically possess an X-shaped mark on their backs, which also explains the specific epithet of their scientific name, *crucifer* (cross bearer). Spring Peepers are widespread in Wisconsin and documented in all but a handful of counties. They are active from late March to November. In the spring, males call to attract females for reproduction by producing a pronounced high-pitched 'peep' every few seconds that echoes loudly into the surrounding landscape. Spring Peepers are one of the first frog species to call during spring in Wisconsin, and their 'peeps' are widely acknowledged to announce the end of winter. Spring Peepers breed in open (often ephemeral) wetlands such as sedge meadows but will also use wooded ponds. Eggs are laid singly or in small clusters, often hatching in two to three days. Larvae typically complete metamorphosis from late May through June. Outside of the breeding season, Spring Peepers are often found in moist woodland habitats, resting on trees or shrubs and below damp leaf litter or fallen logs. During winter, Spring Peepers seek shelter below ground debris on the forest floor. They can survive the cold conditions by production of physiological compounds that protect their cells from damage due to freezing. Spring Peepers eat a wide variety of small arthropods and are consumed by numerous predators, including large aquatic insects, dragonflies, ground beetles, fishes, and snakes. Spring Peepers go by many colloquial names: 'peeper' and 'Pickering's frog' in Wisconsin, while farther east on the Atlantic coast they are sometimes referred to as 'tinkletoes,' 'pinkletinks,' 'pink-winks,' and 'the little frogs from the moon.' They are currently considered a common species by the WDNR, although loss of woodlands and breeding wetlands is a conservation concern.

SYSTEMATICS

Pseudacris crucifer (Spring Peeper) was first described by Wied-Neuwied (1838), who initially placed it in the genus *Hyla*. Since then, it has gone through numerous taxonomic changes. More recently, Hardy and Borroughs (1986) reviewed the results of behavioral, morphological, and molecular studies to justify transferring the species from *Hyla* into the newly described yet largely ignored genus *Parapseudacris*. At roughly the same time, Hedges (1986) used genetic analyses to determine that this species should be placed in the genus *Pseudacris*. Cocroft (1994) combined Hedges's data with additional morphological characters and argued that this species should remain in *Hyla*, although Da Silva (1997) returned it once again to *Pseudacris*, noting that arguments could be made to place it in either *Hyla* or *Pseudacris*. More recent phylogenetic analyses by Moriarty and Cannatella (2004) found that *P. crucifer* was related to *Pseudacris ocularis* (Little Grass Frog, which occurs in the southeastern US), forming a monophyletic "crucifer" clade embedded within *Pseudacris*, thus providing the strongest argument yet for their placement in this genus. In agreement with Austin et al. (2002), Moriarty and Cannatella (2004) also rejected the subspecific designations *Pseudacris crucifer crucifer* (Northern Spring Peeper) and *P. c. bartramiana* (Southern Spring Peeper), which are no longer recognized (Crother 2017). Among anurans in Wisconsin, only one other species shares the generic name *Pseudacris*: *P. maculata* (Boreal Chorus Frog). Both *P. crucifer and P. maculata* are also members of the family Hylidae, as are resident species in the genera *Acris* and *Hyla*.

DESCRIPTION

Adults: Dorsal background color ranges from pinkish brown or light tan to gray or dark brown (Figure 89). The smooth dorsum typically features a dark X-shaped mark that can be distinct but sometimes appears broken or nondescript (see Figures 89 and 90; Knudsen 1958d, Vogt 1981). Regardless, it is this defining mark that is responsible for the specific epithet *crucifer* (cross bearer) (Dodd 2013). The head has a blunt snout with a dusky lateral band on each side that extends posteriorly through the eye and tympanum to the insertion of the hind leg (Figure 89). The diameter of the tympanum is less than that of the eye (Figure 89).

Dark bands or crossbars appear on all four limbs, whereas the undersurfaces of the hind limbs near the groin appear light yellowish or pinkish

Figure 89. Examples of variation in dorsal pattern and coloration in adult Spring Peepers. (a) Individual with dark dorsal coloration that also lacks an X-shaped dorsal mark (Sawyer County; photo by A. Badje). (b) Adult individual with lighter dorsal coloration (Adams County; photo by A. Badje).

Above, top: Figure 90. Adult Spring Peeper with obvious X-shaped mark on the dorsal surface (Taylor County; photo by C. Raimond).

Above, bottom: Figure 91. Spring Peeper eggs (Dane County; photo by J. Kapfer).

(Breckenridge 1944, Vogt 1981). The granular venter ranges in color from white to yellowish or tan, with occasional dark mottling (Pope and Dickinson 1928, Knudsen 1958d, Vogt 1981). The toes on the feet of all four limbs feature slightly enlarged adhesive toe pads that are utilized when climbing. There is no webbing between the toes of the fore feet, while webbing between the toes of the hind feet only extends to the first toe joint (Breckenridge 1944). Spring Peepers are one of Wisconsin's smallest anurans, and the mean SVL of specimens from Wisconsin that were housed in the MPM and UWZM collections was 2.5 cm (1.0 in, $n = 134$; note that only individuals ≥2.3 cm [0.9 in] were measured, following information on size at maturity reviewed by Dodd 2013). The largest Wisconsin specimen measured was 3.2 cm (1.3 in, MPM 33266, collected from Ashland County in 1998).

Preadult stages: When first laid, Spring Peeper eggs are white at one end and black at the other (Knudsen 1958d). Eggs are laid singly or in smaller clusters and are enveloped in a clear gelatinous membrane (Figure 91). In Wisconsin, vitelli diameter is 0.1 cm

(0.04 in), while total egg diameter, which includes the outer membrane, is 0.15 cm (0.06 in; Vogt 1981, Watermolen 1995). Across their range, larvae tend to display great variation in appearance. The dorsal color of larvae ranges from green to brown with a metallic copper flecking, whereas the venter is lighter and more iridescent (Vogt 1981). Tail musculature is uniformly covered with irregular mottling, reticulations, or blotches (Gosner and Black 1957a, Altig 1970, Vogt 1981). There is a clear border immediately surrounding the tail musculature, while the remainder of the tail fin is mottled with a dark or violet pigmentation that is more concentrated along the outer edges (Altig 1970, Vogt 1981). Spring Peeper larvae have laterally positioned eyes similar to larvae of other hylids, with two rows of upper labial teeth and two to three rows of lower labial teeth (Altig 1970, Vogt 1981). Recently hatched larvae from Wisconsin are 0.30–0.35 cm TL (0.12–0.14 in) but can reach 2.4–3.2 cm TL (0.9–1.3 in) prior to metamorphosis (Vogt 1981). Newly metamorphosed juveniles in Wisconsin resemble miniature adults but with less conspicuous patterns and range in TL from 0.8 to 1.5 cm (0.3 to 0.6 in; Knudsen 1958d, Vogt 1981).

Variation: The background dorsal coloration of adults can vary substantially among individuals and by sex (Figures 89 and 90). Males are somewhat smaller and darker than females, possess longitudinal gular folds (which permit the expansion of the single vocal sac), and have an olive-brown vocal pouch that becomes more conspicuous during the breeding season (Smith 1961, Oplinger 1966). Although the two previously identified North American subspecies of Spring Peeper are no longer recognized (see Systematics section), they do reflect phenotypic differences. The southern phenotype, found in southern Georgia and northern Florida, displays dark ventral spotting, broad dorsal stripes, spotting on the upper jaw, and robust coloration, whereas the northern phenotype, which occurs in Wisconsin, has a plain or slightly mottled venter (Harper 1939, Conant and Collins 1998).

Diagnosis: Spring Peepers possess a slender body and have no parotoid glands, large warty tubercles, or keratinized spades on the heel of each hind foot. They have small terminal toe pads that are wider than the toes and a light brown to tan dorsum with a distinct brown (usually X-shaped) dorsal mark, but they do not have bright yellow coloration on the hidden surface of the thighs. In Wisconsin, Spring Peepers are most easily confused with Boreal Chorus Frogs and Blanchard's Cricket Frogs (*Acris blanchardi*) due to similarities in size and color. Yet Boreal Chorus Frogs have a white stripe on their upper lip, generally have a more slender body, and also possess three dorsal longitudinal lines (not an X-shaped dorsal mark) that run the length of the body. Blanchard's Cricket Frogs have rougher skin, a rugged stripe on the inner thighs, and sometimes a dark interorbital triangle (reviewed by Gray and Brown 2005). Spring Peepers could also be mistaken for young treefrogs (*Hyla* spp.) in Wisconsin; however, both treefrog species typically reach a larger adult size, are dorsally emerald-green or mottled gray-green with a bright yellowish coloration on the inside of the thighs, and usually possess a light spot under each eye.

Spring Peeper eggs and larvae are difficult to differentiate from those of other hylids, such as the comparably sized Boreal Chorus Frog. Yet Spring Peeper eggs are often deposited singly or in small clusters, whereas Boreal Chorus Frog eggs are typically deposited in loose irregularly shaped clusters of 10–150 (Watermolen 1995). Spring Peeper larvae are characterized by internal gills and a lack of external forelimbs. They also have laterally or dorsolaterally positioned eyes on the head, a spiracle tube that is short and attached to the body wall, and a light-colored belly. Their tail fins lack black tips and dark mottling but possess an area of clear tissue near the tail musculature. Larvae of Boreal Chorus Frogs are dark brown or black above and bronze below and usually have clear tail fins with two-toned tail musculature (Altig 1970, Vogt 1981). In comparison, Spring Peeper larvae have clear borders immediately surrounding the tail musculature, with dark pigmentation on the outer edges of their tail fins (Altig 1970, Vogt 1981).

DISTRIBUTION AND HABITAT

Global distribution: Spring Peepers are widely distributed across eastern North America. They range from northern Florida west to eastern Texas and Oklahoma, north through eastern Minnesota and into Manitoba, east to New Brunswick and Nova Scotia, and south along the eastern shore of the Atlantic coastal states. Spring Peepers are found from sea level to elevations as high as 1,650 m (5,413 ft; Mathews and Echternacht 1984).

Distribution in Wisconsin: Spring Peepers occur throughout most of Wisconsin, including Chambers Island in Green Bay, islands off the Door County Peninsula, and the Apostle Islands (Suzuki 1951, Long et al. 1978, Vogt 1981, Mossman et al. 1998, Casper 2008c). Published records exist for nearly all counties, although suitable voucher material could not be found for Pierce, St. Croix, Barron, Pepin, Racine, and Kenosha Counties (see the Appendix).

Localities of possible occurrence: Published records do not currently exist for Menominee, Outagamie, Calumet, and Winnebago Counties. Despite this, the Wisconsin Frog and Toad Survey has generated many anecdotal reports of calling males in these locations (see Conservation section; Mossman et al. 1998). These county-level gaps could easily be filled through directed survey efforts in the future.

Habitat: In Wisconsin, Spring Peepers frequently occupy dry to mesic forests when not breeding (Vogt 1981; see Amphibians and Reptiles in Wisconsin Plant Communities chapter). They seem to prefer hardwood and mixed hardwood forest over coniferous forest communities (Gibbs et al. 2005) but still sometimes occur in the latter. Anecdotal reports suggest that Spring Peepers escape high summer temperatures by resting in the shade of flowering plants (e.g., asters and clintonias) or beneath ground debris in forest understories (Breckenridge 1944). Their small adhesive toe pads give them the ability to easily climb into shrubs and trees, where they are also occasionally found (Breckenridge 1944, Allen 1963, Vogt 1981).

During the breeding season, Spring Peepers migrate to various types of permanent or ephemeral wetlands, including sedge meadows (Vogt 1981; see Amphibians and Reptiles in Wisconsin Plant Communities chapter). In Burnett County, Evrard (2000) reported calling males in open wetlands associated with pine barren habitat. They also breed in lake peripheries with dense aquatic vegetation, bogs, and anthropogenic habitats such as ditches and constructed ponds (reviewed by Dodd 2013). Although Spring Peepers use wetlands with a variety of hydroperiods, the results of several studies suggest that they prefer ephemeral wetlands but will use aquatic habitats that hold water for extended periods (e.g., ≥4 months; Kolozsvary and Swihart 1999, Babbitt et al. 2003). Breeding wetlands are generally found adjacent to or within moist woodland communities in eastern North American forests (Vogt 1981, Knutson et al. 2000, Price et al. 2005). For example, Herrmann et al. (2005) determined that Spring Peepers were more inclined to breed in wetlands within 1,000 m (3,281 ft) of forested habitat. Trenham et al. (2003) similarly reported that the presence of breeding activity (i.e., calling males) was positively associated with wetlands surrounded by forest and shrubs in Wisconsin. Despite the apparent preference for wetlands associated with wooded habitat, chosen breeding sites often possess mostly open tree canopies (Skelly et al. 1999) or highly variable canopy cover. For example, Kutka (1998) reported that tree canopy closure at 34 breeding ponds used by Spring Peepers in northern Wisconsin ranged from <1% to 85%. Fishless wetlands are also generally the most successful reproduction habitats for Spring Peepers (Kats et al. 1988). Information on larval habitat preferences is limited, although Smith et al. (2003b) frequently found larvae in the shallower regions of breeding wetlands.

Overwintering occurs in forested upland habitats. Prior to winter, adults and juveniles seek out shallow depressions in the soil or below debris (i.e., rocks, logs, roots, and leaves) on the forest floor (Blanchard 1933b, Froom 1982, Layne and Kefauver 1997). These hibernacula are typically insulated by a layer of snow and ice that forms on the ground in northern climates, such as that found in Wisconsin (see Activity section).

REPRODUCTION AND DEVELOPMENT

Male and female Spring Peepers can successfully overwinter with spermatozoa and eggs in their respective reproductive tracts (Rugh 1941, Dodd 2013), allowing them to engage in reproductive activity soon after emergence from hibernation. In Wisconsin, males begin calling early in the spring and generally continue from late March to early June (Vogt 1981, Mossman et al. 1998, Paloski et al. 2014); however, interannual variability in when males call can be quite pronounced. For example, it is not uncommon to hear the first Spring Peeper calls of the year in mid-March at ponds that are up to 50% covered with ice (see Natural History Box: Enchanting Solos), and they will also call during spring snow showers (Joshua Kapfer, UW–Whitewater, personal observation). In Massachusetts, calling males were even documented in locations with water temperatures as low as −0.6°C (30.9°F; Allen 1868). Yet calling behavior is more typically initiated when water temperatures reach 5–7°C (41–44.6°F; Dodd 2013). Calling males are also occasionally heard well into autumn after the main breeding season has ended (e.g., 6 November 2015; A. Badje, personal observation), but these calls probably do not result in reproduction.

After they arrive at breeding sites, males establish calling territories that generally range from 0.2 to 2 m in size (0.7 to 6.6 ft; Brenowitz et al. 1984, Gerhardt et al. 1989). Males within these territories attract females by calling from the shoreline, on floating vegetation mats, or from low-lying (but elevated) locations on trees, shrubs, logs, branches, twigs, sedges, and other aquatic vegetation (Allen 1963, Vogt 1981). Males also sometimes form small choral groups of three to five individuals that call in the same general location (Ptacek 1984). The bulk of calling occurs at night, although at the height of the breeding season, males sometimes call throughout the day (A. Badje, personal observation). Male advertisement calls, produced by a single vocal sac below the chin, can be described as a high-pitched acute whistle or 'peep' that repeats in one-second intervals (Powell et al. 2016). Hundreds of calling individuals in a chorus can sound (from a distance) like the jingling of sleigh bells and can be heard up to ca. 0.8 km (0.5 mi) away (Knudsen 1958d). Males also produce a secondary aggressive call that resembles a trilled version of

NATURAL HISTORY BOX: Enchanting Solos

Aside from the return of the American Robin (*Turdus migratorius*) to Wisconsin each spring, the familiar 'peep' of the Spring Peeper has signaled the coming of spring for many human generations. One warm spring evening, I set out to catch a glimpse of this tiny and well-concealed hylid in hopes that it would bring extra luck for my upcoming field season. Near a shallow depression in the pond, I singled out a few particularly loud males. As I walked up to the pond's edge to better see the frogs, however, the calls abruptly stopped. As I waited eagerly, one brave male eventually began to call again, so I was able to narrow down his precise location. After a few more rounds of hide-and-seek, I finally located him, perched near the base of a stand of emergent vegetation. Desiring to see for myself how this small frog could produce such an acutely audible sound, I carefully watched every little movement that supervened. First taking in a big gulp of air, he then closed his mouth and sealed his nostrils before inflating the vocal sac beneath his chin. Internally, this male proceeded to drive air from his lungs to his vocal sac in order to vibrate the vocal chords on the floor of his mouth. What came forth was a magnificent and single fortissimo peep, which resounded throughout the pond and nearby forest. In that moment, I continued to crouch, watching in a mesmerized state for a few minutes before allowing him to go back to his nightly business. After I backed away from the water, I chose to sit down on the nearby park bench and delay my departure by a few minutes to hear the chorus accumulate from a few 'peeps' to the harmonious jingling of sleigh bells.

Andrew F. Badje, WDNR

their breeding call. These trills are reportedly used to warn other males against territorial transgression (Rosen and Lemon 1974). Sometimes invading males will reciprocate trills of the resident males with their own, which can lead to direct aggressive interactions between individuals (Vogt 1981).

The arrival of females at breeding sites takes place shortly after males begin to call, although some females can appear up to five to eight weeks later (Lykens and Forester 1987). Females apparently select males to mate with based on the perceived quality of their call characteristics (e.g., duration and frequency; Doherty and Gerhardt 1984, Forester and Czarnowsky 1985). Yet high male reproductive success is not always due to female mate selection. For example, noncalling males (that are younger and smaller) sometimes wait near mature calling males to intercept and mate with females that are attracted to the other males' calls. This behavior reportedly reduces exposure to predators and conserves energy for growth (Fellers 1979a, Forester and Lykens 1986).

Females typically initiate reproduction by nudging the selected male so that he grasps her and amplexus ensues (Figure 92; Dodd 2013). While in amplexus, females deposit eggs, which males fertilize externally. Reported female clutch sizes are large, roughly 900 eggs from a female in Michigan (Delzell 1958) and ranging from 200 to 1,581 eggs in New York and Nova Scotia (Oplinger 1966, Gilhen 1984).

Figure 92. Spring Peepers in amplexus. Note the sexual dimorphism related to size (Juneau County; photo by Sara Viernum).

Although overall clutch sizes per female can be large, eggs are often laid singly or in numerous small bundles of 2–30 (Flores 1978). Eggs are affixed to aquatic vegetation (e.g., grasses, sedges, weed stalks) in shallow water near the bottom of the wetlands, where they develop (Breckenridge 1944, Knudsen 1958d). Eggs in Wisconsin hatch within two or three days of being fertilized (Vogt 1981). Metamorphosis can occur in as few as 45 days (Gosner and Rossman 1960) but more typically occurs from 90 to 100 days (Minton 2001). In northern regions, newly metamorphosed individuals are observed as late as September (i.e., Ontario [MacCulloch 2002], Nova Scotia [Bleakney 1952], and New York [Wright 1914]). In Wisconsin, however, metamorphosis is documented from late May into June, when larvae reach ca. 0.8–1.5 cm TL (0.3–0.6 in; Pentecost and Vogt 1976, Vogt 1981).

By their first winter, Spring Peepers attain sizes of 1.40–1.55 cm (0.55–0.61 in; Dodd 2013). In Michigan and Maryland, males and females reach sexual maturity as early as their first spring after metamorphosis, though maturity is more commonly attained in their second post-metamorphic season (Delzell 1958, Lykens and Forester 1987). Size at sexual maturity for males is ca. 2.2 cm SUL (0.87 in), while eggs are first found in females at ca. 2.3 cm SUL (0.91 in; Dodd 2013). Longevity in the wild is unknown, but Lykens and Forester (1987) determined that some individuals in three Maryland study ponds were four years old.

ACTIVITY

In Wisconsin, Spring Peeper activity usually begins from late March to April (Vogt 1981, Mossman et al. 1998). During the first warm and wet evenings of spring, Spring Peepers migrate from overwintering sites to breeding wetlands, generally between midnight and dawn (Vogt 1981, Pechmann and Semlitsch 1986, Todd and Winne 2006). Migrations to breeding sites can continue for several weeks as more individuals emerge from hibernation (Cummins 1920, Murphy 1963, Pechmann and Semlitsch 1986).

Although limited information exists, adult migrations between and juvenile dispersal from breeding sites and upland habitats likely represent the greatest movements Spring Peepers undertake. For example,

Delzell (1958) found adult Spring Peepers hibernating in uplands at distances of ca. 280 m (920 ft) and 549 m (1,800 ft) from the nearest breeding wetland, which supports that post-emergence migrations at these distances are probable. Following the breeding season, adults move back into adjacent forested uplands and occupy established home ranges throughout the remainder of summer and autumn (Delzell 1958, Ptacek 1984). Few studies have investigated adult movement patterns after migration from breeding ponds. Yet Delzell (1958) estimated that, once in uplands, Spring Peepers occupied small nonbreeding terrestrial home ranges of 1.2–5.5 m (3.9–18.0 ft) in diameter centralized around ground debris and vegetation used for cover. Delzell (1958) further reported that daily movement rates during this time were minimal, ranging from 6.1 to 39.6 m (20.0 to 129.9 ft; Delzell 1958).

Spring Peepers are generally active in summer until air temperatures approach their critical thermal maximum (CTmax) of 34.8°C–36°C (94.6°F–96.8°F; John-Alder et al. 1988). As temperatures drop in autumn, Spring Peepers migrate to overwintering locations. By October and November, they typically prepare for hibernation through physiological production of glucose, a cryoprotectant, from glycogen stores (Churchill and Storey 1996). Glucose accumulates in the blood and in organs such as the heart, kidney, liver, and leg muscles to resist cellular damage that would otherwise occur due to ice crystal formation in the tissues (Storey and Storey 1987). This remarkable ability to resist freezing allows Spring Peepers to survive the harsh winters that typically occur in northern latitudes such as Wisconsin. Individuals can tolerate the freezing of up to 45% of the extracellular water in their bodies (Layne and Kefauver 1997). Research supports that short-term exposure to subzero temperatures is more tolerable than long-term exposure. For example, Storey and Storey (1987) reported full survival of individuals exposed to −2.5°C (27.5°F) for periods from 20 hours to four days. Subsequently, Layne and Kefauver (1997) documented only 52.6% survival after seven days and 0% survival after 28 days at −1.5°C (29.3°F). Once warmer spring and summer temperatures arrive, the production of physiological compounds that aid in freeze tolerance declines (Storey and Storey 1987).

PREY AND PREDATORS

Adults consume primarily small invertebrates that they can swallow whole, such as midges, mites, springtails, mayflies, ants, beetles, arachnids, and mosquitoes (Knudsen 1958d, McAlister 1963a, Oplinger 1967). They also reportedly eat their own shed skin (Johnson 1989), and Redmer (1987) even documented cannibalism in captivity. Most foraging occurs in forested uplands among ground debris but also in thickets of low-lying shrubs and trees (Vogt 1981). Feeding primarily occurs during the day, from early morning into the early evening (Oplinger 1967), and McAlister (1963a) reported that Spring Peepers selected hunting locations near flowers that attract insects, such as Sweet-scented Joe Pye Weed (*Eutrochium purpureum*), elderberry (*Sambucus* spp.), and goldenrod (*Solidago* spp.). Juveniles consume similar, albeit smaller prey, including ants, larval lepidopterans, arachnids, beetles, mites, terrestrial gastropods, springtails, and lacebugs (Oplinger 1967). Unlike other developmental stages, larvae are omnivorous and consume crustacean eggs, protozoans, diatoms, and algae (Munz 1920).

Spring Peeper predators include giant water bugs (Hinshaw and Sullivan 1990), dytiscid diving beetles (Formanowicz and Brodie 1982), odonate larvae (Skelly 1996), ground beetles (Smith 2002), fishes (Cochran and Cochran 2003), and snakes (e.g., gartersnakes and ribbonsnakes [*Thamnophis* spp.]; Test 1958, Klemens 1993). Dodd (2013) stated that "undoubtedly, birds and other predators take large numbers," but specific observations are rarely documented, although Ross (1989a) reported predation by Loggerhead Shrikes (*Lanius ludovicianus*). Spring Peepers can employ several strategies to evade predators. Adults, for example, jump away, climb vegetation, crouch low to the ground, or remain motionless to avoid detection but will also inflate their bodies to deter predation attempts (reviewed by Dodd 2013). Ptacek (1984) noted that during the day adult males retreated into the dense sedges at breeding wetlands for cover and moved into the open as nighttime returned. Furthermore, Spring Peepers can change the color of their skin from light to dark to reflect the substrate and other nearby backgrounds, which likely aids in predator avoidance (Kats and

Van Dragt 1986). Larvae do not possess chemical defenses (Dodd 2013) and are palatable to predators (Kats et al. 1988). Still, they avoid predation by remaining motionless near the bottom of wetlands to better blend into their surroundings. This strategy is reportedly effective and contributes to Spring Peeper larvae being particularly abundant in breeding wetlands despite predation pressure, sometimes outnumbering sympatric larvae of other anuran species (Lawler 1989).

Across their geographic distribution, Spring Peepers are parasitized by a wide variety of protozoans, trematodes, larval flukes, and nematodes (see comprehensive list in Dodd 2013). Yoder and Coggins (2007) identified numerous helminth parasites from Spring Peepers collected in Wisconsin.

CONSERVATION

Status: Globally, the Spring Peeper is listed as a Least Concern species by the IUCN. They were historically considered common in many regions of the US (reviewed by Butterfield et al. 2005). The abundance of Spring Peepers has remained stable or increased over time in many portions of their continental range (reviewed by Dodd 2013). Despite reports from Wisconsin in the 1980s and 1990s of a small yet statistically significant decline (Mossman et al. 1998), recent Wisconsin Frog and Toad Survey data suggest that populations are stable to slightly increasing (Figure 93). The Spring Peeper is currently considered a common species by the WDNR.

Populations: Research on Spring Peeper population characteristics is relatively scarce across its range. Breeding male densities in Missouri wetlands reportedly ranged from 1 to 4.5 individuals/m^2 (0.1 to 0.4 individuals/ft^2; Gerhardt et al. 1989). Over the course of a 16-year study, 2,499 ($\bar{x} = 156.2$ per year) breeding-age females and 9,245 ($\bar{x} = 577.8$ per year) metamorphosed individuals were documented in a 1 ha (2.5 ac) wetland complex in South Carolina (Semlitsch et al. 1996). In a more northern study, Werner (1975) estimated densities that ranged from 1,157 to 1,478/ha (2,859 to 3,652/ac) at a site in Michigan. Little data on Wisconsin populations exist, although many reports refer to Spring Peepers' large state abundances. For example, Knudsen (1958d) wrote that in Wisconsin "there are few outdoorsmen that have not heard the din caused by ponds full of these little fellows during the early spring months." From 1984 to 2014, volunteer calling surveys for the Wisconsin Frog and Toad Survey have consistently recorded calling male Spring Peepers from 70% to 85% of all statewide survey sites and at mean calling intensity indices of 2.5 (out of 3; WDNR, unpublished data). These data indicate that Spring Peepers could currently be the most abundant anuran species detected by calling surveys in the state.

Adult survivorship is not well-studied; however, larval survivorship has been more widely investigated. Survivorship of Spring Peeper larvae is highly dependent on myriad factors, including the species composition of larval anurans that co-occur in wetlands, larval densities, predation pressure, and wetland

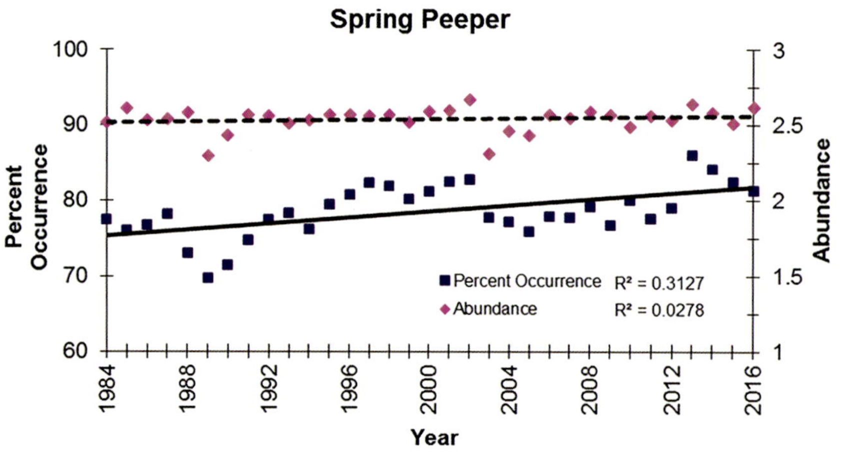

Figure 93. Long-term trend in occurrence (%) and abundance (rank) of Spring Peepers across Wisconsin Frog and Toad Survey statewide monitoring sites. See Introduction to Species Accounts chapter for details on this monitoring program (figure courtesy of WDNR).

permanence (Morin 1986, Skelly 1996). When found with larvae of other anuran species, larval Spring Peepers (which are competitively inferior) attain comparatively low survivorship levels (Morin 1983b). However, survivorship increases when predators such as the Eastern Newt (*Notophthalmus viridescens*) are present and reduce abundances of and competition from other anuran larvae (Morin 1983b).

Most published information on sex ratios supports a bias toward males. Murphy (1963) found that adult sex ratios in the breeding season were skewed toward males at a site in North Carolina (1.43M:1F and 1.2M:1F, respectively, over two study years). Oplinger (1996) found a sex ratio of 9.2M:1F ($n = 416$ males and 45 females) at breeding sites in New York. It is noteworthy, however, that these sex-ratio estimates were calculated during the breeding season, when males are easier to detect. Oplinger (1996) recorded sex ratios of 0.6M:1F ($n = 27$ males and 47 females) and 1.1M:1F ($n = 55$ males and 48 females) based on road surveys in the spring and autumn, respectively.

Threats: Many general threats to amphibians, such as loss and degradation of suitable breeding and nonbreeding habitat, apply to Spring Peepers (see Introduction to Conservation and Management of Wisconsin's Amphibians and Reptiles chapter). Habitat impacts can occur due to intensive mining, agriculture, urbanization, conifer plantations, and clear-cutting for timber harvest (Vogt 1981, Clawson et al. 1997, Waldick et al. 1999, Knutson et al. 2004, Anderson and Arruda 2006). Alteration, loss, and fragmentation of forested upland habitat in urban and industrial landscapes likely play an important role in local Spring Peeper declines. In these human-dominated landscapes, Spring Peepers are found in lower densities and forced to occupy man-made wetlands, narrow strips of undeveloped riparian forests, and highly fragmented woodlots, which are often the only accessible habitat remaining (Vogt 1981, Burbrink et al. 1998). Loss or fragmentation of forested uplands near breeding sites can also result in extirpation of subpopulations (Gibbs 1998b, Kolozsvary and Swihart 1999). Furthermore, the unnatural drainage of small wetlands has undoubtedly diminished reproductive success and led to local Spring Peeper extirpations over the last century.

Habitat degradation also occurs when suitable natural areas are fragmented by roads. The impacts of roads on Spring Peepers are not well-studied, but roads lead to mortality and restrict movement. Carpenter and Delzell (1951) determined that 506 Spring Peepers died during spring migration over one evening along a 1.5 km (0.9 mi) road segment in Michigan. Although probably substantial, the full impact of road mortality is largely unknown because it is likely underreported (Dodd 2013). Many wetland habitats in Wisconsin are degraded due to infestations of invasive vegetation. Although invasive vegetation could negatively affect amphibians, this topic has received little attention. Still, during controlled experiments with anuran larvae, no Spring Peepers survived to metamorphosis when held in enclosures with detritus from the invasive Reed Canary Grass (*Phalaris arundinaceae*) or the nonnative Common Reed (*Phragmites australis*) compared to enclosures with debris from native plants (Maerz et al. 2010).

Habitats used by Spring Peepers experience further degradation due to pollution from chemical contaminants such as fertilizers, herbicides, pesticides, and heavy metals (Westerman et al. 2003, Storrs and Kiesecker 2004, reviewed by Dodd 2013). Several studies have reported on the negative impacts that contaminants have on Spring Peepers. For example, the development and survivorship of larval Spring Peepers were adversely affected by high levels of acidity and low-to-medium-level applications of the herbicide atrazine (Freda 1986, Storrs and Kiesecker 2004). Furthermore, contaminants from passing vehicles and deicing chemicals frequently occur in roadside ponds where Spring Peepers breed. These contaminants can disrupt reproductive activity and lead to inhospitable environments that could negatively affect this species (Langen et al. 2015). Several past studies support that Spring Peepers can act as indicators of habitat quality or environmental disturbance, given Spring Peepers' apparent sensitivity to landscape degradation. For example, work by both Price et al. (2007) and Brazner et al. (2007) determined that Spring Peeper populations showed a negative relationship with disturbed landscapes in the Great Lakes region. Furthermore, Knutson et al. (1999) reported that Spring Peepers were negatively associated with urban land use in Iowa and Wisconsin.

The impacts of several serious amphibian diseases are poorly understood for Spring Peepers in Wisconsin. Spring Peepers infected with iridoviruses, such as ranaviruses, were reported from Maine (Gahl 2007), including an incidence of >200 dead larvae (Green et al. 2002). Yet a small sample of adults tested in Wisconsin were negative for ranaviruses (Standish et al. 2018), so their effect (if any) on state populations is currently unknown. The chytrid fungus *Batrachochytrium dendrobatidis* (Bd) is a disease-causing agent that impacts global amphibian populations (see Introduction to Conservation and Management of Wisconsin's Amphibians and Reptiles chapter) and has been detected in Spring Peepers from Wisconsin (Standish et al. 2018). Mortality due to Bd infection, however, appears to be rare in Spring Peepers. In fact, several studies have reported that among the amphibian species tested for the presence of or mortality due to Bd Spring Peepers were rarely affected (Ouellet et al. 2005, Longcore et al. 2007, Gahl et al. 2012, Phillips et al. 2014). Yet the interaction between Spring Peepers and Bd can be complex and indirectly affected by numerous additional factors. For example, Beyer et al. (2015) reported that annual variation in habitat factors and climate conditions can impact how Bd affects Spring Peepers, which the authors suggested could be further influenced by the onset of climate change.

Management: Many general wetland and forest management recommendations for amphibians apply to this species (see Introduction to Conservation and Management of Wisconsin's Amphibians and Reptiles chapter). Preserving high forest cover across landscapes in close proximity (100 m [328 ft]) to known breeding sites is likely beneficial (Tate 1994, Eigenbrod et al. 2008). Management practices should therefore focus on the preservation or restoration of permanent and temporary wetlands within or adjacent to forested terrestrial habitats. Furthermore, creation or maintenance of upland forest connectivity between sites to protect links among subpopulations at different wetland-upland complexes is important for long-term persistence because it allows for genetic exchange (Kingsbury and Gibson 2012). Habitat connectivity could also be facilitated by the construction of amphibian road underpasses (i.e., tunnels under roads) in areas of high road mortality or where breeding and upland habitats are bisected by roads. The effectiveness of these underpasses can be enhanced if they are accompanied by silt fencing that directs wildlife to underpass entrances (Jackson et al. 2015). The contaminants associated with roads can also be problematic, so the creation and subsequent usage of road salts that biodegrade quickly could be advisable (Langen et al. 2015).

Calling surveys in Wisconsin support that Spring Peepers will use restored wetlands (Nedland et al. 2007), but whether use of these habitats ultimately benefits state populations has not been assessed. Repatriation efforts further offer an enticing solution in areas where this species is declining or extirpated (Cook 2008). In fact, well-meaning Wisconsin citizens have reportedly attempted the unsanctioned relocation of Spring Peepers and Wood Frogs (*Lithobates sylvaticus*) from as far away as Pennsylvania to sites in Wisconsin in an effort to increase local abundances (A. Badje, personal observation). Yet such strategies are largely unvetted for amphibians in the state and could cause great harm. For example, individuals used for repatriation could introduce devastating diseases to local amphibian populations, so they must be quarantined following accepted guidelines prior to release (which requires substantial resources). Although they are the same species, individuals from distant populations differ genetically from local individuals and may not have genotypes adapted to Wisconsin's climate. As such, these strategies should only be considered and carried out with input from specialists and regulatory agencies such as the WDNR.

SPECIAL REMARKS

Male advertisement call surveys are currently used to monitor populations of Spring Peepers throughout Wisconsin. Because they are easily heard and identified via advertisement calls, small population changes can be detected over numerous years of calling data (Mossman et al. 1998). Therefore, citizen-based monitoring projects such as the Wisconsin Frog and Toad Survey could offer an economically efficient and effective way to monitor local, regional, and statewide trends for the Spring Peeper.

Boreal Chorus Frog

Pseudacris maculata (Agassiz 1850)

Andrew F. Badje, Sara E. Viernum, and John D. Peterson

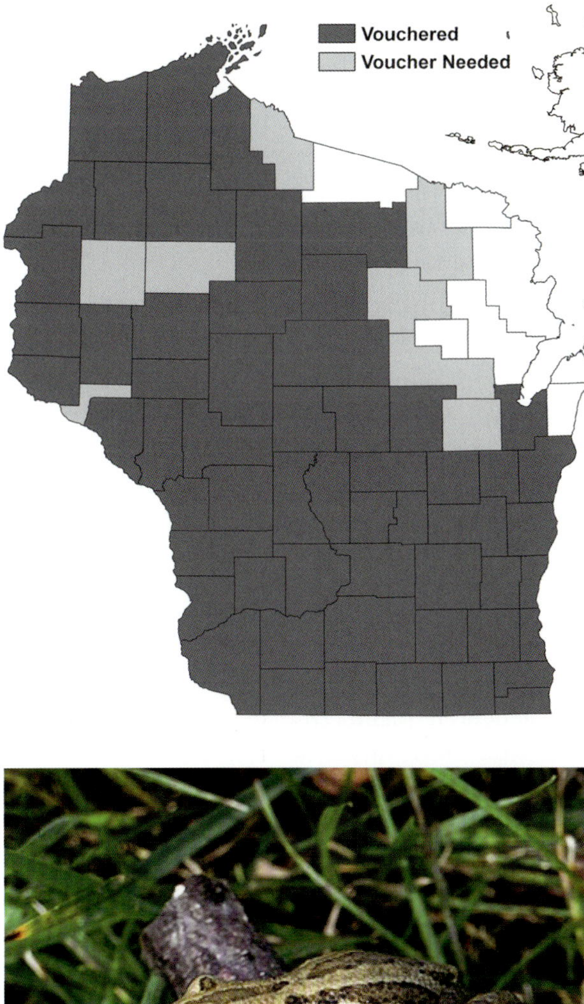

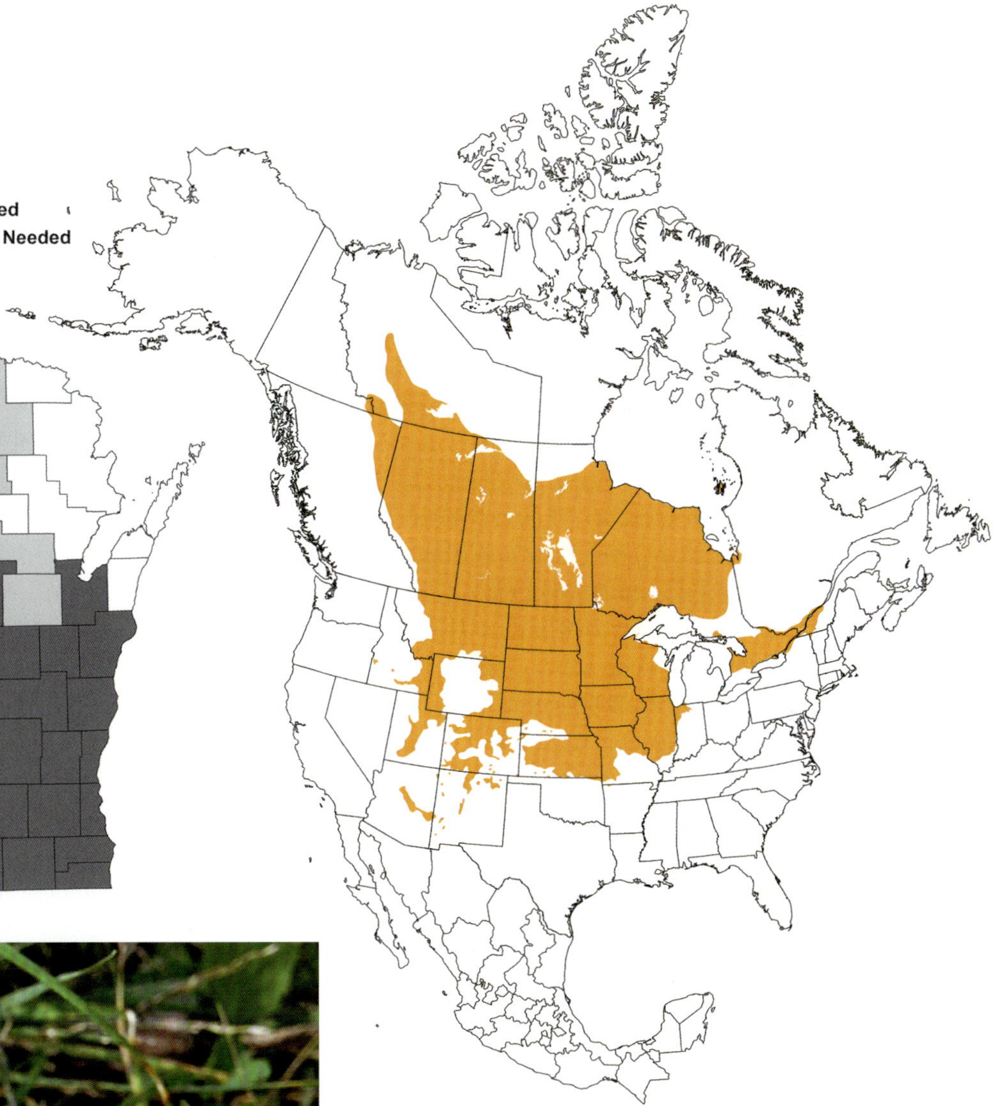

Adult Boreal Chorus Frog (Jefferson County; photo by J. Kapfer).

AT A GLANCE

Boreal Chorus Frogs have gone by several names in Wisconsin, including 'swamp tree frog,' 'striped tree frog,' and 'western chorus frog.' They are small frogs that can be identified by three dark parallel lines composed of stripes and/or spots that run the length of their backs and overlie a background color of green, tan, brown, gray, or even red. Boreal Chorus Frogs are one of Wisconsin's more common frog species and are found throughout much of the state, except several contiguous northeastern counties. They frequent open shallow wetlands during the breeding season, moving to adjacent upland grasslands, old fields, and moist forested areas for the remainder of the year to forage and overwinter. They are one of the first frogs to call in spring, sometimes starting as early as mid-March, and their calls resemble the sound of a finger being dragged across a finely toothed comb. Females typically deposit eggs by late April, and larvae metamorphose from 1.5 to 2 months later. In October, they move to overwintering sites, which are often near breeding wetlands. They hibernate on land below ground debris or in small burrows and survive with the aid of physiological compounds that keep ice crystals from forming within their cells, preventing cellular damage. Adult Boreal Chorus Frogs consume small invertebrates, while larvae scrape algae and detritus off aquatic surfaces. Adults and larvae are prey for larger invertebrates and vertebrates such as snakes and birds. The Boreal Chorus Frog is currently considered a common species in Wisconsin by the WDNR.

SYSTEMATICS

The Boreal Chorus Frog (*Pseudacris maculata*) was initially described by Agassiz (1850) as *Hylodes maculatus*. It has since undergone numerous taxonomic changes that have resulted in the adoption of various scientific names, including *Chorophilus septentrionalis* (Boulenger 1882), *C. nigritus septentrionalis* (Dickerson 1906), *P. septentrionalis* (Stejneger and Barbour 1917), and *P. nigrita septentrionalis* (Wright and Wright 1933). Later studies focused on the distributional patterns of morphological variation in *P. nigrita* (i.e., tibia length to body length ratios; Smith and Smith 1952). Smith (1956) suggested a redetermination of subspecific epithets from *P. n. septentrionalis* to *P. n. maculata*. Reports of intergradation between *P. n. maculata* and what was believed to be the closely related *P. n. triseriata* by Smith (1956) further prompted Schwarz (1957) to suggest that *P. n. maculata* be changed to *P. triseriata maculata*. Platz (1989) performed a multivariate analysis that compared morphological traits and male advertisement call characteristics, which led to the current recognition of *P. maculata* as a species distinct from Western Chorus Frog (*P. triseriata*) rather than separate subspecies (*P. t. maculata* and *P. t. triseriata*). Recent genetic research has upheld the specific status of *P. maculata* and placed it with eight other species into the "Trilling Frog Clade" (Moriarty and Cannatella 2004). Until recently, *P. maculata* and *P. triseriata* were both considered members of Wisconsin's herpetofauna based largely on geographic and morphometric data obtained from Smith (1956, 1957) and Smith and Smith (1952). However, phylogenetic analyses based on molecular data (Moriarty Lemmon et al. 2007a, 2007b, Lemmon and Moriarty Lemmon 2008), as well as statistical analyses of morphological and acoustic behavioral data by Moriarty Lemmon et al. (2008), led to a revision of the geographic ranges of the various chorus frogs and determined that all chorus frogs in Wisconsin were *P. maculata*. Among the other amphibians in the state, this species is most closely related to the Spring Peeper (*P. crucifer*) but also occurs in the same family (Hylidae) as anuran species in the genera *Acris* and *Hyla*.

DESCRIPTION

Adults: Dorsal ground color can be black, gray, dark brown, olive-brown, buff, or bright green and is overlaid with a series of five rows of longitudinal stripes or spots (Knudsen 1958e, Vogt 1981). Typically, there are three parallel dark longitudinal rows that run along the back, extending from the head to the vent (Figure 94). Variation in this pattern is common, however, and some individuals possess three rows of spots or a combination of stripes and spots rather than solid stripes (Figure 94; Ellis and Henderson 1915). Two additional dark lateral stripes, one on each side of the body, run from the nostril through

Figure 94. Variation in coloration and dorsal patterns of Boreal Chorus Frogs. (a) Brown phase adult male with a broken dorsal stripe (Trempealeau County; photo by A. B. Sheldon). (b) Brown phase adult exhibiting a different dorsal stripe pattern (Dane County; photograph by Corey Raimond). (c) Green phase adult with pronounced white upper lip stripe (Price County; photo by Corey Raimond).

the eye and down to the groin (Vogt 1981). A white or cream-colored line also extends along the upper lip on each side of the head (Agassiz 1850). Occasionally, a small brown interorbital triangle or spot occurs on top of the head (Knudsen 1958e). The eyes are small but larger than the tympanum (see Figure 94). Ventral coloration is white or cream-colored, sometimes overlaid with dark speckles on the throat and chest (Knudsen 1958e, Harper 1963). Hind limbs are comparatively short in relation to other species of chorus frogs (see Diagnosis subsection) and generally have transverse bars or spots (Dodd 2013). Forelimbs have four fingers and lack webbing, whereas hind limbs have five toes and can be slightly webbed near the base (Breckenridge 1944). Finger and toe pads are small, expanding slightly at the end of each digit (see Figure 94; Vogt 1981). Boreal Chorus Frogs are one of the smallest species of frog in Wisconsin. Vogt (1981) reported that they ranged in SVL from 2.0 to 3.0 (0.8 to 1.2 in), and the mean SVL of measured Wisconsin specimens in the MPM collection was 2.5 cm (ca. 1.0 in, $n = 26$; note that only individuals ≥2.4 cm [0.9 in] were measured, following information on size at maturity reported by Smith [1987]). The largest specimen measured was 3.0 cm SVL (1.2 in, MPM 2863, collected in 1948 from Clark County).

Preadult stages: Eggs are black and white or tan, although they can be entirely black in some geographic regions (Pettus and Angleton 1967). Egg vitelli are 0.05–0.13 cm (0.02–0.05 in) in diameter (Vogt 1981) and are surrounded by a single indistinct envelope that is ca. 0.30 cm (0.1 in) in diameter (reviewed by Dodd 2013). Boreal Chorus Frog eggs are deposited in loose oval-shaped clusters (1.0–2.0 × 3.0–4.0 cm [0.4–0.8 × 1.2–1.6 in]; Vogt 1981). Boreal Chorus Frog larvae are dark gray, dark brown, or black above and iridescent bronze below (Youngstrom and Smith 1936, Altig 1970). Their eyes are positioned dorsolaterally, and they have two anterior and three posterior labial tooth rows (Youngstrom and Smith 1936, Vogt 1981). Larval tail fins feature a dorsally elevated crest that tapers into an obtuse to rounded tail tip (Youngstrom and Smith 1936). The tail fins are clear but contain brown flecks near the dorsal and ventral crests (Youngstrom and Smith 1936). The tail musculature is bicolored (darker on the

Figure 95. Boreal Chorus Frog tadpole undergoing metamorphosis (Dane County; photo by Corey Raimond).

upper half; Youngstrom and Smith 1936). In Wisconsin, recently hatched larvae measure 0.7–1.0 cm TL (0.3–0.4 in; Vogt 1981). Prior to metamorphosis, larvae may grow to 3.0–3.2 cm TL (1.2–1.3 in; Youngstrom and Smith 1936). Post-metamorphic individuals resemble small adults in color and pattern (Figure 95). In a Michigan study, mean size at metamorphosis was 1.15 cm (0.4 in, range = 0.9–1.5 cm [0.3–0.6 in], $n = 771$; Smith 1987).

Variation: Boreal Chorus Frogs have highly variable coloration and patterning as adults, displaying six to eight possible phenotypes (see Figure 94; Matthews 1971). Knudsen (1958e) further suggested that adult males are generally darker than females in Wisconsin, although robust data to support this are lacking, and both sexes can change background coloration depending on temperature and light conditions (Dodd 2013). Male vocal pouches are usually dark or yellow, though some individuals are described to have plain unpigmented vocal sacs during the breeding season (Harper 1963, Vogt 1981). Vocal sacs also become wrinkled and display transverse folds on the male's throat due to the frequent expansion of the pouch. In addition, males develop nuptial excrescences on the inside of each thumb (i.e., 'thumb pads') that aid in grasping the female during amplexus (Dodd 2013). In a Michigan study, adult females were also generally larger than adult males ($\bar{x}$ SVL = 3.0 and 2.7 cm [1.2 and 1.1 in], respectively, $n = 193$ and 69, respectively; Smith 1987). No polymorphisms have been documented in larvae (reviewed by Moriarty and Lannoo 2005), although amelanistic or albino adults, larvae, and eggs have been documented in Colorado and Illinois (Cahn 1926, Corn 1986).

Diagnosis: Boreal Chorus Frogs are characterized by a slender body without parotoid glands or large warty tubercles and a lack of keratinized spades on the heel of each hind foot. The dorsum has three continuous or broken longitudinal stripes. The head may contain an interorbital triangle or spot, and a white or cream-colored line extends along the upper lip. The terminal toe pads are scarcely wider than the toes, and either the hind limb toes are unwebbed or the webbing is restricted to the base between toes. As adults, Spring Peepers (*Pseudacris crucifer*) and Blanchard's Cricket Frogs (*Acris blanchardi*) are likely the easiest to confuse with Boreal Chorus Frogs, as all can be similar in size. Spring Peepers typically feature an X-shaped dorsal mark rather than the longitudinal stripes of Boreal Chorus Frogs and also have slightly larger toe pads. Blanchard's Cricket Frogs lack dark dorsal stripes and the white upper lip stripe of Boreal Chorus Frogs, have much rougher dorsal skin, and have proportionally longer hind legs. Young Wood Frogs (*Lithobates sylvaticus*) could also be confused with Boreal Chorus Frogs but lack dorsal stripes (or spots) and have dorsolateral folds that run lengthwise down their backs.

Boreal Chorus Frog eggs and larvae are easily confused with those of Spring Peepers, Blanchard's Cricket Frogs, Gray Treefrogs (*Hyla versicolor*), and Cope's Gray Treefrogs (*H. chrysoscelis*). Boreal Chorus Frog eggs are deposited below the water surface in loose clusters of 10–150, whereas similar-sized Spring Peeper, Blanchard's Cricket Frog, Gray Treefrog, and Cope's Gray Treefrog eggs are deposited either individually or in smaller clusters of <40 (Watermolen 1995). Boreal Chorus Frog larvae lack external gills and forelimbs. Their eyes are laterally positioned on the head, while their spiracle tube is short and attached to the body wall. They lack a black tail tip and possess a bronze-colored belly. The tails of Spring Peeper larvae possess a clear area immediately surrounding the tail musculature, unlike the bicolored tail musculature found in Boreal Chorus Frogs (see Preadult stages subsection). Blanchard's Cricket Frog larvae often have black-tipped tails and feature two posterior labial tooth rows, as opposed to three in Boreal Chorus Frogs (Altig 1970). Tail fins in both Gray Treefrog and Cope's Gray Treefrog larvae are extensively reticulated or mottled and thus heavily pigmented; they also frequently possess a reddish coloration (Altig 1970, Dodd 2013).

DISTRIBUTION AND HABITAT

Global distribution: Boreal Chorus Frogs are widely distributed across central North America but primarily occur in north-central regions. Based largely on recent phylogenetic work (see Systematics section) and county-level records, they range from New Mexico and Arizona, north to British Columbia, Yukon Territory, and the Northwest Territories, east to Manitoba and western Ontario, and south to western Indiana, Illinois, Missouri, Kansas, northeastern Oklahoma, and northwestern Arkansas. A disjunct population occurs from southeastern Ontario to southern Quebec (Moriarty Lemmon et al. 2007). Boreal Chorus Frogs have been documented at elevations up to 3,720 m (12,205 ft; Spencer 1971).

Distribution in Wisconsin: Published Boreal Chorus Frog records exist for most Wisconsin counties, although a gap occurs in the northeastern region of the state, which likely reflects low densities rather than overall absence (Mossman et al. 1998). Boreal Chorus Frogs have even been documented on several of the Apostle Islands in Lake Superior (Casper 2008c). A handful of published records currently lack suitable museum vouchers (see the Appendix). Based on data collected through the Wisconsin Frog and Toad Survey, the species appears to be most prevalent in the state's southeastern and central regions (Mossman et al. 1998).

Localities of possible occurrence: Vilas County is an obvious gap in the documented range of this species in Wisconsin. This county is surrounded by counties with published records (although not all are supported by vouchers), so Boreal Chorus Frogs could occur there. In addition, the WDNR's Wisconsin Frog and Toad Survey has generated anecdotal reports of calling males in all counties statewide that currently lack published records (Mossman et al. 1998, WIATRI 2016). Future survey work could expand the documented state distribution of this species even further.

Habitat: Boreal Chorus Frogs are generally associated with open, grassland, and wetland habitats (Knutson et al. 2000, Ouellet et al. 2009), but they are also found in moist forests, ranging from deciduous to mixed boreal-coniferous forests (Marshall and Buell 1955, Roberts and Lewin 1979, Browne et al. 2009). They frequent a variety of wetlands with variable hydroperiods, including marshes, swamps, small lakes, moist prairies, drainage ditches, and wetland-associated edges of deciduous and coniferous forests throughout Wisconsin (Knudsen 1958e, Vogt 1981). Boreal Chorus Frogs attempt to avoid desiccation during hot and dry summer conditions by seeking out refuges in animal burrows and clumps of grass or below terrestrial debris such as leaves, bark, logs, and rocks (Kramer 1973, Roberts and Lewin 1979).

Chorus frogs likely use similar habitat for both breeding and nonbreeding activities, according to the reports of several authors who found that the frogs rarely travel far from breeding sites throughout the active season (Kramer 1973, Roberts and Lewin 1979). It is noteworthy, however, that Vogt (1981) believed that this species spent summers "in moist semiopen areas but not necessarily very close to water." Regardless, Boreal Chorus Frogs in Wisconsin typically breed in shallow emergent wetlands with little tree canopy (Knutson et al. 2000, Trenham et al. 2003; Natural History Box: The Beaver Connection). Breeding wetlands are usually fishless and possess a mosaic of emergent and submersed aquatic plants, such as sedges (*Carex* spp.), rushes (*Juncus* spp.), willows (*Salix* spp.), and other grasses (e.g., *Festuca*, *Hierochloe*, and *Calamagrostis* spp.; Sexton and Phillips 1986, Ouellet et al. 2009). Although breeding wetlands are often ephemeral, some are permanent (Smith 1983, Kolozsvary and Swihart 1999), but specific locations selected for breeding within these wetlands are typically shallow (e.g., <35 cm [14 in] water depth; Ouellet et al. 2009). Larvae, which congregate in shallow warm water, remain in selected breeding

NATURAL HISTORY BOX: The Beaver Connection

In May 2011 I ventured up a small stream in search of Blanchard's Cricket Frogs (*Acris blanchardi*). As I was striking out on that mission, I came across many calling Boreal Chorus Frogs and felt it worthwhile to investigate further. In pursuit of this cryptic species, I waded through a series of five shallow ponds (each ca. ≤0.4 ha [1.0 ac]) that were nestled among willow thickets. To my delight, four of these ponds were created by dams from the North American Beaver (*Castor canadensis*), the largest overlaying an area of ca. 9.1 × 1.5 m (29.9 × 4.9 ft) and rising to 0.9 m (3 ft) above the water's surface. Each dam was made up of the same willows that once stood within ca. 6.1 m (20 ft) of the dams. The dams seemed true to their purpose, pooling up water from the small intermittent stream and allowing the right amount of water to trickle into the pools below. As I approached these frogs with an intent to capture, they evaded my net in every shallow by submerging themselves in water or by hiding within stands of emergent vegetation. Immediately after I left the scene of my half-hearted chase, they continued to taunt me with their calls, all while staying well camouflaged. This complex habitat structure, created by these hardworking furbearers, seems to provide the chorus frog inhabitants a safe haven for breeding and tadpole growth and thus has led to their local proliferation over time. The beavers have additionally become the long-term maintenance crew of these breeding sites, made more ideal for Boreal Chorus Frogs, by keeping woody vegetation from encroaching and becoming too dense. Much later on (and as I've become a more seasoned biologist), I began to make the connection that North American Beavers, with their landscape-scale engineering feats, serve as a valuable species that indirectly affects Boreal Chorus Frogs and the many other anurans and salamanders that rely on the shallow fishless wetlands they create.

Andrew F. Badje, WDNR

wetlands through metamorphosis (Smith 1983). However, individuals will move to deeper water or find cover below aquatic debris if threatened (Moriarty and Lannoo 2005).

Specific observations of overwintering habitat in Wisconsin are rare. Yet reported winter hibernacula elsewhere include cavities within decaying stumps, root systems, and animal burrows and under cover objects (e.g., discarded boards) near breeding sites (Roberts and Lewin 1979, Cochran 1989, Dodd 2013). Carpenter (1953) further found that Western Chorus Frogs from Michigan overwintered in ant mounds at depths ranging from 16.2 to 53.3 cm (6 to 21 in).

REPRODUCTION AND DEVELOPMENT

In Wisconsin, Boreal Chorus Frogs are one of the first anurans to arrive at breeding wetlands in spring (Mossman et al. 1998; Natural History Box: Ecoacoustics), and breeding sites can still be partially covered with ice when males first begin to call (Andrew Badje, WDNR, personal observation). Time of arrival and initiation of calling activity are highly dependent on winter or early spring weather. For example, Vogt (1981) observed that initiation of calling activity ranged from 3 March to mid-April in Dane County and that calling activity in the cooler regions of northern Wisconsin typically began ca. one month after warmer regions of southern Wisconsin. Males generally begin calling when water temperatures approach 2.2°C (36.0°F), increasing in number and intensity as temperatures surpass 4.4°C (40.0°F; WDNR, unpublished data). Smith et al. (2014) also found that calling activity during the breeding season was positively correlated with air temperature. The breeding season can last for several months and in Wisconsin extends

NATURAL HISTORY BOX: Ecoacoustics

Boreal Chorus Frogs are usually the first frogs I hear calling every spring in southern Wisconsin (but see Spring Peeper, *Pseudacris crucifer* species account), though they aren't normally the first species to begin breeding (see Wood Frog, *Lithobates sylvaticus* species account). The first warm nights of every spring, I find myself instinctively migrating toward my neighborhood kettle pond to hear the resident male chorus frogs' signature melody, which sounds like running a finger across the teeth of a fine-toothed comb. In Dane County on one particularly cool spring evening of ca. 7.2°C (45°F), I retrieved a male Boreal Chorus Frog from a minnow trap with a back so dark that his stripes were barely noticeable. After I released him into the water, I stepped back and waited, flashlight in hand, until he began to call two minutes later. The intensity of his call was greatly affected by the cold weather, and he struggled to eke out a few calls at much slower 'creeks' than I typically witness on warmer nights. As spring carried on and nights progressively became warmer, the sunset biophonies of chorus frogs continued to build in intensity, drowning out the calls of local Wood Frogs. Every year, I find that these choruses tend to peak in mid-spring, during which it seems they can reach similar decibel levels to a rock concert. At times, the accumulation and intensity of chorus frog calls can really cause headaches for an unsuspecting homeowner. On a handful of occasions, I've been contacted by concerned citizens living near wetlands who inquire about what, if anything, can be done to stymie these amphibian cacophonies so a person can sleep at night. I give them a little hope and inform them that high-intensity calling periods shouldn't last longer than a few weeks. In the meantime, if one can't learn to enjoy the sounds of nature (of which I'm a biased advocate), then a pack of earplugs in a nightstand should help mitigate the problems these ecoacoustics cause.

Andrew F. Badje, WDNR

until May or June (Vogt 1981, Mossman et al. 1998). Although males occasionally call in autumn (Vogt 1981, Cochran 1989), calling during this period does not likely result in reproduction.

In Wisconsin, males arrive at breeding sites a week or more before females (Vogt 1981), whom males attempt to attract for reproduction via advertisement calls. Advertisement calls, produced by a single vocal sac in the throat region, consist of short separated 'it-it-it-it' notes, similar to the sound produced by running one's fingernail along the teeth of a stiff comb (Vogt 1981). Chorus frog calls can last 0.59–1.16 seconds, with dominant frequencies that range from 2.91 to 3.58 kHz (Platz 1989). Males typically call from within vegetation clumps near the shoreline or in open water while resting on top of submersed vegetation where they are well hidden from observation (Vogt 1981, Whiting 2004). Calling activity can occur night and day (Whitaker 1971), but Bowers et al. (1998) determined that during the height of the breeding season in North Dakota, peak calling activity occurred between dusk and 10:00 p.m.

Although calling to attract females is the primary method most males use to gain reproductive opportunities, males can also engage in alternate mating strategies. For example, some males at breeding sites act as silent 'satellites' of actively calling males. These satellite males wait quietly near calling males and attempt to intercept females that are attracted to the other male's call (Roble 1985). Roble (1985) further determined that this strategy is commonly employed in some populations, estimating that 60.5% of calling Boreal Chorus Frog groups studied in Kansas had at least one male engaged in satellite behavior. During this study, he detected no significant difference in size of satellite vs. calling males ($n = 19$ comparisons)

Figure 96. Boreal Chorus Frog egg mass attached to submerged vegetation (Dane County; photo by Sara Viernum).

and also found that males can switch between use of active calling and satellite strategies during the mating season (Roble 1985). In Colorado, male color phenotypes did not appear to influence female mate choice (Matthews 1971).

In Wisconsin, females normally deposit eggs in mid- to late April, although egg deposition can occur from early March to late May (Vogt 1981, Watermolen 1995). Clutch sizes range from 20 to 100 eggs/mass in Wisconsin (Vogt 1981), and masses are typically attached to emergent or submerged vegetation in shallow water at 7.0–20.0 cm (2.8–7.9 in) below the water's surface (Figure 96; Watermolen 1995).

Although developmental rates of eggs and larvae can be influenced by factors such as water temperature, larval density, and wetland hydroperiod length (Youngstrom and Smith 1936, Whitaker 1971), eggs in Wisconsin typically hatch within 6–18 days, and larvae usually complete metamorphosis in 1.5–2 months (Vogt 1981). Development after metamorphosis is rapid in Boreal Chorus Frogs. Whiting (2004) estimated that they are capable of reaching sexual maturity and breeding the first year following metamorphosis, although developmental rates are not likely uniform across their North American distribution. For example, in northern parts of their range they may not breed until their second or third year (Roberts and Lewin 1979, Muths et al. 2016), and females often take one year longer to reach sexual maturity than males (Smith 1987). Smith (1987) further reported minimum SVL in mature females (i.e., those that were gravid or observed in amplexus) was 2.5 cm (0.98 in), while the mean SVL was 2.96 cm (1.17 in, $n = 193$). Smith (1987) documented minimum SVL of mature males at 2.2 cm (0.87 in), with a mean SVL of 2.66 cm (1.04 in, $n = 193$). Boreal Chorus Frogs are likely a short-lived anuran. The majority of adults might not survive to a second breeding season (Whiting 2004), although Muths et al. (2016) estimated longevity in the wild of up to five to seven years.

ACTIVITY

Spring emergence is likely triggered by increasing air temperature, which induces ice and snowmelt in upland overwintering habitat. In mid-March to April, when air temperatures are consistently above 5°C (41°F; Whiting 2004), Boreal Chorus Frogs thaw (see below), regain bodily functions, and begin to disperse to nearby wetlands to prepare for breeding. After emergence, Boreal Chorus Frogs can remain active over a wide range of air temperatures. For example, Clarke (1958) observed active individuals at air temperatures ranging from 4.4°C to 32.2°C (40.0°F to 90.0°F) in Kansas.

Information on movement patterns is sparse. Following the breeding period, adult Boreal Chorus Frogs move to upland foraging areas adjacent to breeding sites. In Quebec, this post-breeding migration occurred from June through October (Whiting 2004). Yet past studies support that the frogs do not likely stray far from wetlands as the result of such migrations. For example, Roberts and Lewin (1979) found that most individuals remained within a mean distance of 30 m (98 ft) from breeding wetlands in Alberta. Furthermore, most individuals of the closely related Western Chorus Frog remained within 100 m (328 ft) of breeding wetlands in Indiana, although one individual reportedly moved 213 m (699 ft) away (Kramer 1973). Despite this, some authors have recorded even larger movements. For example, Spencer (1964) reported juvenile dispersal distances of >597 m (1,900 ft) from wetlands to meadows. To our knowledge, no studies have estimated home range sizes of Boreal Chorus Frogs in the Upper Midwest.

However, Kramer (1974) tracked the movements of Western Chorus Frogs in breeding wetlands and surrounding upland habitat in Indiana. For the nine individuals from which he obtained at least ten relocations (all male), home range size ranged from 0.064 to 0.602 ha (0.158 to 1.487 ac, $\bar{x} = 0.212$ ha [0.524 ac], MCP estimator). Kramer (1973) further estimated mean daily movements of 3.5 m (11.5 ft), with a maximum daily movement of 42.0 m (137.8 ft, $n = 234$ recaptures of 47 individuals). Using a pitfall trap grid, Spencer (1964) determined that recently metamorphosed Boreal Chorus Frogs typically moved >61 m/day (200 ft/day) when dispersing from a wetland in Colorado.

In Wisconsin, Boreal Chorus Frogs likely enter hibernation in October (Vogt 1981). Edgren (1944) also took two specimens from "the bed of a stream late in September" in Ashland County but did not mention whether they were active or in torpor at the time of collection. They are not known to overwinter while submersed underwater, unlike some anurans in Wisconsin (see Habitat subsection), so they must employ adaptations to survive freezing conditions. To this end, glycogen stores in the liver of Boreal Chorus Frogs synthesize glucose, a cryoprotectant, which is mobilized and accumulates within tissues of the liver and muscles during the winter (Swanson et al. 1996, Jenkins and Swanson 2005, Edwards et al. 2000). These cryoprotectants limit cellular dehydration, and although they allow ice crystals to form in the extracellular fluid, they prevent ice crystal formation (and associated damage) within cells (Storey and Storey 1987). These cryoprotectants thus stabilize cellular structure and function at freezing temperatures, which allows Boreal Chorus Frogs to supercool (i.e., lower their body temperatures below their normal freezing points; MacArthur and Dandy 1982, Storey and Storey 1987). In this state, their breathing, heart rate, and blood flow slow dramatically or suspend entirely (Storey and Storey 1987). Boreal Chorus Frogs can survive several days of exposure to temperatures as low as −3°C to −2.5°C (26.6°F to 27.5°F; MacArthur and Dandy 1982, Storey and Storey 1987). Jenkins and Swanson (2005) further reported that larger Boreal Chorus Frogs show increased overwintering survival because their proportionally larger livers could store more glycogen and subsequently produce higher levels of cryoprotectants.

PREY AND PREDATORS

No robust dietary studies are known for Boreal Chorus Frogs in Wisconsin. Moore and Strickland (1954) identified stomach contents of 37 adults in Alberta. These contained a wide variety of small arthropods (e.g., Diptera and Coleoptera), indicating that Boreal Chorus Frogs are generalist predators of invertebrates. Whitaker (1971) studied their dietary habits across all life stages in Indiana. His findings further support that adult Boreal Chorus Frogs are generalist predators of small arthropods (particularly Arachnida and Formicidae), but, unlike Moore and Strickland (1954), he also documented high amounts of incidentally ingested plant matter during the breeding season. Whitaker (1971) found that larvae primarily consumed algae ($n = 38$ stomachs examined), and individuals undergoing metamorphosis generally ate sparingly ($n = 11$ stomachs examined, 9 of which were empty).

Boreal Chorus Frogs are taken by a variety of predators. In Wisconsin, gartersnakes (*Thamnophis* spp.) and Common Watersnakes (*Nerodia sipedon*) are known predators, while young Eastern Foxsnakes (*Pantherophis vulpinus*) and wading birds are considered likely predators of adults (Vogt 1981). Birds are also probably important predators of larvae and juveniles in Wisconsin, as Gray Jays (*Perisoreus canadensis*) reportedly consume them in Colorado (Tordoff 1980). Larvae are further eaten by numerous additional aquatic and semiaquatic predators, such as various adult and larval mole salamanders (*Ambystoma* spp.), gartersnakes (*Thamnophis* spp.), bass (*Micropterus* spp.), dragonfly naiads (*Anax* spp.), and predaceous diving beetles (*Dytiscus* spp.; Smith 1983, Dodd 2013). Western Chorus Frog larvae are palatable to fishes (Kats et al. 1988). Although this is presumably also true for Boreal Chorus Frogs, under natural conditions larvae should infrequently encounter fishes in the ephemeral wetlands where larvae develop. See Dodd (2013) for a comprehensive list of documented and suspected Boreal Chorus Frog predators.

Adult Boreal Chorus Frogs rely on their minuscule size, reticent behavior, and color patterns to avoid predators (Roberts and Lewin 1979). During the breeding season, males that feel threatened will cease to call and remain still or "seem to throw their voices, as if they were calling from another area," which presumably makes it difficult for predators to detect them (Vogt 1981). Larvae also employ behavioral strategies to avoid predators (i.e., remaining motionless) but further adopt morphological alterations for the same purpose. For example, during aquatic mesocosm experiments, Van Buskirk et al. (1997) documented that Western Chorus Frog larvae responded to the presence of dragonfly naiads (*Anax*) predators by increasing the depth of tail muscles and fins, which likely allowed individuals to swim faster and evade aquatic predators more easily. Bolek and Coggins (1998) recorded protozoan (*Nyctotherus cordiformis*), digenetic trematode (*Glypthelmins pennsylvaniensis*), and nematode (*Cosmocercoides variabilis*) parasites of wild Boreal Chorus Frogs from southeastern Wisconsin. A range-wide comprehensive list of documented parasites is reviewed by Dodd (2013).

CONSERVATION

Status: Globally, the Boreal Chorus Frog is listed as a Least Concern species by the IUCN due to a large number of occurrences, stable trends, and subpopulation viability. However, some populations in Canada (i.e., eastern Ontario and southern Quebec) and the US (i.e., northern New York) have experienced localized declines (Daigle 1997a, Gibbs et al. 2005, Seburn et al. 2008). Historical accounts indicate that Boreal Chorus Frogs have long been considered common in Wisconsin. Elaborating on their midcentury state abundance, Knudsen (1958e) claimed, "Since there is so much water in Wisconsin I would go so far as to say there is hardly a township that does not have a population." The Boreal Chorus Frog is currently considered a common species in Wisconsin by the WDNR. Analyses of Wisconsin Frog and Toad Survey data from 1984 to 1995 found stable populations at the state level but a significant decline in the northern Lake Michigan coastal and central Lake Michigan coastal ecological regions (Mossman et al. 1998). Longer-term trend data (i.e., 1984–2016) from the Wisconsin Frog and Toad Survey support that Boreal Chorus Frogs are stable at the state level (Figure 97).

Populations: Robust population studies in the Upper Midwest are currently lacking. Documented population size estimates at sites in Colorado ranged from 37 to 83, 49 to 262, 471 to 580, and 1,000 to 2,000 individuals (Matthews 1971, Tordoff and Pettus 1977). Corn et al. (2011) further estimated populations of calling males over 15 years at two wetlands in Colorado that ranged from 53 to 1,845 and 18 to 404 individuals. Roberts and Lewin (1979) reported a mean Boreal Chorus Frog minimum density of 23 individuals/ha (9.3 individuals/ac) across two locations in northern Alberta.

Reports of survival and survivorship estimates show substantial variability. In Quebec, estimated survivorship of larvae to metamorphosis ranged from 3.2% to 6.2% (Whiting 2004). In Michigan,

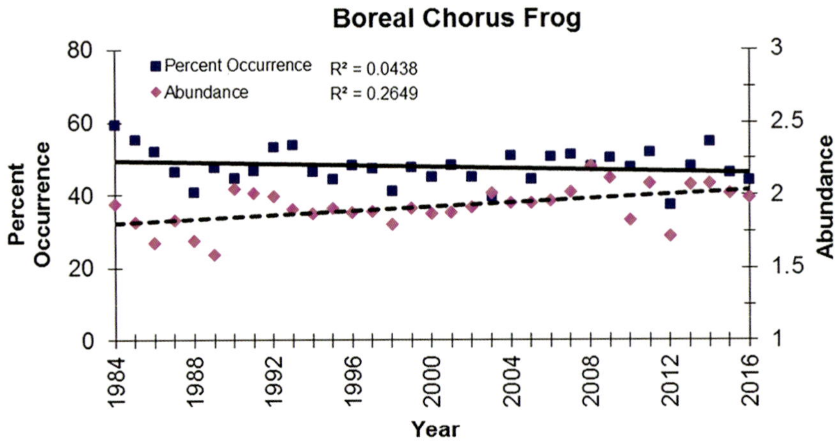

Figure 97. Long-term trend in occurrence (%) and abundance (rank) of Boreal Chorus Frogs across Wisconsin Frog and Toad Survey statewide monitoring sites. See Introduction to Species Accounts chapter for details on this monitoring program (figure courtesy of WDNR).

estimated male survivorship from metamorphosis to maturity was 19%, with early metamorphosis resulting in higher rates of survivorship (Smith 1987). Whiting (2004) investigated adult population dynamics at a wetland in Quebec over the course of two years using a drift fence. This study estimated that survival of adult males during the two breeding seasons was 26.9% and 25.4%, respectively, and estimated survival of adult females was 36.2% and 49.7%, respectively. Of the 304 adults that exited the wetland the first year, 8 returned the following year, resulting in a minimum overwinter survival estimate of 2.6% (Whiting 2004). These survival estimates are lower than probabilities determined for a subalpine population in Colorado, which ranged from 68% to 79% (Muths et al. 2016). Studies in Colorado found that sex ratios were ca. 1M:1F (Matthews 1971, Tordoff and Pettus 1977). However, sex ratios at breeding wetlands are typically skewed toward males (e.g., 3.5M:1F; Matthews 1971).

Threats: Although Boreal Chorus Frogs occur in agricultural and urbanized landscapes (Knutson et al. 2004, Browne et al. 2009, Bartelt and Klaver 2017), the most severe threats they face are likely due to habitat loss or related factors that result from anthropogenic land use. For example, the negative impacts of habitat fragmentation and degradation, or reduction in the quality of available habitat, are important concerns (Ouellet et al. 2005). Several studies have analyzed occurrence data of both the Boreal Chorus Frog and Western Chorus Frog that span several decades in Michigan and New York (e.g., Skelly et al. 1999, Gibbs et al. 2005, respectively). These studies determined that Boreal Chorus Frogs were absent or had disappeared from historically open sites that had experienced degradation due to encroachment of woody vegetation. Habitat quality can also be degraded due to an influx of invasive species and introductions of native species that prey on Boreal Chorus Frogs. For example, Sacerdote and King (2014) found that secondary metabolites of the invasive European Buckthorn (*Rhamnus cathartica*) can cause malformations and significant embryo mortality in Boreal Chorus Frogs. Furthermore, the stocking of fishes into ephemeral wetlands can negatively impact amphibian egg and larvae survivorship (see Introduction to Conservation and Management of Wisconsin's Amphibians and Reptiles chapter). In addition, environmental pollution can degrade Boreal Chorus Frog habitat, and numerous studies have documented the potential impacts that various toxicants can have on this species (reviewed by Dodd 2013).

The potential effects of climate change are not well-studied for most amphibians in Wisconsin (see Introduction to Conservation and Management of Wisconsin's Amphibians and Reptiles chapter). Yet Amburgey et al. (2012) conducted controlled experiments on the influence of hydroperiod in the development and survival of Boreal Chorus Frog larvae. They found that larvae did not undergo metamorphosis earlier in ponds that dried more quickly. Given that climate change is projected to result in hotter and seasonally drier environments than currently exist in Wisconsin (see Past, Present, and Potential Future Landscapes of Wisconsin chapter), the accelerated desiccation of ephemeral wetlands where larvae develop could lead to higher larval mortality rates and population-level impacts.

Numerous diseases that are of conservation concern for many amphibians could also affect Boreal Chorus Frogs. Laboratory research indicates that the chytrid fungus *Batrachochytrium dendrobatidis* (Bd) can negatively impact adult Boreal Chorus Frog survival (Hyman 2012), and Bd infection has been reported from wild populations in various states (Rittman et al. 2003, Ouellet et al. 2005, Hyman 2012, Phillips et al. 2014), including Wisconsin (Standish et al. 2018, but see also Sadinski et al. 2010). However, although Bd can cause Boreal Chorus Frog mortality, whether that mortality alone is substantial enough to cause wild population declines requires further study. This is highlighted by the work of Ouellet et al. (2005). Their study reported widespread Bd infection over time in North America, yet the authors concluded that Boreal Chorus Frog declines in Quebec could actually be attributed to other underlying threats, such as habitat loss and fragmentation, that might exacerbate the impacts of the disease (Ouellet et al. 2005).

Management: Many general wetland management recommendations for amphibians apply to this species (see Introduction to Conservation and Management of Wisconsin's Amphibians and Reptiles chapter). Given that Boreal Chorus Frogs are absent or disappear from sites that become too overgrown with trees (Skelly et al. 1999, Gibbs et al. 2005), management strategies that limit encroachment of woody vegetation in breeding wetlands could benefit this species (see Natural History Box: The Beaver Connection). In addition, the scale of habitat management initiatives is likely important to consider for strategies focused on Boreal Chorus Frogs. For example, the occurrence of this species is reportedly influenced by landscape-scale availability and configuration of suitable habitat (e.g., Kolozsvary and Swihart 1999, Gibbs et al. 2005, Scherer et al. 2012). Bartelt and Klaver (2017) suggested that intact landscapes with conglomerates of numerous and interconnected small (15–130 ha [37–321 ac]) wetlands that are separated by short distances (≤3 km [1.9 mi]) "can provide a network of quality breeding ponds to support metapopulations of anurans," including rapid colonizers such as Boreal Chorus Frogs. Scherer et al. (2012) further found evidence to support that Boreal Chorus Frog presence at a given wetland was positively influenced by the number of neighboring wetlands where this species already occurred. Creation and maintenance of suitable upland habitat adjacent to wetlands can further provide undisturbed areas for nonbreeding activities and possible avenues of connectivity among breeding sites (Werner et al. 2009). Considerations for optimal upland habitat management include maintenance of native forest and grassland land cover but also suitable ground debris (e.g., leaves, logs, rocks) that can provide cool moist refugia during the summer (Kingsbury and Gibson 2012).

Several studies also report that Boreal Chorus Frogs colonize created and restored wetlands without human intervention (Kline 1998, Lehtinen and Galatowitsch 2001, Dodson 2008, Bartelt and Klaver 2017). Work on restored wetlands in Iowa, for example, found that within a year Boreal Chorus Frogs colonized 100% of sites that underwent restoration (Bartelt and Klaver 2017). The authors suggested that this species' small size allowed it to find adequate cover from hot dry conditions around newly restored wetlands, and because these frogs do not need to overwinter underwater, they could quickly colonize restored sites from nearby overwintering locations. Management that maintains the ephemeral and semipermanent nature of wetlands so they remain free of predatory fishes would likely benefit for this species, as such predators can limit larval abundances of Western Chorus Frogs (Werner et al. 2009).

American Bullfrog

Lithobates catesbeianus (Shaw 1802)

Andrew F. Badje and John D. Peterson

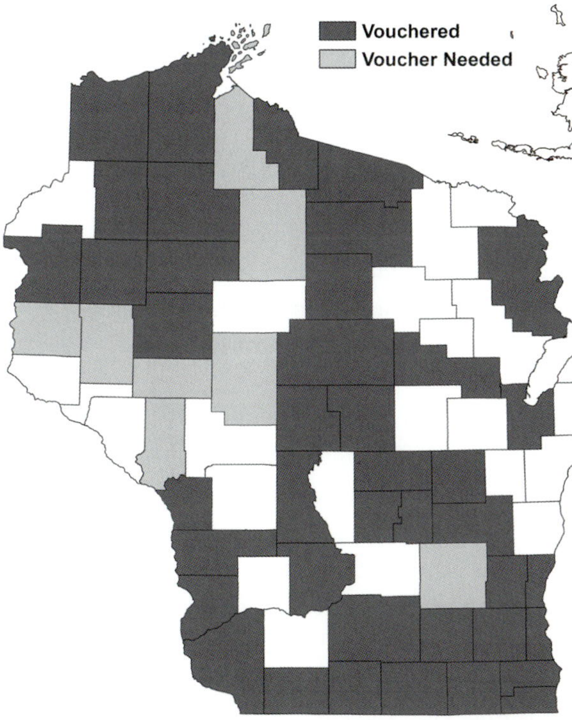

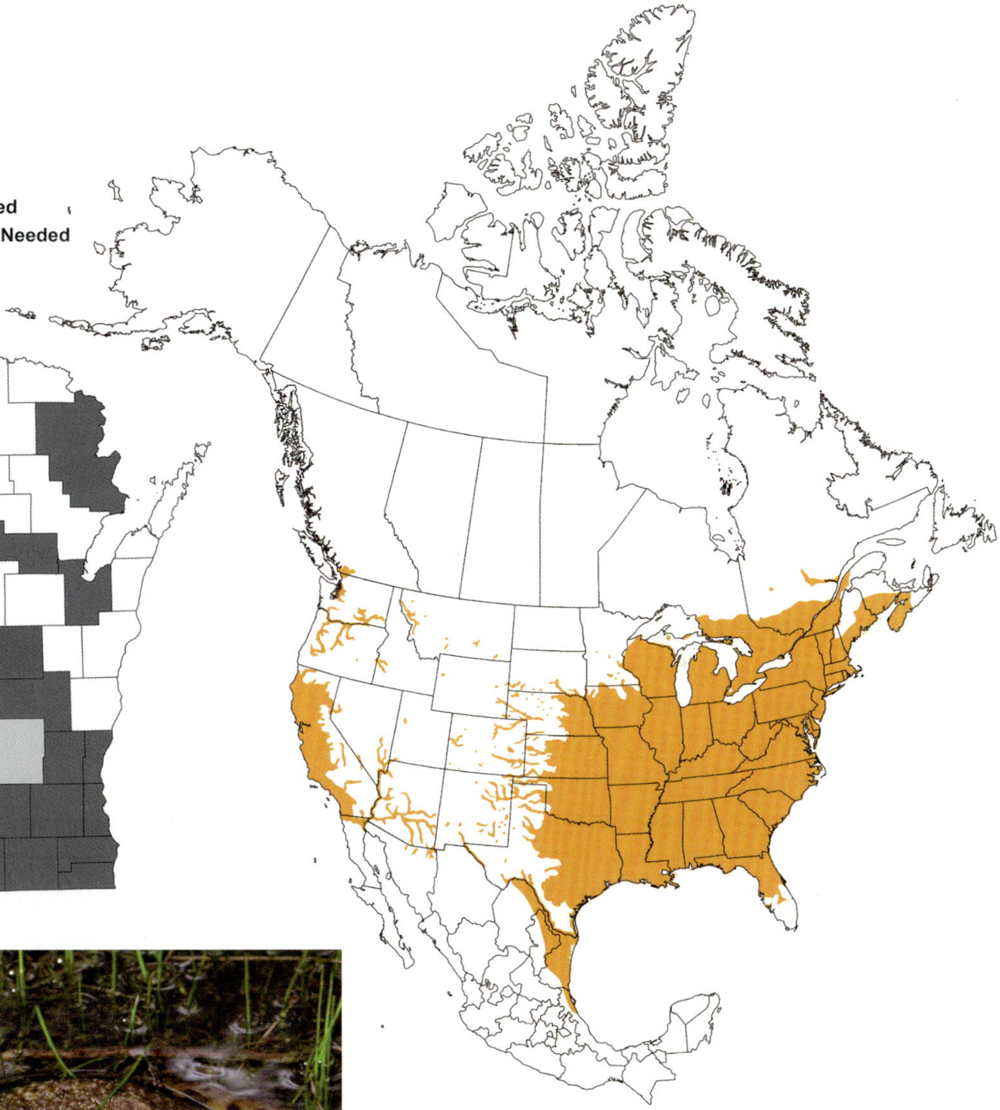

American Bullfrog (Oneida County; photo by D. Feldkirchner).

AT A GLANCE

American Bullfrogs are the largest frog species in Wisconsin and all of North America. Their backs are typically dark gray, green, or brown and often covered in mottling or speckles. In some larger individuals, the darker colors on the back transition to brighter greens around the head (particularly near the mouth) and white to yellow along the lower jaw and throat. American Bullfrogs can also be identified by their very large tympanum and lack of dorsolateral folds along their backs. Although the sexes are similar in appearance, males are generally smaller than females and have a dark yellow throat and tympanums that are larger in diameter than their eyes (whereas the tympanums and eyes of females are of similar diameter). American Bullfrogs occur sporadically throughout much of Wisconsin, although published records do not exist for several contiguous western and northeastern counties. They are typically found in open aquatic habitats with permanent water, such as lakes, rivers, creeks, and deep wetlands. They usually enter hibernation from late September to November and overwinter while submersed in water that does not freeze. Emergence from hibernation occurs later than in other frogs (from mid-May to mid-June), and males do not begin calling to attract mates until early June or mid-July. The breeding call is well known for its reverberating, deep 'jug-o-rum' refrain. Calling males are extremely territorial and will actively defend potential breeding locations from invading males. Larvae usually do not metamorphose in the same summer in which they hatch, so they overwinter as adolescents and transform the following summer (or later). American Bullfrogs can be major predators in aquatic habitats. They eat a wide variety of invertebrates and vertebrates, including small birds and mammals. Larvae are omnivorous, both consuming vegetation (such as algae) and feeding off dead animal material. American Bullfrogs are native to the central and eastern US, but introduced populations exist in western states, where they are considered an invasive species. In their native range, however, past harvest by humans for consumption and bait or by biological supply companies has taken a substantial toll on populations. The WDNR currently classifies American Bullfrogs as a species with Information Needs in Wisconsin, and their harvest is regulated.

SYSTEMATICS

In North America, an eastern and a western lineage of *Lithobates catesbeianus* (American Bullfrog) is recognized, with the western lineage arising out of the eastern one ca. 0.38–1.3 million years ago (Austin et al. 2004, Austin and Zamudio 2008). The geographic boundaries between these two lineages are not entirely distinct. Further complicating matters, it is suspected that irregularities between the boundaries are partially caused by anthropogenic translocation (Austin et al. 2004). Due to a lack of sampling in Wisconsin, it is uncertain which lineage of *L. catesbeianus* occurs in the state (Austin et al. 2004). There are no currently recognized subspecies (Crother 2017). Phylogenetic analyses place *L. catesbeianus* in the same clade as *L. clamitans* (Green Frogs) and *L. septentrionalis* (Mink Frogs; Austin et al. 2003b), to which it is most closely related in Wisconsin.

DESCRIPTION

Adults: Dorsal coloration in Wisconsin ranges from gray or brown to green and yellow, with varying amounts of brown mottling (Knudsen 1958a, Vogt 1981). The majority of the back and legs are dark gray or brown, while the head anterior to the tympanum and below the eyes toward the tip of the nose is brighter green (upper jaw) and white or yellow (lower jaw and throat; Figure 98). The venter is consider-

Figure 98. Adult American Bullfrog (La Crosse County; photo by A. Badje).

ably lighter than the dorsum, ranging from white to cream-colored, although some individuals may have dusky reticulations that partially or completely cover the belly (Vogt 1981, Dodd 2013). Dorsal skin texture is rougher in adults than in subadults due to the presence of skin tubercles (Knudsen 1958a, Vogt 1981). Dark bars occur on the hind legs in some individuals from Wisconsin and Minnesota (Breckenridge 1944, Vogt 1981). Although the front toes are not webbed, the toes on the hind feet are almost fully webbed, except for the distal portion of the fourth and longest toe (Dodd 2013). American Bullfrogs possess large tympanums and prominent supratympanic folds (Knudsen 1958a, Vogt 1981) but lack the dorsolateral folds found in most North American ranids (Bury and Whelan 1984). American Bullfrogs are the largest native frog in North America (adult SVL typically ranges from 10 to 15 cm [3.9 to 5.9 in] in Wisconsin; Vogt 1981). Among Wisconsin museum specimens measured, the largest SVL of an intact individual was 16.3 cm (6.4 in, MPM 761, collected in 1907 from Washington County), although the SVL of a specimen with no other associated data except "Wisconsin" was 17.7 cm (7 in, MPM 63).

Preadult stages: American Bullfrog eggs are typically bicolored (black and white) early in development. Individual eggs have a vitellus diameter of 0.12–0.18 cm (0.05–0.07 in; Knudsen 1958a, Vogt 1981) and a single outer gelatinous membrane that is 0.6–1.0 cm (0.24–0.41 in) in diameter (Dodd 2013). This single outer membrane is difficult to distinguish, as it often merges with the membranes of neighboring eggs (Dodd 2013). Eggs are laid over submerged or emergent vegetation in large sheets (0.6 × 0.3–0.6 m [2.0 × 1.0–2.0 ft]) that float at the water's surface. As the sheet settles, it may remain flat or form a more clustered mass (Vogt 1981, Dodd 2013). Larvae from Michigan and Wisconsin have a brownish-green dorsum with distinct black spots that cover the body and tail (Ruthven et al. 1928, Vogt 1981). Their ventral surfaces are white, cream, or yellow (Dodd 2013). Larvae also possess dorsolaterally positioned eyes, a dextral vent tube, and a sinistral spiracle (Altig 1970). In addition, larvae typically have a labial tooth row formula of two over three (two upper and three lower tooth rows; Breckenridge 1944). American Bullfrog

Figure 99. Juvenile American Bullfrog (Fond Du Lac County; photo by R. Paloski).

larvae can grow very large. The mean SVL of ten larval specimens from Wisconsin in the MPM collection was 5.3 cm (2.1 in), and the TL of the longest specimen measured was 12.4 cm (4.8 in, MPM 11564, collected in 1973 from Vilas County). Immediately following metamorphosis, juveniles in Wisconsin are ca. 5.1 cm SVL (2.0 in; Knudsen 1958a). Subadult American Bullfrogs are similar in color to adults but have smoother skin and small round black spots that fade or disappear with age (Vogt 1981; Figure 99).

Variation: American Bullfrogs exhibit several forms of sexual dimorphism. A four-year study in Michigan found that mean female SUL (range = 12.4–14.3 cm [4.9–5.6 in], $n = 130$) was significantly larger than mean male SUL (range = 11.5–13.2 cm [4.6–5.2 in], $n = 153$) in each year of the study (Howard 1981). The tympanums of adult males are noticeably larger in diameter than their eyes, whereas adult female tympanums are roughly equal in diameter to the eye (Bury and Whelan 1984). This sex-specific difference in tympanum size only becomes evident after sexual maturity, and the larger tympanums likely assist males in projecting their advertisement calls (Schroeder 1966, Purgue 1997). During the breeding season, the throats of male American Bullfrogs become wrinkled and baggy and change from cream to yellow in color (Vogt 1981, Dodd 2013). In addition, mature males have larger forearms and develop swollen thumbs during the breeding season (Dodd

2013). Aside from sexual dimorphisms, albino and mostly yellow-colored individuals are occasionally documented in the wild throughout much of the species' US range (reviewed in Dodd 2013), while blue (axanthic) and melanistic individuals have also been observed in more southern locations (Berns and Uhler 1966, reviewed by Powell et al. 2016).

Diagnosis: American Bullfrogs are characterized by a body without parotoid glands or large warts and without keratinized spades on their heels. The species also lacks terminal toe pads and has no musky odor. Dorsolateral folds are absent, and the supratympanic fold is prominent. Adults can be mistaken for other ranid frog species such as Green Frogs (*Lithobates clamitans*) and Mink Frogs (*L. septentrionalis*), because all three can overlap in size, are sometimes similar in color, and can have large tympanums. However, American Bullfrogs can attain a larger adult size than the other two species. They also have prominent supratympanic folds (absent in the other two species) and lack the dorsolateral folds that are present in Green Frogs and Mink Frogs. Additionally, Mink Frogs produce a musky odor when handled, unlike American Bullfrogs. Male advertisement calls among these species are also very distinct (see Reproduction and Development section).

Green Frog and American Bullfrog egg masses are difficult to distinguish in Wisconsin because both lay their eggs in a large thin mass on the surface of the water (Mink Frogs lay their eggs below the surface of the water). Green Frogs, however, have smaller individual eggs and typically fewer eggs per mass (<5,000 eggs). Furthermore, individual American Bullfrog eggs are encapsulated with a single gelatinous layer, whereas Green Frogs are encapsulated in two layers (Altig and McDiarmid 2015). American Bullfrog larvae have internal gills and lack externally visible forelimbs. Their eyes are dorsally positioned on their heads. Their vent is dextral, and the intestinal coil is at least partially visible. There is no white line from the nostril to the snout, and their tails possess dark spots or mottling (which is more dense along the dorsal than the ventral tail fin) but lack pinkish and fine yellow-colored spots. The larvae of all ranid species are superficially similar in appearance, although those of American Bullfrogs can reach larger sizes than all others in Wisconsin. Other than differences in size, American Bullfrog larvae typically have small distinct black spots on the tail, whereas the tails of Green Frog larvae typically have yellow spots, and the tails of Mink Frogs typically have pinkish-buff spots (Altig 1970, Watermolen and Gilbert 1996).

DISTRIBUTION AND HABITAT

Global distribution: American Bullfrogs naturally occur from southeastern Canada south to Florida and west to the central US and northern Mexico. The historical distribution of American Bullfrogs in North America was traditionally limited in the North by cold temperatures and in the western US by arid climates and a lack of suitable wetlands (Dodd 2013). Yet due to unnatural habitat modifications that lead to permanent water in normally arid regions (e.g., irrigation canals) and human introductions, American Bullfrogs are now established in western states, as well as throughout the world (reviewed by Dodd 2013). Consequently, the current natural western range limits for this species in the US cannot be easily deciphered (Bury and Whelan 1984).

Distribution in Wisconsin: American Bullfrogs are found throughout much of Wisconsin. However, many counties, particularly in the northeastern region, lack confirmed records. Casper (1996) hypothesized that "the many counties lacking documentation indicate local, scattered populations" rather than true absence. Of the published county records, suitable museum vouchers could not be found or have been reported lost for Ashland, Clark, Dodge, Dunn, Eau Claire, Price, and St. Croix Counties (see the Appendix). Furthermore, the voucher specimen reported for Trempealeau County (CHAS HERP 19340) was a misidentified Green Frog (Joshua Kapfer, UW–Whitewater, personal communication) and is not housed at FMNH (as indicated by Casper 2015; Alan Resetar, FMNH, personal communication). Historical data suggest that American Bullfrogs were more abundant in southern Wisconsin until the early to mid-1900s, when collection pressures and conversion of land to agriculture increased (WDNR, unpublished data collected by R. C. Vogt), whereas recent findings indicate that they are now more commonly documented in northern Wisconsin (Vogt 1981, Mossman et al. 1998). A combination of past exploitation and unnatural relocations through the commercial bait industry has likely also influenced their state distribution and made it difficult to determine which counties have historically native vs. introduced populations (Vogt 1981, Casper 1996).

Localities of possible occurrence: The Wisconsin Frog and Toad Survey has recorded American Bullfrogs calling in many counties where they are not officially documented, but no voucher material exists (WIATRI 2016). Given this species' association with the Mississippi and lower Wisconsin Rivers, it is particularly likely that populations occur along the backwaters of these rivers in Pierce, Pepin, Buffalo, Adams, Columbia, Iowa, and Richland Counties.

Habitat: American Bullfrogs occupy permanent and semipermanent waterbodies in Wisconsin such as lakes, large wetlands, oxbows, and bogs but also slow-moving portions or backwater areas associated with rivers and streams (Knudsen 1958a, Vogt 1981, WDNR, unpublished data collected by R. C. Vogt). They are most common in deeper aquatic habitats that have an abundance of undisturbed shoreline vegetation (i.e., floating, emergent, and submersed), including pickerel weed (*Pontederia* spp.), water lilies, sedges, cattails (*Typha* spp.), and willows (*Salix* spp.; reviewed by Bury and Whelan 1984). American Bullfrogs do not often stray far from these aquatic habitats and are rarely found in adjacent upland areas, except on warm, humid, and rainy nights (Currie and Bellis 1969, Dodd 2013).

Breeding in northern regions typically occurs in larger permanent aquatic habitats that are similar to those used by adults during the remainder of the active season (Skelly et al. 1999, Babbitt et al. 2003). Specific oviposition sites often occur in open-canopy shallow-water locations (15.0–60.0 cm [5.9–23.6 in] water depth) associated with emergent and submersed vegetation (Dodd 2013). American Bullfrogs also breed in anthropogenic habitats such as artificial ponds in urban and agricultural areas (Surdick 2005). Little is known about the exact overwintering habitats used by American Bullfrogs in Wisconsin, but adults hibernate underwater in ponds and lakes that do not completely freeze in Ohio, Kansas, and Missouri (Willis et al. 1956, Stinner et al. 1994, Chance 2002). While submersed during hibernation, they likely rest on top of or burrow lightly into aquatic substrates and seem to prefer warmer shallow areas (Ruffner 1933, Emery et al. 1972, Stinner et al. 1994). Bohnsack (1952) further documented a rare case of a terrestrial hibernation in Michigan where the frog successfully overwintered under leaf litter within a 0.9 m (3.0 ft) diameter soil depression and in a 12.7 × 11.4 cm (5 × 4.5 in) soil pocket that was 6.4 cm (2.5 in) deep and 107 m (350 ft) from a permanent marsh. Because larvae in northern latitudes might not undergo metamorphosis for several years (Bruneau and Magnin 1980b), they also typically overwinter in aquatic habitats that are permanent or have long hydroperiods (e.g., Babbitt et al. 2003).

REPRODUCTION AND DEVELOPMENT

Males arrive at breeding sites before females and establish territories (Howard 1978a). Soon after, males begin to call in an effort to attract females for reproduction (Howard 1978a). The distinctive male advertisement call is often described as a deep 'jug-o-rum' or 'charroom' sound that slurs upward and increases in volume as it finishes (Dickinson 1965, Knudsen 1985). The onset of calling by males is significantly affected by water temperatures that exceed 20°C (68°F; Oseen and Wassersug 2002). Males typically call from early June to mid-July in Wisconsin, which is later than other anuran species (Vogt 1981). Despite their tendency to call later in the season, American Bullfrogs are sometimes heard as early as 20 May in the southern half of Wisconsin during unusually warm springs (Andrew Badje, WDNR, unpublished data). Males call in shallow water along the shoreline or while floating on the water's surface up to 30 m (98 ft) from the shoreline (Vogt 1981). Most calling occurs from sunset to sunrise but during the height of the breeding season typically peaks after midnight (Howard 1978a).

Both male and female American Bullfrogs will advertise and defend territories from others of the same sex (Howard 1978a, Dodd 2013). Oviposition usually occurs within the territorial boundaries of the male (Ryan 1980), and Emlen (1968) determined that the mean radius of these territories was ca. 2.7 m (9 ft). When an intruding individual invades a claimed territory, the resident produces a 'hiccup,' 'gronk,' or 'bonk' sound to warn the transgressor (Emlen 1968, Wiewandt 1969). Resident males may further threaten invaders by inflating themselves, sitting high in the water, and exposing their yellow chests (Emlen 1968). If the intruder does not vacate the resident's territory, direct physical conflicts can ensue as residents attempt to force the intruder's retreat (Natural History Box: Jousting for Lily Pad Rights). Males that control territories are typically older and larger than invading males and reportedly 'win' these physical interactions more often, thus achieving greater reproductive success (Emlen 1976, Howard 1978a). Despite this, males that do not hold territories can resort to alternate mating strategies (i.e., opportunism and 'parasitism') in order to gain reproductive opportunities but reduce direct male-male competition (Howard 1978a). For example, opportunistic males will attempt to occupy territories vacated by dominant males and quickly call to attract females, but only until threatened by a larger male (Howard 1978a). Parasitic (or 'satellite') males, on the other hand, remain near actively calling territorial males, silently waiting to intercept and breed with females attracted to the territorial male's call (Howard 1978a).

Although males call over a prolonged period, reproduction occurs during a narrower window of time (reviewed by Dodd 2013). Ovulation in females is asynchronous and generally lasts for a single night (Emlen 1976, 1977). As females become sexually receptive, they venture out from the shoreline to well-advertised and defended male territories, selecting a male based on call characteristics, physical attributes, and the quality of the oviposition sites in his territory (Emlen 1976, Ryan 1980). Eggs deposited in higher-quality territories experience lower mortality rates (Howard 1978b). After a male is selected, as indicated by physical contact from the female, the pair engages in amplexus, and the female moves to a nearby location within the male's territory for egg deposition (Ryan 1980). During amplexus, the pro-

NATURAL HISTORY BOX: Jousting for Lily Pad Rights

In June 2012, while surveying for frogs on the Mississippi River along the Wisconsin-Iowa border, I happened upon a booming foghorn chorus of American Bullfrogs in the midst of the surrounding 'banjo twangs' of Green Frogs. As I neared the water's edge, I spotted an assemblage of American Bullfrogs within ca. 3 m (10 ft) of the boat launch. In a matter of minutes, one male invaded the space of another more vocal, settled, and slightly larger one. Clearly agitated, the defending American Bullfrog wasted no time in confronting the intruder. These prime calling sites are considered high quality due to female preference and are usually correlated with low rates of embryo mortality. The largest and oldest males are generally seen battling over these locations, while smaller and younger males prefer other less aggressive breeding tactics. After a few seconds of strategic positioning, they locked forelimbs and contested their differences, similar to dueling sumo wrestlers. Each male kicked his hind limbs in shallow water, attempting to gain leverage and the much-desired advantage. These heavyweight tussles can also involve males charging, pushing, grabbing, slamming, or dunking the other (Knudsen 1985). An intense, for the two American Bullfrogs at least, 30 seconds later the defending male upended his opponent, pinning him upside down in the water. Instantaneously, their grasp unlocked, and the defeated male retreated. The victor then settled back down on his throne of lily pads and emergent grasses and began to bellow his signature 'jug-o-rum' once again for resident females to hear.

Andrew F. Badje, WDNR

nounced forearms and enlarged thumbs of males aid them in holding on to females for long durations (Dodd 2013). The length of time that mating pairs remained in amplexus in Michigan varied from 17 to 155 minutes ($\bar{x} = 49.3$ minutes, $n = 32$ pairs; Howard 1978a). Larger females also remain in amplexus for longer durations than smaller females, possibly because they require additional time to deposit a greater number of eggs (Howard 1978a).

Clutch sizes of 1,000–5,000 eggs have been reported in Wisconsin (Vogt 1981). Reports from other northern locations indicate that clutch sizes can range from ca. 2,000 to 23,500 eggs and are positively correlated with the age and size of the female (Bruneau and Magnin 1980a, Howard 1988b). Female American Bullfrogs can produce two egg clutches per year, although the second clutch contains smaller individual eggs (Howard 1978b). In Michigan, Emlen (1977) even documented female American Bullfrogs that deposited their first and second clutches three weeks apart. Larger females that are at least two years post-metamorphosis (>13.0 cm [5.1 in] SVL) can produce multiple clutches annually (Howard 1978b).

Proper egg development occurs at water temperatures between 15°C and 32°C (59°F and 86°F; Moore 1942). In northern populations (e.g., Wisconsin), eggs generally hatch three to five days after deposition in mid-summer (Bury and Whelan 1984). Larval growth rates are dependent on several factors, including water temperature, dissolved oxygen levels, food availability, and competition with conspecifics or other anuran larvae (reviewed by Dodd 2013). In Ohio, Bruggers and Jackson (1974) determined that the mean larval SVL was 3.24 cm (1.3 in) by November of their first year. In Wisconsin, American Bullfrogs typically metamorphose in July and August of their second year at 10–16 cm SVL (3.9–6.3 in; Vogt 1981), although in Quebec, transformation can take place as late as their fourth year (i.e., following a third overwintering period; Bruneau and Magnin 1980b). Post-metamorphic American Bullfrogs in northern latitudes reach sexual maturity in two to four years, depending on individual growth rates (Raney and Ingram 1941, Bruneau and Magnin 1980a, Vogt 1981). The minimum SVL of sexually mature females in Missouri was ≥12.3 cm (4.8 in; Willis et al. 1956). Reported longevity in the wild ranged from five to eight years post-metamorphosis in Michigan (Howard 1978a, 1988a) and up to six, possibly eight to nine, years in Quebec (Bruneau and Magnin 1980a).

ACTIVITY

Water temperature (>14.0°C [57.2°F]) and air temperature (20.0°C–24.0°C [68.0°F–75.2°F]) are correlated with American Bullfrog emergence in the spring (reviewed by Willis et al. 1956). Emergence from hibernation typically occurs during mid-May in southern Wisconsin and mid-June in northern Wisconsin (Vogt 1981). Not surprisingly, the timing of emergence varies across the species' broad continental distribution (e.g., February in Texas [Dawson 1953] compared to March or April in Illinois [Durham and Bennett 1963]). See Bury and Whelan (1984) for a summary of spring emergence times.

After spring emergence, activity and movement are dictated in part by thermal state and desiccation rates. American Bullfrogs desiccate at rates similar to those of other frogs; however, being more aquatic, they do not tolerate the loss of body water well and become more easily stressed when desiccating (Thorson 1955). Individuals thermoregulate by basking along wetland edges, but they also frequently move between shorelines and shallow water to prevent desiccation (Lillywhite 1970). Body temperatures more closely reflect surrounding water temperatures as opposed to air temperatures (Brattstrom 1963), although American Bullfrogs can withstand high body temperatures. For example, Lillywhite (1970) found that diurnal activity in California occurred at body temperatures ranging from 26°C to 33°C (79°F to 91°F) and recorded a critical thermal maximum (CTmax) of 38.2°C (100.8°F).

The movement patterns of American Bullfrogs in Wisconsin are not well-understood, although they probably do not travel long distances often. For example, Ingram and Raney (1943) recorded short movements of 122 m (400 ft), 174 m (570 ft), and 206 m (675 ft) over a two-day period in New York. The mean estimate of American Bullfrog activity radii over two years at a study pond in Ontario was only 2.6 m (8.6 ft), but individual activity radius ranged from 0.6 to 11.3 m (2.0 to 37.1 ft, $n = 131$; Currie and Bellis 1969). Furthermore, the size of activity radii was positively related to American Bullfrog population densities and male body size, while males also had larger activity radii than females. Although they tend to remain in or near water, they occasionally use terrestrial habitat for foraging, thermoregulation, and migration to breeding or overwintering sites (reviewed by Dodd 2013). If the threat of desiccation is low, American Bullfrogs are capable of moving substantial distances on occasion. Long-distance movements typically take place during warm rain events, which helps prevent desiccation (Raney 1940, Gibbons and Bennett 1974), and individuals can use celestial cues for navigation (Ferguson et al. 1968). In New York, the greatest recorded distance between recaptures of 237 tagged individuals was 1.6 km (1 mi; Ingram and Raney 1943). Willis et al. (1956) reported that one individual in Missouri moved ≥1.2 km (0.75 mi) over land between ponds. In Arizona, dispersal distances of up to 12.8 km (7.9 mi) were recorded (Suhre 2010). Genetic analyses of nine populations in Ontario indicated that female American Bullfrogs are more likely to disperse than males (Austin et al. 2003a). Homing ability has also been reported for American Bullfrogs. McAtee (1921) documented the return of an individual to its home range one day after being displaced to the opposite shore of a lake and two days after being displaced ca. 0.4 km (0.25 mi) over land to an isolated spring.

American Bullfrogs are typically active until late September to November in the northern states (Raney and Ingram 1941, Ryan 1953). In Missouri, most adults became torpid in preparation for hibernation as water temperatures dropped below 15.5°C (60°F), although activity increased again if water temperatures rose sufficiently prior to the first freeze (Willis et al. 1956). In contrast, juveniles remained active until the first freeze (Willis et al. 1956). American Bullfrogs are not freeze-tolerant and thus must overwinter in habitats that do not completely freeze (e.g., under the surface ice in deep ponds and lakes; see Habitat subsection). Furthermore, they cannot tolerate long periods of anoxic conditions (up to ca. three days; Stewart et al. 2004), although they can reportedly survive prolonged hypoxia (reviewed by Tattersall and Ultsch 2008). In Wisconsin and other northern regions of this species' range, winterkill (due to ice depth penetration and low dissolved oxygen levels) can be a primary source of mortality (Dodd 2013). Stinner et al. (1994) tracked aquatic habitat use of six adults throughout the winter at a lake in Ohio ($N = 208$ total relocations). They found that individuals were mobile throughout winter, likely moving to seek favorable thermal conditions and dissolved oxygen levels or to avoid predation, falling water levels, and freezing conditions.

PREY AND PREDATORS

Adult American Bullfrogs are generalist carnivores that use a 'sit-and-wait' foraging strategy (Bury and Whelan 1984). They eat a wide range of animals in Wisconsin, including aquatic and terrestrial insects, spiders, snails, crayfish, small mammals (e.g., mice), frogs, hatchling turtles, snakes, fishes, and young ducklings (Knudsen 1958a, Vogt 1981; Natural History Box: Batty Appetites). Carpenter and Morrison (1973) determined that adults consumed approximately equal amounts of nocturnal and diurnal prey items in Texas ($n = 124$ stomachs sampled). Like most anurans, they are gape-limited predators, but due to their large adult size they will take prey that most anurans cannot consume. For example, Bolek and Janovy (2007) captured a Wisconsin individual (SVL = 13.5 cm [5.3 in]) that had consumed an adult Meadow Vole (*Microtus pennsylvanicus*) that was nearly 50% of the bullfrog's body length (body length - tail = 6.7 cm [2.6 in]). Furthermore, a Blanding's Turtle (*Emydoidea blandingii*, size unreported) was found in the stomach of an adult American Bullfrog from Nebraska (Lingenfelter et al. 2014). Werner et al. (1995) studied dietary composition of American Bullfrogs during the summer in Michigan ($n = 166$ stomachs sampled). They found that the largest component of adult diet was juvenile frogs, followed by insects (e.g., aquatic hemipterans and adult odonates). These bullfrogs also sometimes reportedly eat large numbers of smaller conspecifics (Stuart and Painter 1993). In Indiana, the most frequently consumed prey items varied by habitat (i.e., a river vs. two ponds), but crayfish were commonly taken at all three sites ($n = 442$ stomachs sampled; Fulk and Whitaker 1969).

NATURAL HISTORY BOX: Batty Appetites

American Bullfrogs are widely considered opportunistic predators, waiting for and ambushing anything capable of fitting in their mouths. Nationally, they've been reported eating vegetation, insects, earthworms, fishes, birds, snakes, salamanders, frogs, and even small turtles and alligators (reviewed by Bury and Whelan 1984). In May 1980 on Pleasant Lake in Iron County, Mr. and Mrs. Ronald Eckstein had witnessed a large male American Bullfrog swallowing a Little Brown Bat (*Myotis lucifugus*) on the lakeshore (Ronald Eckstein, WDNR, personal communication). "When first observed, the bullfrog had half the bat swallowed. The bat appeared dead. The bullfrog eventually did get the bat swallowed." Another instance, as was recalled from the WDNR archives, occurred in 1986 in Menominee Falls, Wisconsin, when an American Bullfrog collector reported "a bullfrog with a Little Brown Bat in its stomach." Since bats aren't normally documented to be on the menu, at least when flying, these outcomes likely took place because a bat was snagged as it swooped down near the water for a drink, splashed frantically (possibly drowning), and was then noticed and seized by an American Bullfrog. It is possible, if unlikely, that these American Bullfrogs also could have lunged at and captured the low-flying bats over the water.

Andrew F. Badje, WDNR

American Bullfrog larvae feed on algae, bacteria, detritus, phytoplankton, and occasionally small invertebrates (reviewed by Bury and Whelan 1984). Larvae also consume eggs and larvae of other amphibians and fishes (Ehrlich 1979, Mueller et al. 2006). Following metamorphosis, the size of prey taken increases as individuals grow (Dodd 2013), but most juveniles likely eat any invertebrate or vertebrate that they can fit into their mouths.

Little is known about predators of American Bullfrogs in Wisconsin. Reports from elsewhere indicate that adults are taken by many larger vertebrate predators such as fishes, snakes, turtles, Raccoons (*Procyon lotor*), various waterbirds, hawks, and owls (reviewed by Bury and Whelan 1984). Eggs and larvae further fall victim to numerous smaller aquatic predators. The American Medicinal Leech (*Macrobdella decora*) was a major predator of eggs at a pond in Michigan (Howard 1978b). A laboratory study in British Columbia documented egg predation by an ambystomatid salamander (Northwestern Salamander [*Ambystoma gracile*]) and fishes (Three-spined Stickleback [*Gasterosteus aculeatus*] and Cutthroat Trout [*Oncorhynchus clarkii*]; Licht 1969). Larvae are also consumed by numerous types of invertebrates and vertebrates found in aquatic habitats (reviewed by Dodd 2013).

Post-metamorphic American Bullfrogs avoid predation when on shorelines or in shallow water by completing a series of rapid hops along the water's surface before diving into dense zones of submerged vegetation (Knudsen 1985). Individuals often chirp as they leap into the water, a presumed tactic to confuse potential predators that can also be useful for identification and population counts (Donald Brown, West Virginia University / USFS, personal observation). When seized, American Bullfrogs produce a distress call that can last up to nine seconds, a 'yeowal' that sounds like the wail of a cat (Hoff and Moss 1974, Vogt 1981). This distressing tone may temporarily distract predators and allow the bullfrog to escape (Hoff and Moss 1974), but it also alerts other individuals to the presence of a possible threat. Unlike some anuran species, American Bullfrog larvae do not produce predator-deterring secretions (Formanowicz and Brodie 1979). However, several studies have found that they are distasteful to some species of fishes and dragonfly naiads, the degree to which changes as they develop (Adams et al. 2011, Szuroczki and Richardson 2011). American Bullfrog larvae also exhibit several morphological and behavioral responses to predation pressure. In a laboratory study of larvae hatched from wild eggs collected in Michigan, larvae responded to the presence of caged Central Mudminnows (*Umbra limi*) and Eastern Tiger Salamander larvae (*Ambystoma tigrinum*) by growing longer tail fins (Relyea 2001a). In another laboratory study using caged predators, larvae reduced their activity in the presence of predators (Smith and Awan 2009).

CONSERVATION

Status: Globally, the American Bullfrog is listed as a Least Concern species by the IUCN because of its wide distribution, large populations, and ability to occupy a wide range of habitats. Although they are native to Wisconsin, invasive American Bullfrog populations have become established throughout much of the western US and many other countries (e.g., France; Ficetola et al. 2007). Lowe et al. (2000) identified American Bullfrogs as among the 100 worst invasive species globally due to their competitive dominance and generalist nature. In Wisconsin, however, American Bullfrogs are listed as a species with Information Needs. They are rare to uncommon throughout much of the state but can be locally abundant where present (Vogt 1981). Long-term monitoring data from the Wisconsin Frog and Toad Survey suggest that populations are currently stable, although slight increases in occurrence and decreases in rank abundance are possible at the state level (Figure 100). Take, or collection, is currently regulated through open/closed seasons for harvest and is limited to five American Bullfrogs per person at any one time (WDNR 2004). In addition, some counties in Wisconsin have enacted regulations that prevent the collection of American Bullfrogs in any season in order to further protect historically degraded populations (WDNR 2004).

Populations: Rigorous estimates of adult abundance and density are rare for purported natural populations, although work on introduced populations indicates that they can reach high densities (reviewed by Dodd 2013). Given this, it is unsurprising that no robust population studies are currently known from Wisconsin. Most state-level population information is inferred from surveys for the presence or absence of calling males during the breeding season. For example, from 1984 to 2016, the Wisconsin Frog and Toad Survey recorded male American Bullfrog calling activity at only 3%–13% of all Wisconsin survey locations (see Figure 100). Elsewhere, existing population density estimates are of the larval stage. In Kentucky, Cecil and Just (1979) documented larval densities of up to 13.2 individuals/m^2 (1.2 individuals/ft^2). Cecil and Just (1979) further estimated mean survivorship to metamorphosis of larvae in three populations ranging from 11.8% to 17.6%. In Michigan, estimated post-metamorphic survivorship per year (up to four years of age) was 42%–52% and 42%–69% for males and females, respectively, while estimated survivorship from year four to five was 23% and 26% for males and females, respectively (Howard 1988a). Throughout an eight-year study on a lake in Illinois, males represented 33%–72% of adults collected ($\bar{x} = 47.5\%$; Durham and

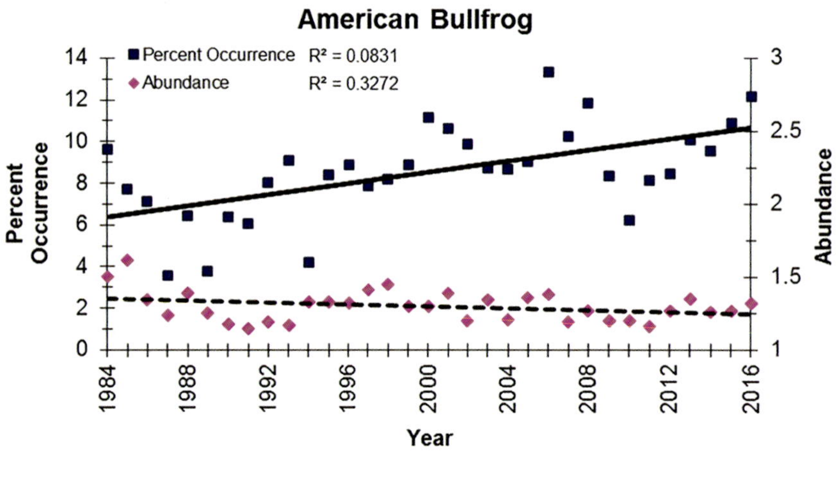

Figure 100. Long-term trend in occurrence (%) and abundance (rank) of American Bullfrogs across Wisconsin Frog and Toad Survey statewide monitoring sites. See Introduction to Species Accounts chapter for details on this monitoring program (figure courtesy of WDNR).

NATURAL HISTORY BOX: Unsustainable Collection

Historical data on American Bullfrog populations in Wisconsin are sparse, and the only relevant published information prior to the 1980s is highly anecdotal. For example, one of the earliest records of the American Bullfrog's state status comes from Hoy (1883), who reported they were "common and well known" in the state. Despite this, American Bullfrogs apparently experienced substantial declines by the early 1900s. Conger (1921) wrote that "it is not common in Wisconsin and is probably only locally distributed." Pope and Dickinson (1928) mentioned that "the final passing of the Bullfrog from the state is probably imminent." Similarly, Cahn (1929) said that in Waukesha County, American Bullfrogs were "still present, but steadily decreasing in numbers. Formerly very abundant; now seldom seen, but sometimes heard." By the mid-1900s, the notion that American Bullfrogs were rare in Wisconsin had persisted, and Knudsen (1958a) stated they were one of the state's "less common species." Knudsen (1985) later validated their perceived decline by proclaiming them "now uncommon and scattered in southern Wisconsin" but stated that "they are still relatively common in many parts of northern Wisconsin." Even as recent as the late 1990s, Mossman et al. (1998) considered them uncommon in Wisconsin.

Overharvest for food and research, as well as broad-scale chemical use in the early to mid-twentieth century, are often cited reasons for their reported population losses. For example, WDNR records from the 1960s documented large harvests of 50 and 26 adult bullfrogs per day at Long and Fish Lakes (Fond du Lac and Dane Counties, respectively). Unsurprisingly, American Bullfrogs are rare or nonexistent at both of these locations today. Furthermore, two of the four major national companies that commercially distributed frogs in the 1960s (accounting directly or indirectly for the supply of six million American Bullfrogs and Northern Leopard Frogs [*Lithobates pipiens*] annually) operated out of Oshkosh, Wisconsin (Gibbs et al. 1971). As highlighted by Knudsen (1958a), past Wisconsin frog hunters effectively harvested frogs through a battery of methods, such as shooting, spearing, netting, clubbing, or capturing them with traditional fishing rigs, which led to "hundreds of thousands" being collected and sold "on the market at fancy prices." The harvesters of American Bullfrogs from Fish Lake, for example, claimed to receive $9.92–$16.53/kg ($4.50–$7.50/lb) for the legs in 1969 (WDNR, unpublished data collected by R. C. Vogt). Culley (1981) reported averages of $6.61–$13.23/kg ($3.00–$6.00/lb) but mentioned that payments could reach upward of $19.84–$22.05/kg ($9.00–10.00/lb).

Andrew F. Badje, WDNR

Bennett 1963). Similarly, males represented 51.8% of collected adults ($n = 426$) across five populations in Missouri (Schroeder and Baskett 1968).

Threats: Many of the general threats that face all anurans in Wisconsin, such as habitat loss and degradation, also probably influence American Bullfrogs (see Introduction to Conservation and Management of Wisconsin's Amphibians and Reptiles chapter). Several sources attribute their declines in southern Wisconsin during the mid-1900s to overcollection by biological supply companies for use in biomedical research or science classes and for human consumption as "frog legs" (Schroeder and Baskett 1968, Gibbs et al. 1971, Culley 1973, Vogt 1981; Natural History Box: Unsustainable Collection). Today, however, American Bullfrogs used for human consumption or scientific research are primarily farmed in private operations that are subject to wildlife trade regulations, a practice that has reduced the exploitation of wild populations and allowed American Bullfrog populations to recover in some areas (reviewed by Dodd 2013).

Historical land conversion and wetland drainage associated with urbanization, agriculture, and silviculture have likely had negative impacts on American Bullfrog populations in Wisconsin (WDNR, unpublished data collected by R. C. Vogt). In addition, practices that destroy or modify shallow and stagnant backwater river habitats, increase the flow of water, and limit vegetation along shorelines can result in population declines (reviewed by Bury and Whelan 1984). Although anthropogenic habitat disturbances can negatively affect American Bullfrogs, they are sometimes found in artificial or heavily modified waterbodies in Wisconsin such as deep ponds in golf courses and fish hatcheries (Joshua Kapfer, UW–Whitewater, personal communication). Road mortality, which is more likely in fragmented landscapes, has contributed to high levels of American Bullfrog mortality in some areas (Ashley and Robinson 1996, Glista et al. 2008).

A plethora of studies have also investigated the influence of environmental contaminants that are often associated with disturbed habitats, determining numerous direct and indirect effects (reviewed by Dodd 2013). Agricultural pesticides alter physiological functions, compromise immune systems, decrease longevity, and limit growth rates associated with size, mass, tibia length, and tympanum size in adults (Marcogliese et al. 2009, Spear et al. 2009). In addition, heavy metals have negative effects on American Bullfrogs, especially in early developmental stages (reviewed by Dodd 2013). Studies of larvae in aquatic systems contaminated with heavy metals report slower response times to stimuli, poor swimming capabilities, and limited abilities to escape predation (e.g., Raimondo et al. 1998).

Several diseases that play important roles in global amphibian declines could impact American Bullfrogs in Wisconsin. For example, the fungus *Batrachochytrium dendrobatidis* (Bd) was detected on 81 of 295 examined American Bullfrog museum specimens taken from locations in the US and Canada from 1954 to 1999 (Ouellet et al. 2005). Limited sampling for Bd has occurred in the state, but Standish et al. (2018) detected it on individuals from Wisconsin. Similarly, Sadinski et al. (2010) detected Bd on American Bullfrogs in nearby northern Iowa, and Phillips et al. (2014) reported it on individuals sampled in Illinois. Daszak et al. (2004) determined that American Bullfrogs show some apparent resistance to Bd infection but can still serve as vectors that distribute Bd to nonresistant anurans. Both wild and captive American Bullfrogs are also susceptible to ranavirus infections and subsequent mortality (e.g., reviewed by Miller et al. 2011, Hoverman et al. 2012). Although few definitive reports of ranaviruses in Wisconsin exist, Standish et al. (2018) detected them on individuals sampled at several sites in the state, and hundreds of infected individuals from Vilas County were documented in 2009 and 2014 (Wildlife Health Information Sharing Partnership 2017).

Management: The primary tactic employed to protect wild American Bullfrog populations in Wisconsin is the regulation of take and the institution of open/closed harvest seasons (WDNR 2015a). Such regulations are designed to prevent overharvest and allow populations enough time to recover before the next harvest season. General habitat management strategies for permanent aquatic systems, such as large wetlands and lakes, could also positively influence this species in Wisconsin (see Introduction to Conservation and Management of Wisconsin's Amphibians and Reptiles chapter). Important management considerations for American Bullfrogs could further include creation or maintenance of wetlands with long hydroperiods and sufficient emergent aquatic vegetation near shorelines. The implementation of this strategy, however, should not include lengthening of hydroperiods in naturally ephemeral wetlands that many other amphibians in Wisconsin require for successful reproduction. Minimizing use of chemicals that are environmental pollutants, such as pesticides, fertilizers, and heavy metals, in and around wetlands would also benefit American Bullfrogs and many other species (Kingsbury and Gibson 2012).

Green Frog

Lithobates clamitans (Latreille 1801)

Patrick C. Owen

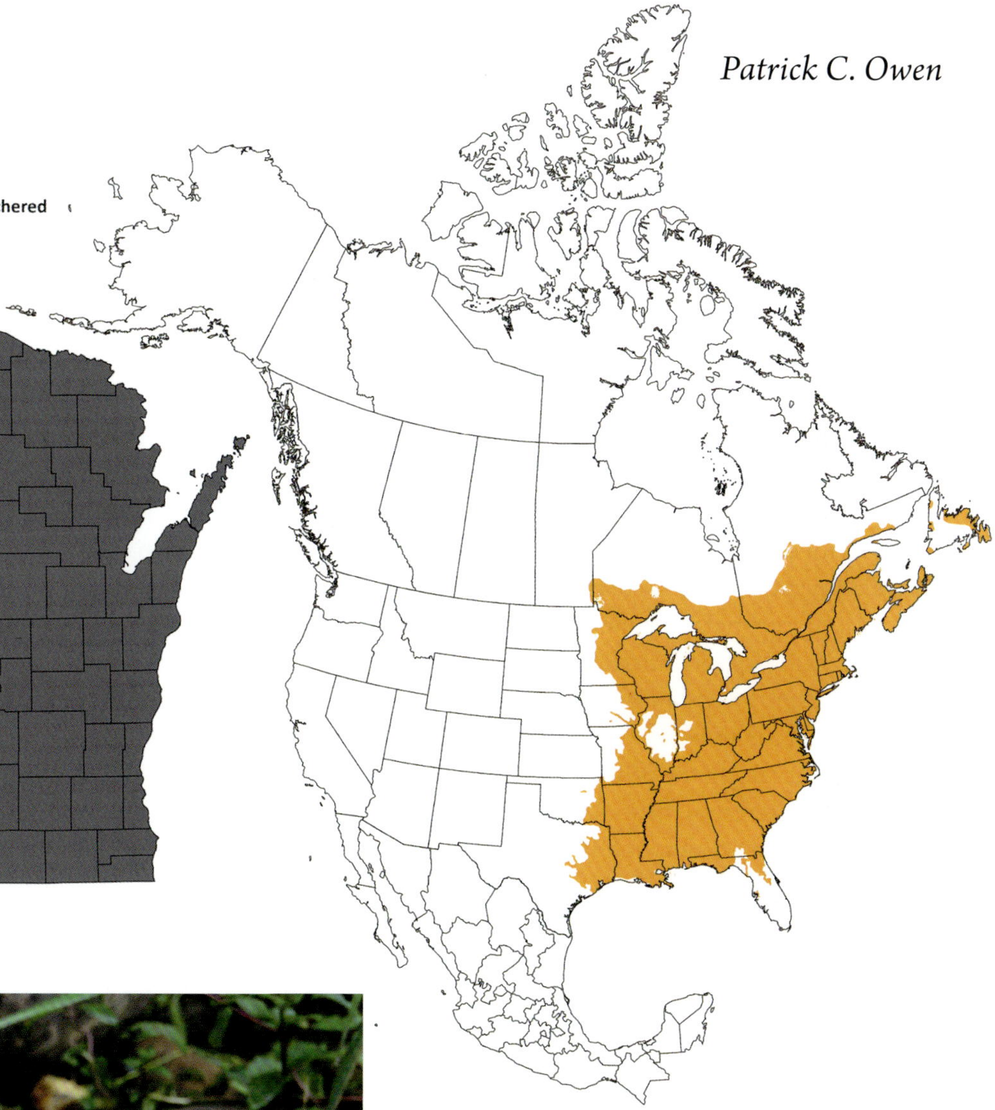

Adult Green Frog (Waukesha County; photo by J. Kapfer).

AT A GLANCE

Green Frogs are one of the state's most widespread amphibians and have been documented in all Wisconsin counties. As their common name suggests, their backs are typically various shades of green that often transition to gray and brown, but they lack large well-defined spots. They also possess large tympanums and two prominent ridges, or dorsolateral folds, running parallel down their backs. Green Frogs are one of the largest frog species in the state, second in size only to American Bullfrogs. Green Frogs frequent the vegetated margins of permanent wetlands and ponds but are also found in lakes and along the edges of streams and rivers. Although they often emerge from hibernation by mid-April, males do not begin calling to attract females for reproduction until summer, and breeding occurs from May to July. Their call is often likened to the sound of a plucked banjo string. Unlike most frogs in the state, Green Frogs can remain in the larval stage through their first winter and undergo metamorphosis the following summer. Adults are opportunistic visual predators. They mainly consume a wide variety of invertebrates, but they will also consume fishes, reptiles, and other amphibians. Larvae, on the other hand, eat mostly vegetation they filter from the water column or scrape off of aquatic surfaces, although individuals have been observed feeding on the eggs of other amphibians. Adults typically remain active through mid- to late October in Wisconsin. They often spend the winter submerged in aquatic habitats that can become covered with ice but do not completely freeze to the substrate. While habitat loss has negatively impacted the populations of many amphibian and reptile species in Wisconsin, Green Frogs seem relatively resilient to habitat disturbance. Because they persist in many types of altered and polluted landscapes, they have become a model organism in the study of threats to amphibian populations. Green Frogs are considered a common species by the WDNR.

SYSTEMATICS

Two subspecies of Green Frogs were previously recognized in North America: *Lithobates c. clamitans* (Bronze Frog) and *L. c. melanota* (Northern Green Frog). The Green Frogs of Wisconsin were considered part of the Northern Green Frog subspecies. However, Crother (2017) rejected these designations based on the work of Austin and Zamudio (2008). The specific epithet reportedly comes from the Latin term *clamito* or *clamitor,* 'to cry loudly' or 'shouter,' which is apparently in reference to the loud nature of male advertisement calls (Dodd 2013). Among Wisconsin anurans, *L. clamitans* occurs in the same clade as *L. catesbeianus* (American Bullfrog) and *L. septentrionalis* (Mink Frog; Austin et al. 2003b).

DESCRIPTION

Adults: Green Frogs have a broad head and a rounded snout (Figure 101). Their hind legs are long and powerful, with webbed feet for jumping and swimming. In contrast, the front legs are short and lack webbing. A pair of dorsolateral folds run parallel from behind the eyes to midway down the back. The skin is somewhat granular. Dorsal coloration ranges from various shades of green to gray or light brown (Figure 101). Dark flecks or small spots can be present dorsally, and the legs may have dark brown spots, striations, or bars. The venter is white, but gray mottling may be present. Green Frogs are one of the largest frogs in Wisconsin (Vogt 1981). The mean SVL of adult Wisconsin specimens housed in the MPM and UWZM collections ranged from 7.1 cm (2.8 in) in males ($n = 155$) to 7.4 cm (2.9 in) in females ($n = 120$, minimum adult SVL based on work by Martof [1956b] in Michigan; see Reproduction and Development section). A collection made by E. R. Wild and colleagues between 1996 and 2006 from locations across the state ($n = 59$, UWSP collection) included juveniles and adults that ranged from 3.0 to 9.6 cm SVL (1.2 to 3.8 in) and weighed 2.4–80.0 g (0.08–2.82 oz). The largest SVL of a Wisconsin specimen examined was 9.1 cm (3.6 in), a female from Kewaunee County collected in 1949 (MPM 2989).

Preadult stages: Typical characteristics of eggs and larvae were described by Altig (1970), Vogt (1981), and Parmelee et al. (2002). Eggs are laid in thin

Figure 101. Color variation in Green Frogs. (a) Green Frog with mostly green coloration and a lack of spotting and striping. The prominent dorsolateral folds positively identify it as a Green Frog (Marquette County; photo by T. Muehlfeld). (b) Green Frog with brown coloration and spotting (Rock County; photo by J. Kapfer).

Figure 102. Green Frog pre-adult life stages. (a) Green Frog egg mass in a pond (Waukesha County; photo by D. Muñoz). (b) Green Frog tadpole (Chippewa County; photo by R. Sajdak).

masses, or sheets, that float at the water's surface (Figure 102), and each egg is surrounded by an inner and outer gelatinous envelope. At hatching, larvae range from 0.74 to 1.2 cm TL (0.3 to 0.5 in), while individuals nearing metamorphosis range from 6.0 to 8.0 cm TL (2.4 to 3.1 in; Figure 102). Larvae have plump bodies that are slightly dorsoventrally compressed, with a vent tube, or anal opening, that is to the right of the tail crest. Larvae have prominent brown to black beaks with a typical labial tooth row formula of two anterior rows and three posterior rows (two over three). Dorsally, larvae are green, olive, or brown in color. They possess dark spots or mottling that is not well defined. The venter of larvae is cream-colored and opaque so that the intestinal coil is not visible through the body wall. The tail is long with a dark tip and low fins. Recently metamorphosed individuals average 3.0 cm SVL (1.2 in) and have many dorsal spots (Harding 1997).

Variation: A number of less common color variants have been described, including very dark (Radaj 1981b, Richards and Nace 1983) and blue (Berns and Uhler 1966) individuals from Wisconsin, as well as yellow (Pinder 2010) and albino (Fowler 1918) individuals from other localities. Specimens collected by E. R. Wild and J. A. Ernst from Sawyer County in 2000 and 2012 (UWSP collection) further illustrate the unusual color patterns that can be found in Green Frogs. One frog had light green to tan blotches

Figure 103. Sexual dimorphism in Adult Green Frogs. (a) Adult male. The tympanum of sexually mature males is large, with a diameter larger than that of the eye. Yellow coloration often extends from the chin down over the forelimbs (Sawyer County; photo by E. Wild). (b) Adult female. The tympanum of females is approximately the same diameter as that of the eye. Yellow coloration is usually not present on the chin or the forelimbs (Jefferson County; photo by J. Kapfer).

running from the rear of the head to the limbs (UWSP 3854). Another frog had a head with turquoise flecks and blotches that extended to the upper lip, eyelid, and tympanum, as well as a faint purple coloration on its flanks and abdomen (UWSP 3852).

Adult Green Frogs are sexually dimorphic. Females attain larger body sizes than males, but males have proportionally larger tympanums. The tympanum of mature males is greater in diameter than the eye (Figure 103), while the tympanum diameter of females (Figure 103) and juveniles (Figure 104) is equal to or smaller than the diameter of the eye (Mecham 1954). Breeding males also have a yellow-pigmented throat, as well as enlarged forelimbs and thumb pads (Walker 1946, Minton 2001).

Figure 104. A juvenile Green Frog. This individual is not yet sexually mature, and the tympanum diameter cannot be used to determine its sex (Oneida County; photo by D. Brown).

Diagnosis: Green Frogs do not possess parotoid glands, large warts, or a keratinized spade on the heel of each hind foot. They also lack terminal toe pads and have no musky odor. Dorsolateral folds are present that extend to the middle of the dorsum, and the webbing on the hind feet extends just beyond the second joint of the fourth digit but does not extend to the tip of the fifth digit. In Wisconsin, adult Green Frogs are most easily confused with American Bullfrogs (*Lithobates catesbeianus*) and Mink Frogs (*L. septentrionalis*; Vogt 1981, Parmelee et al. 2002). American Bullfrogs can attain a larger body size than Green Frogs and lack prominent dorsolateral folds (see Figure 101). Mink Frogs and Green Frogs can be similar in body size and coloration, but Mink Frogs are more extensively covered with larger dark dorsal blotches than Green Frogs. Green Frogs also typically have brown mottling on the green labial area and narrow stripe-like markings on the hind legs (see Figure 101), while Mink Frogs generally do not. Green Frogs reportedly can mimic the coloration of Mink Frogs in the Apostle Islands (Vogt 1981). Unlike Mink Frogs, the webbing on the hind feet of

Green Frogs does not reach the tip of the fifth digit. Adult male Green Frogs have large tympanums not observed in Mink Frogs, while dorsolateral folds of Mink Frogs are sometimes incomplete or absent, unlike Green Frogs. Additionally, Mink Frogs have a distinctive and unique musky odor often likened to the smell of rotten onions.

Preadult stages of Green Frogs are also most easily confused with those of American Bullfrogs and Mink Frogs. Green Frogs deposit eggs in a sheet at the water's surface, whereas Mink Frogs typically attach eggs to vegetation well below the water's surface (e.g., >1 m [3.3 ft] deep; Vogt 1981). Green Frog egg masses resemble those of American Bullfrogs in shape and location of deposition (i.e., a thin sheet of eggs at the water's surface), but individual Green Frog eggs have two gelatinous envelopes, compared to one in American Bullfrogs (Altig and McDiarmid 2015). Green Frog larvae lack external gills and forelimbs, while their eyes are dorsally positioned on their heads, and their vents are dextral. There is no white line from the nostril to the snout. The tail is mottled with dark spots and fine yellow spots, while the intestinal coil is not visible. The determination of whether a larva or an adult is a Green Frog or Mink Frog can consider, in part, geographic location, as the latter is only found in more northern counties. Furthermore, Green Frog larvae lack the pink or buff-colored spots on the tail that are found in Mink Frog larvae. Green Frog larvae do not grow as large as those of American Bullfrogs and also often lack the well-defined spots of American Bullfrog larvae (see Figure 102; Parmelee et al. 2002).

DISTRIBUTION AND HABITAT

Global distribution: Green Frogs are widely distributed in eastern North America, ranging from Ontario, Quebec, and the Maritime Provinces south to the southeastern Coastal Plain. Their western range limit is along a generally north–south line from eastern Minnesota south through eastern Texas. They have been introduced in scattered locations in western North America (Conant and Collins 1998). Despite their broad distribution, they are absent from a large swath of central Illinois, pockets of western to central Indiana, and peninsular Florida.

Distribution in Wisconsin: Published records with vouchers exist for all Wisconsin counties. Green Frogs are even documented on islands in the Great Lakes (Long and Long 1976, Hecnar et al. 2002).

Localities of possible occurrence: Green Frogs have been documented in every Wisconsin county.

Habitat: Green Frogs generally associate with permanent aquatic habitat during both the breeding and nonbreeding periods (reviewed by Pauley and Lannoo 2005, Dodd 2013). They are somewhat general in the habitats and vegetation communities they occupy in Wisconsin, which includes most lentic and lotic types (Vogt 1981; see Amphibians and Reptiles in Wisconsin Plant Communities chapter). Although they are frequently found in open lentic habitats such as lakes and large ponds, Green Frogs also frequent river backwaters, streams, and ditches (Conant and Collins 1998, Minton 2001). Adults are mainly encountered in wetland, riparian, and adjacent upland areas, although the larger adults may range farther away from their aquatic habitats to forage, especially when it is rainy (Martof 1953a, 1953b). Toward the end of the breeding season, Green Frogs may begin to utilize adjacent terrestrial habitats more frequently for foraging (Lamoureax et al. 2002).

When along shorelines, Green Frogs appear to favor specific habitat characteristics. For example, Paloski (2006) determined that in Portage County they were most often found in emergent vegetation, followed by submersed vegetation, but avoided shorelines with open water and bare sand. When adult Green Frogs reach 6.0–7.5 cm SVL (2.4–3.0 in) they move from shallower vegetated regions of aquatic habitats to more open areas with deeper water (Martof 1953a). Interestingly, Vogt (1981) reported that although adults typically remain near water year-round, juveniles wandered "considerable distances in open woodlands, moist fields, and marshes" in Wisconsin. Martof (1953a) further reported that juvenile Green Frogs foraged among vegetative cover along pond edges but also migrated into adjacent fields and woodlands. Church (2008) determined that juveniles were more likely found near ephemeral ponds that had hydroperiods too short for larval development, which suggests they moved to new habitats after metamorphosis.

Habitats used for nonbreeding and breeding do not appear to differ greatly, although more studies in Wisconsin have investigated habitat used during the breeding rather than nonbreeding season. Knutson et al. (2000) found that calling male Green Frogs sampled in Wisconsin were exclusively associated with habitats classified as lakes or those "dominated by water cover types and edges." According to Wells (2007), females also deposit eggs in the backwaters of streams. Kutka and Bachmann (1990) determined that aquatic habitats where males called in northern Wisconsin were commonly at least 0.5 m (1.6 ft) deep and 20 m (66 ft) in diameter. Green Frog larvae develop in the same habitats where breeding occurs, which are open-canopy permanent ponds, where they coexist with fishes and larval dragonfly predators (Werner and McPeek 1994, Skelly et al. 1999, Richardson 2001a). Within these habitats, larvae typically remain in areas of dense aquatic vegetation that can aid in predator avoidance and maximize feeding rates, although larger larvae sometimes venture into open water at night (Warkentin 1992b, Tarr and Babbitt 2002).

Green Frogs frequently overwinter while submerged in aquatic habitats that do not completely freeze in the winter. As such, Lamoureux and Madison (1999) reported that adults primarily used streams and seeps that remained partially ice-free as overwintering sites in New York, but individuals can also overwinter in ponds that are covered in a layer of ice, so long as they do not freeze solid. Although hibernation in aquatic habitats is likely more common, Bohnsack (1951) documented Green Frogs hibernating below forest leaf litter in Michigan, which supports that they may be able to tolerate short bouts of freezing. Furthermore, Badje et al. (2016) found several individuals during late November that were not submerged but apparently overwintering near the water table in crayfish burrows along riparian corridors in southwestern Wisconsin ($\bar{x}$ depth of burrow = 17.6 cm [6.9 in]).

REPRODUCTION AND DEVELOPMENT

Green Frogs breed later than many other frog species. In Wisconsin, sustained calling activity occurs as early as 4 May and continues until as late as 24 July (Vogt 1981, Patrick Owen, University of Cincinnati, personal observation), although sporadic calling may continue into autumn and has been reported as late as mid-October in Michigan (Martof 1953b). Breeding later in spring and summer rather than immediately after emergence from hibernation helps ensure that adults have time to build energy reserves in preparation for reproduction (Jenssen and Klimstra 1966, Wells 2007). During the breeding season, male Green Frogs produce a variety of calls that are used for several purposes, such as attraction of females for reproduction, but also to ward off rival males and maintain spacing in calling groups (Wells 1978, Bee and Perrill 1996). The most familiar vocalization is the advertisement call, which is a short banjo-like 'bonk.' In defense of breeding locations, males also employ a lower-pitch aggressive call to repel intruders (Bee et al. 2000, Owen and Gordon 2005). As a result and depending on choral density, calling males may be spaced from 1 to >6 m (3.3 to 19.7 ft) apart (Wells 1977a, Owen and Gordon 2005). Calling males float on the water's surface with their bodies inflated, displaying their brightly colored yellow throats (Figure 105; Martof 1953b, Wells 1978). In addition to aggressive calls, males may employ splashing displays to repel intruders or engage intruders in wrestling matches if they do not retreat. The breeding season overlaps with that of American Bullfrogs, which have longer and more frequent calls,

Figure 105. Adult male Green Frog calling, floating in his territory with body inflated, displaying his brightly colored yellow throat (Price County; photo by C. Raimond).

so Green Frogs avoid acoustic interference by timing their calling to within the periods of silence between American Bullfrog calls (Herrick et al. 2018).

Green Frogs are prolonged breeders that defend calling sites for up to six to seven weeks during the breeding season (Wells 1977a). Such prolonged territorial activities are energetically taxing, and males that do not feed consistently during the breeding season may lose up to 30% of their mass (Wells 1978, Gordon 2004). In large breeding groups, some males use alternate strategies to gain reproductive opportunities while conserving energy. For example, males that do not actively defend territories sometimes occupy 'satellite' positions in or near the defended territories of other males (Wells 1977a, 1978). These satellite males may intercept females, and they are opportunistic, occupying territories that have been vacated by other males (Wells 2007). Release calls may be given by a male mistakenly amplexed by another male (Wells 1978).

Green Frogs have a resource-defense territorial mating system in which females select males based on the quality of their territories (Wells 1977a, 1978). When females are ready to oviposit, they move from stream areas to ponds and typically spend less than a week visiting the territories of males before selecting a mate (Martof 1953b, Wells 1977a). Higher-quality territories apparently have a higher density of emergent vegetation and warmer water temperature (Brenner 1969, Wells 1977a). Because Green Frog egg masses are laid as a surface film, eggs can absorb oxygen well and thus can survive in warmer water temperatures than the eggs of other ranid species (Moore 1939, 1940, Licht and Grant 1997). Females that oviposit early in the summer are capable of returning three to four weeks later to oviposit again (Wells 1976).

Green Frog eggs hatch within three to seven days of oviposition (Martof 1956a, Licht 2003). Larval development rates are variable and can be influenced by genetics, water temperature, and pond desiccation rates (Martof 1956b, Berven et al. 1979). Larvae sometimes transform into adults the same summer that their egg mass was laid (Martof 1952). Alternately, larvae that hatch in late summer can remain as larvae over the winter and then metamorphose in early summer of the next year (Martof 1956b). Vogt (1981) stated that in Wisconsin, larvae from eggs laid in May metamorphosed that same August, while larvae from eggs laid after 10 July overwintered and transformed in early June of the following year. Overwintering Green Frog larvae metamorphose at larger body sizes than larvae that complete development during a single summer (Smith-Gill and Berven 1979). As modeled in artificial ponds, larvae in ponds of larger area and depth metamorphose at greater mass but are less likely to survive to metamorphosis (Pearman 1993), potentially due to competitive effects. In Michigan, Martof (1952) found that individuals who transformed in late summer ranged from 2.84 to 3.63 cm SVL (1.12 to 1.43 in). In another Michigan study, Werner and McCune (1979) found that during the period of peak larval metamorphosis in July and August, recently transformed Green Frogs consisted of approximately 85% of the anuran biomass.

Peak post-metamorphic growth occurs from mid-May through mid-September. Growth declines after sexual maturity is reached, which takes one to three years (Ryan 1953, Martof 1956b, Shirose and Brooks 1995b). Adult characteristics appear in males that range from 5.9 to 7.7 cm SVL (2.3 to 3.0 in; Martof 1956b). Males as small as 6.3 cm SVL (2.5 in) have been observed calling (Wells 1977a). Martof (1956b) documented gravid females as small as 6.6 cm SVL (2.6 in). Longevity estimates range from five to seven years in the wild (Martof 1956b, Shirose and Brooks 1995a).

ACTIVITY

In Wisconsin, Green Frogs typically emerge from hibernation by mid-April (Vogt 1981). After emergence, adults are primarily diurnal, although during the breeding season they are more nocturnal (Martof 1953a). Green Frog larvae are active at any time of the day, typically feeding around submerged vegetation but sometimes feeding in open water at night (Warkentin 1992b).

Investigations of adult movement patterns support that they move sparingly and exhibit a high fidelity for certain locations within their home ranges. For example, home ranges of adults in Michigan averaged only 60 m^2 (635 ft^2), and most movements during the active season were within a radius of 0.3 m

(1 ft) of previous locations (Martof 1953a). Oldham (1967) further reported that adults returned to the same home range areas that they used in previous years. Green Frogs also show homing ability, as Ingle and Hoff (1990) found that frogs displaced 10–15 m (33–49 ft) from their breeding pond were capable of returning to their approximate original position around the pond. Although movement during most of the active season is reportedly limited, certain activities or behaviors can require larger migrations. If adults move beyond their established home ranges, it is usually related to foraging or migration to breeding and overwintering sites (Martof 1953b, Lamoureux and Madison 1999, Lamoureux et al. 2002). For example, Oldham (1967) observed movement to breeding locations of up to 457 m (1,499 ft) in Ontario. Lamoureux et al. (2002) reported that frogs in New York made late autumn foraging 'forays' of up to 560 m (1,837 ft) from aquatic habitat into adjacent terrestrial habitats where prey biomass was greater. Precipitation can trigger an increase in movement related to foraging activity or breeding migrations. Migrating Green Frogs avoid forested areas and prefer to move through areas with shorter grass (Birchfield and Deters 2005).

Post-metamorphic juveniles likely undertake the greatest movements of all life stages when they disperse from natal ponds. These recently metamorphosed individuals can disperse widely, in some cases up to 4,800 m (15,748 ft), particularly along riparian corridors (Schroeder 1976). Movements coincide with precipitation, and only a small number of individuals return to the natal pond. On the other hand, Whitaker (1961) reported that young Green Frogs captured at a study site in New York were always within 0–18.3 m (0–60 ft) from the water's edge, supporting minimal movement. Active frogs behaviorally thermoregulate to maintain desired body temperatures. In a study by Brattstrom (1963), individuals maintained body temperatures that ranged from 10°C to 29°C (50°F to 84°F) and were slightly warmer than air temperatures by basking in the sun or absorbing heat from the water. Work by Huey and Stevenson (1979) and Knowles and Weigl (1990) on Green Frog jumping performance suggests an optimal activity range between 15°C and 25°C (59°F and 77°F). This wide temperature tolerance probably enables Green Frogs to thrive in varying environmental conditions across a large geographic range.

In Wisconsin, Green Frogs are typically active into October but sometimes early November if environmental temperatures remain warm (Joshua Kapfer, UW–Whitewater, personal observation). Martof (1953a) reported that the active season lasted ca. 220 days in Michigan. During late summer and early autumn, Green Frogs forage in fields and woods near their home ponds to build up lipid reserves for overwintering (Brenner 1969, Lamoureaux et al. 2002). As environmental temperatures drop in mid- to late autumn, however, they move toward overwintering ponds. Although they can apparently tolerate freezing conditions for short periods (Bohnsack 1951), this cannot be maintained throughout the winter, so hibernation typically occurs while submerged. In Michigan, Martof (1953b) observed that when the air temperature fell below 15°C (59°F, often in the last two weeks of October), Green Frogs moved from more exposed positions along the banks of ponds to the water near the shore. By November, as water temperatures further dropped below 15°C (59°F), Green Frogs disappeared into the sediments and leaf litter at the bottoms of the ponds. Environmental temperature drives their activity during cooler late autumn and winter months (Martof 1953a, Brenner 1969). Green Frogs are sporadically active on mild winter days when the water is ice-free (Minton 2001). In Wisconsin, Lannoo et al. (1998) documented mummified Green Frogs that had apparently been active in winter months and killed when entombed in snow during a cold snap.

PREY AND PREDATORS

Adult Green Frogs are opportunistic predators that feed on any prey small enough to fit into their mouths (Hamilton 1948, Roster et al. 1995). They typically ambush prey that comes within reach along the water's edge, but they probably also feed underwater (Hamilton 1948, Martof 1953b). Visual cues are likely important when hunting, and Hamilton (1948) observed a Green Frog capture a crayfish that it stalked from 11 m (36 ft) away. Adult diet composition has been studied in populations from both the Northeast and the Midwest (Hamilton 1948, Whitaker 1961, Jenssen and Klimstra 1966, Stewart and Sandison 1972, Werner et al. 1995). These studies support that Green Frogs are primarily generalists that feed on a plethora of invertebrates. Although they broadly consume many types of invertebrates, dietary composition varies by the habitat that Green Frogs frequent, season, and prey availability. Jenssen and Klimstra (1966) found that Green Frogs in southern Illinois mainly fed on caterpillars and gastropods during the spring. In the summer, their diet shifted to mostly beetles, gastropods, and odonates. By autumn, Green Frogs predominantly consumed moths, beetles, ants, and wasps. Amphipods were the major food item in the winter.

Green Frogs also consume various vertebrates, although vertebrate predation is apparently less frequent than invertebrate predation. For example, documented vertebrate prey items include amphibian larvae, small adult frogs, crayfish, young bullheads (*Ameiurus* spp.), hatchling turtles, and small snakes (Hamilton 1948, Jenssen and Klimstra 1966, Werner et al. 1995, Harding 1997, DeGraaf and Nein 2010). Some additional items reportedly found in Green Frog guts are probably ingested accidentally, such as seeds, pine needles, leaves, duckweed, small rocks, and sand (Hamilton 1948, Stewart and Sandison 1972). Green Frogs also are known to eat their shed skin (Hamilton 1948).

For the most part, adult Green Frogs feed on food items different from those of other ranid frogs with which they share habitats (Stewart and Sandison 1972, McAlpine and Dilworth 1989, Werner et al. 1995). Compared with American Bullfrogs, for example, a smaller proportion of the Green Frog diet consists of aquatic prey items (Stewart and Sandison 1972, McAlpine and Dilworth 1989), and the larger the body size difference between American Bullfrogs and Green Frogs, the less the overlap in types of prey items consumed (Werner et al. 1995). Given these points, adult Green Frogs probably do not directly compete for food with other frogs due to differences in their habitat use and the gape limitations associated with the body sizes of other species.

Green Frog larvae have a generalized feeding morphology that allows them to filter suspended algae out of the water column and to scrape food from substrates (Werner 1994). As larvae become larger, diet may shift from primarily filter-feeding to primarily substrate-scraping (Steinwascher 1979). Cyanobacteria, diatoms, and chlorophyte algae are the predominant organisms consumed by Green Frog larvae (Gosner 1959, Jenssen 1967, Holomuzki 1998). Food items also include other types of algae, nonalgal protists, microcrustaceans, and decomposed plants. Sand and other debris may be consumed inadvertently. Larvae have also been observed feeding on Wood Frog (*Lithobates sylvaticus*) egg masses (Petranka and Kennedy 1999). Larvae apparently feed continuously, not showing a preference for one available food type over another (Jenssen 1967).

Selected food types and feeding behavior can vary based on a variety of factors. Most larvae feed around aquatic vegetation, with higher feeding rates at temperatures above 20°C (68°F; Warkentin 1992a, 1992b). Overwintering larvae are reported to even feed in the water column under ice (Getz 1958). Larvae near transformation have nonfunctional digestive tracts and do not feed (Jenssen 1967). At high densities, Green Frog larvae also have slower growth rates and metamorphose at smaller sizes (Werner 1994, Skelly 1995). Feeding interactions with Spring Peeper (*Pseudacris crucifer*), Gray Treefrog (*Hyla versicolor*), and American Bullfrog larvae have been studied under experimental conditions (e.g., Werner 1994, Skelly 1995, Anholt et al. 2000, Smith et al. 2004). Due to similar morphology and habits, American Bullfrog larvae are potentially one of the most important competitors, although they may avoid competition by partitioning the feeding environment. American Bullfrog larvae spend more time feeding in the water column, while Green Frog larvae spend more time feeding benthically (Werner 1994). High Green Frog larvae consumption rates of phytoplankton, coupled with their ability to overwinter, may have negative effects on available food resources for zooplankton and spring-hatching larval frogs and salamanders (Cortwright and Nelson 1990, Petranka and Kennedy 1999). As far as humans are concerned, a potentially positive effect of the presence of Green Frog larvae is a reduction in the numbers of larval *Anopheles* mosquitos (Petranka and Fakhoury 1991). Post-metamorphic juveniles feed on a greater proportion of terrestrial food items than do adults (Jenssen and Klimstra 1966). Whitaker

(1961) determined that the stomach contents of juveniles at a site in New York were mostly comprised of insect larvae (20.3% of stomach contents by volume), although various additional prey were taken, including a young Gray Treefrog (*Hyla versicolor*).

Adult Green Frogs are apparently palatable to many predators, including American Bullfrogs, snakes, ducks, herons, bitterns, rails, harriers, American Crows (*Corvus brachyrhynchos*), American Minks (*Mustela vison*), Virginia Opossums (*Didelphis virginiana*), and Raccoons (*Procyon lotor*; Carpenter 1952a, Martof 1956a, Brown 1979, Formanowicz and Brodie 1982, Lamoureux and Madison 1999, Bowen 2004a). Defensive behaviors include alarm vocalizations, flight, body inflations, reduced activity, and concealment (Marchisin and Anderson 1978, Formanowicz 1979, Heinen and Hammond 1997). Green Frog eggs and larvae are preyed upon by almost anything frequenting pond habitats (Brodie and Formanowicz 1983, Formanowicz 1986, Wilbur 1997, Moore and Townsend 1998). Early in the spring, important predators include adult Eastern Newts (*Notophthalmus viridescens*) and various turtles. Furthermore, ambystomatid salamanders, dytiscid beetles, hemipterans, and odonates (dragonflies) arrive at newly filled ponds to lay eggs that will develop into predaceous larvae. The most important predator of Green Frog larvae is probably odonate larvae (Relyea and Werner 1999, Relyea 2001a, 2001c). Later in the summer when the ponds begin to dry, widely foraging vertebrates such as Common Watersnakes (*Nerodia sipedon*), herons, and Raccoons feed on the concentrated and easily captured larvae. In more permanent ponds, fishes are also important predators. Six-spotted Fishing Spiders (*Dolomedes triton*) are known to prey on larvae that surface to gulp air (Moore and Townsend 1998). Generally, as larvae develop larger body sizes, predation risk shifts to larger predators (Brodie and Formanowicz 1983, Formanowicz 1986).

Green Frog eggs and larvae apparently possess limited chemical defenses against predators, but their eggs may be somewhat unpalatable to certain species, such as Eastern Newts (Walters 1975). Larvae appear to be palatable to a variety of predators, although several authors support that they are somewhat distasteful to fishes (Martof 1956a, Brodie and Formanowicz 1983, Kats et al. 1988, Relyea 2001a, 2001c, Szuroczki and Richardson 2001a). Although chemical defenses are limited, developmental and behavioral alterations in response to various predators have been documented. For example, the time until eggs hatch will vary when predators such as leeches, dragonfly naiads, and crayfish are present. In some cases, hatching can be delayed in the presence of larval predators or sped up in the presence of egg predators (Schalk et al. 2002, Ireland et al. 2007, Anderson and Brown 2009). Green Frog larvae reduce activity in the presence of visual and chemical cues from potential predators, including fishes (Peacor and Werner 1997, 2000, Relyea and Yurewicz 2002, Fraker 2009, Smith et al. 2011). Green Frog larvae can also modify their body morphology in response to the presence of predators, and this may be useful in improving evasion or foraging ability (Relyea 2000a, 2001a, 2001b).

A wide variety of organisms are parasitic, commensal, or potentially mutualistic symbionts of Green Frogs. Literature on the parasite community of Wisconsin Green Frogs was summarized by Watermolen (2014), and studies on other Green Frog populations were summarized in Owen and Austin (2013). Known symbionts include a diverse range of single-celled eukaryotes, as well as fungi and invertebrate animals. The helminths (trematodes, cestodes, and nematodes) of Wisconsin Green Frogs are well-documented, and this parasite community is similar to that of Michigan Green Frogs (Bolek and Coggins 2001, Yoder et al. 2001). Different species of helminths have adapted to infect different locations or organ systems in Green Frogs (Esch and Fernandez 1994, Muzzall et al. 2001). The trematodes that infect Green Frogs have a life cycle with two intermediate hosts (typically snails and odonates) and a primary host (the frog) harboring the mature adult parasite. For some trematodes, Green Frogs serve as an intermediate host, with birds or mammals as the primary host.

Green Frogs are ideal hosts compared to most other frogs because the larval development period and adult breeding seasons are long, allowing more opportunities for host colonization. In addition, adult frogs have large body sizes, providing more surface area for colonization. Green Frogs can be infected by preying on tadpoles or other frogs (Bolek et al. 2009). They may also serve as accidental alternative hosts for fish parasites (Muzzall et al. 2001). However, the helminth community of Green Frogs is low in diversity compared to that of marine fishes and endothermic vertebrates (Muzzall 1991, McAlpine and Burt 1998, Bolek and Coggins 2001, Sutherland 2005). Furthermore, the Green Frog helminth fauna is not unique, and helminths found in Green Frogs can be found in other frog species, especially American Bullfrogs (Muzzall 1991, McAlpine 1997, Muzzall et al. 2001).

Parasite exposure is dependent upon the local hydrological cycle and terrestrial habitat (Muzzall et al. 2001). For a series of ponds studied in Michigan (Muzzall et al. 2001), core helminth species parasitizing Green Frogs included the trematodes *Fibricola* spp., *Glypthelmins quieta*, *G. attenuata*, *Halipegus eccentricus*, *H. varioplexus*, *Megalodiscus temperatus*, and possibly *Pneumatophilus* spp. Core nematode species included *Raillietnema* spp. and *Spriroxys* spp. Fishless locations had the highest helminth community diversity due to the presence of large numbers of odonate intermediate hosts.

CONSERVATION

Status: Globally, the Green Frog is listed as a Least Concern species by the IUCN. Across their range, Green Frogs are considered a common to abundant species (Hecnar 2004, Pauley and Lannoo 2005). Vogt (1981) believed they were the "most common frog in Wisconsin" at the time of his writing, and there is nothing to support their populations having substantially declined since then. Other than some local declines (see Threats subsection), populations are apparently stable in Wisconsin, and the Green Frog is considered a common species by the WDNR. Not surprisingly, WDNR data from the annual Wisconsin Frog and Toad Call Survey documented a stable pattern of occurrence from 1984 to 2016 (Figure 106).

Populations: Few, if any, studies from Wisconsin exist that focus on population parameters, aside from those inferred through surveys of male advertisement calls (e.g., Trenham et al. 2003). Despite this, studies from other northern locations exist, and these indicate that Green Frog populations can be large. Abundance estimates in Michigan (including both juveniles and adults) ranged from ca. 100 individuals at a 6.1 ha (15 ac) pond, or 16 frogs/ha (7/ac), to 500 individuals at a 2.7 ha (6.7 ac) pond, or 185/ha (15/ac; Martof 1956a). The same study estimated that in August these two populations temporarily reached sizes higher than 4,600 (754/ha or 307/ac) and 1,000 (370/ha or 149/ac) individuals, respectively, with the arrival of newly metamorphosed individuals. These numbers quickly decreased due to dispersal and mortality. Larvae can also reportedly reach high densities. For example, Cortwright (1998) estimated larval densities at numerous sites in Indiana that were as high as 47.67/m^2 (4.43/ft^2). Although larval densities can be high, reported estimates of their survival to metamorphosis are variable.

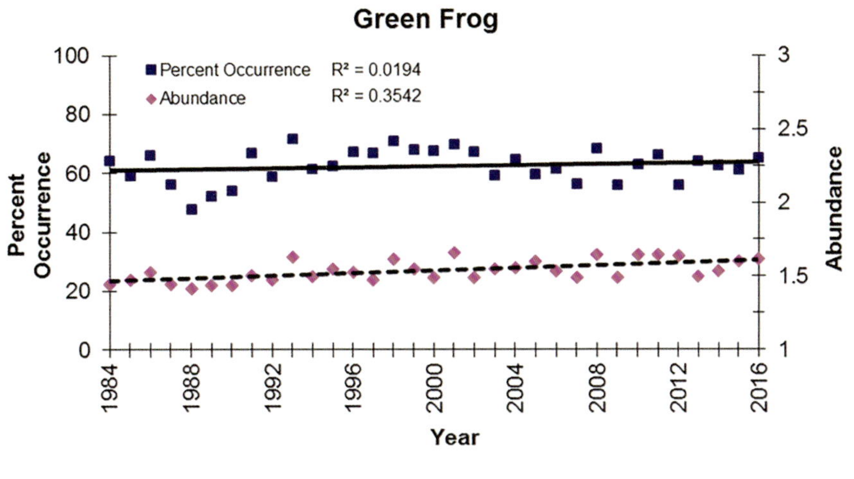

Figure 106. Long-term trend in occurrence (%) and abundance (rank) of Green Frogs across Wisconsin Frog and Toad Survey statewide monitoring sites. See Introduction to Species Accounts chapter for details on this monitoring program (figure courtesy of WDNR).

During the peak of larval metamorphosis in summer, Werner and McCune (1979) counted an average of 0.39 individuals/m (0.12 individuals/ft).

A six-year study in Ontario determined that the percent of larvae surviving to metamorphosis ranged from 32.5% to 85.3% (Shirose and Brooks 1997). Martof (1956a) noted that adult sex ratios were ca. 1M:1F at the beginning and end of the activity season in Michigan. Some studies support that Green Frog populations are naturally dynamic and have high local turnover rates without overall population declines (Hecnar and M'Closkey 1996, 1997a). Populations studied in Wisconsin exhibited losses that were roughly equal to gains over an 18-year period (Trenham et al. 2003).

Threats: Green Frogs are likely affected by many of the general threats that influence other amphibians in Wisconsin (see Introduction to Conservation and Management of Wisconsin's Amphibians and Reptiles chapter). The observed variation in breeding population size and reproductive success can be attributed to factors such as forest succession and changes in wetland hydroperiod, which can be influenced by anthropogenic activities (Semlitsch et al. 1996, Cortwright 1998). Some studies support that Green Frogs are less sensitive to habitat loss around their breeding sites than other amphibian species, and this may allow them to persist in the altered landscapes associated with suburban and urban areas (Rubbo and Kiesecker 2005). Indeed, certain types of development such as golf courses can act as important habitat corridors and even source habitats for Green Frogs (Birchfield and Deters 2005). Green Frogs also readily colonize farm ponds and mitigation wetlands, regardless of fish presence and water depth (Knutson et al. 2004, Porej and Hetherington 2005). Green Frogs can naturally breed in acidic *Sphagnum* bogs (Freda and Taylor 1992), and the ability to tolerate acidic waters allows them to inhabit mine drainage treatment wetlands (Lacki et al. 1992).

Despite their apparent tolerance of anthropogenic habitat changes, several studies support that land development and habitat destruction can threaten Green Frog populations. Knutson et al. (2000) reported that calling males were negatively associated with urban landscapes in Wisconsin. Likewise, Paloski (2006) found them least often on transects in developed areas compared to transects in forested and marshy areas. Gagne and Fahrig (2010) further reported that although Green Frogs persisted in residential landscapes in Canada, their abundances significantly declined in these disturbed landscapes over time. In Wisconsin, development along lake shorelines has had a negative impact on Green Frog populations due to the destruction of vegetation that is important as cover (Woodford and Meyer 2003, Price et al. 2004, Henning and Remsburg 2009). Additionally, silvicultural activities may reduce Green Frog numbers due to the burning and clearing of timber, which also increase the risk of desiccation (McLeod and Gates 1998, Rittenhouse et al. 2008). The impact of habitat degradation due to infestations with invasive vegetation is largely unknown for most amphibians, including Green Frogs. However, Maerz et al. (2005a) found that test plots dominated by an invasive plant were lower-quality habitats for arthropods that Green Frogs prey upon, which indirectly affected foraging success of Green Frogs in the heavily infested plots.

Additional factors that impact Green Frogs and are often associated with disturbed landscapes exist. Roads act as obstacles that separate foraging and breeding habitats, although some authors report that Green Frogs suffer lower road mortality than other anurans (Ashley and Robinson 1996, Carr and Fahrig 2001, Mazerolle 2004). The fact that Green Frogs remain immobile during encounters with vehicles, however, is the major contributor to roadway deaths (Mazerolle et al. 2005). Contamination of breeding ponds from deicing salts, which are often associated with roads in northern locations, could be a concern for amphibians, but the results of some existing research are conflicting. For example, Karraker (2007) found that larval malformation rates were greatest when larvae were exposed to high levels of deicing salts but that survival and growth of eggs and larvae were less affected. Aside from deicing salts, various forms of pollution that are more prevalent in disturbed areas have been linked to amphibian declines (see Introduction to Conservation and Management of Wisconsin's Amphibians and Reptiles chapter) and may have a negative impact on Green Frog populations.

Fertilizer, pesticide, and herbicide pollution presents a threat to Green Frog development, survival, and reproduction. Egg and larval stages are especially susceptible, and effects may vary with larval density, stage of development, type of pollutant, degree of exposure, and presence of predators.

In the Fox River and Green Bay regions of Wisconsin, high heavy metal and polychlorinated biphenyl (PCB) contamination levels from industrial pollution have been associated with Green Frog population declines or absences (Karasov et al. 2005). Heavy metals such as mercury, lead, cadmium, and chromium, as well as PCBs, have negative effects on egg hatching and larval activity (Strickler-Shaw and Taylor 1990, Taylor et al. 1990, McCrary and Heagler 1997, Karasov et al. 2005). Sediment levels of PCBs are correlated with larval tissue levels of PCBs (Karasov et al. 2005), and larval tissue levels are higher than those of adults (Glennemeier and Begnoche 2002). Even so, Green Frogs do not bioconcentrate PCBs at the high levels found in other types of vertebrates (Glennemeier and Begnoche 2002). Other factors that have been examined in Green Frogs that negatively affect their activity levels, reproductive success, and/or development in agricultural and silvicultural settings include ammonium nitrate and phosphorus runoff from fertilizers and animal waste (Hecnar 1995, Knutson et al. 2004, Burton et al. 2009) and components of pesticides and herbicides (e.g., see Edington et al. 2004, Howe et al. 2004, Storrs and Kieseker 2004, Boone et al. 2013).

Both pollutants and parasites can cause deformities in Green Frogs. Some pollutants can disrupt developmental and sex hormones, and abnormal hormonal levels have been associated with limb malformations and intersex Green Frogs (Sower et al. 2000, Skelly et al. 2010). Malformations include ectromelia (the absence of all or part of a limb), ectrodactyly (the absence of all or part of a digit of a limb), and asymmetrical development of limbs. Malformed frogs have been documented in numerous locations in the Midwest, including Wisconsin (Martof 1956a, Green et al. 2002, Lannoo 2008). For example, collections made by Erik Wild and colleagues from Portage and Marathon Counties in 1999 (UWSP collection) included specimens either with missing limbs or with up to five extra limbs and six extra digits. Malformations in Green Frogs are associated with proximity to agricultural land and lawns (Taylor et al. 2005), as well as exposure to various pesticides and herbicides (e.g., see Lowcock et al. 1997, Ouellet et al. 1997, Sessions et al. 1999). Herbicides, as well as synthetic and environmental estrogens, have been found to cause intersex individuals to appear in Green Frog populations (Skelly et al. 2010, Lambert et al. 2015, 2016).

Other population declines and malformations documented in Green Frogs can be tied to parasites, such as the echinostome trematodes *Echinostoma trivolvis*, *E. revolutum*, and *Ribeiroia ondatrae* (Sessions et al. 1999). Cercariae are shed from snail hosts, coinciding with the emergence of Green Frog larvae (Holland et al. 2007). The cercariae of *R. ondatrae* encyst around the base of the hind limbs, causing limb malformation in the metamorphosing larvae (Szuroczki and Richardson 2009). This may facilitate transmission of the trematodes into aquatic bird predators, which are their primary hosts. In contrast, *Echinostoma trivolvis* encysts in kidneys of larvae and can cause up to 40% mortality in early stages, probably due to compromised renal function (Holland et al. 2007, Szuroczki and Richardson 2009). Older larvae can destroy or eliminate metacercariae (Holland 2009). Larvae also become more susceptible to odonate predators as they try to avoid the swimming stages of the parasites (Marino and Werner 2013). Echinostome trematodes may be more prevalent in Green Frog larvae in urban environments, potentially due to larvae being physiologically stressed by herbicides such as atrazine or increases in snail host populations due to eutrophication (Holland et al. 2007, Johnson et al. 2007b, Rohr et al. 2008b).

Fungal parasites, particularly the chytrid fungus *Batrachochytrium dendrobatidis* (Bd), infect both larvae and adults and can cause amphibian population declines (Longcore et al. 2007, Groner and Relyea 2010). In the few studies that sampled amphibians in Wisconsin, Green Frogs were among the species tested with the highest infection rates (Sadinski et al. 2010, Klemish et al. 2012, Standish et al. 2018). Detection rates in some studies were similar to surrounding states, and none of the infected Green Frogs exhibited signs of chytridiomycosis. No population declines were associated with these detections, although sample sizes were often small. Longcore et al. (2007)

suggested that Green Frogs may be more resistant to chytridiomycosis compared to other species of anurans, although habitat and climate modification can influence the prevalence of Bd infection in Green Frogs. For example, in New York, Green Frogs at sites with higher canopy closure and lower water temperature had greater prevalence and intensities of Bd infection (Becker et al. 2012). Other types of important amphibian pathogens, such as ranaviruses, infect Green Frogs in Wisconsin. Although directed surveys for the presence of viruses in state populations have been limited, ranaviruses were identified in several Green Frogs sampled in Wisconsin (Standish et al. 2018) and in a large number of dead Green Frogs sampled from Vilas County in 2009 (WHISPers 2017). The limited reports of Green Frogs infected with ranaviruses in Wisconsin could be the result of limited directed sampling for the pathogen or because some research has found that Green Frogs are poor reservoirs for ranaviruses (Forzan and Wood 2013).

Ultraviolet radiation (UV) may interact with the factors described above and contribute to population declines. Green Frog eggs may not be as adversely affected by UV as other anuran species (Licht and Grant 1997). Green Frog egg masses form a film at the water surface, and the jelly envelopes do not absorb UVB as well as spring-breeding anurans. However, eggs typically hatch within seven days, which reduces the period of UV exposure (Licht 2003). Larvae experimentally exposed to UVA suffered no ill effects, while larvae exposed to UVB suffered arrested development (Grant and Licht 1995). Tietge et al. (2001) found that ambient levels of solar radiation were lethal to larvae if refuges were not available and UVB was not filtered out. The risk for malformations and mortality in Wisconsin wetlands due to natural levels of UV exposure is considered to be low (Diamond et al. 2002). Under most natural lighting conditions, UVA and UVB are probably largely absorbed by suspended particles and vegetation in the water, and larvae are able to behaviorally avoid excessive exposure (Grant and Licht 1995, Crump et al. 1999).

Management: General wetland, stream, and associated upland habitat management considerations for amphibians apply to this species (see Introduction to Conservation and Management of Wisconsin's Amphibians and Reptiles chapter). Green Frogs in Wisconsin could benefit from management strategies that also involve the restoration of degraded wetlands. Nedland et al. (2007) reported that calling male Green Frogs used restored wetlands in Manitowoc County shortly after completion of restoration and were also detected in these wetlands over a decade post-restoration. Because this species has high local turnover rates (Hecnar and M'Closkey 1996, 1997a), preserving or enhancing habitat availability and connectivity among breeding and nonbreeding aquatic habitat patches is important for long-term persistence. This could involve creation or protection of strategically placed wetlands to enhance connectivity and linking isolated aquatic habitats via riparian corridors or upland forest (Schroeder 1976, Price et al. 2004).

Pickerel Frog

Lithobates palustris (LeConte 1825)

Dreux J. Watermolen

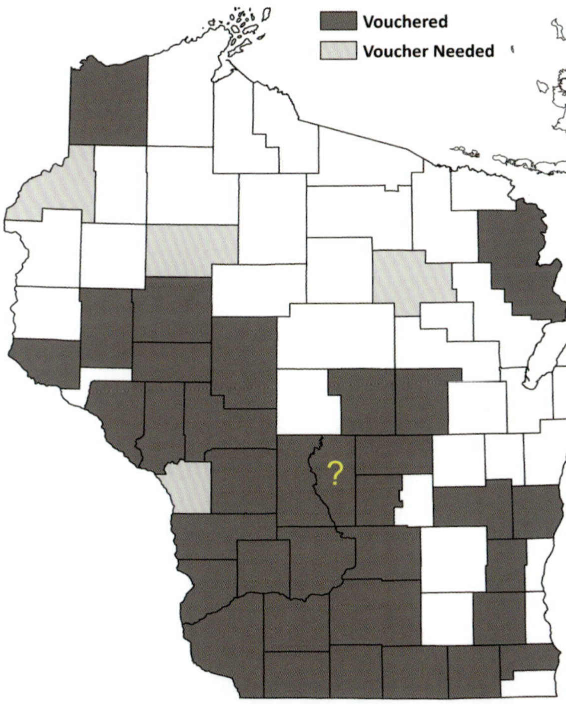

Adult female Pickerel Frog (Dane County; photo by S. Viernum).

AT A GLANCE

Pickerel Frogs are a medium-sized frog with two parallel rows of squarish brown spots that run down the back between the dorsolateral folds and three large spots on the head. Adults also typically possess a yellowish coloration on the inner thigh and groin region. State populations are scattered and isolated but appear to be more concentrated in the southwestern counties. Pickerel Frogs are active from late March through early October. During most of the active season, they typically associate with cool flowing water habitats (such as streams), although they breed in low-current areas of streams or adjacent wetlands in spring. The male's advertisement call is a slow low-pitched snore-like croak that resembles that of the Northern Leopard Frog. Males in Wisconsin call while submerged or while floating at the water's surface from March through May, and larvae metamorphose in 60 to 90 days. Although limited diet information exists, adults appear to be opportunistic carnivores that consume a wide variety of invertebrates. Adult Pickerel Frogs produce skin secretions that could be toxic to some predators, although they are reportedly still consumed by other anurans, reptiles, and raptors. Pickerel Frogs are uncommon in Wisconsin and considered a Special Concern species and Species of Greatest Conservation Need by the WDNR.

SYSTEMATICS

Lithobates palustris (Pickerel Frog) is included in the family Ranidae. Within the genus *Lithobates*, *L. palustris* is most closely related to the *L. areolatus/L. capito/ L. sevosus clade* (Crawfish Frog-Gopher Frog complex of Hillis and Wilcox 2005), a group of frogs with males that use snore-like advertisement calls when mating. Despite its resemblance to *L. pipiens* (Northern Leopard Frog) and related species, the validity of *L. palustris* has never been seriously questioned and has had a stable and simple nomenclatural history. The specific epithet, *palustris*, is derived from the Latin *paluster*, 'of the marsh' (Schaaf and Smith 1971). The common name is said to stem from the frog's use as a popular bait for Pickerels (*Esox americanus*; Dickerson 1969). No subspecies are currently recognized (Crother 2017). Fossil *L. palustris* are known from a Pliocene site in Texas and Pleistocene sites in Arkansas, Pennsylvania, Texas, and Virginia (Holman 2003). *L. palustris* hybridizes with other closely related *Lithobates* species in Wisconsin under laboratory and field conditions (reviewed by Dodd 2013), and mating attempts with other species have occasionally been observed in the wild (e.g., Wright 1914). Hybrids of *L. palustris* and *L. pipiens* can develop normally and metamorphose (Moore 1941, 1946).

DESCRIPTION

Adults: The dorsal background color ranges from light brown or green to olive. Two rows of brownish square spots (often outlined in black) occur on the back between light-colored dorsolateral folds. Typically, there are three large spots on the head, one over each eye and a third on top of the snout (Figure 107). A light line runs along the upper lip. Additional squarish spots occur on the sides of the body, and brown bars cross the hind legs. The belly can be plain white or slightly mottled. Pickerel Frogs have a distinctive bright yellow or orange color on the underside of the hind limbs and in the groin area (Figure 108). The webbing between the hind toes extends nearly to the tip of the toes. Pickerel Frogs are

Figure 107. Adult Pickerel Frog displaying the characteristic two rows of square spots on the back and three large spots on the head, one over each eye and a third on top of the snout (Dane County; photo by T. Muehlfeld).

Figure 108. Adult Pickerel Frog showing its distinctive yellow-orange coloration on the underside of the hind limbs and in the groin area (Dane County; photo by J. Kapfer).

Figure 109. Juvenile Pickerel Frog showing metallic dorsal coloration (Columbia County; photo by A. Badje).

a medium-sized anuran. Johnson (1984) measured 428 Wisconsin individuals and found that adult SVL was 2.2–7.7 cm (0.9–3.0 in), similar to the 3.0–7.0 cm SVL (1.2–2.8 in) reported by Vogt (1981). Schaaf and Smith (1970) determined that the typical length across their geographic range was 4.3–6.5 cm SVL (1.7–2.6 in). The largest Wisconsin specimen measured was a female from Waupaca County (MPM 2896), which had an SVL of 7.4 cm (2.9 in).

Preadult stages: Pickerel Frog eggs are bicolored (brown at one pole and yellow at the other) and are laid in globular masses that contain hundreds of eggs (reviewed by Dodd 2013). The diameter of the entire egg mass at the time of deposition is 3.8–5.0 cm (1.5–2.0 in), which gradually swells to 9.0–13.0 cm (3.5–5.1 in; Livezey and Wright 1947, Johnson 1984) during egg development. Eggs are enclosed in two gelatinous membranes or capsules (not a single capsule, as erroneously indicated by Watermolen [1995]). The diameter of the ovum averages 0.170–0.185 cm (0.05–0.07 in), and the egg and capsules together measure 0.36–0.50 cm (0.14–0.20 in; Livezey and Wright 1947, Dodd 2013).

Larvae are relatively large and full-bodied. Dorsal coloration is green, grayish brown, or brown with scattered black and yellow spots. The venter is iridescent white to cream, and the viscera may be visible through the belly. The tail fins are heavily pigmented with dark spots or blotches. Larvae can reach 7.6 cm TL (3.0 in), with the body measuring 2.7 cm (1.1 in) and the tail measuring 4.9 cm (1.9 in; Wright 1929). Newly metamorphosed individuals with tails completely resorbed measure 1.9–2.9 cm SVL (0.8–1.2 in; Gilhen 1984, Dreux Watermolen, WDNR, unpublished data). Subadult frogs are often metallic in appearance (Figure 109).

Variation: Adults are sexually dimorphic. In Wisconsin, adult females were generally larger (5.2–7.4 cm SVL [2.0–2.9 in], $\bar{x} = 6.2$ cm [2.4 in], $n = 63$) than males (4.9–7.3 cm SVL [1.9–2.9 in], $\bar{x} = 5.3$ cm [2.1 in], $n = 126$; Johnson 1984), an observation similar to results reported throughout the Pickerel Frog's range (Wright 1914, Smith 1961, Gilhen 1984, Hardy and Raymond 1991, Meshaka et al. 2012). Although Pope (1964) reported that males were larger than females (p. 128), elsewhere he gave measurements for males that were smaller than females (p. 130). Males also have enlarged thumbs with inner pads during the breeding season and may be lighter colored than females (Dodd 2013). For a review of geographic variation across the Pickerel Frog's range, see Schaaf and Smith (1970) and Pace (1974).

Diagnosis: Adult and metamorphosed juvenile Pickerel Frogs are characterized by a slender body without parotoid glands or large nodules and no keratinized spade on the heel of each hind foot. They do not possess terminal toe pads or a dark mask, and they do not emit a musky odor. Dorsolateral folds that extend the length of the dorsum are present, as are two rows of large square or rectangular paired spots on the dorsum between the folds. There are also two rows of spots that are ventrolateral to the dorsolateral folds. Pickerel Frogs have a bright yellow or orange coloration on the underside of the hind limbs and in the groin area.

Pickerel Frogs most closely resemble Northern Leopard Frogs in shape, size, general color, and pattern. The dark bordered dorsal spots on Pickerel Frogs tend to be more angular and square or rectangular and are often arranged in two well-defined parallel rows between the dorsolateral folds, whereas those of Northern Leopard Frogs are more rounded or oval, have a light border, and are arranged sporadically. Northern Leopard Frogs also generally lack the bright yellow or orange color in the groin area found in Pickerel Frogs. Some researchers suggest, however, that certain Northern Leopard Frog populations may 'mimic' the spot pattern and yellow groin of the Pickerel Frog, because the latter is somewhat toxic (Vogt 1981, Harding 1997). Although purported mimicry has not been thoroughly investigated, it could complicate differentiating these species. Male advertisement calls of both species also sound similar, but the snore-like call of Pickerel Frogs is higher in frequency and shorter in duration than the call of Northern Leopard Frogs (see Reproduction and Development section). Pickerel Frogs might also be confused with American Bullfrogs (*Lithobates catesbeianus*), although this species lacks dorsolateral folds, as well as Green Frogs (*L. clamitans*), although this species lacks large spots and has only partial dorsolateral folds.

Distinguishing the eggs and larvae of the different ranid frog species in Wisconsin is difficult. The eggs of Pickerel Frogs are reportedly brown on one side and yellowish on the other, whereas those of Northern Leopard Frogs are black on one side and white on the other (Parmelee et al. 2002). Pickerel Frog larvae lack external gills and forelimbs. Their eyes are positioned dorsally on their heads, their intestinal coil is at least slightly visible, and their vent is dextral. There is no white line from their nostrils to their snout, while their body and tail are so heavily mottled or spotted that they are sometimes described as purple in color. According to Harding (1997), Northern Leopard Frog larvae have lighter and more faintly spotted tail fins than Pickerel Frog larvae.

DISTRIBUTION AND HABITAT

Global distribution: Pickerel Frogs occur intermittently throughout much of eastern and midwestern North America from the western end of Lake Superior, northern Wisconsin, and Michigan's Upper Peninsula east to the St. Lawrence River and northern New England, Nova Scotia, and New Brunswick. They extend south from Wisconsin through portions of eastern Minnesota and Iowa to eastern Oklahoma and Texas and east to northern Georgia. They are absent below the Fall Line in the southern Gulf Coast states (Brown and Boschung 1954, Mount 1975, Jensen et al. 2008) and from the prairie regions of, for example, central Illinois and southwestern Indiana (Minton 1972). There are also few records from the Piedmont region in the southeastern US (Brown 1992a, Platt et al. 1999). The similarity between Pickerel Frogs and Northern Leopard Frogs (*Lithobates pipiens*) has led to erroneous records in some localities (Redmer and Mierzwa 1994).

Distribution in Wisconsin: Currently, Pickerel Frogs are sporadically documented throughout the state in scattered local populations. Yet records and reports are most concentrated in the southern half of Wisconsin, particularly in the unglaciated Driftless Area of the southwestern counties (Johnson 1984, Mossman and Hine 1985, Mossman et al. 1998). Of the published county records, suitable museum vouchers could not be found or are reported missing for Burnett, La Crosse, Langlade, and Rusk (see the Appendix). In addition, the record for Adams County is based on a historical specimen (USNM 312213, one of four specimens collected at this location in 1929) with somewhat vague locality data that place it near the border with Marquette County (see discussion of this specimen in Kapfer et al. [2015]).

Localities of possible occurrence: Published county records for this species are currently spotty in southeastern Wisconsin and sparse in the northern half of the state. There are several counties surrounded (or mostly surrounded) by those of known occurrence that could yield records with further survey effort, such as Pepin, Wood, Green Lake, Dodge, Jefferson, Kenosha, Ozaukee, and Milwaukee. Nelson (1915) reported secondhand records but no vouchers of Pickerel Frogs from Green Lake County that he received in 1906 and 1909. Dickinson's (1965) map shows records in Ashland, Winnebago, and Milwaukee Counties, but Vogt (1981) did not include these records, and no vouchers exist to support them. Unverified records generated during the WDNR's Wisconsin Frog and Toad Surveys exist for additional counties (Table 11). If confirmed, records from these counties would greatly expand the Pickerel Frog's known distribution in the state, particularly in the northern region. Although previous records from the Upper Peninsula of Michigan could support this species' presence in additional northern Wisconsin counties (Harding 1997, Holman 2012), few of the reports from the Upper Peninsula are recent (Sargent 2000). Casper et al. (2015b) further stated that the last confirmed record of Pickerel Frogs from the Upper Peninsula was in 1947. Still, survey efforts in northern Wisconsin could substantiate the occurrence of Pickerel Frogs in several additional counties.

Habitat: Few studies have examined habitat associations of Pickerel Frogs. Research in Canada supports that they are positively correlated with wetlands, particularly those that possess deep water

Table 11. Reported counties in Wisconsin with unsubstantiated Pickerel Frog records based on detection of male advertisement calls.

County	Source
Barron	WIATRI 2018
Bayfield	Mossman and Hine 1985, Mossman et al. 1998, WIATRI 2016
Brown	Mossman et al. 1998, WIATRI 2016
Calumet	Mossman et al. 1998, WDNR 2016b, WIATRI 2016
Dodge	Mossman et al. 1998, WIATRI 2016
Florence	WDNR 2016b
Forest	Mossman et al. 1998, WIATRI 2016
Green Lake	Mossman et al. 1998, WIATRI 2016
Iron	Mossman et al. 1998, WIATRI 2016
Lincoln	WIATRI 2016
Marathon	WIATRI 2016
Milwaukee	WIATRI 2016
Oneida	WIATRI 2016
Outagamie	Mossman et al. 1998, WIATRI 2016
Polk	WIATRI 2016
Price	WIATRI 2016
Sawyer	Mossman et al. 1998, WIATRI 2016
Shawano	Mossman et al. 1998, WDNR 2016b, WIATRI 2016
St. Croix	Mossman et al. 1998, WIATRI 2016
Vilas	Mossman et al. 1998, WIATRI 2016
Washburn	WIATRI 2016
Wood	Mossman and Hine 1985, Mossman et al. 1998, WIATRI 2016

tables (Jacobs and Houlahan 2011). In Wisconsin, Johnson (1984) monitored 23 Pickerel Frogs from Sauk County with radio telemetry over the course of their active season. After breeding, these individuals spent most of their time in nearby wet and mesic grasslands and open woods near water, including southern mesic, southern wet, and southern wet-mesic forest communities dominated by Sugar Maple (*Acer saccharum*), White Oak (*Quercus alba*), and American Elm (*Ulmus americana*). Johnson also reported Pickerel Frogs from open-canopy sites with Red-osier Dogwood (*Cornus sericea*), willow (*Salix* spp.), Buckthorn (*Rhamnus catharticus*), and herbaceous vegetation typical of fen and sedge meadow communities. It is noteworthy that Pickerel Frogs reportedly remain close to water in Wisconsin (within 2 m [6.6 ft]) and show a particular affinity for cool streams (water temperature range: 18°C–21.5°C [64.4°F–70.7°F]). Similarly, Pope (1964) indicated that typical habitat in northern Illinois and southern Wisconsin included "clear, spring-fed streams flowing through wet meadows with an abundance of rank grass." Vogt (1981) stated that Pickerel Frogs in Wisconsin required "cold spring water for breeding, usually associated with mesic forests and quartzite formations." Pickerel Frogs in Wisconsin also use waters with a pH range of 5.5–9.6 (Johnson 1984), a broader range than reported elsewhere (4.52–7.80 [$\bar{x} = 5.10$]; Dale et al. 1985). Johnson (1984) found Pickerel Frogs at sites with low to moderate alkalinity (<40–>100 mg $CaCo_3$/L [0.005–0.012 oz/gal]) and high dissolved oxygen levels (94%, $n = 3$ sites).

At the microhabitat scale, Pickerel Frogs in Wisconsin regularly used undercut banks, root tangles, and submerged stick piles as refuges (Johnson 1984). Dense vegetation appears to be another important habitat characteristic for this species, as those "found more than 2 m from the stream were always found in semi-dense vegetation," according to Johnson (1984). Similar observations exist from Illinois, where Redmer and Mierzwa (1994) and Redmer (1998) reported that most adult Pickerel Frogs occupied dense herbaceous vegetation along the margins of cool streams or springs that flowed through or near forests. Elsewhere in their range, Pickerel Frogs are associated with caves and abandoned mines (Smith 1961, Heath et al. 1986, Green and Pauley 1987, Fenoilio et al. 2005, Rimer and Briggler 2010), but similar records are not known from Wisconsin.

Breeding can occur in a variety of stagnant or slow-moving aquatic habitats (either oligotrophic or eutrophic) adjacent to streams that reportedly have high dissolved oxygen levels (97% saturation; Johnson 1984). Johnson (1984) further determined that 6 Wisconsin breeding sites had total phosphorous levels of ≥0.1 mg/L (0.00001 oz/gal) and that 11 sites had total nitrogen levels exceeding 0.4 mg/L (0.00004 oz/gal), levels associated with excessive phytoplankton growth (Bartsch 1972, Mueller and Helsel 1999, Dubrovsky et al. 2010). After hatching, larvae tend to remain in close contact with shoreline vegetation through metamorphosis. In Kentucky streams, larvae were most abundant within 10 cm (3.9 in) of the shoreline, and larval numbers decreased progressively with distance from the shore (Holomuzki 1995). Rittenhouse (2011) found little evidence that diverse vegetation enhanced wetland quality for Pickerel Frog larvae but noted that the quantity of decomposing plant matter had a greater influence on larval performance than did the species of plant matter (see also Conservation section). In Wisconsin, Pickerel Frogs typically overwinter while submersed and under mud, rocks, or debris in spring seeps or pools that receive very little runoff and have temperatures ranging from 6°C to 8°C (43°F to 46°F; Johnson 1984). Dissolved oxygen levels at these overwintering sites were lower than breeding and summer sites (61% saturation, $n = 3$ sites), probably due to inputs of poorly oxygenated groundwater (Johnson 1984).

REPRODUCTION AND DEVELOPMENT

In Wisconsin, mating and reproduction take place from late March through mid-May (Zimmerman 1950, 1951, Vogt 1981). Pickerel Frogs arrive at breeding sites from 7 to 32 days prior to the initiation of mating and typically travel during periods of rain to reach wetlands (Wright 1914). In Dane County, male Pickerel Frogs called from 19 March to 15 May, with greatest calling activity from 2 to 21 April (Zimmerman 1950, 1951). In Sauk County, males called from 7 to 24 April at water temperatures ranging from 9°C to 29°C (48°F to 84°F) and with air temperatures ranging from 3°C to 26.5°C (37°F to 80°F; Johnson 1984). The male advertisement call is a low-pitched snore-like croak (see sonogram in Vogt 1981). Some authors have compared the call to the sound of a door with rusty hinges being slowly opened. Males can call while submerged or while floating at the water's surface. In larger congregations, individual calls merge into a nearly continuous drone. Large choruses can occur throughout the breeding season, with males generally occupying the same calling site on subsequent nights. Males also produce a combination of 'snickers' and 'growls,' which are used in male-to-male interactions during the breeding season (Given 2005, 2009). The dominant frequencies of these additional sounds are similar to those of the advertisement call but of much lower frequency. Snickers and growls are made from both above and below the water's surface and apparently serve to warn intruding males of breeding territory transgressions.

Females that are attracted to males will engage them in amplexus, during which the male mounts the female dorsally and grasps her around the armpits so that eggs can be deposited and externally fertilized. Amplexus begins when mean air temperature reaches 10°C–16°C (50°F–61°F; Wright 1914). Additionally, Johnson (1984) reported that water temperatures at active breeding sites ranged from 6°C to 22°C (43°F to 72°F) in streams and 11°C to 29°C (52°F to 84°F) in ponds. Mating can be nocturnal or diurnal, during which multiple amplectant pairs can occur in a relatively small area. Amplectant pairs move together until the female finds a suitable branch or piece of debris, which she grasps with her hind legs. She then deposits round masses of 1,000–3,000 eggs (reviewed by Dodd 2013). Clutch size is correlated with female body size (Gilhen 1984, Meshaka et al. 2012). Once initiated, oviposition is completed in ca. three minutes (Wright 1914). Egg masses are attached to vegetation or debris submerged up to 10 cm (3.9 in) below the water's surface (Johnson 1984, D. Watermolen, unpublished data). Egg masses deposited in spring pools are often concentrated in areas that receive the greatest sunlight (Johnson 1984). In Kentucky streams, females deposited more egg masses in upstream reaches with lower fish densities than in downstream reaches with higher fish densities. They nearly always oviposited in the stream's main channel but avoided riffles and isolated pools (Holomuzki 1995). Wright (1914) stated that the best place for oviposition in New York was "a quiet back-water of one of our streams." Similar observations have been made at Wisconsin breeding sites (Dreux Watermolen, WDNR, personal observations).

Eggs must remain well-oxygenated, although suitable oxygen concentrations vary based on environmental temperatures. For example, at temperatures of ≥25°C (77°F), eggs in the more hypoxic center of the egg mass may not survive (Moore 1940). Burggren (1985) described how the ciliated epithelia of the embryos are used to create convective ventilation, which may help reduce such mortality. Larvae hatch in 10–24 days (Wright 1914, Johnson 1984). Higher temperatures shorten the incubation period (Harding 1997). Once hatched, larvae congregate in and around vegetation near the shore. Metamorphosis occurs in 60–90 days (Harding 1997, D. Watermolen, unpublished data), but the larval period "can be as short as 54 days under conditions of low density, abundant food, and high temperature" (Wilbur and Fauth 1990). Total energy costs during metamorphosis are positively related to larval body mass and the duration of the metamorphic climax (i.e., the period ranging approximately three to nine days from when the first front leg appears until the tail is completely absorbed; Orlofske and Hopkins 2009). Larvae exhibit a substantial decline in mass during this period, with about half of their available energy used to support maintenance and the other half to support development; larger larvae complete metamorphosis more efficiently than smaller ones because they use proportionally less energy (Orlofske and Hopkins 2009).

Pickerel Frogs reach sexual maturity in the second spring after hatching (Meshaka et al. 2012), but because larvae may overwinter in northern populations, sexual maturity could be delayed for some individuals (reviewed by Dodd 2013). Size at maturity is sexually dimorphic, with females being slightly larger than males (see Variation subsection). Longevity in wild Pickerel Frog populations has not been assessed. Survival in captivity has been reported as less than three years (Snider and Bowler 1992).

ACTIVITY

The activity and movement patterns of Pickerel Frogs in Wisconsin are not well-studied, with the exception of the work by Johnson (1984). Pickerel Frogs emerge from hibernation in late March to early April and remain active until October (D. Watermolen, personal observation). They are mostly diurnal when active but become more nocturnal in the breeding season and during particularly hot periods in the summer (Harding 1997, D. Watermolen, personal observation). Pickerel Frogs have color vision, are sensitive to light in the blue spectrum, and exhibit photopositive responses that suggest they use sunlight and moonlight for orientation, which influences movement patterns (Hailman and Jaeger 1974). In Wisconsin, long-distance movements are likely associated with migrations from breeding habitat to summer habitat, after which movement declines substantially. Johnson (1984) estimated that radio-tracked individuals moved less than once per day in the summer (range = 0.19–0.89 movements/day), while mean distances per move ranged from only 2.38 to 9.56 m (7.8 to 31.4 ft). Despite this, Oldfield and Moriarty (1994) reported a Minnesota individual that was found at least 500 m (1,640 ft) from water. To my knowledge, no studies on home range size have been conducted. Movement rates in autumn can increase as individuals return to hibernacula (Johnson 1984). Pickerel Frogs are one of the last frogs to enter hibernation in Wisconsin and are active well into October (D. Watermolen, unpublished data). Most individuals spend the winter underwater along the bottom of springs or pools (Pope 1964, Harding 1997).

PREY AND PREDATORS

There is little information available regarding the diet and feeding habits of Pickerel Frogs. Although few detailed reports exist, adults are apparently opportunistic generalists that consume spiders, myriapods, beetles, caterpillars, bugs, and other invertebrates (Wright 1920, Gilhen 1984). Adults are active foragers that feed at or above the water's surface and along adjacent shores (D. Watermolen, personal observation). Work in Pennsylvania by Meshaka et al. (2012) showed that fat stores in male Pickerel Frogs are at their lowest after the breeding period and then build monthly until winter dormancy begins. The frequency of food in digestive tracts suggests that females also forage "extensively throughout much of the active season" (Meshaka et al. 2012). Larvae feed primarily on algae and other soft plant material, which they filter from the water column, but they also scrape periphyton and scavenge dead animal matter from the bottom mud (Wilbur and Fauth 1990). Laboratory and field experiments have demonstrated that Pickerel Frog and American Toad (*Anaxyrus americanus*) larvae compete for food at sites where they co-occur (Wilbur 1977a), which generally results in lower survival and growth rates for both species. The effects of this competition, however, vary with the absence or presence of various predators (Wilbur and Fauth 1990).

Few reports of natural predation on Pickerel Frogs exist, particularly for adults. In Michigan, Lagler (1943) found a single adult Pickerel Frog among 323 Snapping Turtle (*Chelydra serpentina*) stomachs examined. Beane (1990) removed two adults from the stomach of an American Mink (*Mustela vison*) in North Carolina. Applegate (1990) reported Bald Eagle (*Haliaeetus leucocephalus*) predation on Pickerel Frogs in Maine, and there are likely other birds that feed on them. Eastern Newts (*Notophthalmus viridescens*) prey on eggs and early-stage larvae (Wright 1914, Wilbur and Fauth 1990). Larvae are also consumed by numerous additional aquatic predators, including Green Darner (*Anax junius*) naiads, Vertical Diving Beetle (*Dytiscus verticalis*) larvae, crayfish (*Cambarus diogenes* and *Orconectes propinquus*), and sunfish (*Lepomis cyanellus* and *L. megalotis*; Formanowicz and Brodie 1982, Wilbur and Fauth 1990, Holomuzki 1995). Although sunfish will consume Pickerel Frog larvae, they are apparently less palatable to fishes than larvae of other anurans (Holomuzki 1995). In New York, newly metamorphosed individuals were reportedly consumed primarily by spiders and various terrestrial insects such as flies, beetles, and hymenopterans (Munz 1920). The palatability of Pickerel Frogs to predators also appears to change as they develop. For example, premetamorphic Pickerel Frogs are fully palatable to Vertical Diving Beetle larvae and crayfish, but post-metamorphic stages are often rejected as food, an observation that is correlated with the development of the granular skin glands (Formanowicz and Brodie 1982).

Pickerel Frogs can deter some predators through chemical defenses. Adults have skin secretions with a complex array of bradykinin-related peptides (Basir et al. 2000, Basir and Conlon 2003, McCrudden et al. 2007) that could protect them from predators. Grant (2001) found that New York Pickerel Frogs also produced an odor "likened to that of alfalfa sprouts or the inside of a latex glove" that could be an olfactory warning to potential predators. Frogs produced this odor while being handled in the field and with electrostimulation in the laboratory. Wright and Wright (1949) suggested that the bright yellow-orange coloration in the groin region could serve to warn predators of their toxicity. Published information on the effects of the noxious secretions, however, is conflicting. For example, Mulcare (1966) could not find strong evidence to support their toxicity in laboratory experiments. Harding (1997) noted that "dogs are said to be repelled by these frogs." Northern Short-tailed Shrews (*Blarina brevicauda*) rubbed their mouths and vocalized or retreated after initial contact with Pickerel Frogs during experiments (Formanowicz and Brodie 1979). Field collectors have also cautioned that other frog species kept in confinement with Pickerel Frogs may die. Despite this, Harding (1997) noted that American Bullfrogs and Green Frogs "eat them without consequence." Pope (1964) speculated that toxicity gives the Pickerel Frogs an immunity that allows them to survive predation by snakes. This is corroborated by the fact that gartersnakes (*Thamnophis* spp.) and watersnakes (*Nerodia* spp.) reportedly reject Pickerel Frogs (Dunn 1935, Babbitt 1937, Walker 1946), although snake predation is still known (Manion 1952, Klemens 1993). The skin toxins probably do not effectively deter fishes, as anglers have used Pickerel Frogs extensively as bait (Cook 1984).

When threatened, Pickerel Frogs can also enact several defensive behaviors. Adults will crouch and remain motionless to avoid detection or flee from a potential predator with a series of rapid and erratic jumps (reviewed by Dodd 2013). Adults sometimes assume an antipredator posture, inflating and tilting the body forward while drawing the chin toward the pectoral region (Marchisin and Anderson 1978, Formanowicz and Brodie 1979). Marchisin and Anderson (1978) noted that this behavior "presents a large flat surface to an approaching snake" and only occurs when a snake is within striking distance. Female Pickerel Frogs tend to oviposit in areas with few or no fish, and larvae responded to fish chemical cues by substantially decreasing activity (Holomuzki 1995).

Few studies have investigated parasites and diseases of Pickerel Frogs in Wisconsin. Metcalf (1923) reported the protozoan *Opalina obtrigonoidea* from Wisconsin Pickerel Frogs. Documented helminth endoparasites include the trematodes *Brachycoelium salamandrae* and *Ribeiroia ondatrae* and the nematodes *Aplectana americana*, *Cosmocercoides dukae*, and *Oswaldocruzia pipiens* (Walton 1929, Coggins and Sajdak 1982, Johnson et al. 2004). Various other bacteria, viruses, protozoans, protozoan-like organisms, and helminthes have been found in Pickerel Frogs elsewhere in their range (reviewed in McAllister et al. 1995, Redmer 2005, Dodd 2013; also see Meyer 1938, 1939). The chiggers *Hannemania dunni* and *Trombicula alfreddugesi* have been reported as ectoparasites (Jenkins 1948, Murphy 1965). Moderate infections of tadpoles with trematodes does not appear to significantly affect larval survival, growth, development, or respiration (Orlofske et al. 2009). Some helminths are associated with malformations (Murphy 1965, Johnson et al. 2004), but malformations have not been observed in Wisconsin Pickerel Frogs. Lee and Franz (1973) reported a "red-leg" bacterial infection in Maryland Pickerel Frogs. Incidences of Pickerel Frog infection with the fungus *Batrachochytrium dendrobatidis* (Bd) are also reported from various locations in North America (see Conservation section).

CONSERVATION

Status: Globally, the Pickerel Frog is listed as a Least Concern species by the IUCN due to its "large extent of occurrence, large number of sub-populations and localities, large population size, and use of a wide range of habitats." However, this species is considered rare in numerous locations throughout its range. Although the frogs have no special status in states that neighbor Wisconsin, they are reportedly declining in Michigan (Eagle et al. 2005) and Iowa (Christiansen

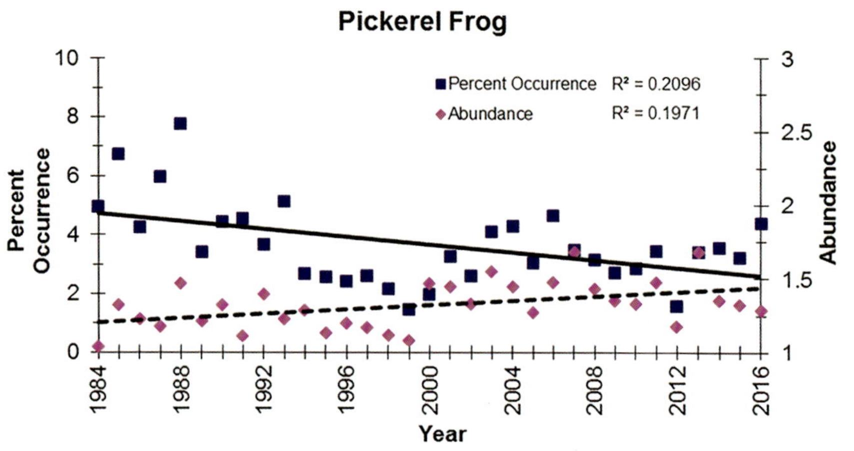

Figure 110. Long-term trend in occurrence (%) and abundance (rank) of Pickerel Frogs across Wisconsin Frog and Toad Survey statewide monitoring sites. See Introduction to Species Accounts chapter for details on this monitoring program (figure courtesy of WDNR).

1981). In Illinois, Redmer (1998) reported Pickerel Frog declines in the Chicago metropolitan area but not elsewhere in that state. In Wisconsin, Pickerel Frogs are currently considered a Special Concern species and a Species of Greatest Conservation Need by the WDNR, although long-term Wisconsin Frog and Toad Survey data suggest slightly increasing abundances (Figure 110). Despite this, some evidence suggests that state populations have declined in recent decades (see Natural History Box: The Pickerel Frog's Status in Wisconsin).

Populations: No rigorous adult abundance or density estimates from Wisconsin exist, and most population information is inferred from surveys of calling males. The Pickerel Frog is rarely considered a common species in Wisconsin, and current estimates suggest that it continues to experience declines in the state (Natural History Box: The Pickerel Frog's Status in Wisconsin: An Unclear Picture). Despite this, Kitchell et al. (2013, 2014) reported that the estimated percent occurrence of Pickerel Frogs among survey sites has increased slightly compared to previous years. Although more recent estimates indicate declines in occurrence of calling males, inferred abundance from these surveys shows increasing numbers (Figure 110). Although population studies focused on this species are few, the most detailed information comes from the southern US. For example, Hardy and Raymond (1991) captured 143 individuals at a single breeding site in Louisiana over a two-year period. They also estimated sex ratios of 1.3M:1F and 5.4M:1F during their investigation. Murphy (1963) further reported a sex ratio of 6.64M:1F at a site in North Carolina. Yet the male bias in these populations could be an artifact of when the work was conducted (i.e., during the breeding season) and not indicative of overall population sex ratios.

Threats: Many Pickerel Frog populations in Wisconsin appear to be isolated in fragmented habitat parcels that are surrounded by agricultural activities or urban development. Johnson (1984) reported no obvious connection between most land use practices and the habitats used or avoided by Pickerel Frogs in Wisconsin. Despite this, he rarely found Pickerel Frogs in mowed fields near fish hatcheries, grazed pastures, or plowed crop fields that "reduced ground cover and possibly insect biomass." Casper (1998) also mentioned a population from the Southern Kettle Moraine State Forest that apparently disappeared "after a shoreline was developed for fishing access." Furthermore, some impacts to habitat are more subtle and cause subdued effects on Pickerel Frog populations. Although roads can destroy a certain amount of habitat when they bisect natural landscapes, they might not completely destroy available habitat within a given area. Still, roads can be problematic because they hinder Pickerel Frog movement such as breeding migrations and juvenile dispersal (Gibbs 1998a) while also causing heightened mortality rates due to vehicular strikes (Glista et al. 1998).

Invasive species could negatively impact Pickerel Frogs in Wisconsin, but the severity of this threat is poorly understood. Some research suggests that the presence of invasive species, particularly vegetation,

NATURAL HISTORY BOX: The Pickerel Frog's Status in Wisconsin: An Unclear Picture

Both historical and modern accounts support that Pickerel Frogs are uncommon in Wisconsin. For example, as far back as the early 1900s, they were referred to as rare but with "a fairly continuous distribution in the state" (Nelson 1915). Decades later, Pentecost and Vogt (1976) again reported they were "rare" in the Lake Michigan drainage basin, while Vogt (1981) even referred to them as "the rarest frog species in Wisconsin." Their rarity was further supported by the infrequent number of reports generated statewide. Vogt (1981) included records from a total of 29 counties but reported that he knew of only 15 active localities. Later work by Johnson (1984) found a greater number of active localities ($n = 61$) but from only 19 counties. During the initial years of the Wisconsin Frog and Toad Survey, volunteers heard them calling at only 26 locations (5% of all surveyed sites; Mossman and Hine 1985).

Unfortunately, not only do they appear to be relatively uncommon in Wisconsin, but several accounts suggest they have experienced recent state-level population declines. For example, over the past three decades Kitchell et al. (2013) estimated a 1.5% decline in percent occurrence of this species along Wisconsin calling routes, which is on top of Mossman et al.'s (1998) report that up to 63% of calling survey routes showed declining trends. The general impression of their rarity in the state was the driver behind the WDNR's decision to list Pickerel Frogs as a Threatened species in the past, although they were later downgraded to Special Concern near the end of the 1980s (Casper 1998). Despite this, Mossman et al. (1998) stated that Pickerel Frogs in Wisconsin are "common (and locally common) in the Driftless Area, rare elsewhere, and uncommon overall."

Dreux J. Watermolen, WDNR

could degrade habitat quality and cause Pickerel Frog population declines. For example, Maerz et al. (2010) investigated how the detritus produced by decaying invasive plants affected larval anurans. They reported that no Pickerel Frog larvae housed in outdoor enclosures with detritus from the invasive Reed Canary Grass (*Phalaris arundinaceae*) or the nonnative Common Reed (*Phragmites australis*) survived to metamorphosis. This, coupled with the extensive invasions of these exotic grasses across Wisconsin, could have contributed to Pickerel Frog declines in the state (see Natural History Box: The Pickerel Frog's Status in Wisconsin: An Unclear Picture). The introduction of fishes into otherwise fishless breeding ponds could also negatively impact Pickerel Frog populations, as some species apparently do not find Pickerel Frogs unpalatable (see Prey and Predators section). Furthermore, Johnson (1984) rarely found Pickerel Frogs associated with cold-water fisheries in Wisconsin, even though these locations contained habitat that superficially appeared suitable. He suggested that the combination of mowed grass near hatcheries and fish predation could have caused population extirpations in these locations.

Substantial evidence exists to support that environmental pollution impacts global amphibian populations (see Introduction to Conservation and Management of Wisconsin's Amphibians and Reptiles chapter), and Pickerel Frogs are no exception. They are, for example, susceptible to heavy metal and pesticide contamination (Devillars and Exbryant 1992, Westerman et al. 2003). Exposure to high levels (600 µg/L [0.00008 oz/gal]) of malathion, a widely used agricultural pesticide, decreased hatching success (80% in experimental treatments compared to 98% in controls) and hatching viability but increased the incidence of malformations (Budischak et al. 2008). Exposure to carbaryl, another widely used insecticide, resulted in death at a single high dose (30 mg/L [0.004 oz/gal]) and a significant reduction in larval activity at a low dose (2.5 mg/L [0.0003 oz/gal]; Bridges and Semlitsch 2000). Although Johnson (1984) generally found no difference in the water quality between Wisconsin sites where Pickerel Frogs were present ($n = 38$) and absent ($n = 4$), susceptibility to some water-quality parameters could vary by age. For example, embryo survival is influenced by inadequate dissolved oxygen levels, but these levels apparently have little effect on adults (see Habitat subsection and Reproduction and Development section). Regardless, unnatural eutrophication due to increased nutrient input and subsequent reduction in dissolved oxygen levels could limit embryo survival and negatively impact populations.

It is noteworthy that some threats have indirect impacts on frogs that are not always easy to identify. High nutrient levels in anthropogenic runoff that enters breeding wetlands can enhance food availability for developing larvae, which could lead to high larval growth rates. Yet subsequent decomposition of excessive organic material and concomitant oxygen depletion could impact larvae later in the season (Rittenhouse 2011). It is of further note that numerous threats can act synergistically, which could exacerbate their negative impacts. High nutrient loads that could occur due to agricultural runoff may also increase the biomass of snails that serve as primary hosts for trematode parasites and result in high larval infection rates (Johnson et al. 2007b). Budischak et al. (2008, 2009) further discussed the complex relationship between pesticide exposure and susceptibility to infection by trematodes. Larvae exposed to pesticides during early development may suffer increased susceptibility and accrue higher parasite burdens. Similarly, exposure to a single dose of atrazine ($\bar{x}$ concentration = 117 µg/L [0.00002 oz/gal]) resulted in immunosuppressed larvae with elevated levels of trematode infections (Rohr et al. 2008a). Understanding the interactive effects of these and other stressors merits additional work.

Pickerel Frogs are also affected by other pathogens that pose substantial risks to global amphibian populations (see Introduction to Conservation and Management of Wisconsin's Amphibians and Reptiles chapter). Yet limited work has been conducted on how these pathogens impact Pickerel Frogs in Wisconsin. Standish et al. (2018) did not detect Bd or a ranavirus on the single Pickerel Frog they tested from Wisconsin, although little can be gleaned from this small sample. Studies completed in other states have investigated a greater number of Pickerel Frogs and found a higher prevalence of infection. Green et al. (2002) identified an iridovirus as the cause of death for numerous larvae or recently metamorphosed individuals from sites in Tennessee, Maine, and New Hampshire. Several studies conducted in various states and Canada also detected Bd on the skin and toe webbing of Pickerel Frogs (Ouellet et al. 2005, Longcore et al. 2007, Rimer and Briggler 2010, Goodman and Ararso 2012). These investigations support that Wisconsin populations could be at risk of infection or already contain infected individuals that have not yet been discovered.

Management: General stream, wetland, mesic grassland, and open forest management recommendations could benefit this species (see Introduction to Conservation and Management of Wisconsin's Amphibians and Reptiles chapter). Protection of suitable breeding and nonbreeding habitats may be the most important step to conserve Pickerel Frogs in Wisconsin. On a landscape scale, Pickerel Frogs could benefit from a diverse assemblage of connected habitats that include streams of various size, open-canopy wetlands, riparian areas with ample herbaceous vegetation, and adjacent forests. The presence of forested stands may moderate the effects of agriculture at landscape scales (Knutson et al. 1999, 2000). Permanent and intermittent streambeds can act as corridors for Pickerel Frog movement (Gibbs 1998a) and aid in dispersal between isolated sites in disturbed landscapes. Furthermore, the frogs' apparent preference for dense vegetation along riparian corridors and in adjacent mesic prairies (Johnson 1984) supports that management activities intended to reduce these habitat features (e.g., frequent mowing or burning) could negatively impact Pickerel Frogs.

Johnson (1984) noted that the creation of small permanent ponds near streams could increase available breeding sites, but there are conflicting accounts of whether Pickerel Frogs successfully use constructed and restored wetlands. Shulse et al. (2010) suggested that Pickerel Frogs do not readily colonize wetland restoration sites, but Brand and Snodgrass (2010) found Pickerel Frogs breeding in stormwater ponds, and Lacki et al. (1992) found that they were abundant in wetlands constructed for treatment of water from surface mines. In Minnesota, Knutson et al. (2004) further reported that constructed agricultural ponds increase available breeding habitat. Creation of suitable physical characteristics in artificial wetlands could increase their success as a management technique. For example, wetland designs that incorporate shallow areas to support emergent vegetation could provide cover for breeding adults and developing larvae. Preventing fish introductions and limiting livestock access can also improve the quality of restoration sites (Knutson et al. 2004). Johnson (1984) noted that although Pickerel Frogs often occupied dense streamside vegetation, they rarely used adjacent areas with short vegetation grazed by cattle.

Individuals who work or recreate in aquatic habitats can take steps to prevent the potential introduction and spread of harmful pathogens and other invasive species by using appropriate disinfection protocols for boats, gear, and equipment (Perdzock 2015). Along these lines, control of invasive wetland plants (such as Reed Canary Grass) around breeding wetlands could positively affect the survival of larval Pickerel Frogs and, thus, population viability.

SPECIAL REMARKS

Monitoring of Pickerel Frog populations across the state is critical for conservation of this species due to its declining trend. Long-term monitoring of most sites has not occurred, however, making it necessary to determine the presence and continued persistence at historical sites. In addition, presence at most sites in the north-central part of the state has never been appropriately confirmed with voucher specimens, photographs, or recordings. The inconspicuous breeding vocalizations and potential habitat specificity make these frogs difficult to detect consistently by common survey methods; detection may require species-specific monitoring techniques. Intensive work at individual field sites, as well as extensive monitoring across the state, is necessary to develop a thorough understanding of the Pickerel Frog's status in Wisconsin.

Northern Leopard Frog

Lithobates pipiens (Schreber 1782)

Rori A. Paloski, Sara E. Viernum, and Tara L. E. Bergeson

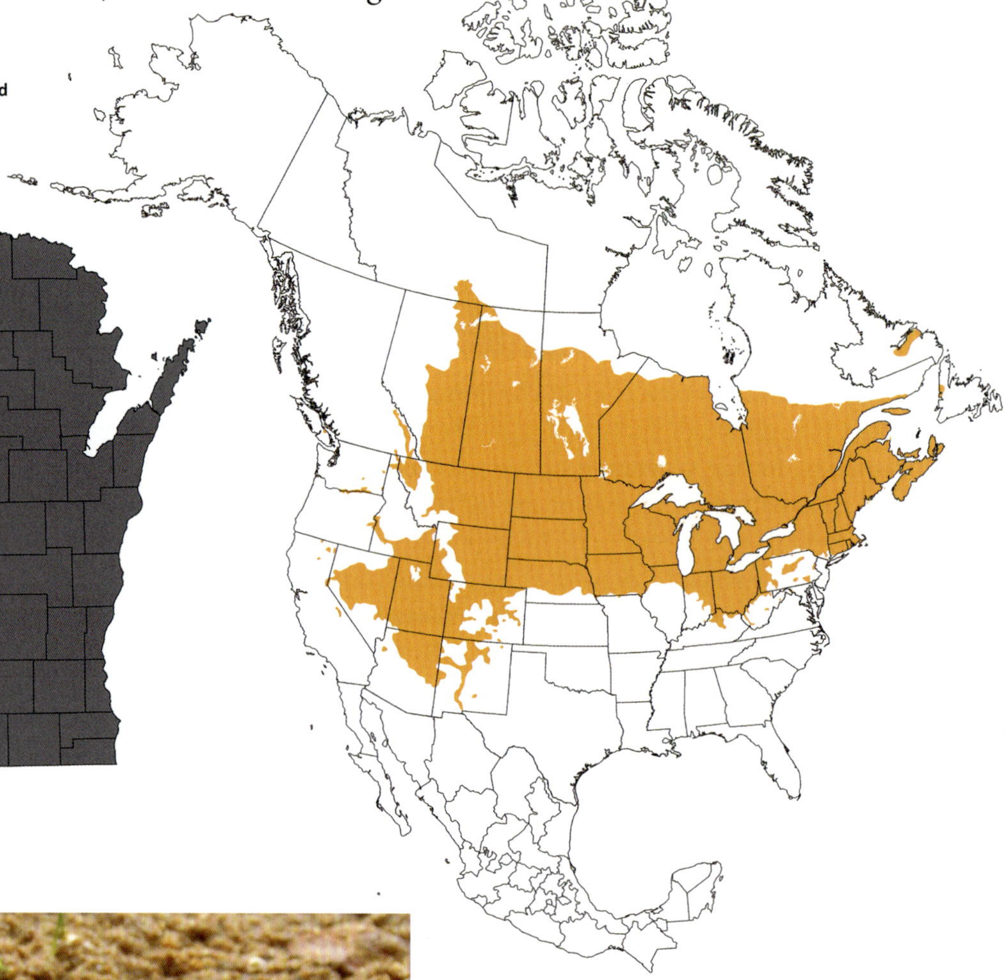

Adult Northern Leopard Frog (Vilas County; photo by C. Raimond).

AT A GLANCE

Northern Leopard Frogs are medium-sized with large dark 'leopard' spots over a green, brown, or greenish-brown background. They are found throughout Wisconsin in a wide variety of semipermanent aquatic and adjacent open- or closed-canopy terrestrial habitats. Northern Leopard Frogs emerge from hibernation in mid-March to mid-April and soon after congregate in open-canopy wetlands for mating. Males attract females with a low-pitched snore-like advertisement call, which some people liken to the sound produced by dragging a finger slowly over the surface of an inflated balloon. Females may lay up to 7,000 eggs during the breeding season, and larvae typically metamorphose in late summer. After breeding, adult Northern Leopard Frogs move to open summer foraging habitats, which can be up to 1 km (0.6 mi) from breeding wetlands. Adults consume a variety of invertebrates, particularly insects, while larvae are omnivorous, feeding on algae and dead animal material with their scraping mouthparts. Adults are prey to many types of vertebrates, such as snakes, birds, mammals, and even other amphibians. Larvae are eaten by similar predators but also fall victim to salamander larvae and larger aquatic invertebrates. In late October, Northern Leopard Frogs begin migrating back to aquatic habitats used for overwintering, where they typically hibernate underwater. Northern Leopard Frogs experienced substantial population declines throughout portions of their range, including Wisconsin, in the 1960s and 1970s. Northern Leopard Frogs are considered common by the WDNR.

SYSTEMATICS

Lithobates pipiens (Northern Leopard Frog) was first described by Schreber (1782). Since then, it has undergone numerous taxonomic changes, sometimes based on phenotypic or regional variants that are not currently recognized as unique species (reviewed by Dodd 2013). The most recent impactful taxonomic revision for *L. pipiens* was the change of its generic epithet from *Rana* to *Lithobates,* an alteration meant to better reflect evolutionary relationships among New and Old World ranid species (Frost et al. 2006). This change has been widely accepted but not without debate (see discussion in Crother 2017). The specific epithet (*pipiens*) is Latin for 'peeping,' possibly assigned because the frogs' describer was confused by the 'peeps' of sympatric *Pseudacris crucifer* (Spring Peepers) heard during collection (Dodd 2013). There are currently no recognized subspecies of *L. pipiens* (Dunlap and Platz 1981, Crother 2017). Based on mitochondrial DNA analysis, Hoffman and Blouin (2004) suggested that geographically isolated eastern and western clades of *L. pipiens* existed, which, interestingly, overlap in Wisconsin. Hybridization between *L. pipiens* and other species of *Lithobates* can occur but does not typically result in viable offspring (reviewed by Hillis 1988). In Wisconsin, Northern Leopard Frogs have been referred to by numerous common names, including Leopard Frog, Grass Frog, Meadow Frog, Common Frog, and Spring Frog (Pope and Dickinson 1928, Knudsen 1958c, Conant and Collins 1998, Dodd 2013). Because there are other leopard frog species in North America (such as *L. sphenocephalus* [Southern Leopard Frog]), the use of 'Leopard Frog' as a common name has led to confusion in the early literature (reviewed by Dodd 2013). Among anurans in Wisconsin, these frogs are closely related to other members of the genus *Lithobates,* such as *L. palustris* (Pickerel Frog).

DESCRIPTION

Adults: Northern Leopard Frogs have numerous dorsal spots over green and/or brown backgrounds (Figure 111). These spots are black, brown, or dark green and often outlined with a cream (see Figure 111), white, or light green border (Vogt 1981, Rori Paloski, WDNR, personal observation). The irregularly shaped and irregularly spaced spots are typically present in two to three rows on the back, as well as on the head, sides, and front legs (Vogt 1981). There is ordinarily one spot over each eye and often one on the snout (Harding 1997), but the total number of dorsal spots across the body varies (reviewed by Dodd 2013). Their rear legs may be spotted or appear barred (Harding 1997, Moriarty and Hall 2014). Their tympanums, which are smaller than their eyes in both males and females (Breckenridge 1944), may or may not have a white spot in the center (Dodd

Figure 111. Examples of color variation in Northern Leopard Frogs. (a) A brown individual (Columbia County; photo by R. Paloski). (b) A green individual (Juneau County; photo by S. Viernum). (c) A brown-and-green individual (Rock County; photo by J. Kapfer).

2013). Cream, tan, bronze, or gold dorsolateral folds are complete and prominent on Northern Leopard Frogs (Knudsen 1958c, Vogt 1981, Johnson 2000b). A light-colored stripe is also present on the upper lip (Dodd 2013). Their ventral surface is uniformly white, cream, or greenish white (Vogt 1981); however, a faint yellowish tint may be present on the inner thighs (Moriarty and Hall 2014). Northern Leopard Frogs are medium-sized among Wisconsin anuran species (SVL = 5–9 cm [2.0–3.5 in]; Vogt 1981). The mean SVL of probable adult specimens (i.e., ≥6.0 cm SVL [2.4 in]; see Reproduction and Development section) from Wisconsin in the MPM and UWZM collections ranged from 6.8 cm (2.7 in; $n = 145$) in males to 7.2 cm (2.8 in; $n = 173$) in females. The largest SVL among Wisconsin specimens measured was 9.5 cm (3.7 in, MPM 831, collected from Milwaukee County in 1909).

Preadult stages: Northern Leopard Frog eggs are bicolored (half black and half white), with the black side typically facing the surface of the water, and they are surrounded by three jelly envelopes (Parmelee et al. 2002, Dodd 2013). In Wisconsin, eggs are reportedly 0.10–0.18 cm (0.04–0.07 in) in diameter immediately after they are deposited by females but enlarge to ca. 0.20 cm (0.08 in) in diameter by the time of hatching (Vogt 1981). Eggs are laid in round globular masses (Figure 112), with overall mass lengths that range from 4.0 to 15.0 cm (1.6 to 5.9 in)

Figure 112. Northern Leopard Frog egg mass (Dane County; photo by R. Paloski).

and widths from 4.0 to 11.0 cm (1.6 to 4.3 in; Hine et al. 1981). Northern Leopard Frog larvae have dorsolaterally positioned eyes and are dark brown, brown, beige, dark green, or olive colored dorsally with a clear, cream, or pinkish belly (Vogt 1981, Parmelee et al. 2002, reviewed by Dodd 2013, Moriarty and Hall 2014, R. Paloski, personal observation). They also sometimes have small gold specks dorsally and slight iridescent coloring on their sides or belly, with the intestinal coil typically visible through the ventral surface (Breckenridge 1944, Vogt 1981, reviewed by Dodd 2013). The spiracle is located on the left side of the body, while the vent tube is located to the right of the ventral fin's midpoint (reviewed by Dodd 2013). Larvae possess three lower labial tooth rows, with the lateral margins of the papillary border folded in toward the corners of the mouth (Vogt 1981). Larvae may reach a maximum total length of 4.5–8.4 cm (1.8–3.3 in; Breckenridge 1944, Harding 1997, Moriarty and Hall 2014). At metamorphosis, the mean TL recorded in Wisconsin is 3.9–4.0 cm (1.5–1.6 in), with a body mass range of 4.0–9.5 g (0.1–0.3 oz; Hine et al. 1981). Juvenile coloration in Northern Leopard Frogs is similar to that of adults (Vogt 1981).

Variation: Adult Northern Leopard Frogs are sexually dimorphic during the breeding season, when male thumb pads (nuptial pads) enlarge, while loose skin becomes visible near their vocal sacs; female body girth expands due to gravidity (Dodd 2013). Several color and pattern variants also occur in upper midwestern populations, although these are not common (or they are absent) in Wisconsin. For example, a spotless color variant, referred to as the 'Burnsi' morph, occurs in Wisconsin, Minnesota, Iowa, North Dakota, and South Dakota (Figure 113; McKinnell and Dapkus 1973, Vogt 1981, reviewed by Dodd 2013, Moriarty and Hall 2014). A second variant, the 'Kandiyohi' morph, is not documented in Wisconsin but occurs as close as Minnesota and possesses the typical large dark dorsal spots, as well as heavy mottling or freckling between the spots (McKinnell and Dapkus 1973, reviewed by Dodd 2013, Moriarty and Hall 2014). The 'Burnsi' and 'Kandiyohi' morphs were once considered distinct species or subspecies but are now generally accepted as phenotypic variants of the Northern Leopard Frog (McKinnell and Dapkus 1973).

Figure 113. 'Burnsi' morph of the Northern Leopard Frog (Crawford County; photo by R. Sajdak).

Diagnosis: Northern Leopard Frogs have smooth skin lacking parotoid glands, warts, and nodules. They do not possess either keratinized spades on the heel of each hind foot or terminal toe pads, and they do not emit a musky odor. The dorsolateral folds are present and extend the entire length of the dorsum. There is no dark 'mask' extending from the tip of the snout to behind the eyes and tympanum. The dorsum is generally covered with large dark spots (except in the 'Burnsi' morph) that are irregular in size and distribution. They do not possess a bright yellow or golden color in the groin area. In Wisconsin, adult metamorphosed juvenile Northern Leopard Frogs are most easily confused with Pickerel Frogs (*Lithobates palustris*). However, the dorsal spots of Pickerel Frogs are more square-shaped and typically form two distinct rows between the dorsolateral folds (Vogt 1981). Pickerel Frogs also have a bright yellow or golden color on the inner surface of the thighs, or groin area, unlike Northern Leopard Frogs (Moriarty and Hall 2014). In addition, while the background dorsal color of Northern Leopard Frogs can be brown, green, or brown and green, the Pickerel Frog is typically only brown. Males of both species have similar advertisement calls, although the Northern Leopard Frog's snore is lower in frequency and longer in duration, with individual clicks that are more widely spaced than Pickerel Frogs (see sonograms in Vogt 1981). The 'Burnsi' morph of the Northern Leopard Frog could be mistaken for a Green Frog (*L. clamitans*) or Wood Frog (*L. sylvaticus*). However, the 'Burnsi' morph contains complete dorsolateral stripes and lacks the green lips of Green Frogs and the dark mask of Wood Frogs (Vogt 1981).

Northern Leopard Frog eggs are probably most difficult to distinguish from Pickerel Frog eggs, although Northern Leopard Frog eggs are half black and half white, while the Pickerel Frog eggs are reportedly half brown and half yellow (Parmelee et al. 2002). Northern Leopard Frog larvae possess internal gills and lack external forelimbs. They have dorsally positioned eyes, a dextral vent, and a white line that extends from the nostril to the snout. Their larvae are superficially similar to larvae of all other ranids in Wisconsin but are particularly difficult to differentiate from Pickerel Frog larvae. However, Pickerel Frog larvae often have dark pigmentation on their fins, whereas Northern Leopard Frogs have fins that are clear or only possess a few small spots (Vogt 1981).

DISTRIBUTION AND HABITAT

Global distribution: Northern Leopard Frogs are one of the most widely distributed members of the genus *Lithobates*. They are found throughout much of the northern, eastern, and central sections of the US and throughout the southern, eastern, and central portions of Canada. Their range extends as far northeast as Newfoundland Island and the Canadian Maritimes, northwest to Alberta and the Northwest Territories of Canada, southwest to Arizona, New Mexico, and into the Rio Grande Valley, south across the Great Plains states, and east to Kentucky, West Virginia, and Pennsylvania. Northern Leopard Frogs have also been introduced to several areas in the US and Canada that are outside their native range (Buckle 1971, Bury and Luckenbach 1976).

Distribution in Wisconsin: Northern Leopard Frogs occur statewide, and published records with vouchers exist for all Wisconsin counties (see the Appendix). They are even reported from the Apostle Islands (Hecnar et al. 2002) and islands off the Door Peninsula in Lake Michigan (Long and Long 1976). Although populations have purportedly declined in the state from historical levels (see Conservation section), the county-level distribution of Northern Leopard Frogs in Wisconsin has not been substantially affected (Casper 1996).

Localities of possible occurrence: Because Northern Leopard Frogs have been recorded from all Wisconsin counties, the possibility exists of finding this species throughout the state wherever suitable habitat occurs.

Habitat: Generally, Northern Leopard Frogs are associated with damp habitats that are open and grassy. These could include not only wet prairies, mesic prairies, and sedge meadows but also open lentic systems of various sizes and depths with suitable vegetation and even lotic systems for overwintering (Vogt 1981; see Amphibians and Reptiles in Wisconsin Plant Communities chapter). Studies conducted in Wisconsin have recorded them in permanent emergent wetlands, ephemeral wetlands, lakes, forested/shrubby wetlands, grasslands, prairies, old fields, and even agricultural landscapes (Knutson et al. 1999, 2000).

Although the list of habitats Northern Leopard Frogs use is broad, they select certain habitat types for specific annual activities: summer foraging, breeding, and overwintering (Pope et al. 2000). Habitats used for summer foraging can be either aquatic or terrestrial but are often open and dominated by herbaceous vegetation (Blomquist and Hunter 2009). Northern Leopard Frogs reportedly spend some portion of the summer foraging in various wetlands, sedge meadows, wet meadows, bogs, waterbody edges, old fields, hayfields, lawns, grasslands, prairies, woodlands, and brushlands (Allen 1868, Hine et al. 1981, Bonin et al. 1997). In Minnesota, Knutson et al. (2018) determined that the frogs preferred to remain in the vicinity of wetlands during the summer, although they sometimes occupied immediately adjacent grassland areas and even forest fragments. Moisture is a major factor influencing summer habitat selection (Pope et al. 2000, Blomquist and Hunter 2009), and Knutson et al. (2018) found that Northern Leopard Frogs in Minnesota remained close to wetlands during periods of drought.

Northern Leopard Frogs primarily breed in open aquatic habitats with emergent and submersed vegetation, including temporary wetlands with clear water and open-water areas in permanent wetlands, as well as shallow vegetated littoral zones in lakes and quiet backwaters of streams (Merrell 1977, Knutson et al. 2004, Sutton 2004). Trenham et al. (2003) determined that calling males in Wisconsin were positively associated with landscapes that contained many diverse emergent wetland types. When natural aquatic habitats are not available, Northern Leopard Frogs will breed in small constructed agricultural ponds (Knutson et al. 2004), urban ponds in residential settings (Gagné and Fahrig 2010), or other artificial waterbodies. Northern Leopard Frogs appear to avoid breeding in large wetlands and deep ponds, instead preferring shallower waterbodies that warm quickly in the spring and are at least 50% covered in submerged and emergent vegetation (Dodd 2013). Larvae remain in the same habitats used for breeding through metamorphosis, after which juveniles appear to occupy similar habitat as adults (R. Paloski, personal observation).

Unlike several other frog species in Wisconsin, such as Spring Peepers and Wood Frogs, Northern Leopard Frogs cannot survive being frozen during hibernation and must overwinter underwater in aquatic habitats that do not completely freeze. Habitats used for overwintering reportedly include lakes, large ponds, and streams where they either lie on the substrate or bury themselves under aquatic vegetation, rocks, logs, mud, and even hibernating turtles (Emery et al. 1972, Ultsch et al. 2000, Kiesow 2006). Cunjak (1986) observed Northern Leopard Frogs overwintering underwater below rocks in riffles, runs, and pools in the Credit River in Canada. The frogs, both adults and juveniles, were in the deepest sections of the stream. Similarly, Breckenridge (1944) made the following observation at an outlet of Martin Lake (Minnesota) in January: "A concrete dam 12 feet wide controlled the outflow from the lake. Immediately below the dam were limestone slabs, and farther downstream the bottom was clean sand. The stream here was 19 to 24 inches deep, and the agitated water was unfrozen for some distance below the dam. The sand bottom for several yards below the dam was literally paved with hundreds of closely crowded leopard frogs." Northern Leopard Frogs can also overwinter in damp refugia where they are not submerged but still remain below the frost line, such as in deep crayfish and mammal burrows, under rocks, and in deep fissures along shorelines (Wright 1914, Wayne 2001). Overwintering locations are not always in breeding wetlands, as radio-tracked Northern Leopard Frogs in Minnesota frequently migrated from breeding ponds to adjacent ponds or streams for the winter (Knutson et al. 2018). Hine et al. (1981) further determined that most aquatic habitats selected for overwintering were within 1.6 km (1 mi) of wetlands used for breeding in Wisconsin.

REPRODUCTION AND DEVELOPMENT

In Wisconsin, Northern Leopard Frogs breed in the spring after emergence from hibernation, which typically occurs from mid-March to late April (Hine et al. 1981, Vogt 1981). Males arrive at the breeding sites from 5 to 14 days before the females and begin calling to attract mates when both water and air temperatures reach 10°C (50°F; Hine et al. 1981, Dodd 2013). In northern populations, calling and reproduction usually take place during the day, when temperatures are warmer, though some males will call at night (Wright 1914, reviewed by Dodd 2013). Males have paired vocal sacs on both sides of the throat that produce a low rumbling snore (dominant frequency: 1.5 kHz) that lasts two to four seconds (Dodd 2013, see sonogram in Vogt 1981). Their snore-like advertisement call is likened to the sound produced by dragging a finger over an inflated balloon (e.g., Vogt 1981) or by a motorboat (Kiesow 2006); the call is punctuated with clucks, chuckles, and croaks. Males typically call with their bodies outstretched, either floating on the water surface or resting below the surface in the shallows (Noble and Aronson 1942, Merrell 1977). Both males and nonreceptive females produce vibrating chuckle release calls that deter over-amorous males from engaging in unwanted advances (Noble and Aronson 1942, McClelland and Wilczynski 1989). Pace (1974) further suggested that males also have a shorter call used to signal aggression toward another male, although potential territorial behavior is not well studied in this species. Both Merrell (1968) and Hine et al. (1981)

Figure 114. Northern Leopard Frogs in amplexus (Brown County; photo by T. Bergeson).

believed that the number of breeding individuals in the population was substantially lower than the total number of individuals at breeding ponds in Minnesota and Wisconsin, respectively.

A receptive female moves toward a calling male so that the male can mount and grasp her around the armpits in pectoral amplexus (Figure 114), aided by his enlarged thumb pads (nuptial pads). Once the pair align themselves into the correct oviposition posture by rotating to bring their cloacae together, the female releases a cluster of eggs, which the male fertilizes externally. Females lay 300–7,000 eggs in one or more gelatinous masses that are attached to emergent or submerged vegetation near the water surface (Wright 1914, Merrell 1977, Corn and Livo 1989, Sutton 2004). Females deposit only one clutch per year, but males can mate more than once during the breeding season (Noble and Aronson 1942).

Hatching usually occurs within 7–20 days after the eggs are deposited, depending on water temperature (Hine et al. 1981, Sutton 2004). Larvae typically metamorphose from late June to mid-July, ca. 71–111 days after the eggs are deposited (Wright 1914, Vogt 1981, Sutton 2004). Development can be rapid, and some individuals may reach sexual maturity within a year, although two years are more typical (Leclair and Castanet 1987). There are differences in developmental rates by sex, with males reaching maturity at a slightly younger age (one to two years) than females (two to three years; reviewed by Dodd 2013). In Wisconsin, males also reportedly reach sexual maturity at a smaller SVL (5.5 cm [2.2 in]) than females (6.0 cm [2.4 in]; Hine et al. 1981). Leclair and Castanet (1987) estimated that longevity is four to five years.

ACTIVITY

Northern Leopard Frogs are generally active from mid-March to late October or early November. Their seasonal activity patterns include (1) spring migration of adults from overwintering sites to breeding sites, (2) summer migration of adults into upland foraging habitat, (3) metamorphosed juvenile dispersal from breeding sites, and (4) autumn migration to overwintering sites (Emery et al. 1972, Merrell 1977, Cunjak 1986, Seburn et al. 1997, Ultsch et al. 2000). After emergence from hibernacula, Northern Leopard Frogs briefly remain along the shores of the waterbodies where they overwinter. When air temperatures exceed 10°C (50°F), adult Northern Leopard Frogs will migrate to breeding sites, which are sometimes in locations different from those of hibernacula (Merrell 1977; Natural History Box: Spring Emergence). The distance of this migration can be large, with documented movements of >1 km (0.6 mi) between overwintering and breeding sites reported (Seburn et al. 1997). During the breeding season, Northern Leopard Frogs move sparingly. Hine et al. (1981) reported that observed movements during this period in Wisconsin were generally <1 m (3.3 ft).

After breeding, Northern Leopard Frogs move to upland foraging areas, where they remain for the majority of the summer. In the Upper Midwest, this migration often begins by early June (Vogt 1981), and individuals can reportedly migrate up to 3 km (1.9 mi; Knutson et al. 2018). Most summer movement is influenced by moisture levels and is greatest after rainfall (Knutson et al. 2018). Mean distance moved per day during summer by radio-tracked individuals in a northern Iowa landscape dominated by agriculture was 21.7 m (71.2 ft), with straight-line movements of up to 857 m (2,812 ft; Bartelt and Klaver 2017). Mean post-breeding distance moved per day in Minnesota was 15 m (49.2 ft; Knutson et al. 2018), while mean 95% fixed kernel home range was 5.3 ha (13.1 ac) and mean 50% kernel core area was 1.05 ha (2.6 ac; Knutson et al. 2018). In Ontario, Blomquist and Hunter (2009) reported a mean home

NATURAL HISTORY BOX: Spring Emergence

It was March 1998, and I was a brand new wildlife ecology master's student at UW–Madison embarking on one of my first research-related field outings. My task was pretty simple that day: find my field sites and get to know them a bit. My fingers were crossed that I would, first, successfully find the sites and, second, maybe even see some Northern Leopard Frogs, the subject of my study. I parked my car at the boat landing parking lot along the Sensiba River and adjacent to the Sensiba State Wildlife Area near Green Bay, Wisconsin. I walked along the road to an area where the water flowed under the road and connected with the wetland to the north. Looking down toward the water, I could see movement and ripples on the water and rocks at the shoreline. As I looked more closely, I was astonished to see hundreds of Northern Leopard Frogs hopping, climbing, and moving about (Figure 115). These frogs had emerged from their overwintering area in the Sensiba River and were getting ready to move to breeding sites. Some were already in amplexus. Many of them simply hopped across the road (at some peril) to the marshy area of the nearby Sensiba State Wildlife Area, where they presumably spent time breeding and foraging. Since that day, I have not seen so many Northern Leopard Frogs congregated in one place. It was a good first day of field research!

Figure 115. Mass spring emergence of Northern Leopard Frogs (Door County; photo by T. Bergeson).

Tara L. E. Bergeson, WDNR

range size of only 0.1 ha (0.3 ac), although the largest home range calculated was 0.84 ha (2.1 ac). Dole (1965) determined that most summer movement by Northern Leopard Frogs at a site in Michigan probably occurred at night (between 8:00 p.m. and 8:00 a.m.) and individuals showed fidelity for certain locations within their home ranges. Merrell (1977) suggested that the distance they moved from water depended on the size of the individual, as larger frogs were better able to withstand desiccation.

After metamorphosing, juvenile frogs remain near their natal aquatic habitat until they have fully absorbed their tails. Once their tails are absorbed (July to mid-August), young Northern Leopard Frogs leave natal aquatic habitat and disperse throughout the surrounding landscape (Merrell 1977, Dodd 2013). These dispersals reportedly occur during warm rains and can span sizeable distances (e.g., 800 m [2,625 ft] in two to three days; Dole 1971). During autumn, Northern Leopard Frogs begin to move from their summer foraging habitats to their overwintering sites. These migrations often occur at night from late October to early November in Wisconsin, after which Northern Leopard Frogs are typically no longer active (Vogt 1981). Although they do not tolerate freezing, they are sometimes observed active under cold conditions. For example, an active individual was reported along the banks of a mostly ice-covered lake in Wisconsin during December (Olson 2010).

PREY AND PREDATORS

Juvenile and adult Northern Leopard Frogs are carnivorous and will consume anything small enough to fit in their mouths. In Wisconsin, Hine et al. (1981) found that adults ate an array of invertebrates, such as isopods (e.g., woodlice, order Isopoda) and annelids (phylum Annelida), as well as insects, including beetles (order Coleoptera), moths and butterflies (order Lepidoptera), wasps, bees, and ants (order Hymenoptera), dragonflies and damselflies (order Odonata), and caddisflies (order Trichoptera), among others. Additional studies have determined that adults and juveniles also prey on spiders (class Arachnida), slugs and snails (class Gastropoda), centipedes and millipedes (subphylum Myriapoda), crayfish (order Decapoda), and earthworms (subclass Oligochaeta; Sutton 2004, Kiesow 2006, Dodd 2013). Small vertebrates are occasionally consumed, including Spring Peepers, Wood Frogs, small birds, small gartersnakes (*Thamnophis* spp.), and even small conspecifics (Breckenridge 1944, Eddy 1976, reviewed by Dodd 2013). Northern Leopard Frog larvae are omnivores that use their scraping mouthparts to feed on plant material such as algae and also to scrape at animal carcasses (Franz 1971, Merrell 1977).

Northern Leopard Frogs are consumed by many types of predators throughout their life stages. In Wisconsin, known predators include Great Horned Owls (*Bubo virginianus*; Hamerstrom and Mattson 1939) and Common Mergansers (*Mergus merganser*; Breckenridge 1944). Additional predators include various wading birds (such as herons), waterfowl, Raccoons (*Procyon lotor*), fishes, turtles, and numerous snakes, such as hognose snakes (*Heterodon* spp.), racers (*Coluber* spp.), gophersnakes (*Pituophis* spp.), watersnakes (*Nerodia* spp.), and gartersnakes (Breckenridge 1944, Vogt 1981, Smith 2003, Dodd 2013). Eggs and larvae are especially vulnerable to predators, including numerous aquatic or semiaquatic insects (e.g., dragonfly naiads, caddisfly larvae, predaceous diving beetles [family Dytiscidae], and giant water bugs [family Belostomatidae]) and vertebrates such as fishes, Eastern Newts (*Notophthalmus viridescens*), Eastern Tiger Salamanders (*Ambystoma tigrinum*), turtles, gartersnakes, watersnakes, and waterfowl (Wright 1914, Merrell 1977, Dodd 2013). Eastern Tiger Salamander larvae will also prey on recently metamorphosed individuals (Natural History Box: Predator vs. Prey; Sara Viernum, Stantec, personal observation).

Adults and juveniles can escape perceived threats by leaping into the water and burying themselves in the muddy substrate (Dodd 2013). If they are not close to water, they will flee by leaping in an erratic, zigzagging fashion until they either reach water or can conceal themselves under vegetation (Hine et al. 1981). Northern Leopard Frogs can employ additional predator-avoidance behaviors, including threat calls when startled and release calls when captured (Vogt 1981, reviewed by Dodd 2013). They can also inflate and arch their bodies, curl their heads down between their front legs, and stand erect on their legs to make themselves difficult to swallow (Dodd 2013). Northern Leopard Frogs with warmer body temperatures are apparently better at avoiding predators than those with cooler body temperatures (Hine et al. 1981).

CONSERVATION

Status: Globally, the Northern Leopard Frog is considered a Least Concern species by the IUCN and a species of concern at the federal level. Despite this, several regions have observed declining populations and even probable extirpations (reviewed by Dodd 2013). Although these frogs are sometimes locally abundant in Wisconsin, numerous mass mortality events were noted in the 1960s and 1970s, and several studies have suggested statewide declines over the last 50 or more years (Hine et al. 1975, Hine et al. 1981, Mossman et al. 1998; Natural History Box: Population Declines). The perceived declining trend in occurrence is supported by long-term monitoring data from the Wisconsin Frog and Toad Survey (Figure 117). The WDNR currently considers Northern Leopard Frogs a common species, while Christoffel et al. (2009) considered them "common but significantly declining" in the state.

NATURAL HISTORY BOX: Predator vs. Prey

During an outing for the 2013 Midwest Partners in Amphibian and Reptile Conservation meeting I helped check funnel traps in a wetland at the Mequon Nature Preserve in Mequon, Wisconsin. A small group of us observed an Eastern Tiger Salamander larva eating a newly metamorphosed Northern Leopard Frog in one of the funnel traps we pulled out of the wetland (Figure 116). The frog must have wandered into the same funnel trap as the salamander and, not being able to escape, soon became prey to the salamander. By the time we had removed the trap from the water and extracted the salamander, all that was left to be seen of the frog were a pair of hind feet sticking out of the salamander's mouth.

Figure 116. Larval Eastern Tiger Salamander consuming a recently metamorphosed Northern Leopard Frog (Ozaukee County; photo by S. Viernum).

Sara E. Viernum, Stantec

NATURAL HISTORY BOX: Population Declines

Northern Leopard Frogs were reportedly once very populous in Wisconsin. In fact, Knudsen (1958c) stated that they were "found abundantly all over Wisconsin" and that they were "seemingly countless" in some areas of the state. During the 1960s and 1970s, however, researchers, field collectors, and biological supply houses reported declines in Northern Leopard Frogs in Wisconsin (Hine et al. 1981, Vogt 1981, Harding 1997). A reduction in the length of time that frogs could be kept alive in-house was noted by the biological supply houses, and reports from field collectors indicated that there were outbreaks of Northern Leopard Frog mortality at locations around the state (Hine et al. 1981). Similarly, Lannoo et al. (1994) reported a drastic decline of Northern Leopard Frogs in Dickinson County, Iowa, in the 1900s. Gibbs et al. (1971) estimated that the leopard frog population in the US had declined by 50% in the 1960s.

Rori A. Paloski, WDNR

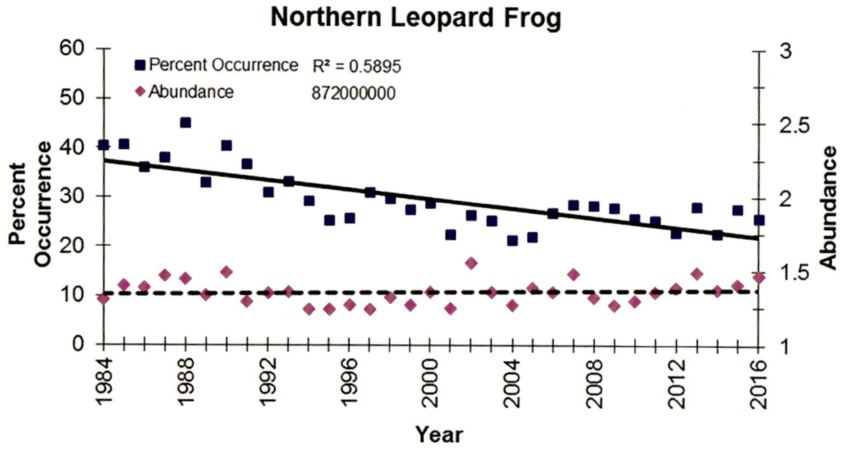

Figure 117. Long-term trend in occurrence (%) and abundance (rank) of Northern Leopard Frogs across Wisconsin Frog and Toad Survey statewide monitoring sites. See Introduction to Species Accounts chapter for details on this monitoring program (figure courtesy of WDNR).

Populations: Few robust population studies have been conducted in the Upper Midwest. Merrell (1968) estimated the number of breeding individuals at six sites in Minnesota, and his results indicated that large variation in population sizes is possible, with estimates from six ponds ranging from 124 to 1,568 breeding individuals. Merrell's (1968) estimates are also substantially larger than those derived from similar attempts at several breeding ponds in Wisconsin, which ranged from 2 to 76 individuals across two years (Hine et al. 1981). Survival of young age classes, especially eggs and larvae, is low. Merrell (1968) estimated that only ca. 5%–6% of fertilized eggs survived to metamorphosis ($n = 20{,}000$ of 346,000 fertilized eggs). Adult mortality rates, on the other hand, apparently vary. For example, Northern Leopard Frog mortality observed during multiyear statewide surveys in Wisconsin ranged from 4% to 23% (although 100% mortality was observed along some survey transects; Hine et al. 1981). Hine et al. (1981) estimated that the sex ratio of populations they investigated in Wisconsin were ca. 1:1.

Threats: Northern Leopard Frogs are likely affected by many threats that face all amphibians in Wisconsin, such as habitat loss, road mortality, pollution, disease, overharvest, species introductions, and climate change (see Introduction to Conservation and Management of Wisconsin's Amphibians and Reptiles chapter). Habitat destruction and degradation, in particular, have serious negative impacts on Northern Leopard Frogs (Knutson et al. 1999, Johnson et al. 2011). Knutson et al. (1999) found a negative correlation between frog populations, including Northern Leopard Frogs, and urban areas in Wisconsin and Iowa. Northern Leopard Frogs also avoid ground surfaces that are barren (e.g., due to agricultural disturbance), which can result in dehydration (Mazerolle and Desrochers 2005). Further studies from the Upper Midwest support that they avoid row-cropped areas, such as cornfields, in agricultural landscapes and are prone to mortality due to equipment used to harvest hay (Bartelt and Klaver 2017, Knutson et al. 2018).

Anthropogenically disturbed landscapes possess higher road densities than undisturbed regions, and these roads can act as barriers to movement and result in substantial Northern Leopard Frog mortality (Ashley and Robinson 1996, Linck 2000, Carr and Fahrig 2001, Mazerolle 2004, Eigenbrod et al. 2008, Bouchard et al. 2009). Knudsen (1958c) anecdotally reported that after some rainy nights "a person may see hundreds of dead [leopard] frogs that tried to cross, only to be crushed by automobiles" along certain roadways in Wisconsin. In Ontario, Ashley and Robinson (1996) reported a staggering 27,846 deceased Northern Leopard Frogs along a 3.6 km (2.2 mi) stretch of road over a four-year period. An additional stressor associated with urban and agricultural landscapes can be greater exposure to environmental contaminants, including metals, UV light, pesticides, and excess nutrients such as nitrogen (Hine et al. 1981, Christen et al. 2003, Hayes et al. 2003, 2006, Howe et al. 2004, reviewed by Rorabaugh 2005, Rohr et al. 2008a, reviewed by Dodd 2013). The impacts of these potentially damag-

NATURAL HISTORY BOX: Malformations

Minnesota was one of the first states in the country to report high numbers of frog malformations when a group of grade-school students found malformed leopard frogs at an agricultural pond in 1994 in the south-central portion of the state. Over the next few years, the Minnesota Pollution Control Agency surveyed sites throughout the state, predominantly responding to citizen reports of malformations. While working for the Pollution Control Agency in the late 1990s and 2000, we conducted malformed frog surveys throughout Minnesota, primarily surveying Northern Leopard Frogs, Mink Frogs (*Lithobates septentrionalis*), and Green Frogs. In between navigating floating bog habitats, dodging female black bears with cubs, and avoiding the occasional tornado in western Minnesota grasslands, we surveyed thousands of frogs each summer. At some sites, we recorded 0%–2% malformation rates in the frog population (typically reported to be the baseline malformation rate), while at other sites we recorded 25%–>50% malformation rates. Malformations ranged from a missing digit, missing limb, or eye malformation to multiple additional limbs. (One juvenile frog in northern Minnesota had seven rear legs!) While the causes of amphibian malformations and declines are still being investigated, our surveys throughout Minnesota provided a substantial amount of data to the state and toward the investigation. The surveys also allowed us to gain a greater understanding of Minnesota's frogs and included many unique amphibian observations, including both 'Burnsi' and 'Kandiyohi' morphs of the Northern Leopard Frog.

Jeff Canfield and Phoebe Vanselow, Minnesota Pollution Control Agency, and Rori A. Paloski, WDNR

ing factors on Northern Leopard Frogs have received substantial attention from researchers (reviewed extensively by Rorabaugh 2005 and Dodd 2013), including numerous studies conducted on wild Northern Leopard Frogs in Wisconsin and wild-collected Wisconsin individuals brought into the laboratory (e.g., Edblom 2001, Chen et al. 2006, 2007, Huang et al. 2007, Jofré and Karasoz 2008). The results of these studies are varied and too extensive to review in detail here, but they support that contaminants and toxicants can result in mortality at various exposure concentrations and under both laboratory and field conditions. In addition, they can reportedly cause various sublethal effects, such as developmental abnormalities (reviewed by Lannoo 2008), that can reduce Northern Leopard Frog reproductive success and fitness. In Wisconsin, Northern Leopard Frogs have had the highest number of malformation reports of any amphibian species (WDNR, unpublished data; see Introduction to Conservation and Management of Wisconsin's Amphibians and Reptiles chapter and Natural History Box: Malformations for additional information on amphibian malformations).

Northern Leopard Frogs fall victim to various diseases that are serious conservation concerns (reviewed by Rorabaugh 2005 and Dodd 2013). Despite this, the chytrid fungus *Batrachochytrium dendrobatidis* (Bd), which affects many amphibians globally (see Introduction to Conservation and Management of Wisconsin's Amphibians and Reptiles chapter), is rarely detected in Northern Leopard Frogs sampled from Wisconsin (Sadinski et al. 2010, Klemish et al. 2012, Standish et al. 2018). Other studies have found evidence that leopard frogs possess characteristics (e.g., structural skin characteristics or antifungal skin peptides) that could make them less susceptible to Bd infection than other anurans (Rollins-Smith et al. 2002, Ortiz-Santaliestra et al. 2013). The bacteria *Aeromonas hydrophila* has often been reported as a possible culprit behind deceased and dying Northern Leopard Frogs encountered in the wild (reviewed by Rorabaugh 2005). In Wisconsin, Vogt (1981) observed "hundreds" of deceased individuals near hibernacula along streams in late autumn with signs of "red leg, a bacterial infection," but indicated that definitive identification of the causative agent was unknown. Sekowska et al. (2014) did not detect ranaviruses on a limited number of Northern Leopard Frogs sampled from Wisconsin but did detect it on individuals from nearby sites in Illinois. Standish et al. (2018) did detect ranaviruses on individuals from Wisconsin, although prevalence was low. Some research supports that the effects of ranaviruses on Northern Leopard Frogs could vary based on context. For example, Echaubard et al. (2010) found that infected Northern Leopard Frog larvae kept in high densities experienced higher mortality rates but lower development and growth rates than larvae kept at lower densities.

Predatory fishes and American Bullfrogs (*Lithobates catesbeianus*) prey on amphibians and can become problematic if introduced into aquatic habitats where Northern Leopard Frogs occur (Bovbjerg 1965, Hammerson 1982, Lannoo 1996, Lannoo et al. 1994, Panik and Barrett 1994, Smith 2003, Germaine and Hays 2007, 2009, Johnson et al. 2011, Herwig et al. 2013). Likewise, human manipulation of hydrology that converts wetlands from ephemeral to permanent creates a more suitable environment for the persistence of predatory fishes, which is also a concern for resident amphibians. The collection of Northern Leopard Frogs for bait or food or by biological supply companies has also negatively impacted their populations in the Upper Midwest (Bider and Matte 1996, Lannoo et al. 1994, Moriarty and Hall 2014). Knudsen (1958c), for example, specifically commented on their frequent use as recreational fishing bait in Wisconsin. In Minnesota, it was estimated that over 45,359 kg (100,000 lb) of Northern Leopard Frogs were commercially harvested per year in the early 1970s, with 454–907 kg (1,000–2,000 lb) harvested per year more recently (Moriarty 1998). Lannoo et al. (1994) indicated that massive numbers were harvested in Dickinson County, Iowa, during the twentieth century, which led to substantial population declines. It is unlikely that the impacts of such harvest on upper midwestern populations can be fully appreciated due to a lack of historical Northern Leopard Frog abundance data, but those impacts almost certainly were profoundly negative. Changes in climate or extreme weather events might also affect Northern Leopard Frogs, and these effects can sometimes be subtle, such as alteration of calling and breeding phenology (Hine et al. 1981, Corn and Fogleman 1984, Johnson et al. 2011, Walpole et al. 2012).

Management: Numerous management techniques that are generally applicable to amphibians and their habitats in Wisconsin could also benefit Northern Leopard Frogs (see Introduction to Conservation and Management of Wisconsin's Amphibians and Reptiles chapter). These include protection and management of preferred upland foraging areas and aquatic habitats used for overwintering and breeding (Lehtinen and Galatowitsch 2001, Stevens et al. 2002, Brodman et al. 2006, Balas et al. 2012). Because Northern Leopard Frogs prefer open-canopy habitats, mechanical removal of woody vegetation or occasional mowing of grass to combat succession around wetlands might also be necessary. However, research in Minnesota supports that Northern Leopard Frogs are frequently killed by equipment used when mowing or during hay harvest (Knutson et al. 2018). Thus, restricting these activities until late autumn or winter could reduce potential for mortality (Maisonneuve and Rioux 2001).

In landscapes dominated by agricultural or urban land, creation of large undisturbed buffers around aquatic breeding habitats and maintenance of connections between adjacent aquatic overwintering habitats are likely beneficial. Bartelt and Klaver (2017) believed that recently metamorphosed individuals could use fencerows and vegetated drainage ditches as corridors to negotiate landscapes disturbed by agriculture. Yet Knutson et al. (2018) determined that Northern Leopard Frogs that migrated between suitable habitats sometimes crossed directly through disturbed landscape features (e.g., roads, row-cropped fields) and did not necessarily follow natural corridors. Therefore, the best strategies for protection and management of landscapes for Northern Leopard Frogs should consider their established movement patterns to maintain large-enough areas of undisturbed natural habitat (see Activity section). Along these lines, Knutson et al. (2018) concluded that "if wetlands are scarce and pond buffer narrow, Northern Leopard Frogs will use an area of up to 6 ha in size and may travel as far as 3 km away from the breeding site."

Maintenance of connectivity among suitable habitats can be realized if the potential for road mortality is minimized or eliminated. Northern Leopard Frogs will use underpasses of at least 0.5 m (1.6 ft) in diameter to bypass crossing roads (Woltz et al. 2008). Exclusion fencing with a height of at least 0.6 m (2 ft) has also been successfully implemented to deter frogs from crossing roads or guide them toward road underpasses (Hoffman 2003, Woltz et al. 2008). In Minnesota, Linck (2000) used a combination of exclusion fencing along roads and volunteers physically moving frogs to reduce Northern Leopard Frog mortality.

Studies from the Upper Midwest support that Northern Leopard Frogs will colonize and successfully reproduce in restored wetlands (Lehtinen and Galatowitsch 2001, Nedland et al. 2007). However, the characteristics of restored wetlands, as well as their position and orientation on the landscape, apparently influence their effectiveness as a management tool. Restored or created wetlands and ponds should be at least 0.9–1.2 m (3–4 ft) deep to prevent the waterbody from freezing solid in the winter but also should not be deep enough to allow survival of fishes (Bovbjerg 1965, Lannoo 1996, Germaine and Hays 2009, Johnson et al. 2011, Herwig et al. 2013). According to Bartelt and Klaver (2017) Northern Leopard Frogs used restored wetlands but took two and three years to colonize wetlands that were >2.5 and >3.5 km from source wetlands, respectively, in a disturbed northern Iowa landscape.

In addition to strategies for habitat management, implementation of tactics that offset the negative impacts of various other threats (e.g., contaminants, introduced species, and harvest) is worthy of consideration. Because of the frogs' sensitivity to contaminants, application of pesticides and fertilizers must be carefully considered. In agricultural areas, it is likely beneficial to limit cattle access to wetlands that Northern Leopard Frogs use in order to keep nitrogen levels <0.5 mg/L (0.00006 oz/gal; Knutson et al. 2004). The harvest of Northern Leopard Frogs for any commercial purpose, including as bait for recreational fishing and for biological supply companies, should be regulated. Currently, best practices for sustainable harvest (based on robust population data) do not exist for Wisconsin, but there are regulations in place that limit collection of Northern Leopard Frogs. There is currently an open season for the collection of frogs in Wisconsin that runs from the Saturday nearest to 1 May through 31 December (Wisconsin Administrative Code NR 19.26). Frogs cannot be collected from the wild during the closed season. An individual can collect or possess up to five individual Northern Leopard Frogs, with the exception that an unlimited number and size of atypically colored or patterned frogs may be possessed (Wisconsin Administrative Code NR 16.12). Up to one full clutch of eggs and their hatched larvae can be collected and possessed up until metamorphosis. Only five adults can be retained in possession. The rest cannot be sold, and they must be returned to the same body of water that they were taken from immediately after transforming to the adult form (Wisconsin Administrative Code NR 16.12). See WDNR (2004) for a summary of additional state regulations related to Northern Leopard Frogs.

Mink Frog

Lithobates septentrionalis (Baird 1854)

Steven J. Price

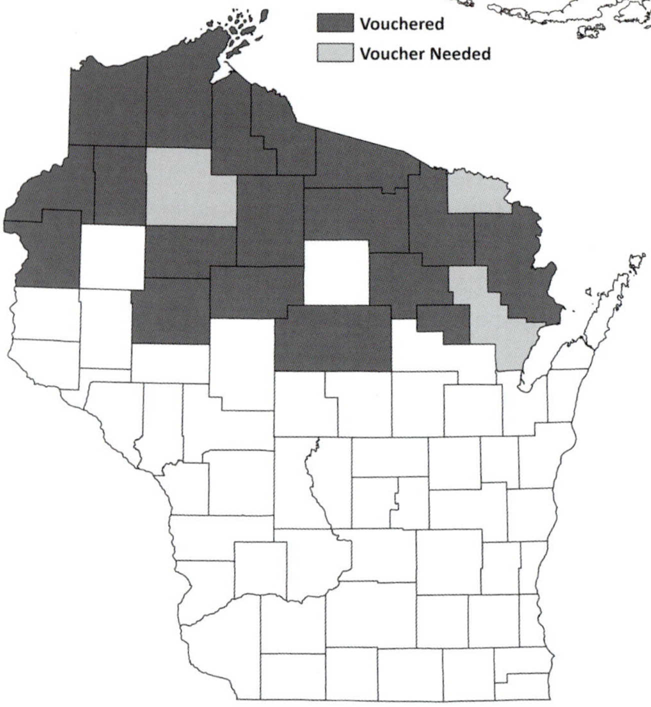

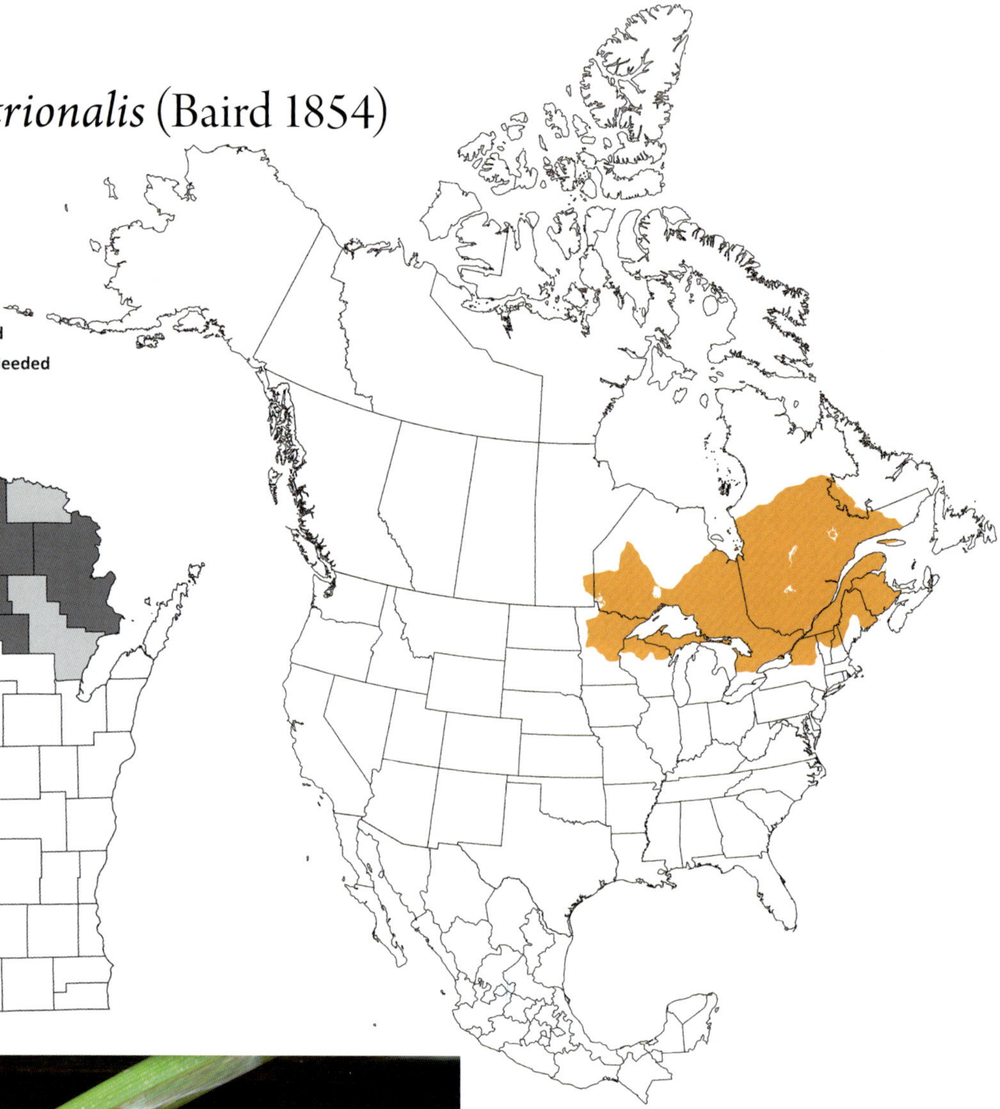

Adult Mink Frog (Iron County; photo by J. White).

AT A GLANCE

Mink Frogs are a medium-sized, highly aquatic frog of the Wisconsin Northwoods. The specific epithet, *septentrionalis,* translates to 'of the North,' a reflection of their northern distribution. The backs and legs of Mink Frogs are typically brown or gray but covered with extensive dark mottling; areas along their upper lips, around their tympanums, and sometimes along their dorsolateral folds are often a clean bright green. They are also frequently described as possessing an odor similar to rotten onions, and their common name is derived from this musky, mink-like odor. Mink Frogs are frequent inhabitants of muddy cold-water bogs and beaver ponds with abundant floating and emergent vegetation. There is little distinction between habitats used for breeding, foraging, and overwintering. They emerge from hibernation in May, but males do not began calling to attract females for reproduction until June through early August. The mating call of males sounds like a series of sharp knocks. Females attach eggs to underwater vegetation and debris at greater depths than other frogs in Wisconsin (e.g., up to 1 m [3.3 ft] or more below the water's surface). Larvae often take a year to transform and spend the long winter months below the ice, along with adults and juveniles, and typically undergo metamorphosis the following spring or summer. Adult Mink Frogs eat an array of aquatic invertebrates, while larvae consume algae that they scrape off aquatic surfaces with their specialized mouthparts. The Mink Frog is currently considered a species of Special Concern and a Species of Greatest Conservation Need by the WDNR.

SYSTEMATICS

Lithobates septentrionalis (Mink Frog) is a New World true frog (family Ranidae) in the *L. catesbeianus* (American Bullfrog) species group (i.e., subgenus) known as *Aquarana* (Hillis and Wilcox 2005). *Aquarana* includes seven species that diverged from a common ancestor approximately 11 mya (Austin et al. 2003b), three of which occur in Wisconsin (i.e., *L. catesbeianus*, *L. clamitans* [Green Frog], and *L. septentrionalis*). Despite substantial variation in adult size and coloration throughout its range (see Variation subsection), there are currently no recognized subspecies of *L. septentrionalis* (Crother 2017).

DESCRIPTION

Adults: Dorsal coloration is light green, brownish olive, or tan and overlain with light tan to almost black spots and/or irregularly shaped blotches (Figure 118). In New York, Kramek and Stewart (1980) found that older female Mink Frogs tended to have dark reticulated dorsal patterns. Bright yellowish-green coloration is prominent below the eyes and along the lips in most individuals, and that coloration sometimes also surrounds the tympanums and runs down the dorsolateral folds (Figure 119). Often, these

Above, top: Figure 118. Adult Mink Frog (Vilas County; photo by C. Raimond).

Above, bottom: Figure 119. Adult Mink Frog (Bayfield County; photo by M. Roth).

greenish areas are devoid of mottling. Ventral coloration ranges from white to yellow with some gray mottling. Dorsolateral folds are often incomplete and not readily apparent. Webbing on the feet extends to the second joint of the fourth toe and to the tip of the fifth toe. Mink Frogs have slightly rough skin, which produces a musky odor, sometimes likened to the scent of American Mink (*Mustela vison*), rotting onions, or garlic.

Mink Frogs are a medium-sized frog, with reproductive adults typically measuring 5.0–6.5 cm SVL (2.0–2.6 in) in Wisconsin (Vogt 1981). The mean SVL of adult specimens from Wisconsin in the MPM and UWZM collections ranged from 5.2 cm (2.1 in) in males ($n = 32$) to 5.6 cm (2.2 in) in females ($n = 14$, based on minimum size at maturity reported by Hedeen 1972c). In north-central Minnesota, adult and subadult males and females ranged from 4.5 to 7.2 cm SVL (1.8 to 2.8 in) and 5.4 to 7.2 cm SVL (2.1 to 2.8 in), respectively ($n = 220$; Hedeen 1972c). The largest SVL of a Wisconsin specimen measured was 7.1 cm (2.8 in, MPM 33258, collected from Ashland County in 1998).

Preadult stages: Mink Frog eggs are brown to black dorsally and yellowish ventrally and are laid as a globular mass of ca. 500–1,700 eggs that measures 7.5–12.5 cm (3.0–4.9 in) in diameter (reviewed by Dodd 2013, Altig and McDiarmid 2015). Two jelly envelopes surround the vitellus; the outer and inner envelopes average 6.3 cm (0.3 in) and 2.7 cm (0.1 in), respectively (Livezey and Wright 1947). Larval dorsal coloration ranges from dark brown to yellowish green with small dark spots. Pinkish or buff-colored spots are often found on the tail. Garnier (1883) noted that the larvae produce a less powerful version of the musky odor emitted by the adult. According to Harding (1997), larvae may reach up to 10.0 cm TL (3.9 in) before undergoing metamorphosis. In Minnesota, the mean size of recently metamorphosed Mink Frogs was 3.72 cm TL (1.5 in, $n = 54$; Hedeen 1971b). Juveniles have a color pattern similar to that of adults.

Variation: Male Mink Frogs have a larger tympanum than females, and in most populations, adult females generally obtain a larger mean body size than males (Schueler 1975, Kramek and Stewart 1980). Additionally, studies by Schueler (1975) and Leclair and Laurin (1996) in Canada revealed geographic variation in size; average body size was larger in northern populations than in southern populations.

Diagnosis: Mink Frogs are characterized by a body without parotoid glands or large warts, without a keratinized spade on the heel of each hind foot, and without terminal toe pads. Their upper jaw is green, without marks, and a musky odor is present. Mink Frogs are similar in appearance to American Bullfrogs and Green Frogs but can be differentiated by reviewing several characteristics. Both American Bullfrogs and Green Frogs reach larger adult sizes than Mink Frogs. They also generally lack the prominent dorsal spots and/or irregularly shaped blotches present on Mink Frogs. These frogs can be further distinguished by the characteristics of their dorsolateral folds, if present. American Bullfrogs lack dorsolateral folds, and Green Frogs have a prominent dorsolateral fold that extends to the midlength of the dorsum, whereas Mink Frogs have incomplete or not readily apparent dorsolateral folds. The extent of webbing between the toes on the hind feet is also particularly useful in discriminating between Mink Frogs and Green Frogs. The hind foot webbing of Green Frogs fails to reach the tip of the fifth toe and rarely extends beyond the second joint of the fourth toe. Furthermore, only the Mink Frog produces a musky odor when handled.

Distinguishing among preadult stages of ranid species can be more challenging than in adults. Although the individual eggs of Mink Frogs, American Bullfrogs, and Green Frogs all look superficially similar, the shape and deposition location of egg masses differ. Mink Frogs often attach globular egg masses to vegetation at depths of 1 m (3.3 ft) or more below the water's surface, whereas Green Frogs and American Bullfrogs lay eggs in a thin sheet at the surface (Vogt 1981). Mink Frog larvae possess internal gills and lack forelimbs. Their eyes are positioned dorsally, while their vent is dextral. They lack a white line from their nostril to their snout, and their mottled tail includes pinkish spots. Larvae can be partially distinguished by the geographic location of their observation. Because Mink Frogs are found in northern counties, anuran larvae encountered in central or southern Wisconsin are unlikely to belong to Mink Frogs. Beyond this distinction, Mink Frog larvae reportedly possess pink or buff-colored spots on the tail, unlike Green Frog and American Bullfrog larvae (Watermolen and Gilbertson 1996, Parmelee et al. 2002).

DISTRIBUTION AND HABITAT

Global distribution: Mink Frogs are found from north-central Minnesota and southeastern Manitoba eastward through the Great Lakes region to Labrador and southward to northern New York. Populations have been introduced into Newfoundland. Mink Frogs are absent from extreme southern Ontario and the Lower Peninsula of Michigan.

Distribution in Wisconsin: Mink Frogs are found throughout northern Wisconsin, extending as far south as Marathon County. Of the published county records, suitable museum vouchers could not be found for Florence, Oconto, and Sawyer (see the Appendix).

Localities of possible occurrence: There are several northern counties with no published records that are obvious gaps in the Mink Frog's state distribution, including Barron, Lincoln, and Shawano Counties. It is noteworthy that Wisconsin Frog and Toad Surveys of calling males conducted through the WDNR have generated several unconfirmed reports of Mink Frogs in Barron County since 2000 (WIATRI 2016). Although the Mink Frog is currently considered a more northern species, several past authors have stated that it was more widespread in Wisconsin. Knudsen (1958f) believed its distribution was "nearly state-wide" but more common in the northern counties. Similarly, Suzuki (1951) and Dickinson (1965) reported Mink Frogs from numerous counties in southern Wisconsin where they have not been observed in recent years. These include Waukesha, Washington, Columbia, Sauk, Richland, and Crawford, but also central counties such as Monroe, Adams, Portage, and Clark. Mink Frogs can be confused with Green Frogs and American Bullfrogs (see Diagnosis subsection), so it is likely that these southern/central records represent misidentifications. Despite this, Mossman et al. (1998) also stated that the species was observed by a reliable source in southern Clark County. Further surveys could, therefore, expand the confirmed range of Mink Frogs in Wisconsin.

Habitat: In Wisconsin, Mink Frogs are generally associated with permanent aquatic habitats in forested landscapes typical of the Wisconsin Northwoods (Knutson et al. 2000). Aquatic habitats preferred by Mink Frogs have an abundance of emergent or floating vegetation and silty substrates such as bogs, wetlands that form in North American Beaver (*Castor canadensis*) impounded streams, lakes, and ponds (Figure 120; reviewed by Dodd 2013). They can also inhabit slow-moving stretches of streams (Schueler 1973). Vogt (1981) reported that Mink Frogs in Wisconsin were abundant in the boggy lake inlets of Vilas and Douglas Counties, while Hecnar et al. (2002) found them in bogs on the Apostle Islands. They were also common in a large North American Beaver–impounded wetland in Vilas County (Steven Price, University of Kentucky, personal observation). However, they were rarely detected in an extensive study of anuran communities in coastal wetlands of Lake Superior and northern Lake Michigan (Price et al. 2007). Apparently, Mink Frogs avoid large lakes with significant wave activity, rivers with a rapid

Figure 120. Adult Mink Frog (Vilas County; photo by C. Raimond).

current, and aquatic habitats with rocky substrates (reviewed by Hedeen 1986, Dodd 2013). Although information on habitat used by post-metamorphic juveniles is rare, Hedeen (1972c) noted that immature males and nonbreeding females were found in shallow temporary wetlands in north-central Minnesota.

Studies on microhabitats selected by Mink Frogs are lacking in Wisconsin. Vogt (1981) anecdotally mentioned that they are "typically found on and along the edge of *Sphagnum* mats and among emergent vegetation throughout the summer" in the state. Research in the Upper Midwest suggests that adult Mink Frogs often inhabit open-water areas and riparian habitat along the shoreline rather than more terrestrial locations (Marshall and Buell 1955, Hedeen 1972c). Their confinement to wetland edges and open water is a result of high vulnerability to desiccation (Schmid 1965).

There is little reported separation between the general aquatic habitats that Mink Frogs use during the breeding season compared to the rest of the active season. In Wisconsin, males call from beds of floating and emergent vegetation within wetlands, bogs, and lakes (see Reproduction and Development section). Mink Frogs often occupy and breed in cold-water habitats (i.e., <21°C [69.8°F]), which provide greater oxygen levels for developing embryos (Hedeen 1986). A small sample of sites in Canada further suggests that selected breeding locations are also slightly acidic (pH 5.2–6.4; Dale et al. 1985). Egg masses are attached to submerged vegetation. Hedeen (1972c) described an egg mass encircling a Bur-reed (*Sparganium*) leaf at a depth of 10 cm (3.9 in) in a Minnesota lake, but some reports indicate that they typically deposit eggs at greater water depths. Reports of specific microhabitats used by Mink Frog larvae are rare. However, Hoopes (1938) described larvae, along with juvenile and adult frogs, inhabiting a clump of partially submerged moss (*Sphagnum* and *Fontinalis*) at the margin of a pond in Maine. Adults, juveniles, and larvae overwinter in the silt and mud at the bottom of aquatic habitats (reviewed by Dodd 2013).

REPRODUCTION AND DEVELOPMENT

In Wisconsin, male Mink Frogs typically call to attract females for reproduction from early June through July (Vogt 1981). Similarly, Hedeen (1972c) recorded Mink Frog vocalizations from 30 May through 9 August in north-central Minnesota. Bevier et al. (2004) studied the calling behavior of Mink Frogs in Maine. They described the male advertisement call as a series of two-note types, 'cuks' and 'rumbles,' while others describe it as a quick 'knock-knock-knock' sound (Vogt 1981) that vaguely resembles rapid beats on a small tight-skinned drum. Mink Frogs reportedly use vocalizations for purposes other than mate attraction. For example, Bevier et al. (2004) suggested that 'cuks' were primarily used by males as aggressive calls to other males. Mink Frogs also produce a warning or distress call when handled or startled (Vogt 1981, but see Casper 2005). Mink Frogs can call day and night. Males observed in Vilas County called sporadically throughout the daylight hours on 8 June 2001 (S. Price, personal observation). Yet during the peak of their calling season (i.e., early July), males mostly call at night (Patrick et al. 2012).

Mating behavior has been studied in Minnesota and Maine populations, as well as in laboratory settings. Males sometimes call when floating in open water, but they also call near aquatic vegetation (Hedeen 1972c). For example, males in a New York population primarily called from floating vegetation, especially lily pads (i.e., *Nymphaea odorata* and *Nuphar microphyllum*; Patrick et al. 2012). The calling locations selected by males also vary in their distance from the water's edge. Hedeen (1972c) found that some individuals called near floating sedge mats on the shoreline, whereas the calling locations selected by other males were in vegetation beds >30 m (98 ft) from the shoreline. Yet calling males are apparently not limited to fixed areas within a wetland, and their position changes throughout the breeding season (Hedeen 1972c, Bevier et al. 2006). Additionally, the chorus tenure of most males is limited to a few nights (Bevier et al. 2006), with a small proportion of adult females active at a breeding site each night (Hedeen 1972c). Although they do not remain in fixed locations during the calling season, males are aggressive toward conspecific males encountered during this time (Bevier et al. 2006).

Oviposition likely occurs during the peak calling period, from late June through July (Hedeen 1972c, Patrick et al. 2012). In Minnesota, nine adult females collected from 25 June through 30 July contained mature eggs (Hedeen 1972c). Prior to oviposition, females locate calling males, and amplexus ensues. The amplexed pair undergo periods of active swimming interspersed with periods of backward shuffling movements, which can continue for up to three hours before oviposition (Aronson 1943). After a series of undulating body movements (termed 'upstroke' and 'downstroke') by both sexes, oviposition occurs underwater until all eggs are released and males have broadcast their sperm over them (Aronson 1943).

Unlike those of other anurans in Wisconsin, eggs are not typically laid near the water surface (Vogt 1981). In New York, egg masses were found at a water depth of 1.5 m (4.9 ft); one clutch was attached to the slim branch of a submerged tree, and another was attached to the stem of a Yellow Pond Lily (*Nuphar lutea*; Patrick et al. 2012). Eggs hatch within 4–13 days, which is likely dependent on water temperature, and reports from several authors indicate that a specific range of temperatures is optimal (Wright 1932; Patrick et al. 2012). Moore (1952) found that eggs died when exposed to temperatures above 30.4°C (86.7°F) and did not develop at temperatures <7.8°C (46°F). These developmental limitations likely influence the geographic distribution of Mink Frogs (Hedeen 1986). Heeden (1971a) studied larval development in north-central Minnesota. In this population, the larval period lasted for ca. one year, and metamorphosis occurred in July and August at 3.28–4.19 cm TL (1.3–1.6 in, $\bar{x} = 3.72$ cm [1.5 in]; Hedeen 1971b). In the same Minnesota population, males reached sexual maturity by one year post-metamorphosis at 4.5–5.0 cm SVL (1.8–2.0 in), while females reached maturity at one to two years post-metamorphosis at 5.4–5.9 cm SVL (2.1–2.3 in; Hedeen 1972c). Mean life expectancy of wild Mink Frogs in Ontario ranged from 1.7 to 4.0 years post-metamorphosis, and estimated longevity ranged from 5 to 6 years post-metamorphosis (Shirose and Brooks 1995).

ACTIVITY

No studies have carefully determined the annual activity period for Mink Frogs in Wisconsin. According to Harding (1997), in the Great Lakes region, they emerge from hibernation in May. Although available information is limited, they appear to actively thermoregulate after emergence from hibernation. Hedeen (1971a) found that Mink Frogs in Minnesota selected microhabitats that maximized exposure to solar radiation, and as a consequence, body temperatures typically exceeded ambient water and air temperatures. In that study, conducted from 25 to 28 May, body temperatures ranged from 16.2°C to 27.1°C (61.2°F to 80.8°F, $n = 27$). In the Upper Peninsula of Michigan, body temperatures ranged from 25.7°C to 30.5°C (78.3°F to 86.9°F, $n = 4$), which was also greater than the air (22°C [71.6°F]) and substrate (25°C [77°F]) temperatures (Brattstrom 1963).

The movement patterns of Mink Frogs are not well studied and largely unknown for Wisconsin populations. Because the frogs rarely leave their aquatic habitats, most movements probably occur within these environments and are related to foraging and reproduction. In a Minnesota lake, Hedeen (1972c) reported movements of 46 m (150 ft) and 72 m (236 ft) by male Mink Frogs. These movements were associated with shifts in the location of breeding choruses. The mean activity area for male Mink Frogs in Maine was 410 m^2 (1,345 ft^2, SD = 112 m^2 [368 ft^2], MCP estimator; Bevier et al. 2006). Although rare, overland movements occur but are reportedly limited to rainy nights (Hedeen 1986).

As temperatures drop in autumn, Mink Frogs presumably prepare for winter dormancy, although detailed information on when the active season ends is sparse. Harding (1997) anecdotally reported that in the Great Lakes region they can enter hibernation by late September. Mink Frogs do not likely possess the ability to resist freezing conditions because they reportedly overwinter while submerged in aquatic habitats that do not freeze (see Habitat subsection).

PREY AND PREDATORS

Adult Mink Frogs eat a large variety of invertebrates. Stomach contents of post-metamorphic individuals contained springtails (Collembola), adult and larval dragonflies (Odonata), beetles (Coleoptera), caddisflies (Trichoptera), true flies (Diptera), sawflies, ants, bees, and wasps (Hymenoptera), and several families of Hemiptera, particularly water boatmen (Corixidae), water striders (Gerridae and Veliidae), and leafhoppers (Cicadellidae; Hedeen 1972b). Other studies have documented that Mink Frogs will consume small fish (Garnier 1883) and various spiders, snails, and leeches (Kramek 1972). Work in New York found that although Mink Frogs consume a variety of invertebrates, the frogs tended to focus more on aquatic prey species, compared to sympatric Green Frogs and American Bullfrogs (Stewart and Sandison 1972). Knudsen (1958f) noted that captive Mink Frogs consumed small frogs (i.e., Spring Peeper [*Pseudacris crucifer*]); however, Steward and Sandison (1972) did not find frogs in stomach contents of Mink Frogs in New York. Plant material has also been documented in the stomachs of adult Mink Frogs, which presumably was accidentally ingested (Hedeen 1972b, Steward and Sandison 1972). In Minnesota, larvae fed primarily on algae (Hedeen 1972b).

Adult Mink Frogs have several known predators. Hedeen (1967) documented Great Blue Herons (*Ardea herodias*) consuming adults and larvae, while Mallory and Lariviere (1998) found three Mink Frogs (each ca. 4.5 cm SVL [1.8 in]) in the esophagus of a Wood Duck (*Aix sponsa*). Hoopes (1938) stated that "snakes feed on the species with great relish." Hedeen (1972a) further observed Raccoons (*Procyon lotor*) attempting to capture Mink Frogs. Predators of larvae include larval Eastern Tiger Salamanders (*Ambystoma tigrinum*), Brook Sticklebacks (*Culaea inconstans*), leeches (e.g., American Medicinal Leech [*Macrobdella decora*]), and Giant Water Bugs (*Lethocerus americanus*), which were found sucking fluids from metamorphosing tadpoles (Hedeen 1972c). Eastern Newts (*Notophthalmus viridescens*) will consume eggs and recently hatched larvae (Wright 1932). Stewart and Sandison (1972) reported predation by American Bullfrogs, and Hedeen (1972a) noted that captive Green Frogs (*Lithobates clamitans*) readily consume recently metamorphosed Mink Frogs. Like other ranids, such as Green Frogs and American Bullfrogs, startled Mink Frogs avoid predation by quickly jumping across the water and eventually diving below the water surface, concealing themselves in vegetation or mud (Hedeen 1972c). Their strong odor is supposedly distasteful to some predators, although accounts of these secretions thwarting predation are lacking.

In Minnesota, Mink Frogs were parasitized by adult and larval digenean trematodes, nematodes, and *Trypanosoma* spp. (Schotthoefer et al. 2009). Bevier and Gorman-Gelder (2018) reported that Mink Frogs parasitized by a digenean trematode (*Halipegus eccentricus*) were less likely to respond to calls of conspecific males.

CONSERVATION

Status: Globally, the Mink Frog is listed as a Least Concern species by the IUCN. Moriarty (1998) and Weir et al. (2009) suggested that populations were stable in Minnesota and Maine, respectively. However, Mink Frogs are considered a species of Special Concern and a Species of Greatest Conservation Need by the WDNR. Data from Wisconsin Frog and Toad Surveys conducted from 1984 to 2016 indicate a declining trend in call survey site occupancy (Figure 121). However, Mink Frogs are not effectively sampled using standard manual calling survey methods (Bevier et al. 2006, Popescu and Gibbs 2009, Shearin et al. 2012), and thus a Mink Frog–specific monitoring effort is likely needed to confidently assess the status and trend of the species in Wisconsin.

Populations: Limited information on Wisconsin Mink Frog populations exists, other than that inferred from data on calling males collected through the WDNR's Wisconsin Frog and Toad Survey. For example, Mossman et al. (1998) reported that Mink Frogs were most commonly detected during call surveys in the northwestern forest region (detected at 33% of call survey stations), perhaps indicating higher abundance there. Mark-recapture estimates of population density in other upper midwestern locations are highly variable. For example, Werner (1975) reported Mink Frog densities of 39–206/ha (15.8–83.4/ac) at a site in Michigan. McAlpine et al. (1998) estimated densities of 70–88/ha (28.5–35.8/ac) across lakes in Canada. In southern Quebec, adult males had lower survivorship than adult females (Leclair and Laurin 1996). Sex-ratio estimates range from 1.4 to 1.8F:1M (reviewed by Dodd 2013).

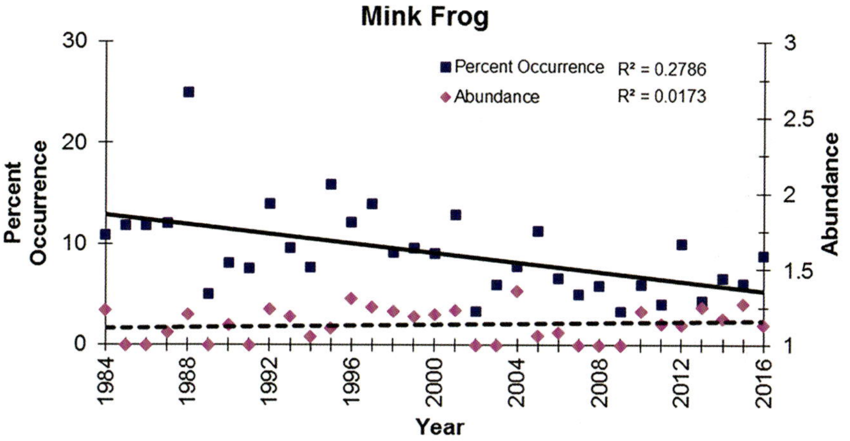

Figure 121. Long-term trend in occurrence (%) and abundance (rank) of Mink Frogs across Wisconsin Frog and Toad Survey statewide monitoring sites. See Introduction to Species Accounts chapter for details on this monitoring program (figure courtesy of WDNR).

Threats: General threats to northern forests and wetlands apply to Mink Frogs in Wisconsin (see Introduction to Conservation and Management of Wisconsin's Amphibians and Reptiles chapter). Specific threats to Mink Frog populations likely center on the drainage and alteration of wetlands, shoreline development, removal of riparian vegetation, and increased road densities near wetlands (Findlay et al. 2001, Maisonneuve and Rioux 2001, Mazerolle 2003, Woodford and Meyer 2003, Jacobs and Houlahan 2011). Because Mink Frogs favor cold aquatic habitats, alterations to water temperature associated with climate change could negatively impact Mink Frog populations, especially those at their southern distributional limit (Popescu and Gibbs 2009). Several diseases that are of conservation concern reportedly affect Mink Frogs. For example, Green et al. (2002) reported mortality due to iridovirus and fungus in Mink Frogs from the Upper Midwest. Sadinski et al. (2010) detected infection of Mink Frogs by the amphibian chytrid fungus (*Batrachochytrium dendrobatidis* [Bd]) in seven of nine sites sampled in the Upper Midwest. Ouellet et al. (2005) found evidence of the fungus in historical and recently collected Mink Frog specimens from Canada. Ruthig et al. (2020) confirmed both ranaviruses and Bd in 6 of 11 individuals tested from Minnesota (an additional 2 were infected only with Bd). In Minnesota, Hedeen (1972a) found larvae and adults with symptoms consistent with redleg disease, likely caused by the bacterium *Aeromonas hydrophila*. In addition, Mink Frogs with limb malformations were reported in Minnesota (Gardiner and Hoppe 1999).

Management: Few, if any, management strategies that specifically target Mink Frogs in Wisconsin exist. Yet many general wetland and riparian management recommendations for amphibians apply to this species, particularly strategies associated with lakes (see Introduction to Conservation and Management of Wisconsin's Amphibians and Reptiles chapter). Little research exists on the effects of specific management practices on Mink Frogs. Presumably, management efforts that maintain or restore vegetation both within the littoral zone and along the shoreline would benefit this species. Also, the use of underpasses along roads near known Mink Frog sites could reduce road mortality, although the effectiveness of underpasses as a conservation tool for this species is not well-studied.

Wood Frog

Lithobates sylvaticus (LeConte 1825)

Jaimie L. Klemish and Tracy A. G. Rittenhouse

Wood Frog (Price County; photo by C. Raimond).

AT A GLANCE

Wood Frogs are medium-sized frogs that range in color from tan to brown, sometimes with a gray, red, pink, or bronze tint. They also have a characteristic dark brown or black mask that extends from their snout backward through their eyes and tympanums. Wood Frogs are aptly named because they are frequently found in woodlands among the debris on the forest floor (their specific epithet, *sylvaticus,* is Latin for 'of the woods'). In Wisconsin, Wood Frogs occupy woodland habitats throughout much of the state but are most abundant in the northern half. They are among the first frogs to breed in Wisconsin each spring, sometimes migrating to wetlands during the early warm rains that melt the snow cover (e.g., mid-March to early April). Upon arrival at breeding sites, males call to attract females with an advertisement call that is likened to quacking ducks. Female Wood Frogs typically deposit their eggs in shallow water next to egg masses laid by other females, forming a large communal cluster of egg masses. Larvae in Wisconsin usually undergo metamorphosis in mid-May to late June, after which juveniles disperse into the surrounding woodland habitat for the remainder of the summer. Wood Frogs are typically active into October, at which time they move to terrestrial overwintering sites below the ground debris in woodlands. They are freeze-tolerant, meaning that water within their tissues can freeze during the winter but thaw again before the frog emerges in the spring. Adult Wood Frogs eat mostly small invertebrates found in damp forest floors. Wood Frogs are considered a common species by the WDNR; the presence of the species can serve as an indicator of high-quality forests and wetlands.

SYSTEMATICS

LeConte (1825) originally described this species as *Rana sylvatica,* although later authors proposed alternate scientific names, including *R. cantabrigensis* (Baird 1854) and *R. temporaria silvatica* (Yarrow 1882). Despite this, *R. sylvatica* was widely accepted as the standard scientific name for many years until Frost et al. (2006) moved this species and other 'true frogs' of the Americas from the predominantly Eurasian genus *Rana* to the genus *Lithobates,* which currently contains North, Central, and South American ranid frogs (Crother 2017; note the specific epithet of *sylvatica* was also changed to *sylvaticus*). There is still controversy over whether *Rana* or *Lithobates* is the appropriate generic name for Ranidae in the Americas (Yuan et al. 2016, reviewed by Crother 2017). Hillis and Wilcox (2005) and Yuan et al. (2016) placed *L. sylvaticus* as a sister group to *Aquarana* (the *L. catesbeianus* [American Bullfrog] species group [i.e., subgenus]).

There are no recognized subspecies of *Lithobates sylvaticus* (Crother 2017). However, molecular data support two clades, eastern and western, that are suggestive of separation during the most recent glacial maximum (Lee-Yaw et al. 2008). *L. sylvaticus* that inhabit the Upper Midwest (including Wisconsin), the Great Lakes region, all the western states, and most of Canada constitute the western clade, while those from the southern Midwest, Appalachian Mountain region, maritime US states, and extreme eastern Canada are within the eastern clade (Lee-Yaw et al. 2008). The Driftless Area of Wisconsin likely served as a glacial refugia for Wood Frogs, and modest phylogenic structure indicates that Wood Frogs rapidly expanded northward and then westward across Canada from the Driftless Area following the retreat of glacial ice (Lee-Yaw et al. 2008).

DESCRIPTION

Adults: Wood Frogs have a dark triangular-shaped 'mask' that extends from the tip of the snout back through the eyes and tympanums, accented by a white line along the upper lip (Figure 122). In Wisconsin, the dorsal color ranges from tan to brown with a gray, red, pink, or bronze tint (Vogt 1981, Christoffel et al. 2009), which differs by sex (see Variation subsection). Some individuals may have black spots along the dorsolateral folds (Martof and Humphries 1959) and a white middorsal stripe from the snout to the vent (Martof and Humphries 1959, Browder et al. 1966). The ventral color is white to cream and may be faintly mottled.

Wood Frogs are a medium-sized species that typically measures 3.0–5.0 cm SVL (1.2–2.0 in; Vogt 1981). Adult specimens from Wisconsin housed in the MPM and UWZM collections had a mean SVL of 3.9 cm (1.5 in, $n = 289$ specimens ≥3.0 cm SVL;

Above, top: Figure 122. Adult Wood Frog (Vilas County; photo by C. Raimond).

Above, bottom: Figure 123. Wood Frog egg masses (Sawyer County; photo by E. Wild).

see Reproduction and Development section for size at sexual maturity). Two specimens in the UWSP collection exceeded the SVL range given in Vogt (1981), with SVLs of 7.7 and 8.8 cm (3.0 and 3.5 in, UWSP 3973 and 3972, respectively).

Preadult stages: Eggs are bicolored (black above and cream below). They measure 0.15–0.33 cm in diameter (0.06–0.13 in), with an inner jelly envelope around each egg that measures 0.30–0.72 cm (0.12–0.28 in) and an outer jelly envelope that measures 0.43–1.73 cm (0.17–0.68 in; reviewed by Dodd 2013). Females deposit the eggs in a globular mass that is 6.0–10.0 cm in diameter (2.36–3.94 in; Vogt 1981; Figure 123). Larvae have brown to dark gray or black dorsal surfaces that are flecked with black and gold (Vogt 1981, Dodd 2013, Altig and McDiarmid 2015). They have no stripes on their tail, but late-stage larvae may develop a white or golden stripe on the upper jaw that will become the light stripe on the upper lip of the adult (Dodd 2013, Altig and McDiarmid 2015). Larvae have three (sometimes two) upper tooth rows and four lower tooth rows (Wright 1914, Altig and McDiarmid 2015). In Wisconsin, larvae are ca. 1.0 cm TL (0.39 in) at hatching and 3.0–4.0 cm TL (1.18–1.57 in) at metamorphosis (Vogt 1981).

Variation: Wood Frog size and color can vary by sex and by geographic location of origin (Martof and Humphries 1959, Berven 1982a, reviewed by Dodd 2013). Females are larger (Noble and Farris 1929, Martof and Humphries 1959, Berven 1981, Howard and Kluge 1985) and reportedly redder in color than males, which are darker and more yellow (Banta 1914, Schueler 1973, King and King 1991, Lambert et al. 2017b). In addition, males have paired vocal sacs and swollen thumbs with dark pads for clasping females during amplexus (Wright 1914, reviewed by Dodd 2013). Color differences between the sexes may not be readily apparent in recently metamorphosed individuals and juveniles. Wood Frogs in some populations possess a white middorsal stripe that extends from the snout to the vent, accompanied by a white stripe on the dorsal surface of each hind leg (Martof and Humphries 1959, Browder et al. 1966, Fishbeck and Underhill 1971). According to Martof and Humphries (1959), striped individuals only occur in a small region of northeastern Wisconsin, but there are no recent assessments of this.

Diagnosis: Wood Frogs lack the parotoid glands, large warts, and keratinized spades on the heels of each hind foot that characterize toads (*Anaxyrus* spp.). They also lack terminal toe pads and have no musky odor. Dorsolateral folds extend the length of the dorsum, and a dark mask passes through the eyes and tympanums. Adult Wood Frogs are similar in shape to other medium-sized *Lithobates* species in Wisconsin but are readily distinguishable by their characteristic dark mask.

Wood Frog eggs and larvae are difficult to differentiate from those of other anurans. In Wisconsin, egg masses that are laid in communal clusters during early spring often belong to Wood Frogs. When a single Wood Frog egg mass is found, however, it could be easily confused with that of a Pickerel Frog (*Lithobates palustris*) or Northern Leopard Frog (*L. pipiens*), and there are few, if any, reliable methods to easily distinguish these in the wild. Wood Frog larvae are characterized by a lack of external gills and forelimbs. Their eyes are dorsally positioned, while their vent is dextral. There is no white line extending from the nostril to the snout, and although the tail mostly lacks mottling or spots, some faint marks are present. Wood Frog larvae are sometimes the only dark larvae of their size found in breeding wetlands (Altig and McDiarmid 2015), which could distinguish them from other species. Despite this, they could still be confused with American Toad (*Anaxyrus americanus*) larvae, which are also dark with iridophores. Yet larval Wood Frogs can be distinguished from larval American Toads by the position of the former's spiracle, which is sinistral, unlike the longitudinally positioned spiracle of American Toad larvae (Altig 1970, Altig and McDiarmid 2015). Wood Frog larvae are also reportedly more active in the midwater region of breeding wetlands than larvae of most other *Lithobates* species (Altig and McDiarmid 2015).

DISTRIBUTION AND HABITAT

Global distribution: Wood Frogs are the most widely distributed anuran in North America (Martof and Humphries 1959, Martof 1970) and the only North American amphibian to occur north of the Arctic Circle (Lee-Yaw et al. 2008). Their range extends from the Pacific coast of Alaska, eastward across Canada to the Atlantic coast in Quebec, and from north of the Arctic Circle southward through the Great Lakes region and the Appalachian Mountains into northern Georgia.

Distribution in Wisconsin: Wood Frogs are documented throughout much of Wisconsin (Vogt 1981, Casper 1996). Yet they are reportedly more common in the northern half of the state and have not been confirmed in several counties along the southern border (Vogt 1981, Casper 1996, Mossman et al. 1998). Among southeastern counties, published records are either sparse (e.g., Walworth and Waukesha), missing vouchers (e.g., Rock), or based on old specimens (e.g., Milwaukee and Racine; see review of the Racine County specimen in Kapfer et al. 2015). Although they might currently be extirpated from some of these southeastern counties (e.g., Milwaukee; Casper 2008c), WIATRI (2016) has received anecdotal reports of calling males in all of them since 2005. Of the published county records, suitable museum vouchers could not be found for Crawford, Grant, Rock, and St. Croix (see the Appendix). The MPM once had a specimen from Grant County (MPM 4997), but it is currently lost (Julia Colby, MPM, personal communication). Furthermore, the Walworth County record is currently based on a photographic voucher of egg masses (MPM VZP538), and it is difficult to use for independent confirmation of the species. The museum notes associated with this photographic specimen also indicate that the eggs belonged to Northern Leopard Frogs, not Wood Frogs (Joshua Kapfer, UW–Whitewater, personal communication). Therefore, collection of a new voucher specimen for Walworth County is warranted.

Localities of possible occurrence: Wood Frogs could feasibly occur within most southern counties that currently lack published records (i.e., Lafayette, Green, Jefferson, and Kenosha), given their proximity to counties with confirmed presence (including several counties in northern Illinois; Phillips et al. 1999) and anecdotal reports of calling males (WIATRI 2016). Yet the restricted and fragmented nature of woodland habitats in these counties could limit the potential for current Wood Frog occurrence.

Habitat: Wood Frogs primarily inhabit forests, requiring a mosaic of wooded upland habitat and appropriate breeding wetlands to support their annual life cycle (Knutson et al. 2000). Their close association with wooded landscapes was highlighted by the work of Gibbs (1998b), who reported that Wood Frogs did not occur in landscapes with <30% forest cover. Similarly, Wood Frogs in Wisconsin were detected more often in the northwestern, north-central, and eastern regions where forested habitats are more abundant than in the Driftless Area and southeastern region

(Mossman et al. 1998). They are often found in both coniferous and deciduous forest types (Gibbs et al. 2005), including mesic forests and northern conifer swamps in Wisconsin (see Amphibians and Reptiles in Wisconsin Plant Communities chapter). Within the wide forest–prairie transition zone that occurs across portions of several midwestern states, including Wisconsin, Minnesota, and parts of the Dakotas, Wood Frogs are sometimes found in wooded riparian areas that can develop in prairies (Tracy Rittenhouse, University of Connecticut, personal observation). Adults select microhabitats based on the presence of ground debris, which acts as cover, but also acceptable microclimates and preferred tree canopy closure (Baldwin et al. 2006a, Blomquist and Hunter 2010, Rittenhouse and Semlitsch 2007). For example, adults often use deep leaf litter and moist hummocks of *Sphagnum* moss as refugia (Baldwin et al. 2006a, Rittenhouse and Semlitsch 2007). Recently metamorphosed Wood Frogs in Missouri experienced greater growth and survival if they had access to sites with downed logs, deep leaf litter, and stable temperatures (Earl and Semlitsch 2015).

Wood Frogs breed in semipermanent wetlands and seasonal/vernal pools in forested habitat (Kutka 1988, Baldwin et al. 2006a, 2006b). Most breeding ponds assessed in northern Wisconsin were small to medium in size, ranging from <20 to 40 m (66 to 131 ft) in diameter and <0.5 to 1 m (1.6 to 3.3 ft) in depth, with tree canopy cover ranging from 0.7% to 85% (Kutka 1988). These ephemeral wetlands are considered ideal breeding habitat for Wood Frogs because they provide the appropriate hydroperiod for larval development but do not support fishes or overwintering larvae of amphibians that prey on Wood Frog eggs and larvae (Hopey and Petranka 1994, Knutson et al. 1999, Porej et al. 2004, Baldwin et al. 2006b, Vasconcelos and Calhoun 2006, Skidds et al. 2007, Karraker and Gibbs 2009, Herwig et al. 2013). Despite this, Wood Frogs also breed in more permanent waterbodies, such as large lakes, in Minnesota and Wisconsin (Herwig et al. 2013). However, these permanent aquatic habitats often possess extensive shallow areas with emergent vegetation that provides cover from fishes (Herwig et al. 2013).

Wood Frogs overwinter in upland forests under leaf litter, logs, and tree roots (Wright 1914, Wright and Wright 1949, Vogt 1981, Schmid 1982, Storey and Storey 1984, 1987, Story 1987, 1990, Larson et al. 2014). Terrestrial hibernacula are likely found in habitats similar to those used during the nonbreeding season (see Activity section). In these terrestrial overwintering locations, Wood Frogs need warm and moist subsurface microhabitats (relative to the ground surface) to survive. Selected microhabitats typically occur in areas with abundant ground debris (leaf litter and woody debris) and also a more open tree canopy that allows for greater snow depth on the forest floor (Churchill and Storey 1993, Groff et al. 2016). Ground debris and a greater snow depth create a sufficient subnivean zone (i.e., the zone in and underneath snowpack), which helps protect frogs from desiccation and insulates frogs and maintains temperatures just above freezing (O'Connor and Rittenhouse 2016a).

REPRODUCTION AND DEVELOPMENT

In Wisconsin, Wood Frogs migrate to wetlands for reproduction soon after emergence from hibernation (see Activity section). They are among the first anurans to call in spring, along with Spring Peepers (*Pseudacris crucifer*) and Boreal Chorus Frogs (*Pseudacris maculata*; Vogt 1981, Mossman et al. 1998). Males in Wisconsin primarily call from late March to mid-April (Vogt 1981) but can continue into mid-May (Mossman et al. 1998). Wood Frogs are 'explosive breeders,' and the majority of females that use a particular breeding site will often deposit their eggs over a span of only a few days (Banta 1914, Wright 1914, Wells 1977b, Howard 1980, Waldman 1982).

Wood Frogs exhibit strong philopatry for specific breeding sites, with ca. 80% of individuals returning to their natal wetland and up to 100% returning to the wetland where they first bred (Berven and Grudzien 1990). Males reach breeding wetlands slightly before females, although the arrival of both sexes greatly overlaps (Banta 1914, Howard 1980, Berven 1981, Howard and Kluge 1985, Semlitsch and Drake 2015). Upon arrival, males produce their guttural quacking call to attract females for reproduction (Banta 1914, Howard 1980, Vogt 1981). Males typically call between sunset and sunrise (Oseen and Wassersug 2002) but may also call both day and night during

NATURAL HISTORY BOX: Breeding Migration

In northeastern Alberta, Canada, I was seated on the bank of a closed-canopy wetland just before dusk. Right at dusk, hundreds of Wood Frogs (females and males) arrived in large numbers from the forested upland habitat to the wetland to breed. As the Wood Frogs hopped through the leaf litter, the sound was loud enough to make me think there was a larger animal (e.g., bear or deer) moving around the wetland. Males positioned themselves in one area of the wetland and immediately began to call. As they called, more males and females arrived. The minute a frog (male or female) entered the water, males would actively pursue the newcomer. On numerous occasions, mating balls formed with more than five males to a single female. Pairs in amplexus began to lay their eggs on the submerged branches of a downed tree. By the time I left the wetland that night, more than 100 egg masses had been laid in a communal area.

Jaimie L. Klemish, University of Lethbridge

the peak of the breeding season (Howard 1980). Females and males that arrive later in the night are drawn to the breeding aggregate by the sound of chorusing conspecifics (Bee 2007a). Within a breeding congregation, there are often more males than females (Meeks and Nagel 1973, Howard 1980, Berven 1981, Howard and Kluge 1985; Natural History Box: Breeding Migration). For example, Howard (1980) reported 5.6 males per female at a breeding site in Michigan, while Semlitch and Drake (2015) reported 2.4 males per female in Missouri. The larger proportion of males than females at breeding sites means that many males in the population do not gain reproductive opportunities in a given year (e.g., 61% in a Michigan study; Howard and Kluge 1985), although larger males mate more frequently than smaller males (Howard 1980, Berven 1981, Howard and Kluge 1985).

Calling males float on the surface of the water with their hind limbs outstretched, actively searching for females (Banta 1914, Noble and Farris 1929, Wells 1977b, Howard 1980, Berven 1981). Females, on the other hand, often stay submerged at the bottom of the wetland and only surface occasionally; if approached by a male upon surfacing, the female will dive back underwater (Banta 1914, Berven 1981). Males aggressively pursue females even when they dive to escape (Banta 1914). Males that follow females are attracted to the movement at the water's surface and may attempt to mate with any amphibian they encounter, including other male Wood Frogs, as well as male or female Northern Leopard Frogs, Spring Peepers, and Spotted Salamanders (*Ambystoma maculatum*; Banta 1914, Wright 1914, Cummins 1920, Berven 1981, Vogt 1981, Kapfer and Muehlfeld 2006, Höbel and Kolodziej 2013).

Males that encounter females will attempt to mount them in a position called amplexus (Banta 1914; Figure 124). In some instances, numerous males will attempt amplexus with a single female, forming a 'mating ball' of up to 15 males per female (Howard 1980, Trauth et al. 2000), all vying for sole reproductive access. Females at the center of these mating balls will often dive below the water's surface, while the males attempt to dislodge rivals by kicking them with their hind limbs (Banta 1914, Wells 1977b, Howard 1980, Berven 1981). However, males are not easily extricated, and females entangled in mating balls sometimes drown (Banta 1914, Howard 1980, Trauth et al. 2000).

The amount of time that an individual pair spends in amplexus varies greatly, reportedly ranging from 1.6 to 96 hours (Howard 1980, Wright 1984). During this time, the pair moves around the wetland searching for a suitable location to deposit eggs (Howard 1980). Females attach eggs to submerged vegetation near the surface of the water, a process that takes ca. 6–20 minutes (Meeks and Nagel 1973, Howard 1980, Waldman 1982, Waldman and Ryan 1983, Egan and Paton 2004). In Wisconsin, females lay 500–800 eggs per

Above, top: Figure 124. A pair of adult Wood Frogs in amplexus (Vernon County; photo by J. Kapfer).

Above, bottom: Figure 125. A communal egg-laying location for Wood Frogs. Note that all of the gelatinous balls are egg masses from different females (Sawyer County; photo by E. Wild).

clutch (Vogt 1981), with larger females laying more eggs than smaller females (Howard and Kluge 1985). Furthermore, females often lay their eggs in communal clusters of numerous egg masses that are located in a warm protected area of the breeding wetland (Banta 1914, Wells 1977b, Seale 1982, Crouch and Paton 2000; Figure 125). In Wisconsin, Vogt (1981) counted 153 individual egg masses in an aggregation that covered a 1 m^3 (35 ft^3) area, while Seale (1982) counted 963 congregated egg masses at a site in Pennsylvania. These communal egg mass aggregations are generally found in the same region of the breeding wetland across years (Berven 1981, Seale 1982).

The practice of communal egg laying may reduce predation on eggs (Berven 1981), help eggs withstand desiccation as ephemeral wetlands dry (Forester and Lykens 1988), or increase developmental rates because warmer temperatures occur in the middle of egg mass aggregations (Berven 1981, Seale 1982, Waldman 1982, Waldman and Ryan 1983). For example, Waldman (1982) reported that eggs in the warmer center of a communal egg mass cluster experienced higher survival (86.3%) than those on the cooler peripheral locations (41.6%). This could explain why females reportedly push to the center of the egg mass aggregates to deposit their eggs (Berven 1981). However, Waldman (1982) also found that females that bred early often deposited their eggs at what eventually became the center of large egg mass aggregations, while late breeders deposited their eggs on the periphery, suggesting it could be advantageous to breed early. After females deposit eggs, they leave the wetland and migrate to forested upland habitats, while males typically remain at wetlands throughout the breeding period to gain additional reproductive opportunities (Banta 1914, Howard and Kluge 1985).

In Wisconsin, eggs hatch in 15–20 days (Vogt 1981). Egg survival to hatching varies greatly by year, wetland conditions, and location of egg mass within the wetland (Seigel 1983, Herreid and Kinney 1966, Berven 1990). Because Wood Frogs generally deposit eggs early in spring, the eggs are sometimes subjected to freezing temperatures that could damage developing embryos (<0°C [32°F]; Wright and Wright 1949). Wood Frogs are more tolerant of colder temperatures during the early stages of development than other northern anurans (Moore 1939), although prolonged exposure to ice and subzero temperatures can kill eggs and delay the embryonic development of survivors (Moore 1939, Frisbie et al. 2000). Even after hatching, larval survival and development are temperature dependent (Berven 1982b, Skelly et al. 2002). For example, survival of larvae monitored during a laboratory investigation was highest at temperatures ranging from 7.5°C to 24.0°C (45.5°F to 75.2°F; Herreid and Kinney 1967). Moore (1939) further determined that larvae housed in water at 23.7°C (74.7°F) developed from Gosner (1960) stage 3 to 20 more rapidly (52 hrs) than larvae housed at 10.0°C (50.0°F, 327 hrs). Even the cooler temperatures that occur in the shaded portions of wetlands can be enough to slow embryonic development. In fact, Skelly (2004) found that embryonic development declined by up to 50% in closed-canopy wetlands vs. open-canopy wetlands in Connecticut.

In addition to temperature, larval developmental rates are affected by interspecific competition with other amphibian larvae (Wilbur and Collins 1973, Wilbur 1977b, Berven and Charda 1988, Peacor and Pfister 2006) and parasite load (see Prey and Predators section). Typically, high larval densities that result in high competition for resources lower rates of survival and development, but they also reduce the size of larvae at metamorphosis (Wilbur and Collins 1973, Wilbur 1977b). Yet, due to their cold tolerance, Wood Frog larvae can reportedly outcompete other larvae species in shaded wetlands with dense tree canopy (Skelly et al. 2002, Williams et al. 2008). In Wisconsin, larvae metamorphose in 45–60 days (Vogt 1981), similar to reports from New York (44–85 days; Wright and Wright 1949) and Alaska (53–78 days; Herreid and Kinney 1967).

Juveniles grow rapidly after metamorphosis, although growth can vary by season. Recently metamorphosed Wood Frogs in Minnesota grew quickly at the beginning of the summer ($\bar{x} = 0.028$ cm/day [0.011 in/day], SD = 0.013 cm/day [0.005 in/day]), but this growth rate slowed in autumn (mean $\bar{x} = 0.007$ cm/day [0.002 in/day], SD = 0.002 cm/day [0.001 in/day]) as temperatures decreased and individuals reduced their feeding rates (Bellis 1961). By late August and early September, most metamorphosed Wood Frogs had reached ca. 3.0 cm SVL (1.18 in; Bellis 1961). Although growth slows even further during their first winter, Wood Frogs continue to grow after emergence from hibernation the following spring. Females grow faster and obtain larger adult sizes than males (maximum of ca. 6.0 and 5.0 cm [2.4 and 2.0 in], respectively; Bellis 1961, Howard and Kluge 1985). The majority of males breed in their first spring after metamorphosis, while females breed in their second spring (Bellis 1961, Berven 1982a, 1990). In Quebec, sexually mature males ranged from one to four years of age ($\bar{x} = 2.49$ yrs, $n = 98$), and sexually mature females ranged from two to five years of age ($\bar{x} = 2.76$ yrs, $n = 33$; Sagor et al. 1998; see Bastien and Leclair 1992 for similar results in Quebec).

ACTIVITY

Wood Frogs are one of the first anurans in Wisconsin to emerge from hibernation in spring, which typically occurs from mid- to late March (Vogt 1981). After emergence, adults migrate to wetlands for reproduction (Baldwin et al. 2006a, Rittenhouse and Semlitsch 2007). Following reproduction, adults move into the moist woodland habitats that surround breeding sites, which can involve numerous long-distance movements that coincide with rainfall (Baldwin et al. 2006a). In Missouri, Wood Frogs moved an average of 76.2 m (250.0 ft, SD = 62.3 m [204.4 ft], maximum = 248.9 m [816.6 ft]) from breeding wetlands to nonbreeding habitat (Rittenhouse and Semlitsch 2007). In July and August, newly metamorphosed Wood Frogs disperse from their natal wetland and establish terrestrial home ranges (Bellis 1961, 1965, Patrick et al. 2006). Bellis (1965) estimated that home range size of young Wood Frogs ranged from 2.9 to 368.3 m^2 (31.2 to 3,964.3 ft^2, $\bar{x} = 64.5$ m^2 [694.3 ft^2], MCP estimator).

Aside from adult breeding migrations and juvenile dispersal from natal wetlands, Wood Frogs generally move sparingly. In Missouri, for example, Wood Frogs typically spent 6–11 days and up to 24 days in a single nonbreeding location (Rittenhouse and Semlitsch 2007). When individuals moved, 56.5% of the distances were <5 m (16.4 ft, $n = 42$ individuals tracked with radio telemetry). Adults and postmetamorphic juveniles are typically diurnal during the nonbreeding season, with daily activity peaks in midmorning (8:00–10:00 a.m.) and late afternoon (ca. 4:00 p.m.; Heatwole 1961, Bellis 1962). Despite these findings, diel activity patterns can fluctuate with temperature, humidity, and light intensity (Heatwole 1961, Bellis 1961, 1962).

Wood Frogs move to overwintering locations from mid-August to early October. Groff et al. (2016) reported that the mean distance migrated from summer habitat to winter hibernacula was 192.5 m (631.6 ft, range = 0–550 m [0–1,804 ft]). Similarly, 40%–46% of overwintering adults observed in Massachusetts were >100 m (328 ft) from breeding wetlands (Regosin et al. 2003, 2005). Wood Frogs use terrestrial hibernacula (O'Connor and Rittenhouse 2016a) and survive extracellular freezing through production of physiologically synthesized cryoprotectants (Schmid 1982, MacArthur and Dandy 1982, Storey and Storey 1987, Constanzo et al. 1992). When the temperature drops below 0°C (32°F), their livers begin to produce glucose from glycogen within five minutes (Storey and Storey 1984, Storey 1987). Concurrently, the Wood Frog's heart rate increases for a period of time to distribute the glucose to body tissues (Layne et al. 1989). Once frozen, the frog's breathing and cardiovascular circulation cease until it thaws (Storey and Storey 1984, Layne et al. 1989). Freezing in this fashion is energetically costly and can take ≥20 hours, depending on the temperature (Layne et al. 1989, Lee et al. 1992, Sinclair et al. 2013). During winter, Wood Frogs frequently undergo several freeze-thaw cycles as environmental temperatures fluctuate (Sinclair et al. 2013) but can survive being frozen for 193 days (Larson et al. 2014). Continuous freeze-thaw cycles during the winter can deplete carbohydrate stores quicker than if individuals remain frozen (Sinclair et al. 2013). When Wood Frogs thaw completely, glucose is converted back into glycogen by the liver, while cardiac function, breathing, and muscle control resume, but this entire process can take several hours (Layne and First 1991, Sinclair et al. 2013). Due to the energetic cost, Wood Frogs are easily fatigued for at least 96 hours after thawing (Storey and Storey 1984, Irwin et al. 2003, Sinclair et al. 2013).

PREY AND PREDATORS

Wood Frogs are generalist predators that likely forage on many types of invertebrates based on their availability (reviewed by Dodd 2013). Analysis of adult and juvenile stomach contents in Alberta revealed mollusks, annelids, arachnids, chilopods, diplopods, and insects (Moore and Strickland 1955). Wood Frogs are ambush predators that can sit at a location in the same posture for multiple days (Rittenhouse and Semlitsch 2007). Wood Frog larvae are more omnivorous than the larvae of closely related species (Janes 1939) and primarily consume detritus (Quammen and Durtsche 2003, Schiesari et al. 2009). Wood Frog larvae are also opportunistic predators of small aquatic invertebrates (Petranka and Kennedy 1999), salamander eggs (Petranka and Thomas 1995, Petranka et al. 1998, Burley et al. 2006), anuran eggs (Petranka et al. 1994), and smaller conspecific larvae (Bleakney 1958, Petranka and Thomas 1995, Jefferson et al. 2014). In a field experiment, the gut contents of larvae in closed-canopy ponds contained primarily detritus and small invertebrates (ostracods, protozoans, rotifers, copepods, and mollusks), whereas the gut content of larvae in open-canopy ponds contained primarily detritus and sediment (Schiesari et al. 2009).

A wide variety of predators eat adults and older juveniles, including snakes (e.g., Common Gartersnakes [*Thamnophis sirtalis*]), birds (e.g., Wild Turkeys [*Meleagris gallopavo*], hawks, and owls), and mammals (e.g., Raccoons [*Procyon lotor*], skunks, and American Minks [*Mustela vison*]; Vogt 1981, Rosenfield et al. 1984, Welch 1987, Berven 1990, Rittenhouse et al. 2009, reviewed by Dodd 2013). Recently metamorphosed individuals are consumed by spiders (Araneae), water snakes (*Nerodia* spp.), and gartersnakes (*Thamnophis* spp.; T. Rittenhouse, personal observation). Adult Wood Frogs are most vulnerable to predators when they are at breeding wetlands (Rittenhouse et al. 2009).

Eggs and larvae are reportedly palatable to many predators (Walters 1975, Formanowicz and Brodie 1982, Szuroczki and Richardson 2011). Eggs are consumed by invertebrates (e.g., Tricoptera larvae and leeches), other anurans (e.g., Green Frog [*Lithobates clamitans*] larvae), and caudates (e.g., *Ambystoma* larvae, adult Eastern Newts [*Notophthalmus viridescens*]; Walters 1975, Stein 1985, Petranka and Kennedy 1999, Trauth and Neal 2004). Similarly, predators of larvae include invertebrates (e.g., dragonfly naiads, damselfly naiads, Dytiscidae larvae, Belostomatidae, and arachnids) and other amphibians (Green Frog larvae, *Ambystoma* larvae, adult Eastern Newts), as well as birds and mammals (Walters 1975, Wilbur 1972, Formanowicz and Brodie 1982, Cortwright and Nelson 1990, Trauth et al. 2000, reviewed by Dodd 2013). Fishes are another potentially important predator of larvae, as Eaton et al. (2005) determined that 7.7 times more Wood Frogs metamorphosed from waterbodies that had experienced a winterkill of fishes in Alberta (Eaton et al. 2005). Larvae concentrated within receding wetlands are more vulnerable to aquatic predators and become available to terrestrial predators such as Raccoons and Wild Turkeys (T. Rittenhouse, personal observations). Not all attacks by predators result in death; often, larvae survive with injuries such as damaged tails or missing limbs (Blair and Wassersug 2000).

Adults and juveniles likely depend on their cryptic coloration to avoid detection by predators. However, when threatened, adults can display many antipredatory behaviors that include fleeing, crouching, chin-tucking, hiding, and walking (Blomquist and Hunter 2007). Controlled experiments further indicate that defense behaviors differ based on predator type. Adult Wood Frogs remained motionless more often during attacks by simulated aerial predators and fled more often during attacks by simulated ground predators (Blomquist and Hunter 2007). When attacked, Wood Frogs also emit a release call that may serve to startle predators (Formanowicz and Brodie 1979). There is evidence that Wood Frogs further emit noxious skin secretions that irritate small predators such as Northern Short-tailed Shrews (*Blarina brevicauda*; Formanowicz and Brodie 1979).

Larvae avoid predators in several ways; use of refugia to avoid detection, avoidance of areas where predators occur, and reduction in activity levels (Petranka and Hayes 1998, Relyea 2000b, Thiemann and Wassersug 2000, Chivers and Mirza 2001, Richardson 2001b, Awan and Smith 2007, Walston and Mullin 2007a, Fraker 2010, Szuroczki and Richardson 2012). In a laboratory study, immobilized larvae were almost four times less likely to be preyed on by dragonfly naiads (*Anax junius*) than were active larvae (Skelly 1994). However, immobility can increase susceptibility to parasitic infections (Szuroczki and Richardson 2012, but see Thiemann and Wassersug 2000), so larvae increase their activity levels when trematodes (e.g., *Echinostoma trivolvis*) contact their skin in an attempt to avoid the parasite (Taylor et al. 2004, Koprivnikar et al. 2014). The conflicting demands of low activity to avoid predators but higher activity to avoid parasites necessitate that a fine balance be struck by larvae to achieve high survival and reproductive success. Wood Frog larvae can also exhibit phenotypic plasticity in the presence of predators.

Variation in physical characteristics can increase the probability that larvae survive predatory attacks, but at the cost of growth. For example, in the presence of predators, larvae can develop deeper tail fins, shorter bodies, and a lower body mass than those reared without predators (Relyea 2001b, Relyea and Auld 2004). These morphological changes are the result of reduced food consumption, which slows development in the presence of predators. Furthermore, such morphological alterations in larvae can carry over into morphological differences in post-metamorphic frogs (i.e., larger hind limbs and forelimbs and narrower bodies; Relyea 2001b). Research is needed to understand how these morphological differences influence adult survival.

Many types of parasites infect Wood Frogs, including nematodes, trematodes, cestodes, leeches, flies, protists, and fungi (reviewed by Muzzal 2005 and Dodd 2013). Parasitic infections can reduce survival, growth, and development rates and increase malformation rates of Wood Frogs (Berven and Boltz 2001, Johnson and Sutherland 2003, Bolek and Janovy 2004, Taylor et al. 2004, Belden 2006, Koprivnikar et al. 2006a, Orlofske et al. 2013, 2017). Although Wood Frogs share many of the same parasites as other *Lithobates* species, their terrestrial habits as adults spare them from the intensity of infection from some parasites that favor more aquatic amphibians (Barta and Desser 1984, Desser and Barta 1984). Exposure to other stressors (e.g., contaminants and predators) in the presence of parasites can intensify the parasitic infection in Wood Frogs (Kiesecker 2002, Koprivnikar et al. 2007, Belden and Wojdak 2011, Szuroczki and Richardson 2012).

In Wisconsin, the following species of parasites have been documented in Wood Frogs: nematodes (*Cosmocercoides dukae* and other *Cosmocercoides* spp., *Oswaldocruzia pipiens*, *Rhabdias ranae*), trematodes (*Alaria mustelae*, *Fibricola texensis*, *Glypthelmins pennsylvaniensis*, *Haematoloechus medioplexus*, *H. varioplexus*), cestodes (*Mesocestoides* spp.), and blowflies (*Bufolucilia silvarum* and *B. elongata*; Williams and Taft 1980, Yoder and Coggins 1996, 2007, Bolek and Janovy 2004). Many additional Wood

Frog parasites are reported in the US and Canada (see Rankin 1945, Bouchard 1951, Cory and Manion 1953, Odlaug 1954, Najarian 1955, Baker 1979, Berven 1982a, Barta and Desser 1984, Nyman 1986, Muzzall and Peebles 1991, Delvinquier and Desser 1996, Mikaelian et al. 2000, Green et al. 2002, Tkach et al. 2006, Dare and Forbes 2009, Pulis et al. 2011, Gustafson et al. 2013). Although Wood Frogs infected with ranaviruses are documented in Wisconsin (Sekowska et al. 2014), mass mortality events are not yet known in the state. Infection of Wood Frogs by the chytrid fungus *Batrachochytrium dendrobatidis* (Bd) has not been documented in Wisconsin, but it has been documented in Minnesota (Rodriguez et al. 2009) and Michigan (Zellmer et al. 2008; see Conservation section).

CONSERVATION

Status: Globally, the Wood Frog is listed as a Least Concern species by the IUCN. It is considered common in Wisconsin by the WDNR. Long-term monitoring data from the Wisconsin Frog and Toad Survey of calling males indicate that the species is stable in the state. For example, between 1984 and 2016, Wood Frog occurrence at Wisconsin monitoring sites ranged from ca. 10% to 45% annually (Figure 126).

Populations: Most information on Wisconsin populations is inferred from surveys of calling males. While calling data can help elucidate the distribution of breeding populations or changes in male calling intensity over time (Mossman et al. 1998, Trenham et al. 2003), these data might not accurately represent population size or density. Several long-term population studies from other states have estimated population size (i.e., Maryland, Michigan, Missouri, New York, and Virginia; Berven 1990, 1995, 2009, Drake et al. 2015). Wood Frog population sizes in the Saginaw Forest of the University of Michigan were studied from 1980 to 2005 (with the exception of 1984; Howard and Luge 1985, Berven 2009). Howard and Kluge (1985) reported that the number of adults in this study area varied from 5,202 to 8,955 over four years (1980–1983). Berven (2009) reported a mean of 3,030 males (SE = 325) and 1,082 females (SE = 206) in the breeding population over 21 years (1985–2005).

Available research suggests that Wood Frog populations are dynamic and can experience substantial fluctuations over time (Berven 1995, 2009). For example, populations in Michigan (over 21 yrs), New York (over 7 yrs), and Missouri (over 6 yrs) varied by a factor of 10 (Berven 1990, 2009, Drake et al. 2015). Furthermore, juvenile production within these populations varied by a factor of 100. There is also a high degree of irregularity in survival among different age classes. Only a small fraction of larvae typically survives to metamorphosis (Seigel 1983, Semlitsch and Drake 2015). Similarly, Herreid and Kinney (1966) estimated that a mean of 96.3% of eggs and larvae died before metamorphosis at four ponds in Alaska, with a mean survival time of ca. 20 days. Despite this, some studies have reported that larvae survival can be high in the absence of predators (e.g., 65.7%; Seigel 1983). Wide variation in juvenile production likely contrib-

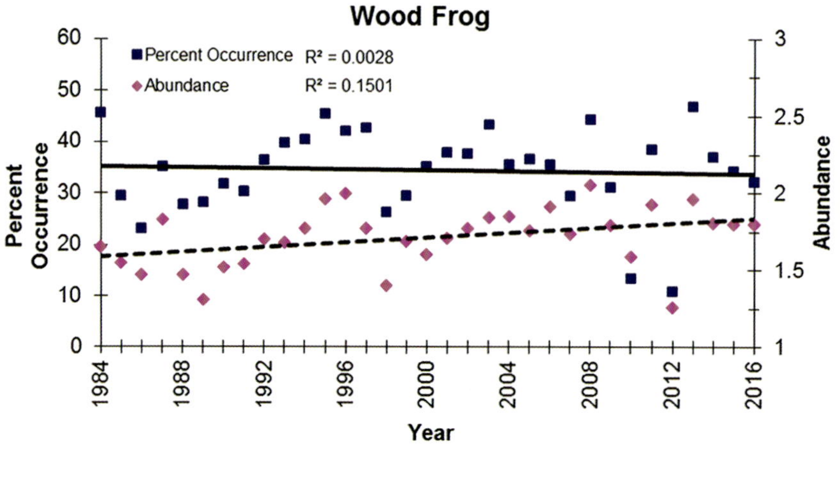

Figure 126. Long-term trend in occurrence (%) and abundance (rank) of Wood Frogs across Wisconsin Frog and Toad Survey statewide monitoring sites. See Introduction to Species Accounts chapter for details on this monitoring program (figure courtesy of WDNR).

utes to fluctuating adult numbers (Berven 1990, 1995, 2009, Berven and Grudzien 1990).

NATURAL HISTORY BOX: Sink Habitats

Female Wood Frogs will lay their eggs in many anthropogenic wetlands that may not meet the species' habitat requirements. These wetlands include cattle ponds, stormwater ponds, feral hog wallows, tire ruts, and roadside ditches. One common shortcoming of these anthropogenic wetlands is an inadequate hydroperiod for the complete development of offspring. In many places, including Wisconsin, snowmelt and rain fill roadside ditches in the spring. This was the case in northeastern Alberta, Canada, where I found more than 100 egg masses in a roadside ditch. In early spring the ditch was connected to a semipermanent wetland. During the summer, the water in the ditch receded, eventually cutting the developing larvae off from the wetland. Unfortunately, the ditch dried completely before any tadpoles were able to metamorphose that season, while the adjacent wetland did not dry. This is an example of how an anthropogenic wetland can serve as a sink habitat with the potential to compromise a population.

Jaimie L. Klemish, University of Lethbridge

Threats: Wood Frogs are affected by many general threats that influence other amphibian species in Wisconsin (see Introduction to Conservation and Management of Wisconsin's Amphibians and Reptiles chapter). Among these, impacts to preferred habitats are particularly important to consider. Studies suggest that Wood Frogs are good bioindicators of forest habitat quality (Knutson et al. 2000). Therefore, degradation and loss of forests due to anthropogenic activities such as logging and exurban development are of great concern to this species' persistence. Not only can these activities outright destroy large swaths of Wood Frog habitat, they can also fragment large habitat areas into smaller parcels. Fragmentation of habitat over many decades in Wisconsin has probably split Wood Frog populations into many small subpopulations that associate with individual wetlands. Each local subpopulation experiences high annual turnover (Trenham et al. 2003) and likely depends on immigration from neighboring subpopulations for long-term persistence (Harper et al. 2008). The ability of Wood Frogs to successfully disperse among wetlands is therefore critical. Yet the loss of suitable upland habitat surrounding wetlands can limit movement and greatly increase the risk of local population extinction (Harper et al. 2008). For example, juvenile and adult Wood Frogs show strong avoidance of logged clear-cut areas as they migrate from wetland to upland habitats (Patrick et al. 2006, Popescu and Hunter 2011, Rittenhouse and Semlitsch 2009). Extensive disturbance through clear-cutting of forested habitats could, therefore, affect the persistence of fragmented Wood Frog populations.

Although they avoid clear-cut areas in forests, Wood Frogs will sometimes use impacted landscapes and occasionally breed in anthropogenic waterbodies (e.g., human-made ponds, stormwater ponds, and ditches; Bishop et al. 2000, DiMauro and Hunter 2002, Vasconcelos and Calhoun 2006, Brand and Snodgrass 2010, Scheffers and Paszkowski 2016). However, these impacted habitats can act as ecological traps that do not support high offspring survival due to inadequate hydroperiods and the presence of predators or contaminants (DiMauro and Hunter 2002, Sanzo and Hecnar 2006, Vasconcelos and Calhoun 2006, Snodgrass et al. 2008, Dananay et al. 2015; Natural History Box: Sink Habitats).

Pollution from anthropogenic activities (e.g., mining, agriculture, manufacturing, and urbanization) is a significant threat to amphibians (Alford and Richards 1999, Blaustein et al. 2003, Croteau et al. 2008, Alfrod 2010). Wood Frog survival, growth, morphology, metamorphosis, sex determination, reproduction, occupancy of contaminated habitat, susceptibility to parasites, and ability to compete for resources and avoid predation are adversely affected by many contaminants that can occur in Wisconsin. Examples include pesticides (Berrill et al. 1995, 1998, Pauli et al. 1999, Relyea 2005a, 2005b, 2005c, 2012, Higley et al. 2013, Lanctôt et al. 2013, Christenson et al. 2014, Navarro-Martin et al. 2014), nitrogenous compounds (i.e., nitrates, nitrites, ammonia, ammonium nitrate; Marco et al. 1999, Griffis-Kyle 2005, Griffis-Kyle and Ritchie 2007, Burgett et al. 2007), road deicing salt (Sanzo and Hecnar 2006, Langhans et al. 2009, Dananay et al. 2015, Hall et al. 2017, Lambert et al. 2017a), metals and metalloids (Horne and Dunson 1995, Laposata and Dunson 1998, Loftin et al. 2012, Peles 2013, Boczulak et al. 2017), nanoparticles (Fong et al. 2016, Thompson et al. 2017a), polychlorinated biphenyls (PCBs; Savage et al. 2002), estrogenic and antiestrogenic compounds (Mackenzie et al. 2003, Tompsett et al. 2013), a pharmaceutical (e.g., sertraline; Carfagno and Fong 2014), cleaning solvent (e.g., tetrachloroethylene; McDaniel et al. 2004), and chemical reagents (e.g., thiosemicarbazide; Riley and Weil 1986). Furthermore, as habitats are often polluted by more than one contaminant, amphibians are frequently exposed to a mixture of impurities (Alford 2010).

Invasive and introduced species can negatively impact Wood Frogs if they are predators, competitors, or plants that degrade the quality of available habitat. For example, introduced fishes will consume Wood Frog larvae (Eaton et al. 2005), while adults avoid breeding in wetlands inhabited by fishes (Hopey and Petranka 1994). The negative impacts of predators such as fishes can also be enhanced when coupled with other stressors such as pollutants (Relyea 2005b). Besides fishes, introduced and invasive plant species that occur in Wisconsin adversely affect Wood Frogs. Dense stands of the nonnative Common Reed (*Phragmites australis*) can slow the development of sympatric larvae (Perez et al. 2013). Furthermore, Maerz et al. (2010) determined that detritus produced from decaying invasive plants, including Reed Canary Grass (*Phalaris arundinacea*), negatively impacted larval growth and development. Red Clover (*Trifolium pretense*), a cultivated and naturalized plant, exudes phytoestrogens from its roots that alter the sex ratio of Wood Frog populations and accelerate metamorphosis of male larvae (Lambert 2015).

Wood Frogs are also affected by several diseases that impact global amphibian populations (see Introduction to Conservation and Management of Wisconsin's Amphibians and Reptiles chapter). Mass Wood Frog mortality due to ranavirus infections has been documented in several states (Connecticut, Maine, Massachusetts, North Carolina, and North Dakota) and southern Ontario (Green et al. 2002, Greer et al. 2005, Duffus et al. 2008, Wheelwright et al. 2014). Some studies have further determined that a high portion of wetlands across a region can contain infected individuals. A study in Connecticut found infected Wood Frog larvae at 27 out of 28 study wetlands (O'Connor et al. 2016b). Similarly, Crespi et al. (2015) found infected Wood Frogs at 25 out of 27 breeding wetlands ($\bar{x}$ within-pond infection rate = 39%). Although the presence of infected frogs in a population is not always a precursor to mass mortality, some models indicate that isolated Wood Frog populations are particularly prone to extinction due to ranavirus-related mortality (Earl and Gray 2014). The prevalence and impacts of ranavirus infections on Wood Frogs in Wisconsin remain largely unknown. Although ranaviruses were identified as responsible for Wood Frog mortalities in Dane County during 2016 (WHISPers 2017) and detected on other amphibians in the state (Sekowska et al. 2014), they were not found on Wood Frogs assessed by Standish et al. (2018). More work is necessary to determine the level of threat that ranaviruses pose to Wisconsin populations of Wood Frogs.

The chytrid fungus *Batrachochytrium dendrobatidis* (Bd) is another amphibian pathogen of serious conservation concern. Past studies have detected Bd on other amphibians in Wisconsin but have not identified it on Wood Frogs in the state (Ouellet et al. 2005, Klemish et al. 2012, Standish et al. 2018). Despite this, the fungus has been found on Wood Frogs in nearby Minnesota (Rodriguez et al. 2009) and Michigan (Zellmer et al. 2008). During laboratory studies, Wood Frogs were more sensitive to Bd than some other amphibian species that occur in Wisconsin, including American Bullfrogs, Northern Leopard Frogs, Gray Treefrogs (*Hyla versicolor*), Spring Peepers, and Blue-spotted Salamanders (*Ambystoma laterale*; Searle et al. 2011, Gahl et al. 2012). Although Bd has not been detected on Wisconsin Wood Frogs yet, these less sensitive species could serve as reservoirs for Bd and facilitate its spread to sympatric Wood Frogs (Woodhams et al. 2008, Greenspan et al. 2012). Yet the threat of Bd infection appears to vary by season and even age. Among the age classes investigated, post-metamorphic frogs were reportedly more susceptible to infection than larvae (Greenspan et al. 2012, Ortiz-Santaliestra 2013). In New York and Maine, the highest infection rates in Wood Frogs were observed during the spring (Longcore et al. 2007, Lenker et al. 2014). Such variability makes it difficult to develop broad conservation strategies that protect Wood Frogs against Bd infection. Although Bd infection of Wood Frogs is documented, large mortality due to chytridiomycosis has not yet been confirmed (Longcore et al. 2007).

The potential impacts of climate change on Wood Frogs in Wisconsin have received limited attention. However, population models developed by Amburgey et al. (2018) suggest that Wood Frog population growth rates were affected by changes in temperature and temperature-precipitation interactions. Furthermore, successful migration and dispersal of amphibians among wetlands, which is required for the persistence of fragmented populations, can be affected by annual weather patterns (Peterman et al. 2013). Given that the progression of climate change is predicted to cause increasingly dramatic environmental shifts, including changes to temperature and moisture patterns, it has the potential to negatively impact Wood Frogs (see Past, Present, and Potential Future Landscapes of Wisconsin chapter).

Management: Many general management recommendations for wetland and woodland habitats could also benefit Wood Frogs (see Introduction to Conservation and Management of Wisconsin's Amphibians and Reptiles chapter). Wood Frogs breed in forested wetlands that are temporary or semipermanent but that also possess submersed vegetation and house few predators (Hopey and Petranka 1994, Knutson et al. 1999, DiMauro and Hunter 2002, Porej et al. 2004, Baldwin et al. 2006b, Vasconcelos and Calhoun 2006, Skidds et al. 2007, Karraker and Gibbs 2009, Herwig et al. 2013). Thus, it is likely important to implement management strategies that maintain the ephemeral nature of small wooded wetlands to discourage survival of introduced fishes. Construction of artificial wetlands and restoration of degraded wetlands could also benefit Wood Frogs. For example, Nedland et al. (2007) found that Wood Frogs used restored wetlands in Wisconsin. Despite this, the long-term breeding success and viability of populations that use restored wetlands in Wisconsin have not been assessed and require further investigation. Maintenance of connectivity among suitable breeding wetlands is likely also critical to the conservation of this species. Thus, management strategies should protect upland forests near breeding sites, ensure that forests possess a suitable tree canopy cover (≥30%), and maintain ground debris to promote proper moisture levels for Wood Frogs (see Habitat subsection; Churchill and Storey 1993, Gibbs 1998b, Baldwin et al. 2006a, Rittenhouse and Semlitsch 2007, 2009, Blomquist and Hunter 2010).

Caudata
Salamanders

Salamanders are members of the class Amphibia and order Caudata (Urodela). All living members of this order have a tail and teeth on both jaws (see Larson et al. 2006 for additional monophyletic characters). The earliest fossil species discovered that contains characteristics representative of modern salamanders existed ca. 170 mya (de Buffrénil et al. 2015). There are 766 recognized salamander species in the world as of October 2021 (https://AmphibiaWeb.org), which are placed in 10 families and 68 genera. The US has 200 salamander species, most of which occur in the southern and central Appalachian region (Kozak 2017). Four families, five genera, and seven species occur in Wisconsin, including Ambystomatidae (*Ambystoma*, three species), Plethodontidae (*Hemidactylium* and *Plethodon*, two species), Proteidae (*Necturus*, one species), and Salamandridae (*Notophthalmus*, one species).

Figure 127. Blue-spotted Salamander (*Ambystoma laterale*) (illustration by E. Wild).

Blue-spotted Salamander

Ambystoma laterale (Hallowell 1856)

Joshua M. Kapfer and Patrick C. Owen

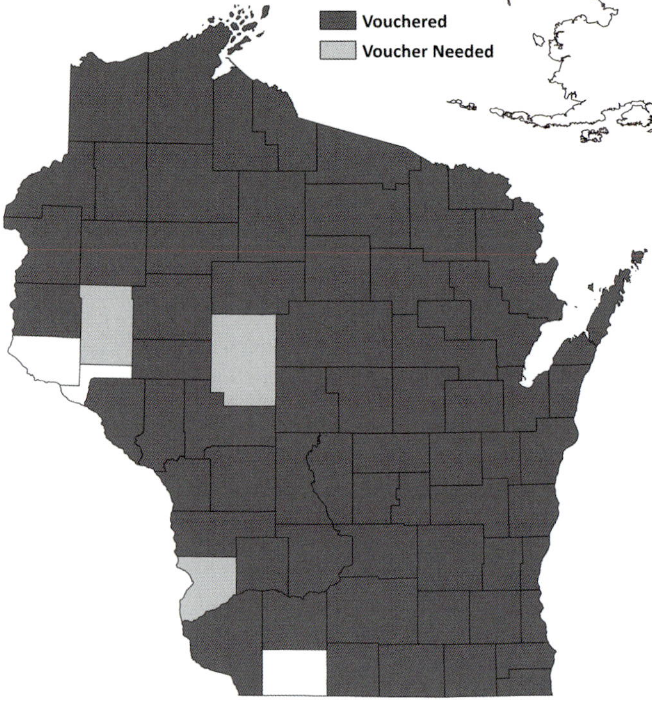

Blue-spotted Salamander (Dane County; photo by C. Raimond).

AT A GLANCE

Blue-spotted Salamanders are a small-to-midsized salamander that are easily identifiable by their blue spots or speckles over a dark gray to black background. They are one of Wisconsin's most widely distributed salamander species and are documented in all but a handful of counties. Blue-spotted Salamanders are found in a variety of habitats associated with ephemeral or semipermanent wetlands, where breeding occurs. These often include woodlands, where adults spend the nonbreeding season resting beneath downed logs, and occasionally drier woodland edges located adjacent to prairies. Blue-spotted Salamanders emerge from hibernation from March to April and shortly thereafter migrate to breeding wetlands where reproduction occurs. Eggs in Wisconsin reportedly hatch in three to four weeks, and larvae typically complete metamorphosis by mid- to late August. As adults, Blue-spotted Salamanders eat numerous types of invertebrates encountered beneath the moist logs where they reside, such as pill bugs, slugs, and earthworms. Larvae are voracious predators of aquatic invertebrates such as mosquito larvae but also fall victim to numerous predators, including large predaceous diving beetles and dragonfly naiads. During the breeding season, adults are likely consumed by various predators that hunt along wetland edges, such as snakes, wading birds, and Raccoons (*Procyon lotor*). During the rest of the year, when adults are beneath woody debris, they could be preyed upon by larger salamanders or small insectivorous mammals with which they share habitat (e.g., shrews). When threatened, this species will sometimes slowly wave its tail and secrete a white sticky substance from the base to deter potential predators. Specific observations of when Blue-spotted Salamanders return to their terrestrial hibernacula in Wisconsin are sparse, but this likely occurs by mid- to late October. Blue-spotted Salamanders are considered a common species by the WDNR.

SYSTEMATICS

Ambystoma laterale (Blue-spotted Salamander) is a close relative of *A. jeffersonianum* (Jefferson Salamander), which is not found in Wisconsin. The two species likely diverged during or prior to the Wisconsin Glacial Episode (Uzzel 1964, Menzel and Goellner 1976), although the genetic data of Demastes et al. (2007) indicate that they could have diverged several million years ago. There are two main *A. laterale* lineages, one represented by populations in eastern and central North America (as far west as Michigan and Illinois) and another represented by populations across portions of the Upper Midwest (Wisconsin, Iowa, Minnesota, and Manitoba; Demastes et al. 2007, Bi et al. 2008). Many individuals that outwardly resemble *A. laterale* encountered throughout their range are likely, in fact, members of a unisexual *Ambystoma* lineage (Natural History Box: Tremblay's Salamander, the Unisexual *Ambystoma* Complex, and Its Relationship to the Blue-spotted Salamander) that overlaps *A. laterale* in its suspected distribution. In Wisconsin, two other members of the genus *Ambystoma* occur: *A. maculatum* (Spotted Salamander) and *A. tigrinum* (Eastern Tiger Salamander).

DESCRIPTION

Adults: Blue-spotted Salamanders possess light blue to bluish-white spots or speckles over a dark gray to black background. Spot size and density across the head and body can vary dramatically among individuals and locations (Figure 128). They are small to medium in size relative to other salamander species in Wisconsin. Mean SVL of metamorphosed Wisconsin specimens housed in the MPM and UWZM was 5.9 cm (2.3 in, $n = 123$), while mean tail length was 4.4 cm (1.7 in, $n = 114$; note that only individuals with SVL of ≥5.0 cm [2.0 in] were considered adults based on reports of minimum size at maturity; see Reproduction and Development section). Living adults from an Ozaukee County population sampled annually from 2004 to 2016 had a mean SVL of 5.9 cm (2.3 in, $n = 32$) and a mean weight of 5.5 g (0.2 oz, $n = 25$; Joshua Kapfer, UW–Whitewater, unpublished data). The largest SVL among Wisconsin museum specimens measured was 8.2 cm (3.2 in, MPM 33267, collected on Basswood Island off of Ashland County in 1998).

NATURAL HISTORY BOX: Tremblay's Salamander, the Unisexual *Ambystoma* Complex, and Its Relationship to the Blue-spotted Salamander

Tremblay's Salamander "(*Ambystoma tremblayi*)" is a poorly understood but fascinating component of Wisconsin's herpetofauna. Unlike most vertebrates in the state, it is not a diploid organism (i.e., it does not possess only two sets of chromosomes). In fact, it is currently best described as a unisexual 'biotype' or 'genomotype' of the *Ambystoma* lineage that is triploid (i.e., possessing three sets of chromosomes; Lowcock et al. 1987, Bogart et al. 2007). It is not considered a distinct species, nor even a hybrid, but rather is referred to as the biotype "LLJ" because it often possesses two Blue-spotted Salamander genomes (LL) and one Jefferson Salamander genome (J) within the cell nucleus (reviewed by Petranka 1998).

Because it is unisexual (meaning populations have only one sex, and in this case typically female), its strategies for reproduction and propagation are not the same as those of bisexual salamander species (i.e., those with both males and females). On the contrary, females of the LLJ biotype incorporate new genomes by picking up spermatophores from males of other genetically compatible bisexual *Ambystoma* species (Bogart et al. 2007, Bogart et al. 2009). Sperm activate egg development with or without the addition of the sperm genome to the egg. Although often considered most compatible with the bisexual Blue-spotted and Jefferson Salamanders, unisexual *Ambystoma* may also possess genomes from the Streamside Salamander (*Ambystoma barbouri*), the Small-mouthed Salamander (*A. texanum*), and the Eastern Tiger Salamander in varying combinations. As such, their ploidy may range from diploid to pentaploid (five sets of chromosomes!), although all unisexuals possess at least one copy of the Blue-spotted Salamander genome. Overall, unisexual *Ambystoma* are currently best thought of as a distinct (although unnamed) *Ambystoma* that has also lost its original nuclear genome and replaced it by parasitizing nuclear genomes from other *Ambystoma* species (Bogart 2019). This genomic exchange with sexual species likely allows unisexuals to persist by preventing the accumulation of lethal mutations that would occur in an exclusively asexual population (Gibbs and Denton 2016).

The ecological interactions between unisexuals and their bisexual relatives are complex. The presence of LLJ biotypes may induce fitness costs to Blue-spotted Salamanders in several respects. In many communities, unisexuals can make up the majority of the *Ambystoma* individuals (Bogart and Klemens 2008). The LLJ biotype often parasitizes the spermatophores of male Blue-spotted Salamanders, which impacts the latter's reproductive success (Downs 1989a). Male Blue-spotted Salamanders may be able to detect unisexuals and are more likely to court and produce spermatophores in the presence of conspecifics presumably to reduce parasitism (Uzzell 1969), although unisexuals may counter this by having a lower motivational threshold at which they pick up spermatophores (Wilbur 1971). Unisexuals also occupy slightly different niches and may be able to use more marginal habitats than their bisexual relatives (Greenwald et al. 2016). While unisexuals regenerate damaged tissue more rapidly than bisexual *Ambystoma* (Saccucci et al. 2016), unisexuals have poorer dispersal ability (Denton et al. 2017). Unisexual larvae could even represent a competitive threat to bisexual larvae, particularly those of Blue-spotted Salamanders (Brodman and Krouse 2007), although Blue-spotted Salamander larvae have higher survivorship and a shorter larval period than LLJ larvae (Wilbur 1971).

While there are over 20 different unisexual *Ambystoma* biotypes, LLJ is probably the only biotype in Wisconsin. The physical appearance of an individual unisexual typically reflects the combination of nuclear genomes that it possesses, and this makes visual identification of unisexuals difficult (Petranka 1998). For the LLJ biotype, larvae and adults resemble Blue-spotted Salamanders (Downs 1989c, Petranka 1998), and this will obviously create problems when visually distinguishing a triploid LLJ unisexual from a true bisexual 'LL' Blue-spotted Salamander in Wisconsin. Generally, adult LLJ unisexuals are somewhat larger and longer than Blue-spotted Salamanders, with most mature unisexuals exceeding the maximum length and head width of Blue-spotted Salamanders (Downs 1989c, Lowcock et al. 1992). LLJ unisexuals are lighter and grayer in color than Blue-spotted Salamanders, with fewer and more widely scattered blue markings. Eggs and larvae of LLJ unisexuals are also larger than those of Blue-spotted Salamanders (Wilbur 1977c, Licht and Bogart 1987). Other signs that unisexuals may be present are large numbers of 'dead' egg masses and a sex ratio strongly skewed toward females (Petranka 1998, Bi et al. 2007). Beyond the morphological and ecological evidence, unisexuals of triploid or greater ploidy traditionally were identified by virtue of their larger cell sizes, which accommodate the greater amount of DNA, and it is possible to identify them through comparisons of red blood cell sizes (Homan et al. 2007). However, only genetic analysis can definitively provide the exact genomic identity of a biotype (e.g., Greenwald and Gibbs 2012, Bogart 2019).

Although unisexual *Ambystoma* salamanders occur in Wisconsin, they have received little attention. Several authors have published records of purported "Tremblay's Salamanders" (presumably the LLJ biotype) from scattered localities across the state (Briggs 1972, 1974, 1977, Sajdak and Buday 1979, Vogt 1981, Cochran and Lyons 1986). A handful of reported voucher specimens, identified by red blood cell measurements, exist from Ashland, Bayfield, Oneida, and Price Counties (Vogt [1981] also later reported this salamander from those counties) but also Florence, Fond du Lac, and Sheboygan Counties. Whole fluid specimens labeled as Tremblay's Salamanders from many of these locations are currently housed at MPM (e.g., MPM 6663, 6679, 6689, 6991, 13226, 16324, 16325, 16311). Cochran and Lyons (1986) further identified this salamander at a site in Vilas County based on red blood cell measurements and reported a photographic voucher (UWZM accession number 1984–147). Briggs (1972) referenced a Washburn County record but made no mention of associated voucher specimens. More recently, unisexual individuals were documented through genetic sampling in Calumet County (Rob Denton, Marian University, personal communication). Unisexuals in Wisconsin likely have the same L genome haplotype as a population of Blue spotted Salamanders and unisexuals sampled in northeastern Minnesota (Bi et al. 2008), and this haplotype is shared with individuals sampled in western Ontario, Indiana, southern Michigan, and western Ohio. However, the exact distribution and genomic composition of unisexuals in Wisconsin remain relatively unexplored.

Patrick C. Owen, University of Cincinnati–Blue Ash College

Figure 128. Variation in spot pattern of Blue-spotted Salamanders from four counties in Wisconsin. (a) Example of isolated spots with minimal speckling on the head (Marinette County; photo by T. Muehlfeld). (b) Example of an individual with greater speckling on the head, particularly around the mouths (Oneida County; photo by D. Brown). (c) Example of an individual with heavy blue speckling on the head and mouths. Note also the presence of larger blotches rather than isolated spots (Waukesha County; photo by J. Kapfer). (d) Example of an individual with heavy and extensive blue speckling across the body. Note the prevalence of fine speckling as opposed to larger spots or blotches (Bayfield County; photo by E. Wild).

Preadult stages: Eggs range from 0.15 to 0.25 cm in diameter (0.06 to 0.1 in), while the diameter of the gelatinous envelope that surrounds each egg ranges from 0.5 to 0.6 cm (0.2 to 0.24 in; Watermolen 1995). Eggs are generally laid in small clusters of four or less (but sometimes 7–40; Figure 129; Watermolen 1995). Larvae are fully aquatic, with bushy external gills. They possess a greenish-brown background color and are lightly mottled, with a broken yellow lateral stripe on both sides of the body and a tail fin that extends up onto the body almost to the head (Figure 130). Larvae also have five toes on the back feet (if present), digits that are rounded at the

Figure 129. Blue-spotted Salamander eggs (Fond du Lac County; photo by C. Raimond).

Figure 130. Larval Blue-spotted Salamander (Chippewa County; photo by R. Sajdak).

tips, <12 gill rakers on the third gill arch, and a head that is noticeably wider than the trunk (Petranka 1998). Blue-spotted Salamander larvae are small to medium in size relative to the larvae of other Wisconsin salamanders. Wisconsin specimens identified as larvae in the MPM and UWZM collections had a mean SVL of 2.1 cm (0.8 in), ranging up to 3.2 cm (1.3 in, $n = 75$), and a mean tail length of 1.4 cm (0.6 in), ranging up to 2.2 cm (0.9 in, $n = 64$). Recently metamorphosed individuals and slightly older juveniles appear similar to adults, although they often have brighter and proportionally larger blue spots.

Variation: The distribution, size, and density of the blue spots or speckles can vary across adults (see Adults subsection). Anecdotal observations further suggest that juveniles frequently have brighter and proportionally larger blue spots than adults (J. Kapfer, personal observation). There is no obvious difference in color or pattern by sex, but females may be slightly longer than males on average (Lowcock et al. 1992).

Diagnosis: Adults and metamorphosed juvenile Blue-spotted Salamanders have no dorsal or ventral fins along the tail musculature, no external gills or nasolabial grooves, five toes on the back feet, a stout body, and smooth dark skin overlaid with light blue or bluish-white markings. Adult Blue-spotted Salamanders are easily discernible from other salamander species in Wisconsin due to their blue to bluish-white spots or speckles. However, there are reportedly populations of unisexual salamanders in some counties of Wisconsin (i.e., "*Tremblay's Salamander*" [*Ambystoma "tremblayi"*]; Vogt 1981). It is particularly difficult to distinguish these unisexual *Ambystoma* from Blue-spotted Salamanders (see Natural History Box: Tremblay's Salamander, the Unisexual *Ambystoma* Complex, and Its Relationship to the Blue-spotted Salamander).

The eggs and larvae of Blue-spotted Salamanders can be difficult to differentiate from those of other ambystomatid salamanders. Yet the diameters of Blue-spotted Salamander eggs and gelatinous membranes combined are often smaller than eggs of Eastern Tiger Salamanders and Spotted Salamanders (Watermolen 1995). Blue-spotted Salamander eggs are also laid in clusters of fewer eggs than those of these other two species. Blue-spotted Salamander larvae are characterized by the presence of external gills and forelimbs. The dorsal fin of their tails extends onto their bodies, which are 3–7 cm TL (1.2–2.8 in), while their ground color is gray or greenish brown. There are no rows of distinct yellow spots visible and no prominent dark stripe through the eyes, although larvae are dark brown with faint yellow transverse bars, yellow midline blotches, and yellow lateral stripes. Larvae have five toes on their hind feet. Toes are rounded at the tip and do not possess a broad base. Generally, larvae of the Eastern Tiger Salamander can grow much larger (TL ≥7 cm [2.8 in]), with digits that have flattened bases and pointed tips and 14–24 gill rakers on the third gill arch (Petranka 1998). Larval Spotted Salamanders also have rounded digits and are <7 cm (2.8 in) in TL. However, older larvae have a row of small yellow spots and light lateral stripes on the dorsal surface (Petranka 1998).

DISTRIBUTION AND HABITAT

Global distribution: Blue-spotted Salamanders have one of the most northern distributions of any North American salamander and occur throughout much of the upper midwestern and northeastern parts of the continent (Petranka 1998, Demastes et al. 2007). Their distribution is centered within the Great Lakes region and Gulf of St. Lawrence but also extends south through New England and eastern New York and into New Jersey. In the Upper Midwest, Blue-spotted Salamanders are found in Wisconsin and northeastern Minnesota and across much of Michigan. A disjunct population occurs in northeastern Iowa. The Blue-spotted Salamander's continental distribution has been greatly influenced by recent glaciation events (Demastes et al. 2007). Their current range likely represents rapid colonization following the retreat of the Laurentide Ice Sheet, which began ca. 18,000 years ago. Subsequently, Blue-spotted Salamanders dispersed into their current range either from a single eastern refugium or from multiple refugia across the unglaciated portions of their distribution.

Distribution in Wisconsin: Blue-spotted Salamanders have the broadest distribution of any *Ambystoma* in Wisconsin. Of the published county records, voucher specimens could not be found for Clark, Crawford, and Dunn Counties (see the Appendix). Despite this, a possible voucher specimen for Clark County at the Field Museum of Natural History (FMNH 13036) has been identified as a member of the "*A. jeffersonianum* complex," and it possesses characteristics consistent with a small Blue-spotted Salamander (Alan Resetar, FMNH, personal communication). The reported voucher for Shawano County (MPM VZP308) is a poor-quality photograph that is difficult to use for independent confirmation of the species. Regardless, there is no reason to doubt the historic or current existence of Blue-spotted Salamanders in the counties with missing vouchers. The reported voucher for Wood County (MPM 2347; Suzuki 1951) is an old specimen in particularly poor shape, but it is noteworthy that at least one other specimen from this county exists (UWSP 4253).

Localities of possible occurrence: Currently, only three counties in Wisconsin have not produced published records of this species: Pierce, Pepin, and Lafayette. There is a high likelihood that Blue-spotted Salamanders occur in these counties given this salamander is documented in numerous adjacent locations.

Habitat: Few studies on Blue-spotted Salamander habitat selection in Wisconsin exist. Anecdotal observations suggest that Blue-spotted Salamanders associate with smaller ephemeral wetlands in wooded or semi-open habitats. Vogt (1981) observed them in northern mixed and boreal forests (Nicolet and Chequamegon National Forests), mixed hardwood forests in eastern Wisconsin (Manitowoc County), and the mesic forests of Northern Kettle Moraine State Forest. The salamanders also occur in semi-wooded pothole wetlands, even those adjacent to sandy oak-hickory savannas, in the Southern Kettle Moraine State Forest, and in floodplain forests associated with the Mississippi and Wisconsin Rivers (J. Kapfer, personal observation).

Detailed research on habitat preferences is more common from other states, and these generally corroborate observations from Wisconsin. It is noteworthy that many of these studies have likely focused on populations with various combinations of pure Blue-spotted Salamanders and polyploid unisexual individuals. Although most studies did not distinguish between these groups, our summary excludes information that specifically pertains only to unisexual *Ambystoma*. In southeastern Michigan, Blue-spotted Salamanders are typically found associated with ephemeral wetlands that also possess high tree canopy cover (Skelley et al. 1999). They are likewise more common in ponds surrounded by extensive woodlands in Maine (Guerry and Hunter 2002), but they also occupy upland forests and wet meadows (Ryan and Calhoun 2014). Ryan and Calhoun (2014) further reported that critical post-breeding terrestrial habitat for this species was concentrated within 152 m (499 ft) of wetland edges.

Blue-spotted Salamanders appear to select for certain microhabitats that are possibly identified through cues left by members of their own species (Belasen et al. 2013). In Maine, individuals used areas with greater coverage of ground debris, deeper leaf litter, and moister soil than other available locations (Ryan and Calhoun 2014). In southeastern Michigan, they preferred forest patches dominated by Red Maple trees (*Acer rubrum*) and associated leaf litter over patches dominated by Black Cherry trees (*Prunus serotina*; Belasen et al. 2013). Blue-spotted

Salamanders are apparently somewhat tolerant of sandy conditions and are sometimes found under logs in typically dry habitats in Wisconsin (savanna or deciduous hardwood edges; J. Kapfer, personal observation). Vogt (1981) frequently found them at the soil surface beneath rotting woody debris when active.

Blue-spotted Salamanders breed in ephemeral wooded wetlands that lack fish. Kutka and Bachmann (1990) found that smaller wetlands were selected for breeding in northern Wisconsin (depth <0.5 m [1.6 ft], diameter <20 m [65.6 ft]). After hatching, larvae remain in wetlands through metamorphosis, but the details of their habitat use in Wisconsin are not well-studied. Yet work in other states supports that they select microhabitats within wetlands based on the presence or absence of potential predators and competitors. For example, larval Blue-spotted Salamanders used vegetation as cover more often when larger predatory Eastern Tiger Salamander larvae were present (Brodman and Jaskula 2002). Blue-spotted Salamander larvae also modified activity levels and microhabitat used to avoid competition with other salamander larvae (Brodman and Krouse 2007). The habitat preferences of recently metamorphosed juveniles are largely unknown but presumably similar to adults. Adults and juveniles overwinter in upland habitats near breeding ponds. Vogt (1981) reportedly observed them 'hibernating' under rocks and logs near wetlands or at wetland peripheries in Wisconsin. Because they do not dig their own burrows (Parmelee 1990), they must hibernate in underground structures or holes that already exist. In fact, Carpenter (1953) found a Blue-spotted Salamander overwintering within an ant mound at a depth of 45.7–48.2 cm (18–19 in) in Michigan.

REPRODUCTION AND DEVELOPMENT

After emergence from hibernation in March and April, Blue-spotted Salamanders migrate in groups to ephemeral wetlands for reproduction (reviewed by Brodman 2005b). Courtship begins soon after arrival at wetlands, which Vogt (1981) observed on April 15 and April 24 in Wisconsin. Research in Illinois suggested that adults locate breeding partners via chemical and, perhaps, tactile cues (Storez 1969). These cues also help ensure that Blue-spotted Salamanders do not accidentally hybridize with similar species that use the same wetlands for reproduction (Storez 1969). Courtship behavior is complex (Storez 1969) and in Wisconsin was described as follows: "The male clasps the female with his forelegs just behind her forelimbs and proceeds to rub his snout over her head and the anterior portions of her body. At the same time there is much tail thrashing and squirming, with the pair often breaking the surface of the water. This alternates with periods of quiescence, when they will lie motionless on the bottom of the pond. This usually lasts for an hour or two. . . . The male eventually dismounts and crawls over the head of the female. She follows him with her nose close to his cloaca. He then deposits 1 to 3 spermatophores on the bottom of the pond. The female picks up the tip of the spermatophore containing the spermatozoa in her cloacal lips, and the eggs are fertilized internally" (Vogt 1981). Uzzell (1969) concluded that male reproductive success could be heavily influenced by intense male-male competition during the breeding season.

Blue-spotted Salamanders are explosive breeders, and the annual reproduction period can last from several days to several weeks (reviewed by Brodman 2005b). Mean number of ova per female ranges from 196 to 250 (reviewed by Brodman 2005b), and the eggs are laid singly or spread among small clusters. Vogt (1981) recorded that the number of eggs per cluster ranged from 1 to 35 in Wisconsin, while Watermolen (1995) reported up to 40 ova per cluster. Sacerdote and King (2009) suggested that deposition of single eggs or eggs in small clusters provides the surface area to volume ratio necessary for effective absorption of oxygen under hypoxic conditions. Eggs are usually adhered to vegetation along the bottom of ponds (Stille 1954b; Figure 129). Eggs found in Wisconsin hatch in three to four weeks, depending on environmental conditions, and larvae metamorphose in mid- to late August (Vogt 1981).

The embryos and larvae of many amphibian species, including Blue-spotted Salamanders, can exhibit dramatic contextual plasticity in morphology, developmental rates, and behavior. In Massachusetts,

Charney et al. (2014) found a gravid female Blue-spotted Salamander during a breeding migration that contained embryos that were developmentally similar to embryos of other species at 60–65 hours post-fertilization. This suggests either that embryo development initiated within the female prior to fertilization or that the female had stored sperm. Regardless, such rapid development of amphibian larvae in ephemeral ponds has adaptive significance (e.g., larvae develop faster to outgrow potential competitors and metamorphose quickly before the pond dries). Wilbur (1976) conducted laboratory experiments and found that Blue-spotted Salamander larvae may also undergo metamorphosis at various sizes, which is influenced by the larval density in the ponds where they live. He suggested that high larval densities lead to fewer resources available, resulting in slower growth and smaller size at metamorphosis. He also suggested that the opposite would be true for larval growth and size in lower-density populations (i.e., larvae would grow faster and achieve a larger size at metamorphosis). In Wisconsin, male Blue-spotted Salamanders reportedly mature at 4–7 cm SVL (1.6–2.8 in), and females mature at 5–7 cm SVL (2.0–2.8 in; Vogt 1981). According to a review by Brodman (2005b), Blue-spotted Salamanders become sexually mature in two yrs, when males attain a minimum SVL of 4.2–4.5 cm (1.65–1.77 in) and females reach a minimum SVL of 5.1–5.2 cm (2.01–2.05 in).

ACTIVITY

Spring emergence, which may be stimulated by warm spring rains, typically occurs in March or April in Wisconsin (Vogt 1981). Limited information exists on Blue-spotted Salamander spatial ecology because they are difficult to track for extended periods (Ryan et al. 2014a). However, it is probable that they move sparingly during most of the year, except during migrations for breeding. For example, most Blue-spotted Salamanders studied in Massachusetts migrated >100 m (328.1 ft) from winter habitat to breeding wetlands (Regosin et al. 2005). The median straight-line distance between tracked Blue-spotted Salamanders and their breeding ponds in a Maine population was 67 m (219.8 ft), but large variation among individuals was observed (range = 7–281 m [23–922 ft]; Ryan and Calhoun 2014). Lehtinen and Galatowitsch (2001) used data from Semlitsch (1998) to estimate that the greatest movement distance of this species was 284 m (931.8 ft). The exact time when Blue-spotted Salamanders enter hibernation in Wisconsin is unreported, although this presumably occurs in mid- to late autumn. Regosin et al. (2005) determined that 52%–55% of Blue-spotted Salamanders likely overwintered at a distance of >100 m (328 ft) from breeding ponds in Massachusetts. They also found that as distance from breeding pond decreased, sex ratios leaned toward males (i.e., a greater proportion of Blue-spotted Salamanders near breeding ponds were males).

PREY AND PREDATORS

In Wisconsin, adults and metamorphosed juveniles eat a variety of invertebrates. The stomach contents of individuals examined from Manitowoc County included "snails, earthworms, beetles, beetle larvae, springtails, and spiders" (Vogt 1981). Bolek (1997c) reported that the stomach contents of adults examined from Waukesha County were dominated, in order of decreasing abundance, by terrestrial isopods (i.e., "pill bugs"), gastropods (i.e., slugs), and oligochaetes (i.e., earthworms). Gilhen (1974) determined that slugs were the most common food item in adults but that aquatic insects were found in the stomachs of individuals from Nova Scotia when they were collected near breeding ponds. Brodman (2005b) observed that larval Blue-spotted Salamanders eat small aquatic invertebrates such as cladocerans, copepods, and larval mosquitoes. When they are larger, larvae also eat anuran tadpoles (Smith 1983) and smaller salamander larvae (Wilbur 1972, Yurewicz 2004).

Adult Blue-spotted Salamanders are vulnerable during the breeding season and could fall victim to numerous species that forage along wetland edges, such as snakes, wading predatory birds (e.g., herons and egrets [*Ardea* spp.]), and Raccoons (*Procyon lotor*). When not breeding, adult Blue-spotted Salamanders are typically found beneath ground debris and could presumably fall victim to small woodland mammals that use the same habitats (e.g., shrews [*Sorex* spp. and *Blarina* spp.]), as well as large invertebrates (e.g., spiders and crayfish). Terrestrial juveniles and adults employ defensive behaviors that include elevation, undulation, and/or thrashing of the tail (Brodie 1977). They may concurrently secrete a sticky white substance that is reportedly distasteful to predators (Brodie 1977). Larval individuals are taken by dragonfly naiads (*Anax* spp.), predaceous diving beetles (family Dytiscidae), and even larger salamander larvae, such as those of the Eastern Tiger Salamander (Jaskula and Brodman 2000, Yurewicz 2004). Larvae can avoid predation through modification of activity (e.g., reducing activity to be less noticeable by predators). Larvae can also modify their morphology, such as developing shorter bodies and deeper tail fins, in the presence of predatory dragonfly naiads (Yurewicz 2004). The parasite fauna associated with this species in Wisconsin has also been investigated by several authors (Coggins and Sajdak 1982, Bolek 1997c, Bolek and Coggins 2003, Yoder and Coggins 2007).

CONSERVATION

Status: Globally, the Blue-spotted Salamander is listed as a Least Concern species by the IUCN. It is considered a common species by the WDNR.

Populations: Extremely limited information is available on adult abundances and densities. Regosin et al. (2005) estimated that winter densities of adult Blue-spotted Salamanders in terrestrial habitats were 0.17–0.36/m^2 (0.02–0.03/ft^2). Estimates of larval densities are more common. Past work indicates that larval densities can be high, but large variability is reported based on habitat, environmental conditions, and species assemblages in a given area. For example, research on Isle Royale discovered that the density of larvae immediately after hatching can range from <20 to 158/m^2 (1.9 to 15.7/ft^2; Van Buskirk and Smith 1991). Over time, however, the competition among larvae apparently plays an important role in structuring populations and communities of salamanders, including Blue-spotted Salamanders (Wilbur 1971, Van Buskirk and Smith 1991). For example, Van Buskirk and Smith (1991) suggested that larval Blue-spotted Salamander populations experience density-dependent regulation (i.e., population size increases or decreases based on the density of individuals that must compete for limited resources). Interestingly, Van Buskirk and Smith (1991) did not actually find that increases in larval density resulted in decreased food availability during their research and suggested that larval densities were more heavily regulated through interference competition. In other words, individuals aggressively interfered with the ability of others to obtain resources (i.e., some larvae chased others away and did not allow them access to resources, despite the resources being plentiful). Other research on *Ambystoma* larvae has also found that larval competition within or between species occurs (including the Blue-spotted Salamander) and can influence populations (e.g., Wilbur 1971, Petranka 1989a). However, the effect of competition is highly variable and depends on the species present, their abundances, and the relative proportion of each species making up the community.

Threats: Many of the general threats that face numerous amphibians and reptiles are also applicable to Blue-spotted Salamanders, such as habitat loss and fragmentation, invasive species, disease, and ecotoxicology (see Introduction to Conservation and Management of Wisconsin's Amphibians and Reptiles chapter). Interestingly, this species appears somewhat tolerant of human disturbance and is even reported from suburban yards on occasion (Holman 2012). Yet Blue-spotted Salamanders require access to suitable breeding and overwintering habitats, which human disturbances can limit. For example, roads associated with disturbed landscapes likely have heavy impacts on this species. Houlahan and Findlay (2003) found that Blue-spotted Salamander presence was negatively correlated with road density at distances of 750 m and 3,000 m (2,461 and 9,843 ft) from wetlands in Canada. High mortality has also been documented in adults that cross roads during spring migrations to breeding ponds. Mazerolle (2004) encountered 234 Blue-spotted Salamanders on roads during an eight-year survey, and 39.8% of these were dead, presumably due to vehicular strike. This species also requires environmental conditions and habitat structure that can be inadequate in disturbed landscapes. For example, they are frequently found in woodlands with high canopy cover and ample downed woody litter. Accordingly, timber harvest or silvicultural practices that alter these characteristics or increase light penetration and leave bare soil can negatively affect Blue-spotted Salamanders (deMaynadier and Hunter 1998, Thompson et al. 2003). Not surprisingly, Windmiller et al. (2008) also found drastic population declines immediately after removal of woodland habitat adjacent to breeding ponds in Massachusetts (although these populations had rebounded by four years after tree harvest). Blue-spotted Salamander abundance also increased as distance from harvested forest edge increased (deMaynadier and Hunter 1998).

The negative impact of environmental toxicants is an important amphibian conservation issue, but few studies have investigated the impacts on Blue-spotted Salamanders. Relyea and Jones (2009) reported laboratory mortality of larvae exposed to concentrations of glyphosate-based herbicides ($LC90_{96\text{-}h} = 3.3$ mg acid equivalents/L [0.0004 oz acid equivalents/gal]; i.e., 3.3 mg acid equivalents per liter of the herbicide was lethal to 90% of exposed larvae after 96 hours). In addition, Kutka (1994) found that Blue-spotted Salamander larvae were affected by alterations in pH levels. For example, they increased swimming activity levels in water with a pH that ranged from 4.1 to 6.5, but larvae exhibited "near inactivity" at a pH of 4.0. Excessive or reduced activity could affect the ability of larvae to avoid predators. Kutka (1988) also reported that Blue-spotted Salamanders and Spotted Salamanders were less sensitive to low pH than Eastern Tiger Salamanders, which could partially explain why the former two species are found in slightly acidic woodland wetlands of northern Wisconsin. Impacts of exposure to UVB radiation are reported to be minimal for Blue-spotted Salamander eggs and larvae (Crump et al. 1999).

The pathogenic chytrid fungus *Batrachochytrium dendrobatidis* (Bd) is a threat to amphibian populations globally, but limited information exists on how it affects Blue-spotted Salamanders. A study investigating the historic presence of this fungus among amphibians found that none of the 120 Blue-spotted Salamanders sampled from Canada in 1999 and 2000 carried the fungus (Ouellet et al. 2005). Yet Standish et al. (2018) detected the presence of both Bd and ranaviruses on a handful of Blue-spotted Salamanders sampled in Wisconsin. Gahl et al. (2012) determined that Blue-spotted Salamanders experimentally exposed to Bd experienced no mortality and suggested that this species is a possible vector of the disease (i.e., it does not often fall victim to the disease but facilitates spread of the disease to susceptible species). Climate change is another critical conservation concern for most amphibians, although studies specific to Blue-spotted Salamanders are limited. A multidecade comparison of breeding periods in northern Wisconsin found that peak breeding activity occurred approximately 12 days earlier in the 2000s than in the 1990s, which the authors attributed to climate change (Donner et al. 2015). A 14-year study in Indiana indicated that temperature and precipitation influenced this species and that it responded favorably to cooler and wetter conditions (Brodman 2009). Thus, climate change that leads to warmer and drier conditions in Wisconsin could influence activity patterns and local habitat suitability for Blue-spotted Salamanders.

Management: Presumably, general management concepts relevant to wetlands and woodlands apply to Blue-Spotted Salamanders, including strategies that maintain habitat connectivity and diverse habitat structure (see Introduction to Conservation and Management of Wisconsin's Amphibians and Reptiles chapter). Regosin et al. (2005) suggested that management of "pool breeding, forest dwelling amphibians" such as this species should focus on maintenance of woodlands adjacent to breeding wetlands (in a 200 m [656 ft] buffer around wetlands) and in uplands acting as corridors that connect wetlands. The authors did, however, suggest flexibility in management recommendations due to great variability in the suitability of conditions across species, locations, and time. In other words, sites with Blue-spotted Salamanders are rarely the same across space and time, which makes it difficult to create effective conservation guidelines at the regional scale. The adoption of appropriate urban land use practices by local and state planning agencies may be particularly helpful in conserving this and many other species (Calhoun et al. 2005).

Rigorous investigations of the best habitat management practices for Blue-spotted Salamanders in

Wisconsin are extremely sparse. Brodman (2005b) cited a personal communication with J. M. Jaskula that indicates that 147 m (482 ft) of "core upland habitat" around a breeding pond were required for a Wisconsin population under investigation. Protection of both suitable upland woods and breeding wetlands is likely critical. Brodman (2009) recommended that wetland complexes of 15 or more wetlands per cluster and undisturbed connecting upland habitat are important to the conservation of amphibian communities in Indiana, which included the Blue-spotted Salamander. He also suggested that greater variation in wetland hydrology and landscape-level habitat heterogeneity are important considerations. For example, maintenance of wetland clusters with a mix of permanent, semipermanent, and ephemeral wetlands (with intact native upland vegetation) resulted in higher amphibian abundance and diversity at a restoration site in Indiana (Brodman et al. 2006). The apparent sensitivity of Blue-spotted Salamanders to changes in tree species composition (Belasen et al. 2013) indicates that problems can arise when silvicultural practices alter woodland tree communities. Therefore, while general forest protection is important to salamanders, it is also important to manage tree community composition (see Habitat subsection).

Although wetland restoration and creation are sometimes touted as viable conservation strategies, more research is required to assess their effectiveness. Work in Minnesota that compared the amphibians present in reference (i.e., natural) wetlands to restored wetlands found that although some species colonized restored locations, Blue-spotted Salamanders did not (Lehtinen and Galatowitsch 2001). The authors suggested that distance separating restored wetlands from natural wetlands impacted the colonization ability of species with low mobility, such as Blue-spotted Salamanders. However, Blue-spotted Salamander eggs reportedly have the ability to survive hypoxia better than the eggs of some other salamanders, such as the Spotted Salamander (Sacerdote and King 2009). Therefore, Blue-spotted Salamanders may be able to persist in disturbed wetlands or newly created wetlands if they can be colonized.

Spotted Salamander

Ambystoma maculatum (Shaw 1802)

Robert D. Brodman

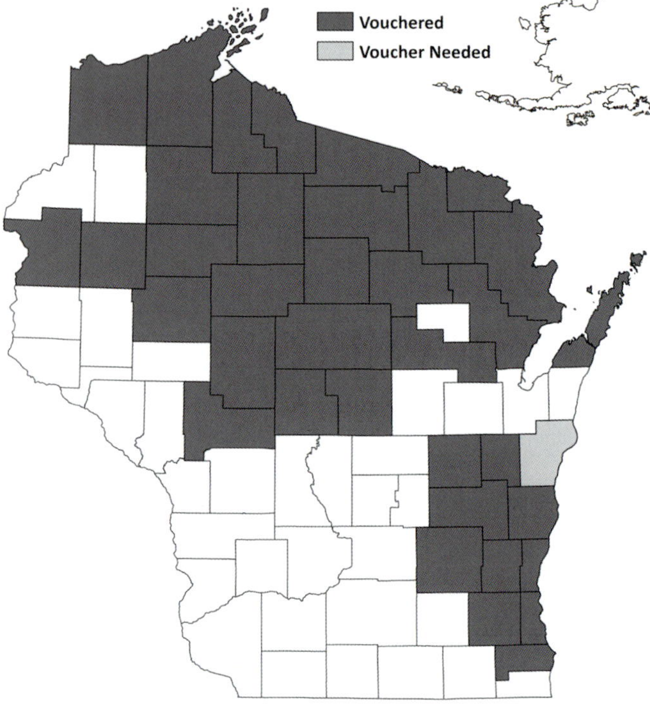

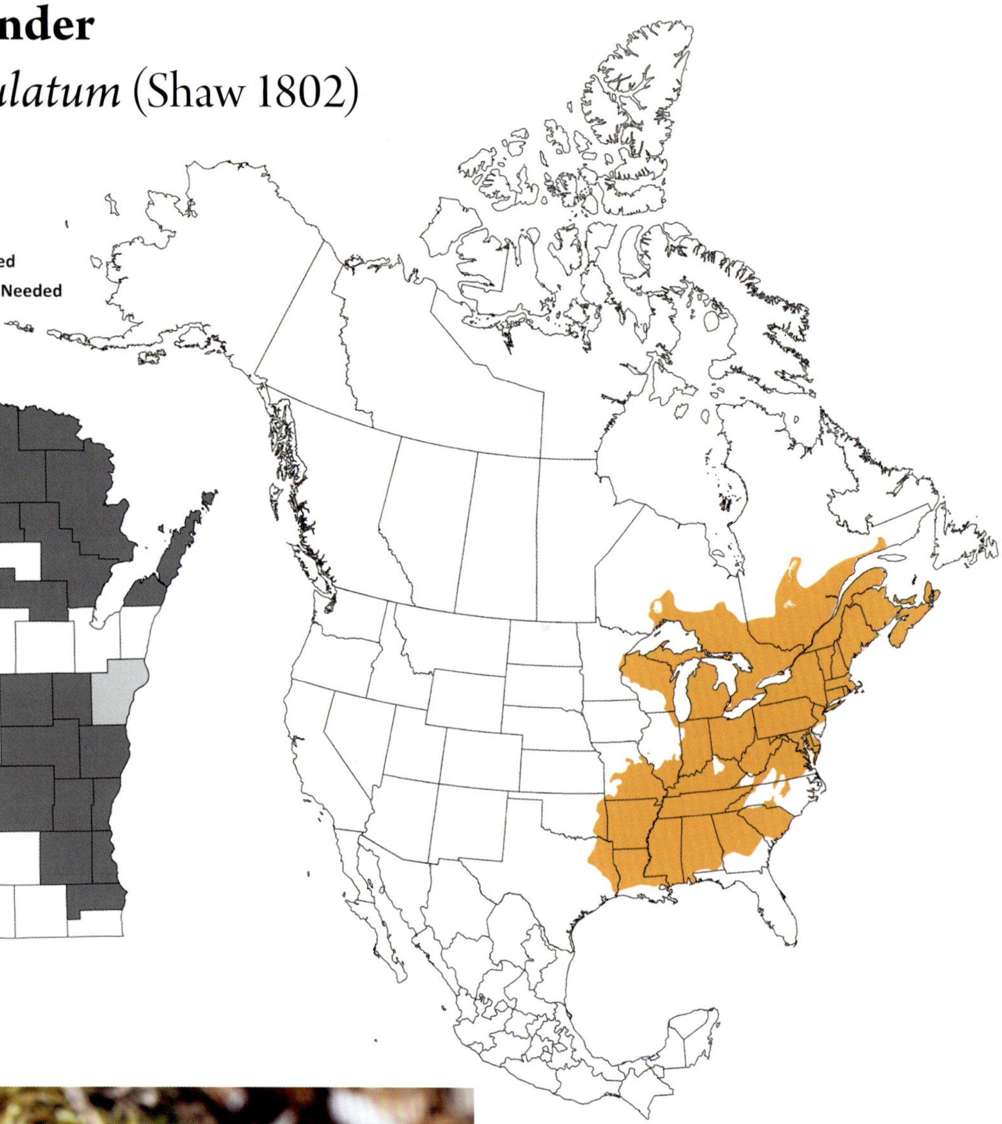

Adult Spotted Salamander (Portage County; photo by T. Muehlfeld).

AT A GLANCE

Spotted Salamanders are a medium-to-large-sized species. Adults are dark, with two rows of distinct isolated yellow to orange spots along the back. Spotted Salamanders are often found in northern Wisconsin but also occur in some southeastern counties. Adults typically occupy woodlands, where they spend much of their time below fallen woody debris. Breeding occurs in wooded wetlands shortly after emergence from hibernation in the spring (March and April), sometimes when ice is still present around wetland peripheries. Preferred breeding wetlands are ephemeral, which ensures that they cannot be colonized by fishes, a major predator of salamander larvae. Spotted Salamander eggs form an interesting symbiotic relationship with algae, whose photosynthetic activity increases oxygen availability to eggs and thus enhances the developmental rates of embryos. Larvae are fully aquatic, with bushy external gills, but they typically metamorphose in late summer and disperse into adjacent uplands. Spotted Salamanders eat a wide range of invertebrates as both larvae and adults, but they also fall victim to a host of invertebrate and vertebrate predators, ranging from caddisflies to birds, depending on developmental stage. From late August through October, adults and metamorphosed juveniles migrate to terrestrial overwintering sites and hibernate in burrows abandoned by other organisms. Spotted Salamanders are one of the iconic amphibians associated with vernal pools and ephemeral wetlands in mature forests throughout midwestern and eastern North America. They serves as a bioindicator of ecosystem health and an umbrella species for the protection of ephemeral and seasonal wetlands. Spotted Salamanders are currently considered a common species by the WDNR, but loss of suitable woodland habitat in Wisconsin is a serious conservation concern.

SYSTEMATICS

There has been little (if any) debate about the taxonomic position of *Ambystoma maculatum* (Spotted Salamander) for many decades (reviewed by Anderson 1967), and it has been the subject of limited phylogenetic investigation. Molecular phylogenetic studies suggest that there is a coastal clade in the eastern US and an interior continental clade that includes Wisconsin populations (Phillips 1994, Zamudio and Savage 2003). There are no subspecies of *Ambystoma maculatum*, but two other members of the genus *Ambystoma* occur in Wisconsin: *A. laterale* (Blue-spotted Salamander) and *A. tigrinum* (Eastern Tiger Salamander).

DESCRIPTION

Adults: Spotted Salamanders are so named because two haphazard rows of rounded yellow to orange spots run along their dark dorsal surface (Pope 1944, Downs 1989b). The number of spots can range from 8 to 45 (Bishop 1943, Minton 2001). The sides and ventral surface are slate gray, and the number of costal grooves varies from 11 to 13 (Smith 1961, Downs 1989b, Minton 2001). They are a medium-sized to large species that is robust, with a wide head (Downs 1989b). The TL of most adults is 15–20 cm (6–8 in), with the tail length measuring 44%–52% of the TL (Downs 1989b, Minton 2001). The mean SVL of 38 mature Wisconsin specimens in the MPM collection was 7.1 cm (2.8 in; note that only specimens ≥6.0 cm SVL were included, following Windmiller 1996).

Preadult stages: Eggs are spherical, generally bicolored (brown to cream), and each egg is enclosed in two gelatinous membranes (Figure 131; Watermolen

Figure 131. Spotted Salamander eggs (Fond du Lac County; photo by C. Raimond).

Figure 132. Spotted Salamander larvae and juveniles (Blount County, Tennessee; photo by R. Brodman). (a) Larva one month after hatching. (b) Recently metamorphosed individual. (c) Juvenile forming spots three months after metamorphosis.

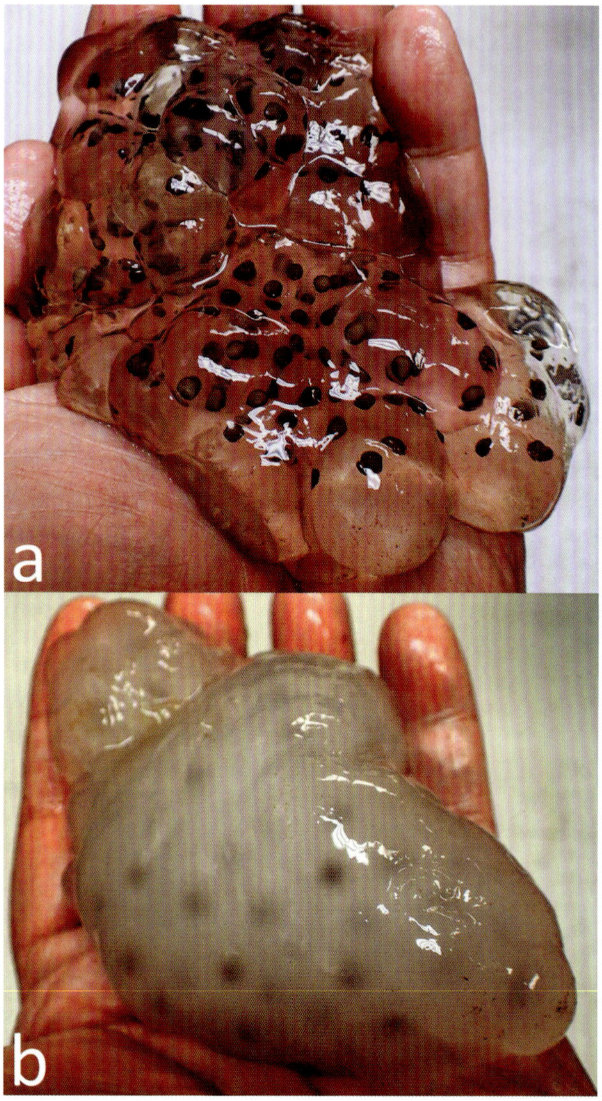

Figure 133. Variation in Spotted Salamander egg masses (Blount County, Tennessee; photo by R. Brodman). (a) An egg mass surrounded by clear jelly. (b) An egg mass surrounded by milky jelly.

1995). They range from 0.18 to 0.28 cm (0.07 to 0.11 in) in diameter and are frequently deposited in larger ovoid-shaped masses of several hundred eggs, although they are sometimes deposited individually (reviewed by Petranka 1998). After hatching, the larvae are uniformly reticulated, with dark mottling on the body, a light belly, and lightly spotted fins (Brandon 1961, Minton 1972; Figure 132). Hatchlings are ca. 1.2–1.7 cm TL (0.5–0.67 in) and have balancers and rudimentary forelimbs (DuShane and Hutchinson 1944, Brandon 1964). At metamorphosis they have greenish-yellow lichen-like markings (Figure 132), which will develop into yellow spots during the first year after metamorphosis (Minton 1972, Petranka 1998; Figure 132). Metamorphosed juveniles and subadults appear similar to adults but have proportionally longer legs and shorter tails (Smith 1961).

Variation: Spotted Salamanders can produce either clear or milky egg masses (Figure 133), depending on the glycoprotein present in the outer jelly layer

Figure 134. Spot pattern variation in adult Spotted Salamanders (Blount County, Tennessee; photo by R. Brodman). (a) An adult with yellow spots on its head. (b) An adult with orange spots on its head.

(Hardy and Lucas 1991, Ruth et al. 1993). The spots on the head of adults can range from yellow or orange (Figure 134), but the variation among populations does not have distinct geographic patterns (Petranka 1998). Some variation in adult size and color/pattern reportedly occurs by sex. Mean values calculated from published literature indicate that adult females have larger SVLs (9.2 cm [3.6 in], $n = 11$) than adult males (8.3 cm [3.3 in], $n = 23$; Downs 1989b, Minton 2001, Blackwell et al. 2003). These sources further reported that adult females have relatively shorter tails than adult males. In western North Carolina (not confirmed elsewhere), males have a higher proportion of their back covered in spots than females and that the spots are more chromatic in males but brighter in females (Morgan et al. 2014).

Diagnosis: Adult Spotted Salamanders are characterized by a lack of dorsal and ventral fins along the tail musculature; smooth skin, with a stout robust body and no nasolabial grooves; and a body and tail covered with a series of dorsolaterally symmetrical yellow to orange spots. Adult Spotted Salamanders can be confused with Eastern Tiger Salamanders in Wisconsin, but the latter has large yellow spots or blotches that are irregular in shape, are not arranged in rows, and also appear on the ventral surface. Furthermore, Eastern Tiger Salamanders have a yellow lower jaw, a broader snout than that of Spotted Salamanders, and flat pointed toes.

The eggs and larvae of salamanders in the genus *Ambystoma* can be difficult to differentiate in the wild. Spotted Salamander eggs are surrounded with two gelatinous envelopes, similar to those of Blue-spotted Salamanders, while Eastern Tiger Salamander eggs possess three (reviewed by Watermolen 1995). The diameter of an individual Spotted Salamander egg with its gelatinous envelopes included is larger than that of Blue-spotted Salamanders, while the former also deposits clusters with a greater number of eggs than the latter (Watermolen 1995). Unlike the eggs of other salamanders, some Spotted Salamander egg clusters can appear milky or greenish if covered in symbiotic algae (see Reproduction and Development section).

Spotted Salamander larvae possess external gills and forelimbs. Their dorsal tail fin does not extend up onto their back, which has a gray, greenish, or brown ground color and lacks distinct rows of spots (on younger individuals). They have no prominent dark stripe through their eyes, and the five toes on each hind foot are rounded at the tip and not broad at the base. They are generally 4.5–7.0 cm TL (1.77–2.76 in), with dark mottling on the body and fins. The tail fins are not transparent. Rows of small distinct and yellowish dorsal spots are present in older larvae (Petranka 1998). Larval Spotted Salamanders can reach larger sizes than larval Blue-spotted Salamanders but do not grow as large as Eastern Tiger Salamanders. Furthermore, Spotted Salamander larvae have rounded digits when viewed in cross section, which are flattened at the base in Eastern Tiger Salamanders, and mottled (not transparent) tail fins (Petranka 1998).

DISTRIBUTION AND HABITAT

Global distribution: Spotted Salamanders are distributed throughout the lowland hardwood and mixed coniferous-deciduous forests of the eastern US and southeastern Canada from Nova Scotia and central Ontario to extreme eastern Minnesota and south to southern Texas and eastern Georgia.

Distribution in Wisconsin: Spotted Salamanders are primarily found in the forested regions of Wisconsin, from Polk and Clark Counties in the northwest to Washington and possibly Racine Counties in the southeast (see the Appendix). Of the published county records, a voucher specimen is missing for Manitowoc County (note: specimens USNM 260904-07 have been identified as both Spotted Salamanders and Blue-spotted Salamanders [*Ambystoma laterale*] but are currently considered the latter; Kenneth Tighe, USNM, personal communication). Although widespread in northern Wisconsin north of the vegetative tension zone (Curtis 1959), they are less common in the southeast, where populations appear to be isolated and threatened by development (Casper 1998). Furthermore, several county records in southeastern Wisconsin are based on old specimens, such as Milwaukee County (MPM 2470, collected in 1934; Casper 1996) and Racine County (USNM 3938, collected in 1858; Kapfer et al. 2015), where the current presence of this species requires additional confirmation (see discussion in Kapfer et al. 2015).

Localities of possible occurrence: Given current records and gaps in the known state distribution, Spotted Salamanders could occur in Burnett, Washburn, St. Croix, and Eau Claire Counties in the northwest and Kewaunee, Brown, Outagamie, Waupaca, and Menominee Counties in east-central Wisconsin. They likely had a more widespread distribution in southeastern Wisconsin prior to the historical disturbance of deciduous forests in that region. Cahn (1929) said it was once "not uncommon locally, though relatively seldom seen" in Waukesha County, where it is currently known to occur but does not appear widespread. This, in conjunction with the existence of old specimens (Milwaukee and Racine Counties) and more recent documentation (Dodge, Washington, and Ozaukee Counties) in the region, lends support to its past (and possibly modern) presence in adjacent counties, such as Jefferson, Walworth, and Kenosha (see also review of Racine County occurrence in Kapfer et al. 2015). Past documentation immediately across the Illinois border in Cook, Lake, and Will Counties (Necker 1939b) and additional recent confirmation in Will County (Anton 1999) further support the possibility that Spotted Salamanders were historically more widespread in southeastern Wisconsin than current records indicate.

Habitat: In Wisconsin, this species associates with woodlands and forests, including mesic forests and northern conifer swamps (Vogt 1981; see Amphibians and Reptiles in Wisconsin Plant Communities chapter). Optimal habitats likely possess a mixed composition of mature hardwood tree species, abundant fallen logs, deep leaf litter, mammal burrows, and vernal pools (Downs 1989b, Guerry and Hunter 2002, Babbitt et al. 2003, Faccio 2003, Homan et al. 2004, Porej et al. 2004, Montieth and Paton 2006). Adults and juveniles prefer forest soils with leaf litter more than grassland soils (Rittenhouse et al. 2004). Outside of the breeding season, Spotted Salamanders spend 72%–91% of their time underground, 7%–21% in or under decaying logs, and up to 7% of their time under wet leaf litter (Kleeberger and Werner 1983, Faccio 2003). They do not dig their own tunnels; instead, adults and juveniles use existing mammal burrows and runways (Gordon 1968, Semlitsch 1983a, Faccio 2003). For example, Spotted Salamanders in New York frequently occupied shallow burrows and leaf litter runways created by Northern Short-tailed Shrews (*Blarina brevicauda*) during the active season (Madison 1997). Adults appear to show fidelity for specific summer burrows across years (Whitford and Vinegar 1966, Shoop 1968, Williams 1973, Downs 1989b). They will aggressively defend their burrows against intruding salamanders (Ducey and Ritsema 1988, Walls 1990, Ducey and Heuer 1991, Ducey et al. 1994). Recently metamorphosed juveniles also defend territories and attack intruders (Nussbaum et al. 2016).

The preferred breeding habitats of Spotted Salamanders are wooded pools that remain fishless.

These wooded wetlands are often ephemeral and typically only hold water from spring through early August but possibly into November (Baldwin et al. 2006b, Petranka et al. 2007, Skidds et al. 2007). Casper (2003) conducted habitat association analyses and found that greater snowfall and mean snow depth in March were related to Spotted Salamander presence in Wisconsin. He suggested that spring snowmelt was an important contributor to water availability and hydroperiod length in ephemeral wooded wetlands that Spotted Salamanders use for breeding. Breeding wetlands assessed in Wisconsin by Kutka and Bachmann (1990) were also small, with a depth of <0.5 m (1.6 ft) and a diameter of <20 m (65.6 ft; Kutka and Bachmann 1990). After hatching, the aquatic larvae remain in breeding ponds but are apparently more successful in those that are open canopy vs. closed canopy (Yurewicz 2004). Larvae further require underwater debris in their aquatic habitats that provides cover from intraguild predators, such as larvae of Eastern Tiger Salamanders (Stenhouse et al. 1983, Stenhouse 1987, Brodman and Jaskula 2002, Brodman 2004). During winter, Spotted Salamanders hibernate on land in mammal burrows that extend below the frost line, such as those of rodents (e.g., *Peromyscus* spp.; Madison 1997, Faccio 2003, Savage and Zamudio 2005).

REPRODUCTION AND DEVELOPMENT

Spotted Salamanders breed in March and April after emergence from hibernation and migration to breeding sites on humid or rainy nights when air temperature exceeds 10°C (50°F; Blanchard 1930b, Bauldauf 1952, Seibert and Brandon 1960, Shoop 1965b, Vogt 1981, Stangel 1988, Downs 1989b, Brodman 1995, Minton 2001). Adults often use the same breeding wetlands from year to year (Vasconcelos and Calhoun 2004), and individuals migrate considerable distances from overwintering to breeding sites in the spring (see Activity section). Males usually arrive at breeding sites one to six days prior to females (Hillis 1977, Douglas 1979), and males that arrive earliest tend to sire more offspring (Tennessen and Zamudio 2003). Females skip reproduction in some years, and it is estimated that 30%–90% of a population's adult females breed in a given year (Phillips and Sexton 1989, Kinkead and Otis 2007).

Spotted Salamander courtship occurs in shallow water (Downs 1989b) and often leads to intense male-male competition. For example, groups of up to 50 reproductive males will nuzzle a single female and deposit spermatophores, which females pick up with their cloacal lips for internal egg fertilization (Wright and Allen 1909, Breder 1927). Males attempt to outcompete other males by producing more spermatophores and by capping the spermatophores of rival males with their own (Arnold 1976, 1977). A single female may pick up spermatophores from several males, and embryos within egg clutches can be sired by two to eight males (Tennessen and Zamudio 2003, Myers and Zamudio 2004).

Spotted Salamanders in northern locations are explosive breeders, and single populations can produce thousands of eggs in as few as three days (Wilson 1976, Talentino and Landre 1991, Brodman 1995). The intensity of breeding activity is so great that most individuals do not feed until after reproduction is complete (Pope 1944, Smallwood 1928). Females take 30–60 minutes to deposit each egg mass and may deposit one to four egg masses over a 48-hour period (Bishop 1941b, Wilbur 1977b, Brodman 1995). Davis et al. (2018) found a strong correlation between adult abundance and egg mass abundance, suggesting one egg mass per adult female. Egg masses are globular and typically contain 50–150 eggs (Trauth et al. 1995, Brodman 2002), although eggs are sometimes deposited singly (reviewed by Petranka 1998).

Females prefer to lay eggs in water with dense submerged vegetation and fallen branches, which are often used for egg mass attachment (Breder 1927, Bishop 1941b, Kern et al. 2013), at water depths of 4–25 cm (1.6–10.0 in; Brodman 1995, Gibson and Merkle 2005). Spotted Salamanders also select locations for egg deposition based on environmental factors. For example, Davis et al. (2018) reported that wetland hydrology and peak mean water temperature of 15°C (59°F) best explained egg mass abundance in breeding ponds. Davis et al. (2018) also found a positive association between the abundance of Spotted Salamander egg masses and the abundance of two co-occurring species, Jefferson Salamander (*Ambystoma jeffersonianum*) and Wood

Frog (*Lithobates sylvaticus*). This suggests shared habitat preferences. After deposition, the gelatinous layers around the eggs absorb water until the mass is 4–15 cm (1.6–5.9 in) wide (Bishop 1941b, Downs 1989b) and firm (Hardy and Lucas 1991). Egg masses sometimes take on a green hue due to colonization by a chlamydomonad algae (*Oophila amblystomati*; Gilbert 1944, Gatz 1973) that increases the oxygen and fixed carbon available to developing embryos (Bachmann et al. 1986, Pinder and Friet 1994, Graham et al. 2013). Embryos in egg masses colonized by these symbiotic algae hatch sooner than embryos in egg masses without algae (Gilbert 1942).

Spotted Salamander eggs typically hatch in four to seven weeks (Bishop 1941b, Worthington 1968). Time to hatching and larval developmental rates depend on shade and water temperature (Worthington 1969, Voss 1993), hydroperiod (Bishop 1941b, Brodman 1996), and intensity of competition with other larvae (Wilbur 1972). Hydroperiod, percent forest coverage, and pond area can also influence larval density in breeding ponds (Ousterhout et al. 2015). High larval densities can lead to greater mortality, prolonged developmental periods, or smaller size at metamorphosis (Brodman 1996, Purrenhage and Boone 2005, Ousterhout et al. 2015). Metamorphosis usually occurs in late summer ca. 57–144 days post-hatching (Talentino and Landre 1991, Parmelee 1993, Trauth et al. 2004, Timm et al. 2007a) when larvae reach a minimum of 2.3–3.2 cm SVL (0.90–1.26 in; Worthington 1968, Talentino and Landre 1991), but mean size at metamorphosis ranges from 2.7 to 6.0 cm TL (1.1 to 2.4 in; Ball 1937, Dundee 1947). Neotenic adults are not known in this species (Savage and Zamudio 2005).

Spotted Salamanders grow rapidly in the first two to three years post-metamorphosis, after which growth rate slows (Newcomb-Homan et al. 2003). Spotted Salamanders typically mature at two to three years in southern locations (Bishop 1941b, Windmiller 1996). However, females in northern populations take longer to mature (e.g., three to five years in Michigan [Wilbur 1977c] and seven years in Quebec [Flageole and LeClair 1992]). The minimum size at maturity for males is 5.7 cm SVL (2.2 in) and for females is 6.0 cm SVL (2.4 in; Windmiller 1996). The mean age of a population in Massachusetts was estimated at 5.2 years, with males ranging from 2 to 13 years and females from 2 to 10 years (Homan et al. 2007). Most adults in a Quebec population were 2–18 years old, but the oldest individual encountered was reportedly 32 years old (Flageole and LeClair 1992).

ACTIVITY

Emergence from hibernation typically occurs in late March to mid-April (Vogt 1981). Soon after, adults migrate to breeding sites, which is typically the most substantial movements they will make annually. Breeding migrations can range from 25 to 249 m (82 to 817 ft; Wacasey 1961, Kleebeger and Werner 1983), with a mean distance of 116 m (381 ft; Savage and Zamudio 2005). During these migrations, Spotted Salamanders are more likely to traverse forested habitat than fields or roads (Regosin et al. 2005) and will even avoid grasslands near the forest edges (Rittenhouse and Semlitsch 2006). Spotted Salamanders also use the same migratory route to breeding ponds across years (Douglas and Monroe 1981, Phillips and Sexton 1989). From late August through October, adults again migrate long distances to overwintering sites. Regosin et al. (2005) determined that 60%–68% of overwintering locations in Massachusetts were >100 m (328 ft) from breeding ponds. Similarly, the mean distance from overwintering locations to the edge of the nearest breeding pond in New York was 118 m (387 ft) but ranged from 15 to 210 m (49 to 690 ft; Madison 1997).

Outside of migrations to breeding and overwintering locations, Spotted Salamanders move sparingly. Most activity consists of short movements to burrow entrances on rainy nights or to forage on the forest floor within the vicinity of their burrow (Gordon 1968). Madison (1997) reported that the mean total movement of radio-tracked individuals per season was only 7.6 m (24.9 ft) in summer, compared to 153.6 m (503.9 ft) and 86.4 m (283.5 ft) in spring and autumn, respectively. Mean home range outside the breeding season is reportedly as low as 4–15 m^2 (43–194 ft^2; Douglas and Monroe 1981, Kleeberger and Werner 1983, Ledford 2011). In summer, Spotted Salamanders often retreat into burrows to avoid air temperatures that exceed their critical thermal maximum (CTmax) of ca. 35°C (90°F; Gatz 1971, Faccio 2003).

Recently metamorphosed individuals usually disperse from breeding ponds into immediately adjacent uplands during rain (Hardy 1952, Petranka 1998, Rothermel 2004). Juvenile dispersal is more sporadic than the migrations of breeding adults, but it also requires moist conditions to avoid desiccation (Shoop 1974, Wilson 1976, Windmiller 1996). When recently metamorphosed Spotted Salamanders emerge from ponds they move through debris to underground locations near pond margins (Pough and Wilson 1970, Windmiller 1996). Windmiller (1996) found that most recently metamorphosed Spotted Salamanders overwintered within 40 m (131 ft) of their natal ponds, with only 14% dispersing farther than 100 m (328 ft) from their natal ponds. The movement of older juveniles is affected by precipitation, minimum air temperature, and degree of habitat disturbance. For example, Gravel et al. (2012) found that juveniles near roads moved shorter distances than juveniles farther from roads. By late October, Spotted Salamanders in Wisconsin enter hibernacula for the winter (Vogt 1981).

PREY AND PREDATORS

Adult and juvenile Spotted Salamanders are generalist carnivores that consume mostly earthworms, mollusks, and terrestrial arthropods (Bishop 1941b, Petranka 1998), although small vertebrates, such as Eastern Red-backed Salamanders (*Plethodon cinereus*), are occasionally taken (Ducey et al. 1994). Larvae are also carnivorous and may consume any prey that fits into their mouths (Petranka 1998). Accordingly, small larvae mostly eat small aquatic crustaceans, such as cladocerans and copepods (Branch and Altig 1981, Nyman 1991), while larger larvae eat larger prey items, including aquatic insects, isopods, mollusks, and amphibian larvae (Freda 1983, Harris 1995, Rowe and Dunson 1995). It is also noteworthy that Spotted Salamander larvae contribute to mosquito control through consumption of mosquito larvae (Rubbo et al. 2011).

Predation of Spotted Salamanders is rarely documented in Wisconsin, although Welch (1987) reported consumption by Red-shouldered Hawks (*Buteo lineatus*). Reports from outside of Wisconsin support that a variety of additional vertebrate predators also consume adults, including Common Gartersnakes (*Thamnophis sirtalis*; Klemens 1993), Eastern Hog-nosed Snakes (*Heterodon platirhinos*; Babbitt 1932, Trauth 1982), Brown Trout (*Salmo trutta*; Bishop 1941b), and Raccoons (*Procyon lotor*; Beachy 1991, Huheey and Stupka 1967). Kuzmin et al. (2001) described a new parasitic nematode species (*Rhabdias ambystomae*) collected from the lungs and body cavity of Spotted Salamanders in Bayfield County.

Younger age classes are particularly susceptible to predators. Eggs are eaten by Eastern Newts (*Notophthalmus viridescens*; Hamilton 1932, Wood and Goodwin 1954), certain caddisflies (Murphy 1961, Stout et al. 1992), midge larvae (LeClair and Bourassa 1981), leeches (e.g., American Medicinal Leech [*Macrobdella decora*]; Cargo 1960), Snapping Turtles (*Chelydra serpentina*; Spohr 2005), centrarchid fishes (Ireland 1989, Boone et al. 2007), Common Grackles (*Quiscalus quiscala*; Duncan 2005), and even Wood Frog larvae (*Lithobates sylvaticus*; Petranka et al. 1998, Baldwin and Calhoun 2002). Caddisflies reportedly prey more on Spotted Salamander eggs in wetlands with low pH compared to wetlands with neutral pH (Rowe and Dunson 1993, Rowe et al. 1994). Although intraguild predation on eggs by Wood Frog larvae also occurs, this reportedly increases Spotted Salamander larval growth rates due to release from intraspecific competition (Burley et al. 2006). Predators of Spotted Salamander larvae include the larvae of other salamander species (Stewart 1956, Walters 1975, Stenhouse 1985, Brodman 1996), sandpipers (Stangel 1983), Great Blue Herons (*Ardea herodias*; Brodman and Pfingsten 2010), and a variety of aquatic insects and fishes (Petranka 1998, Cunnington and Brooks 2000). Fishes can be particularly important predators and can eliminate Spotted Salamanders from wetlands where they co-occur (Ireland 1989, Petranka and Holbrook 2006). Predators of post-metamorphic juveniles reportedly include Wolf Spiders (*Gladicosa pulchra*; Rubbo et al. 2003).

To deter predators, adult Spotted Salamanders secrete a sticky white fluid from skin glands in the tail and head (Brodie 1971, Brodman 1995) that is effective against shrews (Brodie et al. 1979) and reptiles (Barach 1951, Howard 1971). Downs (1989b) even indicated that mammalian predators often avoid eating the outer skin layers of Spotted Salamanders because these contain the glands that produce noxious secretions. The bright spots on adults may act to warn potential predators of this salamander's poisonous secretions (reviewed by Savage and Zamudio 2005). When disturbed, Spotted Salamanders also lower their head, elevate their hips, and undulate their tail (Brodie 1977, Dodd 1977; Figure 135). This position most likely protects the head (Brodie 1977), makes the dorsal spots more visible, and better orients the noxious secretions in the tail region toward the perceived threat (Petranka 1998).

As eggs and juveniles, Spotted Salamanders employ various strategies to deter predators. Their eggs are coated in a thick dense outer jelly matrix that protects them from desiccation (Nyman 1991) and from some predators (Ward and Sexton 1981, Semlitsch 1988). Larvae reduce overall activity, take cover in the leaf litter at the bottom of wetlands, and modify their tail morphology to avoid predation (Semlitsch 1988, Figiel and Semlitsch 1990, Brodman 1996, 1999, Brodman and Jaskula 2002, Yurewicz 2004). Post-metamorphic juvenile Spotted Salamanders responded to attacks from Wolf Spiders by continuously raising and lowering their head and escaped by flipping onto their back (Rubbo et al. 2003).

Figure 135. Spotted Salamander defensive postures (Blount County, Tennessee; photos by R. Brodman). (a) Newly metamorphosed Spotted Salamander with elevated tail and lowered head. (b) Juvenile Spotted Salamander with lowered head and tail-whipping motions.

CONSERVATION

Status: Globally, the Spotted Salamander is listed as a Least Concern species by the IUCN. Recent estimates indicate that Spotted Salamanders are uncommon in several midwestern states compared to other regions. Occupancy rates based on work done in Wisconsin, Illinois, Indiana, Michigan, and Ohio ranged from 0% to 47% (Downs 1989b, Mierzwa 1998, Pfingsten 1998, Brodman 2003, 2010, Porej et al. 2004, Porej and Hetherington 2005). On the other hand, occupancy rates in New England ranged from 48% to 90% (Baldwin and Vasconcelos 2003, Skidds and Golet 2005). Their distribution in Wisconsin is spotty compared to other midwestern states (Zamudio and Savage 2013). Spotted Salamanders were listed as Threatened in Wisconsin until surveys in the 1980s revealed that they were more abundant in the state than previously known, and they were delisted (Casper 1998). Their current status in Wisconsin is apparently secure, and they are considered common by the WDNR.

Populations: Population data, including basic estimates of abundance and density, are sparse for the northern US. Estimates from southern populations provide some context and indicate that population size can vary over time and space. For example, Kinkead and Otis (2007) reported that the number of breeding individuals in populations from two South Carolina ponds ranged widely, from 40 to 507 individuals, while total population estimates were 43–639 individuals (wetland size = 0.6–3.2 ha [1.6–9.2 ac]). Additionally, Blackwell et al. (2004) found that the number of breeding adults in a population from Alabama, estimated across six seasons at a 250 m^2 (2,691 ft^2) wetland, ranged from 189 to 260 individuals.

Several studies have investigated population characteristics beyond abundance, such as survival. For example, annual adult survival rates are high (63%–100%; Douglas and Monroe 1981, Blackwell et al. 2004), with males experiencing greater survival than females (Harris 1980, Flageole and LeClair 1992). Estimates of survival prior to metamorphosis, however, are typically low (<10%) but can be variable. Reported embryonic mortality ranges from 0% to 100% (Pough 1976, Stenhouse 1987) but is commonly ca. 40%–60% (Clark 1986, Brodman 1995, 2002). Larval survival is also generally low, often less than 13%, due to pond desiccation (Stangel 1988, Petranka et al. 2007) or intraguild predation (Stenhouse 1987, Regester et al. 2006). The low survival in young age classes limits population recruitment, as indicated by the work of Homan et al. (2007), which found that on an annual basis, only 5.7 juveniles were added to populations per reproductive female. Dispersal from surrounding populations, long life spans, and low adult mortality help to offset low recruitment (Petranka 1998). Sex ratios at breeding ponds are typically skewed toward males (1.5–4M:1F; Downs 1989b, Blackwell et al. 2004). Despite this, populations assessed outside of the breeding season reportedly have ca. 1M:1F sex ratios (Parmelee 1993).

Threats: Spotted Salamanders are sensitive to many human disturbances, particularly those associated with woodlands (see Introduction to Conservation and Management of Wisconsin's Amphibians and Reptiles chapter). For example, loss of wetlands and adjacent forested habitat is a major threat to this species (Baldwin et al. 2006b). They are rare in agricultural landscapes, including those that are intensively cultivated (Downs 1989b, Minton 2001), and they experience population declines in urbanized areas (Windmiller 1996, Rubbo and Kiesecker 2005). There was also a negative association detected between Spotted Salamander presence and human population density in Massachusetts at several spatial scales (Clark et al. 2008a).

Habitat disturbance associated with timber harvest and removal of fallen logs (or other forms of coarse woody debris) are additional conservation concerns. For example, juvenile and adult numbers decline in timber harvested areas, such as recently clear-cut woodlands, or areas with partial tree canopy removal, especially when coarse woody ground debris is also eliminated (deMaynadier and Hunter 1999, Renken et al. 2004, Patrick et al. 2006, Perkins and Hunter 2006, Todd et al. 2014). Juveniles are less likely to settle in clear-cut areas than closed-canopy forests and more likely to move into areas with coarse woody debris on the ground (Osbourn et al. 2014). Loss of forested habitat is further linked to higher production of stress-induced hormones in Spotted Salamanders, indicating they experience chronic stress, or reduced ability to respond to stress, in disturbed habitats (Homan et al. 2004). Recovery of Spotted Salamander populations in clear-cut areas is slow and can take up to 11–12 years (deMaynadier and Hunter 1999, Morris and Maret 2007).

In addition to habitat loss, habitat fragmentation and degradation are conservation concerns for this species. Gibbs (1998b) found that Spotted Salamanders were absent from fragmented woodlands with less than 30% forest coverage. Although salamanders are able to disperse from clear-cut areas if adjacent forest is kept intact (Semlitsch et al. 2008), the dispersal barriers that arise due to woodland fragmentation can limit movement. Dispersal barriers can range from pastures and grasslands (Rothermel 2004, Rittenhouse and Semlitsch 2006) to roads (Gibbs 1998b, Savage and Zamudio 2005, Richardson 2012). Several studies have reported that Spotted Salamander populations decrease with increasing road density and closer proximity to roads (Fahrig et al. 1995, Trombulak and Frissel 2000, Mazerolle 2004, Gibbs and Shriver 2005, Richardson 2012). Furthermore, Newcomb-Holman et al. (2003) found that movement over paved surfaces significantly increased stress hormone levels in migrating Spotted

Salamanders. Even simple logging roads can limit movement and further reduce occupancy of potential breeding sites (Chambers 2008). Degradation of habitat due to invasive species is an additional concern. Although habitat characteristics preferred by Spotted Salamanders can be directly altered by invasive plants, some impacts of invasive vegetation are more subtle. For example, Martin and Blossey (2013) found that leachates from an invasive wetland plant, Common Reed (*Phragmites australis*), reduced the survival and developmental rates of Spotted Salamander larvae.

Spotted Salamanders are apparently susceptible to extreme environmental variation. Winter mortality is documented in ponds that unexpectedly freeze during the breeding season (Brodman 1995, Gibson and Merkle 2005), and females are reportedly more prone to this type of mortality than males (Kornilev et al. 2006). Along these lines, the unpredictable weather shifts associated with climate change could have substantial impacts on Spotted Salamanders. For example, although higher air and water temperatures recorded between the 1990s and early 2000s did not affect the abundance of Spotted Salamanders at sites in Wisconsin, the shift in temperature was correlated with the earlier initiation of breeding seasons in the state (Donner et al. 2015). If the overall environmental temperature trend continues to increase in Wisconsin, the result could be a permanent shift in the Spotted Salamander breeding season to earlier in the spring, when unpredictable weather swings can occur more frequently. Casper (2003) further suggested that lower mean annual snow depths, which might occur due to climate change, could negatively influence Spotted Salamander presence and persistence at sites in Wisconsin.

Several diseases that impact amphibians are potentially significant conservation concerns for Spotted Salamanders in Wisconsin (see Introduction to Conservation and Management of Wisconsin's Amphibians and Reptiles chapter). The chytrid fungus *Batrachochytrium dendrobatidis* (Bd) was confirmed in wild Spotted Salamanders (Ouellet et al. 2005), and mortality due to ranaviruses was also documented (Green et al. 2002). No reports of infection with the salamander chytrid fungus (*B. salamandrivorans* [Bsal]) are known, but this is a potentially serious threat to all Wisconsin salamanders. Spotted Salamanders are also sensitive to various environmental contaminants. Acid rain results in greater than 50% egg mortality when pH is less than 6.0 (Saber and Dunson 1978, Sadinski and Dunson 1992). Larvae at low pH also experience reduced growth and development (Gosner and Black 1957b, Ireland 1991, McCoy and Harris 2003) and are more susceptible to predation (Brodman 1993). Cations from road salt (Turtle 2000, Karraker et al. 2008, Brady 2012) and other sources (Clark and Hall 1985, Ireland 1991) reduce hatching success, and Spotted Salamander egg masses are smaller in ponds that are closer to highways instead of in forests (Karraker and Gibbs 2011). Yet there is evidence that roadside Spotted Salamander populations can develop a resistance to road salts (Brady 2012). Chlorinated solvents (tetrachloroethylene and its major degradation products) cause developmental malformations in Spotted Salamander embryos (McDaniel et al. 2004). Calfee et al. (2006) found that larvae exposed to UVB radiation experienced reduced growth and increased mortality, but egg mortality was unaffected by UVB exposure. Insecticides and herbicides reduce larval survival, growth, development, and body condition (Metts et al. 2005, Boone and James 2003, Boone et al. 2007, Robert Brodman, unpublished data).

Management: Many general habitat management considerations, particularly those related to wetlands and forests, are applicable to Spotted Salamanders (see Introduction to Conservation and Management of Wisconsin's Amphibians and Reptiles chapter). Spotted Salamanders have been subjects in several long-term population studies spanning 12–30 years (Sexton et al. 1998, Brodman 2002, 2010, Petranka et al. 2003, 2007, Spencer Cortwright, Indiana University Northwest, and R. Brodman, unpublished data). These studies support that populations will persist and thrive in landscapes with high coverage (85%) of intact forests and many shallow fishless ponds that vary in hydroperiod. Protection of a 200–250 m (656–820 ft) mature deciduous forest buffer around breeding ponds could be an effective conservation strategy for Spotted Salamanders (Semlitsch 1998, Porej et al. 2004). They will also likely benefit from the preservation of small wooded ephemeral wetlands (0.05–0.2 ha [0.12–0.49 ac]; Semlitsch 1998, Egan and Paton 2004) and complexes of numerous wetlands with various hydroperiods (Petranka et al. 2004, Rothermel 2004, Brodman 2010, 2013).

Maintenance of connectivity between the habitats required to meet life history needs is of further importance (Compton et al. 2007, Brodman 2013). Habitat linkages to enhance landscape connectivity can be achieved via wooded corridors that span disturbed landscapes between existing populations (Semlitsch and Bodie 1998, Rittenhouse and Semlitsch 2006). Underpasses are also a potentially effective strategy to promote safe migration in habitats fragmented by roads. However, these underpasses are most effective when they are not inundated with flowing water and barriers are used to direct dispersing salamanders toward tunnel entrances (Jackson and Tyning 1989, Gibbs and Shriver 2005, Patrick et al. 2010). Models to predict salamander migration patterns can aid in timing road closures to minimize road mortality (Timm et al. 2007b). A riparian buffer alongside a stream (11–35 m [36–115 ft]) was found to partially mitigate the effects of timber harvest on several species of amphibians by acting as a migration corridor, but there was an 89% reduction of Spotted Salamanders in buffer treatments compared to control plots (Perkins and Hunter 2006). Strategies to manage habitats must be considered carefully, as some techniques could have damaging effects. For example, populations in Indiana declined after application of prescribed fire to manage a prairie and took an average of four to five years post-burn to recover (Brodman 2010).

The construction of breeding ponds within forested habitat is a potentially useful tool for conserving Spotted Salamanders, but its effectiveness requires further investigation. Some research supports that salamanders rapidly colonize new breeding ponds, and long-term studies indicate that such ponds can support population growth over time (Vasconcelos and Calhoun 2006, S. Cortwright and R. Brodman, unpublished data). Constructed ponds are reportedly most effective if they vary in hydroperiod, remain fishless, and occur within 300 m (984 ft) of forested habitat with high tree canopy cover (Peterman et al. 2014). Although conservation strategies that include repatriation via translocations are alluring, many questions remain about the appropriateness of this technique. For example, although this strategy is reportedly effective in some instances with other species, repatriation or translocation studies focused on Spotted Salamanders have produced mixed results (Sexton et al. 1998, Cook 2008; see further review in Introduction to Conservation and Management of Wisconsin's Amphibians and Reptiles chapter).

Eastern Tiger Salamander

Ambystoma tigrinum (Green 1825)

Noah J. Anderson and Joshua M. Kapfer

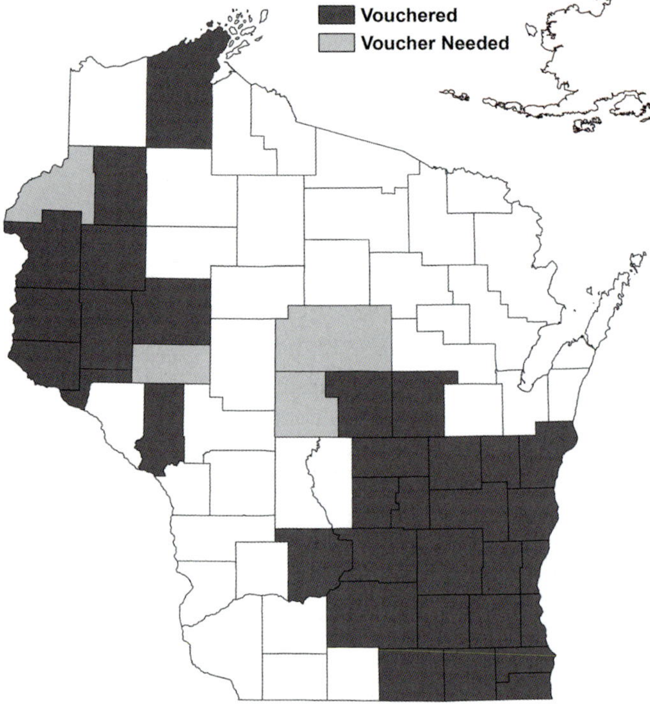

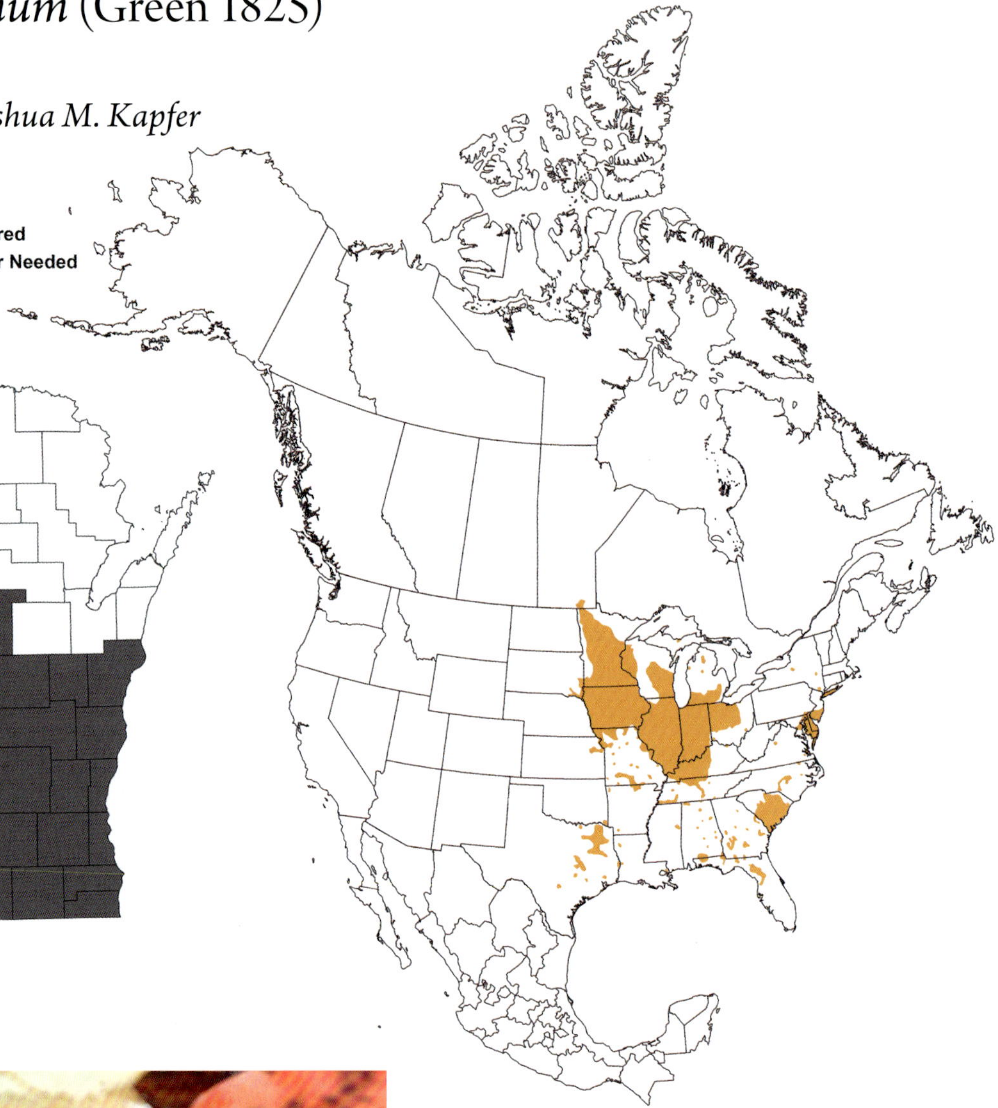

Terrestrial adult Eastern Tiger Salamander (Racine County; photo by J. DeBoer).

AT A GLANCE

Eastern Tiger Salamanders are Wisconsin's largest terrestrial salamander species. Typical adults are dark with large irregular yellow blotches on their back and a solid yellow lower jaw. Eastern Tiger Salamanders occur across Wisconsin in a swath that roughly extends from the west-central through the southeastern regions, although this swath is broken by a narrow strip of counties with no published records in the central part of the state. There are also no published records from much of northern and southwestern Wisconsin. Eastern Tiger Salamanders emerge from terrestrial hibernation sites in March or April and migrate large distances to breed in open wetlands. After the breeding season, adult Eastern Tiger Salamanders move into adjacent upland habitats, such as prairies, grasslands, oak savannas, or woodland edges, where they seek refuge in burrows and beneath ground debris. Larvae are fully aquatic, with large bushy external gills, so they remain in breeding wetlands until they undergo metamorphosis from July through September. In some situations, larvae do not metamorphose but instead develop into fully aquatic adults that retain their larval characteristics (e.g., large external gills). Adult Eastern Tiger Salamanders eat many types of invertebrates, from insects to earthworms, as well as some vertebrates that are small enough to overpower and swallow (e.g., young rodents). Larvae eat large quantities of aquatic invertebrates, including the larvae of some pest insect species, and vertebrates such as juvenile amphibians. Adults are consumed by various types of vertebrates, from snakes to birds and even small mammals such as shrews; larvae also fall victim to fishes and large predatory aquatic insects. From September to October, adults and recently metamorphosed juveniles move to terrestrial overwintering locations, where they hibernate in abandoned or self-excavated burrows that extend below the frost line. Eastern Tiger Salamanders are considered common by the WDNR.

SYSTEMATICS

The evolutionary relationships and taxonomy of the *Ambystoma tigrinum* (Eastern Tiger Salamander) species complex has been the source of much debate and controversy (Shaffer and McKnight 1996, Irschick and Shaffer 1997, Petranka 1998). Some biologists have treated *A. tigrinum* as one morphologically diverse species (Gehlbach 1965, Petranka 1998). However, genetic analyses have revealed a phylogenetic structure that supports a complex of multiple species, although the number of species and the evolutionary relationships among them remain unresolved (Shaffer and McKnight 1996, Irschick and Shaffer 1997, Church et al. 2003, O'Neill et al. 2012). There are currently no recognized subspecies of *A. tigrinum*, although the Wisconsin form was previously considered the subspecies *A. t. tigrinum* (Vogt 1981, Crother 2017). Among Wisconsin salamanders, other species in the genus *Ambystoma* include *A. laterale* (Blue-spotted Salamander) and *A. maculatum* (Spotted Salamander).

DESCRIPTION

Adults: Terrestrial adults are generally black or dark gray, with large light blotches that are haphazardly distributed across their back (Figure 136). These blotches range in color from bright yellow to a duller grayish but are never orange (Figure 136). The lower jaw and often the throat of adults in Wisconsin are also mostly yellow or grayish yellow (Figure 136). Eastern Tiger Salamanders possess four toes on the front feet and five toes on the hind feet. Adults have a chunky and robust appearance, with 11–14 (often 12–13) costal grooves (Vogt 1981). Eastern Tiger Salamanders are the largest terrestrial salamander that occurs in Wisconsin. The mean SVL of post-metamorphic Wisconsin specimens housed in the MPM and UWZM collections was 8.0 cm (3.2 in, $n = 180$), while mean tail length was 6.4 cm (2.5 in, $n = 171$). Vogt (1981) reported that they reach sizes of 34 cm TL (13.4 in) in Wisconsin; the largest SVL among Wisconsin specimens measured was 11.6 cm (4.6 in, UWZM 4920, collected in Dane County in 1927, tail not intact).

Preadult stages: Eggs are brown or cream-colored and small, ranging from 1.0 to 1.2 cm (0.4 to 0.5 in) in Wisconsin (Vogt 1981). Each egg is encased in three gelatinous envelopes, and eggs are often deposited in clusters of several to many eggs (Figure 137; reviewed by Petranka 1998). Larvae possess three pairs of bushy

Above, top: Figure 136. Terrestrial adult Eastern Tiger Salamanders. (a) Portage County (photo by L. Huse). (b) Dane County (photo by C. Raimond).

Above, bottom: Figure 137. Eastern Tiger Salamander egg cluster (Washington County; photo by C. Raimond).

Figure 138. Larval Eastern Tiger Salamanders. Note the bushy external gills on the sides of the head. (a) Ozaukee County (photo by S. Viernum). (b) Waukesha County (photo by J. Kapfer).

external gills, with 14–24 gill rakers on the third gill arch (Petranka 1998). The tail is deep, with dorsal fins that extend onto the body nearly to the head. Later in development, larvae have four legs (Figure 138) and five toes on each hind foot with digits that are flattened (not rounded) on the bottom (Petranka 1998). The overall coloration of larvae is olive green, with darker or lighter mottling that is particularly evident on the tail fins. Larvae possess 12–13 costal grooves (Petranka 1998) and can grow larger than the juveniles of all other salamanders that typically metamorphose into terrestrial adults. The mean SVL of larval Wisconsin specimens housed in the MPM and UWZM collections was 5.4 cm (2.1 in, $n = 129$), and mean tail length was 3.8 cm (1.5 in, $n = 129$). Among these, the largest specimen had an SVL of 11 cm (4.3 in, MPM 30312, collected from Walworth County in 1998).

Variation: The background color of adults can range from gray to black, and the distribution of yellow blotches varies considerably. Albinism has also been documented in Wisconsin individuals (Vogt 1981). Williams et al. (2009b) found that mean SVL is longer in females than in males, but males have greater tail lengths and tail heights than females.

The most obvious examples of morphological variation in this species, however, occur due to the various developmental strategies that it can employ. For example, neotenic aquatic adults (see Reproduction and Development section) in Wisconsin do not have terrestrial adult coloration and tend to be brighter yellow green with more defined black dots than typical larvae (Noah Anderson, Winona State University, personal observation). They also have a longer and narrower body than larvae, a shallower and shorter head, shorter gills, and a less pronounced tail fin that does not extend onto their body as far as in larvae (N. Anderson, personal observation). There appears to be some seasonal variation in the color of neotenic adults, as males possess a bright yellow venter and throat in early spring (presumably to attract females) rather than the typical cream or white coloration (N. Anderson, personal observation). Furthermore, when neotenic Eastern Tiger Salamanders transform into terrestrial adults, they retain this color and pattern (i.e., mostly yellow-green with scattered black dots) and highly resemble their western counterparts, particularly the Gray Tiger Salamander (*Ambystoma mavortium diaboli*).

Eastern Tiger Salamander larvae can also develop into a 'cannibal morph' under some circumstances (see Reproduction and Development section), although this morph is rarely (if ever) documented in Wisconsin. Cannibal morphs are often larger than their noncannibal counterparts (SVL >5 cm [2 in]) and possess a broader flattened head with much larger vomerine teeth than typical larvae (Lannoo and Bachmann 1984).

Diagnosis: Terrestrial adults and recently metamorphosed juvenile Eastern Tiger Salamanders do not have dorsal or ventral fins along the margins of the tail musculature, and they lack external gills and nasolabial grooves. The body and tail are black, with irregular yellow blotches arranged in no obvious pattern. Adults and recently metamorphosed juveniles are most easily confused with Spotted Salamanders (*Ambystoma maculatum*) in Wisconsin. Yet Eastern Tiger Salamanders generally have irregular yellow blotches that are not in obvious rows, rather than the well-defined spots and rows of Spotted Salamanders. Adult Eastern Tiger Salamanders also have a yellow lower jaw, compared to the dark gray jaw in Spotted Salamanders, and reach larger sizes.

Amphibian eggs and larvae are difficult to distinguish in the field, although eggs of other Ambystomatid salamanders in Wisconsin have only two gelatinous envelopes, while Eastern Tiger Salamander eggs have three (reviewed by Watermolen 1995). Larval Eastern Tiger Salamanders have external gills and forelimbs. They are large, often exceeding 7 cm TL (2.8 in). Their dorsal tail fin extends up to their back, and their background color is gray to greenish brown, with no rows of distinct yellow spots and no obvious dark stripes running through the eyes. There are five toes on each hind foot, which are pointed at the tip and broad (or flattened) at the base. Eastern Tiger Salamander larvae can grow to much larger sizes than those of Blue-spotted and Spotted Salamanders, but they also have digits that are flattened on the bottom, unlike the latter two (Petranka 1998). Large aquatic Eastern Tiger Salamanders (both larvae and neotenic adults) could be confused with Mudpuppies (*Necturus maculosus*), as both have large external gills. However, Mudpuppies have proportionally smaller eyes than Eastern Tiger Salamanders and four toes rather than five on their hind feet.

DISTRIBUTION AND HABITAT

Global distribution: Eastern Tiger Salamanders have a scattered distribution across nearly the entire eastern half of the US and extreme southern Canada. However, their distribution is most concentrated in the Midwest (Natural History Box: Evolutionary History and Global Distribution of Tiger Salamanders) and includes large swaths of Minnesota, Wisconsin, Iowa, Illinois, Indiana, Arkansas, Ohio,

NATURAL HISTORY BOX: Evolutionary History and Global Distribution of Tiger Salamanders

The evolutionary story and modern distribution of Eastern Tiger Salamanders in the Upper Midwest is intrinsically tied to glaciations of the Pleistocene Epoch (ca. 2.5 mya to 12,000 ybp). Evidence supports that glacial ice advanced from the north into the midwestern US on numerous occasions throughout the Pleistocene, with the best-documented advance being the most recent, named the Wisconsin Glaciation. During this glaciation, the Laurentide Ice Sheet, which covered vast portions of Canada and the US, reached its maximum extent into Wisconsin by about 30,000 years ago and began to melt by 18,000 years ago (Attig et al. 2011). It was during the Pleistocene that many of our state's native reptiles and amphibians retreated south to avoid these advancing glaciers, and Eastern Tiger Salamanders were probably no exception. Recent studies show that the likely refugium for Eastern Tiger Salamanders at this time was the mid-Atlantic Coastal Plain of Georgia, parts of Florida, and Virginia (Church et al. 2003). It was here that Eastern Tiger Salamanders hunkered down, separated from their western counterparts (current-day Blotched Tiger Salamanders [*Ambystoma mavortium melanostictum*]) for hundreds of thousands, possibly millions, of years.

When the earth began its most recent warming trend at the end of the Pleistocene, Eastern Tiger Salamanders from the southern Coastal Plain spread back north into the Midwest (Church et al. 2003). The timing of their migration event was crucial. The Lower Midwest, particularly central parts of Missouri and Illinois, had become comparatively hot and dry from 9,000 to 6,000 years ago (a period called the Holocene Thermal Maximum), creating more xeric conditions that many species of salamander might not prefer (Templeton 1996). During this time, a "Prairie Peninsula," which supported species that associate with the warmer and drier prairies of the Midwest, such as Eastern Tiger Salamanders, extended across southwestern Wisconsin eastward into parts of Ohio and Pennsylvania (Bailey 1949, Webb et al. 1983). Vogt (1981) hypothesized that it was during the Holocene Thermal Maximum that Eastern Tiger Salamanders and a host of species currently considered more Western Prairie endemics followed the Prairie Peninsula and made their way to Wisconsin. As the Midwest cooled after the Holocene Thermal Maximum, the prairie environments that remained in the state were ideal for Eastern Tiger Salamanders, and the current eastern boundary of the species' core distribution correlates well with the eastern extent of the Prairie Peninsula.

Noah J. Anderson, Winona State University

and the southern portion of the Lower Peninsula of Michigan.

Distribution in Wisconsin: Published Eastern Tiger Salamanders records are concentrated in two large areas that each encompass several counties but are separated by a strip of counties without published records: one in northwestern/west-central Wisconsin and the other in central to southeastern Wisconsin. Of the published county records, suitable museum vouchers could not be found for Burnett, Eau Claire, Marathon, and Wood (see the Appendix). It is noteworthy that although vouchers were reported for Eau Claire and Marathon Counties by Suzuki (1951), the specimens are currently missing, while museum records indicate that MPM 26487, purportedly from Marathon County, was lost due to desiccation.

Localities of possible occurrence: New county records for Eastern Tiger Salamanders are rare. Only a hand-

NATURAL HISTORY BOX: Living with Pocket Gophers

I grew up on a farm, and one of our primary 'varmints' was the Plains Pocket Gopher (*Geomys bursarius*). One of my chores was to trap this small mammal and remove it from our hayfields. Pocket gophers dig long burrows that can go for hundreds of feet and deposit their excavations on the surface of the ground in mounds. These mounds are usually plugged with loosely packed dirt. This plug helps to prevent any other animals, such as the gophers' top predator, the Gophersnake (*Pituophis catenifer*), from entering their burrow. On two separate occasions, I found Eastern Tiger Salamanders cohabitating with Plains Pocket Gophers in their underground burrows or otherwise associated with the mound system. This behavior has also been observed in Minnesota in several localities (Breckenridge 1944, Moriarty and Hall 2014). One salamander was found in a 'burrow plug' ca. 0.8 m (2.5 ft) into the tunnel system. Burrow plugs are formed by Plains Pocket Gophers when there is an intruder in the burrow or the burrow system is opened up. Plains Pocket Gophers rapidly shovel a plug of dirt into the tunnel, effectively sealing off the intruder. In this case, the Eastern Tiger Salamander had been shoved into the burrow plug with the rest of the dirt. I found the other Eastern Tiger Salamander as I dug through a mound in an attempt to find the entrance to the burrow. The soil in these areas was hard-packed ground (i.e., hayfield or pasture), and the Plains Pocket Gopher burrows were likely some of the easiest routes to an underground refuge. What is the nature of this association in relation to the looseness of the soil? How long will Eastern Tiger Salamanders live underground with Plains Pocket Gophers? Are the gophers passive hosts in this association, or was the salamander plugged deliberately? Are the invertebrates in the gopher tunnels the types that the salamander typically eats? In areas where Plains Pocket Gophers and a variety of other burrowing mammals are found, do Eastern Tiger Salamanders exhibit a preference as to burrow maker?

Noah J. Anderson, Winona State University

ful of records have been published since Vogt (1981), and most of these from prior to 2000. Despite this, it is possible that Eastern Tiger Salamanders occur in several undocumented counties that are adjacent to those with published occurrences. Examples include Buffalo, Taylor, Clark, Jackson, Juneau, Adams, Outagamie, and Green. Dickinson (1965) reported them from Door County, but no known voucher exists, and Vogt (1981) ignored this record.

Habitat: In the Midwest, terrestrial adult Eastern Tiger Salamanders often associate with open grassy habitats, such as prairie, oak savanna, pasture, and open fields, but individuals also inhabit woodland edges and occasionally occur within forests (reviewed by Lannoo and Phillips 2005). Most information on habitat use in Wisconsin is anecdotal, but Vogt (1981) reported that they were "most often found in prairie ponds, marshes, and kettle potholes." Outside of the breeding season, they live below rocks and woody debris or in burrows. Eastern Tiger Salamanders will occupy burrows of other animals, including mammals (Madison and Farrand 1998; Natural History Box: Living with Pocket Gophers), but can also dig their own burrows to avoid unfavorable conditions (Semlitsch 1983a).

In spring, Eastern Tiger Salamanders migrate from terrestrial overwintering sites to aquatic habitats for breeding. Breeding sites are typically open ponds and marshes, but Vogt (1981) also reported that they will breed in "lakes, woodland ponds, fish-rearing ponds, and farm ponds" in Wisconsin. Breeding sites are frequently ephemeral, so they lack predatory fishes. Madison and Farrand (1998) found that adult Eastern Tiger Salamanders at breeding ponds showed significant variation in the habitats used, based on pond characteristics and the presence of predators. For example, in ponds with no tree canopy cover, adults mostly occupied deeper water with more dense vegetation. Yet in a monitored pond that contained Bluegill (*Lepomis macrochirus*), adults remained in shallower water, where fish access was restricted. Habitats used by larvae are also variable and shift based on environmental conditions, pond characteristics, and threat of predation. For example, larvae made small-scale movements to various regions and microhabitats within wetlands (e.g., near shore, open water, near bottom) to thermoregulate (Heath 1975), as well as to follow prey and avoid predators (Holomuzki 1986). Most adult tiger salamanders overwinter in upland terrestrial burrows that are either self-excavated or the existing burrows of other animals that extend below the frost line (reviewed by Lannoo and Phillips 2005).

Figure 139. Adult Eastern Tiger Salamander below ice in a breeding pond (Sheboygan County; photo by C. Raimond).

REPRODUCTION AND DEVELOPMENT

Eastern Tiger Salamanders emerge from hibernation after the ground thaws in spring and, often stimulated by warm rains, migrate to breeding sites from early March to mid-April (Vogt 1981). Often hundreds of individuals undertake these migrations during short time periods (Peckham and Dineen 1954, Sever and Dineen 1978). Migrations sometimes occur during cold and inclement weather. For example, a Minnesota individual was found migrating to a breeding wetland on 15 April at the onset of a snowstorm (N. Anderson, personal observation), and they are sometimes observed in breeding ponds beneath thin layers of ice (Figure 139).

Males and females often arrive at breeding sites concurrently, although males sometimes arrive first (Sever and Dineen 1978) and remain until the end of June (N. Anderson, personal observation). Breeding behavior is complex and involves males nudging females with their tail, swinging their head, using their nose to shove females sideways, and tapping female dorsal surfaces with their tail (described in detail by Arnold 1976). Fertilization is internal. Males transfer sperm to females by depositing spermatophores (i.e., small gelatinous packets of sperm) on branches or leaves, which females then pick up with their cloacal lips. This strategy does not ensure that a single male-female pair sires all offspring in a given egg clutch. According to DNA analyses on an Indiana population, females can pick up more than one spermatophore to fertilize eggs, so clutches may be sired by multiple males, who in turn deposit many spermatophores that can be used for fertilization by multiple females (Gopurenko et al. 2006).

Adult breeding success can be influenced by a complex combination of energy costs associated with production of gametes, reproductive behavior, competition, and environmental conditions. For example, because Eastern Tiger Salamanders breed in unstable environments (e.g., ephemeral wetlands that experience periodic drought), their reproductive success is limited even if fertilization occurs (i.e., eggs or larvae could die if the breeding pond dries; Church et al. 2007). Reproduction also incurs a high energy cost for females (e.g., energy devoted to production of eggs). As such, Church et al. (2007) suggested that females only attempt reproduction during years when favorable environmental conditions prevail and the risk of adult mortality is low. Male reproductive success, on the other hand, is more typically hampered by male-male interference competition. Males can place spermatophores on top of those deposited by earlier rival males, push females away from rival males, and lead females away from rivals to ensure their own successful reproduction (Arnold 1976, Petranka 1998).

The mean clutch size estimated for females in a Michigan population was 421 eggs (Wilbur 1977c). Yet females can partition their eggs among several clusters that can be deposited within 5 m (16 ft) of each other (Gopurenko et al. 2006). The mean number of eggs per cluster is 38–59 (reviewed by Lannoo and Phillips 2005). In Wisconsin, eggs hatch in three to four weeks (Vogt 1981). There appears to be up to three waves of egg deposition annually for some Wisconsin populations, resulting in as many as three distinct size classes of larvae residing in a given breeding pond at once (N. Anderson, personal observation).

The larvae of Eastern Tiger Salamanders grow rapidly, and development to metamorphosis takes a minimum of ten weeks (reviewed by Lannoo and Phillips 2005). Various factors can influence development and growth rates, including water quality. Kiesecker (1996) found that low pH can reduce overall size of larvae but increase the rate of larval development. Wisconsin larvae metamorphose into terrestrial forms from July through September (N. Anderson, personal observation). However, larvae show much plasticity in developmental strategies, developmental rates, and the resultant larval morphology. In stable but fish-free environments, Eastern Tiger Salamanders can metamorphose at close to their maximum adult size; in shallow waterbodies or systems with high predator densities, larvae can metamorphose at a relatively small body size and finish growth as a terrestrial juvenile (reviewed by Petranka 1998). Some will remain in the pond throughout winter, providing the opportunity for continued growth prior to metamorphosis if food availability in the subsequent spring and summer remains high. However, this is a risky strategy, as some wetlands freeze solid or become anoxic in winter, while others could desiccate before the following spring. In wetlands with high larval densities or those that dry early, some individuals can develop into "cannibal morphs" (Collins and Cheek 1983, Lannoo and Bachmann 1984). These morphs are particularly adept at eating conspecifics that occupy the same ponds (Reilly et al. 1992). Larvae that develop into cannibal morphs likely benefit from the ability to eat larger prey (i.e., other Eastern Tiger Salamander larvae), which provide high nutrients for rapid growth and quick development to metamorphosis, an advantage in ephemeral wetlands that desiccate (Lannoo and Bachmann 1984). In some deep stable ponds where larvae do not have easy options for pond egress after metamorphosis, they can develop into neotenic adults (i.e., sexually mature adults that retain larval characteristics; Natural History Box: The Neotenic Tiger Salamanders of the Badger Army Ammunition Plant). Neotenic Eastern Tiger Salamander populations are relatively rare in the Upper Midwest (Routman 1993). In Wisconsin, there is only one well-documented neotenic population and one other population where biologists suspect neoteny. Another population in Michigan is also reportedly neotenic (Petranka 1998).

Although data on size at sexual maturity in Wisconsin are lacking, studies in South Carolina found that the smallest terrestrial adults were 8.1–8.5 cm SVL (3.2–3.3 in) and that breeding first occurred at two years of age (Semlitsch 1983b). Longevity records reviewed by Sever and Dineen (1978) included a captive individual that survived from an egg to 11 years of age and another captive neotenic adult that survived 25 years in an aquatic stage and then 16 years after transforming to an adult.

ACTIVITY

Adult Eastern Tiger Salamanders in Wisconsin emerge from hibernation in March or April and migrate to breeding wetlands (Vogt 1981). Most movement during the active season is likely associated with migrations to and from breeding ponds, which can cover substantial distances. Semlitsch (1983c) tracked a male with radio telemetry in South Carolina that moved 162 m (531 ft) from its breeding pond to summer habitat. Madison and Farrand (1998) also followed adults with radio telemetry and found that large migrations were not frequently undertaken at once but instead broken into smaller bouts of movement, interspersed with periods when salamanders remained within burrows. Movements are likely correlated with the availability of suitable habitat, prey, environmental conditions, and a low threat of predation (Madison and Farrand 1998). As

NATURAL HISTORY BOX: The Neotenic Tiger Salamanders of the Badger Army Ammunition Plant

High on the flank of the Baraboo Hills, overlooking the defunct Badger Army Ammunition Plant, a 6-million-gallon, 16-foot-deep water reservoir supports a unique fishless ecosystem. Fed by leaves and limbs fallen from the adjacent woods, supported by the influx of groundwater, and driven by the sun, the reservoir supports abundant aquatic invertebrates and at the top of the food chain a population of some 1,200 Eastern Tiger Salamanders. These salamanders are neotenic, in that they forgo the normal terrestrial phase, reaching breeding condition while maintaining an active aquatic lifestyle and larval characteristics such as tail fins, wide heads, spindly legs, large red external gills, and black-spotted olive skin. This 'badger' population demonstrates the plasticity of this species, unlike most Wisconsin salamanders, which do not entertain the neotenic option, or the Mudpuppy, which always does. Lab experiments and field observations suggest that individuals can become neotenic in ponds that are at least ca. 1 m (3.3 ft) deep and unlikely to winterkill and where food is abundant and fish predators (even Bluegills) are absent or at low densities. Yet whereas it may not be uncommon for some slow-developing 'tiger' larvae to enter and perhaps survive their first winter before transforming, true neotenic populations are rare.

Apparently at the ammunition plant, adults fell into the reservoir, probably during the spring en route to or from breeding ponds, and couldn't get out. Before drowning, some bred, and their larvae found suitable conditions to become neotenic. First documented by Gary Casper in 1993, this population may have begun as early as 1942, and they may have recolonized after regular drainings and repairs. This scenario played out again recently in an adjacent reservoir, which was formerly treated and has subsequently produced neotenic individuals as the chlorine disappeared and a pond ecosystem developed. The future of the 'badger' salamanders is uncertain, for with the closing of the plant, the reservoirs have no longer received their careful and costly maintenance and are likely to fail. In addition, the new owners of this property, the WDNR, considers the reservoirs a safety liability. Thus, it was originally decided that the neotenic individuals would be removed and translocated to suitable nearby habitat in 2013. But when WDNR and USGS biologists tested the population for health issues according to standard protocol prior to release, they discovered an apparent viral infection previously unknown in the species. Without a better understanding of the disease, the populations may be unreleasable, and their future is problematic.

The importance of this unique ecosystem and its rare neotenic salamanders may seem obvious to the many people who have visited the reservoir on guided field trips and to those of us who have studied these salamanders. We are taken by their odd yet compelling beauty and enlightened by their story of adaptation and survival, a story that perhaps would not have been told but by the unintended effects of an ammunition plant. Indeed, their educational and scientific value has proved substantial both in situ and in the lab in studies not only of the nature and triggers of neoteny and of amphibian diseases but also of topics with more immediate human impact such as gene expression and tissue regeneration.

And as we grapple with the future of this remarkable population, there are lessons to be learned about resource management as well.

Michael J. Mossman, WDNR (retired)

such, any combination of these factors may drive an individual's decision to leave or remain within the vicinity of the breeding wetland for prolonged periods after reproduction. For example, movement during migration from breeding sites typically occurred when environmental conditions were favorable, at night in association with rain, and primarily through wooded habitats (Madison and Farrand 1998). Little is known about the movement patterns of Eastern Tiger Salamanders outside of breeding migrations, but they presumably move sparingly.

Although information on larval movement is also sparse, Holomuzki and Collins (1983) found that larvae exhibited distinct daily variation in movement patterns related to vertical space in the water column. For example, larvae remained near the water surface during the day but moved closer to the bottom of the water column in the evening. The daily movement patterns of larvae appear to be partially driven by temperature, prey densities, and risk of predation. Holomuzki (1986) found that larvae occupy shallow water during the day due to the presence of greater food densities and higher temperature but avoided these areas at night when the threat of diving beetle predation increased. Temperatures selected by larvae in laboratory experiments were warmer (23.0°C–25.5°C [73.4°F–77.9°F]) than those selected by metamorphosed individuals (18.4°C–23.7°C [65.1°F–74.7°F]; Heath 1975).

In autumn, adult and metamorphosed juvenile Eastern Tiger Salamanders move to overwintering locations. Sometimes these migrations and dispersals include large numbers of individuals. Vogt (1981) captured 526 migrating individuals over a 45-minute period during September in Dane County. In addition, Duellman (1954) found hundreds of individuals moving across a road in Michigan from mid-October through early November, with the latest active individual observed on 20 November.

PREY AND PREDATORS

Relatively little is known about adult Eastern Tiger Salamander dietary preferences, trophic interactions, or community function as predators (reviewed by Davic and Welsh 2004). This paucity of information is likely due to the secretive and fossorial nature of adults, which makes them difficult to study. Despite this, wild adults reportedly consume numerous types of invertebrates that are found among the ground debris, such as insects, earthworms, and snails; they will also consume vertebrates, including young field mice and lizards (reviewed by Petranka 1998).

Compared to those of adults, the diet and predatory behaviors of larvae are better understood. Larvae consume an array of small aquatic organisms, ranging from zooplankton to amphibian larvae and crayfish, depending on developmental stage (reviewed by Petranka 1998). They also prey heavily on larval insects, including large numbers of juvenile mosquitoes (up to a mean of 114 mosquito larvae per day by a single Eastern Tiger Salamander larva; Brodman and Dorton 2006). Larvae detect prey through visual and olfactory cues but also via lateral line systems and electroreception (reviewed by Lannoo and Phillips 2005). The size of prey selected by larval Eastern Tiger Salamanders is only limited by how wide they can open their mouths, and past research has noted shifts in the prey that larvae consume as they grow. For example, Brophy (1980) reported that in southern Illinois younger larvae consumed mostly small ostracods, but as larvae grew they consumed more mollusks that were larger. Cannibal morph larvae generally consume larger prey than typical larvae because they often eat juvenile conspecifics (Collins and Holomuzki 1984). In addition to details of larval dietary composition, numerous studies have focused on the community-level impacts of their trophic interactions. This work suggests that Eastern Tiger Salamander larvae have great influence over the community structure within ephemeral wetlands (Natural History Box: Larval Tiger Salamanders and Trophic Cascades).

The potential predators of Eastern Tiger Salamanders are numerous and vary substantially depending on life stage. As adults, Eastern Tiger Salamanders are reportedly consumed by Eastern Hog-nosed Snakes (*Heterodon platirhinos*), gartersnakes, American Badgers (*Taxidea taxus*), and a variety of birds (Collins 1993). Madison and Farrand (1998) also believed that Northern Short-tailed Shrews (*Blarina brevicauda*) frequently consumed adults encountered in burrows or beneath ground debris. Younger age classes

NATURAL HISTORY BOX: Larval Tiger Salamanders and Trophic Cascades

Eastern Tiger Salamanders appear to occupy a fascinating niche in the wild, but one that is not fully understood. For example, the lives of adults are somewhat mysterious. Yet conclusions can be drawn about their role in terrestrial communities if we compare the ecology of adult Eastern Tiger Salamanders to what is known about other salamander species. For instance, studies support that in some locales the biomass of Eastern Red-backed Salamanders (*Plethodon cinereus*) during the peak of the breeding season (1.7–16.5 kg/ha [1.5–14.7 lb/ac]) can be double that of birds and equal to that of small mammals (Burton and Likens 1975, Petranka and Murray 2001). Follow-up studies suggest that these numbers could be conservative, and peak abundance may actually be up to 14 times that of mammals (Petranka and Murray 2001, Davic and Welsh 2004). Although similar studies have investigated the biomass of ambystomatid salamanders in forests (Raymond and Hardy 1991), there do not appear to be any focused on the typically prairie-dwelling adult Eastern Tiger Salamanders.

Larval Eastern Tiger Salamanders, on the other hand, have received more attention. Aquatic larvae (and gilled neotenic adults) can be top predators in fishless ponds, exhibiting dramatic control over the structure and function of their aquatic communities (Holomuzki et al. 1994). In North Dakota, aquatic larvae can reach densities of 5,000/ha (2,023/ac) and a biomass of 180 kg/ha (161 lb/ac), with a maximum production of 565 kg/ha (504 lb/ac; Deutschman and Peterka 1988)! In Arizona, larvae can become so dense that their voracious appetites can dramatically lower the abundance of their main prey, tiny aquatic invertebrates collectively called zooplankton. Because predation by tiger salamander larvae lowers the abundance of the zooplankton, the phytoplankton, or 'algae,' normally consumed by zooplankton can become very abundant. If so, the water in wetlands packed with tiger salamander larvae can even appear green because of the high abundance of photosynthetic phytoplankton in the pond. Furthermore, the abundant phytoplankton use nutrients from the water that they require for metabolism, which can ultimately lower the water's nitrogen and phosphate levels (Holomuzki et al. 1994). This type of dynamic is known as a top-down control of a biological community or a 'trophic cascade,' where the influence of predators at the top of the food chain cascades down through numerous lower trophic levels. (The most famous example involves Gray Wolves [*Canis lupus*] in Yellowstone National Park. The wolves influence herbivore populations and, thus, indirectly affect the plant communities that herbivores feed on [Ripple and Beschta 2011].) In other words, not only do Eastern Tiger Salamander larvae directly influence populations of their prey (e.g., zooplankton) through predation, they also indirectly influence lower trophic levels that their prey, in turn, feed on (e.g., phytoplankton) and the available nutrients in the water.

Tiger salamander larvae also impact pond communities by indirectly influencing interactions among their prey. For example, in subalpine ponds, predation by tiger salamander larvae modulates competition between caddisflies, which in turn facilitates greater caddisfly diversity because one species of caddisfly cannot outcompete others and become dominant (Wissinger et al. 1999). Thus, it is possible that seasonal changes in larval density or yearly changes in larval recruitment of tiger salamanders could have major influences on wetland ecosystems. These fascinating interactions deserve more attention from researchers of midwestern ecosystems.

Noah J. Anderson, Winona State University, and Joshua M. Kapfer, UW–Whitewater

typically fall victim to smaller predators. Aquatic insects, other amphibian species, and conspecifics will feed on eggs. Morin (1983a) reported heavy predation of eggs by Eastern Newts (*Notophthalmus viridescens*), pressure from which can even exclude Eastern Tiger Salamanders from breeding ponds. Furthermore, dragonfly naiads, diving beetles, gartersnakes (*Thamnophis* spp.), Great Blue Herons (*Ardea herodias*), and other birds are common predators of the larvae (Holomuzki 1986, reviewed by Petranka 1998). Coggins and Sajdak (1982) identified numerous helminth parasites in Wisconsin individuals.

In response to perceived threats, Eastern Tiger Salamanders can enact various defense strategies. Terrestrial adults raise their hindquarters, wave their tails, and secrete a sticky milky-white substance (note that both terrestrial and aquatic adults can secrete this substance; N. Anderson, personal observation). Several authors have reported that this secretion functions as a noxious agent that deters predators and cited its stickiness as a physical impediment to consumption (reviewed by Lannoo and Phillips 2005). Terrestrial Eastern Tiger Salamanders also sometimes adopt an 'unken reflex,' which includes spreading the limbs, arching the back, and elevating the tail vertically over the body, exposing their noxious skin secretions to deter potential predators (Carpenter 1955). Larvae reportedly modify their movement patterns and habitat use to avoid predators (see Activity section).

CONSERVATION

Status: Globally, the Eastern Tiger Salamander is listed as a Least Concern species by the IUCN. While many eastern populations of the Eastern Tiger Salamander are threatened, endangered, or extirpated (especially in the northern Coastal Plain region), its status in Wisconsin is apparently secure, and the species is considered common by the WDNR. However, a complete understanding of its state status is hampered by its reclusive nature. Anecdotal observations in nearby Iowa and Minnesota suggest that Eastern Tiger Salamanders are currently much less common than they were historically (LeClere 2013, Moriarty and Hall 2014). Although these reports could provide insight into Wisconsin populations, directed population research is warranted to confirm possible declines.

Populations: Robust published data on populations are extremely limited. Sever and Dineen (1978) estimated the number of reproducing adults on sampling dates across two years at a breeding pond in Indiana. These estimates ranged from 245 to 1,132 individuals, and the authors believed that the size of their study populations was as high as 1,500–2,000 individuals. Estimates indicate that larval densities can also be extremely high (e.g., 5,000/ha [2,023/ac]; Deutscham and Peterka 1988). Adult sex ratios at breeding sites are often skewed toward males (e.g., Sever and Dineen 1978, Williams et al. 2009b). Survival varies by life stage, with highest mortality in young age classes. For example, estimates of survival from egg to metamorphosis range from 0% to 8.7%, with factors such as temperature, low pH, pond desiccation, predation, and disease taking heavy tolls on larvae (Anderson et al. 1971, reviewed by Petranka 1998). However, estimates suggest that survival could substantially increase with age. Over a two-year period, Sever and Dineen (1978) estimated that ca. 72% of adults that they documented at a breeding pond in year one survived to the second year.

Threats: Several general conservation concerns are relevant to Eastern Tiger Salamanders, including habitat loss and degradation, road mortality, and disease (see Introduction to Conservation and Management of Wisconsin's Amphibians and Reptiles chapter). Loss of breeding ponds and appropriate upland habitat are likely threats of particular importance, although some information on this topic appears contradictory. For example, Vogt (1981) suggested that Eastern Tiger Salamanders were resilient to the impacts associated with agriculture and urbanization in Wisconsin, as he had observed large numbers in both types of disturbed habitats. He further recounted capturing "526 tiger salamanders in the vicinity of an apartment complex near a marsh on the east side of Madison" (Dane County) in September 1966. Yet recent evidence supports that habitat loss, degradation, and fragmentation have negative impacts on this species, although the effects are sometimes subtle. Eastern Tiger Salamanders studied in Ohio, for example, did not associate with sites that possessed high woodland edge-to-area ratios (i.e., woodland fragments that were small or had convoluted borders in landscapes dominated by agriculture; Weyrauch and Grubb 2004).

The presence of roads can cause substantial mortality due to vehicular strikes (e.g., Duellman 1954, Clevenger et al. 2001) and influence movement patterns. Madison and Farrand (1998) reported that adults migrating for reproduction in New York avoided paved roads, grassy fields, and "commercial areas" in favor of woodlands. In addition, Porej et al. (2004) determined that Eastern Tiger Salamanders in Ohio were negatively associated with total length of paved roads that existed within a 1 km (0.6 mi) buffer around potential breeding wetlands. Other factors exist that often accompany human disturbance and can indirectly make habitat unsuitable for this species. For example, unnatural colonization of breeding wetlands by predators of larvae such as American Bullfrogs (*Lithobates catesbeianus*) and fishes can negatively impact populations (Maret et al. 2006). Stiles et al. (2016) further reported that Eastern Tiger Salamanders did not occupy restored wetlands where fishes or introduced American Bullfrogs became established.

Several important amphibian diseases pose potentially serious threats to Eastern Tiger Salamanders in Wisconsin. Viral diseases, including ranaviruses, affect amphibians globally (Jancovich et al. 1997), and strains exist that reportedly target tiger salamanders, such as the *Ambystoma tigrinum* virus (ATV). To date, ATV has been implicated in several cases of mass tiger salamander mortality (reviewed by Jancovich et al. 2001). Although currently no confirmed reports of ATV in Wisconsin populations exist, it could be difficult to detect if its action and epidemiology are subtle or work synergistically with other stressors. For example, ecologically relevant concentrations of environmental pollutants such as nitrates and the herbicide atrazine suppressed the immune response of laboratory-reared Eastern Tiger Salamander larvae, making them more susceptible to ATV (Forson and Storfer 2006). Studies also support that temperature can influence susceptibility to the disease and the likelihood that it will cause mortality. Laboratory experiments found that most larvae survived when exposed to ATV at 26°C (79°F), while most died when exposed at 18°C (64°F; Rojas et al. 2005). Work by Kerby et al. (2011) further suggested that multiple natural (e.g., presence of predators) and unnatural (e.g., pesticides and ATV) stressors can act synergistically to exacerbate the negative effects of individual stressors, leading to reduced body size or higher mortality. Unfortunately, ranaviruses that affect tiger salamanders can also be spread by the activities of humans (Jancovich et al. 2005). In addition to viruses, the chytrid fungus *Batrachochytrium dendrobatidis* (Bd) causes the disease chytridiomycosis, which is a global threat to amphibians. Despite this, cases of Bd infection are rarely reported from wild-caught Eastern Tiger Salamanders. Davidson et al. (2003) detected Bd in an Arizona population, but the infection did not cause mortality in monitored individuals by 60 days post-infection. The more recently discovered salamander chytrid fungus *Batrachochytrium salamandrivorans* (Bsal) is also a potentially serious threat (Martel et al. 2013), but it has not been identified in tiger salamanders yet.

Management: Many general management considerations that are relevant to wetland, woodland, and open grassland habitats are applicable to Eastern Tiger Salamanders. Strategies to offset road mortality, control the spread of disease, and limit environmental pollution could also benefit this species (see Introduction to Conservation and Management of Wisconsin's Amphibians and Reptiles chapter). Due to these salamanders' tendency to migrate long distances for reproduction, an important management consideration could be the protection of suitable upland habitat around known breeding sites so that connections between overwintering and breeding habitats are maintained. For example, Semlitsch (1998) estimated that buffers of protected vegetation out to 164.3 m (534 ft) from the wetland edge would protect 95% of salamander populations associated with breeding ponds. Given the findings of Porej et al. (2004), roads or road traffic within protected buffers should also be kept to a minimum (or effective road underpasses implemented), particularly in areas where Eastern Tiger Salamanders migrate en masse to breeding or overwintering sites.

Control of tiger salamander predator introductions to breeding wetlands, such as American Bullfrogs and predatory fishes, could be a critical component of management plans for this species. Accordingly, maintenance of ephemeral wetland hydrology to ensure that seasonal desiccation occurs could help restrict colonization of breeding ponds by these aquatic predators. Wetland restoration could be another viable management strategy, given that Eastern Tiger Salamanders colonized and bred in 83% of 118 restored wetlands monitored in Iowa (Stiles et al. 2016). While the overall effectiveness of wetland restoration as a management strategy is debatable and requires further assessment, Stiles et al. (2016) only found Eastern Tiger Salamanders in the smallest restored wetlands sampled ($\bar{x}$ surface area = 0.67 ha [1.66 ac]), which were devoid of fishes and American Bullfrogs. In Minnesota, the widespread construction of cattle stock ponds to help prevent erosion has allowed Eastern Tiger Salamanders to breed in areas that formerly had limited breeding habitat (N. Anderson, personal observation).

Four-toed Salamander

Hemidactylium scutatum (Temminck 1838)

Erik R. Wild and Donald J. Brown

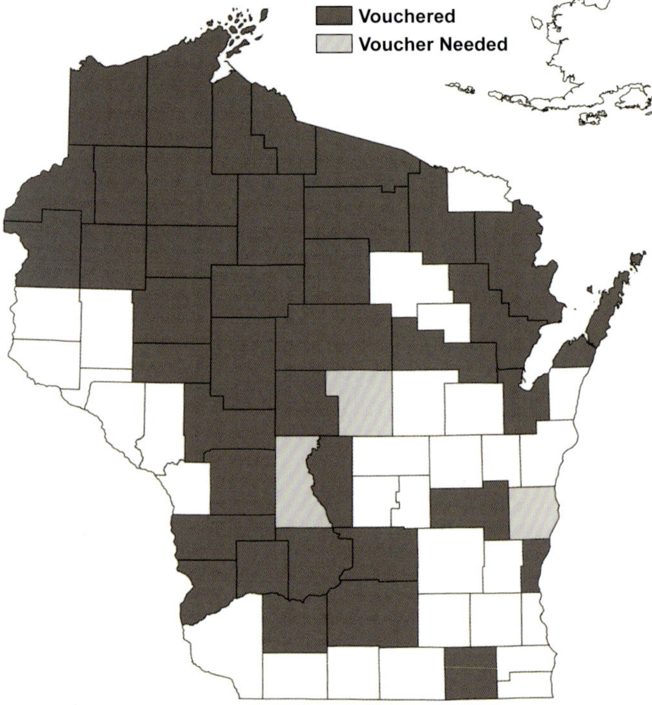

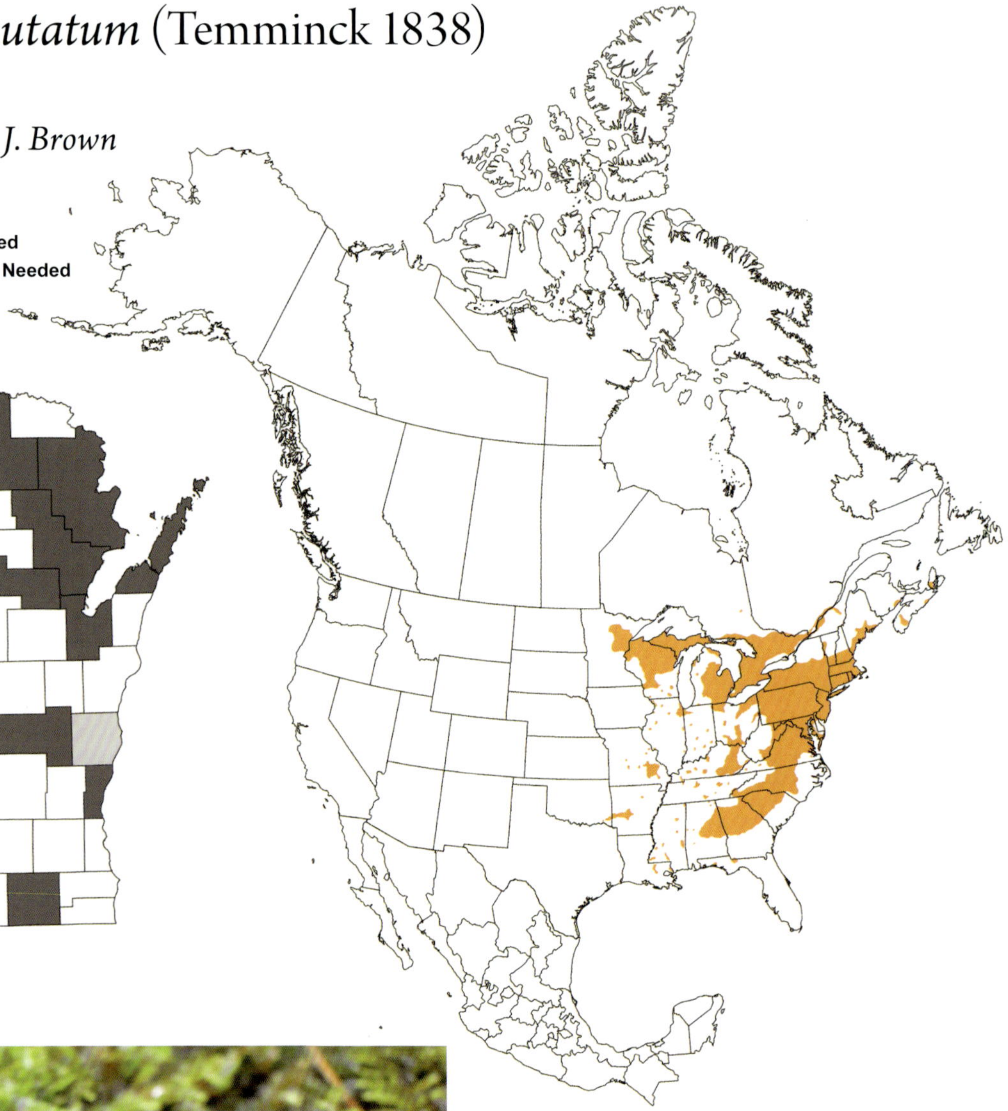

Adult Four-toed Salamander (Iowa County; photo by C. Raimond).

AT A GLANCE

As its common name suggests, the Four-toed Salamander has only four toes on its hind feet, which distinguishes it from nearly all other salamanders in Wisconsin. It is the state's smallest salamander and is rarely seen, due to its secretive nature. Like all members of the family Plethodontidae, the Four-toed Salamander lacks lungs, but it differs from many in having aquatic rather than terrestrial larvae. The tail of the adult narrows at its base and can be curled upward in a curious defensive display. Scattered occurrence records for the species exist throughout Wisconsin, but it seems to be more common in the northern third of the state. Four-toed Salamanders are associated with mesic forests that contain bogs, springs, or spring-fed creeks. They primarily breed in autumn (mid-September through October), but females do not lay their eggs until the spring. Egg clutches are often deposited in *Sphagnum* moss overhanging water in small wooded wetlands. Nests may contain a single clutch or clutches from several individuals, which are attended by females. After about five weeks, larvae that are small and fully aquatic with external gills hatch from the eggs. Larvae metamorphose into terrestrial juveniles in three to eight weeks. Adult Four-toed Salamanders primarily consume arthropods and are preyed upon by a variety of amphibians, fishes, reptiles, and insects. Populations are vulnerable to the loss and degradation of their specialized breeding habitat. The Four-toed Salamander is considered a species of Special Concern and a Species of Greatest Conservation Need in Wisconsin.

SYSTEMATICS

Recently, there has been little question of the placement of the monotypic *Hemidactylium* within the monophyletic Plethodontidae. Historically, however, the position of *Hemidactylium* within Plethodontidae was contentious (reviewed by Vieites et al. 2011). The most thorough taxonomic analysis to date is that of Vieites et al. (2011), who analyzed complete mitochondrial genomes and nuclear genes and found that *Hemidactylium* is most closely related either to the North American *Batrachoseps* (slender salamanders) or to a clade of *Batrachoseps* and the tropical Bolitoglossini. The proposed taxonomy of these authors, formalized in Wake (2012b), is consistent with either topology, with *Hemidactylium* in a monotypic Hemidactyliini (tribe) within the Hemidactyliinae (subfamily), which contains >300 other species. No subspecies are currently recognized (Crother 2017), despite this group's age, disjunct distribution, and extensive genetic variation (Herman and Bouzat 2016). Among Wisconsin salamanders, only one other species occurs in Plethodontidae: *Plethodon cinereus* (Eastern Red-backed Salamander).

DESCRIPTION

Adults: The body of Four-toed Salamanders is dark grayish or rusty reddish brown, mottled with bronze and black, and flecked with blue on the sides (Figure 140). The dorsum of the tail, limbs, snout, and eyelids are dull orange finely mottled with brown. The venter was best described by Vogt (1981) as "porcelain white," but it may have a bluish tinge and has irregularly spaced small black blotches that are usually distributed toward the sides (Figure 141). The body of Four-toed Salamanders has 13–14 costal grooves (Bishop 1919, Vogt 1981) and is shorter than the tail in adults (Blanchard and Blanchard 1931). The tail has a distinct constriction at its base. Individuals have four digits on both the front and

Figure 140. Dorsal view of a Four-toed Salamander. Note the mottled rusty-reddish-brown background coloration and blue flecks on the sides (Sheboygan County; photo by J. DeBoer).

Figure 141. Ventral view of a Four-toed Salamander. Note the porcelain-white color with irregularly spaced small black blotches distributed mostly toward the sides (Iron County; photo by S. Viernum).

hind feet. This likely represents an evolutionary loss of the fifth (outer or postaxial) toe (Alberch and Gale 1985). A distinct median dorsal groove originates near the base of the tail and extends onto the head, where it bifurcates with a branch to each eye (Bishop 1919); this dorsal groove is met by the upper extension of the costal grooves (see Natural History Box: What's in a Name?).

The Four-toed Salamander is Wisconsin's smallest salamander, with individuals reaching up to 5–7 cm SVL (2.0–2.8 in) and 10 cm TL (3.9 in; Vogt 1981). Live individuals sampled from throughout the northern half of Wisconsin ranged from 3.30 to 3.90 cm SVL (1.3 to 1.5 in, $\bar{x} = 3.57$ cm [1.4 in], $n = 13$), from 7.40 to 9.40 cm TL (2.9 to 3.7 in, $\bar{x} = 8.23$ cm [3.2 in], $n = 9$), and from 0.7 to 1.2 g (0.02 to 0.04 oz, $\bar{x} = 0.9$ g [0.03 oz], $n = 10$; Erik Wild, UW–Stevens Point, unpublished data). Preserved Wisconsin specimens in the MPM collection ranged from

NATURAL HISTORY BOX: What's in a Name?

The standard English name, Four-toed Salamander (Crother 2017), is pretty straightforward in referring to the four toes on the hind feet, a characteristic uncommon among salamanders. The scientific name, *Hemidactylium scutatum*, on the other hand, is a bit more interesting. Since the species is placed in its own genus (called a monotypic genus), the generic name, *Hemidactylium*, refers only to that species. Originally established by Tschudi (1838), *Hemidactylium* is formed by the root *hemi-*, from the Greek *hemisus*, meaning 'half,' and *-dactylium*, meaning 'possession of a dactyl-' with 'dactyl-' from the Greek word *daktulos*, meaning 'digit, finger, toe.' Thus, 'half toe' refers to the fewer number of toes on the hind feet (Neill 1963b). The specific epithet, *scutatum*, means 'covered with shield-like plates,' from the Latin *scutatus*, 'armed with a shield' (Neill 1963c). At first this seems to be a strange name for a small, moist, smooth salamander. However, upon close inspection of an individual Four-toed Salamander, one can notice a distinct median dorsal groove in the skin originating near the base of the tail and extending onto the head, where it bifurcates with a branch to each eye (Bishop 1919). This dorsal groove is met by the upper extension of the costal grooves. The shapes outlined by these grooves give a superficial resemblance to overlapping plates or 'shields'; thus, Temminck originally assigned it the name *scutata* (feminine of *scutatus*) to agree in gender with the original genus name, *Salamandra* (Temminck and Schlegel 1838). *Hemidactylium* is masculine, and so when the species was placed in that genus the specific name changed to the masculine *scutatum*. So we have the 'half-toed covered with shield-like plates salamander.'

Erik R. Wild, UW–Stevens Point

Figure 142. Some of the eggs of a Four-toed Salamander nest in Wisconsin uncovered from ca. 8 cm (3 in) deep below the surface of a *Sphagnum* hummock overhanging the edge of open water. Note that the eggs are early in development, and the development of the neural fold is apparent (Wisconsin; photo by E. Wild).

3.3 to 4.4 cm SVL (1.3 to 1.7 in, $\bar{x} = 3.5$ cm [1.4 in], $n = 50$), from 3.2 to 5.3 cm in tail length (1.3 to 2.1 in, $\bar{x} = 4.0$ cm [1.6 in], $n = 41$), and from 6.5 to 9.7 cm TL (2.6 to 3.8 in, $\bar{x} = 7.6$ cm [3.0 in], $n = 41$). Note that minimum SVL for museum specimens included in this size range was 3.3 cm (1.3 in) (see Reproduction and Development section). Furthermore, measured specimens were not dissected to determine sex or sexual maturity and likely included some immature second-year individuals.

Preadult stages: The eggs of Four-toed Salamanders in Wisconsin are cream colored (Figure 142) and measure 0.4–0.5 cm (0.16–0.20 in) in diameter (Watermolen 1995). The aquatic larvae measure 1.1–1.5 cm TL (0.43–0.59 in) at hatching (Figure 143) and 1.6–2.5 cm TL (0.63–0.98 in) at metamorphosis (Vogt 1981, Parmelee et al. 2002). By comparison, Blanchard (1923) reported that larval TL ranged from 1.73 to 2.14 cm (0.67 to 0.85 in) in Michigan. External gills are present. The dorsal tail fin extends onto the body nearly to the head. There are four toes on the hind limbs (when developed). The dorsal background color is gray or greenish brown, and a light median dorsal band with an irregular dark border is present (Watermolen and Gilbertson 1996; Figure 144). A dark band or bar is present behind the eye that becomes more prominent in older larvae and newly metamorphosed individuals (Harding 1997). Blanchard (1923) described this as a horizontal black line through the eye and considered it "characteristic of the larva."

Above, top: Figure 143. Eggs of a Four-toed Salamander nest just prior to hatching (Wisconsin; photos by E. Wild).

Above, bottom: Figure 144. A newly hatched larval Four-toed Salamander (Wisconsin; photo by E. Wild).

Large juveniles of this diminutive species are difficult to distinguish from adults due to similarities in appearance and overlapping size classes (Blanchard and Blanchard 1931). Juvenile Wisconsin specimens in the MPM collection ranged from 2.40 to 3.20 cm SVL (0.9 to 1.3 in, $\bar{x} = 2.95$ cm [1.2 in], $n = 37$), from 1.8 to 3.8 cm tail length (0.7 to 1.5 in, $\bar{x} = 3.2$ cm [1.3 in], $n = 27$), and from 4.2 to 7.0 cm TL (1.7 to 2.8 in, $\bar{x} = 6.1$ cm [2.4 in], $n = 27$). Field samples collected from northern Wisconsin ranged from 1.80 to 3.20 cm SVL (0.7–1.3 in, $\bar{x} = 2.67$ cm [1.1 in], $n = 17$) and from 3.0 to 7.2 cm TL (1.2 to 2.8 in, $\bar{x} = 6.1$ cm

[2.4 in], $n = 14$; E. Wild, unpublished data). Only a small proportion of these individuals were weighed; the weights ranged from 0.5 to 0.8 g (0.02 to 0.03 oz, $\bar{x} = 0.6$ g [0.02 oz], $n = 4$; E. Wild, unpublished data).

Variation: Substantial variation in color or pattern is not known among Wisconsin populations. Current genetic evidence indicates that populations in the state are part of a discrete monophyletic clade that recolonized the entirety of the glaciated region of the Upper Midwest (Herman and Bouzat 2016). Males differ from females in having two nasolabial grooves, which give the snout a more square shape when viewed from above, compared to the more rounded snout of females (Harding 1997). In addition, males have an overhanging upper lip and a mental gland on the chin that becomes enlarged in reproductively active individuals (Petranka 1998). Adult females are longer than adult males (Neill 1963c). In Michigan, adult females ($\bar{x}$ TL = 7.47 cm [2.9 in], range = 6.20–8.77 cm [2.4–3.5 in], $n = 64$) were larger than adult males ($\bar{x}$ TL = 6.51 cm [2.6 in], 4.98–7.66 [2.0–3.0 in], $n = 93$; Blanchard and Blanchard 1931). Furthermore, Blanchard and Blanchard (1931) reported that the length difference between adult females and males increased with age. Blanchard and Blanchard (1931) also found evidence that mean size of adults varied between two sites, indicating that habitat quality may affect growth rates. From a range-wide perspective, Neill (1963c) stated that Georgia specimens had relatively larger black spots on the venter, as compared with more northerly specimens, and Herman and Bouzat (2016) pointed out that morphology and nesting behavior varied across their distribution.

Diagnosis: Adult Four-toed Salamanders are characterized by a lack of dorsal and ventral fins along the tail musculature and smooth skin. The body is long and thin, and nasolabial grooves are present in mature males. Tail coloration is distinct from the body coloration, and the ventral coloration is milky white with black flecks. Individuals have four toes on each hind foot. The only other salamander in Wisconsin with four toes on the hind feet is the much larger and fully aquatic Mudpuppy (*Necturus maculosus*). However, the Mudpuppy rarely overlaps in habitat with the Four-toed Salamander and has external gills as an adult. The salamander in Wisconsin that is most likely to be sympatric and similar in appearance to the Four-toed Salamander is the Eastern Red-backed Salamander. However, Eastern Red-backed Salamanders have five digits on the hind feet, tail coloration that does not distinctly differ from that of the body, and typically a broad middorsal longitudinal red stripe.

Four-toed Salamander eggs can be differentiated from other amphibian eggs by where they are deposited: in *Sphagnum* moss at 2–12 cm (0.8–4.7 in) above the water's surface (Watermolen 1995). Larval Four-toed Salamanders can be distinguished by the presence of external gills and forelimbs. They possess a dorsal fin that extends far onto the body and a light median dorsal band but lack prominent dark stripes through the eyes. Like adults, larval Four-toed Salamanders have four toes on the hind feet, which separates them from most other larval salamanders, except Mudpuppies (with which they rarely share habitat).

DISTRIBUTION AND HABITAT

Global distribution: The Four-toed Salamander has one of the broadest geographic ranges of any plethodontid salamander (Herman and Bouzat 2016). Its scattered distribution extends from extreme southern Maine east to northeastern Minnesota, south to southeastern Oklahoma, and southeast to northwestern Florida. Even in portions of the range considered contiguous, Four-toed Salamander populations tend to be highly localized (Herman and Bouzat 2016).

Distribution in Wisconsin: Published county records for the Four-toed Salamander are scattered throughout Wisconsin, although they appear to be more common in the northern third of the state. Individuals have even been documented on the Apostle Islands (Ashland and Bayfield Counties) in Lake Superior (Hecnar et al. 2002, Casper 2008c). Of the published county records, suitable museum vouchers could not be found for Juneau, Portage, and Sheboygan (see the Appendix).

Localities of possible occurrence: Given the broad distribution of potential habitat in the state and the widely scattered observation records, it seems likely that this species has a wider distribution in the state than is currently documented (WDNR 2018b). Given the confirmed record from Walworth County, this species could be present in other adjacent counties where suitable habitat occurs in the Southern Kettle Moraine State Forest (e.g., Jefferson and Waukesha Counties). In support of this, there are old (but unsubstantiated) reports from Winnebago, Waukesha, and Racine Counties (Cahn 1929, Dickinson 1965), which Vogt (1981) disregarded. Yet even the Walworth County record is based on an old specimen (CHAS HERP 2560, collected in 1940). Perhaps these reports and specimen are representative of a time when woodland habitat in southeastern Wisconsin was more intact and conducive to Four-toed Salamanders (see Past, Present, and Potential Future Landscapes of Wisconsin chapter).

Habitat: No robust studies of Four-toed Salamander habitat preferences have been conducted in Wisconsin. Four-toed Salamanders require both forest and aquatic habitats for their annual cycle; thus, they are limited to areas that provide both of these habitats in close proximity (Blanchard 1923, Neill 1963c, reviewed by Kingsbury and Gibson 2012). In Wisconsin, they are associated with both northern and southern mesic forests (Vogt 1981). Pentecost and Vogt (1976) listed them as a characteristic species of southern mesic (maple-beech) forests in the Lake Michigan drainage basin, although they noted their presence in northern lowland forests, boreal forests, and northern and southern mesic forests of Wisconsin. Kent (1983) described the hardwood forest associated with his study population in Bayfield County as composed of Balsam Fir (*Abies balsamea*) and old second-growth northern hardwoods, including Sugar Maple (*Acer saccharum*), Northern Red Oak (*Quercus borealis*), and Paper Birch (*Betula papyrifera*). Common ground cover species included Large-leaved Aster (*Aster macrophyllus*) and Bracken Fern (*Pteridium aquilinum*). There have also been occasional reports of Four-toed Salamanders in drier habitats in Wisconsin. For example, Suzuki (1951) found them in relatively dry maple–Yellow Birch (*Betula alleghaniensis*) forest and witch hazel (*Hamamelis* spp.) stands. Cochran and Lyons (1986) observed them in an upland forest with no boggy habitat in the vicinity, and Pope (1931) described them as "easily found under both boulders and logs on rather dry hillsides."

Males, juveniles, and nonbreeding females are often associated with specific microhabitats within upland forests. For example, they are found beneath various types of ground cover, such as moss, decaying woody debris, leaf litter, and stones (Neill 1963c, Briggs and Young 1976, Cochran and Lyons 1986). They can estivate in these microhabitats and also within small ground burrows or decaying root systems (Harding 1997). Within suitable forest types, populations of the Four-toed Salamander are found near appropriate nesting habitats. Blanchard (1923) described suitable nesting habitats as ponds, pools, and swamps with moss or soft rotting wood around the shoreline. Likewise, Epstein et al. (1999b) stated that they nest in woodland ponds or seeps with abundant mosses. They also use emergent moss hummocks (Erik Wild, UW–Stevens Point, personal observation). Furthermore, although Vogt (1981) and Kent (1983) stated that they are not characteristic of open *Sphagnum* bogs, individuals have been commonly found nesting in these habitats in Wisconsin (i.e., Bayfield and Forest Counties; E. Wild, personal observation). Although adult Four-toed Salamanders have rarely been observed swimming, they are presumed to do so, since numerous nesting sites have been on *Sphagnum* islands (Blanchard 1934a, E. Wild, personal observation).

Within appropriate wetlands, nesting microhabitats are most commonly associated with hummocks or banks of *Sphagnum* moss that offer a nearly vertical surface (Chalmers and Loftin 2006). Eggs are deposited so that they are placed just under the surface of the moss but positioned mostly above open water (Epstein et al. 1999a, E. Wild, personal observation). Although *Sphagnum* moss seems to be the preferred nesting substrate in Wisconsin, elsewhere in their range individuals have been found nesting in other moss species (Gilbert 1941, Wood 1955, Easterla 1971, Petranka 1998). For example, Casebere and Lodato (2011) found them nesting in nine different species of moss in Indiana, of which *Thuidium delicatum* was the most commonly used, followed by *Sphagnum* species. In addition, Four-toed Salamanders will nest in other substrates such as grasses and sedges and inside rotted logs (reviewed by Thurow 1997, reviewed by Harris 2005, Casebere and Lodato 2011). For example, of 36 nests found at one site in Michigan, 32 were in clumps of *Sphagnum* moss, 2 were within rotten wood, 1 was in a grass hummock, and 1 was in leaf litter (Breitenbach 1982). In Virginia, Wahl et al. (2008) found that nesting Four-toed Salamanders selected specific locations with more live vegetation (e.g., mosses and grasses), steeper bank angles, and lower surface moisture pH at nests than random locations. Furthermore, they found that nesting locations were generally common on north-facing slopes with lower ambient air temperatures. Chalmers and Loftin (2006) found similar results in Maine, as well as a positive association with water depth. Larvae occupy the aquatic habitats associated with the nest location (Neill 1963c). Four-toed Salamanders overwinter in holes, channels, or crevices in the ground or under forest debris (reviewed by Petranka 1998).

REPRODUCTION AND DEVELOPMENT

In the Upper Midwest, Four-toed Salamanders breed in autumn (mid-September to late October; Blanchard 1933c) and deposit eggs in the spring, although they have been reported to breed in the spring as well (Minton 1972). The earliest reported Wisconsin nesting dates are late April in the south (i.e., Vernon County) and mid-June in the north (i.e., Ashland County; Vogt 1981, Watermolen 1995). Nest-site fidelity has been documented for Four-toed Salamanders (Breitenbach 1982). Laboratory observations suggest that courtship occurs at night (Blanchard 1933c, Branin 1935). Blanchard (1933c) speculated that there was no congregation of the sexes for mating (but see Branin 1935). Branin (1935) described the courtship as beginning with a period of excitement, especially with the males, when considerable 'nosing' occurred, occasionally accompanied by males moving briskly about with the tail turned sharply at 90° from the body while the salamander circled once or twice. This was followed by a straddle walk, where the female straddles the tail of the male and presses her chin against the dorsal surface of the base of his tail, maintaining her position as the male moves forward, following his undulating tail with her chin such that constant contact with the dorsal surface of the male is maintained for at least twenty minutes. The male leads the female over the spermatophore, which she picks up in her cloaca (Branin 1935). Observations in the laboratory indicate that water may serve as the stimulus for spermatophore deposition by males (Branin 1935), which presumably deposited two at a time (Blanchard 1933c). Blanchard (1933c) described the spermatophore as being approximately 0.2 cm (0.08 in) in height and having an expanded gelatinous disk (0.2 cm [0.08 in]) at its base that tapers upward into a stalk (0.1 cm [0.04 in]) and ends in a pale yellowish top containing sperm.

Females oviposit while turned upside down (Wood 1955, Duellman and Trueb 1986). Eggs are individually attached to rootlets, rhizoids, moss strands, or other substrates at depths of ca. 5.1–7.6 cm (2.0–3.0 in) and are deposited 2.0–17.8 cm (0.79–7.00 in) above water (Blanchard 1923, Watermolen 1995, Petranka 1998). The eggs are often laid directly over the water, presumably so that hatching larvae drop directly into the aquatic habitat (reviewed by Harris 2005). Wild (2002) reported nests with 10, 42, and 52 eggs in northeastern Wisconsin (Forest and Langlade Counties). Reported clutch sizes in the wild vary tremendously, likely because clutches are sometimes deposited communally in *Sphagnum* nests, which confounds the determination of individual clutch size. Some authors follow the general rule that if eggs are all at the same stage and number fewer than 40, it should be considered a single brood (Blanchard 1934a, Harris and Gill 1980, but see Breitenbach 1982). Watermolen (1995) reported 15–25 (and possibly as many as 64) eggs per clutch. Harding (1997) reported 15–64 eggs per clutch in the Great Lakes region.

Not all females reproduce annually. In a ten-year laboratory/mesocosm study, most females bred in consecutive years (55.6%), but some females skipped up to three years of reproduction ($n = 257$ observations; Harris and Ludwig 2004). In the same study, a feeding experiment prior to winter hibernation showed that body condition the following spring significantly differed between high and low food level treatments, and as a result individuals from the high food level treatment were twice as likely to breed the following year (Harris and Ludwig 2004). Females do not appear to eat while brooding (Wood 1953, Harris 2005).

Communal nesting has been frequently documented for this species. Females may lay and attend single clutches, lay and attend eggs in joint nests, or lay eggs in joint nests and leave them to be brooded by another female (Harris et al. 1995). Brooding females are typically solitary (Blanchard 1934a, Harris et al. 1995), but multiple adult females and even juveniles have been observed in nests in Wisconsin (E. Wild, personal observation) and elsewhere (Wood 1953). In Michigan, Breitenbach (1982) found 13 of 109 nests (12%) to be communal, and Blanchard (1934a) reported communal nesting for 61% of nests, one of which contained 1,110 eggs. Gilbert (1941) reported that 2 of 32 nests were communal in New York. Studies in Virginia found that brooding enhanced embryonic survival, but survival did not differ between solitary and jointly brooded nests (Harris and Gill 1980, Harris et al. 1995). Communal nesting may facilitate the transmission of antifungal bacteria to eggs (Banning et al. 2008). Petranka (1998) reasoned that the brooding of joint nests by a single female is not due to brood parasitism or aggressive usurpation but instead the result of kin selection.

The most thorough estimate of time from laying to hatching of eggs is that of Blanchard (1923), who approximated 38 days in Michigan. Humphrey (1928) described cleavage and gastrulation, and Bishop (1919) thoroughly described and illustrated embryonic development prior to metamorphosis. Upon hatching, the larvae either drop directly into the water or, if the nest is not positioned directly over the water, incorporate vigorous body movements (e.g., flips, violent wriggling) to reach the water (Blanchard 1923). Larvae metamorphose in three to eight weeks in Wisconsin (Vogt 1981). In Michigan, Blanchard (1923) reported that metamorphosis occurred in ca. six weeks. O'Laughlin and Harris (2000) estimated a linear growth curve for the first two-thirds of the larval period, with growth slowing thereafter. In that study, individuals increased in weight from ca. 0.015 g (0.0005 oz) to ca. 0.050–0.080 g (0.0018–0.0028 oz) over a ca. 35-day development period. Blanchard and Blanchard (1931) reported that ratio of tail length to SVL increased with age. The smallest female found attending eggs (and, therefore, presumably an adult) was 3.3 cm SVL (1.3 in). In Michigan, Blanchard and Blanchard (1931) found mature males as small as 5.0 cm TL (2.0 in) and females as small as 6.2 cm TL (2.4 in). Sexual maturity was obtained at 22 months in a population from lowland Virginia (Wood 1953) and by ca. 27 months (near autumn of the third summer) in a Michigan population (Blanchard and Blanchard 1931). Harris (2005) used the recapture history of a single breeding female and, under the assumption that age at first reproduction was two years, estimated that females can live to be at least nine years old.

ACTIVITY

According to Blanchard (1934b), Four-toed Salamanders "emerge from hibernation as soon as there has been sufficient warm weather to thaw the ground well." In Wisconsin, they likely emerge from hibernation between late March and early May (Christoffel et al. 2009). After spring emergence, males and juveniles proceed with normal feeding activities, whereas females migrate to nesting habitat (Blanchard 1934b). However, Bishop (1919) reported that juveniles were found in bogs in New York (as they have been in Wisconsin; E. Wild, personal observation), so he speculated that they too may migrate. In Michigan, peak migrations of females to a nesting site occurred in late March and early April (Breitenbach 1982). Blanchard (1934b) speculated that temperature rather than rainfall stimulates migration to nesting sites. However, a more robust study by Vitale (2013) found that lunar phase (<22%) and precipitation in the previous 24 hours (≥5.2 cm [2.0 in]) were the best predictors of movement nights in Massachusetts.

No intensive studies have been conducted on patterns of nonbreeding space use for this species. Grant (1955) described Four-toed Salamanders as weakly territorial based on laboratory observations. However, no studies have documented individuals defending resources in the wild. Furthermore, Carreno and Harris (1998) did not observe aggressive defense of nests against conspecifics by brooding females in Virginia. Vogt (1981) found eight individuals in a rotten log that he presumed to be a communal hibernaculum on Stockton Island, Wisconsin. In Michigan, Blanchard (1933b) reported a congregation of nearly 200 on 9 November within "an area of a very few square feet . . . snuggled together in little groups or clumps in small depressions or cavities in the ground under the leaves." Moreover, the event was a communal aggregation of Four-toed Salamanders, Spring Peepers (*Pseudacris crucifer*), Western Chorus Frogs (*P. triseriata*), Wood Frogs (*Lithobates sylvaticus*), Jefferson Salamanders (*Ambystoma jeffersonianum*), Eastern Red-backed Salamanders (*Plethodon cinereus*), and Eastern Newts (*Notophthalmus viridescens*) (Blanchard 1933b).

PREY AND PREDATORS

No dietary studies for Four-toed Salamanders have been conducted in Wisconsin. In the Great Lakes region, post-metamorphic individuals have been reported to eat small arthropods, their larvae, and other invertebrates, including beetles, ants, springtails, flies, spiders, snails, and earthworms (Pentecost and Vogt 1976, Harding 1997). Bishop (1919) found that the stomach of a specimen in New York contained Lepidoptera larvae (Crambinae) and rove beetles (Staphylinidae). Larval Four-toed Salamanders are reported to eat plankton and small worms (Harris 2005).

There are no reports of predation on Four-toed Salamanders in Wisconsin. Harris (2005) reported that skunks might prey on adults, and Petranka (1998) suggested that numerous vertebrates might prey on adults, such as shrews, small woodland snakes, and birds. Predation by the Pigmy Rattlesnake (*Sisturus miliarius*) was observed in Arkansas (Trauth and Cochran 1991). Larvae are known to be preyed upon by adult Eastern Newts and some fish species (Kats 1988, Wells and Harris 2001, Harris et al. 2003). Furthermore, Harris (2005) speculated that larval *Ambystoma* species and odonate naiads are among the many potential predators of larvae.

Numerous observations have been made on the defensive behavior of adult Four-toed Salamanders. Harding (1997) described that a threatened adult will "curl its body and tuck its head under its tail. The tail may be raised exposing the bright white underside, perhaps to attract the predator's attention away from the head and body." Additional antipredator mechanisms observed in adult Four-toed Salamanders include noxious skin secretions, particularly from glandular concentrations on the tail dorsum, and several types of behavioral responses exhibited in response to predators (Bishop 1919, Brodie 1977, 1983), for example, raising and undulating the tail, flipping or coiling the body when attacked, exposing the white venter, or remaining motionless (Brodie 1977). Induced and voluntary caudal autotomy to avoid predators has also been observed (e.g., Bishop 1919, Brodie 1977). Cupp (2001) also obtained evidence that the Four-toed Salamander can detect odors of predatory Northern Red Salamanders (*Pseudotriton ruber*).

Larval Four-toed Salamanders exhibit a variety of responses when faced with the threat of predation. Wells and Harris (2001) found that presence of predatory Eastern Newts caused larvae to lower their activity level, presumably to avoid being noticed by predators. However, the authors suggested that the cost associated with this behavior was a reduction in growth rate, as larvae also reduced foraging activity. Hess and Harris (2000) demonstrated that the eggs of Four-toed Salamanders are unpalatable to some insect predators, such as carabid beetles.

In Michigan, ca. 70% of the Four-toed Salamanders examined were infected with the ciliated protozoan *Cepedietta michiganensis* (Woodhead 1928), which are consumed by larvae and are carried through metamorphosis into the adult stage (Woodhead and Kruidenier 1936). This parasite may interfere with the ability to produce eggs (Blanchard and Blanchard 1931). Bishop (1919) found the gelatinous envelopes of several eggs to be infested with small nematodes in New York, and Harris and Gill (1980) reported that species of *Gliocladium*, an ascomycete fungus genus, is pathogenic to the eggs (Harris and Gill 1980). Four-toed Salamanders are known to possess the cutaneous bacterium *Janthinobacterium lividum*, which has shown promise in development of pharmaceuticals to treat a variety of pathologies (i.e., fungal, bacterial, viral, protozoal, parasitic, and oncological) in a number of organisms and has already shown therapeutic effects in treating athlete's foot in humans (United States Patent US 8,506,952 B2) and chytridiomycosis in amphibians (Harris et al. 2006).

CONSERVATION

Status: Globally, the Four-toed Salamander is listed as a Least Concern species by the IUCN. It is considered a species of Special Concern and a Species of Greatest Conservation Need in Wisconsin by the WDNR. However, a thorough understanding of the distribution and abundance of the species in Wisconsin is lacking.

Populations: With the exception of reproductive ecology (see Reproduction and Development section), very little research has been conducted on Four-toed Salamander population dynamics in the wild. The estimated sex ratio for a population in Michigan was 1.35M:1F (Blanchard 1935). King (2012) compared the usage of natural vs. constructed wetlands by Four-toed Salamanders in Kentucky and found that constructed wetlands were used but that nest abundance was higher in natural wetlands ($n = 133$ and 74 for natural and constructed wetlands, respectively). Corser and Dodd (2004) studied metapopulation dynamics of nesting females at 11 wetlands in Great Smoky Mountains National Park over a five-year period. They estimated that ≥730 females oviposited during their study (annual $\bar{x} = 146$, annual range = 104–219). At the site level, they found that the number of nesting females varied dramatically among wetlands and years and that wetland isolation was negatively correlated with relative abundance of females.

Threats: The disjunct occurrence of Four-toed Salamander populations in Wisconsin, coupled with their limited dispersal ability and specialized habitat requirements, makes them vulnerable to local extirpation (Harding 1997). Four-toed Salamanders are dependent on both specialized aquatic nesting habitats and surrounding moist forest; thus, loss or degradation of either required habitat will negatively impact populations (Figure 145). Loss of nesting habitat is probably the greatest threat to Four-toed Salamander populations in the Upper Midwest (Hall and Carlson 2004). This has been a concern in other parts of its range for nearly a century. For example, Wright (1918) stated that "the Four-toed salamander . . . is vanishing with the drying up or draining of *Sphagnum* bogs and feather bed swamps."

Figure 145. A recently harvested forest surrounding Four-toed Salamander nesting habitat. Such practices almost certainly result in extirpation of local breeding populations (Iron County; photo by E. Wild).

Bishop (1919) further stated that it "is certain that the final conversion of *Sphagnum* areas will mark the extirpation of the species in such localities."

Epstein et al. (1999a, 1999b) identified aquatic and terrestrial threats to Four-toed Salamanders in Wisconsin. They reported that wetland modification (i.e., draining, impoundment, and dredging) and water quality degradation were the primary aquatic threats. Forest habitat degradation was identified as the primary terrestrial threat, including reduced canopy cover and reduction in downed woody debris through timber harvest. The Four-toed Salamander is entirely absent from agricultural lands (Pentecost and Vogt 1976); thus, much of southern Wisconsin currently represents unsuitable habitat for the species. In addition, larval survivorship is low in wetlands containing fish (Thurow 1997, Harris 2005).

Four-toed Salamanders harbor cutaneous bacteria that have been shown to inhibit the growth of the fungal pathogen *Batrachochytrium dendrobatidis* (Bd; Harris et al. 2006), indicating that their vulnerability to this pathogen may be low. However, the susceptibility of the species to ranaviruses and the recently discovered fungal pathogen *Batrachochytrium salamandrivorans* (Bsal; Martel et al. 2013) remain unknown.

Management: General recommendations for management of wetland and forest habitat and connectivity are applicable to this species (see Introduction to Conservation and Management of Wisconsin's Amphibians and Reptiles chapter; Casebere and Lodato 2011, Kingsbury and Gibson 2012). See Hall and Carlson (2004) for forestry management guidelines specific to this species in Minnesota. Petranka (1998) suggested that in the construction of vernal pools for Four-toed Salamanders, raised hummocks of earth or raised hills along margins should be added to ponds to encourage growth of moss mats and sedges. Wahl et al. (2008) suggested that banks should be steep and face north. In their evaluation of a state forest in Oneida County, Epstein et al. (1999b) recommended the establishment of an old-growth core area that would maintain or permit the development of a limited successional stage to maintain or increase Four-toed Salamander habitat.

Since information on the distribution and population sizes of Four-toed Salamanders in Wisconsin is so limited, additional survey efforts are needed to better understand the conservation status of the species. Harris (2005) recognized aspects of their biology that researchers should consider in order to minimize sampling biases: their short larval period, the secretive nature of adults, and the propensity of adult females to skip years of reproduction. In addition, nest-site surveys should attempt to minimize nest disturbance, since, as Breitenbach (1982) found, researchers can have a significant negative effect on the presence of brooding females.

Eastern Red-backed Salamander

Plethodon cinereus (Green 1818)

Deahn M. Donner

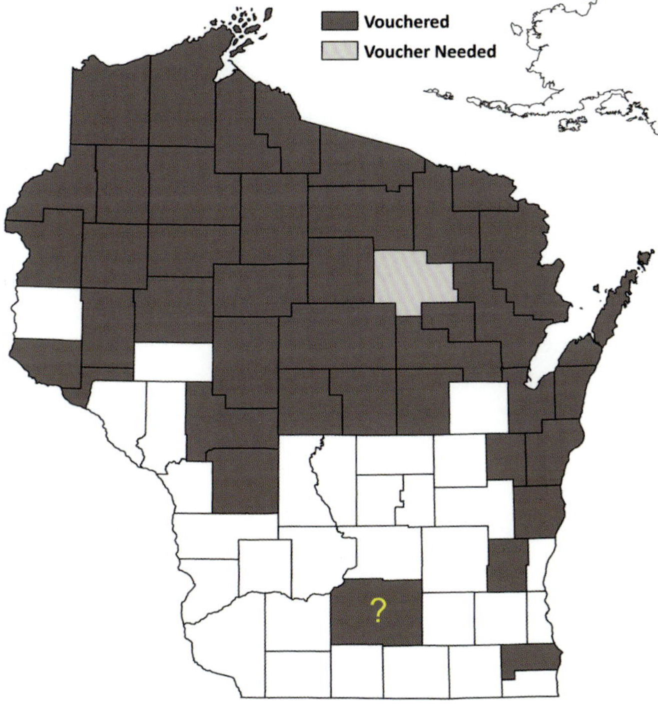

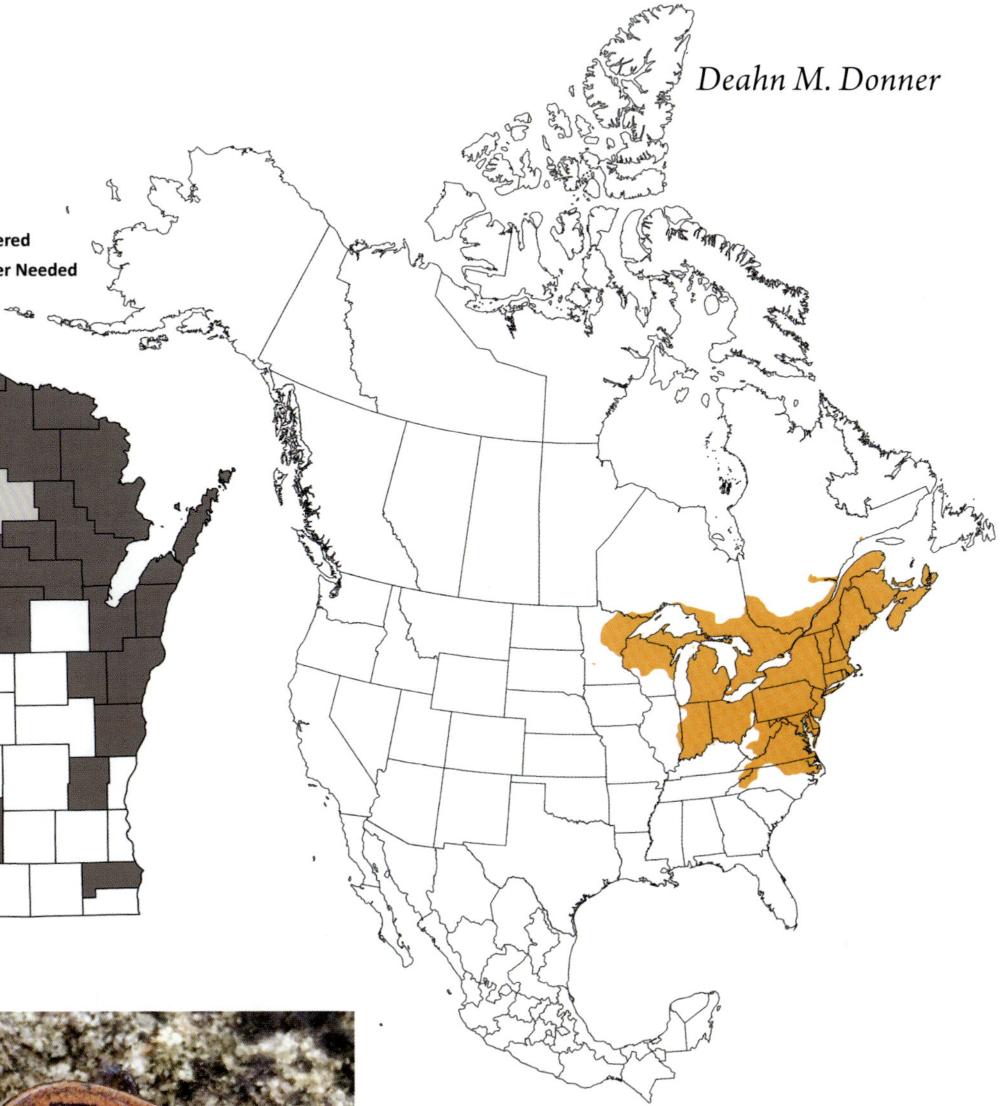

Eastern Red-backed Salamander (Sheboygan County; photo by N. Melhorn).

AT A GLANCE

Eastern Red-backed Salamanders are a woodland species found primarily in northern Wisconsin, with isolated records from southeastern counties. This small slender salamander typically has a reddish dorsal stripe that extends from the head to the tail over a generally dark body, with light flecks on the belly, and five toes on the hind feet. Being lungless, they respire entirely through the skin and thus depend on moist conditions to breathe and are susceptible to desiccation. As such, they closely associate with forested habitats, particularly mesic to wet-mesic hardwoods, where they occur below ground debris. Eastern Red-backed Salamanders are active from April to mid-October. They have complex social systems and exhibit many interesting reproductive and territorial behaviors. Reproduction primarily occurs in autumn, and females store sperm internally, which is used to fertilize the eggs that are laid the following June. Females lay eggs on land in moist decaying logs, and they defend eggs from predators. Unlike other salamanders, there is no aquatic larval phase. Terrestrial hatchlings emerge in August, appearing similar to adults. Eastern Red-backed Salamanders consume a variety of small invertebrates and forage mostly at night by following chemical trails left by prey. This species is consumed by many birds, small mammals, and snakes. It is listed as a common species by the WDNR but is vulnerable to loss of suitable woodland habitat.

SYSTEMATICS

No subspecies of *Plethodon cinereus* (Eastern Red-backed Salamander) are currently recognized (Crother 2017). Highton (1995) reported 4 genetically differentiated geographic groups of the 35 species within the genus *Plethodon* across the eastern US, and 7 of these are part of the eastern *P. cinereus* clade. The majority of the seven species in this clade have highly restricted distributions (Highton 1999). Among these, *Plethodon serratus* (Southern Red-backed Salamander) is morphologically the most similar to *P. cinereus*, but *P. shenandoah* (Shenandoah Salamander) and *P. nettingi* (Cheat Mountain Salamander) are the most similar genetically (Highton 1999). Of the seven species within this eastern clade, only *P. cinereus* is found in Wisconsin. Furthermore, only one other salamander species in Wisconsin is a member of the same family as *P. cinereus* (Plethodontidae): *Hemidactylium scutatum* (Four-toed Salamander).

DESCRIPTION

Adults: Eastern Red-backed Salamanders are so named because of the reddish dorsal stripe that typically extends from the head onto the tail (Figure 146). The rest of the body is generally dark gray or black, often with small light-colored speckles, with the base coloration becoming lighter on the white-flecked belly. They are slender with a long rounded tail (in cross section) and small legs that have four digits on the front feet and five slightly webbed digits on the back feet (Petranka 1998). The number of costal grooves ranges from 17 to 20, and the gular fold is prominent (Petranka 1998).

Eastern Red-backed Salamanders are one of Wisconsin's smaller salamanders, but their tail is long (i.e., nearly as long as the rest of their body). For example, the mean SVL of mature specimens from Wisconsin in the MPM and UWZM collections was 4.4 cm (1.7 in, $n = 145$; to ensure specimens were mature, only those with an SVL ≥3.7 cm

Figure 146. Dorsal view of an Eastern Red-backed Salamander showing the red stripe down the back from head to tail and the white flecking on the sides. The dorsal stripe can be salmon or dark brown in color as well (Sheboygan County; photo by N. Melhorn).

[1.5 in] were included). The mean intact tail length of mature specimens in these collections was 4.1 cm (1.6 in, $n = 102$). These measures are similar to those reported by Liebgold (2014), who determined that the mean SVL of adult males sampled in Michigan ($n = 76$) and Virginia ($n = 88$) was 4.5 cm (1.8 in, SD = 0.3cm [0.1 in]). The largest Wisconsin specimen measured had an SVL of 5.3 cm and a tail length of 4.4 cm (2.1 in and 1.7 in, respectively, UWZM 9429, collected in 1903 from Iron County).

Preadult stages: The eggs are uniformly yellowish white and range in diameter from 0.3 to 0.4 cm (0.1 to 0.2 in). Eggs are laid in small clusters held together with a gel-like substance (Figure 147; reviewed by Petranka 1998). Unlike most Wisconsin amphibians, this species undergoes direct development, and there is no aquatic larval phase (see Reproduction and Development section). As such, hatchlings and juveniles are similar in appearance to adults. In Wisconsin, Vogt (1981) reported that hatchling TL was 2.2 cm (0.9 in), which was also the hatchling size reported in an Ohio population (Pfingsten 1989). The mean SVL from a small sample of Wisconsin hatchling specimens housed at MPM was 1.5 cm (0.6 in, $n = 8$), while mean tail length was 0.8 cm (0.3 in, $n = 8$). In Maryland, Sayler (1966) reported a mean SVL of 2.1 cm (0.8 in, $n = 47$) for hatchlings.

Figure 147. A clutch of Eastern Red-backed Salamander eggs (Door County; photo by C. Raimond).

Variation: There are several color variants of this species documented across its North American range (Moore and Ouellet 2014). For example, in some parts of its distribution, a "lead-back" phase commonly occurs that has no red stripe but often a dark and speckled back (Petranka 1998). The reasons for this color polymorphism have been attributed to differential predation, which influences dominance of the red-backed phase (Venesky and Anthony 2007, Fitzpatrick et al. 2009), and the genetics that control the presence of stripes along the back (Highton 1975). Detailed information on the prevalence of these variant color morphs in Wisconsin is sparse. Knudsen (1958b) anecdotally reported that the dorsal stripes of Wisconsin individuals ranged in color from "salmon-pink to brick-red, though sometimes . . . grayish-yellow, and more rarely dark, grayish-black." Vogt (1981) stated that the lead-backed phase was not documented in Wisconsin and instead reported a "melanistic" population on Stockton Island in Lake Superior (Ashland County). Museum notes identify the specimen MPM 33208 as a "leadback" collected from Basswood Island (Ashland County). In addition, a dark-colored specimen collected from Forest County is listed as a "slate gray morph" (MPM 26484). Interestingly, Moore and Ouellet (2014) indicated the presence of lead-backed morphs in Wisconsin but did not report the source of their information. Although lead-backed morphs have been poorly documented in the state, it is reasonable that they exist in northern Wisconsin, considering Holman (2012) reported that they occurred (albeit infrequently) in the southern reaches of Michigan's Upper Peninsula.

Further variation in appearance has been documented. Normal intact tails are round in cross section for the full length, but regenerated tails can be laterally compressed and dark gray, with the red stripe missing (Petranka 1998). Adult males are smaller than females, but the difference is minimal (reviewed by Petranka 1998). For example, in southwestern Michigan, Werner (1971) noted that the mean SVL of females was 4.5 cm (1.8 in) and that of males was 4.2 cm (1.7 in). In Maryland, Sayler (1966) reported that the mean SVL for adult males was 4.0 cm (1.6 in, $n = 262$) and 4.03 cm (1.6 in, $n = 85$) for females. During the breeding season, differences in external features can be used to distinguish males from females.

Figure 148. An adult male showing enlarged nasolabial grooves that extend below the nares on the front of the upper jaw (Oneida County; photo by C. Raimond).

Figure 149. View of the abdominal cavity showing white eggs near the abdominal walls of a mature female. Also note the presence of five toes on the hind feet, a character used to distinguish this species from the Four-toed Salamander in Wisconsin (Bayfield County; photo by E. Wild).

For example, sexually active males possess a whitish area on the chin called a mental gland, as well as a whitish area lateral to the cloacal opening that is also slightly enlarged, and the snout becomes square due to enlarged premaxillary teeth (Gillette and Peterson 2001). Breeding males also possess a nasolabial groove that extends downward from each of the nares on the front of the upper jaw (Figure 148) and swollen hedonic glands at the base of the tail (reviewed by Petranka 1998). Sex of mature individuals can also be determined by shining a strong light behind the abdominal wall to show black testes in males; if they are absent, then the individual is female (Gillette and Peterson 2001, Jaeger et al. 2002). In gravid females, whitish circles (eggs as small as 0.1 cm [0.4 in]) can sometimes be seen through the skin along either side of the abdomen (Figure 149; Gillette and Peterson 2001).

Diagnosis: Eastern Red-backed Salamanders do not have tail fins that extend off the dorsal or ventral surface of the tail muscles. They do not have bushy external gills, the skin is not rough or granular in appearance, and there are no red dorsolateral spots bordered in black. Their body is long and thin, the tail and body coloration are normally similar (typically with a red dorsal stripe, but see Variation subsection), and there is no constriction at the base of the tail. The ventral surface is light gray (anteriorly) to dark gray or black (posteriorly), and there are five toes on each hind foot (Figure 149). Four-toed Salamanders have an orange-brown to red-brown dorsal stripe and could be confused with Eastern Red-backed Salamanders. However, the former has a constriction at the base of the tail and only four toes on the hind feet rather than five (Figure 149). Four-toed Salamanders also have a bright white belly with bold black spots or blotches (reviewed by Petranka 1998). Mole salamanders (*Ambystoma* spp.) are more stout-bodied, lack nasolabial grooves, and are generally larger as mature adults (reviewed by Petranka 1998).

Eastern Red-backed Salamander eggs look much different from those of other salamanders (uniformly yellowish or whitish in color as opposed to bicolored and encased in gelatinous envelopes). The eggs are also laid terrestrially rather than in the water or below moist *Sphagnum* moss as in other salamanders from Wisconsin. Juveniles should be easily distinguished from most other salamanders in Wisconsin because there is no aquatic larval phase, so they resemble smaller versions of the adults. Although members of the juvenile 'eft' stage in Eastern Newts (*Notophthalmus viridescens*) are small and reddish and could be mistaken for juvenile Eastern Red-backed Salamanders, the red in efts covers most of the body (it is not constrained to a stripe down the middle of the back) and is not bright. Eastern Newts also do not possess costal grooves.

DISTRIBUTION AND HABITAT

Global distribution: Eastern Red-backed Salamanders range from southern Quebec and the Canadian Maritime Provinces, west to northeastern Minnesota, and south to northern and eastern North Carolina. Isolated populations occur in western Minnesota and southern North Carolina (Petranka 1998).

Distribution in Wisconsin: Eastern Red-backed Salamanders are mostly distributed in forested portions of the state north of the vegetative tension zone (Curtis 1959), including islands on the Door Peninsula (Long and Long 1976), although spotty records also occur in eastern to southeastern Wisconsin. The record from Racine County is based on an old specimen, and recent confirmation of Eastern Red-backed Salamanders in that county is desirable (see discussion in Kapfer et al. 2015). The species appears to be absent from counties with well-drained soils in central, south-central, and southwestern Wisconsin (Vogt 1981). Although a vouchered specimen exists for Dane County, this record remains questionable (see discussion in Kapfer et al. 2015). Of the published county records, a voucher could not be found for Langlade County (see the Appendix).

Localities of possible occurrence: This species is not currently documented in St. Croix, Eau Claire, and Outagamie Counties, but its presence in these locations seems likely due to confirmed records in surrounding counties. Given the historical distribution of mesic forests in Wisconsin, Fond du Lac and Winnebago Counties may also have areas of potentially suitable habitat for Eastern Red-backed Salamanders (Curtis 1959). Casper (1996) noted that Eastern Red-backed Salamanders could occur farther south along Lake Michigan in eastern Wisconsin than is currently recorded based on the historical presence of southern mesic forests in that region. Given the existence of old specimens from Racine County, recent confirmation in Washington County, and documentation of this species immediately across the Illinois border (Kapfer et al. 2015), all coupled with the more extensive past distribution of mesic forests in southeastern Wisconsin, this species could also occur (or historically occurred) in Dodge, Jefferson, Ozaukee, Waukesha, Milwaukee, and Kenosha Counties (note that an old, unverified report from Waukesha County exists in Cahn [1929]). In addition, Eastern Red-backed Salamanders could have once been found in portions of Vernon, Crawford, Richland, and Sauk Counties where mesic forests were historically present, and perhaps remnant populations remain there in suitable locations. Vogt (1981) mapped a record from Sauk County that he had not confirmed but that he considered reliable; Casper (1993c) noted unconfirmed reports from the Baraboo Hills that further support its occurrence in Sauk County.

Habitat: Eastern Red-backed Salamanders are completely terrestrial and closely associated with forested habitats in Wisconsin such as mesic to wet-mesic northern forests (hardwood or mixed deciduous-coniferous forest) and conifer swamps (Vogt 1981, Petranka 1998; see Amphibians and Reptiles in Wisconsin Plant Communities chapter). They are lungless and respire entirely through the skin; thus, they rely on moist conditions to facilitate oxygen absorption and avoid desiccation. Therefore, Eastern Red-backed Salamanders in Wisconsin are frequently found in mature forests with dense tree canopies and well-developed moist ground litter (Vogt 1981). However, some sources report that they tolerate a broad range of moisture, temperature, and light levels (Wyman and Hawksley-Lescault 1987, Sugalski and Claussen 1997).

Given the probable importance of adequate moisture and humidity (Heatwole 1962), preferred microhabitats typically include the underside of damp ground cover objects such as old logs, bark, moss, leaf litter, humus substrates, and rocks. These provide a humid environment with a large supply of invertebrate prey and refugia from predators (Gergits and Jaeger 1990a). Accordingly, Bergeson (2001) determined that the volume of downed woody debris, particularly from Eastern Hemlock (*Tsuga canadensis*) and Sugar Maple (*Acer saccharum*), was positively correlated with Eastern Red-backed Salamander abundance at sites across northern Wisconsin (Vilas and Forest Counties) and Michigan's Upper Peninsula (Gogebic County).

The type and texture of cover objects can further influence local abundance and distribution of this salamander. For example, Eastern Red-backed Salamanders at sites in northern Wisconsin and the Upper Peninsula of Michigan were found below logs on 96.9% of captures (385 of 397 total captures; Bergeson 2001). Bergeson (2001) also reported that salamander abundance was positively related to the presence of logs with diameters ranging from 51 to 107 cm (20 to 42 in) that were in advanced stages of decay. On the other hand, work in Vermont (Richmond and Trombulak 2009) and Pennsylvania (Moore et al. 2001) found that they were more common under rocks than woody branches, logs, or leaf litter in hardwood forests. Rocks further housed larger individuals (>3.47 cm SVL [1.37 in]), with smaller individuals more common under woody objects (Richmond and Trombulak 2009). The size of cover objects could also be correlated with the body size and mass of individual salamanders found beneath them (Mathis 1990, Moore et al. 2001), but some studies suggest that this could depend on season (Moore et al. 2001) or geography (Faragher and Jaeger 1997).

Eastern Red-backed Salamanders typically occupy fibrous woody substrates and humus ground layers, in part because they are moist but also because they provide easier movement through subsurface environments than do solid substrates (Heatwole 1962, Richmond and Trombulak 2009). When soil surface layers are dry, salamanders must move vertically below the soil surface to find adequate moisture (Vernberg 1953), which is more challenging in solid substrates. Heatwole (1960) reported that while this species is effective at penetrating loose layers of humus, it does not dig but rather "uses its head as a wedge and forces its body into cracks or small openings in the substrate by pushing with its feet, and in some instances, its tail." Kleeberger and Werner (1982) found 61% of all capture locations ($n = 758$) were at the soil surface, primarily in litter layers and decaying logs in a northern hardwoods system of Michigan's Upper Peninsula, while the remaining locations were in the soil below humus and leaf litter associated with root systems of trees or in raised mounds of soil, roots, and rocks. Eastern Red-backed Salamanders can enlarge existing burrows (Heatwole 1960) and will also often use earthworm (*Lumbricus terrestris*) burrows as refugia (Cáceres-Charneco and Ransom 2010, Ransom 2011).

Females lay eggs in moist locations, such as within or below decaying logs, beneath rocks, underground, or in other damp substrates (Vogt 1981). Because the eggs and larvae are not aquatic, habitat used for breeding and by hatchlings is similar to that occupied during the remainder of the year by adults. Specific data on overwintering habitat in Wisconsin are sparse, and most information comes from anecdotal observations in other states. Laboratory investigations support that Eastern Red-backed Salamanders do not tolerate freezing (Storey and Storey 1986), and studies from west-central Indiana found that they descend below the soil surface to avoid harsh winter conditions (Vernberg 1953). For example, they reportedly use root systems, animal tunnels, and natural crevices to avoid harsh conditions (Jaeger 1980b). Furthermore, Caldwell (1975) found 30 individuals during January in an ant mound (Formicidae) at depths of 31–75 cm (12.0–29.5 in) in Indiana. In Maryland, Grizzell (1949) uncovered a single individual overwintering at a soil depth of 88.9 cm (35 in).

REPRODUCTION AND DEVELOPMENT

In the Great Lakes region, mating typically occurs from October through December but can continue into the winter and early spring (Blanchard 1928). Males employ a "nose-tapping" behavior to detect chemical cues left on the ground by females and locate them for mating (Gergits and Jaeger 1990a). Once the male encounters a female, the male positions himself in front of her, blocking her movement, while arching his tail upward and wiggling it with the tip still on the ground (Gergits and Jaeger 1990a). As courtship progresses, the male approaches the female's side with his body in a low submissive posture. He then rubs glands located on his head and chin along her body until he reaches her snout, while she assumes a flat or submissive posture. The pair then walks forward with the female straddling the male's tail and pressing her head against the base of his tail

(Gergis and Jaeger 1990a). While she is in this posture, the female rubs her cloacal opening over the male's dorsum (Dyal 2006). The male next deposits a spermatophore on the substrate, and the female picks it up with her cloaca. Females store sperm internally, and eggs are fertilized during oviposition the following summer (reviewed by Sever 1997), which generally occurs from early to late June in Wisconsin (Vogt 1981, but see Bergeson 2001).

Eggs are laid in a cluster suspended by a gelatinous stalk attached to the undersurface of logs, rocks, or roots; in the absence of these larger cover objects, eggs are oviposited on the forest floor below ground litter or within small tunnels made by other animals (Test and Heatwole 1962). Reports of clutch size vary. Homyack and Haas (2009) found that the mean number of eggs per female ranged from 6.7 to 7.3 in central Appalachian oak forests ($n = 277$ females). The mean clutch size associated with 13 brooding females captured at sites in northern Wisconsin and adjacent Michigan was 7.8 (range = 4–12; Bergeson 2001). Leclair et al. (2008) found that the mean number of eggs per female was 9.4 (range = 4–15, $n = 21$ females) in Quebec. In Michigan, Blanchard (1928) reported 5–13 eggs per female ($n = 91$ females). In Ohio, the mean number of eggs across eight examined clutches was 7.9 (range = 4–10; Anthony and Pfingsten 2013). Several sources support that mature females do not produce eggs every year. In fact, due to the energy required to produce and subsequently guard eggs, females may only breed biennially (Sayler 1966, Ng and Wilbur 1995). At more northern latitudes, such as Canada, females may reproduce once every three years (LeClair et al. 2008).

Young undergo direct development, and females typically spend a significant amount of time in association with eggs (or 'brooding' the eggs) after they are laid. A brooding female wraps her body around the eggs to help them retain moisture and occasionally vibrates the eggs to maintain aeration and uniform distribution of eggs within clutches (Bachmann 1984, Tornick 2010). Females also guard eggs against invertebrate predators and other larger salamanders until the eggs hatch at six to nine weeks (often during August in Wisconsin; Vogt 1981), after which females remain with the young for up to three weeks until remnant gill buds are absorbed (Harding 1997). When an intruder approaches, the female adopts defensive behaviors such as nudging, chasing, and snapping at intruders and will exhibit heightened aggression when defending late-stage eggs (Tornick 2010). Brooding females do not actively forage while eggs develop, and they only eat opportunistically (Bachmann 1984).

Eastern Red-backed Salamanders, like many species with broad distributions, exhibit geographically variable growth rates due to differences in associated environmental conditions. Slower growth rates in northern latitudes translate to a more gradual increase in size until maturity and later age at first reproduction (Leclair et al. 2006). In Quebec, growth rates for the Eastern Red-backed Salamander were lower than those reported for populations at more southern latitudes in the US (Leclair et al. 2006). In this study, males reached sexual maturity at three to five years of age, and females reached sexual maturity at five to seven years of age; the minimum size at sexual maturity was 3.6 cm SVL (1.4 in) and 4.1 cm SVL (1.6 in) in males and females, respectively ($n = 775$ total salamanders assessed). In comparison, Sayler (1966) found that individuals from Maryland had matured by the end of their second year at a minimum SVL of 3.1 cm (1.2 in) in males and 3.4 cm (1.3 in) in females ($n = 1{,}272$ total salamanders assessed); mean adult SVL was 4.0 cm (1.6 in, range = 31–4.9 cm [1.2–1.9 in], $n = 347$). In Michigan, Blanchard (1928) documented sexual maturity at ca. two years old but noted that females did not successfully produce eggs until their fourth season ($n = 110$ females).

Annual growth rates decline with age. For example, estimates of growth during the first year was 1.5 cm (0.6 in), which declined to 0.8 cm (0.3 in) during year two, with greatly reduced growth rates after reaching sexual maturity (Nagel 1977). Within-year variation in growth also occurs. Sayler (1966) determined that individuals grew rapidly in April and May (0.4 cm [0.16 in], $n = 8$), and from September through December (1.0 cm [0.4 in], $n = 77$). Leclair et al. (2006) reported that female longevity in a Quebec population was eight years and male longevity was nine years but stated that most members of the population were two to four years old. Of these, adult males made up only 25% of captured male three-year-olds, 57% of four-year-olds, and 100% of five-year-olds ($n=28$, 7, and 19, respectively); the oldest male was nine years old.

ACTIVITY

In Wisconsin, Eastern Red-backed Salamanders are typically active from April through October (Vogt 1981), but seasonal variation in activity apparently occurs. In Quebec, Leclair et al. (2008) captured 63.6% of 775 Eastern Red-backed Salamanders in drift fences during the first two weeks of October and <7.8% during spring and early summer (Leclair et al. 2008). Seasonal differences in activity are likely linked to seasonal variation in humidity. Eastern Red-backed Salamanders are nocturnal, and soil surface activity peaks during humid conditions (Keen 1984, Grover 1998, Grover and Wilbur 2002), which are often most prevalent at night in spring and autumn. Accordingly, surface activity declines during hot dry summer days, especially when humidity drops below 85% (Heatwole 1962, Jaeger 1980b). Leclair et al. (2008) captured far fewer individuals during mid-summer surveys in Quebec than during spring or autumn, which the authors linked to high air temperatures. Jaeger (1980b) found that this species modified its behavior during dry periods and moved from leaf litter to under rocks or logs, where conditions were presumably more favorable. Foraging activity also declines under dry conditions and can result in heightened intraspecific competition due to scarcity of prey (Jaeger 1980a, Jaeger et al. 2002).

Eastern Red-backed Salamanders are not particularly mobile, and studies of population genetics confirm that limited migration and gene flow occur among neighboring populations. For example, Cabe et al. (2007) reported that eight subpopulations separated by distances as little as 200 m (656 ft) in unfragmented habitat were genetically differentiated (suggesting that there is little movement between subpopulations). Similarly, Sites et al. (2004) found that populations separated by several kilometers were genetically distinct from each other. Accordingly, reported home range sizes and movement rates for this species are generally small. Mathis (1991) estimated home ranges of only 0.16–0.34 m^2 (1.7–3.7 ft^2, $n=25$) and found no difference in movement among adult males, adult females, and juveniles in a Virginia population. However, these findings are in contrast to other reports. For example, larger home range sizes were estimated from a mark-recapture study in Michigan that also found that males had slightly smaller home ranges than females ($\bar{x}$ male home range $=3.85$ m^2 [41.4 ft^2], $n=15$, $\bar{x}$ female home range $=4.82$ m^2 [51.9 ft^2, $n=13$], modified MCP estimator; Kleeberger and Werner 1982). The authors further found that adult home ranges were larger than those of juveniles ($\bar{x}=2.87$ m^2 [30.9 ft^2], $n=5$). Daily movement rates are also reportedly small and in a Michigan study averaged only 0.43 m/day (1.4 ft/day, $n=34$; Kleeberger and Werner 1982). In Virginia, recaptures between years indicated that adult dispersal movements were smaller, but not significantly (range $=0.85$–0.88 m [2.8–2.9 ft], $n=85$), compared to juveniles (range $=1.14$–1.22 m [3.7–4.0 ft], $n=83$; Ousterhout and Liebgold 2010).

Although they generally do not move great distances, Eastern Red-backed Salamanders are capable of larger mobility if required. Marsh et al. (2004) found that young adults moved up to 55 m (180 ft) to colonize experimental plots in open fields in Virginia. Migration during spring and autumn can also lead to larger bouts of movement. Woolbright and Martin (2014) observed adults moving uphill in autumn and downhill in the spring to wetland areas, most likely due to habitat availability for nesting; young salamanders remained near wetlands year-round.

Given Eastern Red-backed Salamanders do not typically move great distances, it is unsurprising that some authors have reported these salamanders exhibit high fidelity for specific locations. For example, Kleeberger and Werner (1982) indicated that individual bouts of movement were often restricted to beneath the same log or cover object. Furthermore, Gergits and Jaeger (1990b) marked individuals from Virginia and reported that 36% of recaptures occurred at nearly the original capture location, 35% of recaptures occurred within 50 cm (19.7 in), 20% occurred between 50 and 100 cm (19.7 and 39.4 in), and only 1% of recaptures occurred at greater distances ($n=90$ captures and 55 recaptures). Cabe et al. (2007) found a mean intergenerational dispersal distance of 3.5 m (11.5 ft), with no difference detected between sexes, in Virginia. Liebgold et al. (2011) found that 44% of adult females in Virginia were recaptured within 1 m (3.3 ft) of natal locations, and 69% were recaptured within 2 m (6.6 ft). Homing ability is also documented in this species. Kleeberger and Werner (1982) reported that 90% ($n = 30$) of salamanders displaced 30 m (98 ft) from capture location returned and that 25% of individuals displaced up to 90 m (295 ft) also returned. Eastern Red-backed Salamanders exhibit complicated territorial behaviors related to reproduction and food acquisition, which influence their movement patterns (Natural History Box: Complex Behaviors).

PREY AND PREDATORS

Eastern Red-backed Salamanders eat a wide variety of small invertebrates that inhabit the forest leaf litter, such as insects and their larvae, earthworms, sow bugs, centipedes, springtails, mites, snails, slugs, and spiders (Jaeger et al. 1995, Lang and Jaeger 2000, Mitchell and Woolcott 1985). From October to November in New York, insects comprised 73.2% (total bulk) of all stomach contents assessed ($n = 149$ stomachs assessed; Jameson 1944). Of the insects consumed, 20% were adult and larval beetles (order Coleoptera), 10.5% were true flies (order Diptera), and ca. 10% were hymenopterans, of which half were ants (Formicidae). Although comprising only 6.7% of total bulk consumed, earthworms were often the only food in the stomachs of some individuals, indicating their dietary importance to Eastern Red-backed Salamanders (Jameson 1944). In Indiana, ants were also a major part of their diet during December (18.2% by volume), although March Fly larvae (family Bibionidae) were the most abundant food items detected (38% by volume; Caldwell 1975). Gabor and Jaeger (1995) found that Eastern Red-backed Salamanders preferred termites to ants, likely because they were larger and had a softer and easier-to-digest exoskeleton. Furthermore, they are considered pulse feeders that also alter their preferred prey based on what is most readily available (Jaeger and Barnard 1981).

Eastern Red-backed Salamanders will forage throughout the day but most often hunt at night (Jaeger 1978). Yet they can modify foraging behavior with changing weather patterns. After rain, for example, prey are easier to obtain, and salamanders can forage for longer periods with little risk of desiccation, compared to drier periods, when they are restricted to moist patches of the forest floor and so encounter fewer prey items (Jaeger 1980a, Jaeger and Barnard 1981, Placyk and Graves 2001). Most active foraging occurs on the forest floor within the leaf litter and ground debris. They occasionally climb shrubs and tree trunks in search of prey (Jaeger 1978), although some authors suggest that this behavior is primarily adopted by satiated individuals that seek these higher locations to avoid predators that hunt on the ground (Roberts and Leibgold 2008). Chemoreception is the primary sensory modality used while hunting (David and Jaeger 1981), and Eastern Red-backed Salamanders reportedly tap their snout on substrates because it aids in detection of the chemical cues left by prey (reviewed by Jaeger 1984). Placyk and Graves (2001) found that salamanders lowered activity with increasing light intensity, suggesting they switched from active foraging with chemical cues to an ambush behavior via visual cues for prey detection with increasing light. Furthermore, Eastern Red-backed Salamanders are more likely to actively pursue food when prey densities are low, but they otherwise use ambush tactics (Jaeger and Barnard 1981).

Eastern Red-backed Salamanders are significant predators within forest floor communities and play a regulatory role within terrestrial detrital food webs (Davic and Welsh 2004, Walton 2013, Hickerson et al. 2017). During a six-year study in Ohio, Walton (2013) found the density of Eastern Red-backed Salamanders influenced forest floor invertebrate density and composition, especially for invertebrate taxa important to forest floor decomposition and nutrient cycling.

Eastern Red-backed Salamanders are important prey for many birds, small mammals (e.g., shrews), snakes (Brodie 1977, Brodie and Brodie 1980, Ducey and Brodie 1983), and larger salamanders such as the Spotted Salamander (*Ambystoma maculatum*;

NATURAL HISTORY BOX: Complex Behaviors

Eastern Red-backed Salamanders have a complex social system best documented by a series of natural behavioral studies conducted in Virginia (see Mathis 1991, Jaeger and Forester 1993, Gillette et al. 2000, Liebgold 2014). These studies determined that male-gravid female pairs of Eastern Red-backed Salamanders often cohabit and defend territories during both the noncourting and courting seasons. These pairings presumably occur because gravid females select males for reproduction based on male 'quality,' which is inferred through olfactory cues that indicate whether males have obtained high- or low-quality diets (Chouinard 2012), which in turn affects fecundity (Fraser 1980). Males, for their part, compete for and defend high-quality foraging areas that then attract females, who are allowed to share the male's territory (Walls et al. 1989, Mathis 1991, Gabor 1995, Gabor and Jaeger 1995). Lang and Jaeger (2000) further found that males and females defend feeding areas under cover objects, and the paired residents were more aggressive to same-sex intruders. Kohn et al. (2005) determined that females spent more time in threat behavior with one male intruder but more time displaying escape behavior with more than one male.

Territory-holding adults not only engage in aggressive interactions with other adults but also act aggressively toward younger age classes. Presumably, this is because juveniles eventually represent a competitor to adults, as both age classes eat similar food. As such, juveniles and young adults are sometimes forced to modify their behavior or even vacate natal locations, moving long distances but still failing to breed in areas already near carrying capacity (Jaeger 1980a, Mathis 1991, Liebgold 2014). Juveniles are often found at the peripheries of adult home ranges (Liebgold and Jaeger 2007) but will sometimes co-occur with unrelated adults under rocks and logs (Jaeger et al. 1995, Liebgold and Cabe 2008). Jaeger et al. (1995) further found that juveniles were more likely to risk entering adult home ranges during dry conditions but typically left the area under wet conditions. Several studies also suggest that familiarity between young and mature Eastern Red-backed Salamanders can dictate the potential for aggressive interactions. Juveniles decreased nocturnal foraging activities when housed with unfamiliar adult females compared to familiar adults (Liebgold and Dibble 2011). Furthermore, juveniles forced to enter adult home ranges when conditions are dry experienced lower aggression from the resident adult if they had previously encountered each other (Jaeger et al. 1995). Accordingly, adults behaved more aggressively toward juveniles that they were unfamiliar with.

Territorial and aggressive behavior can take several forms in Eastern Red-backed Salamanders. For example, direct aggression usually involves threatening behavior and postures or physical attacks (Jaeger and Schwarz 1991). In a field setting, Gergits and Jaeger (1990a) documented the aggressive interactions between territorial residents and intruders. They determined that residents use 'nose-tapping' to investigate chemical cues left on substrates by intruders. Prior to employing aggressive postures or physical attacks, however, residents often threatened intruders by simply looking directly at them, which appeared to settle many territorial disputes. An encounter that escalates to a physical attack involved the resident biting the intruder's snout, a behavior that continued until the intruder retreated. The bitten and retreating intruder then moved away using quick, jerky undulations of the body to 'skip' across the ground rather than crawling. The aggressor sometimes followed a short distance but eventually returned to the location of the initial interaction. Insight into territorial interactions can also be gleaned from the movement patterns and spatial arrangement of salamanders on the forest floor. Although Mathis (1991) found no difference in home range size among adult males, adult females, and juveniles (which ranged from 0.16 to 0.34 m^2 [1.72 to 3.66 ft^2], $n = 25$, MCP estimator), she did determine that home ranges were spatially segregated within each sex. In other words, home ranges of adults of the same sex were farther apart than home ranges occupied by male and female pairs (based on nearest neighbor distances from center; Mathis 1991). A small number of different territories occupied by male-female pairs overlapped, but no same-sex territories overlapped. Liebgold et al. (2011) found that territorial females ($n = 41$) were more closely related to females separated by 0–2 m (0–6.6 ft) than to females separated by 8–10 m (26.2–32.8 ft), while males were significantly more related to males separated by 6–8 m (19.7–26.2 ft).

Although male-female pairs often occur within a given area, a large proportion of resident males and females can also be nonterritorial 'floaters' that are found singly under cover objects (Mathis 1991). For example, Mathis (1990) found that only 4.6% of various-sized cover objects ($n = 192$) contained cohabiting salamanders during summer and autumn months, with a maximum of three salamanders under a cover object. After transferring known pairs from the field into the laboratory, Gillette et al. (2000) further determined that two-thirds of the captured salamanders ($n = 410$) were alone during the noncourtship period in summer. These individuals showed little interest in associating with their partners at this time, which suggests a socially monogamous system. Toll et al. (2000) reported that unoccupied cover objects and vacated cover objects were more quickly colonized by single females than males.

Deahn M. Donner, USFS

Ducey et al. 1994, Grover and Wilbur 2002, Ransom and Jaeger 2006). If threatened, Eastern Red-backed Salamanders remain motionless when first detected (Brodie et al. 1974, Venesky and Anthony 2007), but once they have been touched by a perceived threat they will adopt a "swimming-running" motion to quickly escape into leaf litter and can apparently use their tail to jump (Harding 1997). If attacked, they release a sticky skin secretion that is distasteful to mammalian predators such as shrews (Brodie et al. 1979). Restrained Eastern Red-backed Salamanders writhe violently and may detach a portion of their tail (Jaeger et al. 1998). The dislocated tail continues to wiggle and distract the predator's attention while the rest of the salamander escapes. Although the broken tail will regenerate, it does not grow as long or possess the same color pattern as the original. Eastern Red-backed Salamanders may also alter their behavior in response to predatory threats. For example, Sullivan et al. (2002) found that they remained within refugia for up to two hours after exposure to Common Gartersnake (*Thamnophis sirtalis*) scent.

CONSERVATION

Status: Globally, the Eastern Red-backed Salamander is listed as a Least Concern species by the IUCN. It is considered a common species in Wisconsin by the WDNR.

Populations: Over 100 years ago, Cope (1889) stated that this was "the most abundant salamander in the northern and central eastern United States." Despite this, it has appeared to experience substantial population losses across parts of its North American distribution. For example, long-term monitoring in the eastern US by Highton (2005) revealed that 79%

of 24 populations sampled prior to and after 1990 declined in abundance. Still, some studies suggest that Eastern Red-backed Salamanders can be one of the most abundant terrestrial vertebrates in eastern North American woodlands. Reported soil surface density (i.e., density of individuals not below ground) estimates for a Virginia population of this salamander ranged from 2.2 to 2.82/m^2 (0.20 to 0.26/ft^2; Jaeger 1980b, Mathis 1991). Within the Hubbard Brooks Experimental Forest in New Hampshire, Eastern Red-backed Salamanders comprised 93.5% of total salamander biomass, with a soil surface density of 2,950/ha (1,200/ac), which was greater than the density of birds and mammals combined (Burton and Likens 1975). Bergeson (2001) estimated mean soil surface densities of 0.068/m^2 (0.006/ft^2) and 680/ha (275.2/ac) at sites spread across Vilas and Forest Counties in northern Wisconsin and Gogebic County in Michigan's Upper Peninsula. Additional studies in Michigan calculated soil surface densities of 838/ha (339/ac; Werner 1975) and up to 0.89/m^2 (0.08/ft^2; Heatwole 1962).

Several studies have investigated aspects of Eastern Red-backed Salamander populations beyond density. Leclair et al. (2008) reported an adult sex ratio of 1M:1.59F ($n = 342$) and a juvenile sex ratio of 1M:2.28F ($n = 154$) for a population in Quebec. Alternately, Test and Bingham (1948) found a sex ratio of 1M:0.36F ($n = 284$) for a population in Michigan. Eastern Red-backed Salamander population growth is likely density-dependent (Bruce 2008), and the factors limiting population growth probably vary from population to population. Important limiting factors include density of suitable substrate and cover objects, food availability, and predation pressure.

Threats: Eastern Red-backed Salamanders are potentially susceptible to numerous conservation concerns that also threaten most Wisconsin amphibians, such as habitat loss, fragmentation, and degradation (see Introduction to Conservation and Management of Wisconsin's Amphibians and Reptiles chapter). Destruction of wooded landscapes has substantial impacts on many salamander species, including Eastern Red-backed Salamanders (e.g., Pough et al. 1987, MacNeil and Williams 2014). The loss of available downed woody debris in forests further affects this species at various life stages. For example, adults seek refuge below damp fallen logs within woodlands, but these substrates are also important for egg deposition. Riedel et al. (2012) found fewer juveniles in disturbed habitats with less available ground cover than in undisturbed habitats (although adult abundances did not differ between disturbed and undisturbed locations) and suggested that lack of cover may specifically impact reproductive success or survival of juvenile age classes. Limited information on the specific effects of habitat loss in Wisconsin exists. Yet woodlands and forests have experienced dramatic declines in the state since the time of European settlement (see Past, Present, and Potential Future Landscapes of Wisconsin chapter), which have probably led to substantial undocumented population losses. Knudsen (1958b) even suggested that Eastern Red-backed Salamanders once occurred across much of Wisconsin but were extirpated from the intensively cultivated portions due to the concomitant degradation and loss of moist woodland habitats.

Eastern Red-backed Salamanders are not well adapted for long-distance movement (see Activity section), so it is doubtful that they will experience much success in crossing large disturbed areas while searching for suitable habitat in fragmented landscapes. Given this, habitat fragmentation that creates unnatural barriers to dispersal is of special concern, as it can limit gene flow and population viability. Key anthropogenic features that often result in habitat fragmentation are roads, which several studies have determined negatively impact Eastern Red-backed Salamanders. Marsh and Beckman (2004) found lower salamander abundance within 20 m (66 ft) of gravel roads that bisected mature forests in Virginia, and edge effects were detected out to 80 m (262 ft) from roads during a dry period. At large scales, Gibbs (1998c) used molecular genetic markers to detect greater genetic differentiation in fragmented populations than in contiguous forest populations from Connecticut ($n = 319$ individuals and 17 populations). Despite this, the impacts that roads have on this species can apparently be influenced by road width. For example, individuals in populations that were fragmented by smaller two-lane paved roads were similar genetically (indicating a lack of isolation due to roads fragmenting the population), but populations separated by a divided four-lane interstate highway existed in genetically distinct subgroups, supporting that these road systems presented a barrier to movement and gene flow (Marsh et al. 2008). Furthermore, whether or not roads are paved also apparently results in different impacts to this salamander. Marsh et al. (2005) found that wide unpaved logging roads and narrow paved roads presented partial barriers to migrating salamanders, leading to a 25%–75% reduction in movement.

Habitat can also be degraded by the presence of invasive species. For example, northern hardwood forests in Wisconsin are susceptible to invasion by exotic earthworms (Gundale et al. 2005). These exotics could alter forest floor trophic interactions and negatively impact Eastern Red-backed Salamander populations, although the magnitude of this threat remains largely unknown. Maerz et al. (2009) reported that abundance of Eastern Red-backed Salamanders was negatively related to earthworm abundance because earthworms decrease forest leaf litter, which serves as habitat for the small arthropods that Eastern Red-backed Salamanders consume. Yet additional research found that this interaction is complicated (Cáceres-Charneco and Ransom 2010). For example, Ransom (2011) reportedly encountered Eastern Red-backed Salamanders in earthworm burrows on 50% of their observations and further suggested that the refuges created by earthworms benefited the salamanders because they facilitated predator avoidance and winter survival.

There are other important threats to Eastern Red-backed Salamanders worthy of consideration, including the potential impacts of climate change. Habitat suitability for and the activity patterns of this species are linked to acceptable temperature and moisture levels. Current projections suggest that Wisconsin will become warmer and drier as climate change progresses (Janowiak et al. 2014), which could reduce the suitability of local habitats for Eastern Red-backed Salamanders. Muñoz et al. (2016) further reported evidence that higher summer temperatures would alter salamander activity patterns and reduce growth rates, prolonging time to maturity and thus lowering fecundity and reproductive success. Emerging amphibian diseases can also pose serious threats to many species, and several infectious agents are of concern, including the fungi *Batrachochytrium dendrobatidis* (Bd) and *B. salamandrivorans* (Bsal), as well as ranaviruses. Yet infections by these agents and their associated diseases are rarely reported from *Plethodon* salamanders such as the Eastern Red-backed Salamander (Muletz et al. 2014). Evidence currently supports that the normal bacterial communities present on the skin of Eastern Red-backed Salamanders could help protect them against a variety of pathogenic infections (e.g., Becker and Harris 2010, Loudon et al. 2014). Despite this, Standish et al. (2018) found a ranavirus on the single individual they sampled from Wisconsin (but they did not detect Bd or Bsal).

Management: Given their reportedly high densities and influential role in forest floor trophic interactions (see Prey and Predators section), Eastern Red-backed Salamanders could be an important species to monitor when assessing the integrity of ecosystems in managed properties (Welsh and Droege 2001). Many general management considerations relevant to woodland habitats could apply to Eastern Red-backed Salamanders, particularly maintenance of the ground debris that this species uses as cover (see Introduction to Conservation and Management of Wisconsin's Amphibians and Reptiles chapter). Habitat modification due to timber harvest, such as clear-cutting, alters tree canopy density and thus forest floor climate. Such habitat alterations are a major concern for *Plethodon* salamanders due to increased risk of desiccation and soil compaction (Ash and Bruce 1994, deMaynadier and Hunter 1995). Bergeson (2001) reported that significantly more Eastern Red-backed Salamanders were captured in old-growth forests vs. even-aged and uneven-aged managed forests in northern Wisconsin and adjacent forests in the Upper Peninsula of Michigan. Macneil and Williams (2014) found that clear-cut areas within oak-hickory and beech-maple forest systems negatively affected the relative abundance of associated Eastern Red-backed Salamanders on the local scale during years immediately following harvest, but shelterwood method treatments that left forest canopies intact had no effect on abundance. Leaving some forest tree canopy cover intact and maintaining the structural complexity on the forest floor (including the presence of downed woody debris) can help minimize impacts following forest harvesting (McKenny et al. 2006). Yet the duration of negative impacts due to timber harvest could be relatively short-lived if the impact area is small and woodlands are allowed to regenerate after disturbance. For example, Brooks and Kyker-Snowman (2008) found that forest clearing had little lasting impact on the forest floor microclimate conditions between cut and uncut northern hardwood forests undergoing regeneration in Massachusetts.

It is noteworthy that recent studies suggest that recreational hiking trails within Eastern Red-backed Salamander habitat can negatively influence abundance of this species. For example, Burger et al. (2017) reported that Eastern Red-backed Salamanders in a recreational nature reserve in Ohio were more abundant near low-traffic trails than high-traffic trails or wooded control sites. Interestingly, Fleming et al. (2011) found more Eastern Red-backed Salamanders near trails than in wooded areas away from trails. It is possible that a higher salamander density near trails occurs because trees that fall across trails must be removed, and the woody debris generated from this removal is often deposited near trail margins (Davis 2007, Burger et al. 2017). Habitat managers should consider these possible outcomes during design and maintenance of trail systems on public land.

Mudpuppy
Necturus maculosus (Rafinesque 1818)

Richard A. Sajdak and Dreux J. Watermolen

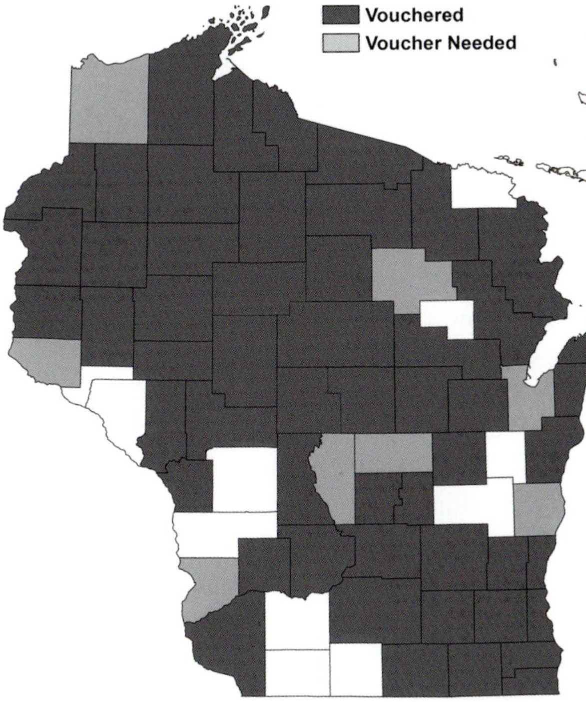

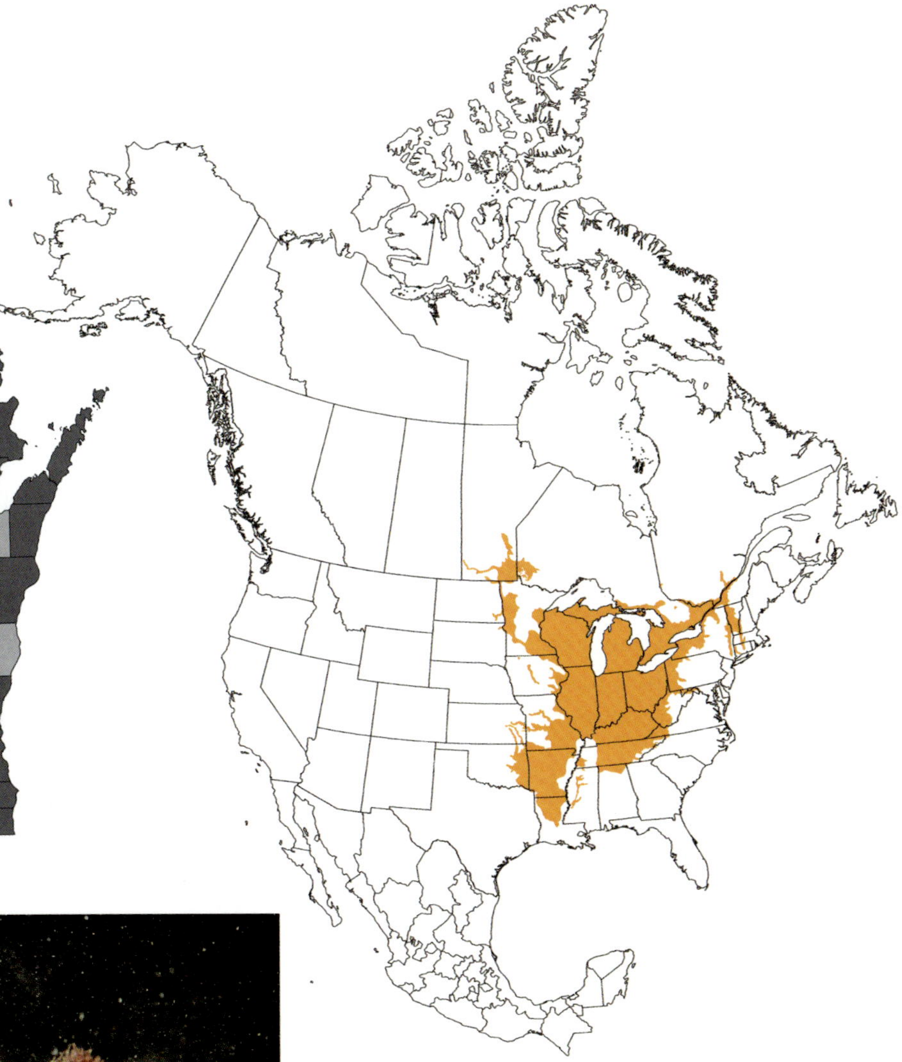

Mudpuppy (Waukesha County; photo by R. Sajdak).

AT A GLANCE

Mudpuppies, or 'Waterdogs,' are large and totally aquatic salamanders. They are brown to gray, with light speckling or scattered spots, but have conspicuous reddish bushy gills on the sides of their head and a tail that is flattened from side to side. Mudpuppies are found statewide in most permanent bodies of water, such as streams, rivers, ponds, and lakes, including Lakes Michigan and Superior. They are active year-round (even beneath the ice in winter) but often hide under rocks, logs, or boards or in holes in streambanks by day. At night, Mudpuppies emerge to forage for small fish, crayfish, and almost any other aquatic animal small enough to swallow. Mudpuppies typically breed in autumn, but fertilization of eggs is delayed until the following spring, and larvae hatch from July to August. Mudpuppies have a reputation as being if not poisonous, then at least distasteful, and some fishermen will cut their lines rather than handle these salamanders if they are accidentally caught. In truth, the animals are harmless. Their slimy secretions are innocuous, and if they can be induced to bite at all, the bite of even a large specimen is a barely painful pinch that is unlikely to draw blood. In Wisconsin, the WDNR considers the Mudpuppy a common species.

SYSTEMATICS

Necturus is considered an ancient genus (Salthe and Kaplan 1966), with known fossil remains that date back to the Paleocene (Hecht 1958). Distinguishing species within this genus can be somewhat difficult due to the lack of derived characters necessary for accurate phylogenetic resolution. All members of the genus are neotenic, and adults retain several ancestral larval characters, including external gills. Most systematic work on the genus has occurred in the southeastern US, where concordant karyological (Sessions 1980, Sessions and Wiley 1985), immunological (Maxson et al. 1988), and electrophoretic (Ashton et al. 1980, Guttman et al. 1990) studies have supported that *Necturus maculosus* is a valid species. People living in the southern US often call them Waterdogs, whereas in Wisconsin they are commonly referred to as Mudpuppies. This species is the only known member of the family Proteidae and genus *Necturus* in the state.

Currently, two subspecies are recognized globally: *Necturus maculosus* (Common Mudpuppy) and *N. m. louisianensis* (Red River Mudpuppy; Crother 2017). The latter is restricted to the southern US (Arkansas, Louisiana, Missouri, and adjacent areas), and available evidence currently recognizes all Wisconsin forms as *N. m. maculosus*. Previously, Bishop (1941a) described the subspecies *N. m. stictus* (Lake Winnebago Mudpuppy) from northeastern Wisconsin, but this poorly defined subspecies actually resembles *N. m. maculosus* from elsewhere in its range (Breckenridge 1944, Hecht 1958, Smith 1961) and generally is not considered valid (Bishop 1948, Hecht 1958, Dickinson 1965, Petranka 1998).

DESCRIPTION

Adults: The body and tail are brown to gray with heavy light speckling and scattered black blotches, while there is sometimes a vague light-colored dorsolateral stripe that runs along the sides of the body and onto the tail (Figure 150). The legs are gray and short but well-developed, with four toes on each foot and pale speckles on the dorsal surface. The head is brown to gray with heavy pale speckling and a dark line with light edges that extends from the nose through the eye to the corner of the jaw, just ahead of the gills. The head is also flattened, with a

Figure 150. Adult Mudpuppy (Sauk County; photo by R. LaFleur).

squared-off nose. Some individuals are dark (almost black), with scattered black flecks (see Variation subsection). As permanently aquatic salamanders, Mudpuppies have prominent red external gills, which are larger in warmer and less oxygenated water (Powell et al. 2016). The body is somewhat dorsolaterally flattened, and the tail has narrow dorsal and ventral fins that extend off of the musculature.

Mudpuppies are the largest salamander in Wisconsin, reaching 30–40 cm TL (11.8–15.5 in; Vogt 1981). Adult males and females are generally similar in size. For example, the mean TL of adult Wisconsin specimens ≥20 cm (7.9 in) housed in the MPM collection only ranged from 28.9 cm in males (11.4 in, $n=51$) to 29.1 cm in females (11.5 in, $n=38$). Live Mudpuppies from the Mukwonago River (Walworth County) and preserved specimens from Rock Lake (Jefferson County) also did not exhibit variation in SVL by sex, regardless of the season when captured (Sajdak 1982). In the autumn of 1976, mean male SVL from the Mukwonago River was 18.9 cm (7.5 in), and mean female SVL was 19.2 cm (7.6 in). Likewise, mean SVL of those collected in Rock Lake during the spring of 1977 was 21.4 cm (8.4 in) for males and 21.5 cm (8.5 in) for females (Sajdak 1982). Despite these findings, females are reportedly larger than males in New York (Bishop 1926) and Michigan (Lagler and Goellner 1941). Some variation in weight was also evident among Mudpuppies captured in the Mukwonago River by season and sex. Both sexes were heavier in the spring of 1977 than in the autumn of 1976. Yet females in the spring of 1977 were heavier than females in the autumn and heavier than males regardless of season ($\bar{x}$ male weight in autumn = 111.8 g [3.9 oz]; $\bar{x}$ female weight in autumn = 112.1 g [4.0 oz]; $\bar{x}$ male weight in spring = 136.4 g [4.8 oz]; $\bar{x}$ female weight in spring = 148.7 g [5.2 oz]; Sajdak 1982). Bishop (1948) reported a record size Mudpuppy from Lake Winnebago that measured 32.7 cm SVL (12.9 in) and 48.6 cm TL (19.1 in).

Preadult stages: Eggs are cream to yellowish in color and 0.50–0.65 cm (0.2–0.3 in) in diameter when freshly laid (Bishop 1941b). Females attach eggs to the roof of nesting cavities beneath submerged aquatic debris with a short stalk that is often easily visible. Eggs are covered by three jellylike envelopes; the entire package (egg, envelopes, and stalk) measures 0.8–1.1 cm (0.3–0.4 in) in diameter and 1.2–2.1 cm (0.5–0.8 in) in length (Bishop 1941b; Figure 151). According to Smith (1911), hatchling Mudpuppies in Wisconsin are ca. 1.8 cm TL (0.7 in). At hatching, the larvae have a large yolk sac and poorly developed limb buds with a shallow tail fin. A light dorsolateral stripe begins to develop at ca. 2.5 cm TL (1.0 in). The yolk is completely absorbed by late November when larvae are ca. 3.4 cm TL (1.3 in). At 3.4 cm TL (1.3 in), the larvae have a dark brown coloration on the head, body, and tail, with a distinct yellow stripe that runs along the sides of the body from the tip of the nose to the tip of the tail (Figure 152). The venter is a pale yellow.

Above, top: Figure 151. Mudpuppy egg and embryo (Richland County; photo by C. Raimond).

Above, bottom: Figure 152. Larval Mudpuppy (Rusk County; photo by C. Raimond).

The juvenile pattern fades with age, and Mudpuppies attain adult coloration and morphology at ca. 15 cm TL (5.9 in). Limited size data have been recorded on larvae in Wisconsin. A larval Mudpuppy found in the Mukwonago River on 14 November 1976 was 3.9 cm TL (1.5 in) and weighed 0.3 g (0.01 oz). It was dark gray on the back with light speckling, a trace of lateral light stripes, and a light venter. The head was dark, with light eye stripes (Sajdak 1982).

Variation: Adult Mudpuppies in Wisconsin can vary in color. For example, many specimens collected in the northeastern and eastern regions of the state exhibit coloration and pattern consistent with the 'Lake Winnebago Mudpuppy,' a subspecies that is no longer recognized (see Systematics section). These individuals have a dark gray, almost black ground color with scattered black specks or dots. Their venter is also dark, but it is lighter than the dorsum. These individuals further lack the black blotches, lighter markings, and eye stripes of more typical Mudpuppy color morphs. Variation in larvae is rarely reported in detail, although Cahn and Shumway (1926) hatched egg masses in captivity that produced a small proportion of melanistic and albino individuals. Hutt (1945) also documented two wild albino specimens collected from adjacent locations in Winnebago County (Lake Butte des Morts near Menasha and presumably Lake Winnebago near Oshkosh). It is noteworthy that Mudpuppies from the Mukwonago River, a small stream environment, were markedly smaller than those from the much larger and lentic Rock Lake (Sajdak 1982). Morphological variation by sex occurs. For example, the vent of females is a simple longitudinal slit, while the male vent possesses a pair of papillae and a transverse groove at the posterior end (Figure 153).

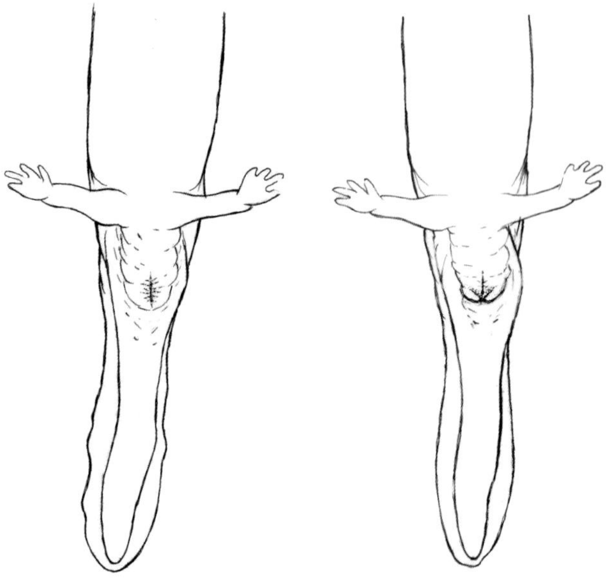

Figure 153. Sexual variation in vent characteristics of Mudpuppies. Note that males (*right*) possess a pair of papillae near the posterior end of their vent (illustration by E. Wild).

Diagnosis: Adult Mudpuppies are characterized by the presence of dorsal and ventral tail fins that extend off of the tail muscles but do not extend anteriorly of the hind legs onto the body, bushy external gills, small eyes, a countersunk lower jaw, and four toes on each hind foot. As the largest and only totally aquatic salamander in Wisconsin, adult Mudpuppies are difficult to confuse with other species. Although adult salamanders of other species are often found in water when breeding, they typically lack external gills.

Mudpuppy eggs can generally be distinguished from the eggs of other salamanders by the habitat where they are laid (typically submerged and attached to the roof of nest cavities beneath underwater debris in lakes, rivers, and streams), the visible stalk that connects them to the cavity roof, and their yellow to cream color. Aquatic larvae of most salamander species typically have external gills and could be confused with Mudpuppies in Wisconsin. Yet the larvae of other species are generally smaller and often found in fishless ephemeral ponds, which Mudpuppies usually do not inhabit. Furthermore, in adult and larval Mudpuppies, tail fins do not extend onto the body, unlike the larvae of other salamanders in Wisconsin (Watermolen and Gilbertson 1996). Larvae of most other salamanders also have five toes on the front and rear feet, whereas Mudpuppies have four toes. The only other Wisconsin salamander with four toes on the hind feet, the Four-toed Salamander (*Hemidactylium scutatum*), is a small species, with terrestrial adults that lack external gills. Neotenic

Eastern Tiger Salamanders (*Ambystoma tigrinum*) appear superficially similar to Mudpuppies because they retain larval morphology and an aquatic lifestyle as adults but have proportionally larger eyes, five toes, and tail fins that extend onto the body.

DISTRIBUTION AND HABITAT

Global distribution: Mudpuppies occur throughout midwestern and eastern North America from southeastern Manitoba to southern Quebec and south to southern Missouri and northern Georgia. They are absent from the Adirondack Mountains and Atlantic Coastal Plain, parts of northern and far southern Minnesota, all but extreme eastern Iowa, and parts of Missouri and Tennessee (Petranka 1998, Matson 2005). Mudpuppies have also been introduced into some states (NatureServe 2020).

Distribution in Wisconsin: Mudpuppies essentially occur statewide in all major watersheds. This includes both Lakes Michigan and Superior, as well as in the Mississippi River (Eckblad 1986) and all major tributary streams. Hoy (1882) considered them "especially numerous in Lake Michigan," and fisheries managers continue to encounter them regularly in the lake proper and in Green and Sturgeon Bays (Dreux Watermolen, WDNR, unpublished data). Of the published county records, vouchers could not be found for several (see the Appendix). In addition, some county records are based on photographs that are of poor quality (e.g., Iron, MPM VZP576) or on photographs of a newspaper article (La Crosse, MPM VZP531). These either rely heavily on the accuracy of anecdotal third-party information or are difficult to use for independent confirmation of species identification. Better vouchers (either photographic or specimens) should be obtained for these counties.

Localities of possible occurrence: With further investigation, Mudpuppies could be found in the larger rivers and lakes of all counties where they are not currently documented. They also certainly exist in locations where Salamander Mussels (*Simpsonaias ambigua*), an obligate Mudpuppy parasite, occur (see Natural History Box: Intimate Relationships). In Wisconsin, Salamander Mussels exist in both the Mississippi River and Lake Michigan drainages and have been reported from the Black, Chippewa, Embarrass, Eau Claire, Flambeau, Lemonweir, Mississippi, St. Croix, Wisconsin, and Wolf river systems (Lisie Kitchel, WDNR, personal communication). These river systems flow through Buffalo, Iowa, Menominee, Monroe, Pepin, and Vernon Counties, where Mudpuppies have not yet been documented.

Habitat: Mudpuppies occupy many types of permanent aquatic habitats, including rivers, perennial streams, lakes, and reservoirs. Hoy (1883) reported that Mudpuppies were "found in abundance in most streams and lakes of the state," and Cahn (1929) considered Mudpuppies "abundant in practically all the lakes of the region, and far less common in the rivers and streams." Vogt (1981) thought that "any large permanent river or lake in Wisconsin" would have Mudpuppies.

Some studies suggest that Mudpuppies are not particular about water depth or velocity. In southeastern Wisconsin, Sajdak (1982) observed Mudpuppies in all available water velocities of the Mukwonago River (Walworth County) "from the dam spillway itself to shallow areas with no perceptible current." They are also found at various water depths (Sajdak 1982), down to 27–30 m (89–98 ft; Hacker 1957, Reigle 1967). Although these reports provide evidence that Mudpuppies are habitat generalists, work by others indicates that they exhibit distinct preferences for certain conditions. Cochran and Borash (2014b) found that in Minnesota, Mudpuppies occupied stream reaches that were relatively wider and deeper than unoccupied reaches in tributaries of the Mississippi River. Mudpuppies also occupied regions that were dominated by riffles and runs, with fish species typical of cool- and warm-water streams. Within Lake Superior, Mudpuppies apparently preferred river mouths and harbors where water was warmer and biological productivity was greater (Casper 2002). There is also evidence for seasonal variation in habitat use. For example, Reigle (1967) collected Mudpuppies from the deep waters of Green Bay in October and December but

NATURAL HISTORY BOX: Intimate Relationships

Mudpuppies share an intimate relationship with the Salamander Mussel (*Simpsonaias ambigua*), a species currently listed as Threatened in Wisconsin and Minnesota and considered to be a Species of Concern by the US Fish and Wildlife Service (Figure 154). Working in the Des Plaines River near Joliet, Illinois, Howard (1915, 1951) first documented Mudpuppies as the host for the larvae of this mussel, the only known obligate host-parasite relationship that involves an amphibian and a freshwater mussel. No other hosts are known.

The mussel has a complicated reproductive cycle. During the breeding season, female mussels produce eggs, which the females brood inside specialized chambers in their gills. Male mussels release sperm into the open water. Females draw the sperm in through their siphons, and the eggs are fertilized in the brooding chamber. Fertilized eggs develop into microscopic larvae (glochidia). Most mussels need to 'infect' a host fish with the glochidia in order to complete development, but Salamander Mussels infect Mudpuppies. How the Mudpuppies become infected remains a bit of a mystery, but Howard (1951) speculated that Mudpuppies are exposed to the larvae when they prey on the adult mussels. Once on the gills, the glochidia become encysted as the host tissue grows over them. Metamorphosis takes place within days or weeks, depending on temperature. The glochidia transform into microscopic juveniles, which then drop off of the host. Glochidia have only been found on Mudpuppies from October through May (Howard 1915, Lisie Kitchel, WDNR, personal communication).

Figure 154. Mudpuppies serve as the host for the rare Salamander Mussel's (*Simpsonaias ambigua*) larvae, the only known obligate host-parasite relationship that involves an amphibian and a freshwater mussel (photo by R. Piette).

In Wisconsin, Salamander Mussels are only found in mud, silt, or sand substrates beneath medium to large flat rocks and undercut ledges, places Mudpuppies frequent. The mussel occurs in both the Mississippi River and Lake Michigan drainages (Baker 1928, van der Schalie and van der Schalie 1950, Mathiak 1979, Clarke 1985). It can be very abundant locally but extremely rare otherwise and is rarely collected (Clarke 1985). Because of the close relationship, the occurrence of Salamander Mussels is a good indicator of the presence of Mudpuppies.

Dreux J. Watermolen, WDNR

took none from deep water during his trawling efforts in August.

Mudpuppies need habitat with an abundance of underwater shelter, as they generally reside beneath logs, rocks, undercut banks, tree roots, or aquatic vegetation during the day. These structures act as both important cover for adults and critical substrate on which females can lay eggs (see Reproduction and Development section). Mudpuppies also occasionally occupy the burrows of other organisms, such as crayfish. Sajdak (1982) found that the majority (80%) of Mudpuppies encountered in the Mukwonago River occurred under some type of hard cover object large enough to conceal the animal. Matson (2005) asserted that a layer of mud along aquatic substrates was an important habitat feature for burrowing beneath underwater cover and that Mudpuppies were generally absent from streams that possessed apparently suitable rock slabs or other retreats but lacked a mud or mud-detritus substrate. However, Mudpuppies in the Mukwonago River were frequently found under cover objects with small gravel or sand substrates (Richard Sajdak, Milwaukee County Zoo [retired], personal observation). Also, Casper (1998) stated that siltation is a problem for Mudpuppies because it degrades substrate quality. Smith (1911) found Mudpuppies nesting in shallow areas (<1.5 m [4.9 ft]) of Lake Monona (Dane County) where the "bottom was strewn with loose flat stones of various sizes." Pentecost and Vogt (1976) claimed that stone riprap used to secure shoreline marshes in Wisconsin "provides excellent nesting habitat." Although limited information on habitat requirements of young Mudpuppies exists, larvae and juveniles apparently congregate in pools where muck or detritus accumulates but are also sometimes found in riffles (reviewed by Matson 2005, R. Sajdak, personal observation). Because Mudpuppies are active year-round and do not seasonally hibernate, it is unlikely that they associate with a specific habitat only during the winter. Yet Mudpuppies may select slightly different microhabitats in winter than during the rest of the year and in Lake Michigan reportedly congregated in shallower water during winter (Pope 1944).

REPRODUCTION AND DEVELOPMENT

Mudpuppies primarily breed in autumn, and mating congregations can form as early as August in Ohio and Wisconsin (Bishop 1941b, Sajdak 1982, Matson 2005). Throughout the breeding season, males and females may share communal retreats beneath rock slabs and other cover objects where mating occurs (Bishop 1941b). Although Mudpuppies breed in autumn, fertilization and egg deposition are delayed until the spring and summer. Egg deposition occurred from April to June, depending on water temperature, in Wisconsin and New York (Bishop 1926, Smith 1911). Eggs are laid in cavities under rocks, logs, or other hard objects (Bishop 1926, Sajdak 1982, Smith 1911). Females within selected nesting cavities turn upside down to adhere egg clusters to the roof (Eycleshymer 1906; Figure 155). Eycleshymer (1906) reported Mudpuppy nests in cavities at water depths as shallow as 10.1 cm (4 in) and as deep as 3.0 m (10 ft), but most were at depths of 0.6–1.2 m (2–4 ft). Data on clutch sizes from Wisconsin are sparse. However, in the Mukwonago River (Walworth County), three nests were found from late May to mid-June on the undersides of a plywood sheet, a concrete slab, and a rock. Female Mudpuppies were in attendance at all three nests. Two of these nests contained 82 and 118 eggs, respectively (Sajdak 1982). In Lake Monona (Dane County), nests were found on the underside of stones that ranged in dia-

Figure 155. Mudpuppy eggs (Waukesha County; photo by R. Sajdak).

meter from 0.4 to 0.6 m (1.5 to 2 ft) in late June and early July (Smith 1911). These nests were at depths of 1–1.5 m (3–5 ft), and a sample of five nests contained 18–87 eggs ($\bar{x} = 66$; Smith 1911). In Michigan, 11 nests contained 28–101 eggs per nest ($\bar{x} = 60$; Fitch 1959). Three nests in Pennsylvania contained 87–140 eggs per nest ($\bar{x} = 107$; Bishop 1941b). The estimated incubation period for eggs is 38–63 days, and hatching occurs from July to August (Bishop 1941b).

Females attend nests and possibly defend the eggs from predators (Bishop 1941b). Females may also attend to larvae after they hatch: Cochran and Lyons (1985) observed a female Mudpuppy near 30–40 juveniles in the Namekagon River (Sawyer County) on 16 September 1982. The juveniles had a mean length of 3.1 cm (1.22 in; range = 2.9–3.2 cm [1.14–1.26 in]). Cochran and Lyons (1985) suggested that the presence of such small juveniles so late in the season indicated either a later hatching date, slower growth, or both compared to the mid-July hatching dates reported from New York (Bishop 1926). Cochran and Lyons (1985) further suggested that the closely associated adult female they discovered (which had an empty gastrointestinal tract, so was not feeding on larvae) might represent maternal guarding of the juveniles.

As described from a series of developing larvae collected in Lake Monona (Dane County), hatching occurs at ca. 1.8 cm (0.71 in; Smith 1911). Work in New York and western Pennsylvania indicated that Mudpuppies develop slowly, reportedly attaining sexual maturity by six years and at a minimum TL of 20 cm (7.9 in; Bishop 1926). McDaniel et al. (2009) determined that the mean estimated age of three populations sampled in Ontario ranged from 7.7 to 12 years and that the oldest individual assessed was ca. 15 years old. Maximum documented longevity is 23 years (Nigrelli 1954).

ACTIVITY

Mudpuppies are active year-round, but their activity appears to peak during autumn and winter (Sajdak 1982). Pearse (1921) stated that they were commonly caught by fishermen through the winter ice of Lakes Mendota and Monona (Dane County), while Cochran (1986a) also reported that scuba divers observed them active below the ice at this time in Lake Mendota. Sajdak (1982) believed that activity in the Mukwonago River (Waukesha County) was greatest in autumn and winter, in part because Mudpuppies congregated for reproduction in autumn. In Vermont, capture probabilities were highest following rain in spring during periods of higher river flow when water temperatures were ca. 3°C–6°C (37°F–43°F; Chellman et al. 2017).

When active, Mudpuppies commonly crawl along the bottom searching for food, but they can also fold their legs against their body and make short swimming spurts using their flattened tail. They are nocturnal in clear-water habitats (Cagle 1954, R. Sajdak, personal observation) but are active by day in murky water and under ice in winter (Bishop 1941b). Information on thermal preferences in the wild is lacking. Sajdak (1982) had much higher Mudpuppy capture rates when water temperature was below 18°C (64°F) in Walworth County. Similarly, Beattie et al. (2017) obtained significantly more captures when water temperature was below 14.1°C (57°F) in northeastern Illinois. Hutchison and Rowlan (1975) conducted laboratory experiments on individuals collected from "northern Wisconsin" and found that their critical thermal maximum (CTmax) ranged from 32.5°C to 35.7°C (90.5°F to 96.3°F). Laboratory studies of Mudpuppies collected from "lakes in central Wisconsin" indicated that thermal tolerances vary by season, ranging from 32.8°C (91.0°F) in January to 34.0°C (93.2°F) in June (Hutchison and Ritchart 1989).

Detailed studies of Mudpuppy movements are rare. Based on recapture data from 53 marked Mudpuppies along a river in Ohio, Matson (1998) estimated a "pooled mean activity range" in summer of 136.1 m^2 (1,465 ft^2), with considerable overlap in activity ranges among individuals. He further found evidence that some Mudpuppies showed fidelity for certain river segments. Seasonal movements or migrations are also occasionally reported. Sajdak (1982) found that when individuals left the autumn breeding congregation they moved up to 1,000 m (3,280 ft) over a 24-hour period, while 24-hour movements in spring were shorter (<100 m [328 ft]). In contrast, Gibbons and Nelson (1968) reported apparently long spring migrations from deep to shallow water in a Michigan lake. Pope (1944) further stated that the "large Lake Michigan population seems to make annual migrations up the inflowing streams and rivers" in the Chicago region.

PREY AND PREDATORS

Mudpuppies consume a wide variety of animals, both vertebrates and invertebrates. In Michigan, Lagler and Goellner (1941) found that insects, earthworms, crustaceans, and fishes made up most of the diet of 105 Mudpuppies examined in late April, when many fish species spawn. Hamilton (1932) reported similar dietary components in the stomachs of 340 New York Mudpuppies that he examined throughout the year but also noted a possible preference for sculpin (family Cottidae). Mudpuppies examined from Minnesota and Wisconsin consumed earthworms, slugs, aquatic insects, crayfish, ranid frogs (e.g., Green Frog [*Lithobates clamitans*] larvae), and fishes such as Rainbow Darters (*Etheostoma caeruleum*; Cochran 1991b, Cochran and Borash 2014a). Several authors have discovered that Mudpuppies also prey on a variety of introduced or invasive aquatic species. For example, Reigle (1967) documented predation of an introduced Alewife (*Alosa pseudoharengus*) in Green Bay (Lake Michigan). In Ohio, shiners (*Notropis* spp.) and the invasive Round Goby (*Neogobious melanostomus*) were reported prey items (Thomas et al. 2007), while Beattie et al. (2017) also documented consumption of Rusty Crayfish (*Orconectes rusticus*) and Zebra/Quagga Mussels (*Dreissena* spp.).

Reports that Mudpuppies in Wisconsin consume various sport fish species and their eggs extend back to the 1800s. For example, Hoy (1883) claimed that Mudpuppies "do considerable damage by eating the spawn of various fishes." Hacker (1957) found that 70% of the Mudpuppies he trapped over artificial fish-spawning structures in Green Lake (Green Lake County) had consumed Lake Trout (*Salvelinus namaycush*) eggs, although it is noteworthy that these Lake Trout were introduced. He attributed the lack of reproduction by this introduced fish to Mudpuppy predation of their eggs. Kempinger (1988) and Caroffino et al. (2010) observed Mudpuppies consuming Lake Sturgeon (*Acipenser fulvescens*) eggs in Lake Winnebago (Winnebago County) and the Peshtigo River (Marinette County), respectively, but neither quantified the impact. On the other hand, Pearse (1921) examined 33 adult Mudpuppies collected from lakes in Dane County, primarily in late November and early December. He determined that Mudpuppies consumed mostly crayfish, insect larvae, small fishes, snails, leeches, and small crustaceans. He found fish eggs (1% of the food items taken) in only a single Mudpuppy that was collected in a gill net along with spawning Yellow Perch (*Perca flavescens*) in early May. Even during this spawning period, crayfish and snails comprised the majority of the Mudpuppy's diet. In spite of these data, Pearse (1921) somewhat surprisingly concluded that the Mudpuppy "doubtless does considerable damage by eating the eggs of fishes." Interestingly, Vogt (1981) cited Pearse (1921) as finding "whitefish" eggs in the stomach of Mudpuppies from Lake Mendota (Dane County), apparently misinterpreting Pearse's (1921) quote of Milner's (1874) report from Michigan. It is worth noting that Whitefish (*Coregonus clupeaformis*) are not native to Lake Mendota (Lathrop et al. 1992), and Cochran (1986a) suggested that Ciscoes (*Coregonus artedii*) were the intended target of this reference. Later investigations corroborated Pearse's (1921) principal findings that Mudpuppies consume a variety of prey opportunistically and do not target fish eggs.

The diet of juvenile Mudpuppies has received less attention. Cochran and Lyons (1985) examined the stomach contents of 16 juvenile Mudpuppies collected in the Namekagon River (Bayfield County). Early instars of aquatic insects were the "most important prey in terms of relative size, frequency of occurrence in stomachs, and numbers." The small grains of sand and algal filaments they discovered were probably ingested incidentally. Beattie et al. (2017) reported that smaller Mudpuppies from a lake in northeastern Illinois consumed more isopods and amphipods than larger Mudpuppies (>20 cm [7.9 in]).

Specific foraging behaviors are not well-studied (see early reviews by Harris 1959), although Mudpuppies primarily hunt at night in clear waters and during the day in muddier or weed-choked waters. Mudpuppies in the Mukwonago River (Walworth County) were seen at night moving through areas of thick weeds feeding on caddis fly larvae (R. Sajdak, personal observation). Mudpuppies were similarly observed in shallow waters along the rocky shore of

Sturgeon Bay (Door County), where they appeared to feed on mayfly and midge larvae (Dreux Watermolen, WDNR, personal observation). Pearse (1921) inferred that foraging Mudpuppies rely primarily on tactile senses, quoting an observation that "whenever the water is disturbed in the vicinity of the snout they snap viciously." Mudpuppies also have well-developed chemosensory abilities that likely aid in prey capture (Duellman and Trueb 1986). Captive Mudpuppies housed in a glass aquarium appeared to visually stalk larval insects (D. Watermolen, personal observation).

Mudpuppies are consumed by a wide variety of organisms, although fishes appear to be particularly important predators. In Wisconsin, predatory fishes that feed on Mudpuppies include Muskellunge (*Esox masquinongy*) and Burbot (*Lota lota*; Fratt 1991, Burri 1997, Bozek et al. 1999). Cochran and Lyons (1985) observed predation of juvenile Mudpuppies by Common Shiners (*Luxilus cornutus*) and Hornyhead Chubs (*Nocomis biguttatus*). A variety of other fishes likely consume Mudpuppies as well, but temporal factors associated with Mudpuppy reproduction and growth may help minimize this predation. Specifically, Mudpuppy eggs hatch from July to August. Thus, juvenile Mudpuppies, which are smaller and presumably subject to a wider range of predators, may not be abundant at times when fishes are feeding most actively (e.g., in Wisconsin, the most active feeding period for many fishes occurs in March to early May just after the breakup of lake ice; Greg Sass, WDNR, personal communication).

Predation by larger amphibians and cannibalism, including conspecific egg consumption, has been reported (Eycleshymer 1906, Harris 1959, Irwin et al. 2014, other reports summarized in Lockhart 2000). Reptile predators include Snapping Turtles (*Chelydra serpentina*) and snakes (family Colubridae; Lagler and Salyer 1947, Collins 1993, Himes 2004, Gibbs et al. 2007). Long et al. (1978) reported predation by a Common Watersnake (*Nerodia sipedon*) along the shoreline of Washington Island (Door County).

Several reports of predation by birds in Wisconsin exist. Watermolen (1996) described a case of attempted Osprey (*Pandion haliaetus*) predation on the lower Menominee River near the Wisconsin-Michigan border. In that instance, a pair of American Crows (*Corvus brachyrhynchos*) harassed the Osprey so that it dropped the Mudpuppy, which was later scavenged by one of the crows. Fassbender and Watermolen (2002) reported a Ring-billed Gull (*Larus delawarensis*) scavenging a partially decomposed Mudpuppy and a Herring Gull (*Larus argentatus*) ingesting a live Mudpuppy on an ice floe in Sturgeon Bay (Door County). They also described a case of a Herring Gull robbing a Common Merganser (*Mergus merganser*) of a recently caught Mudpuppy. Double-crested Cormorants (*Phalacrocorax auritus*) feed on Mudpuppies in New York (Coleman et al. 2012, DeBruyne et al. 2013) and probably do in Wisconsin as well.

Predation of Mudpuppies by mammals is not well documented in Wisconsin, but Jackson (1961) mentioned them as prey of North American River Otters (*Lontra canadensis*). Elsewhere in the Mudpuppy's range, known mammalian predators include North American River Otters (Michigan; Lagler and Ostenson 1942) and American Mink (*Mustela vison*, Maine; Little 2002). Predation by invertebrates is reported, but supporting data are sparse. Without citing sources, several authors state that crayfish consume adults or eggs (Petranka 1998, Matson 2005). Yet in his extensive review of the trophic relationships of North American freshwater crayfishes and shrimps, Hobbs (1993) found no reports where crayfish had consumed Mudpuppies. He did, however, make general statements that crayfish consume amphibians, so predation of Mudpuppies seems reasonable. Mudpuppies primarily rely on cryptic coloration and secretive behaviors to avoid predators. For many years, a popular belief prevailed that Mudpuppies were "poisonous" and somehow dangerous to people (e.g., Milner 1874), which is false.

Parasites taken from Mudpuppies in Wisconsin include the monogenetic trematode *Sphyranura osleri* (found on the gills and skin) and the cestode *Proteocephalus loennbergii* (found in the intestines) of Mudpuppies in Dane and Waukesha Counties (Pearse 1921, Coggins and Sajdak 1982). Cochran et al. (2002) found cestodes that they believed to be the latter species in the intestine of a juvenile Mudpuppy collected in the Iron River in Ashland County, and this tapeworm appears to be common in Michigan as well (Lagler and Goellner 1941). Hughes (2008) observed immature forms (metacercariae) of a digenean trematode, *Clinostomum*, in Mudpuppies collected in Adams Lake (Portage County). Amin (1985) found the acanthocephalan *Acanthocephalus dirus* in Mudpuppies in Racine County. Wisconsin Mudpuppies are also likely hosts for ectoparasitic leeches of the genus *Placobdella*. Briggler and Moser (2008) reported *Placobdella cryptobranchii* feeding on Mudpuppies in Missouri.

CONSERVATION

Status: Globally, the Mudpuppy is listed as a Least Concern species by the IUCN because of its wide distribution, tolerance of a degree of habitat modification, and presumed large and apparently stable populations. The Mudpuppy is listed as a species of Special Concern in Minnesota and Indiana and as a Threatened species in Illinois and Iowa. Mudpuppies are considered a Species of Greatest Conservation Need in the wildlife action plans developed for Illinois, Iowa, Michigan, and Minnesota. They have reportedly declined in Lake Erie (Pfingsten and White 1989) and other parts of the Great Lakes region (Harding 1997). They are further considered endangered and likely extirpated in Maryland and South Dakota (Maryland Natural Heritage Program 2010, South Dakota Game, Fish, and Parks 2018) and a Species of Greatest Conservation Need in Vermont (Vermont Fish and Wildlife 2015). The Mudpuppy is currently considered a common species in Wisconsin by the WDNR. Despite this, some Mudpuppy populations in the state are likely declining or extirpated. For example, the Waukesha County population studied by Sajdak (1982) has disappeared, according to Casper (1998). Mudpuppies have also been nearly or completely eliminated from streams with a history of water quality issues, such as the Sheboygan River (Galarneau and Masterson 2001).

Populations: Annual abundance estimates for a population in Vermont over two years ranged from 132 to 335 adults, although CIs were wide (Chellman 2011, Chellman et al. 2017). Matson (1990c) estimated that population densities along a 600 m (1,969 ft) stretch of river in Ohio were 805/km (1,296/mi; 95% CI = 513–1,188/km [826–1,912/mi]). Unfortunately, additional rigorous population data are limited, and most information must be inferred from anecdotal reports. Some historical accounts indicate that populations are (or were) high in certain locations. For example, Milner (1874) reported that a fisherman who captured 2,000 Mudpuppies with a single haul of a sweep seine estimated the mean number of Mudpuppies "to each square rod to be four" (ca. 1,542/ha [640/ac]). Milner (1874) also recounted a story of a fisherman from Illinois who set 900 hook lines for fish and in one night caught 500 Mudpuppies. In Wisconsin, roughly 150 Mudpuppies were removed from Green Lake (Green Lake County) over an apparently short time period, although exact trapping efforts were not reported (Hacker 1957).

Monthly adult survival rates in Vermont ranged from 0.866 to 0.882 (Chellman 2011, Chellman et al. 2017). The sex ratio of samples from the Mukwonago River (Walworth County) did not differ from 1M:1F (Sajdak 1982), a pattern that is frequently reported in other parts of the Mudpuppy's range (Cagle 1954, Gibbons and Nelson 1968, Shoop 1965a). Despite this, Lagler and Goellner (1941) reported a sex ratio of 1M:0.61F (n = 62M:38F) for a population sampled in a Michigan lake, while McDaniel et al. (2009) found that populations were biased toward females at study locations in Ontario.

Threats: Unfortunately, Wisconsin populations are not rigorously monitored, and limited data exist on trends, serious threats, or impacts. Documentation of inadvertent Mudpuppy captures during fisheries assessment surveys can greatly enhance the collection of distribution and population data (Craig et al. 2015, Watermolen 2015b). Collins (2017) also recently explored the use of environmental DNA (eDNA) collected through water sampling and detected using the quantitative polymerase chain reaction (qPCR). This technique may be a cost-effective and time-saving approach for monitoring Mudpuppies at a large spatial scale.

General threats associated with degradation of rivers and streams are relevant to this species (see Introduction to Conservation and Management of Wisconsin's Amphibians and Reptiles chapter). Casper (1998) stated that silt deposition in disturbed landscapes has negative impacts on Wisconsin Mudpuppy populations. Excessive sedimentation and associated turbidity have buried rock structures and reduced abundance of aquatic vegetation in some rivers, which could eliminate essential cover and nesting habitat. Dams also pose barriers to upstream migration for Mudpuppies (D. Watermolen, unpublished data), which could result in population fragmentation, isolation, and genetic bottlenecks that can compromise population viability. The presence of Mudpuppies above barriers such as waterfalls on some tributary streams suggests they were able to colonize these upriver locations prior to the formation of these barriers (Cochran 1999, Cochran et al. 2002). More research is needed to determine how natural and anthropogenic barriers affect connectivity and interactions among Mudpuppy populations.

Humans impact Mudpuppies in a variety of additional ways, some of which are unintentional. For example, mechanical harvesters deployed to remove aquatic 'weeds' from lakes and other waterbodies can accidentally kill Mudpuppies that become entangled in harvested vegetation (Booms 1999). Yet direct mortality due to humans could be of greater concern. Mudpuppies are occasionally viewed as a major threat to effective game fish management because of their occasional tendency to eat fish eggs (e.g., Milner 1874, Pearse 1921, Hacker 1957). This tendency has, unfortunately, saddled Mudpuppies with a negative reputation among some fisheries managers. Matson (2005) stated that "erroneous information remains pervasive among biologists, fishers, and others who continue to persecute mudpuppies." Similarly, some anglers regularly catch Mudpuppies (Eycleshymer 1906, Pearse 1921) and discard them on land or on the ice to die because they are falsely viewed as destructive (Petranka 1998, D. Watermolen, unpublished data).

Mudpuppies are also harvested by humans for a variety of commercial purposes, including specimens for biological supply companies and to use as bait, which could negatively impact populations. As of 2011, Mudpuppies were still actively collected in Wisconsin for biological supply companies, and this commercial harvest likely continues today (D. Watermolen, unpublished data). Mudpuppies are also occasionally harvested for human consumption. Pearse (1921) described the meat as "fine in quality and very white," going on to state that it "rivals frog legs in flavor and is more desirable because it comes in larger 'packages,' the body muscles of an adult Mudpuppy being more bulky than the legs of the largest frogs." He even went so far as to provide instructions for how to prepare and cook the meat. Vogt (1981) mentioned that some UW–Madison graduate students used to partake of an annual contest to see who could catch the most Mudpuppies from Lake Mendota (Dane County), an endeavor that culminated in a Mudpuppy fry and dinner.

Although research in Wisconsin is limited, environmental aquatic contaminants could be a substantial concern. Because Mudpuppies are distributed widely, relatively sedentary, and long-lived, they could act as biological indicators of aquatic pollution (Bonin et al. 1994, 1995). Mudpuppies bioaccumulate organochlorine pesticides, polychlorinated biphenyl (PCB) congeners, mercury, and other pollutants (Galarneau and Masterson 2001, reviewed by Sparling et al. 2010). Mudpuppies could also be negatively affected by high biological oxygen demand (BOD) due to unnaturally high aquatic nutrient levels from polluted runoff (reviewed by Sparling et al. 2010). McDaniel et al. (2009) reported deformity rates of 9%–11% over a two-year survey in Ontario.

A specific and serious threat to Mudpuppy populations in streams within the Great Lakes basin is the use of TFM (3-trifluoromethyl-4-nitrophenol) to control Sea Lampreys (*Petromyzon marinus*). While TFM is generally not considered harmful to other fish, several authors have found evidence that lampricide can negatively impact Mudpuppies. Gilderhus and Johnson (1980) observed Mudpuppy mortality in 33% of their field observations along lampricide-treated streams feeding into Lake Superior and further found evidence of large die-offs in 18% of observations. This effect was confirmed in 2009, when surveyors of a Vermont river after TFM treatment recovered more than 500 dead Mudpuppies (Chellman et al. 2017). Matson (1990c) reported a 29% decline in an Ohio River Mudpuppy population after exposure to TFM lampricide. Matson (1990c, 1998) also pointed out that for long-lived, slow-maturing species like the Mudpuppy, even treatments at three-to-five-year intervals would adversely affect its ability to recover from lampricide-induced mortality. As such, population declines and even local extirpations could occur unless adjacent source populations exist that can supply migrants to treated river stretches. In spite of these observations, Boogaard et al.'s (2003) experiments led them to conclude that "an adequate margin of safety exists for adult mudpuppies to survive when exposed during stream treatments." The long-term effects that successive lampricide treatments have on Mudpuppy populations are unknown, in part because limited information is available on Mudpuppy reproduction and survivorship and on TFM's impacts on juveniles.

Mudpuppies are likely susceptible to various bacterial, fungal, and viral pathogens. Pearse (1921) stated that "wherever there is an abrasion in a mudpuppy's skin, fish mold is apt to grow" and went on to describe catching Mudpuppies in the Wisconsin River that had "white growths of mold on the body or head." The identity of this fungus was not reported but was possibly a type of *Saprolegnia*. The chytrid fungus *Batrachochytrium dendrobatidis* (Bd) affects many amphibians and is a serious conservation concern (see Introduction to Conservation and Management of Wisconsin's Amphibians and Reptiles chapter). Several authors have detected Bd on Mudpuppies from various locations, including Wisconsin (Standish et al. 2018). Speare and Berger (2013) reported a case of Bd infection in a captive Mudpuppy. Regester et al. (2016) detected Bd infections in wild Mudpuppies from Pennsylvania but observed "no external physical abnormalities associated with infection" and found that "all mudpuppies appeared in overall excellent condition." Ouellet et al. (2005) investigated preserved museum specimens for presence of Bd but did not detect it on a small sample of Mudpuppies collected from Canada in 1998. Chatfield et al. (2012) confirmed the presence of Bd in two related species from the southeastern US: Black Warrior River Waterdogs (*Necturus alabamensis*) in Florida and Gulf Coast Waterdogs (*N. beyeri*) in Louisiana. Although it is uncertain how large a threat this fungus poses to wild populations, Standish et al. (2018) reported a high prevalence of Bd infection (71%) among 62 Mudpuppies sampled in Wisconsin.

In addition to Bd, Martel et al. (2014) recently described the effects of the salamander chytrid fungus *Batrachochytrium salamandrivorans* (Bsal) on European salamander populations, raising concern for species elsewhere should the fungus be introduced to other countries. Yet Standish et al. (2018) did not detect Bsal in a sample of 62 Mudpuppies taken from several locations in Wisconsin. Mass Mudpuppy die-offs in the Detroit River, Lake St. Clair, and Lake Erie were attributed to either an iridovirus similar to largemouth bass virus or type E botulism, caused by *Clostridium botulinum* (Faisal 2006, Robinson 2006). Regester et al. (2016) recently detected a ranavirus in Pennsylvania Mudpuppies but found no outward signs of pathogenesis on infected individuals. Standish et al. (2018) did not detect a ranavirus on Mudpuppies tested from Wisconsin. A unique concern for Mudpuppy conservation initiatives is its intimate relationship with the Salamander Mussel (*Simpsonaias ambigua*), a species currently listed as Threatened in Wisconsin and a species of concern by the USFWS. As an obligate host to the mussel's larval stage, any decline or local extirpation of the Mudpuppy would adversely affect the mussel (see Natural History Box: Intimate Relationships).

Management: Limited research on effective management strategies for this species in the Upper Midwest exists. Presumably, many general management considerations for streams, rivers, and lakes are relevant (see Introduction to Conservation and Management of Wisconsin's Amphibians and Reptiles chapter). Some sources suggest that control of siltation, which smothers suitable locations for egg deposition in streambeds and riverbeds, could be an important consideration (Casper 1998). Mitigation of factors that contribute to aquatic pollution, the use of TFM lampricide, and the spread of emerging diseases is also an important consideration. For example, individuals working in aquatic habitats can take steps to prevent the potential introduction and spread of harmful pathogens and other invasive species by using appropriate disinfection protocols for boats, gear, and equipment (Perdzock 2015). An important component of effective management and conservation of Mudpuppies could be public education to reduce their persecution, particularly among commercial and recreational anglers. For example, dissemination of information that reviews the importance of their trophic roles in aquatic communities rather than focusing on their consumption of fish eggs would put their predation habits in perspective.

Eastern Newt

Notophthalmus viridescens (Rafinesque 1820)

Jaimie L. Klemish, Rochelle M. Stiles, Jonathan W. Swan, and Michael J. Lannoo

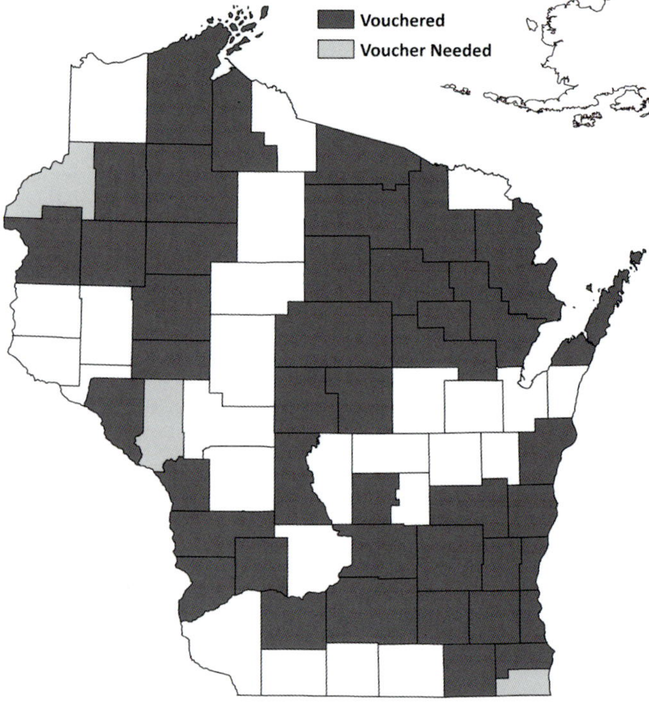

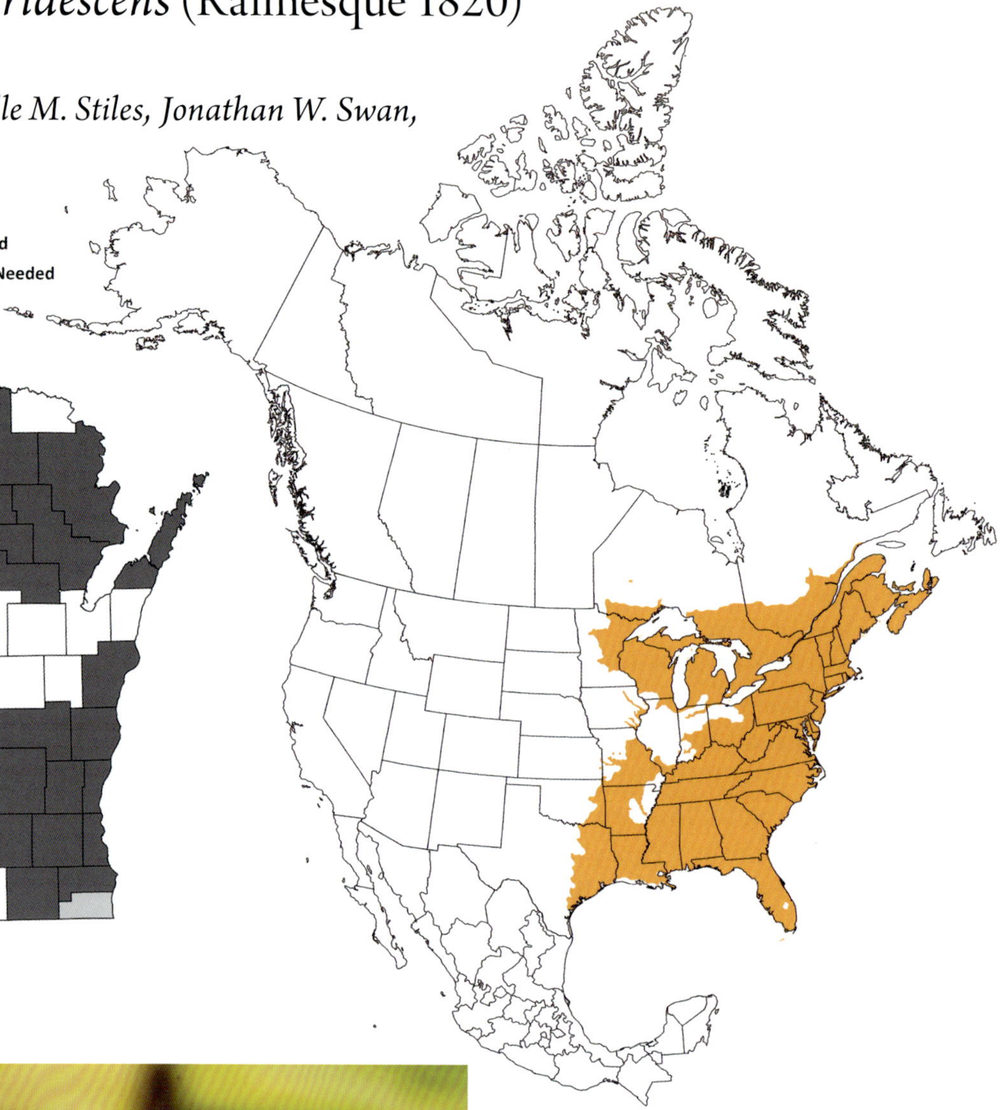

Terrestrial adult Eastern Newt (Sheboygan County; photo by J. DeBoer).

AT A GLANCE

Eastern Newts are highly variable in appearance and have a complex life cycle. Larvae and adults are typically aquatic with deep tail fins and smooth skin. Their dorsal color is olive green to yellow brown, which serves as camouflage, while their bellies are yellow with black spots. Although similar in general coloration to adults, larvae are smaller and have bushy external gills. Larvae eventually develop into terrestrial juveniles, called efts. The efts are more brightly colored than larvae or adults, typically ranging from rusty brown to brick red in Wisconsin. This color is aposematic and serves to warn predators that efts are toxic. Efts remain in terrestrial habitats for several years before returning to wetlands as adults to reproduce. Under certain circumstances, such as pond desiccation or high population densities, aquatic adults can leave the water and live on land. These terrestrial adults develop narrower tail fins and rougher skin than aquatic forms. Eastern Newts occur throughout the state in suitable habitat, such as permanent and semipermanent wooded wetlands with emergent or submersed vegetation. Terrestrial forms seek cover beneath ground debris or vegetation in woodlands adjacent to breeding wetlands. Aquatic adults can be active year-round, although activity decreases in winter, while terrestrial stages seek underground winter retreats in late autumn. Retreats must allow animals access to soil layers that extend below the frost line. Reproduction in Wisconsin probably occurs in autumn, but females store sperm for egg fertilization and deposition the following spring. Eastern Newts eat invertebrates, and their diet composition changes depending on habitat. Predators include numerous types of vertebrates, although their toxic skin secretions deter some fishes, mammals, and birds. Eastern Newts are listed as a common species by the WDNR, but populations are susceptible to loss and degradation of their terrestrial and aquatic habitats.

SYSTEMATICS

Notophthalmus viridescens (Eastern Newt) is the only member of the genus *Notophthalmus* and the family Salamandridae found in Wisconsin. Rafinesque (1820) originally described the eft and adult stages as separate species (*Triturus miniatus* and *T. viridescens*, respectively), which held until Cope (1889) recognized them as different life stages of the same species (Rafinesque 1820, Cope 1859, 1889, Sever 2006). Baird (1850) placed this species in the genus *Notophthalmus*. Today, four subspecies of *N. viridescens* are recognized, including *N. v. dorsalis* (Broken-striped Newt; Harlan 1828), *N. v. louisianensis* (Central Newt, the subspecies in Wisconsin; Wolterstorff 1914), *N. v. piaropicola* (Peninsula Newt; Schwartz and Duellman 1952), and *N. v. viridescens* (Red-spotted Newt; Rafinesque 1820, Crother 2017). However, genetic analyses do not support these subspecific taxa (Gabor and Nice 2004, Whitmore et al. 2013).

DESCRIPTION

Adults: Adults can be aquatic or terrestrial, which leads to variation in appearance (see Variation subsection; Figures 156 and 157). In general, adult Eastern Newts have slightly granular skin (particularly when terrestrial), a laterally compressed body that lacks costal grooves, and a laterally compressed tail (reviewed by Petranka 1998, Mecham 1967, Hunsinger and Lannoo 2005). The dorsal color is dark olive green to pale yellow brown, the ventral color is yellowish, and the whole body is speckled with small dark spots (Mecham 1967, Vogt 1981, Petranka 1998,

Figure 156. A terrestrial adult Eastern Newt. Note the rough appearance of the skin (Portage County; photo by J. Kapfer).

Figure 157. An aquatic adult Eastern Newt. Note the wide tail fins (Chippewa County; photo by R. Sajdak).

Christoffel et al. 2009). In Wisconsin, Eastern Newts (i.e., the Central Newt subspecies) also have two rows of small red spots on either side of the spine from the base of the head to the base of the tail, although these markings can be absent, reduced, or faint (Vogt 1981, Christoffel et al. 2009). Eastern Newts are one of Wisconsin's smallest salamanders, with a TL of 6.5–10.0 cm (2.6–3.9 in) and an SVL of 3.0–5.0 cm (1.2–2.0 in; Vogt 1981). The mean SVL and tail length of adult Wisconsin specimens in the MPM collection were 3.8 and 3.3 cm (1.5 and 1.3 in), respectively ($n = 79$). The largest individual measured had an SVL of 5.1 cm (2 in) and a tail length of 4.4 cm (1.7 in, MPM 9114, collected in 1973 from Bayfield County).

Preadult stages: Eggs measure 0.65–0.80 cm (0.26–0.31 in) in diameter and are deposited singly or in clusters encased in a thick jelly-like envelope (Vogt 1981). Aquatic larvae possess bushy external gills that are similar to but often proportionally smaller than those of *Ambystoma* larvae (Figure 158; Minton 2001). Larvae are grayish to greenish brown (Harding 1997, Mitchell and Gibbons 2010) and have a dark stripe that passes from the snout through the eye on each side of the head (Petranka 1998). Size at hatching is 0.7–0.9 cm TL (0.3–0.4 in; Vogt 1981). Juveniles typically leave wetlands prior to reaching sexual maturity and become a terrestrial form commonly called an 'eft' (see Reproduction and Development section). Efts are stout, with a rounded tail (in cross section) and granular skin; they range from 3.0 to 4.0 cm TL (1.2 to 1.6 in; Figure 159; Vogt 1981). The most notable characteristic is their aposematic coloration, which is a uniform rusty brown to brick red and is slightly darker dorsally, with a spot pattern similar to that of adults (Figure 159; Vogt 1981).

Above, top: Figure 158. Aquatic Eastern Newt larvae with large external gills (Waukesha County; photos by K. McKinney [*top*] and J. Kapfer [*bottom*]).

Above, bottom: Figure 159. A terrestrial juvenile Eastern Newt, called an 'eft.' Note the reddish-orange coloration (Sheboygan County; photo by J. DeBoer).

Variation: Eastern Newts are highly variable in appearance, depending on whether they are aquatic or terrestrial as adults and juveniles (see Reproduction and Development section). The skin of terrestrial adults is rougher than that of aquatic adults, while aquatic adults have a deeper tail to facilitate swimming (Figures 156 and 157). Efts are reddish in color, unlike other life stages (see Preadult stages subsection).

Diagnosis: Aquatic adults have broad dorsal and ventral fins that extend from the tail musculature. They have no external gills, but they do have large eyes, a dark dorsum and yellowish venter with black spots, and five toes on each hind foot. Terrestrial adults and efts do not have broad tail fins, and they have rough granular skin. Adults also possess a dark dorsum and yellowish venter, while the eft's body is reddish with a somewhat darker dorsal surface. Although adult Four-toed Salamanders (*Hemidactylium scutatum*) and Eastern Red-backed Salamanders (*Plethodon cinereus*) are similar in size to terrestrial adult Eastern Newts, the latter do not possess costal grooves (Harding 1997).

Eastern Newt eggs could be confused with those of other amphibians in the wild. Yet their eggs are often laid singly and frequently wrapped in vegetation, unlike eggs of other salamanders (see Reproduction and Development section). Furthermore, the diameter of individual eggs and their gelatinous envelopes is greater than that of other salamander species with which they frequently share habitat in Wisconsin (such as the Blue-spotted Salamander [*Ambystoma laterale*]), with the exception of the Eastern Tiger Salamander (*A. tigrinum*; Watermolen 1995). Juveniles are relatively easy to distinguish from other Wisconsin salamanders. Larvae have a dark stripe on each side of the head that extends from the snout through the eye, have granular skin, and lack costal grooves (Petranka 1998). Juveniles in the terrestrial eft stage could be mistaken for juvenile Eastern Red-backed Salamanders, which are also terrestrial at a small size. Yet efts are somewhat uniformly covered in a rusty or brick color, unlike juvenile Eastern Red-backed Salamanders, which have a bright red dorsal band that is constrained to the back. Efts also have granular skin and lack costal grooves.

DISTRIBUTION AND HABITAT

Global distribution: Eastern Newts are the second most widely distributed North American salamander (following the Eastern Tiger Salamander [*Ambystoma tigrinum*]; Hurlbert 1969). They are found throughout the eastern US in historically forested areas from the Atlantic Ocean to the eastern Great Plains. Eastern Newts occur as far west as eastern Oklahoma and Texas; in east-central and southeastern Kansas; throughout Illinois, eastern Iowa, and northern and central Minnesota; and from southeastern Canada to the Gulf of Mexico (Hunsinger and Lannoo 2005). Analyses of allozyme loci and allele frequencies suggest that population genetics do not fully support the recognized distributions of the four subspecies (Gabor and Nice 2004).

Distribution in Wisconsin: Eastern Newts are widely distributed across Wisconsin, but numerous counties currently lack published records (Vogt 1981, Casper 1996). Of the published county records, voucher specimens could not be found or do not currently exist for Burnett, Trempealeau, and Kenosha Counties (see the Appendix). Although a reported voucher specimen previously existed for Trempealeau County (MPM VZP460; Casper 1996), Casper (2015) reexamined the location associated with this specimen and determined that it was actually collected in La Crosse County. Suzuki (1951) referred to specimens from Kenosha County (MU 1023 and 1024), but these have since been lost, and no alternative voucher is known, which is why Vogt (1981) did not include this county (Richard Vogt, Instituto Nacional de Pesquisas da Amazônia, personal communication). As such, a replacement voucher specimen for Kenosha County is currently needed.

Localities of possible occurrence: There are currently gaps in this species' documented distribution in Wisconsin. Yet many of the counties that currently lack published records are adjacent to counties of confirmed occurrence. Furthermore, Vogt (1981) mapped records from Iron County that he considered reliable, but he did not confirm them, and they are still not corroborated (Casper 1996). There are additional counties where this species likely exists but has not been detected yet.

Habitat: Eastern Newts are generally found in association with wooded landscapes. For example, in New Brunswick, Canada, Eastern Newt occupancy was highest in areas where 'mature' and 'overmature' forests occurred within 1,000 m (3,281 ft) of wetlands (Gibbs 1998a, Jacobs and Houlahan 2011). Use of more specific habitat types within wooded landscapes, however, varies by age and whether individuals are aquatic or terrestrial. Aquatic adults and larvae occupy a variety of permanent or semipermanent waterbodies, such as lakes, reservoirs, ponds, wetlands, ditches, sloughs, canals, and quiet streams in upland and bottomland regions of both woodland and open-canopy areas (Bishop 1941b, Mecham 1967, Vogt 1981, Gates and Thompson 1982, Hunsinger and Lannoo 2005). Vogt (1981) further observed them in floodplain forests of the Wisconsin River but also in open wetlands in oak savanna habitat in Wisconsin (see Amphibians and Reptiles in Wisconsin Plant Communities chapter). Aquatic forms are often found in wetlands with ample emergent or submersed vegetation (Bishop 1941b, Mecham 1967, Gates and Thompson 1982). For example, Burton (1977) determined that presence of this species was related to availability of rooted macrophytes. The water depth that aquatic adults occupy is reportedly broad. Burton (1977) found them frequently at depths of <2 m (6.6 ft), while George et al. (1977) observed them at depths of up to 13 m (42.7 ft) in New York and often at or below the aquatic thermocline.

Adults that abandon aquatic environments will inhabit adjacent moist terrestrial habitats (Hurlbert 1969, Gill 1978b). These terrestrial adults usually remain beneath ground debris (such as rocks or decaying logs) or vegetation or work themselves into mud (Bishop 1941b, Hurlbert 1969, Gill 1978b, Green et al. 2013). Efts are also terrestrial and typically found in woodlands that are adjacent to breeding wetlands, where they rest beneath leaf litter, decaying logs, or humus that provides cover, moisture, and invertebrate prey (Bishop 1941b, Williams 1947, MacNamara 1977, Vogt 1981). Interestingly, efts exhibit preferences for certain types of woodlands over others. Hurlbert (1969) reported that they were more common in beech-maple and beech-maple-hemlock woodlands than oak woodlands. Efts shift microhabitat use throughout the year, perhaps to follow changes in prey availability. Healy (1974b, 1975) found that they occurred near the bases of trees and stumps during the spring, temporary pools in early summer, and areas with decaying mushrooms in late summer. Efts are found less frequently in association with permanent aquatic habitats surrounded by terrestrial environments that are unsuitable or of marginal quality (Wilbur and Collins 1973, Healy 1974a, Smith and Pfingsten 1989, Yoder and Coggins 2007, Takahashi and Parris 2008).

Eastern Newts breed in aquatic habitats that probably do not differ greatly from habitat used by aquatic adults during the rest of the active season. Habitats used for hibernation, however, vary based on whether an individual is aquatic or terrestrial. Aquatic adults overwinter in the water beneath submersed vegetation or large rocks at the bottom of wetlands that do not freeze, while terrestrial individuals overwinter beneath ground debris in adjacent woodlands that allows them to retreat below the frost line (Morgan and Grierson 1932, Bishop 1941b, Cooper 1956, Pitkin and Tilley 1982). These terrestrial overwintering locations are generally <100 m (328 ft) from breeding wetlands (Regosin et al. 2005).

REPRODUCTION AND DEVELOPMENT

Populations of Eastern Newts can consist of numerous developmental forms. Depending on the time of year, these can include (1) embryos, (2) fully aquatic larvae with gills, (3) efts (terrestrial juveniles), (4) aquatic adults with lungs, and (5) terrestrial adults; some populations also include (6) aquatic efts with lungs and (7) aquatic neotenic adults that retain gills (although neotenic adults are not yet documented in Wisconsin; Noble 1926, Bishop 1941b, Brandon and Bremer 1966, Healy 1970, Vogt 1981, Harris 1987, reviewed by Hunsinger and Lannoo 2005). The factors that drive the occurrence of these various developmental states are complex but include unfavorable environmental conditions and high conspecific densities (Healy 1970, Morin et al. 1983, Harris 1987).

Timing and duration of breeding are dependent on latitude and climatic conditions (reviewed by

Petranka 1998, Hunsinger and Lannoo 2005). Eastern Newts can reportedly breed in both autumn and spring (Bishop 1941b, Hurlbert 1969, Harding 1997), only in autumn (Stein 1938, Hurlbert 1969, Healy 1975), or only in the spring (Gill 1978b, Massey 1990, Oldfield and Moriarty 1994). The timing of breeding in Wisconsin is not well documented, although Vogt (1981) reported that in autumn males develop dark horny toe tips and inner-thigh pads, features associated with mating (Petranka 1998). However, on 18 April 2019 a male was found grasping a female behind the head with his hind legs (a courting embrace described below) in Walworth County, which supports that spring breeding could occur in Wisconsin (Joshua Kapfer, UW–Whitewater, personal observation). Females courted in the autumn can store sperm for ten months (Gage 1891, Jordan 1893, Adams 1940, Bishop 1941b, Massey 1990, Mitchell and Gibbons 2010), and regardless of when reproduction occurs, eggs are deposited in the spring. Most adults breed annually, but females can skip reproduction in one or more years (Gill et al. 1983, Gill 1985).

Eastern Newt courtship is aquatic and elaborate (Arnold 1977, Petranka 1998, Mitchell and Gibbons 2010). Males court females either by performing lateral "hula displays" (an undulation or wave-like motion of the body and tail) or by engaging in amplexus (Arnold 1977, Verrell 1982b, 1990). The attracted female subsequently nudges the male's tail with her snout so that he responds by depositing one or more spermatophores on nearby substrates, which the female picks up with her swollen cloacal lips (Bishop 1941b, Arnold 1977, Verrell 1982b, 1990). If the female is unresponsive, the male grasps her behind her head with his hind limbs and begins rubbing her snout with his forelimbs and genial glands on the sides of his head (Schmidt 1930, Arnold 1977). He also fans his tail to waft reproductive pheromones toward her head. After a prolonged period (sometimes over an hour), he increases the fanning and begins violent body contortions, which jerk the female around. The male then again deposits one or more spermatophores, which the female can pick up (Arnold 1977). Eggs are fertilized internally with sperm from the spermatophores, although fertilization can be delayed (see previously). Hula courtship is shorter and requires less energy than amplexus, but sneak inseminations by smaller rival males are more likely to occur when larger dominant males are engaging in hula courtship (Verrell 1982b, 1983, 1985).

Eastern Newts are promiscuous, and both sexes breed several times throughout the mating season (Gill 1978a). Males use visual or olfactory cues to assess female size (Verrell 1982a, 1985) and compete against each other for reproductive opportunities. For example, some males will attempt to dislodge mating pairs that are already engaged in an effort to gain sole access to the female (Verrell 1982a, 1985). Smaller males can also mimic female behavior and induce larger males to deposit spermatophores, even in the absence of a female, so these are not used for fertilization. This behavior allows smaller males to potentially outcompete larger males by increasing the probability that their spermatophore will be one of fewer options available for females that eventually arrive (reviewed by Petranka 1998). If a female is present, smaller males can also sneak in and deposit their own spermatophores between those of the larger male and the female so that she is inseminated with the sneaker male's sperm (reviewed by Petranka 1998).

In Wisconsin, egg laying occurs in or close to April (Schmidt 1930, Vogt 1981). The number of eggs deposited annually per female varies from several dozen to over 300, which are laid in small batches each day over a period of several weeks (Bishop 1941b, Chadwick 1944, Mitchell and Gibbons 2010). Eggs are wrapped in leaves or detritus that is glued together by female cloacal secretions (Schmidt 1930, Bishop 1941b, Morin et al. 1983). The incubation period depends on water temperatures but generally lasts 20–40 days (Gage 1891, Schmidt 1930, Bishop 1941b, Vogt 1981). After hatching, larvae grow rapidly and metamorphose from late summer to autumn (reviewed by Hunsinger and Lannoo 2005). Size at metamorphosis appears to be density-dependent (reviewed by Petranka 1998).

Juveniles that leave aquatic environments, or efts, remain on land for two to three years (Vogt 1981) before they reach sexual maturity and return to breeding wetlands as aquatic adults (Bishop 1941b, Healy 1974b, Harris et al. 1988). The length of the eft stage is variable, because the onset of sexual maturity depends on size rather than age (Healy 1974b, Forester and Lykens 1991, Caetano and Leclair 1996). Females appear to mature later and at a larger body size than males (Caetano and Leclair 1996). Growth and developmental rates are influenced by resource availability and competition. Morin (1983a) reported that adult and larval growth rates were negatively correlated with larval density, suggesting that competition in larvae has fitness effects that carry over into subsequent life history stages (Earl and Whiteman 2015).

Adults are facultatively aquatic. They can transition to a terrestrial lifestyle if wetland conditions become unfavorable (e.g., water levels drop or temperature exceeds tolerances), if high conspecific densities prevail, or if they incur excessive parasite loads (Hurlburt 1969, Gill 1978b, Massey 1990, Grayson and Wilbur 2009). In some populations, adults leave wetlands to overwinter terrestrially and then return to the same wetlands the following spring to breed (Hurlburt 1969, Gill 1978b, 1979). Presumably, newts engage in this behavior to avoid winter mortality. Interannual survivorship is lower in females than males, meaning that females could experience fewer reproductive seasons in their lifetime (Gills 1978a, 1985, but see Caetano and Leclair 1996).

ACTIVITY

Seasonal and daily activity patterns vary by developmental stage. Aquatic adults are active year-round, although activity declines in winter (Morgan and Grierson 1932, Bishop 1941b, George et al. 1977, Vogt 1981, Pitkin and Tilley 1982). Juveniles, on the other hand, exhibit more obvious peaks in activity by season. For example, peak juvenile activity occurs in late summer and autumn as larvae metamorphose and disperse away from wetlands (Chadwich 1944, Hurlbert 1969, Healy 1975). Stein (1938) also observed thousands of efts dispersing in October. The efts reportedly moved en masse from a small stream, up a 4.6 m (15 ft) vertical rock waterfall, through a road drainage tunnel, and over a dam to a wetland. Within 30 minutes, Stein collected over a thousand individuals, describing the streambanks as "literally crawling" with efts. Regarding daily activity patterns, adults are circumdiel, while larvae are nocturnal (Morgan and Grierson 1932, Bishop 1941b, George et al. 1977, Vogt 1981, Pitkin and Tilley 1982, Mitchell and Gibbons 2010). Efts are also mostly nocturnal and restrict activity to when the ground is moist or when conditions are warm and humid. Furthermore, eft activity is greatest at air temperatures >12°C (54°F; Healy 1975).

Information on adult movement patterns is sparse, although some sources suggest that males have larger home ranges than females (reviewed by Hunsinger and Lannoo 2005). Most movement data come from efts (Healy 1974b, 1975, reviewed by Petranka 1998). Healy (1974b, 1975) measured an average home range size of 270 m^2 (2,906 ft^2) for efts. Eastern Newts exhibit strong philopatry. For example, Bellis (1968) reported that Eastern Newts will remain in the same wetlands and even in the same small area within wetlands most of the time (Bellis 1968). Gill (1979) also found that displaced individuals returned to original capture locations.

PREY AND PREDATORS

Eastern Newts are carnivorous and opportunistically feed on a variety of invertebrates and vertebrates. Adults locate prey through visual and chemical cues (Hamilton 1940b, Martin et al. 1974, Mathis and Vincent 2000, Mathis 2003). Larval newts likely use vision in conjunction with lateral line mechano- and electroreceptors to detect invertebrate prey (Bullock et al. 1983, Lannoo 1987, Northcutt et al. 1995), including zooplankton (Ostracoda, *Diaptomus* spp., Cyclops, and Cladocera), mollusks (Planorbidae and Sphaeriidae), insects (Odonata, Diptera, and beetle larvae), arachnids (Hydrachnids), plants (*Wolffia* spp.), and cyanobacteria (*Nostoc* spp.; Hamilton 1940b). Terrestrial adults and efts analyzed in New York consumed primarily arthropods from the following groups (in order of highest to lowest frequency): Psocoptera, Acari, Parasitidae, Sminthuridae, Isotomidae, Poduridae, Araneidae, Thysanoptera, and Cecidomyiidae (MacNamara 1977). Both life stages ate the same prey items; however, adults consumed larger individual prey items than efts (MacNamara 1977). Work in New York and Massachusetts found that aquatic adults consumed insects (Ephemeroptera, Odonata, Plecoptera, Lepidoptera, Diptera, Trichoptera, *Notonecta* spp., Corixidae, *Aedes* spp., and Thysanoptera), zooplankton (Cladocera and *Diaptomus* spp.), amphipods (*Hyalella azteca*), mollusks (snails and *Sphaerium* spp.), arachnids (spiders and Hydrachnidae), leeches, small minnows, and eggs of Spotted Salamanders (*Ambystoma maculatum*) and Wood Frogs (*Lithobates sylvaticus*; Hamilton 1932, Burton 1977). Morin (1983a) also reported heavy predation of Eastern Tiger Salamander eggs by Eastern Newts and speculated that newts were responsible for the disappearance of, on average, 2.1 Eastern Tiger Salamander embryos every day in artificial study ponds ($n = 10$ egg masses monitored). Newts cannibalize smaller conspecifics (Burton 1977, Mathis 2003).

Predators of Eastern Newts include large aquatic insects, fishes (e.g., Smallmouth Bass [*Micropterus dolomieu*] and Pumpkinseed Sunfish [*Lepomis gibbosus*]), other amphibians (e.g., Mudpuppies [*Necturus maculosus*] and American Bullfrogs [*Lithobates catesbeianus*]), reptiles (e.g., Eastern Hog-nosed Snakes [*Heterodon platirhinos*], Common Gartersnakes [*Thamnophis sirtalis*], Common Watersnakes [*Nerodia sipedon*], Painted Turtles [*Chrysemys picta*], and Snapping Turtles [*Chelydra serpentina*]), and mammals (e.g., Raccoons [*Procyon lotor*]; reviewed by Hunsinger and Lannoo 2005).

Eastern Newts employ a suite of antipredator mechanisms, including toxic secretions, warning coloration (efts), cryptic coloration (larvae and adults), and behavioral defenses. Many animals avoid eating Eastern Newts because they produce tetrodotoxin (Brodie 1968, Brandon et al. 1979, Brodie and Formanowicz 1981b). Tetrodotoxin is a sodium channel blocker that affects nervous system function (Wakely et al. 1966, Brodie et al. 1974) and is reportedly effective against predatory fishes and crayfish, but less so against reptiles (e.g., turtles) and some amphibians (e.g., American Bullfrogs; Marion and Hay 2011). Newt toxicity varies with life history stage, and the brightly colored efts contain the highest concentrations of toxins. Eastern Newts couple their chemical defenses and warning colors with a postural defense known as the unken reflex (Figure 160), where individuals bend their head and tail upward until they nearly touch so the body is in the shape of the letter U (Neill 1955, Brodie and Howard 1972, Brodie et al. 1974). Ducey and Dulkiewicz (1994) showed that

Figure 160. An adult Eastern Newt exhibiting the unken reflex (Greene County, Indiana; photo by M. Lannoo).

efts exhibited the reflex more often than terrestrial adults. Following injury, newts have the ability to regenerate toes, limbs, and tail (Vogt 1981).

Parasites of Eastern Newts include Helminthes and leeches (Gill 1978b, Coggins and Sajdak 1982, Muzzall 1991, Yoder and Coggins 2007). In Wisconsin and Michigan, adult Eastern Newts were infected by trematodes (*Clinostomum* spp., *Halipegus* spp., *Megalodiscus temperatus*, and *Telorchis corti*), cestodes (*Bothriocephalus rarus*), and nematodes (*Cosmocercoides* spp., *Falcaustra catesbeianae*, and *Spiroxys* spp.; Coggins and Sajdak 1982, Muzzall 1991, and Yoder and Coggins 2007). In Virginia, leeches are a major source of mortality (Gill 1978b). However, Pough (1974) reported that four leech species avoided Eastern Newts. Mesomycetozoans are fungus-like parasitic protists (Herman 1984, Densmore and Green 2007, Raffel et al. 2008). *Ichthyophonus* spp., which commonly affect marine fishes, also infect Eastern Newts (Herman 1984, Densmore and Green 2007). In Pennsylvania and Vermont, *Ichthyophonus* spp. are common in Eastern Newt populations and likely transmitted by an amphibian leech (*Placobdella picta*; Raffel et al. 2006, Sherman 2009). Raffel et al. (2008) described a new mesomycetozoan, *Amphibiocystidium viridescens*, that infects Eastern Newt populations in Pennsylvania, West Virginia, and Massachusetts. This was the first record of an *Amphibiocystidium* spp. infecting a North American amphibian, but *Amphibiocystidium ranae* infects amphibians in Europe and has been linked to recent amphibian declines in Italy (Raffel et al. 2008).

CONSERVATION

Status: Globally, the Eastern Newt is considered a Least Concern species by the IUCN. It is considered a common species in Wisconsin by the WDNR. Eastern Newts are widely distributed throughout the US, and their historic range is likely similar to their current range, although some local populations have been lost (reviewed by Hunsinger and Lannoo 2005).

Populations: Information on Wisconsin populations is sparse, although reports from other states indicate that variation in population characteristics occurs by age class and geographic location. For example, Burton (1977) estimated that 1,950–2,600 adults occurred in a <1 ha (2.5 ac) area of a lake in New Hampshire. Gill (1978a) estimated breeding population sizes of 6 to more than 2,600 adults per wetland in Virginia ponds that were 10–14 ha (25–35 ac) in size. Healy (1975) estimated that the soil surface density of efts at a site in Massachusetts was $0.03/m^2$ ($0.32/ft^2$). Furthermore, mean terrestrial winter densities near breeding wetlands in Massachusetts ranged from 0.0049 to $0.0055/m^2$ (0.00045 to $0.00051/ft^2$; Regosin et al. 2005). Annual adult survival is also highly variable. While Morin (1983a) reported that aquatic adult survival declined as newt densities increased, Gill (1978a) found that primarily female survival was influenced. While the sex ratio of larvae and efts is ca. 1M:1F, adult sex ratios are male-biased (Gill 1978b).

Threats: General threats to amphibians, including Eastern Newts, are covered in the Introduction to Conservation and Management of Wisconsin's Amphibians and Reptiles chapter. Eastern Newts are affected by habitat disturbances and are particularly susceptible to the loss of forests that are adjacent to wetlands (Guerry and Hunter 2002). For example, Gibbs (1998b) found that Eastern Newts did not occur in areas with <50% forest cover. Porej et al. (2004) suggested that both loss of wooded wetlands and loss of wooded corridors connecting suitable wetlands for migrating efts restricted site occupancy and disrupted the metapopulation dynamics of Eastern Newts. In Pennsylvania, Eastern Newts were found less often in urban areas than in rural and suburban wetlands, likely because of habitat destruction (e.g., deforestation, wetland drainage, and roads) in urban areas (Gibbs 1998a, Rubbo and Kiesecker 2005). Eastern Newts are also negatively impacted by the presence of predatory fishes, which are more often introduced to wetlands in disturbed landscapes (Hecnar and M'Closkey 1997b). Furthermore, contaminants associated with disturbed landscapes (e.g., pesticides and herbicides applied to lawns, gardens, and agricultural fields) impact Eastern Newts. For example, Arias et al. (1989) found that the phenoxyacid herbicide MCPA reduced their ability to regenerate forelimbs.

Ranaviruses and the chytrid fungus *Batrachochytrium dendrobatidis* (Bd) infect Eastern Newts, although the population-level impacts of such infections (if any) remain largely unknown. Eastern Newts infected with a ranavirus have been documented in Pennsylvania and the Tallulah River valley (Glenney et al. 2010, Rothermel et al. 2013). Yet laboratory investigations have reported low mortality rates for infected larvae (Hoverman et al. 2011). Similarly, Bd has been detected in Eastern Newts from Alabama, Connecticut, Georgia, Missouri, New York, North Carolina, Pennsylvania, Virginia, and Canada (Ouellet et al. 2005, Rothermel et al. 2008, Bakkegard and Pessier 2010, Raffel et al. 2010b, Richards-Hrdlicka et al. 2013, Strauss and Smith 2013, Lenker et al. 2014). The fungus was also recently detected on Eastern Newts from Wisconsin, although chytridiomycosis was not observed (Standish et al. 2018).

Management: There is little information on effective management strategies specifically focused on Eastern Newts in Wisconsin. Eastern Newts almost certainly benefit from the protection and restoration of breeding wetlands and nonbreeding habitat (see Introduction to Conservation and Management of Wisconsin's Amphibians and Reptiles chapter). Wetlands should be forested and contain emergent or submersed vegetation but lack fish. Suitable forest canopy cover and other terrestrial habitat requirements of amphibians are often overlooked but are critical habitat components for Eastern Newts and many other amphibians (Semlitsch and Bodie 2003). Because dispersal of the eft phase is important in maintaining viable populations across landscapes, maintenance of wooded corridors that connect wooded wetlands is probably critical (Porej et al. 2004). Habitat management plans that consider Eastern Newts should also promote retention of ground cover for terrestrial life stages, including leaves, woody debris, rocks, grass, and moss (Roe and Grayson 2008). Eastern Newts colonize wetlands constructed in forested habitats (Skelly et al. 1999), but the possible population-level benefits of artificial wetlands have not been investigated for this species.

Squamata
Lizards and Snakes

The order Squamata currently contains over 10,000 species distributed across well over 50 families globally (http://www.reptile-database.org). Present evidence supports that Squamata could have originated as far back as 257 mya, close to the mass extinction event that occurred near the end of the Permian and the beginning of the Triassic period (Hedges and Vidal 2009, Simões et al. 2018). This order is divided into several clades (i.e., groups of related species; see Systematics and Classification chapter), which among resident Wisconsin species include the lizards and snakes.

LIZARDS

Figure 161. Six-lined Racerunner (*Aspidoscelis sexlineata*) (illustration by E. Wild).

Lizards are squamate reptiles that, for the most part, have obvious limbs or at least the skeletal elements to support limbs (i.e., girdles) when limbs are absent.* Evidence suggests that lizards probably originated ca. 257 mya near the beginning of the Triassic period (Simões et al. 2018). Current estimates place the number of living lizard species at over 6,500 globally (http://www.reptile-database.org). Despite this high global diversity, the number of lizard species found in Wisconsin is far lower than any other group of amphibian or reptile: only four species, representing three families (Teiidae, Scincidae, and Anguidae),

*Limb reduction and loss are common themes in squamate evolution, having occurred independently many times (Pianka and Vitt 2003, Sites et al. 2011). As a result, there are lizard species that do not have legs (one legless lizard species occurs in Wisconsin, the Slender Glass Lizard [*Ophisaurus attenuatus*]). Although the superficial similarity between legless lizards and snakes can cause confusion, legless lizards have numerous morphological features, including skeletal elements, that (aside from the lack of legs) make them more similar to lizards than snakes, including external ear openings and moveable eyelids.

and spread across three genera (*Aspidoscelis*, one species, *Plestiodon*, two species, and *Ophisaurus*, one species, respectively).

SNAKES

Snakes are elongated, limbless, and carnivorous species of squamate reptiles. This group likely originated during the mid-Cretaceous period, ca. 108.5 mya, and there are currently over 3,500 recognized species globally (Hsiang et al. 2015). Snakes are the most species-rich group of amphibian or reptile in Wisconsin, with 21 species representing two families and 15 genera. Of the two families found in the state, Colubridae contains 19 of the snake species that occur here, while the family Viperidae contains only 2 species. Due to the large number of species representing the family Colubridae in this region of the US, local field guides typically group snakes first by family and then by subfamily, as this better reflects taxonomic relationships than does simply ordering them alphabetically. We have also chosen to follow this organizational pattern.

Family Colubridae

Subfamily Colubrinae: There are six species across five genera representing this subfamily in Wisconsin: *Coluber* (one species), *Opheodrys* (one species), *Lampropeltis* (one species), *Pantherophis* (two species), and *Pituophis* (one species). Snakes within this subfamily lay eggs (oviparous). Furthermore, most members of this subfamily found in Wisconsin are large constricting snakes, while some species, such as the North American Racer (*Coluber constrictor*) and the Smooth Greensnake (*Opheodrys vernalis*), do not actually constrict their prey.

Subfamily Dipsadinae: There are more snake species in this group worldwide than any other (over 750; Figueroa et al. 2016), but in Wisconsin it contains the fewest: three genera (*Carphophis*, *Diadophis*, and *Heterodon*) represented by one species each. Species in this subfamily lay eggs but can also have teeth that are not uniform in size and shape. The classic example of this is the Eastern Hog-nosed Snake (*Heterodon platirhinos*), which has elongated rear fangs.

Subfamily Natricinae: In Wisconsin, there are more species in this subfamily than in the others. These snakes do not lay eggs but instead give birth to live young (viviparous). The ten

Figure 162. Smooth Greensnake (*Opheodrys vernalis*) (illustration by E. Wild)

Figure 163. Eastern Hog-nosed Snake (*Heterodon platirhinos*) (illustration by E. Wild).

Figure 164. Butler's Gartersnake (*Thamnophis butleri*) (illustration by E. Wild).

Figure 165. Massasauga (*Sistrurus catenatus*) (illustration by E. Wild).

species found in Wisconsin are spread across five genera: *Nerodia* (one species), *Regina* (one species), *Storeria* (two species), *Tropidoclonion* (one species), and *Thamnophis* (five species).

Family Viperidae

The species in this group are venomous. In Wisconsin, there are two genera (*Sistrurus* and *Crotalus*) represented by one species each. Both species have sound-producing structures at the tip of their tails (i.e., 'rattles'), elongated fangs in the front of the mouth on the upper jaw that can inject venom, and give birth to live young.

Slender Glass Lizard

Ophisaurus attenuatus (Cope 1880)

Jeffrey M. Lorch

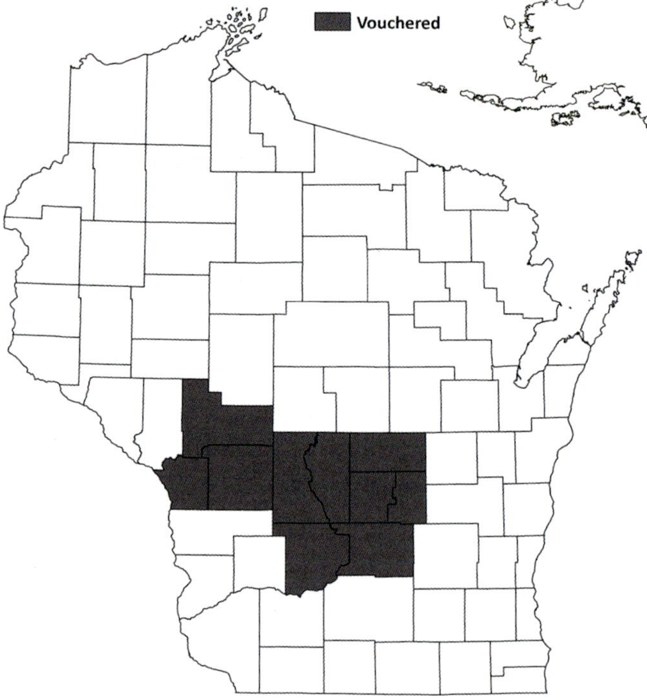

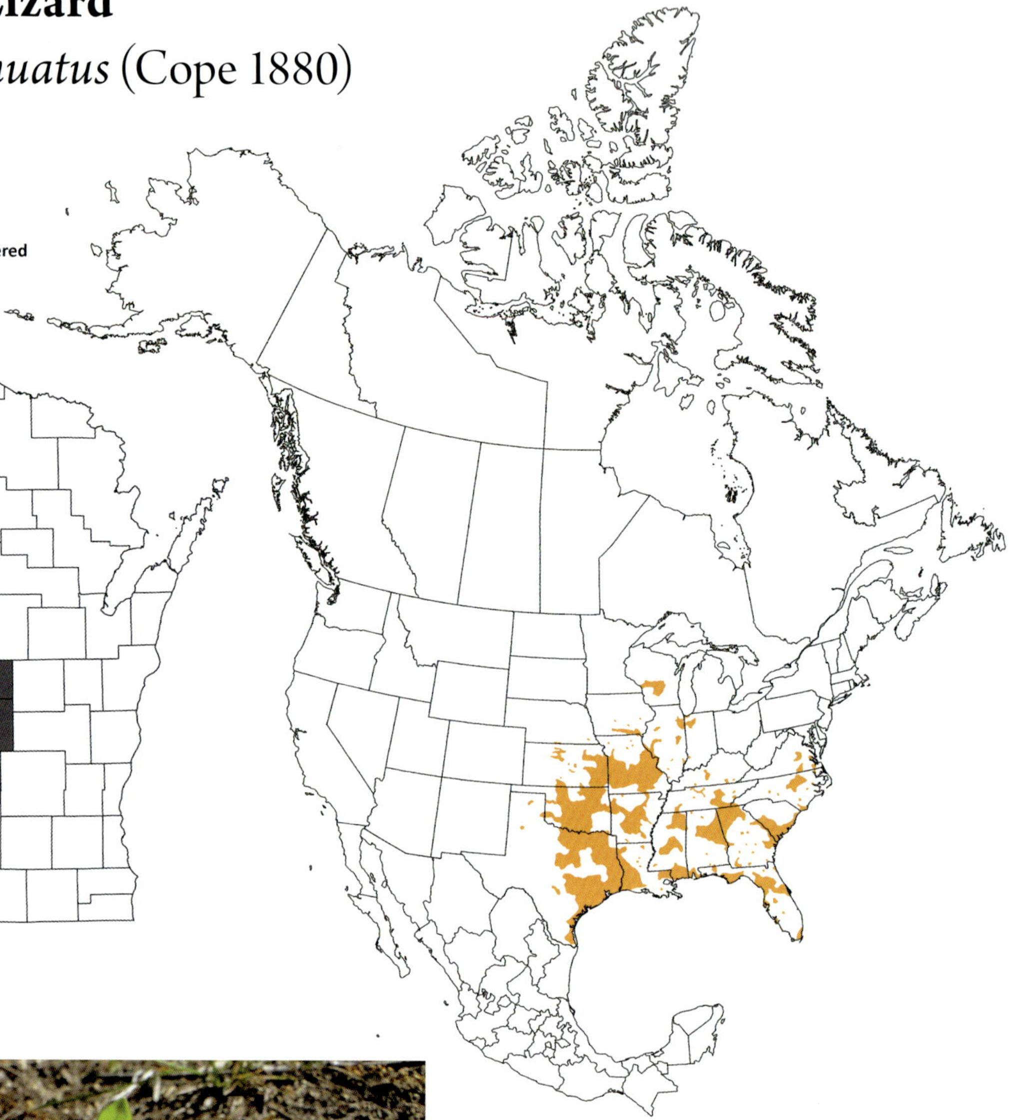

Slender Glass Lizard (Columbia County; photo by C. Raimond).

AT A GLANCE

Slender Glass Lizards are large limbless lizards that can be distinguished from snakes by the presence of external ear openings, movable eyelids, and a lateral fold on each side of the body. The tail makes up nearly two-thirds the length of the animal and is important in locomotion. Slender Glass Lizards live in savannas, in oak and pine barrens, and along the edges of woodlands in the Central Sands region of Wisconsin; here, they are disjunct from their main North American range south of Wisconsin. Slender Glass Lizards spend the majority of their time underground and even during the active season (April through late September) are at the surface for only a few hours a day while they hunt for insects. Although data from Wisconsin are limited, this species likely breeds in April or May after emergence from hibernation. Females presumably lay eggs in June, which hatch from late August to early September. Like the other species of lizards in the state, Slender Glass Lizards can dislocate their tail when threatened. The lizard's common name is derived from the perceived ability of the tail to 'shatter' like glass due to the presence of numerous fracture planes within the tail vertebrae. The secretive nature of Slender Glass Lizards, combined with their rarity and tendency for populations to be localized within suitable habitat, makes them one of the least often encountered of all Wisconsin's amphibians and reptiles. Remaining Slender Glass Lizard populations in the state are threatened by habitat loss and degradation, so it is listed as Endangered and considered a Species of Greatest Conservation Need by the WDNR.

SYSTEMATICS

Glass lizards belong to the family Anguidae, which also includes the galliwasps (genera *Celestus* and *Diploglossus*) and alligator lizards (genera *Elgaria* and *Gerrhonotus*). Within Anguidae, glass lizards belong to the clade Anguinae. The genus *Ophisaurus*, as currently recognized, has representatives in North America, Eurasia, and northern Africa. Some evidence suggests that the genus is polyphyletic and in need of reclassification; however, phylogenies determined on the basis of morphologic and molecular data are somewhat discordant, and reassignment of European and Southeast Asian species to other genera requires further validation (Macey et al. 1999, Conrad et al. 2011).

The species that occurs in Wisconsin (*Ophisaurus attenuatus* [Slender Glass Lizard]) is aptly named, as *Ophisaurus* means 'snake-lizard,' and *attenuatus* refers to its thin or slender body. It was first described by Cope (1880) under the name *Opheosaurus ventralis attenuatus*. However, *O. attenuatus* was not widely accepted as a distinct species until McConkey (1954) reviewed the genus and proposed that *O. ventralis* (Eastern Glass Lizard) was a species complex comprised of several morphologically distinct taxa. Macey et al. (1999) demonstrated that DNA sequence divergence of multiple mitochondrial genes between *O. ventralis* and *O. attenuatus* was nearly as high as the divergence observed between *O. ventralis* and the Asian species *O. harti* (now *Dopasia harti* [Hart's Glass Lizard]), suggesting that *O. attenuatus* is likely distinct from *O. ventralis*. However, more thorough studies that examine genetic relationships across the ranges of North American species of *Ophisaurus* are necessary to better determine species boundaries.

Two subspecies of *O. attenuatus* are currently recognized (*O. a. longicaudus* [Eastern Slender Glass Lizard] and *O. a. attenuatus* [Western Slender Glass Lizard]; Crother 2017), which differ on the basis of tail length, presence of palatine teeth, and geographic distribution (Holman 1971). The subspecies are separated by physical barriers (the Mississippi and Ohio Rivers; Holman 1971), which may have resulted in genetic differentiation between the two taxa, but there have been no genetic studies to verify this hypothesis. The Western Slender Glass Lizard (*O. a. attenuatus*) is the subspecies that occurs in Wisconsin.

DESCRIPTION

Adults: Slender Glass Lizards are a limbless lizard with an elongated body and tail. Like most lizards, Slender Glass Lizards have ear openings and movable eyelids (Figure 166). There is a fold in the skin running along each side of the body (see Figure 166). When held, glass lizards feel stiff due to the presence of boney plates (osteoderms) under the skin.

Figure 166. The Slender Glass Lizard can be differentiated from snakes by the presence of movable eyelids (A), an external ear opening (B), and a lateral fold running the length of the body (C) (Marquette County; photo by J. Lorch).

Figure 167. An intact tail of a Slender Glass Lizard comprises ca. two-thirds of the animal's total length (Columbia County; photo by A. Badje).

The dorsal scales are keeled, and the ventral scales are hexagonal and only somewhat greater in width than in length (unlike the widened ventral scales of snakes). The dorsal background coloration is tan or light brown. There is a dusky to dark brown dorsal stripe and three to seven (usually five) dark brown to black lateral stripes that vary in width and intensity. Most often there are three lateral stripes above the lateral fold (one of which is often narrow and somewhat obscure) and two lateral stripes below the lateral fold (one or both of which are narrow and faint; Figure 167). The ventral surface is white or cream and is usually without markings. The head is tan or brown with varying degrees of black-and-white speckling, especially on the sides of the face. The eye is relatively small, and the iris is gold in color. The tail is exceptionally long (2–2.4 times the length of the body when fully intact; see Figure 167). The tail is patterned similarly to the body, although regenerated tails are usually more or less uniformly brown. McConkey (1954) provided a comprehensive description of this species.

As adults, Slender Glass Lizards are the largest of Wisconsin's four lizard species. The mean SVL of Wisconsin specimens in MPM and UWZM collections was 19.4 cm (7.6 in, $n = 19$). Only nine specimens measured had intact tails, with a mean tail length of 35.6 cm (14 in). Measurements of total length can be difficult to interpret because most adult Slender Glass Lizards lose a portion of their tail at some point in their lives, and the regenerated tail is rarely, if ever, as long as the original (see Prey and Predators section). Knudsen (1956a) reported that the mean TL of Wisconsin individuals was 45.7 cm (18 in), with some exceeding 61 cm (24 in). Holman (1971) stated that maximum SVL was 27.9 cm (11 in) and maximum TL was 106.7 cm (42 in), but no individuals this large have been documented in Wisconsin. The largest SVL of a museum specimen measured, an individual from Juneau County collected in 1986, was 22.9 cm (9 in, UWZM 24122). Pleyte (1975) reported one individual from Waushara County that was 23.5 cm SVL (9.2 in), and a large adult captured in Columbia County was estimated to be ca. 23 cm SVL (9.1 in; Jeffrey Lorch, US Geological Survey, personal observation). Assuming that tail length is approximately two-thirds of TL, these large Wisconsin individuals would have a roughly 70 cm TL (27.6 in).

Preadult stages: When the white elliptical eggs are laid, their mean dimensions are 1.8 × 1.0 cm (0.72 × 0.40 in; Noble and Mason 1933). The eggs

Figure 168. Clutch of seven eggs laid in captivity by a Slender Glass Lizard captured in Monroe County (photo by J. Kapfer).

Figure 169. Pattern variation in the Slender Glass Lizard. *Top*: typical patterning (Columbia County; photo by J. Lorch); *middle*: adult males may have a greater degree of speckling on the head and dorsolaterally (Columbia County; photo by J. Kapfer); *bottom*: juveniles often have fainter striping (Columbia County; photo by N. Schiltz).

absorb moisture during incubation and increase in size to ca. 2.5 × 1.9 cm (0.98 × 0.77 in) just prior to hatching (Noble and Mason 1933; Figure 168). The young are similar in appearance to the adults except that the lateral stripes and the speckling on the head may be faint or lacking altogether (see Figure 169). Fitch (1989) noted that immature Slender Glass Lizards also have relatively wider heads and shorter tails than adults. At hatching, the mean SVL is 5.7 cm (2.2 in, SD = 0.3 cm [0.01 in], $n = 64$), and the mean mass is 1.23 g (0.04 oz, SD = 0.05 g [0.002 oz], $n = 26$; Fitch 1989). Pleyte (1975) reported that the SVL of recently hatched young ranged from 5.5 to 6.5 cm (2.17 to 2.56 in), with a mass of ca. 1 g (0.04 oz) at a study site in Waushara County.

Variation: There is relatively little variation in color or pattern among individuals. While the normal ground color of the dorsum is tan or light brown, a few individuals that were dark reddish brown in color have also been observed in Wisconsin (J. Lorch, personal observation). Vogt (1981) similarly described that color ranged from tan to bronze; however, most specimens are lighter in color. Adult males reportedly have a greater degree of black-and-white mottling on the head and dorsolateral area (Figure 169), a slightly longer tail, and a broader head than females; however, there is some degree of overlap in characteristics between the sexes (Fitch 1989).

Diagnosis: Slender Glass Lizards are characterized by the absence of legs. They also possess a lateral groove along each side of the body, as well as ear openings and movable eyelids. It is unlikely this species can be confused with any other lizard in Wisconsin, due to its lack of legs. Although it superficially resembles a snake, the presence of ear openings, eyelids, and lateral grooves clearly distinguishes it (see Figure 166). Furthermore, ventral scales of Slender Glass Lizards are unlike those of snakes, and no other Wisconsin reptile has a similar stripe pattern.

Slender Glass Lizard eggs are larger than those of other lizards in Wisconsin. Despite this, they could still be confused with the eggs of several snake species that overlap in size. In these instances, egg dimensions, clutch size, and geographic location and habitat where eggs are encountered can provide some guidance (see Preadult stages and Habitat subsections and Reproduction and Development section). Yet it might not be possible to definitively identify Slender Glass Lizard eggs through visual inspection alone.

DISTRIBUTION AND HABITAT

Global distribution: Slender Glass Lizards range from extreme southern Nebraska, central Wisconsin, northern Indiana, and Virginia south to peninsular Florida and the southern tip of Texas. The eastern subspecies (*Ophisaurus attenuatus longicaudus*) occurs east of the Mississippi River (but not in Wisconsin) and south of the Ohio River. The distribution of Western Slender Glass Lizards (*O. a. attenuatus*) roughly traces that of the tallgrass prairie, extending east of the Mississippi River in the former tallgrass prairie peninsula region of Illinois and northwestern Indiana. Throughout most of its range, the Slender Glass Lizard has a spotty distribution (Powell et al. 2016). In Wisconsin, Slender Glass Lizards occur in the central part of the state and are disjunct from the core of their North American range; there are no records from southern Wisconsin, extreme northern Illinois, northern Iowa, or Minnesota. It is unclear how Slender Glass Lizards dispersed into Wisconsin after the last glaciation, but they may have followed prairie corridors on the sand terraces along the Mississippi and Wisconsin Rivers.

Wisconsin distribution: In Wisconsin, Slender Glass Lizards occur primarily in the Central Sands region. Although Breitenbach (2011) suggested that the species may currently be extirpated from many of the counties with published records, recent reliable observations have come from all of these but Green Lake County (J. Lorch, unpublished data; WDNR 2016b). This suggests that while some populations may have declined, the Slender Glass Lizard's overall known range in Wisconsin has remained relatively intact. Despite this, populations appear to be highly localized. Where Slender Glass Lizards occur, they can often be reliably found in a small area but appear to be absent from nearby areas that offer seemingly identical habitat (Knudsen 1956a, J. Lorch, personal observation).

Localities of possible occurrence: Wisconsin represents the northern limit of this species' North American distribution. As such, it might not be able to persist in the climate conditions at latitudes higher than where it is already documented in the state. Casper (1996) believed that its geographic range in Wisconsin was already well-documented, which suggests that the potential for finding new counties of occurrence is low. Yet seemingly suitable habitat for the species occurs in portions of Portage, Waupaca, and Wood Counties, so it could eventually be found in the extreme southern portions of these. For example, in 2013 Sara Viernum (Stantec Consulting) photographed a Slender Glass Lizard in Adams County at a location only ca. 305 m (1,000 ft) from the Wood County line, with continuous habitat extending into Wood County, highlighting the potential for this species to also occur there.

Sandy prairie and oak barrens habitats in the northeastern and southeastern parts of Wisconsin appear superficially suitable for Slender Glass Lizards, but the species' historical range may have never extended into these areas. Hoy (1883) stated that Slender Glass Lizards were once common near Kenosha (Kenosha County) in southeastern Wisconsin, although there are no specimens to support this claim. The species does occur in sandy dune habitats along the southern edge of Lake Michigan in nearby northwestern Indiana and northeastern Illinois (Schmidt and Necker 1935). Thus, it is possible that the species once also extended into the sandy prairies and dunes south of Kenosha. Today, little suitable habitat remains in southeastern Wisconsin, and it is unlikely that past populations in that region have persisted to the present.

The lack of records from the sand terraces along the Lower Wisconsin River, with the exception of two unverified reports from northwestern Dane County (WDNR 2016b), is particularly mystifying. Much habitat in this river valley (and adjacent sand prairies that exist away from the river along eroded sandstone bluffs) appears identical to that used by the species farther north. The paucity of records in the river valley is further perplexing when one considers that Slender Glass Lizards presumably colonized the Central Sands region of Wisconsin by migrating north along the Wisconsin River corridor (Wagner 1922). Hoy (1883) stated that "they occur in the western part of the state as far north as La Crosse," so perhaps Slender Glass Lizards once had a wider distribution in southwestern Wisconsin. Given their cryptic nature, localized populations could still be discovered in the lower Wisconsin River region.

Habitat: The habitat preferences of this species have received little attention in the Midwest and are poorly understood (e.g., Pleyte 1975, Fitch 1989, Breitenbach 2011). Fitch (1989) associated Western Slender Glass Lizards with the former tallgrass prairie ecosystem in North America. In Wisconsin, it has only been found in dry sandy habitats such as sand prairies, oak savannas, and oak and jack pine barrens (see Amphibians and Reptiles in Wisconsin Plant Communities chapter). Glass lizards also utilize anthropogenic habitats such as old fields, grassy roadsides, and cutover areas associated with transmission line rights-of-way (J. Lorch, personal observation). The species appears to avoid wetlands and closed-canopy woodlands (Pleyte 1975, Breitenbach 2011).

Although they do not use woodland interiors, Slender Glass Lizards are strongly associated with edge habitats. At a study site in Monroe County, the mean capture distance from a woodland edge was only 27.1 m (88.9 ft; Breitenbach 2011). Similarly, many glass lizards are observed in open grassy areas that receive shade during at least part of the day from nearby oak woods or pine plantations (J. Lorch, personal observation). Occasionally, individuals are found in open areas farther from habitat edges, but then usually in the vicinity of a tree or other structure that can block the sun and allow for better thermoregulation during the hottest parts of the day (J. Lorch, personal observation). Slender Glass Lizards prefer patches of dense vegetation, such as Pennsylvania Sedge (*Carex pennsylvanica*) or Smooth Brome (*Bromus inermis*) with a well-developed thatch layer (Fitch 1989, Breitenbach 2011). They appear to seek out such plots within larger habitat mosaics that also contain sparsely vegetated areas. Apparently, these microhabitat variations facilitate thermoregulation (glass lizards can often be seen basking on exposed sandy soil on cool sunny days [Pleyte 1975, J. Lorch, personal observation]), hunting, and avoiding predators. Pleyte (1975) stated that Wisconsin individuals spent "a great deal of time hidden under cover, such as logs, brush, or even burrowed into the loose sand itself." Additional cover in the form of brush piles and rotting logs is also favored (Breitenbach 2011, J. Lorch, personal observation). They are documented to use the abandoned burrows of small mammals when inactive (Fitch 1989). The network of tunnels produced by Eastern Moles (*Scalopus aquaticus*) and Prairie Voles (*Microtus ochrogaster*), for example, seem to be particularly well-utilized. Fitch (1989) observed that Slender Glass Lizards shifted their use of hiding places and did not favor a particular shelter over time. Fitch (1989) also pointed out the somewhat paradoxical habitat preferences of this species (see Natural History Box: A Prairie Paradox).

Information on nesting and overwintering habitats in Wisconsin is sparse. Fitch (1989) reported that in Kansas, this lizard laid eggs in abandoned mammal burrows with loose soil (see Reproduction and Development section). Slender Glass Lizards also spend the winter months below the frost line in the abandoned burrows of small mammals. Fitch (1989) believed that the lizards could differentiate between active and abandoned burrows based on the accumulation of soil within the tunnels and that their selection of abandoned burrows was important to avoid predation by moles and shrews during hibernation.

REPRODUCTION AND DEVELOPMENT

Most of what is known about the reproductive habits of Slender Glass Lizards comes from opportunistic encounters with wild individuals or observations of animals held in captivity. Available evidence suggests that breeding occurs in the spring. Examination of gonads from Arkansas specimens revealed physiological changes in testes and ovaries indicative of breeding in animals captured during April and May (Trauth 1984). Fitch (1989) observed an instance of presumptive courtship in Kansas on 6 May. During the event, the male nipped at the female and eventually secured her head in his jaws. The male's vent was brought into close proximity with the female's vent as he rhythmically stroked his body against hers. The female occasionally attempted to escape and eventually did so; actual copulation was never observed.

In Kansas, fertilized eggs are laid between 24 June and 11 July (Fitch 1989). Farther south, egg laying may occur slightly earlier (Noble and Mason 1933). Clutch size is positively correlated with female body size and ranges from 5 to 16 eggs ($\bar{x}$ clutch size = 10.2, SD = 0.47, n = 40; Fitch 1989). In Wisconsin, an individual collected by Vogt (1981) laid a clutch of

NATURAL HISTORY BOX: A Prairie Paradox

Prairies and barrens are ecosystems shaped by fire. The burning of these open landscapes prevents encroachment by woody vegetation (i.e., natural succession) and promotes the growth of grasses and sedges. As a species dependent on prairie and barrens habitats for survival, the Slender Glass Lizard is inevitably dependent on fire as well. Therefore, it would seem logical that the Slender Glass Lizard would be adapted to coping with fire. However, this is apparently not the case. Fitch (1989) summarized the somewhat paradoxical relationship of this reptile with its required habitat at one of his former study sites in Kansas: "The tallgrass prairie vegetation on which the Slender Glass Lizard depended for its natural habitat was a fire-type subclimax, ephemeral without periodic burning. Yet the lizard had no obvious adaptations to resist or escape burning. Even if it survived a grass fire by being underground it was rendered highly vulnerable to predation by the complete lack of surface cover after burning and lacked the capacity to disperse to more favorable areas. It responded slowly to favorable conditions by increasing its rate of reproduction. Indeed, it seemed that the species existed somewhat precariously, i.e., in marginal ecotonal stands of tallgrass prairie, an association dependent on recurrent fire, in which streams, gullies, brush patches, woodland edge, or rock outcrops offered some degree of protection from the burning off of all surface cover."

That the Slender Glass Lizard requires microhabitats that are firebreaks explains its propensity for edge habitats in Wisconsin. It could also explain the lizard's predilection for patches of Pennsylvania Sedge (*Carex pensylvanica*) and Smooth Brome (*Bromus inermis*) grass. Both of these plants green up and provide cover quickly after fire. In contrast, most other prairie grasses in Slender Glass Lizard habitat do not produce adequate cover until late spring or early summer following a burn event. Historically, when barrens covered a large portion of the landscape in central Wisconsin, the Slender Glass Lizard would have had little difficulty finding refugia from burned areas. In today's fragmented landscape, managing the few parcels of critically endangered oak barrens communities that remain in Wisconsin by means of frequent prescribed burning while also attempting to sustain viable populations of the Slender Glass Lizard at those sites poses a significant conservation challenge.

Jeffrey M. Lorch

13 eggs in June. Eggs are usually laid in an abandoned small mammal burrow; the female presses her body against the sides of the burrow to compact the soil and create a nesting cavity (Fitch 1989). After laying, the female coils around the eggs and remains with them until they hatch (Noble and Mason 1933). The reason for this brooding behavior is unclear, as the female neither increases the incubation temperature of the nesting cavity nor defends the eggs against potential predators (Noble and Mason 1933). Fitch (1989) estimated that females reproduced biennially. He observed that reproductive females were often emaciated by the end of the brooding period, indicating that they might not be able to attain sufficient resources to produce clutches on an annual basis. However, he also noted that larger reproductive-aged females were more likely to be gravid, suggesting that at least some older females may have an annual reproductive cycle. The eggs hatch after an incubation period of ca. 50–60 days (Noble and Mason 1933, Fitch 1989). Most clutches hatch during mid- to late August in Kansas, although one clutch did not hatch until 10 October (Fitch 1989). In Wisconsin, hatchlings have been encountered on 18 August, 2 Septem-

ber, and 9 September in Columbia County and on 13 September in northern Adams County (J. Lorch, unpublished data).

Because the young hatch near the end of the active season, little growth occurs during the first year (Pleyte 1975). However, by the autumn of their second active season, juveniles are 9–10 cm SVL (3.54–3.94 in; Pleyte 1975). Growth in Wisconsin is apparently slower than at lower latitudes. For example, Fitch (1989) reported that Slender Glass Lizards in Kansas had a mean SVL of 12.5 cm (4.93 in, SD = 0.15 cm [0.06 in]) by one year of age. Fitch (1989) also noted that beyond two years of age, growth substantially slowed, and cohorts became difficult to differentiate on the basis of size. Adult Slender Glass Lizards do not reach sexual maturity until they are three to four years old (Fitch 1989). The long time to sexual maturity and low reproductive rate of Slender Glass Lizards are potentially offset by a relatively long life span. Fitch (1989) recorded a maximum life span of at least ten years, although he believed that most individuals at his study site probably did not survive beyond five years of age. Their long generation time, combined with low reproductive and high mortality rates, apparently contribute to the relatively slow population growth observed by Fitch (1989).

Figure 170. On cold days, Slender Glass Lizards may achieve a high body temperature by basking on sand warmed by the sun (Columbia County; photo by R. Paloski).

ACTIVITY

Slender Glass Lizards are active from late April to late September (Pleyte 1975). In the spring, they gradually move closer to the soil surface, but surface activity is dependent on temperature. An individual in Kansas tracked by radiotelemetry moved to the surface and emerged on 2 April when the air temperature was 22°C (71.6°F), but after two days, a cold snap forced the animal back underground, and it did not reemerge until 22 April (Fitch 1989). Temperature continues to heavily affect activity patterns after emergence from hibernation. Pleyte (1975) found that the optimal range of air temperatures for Slender Glass Lizard activity was 20–26°C (68.0–78.8°F), with peak activity occurring at 23.5°C–25°C (74.3°F–77°F). He further found that the animals were inactive when air temperatures exceeded 30°C (86°F) or dropped below 16°C (60.8°F). In the barrens habitats where Slender Glass Lizards occur in Wisconsin, air temperatures can be a poor representation of the temperatures actually experienced by the lizards at ground level because exposed sand absorbs radiant heat from the sun (Figure 170). Pleyte (1975) captured a Slender Glass Lizard when the air temperature was only 16°C (60.8°F) but estimated that sand temperatures at the basking location reached 49°C (120.2°F). Similarly, Fitch (1989) reported that mean body temperature of glass lizards at the soil surface was 3.9°C (7°F) higher than that of the surrounding air. Pleyte (1975) observed two distinct peaks in soil surface activity at a site in Waushara County, which included the first week of June and then from late July to the end of August. Similarly, Trauth (1984) noted that Slender Glass Lizards were more likely encountered at the soil surface from April to early June during the breeding season and again just prior to the winter dormancy period.

Daily activity patterns vary by season and conform to environmental temperatures preferred by glass lizards. In spring and autumn (or on overcast summer days) when temperatures are relatively cool, Slender Glass Lizards may be observed above the ground surface throughout the day; however, activity during the warmest part of the year is primarily restricted to the morning and late afternoon (Pleyte

NATURAL HISTORY BOX: It's All about Timing

The Slender Glass Lizard can be one of the most challenging reptiles to find. Even in the early 1900s, the species was considered to be rare in Wisconsin (Wagner 1922). This apparent rarity is, at least in part, due to the secretive nature of the species. My first encounter with a Slender Glass Lizard was in 2005, when I saw a juvenile at a site in Columbia County. The animal escaped into a thick patch of Smooth Brome grass before I could get a good look at it. In 2006 I made numerous trips to the exact same location and finally succeeded in finding another glass lizard in July. Over the last ten years, I have seen several dozen Slender Glass Lizards at that particular site in Columbia County. However, the tally does not reflect the success rate, which was poor. Specifically, I encountered Slender Glass Lizards during only one out of every four visits to the site, and I observed only one animal per trip. I always saw them in the same 0.1 ha (0.25 ac) opening, despite the presence of seemingly identical habitat throughout most of the site. Then, on a cool (18.3°C [65°F]), sunny day in early September 2007, I encountered five Slender Glass Lizards in less than an hour. I had a similar experience under nearly identical temperature conditions on 18 August 2012, when I saw seven glass lizards over the course of 30 minutes. Former WDNR wildlife biologist George J. Knudsen (1956a) summarized activity patterns of Slender Glass Lizards well when he wrote, "On certain days when conditions are right the species can be readily collected in certain spots but on the very next field trip to the same spot one can look for hours and not locate a single specimen." Thus, Slender Glass Lizards can be difficult to find, even at sites where they reach higher population densities, and the species may go undetected in an area unless surveys are conducted in the right place at the right time.

Jeffrey M. Lorch

1975, Fitch 1989). In Wisconsin, daily soil surface activity on hot July and August days may be limited to dusk (J. Lorch, personal observation). Fitch (1989) found that dusk was the most reliable time to find Slender Glass Lizards in summer at his Kansas study site and believed that the animals were probably active for a period of time after sunset. Humid weather seems to be particularly conducive to aboveground activity, and animals are frequently observed after rain showers (Trauth 1984, Fitch 1989, J. Lorch, personal observations). Slender Glass Lizards often go unnoticed during cursory survey efforts because of their secretive nature and narrow soil surface activity periods (see Natural History Box: It's All about Timing). Only a small percentage of the population is above ground at any given time (Pleyte 1975).

Published home range estimations for adult Slender Glass Lizards are scarce but range from 0.49 ha (1.2 ac) in a Kansas population (Fitch 1989) to 1.5–2.0 ha (3.7–4.9 ac) at a Wisconsin study site (Pleyte 1975). Space use by this species reportedly changes over time (Fitch 1989); thus, home range estimates might only quantify the size of temporary activity areas. Fitch (1989) found that Slender Glass Lizards typically occupied a small area familiar to them and moved a mean distance of only 17.2 m (56.4 ft) per day within their activity area. Furthermore, these activity areas were ephemeral, and individuals occasionally undertook long-distance movements averaging 490 m (1,608 ft) to locate another area of suitable habitat where temporary residence was established. Although adults use a larger area for foraging than juveniles, both are just as likely to move long distances to establish a new activity area (Fitch 1989). During these movements, glass lizards may cross through habitats that they otherwise would not occupy (e.g., woodlands; Fitch 1989, Corey Raimond, personal observation).

Surface activity gradually declines as temperatures drop in autumn and glass lizards enter underground overwintering locations (Fitch 1989; see Habitat subsection). In Wisconsin, concentrations of Slender Glass Lizards are sometimes observed in autumn, suggesting that the animals might congregate in areas that are conducive to hibernation (J. Lorch, personal observation). However, this may be biased by greater surface activity in the autumn, giving the impression of increased density. Fitch (1989) reported that there was no autumn migration to hibernacula and that Slender Glass Lizards hibernated in the same areas they used during the summer.

PREY AND PREDATORS

Slender Glass Lizards are primarily insectivorous. Based on fecal analyses, orthopterans (grasshoppers, crickets, and katydids) are the dominant prey item, followed by beetles (Pleyte 1975, Fitch 1989). Other invertebrates, such as ants, caterpillars, cicadas, spiders, and snails, are also consumed, and vertebrates such as small frogs, small snakes, and the young of small mammals (i.e., mice, voles, and shrews) make up less than 2.7% of the diet (Fitch 1989). Pleyte (1975) observed a large Slender Glass Lizard with an adult mouse (*Peromyscus* spp.) in its jaws but was unsure whether the lizard had captured the mouse or had encountered it as carrion. Vogt (1981) reported that captives were fond of raw eggs, but it is unclear how frequently the eggs or young of ground-nesting birds are eaten in the wild. Dietary preferences appear to vary by season, according to the availability of prey. For example, beetles, caterpillars, and spiders are more commonly eaten in the spring, but orthopterans are the predominant food item later in the summer (Fitch 1989). When foraging, Slender Glass Lizards alternate between crawling slowly and remaining motionless, taking most prey by stalk-and-ambush techniques (Fitch 1989). Both vision and smell are likely utilized to locate prey. Olfaction is probably particularly important when foraging near dusk. Although it has not been documented, some prey are probably captured underground.

There are few documented cases of predation upon Slender Glass Lizards. Pleyte (1975) believed that a pair of resident Broad-winged Hawks (*Buteo platypterus*) at his study area probably preyed upon Slender Glass Lizards. Fitch (1974) recorded Broad-winged Hawks and Red-tailed Hawks (*B. jamaicensis*) taking Slender Glass Lizards, but the lizards made up only 1.4% and 2.4% of their diets, respectively. Additional studies have suggested that Slender Glass Lizards usually constitute an even smaller component (0.1%–0.5%) of the diet of Red-tailed Hawks (Fitch and Bare 1978, Fitch 1982). Other raptors such as owls and American Kestrels (*Falco sparverius*) likely also take some Slender Glass Lizards in Wisconsin. Many species of carnivorous mammals occur at sites occupied by Slender Glass Lizards in Wisconsin and probably prey on them opportunistically. Fitch (1989) noted that harvest mice (*Reithrodontomys* spp.) and White-footed Deer Mice (*Peromyscus* spp.) captured in the same funnel traps as Slender Glass Lizards would often kill the reptiles, and he suspected that small rodents, moles, and shrews probably take the occasional glass lizard while they are in underground burrows, especially during hibernation. North American Racers (*Coluber constrictor*) consume Slender Glass Lizards but are infrequent predators (Fitch 1989) and are not common within the range of the Slender Glass Lizard in Wisconsin.

When first encountered by a predator, Slender Glass Lizards will usually attempt to flee. Movement occurs primarily by side-to-side undulation of the tail; the tail pushes off of objects, propelling the animal forward (or sometimes upward; see Natural History Box: Leapin' Lizards). Using this method, Slender Glass Lizards can move with surprising speed and will often crawl to the nearest burrow, thicket, or brush pile or dive below the thatch layer. When there is no immediate cover nearby, Slender Glass Lizards will often move for several meters and then lie motionless on the ground in an apparent attempt to evade further detection (Fitch 1989, J. Lorch, personal observation). When captured, Slender Glass Lizards may wrap their tails around vegetation and, once extracted, will violently thrash about for a minute or so before becoming placid. Fitch (1989) suggested that this calm behavior after capture was a form of death-feigning, but it more likely represents fatigue resulting from physiological processes that occur during strenuous muscle activity (Gleeson and Bennett 1982).

NATURAL HISTORY BOX: Leapin' Lizards

On a warm summer evening in early September 2007, I set out with fellow naturalist Corey Raimond to look for Slender Glass Lizards at a site in Columbia County. Corey had never seen the species before, and I was hoping that conditions would be ideal for finding one. When we reached the opening where I would typically see the lizards, I immediately began walking through the patches of Smooth Brome grass, carefully watching for movement in the grass 1–2.5 m (3.2–8.2 ft) ahead of me. Meanwhile, Corey was trying his luck by looking under logs—a technique that I have found to be minimally successful, as Slender Glass Lizards do not utilize cover objects as frequently as other lizards. Suddenly I heard Corey shout, "A glass lizard! And it's jumping!" He failed to capture it, and I thought that he must have mistaken some other critter for a Slender Glass Lizard. I rushed over to where he had last seen the animal, and soon I flushed it myself. Indeed, the animal leaped into the air three or four times before crawling toward a thick patch of Smooth Brome. Upon capturing the animal, we determined that it was a recent hatchling with a fully intact tail. Since that initial experience, I have observed Slender Glass Lizards perform this 'bouncing' behavior on several other occasions. Of course, something without legs cannot really 'jump.' Instead, the lizards push against the ground with their tails, launching themselves upward. The method is essentially the same that the Slender Glass Lizard uses for moving across the ground, except that the force is generated downward instead of laterally. The reason for this behavior is not clear. Pleyte (1975) noted that Slender Glass Lizards would "jump" across roadside ditches when fleeing, and Cliburn (1957) recounted an instance of a Slender Glass Lizard leaping when the animal was moved from a road onto adjacent grass. I have observed Slender Glass Lizards propel themselves upward when they cannot get traction on surfaces such as pavement and loose sand. In doing so, the lizards also make a small amount of forward progress and can eventually get to an area that facilitates their more typical form of locomotion. I also suspect that Slender Glass Lizards jump to gain a better vantage point, orienting themselves with respect to their favored hiding areas. Every Slender Glass Lizard I have seen bounce along the ground subsequently made a beeline to the nearest burrow, brush pile, or other form of cover.

Jeffrey M. Lorch

The best-known defensive strategy of Slender Glass Lizards and the origin of the name of this group of lizards is the ability to purposely detach their tail (i.e., caudal autotomy). The tail can break into one or more pieces. Each vertabra in the tail has a fracture plane that runs through the center of the bone, and the muscle tissue is only weakly attached to the preceding vertebrae, allowing for easy detachment of the portion of the tail distal to the respective fracture plane. Once detached, the tail wriggles vigorously at first, which progressively slows over the course of several minutes; touching the detached tail causes it to thrash more violently (Fitch 1989). The movement of the detached tail (or tail pieces) likely distracts predators and allows the lizard an opportunity to escape.

Caudal autotomy in Slender Glass Lizards can be the result of pressure being applied to the tail, such as when a predator grasps the tail or pins the tail against the ground. However, the detachment of the tail is not always simply due to force but can be spontaneous and apparently under the lizard's control. In a sample consisting of dozens of Slender Glass Lizard capture events, animals that were pinned by the tail for a fraction of a second, or whipped their tail, or wrapped their tail around a human captor's wrist and arm did not detach their tail (J. Lorch, personal observation). Fitch (1989) further reported several instances of Slender Glass Lizards that detached their tail in the absence of any pressure being applied to the appendage. Fitch (1989) stated that Slender Glass Lizards could lose nearly half of their body mass when the tail detached. He noted that individuals that had lost large sections of their tail were slower, were easier to capture, and had less area in which to store fat reserves. For these reasons, he believed that tail loss may subsequently affect fitness and lead to higher mortality rates. Despite the negative consequences of losing a portion of the tail, few Slender Glass Lizards make it to adulthood with an intact tail (Fitch 2003a). Given the tradeoffs involved between losing a portion of the tail and survival, it makes sense that caudal autotomy involves a conscious decision on the part of the lizard. Contrary to popular belief, the tail rarely, if ever, regenerates to its normal length after a portion detaches. Tissue of the new tail grows to form an abruptly tapered tip with no new vertebrae, resulting in a tail that is, on average, only 15.3% the length of the original tail (Fitch 1989, 2003a).

CONSERVATION

Status: Globally, the Slender Glass Lizard is listed as a Least Concern species by the IUCN due to its "extensive geographic distribution and many relatively stable populations." However, the species is localized and particularly rare at the northern edge of its range (i.e., Illinois, Indiana, Iowa, and Wisconsin). The Slender Glass Lizard has been listed as Endangered in Wisconsin since 1979 and is currently a Species of Greatest Conservation Need, due to the small number of recorded occurrences in the state and the loss of important habitat (WDNR 2017b).

Populations: Limited information on Slender Glass Lizard populations exists. The work of Pleyte (1975) in Waushara County represents the only data available on population parameters from anywhere in the Upper Midwest. Throughout the course of two active seasons in the early 1970s, he captured 165 individuals (not including ca. 40–50 others that escaped capture), which resulted in annual density estimates ranging from 1.2 to 1.7/ha (0.5 to 0.7/ac; Pleyte 1975). Work from Kansas demonstrated that Slender Glass Lizard populations can be much larger than estimates by Pleyte (1975) indicate. Fitch (1989) completed "census estimates" for this species on a 7 ha (17.3 ac) study site in Kansas, which ranged from 128 to 780 individuals (density estimates from 18.3 to 111.4/ha [7.4 to 45.1/ac]). Additional demographic details of Slender Glass Lizard populations are also sparse. The population studied in Waushara County was dominated by older age classes (three-to-four-year-olds), which could result in limited viability (Pleyte 1975). Yet Pleyte (1975) further suggested that a long life span could offset a population structure dominated by old lizards and help populations withstand variable climatic conditions at northern latitudes (see Reproduction and Development section).

Threats: Slender Glass Lizards have a limited distribution in Wisconsin, but they may be locally common. For example, Pleyte (1975) stated that Slender Glass Lizards were present at his Waushara County study site "for at least ten to fifteen years that the author has been in this area and during that time, no major changes have been observed, [which] would indicate that the presence of the lizard is relatively stable." Pleyte (1975) even found that this species was substantially more abundant than the six other reptile species he encountered, including the Common Gartersnake (*Thamnophis sirtalis*).

Nonetheless, their restricted and localized range leaves Slender Glass Lizards vulnerable to numerous anthropogenic influences. Habitat loss and fragmentation likely have substantial impacts, and conservation concerns associated with open habitats apply to this species (see Introduction to Conservation and Management of Wisconsin's Amphibians and Reptiles chapter). Along these lines, one of the greatest threats to Slender Glass Lizards in Wisconsin could be the unchecked succession of plant communities, as woody vegetation overtakes otherwise open habitats. The open structure of the barrens and savanna communities occupied by this species was formerly maintained by wildfires. In recent times, fire suppression has transitioned these communities to dry and dry-mesic forests, which Slender Glass Lizards do not associate with. For example, at a study site in Kansas, Fitch (1989) noted that Slender Glass Lizards avoided wooded areas and that in the absence of habitat management activities, Slender Glass Lizard populations declined as a result of reforestation. Like vegetative succession, the establishment of pine plantations in the Central Sands portion of Wisconsin may also shade out otherwise suitable glass lizard habitat, reducing its availability.

The impacts of invasive plant species on Slender Glass Lizards have not been characterized but presumably threaten the quality of critical habitat (see Introduction to Conservation and Management of Wisconsin's Amphibians and Reptiles chapter). Spotted Knapweed (*Centaurea maculosa*), for example, often invades sites that harbor Slender Glass Lizard populations in Wisconsin (WDNR 2017b). This plant displaces native vegetation and reduces the thatch layer that provides important cover for Slender Glass Lizards (Lacey et al. 1989), so it could threaten this species. The effects of another invasive plant, Smooth Bromegrass (*Bromus inermis*), are difficult to understand, because reported interactions between this grass and Slender Glass Lizards are contradictory. In Wisconsin, small patches of this grass are frequently used by Slender Glass Lizards as a source of cover, but glass lizards have not been reported from areas where this grass dominates in large monocultures, which is unlike observations in Kansas (Fitch 1989, J. Lorch, personal observation). It is possible that at the northern limit of its range in Wisconsin, the dense sods of tall grasses like Smooth Bromegrass do not provide the microclimates and microhabitats needed by the lizards to successfully thermoregulate and forage. Invasive exotic woody plants such as buckthorn (*Rhamnus* spp.), honeysuckle (*Lonicera* spp.), Autumn Olive (*Elaeagnus umbellata*), and Black Locust (*Robinia pseudoacacia*) also invade the preferred habitats of Slender Glass Lizards. These woody species are likely problematic because they promote succession and lower thermoregulatory opportunities.

Glass lizards have great difficulty moving on smooth surfaces (Gans and Gasc 1990, Pleyte 1975), which has prompted concern over whether roads act as barriers to dispersal and gene flow (WDNR 2017b; Figure 171). Pleyte (1975) noted that all of the animals recaptured in his study were on the same side of the road that they had been captured on initially and that lizards were never observed on the pavement. However, reports indicate that Slender Glass Lizards will attempt to cross roads on occasion (Trauth 1984), and their method of locomotion does allow for forward movement on smooth or slightly roughened surfaces, but movement is slowed (Gans and Gasc 1990). While roads may not be complete barriers to movement, they are certainly hazardous for Slender Glass Lizards due to vehicular strikes. Specifically, the difficulty in moving expediently across roads leaves Slender Glass Lizards at great risk of falling victim to an automobile or predator.

Figure 171. Slender Glass Lizards cannot move efficiently on smooth surfaces and may have difficulty crossing paved roads (Adams County; photo by J. Lorch).

Management: Control of succession and invasive plant species that threaten glass lizard habitat is often accomplished through the use of prescribed fire (see Introduction to Conservation and Management of Wisconsin's Amphibians and Reptiles chapter). However, this technique must be applied cautiously when managing areas where Slender Glass Lizards might occur. Fitch (1989) believed that fire was a source of both direct (e.g., related to either fire or smoke) and indirect (e.g., microhabitat modification resulting in exposure to predators) mortality. In Florida, Eastern Glass Lizards were reported as a regular casualty of prescribed burns (Kaufmann et al. 2007), and the same would presumably hold true for Slender Glass Lizards. Fire can also lead to a short-term reduction in the density of vegetation and obliteration of the thatch layer, negatively impacting glass lizard locomotion and decreasing available shelter. Cavitt (2000) found fewer Slender Glass Lizards at a recently burned site compared to an unburned site, and Wilgers and Horne (2006) found that the species was more prevalent in long-term unburned areas than those burned annually or at four-year intervals. Such requirements seem enigmatic for a species restricted to habitats that depend on fire for long-term persistence (see Natural History Box: A Prairie Paradox). Because of its potential susceptibility to mortality from fire, management practices should restrict prescribed burning to times of the year when Slender Glass Lizards are inactive and create a prescribed burn plan that designates unburned areas to act as refugia (with microhabitat characteristics necessary for this species to persist). Other activities that reduce woody cover can also benefit Slender Glass Lizards. For example, rotational harvest of planted pines by clear-cutting may create habitat, as appears to be the case at a site in Adams County (J. Lorch, personal observation). Grazing can also effectively control woody vegetation and maintain the open habitats needed by Slender Glass Lizards, but heavy grazing can be detrimental by reducing grassy cover and thatch (Fitch 1989).

Efforts to mitigate mortality along roads through the use of road underpasses or exclusion fences could benefit this species, particularly in areas where high road mortality is documented (see Introduction to Conservation and Management of Wisconsin's Amphibians and Reptiles chapter). The presence of roads, however, could provide some indirect benefit to Slender Glass Lizards in regions where unabated succession has occurred. The species often utilizes grassy areas along roadways that are kept free of woody vegetation by occasional mowing (Pleyte 1975), and such habitat may provide important dispersal corridors between larger patches of natural habitat. Yet caution must be exercised to avoid excessive mowing or cutting the vegetation too short along roadways, which could result in direct mortality or reduce habitat suitability through removal of cover and thermoregulatory options (WDNR 2017a). Also, if suitable habitat is only concentrated along roads, and this draws unusually high numbers of glass lizards to roadsides, greater road mortality could occur. Efforts should be made to maintain open habitats that extend some distance from roads.

Common Five-lined Skink
Plestiodon fasciatus (Linnaeus 1758)

Tony Gamble

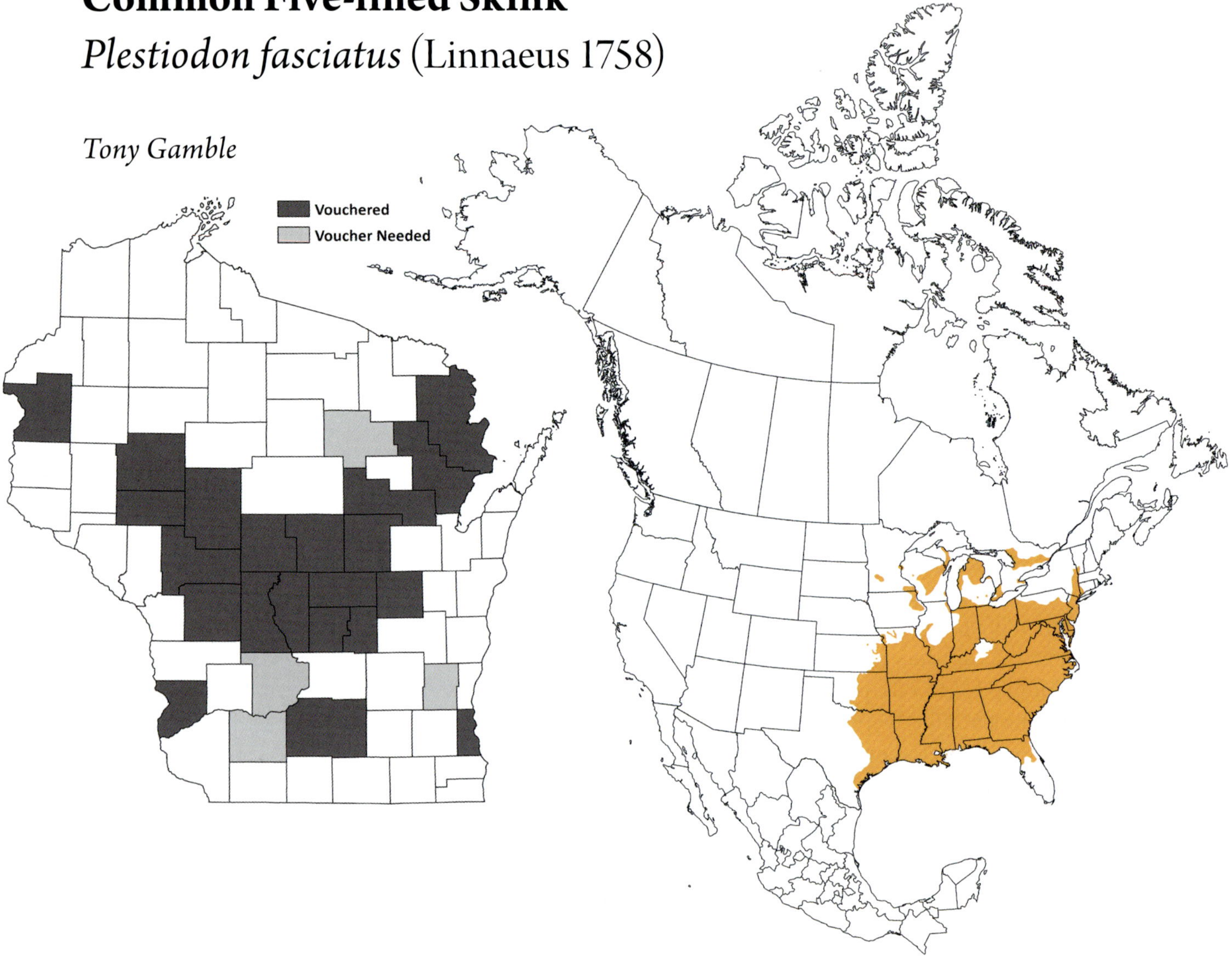

Adult female Common Five-lined Skink (Crawford County; photo by C. Raimond).

AT A GLANCE

Common Five-lined Skinks are small terrestrial lizards primarily found across central Wisconsin in forested areas with partially open canopies and sparse understory vegetation. In these habitats, they are frequently found beneath ground debris such as logs or rocks. Adults are dark brown to black with a grayish tail and have five yellow stripes running the length of their body. The electric blue tail in hatchlings and juveniles, which fades with age, contrasts the alternating dark and light stripes that give the species its name. Common Five-lined Skinks emerge from hibernation from late April to early May and mate soon after. Females lay their eggs in rotten woody debris from mid-May to late June and guard the eggs until they hatch in August. They eat primarily insects and are likely consumed by many mammal and bird species, but shrews may be particularly important predators. Common Five-lined Skinks will readily lose their tail when grasped. The twitching and wriggling of the dismembered tail might distract a predator just long enough to allow a quick escape. The tail will regrow, but it will never regain its original length or color. In October, Common Five-lined Skinks enter hibernacula, which are probably large rotten logs, for the winter. Few investigations on the ecology of this species have been conducted in Wisconsin. Common Five-lined Skinks are listed as a Special Concern species by the WDNR.

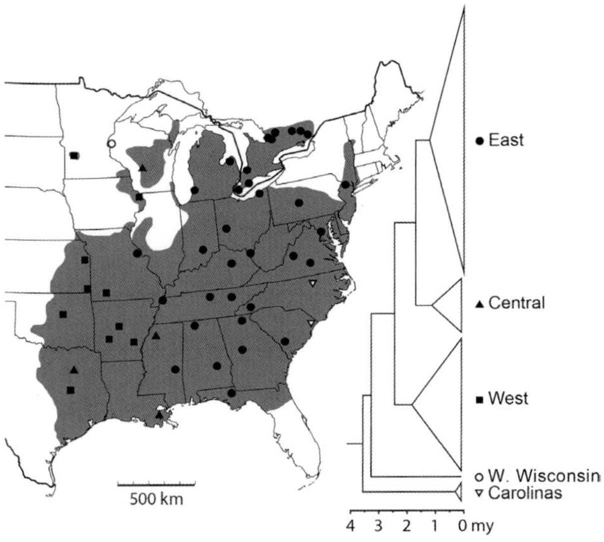

Figure 172. Major genetic lineages of the Common Five-lined Skink (*Plestiodon fasciatus*), modified from Howes et al. (2006; *right*). The map of eastern North America showing the distribution of the Common Five-lined Skink is modified from Conant and Collins (1998; *left*). Symbols represent collecting localities for the phylogenetic analysis. The phylogenetic tree was derived from analysis of the mitochondrial cytochrome *b* gene (Howes et al. 2006) and includes approximate divergence times for major clades in millions of years (my). The deepest divergences delineate samples from the Carolinas, western Wisconsin, and the rest of the species' distribution, comprised of east, west, and central clades.

SYSTEMATICS

Plestiodon fasciatus (Common Five-lined Skink) was formerly part of the cosmopolitan skink genus *Eumeces*. However, *Eumeces* was broken up into multiple genera following a comprehensive molecular phylogenetic analysis and subsequent taxonomic revision (Brandley et al. 2005), with North American and some Asian species placed into the genus *Plestiodon*. While there are no described subspecies of *P. fasciatus*, significant genetic differences exist among populations (Howes et al. 2006). The deepest of these genetic splits distinguishes three distinct lineages that occupy (1) the eastern Carolinas, (2) western Wisconsin, and (3) the rest of the species distribution (Figure 172). These divergent lineages are the result of ancient population fragmentation that predates Pleistocene glacial cycles that began 1.9 mya (Howes et al. 2006). The ancestors to today's disjunct population in western Wisconsin likely survived repeated glacial cycles through persistence in the unglaciated Driftless Area (see Past, Present, and Potential Future Landscapes of Wisconsin chapter). Additional genetic research and a reevaluation of the taxonomic status of the western Wisconsin population are urgently needed.

DESCRIPTION

Adults: The Common Five-lined Skink, like most skink species, has extremely smooth shiny overlapping scales, short limbs, and a long tail. The following combinations of head scales are typical: a single postnasal scale, a divided postmental scale, two postlabial scales, and seven supralabial scales (Smith

Above, top: Figure 173. The head of an adult female Common Five-lined Skink (Crawford County; photo by C. Raimond).

Above, bottom: Figure 174. An example of a subadult or adult Common Five-lined Skink on which color and pattern have not started to fade (Jackson County; photo by C. Raimond).

1946; Figure 173). Adults measure 12.5–21.5 cm TL (4.9–8.5 in) and 6.0–8.6 cm SVL (2.3–3.2 in; Vogt 1981, Conant and Collins 1998). The few adult Wisconsin specimens in MPM and UWZM collections available for measurement had a mean SVL of 6.6 cm (2.6 in, $n = 10$). Because only five of these possessed intact tails, mean tail length was not recorded. The SVL of the largest Wisconsin specimen measured was 7.3 cm (2.9 in), with a tail length of 9.7 cm (3.8 in, UWZM 9914, collected from Waushara County in 1913).

Adults can be quite variable in color and pattern based on age or sex. Young adults retain the juvenile coloration, five cream or yellow stripes on a black background, with a bright blue tail and a middorsal stripe dividing on the head to form a V. These colors often fade as the animals mature (Figure 174). Males become nearly patternless with age, having a uniform gray or light brown color with barely a hint of stripes on the back and sides. The bright blue tail fades to gray or light brown. During the breeding season the male's head, labial scales, and chin become orange or red. Females retain the dorsal stripes into adulthood, although the dark background fades from black to a dark brown, and the cream or yellow stripes become light brown or beige. In Wisconsin and other northern populations, females may retain some blue coloration on the tail (Smith 1882, Vogt 1981, Harding 1997, Moriarty and Hall 2014), although this is rarely observed throughout the rest of the species' range (Smith 1946, Vitt and Cooper 1986).

Figure 175. Eggs of the Common Five-lined Skink found in a rotten log (Jackson County; photo by C. Raimond).

Preadult stages: Eggs are oblong and white with a thin leathery shell (Figure 175). A sample of 36 eggs from Illinois had mean dimensions of 1.27 × 0.89 cm (0.35 × 0.50 in) in late June but swelled to 1.59 × 1.11 cm (0.44 × 0.60 in) by early July, presumably due to water absorption (Cagle 1940). Hatchlings and juveniles have five yellow or cream stripes on a black background and a bright blue tail (Figure 176). The middorsal stripe divides on the head to form a V. Hatchlings measure 5.8–6.5 cm TL (2.3–2.6 in; Smith 1946). The bright blue tail of hatchlings and juveniles was initially thought to limit aggression by adult males (Clark and Hall 1970). However, subsequent studies have found little support for this hypothesis, suggesting instead that a blue tail distracts predators (particularly

Figure 176. A juvenile Common Five-lined Skink with the classic bright blue tail (Adams County; photo by C. Raimond).

birds) away from the head and body (Cooper and Vitt 1985, Vitt and Cooper 1986, Watson et al. 2012, Bateman et al. 2014).

Variation: Common Five-lined Skinks exhibit considerable variation in color and pattern. This is particularly evident when comparing adult males to females. Males tend to have faded dorsal stripes and can have an orange or red head and chin during the breeding season, while females retain the dorsal stripes throughout their lives. Further differences are seen between adults and juveniles, with the juvenile's bold color and pattern fading into adulthood.

Diagnosis: The Common Five-lined Skink is characterized by smooth, large, and overlapping dorsal scales; smooth tail scales and ventral scales that are not plate-like; two subequal postlabial scales; and a dorsolateral stripe that, if present, is no more pronounced than the middorsal light stripe. This species appears superficially similar to the Prairie Skink (*Plestiodon septentrionalis*). The two can be distinguished based on the following characteristics: Common Five-lined Skinks have two postlabial scales, while Prairie Skinks have just one postlabial scale; the dark dorsal stripes of Common Five-lined Skinks are wider than the alternating light dorsal stripes (it is noteworthy that the dark stripes typically fade in adult males), while the light dorsal stripes of Prairie Skinks are wider than the thin black stripes; Common Five-lined Skinks, particularly younger individuals, have a middorsal stripe that divides on the top of the head to form a V, while Prairie Skinks lack this head marking.

It is a challenge to distinguish the eggs of Common Five-lined Skinks from those of some other lizards and snakes. Their eggs are smaller than many reptile species in Wisconsin but could overlap in size with those of Prairie Skinks, Six-lined Racerunners, and perhaps smaller egg-laying snake species such as Western Wormsnakes. Still, it would be possible to narrow down potential options through comparisons of overall egg dimensions and clutch size but also through comparisons of geographic location and habitat where eggs were encountered (see Preadult stages and Habitat subsections and Reproduction and Development section).

DISTRIBUTION AND HABITAT

Global distribution: Common Five-lined Skinks are one of the most widespread lizards in North America, with a distribution that extends from the East Coast, including southern Ontario to southeastern New England south to northern Florida, west to the Great Plains, including southeastern Nebraska south through central Texas. Numerous disjunct populations dot the northern and western edges of this distribution, with isolated populations in South Dakota, Minnesota, Iowa, the Upper Peninsula of Michigan, and Wisconsin.

Distribution in Wisconsin: Common Five-lined Skinks are reported from 25 counties, but voucher specimens could not be found for Iowa, Sauk, and Langlade (see the Appendix and discussion of additional problematic county record vouchers below). The current documented distribution in Wisconsin is also somewhat spotty and restricted (Natural History Box: Factors That Potentially Influenced the Distribution of Common Five-lined Skinks in Wisconsin). This species has been documented throughout central Wisconsin, with remnant populations in the northwestern (Polk County) and northeastern (Marinette, Oconto, and Shawano Counties) parts of the state (Vogt 1981, Casper 1996). Records supported by voucher specimens from southeastern Wisconsin (Milwaukee and Washington Counties) are extremely sparse, very old, and occasionally questionable (Graenicher 1911,

NATURAL HISTORY BOX: Factors That Potentially Influenced the Distribution of Common Five-lined Skinks in Wisconsin

The current distribution of Common Five-lined Skinks in Wisconsin raises many questions about what drives their presence or absence at various locations. Although limited data from Wisconsin exist to make definitive statements on this subject, several key factors could be relevant. Wisconsin is at the periphery of the Common Five-lined Skink's continental distribution, and research suggests that species often have reduced abundances, being more vulnerable to declines and extirpation, at the edge of their range compared to the core (e.g., Bahn et al. 2006). It is, therefore, probable that this species has always existed in lower abundance in Wisconsin compared to the core of its North American range, which could be a major factor explaining its patchy distribution in the state.

Availability of appropriate habitat is also an important consideration. It is reasonable to speculate that loss of woodlands has led to undocumented historical extirpations of this species and contributed to its current range in Wisconsin (see Past, Present, and Potential Future Landscapes of Wisconsin chapter). Common Five-lined Skinks also appear to suffer when open woodlands experience even moderate alteration, such as vegetative succession that leads to overly dense understory vegetation (see Habitat subsection and Conservation section). However, the loss or degradation of a broad vegetative community alone may not be a sufficient explanation for the modern absence of this species from the rather expansive protected lands in Wisconsin with seemingly suitable habitat. The Southern Kettle Moraine State Forest in the southeastern part of the state contains open woodlands, prairies, and savannas that provide qualitatively suitable habitat for Common Five-lined Skinks. However, the species has never been confirmed there and has only been very rarely documented in all of southeastern Wisconsin since the late 1800s (see Distribution in Wisconsin and Localities of Possible Occurrence subsections).

It is further possible that Common Five-lined Skinks require very specific habitat characteristics that are lacking in these locations that appear otherwise suitable. For example, a subtle but critical component of habitat involves appropriate nesting and overwintering sites. Common Five–lined Skinks are frequently reported to nest in downed logs (e.g., Cagle 1940). Fitch (1954) also reviewed reports from other states that described this species as hibernating in rotten woody debris. As such, it is plausible that many Common Five-lined Skinks in Wisconsin would also use downed woody debris to overwinter, if available. Yet evidence supports that this species selects logs of larger diameter when seeking acceptable microhabitat during the active season (Hecnar and M'Closkey 1998). It follows that any such logs used for hibernation in Wisconsin would also need to be of appropriate diameter to ensure that ice does not penetrate deep enough into fallen logs to harm torpid skinks within. If true, the loss of large timber and old-growth forests in Wisconsin could be an important factor explaining the current distribution of Five-lined Skink populations in the state.

However, the loss of large downed woody debris is not completely congruent with the modern presence of several populations in the state. Common Five-lined Skinks that currently occur at sites in Monroe and Polk Counties are not associated with large timber or old-growth forests. They are, however, found in the vicinity of large exposed rocky outcrops that likely possess fissures that extend below the winter frost line (J. Kapfer, personal observation; Breckenridge 1944). In other counties of occurrence, such as Juneau and

Adams along the Wisconsin River and nearby Monroe, appropriate geological features are also present. In accordance, a review of museum records indicates that many of the more recently collected (i.e., post-1970) Common Five-lined Skink specimens come from Juneau, Adams, and Monroe Counties (e.g., MPM 11544–46, 23345–46, 30298, 30319). To exemplify the importance of these specific habitat characteristics, we again consider the Southern Kettle Moraine State Forest. This large protected area possesses habitat that appears superficially suitable, yet, strangely, Common Five-lined Skinks have never been confirmed there. Although there are currently ample fallen logs in the state forest, this woody debris likely comes from smaller-diameter trees due to past human disturbance that removed larger trees. It also lacks ample geological features conducive to overwintering.

When this information is considered together, one might conclude that, given the current paucity of old-growth forests in Wisconsin, modern Common Five-lined Skink populations are now mostly restricted to locations where exposed rock outcrops for overwintering are present (e.g., populations from Monroe and Polk Counties). However, this does not explain the fact that, although rocky outcrops occur in Juneau and Adams Counties, the skink populations in those counties are not typically found at specific locations with the rocky outcrops (Jeffrey Lorch, US Geological Survey, personal communication). It is also puzzling that there are few records from many counties along the Mississippi and lower Wisconsin Rivers, many of which possess exposed rocky outcrops and adjacent open oak woodlands. Perhaps the scarcity of reports from counties along these rivers is simply due to insufficient survey effort, as the skinks have been documented in Crawford County.

The situation in Wisconsin is made even more complex by the findings of Howes et al. (2006). They analyzed mitochondrial DNA across many North American populations and found that the currently disjunct Common Five-lined Skinks in Polk County are genetically divergent from other Wisconsin populations. The authors suggested that this occurred due to glacial activity thousands of years ago. With the exception of the Driftless Area, Wisconsin was covered by the Laurentide Ice Sheet during the last large glacial event. As proposed by Howes et al. (2006), the Driftless Area could have provided a refuge for the ancestors of the Polk County population, so they were not forced to retreat farther south like their counterpart populations in other parts of Wisconsin (see Systematics section). According to Howes et al. (2006), Wisconsin's other modern populations (i.e., those not in Polk County) descended from a different lineage, and they are more closely related to populations in the south-central US. This suggests that the ancestors of the non–Polk County populations in Wisconsin expanded northward after the glaciers retreated. Thus, the modern spotty distribution could be partially attributed to the state's glacial history.

It is probable that a complicated interaction of factors is responsible for the modern distribution and abundance of Wisconsin's Common Five-lined Skinks. These factors include naturally lower population sizes due to Wisconsin's position at the species' range periphery, the fact that this species could have had a fragmented state distribution for thousands of years due to glacial activity, and long-term anthropogenic habitat disturbance in the state.

Joshua M. Kapfer, UW–Whitewater, and Tony Gamble, Marquette University

Casper 1996; see also Localities of possible occurrence subsection). For example, museum records associated with the Washington County voucher specimens (MPM 489 and 490; Casper 1996) state only that they were collected at "Cedar Lake" in 1888, and there are several lakes with this name located in various counties. Despite this, publications from as long ago as the early 1900s referenced this species in Washington County. Pope and Dickinson (1928) reported it there but provided no specimen numbers. Graenicher (1911) reported that specimens MPM 489 and 490 were taken by W. M. Wheeler at "Cedar Lake, Washington Co., Wis." Unfortunately, Graenicher's work was published over 20 years after the specimens were collected, and it is unclear if he confirmed the specific location with Wheeler, as no county-level information currently accompanies the specimens (Julia Colby, MPM, personal communication). Pope and Dickinson (1928) also mentioned MPM specimens from Milwaukee County but again reported no voucher numbers. Graenicher (1911) reported two specimens from New Coeln in Milwaukee County (MPM 488 and 491). Although no dates accompanied them, and current museum records indicate that MPM 488 accidentally had its information tag switched with specimen MPM 493 from Florida, Graenicher (1911) speculated that they were collected before 1890. No other confirmed records or specimens from southeastern Wisconsin are currently known.

The Chippewa County record is based on a photographic voucher (UTADC 8161), which superficially appears to be a misidentified Prairie Skink, although the orientation of the photograph does not allow definitive determination. Regardless, given this record's proximity to counties of documented occurrence, Common Five-lined Skinks are likely present in Chippewa County. The Dane County record is based on a specimen collected from the UW–Madison arboretum in 1958 (UWZM 17589), but no additional specimens from this county are known. Given the lack of vouchers from nearby Sauk and Iowa Counties and the possibility that arboretum staff in the 1950s had purchased landscaping and gardening materials from out of state, it is feasible that the Dane County specimen was accidentally translocated to Wisconsin from elsewhere in the species' range.

Localities of possible occurrence: It is plausible that this species is present in undocumented counties such as Menominee and Outagamie because they are adjacent to counties with documented occurrences. Given reports of this species in various locations from southeastern Wisconsin, it could still occur in several counties in this region where habitat is appropriate (see Distribution in Wisconsin subsection). Higley (1889) and Pope (1944) mentioned Common Five-lined Skinks in Walworth County, but no associated voucher specimens are known. Cahn (1929) reported observing a young male and two adult females near Golden Lake in Waukesha County in 1913 and 1915, but no specimens exist. Graenicher (1911) made specific mention of anecdotal reports he received from Sheboygan County. Hoy (1883) did not include Common Five-lined Skinks but stated that Prairie Skinks were common in areas of southern Wisconsin "as far north as Lake Winnebago." However, this would represent a substantial range expansion for Prairie Skinks in Wisconsin. It is plausible that his report was actually based on misidentified Common Five-lined Skinks. Interestingly, in an earlier publication, he mentioned the presence of the "Blue Tailed Lizard, Sincus Faciata" in the vicinity of Racine (Hoy 1874). Numerous recent surveys at various potentially suitable locations in Walworth and Waukesha Counties have been fruitless (Joshua Kapfer, UW–Whitewater, unpublished data), but many southeastern locations remain to be surveyed.

Habitat: Across their range, Common Five-lined Skinks occupy deciduous forests with abundant ground cover such as logs, rocks, and leaf piles (Smith 1946). In Wisconsin, they may associate with savannas, Jack Pine (*Pinus banksiana*) barrens, and dry woodlands (see Amphibians and Reptiles in Wisconsin Plant Communities chapter). Preferred habitats are typically humid but not wet, with sparse understory vegetation, and have breaks in the canopy or are adjacent to open areas to facilitate basking (Smith 1946, Vogt 1981; Natural History Box: A Long-Term Perspective of Common Five-lined Skink Habitat). Common Five-lined Skinks that occur in northern populations are often closely associated with rock outcrops (Breckenridge 1944, Fitch 1954, Lang 1982,

NATURAL HISTORY BOX: A Long-Term Perspective of Common Five-lined Skink Habitat

The most detailed description of Common Five-lined Skink habitat comes from the work of Henry S. Fitch in Kansas. In general, he reported that they were fond of areas with "deep alluvial soil, alternate with flat or rolling upland some two hundred feet higher in elevation, and having shallow, rocky soil. Where the uplands slope to the valley floors, there are steep hillsides, usually with extensive limestone outcrops along their upper slopes" (Fitch 1954). Interestingly, in an effort to outline their specific habitat needs, Fitch spent a substantial amount of time discussing the specific locations he searched and *did not* find skinks. For example, he did not find this species in bottomlands at the base of the hills referenced in the previous quote and suggested that floods and a lack of ground cover deterred them. He also did not find them on the open hilltops that were once prairie and later used for various agricultural purposes. Where Fitch *did* encounter this species was in the woodlands that existed along slopes between adjacent bottomlands and hilltops. Yet this skink did not use all hillside woodlands equally. For example, Fitch did not observe them on hillsides overgrown with aggressive woody plants (e.g., honey locust [*Gleditsia triacanthos*]) or on shaded north-facing hillsides with "thick second-growth woods" and "dense understory vegetation." Interestingly, Common Five-lined Skinks at his study site were heavily associated with wooded hillsides that were grazed by livestock. These specific habitats possessed well-cropped understory vegetation, "more open ground, and patches of sunlight."

Over 50 years later, Fitch (2006a) again reviewed the habitat associated with this species at the same study sites but with the advantage of a long-term perspective: "The Five-lined Skink was abundant on the Reservation during the early years. It was especially common on two wooded hillside tracts that were accessible to livestock and connected the hilltop pastures with the bottomland pastures. Browsing and grazing in Horse Woods and Skink Woods altered the aspect of these areas. They were more open than other parts of the woodland, with less ground vegetation. Branches that were within reach of cattle had been eaten back resulting in more sunlight reaching the ground." These moderately grazed locations possessed a "sunnier habitat favored [by] the skinks, but when the area became a reservation, livestock were removed and both areas developed thicker canopies and thicker ground vegetation. As a result the skink's habitat deteriorated, and over a period of years they underwent drastic reduction in numbers. By the year 2000 they were seen regularly only at the laboratory building, the residence, and the 60 foot sidewalk between them." Although this information might prompt some individuals to hastily tout the benefits that grazing cattle has on wildlife habitat, Fitch (1954) also outlined that a delicate balance must be met between under- and overgrazing: "They are absent or scarce in woods that have been subjected over periods of years to browsing, by sheep or goats, so heavily that hardly any herbaceous vegetation remains and so heavily that the soil is packed from trampling."

Joshua M. Kapfer, UW–Whitewater

Howes and Lougheed 2004, LeClere 2013, Moriarty and Hall 2014). For example, Wisconsin populations from Polk County, near Taylor's Falls, are found on and near rocky basalt outcrops (Breckenridge 1944).

Appropriate microhabitats appear to be particularly important to this species and include suitable downed woody debris, rock piles, or large rocks with crevices. For example, Wisconsin individuals are reportedly found beneath woody ground debris such as rotten logs and sawdust piles or "humus" (Vogt 1981). Selected ground cover may also possess specific characteristics. For example, Hecnar and M'Closkey (1998) found that Common Five-lined Skinks in Ontario preferred larger downed logs of a moderate decay class (e.g., the presence of soft damp wood, hollows, and moderate to substantial decay). Rock walls and rock piles could likewise be important habitat features (Fitch and von Achen 1977), possibly because open rocks offer increased basking opportunities (Fitch 1954, Lang 1982, Howes and Lougheed 2004). Common Five-lined Skinks in Ontario preferred to hide beneath thin (≤10 cm [3.9 in]) rocks that were longer ($\bar{x} = 55.2$ cm [21.7 in] long) on average than unoccupied rocks (Seburn 1993, Quirt et al. 2006). Occupied rocks were located in areas with more open tree canopies than unoccupied rocks, while the skinks selected locations beneath the rocks with more stable thermal conditions (Quirt et al. 2006). This species readily climbs trees (Fitch and von Achen 1977), and Vogt (1981) stated that Wisconsin individuals occasionally escaped him by "running up the trunk of a jack pine."

Nesting and overwintering habitats in the Upper Midwest are poorly documented. Fitch (1954) reported that of over 100 natural nests found in Kansas, all but one were beneath rocks. Alternately, studies from Illinois (Cagle 1940) and Ontario (Hecnar 1992) observed that nests were primarily located within or beneath rotten logs. Nesting logs in Illinois were internally soft with a hard outer layer and ranged in diameter from 12.7 to 61 cm (5 to 24 in), receiving at least partial shade for most of the day (Cagle 1940). All nesting sites at this location were located on the dry upland areas of an oak-hickory woodland with limited underbrush, high sunlight penetration to the woodland floor, and a "heavy mat of dead leaves" (Cagle 1940). Vogt (1981) anecdotally reported that they spent the winter "deep inside rotten logs or sawdust piles, or . . . underground." In Kansas, they overwinter in sawdust piles (Linsdale 1927), while Fitch (1954) suspected that they overwintered within underground chambers beneath large rocks. Minnesota populations depend on deep rock crevices for overwintering (Lang 1982), and this may also apply to Wisconsin populations associated with rock outcrops (e.g., Polk County).

REPRODUCTION AND DEVELOPMENT

Breeding in northern latitudes occurs soon after emergence from hibernation in late April and May (Minton 1972, Vogt 1981, Moriarty and Hall 2014), but peak breeding season in Kansas was May 3–16 (Fitch 1954). Approximately one month after emergence from hibernation, the heads of sexually mature males become more reddish in color (Fitch 1954). Males soon become aggressive toward each other and also begin chasing females (Fitch 1954). Pursued females often attempt escape, but if captured, they are grasped in the male's mouth so that they cannot wriggle free, after which time copulation typically occurs (Fitch 1954). Bateson et al. (2011) found evidence that Common Five-lined Skinks are polyandrous, and offspring of a single female are frequently sired by multiple males per year. This species is oviparous and lays 2–18 eggs (Smith 1946; $\bar{x} = 8.82$ [Fitch 1954]) in warm and humid locations (see Habitat subsection). Eggs are laid from late May through mid-June and hatch from late July through August (Minton 1972, Vogt 1981).

Unlike most lizard species, female Common Five-lined Skinks guard their eggs until they hatch (Ruthven 1911, Noble and Mason 1933). Females coil around the eggs or lie next to them during the six-week incubation period, leaving regularly to feed or bask (Noble and Mason 1933, Fitch 1954). The brooding female will use her snout to gently nudge any eggs that are displaced back into position (Ruthven 1911, Noble and Mason 1933, Vitt and Cooper 1989). The purpose of egg brooding in skinks is not fully understood, but it may be to protect the eggs from predators or to mitigate environmental changes during incubation (Vitt and Cooper 1989). For example, Hecnar (1992) reported that during low moisture levels, female body contact with incubating eggs was greater than when moisture levels were higher, which suggests that females brood eggs to help maintain moisture levels for appropriate embryo development. Multiple females will nest in the same location (Cagle 1940, Fitch 1954). As reviewed by Hecnar (1992), such nesting aggregations could result in better protection from predators (i.e., multiple females to guard eggs) and better body condition for brooding females (i.e., females can leave to bask or forage because their eggs receive protection from other females), or these aggregations simply occur when several individuals take advantage of limited suitable nesting locations. Oophagy by females on their own unviable eggs has been documented (Groves 1982). The female loses interest in the hatchlings once the eggs hatch and the offspring leave the nest (Fitch 1954).

There is little, if any, available information on the growth rates of individuals from Wisconsin. Extensive multiyear mark-recapture studies from Kansas found that some hatchlings can grow up to 0.05 cm/day (0.02 in/day) during their first spring, and minimum adult size was attained at 6.5 cm SVL (2.6 in; Fitch 1954). The author also found that most individuals reach maturity after emerging from their second winter dormancy period, but some reached maturity by nine or ten months (Fitch 1954).

ACTIVITY

In Wisconsin, Common Five-lined Skinks emerge from hibernation from late April to May and are active throughout the summer (Vogt 1981). Fitch (1954) found them active under very cool spring conditions, such as sunny days with air temperatures between 10°C and 15.6°C (50°F and 60°F). During these cooler spring periods, this species spends much time openly basking in woodlands where deciduous trees have not experienced full leaf-out (Fitch 1954) or under warm ground debris, such as rocks exposed to the sun (reviewed by Fitch and von Achen 1977). Common Five-lined Skinks spend most of their time hidden under rocks or logs but emerge regularly throughout the day to bask and forage for food (Smith 1946, Vogt 1981). The mean body temperature of active individuals found openly basking in Kansas was 31.5°C (88.7°F, $n = 41$; Fitch 1954). Although primarily found on the ground surface, Common Five-lined Skinks will bask on tree trunks (Minton 1972, Fitch and von Achen 1977).

The movement patterns of this species are somewhat erratic and unpredictable but may cluster around areas with suitable ground cover (Fitch and von Achen 1977). Fitch (1954) reported that "recorded movements varied from a few inches to hundreds of yards, the most noteworthy feature in general was the short distance between points of capture . . . after days, weeks, months or years." Mean distance moved per day by adult individuals in a Kansas population was 5.1 m (16.7 ft; Fitch and von Achen 1977), with the largest single movement recorded being 207.3 m (680 ft; Fitch 1954). Fitch (1954) further estimated that the home range of adult males and juveniles was "approximately 90 ft across" but was "little more than 30 ft across" for females. Movement may be greater in spring during the mating season, and activity could also be related to seasonal variation in environmental conditions (e.g., moisture levels; Fitch 1954). For example, Fitch (1954) reported that "after rainless periods in August and September, when morning dew is no longer available these skinks, especially the adults, are no longer regularly seen in the open. They have retreated to underground shelters where they spend nearly all their time."

Hibernation typically begins in October, and skinks likely remain inactive through the winter until the following spring (Vogt 1981), although few specifics are known about their overwintering habits in Wisconsin. Duvall et al. (1980) found evidence to support that Common Five-lined Skinks respond to the olfactory cues of conspecifics when selecting hibernacula. Because they conducted their work on individuals that were past annual breeding condition, they suggested that this olfactory response could aid Common Five-lined Skinks in finding hibernation sites and explains why small aggregations of overwintering skinks are occasionally reported (Duvall et al. 1980).

PREY AND PREDATORS

Common Five-lined Skinks are active foragers that rely primarily on visual cues when hunting (Burghardt 1964). Fitch and von Achen (1977) reported that wild adults held in confinement during the active season consumed (on average) ca. 3% of their body weight per day, but food consumption was generally higher in early summer. Prey items include a variety of invertebrates, such as spiders, caterpillars, crickets, roaches, snails, beetles, and grasshoppers (Vogt 1981, Lang 1982). Of 738 food items recorded from Kansas individuals, 48.7% were arachnids and 43.3% were insects (Fitch 1954). Occasional reports of cannibalism and even ingesting their own tails also exist (reviewed by Fitch 1954).

Common Five-lined Skinks are likely preyed upon by many species, such as domestic Cats (*Felis catus*), Virginia Opossums (*Didelphis virginiana*), Raccoons (*Procyon lotor*), shrews (Soricidae), Striped Skunks (*Mephitis mephitis*), hawks, Eastern Milksnakes (*Lampropeltis triangulum*), Eastern Hog-nosed Snakes (*Heterodon platirhinos*), and Common Gartersnakes (*Thamnophis sirtalis*; reviewed by Fitch 1954, Vogt 1981, Mitchell and Beck 1992). Fitch (1954) observed encounters between Common Five-lined Skinks and Northern Short-tailed Shrews (*Blarina brevicauda*) placed together in terraria. In these interactions, the shrews either attacked or killed the lizards, and one shrew even consumed part of a skink's detached tail. The author postulated that Northern Short-tailed Shrews were important predators because the wounds received by individuals that escaped shrew attacks in terraria resembled scars noted on wild lizards.

Common Five-lined Skinks are secretive and likely avoid danger by remaining hidden beneath cover. Like most skink species, the Common Five-lined Skink is extremely wary and will flee if encountered in the open, in some cases even climbing trees or occasionally entering nearby waterbodies to escape (reviewed by Fitch 1954). Skinks are capable of caudal autotomy and easily lose their tail when grasped by a predator. Once removed, the tail will continue to wriggle, keeping the predator occupied while the skink escapes. In juveniles, the bright blue tail color is thought to act like a decoy that distracts predators away from the more vital head and body (Cooper and Vitt 1985, Vitt and Cooper 1986, Watson et al. 2012, Bateman et al. 2014). Fitch (1954) noted that only 16.5% of adult females over 7.5 cm SVL (3.0 in) possessed tails that had never been broken, suggesting that this is a commonly employed escape tactic. Although tail autotomy can aid in predator avoidance, a missing tail might also lead to greater difficulty in escaping subsequent predator attacks. For example, hatchling Common Five-lined Skinks with missing tails have reduced maximum sprint speeds for several weeks after the tail is detached, which could affect their ability to avoid predation (Goodman 2006). A regrown tail is not as long or as colorful as the original tail (Vogt 1981). Mean rate of tail regrowth of hatchlings studied in the laboratory was 0.6 cm/wk (0.2 in/wk; Goodman 2006).

CONSERVATION

Status: Globally, the Common Five-lined Skink is listed as a Least Concern species by the IUCN. It is considered a species of Special Concern in Wisconsin by the WDNR.

Populations: Hecnar and M'Closkey (1998) conducted annual population censuses over six years at two locations in Ontario that were 2.55 and 2.85 ha (6.30 and 7.04 ac) in size. They reported that populations from these survey sites ranged from 124 to 300 individuals. Fitch (1954) calculated what he believed were reasonable estimates of population density at a study site in Kansas over three years, which ranged from 19.8 to 24.7/ha (8 to 10/ac). The author also found evidence that Common Five-lined Skink populations experienced rapid turnover, and individuals had short life expectancies, with few living past five years. He further suggested that populations could be heavily affected by random fluctuations in weather and that repeated years of poor climatic conditions could result in low brood survival. Populations in southeastern Wisconsin are likely extirpated (Casper 1996), highlighting the fragile nature of isolated disjunct populations that frequently occur at the northern edge of their distribution (Lesica and Allendorf 1995, Curnett et al. 1996, Sagarin et al. 2006).

Threats: Conservation of habitat is likely a primary concern for this species in Wisconsin, and general concepts associated with habitat loss and degradation are likely relevant to Common Five-lined Skinks (see Introduction to Conservation and Management

NATURAL HISTORY BOX: Five-lined Skinks in Portage County, Wisconsin

Growing up on a 64.7 ha (160 ac) farm in the eastern edge of Portage County during the mid- to late 1950s gave me the opportunity to capture, keep, and/or observe many interesting small animals. The property was one-third open woods and two-thirds hayfield or row crop fields. Some of the fields were rented to a local farmer, but most of the old farm buildings were not used. There was a small population of Common Five-lined Skinks that lived on the southern sides of the hay barn and small granary. During summer months, adult skinks could be seen on sunny mornings along the cement foundation, on the lower part of the buildings' unpainted wooden walls, and under old boards strewn about on the ground. Young Common Five-lined Skinks were fun to see during mid-summer with their bright cobalt-blue tail. Males were easily spotted because of the orange red on their head. Even at the age of 15, I was already deeply interested in amphibians and reptiles and kept a wide variety of local species, including Common Five-lined Skinks. I kept them in a wooden cage with a glass front, a screened top, and gravel on the bottom. I placed a small tree branch in the cage and provided a light bulb so the lizards could bask and be warm. I fed captive individuals mostly crickets, grasshoppers, and small 'cutworms' (moth larvae). Sad to say, the Common Five-lined Skink population on our farm seemed to disappear during the early 1960s. My sister wanted a pet cat, which quickly became a family of cats, and they spent most of their time outside. Although I never saw a cat capture any lizards, the cats did spend a lot of time behind the barn. It was likely our skinks were targeted by the cats and their population eliminated.

Tom R. Johnson, State Herpetologist (retired), Missouri Department of Conservation

of Wisconsin's Amphibians and Reptiles chapter). Fitch (2006a, 2006b) observed that Common Five-lined Skinks were among the first group of amphibians and reptiles to disappear at a site he monitored for 58 years in Kansas. He reported that this rapid decline coincided with the progression of vegetative succession over much of the site, and in later years he only found this species in clearings associated with buildings (see Natural History Box: Factors That Potentially Influenced the Distribution of Common Five-lined Skinks in Wisconsin). Disturbance of appropriate microhabitat characteristics could also threaten this species. For example, anthropogenic removal of suitable downed woody debris at a site in Ontario was correlated with a reduction in Common Five-lined Skink abundances (Hecnar and M'Closkey 1998). The loss of logs also appeared to influence the structure of this population, with few subadults observed, presumably due to loss of woody debris for egg laying (Hecnar and M'Closkey 1998).

Numerous other potential threats exist that are not currently well-understood. For example, Hecnar and M'Closkey (1998) identified collecting for the pet trade as a concern and found that local pet shop employees admitted to selling wild skinks from the region. Unfortunately, population-level impacts of this practice were difficult to quantify. It is also possible that various problematic species, such as feral cats, impact populations in ways that are not fully appreciated (Natural History Box: Five-lined Skinks in Portage County, Wisconsin).

Management: Extremely limited data on management strategies for this species in the Upper Midwest exist. Furthermore, this lizard's apparently specific habitat needs make implementation of truly effective management practices a challenge. However, general concepts associated with appropriate timing of prescribed fire application, management of open bluffsides and savannas, and maintenance of habitat structural variability and ground cover are considerations applicable to Common Five-lined Skinks (see Introduction to Conservation and Management of Wisconsin's Amphibians and Reptiles chapter). Moderate application of prescribed fire or light grazing in occupied woodlands could help maintain an appropriate density of trees and reduce encroachment of woody vegetation for optimal light penetration (see Natural History Box: A Long-Term Perspective of Common Five-lined Skink Habitat), although no research on the effectiveness of these tactics for Common Five-lined Skinks in Wisconsin exists. Work on the Southeastern Five-lined Skink (*Plestiodon inexpectatus*) in Florida suggested that application of prescribed fire every one or two years could be too intense and that a five-to-seven-year fire rotation resulted in higher abundances (Mushinsky 1992). Common Five-lined Skinks may also benefit from the selective cutting and burning of vegetation on and near rock outcrops. Eastern Red Cedar (*Juniperus virginiana*) and other plants can invade open rock outcrops and reduce the availability of basking areas (Lang 1982). Work in Illinois supports that this species will use corridors associated with riparian zones for dispersal in disturbed landscapes (Burbrink et al. 1998).

It could be particularly difficult for Common Five-lined Skink populations to rebound from perturbations (either random natural events or anthropogenic impacts) that result in large fluctuations in numbers or rapid population declines in a short period (see Populations subsection). As such, management that includes protection of suitable ground debris that Common Five-lined Skinks use as cover from predators, for overwintering, and for nesting could be especially important. For example, Hecnar and M'Closkey (1998) suggested that park maintenance staff at their study site removed downed logs to keep beaches free from debris for recreationalists. They believed this had negative impacts on the Common Five-lined Skink populations they studied. Given the importance of logs to this species, habitat managers at Common Five-lined Skink sites should promote the persistence of sufficiently large downed woody debris. Furthermore, this species would possibly benefit from the presence of rocks with large surface areas as cover, and such microhabitat features should be allowed to persist.

SPECIAL REMARKS

Explorer Jonathan Carver traveled through Michigan, Iowa, Wisconsin, and Minnesota from 1766 to 1768 and encountered what is most certainly *Plestiodon fasciatus*. Carver (1781) called it the "swift lizard" and wrote, "It is so remarkably agile that in an instant it is out of sight, nor can its movement be perceived by the quickest eye, so that it might more justly be said to vanish, than to run away." He went on to say that the "swift lizard" was said to have a "poisonous" bite, although it was not dangerous to people.

Prairie Skink

Plestiodon septentrionalis (Baird 1858)

Rori A. Paloski

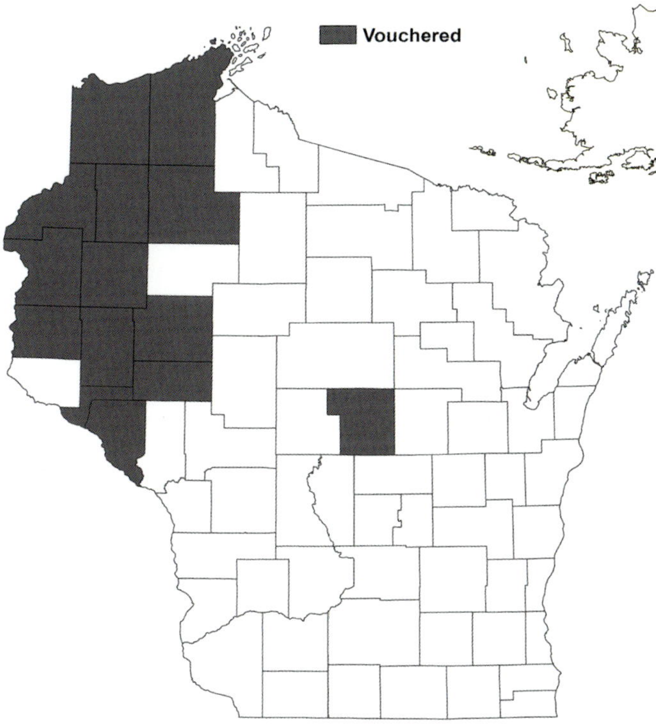

Adult Prairie Skink (Bayfield County; photo by J. Kapfer).

AT A GLANCE

Prairie Skinks are small, semifossorial lizards typically found in grasslands, savannas, barrens, and old field habitats in northwestern Wisconsin. They have a series of dark and light longitudinal stripes running down their back, and juveniles are easily recognized by their bright blue tail. Prairie Skinks are active throughout the day but spend a substantial amount of time under cover objects such as rocks and logs or underground during unfavorable weather conditions. They emerge from hibernation in May, and mating ensues shortly thereafter. Females lay their eggs in early July. Skinks are one of the few families of lizards that provide maternal care to incubating eggs, which typically hatch from late July to early August. Their diet consists of spiders, grasshoppers, and crickets, as well as a variety of other terrestrial arthropods. Prairie Skinks have a tail that is longer than their body when intact, but they are able to detach their tail to escape predation. The tail regenerates, but never to exactly the same length or coloration. In September, Prairie Skinks return to underground burrows, which act as winter refugia. Habitat degradation due to natural succession and the removal of ground cover is likely the largest threat to Prairie Skink populations. Prairie Skinks are currently a species of Special Concern and a Species of Greatest Conservation Need in Wisconsin.

SYSTEMATICS

Plestiodon septentrionalis (Prairie Skink) was originally described by Baird (1858) and reclassified as *Eumeces septentrionalis* by Cope (1875). Later work reverted *Eumeces* to *Plestiodon*, which changed the Wisconsin form back to *P. septentrionalis* (Brandley et al. 2005). Three subspecies are currently recognized: *P. s. septentrionalis* (Northern Prairie Skink), *P. s. obtusirostris* (Southern Prairie Skink), and *P. s. pallidus* (Pallid Skink), with the subspecies *P. s. septentrionalis* occurring in Wisconsin (Crother 2017). Based on mitochondrial DNA analyses, Fuerst and Austin (2004) suggested that *P. s. septentrionalis* and *P. s. obtusirostris* may be on separate evolutionary trajectories. *P. septentrionalis* is commonly recognized by Crother (2017) as the Prairie Skink but has been alternately and/or previously referred to as the Northern Prairie Skink, Black-banded Skink, Western Skink, Northern Skink, and Northern Lizard. Only one other lizard species in Wisconsin is a member of the genus *Plestiodon*, and that is *P. fasciatus* (Common Five-lined Skink; see Systematics and Classification chapter).

DESCRIPTION

Adults: Adults have smooth shiny scales and three wide beige or tan dorsal stripes separated by thin black stripes (Figure 177; Vogt 1981). These stripes extend from the head to the tail but become less pronounced on the tail (Breckenridge 1943). Three black lateral stripes separated by two thin white or cream-colored stripes are also present (Vogt 1981). The ventral surface of the Prairie Skink is a uniform beige, tan, or gray (Vogt 1981). Prairie Skink limbs are brown in color above and lighter in color below (Rori Paloski, WDNR, personal observation, Breckenridge 1943). Prairie Skinks in the Upper Midwest reportedly reach a maximum TL of 13–22 cm (5.1–8.7 in), with a maximum SVL of 7.0–8.2 cm (2.8–3.2 in; Breckenridge 1943, Vogt 1981). The mean SVL from a small number of Wisconsin specimens in the MPM and UWZM collections was 6.4 cm (2.5 in; $n = 15$). When intact, Prairie Skink tails are longer than the rest of the body. Although the tails of most museum specimens assessed were damaged and were not used to create a mean measure, the longest encountered was 11.9 cm

Figure 177. Adult Prairie Skink (Portage County; photo by R. Paloski).

Figure 178. Female Prairie Skink with eggs and nest under a limestone rock (Eau Claire County; photo by R. Hay).

Above, top: Figure 179. Juvenile Prairie Skink (Portage County; photo by R. Paloski).

Above, bottom: Figure 180. Male Prairie Skink with orange chin and lips during the breeding season (Yellow Medicine County, Minnesota; photo by A. B. Sheldon).

(4.7 in; UWZM 20388). The largest SVL of a museum specimen from Wisconsin measured was 7.6 cm (3 in, MPM 13559, collected in 1976 from Bayfield County).

Preadult stages: Eggs are small and elliptical in shape. They are a whitish or buff color and have a soft leathery shell (Figure 178; Breckenridge 1943, R. Paloski, personal observation). In Wisconsin, Vogt (1981) reported a mean egg size, presumably recorded immediately after egg deposition, of 0.8 × 1.3 cm (0.3 × 0.5 in). In Minnesota, Breckenridge (1943) determined that mean egg dimensions at the time of deposition were 0.8 × 1.34 cm (0.31 × 0.53 in), while Clarke (1955) reported slightly smaller mean egg dimensions from Kansas (0.73 × 1.20 cm [0.29 × 0.47 in]). Skink eggs absorb moisture and so become larger during incubation, and Breckenridge (1943) reported mean dimensions of 1.1 × 1.8 cm (0.42 × 0.71 in) for eggs immediately prior to hatching. Juvenile Prairie Skinks have stripe patterns similar to adults, but with an overall darker coloration and seven thin light dorsal and lateral stripes that become less pronounced with age. Hatchlings have a bright blue or bluish-gray tail (Figure 179), which fades with age and is typically absent by the time the skink reaches a body length of 5 cm (2 in; Breckenridge 1943).

Variation: No significant geographic variation in color, pattern, or size among Wisconsin Prairie Skink populations is known. Prairie Skinks exhibit sexual dimorphism during the breeding season, when the chin, lips, and throat of adult males turn orange or reddish orange (Figure 180; Breckenridge 1943, Nelson 1963, Vogt 1981). In Minnesota, reports indicate that the head size of adult male Prairie Skinks averaged 0.12 cm (0.05 in) wider and 0.11 cm (0.04 in) longer than adult females (Breckenridge 1943).

Diagnosis: Prairie Skinks have smooth dorsal and tail scales with large overlapping ventral scales that are not plate-like. They also possess one postlabial scale and light pronounced dorsolateral stripes. Among lizards in Wisconsin, Prairie Skinks are easily distinguished from Slender Glass Lizards (*Ophisaurus attenuatus*), which have no legs. Prairie Skinks could be confused with either Six-lined Racerunners (*Aspidoscelis sexlineata*) or Common Five-lined Skinks. Prairie Skinks occupy similar habitat types and overlap in range with Six-lined Racerunners in portions of Wisconsin. Although Six-lined Racerunners have a vaguely similar TL and stripe pattern as Prairie Skinks, Six-lined Racerunners have duller rougher scales and possess thin light-colored stripes that are typically yellow or yellow green in color (Vogt 1981). Furthermore, while Prairie Skinks have long tubular bodies with short legs, Six-lined Racerunners have proportionally larger hindlegs and feet and a thinner and more whip-like tail.

Prairie Skinks most closely resemble Common Five-lined Skinks in pattern, size, and scale type. However, their known ranges currently only overlap in Polk, Chippewa, Eau Claire, and Portage Counties (although see discussion of the Chippewa County Common Five-lined Skink voucher in the Common Five-lined Skink [*Plestiodon fasciatus*] species account). Although Common Five-lined Skinks can occupy similar habitat as Prairie Skinks, they prefer locations where tree canopies are slightly more enclosed than Prairie Skinks (Vogt 1981). The two species have similar stripe patterns as adults, but unlike adult Prairie Skinks, adult Common Five-lined Skinks have two rather than three dark lateral stripes, thin light dorsal stripes rather than broad dorsal stripes (except in older adults), and a central light dorsal stripe that typically, in younger adults, forms an obvious V or Y on the head, just anterior to the eyes (Vogt 1981, LeClere 2013). In addition, Prairie Skinks have only one postlabial scale, while Common Five-lined Skinks have two (Vogt 1981).

Eggs and juveniles of these species can be particularly difficult to differentiate. Because the eggs are very similar in appearance, the best method to differentiate them could be the geographic location and habitat where the eggs are encountered (see Distribution and Habitat section). Hatchlings and juveniles of both species have a dark background color, thin light stripes, and a blue tail. However, Common Five-lined Skinks typically only possess five light stripes that become blue stripes on the tail, whereas Prairie Skinks have seven light stripes that merge together into a uniform blue tail.

DISTRIBUTION AND HABITAT

Global distribution: Prairie Skinks are primarily found in the central US, ranging from northern Minnesota and northwestern Wisconsin through portions of the Great Plains to eastern Texas. A disjunct population also occurs in southern Manitoba, Canada (COSEWIC 2004).

Distribution in Wisconsin: Published Prairie Skink records are primarily concentrated in the northwestern quadrant of the state. A disjunct population has also been reported from Portage County in central Wisconsin (Orlofske and Anderson 2005). Adult and juvenile skinks have been observed regularly at the Portage County site since the initial sighting in 2004, indicating that a breeding population is present (Natural History Box: Prairie Skinks in Central Wisconsin).

Localities of possible occurrence: Hoy (1883) described this species as "not uncommon as far north as Lake Winnebago" but provided no additional locality or specimen information, and no confirmed records south of Lake Winnebago exist. Regardless, a Walworth County record was reported by Higley (1889), although subsequent authors have ignored this (Walley 1962). Some publications also include Clark County as part of the Prairie Skink's range (Pope and Dickinson 1928, Dickinson 1949). However, this might be an erroneous interpretation of a species list for Clark County compiled by Schmidt (1926), who referenced Prairie Skinks as a hypothetical species in Clark County because they occurred in adjacent Chippewa County. Walley (1962) also reported Prairie Skink records for Marinette and Waukesha Counties. Yet Vogt (1981) examined the Marinette and Waukesha specimens and determined that the Marinette County specimen, which Walley (1962) listed as MPM 2653 (note that a recent museum search revealed that this specimen no longer exists), was a Common Five-lined Skink. Vogt (1981) further determined that although the purported Waukesha County specimens were Prairie Skinks, they were actually collected in Washburn County (UWZM 9800–01).

NATURAL HISTORY BOX: Prairie Skinks in Central Wisconsin

In Wisconsin, the typical range of the Prairie Skink consists of swaths along the St. Croix and Mississippi Rivers in the northwestern portion of the state where dry, sandy soils exist in barrens and grasslands. Knowing the typical habitat and range for the species made it unusual that a breeding population of Prairie Skinks had been discovered in Portage County several years ago. On a sunny day in late August, several of us explored along a river where the Prairie Skinks had been documented. Not ten minutes into the walk, we caught a glimpse of a small dark body on the riprap. This was the first of about ten individuals observed that day. Later in the morning we began to see both juveniles (identifiable by the blue tail) and adults running along the warm rocks and cement. One or two individuals had broken tails, indicating escape from either a predator or an inexperienced human lizard hunter. We were not the only skink hunters out that day, as we were approached by a father and son with nets and a few pointers about the best spots to look. We observed several species of snakes in addition to the skinks, suggesting that this small strip of land near the river may be a haven for reptiles and amphibians.

Heather M. Kaarakka, WDNR

In Wisconsin, Prairie Skinks are commonly associated with the northwest sands ecological landscape, as well as portions of the forest transition, western prairie, and western coulees and ridges ecological landscapes (WDNR 2012). Given the geographic distribution of these ecological landscapes, Prairie Skink records in adjacent Minnesota counties (i.e., Dakota, Washington, Goodhue, Wabasha, and Winona; Moriarty and Hall 2014), and a recent discovery in Pepin County (White et al. 2020b), it is possible that this species could be encountered in Pierce and even Trempealeau Counties. It is also interesting to note that a pre-1950 record exists from extreme northeastern Iowa in Allamakee County, which is adjacent to Vernon and Crawford Counties in Wisconsin (LeClere 2013).

Habitat: Prairie Skinks typically associate with earlier successional habitats (Fitch 2006b), which can include dry cliffs, prairies, savannas, barrens, and old fields (Vogt 1981, Pitt 2001; see Amphibians and Reptiles in Wisconsin Plant Communities chapter). They also use open areas along streambanks, riverbanks, lakeshores, road embankments, and excavation areas (Clarke 1955, Nelson 1963, Vogt 1981). In Nebraska, Prairie Skinks were reported from woodland edges and near major river systems (Ballinger et al. 2010, Geluso and Harner 2013). This pattern appears to hold true in some regions of Wisconsin, where Prairie Skinks are locally common along portions of the St. Croix and Chippewa Rivers, as well as their tributaries. Prairie Skinks are further reported from urban/suburban locations in Wisconsin, Kansas, and Manitoba (Figure 181; Clarke 1955, Krause Danielsen et al. 2014, R. Paloski, personal observation). A Minnesota study found that the highest densities of Prairie Skinks occurred in old field habitats and were positively correlated with field age (i.e., the number of years after the field was abandoned by humans; Pitt 2001).

Prairie Skinks are semifossorial lizards that require sandy or loamy soils for burrowing, nesting, and overwintering (Breckenridge 1943). Adequate ground cover for shelter and thermoregulation also appears to be an important habitat component (Krause Danielsen et al. 2014). In Minnesota, Prairie Skinks have been reported from under a variety of cover objects, including sticks, stones, rocks, bark, boards, tin, and paper (Breckenridge 1943). In Wisconsin they have also been observed under pieces of plastic and sheets of plywood (R. Paloski, personal observation). In fact, research in Manitoba found that the presence of artificial

Figure 181. Edge of a suburban property used by Prairie Skinks in Eau Claire County (photo by R. Paloski).

ground cover was an important microhabitat predictor for this species (Larkin 2011, Krause Danielsen et al. 2014), while the presence of leaf litter was also important (Krause Danielsen et al. 2014). Furthermore, Prairie Skinks showed no preference for specific vegetation types during habitat research, which suggests that habitat structure is more important than vegetation type (Krause Danielsen et al. 2014).

Few reports of nesting and overwintering habitats in Wisconsin exist. Authors from other states have determined that nest cavities are typically constructed in sandy or loamy soils and associated with various types of cover, including rocks, wood, bark, cement, and scraps of tin (Breckenridge 1943, Nelson 1963, Somma 1990, Frese 2003). Nest cavities may be located directly under cover objects (Figure 178; Breckenridge 1943, Frese 2003) or at depths of 3–9 cm (1.2–3.5 in; Breckenridge 1943, Nelson 1963) with cavity diameters of 3–10 cm (1.2–3.9 in; Somma 1987a, Somma 1990, Frese 2003). Prairie Skinks reportedly overwinter in self-excavated or abandoned burrows that extend below the frost line in well-drained soils (see Activity section). Strong nesting and overwintering site fidelity has been observed in some individuals, while other individuals that were relocated within the same study site did not return to their original capture location (Nelson 1963).

REPRODUCTION AND DEVELOPMENT

Prairie Skinks breed once per year in the spring (Fitch 1985). Breeding in Wisconsin typically occurs in May, shortly after emergence from overwintering (Vogt 1981). In nearby Minnesota, breeding has been recorded in late May and early June (Breckenridge 1943). Males actively pursue females prior to copulation but do not appear to defend breeding territories (Breckenridge 1943, Nelson 1963). In Wisconsin, egg deposition has been observed in early July (Vogt 1981), while in Minnesota egg deposition has been observed in late June and early July (Breckenridge 1943). Vogt (1981) reported a mean clutch size of 4–6 eggs in Wisconsin. Clutch sizes documented from other states include 3–13 eggs in Minnesota (Breckenridge 1943, Nelson 1963) and 4–18 eggs for captive individuals from Nebraska (Somma 1987b). The number of eggs laid by a female may be correlated with the female's SVL (Somma 1987b) or related to annual environmental conditions (COSEWIC 2004). Fitch (1985) also suggested that an increase in clutch size occurs along a north–south gradient across the Prairie Skink's North American range (see Description section for information on egg size). Female Prairie Skinks exhibit strong nest-site fidelity (Nelson 1963) and may nest communally (COSEWIC 2004).

Skinks are in one of the few lizard families that provide maternal care to their eggs during incubation (Figure 178; Breckenridge 1943, Somma 1987a). In addition to aggressive defense against would-be predators, brooding females regulate moisture levels, an important factor in Prairie Skink egg development, by changing the size of the nest cavity, moving eggs to a slightly different substrate, and maintaining physical contact with the eggs (Breckenridge 1943, Somma 1985, Somma 1989, Somma and Fawcett 1989). Much like clutch size, the duration of egg incubation appears to vary, which could reflect geographical differences in environmental conditions along a north–south gradient (Nelson 1963). In Minnesota, Breckenridge (1943) estimated that the incubation period ranged from 40 to 52 days (hatching occurring in late August). However, Nelson (1963) reported a slightly shorter incubation period of ca. 30–35 days (hatching occurring in late July or early August). Frese (2003) observed an incubation period of <30 days in southern Iowa (hatching occurring in mid-July). In Wisconsin, hatching likely occurs in late July or August. After eggs hatch, most females remain with the hatchlings for two to three days (Somma 1985, 1987a).

Although Fitch (1985) stated that the time from hatching to sexual maturity is likely three years in northern populations, the time to maturity recorded in Minnesota was only two years (Breckenridge 1943, Nelson 1963). Prairie Skink longevity is not well-studied. Nelson (1963) estimated a mean life span of 2–4 years, although individuals of at least 4.75 and 7 years have been recorded in the wild (Nelson 1963, COSEWIC 2004).

ACTIVITY

The annual activity period for Prairie Skinks in Wisconsin is typically May through September (Vogt 1981). However, in warmer years they are likely active as early as mid-April and as late as October, similar to findings from Nebraska (Ballinger et al. 2010) and Minnesota (Breckenridge 1943, Nelson 1963). There is apparently variation in when emergence from winter dormancy occurs by sex and age, with males emerging before females and juveniles (Breckenridge 1943, Nelson 1963).

During the active season, Prairie Skinks are diurnal but spend portions of the day under cover or underground when weather conditions are not suitable (Nelson 1963, Vogt 1981). The majority of Prairie Skink sightings in Minnesota occurred on warm sunny days from 10:00 a.m. to 3:00 p.m., with peak observations from 2:00 to 3:00 p.m. Nelson (1963) reported that body temperatures ranged from 22.4°C to 35.0°C (72.3°F to 95.0°F) in active Prairie Skinks ($\bar{x}$ = 29.7°C [85.5°F]), noting that the body temperatures were typically several degrees warmer than the surrounding air or substrate temperature. Prairie Skinks spend nights below cover or underground (Nelson 1963).

Data on Prairie Skink movement patterns are sparse, although the few existing reports indicate high variability in daily movements (Nelson 1963) and annual movements (Breckenridge 1943). For example, Nelson (1963) determined that some individuals remained within a 31 m (100 ft) diameter area over the course of several years, while others traveled throughout a 116 m (380 ft) diameter area in only a few weeks. Nelson (1963) also recorded an individual movement of 47.2 m (155 ft, straight-line distance) in four hours, whereas other lizards moved only minimal distances and then only on occasion. Territoriality, or defense of space, is not documented, and Breckenridge (1943) observed multiple skinks occupying a single burrow.

Prairie Skinks overwinter singly or communally below the frost line in self-constructed or existing but abandoned burrows (Scott and Sheldahl 1937, Breckenridge 1943, LeClere 2013). Skinks may lengthen their summer burrows for overwintering (Vogt 1981) or choose a different location in which to dig hibernacula. Distances of 1.5–112.8 m (5–370 ft) have been recorded between summer and overwintering burrows (Nelson 1963). The depth at which Prairie Skinks overwinter is variable, and they will move vertically in the soil profile during winter to find suitable conditions. Nelson (1963) determined that some individuals initially created shallow burrows but moved deeper as the winter progressed, some created a deep burrow initially and gradually moved closer to the soil surface as the end of winter approached, while others remained at a constant moderate depth throughout the winter. A maximum overwintering depth of 0.7 m (2.3 ft) was recorded in the Nelson (1963) study, but other reports indicate that they will hibernate at greater depths. Scott and Sheldahl (1937) reported that workers in Iowa uncovered 52 communally overwintering skinks in mid-February at a depth of 1.37 m (4.5 ft), noting that a "soft web-like material surrounded the mass and adhered to the bodies of the animals." Breckenridge (1943) also provided accounts of overwintering Prairie Skinks exposed by work crews in Minnesota. Groups of three and eight skinks were uncovered at the end of October at a depth of 0.9 m (3 ft), and a single Prairie Skink was uncovered mid-January at a depth of 0.6 m (2 ft).

PREY AND PREDATORS

Prairie Skinks eat a variety of arthropods. In Minnesota, their diets consisted of primarily spiders

(29.5%) and grasshoppers or crickets (27.3%), as well as a variety of other terrestrial arthropods (28%), including leafhoppers, beetles, and caterpillars (Breckenridge 1943). The diet of juvenile Prairie Skinks differs only slightly from that of adults (Nelson 1963), which is likely related to smaller juvenile head size limiting the size of prey consumed. Nelson (1963) reported that the juvenile diet contained more spiders than that of adults (46%), while the remaining prey were divided among grasshoppers or crickets (15%), leafhoppers (13%), and flies (15%). Breckenridge (1943) and Nelson (1963) both remarked on the absence of ants in the diet of the Prairie Skinks they studied. Vertebrate prey are reported infrequently, but consumption of fish was documented in Nebraska (Somma 1991), as was a single case of cannibalism in Minnesota (Breckenridge 1943). Prairie Skinks continuously search for prey when active (Vogt 1981). Investigations of captive individuals support that they can simultaneously handle and search for multiple prey items (Cochran 2008c; Natural History Box: Simultaneous Handling and Searching for Prey). Antagonistic behavior has also been reported between feeding individuals in a captive situation but has not been observed in the wild (Cochran 1983b).

In Minnesota, Plains Hog-nosed Snakes (*Heterodon nasicus*), Loggerhead Shrikes (*Lanius ludovicianus*), Northern Harriers (*Circus cyaneus*), American Kestrels (*Falco sparverius*), Barred Owls (*Strix varia*), Thirteen-lined Ground Squirrels (*Ictidomys tridecemlineatus*), Raccoons (*Procyon lotor*), and American Badgers (*Taxidea taxus*) are documented predators (Blakemore 1940, Giles 1940, Breckenridge 1943, Lampe 1982). A variety of other birds, mammals, and snakes likely also prey on Prairie Skinks (Johnson 2000b, Ballinger 2010).

Prairie Skinks can dislocate their tail to distract and avoid predators (Figure 182; Breckenridge 1944). This tactic is apparently employed frequently, as Nelson (1963) noted seeing a much smaller number of skinks with intact tails than those with tails in some stage of regeneration. The tail will regenerate over several weeks but will be shorter in length and duller in color, and it will lack the pattern and scalation of the original tail (Breckenridge 1944, Vogt 1981, LeClere 2013). In addition to caudal autotomy, Prairie Skinks will often attempt to quickly flee or escape into underground refugia when confronted by perceived threats (Nelson 1963, R. Paloski, personal observation). Prairie Skinks also occasionally retreat into the water, if available, and swim to escape predation (Nelson 1963).

NATURAL HISTORY BOX: Simultaneous Handling and Searching for Prey

"As predators of small invertebrates that typically consume multiple prey during the course of the day, skinks of the genus *Eumeces [Plestiodon]* may provide an example of a group for which this issue [simultaneous handling and searching for prey] is relevant. I have observed simultaneous handling and search by captive specimens of two species of *Eumeces [Plestiodon]*. A Northern Prairie Skink . . . captured a small moth, then noticed and captured a larger moth without dropping the first. . . . Because they consume prey of a wide range of sizes and a wide range of handling times and they often consume multiple prey during the course of a feeding bout, the Five-lined Skink and Northern Prairie Skink may be good examples of predators for which the ability to simultaneously handle and search for prey might be especially beneficial."

Excerpt from Cochran (2008c)

CONSERVATION

Status: Globally, the Prairie Skink is listed as a Least Concern species by the IUCN. It is considered a species of Special Concern and a Species of Greatest Conservation Need in Wisconsin by the WDNR. While this species is not listed federally in the US, it is listed as Endangered in Canada (COSEWIC 2004).

Figure 182. Juvenile Prairie Skink with detached tail (Bayfield County; photo by R. Hay).

Populations: Due to their semifossorial nature, Prairie Skinks are often difficult to detect (Powell and Russell 2007). As a result, limited population information exists for this species, and none is known from Wisconsin. Vogt (1981) anecdotally reported that Prairie Skinks are locally abundant and found in high densities at sites where they occurred in the state. Nelson (1963) provided a conservative density estimate of 26/ha (10.5/ac) at his Minnesota study site. Pitt (2001) further estimated adult densities of 58–206/ha (23.5–83.4/ac) in Minnesota. In addition to determining density, Nelson (1963) determined that the sex ratio of his study population was ca. 1M:1F. He also reported that juveniles outnumbered adults, although only a small number of marked hatchlings were ever recaptured, possibly indicating a high juvenile mortality rate.

Threats: Habitat loss and degradation are likely the largest threats to Prairie Skinks. As such, the conversion of natural open and dry habitats (such as prairies) to residential developments and agriculture has likely led to population declines in Wisconsin (see Past, Present, and Potential Future Landscapes of Wisconsin chapter). In addition to habitat loss, degradation of habitat due to factors that range from unchecked vegetative succession to ingrowth of invasive plant species are conservation concerns (see Introduction to Conservation and Management of Wisconsin's Amphibians and Reptiles chapter). Given their association with open habitats dominated by short herbaceous vegetation, encroachment of these landscapes by woody vegetation is a serious concern. While a small to moderate number of shrubs and trees can provide cover for thermoregulation and shelter, a substantial increase in these woody plants due to vegetative succession can negatively impact Prairie Skink populations. For example, the Fitch Natural History Reservation in Kansas experienced a noticeable increase in woody vegetation in the late 1900s, with the Prairie Skink being noted as one of the first species to disappear as this succession progressed (Fitch 2006b). In Minnesota, Pitt (2001) found 77% higher Prairie Skink densities in old fields that were managed with prescribed fire used to offset woody vegetation encroachment than in those areas where habitat was not actively managed. Exotic plant species that often invade prairies in Wisconsin, such as Spotted Knapweed (*Centaurea maculosa*), could also have negative impacts on Prairie Skinks, but their effects on skinks are largely unknown in the Upper Midwest. In Canada, Prairie Skink declines were documented in areas colonized by the invasive Leafy Spurge (*Euphorbia esula*; COSEWIC 2004). However, additional research has determined that if ground cover objects are present, abundances of Prairie Skinks do not differ between habitat patches dominated by Leafy Spurge compared to patches dominated by native vegetation (Larkin 2011).

Management: Prairie Skinks will likely benefit from the application of techniques that are employed to manage open habitats such as prairies (see Introduction to Conservation and Management of Wisconsin's Amphibians and Reptiles chapter). Often, these techniques are used to offset the excessive ingrowth of woody vegetation in naturally open habitats, such as appropriately timed application of prescribed fire, mowing, selective brush or tree cutting, grazing, and spot herbicide treatments. Restriction of these activities to the lizard's inactive season, typically between October and April in Wisconsin, will reduce the likelihood of management-related mortality (e.g., accidentally killing lizards with prescribed fires or due to lawnmower strikes). If work must be conducted during the active season, restricting activities to cold or cool cloudy days could minimize management-related mortality. While the impacts of herbicides on reptiles is not well studied, herbicides that are considered safe for amphibians may also be safe for reptiles and are recommended.

In addition to cessation of woody vegetation encroachment, promotion of ground cover objects is likely a beneficial habitat management strategy. This ground cover, including logs, rocks, smaller woody debris, and artificial cover objects, is critical for Prairie Skinks, as it provides refuge and thermoregulatory opportunities, but it is also important for females to brood their eggs (Breckenridge 1943). In Manitoba, the presence of artificial cover objects was the most important microhabitat predictor of Prairie Skink abundance during the active season (Larkin 2011). Krause Danielsen et al. (2014) also found that Prairie Skinks in Manitoba were significantly more abundant in areas with artificial ground cover and leaf litter than in areas without them. Given that the presence of specific types of vegetation does not appear important to this species (Krause Danielsen et al. 2014), management of habitats to promote a suitable structure of associated vegetation communities and ground cover may be of primary importance. Although the impact that invasive vegetation has on this species is not fully understood, the removal of invasive vegetation may benefit many plant and animal species that use open habitats, including the Prairie Skink.

Prairie Skinks appear to display some level of tolerance for suburban environments (Krause Danielsen et al. 2014), perhaps because these areas are often open (due to the presence of manicured lawns) and possess anthropogenic ground debris (e.g., wood piles, discarded metal sheets, rocks around campfire pits). In Wisconsin, for example, Prairie Skinks are often observed in suburban areas (R. Paloski, personal observation). In Manitoba and Kansas, suburban areas have also been reported as suitable habitat for this lizard (Clarke 1955, Krause Danielsen et al. 2014). Prairie Skinks could persist in suburban or exurban landscapes where impervious surfaces are limited, suitable natural buffers are maintained, and homeowners make an effort to avoid accidentally killing lizards during yardwork. Regardless, it is important to note that limited data exist on the population-level impacts that Prairie Skinks experience in suburban or exurban landscapes. Therefore, it is difficult to assess if Prairie Skinks can flourish in these landscapes or if individuals can simply survive there while experiencing limited reproductive success that leads to low population viability.

Six-lined Racerunner

Aspidoscelis sexlineata (Linnaeus 1766)

Joshua M. Kapfer and Donald J. Brown

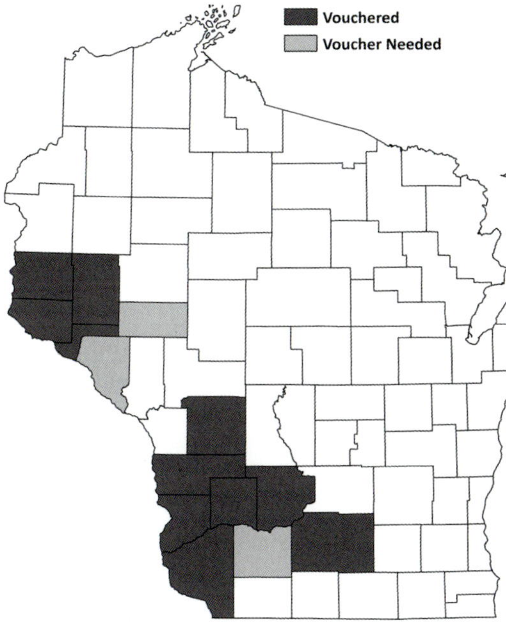

Adult Six-lined Racerunner (Sauk County; photo by T. Muehlfeld).

AT A GLANCE

Six-lined Racerunners are small lizards with six yellowish stripes running longitudinally along the length of their body. They have a long and thin tail, which can be voluntarily broken off while still wriggling to distract would-be predators and afford the lizard a chance at escape. Six-lined Racerunners are most common along rocky open bluffsides and in sandy savannas and prairies in western Wisconsin. In most years, they are only active from mid-May to early September. Adults breed in May, and females lay their eggs from mid- to late June, which hatch in August. Occasionally referred to as Prairie Streaks, Six-lined Racerunners are quick and capable of attaining speeds up to 29 kph (18 mph) when pursuing insect prey or fleeing from threats. George Knudsen (1956b) explained this best when he described their response to a potential human captor: "The small head tilts when you are yet a long way off and a beady eye trains on your every move. You are allowed to approach quite closely and just about the time you think you'll make your grasp Mr. Lizard takes off." They are considered a species of Special Concern and a Species of Greatest Conservation Need by the WDNR.

SYSTEMATICS

Aspidoscelis sexlineata (Six-lined Racerunner) was historically included in the genus *Cnemidophorus* (Duellman and Zweifel 1962). Based on mitochondrial DNA, allozymes, and morphology, Reeder et al. (2002) determined that North American *Cnemidophorus* species were a distinct group (i.e., clade). They proposed using the genus name *Aspidoscelis* for this clade, which has been accepted (Crother 2017). Three *A. sexlineata* subspecies are currently recognized: *A. s. sexlineata* (Eastern Six-lined Racerunner), *A. s. stephensae* (Texas Yellow-headed Racerunner), and *A. s. viridis* (Prairie Racerunner; Crother 2017), which is the subspecies found in Wisconsin. This species is known to hybridize with other *Aspidoscelis* species across its range (e.g., Manning et al. 2005, 2009, Walker et al. 2013). It is the only member of the genus *Aspidoscelis* found in Wisconsin.

DESCRIPTION

Adults: Six-lined Racerunners have a green or brown base color overlaid by six yellow longitudinal stripes, which alternate with dark bands that run down their back (Figure 183). The subspecies present in Wisconsin has an additional thick brown vertebral stripe that may be single or narrowly divided (Trauth and McAllister 1996). The underside is gray or white. Males develop a blue chin during the breeding season, and their underside may have a blue tinge throughout the year (Figure 184). The tail is brown or gray and has rough scales (see Figure 183).

Above, top: Figure 183. The large back legs and proportionally long thin tail of the Six-lined Racerunner (Sauk County; photo by T. Muehlfeld).

Above, bottom: Figure 184. Adult male Six-lined Racerunner displaying a blue-tinged chin (Sauk County; photo by J. Kapfer).

Six-lined Racerunners are small lizards, with a slim body, but proportionally large hindlegs and feet (see Figure 183). They also possess a long tail that accounts for ca. 70% of TL (Ballinger et al. 1979). Vogt (1981) reported that adult SVL in Wisconsin was 6–8 cm (2.4–3.1 in), and TL was up to 24 cm (9.4 in). This is corroborated by Stein (1995), who reported that mean adult SVL was 6.5 cm (2.6 in, $n = 163$) for a population in Wisconsin. Brewster (1984), on the other hand, reported a mean SVL of only 5.4 cm (2.1 in, $n = 30$) but did not distinguish among age classes. Only a handful of museum specimens from Wisconsin were available for investigation, and only individuals with intact tails were included in tail length measures. Of these, mean adult SVL was 6.6 cm (2.6 in, $n = 16$), and mean adult tail length was 13.0 cm (5.1 in, $n = 11$; note that only specimens with SVL of 5.9 cm (2.3 in) or higher were considered adults when taking measurements, based on findings in Brewster (1984). Furthermore, mean adult head length and width were 1.50×0.97 cm (0.60×0.38 in, $n = 16$). The only published report of weight for wild-caught Wisconsin individuals is from Brewster (1984), who reported that mean weight was 4.9 g (0.2 oz, $n = 30$).

Preadult stages: Eggs are elliptical and ca. 1.7×0.9 cm (0.7×0.4 in; Vogt 1981). Hatchlings appear similar to adults, but their tail is blue, and their stripes are more vivid (Figure 185; Fitch 1958b, Duellman and Zweifel 1962). These features fade rapidly as juveniles grow (Fitch 2003a). Harding (1997) stated that hatchlings have an SVL of ca. 3 cm (1.2 in), and Fitch (1958b) found that hatchlings in Kansas had a mean SVL of 3.2 cm (1.3 in). One-year-old individuals in South Carolina were ≤6 cm SVL (2.4 in; Bellis 1964). Mean subadult or juvenile SVL of measured museum specimens from Wisconsin was 3.7 cm (1.5 in, $n = 16$), and mean tail length was 6.3 cm (2.5 in, $n = 16$), while mean head length and width were 1.0×0.6 cm (0.4×0.2 in, $n = 16$).

Figure 185. Vivid stripes and bright blue tail of a juvenile Six-lined Racerunner (Sauk County; photo by J. Kapfer).

Variation: Adults are sexually dimorphic, with males having a larger head than females and a blue chin and belly during the breeding season (see Figure 184). However, males and females do not differ substantially in SVL (Fitch 1958b, Bellis 1964, Warner 2000). Variation exists for many of the species' morphological characteristics across its continental distribution (see Duellman and Zweifel [1962] and Trauth [1980] for detailed information on morphological variation).

Diagnosis: Six-lined Racerunners can be identified by the presence of granular dorsal scales, large rectangular ventral scales, and keeled tail scales. They could be confused with Common Five-lined Skinks (*Plestiodon fasciatus*) and Prairie Skinks (*P. septentrionalis*) in Wisconsin. However, both skinks have smooth shiny scales and dark longitudinal stripes. Both skinks also have short limbs, whereas the hind limbs and feet of the Six-lined Racerunner are comparatively large (see Figure 183). Six-lined Racerunners should be easily distinguished from Slender Glass Lizards (*Ophisaurus attenuatus*) because the latter do not possess legs. The eggs of Six-lined Racerunners can be similar in size to those of both skink species in Wisconsin, as well as smaller egg-laying snakes. Review of the geographic location and habitat where eggs were encountered, as well as egg shape and clutch size, could help distinguish the eggs of this species from others.

DISTRIBUTION AND HABITAT

Global distribution: Six-lined Racerunners occur sporadically in the northern US from Maryland west to southeastern Wyoming but are distributed throughout the southeastern US as far west as Texas and into extreme northeastern Mexico (Pérez-Ramos et al. 2010). Populations in Minnesota and Wisconsin are potentially connected to the broadly distributed southern populations through a fairly narrow band of suitable habitat along the Mississippi River (Trauth and McAllister 1996). There is currently a small isolated population in the Lower Peninsula of Michigan (Harding 1997).

Distribution in Wisconsin: Published county records are mostly concentrated in western Wisconsin along the Mississippi and Wisconsin Rivers. Of these, vouchers could not be found for Buffalo, Eau Claire, and Iowa Counties. Most published county records are at least partially within the Driftless Area of Wisconsin (Paull and Paull 1977), although Pierce and St. Croix Counties are exceptions. Like that of several Wisconsin species, the Six-lined Racerunner's current state distribution may be the result of historic migrations north, following the rivers and receding glaciers into Wisconsin from the south (see review in Past, Present, and Potential Future Landscapes of Wisconsin chapter). Its current concentration in the Driftless Area may be further explained by the abundance of crevices in the exposed rocky hillsides of the region that can provide desirable winter refugia. This species also prefers sandy soils for burrowing and open prairie or savanna (see Habitat subsection), which are common in the Driftless Area and many adjacent counties.

Localities of possible occurrence: Trempealeau and La Crosse Counties are obvious gaps in this lizard's Wisconsin distribution. Although Six-lined Racerunners are often associated with sites immediately adjacent to the Mississippi and Wisconsin Rivers, they are also documented in several counties that are removed from the rivers, including Dunn, Eau Claire, and Monroe (note that a voucher does not currently exist to support the Eau Claire County record and also see Natural History Box: Six-lined Racerunners in Monroe County). Given this, they may yet be found in additional counties with appropriate habitat that are not along the rivers. For example, no records exist for Jackson County, which is in the Driftless Area and lies between Eau Claire and Monroe Counties. This species is also undocumented in Lafayette County, portions of which are in the Driftless Area and which is adjacent to the occupied counties of Dane, Grant, and Iowa (although the latter does not have an associated vouchered specimen).

Several prairie reptile species that coexist with the Six-lined Racerunner in other locations are known in either Jackson or Lafayette (e.g., the Gophersnake [*Pituophis catenifer*] and the North American Racer [*Coluber constrictor*]), which supports that suitable habitat for this lizard could be present in both counties. Along this line of reasoning, it is plausible that Six-lined Racerunners are present in the open sandy habitats of southern Green and Rock Counties. This is particularly true given the documented populations of other rare prairie reptiles in Rock County (e.g., the Ornate Box Turtle [*Terrapene ornata*] and the North American Racer). However, there is a lack of exposed rocky bluffs in south-central Wisconsin, and repeated surveys of several large protected wildlife areas in Rock County during 2013 did not yield Six-lined Racerunners (Joshua Kapfer, UW–Whitewater, unpublished data).

Habitat: Information on Six-lined Racerunner habitat in Wisconsin is limited and mostly observational. Dickinson (1949) and Knudsen (1956b) mentioned that racerunners frequently used the steep hillsides near the Mississippi River in western Wisconsin. Vogt (1981) reported them on hillsides "above the river floodplain" and also noted them in dry or mesic prairie habitats that are open, grassy, or sandy. Kapfer and Pauers (2006) frequently observed this species in oak savanna (58.9% of observations), despite this habitat type comprising only 7% of their study site in western Wisconsin. Observed habitat associations in Wisconsin are similar to those from western and southern states (reviewed by Fitch 1958 and Etheridge 1983). Reports from Kansas (Fitch 1958b) and Arkansas (Paulissen 1988b) indicate that this species uses microhabitats of limited underbrush, short vegetation, and exposed sand. Similarly, Brewster (1984) reported that individuals from a Wisconsin population were often found in prairie grasses. In southeastern Minnesota, Six-lined Racerunners were documented using overhanging rock ledges as thermal refuges (Cochran and Schmitt 2016).

NATURAL HISTORY BOX: Six-lined Racerunners in Monroe County, Wisconsin

The factors that influence Six-lined Racerunner biogeography in Wisconsin could be more complex than the presence of certain soils and vegetation. For example, to our knowledge only one vouchered specimen exists from Monroe County (MPM VZP266), where suitable habitat and soils appear to be somewhat widespread. This photographic specimen was collected on the grounds of the Fort McCoy US Army Installation, but the circumstances surrounding its collection bring the specimen's origin into question (Timothy Wilder, US Army Fort McCoy wildlife biologist, personal communication). The lizard was captured on a paved road among military warehouses that held shipments of materials from around the state and country (T. Wilder, personal communication). This, coupled with the fact that this species has not been vouchered since then in Monroe County, suggests the specimen could have been an accidental translocation from outside of the area. Regardless of whether this species occurs elsewhere in Monroe County, the fact that they are not often observed on the grounds of Fort McCoy, and similar areas nearby, is puzzling. This property is located within the Driftless Area, it is a large, mostly undeveloped site (>22,257 ha [55,000 ac]), and it contains an abundance of seemingly suitable habitat. This information highlights the need for more data to further clarify the extent of this species' distribution in Wisconsin. Future surveys focused on counties with suitable habitat adjacent to known populations may shed further light on this subject.

Joshua M. Kapfer, UW–Whitewater

Wisconsin individuals are typically encountered on the ground, but in Arkansas they have been observed in bird nest boxes at heights of 3.0–4.5 m (9.8–14.8 ft; Caster 1994). Six-lined Racerunners are rarely reported in aquatic habitats, although they may seek refuge from human surveyors in wetlands, ponds, or puddles within preferred savanna and prairie (Stille 1947, Trauth et al. 1996, Gossen and Cochran 2013).

Past authors have noted an association with sandy soils in Wisconsin (Vogt 1981, Brewster 1984, Kapfer and Pauers 2006). Furthermore, Fitch (1958b) indicated that their "well developed and powerful forelegs with long, powerful claws" made them well-suited for digging (Figure 186). The combination of limbs adapted for digging and an inclination for sandy soils facilitates burrowing to create temporary shelters, nests, and winter hibernacula (Hardy 1962, Etheridge et al. 1983). Edgren (1955a) inspected temporary Six-lined Racerunner burrows in the sand dune region of northern Illinois that were short, shallow, and tubular in shape. These burrows were nearly straight and about 1.5 times the length of the lizard that dug them. Although Six-lined Racerunners frequently excavate new burrows, apparent fidelity for existing burrows has been documented (Hardy 1962, Trauth et al. 2013). Fitch (1958b) observed burrow defense by an individual in Kansas, but Hardy (1962)

Figure 186. An adult Six-lined Racerunner in a self-excavated burrow (Sauk County; photo by J. Kapfer).

reported that this behavior was rare. In addition to excavating new burrows, Six-lined Racerunners use the existing tunnel systems of Baird's Pocket Gophers (*Geomys breviceps*) in Arkansas (Connior and Chordas 2012) and Colorado (several *Geomys* and *Thomomys* spp.; Vaughan 1961), but they also inhabit burrows of other small mammals (*Microtus* and *Scalopus* spp.) in Kansas (Fitch 1958b). It is likely that Wisconsin individuals use mammal burrows in a similar fashion.

REPRODUCTION AND DEVELOPMENT

Breeding in Wisconsin occurs during May (Vogt 1981, Stein 1995) but in the southern US can extend into late July (Brackin 1979). When males encounter potential mates they often perform a behavior termed "cloacal rubbing," where the male wags his posterior region from side to side while rubbing his cloaca on the ground (Noble and Bradley 1933, Carpenter 1962). Males also display their colored chins and throats by lifting themselves up on their forelimbs, but this seems to be more common in male-male dominance interactions than courtship (Noble and Bradley 1933). However, dominance displays appear to affect mating frequency, as Brackin (1978) found that dominant males had higher sexual activity than subordinate males in mesocosm experiments. Males chase, capture, and then secure females by using their jaws to grip females by the neck (Noble and Bradley 1933, Fitch 1958b, Carpenter 1962; Natural History Box: Tough Love). After a female has been restrained, the male slides his tail underneath the female to allow cloacal contact and initiation of copulation. Copulation can last for less than two minutes but sometimes lasts up to fifteen minutes (Noble and Bradley 1933, Fitch 1958b, Carpenter 1962).

Oviposition in Wisconsin occurs in mid- to late June (Vogt 1981, Stein 1995). Females often create cavities in the soil that are two to three times larger than the enclosed eggs and seal the cavities with soil after egg deposition (Brown 1956). Limited information on the nest characteristics of upper midwestern populations exists. Of 55 nests observed in sawdust piles in North Carolina, the mean depth was 14.2 cm (5.6 in; Brown 1956). Trauth (1983) studied nesting activity over eight years in Arkansas and found that nests were mostly located in well-drained embankments with eroded soils and sparse vegetation. He also reported that females will return to suitable nesting sites repeatedly over time and further observed colonial nesting, where as many as ten egg clutches were found in a "few square meters of topsoil." Vogt (1981) found clutch sizes of three to eight eggs, and Stein (1995) found clutch sizes of one to five eggs ($\bar{x} = 2.8$, $n = 18$) in Wisconsin. First-year breeders in

NATURAL HISTORY BOX: Tough Love

Amphibians and reptiles exhibit some of the most interesting, strange, and in some cases arguably disturbing mating activities and behaviors one can imagine. I distinctly remember the first time I observed a Six-lined Racerunner mating interaction. I was a new graduate student and budding herpetologist studying freshwater turtles in South Texas. While setting up hoop net traps next to a pond I noticed a couple of individuals scurrying around about 2 m (6.5 ft) away from me. A closer look revealed one individual moving erratically (i.e., in rapid zigzags and circles) while carrying the other individual on its back. Furthermore, the individual on top was tightly grasping the neck of the individual on the bottom with its jaws while being swung violently from side to side. This 'battle' went on for several minutes; then the individual on top released its grasp, and both lizards scurried into the brush. Having little knowledge of lizard behavior at the time, I assumed I was watching two males fighting over a favorable patch of habitat. Later that day I mentioned the event to my advisor, at which point he informed me that I had just been introduced to lizard sex!

Donald J. Brown, West Virginia University / USFS

Kansas laid a mean of 2.0 eggs per clutch, while older individuals laid a mean of 4.4 eggs per clutch (Fitch 1958b). Trauth (1983) determined that clutch size was positively correlated with SVL, and mean clutch size based on 401 clutches in the southern United States was 3.19. Stein (1995) also found a positive correlation between SVL and clutch size in a population from southwestern Wisconsin. Fitch (1958b) hypothesized that females in Kansas often laid two or more clutches per year, and this was confirmed in Texas (Hoddenbach 1966, Clark 1976). In Wisconsin, egg incubation time ranges from 50 to 64 days, so hatchlings emerge in August and are active until early October (Vogt 1981, Stein 1995, Kapfer and Pauers 2006). Cochran and Schmitt (2016) documented the appearance of hatchlings as early as 14 August and as late as 21 September in southeastern Minnesota. Furthermore, Trauth (1977) found evidence of successful egg overwintering in Arkansas.

Younger age classes grow rapidly. Fitch (1958b) reported that hatchlings grew at a rate of ca. 0.05 cm (0.02 in) per day, reaching approximately half of their adult size prior to their first winter hibernation in Kansas. Following hibernation, the individuals grew ca. 0.03 cm (0.01 in) per day until they approached maturity. Stein (1995) documented growth rates between May and September in Sauk County, reporting that yearlings grew at a rate of 0.023 cm (0.009 in) per day ($n = 90$). Growth slows as lizards reach maturity, and adult males from Sauk County grew at a rate of 0.0067 cm (0.003 in) per day ($n = 106$), while adult females grew at a rate of 0.0033 cm (0.001 in) per day ($n = 57$; Stein 1995). Carpenter (1959) studied growth rates of subadults and adults in Oklahoma and failed to detect monthly TL increases once males and females reached 7.5 cm (0.3 in) and 8.0 cm (0.3 in), respectively. Fitch (1958b) found that mean SVL differed by only 0.28 cm (0.1 in) between the second year and the third year but that individuals continued to grow slowly after reaching sexual maturity. Pope (1947) reported that individuals from Illinois were capable of mating by their second spring. The smallest mature females found during a study in Kansas were 6.2 cm SVL (2.4 in; Fitch 1958b), and the smallest in Texas were 6.8 cm SVL (2.7 in; Clark 1976). However, Stein (1995) found gravid females as small as 5.9 cm SVL (2.3 in) in Wisconsin. Six-lined Racerunners can live to be at least six years old (Fitch 1958b), but annual population turnover is high (Fitch 1958b, Clark 1976).

ACTIVITY

Several sources have noted that the annual activity period for Wisconsin adults is limited to late spring and summer (i.e., early May through late August or early September; Stein 1995, Kapfer and Pauers 2006). Similarly, long-term observations (2000–2012) in southeastern Minnesota indicated that the species is generally active from mid- to late May to mid-September (Cochran and Schmitt 2016). Research in Alabama and Georgia indicates that rising environmental temperatures are a likely trigger for emergence from hibernacula in the spring (Etheridge et al. 1983). Brewster (1984) rarely found them aboveground during surveys in July and August, suggesting that they avoid high summer temperatures in Wisconsin.

Like seasonal activity, temperature might be an important influence on daily activity patterns. For example, although generally diurnal, this lizard concentrates its activity to the slightly cooler periods of the day, such as mornings, late afternoons, and early evenings (Barden 1942, Bellis 1964, Cochran and Schmitt 2016). Barden (1942) conducted laboratory experiments which supported that peak daily activity occurred between 10:00 and 11:00 a.m. Bellis (1964) reported that during midmorning activity, lizards alternated behavior between thermoregulation and foraging by covering "several square yards in moving slowly about, searching, and feeding occasionally." Work in South Carolina and Arkansas found that Six-lined Racerunners were also less active during cooler overcast (Bellis 1964) or rainy (Ransom and Plummer 1999) conditions. Interestingly, several studies have noted seasonal variation in the importance of ground temperature as an activity trigger. For example, in Arkansas, temperature has greater influence on daily activity during cooler parts of the active season (Paulissen 1988a). However, in the warmer summer months, ground temperature may have less influence on daily activity than the intensity of solar radiation or factors related to foraging (Paulissen 1988a).

This species has a high thermal tolerance, and Paulissen (1988c) determined that their critical thermal maximum (CTmax) was 51°C (123.8°F) in Arkansas. Furthermore, Werth (1972) noted that Six-lined Racerunners had the highest thermal tolerance of the lizard species he studied in Kansas. Stein (1995) found that mean body temperature in active lizards from Wisconsin was 38.1°C (100.6°F, $n = 16$), while Kapfer and Pauers (2006) reported that ground temperature associated with active lizards in Wisconsin was 34.7°C–39.2°C (94.5°F–102.5°F). The mean substrate temperature near observed Six-lined Racerunners on a sand prairie along an artificial dyke in Minnesota was 32.8°C (91°F; Gossen and Cochran 2013). Although racerunners prefer warm temperatures, Oldfield and Moriarty (1994) found active adults in Minnesota at air temperatures as low as 22°C (72°F) in mid-May.

Research on several congeners (i.e., lizard species in the genus *Aspidoscelis*) determined that body temperature fluctuations were minimal, indicating that whiptail lizards are generally capable of high thermoregulatory precision (Bowker and Johnson 1980). Asplund (1970) reported that whiptail lizards are capable of thermogenesis (i.e., generation of heat), although this was through muscular activity rather than the physiological processes mostly associated with endotherms. For example, 100 g (3.5 oz) whiptail lizards were able to increase their body temperature as much as 1°C–2°C (1.8°F–3.6°F) above ambient air temperature via thermogenesis (Asplund 1970).

Extremely limited data on movement or spatial patterns in the Upper Midwest exist, and most research on spatial ecology is from western and southern states. The only data for Wisconsin are from Vogt (1981), who used capture-recapture methods to estimate a home range of "less than 160 m^2" (1,722 ft^2), which is substantially smaller than other findings. For example, Fitch (1958b) estimated that the home range of a single male in Kansas was 1,266 m^2 (13,635 ft^2). Clark (1976) employed capture-recapture methods in Texas to measure seasonal home ranges (MCP estimator). He reported mean estimates that ranged from 12,603 m^2 (135,657 ft^2) for females to 13,482 m^2 (145,119 ft^2) for males after application of Jennrich and Turner's (1969) correction factor.

The mean distance between captures for a population in South Carolina was 14.2 m (46.6 ft) for males and 13.5 m (44.3 ft) for females (Bellis 1964) and in Kansas was 11.6 m (125 ft) for males and 15.9 m (171.3 ft) for females (Fitch 1958b). Hardy (1962) found aggression among individuals in Kansas, noting that it was common for adults to chase yearlings and for dominant males to chase subordinate males from favorable basking sites and shelters. She also observed a social hierarchy among captive individuals (Hardy 1962). This interaction manifested as an "approach order," where dominant individuals could approach others without eliciting an aggressive response. Hardy (1962) suggested that this could reduce competition for food and active conflict, but she also discovered that more aggressive individuals obtained more food and gained weight. Adults in Wisconsin are generally active until late August or early September, although hatchlings are sometimes observed into early October (Kapfer and Pauers 2006). Extremely little is known about the overwintering physiology or winter survival of this species in the Upper Midwest.

PREY AND PREDATORS

Six-lined Racerunners are insectivores. An investigation of 15 stomachs in Kansas found that cumulative stomach contents consisted primarily of Orthoptera (grasshoppers and crickets, 41.5%), Arachnida (spiders, 24.5%), Lepidoptera (moth adults and pupae, 19.9%), and Mollusca (snails, 7.0%; Burt 1928). An investigation of fecal samples from 25 individuals in northern Illinois found that 52% contained Orthoptera, 24% contained Hymenoptera (probably ants), and 20% contained Coleoptera (beetles; Warner 2000). An investigation of stomach contents from 18 individuals in Oklahoma found that 52% contained Locustidae (short-horned grasshoppers and locusts), 16% contained Formicidae and Myrmicidae (ants), and 15% contained Arachnida (Force 1925). Punzo (1990) found Isoptera (termites) in 97.8% of stomachs examined in Florida ($n = 92$). Diptera (flies) species are also sometimes consumed (Burt 1928, Warner 2000), but in general this species avoids flying insects (Paulissen 1987a, 1987b). Although ants and termites are probably consumed frequently, they do not appear to be a major dietary component by mass (Hamilton and Pollack 1961). Cannibalism and oophagy have also been documented (Etheridge and Wit 1982, Hatch and Stein 1994).

Six-lined Racerunners are active foragers (Paulissen 1987a). Fitch (1958b) hypothesized that prey is located primarily by olfaction, which allows detection of prey underground. Alternately, Hardy (1962) found that captive individuals used sight to hunt larger prey and olfaction to hunt smaller prey. Witz (1996) found that foraging lizards attacked prey at a mean rate of 0.30, 0.04, and 0.08 attacks per minute for free-ranging, field enclosure, and laboratory enclosure individuals, respectively. Occasionally, juveniles will climb shrubs while foraging (Paulissen and Harvey 1985). Diets were not found to differ between sexes in Florida (Punzo 1990). Paulissen (1987a) determined that diets differed between adults and juveniles in Oklahoma, with adults consuming more Dictyopharidae (planthoppers) and Acrididae (grasshoppers) and juveniles consuming more Cicadellidae (leafhoppers), Salticidae (jumping spiders), and juvenile Hemiptera (true bugs). Adults typically consume more energy per unit effort when foraging and spend less time foraging than juveniles (Paulissen 1987b).

Despite their speed (29 kph [18 mph]; Hoyt 1941), agility, and ability to dislocate their tails, Six-lined Racerunners are common prey for many vertebrates, particularly snakes and carnivorous birds (Fitch 1958b, Lamore 1963, Heintzelman 1964, Ross 1989a, Halstead et al. 2008). Even hog-nosed snakes (*Heterodon* spp.), which are anuran specialists, will sometimes prey on this lizard (Beane et al. 1998). Small mammals such as Thirteen-lined Ground Squirrels (*Ictidomys tridecemlineatus*; Bridgwater and Penny 1966) and medium-sized predators such as feral hogs (Taylor and Hellgren 1997) and Swift Foxes (*Vulpes velox*; Cutter 1958) will also consume Six-lined Racerunners. In addition, Eastern Tiger Salamander (*Ambystoma tigrinum*) predation has been documented (Camper 1986).

Invertebrates also prey on Six-lined Racerunners. Whiting et al. (1991) documented a spider (*Geolycosa* sp.) pulling a recently killed juvenile (SVL = 5.25 cm [2.1 in]) into its burrow. Red Imported Fire Ants (*Solenopsis invicta*) are probably a major predator in southeastern states (Etheridge et al. 1983, Stevenson and Dyer 2002). In addition, Six-lined Racerunners are reportedly a favored host of Chiggers (*Eutrombicula alfreddugesi*; Fitch 1958b), although they likely do not significantly impact survival. However, racerunners do not appear to be a common host for Deer Ticks (*Ixodes scapularis*; Giery and Ostfeld 2007).

Although flight often appears to be the primary means of predator avoidance, Six-lined Racerunners can also employ caudal autotomy, or the voluntary dislocation and amputation of their tail, to avoid predators (reviewed by Bateman and Fleming 2009). The dislocated tail continues to move, which draws the attention of a would-be predator and allows the lizard a chance to escape. Numerous authors have reviewed the general phenomenon and costs vs. benefits of this adaptation in lizards (e.g., Clause and Capaldi 2006, Bateman and Fleming 2009, Gilbert et al. 2013). Although caudal autotomy is an effective predator-avoidance strategy (reviewed by Bateman and Fleming 2009), Six-lined Racerunners apparently employ this tactic less often than other sympatric lizard species. For example, Fitch (2003a) found that the percent of Six-lined Racerunners missing tails was lower in most age classes than in the Great Plains Skink (*Plestiodon obsoletus*) and the Slender Glass Lizard (*Ophisaurus attenuatus*), both of which are more likely to employ caudal autotomy than flight when avoiding predators. Furthermore, Ballinger et al. (1979) reported that Six-lined Racerunners with missing tails were 36% slower than those with intact tails, indicating that tail loss could impact a racerunner's ability to quickly escape from threats. These findings further support the importance of flight over tail dislocation for predator avoidance in Six-lined Racerunners (Ballinger et al. 1979).

CONSERVATION

Status: Globally, the Six-lined Racerunner is listed as a Least Concern species by the IUCN. It is considered a species of Special Concern and a Species of Greatest Conservation Need in Wisconsin by the WDNR.

Populations: Substantial variation exists in estimates of Six-lined Racerunner abundance across their North American range, but there are only sparse data from the Upper Midwest. Anecdotal observations suggest that they are abundant at several locations in Wisconsin (Vogt 1981). Brewster (1984), for example, believed that Six-lined Racerunners were more abundant than all other reptile species at a site he surveyed in Sauk County but indicated that high encounter rates were possibly biased by the lizard's higher activity levels. Stein (1995) captured 218 unique individuals across three years in the same study area but did not estimate population size. Maximum population size estimated across two years at a site in northwestern Illinois was 200 individuals, with a density estimate of 3.33/ha (1.35/ac; Warner 2000). However, recapture rates during this two-year study were low ($n = 14$), a phenomenon also observed by researchers in other states (e.g., Werth 1972). Gossen and Cochran (2013) estimated densities of 4.4–37.8/ha (1.8–15.3/ac), which the authors extrapolated to abundances of 26–229 racerunners, along two 50 × 2 m (164.0 × 6.6 ft) transects in Minnesota. Density estimates from western and southern states vary dramatically, ranging from 15 to 2,196/ha (6.1 to 889.0/ac; Fitch 1958b, Carpenter 1959, Bellis 1964, Clark 1976).

Several studies have estimated additional demographic parameters, such as survival and sex ratios. Fitch (1958b) found that 41% of marked individuals survived their first year in Kansas, while Clark (1976) found that only 16% survived their first year in Texas and that males had a shorter life expectancy than females. Brewster (1984) reported sex ratios of ca. 1M:1F for a sample of individuals from Wisconsin ($n = 30$) that was similar to a population studied in northwestern Illinois (Warner 2000).

Threats: Since the mid-1800s, there has been substantial degradation of open habitats in Wisconsin due to the expansion of anthropogenic landscapes and ingrowth of woody vegetation (see the Past, Present, and Potential Future Landscapes of Wisconsin and Introduction to Conservation and Management of Wisconsin's Amphibians and Reptiles chapters). This lizard has low dispersal ability, so it cannot easily migrate through degraded landscapes to new locations, which can limit genetic exchange among populations (Hokit et al. 1999). As a result, Six-lined Racerunners in the state have most likely suffered and experienced population declines due to loss of suitable habitat.

Despite this, it is noteworthy that Fitch (1958b) often observed Six-lined Racerunners in disturbed locations in Kansas, specifically in areas with exposed soil that resulted from excessive agricultural activity (e.g., in heavily grazed pastures and along the edges of crop fields). Six-lined Racerunners will also use discarded anthropogenic debris such as scrap wood, sheet metal, or old structural foundations as cover (Fitch 1958b, Joshua Kapfer, UW–Whitewater, personal observation). However, lizard use of these unnatural habitat features is likely related to limited availability of suitable natural features and because these disturbed locations happen to possess microhabitat characteristics that fulfill an individual's immediate needs (i.e., thermoregulation and cover). It is unlikely that Six-lined Racerunners prefer anthropogenically disturbed landscapes, which probably cannot support stable populations over the long term. Kapfer and Pauers (2006) never found the species in disturbed crop fields, despite an abundance of agriculture at their Wisconsin study location (33.7% of the area surveyed). The apparent reluctance of Six-lined Racerunners to use agricultural fields in Wisconsin, coupled with low dispersal ability, likely limits genetic interchange among populations, which could further threaten this species.

Other potential threats could impact this lizard, but they are poorly understood due to a lack of information. For example, Six-lined Racerunners could be affected by residual or bioaccumulated pesticides in their prey. For example, Clark et al. (1995) found residual agricultural pesticides in the tissues of Six-lined Racerunners from Florida that were collected in areas of heavy pesticide use. Although the authors did not connect detected pesticides to population-level impacts, they recommended that future work be conducted on this subject.

Management: General concepts associated with managing prairie and oak savanna habitats are relevant to Six-lined Racerunners (see Introduction to Conservation and Management of Wisconsin's Amphibians and Reptiles chapter). Given the likelihood of population isolation due to habitat degradation, conservation plans promoting landscape-level connectivity among occupied sites might be beneficial. This would likely require strategic purchase of land and management of habitat in protected areas to create linkages among isolated populations for promotion of genetic exchange. Fire is an important management tool for creating and maintaining the more open habitats preferred by this species (Dodd et al. 2007, Halstead 2007, Ashton and Knipps 2011, Steen et al. 2013). In Oklahoma, Jones et al. (2000) found Six-lined Racerunners more often in areas treated by fire, although sample sizes were small. Mushinsky (1985) also found that parcels in the Florida sandhills burned annually had higher densities of Six-lined Racerunners than those with longer burn intervals, indicating that management through fire is immediately beneficial for this species. Reports from other states indicate that even high-intensity fires during the summer months when the species is active on the surface probably do not result in substantial direct mortality (Ruthven et al. 2008, Brown et al. 2014). However, other sympatric species could be impacted by prescribed fire in the summer, and no studies to date have examined fire-related mortality of this lizard in Wisconsin. Thus, in the absence of regional studies, it is prudent to plan application of prescribed fire carefully at locations where Six-lined Racerunners are present. Possible direct mortality impacts might be mitigated by scheduling prescribed fire application for outside of the active season. Furthermore, planning the annual frequency and patterns of prescribed fire application to ensure the presence of intact vegetation patches (for refugia from the elements and predators) may further reduce potential impacts on populations.

The removal of encroaching woody vegetation from landscapes that normally possess short vegetation and exposed soil (e.g., sand blowouts or dunes) is also important, given the heavy use of exposed soils by this species (see Habitat subsection). Six-lined Racerunners could also benefit from hand removal and clearing of later successional vegetation (particularly woody species such as Eastern Red Cedar [*Juniperus virginiana*]) that, in the absence of fire, encroaches on the normally exposed hillsides or bluffsides they use (Pierce and Reich 2010).

North American Racer

Coluber constrictor (Linnaeus 1758)

Joshua M. Kapfer and Jeffrey M. Lorch

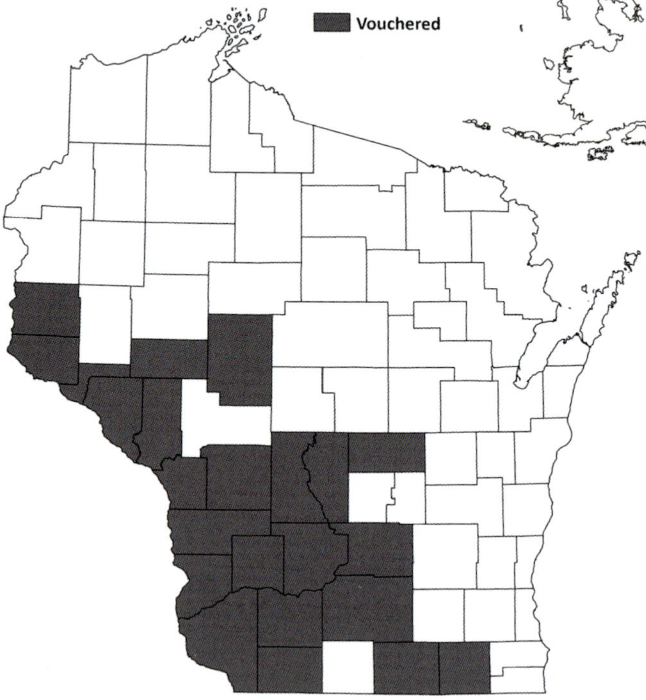

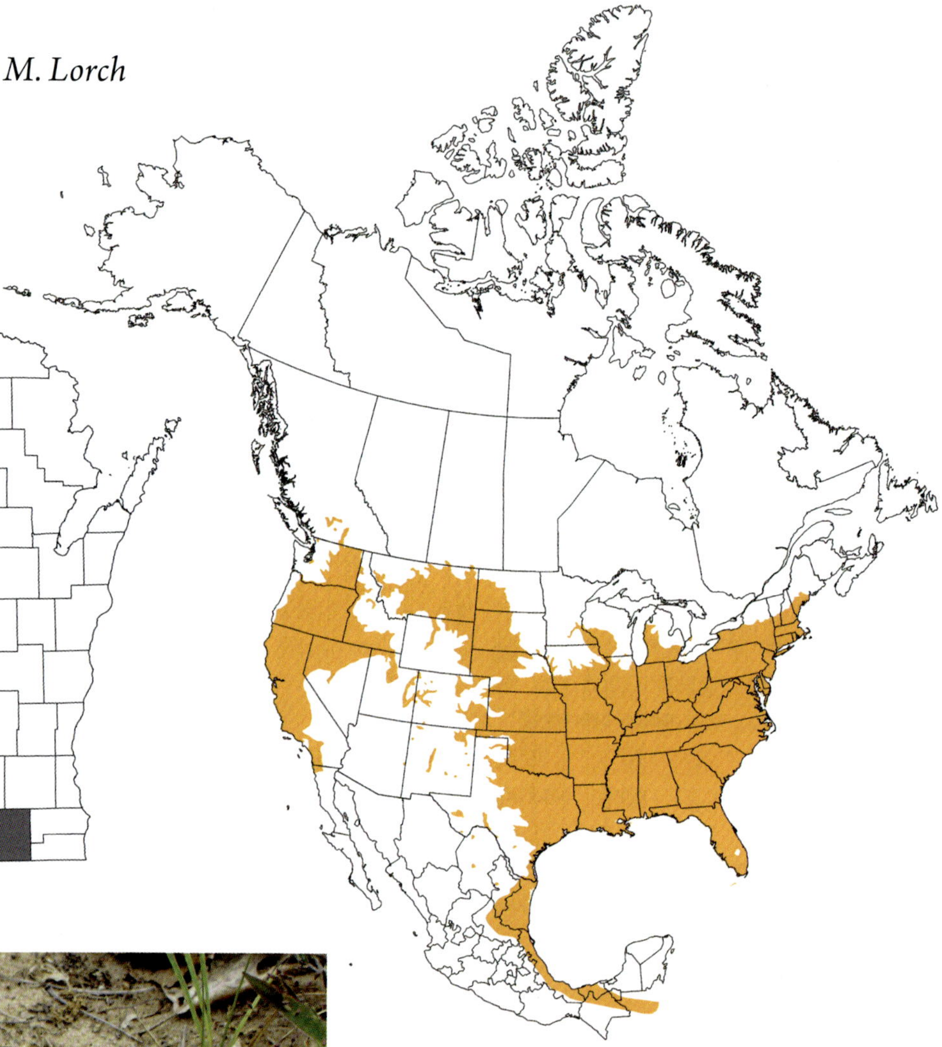

Adult female North American Racer (Sauk County; photo by J. Kapfer).

AT A GLANCE

Adult North American Racers are solid slate gray to olive brown, often with hints of blue and yellow along their sides, making them distinct among the larger snake species in Wisconsin. Juveniles appear much different from adults, typically with a heavy pattern of rusty or brown blotches over a cream to gray background. North American Racers inhabit savannas, prairies, and open bluffsides in the western part of the state and are most common in the Wisconsin and Mississippi River valleys. They are active from mid-April through mid-October in Wisconsin. Reproduction occurs soon after emergence from hibernation. Females lay eggs from early to mid-June, and they hatch from late August to early September. The North American Racer is a harmless species that preys on insects, small mammals, snakes, and lizards. Despite its scientific name (*Coluber constrictor*), it is not a true constrictor and mostly swallows its prey live. Although information on predators in Wisconsin is sparse, North American Racers are likely consumed by various vertebrates, such as larger birds and mammals. The common name 'racer' is apt, as this snake is sleek in appearance and has a tendency to quickly dart for cover when threatened (it is reportedly able to slither as fast as 6.4 kph [4 mph]). Much like its preferred habitat, North American Racers are uncommon in Wisconsin and currently considered a species of Special Concern and a Species of Greatest Conservation Need by the WDNR.

SYSTEMATICS

Coluber constrictor (North American Racer) is a member of the subfamily Colubrinae, which among Wisconsin snakes also includes the true constrictors (*Lampropeltis* spp., *Pantherophis* spp., and *Pituophis* spp.), as well as *Opheodrys vernalis* (Smooth Greensnake). *Coluber constrictor* was among the first New World reptiles to be given a scientific name, having been described by Linnaeus (1758). Based on morphological characteristics, North American Racers were once considered closely related to several Old World snake species that were also placed in the genus *Coluber*. Subsequent phylogenetic analyses demonstrated that the genus was polyphyletic, with the Old World species being only distantly related to the New World species (Nagy et al. 2004, Utiger et al. 2005). The Old World species have since been reassigned to different genera, while *Coluber* is restricted to only some of the racer and whipsnake species that occur in North and South America (Nagy et al. 2004, Utiger et al. 2005). The genus *Coluber* belongs to a larger clade that includes species from the western US and Mexico, *C. taeniatus* (Striped Whipsnake) and *C. flagellum* (Coachwhip), as well as *Salvadora mexicana* (Mexican Patchnose Snake) and *Tantilla melanocephala* (Black-headed Snake), which occur from the western US into Central and South America (Pyron et al. 2013).

Eleven subspecies of *Coluber constrictor* are recognized (Crother 2017), but little genetic data have been presented to support these designations. Burbrink et al. (2008) examined a single mitochondrial DNA gene and found evidence of six distinct lineages of *C. constrictor* in the US. Of these, the central lineage had the largest geographic distribution, occupying most of the US between the Rocky Mountains and Mississippi River, including the Upper Midwest. The authors suggested that racers found in Illinois, Indiana, Iowa, Michigan, Minnesota, Wisconsin, and most of Ohio all belonged to this central lineage. Casper (1996) pointed out inconsistencies in the subspecific designation of racers in Wisconsin and adjacent states, specifically in reference to *C. c. foxii* and *C. c. flaviventris*. However, the work by Burbrink et al. (2008) indicates that these two taxa may not be genetically distinct. Whether the central lineage represents a species distinct from lineages that occur in the eastern, southern, and western US is unclear and will require additional phylogenetic analyses.

DESCRIPTION

Adults: The dorsal surface has no pattern and is a solid gray, slate, or olive brown, sometimes with a bluish tint that is most noticeable on the lateral scales (Figure 187b). Occasionally, the blue is more vibrant and covers much of the dorsal surface. The ventral surface is also without pattern and mostly light gray or cream. There is often a yellow tint on the lateral scales that extends down to the edges of the ventral scales. This yellow coloration is frequently concentrated on the

Figure 187. Examples of North American Racer coloration. (a) An example of the slate-gray-colored dorsal surface, cream-colored labial scales, and lighter preocular scale of adults (Sauk County; photo by J. Kapfer). (b) An example of the blue-and-yellow tint along the lateral scales of the anterior fourth of the body (Sauk County; photo by T. Muehlfeld).

anterior quarter of the body (Figure 187b) and particularly pronounced around the throat. Labial scales are primarily cream or cream yellow in color but may also be gray along the upper edge. The preocular scale is lighter than the surrounding head scales. The body scales are not keeled, giving the snake a very smooth appearance, and the anal plate is divided. Further details of morphological characteristics unique to this species can be found in Ernst and Ernst (2003).

North American Racers are slender-bodied but one of Wisconsin's longer snakes. Measurements of museum specimens housed in the MPM and UWZM collections that were taken from various locations in-state revealed that mean adult SVL was 82.7 cm (32.6 in, $n=13$) for females and 75.9 cm (29.9 in, $n=23$) for males. Similarly, the mean SVL of live adults captured over several years at a site in Sauk County was 79.9 cm (31.5 in, $n=21$) for females and 72.3 cm (28.5 in, $n=10$) for males (Joshua Kapfer, UW–Whitewater, unpublished data). The longest museum specimen measured was a female collected in 1918 from Sauk County (MPM 2104) that had an SVL of 105.8 cm (41.7 in). The tail is proportionally longer than in other large snake species from Wisconsin. Mean tail lengths of adult museum specimens from Wisconsin ranged from 24.1 cm (9.5 in, $n=12$) for females to 22 cm (8.7 in, $n=24$) for males, while mean tail lengths of live snakes captured in Sauk County were 21.3 cm (8.4 in, $n=18$) for females and 23.5 cm (9.3 in, $n=10$) for males. Mean live weights recorded for snakes captured in Sauk County were 191.4 g (6.8 oz, $n=19$) for females and 151 g (5.3 oz, $n=10$) for males. Note that individuals were considered adults and included in the previous measures if they were at least the minimum size for maturity reported by Fitch (1999): 50 cm SVL (19.7 in) for males and ca. 60 cm SVL (23.6 in) for females.

Preadult stages: North American Racer eggs are elliptical and roughly 3.0 × 1.7 cm (1.18 × 0.67 in) when initially laid (Fitch 1963b). Egg shells are leathery and nonadherent, and the surfaces are granular (i.e., sandpaper-like). Eggs absorb moisture during the incubation period and may swell to nearly twice their original size (Fitch 1963b). The color and pattern of young individuals are substantially different from those of adults (Figure 188). The anterior two-thirds of their body possesses distinct dorsal blotches that are brown or rusty colored over a cream, brown, or gray-colored background. The blotches may fade in the posterior third of the body, which results in the tail appearing more solid brown. The sides of the body contain many brown or rusty-colored speckles, and the head is patterned with bars along the posterior edges of the labial scales. There may also be mottling on the head scales that occur between and posterior to the eyes. The ventral surface is cream colored, with two to four rust-colored spots per ventral scale (Vogt 1981). Juvenile patterns fade with age (Figure 189). Vogt (1981) stated that adult col-

Figure 188. Two examples of the color and pattern typical of juvenile North American Racers in Wisconsin. (a) Sauk County (photo by J. Kapfer). (b) Crawford County (photo by R. Sajdak).

Figure 189. A North American Racer in transition from the juvenile to adult color pattern. Note the faint reddish-brown spotting (Lafayette County; photo by J. Lorch).

oration is attained by the "end of the third summer when the snake is 80–90 cm [31.5–35.4 in] long." Limited data on the size of Wisconsin hatchlings and juveniles could be found. Vogt (1981) reported that mean juvenile TL was 21 cm (8.27 in). Two hatchling specimens from Sauk County (MPM 23461 and 2121) had SVLs of 16.6 cm (6.54 in) and 20.4 cm (8.03 in), with tail lengths of 5.1 cm (2.01 in) and 6.3 cm (2.48 in), respectively. In Kansas populations, young were 18.6–24.4 cm SVL (7.32–9.61 in) at hatching, with a mean weight of 4.16 g (0.15 oz; Fitch 1963b). In Michigan, hatchlings were larger, averaging 26.1 cm SVL (10.28 in) and 6.20 g (0.23 oz; Rosen 1991a).

Variation: The most striking variation in this species is the difference in appearance between juveniles and adults. Among hatchlings, the ground and speckling coloration is also somewhat variable. For example, some individuals possess a light dorsal ground color, with brighter reddish-brown blotches and speckles, while other juveniles have a brown wash over much of their body and are duller in appearance. Some variation in adult dorsal coloration also occurs, which can range from slate gray with bright bluish tints to a more drab olive brown. Adults are also sexually dimorphic in that females are typically longer and heavier than males in Wisconsin.

Diagnosis: North American Racers can be distinguished by their smooth dorsal scales that exist in 15–17 rows at midbody, 151–193 ventral scales, divided anal plate, and divided subcaudal scales. They also possess two preocular and postocular scales, seven to eight supralabials, eight to nine infralabials, and two to three temporal scales on each side of the head. There is no light-colored ring around the neck, and in adults the dorsal color is a uniform blue, gray, or light brown (never green). Adult North American Racers are easily distinguished from most other snakes by the combination of their large size and unpatterned dorsal surface. They could possibly be confused with adult Gray Ratsnakes (*Pantherophis spiloides*), as both may have a grayish dorsal color. However, unlike on adult North American Racers, faded patterns are frequently visible on the dorsal surface of Gray Ratsnakes. Adult North American Racers also do not have any markings on their ventral surface, whereas the belly of Gray Ratsnakes is checkered with black and white. Smooth Greensnakes (*Opheodrys vernalis*) take on a blue coloration soon after death, which could cause confusion with North American Racers. However, adult racers that have the solid blue coloration are larger than Smooth Greensnakes. There is little reason why adult North American Racers should be confused with either of the state's venomous snakes. Despite this, Vogt (1981) recounts an instance of a boy from Cross Plains (Dane County) who was bitten by a racer that was misidentified as a venomous "water moccasin." Although Vogt was able to correctly identify the snake, it was only after the boy had received needless treatment for venomous snakebite. It is noteworthy that water moccasins (i.e., Cottonmouths [*Agkistrodon piscivorus*]) do not occur naturally in Wisconsin.

Snake eggs are generally difficult to differentiate. However, the outer surface of North American Racer eggs is granular and rough, unlike the eggs of other snakes. Juvenile North American Racers are easily confused with juveniles of several other snake species in Wisconsin, including Common Watersnakes (*Nerodia sipedon*), Eastern Milksnakes (*Lampropeltis triangulum*), Eastern Foxsnakes (*Pantherophis vulpinus*), Gray Ratsnakes, Gophersnakes (*Pituophis catenifer*), and Massasaugas (*Sistrurus catenatus*), due to a degree of similarity in color and pattern. Unlike North American Racers, however, Eastern Milksnakes and Gophersnakes have single anal plates, while the other species listed possess keeled scales.

DISTRIBUTION AND HABITAT

Global distribution: The North American Racer is one of the few North American reptiles with a transcontinental distribution. It ranges from Maine west to Washington and from southern Canada south to Guatemala and Belize (Wilson 1978). Although there are significant gaps in its distribution associated with the Rocky Mountains, the arid Southwest, and the heavily forested regions of the Upper Midwest and Northeast, the North American Racer occurs (or at least occurred historically) in nearly all of the lower 48 states. Rosen (1991a) suggested that the northern range of this species paralleled a geographic contour line that delineated a mean winter frost depth of 101.6 cm (40 in), but that in the Rocky Mountains and Pacific Northwest the range was further influenced by mean summer air temperature. Rosen (1991a) further indicated that winter conditions faced in Michigan had a strong influence on the survivorship of this species and that such effects had greater impacts than those experienced by overwintering populations studied in Kansas and Utah.

Distribution in Wisconsin: Wisconsin is along the northern periphery of the North American Racer's midwestern distribution. Published county records are concentrated in the southwestern region of the state, primarily along the Mississippi and Wisconsin River valleys in the Driftless Area. These records extend east to western Rock and Waushara Counties and north to St. Croix and extreme southern Clark Counties. In the Driftless Area, the steep south- and west-facing bluff slopes that are exposed to the sun and facilitate basking presumably provide temperature conditions conducive to spring and autumn activity, and they allow access to deep cracks in the bedrock suitable for hibernation. Outside of such habitats, North American Racers occur sporadically in the Central Sands region within the former bed of Glacial Lake Wisconsin (see Past, Present, and Potential Future Landscapes of Wisconsin chapter). A single juvenile specimen collected from Walworth County represents the only known confirmed record from southeastern Wisconsin (CHAS 5977, Peggy Notebaert Nature Museum, Chicago, Illinois). There is no date associated with the specimen, but it was taken by "D. C. Lowrie," who collected other specimens in the area during the late 1930s. Interestingly, other old anecdotal reports of this species in southeastern counties exist, but these are unverified, and there are no new recent reports from that region (see Localities of possible occurrence subsection).

Localities of possible occurrence: Observational reports that lack supporting vouchers exist from Jackson (WDNR 2005a) and Wood Counties (Vogt 1981), which are adjacent to counties with documented occurrences. North American Racers could have also occurred (or still occur) in Green and Marquette Counties, which are located between other counties with voucher specimens.

In addition to the confirmed (but old) specimen from Walworth County (see Distribution in Wisconsin subsection), historical reports exist from other counties in southeastern Wisconsin that are unverified. For example, Cahn (1929) stated that he was presented with "three specimens" that were purportedly from Waukesha County. Also, the original species list of amphibians and reptiles created for Wisconsin (Lapham 1853) included *Bascanion constrictor* (a synonym of *Coluber constrictor*), stating only that it occurred in Milwaukee County. Racers are documented in sandy regions on the nearby southern shore of Lake Michigan in Illinois and Indiana (Anton 1999, Phillips et al. 1999, Minton 2001), and they occur in the Lower Peninsula of Michigan (Rosen 1991a). These reports provide possible corroborating evidence that the Walworth County specimen along with anecdotal observation from the region were not aberrations and that this snake once occurred in southeastern Wisconsin. Regardless, due to a lack of modern reports, it seems unlikely that North American Racers are currently found in this region of the state.

Some published range maps include an isolated occurrence in the Upper Peninsula of Michigan's Menominee County (e.g., Conant and Collins 1991), which Ernst and Ernst (2003) appear to have mapped in northeastern Wisconsin. No confirmed specimens or anecdotal records from northeastern Wisconsin are known. Potentially suitable barrens habitat occurs adjacent to the Michigan report in Marinette County (Wisconsin), which lends some support to the notion that North American Racers could exist in the state's northeastern region. However, further investigation of the Menominee County (Michigan) report revealed that it is not supported by a voucher specimen and may represent a mapping error that has been perpetuated in the literature (James Harding, Michigan State University, personal communication, David Misfud, Michigan Herpetological Atlas, personal communication, Casper et al. 2015b).

Habitat: North American Racers are generally associated with open dry prairies (including those on bluffsides), oak savannas, and sand barrens (see Amphibians and Reptiles in Wisconsin Plant Communities chapter). Several studies from the Midwest suggest variation in the habitat preferred by North American Racers. For example, research in southern Illinois found preferences for edges between forested and open habitats and for successional fields (i.e., fields dominated by "woody species including shrubs, vines, and dense saplings" [Carfagno and Weatherhead 2006]). On the other hand, a study in central Illinois found greater use of more open habitats and little association with forest edges (Keller and Heske 2000). Carfagno and Weatherhead (2006) further recorded differences in habitat use based on sex, with females found closer to habitat edges (with fewer trees and shorter vegetation) than males.

Information on habitat use in Wisconsin is observational. Three years of surveys during the active season at a site in Sauk County resulted in 37 North American Racer observations for which habitat was assessed. The majority of these occurred on open west- and south-facing bluffs or hillsides (31.5% of observations), followed by oak savannas (28.9%), open grasslands or sand barrens (26.3%), and grassy edges between grassland or agricultural fields (13%; J. Kapfer, unpublished data). North American Racers were not found in closed-canopy forest or agricultural habitat, even though these habitat types were common at the study site (for availability of habitats on-site, see Kapfer et al. 2008e). Furthermore, convex densiometer readings revealed no associated tree-canopy closure at 19 of 21 snake locations and <40% closure for the remaining two (J. Kapfer, unpublished data). Research in Kansas and southern Illinois found a similar preference for open habitats, including tallgrass prairie, pasture (Fitch 1963b), or areas with only young trees (Carfagno and Weatherhead 2006). Fitch (1963b) further mentioned a fondness for rocky "hilltop ledges" in spring and autumn, which is similar to observations in Wisconsin (J. Kapfer, personal observation).

North American Racers appear to associate with specific microhabitat qualities, particularly related to structural vegetation characteristics (i.e., percent ground cover of vegetation and mean vegetation height). For example, in Wisconsin they are frequently found in herbaceous (nonwoody) vegetation, and mean estimate of herbaceous vegetation cover within a m^2 area centered on snake locations in Sauk County was 66% (SE = 6.8, n = 26; J. Kapfer, unpublished data). In contrast, coverage of woody vegetation associated with snakes was very low ($\bar{x}$ = 5.2%, SE = 2.9, n = 26), as was woody ground debris ($\bar{x}$ = 2.6%, SE = 1.5, n = 25) and leaf litter ($\bar{x}$ = 7.3%, SE = 4, n = 27; J. Kapfer, unpublished data). This is similar to findings from southern Illinois, where North American Racers preferred ample herbaceous ground cover, as well as fallen logs (Carfagno and Weatherhead 2006). The estimates of woody vegetation and ground debris coverage in Wisconsin are substantially lower than similar measures in Kansas ($\bar{x}$ "shrub" coverage = 18.7%, SE = 1.7, $\bar{x}$ litter coverage = 58.5%, SE = 1.9; Klug et al. 2010b). Measures of vegetation height associated with Wisconsin observations suggests that shorter vegetation is preferred ($\bar{x}$ = 25.7 cm [10.1 in], SE = 11.6, n = 24), although this differs from associated vegetation height recorded in Kansas ($\bar{x}$ = 77.4 cm [30.5 in], SE = 1.8, n = 155; Klug et al. 2010b).

Factors other than the physical structure and composition of vegetation may affect habitat selection. Carfagno and Weatherhead (2006) found that season affects habitat use, with open successional fields used frequently in May and again late in the active season. However, during late spring and summer snakes transitioned to habitats with a denser tree canopy and greater leaf litter (Carfagno and Weatherhead 2006). Carfagno et al. (2006) investigated the influence of prey availability on habitat selection but could not correlate use of habitat edges in southern Illinois with mammalian prey abundance. They suggested that perhaps thermoregulatory opportunities or abundance of other potential prey species (i.e., birds, reptiles, amphibians, insects) were more important in dictating the selection of habitat.

Information on habitats used for nesting in Wisconsin is largely anecdotal. Old eggshells from North American Racers have been found along the edge of open sand blows in dry prairies (Sauk County; J. Kapfer, personal observation). In such dry habitats, most egg laying occurs underground, likely in the burrows of various small mammal species (Fitch 1963b, Rosen 1991a). In Wisconsin, North American Racers hibernate below the frost line in rock crevices along open south- and west-facing bluff slopes, which is similar to the observations of Fitch (1963b) in Kansas. Wisconsin individuals have also been observed near mammal borrows in flat open prairies during early spring (J. Kapfer, personal observation). These observations, coupled with reports of hibernation in mammal burrows in Illinois (Schroder 1950), suggest that burrows are sometimes used as hibernacula in Wisconsin. In Utah, Brown et al. (1974) excavated North American Racers from a hibernaculum in late January at soil depths of 100–110 cm (39.4–43.3 in) in a mix of loose soil and debris or natural tunnels and cavities. Cloacal temperatures of extracted individuals ranged from 4.1°C to 6.5°C (39.4°F to 43.7°F; Brown et al. 1974). Schroder (1950) reported that this species overwintered in northwestern Illinois at depths from 91.4 to 106.7 cm (36 to 42 in, frost depth of soil was 22.3 cm [8.3 in]). Rosen (1991a) suggested that finding suitable overwintering locations affected the distribution of this species (see further discussion in the Global distribution subsection).

REPRODUCTION AND DEVELOPMENT

In Wisconsin, breeding likely takes place in April and May shortly after spring emergence. Yet there is little detailed information on the reproductive habits of these snakes in the state, and most of what is known across the entire Midwest comes from extensive observations by Henry S. Fitch in Kansas and Philip C. Rosen in Michigan. Fitch (1963b) observed copulation on numerous occasions in May; similarly, females he captured in May often contained active sperm in the cloaca, whereas no sperm was found in females captured at other times of year. Brown et al. (1974) also found no evidence that overwintering females had been inseminated prior to hibernation, suggesting that copulation does not commonly occur in autumn, if at all.

Fitch (1963b) described courtship behaviors that involved spasmodic movement of the male's body against the female for several hours prior to copulation (Figure 190). He also noted that the female

Figure 190. A male and female North American Racer encountered in the field that appeared to be courting (Richland County; photo by H. Kaarakka).

would occasionally attempt to crawl away from the male (even during copulation), which seemed to stimulate the male's interest in the mating process. Lillywhite (1985) documented similar behavior, which included apparent olfactory trailing by the male. The dragging of one partner by the other during copulation (even while climbing trees, so that an entire male snake's weight is suspended by his genitalia) has been reported on more than one occasion (Fitch 1963, Lillywhite 1985). It is unclear if such behaviors represent a normal part of the courtship process or if inadvertent disturbance by the observers may have elicited these responses. Mating in racers apparently occurs in pairs or in trios of one female and two males (Fitch 1963b, Lillywhite 1985).

From 2003 to 2005, six females with visibly distended abdomens were found in Sauk County between 3 June and 11 June (J. Kapfer, personal observation), and a gravid female was observed on 7 June in Richland County (Heather Kaarakka, WDNR, personal communication). A female captured in Sauk County on 12 May was brought into captivity and laid a clutch of 17 eggs on 7 June (J. Kapfer, personal observation). These observations suggest that oviposition in Wisconsin occurs primarily in early to mid-June, consistent with peak egg laying reported from Michigan (Rosen 1991a). Fitch (1963b) found parturient females from mid-June to mid-July in Kansas. Females at northern latitudes (such as Wisconsin and Michigan) may lay eggs slightly earlier in the year to compensate for cooler temperatures and a shorter active season.

Clutch sizes ranging from 6 to 24 eggs (means = ca. 12 to 15) have been reported in the Midwest, with older and larger females generally producing more eggs (Fitch 1963b, Rosen 1991a). There is little information on clutch size and incubation periods in Wisconsin. Eggs laid in Kansas and maintained in captivity had an incubation period of about 51 days, which was consistent with the appearance of wild young in late August and early September (Fitch 1963b). In Grant County, young of the year were observed on 1 September, 24 September, and 5 October (Jeffrey Lorch, USGS, personal observation). The individual captured on 1 September was presumably a recent hatchling undergoing its first molt, but the other animals may have hatched weeks earlier.

Juvenile individuals grow rapidly after hatching. By the end of the first active season, young will double or triple in weight and increase in length by ca. 50% (Fitch 1963b). In Michigan, both sexes exceeded 60 cm SVL (23.6 in) at the end of the first full season of growth, with some attaining lengths of >130 cm SVL (51.2 in; Rosen 1991a). North American Racers reach sexual maturity at a minimum of one year of age in males and two years of age in females; however, only a small percentage of females actually produce a clutch in their second year (Fitch 1963b). In Kansas, a significant proportion of females under the age of five were not gravid, which supports either that some females take longer to reach sexual maturity or that young females might not produce a clutch on an annual basis (Fitch 1963b). Rosen (1991a) reported that most two-year-old females assessed were gravid, but they produced fewer eggs per clutch than older females. That females take several years to achieve peak reproductive status may be compensated for by the relatively long life span of the species, which can exceed eight years in the wild (Fitch 1999).

ACTIVITY

North American Racers can be active in Wisconsin from early or mid-April through mid- to late October (J. Kapfer, personal observation). Early- and late-season adult activity is focused around hibernacula on open bluffsides, but much time is spent in adjacent grasslands, prairies, and savannas during the warmer spring and summer months (J. Kapfer, personal observation, Vogt 1981). They are primarily diurnal in Wisconsin (Vogt 1981), and the presence of predominantly cones in the retina of North American Racers reportedly makes them well-suited for daytime activity (Walls 1934, Underwood 1967). Ernst and Ernst (2003) found that North American Racers can remain active at higher air temperatures than many other snakes and recorded cloacal temperatures of >32°C (89°F) from individuals that were "prowling about." Similarly, they were often found aboveground on very warm days during a three-year study in Sauk County, while other sympatric reptiles were within cooler refugia (J. Kapfer, personal observation). The mean surface soil temperature recorded at North American Racer locations in these instances was 33.9°C (93°F, SD = 7.9°C [14.2°F], n = 20; J. Kapfer, unpublished data). Fitch (1963b) reported that the majority of the snakes captured in the field were active when cloacal temperatures ranged from 29°C to 35°C (84.2°F to 95°F, n = 39) and at air temperatures ranging from 26°C to 27°C (78.8°F to 80.6°F). Keller and Heske (2000) reported that radio-tracked North American Racers in Illinois were most often found on the ground surface (ca. 70% of locations), as opposed to underground or in trees or shrubs. Fitch and Shirer (1971) occasionally observed radio-tracked individuals in trees at a maximum height of 3 m (9.8 ft), while Ernst and Ernst (2003) report observing them at heights of 10 m (32.8 ft).

No scientific literature was located that described movement patterns of North American Racers in Wisconsin. Anecdotal observations suggest that they are mobile and wide-ranging (J. Kapfer, personal observation), which is supported by studies conducted in other states. A review of research from more northern locations indicates that home range sizes are large and movement rates are high, but estimates varied greatly among studies (Table 12). The variable results reported may be related to differences in sample size and duration of studies, but they could also suggest that this species exhibits substantial regional variation in movement patterns. Klug et al. (2011a) suggested that the high movement rates recorded (67.3 m/day [220.8 ft/day]) for North American Racers could be due to the active (as opposed to ambush-type) foraging strategy of this species. Variation in movement also occurs by age class. For example, Rosen (1991a) determined that the mean distance between random recaptures during the active season was 424 m (1,391 ft) for adults, 188 m (607 ft) for yearlings, and 62 m (203 ft) for small juveniles in Michigan. North American Racers in Utah and Kansas reportedly make substantial migrations from communal hibernacula to established summer home ranges ($\bar{x}$ movement distance = 739 m [2,225 ft], n = 32, and $\bar{x}$ = 339 m [1,112 ft], n = 124,

Table 12. Comparison of spatial data reported for North American Racers from studies conducted in three states. Home range estimates are all 100% minimum convex polygons (MCP; Fitch 1963b refers to "minimum home range size").

State	Sex	Movement Rate, m/day (ft/day)	Home Range Size, ha (ac)	Sample Size	Source
Illinois	Males	71.9 (235.9)	21.9 (54.1)	7	Carfagno and Weatherhead (2008)
	Females	54.1 (177.5)	19.6 (48.4)	8	Carfagno and Weatherhead (2008)
Kansas	Adults	28.5 (93.5)		12	Fitch and Shirer (1971)
Kansas	Adults	67.3 (220.8)	11.5 (28.4)	12	Klug et al. (2011a)
	Males		21.6 (53.4)	4	Klug et al. (2011a)
	Females		6.4 (15.7)	8	Klug et al. (2011a)
Kansas	Adults		2.7 (6.6)	20	Fitch (1963b)
	Males		3.0 (7.3)	15	Fitch (1963b)
	Females		1.8 (4.5)	5	Fitch (1963b)
Utah	Adults	30.6 (100.4)	1.5 (3.7)	7	Brown and Parker (1976)

respectively; Brown and Parker 1976, Fitch 1963b). Brown and Parker (1976) also found that North American Racers made migrations to the same home range location over multiple years in Utah, indicating strong fidelity for activity areas.

By mid- to late October, most North American Racers have likely moved back toward overwintering sites. Although this could result in substantial movement for individuals that migrate between hibernacula and summer ranges, members of a Sauk County population were frequently captured within the vicinity of known hibernacula throughout the active season (J. Kapfer, personal observation). This suggests that migration for overwintering does not always occur, although the need to migrate could vary based on the composition of the surrounding habitat and prevailing environmental conditions near hibernacula. Young individuals apparently follow scent trails left by adults, which may orient naive hatchlings toward hibernacula in autumn (Brown and Parker 1976). Although few details about behavior and physiology during hibernation are known, hibernacula likely possess underground crevices and structures that allow snakes to descend below the frost line during winter (see Habitat subsection). In Utah, adults return to the same overwintering location annually (Brown and Parker 1976), and casual observations suggest that this also occurs in Wisconsin (J. Kapfer, personal observation). Adult North American Racers in Wisconsin will frequently co-occupy the same hibernaculum with Gophersnakes, Eastern Milksnakes, and Timber Rattlesnakes (*Crotalus horridus*; J. Kapfer, personal observation). Intra- and interspecific communal overwintering is also reported from Kansas (Burt 1935), Illinois (Pope 1944, Schroder 1950), and Michigan (Rosen 1991a).

PREY AND PREDATORS

When hunting, North American Racers rely on visual cues, particularly prey movement, in conjunction with olfactory cues (Herzog and Burghardt 1974, Cooper et al. 2000). They actively search for food rather than relying solely on ambush tactics. Vogt (1981) recounted watching Wisconsin individuals "in ardent pursuit of darting racerunners," and other reports of active prey pursuit exist (e.g., Fitch 1963b). Despite its scientific name, *Coluber constrictor*, this species does not constrict its prey. Instead, captured prey is overpowered and swallowed, although the racer will press a body loop over struggling prey if necessary (Vogt 1981). Interestingly, Knudsen (1955c) made a point to mention that the "common supposition that this snake is able to hypnotize its prey and draw it to its waiting jaws is unfounded."

No detailed dietary analyses have been published for Wisconsin. However, Vogt (1981) anecdotally reported that North American Racers in Wisconsin consume lizards, snakes, amphibians, insects, and small rodents. Analysis of a scat sample from a female (SVL = 71.8 cm [28.3 in]) captured in Sauk County revealed the remnants of several large field crickets (*Gryllus* sp.; Kerry Katovich, UW–Whitewater, unpublished data). Detailed studies from other states, however, provide better insight into important food items. Klimstra (1959) examined the stomach contents of 115 individuals from Illinois and found that by percent volume, prey was primarily insects (39.1%) and mammals (32.9%). Fitch (1963b) examined 479 scats from a Kansas population and determined that the most frequent prey were insects (>75%), but the largest percent consumed by weight were mammals (65.7%). The stomach contents of 55 individuals from another location in Kansas consisted mostly of lizards (61.5%) and mice (15.4%) by weight (Fitch 1963b).

North American Racers are sometimes casually discussed as being decidedly ophiophagous. Yet at both Kansas locations studied by Fitch (1963b), snakes comprised <5.5% of racer diet (including cases of cannibalism) by weight. However, there appears to be substantial regional variation in the importance of snakes as a dietary component. Uhler et al. (1939) examined the stomach contents of 16 individuals from Virginia and found that the largest dietary constituent by volume was snakes (25.6%). These findings are consistent with studies in New Hampshire and Pennsylvania in which the ophiophagous tendencies of North American Racers were thought to influence snake community composition and structure (Kjoss and Litvaitis 2001, Meshaka and Delis 2014, respectively; see also Natural History Box: Complementary Ranges). A road-killed

NATURAL HISTORY BOX: Complementary Ranges

For many reptiles and amphibians, the factors that limit a species' distribution are poorly understood. In the case of the North American Racer, geographic distribution is strongly correlated with frost depth (Rosen 1991a), suggesting that winter temperatures or length of the active season may influence the species' range. In Wisconsin, the presence of hibernacula is another limiting factor and may explain why the North American Racer occurs mostly in the Driftless Area, where deep crevices in bedrock provide suitable hibernacula. However, the distributions of some other Wisconsin reptiles are not so easily explained. Of particular note is the rarity or absence of species such as the Slender Glass Lizard (*Ophisaurus attenuatus*), Smooth Greensnake (*Opheodrys vernalis*), and Red-bellied Snake (*Storeria occipitomaculata*) in parts of the Driftless Area, especially along the Lower Wisconsin River where the North American Racer was once abundant. Suitable habitat for all of these species is present within that region, and there are no obvious dispersal barriers that would prevent colonization.

Could the North American Racer have actually limited population and range expansion of other reptile species in that area? The mechanism(s) could be related to predation (North American Racers frequently prey on other reptiles), direct competition for resources, or transmission of disease agents that some reptiles might be more susceptible to. Several studies in the northeastern US have demonstrated that small-bodied snake species (and smaller individuals of larger snake species) are less abundant in areas where North American Racers are present; the authors in each case hypothesized that racers directly limited the abundance of smaller snakes through predation (Kjoss and Litvaitis 2001, Meshaka et al. 2009, Meshaka and Delis 2014). Yet North American Racers do co-occur at sites with Slender Glass Lizards, Smooth Greensnakes, and Red-bellied Snakes in some regions, and in Wisconsin, Ring-necked Snakes and Dekay's Brownsnakes (*Storeria dekayi*) are frequently found in association with the racer. Thus, the ability for smaller snake species to cohabit sites with North American Racers may depend on a variety of factors such as adaptations possessed by smaller species to resist or cope with pressures exerted by the racer, population densities of the species involved, the assemblage of the community (including abundance of other prey items for the racer), landscape features that allow for habitat partitioning and spatial or temporal avoidance of racers, and the order in which the species colonize a site. While there is no direct evidence to prove that North American Racers influence the abundance or distribution of other reptile species, the correlations that have been observed have led to several interesting hypotheses worthy of further examination.

Jeffrey M. Lorch

individual encountered in Crawford County had swallowed a large Common Gartersnake (*Thamnophis sirtalis*; Figure 191; J. Kapfer, personal observation), and another individual captured in Crawford County regurgitated a live Ring-necked Snake (*Diadophis punctatus*; Nick Schiltz and Jason Dare, Dare Ecosystem Management, LLC, personal communication). These observations could indicate that Wisconsin populations frequently engage in ophiophagy, but further dietary analyses in the state are needed.

Prey selected by juveniles is not well-studied. Vogt (1981) reported that Wisconsin juveniles "favor small moths and other insects, as well as lizards and small

Figure 191. A road-killed North American Racer with the tail of a consumed Common Gartersnake (*Thamnophis sirtalis*) protruding from a wound on the anterior third of the snake (Crawford County; photo by J. Kapfer).

snakes." Brodie et al. (1969) examined the stomach contents of seven hatchlings from Oregon and found that six had consumed a mean of 1.8 "crickets or grasshoppers." Tucker (2000) examined the stomach contents of three young-of-the-year individuals from Illinois (30.5–32.8 cm SVL [12.01–12.91 in]), which included a field cricket and several Six-lined Racerunners (*Aspidoscelis sexlineata*).

North American Racers are consumed by many species of vertebrates and some invertebrates. Cooper et al. (2000) found that they responded strongly to chemical stimuli of milksnakes (presumably *Lampropeltis polyzona* based on Ruane et al. 2013). Due to this, the authors suggested that adult milksnakes are potential predators and that juvenile milksnakes are possible prey of North American Racers. A lengthy list of known predators throughout North America, ranging from ground beetles (*Calosoma* spp.) to American Badgers (*Taxidea taxus*), can be found in Ernst and Ernst (2003). Published reports of documented predators in Wisconsin are unknown, although various mammals, birds, and other snakes are likely predators.

When threatened, most North American Racers first attempt to flee, which is a very effective tactic (Natural History Box: Hasty Retreat). Vogt (1981) reported a maximum speed of 6.5 kph (4 mph). If escape is not possible, racers quickly employ various defensive behaviors. These include vibrating their tail and preparing to strike by coiling the anterior of their body into an S-shaped posture. If restrained, North American Racers will defecate and use their tail to smear foul-smelling cloacal excretions on their captor. They are also likely to strike and bite, although the resultant damage is minimal (J. Kapfer, personal observation). The intensity of aggression employed during defensive behaviors increases with increasing air temperature (Keogh and DeSerto 1994). Death-feigning has been reported (Lynch 1978). Interestingly, the type of response to perceived threats also changes as snakes develop, with juveniles being more defensive and adults relying more on flight (Creer 2005). This shift in behavior was reportedly related to the developmental change in color pattern that occurs as they mature from juveniles to adults (i.e., juveniles can rely on crypsis and defensive behavior due to their patterns, while adults, with no pattern, are more conspicuous and rely primarily on flight [Creer 2005]).

CONSERVATION

Status: Globally, the North American Racer is listed as a Least Concern species by the IUCN. It is considered a Species of Greatest Conservation Need, a species of Special Concern, and a Protected Wild Animal in Wisconsin by the WDNR.

Populations: Published abundance and density estimates for this species at northern latitudes are rare, but those estimates that exist show variation. Fitch and Echelle (2006) used a multiyear dataset to estimate a density of 13.2/ha (5.3/ac) for a population in Kansas. This is substantially greater than the density estimates of 1–3 adults/ha (0.4–1.2 adults/ac) for a

NATURAL HISTORY BOX: Hasty Retreat

For several years, I spent a substantial amount of time in North American Racer country conducting snake research. During this time, I was frequently standing at awkward angles on bluffside prairies, searching for snakes. It was tricky business and hard to stay upright, given that the substantial slopes I was traversing were also often littered with loose rocks. However, on more than one occasion I encountered North American Racers that made good use of these steep hillsides as they undertook hasty retreats upon my approach. They were particularly adept at remaining motionless, slightly obscured in the vegetation I was searching through, until the very last minute. When they finally bolted, I often had little hope of catching them, particularly given my precarious position on the slopes I was searching. On one such occasion, I recall 'scaring up' a racer that shot like an arrow down the slope. I took a few steps forward in a half-hearted attempt to catch the snake, which only made the animal move faster. When the snake reached the crest of a small bump on the hillside below me, it was moving so quickly that it became almost completely airborne as it launched itself off the earthen ramp. I never caught a second glimpse of the animal.

Joshua M. Kapfer, UW–Whitewater

population in northern Virginia (Ernst et al. 1997). North American Racers made up a relatively small proportion of the total number of 12 snake species captured at a location in Kansas (0.63%), suggesting a small population, although this proportion was greater than that of other large snakes on-site (Fitch and Echelle 2006). Likelihood of survival increases with age, as larger snakes are less often taken by predators. For example, annual survivorship of first-year snakes was only 15%–20%, while adult survivorship was 78% during a study in Utah (Brown and Parker 1982). Although survival is higher in adults than juveniles, differences are reported by geographic locations. Rosen (1991a) studied this species for over six years in the southeastern Lower Peninsula of Michigan and estimated that adult annual survivorship was low ($\bar{x} = 54\%$). Rosen speculated that this was due to variable and harsh winter conditions.

Threats: Habitat destruction and fragmentation probably represent substantial conservation concerns for North American Racers, and the general threats facing prairies, grasslands, and savannas apply to this species (see Introduction to Conservation and Management of Wisconsin's Amphibians and Reptiles chapter). Intense agriculture could have negative impacts in Wisconsin, as evidenced by anecdotal observations. For example, at a study site in Sauk County, few North American Racers were observed in agricultural fields compared to frequent observations in adjacent prairies and savannas (J. Kapfer, personal observation). They were also rarely encountered in woodland habitat on-site compared to open habitat (J. Kapfer, personal observation), suggesting that uncontrolled vegetative succession also negatively affects this species. Similarly, Fitch (2006b) studied snake communities at a location in northeastern Kansas for 58 years and reported that North American Racer numbers dwindled over time as survey locations transitioned from early to later successional vegetation community types with more woody vegetation. This decline was dramatic, with over 600 racers captured in the 1950s (when open habitats dominated by a "dense grass-weed mixture" and many "broad-leafed forbs" occurred on-site) and only 31 captured from 1990 to 2006 (by which time the open habitats had been almost completely replaced with "shrubs and trees" [Fitch 1999, 2006]). Klug et al. (2011b) found strong evidence that this species will maintain high genetic diversity in large undisturbed tracts of prairie in Kansas. Given that much traditional prairie and savanna habitat in Wisconsin has been eliminated, research on the genetic diversity and viability of North American Racer populations in the state (and across the Upper Midwest) is warranted.

The effects of roads are largely unknown in Wisconsin. Yet research in Georgia found that North American Racers were more likely to cross roads than expected and also exhibited the greatest speed in crossing roads among snake species tested (Andrews and Gibbons 2005). Despite this, passing vehicles elicited an "immobilization" behavior in racers during 66% of observations, and in 55% of these trials they remained motionless for up to a minute after the vehicle had passed (Andrews and Gibbons 2005). Unfortunately, the tactic of 'freezing' on roads in response to vehicles could make racers vulnerable to vehicular strikes along stretches with high traffic volume.

North American Racers are also potentially susceptible to diseases that impact reptiles (see Introduction to Conservation and Management of Wisconsin's Amphibians and Reptiles chapter). For example, Lorch et al. (2016) detected *Ophidiomyces ophiodiicola* and found clinical signs of snake fungal disease (SFD) on individuals from other regions of the US. Baker et al. (2018) also detected *O. ophiodiicola* on North American Racers in Indiana. Although the prevalence of this disease in-state is largely unknown, wild North American Racers from Wisconsin are frequently captured with external symptoms that appear consistent with SFD infections (J. Kapfer and J. Lorch, personal observations).

Illegal collection for the pet trade may negatively impact North American Racer populations, despite the species' status as protected in Wisconsin. Harvest of this snake with the intent of making it a pet seems a strange choice, given that it does not appear to tolerate captivity well. As Ernst and Ernst (2003) point out, captive individuals "normally remain nervous and high strung," and "many refuse to eat anything but their keeper's hands" (Natural History Box: Surly Snake).

NATURAL HISTORY BOX: Surly Snake

North American Racers are notorious for being defensive and ill-tempered when they perceive something as a threat. Whenever you attempt to capture or handle a wild racer, you will almost inevitably get bitten and musked—something that undoubtedly is effective against many predators. However, one adult North American Racer I encountered in the spring of 2006 had its own idea of how to deal with this annoying human. While I was extricating the snake from a 1.2 m (4 ft) oak sapling it had climbed in an attempt to escape, the snake suddenly struck the side of my thumb. However, instead of recoiling and striking again, the snake held on and began to manipulate its jaws on my thumb until its mouth was positioned over the tip of the appendage. The racer then proceeded to swallow my thumb! I watched in disbelief (thankfully, a racer's bite is essentially painless) until the snake's jaws reached the final knuckle and could not accommodate the width of my hand. With a little coaxing, I was able to retrieve my saliva-covered thumb.

Interestingly, snake keepers frequently use a technique called 'slap feeding' to get fussy snakes to eat. The premise of this feeding method is that the keeper annoys the snake with the food item until the snake strikes the prey in defense, recognizes that it is food, and then proceeds to consume it. In using the technique with captive Corn Snakes (*Pantherophis guttatus*), I have noticed that some animals would not defensively strike prey items offered in this manner but would reluctantly (and rather gently) bite and consume the food when the prey item was repeatedly used to block the snake's escape route. In such instances, it did not seem as though the snakes suddenly recognized the prey as food; instead, they were 'eating' their way out of an annoying situation—perhaps the same strategy that the North American Racer was attempting with my hand.

Jeffrey M. Lorch

Management: Maintenance of dry prairie (including open bluffsides) and savanna habitats will likely benefit North American Racers (see Introduction to Conservation and Management of Wisconsin's Amphibians and Reptiles chapter). These habitats often only persist through management strategies that employ prescribed fire. In fact, some authors report that North American Racers are more abundant in burned areas than nonburned areas (McLeod and Gates 1998). However, research in Kansas suggested that racers avoid recently burned prairie parcels, perhaps due to the initial lack of vegetative cover (Cavitt 2000, Setser and Cavitt 2003). Mosaic burn patterns and variable frequencies of fire application (rather than burning large continuous expanses of prairie annually) are recommended so that refugia from fire and cover for snakes after management remain intact. Although most reports and observations suggest that this species prefers open areas, several studies also indicate a preference for shrubby habitat (Klug et al. 2010) and woodland edges (Carfagno and Weatherhead 2006). These examples highlight the likely importance of habitats with structurally diverse vegetation for North American Racers in the midwestern US. The presence of suitable hibernacula is also critical to the persistence of this species in northern locations such as Wisconsin. Therefore, it could be beneficial if management strategies consider maintenance of more open vegetation communities around known overwintering sites. Cook (2008) reported successful translocations of this species to study sites in New York, although such strategies must be considered carefully before implementation. Because of their apparently high mobility and tendency to cross roads, North American Racers may be able to recolonize reclaimed sites if suitable habitat is made available through management.

Eastern Milksnake

Lampropeltis triangulum (Lacépède 1789)

Joshua M. Kapfer

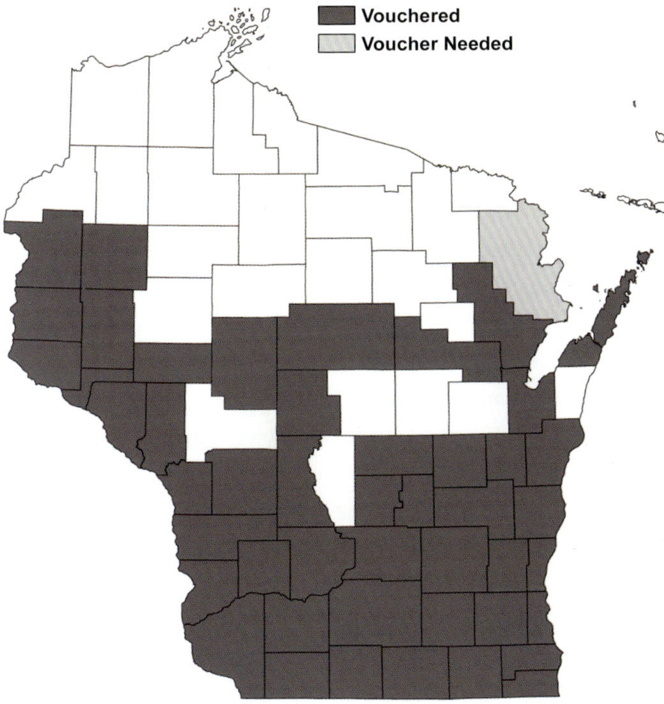

Adult Eastern Milksnake (Columbia County; photo by J. Kapfer).

AT A GLANCE

The Eastern Milksnake is Wisconsin's smallest constrictor. It can be identified by the large saddle-shaped blotches along its back, which are usually overlaid on a gray or tan background. These blotches range in color from brown to reddish or rusty and are outlined with a thin black border. Eastern Milksnakes are primarily found in the southern two-thirds of Wisconsin, where they occupy open and semi-open habitats. These include various types of prairies and shrubby woodland edges, although they are also found adjacent to open wetlands and streams. In addition, they are a well-known denizen of agricultural buildings, and an early myth surrounding their supposed desire to suckle milk from livestock is the basis for their common name. Eastern Milksnakes are active from mid-April to late October in most years. During the late autumn and winter, they sometimes find their way into the basements of houses with old foundations. Reproduction occurs shortly after emergence from hibernation in spring, and females often lay their eggs in warm humid substrates around late June. The eggs hatch by late July or early August. These harmless snakes eat a variety of small vertebrates, particularly rodents, but also have a fondness for consuming ectotherms (including other snakes). Eastern Milksnakes are considered common by the WDNR.

SYSTEMATICS

A recent phylogenetic study of the 'milksnake' group comes from Ruane et al. (2013), who included an assessment of evolutionary relationships and also delineated species distributions. This work incorporated analyses of both mitochondrial and nuclear DNA and evaluated most forms that were previously described as subspecies of *Lampropeltis triangulum* throughout North America. According to Ruane et al. (2013), the Wisconsin form, once considered *L. t. triangulum* (Crother 2012), should be classified simply as *L. triangulum* (Eastern Milksnake; Crother 2017). This species is a member of the subfamily Colubrinae, along with five other Wisconsin snakes: *Opheodrys vernalis* (Smooth Greensnake), *Coluber constrictor* (North American Racer), *Pantherophis spiloides* (Gray Ratsnake), *P. vulpinus* (Eastern Fox snake), and *Pituophis catenifer* (Gophersnake).

DESCRIPTION

Adults: The dorsal surface is covered with brown, rusty, or reddish blotches, which all have a black or dark border and are overlaid on a gray to tan background (Figure 192). The mean dorsal blotch count of 28 adult MPM specimens assessed was 38.6 (range = 31–51). The ventral surface is light gray or cream with numerous small rectangular blotches of black or brown that are haphazardly distributed. There is typically a distinct lightly colored Y- or V-shaped mark on top of the head located roughly between the eyes but posterior to the back corner of the mouth (Figure 193). A dark band runs from each eye to the posterior of the mouth, and thin dark bands are present on the posterior edges of most supralabial scales (Figure 193). The anal plate is undivided, scales are not keeled, and the snout is blunt.

This is the smallest constrictor in Wisconsin, and Vogt (1981) stated that TL was 60–90 cm (23.6–35.4 in). Measurements of adult specimens from Wisconsin housed in the MPM and UWZM collections support this, with mean SVLs that ranged from 67 cm (26.4 in, $n = 25$) in males to 68 cm (26.8 in, $n = 21$) in females. The mean tail length of these specimens was 10.7 cm (4.2 in, $n = 26$) in males and 10.1 cm (4 in, $n = 20$) in females. Live adults from

Figure 192. General body pattern of the Eastern Milksnake. Note the black border around brown elongated blotches (Sauk County; photo by J. Kapfer).

Figure 193. Head detail of the Eastern Milksnake. Note the Y-shaped marking on the top of the head, as well as the labial bars and dark band posterior to the eye (Sauk County; photo by J. Kapfer).

Above, top: Figure 194. Hatchling Eastern Milksnake with associated eggs (Ozaukee County; photo by J. Kapfer).

Above, bottom: Figure 195. Hatchling Eastern Milksnake with reddish blotches (Sauk County; photo by C. Raimond).

a population in Ozaukee County sampled from 2004 to 2018 showed greater variation by sex, with a mean male SVL of 62.4 cm (24.6 in, $n = 30$) but a mean female SVL of 56.3 cm (22.2 in, $n = 21$; Joshua Kapfer, UW–Whitewater, unpublished data). Mean tail length of snakes sampled from Ozaukee County was slightly larger for males (9.4 cm [3.7 in], $n = 28$) than for females (7.9 cm [3.1 in], $n = 21$). Males at this same location were also heavier ($\bar{x} = 78.4$ g [2.8 oz], $n = 30$) than females ($\bar{x} = 64.6$ g [2.3 oz], $n = 21$). The largest specimen measured was an adult female from Dodge County collected in 1974 (SVL = 92.6 cm [36.5 in], MPM 11718). Note that measured specimens were considered adults based on minimum size at maturity reported by Ernst and Ernst (2003).

Preadult stages: Eggs are grayish white and oblong and measure 2.0 × 4.5 cm (0.8 × 1.8 in; Vogt 1981; Figure 194). Remnants of previously hatched eggs found on two occasions in Wisconsin averaged 4.3–4.5 cm in length (1.7–1.8 in, $n = 32$; J. Kapfer, unpublished data). Eggs sampled from northern Illinois were smaller (1.6 × 3.2 cm [0.6 × 1.3 in]; Dyrkacz 1977). Hatchlings and juveniles are generally similar in appearance to adults, although blotches are often redder in coloration (Figure 195). A handful of hatchlings captured in Ozaukee County had SVLs from 18.5 to 25.2 cm (7.3 to 9.9 in) and weighed 1–6 g (0.04–0.21 oz, $n = 5$; J. Kapfer, unpublished data). The mean SVL of juvenile specimens from Wisconsin in the MPM and UWZM collections was 20.9 cm (8.2 in, $n = 49$).

Variation: No significant variation in pattern or size is known among Wisconsin individuals. There are slight differences in the blotch color, which can range from light brown to more reddish or rusty; the background color of the dorsal surface can be tan or gray.

Diagnosis: Eastern Milksnakes possess 21 dorsal scale rows at midbody, which are mostly smooth (unkeeled); ventral scale counts range from 135 to 244. Head scalation includes one loreal, one preocular, and two postocular scales, as well as seven supralabials and eight or nine infralabials on each side. Eastern Milksnakes also possess an undivided anal plate but divided subcaudal scales. Eastern Milksnakes are easiest to confuse with Eastern Foxsnakes and perhaps Gophersnakes. Although the adults of the latter two grow substantially larger than most adult Eastern Milksnakes, they are particularly difficult to differentiate when young. All three species have blotch patterns on their dorsal surfaces, although the large blotches of Eastern Milksnakes and Eastern Foxsnakes are more isolated than those of Gophersnakes. Furthermore, the blotches of Eastern Milksnakes are lighter brown, rusty, or reddish, and each blotch possesses a thin black border. The blotches of Eastern Foxsnakes and Gophersnakes, on the other hand, are dark and lack borders. The head of an adult Eastern Foxsnake is frequently without pattern, so it rarely possesses the obvious Y or V marking found on the head of an Eastern Milksnake. Once in hand, these species can be distinguished by scale characteristics. Eastern Milksnakes have an undivided anal plate, while Eastern Foxsnakes have a divided anal plate. Although both Eastern Milksnakes and Gophersnakes have an undivided anal plate, the latter possesses keeled dorsal scales and an enlarged and somewhat pointed rostral scale.

When threatened, Eastern Milksnakes will vibrate their tail, which sometimes results in their being mistaken for rattlesnakes. Such cases of mistaken identity are due to a lack of experience on the part of the observer, frequently coupled with the associated unease of encountering an unfamiliar snake in the wild. Both Wisconsin rattlesnakes possess a tail tipped with an obvious rattle, whereas nonvenomous snakes have sharply pointed tails. Wisconsin rattlesnakes also have enlarged supraocular scales and elliptical pupils, compared to the round pupils found in nonvenomous species. There are superficial similarities between Eastern Milksnakes and Massasaugas (*Sistrurus catenatus*; e.g., both can have isolated brown or rusty blotches over a gray background), which may lead to further confusion. But Massasaugas are much more thick-bodied, or stout in appearance, than Eastern Milksnakes. The similarities in appearance between Eastern Milksnakes and Timber Rattlesnakes (*Crotalus horridus*) are negligible at best, and it seems unlikely that these species could be confused.

It is difficult to distinguish Eastern Milksnake eggs encountered in the wild from the eggs of other snakes. Some level of differentiation could be possible through comparison of egg dimensions, clutch size, location in Wisconsin where eggs are encountered, and habitat in which eggs are encountered. For example, Eastern Milksnake eggs are smaller than those of Wisconsin's larger egg-laying species (e.g., Eastern Foxsnake, Gophersnake) but generally differ in shape from eggs of smaller species (e.g., Ring-necked Snakes [*Diadophis punctatus*]). Eastern Hog-nosed Snake (*Heterodon platirhinos*) eggs could pose a challenge due to similarity in size and shape, although they are possibly distinguishable by the habitat where eggs are encountered (i.e., Eastern Hog-nosed Snakes tend to dig burrows in sandy soils for nesting). Juveniles can be differentiated by the same characters used for adults and by the fact that young Eastern Milksnakes have reddish-colored blotches.

DISTRIBUTION AND HABITAT

Global distribution: The Eastern Milksnake occurs throughout the midwestern US (west into Iowa, Missouri, and Arkansas, north into the eastern portion of Minnesota, southern Wisconsin, and the Lower Peninsula of Michigan) and northeastern US (southern Maine through New England and New York), extending into Canada (parts of Ontario, southern Quebec, and east of Lake Huron). It also ranges into portions of the southeastern US (North Carolina and the northern edge of southern states such as Alabama and Georgia). There is apparently some ambiguity regarding the true taxonomic affiliation of milksnakes previously considered *Lampropeltis triangulum amaura* in northeastern Louisiana, but these could also be *L. triangulum*, according to Ruane et al. (2013).

Distribution in Wisconsin: Published county records for Eastern Milksnakes exist for much of the southern two-thirds of Wisconsin. They are even documented on Chambers Island off the Door Peninsula in Green Bay (Long et al. 1978). Of the published county records, a voucher specimen could not be found for Marinette County (see the Appendix). A review of Wisconsin records and specimens indicates that the highest concentration of reports comes from the south, particularly Sauk, Dane, Ozaukee, Milwaukee, and Walworth Counties.

Localities of possible occurrence: Several counties represent obvious gaps in the Eastern Milksnake's documented state range, including Jackson, Menominee, Portage, Waupaca, Outagamie, Kewaunee, and Adams. Eastern Milksnakes could exist in all of these counties where suitable habitat occurs. Their tendency to remain hidden beneath ground debris makes them hard to detect and has probably influenced the present gaps in their known state distribution. Eastern Milksnakes are also currently undocumented in much of the northwestern and north-central forested portions of Wisconsin. This may seem odd, given that Eastern Milksnakes appear to occupy a broad range of habitat types, including open woodlands. However, existing studies indicate that this snake prefers more open-canopy habitats, despite an occasional association with woodlands and woodland edges (see Habitat subsection). As such, it is of little surprise that Eastern Milksnakes are rare or absent from the more heavily forested northern regions of the state.

Habitat: Little research exists on the habitats used by Eastern Milksnakes in Wisconsin (e.g., Henderson et al. 1980). Anecdotal observations in the state support that they occur in a variety of vegetative communities and soil types in conjunction with both upland and wetland habitats. These include mesic to dry (and sandy) prairies and oak savannas and among rocky bluffs along the Mississippi and Wisconsin Rivers but also in sedge meadows, open riparian habitats, and sometimes open woodlands (J. Kapfer, personal observation; see Amphibians and Reptiles in Wisconsin Plant Communities chapter). Eastern Milksnakes sometimes occur in disturbed landscapes (Dyrkacz 1977). For example, they are frequently observed in agricultural settings and around the foundations of old homes (J. Kapfer, personal observation; Natural History Box: The Old Wives' Tale).

The need to regulate body temperature plays an important role in selection of habitat at both broad and fine scales. Research in Ontario revealed that Eastern Milksnakes selected habitat based on thermal properties, preferring warmer open habitats (i.e., fields, rocky outcrops, and open marshes and along woodland edges with rocky soil) over cooler forested habitats (Row and Blouin-Demers 2006a, 2006b). This could explain why Eastern Milksnakes are rarely documented in the northern forested habitats of Wisconsin. Eastern Milksnakes also select microhabitats with thermal properties that best align with preferred body temperatures (Row and Blouin-Demers 2006a, 2006b). They spend much of their time thermoregulating under cover such as old logs, rocks, and concealed in grass thatch (J. Kapfer, personal observation). Henderson et al. (1980) tracked the movements of radio-tracked females in Walworth County and almost never found them exposed on the ground surface. During this study, radio-tracked individuals were always located under ground debris or in burrows and often returned to the same shelter (a phenomenon also observed by Fitch [1999] for Western Milksnakes [*Lampropeltis gentilis*] in Kansas).

Some studies suggest that Eastern Milksnakes are not selective about the type of ground debris they use as cover, which ranges from natural debris (e.g., fallen logs and brush piles) to artificial wooden or metal objects (Dyrkacz 1977, Row and Blouin-Demers 2006a). Of note is that although Eastern Milksnakes are rarely aboveground, Gregory (2015) found that larger individuals exhibited a greater tendency to be in the open than smaller individuals. Nothing is known about the habitat preferences of neonate or juvenile Eastern Milksnakes in Wisconsin, although they are often observed in similar locations as adults and presumably use similar habitats (J. Kapfer, personal observation).

Habitats used for nesting and overwintering are not well-studied in Wisconsin. Based on occasional observations in-state and reports from other regions, they likely lay eggs in warm humid locations, such as within rotten logs and stumps, within rock piles or rock crevices on bluff faces, within compost or manure piles, and beneath discarded pieces of wood or tin (Pope 1944, Ernst and Ernst 2003, J. Kapfer, personal observation). In late autumn and early spring, Eastern Milksnakes are sometimes observed

NATURAL HISTORY BOX: The Old Wives' Tale

Most herpetologists are familiar with the story behind the common name 'milksnake.' This species' tendency to frequent barns and other agricultural buildings gave earlier European inhabitants of North America some discomfort. Strangely, the apparent consensus reached by our predecessors was that the snakes they found near barns were there for one insidious purpose: to steal milk from the cows housed therein. So strongly was this belief held that one of Wisconsin's earliest famous naturalists, P. R. Hoy, even referred to this species as the "milk sucker" (Hoy 1883). Yet, even more bizarre than early explanations for why Eastern Milksnakes were found in barns were the purported results of their attempted thievery. Descriptions I have read range from the cow's milk turning sour to the cow being drained of milk entirely by the gluttonous serpent. This is, of course, ridiculous. Vogt (1981) described the absurdity best when he stated, "This is as improbable as it is impossible. A cow would hardly allow a snake less than 1 m long to suck several gallons of milk from its udder. Even if this arrangement were physically possible, the snake's gut could only hold several tablespoons of milk." The likely reason why milksnakes were (and are) found around agricultural buildings has to do with shelter and food, just not in the way these early myths implied. An important food source for Eastern Milksnakes is rodents (not milk). Rodents are attracted to the grain and straw frequently stored in barns. Eastern Milksnakes are, in turn, attracted to the rodents. Barns and other outbuildings in rural settings, with their old foundations and various forms of stacked materials stored inside, could also provide shelter for snakes year-round. So there is no reason to fear that milk yields will be impacted by this rather shy little serpent.

Joshua M. Kapfer, UW–Whitewater

near the crevices of rocky bluff outcroppings in the Driftless Area and near large mounds of rocky glacial deposits (i.e., moraines) in southeastern Wisconsin (J. Kapfer, personal observation), which are presumably used for overwintering. They probably also overwinter in other natural underground crevices, such as channels that follow old tree root systems and abandoned animal burrows. They will sometimes enter the basements of houses with old cracked foundations in late autumn and winter (Natural History Box: Surprise Visitors).

REPRODUCTION AND DEVELOPMENT

The courtship and breeding behaviors of wild Eastern Milksnakes are seldom observed (or at least seldom reported) in Wisconsin. Vogt (1981) suggested that most breeding occurred soon after emergence from winter dormancy (April and early May), when snakes congregate near hibernacula. At these times, numerous males could attempt to mate with the same female, which might lead to male-male competition. Although information on competitive behavior during the breeding season is extremely limited, aggressive wrestling bouts for breeding opportunities between captive males have been documented (Shaw 1951).

Ernst and Ernst (2003) included a detailed description of reproductive behavior from thorough observations of captive individuals on two occasions. In both instances, a male used his mouth to grasp a female by the neck and also placed several body coils over the female's dorsal surface. The male then concurrently performed spasmic body undulations down the female's lateral sides. This behavior continued for 53 and 68 minutes, respectively, at which time the male twisted his vent toward the female's and inserted his hemipenis. Further reports

NATURAL HISTORY BOX: Surprise Visitors

There is an old farmhouse at one of the study sites I frequent in southern Wisconsin. This building was converted to housing for researchers and students a number of years ago. Although the windows were replaced and the plumbing and electricity were updated, the foundation was still old with many crevices. Eastern Milksnakes are fairly common on that site and at the right time of year can frequently be found under artificial coverboards and other debris that are scattered throughout the property. Eastern Milksnakes are also regularly found near the renovated farmhouse, and on several occasions we have encountered them moving through the grass along the house's foundation. I once mentioned this to a senior researcher and ornithologist who worked on the property. He told me that on at least one occasion his graduate students had awoken early to prepare for a day of fieldwork, headed down the steps of the old farmhouse toward the bathroom, and received a shock: an Eastern Milksnake sitting placidly on the floor at the base of the stairwell. Interestingly, the adeptness that Eastern Milksnakes show for finding small foundational crevices and accidentally gaining access to basements during the winter is well-known. In fact, I frequently field questions from concerned members of the public regarding 'home invasions' by Eastern Milksnakes, a phenomenon also mentioned by Vogt (1981). Such behavior occurs so frequently that Eastern Milksnakes were once referred to as "house snakes" in Wisconsin (Knudsen 1955f). As an example, see Figure 196, which shows a picture that I received via email in late September with the following comment: "This snake got caught in a glue trap in my friend's basement."

The following questions posed by an interested (and concerned) individual are representative of those I receive from homeowners who encounter these snakes in their residences. This particular email was sent to me in November by someone who had found several small Eastern Milksnakes in their basement: "Here are my questions: 1. Why are they so small? Does this mean they just hatched? 2. What is the best thing to do with the one I have? I have him in a container with dead wet leaves for the night. 3. If I have found 2 in my basement, should I assume there are more? 4. Any suggestions about getting rid of them?"

Figure 196. An Eastern Milksnake accidentally captured by a glue trap in a homeowner's basement (Dane County; photo by D. Levy).

Unfortunately, the answers to these questions are not always straightforward. In response to question 1: they could be juveniles, and perhaps as the result of naivete, they decided that your home was a suitable hibernaculum. It is also possible that they followed the scent trails of older 'experienced' snakes that used the same foundation as a hibernaculum. Snakes find cracks and crevices in the foundations of old houses, just as they seek such fissures underground in natural settings, to get below the frost line in winter. If their overwintering location is adjacent to a home foundation, this could result in them ending up in a basement. One could also assume that the snakes have used the foundation or areas near the foundation to overwinter for years. (Some snakes often return to the same overwintering spot over time.) Yet if this is true, then why do they end up in basements some years and not others? That question is, unfortunately, harder to answer. Perhaps in very cold winters or in years when the frost line extends deeper than normal, they move farther down and accidentally go so far as to end up in the basement. Alternatively, the fact that the homeowner sees them might just be 'luck.' In other words, maybe there are snakes in the basement every winter, but the homeowner only happens to encounter them in some years.

As far as question 2 goes, the best course of action will vary. If they are found in the basement during the summer or early to mid-autumn, one could release them outside and hope they move on. Unfortunately, this strategy comes with a host of unknown outcomes. If the snake is released too far from the property of capture in the late autumn, it is unlikely they will find a hibernaculum they are familiar with in time, and they could die in the winter. On the other hand, if released in very close vicinity to the house where they were discovered, they may end up back in the same basement again. If the snakes are found in the late autumn or winter, then releasing them outside will surely result in their mortality. In this situation, the best option for the snake might be to place it back in the basement where you found it. It is possible the snake will find the crevice it came from, crawl back inside it, and wait until spring. Another potential solution is to keep the snake in a terrarium in the basement while it hibernates and release it when spring arrives, but this will only succeed with adequate knowledge of how to effectively keep snakes under such conditions.

The answer to question 3 is easy: yes, there probably are more in the foundation. Regarding question 4: the options are limited. If you are not fond of snakes, then you might need to fix cracks in the building's foundation. You could try resealing these with mortar or something akin to expandable foam, but finding every potential crevice that a snake could use would be a challenge. You could also go through the effort to have your basement professionally resealed or new foundation walls constructed, which are expensive options. Alternatively, you could just do nothing and consider yourself fortunate. Most snake lovers have to travel considerable distances and put in long hours of searching to find Eastern Milksnakes in the wild. You need only walk down a flight of stairs!

Joshua M. Kapfer, UW–Whitewater

that outline reproduction in captive individuals exist in the literature (e.g., Miller and Grall 1978). Eastern Milksnakes are oviparous. Henderson et al. (1980) determined that egg deposition at a southern Wisconsin location likely occurred in early July and documented communal nesting. Vogt (1981) reported that females in Wisconsin lay 6–18 eggs per clutch. Similarly, females at two sites in northern Illinois laid an average of 8.9 eggs per clutch ($n = 80$; Dyrkacz 1977). Eggs take several weeks to develop, and Dyrkacz (1977) reported that they hatched by 56.6 days post-deposition in northern Illinois. During a two-year drift fence survey in Ozaukee County, hatchling Eastern Milksnakes were only captured after 10 August, which suggested that eggs hatched in late July or early August (J. Kapfer, unpublished data).

Ernst and Ernst (2003) reviewed available developmental literature on this species from across the US and Canada. They reported that Eastern Milksnakes mature by two to five years of age, when males reach 39.7–61.0 cm TL (15.6–24.0 in) and females reach ca. 40 cm TL (15.7 in). The limited information that exists suggests that growth is rapid in younger age classes. For example, Fitch and Fleet (1970) found that hatchlings added 15% to their length in the first two months of life. Estimates of longevity in the wild are lacking, although Fitch (1999) estimated that the oldest Western Milksnakes in his study were likely over ten years of age. A male from Wisconsin originally marked as an adult in 2006 (SVL = 71 cm [28 in]) was recaptured eight years later in 2014 (SVL = 84.8 cm [33.4 in]), which supports that adults can be long-lived (J. Kapfer, unpublished data).

ACTIVITY

In Wisconsin, Eastern Milksnakes typically emerge from hibernation in mid-April (Vogt 1981), and activity peaks in late spring and early summer. During a two-year drift fence survey in Ozaukee County, the highest rate of adult captures occurred from late May to late June (J. Kapfer, unpublished data), which is similar to findings from Ohio (Meshaka et al. 2008). Work on radio-tracked individuals in Ontario further supported that activity varies by season and sex (Row and Blouin-Demers 2006a). For example, peak movement for males occurred in late spring (ca. 80 m/day [262 ft/day]), perhaps because males are searching for mates during this time. On the other hand, the longest bouts of movement for females occurred in summer (ca. 60–65 m/day [197–213 ft/day]). Although information on the movement patterns of Eastern Milksnakes is generally sparse, several authors have estimated home range size. Maddalena (2018) estimated that mean home range size at two sites in Ontario was 7.02 ha (17.4 ac, SD = 3.02 ha [7.5 ac], $n = 12$) and 11.84 ha (29.3 ac, SD = 3.26 ha [8.1 ac], $n = 22$), respectively, with the largest estimated home range of 30.79 ha (76.1 ac, 95% MCP estimator). Fitch (1999) reported that the mean distance moved between captures of marked Western Milksnakes was 127 m (417 ft) for males ($n = 22$) and 106 m (348 ft) for females ($n = 16$), although recapture data were extremely limited. He further reported maximum movements that ranged from 337 to 504 m (1,106 to 1,654 ft). Because this snake spends much of its time underneath ground debris or vegetation thatch, it is probable that most movement occurs below this cover rather than while traveling exposed on the ground surface (Henderson et al. 1980, Row and Blouin-Demers 2006a). Eastern Milksnakes likely return to hibernacula from mid-October to early November in Wisconsin. The information available on their overwintering behavior and physiology in northern locales is extremely sparse. They will sometimes overwinter communally with conspecifics and other species, such as gartersnakes (*Thamnophis* spp.) and Gophersnakes (J. Kapfer, personal observation).

PREY AND PREDATORS

Detailed dietary analyses for Wisconsin (and Upper Midwest) populations do not exist. Eastern Milksnakes reportedly eat a variety of small vertebrates (Vogt 1981, reviewed by Ernst and Ernst 2003). These include rodents, shrews, birds, lizards, and snakes. Shelley (1938) reported that Eastern Milksnakes are adept at climbing to reach bird nests, presumably for feeding (in one case scaling the side of a house and spanning a foot of open space with outstretched body to reach the nest of a Chipping Sparrow [*Spizella passerina*]). However, it is important to note that birds do not represent a large portion of the snakes' diets. For example, an analysis of 180 prey items taken by milksnakes in the western US found that birds accounted for only 0.6% of all prey consumed (Hamilton et al. 2012). On the other hand, mammals accounted for 33.9% of prey consumed, while ectotherms (65.5%) made up the majority of their diets (Hamilton et al. 2012). Radio-tagged Eastern Milksnakes in Ontario were believed to eat almost exclusively mammals (Row and Blouin-Demers 2006a).

Eastern Milksnakes are ophiophagus and could be important snake predators in some regions. There is no doubt that Eastern Milksnakes consume other snakes in Wisconsin (Figure 197), and it is not uncommon to find snake scales in the scat of wild adults (J. Kapfer, personal observation). It is of further note that snakes in the genus *Lampropeltis* even consume rattlesnakes, a feat made possible by their resistance

Figure 197. An Eastern Milksnake consuming a Dekay's Brownsnake (Green County; photo by B. Richards).

Figure 198. Typical defensive posture exhibited by an Eastern Milksnake when threatened (Ozaukee County; photo by J. Kapfer).

to rattlesnake venom. This resilience is facilitated by venom-neutralizing properties that exist in their blood serum (Weinstein et al. 1992). To date, however, no verified reports of Eastern Milksnakes consuming rattlesnakes in Wisconsin exist. Behavioral studies investigating snake responses to chemical cues further support that Eastern Milksnakes are important predators of other snake species. For example, Cooper et al. (2000) found that North American Racers (*Coluber constrictor*) exhibited a strong response to the scent of Eastern Milksnakes that was consistent with predator-avoidance behavior. Ontogenetic shifts in prey consumption reportedly occur, with juveniles more frequently consuming smaller ectotherms and adults more frequently consuming endotherms (Hamilton et al. 2012, Gregory 2015). Reports of intraspecific oophagy (Mitchell and Groves 1993), dermatophagy (Weldon et al. 1993), and scavenging (Ellison and Ribic 2012) also exist.

Eastern Milksnakes are constrictors that actively forage, but they also ambush prey that are targeted via olfactory and visual cues (Ernst and Ernst 2003). This snake has a somewhat short and narrow mouth, which limits the diameter of food it can swallow. Such gape limitations could in part be what drives the snake's consumption of small or slender prey, such as other snakes (reviewed by Ernst and Ernst 2003). Despite this, Eastern Milksnakes in Ontario can consume Dekay's Brownsnakes (*Storeria dekayi*) that are nearly as long as they are (Gregory 2015).

There are a variety of vertebrates that likely consume Eastern Milksnakes, although information from Wisconsin is limited. Red-tailed Hawks (*Buteo jamaicensis*) are known predators in the state (Ross 1989a). Furthermore, numerous ground-foraging mammals, such as Raccoons (*Procyon lotor*) and Striped Skunks (*Mephitis mephitis*), could frequently encounter adults, juveniles, or eggs. It is even plausible that Wild Turkeys (*Meleagris gallopavo*), which forage on the ground, consume this snake, as could sympatric ophiophagus snakes, such as North American Racers. Eastern Milksnakes are generally cryptic, an effective strategy to avoid detection by predators, but they will also attempt to escape below vegetation thatch when threatened (J. Kapfer, personal observation). Like other constrictors, however, they will quickly adopt an aggressive posture when cornered. This includes poising their necks, apparently ready to strike, and vibrating their tails, which may result in them being mistaken for rattlesnakes (Figure 198). Although they will sometimes bite human captors if handled, their relatively small gape size and teeth cause minimal damage.

CONSERVATION

Status: The Eastern Milksnake is listed as a Least Concern species by the IUCN. It is considered common in Wisconsin by the WDNR.

Populations: Although little data exist on the ecology of this species in the Upper Midwest, one of the most detailed studies from this region is a long-term analysis of a population in Ozaukee County (Hileman et al. 2015b). This study reported an abundance estimate of 85 individuals, with a density of 0.96/ha (0.39/ac). For comparison, Fitch and Echelle (2006) estimated that Western Milksnake density was 0.6/ha (0.24/ac) at a site in northeastern Kansas. The Ozaukee County study further estimated annual survival rates of only 0.159 for juveniles but 0.72 for adults (Hileman et al. 2015b). A ten-year mark-recapture study of this species found that marked individuals were rarely captured in subsequent years. For example, of 83 Eastern Milksnake capture-recapture events that occurred during this study, only 5 recaptures occurred at one year after the original capture, and even fewer recaptures occurred after two and three years (both $n = 2$; J. Kapfer, unpublished data).

Threats: As they do with most snakes, humans represent the greatest threat to Eastern Milksnakes either through habitat destruction, collection for the pet trade, or direct killing (see Introduction to Conservation and Management of Wisconsin's Amphibians and Reptiles chapter). Loss and degradation of open habitats (in particular, prairies, grasslands, and sedge wetlands) have likely led to population declines in Wisconsin. Similarly, Maddalena (2018) found that at large spatial scales, Eastern Milksnakes in Ontario avoided agricultural and urban areas. Despite this, Eastern Milksnakes are often encountered in disturbed areas, such as those dominated by agriculture. It seems likely that this association could be the result of limited habitat options due to the dominance of agriculture in some regions of Wisconsin. Habitat fragmentation also has negative impacts, as milksnakes favor landscapes with large intact suitable habitat patches (Maddalena 2018). Roads are features that greatly enhance habitat fragmentation and can have various direct or indirect effects on Eastern Milksnakes. For example, Eastern Milksnakes experience direct mortality from vehicular strikes along roads (J. Kapfer, personal observation), but the presence of roads also restricts their movement and results in behavioral modifications (Maddalena 2018).

Direct human persecution is a threat, and the potential for encounters with people is elevated in agricultural and urban areas. Unfortunately, human persecution does not only come from citizens who simply lack awareness of the Eastern Milksnake's important ecological role. Past accounts detail how even wildlife advocates, presumably educated on ecological phenomena, have directly killed Eastern Milksnakes for unsatisfactory reasons. For example, Shelley (1938) reported killing several Eastern Milksnakes (which he referred to as both "milk snakes" and "milk adders," with no accompanying scientific name) for the offense of threatening songbirds. He recounted that one of the snakes, which came too near a Chipping Sparrow nest, "had to be shot, using craft to get it," and that he killed other Eastern Milksnakes found near nesting Eastern Phoebes (*Sayornis phoebe*). Although attitudes toward snakes have changed since Shelley's time, this is sometimes the general opinion of how to treat them. Furthermore, the tendency of Eastern Milksnakes to find their way into human domiciles probably has not endeared them to many (see Natural History Box: Surprise Visitors).

Wildlife diseases can be another important conservation issue for many snake species, although few consistent efforts to screen and monitor Eastern Milksnakes for diseases have occurred. Snake fungal disease (SFD) was recently confirmed in a single Eastern Milksnake from the Lower Peninsula of Michigan (Ravesi et al. 2016). Although the population-level impacts of this disease on Eastern Milksnakes are not well-understood, and it is not yet clinically confirmed in Wisconsin populations, numerous individuals with physical symptoms consistent with SFD have been observed in the state (J. Kapfer, personal observation).

Management: Strategies focused on management of open grassy areas and wetlands would likely benefit Eastern Milksnakes (see Introduction to Conservation and Management of Wisconsin's Amphibians and Reptiles chapter). Given the snakes' reported predilection for large intact tracts of suitable habitat (Maddalena 2018), conservation strategies for Eastern Milksnakes in Wisconsin should consider the size of available habitats and the potential connectivity among protected lands. Maintenance of the open habitats that Eastern Milksnakes use often requires an appropriate level of disturbance to offset vegetative succession. When implemented by people, this disturbance could be replicated by brush removal and application of prescribed fire in areas that Eastern Milksnakes frequent. Yet past studies suggest that it is important to consider the intensity and frequency of fire application to address the Eastern Milksnake's needs. For example, although Steen et al. (2015) reported that Eastern Milksnakes in New York responded favorably to prescribed burns, Western Milksnakes in Kansas showed a preference for sites that had not been subjected to prescribed fire for over four years (Wilgers and Horne 2006; note that this study had a small total sample size of 14). Therefore, intense management (annual application of mechanical vegetation removal or fire) could be most effective at sites that have become heavily overgrown with dense herbaceous or woody vegetation. It follows that use of intense management techniques could, therefore, be implemented more infrequently after the initial removal of dense vegetation. Implementation of a heterogeneous burning pattern is further warranted to create a structurally variable habitat. Because this snake also favors substantial ground debris for cover (Maddalena 2018), habitat managers should plan to leave woody or rocky debris of various sizes in place. It is likely beneficial to restrict intense management techniques to the Eastern Milksnake's inactive period (31 October–31 March), which would reduce the probability of accidental snake mortality during management. If seasonal restrictions are not possible, then intense management should be conducted on unusually warm or cold days when snakes are likely in refugia.

Additional management techniques exist that could benefit Eastern Milksnakes, but they require further investigation. For example, the effectiveness of translocation as a conservation and management tool is questionable, but Cook (2008) reported some success with translocations of Eastern Milksnakes in New York. Continued efforts to educate the general public about the value of snakes and the harmless nature of Eastern Milksnakes could also help reduce mortality due to human persecution. These educational efforts could highlight the fact that although Eastern Milksnakes consume birds on occasion, their diet is mostly comprised of ectothermic organisms and mammals (see Prey and Predators section).

Smooth Greensnake
Opheodrys vernalis (Harlan 1927)

Allison B. Sacerdote-Velat

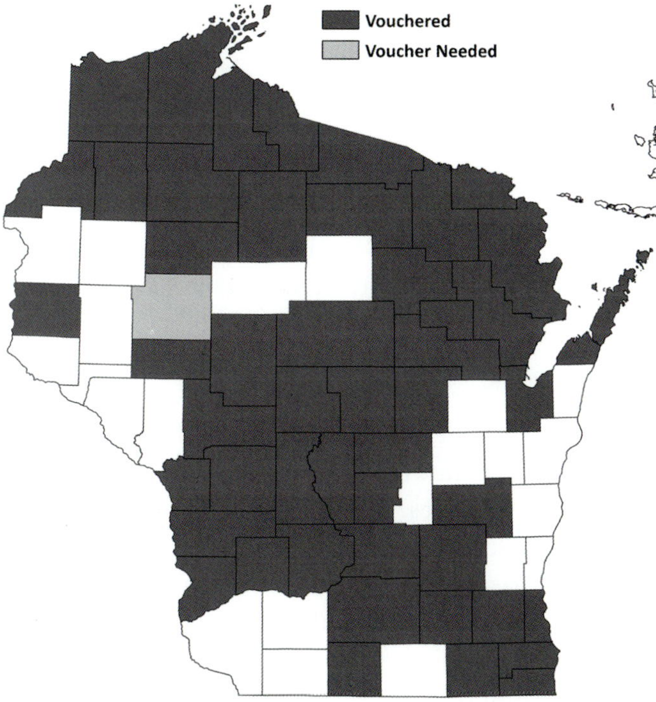

Adult Smooth Greensnake (Portage County; photo by C. Raimond).

AT A GLANCE

Smooth Greensnakes are small egg-laying snakes found throughout much of the state, although reports from some western and eastern border counties are relatively uncommon. Bright green in color and active during daytime, they are well-camouflaged in vegetation as they move through prairies, old fields, savannas, and wet meadows. Smooth Greensnakes are slender (about the width of a pen in diameter) and may reach ca. 46 cm (18.1 in) in length. These snakes are quite docile, rarely attempting to bite or musk. However, when handled they will generally attempt to escape by flailing their tail. Smooth Greensnakes are typically active from late April or early May until late September or October. They primarily feed on crickets and other orthopterans, along with smooth-bodied caterpillars, small spiders, moths, and other small arthropods. They tend to coil around the stems of forbs and grasses, moving from stem to stem above the substrate while foraging for prey. Mating generally occurs in late summer or early autumn prior to overwintering, but spring mating has also been reported. During the active season, females are frequently more conspicuous than males, as they are larger and stouter and may bask along more exposed mowed paths or prairie edges prior to laying eggs in late June or early July. Depending on when eggs are laid and latitude, eggs can hatch from late July to early September. Smooth Greensnakes are thought to be declining range-wide due to loss and fragmentation of grassland habitats but are currently considered a common species in Wisconsin by the WDNR.

SYSTEMATICS

Two clades of *Opheodrys vernalis* (Smooth Greensnake) have been identified in North America (Blais 2015). The eastern clade includes all populations east of the Appalachian Mountains and northeast and southeast of Lake Michigan, but it does not include populations in Ohio (Blais 2015). The western clade includes the Ohio populations and all populations west of Lake Michigan (Blais 2015). However, a contact zone exists within the Upper Peninsula of Michigan in which both clades occur (Blais 2015). While Grobman (1992) described three subspecies based on regional differences in scutellation (an eastern subspecies [*Opheodrys v. vernalis*], a western subspecies [*O. v. blanchardi*], and a boreal or maritime subspecies [*O. v. borealis*]), recent phylogeographic analyses do not support Grobman's morphological definition for eastern and western subspecies. Investigation of the proposed boreal subspecies was limited in sample size (Blais 2015), and the extent of its differentiation from the eastern clade is currently unknown. Relict populations of *O. vernalis* in high-elevation areas could be the result of vegetation and climate change during the recent glacial retreat (Worthington 1973, Hammerson 1981, Holman and Richards 1981, Blais 2015). *Opheodrys vernalis* is the only member of its genus found in the state. Yet it is placed within the subfamily Colubrinae, which also includes Wisconsin snakes in the genera *Coluber*, *Lampropeltis*, *Pantherophis*, and *Pituophis*.

DESCRIPTION

Adults: Smooth Greensnakes have a solid bright green dorsum and white to cream or pale yellow venter that lacks any stripes, blotches, mottling, or pattern. The labial scales are typically yellow to white, with a white throat (Figure 199a). Yellow coloration along the edges of the ventral scales may be observed in adults but seldom in juveniles or neonates (Figure 199b; Allison Sacerdote-Velat, Chicago Academy of Sciences, personal observation). Coloration becomes drab as individuals begin to shed, but coloration does not differ greatly between males and females (Figure 200). The mean SVL and tail length of presumed adult specimens from Wisconsin in the MPM, UWZM, and CHAS collections showed little variation by sex. Mean female SVL of specimens where sex was determined was 26.7 cm (10.5 in, $n = 38$), and mean tail length was 11.3 cm (4.5 in, $n = 35$). Mean male SVL was 24.8 cm (9.8 in, $n = 32$), and mean tail length was 12.2 cm (4.8 in, $n = 30$). The largest Wisconsin specimen measured was a female with an SVL of 39.1 cm (15.3 in), collected by A. E. Kurth in Dodge County in 1893 (MPM 645). In northern Illinois populations, SVL ranged from 20.0 to 37.9 cm (7.8 to 14.9 in, $\bar{x} = 24.9$ cm [9.8 in], $n = 58$), with the SVL of females ranging from 22.5 to 37.9 cm (8.8 to 14.9 in, $\bar{x} = 28.9$ cm [11.4 in],

Figure 199. Coloration patterns of adult Smooth Greensnakes. (a) The head and labial scales of an adult female (Columbia County; photo by J. Kapfer). (b) The bright green dorsum and white venter of an adult. Yellow coloration is visible along the edges of the white ventral scales (Columbia County; photo by J. Kapfer).

Figure 200. Male and female Smooth Greensnakes do not differ in coloration. However, here the male Smooth Greensnake is close to the onset of shedding, as is apparent from the drab coloring, while the female has recently shed and is brighter in coloration (McHenry County, Illinois; photo by A. Sacerdote-Velat).

$n = 26$) and the SVL of males ranging from 20.0 to 29.4 cm (7.8 to 11.6 in, $\bar{x} = 21.6$ cm [8.5 in], $n = 32$; A. Sacerdote-Velat, unpublished data). Total length of adults ranged from 29.5 to 52.0 cm (11.6 to 20.4 in, $\bar{x} = 35.7$ cm [14 in], $n = 58$). In northern Illinois populations, body weight of adults ranged from 6.4 to 42.0 g (0.22 to 1.48 oz, $n = 59$), with females ranging from 8.2 to 31.9 g (0.30 to 1.12 oz, $\bar{x} = 16.5$ g [0.58 oz], $n = 26$), and males ranging from 6.4 to 14.8 g (0.22 to 0.52 oz, $\bar{x} = 7.1$ g [0.25 oz], $n = 32$; A. Sacerdote-Velat, unpublished data).

Preadult stages: Smooth Greensnake eggs are elongate and white (Figure 201), with mean dimensions of 2.19 × 1.04 cm (0.86 × 0.40 in) in Illinois (Stille 1954a) and 2.48 × 1.20 cm (0.97 × 0.47 in) in Michigan (Blanchard 1933a). The eggs are cohesive and are typically deposited in a somewhat linear fashion or a small clump beneath natural or artificial cover. Upon emerging from the eggshell, hatchlings have a solid dark green or olive to tan dorsum that lacks any pattern and a white venter (Figure 202). Wisconsin neonates that were offspring of a buff-colored morph were described as slate gray to brown with a bronze sheen (Radaj 1981a; Natural History Box: A Snake of a Different Color). The mean SVL of Wisconsin neonates from MPM, UWZM, and CHAS collections was 8.5 cm (3.4 in, $n = 33$), and mean tail length was 3.8 cm (1.5 in, $n = 33$). Milwaukee County neonates collected from the same nest

Figure 201. Smooth Greensnake nests with cohesive eggs. Clutch sizes vary, and clutches exceeding ten eggs are likely communal nests where multiple females deposit clutches together (Lake County, Illinois; photo by A. Sacerdote-Velat).

Figure 202. Hatchling Smooth Greensnakes. (a) Two hatchlings emerging from eggs. A small egg tooth is used to slice open the eggshell. The tooth falls off during emergence (Lake County, Illinois; photo by A. Sacerdote-Velat). (b) Hatchlings and juveniles are much darker or drab in color compared to adults (Portage County; photo by E. Wild).

ranged in SVL from 9.5 to 12.7 cm (3.7 to 5.0 in), and tail length ranged from 2.5 to 3.8 cm (0.9 to 1.4 in, $n = 8$; Radaj 1981a). In a communal nest from Lake County, Illinois, neonate SVL ranged from 9.0 to 11.4 cm (3.5 to 4.4 in, $\bar{x} = 10.2$ cm [4.0 in], SD = 0.6 cm [0.2 in], $n = 83$; Sacerdote et al. 2012). Neonate weight ranged from 0.97 to 2.0 g (0.03 to 0.07 oz, $\bar{x} = 1.17$ [0.040 oz], SD = 0.04 g [0.001 oz], $n = 83$; Sacerdote et al. 2012). Juveniles have a solid pale green to bright green dorsum and a white to cream-colored venter. In northern Illinois populations, juvenile SVL ranged from 8.6 to 21.8 cm (3.3 to 8.5 in, $\bar{x} = 14.5$ cm [5.7 in], $n = 25$). Juvenile weight ranged from 0.7 to 5.7 g (0.02 to 0.20 oz, $\bar{x} = 2.6$ g [0.09 oz], $n = 25$; A. Sacerdote-Velat, unpublished data). The green color of typical hatchlings is somewhat muted. After ca. three shedding cycles, the green coloration becomes brighter but is still not the vibrant grass-green hue of the adults. The muted green coloration may persist until juveniles reach approximately two to three years of age (A. Sacerdote-Velat, personal observation).

NATURAL HISTORY BOX: A Snake of a Different Color

In 2014 during Smooth Greensnake surveys in northern Illinois, I found two male snakes beneath a plywood board in matted-down vegetation. At first glance, one appeared to be a bright green, freshly shed Smooth Greensnake and the other a drab Smooth Greensnake that was dull in color because it was close to shedding. On closer inspection, the drab individual was actually a 'buff morph' Smooth Greensnake (Figure 203). It was the only buff morph individual I had encountered since beginning surveys for the species in 2010. It was a treat to find one occupying a cover object, side by side with the bright green morph, so that I could closely examine their differences and similarities (Figure 203). The buff morph has been encountered in Wisconsin, Illinois, Indiana, and Minnesota (Schmidt and Necker 1935). A buff morph from Cook County, Illinois, was erroneously vouchered as the Smooth Earthsnake (*Virginia valeriae*), but the error was subsequently corrected by Weed (1922). However, Weed (1922) described the coloration as similar to Dekay's Brownsnake (*Storeria dekayi*).

Figure 203. Smooth Greensnake color morphs. (a) Note the difference in dorsum and venter color of this morph (DuPage County, Illinois; photo by A. Sacerdote-Velat). (b) A buff morph and green morph encountered together in a prairie adjacent to an oak savanna (DuPage County, Illinois; photo by A. Sacerdote-Velat).

Allison B. Sacerdote-Velat, Chicago Academy of Sciences

Variation: In addition to the brilliant green color that is characteristic of Smooth Greensnakes, a drab or pale brown color mutation, referred to as a buff morph (see Figure 203), is known from several Wisconsin counties, including Milwaukee (Radaj 1981a), Jackson (MPM 11560 and 11562), and Juneau (MPM 30320; see Natural History Box: A Snake of a Different Color). Aside from color variation, scutellation differences have been examined by Stille (1954a), Fritts (1968), and Grobman (1992). Female SVL also tends to exceed male SVL, while tail length of males exceeds that of females. Females are generally stouter and heavier than males, especially when gravid.

Diagnosis: Smooth Greensnakes possess one loreal scale, one preocular scale, two postocular scales, seven (or fewer) supralabial scales, seven or eight infralabial scales, and one anterior temporal scale on each side of the head. The dorsal scales are smooth (i.e., unkeeled) and occur in 15 rows at midbody. Smooth Greensnakes have 106–154 ventral scales, divided subcaudal scales, and a divided anal plate. They have a uniform dorsal coloration that is green, buff, or bronze and no bright ring around their neck. Unlike several neighboring states that harbor populations of both Smooth Greensnakes and Rough Greensnakes (*Opheodrys aestivus*), only the former is found in Wisconsin (but see Hypothetical and Enigmatic Species of Wisconsin chapter). As such, no other snake in the state is easily confused with the typical green morph adult Smooth Greensnake. Another source of misidentification results from the rapid breakdown of the pigment in the scales of Smooth Greensnakes after death. Deceased individuals appear blue in color (Necker 1939a) and could be confused with North American Racers (*Coluber constrictor*) in Wisconsin. However, North American Racers of comparable size to Smooth Greensnakes would be juveniles, and juvenile North American Racers possess a pronounced pattern that easily distinguishes them from Smooth Greensnakes. Adult coloration will be less vibrant close to the onset of shedding. However, they remain distinct from any other Wisconsin snake species.

Eggs of the Smooth Greensnake could be distinguished from those of many other oviparous Wisconsin species by their elongate shape and small size. The eggs also are not tapered at one end, as are the eggs of Ring-necked Snakes (*Diadophis punctatus*), and Smooth Greensnakes are not currently known to co-occur with Western Wormsnakes (*Carphophis vermis*) at sites in Wisconsin. When first deposited in the environment, the eggs have a pinkish hue but become solid white within 24 hours. Unlike many of the other oviparous species that bury their eggs, Smooth Greensnake eggs are often on top of the soil but usually beneath surface cover such as plant roots, moss, and small mammal nests or within sedge hummocks. The drab-colored hatchlings may appear dark brown (Figure 202b) and could be initially mistaken for neonate Dekay's Brownsnakes (*Storeria dekayi*). However, hatchling Smooth Greensnakes lack the dorsal stripes and dark subocular spot of Dekay's Brownsnakes, and they are also darker in color. The brown or buff morph Smooth Greensnakes (see Natural History Box: A Snake of a Different Color) serve as an additional source of confusion in identifying both neonates and adults of the species. Buff-colored morph adults have been initially mistaken for other species of snakes that possess brown or tan coloration, including the Dekay's Brownsnake and the Smooth Earthsnake (*Virginia valeriae*), which does not occur in Wisconsin (but see Hypothetical and Enigmatic Species of Wisconsin chapter). Buff morphs have a solid tan to drab brown dorsum without any lines or patterning and white to cream-colored ventral scales (see Natural History Box: A Snake of a Different Color).

DISTRIBUTION AND HABITAT

Global distribution: Smooth Greensnakes have a wide distribution in the northeastern US and Canada, from the Canadian Maritime Province of Nova Scotia west into Saskatchewan. Along the Atlantic seaboard, the species is distributed from Maine south through the Appalachian Mountains in Virginia. They exhibit a restricted midwestern distribution in close proximity to the Great Lakes in Ohio, Indiana, and Illinois, with Michigan and Wisconsin as the core of the range. In Illinois, they historically occurred throughout the Grand Prairie region, but currently they are principally found in the northeastern counties of Illinois because of agricultural conversion of the Grand Prairie region. Disjunct populations occur in the Great Plains region of Iowa, Nebraska, Minnesota, and North and South Dakota, including the Black Hills, in high-elevation meadows in the Rocky Mountains of Colorado, and in other high-elevation ranges in New Mexico and Utah. Isolated populations have been reported from coastal Texas (Worthington 1973, Grobman 1992) and the northern Sierra Madre in Chihuahua, Mexico (Grobman 1992, Redder et al. 2006).

Distribution in Wisconsin: Smooth Greensnakes are widespread throughout Wisconsin. Of the published county records, a reliable voucher could not be found for Chippewa (note that specimen FMNH 8936 was collected in Chippewa County, but it is only a skull and difficult to use for independent confirmation of identity; Alan Resetar, FMNH, personal communication; see the Appendix). Although the known distribution of Smooth Greensnakes covers much of Wisconsin, records are sparse in the west along the Mississippi River and the east near Lake Michigan.

Localities of possible occurrence: This species could occur in Rock County (south-central Wisconsin), as it is documented in several adjacent Wisconsin and Illinois counties. However, agricultural conversion and development in the southern part of the state has impacted historic grasslands that this species requires and could explain the paucity of records there. Furthermore, intense surveys in 2013 and 2014 at various Rock County locations did not yield Smooth Greensnakes (Joshua Kapfer, UW–Whitewater, unpublished data). Taylor, Lincoln, and Green Lake Counties are also mostly surrounded by counties with published records, so Smooth Greensnakes could occur in these counties where suitable grasslands and wet meadows persist. As noted by Casper (1996), there is a lack of records in the western counties associated with the Driftless Area, but this may be due to limited survey effort, as this species was recently documented in Crawford (Raimond and Thomsen 2016), La Crosse (Casper 2015), and Vernon (Beck et al. 2015) Counties. Given the presence of Smooth Greensnakes in Door County at the extreme eastern end of the Door Peninsula and in Brown County (Vogt 1981), they could occur in neighboring counties, such as Kewaunee, Mantiwoc, Calumet, Outagamie, and Ozaukee.

Habitat: Smooth Greensnakes are apparently a grassland-dependent species. This, coupled with their bright green coloration, likely contributed to the colloquial name 'grass snake' that is sometimes applied to them (Vogt 1981). Unfortunately, robust studies that investigate the habitat preferences of this species are few, particularly in the Upper Midwest. In fact, most reviews of Smooth Greensnake habitat are based on reports of habitat associations or observational notes. For example, Grobman (1941) reviewed many early anecdotes to conclude that Smooth Greensnakes required a "relatively moist grassy situation." Gregory (1977a) reported that individuals captured at a site in Manitoba ($n = 41$) were found "equally in wooded areas and in meadows adjacent to marshes." In Ontario, Judd (1960) encountered them frequently in an open grassy location. In Wisconsin, Vogt (1981) stated that they primarily occur within wet to mesic prairies, oak savannas with abundant grasses and sedges, and open woodlands such as northern or southern dry forests and jack pine barrens (see Amphibians and Reptiles in Wisconsin Plant Communities chapter).

There are a handful of sources that outline specific, small-scale habitat characteristics associated with this species. These sources report that Smooth Greensnakes frequently seek short-term refuge beneath natural (i.e., rocks and rotting logs; Fowler 1966) and anthropogenic ground debris on sandy loam soils (Stille 1954a, Judd 1960). In a northern Illinois radio telemetry study of wild resident and captive-reared Smooth Greensnakes, all foraging locations occurred within patches dominated by goldenrods (*Solidago* spp.) and wild Bergamot (*Monarda didyma*; Sacerdote-Velat et al. 2014). In areas where grassland habitat has been altered, destroyed, or fragmented, they may be found in fallow grazing fields (Knudsen 1955e) and in vacant lots with old field vegetation with forbs such as clover (Seibert and Hagen 1947, Seibert 1950). Baldwin (2007) reported that Smooth Greensnakes were found in habitats dominated by asters and composites in West Virginia.

Nesting reportedly occurs in open-canopy grass or sedge-dominated areas beneath logs, moss (Blanchard 1933a, Cook 1964, Fowler 1966), or artificial cover (Knudsen 1955e, Sexton and Claypool 1978, Sacerdote et al. 2012; see Reproduction and Development section). In drier years, eggs may be found within cracks in the soil, in Prairie Crayfish (*Procambarus gracilis*) burrows, or within ant nests beneath natural or artificial cover. Overwintering locations include rotten logs of large diameter or beneath rocks, gravel banks, and rocky crevices; structures made by other animals are also utilized, including ant mounds and crayfish burrows, that allow snakes to avoid frost penetration (Criddle 1937, Lachner 1942, Carpenter 1953, Lang 1971, Redder et al. 2006, Baldwin 2007). Upon excavation of ant (*Formica ulkei*) mounds in Minnesota, Lang (1971) found that overwintering snakes were spatially partitioned. He reported that while some Smooth Greensnakes were located from 0 to 264 cm (0 to 104 in) below the surface, the majority occupied the deeper portion of the ant mound (76–229 cm [30–90 in] below the surface).

REPRODUCTION AND DEVELOPMENT

Smooth Greensnakes may mate from late summer through early autumn, prior to hibernation, or in spring upon emergence from hibernation (Fitch 1970). For example, Dymond and Fry (1932) reported an initial bout of mating in June, egg deposition in mid- to late July, hatching in early August, and a second bout of mating in mid- to late August in an Ontario population. Similarly, Aldridge et al. (2009) detected the presence of active sperm in female oviducts during spring, late summer, and early autumn. In a captive breeding colony at Lincoln Park Zoo in Chicago, Illinois, a wild-caught female from northern Illinois that was gravid at the time of capture produced a clutch of seven eggs in summer 2010. After completion of nesting, she was individually housed and overwintered but still produced a viable egg the following May. This observation supports that female Smooth Greensnakes are capable of sperm storage and delayed fertilization (Sacerdote-Velat et al. 2014).

Gravid females are frequently encountered during June and July (A. Sacerdote-Velat, personal observation). Smooth Greensnakes are oviparous, and Vogt (1981) stated that eggs are laid in "rotting logs and under boards." Nests have also been found beneath old Masked Shrew (*Sorex cinereus*) nests (Jay Vecchiet, Northern Illinois University, personal observation) and within crayfish burrows and cracks in the soil (A. Sacerdote-Velat, personal observation). Nests are encountered from late June to early August (Cook 1964, Fowler 1966, Fritts 1968, Sacerdote et al. 2012). For example, the earliest nests observed at a survey location in Dane County were on 8 July in 2013 (J. Kapfer, unpublished data). In Wisconsin, mean clutch size is six eggs (Vogt 1981), and clutches of four, six, and eight eggs have been observed (Dane County; J. Kapfer, unpublished data). Reported clutch sizes from other locations are variable, ranging from 1 to 18 eggs per female, with a mean clutch size of 7 (Fowler 1966, Fritts 1968, Sexton and Claypool 1978, Radaj 1981a). However, reports of large clutch sizes may in fact be observations of small communal nests (Cook 1964, Fowler 1966, Sacerdote et al. 2012). The largest clutch size confirmed from a single female in northern Illinois was 13 eggs (A. Sacerdote-Velat, unpublished data), whereas confirmed communal nests may contain as many as 85 eggs (Sacerdote et al. 2012). Smooth Greensnake eggs exhibit great plasticity in incubation time, which shows variation by latitude (Stille 1954a). Incubation can generally last 20–30 days (Sexton and Claypool 1978, Sacerdote et al. 2012), although females can retain eggs internally until the eggs are approximately 4 days from hatching (Blanchard 1933a). In 2014, two wild nests discovered beneath a plywood board in northern Illinois hatched after a 40-day incubation period (A. Sacerdote-Velat, unpublished data). Hatching generally occurs during the last two weeks of July into the third week of August in northern Illinois (A. Sacerdote-Velat, personal observation), but it may occur later in more northern portions of this snake's range (Fowler 1966, Sexton and Claypool 1978).

Growth rates are rapid. Hatchling length increases by 0.71 cm (0.28 in) per week during the active season of their first year. Further, the length of older snakes increases from 0.83–1.11 cm (0.33–0.42 in) per week during the active season until they attain their maximum SVL (Seibert and Hagan 1947). Wild-caught Smooth Greensnakes reach reproductive maturity prior to their third hibernation period (Seibert and Hagan 1947), while juveniles reared in ex-situ facilities matured prior to their second hibernation period (Sacerdote-Velat et al. 2014). Estimates of longevity in the wild are lacking, but a captive individual reportedly lived over six years (Snider and Bowler 1992).

ACTIVITY

Smooth Greensnakes typically emerge from hibernation when the air temperature reaches 21°C–23°C (70°F–75°F) in late April and early May (Seibert and Hagen 1947, Lang 1971, Redder et al. 2006). While investigating ant mound hibernacula enclosed by a drift fence, Lang (1971) found that male Smooth Greensnakes emerged earlier than females, and young of the previous summer emerged later than adults or older juveniles. After emergence, activity is greatest during daytime at air temperatures between 21°C and 30°C (70°F and 85°F; Seibert and Hagen 1947). Limited data on the body temperature of active snakes exist, but Brattstrom (1965) reported that body temperatures ranged from 18.0°C to 31.2°C (64.4°F to 88.2°F, $\bar{x} = 25.6$°C [78.1°F]). Smooth Greensnakes generally move little while foraging, and Sacerdote-Velat et al. (2014) found that mean daily movements ranged from 1 to 4 m (3.2–13.1 ft) in northern Illinois, although movements to find nesting locations can be much larger (see Reproduction and Development section).

Ingress to hibernacula typically occurs from September through October (Lang 1971, Redder et al. 2006). Lang (1971) reported that Smooth Greensnakes arrived at an ant mound hibernaculum during September in Minnesota, when the air temperature reached a minimum of 0.6°C (33°F). In a five-year mark-recapture study in Lake County, Illinois, the latest and coldest date of capture for a Smooth Greensnake (young of the year) was 31 October 2011, with an air temperature of 5.5°C (42°F; A. Sacerdote-Velat, unpublished data). Smooth Greensnakes will communally overwinter in ant mounds with many other small snake species, ranging from Dekay's Brownsnakes to Eastern Milksnakes (*Lampropeltis triangulum*; Criddle 1937, Lachner 1942, Carpenter 1953, Lang 1971). Lang (1971) observed that Smooth Greensnakes and Common Gartersnakes (*Thamnophis sirtalis*) arrived at ant mounds earlier than Red-bellied Snakes (*Storeria occipitomaculata*). Among snakes communally hibernating in ant mounds studied by Carpenter (1953), those arriving at the ant mound earlier occupied deeper overwintering depths than individuals arriving later in the season. Smooth Greensnakes in Minnesota showed variation in fidelity to specific ant mounds used for hibernation across years that ranged from 13% to 100% (Lang 1971).

PREY AND PREDATORS

As small-bodied grassland specialists, Smooth Greensnakes are generally insectivorous. In midwestern populations, major dietary components include crickets, smooth-bodied caterpillars (especially noctuid moth larvae), spiders, harvestmen, ants, moths, worms, slugs, and snails (Judd 1960, Waters 1993, Redder et al. 2006). In Michigan, Waters (1993) found that prey items shifted with seasonal changes in abundance. Burghardt (1967) observed that naive neonate Smooth Greensnakes responded to crickets with the greatest number of tongue flicks compared to other prey species scents, including worms, leeches, fishes, larval and adult amphibians, and a control scent of water. In West Virginia, Baldwin (2007) assessed dietary composition and determined that spiders (Araneae) were the most frequent prey item (38.18% of diet), followed by lepidopteran larvae (27.27%) and orthopterans (18.18%). However, with regard to volume, lepidopteran larvae were 33.17%, orthopterans were 28.64%, and Araneae were 26.63% of the diet (Baldwin 2007). Smooth Greensnakes forage by actively hunting insects and often coil their bodies around graminoids or forbs while striking at potential prey (Baldwin 2007).

Predators are numerous, particularly avian species (e.g., diurnal raptors such as Broad-winged Hawks [*Buteo platypterus*; Rosenfield et al. 1984] and Red-shouldered Hawks [*B. lineatus*; Welch 1987], as well as domestic fowl [Guthrie 1932]). Smooth Greensnakes are also consumed by other snake species, including the Western Gartersnake (*Thamnophis elegans*; Hammerson 1999) and the Massasauga (*Sistrurus catenatus*; Riley et al. 2015). Some Smooth Greensnakes have been encountered with scars and lacerations consistent with attacks from nonnative mantis species and shrews (A. Sacerdote-Velat, personal observation), along with other injuries (Natural History Box: A Flawed Escape). The bright green dorsal coloration of this species allows it to move undetected through the vegetation, which is likely an effective defense against predators. Smooth Greensnakes will release mildly pungent cloacal secretions when captured, although they rarely bite (but see Schlauch 1975). Should an individual happen to bite a human captor, the damage is negligible and best described by Knudsen (1955e) as akin to "pinching your finger between the folds of a wet Kleenex."

Parasite fauna of Smooth Greensnakes have been reported by Chu (1936), Barrow and Stockton (1960), Ernst and Ernst (2006), and Watermolen (2014). The nematode *Rhabdias fuscavenosa* is also known

NATURAL HISTORY BOX: A Flawed Escape

In September 2011 we encountered an adult female Smooth Greensnake by a prairie pond adjacent to an oak savanna. Much to our dismay, her body was in a complete knot, although she was still attempting to move through vegetation (Figure 204). I believe that her condition was the result of a flawed attempt to escape predation. Smooth Greensnakes often coil their bodies around stems of forbs and grasses while basking. While in this position, they loosely knot or coil the last third of their bodies around the plants. My hypothesis is that a predator may have encountered the snake and tried to pull her from the vegetation, and to resist she may have wrapped herself more tightly around the plant and became knotted as a result. In addition to the knot, she had several small abrasions and cuts on her body (see Figure 204). Although she was still moving when we found her, she endured several broken vertebrae and did not survive.

Figure 204. A Smooth Greensnake encountered with a knot in her body following a presumed predation attempt (vertebrae are broken). Several lacerations and abrasions are visible closer to the head of the snake, and blue coloration is visible where scales have been damaged (Lake County, Illinois; photo by A. Sacerdote-Velat).

Allison B. Sacerdote-Velat, Chicago Academy of Sciences

to cause sublethal infections in Smooth Greensnakes (Chu 1936). However, nematodes in the genus *Rhabdias* were found throughout the gastrointestinal tract of three Smooth Greensnakes from a northern Illinois population, and the enteric infection caused subsequent emaciation (Cummins et al. 2012).

CONSERVATION

Status: Globally, the Smooth Greensnake is listed as a Least Concern species by the IUCN. It is considered common in Wisconsin by the WDNR. A conservation assessment was completed for the Western Smooth Greensnake by the USFS (Redder et al. 2006). The assessment found that many range states have reported declines in distribution, population sizes, or a general lack of knowledge regarding the population trends (Redder et al. 2006). Smooth Greensnakes are currently listed as Endangered in Iowa, Ohio, and Indiana and considered a Species of Greatest Conservation Need in Illinois, Michigan, and Connecticut, and they are considered critically imperiled in Wyoming. The species is protected by law in Wyoming, Nebraska, Colorado, Iowa, Indiana, Ohio, North Carolina, and Montana, with prohibited commercial collection (reviewed by Ernst and Ernst 2003, Redder et al. 2006).

Populations: Few rigorous studies of Smooth Greensnake populations exist. Seibert (1950) used two different techniques to estimate population size and density at a site in northeastern Illinois. He concluded that, depending on estimation method, Smooth Greensnake abundance ranged from 142 to 237, while density ranged from 109 to 183/ha (44 to 74/ac). These estimates support that Smooth Greensnake abundances can be high in optimal habitat. Some sparse data on sex ratios exist, but results are variable, with some studies reporting more males (Gregory 1977a), and others reporting more females (Grobman 1992).

Threats: Smooth Greensnakes have reportedly declined in several midwestern states. For example, Brodman et al. (2002) investigated historic vs. current occupancy of sites in Indiana by comparing past information to more recent survey efforts. Their modern surveys resulted in no Smooth Greensnake captures, and the authors believed that this species was perhaps extirpated from their current survey locations. Despite this example, there are limited data available on historic vs. current population trends for this species in the Midwest. Out of necessity, declines reported from certain midwestern states are often based on anecdotal statements in field guides or historic notes that do not always paint a clear picture. Schmidt and Necker (1935) stated that it was "common even within the city limits of Chicago but strangely restricted in its range within our region," which, aside from being purely observational, provides somewhat conflicting views on the historic status of this species in northeastern Illinois. Despite its large geographic range and listing as a common species in Wisconsin, no population-level or regional studies exist that focus on the current conservation status of this species in the state. As such, local or regional population declines may have occurred but have gone unnoticed. These examples also highlight the need for population studies in Wisconsin, where this species is apparently more secure, to help guide conservation efforts in other states.

Given this species' dependence on grasslands, oak savannas, and meadows (see Habitat subsection), Smooth Greensnake population declines are likely to occur in areas that are heavily impacted by conversion of natural grassy habitats to row crop agriculture and urbanization (Redder et al. 2006). Conservation threats related to prairies, grasslands (particularly those adjacent to wetlands), and open dry woodlands are relevant to this species (see Introduction to Conservation and Management of Wisconsin's Amphibians and Reptiles chapter). Fragmentation of habitat due to trails and roads, as well as the associated mortality along them, can further impact this species. Smooth Greensnakes often bask along or attempt to cross roads or trails early in the morning and are frequently hit by motor vehicles or bicyclists (A. Sacerdote-Velat, personal observation). The apparent disparity in abundance of Smooth Greensnakes between Wisconsin and other range states may relate to land use history, specifically to the forms of agricultural pressures across states. While Wisconsin agriculture has historically focused on dairy production (Jackson-Smith and Barham 2000), with livestock grazing as an environmental pressure on native habitats, other midwestern states that appear to have experienced declines of Smooth Greensnakes tend to be corn-producing states where row crop agriculture has altered grassland habitat to the extent that Smooth Greensnakes cannot persist (A. Sacerdote-Velat, personal observation). In grazed pastures, Smooth Greensnakes could still remain hidden from predators and find insects to feed on. For example, Smooth Greensnakes persisted in active cow pastures, despite grazing activity, in West Virginia (Baldwin 2007). With the intensification of row crop agriculture, pesticide application has increased (reviewed by Redder et al. 2006), which potentially impacts some of the arthropods that compose the prey base for Smooth Greensnakes. It is further possible that pesticide application causes direct Smooth Greensnake mortality, as suggested by anecdotal observations of Minton (1972).

Other threats to persistence may include disease and parasites (Judd 1960). While snake fungal disease (SFD) has not yet been reported from Smooth Greensnakes, an infected Common Gartersnake and an infected Eastern Foxsnake (*Pantherophis vulpinus*) were found in grasslands that support populations of Smooth Greensnakes in northern Illinois (A. Sacerdote-Velat, unpublished data). Reports of SFD exist in Wisconsin (see Introduction to Conservation and Management of Wisconsin's Amphibians and Reptiles chapter), so infection remains a possibility. Collection for the pet trade is an additional concern for this species. It is noteworthy that Vogt (1981) specifically mentioned that Smooth Greensnakes "do poorly in captivity" because most refused to eat once removed from the wild.

Management: Few habitat management recommendations specifically target Smooth Greensnakes. However, best management practices for grassland or prairie habitat, particularly strategies that maintain habitat structural complexity, are likely applicable to this species (see Introduction to Conservation and Management of Wisconsin's Amphibians and Reptiles chapter). Smooth Greensnakes could benefit from implementation of habitat corridors to offset loss of grasslands in agricultural landscapes. Knoot and Best (2011) found that Smooth Greensnakes in Iowa frequently used grassy buffer strips along waterways in agricultural habitats as refugia. The authors also found that Smooth Greensnakes associate with plant 'litter' or debris that can potentially act as cover in these disturbed landscapes.

Like that of most grassland snakes, Smooth Greensnake habitat must be managed with prescribed fire or mechanical removal of vegetation to offset vegetative succession and maintain basking and nesting sites. Unfortunately, these management practices can negatively impact Smooth Greensnakes if implemented inappropriately. For example, given the snake's association with plant litter in Iowa (Knoot and Best 2011), overly intense management practices that remove this small-scale habitat characteristic could reduce overall habitat suitability for this species. To avoid direct mortality of Smooth Greensnakes during prescribed burning, appropriately scheduled application to avoid the active season is among the important factors to consider (see further information in the Introduction to Conservation and Management of Wisconsin's Amphibians and Reptiles chapter). In regard to mechanical removal of vegetation, the individuals most often struck by mowing equipment tend to be large reproductive females, so this management activity could take a severe toll on small populations (A. Sacerdote-Velat, personal observation). Therefore, if habitat edges are mowed for trail maintenance or invasive species control, this activity should occur after 10:00 a.m., as gravid females are often found basking along the habitat edge between 8:00 and 10:00 a.m. during the active season (A. Sacerdote-Velat, personal observation). Implementation of strategies to offset direct mortality and habitat fragmentation due to roads and trails should also be considered. A possible solution is the use of wildlife underpasses or overpasses (see Introduction to Conservation and Management of Wisconsin's Amphibians and Reptiles chapter). In Indiana, a bicycle trail was built in 2011 that incorporated a wooden bridge designed to span a prairie and wetland complex where Smooth Greensnakes occurred to minimize habitat impacts and reduce snake mortality along trails (Franklin 2010). No information has yet been reported regarding success of this management effort.

Translocation as a management strategy has been attempted with this species. For example, Johnson (1991b) collected snakes from seven sites in Portage County, Wisconsin, and released them at a location in Atchison County, Missouri, where Smooth Greensnakes were believed extirpated. Although this was an isolated attempt that lacked intensive follow-up monitoring, the species did not reestablish, and the translocations were not considered successful. In the 2010s, Lake, McHenry, and DuPage Counties (Illinois) began working with Lincoln Park Zoo to monitor extant populations, incubate nests, headstart young, and establish zoo breeding colonies to augment existing populations (Sacerdote-Velat et al. 2014). At present, follow-up monitoring indicates 78%–83% survival of headstarts in the augmentation program (Sacerdote-Velat et al. 2014).

Gray Ratsnake

Pantherophis spiloides (Duméril, Bibron, and Duméril 1854)

Richard A. Sajdak

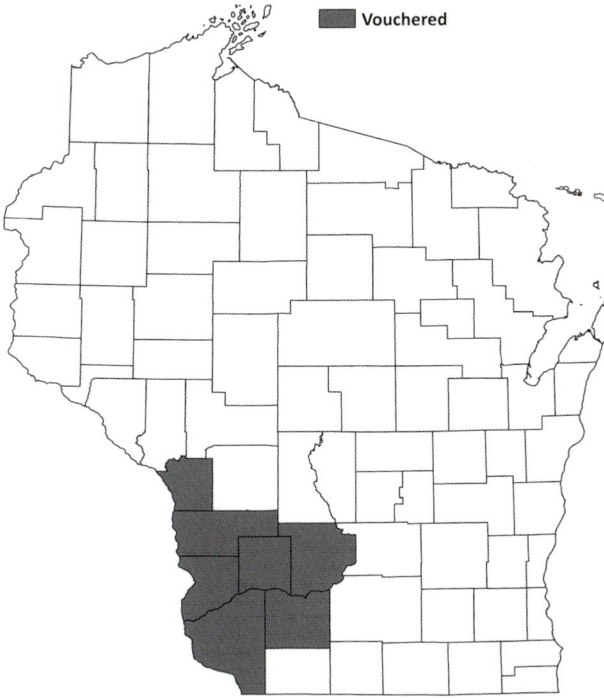

Adult Gray Ratsnake (Sauk County; photo by T. Muehlfeld).

AT A GLANCE

Gray Ratsnakes, previously known in Wisconsin as Black Ratsnakes, are one of the largest snake species in the state. The adult dorsal surface is dark gray to black, with a subtle pattern of blotches occasionally visible, and a cream or white chin and belly. Gray Ratsnakes are found in oak savannas, oak woods, wooded ravines, hollows, and coulees along the Mississippi and lower Wisconsin Rivers in southwestern Wisconsin. They emerge from winter dormancy in late April and breed shortly thereafter. Females lay eggs from mid-June to early July, which hatch from late August to early September. In late September Gray Ratsnakes migrate back to open bluff prairies, where they spend the winter in rock crevices that extend below the frost line. Adults, which are powerful constrictors, feed on rodents, birds, and their eggs. They are strong climbers and are often observed in trees or sometimes in the rafters of buildings. Because Gray Ratsnakes are rather slow-moving, their first response to danger is to freeze and rely on camouflage for protection. When threatened, a Gray Ratsnake will vibrate its tail and raise its head with an open mouth while coiling the forepart of the body. The vibrating tail may cause a buzzing or rattling sound, which, along with the snake's bold defensive display, may cause it to be confused with a rattlesnake. Gray Ratsnakes are not venomous, but they do not hesitate to bite. Bites are superficial but often bleed profusely. This species was once commonly called the Pilot Blacksnake due to the strange misconception that it led other snake species to communal hibernacula in autumn. In Wisconsin, the Gray Ratsnake is considered a Protected Wild Animal, a species of Special Concern, and a Species of Greatest Conservation Need.

SYSTEMATICS

Gray Ratsnakes in Wisconsin are part of a species complex that traditionally included *Elaphe obsoleta* (five subspecies) and *E. bairdi* (Baird's Ratsnake). The Wisconsin form was previously identified as *E. obsoleta* (Black Ratsnake; Ernst and Ernst 2003). Using a combination of morphological and genetic data, Burbrink (2001) recognized three clades (eastern, central, and western) plus *E. bairdi*. These clades did not correspond with the originally described subspecies, and Burbrink (2001) recommended that the use of subspecific epithets be discontinued. He further recognized the three clades as full species: the eastern clade as *E. alleghaniensis* (Eastern Ratsnake), the central clade as *E. spiloides* (Gray Ratsnake), and the western clade as *E. obsoletus* (Western Ratsnake), while retaining *E. bairdi* (Baird's Ratsnake) as a separate full species. Shortly thereafter, Utiger et al. (2002) used the results of genetic analyses to separate the American species of *Elaphe* from the Eurasian species and resurrected the generic name *Pantherophis* for the American clade. The studies by Utiger et al. (2002) and Burbrink (2001) have been mostly accepted, although Gibbs et al. (2006) reported interbreeding between *P. alleghaniensis* and *P. spiloides* in Ontario, casting some doubt on their status as full species. Regardless, Crother (2017) accepted *P. spiloides* (Gray Ratsnake), which is currently applied to the Wisconsin form. There is one other snake species in the genus *Pantherophis* in the state: *P. vulpinus* (Eastern Foxsnake). Both *Pantherophis* species occur in the subfamily Colubrinae, which among snakes in Wisconsin also includes species in the genera *Opheodrys*, *Coluber*, *Lampropeltis*, and *Pituophis*.

DESCRIPTION

Adults: The dorsal surface of Gray Ratsnakes is generally black to gray. They have a muted series of 24 darker middorsal blotches, with one or two rows of smaller alternating dark lateral blotches against a dark ground color (Vogt 1981). The dorsal surface of the tail is black. The head is black, while the chin, throat, and labial scales are white (Figure 205). The ventral surface has a somewhat distinct black-and-white checkerboard pattern. The dorsal scales are weakly keeled, and the sides of the body are straight rather than rounded as in most other snakes, giving the body a 'bread loaf' cross-sectional profile. The ventral scutes have a pair of strong keels running along the sides of the body. These keels and body shape are thought to contribute to the snake's excellent climbing ability (Saenz et al. 1999), allowing it to insinuate the sharp edge of the body into cracks in bark to gain purchase, much as a rock climber will lock his fingers into a rock crevice (Figure 206).

Figure 207. Juvenile Gray Ratsnake (Crawford County; photo by R. Sajdak).

Above, top: Figure 205. Adult Gray Ratsnake (Crawford County; photo by R. Sajdak).

Above, bottom: Figure 206. Gray Ratsnake climbing a tree (Crawford County; photo by R. Sajdak).

Gray Ratsnakes are the second largest species of snake in Wisconsin (TL = 100–180 cm [39.4–70.9 in]; Vogt 1981). In a small sample of wild adult Gray Ratsnakes taken from Crawford County, males were larger than females ($\bar{x}$ male SVL = 132.6 cm [52.2 in], $n = 4$, $\bar{x}$ female SVL = 122.4 cm [48.2 in], $n = 8$; Richard Sajdak, unpublished data). Males were also heavier ($\bar{x}$ = 788.3 g [27.8 oz], $n = 4$) than females ($\bar{x}$ = 740.3 g [26.1 oz], $n = 7$; R. Sajdak, unpublished data). A small number of adult Wisconsin specimens housed in the MPM and UWZM collections had a mean SVL of 115.3 cm (45.4 in, $n = 9$). All but two of these specimens were male, and only six had intact tails ($\bar{x}$ tail length = 20.1 cm [7.9 in]). The largest specimen measured was a female collected from Crawford County in 1926 (SVL = 134.5 cm [52.9 in], MPM 2601).

Preadult stages: Eggs are white, leathery, and nonadherent. Eggs of *Pantherophis spiloides* from Ohio were 38–50 × 2.0–2.4 cm (1.50–1.97 × 0.79–0.94 in; Conant 1951). Hatchlings and juveniles are much more distinctly patterned than adults (Figure 207). There is a middorsal row of dark brown blotches against a light gray ground color on the body and tail, and the venter has a black-and-white checkerboard pattern. The head is light gray with dark spots and a dark bar between the eyes across the prefrontal scales and extending through each eye posteriorly to the corner of the jaw. In Indiana, hatchlings are ca. 35 cm TL (13.8 in; Minton 2001). In Ontario, the mean SVL of 83 female hatchlings was 28.43 cm (11.19 in, SD = 1.6 cm [0.06 in]), and the mean SVL of 77 male hatchlings was 28.55 (11.24 in, SD = 1.6 cm [0.06 in]; Blouin-Demers et al. 2002). The mean SVL of 57 hatchlings (of the ratsnake species complex) was between 17.7 and 50.1 cm (7.0 and 19.7 in, $\bar{x}$ = 38.6 cm [15.2 in]), and hatchlings weighed 7.7–18.3 g (0.27–0.65 oz, $\bar{x}$ = 11.2 g [0.4 oz]; Ernst and Ernst 2003). The only Wisconsin specimen in the MPM collection categorized as a juvenile or subadult had an SVL of 75.6 cm (29.8 in, MPM 2035).

Variation: There is no significant geographic variation or sexual dimorphism in color or pattern in Gray Ratsnakes from Wisconsin, although males appear to be longer and heavier than females (see Adults subsection).

Diagnosis: Gray Ratsnakes can be distinguished by the presence of 2 prefrontal scales, while each side of the head possesses 1 (rarely 0 or 2) loreal scale, 2 postocular scales, 1 (rarely 2) preocular scale, 8 (7–9) supralabials, and 11 (9–14) infralabials. There are 25–30 dorsal scale rows at midbody and over 220 ventral scales. The anal plate and subcaudal scales are all divided. In Wisconsin, the only large snakes that could be confused with adult Gray Ratsnakes are Gophersnakes (*Pituophis catenifer*), Eastern Foxsnakes, and North American Racers (*Coluber constrictor*). Gophersnakes differ in having a yellow head with a pointed rostral shield, four prefrontal scales (vs. two in Gray Ratsnakes), strongly keeled dorsal scales, and a single anal plate. Adult Eastern Foxsnakes have a distinct series of dark blotches along the back on a straw-colored background. Minton (2001) further suggested that adult Gray Ratsnakes have a more "silvery"-colored iris, while the iris in Eastern Foxsnakes was more "golden brown." However, the most reliable difference between these two is the number of ventral scales. Eastern Foxsnakes have fewer than 220 ventral scales, while Gray Ratsnakes have more than 220 (Vogt 1981). North American Racers have smooth scales and an unmarked yellow belly. Mostly due to their strong defensive posture and tail vibrating, Gray Ratsnakes are sometimes mistaken for rattlesnakes. Gray Ratsnakes are very different from the two species of rattlesnakes in Wisconsin, the Timber Rattlesnake (*Crotalus horridus*) and Massasauga (*Sistrurus catenatus*). Both rattlesnakes differ from harmless snakes in having a tail with an obvious rattle, as well as cat-like elliptical pupils and a pit-like sensory organ on the side of the head between the eye and nostril. Unlike the mostly solid black dorsum of adult Gray Ratsnakes, both rattlesnakes have distinct patterns.

Snake eggs are difficult to distinguish in the wild. However, Gray Ratsnake eggs are larger than those of most other Wisconsin snakes. Furthermore, other snakes that also have larger eggs, such as Gophersnakes and Eastern Foxsnakes, typically lay their eggs in different habitat types. Hatchling and juvenile Gray Ratsnakes can be confused with several other Wisconsin snakes. Juvenile North American Racers and Eastern Milksnakes (*Lampropeltis triangulum*) have blotched patterns like juvenile Gray Ratsnakes, but these species have smooth, unkeeled scales. Furthermore, young North American Racers lack the band between the eyes and the checkerboard ventral pattern. Eastern Milksnakes have a distinctive V- or Y-shaped mark on the back of the head and a single anal plate. Juvenile Gophersnakes have strongly keeled scales and a pointed rostral scale and lack distinct separated blotches on the dorsum and a checkerboard ventral pattern. For juveniles, the most easily confused species is the Eastern Foxsnake. Juvenile Eastern Foxsnakes have a very similar pattern to that of juvenile Gray Ratsnakes. However, Eastern Foxsnakes are somewhat yellowish in color, while Gray Ratsnakes tend toward cooler gray tones. As mentioned previously, they also differ in ventral scale count and in the color of their irises.

DISTRIBUTION AND HABITAT

Global distribution: As currently defined, the Gray Ratsnake is found east of the Mississippi River to the Appalachian Mountains. The northern extent of its range is from southwestern Wisconsin to southern Ontario and southern New York, while the southern extent is from eastern Louisiana to Florida west of the Apalachicola River. Along East Coast states, there is currently a broad area of uncertainty regarding presence of the Gray Ratsnake vs. Eastern Ratsnake (Powell et al. 2016). Thus, in this account Maryland populations will be referred to as Eastern/Gray Ratsnakes.

Distribution in Wisconsin: Published county records for Gray Ratsnakes are concentrated in the extreme southwestern corner of the state (Casper 1996). Most populations are found in association with the

Driftless Area, in or near bluffs and ravines along the lower Mississippi and Wisconsin River valleys through the Baraboo Hills (see Past, Present, and Potential Future Landscapes of Wisconsin chapter).

Localities of possible occurrence: It is likely that the published county records represent the entire range of the Gray Ratsnake in Wisconsin. Although a 1994 photographic specimen of an adult Gray Ratsnake in Langlade County exists (MPM P513), there is no specific locality information associated with it. It is unclear whether the snake in the photograph was a waif or an accidental or deliberate translocation, but it almost certainly does not represent an actual population, and Langlade County is not included in the Gray Ratsnake's Wisconsin distribution.

Habitat: Gray Ratsnakes are typically found along bluffs in the Driftless Area of southwestern Wisconsin. They occur in open hillside prairie remnants and southern dry forests such as oak openings and on ridge tops and south- or west-facing slopes in spring, then move into oak woods and wooded ravines on east- or north-facing slopes in summer (Sajdak 2000, Vogt 1981). They are frequently observed around farm buildings and other anthropogenic structures, often in rafters or attics (Natural History Box: Chance Encounters; Figure 208). Gray Ratsnakes are the most

NATURAL HISTORY BOX: Chance Encounters

Gray Ratsnakes are frequently observed in or around anthropogenic structures, particularly in rural areas within their range. I have, for example, observed them in the rafters of old sheds on state land and even picnic shelters within state parks. Yet the best stories I have of their tendency to use anthropogenic structures come from others. For three years I spent my springs and summers working in snake habitat adjacent to the lower Wisconsin River. Although most of my work was conducted on publicly accessible property, it was often necessary to gain permission for access to private land so that I could complete my research. The landowners in this area were incredibly gracious and always granted me access to their property. They also had great stories to impart once they discovered I was working with snakes. Several such stories involved Gray Ratsnake encounters around their houses or outbuildings. Two stories came from the same landowner, who lived at the base of a wooded hillside immediately adjacent to a communal snake hibernaculum. He recounted to me one instance when he and his wife were walking around the side of their home on an early summer evening. Both were surprised to find a large Gray Ratsnake hanging off a window shutter on the house, constricting and consuming a recent meal. The snake apparently paid them little mind, and they were able to discern that its prey was a bat of unknown species. The landowner then explained how he had once forgotten to shut his front garage door on a summer evening. The next day, when he finally lowered the door, he was surprised to see a large black snake slither from its resting place atop the descending door, move out over his driveway, and disappear across the adjacent lawn. The last story that I recall came from a farmer whose barn and silo were near fields at the base of a hill supporting a snake hibernaculum. He explained to me that, years prior, he was climbing the ladder on the outside wall of the silo and had made it to a considerable height when he received a shock. While reaching for the next rung, he looked up to see a large Gray Ratsnake resting in a coil about where he was going to place his hand. Of course, he was in no danger from the snake, but the encounter unnerved him. I can only imagine he was happy that the shock hadn't caused him to lose his grip.

Joshua M. Kapfer, UW–Whitewater

Figure 208. Adult Gray Ratsnake in the rafters of a shed (Grant County; photo by P. Pieper).

arboreal of Wisconsin's snakes. For example, a pair of adult Gray Ratsnakes from Crawford County was implanted with radio transmitters and tracked during the active season of 2000. During this time, the female snake was found in arboreal locations on 70.4% of observations and the male on 87.5% of observations (R. Sajdak, unpublished data; Natural History Box: At Home in a Tree). Vogt (1981) further reported that Wisconsin individuals are often found at heights of 6–12 m (20–39 ft) in oak (*Quercus* spp.) and hickory (*Carya* spp.) trees. In Illinois, Gray Ratsnakes were also primarily arboreal, with ca. 50% of observations in trees >3 m (9 ft) aboveground.

Gray Ratsnakes appear to prefer edge habitat, or the boundary between two habitat types such as prairie and forest (e.g., Blouin-Demers and Weatherhead 2001b). Habitat selection, particularly selection for habitat edges, is apparently influenced by food availability, foraging behavior, and predation risk (Blouin-Demers and Weatherhead 2002). For example, Weatherhead and Charland (1985) reported that Gray Ratsnakes in Ontario preferred field habitat, especially edge or ecotone habitat around fields, during the bird breeding season but used more forested habitat later in the year. Furthermore, McAllister (1995) reported that Gray Ratsnakes at another Ontario site spent most of their time in marsh habitat during the bird breeding season. In contrast, Gray Ratsnakes rarely, if ever, use wetlands in Wisconsin (Richard Sajdak, Milwaukee County Zoo [retired], personal observation). Carfagno et al. (2006) found that small mammal populations in edge habitat in southern Illinois were not higher than in woodland areas. The authors concluded that some other factor, such as an abundance of nesting birds or thermoregulation, was responsible for the snakes favoring edge habitat. Gray Ratsnakes tracked with radio telemetry in Ontario were also more likely to use forest edges after feeding, presumably because these habitats provided appropriate thermoregulatory opportunities for digestion (Blouin-Demers and Weatherhead 2001a). Use of edge habitat is also reportedly related to available vegetation structure, regardless of whether edges are between two natural or natural/anthropogenic habitats (Blouin-Demers and Weatherhead 2001b).

There is documented variation in habitat use based on snake age, season, and geography. Blouin-Demers et al. (2007) found evidence that in Ontario smaller individuals showed no distinct preference for any habitat type, while larger individuals preferred forest edges. In Ontario, Weatherhead and Charland (1985) found that Gray Ratsnakes used old field habitat early in the year but moved into forest habitat later in the year. Carfagno and Weatherhead (2006) reported that edge habitat was used more by Gray Ratsnakes in Ontario, while in the more moderate climate in southern Illinois, Gray Ratsnakes favored woodland habitat over edge habitat. Keller and Heske (2000) also detected a similar lack of edge preference in Illinois, where Gray Ratsnakes used both upland and bottomland forest but did not favor forest edges. If this pattern is accurate, then Wisconsin individuals, being found in a location with intermediate climate, should favor edges more than Illinois snakes but perhaps less than Ontario snakes. Data from the pair of radio-tracked Wisconsin snakes support this hypothesis (Sajdak 2000), although more research on this topic is warranted.

Nesting likely occurs in open forested habitats, where females reportedly lay eggs in hollow logs, stumps, tree holes, sawdust piles, and rotting vegetation (reviewed by Ernst and Ernst 2003). Although observations of nesting in Wisconsin are rare, a female from Crawford County likely oviposited in a hollow log in a ravine dominated by blackberry bushes and other shrubs with scattered large oaks (R. Sajdak, personal observation; see Reproduction

NATURAL HISTORY BOX: At Home in a Tree

The pair of Gray Ratsnakes that I radio-tracked in 2000 had a particular attachment to one favored tree. I found the male in a large Red Oak (*Quercus rubra*) in open prairie habitat on a southwest-facing slope on 16 June, and I located him in the same tree on 12 subsequent observations through 9 August (Figure 209). I occasionally saw him with his head sticking out of a hole 10–12 m (30–35 ft) up in a hollow dead stub branch attached to a live branch. On one occasion I saw him move out to the end of the branch and defecate. I found several old stools beneath the branch, indicating that he used the end of the branch as a latrine on at least a few occasions.

On 29 June I found the male Gray Ratsnake lying exposed ca. 15 m (45 ft) high on a side branch of the same main branch he favored. In my field notes, I wrote the following entry: "The male black rat was still in the same tree. Walked up the hill until I was almost level with the branch he was last on, and scanned with binoculars. Found him higher up on a side branch of the same main branch he had been on before. He was about twelve feet higher, loosely coiled in shade. While I was taking data and photos, I began to notice a lot of bird calls nearby. Three or four orioles, a male and several females or juveniles were in a small tree next to the snake's big oak, chattering away. Over a few minutes, one or more would fly into the oak and land high above the snake, then fly back. Other birds landed in the small tree, including a Rose-breasted Grosbeak and Indigo Bunting. I can't make up my mind if they were mobbing the snake or not. They didn't seem focused on the snake at all, never got closer than twenty feet from him and spent most of their time in the smaller tree. I checked around and couldn't see anything else that would attract them, so maybe it was the rat snake."

He was joined in the tree on 6 July by the female after she had shed her skin and oviposited in a hollow fallen log. At the time she joined the male in the tree, several shed skins were visible on the branches, suggesting that at least one additional Gray Ratsnake had also spent time in the tree. The female had left the tree between 10 and 15 July. She had returned to the male's tree on 18 July, leaving it again by 21 July. On 15 August the male had left the tree, and I found him moving on the ground in oak-opening habitat. Other researchers have noted this tree-loving behavior in other ratsnake populations. In Maryland, radiolocated Eastern or Gray Ratsnakes were found to focus on a single tree, which "proved to be a focal point in the home ranges of many different snakes." The snakes retreated to hollow tree branches and used the tree for pre-molt sunning. Researchers also observed, as in the Wisconsin snakes, that the ratsnakes defecated in one spot under the tree (scat station). Observers also noted courtship in the favored tree (Stickel et al. 1980).

Figure 209. The male Gray Ratsnake in his tree (Crawford County; photo by R. Sajdak).

Richard A. Sajdak

and Development section). Many types of hibernacula are used by members of this species complex throughout their continental range (reviewed by Ernst and Ernst 2003), but work on northern Gray Ratsnake populations in Ontario suggests that they overwinter in fissures that exist along rocky and exposed south-facing dry cliffs or bluff slopes (Prior and Weatherhead 1996). This corroborates observations of Gray Ratsnakes tracked with radio telemetry in Wisconsin (R. Sajdak, unpublished data). Furthermore, according to Vogt (1981), they can frequently be found in late April "sunning on rocky outcrops near bluff tops" that act as hibernacula. Prior and Weatherhead (1996) also found that larger decaying or dying trees near hibernacula were used by Gray Ratsnakes as basking habitat after emerging in the spring (see Activity section).

REPRODUCTION AND DEVELOPMENT

Very little information exists on Gray Ratsnake reproduction in Wisconsin. A large pair of adult Gray Ratsnakes in Crawford County was observed in close proximity to each other on 16 May 2000 (Sajdak 2000). Although coitus between these individuals was not observed, this pairing could have represented premating behavior. Minton (2001) reported that mating in Indiana occurred from late May to early June. Male-male conflict, presumably for breeding opportunities, has been described for closely related Eastern Ratsnakes from Louisiana and Eastern/Gray Ratsnakes from Maryland (Gillingham 1980 and Stickel et al. 1980, respectively) but has not yet been documented for Gray Ratsnakes. Fitch (1999) found active sperm in the cloaca of a female Western Ratsnake during mid-October in Kansas, which suggests that autumn mating could also occur.

Ernst and Ernst (2003) reported that the gestation period for members of this species complex ranged from 37 to 51 days. Oviposition in Wisconsin could occur from mid-June to July, but data are sparse. An egg-carrying female Gray Ratsnake, one of the Crawford County pair mentioned previously, moved to a large hollow log on 26 May 2000 and remained in, on, or near the log, shedding between 29 and 30 June. She stayed in or on the log until 6 July, by which time she had laid her eggs. Egg laying in late July, with clutches of 6–22 eggs, was reported from Ohio (Conant 1951). Fitch (1970) documented egg clutches of 4–44 ($\bar{x} = 15$) in Western Ratsnakes from Kansas. Communal nesting was reported in Indiana (Lynch 1966), where 76 eggs were discovered under boards and tar paper, along with the remains of shells from previous years. Fitch (1999) estimated that only 33% of females in a Kansas population of Western Ratsnakes were gravid in a given year.

Eggs in Wisconsin likely hatch during August and September, based on a mean incubation period of 62.4 days ($n = 18$ clutches; Ernst and Ernst 2003). As it is with many snakes, growth is rapid in early years and slows considerably with age (Stickel et al. 1980). The estimated age of maturity for individuals in Ontario was ca. 9–10 years, and some might be capable of reaching ≥30 years of age (Blouin-Demers et al. 2002), although ratsnakes from warmer, more southern climates mature earlier (i.e., 4 years in Kansas [Ernst and Ernst 2003]). Gray Ratsnakes in Ontario are larger than Gray Ratsnakes in Illinois or Wisconsin, and due to the short active season (ca. 135 days in Ontario vs. 190 days in Maryland), Ontario Gray Ratsnakes of both sexes had slower growth rates and matured later than Eastern/Gray Ratsnakes from Maryland (ca. nine years in Ontario vs. four years in Maryland). However, the rapid growth and early maturation in Maryland individuals apparently occurred at the expense of longevity (maximum longevity of ca. 30 years in Ontario vs. 20 years in Maryland [Blouin-Demers et al. 2002]).

ACTIVITY

Gray Ratsnakes in Wisconsin typically emerge from hibernation in late April (Vogt 1981). Blouin-Demers et al. (2000) found that emergence at a study site in Ontario happened over a period of 40 days and that larger individuals emerged earlier than smaller individuals. This snake is primarily diurnal during the active season in Wisconsin (Vogt 1981, R. Sajdak, personal observation). Yet research from southern states found that there is a positive correlation between air temperature and nocturnal activity (i.e., at higher temperatures they are active at night), and Wisconsin individuals could also become more nocturnal in warm summer months (Sperry et al. 2013).

Mean body temperature of active adults in Ontario, based on thousands of readings collected over several years, was 28.1°C (82.6°F; Blouin-Demers and Weatherhead 2001c). Furthermore, gravid females had slightly higher body temperatures prior to oviposition than males and nongravid females. Landreth (1972) reported that a body temperature of 5°C (41°F) was near the critical thermal minimum for this species but that several individuals exposed to −4°C (24.8°F) in the laboratory for two hours survived.

Limited information exists on the movement patterns of Gray Ratsnakes in northern latitudes. The pair of radio-tracked individuals from Wisconsin had largely overlapping home ranges. The female's home range was 7.8 ha (19.3 ac); the male's home range (excluding a 650 m movement to its hibernation site) was 6.9 ha (17.0 ac, MCP estimator; R. Sajdak, unpublished data). In Illinois, male Gray Ratsnakes had larger home ranges than females (6.2 ha [15.3 ac] vs. 3.5 ha [8.6 ac]; Carfagno and Weatherhead 2008). Given the limited estimates of home range for this species in the Upper Midwest, information from other regions and other ratsnake species is valuable. Durner and Gates (1993) estimated that mean home range size of Eastern/Gray Ratsnakes in Maryland was 9.5 ha (23.5 ac, MCP estimator), while Fitch (1999) estimated a home range area of 9.84 ha (24.3 ac) for Western Ratsnakes in Kansas. Blouin-Demers et al. (2007) reported that juveniles moved more often than adults, but when adults moved, it was for greater distances. The home range size of juveniles was also smaller than that of adults (Blouin-Demers et al. 2007). Habitat influences movement rates, and snakes in forested areas moved more often over larger distances than when along forest edges in Ontario (Blouin-Demers and Weatherhead 2001b). Past work from Maryland and Kansas determined that members of this species complex continue to use the same established home range across many years (Stickel et al. 1980, Fitch 1999, respectively).

In Wisconsin, the active season is ca. five months, as Gray Ratsnakes move back to their hibernacula in mid- to late September (R. Sajdak, unpublished data). Some investigations at the far northern edge of their North American distribution in Ontario indicate that the active season is also limited to five months (Blouin-Demers et al. 2000). However, Weatherhead and Hoysak (1989) reported that the active season was more than six months in eastern Ontario. Such short active seasons result in prolonged periods of hibernation, which can pose significant challenges for this species. For example, research in Ontario found that an extended period of winter dormancy (seven to eight months) contributed to or was the cause of post-emergence mortality of several radio-tracked snakes (Prior and Shilton 1996).

In Wisconsin, Gray Ratsnakes overwinter communally with Timber Rattlesnakes, North American Racers, and other snakes (Vogt 1981, R. Sajdak, unpublished data) but will also overwinter separately. The pair of radio-tracked Gray Ratsnakes from Crawford County was initially found together on 16 May within 0.140 km (0.09 mi) of an active communal snake hibernaculum used by Timber Rattlesnakes, North American Racers, and other Gray Ratsnakes (R. Sajdak, unpublished data). That autumn they chose to ignore this active den site. The snakes did not overwinter together, instead choosing hibernacula separated by 0.67 km (0.42 mi). The female overwintered within their shared home range under a school bus–sized rock on a southwest-facing slope, while the male moved past the communal snake den site >0.65 km (0.40 mi) from the summer range to a south- to southwest-facing sandstone outcrop (Sajdak 2000). A similar pattern was observed in eastern Ontario, where snakes moved large distances to hibernacula that were located outside their active season home ranges (Weatherhead and Hoysak 1989).

PREY AND PREDATORS

The diet of Gray Ratsnakes in Wisconsin has not been well-studied but is likely dominated by birds and mammals, similar to diets recorded elsewhere. For example, species in this complex will seemingly take any suitably sized mammal or bird, including such unusual items as bats, weasels, Virginia Opossums (*Didelphis virginiana*), and adult American Kestrels (*Falco sparverius*; reviewed by Ernst and Ernst 2003). In Indiana, the diet of Gray Ratsnakes included 64% mammals, such as mice (*Peromyscus* spp. and *Mus* spp.), voles (*Microtus* spp.), young rabbits (*Sylvilagus* spp.), and squirrels (*Sciurus* spp.). Birds and eggs comprised 36% of the sample, including a Mourning Dove (*Zenaida macroura*), an American Robin (*Turdus migratorius*) and its eggs, a Red-winged Blackbird (*Agelaius phoeniceus*), domestic chicks and eggs, a House Wren (*Troglodytes aedon*) brood, and unidentified birds and eggs (Minton 2001). A pair of Gray Ratsnakes captured on 16 May 2000 in Crawford County both defecated crushed Wild Turkey (*Meleagris gallopavo*) eggshells (Sajdak 2000). Juveniles also take lizards and frogs (Minton 2001). See Ernst and Ernst (2003) for a comprehensive list of recorded prey items for the ratsnake species complex.

Birds, especially nestlings and eggs, are a tempting prize for foraging arboreal snakes. Both of the radio-tracked Wisconsin Gray Ratsnakes were strongly attracted to hollow trees and branches, even occasionally returning to the same hole from as much as hundreds of meters away, perhaps investigating bird nests for possible prey (R. Sajdak, unpublished data; see Natural History Box: At Home in a Tree). In more southern locations, ratsnakes were the most commonly detected nocturnal predators at bird nests (Stake et al. 2005). Similarly, Gray Ratsnakes in Ontario repeatedly homed to bird nests (Weatherhead and Robertson 1990). Although Gray Ratsnakes can be substantial predators of nesting birds, evidence from Ontario suggests that Gray Ratsnakes do not focus primarily on avian prey but instead take birds and mammals opportunistically (Weatherhead et al. 2003). Although mammals were always taken with greater frequency than birds, Weatherhead et al. (2003) observed seasonal variation in bird predation, with highest rates of bird consumption from May to August.

Vogt (1981) believed that Wisconsin individuals foraged both on the ground and in the tree canopy within woodland habitat. Snakes in this species complex hunt with a combination of sight and olfaction (Ernst and Ernst 2003). Sight is particularly important during arboreal foraging, and controlled experiments in Tennessee have found that prey could first be detected through olfaction but visually confirmed from a distance prior to pursuit (Mullin and Cooper 2000). Large active prey are constricted and then swallowed, while smaller helpless prey are swallowed alive. Eggs are swallowed whole and then broken in the esophagus by contracting the body and forcing the egg against the backbone. Some members of the Ratsnake species complex have enlarged processes on the ventral aspect of vertebrae along the esophagus, which facilitates the cracking of swallowed eggs (Mullen 1996), but such vertebral processes have not been reported in northern populations of the Gray Ratsnake, including those in Wisconsin.

Large adult Gray Ratsnakes probably have relatively few enemies, but juveniles are vulnerable to snakes (e.g., Eastern Milksnakes, North American Racers), hawks, owls, and small mammalian predators (e.g., Raccoons, Virginia Opossums; reviewed by Ernst and Ernst 2003). Fitch (1963a) reported that 51% of Red-tailed Hawk (*Buteo jamaicensis*) pellets from eastern Kansas contained Western Ratsnake remains, with only Prairie Voles (*Microtus ochrogaster*) being found more often. A majority of the snakes taken were juveniles, but adults were taken as well. Fitch (1963a) further believed that Red-tailed Hawks were an important factor affecting Western Ratsnake demography. Broad-winged Hawks (*Buteo platypterus*) and Great Horned Owls (*Bubo virginianus*) were occasional predators. Copperheads (*Agkistrodon contortrix*) and North American Racers were also reported to take ratsnakes (Fitch 1963a). Residents of southwestern Wisconsin commonly believe that Wild Turkeys are a substantial predator of juvenile snakes and could consume many young Gray Ratsnakes, but only anecdotal reports exist to date.

When it is threatened, a Gray Ratsnake's first response is to freeze, relying on camouflage for defense. If closely approached, the snake will attempt to flee. If the snake is cornered, it will coil up, raise its head into the air, open its mouth, and vibrate its tail. If further provoked, the snake will strike out and bite. Gray Ratsnake bites are superficial but prone to profuse bleeding (R. Sajdak, personal observation). Like many snakes, when restrained, Gray Ratsnakes will void a combination of feces, urates, and a rather foul-smelling musk from its cloaca, smearing the mixture on the offender (R. Sajdak, personal observation).

CONSERVATION

Status: Globally, the Gray Ratsnake is listed as a Least Concern species by the IUCN due to its wide distribution and tolerance of some habitat modification and because it does not seem to be in significant decline. It is considered a species of Special Concern, a Species of Greatest Conservation Need, and a Protected Wild Animal in Wisconsin by the WDNR.

Populations: Published data on Gray Ratsnake population parameters are incredibly sparse, and none exist from upper midwestern locations. Studies of Western Ratsnakes in Kansas found a mean density of 0.9/ha (0.4/ac; Fitch 1963a), while Eastern/Gray Ratsnake densities of ca. 0.23/ha (0.09/ac) were documented in Maryland (Stickel et al. 1980). Stickel et al. (1980) reported that the adult sex ratios in populations from Maryland were skewed toward females.

Threats: Several potentially important conservation threats exist in Wisconsin, including habitat loss due to residential development and agriculture in bluffland habitats (WDNR 2018d). Habitat degradation likely also has serious impacts on this species. For example, encroachment of woody vegetation on normally open bluff prairies could create excess shade that alters thermoregulatory behavior and typical patterns of ingress or egress from hibernacula (see Introduction to Conservation and Management of Wisconsin's Amphibians and Reptiles chapter). Habitat quality is further degraded as road densities in disturbed landscapes increase, which can result in snake mortality due to vehicular strikes. In a study of Gray Ratsnakes from Ontario, a total of 115 road crossings by radio-tracked snakes resulted in three instances of vehicular strikes and a mortality rate of 0.026 deaths per crossing (Row et al. 2007). A population viability analysis revealed that this level of road mortality would be enough to increase the extinction probability for the population from 7.3% to 99% over 500 years, strengthening the view that road mortality can have a pronounced negative effect on populations of long-lived species (Row et al. 2007). In Crawford County, Gray Ratsnakes (mostly juveniles) were the most commonly found snake species crossing paved and gravel roads atop bluff ridges along the Mississippi River and in coulees and ravines near the river (R. Sajdak, unpublished data). Individuals were sometimes found crossing roads in areas surrounded by farm buildings and farmland but were more commonly found along roads through ravines surrounded by oak woodlands. These data support the hypothesis that road mortality could be a serious threat to Gray Ratsnakes in Wisconsin, but further research on the population-level impacts is warranted.

Gray Ratsnakes likely face various additional threats that could impact populations in Wisconsin but are not well-studied. For example, direct interactions with humans are probably an important conservation concern. Gray Ratsnakes are occasionally encountered near anthropogenic structures, which could lead to human persecution and subsequent mortality (see Natural History Box: Chance Encounters). Documented cases of diseases such as snake fungal disease (SFD) exist for this species complex in eastern states (Sleeman 2013; see Introduction to Conservation and Management of Wisconsin's Amphibians and Reptiles chapter). Given that SFD has been detected in species that communally overwinter with Gray Ratsnakes in Wisconsin, the epidemiology of this fungal pathogen should be closely monitored. The effects of regional climate patterns on local populations could also indirectly affect Gray Ratsnakes in Wisconsin. For example, Blouin-Demers et al. (2002) suggested that Ontario's cooler climate resulted in slower growth and later maturation of ratsnakes. Their results support that populations from northern regions could experience prolonged recovery periods after declines. As Wisconsin has a cool climate that is somewhat similar to Ontario's, Gray Ratsnake populations in the state might also be slow to recover after disturbance.

Management: Ernst and Ernst (2003) aptly described this species' benefit to humans and the importance of its effective conservation and management when they stated that it "is of positive economic importance" because it is a very "efficient rodent catcher and consumes many destructive mice and rats about farms each year." Conservation strategies for Gray Ratsnakes should likely start with protection and management of its associated habitats. Management to maintain appropriate vegetation and cover around hibernacula on south-facing oak openings along the Mississippi and Wisconsin Rivers can be beneficial (see Introduction to Conservation and Management of Wisconsin's Amphibians and Reptiles chapter). Referring to Ontario populations,

Prior and Weatherhead (1996) also recommended that some large basking trees near hibernacula be preserved. Given this species' preference for forest edges, Blouin-Demers and Weatherhead (2001b) suggested that breaking continuous forest habitat into a mosaic of smaller fragments may benefit Gray Ratsnakes in Ontario by facilitating thermoregulation opportunities. However, this tactic would need careful consideration, as a strategy that actively fragments forests to manage snake populations would almost certainly have detractors in Wisconsin.

Other relevant management strategies exist, although their benefits to Gray Ratsnakes are poorly understood. For example, efforts to mitigate road mortality with road underpasses could positively affect this snake (see Introduction to Conservation and Management of Wisconsin's Amphibians and Reptiles chapter). While translocation of wildlife is sometimes identified as a possible strategy for management, the effectiveness of this technique is largely unknown. Laboratory investigations of Eastern Ratsnakes determined that time in captivity impacts effective foraging behavior, which could have implications for conservation strategies that involve translocation or reintroduction of captive snakes (DeGregorio et al. 2013).

Eastern Foxsnake

Pantherophis vulpinus (Baird and Girard 1853)

Robert W. Hay and Joshua M. Kapfer

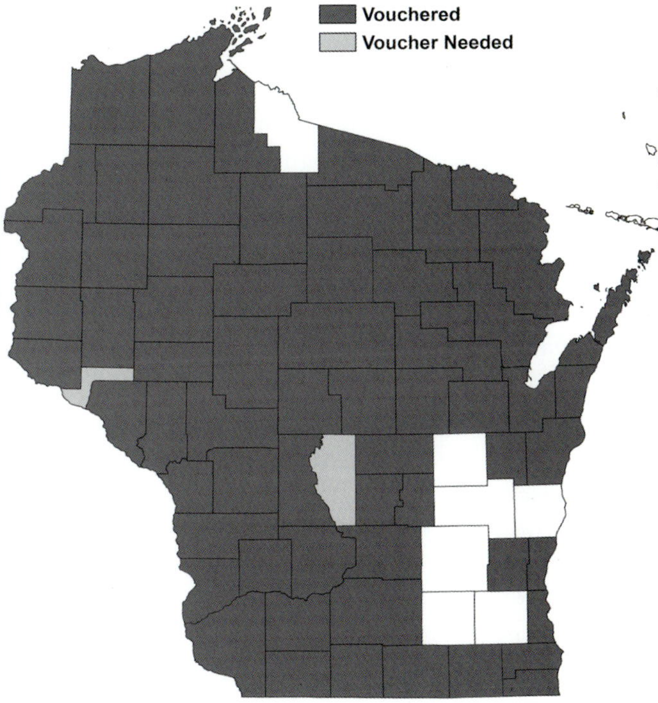

Adult Eastern Foxsnake (Washburn County; photo by E. Wild).

AT A GLANCE

Eastern Foxsnakes are one of the longest snake species in Wisconsin. Most adults have a light brown to copper-colored head that is largely without pattern, although head patterns are more discernible in juveniles. Adults and juveniles have a series of large dorsal blotches over a tan or beige background that runs the length of their bodies. Eastern Foxsnakes occur throughout much of the state, but records are sparse in portions of north-central and southeastern Wisconsin. They occupy a wide range of habitats but appear particularly fond of 'lowland' areas such as grasslands, open woodlands, and woodland edges that are close to river bottoms or wetlands. They are active from mid-April to late October. Breeding occurs in spring, and eggs are laid from June to early August, which hatch from late August to early September. Eastern Foxsnakes are nonvenomous constrictors that feed primarily on rodents, bird eggs, and nestling birds. They are sometimes mistaken for venomous species such as rattlesnakes and Copperheads due to a tendency to vibrate their tail when threatened and adult head color. However, the tails of Eastern Foxsnakes do not possess a rattle, and Copperheads do not exist in Wisconsin. They are also incorrectly referred to as 'pinesnakes' by some Wisconsin citizens. The origin of the common name 'foxsnake' is difficult to determine but may be due to the snake's tendency to emit a musky odor that is similar to that of a fox when threatened. Eastern Foxsnakes are considered a common species by the WDNR.

SYSTEMATICS

Foxsnakes are members of the North American Ratsnake group (Crother 2017) and were previously placed in the genus *Elaphe*. A recent taxonomic revision assigned them to the resurrected genus *Pantherophis* (Utiger et al. 2002), and the specific epithet of the Wisconsin form changed from *vulpina* to *vulpinus*. Until recently, foxsnakes in Wisconsin, Illinois, northwestern Indiana, the Upper Peninsula of Michigan, Minnesota, Iowa, northeastern Missouri, and the eastern regions of South Dakota and Nebraska were grouped together as *Pantherophis vulpinus* (Western Foxsnake; Conant 1940). In addition, geographically separated populations in the eastern portion of Michigan's Lower Peninsula, north-central Ohio, and Ontario were previously classified as *P. gloydi* (Eastern Foxsnake; Conant 1940). Yet analyses of mitochondrial DNA by Crother et al. (2011) supported previous hypotheses that the Mississippi River and glacial lobes in Wisconsin during the Pleistocene were important barriers that divided foxsnakes in the Midwest. They maintained that *P. gloydi* was a junior synonym of *P. vulpinus* and that all foxsnakes east of the Mississippi River were actually *P. vulpinus*. They also erected the scientific name *P. ramspotti* (Western Foxsnake) for foxsnakes west of the Mississippi River. Although some sources do not follow this classification or associated distribution for foxsnakes (e.g., Holman 2012), Crother (2017) recognized the Wisconsin form as *P. vulpinus*. The only other species in this genus that occurs in the state is *P. spiloides* (Gray Ratsnake). *Pantherophis* species are placed in the subfamily Colubrinae, along with snakes in the genera *Opheodrys*, *Coluber*, *Lampropeltis*, and *Pituophis*, all of which are represented by species in Wisconsin.

DESCRIPTION

Adults: Adult head color ranges from light brown to dark orange or copper and typically lacks a prominent pattern (Figure 210). Large dark brown saddle-shaped blotches (three to four scales long; Crother et al. 2011) run the length of the dorsal surface and occur over a yellow to beige background (see Figure 210). In some Wisconsin specimens, several middorsal blotches connect to form a V or W shape. Review of 72 museum specimens collected from various locations in Wisconsin revealed that the number of dorsal botches ranges from 30 to 50 ($\bar{x} = 37$), which is similar to the number reported by Vogt (1981; range = 34–43). The lateral areas of the dorsal surface further possess a series of smaller dark blotches. The ventral surface is pale gray to yellow with scattered dark, rectangular marks in a 'checkerboard' or 'maize' pattern.

Eastern Foxsnakes are one of the longest snakes in Wisconsin. The mean SVL of adult museum specimens from Wisconsin housed in the MPM and UWZM collections ranged from 107.7 cm (42.4 in)

Figure 210. An example of the copper-colored head and distinct dorsal blotches common in adult Eastern Foxsnakes (Portage County; photo by J. Kapfer).

Above, top: Figure 211. Eastern Foxsnake eggs. Note that some eggs in the clutch are adhered to each other (Dane County; photo by J. Kapfer).

Above, bottom: Figure 212. A juvenile Eastern Foxsnake. Note the head pattern and gray-brown background color of the dorsal surface (Richland County; photo by C. Raimond).

in males ($n = 27$) to 92.9 cm (36.6 in) in females ($n = 12$). Mean tail length of these specimens ranged from 20 cm (7.9 in) in males to 16.3 cm (6.4 in) in females. The largest known Wisconsin individual was a wild-caught male from Sauk County (TL = 183 cm [72 in]; Robert Hay, Turtles for Tomorrow, unpublished data). This snake was displayed by Chuck Naidl at his roadside reptile exhibit near Baraboo in 1972 (for more information about C. Naidl, see Lange 2004). The largest museum specimen measured was a male collected from Door County in 1910 (SVL = 141.4 cm [55.7 in], tail = 21.8 cm [8.6 in], MPM 1530). Weights for a small sample of wild-caught Wisconsin adults ranged from 575 to 790 g (20.3 to 27.9 oz, $n = 5$; Joshua Kapfer, UW–Whitewater, and Erik Wild, UW–Stevens Point, unpublished data).

Preadult stages: Eastern Foxsnakes lay oblong eggs that are light gray to white. The eggs have leathery pliable shells and frequently adhere to each other (Figure 211). The mean dimensions of seven eggs measured from two Dane County clutches were 3.9 × 2.6 cm (1.5 × 1.0 in; J. Kapfer, unpublished data), which is slightly smaller than the dimensions reported by Vogt (1981; 4.5 × 3.8 cm [1.8 × 1.5 in]).

Hatchlings and juveniles often have a patterned head, unlike that of adults, with thick bars running from the eyes to the posterior edge of the mouth and spanning the top of the head, slightly anterior to the eyes (Figure 212). The upper labial scales that are below and anterior to the eyes are shaded along the posterior edge. Paired blotches are also often visible on top of the head. The darker bars and spots on the head are typically over a gray, beige, or light brown background color. As juveniles mature, their head patterns fade, and their head color starts to resemble that of adults. Eggs laid by a Wisconsin individual that were incubated in captivity produced hatchling specimens currently housed at the MPM with a mean SVL of 24.4 cm (9.6 in) and mean tail length of 4.4 cm (1.7 in, $n = 11$, MPM 11666–11676). The mean SVL of 19 juvenile and subadult museum specimens in the MPM and UWZM that were collected in Wisconsin was 56.4 cm (22.2 in), with a mean tail length of 11 cm (4.3 in).

Variation: Aside from the differences in juvenile and adult pattern noted already, some variation exists in head color and head pattern intensity among adults. Variation also occurs in dorsal blotch pattern and background dorsal color (see Adults subsection). Albinism is documented in wild individuals but is apparently uncommon and rarely reported (Keegan 1944).

Diagnosis: Eastern Foxsnakes possess lateral dorsal scales that are smooth and central dorsal scales that are weakly keeled, as well as a divided anal plate and subcaudal scales. Head scales include one loreal scale, one (rarely two) preocular scale, and two postocular scales. The number of labial scales on each side of the head is typically 8 (7–9) for supralabials and 11 (9–13) for infralabials. The dorsal scale rows at midbody number 25–30, and ventral scale counts range from 190 to 219. Eastern Foxsnakes are sometimes confused with Eastern Milksnakes (*Lampropeltis triangulum*). Both have isolated dorsal blotches and blunt rostral scales. However, the dorsal blotches of adult Eastern Foxsnakes are solid brown, while those of Eastern Milksnakes are reddish to rusty and outlined in black. Both species possess a heavily patterned head as juveniles, but Eastern Milksnakes have a V- or Y-shaped mark on top of the head, regardless of age. Eastern Foxsnakes also have a divided anal plate, whereas Eastern Milksnakes do not. Eastern Foxsnakes are occasionally referred to as 'pinesnakes' in Wisconsin, although true Pinesnakes (*Pituophis melanoleucus*) only occur in the eastern US. Wisconsin is home to a close Pinesnake relative, the Gophersnake (*P. catenifer*), which could be confused with Eastern Foxsnakes. Yet unlike that of Eastern Foxsnakes, the anterior third of the Gophersnake's body, as well as its head and neck, is heavily patterned with small white, beige, and black flecks or mottling. Furthermore, Eastern Foxsnakes possesses a blunt (rather than pointed) rostral scale and a divided anal plate, unlike Gophersnakes. It is noteworthy that wild hybridization between these two species has been documented in Minnesota, although the incidence is probably low (Le Clere et al. 2012). Regardless, hybrid individuals are difficult to identify in the wild because they can possess characteristics reminiscent of both species (see details of hybrid characteristics in Le Clere et al. 2012).

Eastern Foxsnakes are frequently mistaken for venomous species. Because adults can have a copper-colored head, this snake is often confused with the Copperhead (*Agkistrodon contortrix*), which does not occur in Wisconsin. Eastern Foxsnakes also vibrate their tail when threatened, so they are sometimes misidentified as one of Wisconsin's rattlesnake species: the Timber Rattlesnake (*Crotalus horridus*) or the Massasauga (*Sistrurus catenatus*). Yet rattlesnakes possess obvious rattles on their tail tips, unlike the sharply pointed tail of Eastern Foxsnakes. Like all nonvenomous snakes in Wisconsin, Eastern Foxsnakes do not have heat-sensing pits located between the nostrils and eyes. Eastern Foxsnakes also do not have the protruding supraocular scales (i.e., scale above the eye) typical of Wisconsin's venomous snakes, and they possess round pupils rather than the elliptical pupils of rattlesnakes. Timber Rattlesnakes further have black dorsal bands (not blotches) and a solid black tail anterior to the rattle. While Eastern Foxsnakes and Massasaugas can both have dark isolated blotches over a gray background, the latter is discernible by its two thick parallel bands running from each eye to the back corner of the mouth.

In Wisconsin, Eastern Foxsnakes are easy to confuse with other snakes as eggs and in juvenile stages. Eggs could be partially differentiated by size, as Eastern Foxsnake eggs are among the larger of those produced by Wisconsin snakes. Their egg dimensions still might overlap with those of other large snake species such as the Gophersnake and Gray Ratsnake. Despite this, these other large snake species do not often occupy the same sites or use the same nesting habitats as Eastern Foxsnakes in Wisconsin. As juveniles, Eastern Foxsnakes and Gray Ratsnakes, which both possess a divided anal plate, are particularly difficult to differentiate due to similarities in color and pattern. Much like with eggs, however, they can be coarsely distinguished by geography. In areas of range overlap, juvenile Gray Ratsnakes usually have square or rectangular dorsal blotches, whereas the blotches of Eastern Foxsnakes are more rounded. It is also rare for Gray Ratsnakes to have the connected dorsal blotches sometimes found in Eastern Foxsnakes. Furthermore, Minton (2001) suggested that Eastern Foxsnakes have more heavily keeled scales and a "golden brown" iris compared to the "silvery" iris of Gray Ratsnakes. The most reliable method for differentiating them, however, is to closely inspect and count their ventral scales. Eastern Foxsnakes have <220 ventral scales compared to ≥220 ventral scales typical of Gray Ratsnakes.

DISTRIBUTION AND HABITAT

Global distribution: Eastern Foxsnakes primarily occur east of the Mississippi River in portions of Wisconsin, Illinois, Missouri, Indiana, Michigan, Ohio, and Ontario. An obvious geographic gap exists in this distribution, and they are not currently known in eastern Indiana, western Ohio, and central Michigan. Crother et al. (2011) reportedly agreed with J. Alan Holman's (Michigan State University) communication that Eastern Foxsnakes likely once occurred within this geographic gap but were extirpated due to habitat loss.

Distribution in Wisconsin: Eastern Foxsnakes are probably the most widely distributed constrictor in the state. Yet sightings are concentrated in certain regions of Wisconsin based on a review of over 200 confirmed records compiled from various specimen collections, WDNR reports, and the authors' personal field notes. A large number of records occur in northeastern counties, such as Oconto and Door, including several islands in Lake Michigan (Long and Long 1976, Long et al. 1978). Numerous records also exist from northwestern (Polk, St. Croix, and Buffalo Counties), central (e.g., Portage County), and south-central (Iowa, Sauk, and Dane Counties) portions of Wisconsin. This species is apparently rare or absent in parts of north-central and southeastern Wisconsin. It is also noteworthy that some of the limited existing records from the southeastern region are old or questionable (Natural History Box: Troublesome Specimens).

Of the published county records, museum vouchers could not be found for Adams and Pepin (see the Appendix). Although an Adams County specimen was reportedly present at MPM (MPM 3099), museum records indicate that it is currently "presumed missing." Walley and Walley (2004a) first reported Eastern Foxsnakes from Crawford County, but the associated voucher specimen is a misidentified juvenile Gray Ratsnake (HDW-NIU 1937, now FMNH 284199; Alan Resetar, FMNH, personal communication), invalidating the original county record. More recently a photographic voucher (MPM VZP1035a-b) was obtained to support that Eastern Foxsnakes occur in Crawford County (LaFleur and Kapfer 2022).

Localities of possible occurrence: There are counties in north-central and southeastern Wisconsin that have generated no records but are adjacent to counties of known occurrence. The reason for the paucity of sightings in these regions is not clear (Natural History Box: Perplexing Distribution), although additional survey effort could yield new records.

Habitat: Eastern Foxsnakes could be classified as adaptable habitat generalists in Wisconsin. They are found in both disturbed and undisturbed habitats, from bluff tops to lowland marshes and sometimes even forests (Knudsen 1954a; but see Threats subsection related to agricultural impacts). Despite this, Vogt (1981) specifically linked their occurrence in Wisconsin to the presence of oak savannas, pine barrens, and woodland edges. Studies from other states support this, indicating preferences for open habitats and habitat edges near wetlands or open water (DeGregorio et al. 2011b, Shew et al. 2012). For example, Eastern Foxsnakes reportedly occurred in grassland, prairie, and old field habitats in Illinois (Heske and Keller 1999, Keller and Heske 2000, Griesbaum 2012) and the Lower Peninsula of Michigan (DeGregorio et al. 2011b). The open habitats occupied by Eastern Foxsnakes contained dense herbaceous vegetation (e.g., grasses and forbs) and downed woody debris for cover (DeGregorio et al. 2011b, Shew et al. 2012). Eastern Foxsnakes also associate with woodlands adjacent to creeks, streams, and rivers in Wisconsin (Vogt 1981). For example, they are sometimes found in heavily forested areas of

NATURAL HISTORY BOX: Troublesome Specimens

Reports of Eastern Foxsnakes in southeastern Wisconsin are rare, and some of the catalogued institutional specimens from that region are particularly troublesome. The few existing specimens from Milwaukee County provide prime examples. Only three such specimens are known, and all were collected decades ago. The tag tied to one purported Milwaukee County specimen (FMNH 95163) does not mention Milwaukee County but only "Greenfield, Tank Co., Wisconsin" (Alan Resetar, FMNH, personal communication). "Tank Co." could abbreviate either 'tank company' or 'Tank County.' Yet there is no Tank County in Wisconsin, and Greenfield is in Milwaukee County, which may be why the specimen was assigned to that location. The two other specimens apparently from Milwaukee County were collected in 1973 and have no specific location information below the county level (LSU 27725–6; Jeff Boundy, Louisiana State University Museum of Natural Sciences, personal communication). The collector (Jim McLean) also did not collect extensively in Wisconsin (J. Boundy, personal communication). Thus, it is plausible that Milwaukee County was accidentally assigned as the collection location to these specimens as well.

There are precious few additional museum specimens collected in other southeastern counties, such as Ozaukee, Washington, Racine, and Kenosha. To my knowledge, Ozaukee and Washington counties have each yielded only one catalogued specimen to date (MPM 30350 and MPM 26488, respectively) and minimal (if any) additional anecdotal reports. Although specimens from Racine County are more numerous, they are also the oldest (e.g., National Museum of Natural History, USNM 7269 and 9969, Museum of Comparative Zoology at Harvard MCZ 347, 4857; J. Rosado, Harvard University, personal communication). Some are missing contextual information beyond collector, date, and simple location, while others are suspect for different reasons. For example, museum records indicate that MPM 13489–90, 18197 (collected in 1935) are juveniles and an egg from a female purportedly taken near Wind Lake in Racine County. Although this adult female was also preserved and once part of the museum's specimen collection (listed as MPM 2515), records indicate she is currently missing, so additional scrutiny is not possible. It is difficult to determine if these offspring were incubated in captivity or somehow linked to the female in the wild. If they were laid or hatched in captivity, then the adult female specimen is a more robust wild representative of this species in Racine County (a point that is, alas, moot because the adult specimen is lost). Kenosha County has likewise yielded few specimens over time (e.g., MPM 33158 and Illinois Natural History Survey [INHS] 2014ad), although these are more recent, and at least one population is currently known in the southern half of Kenosha County (Tim Muehlfeld, WE-Energies, personal communication).

It is problematic that currently many of the limited specimens from these counties are old, missing context, or questionable in some way. When combined with the lack of Eastern Foxsnake reports from adjacent counties, these troublesome specimens raise questions about why this species is so seldomly observed in southeastern Wisconsin (see Natural History Box: Perplexing Distribution).

Joshua M. Kapfer, UW–Whitewater

NATURAL HISTORY BOX: Perplexing Distribution

The distribution of Eastern Foxsnakes in Wisconsin is somewhat perplexing. Although they are commonly documented across several regions of the state, they are notably rare or absent from others. For example, they are rarely reported in a number of north-central counties, including Iron, Lincoln (only recently documented; Millis and Badje 2019), Ashland (recently documented; Casper 2015), Vilas, Price, and Oneida (rarely documented in these three). Similarly, there are no published records from the contiguous southeastern counties of Winnebago, Fond du Lac, Sheboygan, Dodge, Jefferson, and Waukesha (but see an anecdotal report from Waukesha County in Cahn 1929). Although they are documented in northeastern Illinois (Phillips et al. 1999) and records exist from several adjacent southeastern Wisconsin counties (i.e., Ozaukee, Washington, Milwaukee, Racine, and Kenosha), confirmed reports in most of these are rare, while some of the associated specimens are problematic (see Natural History Box: Troublesome Specimens). Furthermore, a more recent Walworth County record is from the far southeastern corner of the county and relatively close to known populations in Kenosha County (Badje 2020f).

Unfortunately, possible explanations for these apparent distributional gaps require speculation. On the one hand, the paucity of reports from north-central counties could be due to a lack of survey effort. However, this cannot explain the dearth of reports from the southeastern region, where many snake researchers and snake enthusiasts have been very active for decades. Human influences, such as habitat destruction, might be another factor of significance, as human densities and anthropogenic disturbance are extensive in the southeast. Yet if lack of disturbance is critical, why aren't Eastern Foxsnakes abundant in the large tracts of apparently suitable and less disturbed habitat in the Southern Kettle Moraine State Forest? Furthermore, why aren't foxsnakes more common in north-central Wisconsin, where human densities are lower, than in, for example, the central part of the state, where foxsnakes are somewhat prevalent?

It is possible that the state's glacial history could provide some insight into the current distributional pattern observed. At the end of the last glacial period, Eastern Foxsnakes could have quickly recolonized Wisconsin by following the northward retreat of melting glaciers (reviewed by Crother et al. 2011). If they were particularly fond of habitats that existed immediately after glacial retreat and continued north to follow the disappearing ice, it could possibly explain their current stronghold in counties such as Oconto and Door (compared to, say, the southeastern counties). Unfortunately, this does not adequately address why they are somewhat common in certain south-central counties (e.g., Dane), nor does it explain the limited records in the north-central region.

Perhaps, then, a comparison of available habitat features in areas where Eastern Foxsnakes are present and absent can provide insight into their Wisconsin distribution. Considering that Eastern Foxsnakes are arguably habitat generalists during the active season, sufficient suitable habitat for active foxsnakes should be present in north-central and southeastern Wisconsin. Yet it is plausible that the critical missing habitat component is related to some specific aspect of their ecology. Knudsen (1954a) commented that Eastern Foxsnakes were more prevalent in areas that possessed geological features likely to provide suitable overwintering conditions. This concept is easy to grasp when one considers the geology of several northeastern counties that have produced many foxsnake records, such as Door, eastern Marinette, and eastern Oconto. These counties possess ample exposed bedrock with many fissures near the surface that extend below the frost line (Paull and Paull 1977). Likewise, similar geology exists in some south, central, and northwestern counties that also produce many Eastern Foxsnake records (St. Croix, Polk, Buffalo, Iowa, Sauk, western Dane, and Columbia; Paull and Paull 1977).

That Eastern Foxsnakes are limited by available overwintering opportunities gains further traction if one also considers the geology of regions where this species is rare. Both the north-central and southeastern regions possess more thick clay deposits in lowland areas or sand with gravel-rich glacial deposits in uplands that could limit the presence of or access to subterranean overwintering cavities (Paull and Paull 1977). There is also little exposed surface bedrock in these locales, and that which exists is not known for abundant crevices that extend below the frost line (Paull and Paull 1977, Peter Jacobs, UW–Whitewater, personal communication). This, in conjunction with some combination of other factors, such as human disturbance and limited survey effort, could be responsible for the current dearth of records in these regions. Unfortunately, this explanation is not fully satisfying, and further investigations on this subject could produce interesting results.

Joshua M. Kapfer, UW–Whitewater

northern counties, such as Forest (Dickinson 1972) and Florence (Robert Hay, Turtles for Tomorrow, personal observation).

Anecdotal observations further suggest that Eastern Foxsnakes show a general affinity for aquatic habitats, such as wetlands and riparian corridors, but are sometimes even found in open water. Conant (1938) caught two individuals swimming, "respectively, several hundred yards and nearly a mile off shore in Lake Erie." Their association with water and ability to swim are further highlighted by several specimens at MPM, one of which was encountered on the Strawberry Islands in Green Bay (MPM 1043) and another apparently collected at a depth of "9 fathoms" in Lake Michigan (Door County, MPM 98). Radio telemetry research in Ontario may help explain how this species copes with long overwater movements to islands. MacKinnon et al. (2006) found that Eastern Foxsnakes sometimes subject themselves to large drops in body temperature (as much as 22.6°C [72.7°F]) that result from swimming in cold water. They typically engage in such activity at midday after actively basking to raise their body temperatures (MacKinnon et al. 2006). The authors suggested that this behavior helped minimize the amount of time spent swimming in cold water when necessary (i.e., warm body temperatures prior to entering water resulted in greater swimming performance, quicker arrival at destinations, and less time spent in the water before body temperature dropped).

The apparently general use of numerous habitat types by Eastern Foxsnakes could be related to differences in how they select habitat based on geographic scale. For example, work by Rowe et al. (2012) found that when habitat preferences were assessed at larger scales (i.e., the entire study site where Eastern Foxsnakes occurred), this species was associated with aquatic habitat such as marshes. On the other hand, they determined that when habitat preferences were assessed at smaller scales (i.e., constrained to the snake's home range within the overall study site), this species associated with mostly open dry habitat.

A variety of nesting habitats are reported in Wisconsin, including old stumps, piles of humus, and beneath fallen logs (Vogt 1981). Eastern Foxsnakes also deposit eggs in sawdust piles (Minnesota; Oldfield and Moriarty 1994), within rocky or gravel berms (Ontario; Row et al. 2012), and below riprap or similar locations (Missouri; Shew et al. 2012). Definitive observations of overwintering habitat in Wisconsin are sparse, but they will apparently use either natural or anthropogenic locations for hibernation. Radio-tracked individuals in Missouri overwintered near road embankments, railroad berms, and water con-

NATURAL HISTORY BOX: Communal Denning and Dug Well Destruction

In Richard Vogt's book on Wisconsin herpetofauna (Vogt 1981), a photo on page 140 shows a pile of 68 Eastern Foxsnakes that were extracted from an inactive rock-lined dug well acting as a snake hibernaculum. These snakes represented only a percentage of the 166 snakes that Vogt captured at the den. I had the privilege of visiting this site on several occasions and found that many Common Gartersnakes were also communally overwintering with the Eastern Foxsnakes on-site. In the early 2000s, this well was discovered by private citizens, who reported to the WDNR that it had not been properly filled in when abandoned (following Wisconsin Administrative Code NR 811.01). As the Wisconsin state herpetologist with the WDNR at the time, I was alerted to the situation and knew that a lot of snakes would likely die if attempts to fill the well were made during the winter, prior to snake emergence from hibernation. Although I attempted to sufficiently delay the procedure, my efforts unfortunately failed. Therefore, we attempted to rescue as many snakes as we could before the well was destroyed but were only able to collect 3 adult Eastern Foxsnakes and 14 Common Gartersnakes. That was a very sad battle to lose.

Since that time there have been other instances where construction of an alternative and WDNR-approved den site was possible. Constructed hibernacula were designed to mimic the characteristics of the dug wells that were filled and were installed adjacent to wells prior to well destruction. Follow-up surveys have provided evidence that some of these structures now support overwintering snakes. See Hastings (2009) for information on how to construct a legal den site that reduces the threat of groundwater contamination.

Robert W. Hay, Turtles for Tomorrow

trol canals, while others used natural banks and animal burrows near wetlands and wet prairies (Greisbaum 2012, Shew et al. 2012). Research in Illinois documented an old railroad berm/trestle with large irregular rocks at the base that served as a hibernaculum (Heske and Keller 1999, Keller and Heske 2000). In Wisconsin, they occasionally overwinter in the basements of older residential structures (R. Hay and J. Kapfer, personal observation). Vogt (1981) and Zaremba (1978) further reported that Eastern Foxsnakes overwintered in abandoned wells or cisterns in Portage and Monroe Counties, respectively (Natural History Box: Communal Denning and Dug Well Destruction).

REPRODUCTION AND DEVELOPMENT

Mating occurs soon after spring emergence in Wisconsin (from late April to early May; Gillingham 1974), although later reproductive activity has been documented (Erik Wild, UW–Stevens Point, personal communication). Eastern Foxsnakes often mate near hibernacula and reportedly exhibit "Coluber type" courtship and mating behavior that includes three phases: chasing, tactile and mounting, and biting and intromission (Davis 1936, Gillingham 1974; Figure 213). Gillingham et al. (1977) later defined these three phases as tactile-chase, tactile-alignment, and intromission-coitus. Occasionally, unusual behaviors associated with reproduction are reported. For example, Davis (1936) described a unique observation made near Chicago that was apparently related to reproduction in which one snake "had swallowed some four inches of the anterior end of the second." Despite the situation, both snakes continued to undertake the undulating motions associated with the tactile and mounting phase. The author returned later to find that the second snake had somehow dislodged its head from the throat of the first. Yet rather than fleeing, both snakes remained

Figure 213. Eastern Foxsnakes during a reproductive event on 12 June 2014. Note that the male is on top of the female and has grasped her head in his jaws (Trempealeau County; photo by. E. Wild).

together and continued the undulating movements, although the author did not witness coitus.

The gestation period is short. In fact, Gilligham (1974) found that egg deposition occurred from late June to early August, although observations in Wisconsin indicate that most eggs are laid from late June to early July (Vogt 1981, R. Hay, personal observation). Although gestation generally does not last long, the duration of this period likely varies based on the environmental conditions that females are exposed to while gravid. Furthermore, females apparently do not actively elevate their body temperatures during embryo development, as research in Ontario found that gravid females maintained body temperatures similar to those of nongravid females and males (Willson and Brooks 2006). It is possible that gravid females do not elevate their body temperature because the gestation period is so short. As such, the energy gained by actively thermoregulating is not advantageous to developing embryos, so gravid females might not be justified in spending time on this endeavor over other activities such as foraging (Willson and Brooks 2006).

Eggs are deposited within warm humid debris (see Habitat subsection). Vogt (1981) reported clutch sizes in Wisconsin that ranged from 8 to 27 eggs. Clutches examined from Marathon County contained 18–21 eggs ($\bar{x} = 19.6$, $n = 7$ clutches; R. Hay, unpublished data), while two clutches examined from Dane County contained 12 and 13 eggs (J. Kapfer, unpublished data). Communal nesting is known in Wisconsin but infrequently documented (Natural History Box: Communal Nesting). Eggs hatch after ca. 60 days of incubation in late August or early September (Vogt 1981). Maturity is apparently reached when snakes attain ca. 91 cm TL (35.8 in; Conant and Collins 1998), while mean age of maturity in Ontario was 3.6 and 5.2 years for females and males, respectively. Longevity in the wild is reportedly 12–15 years (Row and Lougheed 2006).

ACTIVITY

Eastern Foxsnakes in Wisconsin emerge from hibernacula in mid- to late April. Vogt (1981) reported capturing numerous individuals from 20 to 25 April as they emerged from a hibernaculum in Portage County. Row and Lougheed (2006) determined that air temperatures of ≥16°C (61°F) were the best predictor of spring emergence for Eastern Foxsnakes in Ontario. During the majority of the active season, Eastern Foxsnakes are mostly diurnal but occasionally nocturnal, especially on warm summer nights with light precipitation (R. Hay, personal observation). They focus their activity at ground level (Ernst and Ernst 2003), although there is evidence that they will climb to forage (e.g., Brown and Brown 1975, Griesbaum 2012).

Published data on movement rates and home range sizes do not exist for Wisconsin populations and are extremely limited for populations elsewhere. Yet the few reports that exist indicate that Eastern Foxsnakes can move appreciable distances. Row et al. (2012) estimated home range sizes in Ontario that excluded migrations to and from hibernacula. They determined that the mean home range sizes of foxsnakes at two sites were 34 and 53 ha (84 and 131 ac, $n = 14$ and 13 snakes tracked, respectively, MCP estimator; Row et al. 2012). Griesbaum (2012) reported a mean home range size of 7.36 ha (18.2 ac) for males ($n = 6$) and 5.35 ha (13.2 ac) for females ($n = 3$, kernel density estimator). Row et al. (2012) further reported that mean home range size did not differ based on snake sex or whether females were gravid. Despite this, others have detected differences in home range

NATURAL HISTORY BOX: Communal Nesting

I have seen and received reports of communal nesting of Eastern Foxsnakes on several occasions. In the summer of 1986, I found a single clutch of Eastern Foxsnake eggs in a crevice under a large flat rock in western Dane County. Further investigation revealed that Gophersnakes had also deposited three clutches of eggs in the same crevice. I revisited the site in early September of that year and saw three hatchling Gophersnakes near the nest rock but no Eastern Foxsnakes. I reached into the crevice to see what was there and extracted shell remnants from dozens of hatched eggs. The shells appeared to be a mix of freshly hatched eggs and older eggs, likely from previous years. In almost 50 years of field surveys, this was the only time I have observed multiple snake species depositing eggs in the exact same location.

In late June 2004, a husband and wife from Marathon County contacted me to report that many snakes had nested in a compost pile on their property. They had pulled a tarp off the large compost pile, inadvertently unveiling the nesting serpents. The husband informed me over the phone that the encounter nearly gave him a heart attack. He commented that the number of snakes below the tarp could have "filled a 55-gallon drum," which apparently included hundreds of ca. 1.2 m (4 ft) long gartersnakes (*Thamnophis* spp.) and several ca. 1.8–2.1 m (6–7 ft) "pinesnakes." I instructed him to lay the tarp back over the compost pile and mentioned I would come up for a look over the weekend. When I arrived and we removed the tarp, we found roughly 40 gravid Common Gartersnakes (*Thamnophis sirtalis*), 2 adult Eastern Foxsnakes (both gravid females), and 5 clutches of Eastern Foxsnake eggs. Fortunately, the landowners appreciated the value of snakes, although the husband did not particularly like them. Because they had planned to use the compost for a large garden, we removed all the eggs, which were hatched in captivity and returned to the site that autumn.

In 2008 road repair workers accidentally discovered two communal nesting locations in Buffalo County. Both were located at bridges below undercut blacktop where the road met the concrete bridge abutments. Eroding soils under the pavement had created an access point for the nesting snakes. The person who reported these incidents brought five eggs into the WDNR office in Alma (Buffalo County) for identification. He told the WDNR wildlife biologist on-staff that there were probably 1,000 eggs combined between these two sites. The wildlife biologist visited the locations and unfortunately discovered that none of the eggs were salvageable by the time he arrived. Most either were buried in deep soil piles that had accumulated due to road repairs or had sloughed into the nearby river. Still, the eggs could be identified by their shape and size. It is highly probable that the original number of eggs reported was greatly exaggerated and that many of the eggs observed could have represented shells from previous years of hatching (Tom Schultz, Buffalo County Land Conservation Department, personal communication to Kris Johansen, WDNR). Regardless, this occurrence provides another example of multiple Eastern Foxsnakes nesting at the same location.

Interestingly, these reports of communal nesting from Wisconsin are corroborated by another communication I received from Ontario in 1992. A regional herpetologist there had stopped to check what looked like a cover object along a roadside. When he approached the object he could see that it was a discarded mattress. He lifted it up and found nothing underneath, but when he dropped the mattress, a clutch of Eastern Foxsnake eggs rolled out from one of the sides. Upon further investigation, he found a total of 220 eggs that snakes had deposited in the mattress. Because this was a Threatened species in Canada, the eggs were taken to the Metro Toronto Zoo, where they hatched, and were then released back into the area where they had been discovered (Kent Prior, Parks Canada, personal communication).

Robert W. Hay, Turtles for Tomorrow

size by sex. In Missouri, for example, mean home range size of males (44.3 ha [109.5 ac], $n=6$) was much larger than that of females (22.8 ha [56.3 ac], $n=3$, MCP estimator; Shew et al. 2012). Interestingly, this study found little difference in movement rates by sex: 44.3 m/day (145.3 ft/day) for males and 53.5 m/day (175.5 ft/day) for females.

Eastern Foxsnakes return to overwintering sites in early to late October, although they might not immediately enter hibernation upon arrival. For example, rare reports from Wisconsin exist of active Eastern Foxsnakes near hibernacula well into November when conditions were unusually warm (Thomas and Richard LaFleur, Turtles for Tomorrow, personal communication). As winter approaches and temperatures drop, aboveground activity normally ceases, although unusual reports of winter activity exist. For example, an adult male was observed crossing a frozen pond in central Illinois on 17 February (Brown and Brown 1995). Behavior and physiology during winter are largely unknown. However, Vogt (1981) and Zaremba (1978) reported instances of Eastern Foxsnakes hibernating while fully submerged in abandoned cisterns or wells. They will communally overwinter with other foxsnakes (e.g., Vogt [1981] counted 166 individuals at a Wisconsin hibernacula) and other snake species (J. Kapfer and R. Hay, personal observations; Figure 214). Reports from Wisconsin and Ontario support that Eastern Foxsnakes typically return to the same hibernacula annually (Vogt 1981, Watson 1995).

Figure 214. One of the authors (J. Kapfer) and an assistant (J. Hay, wife of R. Hay) with several of many Eastern Foxsnakes collected near a communal hibernaculum in April (Portage County; photo by T. Muehlfeld).

PREY AND PREDATORS

Eastern Foxsnakes are active diurnal foragers (reviewed by Ernst and Ernst 2003). They rely heavily on visual stimuli when hunting (Saviola et al. 2012), although olfaction still plays a role. Prey items are typically restrained with the jaws, then quickly immobilized and constricted with body coils. To date, detailed analyses of Eastern Foxsnake diet have not been published, but sources often report that they generally consume small mammals (Vogt 1981, reviewed by Ernst and Ernst 2003). Despite this, published accounts of mammal predation are somewhat sparse and often based on isolated observations (e.g., Hay 1892, Logier 1925, Conant 1938, Minton 2001). It follows that reports from Wisconsin are also quite rare, although Grange (1948) apparently documented individuals that had eaten young "gray squirrels" and "young rabbits" in the state.

Eggs of birds are also apparently an important dietary component (Logier 1925, Conant 1938, Rodríguez-Robles and De Jesús-Escobar 1999, Klug et al. 2010a). Examples of egg predation in Wisconsin include "pheasant" eggs (Kewaunee County; Vogt 1981), Common Tern (*Sterna hirundo*) eggs (Door County; Lyon 1927), and the eggs of passerines (Green and Lafayette Counties; Renfrew and Ribic 2003). Grange (1948) further reported that Eastern Foxsnakes were once an important predator of Ruffed Grouse (*Bonasa umbellus*) eggs in Wisconsin, having destroyed 5 of the 37 nests he monitored across Jackson, Wood, Portage, Juneau, and Adams Counties. Wheeler (1984) also observed several incidences of Eastern Foxsnakes in the act of either consuming waterfowl eggs or investigating waterfowl nests in Green Lake County. The process of egg consumption is rarely reported in detail, although Grange (1948) offered an Eastern Foxsnake from Wisconsin a "grouse egg" in the laboratory and monitored the outcome. According to his observations, the snake engulfed the egg in ca. two minutes, then contorted its body slightly, which caused the egg to burst "with a noise loud enough to be audible in an adjoining room." In addition to eggs, Eastern Foxsnakes in Wisconsin eat nestling and juvenile birds (e.g., Grange 1948, Wilson 1985, Kenow et al. 2009). In fact, Kenow et al. (2009) found four half-day-old radio-tagged ducklings of the same brood in the stomach of an Eastern Foxsnake from Vernon County. They will also scavenge dead nestlings (Ellison and Ribic 2012). Despite the many observations of bird egg and juvenile predation in Wisconsin, it is noteworthy that Eastern Foxsnakes generally consume far fewer young birds than mammalian predators (Wheeler 1984, Kenow et al. 2009, but see Grange 1948). Reports of reptile or amphibian consumption are sporadic (e.g., Logier 1939, Breckenridge 1944), although a captive Eastern Foxsnake that was force-fed an American Toad (*Anaxyrus americanus*) died shortly thereafter (Karstad 1967). Information on juvenile dietary habits is limited. Logier (1931) mentioned a young individual captured in Ontario that regurgitated "a bundle of undigested earthworms."

Numerous vertebrates consume Eastern Foxsnakes. Raptors, including Red-Tailed Hawks (*Buteo jamaicensis*, Brown County; Watermolen 1992a) and Bald Eagles (*Haliaeetus leucocephalus*, Marathon County; Pierce and Ross 1989a) are documented predators in Wisconsin. Wild Turkeys (*Meleagris gallopavo*) will also occasionally attack and kill adult Eastern Foxsnakes (Carlisle et al. 2008). Ernst and Ernst (2003) suggested many potential mammalian predators, but confirmed reports in Wisconsin are few. Eastern Foxsnakes can employ several defensive strategies when threatened. For example, they can emit musky-smelling cloacal fluid onto their captors if handled. They will also vibrate their tail, quickly coil into a defensive posture (Figure 215), and strike toward the perceived threat (J. Kapfer and R. Hay, personal observations). Yet the level of aggression displayed in response to being threatened varies among individual Eastern Foxsnakes, and some are relatively placid when handled (J. Kapfer and R. Hay, personal observations).

CONSERVATION

Status: Globally, the Eastern Foxsnake is listed as a Least Concern species by the IUCN. It is considered a common species in Wisconsin by the WDNR.

Figure 215. Typical Eastern Foxsnake defensive posture (Washburn County; photo by E. Wild).

Populations: Limited information on Eastern Foxsnake populations exists. Because they are difficult to capture and recapture, it is challenging to obtain sufficient data for robust abundance estimates. Much of the data available are based on very small sample sizes or investigations of small populations in eastern Michigan and Canada. To our knowledge, only Freedman and Catling (1978) have published abundance or density estimates for this species to date. Although the authors obtained a limited number of captures during their two-year study in Ontario (including only one recapture), they reported that annual abundance of Eastern Foxsnakes ranged from 52 to 128 individuals, with densities ranging from 1.3 to 3.2/ha (0.53 to 1.30/ac). Later research from Ontario estimated a mean adult survival rate of 0.75 (Row and Lougheed 2006).

Given how difficult Eastern Foxsnakes are to capture, some authors have reported data on encounter rates and survey effort, from which population information was inferred. For example, Lee (2004) calculated combined detection frequency estimates (i.e., number of snakes detected per hour of survey) for transect and time-constrained visual encounter surveys in the Lower Peninsula of Michigan. The resultant detection frequency estimates for two years were low, at 0.10 ($n = 7$ individuals over 70 survey hours) and 0.04 ($n = 4$ individuals over 92 survey hours), perhaps indicative of low foxsnake abundance. Tucker (1995) used road surveys in Illinois to report percent encounters of Eastern Foxsnakes relative to all snake species observed and determined that only 5.3% of his snake observations were Eastern Foxsnakes (i.e., 6 of 113 observed snakes).

Threats: Habitat loss and fragmentation are serious conservation concerns for Eastern Foxsnakes (see Introduction to Conservation and Management of Wisconsin's Amphibians and Reptiles chapter). For example, loss of suitable nesting and overwintering habitats could limit the distribution of this species in Wisconsin. Furthermore, extensive agricultural activities can destroy or degrade habitat, creating conditions that Eastern Foxsnakes avoid. Row et al. (2012) never found Eastern Foxsnakes on active agricultural land, despite the substantial availability of agriculture at their study sites. Shew et al. (2012) only located radio-tracked individuals in agricultural habitat on one occasion in Missouri, while Heske and Keller (1999) and Griesbaum (2012) reported avoidance of agricultural fields in Illinois. Research in Ontario also indicates that Eastern Foxsnake dispersal is inhibited by the presence of unsuitable habitat (primarily agriculture) and roads that do not permit easy passage (Di Leo et al. 2010, Row et al. 2010, 2011). The authors found that such barriers to movement isolate foxsnake populations and reduce their genetic diversity (Row et al. 2010, 2011), a problem that is also exacerbated by their small effective population sizes (i.e., a small number of actively reproducing members of a given population; Di Leo et al. 2010).

The impact of roads on Eastern Foxsnake populations is likely a serious concern that has not been quantified in Wisconsin. Roads inhibit dispersal capabilities and genetic exchange between populations, but they also increase the likelihood of mortality due to vehicular strike (Ashley and Robinson 1996, Row et al. 2010). For example, over half of the 489 reptiles brought into a large wildlife rehabilitation center in DuPage County, Illinois, were the victims of automobile collision (Eastern Foxsnakes represented 78 [16%] of this total; Hartup 1996). In Wisconsin, this is problematic because Eastern Foxsnakes have historically been observed to frequent roads and roadsides. Grange (1948) anecdotally reported that it was "not uncommon to see thirty or fifty" Eastern Foxsnakes a day while driving along the back roads of some central Wisconsin counties. Knudsen (1954a) raised concerns in the state decades ago when he proclaimed that "many drivers will practically run their cars into a ditch to kill any snake that seems to be escaping over the shoulder of the road and the fox snake, being large and readily seen, really 'catches it in the neck.'" He further recounted observing 16 road-killed individuals during a single automobile ride from Black River Falls to Madison in 1949.

A handful of studies have reviewed the potential impacts of pathogens on Eastern Foxsnakes. Snake fungal disease (SFD), caused by *Ophidiomyces* (formerly *Chrysosporium*) *ophiodiicola*, has been identified as an issue of increasing importance in the conservation of global biological diversity (Sutherland et al. 2014; see Introduction to Conservation and Management of Wisconsin's Amphibians and Reptiles chapter). Cases of SFD have been documented in Wisconsin (Lorch et al. 2016), and a Brown County Eastern Foxsnake population surveyed in 2014 appeared to have a high incidence of infection (Jeffrey Lorch, USGS, personal communication). Xuereb et al. (2012) found a greater prevalence of *Hepatozoon* infection in Ontario populations with more northern distributions. They suggested that Eastern Foxsnakes subjected to greater environmental stress, such as those found in colder northern climates, are more susceptible to parasites. Perhaps a similar pattern related to SFD infection will be uncovered in the future.

Overcollecting of snakes for the expanding worldwide pet trade has raised concerns for populations of many amphibian and reptile species (see Introduction to Conservation and Management of Wisconsin's Amphibians and Reptiles chapter). This is particularly true for species such as the Eastern Foxsnake that are popular among hobbyists (R. Hay, personal observation) and easy to collect based on their communal overwintering behavior. Vogt's (1981) encounter of many individuals congregated near a single hibernaculum demonstrates how easily this species could be exploited by collectors.

A possible area of concern is the accidental entrapment or entanglement of snakes in plastic mesh de-

Figure 216. Eastern Foxsnakes entangled in plastic netting (Rusk County; photo by A. Coppens).

ployed to control erosion and for wildlife exclusion (Cochran 2013a, reviewed by Kapfer and Paloski 2011; see Introduction to Conservation and Management of Wisconsin's Amphibians and Reptiles chapter). Limited data are available on the impact that entanglement has on snake populations, but Eastern Foxsnake entrapment has been documented on several occasions in Wisconsin. Examples include Marinette County in 2002 ($n = 22$ individuals, 16 released alive, 6 dead on arrival), Wood County in 2010 ($n = 3$ individuals released alive; Kapfer and Paloski 2011), and Rusk County in 2014 ($n = 2$ individuals released alive; Figure 216; Ami Thompson, Dragonfly Environmental Education, personal communication). The degraded remains of other dead snakes are often reported near entangled individuals, and there are likely many more unreported cases. Photographs of the entangled Marinette County snakes suggested that the majority were gravid females (R. Hay, personal observation). Further research that investigates the population-level impact of mesh entanglement is warranted.

Management: Eastern Foxsnakes could benefit from management strategies that promote protection of undisturbed habitats and reclamation of disturbed agricultural lands, particularly in areas with a mix of lowland and open upland habitats. Management strategies should also promote necessary microhabitat features, such as downed woody debris and brush piles for cover. Protection of habitat adjacent to overwintering sites may also help reduce the likelihood that Eastern Foxsnakes will disperse substantial distances to find suitable summer habitat, which could lead to fewer risky encounters (e.g., Durbian et al. 2008). The restoration and maintenance of habitat corridors (i.e., "grassed waterways") that connected suitable habitat in an agriculture-dominated landscape were effective at improving connectivity among Eastern Foxsnake populations in Iowa (Knoot and Best 2011).

Active management is necessary to maintain the habitats that Eastern Foxsnakes require. Yet certain habitat management practices, though useful, can accidentally kill snakes and should be scheduled to avoid their active periods (see Introduction to Conservation and Management of Wisconsin's Amphibians and Reptiles chapter). For example, a tractor and disc harrow were employed to break up undesirable vegetation in an abandoned agricultural field during habitat restoration (Dane County). This was conducted in mid-spring, after Eastern Foxsnakes typically disperse from hibernacula. A post-management survey of the property revealed that 16 Eastern Foxsnakes had been killed or mortally wounded during these activities (Michael Foy, WDNR [retired], personal communication). Similarly, the use of prescribed fire to slow the encroachment of woody vegetation should also be scheduled appropriately. Although prescribed burns in managed prairies can benefit Eastern Foxsnakes (Shew et al. 2012), they can also lead to mortality if active snakes are inadvertently trapped in areas where fire lines converge and underground retreats are inaccessible (Carlisle et al. 2008).

General management concepts related to mitigation of road mortality could also be beneficial, particularly in areas where this phenomenon is frequently documented (see Introduction to Conservation and Management of Wisconsin's Amphibians and Reptiles chapter). Furthermore, careful consideration of appropriate use of erosion control mesh is advisable (see review of products by Kapfer and Paloski 2011). To address threats posed by overcollecting, the WDNR developed regulations in 2000 for collection or possession of native amphibians and reptiles

(Wisconsin Legislature 2010, Wisconsin Administrative Code NR 16).

SPECIAL REMARKS

Surveys are needed to identify and better characterize hibernacula and oviposition sites for this species, especially communal sites. These data would help guide best practices to reduce pressure from snake collectors and outline appropriate management strategies, particularly the best time to burn, mow, and disk. A study in Ontario found that the hourly rate of Eastern Foxsnake emergence from hibernacula increased significantly as air temperature increased (Row and Lougheed 2006). The study further suggested that monitoring of hibernacula in spring could be most effective when air temperature is ≥16°C (61°F).

Gophersnake

Pituophis catenifer (Blainville 1835)

Joshua M. Kapfer

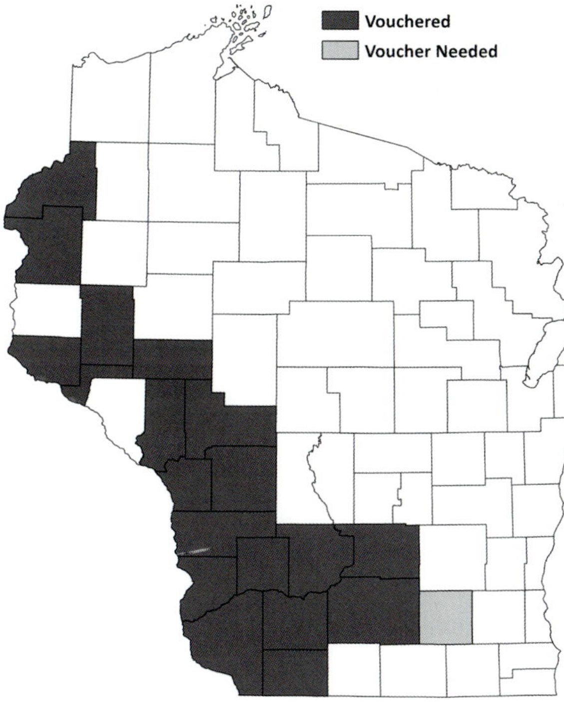

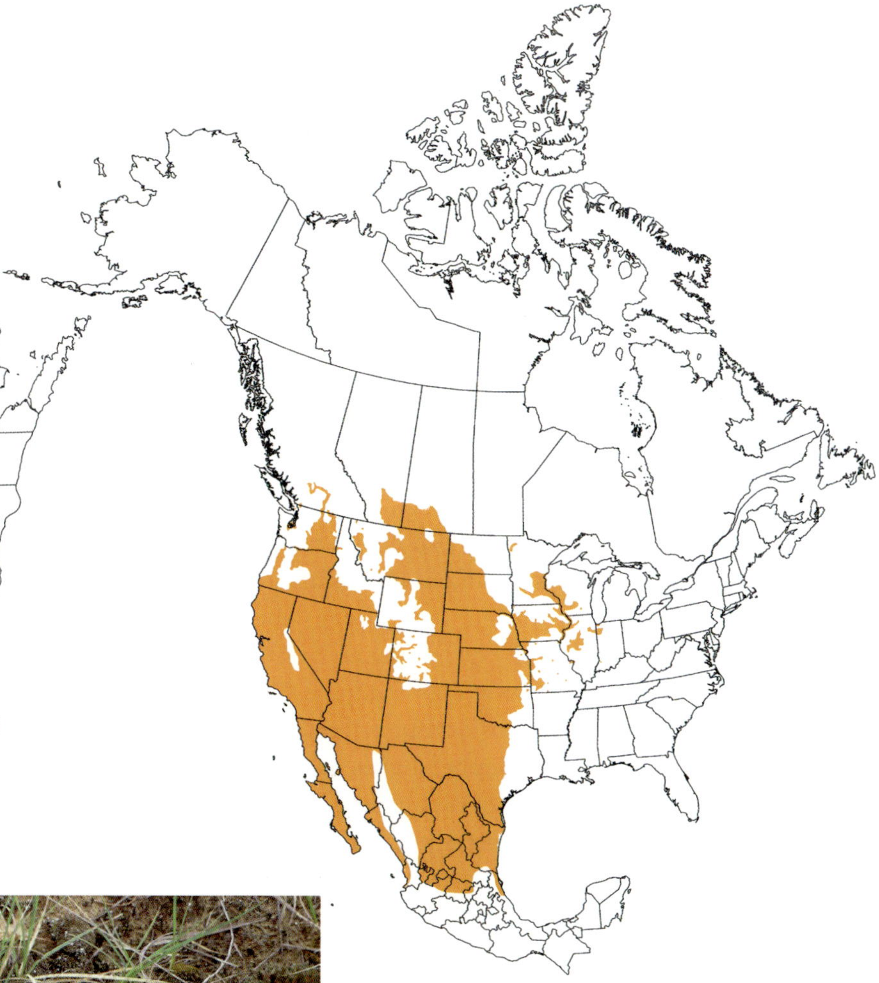

Adult male Gophersnake (Sauk County; photo by J. Kapfer).

AT A GLANCE

Gophersnakes are Wisconsin's largest snake and can exceed 1.5 m (5 ft) in total length. They possess a series of dark blotches that run down their back and that are overlaid on a cream to yellow-beige background. The spaces between these dark blotches and along the snake's flanks are typically filled in with dark mottling. Their head is also heavily patterned, and there is an enlarged and somewhat pointed (forward-facing, not upturned) scale at the tip of their nose. Gophersnakes prefer prairies, oak savannas, and associated open bluff faces in the Wisconsin and Mississippi River valley regions of the state. They emerge from hibernation in April, and reproduction occurs shortly thereafter. Females nest from mid-June to early July, and eggs hatch from late August to early September. They are powerful constrictors that primarily eat warm-blooded prey, such as rodents and other small mammals. By late October, Gophersnakes return to their overwintering sites, which are typically found along bluff prairies with exposed rock outcroppings and deep underground crevices. Gophersnakes can enact several defensive behaviors when threatened. If flight does not work, they can hiss very loudly by forcing air over unique structures associated with their windpipe. They will also coil into a striking position and vibrate their tail. If the threat persists, Gophersnakes will not hesitate to bite. Although they can make a considerable ruckus if threatened, and they are even sometimes misidentified as rattlesnakes, Gophersnakes are completely harmless when left alone. They are considered a Protected Wild Animal, a species of Special Concern, and a Species of Greatest Conservation Need in Wisconsin. In the Upper Midwest, they are often called 'bullsnakes' (the common name associated with the regional subspecies) and sometimes incorrectly referred to as 'pinesnakes.'

SYSTEMATICS

Pituophis catenifer (Gophersnake) is the only member of the genus *Pituophis* that occurs in the Upper Midwest. The most thorough phylogenetic studies of *P. catenifer* are those of Stull (1940) and Rodríguez-Robles and De Jesús-Escobar (2000). The Wisconsin form was once considered a subspecies of *P. melanoleucus* (Pinesnake, found in the eastern US), until Rodríguez-Robles and De Jesús-Escobar (2000) showed otherwise. After this rearrangement, there has been little confusion regarding the taxonomic position of *P. catenifer* in the state. Six subspecies of *P. catenifer* are currently recognized globally (Crother 2017), and although no recent efforts have been made to evaluate subspecific validity, available evidence indicates that the Wisconsin form is *P. c. sayi* (Bullsnake). *P. catenifer* is a member of the subfamily Colubrinae, along with five other Wisconsin species that occur in the genera *Coluber*, *Lampropeltis*, *Opheodrys*, and *Pantherophis*.

DESCRIPTION

Adults: The dorsal surface is distinctly patterned with a combination of 38–53 larger dark blotches and dark mottling, the latter of which is often concentrated along the flanks (Figure 217; Vogt 1981). Along the anterior third of the body, the blotches and mottling are black but overlaid on a light cream or beige background. The blotches along the middle third of the body are typically chocolate brown over a pale yellow or beige background. Occasionally, lighter and sometimes interrupted bars are present between these large blotches on the dorsum (see Figure 217a). The tail contains 8–15 dark rings (Vogt 1981) on a yellow or golden background (see Figure 217). The venter is cream colored with smaller black or brown rectangular blotches that are haphazardly distributed. The head is heavily patterned, which includes a dark bar between the eyes and often another running from each eye to the posterior corners of the mouth (Figure 218a). Thin dark bars are also present on the posterior edges of most supralabial scales (Figure 218b). Gophersnakes have mostly keeled scales, and the snout is pointed, with an enlarged rostral scale (see Figure 218). There is a single anal plate, and the subcaudal scales are all divided.

Gophersnakes are one of the largest snake species native to North America (Devitt et al. 2007) and are Wisconsin's largest snake (TL = 127–200 cm [50.0–78.7 in]; Vogt 1981; Figure 219). The mean SVL of adults and subadults sampled from a Sauk County population ranged from 119 cm (46.9 in, $n = 24$) in

Figure 217. Adult Gophersnakes that highlight the color variation between the anterior, middle, and posterior thirds of the body. (a) Note the light bars that occasionally show in between the large dorsal blotches (Sauk County; photo by J. Kapfer). (b) An alternative example of adult Gophersnake patterns, with a lack of light bars between the dorsal blotches (Sauk County; photo by J. Kapfer).

Figure 218. Head patterns and scales typical of Gophersnakes in Wisconsin. (a) An adult female. Note the dark bars between the eyes and running from the eye to the posterior corner of the mouth (Sauk County; photo by J. Kapfer). (b) An adult male. Note the dark bars on the posterior edges of each supralabial scale and the enlarged rostral scale (Sauk County; photo J. Kapfer).

females to 131.5 cm (51.8 in, $n = 40$) in males (Joshua Kapfer, UW–Whitewater, unpublished data). Males from this population also had longer mean tail lengths (16.6 cm [6.5 in], $n = 37$) than females (14.5 cm [5.7 in], $n = 24$) and were heavier ($\bar{x} = 958$ g [33.8 oz], $n = 39$) than females ($\bar{x} = 844$ g [29.7 oz], $n = 23$; J. Kapfer, unpublished data). Despite this, mean SVL of Wisconsin specimens housed in the MPM and UWZM collections was similar by sex, although females (133 cm [52.4 in], $n = 7$) were slightly larger than males (127.2 cm [50.1 in], $n = 6$). The tail length of preserved specimens was also similar by sex, although mean male tail length (17.9 cm [7.1 in]) was slightly greater than that of females (16.1 cm [6.4 in]). The longest museum specimen measured was an adult female collected in 1926 from Vernon County, which was 159.7 cm SVL (62.9 in, MPM 2092).

Preadult stages: Eggs are oblong and grayish white. Vogt (1981) reported egg dimensions of 5×4 cm (2.0×1.6 in) in Wisconsin. Remnants of previously hatched eggs found in Sauk County on two occasions had lengths that ranged from 4.3 to 4.5 cm (1.7 to

Above, top: Figure 219. Large adult Gophersnakes (Sauk County; photo by J. Kapfer).

Above, bottom: Figure 220. Hatchling Gophersnake (Sauk County; photo by J. Kapfer).

1.8 in, $n = 32$; J. Kapfer, unpublished data). Hatchlings and juveniles are similar in appearance to adults (Figure 220). Although there are limited data on the size of young Wisconsin individuals, SVL measures of individual neonates from Sauk County captured in each of four years were 35.8 cm (14.2 in, 2003), 48.0 cm (18.9 in, 2004), 40.8 cm (16.1 in, 2017), and 49.1 cm (19.3 in, 2018; J. Kapfer, unpublished data). Other wild Wisconsin individuals, estimated at ca. two to three years of age ($n = 5$), ranged from 91 to 99 cm SVL (35.8 to 38.9 in) and weighed 237–403 g (8.4–14.2 oz; J. Kapfer, unpublished data).

Variation: No significant variation in the color or pattern of Wisconsin individuals is known. There appears to be sexual size dimorphism in at least some Wisconsin populations, where males are significantly larger than females (Kapfer 2009).

Diagnosis: Gophersnake dorsal scales are mostly keeled and exist in 27 (or more) rows at midbody. Gophersnakes possess a slightly adpressed forward-pointed thick rostral scale. In most cases, one loreal scale, three to four postocular scales, one to two preocular scales, and often four prefrontal scales occur on each side of the head. Each side of the mouth is typically lined with 8–9 supralabial scales and 12–13 infralabial scales. There are 207–260 ventral scales, and while the anal plate is not divided, the subcaudal scales are all divided. Gophersnakes are easily distinguishable from other snakes in Wisconsin because most species do not possess the thickened rostral scale. Although Eastern Hog-nosed Snakes (*Heterodon platirhinos*) have a large rostral scale, it is strongly upturned toward a sharp point, while the Gophersnake's is more blunt and points forward. Eastern Foxsnakes (*Pantherophis vulpinus*) and Eastern Milksnakes (*Lampropeltis triangulum*) could cause some confusion due to superficial similarities in color and pattern. Yet Gophersnake blotches are less defined than in the other two species due to the mottling that occurs between larger dorsal marks (particularly along the flanks) so that blotches do not appear isolated. The Gophersnake's head is heavily patterned with dark speckles, interorbital bars, and labial bars, unlike that of most adult Eastern Foxsnakes. Although Eastern Milksnakes possess an obvious head pattern, they have a unique V- or Y-shaped mark on top of the head just posterior to the eyes. Gophersnake dorsal scales are strongly keeled, unlike those of the other two species, and their anal plate is undivided, unlike that of Eastern Foxsnakes. There is confirmed wild hybridization between Gophersnakes and Eastern Foxsnakes in Iowa and Minnesota (LeClere et al. 2012, Pinto et al. 2018). Although the potential presence of hybridized individuals can confuse identification, the incidence of such outbreeding in the wild is likely low and is not yet confirmed in Wisconsin. Gophersnakes are sometimes confused with Wisconsin's two venomous snakes, Timber Rattlesnake (*Crotalus horridus*) and Massasauga (*Sisturus catenatus*). This is likely because Gophersnakes, like many other large nonvenomous snakes in Wisconsin, will vibrate their tail in response to a perceived threat. However, both venomous species possess a tail tipped with an obvious rattle, compared to the sharply pointed tail of nonvenomous snakes. Wisconsin's venomous snakes also have elliptical pupils, unlike the round pupils that are characteristic of nonvenomous species in the state.

Although their eggs appear superficially similar to those of other reptiles, the size of Gophersnake eggs is larger than most. Despite this, their eggs could overlap in size with those of Gray Ratsnakes and Eastern Foxsnakes, although these species often separate by geographic distribution or habitat where nesting occurs in Wisconsin. Of the larger snakes in Wisconsin, North American Racers probably most frequently overlap in geographic range, nesting habitat, and egg dimensions with Gophersnakes, so their eggs could be confused. Yet North American Racer eggs have rough shells, with a texture similar to fine sandpaper, unlike those of Gophersnakes (J. Kapfer, personal observation). Juvenile Gophersnakes can be confused with juvenile Eastern Foxsnakes and Eastern Milksnakes because young of the latter two species often have a heavily patterned head that is vaguely reminiscent of Gophersnakes. Differentiating among the juveniles, therefore, requires attention to similar characteristics as used to distinguish adults: geographic range, the shape of the rostral scale, the degree of isolation among body blotches, and whether the anal plate is divided.

DISTRIBUTION AND HABITAT

Global distribution: Gophersnakes are the most widely distributed member of the genus *Pituophis* in North America. They are found throughout much of the Great Plains, although they are absent in portions of the Rocky Mountains. Their range extends west to the coast of the Pacific Ocean and covers most of California, Oregon, and Washington. They occur north into Canada, south into Mexico, and east of the Mississippi River into parts of Wisconsin (the northeastern extreme of their range) and Illinois, with isolated pockets of occurrence in Indiana (Rodríguez-Robles and De Jesús-Escobar 2000).

Distribution in Wisconsin: Most published county records are adjacent to the Mississippi and Wisconsin Rivers. Of the published county records, a voucher specimen could not be found for Jefferson (see the Appendix). The current presence of this snake in Jefferson County seems unlikely, although it could have previously occurred in the western portion if suitable habitat existed.

Localities of possible occurrence: Gophersnakes could occur in Green County, which is partially in the Driftless Area, is adjacent to other counties with confirmed records, and appears to possess suitable habitat in some locations. There are also several counties in northwestern Wisconsin with no published records but where Gophersnakes could occur. For example, St. Croix and Buffalo Counties represent obvious gaps in the known range of this snake. The western half of St. Croix County has soils and historic vegetation similar to nearby counties of known occurrence (i.e., Pierce, Pepin, and Dunn). Furthermore, recent unsubstantiated reports exist from Buffalo County (Richard Staffen, WDNR, personal communication). Given their confirmed presence in Polk and Burnett counties, it seems plausible that Gophersnakes could exist in Bayfield, Douglas, and Washburn Counties (which have appropriate habitat and soils). Casper (1996) mentioned an observational report from Bayfield County, but this remains unconfirmed.

Several unusual Gophersnake literature records and specimens exist from northeastern Wisconsin. For example, a specimen reportedly collected in 1919 from Outagamie County (MPM 1823) exists but is considered "enigmatic" (Casper 1996). Several observational accounts from Shawano and Forest Counties occur in Pope (1930) and Dickinson (1972). Pope and Dickinson (1928) also referred to observations and specimens (but did not report specimen numbers) from Door County. Edgren (1944) further reported the observation of several road-killed "*Pituophis sayi sayi*" in the Rib Hill area of Marathon County. These records lack supporting evidence and modern corroboration. They are also mostly isolated from the Gophersnake's currently documented range, so they remain dubious. Thus, it seems more likely that these northeastern reports and specimens were misidentified Eastern Foxsnakes. No confirmed Gophersnake reports exist from southeastern Wisconsin, despite Vogt's (1981) inclusion of Jefferson County in their range. Several additional accounts of this snake are known from other southeastern locations, but these are old and unsubstantiated (Natural History Box: Gophersnakes in Southeastern Wisconsin?).

Habitat: Gophersnakes are generally found in open dry habitats, including prairies and savannas (see Amphibians and Reptiles in Wisconsin Plant Communities chapter; Kapfer et al. 2008e). Research in Wisconsin indicates that Gophersnakes also prefer south- to west-facing bluff slopes with open tree canopies and exposed rock outcroppings (Kapfer et al. 2008e). These habitats are used throughout the active season and for most ecological functions (e.g., foraging, breeding, egg laying).

Within preferred habitats, Gophersnakes select microhabitats that possess vegetation and soils with

NATURAL HISTORY BOX: Gophersnakes in Southeastern Wisconsin?

It is easy to associate Gophersnakes with the Driftless Area of western Wisconsin because most records of this snake are from the counties that occur along the Wisconsin and Mississippi Rivers. Yet there are a number of old Gophersnake reports and specimens from more eastern locations that can pique one's interest. For example, museum records indicate that several purported Gophersnake specimens from Milwaukee County once existed (MPM 811, MPM 812, MPM 1833). Furthermore, Cahn (1929) mentioned that "*Pituophis sayi*" was "formerly common; now very rare indeed" in Waukesha County. Of additional interest is that Vogt (1981) included Jefferson County in this snake's state range based on a reliable report from the county's western border near Dane County.

The idea of Gophersnakes in south-central and southeastern Wisconsin is not completely without merit. Several counties in these regions, including Green, Rock, Walworth, Racine, and Kenosha, have certain natural characteristics that could appeal to Gophersnakes. First, Green County lies at least partially within the boundaries of the Driftless Area of Wisconsin. Considering that the Driftless Area is where most Gophersnake records occur in the state, Green County seems a likely location for this snake. The other counties are mostly outside of the Driftless Area, although they still appear superficially suitable for Gophersnakes in several ways. They possess areas of sandy soils that a cursory review would suggest are similar to the sandy substrates associated with Gophersnakes in western Wisconsin. There is the additional fact that Gophersnakes in western Wisconsin use oak savanna and prairie habitats (see Habitat subsection). Considering that savannas and prairies were historically common throughout the south-central and southeastern counties, it is easy to imagine that suitable habitat for this snake was once ample there. Of further relevance is that Gophersnakes are documented from just across Wisconsin's southern border in north-central and northeastern Illinois. This includes confirmed records from Ogle, Kane, Will, and Kankakee Counties (Phillips et al. 1999) and an observational record from Winnebago County (Olson 1995). Given these points, one can imagine the Gophersnake's historical range extending north from Illinois into eastern Wisconsin, where possibly suitable soils and habitat occurred. If so, the past presence of Gophersnakes in south-central and southeastern

Wisconsin seems slightly more plausible, which perhaps corroborates the old sightings already mentioned.

However, closer inspection of these historical southeastern Wisconsin records suggests that they are dubious. For example, museum records indicate that the purported Gophersnake specimens from Milwaukee County are now "presumed missing." The reliability of one (MPM 1833) is also particularly suspect for several reasons. First, Pope (1931) originally described this specimen as a "yellow-bellied kingsnake" (*Lampropeltis calligaster*), not a Gophersnake, and because it is currently lost, the specimen's identity remains uncertain (see Hypothetical and Enigmatic Species of Wisconsin chapter). Furthermore, although this specimen was reportedly collected in Milwaukee County in 1920, the collector was not Pope but someone who brought the snake to him. Although museum notes indicate that the specimen was later reidentified as a Gophersnake, its location of collection cannot be truly confirmed. The historically "common" Gophersnakes from Waukesha County that Cahn (1929) mentioned are also hard to fathom. Although such a supposedly common species would have certainly, at some point, allowed Cahn opportunity for correct identification, he provided no specimens as evidence to support his claim. Therefore, one cannot help but wonder if he simply mistook Eastern Milksnakes (currently found in Waukesha County) or some other species for Gophersnakes. In addition, although Vogt's (1981) Jefferson County report could be considered reliable, there are no corroborating voucher specimens and no additional recent reports from that county.

Still, if neither soils nor vegetative communities provide obvious reasons for the snakes' absence, the next reasonable question could be, Why are there no confirmed records of Gophersnakes in far south-central and southeastern Wisconsin? There are several potential explanations for this. First, these regions currently contain high levels of human disturbance in the form of urban development and agriculture, which could restrict modern Gophersnake presence. There is also a possible 'natural' explanation for their apparent absence: a lack of geological features that provide suitable overwintering locations. Although, as previously mentioned, sandy soils exist in many south-central and southeastern locations, these are interrupted by large swaths of soils with silty mantles that rest on top of dense glacial sediments (Jacobs et al. 2011b). Such geological features could make it difficult for Gophersnakes to effectively burrow far enough below ground to escape frost during winter. While it is true that Gophersnakes can hibernate in southwestern counties that also possess silty soils (i.e., Grant, Crawford, Richland, and Iowa), these counties are in the Driftless Area, where exposed rocky bluffs and outcroppings occur along the Mississippi and Wisconsin Rivers. An important characteristic of these exposed rocks is that they contain deep crevices that extend from the soil surface to below the frost line and appear to create suitable winter refugia for Gophersnakes. Exposed rock outcroppings like these are not common in the glaciated regions of the south-central and (in particular) southeastern counties. An absence of these features would require that Gophersnakes instead find other suitable natural cavities and fissures or abandoned mammal burrows that are deep enough to help them survive hibernation. While natural crevices and burrows surely exist in the regions, it is easy to wonder if these are a numerous and reliable-enough resource in what little suitable habitat remains to support populations of such a large snake. Thus, access to suitable hibernacula could largely explain why the Gophersnake's state range is more western in Wisconsin, even extending far north along the Mississippi River into Burnett County.

Joshua M. Kapfer, UW–Whitewater

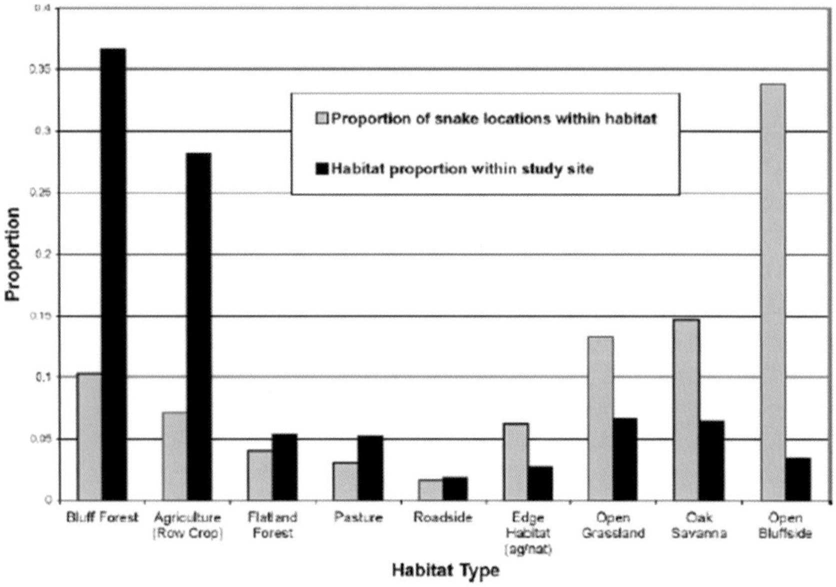

Figure 221. Proportion of habitat available vs. habitat used by Gophersnakes at a site in Wisconsin. Re-created from Kapfer et al. (2008e) with permission from the American Society of Icthyologists and Herpetologists.

certain characteristics. For example, Kapfer et al. (2009a) found that this species preferred to rest in locations dominated by herbaceous vegetation (41%–47.6% coverage) that ranged in height from 35.5 to 37.3 cm (14.0 to 14.7 in). Yet the authors also detected sexual variation in the small-scale habitat features that these snakes preferred, with males more often found below woody debris or associated with stands of woody vegetation than females. Gophersnakes further prefer loose soils (sandy or loamy), which they are adept at burrowing through. Carpenter (1982) estimated that this snake could move 3,400 cm^3 (207.5 in^3) of soil per hour. Such behavior is important when excavating prey animals from their underground tunnel systems or when modifying existing burrows for cover and nesting. Yet loose soils are not exclusively used, as preferred habitats may also possess rocky substrates, such as those associated with exposed rock outcroppings and rock slabs on bluffsides. Radio-tracking of translocated individuals in Minnesota further found a strong association between this snake and "gopher mounds" (Moriarty and Linck 1997). Little is known about the habitat preferences of neonate or juvenile Gophersnakes in Wisconsin or elsewhere. It is reasonable to assume that they prefer habitats similar to those of adults.

Gophersnakes occasionally make use of several presumably suboptimal habitat types in Wisconsin, including closed-canopy forest (Figure 221; Kapfer et al. 2008e). They are also occasionally found on large vegetated dykes running through backwaters of the Wisconsin River, although such sites are often adjacent to dry oak savannas (J. Kapfer, personal observation). Gophersnakes will further occupy anthropogenically altered landscapes, such as those dominated by active agriculture. They will readily seek shelter near rural houses or outbuildings and have been observed among piles of discarded wood and metal debris in rural settings (Natural History Box: Surprising Places). Although these disturbed or suboptimal habitats are not preferred, their use could be due to limited availability of optimal natural habitat.

Females nest in open sandy habitats with exposed soils that receive high sun penetration and are sometimes associated with abandoned mammal burrows (e.g., Plains Pocket Gophers [*Geomys bursarius*]; J. Kapfer, personal observation). They also lay eggs in crevices or chambers below the rock slabs on outcroppings that exist on open bluff prairies in Wisconsin (J. Kapfer, personal observation). Furthermore, Gophersnakes typically use the south-facing slopes on bluffside prairies for overwintering (Kapfer et al. 2008e). The rocky outcroppings in these habitats possess crevices that presumably allow Gophersnakes to descend below the frost line, a critical feature for overwintering success in Wisconsin.

NATURAL HISTORY BOX: Surprising Places

Gophersnakes seem attracted to old discarded junk piles and farmhouses. During radio telemetry work on this snake in Wisconsin, I tracked an individual that frequented a heap of decades-old metal and wood debris that a past landowner had left to decay in the middle of nowhere. I followed this snake on several occasions to the vicinity of a discarded refrigerator lying on its back amid the rubble. I typically observed him only after opening the refrigerator's door, which would reveal the snake sitting peacefully among the nesting material of rodents that had colonized the old appliance and that likely provided the snake with its most recent meal. On another occasion, I radio-tracked a snake from its previous location on a wooded hillside to the attic of a nearby farmhouse. After gaining permission from the concerned homeowner to enter the house and removing our shoes to avoid soiling the carpet, one of my colleagues and I tracked the snake through the main living spaces within the home to no avail. We eventually found it, however, in the attic. There it sat coiled on the insulation between the rafters near a crack in the roof where a number of European Starlings (*Sturnus vulgaris*) had built their nests. I suspected it was attracted to this spot due to its associated food source (i.e., nestling birds).

Joshua M. Kapfer, UW–Whitewater

REPRODUCTION AND DEVELOPMENT

Mating and reproduction in Wisconsin often occur near hibernacula soon after emergence from winter dormancy (April and May; Vogt 1981), although few detailed descriptions of these activities in the region exist (Natural History Box: Romance). Males likely find females by following olfactory cues (Smith and Iverson 1993); sometimes several males will congregate near the same female. Observations from other regions support that congregated males will engage in ritualized combat to compete for access to females (Shaw 1951, Bogert and Roth 1966, reviewed by Shine 1994). Such male-male interactions reportedly involve two parallel males pushing against each other (or 'wrestling'), although credible reports of this behavior in Wisconsin are very limited (Richard Sajdak, Milwaukee County Zoo [retired], personal communication). Regardless, Kapfer (2009) found that males from a Wisconsin population were significantly larger than females, which could be an indication in snakes that male-male combat occurs (Shine 1994).

Gophersnakes are oviparous. Gravid females select nesting locations and deposit eggs from mid-June to early July in Wisconsin, which is similar to reports from Utah and British Columbia (Parker and Brown 1980, Vogt 1981, Shewchuk 1996). Females will return to the same nesting location over several years and even purge the old shells from past clutches while modifying new nest cavities (J. Kapfer, personal observation). In contrast, reuse of past oviposition sites was uncommon in British Columbia (Williams et al. 2015). Communal nesting is reported from Alberta (Wright 2008) and has also been observed in Wisconsin. The latter observation involved multiple females nesting along the grassy edge of an open sand dune among numerous Pocket Gopher mounds (J. Kapfer, personal observation). Kapfer et al. (2008e) further determined that females in Wisconsin nested in locations that were within their normal home ranges, unlike a study in Canada that found that females will leave their normal activity areas and migrate to nesting sites (Shewchuk 1996).

Females in Wisconsin lay an average of 15 eggs per clutch (Vogt 1981). Iverson et al. (2012) reported a smaller mean clutch size in Nebraska (9.5 eggs) but also estimated that clutch mass was 37% of gravid female body mass. Reproductive output can be high, although the survival rates of eggs and hatchlings are unknown. Most adult females in a Wisconsin

NATURAL HISTORY BOX: Romance

In 2004 I happened across a pair of intertwined Gophersnakes in late April at around 11:00 a.m. (temperature: ground = 33°C [91°F], air = 25°C [77°F]). The male was lying on top of the female, pinning down the front third of her body. He concurrently moved the posterior half of his body in long undulations down the length of the female in quick, jerking spasms. He also firmly grasped her behind the head with his jaws. Occasionally, the male vibrated his tail and attempted to bring his cloaca in contact with hers, but in response she typically moved her tail away. This continued for about 20 minutes, during which no copulation took place, before I moved on. In late April 2005 at 10:31 a.m. I observed another reproductive event. In this instance, I again found a male firmly holding a female immediately behind her head with his mouth. The anterior half of the male was either on top of the female (pressing down) or next to her on the ground. He also made quick, jerky undulations against her body with his anterior half, which were accompanied with quick spasms of the ventral muscles just past the midbody region. Every 30–40 seconds, the male vibrated his tail (in what appeared to be an effort to entice the female) for ca. 5–10 seconds before quickly attempting to touch his vent to hers. He repeated this behavior several times, but mostly the female appeared unreceptive, quickly moving her tail away from his while pressing her vent against the ground. Despite this, she appeared to make little effort to escape, but his grip seemed firm around her neck. After several unreceptive responses, the male slightly released the pressure exerted by his jaws and exhaled loudly but then regrasped her and continued his efforts. By 11:14 a.m. the male was able to touch his vent to hers and extend a single hemipene into her cloaca. In response, the female appeared to make attempts to disengage by pulling her tail in the opposite direction. Her actions resulted in the male rolling twice to maintain his hemipene inside her cloaca. After ca. two minutes the male withdrew his hemipene, and I observed a small amount of semen near his vent. The male unhooked his teeth from her neck at about three minutes post-copulation and proceeded to pass his head over the length of her body with no tongue flicking. By 11:21 a.m. the male had left the female and entered a nearby burrow. The entire observation lasted 51 minutes. Interestingly, on this same date I encountered a second female nearby that appeared to have undergone a similar ordeal: the side of her neck and vent were caked with a small amount of dried blood.

Joshua Kapfer, UW–Whitewater

population were gravid in each year of a three-year study (Kapfer et al. 2008a). Likewise, Diller and Wallace (1996) estimated that 97% of adult females in Idaho were reproductive each year, while Williams et al. (2014) observed high oviposition rates among females in British Columbia. Despite this, there is apparently some variation in annual gravidity rates by geographic location, as a higher proportion of females did not successfully reproduce annually in a Nebraska population (Iverson et al. 2012). Egg development takes two to three months, and hatchlings typically emerge from late August through mid-September (Vogt 1981, reviewed by Ernst and Ernst 2003).

Growth rates are largely unknown for wild Wisconsin Gophersnakes, but studies from other regions indicate that substantial age-based variation in growth occurs. Platt (1984) estimated that Kansas individuals experienced rapid growth in their first year (9.4% growth) but found that growth slowed dramatically by the third year (1.2% growth per year). In Utah, Parker and Brown (1980) also estimated much higher growth rates during the first year

(15.6%–24.2%) than in the third year (9.3%–11.2%). Males reach maturity at 90–96 cm TL (35.4–37.8 in; Parker and Brown 1980, Diller and Wallace 1996), while females reach maturity at 90 cm SVL (35.4 in), which was attained after two seasons (Iverson et al. 2012). Estimates of longevity in the wild are lacking. However, a male from Sauk County originally marked as a large adult in 2005 was recaptured nine years later in 2014, which supports that adults can be long-lived (J. Kapfer, unpublished data).

ACTIVITY

Three years of radio telemetry work in southern Wisconsin revealed that peak emergence from winter dormancy occurred from early to late April (Kapfer et al. 2008e). Seasonal activity patterns and behavior are closely related to individual thermal state and environmental temperature. For example, Landreth (1972) found that Gophersnakes awoke from cold narcosis (i.e., hypothermia), and their activity levels increased at a body temperature of 8°C (46.4°F). Furthermore, active Wisconsin snakes that were aboveground (coiled or outstretched) had body temperatures that ranged from ca. 27°C to 28°C (80.6°F to 82.4°F), while snakes resting underground or in dense cover had cooler body temperatures of ca. 24.5°C (76.1°F; Kapfer et al. 2008c). Similarly, Fitch (1999) reported that the mean body temperature of active snakes in Kansas was 28.2°C (82.8°F), while the mean body temperature of active snakes in Idaho was 28.6°C (83.5°F; Diller and Wallace 1996).

Gophersnakes in Wisconsin move extensively during the active season and, as a result, have large home ranges ($\bar{x}$ = 34.4 ha [85 ac], 100% MCP; Kapfer et al. 2008e). Although high mobility is probably the norm for this species, there is still variation in movement patterns by sex. For example, mean male home range size (40.8 ha [100.8 ac]) exceeded that of females (23.4 ha [57.8 ac]; Kapfer et al. 2008e). Similarly, the mean distance traveled per day by Wisconsin individuals also varied by sex (male = 36.7 m/day [120.4 ft/day], female = 31.7 m/day [104 ft/day]; Kapfer et al. 2008e). These differences could occur because females reduce their movement when heavily gravid (Kapfer et al. 2008e). Males, on the other hand, leave the vicinity of hibernacula after breeding but often make intermittent return trips (Kapfer et al. 2008e), which requires higher movement rates. Movement patterns also apparently differ across geographic locations. For example, radio-tracked Gophersnakes in Nebraska (Fox 1986) and British Columbia (Williams et al. 2012b) had smaller MCP home ranges than snakes studied in Wisconsin (range = 3.9–16.9 ha [9.6–41.8 ac]). Snakes studied in Saskatchewan, on the other hand, had a much larger mean home range size than those in Wisconsin ($\bar{x}$ = 86.8 ha [214.5 ac], MCP estimator; Gardiner et al. 2013).

Gophersnakes exhibit high fidelity for certain locations and specific refugia within their home ranges, so they often return to these after long forays into the surrounding landscape (Kapfer et al. 2008e). Their revisiting of refugia appears to coincide with the need to thermoregulate or undergo ecdysis (J. Kapfer, personal observation). In Wisconsin, fidelity is even exhibited for certain overwintering locations across years. For example, 19 of 20 individuals radio-tracked to hibernacula over multiple years in Sauk County did not change their overwintering location between years (Kapfer et al. 2008e). In contrast, only 58.3% of 12 monitored individuals from northern populations in British Columbia reused hibernacula across years, although reuse of hibernacula occurred in all Gophersnakes that overwintered in rock outcrops (Williams et al. 2015). Gophersnakes in Wisconsin return to hibernacula in mid- to late October (Kapfer et al. 2008e). They also frequently overwinter communally with other Gophersnakes, Timber Rattlesnakes, North American Racers (*Coluber constrictor*), Gray Ratsnakes (*Pantherophis spiloides*), and Eastern Milksnakes (J. Kapfer, personal observation).

PREY AND PREDATORS

Gophersnakes primarily consume mammals. Rodríguez-Robles (2002) investigated the stomach contents of >2,600 individuals and found mammalian prey in 75% of them. In Wisconsin, documented prey species include juvenile Eastern Cottontails (*Sylvilagus floridanus*) and White-footed Mice (*Peromyscus leucopus*; Natural History Box: Dining). Fitch (2006c) further reported that Gophersnakes

NATURAL HISTORY BOX: Dining

On one occasion, upon investigating a distress call, I encountered an individual Gophersnake (SVL = 142.5 cm [56.1 in], head circumference = 13 cm [5.1 in], body circumference = 13.3 cm [5.2 in], weight = 917 g [32.4 oz]) that had just constricted and killed a relatively large Eastern Cottontail (snout-to-rump length = 23 cm [9.1 in], diameter at midbody = 17.5 cm [6.9 in], weight = 263 g [9.3 oz]). Although I obviously knew snakes often ate large prey items, the size of this cottontail relative to the snake still felt impressive to behold. I was dismayed that my approach startled the snake and disrupted its attempt to swallow this large meal. Yet because of that, I confirmed the cottontail was 28.6% of the snake's body weight, 134.6% of the snake's head diameter, and 131.5% of the snake's body diameter at the widest point (Figure 222)!

I've had other serendipitous opportunities to learn about Gophersnake feeding habits in Wisconsin. In the early spring of 2003, I captured a male Gophersnake that quickly regurgitated four baby cottontails (each ca. 10.3 cm [4.1 in] snout to rump). In the summer of 2004, I observed a Gophersnake moving through the boughs of a cedar tree at ca. 2.5 m (8.2 ft) off the ground. Further inspection revealed that the snake's head was buried in a mass of leaves and dead vegetation that appeared similar to a squirrel's nest. As I watched, a commotion erupted, and two White-footed Mice (an adult and a juvenile) fell to the ground near my feet. Overhead, I heard a cry and saw that the snake had begun to constrict another juvenile mouse while simultaneously consuming a third (alive and tailfirst). The juvenile mouse that originally fell to the ground near the adult was only ca. 5 cm (2.0 in) snout to rump in length. The concerned mother mouse proceeded to run a perimeter around where I stood in search of her offspring that lay at my boot. At that point, having confirmed the predation event, I decided to leave so that she could reclaim the juvenile and move it to some other location of relative safety.

Figure 222. Adult Gophersnake with an Eastern Cottontail that it had recently constricted and killed (Sauk County; photo by J. Kapfer).

Joshua M. Kapfer, UW–Whitewater

favored Pocket Gophers in Kansas, and they likely consume other fossorial mammals, such as Thirteen-lined Ground Squirrels (*Ictidomys tridecemlineatus*). Although mammals are apparently preferred prey items, Gophersnakes also prey on other vertebrates (for an exhaustive list, see Ernst and Ernst 2003). They are diurnal foragers that actively search for and capture preferred prey through a combination of scent and visual cues (Chiszar et al. 1980, Saviola et al. 2012). They typically hunt by moving through ground vegetation and investigating any burrows or crevices encountered (J. Kapfer, personal observation), but they also forage in trees (Iverson 2014). Gophersnakes are powerful constrictors that typically kill prey animals by wrapping coils around them, although prey encountered in narrow underground burrows can be pinned against the floors and walls to constrict them (de Queiroz 1984).

Gophersnakes have many potential predators, including various birds and mammals (reviewed by Hisaw and Gloyd 1926, Ernst and Ernst 2003). Information on predators in Wisconsin is limited, but Coyotes (*Canis latrans*; Kapfer and Bennell 2005, Kapfer et al. 2008a) and numerous birds of prey are documented (Ross 1989a). Other mammals that share habitat with Gophersnakes presumably also consume them, such as Raccoons (*Procyon lotor*) and American Badgers (*Taxidea taxus*). Hatchling and juvenile snakes are likely more vulnerable to predation than adults. Several ophiophagus snake species, including North American Racers and Eastern Milksnakes, are frequently found sympatrically with Gophersnakes and could be important predators of younger individuals. When threatened, Gophersnakes coil the anterior portion of their body into an S-shaped curve and vibrate their tail. They also forcibly exhale so that air is pushed over anatomical elements of the trachea, larynx, and epiglottis to produce a loud hissing sound, which is characteristic of this species (reviewed by Ernst and Ernst 2003). Although Gophersnakes will often strike at potential threats without biting, they will bite if not left alone. The bite is only marginally painful, but it often bleeds profusely for a short period of time (J. Kapfer, personal observation).

CONSERVATION

Status: Globally, the Gophersnake is listed as a Least Concern species by the IUCN. The subspecies *Pituophis catenifer sayi* has no IUCN ranking. The Gophersnake is considered a species of Special Concern, a Species of Greatest Conservation Need, and a Protected Wild Animal in Wisconsin by the WDNR.

Populations: A three-year study of a Wisconsin population estimated adult abundance at 28.8 individuals, with a density estimate of 0.42 adult/ha (0.169 adult/ac; Kapfer et al. 2008a). This is slightly less than the density estimate for a Kansas population (0.60/ha [0.24/ac]; Fitch and Echelle 2006). Although the ratio of males to females in the Wisconsin study population was equal, and most female snakes encountered during the nesting season were gravid every year, few juveniles or subadults were encountered (i.e., 7.7%–15.4% of all snakes captured). This population further experienced an adult mortality rate of 26.6% (mean over three years; Kapfer et al. 2008a), although the mortality rate was substantially higher for females (43.9%) than for males (14.4%), which could affect long-term population viability.

Threats: Gophersnakes are found in prairies and oak savannas, which have experienced extensive acreage losses in Wisconsin, so general threats to such habitats indirectly affect this snake (see Past, Present, and Potential Future Landscapes of Wisconsin chapter). Habitat disturbance due to agriculture in particular presents a host of risks. Aside from the general habitat loss that occurs when natural landscapes are converted to crops or to contain domestic animals, agricultural equipment can often lead to direct snake mortality. Kapfer et al. (2008a) found that the harvest of crops and the use of heavy machinery for other purposes in agricultural fields were primary sources of snake mortality, as snakes inhabiting harvested fields were sometimes inadvertently run over and killed (Figure 223). It is noteworthy that substantial mortality in agricultural fields occurred despite the fact that local farmers often considered Gophersnakes a beneficial rodent predator (J. Kapfer, personal observation). However, most mortality due to equipment was accidental, because farmers tilling or harvesting fields could not see Gophersnakes hidden belowground in shallow rodent burrows. Unintentional mortality in agricultural landscapes can also occur when Gophersnakes become accidentally entangled in nylon mesh used to exclude wildlife from crops (Kapfer and Paloski 2011).

In addition to habitat destruction, the erosion of habitat quality due to the encroachment of woody vegetation on prairies and open bluffsides likely affects Gophersnakes (Fitch 1999). Unchecked growth of woody vegetation reduces the suitability of habitat by overshading critical nesting, thermoregulatory, and overwintering sites, which could alter normal

Figure 223. A radio-tracked Gophersnake that was accidentally killed by agricultural equipment. Note that the red object in this photograph is the radio transmitter that had been surgically implanted into this individual as part of a research project (Sauk County; photo by J. Kapfer).

patterns of emergence from winter dormancy and also retard embryo development. Another common feature of anthropogenically degraded landscapes is a higher density of paved roads. Paved roads can fragment habitats, restrict snake movement, and result in direct mortality due to vehicular strikes (e.g., Collins et al. 2006, Williams et al. 2014). Although Gophersnake mortality has been anecdotally documented along Wisconsin roadways (Knudsen 1955a), there are little quantifiable data on population-level impacts. It is further important to recognize that although habitat loss and degradation can lead to direct mortality, they can also have subtle effects that are hard to quantify (e.g., altering normal behavior or movement patterns in a way that impacts population viability). These subtle impacts are easily overlooked but no less important. For example, Kapfer et al. (2010) reported that Gophersnake home range size in Wisconsin increased when snakes were faced with a higher prevalence of unsuitable habitats such as agriculture and closed-canopy forest. Because snakes with large home range sizes are more mobile, they are more likely to encounter roads or humans, so their populations tend to experience higher mortality (Bonnet et al. 1999).

The impacts of direct interactions with humans, although rarely quantified, can also negatively impact Gophersnake populations. Breckenridge (1944) outlined a historical example that epitomized a typical outcome of direct human-Gophersnake interactions. This account, published in the *Minneapolis Journal* (27 July 1937), recounted the story of a woman who killed a Gophersnake that she encountered with "about a dozen blows with a long section of pipe. The snake, which was 6½ feet long and about four inches in diameter, did not fight back, she said." The reason for the woman's harsh treatment of the snake was not provided, but it is difficult to imagine that her brutality was justifiable. Although humans often persecute large snakes out of fear, it is noteworthy that direct human interactions can have negative outcomes for entirely contrary reasons (e.g., when Gophersnakes are viewed positively and not feared). A particularly important example is the removal of snakes from the wild for the pet trade or by snake collectors (Ernst and Ernst 2003). The popularity of this snake among herpetoculturists in Wisconsin has likely taken a heavy toll on local populations. For example, in the 1970s and 1980s, numerous well-known populations in the state were repeatedly plundered by collectors, who caused either extirpations or near extirpations at various locales (Robert Hay, Turtles for Tomorrow, personal communication). Removal of individuals from the wild, particularly given the isolated nature of most Gophersnake populations in Wisconsin, could have severe negative impacts. Snake collectors and even enthusiasts can also damage preferred microhabitats by employing careless or destructive searching practices (Goode et al. 2004). This is likely exacerbated if searching is undertaken by large groups of individuals.

Management: Data that support the best management practices for Gophersnakes are limited, and many recommendations are based on knowledge of natural history or anecdotal information. Yet any conservation strategy for Gophersnakes in Wisconsin should surely emphasize the protection of existing suitable habitats. These protections could include attempts to guard knowledge of known Gophersnake populations and hibernacula from the general public.

This could help ensure that snake populations are not easily found and pillaged for the pet trade, but it would also limit the potential for human persecution. Educational programs geared toward the public could also improve awareness of the Gophersnake's important role as a rodent predator and reduce persecution.

Conservation plans for this species should further address its relationship to preferred and disturbed habitats at multiple spatial scales. At the landscape scale, even Gophersnake populations on protected sites are potentially at risk if optimal habitat patches are immediately adjacent to intense agricultural operations or urban/suburban areas. Plans for land purchase and land protection by regulatory agencies should, therefore, attempt to create larger undisturbed buffers around known Gophersnake sites. Unfortunately, due to the Gophersnake's highly mobile nature (Kapfer et al. 2008e, Kapfer et al. 2010), the minimum distance required to keep human activities well-separated from snakes could be large. Yet if adjacent landowners are also cognizant of basic Gophersnake ecology and behavior, the potential for accidental mortality could be reduced. For example, understanding the time of year Gophersnakes are active or when they might enter disturbed locations to forage could guide daily routines for crop harvest or residential lawn care.

At smaller spatial scales, habitat management strategies could contain plans to control the rampant growth of woody vegetation in dry prairies and savannas and along open bluff prairies (see Introduction to Conservation and Management of Wisconsin's Amphibians and Reptiles chapter). Such techniques can include application of prescribed fire and removal of woody plants by hand or with machinery. Yet these management tactics must be considered carefully, because they can result in Gophersnake mortality (e.g., Erwin and Stasiak 1979). Wilgers and Horne (2006) reported that the duration between applications of prescribed fire can have variable effects on different snake species and that Gophersnakes show a preference for sites that are burned annually. However, very few individuals were captured during their study. Furthermore, the frequency with which prescribed fire is applied to a location should depend on how overgrown the site is with woody plants at the time of management (i.e., some sites are heavily overgrown so require frequent burns at first, while fire can be applied less often to sites that are not heavily overgrown). Gophersnakes could also benefit from flexible burn plans, where subplots within larger areas are burned on a variable interannual schedule to create a mosaic-like impact footprint (Wilgers and Horne 2006). Just as with burning regimens, mowing or mechanical brush removal should not impact the entirety of a large area at once. Instead, a rotational mowing regimen could be implemented so that patches of undisturbed vegetation remain to provide refuge for Gophersnakes. Habitat management plans could also implement an appropriate intra-annual schedule for intense management activities. For example, all management activities with substantial impacts to the landscape (i.e., prescribed fire and use of heavy machinery) should be restricted to the snake's inactive period (i.e., 31 October through 31 March). If conducted during the active season, intense activities should at least be restricted to very hot or very cold periods, when snakes will likely be in refugia.

Small-scale habitat management strategies should also consider the 'structure' of vegetation communities on-site. Overly intense application of management techniques, such as prescribed fire, can homogenize habitat structure (see further review of this topic in Introduction to Conservation and Management of Wisconsin's Amphibians and Reptiles chapter). As such, it is important to design management plans that promote creation of structurally heterogeneous habitats for Gophersnakes, with many opportunities for cover and thermoregulation. This can be achieved through management plans that not only schedule activities with appropriate frequency and intensity but also create pockets or corridors of brushy native vegetation, occasional brush or wood piles, or small oak tree stands in prairies. Be aware that the WDNR has regulations for managing habitat (which include restrictions on scheduling such activities), and activities should adhere to those regulations.

Western Wormsnake

Carphophis vermis (Kennicott 1859)

Richard A. Sajdak

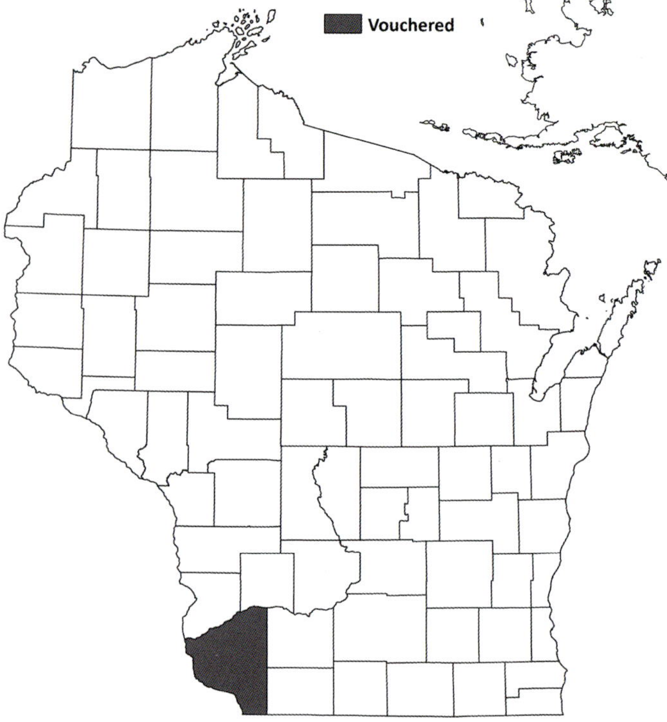

Adult Western Wormsnake (Carter County, Missouri; photo by C. Raimond).

AT A GLANCE

The Western Wormsnake is the smallest snake species in Wisconsin and one of the rarest as well, known only from the extreme southwestern region of the state. As its name suggests, it resembles an earthworm due to its relatively stout, shiny, and cylindrical body, small flattened head, and short sharply pointed tail. Because it is so rare here, little is known about its biology in Wisconsin. It is a burrowing species, most often found under rocks or other objects on bluffside prairies. Although information from Wisconsin is lacking, Western Wormsnakes presumably emerge from hibernation in April or early May. They probably lay eggs in June, which would most likely hatch in August. Based on information from other states, they likely begin to overwinter in late September or October. They presumably eat soft-bodied invertebrates such as earthworms, and little is known about their predators. Western Wormsnakes are completely harmless and rarely observed to bite in defense. They may attempt to make themselves distasteful by defecating and thrashing their bodies, thereby smearing the excrement onto a potential threat. They will also jab their sharply pointed tail into their captor, although the force generated is minimal. Western Wormsnakes are currently considered a species of Special Concern and a Species of Greatest Conservation Need in Wisconsin.

SYSTEMATICS

The most important phylogenetic studies of *Carphophis vermis* (Western Wormsnake) are Clark (1968) and Rossman (1973). Formerly considered a subspecies of *C. amoenus* (Common Wormsnake), the Western Wormsnake was raised to species rank by Clark (1968), while Rossman (1973) presented evidence of intergradation between *C. amoenus* and *C. vermis*. Crother (2017) regarded it as a full species and did not recognize any subspecies. Current phylogenetic analyses of extant snakes place *C. vermis* in the subfamily Dipsadinae, along with two other snake species found in Wisconsin: *Heterodon platirhinos* (Eastern Hog-nosed Snake) and *Diadophis punctatus* (Ring-necked Snake; Figueroa et al. 2016).

DESCRIPTION

Adults: Western Wormsnakes have a relatively stout, shiny, and cylindrical body (Figure 224). Their head is small and flattened, with a countersunk jaw and a thick neck (Figure 225). The tail is short and ends in a spine-like scale tip (Figure 225). There are 13 rows of smooth scales across the dorsum and a divided anal plate. Their ventral surface (including the chin and underside of the tail) and first few scale rows along the sides are bright pink, whereas the dorsum is dark gray to purple gray (see Figures 224 and 225). Unlike most snakes, male Western Wormsnakes have only a single hemipenis (Ernst and Ernst 2003).

Above, top: Figure 224. Adult Western Wormsnake (Grant County; photo by R. Sajdak).

Above, bottom: Figure 225. Head and tail of an adult Western Wormsnake (Grant County; photo by J. Lorch).

Western Wormsnakes are likely the smallest snake species in Wisconsin. The only measurements of Wisconsin individuals available are of a single male (SVL = 25.5 cm [10.04 in], tail length = 5.3 cm [2.09 in]) and single female (SVL = 24.7 cm [9.72 in], tail length = 3.6 cm [1.42 in]). Across the rest of its North American distribution, adult TL ranges from 17.7 to 39.1 cm (7.0 to 15.4 in; Ernst et al. 2003).

Preadult stages: In Wisconsin, the only information on eggs and clutch size is from a single female captured in 2006, which laid three eggs, two of which hatched in captivity (Jeffrey Lorch, unpublished data). Egg dimensions were 3.2 × 0.8 cm (1.3 × 0.3 in; Robert Hay, Turtles for Tomorrow, unpublished data). Eggs from Kansas were oblong (length = 1.9–3.6 cm [0.7–1.4 in], $\bar{x}$ = 2.7 cm [1.1 in]; width = 0.65–0.85 cm [0.26–0.33 in], $\bar{x}$ = 0.79 cm [0.31 in]) and weighed 0.62–1.41 g (0.02–0.05 oz, $\bar{x}$ = 1.05 g [0.04 oz]; Clark 1970). Hatchlings and juveniles are similar in appearance to adults, but no data exist on size of hatchlings or subadults in Wisconsin. In Kansas, mean size of male hatchlings (SVL = 11.2 cm [4.41 in], TL = 2.01 cm [0.79 in], *n* = 10) was similar to that of female hatchlings (SVL = 11.7 cm [4.60 in], TL = 1.52 cm [0.60 in], *n* = 12; Clark 1970).

Variation: No substantial variation in color or pattern among Wisconsin individuals is known. This species appears to exhibit sexual size dimorphism, based on the work of Clark (1970) in Kansas, where females were significantly larger than males. He reported that the mean SVL for the largest 10% of individuals sampled in Kansas was 30.6 cm (12.05 in, *n* = 29) for females and 27.2 cm (10.71 in, *n* = 29) for males.

Diagnosis: Western Wormsnakes can be characterized by the presence of smooth dorsal scales in 13 rows at midbody, an anal plate and subcaudal scales that are divided, and 120–150 ventral scales. They further possess no preocular scales, one loreal scale, one (sometimes two) postocular scale, and five supralabial and six infralabial scales on either side of their head. The only other snake in Wisconsin likely to be confused with the Western Wormsnake is the Ring-necked Snake. Both have shiny smooth scales and a boldly two-toned body. Both also have dark gray to blue-black dorsal surfaces and pink or yellow to red ventral surfaces. However, the Ring-necked Snake has a distinct yellow band across the dorsum of the neck and paired bold black spots on each belly scale. Although other small semifossorial snakes occur in Wisconsin (i.e., Lined Snakes [*Tropidoclonion lineatum*], Brownsnakes [*Storeria dekayi*], and Red-bellied Snakes [*S. occipitomaculata*]), these have stripes or markings on the dorsum and keeled scales.

Western Wormsnake eggs are smaller and more elongated in shape than those of most other snakes in Wisconsin (see Preadult stages subsection). It could be possible to confuse their eggs with those of Smooth Greensnakes (*Opheodrys vernalis*), although current evidence suggests that these species rarely (if ever) co-occur at sites in Wisconsin (see Distribution in Wisconsin subsection).

DISTRIBUTION AND HABITAT

Global distribution: Western Wormsnakes are primarily found on the western side of the Mississippi River, from northern Louisiana and extreme northeastern Texas through Missouri, eastern Oklahoma, and Nebraska to extreme southern Iowa. The only known sites along the eastern side of the Mississippi River are a handful of scattered locations in far western Illinois and Wisconsin.

Distribution in Wisconsin: The only published (and vouchered) county record is for Grant County in extreme southwestern Wisconsin (Sajdak 1978), where it is only known from a handful of locations (see Natural History Box: An Up-and-Down History).

Localities of possible occurrence: As a small secretive snake, Western Wormsnakes are difficult to detect without a directed search. Given that limited surveys for this species have occurred in Wisconsin, it is quite possible that the Western Wormsnake is more widely distributed in the state. It could occur in suitable rocky prairie openings of southwestern Wisconsin along the Mississippi and Wisconsin Rivers, such as in Crawford, Vernon, and Richland Counties. It is often found associated with the Ring-necked Snake

NATURAL HISTORY BOX: An Up-and-Down History

On a warm sunny afternoon in June 1977, my companions were startled to watch me suddenly begin whooping and leaping into the air. We were high on a blufftop prairie overlooking the Mississippi River and had been lifting rocks to enjoy watching disturbed Ring-necked Snakes wave their tightly coiled tails at us. The last rock I had lifted uncovered something different. Once I had it in hand, I recognized the snake as a wormsnake, a species not known from Wisconsin. As a budding herpetologist, I found this to be cause for exuberant celebration, and it became my first herpetological publication, a 14-line note in the Society for the Study of Amphibians and Reptiles' bulletin, *Herpetological Review* (Sajdak 1978).

The next day we started back home to Milwaukee, stopping in Madison for a Wisconsin Herpetological Society meeting. The speaker at the meeting was Dick Vogt, who was in the process of writing his book *The Natural History of Amphibians and Reptiles in Wisconsin* (Vogt 1981). I caught two individuals on this first trip, one of which I brought with me and Vogt photographed for inclusion in his book as part of the account for this species, recognizing it as part of Wisconsin's fauna. The snake was preserved and deposited in the Milwaukee Public Museum (MPM 14754). To this day it carries the tag tied around its body and rests in a jar of alcohol in the museum's collection, a voucher available for future examination and confirmation of its identification. The specimen was a significant record, as the nearest published localities were 235.3 km (146.2 mi) south in Iowa and 255 km (158.5 mi) south in Illinois.

Some years later, Gary Casper began compiling data from the Wisconsin Herpetological Atlas, a database of state distribution records. The wormsnake record was included in the database, but Gary confided to me that he was uncertain that the snake was an actual member of the state's fauna. He had searched the original location without success and suggested that the specimens I found were waifs, perhaps arriving via freight train from the tracks immediately below the bluff. He did, however, later include the species in his book *Geographic Distributions of the Amphibians and Reptiles of Wisconsin* (Casper 1996).

In the interim, the record *was* included in Conant and Collins's field guide (Conant and Collins 1991), but the species *was not* included in the WDNR's original *Snakes of Wisconsin* booklet (Christoffel et al. 2000). However, future discussion on the matter eventually resulted in the wormsnake's inclusion in an updated version of the booklet (Christoffel et al. 2008). Despite the original record going back to 1977, the Western Wormsnake wasn't given protection in Wisconsin as a species of Special Concern until ca. 1985, some eight years later (Rori Paloski, WDNR, personal communication).

At present, more specimens have been found at the original location, and a second location was identified some years later. The species is now a firmly established part of the Wisconsin fauna and protected, but it has been a bumpy ride along the way.

Richard A. Sajdak

in both Wisconsin (J. Lorch, unpublished data, R. Sajdak, unpublished data) and Kansas (Clark 1970). Thus, focusing searches in locations with Ring-necked Snake populations could be fruitful.

Habitat: Although specifics of habitat use in Wisconsin are lacking, Western Wormsnakes are known from hillside prairie remnants with scattered rocks near dry southern forests, such as oak woodlands, along bluff ridges that overlook the Mississippi River (J. Lorch, unpublished data, R. Sajdak, unpublished data; see Amphibians and Reptiles in Wisconsin Plant Communities chapter). Perhaps strangely, they have not been found in the wooded areas adjacent to their prairie locales (J. Lorch, unpublished data, R. Sajdak, unpublished data). Elsewhere, the Western Wormsnake is often reported in flat grassy or wooded areas, as well as along hillsides, such as bluffs and other shaded slopes (Wright and Wright 1957, Clark 1970).

Western Wormsnakes are fossorial and in Wisconsin have been found primarily under rocks embedded in prairie soil, but they are also found under logs, bark, and leaves in other parts of their range. Elick and Sealander (1972) noted that in Arkansas, Western Wormsnakes were most often found under rocks tightly seated in soil, while sympatric Ring-necked Snakes were found under loose rocks. In Kansas, they prefer a loose clay-loam soil with abundant earthworms; being poor burrowers, they cannot penetrate dry packed soil (Clark 1970). The closely related Common Wormsnakes in Maryland and Virginia were documented in active ant nests during summer months. The ants did not show aggressive behavior toward the snakes, and the snakes used the ant tunnels as escape routes in areas with compacted soil (Jones 2014). Elick and Sealander (1972) determined that Western Wormsnakes had high rates of integumentary water loss, which they suggested influenced habitat selection and seasonal activity. They further found Western Wormsnakes in soil with a mean moisture level of 20.4% in Arkansas, while Clark (1967) conducted experiments in captivity and found them in soil with a mean moisture level of 25% in Kansas.

Information on nesting and overwintering habitats in Wisconsin is lacking. Ernst and Ernst (2003) stated that this snake lays its eggs in "depressions under rocks, inside rotting stumps and logs, old sawdust piles, and possibly ground squirrel burrows" but that it spends the winter in "rodent burrows or in rotting logs or stumps." Western Wormsnakes also use active ant nests as hibernacula in Kansas (Pisani 2009a).

REPRODUCTION AND DEVELOPMENT

Little is known about the reproduction and development patterns of Western Wormsnakes in Wisconsin. Mating apparently occurs in two periods, April to May and September to October, with the greatest mating activity in the autumn (based on male reproductive condition; Clark 1970). Adult males possess keeled scales in the anal region, which Blanchard (1931) speculated may serve to stimulate females during courtship. In Wisconsin, single gravid females were captured on 27 May 2006 and 2007 at the same location, which supports that breeding had already occurred (J. Lorch, personal communication). Eggs are laid in late June in Missouri, which could be similar to Wisconsin (Aldridge and Metter 1973). In Kansas, mean clutch size was 2.59 (range = 1–4 eggs; Clark 1970). The Wisconsin female captured in 2006 laid three eggs under captive conditions, which hatched in August (R. Hay, personal observation). Based on the work of Clark (1970) in Kansas, most male Western Wormsnakes attained sexual maturity in the second year and females in the third.

ACTIVITY

Virtually nothing is known of Western Wormsnake activity or behavior in Wisconsin. The earliest individuals have been found under rocks and cover objects in the state in May, but presumably they have similar active seasons to other small fossorial snake species in the state. In Kansas, Western Wormsnakes are most commonly encountered in April and May, moving underground during hotter and drier months except after periods of rainy weather (Fitch 1958b). Clark (1970) reported that the active season was April through October in Kansas. In South Carolina, Common Wormsnakes were active year-round (Russell and Hanlin 1999).

Western Wormsnakes are fossorial and only rarely found aboveground, although Clark (1970) stated that "movements over the surface probably occur every few days when the snakes are active" in Kansas. Flat rocks are often utilized as shelter (see Habitat subsection), and the snakes are able to thermoregulate through contact with the sun-warmed rocks (Fitch 1999). Fitch (1999) reported that mean cloacal temperature was 25.7°C (78.3°F) for a sample of 31 active individuals from Kansas, while Clark (1967) recorded an associated mean substrate temperature of 23°C (73.4°F) during captive experiments.

Based on mark-recapture data, Western Wormsnakes in Kansas are highly sedentary (Clark 1970). Mean distance between captures for all snakes, regardless of sex or age class, was less than 15 m (50 ft). The longest movement recorded in Kansas was a female that moved 125 m (410 ft) in 19 days (Fitch 1958b). Clark (1970) calculated home ranges for individuals in Kansas, reporting home range sizes of 700.8 m^2 (7,543 ft^2) for adult males, 212.7 m^2 (2,290 ft^2) for juvenile males, 74.7 m^2 (804 ft^2) for adult females, and 65.7 m^2 (707 ft^2) for juvenile females (MCP estimator). These estimates were less than one one-hundredth the size of reported home ranges of other snake species in Kansas (Clark 1970).

PREY AND PREDATORS

Nothing is known about the diet of the Western Wormsnake in Wisconsin. The reported diet in Kansas consisted of earthworms and perhaps insect larvae (Clark 1970). Similarly, no information is available on the predators of Western Wormsnakes in Wisconsin. Snakes such as Eastern Milksnakes (*Lampropeltis triangulum*) and North American Racers (*Coluber constrictor*) and mammals such as Virginia Opossums (*Didelphis virginiana*) are reported predators in Kansas (Clark 1970). Clark (1970) further suspected moles and shrews of preying upon this species. Western Wormsnakes are particularly docile when captured. They rarely bite but are known to emit a musky odor when threatened and probe their captor with the pointed spine at the tip of their tail. This probing is not painful and generally does not break the skin. The closely related Common Wormsnake reportedly wags its tail in an arcing serpentine fashion when threatened (Kapfer and Sloane 2011), a behavior also observed in a Western Wormsnake from Wisconsin (R. Sajdak, personal observation; Figure 226). The effectiveness of this tactic in predator deterrence is unknown.

Figure 226. Western Wormsnake performing the tail-wagging defensive behavior (Grant County; photo by R. Sajdak).

CONSERVATION

Status: Globally, the Western Wormsnake is listed as a Least Concern species by the IUCN. It is considered a species of Special Concern and a Species of Greatest Conservation Need in Wisconsin by the WDNR.

Populations: Nothing is known of Western Wormsnake population characteristics in Wisconsin. In a Kansas study, they were the second most abundant species (729/ha [295/ac]) after the Ring-necked Snake ($\bar{x} = 1{,}275$/ha [515/ac]), with an estimated biomass of 5,103 g/ha (73 oz/ac; Clark 1970, Fitch and Echelle 2006).

Threats: The largest threat to the Western Wormsnake in Wisconsin is likely habitat loss and degradation, although further research on this species is sorely needed.

Management: Presumably, habitat management efforts to preserve bluffside prairies and associated oak-hickory forests will benefit Western Wormsnakes, including implementation of prescribed burns and hand removal of woody vegetation in bluff prairies that experience succession (see Introduction to Conservation and Management of Wisconsin's Amphibians and Reptiles chapter). Habitat management strategies that preserve ground cover for this species, such as rocks and downed woody debris, could also be important. Disturbance of rocks that act as shelter for this species should be avoided, as Western Wormsnakes prefer rocks that are well-seated in the soil, and past studies support that rock disturbance negatively impacts reptiles (e.g., Goode et al. 2004). As new populations are identified, protection of associated land should be arranged.

Ring-necked Snake

Diadophis punctatus (Linnaeus 1766)

Rebecca A. Christoffel

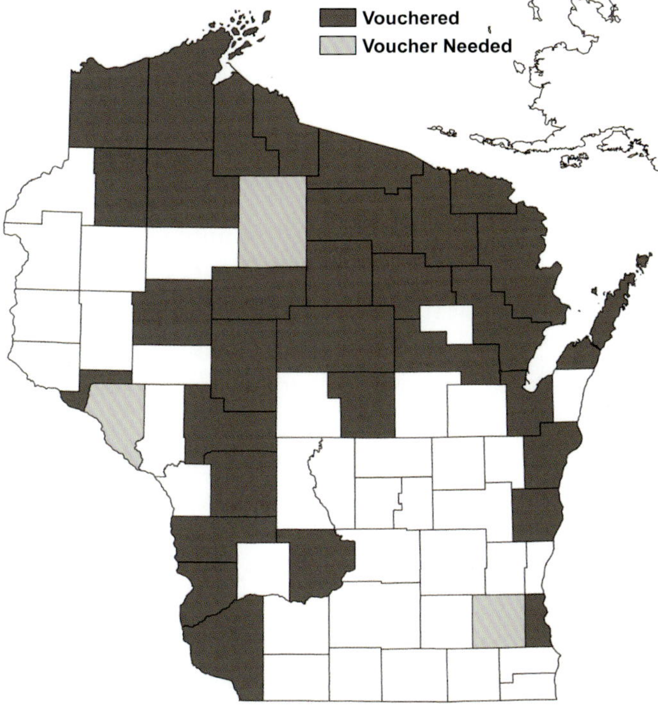

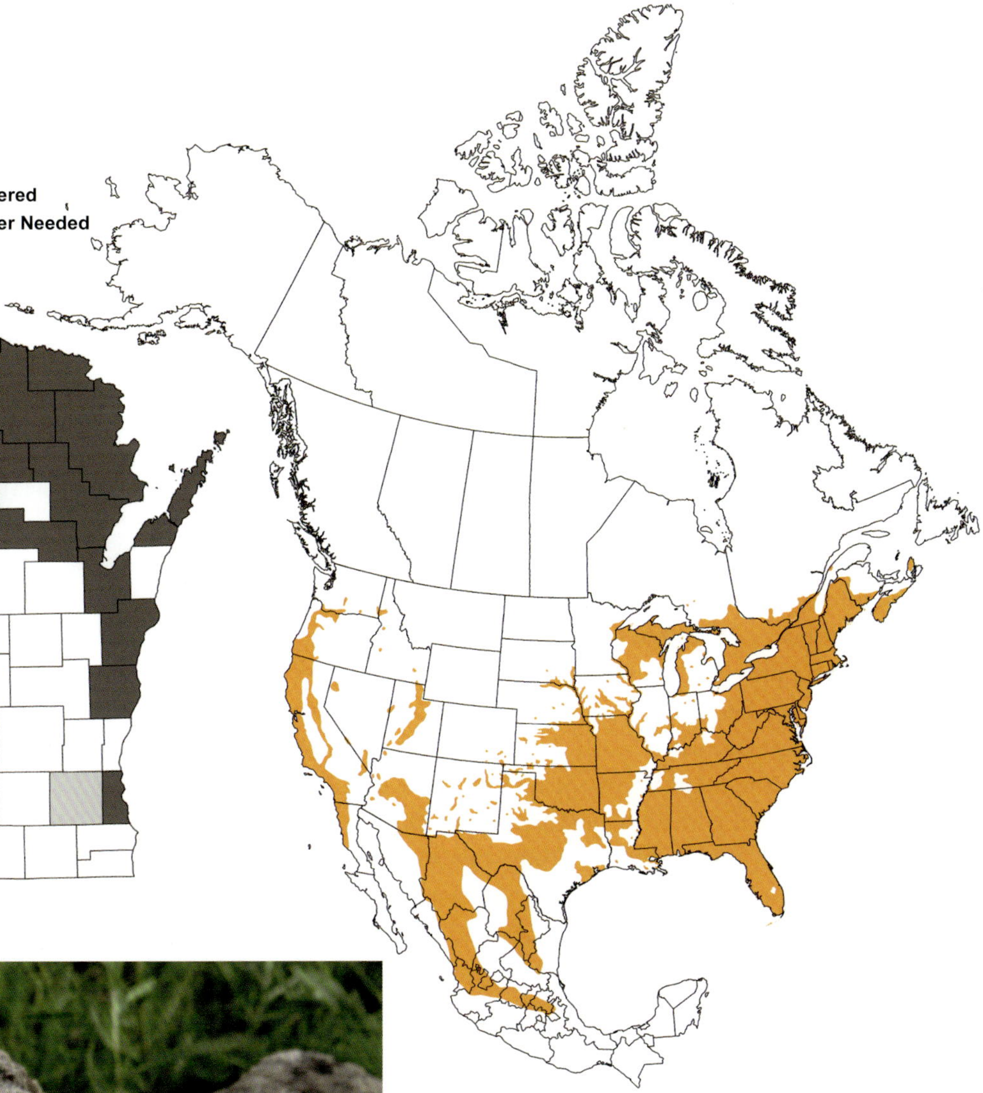

Adult Prairie Ring-necked Snake (Grant County; photo by C. Raimond).

AT A GLANCE

Ring-necked Snakes are among the smallest snakes found in Wisconsin. Two subspecies reside in the state, the Prairie Ring-necked Snake and Northern Ring-necked Snake. Both are easy to distinguish from other Wisconsin snakes by the bright yellow to orange ring around their necks. They appear to be most prevalent in the northern (Northern Ring-necked Snake) and western (Prairie Ring-necked Snake) portions of the state. The distributions of the two subspecies overlap in a small region of Wisconsin, and occasionally mating occurs between subspecies. These matings result in offspring that display coloration and pattern that are intergradations of the two subspecies. Ring-necked Snakes are active from May through September. During the active season, they inhabit deciduous woodlands (Northern Ring-necked Snake) or prairies and semi-open woodlands, including open bluffsides (Prairie Ring-necked Snake). Breeding reportedly occurs in May, and females lay eggs below damp warm ground debris during late June, which hatch in late August. These foot-long snakes dine primarily on invertebrates and amphibians. Their oral secretions are reportedly toxic to some prey, but they are not considered a danger to humans because the negative effects of their bite are apparently mild at worst. They are quite secretive, spending the majority of their time under logs and rocks and in leaf litter. When threatened, Ring-necked Snakes will emit a foul-smelling odor and flip their tails over to expose their colorful undersides—yellow in the Northern Ring-necked Snake and yellow orange grading to bright orange red in the Prairie Ring-necked Snake. The Prairie Ring-necked Snake is considered a Species of Greatest Conservation Need in Wisconsin. The Northern Ring-necked Snake is considered a common species in Wisconsin.

SYSTEMATICS

Blanchard (1942) recognized four species within *Diadophis* based on a combination of morphological characters and geographic distribution. Later studies, which found continuous clinal variation in morphological characters used by Blanchard (1942), suggested that there was gene flow among the populations of *Diadophis*. As a result, the four species recognized by Blanchard (1942) were synonymized into one species (*Diadophis punctatus*) with 13 subspecies. Collins (1991) proposed that two subspecies be raised to species status based on their allopatric distributions and presumed morphological uniqueness. Pinout et al. (1995) examined immunological distances for a small sample of *D. punctatus* and concluded that *Diadophis* may not be a monotypic genus but instead contained at least two genetically distinct species. They argued that their results did not support placement of *D. punctatus arnyi* as a subspecies of *D. punctatus*. Feldman and Spicer (2007) conducted thorough phylogeographic research on *D. punctatus* but restricted their analysis of mitochondrial DNA sequences to *D. punctatus* in California. A later comprehensive phylogeographic analysis of *D. punctatus* was published by Fontanella et al. (2008). Their analyses supported 14 mitochondrial DNA lineages occupying specific habitats separated by putative physical, biotic, or ecological barriers. As a result of their analyses, the authors concluded that species-level diversity is underestimated and a full taxonomic review is warranted, but further sampling, particularly in Mexico, must be undertaken before such a review and revision are conducted. They found evidence for range-wide expansion of the species during the late Pleistocene as the ice sheets retreated (ca. 16,000 yrs ago).

Currently, there are 12 recognized subspecies of *Diadophis punctatus* in North America (Crother 2017). Two *D. punctatus* subspecies occur in Wisconsin. *D. p. arnyi* (Prairie Ring-necked Snake) was named for Samuel Arny, who collected the type specimen (LeClere 2013). *D. p. edwardsii* (Northern Ring-necked Snake) was named in honor of the English ornithologist George Edwards, who collected the type specimen (Beltz 2006). Among Wisconsin snakes, *D. punctatus* occurs in the same subfamily (Dipsadinae) as two other species: *Carphophis vermis* (Western Wormsnake) and *Heterodon platirhinos* (Eastern Hog-nosed Snake).

DESCRIPTION

Adults: Regardless of subspecies, Ring-necked Snakes have brown or black to blue-gray dorsal coloration

Figure 227. Ventral surface of a Prairie Ring-necked Snake showing black spots on the ventral scales and a gradation of ventral color from yellow anterior to orange or orange-red under the tail (Vernon County; photo by J. Kapfer).

Figure 228. An adult Northern Ring-necked Snake displaying a yellow band around its neck and a yellow venter without black spots (Gogebic County, Michigan; photo by J. Kapfer).

with yellow to orange rings around the neck. The ventral surface ranges from yellow to red orange and may include spots on each ventral scale, depending on subspecies. Due to the variation in color and pattern that exists between the Prairie Ring-necked Snake and Northern Ring-necked Snake subspecies, they are described separately here. Prairie Ring-necked Snakes have a dorsal color that varies from dark brown to almost black but is typically a dark bluish gray (Fitch 1999), with a bright yellow or orange-yellow ring around their neck (Collins 1993). Their ventral surface is yellow orange from the throat to the vent and bright orange red on the tail, with many small black half-moon spots scattered across the ventral scales (Vogt 1981; Figure 227). Northern Ring-necked Snakes have a dorsal color that varies from gray or slate to blue black, with a light yellow ring around the back of their neck (Vogt 1981). Their ventral surface is yellow or yellow orange and often has few or no black spots (Vogt 1981; Figure 228) or a single row of black spots down the middle (Knudsen 1955d). Their heads are small and flattened, and their bodies are slender (Conant 1938). Additional morphological characteristics of this species and its subspecies are reviewed by Ernst and Ernst (2003).

The mean SVL of Wisconsin Ring-necked Snake specimens housed in the MPM and UWZM collections was 25.2 cm (9.9 in, $n = 15$) for males and 26.1 cm (10.3 in, $n = 18$) for females, while mean tail length was 7.4 cm (2.9 in, $n = 15$) for males and 5.9 cm (2.3 in, $n = 16$) for females. Larger samples of Prairie Ring-necked Snakes from Kansas revealed similar sizes to those from Wisconsin. For example, Fitch (1999) compared SVLs and weights of adult Prairie Ring-necked Snakes over a 46-year period. He explained the small differences recorded in SVLs across years as being due to chance sampling, minor changes in the age structure of the population, and variation in food availability. As an example, in 1985, adult males averaged 24.9 cm SVL (9.8 in, SD = 0.4 cm [0.2 in]), with a mean weight of 6.13 g (0.22 oz, SD = 0.20 g [0.01 oz], $n = 121$); females averaged 27.6 cm SVL (10.8 in, SD = 0.5 cm [0.2 in]), with a mean weight of 7.35 g (0.26 oz, SD = 0.41 g [0.01 oz], $n = 42$). Females were consistently heavier and longer than males in his population. Male Northern

Figure 229. An egg from a wild female Northern Ring-necked Snake that laid eggs while briefly held in captivity (Wisconsin; photo by J. Kapfer).

Ring-necked Snakes averaged 24.6 cm SVL (9.7 in, SD = 1.5 cm [0.6 in], $n = 6$), while females averaged 27.8 cm SVL (10.9 in, SD = 3.3 cm [1.3 in], $n = 13$) on a military site in Pennsylvania (Meshaka and Delis 2014). The longest Wisconsin specimen measured was an adult male Northern Ring-necked Snake collected in 1917 from Door County (SVL = 33.3 cm [13 in], UWZM 2153).

Preadult stages: The eggs of Prairie Ring-necked Snakes in Wisconsin are yellow white and elongated but often tapered at one end with irregular lumps, and they measure 2.8 × 1.3 cm (1.1 × 0.5 in; Vogt 1981). In Kansas, mean egg dimensions were 2.5 × 0.7 cm (1.0 × 0.3 in), and mean weight was 0.89 g (0.03 oz, $n = 251$; Fitch 1999). Northern Ring-necked Snakes lay eggs that are thin-shelled, elongate, and sometimes tapered, measuring 2.2–3.6 cm (0.9–1.4 in) in length (Figure 229; Harding 1997). Knudsen (1955d) described the eggs as humped on one side and flat or concave on the other. In northern Michigan, Blanchard (1937b) examined 202 egg clutches from Northern Ring-necked Snakes and found that eggs measured 1.8–4.3 × 0.6–1.2 cm (0.7–1.7 × 0.2–0.5 in). In a clutch examined in Ohio, the eggs were adherent and white in color but stained yellow in places (Conant 1938). The eggs of both Prairie Ring-necked Snakes and Northern Ring-necked Snakes gain size and weight during incubation due to water uptake (Ernst and Ernst 2003). In Wisconsin, newly hatched and juvenile Ring-necked Snakes appear similar to adults (Vogt 1981). A sample of 12 juvenile specimens from Wisconsin in the UWZM and MPM collections had a mean SVL of 10.9 cm (4.3 in) and a mean tail length of 2.4 cm (0.9 in). Vogt (1981) reported that the length (presumably TL) of hatchling Prairie Ring-necked Snakes from Wisconsin ranged from 12.5 to 13.5 cm (4.9 to 5.3 in), and TL of hatchling Northern Ring-necked Snakes from Wisconsin was 14 cm (5.5 in). Hatchling Northern Ring-necked Snakes in Michigan measured 10.5–14.3 cm TL (4.1–5.6 in, $\bar{x} = 12.4$ cm [4.9 in]; Blanchard 1930a). In the laboratory, hatchlings measured an average of 12.7 cm TL (5 in, range = 11.1–14.5 cm [4.4–5.7 in], $n = 35$; Blanchard et al. 1979). In Kansas, hatchlings are relatively large compared to adults (mean hatchling weight was 21% of mean adult weight; Fitch 1982). Age for the first and potentially second and third years can be estimated by size, but after that, there is too much variation in individual growth rates to estimate age from size (Blanchard et al. 1979).

Variation: There is variation in the appearance between subspecies of the Ring-necked Snake in Wisconsin (see Description section). Furthermore, in the region where Prairie Ring-necked Snakes and Northern Ring-necked Snakes occur together and potentially intergrade, the belly pattern of individuals can be intermediate between the two subspecies. Females are slightly longer than males, but males have longer tails (Blanchard et al. 1979, Collins 1993). Fitch (1999) measured hundreds of individuals from Kansas and found that female SVL was 11% longer than male SVL. Males who have reached sexual maturity also possess keeled scales in the anal region not found in females (Blanchard 1930a). Partial albino and leucistic Northern Ring-necked Snakes have been documented in Nova Scotia (Gilhen et al. 2012).

Diagnosis: Ring-necked Snakes can be identified by their smooth dorsal scales in 15 rows at midbody, divided subcaudal scales and anal plates, and a yellow or orange neck ring just behind their head. They also possess 1 loreal scale, 2 postocular scales, 2 preocular scales, 7–8 supra- and infralabial scales, and 126–238 ventral scales. They should be easily differentiated from other snakes by the presence of their bright neck ring. However, young Dekay's Brownsnakes (*Storeria dekayi*) and Red-bellied Snakes (*Storeria occipitomaculata*) also have lighter markings around their neck, but these markings are less bright than those of Ring-necked Snakes. Young Dekay's Brownsnakes have keeled scales, and their dorsal color is light brown and spotted. Red-bellied Snakes also have keeled dorsal scales and usually some dorsal markings. In Ohio, many rural residents apparently once thought the Northern Ring-necked Snake was the juvenile form of the North American Racer (*Coluber constrictor*), although the dorsum of juvenile racers is heavily patterned, unlike the solid brown or gray of Ring-necked Snakes (Conant 1938). Ring-necked Snake eggs are uniquely shaped and relatively easy to differentiate from the eggs of other reptiles.

DISTRIBUTION AND HABITAT

Global distribution: Ring-necked Snakes occur only on the North American continent but are one of the few reptiles with a transcontinental distribution. Although they are broadly distributed, their range is far more widespread in the eastern US while only spotty and sporadic in the western US. In northern regions, they occur from New Brunswick and Nova Scotia west to Minnesota. Although they are not found throughout large portions of the Great Plains and Rocky Mountains, they occur again in the far western US from southern Washington south to southern California and northern Baja California. In the South, they are found down to the Florida Keys, west across portions of the Gulf Coast, and south to San Luis Potosí in eastern Mexico.

Distribution in Wisconsin: Published county records for Prairie Ring-necked Snakes are primarily concentrated in the southwestern Driftless Area, extending north to Pepin County and east to Sauk County. Published Northern Ring-necked Snake county records are mostly from north of the vegetative tension zone in Wisconsin (Curtis 1959) but occasionally from counties just to the south, such as Monroe. Among published county records of both subspecies, voucher specimens could not be found for Buffalo, Price, and Waukesha (see the Appendix).

The zones of overlap and potential intergradation for these subspecies in Wisconsin are not particularly well-defined. According to Casper (1996), both subspecies were reported from Sauk and Pepin Counties. A specimen that had a ventral pattern intermediate between the Prairie Ring-necked Snake and Northern Ring-necked Snake was collected in Jackson County (Casper and Burnham 2001).

Localities of possible occurrence: Ring-necked Snakes likely occur in other counties but have not yet been encountered due to their small size and fossorial nature. It is likely that Prairie Ring-necked Snakes occur in La Crosse and Trempealeau Counties, given records to the north and south (although no voucher could be found for the Buffalo County record). They may also exist farther north in the counties along the Mississippi River and into south-central counties with appropriate habitat, such as Richland, Iowa, Lafayette, Rock, and western Dane. Northern Ring-necked Snakes probably occur in several additional counties north of the vegetative tension zone in Wisconsin where they have not yet been documented (i.e., Barron, Rusk, Menomonee). Reports from localities in eastern Wisconsin are particularly sparse, and records either lack vouchers (i.e., Waukesha County) or are supported by old specimens (e.g., Sheboygan and Milwaukee Counties). Historically, Ring-necked Snakes could have been more widespread in eastern counties, particularly when deciduous hardwood forests were more prevalent along the shores of Lake Michigan (see Past, Present, and Potential Future Landscapes of Wisconsin chapter). Given the existence of anecdotal reports from other eastern counties (i.e., Ozaukee County; reviewed by Casper 2007), this snake could still be discovered in additional eastern locations with appropriate habitat (e.g., Washington County).

Habitat: There are no detailed analyses of Ring-necked Snake habitat preferences in Wisconsin, and much information from the Upper Midwest is anecdotal. Vogt (1981) stated that Prairie Ring-necked Snakes are found on hill prairies and open dry cliff or bluff prairies along the Mississippi and lower Wisconsin Rivers (see

Amphibians and Reptiles in Wisconsin Plant Communities chapter) and that they spend the spring and autumn on warm southwest-facing slopes but move to "moister places near springs or talus piles" in summer. Numerous additional sources from throughout the Midwest report this subspecies in habitats that range from forested areas to prairies and pastures, including both wooded and open south- to west-facing bluffsides or hillsides (Smith 1961, Johnson 1987, Oldfield and Moriarty 1994, Le Clere 2013). More detailed work has been conducted on suitable habitat in Kansas, which found that Prairie Ring-necked Snakes were not limited to a specific habitat type, so long as certain habitat features were present. For example, Fitch (1975) found that they required soil that was slightly damp (but not saturated), abundant cover in the form of litter (dead vegetation) or loose objects such as flat rocks, and screening shrubs or trees with leaf canopies that were open enough to allow substantial sunshine to reach the ground. Fitch (1982) further mentioned their strong association with edge habitats that exist between woodlands and grasslands. In his study of Prairie Ring-necked Snakes from northern Missouri, Riedle (2012) believed that these snakes could thrive in urban environments because of their ability to use an abundant food resource (i.e., earthworms) coupled with their small size and secretive nature. Northern Ring-necked Snakes, on the other hand, reportedly occupy moist deciduous forests, which makes them somewhat unique among snakes native to Wisconsin (Vogt 1981). Harding (1997) further suggested that they use moist shady upland woods, avoiding flood-prone bottomlands, but they may also opt for more open habitats adjacent to woodlands, including clear-cuts, old fields, grassy dunes and beaches, and trash dumps. In Michigan, Blanchard et al. (1979) documented a population of Northern Ring-necked Snakes on a ca. 30.5 m (100 ft) wide clearing that extended along the shore of Lake Michigan next to a sand and pebble beach and backed by low damp evergreen woods.

Ring-necked Snakes spend most of their time under numerous types of ground debris. Objects used as cover include flat rocks that are piled on top of each other (Figure 230), within logs, under fallen bark, and below woody debris in woodlands, but Ring-necked Snakes will also find refuge underground on dry hillside prairies and within mammal or insect tunnels (Vogt 1981, Smith 1961, Fitch 1975, Harding 1997). Conant (1938) found no apparent preference for the types of rocks used as cover by Northern Ring-necked Snakes in Ohio, observing them under limestone, sandstone, and shale, so long as they were sufficiently damp. Ring-necked Snakes seem to require moist soil to balance evaporative water loss (Myers 1965, Henderson 1970). Clark (1967) estimated that mean soil moisture associated with snakes found beneath ground debris was 29%.

Figure 230. An example of a rock pile, with large flat rocks on top of other rocks, commonly associated with Prairie Ring-necked Snakes (Monroe County; photo by J. Kapfer).

Little is known about nesting habitat in Wisconsin. A handful of observations suggest that they lay eggs in humid locations under ground debris, such as rocks and fallen wood (see Reproduction and Development section), likely favoring areas that receive enough sunlight for adequate incubation. Ring-necked Snakes potentially overwinter in available fissures or holes in the ground that descend below the frost line. Anecdotal reports of potential hibernacula exist, ranging from rock crevices to mammal and insect burrows, but few observations have been documented (Johnson 1987, Oldfield and Moriarty 1994, Harding 1997, LeClere 2013). Fitch (1999) believed that snakes in his Kansas study area used the tunnels of small mammals or insects within their summer activity ranges as hibernacula because he found them in these locations in early spring. Pisani (2009a) made observations that supported Prairie Ring-necked Snakes and several other snake species used an ant tunnel system as a hibernaculum in Kansas. In Michigan, Blanchard et al. (1979) found Northern Ring-necked Snakes emerging from a pile of brush on a ridge and posited that the snakes had overwintered under the ground below the pile. In Maryland, a Northern Ring-necked Snake was found overwintering at a depth of 81.2 cm (32.0 in) in sandy loam (soil temperature = 9.5°C [49.1°F]) in February (Grizzell 1949).

REPRODUCTION AND DEVELOPMENT

As it is with many aspects of this species' ecology in Wisconsin, information on reproduction and development is largely anecdotal. Vogt (1981) reported that breeding primarily occurred in May after emergence from hibernation in Wisconsin. Reports of late summer and autumn breeding exist from Illinois (Phillips et al. 1999), Michigan (Blanchard et al. 1979), and Kansas (Fitch 1999). The anatomy of female Ring-necked Snakes allows for delayed pregnancy after autumn breeding events occur because the vagina contains long furrows in the mucosa that serve as receptacles that store sperm throughout hibernation (Perkins and Palmer 1996). Pheromones released from the skin of female Ring-necked Snakes attract males and aggregate them for mating in spring and autumn (Dundee and Miller 1968, Noble and Clausen 1936). According to Fitch (1975), when a male Prairie Ring-necked Snake has located a receptive female, he rubs his closed mouth up and down the female's side and moves forward beside her. When the male encounters the female's neck ring he bites it and lines up the lower part of his body against hers, wraps his tail around her tail, and achieves intromission (Fitch 1975). Ernst and Ernst (2003) stated that the "female's neck ring appears to serve as a releaser for male aggressive biting behavior."

Aside from isolated observations, little is known about the schedule of egg deposition in Wisconsin. Two gravid Prairie Ring-necked Snakes collected in Grant County on 29 May laid eggs on 21 and 23 June in captivity (Vogt 1981). Female Northern Ring-necked Snakes collected in Florence County on 20 June laid eggs on 28–29 June (Vogt 1981). In his 50-year study of snakes in Kansas, Fitch (1999) found that Prairie Ring-necked Snakes laid eggs primarily in the first week of July. Northern Ring-necked Snakes deposited eggs mostly in late June or early July in Kentucky (Barbour 1950) and Michigan (Blanchard et al. 1979). In Maryland, eggs were laid from 17 June to 21 July (median = 2 July; Clark et al. 1997). Locations that Northern Ring-necked Snakes selected to lay eggs in Michigan included moist areas under the bark of rotting logs or within decaying logs lying in full sunshine (Blanchard 1937b, Blanchard et al. 1979). Blanchard et al. (1979) also found several clutches on the beach shore of Lake Michigan under cover of a log or board, which kept them moist. Communal nesting has been reported in Prairie Ring-necked Snakes (Fitch 1999) and Northern Ring-necked Snakes, where as many as 55 eggs were found in one nest from Michigan (Blanchard 1937b). In Wisconsin, three gravid female Northern Ring-necked Snakes were found under the same piece of tar paper prior to nesting (Vogt 1981).

Documented records of clutch size in the Upper Midwest are sparse. Two gravid female Prairie Ring-necked Snakes captured by Vogt (1981) in Grant County each laid four eggs. Prairie Ring-necked Snake egg clutches in Kansas varied from one to ten eggs ($\bar{x}$ = four; Fitch 1975), although great annual variability was detected. Over a 30-year period of monitoring reproductive output in Kansas, the mean clutch size in a low reproduction year (1976) was 3.11 (SD = 0.60, $n = 9$), while mean clutch size in a high reproduction year (1963) was 5.17 (SD = 2.04, $n = 12$; Seigel and Fitch 1985). Clutch size is positively correlated with female body size, as Fitch (1999) reported mean number of eggs per clutch by third-year females was 2.7, while the oldest and largest females laid >5 eggs per clutch in Kansas. Females in Kansas also experienced an increase in the number of eggs per clutch during a breeding season that immediately followed an autumn with ample precipitation, which presumably resulted in greater energy for reproduction due to an increase in available prey (e.g., earthworms; Seigel and Fitch 1985).

Female Northern Ring-necked Snakes collected in Forest County by Vogt (1981) laid clutches of four eggs in captivity. Also, fewer than four eggs or sometimes four eggs were found in the few Wisconsin clutches of Northern Ring-necked Snakes examined by Knudsen (1955d). This is similar to findings from more in-depth reviews of clutch size from Michigan (Blanchard 1937b), Kentucky (Barbour 1950), and Maryland (Clark et al. 1997). As it does in Prairie Ring-necked Snakes, clutch size of Northern Ring-necked Snakes increases with female SVL (Clark et al. 1997). Clutches from females collected in Grant and Florence Counties hatched in ca. two months (late August; Vogt 1981). Reported incubation times from various northern locations outside of Wisconsin range from 42 to 56 days (Blanchard 1930a, Conant 1938, Clark et al. 1997, Fitch 1999). Several authors reported that Ring-necked Snake embryos are already partly developed by the time eggs are laid (Fitch 1999), which Knudsen (1955d) postulated results in a shorter egg incubation period compared to many other egg-laying snakes.

Juvenile Ring-necked Snakes grow rapidly, although variation in growth rates by sex occurs. At the end of their first summer (i.e., August or early September), young female Prairie Ring-necked Snakes are larger (SVL = 11.4 cm [4.5 in], $n = 33$) than males (SVL = 10.9 cm [4.3 in], $n = 50$), and they continue to grow faster than males in Kansas (Fitch 1999). Growth rates for Prairie Ring-necked Snakes are similar to those of Northern Ring-necked Snakes. A multiyear study of Northern Ring-necked Snakes in Michigan revealed that females increased their length by ca. 60% at the end of their first year, but this growth rate decreased steadily with age (second year = 23% growth, third year = 18% growth, fourth year = 17% growth, fifth year = 15% growth; Blanchard et al. 1979). Changes in growth rates by age were more extreme in males, which ranged from a 68% size increase during year one to an 11% size increase during year four (Blanchard et al. 1979). In Kansas, male Prairie Ring-necked Snakes become sexually mature during their second active season, while females required three seasons to reach sexual maturity (Fitch 1975) but did not reproduce until their fourth year (Fitch 1999). In Pennsylvania, estimated growth trajectories for Northern Ring-necked Snakes were suggestive of sexual maturity by the beginning of their second spring at 20–21 months of age (Meshaka and Delis 2014). Males reach sexual maturity by their third or fourth years at a length of ca. 28.0 cm (11.0 in), a year earlier than females (Blanchard et al. 1979). Maturity in females is determined by the presence of discernible eggs in early summer at a length of about 31.5 cm (12.4 in; Blanchard et al. 1979).

As they are with most snakes, records of longevity in the wild are sparse for Ring-necked Snakes. Prairie Ring-necked Snakes can apparently live for >20 years (Fitch 1999). A male Northern Ring-necked Snake repeatedly recaptured over the course of 15 years in Michigan was estimated to be well over 20 years old (Blanchard et al. 1979). In this study population, females >36.5 cm TL (14.4 in) and males >33.5 cm TL (13.2 in) often failed to grow during a year, and those snakes that had ceased to grow were estimated to be at least 20 years old (Blanchard et al. 1979). The record for Ring-necked Snake longevity in captivity is six years and two months (Snider and Bowler 1992).

ACTIVITY

Ring-necked Snakes are primarily active in Wisconsin from May through September (Vogt 1981), although reports of activity in late April and October exist from other states in the Upper Midwest (Blanchard et al. 1979, Oldfield and Moriarty 1994, LeClere 2013). Fitch (1975) reported that emergence from hibernation in Kansas is controlled by environmental temperature, which depends on weather and location or depth of the hibernaculum. After emergence from hibernation, the intensity of activity varies by time of day, season, and sex. Overall, the greatest activity occurs in the spring (e.g., May), with a second lesser spike in activity during autumn (reviewed by Ernst and Ernst 2003, Meshaka and Delis 2014). Furthermore, Fitch (1999) encountered male Prairie Ring-necked Snakes three times as often as females in the spring, but females were found more often than males in the summer. He attributed this pattern to the greater activity of males during the spring breeding season. Even when active, Ring-necked Snakes are primarily nocturnal in their habits and spend most of their time underneath ground debris (Fitch 1975, Vogt 1981, Oldfield and Moriarty 1994, reviewed by Ernst and Ernst 2003). In a long-term study from Michigan, only one individual was observed on the surface of open ground during daylight, coiled in a sunny spot near a brush pile on a windy day in May (Blanchard et al. 1979).

Fitch (1956) studied the relationship between environmental temperature, body temperature, and activity of Prairie Ring-necked Snakes from Kansas. He concluded that environmental temperature was the most important factor controlling the activity of this snake and that its optimal body temperature range was 27°C–29°C (80.6°F–84.2°F; $n=145$). Overall, the mean body temperature of snakes found under cover was 26.1°C (79°F, $n=129$), and the mean body temperature of snakes found in the open was 26.6°C (79.9°F, $n=37$). He further reported that snakes found when air temperatures were below 16.2°C (61.2°F) were always under some form of ground debris. Based on laboratory observations, Clark (1967) reported a mean preferred body temperature of 26.43°C (79.6°F). Henderson (1970) found that because water requirements increased with rising environmental temperatures, Ring-necked Snakes were susceptible to dehydration. At temperatures higher than 30°C (86°F), test subjects showed duress from the heat and died at extended exposures to 35°C–36°C (95°F–96.8°F; Henderson 1970).

The secretive nature of this snake makes it difficult to investigate its movement patterns. In a Kansas population, the normal home range diameter of Prairie Ring-necked Snakes was 70 m (230 ft; Fitch 1970), and the mean area occupied was ca. 0.65 ha (1.6 ac; Fitch 1975). Individuals usually move only a few meters over the course of a day, and that movement is very slow, with many long pauses (Fitch 1975). Tennant and Bartlett (2000) found marked Prairie Ring-necked Snakes that were recaptured a year or more later near the spot the animals were originally marked, which prompted them to estimate that individual home ranges were no more than 121.9 m (400 ft) in diameter. Fitch (1999) further found evidence of philopatry for the home range locations of certain snakes in Kansas. He tracked 151 individuals over periods ranging from a few days to more than six years. In 29 instances, he recorded snakes returning to approximately the same spot after long periods of time. Northern Ring-necked Snakes in Michigan also showed evidence of philopatry, with several individuals displaced 0.8–1.6 km (0.5–1.0 mi) by researchers returning to their original capture location (Blanchard et al. 1979). Blanchard et al. (1979) suggested that Northern Ring-necked Snakes in Michigan make their longer-distance movements after dark.

Despite their small overall home range sizes and philopatric tendencies, there is variation in movement based on sex, age class, and season. Fitch (1975) found that females were more vagile than males, while adults were more vagile than juveniles. Several sources also report that Ring-necked Snakes make substantial migrations between overwintering locations and areas where they spend most of the active season (Blanchard et al. 1979, LeClere 2013). Similarly, Fitch (1975) found that some of his study animals completed relatively large autumn movements from grassy summer habitat to their hibernacula, migrating 100 m (328 ft) or up to 380 m (1,246 ft). Anecdotal reports suggest that numerous Ring-necked Snakes will overwinter communally with other species, as well as with conspecifics (Vogt 1981, Harding 1997, LeClere 2013).

PREY AND PREDATORS

Ring-necked Snake diet apparently varies by geographic location (Rossi and Rossi 1995). Numerous sources have reviewed Ring-necked Snake prey in the Upper Midwest, which purportedly ranges from earthworms and various other invertebrates to vertebrates such as salamanders, small snakes, and newborn rodents (Oldfield and Moriarty 1994, Phillips et al. 1999, LeClere 2013). In Kansas, Fitch (1975) examined the stomach contents of Ring-necked Snakes ($n=182$) and found 221 earthworms, 2 insect larvae, and 1 hatchling Slender Glass Lizard (*Ophisaurus attenuatus*). Harding (1997) suggested that Northern Ring-necked Snakes may actually specialize in certain prey species, such as Eastern Red-backed Salamanders (*Plethodon cinereus*), while Cathro and Lindquist (2016) reported that 88% of those examined from Pennsylvania ($n=58$) contained plethodontid salamanders in their stomachs. Blanchard et al. (1979) found that Ring-necked Snakes held in captivity refused many types of vertebrate and invertebrate prey while eating almost exclusively

Red-backed Salamanders (and their eggs) and the occasional small squamate. Lancaster (1994) reported that Northern Ring-necked Snakes detected and potentially identified substrate odors from Eastern Red-backed Salamanders. In subsequent work, Lancaster and Wise (1996) demonstrated that Northern Ring-necked Snakes increased their tongue-flicking when exposed to swabs containing chemicals from the ventral surface of the tail of Eastern Red-backed Salamanders, as opposed to the salamander's body.

Fitch (1975) estimated that Prairie Ring-necked Snakes consume, on average, 12% of their body weight per feeding and three times their body weight in earthworms annually. Henderson (1970) found that snakes could consume up to 45% of their body weight in a single feeding. Feeding frequency was estimated at once every eight days in Kansas, with about half that time interval being needed to digest and assimilate the last meal (Fitch 1975). Henderson (1970) observed that the optimal environmental temperature for Prairie Ring-necked Snakes to feed in Kansas was 25°C–35°C (77°F–95°F).

Prey are apparently located through chemical cues (Henderson 1970, Lancaster and Wise 1996, Weaver 2010) and are often overpowered and swallowed whole (Vogt 1981). Some authors reported that prey are constricted, although this is not well-documented in eastern subspecies (reviewed by Ernst and Ernst 2003). Ring-necked Snakes apparently produce toxic oral secretions to help subdue prey, but research investigating the toxicity of these secretions and their effectiveness when hunting is relatively sparse (e.g., O'Donnell et al. 2007). There is evidence that toxicity varies by Ring-necked Snake subspecies and geographic location, but this is not well-studied in the Upper Midwest. As reviewed by McKinstry (1978), Ring-necked Snakes' oral secretions may be particularly toxic to specific prey types in regions where that prey is more often consumed (e.g., subspecies in the southwestern US that often consume snakes have toxic secretions that immobilize snakes). Although Prairie Ring-necked Snakes (and many other *Diadophis punctatus* subspecies) have elongated rear fangs to potentially administer these toxins, the Northern Ring-necked Snake either is fangless or has reduced fangs (reviewed by McKinstry 1978, O'Donnell et al. 2007).

Information on the predators of Ring-necked Snakes in the Upper Midwest comes mostly from statements in field guides. These identify a variety of predators, including small mammals, birds, and other snakes (Phillips et al. 1999, Oldfield and Moriarty 1994). In Wisconsin, Broad-winged Hawks (*Buteo platypterus*) are documented predators (Rosenfield et al. 1984). Vogt (1981) also specifically mentioned that North American Racers "feed readily" on Prairie Ring-necked Snakes in Wisconsin, and North American Racer predation has been documented in Crawford County (Nick Schiltz and Jason Dare, Dare Ecosystem Management, LLC, personal communication). Harding (1997) postulated that potential predators are primarily those willing to enter burrows or dig through ground debris.

Several sources mention this species' tendency to discharge fecal matter and foul-smelling musk from glands at the base of the tail when threatened (Vogt 1981, Johnson 1987, Tennant and Bartlett 2000). In Kansas, Fitch (1999) observed that when encountered in the open, Prairie Ring-necked Snakes will remain still until actually touched before making a hasty retreat. In Ohio, Conant (1938) found that Northern Ring-necked Snakes suddenly exposed to light tried to crawl under nearby objects. He further observed that captured snakes rarely bit and only sometimes attempted to wriggle free, although most were docile, similar to reports from Wisconsin (Vogt 1981). Although Ring-necked Snakes likely have toxic oral secretions, past authors suggest that these secretions are mostly only effective against prey, such as other reptiles (reviewed by Brock and Camp 2018) and presumably of little use for defense. This, coupled with their docile temperament, supports the claim that Ring-necked Snakes pose a minuscule threat to humans, with described effects of bites ranging from mild (i.e., a burning sensation and swelling) to nonexistent (reviewed by Ernst and Ernst 2003, Brock and Camp 2018). Death-feigning behavior, supposedly in response to being threatened, was reported in Minnesota (Oldfield and Moriarty 1994).

A tail-spiraling or coiling defensive behavior is well-known in this snake. For example, when threatened, Prairie Ring-necked Snakes often bury their heads under their coils and raise their tails to expose their brightly colored ventral surface, which is then

Figure 231. Adult female Prairie Ring-necked Snake displaying the tail-spiraling or 'corkscrew' defensive posture (Vernon County; photo by J. Kapfer).

spiraled like a corkscrew (Figure 231). Fitch (1975) tested 100 Prairie Ring-necked Snakes (including 48 adult males) for spiraling behavior by holding snakes down momentarily after uncovering them. Of these individuals, 95 spiraled their tails, 1 held its tail in a loop, and only 4 individuals failed to show any spiraling. Smith (1975) demonstrated that large male Prairie Ring-necked Snakes exhibit tail-spiraling behavior less often than smaller individuals and females. Gehlbach (1970) posited that the use of "erratic behavior" preceding defensive postures by some Ring-necked Snake subspecies appears to be a stress response to the precursors of predation (i.e., attack and capture). He averred that the snakes may advertise their distastefulness and startle predators effectively with their stereotypical defense displays.

Ernst and Ernst (2003) stated that this display could either startle a predator or act as a decoy (i.e., direct a predator to attack the tail rather than the head). Vogt (1981) speculated that tail-coiling might function as a distraction to avian predators. Fitch (1999) also postulated that birds were likely important predators, as the tail-coiling display should evolve under pressure from predators that hunt largely through visual cues, such as birds, as opposed to those that lack color vision or those encountered in dark places (i.e., below leaf litter or in tunnels). Fitch (1999) reported that 6.5% of Prairie Ring-necked Snakes examined from Kansas were missing part of their tail ($n = 13{,}927$ individuals examined). In first-year snakes, the incidence was slightly greater than 1%, but in those snakes judged to be more than 12 years old, the incidence had risen to >20%. He estimated that about 1.3% of the Ring-necked Snakes in the local population lost their tail tip each year. Blanchard et al. (1979) also found that Northern Ring-necked Snakes in Michigan often had the end of their tail tip missing. McCallum et al. (2006) determined that significantly more Ring-necked Snakes had a damaged tail than other small fossorial species analyzed, which supported that tail-coiling behavior is employed to distract predators and direct them toward the less vital tail tip rather than act as a startle display. Vogt (1981) reported that Northern Ring-necked Snakes in Wisconsin did not exhibit tail-coiling when threatened, but this behavior has been observed in populations from the Upper Peninsula of Michigan (J. Kapfer, personal observation; see Figure 228). The tail-coiling behavior of the Ring-necked Snake has also led to the colloquial name 'corkscrew snake' (Oldfield and Moriarty 1994).

CONSERVATION

Status: Globally, the Ring-necked Snake is listed as a Least Concern species by the IUCN. Neither subspecies is currently listed as threatened or endangered in Wisconsin. The Prairie Ring-necked Snake is considered a Species of Greatest Conservation Need, and the Northern Ring-necked Snake is on the state 'watch list' as a potential future species of concern (WDNR 2016a).

Populations: Ring-necked Snakes are small, secretive, and primarily fossorial, which makes it challenging to study population parameters. Furthermore, little published work exists, in part because recapture rates are low, and estimating population size via mark-recapture techniques is difficult (Turner 1977, Parker and Plummer 1987, Vitt 1987, Riedle 2012). Yet casual observations suggest that this snake can exist in high numbers at certain locations. For example, according to a personal communication to George Wagner (UW–Madison), A. I. Ortenburger, who did extensive collecting for the UWZM, found "'bushel baskets full' in a communal hibernating area on the northeast side of Rib Hill, Marathon County, in 1918" (Vogt 1981). Although Ortenburger's observation might suggest a large population, actual quantitative information from Wisconsin is limited. Intense work from other states provides potentially relevant data that support that this snake can exist in high abundances. Fitch (1999) studied Prairie Ring-necked Snakes for more than 50 years in Kansas. He concluded that his study site had "some 10,000 snakes (1040 per ha)" (421/ac). Of 12 species of snake he studied at the University of Kansas Natural History Reservation, the Prairie Ring-necked Snake accounted for more than twice as much biomass as all other snake species combined (Fitch 1982). Of the 33,117 snakes in 44 samples taken from localities across Kansas, Prairie Ring-necked Snakes made up 30.3% of the total sample (Fitch 1993). In three northeastern Kansas counties, Prairie Ring-necked Snakes made up 59.2% of the 7,062 snakes captured from 1981 to 1991 (Fitch 1993). Riedle (2012) studied an urban population of Prairie Ring-necked Snakes in northern Missouri. He calculated a density of 188 ± 29 individuals for his 0.05 ha (0.12 ac) study area (i.e., ca. 3,760 individuals/ha [1,567/ac]).

Several studies have estimated other population parameters. Researchers of both Prairie Ring-necked Snakes (Fitch 1975) and Northern Ring-necked Snakes (Blanchard et al. 1979) report a rather constant mortality rate of 21%–29% among each age or size class. On the other hand, Riedle (2012) found low recapture rates for all age classes studied (i.e., year one, year two, and adult), but particularly so for year two individuals. Less than half of his study population was predicted to survive to three years of age (i.e., adult), although his study population showed signs of increasing. He posited that the persistence of Ring-necked Snake populations may be most dependent on two factors: the survival of juveniles as they transitioned from their second year of life to sexual maturity at three years of age, followed by adult females achieving a relatively long life span.

In his sample of 1,598 Prairie Ring-necked Snakes captured from 1971 to 1973, Fitch (1982) recorded a sex ratio of 1.71M:1F and an adult-to-juvenile ratio of 1.21A:1J. Riedle's (2012) study population had a ca. 1M:1F sex ratio for all age classes. Blanchard et al. (1979) reported an adult-to-juvenile ratio of 3A:1J in their northern Michigan population of Northern Ring-necked Snakes. They attributed the scarcity of juveniles in their sample in part to the difficulty in finding them and an underestimate of how long the snakes lived. They believed juvenile survivorship must be high for the study population to persist despite recapturing only 18% of their marked juvenile snakes as adults. They concluded that Northern Ring-necked Snakes are long-lived, produce few eggs annually, and have low mortality of young. Fitch (1999) found that the percentage of first-year young in his annual samples fluctuated drastically (10%–>50%) due to variations in weather, which affected the success of reproduction.

Threats: Given their propensity to remain under ground debris, Ring-necked Snakes could be less vulnerable to direct human persecution than other more noticeable snake species (Blanchard et al. 1979). However, Ernst and Ernst (2003) noted that many populations have been lost or greatly reduced because of habitat destruction. Minton (1972) reported that habitat for Ring-necked Snakes was being lost to deforestation and intensive agriculture. In addition to habitat loss, this snake suffers when the quality of existing habitat is degraded. For example, research on the impacts of habitat fragmentation in Canada suggested that Northern Ring-necked Snakes could be sensitive to the amount of suitable habitat available, being absent from smaller habitat islands (Hager 1998). Andrews and Gibbons (2005) determined that Ring-necked Snakes could be indirectly impacted by the presence of roads that fragment habitat. They found that although Ring-necked Snakes often made initial efforts to cross roads in experimental trials, monitored snakes frequently abandoned these attempts and left the roadside. This tendency can restrict dispersal from currently occupied habitat fragments, deterring colonization of new sites and genetic exchange over large geographic areas. In his 50-year study of snakes in Kansas, Fitch (1999) found that reproductive success for Prairie Ring-necked Snakes declined during the latter decades of study. He attributed this to shading of previously open habitats caused by vegetative succession, which may have been responsible for a delay in hatching and a reduction in hatching success. Therefore, the impacts of habitat degradation due to vegetative succession and fragmentation, as well as habitat loss, must be considered (see Introduction to Conservation and Management of Wisconsin's Amphibians and Reptiles chapter).

In Wisconsin and Minnesota, populations of Prairie Ring-necked Snakes are locally abundant on certain hillsides but absent from nearby hillsides of comparable quality (Vogt 1981 and Oldfield and Moriarty 1994, respectively). It is unclear why the snake's distribution is spotty, although proximity to suitable overwintering areas may be one factor. Harding (1997) believed that the colonial nature of the Northern Ring-necked Snake makes the species particularly vulnerable to local extirpations when its habitats are disturbed.

Because Ring-necked Snakes live and eat prey found in soil, they could be exposed to toxins in landscapes where pesticide use is heavy, which could also bioaccumulate through feeding on contaminated prey. In the Great Lakes region, Harding (1997) believed that the widespread use of persistent pesticides during the twentieth century could have played a role in regional Northern Ring-necked Snake population declines. They could also be susceptible to various diseases or pathogens that can cause high mortality, such as snake fungal disease (SFD; see Introduction to Conservation and Management of Wisconsin's Amphibians and Reptiles chapter). Although anecdotal, blistering under the skin that appeared consistent with SFD has been reported in both Prairie Ring-necked Snakes (Fitch 1975) and Northern Ring-necked Snakes (Blanchard et al. 1979). In captivity, if this condition is left untreated, it usually results in death (Fitch 1975).

Management: Although limited data on best habitat management practices for this snake exist, general management considerations associated with its preferred habitats are likely appropriate (see Introduction to Conservation and Management of Wisconsin's Amphibians and Reptiles chapter). Despite a lack of rigorous research on this topic, Ring-necked Snakes would also probably benefit from management plans that promote the persistence of ground debris for cover and nesting (e.g., rocks and damp woody debris). It is also important to consider that some habitat management techniques can pose threats to snake populations. For example, mortality due to prescribed fire used to manage vegetation is documented for several snake species (Erwin and Stasiak 1979). If prescribed fire is employed, research in Kansas supported that Ring-necked Snakes may benefit from a schedule where there are several years between fire application, likely due to their preferences for greater soil moisture and earthworm prey (Wilgers and Horne 2006). Annual timing of habitat management activities appears to be key in reducing or preventing such mortality, and these should be scheduled for the inactive season.

Eastern Hog-nosed Snake

Heterodon platirhinos (Latreille 1801)

Gabriel J. Miller

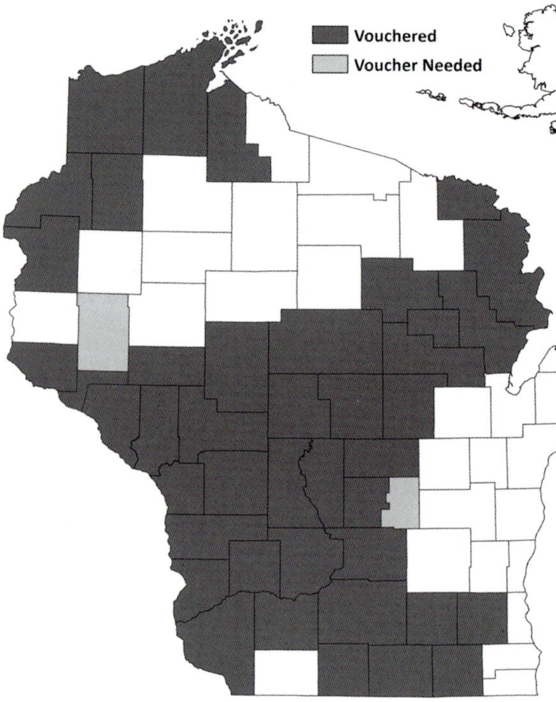

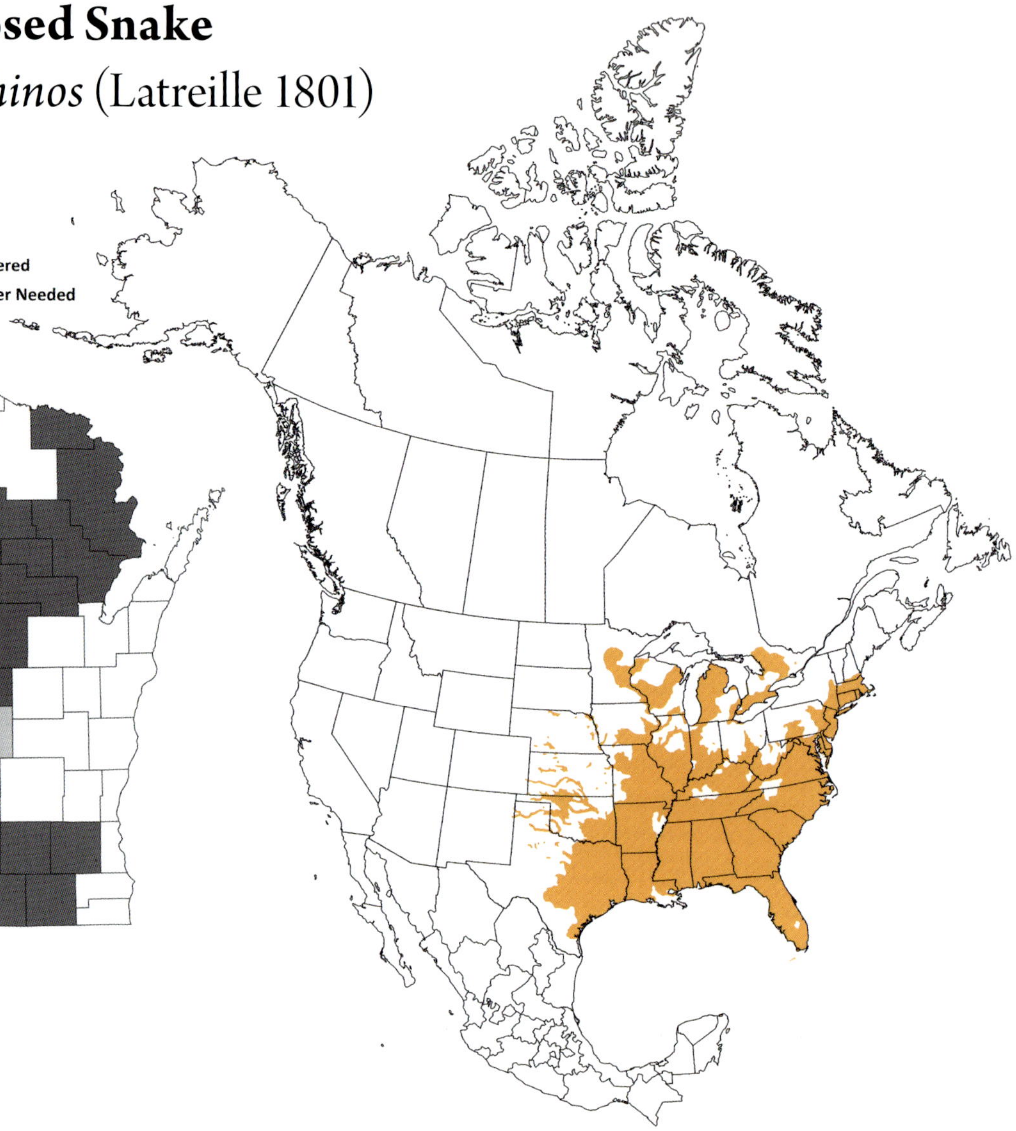

Adult Eastern Hog-nosed Snake (Jackson County; photo by A. Badje).

AT A GLANCE

The Eastern Hog-nosed Snake is a unique and interesting Wisconsin snake. A denizen of open to semi-open habitats such as prairies, savannas, open woodlands, and floodplains with sandy soils, it has been documented somewhat broadly in southern Wisconsin, but only in extreme eastern and western counties in the northern part of the state. This snake possesses a distinctive upturned spade-shaped rostrum that is its namesake and aids in burrowing through sandy soils. Eastern Hog-nosed Snakes emerge from hibernation in April and probably engage in reproduction shortly thereafter. Females nest and lay eggs primarily in late June and early July, while juveniles usually hatch from eggs in late August and September. This snake likely returns to hibernacula for winter by late October or early November, although little confirmed information exists regarding its overwintering habits in Wisconsin. Although typically considered a snake of dry habitats, it is sometimes found adjacent to aquatic environments that are home to its preferred prey: amphibians such as American Toads. Eastern Hog-nosed Snakes are well known for their defensive displays, which include flattening their neck like a cobra while hissing loudly and ultimately 'playing dead.' Although mostly harmless, these theatrics have likely contributed to several of the colorful colloquial names used to describe this snake, such as 'blow snake,' 'puff adder,' and 'spreading adder.' It is currently considered a common species in Wisconsin by the WDNR.

SYSTEMATICS

Heterodon platirhinos (Eastern Hog-nosed Snake) belongs to the Dipsadinae subfamily of the family Colubridae, which includes the rear-fanged snakes of the genus *Heterodon* (Crother 2012). Although the currently accepted scientific name for this species is *H. platirhinos*, a variety of outdated synonyms exist, such as *Coluber heterodon* (Daudin 1803) and *H. contortrix* (Allen 1932). In the mid-1800s, it was reclassified as *H. platirhinos*, based on the description by Latreille (1801). Baird and Girard (1853) altered the specific epithet spelling to *platyrhinos* (Platt 1985). Despite this, however, most current sources use the original spelling by Latreille (1801), which we follow here. Despite the broad continental distribution of this species and some variability in appearance, no subspecies are currently recognized (Crother 2017). No other species in the genus *Heterodon* occur in Wisconsin, although snakes in the genera *Diadophis* (ring-necked snakes) and *Carphophis* (wormsnakes) are also placed in the subfamily Dipsadinae.

DESCRIPTION

Adults: The dorsal coloration and pattern of Eastern Hog-nosed Snakes are variable across their geographic distribution. Adults in the Midwest are often a drab olive, dark brown, or beige color, with only faint outlines of a dorsal pattern but with large dark 'eyespots' on either side of the neck (Figure 232). These 'eyespots' are prominently displayed when

Figure 232. Dorsal color pattern commonly found in many Eastern Hog-nosed Snakes from Wisconsin, including 'eyespots' on either side of the neck (Columbia County; photo by J. Kapfer).

Figure 233. An Eastern Hog-nosed Snake with a slightly more orange to rusty tint to its dorsal pattern (Sauk County; photo by R. Paloski).

Figure 234. The head characteristics of a typical Eastern Hog-nosed Snake in Wisconsin. Note the upturned rostral scale and 'eyespots' on either side of the neck (Sauk County; photo by J. Kapfer).

Eastern Hog-nosed Snakes enact typical defensive postures (see Prey and Predators section). Head markings include a darker band that passes primarily across the prefrontal scales, supraocular scales, and through the eyes, terminating at the last supralabial scale. Although the dorsal pattern is only faintly visible in many adults, some individuals possess more distinct markings. In such instances, 20–30 darker squarish blotches are visible over a gray, tan, brown, rust, or pinkish background along most of the dorsum (Figure 233). The flanks exhibit a pattern of dark spots that are offset between the dorsal blotches. In those snakes with a more obvious pattern, the posterior portion of the dorsum can also possess 6–11 obvious tail bands (Vogt 1981, Moriarty and Hall 2014). The ventral surface is typically a mottled cream to gray. The most notable characteristic of the Eastern Hog-nosed Snake (and its namesake) is the upturned spade-shaped rostral scale (Figure 234). The rostral scale is sharply pointed at the tip and smooth on the ventral surface, with a strong keel down the center of the dorsal surface. This shape aids the snake in fossorial activities such as excavating nests and burrows or unearthing prey. Their dorsal scales are mostly keeled, and the anal plate is divided (see additional scale count information in Diagnosis subsection).

Eastern Hog-nosed Snakes are a medium-sized heavy-bodied snake. Across their North American distribution, adults typically range from 51 to 84 cm TL (20 to 33 in) but can be as large as 115 cm TL (45 in; reviewed by Ernst and Ernst 2003). Similarly, Vogt (1981) reported that Wisconsin individuals ranged from 50 to 90 cm TL (20 to 35 in). The mean SVL of adult Wisconsin specimens housed in MPM and UWZM collections was 58.5 cm (23 in, $n = 15$) in males and 64.7 cm (25.5 in, $n = 6$) in females (minimum size of mature individuals determined from Platt [1969]; see Reproduction and Development section). The mean tail length of these adult Wisconsin specimens was 12.6 cm (5 in) in males ($n = 15$) and 11.5 cm (4.5 in) in females ($n = 6$). The largest SVL of a measured Wisconsin specimen was 77.6 cm (30.6 in, MPM 23668, collected from Shawano County in 1991), but midwestern individuals can grow larger than the typical range reported by both Vogt (1981) and Ernst and Ernst (2003). For example, Moriarty and Hall (2014) reported a female from Minnesota with an SVL of 109 cm (43 in).

Preadult stages: The eggs of Eastern Hog-nosed Snakes are cream to white in color, linearly oblong, and 2.0–4.3 cm (0.78–1.69 in) long by 1.3–2.4 cm (0.47–0.94 in) at maximum width (reviewed by Ernst and Ernst 2003). A review by Moriarty and Hall (2014) of eggs from Minnesota yielded similar measurements of 2.1–3.9 cm (0.83–1.54 in) long by 1.3–2.8 cm (0.51–1.10 in) at maximum width. Hatchling and juvenile dorsal pat-

Figure 235. A hatchling Eastern Hog-nosed Snake with a more striking pattern than is found in many adults (Waukesha County; photo by J. Kapfer).

tern is similar to that of the adults but often more vivid and contrasting (Figure 235). The ventral surface of juveniles is predominantly dark gray to black, though the neck and tail are white or yellowish (Moriarty and Hall 2014). Mean hatchling SVL for a small number of museum specimens collected from Wisconsin was 13.6 cm (5.4 in, $n = 5$). This is within the range of 12.7–29.4 cm (5.0–11.5 in) hatchling TL reported by Ernst and Ernst (2003).

Variation: Eastern Hog-nosed Snakes in Wisconsin show much variation in dorsal color and pattern intensity. Some individuals are a drab olive with little discernible pattern, while others exhibit a more distinct pattern over a tan to rusty background (see Figures 232 and 233). Females generally have longer SVLs than males, but male tail lengths typically exceed those of females (see Adults subsection). Melanism appears to be common in populations from the southern US, such as in Georgia (Gabriel Miller, Prairie Island Indian Community Department of Land and Environment, personal observation), but is rarely reported in Wisconsin (e.g., Edgren 1957). Where melanism occurs, it is more often observed in mature individuals (Platt 1969), although melanistic hatchlings have been documented (Lazell 1993).

Diagnosis: Eastern Hog-nosed Snakes are characterized by the presence of a pointed and upturned rostral scale, a divided anal plate, and divided subcaudal scales. Some or all of the dorsal scales are keeled and occur in 25 rows at midbody. On each side of the head there are 2 prefrontal scales, 1 loreal scale, 8 supralabial scales, and 10–11 infralabial scales, while ventral scale counts range from 109 to 156 (reviewed by Ernst and Ernst 2003). The sharply upturned rostral scale should distinguish adult and juvenile Eastern Hog-nosed Snakes from all other snakes in Wisconsin. Most Wisconsin snakes encountered in the wild also do not exhibit the elaborate death-feigning behavior frequently employed by Eastern Hog-nosed Snakes (see Prey and Predators section).

There are several species with superficially similar body patterns and scale characteristics that could cause confusion. Common Watersnakes (*Nerodia sipedon*), for example, possess keeled scales and dorsal blotches. Despite this, they typically have a greater number of blotches along the body than Eastern Hog-nosed Snakes and <25 scale rows at midbody. Furthermore, Common Watersnakes possess an array of black-bordered rusty crescents on each of the ventral scales. In addition to differences in physical traits, these two snakes can often be distinguished by preferred habitat. For example, Common Watersnakes are almost exclusively associated with water, whereas Eastern Hog-nosed Snakes are most often found in drier uplands. Eastern Milksnakes (*Lampropeltis triangulum*), Eastern Foxsnakes (*Pantherophis vulpinus*), and juvenile Gray Ratsnakes (*P. spiloides*) can also have similar markings to Eastern Hog-nosed Snakes, but the latter two have a greater number of ventral scales (≥190). Eastern Milksnakes are a slimmer and typically shorter species with an undivided anal plate and fewer scale rows at midbody (21). Eastern Milksnake blotches are also brown or reddish and outlined with a black border over a grayish background. Gophersnakes (*Pituophis catenifer*) sometimes frequent the same habitats as Eastern Hog-nosed Snakes and also possess a specialized rostral scale. However, rather than the rostral scale being upturned like a shovel, the Gophersnake's rostral scale is more appressed against the snout. Furthermore, adult Gophersnakes are larger but proportionately slimmer than adult Eastern Hog-nosed Snakes. Gophersnakes also possess an undivided anal plate, more scale rows at midbody (27), a greater number of ventral scales (207–260), and 4 prefrontal scales.

Distinguishing the eggs of different snake species in the wild can be difficult, particularly for those that overlap in size. However, Eastern Hog-nosed Snake eggs could be differentiated from other snake eggs based on individual egg dimensions, clutch size, and the habitat where they are deposited (see Preadult stages subsection and Reproduction and Development section; also reviewed by Ernst and Ernst 2003).

DISTRIBUTION AND HABITAT

Global distribution: Eastern Hog-nosed Snakes are the most widely distributed species in the genus *Heterodon*, occurring throughout most of the eastern continental US. Their range also extends northward into the Great Lakes region and Canada, from southern Ontario to the northeast shore of Lake Huron. Despite their widespread distribution throughout the eastern US, they occur more sporadically along the peripheries of their continental range in the Midwest and Northeast, north of the Ohio River valley. Their distribution is also more sporadic in the western portions of their range (Platt 1969).

Distribution in Wisconsin: Of the published county records, suitable museum vouchers could not be found for Dunn and Green Lake Counties (see the Appendix). They are reported throughout most of the southern two-thirds of the state, but published records do not exist for extreme southeastern counties. The range in the northern tier of the state appears to be more restricted, and they are confirmed only in the far northwestern and far northeastern counties, perhaps associated with sandy soils or river watersheds (such as the Mississippi and St. Croix Rivers) in those regions.

Localities of possible occurrence: There are a number of counties that lack Eastern Hog-nosed Snake records but possess habitat and, in some cases, soils that this species prefers (see Habitat subsection). In northwestern Wisconsin, St. Croix County is surrounded by counties with published records, including several on the Minnesota side of the border (reviewed by Moriarty and Hall 2014). While preferred soils occur in some areas of St. Croix County, these are often overlaid with silts and loess deposits from the last glacial event (Gassen et al. 1987), which could limit Eastern Hog-nosed Snake access. Other counties that lack records but are connected to potentially appropriate historic habitats include Winnebago, Chippewa, Lafayette, Sawyer, and Barron Counties. Despite this, suitable soils in many of these counties are either unavailable (due to silt caps) or widely dispersed (WDNR 2015b). Regardless, a skeletonized specimen listed as an Eastern Hog-nosed Snake exists that was reportedly captured in Sawyer County (UWZM 21499). Unfortunately, it is difficult to independently confirm the species identity of this specimen from disarticulated bones alone. Though sandy soils exist in the counties south of Green Bay along Lake Michigan, the current lack of records from many of them, such as Milwaukee, Kenosha, and Racine, could be due to the presence of high human densities and development.

Habitat: Eastern Hog-nosed Snakes frequently occur in semi-open to open upland habitats with sandy or sandy loam soils (Platt 1969, Ernst and Ernst 2003, Lagory et al. 2009, Goulet et al. 2015). Platt (1969) highlighted that Eastern Hog-nosed Snakes in Wisconsin used oak savannas but also oak-hickory, beech-maple, northern hardwood, and pine forests. They are additionally known from brushy woodland edges, woodland clearings, sand prairies, and open bluffsides in the state (J. Kapfer, personal observation). Though not often considered a snake of wet environments, they will use habitats with well-drained soils adjacent to wetlands, lakes, and rivers, perhaps because these aquatic habitats are home to preferred amphibian prey (Platt 1969, Goulet et al. 2015). During survey work along rivers in Portage County from 2000 to 2003, Eastern Hog-nosed Snakes were found with some regularity in sandy open woodlands, as well as secondary growth forest, marginal disturbed lands, and oak savanna along the river bottoms in the Wisconsin River watershed (G. Miller, personal observation). Eastern Hog-nosed Snakes sometimes occur within disturbed habitats, such as the ecotones between woodland edges and cultivated areas (Ernst and Ernst 2003, Lagory et al. 2009), old fallow fields (Plummer and Mills 2000), and developed lands (Goulet et al. 2015).

Several studies indicate that Eastern Hog-nosed Snakes seek out specific features within their selected habitats, but there is reportedly variation in which particular characteristics are preferred. For example, Goulet et al. (2015) found they use locations with an abundance of shrubs, ground debris, and rock cover. Yet Lagory et al. (2009) reported that they often used locations with a higher percentage of bare ground and graminoid coverage but with less ground debris than random sample points. Past authors have surmised that microhabitat selection by this snake is primarily driven by the need to thermoregulate (Lagory et al. 2009, Goulet et al. 2015). In fact, even in semiwooded habitats, they still reportedly seek specific locations where sunlight easily penetrates to the ground surface and temperatures are higher (Goulet et al. 2015). Ernst and Ernst (2003) suggested that Eastern Hog-nosed Snakes spend much of the active season buried in loose soil or using excavated burrows. However, several authors have reported that Eastern Hog-nosed Snakes rarely seek cover beneath ground surface debris (e.g., Platt 1969, Plummer and Mills 2000, 2010). Plummer and Mills (2000) found radio-tracked Eastern Hog-nosed Snakes under cover objects on only 17% of locations and rarely observed them in underground burrows (5% of locations). Instead, the authors frequently found snakes partially concealed near the base of grass clumps, where thermoregulatory opportunities were presumably superior.

In the Upper Midwest, Eastern Hog-nosed Snakes tend to nest in sandy habitats at locations with no tree canopy cover to maximize sun exposure and optimize thermal conditions for egg development (Peet-Paré 2010). In Wisconsin, observations of nesting activity are reported from open sand prairies, savannas, and woodland clearings (Koch 2009, J. Kapfer, personal observation). Females select specific oviposition sites with sparse vegetation coverage along southeast aspects of small rises within open areas (reviewed by Cunningham and Cebek 2005). Peet-Paré and Blouin-Demers (2012) further found that females choose nesting sites with less woody vegetation cover, more grasses, and warmer temperatures than random locations adjacent to nest sites. Eastern Hog-nosed Snakes often excavate nest chambers in the soil (Koch 2009), although they also reportedly lay eggs in rotting logs (Degraaf and Rudis 1983). Limited information on overwintering habitat in Wisconsin exists. Yet in the Wisconsin and Mississippi River valleys, they are sometimes found along open bluff faces associated with hibernacula used by Gophersnakes and Timber Rattlesnakes (*Crotalus horridus*). Ernst and Ernst (2003) stated that they overwintered in burrows abandoned by mammals or self-excavated. In Wisconsin, these burrows would likely occur in open habitats with loose soils so that overwintering snakes could easily descend below the frost line.

REPRODUCTION AND DEVELOPMENT

Eastern Hog-nosed Snakes reportedly breed in spring, shortly after emergence from hibernation, but also in autumn (Platt 1969, Plummer and Mills 1996, Cunnington and Cebek 2005). Vogt (1981) recounted a secondhand report of copulation observed on 23 September in Wisconsin, but breeding is probably more common in the spring. Like that of many other larger snake species, the entire process of mate-finding and copulation is prolonged and complicated for Eastern Hog-nosed Snakes. Males tend to search large areas in their pursuit of females (Platt 1969). Once males encounter female scent cues, they may spend long periods of time following them. If a receptive female is encountered, the male moves alongside her and typically coils around her. Cunningham and Cebek (2005) described courtship behavior as follows: "A 59 cm male and 62 cm female were observed in close proximity to each other. The male proceeded alongside the back end of the female until she twitched her tail rapidly and moved quickly away. The female was observed to change directions quickly, moving in a zigzagging manner. The male would consequently pause and begin to follow her again. The two snakes were observed to continue this behavior over a 2 hour period." The authors further described the act of copulation: "His entire body was positioned on top of hers with the exception of the tail, which was extended along the side of her body in such a manner as to allow for copulation. The snakes remained linked in copulation for a period of 3 days."

Female Eastern Hog-nosed Snakes reportedly breed with multiple males. Cunningham and Cebek

Figure 236. A female Eastern Hog-nosed Snake excavating a nest cavity (Sauk County; photo by J. Kapfer).

(2005) tracked a female over the course of 43 days, during which she was seen within 5 m (30 ft) of at least nine different males, and she copulated with five of them. It is unclear if multiple copulations in this snake provide a genetic benefit to offspring or if the female is able to select which male's sperm will fertilize her eggs after copulation. Shine (2003) suggested that some snake species with multiple mates may be able to reject sperm from genetically similar males, thereby maximizing genetic variability in the brood, but this has not been confirmed for Eastern Hog-nosed Snakes.

Eastern Hog-nosed Snakes are oviparous. Clutch size can range from 4 to 61 eggs and is often positively correlated with female SVL (reviewed by Ernst and Ernst 2003). Females in Wisconsin typically excavate nest chambers in June (Vogt 1981; Figure 236), which was described in detail by Koch (2009). After a satisfactory chamber is excavated, eggs are deposited "a few inches" below the soil surface (Platt 1969). Females appear to take great care in selecting a desirable nesting site and often dig multiple test nests over many days before finding a suitable location (Cunnington and Cebek 2005). Females can also travel great distances in search of nesting sites. According to observations from Rouse et al. (2011), at least 50% of observed gravid female Eastern Hog-nosed Snakes traveled "considerable" distances to oviposit in southern Ontario. After a satisfactory oviposition site is found, females reportedly return to that location across years for egg deposition (Cunnington and Cebek 2005). Peet-Paré (2010) also noted that Eastern Hog-nosed Snakes sometimes nest communally, particularly in disturbed areas where suitable nesting habitat is limited.

Peet-Paré (2010) found that wild nests at their Ontario study site maintained a mean incubation temperature of 23.81°C (74.86°F), which was higher than soil temperatures recorded at random locations nearby (21.73°C [71.11°F]). Selection of nesting locations with optimal thermal properties is likely important to offspring success. In a controlled experiment on the factors influencing neonate fitness, Pete-Paré and Blouin-Demers (2012) found that nests incubated at 24°C (75.2°F) produced larger neonates with fewer developmental anomalies than those incubated at 22°C (71.6°F). Despite the importance of high nest temperatures in offspring fitness, the nest sites selected by Eastern Hog-nosed Snakes are cooler, on average, than those selected by other colubrid species in the northern portions of their range. According to a review by Peet-Paré (2010), most other colubrid nests have an optimal incubation temperature of 28.7°C (83.66°F).

The incubation period for eggs reportedly ranges from 39 to 65 days, depending on mean daily incubation temperature, and eggs typically hatch from August to September (reviewed by Ernst and Ernst 2003). Koch (2009) found recently hatched eggs on 13 September and a hatchling on 24 September at a site in southern Wisconsin. According to Platt (1969), young individuals experience rapid growth, but this slows considerably with age. He further estimated that growth rates for Eastern Hog-nosed Snakes in Kansas were 4.3 cm (1.69 in) per month for males at 9 months of age and 1.1 cm (0.43 in) per month at 21 months. Females grew slightly faster than males with 4.5 cm (1.77 in) per month at 9 months of age and 1.2 cm (0.47 in) per month at 21 months of age. Eastern Hog-nosed Snakes reach sexual maturity at ca. 50 cm SVL (19.7 in) for females and 45 cm SVL (17.7 in) for males (Platt 1969). They can reportedly mature in as little as two years (Platt 1969); however, in the northern reaches of their range, including Wisconsin, this may take up to five years (reviewed by Seburn 2009). Average longevity in the wild is not known, but a captive male lived for seven years and eight months (Snider and Bowler 1992).

ACTIVITY

Emergence from hibernation in Wisconsin begins in April (Vogt 1981), but there is likely annual variation in the exact date of egress due to annual variation in environmental temperatures. Patterns of activity and behavior after emergence are also often driven by temperature. According to Platt (1969), voluntary activity in Kansas occurred at ambient air temperatures ranging from ca. 22°C to 34°C (71.6°F to 93.2°F), although snakes were occasionally active outside of this range. Reports of mean body temperature for active individuals suggests that variation exists by latitude, from 30.5°C (86.9°F) in a southern location (Arkansas; Plummer and Mills 2010) to 29.2°C (84.6°F) farther north (Massachusetts; Buchanan et al. 2016). It is reasonable to suggest that Eastern Hog-nosed Snakes from Wisconsin operate at mean body temperatures similar to those of the Massachusetts population, although this has not been confirmed.

Overall, annual activity levels are apparently higher in spring and autumn than during the summer months, when Eastern Hog-nosed Snakes must frequently aestivate to avoid hot conditions. Plummer and Mills (2000) noted that one study animal aestivated at the same location for 68 consecutive days in summer to avoid an extended period of high temperatures. Plummer and Mills (2000) also found that males exhibited greater individual movements during spring and autumn, which can be attributed to mate searching, but that females exhibited more consistent activity than males throughout the active season.

Eastern Hog-nosed Snakes are almost exclusively diurnal when active, with the majority of movements occurring from morning to midday. Plummer and Mills (2000) observed that from April to October, peak diel activity began at ca. 8:00 a.m. but decreased significantly after 2:00 p.m. Similar findings were reported by Buchanan et al. (2016), who determined that activity peaked at two hours past sunrise and then declined from ca. 11 to 14 hours after sunrise. Plummer and Mills (2000) observed snakes tracked with radio telemetry equipment aboveground on nearly 60% of active season days, likely due to their foraging habits (see Prey and Predators section). Nearly half of inactive days were associated with ecdysis, with the remainder occurring because environmental temperatures were near high or low thresholds of preference.

On active days, Eastern Hog-nosed Snakes can be quite mobile. For example, Akresh et al. (2017) reported daily movements of up to 455 m (1,493 ft). Despite this, 74% of recorded movements across all tracked snakes were <50 m/day (164 ft/day), with a mean movement rate of 39 m/day (128 ft/day, SD = 57 m/day [187 ft/day]; Akresh et al. 2017). Reported home range sizes are variable across studies. In Massachusetts and New York, mean home range sizes were only 18.6 ha (46.0 ac, SD = 19.3 ha [47.7 ac], *n* = 12, MCP estimator; Akresh et al. 2017) and 23.7 ha (58.6 ac, SD = 21.2 ha [52.4 ac], *n* = 6, MCP estimator; Vanek and Wasko 2017), respectively. In Ontario, Robson (2011) estimated that mean home range size was 39.4 ha (97.4 ac, SD = 6.3 ha [15.6 ac], *n* = 17, MCP estimator). In Arkansas, mean home range size was 50.2 ha (124.0 ac, SD = 6.4 ha [15.8 ac], *n* = 8, MCP estimator; Plummer and Mills 2000), with significant overlap among individual snake home ranges. Goulet et al. (2015) reported a large mean home range size of 72.7 ha (179.6 ac, SD = 35.3 ha [87.2 ac], *n* = 5, MCP estimator) for snakes tracked in New Hampshire. The differentiation in home range sizes for Eastern Hog-nosed Snakes throughout the various regions could be attributed to spatial and temporal differences in resource availability (reviewed by Plummer and Mills 2000). Regardless of variation in home range size by geographic location, several studies have found that no substantial difference exists between the home range sizes of males and females in the same population (Plummer and Mills 2000, Robson 2011, Akresh et al. 2017).

Eastern Hog-nosed Snakes return to their hibernacula in early to mid-autumn, depending on weather conditions and latitude. In Wisconsin, this likely occurs by the beginning of November (Vogt 1981). Platt (1969) indicated that their return to hibernation is triggered when soil temperatures fall below 15.5°C (60°F); thus, soil temperature appears to be the more reliable indicator of when dormancy begins than air temperature or date. Little information exists regarding physiological condition during winter dormancy. Arndt (1980) recorded a body temperature of 9.2°C (48.6°F) for a hibernating individual that buried itself 58 cm (22.8 in) below the soil surface, with a soil temperature of 8.5°C (47.3°F).

PREY AND PREDATORS

Eastern Hog-nosed Snakes are well-documented predators of anurans, particularly American Toads (*Anaxyrus americanus*), but Vogt (1981) indicated that Wisconsin individuals also consumed Northern Leopard Frogs (*Lithobates pipiens*), Green Frogs (*Lithobates clamitans*), and Prairie Skinks (*Plestiodon septentrionalis*). In other states, salamanders are taken (Trauth 1982), which could also be the case in Wisconsin. Additional identified prey includes insects, snakes, birds, small mammals, and reptile eggs (Edgren 1955b, Platt 1969). Investigation of scat from a midsized Wisconsin male (SVL = 46.9 cm [18.5 in]) from Sauk County revealed various insects, such as crickets, and remnants of a nestling bird (Kerry Katovich, UW–Whitewater, unpublished data). Cooper and Secor (2007) conducted controlled experiments to determine if this species responded to chemical cues of different prey. Using ingestively naive hatchling Eastern Hog-nosed Snakes, they found that chemosensory response was significantly greater for toads (i.e., Southern Toad [*Anaxyrus terrestris*]) than other food items, though other anurans, specifically Green Frogs, also showed elevated responses compared to control stimuli. It is unique that Eastern Hog-nosed Snakes are able to ingest toads, since toads produce an alkaloid-containing toxin that is typically distasteful or poisonous to and avoided by other species. In addition to their ability to eat a toxic animal, Eastern Hog-nosed Snakes possess specialized dentition that is important when killing prey (Kardong 1979). Prey are subdued, in part, by serous secretions from the snake's elongated rear fangs (Datta and Tu 1993; Figure 237) and swallowed whole. Although these secretions are mildly toxic, Eastern Hog-nosed Snakes are generally not considered a threat to humans, as their toxin is selective for their preferred prey (Natural History Box: A Venomous Colubrid?). The rear fangs of hog-nosed snakes can also be employed to puncture and 'deflate' toads, which sometimes attempt to increase their body size by inhaling air into the lungs as a defense mechanism (Weaver 1965). Eastern Hog-nosed Snakes are active foragers, moving often in search of food (Rouse et al. 2011).

Figure 237. The elongated rear fang of an adult Eastern Hog-nosed Snake (Sauk County; photo by J. Kapfer).

Predators of Eastern Hog-nosed Snakes include other snakes and various birds (such as crows and raptors; reviewed by Ernst and Ernst 2003). Though mammalian predators are infrequently reported, it is plausible that Raccoons (*Procyon lotor*), Coyotes (*Canis latrans*), and other mammals will opportunistically prey upon all age classes, including eggs. Plummer and Mills (2000) determined that mammal, bird, and even fish (though no species were identified) predation was responsible for the mortality of several transmitter-fitted Eastern Hog-nosed Snakes in their study. Eastern Hog-nosed Snakes are likely put into frequent contact with predators due to their active foraging behavior (Plummer and Mills 2000, Rouse et al. 2011). To deter potential predators, this species has evolved a suite of elaborate defensive behaviors (Natural History Box: Thespian Death Throes; Figure 238). Rouse et al. (2011) noted that Eastern Hog-nosed Snakes must choose between safety and satiety, so they prefer taking the risk to forage over remaining below cover to avoid detection. As such, the need for elaborate defensive behaviors is reasonable.

Very little information exists regarding parasites of Eastern Hog-nosed Snakes, but Dyer (1994) found that they were host to at least two parasites: the ochetosomatid digenean (fluke) *Ochetosoma elongatum* and

NATURAL HISTORY BOX: A Venomous Colubrid?

The Eastern Hog-nosed Snake has long been considered a mildly venomous species (McAlister 1963). As a member of the subfamily Dispadinae, it is opisthoglyphous (with elongated rear fangs) and possesses Duvernoy's glands, which are homologous to the venom glands of well-known toxic snakes, such as rattlesnakes (Weinstein and Kardong 1994). The Duvernoy's gland is responsible for production of a unique serous secretion in Eastern Hog-nosed Snakes that is often miscategorized as venom. In comparison, the Timber Rattlesnake is solenoglyphous; its special dentition includes an articulating hollow fang at the front of the maxillae that injects true venom from the primary venom gland to its prey, much like a hypodermic needle. The elongated back fangs of Eastern Hog-nosed Snakes, on the other hand, are not hollow. Instead, some reports support that their rear fangs include an external shallow groove that serves as the pathway for the delivery of the snake's specialized toxic secretions.

Although they possess anatomical structures used for administering toxins to prey, several lines of evidence support that Eastern Hog-nosed Snakes are not efficient at delivering their specialized secretions. For example, some authors have reported that only the faintest trace of an anterior groove exists in Eastern Hog-nosed Snake teeth (Ditmars 1912). Other researchers (Kapus 1964, Weaver 1965) claim that this species possesses no teeth grooves at all. More definitive research on Eastern Hog-nosed Snake dentition is apparently needed. Whether or not their teeth possess a viable toxin pathway for prey inoculation, there is no current debate that the Eastern Hog-nosed Snake possesses the Duvernoy's gland (Taub 1967), which produces the aforementioned serous secretion (Datta and Tu 1993). This secretion's effectiveness, however, is targeted at certain organisms. Salivary extracts of Eastern Hog-nosed Snakes were tested on mice, toads, and frogs to determine toxic effect, which yielded no damaging effect on mice but resulted in an 80% mortality rate in tested amphibians (McAlister 1963). Young (1992) showed that in lab preparations, secretions from the Eastern Hog-nosed Snake's Duvernoy's gland induced neuromuscular blockade in isolated frog sciatic nerve–gastrocnemius muscle, whereas mice exhibited only minor transient signs of toxicity.

These findings then beg the question: Is the Eastern Hog-nosed Snake a danger to humans? To answer this, it is first important to note that few reports exist of humans who were actually bitten by Eastern Hog-nosed Snakes (reviewed by Weinstein and Keyler 2009). In fact, Vogt (1981) noted that despite sometimes striking repeatedly, this species "strikes with mouth closed" and that he had "never known one of these gentle creatures to bite." Reports of clinical symptoms from the toxic secretions of hog-nosed snake bites are also rare and associated with unlikely circumstances. Some incidents appear to involve captive snakes that bit their handlers while being fed but where handlers (for some reason) allowed snakes to maintain their bite for numerous minutes (e.g., Bragg 1960, Weinsten and Keyler 2009). Another involved an accidental scratch from the teeth of a captive snake during its death-feigning behavior (Grogan 1974). In some cases of apparent envenomation, the resultant swelling could be attributable to residual toad toxins on the teeth of snakes that had recently fed (Datta and Tu 1993). In reported bite cases, symptoms included swelling, pain, and a tingling sensation that lasted from several days to weeks, but no extensive tissue damage or systemic effects were noted (reviewed by Weinstein and Keyler 2009). This information supports that Eastern Hog-nosed Snakes exhibit a specific toxin that works almost exclusively on their preferred prey, do not bite people in the wild, and pose (at most) a minimal risk to humans.

Gabriel J. Miller, Prairie Island Indian Community Department of Land and Environment

NATURAL HISTORY BOX: Thespian Death Throes

In response to perceived threats, Eastern Hog-nosed Snakes can enact several interesting defensive behaviors (Sexton 1979). When the snake is initially threatened and escape is not possible, the snake's first response is to elevate its head off the ground and flatten its neck while tilting its head toward the antagonist. In this posture, Eastern Hog-nosed Snakes vaguely resemble cobras spreading their hood, which can lead to misplaced concern among Wisconsin residents who are not familiar with this species. It is noteworthy that wild cobras do not naturally exist in the US. Flattening of the neck enhances the eye spot pattern along the flanks of its neck, which presumably makes the snake look bigger or more dangerous to its adversary. During this display, the snake also often hisses in a loud and intimidating fashion. If this does not work, the Eastern Hog-nosed Snake will engage in more intense acts of aggression, such as striking at the provoker. Although its strikes rarely land and the snake almost never bites (Vogt 1981), these actions resemble those of a dangerous animal and help to deter would-be antagonists. If its efforts fail, however, the Eastern Hog-nosed Snake will then 'play dead.' In this climax of theatrics, the Eastern Hog-nosed Snake writhes on the ground, rolls onto its back, occasionally regurgitates its last meal, and emits a foul-smelling musk. The snake soon slows its writhing and eventually stops moving altogether. At the end of this final act, the snake lies on its back with mouth open and tongue extended to best resemble something dead. However, if a snake in this position is rolled onto its belly by an amused onlooker, it will quickly flip back over again, and the ruse is up (Moriarty and Hall 2014).

Gabriel J. Miller, Prairie Island Indian Community Department of Land and Environment

the cosmocercoid nematode (roundworm) *Cosmocercoides variabilis*. The fluke infects the lungs of the snake and the roundworm infects the oral cavity.

CONSERVATION

Status: Globally, the Eastern Hog-nosed Snake is listed as a Least Concern species by the IUCN due to its wide distribution and predominantly stable populations. NatureServe lists Eastern Hog-nosed Snakes as G5 (globally secure); however, in Wisconsin the species is listed as S4 or S3 (apparently secure to vulnerable). Because it is considered a common species in Wisconsin, it has no special protective status. However, information from regions near Wisconsin might suggest that more protection is necessary. In Minnesota, for example, the Eastern Hog-nosed Snake is recognized as a Species of Greatest Conservation Need (Minnesota Department of Natural Resources 2016). In southern Ontario, this species is recognized as threatened by the Committee on the Status of Endangered Wildlife in Canada (COSEWIC), with anthropogenic factors having the most significant impact (COSEWIC 2007).

Populations: There are limited data available on Eastern Hog-nosed Snake population parameters. Platt (1969) estimated mean population size based on capture-recapture data at two sites in Kansas, which ranged from 10 to 43 individuals in a 20.2 ha (50 ac) park, but indicated that his estimates should be viewed cautiously. The few published estimates of population size and density that exist could reflect the fact that marked Eastern Hog-nosed Snakes are rarely recaptured (Akresh et al. 2017).

Platt (1969) found that populations have roughly equal sex ratios and that his study population was dominated by hatchlings and yearlings (suggesting low survival to adulthood). Plummer and Mills (2000) estimated survivorship in their study population and found that when discounting individuals lost due to presumed transmitter failure, Eastern

Figure 238. A subadult Eastern Hog-nosed Snake in the middle of a death-feigning display (Adams County; photo by T. Muehlfeld).

Hog-nosed Snake survivorship was ca. 50% during the active season. This is similar to a more recent estimate of active-season survival for a Massachusetts population of 61% (Akresh et al. 2017).

Threats: Eastern Hog-nosed Snakes in Wisconsin face numerous threats, but the destruction and degradation of habitat are particularly relevant (reviewed by Lagory et al. 2009; see Introduction to Conservation and Management of Wisconsin's Amphibians and Reptiles chapter). Loss of their preferred open to semi-open habitats due to unabated vegetative succession and extensive conversion for anthropogenic use (such as timber production or residential development) has likely had a substantial impact on this species. Although the effects of habitat disturbance can be obvious, such as when populations decline due to the loss of suitable nesting or overwintering sites, some impacts are indirect and poorly understood. For example, extensive exotic plant invasions, such as those that can occur due to European Buckthorn (*Rhamnus cathartica*) in open woodlands and Spotted Knapweed (*Centaurea maculosa*) in prairies, are unlikely to cause direct snake mortality but could eliminate thermoregulatory opportunities and restrict movement. Although Eastern Hog-nosed Snakes do not typically prefer aquatic habitats, loss of wetlands can reduce habitat available for their preferred amphibian prey, which could indirectly affect this snake. Adding further confusion, the observed effects of habitat degradation are sometimes counterintuitive. Goulet et al. (2015) noted that Eastern Hog-nosed Snakes actually selected for disturbed areas at a site in New Hampshire, which could imply that such habitats are suitable. Yet the authors surmised that these developed areas (with maintained lawns) created pockets of preferred open-canopy conditions in landscapes where succession of woody vegetation and shade was otherwise extensive. Eastern Hog-nosed Snakes in such situations are likely forced to use what limited open-canopy habitat remains and end up in residential areas. Despite this, disturbed landscapes can be ecological traps (Battin 2004). Although some aspects of disturbed landscapes attract Eastern Hog-nosed Snakes, individuals that use developed areas can experience higher mortality rates than those that use undisturbed locations (reviewed by Seburn 2009).

Habitat fragmentation, which occurs when natural landscapes are divided by anthropogenic disturbances, also negatively impacts Eastern Hog-nosed Snakes. Fragmented habitats possess greater opportunities for mortality, so they could lead to smaller populations and higher risks of extirpation while also isolating populations by restricting dispersal, which limits gene flow and genetic diversity (Young and Clarke 2000, Robson and Blouin-Demers 2013). Xuereb et al. (2015) determined that Eastern Hog-nosed Snake populations in fragmented landscapes of Ontario experienced population declines and isolation that likely limited the genetic exchange among groups. Roads are frequently responsible for fragmenting habitats, which then affects Eastern Hog-nosed Snakes (Robson 2011, Rouse et al. 2011, Robson and Blouin-Demers 2013). Yet disentangling the effects that roads have on this snake can be challenging. For example, although Andrews and Gibbons (2008) frequently found Eastern Hog-nosed Snakes along roads in Georgia, the individuals encountered were more often alive than dead. On the other hand, Plummer and Mills (2006) reported that individuals tracked with telemetry equipment avoided paved roads and rarely crossed the roads they encountered. Robson and Blouin-Demers (2013) further found that the frequency of road crossings was affected by road substrate type and that snakes avoided crossing paved roads, though low-traffic sand roads were often traversed.

Other factors could also threaten Eastern Hog-nosed Snakes but remain poorly understood. For example, Eastern Hog-nosed Snakes experience direct human persecution because they are often misidentified as a dangerous species. The colloquial names 'blow snake,' 'spreading adder,' and 'puff adder' are sometimes applied to Eastern Hog-nosed Snakes due to elaborate defensive displays that are reminiscent of venomous snake behavior (see Natural History Box: Thespian Death Throes). Although many individuals are likely killed by people as a result of these misidentifications, the population-level effects of such persecution are unknown. Entanglement in erosion control mesh is another potentially significant threat to snakes (Kapfer and Paloski 2011), including Eastern Hog-nosed Snakes (G. Miller, personal observation). Furthermore, Vogt (1981) reported the death of a captive Eastern Hog-nosed Snake that was fed live American Toads captured in a recently sprayed cornfield. While it seems likely that wild snakes would face similar threats, mortality from uptake of various biocides through trophic interactions has not been thoroughly investigated in this species.

Management: General management considerations for open sandy habitats (e.g., prairies, oak woodlands, savannas, pine-oak forests, dry-mesic forests) are relevant to Eastern Hog-nosed Snakes (see Introduction to Conservation and Management of Wisconsin's Amphibians and Reptiles chapter). Management strategies designed to maintain a mix of open and semi-open vegetation communities across the landscape could benefit this species (Lagory et al. 2009). Akresh et al. (2017) found that Eastern Hog-nosed Snakes in Massachusetts preferred habitats managed specifically to maintain open canopies. Prescribed fire can be an effective tool for controlling the extent of woody and invasive vegetation on normally open properties. As such, management plans that include periodic application of fire could benefit Eastern Hog-nosed Snakes. It is noteworthy, however, that prescribed fire can cause mortality if used carelessly. It must be applied appropriately so that it does not negatively impact this and other species. Restriction of prescribed fire application to the inactive season while also dividing burn units into smaller management plots with refugia interspersed could limit accidental snake mortality (Kingsbury and Gibson 2012). Programs that promote fallow agricultural land, such as that created by the USDA's Conservation Reserve Program (CRP), could also provide important open habitat, particularly in landscapes that are otherwise overgrown and unsuitable (Vanek and Wasko 2017).

Conservation plans should further address the management of specific habitat components that support particular ecological needs. For example, although open to semi-open habitats are often preferred, these must still be structurally variable, with ample cover for snakes. Maintenance of patchy ground cover that includes clump grasses or ground debris helps create heterogeneity in habitat structure, an important microhabitat feature for this species (Plummer and Mills 2000). Protection and maintenance of open nesting habitat are also likely critical components of effective Eastern Hog-nosed Snake conservation (reviewed by Robson 2011, Vanek and Wasko 2017). Protection of wetlands used by prey species could be important to maintain suitable resources for Eastern Hog-nosed Snakes. Seburn (2009) further suggested that understanding seasonal migrations and population changes in *Anaxyrus* species, which Eastern Hog-nosed Snakes often consume, may be important when developing conservation tactics for this snake. Conservation plans that attempt to offset the effects of habitat fragmentation, for example, by alleviating road mortality, should also be considered. These could include infrastructure plans that limit road densities in areas adjacent to known Eastern Hog-nosed Snake populations or implementation of wildlife road underpasses that create safe passageways in areas where road mortality is high. Unfortunately, little research exists that focuses on whether these measures are effective for Eastern Hog-nosed Snakes, and such investigations are warranted.

The efficacy of additional management techniques for Eastern Hog-nosed Snakes has received some attention by researchers. For example, translocation of individuals to supplement or reintroduce populations might be a useful strategy for this species under extreme conditions. Yet snake translocations are often unsuccessful (Reinert and Rupert 1999), and specific efforts to translocate Eastern Hog-nosed Snakes have not yielded encouraging results. Cook (2008) reported several translocation efforts at three properties in New York that appeared to result in only "probable" success at one site and uncertain results or failure at others. Plummer and Mills (2000) determined that translocated Eastern Hog-nosed Snakes did not survive as long as resident individuals did. The authors further found that translocated snakes exhibited aberrant movement patterns reminiscent of snakes unfamiliar with their surroundings, which resulted in higher predation rates. Therefore, translocation of Eastern Hog-nosed Snakes should only be considered if thoroughly vetted and after consultation with experts or natural resource agency specialists.

Common Watersnake

Nerodia sipedon (Linnaeus 1758)

Elizabeth Mack McKeag and Richard B. King

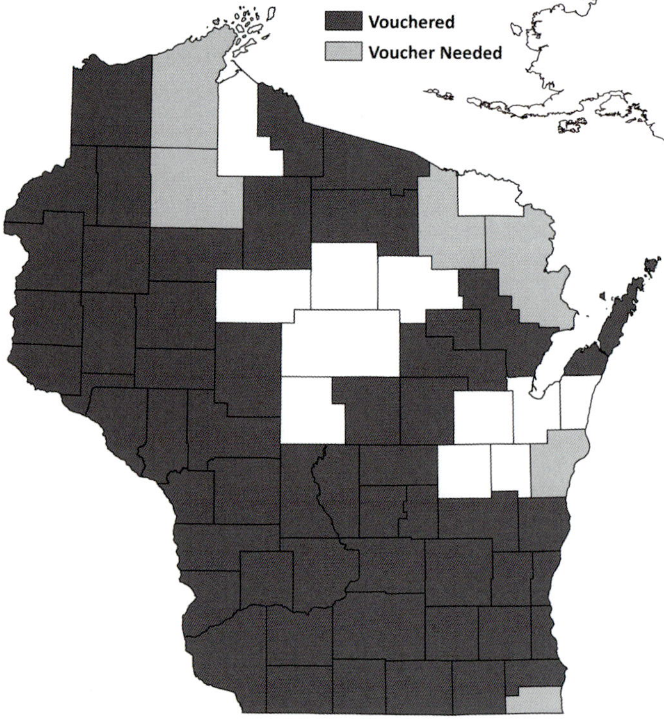

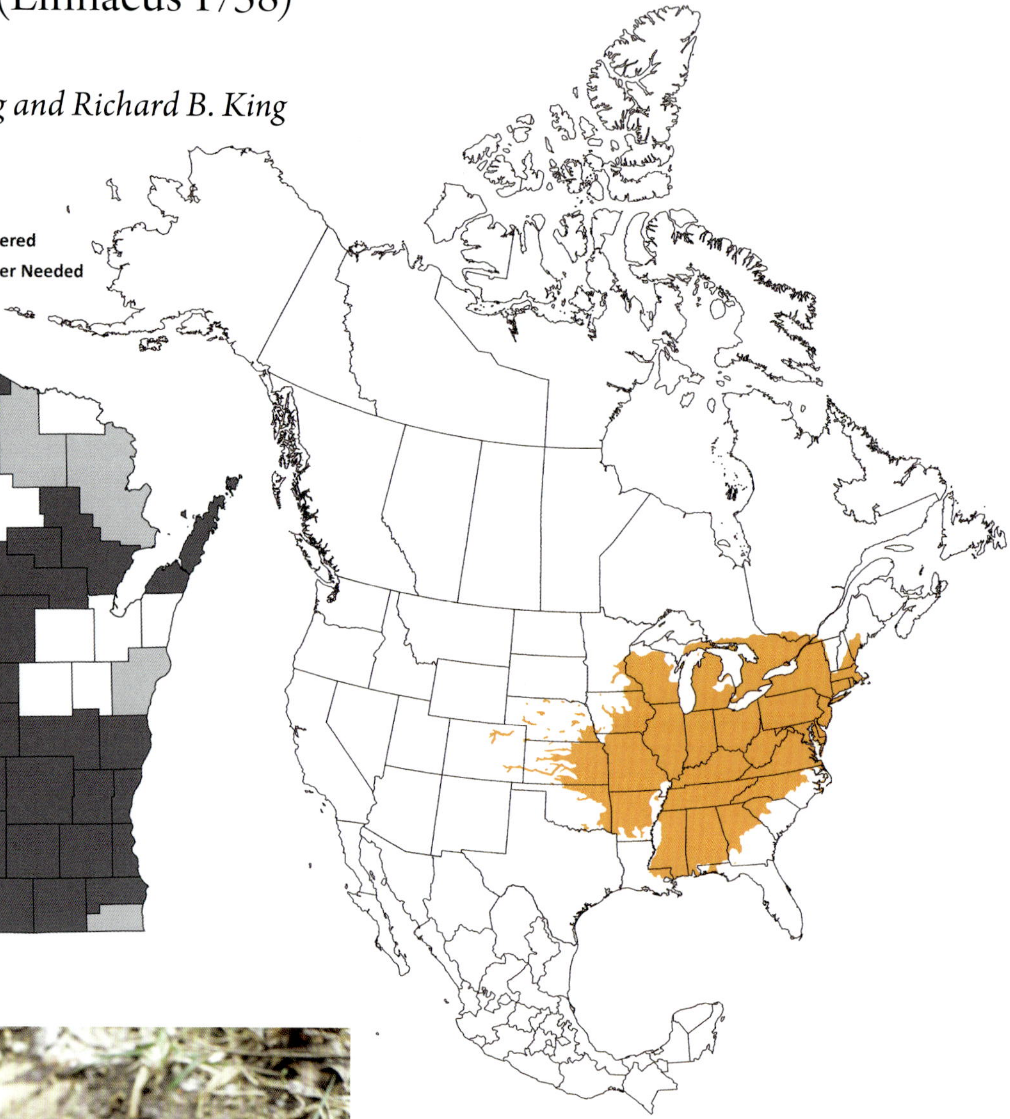

Juvenile Common Watersnake (Fond du Lac County; photo by C. Raimond).

AT A GLANCE

Common Watersnakes are a medium-sized, nonvenomous species that can be somewhat variable in appearance. When adults are dry, their dorsal surface is often drab gray or olive in color. When they are wet, however, the series of starkly contrasting brown to rust-colored bands that run the length of their back become evident. This pattern helps Common Watersnakes blend in with their surroundings so that most passersby are unaware of their presence, which can allow these snakes the opportunity to drop into the water and escape unnoticed. Common Watersnakes occur in many aquatic habitats throughout Wisconsin, where they spend their time sunning on rocks, logs, or tree limbs that occur along the shorelines of wetlands, lakes, rivers, and streams. They are typically active from April through October in Wisconsin. Breeding occurs soon after emergence from hibernation, and females give birth to live young from late August to early September. Common Watersnakes consume primarily amphibians and fishes, although they occasionally eat other vertebrates and invertebrates. When threatened, watersnakes will emit a foul-smelling musk and are quick to bite, but the bite does no lasting damage. Unfortunately, these snakes are often mistaken for the venomous Cottonmouth (*Agkistrodon piscivorus*), a species not found in Wisconsin, and killed unnecessarily. Despite this, Common Watersnake populations in the state appear to be secure, and they are considered common by the WDNR.

SYSTEMATICS

Nerodia sipedon (Common Watersnake) is a member of the widespread New World natricine group, which includes species such as watersnakes, gartersnakes, and crayfish-eating snakes that diversified during the Miocene ca. 14–11 mya (Lawson 1987, McVay et al. 2015). The natricines are a diverse group of snakes with highly variable habitats, diets, and reproductive forms that are distributed across North America (Burbrink and Crother 2011). There are four recognized subspecies of *N. sipedon*: *N. s. insularum* (Lake Erie Watersnake), *N. s. pleuralis* (Midland Watersnake), *N. s. williamengelsi* (Carolina Watersnake), and *N. s. sipedon* (Northern Watersnake), which is the form found in Wisconsin (Crother 2017). *N. sipedon* is known to hybridize with *N. fasciata* (Southern Watersnake; Crother 2017), but one phylogenetic study suggested that its closest relative is actually *N. harteri* (Brazos River Watersnake; Lawson 1987), which is found only in Texas (Gibbons and Dorcas 2004). A more recent study found the closest relatives of *N. sipedon* to be *N. fasciata* and *N. clarkii* (Saltmarsh Watersnake); however, *N. harteri* was not included in this analysis (McVay et al. 2015). There are no other *Nerodia* species in Wisconsin, but *N. sipedon* is placed in the same subfamily (Natricinae) as regional snake species in the genera *Regina*, *Storeria*, *Tropidoclonion*, and *Thamnophis*.

DESCRIPTION

Adults: Common Watersnakes possess ≥30 bands that run the length of the dorsal surface and alternate between red or brown and gray or tan vertical stripes or blotches (Figure 239; Vogt 1981). Bands on older individuals may be faded or barely discernible (Vogt 1981), especially when they are dry, but bands can appear more vibrant when snakes are wet or submerged. In many adults, the banding pattern is more pronounced anteriorly and more faded or blurry posteriorly (Elizabeth Mack McKeag, Northern Illinois University, personal observation). The head is mostly brown or gray, but the lower jaw is a lighter buff color with brown or rusty bands on the posterior edge of each labial scale (Figure 240). The venter is white

Figure 239. Adult Common Watersnake (Walworth County; photo by N. Melhorn).

Figure 240. The head of an adult Common Watersnake that highlights the light-colored lower jaw and labial bands (Sauk County; photo by J. Kapfer).

Figure 241. The typical ventral pattern of Common Watersnakes in Wisconsin (Waupaca County; photo by N. Melhorn).

with dark gray speckling, and the ventral scales are marked with red 'half moons' (Figure 241).

Common Watersnakes are a medium-sized snake, ranging from 60 to 100 cm TL (23.6 to 39.4 in) in Wisconsin (Vogt 1981). Measures of Wisconsin specimens in the MPM and UWZM collections revealed that mean SVL of adult females was 63.8 cm (25.1 in, $n = 60$), and mean tail length was 17.6 cm (6.9 in, $n = 55$). Mean SVL of adult males was 56.7 cm (22.3 in, $n = 38$), and mean tail length was 17.9 cm (7.1 in, $n = 32$; see Reproduction and Development section for size limits used to estimate maturity). The largest SVL among Wisconsin museum specimens measured was 93.1 cm (36.7 in) for a male from Vernon County taken in 1974 by Richard Vogt (MPM 11665).

Preadult stages: Neonates and juveniles exhibit the same color pattern as adults, but it tends to be brighter, with more pronounced banding (Vogt 1981). There are no published size data for juvenile watersnakes in Wisconsin. Studies in Ohio and Ontario have found that neonates range from 11.3 to 21.0 cm TL (4.4 to 8.3 in) and weigh 1.5–6.1 g (0.1–0.2 oz; King 1986, Weatherhead et al. 1999).

Variation: Common Watersnakes in Wisconsin exhibit a high level of individual variation in color pattern. The dorsum can have a base color of gray, brown, or tan, while banding may be brown to rusty and more vibrant or faded depending on the individual and whether its skin is wet (Vogt 1981). Throughout this species' range, there is pronounced sexual size dimorphism, with females being significantly longer and heavier than males (Brown and Weatherhead 1999a). For example, females sampled in Walworth County were markedly longer and heavier (SVL = 63.5–83.8 cm [25–33 in], weight = 237–465 g [8.4–16.4 oz], $n = 10$) than males (Tiebout and Cary 1987).

Diagnosis: Common Watersnakes possess obviously keeled dorsal scales (Figure 242), divided subcaudal scales, and a divided anal plate. Their rostral scale is not upturned or keeled. There is 1 (sometimes 2) loreal scale, 1–2 preocular scales, and 2–3 postocular scales on each side of the head. Typically, 8 (9–12) supralabial scales and 10 (9–12) infralabial scales are present. The

Figure 242. The dorsal surface of a Common Watersnake is covered with scales that possess an obvious keel (Dane County; photo by S. Viernum).

dorsal scale rows at midbody number <25 (21–25), while there are 123–155 ventral scales that each contain two red half-moon shapes over a light background.

Common Watersnakes in Wisconsin are most easily confused with other 'blotched' species, including Eastern Milksnakes (*Lampropeltis triangulum*), Eastern Foxsnakes (*Pantherophis vulpinus*), and Massasaugas (*Sistrurus catenatus*). Watersnakes are distinguishable from Eastern Milksnakes by the absence of black borders on the dorsal blotches and lack of a distinctive V- or Y-shaped blotch on the head. Milksnakes also have smooth scales and an undivided anal plate (Vogt 1981). Watersnakes are smaller and more aquatic than Eastern Foxsnakes, but the two can be found together along the edges of aquatic habitats (E. Mack McKeag, personal observation). Eastern Foxsnakes have more ventral scales (190–219) and sometimes a more yellow base color than watersnakes, with blotches (not bands) that are typically darker brown than the bands of watersnakes. Adult Eastern Foxsnakes also often have a patternless copper-colored head, unlike the typically darker head of Common Watersnakes. Foxsnakes also possess dark ventral spots arranged in a 'checkerboard' or 'maize' pattern, not the red half moons on the ventral scales of Common Watersnakes. Massasaugas are venomous and occupy semiaquatic habitats similar to those of the Common Watersnakes. The most defining differences between Massasaugas and watersnakes are that the former possesses a rattle at the end of the tail and a heat-sensing pit between the eye and the nostril. Massasaugas also have a white stripe and thick dark brown bars that extend from the eyes and heat-sensing pits diagonally across the side of the head, while watersnakes have less distinct head markings, with the exception of thin rust-colored bars on each labial scale, which Massasaugas lack (Vogt 1981).

Throughout their range, Common Watersnakes are often mistaken for the venomous Cottonmouth, which does not occur in Wisconsin. Watersnakes can be distinguished from Cottonmouths by appearance and swimming behavior. Cottonmouths tend to have a darker base color, sometimes approaching black, and a thick dark stripe extending from the heat pits to the articulation of the jaw. Although many people cite a triangular-shaped head as a distinguishing feature of the Cottonmouth, defensive behavior of watersnakes includes flattening and widening the head into a similar shape when threatened. When swimming, watersnakes submerge their bodies, while Cottonmouths tend to float along the surface (Figure 243).

DISTRIBUTION AND HABITAT

Global distribution: Common Watersnakes are widely distributed across eastern and central North America. Populations occur from southern Ontario to Mississippi and west to eastern Colorado. Common Watersnakes have successfully colonized several island chains in the Great Lakes region, including the Georgian Bay Islands in Lake Huron, the St. Lawrence Islands in the St. Lawrence River, the Beaver Island archipelago in Lake Michigan, and the Lake Erie Islands (Long and Long 1976, King 1988a, Hager 1998). However, there are no records of the species on the Apostle Islands of Lake Superior in northern Wisconsin (Hecnar et al. 2002).

The fossil record indicates that Common Watersnakes have occupied the Great Lakes region since at least the Wisconsin period (ca. 25,000–10,000 yrs ago; Holman 2001). They likely originated in the deciduous forests of the eastern US and then expanded westward during periods of changing climate (Vogt 1981). Following glacial retreat, watersnakes dispersed into Wisconsin either through the Upper Peninsula

Common Watersnake
Nerodia sipedon

Cottonmouth
Agkistrodon piscivorus

Figure 243. A Common Watersnake (*left*; Walworth County) compared to a Cottonmouth (*right*; Collier County, Florida). Note the more obvious banding pattern and lack of a facial stripe on the Common Watersnake. Both Common Watersnakes and Cottonmouths swim with their heads above water. However, Common Watersnakes tend to keep their bodies submerged (*lower left*; Walworth County), while Cottonmouths often swim with their bodies floating on the surface (*lower right*; Union County, Illinois; photos by N. Melhorn).

connecting Michigan and Wisconsin, around the southern end of Lake Michigan, or both (Vogt 1981).

Distribution in Wisconsin: Published county records exist for much of Wisconsin, including several far northern counties such as Iron and Douglas (see the Appendix). Despite this, Common Watersnakes are apparently more abundant in the southern half of the state (Knudsen 1954c, West and Dallman 1980, Vogt 1981). This distributional pattern could have occurred because Common Watersnakes were able to find refuge from growing Pleistocene glaciers in the unglaciated Driftless Area of southwestern Wisconsin over ca. 12,000 years ago (West and Dallman 1980, Vogt 1981).

Voucher material is missing for several published county records (see the Appendix). Gibbons and Dorcas (2004) attempted to list the voucher specimen numbers for all existing county records published for each watersnake species in the US. They reported specimen numbers to support most counties of occurrence listed by Vogt (1981), who had not provided information on vouchers. Despite their efforts, a handful of these counties (e.g., Forest and Sawyer) still lacked suitable voucher material. Gibbons and Dorcas (2004) further included previously unpublished records for Bayfield, Manitowoc, and Marinette Counties based on observational reports that did not reference voucher material. In addition, they reported Kenosha County as a new record, listing MPM 1079 for the supporting voucher, although museum notes associated with this specimen indicate it was disposed of (J. Kapfer, personal communication). Casper (2015) published a Chippewa County record based on a specimen at UMMZ (UMMZ 64740), but Gibbons and Dorcas (2004) had already published a record for this county using the same specimen as a voucher.

Localities of possible occurrence: To our knowledge, no published records exist from Ashland County or the Apostle Islands. Yet Common Watersnakes could occur there, as they are documented in nearby Douglas and Iron Counties, while an observational record exists from Bayfield County (Gibbons and Dorcas 2004; see the Appendix). Although Common Watersnakes are currently undocumented on the Apostle Islands, they occur on other Great Lakes islands in locations that are farther south (Long and Long 1976, Long et al. 1978, Hecnar et al. 2002). Despite this, their absence from the Apostle Islands could be due to high latitudes and concomitant harsh climates (Vogt 1981). There are also currently no records or documented observations in two blocks of contiguous counties located in central and east-central Wisconsin. Yet counties of known occurrence surround both of these locations, so it is plausible that further effort could uncover Common Watersnakes in these distributional gaps.

Habitat: Common Watersnakes associate with a broad array of aquatic habitats in Wisconsin. These range from habitats that simply possess moist soil (e.g., sedge meadows) to most other aquatic communities found in the state, such as wetlands, lakes, streams, rivers, and river backwaters (Vogt 1981, reviewed by Ernst and Ernst 2003; see Amphibians and Reptiles in Wisconsin Plant Communities chapter). Although Common Watersnakes generally occur in many aquatic habitat types, they apparently prefer those that are open and possess limited overhead tree canopy cover. For example, Vogt (1981) stated that "waters in dense forests are not usually inhabited by water snakes" in Wisconsin.

Within their preferred aquatic systems, Common Watersnakes rarely stray far from the shoreline. For example, the only robust study of Common Watersnake habitat preference in Wisconsin found that active watersnakes mainly utilized shoreline locations and areas of emergent vegetation (Tiebout and Cary 1987). Despite a wider variety of habitats available, individuals used several functional types most of the time, which included emergent vegetation such as cattails (*Typha* spp.) and shorelines with open water (Tiebout and Cary 1987). Roe et al. (2003) further reported them in upland habitats on <3% of observations, and these terrestrial encounters occurred within 30 m (98.4 ft) of adjacent wetlands. When at the water's edge, Common Watersnakes spend most of their time on debris or hiding among emergent vegetation (Tiebout and Cary 1987). They frequently rest on rock piles, logjams, overhanging trees, docks, dams, and piles of scrap metal or concrete that occur along shorelines (Figures 244 and 245; Knudsen 1954c, Pattishall and Cundall 2009).

Figure 244. An example of typical Common Watersnake habitat in Wisconsin. Can you spot the snake in this photo? (Washington County; photo by N. Melhorn).

Figure 245. A Common Watersnake basking on a tree branch along the water's edge. Utilizing low-hanging tree branches for basking allows Common Watersnakes to quickly drop into the water and escape when approached by a predator (Washington County; photo by N. Melhorn).

Common Watersnake use of habitat structure at the water-land interface facilitates thermoregulation, which is further influenced by annual or daily environmental conditions (see the Activity section). Tiebout and Cary (1987) found that snakes exploited different microhabitat locations based on the month and time of day to optimize thermoregulatory opportunities. As such, they used clumps of dead cattails for thermoregulation and cover in April (Teibout and Cary 1987). Yet from May to July, snakes shifted their behavior to use thick stands of living cattails, which provided better refugia from rising summer temperatures (Tiebout and Cary 1987). Robertson and Weatherhead (1992) further found that while Common Watersnakes selected microhabitats with preferred thermal conditions, they also occupied areas where prey species such as Northern Leopard Frogs (*Lithobates pipiens*) congregate.

Common Watersnakes overwinter in natural fissures and holes, mammal and crayfish burrows, and crevices in man-made structures such as building foundations, dams, and cellars (Ernst and Ernst 2003, Stanford et al. 2010). Hibernacula are often located directly along shorelines, but Common Watersnakes can also migrate from aquatic habitats to rocky hillsides and overwinter in rock fissures that extend below the frost line (Fitch 1982).

REPRODUCTION AND DEVELOPMENT

Few researchers have studied the reproductive biology and development of Common Watersnakes in Wisconsin. Yet these subjects have received more attention elsewhere in the Great Lakes region, and the results of these studies are applicable. Throughout their range, Common Watersnakes begin breeding soon after emergence from hibernation (April in Wisconsin; Tiebout and Cary 1987). The length of the breeding season varies across years, but it generally extends from mid-May to early June. Mating can occur as early as 23 April in Missouri (Weil and Aldridge 1981) and as late as 11 June in Ohio (King 1986).

Females become receptive for mating multiple times during the breeding season, causing their "attractiveness" and the concomitant intensity of male courtship to vary during the reproductive period (Aldridge et al. 2005). Given that females do not make large overland movements and are frequently submerged or near water, males often have limited opportunities to find them by following scent trails on substrates. Despite this, males are reportedly able to locate females via volatile pheromones in the air, particularly when females are basking in elevated positions along shorelines (Aldridge et al. 2005). Watersnakes are well known for their mating aggregations, often referred to as 'mating balls,' which occur on open habitat patches such as cattail mounds, beaver dams, and docks, where many males compete for access to females (Mushinsky 1979, Tiebout and Cary 1987, Brown and Weatherhead 1999b, E. Mack McKeag, personal observation). Successful males will tap the female with their head until she becomes receptive and allows copulation (Brown and Weatherhead 2004). Despite participation in multiple mating aggregations each breeding season, male mating success can be quite low (Weatherhead et al. 1995, Prosser et al. 2002). Females are viviparous, and parturition occurs in late August to early September (Tiebout 1983). Litter size varies, often increasing with the female's age and size (Bauman and Metter 1977, King 1986, Prosser et al. 2002, Kissner et al. 2005). Across the Great Lakes region, litters can reportedly contain as few as 5 (Canada) and as many as 50 (Ohio) neonates (King 1986, Prosser et al. 2002). Mean litter size appears to be ca. 20 live neonates (King 1986, Barry et al. 1992, Prosser et al. 2002).

An interesting facet of Common Watersnake reproductive biology is the frequent occurrence of multiple paternity (Barry et al. 1992, Prosser et al. 2002, Kissner et al. 2005). For example, research on an Ontario population determined that 54.3%–85.7% of individual litters assessed during a two-year study were likely fathered by multiple males (Barry et al. 1992, Kissner et al. 2005). Although most litters have genetic contributions from one to three males, analyses in Ontario verified that five males had contributed to a single litter (Prosser et al. 2002, Kissner et al. 2005). Multiple paternity is reportedly more common in litters produced by larger females. Furthermore, Kissner et al. (2005) found that litters fathered by multiple males contained a mean of six more offspring than litters from single males, but the litters sired by multiple males also had higher rates of stillbirth. Some researchers have suggested that multiple paternity is a function of female choice and that it may provide benefits such as safeguards against male sterility and promotion of genetic diversity or competition among sperm of different males after insemination (Barry et al. 1992).

The developmental strategies employed by watersnakes to maximize offspring success are varied. Neonate size at birth differs greatly both between and within litters (Kissner and Weatherhead 2005). This variation could be a function of birth date and energy investment into offspring by females, but different studies have found contradictory evidence, with birth size both increasing and decreasing in neonates born later in the year (Weatherhead et al. 1999, Kissner and Weatherhead 2005). This has also led to contradictory conclusions about how birth date can affect winter survival of neonates. It is widely accepted that larger and faster-growing neonates have a higher chance of surviving their first hibernation (Jayne and Bennett 1990, Kissner and Weatherhead 2005). Thus, earlier birth dates could give neonatal watersnakes more time to forage before hibernation, allowing them to increase their size and chances of survival (Weatherhead et al. 1999). In contrast, being born later could give females more time to invest in the condition of their developing offspring and increase their chances of survival (Kissner and Weatherhead 2005). Kissner and Weatherhead (2005) further suggested that females who withhold offspring for longer periods during development decrease the length of time that neonates are exposed to predators between parturition and their first hibernation. In Ohio, the majority of neonatal Lake Erie Watersnakes, a subspecies of the Common Watersnake, did not feed in the period between birth and their first hibernation. Females invested heavily in their offspring's fat stores (yolk and fat bodies), allowing them to delay feeding until the spring while still maintaining a high condition for surviving hibernation (Mack et al. 2017). This is similar to the pattern observed in Common Watersnakes in Ontario, where the majority of neonates studied did not feed (evidenced by lost weight) during the period before hibernation (Kissner and Weatherhead 2005).

Yet high investment of energy into offspring by gravid females has costs. Brown and Weatherhead (1997) found that after undergoing parturition, females were emaciated and had little time remaining to restore their own energy reserves, which lead to high winter mortality. Furthermore, laboratory experiments supported that the amount of food consumed by females prior to hibernation influenced reproductive output the following spring (i.e., greater food consumption in autumn leads to more follicles that complete primary vitellogenesis during hibernation; Barron and Andraso 2001).

Although growth in young snakes is fastest prior to reaching sexual maturity, rapid growth does not always begin immediately after parturition (King et al. 2016, Mack et al. 2017). In Ohio, neonatal Lake Erie Watersnakes, which do not feed prior to their first hibernation, also do not experience substantial growth until the following spring (E. Mack McKeag, personal observation). It is unknown if young watersnakes in Wisconsin experience the same delay in growth immediately after parturition. The growth rates of neonates in their first year are also highly variable. In one Ohio population, growth rates of snakes during their first 12 months ranged from 0.08 to 0.14 cm per day (0.03 to 0.06 in) across ten years (Mack 2015). Work in Ontario found that growth rates can be affected by environmental conditions (e.g., snakes grew faster in years with higher air temperature during the active season; Brown and Weatherhead 1999c). If such variation in growth during early development continues past year one, it could lead to variation in when juveniles reach maturity (i.e., faster-growing neonates could reach sexual maturity up to two seasons earlier than slower-growing neonates) (Mack 2015). This could cause marked differences in survival and lifetime fitness among cohorts of young snakes. There is also apparently variation in growth rates by sex, with females typically growing at a faster rate than males (Brown and Weatherhead 1999c). Males generally become reproductively mature at two to three years of age and females at three to four years (Bauman and Metter 1977, Feaver 1977, King 1986, Weatherhead et al. 1995, Brown and Weatherhead 1999a). For example, the minimum SVL at sexual maturity reported for snakes in northern locations ranged from 47.6 cm (18.7 in, Michigan; Feaver 1977) to 55.4 cm (21.8 in, Ontario; Weatherhead et al. 1995) for females. On the other hand, the SVL of most sexually mature males in these studies was ≥42.5 cm (16.7 in; Feaver 1977, Weatherhead et al. 1995). During a comprehensive demographic study in Canada, female and male Common Watersnakes reportedly attained asymptotic SVLs of 92.9 cm (36.6 in, $n = 123$) and 62.0 cm (24.4 in, $n = 143$), respectively (Brown and Weatherhead 1999c). Similar asymptotic sizes have been reported for the closely related Lake Erie Watersnake in Ohio (population mean female asymptotic SVL = 101.8 cm [40.1 in], $n = 730$; population mean male asymptotic SVL = 74.0 cm [29.1 in], $n = 1{,}152$; King et al. 2016). Information on longevity in the wild is sparse, but Snider and Bowler (1992) reported an individual that lived to >9.5 years in captivity.

ACTIVITY

Common Watersnakes in Wisconsin are active from April to October (Vogt 1981, Tiebout and Cary 1987). During the active season, they spend a substantial amount of time regulating body temperature (Robertson and Weatherhead 1992). Many factors can influence their thermoregulatory behaviors, including time of day, associated habitat, and environmental conditions. For example, although Common Watersnakes will rest on shoreline debris to bask during midday, they will commonly remain fully or partially submerged in the late afternoon and at night to modulate temperature (Robertson and Weatherhead 1992, Brown and Weatherhead 2000). Insolation, or an individual's exposure to sunlight, is greatest during midday (ca. 9:00 a.m. to 3:00 p.m.; Tiebout and Cary 1987, Burger et al. 2004). As water temperatures warm and approach air temperatures later in the day, watersnakes frequently remain in the water and avoid terrestrial activity (Robertson and Weatherhead 1992). Burger et al. (2004) found that Common Watersnakes along an urban canal in New Jersey basked most often when water temperature reached 19°C (66.2°F), while basking ceased when water temperature exceeded 30°C (86°F) and air temperature exceeded 32°C (89.6°F). Brown and Weatherhead (2000) reported that the mean preferred body temperature for captive and wild Common Watersnakes at the northern extent of their geographic distribution (Ontario) was 27.1°C (80.8°F), with a range of preferred body temperatures of 25°C–30°C (77°F–86°F). This study also determined that feeding did not stimulate snakes to raise their body temperature but that females, particularly gravid females, exhibited more thermoregulatory precision than males. The authors suggested that this could be because males choose to remain hidden later in the summer to avoid predation, while females must maintain preferred body temperatures during gestation (Brown and Weatherhead 2000).

Common Watersnake movement patterns are heavily influenced by the presence of water and the distribution of waterbodies on the landscape. Individuals found in lentic systems, such as large wetlands or lakes, primarily associate with a single waterbody or a few closely located waterbodies during the active season. For example, Fraker (1970) recaptured the same individuals along the shore of the same ponds 61%–80% of the time during a single active season in Indiana. Because they rarely stray from the water's edge, there is some debate as to whether Common Watersnakes move throughout a true 'home range' during the active season (Tiebout and Cary 1987, Roth and Greene 2006, Patishall and Cundall 2008). For example, Tiebout and Cary (1987) determined that snakes in Walworth County made mostly linear movements as individuals followed shorelines.

They further found that Common Watersnakes occupied shifting activity areas throughout the active season instead of stationary home ranges (Tiebout and Cary 1987). The sizes of activity areas ('95% isopleths,' or areas that contained 95% of an individual's locations) in Wisconsin were highly variable and spanned four orders of magnitude, ranging from 0.0007 to 15.4 ha (0.002 to 38.1 ac, $\bar{x} = 5.4$ ha [13.3 ac]). Core movement areas ('50% isopleths,' or areas that contained 50% of an individual's locations) used by snakes averaged only 7.7% of their total activity area. The same patterns of linear movement and shifting activity areas were found in Common Watersnakes from a Missouri population, but home ranges and core areas estimated using the same technique were significantly smaller (95% isopleth = 0.1–1.1 ha [0.3–2.6 ac], 50% isopleth = 0.01–0.3 ha [0.02–0.8 ac]; Roth and Greene 2006). Sexual differences in the size of activity areas and rates of movement have not been investigated in Wisconsin, but these do not seem to differ between males and females based on populations studied in Ohio, Michigan (Roe et al. 2004), and Missouri (Roth and Greene 2006). Mean movement rate of individuals tracked in Wisconsin was 18.7 m/hr (61.4 ft/hr). Movement rates were higher in open water and open vegetated habitats (Tiebout and Cary 1987).

Figure 246. A Common Watersnake consuming an American Toad (Columbia County; photo by G. Braatz).

PREY AND PREDATORS

Most studies of Common Watersnake diet, including investigations of their responses to chemical cues from prey, support that this species feeds primarily on fishes and amphibians (Burghardt 1968, Vogt 1981). For example, they reportedly eat a variety of fishes and their eggs in Wisconsin (Vogt 1981). Anderson and Rozmarynowski (2011) even observed several Largemouth Bass (*Micropterus salmoides*) mobbing a Common Watersnake in Sauk County, which the authors suggested was an antipredator tactic that inferred that Largemouth Bass are a prey source. Common Watersnakes in Wisconsin also frequently consume adult and larval amphibians (Vogt 1981; Figure 246).

Numerous factors can potentially influence the types of prey that are commonly consumed by Common Watersnakes, including prey availability, habitat, and snake developmental state. An exhaustive dietary analysis of Common Watersnakes on the Beaver Islands of Lake Michigan showed that snakes consumed a variety of fish and amphibian species, but each population's dominant prey was dictated by its foraging habitat. Shore-dwelling snakes consumed mostly fish, but inland snakes ate amphibians almost exclusively. A diet comparison of island-dwelling Lake Erie Watersnakes and mainland Common Watersnakes in Ohio produced similar results. Only island snakes ate amphibians, and while both island and mainland snakes ate fishes, the species composition of fish taken varied between island and mainland populations (King 1993b). There is evidence of an ontogenetic shift in Common Watersnake diet related to snake size. Prey length and weight tend to increase with snake size, and larger snakes tend to drop smaller prey items from their diets (Bowen 2004a). This can lead to differences in the dietary composition of small (young) and large (mature) watersnakes, as larger watersnakes eat prey that attain larger sizes. Consequently, sexual size dimorphism in Common Watersnakes facilitates the consumption of larger meals by females (King 1993b). Prey availability can also affect the type of habitat that Common Watersnakes select and their distributional patterns. For example, leopard frog abundances can contribute to the spatial distribution and microhabitat preferences of Common Watersnakes within wetlands (Robertson and Weatherhead 1992).

Common Watersnakes will frequently forage during the day but also sometimes at night, employing both active and passive strategies to capture prey (Raney and Roecker 1947, Drummond 1983, Balent and Andreadis 1998, Cundall and Patishall 2011). During active foraging, Common Watersnakes swim open-mouthed along the bottom of a pond or river close to shore and inquisitively probe between rocks and crevices, snatching unsuspecting fish where they hide. Active underwater foraging is facilitated by the fact that watersnakes can remain submerged for up to two minutes (Drummond 1983). While passively foraging, Common Watersnakes position themselves along the water's edge, sometimes anchored on dry land by their tail, waiting to strike at prey as it passes by (Drummond 1983, Balent and Andreadis 1998). The use of both active and passive foraging techniques, in conjunction with the large differences reported in dietary composition between populations, has led to some debate over whether Common Watersnakes are dietary specialists or generalists (Drummond 1983, King 1993b, Bowen 2004a). For example, Common Watersnakes in Wisconsin eat a wide variety of prey items, but the Lake Erie Watersnake in Ohio preys almost exclusively on the invasive Round Goby (*Neogobius melanostamus*; King et al. 2006). Common Watersnakes are also documented scavengers that reportedly consume dead fish in New York (Raney and Roecker 1947).

Known and suspected predators of Common Watersnakes throughout their North American range include small and medium-sized predatory mammals such as Raccoons (*Procyon lotor*) and weasels (Mustelidae), large birds of prey such as hawks (Accipitridae) and Bald Eagles (*Haliaeetus leucocephalus*), snapping turtles, and bullfrogs (Pierce and Ross 1989, Brown and Weatherhead 1999a). On several occasions, the authors of this account have witnessed Great Blue Herons (*Ardea herodias*) carrying away adult watersnakes in Ohio. When threatened, this species' first instinct is to quickly drop or slither into the nearest body of water (Cooper et al. 2008). Sometimes this involves falling from substantial heights, as Common Watersnakes in Wisconsin have been recorded perching at heights of up to 213 cm (83.9 in; Tiebout and Cary 1987). If escape is impossible, the Common Watersnake will coil and strike, biting repeatedly, although the resultant damage is minimal. Watersnakes are also notorious (even infamous) for the foul-smelling musk they emit from glands in the cloaca when perturbed. The authors of this account can attest to the lingering effects (i.e., days!) of this musk and the effect it has on dinner guests unfortunate enough to take a seat next to a watersnake researcher.

CONSERVATION

Status: Globally, the Common Watersnake is listed as a Least Concern species by the IUCN. It is considered a common species in Wisconsin by the WDNR.

Populations: No known research that is focused on the characteristics of Wisconsin populations exists, and studies from the Upper Midwest are apparently limited. Feaver (1977) estimated that the mean annual size of a Common Watersnake population in nearby Michigan ranged from 64.7 (± 12.4) to 105.7 (± 16.6) over a four-year study. Fitch (1982) sampled Common Watersnakes for five years in Kansas and captured 206 snakes, which he used to estimate a density of 0.13/ha (0.32/ac). Yet due to limited recaptures, his estimate was determined by comparing Common Watersnake captures to a more common species, Common Gartersnake (*Thamnophis sirtalis*). Intense population research has been conducted on the Lake Erie Watersnake (a subspecies of *Nerodia sipedon*; King 1986). This work has focused primarily on island populations in Lake Erie and resulted in an estimated population size of 1,262 snakes for all sampled islands, although it is noteworthy that estimates for individual islands varied dramatically. Estimated linear (shoreline) density for this population was high at 90/km (56/mi; King 1986). It is noteworthy that some reports suggest that Common Watersnakes can exist at higher than normal densities under certain conditions, including unnatural situations where a potential food source is highly concentrated and easy to obtain (e.g., fish hatcheries; Bauman and Metter 1975).

Some studies have reported data on Common Watersnake population parameters other than estimates of abundance and density, such as sex ratios. For example, a slight male bias was discovered in the operational sex ratio of Ontario Common Watersnake populations, potentially due to a combination of higher female mortality and the fact that some females have more female offspring than male offspring (Weatherhead et al. 1995, 1998). Despite this, it is uncommon for reported sex ratios to differ significantly from 1M:1F (e.g., Brown and Weatherhead 1999a). Information on survival indicates high variability, although juvenile survival appears to be lower than that of adults. Work by Feaver (1977) in Michigan found that juvenile survivorship ranged from 0.193 to 0.235, while adult survivorship ranged from 0.296 to 0.382. Feaver (1977) also reported that adult males had higher survivorship than adult females. Yet juvenile survival also appears to shift seasonally. For example, survival of neonates during the period between birth and their first hibernation in Ontario was reportedly ca. 65%, while survival during hibernation was ca. 47%. Therefore, only 31% of neonates survived to their first spring (Kissner and Weatherhead 2005). Other studies from Ontario have also found high variability in annual neonatal survival, ranging from 4% to 42% (Brown and Weatherhead 1999a). Feaver (1977) further estimated that a study population in Michigan experienced turnover every 2.8 years, on average.

Threats: Many of the general threats that face aquatic habitats, from streams and rivers to wetlands and lakes, are potentially applicable to this species (see Introduction to Conservation and Management of Wisconsin's Amphibians and Reptiles chapter). Indeed, Common Watersnakes face many of the same threats that impact other snake species, including habitat loss and fragmentation, emergent diseases, environmental contaminants, entanglement in erosion netting, and direct mortality due to human persecution (Vogt 1981, Bishop and Gendron 1998, Attum et al. 2007, Kapfer and Paloski 2011, Chin et al. 2013a, 2013b, Baker et al. 2018). The most severe threat to Common Watersnakes is likely the loss and fragmentation of aquatic habitats. By the early twenty-first century, more than half of Wisconsin's natural wetlands and ponds had been drained (Dodson and Lillie 2001). Common Watersnake abundance reportedly correlates positively with wetland size and connectivity, reiterating the species' need for large wetland complexes instead of small isolated ponds (Attum et al. 2007). Not only has the number of available wetlands in the state declined over time, but the distance between aquatic habitats has also increased, potentially forcing watersnakes to travel long distances to find suitable habitat. Common Watersnakes that undertake these large migrations are more likely to engage in risky behaviors, such as crossing roads, which increases the probability of mortality. In Indiana, the majority of road mortality risk (94% of crossings) for Common Watersnakes was at locations where roads bisected snake migration corridors between wetlands during the active season, and road mortality was estimated at 3%–5% (Roe et al. 2006). This species also uses urban habitats, which could indirectly put it at higher risk of mortality due to greater contact with humans (Burger et al. 2004).

Among Wisconsin's snakes, Common Watersnakes appear to be particularly susceptible to threats that have minimal influence on more terrestrial species, such as waterborne contaminants (Bishop and Gendron 1998, Chin et al. 2013a, 2013b). A study of Great Lakes reptiles determined that among examined reptile specimens, the individual with the highest concentrations of DDE (dichlorodiphenyldichloroethylene, a compound created by the breakdown of DDT) and Aroclor (a toxic PCB banned from manufacturing in 1979) was a Lake Erie Watersnake (Bishop and Gendron 1998, US Environmental Protection Agency 2018). Furthermore, Common Watersnakes at a disturbed site in Tennessee had significantly higher levels of several heavy metals in their tissues than snakes sampled from an unimpacted reference site (Campbell et al. 2005). Similarly, female Common Watersnakes from a contaminated river in Virginia transferred mercury, a neurotoxin, through reproductive tissues to their young. Although this maternal transfer of mercury did not affect litter size or offspring survival, neonates from litters with higher mercury concentrations demonstrated sublethal effects, such as decreased feeding and strike efficiency (Chin et al. 2013a, 2013b). In addition to contaminants, watersnakes are reportedly susceptible to entanglement in refuse, such as fishing line, and erosion control mesh, often used to stabilize banks along aquatic systems (Natural History Box: Aquatic Debris; Kapfer and Paloski 2011).

Unfortunately, direct persecution by humans has likely had a substantial but unquantified impact on this species. A large number of watersnakes are

NATURAL HISTORY BOX: Aquatic Debris

Although being able to quickly swim away from terrestrial predators is an advantage of being a watersnake, aquatic habitats come with their own unique threats. When out doing research on the islands of western Lake Erie, we often encounter snakes entangled in different types of aquatic debris. Hazards include old aluminum cans, fishing line, lures, and all kinds of plastic wrapping (Figure 247). In one instance, Elizabeth W. Mack McKeag had to cut away some fishing line that had knotted around the middle of an adult male watersnake (Figure 248). The line had been there long enough to constrict the body, causing the skin to actually start growing around the fishing line. Thankfully, it was removed before it could cause any lasting harm. Unfortunately, none of the snakes we have found with their heads stuck in rusty cans have been so fortunate (Figure 249).

Above, left: Figure 247. An adult watersnake snagged by a fishing leader (Ottawa County, Ohio; photo by K. Stanford).

Above, right: Figure 248. An adult Common Watersnake with fishing line knotted around the midsection (Ottawa County, Ohio; photo by E. W. Mack McKeag).

Left: Figure 249. Two adult Lake Erie Watersnakes (*Nerodia sipedon insularum*) that were found dead with their heads stuck inside aluminum cans (Ottawa County, Ohio; photo by K. Stanford).

Elizabeth Mack McKeag and Richard B. King, Northern Illinois University

senselessly killed each year because they are mistaken for venomous Cottonmouths, which do not occur in Wisconsin (Vogt 1981). In addition, the belief that watersnake predation has detrimental impacts on sport fish populations has also sometimes led to heavy persecution by humans. For example, Bauman and Metter (1975) estimated that Common Watersnakes consumed >$80,000 worth of goldfish at a commercial hatchery in Missouri. To offset monetary losses, the hatchery instituted a snake bounty program that (in conjunction with the article authors' own efforts) led to the mortality of over 6,300 snakes (Baumann and Metter 1975). They apparently believed this was justified by their statement: "It is obvious that killing the snakes represents a substantial savings to the hatchery, and that the control measures are not an unwarranted persecution of an innocent species." Despite this example, quantitative data on the impacts (if any) that Common Watersnakes have on fish populations, particularly game species, in natural settings are lacking (reviewed by Vogt 1981).

Management: Many management strategies that generally benefit herpetofauna in aquatic habitats likely apply to Common Watersnakes. These include the maintenance of upland vegetation buffers adjacent to aquatic habitats and minimization of road mortality for the promotion of connectivity among aquatic habitats (see Introduction to Conservation and Management of Wisconsin's Amphibians and Reptiles chapter). Of particular importance to this species is likely the preservation of shoreline habitat that includes ample logs, snags, and other natural debris used for thermoregulation and as cover. As such, habitat management strategies that regulate shoreline development and allow for the persistence of woody debris and snags in aquatic systems will likely benefit this species (Burger et al. 2004). Furthermore, past researchers have suggested that mowing of grass along shorelines should be kept to a minimum (Burger et al. 2004), as snakes use longer vegetation for cover and could accidentally be struck by mowing equipment. Entanglement in erosion control netting can be prevented by using mesh with large openings or choosing more natural materials such as wood mulch (Kapfer and Paloski 2011).

SPECIAL REMARKS

Although there are suites of studies on the Common Watersnake throughout the rest of its range, much of the species' biology and ecology is still unknown in Wisconsin. For instance, it would be simple to investigate the feeding patterns of neonates and compare them to watersnakes in Ohio or to look at habitat and prey availability in counties where watersnake records currently do not exist. Since this species is so dependent upon permanent wetlands, studies of climate change and changing water regimes would also be valuable and could possibly inform future management decisions. So for you burgeoning snake ecologists with an itch to do some research, look no further than the wetlands of Wisconsin!

Queensnake

Regina septemvittata (Say 1825)

Joshua M. Kapfer and Richard A. Staffen

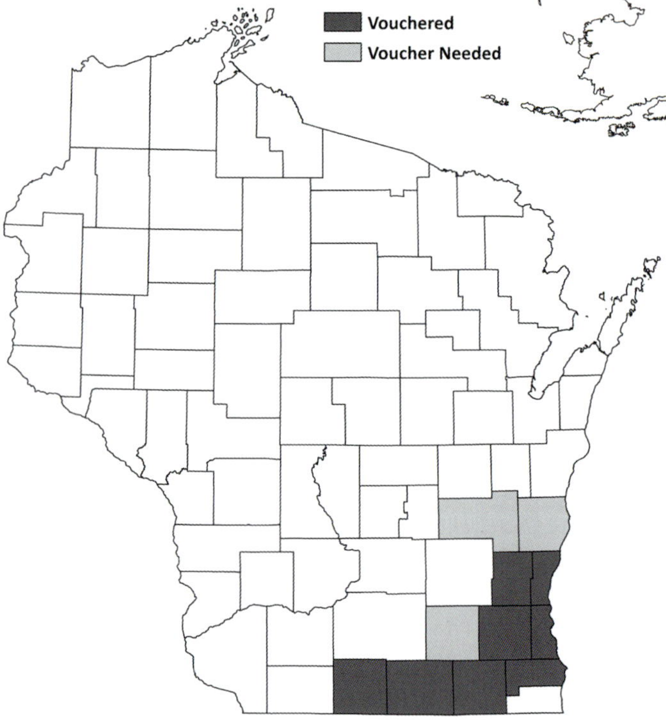

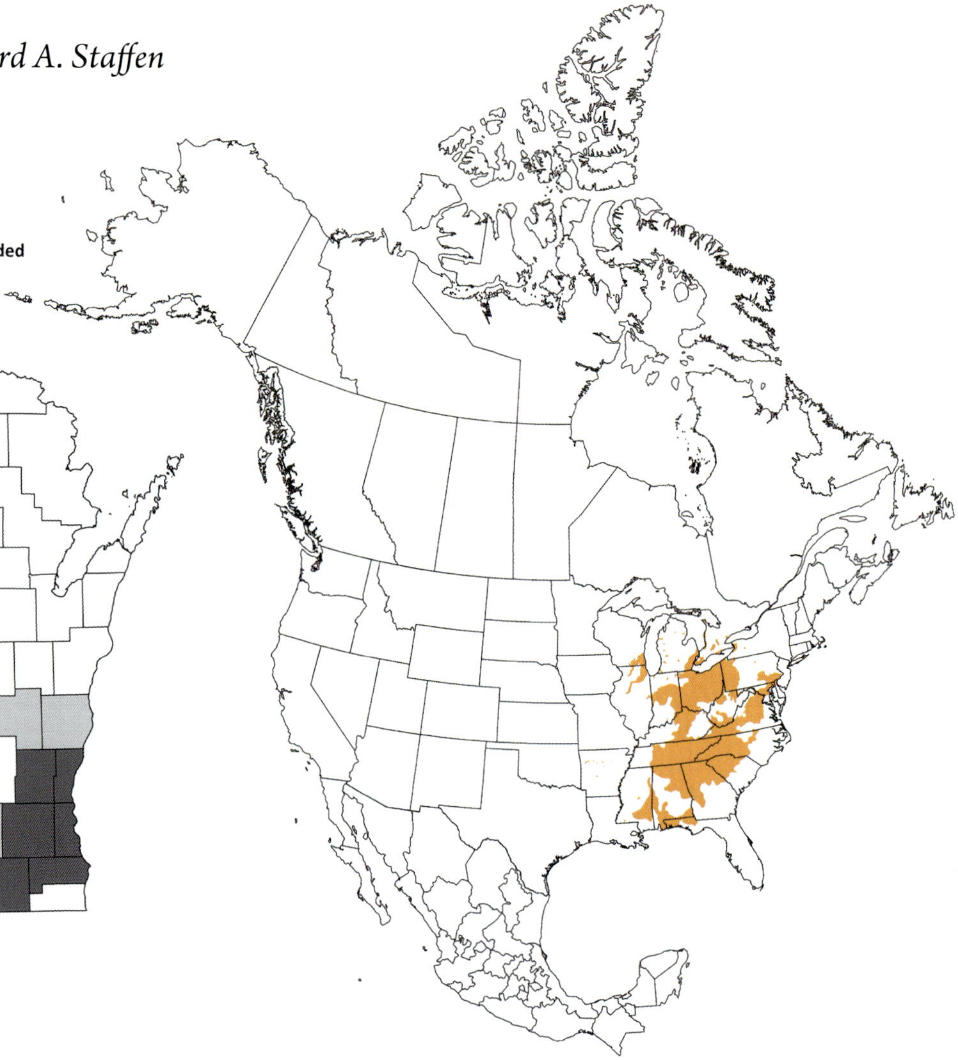

Adult female Queensnake (Washington County; photo by L. Huse).

AT A GLANCE

Queensnakes are one of the rarest and least often encountered snakes in Wisconsin. They are currently documented in only a handful of southeastern counties, and several historical locations have not produced confirmed observations in decades. This species is small to midsized, with a dark brown to olive back. The back also has three darker lines that run the length of the body and that are not always distinct, so that the dorsal surface sometimes appears uniform in color. The scales around the mouth are cream or beige, and this color extends to lighter lines that run down the lateral sides of the body; these lines further highlight single dark lateral stripes on each side. The belly is cream to beige colored, with two more dark parallel lines that also run the length of the body, so there are seven total dark parallel stripes. Queensnakes are typically encountered near open canopy clear-water streams, and observations in the wild indicate that they require a substantial amount of rocks and cobble along the streambeds and streambanks for cover. They appear to remain close to the shoreline under rocks or basking in woody vegetation that hangs over the water. In Wisconsin, Queensnakes probably emerge from winter dormancy in April, and breeding occurs shortly thereafter. They do not lay eggs but give birth to live young from late August to early September. Queensnakes eat primarily freshly molted crayfish, although other prey are taken. They are likely consumed by a variety of vertebrates that frequent stream habitats. Yet when threatened, this placid species rarely, if ever, bites but instead attempts to escape by fleeing into the water. In October, Queensnakes move back to overwintering locations, which are probably crevices in anthropogenic structures or crayfish burrows along streambanks that allow them to descend below the frost line. Because this snake is so rare, few details are known about its ecology in the state. It is considered a Species of Greatest Conservation Need and listed as an Endangered species by the WDNR.

SYSTEMATICS

As Gibbons and Dorcas (2004) stated, *Regina septemvittata* (Queensnake) "has provided minimal taxonomic confusion" over time. This is likely because there are no accepted subspecies, although several have been proposed (e.g., Neill 1963a), and because it has reportedly never been accidentally described as a different watersnake species (Gibbons and Dorcas 2004). Several recent studies have demonstrated that *Regina* is a paraphyletic genus (Alfaro and Arnold 2001, Pyron et al. 2011, Guo et al. 2012, McVay et al. 2015). Guo et al. (2012) also concluded that if additional sampling and analyses continue to suggest paraphyly, a massive taxonomic overhaul of the entire watersnake group would be necessary to rectify the situation. The translation of its Latin scientific name is 'queen' (*Regina*) and 'seven striped' (*septemvittata*), the latter of which refers to the dark stripes that run down the snake's body (see Description section). The reason behind selection of the generic name *Regina* is less obvious, however. Gibbons and Dorcas (2004) explained that it "was presumably chosen because of the attractive appearance of the slender, striped snakes in this genus." Minton (2001) seemed less enthusiastic about *R. septemvittata*'s appearance when he said, "Why this name was proposed for a drab, unimpressive serpent is uncertain." Currently, no other snakes in the genus *Regina* are known from Wisconsin, and *R. septemvittata* is most closely allied with *Nerodia sipedon* (Common Watersnake; Guo et al. 2012), although it also occurs in the same subfamily (Natricinae) as snakes in the genera *Tropidoclonion*, *Storeria*, and *Thamnophis*.

DESCRIPTION

Adults: The dorsal surface is dark brown, gray, or olive, with three darker brown or black stripes along the length of the body (Figure 250). These dark stripes are not always well-defined, which sometimes gives the dorsal surface a uniform brown or gray appearance. The ventral surface is tan or cream, with two parallel dark bands down the center. The ventral surface from midbody to the tail is sometimes filled in with dark flecking that partially obscures the dark bands down the center (Figure 251). The labial scales are cream, which connects to a similarly colored band that extends from the labial scales down the length of the body on both sides (usually involving part of scale rows one and two; Figure 252). Queensnakes

Figure 252. Queensnakes have cream-colored labial scales that connect to a lateral stripe or band extending the length of the body (Rock County; photo by J. Kapfer).

Above, top: Figure 250. The dorsal surface of an adult Queensnake is often a brown to gray color, and occasionally the dark longitudinal stripes are difficult to discern (Rock County; photo by J. Kapfer).

Above, bottom: Figure 251. Examples of ventral surfaces in Queensnakes. (a) Adult (Rock County; photo by J. Kapfer). (b) Juvenile (Rock County; photo by J. Kapfer).

possess keeled scales and a divided anal plate (Ernst and Ernst 2003).

Extremely limited measures of size exist for Wisconsin individuals. Measures of SVL from wild-caught Wisconsin adults listed in technical reports to the WDNR ranged from 46.0 to 58.5 cm (18.1 to 23.0 in), the largest of which was a female from Washington County (Casper et al. 2011). A total of only nine adult specimens from Wisconsin were found in the MPM and UWZM specimen collections. The mean SVL of these specimens ranged from 46.6 cm (18.4 in, $n = 3$) in females to 39.9 cm (15.7 in, $n = 6$) in males, while mean tail length ranged from 14.5 cm (5.7 in) in females to 15.1 cm (5.9 in) in males. This is similar to findings of Baker (1964) in Kentucky, who determined that mean adult SVL ranged from 50.9 cm (20 in, $n = 89$) in females to 44.9 cm (17.7 in, $n = 90$) in males. The largest reported individual from across the range of this species was from Ohio, with an SVL of 71.6 (28.2 in) and a tail length of 20.6 cm (8.1 in; Triplehorn 1949).

Preadult stages: Adults and juveniles are similar in appearance (reviewed by Ernst and Ernst 2003), although limited observations of neonates or juveniles from Wisconsin exist. A handful of neonates observed in Rock County appeared similar to adults, although ventral and dorsal stripes were slightly more distinct (J. Kapfer, personal observation). Size data on younger Queensnakes from Wisconsin are extremely sparse.

A young juvenile captured in Rock County had an SVL of 14.5 cm (5.7 in), a tail length of 5 cm (2 in), and a weight of 5 g (0.2 oz). Furthermore, the SVL of a male juvenile or subadult captured at the same location was 24.9 cm (9.8 in), while tail length was 8.8 cm (3.5 in), and weight was 11 g (0.4 oz; J. Kapfer, unpublished data). The SVL of two juvenile Wisconsin specimens in the MPM collection measured 15.7 cm and 18 cm (6.2 and 7.1 in), respectively. Burghardt (1968) reported a mean SVL of 16.1 cm (6.3 in) and a weight of 2.9 g (0.1 oz, $n = 20$) for neonates from Illinois. Baker (1964) reported that mean SVL of 128 neonates from Kentucky was 17.7 cm (7 in), and mean weight was 3.3 g (0.1 oz). The mean TL of recently born individuals from New York was 21.8 cm (8.6 in, $n = 17$; Raney and Roecker 1947).

Variation: There is no substantial variation in color known among individuals, sexes, or age classes in Wisconsin, although limited size data recorded in the state suggest that females tend to be larger than males (see Adults subsection). Some Queensnakes also exhibit slightly more or less flecking on the ventral surface, while others have slightly more distinct dorsal or ventral stripes.

Diagnosis: Queensnakes have heavily keeled dorsal scales that occur in <25 (often 19) rows at midbody. Ventral scale counts range from 118 to 159, while the anal plate and all subcaudal scales are divided. They lack a sharply upturned rostral scale, but each side of the head possesses 1 loreal scale, 2 (sometimes 1) preocular scales, 2 (1–3) postocular scales, 7 (6–8) supralabials, and 9–10 (8–11) infralabials. Adult and juvenile Queensnakes could be confused with various gartersnakes, ribbonsnakes (*Thamnophis* spp.), and Common Watersnakes (*Nerodia sipedon*). However, Queensnakes can be distinguished from members of the genus *Thamnophis* by their divided anal plate. Queensnakes also do not have the yellow middorsal stripe or unmarked ventral scales found in *Thamnophis*. Basking Common Watersnakes often appear a drab olive color when dry and so can be confused for Queensnakes at a distance. However, close inspection will reveal that although Common Watersnakes also have a divided anal plate, they possess dark reddish bands on their dorsal surface, reddish half-moon shapes on their ventral surface, and dark bands across their labial scales, unlike Queensnakes.

DISTRIBUTION AND HABITAT

Global distribution: The core of the Queensnake's continental distribution lies east of the Mississippi River, where it is most commonly found in the midwestern and southeastern US. Its range there extends from the upper Great Lakes (Wisconsin, Lower Peninsula of Michigan, southern Ontario, and western New York), south to the Gulf of Mexico, and east to southeastern Pennsylvania, western New Jersey, and the Delmarva Peninsula. The southern portion of the snake's range includes Mississippi, Alabama, Georgia, and the panhandle of Florida, extending eastward into the western Carolinas. A few disjunct populations are known from north-central Arkansas, southwestern Missouri (extirpated), and an island in western Lake Huron (Conant 1960, Ernst and Ernst 2003, NatureServe 2020).

Distribution in Wisconsin: Wisconsin is at the northwestern limit of the Queensnake's North American range. Reports of this species are mostly concentrated in the far south-central to southeastern regions of the state, and published records currently exist for only a handful of counties (see the Appendix). Among the counties listed in Gibbons and Dorcas (2004), no vouchers could be found for Fond du Lac or Sheboygan. Although both counties are adjacent to recently confirmed records in Washington County, these records are currently based on anecdotal observations and remain questionable (Table 13). Furthermore, the voucher specimen reported for Jefferson County (UWZM 9709) is missing (Table 13). Several other existing county records are based on historical or missing specimens, which begs the question of whether populations are extant in these locations (Table 13; see Localities of possible occurrence subsection).

Localities of possible occurrence: Undocumented Queensnake populations could occur in additional counties, given proximity to published records, pending the existence of enough suitable habitat. For example, this species could occur in eastern Dodge

Table 13. Counties of known or hypothetical occurrence for Queensnakes in Wisconsin, based on published reports and museum specimens. Included are the last known voucher specimens, the publication that references or confirms the record, and the currently known status of this species in each county.

County	Voucher #	Publications	Status (Year of Last Observation)[b]
Dane	UWZM 1907	N/A	Unknown; specimen disposed of and associated data suspect (early 1900s)
Fond du Lac	N/A	Gibbons and Dorcas (2004)	Unknown (unconfirmed report, G. Casper)[c]
Green	JFBM Herp Photo 471	Casper (2015)	Extant (2016)
Jefferson	UWZM 9709	Gibbons and Dorcas (2004), Vogt (1981)[a]	Unknown; UWZM 9709 lost; presumed extirpated, but potentially suitable habitat exists in county (1894)
Milwaukee	MPM 99	Casper (1996)	Historical and presumed extirpated, but potentially suitable habitat exists in county (1890)
Ozaukee	MPM 2225	Casper (1996)	Historical and presumed extirpated, but potentially suitable habitat exists in county (1928)
Racine	MPM 4547	Gibbons and Dorcas (2004), Vogt (1981)[a]	Historical and presumed extirpated, but potentially suitable habitat exists in county (1971)
Rock	INHS 2008i	Kapfer et al. (2009c)	Extant (2019)
Sheboygan	N/A	Gibbons and Dorcas (2004)	Unknown (unconfirmed report, G. Casper)[c]
Walworth	CHAS HERP 501	Schmidt and Necker (1935), Vogt (1981)[a]	Extant (2021)
Washington	MPM VZP34	Casper (1996)	Extant (2020)
Waukesha	UWZM 9707	Vogt (1981)[a]	Historical and presumed extirpated, but potentially suitable habitat exists in county (1903)

[a] Vogt (1981) did not report vouchers specimens for this county, so reference to voucher material herein was reported by Schmidt and Necker (1935), added post hoc by Gibbons and Dorcas (2004) or the authors (see the Appendix).

[b] Last year of observation determined either through direct observation by authors, dates of specimen collection, most-recent published reports, or personal communication and associated photographs from others.

[c] Although Gibbons and Dorcas (2004) include this county based on a personal communication with G. Casper, Casper et al. (2011) later reported that surveys in this county did not yield Queensnakes. No additional observations in this county are known.

County because it was recently documented in nearby Washington County. Kenosha County is also surrounded by counties of known occurrence (including numerous adjacent counties in northern Illinois; Gibbons and Dorcas 2004), although records from neighboring Racine County in Wisconsin are old (see below). A single specimen purportedly collected from Dane County in the early 1900s once existed in the UWZM collection (UWZM 1907). Museum records indicate that the only specific location connected with this specimen was "Madison," but it was apparently disposed of decades ago due to desiccation (Laura Monahan, UWZM, personal communication). Interestingly, Vogt (1981) did not include Dane County in the range of this species, perhaps because this specimen was lost before his time. Casper et al. (2011) speculated that this record came from the Madison chain of lakes associated with the Yahara/Rock River system. Queensnakes were documented in 2005 along the Sugar River in Green County (Casper et al. 2011). The Sugar River extends north into Dane County, which further supports that Queensnakes could occur there in suitable habitat. Yet because a record for Dane County was never published, and no new specimens exist, the presence of Queensnakes in this county remains hypothetical.

Several published county records are supported by very old voucher specimens or missing voucher material altogether, with no additional recent observations. As such, it might be reasonable to currently consider these counties locations of only hypothetical occurrence (Table 13). For example, the only known specimens from Racine County were collected along the Root River in 1958 and 1971 (UWZM 22760 and MPM 4547, respectively). The single existing specimen from Ozaukee County (MPM 2225) was collected in 1928, with no specific locality information aside from township and range (T10N, R21E) and the notes "Cedarburg only data. Possibly Cedar Creek or Milwaukee River." The two existing specimens from Waukesha County (UWZM 9707, 9708) were collected along the Bark River in 1903 and 1900, respectively. Although these specimens are old, there are more recent unsubstantiated Queensnake reports associated with the Scuppernong River and Paradise Springs in Waukesha County (reviewed by Casper et al. 2011, WDNR 2016b). The only known specimens from Milwaukee County (MPM 99 and 13515) were collected in 1890 from the Bayview area in the current city of Milwaukee. Extensive development in Bayview since the late 1800s has likely caused the extirpation of this historical Milwaukee County population (see further review by Casper et al. 2011). There are no additional locations in Milwaukee County where Queensnakes are confirmed. Jefferson County has produced only one known specimen, which was apparently collected at Lake Koshkonong in 1894 by the eminent Wisconsin naturalist Ludwig Kumlien (UWZM 9709; Taylor 1937). According to UWZM museum records, this specimen was sent to Karl Schmidt at the Field Museum of Natural History in the early 1900s, where identity was confirmed. Unfortunately, recent investigations of museum collections and discussion with museum staff have revealed that this specimen is likely lost (L. Monahan, personal communication, Alan Resetar, FMNH, personal communication). Lake Koshkonong has experienced heavy disturbance since this specimen was collected in the late 1800s (reviewed by Casper et al. 2011), and suitable habitat at this historical location may be limited today, if present at all. Despite this, habitat that appears superficially suitable for Queensnakes exists in other areas of Jefferson County, such as near the border with Waukesha County in the Southern Kettle Moraine State Forest, although recent survey efforts in this area have been fruitless (J. Kapfer, unpublished data).

Habitat: Queensnakes are semiaquatic, spending much of their lives along the edges of permanent flowing water sources such as streams or rivers (Raney and Roecker 1947, Ernst 2003b; see Amphibians and Reptiles in Wisconsin Plant Communities chapter). They appear to strongly associate with small to moderate-sized streams and rivers that are shallow, with moderate flow rates (Wood 1949, Ernst 2003b). The flowing aquatic systems they favor in Wisconsin have predominantly open-canopy shorelines, with only partial to moderate tree canopy cover and shrubs that overhang the water (Figure 253). Similar habitat characteristics are reported from Kentucky, where mean tree canopy cover associated with Queensnakes was only 15%, and every snake encountered was below <50% tree canopy cover (Branson and Baker 1974). Although this snake does not prefer

Figure 253. Example of a Wisconsin stream with potentially suitable Queensnake habitat, including open tree canopy, the presence of shrubs that overhang the water, and substantial cobble along the shore. Noteworthy is that Queensnakes have not been confirmed at the location in this picture, despite the fact that it contains apparently suitable habitat (Jefferson County; photo by J. Kapfer).

Figure 254. Queensnakes resting among branches of young willows (*Salix* spp.) along the shoreline (Washington County; photo by L. Huse).

shorelines with larger trees, shorter woody vegetation is apparently important, as Queensnakes often climb into shrubs at the water's edge to bask. In fact, Queensnakes from Wisconsin have been observed basking in Red-osier Dogwood (*Cornus sericea*) and willow (*Salix* spp.) growing along shorelines (Richard Staffen, WDNR, personal observation; Figure 254). They also have a relatively long tail compared to other *Regina* species, which likely assists in climbing among the branches of woody vegetation along shorelines (Rossman 1963a). From these elevated thermoregulatory perches, Queensnakes can also easily slip into the water when threatened.

Queensnakes do not stray far from the water's edge, possibly because their skin is highly water permeable, making them more prone to desiccation (Stokes and Dunson 1982). Of 11 individuals captured or observed by Casper et al. (2011) in Washington County, 7 were basking either along the water's edge, over the water, or within 2 m (6 ft) of the shoreline. Recent additional observations at this same Washington County site yielded similar findings: basking individuals were encountered on a small island in the river or on shoreline shrubs within 1 m (3 ft) of the water's edge (R. Staffen, personal observation, Logan Huse, personal observation). This behavior appears to be common among Queensnakes found across a broad geographic area, as all individuals encountered during a Kentucky study were within 3 m (9.8 ft) of the water (Branson and Baker 1974), and all encountered during a study in Pennsylvania were within 6 m (19.7 ft) of the water (Ernst 2003b). Few details are known about juvenile habitat use in Wisconsin, but minimal anecdotal information suggests that they also remain close to water. For example, young individuals captured during drift fence surveys in Rock County were all at the water's edge, but sample size was small ($n = 2$; J. Kapfer, unpublished data).

Occupied streams reportedly have cool to warm water temperature (≥18.3°C [65°F]) and a substrate dominated by flat rocks for cover and foraging (Wood 1949, Ernst 2003b). Branson and Baker (1974) reported that >95% of 229 Queensnakes encountered in Kentucky were under rocks on the streambed. The presence of crayfish, the snake's primary prey item, has been further identified as a major determinant of suitable habitat (Raney and Roecker 1947, Wood 1949, Godley et al. 1984, Ernst 2003b, Ernst and Ernst 2003, Gibbons and Dorcas 2004). Although often associated with streams, Queensnakes are known from other habitats in the Upper Midwest, such as abandoned quarries southwest of Chicago (Weed 1922).

Queensnakes overwinter within or adjacent to the aquatic habitats that they use during the active period (Ernst 2003b). Vogt (1981) suggested that in Wisconsin, Queensnakes may overwinter under rocks in or near streambeds. It is also plausible that crayfish burrows along the shore could make sufficient hibernacula, as other snake species in Wisconsin hibernate in such microhabitats (e.g., Common Watersnakes and Massasaugas [*Sistrurus catenatus*]). Several authors further report the use of small mammal tunnels or rock shelters to overwinter (Vogt 1981, Harding 1997, Ernst and Ernst 2003). There are only two known overwintering sites in Wisconsin: one is associated with a bridge abutment, and the other is associated with a dam (Casper et al. 2011, WDNR 2016b). In these instances, crevices in the concrete structures apparently created fissures that allowed snakes to escape the frost during winter. Use of similar anthropogenic structures to hibernate was also documented in Pennsylvania (Ernst 2003b). Congregations of individuals during the Pennsylvania study hibernated in cracks at depths ranging from 16 to 21 cm (6.3 to 8.3 in), although other snakes likely also overwintered at greater depths (Ernst 2003b).

REPRODUCTION AND DEVELOPMENT

Breeding in Wisconsin is likely concentrated in the spring (Vogt 1981). The latest observation of copulation in Indiana was 28 May (Minton 2001). Vogt (1981) mentioned that autumn breeding occurs in Wisconsin, although to our knowledge there are no additional reports of late-season mating activity from the state. Branson and Baker (1974) believed that autumn mating was a possibility in Kentucky. Gibbons and Dorcas (2004) recounted a personal communication with "A. James," who made numerous observations of captive mating behavior in November between Queensnakes collected from North Carolina.

Male Queensnakes reportedly find females for reproduction by following pheromone trails (Ford 1986). Ford (1982a) described the mating activities of wild Queensnakes held briefly in captivity. Although many aspects of their breeding behavior resembled that of other natricine snakes, he reported that males did not employ the typical lateral undulations against the body of females, a behavior common in many related snake species. Instead, males "bounced" (Ford 1982a) or produced vertical body waves against the female in an apparent attempt to initiate copulation. Ford (1982a) further suggested that this unique behavior could "convey species specific identification designed to bring about female passivity" but that more research on the topic was necessary. Multiple paternity has been documented in Queensnake offspring from Ontario (Wusterbarth et al. 2010). This supports a polygynandrous mating system and suggests that post-copulatory competition among sperm occurs (Wusterbarth et al. 2010).

Queensnakes are viviparous, so they do not lay eggs, and young are typically born from early August to early September in Wisconsin (Vogt 1981). For example, a gravid wild Wisconsin female was briefly kept in captivity and released after parturition in late August (J. Kapfer, unpublished data). Similarly, Trauth (1991) found gravid females in July and presumed that parturition occurred in August or early September in Arkansas. High variability in litter size is reported (5–23 offspring per litter across 39 instances; reviewed by Gibbons and Dorcas 2004). Research in Kentucky found a correlation between female size and number of offspring, with larger females having larger litter sizes (Baker 1964, Branson and Baker 1974).

Limited data exist on growth and development in this species. A review of several studies suggests that the growth rate of Queensnakes in their first year varies by location, from a 35.6% size increase in northern Ohio (Bekker 2007) to a 79% increase in New York (Raney and Roeker 1947). Growth rates appear to decline substantially from their first year (75% increase) to their second year (44.8% increase; Branson and Baker 1974). Similarly, Wood and Duellman (1950) reported that Queensnakes from Ohio increased their size by only ca. 50% during their second year. Females in Kentucky reach sexual maturity by two to three years of age (minimum size at maturity = 34.4 cm SVL [13.5 in]), while males reach maturity by year two (30 cm SVL [11.8 in]; Branson and Baker 1974). These reports of maturity are similar to what Vogt (1981) estimated for Wisconsin. Longevity in the wild is unknown, but Snider and Bowler (1992) found a record that a wild-caught individual survived for >19 years in captivity at the Racine Zoo.

ACTIVITY

Vogt (1981) observed that this species was active from May through September in Wisconsin, similar to accounts from New York (Raney and Roecker 1947). During the active period, however, no seasonal variations in activity levels are reported (Ernst 2003b). Although specific investigations are lacking, the snakes are likely diurnal in Wisconsin, which is the case elsewhere (Wood 1949, Branson and Baker 1974, Ernst 2003b). In Pennsylvania, Queensnakes exhibit two main peaks in daily activity, one from 9:00 to 11:00 a.m., and another less pronounced peak from 2:00 to 6:00 p.m. (Ernst 2003b).

Reports of mean active body temperatures are similar across geographic locations. Examples include 25.6°C (78.1°F, $n = 205$) for active snakes in Kentucky (Branson and Baker 1974), 27.2°C (80.9°F, $n = 38$) for basking snakes in Ohio (Butler 1978), and 26.3°C (79.3°F, $n = 59$) for basking snakes in Pennsylvania (Ernst 2003b). Similarly, Butler (1978) estimated a mean preferred body temperature of

23.6°C (74.5°F) for snakes that he provided a gradient of temperatures to choose from in the laboratory ($n = 18$). The mean critical thermal maximum (CTmax) of this species in the laboratory ranged from 36.4°C to 37.9°C (97.5°F to 100.2°F, $n = 7$), and the mean critical thermal minimum ranged from 4.2°C to 9.7°C (39.6°F to 49.5°F, $n = 6$), depending on acclimation temperature (Butler 1978). There also appeared to be age-specific variation in thermal tolerances detected by Branson and Baker (1974), as adults had a higher mean CTmax (43.9°C [111°F]) than juveniles (40.3°C [104.5°F]).

Data on observed basking behavior suggest that substantial geographic variation occurs. For example, most of the 38 wild snakes encountered by Butler (1978) in Ohio were openly basking, and Cliburn (1973) found many individuals basking along a stream in Mississippi, but Branson and Baker (1974) found that most individuals were under rocks during surveys in Kentucky. Ernst (2003b) reported that most of the Queensnakes he observed were foraging (56.8% of observations), while far fewer were basking (21.1%), and that mean duration of the basking period was 2.3 hours for females and 1.5 hours for males. Gravid females are less active than males and nongravid females, spending more time basking than foraging (Ernst 2003b). Several studies have found additional relationships between Queensnake body temperature and behavior. For example, both crawling and swimming speeds increase with higher body temperature (Finkler and Claussen 1999).

Few estimates of Queensnake home range exist, although Ernst (2003b) suggested that home ranges were small and largely linear in shape, conforming to stream shorelines. Movement data are mostly based on recorded distances between recaptures of marked individuals. For example, although recaptures were sparse, the mean distanced between recaptures recorded in a Kentucky population was 22.6 m (74.2 ft; Branson and Baker 1974). In this same study, the greatest distance between recaptures was 137.2 m (450 ft), but most locations were separated by <30.5 m (100 ft; Branson and Baker 1974). In Pennsylvania, the distance between recaptures was never greater than 101 m (331.4 ft), and mean distance between recaptures was only 4 m (13.1 ft) for gravid females and 12 m (39.4 ft) for juveniles (Ernst 2003b). Furthermore, most of the snakes first captured at a hibernaculum in the same research project remained within 25 m (80 ft) of the original capture location for the duration of the study (Ernst 2003b). Branson and Baker (1974) attempted translocations of Queensnakes to investigate homing ability but recorded no recaptures at the release site or original capture site. They believed that snakes dispersed from the location of release, but homing ability could not be assessed. Newcomber et al. (1974) suggested that Queensnakes used celestial cues (i.e., the location of the sun) for spatial orientation, but this has not been confirmed by others.

The factors that initiate egress from or ingress to overwintering locations in Wisconsin are unknown, but the latter could be related to the cooler environmental conditions in autumn. Several authors have reported autumn aggregations of Queensnakes prior to hibernation (Wood 1944, Wood and Duellman 1950, Ernst 2003b), a behavior that might also occur in Wisconsin, but to our knowledge no published records exist. Aggregations can include both conspecifics and other species, such as Common Watersnakes (Branson and Baker 1974). The mean body temperature of hibernating Queensnakes in Pennsylvania was 2.8°C (37°F, $n = 10$), compared to a mean winter air temperature of 3°C (37.4°F) and mean surface temperature of 2.6°C (36.7°F; Ernst 2003b). Conant (1938) encountered an individual on the ice of a frozen creek in January that reportedly survived when brought into the lab and warmed. Interestingly, Ernst (2003b) observed juveniles with damaged tails and believed that >50% of these cases were due to frostbite during the winter.

PREY AND PREDATORS

Queensnakes, along with other members of the genus *Regina*, are reputed to be dietary specialists, with crayfish comprising the vast majority of consumed prey (Wood 1949, Branson and Baker 1974, Jackrel and Reinart 2011, reviewed by Gibbons and Dorcas 2004). Ernst (2003b) found that all individuals captured in southeastern Pennsylvania were either foraging for crayfish (94.6%) or ingesting crayfish (5.4%). Although crayfish are, in general, a dietary staple, Queensnakes do not seem to favor a particular species of crayfish as prey. As reviewed by Gibbons and Dorcas (2004), they consume Rusty Crayfish (*Orconectes rusticus*), Allegheny Crayfish (*O. obscurus*), Kentucky River Crayfish (*O. juvenilis*), and Devil Crayfish (*Cambarus diogenes*), although additional species found in Wisconsin are likely eaten. Godley et al. (1984) and Meyer et al. (2019) found evidence to support that Queensnakes, rather than preferring to consume certain crayfish species, favor freshly molted individuals, a preference implied by observations in Illinois (Weed 1922) and observations of regurgitated meals in Wisconsin (J. Kapfer, personal observation). Godley et al. (1984) even estimated that Queensnakes consumed most crayfish within six hours of molting. Laboratory experiments support that Queensnakes are highly attracted to chemicals in the water associated with molting crayfish, such as ecdysone (Jackrel and Reiner 2011). Queensnakes also apparently have much difficulty capturing and securing hard-shelled crayfish, which can injure snakes during predation attempts (Wood 1949, Branson and Baker 1974).

Although uncommonly reported, additional prey types are consumed. These include various other invertebrates, such as dragonfly naiads, fairy shrimps, and snails, as well as vertebrates, including fishes such as the Mud Minnow (*Umbra limi*) and several species of amphibians and their larvae (Raney and Roecker 1947, Adler and Tilley 1960, Conant 1960, Branson and Baker 1974, reviewed by Gibbons and Dorcas 2004). In addition, Wood (1949) reported consumption of carrion.

Based on existing reports, Queensnakes regularly forage underwater. Ernst (2003b) observed that they "crawled or swam along the bottom of the waterway and explored possible crayfish retreats, such as beneath or on the downstream side of rocks." Olfaction is important while searching for prey, and several authors observed that this species increases its rate of tongue flicking when foraging underwater (Burghardt 1968, Pinder 1996, Jackrel and Reinert 2011). Queensnakes seize appropriate prey by the tail and swallow it tail-first, consuming mostly small to medium-sized items (Wood 1949, reviewed by Ernst and Ernst 2003). Most feeding is done in the morning from 8:00 to 11:00 and to a lesser extent the late afternoon and early evening from 3:00 to 6:00 (Ernst 2003b). In the southern parts of their US range, Queensnakes may forage both day and night (Gibbons and Dorcas 2004).

Past studies have noted that several vertebrates will eat Queensnakes. These include large fishes, large Common Watersnakes, Bullfrogs (*Lithobates catesbeianus*), and Great Blue Herons (*Ardea herodias*; Wood 1949, reviewed by Ernst and Ernst 2003). Other wading birds, Raccoons (*Procyon lotor*), North American River Otters (*Lontra canadensis*), and raptors such as Red-shouldered Hawk (*Buteo lineatus*) also probably take Queensnakes (Vogt 1981, Ernst and Ernst 2003). The young are particularly vulnerable and, ironically, may sometimes be preyed upon by large crayfish. For example, Branson and Baker (1974) observed a crayfish that captured and secured a neonate Queensnake under the water, drowning the snake. While dormant in the hibernacula or after exposed to extremely cold weather, Queensnakes are also apparently vulnerable to attacks from crayfish and rodents (Wood 1949, Branson and Baker 1974).

Queensnakes are generally mild-natured (reviewed by Ernst and Ernst 2003). Several studies reported that larger adults may bite or strike repeatedly, although they are mostly docile (Wood 1949, Brust 2013). Queensnakes also commonly expel musk or feces when captured and may squirm or roll (reviewed by Ernst and Ernst 2003 and Gibbons and Dorcas 2004), possibly dislocating their tail in the process (see Natural History Box: Distraction). Queensnakes are frequently found basking along shorelines (see Habitat subsection) and can easily attempt to flee toward the safety of water when confronted or approached. Baker (1964) observed that "in practically every instance, specimens attempting to avoid capture crawled into the water and hid beneath rocks and debris rather than remaining on land." This tendency may be partially explained by the fact that Queensnakes move at a greater velocity in the water than on land (Finkler and Claussen 1999). Body temperature also plays an important role in whether Queensnakes attempt to escape or remain motionless when approached. For example, past studies supported that snakes with higher body temperatures were more likely to flee if threatened, while the use of flight behavior steadily increased with higher air temperatures (Layne and Ford 1984, Brust 2013).

NATURAL HISTORY BOX: Distraction

After being captured, a large adult Queensnake displayed a unique behavior to attempt to escape captivity. While in hand, the snake performed a rolling and twisting behavior. This activity exposed the ventral surface and the beautiful pale striped markings on this snake. The behavior was most evident during the initial stages of capture and handling and seemed to dissipate after the snake was held for a period of time. Perhaps the individual realized we did not mean it any harm. I have not observed any similar behavior in other snakes I have captured. This behavior has only been somewhat vaguely described in the literature: Ernst and Ernst (2003b) noted that the Queensnake will often "squirm violently" when captured, while Wood (1949) described collected individuals that would "actively writhe and twist" in Ohio. All other literature describes their defense strategy as fleeing to water or emitting a strong musky odor. What is the benefit to the snake of this type of activity? The captured Queensnake never made an attempt to bite, but, as described in multiple sources, it did emit a strong musk and odor. Because of the snakes' docile disposition, this adaptation may be similar to the elaborate distraction display of hog-nosed snakes (*Heterodon* spp.), during which they roll over on their back and play dead. In lieu of aggression, the Queensnake seems to prefer to roll or twist to escape capture. Is it a successful strategy? In this case it did not lead to escape, but in the wild one could hypothesize that this could be a method to elude the grasp of a fish or wading bird or a distraction technique to have a predator focus on the brighter colors and markings on the snake's ventral surface to facilitate escape. Another hypothesis is that this strategy could be utilized to enhance capture of crayfish, enabling them to work into and out of tight crevices where crayfish reside.

Richard A. Staffen, WDNR

CONSERVATION

Status: Globally, the Queensnake is listed as a Least Concern species by the IUCN. It is currently listed as Endangered in Wisconsin by the WDNR. The WDNR receives very few Queensnake reports, and new records are scarce in the state, which speaks to their rarity. For example, in the past five years only two Wisconsin populations have consistently produced reports, although occasional records exist from a small number of other sites in the state.

Populations: Limited population data exist for this species throughout its range, including Wisconsin, where it is so rare that sufficient sample sizes to estimate population size are difficult to obtain. For example, a two-year drift fence study in southern Wisconsin yielded fewer than five Queensnakes (even though the total number of adults across all snake species captured = 130; J. Kapfer, unpublished data). A review of technical reports from Ontario outlined the challenges associated with detecting this species during surveys (COSEWIC 2010b), while studies have also encountered limited recapture potential for marked individuals (Bekker 2007). An intense three-year mark-recapture study along a 200 m (656 ft) stretch of river in Ontario estimated an adult population of only 50 snakes (0.25 snake/m [0.08 snake/ft] of shoreline; note: these data are from a review of technical reports in COSEWIC [2010b], and individual reports could not be released by the authors due to the sensitive nature of these snakes in Ontario). Furthermore, a three-year mark-recapture study on 2,000 linear m (6,562 ft) of shoreline in northern Ohio revealed adult population estimates that ranged from 23 to 359 (density = 0.01–0.18 snake/m [0.003–0.6 snake/ft] of shoreline; Bekker 2007). Branson and Baker (1974) estimated populations of

35 (density = 0.018 snake/m [0.06 snake/ft] of shoreline) and 62 (density = 0.024 snake/m [0.08 snake/ft] of shoreline) at two locations on the same creek in Kentucky. A population in Pennsylvania was estimated at 216 individuals (SE = 19) at a location with 3 ha (7.4 ac) of suitable habitat, resulting in a density estimate of 68.3/ha (27.6/ac; Ernst 2003b). Ernst (2003b) further estimated an autumn density of 59 individuals / 100 m (59 individuals / 328 ft) of shoreline at a site where Queensnakes aggregated prior to hibernation. Reports of additional population characteristics are rare. Sex ratios are ca. 1M:1F in both adults and neonates or juveniles (Branson and Baker 1974, Ernst 2003b). A population in Pennsylvania was comprised of far more adults (68.3%) than juveniles (31.7%; Ernst 2003b).

Threats: According to Gibbons and Dorcas (2004), Queensnakes are of conservation concern at the peripheries of their continental distribution, which applies to Wisconsin. Habitat loss and degradation are likely key threats in the state, and many conservation concerns that generally apply to flowing aquatic habitats are likely relevant to Queensnakes (see Introduction to Conservation and Management of Wisconsin's Amphibians and Reptiles chapter). Unfortunately, their overall ecological specificity (i.e., specific habitat and food requirements) make Queensnakes particularly challenging to conserve in Wisconsin without active habitat management for the species.

Although Queensnakes are likely directly impacted by loss of habitat, the indirect impacts of habitat degradation could also be problematic. Given that Queensnakes reportedly forage by searching among the cobble in streambeds and on streambanks (see Prey and Predators section), excessive siltation of such benthic structure is a potential concern. Gibbons and Dorcas (2004) cited a personal communication with Gary Casper (UW–Milwaukee Field Station), who suggested that siltation buries cobble in streams that are adjacent to "development and agriculture," which makes it more difficult for Queensnakes to search for food beneath cobble. In conjunction with stream siltation, cattle access to streams negatively impacted the abundance of Queensnakes over a two-year study in Pennsylvania (Homyack and Giuliano 2002). However, Wood (1949) stated that "drastic changes in the character of the streams due to the erosion of soil, change in water run-off, and change in water temperature resulting from the absence of forest do not appear to have had an unfavorable effect on the [*Regina*] *septemvittata* population." Indeed, current Queensnake populations in Wisconsin are sometimes found near streams that possess silty or mucky bottoms (J. Kapfer, personal observation). Yet these could represent relict populations that have persisted in low quality conditions and will eventually be extirpated.

Indirect effects of impacts to prey species could also be an important concern. Godley et al. (1984) reviewed past studies and reported that freshly molted crayfish make up a small portion of the populations of several crayfish species within the range of the Queensnake (annual $\bar{x}$ = <15% of populations). They further reviewed that these crayfish species often engage in a "mass, synchronized molt," which places temporal limitations on food availability (i.e., molted crayfish are only periodically available). Furthermore, any environmental perturbations that influence crayfish populations could have a strong impact on Queensnakes in Wisconsin. Given the importance of prey chemical cues in hunting (e.g., ecdysone; Jackrel and Reinert 2011), environmental contaminants that interfere with the ability of arthropods to molt could unnaturally alter food availability for Queensnakes. Indirect impacts from invasive plants and animals have not been thoroughly assessed, but Bekker (2007) reported declines in the estimated population size at a site along the shore of Lake Erie in Ohio. Although confounding factors related to variability in captures by year and low recaptures impacted the study, the author suggested that this decline might be related to the introduction of the Round Goby (*Neogobius melanostomus*) into Lake Erie, which could have affected the crayfish populations that Queensnakes depended on.

Various additional conservation concerns also potentially threatened this species, but their impacts in Wisconsin remain a mystery. For example, although the effects of disease on Queensnakes have not been investigated in the state, Baker et al. (2018) detected *Ophidiomyces ophiodiicola* (the causative agent of snake fungal disease [SFD]) on one of five Queensnakes sampled in Indiana. Furthermore,

Lorch et al. (2016) reported *O. ophiodiicola* from Queensnakes in Kentucky. Direct human persecution could affect this and all snake species. An instance of wanton killing by humans was documented in Pennsylvania, where 107 basking individuals (>50% of the estimated population) were shot in a short time period by teenage boys (Ernst 2003b). The open basking behavior exhibited by this species could facilitate human persecution and, therefore, create conservation challenges. A severe storm and subsequent flood also displaced many individuals from the Pennsylvania population studied by Ernst (2003b). As such, Queensnakes may be susceptible to other forms of extreme weather that have not currently been documented or investigated but that could have substantial impacts on small populations.

Management: Management of vegetation along streams with documented Queensnake populations should consider this species' association with mostly open canopy but also shrubby shorelines (see Habitat subsection). Specifically, maintenance of shrubby vegetation that overhangs the water is probably beneficial, but management to limit succession and ensure that larger woody species do not shade riparian corridors should be considered. Control of erosion and runoff to limit the siltation that buries cobble in streambeds may be particularly important, given the specific habitat and diet needs of this species. Due to the snakes' use of rocks for foraging and as cover, habitat or fisheries managers could implement strategies to promote or maintain silt-free rocky substrates in riparian corridors that positively influence Queensnakes and game fish species such as trout. However, caution should be exerted and careful consideration occur before installation of riprap (such as is sometimes deposited along the water's edge in aquatic systems to control erosion), as this can restrict movement of species that must migrate between water and land (e.g., amphibians and turtles). It could instead be more prudent to focus on use of flat rocks for bank stabilization.

Protection of overwintering locations where snakes congregate could be particularly beneficial. Unfortunately, several suspected overwintering locations in Wisconsin are associated with anthropogenic structures (e.g., dams and road crossings; Casper et al. 2011, J. Kapfer and R. Staffen, personal observations). Any modifications or maintenance that alter conditions within or access to these structures could negatively impact Queensnakes that use them for overwintering. For example, Ernst (2003b) studied a population associated with a Pennsylvania dam in the 1960s. He reported that improvements to the dam and adjacent road altered both active and inactive season habitat available to Queensnakes. This, coupled with numerous additional impacts to the population over several years, resulted in the snake's eventual extirpation from the location. However, a thorough understanding of overwintering habits and selected hibernacula in the Upper Midwest is lacking, and further data on this subject would be very useful. Important examples include the relative use of anthropogenic vs. natural overwintering locations, fidelity to overwintering locations, and susceptibility to mortality during hibernation in Wisconsin.

The lack of data on Queensnake populations in Wisconsin could represent the largest challenge to enacting effective protection. As reviewed by Bekker (2007), data on demographic characteristics and growth rates are extremely important when developing effective conservation strategies. For example, the slow growth of this species in some locations results in snakes that are small and more likely of appropriate prey size for various predators over extended periods of time compared to larger snakes. Furthermore, delayed development leads to snakes that take longer to mature and perhaps mature at smaller sizes, which is problematic because smaller females also produce litters with fewer offspring (Branson and Baker 2007). Unfortunately, these variables and their impacts are not well-understood for Wisconsin (or Upper Midwest) populations, highlighting the need for continued research and monitoring of Queensnakes.

Dekay's Brownsnake

Storeria dekayi (Holbrook 1836)

Richard B. King

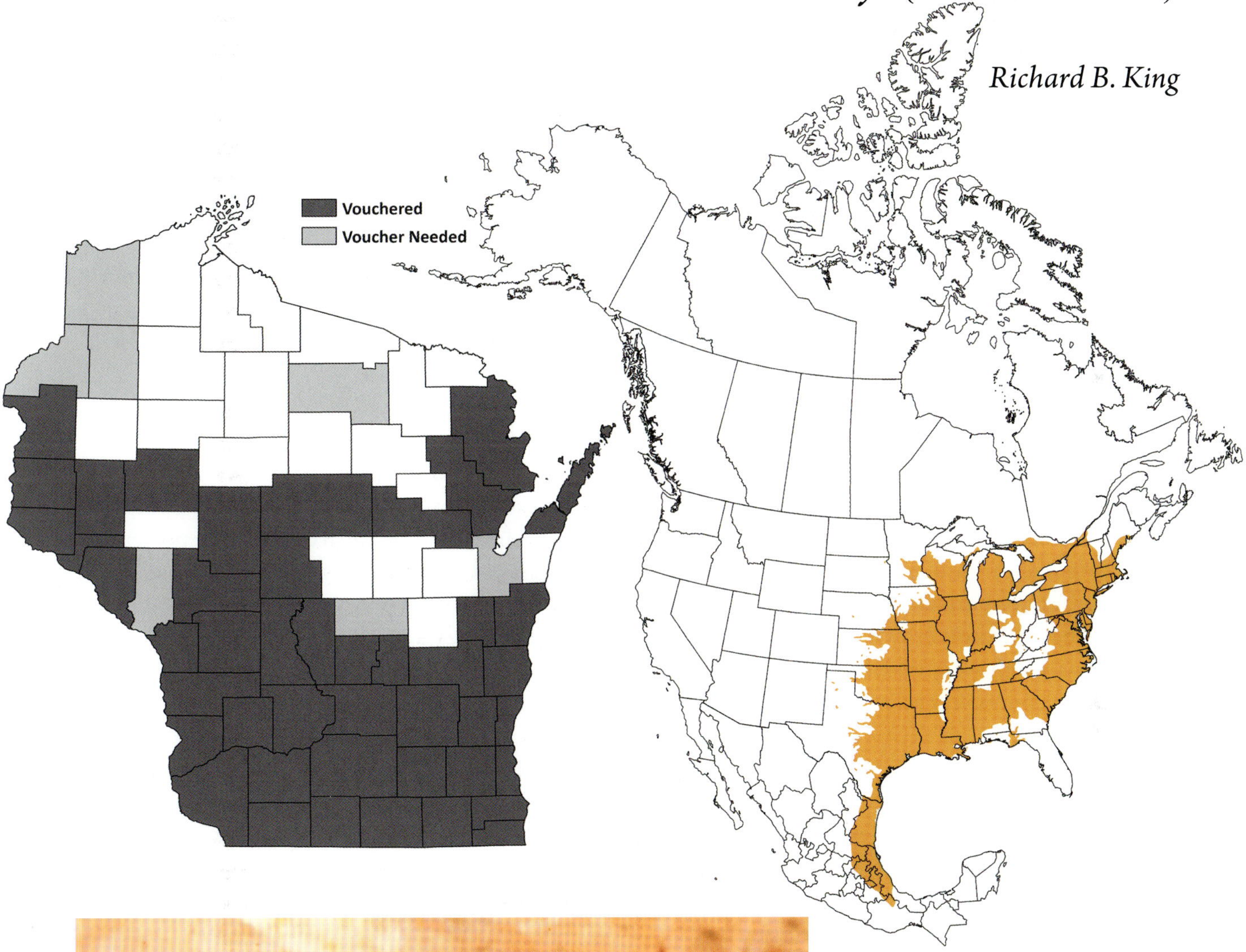

Dekay's Brownsnake (Walworth County; photo by J. DeBoer).

AT A GLANCE

"A more inoffensive snake is difficult to imagine"—this is how Roger Conant described Dekay's Brownsnakes in his seminal *Reptiles of Ohio* (Conant 1938). Dekay's Brownsnakes are among the most diminutive of Wisconsin's snakes, rarely exceeding 30 cm (12 in) from snout to vent. Dekay's Brownsnakes are gray, brown, rust, or pink in body color, with dark dorsal spots or crossbars. They are a widespread resident of grassland, prairie, wet meadow, savanna, open woodland, and old field habitats, including suburban parks and overgrown vacant lots. However, their small size and secretive behavior mean that Dekay's Brownsnakes are most frequently found under cover objects; only occasionally are active individuals encountered. Dekay's Brownsnakes emerge from hibernation in late April, and mating probably occurs shortly thereafter. They feed on slugs, snails, and earthworms and, in turn, may fall prey to a variety of small to medium-sized mammals, birds, and sometimes other snakes. Females produce large litters of live-born offspring in late summer. By mid-October, Dekay's Brownsnakes move toward hibernacula for the winter. Dekay's Brownsnake is considered a common species by the WDNR.

SYSTEMATICS

Phylogenetic analyses of New World natricine snakes consistently place *Storeria dekayi* (Dekay's Brownsnake) among the semifossorial group, which includes, among species found in Wisconsin, *S. occipitomaculata* (Red-bellied Snake) and, among species elsewhere, *Clonophis kirtlandii* (Kirtland's Snake), *Regina alleni* (Striped Crayfish Snake), *R. rigida* (Glossy Crayfish Snake), *Seminatrix pygaea* (Black Swampsnake), *Storeria storeriodes* (Mexican Brownsnake), *Virginia valeriae* (Smooth Earthsnake), and *V. striatula* (Rough Earthsnake; Alfaro and Arnold 2001, McVay and Carstens 2013, McVay et al. 2015). Eight subspecies have been recognized, two of which occur in Wisconsin: *Storeria d. texana* (Texas Brownsnake) and *Storeria d. wrightorum* (Midland Brownsnake) (Christman 1982). However, subspecies designations are based on subtle differences in color pattern and scalation. For example, the subspecies found in Wisconsin differ in the coloration of the anterior temporal scale (Trapido 1944). Furthermore, broad areas of overlap exist between these and other subspecies, and range-wide molecular-based phylogeographic analyses have not been conducted (Trapido 1944, Christman 1982, McVay et al. 2015). Consequently, subspecies designations are not considered further in this account. In addition to the standard common name, Dekay's Brownsnake (Crother 2017), colloquial names include 'Dekay's snake,' 'little brown snake,' 'ground snake,' 'brown snake,' and 'rock snake' (Surface 1906, Pope and Dickinson 1928, Vogt 1981). *Storeria* is in the subfamily Natricinae, which, among Wisconsin snakes, also includes species in the genera *Regina*, *Nerodia*, *Tropidoclonion*, and *Thamnophis*.

DESCRIPTION

Adults: Dekay's Brownsnakes may be gray, brown, rust, or pink in body color, with dark dorsal spots or crossbars that decrease in prominence posteriorly (Trapido 1944, reviewed by Ernst and Ernst 2003). Middorsal scale rows are usually lighter in color, giving the appearance of a light dorsal stripe. Dark lateral spots typically alternate with dorsal markings (Figure 255). Body coloration extends to the dorsal aspect of the head, but labial scales are lighter. The venter is typically unpatterned and lighter than the dorsal body color (Trapido 1944, Ernst and Ernst 2003).

The dorsal scales of Dekay's Brownsnakes are clearly keeled, and the anal plate is divided. Subventral scales vary from ca. 110 to 140, with females exceeding males, and subcaudal scales vary from ca. 40 to 70, with males exceeding females (Trapido 1944, Ernst and Ernst 2003). Supra- and infralabial scales typically number seven per side (Trapido 1944). Preocular and anterior temporal scales are usually single; postocular and posterior temporal scales are usually paired (Trapido 1944).

At study sites in northern Illinois (DeKalb, Lee, and Ogle Counties), adult SVL ranged from 17.5 to 30.9 cm in males (6.9 to 12.2 in, $\bar{x} = 22.9$ cm [9.0 in], $n = 215$) and 22.5 to 37.8 cm in females (8.9 to 14.9 in, $\bar{x} = 28.3$ cm [11.1 in], $n = 395$; Richard King, Northern Illinois University, unpublished data; Figure 256). Adult weight ranges from 2.2 to 12.4 g in males (0.1 to 0.4 oz, $\bar{x} = 6.2$ g [0.2 oz], $n = 212$) and

Figure 255. The dorsal (*top*) and ventral (*bottom*) surfaces of an adult female Dekay's Brownsnake (Rock County; photos by J. Kapfer).

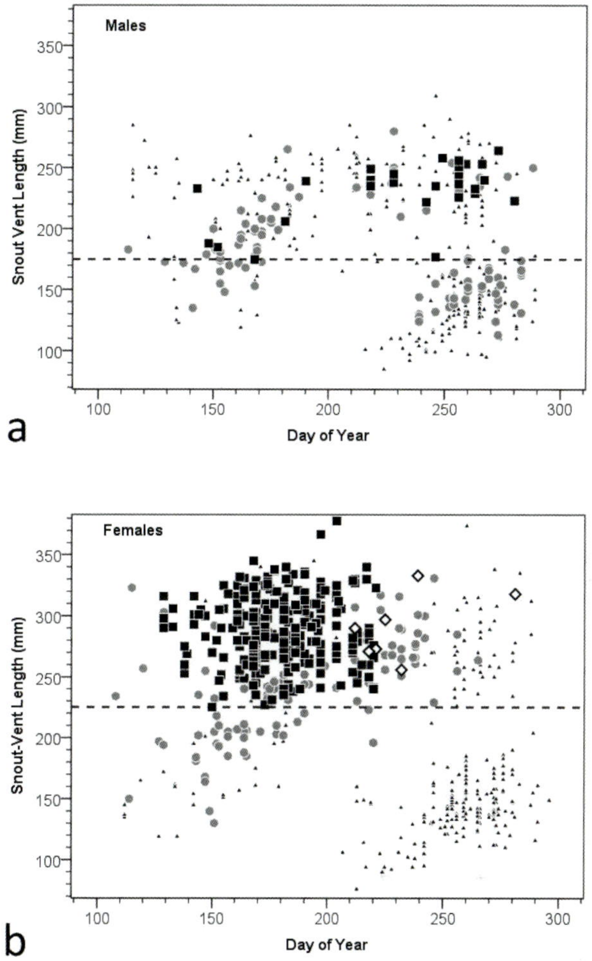

Figure 256. Plots of SVL vs. day of year for male (a) and female (b) Dekay's Brownsnakes captured at study sites in Northern Illinois (Dekalb, Lee, and Ogle Counties). For males, black squares and gray circles denote animals for which sperm were or were not detected in cloacal smears; small triangles denote animals from which cloacal smears were not collected. The dashed line represents size at maturity (17.5 cm SVL [6.9 in]). For females, black squares and gray circles denote animals that were gravid or not gravid based on palpation, diamonds denote animals that were postpartum, and small triangles denote animals that were not palpated. The dashed line represents size at maturity (22.5 cm SVL [8.9 in]).

4.3 to 32.4 g in females (0.2 to 1.1 oz, $\bar{x}$ = 13.8 g [0.5 oz], n = 391; R. King, unpublished data). Tail length is greater in males than females, equaling ca. 32% of SVL in males and 24% of SVL in females (King 1997b).

Body size varies geographically. Adult females range in size from 24.1 to 29.5 cm SVL in Wisconsin (9.5 to 11.6 in; Kapfer 2010), 22.5 to 37.0 cm SVL in Illinois (8.9 to 14.6 in; R. King, unpublished data), 21.4 to 31.9 cm SVL in Indiana (8.4 to 12.6 in; Minton 1972), 23.4 to 31.0 cm SVL in Ohio and Ontario islands in Lake Erie (9.2 to 12.2 in; King 1993b), 25.0 to 34.0 cm SVL in Ohio (9.8 to 13.4 in; Meshaka et al. 2008), 22.2 to 30.8 cm SVL in Pennsylvania (8.7 to 12.1 in; Hulse et al. 2001, Gray 2014a), 19.2 to 31.2 cm SVL in Connecticut (7.6 to 12.3 in; Klemens 1993), 17.5 to 29.6 cm SVL in Virginia (6.9 to 11.7 in; Mitchell 1994), 18.3 to 35.8 cm SVL in North Carolina (7.2 to 14.1 in; Palmer and Braswell 1995), 21.8 to 33.0 cm SVL in Florida (8.6 to 13.0 in; Meshaka 1994), 17.0 to 32.0 cm SVL in Louisiana (6.7 to 12.6 in; Kofron 1979), and 18.5 to 28.5 cm in Texas (7.3 to 11.2 in; Ford et al. 1990).

Preadult stages: Dekay's Brownsnakes are viviparous, so they do not lay eggs. Mean SVL among offspring from litters born to 25 females from Ohio and Ontario islands was 7.7 cm (3 in, range of litter means = 6.5–8.6 cm [2.6–3.4 in]), with a mean weight of 0.35 g (0.012 oz, range of litter means = 0.26–0.46 g [0.009–0.016 oz]; King 1993b). Newborns tend to be darker (brownish, grayish) and less strongly patterned than adults, and they possess a cream-colored nuchal band that fades with age (Trapido 1944, Jackson and Pounds 1980, Rowell 2012).

Variation: No significant geographic variation among Wisconsin individuals has been described. There appears to be no sexual dimorphism in color or pattern. Females are larger on average than males, but males have relatively longer tails and larger heads (King 1997b, 2002b, King et al. 1999).

Figure 257. The head of the Dekay's Brownsnake has dark spots below the eyes that extend onto adjacent labial scales (Columbia County; photo by J. Kapfer).

Diagnosis: Dekay's Brownsnakes are characterized by keeled dorsal scales in 17 (range = 15–17) rows at midbody, a divided anal plate, and divided subcaudal scales. They lack an upturned rostral scale and lack loreal scales and possess 2 postocular scales, 1 (rarely 2) preocular scale, 7 supralabial scales on each side of the head, and 112–149 ventral scales.

Dekay's Brownsnakes could be confused with other small-bodied snakes in Wisconsin, including Red-bellied Snakes, Ring-necked Snakes (*Diadophis punctatus*), Lined Snakes (*Tropidoclonion lineatum*), Western Wormsnakes (*Carphophis vermis*), juvenile gartersnakes and ribbonsnakes (*Thamnophis butleri, T. proximus, T. radix, T. saurita*, and *T. sirtalis*), and the buff morph of Smooth Greensnakes (*Opheodrys vernalis*). Dekay's Brownsnake can be distinguished from its congener, the Red-bellied Snake, by the presence of one to two dark spots extending from the bottom of the former's eyes to the adjacent labial scales (Figure 257) and the latter's orange-red belly. Scale counts also differ between these two species, with Dekay's Brownsnakes typically possessing 7 supralabial scales (6 in Red-bellied Snakes) and 1 preocular scale on each side of the head (2 in Red-bellied Snakes), while scale rows at midbody are usually 17 (15 in Red-bellied Snakes; Trapido 1944, reviewed by Ernst and Ernst 2003). The distinctive neck ring distinguishes the Ring-necked Snake from adult Dekay's Brownsnake, but the light-colored nuchal band possessed by neonatal Dekay's Brownsnakes may cause confusion. However, Ring-necked Snakes possess smooth (vs. keeled) scales, are uniform in dorsal body color, and have yellow, orange, or red venters. Although Dekay's Brownsnakes sometimes appear to have a dorsal stripe, they lack lateral stripes and thus can be readily distinguished from Lined Snakes, gartersnakes, and ribbonsnakes. Dekay's Brownsnakes can be distinguished from the Western Wormsnake by the latter's narrow head, small eyes, pointed tail tip, smooth scales, and uniform body color. Similarly, it can be distinguished from the Smooth Greensnake (buff morph) by the latter's smooth scales and uniform dorsal body color.

DISTRIBUTION AND HABITAT

Global distribution: Dekay's Brownsnakes are broadly distributed across eastern North America in an area bounded by the US and Mexico gulf states from Florida to Veracruz, the Atlantic coast states from Florida to Maine, west through southern Ontario to Minnesota, and south to Texas, with disjunct populations in Kansas, Texas, and Central America (spanning parts of Mexico, Guatemala, and Honduras; Christman 1982, Ernst and Ernst 2003).

Distribution in Wisconsin: Published Dekay's Brownsnake records exist across much of the southern two-thirds of the state (see the Appendix), a number of more northerly counties (e.g., Polk, Marinette, Door), and islands in the Great Lakes (Hecnar et al. 2002). Of the reported counties of occurrence, vouchers are currently lacking for Douglas, Burnett, Oneida, Trempealeau, and Waushara. In addition, the voucher listed for Washburn County (UWZM 24065, collected in 1971; Casper 2015) possesses characteristics that are consistent with Red-bellied Snakes (2 preocular and 6 supralabial scales on each side of the head, 15 scale rows at midbody; J. Kapfer, personal communication). Furthermore, the Brown County voucher specimen (BBWS RE08; Cochran et al. 1987) was part of a collection at the Bay Beach Wildlife Sanctuary (Green Bay) that no longer exists, according to recent communications with sanctuary staff (J. Kapfer, personal communication).

Localities of possible occurrence: Dekay's Brownsnakes likely occur in a number of additional counties across central Wisconsin that are surrounded by or adjacent to counties of known occurrence (e.g., Eau Claire, Portage, Waupaca, Outagamie, Winnebago, Kewaunee). Casper (1996) suggested that their distribution potentially includes all but the very northern tier of counties in the state (Bayfield, Ashland, Iron, Vilas, Forest, Florence). Further work will be required to better define its range in northern Wisconsin, given the current lack of suitable museum vouchers for several existing northern county records, such as Douglas, Burnett, Washburn, and Oneida (see Distribution in Wisconsin subsection). Casper (2001) further reported a verified (but apparently unpublished) record from Rusk County for which a voucher is unknown.

Habitat: Dekay's Brownsnakes are a common resident of grasslands, mesic and wet prairies, savanna, open forests, and anthropogenically altered habitats, including suburban parks and overgrown vacant lots (Klemens 1993, Ernst and Ernst 2003, Trauth et al. 2004, Gray and Lethaby 2008b, Rowell 2012, Moriarty and Hall 2014; see Amphibians and Reptiles in Wisconsin Plant Communities chapter). Vogt (1981) stated that in Wisconsin they were often "found only in fairly moist situations, although they are in no way aquatic." Quantitative studies of habitat use in Ontario and Texas indicate a preference for areas with higher moisture and dense grass and forb cover, including moist ditches, compared to drier upland areas with less vegetation (Catling and Freedman 1980b, Ford et al. 1991). In Kansas, coverboards placed in grass, edge, and woods habitats were used by the snakes in proportion to coverboard placement, but there was a seasonal pattern of movement between woods and edge habitats (used in spring and autumn) and open grassy habitat (used in summer; Pisani 2009b). In a study restricted to patches of old field in New York, Dekay's Brownsnake presence under coverboards increased with increasing distance to the forest edge but was unaffected by distance to roads, soil moisture, vegetation height, litter depth, or their interactions (Patrick and Gibbs 2009). In a study using drift fences and funnel traps in tallgrass prairie habitat in Illinois, Dekay's Brownsnake abundance was positively associated with distance to grassland edge, and occurrence was positively associated with distance to grassland edge and road density within 1 km (0.6 mi; Cagle 2008).

Due to their small size and secretive behavior, Dekay's Brownsnakes are most frequently found under ground cover and only occasionally encountered in the open (mostly in spring and autumn). Consequently, studies of the species typically focus on the use of natural cover objects (Catling and Freedman 1980b, Hecnar and Hecnar 2011), human-deposited debris (wood, carpet, tin, cardboard; Ernst 2003a, Gray 2013, 2014a), or cover objects placed for the purpose of increasing snake detection and capture probability (Fitch 1999, Pisani 2009b, Patrick and Gibbs 2009). This work suggests that Dekay's Brownsnakes select ground cover with specific characteristics. For example, in Ontario, this species selected larger logs and larger boards that were at moderate to advanced stages of decay compared to available woody debris (Hecnar and Hecnar 2011). Individual snakes also show high fidelity for ground-cover items at specific locations for gestation (Richard King, Northern Illinois University, personal observations) and ecdysis (Gray 2008, 2014, Pisani 2009b). When beneath ground debris, Dekay's Brownsnakes are frequently encountered in aggregations of two or more individuals. This phenomenon was well documented by Noble and Clausen (1936), who demonstrated that aggregation reduced water loss and was mediated through visual and olfactory cues. Aggregations occur more frequently in Dekay's Brownsnakes than in sympatric species and often consist of similarly sized individuals, especially gravid females (Gregory 2004). Dekay's Brownsnakes are also sometimes found in multispecies aggregations that include Common Gartersnakes (*Thamnophis sirtalis*), Common Watersnakes (*Nerodia sipedon*), and Red-bellied Snakes (Burghardt 1983, Gregory 2004, R. King, personal observation).

Mammal burrows and ant mounds, including active ant mounds, are sometimes used as hibernation sites (Noble and Clausen 1936, Carpenter 1953, Pisani 2009a). Hibernation also occurs under logs, in rock crevices, and in human-made structures (Rowell 2012). Multiple individuals, including other snake species, sometimes hibernate together (Lachner 1942, Carpenter 1953). Mortality during hibernation was described by Bailey (1948).

REPRODUCTION AND DEVELOPMENT

Dekay's Brownsnakes have figured prominently in studies of courtship and mating of New World natricine snakes due to the work of G. K. Noble (1937). As it does in many North American natricines, mating occurs annually in spring, and parturition occurs in summer after a ca. 75–90-day gestation period (Kofron 1979, Ernst 2003a). Vogt (1981) presumed that mating in Wisconsin occurred shortly after spring emergence in late April, but specific observations in the state do not appear to exist. Wild-caught Dekay's Brownsnakes from Pennsylvania that were maintained in captivity mated between 20 April and 15 May, and parturition occurred between 20 August and 15 September (Ernst 2003a). Field observations of mating, while rare, mostly conform to these dates (Ernst 2003a), although Clausen (1936) reported mating activity in February and March, while other authors have reported autumn mating (Johnson 1987, Collins 1993), which could contribute to the occurrence of multiple paternity within litters (Wusterbarth et al. 2010).

Males have a single hemipenis and use olfactory cues to find females for reproduction (Ernst and Ernst 2003). Copulatory behavior was described by Noble (1937). In southern Wisconsin, all adult females encountered ($n = 21$) across several sites during late spring and early summer were gravid (Kapfer 2010). At study sites in northern Illinois (DeKalb, Lee, and Ogle Counties), 167 of 198 adult females (86%) were gravid (R. King, personal observation); in Kansas, 49 of 52 adult females (94%) were gravid (Fitch 2003b); and in Louisiana, 21 of 22 adult females (95%) were gravid (Kofron 1979). However, Gray (2014a) reported that just 10 of 19 adult females (53%) were gravid at a site in Pennsylvania.

Dates of mating and parturition likely vary geographically and among years, depending on climate and weather. For example, 25 females from Ohio and Ontario islands in Lake Erie gave birth from 6 to 23 August (King 1993a), while 10 females from a more southern location in Kansas gave birth from 26 July to 13 August (Fitch 1999). Females from Louisiana gave birth as early as 11 June (Kofron 1979), and 10 females from Big Pine Key, Florida, gave birth from June to September (Dalrymple et al. 1991). Dekay's Brownsnakes are viviparous and can produce large numbers of offspring. Litter sizes ranged from 4 to 24 in Kansas (Fitch 2003b), 5 to 28 on Ohio and Ontario islands in Lake Erie (King 1993a), 6 to 19 in Pennsylvania (Ernst 2003a), 9 to 20 in New York (Clausen 1936), 3 to 26 in Virginia (Mitchell 1993), 4 to 25 in North Carolina (Palmer and Braswell 1995), 5 to 17 in Florida (Iverson 1978, Dalrymple et al. 1991, Meshaka 1994), 5 to 24 in Louisiana (Kofron 1979), and 4 to 15 in Texas (Ford et al. 1990). An unusually large litter, numbering 41 offspring, was born to a large female in Illinois (39 cm SVL [15.4 in]; Morris 1974). Offspring number increases with increasing female size (Kofran 1979, King 1993a, Ernst 2003a), but a trade-off exists between offspring number and offspring size, such that females producing more offspring, given their size, have young that are smaller (Natural History Box: More Small or Fewer Big Offspring? Trade-offs between Offspring Size and Number).

Neonate Dekay's Brownsnakes are tiny (see Description section). Because they are born over a short time period in late summer, neonatal Dekay's Brownsnakes can be distinguished from other age classes by

NATURAL HISTORY BOX: More Small or Fewer Big Offspring? Trade-offs between Offspring Size and Number

Reptiles such as Dekay's Brownsnakes are 'capital breeders,' meaning that energy for reproduction comes from stored energy (capital) accrued prior to ovulation (Bonnet et al. 1998). In contrast, 'income breeders' (e.g., birds and mammals) fuel reproduction from current energy consumption (income; Bonnet et al. 1998). As a consequence, capital breeders have a limited amount of energy for reproduction and are expected to exhibit a trade-off between offspring size and number: as offspring number increases, offspring size should decrease. Detecting such a trade-off in snakes is complicated by the fact that females are highly variable in size, and female size also influences offspring number. Consequently, characterizing the trade-off between offspring size and number requires statistical techniques that control for female size (e.g., path analysis, part or partial correlation). In Dekay's Brownsnakes, this trade-off is unusually strong, with offspring size decreasing by 0.84 standard deviations for each standard deviation increase in offspring number (King 1993a; Figure 258).

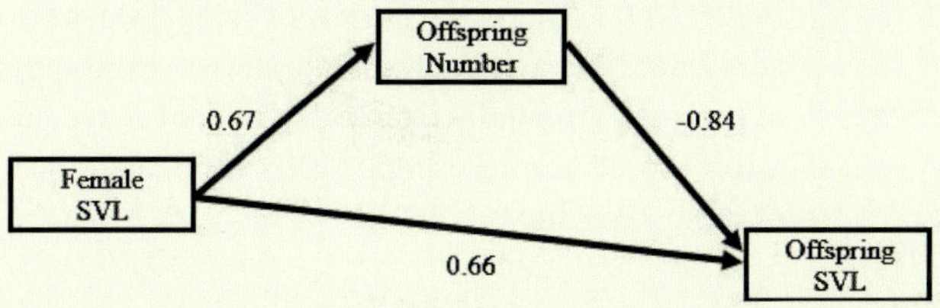

Figure 258. Path diagram showing direct effects among female SVL, offspring number, and offspring size in Dekay's Brownsnake (standard deviation units, adapted from King 1993a). A trade-off between offspring number and offspring size is reflected in the negative path coefficient connecting these two variables. Because the effects of female SVL and offspring number on offspring SVL are of opposite sign, the overall effect (considering both direct and indirect effects) of female SVL on offspring number is close to zero (King 1993a).

Richard B. King, Northern Illinois University

plotting size (SVL) vs. day of capture. Sampling this age class regularly from birth until the onset of hibernation in northern Illinois (DeKalb County) demonstrated that growth of neonates is rapid, ca. 0.077 cm/day (0.03 in/day; Figure 259). However, the size neonates achieved at this site by the onset of hibernation varied annually, ranging from an average of 12.5 to 18 cm (4.9 to 7.1 in) among years (O'Brien 2014). This year-to-year variation was positively correlated with air temperature in April and May, presumably because ovulation and parturition occur earlier in years with higher spring temperatures, leaving young snakes with a longer period for growth (O'Brien 2014).

Among females, size at sexual maturity can be determined by palpating for enlarged follicles or developing young (Figure 256). Among males, size at sexual maturity can be determined by microscopic examination of cloacal smears (Figures 256 and 260). In northern Illinois, female sexual maturity occurred at ca. 22.5 cm SVL (8.9 in), but as noted above, female size at sexual maturity varies geographically. The sexual maturity of males from Illinois occurred at 17.5 cm SVL (6.9 in; Figure 256). In Kansas, motile sperm was obtained from 9 of 17 males ranging in size from 15.9 to 18.4 cm (6.3 to 7.2 in) and captured between 13 April and 5 May (Fitch 1999). Based on

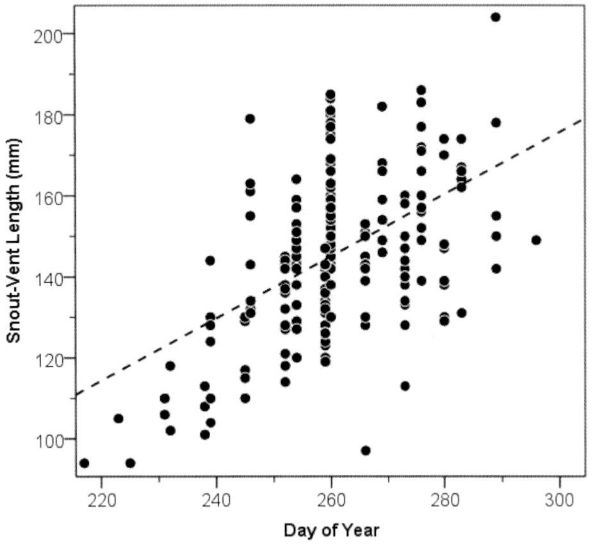

Figure 259. Plot of SVL vs. day of year for neonatal Dekay's Brownsnakes showing rapid growth following birth. Data are from northern Illinois study sites (Dekalb, Lee, and Ogle Counties) and were collected in 2014. Based on the regression of SVL on day of year, growth averaged 0.077 cm/day (0.03 in/day).

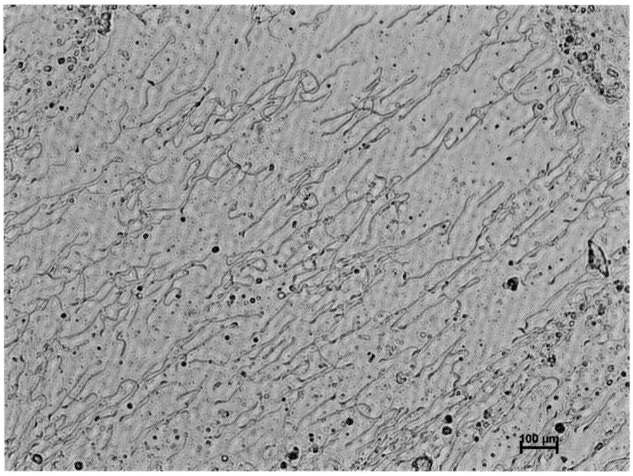

Figure 260. Microscopic view of cloacal smear from an adult male Dekay's Brownsnake showing the appearance of sperm in a sexually mature individual as seen using a 10× objective microscope (100× total magnification; photo by R. King).

neonatal growth trajectories (Figure 259), males may reach sexual maturity during the summer following their first hibernation, while females likely reach sexual maturity by spring following their second hibernation (Fitch 1999).

Information on longevity is scant. Ernst (2003a) reported that animals marked as juveniles were recaptured 1–7 years later (mean time between first and last capture = 4.2 yrs). Snider and Bowler (1992) reported a wild-caught individual, captured as an adult, that lived to slightly over seven years in captivity (Jeff Beane, North Carolina Museum of Natural Sciences, personal communication).

ACTIVITY

In southern Wisconsin, spring emergence likely occurs in late April (average date of first sighting = 26 April in Dane County; Cochran 2013b). Farther south, spring emergence happens earlier. In Kansas and Florida, emergence typically occurs in mid-March (Fitch 1999, Dalrymple et al. 1991, respectively). Furthermore, records of very early emergence (e.g., January and February) also exist from Kansas and Pennsylvania (Pisani and Pittman 2009, Gray and Lethaby 2008b).

Incidental observation or quantification of road mortality suggests that peaks in activity occur during spring and autumn, associated with moves away from or toward hibernation sites (Ernst 2003a, Gaul 2008, Cochran 2013b, Gray 2014a). By following a standard driving route almost daily, Cochran (2013b) observed snakes on the road only before 1 June and after 1 September in Wisconsin. Drift fence surveys in South Carolina indicated that activity not associated with migration to and from hibernacula peaks in July and decreases over successive months to a low in January (Gibbons and Semlitch 1987). Data on body temperature of active individuals are extremely limited, but Brattstrom (1965) reported that the body temperature for a single snake encountered in the field was 27°C (80.6°F). Fitch (1999) measured body temperature for three active individuals in Kansas, which were 20.4°C, 23.8°C, and 24.5°C (68.7°F, 74.8°F, and 76.1°F).

Given these snakes' small size, typical movements by Dekay's Brownsnakes during the active season tend to be quite restricted. For example, recaptures of marked animals in Pennsylvania ($n = 41$ males and 63 females) mostly occurred within 40 m (131 ft) of the prior capture site (Ernst 2003a). A similar pattern was observed by Gray (2014a), also in Pennsylvania, who reported that the distances between recaptures of 37 males and 27 females were usually <50 m (164 ft), and all were <100 m (328 ft). In Kansas, mean recorded movement was ca. 50 m (164 ft), and all movements were <125 m (410 ft, $n = 19$; Pisani 2009b), while in Ontario 8 of 10 moves recorded were <100 m (328 ft; Freedman and Catling 1979). Although they are found on islands in the Great Lakes (see Distribution in Wisconsin subsection), overwater movements by Dekay's Brownsnakes are likely uncommon compared to other larger-bodied North American natricines, as indicated by molecular marker-based estimates of gene flow (King and Lawson 2001).

Dekay's Brownsnakes are cold tolerant. They can be active at temperatures close to freezing (Fitch 1999, Rowell 2012, Gray 2014b) and are among the first snakes to emerge from hibernation, but they are likely among the last to enter hibernation (Pisani and Pittman 2009, Pisani 2009b). By early to mid-October in Wisconsin, this species begins traveling back to hibernacula. For example, Vogt (1981) observed "hundreds" crossing roads while migrating to overwintering locations in mid-October, while Cochran (2013b) reported that the mean date of last sighting for this species in Dane County was 7 October. Long-distance movements to hibernacula are occasionally recorded. Moves of 555, 605, and 670 m, respectively (1,821, 1,985, and 2,198 ft, respectively) were listed by Ernst (2003a), and moves of 1,200 m (3,937 ft) between summer ranges and hibernacula were reported by Noble and Clausen (1936). Fidelity for hibernacula was evident from recaptures of animals returning to the same hibernation site across years (Ernst 2003a).

PREY AND PREDATORS

Quantitative analyses of diet composition demonstrate a preponderance of slugs and earthworms among the prey of Dekay's Brownsnakes (reports based on fewer than ten individual snakes include Judd 1954, Hamilton and Pollock 1956, Catling and Freedman 1980a, Gray 2013, Thomas 2014). In the most extensive investigation of Dekay's Brownsnake diet to date, fecal samples from 84 individuals were analyzed (Gray 2013). This study found that 47 samples consisted primarily of slugs, 6 samples consisted primarily of earthworms, and 7 samples contained both prey types (Gray 2013). Furthermore, by using the slug shell length as a proxy for slug size, Gray (2013) cleverly demonstrated that prey size increases with SVL for female but not male Dekay's Brownsnakes.

Snails are also consumed (Surface 1906), and Dekay's Brownsnakes, as well as Red-bellied Snakes, display morphological and behavioral adaptations that allow them to extract snails from their shells (Rossman and Myer 1990). Specifically, the relatively shallower maxillae and longer teeth of *Storeria* compared to other New World natricines allow snakes to grasp snail soft body parts and apply torsion until the columnar muscle of snail prey becomes fatigued and the ligament securing snails to their shell is broken (Rossman and Myer 1990). Despite these adaptations, available data suggest that snails make up a small proportion of the diet of Dekay's Brownsnakes. At study sites in Northern Illinois (DeKalb, Lee, and Ogle Counties), 115 prey were recovered from 49 Dekay's Brownsnakes and included 99 slugs (*Deroceras* spp.), 13 snails (unidentified), and 3 earthworms (likely *Lumbricus* or *Apporectodea* spp.; Emily Virgin, Northern Illinois University, and R. King, unpublished data; Figure 261). Possibly, snails extracted from their shells have been misidentified as slugs in other studies (Hulse et al. 2001). However, when recently consumed prey were immediately preserved in ethanol, snails could be distinguished from slugs by the characteristic spiral body shape retained by the former (Figure 261). A recent investigation of diet in northern Illinois recovered multiple prey items from individual Dekay's Brownsnakes on 19 occasions (E. Virgin and R. King, unpublished data). Usually, these consisted of a single prey type, but on two occasions, both slugs and snails were recovered from the same snake. In one case, 15 slugs were recovered from a single snake, a female measuring 23.2 cm SVL (9.1 in). Head rubbing and mouth gaping frequently follow consumption of earthworms by Dekay's Brownsnakes (Cunningham and Burghardt 1999).

Responses of Dekay's Brownsnakes to prey chemical extracts are consistent with prey consumption data summarized previously. Ingestively naive newborn

Figure 261. Snail (*top*) and slug (*bottom*) recovered from Dekay's Brownsnakes captured in northern Illinois. The snail's foot is to the right; tissues contained within the shell apex are to the lower left and spiral out of the plane of the photograph (photos by R. King).

Dekay's Brownsnakes direct elevated tongue flicks and attack scores toward slug and earthworm extracts when compared to distilled water; in contrast, responses to fish extracts were no different from distilled water (Burghardt 1967). Positive responses to earthworms but not aquatic snails or mealworms were reported by von Achen and Rakestraw (1984). However, terrestrial snail, slug, and earthworm extracts all elicited positive responses by wild-caught neonates from northern Illinois (E. Virgin and R. King, unpublished data). Few studies of diet or responses to chemical extract have identified prey more specifically than statements of 'earthworm,' 'slug,' or 'snail.' Where further identification is provided (e.g., Gray 2013), species consumed are mostly nonnative (e.g., *Derocoras* slugs, *Lumbricus* earthworms), raising questions of how Dekay's Brownsnake diet, distribution, abundance, and habitat utilization may have changed since the arrival of these species.

Dekay's Brownsnakes reportedly also consume a number of other prey taxa, including insects, mites, isopods, frogs, amphibian eggs, and fishes (e.g., Conant 1938, Smith 1961, Harding 1997, Tekiela 2004, Trauth et al. 2004, Moriarty and Hall 2014), but the importance of these prey and the veracity of these records are unclear (Rossman and Myer 1990). For example, Surface (1906) reported that prey in 4 of 21 specimens included earthworms (in 1), slugs (in 3), snails (in 2), and insect larvae (in 2) but presented these results as a pie chart showing insect larvae as constituting 17% of prey consumed, a number later reported by Conant (1938) without qualification. Consumption of toads and fish is based on observations of captive snakes (Rossman and Myer 1990). Insect parts and mites were recovered from two of ten animals analyzed by Judd (1954), but whether these represent prey or were consumed accidentally is unknown. Recovery of millipedes, mites, an insect, plant material, and inert objects (i.e., soil particles, glass) was interpreted by Gray (2013, 2015) as incidental ingestions that occurred while more typical prey (i.e., slugs, earthworms) were being consumed. In a largely overlooked study, Harwood (1945) repeatedly offered neonatal Dekay's Brownsnakes insects (ten species including aphids, cabbage worms, cutworms, beetle larvae, ant pupae, and crickets), millipedes, slugs, and earthworms. Arthropods were ignored, but slugs and earthworms were taken eagerly (Harwood 1945).

Dekay's Brownsnakes are likely consumed by a wide range of predators, including small to medium-sized mammals, from shrews (*Sorex* and *Blarina* spp.) to Raccoons (*Procyon lotor*), and birds, from small ground-feeding species to larger birds of prey (reviewed by Ernst and Ernst 2003, Rowell 2012). Other snakes also feed on Dekay's Brownsnakes, including Eastern Milksnakes (*Lampropeltis triangulum*; Gray 2014a), Massasaugas (*Sistrurus catenatus*; Johnson 1987), North American Racers (*Coluber constrictor*; Thomas 2014), and Common Gartersnakes (Ernst 2003a). Predator avoidance strategies apparently vary, but this species' small size and cryptic behavior likely play an important role in avoiding many potential threats. Among 82 free-ranging snakes found under ground cover, most (65) remained still when exposed, but a small number attempted to flee (16) or hid their head (1; Gray 2015). Upon contact, snakes often became defensive, flattening (42), adopting defensive postures (22), or exhibiting closed-mouth (5) or open-mouth (11) strikes (Gray 2015). Furthermore, snakes exhibiting defensive behaviors were significantly cooler (16.1°C [61.0°F], SD = 2.2°C [4.0°F]) than snakes that attempted to flee (22.3°C [72.1°F], SD = 1.3°C [2.3°F]; Gray 2015). Upon capture, most (76) discharged cloacal contents (Gray 2015). Temperature dependence of predator-avoidance behavior is also evident in laboratory studies. Among neonatal Dekay's Brownsnakes, the probability and duration of aquatic death-feigning increased with increasing water temperature (Gerald 2008). Furthermore, swimming velocity increased with increasing water temperature and increasing SVL, whereas death-feigning duration decreased with increasing SVL, suggesting that a snake's choice of predator-avoidance behavior (death-feigning vs. flight) is modulated by potential flight speed (Gerald 2008).

Ectoparasitic mites sometimes infest Dekay's Brownsnakes. In Pennsylvania, mites were noted on 41 of more than 700 snakes captured, primarily in spring (Ernst 2003a). In contrast, in Illinois, mite parasitism occurs in summer (typically August), coincident with the appearance of neonatal Dekay's Brownsnakes (R. King, personal observation). Endoparasitic nematodes have been reported by Judd (1954), Rau and Gordon (1980), Ernst (2003a), Pisani (2009b), and Gray (2013).

CONSERVATION

Status: Globally, the Dekay's Brownsnake is listed as a Least Concern species by the IUCN. It is considered a common species in Wisconsin by the WDNR.

Populations: Dekay's Brownsnakes can be locally abundant, with populations estimated from capture-recapture data numbering in the hundreds and density estimates equaling 13/ha (5.3/ac; Fitch 1993) and 32/ha (13/ac; Pisani 2009b) in Kansas, 66/ha (26.7/ac; Ernst 2003a), 244/ha (98.7/ac; Gray 2014a), and 300/ha (121.4/ac; Ernst 2003a) in Pennsylvania, and 14/ha in Ontario (5.7/ac; Freedman and Catling 1978). Based on limited observations, mortality rates can be high, particularly in winter (Bailey 1948), but the overall effect this has on populations is unknown. Information on several additional key population parameters, such as sex ratios and annual survival, is lacking. The significant gaps in population data prevent characterization of Dekay's Brownsnake population dynamics, despite detailed information on other key demographic parameters (e.g., age and size at maturity, reproductive frequency, size-specific fecundity).

Threats: Several general conservation concerns are potentially relevant to Dekay's Brownsnakes, including threats to wetlands or adjacent upland habitats (see Introduction to Conservation and Management of Wisconsin's Amphibians and Reptiles chapter). Given this species' propensity to select relatively open habitat (see Habitat subsection), successional change (e.g., from grassland to forest) could be an important threat if it alters habitat suitability. The influence of succession is evident in capture numbers reported by Fitch (2006b) from his long-term study of snake ecology at the Kansas University Fitch Reservation. There, succession from agricultural plots to old field and then to forest brought an initial increase (with the transition from agriculture to old field) and subsequent decrease (with the transition to forest) in Dekay's Brownsnake captures per decade from 23 to 164, 32, 23, 15, and 0 in the 1950s, 1960s, 1970s, 1980s, 1990s, and 2000s, respectively (Fitch 2006b).

This species' use of urban environments (e.g., Zappalorti and Mitchell 2008) could increase the likelihood for negative interactions with humans, but the impacts of such interactions have not been rigorously investigated. In some locations, motor vehicles are a significant cause of mortality (Gibson and Merkle 2004, Cochran 2013b), which may be particularly pronounced in spring and autumn when Dekay's Brownsnakes disperse from or return to hibernacula. Vogt (1981) observed many Wisconsin individuals crossing roads during autumn migration to overwintering sites, supporting this species' vulnerability during such activities. At Fox Ridge State Park

NATURAL HISTORY BOX: Grassland Snakes and Prairie Restoration

Ecological restoration efforts frequently focus on reestablishing native plant communities with the assumption that animal species will subsequently colonize them. Long-term restoration projects provide an opportunity to test this assumption. At Nachusa Grasslands, a large-scale tall-grass prairie restoration project undertaken by the Nature Conservancy in northern Illinois, restoration units vary in age from 1 to 26 years. Systematic sampling of 12 of these units using arrays of coverboards during 2013 and 2014 revealed a pattern of increasing capture number with time since restoration for both DeKay's Brownsnake and the Common Gartersnake, the two species encountered most frequently at the site (Figure 262). Over two years, 6,989 coverboard checks were conducted at these 12 locations, resulting in 231 DeKay's Brownsnakes and 205 Common Gartersnake captures. In both species, captures per coverboard check were positively correlated with years since restoration (Figure 262). While Common Gartersnakes were absent only from the youngest restoration unit, Dekay's Brownsnakes were absent from five units, including the four most recent restorations (Figure 262). The observation that Dekay's Brownsnakes were only present in units that had been restored for more than six years, whereas Common Gartersnakes were present as little as two years post-restoration, suggests a difference in colonization rate between species possibly related to body size (King and Vanek 2020).

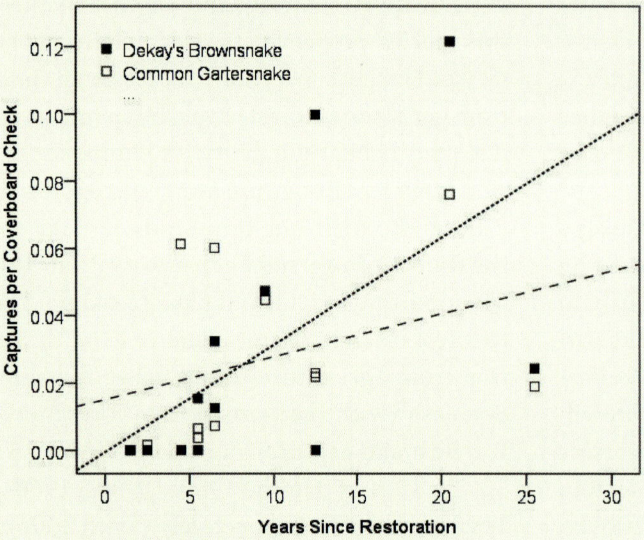

Figure 262. Relationship between number of snake captures per coverboard check and years since restoration for Dekay's Brownsnakes (filled squares, dotted line) and Common Gartersnakes (open squares, dashed line) in 12 restoration units at the Nature Conservancy's Nachusa Grasslands in northern Illinois. For Dekay's Brownsnakes, $R_{Spearman} = 0.69$, $n = 12$, $p = 0.013$; for Common Gartersnakes, $R_{Spearman} = 0.55$, $n = 12$, $p = 0.063$. Lines are shown for heuristic purposes only.

Richard B. King, Northern Illinois University

in central Illinois, where a road separates lowland summer habitat from upland hibernation sites, 63% mortality was observed (n = 1,184 of 1,885 individuals) among animals encountered using drift fences, pitfall traps, and twice-daily surveys of 2.4 km (1.5 mi) of road (Gross 2013). Mortality was greater for males than females and in autumn when traffic was heavier. Similarly, in a residential section of Greenville, North Carolina, 63% morality was observed (n = 212 of 337 individuals) among Dekay's Brownsnakes encountered on neighborhood streets and adjacent vacant lots (Gaul 2008). Individuals that exhibit lesions and blisters consistent with snake fungal disease (SFD) have been encountered in Wisconsin (J. Kapfer, personal communication), but the cause of these symptoms has not been definitively confirmed.

Management: Given its habitat requirements, Dekay's Brownsnake may serve as an indicator species of habitat restoration and management success (see Natural History Box: Grassland Snakes and Prairie Restoration). General management considerations related to combating successional change, promoting structural variability in habitats, and building road underpasses to mitigate road mortality are applicable to this species (see Introduction to Conservation and Management of Wisconsin's Amphibians and Reptiles chapter). Habitat management strategies that offset successional change, such as prescribed fire and mowing, can also have negative impacts if they cause direct snake mortality; thus, they must be timed and implemented appropriately. For example, Dekay's Brownsnake captures in upland pine forests in Maryland were significantly lower in burned compared to unburned units (McCleod and Gates 1998; but note that only ten individuals were captured in unburned pine forest). Dekay's Brownsnakes, along with three other small-bodied earthworm-eating snakes (i.e., Ring-necked Snake, Lined Snake, and Western Wormsnake) occurred at lower than expected abundance in annually and four-year burned vs. long-term unburned prairie units (Wilgers and Horne 2006). Furthermore, given the snakes' substantial use of ground debris, management strategies that promote downed woody debris will likely benefit Dekay's Brownsnakes. Culverts may represent one method to mitigate road mortality (Dodd et al. 2004), but their effectiveness as a management strategy for this species requires further assessment. Cook (2008) attempted translocations of this species to sites in New York that reportedly lead to successful repatriation. Despite this, such a strategy should be considered carefully (and in conjunction with experts and regulatory officials) before implementation.

Red-bellied Snake

Storeria occipitomaculata (Storer 1839)

Dreux J. Watermolen

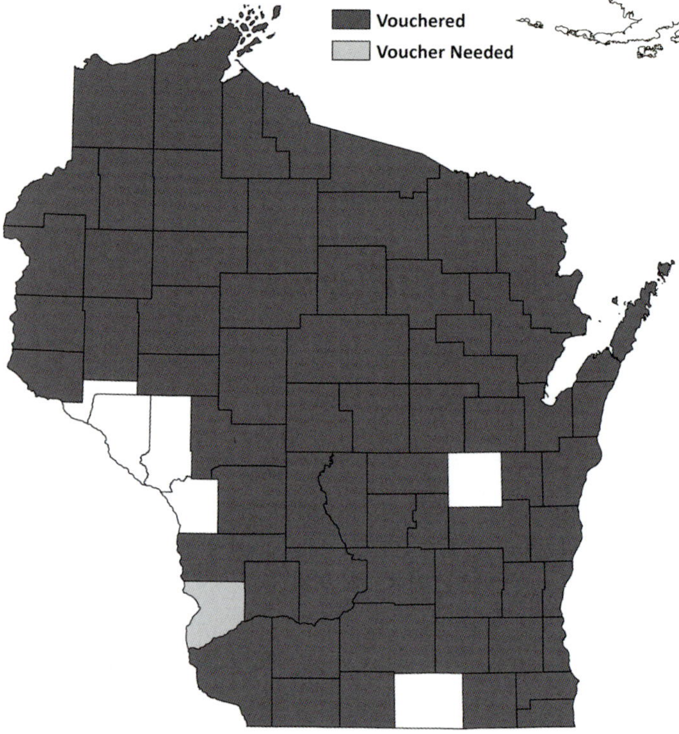

Adult Red-bellied Snake (Waukesha County; photo by J. DeBoer).

AT A GLANCE

The Red-bellied Snake, Wisconsin's smallest snake, is a secretive inhabitant of woodlands that feeds mostly on slugs. This attractive and docile species is likely one of the most widespread and common snakes in Wisconsin, but its small size, inconspicuous dorsal color, and semifossorial habits make it difficult to detect. Nonetheless, Red-bellied Snakes are easily identified by their red to yellowish-orange bellies and three small light yellow spots on the back of the head, a single larger one situated middorsally, and smaller ones on each side. The chin and throat are usually whitish. Red-bellied Snakes are active from mid-April through October in Wisconsin. They mate from mid-summer to early autumn, but females give birth to live young (rather than lay eggs), which are not born until the following spring or summer. Red-bellied Snakes are completely harmless, rarely, if ever, attempting to bite, but they display interesting defense behaviors when disturbed. The Red-bellied Snake is considered a common species by the WDNR.

SYSTEMATICS

Storeria occipitomaculata (Red-bellied Snake) belongs to the family Colubridae, a diverse group of serpents that includes the larger constrictors, watersnakes, and gartersnakes, many of which have wide North American distributions (reviewed by Ernst and Ernst 2003). Within this family, the genus *Storeria* includes relatively short slender live-bearing snakes that feed primarily on earthworms and slugs. There are currently four recognized species in the genus *Storeria*, including *S. occipitomaculata*, and all are restricted to eastern North America (Ernst 2012). Some early authors hyphenated the specific epithet, but aside from this minor variation, the nomenclature for *S. occipitomaculata* has been stable since Baird and Girard (1853) placed this species in the genus *Storeria*. Currently, three subspecies of *S. occipitomaculata* are recognized (Ernst 2002, Crother 2017), but only *S. o. occipitomaculata* (Northern Red-bellied Snake) occurs in Wisconsin. Fossils of *S. occipitomaculata* date back to the Pleistocene (Holman 1995, 2000).

In general, snakes in the genus *Storeria* are most closely related to other semifossorial North American species, such as Kirtland's Snake (*Clonophis kirtlandii*) and the earthsnakes (*Virginia* spp.), which have similar diets and natural histories (Alfaro and Arnold 2001, McVay et al. 2015). Among snakes native to Wisconsin, *S. occipitomaculata* is most closely related to its congener, *S. dekayi* (Dekay's Brownsnake). The genus name *Storeria* honors David Humphreys Storer, an eighteenth-century zoologist. The species name *occipitomaculata* is derived from the Latin *occiput*, 'back part of the head,' and *macula*, 'spot.' For more information, see Ernst (2002, 2012), who provided complete nomenclatural histories for the genus and species.

DESCRIPTION

Adults: The dorsal body color is dark or light brown, gray, or olive black and extends onto the venter. The dorsum may have a narrow pale vertebral stripe or narrow black longitudinal stripes (Figure 263). The venter ranges from vermilion to bright red to yellowish orange (Figure 264). The obtusely shaped head is distinct from the neck and is brown or reddish brown and generally darker than the body. Three small whitish-yellow spots are present on the back of the head, a single larger one situated middorsally and smaller ones on each side. The chin and throat are usually whitish, and the lower jaw may be dark spotted. The eyes are relatively large and dark (Ernst 2012).

The Red-bellied Snake is Wisconsin's smallest snake. Measurements of Wisconsin specimens in museum collections revealed that mean adult SVL was 18.9 cm (7.4 in, $n = 156$), and mean tail length was 5.2 cm (2.1 in, $n = 142$). Specimens with SVL ≥12 cm (4.7 in) were considered adults based on reports of size at first reproduction (see Reproduction and Development section). Given the presence of a single hemipenis in males, sex of preserved specimens was not determined via probe due to the potential for associated error. Live weights are unavailable for Wisconsin specimens. The largest Red-bellied Snake reported was a gravid female found in Pennsylvania that measured 34.4 cm SVL (13.5 in) and 42.2 cm TL (17 in) and weighed 18 g (0.6 oz; Gray and Lethaby 2008a).

Figure 263. Examples of the variation in dorsal body color of the Red-bellied Snake, ranging from more brown (*top*) to more gray (*bottom*). Note the light dorsal stripe evident on the top individual (Walworth County; photos by S. Viernum).

Preadult stages: Juveniles are similar in appearance to adults but typically darker, with a paler venter and more prominent occipital spots (Vogt 1981). Neonates in Michigan measured 6.7–9.8 cm TL (2.6–3.9 in, $\bar{x}=8.6$ cm [3.4 in]; Blanchard 1937a). In Minnesota, juveniles ranged from 8.0 to 10.1 cm TL (3.1 to 3.9 in, $\bar{x}=9.1$ cm [3.6 in]; Nelson 1969). Juveniles taken in August and September by Meshaka et al. (2008) in Ohio had SVLs that ranged from 9.5 to 10.0 cm (3.7 to 3.9 in). Juveniles from Wisconsin are similar in size to reports from other states (7–10 cm [2.75–3.94 in]; Vogt 1981). Semlitsch and Moran

Figure 264. An adult Red-bellied Snake with the colorful venter visible (Walworth County; photo by S. Viernum).

(1984) reported that juveniles were indistinguishable in SVL from adults by the end of their first year.

Variation: Distinctly light (gray) and dark (brown) color morphs occur, with populations varying in the proportion of each. These morphs apparently share a common gene pool, as Grudzien and Owens (1991) noted that the "polymorphism . . . occurs in sympatric populations, not as geographically distinct variation," and pointed out that "no selective advantage is known for either the gray or brown color" morph. Both color phases are documented in Wisconsin, and the two occur together on the Apostle Islands (data from MPM specimens 11735–11738, 11740–11742). Interestingly, there is no color variation within the light and dark morphs and no intergradation between them. The pale spots on the neck are sometimes fused, suggesting a ring, which is most often enlarged at the normal locations of the spots. Occasionally, one or more of the pale neck markings are absent (Ernst 2002). No published reports of albinism in Wisconsin are known, although a partial albino individual was collected in Vermont (Watkins-Colwell 2002).

Some reports support that adult females are generally larger than males (female maximum SVL = 30.2 cm [11.9 in]; male maximum SVL = 28.3 cm [11.1 in]; Meshaka and Klippel 2011). Yet Semlitsch and Moran (1984) found no significant differences in SVL between females and males in South Carolina specimens, but they noted that females were heavier (females = 2.37 g [0.08 oz], males = 1.79 g [0.06 oz]). Range-wide, the tail of males is proportionally longer (tail length = 20%–27% of TL) than the tail of females (tail length = 17%–24% of TL; Ernst 2002, Ernst and Ernst 2003). This is true in Wisconsin as well (17.2%–38.2% of TL [$\bar{x}$ = 23.4%] in males and 17.2%–25.4% of TL [$\bar{x}$ = 20.0%] in females; data from 33 male and 42 female MPM specimens).

Diagnosis: Red-bellied Snakes possess keeled scales that typically occur in 15 rows at midbody, an anal plate that is divided, and subcaudal scales that are divided. The loreal scale is absent (Trapido 1944, Ernst and Ernst 2003). There are typically six supralabial scales, the third and fourth of which enter the orbit, seven infralabial scales, two nasal scales, two preocular scales, and two postocular scales on each side of the head. A pale spot is often present on the fifth supralabial scale (Figure 265). The ventral surface is orange to red in life, and ventral scale counts range from 107 to 136.

Adult and juvenile Red-bellied Snakes might be confused with other small snakes. Yet Dekay's Brownsnakes (*Storeria dekayi*) have 17 scale rows and a single preocular scale. In addition, the light spots on the nape are nearly always present in Red-bellied Snakes in Wisconsin but are generally lacking in Wisconsin Dekay's Brownsnakes, as is the red venter. Red-bellied Snakes may be confused with young Ring-necked Snakes (*Diadophis punctatus*), because both have orange to red venters and because when the Red-bellied Snake's neck spots appear fused, they resemble a neck ring. However, the scales of Ring-necked Snakes are not keeled. In addition, the light marks on the nape completely encircle the neck of Ring-necked Snakes. Western Wormsnakes (*Carphophis vermis*) superficially resemble Red-bellied Snakes but have smooth scales.

Figure 265. The head of an adult Red-bellied Snake that shows the pale spot on the fifth supralabial scale (Oneida County; photo by D. Brown).

DISTRIBUTION AND HABITAT

Global distribution: Red-bellied Snakes occur throughout eastern and midwestern North America, from southeastern Saskatchewan east to northeastern Nova Scotia in the North and from southeastern Texas along the Gulf Coast to central Florida in the South. They reach their northernmost distribution west of Lake Winnipeg in Manitoba. They are absent from much of Iowa, northern Missouri, and parts of Illinois, northern Indiana, and western Ohio (Ernst 2002, Ernst and Ernst 2003). Relict populations occur in the Black Hills of South Dakota and adjacent Wyoming (Smith 1963), Nebraska (Lynch 1985, Peyton 1989) and possibly western Kansas, based on historic records (Peyton 1989, Collins 1993).

Distribution in Wisconsin: Red-bellied Snakes likely occur throughout Wisconsin, and published records exist for most counties. They are even documented on various islands in the Great Lakes. For example, a population of Red-bellied Snakes occurs on Chambers Island in northern Green Bay (Long et al. 1978), and they have been observed on the islands off the tip of Door County (Long et al. 1987, Dreux Watermolen, WDNR, unpublished data) and on islands in Lake Superior (Corin 1976, Vogt 1981, Casper 2008a). Of the published county records, suitable museum vouchers could not be found for Crawford (see the Appendix). Furthermore, the Menominee County record is based on a specimen collected in 1923 by T. E. B. Pope (MPM 1903). The only location information that accompanied this specimen was "Big Eddy Falls," from which county of occurrence was apparently extrapolated (J. Kapfer, personal communication; see the Appendix).

Localities of possible occurrence: There are currently no published records of Red-bellied Snakes from Winnebago and Rock Counties, which are both surrounded by counties of documented occurrence. They are also not documented from a small number of contiguous west-central counties along the Mississippi River. With further investigation, they will likely be found in suitable habitats in all of these counties, although perhaps in relatively small populations.

Habitat: Detailed investigations of this species' habitat relationships in Wisconsin do not exist. Yet anecdotal reports indicate that Red-bellied Snakes are frequently encountered in moist forests (both conifers and mixed conifer-hardwoods), and some authors consider them a woodland species (e.g., Vogt 1981). However, they appear to be somewhat of a habitat generalist and often occur along the edges of wetlands, bogs, and wet prairies (D. Watermolen, unpublished data; see Amphibians and Reptiles in Wisconsin Plant Communities chapter). Populations in the north-central US, including Wisconsin, can also occur in more xeric upland habitats (reviewed by Ernst and Ernst 2003, D. Watermolen, unpublished data). For example, recent work supports that Red-bellied Snakes are not restricted to forests but found in drier prairies and prairie-like habitats throughout Illinois (Brown and Phillips 2012). Similarly, Meshaka et al. (2008) collected specimens in old field habitats in Ohio, while Diaz (2016) captured significantly more Red-bellied Snakes in open vs. wooded habitats in Ontario. Red-bellied Snakes appear to tolerate minor anthropogenic habitat alterations, as they can be encountered along road embankments and in moderately disturbed woodlots in agricultural and urban areas (Kjoss and Litvaitis 2001, Zappalorti and Mitchell 2008, Brown and Phillips 2012). Cahoe and Troelstrup (2004) collected a large percentage of their specimens on unpaved roads consisting of 90% or more sand and gravel. Similarly, in Door County, active Red-bellied Snakes were documented along roadways in spring and autumn during midmorning and early evening (D. Watermolen, unpublished data).

Red-bellied Snakes are secretive and spend much time underneath ground debris in their chosen habitats. Most field investigators report encountering Red-bellied Snakes primarily under rocks or logs, and the presence of such cover objects creates presumably important microhabitat for this species. In addition, the accumulation of leaf litter on the forest floor provides a matrix through which Red-bellied Snakes can move when foraging, searching for mates, and dispersing. For example, they were frequently captured in exposed leaf litter on cloudy summer afternoons in Door County (D. Watermolen, unpublished data). In grassland habitats, tussocks can also provide suitable cover and thermoregulatory opportunities for gravid females (Meshaka et al. 2008). Although typically found under ground cover, Red-bellied Snakes are occasionally observed climbing vegetation at heights of up to 1 m (3.3 ft) above the ground (Smith 1961, Barbour 1971, Minton 1972, Cupp et al. 2012). This could be related to differences in thermoregulatory behaviors between the sexes, with gravid female snakes seeking sites that provide optimum temperatures for development of their embryos.

Hibernation occurs below the frost line in loose soil and gravel, rock crevices, or existing burrows. For example, Lachner (1942) reported that Red-bellied Snakes were excavated by construction workers from a gravel bank in Pennsylvania on 3 January 1939. The snakes were ca. 0.8 m (2.5 ft) below the surface within "soil that contained some gravel, from small pebbles to rocks averaging 2 to 3 inches in diameter. . . . A small partly decayed root nearly 2 inches in diameter which led from the surface of the ground to the hibernation area probably served as an entrance." In Minnesota and Michigan, they are reported to use ant mounds as hibernacula (Carpenter 1953, Lang 1969), but similar observations have not yet been reported in Wisconsin. Vogt (1981) indicated that Red-bellied Snakes may hibernate in and around building foundations as well (see Activity section). In areas of high groundwater, they may remain partially submerged during hibernation (Criddle 1937).

REPRODUCTION AND DEVELOPMENT

Throughout their range, Red-bellied Snakes mate from late July to September and give birth the following spring or summer (Trapido 1940, Smith 1950, Anderson 1965, Barbour 1971, Gregory 1977a, Martof et al. 1980, Willson and Dorcas 2004, Cupp et al. 2012), but Ernst and Ernst (2003) observed a pair copulating in May in Minnesota. Males possess a single rather than bifurcated hemipenis, which is inserted into the female's vent during coitus (Ernst and Ernst 2003). All female Red-bellied Snakes captured during June and July in Ohio were gravid (Meshaka et al. 2008). A similar schedule likely occurs in Wisconsin, with most females reproducing annually. Gestating gravid females may seek microhabitats that provide both optimum temperatures for development of their embryos and shelter from potential predators (Meshaka et al. 2008; see Habitat subsection). Gregory (1975) and Gordon and Cook (1980) observed aggregations of gravid females in Manitoba and Quebec, respectively, and speculated that the aggregations could be due to limited availability of sites with these characteristics. Similar aggregations are not reported in Wisconsin, perhaps due to a greater abundance of suitable habitat.

Red-bellied Snakes are viviparous (Blackburn 2000). In the Upper Midwest, most young are born in mid- to late August of the year after copulation, although parturition sometimes occurs as early as late July or as late as September (Ruthven 1906, Blanchard 1937a, Vogt 1981). Brood size is highly variable and ranges up to 21 young per year, with a mean of 6–8 young (Smith 1950, Oldfield and Moriarty 1994, Meshaka and Klippel 2011). Larger females tend to produce greater numbers of offspring, and variation in neonate TL is loosely related to the number in the brood (Blanchard 1937a, Brodie and Ducey 1989). Multiple paternity has also been documented in this species (Wusterbarth et al. 2010).

The young grow and develop rapidly. In Pennsylvania and South Carolina, Red-bellied Snakes approach sexual maturity at ca. one year (male SVL = 11.8 cm [4.6 in], female SVL = 12.6 cm [5.0 in]), and first reproduction takes place the following spring or summer when females are two years of age (Semlitsch and Moran 1984, Meshaka and Klippel 2011). Longevity of four to five years in captive individuals has been reported (Conant and Hudson 1949, Snider and Bowler 1992).

ACTIVITY

Red-bellied Snakes in Wisconsin emerge from winter dormancy in mid-April. Lang (1969) reported moisture and temperature as important factors affecting spring emergence of Red-bellied Snakes in Minnesota, and this seems to be the case in Wisconsin as well (Dreux Watermolen, WDNR, personal observation). In Ohio, Meshaka et al. (2008) observed a bimodal seasonal activity pattern, with most observations made in June (females) and a lesser peak in September (juveniles).

Red-bellied Snakes may be active throughout the day, but Ernst and Ernst (2003) and Cahoe and Troelstrup (2004) reported that most spring and autumn activity occurred during the morning or afternoon and early evening, with summer activity being crepuscular or nocturnal. Red-bellied Snakes in Door County were often encountered along roadways in midmorning and early evening on spring and autumn days (D. Watermolen, unpublished data). A tendency toward nocturnal activity in summer could be because slugs and earthworms, the snakes' primary prey, are more active at night and thus more easily encountered. Semlitsch and Moran (1984) found that seasonal activity patterns were related to the drying cycle of the snakes' habitat in South Carolina, and Cahoe and Troelstrup (2004) reported a positive correlation between air and ground temperatures in places where they encountered individuals. Brattstrom (1965) collected body temperature data from two active individuals in Minnesota, which were 24.3°C and 28.2°C (75.7°F and 82.8°F), and reported that critical thermal maximum (CTmax) for this species was 37.8°C (100°F).

Extremely limited data exist on the movement or spatial ecology of Red-bellied Snakes. Recaptures of marked snakes at sites relatively close to their original capture points and near hibernacula suggest that Red-bellied Snakes have relatively small home ranges. For example, Blanchard (1937a) recaptured a large female that had moved ca. 402 m (1,319 ft) within 24 hours of being marked and released at a site in Michigan. In Minnesota, Lang (1969) reported that most summer captures were within 152–305 m (500–1,000 ft) of hibernacula. Cahoe and Troelstrup

(2004) recaptured South Dakota snakes within 50 m (164 ft) of initial capture sites.

In Wisconsin, migration to hibernacula begins in September (Vogt 1981), and Red-bellied Snakes probably enter winter dormancy by October. However, those snakes in the northern parts of the state probably have a slightly shorter period of activity (D. Watermolen, unpublished data). Work in Minnesota found that Red-bellied Snakes exhibited a strong tendency to return to the same hibernacula across successive winters (Lang 1969). Hibernation may be solitary or occur communally, often with other species using the same hibernaculum (Criddle 1937, Carpenter 1953, Lang 1969). In the Madison area, Vogt (1981) mentioned finding Red-bellied Snakes, Dekay's Brownsnakes, gartersnakes (*Thamnophis* spp.), and juvenile Eastern Foxsnakes (*Pantherophis vulpinus*) "sunning together near foundations in early May. Presumably they had just emerged from hibernation."

PREY AND PREDATORS

The scant literature concerned with the feeding habits of Red-bellied Snakes contains primarily a few reports of stomach contents, statements about food items that captive individuals accepted, and apparently secondhand reports in field guides. Despite limited information, slugs are the principal prey identified by authors who have analyzed gut contents (Hay 1892, Surface 1906, Wright and Bishop 1915, Brown 1979, Gilhen 1984). Semlitsch and Moran (1984) reported that slugs from the families Limacidae and Philomycidae comprised 100% of the stomach contents of ten Red-bellied Snakes they examined in North Carolina. Gilhen (1984) noted that Red-bellied Snakes in Nova Scotia prey upon the native Carolina Mantleslug (*Philomycus carolinianus*), as well as members of the nonnative genera *Arion* and *Deroceras*. Gray et al. (2013) discussed a Red-bellied Snake that disgorged a Brown-banded Arion (*Arion circumscriptus*) in Pennsylvania. There are also a few reports of earthworms in stomachs (Hamilton and Pollack 1956, Minton 1972), miscellaneous reports of insect remains in Red-bellied Snake "rectums" (Wright and Bishop 1915), soft-bodied beetle larvae (Ditmars 1936), and "50 per cent slugs, 10 per cent *Gryllus*, and 40 per cent unidentified animal matter, by volume" in the stomachs of two snakes (Barbour 1950). Red-bellied Snakes apparently will not prey on introduced flatworms (Ducey et al. 1999). From the available information, it can be concluded that Wisconsin Red-bellied Snakes likely feed on slugs and perhaps earthworms. It is worth noting, however, that relatively few earthworms and slugs are native to Wisconsin (Jass 2004, Hale 2007, Reynolds 2012, Reynolds and Wetzel 2012). So it is reasonable to conclude that in the past, this species likely relied on other prey, probably terrestrial snails, as the snake's jaw and tooth adaptations could aid in snail extraction (Rossman and Myer 1990, do Amaral 1999). Prey ingestion and digestion may be aided by secretions of venom-like compounds from the Duvernoy's glands, which open near the maxillary teeth (do Amaral 1999).

A wide variety of predators likely feed on Red-bellied Snakes, but detailed field observations in Wisconsin are sparse. Regardless, many small to midsized mammals that consume snakes are probable candidates, ranging from Northern Short-tailed Shrews (*Blarina brevicauda*) to Raccoons (*Procyon lotor*) and Virginia Opossums (*Didelphis virginiana*; Hamilton 1951, Jackson 1961, Kurta 1995). Several small mammal species that are not typically considered predators of other animals apparently also consume Red-bellied Snakes. Thirteen-lined Ground Squirrels (*Ictidomys tridecemlineatus*) occasionally feed on small snakes (Wistrand 1972), while Harriot (1940) observed a Chipmunk (*Tamias striatus*) eating a Red-bellied Snake in New York. Reported avian predators include Barn Swallows (*Hirundo rustica*), American Robins (*Turdus migratorius*), and Ruffed Grouse (*Bonasa umbellus*) in South Dakota (Smith and Stephens 2003, Cahoe and Troelstrup 2004). Various raptors also consume this species across its range (Breckenridge 1944, Conant 1951, Sherrod 1978, Ross 1989a, Barrett and Villarroul 1994), but currently only Red-shouldered Hawks (*Buteo lineatus*) are documented predators in Wisconsin (Welch 1987).

Other snakes may also feed on Red-bellied Snakes, including Eastern Milksnakes (*Lampropeltis triangulum*; Medsger 1922). Vogt (1981) fed young Red-bellied Snakes to captive Eastern Milksnake hatchlings from Wisconsin. Hodgson and Hansen (2005) further reported occasionally finding them in the stomachs of Largemouth Bass (*Micropterus salmoides*) from lakes in Michigan's Upper Peninsula. Juvenile snakes might fall victim to a wider range of predators than adults due to their smaller size. For example, Swanson (1952) described a case of a captive newborn Red-bellied Snake being preyed upon by a spider. Although a primary method of avoiding predation appears to be small size and secrecy, several interesting predator-avoidance behaviors have been noted (Natural History Box: Curl Your Lips or Play Dead? Defense Behaviors). This species shows very little, if any, aggression when handled by humans.

No one has studied the parasites of Red-bellied Snakes in Wisconsin (reviewed by Watermolen 2014), but a number of endoparasites have been reported from elsewhere in North America. These include apicomplexans (*Hepatozoon eurytopis*), trematodes (*Distomum* spp., *Stomatrema pulsillum*), and nematodes (*Cosmocercoides dukae*, *Ophidascaris ashi*, *Rhabdias fuscovenosa*; Sprent 1988, Ernst and Ernst 2006, Telford 2010).

CONSERVATION

Status: Globally, the Red-bellied Snake is listed as a Least Concern species by the IUCN, and populations are considered secure by NatureServe. Yet the species is still characterized as uncommon to rare in many states (Ernst and Ernst 2003). Ernst and Ernst (2003), however, point out that Red-bellied Snakes are secretive and mostly nocturnal and as such are likely more common than the available records might suggest. Breckenridge (1944) stated that "this tiny snake is probably neither as nocturnal nor as secretive in its habits as it is often said to be, but its small size and inconspicuous dorsal color make it difficult to detect among the leaves and dead grass." Pentecost and Vogt (1976) referred to Red-bellied Snakes as "common" in the Lake Michigan drainage basin, and Harding (1997) considered them "locally common" throughout the Great Lakes region. The Red-bellied Snake is currently considered common in Wisconsin by the WDNR.

Populations: Hoy (1883) considered Red-bellied Snakes "abundant" in Wisconsin, Knudson (1955) noted "local hot-spots" where the snake was very common, and Casper (1996) thought this might be the most abundant snake in the state. Yet little robust information exists on Red-bellied Snake populations. Low rates of recapture (3%–13%) for marked Red-bellied Snakes in studies from other states suggest that they are very abundant and short-lived or that their populations experience rapid turnover (Blanchard 1937a, Semlitsch and Moran 1984, Cahoe and Troelstrup 2004). Reported sex ratios are generally close to 1M:1F, but this could be due to collection bias (i.e., females are encountered more frequently than males), especially in the summer (Blanchard 1937a, Semlitsch and Moran 1984, Cahoe and Troelstrup 2004, Meshaka et al. 2008). Although no published estimates of population survival and mortality rates were found, it is possible that these vary by season. For example, winter mortality rates might be relatively high if temperatures within selected hibernacula drop below freezing for extended periods (Lang 1969).

Threats: No major threats specific to Red-bellied Snakes are recognized in the Upper Midwest, and sources of mortality are neither well understood nor adequately documented. General threats that impact wooded habitat and mesic prairies or impact uplands immediately adjacent to wetlands could affect this species (see Introduction to Conservation and Management of Wisconsin's Amphibians and Reptiles chapter). Furthermore, threats to habitat connectivity, habitat structural diversity, and loss of ground debris for cover are important considerations. Red-bellied Snakes use urban environments (Zappalorti and Mitchell 2008), which could increase contact with humans and lead to greater mortality related to anthropogenic activities. According to Ernst and Ernst (2003), "habitat destruction and road mower deaths take a heavy toll." Like many snakes, Red-bellied Snakes occasionally seek warmth on roads, particularly in summer and after rains (Vogt 1981), and as a result could fall victim to road mortality (Gibson and Merkle 2004). Factors that alter normal microclimates necessary for thermoregulation could also have negative consequences for this species and its prey. Semlitsch and Moran (1984) found that Red-bellied Snakes move to moist areas following

NATURAL HISTORY BOX: Curl Your Lips or Play Dead? Defense Behaviors

Red-bellied Snakes are docile creatures. Although I have handled several hundred over the past 30 years, not a single one ever attempted to bite me. They do, however, exhibit a variety of interesting defense behaviors when handled. Most often, Red-bellied Snakes expel musk from glands at the base of the tail and fecal matter, a defensive behavior common to several species of small snakes. The unpleasant odor and texture are effective deterrents for some would-be predators. Sometimes Red-bellied Snakes will flatten their bodies and curl their upper lips to expose the maxillary teeth, resulting in a somewhat startling 'grin' (Gosner 1942, Jordan 1970). This lip curling and head distortion behavior resembles head shape changes during feeding. Lateral writhing and jerking of the anterior portion of the body and the tail may accompany the lip-curling display. During the display, the side of the head may be pushed against the snake's captor so that the tiny teeth snag the skin. This behavior may aid in the delivery of mouth secretions (do Amaral 1999; see Prey and Predators section). Some Red-bellied Snakes may go into convulsions, roll onto their back, gape open the mouth, protrude the tongue, and then become motionless and contorted (Jordan 1970), a behavior more typically thought of in association with the Eastern Hog-nosed Snake (*Heterodon platirhinos*). This activity exposes the bright red belly and may startle a predator momentarily, allowing the snake to escape. Although he never observed the death-feigning behaviors, Vogt (1981) described an individual caught in Portage County that extended its body straight out, became rigid, and was able to be held erect like a matchstick. His observations were made on a cold day in May. After a few minutes in the stiffened position, the snake became flaccid and attempted to wriggle away. In Ontario, Warburton (1951) described behaviors resembling an "extremely venomous snake" in which the Red-bellied Snake raises the head and simulates strikes.

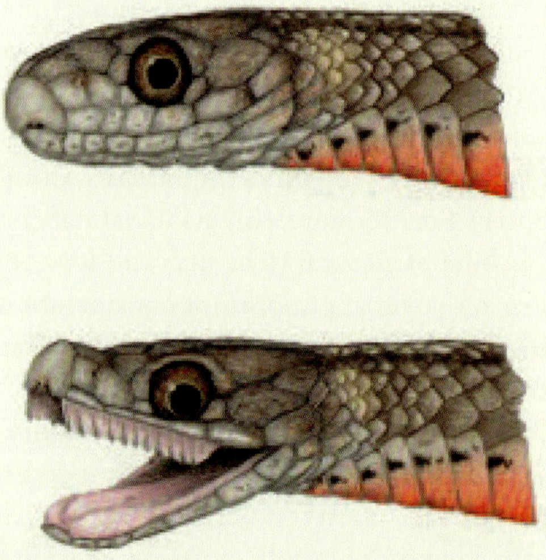

Figure 266. An example of the lip-flaring behavior in Red-bellied Snakes (illustration by E. Wild).

I observed Red-bellied Snakes display a different defense behavior in Door County (Watermolen 1991b). On a warm day in May, I captured an individual on a gravel road. When I picked the snake up, it simply ceased moving, and its body became flaccid. It rolled its head to the side and exhibited the lip-flaring display but otherwise remained motionless (Figure 266). Several additional individuals exhibited this same behavior when captured. In all cases, when the snakes were placed on the ground, they crawled away after a short period of immobility. Given the potentially large number of predators that these small snakes might encounter, it is not surprising that they have developed multiple, varied defense behaviors.

Dreux J. Watermolen, WDNR

seasonal drying of the soil. Along these lines, work in the southeastern US found fewer Red-bellied Snakes in clear-cut vs. unharvested forests (Todd and Andrews 2008). Higher temperatures and habitat desiccation may occur after forest fragmentation, which could be a particular concern in small isolated habitat fragments. For example, Kjoss and Litvaitis (2001) found fewer Red-bellied Snakes in small habitat patches, which had shorter vegetation, lower soil moisture, and possibly less gastropod prey, than in habitat patches of intermediate to large size. On a landscape scale, however, environmental variables (e.g., drought) may have greater impacts on populations than forest management actions (Renken et al. 2004). The significant reduction of leaf litter caused by the feeding activity of nonnative earthworms could further eliminate an important microhabitat component for this species (Burtelow et al. 1998, Ziemba et al. 2015).

Management: Management considerations related to woodlands, mesic prairies, and open wetlands are likely applicable to this species, as are concepts associated with timing of management and mitigation of potential road mortality (see Introduction to Conservation and Management of Wisconsin's Amphibians and Reptiles chapter). Management of suitable habitats that promotes adequate ground moisture year-round could help ensure the continued presence of this species. Maintenance of leaf litter and ground debris (e.g., downed woody debris) during habitat management can further provide suitable microhabitats. Although application of fire is useful in managing open habitats, the potential effects of prescribed fire on Red-bellied Snakes have not been assessed in Wisconsin. Still, this practice presumably leads to mortality if conducted during the snake's active season.

Interestingly, research in Pennsylvania suggests that existing transmission line rights-of-way can provide suitable habitat for this species. For example, Yahner et al. (2001) found Red-bellied Snakes within transmission line rights-of-way that were managed to control growth of woody vegetation but not in adjacent woodlands. This is in contrast to work in New Hampshire that suggested that smaller plots of habitat that received frequent mowing, which reduced vegetation height and soil moisture, had fewer Red-bellied Snakes than sites mowed infrequently, if at all (Kjoss and Litvaitis 2001). Conservation of habitat with suitable overwintering locations (e.g., ant mounds and animal burrows) is also likely important for the Red-bellied Snake, as winter mortality rates have been suggested to be relatively high in some hibernacula (Lang 1969).

SPECIAL REMARKS

Blanchard (1937a) conducted the first detailed life history study of Red-bellied Snakes at the University of Michigan's Biological Station on Douglas Lake. His work remains one of the most frequently cited studies of this species, and no similar detailed work has addressed the Red-bellied Snake in Wisconsin. Most of the recent publications involving this species in Wisconsin are geographic distribution notes (Watermolen 1992b, 2013). As a result, much of what is known about this species' natural history in-state is derived from studies conducted elsewhere.

Butler's Gartersnake

Thamnophis butleri (Cope 1889)

Eric T. Hileman and Danielle R. Bradke[†]

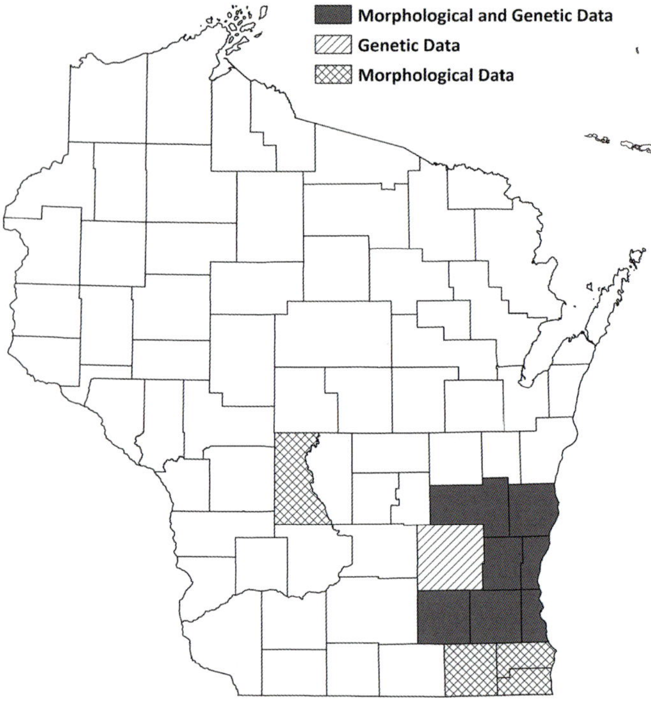

Adult female Butler's Gartersnake (Waukesha County; photo by J. Kapfer).

[†] We acknowledge J. Kapfer, J. Lorch, G. Schuurman, and R. Paloski for assistance with the Description and Distribution and Habitat sections.

AT A GLANCE

Butler's Gartersnakes are the smallest of five gartersnakes that occur in Wisconsin. They are largely restricted to the southeastern part of the state and are often found in open moist habitats such as prairies, wet meadows, wetlands, old fields, and vacant lots. This species is stout with a small narrow head and almost no discernible neck. The characteristic most often used to distinguish Butler's Gartersnakes from sympatric gartersnake species is the position of the lateral stripes. Anteriorly, the lateral stripes cover the third dorsal scale row and typically a portion of scale rows two and four. Butler's Gartersnakes are potentially active from late March to the beginning of November in Wisconsin, and reproduction probably occurs soon after egress from hibernation. Females do not lay eggs but instead give birth to live young from July to early September. This snake feeds primarily on earthworms. However, before European settlers introduced earthworms in the mid-1800s, the snake was likely restricted to habitats supporting leeches. A variety of animals probably prey on Butler's Gartersnakes, including crayfish, snakes, birds, shrews, and other small to midsized carnivorous mammals, ranging from weasels to Red Foxes. When threatened, Butler's Gartersnakes are more prone to flee than strike. This species is closely related to the Plains Gartersnake, and the two species are known to hybridize at some sites where they co-occur in southeastern Wisconsin. Butler's Gartersnake is considered a species of Special Concern and a Species of Greatest Conservation Need by the WDNR.

SYSTEMATICS

Thamnophis butleri (Butler's Gartersnake) is closely related to *T. radix* (Plains Gartersnake) and *T. brachystoma* (Short-headed Gartersnake). Morphological and molecular evidence supports *T. butleri* and *T. radix* as sister species that diverged from *T. brachystoma* (de Queiroz and Lawson 1994, Alfaro and Arnold 2001, de Queiroz et al. 2002, Pyron et al. 2013, McVay et al. 2015). Rossman et al. (1996) hypothesized that *T. butleri* is a reduced derivative of *T. radix*.

According to mitochondrial DNA sequence and amplified fragment length polymorphism data (Burghardt et al. 2006, Placyk et al. 2012), Wisconsin *Thamnophis butleri* populations are genetically distinct from those in Ohio, Michigan, and Ontario. Genetic analyses also support hybridization between *T. butleri* and *T. radix* along the Wisconsin southern range limit (Burghardt et al. 2006, Fitzpatrick et al. 2008, Placyk et al. 2012). Fitzpatrick et al. (2008) demonstrated that the two species have maintained distinctive species-specific characteristics in Wisconsin despite hybridization. Placyk et al. (2012) suggested that this hybrid zone predates European settlement and is not a consequence of anthropogenic factors (e.g., habitat modification). In Wisconsin, *T. butleri* may also hybridize with *T. sirtalis* (Common Gartersnake), but this phenomenon appears to be rare (see Distribution in Wisconsin subsection; Kapfer et al. 2013a).

DESCRIPTION

Adults: This species is stout with a small narrow head and almost no discernible neck (Figure 267). It has keeled scales with a black, brown, reddish-brown, or olive dorsum and three stripes varying from pale yellow to orange (Figure 268). The stripes run longitudinally, one along the dorsal midline and one on either side of the body. The lateral stripes cover the third scale row, bleeding onto scale rows two and four (Figure 269). Occasionally, a checkerboard pattern

Figure 267. Adult Butler's Gartersnake. Note the small head in comparison to the body (Ozaukee County; photo by E. Hileman).

Above, top: Figure 268. Color variation among Butler's Gartersnakes in Ozaukee County, Wisconsin. (a) Adult (Ozaukee County; photo by G. Mayer). (b) Adult (Ozaukee County; photo by E. Hileman).

Above, bottom: Figure 269. Anterior body pattern typical of Butler's Gartersnake. Note the placement of the lateral stripe and only thin dark bars on few supralabial scales, which are important characteristics used in recognizing this species (Ozaukee County; photo by E. Hileman).

Figure 270. Butler's Gartersnake littermates less than one week old. Note similarities of pattern and coloration to adults (Waukesha County; photo by R. Paloski).

of alternating spots is visible between the stripes on either side of the body. The yellow labial scales may be finely speckled with brown or orange, with occasional thin short bars on the back edge of some posterior labial scales, and there are usually two small yellow markings on the parietal scales. The venter varies from grayish yellow to pale green and commonly has black spots present along the outer edges; occasionally some medial spotting is also present. The Butler's Gartersnake is the smallest of five gartersnakes (*Thamnophis* spp.) that occur in Wisconsin (Vogt 1981). In a study of three Wisconsin populations, one in Ozaukee County and two in Waukesha County, females were typically longer ($\bar{x}$ SVL = 39.9 cm [15.7 in], n = 605) than males ($\bar{x}$ SVL = 35.2 cm [13.9 in], n = 193; Eric Hileman, Northern Illinois University, personal data). Females (nongravid) were also heavier ($\bar{x}$ = 38.7 g [1.4 oz], n = 206) than males ($\bar{x}$ = 24.0 g [0.8 oz], n = 192; E. Hileman, unpublished data). See Rossman et al. (1996) for an exhaustive species description.

Preadult stages: Neonates and juveniles look similar to adults (Figure 270). Neonates born to 11 females captured in Milwaukee County averaged 12.1 cm SVL (4.8 in) and 1.2 g (0.04 oz, n = 112; Albright 2001). However, some females were maintained in the lab prior to pregnancy, potentially affecting maternal

Figure 271. Albino female neonate Butler's Gartersnake (15.0 cm SVL [5.9 in]) found underneath a plywood board on 13 September 2009 (Waukesha County; photo by E. Hileman).

and offspring body condition. Additionally, some of the individuals in the study may have been hybrids, as they were captured in southern Milwaukee (Albright 2001) near a known area of hybridization (Fitzpatrick et al. 2008, Placyk et al. 2012). In southeastern Michigan, mean neonate sizes for 28 litters ranged from 10.1 to 12.5 cm SVL (4.0 to 4.9 in) and 0.9 to 1.7 g (0.03 to 0.06 oz; Ford and Killebrew 1983). In Ohio, 22 neonates born to 3 females ranged from 12.7 to 18.1 cm TL (5.0 to 7.1 in; Conant 1938). In Ozaukee and Waukesha Counties, young of the year measured 10.4–23.3 cm SVL (4.1–9.2 in) and 0.8–8.0 g (0.03–0.3 oz, $n = 1{,}489$; E. Hileman, unpublished data). Juveniles from these counties measured 14.0–40.7 cm SVL (5.5–16.0 in, $n = 861$) and 2.0–42.0 g (0.07–1.5 oz, $n = 862$; E. Hileman, unpublished data).

Variation: Albinism is rare but was observed in one individual collected from Milwaukee County on 11 October 1938 (Dyrkacz 1981) and in one neonate captured on 13 September 2009 in Waukesha County (E. Hileman and Julia Robson, Milwaukee County Parks, unpublished data; Figure 271). Melanistic individuals were found in Amherstburg, Ontario (Catling and Freedman 1977), but have not been reported elsewhere. In Wisconsin, males of all age classes have a greater tail-to-body-length ratio ($\bar{x} = 23.3\%$, $n = 165$) than females ($\bar{x} = 21.2\%$, $n = 221$; E. Hileman, personal data). However, adult males tend to be more slender and shorter than adult females (E. Hileman, personal observation).

Diagnosis: Butler's Gartersnakes possess keeled scales that occur in 19 or fewer scale rows at midbody, an undivided anal plate, and divided subcaudal scales. There are 2 prefrontal scales, while each side of the head contains 1 loreal scale, 1 preocular scale, 2–3 postocular scales, 7 (sometimes 6–8) supralabials, and 8–9 infralabials. There are 129–154 ventral scales that possess no obvious pattern. The lateral stripe either partially or fully covers scale rows two, three, and four. Butler's Gartersnakes are similar in appearance to other gartersnake species, and the characteristic generally used to separate them is the scale row position of the lateral stripe. In Butler's Gartersnakes, lateral stripes typically cover part of scale row two, all of three, and part of four. The Common Gartersnake's lateral stripes typically cover part (or all) of scale row one and all of scale rows two and three. The Plains Gartersnake's lateral stripes are usually narrow and cover only scale row three and some proportion of row four. Plains Gartersnakes also typically have thicker bars along the posterior edge of the labial scales than Butler's Gartersnakes. The Eastern Ribbonsnake (*Thamnophis saurita*) and Western Ribbonsnake (*T. proximus*) are much more slender, with white labial scales and white preocular scales, and their lateral stripes include only scale rows three and four. The head of Butler's Gartersnake also appears superficially shorter and narrower (i.e., similar in width to the neck) than the head of other gartersnake species in Wisconsin. Unfortunately, distinguishing Butler's Gartersnakes from Plains Gartersnakes is extremely difficult in areas of geographic overlap and possible hybridization. In these regions, the use of typical characteristics could be unreliable for rapid field identification.

DISTRIBUTION AND HABITAT

Global distribution: Butler's Gartersnakes have a small global geographic distribution and are found only in the Great Lakes region of the US and southern Canada. This species ranges from central Indiana to central Ohio, northward through eastern Michigan and the southernmost edge of Ontario, with isolated populations in central southern Ontario and southeastern Wisconsin (Rossman et al. 1996).

Distribution in Wisconsin: The distribution of Butler's Gartersnakes in Wisconsin is difficult to delineate because they hybridize with other gartersnake species (*Thamnophis* spp.; see Systematics section). Recognized counties of occurrence for this snake have changed over time based on the results of genetic and morphological investigations. Vogt (1981) used morphology to report Butler's Gartersnakes from Fond du Lac, Washington, Waukesha, Milwaukee, Walworth, Racine, and Kenosha Counties. Casper (1996) also reported Butler's Gartersnakes in Ozaukee County based on morphological assessments but questioned records from Fond du Lac, Washington, Walworth, Racine, and Kenosha Counties. Later mitochondrial DNA analyses by Burghardt et al. (2006) supported the occurrence of Butler's Gartersnakes in Fond du Lac, Washington, and Ozaukee Counties while also identifying this species in Sheboygan County. Their analyses further illustrated a lack of concordance between morphology and genetics in snakes sampled from Racine and Kenosha Counties, providing evidence that hybridization with the Plains Gartersnake may be common in that area. Kapfer et al. (2013a) provided genetic and morphological evidence to support the presence of Butler's Gartersnakes in Jefferson County. Recent analysis of nuclear DNA also identified genetic markers of Butler's Gartersnakes in a population of snakes from Dodge County; however, all individuals examined in this population had morphological characteristics suggestive of Plains Gartersnakes (Sloss 2011; Rori Paloski, WDNR, personal communication). In total, genetic analyses have revealed suspected hybrid Butler's × Plains Gartersnakes, or lack of concordance between morphological and genetic data from snakes sampled, in Dodge, Dane, Jefferson, Waukesha, Milwaukee, Walworth, Racine, and Kenosha Counties (Burghardt et al. 2006, Sloss 2011, Placyk et al. 2012, Paloski et al. 2017; R. Paloski, unpublished data). Putative Butler's × Common Gartersnakes have been identified through analysis of nuclear DNA from snakes in Dane, Dodge, Washington, Ozaukee, Waukesha, and Milwaukee Counties (R. Paloski, unpublished data see Common Gartersnake, *Thamnophis sirtalis* species account).

Numerous studies and reports have contributed to the understanding of this species' distribution in Wisconsin (Vogt 1981, Casper 1996, Burghardt et al. 2006, Casper 2008d, Fitzpatrick et al. 2008, Sloss 2011, Placyk et al. 2012, Kapfer et al. 2013c, WDNR 2014b, also R. Paloski, personal communication). However, there is apparent discordance between the morphological traits and genetic markers proposed to differentiate Butler's Gartersnakes and Plains Gartersnakes, with no standardized criteria for distinguishing them when encountered in areas of range overlap. As such, the creation of a definitive county-level range map for Wisconsin is challenging. County records supported by both morphological and genetic data (i.e., counties from which assessed specimens had both morphological characteristics and genetic markers consistent with Butler's Gartersnake) include Fond du Lac, Sheboygan, Washington, Ozaukee, Jefferson, Waukesha, and Milwaukee (shaded black on the range map). Reports from Dodge County are based on genetic data only (i.e., snakes sampled in that population were morphologically consistent with Plains Gartersnakes despite having genetic markers indicative of Butler's Gartersnakes; shaded with black diagonal bars). Several counties have produced records or specimens not corroborated via genetic or morphological analyses, although some specimens are morphologically suggestive of Butler's Gartersnakes, including Juneau, Walworth, Racine, and Kenosha (shaded with crosshatching; Vogt 1981, Casper 1996; Natural History Box: Unusual Butler's Gartersnake Records in Wisconsin: A Historical Perspective).[‡] It is noteworthy that both 'pure' Butler's Gartersnakes and putative hybrid Butler's × Plains Gartersnakes have been reported in Dodge, Jefferson, Waukesha, and Milwaukee Counties based on genetic analyses (Fitzpatrick et al.

[‡] Note that Dodge and Sheboygan County records have not been published or have been published only in technical government reports (e.g., Burghardt et al. 2006, Casper 2008d, Sloss 2011). It is also noteworthy that not all reported voucher specimens for published county records have been confirmed through genetic analysis (see the Appendix).

NATURAL HISTORY BOX: Unusual Butler's Gartersnake Records in Wisconsin: A Historical Perspective

Recent genetic and morphological studies have not addressed Butler's Gartersnake specimens collected from Juneau County in the 1960s (e.g., UWSP 546, MPM 11791). Casper (1996) suggested that these were part of an "introduced" population. However, the collector of the Juneau County specimens believed the associated population was native, and aside from its disjunct position relative to the known range in Wisconsin, no evidence to support introduction was known (Tom Johnson, Missouri Department of Conservation [retired], personal communication). Regardless, additional work is necessary to both reconfirm the presence of this Juneau County population and determine its genetic relationship to populations in southeastern Wisconsin. If nontranslocated populations are documented in Juneau County, then Butler's Gartersnakes could further occur in adjacent counties (e.g., Adams, Marquette, Green Lake, Columbia).

This species was also reported in Dane County by Threlfall et al. (1973), but the record is suspect because no corroborating specimens or additional observations are known. Still, recent genetic analyses of closely related Plains Gartersnakes from Dane County found evidence of Butler's Gartersnake genetic markers (Rori Paloski, WDNR, personal communication). These findings could support the proposition that the distribution of Butler's Gartersnakes in Wisconsin was historically larger but that perhaps past populations in peripheral locations were genetically overwhelmed through hybridization with the Plains Gartersnake. This could also explain the presence of putative hybrid and pure Butler's Gartersnakes identified in Dodge County and the old specimens collected from Juneau County.

Joshua M. Kapfer, UW–Whitewater

2008, R. Paloski, personal communication). Further work is necessary to clarify this species' distribution in Wisconsin.

Localities of possible occurrence: Reports of disjunct populations in Wisconsin may support that this species' state distribution was once broader, encompassing more counties than are currently known (Natural History Box: Unusual Butler's Gartersnake Records in Wisconsin: A Historical Perspective). Kapfer et al. (2013a) postulated that 'pure' Butler's Gartersnakes could occur in areas of Walworth County based on the presence of genetically and morphologically confirmed individuals found ca. 1 km (0.62 mi) north of the county border into Jefferson County. This is consistent with reports of the species from Walworth County by Vogt (1981). Furthermore, Butler's Gartersnakes could occur in Calumet and Manitowoc Counties, given the proximity of known populations in Sheboygan County. 'Pure' Butler's Gartersnakes may also be present in some populations from Racine and Kenosha Counties. Confirmation of additional county records would require a combination of morphological and genetic analyses due to the potential for hybridization in Wisconsin.

Habitat: In Wisconsin, Butler's Gartersnakes can be found in mesic to wet prairies, sedge meadows, open wetlands, old fields, and vacant lots (Vogt 1981, Hileman 2010; see Amphibians and Reptiles in Wisconsin Plant Communities chapter). Open upland habitat adjacent to wetlands is particularly important for this species during the active season (Joppa and Temple 2005). For example, at three study sites in Wisconsin, Butler's Gartersnake captures were highest in open grassy upland areas near wetlands when compared to other upland and wetland habitat types (Kapfer et al. 2013b). Similarly, in southwestern Ontario, these snakes were most common in dry upland habitat (Catling and Freedman 1980b). Wetlands provide foraging and overwintering locations, and those dominated by native sedges (e.g., *Carex* spp.) are used more heavily than those with large stands of cattail (e.g., *Typha* spp.) or invasive species (Kapfer et al. 2013a). Closed canopy and shrubby habitats appear to be avoided in Wisconsin (Kapfer et al. 2013b, E. Hileman, personal observation) and Michigan, which suggests that wooded habitats could serve as a dispersal barrier (Carpenter 1952a). This species is frequently found beneath debris such as old logs, bark, or even discarded anthropogenic structures of wood or metal (E. Hileman, personal observation). Because its distribution includes several counties with extensive developed lands in southeastern Wisconsin, Butler's Gartersnakes are sometimes found in green space or undeveloped plots in disturbed landscapes (see Conservation section). For example, Rossman et al. (1996) collected this species "in old house foundations near urban areas" in Wisconsin, while Shonfield et al. (2019) frequently found radio-tracked snakes from Ontario in semi-open to open human-altered habitats.

Butler's Gartersnakes overwinter in a variety of hibernacula. Shonfield et al. (2019) followed four snakes to winter hibernacula and found two in crayfish burrows, one below a wood pile, and another along a "creek drain." In Wisconsin, they are more abundant at sites with crayfish burrows (Sarah Orlofske, UW–Stevens Point, personal communication) and may use crayfish burrows for overwintering and shelter during the active season (Joppa and Temple 2005, WDNR 2014b). Rossman et al. (1996) recounted collecting them "on 5 April 1955 in an old dump near Waukesha, Wisconsin. Most of the snakes were taken on the south face of a dirt bluff that appears to be their overwintering site. Others were basking on piles of bricks or matted grass." In Michigan, Butler's Gartersnakes were found communally overwintering in an ant mound (depth 35.6–68.6 cm [14–27 in]; Carpenter 1953). On another occasion, Carpenter (1953) excavated a number of amphibians and reptiles, including one Butler's Gartersnake, hibernating in a probable Meadow Vole (*Microtus pennsylvanicus*) burrow at a depth of 15.2–38.1 cm (6–15 in).

REPRODUCTION AND DEVELOPMENT

Copulation in wild individuals has not been observed in Wisconsin, but a single observation was recorded in Ohio on 4 April 1930 (Conant 1938). Finneran (1949) described an aggregation comprised of several males courting a female in Michigan on 19 April 1948. Other instances of multiple males pursuing a single female have been documented in captivity (Ruthven 1912, Noble 1937), and captive mating has been observed between late March and late April (Pope 1944). The typical breeding period for wild Butler's Gartersnakes is likely consistent with these observations, since other gartersnake species in the northern US and Canada typically mate soon after spring emergence; however, autumn mating also reportedly occurs in some species (reviewed in Rossman et al. 1996).

Males purportedly find females for reproduction via pheromone trails and exhibit preference for trails left by conspecifics over those produced by Common Gartersnakes (Ford 1982b). When given the option between conspecific and Plains Gartersnake female scents, male Butler's Gartersnakes could not differentiate between the two (Ford 1982b). However, individuals included in the latter experiment may have been hybrids, limiting interpretation of the results (Neil Ford, University of Texas–Tyler, personal communication). In courtship prior to copulation, the male climbs atop the female, moving over her in an anterior direction while contacting her dorsum with his chin and tongue. Once his head reaches her neck, he pushes his cloaca under her and produces high-frequency caudocephalic waves before inserting one of his hemipenes into her cloaca (see Noble [1937] for a more detailed account). After ejaculation, the male deposits a copulatory plug, which likely prevents other males from successfully mating with that female until the plug is expelled (Devine 1975). Additionally, Devine (1977) found that females with a copulatory plug were less likely to be pursued by males. Despite these deterrents, Albright (2001) documented multiple paternity in Milwaukee County.

Among three sites studied in Ozaukee and Waukesha Counties, the proportion of adult females gravid at each site in 2008 and 2009 ranged from 50.4% to 95.6% (E. Hileman, unpublished data). However, these reproductive rates are likely positively biased, as the coverboards used in this study created a favorable gestation environment (E. Hileman, personal observation). Reported reproductive rates from Michigan may be less biased, where visual searches were conducted without coverboards, and 67% of captured females were gravid (Carpenter 1952a). In Waukesha County, 40.0% of individual adult females were gravid consecutively in 2008 and 2009, 55.6% were gravid only one year, and 4.4% were not gravid either year ($n = 45$; E. Hileman, unpublished data), demonstrating that reproduction is not strictly biennial. In contrast, Carpenter (1952a) found that only 7% of Butler's Gartersnakes recaptured over time were gravid for consecutive years at a site in Michigan.

Butler's Gartersnakes are viviparous and have the lowest fecundity among the five gartersnake species that occur in Wisconsin (Carpenter 1952a, Vogt 1981). Albright (2001) reported litter sizes for captive and wild-bred females from Milwaukee County of 5–20 ($\bar{x} = 11.3$, $n = 9$), and Rossman et al. (1996) reported a mean litter size of 11.9 for Wisconsin. However, some individuals sampled in these studies may have been hybrid Butler's × Plains Gartersnakes (Placyk et al. 2012; N. Ford, personal communication). Despite this, similar litter sizes have been documented in Michigan. In Washtenaw County (Michigan), four females produced a mean of 11 offspring (Carpenter 1952a). Additionally, a range of 6–20 ($\bar{x} = 11.4$) offspring were produced by 28 females in southeastern Michigan (Ford and Killebrew 1983). Smaller litters were reported in Ontario, where six females produced 4–11 neonates ($\bar{x} = 8.5$; Freedman and Catling 1978).

In Wisconsin, both female weight and length were positively correlated with number of offspring produced (Albright 2001, Kirby 2005), but individual neonate weight decreased as litter size increased (Albright 2001). The findings of Albright (2001) and Kirby (2005) corresponded with those of Ford and Killebrew (1983; southeastern Michigan), who also noted this negative relationship and demonstrated that longer females have more offspring and heavier litters. Additionally, Ford and Killebrew (1983) provided evidence that relative clutch mass, as quantified by total litter mass / female mass, does not vary with female size. However, relative clutch mass per offspring decreases as female size increases. Thus, as adult female size increases, the influence of litter size on neonate mass decreases (Ford and Killebrew 1983). Vogt (1981) suggested that parturition occurs from 2 July through 18 September. Of three populations studied in Wisconsin, the earliest dates of neonate observation ranged from 18 July to 12 August (E. Hileman, unpublished data).

In Michigan, Carpenter (1952b) observed that growth rates slow with age and that male growth declined more rapidly than female growth. Due to this decrease in growth rates and variation in individual growth, young of the year and juveniles are easier to age than adults (Carpenter 1952b). In Wisconsin, females generally obtain adult size after the breeding period of their second active season (E. Hileman, unpublished data). Therefore, the earliest mating opportunity is likely in their third growing season (i.e., second spring). Carpenter (1952b) found a sexually mature male and three sexually mature females in Michigan that were comparable in length to snakes in their second spring, but he speculated that many individuals likely remain sexually immature until their third spring. In Wisconsin, the smallest gravid female found in Ozaukee and Waukesha Counties was 30.8 cm SVL (12.0 in; E. Hileman, unpublished data). In comparison, the smallest gravid female found in eastern Michigan was slightly longer (34.5 cm SVL [13.6 in]; Carpenter 1952b). Longevity in the wild has not been reported for this species.

ACTIVITY

Spring emergence is likely triggered after shallow soil temperatures (10 or 30 cm [3.9 or 11.8 in] belowground) exceed deeper soil temperatures (60 or 100 cm [23.6 or 39.4 in] belowground) for several days (Smith 2009, King and Hileman 2012, 2013). In Michigan, Butler's Gartersnakes were observed emerging from overwintering in an ant mound between 27 March and 24 May (Carpenter 1953). Carpenter (1952a) estimated an active season of 215 days in Michigan. Minton (1972) reported that

Butler's Gartersnakes are active from mid-March through the first week of November in northern Ohio and southern Michigan. Vogt (1981) suggested that Wisconsin populations likely have a similar phenology.

Cloacal temperatures of individuals in a Michigan population increased over the active season from 25.7°C (78.3°F) from April to June to 29.5°C (85.1°F) from July to September, with an overall active season mean cloacal temperature of 26.1°C (79.0°F; Carpenter 1956). Cloacal temperatures were typically higher than environmental temperatures in cooler weather and lower than environmental temperatures in warmer weather, and Carpenter (1956) attributed this to behavioral thermoregulation. Doughty (1994) demonstrated under laboratory conditions that critical thermal minimum (CTmin, defined as the inability of a snake placed on its back to right itself) generally decreased with latitude in five gartersnake species, including Butler's Gartersnake. Excluding neonates, the CTmin for Butler's Gartersnakes was 5.8°C (42.4°F; Doughty 1994).

Data on Butler's Gartersnake movement patterns are particularly sparse. Using 2009 capture-recapture data from a population in Waukesha County, adult females had a mean home range of 0.0654 ha (0.1616 ac, *n* = 29, MCP estimator), and individual snakes moved a mean of 2.2 m/day (7.1 ft/day, *n* = 82; E. Hileman, unpublished data). Based on typical distances traveled, Carpenter (1952a) estimated that the 'activity range' of Butler's Gartersnakes at a study site in Michigan was ca. 0.8 ha (2 ac). A radio telemetry study in Ontario reported that mean female home range size was 0.9 ha (2.23 ac, *n* = 12), while the home range size of the single male tracked was 0.26 ha (0.64 ac, MCP estimators; Shonfield et al. 2019). Carpenter (1952a) estimated minimum travel distances of 120.0 m (393.6 ft, *n* = 11) in >200 days and 114.8 m (376.7 ft, *n* = 15) in <200 days. One individual traveled a minimum distance of 121.9 m (400 ft) in two hours (Carpenter 1952a). In Amherstburg, Ontario, minimum travel distances over 70 days ranged from 10 to 433 m (33 to 1,421 ft, *n* = 5) for males and 10 to 517 m (33 to 1,696 ft, *n* = 19) for females (Freedman and Catling 1979). However, >50% of all snakes captured in this study moved <50 m (164 ft) from their initial capture location. Shonefield et al. (2019) reported movement rates of ca. 14 m/day (46 ft/day) for individuals tracked with radio telemetry equipment in Ontario.

Autumn ingress is likely initiated after late-season soil temperatures invert once again, and deeper soil temperatures (60 or 100 cm [23.6 or 39.4 in] belowground) become warmer than shallow soil temperatures (10 or 30 cm [3.9 or 11.8 in] belowground; Smith 2009, King and Hileman 2012, 2013). Carpenter (1952a) estimated a hibernation period of 150 days in Michigan.

PREY AND PREDATORS

Butler's Gartersnakes are well adapted to feed on earthworms, which are their primary food source. In Ohio, earthworms were the only prey item disgorged upon capture (Conant 1938), and in Ontario earthworms comprised 96% of regurgitated meals (Catling and Freedman 1980a). In Michigan, forced regurgitation revealed a diet composed of earthworms (83%) and leeches (10%), with the remaining material unrecognizable (Carpenter 1952a).

Dietary preferences may have influenced the North American distribution of this species. Relevant to this discussion is the fact that the earthworms (Lumbricidae) commonly encountered in the Great Lakes region today are of European origin and presumably introduced by settlers in the mid-1800s (Hale 2007). Historically, Butler's Gartersnakes likely persisted along the periphery of the Wisconsin Ice Sheet and invaded the Great Lakes region soon after the Quaternary glaciation (Conant et al. 1945, Smith and Minton 1957, Minton 1972). If native earthworms (Megascolecidae) occurred in this region, they were probably extirpated prior to the post-glacial invasion of the region by Butler's Gartersnakes (Hale 2007). Therefore, this annelid specialist could have been restricted to an alternate food source, such as leeches, which limited the snake to wetland habitats, until the introduction of European earthworms (Lumbricidae). The expansion of this new and eventually widespread food source then presumably expanded the range of the Butler's Gartersnake.

Butler's Gartersnakes likely use scent trailing as their principal means to locate prey and are able to capture earthworms while foraging underground

(Catling and Freedman 1980a). Catling and Freedman (1980a) also observed this species foraging for leeches underwater. The vomeronasal system used to detect chemical cues left by prey is likely functional at birth (see Special Remarks section), and Carpenter (1952a) observed that neonates in Michigan consume their first meal within hours of parturition.

Despite a clear specialization in annelid prey, Test (1958) documented consumption of a Spring Peeper (*Pseudacris crucifer*) by a wild Butler's Gartersnake in Michigan, and forced regurgitation yielded the first record of a Western Chorus Frog (*Pseudacris triseriata*) as prey in Ontario (Catling and Freedman 1980a). Additionally, a variety of laboratory experiments have demonstrated that Butler's Gartersnakes opportunistically consume amphibians and minnows in captivity (e.g., Carpenter 1952a, Halloy and Burghardt 1990) and can exhibit plasticity in prey preferences (Lyman-Henley and Burghardt 1995, see Special Remarks section for more detail). These phenomena may be a consequence of the species' close phylogenetic relationship with the Plains Gartersnake, which is more of a generalist predator (Burghardt 1993; see Systematics section).

A variety of animals likely include Butler's Gartersnakes in their diets. In three Butler's Gartersnake populations studied in Wisconsin, Eastern Milksnakes (*Lampropeltis triangulum*, Ozaukee County) and Northern Short-tailed Shrews (*Blarina brevicauda*, Waukesha and Ozaukee Counties; Natural History Box: Predation Event) were the most common predators found under coverboards (Hileman 2010). Carpenter (1953) provided evidence of crayfish predation on unidentified species of gartersnakes. Crayfish likely consume Butler's Gartersnakes, as this snake is known to use crayfish burrows as refugia in Wisconsin (Joppa and Temple 2005, WDNR 2014b). Throughout the Great Lakes region, additional predators probably include birds of prey and numerous small to midsized carnivorous mammals, ranging from weasels (*Mustela* spp.) to Red Foxes (*Vulpes vulpes*; Vogt 1981, Harding 1997).

When threatened, Butler's Gartersnakes are more prone to flee than strike (Herzog and Burghardt 1986, Herzog et al. 1992, Bowers et al. 1993, Albright 2001). In fleeing, they often thrash their body laterally while making limited forward movement. In response to simulated predator threats in the lab, Butler's Gartersnakes also exhibit tail waving, which is likely intended to redirect attacks away from the head (Bowers et al. 1993). If handled, this snake may secrete a foul-smelling musk containing the same volatile compounds found in musk of other North American gartersnakes (Wood et al. 1995).

Despite the overall tendency of this species to flee, neonates more frequently struck defensively at moving stimuli than stationary stimuli in laboratory experiments (Herzog et al. 1989a). Additionally, neonate males struck more often than females regardless of whether stimuli were moving (Herzog and Burghardt 1986, Herzog et al. 1989a). Juveniles, on the other hand, were equally unlikely to strike whether or not a stimulus was moving (Herzog et al. 1989b). Kirby (2005) conducted antipredator tests and compared Butler's Gartersnakes from Wisconsin, Michigan, and Ohio to Plains Gartersnakes from Illinois at 2 days old and 22–25 days old. Overall, she found that Wisconsin Butler's Gartersnakes (many from counties now known to have hybrids; Placyk et al. 2012) displayed antipredator behaviors more similar to those of Plains Gartersnakes from Illinois than to those of Butler's Gartersnakes from Michigan and Ohio. Specifically, at two days old, Wisconsin Butler's Gartersnakes were less likely to flee or tail wave when exposed to simulated threats than individuals from Michigan and Ohio. Similarly, at 22–25 days old, Wisconsin Butler's Gartersnakes were more prone to strike and less likely to flee than snakes from Michigan and Ohio. Additionally, Kirby (2005) found that Butler's Gartersnakes from Racine County behaved more aggressively (similar to Plains Gartersnakes) than Butler's Gartersnakes from Ozaukee County. For instance, individuals from Racine County were more likely to strike and less prone to flee than those from Ozaukee County. As previously mentioned, hybrids have since been identified in Racine (see Systematics and Distribution and Habitat sections).

CONSERVATION

Status: Globally, Butler's Gartersnake is listed as a Least Concern species by the IUCN. In Ontario and Indiana, it is listed as Endangered (COSEWIC 2010a, Indiana Legislature 2011). It is currently

NATURAL HISTORY BOX: Predation Event

On 14 March 2010, while removing plywood coverboards used for a capture-recapture study of Wisconsin Butler's Gartersnakes, I discovered a shrew cache of ten young Butler's Gartersnakes under one of the boards (Figure 272a). Seven of the ten snakes were fresh kills, with their heads chewed off but their bodies still writhing (Figure 272b). The shrew was not present at the time, but based on the nest, prey cache, and ubiquity of the species, it was most likely a Northern Short-tailed Shrew (*Blarina brevicauda*) or, less likely, a Masked Shrew (*Sorex cinereus*). These voracious predators likely pose a threat year-round, but hibernating snakes may be at greatest risk, since shrews remain active throughout winter (Jackson 1961). Moreover, the Northern Short-tailed Shrew has the added advantage of subduing its prey with venom secreted from its submaxillary salivary glands (Pearson 1942).

Figure 272. (a) Shrew cache of young Butler's Gartersnakes found under a plywood coverboard (Ozaukee County; photo by E. Hileman). (b) Photograph documenting the carnage inflicted by the shrew on individual snakes (Ozaukee County; photo by E. Hileman).

Eric T. Hileman, Mississippi State University

considered a species of Special Concern and a Species of Greatest Conservation Need in Wisconsin by the WDNR (Natural History Box: A Brief History of Butler's Gartersnake Protection in Wisconsin).

Populations: Little is known about the population ecology of Butler's Gartersnake in Wisconsin. Hileman (2010) provided population estimates for three study sites in Ozaukee and Waukesha Counties; however, the methods used to make these estimates are unreliable. Carpenter (1952a) estimated a population size of 121 individuals on a 19.4 ha (48 ac) study plot in Washtenaw County, Michigan. Freedman and Catling (1978) conducted a capture-recapture study from 14 May to 24 July 1976 on a 40 ha (99 ac) abandoned quarry in Amherstburg, Ontario. They pooled age classes and sexes (excluding neonates) to estimate population size (900 snakes) and density (23/ha [57/ac]) with traditional closed-capture models. Using data provided by Catling and Freedman (1980b),

NATURAL HISTORY BOX: A Brief History of Butler's Gartersnake Protection in Wisconsin

The history of Butler's Gartersnake conservation in Wisconsin is rife with conflict and controversy. Originally listed as state Endangered in 1972 (Wisconsin Administrative Code NR 27.01), Butler's Gartersnake was delisted in 1975 after significant populations were found in the greater Milwaukee area. Habitat loss and fragmentation caused by human development and encroachment of woody vegetation, in addition to concerns of hybridization with the Eastern Plains Gartersnake, resulted in relisting the species in 1997, this time as state Threatened (Hyde et al. 2007).

The protection afforded to Butler's Gartersnake as a Threatened species caused intense debate and conflict between conservationists, politicians, landowners, and land developers over how to balance economic development with preservation of the species and its habitat (Bergquist 2005, Schultze 2005, Enriquez 2006, Lydersen 2006, Sheeley 2007). This is not surprising, given that the core of the snake's range falls within the greater Milwaukee metropolitan area, the most densely populated and urbanized region in Wisconsin. The conflict came to a head in 2006 when a state legislative committee voted to delist the snake (Bergquist 2006b). According to Burghardt et al. (2009), "Only political change through recent elections prevented politicians from, for the first time in U.S. history, delisting an endangered species without even a superficial reliance on scientific data." This political change resulted in the vote being rescinded by the Joint Committee for Review of Administrative Rules (by a vote of seven to two), and the snake's Threatened status was preserved (Bergquist 2006a).

Protection for Butler's Gartersnake was threatened again on 16 May 2008, when the Wisconsin Department of Natural Resources (WDNR) issued a proposed regulatory framework regarding authorization and the definition of 'incidental take' (hereafter, 'take') under Wisconsin's Endangered and Threatened Species Law (Wisconsin Statutes 29.604). (The WDNR currently considers 'incidental take' to be the unintentional loss [or take] of species that are listed as endangered or threatened at the state level [Rori Paloski, WDNR, personal communication]. 'Take,' then, is defined as "shooting, shooting at, pursuing, hunting, catching or killing any wild animal" [Wisconsin Administrative Code NR 27.01].) Presidents of the American Society of Ichthyologists and Herpetologists, the Herpetologists' League, the Society for the Study of Amphibians and Reptiles, and the Canadian Association of Herpetologists argued that the proposed regulatory framework reinterpreted the WDNR's authorization and the definition of 'take' and substantially weakened protection of Butler's Gartersnake as a Threatened species. This interpretation authorized agency programs to approve development projects on the snake's critical upland habitat as long as mortality would likely not occur. Recognizing that destruction of overwintering habitat would result in indirect mortality, society presidents urged officials to reconsider its new interpretation of the state statute in a letter on 1 October 2008. The WDNR contended that agency programs already had this authority and that the regulatory framework adhered to the definition of 'take' as defined in the state statute. It therefore maintained its position. Citing species stability in Wisconsin based on new information on abundance, range, and hybridization, the WDNR downgraded Butler's Gartersnake from Threatened to Special Concern per administrative rule ER–27-11 on 1 January 2014 (Hyde et al. 2012; WDNR 2014b).

Eric T. Hileman, Mississippi State University

Parker and Plummer (1987) estimated density for this same site as 13–40 adults/ha (32–99/ac) and reported SE (26–42/ha [64–104/ac]) as a percentage of the population estimate. Using Carpenter's (1952a) population estimate, Parker and Plummer (1987) also estimated a density of 6.2 individuals/ha (15.3/ac). However, this is likely an underestimate, as the area used for density estimation incorporated all habitat within the plot, including a wooded area where snakes were never found (Carpenter 1952a). Hileman (2014) also estimated adult density and abundance at a Wisconsin site from 2011 to 2013 by employing spatially explicit capture-recapture models for two trapping areas within this same site. Density ranged from 291 to 609/ha (719 to 1,505/ac). Based on SE, the most reliable density estimates from these two study areas in 2011 were 309/ha (764/ac, 95% CI = 257–362/ha [635–895/ac]) and 437/ha (1,080/ac, 95% CI = 316–558/ha [781–1,379/ac]). Combining the two trapping areas (0.4 ha [1.0 ac]), the 2011 adult abundance estimate was 439 (95% CI = 409–485; Hileman 2014). Density estimates reported in Wisconsin are much higher than those reported for Michigan and Ontario; however, because estimation methods varied widely between studies, they may not be comparable. Using Cormack-Jolly-Seber models and a composite of males and females age one year or greater, Hileman (2014) estimated apparent survival for a Milwaukee County population. Apparent survival ranged from 0.42 to 0.55 across years (2007–2013), but the annual estimate of 0.55 (95% CI = 0.38–0.70) had the lowest SE. Annual recapture probabilities varied from 0.04 to 0.21 across years.

Threats: Habitat loss and fragmentation from agriculture, development, natural succession, and proliferation of invasive plant species pose significant threats to Butler's Gartersnakes (see Introduction to Conservation and Management of Wisconsin's Amphibians and Reptiles chapter). Interestingly, over 20 years ago, Rossman et al. (1996) reported that Butler's Gartersnakes were a suburban species with relatively stable populations. Citing Minton (1968) and Vogt (1981), Ernst and Ernst (2003) also suggested that urbanization may be beneficial to Butler's Gartersnakes. However, Minton suggested that this benefit might be temporary, and Vogt merely stated that the snakes inhabit city lots and that "excellent populations of Butler's garter snakes can still be found within the city limits." More recently, however, this species' use of anthropogenically disturbed sites has raised concerns. For example, its association with suburban landscapes in Wisconsin is likely a by-product of a restricted distribution that naturally coincides with the heavily populated southeastern counties. Consequently, Butler's Gartersnake populations in southeastern Wisconsin are often limited to small disturbed plots and green spaces within urban landscapes, which perhaps give the false impression that they are urbanophiles. Populations in urban areas are likely more susceptible to the effects of habitat degradation, vehicular strike, and human persecution. Furthermore, Freedman and Catlin (1979) pointed out this species' tendency to avoid roads, which could fragment urban populations. Rossman et al. (1996) noted that fragmented Butler's Gartersnake populations could be more heavily impacted if isolated subpopulations occur on landscapes faced with habitat loss due to heavy development pressure. Conversely, Rossman et al. (1996) cited evidence via personal communication with J. Harding that indicates that this species does not appear to avoid roads in Michigan. These conflicting reports suggest that the impact of roads on this species is complex and warrants further investigation.

Research in Wisconsin supports the negative impacts of invasive vegetation. For example, Kapfer et al. (2013b) captured fewer Butler's Gartersnakes in wetland areas dominated by the invasive Reed Canary Grass (*Phalaris arundinacea*) than in areas dominated by native wetland sedges. Given the pervasiveness of Reed Canary Grass, extensive degradation of habitat will likely continue if this invasive grass is not controlled. While hybridization with Plains Gartersnakes could be a historical and nonanthropogenic phenomenon (Placyk et al. 2012), the long-term conservation impacts of this are unknown. Recent molecular work provides evidence that Wisconsin Butler's Gartersnake populations have unique haplotypes from populations in other states and are especially important in maintaining overall genetic diversity for this species (Burghardt et al. 2006, Placyk et al. 2012; see Systematics section).

Climate change poses a threat to reptiles and amphibians worldwide. Using maximum entropy ecological niche modeling, King and Niiro (2013) evaluated the effects of climate change on 11 reptile species of conservation concern in the midwestern United States. Butler's Gartersnake was predicted to be one of the least affected species as measured by reductions in climatically suitable habitat. In addition, their results suggest that climatically suitable habitat for Butler's Gartersnake might actually expand into new localities. However, this expansion would not necessarily lead to a range expansion for Butler's Gartersnakes due to the widespread threats of habitat loss and fragmentation described above.

Management: Preserving open wetlands and adjacent open uplands is critical to maintaining suitable habitat for this species. As such, general management concepts associated with these habitat types are relevant, as are considerations of habitat connectivity and timing of management activities (see Introduction to Conservation and Management of Wisconsin's Amphibians and Reptiles chapter). Protection of upland buffer zones around wetlands is important (Kapfer et al. 2013a), although the ideal width of these zones remains in question. At six sites studied in southeastern Wisconsin, Joppa and Temple (2005) found snakes in adjacent uplands as far as 122 m (400 ft) from the wetland edge. Additionally, Kapfer et al. (2013a) cited R. Hetzel (via personal communication) as finding Butler's Gartersnakes up to 182 m (597 ft) from wetland habitat. These reports and other information on movement (see Activity section) could help in the design and implementation of appropriate wetland buffers to protect this species. It is also important to preserve suitable overwintering sites for population persistence. Therefore, potential hibernacula (e.g., crayfish burrows) should be identified prior to land development or management activities. Given that Butler's Gartersnakes do not frequently use shrubby habitats (Kapfer et al. 2013b), mitigating the spread of woody vegetation in preferred wetlands is an important consideration. Management strategies for this species should also consider strategies to control the invasive plants, such as Reed Canary Grass (Kapfer et al. 2013b). Road underpasses have been used in attempts to mitigate road mortality at several sites in Wisconsin (R. Paloski and Lisie Kitchel, WDNR, personal communication). However, the effectiveness of road underpasses as a management tool for Butler's Gartersnakes has not been demonstrated.

SPECIAL REMARKS

In laboratory conditions, Butler's Gartersnakes voluntarily consumed a Northern Leopard Frog (*Lithobates pipiens*), a Spring Peeper, an American Toad (*Anaxyrus americanus*), an Eastern Red-backed Salamander (*Plethodon cinereus*), and small minnows (Carpenter 1952a). Snakes as young as two months old were able to capture and ingest fish in the lab but exhibited difficulty doing so (Halloy and Burghardt 1990). Others have also observed captive Butler's Gartersnakes consuming small frogs and fishes in addition to earthworms (Ruthven 1904, Ruthven 1908, Conant 1938, Lyman-Henley and Burghardt 1995). Burghardt (1993) speculated that the ability of this species to recognize fishes as prey items is a consequence of its ancestral lineage to the Plains Gartersnake. Additionally, he hypothesized that while this prey recognition is still intact, Butler's Gartersnake has evolved morphological traits better suited to annelids. Cunningham and Burghardt (1999) noted that captive individuals provided with earthworms groomed themselves after 95% of feedings by rubbing their labial and rostral scales against various surfaces.

Laboratory experiments have also revealed details about dietary preferences and sensory function in neonates. Inexperienced captive-born individuals from a wild-caught Michigan female displayed significant tongue flicking and attack responses when presented with surface chemicals extracted from worms, leeches, fishes, and amphibians but not slugs, mice, or a water control (Burghardt 1967, 1969). Additionally, Lyman-Henley and Burghardt (1995) demonstrated plasticity in prey preference. In their experiments, naive ten-day-olds born to wild-caught Michigan females showed preference for worm surface extracts over fish extracts. However, after 149 days, neonates that were fed a strict Mosquito Fish (*Gambusia affinis*) diet shifted their preference to fish extracts, while neonates fed only earthworms (*Lumbricus terrestris*) did not. Burghardt and Pruitt (1975) reported that neonate Common Gartersnakes with tongues completely removed did not attack swabs containing prey extracts. Moreover, in comparison to snakes with tongues intact, those without tongues showed significantly lower interest in the extract swabs (Burghardt and Pruitt 1975). Based on these results, the vomeronasal system seems to be functional at birth and likely mediates chemical cue responses (Burghardt and Pruitt 1975, Holtzman 1993). These findings extend to Butler's Gartersnakes because the vomeronasal system is structurally identical to the Common Gartersnake's throughout development (Holtzman and Halpern 1990). Olfaction may also contribute to chemical cue recognition by newborns (Holtzman 1993).

Kirby (2005) performed a series of experiments comparing neonate Butler's Gartersnakes from Wisconsin (Milwaukee, Ozaukee, Racine, and Waukesha Counties), Michigan, and Ohio to neonate Plains Gartersnakes from Illinois. In chemical cue choice tests, Butler's Gartersnakes from Wisconsin were more likely to respond to fish extracts than Butler's Gartersnakes from Michigan and Ohio. Based on these results, Kirby (2005) concluded that Wisconsin Butler's Gartersnakes share greater feeding response similarities to Plains Gartersnakes than to other Butler's Gartersnakes across their range. Additionally, when compared to Ozaukee County, Butler's Gartersnakes from Racine County were more prone to attack fish extracts. These results are consistent with recent genetics work, which has identified hybrids in Racine and pure Butler's Gartersnakes in Ozaukee (see Systematics and Distribution and Habitat sections).

Plains Gartersnake

Thamnophis radix (Baird and Girard 1853)

Collin P. Jaeger and Joshua M. Kapfer[§]

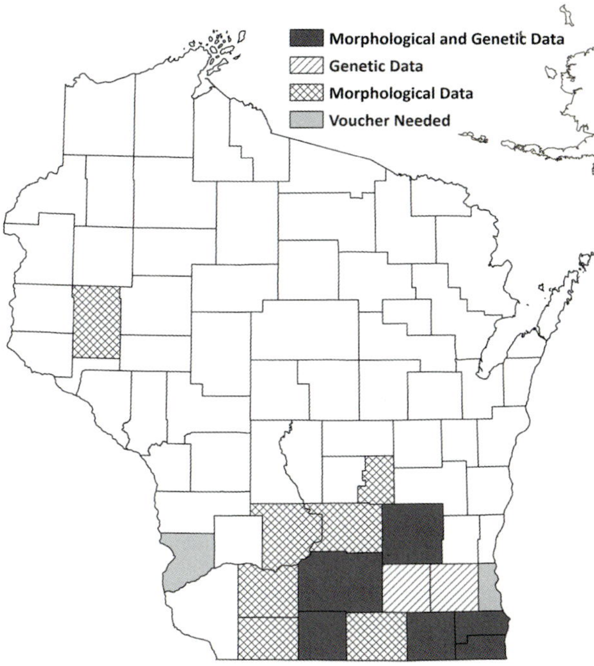

Adult Plains Gartersnake (Dane County; photo by C. Raimond).

[§] We acknowledge J. Lorch, G. Schuurman, and R. Paloski for assistance with the Systematics, Description, and Distribution and Habitat sections.

AT A GLANCE

Plains Gartersnakes are one of several members of the genus *Thamnophis* found in Wisconsin. They primarily occur in the southern counties and have been most often documented in the south-central part of the state. Although Plains Gartersnakes also occur in southeastern Wisconsin, these populations overlap with those of Butler's Gartersnakes, and hybridization can occur. As such, definitive identification of Plains vs. Butler's Gartersnakes (and hybrids between the two) in the field can be particularly challenging in southeastern Wisconsin. Generally, the lateral yellow stripes on Plains Gartersnakes occur higher up on the body than on other gartersnakes, covering primarily scale rows three and four. Plains Gartersnakes also have distinct black bars on their supralabial (upper jaw) scales, unlike other gartersnakes in the state. Plains Gartersnakes are active in Wisconsin from early April to September or October. They are primarily associated with grassy areas or 'meadows' adjacent to aquatic habitats. This species could occupy locations that are slightly warmer and drier than the locations occupied by other sympatric gartersnakes, such as the Common Gartersnake, and may therefore occur farther from wetland edges than their relatives. Breeding occurs soon after emergence from hibernation, and females give birth to live young in August. Although somewhat of a dietary generalist, they reportedly prey heavily upon earthworms and frogs. Plains Gartersnakes are currently considered a species of Special Concern and a Species of Greatest Conservation Need by the WDNR.

SYSTEMATICS

Thamnophis radix (Plains Gartersnake) is one of more than 30 species currently recognized within the genus *Thamnophis* (Rossman et al. 1996). Multiple phylogenetic analyses based on a variety of molecular and morphological characters position *T. radix*, *T. butleri* (Butler's Gartersnake), and *T. brachystoma* (Short-headed Gartersnake) in a well-supported clade, with *T. radix* and *T. butleri* as sister taxa (de Quieroz and Lawson 1994, Alfaro and Arnold 2001, Pyron et al. 2013, McVay et al. 2015).

Analyses of genetic variation at mitochondrial DNA loci and amplified fragment length polymorphisms have revealed evidence of potential hybridization between *Thamnophis radix* and *T. butleri* in the Milwaukee area where both species co-occur (Fitzpatrick et al. 2008, Placyk et al. 2012). Interestingly, the genetic variation appears to be bimodally distributed, and the distinctiveness of each species has been maintained, suggesting that this may be an example of speciation with gene flow (Nosil 2008). This pattern of gene flow within the hybrid zone is probably natural and historical rather than the result of anthropogenic factors such as habitat loss and fragmentation (Placyk et al. 2012). The degree to which individuals can be accurately identified in the field as *T. radix*, *T. butleri*, or possible hybrids remains to be examined.

DESCRIPTION

Adults: The dorsum is dark, from black to dark olive brown, with three longitudinal stripes of light yellow to orange (Figure 273). In addition to a single middorsal stripe, a lateral stripe runs down each side of the body that typically contacts scale rows three and four (Figure 274). Ventral coloration varies from light yellow to grayish green, and black bars are present on the supralabial scales (Figure 273).

Figure 273. Plains Gartersnake displaying the yellow-orange coloration of the middorsal stripe and black bars on the supralabial scales (Dane County; photo by C. Raimond).

Figure 274. A close-up of the lateral stripe primarily covering scale rows three and four on Plains Gartersnakes (Dane County; photo by C. Raimond).

Plains Gartersnakes are medium-sized, with adult SVLs that typically range from 38 to 71 cm (15–28 in; Vogt 1981, Ernst and Ernst 2003). Stanford and King (2004) modeled individual growth in a Plains Gartersnake population in northern Illinois and reported that the mean SVL of the largest snakes was 48.3 cm (19.0 in) for males and 58.5 cm (23.0 in) for females. However, some individuals reportedly exceed 100 cm SVL (39.3 in; Breckenridge 1944).

Preadult stages: Hatchlings and juveniles appear similar to adults in color and pattern. Stanford (2002) reported a mean SVL at birth of 13.8 cm (5.4 in, $n = 557$ individuals from 37 families). Estimates of mean SVL for juveniles captured during their first year of life are 17.1 cm for males (6.7 in, range = 12.9–24.8 cm [5.1–9.8 in], $n = 25$) and 19.7 cm for females (7.8 in, range = 13.0–45.5 cm [5.1–17.9 in], $n = 40$; Stanford 2002).

Variation: Reports of albino (May 1952, Hensley 1959, Dyrkacz 1975, 1981, Hallmen and Chlebowy 2001, Hallmen 2002) and melanistic (Morris 1975, 1976) individuals exist. Furthermore, Redmer and Zaworski (1987) encountered two red Plains Gartersnakes in Illinois. Females grow more quickly and attain larger sizes than males (Seibert and Hagen 1947, Stanford 2002).

Diagnosis: Plains Gartersnakes can be identified by their keeled scales in 21 or fewer rows at midbody, divided subcaudal scales, and a single anal plate (Ernst and Ernst 2003). They have 2 prefrontal scales, while each side of the head possesses 7–8 (rarely 6) supralabial scales, 9–10 (rarely 8 or 11) infralabial scales, 1 (rarely 2) preocular scale, and 1–3 (typically 3) postocular scales. Their ventral scale count ranges from 138 to 175 ($\bar{x} = 156.5$). Their light lateral stripe covers the third and fourth scale rows.

Plains Gartersnakes could be confused with any other gartersnake species (*Thamnophis* spp.) in Wisconsin. They are especially similar in appearance to Butler's Gartersnakes, and potential hybridization could prevent accurate diagnosis in the field where these species co-occur (see Systematics section). In locations where they do not co-occur, the position of the lateral stripes on the third and fourth scale rows can generally distinguish Plains Gartersnakes from Butler's Gartersnakes and Common Gartersnakes (*Thamnophis sirtalis*) because the lateral stripe in the latter two species contacts lower scale rows. Plains Gartersnakes also have thick black bars on the posterior edges of the supralabial scales, which are completely absent in Eastern Ribbonsnakes (*T. saurita*) and Western Ribbonsnakes (*T. proximus*). While supralabial bars are present in Butler's Gartersnakes and Common Gartersnakes, they are generally thinner than those of Plains Gartersnakes.

DISTRIBUTION AND HABITAT

Global distribution: Plains Gartersnakes are found throughout the North American Great Plains region, westward to the Rocky Mountains, northward into central Canada, and southward to the northern reaches of New Mexico, Texas, and Oklahoma. They also occur eastward into the north-central plains regions of Minnesota, Wisconsin, Illinois, Indiana, and Missouri. An isolated population occurs in central Ohio.

Distribution in Wisconsin: Plains Gartersnakes hybridize with Butler's Gartersnakes in southern Wisconsin (see Systematics section), which makes delineation of their state range problematic (Vogt 1981, Casper 1996, Burghardt et al. 2006, Fitzpatrick et al. 2008, Sloss 2011, Placyk et al. 2012, WDNR 2016a, Rori Paloski, WDNR, unpublished data). The list of counties where Plains Gartersnakes reportedly occur has changed over time as genetic analyses investigating hybridization have either confirmed or challenged existing county records (see also Butler's Gartersnake, *Thamnophis butleri* species account).

Based on concurrence between morphological characteristics and genetic markers (nuclear or mitochondrial DNA), Plains Gartersnakes occur in Dodge, Dane, Green, Walworth, Racine, and Kenosha Counties (shaded dark on range map; Vogt 1981, Casper 2003, Burghardt et al. 2006, Sloss 2011, Placyk et al. 2012, R. Paloski, unpublished data). Plains Gartersnakes also reportedly occur in Jefferson and Waukesha Counties, according to investigations of genetic markers alone (nuclear DNA), with limited to no corroborating morphological information (black diagonal bars on range map; Sloss 2011, Paloski et al. 2020).** Furthermore, records from Columbia, Dunn, Green Lake, Iowa, Lafayette, Rock, and Sauk Counties exist that are based only on morphological characteristics, with no corroborating genetic information (cross-hatched on range map; Vogt 1981, Casper 2003, 2015). Several previously published county records are also currently missing suitable museum vouchers. For example, the voucher reported for Crawford County (shaded gray) is apparently lost (HDW-NIU 1936; Richard King, Northern Illinois University, personal communication), and no additional supportive morphological or genetic data are known. In addition, a suitable voucher specimen could not be found to support Milwaukee County records (although genetic investigations have confirmed Plains Gartersnakes in Milwaukee County). Potential Milwaukee County voucher specimens exist in museum collections, such as MPM 552 and 2518, reported by Smith (1949), but many have not been rigorously analyzed, or they were later reidentified as either Butler's Gartersnakes or possible hybrids, according to museum records. Genetic investigations have identified both Plains Gartersnakes and putative Plains × Butler's Gartersnake hybrids in Dane, Dodge, Jefferson, Milwaukee, Racine, Walworth, and Waukesha Counties (Fitzpatrick et al. 2008, R. Paloski, unpublished data). Additional genetic investigations could lead to further refinement of this species' state distribution, particularly in counties where hybridization with other gartersnakes occurs.

Localities of possible occurrence: The currently documented distribution of Plains Gartersnakes in the state is likely incomplete given the disjunct record from Dunn County. Casper (1996) also mentioned a "single sight record" from Eau Claire County (adjacent to Dunn) and suggested that northwestern Wisconsin populations could be "contiguous with Minnesota populations." Given the missing Crawford County voucher (see Distribution in Wisconsin subsection), Plains Gartersnakes cannot be confirmed in any county immediately adjacent to the Mississippi River. Yet this species' presence in Minnesota, Iowa, and Illinois (Ernst and Ernst 2003) supports that it could occur in counties bordering the Mississippi River in Wisconsin. It is also possible that Plains Gartersnakes could exist in Richland and Grant Counties, given their presence in adjacent counties (e.g., Sauk and Iowa). Future work to obtain additional records in western Wisconsin is warranted.

Habitat: Most information on the habitats used by this species comes from anecdotal statements in field guides or accounts of observed habitat associations in the field with small sample sizes. In Wisconsin, they reportedly occupy open grassy areas associated with wetlands, such as mesic and wet prairies, but also drier prairies and oak savannas (Vogt 1981; see Amphibians and Reptiles in Wisconsin Plant Communities chapter). Vogt (1981) further found them in open grassy habitats adjacent to rivers but "not in bogs or marshes." In Minnesota, Jordan (1967) reported 21 individuals captured in "typical prairie situations" that were closely associated with aquatic habitats. Breckenridge (1944) surveyed a 17.7 km (11 mi) segment of road that crossed "low, marshy prairie land" in Minnesota and encountered 93 individuals. In the southern Minnesota River valley, Jacobs (1950) reported them from a grassland-deciduous forest ecotone but collected few observations. Gregory (1977a) found ca. 51 active Plains Gartersnakes "in or near marshes or in meadows adjacent to marshes" in Canada.

** It is noteworthy that not all reported voucher specimens for published county records have been analyzed and confirmed genetically (see the Appendix).

Hart (1979) analyzed a much larger sample of Plains Gartersnakes from Canada ($n = 210$) and determined that this species is strongly associated with wetlands that possess open water and a well-defined wetland-upland transition zone. Plains Gartersnakes frequently co-occur at sites with Common Gartersnakes. Both have similar habitat associations, but Ernst and Ernst (2003) believed that when sympatric with Common Gartersnakes, Plains Gartersnakes tended to occupy more open, drier locations. Hart (1979) also found evidence for niche discrimination between these species when sympatric. For example, he reported that Plains Gartersnakes often occurred in "meadows" with higher light intensity, high air temperatures, and relatively lower substrate temperatures than did Common Gartersnakes.

Direct observations of overwintering habitats are particularly sparse. In Wisconsin, Vogt (1981) reportedly found Plains Gartersnakes using old house foundations and ant mounds as hibernacula. They also overwintered in "postholes and under sidewalks of defunct real estate developments in the Chicago region," according to Pope (1944). Dalrymple and Reichenbach (1981) encountered active individuals during the early spring (first week of April) that escaped down nearby crayfish burrows in Ohio, and they surmised that these burrows were used as hibernacula. The burrows of small mammals (i.e., gophers, ground squirrels, voles) have also been suggested as hibernacula (Ernst and Ernst 2003). The exact soil depth preferred by hibernating snakes is unknown. Bailey (1949) conducted manipulative experiments in Iowa that involved burying individuals in containers at various depths during the winter while concurrently monitoring soil temperature. He found that snakes kept at depths of 15.2 and 30.5 cm (6 and 12 in) died, while those housed at soil depths of 45.7–91.4 cm (18–36 in), with minimum temperatures of −2.2°C to −1.5°C (28.0°F to 29.3°F), survived, although this was above the maximum frost depth (see further discussion in Activity section).

REPRODUCTION AND DEVELOPMENT

Mating and courtship occur in spring. Stanford (2002) observed mating activity between late April and early May but found sperm in the cloacal smears of males from early April to late August. Males are able to locate females for reproduction through olfactory cues (i.e., sex pheromones; Ford and Schofield 1984). Breeding aggregations of numerous individuals are reported, and Vogt (1981) found two males courting a female "in juniper bushes about ½ m off the ground near an apartment complex in downtown Madison. At least six other snakes were around the bushes." There also appears to be substantial competition among males to achieve high reproductive success and interfere with rival males during the breeding season. For example, males secrete mating plugs, which they adhere to female cloacae after copulation, making them less attractive to other males (Ross and Crews 1978). Despite this, Wusterbarth et al. (2010) documented multiple paternity in the offspring produced by individual females. A single case of apparent parthenogenesis in a captive female has been reported (Murphy and Curry 2000a, 2000b).

Females are viviparous (Blackburn 2000), and parturition in the Upper Midwest occurs from late summer to early autumn, typically during August (Smith 1961, Stanford 2002, Stanford and King 2004). The number of offspring per litter varies annually and geographically but is often positively correlated with the SVL of the mother (Seigel and Fitch 1985, Stanford 2002, Tuttle and Gregory 2014). Work in Canada also found that litter size can be influenced by environmental conditions, such as weather patterns in the previous year (Tuttle and Gregory 2014), which the authors suggested was because "warm, sunny weather facilitates the acquisition of capital resources for the subsequent year's reproduction."

Vogt (1981) observed that the mean litter size from Wisconsin females was "around 25." In Illinois, reported mean litter sizes include 18.4 (range = 9–31, $n = 46$ litters; Stanford 2002), 14.6 (range = 10–27, $n = 7$ litters; Smith 1961), 13.3 ($n = 43$ litters; Cieslak 1945), and 17 ($n = 12$ litters; Rossman et al. 1996). In Ohio, Fitch (1970) uncovered a mean litter size of 29.2 ($n = 16$ litters), and Reichenbach and Dalrymple (1986) determined that mean litter size was 15.2 ($n = 39$ litters). The work of Seigel and Fitch (1985) in Missouri revealed litter sizes of ca. 10 ($n = 52$ litters). In the western US, Rossman et al. (1996) reported a mean litter size of 11.6 ($n = 8$ litters) from Colorado, and Lesch and Fawcett (1978) reported a mean litter size of 20.6 ($n = 16$ litters) based on data from Nebraska, Kansas, and Colorado. In South Dakota, Smith and Zimmer (2013) documented a litter of 90 apparently healthy offspring from a single female, along with 1 unfertilized ovum and 6 stillborn offspring. Further north, Gregory (1977) reported a mean litter size of 29.5 ($n = 6$ litters) in Manitoba, and Tuttle and Gregory (2014) reported a mean litter size of 22.7 (range = 7–41, $n = 84$ litters) from a population in Alberta.

Growth rates vary by snake age, size, and season. For example, within their first year, Plains Gartersnakes in Illinois grew a mean of 1.57 cm/wk (0.62 in/wk) from May through July but only 0.9 cm/wk (0.36 in/wk) from August through the end of October (Seibert and Hagen 1947). Standford (2002) further found that growth rate decreased as size increased, leading to an asymptotic growth curve typical of most snakes. Size at sexual maturity was 36 cm SVL (14.2 in) for males and 38 cm SVL (15.0 in) for females in Illinois (Stanford 2002). Individuals likely mature during their second year (reviewed by Gregory 1977a). Females exhibit a higher rate of growth and attain larger adult sizes than males (Siebert and Hagen 1947, Stanford 2002). Maximum reported longevity was slightly greater than eight years and five months in a captive individual (Snider and Bowler 1992).

ACTIVITY

Plains Gartersnakes emerge from overwintering sites in early April and disperse soon after (Gregory 1977a, Dalrymple and Reichenbach 1981, Tuttle and Gregory 2009, Moriarty and Hall 2014). When active, this species is primarily diurnal. Vogt (1981) stated that, unlike Common Gartersnakes, he never observed Plains Gartersnakes active at night in Wisconsin. Activity is strongly influenced by temperature, with increased activity associated with ambient air temperatures from 21°C to 29°C (69.8°F to 84.5°F; Seibert and Hagen 1947). In the spring, activity is greatest during midday, but when air temperatures exceed acceptable levels during the summer months, activity becomes mainly bimodal (i.e., greatest earlier and later in the day; Dalrymple and Reichenbach 1981). The mean cloacal temperature of free-ranging Plains Gartersnakes in Ohio was 27.9°C (82°F, 95% CI = 27.0°C –28.8°C [80.6°F–83.8°F], $n = 65$; Dalrymple and Reichenbach 1981). Lueth (1941) reported that the upper thermal tolerance of Plains Gartersnakes in captivity was between 40°C and 42°C (104°F and 107.6°F).

Published data on the movements and spatial ecology of this species in Wisconsin do not exist. Work from other Upper Midwest locations indicates that the activity centers of Plains Gartersnakes are small. Seibert and Hagen (1947) found that most movements recorded for a population in Illinois were <2 m/day (6.6 ft/day), and the majority of individual movements recorded were between 9.1 and 27.3 m (29.9 and 89.6 ft). Although snakes generally did not move great distances, the authors recorded several individual movements of >90 m (295 ft). On the other hand, Macartney et al. (1988) used data from Canada presented by Dukart (1984) to calculate a mean distance moved per day that ranged from 11.3 m (37.1 ft) for gravid females to 28.6 m (93.8 ft) for males. They also determined that mean MCP home range size ranged from 0.7 ha (1.7 ac) for gravid females to 3.5 ha (8.6 ac) for males.

Beginning in September and October, snakes return to their overwintering sites (reviewed by Ernst and Ernst 2003). Plains Gartersnakes are quite cold-tolerant relative to other snakes and have been observed basking on warm sunny winter days. For example, Oldfield and Moriarty (1994) witnessed an active individual on a warm day in mid-November. Snakes may use olfactory cues from conspecific skin extracts (Graves and Halpern 1988) or solar cues (Lawson and Secoy 1991) to locate hibernacula. As a result of experiments in which snakes that overwintered above the frost line survived, Bailey (1949) suggested that Plains Gartersnakes could supercool during hibernation (i.e., body tissues do not crystalize even when exposed to temperatures below the freezing point of water).

PREY AND PREDATORS

Although Plains Gartersnakes are generalists, numerous reports indicate that they often feed heavily on earthworms, small amphibians, and other prey such as minnows (Jordan 1967, Gregory 1977a, Hart 1979, Dalrymple and Reichenbach 1981). In Alberta, the majority of juvenile snakes captured and analyzed had consumed small frogs (Tuttle and Gregory 2009). In contrast, the authors found that adults consumed a wider variety of prey, including earthworms, leeches, frogs, and small rodents. Research in Ohio further determined that the primary type of prey consumed changed as seasons progressed, with more earthworms taken in the spring but more young frogs consumed in summer (Dalrymple and Reichenbach 1981). Ophiophagy was also reported in Illinois (Vecchiet et al. 2018). Prey are located through chemical trails (Kubie and Halpern 1979) and once encountered are grasped with the jaws and swallowed (reviewed by Ernst and Ernst 2003).

Numerous predators are reported, including birds of prey, from American Kestrels (*Falco sparverius*) to Red-shouldered Hawks (*Buteo lineatus*), and carnivorous mammals, from Striped Skunks (*Mephitis mephitis*) to Coyotes (*Canis latrans*; reviewed by Ernst and Ernst 2003). Pope (1944) reportedly had pet Raccoons (*Procyon lotor*) that showed "an unfailing interest in them and a marked willingness to eat them" and believed that wild predation by Raccoons was likely. Eastern Milksnakes (*Lampropeltis triangulum*) are also reported predators (Ernst and Ernst 2003), and Vogt (1981) further suggested that North American Racers (*Coluber constrictor*) would consume Plains Gartersnakes. Experimental work supports that this species employs several defensive behaviors, such as tail-waving (Figure 275), which become more intense as the perceived level of threat increases (Arnold and Bennett 1984). Despite this, several authors have commented on this species' more docile temperament compared to that of Common Gartersnakes but noted that they often release foul-smelling cloacal fluid when threatened (e.g., Pope 1944, Vogt 1981). Jordan (1967) observed that individuals encountered in Minnesota frequently took refuge in water when threatened.

Figure 275. A juvenile Plains Gartersnake exhibiting the tail-waving antipredator display (Rock County; photo by J. Kapfer).

CONSERVATION

Status: Globally, the Plains Gartersnake is listed as a Least Concern species by the IUCN. It is considered a species of Special Concern and a Species of Greatest Conservation Need in Wisconsin by the WDNR.

Populations: Robust population data do not exist for Wisconsin, and most information comes from anecdotal observations. Relatively few individuals were encountered during recent surveys (2008–2010) at numerous sites across southern Wisconsin, including locations where this species was previously confirmed by Vogt (1981; Joshua Kapfer, UW–Whitewater, and R. Paloski, unpublished data). As such, it seems unlikely that most modern populations in the state achieve high densities. Past reports from neighboring states indicate that this snake can be very abundant, although many of these reports are old. Pope (1944) stated that in Chicago it was once "abundant locally, and individuals often turn up in storm sewers near populous business districts of Chicago's outskirts." Seibert (1950) used two estimation methods to also report high abundances at a small disturbed site in the Chicago area (1,093–1,152 individuals), with densities of 845–889/ha (342–360/ac). An abundance estimate from Ohio by Dalrymple and Reichenbach (1981) was 219.6 ± 64.6, with density estimates of 51/ha (20.6/ac). More recently, Stanford and King (2004) reported abundance estimates ranging from 64 to 173 adults and density estimates ranging from 40 to 107/ha (16.2 to 43.3/ac) in northern Illinois.

Pilgrim et al. (2011) found roughly equal sex ratios among broods of neonates, but studies from Illinois and Minnesota reported that adult sex ratios were skewed toward females (Seibert and Hagen 1947, Jacobs 1950, respectively). Illinois populations studied by Seibert and Hagen (1947) were comprised mostly of first-year snakes (52.7%), followed by second-year snakes (34.6%) and third-year snakes (12.6%). King and Stanford (2006) estimated that survival of free-ranging young Plains Gartersnakes in Illinois was 16% (year one) to 40% (year two) and that expected survival to reproductive maturity was 6%. Survival of adults varied by sex, with female survival (45%) being higher than male survival (35%; Stanford and King 2004).

Threats: General concepts associated with disturbance or loss of open grassy areas, prairies, and wetlands, as well as the impacts of habitat fragmentation, are applicable to this species (see Introduction to Conservation and Management of Wisconsin's Amphibians and Reptiles chapter). Given its association with open and warm conditions, it may prefer to occupy locations that occur farther from wetland edges than other gartersnakes. If true, this species could experience greater impacts due to disturbance of upland habitat that is immediately adjacent to wetlands. Road mortality could also be a concern for some populations (Breckenridge 1944).

Management: Protection and management of upland habitat immediately adjacent to wetlands could be an important part of any conservation strategy for Plains Gartersnakes. General concepts related to managing prairies and wetlands (including appropriate implementation of techniques such as prescribed fire) but also managing for habitat connectivity and structural heterogeneity are relevant to Plains Gartersnakes (see Introduction to Conservation and Management of Wisconsin's Amphibians and Reptiles chapter). Management to control vegetative succession is likely of particular importance, considering that Plains Gartersnakes typically associate with open habitats. Strategies to offset succession, such as application of prescribed fire, should be conducted during the inactive period, as post-burn mortality of Plains Gartersnakes has been documented (Seibert and Hagen 1947, Erwin and Stasiak 1979). Furthermore, Seibert and Hagen (1947) reported that Plains Gartersnakes did not recolonize recently burned areas studied in Illinois until post-burn vegetation began to return. Dalrymple and Reichenbach (1981) found that this species likely fell victim to mowing of open grasslands to improve Canada Goose (*Branta canadensis*) habitat in Ohio, where they estimated that 5% of all Plains Gartersnakes they observed had been killed by mowing machinery. Management strategies to offset road mortality, particularly in areas where roads bisect suitable habitat, are also likely important to consider.

King and Stanford (2006) conducted research on the effectiveness of headstarting this species as a management strategy, with promising results. For example, the estimated survival of headstarted snakes to reproductive maturity after release into the wild under a 'best-case scenario' was 33%, which is substantially greater than for wild-born snakes (see Populations subsection). They also reported that adult females reared in captivity to use as breeding stock produced fewer 'stillbirth' offspring than those collected from the field shortly before parturition. As such, attempts to supplement populations via this technique may experience greater success with snakes that are acclimated to captivity, if enough individuals are available to avoid low genetic diversity in captive populations. They also found that mortality during headstarting of young snakes was reduced when captive snakes were provided an appropriate thermal gradient, kept at higher humidity to avoid desiccation (≥50% relative humidity), encouraged to feed shortly after birth (hand-feeding, if necessary, in those that initially refuse food), and fed every two days to encourage quick growth to reach sexual maturity in their second year. King and Bowden (2011) determined that male Plains Gartersnakes with better body condition also produced higher testosterone levels, which implies that body condition could positively relate to reproductive success. Although further work is needed, these results suggest that headstarted snakes that are larger at the time of release may have greater reproductive success in the wild.

Eastern Ribbonsnake

Thamnophis saurita (Linnaeus 1766)

Western Ribbonsnake

Thamnophis proximus (Say 1823)

Richard A. Staffen and Jeffrey M. Lorch

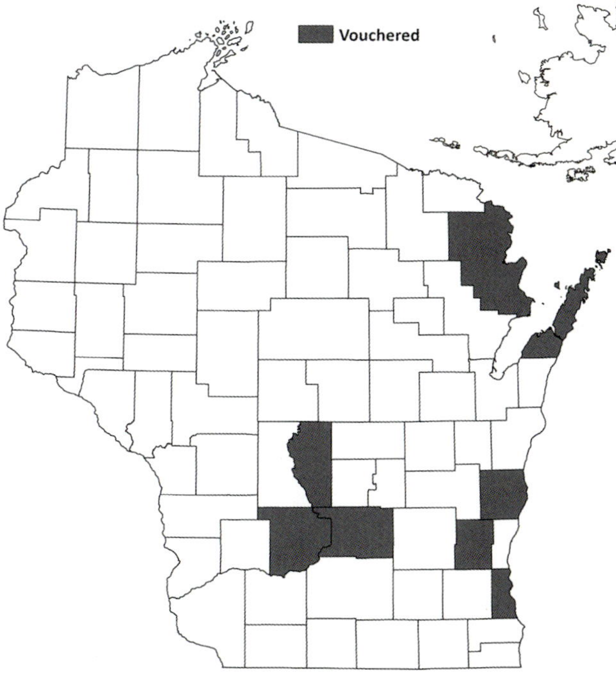

Adult Eastern Ribbonsnake (Sheboygan County; photo by J. Lorch).

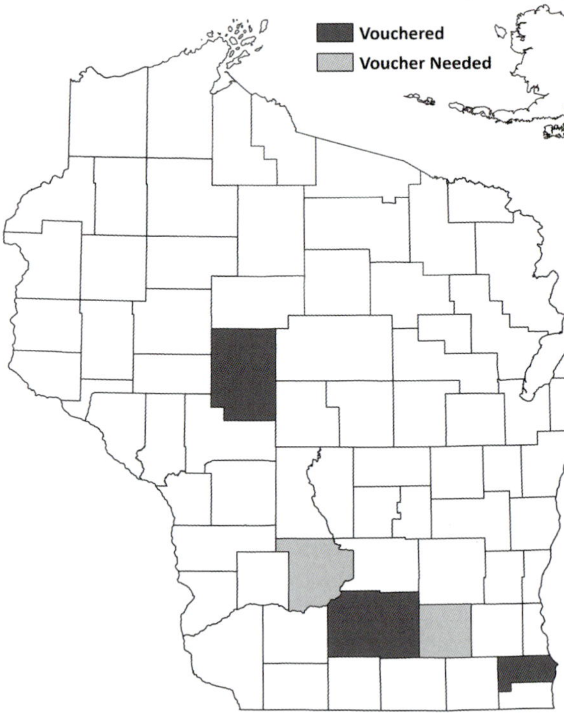

Adult Western Ribbonsnake (St. Martin Parish, Louisiana; photo by B. Glorioso).

AT A GLANCE

Two species of ribbonsnake occur in Wisconsin: the Eastern Ribbonsnake and the Western Ribbonsnake. Because they have similar life histories and because many museum specimens from Wisconsin cannot be definitively identified as one or the other, we review both ribbonsnakes in a combined account. Ribbonsnakes are slim-bodied gartersnakes with a small head and very long tail. They appear similar to Common Gartersnakes at first glance but have a much longer tail, unmarked white upper labial scales, and lateral stripes on dorsal scale rows three and four. In Wisconsin, they are only known from a handful of locations, and very few legitimate records exist. Given their rarity, extremely little is known about their ecology in the state. Despite this, most observations suggest that ribbonsnakes spend much of their life in and around aquatic habitats (such as wetlands) where their chief prey, frogs and other amphibians, are most abundant. They are often found climbing up into vegetation, where it is likely easier for them to regulate body temperature and forage. Ribbonsnakes are presumably active from April to October in Wisconsin, although data are sparse. They probably reproduce in the spring (late March through April) after emergence from hibernation. This species gives birth to live young, which could occur from late July through early September. Both ribbonsnakes are listed as Endangered and considered a Species of Greatest Conservation Need in Wisconsin. After two decades with no reliable sightings, Eastern Ribbonsnakes were rediscovered in Wisconsin in 2008; Western Ribbonsnakes have not been reported in Wisconsin since the 1970s and may be extirpated.

SYSTEMATICS

Confusion regarding the original description of *Thamnophis saurita* (Eastern Ribbonsnake) exists, although Linneaus (1758) was ultimately credited (Natural History Box: The Eastern Ribbonsnake's Convoluted Description Story). *T. proximus* (Western Ribbonsnake) was first described by Say (1823) but was later reclassified as a western subspecies of *T. saurita* (*T. saurita proximus*). Rossman (1962) and Gartside et al. (1977) demonstrated that the eastern and western subspecies of *T. saurita* retained discrete morphological and ecological traits where they were sympatric, suggesting that they indeed represented two distinct species, and elevated *T. saurita proximus* to *T. proximus*.

Some additional evidence exists to further support that *Thamnophis saurita* and *T. proximus* are distinct but closely related species. Gartside et al. (1977) noted molecular differences between these species based on protein electrophoretic analyses but hypothesized that their divergence was relatively recent. Maximum likelihood analyses conducted on all DNA sequences of four mitochondrial loci (cytochrome b and NADH dehydrogenase subunits 1, 2, and 4) available in GenBank at the time of writing this account corroborated that *T. saurita* and *T. proximus* are sister taxa (Jeffrey Lorch, unpublished data). Furthermore, analyses support that the genetic distance between the two ribbonsnakes was generally greater than that exhibited between disparate populations of *T. sirtalis*, the species to which both ribbonsnakes are most closely related (de Queiroz and Lawson 1994, de Queiroz et al. 2002). Yet these results were based on small sample sizes with poor representation of the ribbonsnakes across their geographic ranges. Additional studies (including those that take into account chromosomal markers) are necessary to more definitively demonstrate that *T. saurita* and *T. proximus* are two genetically isolated species.

There are four subspecies of *Thamnophis saurita* currently recognized in North America (Rossman 1970b, Crother 2017). Specimens from Wisconsin are referable to the *T. s. septentrionalis* (Northern Ribbonsnake). Six subspecific designations of *T. proximus* were described by Rossman (1970a), with *T. p. proximus* being the subspecies that occupies the Midwest (including Wisconsin). Additional phylogenetic studies on the described subspecies of each ribbonsnake have not been attempted, making it difficult to assess their validity. There are three other species in Wisconsin that occur in the genus *Thamnophis*: *T. butleri* (Butler's Gartersnake), *T. radix* (Plains Gartersnake), and *T. sirtalis* (Common Gartersnake). Furthermore, all *Thamnophis* species are members of the subfamily Natricinae, which also includes members

NATURAL HISTORY BOX: The Eastern Ribbonsnake's Convoluted Description Story

Linnaeus (1766) is ultimately credited with the description of *Thamnophis saurita* (Eastern Ribbonsnake), but this came to pass in a somewhat convoluted manner. Linnaeus (1758) had originally described a snake specimen collected in Canada as *Coluber sirtalis*, a scientific name that was subsequently placed in synonymy with *Thamnophis sirtalis* (Common Gartersnake; Klabuer 1948, Dowling 1951, Rossman 1970b). Later, Linnaeus (1766) described a specimen from "Carolina" as *Coluber saurita*, which was referred to as the "southern ribbon snake" (Klauber 1948, Dowling 1951). These two descriptions by Linnaeus, unfortunately, caused subsequent confusion among zoologists. For example, Harlan (1827) counted the scales of Common Gartersnake specimens available to him and realized that these were not as Linnaeus (1758) had described for *Coluber* [*Thamnophis*] *sirtalis* (Klauber 1948). Furthermore, Klauber (1948) reviewed Linnaeus's (1758) description and determined that the scale counts he had used to describe *sirtalis* were actually in line with the scale counts of a ribbonsnake and not the snake species that most knew as the Common Gartersnake. The conclusion reached by Klauber (1948) was that Linnaeus had accidentally described the same species (which is currently known as the Eastern Ribbonsnake) as two different species: the Common Gartersnake and the Eastern Ribbonsnake. Furthermore, the problem became exacerbated because this nomenclatural quandary was not addressed for some time after Linnaeus's descriptions.

Explanations for the lack of early correction appear speculative. Yet perhaps other European herpetologists who encountered or received reports of any snake that generally resembled a gartersnake (i.e., dark with lighter longitudinal lines down its back) simply assumed it was the snake Linnaeus had described but did not confirm this (Klauber 1948). As Dowling (1951) stated, application of the name *sirtalis* to the Common Gartersnake could have persisted because of "continued usage through the years, [with] everyone apparently basing his work upon that just previous, rather than upon the original descriptions." The first attempt to correct this issue appears to be Klauber (1948). He argued that the specific epithet of *sirtalis* should be used for the ribbonsnake rather than the Common Gartersnake. Yet *sirtalis* was already widely recognized as the specific epithet for the Common Gartersnake, and Klauber's suggested change quickly caused its own problems (see examples in Dowling 1951). As such, the scientific name *sirtalis* was officially retained for the Common Gartersnake to avoid further confusion, while *saurita* was assigned to the Eastern Ribbonsnake (reviewed by Rossman 1970b, Crother 2017).

Jeffrey M. Lorch, Richard A. Staffen, WDNR, and Joshua M. Kapfer, UW–Whitewater

of the genera *Nerodia, Regina, Storeria,* and *Tropidoclonion* that are found in the state.

DESCRIPTION

Adults: Ribbonsnakes are slim-bodied striped snakes with relatively large eyes and a long tail. Their tail is often about one-third of their TL, while that of other gartersnakes is usually less than one-fourth of their TL (Figure 276; Rossman et al. 1996). Their scales are keeled, and their anal plate is undivided. Ribbonsnake color pattern is superficially similar to that of Wisconsin's other *Thamnophis* species, but with a distinctively whiter ventral surface, unmarked white to cream-colored labial scales, and white preocular scales on each side of

Figure 276. Ribbonsnakes, such as this Eastern Ribbonsnake, have a tail that is up to one-third of the total body length. The arrow indicates the approximate location of the vent and start of the tail. (Sheboygan County; photo by R. Paloski).

the head. The lateral stripes are on dorsal scale rows three and four.

Eastern Ribbonsnakes have dark brown or black dorsal surfaces overlaid with three yellow, greenish-yellow, or white stripes (one down the middle of the back and one on each side). The middle (or vertebral) stripe may be different in coloration from the side (or lateral) stripes, often dull gold to orange yellow in Wisconsin individuals. Below each lateral stripe is a broad chestnut to dark brown ventrolateral stripe that is noticeably different in color from the dorsal surface and extends well onto the ventral surface, often covering nearly one-fourth of the ventral scales' left and right edges (Rossman 1962). Aside from their darker edges, the ventral scales are mostly white, although this slowly transitions to yellow when nearing the tail (Vogt 1981). Eastern Ribbonsnakes typically have seven supralabial scales but occasionally have eight (Figure 277a) on each side of the head, which is short for a gartersnake (the head constitutes only 3.9% of the SVL; Rossman 1963b). There are sometimes white to yellowish parietal spots (which are rarely fused) on the back of the head, although if present they can be small or faint (Rossman 1962; Figure 277b). Eastern Ribbonsnakes reach a maximum TL of 101.8 cm (40.1 in; Ernst and Ernst 2003). Small samples of adults measured from Wisconsin and Nova Scotia had SVLs that ranged from 38 to 42 cm (15 to 16.5 in; Table 14). The tail is long, up to 28.8%–38.8% of the TL (Rossman 1963b). In the adult ($n = 6$) and neonate ($n = 11$) Wisconsin specimens examined, the tail was nearly always slightly less than one-third of the TL (J. Lorch, unpublished data).

Western Ribbonsnakes usually have a jet-black dorsal surface with yellowish lateral stripes and an orange-yellow dorsal stripe; the venter is white. A narrow black ventrolateral stripe occupies scale rows one and two below the lateral stripe but does not (or barely) extend onto the ventral scales (Rossman 1962). According to Rossman (1962), Western Ribbonsnakes have longer snouts than Eastern Ribbonsnakes, almost always have eight supralabial scales, and possess parietal spots that are bright and fused across the interparietal suture. However, these features may not be consistent with Wisconsin specimens (see below). Western Ribbonsnakes attain a greater size than Eastern Ribbonsnakes, reaching up to 126.8 cm TL (49.9 in) but averaging 51–76 cm TL (20.1–29.9 in; Ernst and Ernst 2003). The tail is relatively short compared to Eastern Ribbonsnakes, usually being less than one-third (ranging from 25.2% to 33.6%) of the TL (Rossman 1963b). The largest Wisconsin museum specimen measured was an adult male from Dane County (UWZM 21232), whose TL was 49 cm (19.3 in).

Distinguishing between Eastern and Western Ribbonsnakes in Wisconsin can be challenging. Widely utilized references (e.g., Conant and Collins 1991, Ernst and Ernst 2003) emphasize supralabial count and the intensity and fusion of parietal spots as useful characteristics for identification. However,

Table 14. Mean and ranges of SVL and weights for Eastern Ribbonsnakes of different age classes and sexes from Wisconsin (J. Lorch, unpublished data, WDNR 2016b) and Nova Scotia (Bell et al. 2007).

SVL, cm (in)	Weight, g (oz)	Age Class	Sex	Location	Sample Size
15.9, 13.5–17.9 (6.3, 5.3–7.0)	Unknown	Neonate	Unknown	Wisconsin	11
16.8, 14–21 (6.6, 5.5–8.3)	4.6, 3.5–6 (0.16, 0.12–0.21)	Neonate	Female	Nova Scotia	8
15 (5.9)	4.5 (0.16)	Neonate	Male	Nova Scotia	1
32.8, 23–41.5 (12.9, 9.1–16.3)	14.6, 7–26.5 (0.51, 0.25–0.93)	Juvenile	Female	Nova Scotia	13
31, 18–37 (12.2, 7.1–14.6)	14, 6–22 (0.49, 0.21–0.78)	Juvenile	Male	Nova Scotia	31
38.5, 35–43.5 (15.2, 13.8–17.1)	Unknown	Adult	Unknown	Wisconsin	4
41.9, 34–49 (16.5, 13.4–19.3)	31.7, 19.5–55.5 (1.12, 0.69–1.96)	Adult	Female	Nova Scotia	31
41.1, 35.5–50 (16.2, 14.0–19.7)	25.6, 16.5–37.5 (0.90, 0.58–1.32)	Adult	Male	Nova Scotia	21

a review of 32 live Eastern Ribbonsnakes from Sheboygan County suggests that there could be substantial overlap in these traits between the two species in the state. For example, over 20% of Eastern Ribbonsnakes captured in Sheboygan County had eight supralabials on at least one side of the head (Figure 277a). Parietal spots on these individuals were highly variable, ranging from absent to very bright and from well-separated to nearly touching (separated only by the thin suture line between the parietal scales). In addition, some presumed Western Ribbonsnake museum specimens from Wisconsin do not have the fused parietal spots that are often considered an identifying feature (Figure 277b). The characteristics of the parietal spots are also somewhat subjective, cannot be easily interpreted on many faded museum specimens, and can be difficult to assess without side-by-side comparison of the two species. This is corroborated by Rossman (1962), who did not consider the number of supralabials a reliable trait for identifying Western Ribbonsnakes (since ca. 20% of Eastern Ribbonsnakes he examined had eight supralabials on at least one side of the head) and stated that there were exceptions to the parietal spot characterizations. However, the presence of seven supralabials is a reliable indication of the Eastern Ribbonsnake (Rossman 1962).

The lack of conformity among Wisconsin specimens regarding characteristics typically used to differentiate Eastern and Western Ribbonsnakes is problematic. Since Wisconsin was represented in Rossman's study by a single Western Ribbonsnake specimen and no Eastern Ribbonsnake specimens, such overlap in characteristics may have been overlooked. Wisconsin lies on the edge of both species' ranges. Factors such as founder effect, genetic drift (to which small populations such as those found in Wisconsin are more susceptible), selection for similar morphological traits, or possible hybridization between the two species could be responsible for the differences in characteristics between Wisconsin specimens and those from other states. Alternatively, the differences in some of the observed characteristics may be an artifact of the small amount of Wisconsin material available for examination.

Although the previously outlined characteristics for differentiating snakes with eight supralabials are unreliable in Wisconsin, other physical traits are apparently more useful. Rossman (1962) described the degree to which the dark ventrolateral stripe extended onto the ventral scutes as being the most defining criteria for distinguishing the two species. As corroborating evidence, all live Eastern Ribbonsnakes examined from Wisconsin had dark ventrolateral stripes that covered one-sixth to one-fourth the width of the ventral scutes at each lateral margin (Figure 277c; J. Lorch, unpublished data). On museum specimens of the Western Ribbonsnake for which the extent of the dark ventrolateral stripe was discernible, the stripe either did not touch or barely extended onto

Figure 277. Variation in labial scale counts (a), parietal spots (b), and ventrolateral stripes (c) in the ribbonsnakes. (a) The Western Ribbonsnake nearly always has eight supralabials (*top*; Racine County; USNM 776; photo by J. Poindexter II, Smithsonian Institution National Museum of Natural History). The Eastern Ribbonsnake typically has seven supralabials (*bottom*; Sheboygan County; photo by J. Lorch), but many Wisconsin individuals have eight supralabials (*middle*; Sheboygan County; photo by J. Kapfer). (b) In the Western Ribbonsnake, the parietal spots are typically bright and fused (*upper left*; Vermilion Parish, Louisiana; photo by B. Glorioso). However, in Wisconsin specimens, the spots are often only partially fused (*upper middle*; Dane County, UWZM H21232; photo by J. Kapfer) or separated (*upper right*; Clark County, FMNH 14580; photo by R. Grills, Field Museum of Natural History). In the Eastern Ribbonsnake, the parietal spots are usually faint and distinctly separated by the scale suture between the parietals (*lower left*; Sheboygan County; photo by J. Lorch). However, the spots may be absent (*lower middle*; Sheboygan County; photo by J. Lorch) or bright and partially fused (*lower right*; Sheboygan County; photo by J. Lorch). The degree of overlap for this trait in Wisconsin specimens makes this characteristic of limited value in distinguishing the two species. (c) In the Western Ribbonsnake (*upper*; Vermilion Parish, Louisiana; photo by B. Glorioso), the dark brown to black ventrolateral stripe does not (or barely) touches the ventral scales. In the Eastern Ribbonsnake (*lower*), the chocolate-brown to chestnut ventrolateral stripe extends well onto the ventral scales. This trait appears to be the most reliable way of distinguishing the two ribbonsnake species in Wisconsin.

Figure 278. Neonate Western Ribbonsnake (St. Martin Parish, Louisiana; photo by B. Glorioso).

the ventral scutes. These findings support that the dark ventrolateral stripe may be the best way of distinguishing Wisconsin specimens with eight supralabials. Unfortunately, this characteristic is not useful for museum specimens with faded coloration.

Preadult stages: Neonates and juveniles are similar in appearance to adults (Figure 278). Measurements of neonatal Eastern Ribbonsnakes are presented in Table 14. Newborn Western Ribbonsnakes are 15.0–21.4 cm SVL (5.9–8.4 in) and weigh 1.4–2.5 g (0.05–0.09 oz; Clark 1974, Ernst and Ernst 2003). Ribbonsnake neonates have a disproportionately large head (Hampton 2011, Jeffrey Lorch, US Geological Survey, personal observation). Morphological characteristics used to differentiate the two ribbonsnake species should be applied cautiously to immature animals; the degree to which the ventrolateral stripe extends onto the ventral scales is the best trait for distinguishing juveniles of the two taxa.

Variation: No significant geographic variation among individuals of either species is known. There appears to be no sexual dimorphism in color or pattern, although Rossman (1963b) reported that males and females varied in the number of ventral and subcaudal scales. The sexes are very similar in length across age classes, but females from Nova Scotia were significantly heavier than males regardless of whether snakes were gravid (Bell et al. 2007). Rossman (1963b) also found proportionally longer tails in males compared to females.

Diagnosis: Although it is difficult to distinguish between the two species in Wisconsin, ribbonsnakes can generally be differentiated from other snakes by various characteristics. They possess keeled scales in fewer than 27 rows at midbody and ventral scale counts ranging from 141 to 181. They also possess a tail that is 27% or more of their total body length, and although the anal plate is not divided, subcaudal scales are all divided. Their supralabial scales are white with no dark marks, and they have white preocular scales. Sometimes a pair of parietal spots (which can be fused or unfused) occurs on top of the head. Their ventral surface is unmarked, and their lateral stripes contact the fourth scale row. They also possess two prefrontal scales, one loreal scale, two to four preocular scales, a rostral scale that is not pronounced, seven to eight supralabials, and more than eight infralabials. Three other species of *Thamnophis* occur in Wisconsin that could be confused with the ribbonsnakes, including the Common Gartersnake (*Thamnophis sirtalis*), Plains Gartersnake (*T. radix*), and Butler's Gartersnake (*T. butleri*). None of these species have immaculate white supralabials or light preocular scales or such a long tail (the tail is almost always less than one-fourth of the TL in these other species).

DISTRIBUTION AND HABITAT

Global distribution: Eastern Ribbonsnakes are found east of the Mississippi River, extending from Maine, Nova Scotia, and southern Ontario west to Michigan's Lower Peninsula and south to Florida and Louisiana (reviewed by Rossman 1970b). They are not reported from the more mountainous parts of the eastern US (reviewed by Ernst and Ernst 2003) or from areas that were historically tallgrass prairie. Wisconsin populations are disjunct and represent the northwestern limit of the species' range.

The core distribution of Western Ribbonsnakes lies in the lower Midwest and south-central US. They range north to Nebraska, Iowa, and Wisconsin and south to Costa Rica (Rossman 1970a). In the US, the species is found primarily west of the Mississippi River in the southern plains region west to Colorado and New Mexico. However, Western Ribbonsnakes range eastward into the former tallgrass prairie 'peninsula' of the Midwest, reaching extreme western Ohio (Rossman 1970a, Ernst and Ernst 2003). In general, Eastern and Western Ribbonsnakes have complementary ranges. In areas where the two species overlap, populations are often spatially separated by habitat (Rossman 1962). Wisconsin is one of the few states where their ranges overlap.

Distribution in Wisconsin: Substantiated Eastern and Western Ribbonsnake records combined exist from only a handful of counties in Wisconsin (Table 15). There is also little material available from Wisconsin for review, and many preserved specimens are ambiguous so cannot be definitively identified. Specifically, color patterns in many museum specimens are faded and undiscernible. There are also no detailed written descriptions of the morphological characteristics of the specimens at the time of collection, making key features related to the ventrolateral stripes and parietal spots difficult to interpret. Without supporting

Table 15. Wisconsin counties with published records, existing specimens, or anecdotal observations of ribbonsnakes. Note that substantiated records include those that have been verified by examination of photographs or museum specimens.

Species Name	County Name	Substantiated	Source and Catalog Number
Thamnophis proximus	Clark	Yes	FMNH 14580
Thamnophis proximus	Dane	Yes	UWZM 21232
Thamnophis proximus	Grant	No[a]	Hoy (1883)
Thamnophis proximus	Jefferson	No[b]	BYU 40420
Thamnophis proximus	Racine	Yes	USNM 776
Thamnophis proximus	Sauk	No[c]	CM 70518 (see below)
Thamnophis proximus	Waukesha	No[d]	WDNR (2015)
Thamnophis saurita	Adams	Yes	USNM 307855
Thamnophis saurita	Columbia	Yes	UMMZ Photo 1262–4
Thamnophis saurita	Door	Yes	UMMZ Photo 1257–8
Thamnophis saurita	Marinette	Yes	MPM 1829
Thamnophis saurita	Milwaukee	Yes	MPM 2605
Thamnophis saurita	Oconto	No[a]	Vogt (1981)
Thamnophis saurita	Sauk	Yes	UWZM 21231
Thamnophis saurita	Sheboygan	Yes	MPM 18295
Thamnophis saurita	Washington	Yes	MPM 1007
Thamnophis sp.	Sauk	Yes[c]	CM 70518
Thamnophis sp.	Washburn	Yes[c]	UWZM 2206

[a] Record not mapped, although it was from a published source, because source either referenced record as second-hand report or source was very old and no additional records from the location exist.

[b] Museum specimen has apparently been lost and could not be examined by the authors.

[c] Museum specimens examined are either *T. proximus* or *T. saurita*, but lack definitive characteristics for species determination.

[d] Specimen USNM 731 is reportedly from the "*Fox, River, Wisc.*" (no county listed), but could have come from Waukesha County.

information, it seems plausible that most species determinations for these specimens were made on the basis of supralabial scale counts, which is not always a reliable method for distinguishing these species (see Description section).

The currently documented range of Eastern Ribbonsnakes in Wisconsin is primarily north of the vegetative tension zone (Curtis 1959), with the majority of occurrences concentrated in the eastern part of the state. The main exceptions are a few records from the vicinity of the Wisconsin River in south-central Wisconsin. On the basis of suitable museum specimens and photographic vouchers, Eastern Ribbonsnakes have been documented in only Adams, Columbia, Door, Marinette, Milwaukee, Sauk, Sheboygan, and Washington Counties. Casper (1996) considered Eastern Ribbonsnakes extirpated from Milwaukee County, and the only recent records known are from Columbia, Door, and Sheboygan Counties.

Wisconsin marks the Western Ribbonsnake's northern range limit in North America. Its distribution is spotty in Wisconsin and in central and northern Illinois, perhaps indicating that this snake was never widespread in the Upper Midwest. Confirmed occurrences (after reevaluation of all known Wisconsin specimens in museum collections) currently only exist for Clark, Dane, and Racine Counties (see Table 15). Existing specimens collected from additional counties (Sauk and Washburn) cannot be definitively identified to species based on currently accepted characteristics of identification. Yet Vogt (1981) mapped what he considered to be a reliable report of Western Ribbonsnakes from along the Wisconsin River in Sauk County. A museum specimen (UWZM 21231) from roughly the same area that may have been collected by the observer who relayed the report to Vogt is actually an Eastern Ribbonsnake. However, an additional specimen currently labeled as a Western Ribbonsnake collected farther south in Sauk County (CM 70518) has eight supralabials. As such, this specimen might be a Western Ribbonsnake, but definitive identification was not possible from external features alone. Vogt (1981) also mentioned that Chuck Naidl (a well-respected herpetoculturist who owned a famous reptile farm in Sauk County for decades [Lange 2004]) found them near "the Wisconsin River" in the 1970s, but specific locations were unreported. The Dane County record listed by Vogt (1981) is possibly based on a UWZM specimen (UWZM 21232) that was originally identified as an Eastern Ribbonsnake. Reevaluation of the specimen indicates that it is actually a Western Ribbonsnake based on characteristics of the dark ventrolateral stripe (J. Lorch, unpublished data). Casper (2015) reported a Jefferson County specimen (BYU 40420) collected in 1963 that is currently lost and could not be evaluated any further (Jack Sites, Brigham Young University, personal communication). There are no confirmed reports or collected specimens of Western Ribbonsnakes from Wisconsin after 1972.

Localities of possible occurrence: The dearth of ribbonsnake reports, vouchered or otherwise, speaks to the need for continuing survey work focused on both species. In recent years, several populations of Eastern Ribbonsnakes have been discovered in Wisconsin (e.g., Sheboygan and Columbia Counties). Some of these populations are in areas frequently visited by outdoor enthusiasts, but ribbonsnakes had gone unreported there for years, demonstrating that these snakes can be easily overlooked, even by experienced naturalists. Considering this, and given the anecdotal information outlined below, ribbonsnakes could occur in additional counties. For example, Vogt (1981) mapped a record of Eastern Ribbonsnakes in Oconto County based on a secondhand report that he considered reliable but had not confirmed himself (see Table 15). This observation could be legitimate, as populations are known from nearby Marinette and Door Counties. Furthermore, Calumet, Fond du Lac, Manitowoc, and Ozaukee Counties are adjacent to documented Eastern Ribbonsnake populations. These counties also contain large boggy wetland complexes, a habitat type frequently associated with ribbonsnakes.

Historically, reports and records of Western Ribbonsnakes in Wisconsin are less common than those of Eastern Ribbonsnakes. Even Vogt (1981) observed only three live individuals during his time in the state. Despite this, a handful of unverified reports and snippets of information exist that could point to currently unconfirmed populations and warrant further investigation. For example, Smith (1961) believed that the range of Western Ribbonsnakes in Illinois extended north along the Mississippi River to the Wisconsin border. Recent reliable reports from along the border between Carroll and Whiteside Counties (Illinois; Matt Ignoffo, personal communication) suggest that the species still occurs in northwestern Illinois. Along these lines, Hoy (1883) listed *Eutaenia faireyi* (a synonym of *Thamnophis proximus*) in Grant County, but no supportive vouchers are known. Given this information, the possibility that they currently exist in southwestern Wisconsin remains reasonable. Outside of southwestern Wisconsin, unsubstantiated occurrences also exist from Waukesha County, including a reported observation from 1973 (WDNR 2016b), and a specimen collected from along the "Fox [River], Wisc." (unknown county, but possibly Waukesha, USNM 731) that could not be located for examination.

It is noteworthy that a museum specimen from Washburn County (UWZM 2206) exists that is clearly a ribbonsnake but cannot be accurately assigned to species. If confirmation can be obtained to supplement this specimen, it would expand the known range of either species in Wisconsin.

Figure 279. Typical habitat utilized by Eastern Ribbonsnakes at a wetland in Sheboygan County. Note the mixture of shrubs and emergent vegetation that provide hunting, basking, and hiding areas (photo by J. Kapfer).

Habitat: Most information on the habitats used by ribbonsnakes in Wisconsin is anecdotal. Both species often occur in aquatic habitats (Rossman et al. 1996), likely because they feed heavily on amphibians (see Prey and Predators section). It is also possible that their habitat use is influenced by the presence of other snakes. For example, habitat partitioning was noted among sympatric populations of *Thamnophis* species, with ribbonsnakes occupying a more amphibious niche (Carpenter 1952a, Fouquette 1954). Furthermore, Tinkle (1957) noted that Western Ribbonsnakes avoided competition by selecting wetland edge habitats that were not utilized as frequently by either the more terrestrial or more aquatic snake species found in the same area.

Of the two species, Eastern Ribbonsnakes are more frequently associated with wetlands in wooded landscapes rather than prairies. Vogt (1981) reported them only from *Sphagnum* bogs in Wisconsin, but recent observations in the state suggest that they are not limited to this habitat type. For example, at a Sheboygan County site this species utilized ephemeral and permanent kettle ponds, as well as a large marsh (Figure 279; J. Lorch, personal observation), while a museum specimen from Sauk County (see Distribution in Wisconsin subsection) was collected in a backwater slough along the Wisconsin River. Regardless of the aquatic habitat type used, adult Eastern Ribbonsnakes reportedly stay close to the water's edge. In Nova Scotia, they were rarely found beyond 5 m (16.4 ft) from water during the active season (May through September; Bell et al. 2007). Similarly, all adult Eastern Ribbonsnakes encountered by the authors in Wisconsin have been in standing water or on the shoreline within a few meters of the water's edge. In contrast, over 95% of young of the year ($n = 22$) were encountered in upland grassy areas ca. 50–100 m (164–328 ft) from the wetland; only one neonate was found along the shoreline (J. Lorch, unpublished data). Young from the previous year were also observed in the same upland areas the following summer (J. Lorch, personal observation). There are likely seasonal changes in habitat use as well. For example, ephemeral ponds, which attract amphibians that breed early in the season, may be utilized by Eastern Ribbonsnakes in spring and early summer when food is abundant at such locations but could be abandoned in favor of more permanent waterbodies later in the year.

Shrubby vegetation is a specific habitat feature that is apparently important to Eastern Ribbonsnakes, more so than for most other gartersnake

NATURAL HISTORY BOX: A Shrub-Loving Snake

Few Wisconsin residents have knowingly encountered ribbonsnakes in the state. Even among local herpetological enthusiasts, experience with these snakes in the wild is rare. I did not capture my first ribbonsnake in the state until 2013. In July of that year, I visited a site with three UW–Whitewater students (Phil Pieper, Karl Rutzen, and Sarah Wyrick) and Bill Poole from Stantec, an environmental consulting firm located in Cottage Grove (Dane County). We arrived at a wetland selected for Eastern Ribbonsnake surveys in midmorning. It was not too hot out yet, so I was hopeful that snakes would still be active. However, when we reached the wetland edge, my hopes dwindled. We were standing in a strip of dense Reed Canary Grass intermixed with small pockets of short willows (*Salix* spp.) and dogwood (*Cornus* spp.). Considering that one normally looks on the ground for snakes and that the dense vegetation wasn't very conducive to seeing the ground, I started to lose confidence that we'd catch an Eastern Ribbonsnake. Still, we proceeded to push through the waist- to shoulder-height vegetation in the hopes of figuring out a good strategy. Then I remembered what Vogt (1981) said in his book: "Northern Ribbon Snakes are found above the ground in low bushes." At about the same time, Bill spoke up: "Vogt said that these snakes bask up in bushes a lot." In what seemed like seconds, Phil looked into a clump of young willows nearby and exclaimed, "There's one!" He was right. In a matter of about two hours, we captured 12 Eastern Ribbonsnakes, and we found all of them intertwined in the twigs of shrubs located along the wetland edge.

Joshua M. Kapfer, UW–Whitewater

species. Carpenter (1952a) consistently found Eastern Ribbonsnakes near or in bushes, trees, or similar cover in Michigan. Wisconsin individuals are also often observed climbing among the branches of shrubs, where they presumably bask or hunt (see Natural History Box: A Shrub-Loving Snake). Shrubs commonly found in or near bog and wetland habitats are particularly favored (Carpenter 1952a, Rossman et al. 1996). In fact, all wetlands where Eastern Ribbonsnakes are currently documented in Wisconsin are bordered, at least partially, by forest. Based on limited observations in Sheboygan County (J. Lorch, personal observation), it appears that this species is more common on the east side of wetlands, where vegetative cover is often denser and there are large numbers of larval and metamorphosing amphibians at certain times of the year. Although shrubby vegetation along aquatic peripheries is favored, other types of cover are also reportedly used. For example, Ernst and Ernst (2003) observed ribbonsnakes emerging from bankside Muskrat (*Ondatra zibethicus*) burrows, which suggests that they may use existing mammal burrows as refugia. It is of further note that some level of habitat degradation is apparently tolerated. In Sheboygan County, Eastern Ribbonsnakes are frequently encountered in areas with invasive Reed Canary Grass (*Phalaris arundinacea*), which they utilize for cover and basking (J. Lorch, personal observation).

In Wisconsin, far less is known about the habitats used by Western Ribbonsnakes than Eastern Ribbonsnakes. Most available information comes from limited observational reports of habitats associated with old museum specimens. The Dane County specimen (UWZM H21232), for example, was reportedly found adjacent to a large wetland consisting of Tamarack (*Larix laricina*) swamp, deep-water marsh, shallow marsh with emergent vegetation, and a floating bog mat. The Racine County specimen was taken along the Root River, although no specific habitat information for that specimen or for the Clark County specimen is available. Given that the Western

Ribbonsnake's distribution in the Upper Midwest is primarily restricted to the tallgrass prairie region, it presumably occupies more open landscapes than the Eastern Ribbonsnake does. This is corroborated by reports from other states, such as Indiana, where Minton (2001) typically observed Western Ribbonsnakes in areas that were "more open" than those where he observed Eastern Ribbonsnakes. Kennicott also found them along backwater sloughs that formed in the prairies of northeastern Illinois and believed they were just as aquatic as watersnakes (*Nerodia* spp.; Mierzwa 1985). In the southern US, Western Ribbonsnakes occur in a variety of permanent or semipermanent waterbodies but seem most abundant in swamps along river floodplains. In such areas, they forage in the lowlands and seek shelter during periods of inactivity on drier ridges (Tinkle 1957). Even in wooded swamps, Western Ribbonsnakes seem to prefer sunny patches, such as open and well-vegetated wetland flats; however, shaded areas are also used when they are beneficial for thermoregulation (Tinkle 1957). Like Eastern Ribbonsnakes, this species is frequently found elevated above the ground in shrubs and herbaceous vegetation (Rossman et al. 1996).

Limited information is available on the overwintering habitat preferred by either ribbonsnake species in Wisconsin. In Michigan, Eastern Ribbonsnakes reportedly utilized crayfish burrows, small mammal tunnels, or ant mounds as hibernacula (Carpenter 1953). Eastern Ribbonsnakes may also overwinter in well-drained upland sites, where they locate crevices in subsurface rocks that descend below the frost line. For example, an adult found in mid-October on the side of a bluff in Columbia County was presumably migrating to such a hibernaculum (Lester Doyle, personal communication). Eastern Ribbonsnakes sometimes hibernate underwater in colder regions (Rossman et al. 1996). Carpenter (1953) suggested that Michigan snakes hibernated in moist conditions to offset the threat of dehydration that occurred at low temperatures. Information on hibernacula used by Western Ribbonsnakes in the Upper Midwest is more limited than it is for Eastern Ribbonsnakes. In Wisconsin, it is possible that they overwinter in habitats similar to those utilized by Eastern Ribbonsnakes and other species of gartersnakes. In southern Illinois and Louisiana, Western Ribbonsnakes hibernate in rock crevices with other snake species (Smith 1961, Tinkle 1957). Clark (1974) proposed that, similar to Eastern Ribbonsnakes, desiccation during hibernation was a significant source of mortality, which suggests that Western Ribbonsnakes should seek hibernacula with adequate moisture.

REPRODUCTION AND DEVELOPMENT

There are no reported observations of ribbonsnake breeding activity in Wisconsin. In nearby Michigan, Carpenter (1952a) observed gravid female Eastern Ribbonsnakes in May, which suggests that reproduction occurred earlier in the spring. Although data on Western Ribbonsnakes from the Upper Midwest are virtually nonexistent, studies from southern locations suggest that they also breed in the spring. For example, males of a Texas population reportedly increased their movement rates in April and May (presumably as they sought mates), and inseminated females were found from late April to late June (Clark 1974). Ford (1981) suggested that mating activity in Oklahoma might occur throughout the active season and pointed to the fact that there was no distinct period when male Western Ribbonsnakes followed scent trails left by females. Ford (1981) further noted that a comparison of when sperm were found in the cloaca of females to the time at which gravidity was first observed supported the possibility of delayed fertilization or delayed development of young (i.e., fertilization might not occur immediately after mating). No specific description of ribbonsnake courtship behavior exists, but it is presumably similar to that of other *Thamnophis* species.

Like all gartersnakes, ribbonsnakes are viviparous and give birth to live young in late summer (July–September). Neonate Eastern Ribbonsnakes were observed in Sheboygan County on 30 August 2014 ($n=22$) and 22 August 2015 ($n=1$), which corroborates late summer parturition. Eastern Ribbonsnakes typically have 10–12 young per brood, but brood size can range from 3 to 26 young (reviewed by Ernst and Ernst 2003). Nothing is known about when parturition occurs or mean brood size for the Western Ribbonsnake in Wisconsin. In Louisiana, young were born in July, and mean brood size was 13 (Tinkle 1957). A Texas study reported that mean brood size was 8.4, and young were typically born in late July, although one animal housed in captivity did not give birth until 13 October (Clark 1974). Smith (1961) reported that individual females from Indiana and Illinois gave birth to 11 and 12 young in late July and August, respectively. In a Louisiana study population, 88% of the adult females captured were gravid, which suggests that most mature females produce young annually (Tinkle 1957).

Limited information on growth and development exists for either ribbonsnake species. Carpenter (1952b) reported that growth rates of Eastern Ribbonsnakes in Michigan were greatest in younger snakes, or those ranging from 15 to 24 cm TL (5.9 to 9.4 in), and that female growth rates declined more slowly with age than male growth rates. Female Eastern Ribbonsnakes could become sexually mature in Michigan at two years of age, and the smallest gravid female captured by Carpenter (1952b) had an SVL of 42.1 cm (16.6 in). Young-of-the-year Western Ribbonsnakes in study populations from Texas and Louisiana experienced a nearly 50% increase in length by their first winter and doubled in size by one year of age (Tinkle 1957, Clark 1974). In those populations, female Western Ribbonsnakes reached sexual maturity at ca. 50 cm SVL (19.7 in) and probably bred by their second spring, while males were sexually mature by their first or second spring at ca. 40 cm SVL (15.7 in; Tinkle 1957, Clark 1974). However, time to sexual maturity could be longer in northern populations where the active season is shorter (Tinkle 1957). Longevity in the wild is largely unknown, but Tinkle (1957) suspected that Western Ribbonsnakes lived to ca. three years at his study site. Some information on captive ribbonsnake longevity exists that indicates that Eastern Ribbonsnakes survive to ca. 10.5 years and Western Ribbonsnakes survive to 3.5 years (Snider and Bowler 1992). The seemingly shorter life span of the Western Ribbonsnake may be an artifact of the species' failure to thrive under captive conditions rather than a true reflection of maximum life span.

ACTIVITY

Based on anecdotal observations in Wisconsin, data from other *Thamnophis* species in the state, and information from other states, ribbonsnakes likely emerge from winter dormancy in late March or April. Active Eastern Ribbonsnakes have been found in Sheboygan County on 24 April and 6 May (J. Lorch and Richard Staffen, WDNR, personal observation). Carpenter (1953) monitored egress from an ant mound hibernaculum in Michigan and found no Eastern Ribbonsnakes emerging after 15 April. Specific dates for Western Ribbonsnake emergence from hibernacula in the Upper Midwest are lacking. According to Rossman et al. (1996), they are active by 1 April in northern Missouri. Clark (1974) reported that activity is greatest in spring due to mate-searching and migration from hibernacula to the summering areas. The few specimens taken in Wisconsin were primarily collected in summer.

Daily activity patterns are dependent on local environmental conditions and snake body temperature. Carpenter (1956) determined that the mean body temperature of Eastern Ribbonsnakes from Michigan was 26°C (78.8°F, $n = 123$). Yet work on Western Ribbonsnakes revealed that critical thermal maxima and minima varied based on the air temperatures that individuals were acclimated to (Jacobson and Whitford 1970). During the active season, both ribbonsnake species are likely diurnal in Wisconsin, given information from other states (reviewed by Ernst and Ernst 2003). Although >60% of Western Ribbonsnakes observed during nighttime surveys in southern states were above the ground, these appeared inactive and may have been sleeping on elevated perches to avoid predators (Brad M. Glorioso, USGS, personal communication). Numerous studies and observations indicate that ribbonsnake activity is also heavily influenced by precipitation. For example, Bell et al. (2007) reported that Eastern Ribbonsnake activity often increased immediately after precipitation in Nova Scotia. After a brief heavy downpour in late August, 22 neonatal Eastern Ribbonsnakes were observed in upland grassy areas adjacent to a marsh in Sheboygan County; none were observed prior to the downpour (J. Lorch, personal observation). Similarly, Clark (1974) noticed spikes in trapping success for Western Ribbonsnakes following precipitation, and Tinkle (1957) documented a sudden increase in captures of this species during and after a light drizzle. Such observations could be related to an increase in activity of amphibian prey associated with precipitation.

Movement patterns are poorly understood and mostly estimated via recapture techniques as opposed to radio telemetry. The only known report of home range size for either species comes from Imlay et al. (2015), who estimated that mean home range size (measured as minimum convex polygon) for Eastern Ribbonsnakes in Nova Scotia was 0.43 ha (1.06 ac). Although Tinkle (1957) did not estimate Western Ribbonsnake home range size due to low recapture rates (22%), he suspected that this was because home range size exceeded the size of his study area. More published information exists on ribbonsnake movement rates than home range size, but these are highly variable. For example, a review of several studies suggests that movement rates vary by species, geographic location, sex, and size. Some authors describe Western Ribbonsnakes as moving only short distances during most of the active season (Clark 1974), while others consider them highly mobile (Tinkle 1957). Furthermore, mean distances moved per day by adult Eastern Ribbonsnakes from sites in Nova Scotia were comparatively large, ranging from 4.8 to 17.4 m (15.8 to 57.1 ft; Imlay et al. 2015). Carpenter (1952a) estimated that the greatest distance between Eastern Ribbonsnake recaptures in Michigan was 278 m (912 ft). The longest distanced recorded in a Nova Scotia study was 391 m (1,283 ft; Imlay et al. 2015).

A review of movement rates by sex also shows that variation exists and that results are somewhat contradictory. Langford et al. (2011) reported that the mean distance between Eastern Ribbonsnake recaptures was substantially smaller for females (41 m [134.5 ft]) than for males (201 m [659.5 ft]) in Alabama. Despite this, Clark (1974) estimated that mean distance moved per day by adult Western Ribbonsnakes was slightly larger for females (4.3 m/day [14.1 ft/day]) than for males (3.0 m/day [9.8 ft/day]). It is possible that females move greater distances at certain times of year (e.g., when gravid) as they seek out optimal locations for thermoregulation during gestation or to give birth. The presence of neonatal Eastern Ribbonsnakes in the uplands surrounding a wetland in Sheboygan County suggested that females moved away from the water's edge for parturition (J. Lorch, personal observation). Clark (1974) further found that movement distances of Western Ribbonsnakes were correlated with body size and that larger individuals were more vagile. Little information exists on the movement patterns of young age classes. Imlay et al. (2015) estimated that the mean daily movement rate for juvenile Eastern Ribbonsnakes ranged from 3.3 to 14.6 m (10.8 to 47.9 ft). Furthermore, Clark (1974) reported that juvenile Western Ribbonsnake movement rates ranged from 0.6 to 2.5 m/day (2.0 to 8.2 ft/day).

Ribbonsnakes in Wisconsin probably return to hibernacula in October. An adult Eastern Ribbonsnake in Columbia County that was presumably migrating to a hibernaculum on a rocky upland site was observed on 13 October (L. Doyle, personal communication). In Michigan, Eastern Ribbonsnakes commonly hibernated with conspecifics, other snake species (especially members of the genus *Thamnophis*), amphibians, and various invertebrates (Carpenter 1953). There are no known observations or even anecdotal reports of when Western Ribbonsnakes enter hibernation in Wisconsin. In northern Missouri, they are active until 12 November (Rossman et al. 1996).

PREY AND PREDATORS

There have been no detailed dietary studies focused on ribbonsnakes in Wisconsin. Reports from other states support that amphibians make up the vast majority of prey items taken. For example, it is estimated that amphibians comprise 80%–90% or more of the diets of Western Ribbonsnakes (Fouquette 1954) and Eastern Ribbsonsnakes (Brown 1979, Carpenter 1952a). Of the amphibians eaten, recently metamorphosed anurans make up the bulk of the diet, but anuran larvae are also frequently consumed (Brown 1979, Carpenter 1952a, Fouquette 1954, Clark 1974). Along these lines, stomach contents of an Eastern Ribbonsnake in Sheboygan County (MPM P171–175) were listed as "*Rana clamitans*," and a Western Ribbonsnake from Clark County (FMNH 14580) reportedly contained "3 tadpoles 7 mm long." Unlike the other *Thamnophis* species in Wisconsin, ribbonsnakes do not readily consume earthworms, and dietary partitioning between sympatric populations of ribbonsnakes and other gartersnakes reportedly occurs (Carpenter 1952a, Fouquette 1954).

Consumption of salamanders, fishes, various invertebrates, and reptiles (including captive cannibalism of juveniles) is also documented in ribbonsnakes, but most of these prey items apparently constitute a much smaller portion of the diet (Carpenter 1952a, Fouquette 1954, Smith 1961, Clark 1974, Hampton 2008, further reviewed by Ernst and Ernst 2003). Although amphibians are favored, prey is apparently taken based on availability, which varies due to many factors. For example, Ford and Hampton (2008) documented seasonal shifts in diet, with amphibian larvae being the most abundant prey item in spring and summer, while in autumn, when most larvae had metamorphosed, juvenile anurans made up the bulk of the diet. The species and life stage of anurans eaten can also vary by geographic location and habitat type (Fouquette 1954). Carpenter (1952a) noted that Eastern Ribbonsnake size further influences choice of amphibian prey, with small or young snakes eating smaller anurans (e.g., Spring Peepers [*Pseudacris crucifer*] and Boreal Chorus Frogs [*P. maculata*]) or larvae and adult snakes eating larger metamorphosing or adult amphibians (e.g., Northern Leopard Frogs [*Lithobates pipiens*], Green Frogs [*L. clamitans*], Pickerel Frogs [*L. palustris*], Wood Frogs [*L. sylvaticus*], and Eastern Newts [*Notophthalmus viridescens*]). Ford and Hampton (2008) similarly observed a positive correlation between snake SVL and prey size in the Western Ribbonsnake. That study noted that large females sometimes preyed on adult anurans, whereas males and younger age classes primarily consumed newly metamorphosed anurans or adults of smaller anuran species. Because of their smaller size, juveniles had a more restricted diet that was limited to the smallest amphibians present at a site. Juveniles have a disproportionately large head, a possible adaptation for consumption of amphibians, which are large relative to the snake's body size (J. Lorch, personal observation).

Ribbonsnakes are active foragers that use olfactory and visual cues to hunt (Ernst and Ernst 2003). They are reported to forage primarily in the morning and early evening (Ernst and Ernst 2003), although nighttime foraging has been documented in southern states (Rossman 1963b, Kennedy 1964, Rosen 1991b, Ernst et al. 1997, Ernst and Ernst 2003). Nocturnal hunting is probably less frequent in northern areas such as Wisconsin, where evening temperatures could drop below the thermal optima for ribbonsnakes. Ribbonsnakes do not employ constriction but instead overpower prey by capturing it in their jaws and swallowing it whole (Ernst and Ernst 2003). Carpenter (1952a) provided a detailed observation of a ribbonsnake hunting (see Natural History Box: In Search of Food).

NATURAL HISTORY BOX: In Search of Food

Charles C. Carpenter made the following observation on the efficiency of an adult female ribbonsnake attempting to catch frogs on 9 August 1948 at Sedge Point, Michigan: "I watched through field glasses and saw no indication that the snake was aware of my presence about 25 feet away. The following is quoted directly from my field notes: 'At 3:00 pm I observed this snake foraging along shore. In the next three quarters of an hour it moved back and forth along the shore searching for food. It chased and missed eight frogs in this time. When within 8–10 inches of a frog, it seemed to be able to sense it and would stalk towards it, the frog jumping just as the snake was ready to strike. At times, it would freeze, then after a short time its tongue would start flicking as it moved along, head going from side to side as if to pick up some scent of food. At other times it would lift its head up and peer about. At one time it came within about two to three feet of a group of Rana pipiens at the edge of the pool but did not notice them and turned and went the other way.'"

Excerpt from Carpenter (1952a)

No definitive reports of ribbonsnake predators in Wisconsin are known. Given information from other states, they likely fall victim to many vertebrates and invertebrates found in aquatic habitats (reviewed by Ernst and Ernst 2003). Carpenter (1953) specifically documented the predation of overwintering Eastern Ribbonsnakes in Michigan by crayfish. Ribbonsnakes are alert, and flight is likely their primary strategy for threat avoidance. In fact, Ford and Hampton (2008) proposed that ribbonsnakes rely on flight more frequently than other species of *Thamnophis*, which are more likely to remain motionless or behave defensively. In accordance, Eastern Ribbonsnakes encountered in Wisconsin always darted toward the nearest cover, usually a clump of dense vegetation (J. Lorch, personal observation). There was also no noticeable preference for seeking refuge in the water vs. along the shoreline. As noted by Vogt (1981), Eastern Ribbonsnakes in water usually swim across the surface to reach cover but will occasionally dive. In contrast, Western Ribbonsnakes are apparently more prone to head directly for water when threatened and immediately dive below the surface (Mierzwa 1985, Ernst and Ernst 2003), although Tinkle (1957) stated that in Louisiana they frequently sought cover in dense vegetation.

When handled, ribbonsnakes may smear musk and feces on their captor, but as Vogt (1981) mentioned, they are generally not aggressive. In fact, only 1 of 13 adult Eastern Ribbonsnakes handled by the authors actually followed through with the threat to bite. Eastern Ribbonsnakes also utilize caudal pseudoautotomy to escape capture (Todd and Wassersug 2010, Willis et al. 1982). Of 406 individuals captured in southwestern Nova Scotia, 14.8% were missing a portion of their tail (Todd and Wassersug 2010). The authors of that study observed tail pseudoautotomy three times when capturing snakes, and in one instance the detached tail twitched for ca. four minutes. Rossman (1962) also noted that tails were often truncated in museum specimens, which affected the ability of tail length to serve as a reliable means by which to differentiate between the two species of ribbonsnakes. Tail pseudoautotomy is apparently not as commonly utilized by (or at least not as well described in) the Western Ribbonsnake.

CONSERVATION

Status: Globally, Eastern and Western Ribbonsnakes are considered a Least Concern species by the IUCN. However, in Wisconsin both are considered Species of Greatest Conservation Need and listed as Endangered.

Populations: Information on population parameters, such as abundance and density, is scarce for both ribbonsnakes across their North American distributions. Carpenter's (1952a) work on a 19.4 ha (48 ac) site in Michigan is likely the best attempt at an Eastern Ribbonsnake abundance estimate published to date. However, he reported numerous abundance estimates made with various methods, as well as averages among them, making it difficult to disentangle the most robust number. Regardless, after considering all of his various estimates, he believed that 477 individuals was a "fair estimate" of abundance at his study site. From this, a density of 24.6/ha (9.9/ac) can be extrapolated. Estimates of Western Ribbonsnake abundance, density, and other population characteristics only currently exist for a few southern populations. For example, a multiyear study in Texas reported density estimates that ranged from 16/ha (6.5/ac) to 61/ha (24.7/ac) over time (Clark 1974). In addition, several studies from southern locations estimated that sex ratios were close to 1M:1F (reviewed by Ernst and Ernst 2003). Tinkle (1957) further believed that few young individuals survived their first year, which suggests that populations are dominated by older individuals.

Threats: Because they are often found in wetlands, there are numerous general threats associated with aquatic habitats that likely apply to ribbonsnakes (see Introduction to Conservation and Management of Wisconsin's Amphibians and Reptiles chapter). Although few investigations of threats to ribbonsnakes exist, Ernst and Ernst (2003) suggested that habitat destruction could result in substantial declines throughout their North American distributions. Hammerson (2007) elaborated that the declines mentioned by Ernst and Ernst (2003) were likely referencing populations at the edge of their distributions, where they were rare (such as in Wisconsin). Given that Wisconsin represents the distributional peripheries for both ribbonsnake species in North America, it is possible that state populations were historically limited to small isolated pockets of suitable habitat, even prior to European settlement of the area. Unfortunately, the past state abundances of both species are largely unknown. Hoy (1883) did not even include the Eastern Ribbonsnake in his early list of Wisconsin's reptiles (although he could have confused it with the Western Ribbonsnake, which he reported), and relatively few old specimens exist. However, it seems plausible that substantial populations of Eastern Ribbonsnakes once inhabited Wisconsin where wetlands used to be prevalent, but the snakes were quickly decimated when settlers converted wetland habitat to agricultural fields. Presently, Eastern Ribbonsnakes persist in areas that were historically unsuitable for agriculture and largely spared from human-mediated disturbance. For example, Wisconsin populations are currently known in areas of Sheboygan and Door Counties where glacial till dominates or the soils over bedrock are thin, both of which make effective implementation of agriculture challenging. Populations on public lands are protected for the time being. However, this species' state distribution closely overlaps many of the large human population bases in southeastern Wisconsin, and development pressure could impact unprotected sites in the future.

Western Ribbonsnakes face similar conservation challenges to those faced by Eastern Ribbonsnakes. In addition to its endangered status in the state, the species has declined in Indiana and northern Illinois (Smith 1961, Brodman et al. 2002). Like Eastern Ribbonsnakes, Western Ribbonsnakes were probably never widespread or abundant in Wisconsin, although they could have been locally common in wetlands associated with tallgrass prairies and oak savannas in the southeastern and south-central parts of the state. These areas were rapidly converted to agricultural land soon after European settlement, and little suitable habitat for the Western Ribbonsnake currently exists in Wisconsin (see Past, Present, and Potential Future Landscapes of Wisconsin chapter). Furthermore, that which remains is threatened by development, changes in hydrology, and other landscape alterations. For example, the southern unit of the Kettle Moraine State Forest contains extant wetland mosaics potentially capable of supporting Western Ribbonsnakes, but reforestation via vegetative succession and invasion of exotic plants (especially shrubs) are causing significant habitat degradation (J. Lorch, personal observation). Declines in the abundance of prey for neonatal Western Ribbonsnakes could also play a role in the species' demise in Wisconsin (see Natural History Box: Domino Effect?).

A variety of other potential threats to ribbonsnakes exist, but these are poorly understood. The snakes' alertness, close association with water sources, and tendency to go unnoticed by humans likely preclude direct human persecution as a significant source of mortality. However, the effects of roads could be of concern, given work on this subject in southern states. Research in Florida recorded 211 Eastern Ribbonsnakes as either active on roads or as roadkill over a three-year period (Dalrymple et al. 1991). Although this study was conducted in a state where Eastern Ribbonsnakes are more common than in Wisconsin (which likely contributed to the many roadkills observed), it speaks to the potentially high numbers of snakes that can be killed by vehicles. Clark (1974) attributed Western Ribbonsnake population declines he observed in Texas to an onset of dry conditions that led to fewer amphibians. Therefore, potential environmental alterations due to climate change, which might lead to more frequent droughts, could directly (via loss of wetlands) or indirectly (via loss of important amphibian prey) impact ribbonsnakes.

Management: Given the rarity and lack of research on both ribbonsnake species in Wisconsin, the best management practices for their conservation are difficult to outline. Presumably, many management considerations and practices that generally apply to aquatic habitats could benefit these snakes (see Introduction to Conservation and Management of Wisconsin's Amphibians and Reptiles chapter). Habitat management that focuses on the structure of plant communities, with an additional focus on connecting open-canopy uplands to adjacent wetlands, could benefit ribbonsnakes. Ribbonsnakes appear to strongly associate with shrubs and vegetative cover along wetland peripheries (see Habitat subsec-

NATURAL HISTORY BOX: Domino Effect?

Although many reptiles and amphibians are imperiled in Wisconsin, no herpetofaunal species that has inhabited the state in recent history is officially listed as extirpated. The Western Ribbonsnake, however, is a candidate for that designation. There have been no confirmed records of this snake since the early 1970s. While the decline of the Western Ribbonsnake in Wisconsin was likely multifactorial, one wonders how prey availability may have contributed to the snake's demise. Western Ribbonsnakes prey almost exclusively on small frogs. In the prairies and oak savannas of southern Wisconsin, Blanchard's Cricket Frogs (*Acris blanchardi*) were historically abundant and would likely have been a significant dietary item. Blanchard's Cricket Frogs undergo metamorphosis in August and September, and these recently metamorphosed individuals may have been especially important prey for newborn Western Ribbonsnakes that are too small to eat many of the other amphibians available in late summer. In the 1970s Blanchard's Cricket Frogs underwent catastrophic declines throughout the northern portion of their range and nearly disappeared from Wisconsin (see Blanchard's Cricket Frog, *Acris blanchardi* species account). Whether or not the Western Ribbonsnake was dependent on Blanchard's Cricket Frogs for survival is unclear, but Clark (1974) attributed a decline in Western Ribbonsnakes at his Texas study site to an abnormally dry winter that led to fewer amphibians. This alludes to the strong connection between ribbonsnakes and their prey. Although a link between Blanchard's Cricket Frog and Western Ribbonsnake population declines in Wisconsin is purely speculative, the collapse of Blanchard's Cricket Frog populations in the state coincides with some of the last sightings of this snake. However, there is room for optimism. In recent years, Blanchard's Cricket Frog populations appear to be rebounding. If Western Ribbonsnakes are dependent on Blanchard's Cricket Frogs, then the increase in abundance of these amphibians may allow the snakes to reclaim some of their former range in Wisconsin. This may be especially true along the Mississippi River corridor, as populations of Western Ribbonsnakes have been recently reported from northwestern Illinois.

Jeffrey M. Lorch

tion; Carpenter 1952a, Tinkle 1957, Charland and Gregory 1995). As such, management practices that promote an intact native shrub layer along portions of wetland and bog edges or on uplands adjacent to waterways could be beneficial (although unchecked ingrowth of woody plant species is likely undesirable). The WDNR (2013b) recommends that upland areas within 5–10 m (16.4–32.8 ft) of water be maintained to provide an appropriate balance between a dense ground layer of vegetation and woody vegetation for cover. Active management of invasive plant species (e.g., Purple Loosestrife [*Lythrum salicaria*], Common Buckthorn [*Rhamnus cathartica*], Reed Canary Grass) in wet meadows, bogs, and fens could further benefit this species (WDNR 2013b), although research on how invasive plants affect ribbonsnakes is lacking. The most striking feature of areas where ribbonsnakes are documented in Wisconsin is the abundance and diversity of amphibians at those sites (J. Lorch and R. Staffen, personal observation). Therefore, protecting landscape-wide mosaics of shallow wetlands and surrounding upland habitats that support a sufficient prey base could be important for ribbonsnake conservation.

Common Gartersnake

Thamnophis sirtalis (Linnaeus 1758)

Joshua M. Kapfer, Gregor W. Schuurman, and Rori A. Paloski

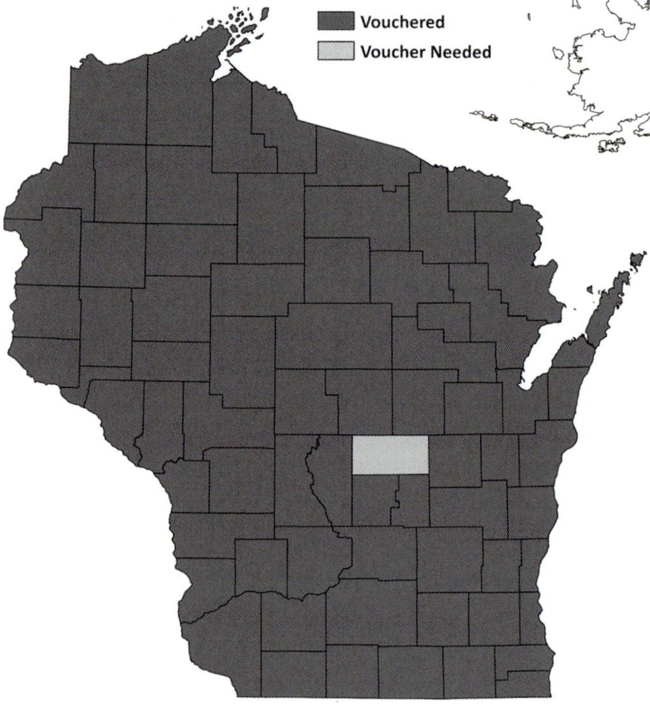

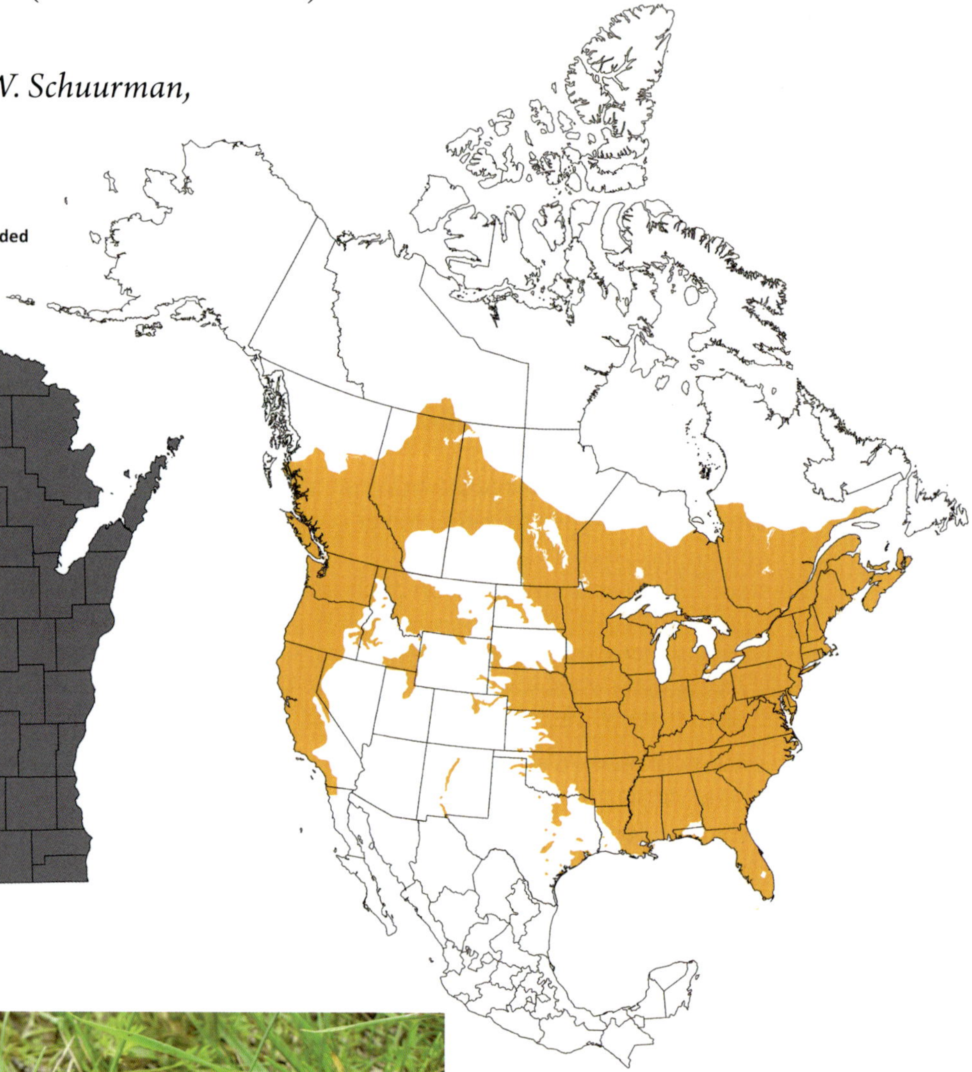

Adult Female Common Gartersnake (Waukesha County; photo by J. Kapfer).

AT A GLANCE

Common Gartersnakes can generally be identified by their dark upper surface with three yellowish stripes that run the length of the body: one down the center of the back and one along each side. Wisconsin is home to several related gartersnake species, which are difficult to differentiate, so careful inspection of these side stripes is necessary to distinguish the species. The side stripes of Common Gartersnakes are usually wider and in a lower position than those of related species. The stripes of Common Gartersnakes partially or completely cover the third, second, and often first scale rows that run longitudinally above the belly scales. Common Gartersnakes are documented in all Wisconsin counties. They are found in a variety of habitats, both natural and anthropogenic. This species is particularly abundant in damp environments such as wetlands and along stream or river corridors, but it is also common in adjacent uplands (e.g., mesic prairies near wetlands). They have a lengthy active season that often lasts from late March through mid-October but can be longer. Common Gartersnakes breed in the spring, sometimes in large courting groups, and give birth to live young from mid-August to early September. Common Gartersnakes are active daytime foragers that consume a wide array of small prey species but are particularly fond of earthworms and amphibians. They are harmless but will sometimes bite if threatened or handled (the bite is no more painful or dangerous than a scratch). The origin of the name "gartersnake" may be related to the snake's stripe pattern, which superficially resembles an undergarment called a garter. Garters consist of narrow black fabric strips that fasten below the knee and were commonly used in the eighteenth and nineteenth centuries to hold up stockings before the advent of elastic. It is noteworthy that gartersnakes are sometimes commonly, and incorrectly, referred to as 'garden snakes' or 'grass snakes.' This species is considered common by the WDNR, and Vogt (1981) stated, "This ubiquitous snake is perhaps the first and often the only snake many Wisconsinites encounter."

SYSTEMATICS

Thamnophis sirtalis (Common Gartersnake) was originally described by Linnaeus (1758) as *Coluber sirtalis*, but Klauber (1948) later determined that Linnaeus had actually described a specimen of *T. saurita* (Eastern Ribbonsnake). The rule of priority would generally require the specific epithet *sirtalis* subsequently be used for *T. saurita*, but because *sirtalis* had already been applied to the Common Gartersnake for so long, it was retained as the scientific name to avoid confusion. *T. sirtalis* has since undergone several taxonomic changes (summarized by Fitch 1980 and Rossman et al. 1996). Twelve subspecies are currently recognized (Crother 2017), two of which occur in Wisconsin (*T. s. sirtalis* and *T. s. semifasciatus*; Rossman et al. 1996). It is of further note that Crother (2017) stated that *T. sirtalis* "may be composed of multiple independently evolving lineages often not concordant with the subspecific taxonomy," in reference to the results of mitochondrial DNA analysis described via personal communication with F. T. Burbrink (American Museum of Natural History). *T. sirtalis* is currently positioned within a clade that also includes Wisconsin's ribbonsnake species (*T. saurita* and *T. proximus*), consistent with the understanding that these three species are more closely related to each other than to the state's other gartersnakes, *T. butleri* and *T. radix* (de Queiroz et al. 2002, Pyron et al. 2013, McVay et al. 2015). Despite the relative genetic 'distance' between *T. sirtalis* and *T. butleri*, hybridization between the two species has been documented in Wisconsin (Kapfer et al. 2013a; see Distribution in Wisconsin subsection). All species within the genus *Thamnophis* are members of the subfamily Natricinae, which, in Wisconsin, also includes members of the genera *Nerodia*, *Regina*, *Storeria*, and *Tropidoclonion*.

DESCRIPTION

Adults: In general, the dorsal surface of Common Gartersnakes in Wisconsin is dark gray, olive, or black; yellow, white, or bluish vertebral stripes and three lateral stripes run the length of the body (Ernst and Ernst 2003; Figure 280). The light dorsal stripes are typically prominent but sometimes indistinct

Above, top: Figure 280. An adult Common Gartersnake showing color and pattern typical in Wisconsin (Ozaukee County; photo by S. Viernum).

Above, bottom: Figure 281. A close-up of the head and labial scale pattern of an adult Common Gartersnake (Sauk County; photo by C. Raimond).

and covered by dark side spots (Ernst and Ernst 2003). One dorsal stripe runs down the center of the back, covering the vertebral scale row and at least half of each paravertebral scale row (i.e., the scale rows adjacent to the vertebral scale row). A single light lateral stripe also generally occurs along each side of the body. These lateral stripes cover scale rows two and three but often also scale row one and sometimes even extend up to the lower edges of row four. The top of the head is dark, but the labial scales and throat are whitish to light yellow, occasionally with a green tint. Dark bars occur on the posterior edge of most supralabial scales directly below and behind the eyes (Figure 281). The ventral surface is gray, greenish, yellow, or cream and often has two rows of black spots (Ernst and Ernst 2003).

Although difficult to distinguish, the two subspecies in Wisconsin differ slightly in appearance. Unlike *Thamnophis sirtalis sirtalis* (Eastern Gartersnake, the most common subspecies in Wisconsin), *T. s. semifasciatus* (Chicago Gartersnake) has broad black vertical bars on the sides of the body that interrupt the lateral stripes. Although the Chicago Gartersnake's range overlaps with that of the Eastern Gartersnake in Wisconsin, it is reportedly restricted to a small pocket in the southeastern portion of the state near Lake Michigan (Rossman et al. 1996).

Following thresholds for adult size in Rossman et al. (1996), we measured adult specimens from Wisconsin in the MPM and UWZM collections. Mean SVL of these specimens was 42.4 cm (16.7 in, $n = 83$) and 54.1 cm (21.3 in, $n = 154$) for males and females, respectively, while mean tail length was 13.7 cm (5.4 in, $n = 66$) and 14.5 cm (5.7 in, $n = 127$) for males and females, respectively. Froom (1972) reported an SVL that exceeded 137 cm (53.9 in) in Canada. The largest Wisconsin specimen measured had an SVL of 85.7 cm (33.7 in, UWZM 17289, collected in 1963 from Wood County, sex undetermined).

Preadult stages: Neonates and juveniles are similar in appearance to adults. The mean SVL of 29 neonates from a Washington County female was 15.9 cm (6.3 in) immediately after birth (Joshua Kapfer, UW–Whitewater, unpublished data). Neonates were sampled from late August through early October for two years in Rock County ($n = 37$). Mean SVL of these snakes was 18.9 cm (7.4 in), mean tail length was 5.2 cm (2.1 in), and mean weight was 4.2 g (0.2 oz; J. Kapfer, unpublished data).

Variation: Adult appearance varies widely at both broad (Rossman et al. 1996, Ernst and Ernst 2003) and local (King 1988b, Krause et al. 2003, Mooi et al. 2011) geographical scales. Across this species' North American distribution, head and body base color ranges from olive to red to black, while dorsal stripes can also range in color from yellow to greenish. Common Gartersnakes in Wisconsin, however, generally possess the darker base color with light dorsal stripes,

Figure 282. A melanistic adult Common Gartersnake from Isle Royale (photo by R. Paloski). Note that melanistic Wisconsin Common Gartersnakes are known on the Apostle Islands off the coast of Ashland and Bayfield Counties. Both Isle Royale and the Apostle islands occur in Lake Superior.

Figure 283. Examples of variation in the lateral stripe position that can occur in Wisconsin Common Gartersnakes. (a) The lateral stripe often covers at least half of the first scale row (Jefferson County; photo by J. Kapfer). (b) The lateral stripe can appear to occur on the second scale row, but note that the first scale row is lighter in color than those above the lateral stripe on the fourth scale row (Rock County; photo by J. Kapfer).

but rare exceptions are known. Melanistic individuals occur in populations from central Canada and the Great Lakes region (Bittner et al. 2002, Mooi et al. 2011), including the Apostle Islands off northern Wisconsin's Lake Superior shoreline (Hecnar et al. 2002; Figure 282). Melanism is a simple Mendelian trait that is recessive to a striped pattern (King 2003) and can result in the dark side pigment extending onto the ventral surface. Variability in the vertical position and width of lateral stripes is also sometimes evident among Wisconsin individuals (Figure 283).

Common Gartersnakes further exhibit sexual size dimorphism, with adult females attaining sizes up to ca. 25% longer than adult male snakes and weighing almost twice as much (Ernst and Ernst 2003; Table 16). Yet differences in size by sex apparently vary with age. For example, King et al. (1999) found the opposite pattern among neonates analyzed, where male SVL exceeded female SVL.

Diagnosis: Common Gartersnakes possess keeled dorsal scales that occur in <27 rows at midbody, and a yellow lateral stripe typically covers the second and third scale rows (although these often include the first and sometimes lower edges of the fourth scale row). They have a single anal plate, divided subcaudal scales, and ventral scale counts that range from 137 to 177. They are further characterized by the presence of two prefrontal scales, and there is a single loreal scale, a single preocular scale, three postocular scales (typically), seven supralabial scales, and ten infralabial scales on each side of the head. Common Gartersnakes are most easily confused with the other species of *Thamnophis* in Wisconsin: Butler's Gartersnake, Plains Gartersnake, Eastern Ribbonsnake, and Western Ribbonsnake. The geographic distribution and morphological features of Common Gartersnakes substantially overlap those of the other Wisconsin snakes in this genus, which can make correct identification a challenge. Yet

Table 16. Mean SVL, tail length, and weight of wild-caught adult Common Gartersnakes sampled at three locations in southern Wisconsin in 2013 and 2014.

County	Sex	SVL cm (in)	Tail cm (in)	Weight g (oz)
Jefferson	Female	45.1 (17.8), $n=51$	11.9 (4.7), $n=45$	46.4 (1.6), $n=49$
Jefferson	Male	42.7 (16.8), $n=54$	12.8 (5.0), $n=52$	35.9 (1.3), $n=52$
Rock	Female	46.1 (18.2), $n=19$	11.8 (4.7), $n=18$	48.9 (1.7), $n=18$
Rock	Male	42.3 (16.7), $n=21$	12.2 (4.8), $n=18$	40.4 (1.4), $n=21$
Walworth	Female	45.2 (17.8), $n=37$	11.5 (4.5), $n=35$	45.4 (1.6), $n=37$
Walworth	Male	41.9 (16.5), $n=48$	12.3 (4.8), $n=47$	33.4 (1.2), $n=48$

Common Gartersnakes are distinguishable from both ribbonsnake species because the latter have a longer tail and are more slender. They also possess unbarred white labial scales, white spots on the preocular scales, and a lateral stripe that only contacts scale rows three and four. Although the subralabial scales of Plains and Butler's Gartersnakes are typically barred (similar to those of Common Gartersnakes), the lateral stripe of Plains Gartersnakes usually covers the third and fourth scale rows, while in Butler's Gartersnakes the lateral stripe often covers the third scale row and part of the second and fourth rows. Analysis of microsatellite DNA shows that Common Gartersnakes hybridize with Butler's Gartersnakes in Wisconsin, but incidence in the wild is likely rare (Kapfer et al. 2013a). However, distinguishing hybrids from 'pure' gartersnakes can be difficult or impossible with only external morphological characteristics.

DISTRIBUTION AND HABITAT

Global distribution: Common Gartersnakes are one of the most widespread reptile species in North America. They occur throughout much of southern Canada and are widespread in the eastern and central US. Their distribution is more restricted in the western US, particularly in the southern Great Plains, Rocky Mountains, and desert regions, but they occur along the West Coast from southern Alaska to southern California. Several disjunct populations also exist in New Mexico and parts of southeastern Texas.

Distribution in Wisconsin: Published records exist for all of Wisconsin's counties, although a suitable voucher specimen could not be located for Waushara County (see the Appendix). Museum records indicate that UWZM 9912 was collected in Waushara County, but this specimen could not be located (Laura Monahan, UWZM, personal communication). Common Gartersnakes are also found on the Apostle Islands in Lake Superior (Ashland and Bayfield Counties; Schulz 1976, Hecnar et al. 2002) and on islands in Green Bay and Lake Michigan (Door County; Long and Long 1976, Long et al. 1978). Recent analyses detected hybrid Common × Butler's Gartersnakes in Dodge, Washington, Ozaukee, Waukesha, and Milwaukee Counties (Rori Paloski, WDNR, unpublished data).

Localities of possible occurrence: Common Gartersnakes are documented in every Wisconsin county.

Habitat: Common Gartersnakes can occupy a wide range of habitats but are particularly common in wetter environments, such as open-canopy wetlands and riparian corridors (Fitch 1965, Jordan 1967, Catling and Freedman 1980b). In Wisconsin, they appear to favor certain wetland types over others. For example, Kapfer et al. (2013b) captured more Common Gartersnakes in wetlands dominated by native sedges, shrubs, and the invasive Reed Canary Grass (*Phalaris arundinacea*) than in wetlands with substantial stands of cattail (*Typha* spp.). Wisconsin and Michigan studies also reported a strong affinity for open upland habitat immediately adjacent to wetlands, even over the species' affinity for wetlands themselves (Kapfer et al. 2013b and Carpenter 1952a, respectively). Common Gartersnakes in Wisconsin are less often encountered in dry open prairies that lack direct access to damp habitats, but they are sometimes observed on dry bluffs adjacent to the Mississippi and Wisconsin Rivers and in sandy oak savannas with pothole wetlands (J. Kapfer, personal observation). They appear somewhat tolerant of human disturbance in Wisconsin. For example, they occupy agricultural and urban landscapes along vegetated buffers and green spaces near ponds or ditches (J. Kapfer, personal observation). The microhabitats selected by Common Gartersnakes are not well-studied in the Upper Midwest. Most (2013) modelled microhabitat characteristics associated with a population in northern Illinois and reported that the presence of "aquatic" and "graminoid" vegetation, as well as distance to the nearest crayfish burrow, partially predicted where Common Gartersnakes existed on the landscape. Carpenter (1952a) found that Common Gartersnakes in Michigan sometimes rested in elevated locations on the branches of shrubs, particularly hawthorne (*Crataegus* spp.). Information on habitats used by juveniles during the active season in Wisconsin is anecdotal, but they presumably occupy similar habitats as adults.

Because Common Gartersnakes cannot survive freezing for prolonged periods (see Activity section), they must spend the duration of winter in hibernacula that protect them from prolonged cold. In Wisconsin, this species likely overwinters under a variety of conditions as long as they can descend below the frost line, although published information is not extensive. Reports from northern Illinois and Ohio indicate the snakes will spend the winter in various burrows or holes, such as the burrows of Devil Crayfish (*Cambarus diogenes*; Most 2013, Dalrymple and Reichenbach 1981). Late autumn and early spring observations in Wisconsin suggest they will overwinter belowground in decaying woody structures (e.g., old tree stumps) near wetlands or in floodplain forests and also fissures within rock outcroppings along open river bluff faces (Figure 284; J. Kapfer and Rori Paloski, WDNR, personal observations). Common Gartersnakes sometimes overwinter in anthropogenic structures, including old railroad embankments and the foundations or basements of old homes (J. Kapfer, personal observation). Costanzo (1985) even reported Common Gartersnakes that overwintered underwater in an abandoned buried cistern in Wisconsin (see also Activity section).

Figure 284. An adult Common Gartersnake exits a hibernaculum at the base of an old tree stump (Waukesha County; photo by J. Kapfer).

Information on hibernacula selected by juveniles in the Midwest comes primarily from Minnesota (Lang 1971) and Michigan (Carpenter 1953). Both sources reported that very young Common Gartersnakes overwintered in abandoned ant mounds (*Formica ulkei*). Carpenter (1953) further recorded young Common Gartersnakes using Meadow Vole (*Microtus pennsylvanicus*) and Devil Crayfish tunnels as hibernacula. Very young members of a population also might not overwinter in the same location as older juveniles and adults. Studies in Portage County and northern Canada found that most snakes captured near communal hibernacula were either in their second year or older (Costanzo 1986, Larsen and Gregory 1989), which suggests that younger snakes overwintered elsewhere. Costanzo (1986, 1989b) provided evidence that younger snakes eventually locate adult hibernacula by following older snakes' pheromone trails.

REPRODUCTION AND DEVELOPMENT

In Wisconsin, Common Gartersnakes breed primarily in spring (March through early May), although some autumn breeding reportedly occurs (Vogt

1981). Gartska et al. (1982) determined that a period of winter dormancy is required for stimulation of male sexual activity. Furthermore, O'Donnell et al. (2004) reported that although males emerged from hibernation with low energy reserves, their post-emergence behavior was centered on reproduction over feeding.

Females attract males via pheromones produced in the liver, and their receptivity to male advances is related to environmental temperature (Gartska et al. 1982). After males have located females, male courtship and female selection of mates occur. The courting behavior of males has been detailed by Fitch (1965) and Blanchard and Blanchard (1942) but is summarized here. It involves a male lying alongside a female and pressing their bodies together. The male then sends a series of ripples down the length of both snakes' bodies while rubbing his chin on the female's back. Copulation only occurs if the female raises her tail (or allows the male to raise her tail) so that intromission can ensue, which then lasts 15–20 minutes (Fitch 1965). Males court heavier and larger females more actively than thinner females (Gartska et al. 1982). This is presumably because larger females produce more offspring, which increases the male's fitness (reviewed by Shine 2003). Although several sources report that females copulate once per season (e.g., Fitch 1965, Whittier et al. 1985), multiple paternity has been documented in the Upper Midwest (King et al. 2001). The courtship season in Wisconsin can be very short (three to five days), according to research on a population in Outagamie County (Clesson et al. 2002).

Common Gartersnakes sometimes breed in congregations that can include hundreds of snakes (Gartska et al. 1982, Shine et al. 2001b). These large 'mating balls' often consist of a few females that are courted by numerous males, and intense competition occurs among males as they vie for access. Mating balls have been documented in Wisconsin, albeit with fewer participants than reported in Canadian populations (Clesson et al. 2002; Natural History Box: Many Amorous Snakes). Interestingly, work in Manitoba revealed that some snakes take advantage of these mating balls for purposes other than reproduction. Mason and Crews (1985) reported that upon emergence from overwintering, some males released pheromones with characteristics that were similar to those of females, which attracted male snakes. The benefit of this strategy was not immediately apparent when first discovered because these female mimics did not experience higher reproductive success than other males (Shine et al. 2000b). However, Shine et al. (2001a) later found that female mimicry likely provided a survival advantage rather than a reproductive advantage. Large breeding congregations of snakes are warm (Shine et al. 2000b). Upon emergence from hibernation, snakes are cold and require heat to warm themselves, which affects survival (Shine et al. 2000a). Interestingly, the pheromones produced by males that mimicked females were only attractive to other males when these female-mimicking males were cold (Shine et al. 2012). Thus, immediately upon emergence, the female mimics attracted other males and took advantage of heat produced by the large group of male snakes that mobbed them. After they warmed sufficiently and their pheromones no longer attracted other males, these female mimics were ignored and able to attend to other post-emergence activities. Although this phenomenon is poorly investigated in Wisconsin, Clesson et al. (2002) believed that males who mimicked females occurred in their Outagamie County study population.

Common Gartersnakes are viviparous (Blackburn 2000), a developmental strategy that is more frequent in snakes from higher latitudes (reviewed by Gregory 2009). This strategy's distributional pattern could be related to the effect that environmental factors have on embryonic development and survival (Shine 1995). For example, embryos held in the female's body during development experience prolonged optimal conditions because gravid females maintain high and precise body temperatures (Charland 1995, Gregory 2009). Oviparous snakes, on the other hand, leave more to chance when they deposit and abandon eggs. For these eggs to successfully develop, nest sites must be selected so that eggs experience consistently optimal environmental conditions (reviewed by Gregory 2009). Given that higher latitudes have lower average temperatures and fewer warm days for egg development than lower lat-

NATURAL HISTORY BOX: Many Amorous Snakes

When I was a boy, a favorite pastime was to search for amphibians and reptiles in the 'wild' areas near my home in Dane County. Frequently, these trips led me and a neighborhood friend to a nearby city park. The park was adjacent to the Yahara River, from which came the nesting turtles and frogs that fueled our interest in wildlife. Although the park was not far from where we lived, it required we walk either under or over a concrete railroad overpass at the end of the street. The railroad tracks were another source of adventure, and we spent many hours walking the tracks looking for Common Gartersnakes during the spring and summer. One spring day in elementary school, we decided to investigate the nearby park for 'critters' and took the trail up the small earthen embankment that led over the concrete railroad overpass at the end of the road. The embankment was forever overgrown with dense vegetation, but a path had been worn down over the years by folks climbing up to walk the tracks. On this particular day we reached the peak of the hill, where the tracks bisected our path, and were suddenly confronted with a writhing pile of snakes at our feet. My memory is that there were hundreds of snakes in the moving mass, but the many years between then and now have likely inflated reality. My friend and I stood there watching the snakes for quite a while, not sure what they were up to. In hindsight, it seems likely that the snakes had overwintered in fissures within the cracked concrete of the railroad track overpass. I vividly recall grabbing one of the larger snakes, likely under the pretense that I could convince my parents it should be my pet for the summer. Yet to my surprise, when I lifted the one snake, another came with it, dangling with its tail connected to the tail of the first! I quickly released the first snake and let them go on about their business. Although I don't know exact numbers, this was the largest breeding group of Common Gartersnakes I've ever seen.

Joshua M. Kapfer, UW–Whitewater

itudes, the greater prevalence of viviparous snakes in the North is sensible.

In Wisconsin, female Common Gartersnakes become noticeably gravid (via palpation) in late May (J. Kapfer, personal observation), a pattern observed in other midwestern studies (e.g., Meshaka et al. 2008). Offspring are typically born from mid-August to early September (Vogt 1981), but some are likely also born in late July. Great variation in litter size among female Common Gartersnakes has been reported. Vogt (1981) found that females produced 6–73 offspring per litter in Wisconsin. A female from Washington County (SVL = 67.5 cm [26.6 in]) produced 29 offspring on 13 August (Figure 285; J. Kapfer, unpublished data). Fitch (1965) determined that mean litter size in Kansas was 14.5 (range = 4–29, *n* = 132 females), but mean litter size across 100 other

Figure 285. Newborn Common Gartersnakes from the same brood (Washington County; photo by J. Kapfer).

NATURAL HISTORY BOX: Frigid Encounter

Among Wisconsin's snakes, Common Gartersnakes appear to be champions at tolerating cold weather. Knudsen (1955b) provided an excellent example of this: "Once while kneeling at a spring in late January the author was astounded to find a medium-size garter snake, too stiff to move, lying in the icy water. A previous warm spell had brought it out of a crevice above the spring and then a quick change to cold caught the snake." In Portage County, Costanzo (1986) also captured two individuals that had recently emerged from hibernation to bask on snowbanks. I have further received photos of active Wisconsin individuals taken by deer hunters in November. I've even found evidence of active Common Gartersnakes late in the year or under very cold conditions. On 17 November 2013 I found two fresh road-killed individuals in Jefferson County during a spell of warm rains (UW–Whitewater Natural History Collections, UWWHERP 0202 and 0203). However, my most astounding experience of cold-weather activity by this species occurred during my childhood. On a late winter day when school had been cancelled due to a snowstorm, a friend and I donned our winter gear and went out to explore a frozen ditch in rural Dane County. As we trudged along, we were both shocked to encounter a Common Gartersnake resting atop the snow and ice in the middle of the ditch. Although the snake was mostly motionless, closer inspection revealed that it was definitely alive. Furthermore, the little serpent had left a trail in the snow that ran across the ditch, showing us the path it had taken. We carefully followed its trail back to a large exposed muddy bank along the ditch. The bank was too steep to allow the accumulation of snow and was oriented to receive substantial direct sunlight over the course of the day. Many holes and crevices dotted the slope, and we assumed one was the entrance to the snake's winter retreat. We surmised that the sun had warmed the exposed mud just enough to draw the snake from its winter's rest. We could not be sure when the snake left the hibernaculum or how long it had been above ground, but its trail on fresh snow led to this general location. We gingerly took the snake and placed its nose into one of the holes on the bank. To our delight, it slowly crawled forward and was gone in a matter of seconds.

Joshua M. Kapfer, UW–Whitewater

North American literature records was 26. Litter size can exceed these reports, however. Martof (1954) documented a Michigan female that gave birth to 85 offspring. The number of offspring a female can produce per litter increases as she increases in size (Gregory and Larsen 1996). There is also a reported inverse relationship between litter size and neonate SVL (i.e., in populations where females produced larger litters, neonates had shorter SVLs; Gregory and Larsen 1996).

Data on the development of neonate Common Gartersnakes indicate that growth rates are highest in younger individuals (12%–30% increase in length per 'growing month'), compared to adults (0.2%–2.0% increase in length per 'growing month'; Carpenter 1952b). Females in Michigan are reportedly capable of producing young when they reach two years of age, while males can reach maturity by their second spring, slightly before the two-year mark (Carpenter 1952b). Rossman et al. (1996) reported that females matured at an SVL of 42–55 cm (16.5–21.7 in), while males matured at an SVL of 36–39 cm (14.2–15.4 in). Longevity data indicate that wild Common Gartersnakes can live for six to nine years (Larsen and Gregory 1989, Fitch 1965, respectively).

ACTIVITY

Common Gartersnakes have long active seasons in Wisconsin that can extend from early March to mid-November, but even mid-winter activity is noted if conditions are appropriate (Vogt 1981; Natural History Box: Frigid Encounter). Costanzo (1986) reported that Common Gartersnakes typically emerged from a Portage County hibernaculum over a period of several days in the spring and found that mean body temperature of emerging snakes was 4.8°C (40.6°F, $n = 23$). When active, Common Gartersnakes are mostly diurnal (Rossman et al. 1996), but reports of peak daily activity show variation. For example, studies of an Ohio population revealed that daily activity was concentrated primarily from late morning through afternoon in spring, while summer activity peaked twice: first in the morning and again in late afternoon (Reichenbach and Dalrymple 1986). Activity patterns are also affected by the shedding cycle, with lower activity documented in snakes that are about to shed their skin (King 1997a). Snakes that are close to shedding typically experience eye opacity as the old skin separates from the new skin beneath. Reduced activity levels immediately prior to shedding could be related to the impairment in vision that occurs at that time (King 1997a).

Carpenter (1956) found that the body temperature of active snakes in Michigan was highly correlated with air temperature and that mean body temperature was slightly lower in the spring (April through June, 25.5°C [77.9°F]) than in the late summer to early autumn (July through September, 28.4°C [83.1°F]). He also determined that the range of body temperatures associated with activity could be broad: 20°C–40°C (68°F–104°F). Others have reported that the body temperatures of active individuals were often kept within a narrower range, which supports that this species can precisely thermoregulate. For example, studies from various North American locations found that active body temperatures ranged from only 22°C to 34°C (71.6°F to 93.2°F; Fitch 1965, Stewart 1965, Rosen 1991b, additional literature reviewed by Lillywhite 1987). Fitch (1965) found that a body temperature of 41°C (105.8°F) was at or near the critical thermal maximum (CTmax) for the individuals he tested in Kansas. Bittner et al. (2002) observed that variation in color pattern and morphology affected the thermal state of adult snakes. They reported that the mean equilibrium body temperature of melanistic color morphs was 1.3°C (33.3°F) warmer than normal color morphs, although no difference in rates of warming were detected between morphs.

There appears to be substantial variation in this species' movement patterns throughout North America. Rossman et al. (1996) suggested that the "variation in movements probably reflects population differences in distance between suitable hibernation sites and summer foraging areas." Larsen (1987) determined that radio-tracked snakes moved along a mostly straight trajectory away from hibernacula during spring and early summer. By mid-summer, these snakes had reached a maximum distance from the hibernacula. They then began return migrations in order to arrive at hibernacula by autumn. Gregory (1984) reviewed the results of several studies and found that dispersal distances from summer activity areas to hibernacula can be substantial (1.0–17.7 km [0.6–10.9 mi]). This distance was more exaggerated in areas where the availability of suitable winter hibernacula was limited (Gregory 2009).

Other factors also reportedly affect movement patterns over both long and short time frames. For example, Larson (1987) linked large migration distances from hibernacula to the amount of prey available to a population from northern Canada. Carpenter (1952a) reported variation by sex in distances moved between the first and last captures of 42 individuals in Michigan. These were 108.9 m (357.44 ft) for females but only 59.9 m (196.75 ft) for males after "more than 200" days of monitoring (overall $\bar{x} = 80.9$ m [265.6 ft]). Mean distance between recaptures for males and females in northern Illinois also varied by season. For example, while distance between recaptures of males and females was 16.4 m (53.9 ft) and 11.7 m (38.5 ft) in the spring, respectively, this distance shifted to 1.2 m (3.8 ft) and 18.7 m (61.4 ft) in summer, respectively (Most 2013). Common Gartersnakes further appear capable of moving large distances over short time periods. For example, Carpenter (1952a) found that one individual moved at least 31.7 m (104 ft) in ca. 30 minutes, while another moved 30.5 m (100 ft) in ca. 42 minutes. Larsen (1987) found that distances moved by radio-tracked snakes over a single day varied substantially, ranging from <50 to 1,000 m (164 to 3,281 ft).

Home range estimates for Common Gartersnakes in Wisconsin do not exist. Although a handful of authors have estimated home range size for other midwestern states, these reports are few and include estimates that are highly variable. Most (2013) determined single-year home range sizes for a small sample of males ($n = 2$) and females ($n = 5$) in northern Illinois. The resultant mean areas were small, only 0.006 ha (0.01 ac) for males and 0.05 ha (0.14 ac) for females, with large standard deviations (MCP estimator). Carpenter (1952a) estimated a similar but slightly larger mean home range size for Common Gartersnakes in a Michigan population (0.8 ha [1.98 ac]). Fitch (1965) reported much larger mean home range sizes for a Kansas population, which ranged from 14.2 ha (35.0 ac) for males to 9.2 ha (22.7 ac) for females. Although these home range estimates were appreciably larger than in the other studies, the author incorporated a different home range estimator based on recapture data used to create circular home ranges.

In autumn, adult Common Gartersnakes often return to the same hibernacula they occupied in past years (Costanzo 1986). Although they may remain in the vicinity of hibernacula for a period, the timing of actual ingress into hibernacula for winter likely varies annually. Costanzo (1986) reported that Wisconsin individuals entered hibernacula when maximum daily air temperatures dropped to 12°C–17°C (53.6°F–62.6°F) in autumn. Yet Common Gartersnakes have a high tolerance for cold, and Wisconsin individuals are even sometimes observed outside of hibernacula in winter (Natural History Box: Frigid Encounter). Laboratory experiments in Michigan found that the mean critical thermal minimum (CTmin) for both adults and neonates was 3.9°C (39°F; Doughty 1994). Fitch (1965) experimented on Kansas individuals and found that they survived even when body temperature dropped to between −3°C and −2°C (26.6°F and 28.4°F) but suggested that the former was near their CTmin. In an extreme example, Costanzo (1988) determined that 62.1% of captive individuals accidentally entombed in ice due to malfunctioning laboratory equipment ($n = 29$) survived the ordeal. Costanzo and Lee (1988) further reported that individuals even survived 48 hours with 34% of their total body water frozen. Interestingly, they also found little evidence to support that snakes employed physiologically produced cryoprotectants in their blood or tissues to resist freezing. They concluded that freeze tolerance is a strategy used in addition to behavioral avoidance of cold temperatures (i.e., overwintering) during winter. Churchill and Storey (1992) concluded that tolerance of freezing for at least short periods is likely beneficial if snakes are aboveground during frost events in early spring or late autumn. Despite their apparent ability to withstand cold conditions, winter mortality can be substantial in some years (see Populations subsection).

Costanzo (1985) studied the overwintering physiology of Portage County snakes in an abandoned cistern. He found that their energetic demands during winter were largely met by metabolism of glycogen and protein stored in the liver and muscle tissue. He also determined that the largest depletion of these energy reserves occurred early in the overwintering period as snakes acclimated to dropping temperatures experienced in hibernacula. Unlike some reptile species, Common Gartersnakes do not metabolize energy from stored fats during winter dormancy, because these resources are used to support reproduction after spring emergence (Costanzo 1985). This species also apparently overwinters with an empty gut, given the lack of food recovered from the stomachs and intestines of 83 snakes removed from a hibernaculum in late October (Costanzo 1986). Dormant individuals can overwinter while submerged in water. In fact, when given the choice to overwinter in wet vs. dry conditions, Common Gartersnakes in Wisconsin elected to remain submerged (Costanzo 1985), which could be related to energy conservation. Costanzo (1989a) determined that submerged individuals used 55.3% less energy than those overwintering while exposed to air. There are suggestions that the higher winter survival while submerged occurs because oxygen absorption through the skin is possible, particularly when coupled with their reduced metabolic demands in cold water (Costanzo 1989a). Common Gartersnakes that remain submerged during winter also benefit from the thermal buffering capacity of water and the reduced potential for dehydration (Costanzo 1985, 1988, 1989c).

PREY AND PREDATORS

Only anecdotal information exists on the diet of Common Gartersnakes in Wisconsin, but earthworms, amphibians, and fishes are known prey (Vogt 1981, R. Paloski, personal observation). Rigorous investigations from other midwestern states corroborate observations from Wisconsin. For example, several studies from Michigan (Carpenter 1952a, Krause et al. 2003) and Ontario (Catling and Freedman 1980a) reported that earthworms and various amphibians were primary dietary components, and work in Kansas also found that anurans were the most common prey consumed (Fitch 1965). This information corresponds with behavioral research that found that olfactory cues from earthworms elicited a strong interest from this species (e.g., Burghardt 1966).

Although earthworms and amphibians appear to be dietary staples, Common Gartersnakes consume a diversity of other prey types. For example, neonates from Bayfield County exhibited the greatest feeding response to fish and leech extract (more so than earthworm extract; Burghardt 1970). Isolated reports of predation on nestling birds were also documented in southwestern Wisconsin (Ellison and Ribic 2012), Michigan (Carpenter 1952a, Krause et al. 2003), Minnesota (Breckenridge 1944), and Canada (Martin 1979), although bird predation appears uncommon. Furthermore, consumption of Meadow Voles (*Microtus pennsylvanicus*) (Carpenter 1952a) and Northern Short-tailed Shrews (*Blarina brevicauda*; Krause et al. 2003) was observed in Michigan. Apparent scavenging behavior has been reported in several Upper Midwest locations, including Wisconsin (i.e., Iowa County [Sajdak and Sajdak 1999] and Forest County [Casper et al. 2015a]). Juveniles typically consume smaller prey than adults, such as slugs, snails, and other invertebrates (reviewed by Ernst and Ernst 2003).

Extensive research on Common Gartersnake feeding behavior exists (see thorough review in Ernst and Ernst 2003). In summary, Common Gartersnakes do not constrict their prey but instead overpower and swallow it whole. They are active predators that rely primarily on olfactory cues to forage, and even temporarily blinded snakes can locate prey by scent (Gillingham et al. 1990). Successful prey capture is further influenced by repeated experience with certain prey items (Burghardt and Krause 1999).

The list of potential Common Gartersnake predators is long and includes many small to midsized mammal species (reviewed by Ernst and Ernst 2003). Numerous raptor species also eat Common Gartersnakes (Ross 1989a), and predation by Broad-winged Hawks (*Buteo platypterus*) and Red-shouldered Hawks (*B. lineatus*) has been documented in Wisconsin (Rosenfield et al. 1984 and Welch 1987, respectively). Other snake species are also likely predators. Vogt (1981), for example, mentioned that Eastern Milksnakes (*Lampropeltis triangulum*) and Massasaugas (*Sistrurus catenatus*) consumed Common Gartersnakes, while North American Racers (see North American Racer, *Coluber constrictor* species account) also prey upon them in Wisconsin and elsewhere (Fitch 1965). Fitch (1965) reviewed information on the parasites of Common Gartersnake in Kansas.

Common Gartersnakes can employ several strategies to deter or avoid perceived threats. These usually involve either remaining motionless to avoid detection or fleeing (J. Kapfer, personal observation). Their color pattern apparently aids in escape because it obscures the outline of their body, while the continuous dorsal stripe makes forward motion difficult to detect (Fitch 1965). Fitch (1965) stated that "motion can be readily detected only if the head is in view. Otherwise, as the snake moves forward the section of the body that is exposed seems to shrink and vanish as the observer watches, unaware that it is in motion." This species will also flatten its body, gape its mouth, and strike if approached. The bite typically causes no more pain or damage than a scratch and often does not break the skin (J. Kapfer, personal observation). In a series of laboratory tests, King (2002a) found that females were more likely than males to strike when threatened. When restrained, Common Gartersnakes typically release a foul-smelling fluid from their cloaca that is smeared on the hands of their captors (J. Kapfer, personal observation). Vogt (1981) aptly described this fluid as "the most noxious of that of any species of Wisconsin snake." If grasped by the tail, Common Gartersnakes can twist their body and break the tail off behind the vent to escape (Fitch 1965). Although this species can employ an array of potential defensive strategies, body temperature plays a role in behaviors selected. For example, individuals with lower body temperature are more likely to rely on crypsis or defensive behaviors (i.e., flattening their body, gaping, or striking), whereas warmer individuals are more likely to flee (Shine et al. 2000a).

CONSERVATION

Status: Globally, the Common Gartersnake is listed as a Least Concern species by the IUCN. It is currently considered a common species in Wisconsin by the WDNR.

Populations: Published abundance and density estimates for Common Gartersnakes in Wisconsin do not exist. Estimates from Canada and other midwestern states show substantial variation (Table 17). For example, Larsen and Gregory (1989) estimated that abundances ranged from 500 to 600 individuals (mostly adults) associated with a den in Canada but with a large confidence interval.

Several authors have noted a high degree of interpopulation variability in demographic parameters (Gregory 1977b, Larsen and Gregory 1989). For example, reported adult sex ratios were female-biased in some locations (e.g., Kansas; Fitch 1965) and male-biased in others (e.g., Michigan and Canada; Carpenter 1952a, Larsen and Gregory 1989). Existing data suggest that Common Gartersnake populations are mostly comprised of young individuals. Fitch (1965) found that nearly 50% of 650 individuals sampled in a Kansas population from 1949 to 1957 were in only their second year, whereas only ca. 10% of the population were five years or older. In Kansas, ca. 50% annual population declines were estimated regardless of age class (Fitch 1965), while annual survival of adult males in Canada was reportedly as high as 98%, although standard errors were large (Larsen and Gregory 1989). Fitch (1965) suggested that population fluctuations in Kansas were due to environmental variation and changes in prey availability. Similarly, winter mortality likely has large impacts on northern populations. Shine and Mason (2004) estimated that >60,000 overwintering snakes died during a study of hibernacula in Canada. In one year, early snow melt due to unseasonable warmth flooded hibernacula and killed snakes. In another year, high mortality occurred because limited snowpack did not effectively insulate hibernacula, which in turn caused temperatures within to drop below the thermal tolerances of hibernating snakes.

Threats: Common Gartersnakes are probably affected by a number of threats that impact many amphibians and reptiles in Wisconsin (see Introduction to Conservation and Management of Wisconsin's Amphibians and Reptiles chapter). Despite being a generalist with an apparent tolerance of disturbed landscapes, the Common Gartersnake has limits as to what it can withstand. Extreme disturbance or isolation of populations can result in the extirpation of even this resilient species, so habitat destruction and fragmentation remain viable concerns. Common Gartersnakes also frequently die along roads in Wisconsin (J. Kapfer, personal observation), and several studies from other regions have observed substantial population-level impacts due to road mortality. Shine and Mason (2001) commented on the extensive road mortality at their study location in Canada: "The highway carnage was so great that the road became slippery due to the crushed bodies of snakes; a bulldozer was brought from Winnipeg to scrape the

Table 17. Estimates of various Common Gartersnake population parameters (including confidence intervals, if available) from several locations in the Midwest and Canada.

Location	Area, ha (ac)	Abundance	Density, ha (ac)	Source
Illinois	6.5 (16)	209	32.2 (13.1)	Most (2013)
Michigan	19.4 (48)	436–529	22.5–27.3 (9.1–11)	Carpenter (1952a)
Kansas	Variable	189–1,427	2.7–10.9 (1.1–4.4)	Fitch (1965)
Ohio	4.3 (10.6)	58 ± 21.2	14 (5.7)	Dalrymple and Reichenbach (1981)
Ontario	40 (98.8)	150	4 (1.6)	Freedman and Catling (1978)

bodies off the road so that cars could safely negotiate the highway." They later determined that road mortality can indirectly influence Common Gartersnake populations through alteration of age class / size distributions, given that the mean size of road-killed snakes they analyzed was smaller than the size of surviving snakes (Shine and Mason 2004). Roads may also indirectly influence Common Gartersnake populations by altering movement patterns and the ability of males to find pheromone trails left by reproductive females (Shine et al. 2004).

Various other threats likely affect Common Gartersnakes in Wisconsin, but their possible population-level impacts are largely unknown. For example, various pathogens, such as snake fungal disease (SFD), could impact this species (see Introduction to Conservation and Management of Wisconsin's Amphibians and Reptiles chapter). Given that extreme weather and environmental fluctuations are projected to become more extreme as the result of climate change (WICCI 2011, Pryor et al. 2014), overwintering mortality could become an increasingly serious issue as well. Common Gartersnakes are also among several documented species in Wisconsin to die from entanglement in erosion control or wildlife exclusion netting (Kapfer and Paloski 2011).

Management: General concepts associated with effective management of wetlands and adjacent uplands, as well as appropriate management of invasive species and roads, are relevant to Common Gartersnakes (see Introduction to Conservation and Management of Wisconsin's Amphibians and Reptiles chapter). In particular, maintenance of vegetative buffers (based on snake movement patterns) and structural variability of vegetation in aquatic habitats and adjacent uplands would likely benefit this species. Adherence to general recommendations for appropriate management schedules (e.g., burning and mowing at times of year when snakes are not active) could minimize the risk of accidental mortality due to management activities. Unfortunately, habitat management strategies for other wildlife species can sometimes conflict with best practices for Common Gartersnakes. For example, 5% of Common and Plains Gartersnakes observed during surveys in Ohio had been killed by lawn mowers employed to improve grazing areas for Canada Geese (*Branta canadensis*; Dalrymple and Reichenbach 1981).

Although control of invasive plants is generally an advisable management strategy, Common Gartersnakes in Wisconsin did not appear to use wetlands dominated by native sedges more often than those overrun with invasive species (i.e., Reed Canary Grass; Kapfer et al. 2013b). Specific efforts to manage habitat around known hibernacula may benefit this species. Enhancement and maintenance of habitat characteristics suitable for crayfish and ants, whose burrows and mounds provide overwintering habitat, could improve winter survival in Wisconsin. This species could also benefit from strategies to protect known communal hibernacula, particularly if hibernacula are anthropogenic structures that could be altered or destroyed (e.g., old foundations, abandoned wells). If erosion control materials must be deployed during management, use of mats with biodegradable jute/twine nets that are not bonded at strand intersections have a lower potential to entangle snakes. In areas with high potential for road mortality, road underpasses and associated drift fences that guide snakes to passageways will help limit mortality (Shine and Mason 2001).

SPECIAL REMARKS

Analysis of mitochondrial DNA and fossil evidence suggests that Common Gartersnakes were among the first group of herpetofaunal species to colonize Wisconsin and Michigan after the last glacial period, migrating north from Illinois and Indiana as ice sheets retreated (Placyk et al. 2007, Holman 1992b). They are also among several species that possibly found a refuge from glacial activity in the Driftless Area of the Upper Midwest (see Past, Present, and Potential Future Landscapes of Wisconsin chapter).

Lined Snake

Tropidoclonion lineatum (Hallowell 1856)

Corey Raimond and Jeffrey M. Lorch

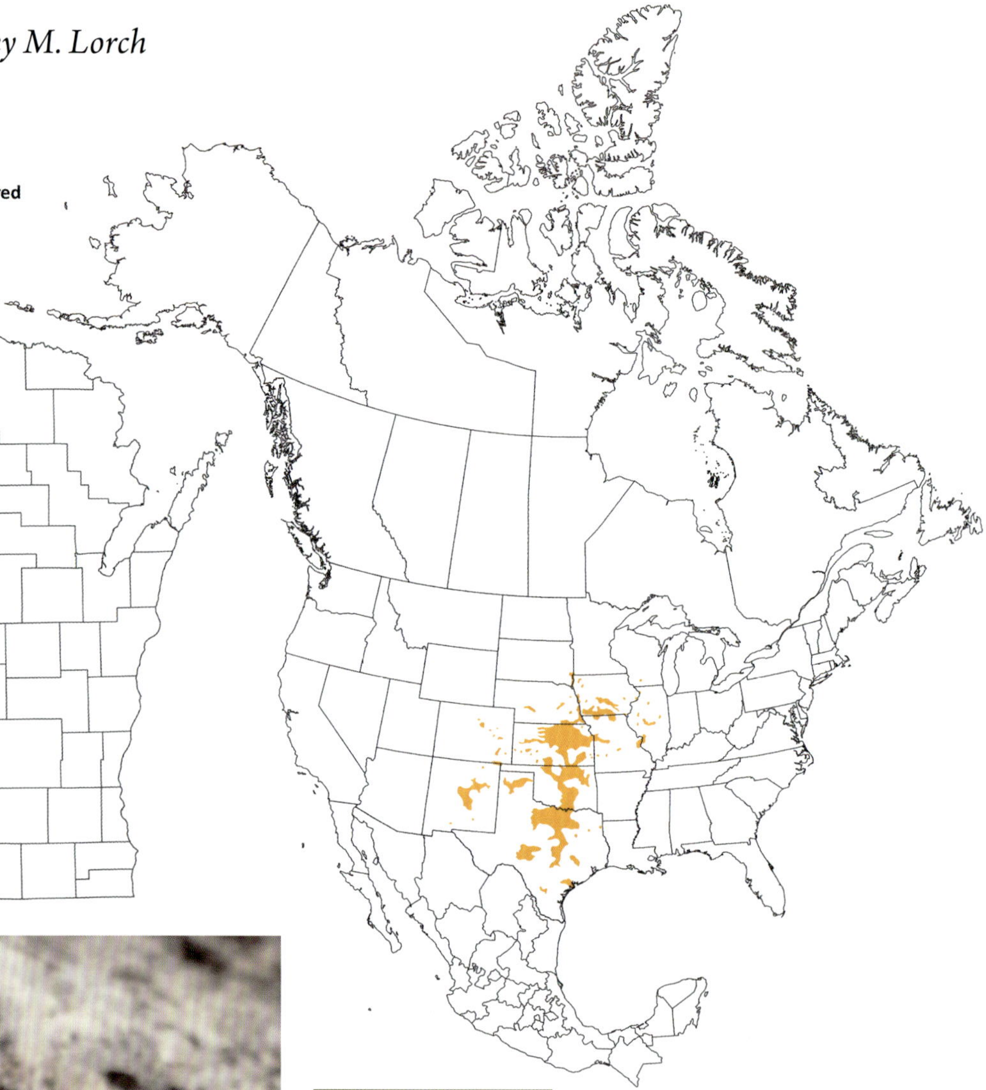

Lined Snake (Iowa County; photo by C. Raimond).

AT A GLANCE

The Lined Snake is a small nonvenomous species most closely related to gartersnakes, Common Watersnakes, and Queensnakes. At first glance, Lined Snakes resemble juvenile gartersnakes, having three light stripes that run the length of their bodies on a grayish-brown dorsum. However, this species can be distinguished from other striped snakes in Wisconsin by having black semicircles on the ventral scales, which form two rows running the length of the ventral surface. Lined Snakes are very secretive and generally remain underground or below rocks and other ground debris in open grasslands. They are potentially active from early April through early to mid-October in Wisconsin, although observations are limited. This species is typically found aboveground only in spring and autumn during short seasonal migrations and after dark, when it leaves daytime retreats in search of earthworm prey. Unlike most Wisconsin snakes, Lined Snakes breed in late summer and autumn. The young are born live the following summer and reach sexual maturity in one year. Lined Snakes were first discovered in Wisconsin in 2011 and are currently known only from a small area in the southwestern part of the state. This species is often associated with the eastern Great Plains region of North America, and the Wisconsin population appears to be disjunct from the nearest populations in Minnesota, Iowa, and Illinois. Because there are few documented locations where they occur, Lined Snakes are considered a species of Special Concern and a Species of Greatest Conservation Need in Wisconsin. Given its rarity in the state, conservation efforts will be essential for securing this esoteric reptile as a member of our fauna.

SYSTEMATICS

Tropidoclonion is a monotypic genus within the family Colubridae. *Tropidoclonion lineatum* (Lined Snake) was originally described by Hallowell (1856) under the name *Microps lineatus* and was erroneously classified as being most closely related to wormsnakes (*Carphophis*) and earthsnakes (*Virginia*). Cope (1891) assigned *T. lineatum* to the clade Natricinae and believed that it was most closely related to members of the genus *Thamnophis* based on its color pattern. Alfaro and Arnold (2001) and Alfaro (2003) performed phylogenetic analyses of the natricine clade Thamnophiini using DNA sequence data from three mitochondrial genes and found strong support for the placement of *T. lineatum* within a clade that included *Nerodia* and *Regina*. However, McVay (2013) was unable to resolve the placement of *Tropidoclonion* within Thamnophiini using a more rigorous multilocus approach that included one mitochondrial gene and nine nuclear loci. Thus, the relationship of *T. lineatum* to other genera within Natricinae remains uncertain, and the hypothesis of Ramsey (1953), that *T. lineatum* diverged from *Nerodia* and *Thamnophis* long ago, remains a possibility.

Ramsey (1953) described three subspecies of *Tropidoclonion lineatum* on the basis of ventral and caudal scale counts, and Smith (1965) proposed a fourth subspecies. Only one of Ramsey's (1953) proposed 'subspecies,' *T. l. lineatum* (Northern Lined Snake), occurs in Wisconsin. No molecular analyses have been conducted to support these subspecific designations, and many authors have chosen not to recognize them (Crother 2017).

DESCRIPTION

Adults: Lined Snakes have a short pointed head that is scarcely wider than the neck. Their body is slender, although it may appear to be thick due to the small size of the head. Their dorsum is typically gray to olive brown in coloration. A cream to light gray dorsal stripe covers the vertebral scale row and half of each adjoining scale row, and similarly colored lateral stripes occur on portions of the third and second dorsal scale rows (sometimes extending onto the first). Each stripe is bordered on both sides by a row of small black spots. Like other natricine snakes, the dorsal scales (with the frequent exception of the first row) are keeled (Ramsey 1953). Near the tip of each dorsal scale, there is a tiny pit on either side of the keel (Ramsey 1953). The ventral surface is white to cream colored with two rows of black semicircles originating on the neck and extending onto the tail (Figure 286). The two semicircles are located on the anterior portion of each ventral scale and may be fused in the neck and body regions. The

Above, top: Figure 286. Ventral scales of the Lined Snake showing two rows of black semicircles. These semicircles distinguish this species from most other snake species that occur in Wisconsin (Iowa County; photo by C. Raimond).

Above, bottom: Figure 287. Juvenile Lined Snake showing mottled pattern on the head (Iowa County; photo by C. Raimond).

area between the spots is often more yellowish in color than the rest of the venter. The anal plate is undivided. The top of the head is usually uniform gray to brown, but some Wisconsin specimens have substantial mottling (Corey Raimond, personal observation; Figure 287). The labial scales are light in color, often similar in coloration to the lateral stripes.

Lined Snakes are small; Ernst and Ernst (2003) reported a maximum TL of 57.2 cm (22.5 in); however, most individuals are much smaller. In Oklahoma, Force (1930) reported that the majority of Lined Snakes were 24–35 cm TL (9.4–13.8 in). Ramsey (1953) noted that the northern 'subspecies' (see Variation subsection) was even smaller. The largest female he measured was 33.8 cm TL (13.3 in), and most males were less than 27.9 cm TL (11.0 in). This is consistent with measurements of individuals from a Missouri population where the maximum size of males was 25 cm SVL (9.8 in) and females had a mean size of 31.3 cm SVL (12.3 in; Krohmer and Aldridge 1985a, 1985b). No data could be located for the size of adults measured in Wisconsin.

Preadult stages: Neonates are similar in appearance to adults but generally have a gray dorsal surface with lighter gray stripes; the stripes remain faint until after the second molt (Force 1931). As juveniles age, their dorsal surface gradually turns brownish, with gray, yellow, or orange stripes, and eventually attains the same coloration as adults (Force 1931). No data could be located regarding the size of neonate Lined Snakes in Wisconsin or the Upper Midwest. Based on a sample of 145 neonates from Oklahoma, Force (1931) reported that newly born young had a TL of 7–13 cm (2.8–5.1 in). In Texas, the young at birth measure approximately 10–12 cm TL (3.9–4.7 in) and weigh 0.45–1.0 g (0.02–0.04 oz; Ramsey 1946, 1953). Newborn Lined Snakes in Missouri had a mean SVL of 9.5 cm (3.7 in) for males and 8.4 cm (3.3 in) for females (Krohmer and Aldridge 1985a, 1985b).

Variation: Lined Snakes apparently exhibit little geographic variation across their North American range. Ramsey (1953) noted slight variations in body size and in ventral and caudal scale counts among populations. There is also little difference between the sexes. Females are generally larger on average and have a lower caudal scale count than males, but there is considerable overlap (Ramsey 1953). The mean ratio of head length to SVL in females is reportedly slightly greater than in males (Shine 1991).

Diagnosis: Lined Snakes are characterized by the presence of scales that are mostly or entirely keeled, an undivided anal plate, and fewer than 27 scale rows at midbody. They also possess two prefrontal scales, a rostral scale that is not pronounced, fewer than eight infralabial scales, and a double row of black spots on the ventral surface. Lined Snakes could be confused with several other snake species that occur in Wisconsin. Their light dorsal and lateral stripes contrasting with a darker ground color superficially resemble gartersnake and ribbonsnake species (*Thamnophis* spp.) in the state. However, the two midventral rows of black semicircles separate the Lined Snake from gartersnakes and ribbonsnakes, which lack distinctive markings near the middle of the ventral scutes. The only other Wisconsin snake with a similar ventral pattern to that of the Lined Snake is the Queensnake (*Regina septemvittata*). However, Queensnakes can easily be distinguished by the lack of a light dorsal stripe (the Queensnake's dorsal stripe, if visible, is darker than the dorsum ground color) and a divided anal plate. Common Watersnakes (*Nerodia sipedon*) have half-moon patterning on their venter as well, but the half-moon shapes are typically brown or red and generally not arranged in rows; furthermore, the dorsal surface of Common Watersnakes is banded, not longitudinally striped.

DISTRIBUTION AND HABITAT

Global distribution: Lined Snakes are endemic to the central portion of the US. The core range is from south-central Texas north to extreme southeastern South Dakota and southwestern Minnesota (Conant and Collins 1991, Ernst and Ernst 2003) in a relatively narrow band stretching roughly from western Missouri to western Kansas and central Texas. There are many disjunct populations on the western (Colorado and New Mexico) and northeastern (Iowa, Illinois, and Wisconsin) edges of its range. The distribution closely mirrors that of the historic tallgrass and mixed-grass prairies; its range becomes more discontinuous in the shortgrass prairies to the west and in the 'prairie peninsula' of the Upper Midwest (i.e., Illinois and Wisconsin). Ramsey (1953) believed that the apparent discontinuous distribution of the Lined Snake was the result of a recent range contraction. He reviewed many distribution records and rejected previous reports of the species' occurrence in Arkansas, Ohio, Pennsylvania, and Tennessee. However, the Lined Snake has since been reconfirmed in Arkansas (Kory Roberts, personal communication).

Distribution in Wisconsin: Lined Snakes are currently known from only two counties in Wisconsin: Iowa County (Natural History Box: An Unusual Discovery) and Dane County (Raimond and Lorch 2012, Casper and Kleinhenz 2016). Most Lined Snakes have been found within an extremely small area, less than 0.5 ha (1.2 ac) in size. However, individuals have occasionally been found as far apart as 1.6 km (1.0 mi) within the same site with fairly continuous grassland habitat in between.

Localities of possible occurrence: Lined Snakes were only recently discovered in Wisconsin (Raimond and Lorch 2012) and northwestern Illinois (Bowen 2004b). The lack of historical reports could be the result of their secretive nature and localized populations and because they were never widespread in the state. Still, additional survey efforts could uncover Lined Snakes in more localities than are currently documented. The nearest known population to those in Wisconsin is in the sand prairies of northwestern Illinois (Jo Daviess County) ca. 95 km away (59 mi; Bowen 2004b), which is also believed to be isolated from the snake's core distribution. It is plausible that Lined Snakes once occupied other southwestern counties in Wisconsin that lie between currently known populations in the state and northwestern Illinois. For example, Lined Snakes may have occurred (or perhaps still persist) in portions of Grant, Green, Iowa, and Lafayette Counties, where vast prairies and savannas once dominated the rolling terrain (Curtis 1959). In addition, sand prairies similar to those found at the northwestern Illinois site were historically common along the Mississippi and lower Wisconsin Rivers in Wisconsin, providing potential habitat for Lined Snakes. Additional survey work by one of the authors (C. Raimond) and WDNR employees has not identified additional Wisconsin populations to date. However, this species can easily evade detection during cursory surveys. For example, in South Dakota, Lined Snakes could not be located during survey efforts even though a population was

NATURAL HISTORY BOX: An Unusual Discovery

On a September day in 2011, I set out to explore a prairie in Iowa County that I had not previously visited. I was hoping to see blooming wildflowers and perhaps catch a glimpse of an Upland Sandpiper (*Bartramia longicauda*) before its long autumn migration to Argentina. It had rained the previous day, which brought with it a cold front. The air was cool and the sky was overcast—not the best weather for observing reptiles. On some of the hillsides, the bedrock was close to the surface, and I observed many plant species indicative of a remnant prairie, suggesting the site had escaped the plow. Despite the cool temperatures, I thought I would try my luck in looking for snakes. Perhaps I would be fortunate enough to find something unusual such as a Plains Gartersnake (*Thamnophis radix*). I lifted and peered under some of the rocks lying on the surface of the hillside (carefully replacing them to their original positions afterward), and within around 15 minutes I had found a small gartersnake. I crouched down to take a photo, but as I examined the animal more closely, I realized that something seemed unusual about it. The head was too small for a Plains Gartersnake, and the coloration was too dull. Suddenly I realized . . . it looked kind of like a Lined Snake! I had read about this species and seen pictures of it but had never seen one in person. Astonished, I picked up the snake and turned it over, which exposed the rows of black half-moon-shaped spots on the ventral surface and confirmed the identification. Not only was this a species I had never seen before, but it was apparently the first time one had ever been found in Wisconsin!

The find was unexpected for several reasons. First, it was in Iowa County and not particularly close to the Illinois or Iowa border (i.e., it seemed to represent a population disjunct from other known Lined Snake sites in neighboring states). Second, this was likely a native relict population that had somehow gone undetected for hundreds of years. In contrast, most new vertebrate species discovered in Wisconsin are the result of an introduction or recent range expansion. Third, of all the amphibian and reptile species found in adjacent states (but not in Wisconsin) that were 'on the radar,' the Lined Snake was low on the list. Many others, such as the Kirtland's Snake (*Clonophis kirtlandii*), Plains Hog-nosed Snake (*Heterodon nasicus*), Graham's Crayfish Snake (*Regina grahamii*), Smooth Earthsnake (*Virginia valeriae*), Fowler's Toad (*Anaxyrus fowleri*), and Plains Leopard Frog (*Lithobates blairi*), seemed like more plausible candidates to be discovered in the state. This experience is testament to the point that there are still great discoveries to be made in Wisconsin by enthusiastic naturalists.

Corey Raimond

known to occur at a site based on the presence of roadkill specimens (City of Sioux Falls 2002).

Habitat: Lined Snakes live primarily in open-canopy grasslands, although robust data on habitat preferences in Wisconsin do not currently exist. In Texas, Ramsey (1953) reported them from open prairie and lightly wooded habitats with thin soil overlying bedrock. This is consistent with the habitat occupied by the known Wisconsin population, which consists of dry-mesic prairie remnants embedded within a larger grassland landscape that possess relatively thin soil on hilltops and hillsides with underlying limestone bedrock (J. Lorch and C. Raimond, personal observation). In South Dakota and Minnesota, Lined Snakes also inhabit mesic to dry prairies on gently rolling terrain (City of Sioux Falls 2002, Moriarty and Hall 2014). Although the snake is known from only prairie remnants in Wisconsin, numerous additional surveys of various habitats adjacent to the known Lined Snake population have been conducted. Habitats surveyed included heavily grazed cattle pastures, weedy fields, areas near human dwellings, unmanaged shrubby habitats, and woodland edges, none of which yielded Lined Snakes (C. Raimond, unpublished data). However, the failure to detect Lined Snakes could be due to limitations of traditional survey methods (e.g., low density of appropriate cover objects to survey). For example, Lined Snakes at the known Wisconsin sites were eventually detected in a Smooth Brome (*Bromus inermis*) grass field with few natural cover objects, a specific location where they were not detected before but only after coverboards were placed in the field, creating refugia (Andrew Badje, WDNR, personal communication).

In contrast to what is reported in other literature, Ernst and Ernst (2003) stated that Lined Snakes often live in the vicinity of wetlands such as springs, ponds, marshes, and streams. However, Ernst and Ernst (2003) provide no references to indicate which regional populations they were referring to or how far from such wetlands Lined Snakes are found. Ramsey (1953) stated that Lined Snakes actually avoided damp areas in Texas. Similarly, Lined Snakes were three times more likely to be captured in uplands compared to wetland sites in Nebraska (Harner and Geluso 2011). There are few wetlands in the vicinity of the known Lined Snake population in Wisconsin. It is possible that this species occurs closer to water in arid regions or in areas where vegetated buffers along wetlands represent the only available habitat (e.g., landscapes dominated by cropland). Although Lined Snakes are not dependent on standing water, they do seem to require somewhat moist conditions. In Oklahoma, they are found under cover objects where the soil is moist but not wet (Force 1931). Although Lined Snakes occupy dry sand prairies at a site in northwestern Illinois (Bowen 2004b, Charles Tucker, USFWS, personal communication), moister microhabitats are intermixed with the drier areas at that location. In Wisconsin, Lined Snakes are also often found under rocks when the soil is moist in spring and autumn, but they become more difficult to locate as the surface soil dries in summer (C. Raimond, personal observation).

Much attention has been brought to the occurrence of Lined Snakes in urban areas. Force (1931) found that the snakes were common in the city of Tulsa, Oklahoma. She noted that the snakes could frequently be observed under boards, tar paper, cement blocks, and other objects in vacant lots and near dwellings. Lined Snakes are also sometimes found in urban areas in Texas (Hamilton 1947, Ramsey 1953), Iowa (Jeff LeClere, Minnesota Department of Natural Resources, personal communication), and Missouri (Krohmer and Aldridge 1985b). In Wisconsin, Lined Snakes have not been reported from suburban or urban locations. Portions of known Wisconsin sites were subjected to Cattle (*Bos taurus*) grazing until the 1990s (Williams 1996), which could indicate that Lined Snakes are tolerant of certain forms of disturbance (see Natural History Box: Pleistocene Microfauna).

Habitats used for overwintering in Wisconsin are currently unknown. In Texas, Hamilton (1947) excavated Lined Snakes from soil depths of 15–20 cm (5.9–7.9 in) during a January cold snap, consistent with Ramsey's (1953) belief that the species hibernated under objects or underground in loose soil. In the northern part of their range, the active season is shorter, and Lined Snakes probably hibernate deeper underground in bedrock crevices and burrows.

REPRODUCTION AND DEVELOPMENT

Little data are available on reproduction in Wisconsin or the Upper Midwest; therefore, most information presented here comes from southern populations that have been better studied. Copulation has been reported in the laboratory among wild-caught Texas specimens in September and October (Ramsey 1946, Curtis 1949). No observations of mating in spring are known; even when males and females were housed together for the entire active season, mating was only witnessed in late summer (Ramsey 1946). These observations are supported by late summer physiological changes in males that are indicative of mating activity and by the presence of sperm in the oviducts of females in autumn (Krohmer and Aldridge 1985a, 1985b). Courtship rituals in the Lined Snake have not been described. In captivity, a mating pair spent 6.5 hours hidden under shelter, where copulation was suspected to have occurred; no specific courtship behaviors were witnessed (Ramsey 1946).

NATURAL HISTORY BOX: Pleistocene Microfauna

It may be difficult to imagine today, but during the late Pleistocene (ca. 15,000 yrs ago) massive herds of large herbivorous mammals roamed North America. In fact, the scene in parts of Wisconsin might not have looked much different from the plains of Africa. Such large numbers of grazing mammals would have had a profound impact on the landscape. Some of these influences persisted (even after the Quaternary megafauna extinction event) as a result of the remaining American Bison (*Bison bison*) and Elk (*Cervus canadensis*) herds that continued to wander the plains (including parts of Wisconsin) for thousands of years (Jackson 1961). Given that many amphibian and reptile species require more dense vegetation for cover, intense grazing could limit the habitat available to them. A lack of competition among snake species in heavily grazed landscapes would have favored the evolution of those that could withstand such pressure. The Lined Snake appears to be one of those species that thrives in the presence of certain forms of disturbance. It is frequently reported on prairie hillsides, in pastures, and in urban areas where sparsely vegetated soils, manicured lawns, and vacant lots may approximate conditions of a grazed landscape. The spotty distribution of the Lined Snake may also be explained by its preference for such habitats. With the large-scale decline of American Bison and Elk populations in the Upper Midwest due to overhunting and habitat destruction, the availability of 'grazed' prairie was reduced to areas where cattle were maintained, providing only islands of suitable habitat. The site where Lined Snakes occur in Wisconsin is within a large prairie complex on thin soil that was largely unsuitable for agriculture. Thus, the land was used primarily for grazing cattle. The site itself was grazed until the 1990s and supported many native plant species that thrive under the pressure of light to moderate grazing. Like the Lined Snake, many of these plants are now rapidly disappearing from the Midwest due to changes in grazing practices. Former grazing, along with the presence of thinly vegetated slopes, may be the reason that Lined Snakes persisted at the Iowa and Dane County site and, as yet, apparently nowhere else in the state. Although Lined Snakes are often associated with disturbed habitats, they are not entirely dependent on grazing. However, in some parts of the Lined Snake's range, large herbivores could have significantly contributed to maintaining suitable habitat. Such an example may represent the important relationships that existed between the mega- and microfauna that once occupied the open range.

Jeffrey M. Lorch

Lined Snakes are viviparous. Based on the few reported copulation and parturition dates, the gestation period is likely ca. 10–12 months. Females usually fast for a period of 15 days to 3 months leading up to parturition (Force 1931, Ramsey 1946, Krohmer and Aldridge 1985a). Approximately one week prior to parturition, captive females became restless and often laid exposed (i.e., not under cover) within their enclosures (Force 1931). In Texas and Oklahoma, parturition has been reported on 1 August and 31 August (Force 1931, Ramsey 1946). In Missouri, the young are born in mid-August (Krohmer and Aldridge 1985a). Ernst and Ernst (2003) reported litter sizes of 2–17 ($\bar{x} = 7.6$). Ramsey (1953) recorded females giving birth to 3–12 young ($\bar{x} = 6$). The number of embryos held by females (and potentially clutch size at parturition) is positively correlated with female SVL (Krohmer and Aldridge 1985a). After giving birth, females resume eating and are able to accumulate enough body fat to produce clutches on an annual basis (Ramsey 1946, Krohmer and Aldridge 1985a).

Newborn Lined Snakes are very active and grow rapidly. Force (1931) reported that the first molt occurs within an hour of birth and then daily for the first few days of life. The juveniles enter their first hibernation at a mean SVL of 14.5 cm (5.7 in) for males and 12.4 cm (4.9 in) for females, equating to a growth rate of ca. 0.1 cm (0.04 in) per day in their first few months of life (Krohmer and Aldridge 1985a, 1985b). Juveniles continue to grow quickly the following spring and summer. Males reach 80% ($\bar{x}$ SVL = 20 cm [7.9 in]) of their adult size by 10 months of age, and females are approximately 67% ($\bar{x}$ SVL = 24 cm [9.4 in]) of their adult size by the end of their first year (Krohmer and Aldridge 1985a, 1985b). Both males and females mature by ca. one year of age, and females produce their first clutch the following summer at two years of age (Ramsey 1946, Krohmer and Aldridge 1985a). No longevity data are available, but given their rapid development, Lined Snakes may not be particularly long-lived.

Figure 288. A Lined Snake displaying the small pointed head typical of the species. This morphological feature is likely an adaptation for a fossorial lifestyle (Iowa County; photo by C. Raimond).

ACTIVITY

Due to Lined Snakes' secretive nature, little is known about their activity and movement patterns. The activity period in the southern part of their range is primarily from March to November, although they can be found aboveground during warm spells in the winter (Ramsey 1953). At a site in Wisconsin, they were encountered from 3 April to 2 October (J. Lorch and C. Raimond, personal observation). However, very few surveys were conducted earlier or later in the year, and the activity period may be slightly longer than is documented.

Lined Snakes are semifossorial and spend the majority of their time either underground or hidden under debris (Force 1930, Krohmer 1989; Figure 288). Nearly all of the Lined Snakes found in Wisconsin have been under flat rocks. The species is most frequently seen in the spring and autumn. Most observations in Wisconsin have been from April to May and again from September to October. Ramsey (1953) found that Lined Snakes in Texas were most frequently observed in April. In Nebraska, Lined Snake capture rates in drift fence arrays increased substantially from summer to early autumn (from 0.6 individual per day in July, to 2 individuals per day in August, to 3.1 individuals per day in September; Harner and Geluso 2011). Aboveground activity in the spring is likely related to spring migration away from hibernacula and thermoregulation (Krohmer 1989). The increase in late summer and early autumn observations is undoubtedly the result of breeding activity and migration to winter hibernacula. In summer, Lined Snakes become more fossorial and are encountered less often (Krohmer 1989, C. Raimond, personal observation).

Daily activity patterns of Lined Snakes are probably tied to that of their earthworm prey. Ramsey (1953) observed Lined Snakes prowling after dark and even witnessed one consuming an earthworm. He concluded that they are nocturnal in summer and crepuscular in spring and autumn (when nighttime temperatures are too low for activity), corresponding with the time periods when earthworms are most active at the ground surface. Nocturnal activity has also been reported in captive Lined Snakes, which spent most daytime hours under cover objects (Force 1931). The recorded range in body temperatures of active Lined Snakes at a site in Missouri was 22°C–34°C (72°F–93°F, $\bar{x}$ = ca. 29°C [84°F]; Krohmer 1989). Similarly, mean body temperature of field-captured Lined Snakes in Kansas was 30.5°C

(86.9°F, SD = 0.53°C [0.95°F]; Fitch 1956). Despite seasonal changes in air and substrate temperatures, Lined Snakes maintain a similar body temperature throughout the active season by seeking out microclimates near their thermal optimum. Specifically, Krohmer (1989) reported that Lined Snakes used thin cover objects slightly elevated off the ground in spring to increase body temperature above that of the surrounding air and soil but sought cover beneath large rocks or underground in the summer when daytime air temperature exceeded their thermal preference. No estimates of movement rates or home range size are known for Lined Snakes, but they presumably move little during the summer months.

PREY AND PREDATORS

The diet of Lined Snakes is thought to consist almost entirely of earthworms (Megadrilacea). Ramsey (1947) found earthworms in the stomachs of wild individuals and noted that captives refused small orthopterans, insect larvae, snails, and sow bugs (Isopoda) unless they were coated with fluids from chopped earthworms. Despite this, Curtis (1949) observed a Lined Snake preying on a sow bug, and Force (1931) recounted an instance in which a captive female consumed her young shortly after giving birth, although Ramsey (1953) did not believe that cannibalism was a regular phenomenon. In areas where Lined Snakes occur, earthworms are often abundant under the same types of cover objects used by the snakes (Ramsey 1953). Interestingly, earthworms have not been observed under the ground debris surveyed at a Lined Snake site in Wisconsin, although crickets and other small soft-bodied insects are abundant (Figure 289; see Natural History Box: Vermivores). The small head and gape size and unique feeding behavior of the Lined Snake suggest that the species is highly specialized (Ramsey 1947), and Ramsey (1953) believed the species to be the most derived form of any natricine snake found in North America.

Feeding behavior in captive Lined Snakes was described in detail by Ramsey (1947). Although detection was primarily facilitated by olfaction, movement of prey elicited excitement. After first smelling a potential prey item, a Lined Snake approached the prey gingerly and placed its head in contact with the prey before rapidly seizing it with its jaws. Lined Snakes observed while feeding almost always manipulated earthworms in their jaws so that they could swallow them by the middle of the body (with the two halves of the worm's body folded together). Ramsey (1947) suggested that swallowing earthworms in this manner led to a higher success rate in procuring prey because worms swallowed by the end of the body would frequently become severed, allowing a portion of the worm to escape. He noted that Lined Snakes were adept at removing small worms from tight crevices and believed that the snakes' feeding behavior evolved as a mechanism to prey on worms that are partially exposed from their burrows. After consuming a worm, a Lined Snake would typically rub its head against nearby objects to remove slime from its labial scales.

Because of their small size, Lined Snakes could be preyed upon by many vertebrates. Neonates are even small enough to be taken by some predatory invertebrates. Ramsey (1953) believed that ophiophagous snakes were the Lined Snakes' most significant predator. In Wisconsin, Eastern Milksnakes (*Lampropeltis triangulum*) are the only ophiophagus snake known to co-occur with Lined Snakes, although the North American Racer (*Coluber constrictor*) may have been historically sympatric. Numerous species of birds, such as American Kestrels (*Falco sparverius*), owls, hawks, Ring-necked Pheasants (*Phasianus colchicus*), Wild Turkeys (*Meleagris gallopavo*), and larger passerines, probably also consume Lined Snakes in Wisconsin. In Oklahoma, Loggerhead Shrikes (*Lanius ludovicianus*) were reportedly major predators of Lined Snakes, which were often found impaled by shrikes on the needles of yucca leaves and other sharp objects. An American Kestrel, Blue Jay (*Cyanocitta cristata*), and meadowlark (*Sturnella* sp.) were also observed preying on Lined Snakes (Black 1976). Mammalian predators likely include Striped Skunks (*Mephitis mephitis*), Raccoons (*Procyon lotor*), weasels (*Mustela* spp.), American Badgers (*Taxidea taxus*), Virginia Opossums (*Didelphis virginiana*), Thirteen-lined Ground Squirrels (*Ictidomys tridecemlineatus*), and Northern Short-tailed Shrews (*Blarina brevicauda*).

Aside from remaining concealed, Lined Snakes possess few defensive strategies to avoid predation.

NATURAL HISTORY BOX: Vermivores

Earthworms are abundant throughout much of Wisconsin and provide an important food source for a variety of wildlife. However, the earthworms most frequently encountered in the state are invasive and were introduced from elsewhere (James 1995). While some species of native earthworms occur in North America, it is thought that the last glaciation caused the extirpation of native earthworms from Wisconsin ca. 11,000 years ago. These extirpated natives were slow to recolonize by natural dispersal and may not have been present by the time the first Europeans reached the state (James 1995). Yet some of our native snake species, such as the Lined Snake, Butler's Gartersnake (*Thamnophis butleri*), Common Gartersnake (*Thamnophis sirtalis*), Dekay's Brownsnake (*Storeria dekayi*), and Western Wormsnake (*Carphophis vermis*), are decidedly vermivorous (worm-eating). How is this possible? One explanation is that the diets of these snake species shifted once earthworms were introduced. Although gartersnakes and Dekay's Brownsnakes primarily feed on earthworms, they also take other prey items, and the large number of earthworms in their diet may simply reflect availability (i.e., earthworms are more abundant than other prey). However, the Lined Snake and Western Wormsnake are currently considered to be obligate vermivores, and the explanation of greater prey availability does not adequately explain their diet. A second possible explanation is that some species of native earthworms persisted in portions of the Driftless Area that did not experience permafrost during the last glaciation. This could explain why the Lined Snake and Western Wormsnake are currently limited to this region of the state. We very rarely see earthworms in remnant prairies in southwestern Wisconsin and have never observed any at the Lined Snake site. Perhaps the native earthworms are less likely to be found under cover objects, or perhaps these snakes simply have a different diet in Wisconsin than in the rest of their range (e.g., other soft-bodied invertebrates are abundant at the Wisconsin Lined Snake site; Figure 289). Unfortunately, it is difficult to find support for either of the two hypotheses due to the lack of historical data on earthworms and the introduction of earthworms native to the southern United States (in addition to their European cousins) to the Upper Midwest. However, it is interesting to ponder how the introduction of these unassuming invertebrates might have altered the geographic ranges, abundances, and ecologies of the animals that feed on them.

Jeffrey M. Lorch

Figure 289. An adult Lined Snake next to a field cricket (*Gryllus* sp.). Earthworms (the reported staple of the Lined Snake diet) were not observed at the Wisconsin site, but other small invertebrates were abundant (Iowa County; photo by J. Lorch).

Ramsey (1946) described an incident in which a gravid female flattened her body, coiled, and struck at his hand; however, this behavior seems uncommon based on observations in Wisconsin (C. Raimond, personal observation). Most Lined Snakes initially remain motionless when encountered underneath ground debris and then attempt to flee when touched. Often, the snakes are near an invertebrate tunnel, in which they will seek refuge. When grasped, they thrash wildly and smear musk and feces on their captor. A fresh Lined Snake carcass found in northwestern Illinois had tied its body into a knot (Charles Tucker, USFWS, personal communication), a behavior that, although potentially accidental, would certainly prevent small predators from being able to swallow the snake.

CONSERVATION

Status: Globally, the Lined Snake is listed as a Least Concern species by the IUCN. Within the heart of its range, it is widespread and locally abundant. However, in the Upper Midwest it is quite uncommon and extremely localized in distribution. Lined Snakes are listed as Endangered in South Dakota, Protected in Iowa, and Threatened in Illinois. In Minnesota (Moriarty and Hall 2014) and Wisconsin, the Lined Snake is considered a species of Special Concern and a Species of Greatest Conservation Need.

Populations: Due to annual shifts in activity and the species' secretive nature, it can be difficult to assess Lined Snake abundance in a given area. Despite this, Lined Snakes appear to reach high densities at some locations. Force (1931) and Ramsey (1953) reported the species as abundant in Oklahoma and Texas, respectively. During surveys at the Wisconsin site, as many as 15 individuals have been observed in a single visit; at other times, no Lined Snakes were encountered (J. Lorch and C. Raimond, personal observations). Studies using drift fences and pitfall traps vary widely in the numbers of Lined Snakes captured. Cavitt (2000) found that this species accounted for only 0.4% of 550 total snake captures in seemingly ideal habitat in Kansas. However, Lined Snakes represented 12.2% of 444 total snake captures at an upland site in Nebraska (Harner and Geluso 2011). Surveys that involve looking under cover objects may be more effective than drift fences for the species. Using this method, Taggart (1992) reported finding 47 and 72 Lined Snakes per day, respectively, at two sites in Kansas. However, without mark-recapture studies, it is difficult to draw conclusions about population sizes from these reported capture rates. Little information exists on additional population parameters. Funk and Tucker (1978) estimated a nearly 1M:1F sex ratio in an exceptionally large litter of 13 young, but adult sex ratios are unreported. Given their rapid development (i.e., sexual maturity reached at only one year of age) and potentially short life span, Lined Snake populations may have rapid turnover rates and be more prone to fluctuations than slower-maturing, longer-lived snake species. Such a life history strategy may be beneficial for a snake that has a tendency to exploit disturbed habitats such as urban areas and recently grazed grasslands (see Natural History Box: Pleistocene Microfauna).

Threats: Given that Lined Snake ecology is virtually unknown in Wisconsin, the major threats facing it are equally unknown. Presumably, many of the conservation issues that negatively impact prairie and savanna habitats relate to this species (see Introduction to Conservation and Management of Wisconsin's Amphibians and Reptiles chapter). It is highly probable that the largest threat to Lined Snakes in the Upper Midwest is loss of habitat. Less than 1% of the historic midwestern tallgrass prairie remains intact (Mirk 1997). The majority of this land has been lost to high-intensity agriculture and urbanization. The threat of habitat loss will intensify with increasing human populations. Many of the remaining parcels of prairie are small, fragmented, unmanaged, and becoming invaded with exotic plants (C. Raimond, personal observation). Alteration or removal of ground debris, particularly soil surface rocks, for agricultural purposes could negatively affect this species. Reports exist of road mortality during spring and autumn migration periods (City of Sioux Falls 2002, Minnesota Department of Natural Resources 2014). At the known Wisconsin site, the identified areas of quality habitat are not bisected by roads, and trauma from vehicle strikes is probably an insignificant form of mortality. In other parts of its range, direct human persecution can also serve as an important source of mortality for the Lined Snake. Ramsey (1953) reported that Lined Snakes were intentionally killed in urban areas. He recounted an instance of finding over 40 dead individuals on a sidewalk in Fort Worth, Texas.

Management: Managing remaining prairie patches so that suitable habitat persists is likely vital to the long-term conservation of Lined Snakes in Wisconsin. Therefore, many of the management concepts that are generally associated with prairie and savanna habitat apply to the Lined Snake (see Introduction to Conservation and Management of Wisconsin's Amphibians and Reptiles chapter). However, in some cases, the management needs of Lined Snakes may conflict with those of other amphibians and reptiles. For example, Harner and Geluso (2011) recorded high herpetofaunal species richness and abundance in areas managed with prescribed fire in Nebraska. Yet while most other snakes in the study responded well to fire, Lined Snakes were most common in pastures that had not been burned. In contrast, Wilgers and Horne (2006) reported only capturing Kansas Lined Snakes in annually burned prairies (although only three individuals were found during their study). Still, these conflicting reports suggest that Lined Snakes are tolerant of at least the occasional use of prescribed fire to manage their habitat. Unfortunately, little rigorous data exist on how various habitat management strategies affect this species. However, it is reasonable to assume that management actions (e.g., prescribed fire, use of heavy machinery) should be restricted to the inactive season (when snakes are in hibernation) to avoid accidental mortality of active individuals.

Timber Rattlesnake

Crotalus horridus (Linnaeus 1758)

Richard A. Sajdak, Robert W. Hay, and Craig S. Berg

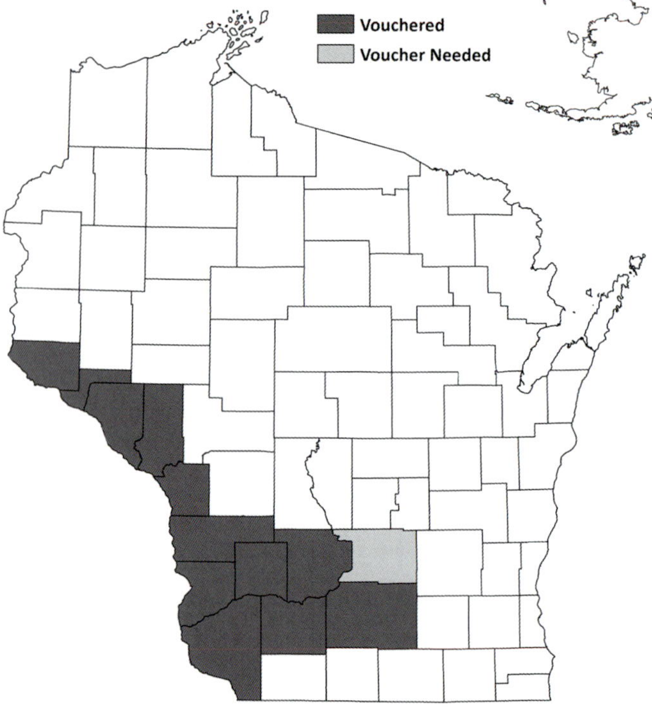

Adult Timber Rattlesnake (Sauk County; photo by J. Kapfer).

AT A GLANCE

Timber Rattlesnakes are one of only two venomous snakes in Wisconsin and among the largest species of rattlesnake in the world. Their head and body are yellow to tan or gray, with a series of thick black crossbands that occur from roughly the neck to near the tail, where the color transitions to mostly black. Although the top of the head is largely without pattern, there is a thick brown band on the side of the head that extends from each eye rearward to the back of the jaw. They also possess elliptical pupils, and adults typically have a distinct segmented rattle at the end of the tail. Timber Rattlesnakes are found along the Mississippi and lower Wisconsin Rivers on open bluff prairies, in oak openings, and in wooded ravines. They are usually active from mid-May to late September. Mating occurs from July through August, but females delay giving birth to their live young until autumn of the following year. Timber Rattlesnakes feed primarily on rodents and other small mammals, from shrews to squirrels and even young rabbits. As a venomous species, the Timber Rattlesnake demands respect but is normally a shy and retiring snake that prefers to rely on camouflage to avoid detection. When surprised or harassed, it will make a run for cover, possibly while rattling. If a retreat site is not available or the snake is threatened at a close distance, it will coil up and rattle loudly. In such situations, it may also strike out and attempt to bite as a last resort. In Wisconsin, the Timber Rattlesnake is considered a Species of Greatest Conservation Need and a species of Special Concern and designated as a Protected Wild Animal.

SYSTEMATICS

Crotalus horridus (Timber Rattlesnake) in Wisconsin was historically considered a member of the subspecies *C. h. horridus* and the northern counterpart to *C. h. atricaudatus* (Canebrake Rattlesnake), which occurs in the southern US (reviewed by Ernst and Ernst 2011). More recent work, however, supports that the variation identified between the proposed subspecies was likely clinal, which substantiated the argument against the use of subspecific designations (Pisani et al. 1972). Brown and Ernst (1986) used morphological characteristics to separate eastern populations into Coastal Plain and montane subspecies. Later analyses of mitochondrial DNA further supported the presence of three lineages that were not arranged along a north–south gradient: one to the west of the Appalachian Mountains, one to the east of the Allegheny Mountains, and a third between the two mountain ranges (Clarke et al. 2003). The most recent review of morphological characteristics across the range of this species found large overlap in characters and concluded that variation within *C. horridus* was too extensive and complex to define subspecies (Allsteadt et al. 2006). As such, the present consensus does not favor the use of subspecific designations for *C. horridus* (Crother 2017). Only one other snake species from Wisconsin is placed in the same family as *C. horridus* (Viperidae): *Sistrurus catenatus* (Massasauga).

DESCRIPTION

Adults: This is a large heavy-bodied snake with a short thick tail tipped with a distinctive segmented rattle. The eyes have elliptical pupils, and there is a heat-sensing pit between the eye and nostril on each side of the head (Figures 290 and 291). The head and body are yellow, tan, or gray (Figures 290 and 291). The head is largely unmarked, although a darker gray to brown band may extend rearward from the eye to the corner of the jaw on both sides of the head (Timber Rattlesnake Cover Photo). A series of dark brown to

Figure 290. Typical yellow phase Timber Rattlesnake (Sauk County; photo by R. Staffen).

Figure 291. Gray phase Timber Rattlesnake. Note the obvious heat-sensing pit located slightly below the nostril, visible on the left side of the head (Sauk County; photo by T. Muehlfeld).

black chevrons or crossbands encircle the dorsal surface and occur at somewhat consistent intervals from the neck to the tail (Figure 291). These dorsal crossbands are irregular in shape and occasionally broken on the anterior half to third of the body. There is often a reddish-brown to cinnamon-brown middorsal stripe three or four scales wide along the midline of the back. The tail is a satiny black, which gives rise to one of this species' colloquial names, the 'velvet tail.' The venter is yellowish or white, with darker speckles. Numerous small scales occur between the eyes on top of the triangular-shaped head, which is distinct from a thin neck.

The largest 16 females from a sample of 81 Wisconsin Timber Rattlesnakes had a mean SVL of 96.2 cm (37.9 in), a mean tail length of 6.2 cm (2.4 in), and a mean weight of 834.6 g (1.8 lb, Crawford and La Crosse Counties; Richard Sajdak, unpublished data). The largest 16 of 77 males sampled from the same counties had a mean SVL of 110.7 cm (43.6 in), a mean tail length of 9.9 cm (3.9 in), and a mean weight of 1,158.2 g (40.9 oz; R. Sajdak, unpublished data). The maximum reported total lengths for Timber Rattlesnakes from throughout their North American range are 244 cm (96 in) in southern states, 188 cm (74 in) in Missouri, and 183 cm (72 in) in Massachusetts (reviewed in Brown 1993). The largest Wisconsin male measured during surveys in Crawford County was 128 cm SVL (50.4 in), 9 cm tail length (3.5 in), and 1,395 g (3.1 lb; R. Sajdak, unpublished data). The largest Wisconsin female, also from Crawford County, was 102.5 cm SVL (40.4 in), 6.0 cm tail length (2.4 in), and 1,000 g (2.2 lb; R. Sajdak, unpublished data). Ernest Messeling, a longtime Wisconsin rattlesnake bounty hunter who reportedly caught over 1,000 Timber Rattlesnakes per year for more than a decade, reported that the longest rattlesnake he had killed was 129.5 cm TL (51 in; Messeling 1953), although no corroborating specimen exists.

The Timber Rattlesnake's tail ends in an elaborately ridged bony tip, which holds a loosely joined string of keratinized rings or segments typically referred to as 'rattles.' Each time the snake sheds its skin, the base of the tail also sheds, leaving a new rattle segment behind. Messeling (1953) reported that the most rattle segments he encountered on a Wisconsin individual was 23, although adults typically have fewer than 15 segments (Fitch 1999). It is a common belief that the age of a rattlesnake can be determined by counting the rattles, with each year of life supposedly represented by a single rattle segment. This belief is false for two reasons: first, the oldest segments become worn and break off, leaving an incomplete rattle string; second, Timber Rattlesnakes in Wisconsin can shed up to three times in a year ($\bar{x}$ sheds/year = 1.74, range = 1–3, $n = 16$; Craig Berg, Milwaukee County Zoo [retired], R. Sajdak, and Armund Bartz, WDNR, unpublished data). Despite this, a general idea of the snake's age can be gained by assessing the characteristics of retained rattle segments. The first shedding cycle produces a smoothly rounded rattle segment, referred to as the 'button.' If the button has not broken off, the snake's age can be roughly estimated by counting the rattles and dividing by 1.85, which is the average number of sheds per year for juvenile snakes (>65 cm SVL [25.6 in] at first capture, Crawford County; C. Berg, unpublished data). If the button is missing, the rattle cannot easily be used to estimate age. Despite this, relative age can be determined, as young snakes tend to possess a tapered rattle string, because early growth is more rapid, and rattle segments become progressively wider with greater size and age. Old snakes, on the other hand, grow much more slowly, and the rattle string will have a uniform and more columnar shape (Figure 292).

Above, top: Figure 292. Comparison of rattles from an old adult (*left*; broken string of seven rattle segments) and juvenile (*right*; six rattle segments plus button) Timber Rattlesnake (Crawford County; photo by R. Sajdak).

Above, bottom: Figure 293. Neonate Timber Rattlesnake (Grant County; photo by S. Viernum).

Preadult stages: Neonates and juveniles are similar to adults in pattern and color (Figure 293). In a series of ten neonates from Crawford County (two distinct litters plus an additional single individual), mean SVL was 26.3 cm (10.4 in, range = 18.5–29.5 cm [7.3–11.6 in]), mean tail length was 2.3 cm (0.9 in, range = 19.9–32.5 cm [7.8–12.8 in]), and mean weight was 22.0 g (0.78 oz, range = 12.7–26.0 g [0.45–0.92 oz]; R. Sajdak, unpublished data).

Variation: Background coloration can range from gray to yellow, although this variation is not substantial across Wisconsin populations (see Figures 290 and 291). The sexes are similar in color and pattern, but adult females are somewhat heavier-bodied with a shorter tail, while adult males can be substantially longer.

Diagnosis: Timber Rattlesnakes can be characterized by the presence of verticle pupils, one or more rattle segments on the end of the tail, a heat-sensing pit between the eye and nostril, and head scales on the top of the head that are small and asymmetrical. They also possess keeled dorsal scales in 23–25 scale rows at midbody, 154–183 ventral scales, an undivided anal plate, and mostly undivided subcaudal scales. On either side of the head, there are 13–15 supralabial scales, 14–16 infralabial scales, 2 loreal scales, 2 preocular scales, and 4 postocular scales. The tail anterior to the rattle segments is typically solid black.

The only other rattlesnake in Wisconsin, the Massasauga, is much smaller, with a series of rounded dark blotches along the back and thick dark bands on the tail and along the sides of the head over a gray background. Massasaugas also have a series of large paired plate-like scales on the top of the head (unlike the small scales on the head of Timber Rattlesnakes), 21–27 scale rows at midbody, 139–160 ventral scales, 11–12 supralabial scales, and 11–13 infralabials. Unfortunately, many larger nonvenomous species in Wisconsin are sometimes confused with a rattlesnake by persons unfamiliar with snakes, particularly if the snake is cornered and defending itself (many nonvenomous snakes vibrate their tails when threatened, so they vaguely resemble a rattlesnake). However, calm inspection reveals that they do not closely resemble the Timber Rattlesnake. For example, Gophersnakes (*Pituophis catenifer*), Eastern Foxsnakes (*Pantherophis vulpinus*), Gray Ratsnakes (*P. spiloides*), Eastern Milksnakes (*Lampropeltis triangulum*), and Eastern Hog-nosed Snakes (*Heterodon platirhinos*) can all make spectacular defensive displays: coiling up with open mouths, hissing, or vibrating their tail. Yet they are easily distinguished from the Timber Rattlesnake because their tail ends in a sharp tip with no rattle. They also have large paired plate-like scales on the top of the head and round pupils (not vertical pupils), and they differ in pattern (e.g., they lack the narrow brown stripe down the middle of the back and the solid black tail).

DISTRIBUTION AND HABITAT

Global distribution: The current range of the Timber Rattlesnake encompasses 30 states, from New Hampshire west to eastern Minnesota in the north; south along eastern Iowa; west to extreme southeastern Nebraska; south through eastern Kansas, eastern Oklahoma, and eastern Texas; east across the Gulf Coast states to northern Florida; and north along the eastern Coastal Plain. This snake has been extirpated from Maine (Hutchinson and Hunter 1992), Rhode Island (Raithel 1992), and Canada (Government of Canada 2018).

Distribution in Wisconsin: Timber Rattlesnakes in Wisconsin occupy the northwestern fringe of their global distribution (Powell et al. 2016). Here they are largely restricted to bluff habitats adjacent to major rivers where suitable hibernacula exist (see Habitat subsection). Although published records exist for a handful of Wisconsin counties, a suitable voucher could not be found for Columbia (see the Appendix). The originally referenced voucher to support the Columbia County record (MPM 2554; Casper 1996) was reportedly a preserved skin, which recent investigations could not locate (Julia Colby, MPM, and Joshua Kapfer, UW–Whitewater, personal communication).

Localities of possible occurrence: Schorger (1967–68) reported Timber Rattlesnakes in four additional counties, based mostly on unverified accounts that he had compiled: St. Croix, Clark, Lafayette, and Adams. Vogt (1981) further listed what he considered trustworthy anecdotal accounts for Dunn, Juneau, and Adams Counties. The WDNR has also received two unverified reports from Monroe County (WDNR 2016b). Of these potential locations, Clark County seems the least likely, as it is one or more counties separated from those with known occurrences. The presence of Timber Rattlesnakes is more likely, however, in undocumented counties where potentially suitable hibernacula exist, such as St. Croix and Lafayette (see Habitat subsection). These counties are also particularly close to confirmed populations near shared borders with Pierce and Grant Counties, respectively (Vogt 1981, Oldfield and Keyler 1989).

It is unlikely that Timber Rattlesnakes naturally occur far outside of their currently documented range in Wisconsin because they rely on habitat features concentrated near the Wisconsin and the Mississippi Rivers. Despite this, individuals have been verified in more disjunct locations under unusual circumstances. For example, a Timber Rattlesnake was confirmed in the village of Windsor (northeastern Dane County) that was likely a translocated individual or an illegally kept pet that escaped (Robert Hay, Turtles for Tomorrow, personal observation). Another was identified in the Cherokee Park in the city of Madison (central Dane County). This snake was apparently released in the park by a "well-meaning but uninformed snake enthusiast" in hopes of establishing a new population (Russ Hefty, City of Madison Parks Department, personal communication to R. Hay). Vertebrae from "a large crotalid, probably the timber rattlesnake," were found at an archaeological site used by the Fox Native American tribe near Oshkosh in eastern Wisconsin (Winnebago County; Parmalee 1963). Parmalee (1963) suggested that the "presence of these vertebrae point to the former occurrence of timber rattlesnakes in this area, a region considerably northeast of this snake's present range in Wisconsin." The disjunct nature of this report and the lack of appropriate habitat in the region make the presence of these vertebrae puzzling. Yet because ancient humans could have transported these rattlesnake remains, the discovery probably does not represent a former population in Winnebago County.

Habitat: In Wisconsin, Timber Rattlesnakes primarily occur along the steep rugged bluffs of the Driftless Area in the Mississippi and Wisconsin River valley watersheds (Vogt 1981). They are often found on south- and west-facing slopes in bluffside prairies and oak savannas with exposed rocky outcroppings (Curtis 1959; see Amphibians and Reptiles in Wisconsin Plant Communities chapter). Timber Rattlesnakes also use the wooded areas in ravines between open bluff faces and on north slopes. These habitats are more shaded but still quite open, allowing for high sun penetration to the substrate.

Brown (1993) provided details of the essential habitats used by this species throughout the year, which he divided into summer habitat, transient habitat, and hibernacula. Summer habitat is used by Timber Rattlesnakes during the peak of the active season. The types of summer habitat selected vary by sex, gravidity, and age class (Reinert 1984a, 1984b, Sajdak 1999, Sajdak and Berg 2005, Hamilton 2009). Gravid females appear to require habitat with high insolation but adequate ground cover to effectively thermoregulate and enhance gestation. As a result, gravid females in summer strongly associate with open or partially open habitats that are warm and typically possess 'shelter rocks' or 'slab rocks' that act as retreat sites (Figures 294 and 295; Keenlyne 1972, Sajdak 1999, Sajdak and Berg 2005, Hamilton 2009). For example, all gravid females monitored with radio telemetry in Crawford County utilized open prairies

Above, top: Figure 294. Example of Timber Rattlesnake bluffside habitat (Sauk County; photo by J. Kapfer).

Above, bottom: Figure 295. Timber Rattlesnake den (Crawford County; photo by R. Sajdak).

on rocky bluffs more often than expected (77% of locations) based on this habitat's availability (<1%; R. Sajdak, unpublished data). Gravid females at this site were also found in contact with large rocks or rock ledges on 43% of observations, which was four times as often as males or nongravid females (Sajdak 1999). After parturition, females often moved from open summer habitat to more shrubby or wooded habitats to forage and regain the fat reserves lost while gravid (Richard Sajdak, Milwaukee County Zoo [retired], personal observation).

Adult males and nongravid females use a broader array of summer habitats compared to gravid females. In Wisconsin, these range from bluff-top prairies to lowland swamps and dense oak forest (*Quercus* spp.), often with ample downed woody debris (Sajdak 1999, 2000). Despite the variety of habitat types used, foraging males and nongravid females in Crawford County were found in secondary woodland, shrub, and prairie more than expected, while mature oak woods, wetlands, and agricultural fields were used less than expected (Sajdak 1999; see further discussion of disturbed habitat use in Conservation section). Wisconsin juveniles appear to rely on a combination of open prairie and secondary forest during the summer, associating with herbaceous ground cover when small (Sajdak 1999) but transitioning to use more wooded habitats as they grow (Hamilton 2009).

Transient habitat is loosely defined as the area immediately adjacent to the den site (i.e., within 200 m [650 ft]), which is used seasonally for thermoregulation as snakes enter and leave winter hibernacula (Brown 1993). This habitat has an open to partially open tree canopy and vegetation that is dominated by grasses. Transient habitats are typically found on bluff slopes that possess scattered rocks and rock outcroppings, used by snakes for cover. Timber Rattlesnakes in the northern parts of their range, including Wisconsin, often used the same shelter rock outcroppings in transient habitat from year to year as they dispersed from and returned to the den (Brown 1993, R. Sajdak, unpublished data). It is noteworthy that gravid females often show little distinction between their transient and summer habitats.

Microhabitat features of summer and transient habitats are structurally diverse, which aids in thermoregulation and foraging. In Crawford County, radio-tracked individuals were most frequently found coiled in locations of dappled shade (46.4% of 367 snake observations; Sajdak 1999), likely because these provided a spectrum of thermal conditions for regulating body temperature. Selected microhabitat features can also vary by sex and gravidity, presumably related to thermoregulatory opportunities (Gibson et al. 2008). For example, three radio-tracked males in Crawford County were associated with a substantially higher mean canopy cover (58.4% of 190 observations) than two gravid females (12.3% of 98 observations; Sajdak 1999), the latter typically maintaining higher body temperatures during gestation (see Reproduction and Development and Activity sections). Similar use of locations with a more open canopy by gravid females was observed in Indiana (Gibson et al. 2008). Cochran and Schmitt (2014) also reported the importance of bluff prairie shelter rocks as refugia from high midday summer temperatures for gravid females in Minnesota. Research in Missouri further suggested that use of habitat edges was related to how microhabitat features affect environmental conditions, particularly for gravid females (Anderson and Rosenberg 2011). The authors typically found gravid females near fences, roads, and field edges, where overhead canopy was more open than in adjacent habitats, which suggested that snakes were selecting features that aid in thermoregulation. It is noteworthy that the structure of vegetation and time of day resulted in variable temperature gradients along roads and field edges, which likely influenced spatial and temporal habitat use patterns along roads and field edges (Anderson and Rosenberg 2011).

Timber Rattlesnakes select microhabitats that aid not only in regulation of body temperature but also in prey capture. They frequently rest near rocks and downed woody debris, such as logs, that can act as effective ambush locations for hunting small mammals (Reinert et al. 1984, Sajdak and Berg 2005). This species is also occasionally encountered in elevated locations among the branches of trees and shrubs, presumably for thermoregulation or to forage (Natural History Box: Rattlesnakes in Trees).

Timber Rattlesnakes in the Upper Midwest appear to use hibernacula with very specific characteristics (Natural History Box: The Importance of Hibernacula). Wisconsin hibernacula often occur in limestone bluff outcrops with deep crevices that allow snakes to access winter retreats below the frost line (C. Berg, unpublished data; see Figure 295). Dens are usually located on south- or west-facing slopes above the transition zone between the sandstone along the lower bluff slope and the overlying limestone near the top. Observations suggest that den entrances often occur below the crest of the bluff where groundwater penetration probably maintains damp to wet conditions within dens during hibernation. Occasionally, hibernacula are found near the top of bluffs or along north- to east-facing bluff slopes. Talus slopes consisting of large dolomite or quartzite rocks piles, such as those found in Wisconsin's Baraboo Range, may also provide suitable overwintering structure (Brown 1993, R. Hay, personal observation; Figure 297). Appropriately structured sandstone hillsides can also support rattlesnake hibernacula in Wisconsin, but these occurrences are apparently rare (R. Hay, personal observation). Unlike for some other snake species in the state (e.g., Eastern Foxsnake), there are no published accounts of Timber Rattlesnakes using artificial structures for overwintering, such as abandoned dug wells. Despite this, use of unnatural hibernacula remains a possibility under appropriate conditions. Timber Rattlesnakes show high fidelity for specific den locations and return to the same hibernacula across years (R. Sajdak, personal observation).

REPRODUCTION AND DEVELOPMENT

The breeding season in Wisconsin begins in July and continues through August (Sajdak 1999, Hamilton 2009), although parturition does not occur until the following autumn. To stimulate reproduction, the male rubs the female with his chin and then rests alongside her. He next employs quick body spasms and attempts to wrap his tail beneath hers for copulation (Ernst and Ernst 2011). Males reportedly compete for reproductive opportunities by 'wrestling' other males (Merrow and Aubertin 2005). These wrestling bouts involve males holding roughly the front one-third of their body in a vertical position, intertwining, and attempting to push each other. To date, this behavior has not been documented in

NATURAL HISTORY BOX: Rattlesnakes in Trees

Finding such heavy-bodied snakes as Timber Rattlesnakes in trees may seem unlikely, but with the advent of radiolocation technology, arboreal activity has been found to be a part of the Timber Rattlesnake's lifestyle. While radio-tracking Timber Rattlesnakes along the Mississippi River in Crawford County, I occasionally found the snakes above ground, resting in saplings, in the branches of a fallen tree, the crotch of a standing dead tree, and even 6 m (20 ft) up clinging to the bark on the trunk of a large oak tree. I have also seen a photo of a Minnesota Timber Rattlesnake on a branch holding a Yellow-bellied Sapsucker (*Sphyrapicus varius*) in its jaws (Sajdak and Bartz 2004).

But one of the most interesting observations I know of took place over a 17-day period beginning on 16 August 2001. On that date, Armund Bartz found an adult male he was tracking coiled up on the ground with a large pre-shed female. Four days later they were together in a tangle of grapevines and brushy trees about 1 m (3.3 ft) above ground and spent the next five days in the same place and posture (see Figure 296). On the morning of the 26th, the male was on the ground 5 m (16.4 ft) from the female, who was a little higher up in the same tree. On the 29th, the male had joined the female, and they remained near each other or together until the afternoon of 1 September. On the 2nd, both snakes were gone, and a neighbor reported that two snakes were seen crossing a nearby road. Two days later the male was found alone (Bartz and Sajdak 2004).

Figure 296. Courting Timber Rattlesnakes in a shrub; the male is the lighter-colored snake (Crawford County; photo by A. Bartz).

Richard A. Sajdak

Wisconsin. Wisconsin Timber Rattlesnakes show great fidelity for particular den sites, and populations from neighboring hibernacula are somewhat isolated from each other. As such, there is the danger of limited genetic exchange between populations. Yet males can follow female scent trails and disperse long distances to locate them for reproduction (Coupe 2002, Clark et al. 2008b). The home ranges of snakes from adjacent dens also sometimes overlap, which could permit the sharing of genetic material between neighboring groups (Sajdak and Bartz 2001).

Females store sperm from summer copulation internally until the following spring, at which time fertilization commences (reviewed by Ernst and Ernst 2011). Gravid females emerge from hibernacula and move to adjacent open bluffsides, where they remain among the warm exposed rock ledges. Timber Rattlesnakes are viviparous, and these ledges provide a more stable thermal environment for regulation of body temperature and development of embryos (Cochran and Schmitt 2014). Parturition occurs in similar open bluff habitat, sometimes beneath shelter

NATURAL HISTORY BOX: The Importance of Hibernacula

The presence of suitable habitat for overwintering could be an important factor limiting the distribution of this species in Wisconsin. Even in areas of supposedly good habitat, known Timber Rattlesnake dens in the state are widely scattered (often separated by 1.6 km [1 mi] or more), despite the fact that unoccupied habitat between known den sites appears otherwise suitable. This species is also apparently more selective of den habitat than other snake species that overwinter along bluffsides in Wisconsin. For example, Gray Ratsnakes that coinhabited a Crawford County site with Timber Rattlesnakes successfully overwintered in various hibernacula that were likewise accessible but not occupied by Timber Rattlesnakes (Sajdak 1999, 2000, R. Sajdak, unpublished data). The reason for this is not fully understood but could be related to on-site habitat composition and associated thermal properties near dens or the location's history of anthropogenic disturbance (e.g., perhaps some den locations once supported this species, but the associated populations are now extirpated). An additional unanswered question is how the underground structure of rock fissures, hydrology, and associated space within potential dens might affect hibernacula use by Timber Rattlesnakes. Observations of wild-caught Timber Rattlesnakes held in captivity during winter support the importance of hibernacula hydrology. For example, for a weeklong period during the autumn, when they would normally be entering their dens, wild-caught snakes expended considerable effort attempting to push through the bottom of water bowls in their enclosures (Craig Berg, Milwaukee County Zoo (retired), and Robert Hay, Turtles for Tomorrow, personal observations). This suggests that water or high subsurface moisture levels may be an important feature of suitable hibernacula.

Richard A. Sajdak

Figure 297. Timber Rattlesnake talus slope habitat (Sauk County; photo by J. Kapfer).

rocks, and these locations are referred to as nursery habitats or birthing rookeries (reviewed by Ernst and Ernst 2011). Most radio-tracked females monitored in Crawford County gave birth in late August and September (Sajdak 1999, 2000, Sajdak and Bartz 2001); however, parturition in Wisconsin has been recorded as late as October (C. Berg and A. Bartz, unpublished data). Vogt (1981) reported that litter sizes in Wisconsin ranged from 6 to 15 neonates. Keenlyne (1978) found a mean of 8.4 ova or young per gravid Wisconsin female, but if atretic ova are removed from the estimate, the mean drops to 7.7 ova or young per gravid female. In other parts of the range, litter sizes from 6 to 10 have been reported, with a maximum litter size of 20 (reviewed by Ernst and Ernst 2011). Females remain near offspring for 7–10 days after parturition, and then all disperse (reviewed by Ernst and Ernst 2011).

Timber Rattlesnakes are capital breeders (Bonnet et al. 1998, Olson et al. 2015), meaning that they gain energy for reproduction by storing energy from food consumed in previous seasons rather than food consumed immediately prior to reproduction. The reproductive output and developmental rates for this species are also low. For example, seven radio-tracked female Timber Rattlesnakes were monitored at several sites in western Wisconsin between 1999 and 2004. These seven snakes produced a total of only 15 litters over this time, with a mean time interval between litters of 3.14 years (range = 2–5 yrs; A. Bartz, C. Berg, and R. Sajdak, unpublished data). This contradicts the findings of Keenlyne (1978), who reported that Wisconsin Timber Rattlesnakes had a biennial reproductive cycle. Keenlyne's results were skewed by an overestimation of shedding frequency, which led to overestimates of growth and frequency of reproductive episodes. He estimated that snakes with up to eight rattle segments were in their third summer of growth and had completed six or more shedding cycles in two years. Recent results, based on 16 adult Timber Rattlesnakes in Crawford County (12 males, 4 females) recaptured after at least one year, demonstrated a much lower shedding rate ($\bar{x}$ sheds/yr = 1.74, range = 1–3; A. Bartz, C. Berg, and R. Sajdak, unpublished data). Adult males shed more frequently than females (1.80 vs. 1.54 sheds/yr), while juvenile snakes (<65 cm SVL [25.6 in] at first capture) shed more frequently than adults (1.85 sheds/yr; A. Bartz, C. Berg, and R. Sajdak, unpublished data). Reanalyzing Keenlyne's (1978) data with this lower annual shedding frequency suggests that females first reproduce when they are seven to nine years old rather than four years old and experience a triennial or greater reproductive cycle rather than a biennial one. This better conforms to more recent estimates of developmental rates and reproductive output from other northern populations. For example, Martin (2002) estimated that the mean time between female parturition events was five years, while Brown (2016) reported that the mean number of lifetime reproductive events was only 1.7 per female in New York. These studies further revealed that females develop slowly and take from 9.6 to 11 years to reach maturity (Brown 2016, Martin 2002, respectively). Males in New York, on the other hand, reportedly reach sexual maturity at a mean age of 5.3 years (Aldridge and Brown 1995).

Growth rates for this species also appear to be slow. For example, an immature Wisconsin female captured in May 1997 (Crawford County) weighed 200 g (7.0 oz) and had a TL of 66.6 cm (26.2 in). It had a string of three rattles (missing the button) and was estimated to be at the start of its fourth growth season (estimated age = 3.6 yrs). It was recaptured 37 months later (June 2000) and was gravid for the first time at an estimated age of 6.6 years. While its weight had increased by 195% (390 g [13.8 oz]) partly due to gravidity, its TL had increased by only 32.4% (21.6 cm [8.5 in]) during that roughly three-year period. Likewise, an immature male captured in May 1999 (Crawford County) at a TL of 73.0 cm (28.7 in) and a weight of 268 g (9.4 oz) had gained only 5.1 cm (2.0 in) and 129 g (4.6 oz) when recaptured in June 2000 (Sajdak and Bartz 2001). Growth is likely influenced by various factors, ranging from food availability to environmental conditions. Beaupre and Zaidan (2012) conducted laboratory studies on the relationship between temperature and growth in Timber Rattlesnakes from several southeastern states. They discovered that snakes housed at 20°C (68°F) allocated more energy to 'bulk,' while those housed at a warmer 30°C (86°F) allocated more to length, which the authors related to annual activity cycles. For example, allocation of energy toward additional mass during cooler seasons (spring and autumn) would help offset weight loss after winter and also help achieve appropriate body condition to enter winter dormancy.

Although they grow slowly, Timber Rattlesnakes are long-lived (Brown 1993). A wild-caught male reportedly survived for over 30 years in captivity (Snider and Bowler 1992). Brown et al. (2007) found that time between recaptures could be up to 18 years for wild individuals in New York. Similarly, in 2004 two adult female Timber Rattlesnakes from Wisconsin (SVL = 91 cm [35.8 in] and 89.5 cm [35.2 in]) were captured in Sauk County, marked, and released. Both individuals were recaptured alive over ten years later (the first in 2015 at SVL = 100 cm [39.4 in] and the second in 2017 at SVL = 107.9 cm [42.5 in], respectively; J. Kapfer, unpublished data).

ACTIVITY

Wisconsin Timber Rattlesnakes are active from mid-May to late September, which is several weeks shorter than other snake species with which they share hibernacula, such as North American Racers (*Coluber constrictor*) and Gophersnakes (Sajdak 2000, Sajdak and Bartz 2001). A three-year radio telemetry study in Crawford County determined that most spring emergence occurred from 7 to 21 May (Craig Berg, Milwaukee County Zoo [retired], personal observation, R. Sajdak, unpublished data), but anecdotal reports exist of individuals active as early as 12 and 16 April (R. Hay, personal observation). Survival may be lower for individuals that emerge from hibernation earlier than what is typical, and several examples exist of Timber Rattlesnakes in Wisconsin that died shortly after premature emergence from hibernacula. A large male in Crawford County emerged prior to 6 April 2000, before other members of the populations, and died of an apparent infection several weeks later (C. Berg, unpublished data). Another large male emerged from hibernation prior to 4 April 2002 and was consumed by an unknown avian predator shortly thereafter (C. Berg, unpublished data). Among Timber Rattlesnakes monitored with radio telemetry in Wisconsin, the earliest record of egress from hibernacula for an individual that also survived the active season was 20 April (C. Berg, unpublished data).

Mean active body temperature and degree of body temperature variation differ among individuals, with gravid females showing less variation than males and nongravid females (Gardner-Santana and Beaupre 2009, Sajdak 1999). For example, the mean active body temperature of two gravid females from Crawford County monitored for several years was 29.7°C (84.5°F), with relatively little variation (20.0°C–35.1°C [68.0°F–95.2°F], $n = 97$; Sajdak 1999). Alternatively, mean body temperature of males at this location was a cooler 24.4°C (75.9°F, $n = 145$), and recorded body temperatures showed greater variation than those of gravid females (range = 12.2°C–35.2°C [54.0°F–95.4°F]; Sajdak 1999). This was also true for the body temperature of two females that were monitored post-parturition ($\bar{x} = 22.4$°C [72.3°F], range = 13.6°C–32.9°C [56.5°F–91.2°F], $n = 29$; Sajdak 1999).

There are large differences in reported movement patterns of this species across its North American distribution, with estimated home range sizes as large as 207 ha (511.5 ac; reviewed by Ernst and Ernst 2011). Although high mobility is documented, gravid females in Wisconsin move less than males and nongravid females (Sajdak 1999, 2000, Hamilton 2009). For example, the mean home range size (MCP estimator) of gravid females in Crawford County was 0.7 ha (1.7 ac), compared to 21.1 ha (52.1 ac) in adult males and 13.3 ha (32.9 ac) in adult nongravid females. Both mean movement per day (DM) and mean total distance moved by individuals over a year (YM) showed a similar pattern, with males (DM = 27.7 m [90.9 ft], YM = 2,981 m [9,780 ft]) and nongravid females (DM = 21.8 m [71.5 ft], YM = 2,082 m [6,831 ft]) moving greater distances than gravid females (DM = 4.9 m [16.1 ft], YM = 483 m [1,585 ft]; Hamilton 2009). Juveniles also have smaller home ranges than adults, but their activity areas appear to increase as they grow (Hamilton 2009, R. Sajdak, personal observation).

Timber Rattlesnakes in Crawford County generally entered hibernacula between 14 September and 1 October (R. Sajdak, unpublished data), a phenology similar to that observed for New York populations (Brown 1993). The latest reported date of activity documented in Wisconsin was for a neonate encountered on 7 November 2008 in Crawford County (Christopher Hamilton, NRCS, unpublished data). This individual was found within 3 m (10 ft) of a hibernaculum (C. Hamilton, unpublished data). Laboratory experiments show that neonates follow the scent trails of adults and may use these as cues to locate hibernacula in the autumn of their first year (Brown and MacLean 1983, Cobb et al. 2005). Although some neonates in eastern populations overwintered in the same den as their mother, others dispersed to other dens as distant as 1.7 km (1.1 mi) from the parturition site (Reinert 2005).

The details of behavior during hibernation are difficult to investigate because wild snakes are typically not accessible when dormant. Over a five-year period, winter body temperatures were recorded with temperature-sensitive radio transmitters in 16 hibernating Timber Rattlesnakes spread across 6 hibernacula in western Wisconsin. The mean body temperature of all hibernating snakes was 10.8°C (51.4°F, range = 5.0°C–18.9°C [41°F–66°F], n = 137 temperature readings; C. Berg, unpublished data). These results are virtually identical to those published by Brown (1982, 1992b, 1993) for northeastern New York, where winter environmental pressures placed on Timber Rattlesnakes are likely similar to those in Wisconsin. The approximate position and depth of radio-tracked Timber Rattlesnakes overwintering in a Crawford County den was estimated by localizing the source of the strongest radio signal through the soil (n = 10 snake locations, 1999–2003). Although the position of most Timber Rattlesnakes remained fixed throughout much of hibernation, there was evidence of small movements by some individuals over the course of the winter. Furthermore, snakes appeared to reach their deepest position within the den during December and January, while probable movements toward the den entrance were first noted in late March and early April as within-den temperatures begin to rise (A. Bartz and C. Berg, unpublished data).

Depletion of water and fat reserves during winter dormancy leads to declines in body weight by the time spring emergence occurs. Mean weight loss during hibernation was assessed for Timber Rattlesnakes at the six sites in western Wisconsin. A total of 11 individuals were monitored across five years, with weight at time of ingress and egress recorded, which revealed a mean weight reduction of 10.2% (range = 3.7%–19.4%, n = 24 records; A. Bartz and C. Berg, unpublished data). There was no evidence that weight loss was related to differences in snake size at ingress (by weight) or sex; females (n = 5 individuals and 11 records) lost an average of 9.9% of their weight, while males (n = 6 individuals and 13 records) lost an average of 10.5% of their weight (A. Bartz and C. Berg, unpublished data).

PREY AND PREDATORS

Timber Rattlesnakes are sit-and-wait (i.e., ambush) predators. As such, they choose promising ambush sites and remain coiled at these locations waiting for suitable prey. Clark (2006) estimated that the mean time spent at ambush locations by foraging Timber Rattlesnakes in New York was 17 hours. Foraging individuals reportedly select optimal ambush locations through chemical cues left by energetically profitable prey and by past hunting experience (Clark 2004a, 2004b). Individuals often choose 'runways' along fallen logs that are frequented by rodents (Douglass and Reinert 1982), where they rest in a coil alongside the log with their head pointed toward the rodent trail (Figure 298). Despite these efforts to select optimal hunting locations, Clark (2006) found that Timber Rattlesnakes in New York successfully struck only 13% of encountered prey (n = 17 individuals and 87 encounters). Clark (2006) also estimated that the Timber Rattlesnakes he studied typically fed 12–15 times throughout the active season.

Figure 298. Timber Rattlesnake in ambush posture (Crawford County; photo by R. Sajdak).

This species appears to feed heavily on small mammals (reviewed by Ernst and Ernst 2011), although limited data from Wisconsin exist. Breckenridge (1944) reported a Gray Squirrel (*Sciurus carolinensis*) in the stomach of a specimen that was probably from Pierce County, while Keenlyne (1972) reported nine mice (*Peromyscus* spp.), seven Meadow Voles (*Microtus pennsylvanicus*), two Gray Squirrels, two Eastern Chipmunks (*Tamias striatus*), and an unidentified bat (Chiroptera) in the stomach contents of 76 snakes from Pepin and Buffalo Counties. They apparently also feed on birds and occasionally lizards, snakes, and frogs (Clark 2002, reviewed by Ernst and Ernst 2011). DeVault and Krochmal (2002) reviewed several documented cases of scavenging.

Timber Rattlesnakes are likely taken by a range of predators, including hawks, owls, Wild Turkeys (*Meleagris gallopavo*), Coyotes (*Canis latrans*), Striped Skunks (*Mephitis mephitis*), and other snakes (reviewed by Ernst and Ernst 2011). Olson et al. (2015) found that 50% of the adult mortality documented in a sample of Timber Rattlesnakes from Indiana was due to predation. Minton (2001) also reported that this species was occasionally trampled by White-tailed Deer (*Odocoileus virginianus*). Interestingly, research on wild American Black Bears (*Ursus americanus*) indicated that they did not consume Timber Rattlesnakes but showed avoidance of them when encountered in the wild (Rogers et al. 2014).

When disturbed but not provoked, this species often remains silently coiled, but it will also rattle to warn a would-be threat (R. Sajdak, personal observation). Yet Timber Rattlesnakes will not always rattle in response to disturbance, so the presence or absence of a rattling sound should not be relied on to avoid encounters. Fitch (1999) recounted accidentally stepping on a Timber Rattlesnake that he was tracking with radio telemetry equipment, noting that "even then it did not always rattle, but uncoiled and slithered away silently and rapidly." If molested, this species will remain partially coiled and raise its forebody into a striking position. Timber Rattlesnakes that are handled roughly or restrained will thrash laterally and possibly bite (R. Sajdak, personal observation). The frequency of Timber Rattlesnake bites in the Upper Midwest is incredibly low and likely represents a minuscule threat to humans (see Snakebite Envenomation in Wisconsin and the Northern Upper Mississippi River Valley chapter).

CONSERVATION

Status: Globally, the Timber Rattlesnake is listed as a Least Concern species by the IUCN. While this snake occurs in 30 states, it is listed as Rare to Imperiled in 16 and designated as Endangered or Threatened in 13, including Illinois and Minnesota. It is a protected species and also a Species of Greatest Conservation Need in Iowa. In Wisconsin, the Timber Rattlesnake is considered a Species of Greatest Conservation Need and also a species of Special Concern. It was legally designated as a Protected Wild Animal in Wisconsin in 1998 (Wisconsin Legislature 2018) after a proposal by the WDNR to list the snake as a Threatened Species was defeated by legislative committee. Protected Wild Animals are protected from hunting or collecting, dead or alive (see Introduction to Conservation and Management of Wisconsin's Amphibians and Reptiles chapter).

Populations: Estimates of population size and density do not exist for Wisconsin, and estimates from across the Timber Rattlesnake's North American distribution are highly variable (reviewed by Ernst and Ernst 2011). A handful of examples from northern latitudes exemplify this. Brown (1993) reported population sizes associated with unexploited hibernacula in New York that ranged from 40 to 200 individuals. Martin et al. (1990) reported that the mean population size per den site surveyed in Pennsylvania ($n = 312$ hibernacula investigated) was only 36 individuals. Fitch and Echelle (2006) estimated very low mean densities at locations in Kansas, which ranged from 0.2/ha to 1.2/ha (0.5/ac to 3.0/ac) over many years of surveys. Despite these low estimates, Timber Rattlesnakes can appear locally abundant if the right location is surveyed at an appropriate time of year. Because they congregate near den sites after spring emergence from hibernation, as many as 15–20 individuals can be encountered during a single spring survey of these locations (R. Sajdak, C. Berg, and R. Hay, personal observations), which can encourage a misleading notion that this species is common.

Oldfield and Keyler (1989) reported a roughly 1M:1F sex ratio in snakes sampled from Wisconsin populations, but their sample size was small ($n = 9$ individuals). A review of other studies suggests large variation in sex ratios across the snake's distribution (reviewed by Ernst and Ernst 2011), which could be due to differences in the threats that each sex faces. Other demographic variables, such as survival, also appear to vary by sex and age. Brown et al. (2007) found that while the survival probability of first-year snakes in New York was 0.65, this increased to ca. 0.90 in snakes five years or older. In Wisconsin, a small number of observations support that adult survival rates and longevity are also high (see Reproduction and Development section), so long as individuals do not experience human persecution or habitat disturbance. Yet despite high rates of adult survival, Timber Rattlesnake populations are likely susceptible to even small perturbations because of their slow development and small reproductive output (Ernst and Ernst 2011). It is also reported that Timber Rattlesnake populations can be indirectly affected by natural environmental variation. For example, Olson et al. (2015) documented oak mast production failures in Indiana that led to a concomitant 50% reduction in rodent populations. This reduction in prey availability was further linked to high female Timber Rattlesnake mortality in those years (Olson et al. 2015).

Threats: Unfortunately, Timber Rattlesnakes have experienced significant human persecution across their range since early European settlement (Brown 1993). Schorger (1967–68) aptly summarized the situation when he stated that "man has been the greatest enemy of the rattlesnake since the first European set foot in Wisconsin." Although no formal population analyses have occurred in the state, records of past persecution can provide insight into the level of human impacts on this species. Schorger (1967–68) reviewed historical accounts and found that numerous Wisconsin hibernacula were destroyed with explosives or filled in with concrete. Furthermore, available information suggests that persecution due to Wisconsin's rattlesnake bounty was disturbingly high (see Natural History Box: Historical Anecdotes Describing Human Interactions with Reptiles in Early Wisconsin in the Introduction to Conservation and Management of Wisconsin's Amphibians and Reptiles chapter). As a result, many historical Wisconsin populations are likely extirpated, and most that remain are probably vulnerable to elimination (Vogt 1981, Oldfield and Keyler 1989).

Unfortunately, high mortality due to humans is not only a thing of the past. Olson et al. (2015) recently reported that 30% of the 47 adult Timber Rattlesnakes radio-tracked for five active seasons in Indiana died due to direct persecution or vehicle strikes. Similarly, direct human persecution also still occurs in Wisconsin, despite this species' protected status (John Buss, WDNR Conservation Warden [retired], personal communication). According to work in New York, the highest probability of encounters between Timber Rattlesnakes and humans occurred during the breeding season (Aldridge and Brown 1995). At this time, male Timber Rattlesnakes move greater distances and move more frequently than females do, a behavior associated with searching for mates. Some research suggests that the more mobile males are also highly susceptible to mortality, which can skew population sex ratios (Aldridge and Brown 1995). For example, Sealy (2002) found a skewed sex ratio that substantially favored females in North Carolina and that was associated with the vulnerability of the more active males to road mortality. On the other hand, the Timber Rattlesnake bounty in Wisconsin may have had opposite unintended impacts on sex ratios in the state. Because partial bounties were paid for juvenile snakes (even unborn individuals) and the fact that gravid females can be reliably encountered year-round at known locations, females were likely targeted by many rattlesnake hunters (e.g., Messeling 1953). That bounty hunters took a heavier toll on gravid females could possibly explain the anecdotal observations that Wisconsin populations appear more male-heavy, although rigorous analyses are lacking (R. Hay and J. Kapfer, personal observation). It is noteworthy that population impacts can also occur due to illegal collection for the pet trade. Unfortunately, their slow development and low reproductive output make Timber Rattlesnake populations susceptible to minor levels of disturbance. Therefore, harvesting or collecting even a handful of individuals can have substantial long-term negative effects.

Habitat loss appears to be a critical threat to Timber Rattlesnakes in Wisconsin. Many general conservation concerns associated with prairies, grasslands, and savannas, as well as habitat loss due to conversion of natural land to agricultural or developed land, are applicable to this species (see Introduction to Conservation and Management of Wisconsin's Amphibians and Reptiles chapter). Open-canopy bluffside habitat is particularly important for Timber Rattlesnakes, and its loss or degradation could also have substantial impacts. For example, suppression of wildfire and lack of management to remove woody vegetation have resulted in unchecked vegetative succession along bluff prairies. This succession has caused shading that can limit appropriate thermoregulatory opportunities and gestation habitat for gravid females. It can also influence the overwintering phenology of snakes whose hibernacula occur in these overgrown habitats.

In recent decades, habitat along bluff ridges in western Wisconsin has been converted from natural or even agricultural land to spacious residential subdivisions. This exurban or rural development poses further risks to rattlesnakes by interrupting the connectivity between rattlesnake populations or through increased potential for negative human-snake interactions (Rudolph and Bergdorf 1997, Bonnet et al. 1999, Driscoll 2004, Andrews and Gibbons 2008). Evidence suggests that Timber Rattlesnakes either avoid disturbed land or use it sparingly. A radio telemetry study in western Wisconsin encountered adult snakes within agricultural fields on less than 1% of 261 telemetry locations (R. Sajdak, unpublished data). Despite this, the driving forces behind use of disturbed habitats in Wisconsin are likely complex and influenced by a variety of factors, including the composition of habitat immediately adjacent to hibernacula. For example, despite the behavior of most snakes in the previously mentioned study, a single radio-tracked male overwintered in a den surrounded by disturbed habitat and was subsequently located in disturbed habitat on 70.6% of active observations. This included use of a tree plantation, roadsides, a barnyard, and a cornfield (Sajdak 2000). Results from other midwestern states also support regional differences in use of disturbed habitat. For example, Timber Rattlesnakes in Nebraska used agricultural fields and field edges frequently, but this use was possibly due to limited availability of preferred habitat (Fogell et al. 2002).

Although some Timber Rattlesnakes use disturbed habitat, those that occupy such landscapes likely face a greater risk of mortality. In 2000 a freshly dead adult Timber Rattlesnake was encountered in a Wisconsin hayfield the day after the field was mown (R. Sajdak, personal observation), a presumably common occurrence. Roads associated with developed landscapes are also frequently identified as important sources of disturbance. Andrews and Gibbons (2005) suggested that Timber Rattlesnakes may be particularly prone to road mortality because they remain immobile (rather than opting to flee) as vehicles approach and pass. In addition to impacting abundance and sex ratios, road mortality can adversely affect the genetic exchange between Timber Rattlesnake populations that are fragmented by heavily used roadways (Clark et al. 2010).

Fungal dermatitis has been documented on Timber Rattlesnakes throughout their North American range (e.g., Clark et al. 2011, McBride et al. 2015). Current research suggests that the likely cause of this dermatitis is the fungal species *Ophidiomyces ophiodiicola*, which results in the disease commonly known as snake fungal disease (SFD; see Introduction to Conservation and Management of Wisconsin's Amphibians and Reptiles chapter). Recent surveys and analyses have confirmed the presence of SFD in Wisconsin Timber Rattlesnake populations (Lorch et al. 2016). Although mortality due to the disease is known in several states, the population-level impacts remain largely indefinite. However, Clark et al. (2011) reported the impacts of an undetermined skin disease with clinical symptoms consistent with SFD as one of several reasons for the substantial decline of an isolated Timber Rattlesnake population in New Hampshire. Given its potential to impact snake populations, this disease should be monitored in Wisconsin. Timber Rattlesnakes also face other threats, such as entrapment in plastic mesh used for garden pest exclusion and erosion control (Kapfer and Paloski 2011, Cochran 2013a), but the magnitude of these impacts is unknown.

Management: Conservation, management, and protection strategies for Timber Rattlesnakes can involve a variety of approaches (Brown 1993). Among them is the critical acquisition and protection of land to provide habitat where this species can remain unmolested. Agencies and organizations such as the WDNR, TNC, and the Mississippi Valley Conservancy have acquired and continue to acquire, manage, and restore important bluffland habitats. These efforts will likely remain instrumental in the future conservation of Timber Rattlesnakes in Wisconsin. Although acquisition of land is an important conservation initiative, appropriate management of these properties is also critical.

Rigorously obtained data on the best management strategies for Timber Rattlesnakes in Wisconsin do not exist. Regardless, general techniques used to manage prairies (including bluffside prairies) could benefit this species (see Introduction to Conservation and Management of Wisconsin's Amphibians and Reptiles chapter). For example, because they appear to maintain high body temperatures and select stable thermal environments (particularly gravid females), management to enhance thermoregulatory opportunities could be beneficial. Therefore, efforts to offset encroachment of woody vegetation on bluffside prairies could be a critical component of habitat management for this species (Gardner-Santana and Beaupre 2009). Along these lines, it is reasonable to suggest that den and transient habitats support a diversity of habitat structure but predominantly open canopy (>70% open canopy; Sajdak 1999).

Several common techniques exist to maintain open habitats in areas that Timber Rattlesnakes frequent. These methods include prescribed fires, hand removal of vegetation (e.g., via chainsaw), and mechanical removal of vegetation (e.g., removal with heavy machinery; see Introduction to Conservation and Management of Wisconsin's Amphibians and Reptiles chapter). However, appropriate implementation of these techniques requires careful consideration, which should be driven by the time of year and rattlesnake activity patterns. Beaupre and Douglas (2012) reported that application of fire to transitional habitats after rattlesnakes emerged from hibernation led to adult mortality and population impacts that persisted for over a decade. Yet the authors found that prescribed burns completed prior to emergence not only reduced mortality but also resulted in enhanced growth and better body condition in snakes. They believed this was due to an increase in prey (small mammal) populations that was presumably the result of appropriate fire application (Beaupre and Douglas 2012). Thus, intense management should be conducted during the inactive period for this species, which could extend from 1 November through 10 April (although activities that disturb soil, such as 'skidding,' should be restricted to when the ground is frozen). In addition to maintenance of open conditions, the promotion of ground cover such as brush piles and downed logs likely benefits Timber Rattlesnakes (Hamilton 2009). Hamilton (2009) found that Wisconsin Timber Rattlesnakes within disturbed habitats used brush piles extensively. He speculated that snakes used these piles as cover to "island hop" across open disturbed areas such as mowed fields or lawns and for foraging, because brush piles attract rodents. Further care should be taken to avoid disturbance of cover in nursery habitats (e.g., moving rocks) so that snake retreat sites are not destroyed and gravid females or neonates are not accidentally crushed.

Some Timber Rattlesnake hibernacula in Wisconsin exist on private property that is not actively managed by state agencies, which must focus their efforts on state land. As a result, hibernacula on private lands can become overgrown with woody vegetation. Yet private landowners interested in promoting Timber Rattlesnake populations could focus on the creation of open conditions through simple measures such as hand removal of woody vegetation. Private landowners are strongly encouraged to develop strategies with input from agency employees or professional habitat managers so that techniques are implemented in a manner that is not only legal but also safe for humans and snakes. It is further critical that land developers consider Timber Rattlesnakes when building in bluffland habitats (see Andren 1994, Driscoll 2004, Zappalorti and Mitchell 2008, Hamilton 2009). This could include making efforts to identify and avoid hibernacula in development plans. Furthermore, residential buildings should be

NATURAL HISTORY BOX: Public Education as a Conservation Strategy for Timber Rattlesnakes

Public education, particularly directed at those who own land that Timber Rattlesnakes use, is extremely important to the long-term persistence of this species. Following its listing as a Protected Wild Animal in Wisconsin in 1998, 51 rare snake educational workshops were conducted for citizens throughout the Timber Rattlesnake's range. The goal of these workshops was to inform people about rattlesnakes, including ecology and appropriate identification, and to help change public attitudes toward venomous snakes. Public education and outreach efforts regarding Timber Rattlesnakes in Wisconsin have had positive impacts. At least two documented examples since 2000 exemplify this. One of these examples is outlined in Sajdak and Berg (2005). The other involves a local landowner from southwestern Wisconsin who was in favor of Timber Rattlesnake conservation but knew that other landowners in the immediate area either were indifferent or opposed or actively killed Timber Rattlesnakes on sight. The landowner had been trained to safely capture and move Timber Rattlesnakes (under the direction of WDNR employees and rattlesnake biologists, following protocols in Upton et al. 2002). As his neighbors became aware of his training, he was granted many opportunities to interact with and educate them during visits to remove 'problem' rattlesnakes from their properties. As a result of these positive interactions with him, many citizens in the region changed their opinions of Timber Rattlesnakes and now often view them as important ecological components of Wisconsin's landscape. The WDNR and nonprofit conservation organizations are urged to work with individual landowners to promote the protection of this species, since human persecution continues to pose a threat (Brown 1993, Dodd 1993).

Robert W. Hay, Turtles for Tomorrow

spaced appropriately to maintain natural corridors that allow snakes passage through developed areas while minimizing human contact.

A critical component of Timber Rattlesnake conservation and management is the restriction of human access to hibernacula. As such, the locations of hibernacula should only be shared with professionals who can aid Timber Rattlesnake conservation. Social media and other internet-based forums are especially problematic in the dissemination of sensitive data. These outlets can rapidly and widely distribute information where it can easily fall into the hands of those with little regard for natural resource conservation. There are many examples of well-intentioned Wisconsin citizens distributing the locations of hibernacula that have led to exploitation of associated rattlesnake populations (R. Hay, personal observation). In addition to exploitation of populations, regular human activity near hibernacula could lead to habitat disturbance and higher probabilities of accidental rattlesnake bite. It is also valuable to recognize that the locations of rare species (including their hibernacula) are currently exempt from Wisconsin's Open Records Law, which helps protect these species from insidious activities. Finally, the importance of public outreach as a conservation strategy for venomous snakes should not be overlooked. For example, public outreach on rattlesnakes in Wisconsin was particularly useful in garnering support for the current protections afforded Timber Rattlesnakes (Natural History Box: Public Education as a Conservation Strategy for Timber Rattlesnakes).

Massasauga

Sistrurus catenatus (Rafinesque 1818)

Richard S. King, Robert W. Hay, Billie C. Harrison, Eric T. Hileman, and Craig S. Berg

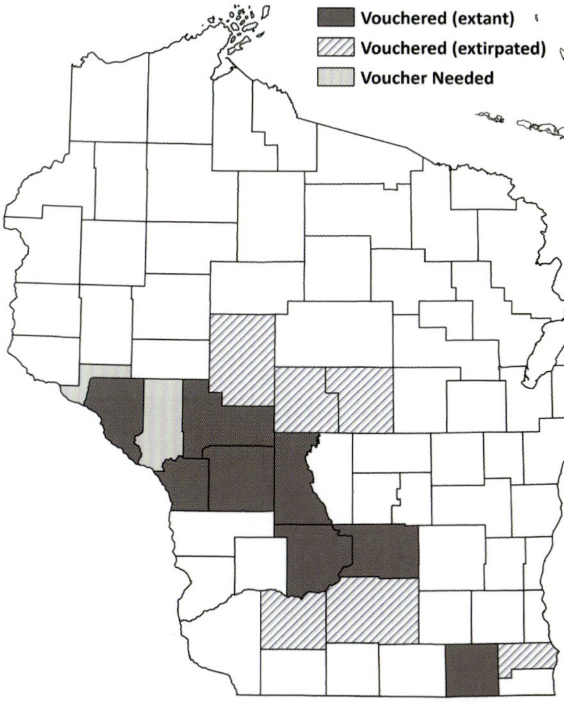

Adult Massasauga (Buffalo County; photo by R. Paloski).

AT A GLANCE

Massasaugas (a.k.a. 'swamp rattlesnakes') are one of two venomous snakes found in Wisconsin. They can be identified as venomous by their elliptical pupils, obvious rattle on the end of the tail, and presence of a heat-sensing pit between the eyes and nostrils on either side of the head. Their dorsal pattern consists of numerous saddle-shaped chocolate-colored blotches over a gray background that run along the back. Two broad brown stripes bordered by white occur on each side of the head: one that runs from the back of the eye to the back corner of the jaw, and another that runs parallel to the mouth. Massasaugas frequent wet or swampy habitats, including damp meadows, fens, and wet prairies. They are active in Wisconsin from mid-April to mid-October. The period when adults breed in the state is somewhat difficult to determine, based on past reports. Some indicate that breeding mostly occurs in summer or autumn, while others suggest that it occurs in spring and summer. Regardless, females give birth to live young from early August to mid-September. Their diet in Wisconsin consists almost entirely of small mammals, while the snakes are likely eaten by various birds and mammals found in wetland habitats. Although Massasauga populations are much smaller and more isolated than those of Wisconsin's other venomous snake, the Timber Rattlesnake, this species was once widespread and locally abundant in southern Wisconsin. Habitat loss and human persecution, including a state-paid bounty for Massasaugas from 1903 to 1973, have contributed to its decline. Massasaugas are now among Wisconsin's rarest vertebrates and only reliably found in a handful of locations from the western and central portions of the state. It remains unclear if the species will persist in Wisconsin. The Massasauga has been listed as Endangered in the state since 1975 and was listed as Threatened in the US under the Endangered Species Act in 2016.

SYSTEMATICS

The phylogenetics of *Sistrurus catenatus* (Massasauga) have been extensively studied. The species belongs to a group of snakes (genus *Sistrurus*) collectively known as pigmy rattlesnakes. There are three species in the genus: *S. catenatus* (Rafinesque 1818), *S. miliarius* (Pigmy Rattlesnake; Linnaeus 1766), and *S. ravus* (Mexican Pigmy Rattlesnake; Cope 1865). *S. miliarius* is widely distributed across the southeastern and south-central United States. *S. catenatus* occurs from Arizona to New York and Ontario. *S. ravus* has a relatively small range, limited to southern Mexico.

Until recently, *Sistrurus catenatus* was thought to encompass three subspecies, including *S. c. catenatus* (Eastern Massasauga), *S. c. tergeminus* (Western Massasauga), and *S. c. edwardsii* (Desert Massasauga). Extensive genetic research (Anderson et al. 2010, Gibbs and Chiucchi 2011, Ray et al. 2013, Ryberg et al. 2015) indicates that only *S. c. catenatus* occurs in Wisconsin, and this subspecies should be considered a separate species. Among Wisconsin snakes, *S. catenatus* is one of only two species in the family Viperidae, the other being *Crotalus horridus* (Timber Rattlesnake).

DESCRIPTION

Adults: In Wisconsin, Massasaugas have 26–40 prominent middorsal blotches that are chocolate brown to very dark brown or nearly black. The blotches are overlaid on a ground color that ranges from light gray to gray brown (Vogt 1981, Oldfield and Moriarty 1994; Figure 299). Each side of the body has two or three

Figure 299. Two adult Massasaugas showing the slight color variation that can occur in Wisconsin (Juneau County; photo by R. Hay).

Figure 300. Massasauga head pattern (Buffalo County; photo by R. Paloski).

Above, top: Figure 301. Photograph highlighting the broad brown stripes bordered by white on the side of the head. Note that the green tint on the rattle is due to paint used as a short-term identifying mark by researchers (Buffalo County; photo by J. Kapfer).

Above, bottom: Figure 302. A close-up of the rattle from a Massasauga (Buffalo County; photo by R. Paloski).

rows of smaller alternating blotches, the lowest often merging with the ventral scales. The blotches can be faintly bordered by a very thin white line (Ditmars 1907). The ventral surface is dark gray to black and mottled with yellow, cream, or white, but it can also be mostly black. The dorsal body scales are keeled.

The head is noticeably wider than the neck and has nine large plate-like scales between the eyes (Figure 300). The top of the head also has two broad dark brown and irregularly shaped bands running from behind the parietal scales on top of the head onto the neck (Figure 300). The sides of the head have two short broad brown stripes bordered with white: one that runs from the eyes to the back of the head, and another that runs along the labial scales (Figure 301). This snake has a pit between the nostril and eye on both sides of the head that detects radiant heat (Bakken and Krochmal 2007). Both of Wisconsin's rattlesnakes have vertical pupils, unlike Wisconsin's nonvenomous snakes, which have round pupils (Ernst and Ernst 2003). The tail has dark dorsal bands that alternate against the ground color, giving it a ringed pattern. Males have a longer tail than females, with five to seven tail rings, whereas females have fewer than five or exactly five rings (Vogt 1981). The anal plate and subcaudal scales are undivided. The rattle consists of horny caudal scales that are loosely connected. After every shedding cycle a new scale or segment is added to the rattle. A special muscle in the tail allows the snake to elevate the rattle vertically while quickly vibrating it from side to side, producing a distinct rattling sound (Friederici 2008). The number of rattle segments is not indicative of age, as they can shed more than once a year, and the rattles can also break off (Steen 2011; Figure 302).

Massasaugas in Wisconsin are medium-sized, stout-bodied snakes ranging from 50 to 75 cm TL (19.5 to 29.5 in; Vogt 1981), although individuals as large as 100 cm TL (39 in) have been reported (Ditmars 1907, Conant and Collins 1998). The largest known wild specimen from Wisconsin came from

Juneau County and measured 85.2 cm TL (33.5 in, UWZM 23946). Wild males sampled from a three-county area of west-central Wisconsin ranged from 56.5 to 73.5 cm SVL (22.2 to 28.9 in), and tail length ranged from 5.3 to 8.5 cm (2.0 to 3.3 in, $n = 24$). Female SVL and tail length ranged from 53 to 71 cm (20.9 to 28.0 in) and 4.0 to 5.8 cm (1.6 to 2.3 in, $n = 24$), respectively. Weights for adult males ranged from 195 to 550 g (6.9 to 19.4 oz), nongravid females ranged from 220 to 380 g (7.8 to 13.4 oz), and gravid females ranged from 225 to 510 g (7.9 to 18.0 oz; Robert Hay, Turtles for Tomorrow, unpublished data).

Preadult stages: Neonates look similar to adults in pattern but usually have a lighter ground color and are born with a distinctly yellow tail tip (ca. the last centimeter of the body; Figure 303). This tail color fades somewhat quickly as the snake grows and becomes dark like the dorsal blotches. Mean TL of neonates from Buffalo County was 22.0 cm (8.66 in, SD = 0.1 cm [0.04 in], $n = 207$) and mean weight was 9.7 g (0.349 oz, SD = 0.1 g [0.004 oz]; Keenlyne and Beer 1973b). King et al. (2004) reported that captive-born Wisconsin neonates from wild-caught females in Juneau and Monroe Counties had a mean weight of 8.3 g (0.29 oz, SD = 0.7 g [0.02 oz], $n = 32$). Hileman et al. (2017) indicated that the weights reported by King et al. (2004) were the lowest for any Massasauga population and that across the species' range, neonate mass was positively correlated with mean annual precipitation.

Figure 303. A neonate Massasauga. Note the yellowish tail (Buffalo County; photo by J. Kapfer).

Variation: No significant variation in appearance is known in Wisconsin, although dorsal blotches can range from light to dark brown (see Figure 299), and juveniles have a yellow tail tip (see Figure 303). There is no apparent sexual dimorphism in color or pattern (R. Hay, unpublished data), but adult males are likely longer and heavier than females.

Diagnosis: Massasaugas can be characterized by the presence of verticle pupils, one or more rattle segments on the end of the tail, a heat-sensing pit between the eye and nostril, and large plate-like scales on the top of the head. They also possess strongly keeled dorsal scales in 21–27 rows at midbody and 139–160 ventral scales. Their tail anterior to the rattle segments is banded, while the anal plate is undivided, and most subcaudal scales are undivided. There are 2 prefrontal scales, and each side of the head has 11–12 supralabial scales, 11–13 infralabial scales, 2 preocular scales, and 3–4 postocular scales. Massasaugas are distinguishable from Wisconsin's other venomous snake, the Timber Rattlesnake, because the latter is beige with dark brown or black bands that cross the dorsal surface, has a mostly unpatterned head, and has a solid black tail. Timber Rattlesnakes possess many small head scales (reviewed by Ernst and Ernst 2003), as opposed to the nine large scales on the head of the Massasauga. The scale counts of Timber Rattlesnakes also differ from those of Massasaugas, with 13–15 supralabial scales, 14–16 infralabial scales, and 154–183 ventral scales. Although Bailey (1942) reported a natural Massasauga × Timber Rattlesnake hybrid, the rate of incidence in the wild is likely low, and no reports of hybridization exist from Wisconsin.

Five Wisconsin snake species are considered 'rattlesnake mimics,' since they often vibrate the end of their sharply pointed tail when disturbed, which confuses differentiation from Massasaugas. These include Eastern Foxsnakes (*Pantherophis vulpinus*), Gophersnakes (*Pituophis catenifer*), Eastern Milksnakes (*Lampropeltis triangulum*), Gray Ratsnakes (*Pantherophis spiloides*), and North American Racers (*Coluber constrictor*). Eastern Foxsnakes and Eastern Milksnakes, for example, have somewhat similar dorsal blotch patterns. Yet adult Eastern Foxsnakes have little to no pattern on an orange or copper-colored head, whereas Massasaugas have a prominently patterned head. Juvenile Eastern Foxsnakes, on the other hand, do have a heavily patterned head but possess round pupils and patterned ventral surfaces, unlike Massasaugas. Eastern Milksnakes possess blotches that are brown or reddish but with a narrow black border, while the Massasauga's blotches are usually bordered by a faint white line. Eastern Milksnakes also typically have a light-colored V- or Y-shaped mark on the nape of the neck, while Massasaugas have two somewhat parallel brown bands behind the eyes that run onto the neck (Vogt 1981). Gophersnakes possess many poorly defined dorsal blotches, with variegated spots between them (unlike the Massasauga's well-defined blotches), while North American Racers and Gray Ratsnakes do not have obvious dorsal patterns as adults. Eastern Hog-nosed Snakes (*Heterodon platirhinos*) are another potentially confusing species, although they do not vibrate their tails when threatened. They are thick-bodied, like Massasaugas, and some adults possess prominent dark dorsal blotches (although many have a more uniform tan or olive dorsal surface). The Eastern Hog-nosed Snake's rostral scale is pointed and upturned. This snake also often flares its neck when approached, a behavior not typical in Massasaugas. As stated previously, all nonvenomous snakes in Wisconsin have round pupils and a sharply pointed tail, unlike venomous species.

DISTRIBUTION AND HABITAT

Global distribution: Massasaugas occur in a southwest to northeastern swath across North America. Their range extends from Arizona and Texas in the South, north through Oklahoma, and into southern Kansas (with an isolated pocket in Colorado). A gap in the distribution occurs across much of Missouri and Iowa, after which they range into the Upper Midwest across Illinois, into Wisconsin and east across Ohio, Indiana, and the Lower Peninsula of Michigan, and into western New York, Pennsylvania, and southern Ontario.

Distribution in Wisconsin: Vogt (1981) reported that Massasaugas could only be found with regularity at a handful of sites through the 1970s. Today they are even rarer in Wisconsin and perhaps the state's most endangered reptile (Christoffel et al. 2008). At best, they occur at fewer than ten locations, and the species' future in the state is uncertain (Faust et al. 2011). Massasaugas are currently reported from 16 counties, although cataloged specimens are either missing or do not exist for two of these. For example, Vogt (1981) included Pepin County in the snake's range, but no vouchered specimen from that county is known. Casper (1993b) included an anecdotal observation from Pepin County by "R. Hoffman" in 1987, but no further anecdotes are documented.

Several additional voucher specimens reported in support of county records are either missing or suspect. Recent inquiries into the existence of the voucher for Trempealeau County (UWEC 1455, presumably the specimen listed as "UWEC Unnumbered" in Casper [1993b]) revealed that it is currently lost (Joshua Kapfer, UW–Whitewater, personal communication). The voucher specimen for Dane County (UWZM 17584) submitted in 1962 by "R. K. Nelson" includes the note "died at birth in captivity," and it is unclear if Dane County is the location of collection or captivity. No other specimens from Trempealeau or Dane County are known. Casper (1993b) submitted an observational report by "J. Ramsey" in Trempealeau County, while R. Hay (personal observation) and R. S. King (personal observation) have also observed them in this county. Several counties with records supported by historic vouchers may no longer represent extant populations. For example, the specimens taken from Racine County were collected in the 1800s by P. Hoy (USNM 524–525), and no recent substantiated reports from that county are known.

Localities of possible occurrence: A review of various sources provides insight into the historical range of this species in Wisconsin. For example, Schorger (1967–68) compiled anecdotal historical reports of Massasaugas from 26 Wisconsin counties, which did not include Portage County, where they have since been documented via museum specimen (CM 70492; Casper 2015). Schorger's review also revealed some surprising locations for purportedly occupied sites, which included several that are now considered major urban centers and unlikely to have extant populations.

For example, although museum specimens from Milwaukee County are not known, the area of the current city of Milwaukee was once apparently home to a substantial Massasauga population. Olin (1930) described a Massasauga site in Milwaukee as "the marsh where Mason Street now strikes the river." This wetland was likely destroyed long ago, as the area is now the site of residential and commercial buildings. Historic accounts also indicate that the Madison area had at least two Massasauga populations: one near the outlet of Lake Monona (Schorger 1967–68) and the other on the marshes that once covered most of the area between the Yahara River and the state capitol building (Vogt 1981). Currently, both locations are largely urban or residential, although a small wetland (Paunack Marsh) remains near the outlet of Lake Monona.

Schorger (1967–68) compiled further historical records from additional counties where this species apparently no longer exists, which include (but are not limited to) Chippewa, Dunn, Grant, Jefferson, and Vernon Counties. The WDNR has also compiled a handful of more recent observational reports of this species from counties without known voucher specimens, including Adams, Green, and Crawford Counties (WDNR 2016b). Casper (1993b) reported reliable observational records from 1991 in additional counties that currently lack voucher specimens, including Rock, Green, and Grant Counties. Casper (1996) also included an observation from Kenosha County in 1987 that additional surveys could not verify.

Habitat: Massasaugas generally inhabit a mosaic of open wetland habitat interspersed with pockets of drier uplands and floodplain forest (e.g., Keenlyne 1968, Marshall et al. 2006, Moore and Gillingham 2006, Harvey and Weatherhead 2006a; see descriptions in Amphibians and Reptiles in Wisconsin Plant Communities chapter). Knudsen (1954b) described Massasauga habitat in Wisconsin as "marshy areas of the southwestern one-quarter of the state." Vogt (1981) identified mesic and lowland areas along rivers, lakes, and marshes as characteristic habitats for this species. Other historical accounts provide a similar picture of the habitats once occupied in Wisconsin. For example, Hoy (1883) wrote: "This species inhabits marshy, grassy prairies where meadow mice live, on which they almost exclusively subsist." Higley (1889) wrote: "Not rare, in grassy marshes throughout the state." Pope (1926) located specimens at a farm near Portage and described the area as the "Massasauga swamp." However, based on a photograph in Pope (1926), this area could be referred to as an open, perhaps mesic, meadow.

These historic descriptions of Massasauga habitat are consistent with locations where it still occurs in Wisconsin. Research on five of the remaining Massasauga sites in the state describe the associated habitat as meadows in low-lying wet areas (such as sedge meadows) and adjacent wet or mesic prairies (McCumber and Hay 2003, King et al. 2004, Durbian et al. 2008). Suitable habitat is frequently described as open canopy, and individuals tracked using radio telemetry in Buffalo County were located in areas with 26.5%–44.9% mean tree canopy cover (McCumber and Hay 2002, 2003). Unfortunately, many of these meadows are now dominated by invasive Reed Canary Grass (*Phalaris arundinacea*) or have succeeded to habitats overrun with woody vegetation that may not be suitable for this species (see Conservation section).

McCumber and Hay (2002, 2003) recorded microhabitat associations for individuals from Buffalo County for two years ($n = 6$ and 10 in 2002 and 2003, respectively). Their findings indicate that Massasaugas are frequently found in leaf litter ($\bar{x}$ litter coverage centered on snake location = 37%–40.8%) and herbaceous vegetation ($\bar{x}$ coverage = 34%–44.9%), compared to rock ($\bar{x}$ = 9.3%–11%) and woody debris ($\bar{x}$ = 14%–17.9%) cover. Harvey and Weatherhead (2006a) reported that Massasaugas in Ontario were always within ca. 0.5 m (1.6 ft) of retreat sites and frequently found near shrubs. The authors suggested this might be due to a combination of the cover from predators and thermoregulatory opportunities provided by both. They also reported variation in microhabitat preference among gravid females, which selected sites that were rockier, with more open canopies than males and nongravid females.

Given the length of hibernation in Wisconsin, habitat used for overwintering is of equal importance to that used during the active season. Wetlands with relatively stable winter water levels are critical for Massasauga hibernation. Massasaugas must descend below the frost line to overwinter and typically hibernate underwater or at groundwater level. A variety of hibernacula are reportedly used, such as below-ground channels associated with tree and shrub roots (Conant 1951, Johnson 1995, King et al. 2004), small mammal and crayfish (Cambaridae) burrows (Maple and Orr 1968, Hallock 1991), and even anthropogenic structures (Ravesi et al. 2015). Several authors have mentioned that crayfish burrows near the water table line, in particular, are important hibernacula for this species in Wisconsin (Keenlyne 1968, Vogt 1981, McCumber and Hay 2003). Maple and Orr (1968) reported that Massasaugas in Ohio overwintered at depths of 30–90 cm (11.8–35.4 in). Research in Ontario found Massasaugas rarely used the same hibernacula in consecutive years but often overwintered within 100 m (328 ft) of a previously selected hibernaculum (Harvey and Weatherhead 2006b).

Without a relatively stable water supply within the hibernacula, snakes can die of exposure if the water level drops and frost penetrates into the soil. This threat could be a defining characteristic that limits the northern edge of their range and may partially explain the species' North American distribution. An examination of locations across the eastern range of this species illustrates the importance of wetlands with stable water levels. For example, all of Canada's extant populations are less than 50 km (31.1 mi) from the Great Lakes. While the Great Lakes water levels do fluctuate, this fluctuation is typically minimal during the course of one winter. The majority of extant populations in the US are found in Michigan (Szymanski 1998), where they often associate with wetlands known as fens (Moore and Gillingham 2006, Bailey et al. 2012). Fens are similar to mesic or wet meadows or marshes but are groundwater driven and therefore have a stable water supply that fluctuates little in the winter (Marshall et al. 2006).

REPRODUCTION AND DEVELOPMENT

There are reported differences in when Massasaugas breed. Vogt (1981) claimed Wisconsin populations mated in spring and late summer but provided no evidence to support this. Spring breeding in Ohio was speculated by Crawford (1936) and recorded in Iowa (Guthrie 1927) based on a single captive pair. Other reports, however, suggest Massasaugas breed during summer or autumn and that reproduction is biennial throughout much of their range, including Pennsylvania (Reinert 1981), New York (Johnson 1995, 2000), Illinois (Aldridge et al. 2008), and Michigan (Degregorio et al. 2011a). In Wisconsin, Keenlyne (1978) indicated that adults reproduced annually, but more recent studies support that females reproduce biennially (McCumber and Hay 2002, 2003). It is noteworthy that Johnson et al. (2016a) concluded that some females in New York reproduced on a triennial schedule, which could also occur in Wisconsin.

Adult males are longer and heavier than females (see Description section). Evolutionarily, this sexual size dimorphism is likely a result of intraspecific male-male competition for breeding opportunities with females (Shine 1978, 1994). Male-male combat (VanDeWalle 2005) and aggression (Shepard et al. 2003, E. McCumber, WDNR, personal observation) preceding copulation has been documented. After summer or autumn copulation, females store sperm in the oviduct over winter (ca. eight months) until the following spring, when fertilization occurs (Aldridge et al. 2008). In a range-wide analysis, Hileman et al. (2017) found the proportion of gravid females in Massasauga populations ranged from 23% to 82%. Communal gestation sites have been documented in Pennsylvania (Reinert and Kodrich 1982), while communal parturition sites have been observed in Wisconsin (R. King, personal observation) and Michigan (Eric Hileman, Northern Illinois University, personal observation). In Wisconsin, parturition occurs from early August to mid-September, and snakes are born live, with an average litter size of 11.1 ($n = 58$; Keenlyne 1978). Multiple paternity is widespread in snakes (Jellen and Aldridge 2010) and was recently identified in Massasaugas from Michigan, Illinois, and Pennsylvania (Stedman et al. 2016).

After parturition, females maintain close proximity to their young until completion of the neonates' first ecdysis cycle, which lasts ca. one to two weeks (Butler et al. 1995). During this time neonates may be particularly vulnerable to predators due to visual impairment and possible impairment of the heat-detecting pit (Greene et al. 2002, Durbian et al. 2008). This form of parental care or 'maternal attendance' is common among pit vipers (Butler et al. 1995) and is documented in Massasaugas from Michigan (Hileman et al. 2015a), Pennsylvania (Reinart and Kodrich 1982), and Wisconsin (R. Hay, personal observation, R. King, personal observation). During this time, neonates actively maintain mother-offspring associations using chemical cues and scent trailing. However, neonates are unable to distinguish between maternal scent and scent of unrelated conspecifics (Hileman et al. 2015a).

In Michigan, individual growth rates are most rapid during the first and second seasons of life and decrease with each passing year thereafter (King and Hileman 2013). Hileman et al. (2017) reported that among nine Massasauga sites (none from Wisconsin), neonate growth rates in their first year ranged from 2.2 to 8.5 cm/year (0.87 to 3.35 in/year). Age of male sexual maturity is largely unknown and likely varies geographically. Based on growth rates of individuals from the Chicago area, Pope (1944) suggested that both sexes reach sexual maturity at three years of age. Motile sperm was detected in one two-year-old male (SVL = 43.7 cm [17.2 in]), but male sexual maturity is probably more common at three years of age in Michigan (E. Hileman, unpublished data). In Wisconsin, females typically show follicular growth in their second season and reproduce for the first time in their third or fourth season (Keenlyne 1978). Although longevity records for free-ranging individuals are lacking, a captive male from the Staten Island Zoo reportedly lived for 20 years and 5 days (Snider and Bowler 1992).

ACTIVITY

Massasaugas in Wisconsin are primarily active from mid-April to mid-October, although variation can occur. Timing of spring emergence varies annually and geographically but is apparently triggered by increasing soil temperature (Smith 2009, King and Hileman 2013). Wisconsin Massasaugas typically emerge from hibernation when soil temperatures at a depth of 15 cm (6 in) reach 10.6°C (51°F; E. McCumber, personal observation). King (1999) observed emergence from hibernation in Wisconsin as early as 4 April, although individuals used for a repatriation experiment in Wisconsin emerged slightly later than this (King et al. 2004).

Both Western Massasaugas (Seigel 1986) and Desert Massasaugas (Hobert et al. 2004) are primarily diurnal except in summer, when they are primarily nocturnal. A similar pattern was documented for Massasaugas in Clinton County, Illinois (Mike Dreslik, Illinois Natural History Survey, unpublished report). Seigel (1986) reported that in northwestern Missouri, spring and autumn activity was greatest from noon to 4:00 p.m., while summer activity was slightly later (i.e., 4:00 to 8:00 p.m.). Ernst and Ernst (2011) indicated that activity increases after summer rain showers. The mean body temperature of ten Buffalo County individuals tracked with radio telemetry was 23.9°C (75°F; McCumber and Hay 2003). Peak active body temperatures for a population in Illinois were high but also differed by sex and gravidity: males (30°C–34°C [86°F–93.2°F]), gravid females (29°C–33°C [84.2°F–91.4°F]), and nongravid females (26°C–28°C [78.8°F–82.4°F]; Dreslik 2005).

Published estimates of home range size for Wisconsin Massasaugas show large variability. Mean home range size (based on minimum convex polygon estimates) was 2.4 ha (5.9 ac) for a Monroe County population ($n = 10$), 135.8 ha (335.6 ac) for a Juneau County population ($n = 4$; Durbian et al. 2008), and 8.9 ha (22 ac) for a Buffalo County population ($n = 10$; McCumber and Hay 2003). Range length (i.e., the distance between the most widely separated locations within an individual's home range) averaged 272.1 m (992.7 ft) and 1,378.6 m (4,523.0 ft) for populations in Monroe and Juneau Counties, respectively (Durbian et al. 2008). In comparison to other midwestern states, Wisconsin home ranges are larger than those reported from Illinois (Dreslik 2005, Dreslik et al. 2016a), Indiana (Marshall et al. 2006), and southern Michigan (Moore and Gillingham 2006) but smaller than those reported from northern Michigan (Degregorio et al. 2011a).

Massasauga movement patterns are apparently affected by the severity of habitat disturbance, particularly in areas between summer foraging grounds and winter hibernacula (Weatherhead and Prior 1992, Johnson 2000a, Durbian et al. 2008). These patterns also appear to vary by sex. Several studies found home range sizes and movements of males were greater than those of females (e.g., Dreslik 2005, Moore and Gillingham 2006, Durbian et al. 2008, Bieser 2008, Degregorio et al. 2011a). This disparity could be related to the mate-searching behavior of males and its influence on reproductive success. In fact, Jellen et al. (2007) demonstrated that travel distance and body mass of males from southern Illinois were positively correlated with mating success.

In Wisconsin, Massasaugas hibernate for ca. six months. They enter hibernacula by mid- to late October when mean surface water temperature exceeds mean air temperature (McCumber and Hay 2003, King et al. 2004). In autumn, individuals move between openly basking under warm conditions and remaining submerged when temperatures drop. As an example, on warm October days, King et al. (2004) noted the midday emergence of Massasaugas from the hibernacula they selected. These snakes quickly became tannin-stained (brown) from the subsurface water and continued this behavior until temperatures dropped enough to trigger hibernation. During October floods, radio-tracked Massasaugas in Wisconsin were observed thermoregulating while submerged at their hibernation sites. They remained coiled in areas of shallow water with high light penetration and surfaced occasionally to breathe (McCumber and Hay 2003). This behavior has also been noted in Canada (Kent Prior, Canada Ministry of Natural Resources, personal communication). The latest observation of active Massasaugas in Wisconsin was 1 November (King 1999). Maple and Orr (1968) reported that mean body temperature of overwintering snakes monitored in Ohio was 2.6°C (36.7°F).

PREY AND PREDATORS

The diet of Massasaugas has received extensive attention. Past reports indicate this species is an opportunistic, generalist feeder and is able to adapt to whatever appropriately sized prey items are available (reviewed by Ernst and Ernst 2011). Examples of prey include insects, crayfish, frogs, toads, birds, mammals, snakes, and lizards (Holycross and Mackessy 2002). They are also reported to occasionally consume carrion (reviewed by DeVault and Krochmal 2002). The Massasauga's venom is highly toxic to small mammals (Gibbs and Mackessy 2009), which consequently make up the majority of adult diet regardless of geographic location. Among small mammals, shrews (*Sorex* and *Blarina* spp.) appear to be an important food item (Hallock 1991, Shepard et al. 2004, Weatherhead et al. 2009). Furthermore, Keenlyne and Beer (1973a) collected and euthanized 323 Massasaugas from a population in Buffalo County in 1967 to analyze their stomach contents. They found that Meadow Voles (*Microtus pennsylvanicus*) constituted more than 85% of all prey items. All other prey species individually represented less than 5% of the total food items collected and included White-footed Mice (*Peromyscus leucopus*), Meadow Jumping Mice (*Zapus hudsonius*), a Masked Shrew (*Sorex cinereus*), gartersnakes (*Thamnophis* spp.), and one fledgling Red-winged Blackbird (*Agelaius phoeniceus*). Shepard et al. (2004) found that wild neonates consumed mostly Southern Short-tailed Shrews (*Blarina carolinensis*), but neonates in laboratory feeding trials preferred snakes, which indicates that both may be important prey types. Neonates in some regions will also wave their bright yellow tail to lure anuran prey, although there is apparently geographic variation in the likelihood that such behavior is employed (reviewed by Shepard et al. 2004). Massasaugas are generally ambush hunters that lie in wait for unsuspecting prey. After potential prey is detected through a combination of olfaction, thermal reception, and vibrations, it is struck and injected with venom (Bakken and Krochmal 2007, Friedel et al. 2008). Although the envenomated prey often leaves the vicinity, the snake follows via olfaction and consumes it.

Predation is an apparently important source of mortality in Massasauga populations (Szymanski 1998, King et al. 2004, Durbian et al. 2008, Baker et al. 2016). Numerous confirmed and suspected predators are reported from Wisconsin (McCumber and Hay 2002, King et al. 2004, Durbian et al. 2008), Michigan (Bailey et al. 2011), Illinois (Anton 1993, Baker et al. 2016), and Pennsylvania (Jellen and Kowalski 2007). In some cases, specific species are identified, such as American Mink (*Mustela vison*; McCumber and Hay 2002) and Coyotes (*Canis latrans*; Harvey et al. 2013). Yet reports of predators are often more general, including "owl" (Durbian et al. 2008), "bird" (Moore and Gillingham 2006), and "mammal"

(Moore and Gillingham 2006, Bailey et al. 2011). Although not Massasauga predators, humans are also a leading cause of mortality both directly and indirectly (e.g., Keenlyne 1968, Dreslik 2005; see Conservation section).

Massasaugas typically rely on their cryptic coloration to avoid threats (Robillard and Johnson 2015) and are generally considered mild-mannered (reviewed by Keenlyne 1968). Avoidance tactics often involve either crypsis or flight (Harvey and Weatherhead 2006a). Massasaugas can also warn potential threats of their presence by producing a rattling sound with their specialized tail scales. Yet Keenlyne (1968), who collected hundreds of Massasaugas in Buffalo County, noted that only two individuals he encountered rattled before they were visually detected. Similarly, <40% of Massasaugas monitored in Ontario rattled in response to human disturbance, but snakes with higher body temperatures were more likely to rattle (Prior and Weatherhead 1994). Massasaugas that feel especially threatened could bite and possibly envenomate perceived threats. Yet some studies suggest that this strategy is rarely employed. Prior and Weatherhead (1994) observed that Massasaugas never struck during 174 trials during which researchers approached snakes to within 0.5 m (1.6 ft). Due to their small size, Massasaugas administer a smaller volume of venom compared to larger rattlesnake species and confirmed incidences of Massasaugas envenomating people are rare (see Snakebite Envenomation in Wisconsin and the Northern Upper Mississippi River Valley chapter).

CONSERVATION

Status: Globally, the Massasauga (including the Eastern, Western, and Desert subspecies) is listed as a Least Concern species by the IUCN. However, this ranking could change if the subspecies are elevated to separate species (see Systematics section). In the US, the Eastern Massasauga was listed as Threatened under the Endangered Species Act in 2016 (*Federal Register* 31, no. 190). Wisconsin listed the Massasauga as a State Endangered species in 1975 and considers it a Species of Greatest Conservation Need (WDNR 2005b).

Populations: Rigorous estimates of population sizes from Wisconsin do not exist (Faust et al. 2011). Annual abundance estimates from 1999 to 2010 in Carlyle Lake, Illinois, varied across years ranging from 18 to 70 individuals, depending on the estimator used (Dreslik et al. 2016b). Population estimates for Beausoleil Island in southeastern Georgian Bay, Canada, ranged from 35 to 77 individuals from 1993 to 2007 (Jones et al. 2017). Johnson et al. (2016a) estimated that the number of gravid females in Cicero, New York, ranged from 9 to 46 individuals from 2006 to 2014. Using the year with the largest estimated number of gravid females and assuming a 1:1 ratio of gravid to nongravid females and a 1M:1F sex ratio, Johnson et al. (2016a) suggested a maximum population size of 164 individuals for their Cicero Swamp Wildlife Management Area study site.

Using conditional likelihood and full likelihood closed population models, Bradke et al. (2013, 2015) estimated adult density for a population in Barry County, Michigan. Density was calculated using the upper and lower 95% CIs of their abundance estimate divided by the total area surveyed (ca. 19.9 ha [49.2 ac]). This resulted in adult female density estimates of 1.4–3.5 snakes/ha (0.57–1.42 snakes/ac) in 2013 and adult male and female density estimates of 1.8–4.5 and 2.5–3.2 snakes/ha (0.73–1.82 and 1.01–1.30 snakes/ac), respectively, in 2015. As this method did not include a boundary strip (i.e., the area just outside of the study area that is periodically occupied), the density estimates may be positively biased. Based on number of captures and confidence interval width, the best estimate for this population was 101 adults (95% CI = 86–143; Bradke et al. 2015).

Local extinction risk has been evaluated multiple times through simulations and population viability analyses (Seigel and Sheil 1999, Brennan and Tischendorf 2004, Middleton and Chu 2004, Bissell 2006, Bailey 2010). Several of these reports found that populations may be particularly sensitive to juvenile mortality (Brennan and Tischendorf 2004, Baily 2010). Bailey et al. (2011) estimated an adult (8 males and 19 females) active season survival rate of 0.9472 (95% CI = 0.8518–1.0000) in Michigan. In the most complete Massasauga survival analysis to date, Jones et al. (2012) characterized range-wide geographic patterns of adult survival for 16 locations across the species' range. This study used known fate models for 499 telemetered snakes from 21 radio telemetry datasets. Although unable to detect differences in survival between sexes, Jones et al. (2012) provided evidence that annual adult survival ranged from 0.35 to 0.95 and increased along a southwest to northeast geographic axis. Adult annual survival estimates in Cicero, New York (females = 0.78, 95% CI = 0.67–0.86; Johnson et al. 2016a), and Beausoleil Island, Georgian Bay, Canada (males = 0.74, 95% CI = 0.67–0.80, females = 0.73, 95% CI = 0.64–0.80; Jones et al. 2017), based on capture-recapture methods fall within the range-wide annual survival estimates based on radio telemetry (Jones et al. 2012). Realized population growth rate (λ) estimates are only available for the Beausoleil Island population and suggested the population was stable from 1992 to 2008 ($\lambda = 1.02$; Jones et al. 2017).

Threats: The large-scale conversion of landscapes from natural to anthropogenic use throughout Wisconsin has almost certainly taken a toll on Massasaugas (see Past, Present, and Potential Future Landscapes of Wisconsin chapter). As such, many general concerns associated with the conservation of wetlands and adjacent prairies potentially apply to this species (see Introduction to Conservation and Management of Wisconsin's Amphibians and Reptiles chapter). The impacts of habitat alteration and degradation are more subtle than those of outright habitat destruction but nonetheless lead to sizable conservation concerns. For example, the effects of habitat degradation via encroachment of woody vegetation into otherwise open areas can be particularly important, given the Massasauga's association with open habitats. Seigel (1986) suggested that extensive habitat degradation during the 1960s at a site in northern Missouri had drastic effects on the resident Massasauga population, which did not "stabilize" until the late 1980s despite continued protection.

Past research further suggests indirect connections between habitat loss and declining Massasauga populations also exist. For example, Durbian et al. (2008) proposed that Massasauga movement patterns were influenced by habitat fragmentation in Juneau County. The authors indicated the large home range sizes they recorded were related to limited and widely dispersed patches of suitable habitat. They also found evidence that individual snakes traveling large distances could be more prone to mortality. The authors regularly located a radio-tracked male in what are often considered unsuitable habitats, such as "closed canopy forests, pastures, and residential areas," the latter being particularly hazardous due to a higher threat of human interactions. As habitat dwindles, populations could also become more concentrated in small remaining pockets of suitable habitat or be forced to use risky open areas (e.g., grassy road shoulders), which increases their vulnerability to mortality (e.g., predation and vehicle collisions) and collection (Szymanski 1998, Durbian et al. 2008, Shepard et al. 2008b, reviewed by Ernst and Ernst 2011). Durbian et al. (2008) provide an example of this from Monroe County, where gravid females were concentrated on the only available open-canopy upland habitat (a railroad corridor) and the remains of at least four Massasaugas (three neonates and one adult) were found in one owl pellet.

Several studies have noted that road mortality is an important anthropogenic factor affecting Massasauga populations (e.g., Seigel 1986, Shepard et al. 2008b). Baker et al. (2016) even determined that vehicle strikes were the most prominent source of mortality (32% of 155 dead snakes observed) for an Illinois population studied from 2000 to 2011. Along these lines, Durbian et al. (2008) found road-killed individuals at a Juneau County study site and suggested the use of open roadsides was caused by the dominance of closed-canopy forests at this location. On the other hand, Shepard et al. (2008a) found that roads were avoided by Massasaugas in Illinois, a phenomenon that could fragment and disrupt populations.

Natural and anthropogenic-induced changes in hydrology, such as water flow and depth, along rivers in Wisconsin likely have negative impacts on Massasaugas (but see the discussion of the potential benefits of dams that are implemented appropriately below). Many populations in Wisconsin were once associated with wetlands along rivers, such as the Black, Yellow, Chippewa, Fox, Yahara, and Mississippi (Vogt 1981). All of these rivers have been greatly altered by dams, and associated Massasauga populations have probably either been extirpated or experienced substantial declines since the mid-1900s. Such impacts are evidenced by anecdotal observations. For example, in Juneau County radio-tracked individuals were observed swimming during flood events to search for dry ground (R. King, personal observation). Although this did not result in direct mortality (King 1999), snakes are more susceptible to mortality when concentrated in small dry areas, as could be the result of high water. As another anecdotal example, a population in Buffalo County was subjected to unusually high water levels that began prior to emergence from hibernation in spring 2003 and lasted for eight to ten weeks. After the flood subsided, no Massasaugas were found at survey locations on-site where they had been previously observed for several years (E. McCumber and R. Hay, unpublished report). Water-level fluctuations during the winter could be particularly problematic. In theory, a dam closed during the winter will lead to a drop in downstream water levels, which facilitates the penetration of ice into newly exposed damp soils. If snakes overwintering at or below the water table are unable to retreat to below the frost line as water recedes, mortality could occur as a result of exposure to freezing conditions. Although the full effect of this is unknown, the impacts to Massasauga populations could be severe. As an example, winter drawdowns to replace water-control structures were conducted in 1989 at a Wisconsin wildlife management area where Massasaugas had, to that point, been well-documented. Yet post-drawdown surveys yielded no Massasaugas for ca. 27 years (R. Hay, unpublished data, R. King, personal observation, Rori Paloski, WDNR, personal communication).

Lack of genetic variation due to restricted migration between Wisconsin populations could be another cause for concern. Chiucchi and Gibbs (2010) investigated the genetic composition of largely separated populations from Ohio, Pennsylvania, and Ontario and (perhaps predictably) found high genetic similarity among individuals within each population. Yet their results indicate that recent habitat fragmentation was not the primary cause of low intrapopulation genetic variation. Rather, the authors believed that Massasaugas historically existed in smaller isolated populations with limited potential for dispersal. As such, they suggested that high genetic similarity among individuals within populations may naturally occur; therefore, this species has adapted and is resistant to long-term impacts of genetic isolation. Still, the possible effects of inbreeding and low genetic variability within Wisconsin populations are worthy of additional investigation.

The influences of climate change on Massasaugas should also be considered. Climate change could impact snake thermoregulatory opportunities and cause changes in available habitat. Using spatially explicit climate change vulnerability assessment models, Pomara et al. (2014) found that drought, flooding, and other extreme events have increased in frequency and severity over time and are contributing to long-term range contractions and population declines in Massasaugas. Moreover, they provided evidence that these climatic events are good predictors of historic extirpations (Pomara et al. 2014). King and Niiro (2013) modeled the future effects of climate change on this species in the Midwest. Their results indicate that by the year 2080, if climate change continues on its current trajectory, suitable conditions for this species will have shifted away from the locations in Wisconsin where they currently persist. Further research on the potential impacts of climate change and mechanisms to offset them would be useful for proactive conservation of Massasaugas.

There is little doubt that direct human persecution has always been a prominent threat to this species in Wisconsin. Historic accounts suggest that since the mid-1800s, Wisconsin residents have killed thousands of Massasaugas, which has almost certainly contributed to its current rare status in the state (see Natural History Box: The History of Massasauga Persecution and Conservation in Wisconsin). For example, reports from the 1800s indicate that both Milwaukee and Columbia Counties had sizeable Massasauga populations, but due to heavy human persecution these populations are either extirpated or so small that they are no longer viable. Although this species is protected by state and federal law so that direct killing is less common than in the past, even the subtle impacts of human presence are an important consideration. Middleton and Chu (2004) reported that populations could be sensitive to even

NATURAL HISTORY BOX: The History of Massasauga Persecution and Conservation in Wisconsin

Early in Wisconsin's history and even prior to achieving statehood, Massasaugas were apparently plentiful in many locations. Although exact numbers are not known, Schorger (1967–68) reviewed numerous Massasauga reports from the 1800s and early 1900s, which conjure an image of a historically abundant species. For example, an important historical location in Wisconsin was the canoe portage between the Fox and Wisconsin Rivers (present-day city of Portage, Columbia County). This famous portage route had been used by French, British, and American traders for hundreds of years and by Native Americans for millennia. In 1826, on his way to sign the Treaty of Fond du Lac, Thomas McKenney walked the short portage between these rivers and wrote of Massasaugas, "This whole country is full of them; and so constant is the noise of their rattles . . . that the ear is kept half the time deceived by what seems to be the ticking of watches, in a watch-makers window" (McKenney 1827). The anxiety of potentially stepping on a rattlesnake must have made carrying supplies and canoes across this three-mile portage even more difficult! Unfortunately, many past Massasauga tales not only indicate abundance but also reveal the heavy persecution this species endured from early Wisconsin residents. In 1849 hundreds of Massasaugas were reportedly killed per day in the Portage area (Schorger 1967–68). Furthermore, Olin (1930) wrote of this species in Milwaukee during the mid-1830s: "The first day we mowed we killed a quantity of rattlesnakes. I will not say a thousand for fear someone will think it a snake story." These early attitudes and practices, which did not abate for many decades, almost certainly contributed to this species' substantial decline and current rarity in the state.

In the mid-1800s, the Daniel Muir family moved from Scotland to a farm on the Fox River not far from the previously mentioned portage. It was on this farm that Daniel's son, John Muir, developed a connection with the natural world (Muir 1913). This connection later inspired him to advocate for the establishment of the National Park System and Sierra Club and challenge people to reconnect with wild places and the creatures found in them. Muir used rattlesnakes to question contemporary utilitarian views of nature. "Nevertheless, again and again, in season and out of season, the question comes up, 'What are rattlesnakes good for?' As if nothing that does not obviously make for the benefit of man had any right to exist; as if our ways were God's ways. Long ago, an Indian to whom a French traveler put this old question replied that their tails were good for toothache, and their heads for fever. Anyhow, they are all, head and tail, good for themselves, and we need not begrudge them their share of life" (Muir 1901).

Unfortunately, Muir's challenge to utilitarian views of nature made no difference for Massasaugas in his lifetime. In the 1900s, perhaps not too long after the time of Muir's writings, Wisconsin established a bounty for rattlesnakes. The bounty system lasted approximately 70 years and resulted in the mortality of innumerable snakes (see Natural History Box: Historical Anecdotes Describing Human Interactions with Reptiles in Early Wisconsin in the Introduction to Conservation and Management of Wisconsin's Amphibians and Reptiles chapter). The true impact of the bounty on Massasauga populations is unknown for various reasons, which includes poor record-keeping. To make matters more difficult, while some counties maintained detailed rattlesnake bounty records, these generally did not distinguish between bounties paid for Massasaugas or Timber Rattlesnakes. However, we can assume that the bounty's effect on Massasaugas was substantial from records in counties where it was likely the only rattlesnake species present. Numbers from a small area in Juneau County, where Timber Rattlesnakes can effectively be ruled out, indicate that more than 4,000 rattlesnakes were killed in 20 years (R. Theil, WDNR, unpublished data). One can only imagine the sizeable negative influence this level of mortality had on populations of a species with relatively low reproductive output and slow development (see Reproduction and Development section).

Although early research on Massasauga populations in Wisconsin was limited (Keenlyne 1968), the likely impacts of human persecution had already become apparent by the 1920s. At this time, Pope (1926) described the substantial effort required to find these snakes in areas where they were previously common. The prevalent view of Massasaugas as a nuisance species lasted well into the latter half of the 1900s. One need only review Keenlyne's (1968) thesis to find support for this contention. He conducted research on a population from Buffalo County, for which he killed 323 wild individuals during a single active season and 207 neonates born to wild females in captivity. His primary justification for conducting the work was that "more ecological information will be necessary on the life history of the Massasauga if for some reason it is found necessary to eradicate it as a menace to human life or to manage it as a sport animal" (Keenlyne 1968).

Attitudes apparently changed dramatically in the 1970s, when the rattlesnake bounty was discontinued in Wisconsin. The bounty's termination probably reduced the intensity of direct killing by Wisconsin citizens, given that rattlesnake hunters could no longer profit by actively harvesting Massasaugas. Yet Massasauga persecution certainly continued. Many landowners and outdoor recreationists surely killed them during incidental encounters, whether the bounty was active or not. Their preferred habitat was also under heavy pressure from Wisconsin's expanding human population, which required conversion of natural landscapes to suit anthropogenic needs (i.e., agricultural and urban/suburban development). Active conservation initiatives and research focused on the Massasauga in the state would also remain uncommon for years after the bounty ended. Although Richard C. Vogt began his master's degree at UW–Madison in the early 1970s with the intention of studying this species, the high human-related mortality he observed, which he knew would influence sample size, caused him to change his mind (see Some Prominent Figures in Wisconsin Herpetology chapter). In 1975 the WDNR listed Massasaugas as an endangered species in Wisconsin, which afforded it legal protection, although as Vogt (1981) noted, this action may have come "too late." Throughout the 1980s and early 1990s, greater effort was expended on conservation of this species, which was primarily channeled into field surveys to better delineate its current range in Wisconsin. During that time a clear picture came into focus; the species was distressingly rare in former locations once considered strongholds, and it had completely disappeared from several others (Szymanski 1998). It appeared this snake was currently affected by a general lack of preferred early successional habitat (i.e., mesic prairies and other open-canopy habitat within and adjacent to wetlands; see Habitat subsection). Research intensified in the latter half of the 1990s, which included radio telemetry investigations of habitat selection, movement, mortality, and the effectiveness of translocation. Unfortunately, the Massasauga's rarity and the small, isolated nature of its populations in Wisconsin hampered the ability of these projects to achieve large sample sizes, a problem that continues to limit the potential for future research endeavors. Generally, despite current efforts, the future of this species in Wisconsin remains uncertain.

Once so abundant to be considered a pest in Wisconsin, the Massasauga has dwindled to a point that conservationists must make difficult decisions regarding preservation of the species. Despite a long conservation history, limited data exist to construct sound management strategies for Massasaugas in the state, and few options for effective conservation are apparent. At this current point, perhaps more than any other, this species' fate in the state could be the best measure of Wisconsin's ability to conserve a nongame species. As eminent Wisconsin conservationist Aldo Leopold wrote, "The last word in ignorance is the man who says of an animal or plant, 'What good is it?' If the land mechanism as a whole is good, then every part is good, whether we understand it or not. If the biota, in the course of aeons, has built something we like but do not understand, then who but a fool would discard seemingly useless parts? To keep every cog and wheel is the first precaution of intelligent tinkering" (Leopold 1972).

Richard S. King, Robert W. Hay, Turtles for Tomorrow, and Joshua M. Kapfer, UW–Whitewater

small incidental levels of mortality caused by vehicular strikes along roads or direct killing by humans. Prior and Weatherhead (1994) reported that Massasaugas also do not habituate to repeated disturbance by humans. In fact, past research indicates this species will alter its movement patterns at sites where disturbance from humans is more frequent (e.g., outdoor recreation at nature preserves or parks). Parent and Weatherhead (2000) reported that Massasaugas in areas where human disturbance was common moved less often and for shorter distances. However, they cautioned that accurately assessing how this disturbance ultimately influences Massasauga life history is complicated and difficult to ascertain.

Massasauga populations are susceptible to diseases such as the keratinophilic fungus *Ophidiomyces ophiodiicola*, the cause of snake fungal disease (SFD; Allender et al. 2011, Lorch et al. 2015, 2016; see Introduction to Conservation and Management of Wisconsin's Amphibians and Reptiles chapter). Massasaugas collected in Wisconsin and temporarily used for a repatriation experiment (King et al. 2004) were transferred to the National Mississippi River Museum and Aquarium, where they were later diagnosed with this fungal disease (L. Jackson, National Mississippi River Museum & Aquarium, personal communication). The disease was also detected in Illinois (Allender et al. 2011, 2016a) and Michigan (Tetzlaff et al. 2015, Allender et al. 2016b, Tetzlaff et al. 2017) populations. Lorch et al. (2016) did not detect SFD in Massasaugas from various Wisconsin locations. Yet, later work found SFD was present in Massasaugas biopsied from Buffalo County (Richard Staffen, WDNR, personal communication). It is noteworthy that Jaeger et al. (2014) identified four putative functional major histocompatibility complex (MHC) loci in Massasaugas. As MHC is important in immune response, future research in this area may provide important insights into this pathogen-host interaction. Continued monitoring of Wisconsin populations for outbreaks of this fungal infection should be considered until more transmission and population-level effects data are gathered.

Management: Because Massasaugas are sensitive to population-level perturbations, continued legal protection that guards against rampant persecution will be critical to promote their persistence in Wisconsin. Additionally, Faust et al. (2011) conducted an expert opinion-based range-wide population viability analysis that suggested that most populations are imperiled and would benefit from active management. General habitat management considerations for herpetofauna in wetlands and mesic or wet prairies apply to Massasaugas, as do strategies to offset road mortality (see Introduction to Conservation and Management of Wisconsin's Amphibians and Reptiles chapter). Although specific recommendations

NATURAL HISTORY BOX: Repatriation and Population Augmentation as Possible Tools for Massasauga Conservation

Massasaugas are currently protected in Wisconsin by the State Endangered and Threatened Species Law (Statute 29.604) and were recently listed as federally Threatened under the Endangered Species Act (see Conservation section). The new protection granted may require the implementation of additional conservation tactics in Wisconsin. For example, as part of a recovery plan it could be necessary to attempt strategies that involve repatriation or augmentation of existing populations. This could even require that remaining populations be brought into captivity to preserve the species and potentially protect unique genes present in populations across its range. As was the case with other federally protected species, including Red Wolves (*Canis rufus*; USFWS 1989) and California Condors (*Gymnogyps californianus*; USFWS 1996), at some point, removing all remaining wild Massasaugas from Wisconsin might represent the best option to achieve safeguards against extinction. In this hypothetical scenario, wild individuals would be placed into facilities and used for captive breeding, with the goal of bolstering numbers for future establishment or augmentation of populations through repatriation or reintroduction. This very difficult conservation decision is typically made only if three criteria are met: (1) if nothing is done all remaining wild individuals will be lost anyway; (2) placing individuals in captivity will preserve their genetics; and (3) captive-bred individuals can effectively be used to restore populations once conditions are met that allow reintroduction.

There are currently no Massasaugas from Wisconsin in captive populations. However, zoos accredited through the Association of Zoos and Aquariums (AZA) have already enacted captive breeding strategies to promote a stable demographic structure (i.e., stable ratios of males to females and juveniles to adults), and strong genetic diversity in zoo populations to ensure captive breeding can continue in perpetuity. Genetic analyses indicate that 17 unique haplotypes exist among captive Massasaugas (Ray et al. 2013). A consortium of facilities within the AZA called the Eastern Massasauga Rattlesnake Species Survival Plan (EMR SSP; Lentini and Earnhardt 2013) are coordinating the captive breeding program, as well as ongoing in-situ research in southwestern Michigan.

The use of wild-born or captive-born individuals to repatriate formerly occupied or restored habitat remains a controversial issue (Burke 1991, Reinert 1991, Germano and Bishop 2009). Part of the reason reptile reintroductions or repatriations remain contentious is that they are widely viewed as ineffective (Dodd and Seigel 1991), are poorly studied (King and Stanford 2006), and often lack enough released individuals to be successful (Germano and Bishop 2009). King et al. (2004) conducted a Massasauga repatriation experiment in Juneau County in 1999 and 2000. This experiment involved headstarting neonates in captivity for one or two years and briefly releasing them at a location where Massasaugas no longer occurred in the state. This project made no attempt to establish a population but instead was focused on assessing if repatriated Massasaugas could find food, avoid predators, and survive a single hibernation period. King et al. (2004) radio-tracked repatriated individuals and found they had smaller home ranges than wild (never captive) Massasaugas (Durbian et al. 2008), and while the mortality rate among one-year-old individuals was very high, most two-year-old snakes survived during the observation period (including hibernation).

Studying repatriated, headstarted Massasaugas in Michigan, Bieser (2008) found very similar results to King et al. (2004) in regard to lower movement rates and high mortality rates of repatriated snakes compared to resident wild individuals. Bieser (2008) also found that older repatriated Massasaugas fared much better than younger individuals. The results of King et al. (2004) and Bieser (2008) support repatriation to augment populations as a potentially useful conservation tool. However, establishing new populations would likely require sustained efforts over many years, including numerous translocations and intense monitoring. Unfortunately, such projects necessitate substantial resources and often come with an unclear likelihood of success. Yet debate regarding repatriation as a viable conservation tool in Wisconsin matters little at this point because the factors that led to the extirpation or decline of the state's populations have not been reversed. Therefore, without additional habitat restoration and management, augmentation of existing populations or attempts to repatriate formerly occupied locations are likely not prudent conservation measures at this time.

Richard S. King, Robert W. Hay, Turtles for Tomorrow, Billie C. Harrison, Milwaukee Public Museum, and Craig S. Berg, Milwaukee County Zoo (retired)

for habitat management strategies exist, these are often based on limited research or anecdotal observations. Therefore, additional rigorous work is warranted to determine the best management practices for Massasaugas, particularly as it pertains to habitat. It is noteworthy that Durbian et al. (2008) reviewed past studies and warned researchers against the use of populations at small study sites to model space and habitat requirements for management. Yet strategies that create and maintain open-canopy habitat by removal of woody vegetation are likely beneficial to Massasaugas (see suggestions by Seigel and Pilgrim 2002, Dreslik 2005, Moore and Gillingham 2006).

Radio telemetry research on Wisconsin populations in Juneau and Buffalo Counties (McCumber and Hay 2002, 2003, King et al. 2004, Durbian et al. 2008) and capture-recapture research on a population in La Crosse and Trempealeau Counties (R. Hay, unpublished report) illustrate the importance of habitat management that appropriately juxtaposes suitable hibernacula and adjacent open-canopy habitat. At these locations, suitable open-canopy habitat is often separated from hibernacula by unsuitable or degraded habitat. Individuals are, therefore, forced to emerge from hibernation and travel through unsuitable habitat in search of open-canopy areas to thermoregulate (King 1999, Durbian et al. 2008). A hypothetically beneficial scenario includes a landscape with open-canopy summer habitat and overwintering locations that are connected by a contiguous swath of suitable open habitat, which can be achieved through active management. Using a robust Missouri population as a model, Durbian et al. (2008) suggested that maintenance of easy travel corridors to hibernacula (less than 400 m [1,312 ft] long) was desirable. They further determined that if habitat restoration is the goal, then a minimum of 100 ha (247 ac) of suitable habitat should be the target for natural resource managers who hope to sustain Massasauga populations.

Despite this, research on how various management strategies affect Massasaugas is limited. Johnson and Leopold (1998) found some evidence that Massasaugas in New York preferred areas where woody vegetation was removed, indicating that this management strategy could be effective (although a small sample size makes their results somewhat inconclusive). Johnson et al. (2016b) found that gravid females in New York preferred gestation sites where brushy vegetation was mechanically shortened. They found that basking was less frequent on shrubs that approached 0.5 m (1.6 ft) in height and recommended they be kept at ≤0.25 m (0.8 ft) tall.

Poorly timed application of prescribed fire or mowing operations to manage vegetation results in Massasauga mortality (Seigel 1986, Durbian 2006, Cross et al. 2015). Several published studies recommend that land managers consider restriction of habitat

NATURAL HISTORY BOX: Massasauga Conservation in Wisconsin through Public Outreach

Public outreach is an important tactic that can be employed to promote conservation initiatives for Massasaugas. It is more difficult to endear this venomous snake species to the average member of the public than is the case for charismatic endothermic organisms. Therefore, outreach initiatives must occur across a broad array of professional institutions and take many forms to be impactful. The Milwaukee County Zoo conducts outreach efforts designed to expand public understanding and appreciation of this very rare snake in an effort to promote its conservation value to Wisconsin's populace. These outreach endeavors include training docents to interact with zoo guests near snake exhibits and educate them, as well as encourage a respectful and positive attitude toward the species. Zoo staff members even don a Massasauga mascot costume during special events to engage the public of all ages and showcase the importance of these animals in the ecosystem (Figure 304).

Through the Partners for Fish and Wildlife Program, US Fish and Wildlife Service staff contacted private landowners with Massasauga populations to offer technical assistance and financial support for habitat management/restoration that would benefit this species (Pultz 2001). Wisconsin (via the WDNR) was one of only two states to develop a Candidate Conservations Agreement with the USFWS for Massasaugas. The goal of this agreement is to manage and restore sufficient habitat to maximize the potential for maintenance and protection of a population in western Wisconsin. With these conservation measures in place, this particular Massasauga population represents Wisconsin's best chance to maintain this rare snake. Research conducted by natural resource regulators at the state and federal levels, as well as by academics, has helped shape these strategies.

Figure 304. Milwaukee County Zoo staff interact with visitors to promote positive attitudes toward Massasaugas (Milwaukee County; photo by C. Berg).

Despite all of the outreach and conservation efforts noted in this account, Massasaugas have all but disappeared from Wisconsin. Only seven extant populations may remain within the state. Among them, six will likely disappear, if they are not already extirpated. The remaining population is not secure despite measures in place to protect it. This population, Wisconsin's most robust, appears to have a 15% chance of disappearing in 25 years (Faust et al. 2011).

Richard S. King, Robert W. Hay, Turtles for Tomorrow, Billie C. Harrison, Milwaukee Public Museum, and Craig S. Berg, Milwaukee County Zoo (retired)

management with prescribed fire or heavy equipment to periods when snakes hibernate (Johnson et al. 2000, Baker et al. 2016). Based on observations in Missouri, Seigel (1986) indicated that Massasaugas were more susceptible to mortality from prescribed burns under warm environmental conditions, when they were likely to be active aboveground. Durbian (2006) also recorded high mortality rates in the summer after completion of mowing regimens to manage habitat in northern Missouri.

Brush piles created as the result of woody vegetation removal are frequently burned on-site by habitat managers as opposed to hauling debris to an ex-situ location. Massasaugas have been observed colonizing brush piles in Trempealeau County (R. King, personal observation). As a result, it is recommended brush piles be left unburned or burned immediately after construction (so that snakes do not have time to occupy them) or during the inactive season (Cross et al. 2015). The removal of brush piles prior to conducting prescribed burns is also recommended (Cross et al. 2015).

Use of dams to regulate water movement should maintain consistent water levels and flow rates in late autumn, winter, and early spring to avoid mortality of overwintering Massasaugas. It is also noteworthy that installation of dams or water control structures to restore wetland habitat might benefit this species, so long as stable water levels are maintained during hibernation. This was the case at Illinois's Carlyle Lake (Shepard et al. 2008b), where dam installation led to flooded fields that were ultimately used by Massasaugas for hibernation. Translocated snakes tracked as part of a repatriation experiment in Juneau County also used impounded areas to successfully hibernate (King et al. 2004). Given the heavy toll road mortality takes on some populations during specific times of year, it may also be wise to periodically close road sections that cause heavy mortality during autumn migration to hibernacula (Seigel 1986, Baker et al. 2016) or create underpass travel corridors for snakes below the roads.

Despite the conservation concerns and population declines noted previously, Wisconsin has been a leader in Massasauga conservation through implementation of various research projects and outreach endeavors over the last two decades. These include studies focused on the habitat use and ecology of the few remaining populations (e.g., King 1999, McCumber and Hay 2002, 2003, Durbian et al. 2008), as well as research on the viability of conservation strategies such as captive breeding and repatriation (Natural History Box: Repatriation and Population Augmentation as Possible Tools for Massasauga Conservation). Furthermore, a critical component for garnering support of effective conservation and management could be landowner and public outreach efforts, which are also ongoing in Wisconsin (Natural History Box: Massasauga Conservation in Wisconsin through Public Outreach).

Snakebite Envenomation in Wisconsin and the Northern Upper Mississippi River Valley

Daniel E. Keyler

The presence of venomous snakes in Wisconsin and the northern upper Mississippi River valley would not seem likely. The extreme seasonal climatic conditions, where winter temperatures dip well below −18°C (0°F) and summers are short, create a demanding environment for rattlesnake survival. Nonetheless, two species of venomous snakes, the Massasauga (*Sistrurus catenatus*) and Timber Rattlesnake (*Crotalus horridus*), are indigenous to the region, and snakebites to humans, although infrequent, do occur (Parrish 1956a, 1956b, Shorger 1967–68, Keyler 2005, 2008). Documentation of snakebites and their epidemiology, clinical manifestations, and medical treatment is minimal to virtually nonexistent in Wisconsin. Yet the thought or talk of a rattlesnake bite comes with a perspective unfavorable for the conservation of these highly unique and misunderstood creatures. The threat of snakebite is also greatly exaggerated, and when bites occur they are highly sensationalized. In recent decades, medical events associated with Timber Rattlesnake and Massasauga envenomations in Wisconsin reveal that the incidence of bites is quite low, and serious medical consequences are extremely rare. Although venomous snakebites should be considered serious medical emergencies, the vast majority of bites are not life-threatening. As such, with proper medical care and treatment, limbs are saved and life is not lost.

Frequently, snakebites to humans are unjustly attributed to rattlesnakes in the absence of the offending snake. Several snake species, such as Gophersnakes (*Pituophis catenifer*), Eastern Foxsnakes (*Pantherophis vulpinus*), and Eastern Milksnakes (*Lampropeltis triangulum*), are frequently mistaken for 'rattlesnakes.' They are blotched in pattern and will vibrate their tail when threatened, which makes a buzzing or rattling sound. As a result, they can resemble rattlesnakes in appearance and behavior, but there is no rattle to be found! So simply because the recipient's skin appears broken following the bite of a snake, one should not assume that the bite was from a rattlesnake. Furthermore, rattlesnakes bite humans out of defense, and such bites may not result in envenomation (the actual adverse result of venom delivered by injection with the bite). Several medical publications have reported

Table 18. Age, gender, and activity of Timber Rattlesnake (TR) and Massasauga (M) bite victims in Wisconsin (1982–2015).

Species	Month and Year	Age	Gender	Activity of Victim at Time of Bite
TR	July 1983	16	F	Walking by woods in short pants
TR	June 1986	29	F[a]	Stepping off back porch onto snake
TR	July 1987	32	F (pregnant)	Mowing grass
TR	June 1988	45	M	Searching for golf balls in rough
TR	June 1993	36	M	Hunting rattlesnakes
TR	June 1995	26[b]	M	Bare handling snake
TR	July 2002	51	M	Professionally processing snake
TR	May 2006	40	M	Reaching under deck of home
M	August 2007	60	F[a]	Professionally containing snake

[a] Dry bite—no envenomation.

[b] Alcohol involved.

that in 20%–25% of rattlesnake bite cases, no venom is injected (Bradley and Lichtenhan 1989, Russell 1990, Rudolph et al. 1995, Kitchens 1996). This percentage is similar to that of the author's experience with Wisconsin rattlesnake bites, in which 33% of the nine bitten patients suffered no effects of envenomation (two Timber Rattlesnake and one Massasauga bites; Table 18).

HISTORY

From a historical perspective, little documented information exists on venomous snakebites in the northern Mississippi River valley region. Minton (2001) mentioned the general rarity of snakebites in the Midwest. He attributed the rarity of Timber Rattlesnake bites to the snakes' secretive behavior and reluctance to bite unless they are quite aggravated. Prior to 1880, 70 cases of venomous snakebite in humans were documented in Wisconsin, with only 12 resulting in death (Shorger 1967–68). Whether death was truly the result of snakebite in many of these cases is questionable, as it may have been the treatment that killed the patient (Shorger 1967–68). Treatment for snakebite in the nineteenth century was frequently provided by an individual with no formal medical training who used antidotes that were anything but a true antidote. Whiskey (in great quantities), alum, gunpowder, salt and egg yolk, black mud and tobacco, all ingested or applied as a poultice to the bite, were just a few of the medical 'cures' of the time (Shorger 1967–68, Russell 1983).

Parrish (1965a) reported an estimated 15 rattlesnake bites annually from 1958 to 1959 in Wisconsin, with no fatalities. Across the Mississippi River, he further reported that Minnesota had the lowest incidence of venomous snakebites west of the river, with only four having occurred from 1958 to 1959 (Parrish 1965b). He implied that three of these were probably the result of Timber Rattlesnake envenomations (no deaths were recorded; Parrish 1965b). However, these numbers could be underestimates, because health professionals were not required to report cases of snakebite during this time. In stark contrast to the

1800s, a time when both species were likely more geographically widespread in the state, the current distribution of Wisconsin's rattlesnakes is highly restricted due to numerous anthropogenic influences (Vogt 1981, Oldfield and Keyler 1989, Keyler and Oldfield 1992; see also Timber Rattlesnake, *Crotalus horridus* species account). As such, the likelihood of encounters with Timber Rattlesnakes or Massasaugas today is low, which makes the chance of being bitten low and far less than historically.

Current rattlesnake bite documentation in all states is fraught with a lack of comprehensive reporting, which includes biased reporting, and other confounders. As such, there is a lack of truly accurate information on the actual incidence of rattlesnake bite in the US. A brief review of rattlesnake bites from US Poison Center records for 1983–2007 revealed a total of 19,151 rattlesnake bite cases, with only 24 fatalities reported (Walter et al. 2009). This calculates to ca. one fatality/year from rattlesnake bite for the entire country. A single death from rattlesnake envenomation in Wisconsin occurred during the latter half of the twentieth century (Keyler 2008). The author of this chapter has consulted on the medical management or been involved in the treatment of only nine Timber Rattlesnake and one Massasauga bite patients over the past 30 years in Wisconsin (see Table 18; Dan Keyler, University of Minnesota, unpublished data). During this same period in Minnesota, consultation was made on 19 Timber Rattlesnake bites and no cases that involved Massasauga envenomation. These findings suggest that rattlesnake bites in Wisconsin and Minnesota are not frequent.

To provide a proper perspective on venomous snakebite in Wisconsin, it should be noted that the chances of death from snakebite are nearly nonexistent compared to other wildlife-related fatalities. In 2013 alone, the Wisconsin Department of Transportation traffic accident statistics show that there were 18,338 car collisions involving deer, with 70 people suffering incapacitating injuries, 226 suffering general injuries, and 8 dying (Wisconsin Department of Transportation 2013). Deaths in the US from bees, hornets, and wasps accounted for 79 human deaths per year between 1999 and 2007 (Forrester et al. 2012). Collectively, these statistics support that venomous snakebite is not a major medical problem for residents of Wisconsin or the country.

VENOM

The venoms of the Timber Rattlesnake and Massasauga differ. However, in general, both venoms are comprised of a complex mixture of varying quantities and types of nonenzymatic proteins, enzymatic proteins, metalloproteins, glycoproteins, metals (magnesium, zinc, iron, cobalt, manganese, nickel), other inorganic substances (calcium, potassium, sodium), and amino acids (Russell 1983). Venom is delivered from the venom gland in the cheek region of the upper jaw through venom ducts to the hollow lumina of retractable fangs, which are located in the front of the upper jaw (Figure 305).

The primary purpose of rattlesnake venoms is for acquiring prey, not for inflicting ill effects on humans. Venom components work collectively in a staged progression to achieve a net effect on certain physiological systems to incapacitate prey. Massasauga and Timber Rattlesnake venoms affect primarily the hematological (i.e., blood) and vascular systems by causing the blood to clot abnormally and ineffectively. They further alter the permeability of vessel walls, which causes blood to leak from vessels and thereby reduces the circulating blood volume. This action leads to a drop in blood pressure and consequent shock. Timber

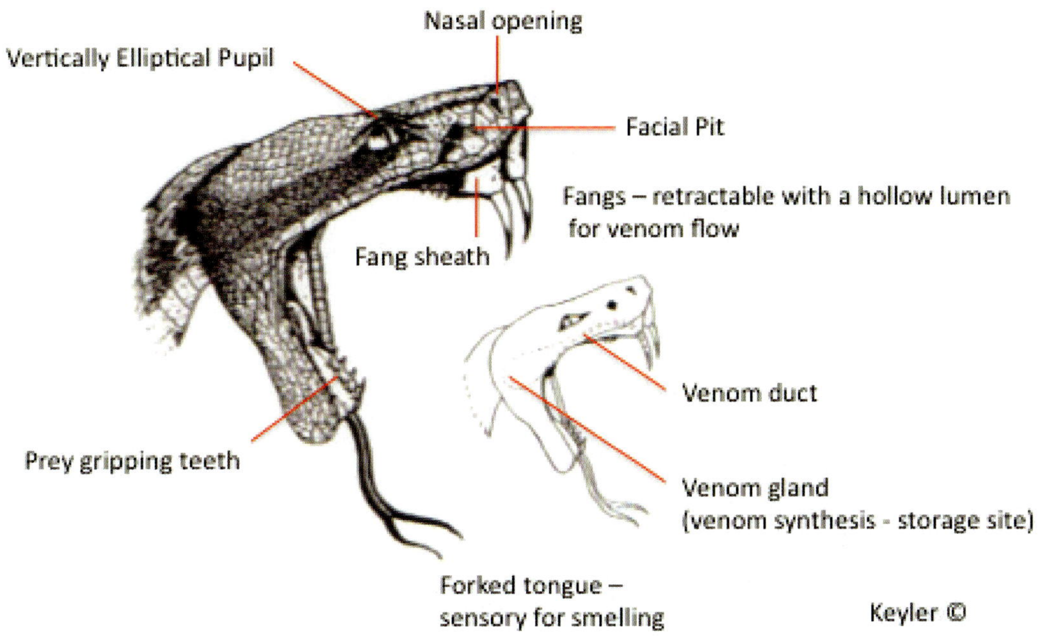

Figure 305. Head characteristics of venomous snakes found in Wisconsin.

Rattlesnake venom contains an active toxin known as *crotalocytin*, which causes platelets to clump together and results in their inability to function in the repair of leaky vessel walls (Schmaier and Coleman 1980). Massasauga venom, on the other hand, contains the proteolytic enzyme (i.e., an enzyme that breaks down protein) fibrinogenase, which prevents effective blood clot formation due to the destruction of the proteins fibrin and fibrinogen (Hankin et al. 1987). The point here is that the venoms of both rattlesnake species cause difficulty in blood coagulation and clotting but via different mechanisms due to their venom chemistry. In addition to the systemic effects on the blood system, significant local tissue damage can occur at the location of the bite from both species, and potential permanent tissue destruction, along with effects on nerves and muscle function, can result (Hankin et al. 1987, Christiansen and Fieselmann 1993, Sing et al. 1994). It is interesting that although the Massasauga is a much smaller snake than the Timber Rattlesnake, the lethality of its venom (based on animal lethality studies) is nearly ten times more potent than that of the Timber Rattlesnake, and Massasaugas produce a considerable quantity of venom, given their smaller size (Glenn and Straight 1982).

RATTLESNAKE BITE AND ENVENOMATION

Venomous snakebite is a medical emergency, and bites from both Wisconsin rattlesnakes have caused fatalities. However, these fatalities were typically the result of a major delay in time to treatment, extremely poor human judgment, or impaired function due to the consumption of alcohol (Keyler 2005, 2008, Forrester et al. 2012). Furthermore, both human medical and veterinary medical complications are commonly manifested as residual damage to tissue or limb function rather than death.

The degree of envenomation following a bite is variable, and several factors determine severity: the quantity of venom injected (the venom dose delivered with bites to humans is highly variable), the health and size of the victim, the venom route of delivery (intravenous > intramuscular > intradermal/superficial), time to medical treatment, allergic complications, and an individual patient's unique susceptibility to venom (Moss et al. 1997). Clinical symptoms associated with North American rattlesnake envenomation are usually evidenced by fang marks, edema or swelling, and significant pain (typically within 30–60 minutes), significant bruising (ecchymosis), and localized tissue effects, followed by systemic symptoms such as hematological effects (including low fibrinogen), prolonged clotting times, and/or low platelets (Kunkel et al. 1984, Russell 1991, Bush et al. 2000, Gold et al. 2002). In cases of Timber Rattlesnake envenomation, including Wisconsin cases, a consistently observed systemic complication has been reduction in the number of platelets (Weaver et al. 1991, Bond and Burkhart 1997, Rao et al. 1998, Keyler 2008). In cases of severe envenomation, cardiovascular and central nervous system effects may occur. Importantly, these manifestations and symptoms may occur at delayed and different times following the initial envenomation. A final consideration in the severity of envenomation relates to the site of the bite. Bites more distal to the body torso are usually less severe. As such, bites to fingers and toes are less life-threatening than a bite to the face, upper thigh, or shoulder. This is partially because blood volume circulated to distal sites is less than to the core of the body (Moss et al. 1997).

Animals such as livestock and pets also suffer the consequences of envenomation. Although their complications are similar to those experienced by humans, they can exhibit greater tolerance of the toxicological effects. Again, the same factors that affect the severity of envenomation in humans apply to other animals. One of the most severe complications with envenomation of domestic dogs, cats, cattle, and horses is airway obstruction. This commonly occurs because of the tendency for these animals to investigate rattlesnakes headfirst, which often results in a face-to-face encounter. Thus, they are typically bitten on the face, jaw, lips, or neck, which leads to extensive swelling, edema, and inflammation, which compromise the airway. As a result, breathing may be impossible, and mortality due to asphyxiation occurs if endotracheal intubation and respiratory support are not provided in a timely fashion (Keyler 2011).

FIRST AID

What is the first thing to do if you are bitten by a rattlesnake? Unfortunately, there is a plethora of poor advice and misinformation on this topic. However, in reality the actions to take are relatively simple. It is important to understand that you will not just drop dead after being bitten by a rattlesnake; time is on your side, despite what you might think. You should immediately ensure that you have distanced yourself from the snake by slowly backing away. Yet only do this after you have checked for other rattlesnakes in the vicinity that you could accidentally step on. Do not try killing or flipping the snake out of the way, which could put you in danger of additional bites. Furthermore, do not apply suction devices or a light constriction bandage. When these actions are taken, time is wasted, and typically they are incorrectly applied, which may lead to further complications.

The single most important action is to get to a medical facility, preferably a hospital, as soon as possible. Almost all victims of snakebite in the US are within an hour of a medical facility (Russell 1990), and it is extremely unusual for severe consequences of envenomation to set in within one hour. Maintain your presence of mind so that you can act rationally, and call or send for help. With cell phones in today's world, there is reception nearly everywhere, and a 911 call can be placed from most locations. If you are with a conscious bite victim and a vehicle is within a reasonable distance but ambulance transport is not readily available, the unbitten individual should drive the victim to a medical facility immediately. Bite recipients should minimize physical activity as much as possible and remove any restrictive clothing (e.g., belts and shoes) or watches and jewelry, depending on the location of the bite. Again, to reiterate, the primary response to having been bitten by a rattlesnake is to seek medical attention immediately.

MEDICAL TREATMENT: ANTIVENOM

The medical management of the patient will be the same, regardless of which Wisconsin rattlesnake species has bitten them, as treatment is based on the symptoms of the presenting patient rather than the species of snake. The keystone of treatment for venomous snakebite in patients who show signs and symptoms of envenoming is antivenom, not surgery. In the past, the use of antivenom was considered risky due to adverse effects that followed its administration, and surgery was frequently performed initially or alternatively (Glass 1991, Dart and McNally 2001). However, antivenoms today (both human and veterinary) are significantly improved biopharmaceutical products that have good track records for medical efficacy when used appropriately. Antivenom is produced by the hyperimmunization of horses or sheep through injecting small amounts of diluted venom (from different species of snakes to make a polyvalent or polyspecific antivenom) beneath the skin. After a period of time, the animal's immune system makes antibodies against the venom toxins. These antibodies are harvested from the plasma without harming the animal and are purified as a pharmaceutical preparation.

Antivenoms for human use in the US are CroFab (Crotalidae Polyvalent Immune Fab [Ovine]; BTG International, Inc., West Conshohocken, Pennsylvania) and Anavip (Crotalidae Immune $F(ab')_2$ [Equine]; Rare Disease Therapeutics, Inc., Franklin, Tennessee). Both are FDA-approved polyspecific antivenoms that are effective treatment for bites from all North American pit vipers. These products are unique because the immunoglobulin fragments that bind the venom components to neutralize them are one-third and two-thirds the molecular size of a complete or whole immunoglobulin (IgG), respectively, that was in the antivenom traditionally used in the US. The smaller antibody sizes reduce the risk of adverse effects such as allergic reactions and allow for efficient administration times in emergent situations.

Antivenom should only be administered in a hospital setting and cannot be safely carried or administered in the field. Veterinary antivenoms are available as three USDA-approved products: Venom Vet (MT Venom, LLC, Canoga Park, California), Antivenin (Crotalidae) Polyvalent (Boehringer Ingelheim Vetmedica, Inc., St. Joseph, Missouri), and Rattler Antivenin (Mg Biologics, Ames, Iowa). All are produced in horses. They are effective for treating North American pit viper envenomations and approved for use in dogs and horses. Venom Vet is a liquid preparation in which the antibody fragments are two-thirds the molecular size of a whole immunoglobulin, while the Antivenin (Crotalidae) Polyvalent is a full-sized immunoglobulin. Rattler Antivenin is a liquid plasma preparation of whole antibodies (IgG).

AVOIDING SNAKEBITE

The ability to avoid snakebite basically involves common sense and some simple measures that can be taken to reduce encounters with snakes. Incidence of rattlesnake bite is often greatest during peak season for human outdoor recreation, as this coincides with the active season of rattlesnakes (see Table 18). Think about watching for snakes when you are out in the wild just as you would be cognizant of traffic when you are in a big city. You would not walk across the street without looking for a traffic light or oncoming vehicles. So when you are in the outdoors, you should use that same conscientiousness. First, it is important to know if you will undertake an outdoor activity where venomous snakes are potentially present. In the summer, Timber Rattlesnakes may rest in shady locations along trail edges, woodland borders, and rocky areas. On the other hand, Massasaugas may hunt in lowland river bottoms and related riparian zones (see Massasauga, *Sistrurus catenatus* and Timber Rattlesnake, *Crotalus horridus* species accounts for further details regarding habitat use). Second, watch where you walk, and inspect areas where you intend to sit, place your feet, and place your hands.

If you live in or adjacent to rattlesnake habitat, try making your yard and immediate areas around your home less attractive to snakes. Make sure there are no cracks or holes where snakes can hide around the foundation of your home, sheds, or other outbuildings. Keep grain and bird seed stored so as to not attract rodents (an important snake prey item). Realize that if you have bird feeders in your yard, the seeds and grain falling to the ground create a feeding area for chipmunks and mice. Rattlesnakes in the area will learn to utilize these locations to find prey and will frequently rest directly below bird feeders (Dan Keyler, University of Minnesota, personal observation). Basically, you should be conscientious of the environment you are in or the environment you have created. Nuisance rattlesnakes that persist on one's property may be translocated to an area away from the home. However, the WDNR should be contacted for assistance in such circumstances.

CONCLUSION

The incidence of Massasauga and Timber Rattlesnake bites in Wisconsin and the upper Mississippi River valley are currently rare and will likely become increasingly uncommon, because remaining rattlesnake populations in this region are declining. As humans expand into rattlesnake environments, the snake's native home is converted to anthropogenic landscapes, leaving few wild places for the snake to remain. Thus, by the mid-twenty-first century the probability of human encounters with rattlesnakes and risk of consequential snakebite in the upper Mississippi River valley will likely decline, albeit not for positive reasons. As a result, rattlesnakes may withdraw to a faded but colorful bit of regional history as they disappear from the blufflands and lowland river bottoms that feed the mighty Mississippi River.

Testudines
Turtles

Turtles are members of the class Reptilia and order Testudines. All members of the order Testudines have a bony shell and lack teeth (see Meylan [2001] for additional monophyletic characters). The earliest fossil species containing characteristics representative of modern turtles existed ca. 260 mya (Bever et al. 2015), and fully shelled species existed at least 200 mya (Szczygielski and Sulej 2016). There are currently 349 recognized extant turtle species in the world, which are placed in 14 families (Rhodin et al. 2017). The US has the highest turtle species richness of any country in the world, with 62 species and 27 subspecies (Rhodin et al. 2017). Four families, eight genera, and eleven species occur in Wisconsin, including Chelydridae (*Chelydra,* one species), Emydidae (*Chrysemys, Emydoidea, Glyptemys, Graptemys, Terrapene,* seven species), Kinosternidae (*Sternotherus,* one species), and Trionychidae (*Apalone,* two species).

Figure 306. Blanding's Turtle (*Emydoidea blandingii*) (illustration by E. Wild).

Snapping Turtle
Chelydra serpentina (Linnaeus 1758)

Deahn M. Donner and Donald J. Brown

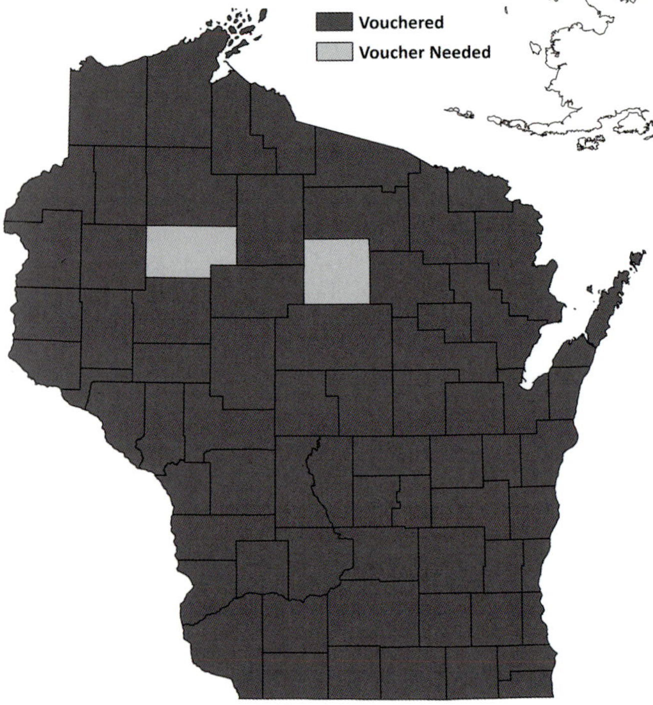

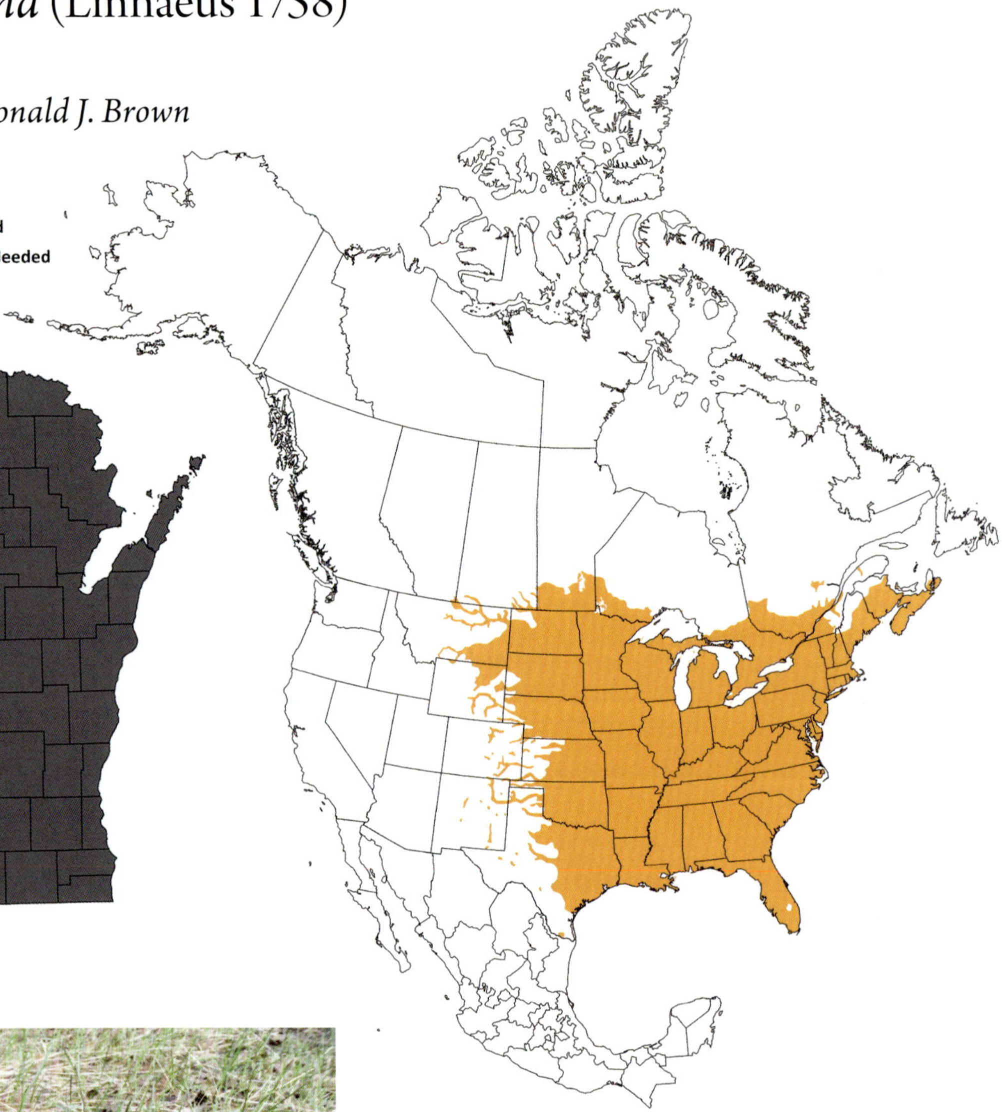

Snapping Turtle (Oneida County; photo by D. Brown).

AT A GLANCE

Snapping Turtles are the largest turtles found in Wisconsin. They can be identified by their large body, wide head (with hooked upper jaw), long thick legs and tail, and brown, tan, or olive-green coloration. Mature males are larger than females. Snapping Turtles are omnivorous and consume a wide variety of food, including carrion. They are ambush predators and scavenge from all sections of the water column. This species can be found in most semipermanent and permanent waterbodies throughout Wisconsin but prefers slow-moving or still waterbodies with high turbidity and soft bottoms. They rarely leave the water and only occasionally bask above the surface to regulate their body temperature. Although they are highly aquatic, female Snapping Turtles can travel long distances over land to nest in early to mid-June. Nests are usually laid in loose sandy or loamy soil and can include anthropogenic habitats such as roadsides, railways, and dams. Hatchlings emerge from the nests in ca. three months. Adults are cantankerous when encountered on land or captured and will not hesitate to 'snap' at a perceived threat. Snapping Turtles are heavily exploited for consumption, educational purposes, and international trade. In response, the species was given CITES Appendix III protection in 2016, allowing for improved regulation of international trade by requiring international exporters to have a permit. The Snapping Turtle is currently listed as a Least Concern species by the IUCN and considered a common but declining species by the WDNR.

SYSTEMATICS

There are currently three accepted New World species in the genus *Chelydra*, although only *Chelydra serpentina* occurs naturally in the US (according to Ernst 2008). No subspecies of *C. serpentina* are currently recognized (Crother 2017). Historically, two North American subspecies were recognized: *Chelydra serpentina serpentina* (Eastern Snapping Turtle), the subspecies whose range covered most of the midwestern and eastern US, and *C. s. osceola* (Florida Snapping Turtle), whose range was restricted to the southeastern US. However, recent genetic work provided evidence against the use of subspecies, indicating that *C. serpentina* represented a "single, virtually invariant lineage" (Shaffer et al. 2008). This species was previously referred to as the Common Snapping Turtle due to its wide distribution, but 'Common' was dropped to avoid misperceptions related to abundance (Crother 2012).

DESCRIPTION

Adults: The carapace generally appears smooth in larger individuals but possesses three lengthwise keels (visible as scute knobs) that are obvious in smaller individuals (MCL <20.0 cm [7.9 in]; Figure 307). The posterior portion of the carapace is wider than the anterior portion, and the back edge is strongly serrated (Figure 308). The carapace of both sexes is brown, tan, or olive green to black, with light narrow lines between each scute (reviewed by Ernst and Lovich

Above, top: Figure 307. Adult Snapping Turtle showing typical coloration, a low lengthwise keel, and a moss-covered back (Polk County; photo by D. Donner).

Above, bottom: Figure 308. Adult Snapping Turtle showing its large head, strong posterior carapace serration, and long tail with elongated scales (Polk County; photo by D. Donner).

2009). Mud and algae are often present on the carapace, which can obscure the true color (Figure 309).

The plastron is yellowish to tan and diamond-shaped (Figure 310). The head, neck, and limbs are rough in appearance but mostly patternless, with occasional whitish flecks and thin black bars on the jaws (reviewed by Ernst and Lovich 2009). The head and limbs grade in color from gray or black at extremities to yellow or tan near the shell. The head is large, with a hooked upper jaw and barbels located on the chin (see Figures 308 and 309). Smooth round tubercles are also present on the dorsal side of the neck and limbs. The legs are large and powerful, and all four feet have webbed toes with long, strong claws. The tail is long (equal to or longer than the MCL), with three longitudinal rows of elongated scales on the dorsal surface (Figure 308).

Above, top: Figure 309. Adult Snapping Turtle with algae on its carapace and displaying its chin barbels (Walworth County; photo by N. Rudolph).

Above, bottom: Figure 310. Diamond-shaped plastron of the Snapping Turtle (Portage County; photo by J. Kapfer).

Knudsen (1956c) reported that large individuals in Wisconsin typically range from 46 to 61 cm CL (18 to 24 in) but can be as large as 71 cm CL (28 in). From 129 captured males along the St. Croix River in Wisconsin and Minnesota, mean MCL was 30.4 cm (12.0 in, range = 20.2–43.1 cm [8.0–17.0 in]), and mean PL was 21.9 cm (8.6 in, range = 9.9–33.3 cm [3.9–13.1 in]; DonnerWright et al. 1999). From 54 captured females along the St. Croix River in Wisconsin and Minnesota, mean MCL was 27.6 cm (10.9 in, range = 20.2–34.5 cm [8.0–13.6 in]), and mean PL was 20.4 cm (8.0 in, range = 12.1–25.8 cm [4.8–10.2 in]; DonnerWright et al. 1999). From the same study area, mean weight of males was 3.0 kg (6.7 lb, range = 0.8–7.8 kg [1.8–17.1 lb]), and mean weight of females was 2.1 kg (4.7 lb, range = 0.5–3.8 kg [1.0–8.4 lb]; Deahn Donner, USFS, unpublished data from the DonnerWright et al. [1999] study).

Preadult stages: Eggs are white and spherical, with leathery pliable shells (Figure 311). Fresh eggs have

Figure 311. Snapping Turtle eggs (Chippewa County; photo by R. Sajdak).

Figure 312. Hatchling Snapping Turtles. (a) Example showing the serrated posterior carapace and long tail (Polk County; photo by D. Donner). (b) Newly hatched individuals with yolk scars on their plastron (Rock County; photo by J. Kapfer).

a pinkish appearance on one end and a white appearance on the other end (Ernst and Lovich 2009). Ernst and Lovich (2009) reported that Snapping Turtle eggs range from 2.3 to 3.3 cm in diameter (0.9 to 1.3 in, $\bar{x} = 2.8$ cm [1.1 in]) and weigh 4.7–17.5 g (0.2–0.6 oz, $\bar{x} = 11.0$ g [0.4 oz], $n = 2{,}004$). From a collection of eggs in Wisconsin, egg diameter ranged from 2.3 to 3.2 cm (0.9 to 1.3 in, $\bar{x} = 2.7$ cm [1.1 in], $n = 84$ clutches), and mean weight of eggs was 10.6 g (0.4 oz, $n = 3{,}926$; Yntema 1970). Hatchlings are about the size of a nickel, and yolk scars are apparent on newly hatched individuals (Figure 312). Hatchlings and juveniles are similar in color to adults, but the three keels on the carapace are more prominent in younger age classes. As it is with adults, sex of hatchlings and juveniles can be determined by their pre-cloaca tail length (Ernst and Lovich 2009).

Variation: No significant geographic variation among Wisconsin individuals is known, although some sexual dimorphism occurs. Adult males reach larger sizes than females (Mosimann and Bider 1960, Hammer 1969, Froese and Burghardt 1975). The tail of adult females is thinner at the base than males, and the cloaca is beneath or just under the carapace rim, whereas in males the cloaca is posterior to the edge of the carapace (Ernst and Lovich 2009).

Diagnosis: Adult and juvenile Snapping Turtles are characterized by a carapace with serrated posterior margins that is covered with keratinized scutes and a narrow plastron that leaves the legs exposed. The body size and tail length of Snapping Turtles make adults easily identifiable among other turtle species found in Wisconsin. From a distance, juveniles and small adults might be confused with Eastern Musk Turtles (*Sternotherus odoratus*) or Wood Turtles (*Glyptemys insculpta*) because of coloration. However, these species are easily distinguished from Snapping Turtles by their carapaces, negating the need for physical handling to identify them. The carapace of Eastern Musk Turtles is highly domed with smooth marginal scutes, and the carapace of Wood Turtles contains raised concentric growth rings, resulting in pyramid-shaped interior scutes, in contrast to the three lengthwise keels in Snapping Turtles (see Figure 307).

Snapping Turtle eggs are spherical, like those of the softshells (*Apalone* spp.) found in Wisconsin (see Preadult stages subsection). Although this similarity in shape makes it difficult to differentiate the eggs of these three species, their eggs can generally be distinguished from the more elliptical eggs of other turtles in Wiconsin.

DISTRIBUTION AND HABITAT

Global distribution: Snapping Turtles range from Texas to southern Canada. The species is distributed throughout the eastern two-thirds of the United States, with the western edge of its distribution in central Montana. In addition, at least four isolated regions are currently occupied in the US and southern Canada, including a large stretch of the Rio Grande in New Mexico (reviewed by Ernst and Lovich 2009).

Distribution in Wisconsin: This species has been reported from every county in Wisconsin, although associated vouchers are missing or suspect for several counties (see the Appendix). For example, Vogt (1981) reported this species in Lincoln and Rusk Counties, but the only known specimens from these counties that existed in local natural history collections were eggs (Lincoln: UWZM 20760, Rusk: UWZM 23952). Because it is difficult to independently confirm species identification from eggs, collection of additional vouchers from these counties is desirable.

The voucher for Outagamie County (MPM VZP450; Casper 1996) is one of several photographs of an artistically modified turtle carapace collected by "J. Povlich" and adorned with a painting of a Flathead Catfish by George S. Bachay. On the underside of the carapace are the handwritten notes "7-4-88", "38 lbs," and "New London," and museum records include additional information ("trib. of Wolf R." with associated township and range). Another photographic specimen reportedly from Outagamie County also exists (MPM VZP451; Casper 1996), which is similarly modified (but with North American River Otters [*Lontra canadensis*] painted on the carapace). These photographic specimens are problematic as vouchers for various reasons. They rely heavily on the assumed accuracy of anecdotal reports and the recollection of associated collection information by a third party. This is exacerbated by the fact that the photograph vouchers do not show a whole (or even nearly whole) specimen in situ at the point of capture. Despite this, Snapping Turtles surely occur in Outagamie County, and new vouchers should be sought.

Localities of possible occurrence: Snapping Turtles are reported from every county in Wisconsin, although several current voucher specimens are missing or questionable (see Distribution in Wisconsin subsection).

Habitat: Snapping Turtles are a highly aquatic species and occupy many types of semipermanent and permanent waterbodies (e.g., Bodie et al. 2000, Swihart et al. 2006, Anthonysamy et al. 2014a; see Amphibians and Reptiles in Wisconsin Plant Communities chapter). In Wisconsin, this species frequently associates with slow-moving or still waterbodies, particularly those with high turbidity and soft bottoms (Vogt 1981, DonnerWright et al. 1999). Paterson et al. (2012) found that at the population level, Common Snapping Turtles showed no apparent preferences for specific aquatic habitat types in Ontario. However, individual variation in habitats used was detected, with some turtles showing a strong preference for marshes and swamps over flowing waterbodies. Similarly, Tucker et al. (2008) found that Snapping Turtles in the upper Mississippi River system used all surveyed habitats but were most frequently captured in stagnant backwater areas (i.e., 174 of 243 total Snapping Turtles captured). Primary productivity of aquatic systems does not appear to influence Snapping Turtle habitat use patterns (Brown et al. 1994), but more productive systems can support larger populations (Galbraith et al. 1988; see Populations subsection).

In Illinois, Snapping Turtles used microhabitats that possessed a broad range of characteristics (Anthonysamy et al. 2014a). Yet Paisley et al. (2009) studied a population in the Mississippi River adjacent to Wisconsin and Minnesota and found that individuals selected specific microhabitats with dense live aquatic vegetation. Braun and Phelps (2016) further found that Snapping Turtles preferred shallow (<1 m [3.3 ft]) slow-moving water with soft or silty substrates, as is often found in river tributaries of the middle Mississippi River. In addition, there appears to be an ontogenetic component to aquatic habitat selection, with hatchlings and juveniles showing stronger preferences for shallow and warm microhabitats than adults (Congdon et al. 1992, Bury et al. 2000). A laboratory experiment also concluded that juveniles preferred covered areas over open water during periods of darkness (Froese 1978). Thermoregulatory behavior does not appear to influence habitat selection in Snapping Turtles to the degree of many other aquatic turtle species. Although they do not bask frequently, individuals will sometimes rest on shorelines, logs, and floating vegetation in the water (Figure 313; see Activity section). Nests are typically located in areas with open tree canopies and

Above, top: Figure 313. Steep sandy bank where Snapping Turtles were observed basking close to the water's edge in the St. Croix River (Polk County; photo by D. Donner).

Above, bottom: Figure 314. Female Snapping Turtle nesting on the side of a road (Dane County; photo by J. Kapfer).

loose sandy or loamy soil (Punzo 1975, Thompson et al. 2017b). In Wisconsin, Mathiak (1966) reported Snapping Turtle eggs in muskrat houses throughout Horicon Marsh (Dodge County), and females will also use beaver lodges as nest sites (Ernst and Lovich 2009). Disturbed habitats such as roadsides, railways, and dams are also often used for nesting (Figure 314).

Snapping Turtles hibernate underwater or buried deep in soggy muck (Lagler 1943, Brown and Brooks 1994). Selection of overwintering habitat is primarily influenced by water temperature and oxygen level, with individuals selecting waterbodies that are cold but that do not freeze completely to both reduce energy use and possibly allow oxygen uptake (Brown and Brooks 1994, Paterson et al. 2012, Strain et al. 2012; see Activity section). Brown and Brooks (1994) identified three primary types of hibernacula used by Snapping Turtles in Ontario: in streams beneath logs and overhanging banks, in lakes beneath logs (but not buried in sediments) 5 m (16 ft) from shore, or buried deep in mud or beneath vegetation in "marshy areas" that froze but were not covered in water. Paterson et al. (2012) further noted that individuals hibernated in marshes, ponds, lakes, and creeks in Ontario. Paisley et al. (2009) found that most individuals hibernated under dense vegetation mats in backwater sloughs associated with the Mississippi River in Wisconsin and Minnesota. Abandoned North American Beaver (*Castor canadensis*) dens, Muskrat (*Ondatra zibethicus*) lodges, and submerged tree root systems are also favored spots for Snapping Turtle hibernacula (Mathiak 1966, Paisley et al. 2009, Strain et al. 2012).

Brown and Brooks (1994) documented 18 interyear recaptures at hibernacula, and 13 of these individuals showed fidelity for specific overwintering locations. Individuals moved underwater during the wintering hibernation period, with 95% of observations separated by <10 m (33 ft, maximum observed movement = 13 m [43 ft]). Less is known about the overwintering habitat of young age classes. However, Ultsch et al. (2007) found that hatchlings in New York successfully overwintered in "spring seeps," while hatchlings in New Hampshire overwintered in a shallow "wetland stream."

REPRODUCTION AND DEVELOPMENT

Mating in Wisconsin is most frequent in spring and autumn right before and after hibernation but can occur throughout the active period (Vogt 1981). Mating behavior was reviewed by Ernst and Lovich (2009), who stated that either the male will mount the female after a brief chase or the male and female will court each other by facing one another and repeatedly moving their heads back and forth. Often, the male will mount the female near the water surface, and then the pair will slowly sink to the bottom. During copulation, the male will grip the female's shell with its claws using all four feet and use his chin to press the female's snout down so that she cannot extend her neck forward while also extending his tail beneath the female to facilitate copulation (Hamilton 1940a). The male will sometimes grip the female's neck with his jaws. The female usually remains passive but may struggle if the male bites her. Sperm can remain viable in the female for several years, and multiple paternity has been detected in clutches of Snapping Turtle eggs (Galbraith et al. 1993). Reproductive physiology of Snapping Turtles has been the subject of much research in Wisconsin (e.g., Mahmoud et al. 1986, Mesner et al. 1993, Mahmoud et al. 2004).

In Wisconsin, females begin nesting migrations in May, with early to mid-June being the peak of the nesting season (see Natural History Box: Use of Camera Traps to Monitor Nesting Behaviors of Snapping Turtles in Wisconsin). Obbard and Brooks (1981) also reported that mid-June was the peak nesting period for Snapping Turtles in Ontario. Iverson et al. (1997) synthesized the Snapping Turtle nesting season literature and found that northern populations began nesting later than southern populations and that duration of the nesting season was variable among studies (ca. two to five weeks). Congdon et al. (1987) studied nesting activity of a Snapping Turtle population in Michigan and found substantial annual variation in both initiation of nesting activity (22 May to 12 June) and duration of nesting season (13–31 days). Congdon et al. (1987) further noted that cooler springs resulted in later initiation of nesting activity.

Female migration distances to nesting locations can be substantial. Maximum observed total migration distances (including in water and over land) range from 1.6 km (1.0 mi) in Michigan (Congdon et al. 1987) to 16 km (9.9 mi) in Ontario (Obbard and Brooks 1980). Maximum observed distances traveled over land during nesting migrations range from 183 m (600 ft) in Michigan (Congdon et al. 1987) to 500 m (1,640 ft) in Ontario (Obbard and Brooks 1980). Variability was observed in the likelihood that females will return to the same nesting sites in Michigan, where some nested within 5 m (16 ft) of previous nest locations and others nested up to 1,000 m (3,281 ft) from previous locations (Congdon et al. 1987). Loncke and Obbard (1977) reported that 34.3%–73.7% of tagged nesting females returned to the same nesting locations in subsequent years.

Nesting activity occurs primarily in the morning and evening hours in Wisconsin (Table 19). In Nebraska and South Dakota, 47% of nests were laid in the morning during a warmer than average June ($n = 36$), but only 20% of nests were laid in the morning during colder than average June temperatures ($n = 30$; Iverson et al. 1997). In Michigan, 60% of nests were laid by 11:00 a.m., and 27% were laid between 8:00 and 11:00 p.m. ($n = 70$; Congdon et al. 1987). The female digs the nest using her hind feet and may test several locations (see Natural History Box: Use of Camera Traps to Monitor Nesting Behaviors of Snapping Turtles in Wisconsin). Heinen (2014) recorded aggregated nesting by four adult Snapping Turtles in a marsh in Michigan. The size of nesting females ranged from ca. 26 to 33 cm MCL (10 to 13 in). Two females nested within 1 m (3.3 ft) of each other, while the other two were ca. 10–15 m (33–49 ft) apart. Nightly visits to the location suggested synchronized nesting behavior. Robinson and Bider (1988) also reported synchronized nesting in Quebec and obtained strong evidence that this behavior can increase overall nest success by decreasing probability of predation of individual nests.

Females deposit eggs in a bowl-shaped nest with a narrow opening that descends at an angle to a large egg chamber 7–20 cm (2.8–7.9 in) belowground (Ernst and Lovich 2009). Ernst and Lovich (2009) reported that clutches range from 4 to 109 eggs (typically 25–45, $\bar{x} = 35.2$, $n = 252$). Vogt (1981) considered ca. 25 eggs to be a typical clutch size in Wisconsin. Mathias (1966) investigated 70 Snapping Turtle egg clutches from muskrat houses in the Horicon Marsh (Dodge County) and reported that the mean clutch size was 44 but that the largest clutch observed included 85 eggs. Clutch size appears to increase with latitude, averaging 30–35 eggs in the northern parts of the turtle's range (Iverson et al. 1997). Larger females lay more eggs and have the greatest clutch mass, but typically only one clutch is produced each year (Iverson et al. 1997). Not all females oviposit annually (Ernst and Lovich 2009).

NATURAL HISTORY BOX: Use of Camera Traps to Monitor Nesting Behaviors of Snapping Turtles in Wisconsin

In 2015 and 2016 we monitored Snapping Turtle nesting activity at four sites across Florence and Marinette Counties using camera traps (Ltl Acorn Model 5210, Reconyx HyperFire HC600). Camera traps were deployed in both years by mid- to late May and programed to snap photographs once per minute from late afternoon (3:00 p.m.) to late morning (10:00 a.m.) through late June. The camera traps also recorded date, time, and air temperature with each photograph.

Most Snapping Turtles investigated nesting locations from early to mid-June, primarily in the morning (prior to 10:00 a.m.), when air temperatures were between 10°C and 21°C (50°F and 70°F). Snapping Turtles also spent substantial time on land exploring possible nesting locations prior to egg deposition, during which they employed several apparent methods for nest site assessment and selection. For example, as females meandered throughout a potential nesting area, they occasionally stopped to rest their chin or nose on the ground for brief periods (a poorly understood behavior referred to as "chin rubbing" and "sand sniffing" by Congdon et al. [2008a]). This was presumably a method used to assess soil surface temperature prior to digging. In addition, females employed what we describe as 'plowing' behavior, which appeared to assist in nest-site selection or perhaps facilitated digging the egg chamber by loosening the soil in the immediate vicinity first. This behavior typically began with the female pushing headfirst into the soil to upturn a small area of earth. These actions were repeated in several directions to expand the disturbed area so that it was slightly larger than the body of the turtle. During the process, some females would sometimes dig down farther into the upturned soil than others, either headfirst or feetfirst, which we presumed was to test subsurface soil conditions with their head or feet. Although it was possible to confuse nesting and plowing behaviors, we identified the latter by the general orientation of the turtle (i.e., a turtle headfirst in soil was obviously not nesting) and the duration of activity. Furthermore, we assumed that females were nesting when they backed into nests and slowly rocked back and forth as their hind legs removed soil for prolonged periods. In most instances, nesting females also ended activities by completely covering nests with soil, which was then smoothed with their plastrons (rather than simply walking away from disturbed soils, which they did at the end of plowing behavior that was not followed by nesting).

Plowing behaviors always occurred prior to nesting (if nesting followed), and often females plowed several locations per visit. The mean number of locations plowed per site visit by female Snapping Turtles was 2 (range = 1–3), regardless of whether they ultimately nested. Plowing events were shorter in duration than nesting, ranging from 2 to 87 min ($\bar{x} = 25.9$ min, $n = 16$). In several instances, cameras recorded complete bouts of actual nesting by female turtles. Bouts of nesting ranged from 111 to 152 min ($\bar{x} = 130.2$ min, $n = 5$, not including one instance where the camera shut off prematurely and another where the presence of people may have deterred normal nesting behavior; Table 19). We also observed that while assessing nesting locations, individuals frequently returned to a location they had previously plowed or investigated a location plowed by other turtles on a prior date. Although we could not be certain if the same individual or different individuals returned on these subsequent visits, this observation suggests that plowing behavior influenced future selection of nest sites. It is noteworthy that not all nesting Snapping Turtles appeared to undertake this behavior, although it could have been performed outside the camera's field of view.

Robert W. Hay, Turtles for Tomorrow, and Joshua M. Kapfer, UW–Whitewater

Table 19. Detailed summary of nesting activity and associated behaviors exhibited by Snapping Turtles at sites in Florence and Marinette Counties.

Site, County	Date	Air Temperature, °C (°F)	Number of Locations Visited by Turtles	Duration of Plowing Behavior	Duration of Nesting	Comments
2, Florence	6/5/2015		1	16:00–16:12	No nesting observed	
2, Florence	6/18/2015	16.1–20.6 (61–69)	2	05:59–07:00	07:42–08:59[a]	Returned to plowed location from 6/5/2015
2, Florence	6/13/2016		2	06:11–06:46; 06:57–08:24	No nesting observed	
2, Florence	6/14/2016	13.3–17.2 (56–63)	3	06:18–06:21; 06:27–06:31; 06:32–07:41	07:42–09:45	Last location plowed on 6/14/2016 was also where nesting occurred and was the same location as the one plowed on 6/13/2016
2, Florence	6/16/2016	11.1–14.4 (52–58)	3	22:20–22:23; 22:37–22:42	Nesting unlikely	Backs in and digs as if nesting from 22:42 to 01:02, then appears to abandon nest; seen in water until 01:13 and then leaves; does not return
1, Marinette	6/5/2015	13.9–16.7 (57–62)	2	06:28–06:30	06:52–09:00	Second location was primarily nesting activity, which the turtle returned to after briefly leaving, and was also the site visited on the previous day by a Snapping Turtle
1, Marinette	6/11/2015	11.7–12.2 (53–54)	1	06:47–07:08	No nesting observed	
1, Marinette	6/12/2015	11.7 (53)	1	05:44–05:56	No nesting observed	
1, Marinette	6/20/2015	6.7–7.2 (44–45)	1	04:41–05:31	No nesting observed	Although an individual wandered the area from 04:09–04:29, it did not plow
1, Marinette	6/3/2016				No nesting observed	On 6/1/2016 and 6/3/2016 turtles entered the area on three occasions from 05:45–07:33, but no plowing or nesting was observed
1, Marinette	6/8/2016	10.6–15.6 (51–60)	1		20:57–23:29	
1, Marinette	6/10/2016	20.0–22.8 (68–73)	1		18:32–19:49	Possibly rushed due to a fisherman being nearby during nesting
1, Marinette	6/15/2016	14.4–19.4 (58–67)	1		06:36–08:27	
3, Florence	6/4/2016	12.8–13.9 (55–57)	2	05:12–05:41	05:51–08:14	Second location was nesting only
4, Florence	6/9/2015	20.6–21.7 (69–71)	2	07:37–07:41; 07:42–08:00	No nesting observed	

[a] Camera turned off before nesting was completed.

Snapping Turtles have temperature-dependent sex determination, with males produced at lower incubation temperatures (ca. ≤26°C [79°F]; Yntema 1979).

Vogt (1981) stated that in Wisconsin "the young emerge in the fall," and a multiyear research project from Michigan found that the average incubation period (from egg laying to hatching) was 93.2 days (Congdon et al. 1987). Vogt (1981) also reported that "overwintering may take place in the nest in some areas" in Wisconsin. Lovich et al. (2014) monitored turtle nest emergence patterns over a 20-year period in Pennsylvania and confirmed that Snapping Turtles can overwinter in the nest (known as a facultative nest emergence strategy). Specifically, they found that 27% of emergences occurred prior to winter (August and September), and 73% occurred the following spring (April and May, $n = 264$). The hatchlings that exhibited delayed emergence were significantly larger (CL) than those that emerged prior to winter. Lovich et al. (2014) reported that precipitation in August and September was positively correlated with the number of hatchlings that delayed emergence until the following spring. In Wisconsin, the extent to which individuals overwinter in the nest and the relationships between emergence strategy and weather patterns are currently unknown. Upon emergence, hatchlings apparently use visual cues and migrate toward open 'illuminated' horizons that may or may not be associated with waterbodies (Congdon et al. 2011a). For a detailed review of hatchling physiology and ecology, see Finkler and Kolbe (2008).

Snapping Turtle size (MCL), growth rates, and age at maturity vary geographically. Females in Nebraska and South Dakota were larger, matured earlier, and had faster growth rates than populations in northern latitudes such as Ontario (Iverson et al. 1997). In Ontario, hatchling growth was ca. 1 cm/yr (0.4 in/yr) from ages 1–3, and females matured at 11–13 years of age; males matured at 17–19 years of age (Galbraith et al. 1989). In Iowa, Christiansen and Burken (1979) reported similar growth rates between males and females until ca. 5 cm PL (2 in), after which point males grew faster than females. Males also matured earlier, at 4–5 years of age, and 14.9–15.5 cm PL (5.7–6.1 in, $n = 27$), compared to females, which matured at 6–7 years of age (or more) and 12.3–17.5 cm PL (4.8–6.9 in, $n = 38$; Christiansen and Burken 1979). In Quebec, males and females matured at 20–21 cm MCL (7.9–8.3 in) and ca. 15 cm PL (5.9 in), but the smallest mature male had a ca. 14–15 cm PL (5.5–5.9 in; Mosimann and Bider 1960). In Tennessee, White and Murphy (1973) reported that both males and females sexually matured at ca. 14.5 cm PL (5.7 in, $n = 28$ males and 25 females). In southeastern Michigan, females matured at 15.9–23.5 cm PL (6.3–9.3 in; Congdon et al. 1987). In Ontario, females showed ovarian follicle enlargement at an estimated 11–13 years of age, but mean age of first nesting was estimated at 17–19 years of age.

ACTIVITY

The active season in Wisconsin starts in mid-April (Vogt 1981) and is stimulated by increasing water temperature at hibernacula (Strain et al. 2012). Obbard and Brooks (1981) found that Snapping Turtles in Algonquin Park (Ontario) emerged from hibernation when water temperatures were 7.5°C (46°F). Mahmoud and Licht (1997) noted that Snapping Turtles in central Wisconsin were most active during the summer when water temperatures were 18°C–26°C (64°F–79°F). In a Minnesota lake, activity levels were highest in June (Schneider et al. 2019). In Ontario, Brown and Brooks (1993) reported seasonal and sexual variation in activity levels throughout the year. For example, they found that male activity was significantly higher than female activity shortly after emergence in May, which was explained by the tendency of males to move frequently while searching for mates. By June, female activity was similar to that of males and then exceeded male activity in July due to onset of the nesting season. Activity is primarily diurnal and greatest during crepuscular hours, although some nocturnal activity has been reported (Obbard and Brooks 1981, Smith and Iverson 2004, Ernst and Lovich 2009, Schneider et al. 2019).

Field and laboratory investigations of Snapping Turtle body temperature support that they actively regulate their body temperature either by selecting water of a desired temperature or (more rarely) by basking above the water level (reviewed by Spotila and Bell 2008). For example, Schuett and Gatten (1980) found that Snapping Turtles in the laboratory selected a preferred temperature ($\bar{x}$ = 28.1°C [82.6°F]) when offered a gradient of temperatures. Yet mean body temperature of active wild Snapping Turtles at various sites in Canada was lower than that recorded in laboratory investigations and ranged from 19.5°C to 22.7°C (67.1°F to 72.9°F; Brown et al. 1990, 1994). Past research supports that when active wild Snapping Turtles are not basking above the water, their body temperature does not vary greatly from water temperature (Punzo 1975), or it is similar to the coolest water temperature available (Brown et al. 1990).

Although basking can increase their body temperature, Snapping Turtles are not as frequently observed basking as other aquatic turtle species in Wisconsin. The possible explanations for this and the advantages of basking in certain situations but not others are complex (reviewed by Spotila and Bell 2008) and could be related to season, latitude, solar radiation, and maximum air temperature (Obbard and Brooks 1979, Obbard and Brooks 1981, Brown et al. 1994). Instead of basking above water, individuals often regulate body temperature while submerged just below the water surface, with only eyes and nostrils exposed to the air (Lagler 1943). Other turtle species, such as the Painted Turtle (*Chrysemys picta*), will sometimes bask on floating Snapping Turtles (Ernst and Lovich 2009). Based on laboratory investigations, Hutchinson et al. (1966) concluded that the mean critical thermal maximum (CTmax) of adult Snapping Turtles was 39.5°C (103.1°F).

Daily movements are likely limited outside of the mating and nesting seasons (see Reproduction and Development section). Snapping Turtles rarely swim in open water, instead usually opting to slowly walk along the bottom of a given aquatic habitat (Ernst and Lovich 2009). They spend much of their time in shallow water buried in mud, hiding beneath muskrat and beaver mounds, or resting below aquatic debris (e.g., stumps, roots, brush; Ernst and Lovich 2009, Paisley et al. 2009). However, Snapping Turtles can migrate several kilometers between their hibernacula and their summer home ranges (Brown and Brooks 1994, Haxton 2000b). In the Upper Midwest, Paisley et al. (2009) found that the average migration distance to hibernacula was 0.62 km (0.39 mi, range = 0.03–3.20 km [0.02–2.00 mi], n = 103 migration distances). Otten and VanDeWalle (2014) further documented strong homing ability in three displaced Snapping Turtles from Michigan, which moved ca. 2 km (1.2 mi) along a river from their released location back to their original capture locations.

Several studies have estimated home range sizes for Snapping Turtles in northern locations. Paisley et al. (2009) used radio telemetry to estimate home ranges on Pool 8 of the upper Mississippi River (adjacent to Minnesota and Wisconsin) from 1997 to 2001 (n = 76 females and 7 males). The overall mean home range size was 11.3 ha (27.9 ac) for females and 9.3 ha (23.0 ac) for males based on a Poly-Buff estimator and 32.3 ha (79.8 ac) for females and 18.3 ha (45.2 ac) for males based on an MCP estimator. Several additional studies have found variation in movement patterns by sex. In a marsh area of Lake Ontario, females traveled farther than males due to nesting activities (Pettit et al. 1995). Ryan et al. (2014b) studied movement patterns of 23 radio-tracked Snapping Turtles for two years in an urbanized canal in Indiana. They found that total range of movement (i.e., maximum distance between two relocation points) significantly differed between years for females only and that mean annual movement (i.e., distance between locations and number of movements) significantly differed between years for males only. In Lake Ontario, home ranges were typically larger for females ($\bar{x}$ = 8.5 and 9.7 ha [21.0 and 24.0 ac], n = 15) than males ($\bar{x}$ = 2.3 and 3.4 ha [5.7 and 8.4 ac], n = 4) over two study years (modified MCP estimator; Pettit et al. 1995). However, Obbard and Brooks (1981) reported that summer aquatic home ranges for a population in Algonquin Park (Ontario) did not significantly differ between sexes, with a mean home range size of 3.4 ha (8.5 ac, n = 10, modified MCP estimator). The variability in movement documented across these studies could indicate that movement patterns are influenced by the aquatic system a population inhabits (e.g., marsh vs. canal) and can vary annually.

Paisley et al. (2009) reported that Snapping Turtles in a Wisconsin population typically moved to hibernacula from mid-September to mid-October and entered hibernation in late October, which is reportedly triggered by decreasing water temperature (Strain et al. 2012). During hibernation, Snapping Turtles rarely (if ever) surface to breathe. Given the duration of the overwintering period (potentially four to six months in Wisconsin), these turtles face many challenges related to insufficient oxygen availability. Their physiological mechanisms to deal with low oxygen conditions are complicated, but we provide a cursory overview based on extensive reviews by Gatten (2008) and Ultsch and Reese (2008). Importantly, the metabolic (and, therefore, oxygen) demands of Snapping Turtles are reduced when they are submerged, regardless of water temperature. Snapping Turtle metabolic rates decline even farther when turtles are submerged in cold water during hibernation. While submerged, Snapping Turtles are capable of some oxygen uptake through the skin. As such, the simple act of selecting well-oxygenated aquatic locations as hibernacula (e.g., those with flowing water that do not freeze at the surface) can permit sufficient oxygen uptake across the skin to help meet lowered metabolic demands in winter. In an Ohio marsh system, for example, Meeks and Ultsch (1990) reported that Snapping Turtles moved to shallow water near mouths of tributaries or into streams presumably to ensure well-oxygenated waters throughout winter months, a pattern also seen by Lagler (1943). However, this species frequently hibernates in locations that become anoxic during the winter (e.g., buried in mud or in stagnant aquatic systems that are completely ice-covered). In these situations, there is no oxygen available from the surrounding water for absorption through the skin to meet even low metabolic demands. Yet in these situations Snapping Turtles can make use of metabolic pathways that do not require oxygen (i.e., anaerobic pathways such as glycolysis). This can be dangerous, because a product of anaerobic metabolism is lactate, an acid that can build to potentially lethal levels in the blood and tissues. However, a Snapping Turtle's shell contains substantial calcium and magnesium carbonates, which can be mobilized to help offset the dangerous buildup of lactic acid. As such, these turtles are capable of withstanding prolonged winter periods in anoxic environments.

PREY AND PREDATORS

Snapping Turtles are considered ambush predators (Feuer 1971, Punzo 1975). They often obtain food by burying themselves in the mud, then swiftly striking with their head and jaws to grab passing animal prey (Lagler 1943). Their ability to quickly extend their neck and snap their jaws during predation (or defense) is well-known. Video analyses have revealed that Snapping Turtles can reach a mean velocity of up to 77.8 cm/sec (30.6 in/sec) during neck extension and generate up to a mean of 136.5 newtons of bite force (Vervust et al. 2011). Snapping Turtles consume small food items whole, while they seize large items with their jaws and then tear those items into smaller pieces using their front feet (Lagler 1943).

Mathias (1966) reviewed the intestinal contents from a "series" of turtles trapped in the Horicon Marsh (Dodge County) and found that "bullheads, carp, crayfish, and white waterlily leaves were the main food items." Based on stomach contents of 173 Snapping Turtles captured in Michigan, vegetation was the principal dietary component (36.2% stomach composition by volume), followed by game and pan fishes (34.2% by volume), and carrion or live birds and mammals (19.6% by volume; Lagler 1943). Snapping Turtles prey on both juvenile and adult waterfowl (Lagler 1943). Abel (1992) even reported Snapping Turtle attacks on reintroduced Trumpeter Swan (*Cygnus buccinator*) cygnets at several sites in Wisconsin. Yet Kenow et al. (2009) found that the consumption of juvenile waterfowl in several pools of the upper Mississippi River adjacent to Wisconsin was low and less intense than that of other vertebrate predators. Smith and Casper (2015) reported an observation of a Snapping Turtle preying on a young live Porcupine (*Erethizon dorsatum*) on the Turtle-Flambeau Flowage in Wisconsin. The observed turtle succeeded in drowning the Porcupine and then proceeded to consume it. In Pennsylvania, several individuals were observed gathering leaves of Skunk Cabbage (*Symplocarpus foetidus*) on land and then retreating to the water to consume them (Hartzell 2016). The diet of juvenile Snapping Turtles is rarely documented, but in Ontario they were observed consuming aquatic insects and amphibian larvae in ephemeral wetlands (Moldowan et al. 2015).

There are probably few predators that regularly consume adult Snapping Turtles. Bald Eagles (*Haliaeetus leucocephalus*), Red-shouldered Hawks (*Buteo lineatus*), and Broad-winged Hawks (*Buteo platypterus*) have been reported as predators (Ross 1989a). North American River Otters also depredated 17 hibernating Snapping Turtles over two winters in Ontario (Brown and Brooks 1994). Eggs and juveniles are more susceptible to predation than adults, with nest predation rates as high as 100% in some locations (e.g., Congdon et al. 1987). In the lower Wisconsin River, Raccoons (*Procyon lotor*) were the primary predators of *Graptemys* species nests (92%, $n = 561$), followed by Striped Skunks (*Mephitis mephitis*), Virginia Opossums (*Didelphis virginiana*), and Coyotes (*Canis latrans*; Geller 2012c). During this study, Snapping Turtles were observed traveling through the nesting sites of these other turtle species, and thus it is likely that Snapping Turtle nests were susceptible to these same nest predators. Review of other studies suggests that similar species are primary nest predators across northern latitude locations (e.g., Hammer 1969, Congdon et al. 1987). Interestingly, no evidence of predation on nests located in muskrat lodges was detected in Horicon March (Dodge County) by Mathiak (1966).

Parasitic protozoans and sometimes leeches can reduce the fitness of females, potentially negatively impacting clutch sizes (Brown et al. 1994). The primary tactics used by Snapping Turtles to avoid threats is to remain underwater, lodged beneath debris or buried in sediments, where they are rarely encountered. Snapping Turtles are very aggressive if cornered on land or captured. When approached or handled, they fiercely snap with their powerful jaws, which can cause appreciable damage. This turtle is not a threat, however, if left alone or observed from a respectable distance. Reports of Snapping Turtle 'attacks' on humans engaged in aquatic recreation are rare and often unsubstantiated.

CONSERVATION

Status: Globally, the Snapping Turtle is listed as a Least Concern species by the IUCN. In Wisconsin, it is considered common but declining (Christoffel et al. 2002). In response to continued exploitation for the international turtle trade, the species was given CITES Appendix III protection in 2016, allowing for improved regulation of international trade by requiring international exporters to have a permit.

Populations: There is little population information available on this species in Wisconsin, and robust population analyses do not exist. The only published density estimate found comes from Pearse (1923), who reported that the density of Snapping Turtles in Lake Mendota (Dane County) was 0.17/ha (0.07/ac) but probably based this on an inadequate sample size. Gailbraith et al. (1988) synthesized the Snapping Turtle density literature across its geographic distribution and found that density ranged from 1.01 to 74.30/ha (0.41 to 30.07/ac) and was positively correlated with habitat productivity. In Ontario, density was estimated to be 30 times higher in a eutrophic effluent pond compared to an oligotrophic lake (Gailbraith et al. 1988).

Snapping Turtles are often reported as less common than other turtles within sampled communities. For example, they comprised only 1%–11% of all turtles captured at six locations in Illinois (Cagle 1942). In Wisconsin, Pearse (1923) reported that of 1,006 total turtles captured in Lake Mendota, only 15 were Snapping Turtles. Many investigations have calculated statistically unbiased sex ratios in study populations (e.g., Mosimann and Bider 1960, Hammer 1969, Froese and Burghardt 1975, Hogg 1975, Obbard 1983), although male-biased sex ratios occur (reviewed in Iverson and Smith 2010). Based on hoop net trap capture data, DonnerWright et al. (1999) found male-biased sex ratios at 10 of 12 sampled sites in the St. Croix River in Wisconsin and Minnesota, with the remaining 2 sites having equal sex ratios (overall sex ratio = 2.4M:1F, total $n = 183$, site-level $n = 3–44$).

Estimates of survival probability are low for eggs but increase with age. Over a 17-year study in Michigan, mean annual survival of Snapping Turtles from egg deposition to hatchling emergence was only 23% ($n = 378$ monitored nests; Congdon et al. 1994). This same study found that the mean estimated annual survival of juveniles between the ages of 2 and 12 was 77% (range = 65%–82%, $n = 997$ original captures), while adult female survival was 88% ($n = 1,121$ original captures). Other studies have also determined that adults have high survival rates. For example, a four-year study of an urban population in North Carolina estimated that annual survival was 91.4% for adult females and 99.4% for adult males ($n = 21$ original captures; Eskew et al. 2010). On Pool 8 of the upper Mississippi River (Minnesota and Wisconsin), annual survival ranged from 85.7% to 100% for adults over four years of age ($n = 89$ radio-tracked adults; Paisley et al. 2009). Based on recaptures 7–13 years after initial captures, mean annual adult female survivorship was estimated to be 96.6% ($n = 75$ original captures) for a population in Ontario (Galbraith and Brooks 1987). In Ontario, annual female recruitment for a population estimated at 63 individuals was 1.8% (Brooks et al. 1991). Snapping Turtle populations appear to be more sensitive to changes in adult or juvenile survival than age of sexual maturity, nest survival, or fecundity (Brooks et al. 1991, Congdon et al. 1994).

Threats: Habitat loss and alteration likely affect this species in various ways, and general threats associated with aquatic habitats, road mortality, and ecotoxicology are applicable to Snapping Turtles (see Introduction to Conservation and Management of Wisconsin's Amphibians and Reptiles chapter). The impacts of habitat disturbance can be obvious (e.g., direct mortality due to habitat degradation) or subtle (e.g., habitat alteration affecting reproductive success). Wirsing et al. (2012) found that Snapping Turtle nests in locations that experience high human disturbance (e.g., heavy human activity, heavy road traffic, routine mowing) were 1.9 times more likely to be destroyed by nest predators. Thompson et al. (2017b) reported heavy nest predation in agricultural fields in Virginia ($n = 23$ of 32 nests). Heavy nest predation can directly influence the number of Snapping Turtles in a population. It can also reduce the recruitment of juveniles and cause shifts in the size and age distribution of impacted populations (Browne and Hecnar 2007). In addition, Kolbe and Janzen (2002a) reported that lower nest temperatures in disturbed landscapes (due to shading from houses and greater tree canopy cover) could negatively impact embryo survival, size, and offspring sex ratios in northern Illinois.

Urbanization is associated with higher road densities, which can impact Snapping Turtle populations (Gibbs and Steen 2008, Patrick and Gibbs 2010). Adult females use roadsides as nesting sites and thus are disproportionately vulnerable to road mortality (Steen et al. 2006). As a consequence, male-biased sex ratios can occur in wetlands adjacent to roads (Steen and Gibbs 2004, Steen et al. 2006, Patrick and Gibbs 2010). Roads can also result in reduction of mean size of individuals (and thus possibly reduced fecundity) in populations (Patrick and Gibbs 2010). Haxton (2000b) performed road searches for Snapping Turtles during the nesting season over a three-year period in Ontario. The percentage of encountered individuals that were dead each year was 35.2% ($n = 37$ killed), 23.9% ($n = 21$ killed), and 32.6% ($n = 28$ killed), respectively, with 80.2% of these individuals found along a major highway.

Overharvesting of Snapping Turtles is a concern, as they are widely exploited for the pet trade, human consumption, and educational purposes (van Dijk 2012). Between 1999 and 2014, reported exportation of live Snapping Turtles from the US increased from 7,279 to 1,324,089 individuals (Colteaux and Johnson 2017). In Wisconsin, large numbers of turtles are sometimes taken by professional trappers who use various methods, ranging from traditional baited hoop net traps to a 'hooking' harvest technique, which employs hooks at the end of long poles to snag and extract Snapping Turtles while they are congregated and inactive at wintering sites (Knudsen 1956c, Paisley et al. 2009). Paisley et al. (2009) reported that four of seven adult Snapping Turtles that died during a radio telemetry study in the upper Mississippi River (Goose Island, Wisconsin, and Lawrence Lake, Minnesota) were from legal harvesting. Furthermore, Paisley et al. (2009) stated that "interviews with turtle trappers and discussions with commercial fishermen indicate that Snapping Turtle populations have declined substantially over the past 20–30 years on Pool 8" of the upper Mississippi River. Interestingly, some studies indicate that Snapping Turtles are less often captured by commercial fishermen than other turtle species. For example, <1% of turtles captured as bycatch by commercial fishing operations in the Mississippi River system near Missouri and Illinois were Snapping Turtles ($n = 11$ of 2,584 turtles; Barko et al. 2004). However, estimated proportion of individuals taken during mock harvests at four sites in Missouri ranged from 0.086 to 0.568 (Shaffer et al. 2017). Common Snapping Turtle populations can be particularly susceptible to even small population impacts, given their unique life histories. Adults are long-lived but slow to mature, and juveniles can experience high but variable mortality (reviewed by Brooks et al. 1991). In an Ontario population, Brooks et al. (1991) found that elevated adult mortality, coupled with poor juvenile recruitment and immigration rates, led to rapid population declines with slow potential for recovery. These findings led the authors to suggest that even low levels of adult "exploitation" would have long-term negative impacts (see also Management subsection). Although typically not sought by those engaged in recreational 'hook-and-line' fishing, Snapping Turtles that accidentally ingest recreational fishing tackle (e.g., hooks and lead sinkers) or monofilament line can experience intestinal perforations and lead poisoning (Borkowski 1997). Substantial research on the ecotoxicology of northern Snapping Turtle populations has been conducted, with numerous studies having detected various contaminants in tissues and impacts to several life stages (e.g., Golet and Haines 2000, de Solla and Fernie 2004, de Solla and Martin 2007, Eisenrich et al. 2009).

Management: Many general management considerations associated with aquatic habitats, road mortality, and overharvesting potentially apply to Snapping Turtles (see Introduction to Conservation and Management of Wisconsin's Amphibians and Reptiles chapter). Snapping Turtles appear to be fairly tolerant of structural modifications to their aquatic and surrounding terrestrial habitats (reviewed by Ernst and Lovich 2009). For example, they are often observed in disturbed wetlands and artificial waterbodies, such as stormwater retention ponds, in Wisconsin (Joshua Kapfer, UW–Whitewater, personal communication). They are also reported to rapidly colonize constructed wetlands (e.g., Palis 2007). However, the survival and reproductive success of turtles in these disturbed or unnatural situations are unknown, and such habitats could act as ecological traps. Due to the potential for skewed sex ratios resulting from female road mortality, construction of road barriers to reduce road mortality may be an effective management action to enhance population viability (Christoffel and Hay 1995b). Widespread implementation of this management tool would undoubtedly benefit Snapping Turtles and many other reptiles and amphibians in the state.

Protection of known nests may offset nest predation and increase population recruitment. Work in Michigan suggests that early detection and protection of nests is important, as 59%–73% of monitored Snapping Turtle nests were destroyed within one to six days of egg deposition (Congdon et al. 1987). Steen et al. (2012b) further analyzed a compilation of many reports ($n = 1,043$) and found that 95% of reported Snapping Turtle nests or gravid females were within 116 m (381 ft) of wetlands. These findings can help generate guidelines for implementation of terrestrial buffers around wetlands that can potentially protect Snapping Turtle nests.

It is currently legal to harvest this species in Wisconsin between 15 July and 30 November (WNDR 2015). The existing harvest regulations include a possession limit of ten individuals on the Mississippi River and three individuals in the remaining waterbodies of Wisconsin (WDNR 2015a). Although these harvest limits protect populations from commercial-level harvest, even minimal removal of individuals could negatively impact populations. Zimmer-Shaffer et al. (2014) estimated that only 2.3% of adult and juvenile Snapping Turtles could be harvested annually without causing population declines based on mean demographic rates. The life-history characteristics of turtles make them difficult to manage as a game species, and thus harvested populations should be closely monitored to ensure that the populations remain viable. Wisconsin regulations currently restrict harvest of Snapping Turtles to individuals that are 30.5–40.6 cm MCL (12–16 in). Implementation of size restrictions on harvested turtles has effectively reduced the total number of turtles harvested in the US (Colteaux and Johnson 2017). However, harvests that target large individuals can lead to shifts in the distribution of population size toward smaller and less fecund individuals, which can negatively impact population viability (reviewed by Zimmer-Shaffer et al. 2014, Colteaux and Johnson 2017).

SPECIAL REMARKS

Intraspecific aggressive behavior has been documented surrounding competition for food, with individuals biting or clawing each other (Raney and Josephson 1954, Froese and Burghardt 1974). Ritualized aggressive interaction between two males was observed in a small pond in Champaign, Illinois (Merrill 2004). The behavior involved the turtles facing each other with their heads submerged while their carapaces remained above the surface of water and periodically lunging at one another.

Painted Turtle

Chrysemys picta (Schneider 1783)

Joshua M. Kapfer

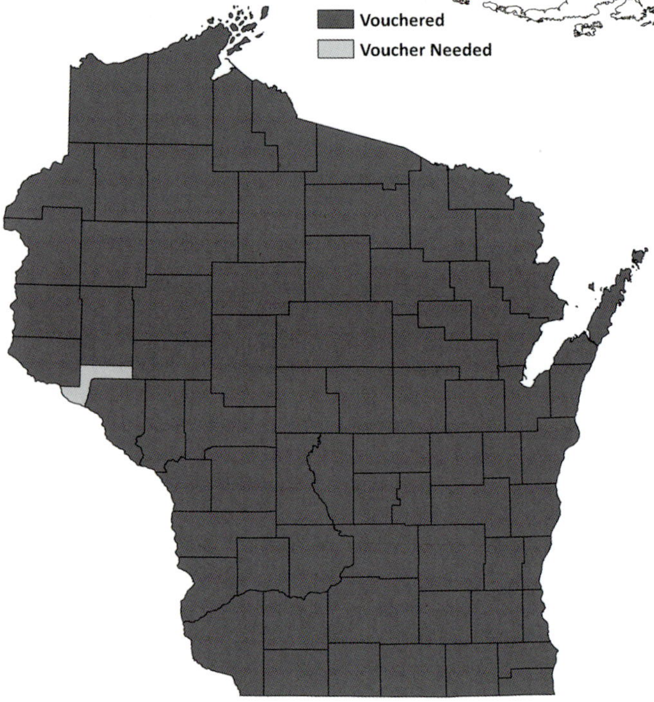

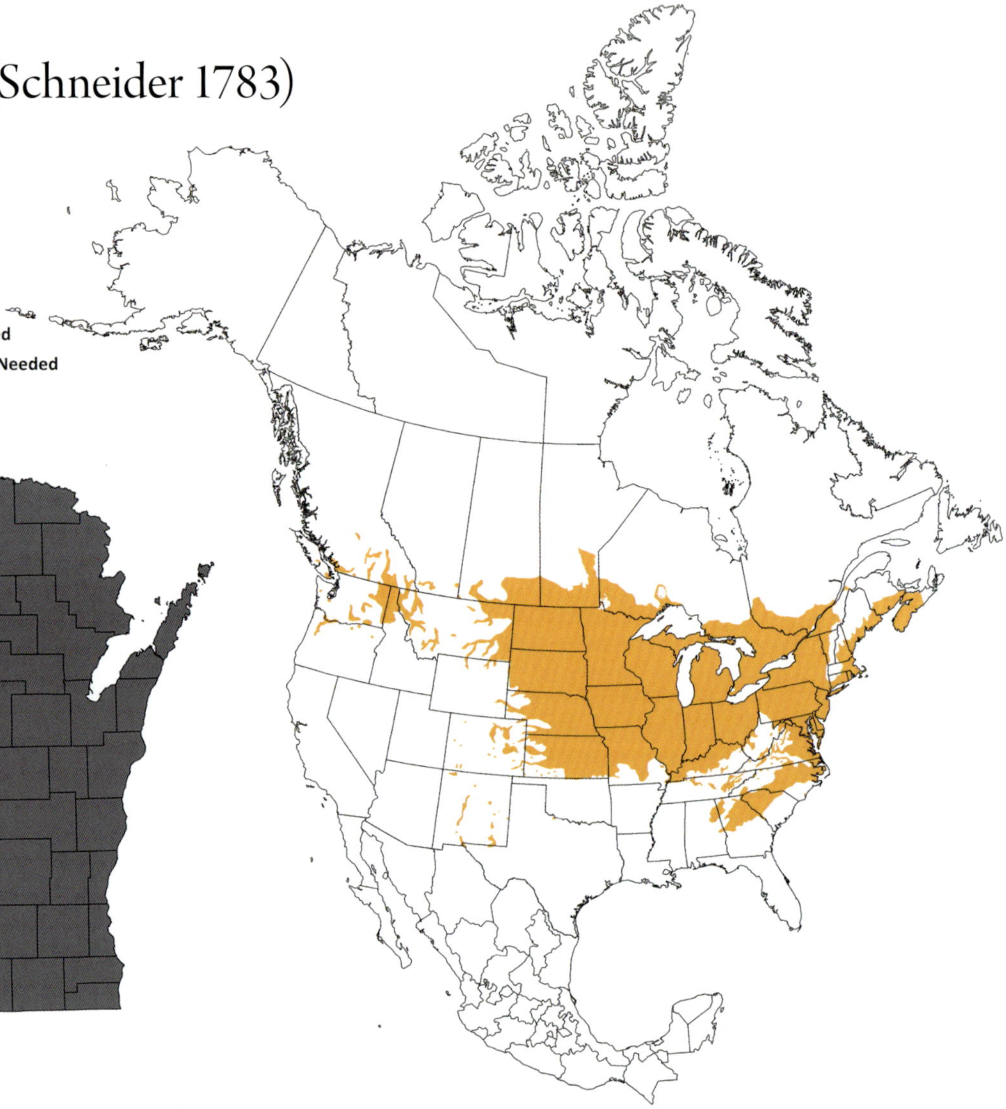

Painted Turtle (Dane County; photo by C. Raimond).

AT A GLANCE

Painted Turtles are probably Wisconsin's most frequently observed and most broadly distributed turtle species. Indeed, records of this turtle are known from every county in the state, and they are found in most available aquatic habitat types, even urban ponds. Painted Turtles are aptly named, with bold yellow and reddish stripes running along the length of their head and legs. Although variation exists, many individuals also have red-orange plastrons with shell margins covered in a striking array of red-orange and dark gray lines. These bright colors and distinct patterns give them the appearance of being painted. Painted Turtles are generalist omnivores, eating a wide range of vertebrates, invertebrates, and plants. They are mostly active from March to late October in Wisconsin and spend winters submerged in waterbodies that do not completely freeze. Breeding typically occurs in early spring, and females leave the water to lay eggs from late May to early June. These eggs often hatch in September, but females will occasionally lay a second clutch of eggs in late summer. This later clutch hatches in late autumn, but the young turtles overwinter in the nest and do not emerge until the following spring. Painted Turtles have provided many children with their formative wild turtle experiences, as they are often encountered crossing roads or attempting to nest in backyards. They are considered common by the WDNR.

SYSTEMATICS

The genus *Chrysemys* is restricted to the painted turtle complex (Vogt and McCoy 1980, Jensen et al. 2015, Crother 2017), although some disagreement exists on the number of species and subspecies in this genus (see Starkey et al. [2003], Ernst et al. [2006], Ernst and Lovich [2009], Weller et al. [2010], and Crother [2012] for further details and information on subspecific systematics and geographic range). There is agreement that only one species, *Chrysemys picta* (Painted Turtle), is found in Wisconsin (Vogt 1981, Starkey et al. 2003, Ernst and Lovich 2009), but reports of subspecies in the state are somewhat conflicting. For example, past authors have indicated that two subspecies and intergrades between them occur in Wisconsin: *C. p. belli* (Western Painted Turtle) and *C. p. marginata* (Midland Painted Turtle; Bishop and Schmidt 1931, Vogt 1981). The distributions outlined in Weller et al. (2010), on the other hand, indicate that only *C. p. belli* and intergraded individuals occur in the state. *C. picta* is a member of the family Emydidae, which includes not only other aquatic turtle species in the genera *Emydoidea*, *Graptemys*, and *Glyptemys* but also the terrestrial *Terrapene ornata* (Ornate Box Turtle).

DESCRIPTION

Adults: Painted Turtles are aptly named. Although the weakly domed carapace is drab (olive, dark gray, or dark green brown; Figure 315), the undersides of the carapace and the shell bridge are covered with alternating dark gray and bright red-orange bands in various patterns (Figure 316). Painted Turtle plastrons can be beige to orange red and often possess a gray pattern that extends down the center and that varies in size and shape. In Western Painted Turtles, the gray pattern is larger, with finger-like projections extending outward along the seams of the plastral scutes. In Midland Painted Turtles, this dark pattern is greatly reduced or absent. Size and shape of plastral patterns vary dramatically among individuals found

Figure 315. Stripe patterns on the head, neck, and legs of a Painted Turtle. Note the red color of stripes near the base of the neck and on the forelegs (Jefferson County; photo by J. Kapfer).

in the intergradation zone (see Variation subsection). The head, neck, and legs are covered with bold stripes that range in color from yellow to reddish (Figure 315). Painted Turtles are a relatively small aquatic species that Vogt (1981) reported ranged in CL from 10 to 25 cm (3.9 to 9.8 in). Individuals captured during multiple years of surveying at various locations in Wisconsin were on the smaller end of this spectrum, with mean MCLs ranging from 12.8 to 15.5 cm (5.0 to 6.1 in; Table 20). The largest MCL among Wisconsin specimens measured that were housed in the UWZM and MPM collections was 19.8 cm (7.8 in, UWZM 2025, a female collected in 1919 from the Apostle Islands off of Ashland County).

Preadult stages: Eggs are oblong in shape, ca. 3.0 × 1.9 cm (1.2 × 0.7 in), and white in color (Moll 1973, Vogt 1981; Figure 317). Hatchlings and juveniles appear similar to adults, although color may be more vibrant (Figure 318). Hatchlings are small ($\bar{x}$ MCL = 2.7 cm [1.1 in], n = 573; Ernst and Lovich 2009), with shells that are more round than oblong when viewed from above.

Figure 316. Variation in the color pattern of the plastron, carapace undersides, and shell bridge among individual Painted Turtles from the same population (Dane County; photo by J. Kapfer).

Variation: The anterior edge of the marginal scutes, just above the head, appear jagged in some individuals. Adults are sexually dimorphic, with males possessing elongated claws on their fore feet (Figure 319). As it is in most pond turtles (family Emydidae), the cloacal opening on the underside of the tail in males is posterior to the back edge of the carapace (when the tail is extended), whereas in females it is not. Males also have a slightly more concave plastron than females.

Substantial variation exists in plastral patterns, which can be especially diverse in areas where the subspecies intergrade (Figure 320). Weller et al. (2010) suggested that in lieu of conducting genetic analyses, plastral pattern length and geometric shape could be used to distinguish intergrades from either subspecies. For example, pattern length, width, and perimeter-to-area ratio of intergrades from the Upper Peninsula of Michigan differed from Western and Midland Painted Turtles.

Diagnosis: Painted Turtles have 12 scutes on their plastron, and their carapace is not highly domed but depressed in cross section. The plastron is not hinged, so it cannot close against the carapace. The carapace is covered with keratinized scutes that do not have strong elevated concentric grooves on each and are not blue black with yellow specks, although marginal scutes are usually marked with red. The

Table 20. Geographic and sexual variation in mean MCL and mean weight of Painted Turtles sampled at several representative locations in Wisconsin between 1999 and 2017.

County	Female			Male		
(Duration)	MCL, cm (in)	Weight, g (oz)	Sample Size	MCL, cm (in)	Weight, g (oz)	Sample Size
Bayfield[a] (7 yr)	15.5 (6.1)	525.6 (18.5)	46	13.4 (5.3)	283.8 (10.0)	48
Ozaukee[b] (11 yr)	13.9 (5.4)	361.6 (12.8)	35	12.8 (5.0)	255.2 (9.0)	39
Jefferson[b] (6 yr)	14.3 (5.6)	416.6 (14.7)	19	13.4 (5.3)	283.0 (9.9)	54

[a] E. Wild, UW–Stevens Point, unpublished data.

[b] J. Kapfer, UW–Whitewater, unpublished data.

Above, top: Figure 317. Painted Turtle eggs (Chippewa County; photo by R. Sajdak).

Above, bottom: Figure 318. Plastron and head pattern of a hatchling Painted Turtle. Note the bright coloration and distinct pattern (Jefferson County; photo by J. Kapfer).

Above, top: Figure 319. Comparison of the foreclaw length in male (*left*) vs. female (*right*) Painted Turtles (Dane County; photo by J. Kapfer).

Above, bottom: Figure 320. Examples of the substantial variation found in Painted Turtle plastral color and pattern in populations from locations across Wisconsin. (a) Bayfield County (photo by E. Wild). (b) Western Dane County (photo by J. Kapfer). (c) Ozaukee County (putative intergradation zone; photo by J. Kapfer). (d) Southern Jefferson County (putative intergradation zone; photo by J. Kapfer).

plastron has a distinct bridge and 12 scutes that are more orange than yellow but without black blotches on the edge of each. There is webbing between the toes. The jaws are not prominently beaked, and the apex of the upper jaw is notched so that one tooth-like structure occurs on each side directly below the nostril. Painted Turtles could be confused with Blanding's Turtles (*Emydoidea blandingii*) and all three map turtle species (*Graptemys* spp.) in Wisconsin. Yet Painted Turtles have a relatively flat carapace, unlike the highly domed carapace found in Blanding's Turtles. Painted Turtles also do not possess the bright yellow lower jaw or the yellow plastron with prominent black blotches on each scute typical of Blanding's Turtles. Painted Turtles and map turtles all have brightly colored lines along the head, legs, and shell margins, which may cause confusion.

However, in Painted Turtles these lines can contain as much orange red as yellow, particularly near the base of the neck and legs. The lines along the head and legs of map turtles, on the other hand, are all yellowish. Painted Turtles also have a more orange plastron, often with a darker pattern down the center, while the map turtle plastron is generally a solid dull yellow. Furthermore, the map turtle carapace is serrated along the posterior edge, and there are keeled scutes along the vertebral ridge (although this keel is not prominent in Northern Map Turtles [*Graptemys geographica*]), unlike Painted Turtles.

The elliptical shape of Painted Turtle eggs is similar to that of numerous species in the families Emydidae and Kinosternidae. Despite this, Painted Turtle eggs are smaller and possibly laid in lower clutch sizes than those of other aquatic turtles with elliptical eggs in Wisconsin (see Preadult stages subsection and Reproduction and Development section), with the possible exception of the Musk Turtle (*Sternotherus odoratus*).

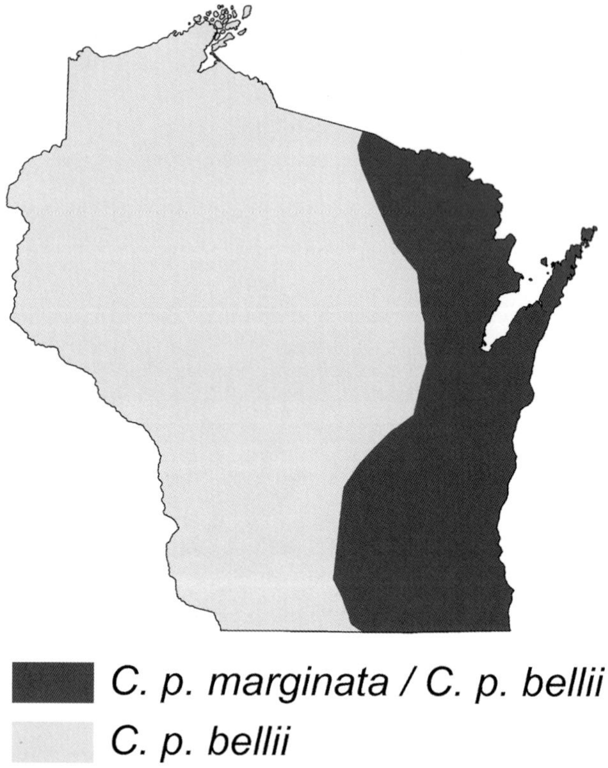

Figure 321. Approximate ranges of the Western Painted Turtle relative to the proposed intergrade zone in Wisconsin. Note that this map is an approximation of distribution maps in Weller et al. (2010) and Conant and Collins (1998). Further research is needed to clarify the distribution of the subspecies and the intergradation zone in Wisconsin.

DISTRIBUTION AND HABITAT

Global distribution: Ernst and Lovich (2009) reported that the Painted Turtle is the only North American turtle species whose distribution extends east to west across the entire continent. Despite this, its range is mostly concentrated in the Midwest and Northeast, including areas of southern Canada. It is less prevalent in the western, southwestern, and southeastern US, although it still occurs in portions of Washington, Oregon, New Mexico, West Texas, Alabama, Georgia, and the Carolinas.

Distribution in Wisconsin: Published Painted Turtle records exist from every county in Wisconsin, although a voucher for Pepin County is currently missing (see the Appendix). The exact range of each subspecies relative to the zone of intergradation, on the other hand, is not well established in the state. Vogt (1981) suggested that "pure" Western Painted Turtles were found in some western and northwestern regions, while individuals that "resembled" the Midland Painted Turtle only occurred in southeastern Wisconsin. He concluded that a "broad area of intergradation in characteristics occurs throughout the remainder of the state." Although he did not provide a detailed distribution map for intergrade individuals, his range maps also showed substantial overlap in the distributions of Western and Midland Painted Turtles, particularly in central and southern Wisconsin. Weller et al. (2010) and Conant and Collins (1998) provided distribution maps that delineated the intergradation zone throughout the Upper Midwest. These maps provide a coarse representation of where this zone occurs in Wisconsin and suggest that only Western Painted Turtles and intergrades exist in the state (Figure 321).

Localities of possible occurrence: Painted Turtles are reported from every county in Wisconsin, but a voucher is needed for Pepin County.

NATURAL HISTORY BOX: The Ambassador

During the nesting season, Painted Turtles are frequently encountered in playgrounds, city parks, and even backyards. Because of this, the first direct contact that many children have with wild turtles (and perhaps wildlife in general) is provided by this unassuming little chelonian (Figure 322). I vividly remember a Painted Turtle that my parents found in our back flower bed one summer evening when I was a child (likely a gravid female searching for a nesting site). I also remember another Painted Turtle I encountered at a friend's birthday party one early autumn day when I was very young. I recall the young partygoers all excitedly gathering around the turtle intently, enamored by its existence. Back then, a 'wild animal' was a very abstract concept to me, something usually encountered on television or in books and zoos. I remember being amazed that these wild creatures were actually available for my own discovery and not just for the likes of Marty Stouffer or Marlin Perkins. Such experiences sharpened my interest in wildlife as I grew up.

When I was slightly older, I began to bring home 'critters' that I encountered while exploring nearby parks and riverbanks to keep as short-term pets. Often these included a small Painted Turtle or two, which I housed in plastic wading pools outside and fed earthworms, crayfish, and the occasional minnow. Although I never kept the turtles for more than a month or so in the summer, these little captives, much like the wild turtles I encountered as a very young child, also had a profound impact on me. Watching them swim, feed, and bask in these backyard wading pools was like peering into my own small bit of the natural world, one that was telling me its secrets. These experiences with Painted Turtles were surely important in honing a childhood interest in wildlife that eventually grew into my chosen adult profession. Although this is but my own example, it's easy to imagine how children's experiences with Painted Turtles and other species like it may affect their attitudes toward wildlife for the rest of their life. If that is true, then Painted Turtles are important ambassadors for wild places and the wild things that live there. For this alone, they deserve our respect.

Figure 322. Two budding turtle enthusiasts holding their first wild Painted Turtle (Dane County; photo by J. Kapfer).

Joshua M. Kapfer, UW–Whitewater

Habitat: Little rigorous research on the habitat preferences of Painted Turtles in Wisconsin exists (e.g., DonnerWright et al. 1999), although habitat associations are well-studied in many other parts of this species' range (reviewed by Ernst and Lovich 2009). General observations support that Painted Turtles are found in a wide range of aquatic habitats throughout the state. For example, they occur in still and slower-moving permanent waterbodies, including most types of wetlands and lakes (Vogt 1981, Joshua Kapfer, UW–Whitewater, personal observation). Yet they will also use creeks and rivers (particularly low-current warm backwaters), but Vogt (1981) further reported them from "fast-moving, spring-fed streams." At large spatial scales, Painted Turtles are more common in landscapes that possess numerous and highly connected waterbodies (Cosentino et al. 2010).

Although they are somewhat general in the type of aquatic habitats they use, Painted Turtles appear to favor specific habitat features within occupied waterbodies. For example, they are often found in aquatic systems that are warm and eutrophic, with dense aquatic vegetation (Hughes et al. 2016). Painted Turtles studied in Illinois preferred waterbodies with open canopies, deeper water (Anthonysamy et al. 2014a), and long hydroperiods (Cosentino et al. 2010). The need to bask for thermoregulation (see Activity section) also strongly influences the microhabitats they select in aquatic systems. DonnerWright et al. (1999) found that Painted Turtle captures along a stretch of the St. Croix River in Wisconsin were positively associated with soft substrates and the number of available basking locations (i.e., 'snags'). These basking sites can include woody debris along the water's edge, as well as vegetation hummocks, exposed banks, and even other vertebrates (Jones and Cochran 2014). Habitat preferences appear to change as Painted Turtles develop, and several studies have found that juveniles use habitats with shallower water and more dense vegetation than other age classes (Congdon et al. 1992, Hughes et al. 2016).

The habitats selected for nesting are mostly open canopy but still variable. For example, females will lay eggs in upland prairies and shrubby fields but also along roadsides and in pastures with soils that are either sand, loam, clay, or dominated by loose gravel (Mahmoud 1968b). Painted Turtles hibernate while submerged in waterbodies that do not completely freeze but are often covered in a sheet of surface ice, such as lakes, large ponds, and deep wetlands. Vogt (1981) further stated that overwintering Painted Turtles "congregate around springs and areas of flowing water and bury in the mud or pile up behind logs or rocks." Although these sites are likely chosen because they maintain suitable oxygen availability, Painted Turtles can withstand very low oxygen levels while hibernating (see Activity section). Sometimes overwintering and summer habitats are separated from each other (e.g., a creek that does not freeze used for hibernation and a nearby wetland occupied during the active season), which requires that individuals undertake migrations at certain times of year (J. Kapfer, personal observation).

REPRODUCTION AND DEVELOPMENT

Male and female reproductive cycles are triggered by a combination of changes in photoperiod and temperature (reviewed by Ernst and Lovich 2009). In Wisconsin, breeding mostly occurs in spring, and Vogt (1981) even reported that he had "seen them breeding in early April amid the crushed ice just after the break-up of Lake Mendota" (Dane County). Despite this, breeding and courtship behavior also occur during the summer through late autumn (Vogt 1981, J. Kapfer, personal observation). According to Ernst and Lovich (2009), courtship involves the male pursuing the female in the water until she gradually submits to his advances. Subsequently, the male faces the female and flutters the back of his long fore claws along her chin and neck. If she is receptive, she responds by stroking the male's forelimbs with her own claws. In between bouts of claw-fluttering, the male swims away from the female, as if enticing her to follow. Eventually, both turtles swim to the bottom, and copulation begins.

Females in Wisconsin nest from late May through early July (Ream 1967, Moll 1973, Vogt 1981). During this time, egg-laying activity is concentrated in early morning and late afternoon (Ream 1967, Mahmoud 1968b) and occasionally at night. Females apparently choose to nest during times when temperature

Figure 323. Female Painted Turtle nesting along the margins of a gravel driveway (Waukesha County; photo by J. Kapfer).

is optimal (Frye et al. 2017) and frequently under rainy or cloudy conditions (Ream 1967). Painted Turtles can travel some distance from the water to select nesting sites in Wisconsin. For example, Mahmoud (1968b) and Ream (1967) reported that nesting occurred at a distance of up to 152 m (499 ft) from water, while Vogt (1981) reported that it could occur >300 m (984 ft) from water.

Nest-site selection apparently involves a substantial amount of trial and error by the female, and mean nest-site search time for females in a Nebraska study was 26 minutes (Frye et al. 2017). Ernst and Lovich (2009) further reported that females will "nuzzle" the ground with their head and neck, presumably to test soil texture, moisture, or temperature, and also excavate several trial holes prior to selecting the final nest location. Nests are dug by scooping out dirt with the hind feet, alternating one foot at a time (Figure 323), which results in a flask-shaped excavation with a mean depth of 7.75 cm (3.1 in) and a mean diameter at the base of 6.10 cm (2.4 in; Mahmoud 1968b). Females frequently moisten the ongoing excavation with water from their bladder and will carry up to 11% of their body mass in water to the nesting location (which exceeds the mass of their eggs; Kinney et al. 1998). The entire nesting process can exceed two hours (Mahmoud 1968b). Females often return to the same nesting locations in subsequent years (Lindeman 1992).

Reported clutch sizes in Wisconsin range from 4 to 20 eggs (Vogt 1981 and Moll 1973, respectively). Painted Turtles exhibit temperature-dependent sex determination, where the temperature at which eggs incubate determines hatchling sex. Bull and Vogt (1981) found that 100% of eggs incubated at 25°C (77°F) resulted in male hatchlings, while incubation at 31°C (88°F) resulted in more females. Thus, variation in the incubation temperatures experienced by wild egg clutches promotes balanced sex ratios.

Female Painted Turtles in Wisconsin can produce two egg clutches per year (Moll 1973). Hatchlings from clutches laid early in the year often emerge in September (Vogt 1981), while hatchling turtles from the second clutch remain in the nest throughout the winter and emerge the following spring (Vogt 1981, J. Kapfer, personal observation). Hatchlings from later clutches, which overwinter in the nest, can be exposed to freezing temperatures, and much research exists on the mechanisms by which these young turtles survive this extreme cold. The following summary is based on Ernst and Lovich's (2009) detailed review of the literature. There appears to be two methods by which hatchlings survive subzero temperatures (−11°C to −2°C [12.2°F to 28.4°F]): freeze tolerance and supercooling. Freeze tolerance is possible because hatchlings have layers of keratin in their epidermis that prevent ice crystals from entering the body and contacting their body fluids. They also produce body fluid compounds that can prevent ice crystals from forming in the intracellular space, even if their extracellular body fluids freeze. Yet the ability to tolerate freezing under these conditions is limited by time (i.e., hatchlings cannot survive freezing of extracellular fluids for long). Hatchlings in the nest can, however, remain in a supercooled state during the extended period of winter inactivity. Supercooling occurs when extracellular fluids remain liquid (rather than forming a solid) even though the temperature of the body fluid drops below freezing. This occurs when turtle body fluids have not been exposed to ice crystals, which is the typical catalyst that initiates transition of fluid from a liquid to a solid form. So long as the body fluids are not exposed to ice crystals when in a supercooled state, ice crystal formation within the body is unlikely to occur, and the turtle will not freeze.

Wisconsin Painted Turtles experience rapid growth in their first year (a 95.7% increase in size; Ross 1989b), which declines dramatically in later years (e.g., a 14.5% size increase during their seventh year; Ross 1989b). Attainment of sexual maturity is related more to size than to age (reviewed by Ernst and Lovich 2009), with males maturing at smaller sizes and earlier ages than females (Congdon et al. 2003). In Wisconsin, males reportedly reach sexual maturity at 8.0–8.5 cm PL (3.2–3.4 in) and three to five years, while females reach sexual maturity at 13.0–15.8 cm PL (5.1–6.2 in) and seven to eight years (Moll 1973, Ross 1989b). Painted Turtles are long-lived, with maximum reported longevity in the wild of >55 years (Congdon et al. 2003).

ACTIVITY

In Wisconsin, Painted Turtles can be active from March through October and have even been observed swimming in partially frozen wetlands during early spring (Vogt 1981, Violet Kapfer, personal observation). Emergence from hibernation in the northern part of their range is triggered by increasing water temperature rather than changes in photoperiod (Crawford 1991). Throughout the active season, they are primarily diurnal, although nocturnal behavior is known in the Upper Midwest (i.e., Michigan; Rowe 2003) and in other regions (reviewed by Ernst and Lovich 2009). Nesting females may also be active at night or at dusk and dawn (see Reproduction and Development section). When active, Painted Turtles frequently bask to thermoregulate, and the average bout of basking reportedly lasts ca. two hours (Ernst and Lovich 2009). Basking behavior typically peaks in the morning for northern populations, although this behavior can occur sporadically throughout the day (Schwarzkopf and Brooks 1985). There is variation in the time spent basking by sex. For example, gravid females bask longer than males because they must maintain body temperatures necessary for adequate embryo development (Carrière et al. 2008).

Painted Turtles can be highly mobile (reviewed by Ernst and Lovich 2009). The largest bouts of movement are typically related to female nesting, migration from drying wetlands or ponds in summer, and perhaps migration between overwintering sites and summer habitats. Home range size is not well reported in Wisconsin, and no published estimates exist for state populations, although data from other midwestern states are applicable. Two years of radio telemetry work in nearby Michigan determined that mean home range size was 0.7–1.8 ha (1.7–4.5 ac, MCP estimator; Rowe 2003). Individuals tracked at this site also often patrolled smaller core areas that existed within their larger overall home range (Rowe 2003, Rowe and Dalgarn 2010). Further investigations on this same Michigan population by Rowe and Dalgarn (2010) found that average home range size did not differ drastically from the previous study by Rowe (2003), nor did it differ substantially by sex (females = 1.8 ha [4.5 ac], males = 2.9 ha [7.2 ac]) or age class (juveniles = 1.5 ha [3.7 ac]). Rowe (2003) calculated mean daily movement rates of 68.1 m/day (223.4 ft/day). Research on a riverine population in Saskatchewan found that the greatest distance moved varied by sex (males moved up to 6.0 km [3.7 mi], females moved up to 3.2 km [2.0 mi]) and age class (juveniles moved up to 0.7 km [0.4 mi]; MacCulloch and Secoy 1983).

Painted Turtles in Wisconsin enter hibernation by mid- to late October (Vogt 1981) but can be active later if environmental conditions are suitable. Overwintering locations are frequently ice-covered during winter, and hibernating turtles can face oxygen-poor or anoxic conditions. Painted Turtles deal with low oxygen levels during hibernation in several ways. For example, because body temperature drops during winter, metabolic and oxygen demands are also greatly reduced (i.e., during hibernation, metabolic rates are 10% of those experienced under conditions with normal oxygen levels; reviewed by Ernst and Lovich 2009). Furthermore, turtles faced with anoxic conditions during winter can use anaerobic glycolysis (metabolic pathways that produce energy in the absence of oxygen; Crawford 1994). However, a product of anaerobic glycolysis is lactate, which is acidic so can lead to a decrease in the pH of body fluids and tissues (also called lactic acidosis) that could cause substantial damage to the turtle. In fact, Painted Turtles that use anaerobic glycolysis during the winter can experience a 551% increase in lactate levels (Crawford 1994), which must be offset for the turtle to survive. In such situations, overwintering Painted Turtles can rely on the buffering capacity of their skeleton (particularly their shell) to mitigate the negative impacts of winter lactic acidosis. Carbonates are released from the skeleton that raise the body pH and help balance internal acid-base levels (reviewed by Ernst and Lovich 2009). Interestingly, metabolic accumulation of lactate occurs at a lower rate and the skeletal buffering capacity is greater in Painted Turtles from the northern part of their range (e.g., Wisconsin) compared to those from the South (e.g., Louisiana, Alabama, Arkansas; Reese et al. 2004).

PREY AND PREDATORS

The list of food items consumed by Painted Turtles is lengthy (reviewed by Ernst and Lovich 2009) and includes many vertebrate, invertebrate, and plant species. Vogt (1981) reported that the ratio of animal to plant material in their diet was 60:40, while stomach content analysis of over 300 Painted Turtles from the Lower Peninsula of Michigan indicated nearly the opposite ratio (slightly more than 60% of their diet was composed of aquatic vegetation; Lagler 1943). Consumption of animal vs. plant material can vary by month, with a higher volume of animal material consumed in May, June, and August but a higher volume of plant material consumed in July (Cooley et al. 2003). Some researchers even suggest that their herbivorous tendencies play an important role in the dispersal of aquatic macrophyte seeds (Padgett et al. 2010). Painted Turtles are visual predators that hunt along the substrate and around the vegetation of the waterbodies they occupy. Although these turtles are visual hunters, experimental work has shown that increased turbidity (up to 40 nephelometric turbidity units) did not substantially increase their search time while foraging (Grosse et al. 2010). They also readily consume carrion (Ernst and Lovich 2009).

Adults are generally less susceptible to predation than eggs and juveniles, although data specific to Wisconsin are lacking. Ernst and Lovich (2009) noted that adults may be taken by Raccoons (*Procyon lotor*), North American River Otters (*Lontra canadensis*), Red-shouldered Hawks (*Buteo lineatus*), Ospreys (*Pandion haliaetus*), and Bald Eagles (*Haliaeetus leucocephalus*). Adult Painted Turtles are wary while basking and often return to the relative safety of the water at the least provocation; thus, they are difficult for predators to capture. They could be most vulnerable to aquatic predators immediately after emergence from hibernation, when they are cold and not operating at peak behavioral efficiency (Ross 1988), or to terrestrial predators during migrations for nesting or between waterbodies. When captured, most adults retreat into their shells or flail their arms and legs, but some will attempt to bite (J. Kapfer, personal observation).

Several mammals are known predators of eggs and young, particularly Raccoons, Virginia Opossums (*Didelphis virginiana*), and Striped Skunks (*Mephitis mephitis*), but birds, small mammals, and snakes are also potentially important predators (see the comprehensive list in Ernst and Lovich [2009]). Ream (1967) estimated that 70%–80% of the nests she observed near Lake Mendota (Dane County) were destroyed by Thirteen-lined Ground Squirrels (*Ictidomys tridecemlineatus*), while other fossorial predators such as moles also consume eggs (Robert Hay, Turtles for Tomorrow, personal communication).

Painted Turtle eggs are particularly vulnerable to predators, and nest predation rates of 95%–100% are sometimes reported (Ernst and Lovich 2009). Nest predation rates can be influenced by nest distribution, immediate habitat composition, and time. For example, Marchand and Litvaitis (2004a) reported that congregated Painted Turtle nests were more at risk of predation than scattered nests. Also, nests in agricultural areas were at greater risk than those along roadsides or in suburban lawns. Several studies report that predation rates increase when nests are placed closer to the water's edge. For example, Marchand et al. (2002) reported that higher predation occurred on nests closer to the water, and Kolbe and Janzen (2002b) estimated that predation was greatest within 25 m (82 ft) of the water's edge but leveled off at distances of 25–40 m (82–131 ft). Kolbe and Janzen (2002b) also found that predation was sometimes high when nests were placed closer to woodland edges, although other studies found no relationship between predation rates and the distance to edges of nonaquatic habitats (e.g., Marchand and Litvaitis 2004a). According to Christens and Bider (1987), the majority of nest predation (85.7%) occurred on the same night eggs were laid.

CONSERVATION

Status: Globally, the Painted Turtle is listed as a Least Concern species by the IUCN. It is considered a common species in Wisconsin by the WDNR.

Populations: Although a substantial amount of published research on the population dynamics of Painted Turtles exists, relatively little is from Wisconsin (but see Pearse 1923, Ream and Ream 1966, Ross 1989b). Despite this, anecdotal observations suggest that it is the most abundant turtle in many Wisconsin waterbodies. For example, Painted Turtles were always the most frequently captured species during multiyear turtle-trapping efforts at various locations in the state (J. Kapfer, unpublished data). However, great variability in abundance and density can occur due to waterbody size and type (lentic vs. lotic), time of year, hydrologic condition, and location (reviewed by Ernst and Lovich 2009). Pearse (1923) estimated that 2,744 individuals inhabited the suitable habitat of his study site in Dane County. Later, Petokas (1981) reanalyzed the data collected by Pearse and estimated a density of 135.5/ha (54.8/ac). Petokas (1981) also analyzed capture data from Ream and Ream (1966) to estimate a density of 32.8–33.3/ha (13.3–13.5/ac) at a site in Dane County. Ross (1989b) reported a density of 104/ha (42.1/ac) at a site in Adams County. Much greater abundances and densities have been recorded at other locations in the midwestern US, supporting the high variability in populations across this species' range. For example, Sexton (1959) estimated that ca. 1,000 individuals existed in a Michigan pond, resulting in a density of 500–555/ha (202.3–224.6/ac).

Despite adult female susceptibility to road mortality and predators during the nesting season, female survivorship was greater than adult male survivorship in Michigan (Congdon et al. 2003, but see Steen and Gibbs 2004). The authors of that study suggested that because males migrate between wetlands more frequently than females they were more exposed to sources of mortality. Regardless, most studies of this species across its North American range, including Wisconsin, indicate that sex ratios are generally 1M:1F (Ross 1989b, reviewed by Ernst and Lovich 2009).

Threats: Many general threats, such as degradation of habitat, overharvesting, and road mortality, apply to this species (see Introduction to Conservation and Management of Wisconsin's Amphibians and Reptiles chapter). Interestingly, Painted Turtles frequently inhabit degraded and artificial environments, including disturbed natural waterbodies and artificial ponds such as agricultural ponds and stormwater retention ponds (J. Kapfer, personal observation). For example, in Connecticut, Lambert and Giller (2015) documented a high density of Painted Turtles in an urban pond (34.5/ha [14/ac]) while employing minimal sampling effort (eight baited hoop nets set for one day). Despite this, apparently not all anthropogenic habitats are of equal suitability for this species. For example, research in North Carolina found that Painted Turtle abundances were higher in golf course ponds than in ponds in more urban landscapes (Price et al. 2013).

Several potential stressors on Painted Turtle populations can be especially prevalent in disturbed landscapes, such as road mortality. Knudsen (1956d) stated that "thousands of painted terrapins are killed annually in Wisconsin by automobiles, especially during their egg-laying excursions. It seems that many people enjoy seeing how many they can run over." Research on Painted Turtles in New York indicated that roads affect population sex ratios (Steen and Gibbs 2004) and that this may be related to higher mortality of nesting females that cross roads. Although some past work supports that high road densities near wetlands skew the sex ratio of Painted Turtle populations toward males (because females experience road mortality during nesting forays; Marchand and Litvaitis 2004b), research in Ontario found no difference in abundances or sex ratios based on the presence of roads (Dorland et al. 2014). Dorland et al. (2014) hypothesized that high mortality along roads may cause natural selection that favors turtles that avoid roads. They also proposed that nests close to roads are less susceptible to nest predators. This higher nest survival may result in greater recruitment of offspring that could offset the road mortality experienced by adults. In Wisconsin, substantial road mortality has also been observed in spring and autumn in situations where individuals migrate between overwintering and summer habitats (J. Kapfer, personal observation). Despite their susceptibility to road mortality, Painted Turtles appear to have a rather remarkable ability to survive serious injury. For example, Cagle (1945) reported a female who survived being ruthlessly shot with a rifle, which created a bullet wound that healed in ca. two years. Similarly, individuals with large healed cracks in their shells (presumably due to motor vehicle impacts) are frequently encountered in Wisconsin (J. Kapfer, personal observation).

Various additional factors related to anthropogenic landscapes also affect Painted Turtle populations. For example, encroachment of woody vegetation into areas where Painted Turtles nest may cause excess shading that can reduce thermal quality and affect normal embryo development. Anecdotal observations also support that hatchling Painted Turtles can become entangled in the dense roots of the invasive Spotted Knapweed (*Centuarea maculosa*) while still in the nest (Fish & Wildlife Compensation Program 2010). The effects of invasive wetland grasses such as Reed Canary Grass (*Phalaris arundinacea*) and *Phragmites* species on Painted Turtles are mostly unknown but are a subject worthy of further research. Gamble and Simons (2004) conducted a two-year study of harvested vs. nonharvested Painted Turtle populations in Minnesota. They found that populations exposed to harvesting had reduced abundances, suggesting that turtle harvests may negatively impact populations. Jergenson et al. (2014) found that Painted Turtles are resilient to extreme environmental disturbances (such as floods), which could support their resistance to the damaging impacts of extreme weather predicted to become more common as the result of climate change (WICCI 2011).

Although predation is a natural ecological process, the prevalence of nest predation could be greater in landscapes disturbed by humans and, thus, negatively impact Painted Turtle populations (Marchand and Litvaitis 2004b). For example, a primary predator of turtle eggs is the Raccoon, whose densities in urban and suburban landscapes can be higher than in rural settings (Prange et al. 2003). As such, large populations of Raccoons associated with areas of high human density, such as in southern Wisconsin, could have profound impacts on Painted Turtle population recruitment. Survival of Painted Turtle eggs to hatching is typically low (21%–33%), even in populations that experience no pressure from nest predators (Lindeman 1991). Given their naturally low hatching success, the added effect of nest predation on Painted Turtle populations is likely substantial. Interestingly, work in Illinois found that as distance to anthropogenic "structures" (e.g., campsites, roads, fish-cleaning stations) decreased, nest predation rates also decreased (i.e., Strickland and Janzen 2010), and the authors speculated that perhaps human activity deterred nest predators. However, more research is needed to better understand relationships between prevalence of mesopredators and human land use.

Management: Many general management recommendations are applicable to Painted Turtles, including those related to aquatic habitats (such as large wetlands, lakes, and rivers), mitigation of road mortality, and regulations for harvest or collection of wild turtles (see Introduction to Conservation and Management of Wisconsin's Amphibians and Reptiles chapter). Landscape-level management considerations are likely important components of conservation plans for Painted Turtles. Studies support that this turtle thrives in landscapes with many connected waterbodies that possess variable characteristics (Cosentino et al. 2010). Furthermore, Hughes et al. (2016) suggested that Painted Turtles benefit when these landscapes contain a mix of shallow and well-vegetated 'nursery ponds' for juveniles that are adjacent to larger and less vegetated ponds for adults. Based on these turtles' tendency to bask, habitat improvements through creation or maintenance of basking sites (e.g., logs or snags placed perpendicular to the shoreline) are likely beneficial.

An often-overlooked aspect of habitat management for Painted Turtles involves upland habitat adjacent to waterbodies. These terrestrial habitats are important to the life cycle of turtles for a variety of reasons (Gibbons 2003), but particularly for nesting. Therefore, control of encroaching woody vegetation and invasive plants in open nesting areas, such as with prescribed burns, could be an important conservation and management strategy for this species. Yet Painted Turtles are occasionally killed during prescribed burns to manage terrestrial vegetation adjacent to wetlands (J. Kapfer, personal observation). Although the frequency and population impact of this mortality are unknown, management activities should be timed appropriately.

In addition to management of existing nesting sites, research supports that creation of artificial turtle nesting sites, particularly in areas where natural nesting sites have become overgrown with woody vegetation, can be a successful conservation strategy (Beaudry et al. 2010, Buhlmann and Osborn 2011). Paterson et al. (2013) reported that more Painted Turtles used artificial nesting mounds (where predators were excluded) than expected and that hatching success of eggs transplanted to artificial mounds was high. It should be noted, however, that discrete nesting habitats frequented by many turtles could also attract nest predators, so the usefulness of artificial nesting sites might be limited if they are not coupled with measures to exclude predators. The use of solar-powered electrified fences to protect known turtle nesting areas in Wisconsin from Raccoons has been found to be effective (Geller 2012c). It is possible that removal of predator species in turtle nesting areas would have a positive influence on Painted Turtle nest survival (as is the case with sea turtle nests; Engeman et al. 2006), but the effectiveness of this strategy has not been investigated in Wisconsin. Although human recreation near potential turtle nesting localities may affect nesting success, a study in Illinois found that intensity of human recreational activities did not influence the decision of Painted Turtles to emerge for nesting or the locations they chose to nest in (Bowen and Janzen 2008).

Blanding's Turtle
Emydoidea blandingii (Holbrook 1838)

Robert W. Hay and Samantha Foster

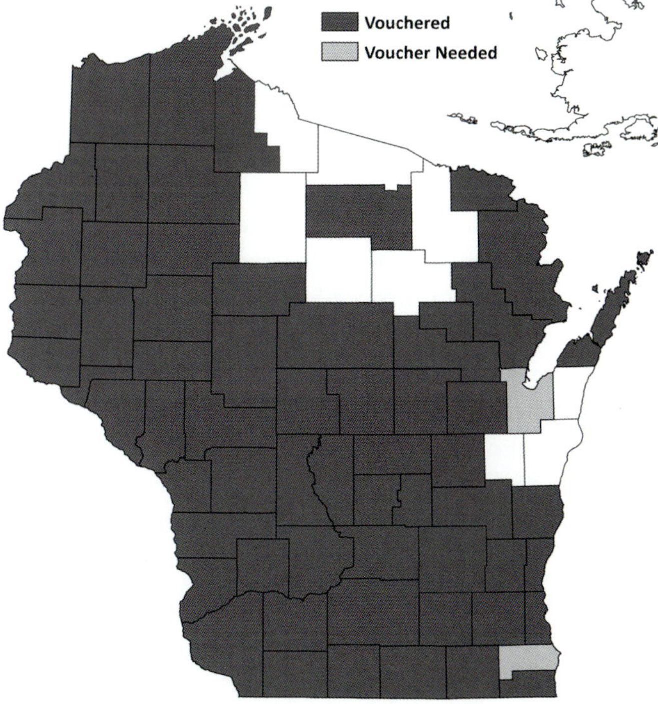

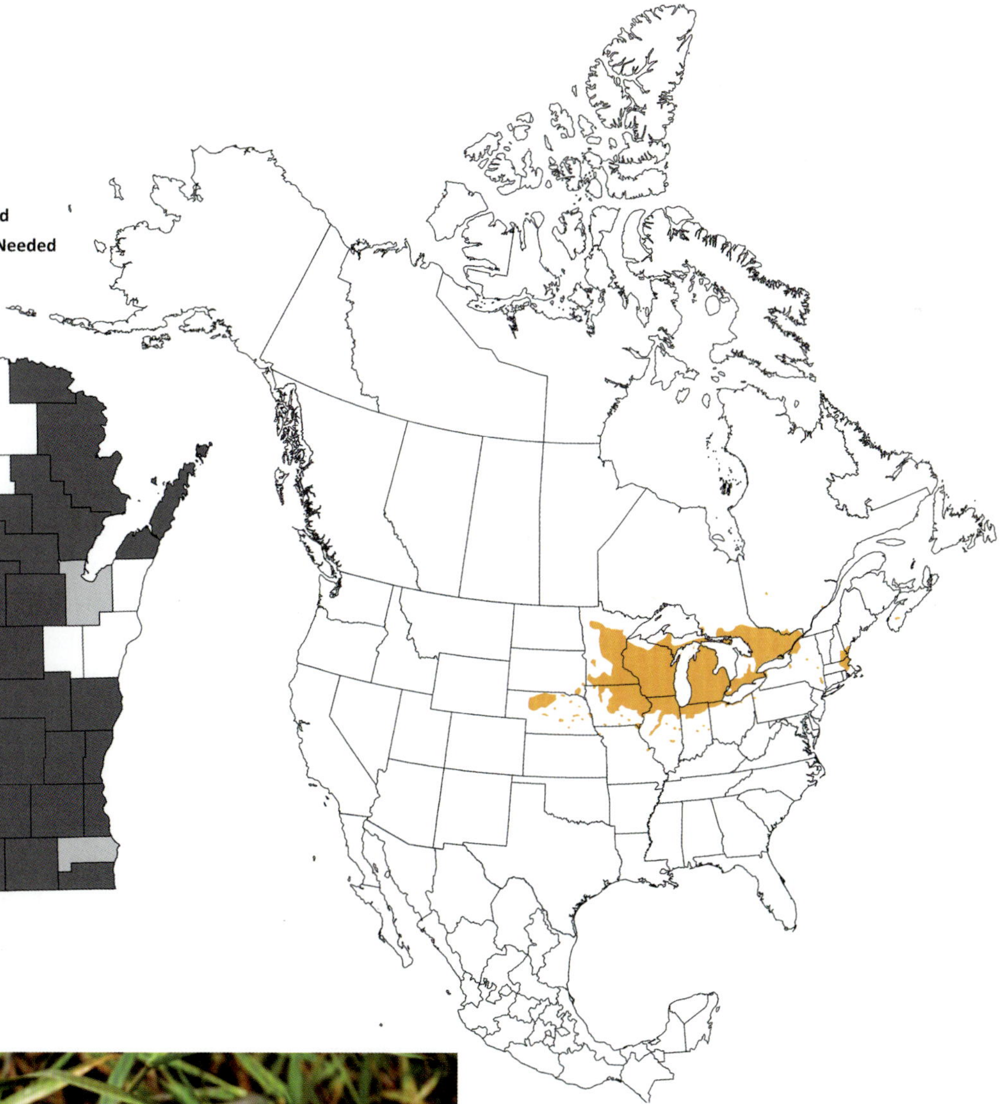

Adult Blanding's Turtle (La Crosse County; photo by L. Huse).

AT A GLANCE

Blanding's Turtles are a medium-sized species with a bright yellow chin and a dark highly domed carapace. They have a somewhat broad distribution in Wisconsin, although reports are less common from the north-central region. Despite this, Blanding's Turtles are generally considered rare and seldom seen outside of the nesting season, when females are sometimes observed crossing roads. Blanding's Turtles typically emerge from winter dormancy in late March, although they can be active earlier. During the active season, this turtle often moves among various habitat types, utilizing the diverse array of resources available within large habitat complexes. As a result, Blanding's Turtles are found in many types of open aquatic habitat with slow-moving or still water, including wetlands, lakes, and stagnant streams or river backwaters. Some individuals also spend a substantial amount of time on land during the summer months, particularly when moving between aquatic habitats. Blanding's Turtles have long life spans and do not reach reproductive maturity until 14 to >20 years of age. This delay in maturity puts the species at high risk of meeting an untimely end before they are able to reproduce. The primary breeding period is April to July, and females will often nest from late May to early July. Although nest survival can be low, hatchlings from successful nests will emerge from late August to mid-September. This species is frequently considered omnivorous but feeds largely on aquatic invertebrates. By late October, Blanding's Turtles move to overwintering locations that are often in deeper aquatic habitats, where they can remain submersed until spring. Blanding's Turtles are listed as globally endangered by the IUCN due to habitat loss and low recruitment. Despite their imperiled status across much of North America, Blanding's Turtles were removed from Wisconsin's Threatened species list in 2014 and are currently considered a species of Special Concern and a Species of Greatest Conservation Need by the WDNR.

SYSTEMATICS

Emydoidea blandingii (Blanding's Turtle) is currently the only known living representative of the genus *Emydoidea* (Crother 2017). Holbrook (1838) originally described *E. blandingii* as *Cistuda blandingii,* named after the early nineteenth-century naturalist William Blanding. Some researchers have disagreed about the evolutionary history of this species. Pritchard (1979) stated that *E. blandingii* was congeneric with *Emys orbicularis* (European Pond Turtle) because of anatomical and functional similarities and the presence of a plastral hinge. Based on molecular data, Feldman and Parham (2002) placed *Emydoidea blandingii, Emys orbicularis,* and *Emys trincaris* (Sicilian Pond Turtle) under the genus *Actinemys* as a congener of *A. marmorata* (Western Pond Turtle). Yet Fritz et al. (2011) reviewed existing information in order to argue instead that *Actinemys, Emydoidea,* and *Emys* were distinct genera. Further review of this topic is available in Crother (2017), who also considered *Emydoidea blandingii* the appropriate scientific name for this species. There are no recognized subspecies of *E. blandingii,* but three distinct evolutionary units apparently exist based on geography: (1) populations west of the Appalachian Mountains; (2) populations east of the Appalachian Mountains (specifically, southern New York and New England); and (3) disjunct populations in Nova Scotia (Mockford et al. 2007). Among Wisconsin turtles, *E. blandingii* occurs in the family Emydidae, which includes *Chrysemys picta* (Painted Turtle), *Glyptemys insculpta* (Wood Turtle), *Terrapene ornata* (Ornate Box Turtle), and three species of *Graptemys* (map turtles). Furthermore, it is a member of the subfamily Emydinae, as are *T. ornata* and *G. insculpta.*

DESCRIPTION

Adults: Blanding's Turtles have a bright yellow chin and neck and an elongated carapace that is more domed than that of most other aquatic turtles in Wisconsin (Knudsen 1957b, Vogt 1981, Ernst et al. 1994; Figure 324). The carapace is dark brown to black with varying degrees of yellow speckles, although some individuals have horn-colored carapaces with brown mottling. The plastron is typically yellow to gold with dark brown blotches toward the lateral third of each scute (Figure 325a) but is occasionally solid red brown to blackish blue in color (Figure 325b). The plastron is also hinged between the pectoral and abdominal scutes (Ernst et al. 1994), allowing the anterior portion to partially close toward the carapace. The limbs and head are usually brown to black, often with pale yellow or light brown speckles, although head patterns can vary considerably (Figure 326).

Blanding's Turtles are a medium-sized turtle. Females sampled in central Wisconsin ranged from 17.6 to 23.8 cm MCL (6.9 to 9.4 in, $\bar{x} = 21.5$ cm [8.5 in], $n = 59$), while males from the same region ranged from 19.5 to 25.5 cm MCL (7.7 to 10.0 in, $\bar{x} = 22.4$ cm [8.8 in], $n = 37$; Jonathan Zellmer, independent researcher, unpublished data). Mean adult weight was also slightly lower in females from this population (1,436 g [3.2 lb], $n = 59$) than in males (1,509 g [3.3 lb], $n = 33$; J. Zellmer, unpublished data). The largest Blanding's Turtle reported in the literature was a wild male from Minnesota whose MCL was 28.4 cm (11.2 in; Blasus et al. 2004).

Preadult stages: The elliptical eggs of Blanding's Turtles are whitish and range from 3.4 to 4.1 cm (1.3 to 1.6 in) long by 2.1 to 2.8 cm (0.8 to 1.1 in) wide in Wisconsin (Vogt 1981; Figure 327). Congdon and van Loben Sels (1991) reported that egg weights in nearby Michigan ranged from 11.0 to 13.9 g (0.4 to 0.5 oz). The carapace of hatchlings is dark gray to brown and is not domed like that of adults (Figure 328). The plastron is primarily black except near the scute margins in most individuals, and the chin color is pale compared to the bright yellow chin of adults (Figure 329). The tail length:CL ratio is also higher in hatchlings than in adults (Figure 328). Vogt (1981) reported that hatchlings ranged in size from 2.8 to 3.5 cm (1.1 to 1.4 in). In northeastern Illinois,

Left, top: Figure 324. The obvious yellow chin of a female Blanding's Turtle (Grant County; photo by J. Kapfer).

Left, bottom: Figure 325. Plastron coloration for subadult and adult Blanding's Turtles. (a) Typical coloration. Note the rust-colored staining on the yellow portions of the scutes (Dane County; photo by R. Hay). (b) A plastron that is stained black from prolonged contact with organic substrates (Grant County; photo by J. Kapfer).

Left, top: Figure 326. Variable head patterns among individual turtles (Dane County; photos by R. Hay). (a) Head with dark base coloration and light-colored spotting. (b) Head with light base coloration and dark-colored spotting and lines.

Left, bottom: Figure 327. Recently excavated Blanding's Turtle eggs (Chippewa County; photo by R. Sajdak).

Above, top: Figure 328. Hatchling Blanding's Turtle. Note the limited pattern, flat carapace profile, and long tail relative to CL (Iowa County; photo by R. Hay).

emerging hatchlings had a mean CL of 3.3 cm (1.3 in) and a mean weight of 8.0 g (0.3 oz, $n = 42$; Glowacki and Kuhns 2010).

Variation: No significant variation in appearance is known among Wisconsin individuals. There is also no strong sexual dimorphism in color or pattern, although males are larger than females (e.g., Pappas et al. 2000). Males also have a more concave plastron and longer tail than females, with the cloacal opening past the back edge of the carapace.

Diagnosis: Blanding's Turtles can be identified by a dark brown to black carapace with yellow specks that is also more domed than most aquatic turtle species, the presence of keratinized scutes that cover the carapace, the presence of a plastral bridge and hinge (although the plastron typically does not close tightly against the carapace), 12 scutes on the plastron, webbed toes, no prominent beak shape to the upper jaw, and a yellow chin and lower neck without dark marks.

Adult Blanding's Turtles are distinct in appearance from most other Wisconsin turtles. Yet biologists that lack experience with turtles sometimes have difficulty distinguishing them from adult Wood

Figure 329. Plastron pattern and pale chin color of a juvenile Blanding's Turtle (Dane County; photo by R. Hay).

Turtles, as the plastrons of both species are similar in color and pattern (Robert Hay, Turtles for Tomorrow, personal observation). Even young Blanding's Turtles also have a flatter carapace than adults and somewhat distinct annual growth rings, which could cause further confusion with Wood Turtles. However, hatchling Blanding's Turtles have a dark gray to brown carapace, whereas the carapace color of hatchling Wood Turtles is light brown to greenish gray (Oldfield and Moriarty 1994). The carapace of young and adult Wood Turtles also has a low middorsal keel, which Blanding's Turtles lack. In addition, the carapace of adult Wood Turtles is usually rougher in texture than that of adult Blanding's Turtles and is not as dome-shaped.

Blanding's Turtles could be misidentified as Ornate Box Turtles (*Terrapene ornata*), because both species have highly domed shells and hinged plastrons. Such confusion could only occur at unique locations where the terrestrial Ornate Box Turtle exists and the aquatic Blanding's Turtle comes ashore to nest. Regardless, Blanding's Turtles are much larger than Ornate Box Turtles as adults, and juvenile Blanding's Turtles do not have the domed shell of younger Ornate Box Turtles. Carapace and plastron color also usually differ substantially between these species. Although Eastern Musk Turtles are aquatic and have domed shells, they do not have plastral colors or patterns, nor do they have the bright yellow chin of Blanding's Turtles.

Turtle eggs encountered in the wild are difficult to identify. Blanding's Turtle eggs are similar in size and general appearance to Northern Map Turtle and Wood Turtle eggs. Yet Blanding's Turtle eggshells are thicker due to a coarse calcified layer that makes them firmer and rough-textured (gently rubbing them together produces a gritty sound) until later during incubation (R. Hay, personal observation).

DISTRIBUTION AND HABITAT

Global distribution: Blanding's Turtles are primarily found in the Great Lakes region of the US and Canada. There are scattered disjunct populations in eastern areas such as southern Maine, southeastern New Hampshire, eastern Massachusetts, and portions of New York but also in western locations such as extreme southeastern South Dakota, parts of Nebraska, and even northern Missouri (Graham et al. 1987, Congdon et al. 2008).

Distribution in Wisconsin: Published county records exist for much of the state, with the exception of some counties in the extreme north-central and northeastern regions. Of the published county records, suitable museum vouchers could not be found for Brown and Racine (see the Appendix).

Localities of possible occurrence: Blanding's Turtles likely occur in several counties where there are currently no published records, including the far northern counties of Forest, Iron, and Vilas. While Blanding's Turtles are not documented in these counties, they have sandy soils, which are frequently used by Blanding's Turtles for nesting (see Habitat subsection). They also contain portions of watersheds that are occupied by Blanding's Turtles in adjacent counties. Plausible reports that are not supported with voucher material also exist from Lincoln, Manitowoc, and Price Counties (WDNR 2016b). The lack of reports in some counties could simply be due to limited survey effort or because resident populations are too small to easily detect.

Habitat: Blanding's Turtles use many types of open-canopy waterbodies with little to no current. These include marshes, ponds, shallow bays of lakes, ditches, backwater sloughs or oxbows associated with rivers, and smaller streams (Vogt 1981, Oldfield and Moriarty 1994, reviewed by Ernst and Lovich 2009). This turtle associates with such a wide variety of aquatic habitats in part because it frequently moves throughout large areas that contain many contiguous or connected waterbodies (Refsnider and Linck 2012).

Although Blanding's Turtles will use a variety of aquatic habitats, they appear to show preferences for certain types. For example, distribution modeling in Ontario supported that Blanding's Turtle presence was best predicted by large wetlands that possessed limited open water because they contained ample aquatic vegetation (Millar and Blouin-Demers 2012). Furthermore, adults often associate with eutrophic aquatic habitats that are shallow (<1.5 m [4.9 ft]) and that possess soft organic substrates (rather than sandy substrates), emergent vegetation along shorelines, dense submersed vegetation (e.g., Spiny Naiad [*Najas flexilis*], Coontail [*Ceratophyllum demersum*]), and floating vegetation (e.g., Pondweed [*Potamogeton pectinatus*], Duckweed [*Lemna minor*]; Ross and Anderson 1990, Pappas and Brecke 1992). A study in Wisconsin further found that Blanding's Turtles selected wetlands with relatively high total nitrogen, total phosphorus, and dissolved oxygen levels (Ross and Anderson 1990). Research in Maine revealed that seasonal shifts occurred in the aquatic habitats used by adult Blanding's Turtles, with higher use of forested wetlands in spring, open wetlands in early summer, and wetlands with deep water during late summer and autumn (Beaudry et al. 2009). Yet not all studies have detected seasonal shifts in habitat use (e.g., Edge et al. 2010), and variability among populations likely occurs. Although often observed in aquatic habitats, Blanding's Turtles sometimes spend a substantial amount of time on land. Terrestrial habitats are used as travel corridors between wetlands and for nesting, as well as for aestivation under certain conditions in the summer (see Activity section).

Less is known about the habitat preferences of juvenile Blanding's Turtles. In Minnesota, they appear to spend much time in aquatic habitats that possess shallow water (10–30 cm [4–12 in]) among scattered clumps of emergent vegetation (Pappas and Brecke 1992). These clumps are often closely spaced (<2 m [6.6 ft]) emergent root hummocks of plant species such as Speckled Alder (*Alnus rugosa*) and Bottlebrush Sedge (*Carex comosa*), which juveniles often bask on (Pappas and Brecke 1992). As they grow, juveniles move closer to the interface between emergent and open water, and subadults eventually move to deeper water with dense submersed vegetation that is more often associated with adults (Pappas and Brecke 1992).

Nesting habitats typically consist of dry sandy soils in open-canopy areas (e.g., dry prairies and old fields; R. Hay, personal observation). However, Blanding's Turtles will also attempt to nest in disturbed habitats, and Beaudry et al. (2010) reported that 84% of observed nests in Maine occurred within anthropogenic locations (see Conservation section). In Wisconsin, they are known to opportunistically nest in sandy agricultural fields in June when row crops are still short and well-spaced (R. Hay, personal observation). In locations where sandy soils are limited or inaccessible, females occasionally nest in richer soils, including clay. Unlike sandy soils, clay soils often have lower oxygen levels, warm up more slowly in spring, are poorly drained compared to sandy soils, and often remain cooler during the growing season. These characteristics potentially result in delayed embryonic development, especially in row-cropped agricultural fields. Nesting success appears to be lower when eggs are deposited in such soils (Gary Casper, UW–Milwaukee Field Station, personal communication).

Blanding's Turtles can overwinter while submersed in a variety of aquatic habitats with organic substrates (Ross and Anderson 1990, Ernst et al.

1994). Ross and Anderson (1990) found that many Wisconsin individuals overwintered in the deeper portions of ponds and streams that were less likely to freeze ($\bar{x}$ overwintering depth = 0.9 m [3.0 ft]). Thiel and Wilder (2010) observed that adult Blanding's Turtles will even overwinter in human-made or human-altered habitats (e.g., ditches, impoundments, and borrow pits).

REPRODUCTION AND DEVELOPMENT

Blanding's Turtles can breed from March to November, but spring and early summer (i.e., April–July) appear to represent the primary mating periods (Conant 1951, Graham and Doyle 1979, Vogt 1981). Elaborate male courtship behaviors have been documented, which include aggressive mounting, gulping of water and expelling it over the female's snout, neck waving that contacts the female's snout, and vigorous side-to-side swaying of the carapace prior to copulation (Graham and Doyle 1979, Vogt 1981, Baker and Gillingham 1983, reviewed in detail by Ernst and Lovich 2009). There appears to be variation in reproductive strategies by geographic location. For example, research on Minnesota (Refsnider 2009) and Michigan (McGuire et al. 2013) populations supports that females mate with several males, as multiple paternity is common. On the other hand, Anthonysamy et al. (2014b) reported that multiple paternity was less common in a population from Illinois, perhaps because of habitat degradation or small population size, both of which likely result in limited options for reproduction.

The duration of gravidity varies depending on when breeding occurred and spring temperatures, particularly during the month of April (Congdon et al. 1983). Nesting in Wisconsin has been observed from late May to early July, similar to observations from other states (Bleakney 1963, Congdon et al. 1983, Rowe and Moll 1991, Rowe 1992b). There is a relationship between environmental temperatures and nesting phenology in several upper midwestern states. For example, a 16-year study of Wisconsin populations determined that nesting phenology was strongly correlated with temperature in March, with warmer years leading to earlier nesting (Byer et al. 2020). Congdon et al. (1983) found that in Michigan, the onset of nesting was significantly correlated with mean minimum and mean daily temperatures in April but not May. In northeastern Illinois, initiation of nesting was negatively correlated with mean daily temperature between April and May (i.e., a degree increase in mean temperature during these months triggered nesting to occur a week earlier; Buckhardt et al. 2020). These authors further reported that females typically nested on evenings when mean daily air temperature exceeded 18.9°C (66°F), during the brightest moon phase, and when wind direction was east or south. In southern Wisconsin, Blanding's Turtles usually begin to nest ca. one week after Painted Turtles and Snapping Turtles (*Chelydra serpentina*; Thomas LaFleur, Turtles for Tomorrow, and R. Hay, personal observations). In northwestern Wisconsin, Blanding's Turtles often nest synchronously with Painted Turtles, Snapping Turtles, and Wood Turtles (Gregory Kessler, WDNR, personal communication). After the nesting season initiates in a given year, it lasts from 16 to 30 days (Congdon et al. 1983). Most females begin the process of nesting in the evening and often finish well after dark (Vogt 1981, Ernst et al. 1994). They can also spend up to several days on land prior to nesting, during which time they search for an appropriate nesting site, but during warm periods they cease activity and take refuge in the shade of trees or under shrubs and leaf litter (R. Hay, personal observation). One or more nesting attempts may be started but abandoned by females before actual egg deposition occurs (Congdon et al. 1983).

Blanding's Turtles can move substantial distances to nest. Ross and Anderson (1990) found that the mean distance of nest sites from the nearest wetland was 168 m (551 ft, $n = 16$). Furthermore, nesting forays that extend well beyond 300 m (984 ft) of the water's edge are quite common in Wisconsin (R. Hay, personal observation). In Maine, distances traveled by females ranged from 100 to 1,620 m (328 to 5,314 ft; Joyal et al. 2001), and J. Congdon observed nesting as far as 2,400 m (7,874 ft) from water (personal communication in Ross and Anderson 1990). In Michigan, the majority of females monitored over time exhibited nest-site fidelity (i.e., 73%, $n = 11$), although some nested up to 1,300 m (4,265 ft) from previous nesting locations (Congdon et al. 1983). Congdon et al. (1983) found that females did not nest every year ($\bar{x}$ = 0.48 clutches per female per year) but that nesting frequency increased in females at least 55 years old ($\bar{x}$ = 0.57 clutches per female per year, compared to 0.35 clutches per female per year for those that were 20–30 years old). Females also do not produce multiple clutches per year (Congdon et al. 1983). Refsnider (2009) suggested that females can store sperm internally between years for delayed fertilization.

Clutch size is positively correlated with adult female length and mass (MacCulloch and Weller 1988, Congdon and van Loben Sels 1991) and varies widely across Wisconsin. The mean clutch size of 39 gravid females collected along the lower Wisconsin River (MCL = 19–21 cm [7.5–8.3 in]) was 10 (range = 8–11; R. Hay, unpublished data). By comparison, 9 clutches encountered in central Wisconsin had a mean clutch size of 17.3 eggs, and females from this population ranged in PL from 20.2 to 23.9 cm (8.0 to 9.4 in, $\bar{x}$ = 21.9 cm [8.6 in], *n* = 60; J. Zellmer, unpublished report to the WDNR). The length of egg developmental periods varies as a function of environmental and nest temperature. In Michigan, days from egg deposition to hatchling emergence ranged from 73 to 104 (Congdon et al. 1983). In Wisconsin, emergence from most nests occurs from late August to mid-September (T. LaFleur, personal communication). The sex of hatchlings is affected by incubation temperature. Clutches artificially incubated at 22.5°C–26.5°C (72.5°F–79.7°F) produce 97%–100% males, and clutches incubated at 30°C–31°C (86.0°F–87.8°F) produce only females (Gutzke and Packard 1987, Ewert and Nelson 1991).

Most nests do not produce viable hatchlings, as they either are eaten by predators or consist of infertile eggs, or the hatchlings do not develop. Congdon et al. (2000) reported that only 18% of the 182 nests they observed in Michigan over 22 years were partially or completely successful. At the same location, nest survival rates averaged 3.3% annually (range = 0%–17%, *n* = 75) from 1985 to 1991 due to an increase in predation by mesopredator mammal species (Congdon et al. 1993; see Prey and Predators and Conservation sections). Congdon et al. (2001) also reported that nests of younger females were more often destroyed by nest predators than nests of older females. The authors suggested that perhaps older females learn to nest in safer locations with experience but noted that sampling artifacts could have confounded their results. Hatchling ability to orient themselves immediately after emerging from nests has received some attention. McNeil et al. (2000) found that upon emergence from nests, hatchlings did not specifically orient toward nor seek water in Nova Scotia. Research in Minnesota suggested that hatchlings orient themselves toward dark horizons up to 325 m (1,066 ft) away, perhaps because dark horizons could represent or be associated with denser vegetation along the edges of aquatic habitats (Pappas et al. 2009). Paterson et al. (2012) found that 16% (*n* = 48) of radio-tracked hatchlings were eaten by predators, 2% died on roads, and 2% were desiccated prior to their first winter. Similarly, the survival of radio-tracked hatchlings in Rock County was high for ca. four weeks after hatching and into autumn (Kastle et al. 2021).

Blanding's Turtle growth rates vary with age. In Wisconsin, Ross (1989b) determined that hatchlings increased their size during the first year of life by 85.9%. He also reported that Blanding's Turtles grew 9.3 cm (3.6 in) in PL from the time of hatching to seven years of age. This is similar to growth reported for the same age range in Michigan (9.3 cm [3.7 in] CL; Congdon and van Loben Sels 1991). In Michigan, juveniles also showed fairly linear growth rates from ages 4 to 13, adding a mean of 1.04 cm (0.4 in) to CL and 75.3 g (2.7 oz) to body weight per year. Growth slowed considerably between 16 and 18 years of age (Congdon and van Loben Sels 1991). Research on populations in Nova Scotia supported that both sexes exhibit indeterminate growth patterns and that adult males grew at a faster rate than adult females (Lefebvre et al. 2011).

Sexual maturity in Blanding's Turtles is determined by size rather than age (Congdon et al. 2001). However, size at maturity varies among populations (Congdon and van Loben Sels 1991). In a Wisconsin population, female Blanding's Turtles matured at 17.2 cm PL (6.8 in) and 18 years of age (*n* = 19 mature and 4 immature females; Ross 1989b). In contrast, Congdon and van Loben Sels (1991) documented a 14-year-old mature female in Michigan with a 15.7 cm PL (6.2 in). Congdon and van Loben Sels (1991) further stated that the age of first reproduction in females ranges from 14 to 20 years of age and that juvenile growth rates and ages at maturity are the primary cause of variation in body size among adults (Congdon and van Loben Sels 1993).

The age of maturity for males has not been studied in Wisconsin. In Massachusetts, Graham and Doyle (1977) suspected that males matured more quickly than females, at 18–19 cm CL (7.1–7.5 in) and 12 years of age. It appears likely that most male Blanding's Turtles take longer to mature in Wisconsin than in Massachusetts based on observations of subadult males that have yet to develop all of their secondary sexual characteristics by ages 12–16 (R. Hay, personal observation). This species is long-lived, and longevity estimates ranging from 45 years in Nebraska to >60 years in Michigan exist (Germano et al. 2000, Congdon et al. 1993, respectively). The oldest known Blanding's Turtle was captured in the wild (Chisago County, Minnesota) in 1922 and recaptured while nesting in 1982 (Brecke and Moriarty 1989). The authors estimated that the turtle was ≥77 years old at the time of last capture.

ACTIVITY

Activity schedules and movement patterns are apparently influenced by many environmental factors (e.g., water temperature, water quality, vegetation structure, substrate composition, prey abundance) and behaviors (e.g., mating, nesting, aestivation, overwintering; Bennett et al. 1970, Buhlmann and Gibbons 2001, Bowne et al. 2006). Emergence from hibernation in particular is heavily affected by environmental conditions. Some turtles begin to emerge from underwater overwintering locations when the ice thaws around wetland and stream edges as air temperatures reach or exceed 10°C (50°F) on sunny spring days (R. Hay, personal observation). Blanding's Turtles in Wisconsin can emerge from hibernation in late March (Samantha Foster and R. Hay, personal observations), but this varies annually based on weather. For example, reports exist of actively basking Blanding's Turtles in southern Wisconsin as early as late February under particularly warm conditions (T. LaFleur, personal observation). Wisconsin individuals have also been observed at the peripheries of mostly frozen wetlands in early spring attempting to emerge from under the edges of ice (G. Casper, personal communication). In Minnesota, emergence tends to begin ca. two weeks later than in southern locations, where spring temperatures increase more rapidly (Piepgras and Lang 2000).

After emergence from hibernation, there are reportedly two peaks in activity: the first from early April to early July, and the second from mid-August to mid-September (Kofron and Schiefer 1985). Research in central Minnesota found that active thermoregulation is most common in May and that variation in body temperature often follows a daily pattern: gradual to rapid heating in the morning, a period of plateau during midday, rapid cooling in the afternoon, and then gradual cooling throughout the evening (Sajwaj and Land 2000). Blanding's Turtles have a lower critical thermal maximum (CTmax, 39.6°C [103.3°F]) than many other turtle species, such as Painted Turtles (40.9°C [105.6°F]; Hutchison et al. 1966). This tolerance of higher temperatures could facilitate this species' tendency to concentrate most basking activity to spring and early summer while also selecting colder water in the warmer summer months (Millar and Blouin-Demers 2011).

During the warmest part of the summer, Blanding's Turtles in Wisconsin sometimes aestivate in either terrestrial or aquatic microhabitats (Ross and Anderson 1990, S. Foster, unpublished data, S. Thiel, personal communication). Ross and Anderson (1990) observed one turtle from Adams County that aestivated for more than two days in an upland location on 15 separate occasions from 21 July to 28 August (ambient air temperature during aestivation ranged from 18°C to 33°C [64.4°F to 91.4°F]). During the same study, two turtles aestivated while partially buried in wet silt under a cattail mat for extended periods (water temperature = 18.0°C–28.5°C [64.4°F–83.3°F], air temperature = 27.0°C–37.5°C [80.6°F–99.5°F]). Furthermore, a radio-tracked subadult from Walworth County remained inactive on land at the same location for nearly five weeks in the summer. It then moved to a dry ephemeral basin, where it buried in mud for the first 24 days of September before moving to deep water within the same wetland (S. Foster, personal observation). In southern Maine, Joyal et al. (2001) also recorded long periods of late summer dormancy in upland forested areas, where turtles spent from 9 to 21 days out of water, buried in leaf litter at distances of up to 110 m (361 ft) from the nearest wetland.

Movement patterns are often driven by this turtle's tendency to travel substantial distances among wetlands during its normal activity season. For example, Edge et al. (2010) reported that the mean annual interwetland movement in Ontario was 231 m (758 ft) for males and 497 m (1,631 ft) for females. However, mean daily movements reportedly differ greatly among populations. Blanding's Turtles in Michigan have been documented traveling >2,000 m (6,562 ft) between wetlands (Sexton 1995). Ross (1989b) used captures in hoop nets to study interwetland movements in Adams County and reported recapture distances that ranged from 212 to 652 m (696 to 2,139 ft, $\bar{x} = 396$ m [1,299 ft], $n = 17$). Large variability in home range size also apparently exists by geographic location throughout the Midwest. For example, the mean summed activity centers for six male and five female Blanding's Turtles in northern Illinois was 1.3 ha (3.2 ac; Rowe and Moll 1991). In south-central Minnesota, mean estimates of home range size were 35.4 ha (87.5 ac, $n = 13$) for females and 38.4 ha (94.9 ac, $n = 6$) for males (MCP estimator; Piepgras and Lang 2000). Even larger home range estimates were reported in southeastern Minnesota, where Hamernick (2001) determined that mean female home range size was 60.8 ha (150.2 ac, $n = 16$) and mean male home range was 94.2 ha (232.8 ac, $n = 8$, MCP estimator). In central Wisconsin, Ross and Anderson (1990) reported that the size of activity centers was 0.57 ha (1.41 ac) and 0.92 ha (2.27 ac) for two radio-tracked males, while the mean for six females was 0.56 ha (1.38 ac). These are substantially smaller than the estimated home range sizes in Illinois and Minnesota, although home range estimators differed among studies.

Variation in movement appears to occur by sex and female gravidity. Ross and Anderson (1990) found that the mean minimum daily distances moved for adult females was 95.1 m (312.0 ft, $n = 56$ observations of 6 turtles) but for adult males was 48.4 m (158.8 ft, $n = 21$ observations of 2 turtles). Gravid females in Ontario had much larger mean home range sizes (20.3 ha [50.2 ac], $n = 4$) than nongravid females (7.3 ha [18.0 ac], $n = 3$) and males (8.2 ha [20.3 ac], $n = 6$, MCP estimator; Millar and Blouin-Demers 2011). The larger mean daily travel distances in gravid females are likely attributable to the fact that females must migrate to and from nesting sites (Ross and Anderson 1990, S. Foster, personal observation).

Numerous studies have reported that movement patterns can further change seasonally (Grgurovic and Sievert 2005) and across years (Schuler and Thiel 2008). This might be explained by the results of studies that show that movement is affected by a host of factors from food availability (Cagle 1944) to habitat variability (Bury 1979), which could change seasonally and annually. Anthonysamy et al. (2013) further found evidence that shifts in movement patterns can be driven by environmental variation (e.g., drought) that forces individuals to migrate from drying wetlands to more stable aquatic habitats (e.g., rivers and deeper marshes). Fewer data exist on hatchling and juvenile movement (see post-hatching dispersal information in Reproduction and Development section). McMaster and Herman (2000) reported that juveniles from Nova Scotia did not travel as widely as adults and that home range size and daily movement distance were positively correlated with age.

In Wisconsin, Blanding's Turtles move to overwintering sites from late September to late October (Ross and Anderson 1990, S. Foster and R. Hay, personal observations). Changes in water temperature, photoperiod, food supply, and rainfall may induce the onset of overwintering behavior in this species (Gregory 1982). Ross and Anderson (1990) hypothesized that Blanding's Turtles chose the waterbodies where they would overwinter in early autumn based on warm water temperatures and stable water levels. Although the turtles' activity declines by late October as they begin to enter hibernation, there are reports of active Blanding's Turtles in November from Indiana and Wisconsin (Evermann and Clark 1916, Ryan Jacques, UW–Whitewater, personal observation). In more southern parts of their North American range, Blanding's Turtles continue to move short distances during winter (seasonal movements <2 m [6.6 ft]; Kofron and Scheiber 1985). In Wisconsin, however, winter movements are apparently limited, and turtles often opt to remain buried in organic aquatic substrates (Thiel and Wilder 2010). Despite this, movements below the ice in late November are still sometimes observed (S. Foster, personal observation).

Thiel and Wildler (2010) found that adults in Wisconsin could successfully survive the winter even after exposure to frozen aquatic substrates for extended periods of time under anoxic conditions. The authors also reported that overwintering individuals experienced periods when their body temperatures were slightly below freezing (i.e., temperatures between 0°C and −1°C [32°F and 30.2°F]) but still survived. Adults that overwinter underwater apparently do not select locations with high dissolved oxygen levels and instead select thermally stable but cold (ca. 0°C [32°F]) water to reduce their metabolic rate and slow their oxygen demand (Edge et al. 2009). Gregory (1982) observed communal overwintering in Nova Scotia, which he suggested occurred because suitable hibernacula were scarce. In Wisconsin, it is not uncommon for multiple turtles to overwinter in the same wetland (Ross and Anderson 1990, Thiel and Wilder 2010, S. Foster, personal observation). Thiel and Wilder (2010) reported that Blanding's Turtles often select different overwintering locations across years.

Winter behavior and physiology of both hatchlings and adults have received some attention. Laboratory research on hatchlings found they can tolerate frozen body fluids and even recover after being frozen at 3.5°C (−37.4°F) for 72 hours (Dinkelacker et al. 2004). These findings support that terrestrial hibernation is also possible, but selected locations must be moist enough to resist dehydration (Dinkelacker et al. 2004). Dinkelacker et al. (2005) further found evidence that hatchlings are not physiologically suited to tolerate extensive submergence in hypoxic water. The authors suggested that, unlike adults, hatchlings that overwinter underwater are likely restricted to well-oxygenated locations or sites where periodic thaws allow access to air at the water's surface.

PREY AND PREDATORS

Blanding's Turtles are primarily carnivorous but have long been considered omnivores due to the frequency of plant matter in the stomachs of analyzed individuals (Lagler 1943, Kofron and Schreiber 1985). More recent studies indicate that although plant matter is often found in the stomach, it comprises only a small percentage of the total gut contents (Rowe 1992a). Furthermore, Rowe (1992a) suggested that much of the ingested plant material is probably consumed incidentally while turtles feed on invertebrates or other prey. Regardless, the list of reported food items is diverse. For example, they consume earthworms (Lumbricidae), leeches (Hirudinea), slugs (*Gyraulus parvus*), grubs, grasses, berries, and succulent vegetation (Cahn 1937, reviewed by Ernst and Lovich 2009).

In Wisconsin, Blanding's Turtles reportedly feed on snails, crayfish, insects, frogs, frog larvae, fishes, fish eggs, and duckweed (Vogt 1981, Ross 1989b). Knudsen (1957b) also specifically observed Wisconsin individuals consuming "dragonfly and mayfly nymphs." Investigations conducted in other states report that crustaceans are a primary dietary component (Penn 1950). Lagler (1943) analyzed the gut contents of 67 individuals from Michigan and reported that crustaceans and insects comprised 56.6% and 21.4% of the dietary volume, respectively. The diet of juvenile Blanding's Turtles is poorly understood because few dietary reports from younger age classes exist. A single fecal sample from a Rock County juvenile contained primarily invertebrates (98% of sample contents; Rutzen et al. 2015). The families of invertebrates identified were mostly those that live at water depths <30 cm (12 in), are good swimmers, and spend significant time attached to the stems of emergent and submersed plants (Rutzen et al. 2015). For a more comprehensive list of recorded Blanding's Turtle prey items, see Ernst and Lovich (2009).

Adult Blanding's Turtles have no documented predators other than humans. Mammals such as North American River Otters (*Lontra canadensis*) can likely remove a turtle's appendages, but whether this causes direct mortality remains unknown. Younger age classes, particularly eggs, are more vulnerable to predators than adults. In fact, nest predation is often the primary cause of egg mortality, with annual nest predation rates that are near 100% in some populations (Congdon et al. 1983). A multidecade monitoring project on the nesting habits of this species in Michigan revealed that 79% of 182 nests were destroyed by mammals (Congdon et al. 2000), which raises important conservation concerns for this slow-to-mature species (see Conservation section). Major nest predators include Raccoons (*Procyon lotor*), Red Foxes (*Vulpes vulpes*), Coyotes (*Canis latrans*), and Striped Skunks (*Mephitis mephitis*; for a more exhaustive list of nest predators, see Ernst and Lovich [2009]). Even if they survive through hatching, young turtles are still vulnerable to predators as they move from nesting sites to aquatic habitats, especially in suburban areas (Jones and Sievert 2012). Ernst and Lovich (2009) suggested that a variety of birds and mammals likely prey on hatchlings and juveniles, and Northern Short-tailed Shrew (*Blarina brevicauda*) predation was documented in Nova Scotia (Standing et al. 2000a). As they mature, the turtles' vulnerability to predators appears to decline. Blanding's Turtles can employ various strategies to offset the threat of mortality due to predators. For example, they often bask near the water and quickly dive below the surface when approached (S. Foster, personal observation). On land, subadult and adult Blanding's Turtles can also use their hinged plastron to partially close their shells and protect against predators (Oldfield and Moriarty 1994). *Placobdella ornata* is a documented parasite of Blanding's Turtles in Wisconsin (Bolek 2001), but there is no evidence that such interactions cause direct mortality. DeFoliart (1967) also documented parasitism of Blanding's Turtles by the mosquito *Aedes canadensis*.

CONSERVATION

Status: Globally, the Blanding's Turtle is listed as Endangered by the IUCN. It has also been proposed for listing under Appendix II of CITES (CITES 2013) and is currently under review for listing under the Endangered Species Act. The WDNR estimated a 30%–50% long-term decline in populations based on habitat loss, as well as recent and ongoing short-term declines ranging from 10% to 30% (WDNR 2011b; see Natural History Box: Status Change for Blanding's

NATURAL HISTORY BOX: Status Change for Blanding's Turtles in Wisconsin

The delisting of the Blanding's Turtle as a Threatened species in Wisconsin in 2014 has many herpetologists, biologists, and nature lovers concerned about the future of this species in the state. It is currently listed in the state as a species of Special Concern, which does not require efforts to avoid or minimize mortality. Blanding's Turtles were first included on the state's Endangered and Threatened Species List in 1979 because biologists from within and outside the WDNR believed that habitat loss and fragmentation, along with low evidence of recruitment, were threatening the long-term persistence of the species in the state. A WDNR status assessment report (WDNR 2011b) supported the original basis for listing, stating that Blanding's Turtles had experienced a 30%–50% decline related to wetland habitat loss and estimated that they may currently be declining at a rate of 10%–30%! Many of the threats identified in the Conservation and Management chapter of this book and the conservation information in this species account have increased in intensity over the past several decades, and Blanding's Turtles clearly need support from conservation biologists and natural resource regulators. The authors of this account recommend its protection as both a state and federal Threatened species and the initiation of conservation actions in Wisconsin to ensure its long-term persistence.

Robert W. Hay, Turtles for Tomorrow

Turtles in Wisconsin). However, in 2014 the Blanding's Turtle was downlisted from a Threatened species to a species of Special Concern in Wisconsin (WDNR 2014a). It is currently considered a Species of Greatest Conservation Need and a Protected Wild Animal in Wisconsin by the WDNR.

Populations: Few robust studies of population characteristics in Wisconsin exist. Evrard and Canfield (2000) determined that the number of individuals at a Burnett County site was possibly between 107.3 and 161.3 but admitted that the estimates were questionable. Ross and Anderson (1990) estimated that the density of individuals at an Adams County site was 27.5/ha (11.1/ac). Elsewhere, estimates of population density show substantial variation. For example, comparatively low densities were estimated by studies in Massachusetts (6.3/ha [2.6/ac]; Graham and Doyle 1977) and Michigan (8.8–10.0/ha [3.6–4.0/ac]; Congdon et al. 1986), while much higher densities were reported by a study in Missouri (55.0/ha [22.3/ac]; Kofron and Schrieber 1985).

Limited data on other demographic parameters are available, and existing information also supports that variability by sex, age, and across studies occurs. For example, Congdon et al. (2001) determined that survivorship was greater in older adults than in middle-aged adults. Similarly, annual survival estimates for older juvenile turtles in Illinois were substantially greater (67%) than survival estimates for hatchlings (0.07%, from egg deposition through year one; Glowacki and Kuhns 2010). Yet more recent work from northern Illinois and southern Wisconsin support that survival of young turtles immediately after hatching can be high (Kastle et al. 2021). Although adult survivorship is also generally high (e.g., >93%; Congdon et al. 1993), data on annual adult survival by sex are contradictory. Research in Nebraska found that annual survivorship of females (59%) was lower than that of males (90%; Ruane et al. 2008), while research in Michigan detected no difference by sex (Congdon et al. 2001). Reports of population sex ratios also vary across studies. Some researchers have found that populations are skewed toward females (Congdon and van Loben Sels 1991, Ruane et al. 2008), although recent work in Wisconsin found that study populations were male-biased (Reid and Peery 2014). Some populations appear to be dominated by adults, and an investigation in Adams County determined that local populations had substantially more adults than juveniles (1A:0.19J; Ross 1989b).

Threats: Destruction and degradation of aquatic habitats are probably the greatest threats to Blanding's Turtles (see Introduction to Conservation and Management of Wisconsin's Amphibians and Reptiles chapter). The historical expansion of agriculture across Wisconsin resulted in widespread loss of the shallow wetlands required by this turtle (Dahl 1990). It is currently impossible to quantify the full impacts of such extensive habitat loss on Blanding's Turtles in the state due to a lack of historical population data. Yet the results of studies from other regions suggest that increasing agricultural acreage in Wisconsin could have had substantial negative effects on Blanding's Turtles. For example, Millar and Blouin-Demers (2012) determined that the distribution of this species in Ontario was negatively associated with large areas of agricultural land. VanDeWalle and Christiansen (1996) further postulated that draining wetlands to facilitate agricultural activities near rivers probably had substantial negative impacts on Blanding's Turtles in Iowa.

Anthropogenic habitat disturbance due to agriculture or other factors can result in either direct or indirect pressures on populations (e.g., causing outright mortality that leads to population declines vs. causing poor reproductive success that lowers population viability). Of these, the effects of indirect pressures can be subtle and difficult to detect but no less damaging. For example, Reid and Peery (2014) found that decreased genetic diversity and male-skewed sex ratios in Wisconsin populations were positively related to the intensity of associated anthropogenic land use. Rubin et al. (2001a) also detected low genetic diversity of a Blanding's Turtle population in Wood County, which they surmised arose from historical wetland declines that prompted a population bottleneck. Along these lines, loss or degradation of nesting habitat in disturbed landscapes is a particularly insidious indirect factor that likely affects population recruitment. Preferred nesting locations typically contain well-drained soils with sparse vegetation and an open tree canopy. In these habitats, unchecked vegetative succession could reduce suitability and drive gravid females into adjacent disturbed habitats that are open but risky for nesting (e.g., roadsides or residential lawns). Agricultural fields could also attract nesting females if the shorter height of crops in late spring or early summer resembles the open habitats preferred for egg deposition (Mui et al. 2016). As crops grow and nests are shaded, the thermal properties within the nest chamber are altered, which can negatively impact embryo development and survival. Embryos and hatchlings in agricultural landscapes could be exposed to pesticides or crushed by agricultural machinery (Mui et al. 2016).

Habitat fragmentation, as opposed to outright habitat destruction, can also have either direct or indirect effects on Blanding's Turtles. When habitats are fragmented by anthropogenic features such as roads, mortality can occur due to vehicular strikes, which could directly result in population declines (van Dijk and Rhodin 2013). In fact, at road densities of >1 km/km^2 (1.6 mi/mi^2) the associated road mortality can have substantial population-level impacts on turtles such as Blanding's Turtles (Gibbs and Shriver 2002). Females are more vulnerable to vehicular strikes than males, which can lead to male-biased populations (Haxton 2000a, Steen and Gibbs 2004, Steen et al. 2006, Reid and Peery 2014). On the other hand, roads might not cause heavy mortality but instead influence movement patterns or fragment populations. For example, Proulx et al. (2014) found that Blanding's Turtles in Ontario avoided crossing roads, whether paved or unpaved. This restricted movement could further limit reproductive opportunities and lead to loss of genetic diversity or reduced population viability (Anthonysamy et al. 2014b). Roads can also influence large-scale distributional patterns, as research in Ohio, Michigan, and Ontario supports that Blanding's Turtles are more likely to occur in wetlands that are farther from roads or in areas with lower road densities (Attum et al. 2008, Millar and Blouin-Demers 2012). In Wisconsin, observations of road mortality suggest that roads do not act as a movement barrier (R. Hay, unpublished data) but also indicate that road mortality could threaten local populations. The substantial amount of time Blanding's Turtles spend on land, especially mature females, increases their odds of experiencing human-caused mortality due to vehicular or agricultural equipment strikes (R. Hay, personal observation).

Another threat to turtle populations that is indirectly linked to disturbed landscapes is higher densities of turtle nest predators. This is of particular concern, given that females are sometimes attracted to residential lawns in developed landscapes for nesting (Jones and Sievert 2012). Yet nest and hatchling mortality rates in such locations are high, and even ostensibly innocuous mammals that are tolerant of humans, such as Eastern Chipmunks (*Tamias striatus*), will consume young turtles (Jones and Sievert 2012). Along these lines, the populations of some nest predators, including Raccoons, have increased across much of the Upper Midwest in recent decades (Gehrt et al. 2002), likely because they appear to thrive in anthropogenic landscapes (Oehler and Litvaitis 1996, Pedler et al. 1997, Dijak and Thompson 2000). Increasing Raccoon abundances have been further correlated with declining populations of their prey (e.g., Wilcove 1985, Robinson and Wilcove 1994, Brown and Litvaitis 1995), and high Raccoon densities could similarly threaten Blanding's Turtles through nest destruction (Congdon et al. 1983, Temple 1987).

Additional conservation concerns exist, but many remain poorly understood. Given their reported sensitivity to habitat alteration (VanDeWalle and Christiansen 1996) and their low critical thermal maxima relative to other freshwater turtles (Hutchison et al. 1966), Blanding's Turtles could be particularly sensitive to extreme climatic variations that raise environmental temperatures and cause droughts or alter phenology (Anthonysamy et al. 2013, Byer et al. 2020). As such, climate change probably poses a substantial threat to this species, although this subject is in need of further investigation. A recent modeling study estimated that only 25%–50% of midwestern locations currently occupied by Blanding's Turtles would possess a suitable climate for this species by the year 2050 (King and Niiro 2013).

Management: Protection and management of large wetland complexes with intact natural corridors and adjacent upland habitats are beneficial for this species (Joyal et al. 2001, Attum et al. 2008, Refsnider and Linck 2012). In particular, conservation strategies should consider the management or restoration of open sandy uplands that turtles use for nesting. These areas are vulnerable to vegetative succession, and the excessive shading caused by encroaching woody vegetation can alter the thermal properties of nests and egg incubation rates. Management to offset succession in nesting areas is likely beneficial, as Reid et al. (2016) determined that female Blanding's Turtles in Wisconsin successfully nested in restored habitats, and associated populations experienced improved recruitment and growth.

In addition to management of nesting sites, conservation strategies should consider the availability of and connections among wetlands with variable hydroperiods. A study in suburban Illinois found that desiccation of ephemeral wetlands in the summer forced Blanding's Turtles to move through disturbed terrestrial habitats in search of new wetlands, which could put them at risk of mortality (Rubin et al. 2001b). Given the importance of permanent wetlands to this species (Anthonysamy et al. 2013), management strategies for Blanding's Turtles should maintain linkages among various types of aquatic habitats so that individuals can easily find inundated sites as necessary throughout the year. Along these lines, Congdon et al. (2011b) estimated that terrestrial habitat buffers from 300 to 450 m (984 to 1,476 ft) around wetlands would protect 90%–100% of Blanding's Turtle nests, while 2,000 m (6,562 ft) habitat buffers around wetlands would protect 100% of adults that move between wetlands.

Artificial wetland creation could be a beneficial conservation tool for this turtle. For example, Kiviat et al. (2000) found that Blanding's Turtles used a created wetland and successfully nested on constructed nesting areas at a restoration site in New York. Hartwig and Kiviat (2007) further reported that constructed wetlands in New York were frequently used for basking in spring and early summer because they were warmer and contained less vegetation cover than natural wetlands in the study area. The authors recommended that designs for artificial wetlands promote ample submersed vegetation and dense mucky substrates, as turtles monitored during their study preferred these features. Although wetland creation and restoration strategies could be useful for Blanding's Turtle conservation, successful re-creation of natural conditions in artificial wetlands is difficult to achieve (see Introduction to Conservation and Management of Wisconsin's Amphibians and Reptiles chapter for further discussion of created wetlands). As such, construction of artificial wetlands is not a failsafe strategy to offset the loss or degradation of natural wetlands that Blanding's Turtles prefer.

Creation of open artificial nesting sites or maintenance of existing nesting sites with sparse vegetation and well-drained soils could benefit Blanding's Turtles. Beaudry et al. (2010) found that 58% of the anthropogenic nesting locations used by Blanding's Turtles in Maine were sites that had only been available to turtles for about five years. This supports that gravid female Blanding's Turtles can find and will use newly created nesting habitat, but continued maintenance (e.g., removal of vegetation) of these sites will help ensure their long-term use. Dowling et al. (2010) determined that radio-tracked females in New York preferred to nest in experimental plots with vegetation removed through tilling as opposed to hand-weeding or mowing. Long-term protection of Blanding's Turtle nests from predators with electrified predator fences and nest protectors to improve hatching rates are apparently useful techniques (see Geller 2012c). Standing et al. (2000b) reported that only 1 of 101 nests covered with protective screens fell victim to predators. Of note is that Congdon et al. (1983) found that 84% of Blanding's Turtle nest predation occurred within five days of egg deposition, and 47% occurred within the first 24 hours. Thus, enacting protective measures for turtle nests shortly after egg deposition will likely increase success.

Measures to reduce road mortality, especially of reproductive-aged females, could increase stability of Wisconsin populations. This can be accomplished through the construction of fences that prevent turtles from accessing roadways where road mortality is known to occur and by funneling turtles into underpasses below roads (e.g., culverts; Lang 2000) in order to maintain connectivity with adjacent habitats (see further discussion in Introduction to Conservation and Management of Wisconsin's Amphibians and Reptiles chapter). Fencing off sandy agricultural fields to exclude nesting females and direct them to natural nest sites can also reduce agriculture-related mortality, especially where natural nesting opportunities exist adjacently.

Augmentation of populations through headstarting could be a useful conservation tool for this species. While this strategy remains somewhat controversial, it has been effective for rebuilding some freshwater turtle populations, including Blanding's Turtles (King et al. 2013). These efforts may be most successful where habitats are already afforded protection (e.g., state or federal lands), where populations are currently or were historically documented, and at sites where the causes for possible declines have been mitigated. The most effective headstarting endeavors for this species are reportedly those conducted over a long period of time and involve the release of many young individuals (e.g., Glowacki and Kuhns 2010). Although protection of hatchlings and higher recruitment, which headstarting strategies promote, are important for the maintenance of populations, Congdon et al. (1993) indicated a strong association between population stability and survival of both adults and juveniles. Thus, efforts to protect all age classes will likely be necessary for this species to persist in Wisconsin.

Some commonly employed habitat management techniques can potentially pose risks for Blanding's Turtles, including prescribed fire, mechanical vegetation removal (e.g., mowing and brush cutting), and autumn or winter water drawdowns of reservoirs to enhance waterfowl habitats (see Introduction to Conservation and Management of Wisconsin's Amphibians and Reptiles chapter). Since drawdowns of impoundments in autumn and winter can devastate local turtle populations (Hall and Cuthbert 2000), they should be avoided except when issues related to potential dam breaches threaten the safety of humans and residences downstream. The WDNR currently has an informal policy that specifies that drawdowns should be completed prior to 1 October. The authors of this account suggest that drawdowns be completed before 15 September in the northern half of Wisconsin and before 25 September in the southern half of the state. The WDNR has also developed a variety of protocols to help avoid the incidental take of Blanding's Turtles due to land management activities (WDNR 2011a). While this species is no longer listed as a Threatened species, the guidelines provide important considerations for habitat management professionals.

SPECIAL REMARKS

Harding and Davis (1999) confirmed hybrid Blanding's Turtle × Wood Turtle hatchlings from egg clutches produced in a large outdoor turtle enclosure in Michigan. They also speculated that hybridization between these species "could occur under natural conditions." An adult male turtle that possessed characteristics of both species, consistent with hybridization, was captured in southern Jefferson County in 2017 (MCL = 22.5 cm [8.9 in]; Figure 330). Although this turtle was in an area of the state where Blanding's Turtles are well documented, it is outside

Figure 330. A possible Blanding's Turtle × Wood Turtle hybrid (Jefferson County; photos by J. Kapfer). (a) Pronounced annuli on carapace scutes (similar to a Wood Turtle). (b) Pronounced yellow spots on the head and bright yellow on the throat and portions of the chin (similar to a Blanding's Turtle).

of the known state range for the Wood Turtle. Unfortunately, genetic confirmation of hybridization was not completed, and the origin of this turtle (wild or captive) was not determinable (J. Kapfer, UW–Whitewater, personal communication).

Ferrara et al. (2018) recently found evidence that Blanding's Turtles in Minnesota emit distinct underwater vocalizations. Although the purpose of these vocalizations is not known, and currently available data are too sparse for the description of a full vocal repertoire, this species appears to be capable of some vocal communication. Further work on this topic could yield important and interesting results on how turtles in Wisconsin interact with each other.

Wood Turtle

Glyptemys insculpta (LeConte 1830)

Donald J. Brown and Carly N. Lapin

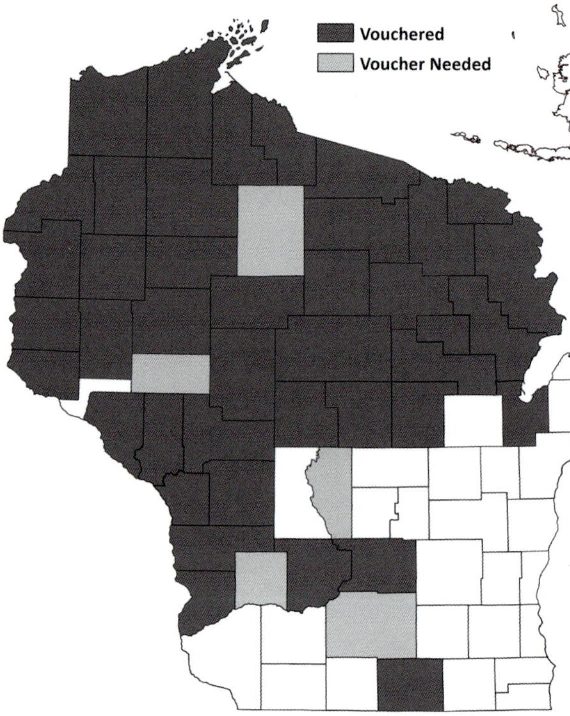

Adult Wood Turtle (Oneida County; photo by D. Brown).

AT A GLANCE

The Wood Turtle is arguably one of the most attractive and interesting freshwater turtles in North America. It is a medium-sized species with yellow or orange coloration on its throat and forelegs and pyramid-shaped rings on its shell. The species is primarily distributed throughout the northern half and the west-central and south-central portions of the state. This mild-mannered species spends the winter hibernating in rivers and streams but spends much of its time feeding and basking on land in late spring, summer, and early autumn. Nesting occurs from late May to early July, and hatchlings typically emerge from nests in September. The Wood Turtle is an omnivorous species that is known for its ability to capture earthworms by rapidly stomping on the ground, which creates vibrations that cause the worms to emerge. Unfortunately, Wood Turtles are negatively affected by a variety of human activities and habitat modifications, and population declines have occurred across the species' distribution. It is currently listed as a Threatened species and considered a Species of Greatest Conservation Need by the WDNR.

Above, top: Figure 331. Adult Wood Turtle displaying its yellow-orange throat and forelegs (La Crosse County; photo by J. Kapfer).

Above, bottom: Figure 332. Adult Wood Turtle displaying its pyramid-shaped interior scutes (Oneida County; photo by D. Brown).

SYSTEMATICS

Glyptemys insculpta (Wood Turtle) was historically included in the genus *Clemmys* (McDowell 1964) with three other species: *C. guttata* (Spotted Turtle), *C. marmorata* (now *Actinemys marmorata,* Western Pond Turtle), and *C. muhlenbergii* (Bog Turtle). Recent phylogenetic studies found this application of *Clemmys* to be paraphyletic, but with strong support for *C. insculpta* and *C. muhlenbergii* as sister species, thus requiring the placement of both species into a separate genus, *Glyptemys* (Bickham et al. 1996, Holman and Fritz 2001, Feldman and Parham 2002, Crother 2017). There are no recognized subspecies of *G. insculpta*.

DESCRIPTION

Adults: The base skin coloration is dark brown, with yellow or orange on the throat and forelegs (Figure 331). Carapace coloration is primarily various shades of brown, often with black or yellow specks or streaks. The carapace has a low middorsal keel, and the scutes contain raised concentric growth rings, resulting in pyramid-shaped interior scutes (Figure 332). The posterior margin of the carapace is serrated due to offset marginal scutes. The plastron has a yellow base with a large black blotch on each scute (Figure 333). Additional information regarding morphology can be found in Ernst and Lovich (2009). The Wood Turtle is a medium-sized species. Mean CL of individuals ten or more years old for a population in La Crosse and Trempealeau Counties was 20.9 cm (8.2 in, $n = 28$) for males and 18.7 cm

Figure 333. Comparison of adult male (*left*) and female (*right*) Wood Turtle plastrons (Oneida County; photo by D. Brown). The concavity in male plastrons is a useful sex identifier for Wood Turtles and many other turtle species. Biologically, it assists with mating by helping males to hold their plastron against the rounded carapace of females (Pritchard 2008).

Figure 334. Hatchling Wood Turtles (Washburn County; photo by A. Badje).

(7.4 in, $n = 48$) for females (Ross et al. 1991). The current record MCL is 24.0 cm (9.4 in) for a male in Maine (Jones and Compton 2010).

Preadult stages: Eggs are elliptical in shape (ca. 3.2 × 2.3 cm [1.3 × 0.9 in]), white in color, and smooth (Vogt 1981). Hatchlings have a gray or black body and plastron and light brown or tan carapace (Harding and Bloomer 1979, Vogt 1981; Figure 334). Juveniles have strongly flared posterior marginal scutes, a morphological characteristic absent in adult males (Ernst and Lovich 2009). Lovich et al. (1990) noted that mean hatchling PL was 2.66 cm (1.05 in, $n = 8$) in Pennsylvania.

Variation: Adults are sexually dimorphic, with males having a deeply concave plastron and higher and narrower carapace (Harding 1997, Ernst and Lovich 2009; Figure 333). In addition, the cloacal opening of males is located posterior to the back edge of the carapace, a useful characteristic for definitive sex determination. Males are typically 7%–10% larger than females (Lovich et al. 1990). There is geographic variation in adult male skin pigmentation, with males in the Great Lakes region typically having light yellow or light orange coloration on their throat, neck, and forelimbs, while males in eastern populations typically have darker orange or red coloration (Harding 1997, Ernst and Lovich 2009).

Diagnosis: The Wood Turtle is characterized by a carapace that is depressed or angular in cross section, has no keel, and has strong elevated concentric grooves on each scute. This species has a yellow plastron with a bridge, 12 scutes, and large black blotches on the lateral edge of each scute. The jaws are not beaked, and the toes are webbed. Adult Wood Turtles can be distinguished from all other turtle species in Wisconsin by their distinct sculptured carapace (Figure 332). However, biologists sometimes misidentify Wood Turtles for Blanding's Turtles and vice versa (Robert Hay, Turtles for Tomorrow, personal communication). Adult Blanding's Turtles (*Emydoidea blandingii*) have a similarly patterned plastron, but their plastron is hinged, and they have a smooth dome-shaped carapace. Juvenile and subadult Blanding's Turtles have a flatter carapace and somewhat distinct annual growth rings, which could cause them to be confused with Wood Turtles. However, they lack a middorsal keel.

Wood Turtle eggs are similar in shape and appearance to numerous other aquatic turtles. However, the size of their eggs probably most overlaps with those of Blanding's Turtles and Northern Map Turtles (see Preadult stages subsection). It is anecdotally possible to distinguish Wood Turtle eggs from Blanding's Turtle eggs because shells of the latter have a rougher texture during earlier incubation periods, so that gently rubbing them together produces a gritty sound (R. Hay, personal communication).

DISTRIBUTION AND HABITAT

Global distribution: The Wood Turtle is restricted to the northeastern and north-central US and southern Canada (Ernst and Lovich 2009). The species ranges from Nova Scotia to eastern Minnesota. On the East Coast, Wood Turtle populations occur as far south as Virginia. An isolated population currently exists in northern Iowa (Spradling et al. 2010). This species occurred as far south as Georgia and Alabama during the last glacial cycle and moved north with the receding glaciers beginning ca. 12,000 years ago (Holman 1976, Parmalee and Klippel 1981, Amato et al. 2008).

Distribution in Wisconsin: County records indicate that the Wood Turtle is primarily distributed throughout the northern half and the west-central and south-central portions of the state. Of the published county records, voucher specimens could not be found for Adams, Dane, Eau Claire, Price, and Richland (see the Appendix). According to WDNR (2016b) records, live individuals collected in Adams County in 1950 were deposited at the Henry Vilas Zoo in Madison, Wisconsin. However, the zoo has no records of these turtles (Jeff Stafford, Henry Vilas Zoo, personal communication). The only known recent observation from Eau Claire County occurred in 2006, but no specimen or photograph was collected (R. Hay, personal communication). Casper (1996) questioned the validity of the Dane County and Rock County records, attributing both to potentially "displaced" individuals. A purported Dane County specimen exists at the UW–Madison Zoology Museum (UWZM 19168). This is a jar of eggs that museum records indicate belong to a Wood Turtle. There is no date of collection, and the only information accompanying the specimen is vague locality data (i.e., "Madison" and "Dane County"). Aside from the ambiguity of the specimen's origin, eggs are difficult to use for species identification, and thus Dane County remains an unconfirmed location. The only known specimen from Rock County was supposedly collected along the Rock River "11 mi. N. of Illinois Line" in 1933 by C. R. Naeser (UIMNH 2287). Based on Curtis (1959), the land associated with the southern Rock River was historically savanna. Wood Turtles are currently associated with primarily forested habitat in Wisconsin, although they are also known to heavily use canopy gaps (see Habitat subsection) and possibly occupied more open areas in Wisconsin in the past (see Natural History Box: Is the Wood Turtle a Forest Species?). In addition, areas of lowland hardwood forest that are potentially more suitable for Wood Turtles currently exist along the nearby Sugar River in southwestern Rock County, and pockets of southern hardwood forest occurred historically in southwestern Rock County (Curtis 1959).

Localities of possible occurrence: There are several counties without confirmed records that represent obvious gaps in the documented range of this species. According to WDNR (2016b) records, unconfirmed sightings exist for Juneau and Outagamie Counties. Future surveys may confirm their presence in these counties. Based on its presence in surrounding counties, the species could also occur in Pepin County. A handful of anecdotal reports also exist from Grant and Iowa Counties (WDNR 2016b), which would represent extensions in the known state distribution for this species. Wood Turtles are occasionally encountered in southeastern Wisconsin, such as Milwaukee County (e.g., MPM VZP833), Kenosha County, and Racine County (MPM VZP216). Pope and Dickinson (1928) mentioned a secondhand report of Wood Turtles near Waukesha (Waukesha County) in the summer of 1925. These locations are far outside of the currently documented range for this species, and the sporadic observations have often been attributed to translocations. Yet the shore of Lake Michigan in eastern Wisconsin was once dominated by forest (Curtis 1959), and habitat could have historically existed for this species in Milwaukee, Kenosha, and Racine Counties. However, given the current level of anthropogenic disturbance in this region, it seems unlikely that these recent observations represent the occurrence of viable populations.

Habitat: Wood Turtles are associated with clear and tannin-stained rivers, perennial streams, and creeks with moderate to fast currents and hard bottoms

NATURAL HISTORY BOX: Is the Wood Turtle a Forest Species?

The terrestrial habitat surrounding rivers and streams that are currently inhabited by Wood Turtles is often primarily forested (see Habitat subsection), which could indicate that the preferred terrestrial habitat of Wood Turtles is forest. However, the scientific literature does not entirely support this assumption. Early reports often indicated that Wood Turtle habitat was nonforested. For example, Allen (1868) stated that in summer Wood Turtles were "generally found in dry fields remote from streams or ponds" in Massachusetts. More recent studies have also observed Wood Turtles occupying distinctly nonforested areas. For instance, Farrell and Graham (1991) noted that their study stream was "surrounded by open, well-cropped pasture, and contained the largest Wood Turtle population known to exist in New Jersey at the time." Several investigations have concluded that individuals select for canopy gaps within a forest matrix (e.g., Compton et al. 2002), have found that individuals preferred young forest to mature forest (Arvisais et al. 2004), and have even indicated that individuals preferred nonforest to forest (Kaufmann 1992a, Dubois et al. 2009).

If Wood Turtles can survive in and possibly even prefer nonforested habitat, it is strange that in Wisconsin the species is currently only found in primarily forested regions. One possibility is that Wood Turtles historically occupied a much broader distribution in the state but have been relegated to forests since European settlement. In particular, the conversion of savanna and prairie habitats to agriculture could have resulted in the extirpation of local Wood Turtle populations in the southern portion of the state (see Conservation section). Several other anthropogenic landscape changes, such as the installation of dams throughout the state, could also have influenced where we find the species today. Unfortunately, we can only speculate on the natural distribution of Wood Turtles in Wisconsin prior to European settlement. Regardless, our knowledge of the ecology of this species indicates that although the Wood Turtle is a forest species, it should not be considered 'strictly' a forest species.

Donald J. Brown, West Virginia University / USFS

(composed primarily of sand and/or gravel), as well as adjacent forested riparian and upland habitat with openings in the streamside canopy and suitable nesting habitat (Compton et al. 2002, Bowen and Gillingham 2004; Figure 335). Vogt (1981) stated that in Wisconsin "they are restricted in habitat to forested areas along fast-moving streams." Within their habitat complex, Wood Turtles follow an annual chronological sequence of aquatic and terrestrial habitat use patterns dictated by seasonal air and water temperature changes and vegetation dynamics (Arvisais et al. 2004, Brown et al. 2016). Thus, suitable aquatic and terrestrial habitats are required for occupancy by Wood Turtles (Bowen and Gillingham 2004).

Wood Turtles are primarily aquatic in the spring and autumn but mostly terrestrial in the summer (Cochrane et al. 2019). However, individuals will return to water even in the summer to move, thermoregulate, and sometimes mate (Kaufmann 1992a, Compton et al. 2002, Dubois et al. 2009). In the transitional periods of spring and autumn, male and female Wood Turtles favor riparian shrub cover, such as Gray Alder (*Alnus incana*), and grassy openings near their overwintering streams (Quinn and Tate 1991, Ross et al. 1991, Compton et al. 2002, Arvisais et al. 2004). The likely reason for this close proximity to water is to assist with thermoregulation by alternating between terrestrial and aquatic habitats depending on the air temperature (Arvisais et al. 2004). During the warmer summer months, male Wood Turtles continue to use these low-elevation riparian areas, while females tend to venture into

Figure 335. Wood Turtle habitat in northwestern Wisconsin (Douglas County; photo by D. Brown).

Figure 336. A female Wood Turtle laying eggs in well-exposed sandy soil (La Crosse County; photo by J. Kapfer).

the uplands, using mixed forest stands in proximity to grassy or scrub-shrub openings (Compton et al. 2002, Arvisais et al. 2004, Tingley et al. 2010). Knudsen (1957e) noted that Wood Turtles use a wide variety of holes and depressions for shelter, such as rabbit holes, hollow logs, and sod overhangs along gullies, and recent population surveys in Wisconsin and Minnesota agree with this statement (Carly Lapin, WDNR, and Donald Brown, West Virginia University / USFS, personal observations).

Wood Turtles likely select for habitat edges or areas with a diverse availability of habitat types rather than large areas of only one habitat type because their habitat selection is driven by both foraging and thermoregulatory needs (Kaufmann 1992a, Compton et al. 2002, Arvisais et al. 2004). Tuttle and Carroll (2005) studied terrestrial habitat selection of 53 hatchling Wood Turtles in New Hampshire and found that hatchlings selected dense herbaceous and woody cover for shelter and concealment from predators. A habitat selection study of headstarted juveniles in Wisconsin revealed that individuals strictly used riparian habitat close to flowing water edges (Todd Thayer, UW–Green Bay, unpublished thesis). In that study, some individuals almost exclusively stayed in the water near the shoreline, while others spent much of their time on land within several meters of the water's edge and used deciduous forest, Gray Alder, and sedge-dominated riparian habitat. Given that these are broad generalizations from a number of studies, we emphasize that individual Wood Turtles display strong associations with a variety of habitats, possibly due to discovery of local food resources over the course of a long lifetime (Kaufmann 1992a, Arvisais et al. 2002, Compton et al. 2002).

Buech et al. (1997) determined that preferred nesting areas in Minnesota included well-exposed sandy substrates within 40 m (131 ft) of flowing waterbodies (ca. 50% of nests were within 10 m [32.8 ft]), with slopes <40° and heights 2–5 m (6.6–16.4 ft) above the water line (Figure 336). These preferences seem to be consistent across the species' distribution (Harding and Bloomer 1979, Farrell and Graham 1991, Hughes et al. 2009), but exceptions have been observed. For example, Wood Turtles have been found nesting >600 m (1,969 ft) from flowing water in Wisconsin (Greg Kessler, WDNR, and R. Hay, personal observations). In addition, nesting is common in gravel along railroad grades in Ashland, Iron, Marathon, Monroe, and Washburn Counties (Vogt 1981, C. Lapin and R. Hay, personal observations). Bougie et al. (2020) documented nests between 0.5 and 330 m (1.6 and 1,083 ft, $n = 120$) from rivers in Oneida and Washburn Counties. Pallas (1960) found an individual nesting in a suburban backyard in New Jersey. Females nesting in open sand patches, where digging is easier than in gravel and compacted soil, will often dig a body pit prior to laying eggs, allowing the female to place the eggs deeper in the soil (R. Hay, personal observation). This may be because the female recognizes that the eggs need to be buried deeper, as these sites may be warmer at the surface than what is preferred, and/or the female may know that these sites could dry out faster than other nesting substrates (R. Hay, personal communication).

Wood Turtles overwinter in flowing water that does not freeze completely (Ernst and Lovich 2009). Individuals have been documented using muskrat burrows and overhanging tree roots along streambanks and resting unprotected on the stream bottom during the winter months (Bishop and Schoonmacher 1921, Farrell and Graham 1991, Quinn and Tate 1991, Neiderberger and Seidel 1999). In contrast, Greaves and Litzgus (2007) studied overwintering behavior in Ontario and reported that individuals rested exposed on the riverbed at a mean distance from shore and depth of ca. 1 m (3.3 ft). Several studies have documented multiple Wood Turtles using the same hibernaculum, as well as overwintering site fidelity (Farrell and Graham 1991, Parren 2013).

REPRODUCTION AND DEVELOPMENT

Breeding can occur at any time during the active season but primarily occurs in spring and autumn (reviewed by Ernst and Lovich 2009). Mating typically occurs in shallow water but sometimes occurs on land, and copulation can last ≥12 hours (Ernst 1986b, Niederberger and Seidel 1999, Ernst and Lovich 2009). Males typically rest their plastron on the carapace of the female, but occasionally plastron-plastron copulation occurs (Tronzo 1993, Mitchell and Mueller 1996). Figueras and Burke (2017) discovered that female Wood Turtles can store sperm for more than two years. Females usually lay one clutch of eggs per year but occasionally lay two (Akre 2002). In addition, females sometimes do not nest in a given year (Ross et al. 1991, Akre 2002). Number of eggs laid per clutch varies widely, with the current record being 20 eggs from an individual in Quebec (Walde and Saumure 2008). Based on clutches in northern Wisconsin collected from 2011 to 2013, clutch size ranged from 3 to 17 ($\bar{x} = 11$, $n = 156$; Table 21). Of the 1,792 eggs observed in 2012 and 2013, 369 (20.6%) were infertile (Table 21).

Nesting in Wisconsin occurs from 20 May to 7 July, although nesting after 30 June is uncommon, and hatchlings emerge between August and October (Carroll and Ultsche 2007, Walde et al. 2007, R. Hay, personal communication). Walde et al. (2007) observed strong nest-site fidelity for a population in Quebec, with 95% of females nesting at the same site in consecutive years. To our knowledge, communal nesting has not been reported for the Wood Turtle. However, multiple females will use the same nesting area. Presumably, this is driven by the availability of preferred (or tolerated) nesting habitat rather than a behavioral or reproductive strategy.

Vogt (1981) stated that eight eggs is the most common clutch size in Wisconsin, but Harding and Bloomer (1979) found that the mean in Michigan was 10.36 eggs per clutch. R. Hay documented a mean clutch size of 10.95 ($n = 195$) in northern Wisconsin (Table 21). Unlike most turtle species in Wisconsin, the Wood Turtle does not have temperature-dependent sex determination (Bull et al. 1985, Ewert and Nelson 1991). Buech et al. (2004) and Walde et al. (2007) found that embryos in Minnesota and Quebec, respectively, must complete development before winter or the nest will be lost. This is supported by recent observations

Table 21. Wood Turtle clutch sizes and infertility rates of naturally laid and artificially incubated eggs in northern Wisconsin from 2011 to 2013 (eggs incubated at 26°C [78.8°F]). Infertility rates for individual clutches ranged from 0 to 100% over the two years of observation. Infertility rates were especially high for clutches with at least five eggs (R. Hay, unpublished data).

Year	Clutches	Total Eggs	Clutch Range	Mean Clutch Size	Infertile Eggs (%)
2011	41	392	4–17	9.57	N/A
2012	78	904	3–17	11.59	197 (21.79)
2013	76	888	5–17	11.68	172 (19.37)

from Oneida and Washburn Counties (C. Lapin, personal observation). In addition, 12 Wood Turtle nests from the 2014 nesting season were monitored in Douglas County (G. Kessler, unpublished data). Of the 12 nests, 2 hatched in September, and the remaining 10 were excavated the following spring after the potential post-winter emergence period, which revealed that individuals died in the eggs as partially developed embryos. Lovich et al. (2014) found no evidence of Wood Turtles overwintering in nests in Pennsylvania over a 21-year period. However, Parren and Rice (2004) found one hatchling in May that likely survived a warm winter in Vermont before hatching.

Wood Turtles are a long-lived species with indeterminate growth (Lovich et al. 1990). Growth rates for juveniles were ca. 1 cm (0.4 in) per year in Pennsylvania, with growth slowing dramatically when individuals neared sexual maturity (ca. 16 cm [6.3 in]; Lovich et al. 1990). Mean MCL for individuals in Jackson County differed by 3.4 cm (1.3 in) between the ages of three ($\bar{x}=9.6$ cm [3.8 in], $n=2$) and five ($\bar{x}=13.0$ cm [5.1 in], $n=2$) and by 2.4 cm (0.9 in) between the ages of six ($\bar{x}=14.0$ cm [5.5 in], $n=5$) and nine ($\bar{x}=16.4$ cm [6.5 in], $n=6$; Table 22). Based on these data, the estimated mean annual growth rate was 1.1 cm (0.4 in) per year, with a corresponding weight increase of 80.1 g (2.8 oz) per year, between the ages of three and nine (Table 22). Sexual maturity is a function of size (male MCL = 19.2–20.0 cm [7.6–7.9 in], female MCL = 15.8–18.5 cm [6.2–7.3 in]; Ernst and Lovich 2009). Both males and females typically reach sexual maturity between the ages of 14 and 18 (Farrell and Graham 1991, Brooks et al. 1992), but according to Harding (1997), individuals in northern populations can take as long as 20 years to mature. The smallest gravid female found during a study in La Crosse and Trempealeau Counties was 17.1 cm CL (6.7 in; Ross et al. 1991).

Plastron growth rings can be used to reliably age individuals up to at least 15 years old, but reliability decreases with age because the distance between rings decreases and plastron wear obscures growth-ring delineation (Harding and Bloomer 1979). Thus, long-term capture-recapture studies are the best way to assess longevity in the wild. The oldest known individuals were part of a study conducted in Minnesota, which recorded adult males that were at least 48 years old in 2013 and adult females that were at least 55 years old in 2014 (Brown et al. 2015). In addition, Jones (2009) estimated that wild individuals in Massachusetts and New Hampshire exceeded 80 years old, based on a shell-wear index.

ACTIVITY

Wood Turtles in Wisconsin typically emerge from hibernation in March (Vogt 1981). However, the annual active period in a given year is influenced by environmental temperature (Merritt 1980, Dubois et al. 2009, Berg 2014); thus, emergence from hibernation varies from year to year. Wood Turtles have even been observed attempting to mate under ice in northern Wisconsin (Kevin Brewster, independent researcher, personal communication to R. Hay). For a population in Pennsylvania, Ernst (1986b) found

Table 22. Observed mean and ranges of juvenile and sub-adult Wood Turtle MCL and weights in Jackson County from 1990 to 1995. Ages were estimated from plastron growth rings (R. Hay, unpublished data).

Age (*N*)	**MCL, cm (in)**	**Weight, g (oz)**
3 (2)	9.6, 9.2–9.9 (3.8, 3.6–3.9)	119, 100–138 (4.2, 3.5–4.9)
5 (2)	13.0, 12.7–13.3 (5.1, 5.0–5.2)	269, 258–280 (9.5, 9.1–9.9)
6 (5)	14.0, 13.2–14.7 (5.5, 5.2–5.8)	356, 292–407 (12.6, 10.3–14.4)
7 (6)	14.7, 13.3–16.8 (5.8, 5.2–6.6)	407, 318–585 (14.4, 11.2–20.6)
8 (6)	15.3, 14.1–17.5 (6.0, 5.6–6.9)	478, 375–618 (16.9, 13.2–21.8)
9 (6)	16.4, 14.7–17.7 (6.5, 5.8–7.0)	603, 455–705 (21.3, 16.0–24.9)

that individuals were active at air temperatures as low as 3°C (37°F) and water temperatures as low as 6°C (43°F), but feeding in water did not occur unless water temperatures were ca. 17°C (63°F). In addition, Dubois et al. (2009) determined that daily terrestrial habitat use patterns were influenced by temperature selection. A more comprehensive review of temperature selection can be found in Ernst and Lovich (2009).

Space use and activity patterns of Wood Turtles are strongly tied to flowing water throughout the year. Pentecost and Vogt (1976) stated that "in Wisconsin, individuals tend to be much more aquatic than are members of eastern US populations," but no study has attempted to test this hypothesis. This species is primarily diurnal (Vogt 1981), although nesting sometimes occurs at night (Ernst 2001b). Brown et al. (2016) studied activity patterns of a population in Minnesota and found that individuals generally remained within 0.1 km (0.06 mi) of flowing water but occasionally moved much farther, with one male observed 0.4 km (0.25 mi) from flowing water. Two juveniles were observed ≥0.5 km (0.3 mi) from flowing water in Douglas and Iron Counties, respectively (R. Hay, personal observation), and an adult male was observed ca. 0.85 km (0.53 mi) from flowing water in Douglas County (Andrew Badje, WDNR, unpublished data). In addition, a gravid road-killed female was observed ca. 0.6 km (0.37 mi) from flowing water in Douglas County (R. Hay, personal observation). Females sometimes travel long distances in search of suitable nesting sites. Behler and Castellano (2005) found a gravid female that moved 3.9 km (2.4 mi) through upland habitat in Delaware, and Ernst (2001b) documented nest-searching movements up to 1 km (0.6 mi) in Pennsylvania. Individuals in Oneida and Washburn Counties have been documented traveling up to 3.2 km (1.9 mi) and 4.1 km (2.5 mi) to nesting habitat, respectively (C. Lapin, personal observation).

Wood turtles often use rivers and streams as long-distance movement corridors, traveling 1 km (0.6 mi) or more (Parren 2013). An unusually long-distance movement event was documented for a radio-tracked female in Vilas County that moved ca. 16 km (10 mi) upstream to a nesting site (K. Brewster, personal communication to R. Hay). In addition, two years after initial capture, an adult female was recaptured 10.1 km (6.3 mi) downstream from its first location in Douglas County (A. Badje, unpublished data). Carroll and Ehrenfeld (1978) studied the homing ability of this species and found that displaced individuals had high return success up to 2 km (1.2 mi), with displacement direction having no influence on homing ability. In addition, Tinklepaugh (1932) performed maze experiments with a Wood Turtle and concluded that it was highly intelligent for a reptile, with a visual learning ability equivalent to that of a rat.

Several studies have investigated the social behaviors of Wood Turtles. The most thorough investigation was conducted on a population in Pennsylvania by Kaufmann (1992b). He classified 27 different social behaviors, with most female-female interactions classified as nonagonistic and most male-male interactions classified as agonistic. Kaufmann (1992b) found that males displayed dominance hierarchies that were associated with age and size of individuals, which he referred to as an "absolute dominance" social structure because agonistic encounters were not associated with territory defense. Male combative behavior is most prominent during the spring and autumn mating seasons, tends to occur more often in water than on land, and often happens during attempted mating events (Farrell and Graham 1991, Kaufmann 1992b). Female aggression has been observed in relation to competition for preferred nesting sites (Rutherford 2012).

Although Wood Turtles clearly have a social system, they are not territorial (Kaufmann 1992b, 1995). However, individuals occupy one or more activity centers that are adjacent to stretches of flowing water (Compton et al. 2002). Because Wood Turtles use rivers and streams as travel corridors, it is challenging to quantify their use of space. If an individual has two or more activity centers separated by hundreds or thousands of meters, traditional home range estimators such as minimum convex polygons (MCP) can dramatically overestimate space use. For example, Quinn and Tate (1991) estimated that activity areas ranged from ca. 1 to 115 ha (2.5 to 284.0 ac) in Ontario, and Remsberg et al. (2006) estimated that activity areas ranged from

0.2 to 390.0 ha (0.5 to 964.0 ac) in Michigan, based on MCP estimators. However, published estimates for Wisconsin indicate that activity areas are generally small. Ross et al. (1991) radio-tracked a total of three males and four females in the summers of 1975, 1985, and 1986 in La Crosse and Trempealeau Counties. Based on MCP estimates, activity areas ranged from 0.08 to 0.41 ha (0.20 to 1.01 ac) and 0.27 to 0.91 ha (0.67 to 2.25 ac) for males and females, respectively. Similarly, Harding and Bloomer (1979) reported activity area estimates of 0.91 ha (2.25 ac) and 0.48 ha (1.19 ac) for two females in Wisconsin (specific location and home range estimator were not reported). Ross et al. (1991) also reported that mean minimum daily movement was 41.9 m (137.5 ft) and 27.4 m (89.9 ft) for males and females, respectively. Woods (1945) observed an individual traveling at a rate of 0.3 km/hr (0.2 mi/hr) and hypothesized that this was close to the maximum terrestrial travel speed for Wood Turtles.

Brewster and Brewster (1991) studied movement patterns of headstarted juvenile Wood Turtles in Vilas County. They found that all individuals remained within 40 m (131 ft) of the river channel where they were released but moved up to ca. 50 m/day (164 ft/day) in the river for the first eight days after release, with minimal movement afterward. Castellano et al. (2008) studied hatchling movement patterns in an agricultural field in New Jersey and found that minimum daily movement was between 7 and 34 m (23 and 112 ft), and individuals spent several days to weeks in the field prior to entering the stream. Tuttle and Carroll (2005) studied hatchling movement patterns in New Hampshire and found that hatchlings took 1–24 days to reach an adjacent stream ($\bar{x} = 6.2$ days), and mean minimum distance traveled was 26.8 m (87.9 ft) per day.

Adult Wood Turtles in the western Great Lakes region begin reducing their terrestrial activity in September and October and enter hibernation in October or November (Vogt 1981, Brown et al. 2016). After emergence from nests in late summer and autumn, hatchlings travel to flowing water within a few weeks and presumably begin hibernation shortly thereafter (Tuttle and Carroll 2005, Castellano et al. 2008). Walde et al. (2008) found that hatchlings could survive short-term bouts of freezing temperatures on land in Vermont. During hibernation, Wood Turtles are thought to be intolerant of anoxic conditions (Graham and Forsberg 1991, Ultsch 2006) and thus overwinter in flowing waterbodies such as rivers, creeks, and streams (Vogt 1981). Greaves and Litzgus (2007) found that individuals made short-distance movements underwater during the winter hibernation period (i.e., 0.1–10.0 m [0.3–32.8 ft] between relocations, which occurred at intervals of 7–12 days).

PREY AND PREDATORS

Wood Turtles are omnivorous and opportunistic feeders that eat a wide variety of plants and animals (Vogt 1981, Walde et al. 2003). Their aquatic diet consists primarily of insects and mollusks (Surface 1908, Lagler 1943). Brewster (1985) studied Wood Turtle behavior in a terrestrial enclosure in Oneida County. He reported that individuals fed on planted raspberries (*Rubus* spp.) and strawberries (*Fragaria* spp., fruit and leaves), Common Dandelions (*Taraxacum officinale*), Common Plantains (*Plantago major*), grasses, earthworms, and a Mallard (*Anas platyrhynchos*) duckling that died in the enclosure. Cochran et al. (2014) collected 7 stomach samples and 11 fecal samples from wild Wood Turtles in Brown County and found that common dietary items included dewberries and blackberries (*Rubus* spp.), White Mulberries (*Morus alba*), mushrooms, and invertebrates. Crayfish claws were observed in the feces of several gravid females from Wisconsin (R. Hay, unpublished data). Jones and Sievert (2009b) studied Wood Turtle foraging behavior in Massachusetts and New Hampshire. They detected individuals feeding on Spotted Salamander (*Ambystoma maculatum*) egg masses and animal carcasses (fishes and birds), in addition to plants, fungi, and invertebrates. Furthermore, an individual was observed feeding on a White-tailed Deer (*Odocoileus virginianus*) carcass in Oneida County (Jim Klosiewski, WDNR, personal communication). Ernst (2001b) observed consumption of Green Frog (*Lithobates clamitans*) adults and tadpoles. Castellano and Behler (2003) found individuals eating Corn (*Zea mays*) in an agricultural

NATURAL HISTORY BOX: Worm Thumping

Wood Turtles consume a notably wide variety of plant and animal foods, but anyone who spends much time observing these turtles, as I have, will certainly witness their fondness for earthworms. They display a unique behavior for obtaining this slimy prey variously called worm 'stomping' or 'thumping.' One cool but sunny spring morning I watched an adult male Wood Turtle emerge from the water and move slowly over the damp ground near shore. He stopped and began to push down with a front foot several times in succession, with increasing tempo, lifting and dropping the front of his plastron with each push. He then moved forward about 5 cm (2 in) and duplicated the process with the alternate foot. After repeating the procedure several times, he suddenly lunged forward, seized a worm in his jaws, yanked it upward, and quickly swallowed it. He continued with this behavior for about 20 minutes, catching and consuming six worms before moving into tall grass to bask. It has been speculated that thumping the ground may cause vibrations similar to those made by tunneling moles, driving worms to the surface to escape. Worm thumping has been observed for Wood Turtles in most parts of their range, but apparently not all populations or individuals do it. Whether it is a mostly learned or inherited behavior is still unclear, but it does seem rather stereotyped; we do know that the alternating limb pushing behavior is used for other things unrelated to feeding. For example, Wood Turtles will sometimes thump the ground during or after aggressive encounters with rival turtles, and courting males often mount a female and then lift and drop onto the female's carapace several times in succession with the same increasing tempo used in worm thumping.

James H. Harding, Michigan State University

field in New Jersey. The most unique Wood Turtle foraging behavior involves rapidly stomping their forelimbs on the ground, which creates vibrations that cause earthworms to emerge (see Natural History Box: Worm Thumping). Kaufmann (1986) noted that this foraging behavior can last for over four hours at a time, with a mean yield of 2.4 worms/hour in Pennsylvania.

Like all turtle species, Wood Turtles are most vulnerable to predation at the egg and hatchling life stages (Paterson et al. 2012). Based on a manipulative field experiment in the Upper Peninsula of Michigan, Rutherford et al. (2016) concluded that predators mostly used soil disturbance cues to locate artificial Wood Turtle nests. Females exhibit high fidelity to nesting sites, and after the discovery of these locations by mesopredators (e.g., Raccoons [*Procyon lotor*]), nests can be subjected to repeated annual plundering (Browne and Hecnar 1995). Predation rates can be very high (e.g., 15 of 17 nests were lost to predators in Ontario [Brooks et al. 1992], 34 of 50 unprotected nests were lost to predators in Oneida and Washburn Counties [Bougie et al. 2020]), which can result in long-term population declines (Daigle and Jutras 2005, Browne and Hecnar 2007). Throughout their range in Wisconsin, once Wood Turtle nesting is under way for two or more days, predation of nests often occurs within 24–48 hours (R. Hay, personal observations). A mail survey was conducted in northwestern Wisconsin in 2013 to obtain landowner feedback regarding individual and nesting Wood Turtle observations. Of the respondents, all of those who had observed Wood Turtle nesting commented that they had not seen a Wood Turtle nest last more than two days before being destroyed by predators (G. Kessler, personal communication to R. Hay). In Wisconsin and Minnesota, nest predation has also been observed late in the nesting season (including just prior to emergence), as well as on unhatched nests the following year (C. Lapin and Madaline Cochrane, University of Minnesota–Duluth, personal observations). It appears that some of the earliest laid nests and those laid very late in the nesting season have the best chance at surviving, especially if nesting begins during a rainy night (R. Hay, personal observations). Nest predation by American Badgers (*Taxidea taxus*), American Black Bears (*Ursus americanus*), Raccoons, Red Foxes (*Vulpes vulpes*), and Striped Skunks (*Mephitis mephitis*) has been observed in Wisconsin and/or Minnesota (Cochrane et al. 2015, Vraniak et al. 2017, C. Lapin and R. Hay, personal observation, Kurt Staab, landowner, personal communication to R. Hay). Parasitic amoeba (*Entamoeba* spp.) have been implicated in captive hatchling deaths (MacNeill et al. 2002), but to our knowledge this is undocumented in the wild. Michell and Michell (2015) radio-tracked ten headstarted Wood Turtles (six one-year-olds and four two-year-olds) for two years following their release in New York and reported that survivorship was 100%.

Adults are less vulnerable to predation, but it does occur. Loss of limbs to Snapping Turtles (*Chelydra serpentina*) and mesopredators, particularly Raccoons, is common (Harding 1985, reviewed by Ernst and Lovich 2009). In Jackson County, ca. 20% of adult Wood Turtles encountered were missing part or all of one appendage (R. Hay, personal observation). Lapin et al. (2019) determined that predation accounted for 75% of known-cause deaths ($n = 12$) of radio-tracked adult Wood Turtles in Wisconsin and Iowa. Carroll and Ultsch (2006) documented limb mutilation of hibernating adults by North American River Otters (*Lontra canadensis*) and suggested that additional individuals were likely killed. Parren (2013) documented predation of an adult male by a Fisher (*Pekania pennanti*). In addition, although leeches probably do not significantly affect survival, leech parasitism is common for Wood Turtles in Wisconsin and elsewhere (Hulse and Routman 1982, Brewster and Brewster 1986).

CONSERVATION

Status: Globally, the Wood Turtle is listed as Endangered by the IUCN. It is considered a Species of Greatest Conservation Need and listed as Threatened in Wisconsin by the WDNR.

Populations: In areas where Wood Turtles are present, documented densities are quite variable due to the species' uneven local distribution. For example, in New Jersey, Harding and Bloomer (1979) recorded an average density of 12.5/ha (5/ac), with local densities as high as 50/ha (20/ac). In Pennsylvania, the density of Wood Turtles for an entire study area (240 ha [593 ac]) was 0.7/ha (0.3/ac), but the density in areas where most Wood Turtles were found was 4.4/ha (1.8/ac; Ernst 2001b). Using replicated surveys and *N*-mixture models, Brown et al. (2017) estimated that population sizes ranged from 5 to 77 across eight sites in northeastern Minnesota (sites = 380–560 m [1,247–1,837 ft] stretches of river). Buech et al. (1997) hypothesized that this uneven distribution was correlated with geologic factors, such as the distribution of soils from glacial outwash plains. Wood Turtle density appears to decrease with increasing latitude (Walde et al. 2003, Bowen and Gillingham 2004). For example, Walde et al. (2003) estimated Wood Turtle density to be 0.4/ha (0.2/ac) in Quebec.

Reported sex ratios and adult-to-juvenile ratios vary widely among studies for the Wood Turtle. Ratner and Anderson (1978) surveyed a population in La Crosse and Trempealeau Counties. The sex ratio was 1M:1.9F when nest-site captures were included in the calculation ($n = 57$) and 1.6M:1F when nest-site captures were excluded in the calculation ($n = 26$). Ross et al. (1991) reported a sex ratio of 1M:0.6F in La Crosse and Trempealeau Counties (after removing nest-site captures; $n = 27$) and 1M:1.9F in Vilas County ($n = 24$). In Jackson County, the adult-to-juvenile ratio was 3.4A:1J during spring surveys ($n = 120$; R. Hay, unpublished data). Based on spring surveys from 1990 to 1995 in Jackson County, the sex ratio (including females found on nest sites) was 1M:1.1F ($n = 93$; R. Hay, personal observations). In Michigan, the sex ratio was 1M:1.2F ($n = 191$; Harding and Bloomer 1979). In Iowa, Berg (2014) found a sex ratio of 1M:1F for a suburban population ($n = 32$) and 1M:1.5F for a rural population ($n = 60$). In addition, the adult-to-juvenile ratio was 11.7A:1J and 61A:1J, respectively. In Quebec, Daigle (1997b) found that the sex ratio was 1M:1.2F and that the adult-to-juvenile ratio was 4.8A:1J ($n = 52$). Walde et al. (2003) found that in Quebec, the sex ratio was 1M:1.6F when nest-site captures were included in the calculation ($n = 138$)

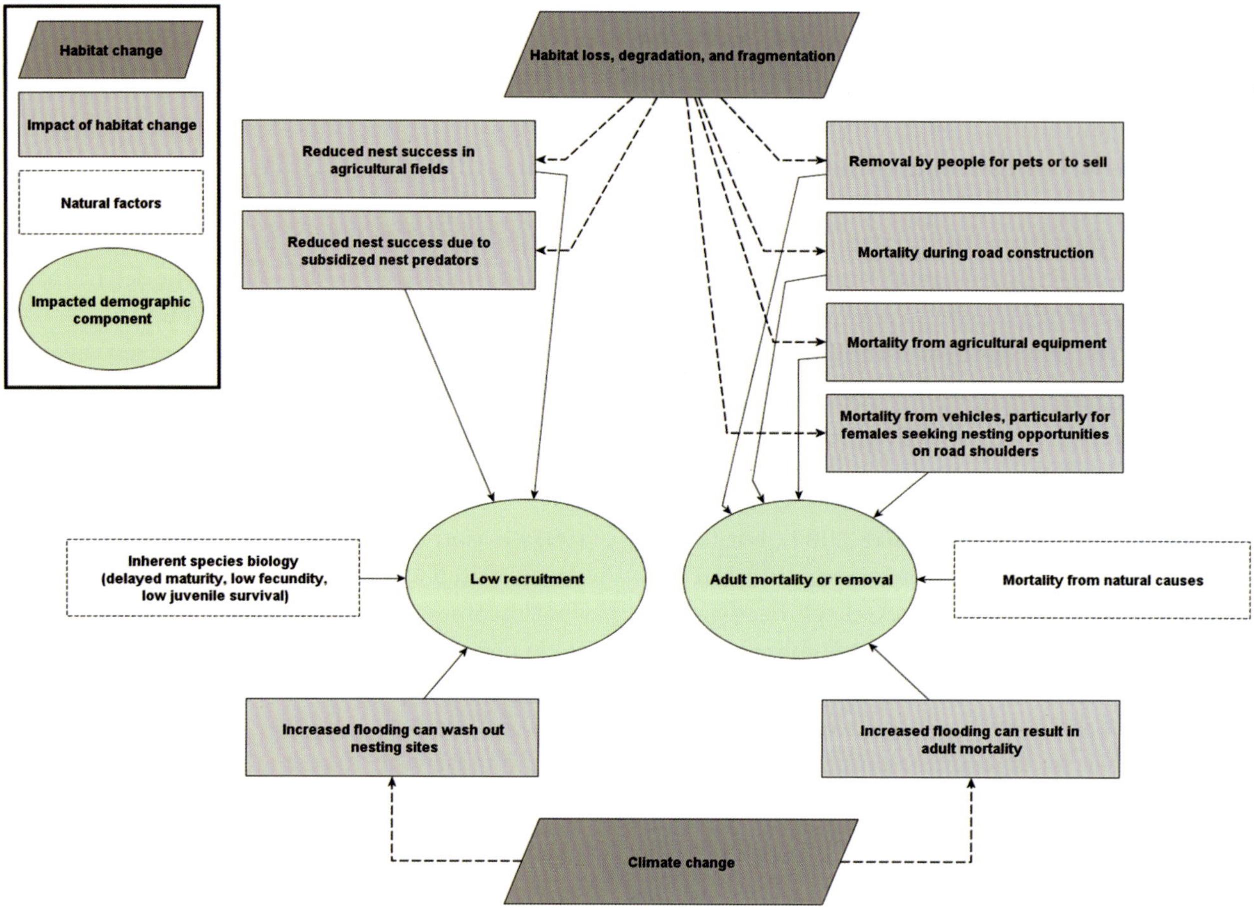

Figure 337. Major threats to Wood Turtle populations across the species' distribution and demographic components directly impacted by the threats. See Conservation section for further discussion and supporting references.

and 1M:1F when nest-site captures were excluded in the calculation ($n = 109$). Furthermore, Walde et al. (2003) documented an adult-to-juvenile ratio of 4A:1J when the population was surveyed in the spring and 2A:1J when the population was surveyed in the autumn ($n = 50$ juvenile captures). In Pennsylvania, the sex ratio was 1.4M:1F, and the adult-to-juvenile ratio was 3.2A:1J ($n = 88$; Ernst 2001b). In West Virginia, the sex ratio was 1.1M:1F, and the adult-to-juvenile ratio was 1.2A:1J ($n = 187$; Niederberger and Seidel 1999). In New Jersey, the sex ratio was 1M:1.6F, and the adult-to-juvenile ratio was 1.2A:1J ($n = 316$; Farrell and Graham 1991).

Threats: Wood Turtle populations are at risk of experiencing declines due to any adult mortality beyond natural levels due to their delayed maturity, low fecundity (small clutch sizes and typically only one clutch per year), and high juvenile mortality rates (see Reproduction and Development section; Figure 337). When these factors are combined with even a slight increase in adult mortality or removal, rapid declines and subsequent extirpations of local populations are expected (Burger and Garber 1995, Gibbs and Shriver 2002). For example, Garber and Burger (1995) observed the extirpation of two Wood Turtle populations over the course of just ten years after a wildlife area, previously closed to the public, was opened for recreation. Garber and Burger (1995) speculated that the cause of the declines was likely due to one or more of the following factors: "(1) removal, (2) road kills, (3) handling by recreationists,

(4) increased predation as a function of food waste and increased number of predators, and (5) disturbance by dogs."

An overall decline of Wood Turtle populations across their distribution has not been documented, but a compilation of available evidence (i.e., site and region-specific scientific literature, expert opinion, and results from temporally repeated site visits) suggests that the species has undergone drastic distribution-wide declines by as much as 50% or more in the last 100 years (Burger and Garber 1995, Bowen and Gillingham 2004, van Dijk and Harding 2011). In addition, average adult survival rates in Wisconsin, Minnesota, and Iowa may be lower than 0.9 (Lapin et al. 2019), which is likely lower than what is required for long-term persistence of populations (Compton 1999). Due to widespread long-term land and water use changes, many of the documented population declines are likely irreversible (van Dijk and Harding 2011). As is the case for most declining species, there are multiple factors contributing to the decline of Wood Turtles across their distribution (Figure 337). Habitat loss, degradation, and fragmentation (of both terrestrial and aquatic habitats) are currently the largest known threats to Wood Turtles (Burger and Garber 1995, Buech 1995, Buech et al. 1997, Ernst 2001a, Bowen and Gillingham 2004), and these issues are discussed in broad context in the Introduction to Conservation and Management of Wisconsin's Amphibians and Reptiles chapter. Below, we expound on two additional major threats for Wood Turtles: mortality from vehicular and agricultural equipment and climate change.

Vehicle traffic, road construction, and agricultural equipment are major threats to Wood Turtle populations (Brooks et al. 1992, Saumure and Bider 1998, Gibbs and Shriver 2002, Steen and Gibbs 2004, Saumure et al. 2007, Roberts et al. 2021). It is likely that a reduction in the availability of natural nest sites caused by development and succession results in increased road exposure for female Wood Turtles as they look for nesting opportunities on gravel road shoulders. Road mortality not only removes individuals from the population but often creates male-biased population sex ratios because proportionally more females are killed (Steen et al. 2006). Daigle and Jutras (2005) documented a decline of 50% for a Wood Turtle population at an agricultural site over a period of seven years in southern Quebec. Saumure et al. (2007) tracked 30 Wood Turtles in an agricultural area in Quebec from 1998 to 2000. Four of the individuals died and 90% were mutilated by agricultural equipment. Jones (2009) found that Wood Turtle mortality rates in New England were higher as a result of agricultural activities than those associated with low-intensity roads, likely because threats associated with agricultural activities were widespread (they affected a large overall area), while threats related to roads were localized.

One of the results of rapid contemporary climate change is an increase in extreme events (e.g., extreme drought and precipitation). Abnormally high precipitation can result in high-water events (i.e., flash floods), which can negatively impact adult Wood Turtle survival and recruitment. In Massachusetts, over 40% of one Wood Turtle population was displaced annually over a four-year period by flooding events, resulting in both adult mortality and reduced mating and nesting for the displaced turtles (Jones and Sievert 2009a). In Iowa, an isolated population of Wood Turtles is currently suffering complete nest losses nearly every year due to flooding (Spradling et al. 2010, J. Tamplin, University of Northern Iowa, personal communication). In addition, Mothes et al. (2020) estimated that 29%–52% of the Wood Turtle's climatic niche in the northeastern US will be lost by 2070.

Management: Given their threatened status in Wisconsin, it is illegal to harm, kill, or remove Wood Turtles from the wild, unless permitted by the WDNR. Furthermore, it is important to consider the impacts of land management activities (e.g., logging, recreational development) on known populations (Burger and Garber 1995, Compton et al. 2002, van Dijk and Harding 2011). Because Wood Turtles are such a mobile species, it is especially important to protect wide continuous corridors of riparian habitat from development and conversion to agriculture. Additional research on the sizes and types of riparian habitat buffers that are most beneficial to Wood Turtles is needed. Identification and protection of

NATURAL HISTORY BOX: Nest-Site Availability

In many areas throughout the range of Wood Turtles in Wisconsin, nesting sites are limited. This has resulted from natural succession and a lack of nest-site management. From 2007 to 2009 funds were provided to the WDNR by the American Transmission Company to survey all suitable Wood Turtle rivers and streams by canoe, walking, and driving roads that closely paralleled or crossed these waters. The surveys were conducted in seven northeastern and north-central Wisconsin counties and part of an eighth county. The goal of this work was to identify nesting sites, evaluate their condition, identify threats to each site (e.g., recreational use, natural succession of vegetation, livestock presence, agricultural use), and collect data on observed nest predation. Very few of the detected nesting sites were classified as 'natural sites' (i.e., sites that were clean sand or sandy, with some degree of sparse vegetation). Rather, the vast majority of sites actively used by turtles (i.e., evidence of at least one Wood Turtle or nest that was predated or intact) were at bridge crossings. In fact, bridge crossings and roadsides close to rivers are conservatively estimated to comprise at least 75% of the actively used nesting sites. This is particularly problematic because Wood Turtle shell fragments from old roadkill or freshly killed Wood Turtles were observed at a number of these bridge crossing sites during surveys (WDNR unpublished data). As a result of these preliminary findings, a nest-site restoration and creation program was developed by Turtles for Tomorrow, a Wisconsin nonprofit organization, to improve nesting opportunities for Wood Turtles, with the goal of reducing adult female road mortality and increasing populations. To date, 16 nesting sites have been restored or created, and 11 of these are equipped with electric fences to deter nest predators. This strategy appears effective, as no predation was detected at 9 of the 11 electrified sites, and the production of hatchlings at these locations has provided further reason for hope. For example, four nesting sites created in northwestern Wisconsin during 2014 produced 125 hatchlings the following year.

Robert W. Hay, Turtles for Tomorrow

known nesting locations from predators and poachers (Levell 2000) while also enacting habitat management strategies to combat succession are efforts that can be made toward the conservation of Wood Turtle populations (Buech et al. 1997; see Natural History Box: Nest-Site Availability).

The WDNR has experimented with several strategies to protect nests and nest sites from predation using nest cages and electric fencing, with positive results (C. Lapin, G. Kessler, and R. Hay, personal observations). Measures that have been used to reduce adult Wood Turtle mortality include installation of roadside fencing and underpasses and raising blade heights on agricultural and roadside mowers (Saumure et al. 2007, Tingley et al. 2009, van Dijk and Harding 2011, WDNR 2015d). The creation of small (≤0.4 ha [1 ac]) stream-side openings in forest canopies, particularly in areas that historically were maintained in an early successional state by natural disturbance (i.e., fire), is a management strategy that could also increase Wood Turtle foraging and nesting habitat (Ernst 2001b, Bowen and Gillingham 2004). However, this strategy should be considered carefully, including consultation with the WDNR Natural Heritage Conservation staff and compliance with WDNR regulations and would only be appropriate in areas where Wood Turtle populations already exist. We also note that Wood Turtle mortality from fire has been observed (Allard 1909). The Wood Turtle is protected by state statute in Wisconsin, so it is illegal to take or possess them, regardless of their origin, without an Endangered and Threatened Species Permit. However, if an individual turtle is encountered on a road, people are encouraged to move it off the road, so long as they do not put themselves at risk while doing so.

SPECIAL REMARKS

Wood Turtles have been reported to breed with Blanding's Turtles (Harding and Davis 1999), but whether or not the hybrid offspring are fertile is currently unknown (Scott Davis, Turtle Survival Alliance, personal communication). An adult male turtle that possessed characteristics of both species, consistent with hybridization, was captured in southern Jefferson County in 2017 (see Figure 330 in the Blanding's Turtle, *Emydoidea blandingii* species account).

Northern Map Turtle

Graptemys geographica (Le Sueur 1817)

Richard C. Vogt

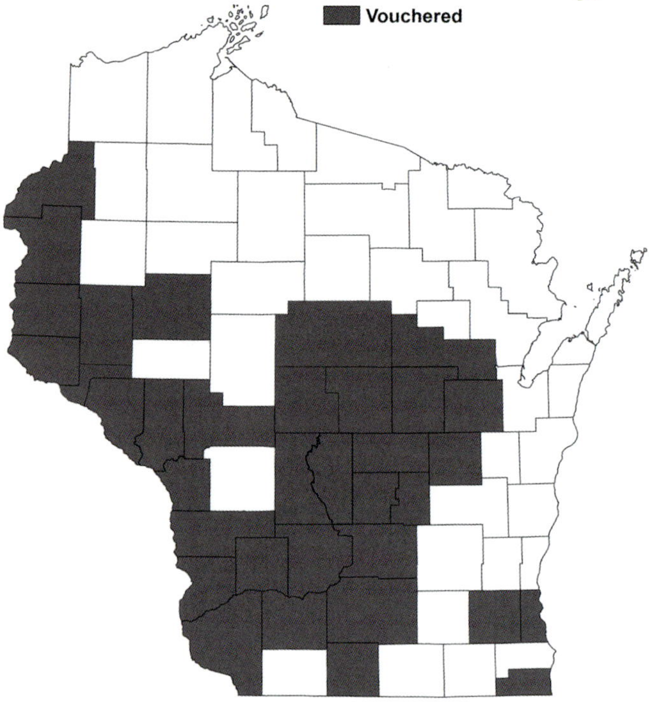

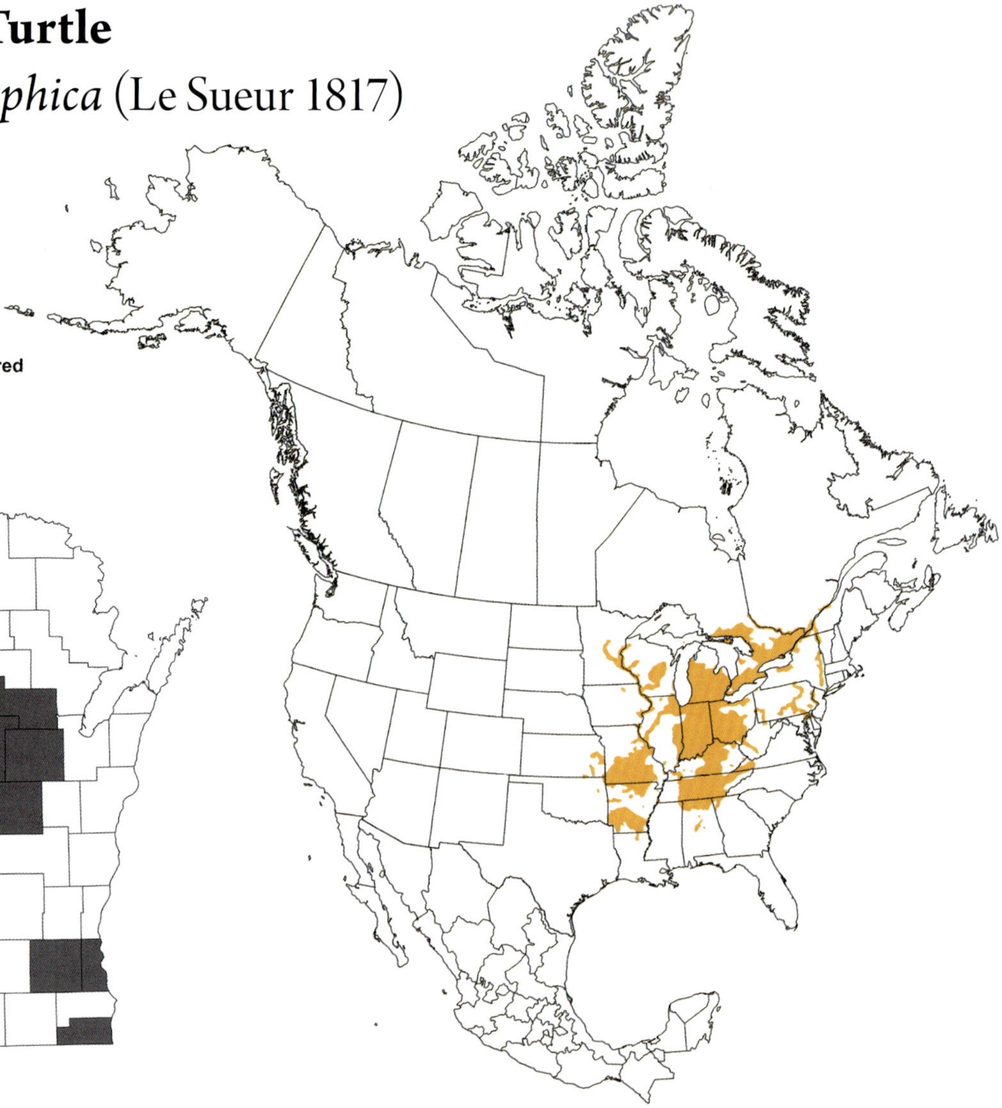

Adult Male Northern Map Turtle (Waukesha County; photo by D. Skryja).

AT A GLANCE

Northern Map Turtles have an olive to brown carapace and a yellow to cream plastron. Their head, limbs, and shell bridge are dark green and covered with a network of thin yellowish stripes. There is a keel down the center of their carapace, and the back edge appears notched or serrated. Northern Map Turtles inhabit large clear rivers and large lakes associated with rivers. Their state range is mostly concentrated in western and central Wisconsin along the Mississippi, St. Croix, and Wisconsin Rivers, but they also occur in eastern locations such as along the Fox River. Northern Map Turtles are active in northern locations such as Wisconsin from approximately late April through late October. They breed in both early spring and autumn. Females nest near the water on sandy riverbanks or sandbars in June, and hatchlings emerge in August or September. They feed principally on snails and clams, but they also devour other prey such as insect larvae. Northern Map Turtles can be distinguished from Wisconsin's other two map turtle species by their solid-colored plastron, broader head, and postorbital head spots, which are more yellow green in color as opposed to bright yellow. They also possess a less pronounced keel down the midline of the carapace than the other map turtle species in Wisconsin. The Northern Map Turtle is currently listed as a common species by the WDNR.

SYSTEMATICS

Graptemys geographica (Northern Map Turtle) was formerly called the Common Map Turtle, but its name was changed because the word 'common' could be construed to indicate that the species was abundant (Crother 2017). Although its new name suggests that it is not found in the southern US, it does occur as far south as Alabama (McCoy and Vogt 1990). Since its description, *G. geographica* has been placed in four different genera: *Testudo, Malacoclemmys, Malaclemys,* and *Graptemys* (the currently accepted genus; Crother 2017). Phylogenetically, this species is the basal group for all other *Graptemys* species (Lamb et al. 1994). No subspecies have been described or proposed.

DESCRIPTION

Adults: The carapace is green, olive, or brown, with a variable network of light green, yellow, or orange lines, and there are sometimes dark ocelli in each scute. The plastron is smooth and usually solid yellow or cream-colored in adults. The head, neck, and limbs are dark green and streaked with lighter green, yellow, or orange stripes. There is a prominent green or yellow dorsal blotch just posterior to the eye that can be oval or irregular in shape (Vogt 1981, McCoy and Vogt 1990; Figure 338). The Northern Map Turtle is a medium-sized turtle. Females are 17–27 cm MCL (7–11 in), and males are 10–16 cm MCL (4–7 in). Live weight of females and males from the Mississippi River in Wisconsin ranged from 900 to 2,000 g (31.7 to 70.5 oz) and 165 to 350 g (5.8 to 12.3 oz), respectively (Vogt 1980). Vertebral spines are low in adult males and absent in adult females. Although a medial keel and serrated posterior marginal scutes are present, they are not as pronounced as in the other map turtle species in Wisconsin (McCoy and Vogt 1990; see Diagnosis subsection). *Graptemys* spp. are members of the family Emydidae, which also includes the genera *Chrysemys, Emydoidea, Glyptemys,* and *Terrapene* in Wisconsin.

Preadult stages: The oblong eggs are 3.4–3.8 × 2.2–2.6 cm (1.3–1.5 × 0.9–1.0 in; Vogt 1981; Figure 339).

Figure 338. Dorsal surface of an adult female Northern Map Turtle's head. Note the oval-shaped spot or blotch posterior to the eye on either side of the head (Grant County; photo by J. Kapfer).

Figure 339. An excavated Northern Map Turtle nest containing 12 elliptical eggs (La Crosse County; photo by J. Kapfer).

Figure 340. The broad head of an adult female Northern Map Turtle (Jackson County; photo by C. Raimond).

Hatchlings appear similar to adults but have a more prominent midline keel on the carapace (Vogt 1981, McCoy and Vogt 1990). The plastral pattern in hatchlings is confined to the area along the scute sutures (Vogt 1981). Hatchlings from two clutches encountered in La Crosse County were 2.58–3.39 cm MCL (1.02–1.33 in, $\bar{x} = 2.96$ cm [1.17 in], $n = 25$; Andrew Badje, WDNR, unpublished data). Nagle et al. (2004) reported that in Pennsylvania hatchling CLs were 2.5–3.3 cm (1.0–1.3 in, $\bar{x} = 3.02$ cm [1.20 in]), PLs were 2.3–3.0 cm (0.9–1.2 in, $\bar{x} = 2.75$ cm [1.10 in]), and weights were 4.2–8.2 g (0.15–0.29 oz, $\bar{x} = 6.40$ g [0.23 oz]).

Variation: The tail of adult males is longer than that of females, with the cloacal opening posterior to the rear edge of the carapace (Vogt 1981). Vertebral spines along the midline of the carapace are also more pronounced in males than females. Furthermore, females have a broader head than males do, with jaws that possess wide crushing surfaces (Figure 340). Bulté et al. (2013) reported that the coloration of postorbital spots differs between the sexes, with adult males and subadult females exhibiting brighter spots than adult females. Adult males and juveniles also have dark lines along the plastral seams, unlike females.

Diagnosis: Northern Map Turtles have 12 scutes on their plastron and a smooth carapace that is not highly domed. The plastron, which is a solid yellow or cream color, cannot close against the carapace, the toes are webbed, and the carapace possesses a low midline keel with no definite knobs on the central scutes and no raised concentric grooves on individual scutes. The carapace is not blue black in color with yellow flecks, and the underside of the neck and chin is not an immaculate yellow color. The upper jaw is notched at the apex, while there is a yellow green or green dorsal spot located posterior to each eye, which is encircled by two or more narrow green or yellow lines.

Northern Map Turtles are most easily confused with the two other map turtle species in Wisconsin, the False Map Turtle (*Graptemys pseudogeographica*) and the Ouachita Map Turtle (*G. ouachitensis*). Yet Northern Map Turtles can be distinguished from the other map turtles by the lower keel on their carapace (compared to the more elevated knobs on the other two species) and the solid-colored plastron (Vogt 1981, McCoy and Vogt 1990). The head of adult female Northern Map Turtles is much wider than that of adult female False Map Turtles and Ouachita Map Turtles. The foreclaws are not elongated in adult male Northern Map Turtles, while they are elongated in adult male False Map Turtles and Ouachita Map Turtles. Head spot patterns also differ between Northern Map Turtles, Ouachita Map Turtles, and False Map Turtles. Ouachita Map Turtles have large yellow head blotches, one behind each eye, one beneath each eye (this may be fused with the head blotch behind the eye), one beneath this on the lower jaw, and one at the symphysis of the lower jaw. False Map Turtles have a narrow yellow postorbital spot and either a lack of or a small light yellow head spot. Northern Map Turtles are also more pugnacious than the other two species of map turtles. As such, they are less likely to stick their head out so that markings can be easily observed but will instead remain retracted into their shell and open their jaws as if attempting to bite (Richard Vogt, Instituto Nacional de Pesquisas da Amazônia, personal observation). The other two Wisconsin map turtles, on the other hand, do not typically attempt to bite (R. Vogt, personal observation). The eggs of different map turtle species in Wisconsin cannot be reliably differentiated in the wild (R. Vogt, personal observation).

DISTRIBUTION AND HABITAT

Global distribution: Northern Map Turtles only occur in North America and are principally found in the central and northeastern US. They are found in northwestern Vermont and northern New York along the St. Lawrence River basin through southern Michigan, Wisconsin, and eastern Minnesota and then south along the Mississippi River basin to northeastern Oklahoma and Louisiana and central Alabama.

Distribution in Wisconsin: Northern Map Turtles are primarily distributed in central and southwestern Wisconsin but also frequently occur along the western border counties and occasionally in southeastern Wisconsin (Kenosha, Milwaukee, and Waukesha Counties; Bolek 1997a, Casper 2015, Skryja 2017). They are currently documented in the Black, Chippewa, Fox, Mississippi, Big Rib, Milwaukee, Namekagon, St. Croix, Sugar, Wisconsin, and Wolf Rivers (Casper 1996, 2002, Badje 2020d). A reported Chippewa County voucher for the False Map Turtle (MPM 6836; Casper 2015) was a misidentified Northern Map Turtle (Joshua Kapfer, UW–Whitewater, personal communication). At present, this represents the only known Northern Map Turtle specimen from Chippewa County. The voucher reported for Outagamie County (MPM VZP449; Casper 1996) is based on photographs of a lacquered carapace from a dead individual with "Wolf River," "1993," "Povlich," "6-12-93," and "found by barn" handwritten on the bottom (J. Kapfer, personal communication).

Localities of possible occurrence: Based on occupancy in adjacent counties and their association with large rivers, Northern Map Turtles could be present in additional counties, such as Eau Claire. Further work in southeastern Wisconsin could better define this species' range in that region. Given records from Waukesha, Milwaukee, and Kenosha Counties, coupled with reports from across the Illinois border in Lake, Cook, and DuPage Counties (Phillips et al. 1999), Northern Map Turtles could be present in additional southern counties, such as Racine, but since they require unpolluted water, this is unlikely. Confirmation of observational reports from the Milwaukee River in Ozaukee County is also warranted (Casper 2015).

Habitat: In Wisconsin, Northern Map Turtles typically associate with medium-sized to large rivers that possess moderate currents (Vogt 1981). Accordingly, Bluett et al. (2013) investigated turtle community composition in relation to stream order in the Sangamon River (Illinois) and found that Northern Map Turtles were only present in the seventh-order stream, the largest order surveyed. In Pennsylvania, they typically occupied deep, slow-moving portions of a river for ca. ten months of the year and migrated to more shallow and heterogeneous sections during the summer months (Pluto and Bellis 1988).

Northern Map Turtles are common in backwater sloughs and oxbows associated with large rivers. During surveys of the upper and middle Mississippi River, Tucker et al. (2008) captured them most often in backwater areas ($n = 136$ of 189 total Northern Map Turtle captures, but note that the sampling design in this study may have weighted captures against fast-flowing sections). Conversely, Northern Map Turtles in Ohio appeared to prefer channel to marsh habitat (Tran et al. 2007). Despite a common association with rivers and backwaters, Northern Map Turtles sometimes use large lakes and occasionally small ponds or marshes (Vogt 1981). The overall landscape features within riparian zones can influence their habitat selection. For example, Northern Map Turtle occupancy in Indiana was negatively associated with the proportion of grassland habitat in the surrounding landscape but positively associated with wetland patch area (Rizkalla and Swihart 2006).

Existing studies report variation in how this species relates to the specific habitat characteristics within occupied waterbodies. For example, Vogt (1981) reported that they used habitats with soft bottoms and ample aquatic vegetation in Wisconsin. DonnerWright et al. (1999), on the other hand, did not detect significant relationships between aquatic habitat characteristics and captures of Northern Map Turtles based on 12 sampled sites along the St. Croix River ($n = 170$ captures). In Ontario, individuals showed preference for shallow water habitats (<2 m [6.6 ft]), as well as "open reeds" and "matted marsh" habitat classes (Carrière and Blouin-Demers 2010). While deep water was not preferred, adult females used deep-water habitat more than adult males and juveniles (Carrière and Blouin-Demers 2010). They also preferred natural shorelines over developed shorelines. Preferences for certain habitat characteristics further appear to differ by turtle size, with larger individuals using deeper water and faster currents than smaller individuals (Pluto and Bellis 1986).

Northern Map Turtles bask on a variety of structures to regulate body temperature. Appropriate basking sites range from partially submerged logs to rocks and sandbars not connected to the shoreline. Studies in Quebec found that this species selected basking logs that were longer, farther from shore, and in deeper water than available logs that turtles did not select (Flaherty and Bider 1984). They also bask on branches overhanging rivers and may fall 2–5 m (6–15 ft) to the water surface when disturbed (Vogt 1981). Preferred basking sites are often used concurrently by numerous individuals of several map turtle species (Figure 341).

Northern Map Turtles nest in open sandy areas near the water's edge, which can also include large sandbars and islands (see Reproduction and Development section). They hibernate while fully submersed in well-oxygenated water, usually behind logs and rock piles, near springs, or within Muskrat (*Ondatra zibethicus*) and North American Beaver (*Castor canadensis*) lodges, but also along anthropogenic structures such as wing dams in the Mississippi and Wisconsin Rivers (Vogt 1981).

Figure 341. Numerous map turtles basking on a log in the Wisconsin River. Note the Northern Map Turtle (*center*, identified by her wider head) is sharing this log with many Ouachita Map Turtles (Dane County; photo by C. Raimond).

REPRODUCTION AND DEVELOPMENT

Vogt (1980) studied the reproductive ecology of a Mississippi River population in Vernon County, and most of the information on courtship and nesting in Wisconsin reviewed here comes from that study. During courtship, males solicited females through head bobs in the vertical plane while turtles faced each other. Unlike the other two species of map turtles in Wisconsin, male Northern Map Turtles did not vibrate the back of their foreclaws along the sides of the female's face. Copulation occurred both in early spring and autumn, and evidence suggests that females store sperm from autumn copulations (Natural History Box: Sperm Storage). Egg laying began in June and lasted through mid-July. Most observed nesting activity occurred just after sunrise but continued throughout the day and night during overcast, humid, and warm conditions. Nests were usually placed in sandy open areas near the water (Figure 342), and clutch sizes typically ranged from 10 to 20 eggs. Additional counts of clutch size from other Wisconsin locations show similarities to Vogt's (1980) findings. A group of females nesting on an island in the Mississippi River laid

NATURAL HISTORY BOX: Sperm Storage

Northern Map Turtles do not appear to feed for nearly a month after emergence from hibernation in the spring. Feeding during this time is apparently not required to assist in egg development or prepare for egg deposition because females enter hibernation with a full set of large vitellogenic follicles. In the late 1970s, I removed five females from a hibernaculum in the Mississippi River (Vernon County) in late November and kept them in a refrigeration chamber until February. When I took the females out of the refrigeration chamber and placed them at a temperature of 30°C (86°F), I detected shelled eggs in the oviducts within 12 days (Vogt 1980). Oviposition was induced by injection of the hormone oxytocin, and the fertile eggs, which were placed in an incubator, successfully hatched later. There was apparently sperm in the oviducts stored from autumn mating or before.

Richard C. Vogt, Instituto Nacional de Pesquisas da Amazônia

Figure 342. A female Northern Map Turtle digging a nest chamber (La Crosse County; photo by J. Kapfer).

9–18 eggs per nest ($\bar{x} = 12$, $n = 20$). The females were 21.2–25.6 cm MCL (8.3–10.1 in, $\bar{x} = 22.3$ cm [8.8 in]) and weighed 1,155–2,138 g (40.7–75.4 oz, $\bar{x} = 1{,}510$ g [53.3 oz]; R. Vogt, unpublished data). Furthermore, eight clutches encountered in La Crosse County in 2018 contained 9–18 eggs ($\bar{x} = 14$; A. Badje, unpublished data). Vogt (1980) found that females can lay one or two clutches each year. Nagle and Congdon (2016) estimated that females in Pennsylvania regularly produced two clutches (12%–21% of females) annually, and some females also produced three clutches annually (0.5%).

Northern Map Turtle hatchlings exhibit temperature-dependent sex determination during development (Bull and Vogt 1979, Bull et al. 1982a, Vogt and Bull 1984). Bull and Vogt (1979) found that eggs incubated in the lab at 25°C (77°F) produced only males, and eggs incubated at 30.5°C (87°F) produced only females. Furthermore, they found that natural nests deposited in the shade produced only males, while those in the sun produced only females. Bull and Vogt (1981) determined that incubation temperature during the middle third of the development period was the most influential for sex determination (for further discussion, see Natural History Box: Temperature-Controlled Sex Determination [TSD] in the Ouachita Map Turtle, *Graptemys ouachitensis* species account).

Most hatchlings emerge in August or September in the Great Lakes region (Harding 1997). Some studies report that hatchlings can overwinter in the nest to emerge the following spring (Nagle et al. 2010) and that these young turtles exhibit tolerance to freezing or have the ability to supercool in the nest chamber (i.e., allow body temperature to drop below freezing without formation of ice crystals in their body; reviewed by Lindeman 2013). However, Vogt (1980) found no evidence that hatchlings overwinter in nests from over 500 map turtle nests excavated from mid-September to November in Wisconsin.

Males become sexually mature in three to four years (Vogt 1980). Females matured during their eighth year in Lake Erie (Lindeman 2013), but they probably do not reach maturity until 12–13 years of age on the Mississippi River in Wisconsin (Vogt 1980). In Quebec, sexual maturity occurred at MCLs of ca. 7.5 cm (3.0 in) and 17.5 cm (6.9 in) for males and females, respectively (Gordon and MacCulloch 1980). Females achieve larger adult sizes than males, because although both sexes grow at the same rate when juveniles, females take longer to reach maturity (Vogt 1980). Furthermore, some studies support that adults obtain larger sizes in northern populations than in southern populations (Iverson 1988, Iverson et al. 2019). Longevity of over 20 years is reported in the wild (reviewed by Lindeman 2013), with females potentially reaching older ages than males (Iverson et al. 2019), and an individual captured in the wild as an adult lived for an additional 18 years in captivity (Harding 1997).

ACTIVITY

In northern locations, this species emerges from hibernation and begins basking by late April (Vogt 1981, Graham et al. 2000), although feeding does not begin until late May (Vogt 1981). They are generally diurnal during the active season, but Smith and Iverson (2004) found that diel activity in an Indiana population was greatest from midmorning to noon. Most of their time when active is spent either foraging or basking. Research from Ontario found that basking abovewater was an important thermoregulatory behavior that allowed juvenile and female turtles to reach net energy retention so that they could maximize energy used for growth and reproduction (Bulté and Blouin-Demers 2010). At night, Northern Map Turtles sleep on the bottom of the river or on logs beneath the water (Vogt 1981).

No studies have focused on movement patterns of this species in Wisconsin. In Pennsylvania, the mean length of river stretches used by males was 2.11 km (1.3 mi, range = 0.17–6.10 km [0.1–3.8 mi], $n = 46$) and by females was 1.2 km (0.8 mi, range = 0–5.3 km [0–3.3 mi], $n = 14$; Pluto and Bellis 1988). In Quebec, the mean length of river used by males was 1.65 km (1 mi, range = 1.19–2.41 km [0.7–1.5 mi], $n = 5$) and by females was 2.3 km (1.4 mi, range = 1.21–3.51 km [0.8–2.2 mi], $n = 5$; Ouellette and Cardille 2011). Ryan et al. (2008) radio-tracked 16 female Northern Map Turtles from May to September in an urban canal in Indiana and found that the mean length of canal used was ca. 3 km (1.9 mi). In Missouri, 76% of individuals were recaptured within 0.184 km (0.1 mi) of their original capture location, based on sampling during the summer months over a three-year period (Pitt and Nickerson 2014).

Although active foraging mostly ceases by mid-September, Northern Map Turtles may still bask until mid-October (Vogt 1981). Hibernation typically begins in Wisconsin by late October or early November, and various hibernation sites are used (Vogt 1981; see Habitat subsection). Northern Map Turtles remain submersed throughout winter dormancy but are intolerant of anoxic conditions, so they do not bury in the mud and must overwinter in well-oxygenated water where cutaneous uptake of water is effective (i.e., rivers or large lakes; Reese et al. 2001). Graham et al. (2000) found that 47% of individuals (i.e., 7 of 15) returned to the same overwintering location between years in Vermont.

PREY AND PREDATORS

According to Harding (1997), this species only feeds in the water. Vogt (1981a) analyzed stomach contents of 21 Northern Map Turtles from Vernon County and found that adult females fed primarily on mollusks (66% by volume). Remaining prey items consisted of insect larvae (caddisfly, mayfly, dragonfly) and fish carrion, with only trace amounts of vegetation. Vogt (1981a) found that males and females primarily consumed mollusks, followed by insect larvae, with little vegetation consumed. Northern Map Turtles also consume crayfish (Lagler 1943, Penn 1950, Vogt 1981). In Ontario, Bulté and Blouin-Demers (2008) estimated that an average-sized female consumed 33–137 kg (73–302 lb) of invasive Zebra Mussels (*Dreissena polymorpha*) a year, while an average-sized male consumed 0.03–0.27 kg (0.07–0.60 lb) a year. In Maryland, adult males fed primarily on a number of small gastropod species (Planorbidae, Hydrobiidae, Physidae), trichopterans, and *Corbicula* spp., while adult females fed primarily on pleurocerid snails, with virtually no overlap in the diets of the two sexes (Richards-Dimitrie et al. 2013). On the contrary, substantial dietary overlap between the sexes was found for a population inhabiting a lake in Ontario (Bulté et al. 2008).

Adults rarely fall victim to predators, and Vogt (1980) did not document any predation of basking adults during his extensive observations in Vernon County. Although midsized mammals occasionally prey on nesting females (Cochran 1987), these predators pose the largest threat to nests and hatchlings. Nest predators include Striped Skunks (*Mephitis mephitis*), Raccoons (*Procyon lotor*), Virginia Opossums (*Didelphis virginiana*), and Red Foxes (*Vulpes vulpes*; Vogt 1981a, Cochran 1987). Geller (2012a) documented a 90% map turtle nest predation rate (*Graptemys geographica* and *G. ouachitensis*; $n = 38$ of 42 nests) over a three-year study period in Iowa County, primarily due to Raccoons. The median time from egg deposition to depredation was 12.5 hours, and 25 of the nests monitored (i.e., 74%) were destroyed within 48 hours of egg deposition.

A variety of birds (e.g., grackles [*Quiscalus* spp.], herons [*Egretta* spp.], American Bitterns [*Botaurus lentiginosus*]), as well as fishes (e.g., catfish, Largemouth Bass [*Micropterus salmoides*], Northern Pike [*Esox lucius*]), also prey on hatchlings once they are in the water (Vogt 1980). Furthermore, Vogt (1981c) found that larvae of the sarcophagid fly *Metoposarcophaga importuna* attacked live hatchlings while they were still in the nests (see Natural History Box: Fly Predation on Hatchling Map Turtles in the Nest in the False Map Turtle, *Graptemys pseudogeographica* species account). Northern Map Turtles are hosts to an array of parasitic species, including a variety of leeches, six coccidians, a trypanosome, one monogenean, eight trematode flukes, one enteric spy-headed worm, and five round worms (Stunkard 1917, Harrah 1922, Hughes et al. 1942, Rausch 1946, 1947, Woo 1969, Pluto and Rothernbacher 1976, Wacha and Christiansen 1976, Brooks 1979), as well as epizoophytic algae (Edgren et al. 1953).

CONSERVATION

Status: Globally, the Northern Map Turtle is listed as a Least Concern species by the IUCN. It is considered a common species in Wisconsin by the WDNR. Across their distribution, they are currently listed as Endangered in Maryland, Threatened in Kansas, rare in Georgia, and as a species of concern in Ontario, Oklahoma, and Vermont.

Populations: No density or abundance estimates exist for Northern Map Turtle populations in Wisconsin. Gordon and MacCulloch (1980) estimated that their study population at a 2,000 ha (4,942 ac) site in Quebec ranged from 322 to 379 turtles. Bulté and Blouin-Demers (2008) estimated a population size of 1,569 (95% CI = 1,487–1,662) individuals in a 788 ha (1,947 ac) lake in Ontario. They further estimated the number of adults by sex at this location, which ranged from 317 females (95% CI = 243–391) to 335 males (95% CI = 281–391; Bulté and Blouin-Demers 2009). Estimated abundance of Northern Map Turtles in a 28 ha (69 ac) Tennessee reservoir was 71 (Weber and Layzer 2011).

The above estimates appear to be much lower than those determined for locations along the Mississippi River in Vernon County during the 1960s and 1970s, as hundreds were often collected per day just as the bycatch of commercial fishermen's gill nets (Vogt 1980). Northern Map Turtles represented ca. 7% of the total number of three map turtle species collected by fyke nets, gill nets, and trammel nets along the Mississippi River in Vernon County (Vogt 1980). Similarly, Northern Map Turtles represented 124 of 1,965 total turtles (6.3%) captured in two northern reaches of the Mississippi River surveyed for four years by Tucker et al. (2008). These capture proportions were less than those reported farther north during work along the St. Croix River, where 170 Northern Map Turtles of 1,204 total turtles (14.1%) were captured, with only Common Snapping Turtles (*Chelydra serpentina*) and Spiny Softshells (*Apalone spinifera*) captured more often (DonnerWright et al. 1999).

Estimates of annual adult survival in Ontario were high, ranging from 82.6% in males to 87.3% in females (Bulté and Blouin-Demers 2009). Hatchling sex ratios can vary widely for Northern Map Turtles due to temperature-dependent sex determination (e.g., Vogt and Bull 1984). Overall, subadult and adult sex ratios appeared to be close to 1M:1F in samples from Wisconsin taken during the 1970s (R. Vogt, personal observation). Pitt and Nickerson (2014) also found that the sex ratio did not differ from 1M:1F for a population in Missouri sampled over three years ($n = 182$ turtles). Weber and Layzer (2011) estimated that the sex ratio was 0.24M:1F in a reservoir in Tennessee but noted that the sample size was small ($n = 21$).

Threats: Habitat degradation due to anthropogenic activities probably has the greatest impact on most Northern Map Turtle populations in Wisconsin (see Introduction to Conservation and Management of Wisconsin's Amphibians and Reptiles chapter). Examples of unnatural habitat alterations can include the removal of snags or logjams that turtles bask on and channelization of rivers to enhance their use for recreation and commercial boat traffic, as well as the construction of impoundments and excessive siltation (VanDeWalle and Christiansen 1996, Lindeman 1999b, Bodie 2001, Smith et al. 2006, reviewed by Ernst and Lovich 2009). Dams can further affect this species by restricting daily movements and unnaturally altering home range sizes (Bennett et al. 2010). Impacts to shorelines and adjacent terrestrial land can also reportedly affect Northern Map Turtles, as research in Ontario determined that they preferred natural shorelines over developed shorelines (Carrière and Blouin-Demers 2010).

Given this turtle's preference for molluscan prey, factors that impact native mollusk populations in Wisconsin could also affect Northern Map Turtles. For example, the high abundance of invasive Zebra Mussels (*Dreissena plymorpha*) and Quagga Mussels (*D. bugensis*) in some Wisconsin rivers could disrupt Northern Map Turtle energetics or physiology and alter the food webs that turtles are members of (reviewed by Lindeman 2013). Lindeman (2017) further reported that invasive mussels can colonize Northern Map Turtle shells. Although the negative impacts of this phenomenon (if any) remain largely unknown, substantial mussel colonies on turtle shells could create excess weight or drag while swimming that would require turtles to use more energy than normal. Water pollution can further exacerbate the negative impacts of invasive mussels on Northern Map Turtles. For example, pollution that reduces native aquatic mollusk populations can allow invasive mollusk species a foothold to reach higher abundances (Neves et al. 1997), which can alter Northern Map Turtle diets. In fact, Moll (1977) found that the native mollusks consumed by a turtle population in Illinois declined after 80 years of habitat degradation that reduced snail and clam abundances, after which only an introduced Asian mollusk (*Corbicula* sp.) was consumed.

The viability of Northern Map Turtle populations could be negatively affected by the unnatural removal of individuals or factors that impact reproductive success. Gravid females can be susceptible to road mortality during overland nesting migrations (e.g., Ashley and Robinson 1996, Nagle and Congdon 2016). Loss of open nesting habitat due to shoreline development or encroachment of open shoreline areas by woody vegetation that shades out nests could also alter egg development and survival (Nagle and Congdon 2016). Increased densities of mammalian nest predators in anthropogenically disturbed landscapes can impact annual reproductive success and population recruitment (Geller 2012a). Northern Map Turtles are sometimes harvested recreationally and accidentally removed as bycatch during commercial fishing operations, which combined could have substantial population-level effects. For example, from 1989 to 1992 Northern Map Turtles made up 25% of 145 turtles harvested that were documented in an Iowa study (Gritters and Mauldin 1994). Although the long-term impacts of harvest in the Upper Midwest have not been quantified, work in Missouri determined that recovery from substantial population declines, such as could occur due to harvest, can take decades (Pitt and Nickerson 2013). Harvest of this species is currently restricted in Wisconsin (see Introduction to Conservation and Management of Wisconsin's Amphibians and Reptiles chapter).

Several studies have found that human boat traffic can have negative impacts on Northern Map Turtles. Bulté et al. (2010) detected frequent boat propeller injuries among individuals from study populations in Ontario and noted that the prevalence of injuries was greater in adult females than adult males or juvenile females. They also determined that the probability of associated population extinction ranged from 63% to 99% if the likelihood of mortality after a propeller strike was >10% (the equivalent of one adult female killed every three years by propellers). In addition, a study in Missouri found that even passing canoes negatively impacted the ability of Northern Map Turtles to bask effectively (Pitt and Nickerson 2012), indicating that even seemingly inconsequential recreational activities can affect this turtle. King and Niiro (2013) investigated potential future shifts in the distribution of Northern Map Turtles based on projected climate changes. They reported that under the high climate emissions scenario created for their models, habitat in western border counties currently occupied by the species could become unsuitable by 2080.

Management: Management strategies that protect rivers, promote heterogeneous aquatic habitats with diverse structures (i.e., preservation of basking logs and overwintering locations), and limit recreational activity and motorboat traffic would likely benefit Northern Map Turtles (see Introduction to Conservation and Management of Wisconsin's Amphibians and Reptiles chapter). Current examples support that protection of rivers in particular can benefit this species in the Upper Midwest. Large Northern Map Turtle populations occur within the 420 km (261 mi) stretch of the Mississippi River that occurs in the upper Mississippi River National Wildlife and Fish Refuge. This 97,000 ha (239,692 ac) reserve encompasses parts of Minnesota, Wisconsin, Iowa, and Illinois and protects a diversity of aquatic habitats, including the main channel, backwaters, and nesting islands for numerous turtle species. Habitat protection is also granted to Northern Map Turtles in the St. Croix River, a 406 km (252 mi) section of river between Minnesota and Wisconsin that is managed by the NPS.

In addition to habitat protection, conservation strategies that protect populations from various sources of mortality are likely important. Although response is slow, Northern Map Turtle populations can reportedly recover from substantial losses if the causes of declines are remedied (Pitt and Nickerson 2013). Accordingly, management strategies that offset mortality can be effective if they are sustained for long periods. Strategies can include harvest regulations (i.e., restrictions on acceptable body size of harvested turtles and acceptable collection yields) and regulations that dictate that commercial fishing practices must limit turtle mortality due to bycatch. Installation of exclusion fences to direct the migration routes of nesting females and reduce road mortality can also benefit this species (Nagle and Congdon 2016).

Although protection of adults is likely important, strategies to enhance recruitment in young age classes could also help ameliorate population losses. Nagle and Congdon (2016) found that gravid females used protected nesting sites managed through removal of vegetation to combat succession and the addition of nesting substrate, which enhanced recruitment potential (but see the considerations these authors outlined). The USFWS recently constructed islands for duck and turtle nesting habitat in Pool 8 of the upper Mississippi River National Wildlife and Fish Refuge (Vernon County). The islands were designed to be permanent, with a riprap base covered with dredge spoil sand, and they are shaped like crab claws when viewed from above to capture sand and silt. In 2011 the islands were only a year old, but Northern Map Turtles and softshells (*Apalone* spp.) were already nesting on some of them (R. Vogt, personal observation). Construction of nesting islands away from the mainland is an attractive management option because it limits easy access by mammal predators such as Raccoons. Additional management options for aquatic turtles can be found in the Introduction to Conservation and Management of Wisconsin's Amphibians and Reptiles chapter.

SPECIAL REMARKS

Northern Map Turtles are of economic value because they consume large numbers of mollusks, which serve as intermediate hosts for many trematode parasites of humans, domestic animals, and game animals. Furthermore, managing for healthy Northern Map Turtle populations could help with control of exotic invasive mussels, such as Zebra Mussels in the Mississippi River. Bulté and Blouin-Demers (2008) estimated that an adult female consumed between 33 and 137 kg (73 and 302 lb) of Zebra and Quagga Mussels a year in Ontario (but see possible drawbacks of consuming invasive mollusks in the Threats subsection). Northern Map Turtles hybridize with both False Map Turtles and Ouachita Map Turtles in the Mississippi River in Wisconsin (Vogt 1978).

Ouachita Map Turtle

Graptemys ouachitensis (Cagle 1953)

Richard C. Vogt

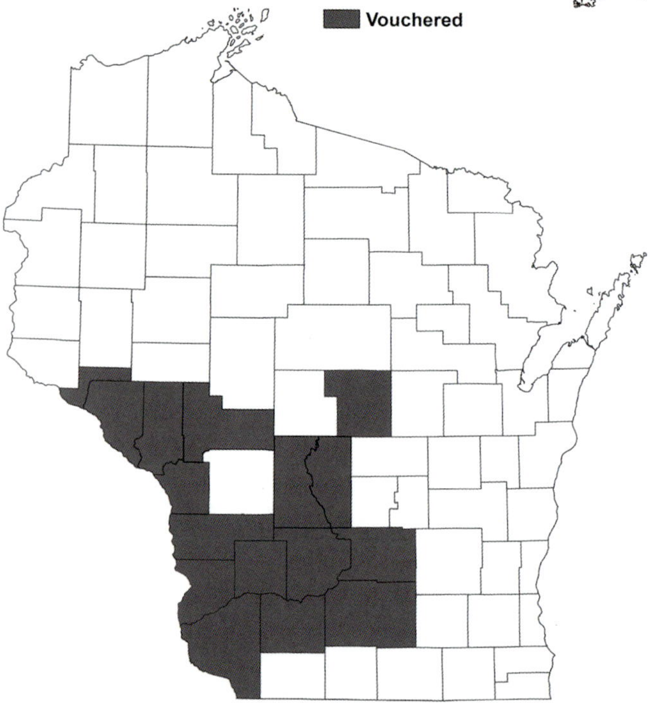

Ouachita Map Turtle (Vernon County; photo by A. Badje).

AT A GLANCE

Ouachita Map Turtles have greenish carapaces with black blotches encircled by thin yellowish lines (but this pattern is often faded in adults). Adult plastrons are often light yellow with dark mottling that follows the creases between individual scutes. Their head, limbs, and shell bridge are dark with thin yellow stripes that create intricate patterns, and there are several large yellow blotches on the top and sides of the head. The midline of the carapace is also keeled, with black knobs on several of the vertebral scutes, while scutes along the back edge of the carapace appear serrated. Ouachita Map Turtles inhabit large rivers, where they are frequently seen basking on logs in the water. Males are typically found in areas with a strong current, while females can often be found feeding in backwaters. Their range in Wisconsin is primarily limited to the western counties along the Mississippi River and areas along the Wisconsin River. They are active from early to mid-April through mid-October or early November. Breeding takes place in both spring and autumn. Females lay eggs on sandy riverbanks and sandbars from late May through early July, while hatchlings typically emerge in late August to early September. They are omnivorous, feeding on a wide variety of aquatic plants and insects. Ouachita Map Turtles can be distinguished from other map turtle species in Wisconsin by the presence of large yellow head blotches: one behind the eye, one beneath the eye, one on the lower jaw, and one at the symphysis in the middle of the lower jaw. The Ouachita Map Turtle is currently listed as a common species by the WDNR.

SYSTEMATICS

Graptemys ouachitensis (Ouachita Map Turtle) was formerly considered a subspecies of *G. pseudogeographica* (False Map Turtle) and described as *G. p. ouachitensis* (Cagle 1953b). For a short time it was listed under the name *Malaclemys pseudogeographica ouachitensis* (Cochran and Goin 1970). Vogt (1978) was the first to recognize *G. ouachitensis* as a species distinct from *G. pseudogeographica* and, following Cagle (1953b), included two subspecies: the widespread subspecies *G. o. ouachitensis* and *G. o. sabinensis* (Sabine Map Turtle), which is restricted to rivers in southwestern Louisiana. Recently, the Sabine Map Turtle was elevated to species status (*G. sabinensis*; Lindeman 2013), as the distributions are allopatric, and genetic studies suggest that *G. sabinensis* is a species distinct from *G. ouachitensis* (e.g., Wiens et al. 2010, Brown et al. 2012a). Phylogenetically, *G. ouachitensis* and *G. sabinensis* are most closely related to *G. pseudogeographica* (Lamb et al. 1994). *Graptemys* spp. are members of the family Emydidae, which also includes the genera *Chrysemys, Emydoidea, Glyptemys,* and *Terrapene* in Wisconsin.

DESCRIPTION

Adults: The carapace is green with one to six (usually one) black blotches on the posterior border of each scute, which are encircled with thin yellow or orange lines. Alternately, a lattice of interconnected circles may be present without the black blotches. The pattern is often faded in adults, and melanism is frequent. The carapace is keeled, with low black knobs on the second, third, and fourth vertebrals (Figure 343). The plastron is marked with dark concentric swirls of alternating yellow and dark green, the green being wider than the yellow (Figure 344). This pattern, which covers less than 60% of the plastron, fades to a brown-yellow mottling in adults and is replaced by dark lines along the sutures. The head

Figure 343. Adult female Ouachita Map Turtle. Note the knobs on the midline of the carapace (Sauk County; photo by C. Raimond).

Figure 344. Plastron of an adult Ouachita Map Turtle (Dane County; photo by C. Raimond).

Figure 345. Hatchling Ouachita Map Turtle. Note the presence of the white pointed egg tooth at the front of the upper jaw (Dane County; photo by C. Raimond).

is dark green with yellow markings. A large blotch extends from under the eye around the posterior border to meet a pair of longitudinal narrow lines running the length of the head. This crescent may be broken behind the eye and may have one to four wide stripes that come in contact with the orbit. The spot below the orbit is large. Directly below this spot is another large yellow spot on the mandible (Figure 343). Four large yellow spots (at the symphysis, on each mandible, and in the center) characterize the underside of the head. The head, neck, and limbs are dark green and streaked with lighter green, yellow, or orange (Vogt 1993, 1994).

The Ouachita Map Turtle is a medium-sized species. Vogt (1980) measured morphological characteristics for a population in Vernon County. He reported that CL of adult females was 16.3–24.2 cm (6.4–9.5 in, $\bar{x} = 20.5$ cm [8.2 in], $n = 265$), while the CL of adult males was 10.9–13.7 cm CL (4.3–5.4 in, $\bar{x} = 12.3$ cm [4.94 in], $n = 68$). Live weight of females was 557–2,300 g (19.6–81.1 oz, $\bar{x} = 1{,}136$ g [40.1 oz], $n = 265$). Live weight of adult males was 161–309 g (5.7–10.9 oz, $\bar{x} = 211$ g [7.4 oz], $n = 68$).

Preadult stages: Eggs are whitish and elliptical in shape. Vogt (1980) reported that eggs measured from nests in Vernon County were 3.02–3.59 × 2.12–2.95 cm (1.19–1.41 × 0.83–1.16 in, $\bar{x} = 2.2 \times 3.4$ cm [0.88 × 1.36 in], $n = 479$). Hatchling color and pattern are similar to those of adults (Figure 345). The dark plastral color pattern of hatchlings usually covers less than 60% of the plastron and is confined to the area along the scute sutures (Vogt 1980). Vogt (1980) recorded morphological characteristics of 116 hatchlings from eggs collected in Vernon County that were incubated in the lab. These hatchlings were 2.71–3.34 cm CL (1.07–1.31 in, $\bar{x} = 3.07$ cm [1.21 in]) and 2.65–3.40 cm CW (1.04–1.34 in, $\bar{x} = 3.01$ cm [1.19 in]).

Variation: This species exhibits marked sexual dimorphism, with females weighing ca. five times more than males (Vogt 1980). The tail of adult males is longer than that of females, with the cloacal opening posterior to the rear edge of the carapace (Vogt 1980). The second, third, and fourth foreclaws of males are disproportionately elongated (Vogt 1980). Head markings vary by individual and are influenced by incubation temperature (see Natural History Box: Incubation Temperature Influences the Characteristic Head Markings in False Map Turtles and Ouachita Map Turtles in the False Map Turtle, *Graptemys pseudogeographica* species account).

Diagnosis: Ouachita Map Turtles have 12 scutes on their plastron, while the carapace is not highly domed and smooth (the scutes do not possess concentric grooves), with a definite keel and knobs on the second, third, and sometimes fourth central scutes. The plastron is not hinged, so it cannot close against the carapace, and the toes are webbed. The carapace is not blue black in color with yellow flecks, nor is the lower neck and chin solid bright yellow. Ouachita Map Turtles further possess large yellow head blotches or spots: one behind the eye, one beneath the eye, one on the lower jaw, and one at the symphysis of the lower jaw. The spot under the eye may be fused with the spot behind the eye to form a crescent-shaped blotch.

Ouachita Map Turtles are most easily confused with the other two map turtle species found in Wisconsin. Ouachita Map Turtles can be readily distinguished from Northern Map Turtles by the latter's wider head, low keel on the carapace, immaculate yellow plastron, and yellow-green postorbital spot (Vogt 1981, McCoy and Vogt 1990). Unlike Ouachita Map Turtles, False Map Turtles in Wisconsin have thinner postorbital head markings, a postorbital blotch that is usually connected to a thin dorsal neck stripe, and three to five thin stripes that pass under the postorbital spot and reach the orbit. In Wisconsin, the False Map Turtle never has a complete postorbital crescent. In addition, the color of the neck stripes and postorbital spots are light yellow in False Map Turtles and bright yellow in Ouachita Map Turtles. No reliable method exists for differentiating the eggs of map turtle species in the wild (Richard Vogt, Instituto Nacional de Pesquisas da Amazônia, personal observation).

DISTRIBUTION AND HABITAT

Global distribution: Ouachita Map Turtles only occur in North America and are principally found in the central US. They occur from the Mississippi River basin in Wisconsin and Minnesota south through Louisiana, west to Lake Texoma, Oklahoma, and east into Indiana, Ohio, and West Virginia.

Distribution in Wisconsin: Ouachita Map Turtles are primarily concentrated in the western border counties along the Mississippi River and in the south-central portion of the state along the Wisconsin River. Although they are mostly found in these two larger rivers, they also occur in the Black River and near the mouth of the Chippewa River.

Localities of possible occurrence: Based on the presence of large rivers and occupancy in adjacent counties, Ouachita Map Turtles could be present in additional counties along the western border of the state (e.g., Pierce and St. Croix).

Habitat: Ouachita Map Turtles are almost exclusively a riverine species in Wisconsin (Vogt 1981) but sometimes can be found in lakes near river mouths (Lindeman 2000). They occur within the faster-flowing segments of larger rivers and usually in areas with ample submersed aquatic vegetation (Vogt 1981). Bluett et al. (2013) investigated turtle community composition in relation to stream order in the Sangamon River (Illinois) and found that Ouachita Map Turtles were only present in fifth-order or higher streams or rivers. In Kansas, this species appeared to select deeper and wider river sections with above average water temperatures (Fuselier and Edds 1994). Tucker et al. (2008) captured Ouachita Map Turtles more often in backwaters of the Mississippi River than in other sampled habitats ($n = 27$ of 44 Ouachita Map Turtle captures).

Vogt (1980) studied basking site selection and behavior in Vernon County along the Mississippi River. Turtles in this population regularly basked communally on logs, stumps, and fallen trees that were usually not connected to the shoreline and only rarely along sandbanks in the river channel. Accordingly, reports suggest that floating woody debris is particularly important to this species. Vogt (1980) observed 47 adult map turtles basking together on a 6 m (20 ft) floating log in Vernon County. Lindeman (1999b) found a positive relationship between basking density of Ouachita Map Turtles and density of downed wood in rivers in Mississippi, Louisiana, and Kentucky (Figure 346).

Nesting habitats used by Ouachita Map Turtles in Wisconsin are often along open beaches or sandbars (Vogt 1980). Geller (2012b) described nesting sites as open areas that are sometimes surrounded by woody vegetation. Furthermore, nesting habitats are in sandy or well-drained soils and sparsely covered with short vegetation (ca. 80% exposed ground surface; Geller 2012b). The two nesting sites in Iowa County characterized by Geller (2012b) were relatively close to a large river (i.e., within 15 m [49 ft] and 52 m [171 ft]),

Figure 346. Numerous Ouachita Map Turtles basking on fallen woody debris in the Wisconsin River. Note the False Map Turtle (*far left*) and the Northern Map Turtle with the characteristically large head (*third turtle from the left*; Dane County; photo by C. Raimond).

while Vogt (1980) indicated that nesting sites in Vernon County could be up to 100 m (328 ft) from the shoreline. Ouachita Map Turtles overwinter underwater behind logs and rock piles, and sometimes in association with Muskrat (*Ondatra zibethicus*) and North American Beaver (*Castor canadensis*) lodges, and near springs (Vogt 1980). On 1 October 1976, 56 Ouachita Map Turtles, 25 False Map Turtles, and 4 Northern Map Turtles were captured behind one rock pile in the Mississippi River (Vernon County; Vogt 1980). Behind the same rock pile in 1977, 131 individuals across all three species were captured (Vogt 1980). Like Northern Map Turtles and False Map Turtles, Ouachita Map Turtles hibernate near anthropogenic structures in Wisconsin, such as wing dams in the Mississippi River (Vogt 1980).

REPRODUCTION AND DEVELOPMENT

Vogt (1980) studied mating behavior of Ouachita Map Turtles along the Mississippi River in Vernon County. Copulation was not observed in the field but probably took place during both spring and autumn, as courtship behavior was documented in October and November. Males probably recognize conspecific females by head markings or cloacal scents (Vogt 1978). Based on the observations made by Vogt (1980), after a male identifies an appropriate conspecific female, he circles around to her head, places his nose against hers, and simultaneously arches both forearms above and in front of their heads. The male then drums his long claws against the tympanic region of the female's head 1–14 times before attempting to mount. The probable purpose of such an act is to identify the male as the proper species to the female. There is no regular number of drumming bouts undertaken before mating, nor is there a typical duration of the drumming behavior, but the number of contacts per bout seems to be species-specific (Vogt 1978, 1980, 1993). The number of drumming contacts by male Ouachita Map Turtles ranges from 4 to 8 ($\bar{x} = 5.2$ contacts per bout), while those of False Map Turtles range from 4 to 14 ($\bar{x} = 9.8$; Vogt 1980). Possibly being drummed for ca. half the time is sufficient for female Ouachita Map Turtles to distinguish the correct males. In addition, Ouachita Map Turtles do not bob their heads in the vertical plane as they drum their claws, but False Map Turtles do (Vogt 1993).

Vogt (1980) reported that nesting occurred from late May through early July on open sandbars and beaches in Vernon County (see Habitat subsection). Nesting was also found to occur from late May through early July in Iowa County (Geller 2012b, Geller et al. 2020b). Geller (2012b) determined that individuals in Iowa County nested throughout the day when temperatures were cool (ca. 21°C–29°C [70°F–84°F]), but on warmer days (ca. 30°C–37°C [85°F–99°F]), nesting activity was principally restricted to mid- to late morning and midafternoon. Similarly, most nesting activity in Vernon County occurred between 6:30 and 10:00 a.m., but nesting occurred throughout the day when temperatures were cool (Vogt 1980). Females laid 8–20 eggs per nest ($\bar{x} = 10.5$, $n = 65$), and there was a positive correlation between female body size and clutch size. Examination of reproductive organs indicated that females typically produced two clutches of eggs per year, but some produced three clutches. A study in Sauk County conducted from 2015 to 2019 documented lower clutch sizes than in Vernon County ($\bar{x} = 7.4$, range = 3–13, $n = 56$; Geller et al. 2020c).

Ouachita Map Turtle hatchlings exhibit temperature-dependent sex determination during

NATURAL HISTORY BOX: Temperature-Controlled Sex Determination (TSD)

The classic definitive study of temperature-controlled sex determination (TSD) was conducted with the three species of map turtles in Wisconsin at the nesting beach on Turtle Island along the main channel of the Mississippi River at Stoddard (Vernon County; Bull and Vogt 1979, 1981, Bull et al. 1982a, 1982b, Vogt and Bull 1984). Although TSD in turtles had been previously suggested in the literature by Pieau (1974) and Yntema (1976), their sample sizes were so small and the mortality so high that it appeared to most people that the effect was merely differential mortality, with males dying at high incubation temperatures and females dying at low temperatures. However, the study of Bull and Vogt (1979), with its large sample size and high survivorship in the experiments, confirmed unequivocally that TSD in fact occurred in many species of turtles and opened the door to a distinctive branch of research covering the spectrum of turtle taxa, crocodilians, and some lizards.

Not all species of turtles exhibit TSD. Some have heteromorphic sex chromosomes, while others have genetic sex determination (GSD) but have homomorphic sex chromosomes (Vogt and Bull 1982). In the study at Stoddard, most nests produced only one sex, with nests placed in the shade producing males and nests placed in the sun producing females (Vogt and Bull 1984). Through experiments in the laboratory, it was determined that the thermal critical period of incubation was the middle third of development; temperatures before or after this period has little influence on the sex of the hatchlings (Bull and Vogt 1981). These results also supported the conclusion that the effect of temperature was additive; more than four hours per day above the threshold temperature (the incubation temperature that produces a 1:1 sex ratio) during the middle third of development was enough to produce female hatchlings.

The exact mechanism that produces TSD is still not known. It appears that there is a genetic factor that produces a hormone or enzyme that changes the production of the primordial germ cells of the gonad from female-producing cells to male-producing cells and that this function is controlled by temperature. Below the thermocritical minimum or above the thermocritical maximum of this reaction the gonad remains female. Within the activity temperature range of this enzyme or hormone it becomes male. This is why females are produced at low temperatures in some species of turtles, crocodilians, and lizards, males are produced at intermediate temperatures, and females are again produced at the highest viable incubation temperatures. However, females are not produced at low temperatures in most species of turtles, since it is too cold for the embryos to develop. Low and high temperature females are the norm in all crocodilians and lizards that display TSD.

Why this system evolved and whether or not TSD is advantageous compared to GSD are unknown and have been open to speculation for over 35 years. In Wisconsin, Wood Turtles (*Glyptemys insculpta*) and both softshell species (*Apalone* spp.) have GSD, while all others have TSD. Most researchers believe that TSD evolved several times from GSD species. TSD is still a fertile field for research and experimentation. For example, is there evolutionary significance of producing low temperature males and high temperature females? Are high temperature males and low temperature females less fit than high temperature females and low temperature males? What is needed is long-term (i.e., from hatchling to maturity) studies on the fitness of males and females produced at different incubation temperatures to determine if there is a selective advantage of producing low temperature males and high temperature females.

Richard C. Vogt, Instituto Nacional de Pesquisas da Amazônia development (Natural History Box: Temperature-Controlled Sex Determination [TSD]). Bull and Vogt (1979) found that eggs incubated in the lab at 25°C (77°F) produced only males, and eggs incubated at 30.5°C (87°F) produced only females. Furthermore, they found that natural nests located in the shade produced almost exclusively males, while those in the sun produced only females, and that temperature during the middle third of the development period was the most influential for sex determination (Bull and Vogt 1981). The developmental environment that embryos are exposed to also has implications beyond sex determination. For example, recent findings support that eggs incubated at lower temperatures (which produce males) also yielded hatchlings with the highest innate immune response (Freedberg et al. 2008).

Hatchlings emerge from mid-August to late September (Geller et al. 2020b). In Wisconsin, no evidence of overwintering in the nest was found from over 500 map turtle nests excavated from mid-September to November (Vogt 1980). Males and females grow at the same rapid rate until maturity is reached, then growth tapers off abruptly (Vogt 1980). Lindeman (1999a) estimated that in Kentucky, mean upper PL was 9.6 cm (3.8 in, $n=26$) for males and 17.5 cm (6.9 in, $n=35$) for females. The smallest gravid female documented by Lindeman (1999a) measured 16 cm PL (6.3 in). Longevity in the wild is unknown. In the laboratory, both sexes have lived over 15 years (Ernst and Lovich 2009).

ACTIVITY

Ouachita Map Turtles emerge from hibernation and begin basking in early to mid-April (Vogt 1980). In Wisconsin, they are diurnal, and initial morning hours are spent basking to raise body temperature, which is followed by foraging throughout midmorning (Vogt 1980). The remainder of daily activity then alternates between basking and feeding until dusk, similar to reports in Kentucky by Lindeman (2000). At night, Ouachita Map Turtles sleep on the bottom of the river (Vogt 1981). Information on preferred environmental and body temperatures in Wisconsin is sparse. In Vernon County, six females and one male were chased from a basking log into a trammel net on 14 June 1977 at ca. 3:45 p.m., and thermal data were recorded. The air temperature was 21.5°C (70.7°F), and the water temperature was 22.5°C (72.5°F). Cloacal temperatures of the six females ranged from 23.8°C to 27.0°C (74.8°F to 80.6°F, $\bar{x}=25.8$°C [78.4°F]); cloacal temperature of the male was 22.5°C (72.5°F; R. Vogt, unpublished data).

Little information exists on Ouachita Map Turtle movement patterns. Geller et al. (2020a) found that hatchlings in southern Wisconsin typically dispersed from the nest toward dense vegetation cover rather than the river. Vogt (1980) documented individuals that moved >4 km (2.5 mi) annually along their selected stretch of river. He also observed that females moved up to 1.6 km (1.0 mi) from river channels to nesting islands in May, while most males remained in the channels until summer, at which point many individuals moved to river backwater areas to forage. By mid-September feeding activity ceases, but Ouachita Map Turtles may still be seen basking until mid-October (Vogt 1980).

In Wisconsin, most individuals enter hibernation from mid-October through November (Vogt 1980), during which they remain submerged and associate with various underwater structures (see Habitat subsection). Although few studies of overwintering ecology in Ouachita Map Turtles exist (reviewed by Ultsch 2006), they are presumably intolerant of anoxic conditions (like other map turtle species). As such, they likely do not bury in the mud and must remain in well-oxygenated water for cutaneous oxygen uptake that also does not completely freeze during winter. Vogt (1980) found large numbers of Ouachita Map Turtles together near presumably suitable overwintering sites during late autumn in Wisconsin, indicating they will congregate to hibernate.

PREY AND PREDATORS

Ouachita Map Turtles are omnivorous. Individuals feed both on and beneath the surface of the water (Vogt 1980). They seem to be opportunistic feeders, 'grazing' on surface vegetation and available insects that are randomly encountered (R. Vogt, personal observation). For example, individuals have been observed floating on the surface indiscriminately scooping duckweed (*Lemna* spp.) and insects into their mouths (R. Vogt, personal observation). Pondweed (*Potamogeton* spp.), duckweed, eelgrass (*Vallisneria* spp.), manna grass (*Glyceria* spp.), arrowhead (*Sagittaria* spp.), and algae made up 40% of the food (by volume) consumed by 58 adult females collected from 26 May to 6 September in Vernon County (Vogt 1981a). The remainder of the diets consisted of insects (caddisfly, mayfly, and damselfly larvae and beetles, 44%), fishes and carrion (6.2%), and mollusks (2.5%). Males had a more carnivorous diet than females, with animal matter accounting for ca. 60% (by volume) of food consumed (Vogt 1981a).

Natural predation of adults is probably minimal, and Vogt (1980) did not witness any predation of basking adults during his extensive observations in Vernon County. However, Geller (2020) documented predation of adult females by Raccoons (*Procyon lotor*) at nesting sites in Sauk County. Eggs and hatchlings are highly vulnerable to predation. Geller (2012a) determined that 90% (i.e., 38 of 42) of the map turtle nests (*Graptemys geographica* and *G. ouachitensis*) he monitored between 2009 and 2011 in Iowa County were destroyed by predators, primarily Raccoons. The median time from egg deposition to predation was 12.5 hours, and 25 of the monitored nests (74%) were predated within 48 hours of deposition. However, if Raccoons disturb female Ouachita Map Turtles while nesting, they may abandon the nest (Geller 2014). Interestingly, while several bird species consume many hatchlings (e.g., Ring-billed Gulls [*Larus delawarensis*], American Crows [*Corvus brachyrhynchos*], Common Grackles [*Quiscalus quiscula*], Red-winged Blackbirds [*Agelaius phoeniceus*]), Vogt (1979) reported a symbiotic relationship between adult map turtles and Common Grackles, in which the Common Grackles removed leeches (*Placobdella parasitica*) from basking adult map turtles (Natural History Box: Cleaning-Feeding Symbiosis between Grackles and Map Turtles). Vogt (1981c) found that larvae of the sarcophagid fly *Metoposarcophaga importuna* preyed on pipped live hatchlings while they were still in the nests (see Natural History Box: Fly Predation on Hatchling Map Turtles in the Nest in the False Map Turtle, *Graptemys pseudogeographica* species account).

CONSERVATION

Status: Globally, the Ouachita Map Turtle is listed as a Least Concern species by the IUCN. It is considered a common species in Wisconsin by the WDNR.

Populations: No density or abundance estimates of this species in Wisconsin exist. For the closely related Sabine Map Turtle, estimated densities ranged from 21 to 61/ha (9 to 25/ac) among four sites in a Louisiana river (reviewed by Ernst and Lovich 2009). Vogt (1980) reported that Ouachita Map Turtles represented ca. 50% of all map turtles he captured over several years (573 of 1,117 total captures) along the Mississippi River in Vernon County. The sex ratios of Ouachita Map Turtles captured were skewed toward females (ca. 1M:1.5F), which may have been attributable to collection method (Vogt 1980).

Threats: Degradation of riparian habitats, road mortality of nesting females, and nest destruction by mammalian predators probably have the greatest impacts on Ouachita Map Turtles in Wisconsin (see Introduction to Conservation and Management of Wisconsin's Amphibians and Reptiles chapter). Given this species' heavy use of woody debris for basking, removal of snags and logjams from rivers to improve recreational use and maintain channels for commercial barge traffic could have substantial impacts. Furthermore, construction of dams, impoundments, and associated siltation negatively impacts habitat suitability for Ouachita Map Turtles (Lindeman 1999b, Bodie 2001). For example, dams in riverine habitats likely restrict natural movements and home range size of this species, as is reported for the Northern Map Turtle (Bennett et al. 2010).

Various factors exist that could impact reproductive success and, thus, population recruitment. For example, encroachment of woody vegetation that shades out nests on normally open sandbars and sandy shorelines could have negative effects on development and hatching success (see also Northern Map Turtle, *Graptemys geographica* account). Anthropogenic alterations to water levels due to lock and dam systems could further impact populations if nests and eggs are flooded prior to hatching (Geller 2012b).

NATURAL HISTORY BOX: Cleaning-Feeding Symbiosis between Grackles and Map Turtles

Many animals have evolved symbiotic cleaning-feeding relationships (e.g., fishes, large terrestrial mammals and oxpeckers, crocodiles and sandpipers, fishes and crocodiles, geckos and tortoises, crabs and finches, finches and marine iguanas). For example, Darwin's finches (Geospizinae) have been documented removing ticks from Galapagos Giant Tortoises (*Geochelone elephantopus*; McFarland and Reeder 1974). On 9 June 1977 I observed a Common Grackle (*Quiscalus quiscula*) pulling leeches from basking map turtles at 11:15 a.m. (Vogt 1979). The mixed species basking log had 5 male and 18 female map turtles basking on it. The log was ca. 150 m (492 ft) from the shoreline of an island just off the main channel of the Mississippi River in Pool 8 (Vernon County). A Common Grackle landed on the log at 11:15 a.m. and began examining the axial and inguinal cavities of two adult female Ouachita Map Turtles; it pulled something from one cavity and flew away. The Common Grackle returned at 11:25 a.m., inspected various leg cavities of 11 more turtles, and removed three more objects before flying away. At 11:30 a Common Grackle returned, inspected two female turtles, and then began pulling something from the left inguinal cavity of the second turtle. As the bird attempted to dislodge the object, the turtle was rocked back and forth on the log. During this process the turtle remained motionless, with all legs extended. After successfully removing the object, the Common Grackle flew ca. 3 m (10 ft) to another log and began pecking at the object while holding it on the log with its right foot. At that time, I was able to identify the object as a leech. The bird continued foraging intermittently on leeches in this manner until 1:30 p.m., when it flew away. I terminated my observations at 2:00. The leeches were presumably *Placobdella parasitica*, as this was the only leech ever found on over 2,000 map turtles examined from this site (Vogt 1979).

I observed a grackle foraging in a similar manner at the same site on 14 June 1977 from 10:30 a.m. to 12:30 p.m. Basking turtles had been observed in the same area for over five years, but this was the first time that this behavior was documented. Given that Lindeman (2013) noted that no one else has ever reported birds feeding on leeches from basking map turtles, these observations may represent the result of a learned behavior by an individual bird. It is interesting that hatchling map turtles are preyed upon by grackles and other birds, so there must be a time in a turtle's life when it recognizes that these birds are no longer a threat but are beneficial in preening the leeches from the turtle's body.

Richard C. Vogt, Instituto Nacional de Pesquisas da Amazônia

Increased densities of medium-sized mammalian nest predators that are concomitant with human development could also impact annual reproductive success (Geller 2012a). Exposure to the widely used pesticide atrazine appears to inhibit the ability of hatchlings to escape nests and results in reduced hatchling activity rates, impacting hatchling survivorship (Neumann-Lee and Janzen 2011).

Research supports that turtle populations are particularly susceptible to harvest of adult individuals (see Introduction to Conservation and Management of Wisconsin's Amphibians and Reptiles chapter). Harvest of Ouachita Map Turtles by noncommercial fishermen is probably minimal, but bycatch in gill nets from commercial fishing operations could be substantial and pose a threat (Vogt 1980). For example, Barko et al. (2004) reported the accidental capture of 2,139 map turtles (Ouachita Map Turtles and False Map Turtles, 83% of all turtle species accidentally captured) by commercial fishing operations in a stretch of the upper Mississippi River. Harvest of Ouachita Map Turtles is currently restricted in Wisconsin.

Management: General management strategies to mitigate road mortality, such as installation of fences to restrict turtle access to roadways near nesting locations, and strategies to maintain suitable riparian habitats (including nesting habitat) are likely applicable to this species (see Introduction to Conservation and Management of Wisconsin's Amphibians and Reptiles chapter). Preservation of existing habitat, such as that found in the Upper Mississippi River National Wildlife and Fish Refuge, is likely a highly effective strategy for this and other aquatic turtle species. Given the potentially substantial impacts that harvest of adult turtles can have on populations, regulations that impose limits on intentional or accidental removal of adults would likely benefit Ouachita Map Turtles.

If facilitation of population recruitment is desired, then strategies to protect or improve nesting habitat are important. For example, Geller (2012c) found that placing electric fences around nests significantly decreased nest depredation in Sauk County; 15% of nests with electric fences were destroyed ($n = 53$), compared to 89% of nests without electric fences ($n = 37$). Furthermore, efforts to offset the encroachment of woody vegetation on nesting beaches and promote warmer nest temperatures and thus development of females (Vogt and Bull 1984) are a potential management strategy (Vogt 1994). Construction of nesting islands to promote population recruitment has been enacted, including islands for duck and turtle nesting in Pool 8 of the Upper Mississippi River National Wildlife and Fish Refuge (Vernon County; see review in Northern Map Turtle, *Graptemys geographica* account). Commercial fishermen using gill nets should be prohibited from setting nets within 2 km (1.2 mi) of nesting beaches to avoid drowning adult female Ouachita Map Turtles.

SPECIAL REMARKS

Ouachita Map Turtles hybridize with both Northern Map Turtles and False Map Turtles in the Mississippi River in Wisconsin (Vogt 1978). However, the only *Graptemys ouachitensis* × *G. geographica* hybrid found nesting in this area did not produce fertile eggs (R. Vogt, unpublished data).

False Map Turtle

Graptemys pseudogeographica (Gray 1831)

Richard C. Vogt

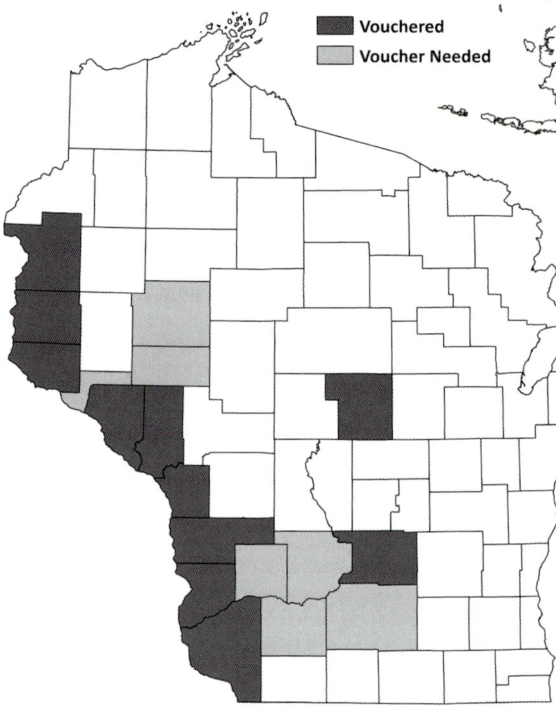

Juvenile False Map Turtle (Wisconsin; photo by R. LaFleur).

AT A GLANCE

False Map Turtles can be recognized by the presence of a typically thin L-shaped yellow blotch behind the head and the presence of many thin yellow lines below the eyes. The midline of the carapace is raised, or keeled, and black knobs occur on each scute down the center of the upper shell. The back edge of the carapace appears serrated. The plastron is yellowish, with dark swirling marks of alternative light and dark stripes, although this pattern can be faded in adult females. False Map Turtles inhabit backwater sloughs and the main channel of large rivers. They are primarily distributed along the southern St. Croix, Mississippi, and Wisconsin Rivers in western and southwestern Wisconsin, where they are frequently observed basking on woody debris. False Map Turtles are active from early to mid-April through mid-October. Reproduction occurs in spring and autumn, while nesting occurs from late May through July. Hatchlings begin to emerge by late August, although emergence can extend to mid-September. They feed principally on insect larvae but will also eat small snails and clams, and females consume aquatic vegetation. The False Map Turtle is currently considered a species of Special Concern in Wisconsin by the WDNR.

SYSTEMATICS

Since its original description, *Graptemys pseudogeographica* (False Map Turtle) has been placed in five different genera: *Emys, Clemmys, Malacoclemmys, Malaclemys,* and *Graptemys.* From 1831 to 1911, *G. pseudogeographica* was referred to as *Emys lesuerii* (Gray 1831), *G. lesuerurii* (Agassiz 1857), *Malacoclemmys lesueuri* (Yarrow 1882), and *Malaclemys lesueuri* (Hurter 1911) while at the same time being referred to as *G. pseudogeographica* by other authors.

There are currently two recognized subspecies of *Graptemys pseudogeographica,* including *G. p. pseudogeographica* (Northern False Map Turtle, the subspecies found in Wisconsin) and *G. p. kohnii* (Mississippi Map Turtle; Crother 2017). *G. p. kohnii* was historically considered a distinct species (Baur 1890, Stejneger and Barbour 1943, Carr 1952, Cagle 1953a, Conant and Collins 1998). Vogt (1978) used multivariate analysis to show that the head patterns, skull characteristics, courtship behavior, and allozymes of *G. p. kohnii* were conspecific with *G. p. pseudogeographica.* Furthermore, he also separated *G. ouachitensis* (Ouachita Map Turtle) as a distinct species from *G. pseudogeographica.* Three additional taxa were often historically considered to be subspecies of *G. pseudogeographica* (*G. oculifera* [Ringed Map Turtle], *G. ouachitensis sabinensis* [Sabine Map Turtle], and *G. versa* [Texas Map Turtle]). *Graptemys* spp. are members of the family Emydidae, which also includes the genera *Chrysemys, Emydoidea, Glyptemys,* and *Terrapene* in Wisconsin.

DESCRIPTION

Adults: The carapace is olive green, usually with one dark blotch on the posterior border of each scute, which is encircled with yellow or orange (Figure 347). However, the number of blotches ranges from zero to six, and the pattern is often faded in adults. The carapace is keeled, with low black knobs on the second, third, and fourth vertebrals. The plastron is marked with dark concentric swirls of alternating yellow and dark green. In adult females, the plastron fades to a yellow-brown mottling. The head, limbs, and shell bridge are covered in many thin yellow lines. The head markings range from a complete yellow-orange

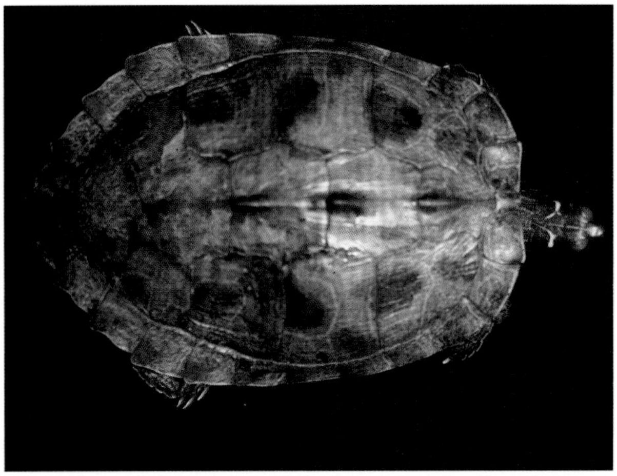

Figure 347. Adult female False Map Turtle displaying the typical carapace pattern, with a dark blotch on the posterior border of each scute that is encircled with yellow or orange. Also note the low black knobs on the second, third, and fourth vertebrals and the yellow crescent on the head (Vernon County; photo by R. Vogt).

crescent posterior to the orbit to a broken crescent with six narrow lines in contact with the orbit. The underside of the jaw is marked with alternate longitudinal yellow and green lines. Anteriorly, the lines break up into a highly variable pattern.

The False Map Turtle is a medium-sized turtle. Vogt (1980) measured morphological characteristics for a population in Vernon County and determined that the CL of adult females was 19.3–27.4 cm (7.6–10.8 in, $\bar{x} = 22.5$ cm [8.1 in], $n = 109$) and of adult males was 11.1–15.1 cm (4.4–5.9 in, $\bar{x} = 13.3$ cm [4.13 in], $n = 68$). Live weight of females was 857–2,502 g (30.2–88.3 oz, $\bar{x} = 1{,}477$ g [52.1 oz], $n = 109$). Live weight of males was 143–364 g (5.0–12.8 oz, $\bar{x} = 251$ g [8.9 oz], $n = 68$).

Preadult stages: The eggs are elliptical in shape and whitish. Vogt (1980) reported that eggs in Vernon County were 2.78–3.88 × 1.79–2.47 cm (1.09–1.53 × 0.70–0.97 in). Eggs from a single clutch measured in La Crosse County on 13 June 2018 averaged 3.28 × 2.34 cm (1.29 × 0.92 in, $n = 12$; Andrew Badje, WDNR, unpublished data). Hatchlings appear generally similar to adults. However, they have a more prominent keel than the adults do, and the plastral pattern is confined to the area along the scute sutures (Vogt 1980). Vogt (1980) recorded morphological characteristics of 99 hatchlings based on eggs collected in Vernon County that were incubated in the lab. Hatchlings were 2.50–3.30 cm MCL (0.98–1.30 in, $\bar{x} = 3.04$ cm [1.20 in]) and 2.66–3.37 cm CW (1.04–1.33 in, $\bar{x} = 3.04$ cm [1.20 in]). Hatchlings from a single clutch in La Crosse County were 2.92–3.14 cm MCL (1.15–1.24 in, $\bar{x} = 3.04$ cm [1.20 in], $n = 4$; A. Badje, unpublished data; Figure 348).

Variation: This species is markedly sexually dimorphic, with females weighing ca. six times more than males (Vogt 1980). The tail of adult males is longer than that of females, with the cloacal opening posterior to the rear edge of the carapace (Vogt 1995). The second, third, and fourth foreclaws of males are disproportionately elongated (Vogt 1980). Head markings vary by individual and are influenced by incubation temperature (Natural History Box: Incubation Temperature Influences the Characteristic Head Markings in False Map Turtles and Ouachita Map Turtles).

Figure 348. Egg and hatchling False Map Turtles (La Crosse County; photo by A. Badje).

Diagnosis: False Map Turtles have 12 plastral scutes and a smooth carapace (no concentric grooves on the scutes) that is not highly domed but with a definite midline keel and knobs on the second, third, and sometimes fourth central scutes. The plastron is not hinged, so it cannot close against the carapace, and the toes are webbed. The carapace is not blue black in color with yellow flecks, nor are the lower neck and chin solid bright yellow. There is a narrow yellow postorbital spot and either a lack of or small light yellow head spots. The plastral pattern of juveniles and males covers 75%–100% of the surface. False Map Turtles are easiest to confuse with the other two map turtle species found in Wisconsin. Northern Map Turtles (*Graptemys geographica*) can be readily distinguished from False Map Turtles by their wider head, low keel on the carapace, and immaculate yellow plastron (Vogt 1981, McCoy and Vogt 1990). The head stripes and blotches on the False Map Turtle are narrower than those on the Ouachita Map Turtle. Furthermore, the color of the head spots on Ouachita Map Turtles is bright yellow, unlike those of False Map Turtles. Ouachita Map Turtles also have large yellow head blotches that False Map Turtles do not possess: one behind the eye, one beneath the eye (this may be fused with the head blotch behind the eye), one on the lower jaw, and one at the symphysis of the lower jaw. The eggs of Map Turtles in Wisconsin appear similar, and no reliable method exists for differentiating them by species in the wild (R. Vogt, Instituto Nacional de Pesquisas da Amazônia, personal observation).

NATURAL HISTORY BOX: Incubation Temperature Influences the Characteristic Head Markings in False Map Turtles and Ouachita Map Turtles

Incubation temperature has a dramatic effect on head patterns, scute anomalies, and sex determination in map turtles. Incubation experiments were conducted with 11 clutches of Ouachita Map Turtles and 6 clutches of False Map Turtles. Eggs were incubated at 25°C (77°F, $N = 66$) and 35°C (95°F, $N = 70$). The eggs incubated at 25°C (77°F) hatched in 66–70 days, with 81.8% hatching success. Six did not show signs of development, and six died during development. The eggs incubated at 35°C (95°F) hatched in 49–54 days, with 24.3% hatching success. Twenty-nine did not show signs of development, and 24 died during the late stages of development. Humidity and high temperature were thought to cause this high mortality (Vogt 1980).

Interestingly, the size of the yellow blotches on the heads of both species of map turtles was affected by incubation temperature. Postorbital wide crescents were formed in Ouachita Map Turtles hatching from eggs incubated at 25°C (77°F) but not at 35°C (95°F). Measurement of the areas of 13 of the yellow head blotches of the hatchlings of both species of map turtles showed a significant difference between hatchlings from the same clutches emerging from the two incubation temperatures (Vogt 1978, 1980). Larger yellow blotches and fewer postorbital stripes were formed at the lower temperature, while the higher temperature produced smaller blotches and more (but thinner) postorbital stripes.

Richard C. Vogt, Instituto Nacional de Pesquisas da Amazônia

DISTRIBUTION AND HABITAT

Global distribution: False Map Turtles are found primarily in large drainages in the Mississippi River basin, occurring from the St. Croix and Wisconsin Rivers in northern and central Wisconsin and the upper Mississippi River in Minnesota south to Louisiana and eastern Texas. They follow the Missouri River in North Dakota and the Ohio and Wabash Rivers east to the western edge of Tennessee, Kentucky, Indiana, and central Ohio (Vogt 1995).

Distribution in Wisconsin: False Map Turtles occur primarily along the western border of the state but are also found in the south-central and southwestern portions. They are often associated with the southern St. Croix, Mississippi, and Wisconsin Rivers. Of the published county records, museum vouchers could not be found or were unsuitable for Chippewa, Dane, Eau Claire, Iowa, Pepin, Richland, and Sauk (see the Appendix).

Localities of possible occurrence: Based on the presence of large rivers and their occupancy in adjacent counties, False Map Turtles could be present farther north along the western border of the state (e.g., Burnett County). Similarly, they could occur in additional counties in central Wisconsin, such as Adams, Juneau, and Wood.

Habitat: False Map Turtles are primarily a riverine species, though they are also found in some lakes and marshes in southern Wisconsin (Vogt 1981). They typically use slow-moving sections of rivers, including backwater sloughs or oxbows, and are particularly fond of open-canopy areas with dense aquatic vegetation (Vogt 1981). For example, in areas of the Mississippi River adjacent to Wisconsin that Tucker et al. (2008) surveyed, most of the 311 False Map Turtles captured were in habitats considered main river channel borders. In the Missouri River, Bodie and Semlitsch (2000) found that this species often moved between river and adjacent floodplain habitats, although radio-tracked females ($n = 10$) were located most often in rivers, and males ($n = 5$) were located most often in adjacent scoured wetlands. Little is known about preferred microhabitat characteristics, but DonnerWright et al. (1999) detected positive relationships between False Map Turtle captures and percent mud in river substrates and bank slope across 12 sampled sites in the St. Croix River ($n = 40$ captures). In Kansas, individuals also appeared to select aquatic habitats with mud substrates, as well as areas with a greater abundance of basking sites (Fuselier and Edds 1994).

Nesting habitats are usually open locations along riverbanks and islands in the main channel with well-drained soils (Vogt 1980). Selected nest locations occurred in a variety of habitats that ranged from beaches with low shrub cover along the peripheries to open sandy areas, and nests were typically placed adjacent to clumps of sedges (*Carex* spp.) or other plants (Vogt 1980). Females also typically placed nests in sandy soil on river islands or within 100 m (328 ft) of the riverbank, often in cultivated fields. False Map Turtles overwinter underwater. Studies of overwintering physiology in Wisconsin do not exist. Yet this species is presumably intolerant of anoxic conditions, similar to Northern Map Turtles (Ultsch 2006), and therefore must overwinter in well-oxygenated locations to facilitate cutaneous uptake of oxygen. They hibernate behind logs and rocks on the bottom of rivers. In the Mississippi River, they congregate in large numbers behind wing dams, not buried in the mud but lying on the bottom behind the rock piles. On 1 October 1976, 56 Ouachita Map Turtles, 25 False Map Turtles, and 4 Northern Map Turtles were captured behind one rock pile in the Mississippi River (Vernon County; Vogt 1980). Behind the same rock pile in 1977, 131 individuals across all three species were captured (Vogt 1980). It is likely that False Map Turtles will also overwinter near natural underwater structures that create similar conditions in the winter, such as underwater woody snags.

REPRODUCTION AND DEVELOPMENT

Vogt (1980) studied the mating behavior of False Map Turtles in the Mississippi River in Vernon County. Mating occurred in spring and autumn, with reproductive behaviors that were similar to those of other map turtle species. Males likely recognize conspecific females through head patterns or olfactory cues (Vogt 1978). Based on observations of Vogt (1980), after the male identifies an appropriate female, he faces her, arches both of his forearms near her head, and 'drums' her tympanic region with the back of his long foreclaws. The purpose of this drumming behavior could be to help the female identify the male as an appropriate mate and may be done from 1 to 14 times before the male attempts to mount for reproduction. The number of contacts per bout seems to be species-specific (Vogt 1978, 1980, 1993), with Ouachita Map Turtles ranging from 4 to 8 contacts per bout ($\bar{x} = 5.2$) and False Map Turtles ranging from 4 to 14 ($\bar{x} = 9.8$).

The nesting behavior of False Map Turtles in Wisconsin was also studied extensively by Vogt (1980) in Vernon County. He reported that nesting occurred from late May into July and that foraging behavior was related to whether females had completed egg deposition. Females do not begin to feed until after they have laid their first clutch of eggs for the year; otherwise, the coelomic cavity is too full of eggs to allow room for the intestines to hold an adequate amount of food (Vogt 1980). The earliest date that Vogt (1980) documented feeding by females along the Mississippi River (Vernon County) was 26 May.

Females were found nesting from 5:45 a.m. to 8:30 p.m. under various weather conditions, but most nesting activity occurred in the morning from 6:30 to 10:00. Cloacal temperatures of nesting females ranged from 24.6°C to 28.2°C (76.3°F to 82.8°F); air temperatures ranged from 21.1°C to 32.0°C (70.0°F to 89.6°F; R. Vogt, personal observation). Presumably, the sand temperature on clear days becomes too hot for nesting in the afternoon, because under overcast conditions nesting activity is observed throughout the day. Other observations in Wisconsin further suggest that nesting behavior is heavily influenced by environmental conditions. For example, on an overcast day, 16 nesting females in the Vernon County study population were observed on land at 2:00 p.m. A rain shower at 2:30 sent all but one female that was covering her nest back to the water, and seven females abandoned the nest cavities they had partially dug (R. Vogt, personal observation). During June, up to 75 females were seen floating at the surface of the water 3–15 m (9.8–49.2 ft) from shore between 4:30 and 10:00 a.m., apparently waiting for preferred environmental conditions to leave the water to nest (R. Vogt, personal observation).

After arrival on the shore, female False Map Turtles in Vernon County wandered from 5 to 150 m (16.4 to 492.1 ft) before attempting to dig a nest cavity (Vogt 1980). Nest cavities are dug entirely with the hind limbs (i.e., no body pit is dug; Vogt 1980). The nest is flask-shaped and 10–16 cm (3.9–6.3 in) deep, and it broadens from 4 to 8 cm (1.6 to 3.1 in) beneath the surface (Vogt 1980). Eggs are positioned and packed into the nest with one of the hind feet. As an egg is laid, the female fully extends her neck and bobs her head slightly in the vertical plane. After the last egg is laid, the hole is filled by scraping and packing sand and nearby debris into the hole with the hind limbs. Once the cavity is filled, the female continues to pack down the soil by alternately raising each hind limb ca. 5 cm (2 in) above the surface and then slapping the sand with it (Vogt 1980).

Fresh nests can be readily distinguished by the striking difference in the sand particle size between the nest and the surrounding sand surface (R. Vogt, personal observation). The surrounding sand has a coarse mosaic of small pebbles created by wind erosion of the lighter sand, while the nest has a rather uniform small-grained appearance. Within one or two days, wind activity obscures the nest locality (R. Vogt, personal observation). Temperatures within nest cavities monitored from 16 July to 17 August 1972 in Vernon County ranged from 2.2°C to 12.2°C (36.0°F to 54.0°F; Vogt 1980). Mean daily fluctuations for two nests monitored for 28 days was 6.7°C (44.1°F).

Vogt (1980) reported that mean clutch size in Wisconsin was 14.1 ($n = 50$), and there was a positive correlation between female body size and clutch size. Similarly, one clutch laid in La Crosse County on 13 June 2018 contained 12 eggs (A. Badje, personal observation). Egg incubation ranged from 60 to 75 days ($n = 14$ nests), and hatching success was 95% in 285 nests excavated from 1972 to 1977 (Vogt 1980). Examination of female reproductive organs indicated that females typically produced two clutches of eggs per year, but some produced three clutches (Vogt 1980).

Sex of hatchlings is dependent on incubation temperature, with males typically produced in cool shaded nests and from nests laid later in the season and females typically produced in warmer nests (Bull and Vogt 1979, Vogt and Bull 1982, 1984). Eggs incubated in the laboratory produced all males at 28°C (82.4°F) and below and all females at 30.6°C (87.1°F) and above. The threshold temperature for sex determination was 29.4°C (84.9°F) for this species in Wisconsin and 29.2°C (84.6°F) in Tennessee (Bull et al. 1982a, 1982b; for more discussion, see Natural History Box: Temperature-Controlled Sex Determination [TSD] in the Ouachita Map Turtle, *Graptemys ouachitensis* species account).

Emergence of hatchlings begins in late August (Vogt 1981). In Wisconsin, no evidence of overwintering in the nest was found from over 500 map turtle nests excavated from mid-September to November (Vogt 1980). Growth is rapid in early age classes but slows dramatically after sexual maturity is reached (Vogt 1980). Maturity in False Map Turtles occurs at ca. four to six years of age in males and ca. eight years of age in females. Estimates of mean upper PL were 11.45 cm (4.51 in, $n = 22$) for males and 21.07 cm (8.30 in, $n = 27$) for females in Kentucky (Lindeman 1999a). Growth rings are useful for estimating age of False Map Turtles up to 6 and 12 years old for males and females, respectively (Vogt 1980). Longevity in the wild is unknown, but a captive individual reportedly lived for 32 years (Bowler 1977).

ACTIVITY

False Map Turtles emerge from hibernation and begin basking in early to mid-April (Vogt 1981). When active, much of this species' time is split between foraging and bouts of basking to thermoregulate. False Map Turtles are gregarious and bask in large numbers on Muskrat (*Ondatra zibethicus*) houses, logs, or stumps with Northern Map Turtles and Ouachita Map Turtles. They rarely bask on shorelines or logs connected to shorelines and prefer basking sites surrounded by water (Vogt 1980; Figure 349). Lindeman (2000) found that basking activity occurred

Figure 349. Adult False Map Turtle basking (Dane County; photo by C. Raimond).

throughout the day in Kentucky but peaked between 11:00 a.m. and 2:00 p.m.

Little research exists on False Map Turtle movement patterns. Vogt (1980) recaptured individuals in the Mississippi River that had moved >2 km (1.2 mi) upstream in less than a month. He also marked a female at a nesting island and recaptured her ca. four months later at her hibernaculum, ca. 8 km (5 mi) away. False Map Turtles stop feeding by mid-September, and hibernation begins in October (Vogt 1980). This species hibernates while fully submersed in well-oxygenated locations. They are not totally dormant throughout the winter, as commercial fishermen catch them in gill nets under the ice (Vogt 1980).

PREY AND PREDATORS

Vogt (1981a) investigated diets of a population in Vernon County and found that False Map Turtles were omnivorous. They fed in the water on the larvae of caddisflies, mayflies, dragonflies, damselflies, flies, beetles, butterflies, and moths. They also fed on snails, clams, crayfish, and dead fish. The vegetative diet consisted of algae, duckweed (*Lemna* spp.), pondweed (*Potamogeton* spp.), eelgrass (*Vallisneria* spp.), and arrowhead (*Sagittaria* spp.). Adult females from the Mississippi River collected throughout the feeding season consumed (by volume) vegetation (41.5%), insects (22.5%), mollusks (15.7%), fishes (10%), and miscellaneous unidentified matter (9.5%, $n = 35$ turtles examined). Based on Vogt (1981a), males have a more carnivorous diet than females. Feeding takes place primarily in 1–2 m (3.3–6.6 ft) of water from the bottom to the surface.

Humans are the only species documented to cause adult mortality in Wisconsin through drowning in gill nets, being caught on setlines, and shooting (Vogt 1980, 1981b). Yet Shivley (2014) reported evidence that Bald Eagles (*Haliaeetus leucocephalus*) had preyed upon "large juveniles or young adult males" that ranged from 13 to 15 cm MCL (5.1 to 5.9 in) in Louisiana, and similar predation could also occur in Wisconsin. Eggs and hatchlings are much more susceptible to predators than adults. Great hordes of gulls, Red-winged Blackbirds (*Agelaius phoeniceus*), American Crows (*Corvus brachyrhynchos*), and Great Blue Herons (*Ardea herodias*) consume hatchlings as they leave the nest (Vogt 1981). Hatchlings and juveniles are also prey for fishes (e.g., bass, catfish, pike; Vogt 1981).

Nesting on sandbars and islands affords protection for the eggs from most mammalian predators. However, Raccoons (*Procyon lotor*) and Red Foxes (*Vulpes vulpes*) often become stranded on islands after the ice melts in the backwaters, which can result in high predation rates of eggs (Vogt 1980). It is rather paradoxical that even though birds prey on hatchlings, Common Grackles (*Quiscalus quiscula*) were observed picking the leeches off of adult map turtles while they were basking on a log on the Mississippi River (Vernon County; Vogt 1979). This is one of the few commensal activities known between birds and freshwater turtles in the world (see Natural History Box: Cleaning-Feeding Symbiosis between Grackles and Map Turtles in the Ouachita Map Turtle, *Graptemys ouachitensis* species account). Larvae of the fly *Metoposarcaophaga importuna* have been observed preying on live hatchlings in Vernon County (Natural History Box: Fly Predation on Hatchling Map Turtles in the Nest).

CONSERVATION

Status: Globally, the False Map Turtle is listed as a Least Concern species by the IUCN. It is considered a species of Special Concern in Wisconsin by the WDNR.

NATURAL HISTORY BOX: Fly Predation on Hatchling Map Turtles in the Nest

Turtle eggs have been reported to be vulnerable to infestation by fly larvae for over a century (e.g., Aldrich [1916] reported them in the eggs of the Diamond-backed Terrapin [*Malaclemys terrapin*]). Muller (1921) reported on the incidence of fly larvae in the Smooth Softshell (*Apalone mutica*) and Spiny Softshell (*Apalone spinifera*). He thought that healthy embryos were not attacked but rather that the fly larvae were focused on dead embryos because he found only two of 20 eggshells packed with fly larvae, and the other eggshells empty, with hatchlings having emerged and dispersed. During a long-term study that I conducted on map turtles in Pool 8 on the Mississippi River (Vernon County), I regularly found that 1–2 eggs per nest were packed with maggots, while the remaining 10–15 eggs had hatched and the hatchlings had dispersed. I observed this in at least 150 of ca. 1,000 nests I opened in late summer to autumn from 1972 to 1977.

I identified the fly as *Metoposarcaophaga importuna* from raising the larvae in the laboratory. I thought that the maggots were consuming dead embryos; however, in 1978 I discovered that the hatchlings were being infested with fly larvae when they pipped, or barely started emerging from the egg, and that the larvae were devouring the living young! I found both natural nests and transplanted nests to be infested with maggots. The only avenue for entering the hatchling was through the yolk plug, and eggs that were transferred to the laboratory before pipping did not develop maggots. Apparently, the fly deposits living maggots on the surface. The maggots burrow into the sand and attack the recently pipped egg, and then the maggots begin feeding on the live hatchling (Vogt 1981b). Flies may be attracted to the odor of the freshly hatched egg, or they may be cueing in on vocalizations of the hatchlings, since it is known that some species of turtles vocalize in the egg before hatching, and hatchlings also vocalize in the nest after hatching, perhaps to synchronize group emergence (e.g., Ferrara et al. 2014).

The sand was more humid than normal in 1978, and on 18 August I excavated some of the marked nests in the sun and noted that some of the eggs had pipped, but there were no maggots present. Since some of the eggs had not yet hatched, I left all of the eggs in the nest. During the following ten days, fly larvae infested 36% of the hatchlings from 23 nests (0%–100% of the hatchlings were attacked among nests). Many hatchlings were still alive, even though they were full of maggots. The presence of maggots in live hatchlings suggests that maggots are important hatchling predators and are not scavenging dead embryos. No maggots were found in unhatched eggs, suggesting that the longer the hatchlings remain in the nest, the longer they are at risk of maggot predation. Northern Map Turtles are known to overwinter in the nest in other parts of their range (Gibbons and Nelson 1978), but none of the three species of map turtles in Wisconsin have been found to overwinter in the nest. This possibly suggests that prompt hatchling emergence may have evolved locally, in part, to avoid predation by fly larvae (Vogt 1981b).

Richard C. Vogt, Instituto Nacional de Pesquisas da Amazônia

Populations: Little research exists on density or abundance of False Map Turtle populations throughout their North American range (reviewed by Ernst and Lovich 2009). Although there are no estimates from Wisconsin, some ideas about populations can be inferred from False Map Turtle capture rates during turtle community studies conducted in the state. For example, this species represented ca. 50% of the over 1,000 map turtles captured in the Mississippi River over several years in Vernon County (Vogt 1980). They also represented 15.8% ($n = 311$) of all turtles captured across seven species ($n = 1,965$) detected in the upper two reaches of the Mississippi River sampled by Tucker et al. (2008). Yet they represented only 6% of over 600 turtles captured in the lower St. Croix River by DonnerWright et al. (1999). Sex ratios vary widely, possibly due to temperature-dependent sex determination. For example, Vogt (1980) found a sex ratio of ca. 1M:6F in Vernon County. In contrast, sex ratios of 2–3M:1F have been reported for some populations in Oklahoma, Arkansas, and Louisiana (reviewed by Vogt 1980).

Threats: Habitat degradation due to anthropogenic modification probably has the greatest impact on most False Map Turtle populations (see Introduction to Conservation and Management of Wisconsin's Amphibians and Reptiles chapter). Removal of snags and logjams from rivers to improve navigability for boat traffic, river channelization, construction of impoundments, industrial pollution, and siltation negatively impact habitat suitability for this species (VanDeWalle and Christiansen 1996, Lindeman 1999b, Bodie 2001). Tourists who camp on beaches or sandbars in the Mississippi and Wisconsin Rivers disturb females attempting to nest, as well as inadvertently crush existing nests buried in the sand during recreational activities (reviewed by Lindeman 2013). Road mortality of females during their migrations to lay eggs could be an important concern in locations where roads occur near shorelines (reviewed by Lindeman 2013). The widely used pesticide atrazine appears to inhibit nest escape behavior of hatchlings and results in reduced activity rates, impacting survivorship (Neumann-Lee and Janzen 2011).

Harvest of False Map Turtles is currently restricted in Wisconsin. Current data support that False Map Turtles are particularly prone to accidental capture and mortality through passive fishing operations (Vogt 1980, Barko et al. 2004). During a study in the upper Mississippi River from 1996 to 2001, a combination of Ouachita Map Turtles and False Map Turtles represented 83% ($n = 2,139$) of all turtles captured in passive fish surveys, and 8% ($n = 161$) of captured individuals died in traps (Barko et al. 2004). Historical commercial harvest may have resulted in substantial population declines for some states (e.g., Missouri; reviewed by Ernst and Lovich 2009). Unregulated harvest of map turtles for human consumption and the pet trade is likely another important threat (reviewed by Lindeman 2013). False Map Turtles were reportedly the fourth most exported turtle species from the US, with 1,149,076 individuals exported from 2001 to 2009 (note that a portion of these were captive-raised, but the percentage is unknown; Herrel and van der Meijden 2014). Although Mali et al. (2014) reported that turtles are often exported out of Louisiana, their original capture locations are frequently unreported.

Management: Several general management concepts are likely applicable to False Map Turtles, including management of river and stream habitats (including nesting habitat), mitigation of possible road mortality, and overharvest (see Introduction to Conservation and Management of Wisconsin's Amphibians and Reptiles chapter). In addition to management of habitat, protection of riverine habitat from development could be an important component of any strategy to conserve this and many other species. For example, large False Map Turtle populations exist within the 420 km (261 mi) stretch of the Mississippi River in the Upper Mississippi River National Wildlife and Fish Refuge (see review in Northern Map Turtle, *Graptemys geographica* species account). Although modifications to river hydrology can negatively affect aquatic biota, the damming of the Mississippi River to maintain the channel may occasionally be beneficial to adult map turtles. For example, in Pools 6–10, huge backwater areas were created by the lock and dam system that provide habitat for feeding, while dredge spoil islands in the area are maintained for nesting sites. Yet alterations in water levels that flood nests prior to hatchling emergence could affect population recruitment rates.

Active management to preserve suitable nesting areas could benefit False Map Turtles. Applicable techniques could include exclusion of potential nest predators and concerted efforts to combat encroachment of woody vegetation that would otherwise shade nests. The USFWS recently constructed islands for duck and turtle nesting habitat in Pool 8 of the Upper Mississippi River National Wildlife and Fish Refuge (Vernon County) that likely benefit this species (see review in Northern Map Turtle, *Graptemys geographica* species account). In 2011 the islands were only a year old, but False Map Turtles and softshells (*Apalone* spp.) were already nesting on some of them (R. Vogt, personal observation). The continual work of the US Army Corps of Engineers has been beneficial for False Map Turtles on the Mississippi River, creating and maintaining many nesting islands through the regular enhancement of islands with fresh dredge spoil. This is beneficial as long as the spoil is not deposited on beaches when nests are still incubating. Fresh dredge spoil is not used for nesting because it is usually too high and dry to support a nest cavity. It usually takes ca. four years for the spoil to be ready for turtle nesting (Vogt 1980).

SPECIAL REMARKS

False Map Turtles are known to hybridize with both Northern Map Turtles and Ouachita Map Turtles in the Mississippi River in Wisconsin (Vogt 1978). *Graptemys pseudogeographica* × *G. geographica* hybrids have produced fertile eggs that hatched in nature and in the laboratory (R. Vogt, unpublished data).

Ornate Box Turtle

Terrapene ornata (Agassiz 1857)

Robert W. Hay and Rebecca A. Christoffel

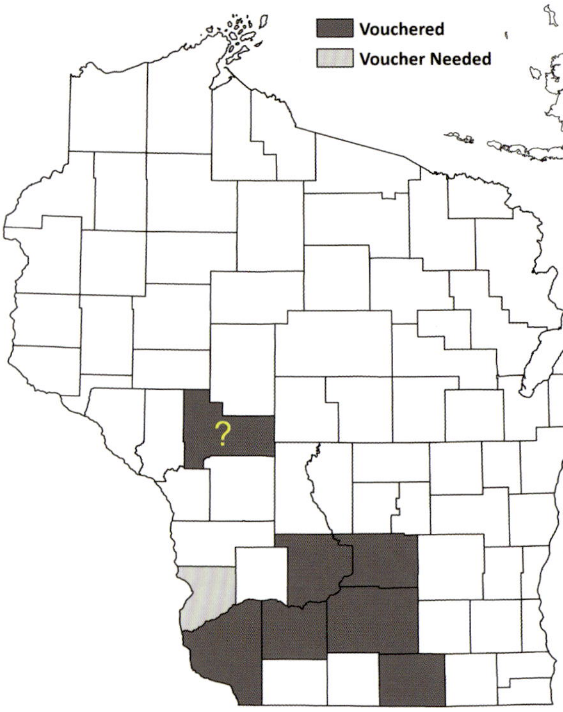

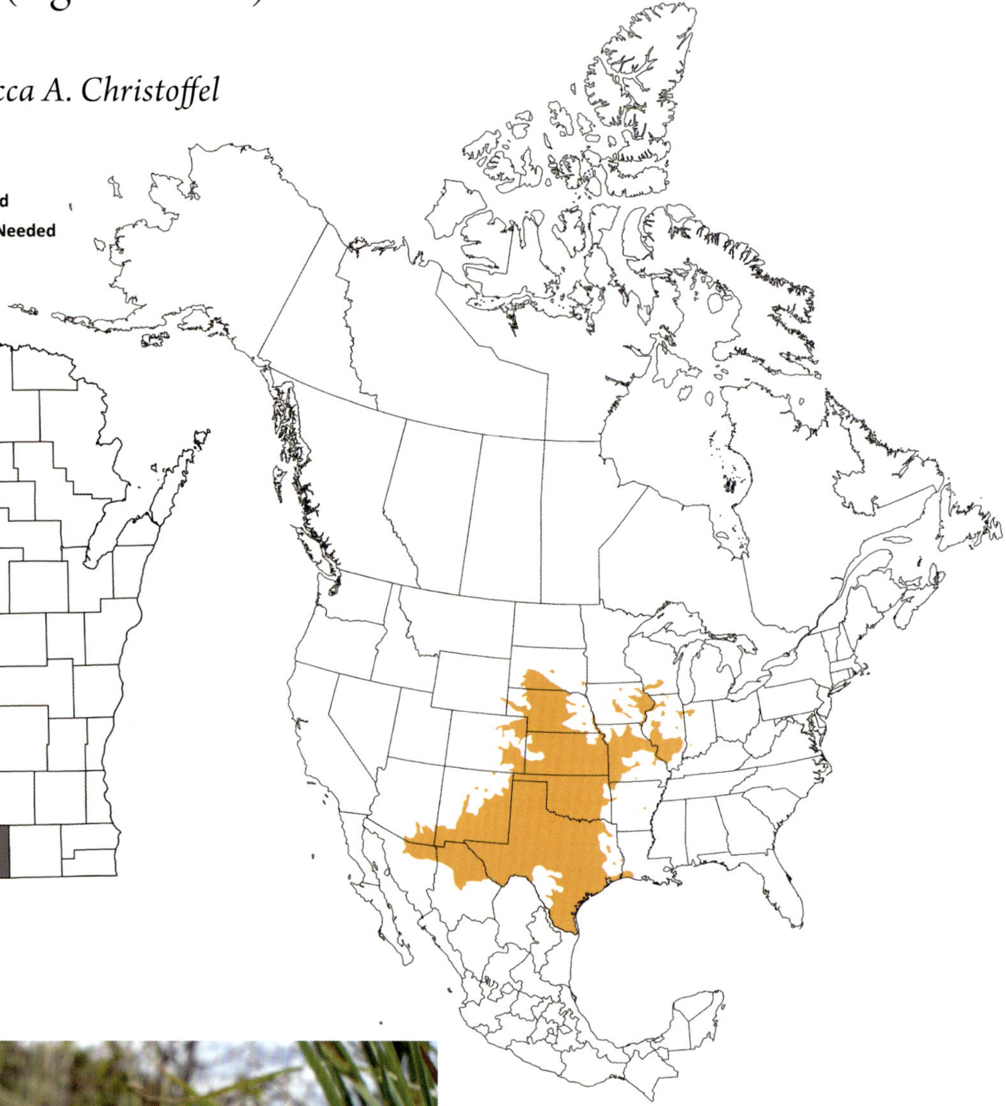

Juvenile Ornate Box Turtle (Sauk County; photo by J. Kapfer).

AT A GLANCE

Ornate Box Turtles are the only fully terrestrial turtles in the state. They prefer the sand prairie and oak savanna habitats in southwestern Wisconsin. The deep sandy soils associated with these habitats are especially critical for overwintering and nesting. Friable sand is also important, as it allows turtles to easily burrow beneath the soil surface for thermoregulation on hot days and for protection at night. Ornate Box Turtles are mostly active from early April into October. Most breeding occurs in late summer, but females typically do not lay eggs until the next year in June. Eggs hatch by late August, but hatchlings often remain in the nest until the next spring and only sometimes emerge in early autumn of the year eggs were laid. Ornate Box Turtles are unique among Wisconsin turtles because they possess a hinged plastron that allows the anterior and posterior portions to close tightly against the carapace, encasing them in what resembles a closed box, thus the name 'box' turtle. The ability to close their shells tightly is their major defense against predators, which include Coyotes (*Canis latrans*), weasels (*Mustela* spp.), and large birds. They feed primarily on succulent plants and invertebrates such as insects and slugs. The Ornate Box Turtle is by far Wisconsin's rarest turtle and is listed as an Endangered species and considered a Species of Greatest Conservation Need by the WDNR.

SYSTEMATICS

The most thorough phylogenetic study that involved Ornate Box Turtles to date was conducted by Martin et al. (2013). The Wisconsin form is currently recognized as *Terrapene ornata* (Ornate Box Turtle; Crother 2017). The uniqueness of *T. ornata* has never been questioned (Dodd 2001), although Martin et al. (2013) suggested that the distinction of two recognized subspecies, *T. o. ornata* and *T. o. luteola*, be disregarded. Some past sources have referred to this species as the 'Painted Box Turtle,' 'Northern Box Turtle,' and 'Sand Turtle' (Pope and Dickinson 1928, Moore 1993). It is the only species in the genus *Terrapene* found in Wisconsin. Further, it is a member of the family Emydidae, as are turtles in the genera *Chrysemys, Emydoidea, Glyptemys,* and *Graptemys* in Wisconsin (see Systematics and Classification chapter).

DESCRIPTION

Adults: Ornate Box Turtles have a highly domed brown to black carapace, with five to nine radiating yellow lines on each scute. The carapace is nearly as wide as it is long when viewed from above, and a weak midline keel is discernible in some individuals (Knudsen 1956e). The plastron is darker brown in color and has a pattern of light radiating lines that varies considerably among individuals (Figure 350). There is a hinge on the plastron that allows the turtle to close the anterior and posterior ends tightly against the carapace. This plastral hinge normally occurs between the fifth and sixth or at the sixth marginal scute, and no shell bridge is present (reviewed by Ernst and Lovich 2009). The front feet have five toes, while the hind feet have four (very rarely three) toes (reviewed by Ernst and Lovich 2009). The color of the neck and front legs ranges from tan or brown to orange or red (see Variation subsection).

The Ornate Box Turtle is one of Wisconsin's smallest turtle species (Vogt 1981). A sample of 144 adults measured at various locations in Wisconsin indicated that males and females are similar in size. The mean adult male CL at these locations was 10.8 cm (4.3 in) and ranged from 9.5 to 13.2 cm (3.7 to 5.2 in, $n = 80$),

Figure 350. Plastron of a subadult Ornate Box Turtle (Dane County; photo by R. Hay).

while mean adult female CL was 10.7 cm (4.2 in) and ranged from 9.5 to 13 cm (3.7 to 5.1 in, $n = 64$; Robert Hay, Turtles for Tomorrow, unpublished data). Furthermore, mean live weights were 304 g (10.7 oz) and 315 g (11.1 oz) for adult males and females, respectively (R. Hay, unpublished data).

Preadult Stages: Eggs of Ornate Box Turtles are hard-shelled and ovoid in shape (Knudsen 1956e). Legler (1960) described 42 newly laid eggs of Ornate Box Turtles in Kansas. Egg length ranged from 3.1 to 4.1 cm (1.2 to 1.6 in, $\bar{x} = 3.6$ cm [1.3 in]), and eggs weighed 8.0–14.3 g (0.3–0.5 oz, $\bar{x} = 10.1$ g [0.4 oz]). Similarly, the mean weight of newly laid eggs from a population in South Dakota was 12.0 g (0.4 oz; Quinn et al. 2014).

Hatchling Ornate Box Turtles in Wisconsin usually have a light to medium brown carapace with poorly defined pale yellowish blotches or spots on each dorsal scute (R. Hay, personal observation). Each lateral scute has a central pale spot with four to seven smaller marks arranged around the edge. By the end of the first full season, the markings exhibit a radial pattern. The plastron has a ground color of cream yellow but can be bright yellow, with a dark area in the center of each scute. The head of hatchlings is dark to lighter brown, with some yellow speckling on the sides (Figure 351). Hatchlings possess a caruncle at the tip of the snout, which is used to assist in hatching. Although the caruncle eventually falls off, it can remain for >30 days after emergence from the egg. Limited data on the weight of wild Wisconsin hatchlings exist, but two that emerged in the spring from Dane County weighed 6.4 g (0.2 oz) and 7.4 g (0.3 oz; Doroff and Keith 1990). In Wisconsin, hatchlings have a CL of 2.3–3.2 cm (0.90–1.25 in, $n > 600$; R. Hay, personal observation). The MCL of hatchlings measured from other states ranged from 2.8 to 3.2 cm (1.1 to 1.3 in, $n = 21$, $\bar{x} = 3.0$ cm [1.2 in]; Ernst and Lovich 2009).

Figure 351. Hatchling Ornate Box Turtle emerging from an egg incubated in captivity (eggs collected from Sauk County; photo by J. Kapfer).

Legler (1960) described older juveniles as patterned similarly to hatchlings, except their yellow head and neck markings were larger and contrasted more sharply with the dark brown ground color. As turtles grow into subadulthood, their plastrons appear to have a pattern of pale radiations on a dark background (rather than a dark pattern on a light background, as in hatchlings), and the markings on the plastron are less sharply defined than those on the carapace.

Variation: Although subtle variation in shell color and pattern exists across individuals, the most striking differences in appearance occur between adult males and females. The head and neck of adult females are usually brown with some pale yellow spots, and the front leg scales are often a light to medium yellow. In contrast, adult males often have an orange or red tinge on the scales of their front legs, and their head color can range from bluish, to various shades of green, to reddish or gold. The irises of adult females are often golden or brown (see Ornate Box Turtle Cover Photo), whereas male irises are usually red (Figure 352; Vogt 1981). The tail of adult males is longer than that of females, with the cloacal opening posterior to the rear edge of the carapace, while females have a short tail with the vent located anterior to the rear edge of the carapace. Shell shape can also differ by sex. Males have a more concave plastron than females (Legler 1960), but the male carapace can be flatter at the top compared to that of females (Knudsen 1956e). When viewed from above, the carapace of adult males can also appear somewhat pear-shaped, being wider near the posterior. Unlike

Figure 352. Adult male Ornate Box Turtle with red eyes and colorful foreleg scales (Rock County; photo by J. Kapfer).

females, adult males have an inner toe on each hind foot that is thickened and usually turned inward (Ernst and Lovich 2009).

Diagnosis: Ornate Box Turtles are characterized by a highly domed carapace that is covered with keratinized scutes, a plastron with 12 scutes, a well-developed hinge that allows the entire plastron to be closed tightly against the carapace, toes that are not webbed, and an upper jaw with a prominent beak. The only other turtle species in Wisconsin that has a functional shell hinge and domed carapace and that can sometimes be found in drier prairies is the Blanding's Turtle (*Emydoidea blandingii*). However, Blanding's Turtles have darker carapace coloration than Ornate Box Turtles, and the carapace is also covered with yellow flecks or spots rather than radiating lines. Furthermore, Ornate Box Turtles have a dark plastron with yellow streaks, whereas the plastron of Blanding's Turtles is yellow with large black blotches near the marginal half of each scute. Blanding's Turtles are also much larger as adults, attaining twice the length of an adult Ornate Box Turtle. Adult and subadult Blanding's Turtles are most easily distinguished from other turtles by their bright yellow throat and notched upper jaw, neither of which occurs in Ornate Box Turtles. Hatchlings and juveniles of these species can be more easily confused, but juvenile Ornate Box Turtles possess a domed shell similar to that of adults, while juvenile Blanding's Turtles have a more streamlined shell profile similar to that of other aquatic turtles.

Turtle eggs encountered in the wild are difficult to identify. The habitat associated with turtle nests can help differentiate Ornate Box Turtle eggs from the eggs of other turtle species. For example, eggs found in sandy prairies in southwestern Wisconsin that are also inaccessible by aquatic turtles most likely belong to Ornate Box Turtles. Furthermore, Ornate Box Turtles lay smaller eggs and have lower clutch sizes than most other turtles in Wisconsin. A possible exception to this, however, could be the Eastern Musk Turtle (*Sternotherus odoratus*). Yet Eastern Musk Turtles rarely stray far from water to nest, so their eggs are not frequently encountered at sites in Wisconsin where Ornate Box Turtles also occur.

DISTRIBUTION AND HABITAT

Global distribution: Ornate Box Turtles only occur in North America and are primarily found in the central US. Their northern range extends from western Indiana to southeastern Wyoming. Their southern range extends from Louisiana west to southwestern Arizona and south to northern Mexico. Legler (1960) posited that the distribution of this species is limited by three factors: (1) access to substrate in which they can dig nests and refugia (primarily deep sandy soils); (2) winter temperatures that do not result in subsoil frost penetration deep enough to kill hibernating turtles; and (3) suitable precipitation levels during summer months to enable turtle emergence from aestivation. It is reasonable to assume that these factors also apply to the distribution of Ornate Box Turtles in Wisconsin.

Distribution in Wisconsin: Ornate Box Turtles have a very limited distribution in Wisconsin that is closely tied to the sandy soil deposits associated with the Wisconsin River floodplain of the southwestern counties. They also occur in sandy soils associated with portions of the Rock River system in south-central Wisconsin. With the inclusion of Jackson County (but see Localities of possible occurrence subsection), they are documented in only eight counties, of which a voucher specimen could not be found for Crawford.

Localities of possible occurrence: Many areas of Wisconsin have not been thoroughly surveyed for Ornate Box Turtles, and their documented distribution in the state might not accurately reflect their true distribution. Although they are not currently confirmed in Juneau County, two individuals were reportedly found there in separate areas with sandy soils during the 1980s, but these were not vouchered (Richard Vogt, Instituto Nacional de Pesquisas da Amazônia, personal communication). Further surveys in appropriate habitats within this county could, therefore, reveal more Ornate Box Turtles. A museum specimen that reportedly originated from Hatfield Township, Jackson County, exists in the UW–Green Bay Richter Museum of Natural History collection (R451, Tom Erdman, UW–Green Bay, personal communication). This individual was likely either a released pet or an unintended stowaway. There are no other documented records for Ornate Box Turtles within 161 km (100 mi) of the site (see further discussion of this specimen in Kapfer et al. 2015).

Habitat: In Wisconsin, Ornate Box Turtles have been described as "strictly a prairie relict" (Vogt 1981). They rely on prairies and oak savannas with loose well-drained soils and favor south- to west-facing exposures with warm dry conditions (R. Hay and Rebecca Christoffel, Snake Conservation Society, personal observation, Legler 1960, Bernstein et al. 2007). The preference for drier conditions is highlighted by Legler (1960), who reported that the average annual precipitation during the warm season was ca. 64 cm (25 in) in the northeastern portion of the Ornate Box Turtle's North American range (which includes Wisconsin; Legler 1960). This species also reportedly occupies a variety of habitats other than prairie and savanna, albeit less frequently. For example, a population studied in Dane County used woodlands and wetlands in addition to dry open uplands (Doroff and Keith 1990). In Sauk County, radio-tracked Ornate Box Turtles were often found on dry open bluffsides (Joshua Kapfer, UW–Whitewater, unpublished data). In northwestern Illinois, Ornate Box Turtles used a riparian area extensively during hot weather (Refsnider et al. 2012). Legler (1960) reviewed reports from Kansas and indicated that this species was most frequently encountered in lightly grazed pastures, as well as open woodlands, fields, and prairies, highlighting the range of upland habitats they can inhabit. Microhabitat use by adult Ornate Box Turtles in Wisconsin is not well-studied. Hatch (1996, 1997) reported that adults in Iowa County ($n = 10$) used patches of Dewberry (*Rubus flagellaris*) and sumac (*Rhus* spp.) on a restored prairie proportionately more often than that habitat's availability (Table 23). Loose soils are a particularly important characteristic of selected microhabitats because they allow Ornate Box Turtles to excavate temporary shallow burrows, or 'forms,' which act as short-term refugia for thermoregulation and cover at night (Figure 353).

Information specifically on hatchling and juvenile habitat use is rarely reported in Wisconsin. However, a 1999 radio telemetry investigation that tracked one-year-old headstarted Ornate Box Turtles in Iowa County ($n = 4$) revealed that 47% of turtle locations were near young trees, mostly Sumac, with an understory dominated by Dewberry (R. Christoffel and R. Hay, unpublished data). In addition, 24% of locations were in prairie, 20% in areas that had been restored as prairie, and 9% in savanna. During this same study, two-year-old headstarted turtles ($n = 4$)

Table 23. Radio-tracked locations for ten adult Ornate Box Turtles ($n = 199$ observations) in eight habitat types compared to the availability of each habitat type within the study area at one site in Iowa County in 1995 (Hatch 1997).

	Woods	**Savanna**	**Dewberry/ Sumac**	**Dewberry**	**Alfalfa**	**Prairie**	**Restored Prairie**	**Grass, Nonprairie**
% Observations	15	8	15	4	4	41	9	4
% Available habitat	11	8.5	2.5	3	23	42	6	4

Figure 353. Ornate Box Turtle outside of a burrow (Dane County; photo by R. Hay).

Figure 354. Ornate Box Turtles mating. Note the position of the male's feet (lodged between the female's plastron and carapace), which ensures she cannot completely close her shell while they are fully engaged. Also note that attached to the back of the female's carapace are a temperature data logger (*left*) and a radio transmitter (*right*) that were used as part of an ongoing research project (Sauk County; photo by J. Kapfer).

tracked with radio telemetry equipment were most frequently located in areas that had been restored as prairie (45%), followed by prairie (27%), savanna (20%), and an area of young trees with Dewberry beneath them (8%).

Nesting and overwintering sites are frequently associated with sandy or loamy soils in open habitats that are sparsely vegetated and dominated by short plant species (ca. <20 cm [18 in] in height; R. Hay, personal observation). These locations also have low root densities, which presumably facilitates digging (R. Hay, Rob and Kim Russell, citizen land stewards, personal observation). In Wisconsin, Ornate Box Turtles reportedly burrow to a mean depth of 0.9 m (3.0 ft) during winter (Doroff and Keith 1990; see Activity section).

REPRODUCTION AND DEVELOPMENT

Ornate Box Turtles primarily breed in late summer, and females deposit eggs the following spring. Although breeding in Wisconsin has been observed from late April to early September, the majority of breeding occurs in early to mid-August (Doroff and Keith 1990, R. Hay, personal observation). Researchers speculate that olfaction plays a role in mating (Davis 1981), while anecdotal observations suggest that visual and olfactory cues trigger breeding activity and aid in mate searching in Wisconsin (R. Hay, personal observation). A male that successfully engages with a female typically places his feet between the female's carapace and plastron near the tail, which helps ensure she cannot completely close her shell when they are in coitus. When in this position, the male frequently rests on the back edge of his dorsal surface (Figure 354). Wild Wisconsin males will often attempt to mount females placed in their proximity, regardless of season, which indicates a strong mating response (R. Hay, personal observation).

Nesting in Wisconsin is frequently observed from 29 May to 26 June, and most gravid females migrate to nesting sites several days prior to oviposition (Doroff and Keith 1990, Vogt 1981, R. and K. Russell, personal observation). Many of the females in a population monitored for over 20 years in Sauk County returned to the same general nesting area across numerous years (R. and K. Russell, personal observation). Furthermore, females within this population were frequently documented to nest in close proximity to each other during a given year (i.e., 1–3 m^2 [11–32 ft^2]; R. Russell, personal observation). Egg deposition often occurs in the evening or at night (R. Hay, R. and K. Russell, personal observations). Females typically burrow so their bodies are completely below the soil surface to deposit eggs and emerge from the nest location the following morning, after which they will spend up to two hours replacing and smoothing the soil on the nest surface (R. Hay, R. and K. Russell, personal observations). Doroff and Keith (1990) described nest cavities as "flask shaped," with a short "neck" toward the soil surface and wider cavity toward the bottom where eggs are deposited. Doroff and Keith (1990) further reported that the bottom of these cavities ranged in depth from 5 to 8 cm (2.0 to 3.2 in), but deeper cavities are often observed in Wisconsin (depths of up to 20 cm [8 in]; R. Hay, personal observation). Although Legler (1960) reported that females nested closer to the soil surface in Kansas (i.e., they did not burrow their entire bodies belowground to oviposit), this behavior has only been observed in 1 of >335 nests documented over ca. 25 years in Wisconsin (R. Hay, personal observation).

Doroff and Keith (1990) found that, on average, 57% of females monitored during a two-year Wisconsin study were gravid during the nesting season (range: 50%–63%). The authors further reported that clutch size ranged from two to seven eggs ($\bar{x} = 3.5$). The mean clutch size of nests investigated from 1993 to 2012 across seven locations in Wisconsin was four (range = 1–6, $n = 335$; Table 24; R. Hay, unpublished data). Interestingly, three of the four clutches containing six eggs were observed at two separate sites within the same year. Thus, typical clutch sizes in Wisconsin are less than the 4.7 mean size reported for Kansas (Legler 1960) but greater than the clutch sizes reported for Nebraska (Converse et al. 2002). Legler (1960) also found that 33% of females double-clutched (i.e., produced two egg clutches in a single year) in Kansas. Double-clutching has not been reported in Wisconsin. Limited data exist on the length of incubation for wild eggs in Wisconsin, but Doroff and Keith (1990) reported a mean duration of 80 days (range: 79–81 days) for three nests monitored in Dane County. They noted that the majority of hatchlings from 13 nests did not emerge in late summer but burrowed below the nest to overwinter. Emergence of hatchlings in autumn was only noted for one nest. The authors excavated two hatchlings that had buried themselves to the subsurface bedrock at a depth of 18 cm (7.2 in) but were found dead the following spring, presumably from freezing (Angela Doroff, UW–Madison, personal communication). Egg fertility rates do not exist in the literature, but a mean fertility rate of 95% (annual range = 90%–97%) for 335 Wisconsin clutches was recorded over a 20-year period (Table 24; R. Hay, unpublished data).

As hatchlings grow, an annual layer of epidermis is deposited along the periphery of each shell scute (Vogt 1981). This new epidermal layer or annual 'growth ring' gives an indication of age for up to ca. 20 years. Because growth slows considerably once maturity is reached, the growth rings from later years are narrower and become difficult to dif-

Table 24. Reproductive output and fertility rates for Ornate Box Turtles at seven locations in southern Wisconsin from 1993 to 2012. Fertility determined by the candle method >21 days post-oviposition. Fertility rate = total number of fertile eggs per total eggs (R. Hay, unpublished data).

Site	Years Sampled	# Clutches	Mean Clutch Size	# Fertile Eggs (All Eggs)	% Fertile Eggs
1	12	68	3.6	222 (248)	0.90
2	12	24	4.0	94 (97)	0.97
3	12	41	3.7	142 (152)	0.93
4	14	91	3.9	334 (351)	0.95
5	7	35	4.9	163 (171)	0.95
6	14	67	4.0	257 (271)	0.95
7	4	9	3.7	32 (33)	0.97
Mean	10.7	47.9	4.0	177.7 (189.0)	0.95

ferentiate (also note that the work by Germano and Bury [1998] brings into question the reliability of this method for aging turtles). Little is known about hatchling and juvenile growth rates for wild Ornate Box Turtles. Legler (1960) determined that plastron lengths of young turtles increased by 68%, 20%, and 18% during the first three years of life, respectively. A small sample of juvenile turtles monitored from ages 4–12 in Wisconsin increased considerably in CL (32.6%) and weight (74.5%) over the period of observation, although the largest increases occurred in turtles ranging from 6 to 8 years of age (R. Hay, unpublished data; Table 25).

Sexual maturity is determined by size rather than age (Ernst et al. 1994), and males mature when smaller and younger than females (Legler 1960, Blair 1976). Legler (1960) reported that the majority of Kansas males matured at 10.0–10.9 cm PL (3.9–4.2 in) during year 8 or 9, while most females matured at 11.0–12.9 cm PL (4.3–5.1 in) during year 10 or 11. Quinn et al. (2014) reported that males and females from South Dakota matured at nine years and noted that adult females were larger than adult males. In Wisconsin, adults mature at ca. 9.5 cm CL (3.7 in; R. Hay, personal observation). Based on size of maturity reported in Kansas (Legler 1960), sexually mature Wisconsin individuals are likely between 10 and 12 years for males and 11 and 13 years for females (R. Hay, personal observation). Little data exist on the average and maximum life span for wild Ornate Box Turtles in Wisconsin. Ernst and Lovich (2009) reviewed several published wild and captive longevity records and reported that maximum ages ranged from 28 to 42. However, recent reports by Lewis and Iverson (2018) indicate that Ornate Box Turtles in northern populations can exceed 57 years of age.

ACTIVITY

In Wisconsin, emergence from hibernation typically begins from the last week of March (R. Russell, personal communication) to the first week of April (Curtin 1997), although it can reportedly occur up to late May (e.g., Vogt 1981). Curtin (1997) correlated emergence from hibernation in Wisconsin with an inversion in soil temperature that persisted for >48 hours. In Missouri, Grobman (1990) studied a population that emerged after five consecutive days where temperatures of ≥7°C (45°F) had occurred at soil depths of 10–20 cm (3.9–7.9 in). The author further reported that the first warm post-emergence rains triggered turtle migrations away from overwintering sites. After emergence, aboveground activity in Wisconsin is greatest during the spring and autumn, when environmental conditions are appropriate (Vogt 1981). It follows that activity declines during the hottest parts of summer (Vogt 1981). In fact, as summer air temperatures exceed 28°C (82°F), turtles frequently retreat into temporary self-excavated burrows or find refuge beneath vegetation to avoid overheating (L. Han, UW–Madison, unpublished data [cited in Doroff and Keith 1990], Refsnider et al. 2012). Both Johnson (1987) and Tucker et al. (2015) further reported that turtles were most often exposed aboveground during or immediately after precipitation, particularly during hot summer months.

Ornate Box Turtles are strictly diurnal, with the exception of nesting females (Ernst and Lovich 2009). Several sources report that two peaks in daily activity occur (i.e., morning and late afternoon); in between these activity peaks, turtles avoid unfavorable conditions by resting in various retreat locations (Converse et al. 2002, Tucker et al. 2015). In Wisconsin, >90% of all box turtle activity occurred when

Table 25. Carapace lengths (CL) and weights for juvenile Ornate Box Turtles from Rock County (R. Hay, unpublished data).

	4-Year-Olds	6-Year-Olds	8-Year-Olds	12-Year-Olds
CL (cm [in])	6.0 (2.4)	6.4 (2.5)	8.4 (3.3)	8.9 (3.5)
Weight (g [oz])	45 (1.6)	71 (2.5)	155 (5.5)	177 (6.2)
N	4	3	4	2

Table 26. Comparisons of mean Ornate Box Turtle home range sizes, sample sizes, and home range estimators reported for several studies in the Midwest.

County (State)	Age Class and Sex	Home Range Size, ha (ac)	Sample Size	Home Range Estimator	Source
Dane (Wisconsin)	Subadults	1.5 (3.7)	5	Dixon Chapman method	Doroff and Keith 1990
Dane (Wisconsin)	Adult males in 1986	8.2 (20.3)	13	Dixon Chapman method	Doroff and Keith 1990
Dane (Wisconsin)	Adult males in 1987	3.4 (8.4)	5	Dixon Chapman method	Doroff and Keith 1990
Dane (Wisconsin)	Adult females in 1986	12.0 (29.7)	17	Dixon Chapman method	Doroff and Keith 1990
Dane (Wisconsin)	Adult females in 1987	6.9 (17.1)	12	Dixon Chapman method	Doroff and Keith 1990
Dane (Wisconsin)	Adults	2.5 (6.2)	24	MCP	Curtin 1997
Iowa (Wisconsin)	Adults	2.6 (6.4)	5	MCP	Hatch 1996
Carroll (Illinois)	Adults	4.0 (9.9)	24	MCP	Refsnider et al. 2012
Johnson (Iowa)	Adult males	5.8 (14.3)	5	MCP	Bernstein et al. 2007
Johnson (Iowa)	Adult females	2.7 (6.7)	5	MCP	Bernstein et al. 2007
Douglas (Kansas)	Adults	2.2 (5.4)	14	Area based on distance between recaptures	Fitch 1958a
Douglas (Kansas)	Adults	2.3 (5.7)	44	Area based on distance between recaptures	Legler 1960

ground temperature was 20°C–28°C (68°F–82°F), primarily from 9:00 a.m. to 6:00 p.m. (Ellner and Karasov 1993). At similar latitudes in northwestern Illinois, Ornate Box Turtles were most active at ambient temperatures of 26°C–27°C (79°F–81°F; Tucker et al. 2015). In periods of clear cool weather at the beginning and end of their active seasons, they may bask for most of the day (Legler 1960). Ellner and Karasov (1993) determined that the mean body temperature of active Wisconsin individuals across season and sex and regardless of behavior was 25.9°C (78.6°F). They further estimated that on cool days Ornate Box Turtles would need to bask for 45–50 minutes to reach their mean active body temperature.

Several studies from the Midwest (including Wisconsin) have published home range estimates for this species (Table 26). Although sample sizes and methods to estimate home range were not consistent, differences in home range size based on age class, sex, and location were evident. For example, Doroff and Keith (1990) radio-tracked hatchlings from a single nest in Dane County, Wisconsin. They found that all hatchling locations (>40) occurred within only a 16 m^2 (172.2 ft^2) area around the nest site. In comparison, the overall mean annual home range size for adults in this population was 8.7 ha (3.5 ac, $n = 47$), although variability across individuals was high (0.2–58.1 ha [0.5–143.6 ac]). In northwestern Illinois, Refsnider et al. (2012) documented philopatry (i.e., a tendency to return to the same area) for annual home range locations, and several studies have documented homing behavior in Ornate Box Turtles that are moved or translocated to other sites (Legler 1960, Metcalf and Metcalf 1970, 1979). Average daily movement rates of Ornate Box Turtles vary seasonally, and greatest movement coincides with peak seasons of activity. One- and two-year-old Ornate Box Turtles that were radio-tracked during a single active season in Iowa County had mean daily movement rates of 3.6 m (11.8 ft) and 10.9 m (35.8 ft), respectively (R. Christoffel and R. Hay, unpublished data). This species does not appear to defend territories, and individuals often exhibit overlap in home ranges (Legler 1960, Bernstein et al. 2007). Although individuals frequently overlap in areas used, Ornate Box Turtles are generally solitary animals except when breeding (Legler 1960).

Ornate Box Turtles in Wisconsin typically return to overwintering locations from September to mid-October, and the adult active season lasts ca. 150–177 days (Doroff and Keith 1990, Curtin 1997). Low air temperatures likely trigger the behavior associated with hibernation (Ernst and Lovich 2009). Ornate Box Turtles in Wisconsin often hibernate alone in self-excavated burrows or abandoned mammal burrows (Vogt 1981), although occasionally several individuals will hibernate in the same tunnel system (J. Kapfer, unpublished data). Overwintering strategies and behavior apparently vary by sex. For example, Wisconsin females hibernated at slightly greater mean depths (1.1 m [3.6 ft], $n=6$) than males (0.7 m [2.3 ft], $n=6$; Doroff and Keith 1990). Furthermore, environmental conditions within chosen hibernacula shift over time; deeper burrows tended to have higher internal temperatures in winter, and the coldest mean temperature recorded for all burrows monitored by Doroff and Keith (1990) during the winter was 1.2°C (34.2°F). The structure of vegetation that surrounds overwintering locations can further impact the environmental conditions within hibernacula. From late February to early April, shading of sites in Wisconsin had the greatest impact on internal hibernacula temperatures, with burrows in complete shade being colder than those with greater sun exposure (Doroff and Keith 1990). In 11 of the 12 hibernacula monitored, temperatures remained above freezing all winter, and all turtles survived to spring. Doroff and Keith (1990) further reported that 8 of 14 individuals monitored with radio telemetry returned to within ca. 1 m (3.3 ft) of overwintering locations from the previous year. Philopatry for overwintering sites, particularly in females, was also evident in a population in northwestern Illinois (Refsnider et al. 2012).

PREY AND PREDATORS

Ornate Box Turtles are omnivorous. Food items are located by scent and sight (Ernst and Lovich 2009). Although omnivorous, adults consume a substantial amount of animal material, and they are effective predators, able to actively chase and deftly capture prey (Ernst and Lovich 2009). Vogt (1981) reported a broad list of food items, including spiderwort (*Tradescantia* spp.), ground cherry (*Physalis* spp.), prickly pear cactus (*Opuntia* spp.), Blackberries (*Rubus allegheniensis*), mulberries (*Morus* spp.), and various grasses, other plants, earthworms, beetles, grasshoppers, and caterpillars. Turtles in a Dane County population were often observed feeding upon prickly pear cactus and the fleshy bases of spiderwort (Doroff and Keith 1990). A population from Columbia County fed on ground cherry, prickly pear cactus, and Blackberries, as well as various invertebrates and vertebrates, from earthworms to fishes (Ramminger 1992). The author further noted consumption of carrion and various plants in anthropogenic settings (i.e., Tomatoes [*Solanumly copersicum*], strawberries [*Fragaria* spp.], and Dandelions [*Taraxacum officinale*]). Turtles in northwestern Illinois feed heavily on raspberries (*Rubus* spp.) when available (Refsnider et al. 2012). In Kansas, insects made up 90% of Ornate Box Turtle diets (Legler 1960). According to Ernst and Lovich (2009), dung beetles (superfamily Scarabaeoidea) are important dietary components. The authors reflected that Ornate Box Turtles could have benefited in "the past, when bison and cattle roamed the plains, [because] there must have been a constant supply of dung beetles in the warmer months." For an exhaustive list of food items, see the review by Ernst and Lovich (2009).

Adult Ornate Box Turtles generally have few predators, and most mortality is due to anthropogenic activities (Legler 1960, Vogt 1981, Doroff and Keith 1990, Ernst and Lovich 2009). However, they are apparently more susceptible to predators during winter dormancy or recently after emergence from hibernacula (e.g., Legler 1960). For example, Coyotes (*Canis latrans*) were observed consuming Ornate Box Turtles during spring emergence at a site in Wisconsin (R. Christoffel, personal observation). When threatened, most juvenile, subadult, and adult Ornate Box Turtles will pull into their shells and seal themselves in tightly, using the flexible shell hinge. However, if handled, they will sometimes bite or flail their limbs in an attempt to escape (Vogt 1981).

Ornate Box Turtle nests experience heavy disturbance from numerous predator species (for a comprehensive list of documented nest predators across this species' range, see the review by Ernst and Lovich 2009). Temple (1987) monitored 22 turtle nests (including 16 of Ornate Box Turtles) in northern Dane County and found that 14 (64%) of the nests were destroyed by mammals, including Striped Skunks (*Mephitis mephitis*), Raccoons (*Procyon lotor*), and Virginia Opossums (*Didelphis virginiana*). In addition, Eastern Moles (*Scalopus aquaticus*) are documented to disturb Ornate Box Turtle nests in Wisconsin (Doroff and Keith 1990). Because of their small size and pliable shells, hatchlings are particularly vulnerable to predation (Dodd 2001). In fact, young Ornate Box Turtles are likely subject to many predators, including birds of prey, snakes, foxes, Virginia Opossums, and Striped Skunks (Vogt 1981).

CONSERVATION

Status: Globally, both subspecies of *Terrapene ornata* are listed as Near Threatened by the IUCN. In addition, all North American box turtles are listed under the Convention on International Trade in Endangered Species of Wild Fauna and Flora (CITES 2013). Within CITES, Ornate Box Turtles are listed in Appendix II (Lieberman 1994, US Fish and Wildlife Service 1995). The Ornate Box Turtle is considered a Species of Greatest Conservation Need and listed as Endangered in Wisconsin by the WDNR. As such, it is protected from collection or killing. However, this unfortunately does not provide protection for the turtle's habitat on private lands.

Populations: Various studies have reported information on Ornate Box Turtle populations throughout their geographic distribution. Doroff and Keith (1990) estimated a population size of 54–56 adult individuals within their 8 km^2 (3 mi^2) study site in Dane County. They also estimated that the density at this site was 2.9–5.0/ha (1.2–2.0/ac; Doroff and Keith 1990). Density estimates for a nearby Iowa population were 5.1–8.0/ha (2.1–3.2/ac; Bernstein et al. 2007), and Legler (1960) reported larger densities of 6.4–15.6/ha (2.6–6.3/ac) for a population in a Kansas pasture.

Annual survival appears to vary based on age class and geographic location. Doroff and Keith (1990) reported that mean annual survival for a Wisconsin population studied from 1977 to 1987 was 0.81 for adults and 0.21 for first-year turtles. The authors further estimated that survival of ages zero to ten years ranged from 0.37 to 0.54, if a population sex ratio of 1M:1F was assumed. Brown et al. (2004) reported that mean annual adult survival over eight years was 0.97 for a "human-impacted" population in northwestern Illinois. Given the recorded variability in survival by age, it is not surprising that Doroff and Keith (1990) determined that adults were the dominant age class in their Dane County study population. Several studies have found that population sex ratios skew toward females. The observed sex ratio of Ornate Box Turtles studied in Dane County was 0.63M:1F ($n = 101$; Doroff and Keith 1990), similar to the 0.59M:1F ($n = 100$) sex ratio reported for a Kansas population (Legler 1960) and the 0.51M:1F ($n = 121$) sex ratio reported for a Nebraska population (Converse et al. 2005). However, at another Wisconsin site, sex ratios were slightly skewed toward males (1.19M:1F, $n = 70$; R. and K. Russell, unpublished data).

Threats: Ornate Box Turtles are a slow-to-mature, long-lived species with low reproductive output and specific habitat needs (e.g., Doroff and Keith 1990, Congdon et al. 1993). As such, their populations are highly susceptible to anthropogenic disturbances. Knudsen (1956e) stated that it was "not inconceivable that [Ornate Box Turtles] could be exterminated from this state." Twenty-five years later, Vogt (1981) stated that Ornate Box Turtle numbers continued to decline, even at sites that once supported sizeable populations. Accordingly, Doroff and Keith (1990) surmised that their study population in Wisconsin, which consisted of ca. 50 turtles in the mid-1980s, had characteristics that made the population prone to declines and extirpation. This was later confirmed when Curtin (1997) reported that Doroff and Keith's (1990) study population had nearly disappeared by the mid-1990s. Currently, there are only six publicly owned sites and one privately owned site that are known to support Ornate Box Turtles in the state. Although further study is necessary, existing evidence suggests that none of these sites contain self-sustaining populations. Given this, rigorous investigation and modeling of these remaining populations, in conjunction with aggressive conservation measures, are warranted to preserve this species in Wisconsin.

The greatest threat to North American box turtles is the loss, degradation, and fragmentation of habitat (reviewed by Dodd 2001). Extensive conversion of natural landscapes to agriculture has historically occurred across much of the Midwest (Bragg and Teuter 1996), which has negatively impacted numerous habitats. In the 1950s and 1960s, for example, the development of more effective agricultural equipment, such as center-pivot irrigation systems (Hays 2010), made row crop agriculture on dry sandy soils financially profitable, which contributed to the loss of dry prairies in the state. In fact, Curtis (1959) estimated that only 1% of Wisconsin's pre-settlement dry sand prairie acreage remained by the 1950s. Today, that percentage is even less. Given that Ornate Box Turtles in Wisconsin prefer these sandy habitats and avoid agriculture (Doroff and Keith 1990), it is easy to make the connection between the loss of habitat due to agriculture and turtle population declines.

Ornate Box Turtle populations can also suffer when the quality of existing natural landscapes is degraded rather than destroyed. Such degradation of habitat can lead to indirect impacts that do not cause immediate mortality but still eventually result in population declines. For example, habitat degradation could force box turtles to increase their home range sizes as they search for additional suitable areas, and the energy required for greater movement could impact growth rates (Doroff and Keith 1990, Curtin 1997). There are various anthropogenic activities that can degrade the quality of box turtle habitat. In the 1940s and 1950s, it became common practice to plant tree species that tolerate the sandy prairie soils of southern Wisconsin (WDNR 1977), including White Pine (*Pinus strobus*) and Red Pine (*P. resinosa*). As these trees grew, the resultant shade could have played an important role in changing the thermal properties of the open-canopy habitats preferred by Ornate Box Turtles. Similarly, unchecked vegetative succession leaves normally open prairie habitats overgrown with woody plant species (see Introduction to Conservation and Management of Wisconsin's Amphibians and Reptiles chapter). If succession in Ornate Box Turtle habitat is not offset with active management, then habitat quality degrades, and local turtle populations could suffer (Moore 1993). The shading of critical open microhabitats used by box turtles for nesting and overwintering is likely of particular concern because it could lead to lower reproductive success or alter normal periods of activity (R. Hay, personal observation). In addition to rampant succession in open areas, habitat degradation can occur due to burgeoning populations of aggressive or invasive plant species. The invasive Spotted Knapweed (*Centaurea maculosa*) is an allelopathic (soil-toxifying) plant (Bais et al. 2003) that can occur in dense stands at some sites with Ornate Box Turtles in Wisconsin. This plant could pose a very serious threat, as anecdotal observations from several sites suggest that Ornate Box Turtles frequently avoid areas with heavy Spotted Knapweed infestations (R. Hay, personal observation).

Habitat can also be fragmented into smaller areas by roads or patches of land that become overgrown with dense woody vegetation. Much like habitat loss and degradation, habitat fragmentation can have serious impacts on Ornate Box Turtles. For example, habitats that are fragmented have a greater proportion of 'edges,' or areas that occur at the boundary between two habitats, such as suitable and unsuitable habitats. High proportions of edge habitats can be problematic for box turtles if the edges possess characteristics similar to the unsuitable habitats they border or if box turtle predators prefer to use these edges. Temple (1987) studied the impacts of habitat fragmentation and edges on turtles, including Ornate Box Turtles. He reported that nests created farther from prairie / oak forest or marsh edge (i.e., >50 m [164 ft]) were less likely to be destroyed by nest predators than nests found closer to an edge (40% vs. 83% nest predation, respectively). It is plausible that high predation rates along edges could intensify as habitat fragmentation becomes more prevalent. The fragmentation of large habitats into isolated patches can also restrict gene flow between populations, which contributes to the loss of genetic diversity. While the genetic diversity of Wisconsin's Ornate Box Turtle populations is not currently well-studied, genetic simplification or bottlenecks were documented in isolated populations from Iowa (Richtsmeier et al. 2008) and Illinois (Kuo and Janzen 2004). Most Wisconsin populations are similarly isolated from one another because the distance between them is too great for turtles to easily traverse. As such, the restriction of gene flow and concomitant declines in genetic diversity are likely important conservation concerns for Ornate Box Turtle populations in the state.

Wild collection of this species for the pet trade occurs throughout its North American distribution, although the population-level impacts of this phenomenon are poorly understood. Still, several authors have outlined past examples where Ornate Box Turtles were collected and sold on international and domestic pet markets (e.g., Zug 1994, Dodd 2001). Although collection for the pet trade is unethical and illegal in Wisconsin (William Smith, WDNR, personal communication), this activity is difficult to track, which makes the effects hard to quantify. Despite this, during the 1970s and 1980s, well-known Wisconsin populations were routinely targeted by local pet collectors (R. Hay, personal observation), which Vogt (1981) suggested was a primary reason for their decline in the state. The results of research from other states further suggest that even minimal removal of individual turtles can jeopardize population stability. Bowen et al. (2004) found that seemingly healthy study populations can actually experience recruitment rates that are near unity (i.e., recruitment is nearly offset by natural mortality). Thus, if the population's removal of individuals due to natural mortality equally offsets recruitment of new individuals, then any additional unnatural removal of turtles (e.g., road mortality, collection for pets) could have substantial negative impacts. Converse et al. (2005) reported similar susceptibility to declines in a Nebraska population. Intense monitoring of and further research on the impact of pet trade collection in Wisconsin are warranted.

Other threats exist that have yet to be thoroughly investigated in the state. These include currently unknown or emerging infectious diseases, road mortality (Knudsen 1956e, Legler 1960, Blair 1976, Doroff and Keith 1990, Dodd 2001) and mortality due to strikes from agricultural equipment and lawnmowers (Doroff and Keith 1990, Kim Reinke, citizen land steward, personal communication). Furthermore, possible direct persecution of Ornate Box Turtles by humans has been observed in Wisconsin (R. Christoffel and R. Hay, personal observation). This incident involved at least 16 adults that were beheaded at a site in south-central Wisconsin. The dead turtles were found in very close proximity to each other over a three-week period, which suggested that they were deposited there by humans. The turtles further had no recent damage to their shells or appendages that would otherwise suggest attempted predation. In addition to these threats, climate change likely impacts this species, but its effects in Wisconsin are currently unquantified (see Past, Present, and Potential Future Landscapes of Wisconsin chapter and WICCI 2011). Yet Converse et al. (2005) found that survival rates in Nebraska were highest in years when the minimum winter temperatures were lowest. On the other hand, a multiyear study by Tucker et al. (2015) in northwestern Illinois reported that Ornate Box Turtles were generally less active in warmer years. Thus, warming trends associated with climate change could alter activity patterns and survival rates and cause population declines, but further research is necessary (reviewed by Ernst and Lovich 2009).

Management: Unfortunately, limited access to funding and resources has restricted the ability of conservation biologists to determine the effectiveness of most conservation strategies for this species. Still, it is reasonable to suggest that conservation initiatives for Ornate Box Turtles will likely require a multifaceted approach that involves monitoring of populations, protection of land, and management of preferred habitats. These strategies have been attempted to some degree at several sites in Wisconsin. For example, a population in Sauk County has been the subject of monitoring for over two decades, and most other state populations have received some level of monitoring or mark-recapture work to establish baseline population data. Land acquisition, protection, and habitat management strategies have also been designed with this species in mind (e.g., Randall Hoffman and Mark Martin, WDNR, personal communications). Furthermore, aggressive conservation tactics such as translocations have been attempted with varying degrees of success (Natural History Box: Translocations from within Wisconsin;

NATURAL HISTORY BOX: Translocations from within Wisconsin

Starting in 1992, a translocation experiment was conducted that primarily involved adult Ornate Box Turtles at a site in Wisconsin. The goal of this work was to determine if adult turtles could survive after being moved from one site in Wisconsin to another, with the hope of preserving the representative genetic diversity of the donor sites (meeting between R. Hay, WDNR, and L. Keith, UW–Madison). Turtles were collected from several very small relict populations in Wisconsin that were expected to shortly disappear, based on various circumstances, and moved to a recipient site where Ornate Box Turtles did not exist at the time. The recipient site had previously been unsuitable for many years due to agricultural activities, but it possessed appropriate soil, and the vegetation had since been allowed to recover from years of disturbance. Twenty adult turtles were initially translocated in 1992 and placed into a 2.5 ha (6.2 ac) enclosure that contained a variety of microhabitats, including those essential for nesting and overwintering. Fourteen more adults were translocated to the enclosure during 1993 and 1994, bringing the number of translocated adults to 34. Research was then conducted to determine if the translocated animals would imprint on the site after being enclosed for at least a year. In 1994 five translocated adults that had been enclosed for a full year were fitted with radio transmitters and released immediately outside of the enclosure in suitable habitat. During this first release, three of the turtles dispersed from the translocation site, while two remained (C. Curtin, WDNR, unpublished report). The two that did not disperse and the three that had dispersed were placed back into the enclosure for another winter. In 1995 11 adults from the enclosure, including 9 that had spent two full years enclosed and the 2 that had been placed back in the enclosure the previous autumn, were released in the adjacent prairie. During this second release, 100% of the turtles remained on-site. One of the 11 turtles was accidentally killed during the summer by farm equipment, and 10 survived until autumn. Nine of these surviving 10 individuals returned to the vicinity of the enclosure and overwintered (Hatch 1996). The tenth turtle remained about 0.8 km (0.5 mi) from the enclosure but did not appear to burrow appropriately for hibernation and was found dead from exposure in mid-November.

Several additional adults were translocated to the experimental site in 1995. In 1996 funding was available to track only four of the nine remaining radio-tagged turtles. All remained on-site, and three returned to the vicinity of the enclosure to overwinter, while the fourth remained on-site but did not return to the area of the enclosure. That autumn, three additional adult turtles from within the enclosure were radio-tagged and released. These individuals also remained on-site. In 1997 26 adults that had been in the enclosure for at least two full years were released on the adjacent prairie. Since that time, several additional adult Ornate Box Turtles were released after spending at least two full years in the enclosure. Periodic monitoring indicates that a number of translocated individuals continue to persist on-site, and evidence of natural recruitment has since been noted, including the capture of a large subadult many years after the initial translocation occurred. This work demonstrated that translocation of Ornate Box Turtles could be an effective strategy under appropriate conditions. We found that a two-year acclimation period within a suitable enclosure improved site acceptance and reduced dispersal into unfavorable adjacent property after release (Hatch 1997). More recently, Germano and Bishop (2009) reviewed the results of numerous herpetofaunal translocation projects and suggested that allowing subjects a period of acclimation on-site (perhaps in an enclosure) prior to release may increase the success of reintroductions, which corroborated our findings. Further work and monitoring are necessary to understand the effectiveness of translocation as a management tool, but the results of our pilot attempt are very encouraging.

Robert W. Hay, Turtles for Tomorrow

NATURAL HISTORY BOX: Translocations from outside Wisconsin

In 1992 an additional translocation project was initiated that involved the relocation of Ornate Box Turtles from northern Texas (*n* = 6), central Kansas (*n* = 7), and northern Nebraska (*N* = 6) to an enclosure at an experimental site in southwestern Wisconsin. This work was primarily conducted to determine if Ornate Box Turtles could be taken from other parts of their North American range and released in Wisconsin to seed new populations should ongoing conservation efforts in the state eventually fail and the species face a greater risk of local extinction. All 19 turtles survived the first season, and all but one turtle survived the first overwintering period. It is noteworthy that the single incident of mortality was an individual that had been caught in a prairie burn prior to translocation from Kansas. This had resulted in significant scute loss and some bone scorching, which may have contributed to this individual's early demise more so than the translocation. All of the remaining 18 turtles survived the second full year, but in the spring of 1994, the 12 Texas and Kansas turtles emerged from overwintering with serious respiratory ailments and died within a month. The six northern Nebraska turtles survived the first three years and appeared to be healthy. We then translocated an additional eight adult turtles from northern Nebraska in 1995 to the same recipient site (total *n* = 14 on-site), with a 1:1 sex ratio, and held all of them for five more years (a total of eight years for the original six Nebraska turtles). During the eight years, the turtles maintained favorable body weights and appeared healthy but failed to produce a single clutch of eggs that we detected. We concluded that Ornate Box Turtles from other, more distant parts of their range would not be suitable candidates for translocation to maintain the species in Wisconsin. Additional details of this work, including an analysis of thermal gradients across the range of the turtle, can be found in Curtin (1998).

Robert W. Hay, Turtles for Tomorrow

Natural History Box: Translocations from outside Wisconsin).

Ongoing efforts to manage habitat for Ornate Box Turtles on public lands and some private lands have occurred for several decades in Wisconsin but have intensified in the past 20 years. Critical to the long-term maintenance of populations, habitats on sites where Ornate Box Turtle exist must provide a variety of optimal conditions to allow behavioral thermoregulation and the protection of all life stages (Adams and DeCarvalho 1984). Conservation plans for this species, therefore, could center around the control of vegetative succession so that open prairies and savannas are not overrun with woody plant species that alter the thermal conditions near the soil surface. Furthermore, intense control of invasive plants such as Spotted Knapweed on these properties could be beneficial.

Prescribed fire is widely used in Wisconsin during restoration or maintenance of prairie habitats, as this method can effectively combat invasive plants and offset vegetative succession. While prescribed fire is useful, overly intense application of this technique could reduce available vegetative cover used by box turtles to hide from predators and for thermoregulation. There are also numerous documented examples of box turtle mortality due to fire (Babbitt and Babbitt 1951, Carr 1952, Smith 1958, Rose 1986, Ernst et al. 1995, Dodd et al. 1997, Larson 2014). Given that the loss of any individuals from populations of this rare species in Wisconsin could be extremely detrimental, intense management strategies should be applied carefully. For this reason, the application of prescribed fire in Wisconsin is restricted to the Ornate Box Turtle's inactive period (WDNR 2018c). The date that turtles leave hibernacula varies from year to year based on environmental conditions. Therefore, annual monitoring of spring emergence in Wisconsin is currently used to dictate when the application of prescribed fire for management should cease at sites where this species occurs. Mowing and mechanical brush removal can also be employed to offset vegetative succession, but they are also restricted to the Ornate Box Turtle's inactive season, with the exception of private agricultural lands.

Eastern Musk Turtle

Sternotherus odoratus (Latreille 1801)

Rebecca A. Christoffel

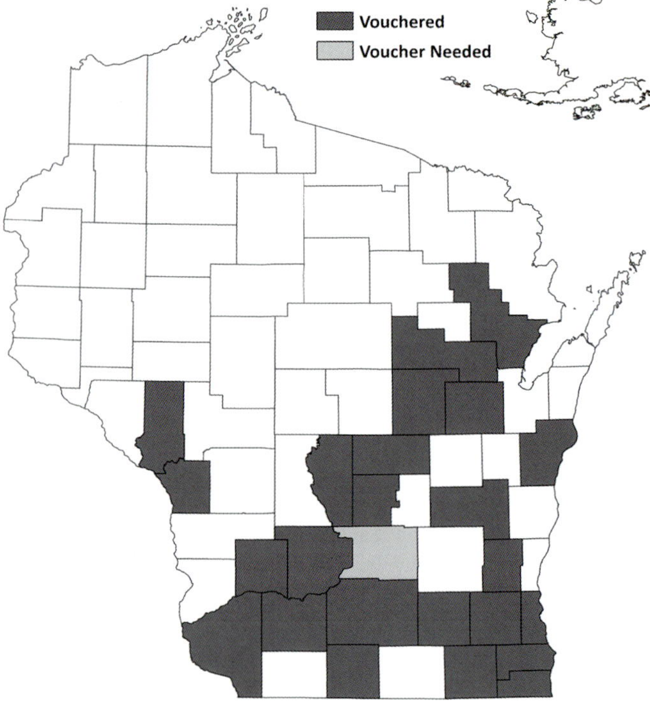

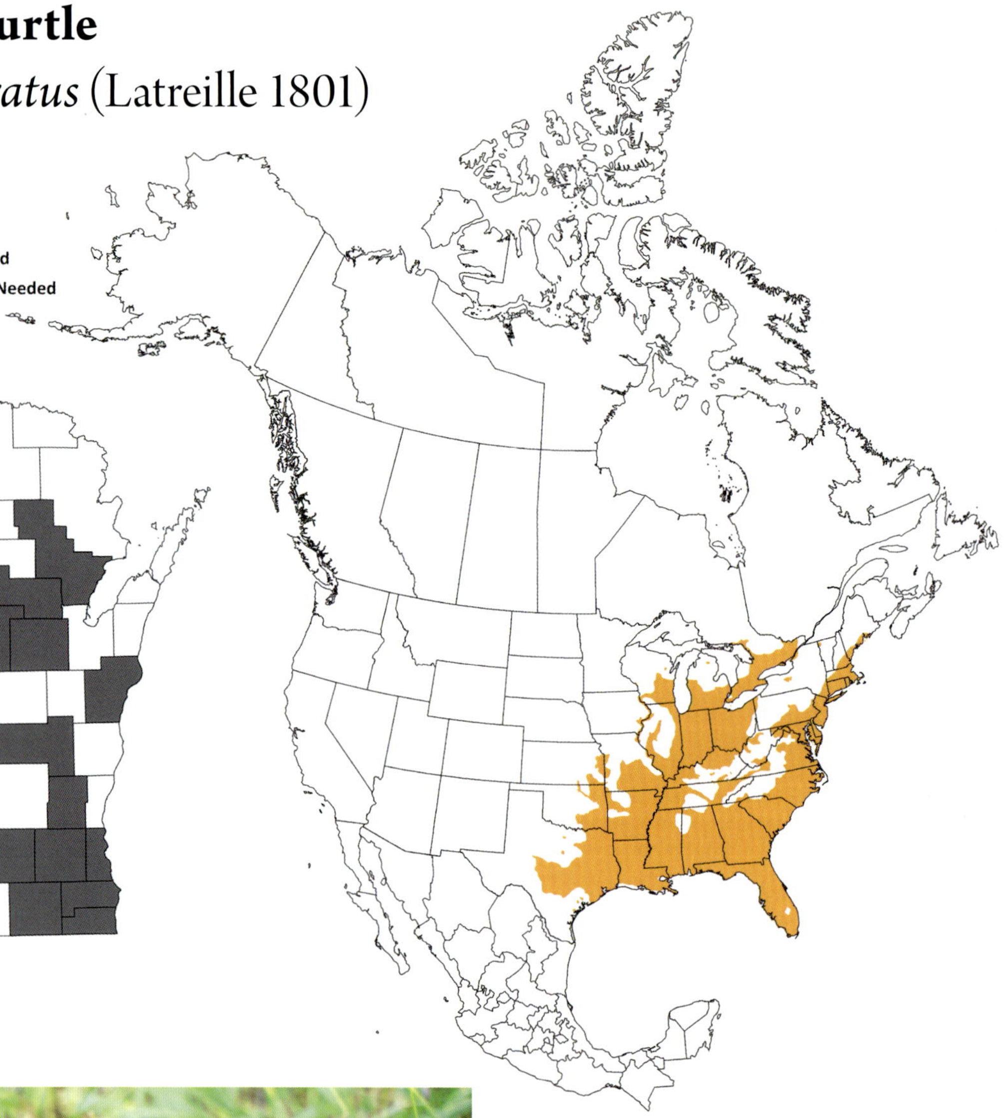

Adult Eastern Musk Turtle (Waukesha County; photo by J. Kapfer).

AT A GLANCE

Eastern Musk Turtles (a.k.a. Stinkpots) are one of Wisconsin's smallest turtles, with a CL<15.0 cm (5.9 in). They are found primarily in the southern half of the state, although they are also reported from some northeastern counties. Eastern Musk Turtles are most readily distinguished from other Wisconsin turtles by the presence of two yellow stripes that run along both sides of the head, one above and one below each eye, and light-colored barbels on the chin. The carapace is highly domed and dark brown to black, with no patterning or ridges. This provides excellent camouflage for the turtle when its head and limbs are pulled in, making it look like a rock sitting on the bottom of a waterbody or on dry land. Yet mind yourself if you are in a canoe under overhanging branches on a large river. Eastern Musk Turtles reportedly sometimes bask high on overhanging branches and, when startled, will drop down and may end up as a stowaway in your vessel. Otherwise, lucky is the person who is fortunate enough to encounter one of these denizens of lakes, ponds, and slower backwaters of rivers. Eastern Musk Turtles are secretive and spend most of their time walking along the bottoms of these waterbodies rather than on dry land. In fact, most of the time, the only opportunity to see these turtles is in June, when females come on land to lay their eggs, which mostly hatch by late August to early September. They reportedly consume many insects and mollusks but also have strong omnivorous tendencies. Eastern Musk Turtles are primarily active from early April through October, after which they can be found overwintering underwater. This turtle's common name is apparently derived from the musky odor it produces from scent glands on shell peripheries. It is considered common by the WDNR.

SYSTEMATICS

Several phylogenetic studies on turtles in the genus *Sternotherus* exist. These include Tinkle and Webb (1955), Tinkle (1958), Killebrew (1975), Iverson (1977), Sites et al. (1979), Seidel and Lucchino (1981), Seidel et al. (1981), Bickham and Carr (1983), Olmo (1984), and Iverson (1998). Iverson (1998) supported that *Sternotherus* was a monophyletic genus and that the Eastern Musk Turtle was a sister taxon to all other *Sternotherus* species. There are no subspecies currently recognized for *S. odoratus* (Eastern Musk Turtle; Crother 2017). This species is also commonly known as the Stinkpot. MacGowan et al. (2005) provide an explanation of the Latin name for the Eastern Musk Turtle. The genus name *Sternotherus* comes from the Greek, meaning 'hinged chest,' in reference to the turtle's hinged plastron, and the specific epithet, *odoratus*, comes from the Latin 'to have a strong odor,' in reference to the turtle's musk glands. The accepted common name was recently changed from Common Musk Turtle to Eastern Musk Turtle to eliminate any current or future confusion about this turtle's status (e.g., the perception that 'common' indicates population size rather than having widespread distribution [Crother 2017]). This is the only member of the genus *Sternotherus* and family Kinosternidae found in Wisconsin.

DESCRIPTION

Adults: The Eastern Musk Turtle's shell is gray, brown, or greenish, often with dark flecks and covered with algae (Vogt 1981, Tyning 1990). It has two yellow or whitish stripes running from the snout onto the neck, one above the eye and one below the eye (Pritchard 2003). These stripes may be absent or faded in older adults (Vogt 1981). The carapace is highly domed and has smooth marginal scutes (Figure 355; Ernst and Lovich 2009). Its legs are short, and there is little webbing between the toes (Christoffel et al. 2002). The musk glands, characteristic of Eastern Musk Turtles, are found on each side of the shell at the back end of the bridge and midway between the bridge and the front edge of the carapace (Ehrenfeld and Ehrenfeld 1973). The Eastern Musk Turtle has a hinge on the plastron that may or may not be functional (Pritchard 2003). The plastron itself is much reduced and consists of 11 scutes separated by wide cartilaginous sutures (Figure 356; Vogt 1981). The reduced plastron allows the legs a great range of motion but also makes the species more vulnerable to predators, particularly on land (Christoffel et al. 2002). Eastern Musk Turtles have obvious barbels (i.e., little fleshy growths found on the chin and throat; Figure 357).

Eastern Musk Turtles are one of Wisconsin's smallest turtles, with MCLs that range from 8 to 14 cm (3.1 to 5.5 in; Vogt 1981). Mean MCL of museum specimens from Wisconsin was 10.5 cm (4.1 in, $n = 14$). The largest individual measured was an adult female collected in the early 1900s, whose MCL was 13.4 cm (5.3 in; note that the metal UWZM specimen tag on this individual had disintegrated, but the label on the jar indicated it was collected in Wisconsin). In Iowa, adult length is reported to be 5.1–11.5 cm CL (2.0–4.5 in; LeClere 2013). Tucker et al. (2008) reported a particularly large specimen captured in Illinois with a 15.4 cm MCL (6.1 in).

Preadult stages: The white eggs of Eastern Musk Turtles are hard-shelled and more blunt-ended than ellipsoidal in shape (Figure 358; Iverson and Ewert 1991). Egg dimensions in Wisconsin are ca. 2.8 × 1.5 cm (1.1 × 0.6 in; Vogt 1981). Furthermore, Edgren (1942) examined 15 eggs from a group of 130 in Racine County. He reported a mean length of 2.69 cm (1.1 in, range = 2.4–2.9 cm [0.9–1.1 in]) and a mean width of 1.58 cm (0.6 in, range = 1.5–1.7 cm [0.6–0.7 in]). In a later study, he reported a mean egg mass of 3.8 g (0.1 oz; Edgren 1949). In Illinois, Phillips (1961) reported that the mean size of eggs was 2.75 × 1.5 cm (1.1 × 0.6 in), and Tucker (2001) found that eggs averaged 2.66 × 1.56 cm (1.1 × 0.6 in) and weighed 3.6–4.5 g (0.1–0.2 oz). Unlike many other turtles' eggs, those of the Eastern Musk Turtle do not increase in size during incubation (Edgren 1956, Palmer and Braswell 1995).

Newly hatched Eastern Musk Turtles are quite small, about the size of large grapes (Rebecca Christoffel, Snake Conservation Society, personal observation). Measured hatchlings from Wisconsin in mu-

Left, top: Figure 355. Adult Eastern Musk Turtle showing the highly domed carapace and smooth marginal scutes (Sauk County; photo by J. Kapfer).

Left, middle: Figure 356. Plastron of an adult Eastern Musk Turtle (Waukesha County; photo by J. Kapfer).

Left, bottom: Figure 357. Light-colored barbels on the underside of an Eastern Musk Turtle's chin (Sauk County; photo by J. Kapfer).

Figure 358. Eastern Musk Turtle eggs produced by a captive individual (Wisconsin; photo by J. Kapfer).

seum collections had a mean MCL of 2.4 cm (0.9 in, $n = 12$). In Illinois, hatchlings ranged from 2.15 to 2.72 cm CL (0.8 to 1.1 in; Phillips 1961). The young have a black carapace, head, and limbs, a noticeable keel along their back, and a lesser lateral keel on each side of their carapace (Phillips 1961, Ernst and Lovich 2009). There is a white or yellow spot on each marginal scute (Vogt 1981). The plastron of juveniles lacks a hinge and is rough to the touch (Ernst and Lovich 2009). Young Eastern Musk Turtles also possess the characteristic light stripes on the head, as described above for adults.

Variation: Male Eastern Musk Turtles differ from females in appearance, although the differences are subtle. Males have a thicker and longer tail than do females (Vogt 1981). Unlike females, males have a blunt tail with a horny end, and their cloacal opening is posterior to the back edge of the carapace (Ernst and Lovich 2009). Males possess two small patches of horny scales on the inner surface of each hindleg (Vogt 1981). The plastron of males is generally shorter than that of females, averaging 70% (63%–76%) and 75% (68%–81%) of the MCL, respectively (Ernst and Lovich 2009). In Illinois, Phillips (1961) reported that PL for 16 males and 21 females averaged 73.3% and 79.7% of the MCL, respectively (range = 67.0%–78.7% and 75.0%–90.3%, respectively). In addition, males have more skin exposed between their plastral scutes than do females (Phillips et al. 1999, Ernst and Lovich 2009).

Diagnosis: Eastern Musk Turtles are characterized by a carapace covered with keratinized scutes, a posterior margin that is not serrated, and a plastron with 11 scutes. Eastern Musk Turtles are unlikely to be confused with any other turtle found in Wisconsin. The only other turtle with visible barbels on the chin is the Snapping Turtle (*Chelydra serpentina*). Unlike the Snapping Turtle, which has a long tail with prominent protuberances, the Eastern Musk Turtle has a short and much smoother tail.

Turtle eggs encountered in the wild are difficult to differentiate, although Eastern Musk Turtles lay eggs that are elliptical in shape and so can be distinguished from the round eggs of Snapping Turtles and softshell turtles (*Apalone* spp.). Although other aquatic turtles in the state also have elliptical-shaped eggs, the eggs of Eastern Musk Turtles are, on average, smaller and blunter.

DISTRIBUTION AND HABITAT

Global distribution: Eastern Musk Turtles are found from New England, Quebec, and southern Ontario in the North to Florida, Alabama, Mississippi, Louisiana, and Texas in the South. The species occurs as far west as central Texas, eastern Kansas, Oklahoma, and Iowa. Wisconsin is on the northern edge of its midwestern distribution.

Distribution in Wisconsin: Published county records are mostly found in southern Wisconsin, where Vogt (1981) stated that populations were locally abundant, particularly along the lower Wisconsin River floodplain. Despite this, Eastern Musk Turtles are also documented in appropriate lacustrine habitat from other locations in the state, including central and northeastern Wisconsin.

Localities of possible occurrence: Eastern Musk Turtles are likely present in Crawford and Vernon Counties, which occur along the Mississippi River between counties where it is already confirmed (i.e., Grant and La Crosse). Given documentation of this turtle in counties not associated with the Mississippi or Wisconsin Rivers, such as Manitowoc (Surendonk and Kapfer 2008), Waukesha (Vogt 1981), and Oconto (Ewert et al. 2007), it seems likely that there may be additional undocumented counties in which this species occurs.

Habitat: In Wisconsin, Eastern Musk Turtles are found in ponds, rivers, backwater sloughs, and lakes with soft mud or marl bottoms (Vogt 1981). Although this species makes use of many aquatic habitat types, research supports that they generally prefer large still waterbodies or slow-moving sections of lotic environments. For example, a variety of research projects conducted in the Upper Midwest supported that these turtles prefer open water ponds and lakes but will also use cattail marshes, rivers, and streams with large backwater areas and adjacent wet meadows (e.g., Dreslik and Phillips 2005, Tucker et al. 2008, Rowe et al. 2009, Anthonysamy et al. 2014a). Small-scale habitat characteristics and thermoregulatory opportunities appear to be important factors that influence microhabitat selection. In southwestern Michigan, most Eastern Musk Turtle activity occurred in water that was <1 m (3.3 ft) deep and in well-vegetated littoral zones (Rowe et al. 2009). In Ontario, individuals selected aquatic habitats based on thermal preferences, which consisted of shallow water covered with lily pads and other macrophytes (Picard et al. 2011). Eastern Musk Turtles reportedly spend much of their time partially or fully submerged (Vogt 1981). Root (1949) found that they can remain underwater for long periods of time, although dives of <20 minutes are most common (Stone et al. 1992). They also use Muskrat (*Ondatra zibethicus*) and North American Beaver (*Castor canadensis*) lodges, as well as bank burrows and undercuts for shelter (Kiviat 1978, Anthonysamy et al. 2014a).

Nesting habitats selected by females are often open locations with loose soil near the water's edge that receive high light penetration (see Reproduction and Development section). Eastern Musk Turtles overwinter in aquatic habitats that do not completely freeze, where they remain submerged until spring emergence. Vogt (1981) reported that Eastern Musk Turtles in Wisconsin hibernate at the bottom of ponds or similar waterbodies in the mud at depths of 15–30 cm (6–10 in) or under submerged logs, rocks, or other debris. In Ohio, Thomas and Trautman (1937) reported a hibernating congregation of ca. 450 individuals in an area of 24.3 m^2 (261.6 ft^2) in the bed of a partly drained canal.

REPRODUCTION AND DEVELOPMENT

Across their North American range, Eastern Musk Turtles court and mate throughout their active seasons, although peak breeding activity is in the spring and autumn (Vogt 1981, Ernst and Lovich 2009). The majority of breeding in the northern US, however, occurs in the spring (April or May), which coincides with ovulation in females. In Pennsylvania, Ernst (1986a) reported that most mating occurred in April and May and that all 12 reproductive events that he observed occurred in the morning between 9:15 and 10:27 in shallow water (<30 cm [12 in] deep). The second period of mating in the autumn, if it occurs, is instigated by decreased day length, which stimulates male sexual activity (Mendonça and Licht 1986, Mendonça 1987). Females that breed in autumn are able to store viable sperm internally until the spring, when fertilization commences (Gist and Congdon 1998).

Laboratory experiments revealed that chemical cues in the musk and urine aid males in mate-searching and determination of female sexual readiness (Fadool et al. 2001, Lewis et al. 2007). Mahmoud (1967) studied the sexual activities of Eastern Musk Turtles in Oklahoma. He reported three phases of reproductive behavior: tactile, mounting, and rubbing/biting. During the tactile phase, a male with its head extended approaches another turtle from behind and touches or smells its tail to determine its sex. Courtship only proceeds if the turtle is female, and then the male moves to her side and nudges the region of her bridge with his nose to detect her musk glands. If the female is not receptive, she moves away, and the male either chases her or goes elsewhere. If receptive, the female remains still while the male gently nudges her behind the eye, and a few seconds later the male assumes the mounted position. The mounting phase consists of the male positioning himself with his plastron directly over the female's carapace by grasping the rim of her carapace with the toes and claws of all four of his feet. The male's tail is looped so that the terminal nail touches the skin at one side of the female's cloaca to bring their vents together for copulation. Once the coital position is attained, the rubbing and biting phase begins.

June is the month when most Eastern Musk Turtle nesting occurs in Wisconsin (Vogt 1981). Eastern Musk Turtles typically do not stray far from the water to nest, although Ernst and Lovich (2009) observed nesting as far as 45 m (147.6 ft) from the water. Nest sites include shallow self-excavated cavities in the soil, as well as amid decaying vegetation or rotting logs in or near pond edges (Vogt 1981). Cagle (1937) reported communal nesting, having found 16 egg clutches under a single log. In Wisconsin, Edgren (1942) also found 130 Eastern Musk Turtle eggs buried within the debris of an old structure used by duck hunters. Vogt (1981) reported that he observed many "uncovered nests" and posited that this happens when females are disturbed during oviposition. Nest excavation and oviposition can occur from early morning into the night (Ernst and Lovich 2009). In Wisconsin, however, numerous anecdotal observations suggest that nesting activity is concentrated in the morning between 9:00 and 11:00, but nesting females are also observed on land in the early evening (Thomas LaFleur, Turtles for Tomorrow, personal communication).

In Wisconsin, females typically lay two to eight eggs (Vogt 1981), although a very large female (12.0 cm CL [4.7 in]) at a fish farm in Washington County laid a clutch of ten eggs (Ewert 2005). The largest reported clutch size in Wisconsin is 11 eggs, encountered at a site in Sauk County (Robert Hay, Turtles for Tomorrow, personal communication). Tucker et al. (2001) found a mean clutch size of 5.8 (range = 5–7) eggs in Illinois. Clutch size is positively correlated with female CL, with larger females producing more eggs per nesting (Mitchell 1985, Ernst 1986a). Relative clutch mass (i.e., the ratio of total clutch weight to turtle body weight) in Illinois averaged 13% (9.8%–20.0%; Tucker 1999), and Ewert (2005) reported relative clutch masses of 13.7%–20.0% in Washington County. Almost all female Eastern Musk Turtles reproduce annually, although occasionally a year is skipped due to adverse weather conditions (Gibbons et al. 1983). In some regions of the US females reportedly produce multiple clutches per year (Mitchell 1985), although this phenomenon has not been documented in Wisconsin. Eastern Musk Turtle egg size appears to decrease with increasing latitude (reviewed by Ernst and Lovich 2009), as do the number of eggs in a clutch and annual nesting frequency (Iverson and Meshaka 2006).

Eastern Musk Turtles have temperature-dependent sex determination, where temperatures ≥28°C (82.4°F) produce all females, and temperatures <25°C (77°F) typically produce males (Vogt et al. 1982). The incubation period ranges from 65 to 86 days ($\bar{x}$ = 74.7 days, n = 11 clutches; Ernst and Lovich 2009). In Pennsylvania, incubation period averaged 75–82 days ($\bar{x}$ = 78.3 days, n = 32; Ernst 1986a). Hatchlings emerge from nests in late August and September in Wisconsin (Vogt 1981). Upon nest emergence, hatchlings orient toward the light of large open areas (such as lakes and rivers) and so find their way to water (Collins 1993). Hatchling Eastern Musk Turtles apparently do not overwinter in the nest cavity (Mitchell 1988, Ernst and Lovich 2009), and there are no reports that they can tolerate freezing (Costanzo et al. 2001).

Young Eastern Musk Turtles grow rapidly (Risley 1933). Yet as their body shape becomes more elongated and flattened with maturity, their growth rate decreases (Ernst and Lovich 2009), so that adult growth rate is very low or nonexistent. Mahmoud (1969) reported that three Eastern Musk Turtles that exceeded ten years of age showed no growth after being held in captivity for two years. Risley (1933) examined Michigan individuals and reported that those with CL >8 cm (3.2 in) were ten or more years old. In Toronto, Edmonds (1998) reported that the average CL for one- to nine-year-old individuals was 3.0 cm (1.2 in), 4.0 cm (1.6 in), 4.7 cm (1.8 in), 5.3 cm (2.1 in), 6.0 cm (2.4 in), 6.6 cm (2.6 in), 7.0 cm (2.8 in), 7.4 cm (2.9 in), and 7.7 cm (3.0 in), respectively. In Ontario, males matured in five to six years, while females took eight to nine years (Edmonds 1998). Risley (1933) speculated that Michigan males matured in three to four years, and females matured in nine to eleven years. This is in contrast to a more southern population (i.e., Virginia), where females reached sexual maturity at four years of age (Mitchell 1985). Tinkle (1961), in his examination of Eastern Musk Turtles across a broad geographic area, reported that no mature northern female had fewer than three growth rings, and most had at least four, while the youngest mature males had three growth rings (but many with this number were still immature). The longevity record in captivity is >54 years at the Philadelphia Zoo (Snider and Bowler 1992). Ernst (1986a) reported that a wild female in the Pennsylvania population he studied was at least 28 years old and a male was at least 27 years old at the time of their last capture.

ACTIVITY

In Wisconsin, Eastern Musk Turtles typically emerge from hibernation in late March or April. They are active for ca. 200 days (from April to October) in Michigan (Risley 1933) and Ohio (Conant 1951), although most activity occurs during the warmer months (June–August; Ernst and Lovich 2009). During the spring and autumn, Eastern Musk Turtles concentrate most activity during the warmer parts of the day but in mid-summer shift to become crepuscular or nocturnal (Mahmoud 1969).

Eastern Musk Turtles are secretive, and chance sightings of them are uncommon (LeClere 2013). Individuals rarely bask openly on structures above the water but instead float at the surface among aquatic vegetation or with just the top of their carapace exposed (Riedle et al. 2009, Rowe et al. 2009). In Missouri, they occasionally bask on logs, rocks, or small horizontal tree trunks (Johnson 1987). On the rare occasions when this species is observed basking outside of water in Wisconsin, it is reportedly along slanting willow trees 1–3 m (3.3–9.8 ft) above the water's surface and "may dive-bomb your canoe if you startle them by paddling underneath!" (Vogt 1981). Older individuals drop into the water when disturbed while basking, but juveniles are reported to hold on tightly and can be difficult to remove (Ernst and Lovich 2009). Although data on preferred body temperature do not exist for Wisconsin individuals, research in Michigan found that mean shell temperature of active radio-tracked turtles was 24.2°C (75.6°F; Rowe et al. 2009), which is similar to the preferred body temperature reported for Oklahoma (24.1°C [75.4°F], range = 21.6°C–27.5°C [70.9°F–81.5°F], $n = 8$; Mahmoud 1969). Additional information on thermal relations for this species can be found in Ernst and Lovich (2009).

Eastern Musk Turtles do not display territorial behavior, but they have a tendency to remain within a relatively small home range (Mahmoud 1969, Ernst 1986a). Some past research suggests that they further exhibit a reluctance to disperse between waterbodies. For example, research in Georgia found that even when threatened by drought, Eastern Musk Turtles were unlikely to abandon their aquatic habitat (Gibbons et al. 1983). Yet variation in the tendency to leave the water exists (Ernst 1986a). Rowe et al. (2009) found that one radio-tracked individual moved 900 m (2,953 ft) from the lake where it was originally captured over land to an adjacent "grass-sedge-willow area." The general reluctance to leave the water may be because the turtles experience rapid evaporative water loss when on dry land due to their high surface area to mass ratio, and they will usually show signs of distress if water is withheld for extended periods (Stone and Iverson 1999).

There is some evidence that home range size increases with latitude (Bennett et al. 2015). In Ontario, mean home range size ranged from 48.9 ha (120.4 ac) for females to 155.4 ha (383.4 ac) for males (100% MCP estimator; Edmonds 1998). Bennet et al. (2015) reported home range sizes in Ontario for two males (430.1 ha [1,062.3 ac] and 7.6 ha [18.3 ac]) and two females (17.2 ha [42.2 ac] and 35.6 ha [87.4 ac], 100% MCP estimator). In Pennsylvania, males had a mean home range size of 1.8 ha (4.3 ac, $n = 12$), while females averaged 0.9 ha (2.3 ac, $n = 9$; Ernst 1986a). Rowe et al. (2009) reported a mean home range size of 2.8 ha (6.9 ac; 95% fixed kernel estimator) for 12 tracked individuals from Michigan with no significant difference in home range size by sex. The mean total distance moved during a 12-hour period by turtles in this study was 27 m (88.6 ft), which did not significantly vary by sex or due to weather (Rowe et al. 2009). Past research indicates that at least some individuals demonstrate fidelity to sites and homing ability (Williams 1952, Andres and Chambers 2006). Otten and VanDeWalle (2014) reported on the homing ability of Eastern Musk Turtles that were rehabilitated following an oil spill in the Kalamazoo River (Michigan). Turtles were removed from the oil spill area, cleaned, and then released at new locations in the river that were unaffected by the spill. Two of five relocated individuals homed in to their original points of capture ca. 12 months later, traveling distances of 6.7 km (4.2 mi) and 5.6 km (3.48 mi), respectively.

Ernst and Lovich (2009) reported that hibernation begins when water temperatures fall below 10°C (50°F). Ernst (1986a) recorded body temperatures as low as 6°C (43.8°F) for dormant individuals in Pennsylvania ($\bar{x} = 8.3$°C [46.9°F], $n = 5$). Very little specific data on selected overwintering locations and overwintering behavior/physiology exist for Wisconsin.

PREY AND PREDATORS

Past authors sometimes indicate that Eastern Musk Turtles are primarily carnivorous (Vogt 1981, Ernst and Lovich 2009) although there are reports of strong omnivorous tendencies (Ford and Moll 2004). Vogt (1981), for example, stated that Wisconsin individuals consumed mostly mollusks and crayfish. Other authors, however, have determined that this turtle's diet is less restrictive. In Michigan, Lagler (1943a) reported the frequency of dietary items, which included mollusks (28.3%), insects (24.2%), plants (15.0%), debris (12.5%), carrion (5.8%), and fish (0.8%), with the greatest volume attributed to carrion. In Oklahoma, Mahmoud (1968a) reported the frequency of dietary items as insects (98.3%), aquatic vegetation (97.4%), mollusks (96.2%), crustaceans (61.1%), carrion (37.4%), and amphibians (5.2%), with the greatest volume attributed to insects. More recently, Ford and Moll (2004) analyzed the diets of Eastern Musk Turtles in Missouri and also recorded frequency of occurrence for food items. For the 249 sampled individuals, plant material was the most prominent food item (95.4%), followed by insects (85.8%), mollusks (81.9%), seeds (62.3%), crayfish (57.7%), algae (44.6%), and fish (12.3%), with the greatest volume from algae (22.0%). Diets vary seasonally, depending on what food items are most readily available (Mahmoud 1968a), as well as seasonal and sexual preferences (Ford and Moll 2004). Individuals with MCL ≤5 cm (2 in) feed mainly on small aquatic insects, algae, and carrion (Ernst and Lovich 2009). Minton (1972) believed that because of the great numbers of snails eaten, the species might be valuable in controlling many fish parasites for which snails are an intermediate host.

Eastern Musk Turtles forage by walking along the bottoms of waterbodies with their necks fully extended or by probing their heads into soft mud or sand to seek out food items (Mahmoud 1968a). It is believed that olfaction plays an important role in prey location (Fadool et al. 2001, Murphy et al. 2001). Ernst (1986a) found that feeding only occurred when water temperature was above 18°C (64.4°F) in Pennsylvania. Prey may be swallowed whole, but large items can be dismantled by securing them in the jaws and tearing off pieces with the claws of the forelimbs (Ernst and Lovich 2009).

Like many turtle species, Eastern Musk Turtles are most susceptible to predators as juveniles. Eggs are eaten by a variety of predators, including Snapping Turtles, snakes, Muskrats, skunks, Raccoons (*Procyon lotor*), foxes, and various herons and crows (reviewed by Ernst and Lovich 2009). Nest predation rates can be substantial; Ernst (1986a) reported that egg hatching success at study sites in Pennsylvania was only 15.4% ($n = 32$ clutches and 104 eggs). Juvenile and adult Eastern Musk Turtles are also consumed by many types of vertebrates (see exhaustive list in Ernst and Lovich 2009). Harding (1997) believed that Muskrats will attack dormant Eastern Musk Turtles in winter and reported that older turtles often show signs of injury from previous encounters with predators. Young Eastern Musk Turtles are occasionally trapped by freshwater clams and drown, although this is accidental and not predation (Plummer and Goy 1997). The behavior of Eastern Musk Turtles when threatened or handled varies. Some will retreat into their shells (R. Christoffel, personal observation), while others will immediately try to bite and scratch (Vogt 1981, Ernst and Lovich 2009). They may also release the contents of two pairs of scent glands located along the shell margins, which is quite unpleasant (Ehrenfeld and Ehrenfeld 1973, Eisner et al. 1977, Vogt 1981).

CONSERVATION

Status: Globally, the Eastern Musk Turtle is listed as a Least Concern species by the IUCN. It is considered a common species in Wisconsin by the WDNR.

Distribution-wide, Eastern Musk Turtle populations appear to be stable or in some rare instances slightly increasing (Ernst and Lovich 2009).

Populations: Estimated population densities have ranged widely among studies. In Pennsylvania, Ernst (1986a) studied a small population that had a mean density of 24/ha (10/ac). Dreslik et al. (2005) reported a mean density of 2.67/ha (ca. 1/ac) in a southern Illinois pond. However, densities as high as 150–194/ha (61–79/ac) were reported from Oklahoma and Virginia (Mahmoud 1969 and Mitchell 1988, respectively). Tucker and Lamer (2004) found that the Eastern Musk Turtle was the fourth most common of seven species sampled at three Illinois ponds. Dreslik and Phillips (2005) reported that the Eastern Musk Turtle had the fourth highest population density and biomass of ten turtle species sampled in Illinois. Adults made up most of the individuals sampled in demographic studies (reviewed by Ernst and Lovich 2009). In Ontario, adults made up 94.5% of the individuals detected ($n = 314$; Edmonds and Brooks 1996), and in Pennsylvania, adults made up 70% of the individuals detected ($n = 204$; Ernst 1986a). However, newly hatched Eastern Musk Turtles are quite small and hard to find; thus, they may be underrepresented in sampled populations (Ernst and Lovich 2009). Annual survivorship estimates were high for juveniles (0.86) and adults (0.84) in Virginia (Mitchell 1988). Male-biased sex ratios are often reported (Edmonds and Brooks 1996, Smith and Iverson 2002, Conner et al. 2005, Reehl et al. 2006), although some studies have found statistically unbiased sex ratios (Ernst 1986a, Mitchell 1988).

Threats: Like many of Wisconsin's amphibians and reptiles, Eastern Musk Turtles are threatened with loss and alteration of aquatic habitats by humans (see Introduction to Conservation and Management of Wisconsin's Amphibians and Reptiles chapter). For example, modifications to streams and rivers are important considerations, as VanDeWalle (1993) reported that alteration of these habitats negatively impacted turtle communities in Iowa (which included the Eastern Musk Turtle). Given the popularity of lakeside housing, Harding (1997) suggested that shoreline development was a serious threat to Eastern Musk Turtles in the Great Lakes region. Furthermore, riprap or retaining walls to stabilize shorelines can be difficult for turtles to traverse and could restrict the movement of nesting females, indirectly impacting populations. The lacustrine habitats that Eastern Musk Turtles prefer can also experience heavy recreational use by humans that often involves motorized watercrafts, which can cause turtle injury or mortality. After finding 3 dead and 77 damaged individuals from propeller strikes and boats, Bancroft et al. (1983) concluded that propeller strikes and other injuries caused by boats were potential population threats.

Because Eastern Musk Turtles usually stay close to water, they are apparently less vulnerable to roadkill than many turtle species (van Dijk 2013). However, some individuals can be put at risk if their aquatic habitat dries up during summer and they are forced to disperse to other locations (Ernst 1986a). Stone et al. (1993) researched turtle communities in Alabama and suggested that the Eastern Musk Turtle was not as effective at colonizing new ponds as the other sympatric species studied. This could indicate that natural colonization of new sites via dispersal would be slow. Anthonysamy et al. (2014a) found that Eastern Musk Turtles used degraded habitats, which could be due to their aversion to leaving the water in order to find less disturbed locations. The potentially higher threat of nest predation in disturbed habitats by synanthropic species (e.g., Raccoons) might also be a conservation concern, given the low nesting success recorded by past studies (e.g., Ernst 1986a). A variety of other important threats exist, including the effects of capture by humans (either accidental or intentional). For example, accidental bycatch by commercial fisheries in fyke nets set below the water surface may accidentally drown this (and other) turtle species (Casper 1997e). Collection for the pet trade has been documented, and reports of potentially damaging collection practices for other *Sternotherus* species exist (Muir 1984). Although diseases of Eastern Musk Turtles are poorly studied, necrotic shell disease can have local impacts on their populations and has been reported from Virginia (Ernst et al. 1999).

Management: Conservation strategies and considerations associated with management of large open wetlands, lakes, and riverine habitats potentially apply to Eastern Musk Turtles (see Introduction to Conservation and Management of Wisconsin's Amphibians and Reptiles chapter). Harding (1997) suggested several strategies related to habitat management that are untested but could benefit this species. For example, shoreline developers and homeowners can promote growth of native aquatic vegetation in the shallow littoral zones of lacustrine habitats that provide useful thermoregulatory and foraging opportunities. Protection of nesting sites and maintenance of egress points for nesting females along shorelines could also be beneficial. Because Eastern Musk Turtles favor habitats that include sluggish water, maintenance of historical water levels and flow rates in such areas is likely important. In fact, VanDeWalle (1993) suggested that simplification of rivers and streams due to anthropogenic modification (e.g., channelization) reduces habitat diversity and negatively impacts associated turtle communities. Thus, maintenance of diverse riverine habitats that allow for variation in water flow rate and depth could benefit this species.

Although they are not a species of concern (i.e., Threatened, Endangered, or Special Concern) in Wisconsin, harvest regulations for common turtle species in the state apply to the Eastern Musk Turtle (see Introduction to Conservation and Management of Wisconsin's Amphibians and Reptiles chapter). Fyke nets and other commercial fishing nets should provide a means of escape for turtles that are inadvertently captured. In waterbodies that are inhabited by Eastern Musk Turtles, it may be appropriate to restrict motorboat use in shallow littoral areas to reduce the incidence of injuries to turtles caused by propellers, such as those found on the trolling motors of fishing boats. Additional management techniques exist that could benefit Eastern Musk Turtles, such as translocation of wildlife from disturbed to undisturbed locations. Attum et al. (2013) assessed the success of translocating Eastern Musk Turtles and found that survival was similar between translocated and resident turtles. They also reported that movement patterns were similar but that translocated individuals exhibited a high fidelity for their release location. They believed this was related to the highly aquatic nature of this species and the lack of alternative habitats available for turtle dispersal (Attum et al. 2013). Yet other studies have found that Eastern Musk Turtles exhibit a strong homing ability (see Activity section), which could nullify translocation efforts.

Smooth Softshell

Apalone mutica (LeSueur 1827)

R. Neal Paisley and Robert W. Hay

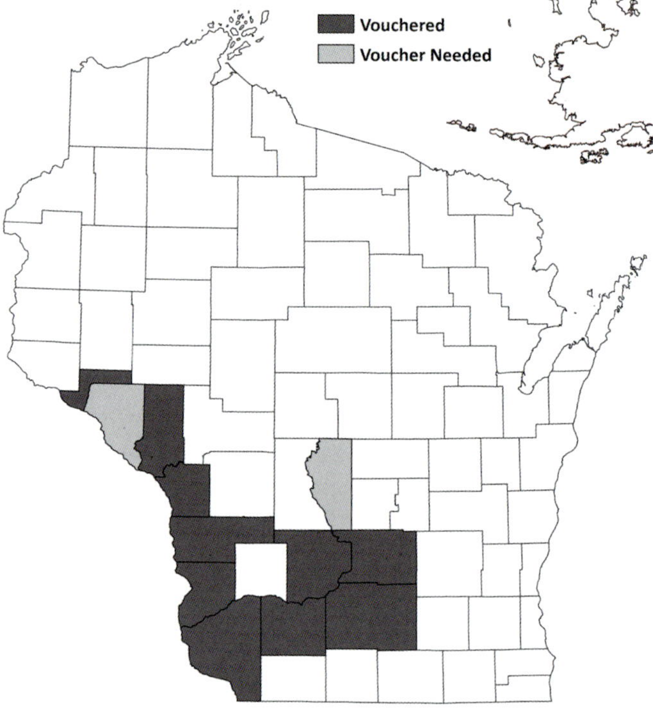

Adult female Smooth Softshell (Dane County; photo by R. Hay).

AT A GLANCE

Smooth Softshells are one of two softshell turtle species in Wisconsin. This medium-sized turtle has a flat carapace with a pancake-like appearance. The carapace is smooth and leathery, lacking the hard scutes found on most turtle species. It is a wary and swift turtle that inhabits almost exclusively large rivers with moderate to fast currents and sandy substrates, such as the Mississippi River and lower Wisconsin River. Smooth Softshells primarily use the main channel borders and side channels of these large river systems. They emerge from hibernation and breed shortly thereafter in early to mid-May. The sandy islands and sandbars that occur within riverine areas are important habitat features for this species, particularly for basking and nesting. Females move to these nesting locations from mid-June to early July. Eggs typically hatch during August or September. Smooth Softshells are largely carnivorous, focusing on aquatic insects, but they will also eat a variety of other animals, including smaller fishes, crayfish, worms, snails, and clams. They likely enter winter dormancy in mid- to late October, and they overwinter while submerged in flowing water. Compared to the more cantankerous Spiny Softshell, this species is relatively placid when captured and handled. Softshells on the Upper Mississippi River are also known as leatherbacks by commercial fishermen and turtle trappers. The species was given CITES Appendix III protection in 2016, allowing for improved regulation of international trade by requiring international exporters to have a permit. The Smooth Softshell is currently listed as a Least Concern species globally by the IUCN and considered a species of Special Concern and a Species of Greatest Conservation Need in Wisconsin by the WDNR.

SYSTEMATICS

In a phylogenetic study of Trionychidae, Meylen (1987) resurrected the generic name *Apalone* for the clade of the three North American species, including *Apalone mutica* (Smooth Softshell), which had previously been included in *Trionyx*, a name that is still used for these turtles by some authors who cite Webb (1990). Meylan (1996) pointed out that softshells in North America are both morphologically and genetically distinct from those elsewhere in the world. Crother (2017) agreed with this view in part by supporting the findings of Webb (1962). Two subspecies of *A. mutica* are currently recognized in the US, one of which is found in Wisconsin (*A. m. mutica*). Among Wisconsin turtles, this species is most closely related to *A. spinifera,* as both occur in the family Trionychidae.

DESCRIPTION

Adults: The carapace has a flat, pancake-like appearance and is flexible because it lacks the hard scutes found on most turtle species. The anterior edge of the carapace sometimes has regularly spaced furrows or wrinkles (but lacks spines or protuberances; Webb 1962). The carapace of adult females is tan or brown, with irregular dark blotches or mottling (see Smooth Softshell Cover Photo). Adult males have a brown or grayish-tan carapace, with scattered small dark brown dots or dashes (Figure 359). The plastron of both sexes is white or gray, with underlying bones partially visible through the highly vascularized skin (Ernst et al. 1994; Figure 360).

The dorsal coloration of the limbs, head, and tail is similar to the carapace (background) color. A black-bordered, peach-colored line extends through each eye and back onto the neck (see Figure 359). Their feet are heavily webbed and function well for powerful swimming. Smooth Softshells have an elongated

Figure 359. Adult male Smooth Softshell (Dane County; photo by R. Hay).

Above, top: Figure 360. Ventral side of a Smooth Softshell showing the highly vascularized skin and shell (Dane County; photo by C. Raimond).

Above, bottom: Figure 361. Hatchling male Smooth Softshell (Dane County; photo by R. Hay).

tubular snout that lacks a septal ridge (see Diagnosis subsection). They range in CL from 11.5 to 35.6 cm (4.5 to 14 in), but females are considerably larger than males (see Variation subsection; Oldfield and Moriarty 1994).

Preadult stages: Eggs are spherical, with a mean diameter of 2.3 cm (0.9 in, range = 2.0–2.4 cm [0.8–0.9 in]) and mean weight of 6.7 g (0.2 oz, range = 3.4–9.0 g [0.1–0.3 oz]; Fitch and Plummer 1975). Hatchlings and juveniles have a carapace pattern that is similar to that of adults (Figure 361) and are sexually dimorphic upon hatching, with males possessing more clearly defined carapace markings than females (Robert Hay, Turtles for Tomorrow, personal observation). Hatchlings from Columbia County measured 3.3–4.1 cm CL (1.3–1.6 in, $\bar{x}$ = 3.8 cm [1.5 in], n = 20) and 2.6–3.9 cm CW (1.0–1.5 in, $\bar{x}$ = 3.3 cm [1.3 in], n = 20).

Variation: Smooth Softshells are sexually dimorphic in color and pattern, although no significant geographic variation among Wisconsin populations is known. Females attain a larger size than males, with a maximum MCL of 35.6 cm (14.0 in), compared to 26.6 cm (10.5 in) for males (reviewed by Ernst and Lovich 2009). Males have a thick tail, with the cloacal opening beyond the posterior margin of the carapace near the tip of the tail. Females have a shorter tail, with the cloacal opening anterior to the back margin of the carapace (Ernst and Lovich 2009).

Diagnosis: Smooth Softshells are characterized by a carapace that lacks keratinized scutes but is instead covered with leathery skin and without spiny protuberances on the anterior edge. Their front feet possess three claws, and there is one peach-colored stripe on each side of the head, while the nasal septum has no horizontal protuberances. Spiny Softshells (*Apalone spinifera*) are the only other species of softshell turtle found in Wisconsin and the only species that could be confused with Smooth Softshells. Smooth Softshells lack the 'spines' (i.e., protuberances) that are found along the anterior edge of the carapace in Spiny Softshells. The carapace of Spiny Softshells also has a rough texture (sandpaper-like) compared to that of the Smooth Softshell. The carapace of the Spiny Softshell is olive green to tan, with dark brown or black markings. Males and very young juvenile females have prominent ocelli and a dark line along the carapace margin. As female Spiny Softshells grow, the ocelli and marginal line fade and are usually missing in adults, whose carapace is covered in irregular light and dark blotches that presumably aid in camouflage. Turtle eggs that are encountered in the wild are difficult to differentiate. Softshell (both Smooth and Spiny Softshells) and Snapping Turtle (*Chelydra serpentina*) eggs are round. As such, the eggs of these three species are difficult to distinguish from each other, but they can be differentiated from those of many other aquatic turtles, which have elliptical eggs.

DISTRIBUTION AND HABITAT

Global distribution: Smooth Softshells naturally occur only in the US. Their northern distributional limit extends from the Dakotas (Missouri River drainage), through Minnesota and Wisconsin (Mississippi and Wisconsin River watersheds), to Ohio (Ohio River drainage; Ernst and Lovich 2009). They are found southward through the eastern half of Texas and east into the state of Mississippi. An isolated population also occurs in eastern New Mexico (Ernst and Lovich 2009). Smooth Softshells once inhabited western Pennsylvania, where they are now likely extirpated (Ernst 1985).

Distribution in Wisconsin: Smooth Softshells occur from the Wisconsin River at Wisconsin Dells downstream to its confluence with the Mississippi River. They are found along the Mississippi River from Pepin County to the state's southern border (Knudsen 1957c, Vogt 1981). Of the published county records, vouchers could not be found for Adams or Buffalo.

Localities of possible occurrence: There are several additional counties in Wisconsin where Smooth Softshells could occur, given proximity to counties of known occurrence and the presence of suitable habitat. For example, Juneau and Richland Counties along the Wisconsin River are obvious gaps in the state distribution of this species. Their state range could extend up the Mississippi and St. Croix Rivers into Pierce County because they have been documented in the St. Croix River on the Minnesota border (Oldfield and Moriarty 1994). Smooth Softshells could also occur in the Chippewa River, although Vogt (1981) conducted extensive trapping for aquatic turtles in both the St. Croix and Chippewa Rivers and captured no Smooth Softshells. It is of further note that Smooth Softshells have been observed but not vouchered from the lower Black River in Jackson County (R. Hay, personal observation).

Habitat: In the Upper Midwest, Smooth Softshells are exclusively a riverine species that associates with large aquatic systems possessing moderate to fast-flowing water, although they are sometimes found in impoundments that possess at least minimal current (Vogt 1981, Williams and Christiansen 1981, Anderson et al. 2002, Dreslik and Phillips 2005, Tucker et al. 2008, Bluett et al. 2013). They prefer aquatic systems with sandy soils and are frequently seen basking along the shores of sandbars or sandy islands (see below). Except for nesting females, Smooth Softshells rarely leave the water and are almost always submersed or partially submersed while basking (Ernst and Lovich 2009).

Although Smooth Softshells are generally associated with rivers, past work supports that they do not distribute themselves evenly throughout riverine habitats. For example, all Smooth Softshells captured or observed during a Snapping Turtle study from 1997 to 2001 on Pool 8 of the Upper Mississippi River (La Crosse and Vernon Counties, Wisconsin, and Houston County, Minnesota) were associated with main channel habitats (Paisley et al. 2009, Neal Paisley, WDNR [retired], personal observation). Furthermore, Tucker et al. (2008) reported that they mostly used the main and side channels of the upper Mississippi River. On the middle Mississippi River adjacent to Missouri, Smooth Softshells were most abundant in open side channels and main channel borders or the area between the main channel and the nearest shoreline that did not possess wing dams (Barko and Briggler 2006). The mean water velocity associated with Smooth Softshell captures in this study was 0.2 m/sec (0.7 ft/sec).

Research from the Midwest further indicates variation in the river habitats used based on age class and sex. Although adult males used all available riverine habitats, they were most frequently found in shallow areas near sandbars. Adult females, on the other hand, used primarily deeper waters (e.g., Anderson et al. 2002), but Ewert (1979) also observed them in "seasonally flushed oxbow ponds" in Indiana. Subadult females are more prone to bask on sandbars than their adult counterparts, who mostly bask during egg gestation (Plummer 1977a). Hatchlings prefer small warmer puddles and shallow slack waters along irregular shorelines, where they rely heavily on sand substrates for burrowing (Plummer 1977a).

Smooth Softshells appear to rely heavily on sandy substrates for various purposes. They frequently bask on elevated sandy banks within meters of the water's edge or in shallow water along the shoreline (Fitch and Plummer 1975, Williams and Christiansen 1981). Sandy river substrates are further preferred because they facilitate burrowing to escape danger and overwinter. They are also used for nesting in adjacent uplands (Knudsen 1957c, Plumber and Shirer 1975; see Reproduction and Development section). In addition, hatchlings require access to sandy substrates for burrowing and because sand appears to play an important role in proper shell development (Plummer et al. 2008). Plummer (1977a) reported that Smooth Softshells specifically congregated along sandbars with ridges that possessed a sharp slope on the lee (or downstream) side, which extended out into the water. These created small "inlets of quiet shallow water . . . which are utilized heavily by turtles for aquatic basking late in the season." Plummer (1977a) collected 73, 48, and 64 individuals (with few recaptures) from such locations on three separate occasions, supporting the importance of these habitat features.

Nesting sites are primarily on sandbars and sandy islands that occur in main and side channel areas but possess little vegetation or debris (Goldsmith 1944, Riley et al. 2005, N. Paisley, personal observation). Limited information on the overwintering habitat used by Smooth Softshells in Wisconsin exists, aside from the fact that they hibernate underwater in deeper locations of rivers but where the current provides suitable oxygen levels (see Activity section).

REPRODUCTION AND DEVELOPMENT

Breeding typically occurs shortly after emergence from hibernation (Plummer and Shirer 1975) during early to mid-May in Wisconsin (Vogt 1981). Males actively search for females in the water and at basking sites, which results in various behavioral interactions depending on whether the turtles encountered were females or other males (Plummer 1977b). For example, Plummer (1977b) observed male turtles swimming along lines of partially submerged basking conspecifics in an effort to investigate their posterior regions. Although other males often ignored these approaches, unreceptive females "would spin around and aggressively charge and bite at the male." He further mentioned that "semicircular wounds from bites, many still bleeding, are common" on the peripheries of male carapaces in spring. Successful copulation occurs in the water, and because males are not known to actively grasp females during coitus, they are required to swim to maintain position (Plummer 1977b). Copulation can last >30 minutes (Ernst and Lovich 2009).

In Wisconsin, gravid females move to nesting sites from mid-June to early July (Vogt 1981, N. Paisley, personal observation). Individuals select nesting sites on sandbars or islands that are 4–90 m (13–295 ft, $\bar{x} = 30$ m [98 ft]) from the shoreline and more than 1.34 m (4.40 ft) above normal river stage (Muller 1921, Goldsmith 1944). Nesting at these elevations can increase egg survival by reducing the chances that eggs will become inundated during floods (Plummer 1976). In Kansas, the mean depth from the surface of the sand to the top of eggs in the nest cavity was 16.2 cm (6.4 in, range = 7.6–25.4 cm [3.0–10.0 in], $n = 95$; Fitch and Plummer 1975). Females often burrow near the nest site (typically within 1 m [3.3 ft] of their nest) immediately following completion of their nests and before returning to the water. This activity creates a shallow trough leading away from the nest (Plummer 1976). Nesting females are wary and may abandon a nesting attempt if disturbed (Oldfield and Moriarty 1994).

Mean clutch size was 10.4 eggs (range = 3–26 eggs, $n = 102$ nests) and 12.6 eggs ($n = 122$ nests) for two studies in Kansas (Plummer 1977c, Fitch and Plummer 1975, respectively). Clutch sizes in more northern populations range from 18 to 22 eggs (Webb 1962, Vogt 1981), although 10–11 eggs are common in Wisconsin clutches (R. Hay, personal observation). While females may lay two or more clutches annually in other midwestern locations (Plummer 1977c), it is unclear if this occurs in the Upper Midwest. Fitch and Plummer (1975) frequently found over 20 eggs in dissected gravid females from Kansas. Because this was higher than the mean clutch size in nests, they suggested that females may divide their clutches between two nests.

Riley et al. (2005) reported that the range of surface soil temperatures for actively incubating nests was 36°C–37°C (96.8°F–98.6°F) along the Mississippi River in western Illinois. According to a review by Ernst and Lovich (2009), the range of incubation temperatures that yielded the highest hatching success rates for Smooth Softshell eggs was 27°C–33°C (80.6°F–91.4°F). In addition to egg hatching success, incubation temperatures could have a substantial influence on various physical and behavioral characteristics of hatchlings, but distinguishing the influence of temperature from other factors, such as genetics, is difficult (reviewed by Ernst and Lovich 2009). The mean incubation period is 63.2 days, but there is high variability (44–82 days; Ernst and Lovich 2009), and Wisconsin hatchlings typically emerge from the nest cavity 8–12 weeks after egg deposition (i.e., they emerge in August or September; R. Hay and Thomas LaFleur, Turtles for Tomorrow, personal observations). Hatchlings do not possess a caruncle and instead use their forelimbs to tear through the eggshell (Vogt 1981). Plummer (2007) reported that hatchlings typically emerged from the nest at sunset over a short period of time. In Wisconsin, asynchronous emergence (i.e., occurring over more than one day) of Smooth Softshell hatchlings from a single nest has been observed on at least three occasions (T. LaFleur, personal communication).

Growth is rapid early in life and slows considerably with age. Hatchlings roughly double in size by the end of their first full season of growth. Plummer (1977c) found that plastron length increased by 0.2 cm (0.08 in) per month for juvenile males and 0.2–0.3 cm (0.08–0.12 in) per month for juvenile females. Although growth rates decrease at sexual maturity, mature females continue to grow at a greater rate than males (Plummer 1977c). In a Kansas population, males became sexually mature in their fourth year (PL = 8.0–8.5 cm [3.1–3.3 in]), although the highest percentage of mature males in the sample population had PL >8.5 cm (3.4 in; Plummer 1977c). Furthermore, females typically became sexually mature in their ninth year (PL = 14.0–15.0 cm [5.5–5.9 in]), although some females attained sexual maturity by their sixth year, and the highest percentage of mature females in the population had PL >15.0 cm (5.9 in; Plummer 1977c). Longevity records are not available for wild Smooth Softshell populations, although Plummer (1977c) suggested that some individuals could survive >40 years.

ACTIVITY

In Wisconsin, Smooth Softshells emerge from hibernation in early to mid-May (Vogt 1981). Activity is primarily diurnal (Vogt 1981), and the turtles spend much of their time in the water either swimming or buried at a depth that just allows them to extend their neck to the surface for air (Ernst and Lovich 2009). However, individuals will emerge from the water occasionally to bask and spend time feeding in shallow water along shorelines (typically within 1 m [3.3 ft] of the shoreline). Lindeman (2000) reported that most basking occurs from 7:00 a.m. to 5:00 p.m. in Kentucky. Smooth Softshells typically bask along sandy shorelines or sandbars that drop steeply into deep and swift water or while in shallow water partially or completely buried (Fitch and Plummer 1975, Lindeman 2001).

Aside from nesting forays, all movements occur in the water (Plummer and Shirer 1975). Work in Kansas determined that home ranges are linear in shape, and movements within home ranges corresponded closely with the contours of the shoreline (Plummer and Shirer 1975). This work further found that female home ranges tended to be larger than those of males and included both sides of the river (linear home range distance = 1.2 km [0.8 mi], range = 0.8–1.7 km [0.5–1.1 mi]), while male home ranges typically only included one side of the river (linear distance = 0.47 km [0.29 mi], range = 0.35–0.62 km [0.21–0.39 mi]). Females also roamed substantially more within their home ranges (165 m/day [541 ft/day]) than males (61 m/day [200 ft/day]) and were able to span the length of their home ranges in a shorter period of time ($\bar{x} = 7.1$ days and 8.7 days for females and males, respectively). Adults will sometimes embark on long movements that take them outside the normal boundaries of their home ranges. For example, Plummer and Shirer (1975) tracked one female that moved 4.1 km (2.6 mi) in a single day. Limited movement data exist for hatchlings, although Plummer (1977a) released 500 marked hatchlings on the Kansas River and recaptured 3 several days later on a sandbar 4 km (2.5 mi) downstream. He suggested that perhaps passive movement with river currents is important to hatchling dispersal.

Smooth Softshells overwinter while submerged and likely enter hibernation by mid- to late October (Ernst and Lovich 2009). Definitive characteristics of overwintering locations have not been reported for this species. Ultsch (2006) suggested that hibernating softshells bury themselves in areas with some current or move into deeper water where ice will not penetrate to the bottom but acceptable oxygen levels are maintained throughout the winter. This notion was supported by a diver in Wisconsin who witnessed Smooth Softshells retracting their heads into the sandy substrates of a deep channel in the St. Croix River during late autumn (Ron Benjamin, WDNR, personal communication). Softshells can extract oxygen from the water column via their buccal cavity (e.g., buccopharyngeal respiration) but also across the tissues of their limbs, cloaca, and leathery shell (Dunson 1960, Jackson et al. 1976, Stone and Iverson 1999). As a result, they can remain submerged during extended periods of hibernation (Jackson et al. 1976, Stone and Iverson 1999). Although they can absorb oxygen in these ways, some research supports that softshell turtles cannot withstand extensive periods of anoxic conditions. A related species, the Spiny Softshell, was the least tolerant of anoxia of all turtle species studied by Reese et al. (2003). Presumably, Smooth Softshells are also particularly sensitive to anoxia, given their affinity for lotic conditions that presumably possess acceptable water oxygen levels.

PREY AND PREDATORS

Smooth Softshells are generally active foragers and feed primarily on aquatic insects (Ernst and Lovich 2009). They will also remain buried in the sediment and ambush prey that comes near them (Ernst and Lovich 2009). According to Ernst and Lovich (2009), the Smooth Softshell is "decidedly insectivorous." Williams and Christiansen (1981) found that their diet consisted of 75% insects, including many that occurred in the open water column, which reflects their ability to aggressively pursue prey. Plummer and Farrar (1981) analyzed stomach contents of individuals collected in Kansas ($n = 105$) and found that females consumed more aquatic prey, particularly caddisfly larvae (e.g., Hydropsychidae), while males consumed more terrestrial prey. This observed sexual variation in diet was likely a result of habitat selection, with females spending more time foraging in deeper water and males spending more time foraging in shallow water near terrestrial-aquatic interfaces (Plummer and Shirer 1975, Plummer 1977a). Smooth Softshells also feed on a range of other aquatic animal matter, including small fishes, worms, snails, isopods, clams, crayfish, and carrion (Ernst and Lovich 2009). Small amounts of plant material are ingested, and males even reportedly consumed mulberry (*Morus* spp.) fruit and Cottonwood (*Populus deltoides*) seeds in Kansas (Plummer and Farrar 1981).

Specific reports of predation on Smooth Softshells in the Upper Midwest are sparse, and adults likely have few natural predators. Bald Eagles (*Haliaeetus leucocephalus*) have been documented to prey on both Smooth Softshells and Spiny Softshells along the lower Wisconsin River (Watermolen 2004). Most other potential predators probably cannot capture Smooth Softshells because they are incredibly wary and will swiftly dash to the water when approached. If captured, they are generally more mild-tempered than their relative in Wisconsin, the Spiny Softshell, and usually retract into their shell for protection (Vogt 1981).

Younger age classes, on the other hand, are vulnerable to predation. Nests are destroyed by American Crows (*Corvus brachyrhynchos*), Raccoons (*Procyon lotor*), Striped Skunks (*Mephitis mephitis*), Red Foxes (*Vulpes vulpes*), and Domestic Dogs (*Canis familiaris*). Hatchlings and juveniles can fall victim to large fishes, turtles, snakes, wading birds, and various mammals (reviewed by Ernst and Lovich 2009).

CONSERVATION

Status: Globally, the Smooth Softshell is listed as a Least Concern species by the IUCN. However, riverine turtles have the highest proportion of threatened species when compared with other vertebrate groups, and softshell turtles have been reported as declining worldwide (Johnson 2000b, Bodie 2001). It is considered a species of Special Concern and a Species of Greatest Conservation Need in Wisconsin by the WDNR. In response to continued exploitation for the international turtle trade, the species was given CITES Appendix III protection in 2016, allowing for improved regulation of international trade by requiring international exporters to have a permit.

Populations: There are reported discrepancies in perceptions of how abundant Smooth Softshells are in Wisconsin. Knudsen (1957c) noted that the species was abundant where suitable habitat features existed. Several decades later, Vogt (1981) reported that the species was not abundant anywhere in the state. Observations by field biologists in recent years support a view closer to Knudsen (1957c), which suggests that this turtle is common, or at least not uncommon, in the lower Wisconsin River and portions of the upper Mississippi River (N. Paisley, R. Hay, T. LaFleur, personal observations).

Aside from these anecdotal reports, however, little information on Wisconsin populations is available. Incidental capture data for the upper Mississippi River exist from which population information could be inferred. For example, the Upper Mississippi River Restoration (UMRR) Long Term Resource Monitoring (LTRM) program records turtle bycatch in its fish sampling activities. From 1995 to 2011, the program recorded 255 Smooth Softshells in its routine sampling activities in navigation Pools 4, 8, 13, and 26, as well as in an open river reach near Cape Girardeau, Missouri, and in the La Grange reach of the Illinois River. Along the Wisconsin border, 16 Smooth Softshells were recorded from Pool 4, and 14 were recorded from Pool 8. Anecdotally, it also seems that Smooth Softshells were more abundant in the 1990s than today, but trend data do not exist to validate this perception (Andy Bartels, WDNR, personal communication).

The most detailed information on Smooth Softshell populations comes from Kansas. Depending on the estimation method employed, Plummer (1977a) calculated a population size of 1,762–2,251 turtles in a 1.5 km (0.9 mi) stretch of river. This turtle was so abundant at his site that he once counted 88 individuals basking along a 100 m (328 ft) stretch of sandbar at one time. He also stated that a "large, but undetermined number of heads were visible in the adjacent waters." Several studies have further reported population characteristics beyond abundance and density, such as sex ratios, proportional sizes of different age classes, and survival rates. For example, adult sex ratio estimated in Kansas heavily favored males (i.e., 1.98M:1F; Plummer 1977a). In contrast, studies from Illinois have found female-skewed sex ratios (Anderson et al. 2002, Barko and Briggler 2006). Plummer (1977a) estimated that his study population in Kansas consisted of ca. 37% immature individuals. Iverson (1991) used data reported by Fitch and Plummer (1975) to estimate that the annual survivorship of Smooth Softshells was 43% in Kansas. Survival of nests can be highly variable but is generally low. Plummer (1976) estimated that across two years, nests failed for various reasons (e.g., predation, flooding) and at distinctly different rates: 98% nest failure in the first year and 51% failure in the second.

Threats: Anthropogenic activities that threaten or alter large riverine habitats could impact Smooth Softshells, including unnatural manipulation of river channels or water levels, degradation of water quality, and introduction of invasive species (see Introduction to Conservation and Management of Wisconsin's Amphibians and Reptiles chapter). VanDeWalle and Christiansen (1996) found that Smooth Softshells were intolerant of anthropogenic modifications to river hydrology in Iowa that included straightening of channels, dredging of sediments, and creation of levees. Smooth Softshells were also intolerant of human activities that eliminated submerged structure from river channel habitats, such as the removal of underwater debris (VanDeWalle and Christiansen 1996). Similar hydrological alterations have occurred along the Upper Mississippi and Wisconsin Rivers. They have likely impacted this species in Wisconsin, but no quantifiable data exist.

Impacts to sandbars and sandy shorelines used for nesting are a threat to all riverine turtles, including Smooth Softshells. Several authors report that Smooth Softshells need clean and level sandy banks that are free of weeds, trash, or silty sediments for nesting (Goldsmith 1944). Shorelines that do not experience natural periodic disturbance can also become overgrown with woody vegetation that shades nesting sites, which could alter the thermal properties experienced by developing eggs. The off-loading of dredge spoil on sandy islands within the Mississippi

NATURAL HISTORY BOX: Turtle Nest Rescue from a Dredge Spoil Pile in Pool 4 of the Mississippi River

A very large and long-established dredge spoil pile in the Mississippi River (Pool 4) near Weaver Dunes, Minnesota, was slated for sand removal to an inland site to make room for newly dredged sand from the river's main commercialization channel. Although the Minnesota Department of Natural Resources and WDNR have no jurisdiction in this matter, they, along with several turtle biologists, were able to obtain permission from the US Army Corps of Engineers to rescue eggs from nests prior to the start of the project. Several hundred turtle nests were moved from this site to other suitable nesting sites in Pool 4 to prevent their loss that season (Mike Pappas, independent researcher, personal communication). While the loss of those clutches alone would not likely have caused a significant problem for turtle populations within Pool 4, the cumulative impacts of dredge spoil removals within the Upper Mississippi is likely contributing to turtle population declines in the Mississippi River.

Robert W. Hay, Turtles for Tomorrow

River during the nesting season can smother nests prior to hatchling emergence (see Natural History Box: Turtle Nest Rescue from a Dredge Spoil Pile in Pool 4 of the Mississippi River). Furthermore, the lock and dam systems on the Wisconsin and Mississippi Rivers have not only restricted turtle movement but also caused water levels to rise, which has permanently inundated countless sandbars and sand islands that once provided nesting habitat for Smooth Softshells (Alexander et al. 2012, Lenhart et al. 2013).

The threats posed by high densities of nest predators are likely substantial, although they have not been quantified in the Upper Midwest. Raccoons, for example, prey heavily on the eggs of all turtle species. Goldsmith (1944) reported that one Raccoon destroyed seven nests along a 60 m (197 ft) stretch of shore over a single night in Iowa. Over two years, Dieter et al. (2014) found 180 destroyed nests of "softshell species" on four sandbars surveyed along the Missouri River in Nebraska and South Dakota. Because they are synanthropic, Raccoons can reach unnaturally high densities in landscapes with high human populations (e.g., southern Wisconsin), which can result in accordingly high nest predation rates.

The impacts of commercial fishing and associated incidental take of turtles, plus legal turtle harvest, have caused significant declines in some turtle populations (Moll and Moll 2000, Ceballos and Fitzgerald 2004, Barko and Briggler 2006, Zimmer-Shaffer et al. 2014), particularly along the Mississippi River. Turtle populations are not amenable to sustained commercial harvest because of their slow development and the generally low survivorship experienced by nests and hatchlings (Congdon et al. 1993, 1994, Galbraith et al. 1997, Heppell 1998, Zhour and Jiang 2008). Demographic studies of softshell species in Missouri revealed that no level of harvest was sustainable if turtles achieved typical population growth rates (i.e., based on average rates of survival and reproductive output; Zimmer-Shaffer et al. 2014). Unfortunately, yields due to accidental and intentional turtle harvest are sometimes substantial and could impact Smooth Softshell populations (see Introduction to Conservation and Management of Wisconsin's Amphibians and Reptiles chapter).

Management: General habitat management strategies that could benefit large flowing aquatic systems are relevant to Smooth Softshells. These can include implementation of practices that consider turtles when manipulating water flow or depth, as well as practices that reduce simplification of river corridors due to anthropogenic structures (i.e., dams), but also control of erosion, pollution, and invasive species (see Introduction to Conservation and Management of Wisconsin's Amphibians and Reptiles chapter).

Riverine management plans that promote more natural flow regimes to preserve variability in the exposure of sandbars and sandy banks could greatly benefit this and other riverine turtle species. Protection of nesting habitat from potential predators and efforts to control encroachment of woody vegetation on open nesting sites could also have positive impacts on Smooth Softshells.

Given their apparent sensitivity to overharvest (Zimmer-Shaffer et al. 2014), there is a need to reduce or eliminate the incidental bycatch and sale of the Smooth Softshell by commercial fisheries on the Mississippi River. This could require the development of more stringent regulations for harvest. It is noteworthy that the sale of any turtles collected in Wisconsin, other than those taken from Minnesota and Iowa boundary waters (Mississippi and St. Croix Rivers), is currently prohibited. Turtle mortality due to bycatch in commercial fishing operations could be reduced through the development and use of turtle excluder devices that lessen the potential for accidental capture in passive commercial fishing gear (Barko et al. 2004). Furthermore, regulations that reduce the length of time that passive fish nets can be deployed so that turtles do not drown are likely needed (Barko et al. 2004).

SPECIAL REMARKS

Smooth Softshells typically use riverine habitats that are difficult to survey. For example, surveys and research on the main channel of large river systems can require expensive watercraft, which can limit access for regular data collection due to budgetary constraints. As such, it is challenging to create practical monitoring and conservation strategies for this turtle. Implementation of long-term monitoring plans for turtles by agencies that have access to suitable resources (e.g., the US Army Corps of Engineers and USGS) could greatly improve knowledge of population dynamics for riverine turtle species in Wisconsin (Barko and Briggler 2006).

Spiny Softshell

Apalone spinifera (LeSueur 1827)

Deahn M. Donner and Donald J. Brown

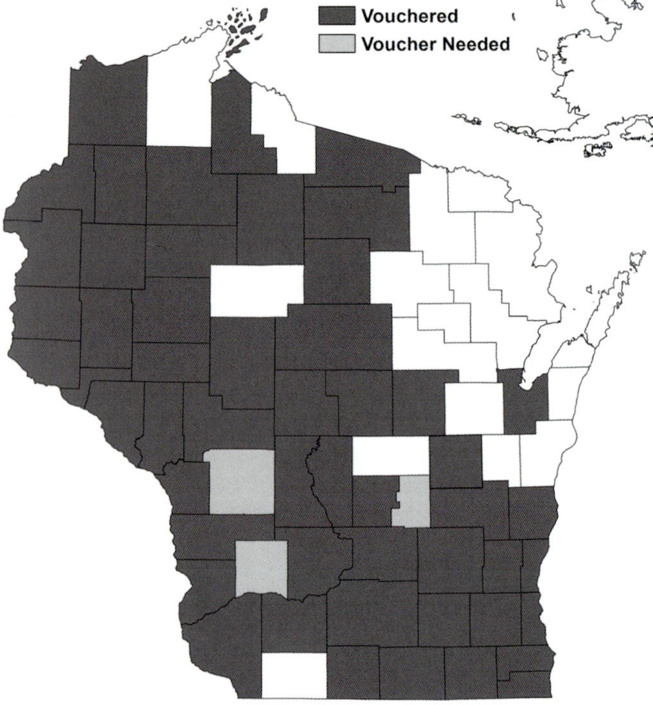

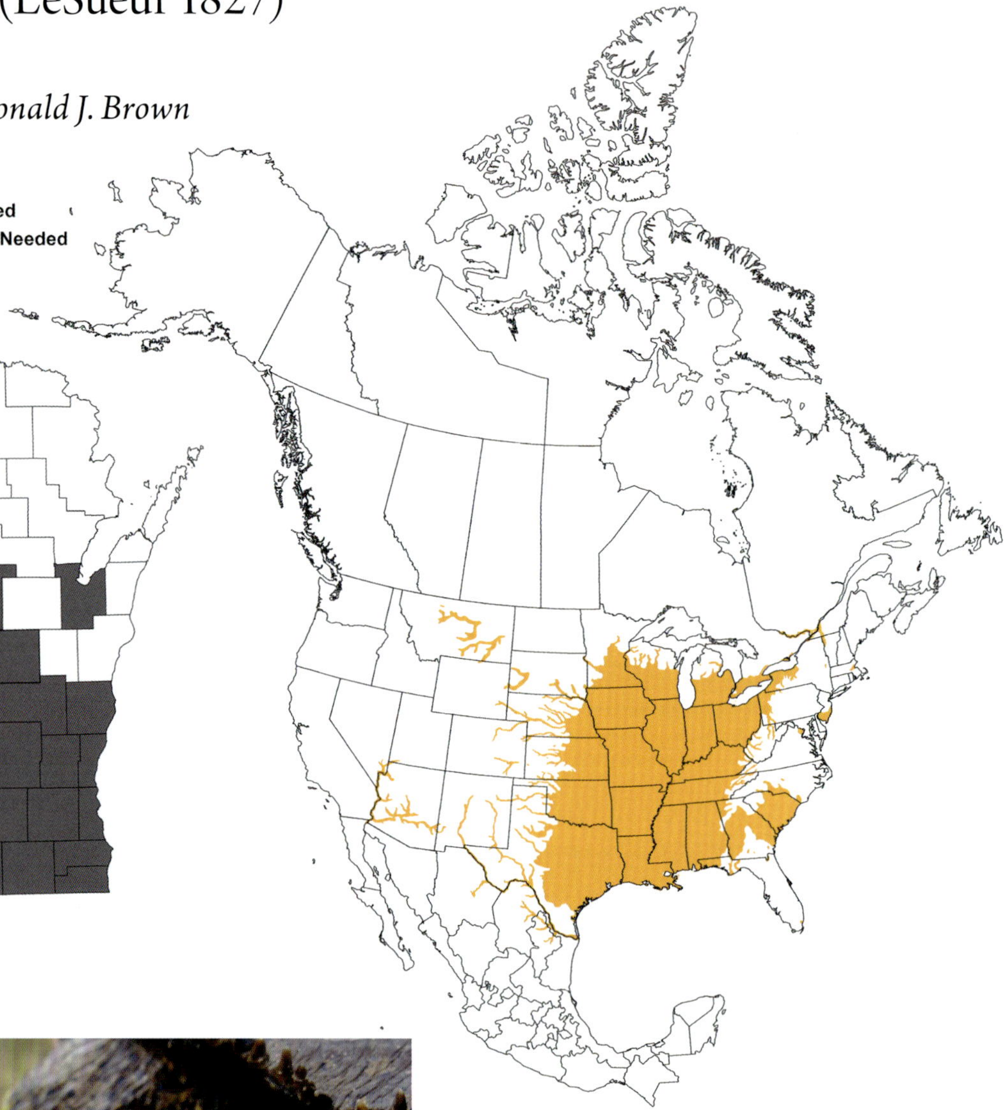

Spiny Softshell (Sauk County; photo by C. Raimond).

AT A GLANCE

The Spiny Softshell is one of two softshell turtles found in Wisconsin and is often found in the southern, western, and central portions of the state. It is a wary aquatic turtle with a narrow pointed snout, thin body, and relatively flat leathery shell. Mature females are significantly larger than males and differ in color and pattern. Spiny Softshells are active from May to October or November in Wisconsin. This species is capable of great speed both in the water and on land. It primarily eats crayfish, insects, and fishes and typically forages by crawling along the bottom of waterbodies searching for prey or by burrowing just below the surface of a soft bottom and ambushing passing prey. Spiny Softshells prefer moderate to large rivers and other large bodies of water in Wisconsin. Individuals are often found in sections of water with floating logs, which they use for basking, and with submerged trees, where they rest among the branches underwater in the evening. They move seasonally between wintering and nesting sites, sometimes traveling long distances in rivers and streams. Mating typically occurs in May, and nesting occurs in June and July. Nests are usually laid on sandy banks close to the water's edge, and hatchlings emerge from the nests in August or September. The Spiny Softshell has a reputation for aggressive behavior when handled; it will often attempt to bite and claw its captors. The species was given CITES Appendix III protection in 2016, allowing for improved regulation of international trade by requiring international exporters to have a permit. The Spiny Softshell is currently listed as a Least Concern species by the IUCN and is considered a common species by the WDNR.

SYSTEMATICS

Apalone spinifera (Spiny Softshell; Crother 2017), a member of the family Trionychidae, was historically included in the genus *Trionyx* (Webb 1962, 1990). A phylogenetic study by Meylan (1987) concluded that North American softshell turtles represented a monophyletic genus (*Apalone*), which is currently accepted (Crother 2017). There are two additional species of *Apalone*, including *A. mutica* (Smooth Softshell), which also occurs in Wisconsin, and *A. ferox* (Florida Softshell). Five subspecies of *A. spinifera* are currently recognized, with the subspecies found in Wisconsin being *A. s. spinifera* (Eastern Spiny Softshell). The remaining subspecies include *A. s. aspera* (Gulf Coast Spiny Softshell), *A. s. emoryi* (Texas Spiny Softshell), *A. s. guadalupensis* (Guadalupe Spiny Softshell), and *A. s. pallida* (Pallid Spiny Softshell). Historically, *A. s. spinifera* and *A. s. hartwegi* (Western Spiny Softshell) were considered to be distinct subspecies, with a zone of intergradation along the Mississippi River (Webb 1962). However, genetic research did not support subspecies separation of *A. s. spinifera* and *A. s. hartwegi* (McGaugh et al. 2008), and both are currently considered *A. s. spinifera* (Crother 2017).

DESCRIPTION

Adults: The carapace is flat, round, and leathery and lacks a keel. It has small conical spiny protuberances along the anterior edge (Figure 362). Females and males have distinct differences in the color and pattern of their carapace (Figure 363). The carapace of females is dark olive or tan with brown and gray mottling, and the carapace of males is olive gray with small black ocelli and is rough like sandpaper (Vogt 1981, Christoffel et al. 2002). There is a single

Figure 362. An adult Spiny Softshell female with small conical projections along the anterior edge of the carapace clearly visible. Note also the presence of numerous leeches in between these projections (Sauk County; photo by C. Raimond).

Figure 363. Carapace color and patterns typical of Spiny Softshells in Wisconsin. (a) A female displaying a mottled pattern on the carapace and neck striping (Polk County; photo by D. Donner). (b) A male displaying black ocelli and one marginal black line on the carapace (Jackson County; photo by T. Muehlfeld).

Figure 364. Plastron color and patterning typical of Spiny Softshells (Polk County; photo by D. Donner). Note the transparency of the underside.

dark line interior to the side and rear margin of the carapace (Figure 363b) that distinguishes this subspecies from the others (Ernst and Lovich 2009). The plastron is small, hingeless, and a transparent white or yellow, which shows the underlying bones (Figure 364). The neck is long when extended; it can be stretched nearly as long as the carapace in small individuals (Donald Brown, West Virginia University / USFS, personal observation). The head, neck, and limbs have heavy yellow-green streaking and black spots against an olive-green background. All four feet are webbed. Two dark-bordered greenish-yellow stripes on each side of the head and neck extend backward from the eye and from the angle of the jaw (see Spiny Softshell Cover Photo). The tail of adult females has a mottled or blotched pattern and is short; the cloaca does not extend past the edge of the carapace. The tail of adult males has black spots and lines and is longer; the cloaca is near the tip of the tail, well beyond the edge of the carapace in fully grown individuals (Ernst and Lovich 2009). The nose is tubular and has a septal ridge.

From 207 females captured along the St. Croix River in Minnesota and Wisconsin, mean MCL was 27.3 cm (10.7 in, range = 12.0–41.2 cm [4.7–16.2 in]), mean weight was 997.9 g (35.2 oz; Deahn Donner, USFS, unpublished data). From 275 males captured in the same study, mean MCL was 17.4 cm (6.9 in, range = 13.0–31.2 cm [5.1–12.3 in]), and mean weight was 226.8 g (8.0 oz; D. Donner, unpublished data). Dunson (1867) created a predictive relationship between weight and MCL in Michigan that did not appear to be affected by sex (weight [g] = 0.102 × length [cm]$^{2.95}$, $n = 34$).

Preadult stages: Eggs are smooth, white, brittle, and spherical (Figure 365). Ernst and Lovich (2009) reported that Eastern Spiny Softshell eggs range from 2.2 to 3.2 cm (0.9 to 1.3 in) in diameter ($\bar{x} = 2.7$ cm [1.1 in]) and weigh 6.7–11.9 g (0.2–0.4 oz, $\bar{x} = 10.3$ g [0.4 oz], $n = 50$). Feltz and Tamplin (2007) recorded mean egg diameter as 2.6 cm (1.0 in) from 28 eggs collected in Iowa.

Figure 365. Spiny Softshell eggs (Chippewa County; photo by R. Sajdak).

Hatchlings and juveniles resemble adults in general shape and coloration. They also display sexual dimorphism in carapace patterns and coloration. Graham and Cobb (1998) observed that carapace spots in male hatchlings were bordered by a dark ring, while carapace spots in female hatchlings "had either an indistinct or absent ring" in Vermont. Laboratory-reared hatchlings in Wisconsin showed a similar pattern, with male carapace spots being larger and more well-defined, often bordered by a dark ring (n >20; Robert Hay, Turtles for Tomorrow, unpublished data; Figure 366). Graham and Cobb (1998) also observed that the base carapace coloration was gray in male hatchlings and brownish olive in female hatchlings. Hatchlings are small, and the mean MCL of recently hatched Wisconsin specimens in the MPM collection was 4.2 cm (1.7 in, $n = 32$).

Variation: Females attain a much larger size than males, with a maximum MCL of 54.0 cm (21.3 in), compared to 21.6 cm (8.5 in) for males (reviewed by Ernst and Lovich 2009). There is also sexual dimorphism in color and patterning (see Adults and Preadult stages subsections), but no significant geographic variation among Wisconsin populations is known.

Diagnosis: Juvenile and adult Spiny Softshells are characterized by a carapace lacking keratinized scutes and covered with leathery skin that has a texture similar to sandpaper, with the presence of spiny protuberances on the anterior edge of the carapace.

Figure 366. Carapace pattern of hatchling Spiny Softshells. (a) A male showing larger and more well-defined carapace spots with dark borders (Wisconsin; photo by R. Hay). (b) A female showing smaller and less well-defined carapace spots (Wisconsin; photo by R. Hay).

Their front feet possess three claws, two yellow-green stripes occur on each side of the head, and the nasal septum has horizontal protuberances (Figure 367).

The Smooth Softshell is the only species in Wisconsin that could be confused with the Spiny Softshell, as the other turtle species in the state all have keratinized shells. Superficially, Smooth Softshells appear similar to Spiny Softshells. However, they lack spines on the anterior border of the carapace and horizontal ridges on the nose septum. Smooth Softshells also have one black-bordered peach-colored neck stripe rather than the two black-bordered yellow-green neck stripes (see Smooth Softshell, *Apalone mutica* species account). Spiny Softshell eggs are round, similiar to the eggs of Smooth Softshells and Snapping Turtles

Figure 367. The horizontal protuberances of the nasal septum of Spiny Softshells (Sauk County; photo by C. Raimond).

(*Chelydra serpentina*), making them difficult to distinguish when encountered in the wild. Yet these round eggs can be differentiated from the more elliptical eggs of most other Wisconsin turtles.

DISTRIBUTION AND HABITAT

Global distribution: The Spiny Softshell is the most widespread softshell species in North America, ranging from east-central Mexico to southern Quebec. In the US, it is primarily found throughout the eastern half of the country, with notable absences in southern Florida and most of the land base within ca. 400 km (250 mi) of the Atlantic Ocean in northeastern states (Ernst and Lovich 2009). In addition, this species currently occupies at least 12 isolated regions in the United States and southern Quebec (Ernst and Lovich 2009). Successful introductions have occurred in New Jersey and the Colorado River system in the western United States (Webb 1962).

Distribution in Wisconsin: County records indicate that the Spiny Softshell is primarily distributed throughout the southern, western, and central portions of the state. The species appears to be absent from numerous counties bordering Lakes Michigan and Superior and is particularly underreported in blocks of contiguous counties from the northeastern region of the state. Of the published county records, vouchers could not be found for Green Lake, Monroe, and Richland (see the Appendix), although Spiny Softshells are likely present in all of them.

Localities of possible occurrence: Several county record gaps exist for the Spiny Softshell in Wisconsin that could be filled with additional survey effort. Based on presence in surrounding counties, this species could occur in Calumet, Lafayette, Outagamie, Shawano, Taylor, and Waushara Counties (and possibly other bordering counties).

Habitat: The Spiny Softshell is a highly aquatic turtle that appears to prefer moderate-sized to large rivers with slow-moving backwater areas, but it will readily use large open lentic waterbodies and sometimes even smaller ponds, streams, and irrigation ditches (Lagler 1943, Knudsen 1957c, Vogt 1981; see Amphibians and Reptiles in Wisconsin Plant Communities chapter). Extensive webbing on both the forelimbs and hind limbs, a streamlined body shape, and physiological and morphological adaptations for life in well-oxygenated water make this species particularly suited for flowing aquatic systems (Pace et al. 2001; see Activity section). DonnerWright et al. (1999) investigated turtle relative abundance along habitat gradients in the St. Croix River in Minnesota and Wisconsin and found that Spiny Softshell relative abundance was positively associated with river depth and velocity but that the species was present across a wide range of habitat characteristics. Radio-tracked individuals in the Missouri River moved from the main stem to tributaries during peak river flow periods, but it was unclear if the individuals were selecting for reduced velocity or warmer temperature (Tornabene 2014).

Some studies indicate that this species avoids locations within large lotic systems that possess the strongest currents (Williams and Christiansen 1981, Barko and Briggler 2006). For example, in the upper Mississippi River near Minnesota and Wisconsin, Spiny Softshells were commonly found in slower-moving open backwater habitats adjacent to the main channel (Cochran 1977, Tucker et al. 2008). Several studies have further determined that individuals often used tributaries and closed side channels adjacent to the

Figure 368. Example of a sandy bank used commonly for nesting and basking by Spiny Softshells in Wisconsin (Polk County; photo by D. Donner).

main channels of the middle Mississippi River in Missouri and Illinois (Barko and Briggler 2006, Braun and Phelps 2016). Spiny Softshells typically nest on open sandbars or shorelines (Webb 1962; Figure 368).

Spiny Softshells are frequently found in aquatic systems with ample woody debris, including submerged or floating logs (Williams and Christiansen 1981, Ernst and Lovich 2009). Rizkalla and Swihart (2006) determined that relative abundances were positively associated with number of woody stems and wetland proximity in an agriculture-dominated landscape of Indiana. Despite an association with woody debris and vegetation, Spiny Softshells may avoid areas with dense stands of emergent plant species (Vogt 1981). They further appear to prefer waterbodies with soft bottoms composed of sand, silt, or mud, as they spend much of their time underwater buried in the sediment (Plummer and Burnley 1997, Plummer et al. 1997, Tornabene 2014). For example, in Lake Champlain (Quebec, Canada, and Vermont, United States), this species spent most of its time immobile and underwater (86.5% and 73.5% of time for males and females, respectively; Galois et al. 2002). Whether the turtles are submerged or not, microhabitat selection is likely influenced by environmental temperatures and thermoregulatory needs (Tornabene 2014). Spiny Softshells typically overwinter in the sediment of well-oxygenated waterbodies (see Activity section).

REPRODUCTION AND DEVELOPMENT

Mating in Wisconsin typically occurs in May following emergence from winter dormancy (Vogt 1981). During this period, males actively search for females in the water and at basking sites. Ernst and Lovich (2009) summarized mating behavior of Smooth Softshells and indicated that Spiny Softshell mating behavior was similar. According to Ernst and Lovich (2009), to gauge whether or not a female is receptive to mating, a male will probe under the edge of her carapace with his snout. If the female is not receptive, she may turn and bite the male. In water, receptive females remain passive as males mount them. On land, receptive females may move to the water followed by several males or may chase after a single male to initiate mating. Copulation occurs at the bottom of the waterbody and typically lasts ≥30 minutes.

According to Vogt (1981), nesting occurs in June and July in Wisconsin, consistent with observations in Quebec (Daigle et al. 2002, Galois et al. 2002, Lazure et al. 2019) and Tennessee (Robinson

Figure 369. An example of the flask-shaped nest with eggs deposited by a female Spiny Softshell (Chippewa County; photo by R. Sajdak).

and Murphy 1978). According to Ernst and Lovich (2009), nesting activity can extend to August in some regions. Individuals rarely make long-distance movements over land for any purpose, including egg deposition (Plummer et al. 1997). As such, nests are usually located within ca. 3 m (10 ft) of water but can be as far as 100 m (328 ft) and placed in sandy or (less frequently) clay-loam soil (Vogt 1981, Daigle et al. 2002, Moler 2006, Bolton and Brooks 2010). Females dig nests in late afternoon or early evening using their hind feet, and they often void their bladder water to help soften the soil for excavation (reviewed by Ernst and Lovich 2009). In Quebec, a female completed the nesting process in less than an hour (Daigle et al. 2002). Females deposit 3–39 eggs ($\bar{x} = 17.9$, $n = 82$) in a flask-shaped nest that is 10–25 cm (3.9–9.8 in) deep, with a 2.5–7.5 cm (1.0–3.0 in) neck, and an egg chamber 7.5–12.5 cm (3.0–4.9 in) wide (Figure 369; Ernst and Lovich 2009). According to Ernst and Lovich (2009), Spiny Softshells commonly lay two and sometimes three clutches each year, but to our knowledge females laying three clutches per year has not been documented in Wisconsin.

Hatchlings typically emerge in August or September after an incubation period of 55–85 days (Vogt 1981). The latest reported emergence date in Wisconsin occurred on 25 November 2016, when ≥12 hatchlings emerged from a single nest in Washburn County (Damian Vraniak, private landowner, personal observation). Sex is determined genetically (Vogt and Bull 1982, Greenbaum and Carr 2001), and hatchling sex ratio was ca. 1M:1F in a laboratory investigation using eggs collected in Vernon County (Vogt and Bull 1982). Breckenridge (1955) estimated size-based growth rates for a Spiny Softshell population in Minnesota. For both males and females, annual MCL growth decreased as MCL increased. For females, annual MCL growth was ca. 3.5 cm (1.4 in) when MCL was ca. 8 cm (3 in) and decreased to ca. 0.7 cm (0.3 in) when MCL was ca. 28 cm (11 in). For males, annual MCL growth was ca. 1.5 cm (0.6 in) when MCL was ca. 13 cm (5 in) and decreased to ca. 0.25 cm (0.1 in) when MCL was ca. 18 cm (7 in). Plummer and Mills (2015) also found that hatchling and young juvenile growth was rapid but slowed as body size increased, with male growth rates slowing at a younger age than female growth rates. Females and males typically reach sexual maturity at a PL of 18–20 cm (7.1–7.9 in) and 8–10 cm (3.1–3.9 in), respectively (reviewed by Ernst and Lovich 2009). In Arkansas, females matured at 12–13 years of age, while males matured at 4–5 years of age (Plummer and Mills 2015). Longevity in the wild is unknown, but a female Spiny Softshell lived to be 25 years old at the Racine Zoo (Snider and Bowler 1992).

ACTIVITY

In Wisconsin, Spiny Softshells become active in May (Vogt 1981). According to Ernst and Lovich (2009), adults are active earlier in spring than juveniles. In Minnesota, activity was highest from June to August, with moderate activity in May and September and very low activity in April and October (Schneider et al. 2019). Diel activity is primarily diurnal, although nocturnal activity has been reported (Lagler 1954, Wade and Gifford 1964, Schneider et al. 2019). Galois et al. (2002) observed activity patterns in Lake Champlain (Quebec, Canada, and Vermont, United States) between April and November between 1996

Figure 370. Female Spiny Softshell basking near the shoreline (Burnett County; photo by E. Wild).

and 1999. They found that while active, individuals frequently remained submerged, and ca. 6% and 13% of male and female time was spent basking out of the water, respectively. Basking primarily occurs in the afternoon, and locations with preferred thermal properties are targeted for this behavior (Smith and Iverson 2004; Figure 370). Galois et al. (2002) observed this species basking primarily on shorelines (56%), followed by logs (20%), rocks (14%), and floating vegetation (10%). The body temperature of Spiny Softshells in Arkansas ranged from 19.9°C to 29.3°C (67.8°F to 84.7°F), depending on turtle behavior (e.g., buried in sediments, swimming, or basking; Plummer et al. 2005). Graham and Graham (1991) found that water temperature near Spiny Softshell burial locations was up to 30.5°C (86.9°F), while mean water temperature at other sample locations was 22°C (71.6°F). Feltz and Tamplin (2007) performed a thermal-selection lab experiment with juvenile Spiny Softshells in Iowa. They found that individuals preferred the warmest temperature available (30°C [86°F], gradient = 15°C–30°C [59°F–86°F]).

In Lake Champlain (Vermont and Quebec, Canada), annual home range was significantly larger for females (3,206 ha [7,922 ac], $n = 11$) than for males (275 ha [680 ac], $n = 4$; Galois et al. 2002). In a stream system in Arkansas, mean linear home ranges for females and males were 1.8 km (1.1 mi, $n = 9$) and 1.4 km (0.9 mi, $n = 7$), respectively (Plummer et al. 1997). In Montana, mean linear home range size did not differ between females and males, but female movement rates were greater than male movement rates (Tornabene et al. 2017). Homing behavior was displayed by five individuals from Michigan after being released 25–45 km (16–28 mi) from their point of capture (Otten and VanDeWalle 2014). Spiny Softshells sometimes undertake long bouts of movement seasonally during migration to nesting sites or overwintering locations. Daigle et al. (2002) documented a female in the Lake Champlain region that moved 7 km (4.3 mi) upstream in three days to nest. Daily activity patterns recorded during this movement included basking and short-distance movements in the morning, long-distance movements in the afternoon, and short-distance movements during the evening. These same activity patterns were exhibited by three females on the Lamoille River in Vermont, but much smaller distances were traveled to reach their overwintering site (3–4 km [1.9–2.5 mi]; Graham and Graham 1997). In a small stream in Arkansas, ten individuals of both sexes made long-distance movements of >0.90 km/day (0.56 mi/day) and returned within a few days to the original point of capture (Plummer et al. 1997).

Galois et al. (2002) found that Spiny Softshells entered hibernation beginning in mid-September in Vermont. In contrast, individuals in Arkansas were active until November (Plummer and Burnley 1997). Spiny Softshells will move to well-oxygenated water to hibernate (Galois et al. 2002, Ultsch 2006), as they are intolerant of anoxic conditions (Reese et al. 2003). However, when they are in water that is adequately aerated, they can remain completely submerged and rely on uptake of oxygen across the skin for extended periods of time (Jackson et al. 2001). This is facilitated in part by the nature of their 'soft' shells, which allows for O_2/CO_2 exchange with their surroundings and creates a large surface area across which this exchange can occur (Stone et al. 1992, Stone and Iverson 1999). Spiny Softshells are also capable of cloacal respiration via water moving over the cloacal tissues, as well as buccopharyngeal respiration, during which they suck water into their mouth and push it out through the nostrils while submerged (Dunson 1960). As the water passes over the delicate cloacal or oral tissues, oxygen is extracted into the blood, allowing Spiny Softshells to remain underwater for extended periods if aquatic oxygen levels are adequate (Dunson 1960, reviewed by Ultsch 2006). During hibernation, Spiny Softshells bury themselves under the water in sediment, most often mud (reviewed by Ultsch 2006). In Arkansas, individuals buried themselves an average of 4.8 cm (1.9 in, $n = 14$; Plummer and Burnley 1997). Tornabene et al. (2017) reported that hibernation locations differed by a mean of 1.70 km (1.06 mi, range = <0.05–25.90 km [0.03–16.09 mi], $n = 41$) between years in Montana, with 46% of individuals hibernating within 0.05 km (0.03 mi).

PREY AND PREDATORS

Spiny Softshells are diurnal foragers that actively search for prey using visual cues, with olfactory cues used secondarily (Webb 1962). Direct observations showed that individuals crawl or swim along the bottom and use their snout to explore under stones or masses of aquatic vegetation, usually swallowing their prey whole rather than tearing off pieces (Webb 1962, Herrel et al. 2002). Spiny Softshells also reportedly burrow just below the surface of soft aquatic sediments in shallow water and wait to ambush passing prey (Breckenridge 1944, Webb 1962). Spiny Softshells are primarily carnivorous but are considered generalists because of the wide variety of prey consumed (Cochran and McConville 1983). Plant material is often incidentally ingested (Williams and Christiansen 1981). Most diet studies for this species have been conducted in riverine systems, primarily the upper Mississippi River and several rivers in Michigan. These studies found that Spiny Softshell diets consisted largely of crayfish, aquatic insect nymphs and larvae, and fishes (Lagler 1943, Breckenridge 1944, Cochran 1977). However, individuals also prey on earthworms, snails, clams, amphibians, carrion, and small snakes (Cahn 1937, Anderson 1965, Williams and Christiansen 1981, Jensen 1998). Cochran and McConville (1983) reported that male stomachs contained mostly dragonfly naiads, while female stomachs contained mostly fish remains, with no differences between sexes for crayfish consumption, based on 28 turtles collected from backwaters of the upper Mississippi River in Minnesota. Knudsen (1957c) observed a Spiny Softshell on land consuming a young Killdeer (*Charadrius vociferus*) it had just captured.

Spiny Softshells are likely most vulnerable to predation when in the egg and juvenile stages. In the lower Wisconsin River, Raccoons (*Procyon lotor*) were the primary predators at nesting sites (92%, $n = 561$), followed by Striped Skunks (*Mephitis mephitis*), Virginia Opossums (*Didelphis virginiana*), and Coyotes (*Canis latrans*; Geller 2012c). Hatchlings and juveniles are also eaten by large fishes, other turtles, snakes, wading birds, and mammals (Webb 1962). As adults, this species' best defense could be its ability to escape predators. Spiny Softshells not only are rapid swimmers but also can move extremely quickly on land (D. Brown, personal observation). In addition, Spiny Softshells can quickly bury themselves in aquatic substrates by plunging headfirst into soft sediments and 'shimmying' back and forth so sand or muck settles on top of the carapace (Graham and Graham 1991). Spiny Softshells are aggressive when captured and readily strike with their long neck and sharp beak (Natural History Box: The "Pugnacious" Turtle). Despite their speed and aggressiveness, predation of adults does occur. Watermolen (2004) documented 36 softshell turtle shells in a single Bald Eagle (*Haliaeetus leucocephalus*) nest in southeastern Wisconsin. American Alligators (*Alligator mississippiensis*) could be important predators of adults in the southern portion of their distribution (Ernst and Lovich 2009; also see Aresco 2009).

CONSERVATION

Status: Globally, the Spiny Softshell is listed as a Least Concern species by the IUCN. It is considered a common species in Wisconsin by the WDNR. In response to continued exploitation for the international turtle trade, the species was given CITES Appendix III protection in 2016, allowing for improved regulation of international trade by requiring international exporters to have a permit.

Populations: No studies exist that have rigorously investigated Spiny Softshell population dynamics in Wisconsin. Furthermore, throughout the Spiny Softshell's range, reported estimates of basic population parameters, such as density, are rare. Dreslik et al. (2005) reported a mean density of 0.39/ha (0.96/ac)

NATURAL HISTORY BOX: The "Pugnacious" Turtle

"Spiny softshells are very pugnacious, readily offering to inflict injury with their sharp beaks. Their agility and long necks pose a definite problem in holding one safely. I have found a firm hold between the anterior portion of the hind leg sockets: by inserting thumb in one side and forefinger in the other you can restrain a very large turtle and keep out of reach of the snapping beak. If not careful, however, you are left at the mercy of the thrashing claws. The only serious bite I have received from a turtle came from a large female spiny softshell. I had climbed inside the hoops of a large fyke net and grasped the turtle in the manner described. After handing it to my assistants at the opening of the net, I bent down again to readjust the bait. Simultaneously, I felt a sharp pain in my lower leg and heard my assistants saying, 'We dropped the turtle!' I removed the turtle from my leg and again dispensed it to my assistants, but this time I didn't turn around until the turtle was out of the net!"

Excerpt from Vogt (1981)

over six years of surveys in a 30 ha (74 ac) lentic system in Illinois, although this estimate was based on few captures. Some information can be extrapolated from studies that reported Spiny Softshell captures relative to other turtle species. For example, this species was among the most frequently captured species from 1992 to 1995 along the Mississippi River near Minnesota and Wisconsin (>300 captures of >1,900 total turtle captures across seven species; Tucker et al. 2008). Spiny Softshells were also the most frequently captured species from 1994 to 1995 along the St. Croix River in Minnesota and Wisconsin (47% of 1,369 total captures across seven species; DonnerWright et al. 1999). Statistically unbiased adult sex ratios have been reported in many studies across the species' distribution (reviewed by Ernst and Lovich 2009). However, DonnerWright et al. (1999) found adult sex ratios at 12 sites in the St. Croix River in Minnesota and Wisconsin varied from 0.36M:1F to 4.05M:1F (total $n = 448$, site-level $n = 13–96$), and 8% of captures in the study were juveniles. In addition, Barko and Briggler (2006) found that sex ratios skewed toward females (1M:2F, $n = 55$) in the middle Mississippi River in Illinois. Survival rates reportedly increased as individuals matured in an Arkansas population, with an average adult and juvenile annual survival of 0.84 and 0.72, respectively (Plummer et al. 2008).

Threats: Many general threats to lotic aquatic systems likely apply to Spiny Softshells, such as habitat loss, pollution, eutrophication, and proliferation of invasive species (see Introduction to Conservation and Management of Wisconsin's Amphibians and Reptiles chapter). River alterations due to dams, pollution, or a change in water flow regimes that impede or restrict movement can result in habitat fragmentation for Spiny Softshells (Shively and Jackson 1985, Dodd 1990, DonnerWright et al. 1999). Control of water levels in lotic aquatic systems that limits the availability of open sandbars or open sandy beaches required for thermoregulation and nesting likely impacts this species. The influence of aquatic pollution could also have serious but largely unknown effects on Spiny Softshell populations. Intolerance of anoxic conditions, for example, makes Spiny Softshells inherently vulnerable to aquatic eutrophication (Jackson et al. 2000, Ultsch 2006). Van Dyke et al. (2013) also found higher concentrations of bioaccumulated heavy metals, such as selenium, in Spiny Softshells than in other turtle species at disturbed sites in Tennessee. Impacts due to human recreation in aquatic systems are also documented, as Galois and Ouellet (2007) reported injuries and mortality due to boat propellers and fishing equipment in Lake Champlain.

Egg predators such as Raccoons can be problematic in areas where their population densities are particularly high and lead to excessive predation of turtle nests. Dieter et al. (2014) recorded 180 "softshell species" nests that were destroyed by predators along the Missouri River over two nesting seasons, which was greater than the nest mortality observed for other turtles. Bolton and Brooks (2010) found that in Ontario, Canada, the presence of the Common Reed (*Phragmites australis*), an exotic invasive species also found throughout Wisconsin, negatively impacted the hatching success of Spiny Softshells. The rapid growth of Common Reed shaded out some of the nests, and the root system grew through the nests, causing nests to either fail or produce fewer hatchlings.

The full impact of accidental or intentional human harvest of Spiny Softshells is largely unknown in Wisconsin. However, Barko et al. (2004) summarized incidental turtle captures along the Mississippi River adjacent to southern Illinois as part of fish surveys for the Upper Mississippi River Restoration (UMRR) Long Term Resource Monitoring (LTRM) program. During this study, 36% of captured Spiny Softshells accidentally died in survey nets from 1996 to 2001. If similar mortality rates occur in commercial fishing operations, the overall impacts to Spiny Softshell populations could be substantial, but further investigation is necessary. Zimmer-Shaffer et al. (2014) performed a demographic modeling study for softshell (i.e., Spiny Softshell and Smooth Softshell) harvest in Missouri and found that no field-estimated harvest levels were sustainable using mean demographic values from the literature. Furthermore, estimated proportion of individuals taken at three sites through a mock harvest were 0.062, 0.219, and 0.792, respectively (Shaffer et al. 2017). The life-history characteristics of turtles (naturally low recruitment but high adult survival and delayed sexual maturity) make them difficult to manage as a game species, and probable harvest impacts on populations have been found in several studies (Warwick et al. 1990, Close and Seigel 1997, Gamble and Simons 2004, Brown et al. 2011a).

Management: Many general concepts associated with management of flowing aquatic ecosystems likely apply to this species, such as control of erosion, pollution, and problematic native or invasive species (see Introduction to Conservation and Management of Wisconsin's Amphibians and Reptiles chapter). Spiny Softshells appear to be fairly tolerant of structural modifications to their aquatic and surrounding terrestrial habitats (VanDeWalle 1993, Rizkalla and Swihart 2006, Plummer and Mills 2008). However, maintaining aquatic connectivity among different landscape features used by Spiny Softshells (e.g., hibernacula and nesting sites) is important, given that they rarely use terrestrial habitat as movement corridors (Plummer et al. 1997, Galois et al. 2002). Adopting aquatic resource management strategies for large rivers that encourage variable flow rates and variation in depths at appropriate times of year (to maintain open sandbars for thermoregulation and nesting) is likely critical to the persistence of many riverine turtle populations (see Introduction to Conservation and Management of Wisconsin's Amphibians and Reptiles chapter). In addition, although Spiny Softshells can inhabit eutrophic environments during their active months, well-oxygenated areas must be accessible for winter hibernation (reviewed by Ultsch 2006).

Habitat management through control of invasive aquatic plants could be beneficial for Spiny Softshell conservation but difficult to achieve. Often, aquatic plants are controlled through application of herbicides, the effects of which are largely unknown for reptiles. Laboratory studies have found that herbicides (i.e., diquat and potassium endothall) used to control invasive aquatic plants such as Eurasian watermilfoil (*Myriophyllum spicatum*) do not have observable toxic effects on Spiny Softshells after short-term exposure (Paul and Simonin 2007). Despite this result, further research is necessary to assess possible sublethal impacts due to long-term exposure to biocides before these chemicals can be deemed safe for Spiny Softshells.

Possible impacts of accidental and intentional harvest can be mitigated through adherence to state regulations on such activities and use of commercial fishing equipment that excludes turtles (e.g., Barko et al. 2004). It is currently legal to harvest Spiny Softshells in Wisconsin between 15 July and 30 November (WDNR 2015a). The current harvest regulations include a possession limit of five individuals on the Mississippi River and three individuals in the remaining waterbodies of Wisconsin. It is illegal to sell any live native species of turtle in Wisconsin (see Introduction to Conservation and Management of Wisconsin's Amphibians and Reptiles chapter).

Hypothetical and Enigmatic Species of Wisconsin

Joshua M. Kapfer

The early published lists of Wisconsin's amphibians and reptiles are interesting to review due to their historical significance (Hoy 1883, Higley 1889).[††] These lists are also intriguing because they include numerous species whose presence in Wisconsin is not currently supported. Although many of these species have not been reported since the previously cited works by Hoy and Higley, their presence was perpetuated by later authors such as Pope (1928, 1929, 1930, 1931) and Dickinson (1949, 1965) without the addition of new information or observations. Later, Vogt (1981) took measures that the authors before him had not and reviewed all enigmatic species in the state. His work identified which species he believed should be considered hypothetical (i.e., possibly present but unconfirmed) and which should be removed from the state species list. However, Vogt's publication is ca. 40 years old, and reports of several enigmatic species listed below continue to trickle in (albeit infrequently). Therefore, it is valuable to reconsider these hypothetical species and unusual accounts and to clearly define Wisconsin's herpetofaunal species list vs. its hypothetical residents.

Although some would argue that past unusual reports should be dismissed outright, Dickinson (1949) eloquently expressed the danger of doing so when he stated, "These older records [are based] on the assumption that the clearing of lands is probably responsible for the scarcity or absence of former native species, and not, uniformly, carelessness or lack of ability on the part of the early workers." Dickinson's point is well-taken. Therefore, the hypothetical and enigmatic species listed as part of the state's herpetofauna at one time or that currently exist in close proximity to Wisconsin's borders (but are not currently confirmed in the state) are addressed here. The species in this chapter were compiled by review of existing literature and through the investigation of museum databases, collections,

[††] It is noteworthy that a review of correspondences between George Wagner (UW–Madison) and Karl P. Schmidt (Field Museum of Natural History) on 2 April 1941 uncovered the following statement by Wagner about W. K. Higley: "Having known Higley personally, and having worked close to him for a year and a half, I am not inclined to take his paper very seriously" (copies of correspondence provided by Armand Esai, Field Museum of Natural History).

and anecdotal reports. New observations and information on several of these species are included. Attempts were made to inspect as many of the controversial specimens in museum collections as possible either in person or via photographs of specimens. Some of these species could have once occurred in Wisconsin and might still be present, given the proximity of known populations in adjacent states. However, it is unlikely that they are currently established in the state; thus, these species are considered either hypothetical or not part of the state's current herpetofauna.

CLASS AMPHIBIA—ORDER ANURA

Great Plains Toad (*Anaxyrus cognatus*)

No known reports of this western species exist from Wisconsin, but it is documented as far east as south-central Minnesota (Nicollet and Blue Earth Counties; Moriarty and Hall 2014). Due to proximity, it is feasible that Great Plains Toads could reach Wisconsin along major watersheds such as the Minnesota and Mississippi Rivers. They typically associate with open flat grasslands and floodplains, as well as agricultural fields (Moriarty and Hall 2014). Although such habitats exist in western Wisconsin, there are also ample deep river valleys and bluffs, which this species does not prefer, so it is currently considered only an unlikely but hypothetical state resident.

Fowler's Toad (*Anaxyrus fowleri*)

Conger (1921) included this species based on a specimen taken from La Pointe, Madeline Island (Ashland County, UWZM 2501). He stated that this was "identified by an expert so there is little doubt that this species occurs in the state." Karl P. Schmidt (past curator of the Field Museum of Natural History) later confirmed that this specimen was actually an American Toad (*Anaxyrus americanus*; Edgren 1944), which was also later reconfirmed by Vogt (1981). There was a specimen in the collection at the Monte L. Bean Life Science Museum at Brigham Young University purportedly taken from Bayfield County identified as "*Bufo woodhousei fowleri*" (BYUH 40549), a synonym of *A. fowleri*. Presumably, this was also a misidentified American Toad, but the specimen is currently lost, and further review is not possible (Jack Sites, Brigham Young University, personal communication). Pope and Dickinson (1928) listed this species as "doubtful" in Wisconsin. Suzuki (1951) indicated that because Fowler's Toads were found along the Lake Michigan sand dunes in Mason County (Michigan), they may exist in similar habitat along the lake in eastern Wisconsin. Vogt (1981) stated that if Fowler's Toads were present, they would likely be found in the extreme southeastern corner of Wisconsin but that he had "diligently scanned Chiwaukee Prairie and other areas in Kenosha County without finding this species." It is possible that Fowler's Toad records just south of the eastern Wisconsin-Illinois border (Lake and Cook Counties) lend credence to its potential presence in southeastern Wisconsin. However, according to Phillips et al. (1999), the Lake County record is old (prior to 1980). Anton (1999) believed that if this species previously existed in Cook County, it is now extirpated. Interestingly, Fowler's Toads were recently identified in Wisconsin, but on the opposite side of the state. Calling males were recorded along the Mississippi River in La Crosse County in 2018 (MPM VZP905; Badje and Dyke 2018). Yet it is currently unknown if this represents a natural occurrence of a previously undetected population, is the result of accidental human translocation, or is a recent northward range expansion of a more southern species due to climate change. The nearest but still quite removed location of occurrence to the La Crosse County record is likely Mercer County in western Illinois, but this is based on a specimen collected in 1948 (Christopher Phillips, Illinois Natural History Survey, personal communication). Further survey effort and investigations into the origin of these individuals from La Crosse County are warranted.

Plains Leopard Frog (*Lithobates blairi*)

There are no known reports or specimens of this species in Wisconsin. The closest confirmed records are likely those from the northern third of Illinois (e.g., Rock Island, Henry, De Kalb, La Salle, and Will Counties; INHS 2014), and this frog could conceivably enter extreme southern Wisconsin due to proximity. Plains Leopard Frogs prefer open prairie wetlands in Illinois (Phillips et al. 1999), which exist in southern Wisconsin but have declined in availability since European settlement. Regardless, the Plains Leopard Frog is considered only a hypothetical inhabitant of the state.

Upland Chorus Frog (*Pseudacris feriarum*)

Conger (1921) reportedly included this species based on eight specimens in the UWZM collection taken from Adams County but did not report specimen numbers. The only chorus frogs found in the UWZM from Adams County were eight individuals listed only as "*Pseudacris* sp." (UWZM 3821, 3856–3862), which were collected in 1918 and are probably those referenced by Conger (1921). It is likely that these specimens are actually Boreal Chorus Frogs (*Pseudacris maculata*), which were simply misidentified because the upper midwestern distributions of chorus frogs were not well-understood in the early 1900s. In support of this notion, Conger (1921) did not include entries for other chorus frog species in the state but did mention that Upland Chorus Frogs were easily confused with Western Chorus Frogs (*P. triseriata*), which until recently were considered Wisconsin's native chorus frog species. Note that more current phylogenetic work supports that the state's native chorus frog is actually the Boreal Chorus Frog. Regardless, Conger's publication was the impetus behind Pope and Dickinson's (1928) inclusion of Upland Chorus Frogs as a Wisconsin species. Vogt (1981) later agreed that this was based on the misidentification of what are currently considered Boreal Chorus Frogs. Upland Chorus Frogs are typically found in the southern US, no farther north than the southern tip of Illinois. Given that no recent evidence supports the presence of this species in Wisconsin, it is unlikely that Upland Chorus Frogs naturally occur in the state.

CLASS AMPHIBIA—ORDER CAUDATA

Jefferson Salamander (*Ambystoma jeffersonianum*)

Hoy (1883) and Higley (1889) both included this species, and there is mention of Wisconsin specimens by Conger (1921), Pope and Dickinson (1928), Suzuki (1951), and Dickinson (1965, 1972). As late as 2016, a handful of old specimens from Wisconsin in various institutional collections still existed that were (almost certainly incorrectly) labeled as *Ambystoma jeffersonianum* (Joshua Kapfer, UW–Whitewater, personal observation). This species generally has a more southern and eastern US distribution, and no known populations occur immediately adjacent to Wisconsin (reviewed by Brodman 2005a). It is also easily confused with the Blue-spotted Salamander (*A. laterale*), which is currently common in Wisconsin but the range of which was not well understood until the 1960s (reviewed by Vogt 1981). Furthermore, the existence of unisexual hybrid *Ambystoma* species populations was not well-known until long after the work of Hoy (1883) and Higley (1889, reviewed by Phillips and Mui 2005). These unisexual hybrids can potentially appear similar to Jefferson's Salamanders, further complicating matters. Thus, without additional evidence, it is likely that references to *A. jeffersonianum* in Wisconsin (either in past publications or on existing specimen labels) are due to misidentification of the Blue-spotted Salamander or unisexual hybrid *Ambystoma* species.

Marbled Salamander (*Ambystoma opacum*)

Hoy (1883) reported this species in the Racine area, and Higley (1889) cited Hoy as the authority. Pope and Dickinson (1928) and Cahn (1929) also included it. The latter stated, "Two specimens both from under rotten logs are all I have seen; one from near Oconomowoc Lake, the other near Dutchman's Lake" (Waukesha County), but none were collected. A specimen that was reportedly collected in "southeastern Wisconsin" exists in the UWZM with no associated date of collection (UWZM 19020; Figure 371). Another specimen was housed in the Smithsonian National Museum of Natural History (USNM 3964) with the collection locality of "Racine," but it has since been destroyed, according to museum records. The presence of this species in Wisconsin is problematic and has been discussed by several past herpetologists in the state as attributable to a mistake by Philo R. Hoy. Hoy (1883) included several species as part of the state's herpetofauna that are now usually associated with the southern or eastern US. Unfortunately, there are no existing specimens to support many of his claims. The Marbled Salamander is an exception, with at least one existing specimen reportedly collected from Wisconsin (the previously referenced UWZM 19020; Figure 371) and at least one more specimen that apparently no longer exists but without additional documentation since Hoy's time. Edgren (1944) believed that many of the eastern species that Hoy considered part of Wisconsin's herpetofauna were erroneously included. He put forth the following explanation for the associated specimens with these unusual reports: "It appears evident that Mr. Hoy brought his personal collection from Ohio with him to Racine; and these Ohio specimens found their way into the United States National Museum after Mr. Hoy's death, with no data except 'Racine.'" Vogt (1981) later revisited Edgren's explanation and considered it reasonable.

Figure 371. A Marbled Salamander (*Ambystoma opacum*) reportedly collected in "Southeastern Wisconsin" (UWZM 19020). No information exists in museum records that indicates who the collector was or the date on which it was collected. Given that P. R. Hoy had extensive collections from southern Wisconsin and reported this species from Racine in the mid- to late 1800s, this specimen could be attributable to his efforts (photo by J. Kapfer).

Yet it is noteworthy that Hoy's own published work (Hoy 1883) lists the Marbled Salamander as "Found at Racine." If he had published a report of Marbled Salamanders in Racine, it seems probable that he would have already accurately labeled the origin of the specimens in his personal collection. That he would have inaccurately assigned the collection location as his city of residence (Racine) or left the specimens unlabeled to be later assigned an incorrect location by someone else seems inconsistent with past reviews of his character. Reports indicate that Hoy was a thorough and well-respected naturalist in his day (see Hoy entry in Some Prominent Figures in Wisconsin Herpetology chapter). Furthermore, the fauna of Racine was likely much different in Hoy's time from what it is today, as evidenced by his collection and observations of numerous bird species near Racine that are no longer found in Wisconsin. Much like these unusual bird species, the Marbled Salamander could have been present in the historical Wisconsin landscape, which has since experienced substantial anthropogenic modification.

Marbled Salamanders are currently considered an inhabitant of the eastern and southern US (Scott 2005). In Indiana they are frequently associated with hardwood forests in both "poorly drained flatlands and forested uplands" (Minton 2001). Historically, hardwood forests were more contiguous and undisturbed across much of the eastern US, and they extended into southeastern Wisconsin from along the shore of the southern tip of Lake Michigan (Curtis 1959, Dyer 2006). In Wisconsin, this habitat type would have extended into the counties of Kenosha, Racine, and Milwaukee (Curtis 1959, Dyer 2006), allowing a historical point of entry for Marbled Salamanders to colonize Wisconsin. This could further explain the other eastern species that Hoy reported in the state.

Documented Marbled Salamander records exist from nearby locations along the southern tip of Lake Michigan not far from Racine and in several adjacent states. There are records from northwestern Indiana (Porter and La Porte Counties; Minton 2001) and records from the southern Michigan counties of Berrien and adjacent Allegan (James Harding, Michigan State University, personal communication). Unfortunately, this species has not been reconfirmed in these Michigan counties for many years and is believed to be extirpated (J. Harding, personal communication). More directly to Wisconsin's south, Marbled Salamanders are documented in central and east-central Illinois (Shelby and Vermillion Counties; Scott 2005). Other amphibians exist in Wisconsin that have current midwestern distributional patterns that would be similar to those of Marbled Salamanders (e.g., Spotted Salamanders [*Ambystoma maculatum*]), if the latter happened to extend into the state from the southeast (i.e., spreading from the eastern US and following a thin strip of woodland along the southern tip of Lake Michigan into Wisconsin). As Anderson (1967) reviewed, Marbled Salamander records from Michigan, Indiana, and Wisconsin (citing Dickinson 1965, who cited Hoy) could represent a disjunct northern population. Even Vogt (1981) mentioned, "There may once have been or still is a relict population of marbled salamanders in the southeastern corner of Wisconsin." Ultimately, the truth behind Hoy's specimens and the historical distribution of this species in Wisconsin will probably never be known. Still, there is a distant possibility that future survey work in southeastern counties could yield records. Suzuki (1951) recommended that Marbled Salamanders be removed from the state species list, but Dickinson (1965) tentatively included them. Vogt (1981) considered this a hypothetical species and did not treat it as part of Wisconsin's herpetofauna. In the absence of any new information, this species remains only a hypothetical inhabitant of the state.

Smallmouth Salamander (*Ambystoma texanum*)

No reports of this species are known from Wisconsin, although it is confirmed in northern Illinois (e.g., Rock Island, Henry, La Salle, and Will Counties; INHS 2014) and northwestern Indiana (e.g., Porter County; Minton 2001). Given the geographic proximity, there is a remote possibility that Smallmouth Salamanders once existed in southern Wisconsin or still exist and could someday be found. They prefer wooded lowlands along larger streams or wooded wetlands (Minton 2001). Both habitats have experienced drastic declines in southern Wisconsin but occur in limited amounts on protected land in the Southern Kettle Moraine State Forest and the Avon Bottoms State Wildlife and State Natural Areas. As such, Smallmouth Salamanders could hypothetically occur (or could have once occurred) in Wisconsin, but without documentation this species is not considered part of the state's herpetofauna.

Northern Two-lined Salamander (*Eurycea bislineata*)

Hoy (1883) reported this eastern species near Racine, although no museum specimens are known. Higley (1889) mentioned that "two or three specimens have been reported from Wisconsin" and specifically referred to Racine, citing Hoy. Pope and Dickinson (1928) reviewed Northern Two-lined Salamanders in their work but seemed hesitant to include them as a Wisconsin species and cited Dunn (1926). There was apparently a specimen present in the Smithsonian National Museum of Natural History from Wisconsin (USNM 3747). Suzuki (1951) corresponded with curators there in the 1950s and discovered that this specimen had been destroyed. Vogt (1981) again believed that Hoy had a memory of this species from his time in Ohio, which he surmised influenced Hoy's decision to include it in his list for Wisconsin. Current research supports that Northern Two-lined Salamanders are found primarily in the northeastern US, with the closest populations to Wisconsin occurring in Ohio (reviewed by Sever 2005). Southern Two-lined Salamanders (*Eurycea cirrigera*), on the other hand, range farther west into Indiana and Illinois, which would put them in closer proximity to Wisconsin than Northern Two-lined Salamanders (reviewed by Pauley and Watson 2005). Southern Two-lined Salamanders prefer rocky streams, seeps, and springs (Phillips 1999). The highest likelihood of occurrence might be southeastern Wisconsin, given that they are vouchered in northeastern Illinois (Will County in 1995 [INHS 12593] and nearby Kankakee County in 1994 [INHS 11270]; INHS 2014). Neither Suzuki (1951) nor Vogt (1981) believed that this species should be considered part of the state's herpetofauna. No confirmed observations exist for either species in Wisconsin, and it is unlikely that they are part of Wisconsin's current herpetofauna.

Long-tailed Salamander (*Eurycea longicauda*)

An eastern species that Hoy (1883) reported was found in Racine, but no supporting specimens exist. Higley (1889) stated that Long-tailed Salamanders occurred in Walworth County, although he considered them rare, and he also indicated Racine (citing Hoy). Pope and Dickinson (1928) later included them, based on the reports of Hoy and Higley, but they also referred to Dunn's (1926) claim that past Wisconsin records were "dubious." Vogt (1981) again suggested that Hoy accidentally included this species based on memories from Ohio. The nearest locations with cataloged specimens are not in close proximity to Wisconsin. For example, specimens exist from the southern half of Illinois along the western and eastern borders, although they are largely unknown throughout much of the state (INHS 2014). Long-tailed Salamanders also range into central Indiana (Minton 2001). Neither Suzuki (1951) nor Vogt (1981) believed that this inhabitant of rocky cool-water streams and springs occurred in Wisconsin. There are no known specimens from Wisconsin, and no new data exist to support that they should currently be considered part of the state's herpetofauna.

Northern Slimy Salamander (*Plethodon glutinosus*)

Five specimens, all confirmed as members of the *Plethodon glutinosus* complex, exist in the Smithsonian National Museum of Natural History collections (USNM 3789, 276840–276843; Kenneth Tighe, Smithsonian National Museum of Natural History, personal communication). The specimens were purportedly taken from "Racine Wis" in the mid-1800s (although no specific collection date exists, museum records indicate that they were cataloged in 1858; K. Tighe, personal communication). Interestingly, Hoy (1883) also reported Northern Slimy Salamanders from Racine but with no reference to voucher specimens. Similarly, Higley (1889) stated that they existed in Wisconsin but were uncommon. Cahn (1929) further mentioned finding them in Waukesha County, although he did not collect vouchers. Pope and Dickinson (1928) also included this species, while Dickinson (1965) later listed it as present in nine counties.

Despite this information, not all authors have agreed that Northern Slimy Salamanders existed in Wisconsin. Neither Suzuki (1951) nor Vogt (1981) considered Northern Slimy Salamanders part of the state's herpetofauna. The latter also discussed how Northern Slimy Salamanders can be easily confused with Blue-spotted Salamanders (*Ambystoma laterale*) and suggested this as the reason why past authors included them. Although the collector of the previously mentioned Smithsonian specimens from Racine is not known, many early specimens cataloged at the museum were apparently from the private collection of Spencer F. Baird. Baird collected widely in the US and donated his collections to the museum when he began as assistant secretary at the Smithsonian (K. Tighe, personal communication). It is, therefore, possible that the specimens came from a different location and that the data were confused during accession at the museum, although no information exists to confirm this (K. Tighe, personal communication). On the other hand, it is documented that Baird spent time collecting natural history specimens in southern Wisconsin (including the Racine area) with Philo R. Hoy in 1853 (see Hoy entry in Some Prominent Figures in Wisconsin Herpetology chapter). Unfortunately, and much like the Marbled Salamander specimens collected by Hoy around the same time (see Marbled Salamander, *Ambystoma opacum* entry), the true origin of the Smithsonian's Slimy Salamander specimens from Wisconsin may never be known.

Wisconsin would be a substantial range extension for this species, whose distribution is more concentrated in the southern and eastern US (reviewed by Beamer and Lannoo 2005). A specimen was apparently collected from central Illinois (Knox County; INHS 2014); however, the accuracy of its identification has not yet been confirmed (C. Phillips, personal communication). Northern Slimy Salamanders are also documented in nearby northwestern Indiana along the southern shore of Lake Michigan (Minton 2001). They are an inhabitant of dense mature hardwood forests, which reports suggest were more widely available in the Racine area during the 1850s and thus could have supported past Northern

Slimy Salamander populations. Regardless, no new evidence exists to support that Northern Slimy Salamanders currently occur in Wisconsin, and we do not consider it part of the state's herpetofauna without collection of further specimens or evidence.

Red Salamander (*Pseudotriton ruber*)

Hoy (1883) reported that Red Salamanders were "found at Racine." Higley (1889) stated that they occurred in swamps in the southern half of the state and were not rare, but neither author provided specimen information that supported these claims. They were included by Pope and Dickinson (1928) based on the previous reports of Hoy and Higley, but the authors also cited Dunn (1926), who stated that it is improbable that this species occurs in Wisconsin. Neither Suzuki (1951) nor Vogt (1981) considered this species as part of the state's herpetofauna. Vogt again believed that Hoy was influenced by his time with this species in Ohio. However, because Higley felt that Red Salamanders were not rare in the southern half of the state, he may have mistaken the reddish terrestrial juveniles (or red efts) of the common Eastern Newt (*Notophthalmus viridescens*) for juvenile Red Salamanders. Red Salamanders are normally found in the eastern US. They are not currently known in Illinois and only confirmed in extreme southeastern Indiana (Minton 2001). Thus, a population in Wisconsin would represent a substantial range extension. No known evidence exists to support that Red Salamanders are or ever were present in Wisconsin.

CLASS REPTILIA—ORDER SQUAMATA—LIZARDS

Prairie Lizard (*Sceloporus consobrinus*)

It is noteworthy that many past sources refer to the *Sceloporus* species in the Midwest as *Sceloporus undulatus* (Eastern Fence Lizard). However, work by Leaché and Reeder (2002) supports that this lizard is more accurately identified as *S. consobrinus* (Prairie Lizard). This is normally considered a southern species, and its known range barely extends into central Illinois (Phillips et al. 1999) and Indiana (Minton 2001), although there are spotty historic reports in Wisconsin. Hoy (1883) mentioned a single specimen that was sent to him from Lafayette County in 1850 (no associated specimen number). Higley (1889) said of the Prairie Lizard: "A few specimens have been reported from this state [no specifics given—JMK]. I have not met it myself." Graenicher (1911) considered the evidence to support its presence in Wisconsin "extremely meager." Pope and Dickinson (1928) and Dickinson (1949) included it, but apparently based this inclusion on Hoy's Lafayette County report. Furthemore, Suzuki collected a specimen from Kenosha County in 1948 (MPM 2771; Figure 372a) with the accompanying remarks of "escaped pet?" Vogt (1981) stated that a "student from Milwaukee" reported finding this species in the bluffs of Crawford County (near the Lynxville Dam) in 1979, but the student provided no verification. In 2009 a young girl presented a WDNR fisheries biologist (Don Bush) with a juvenile Prairie Lizard as he worked along the Rock River near Newville (Rock County; Figure 372b). Bush then forwarded a picture to Rori Paloski at the WDNR Bureau of Natural Heritage Conservation, and its identity was confirmed. The girl apparently indicated that these lizards lived around her house and that she had seen them for several years. Unfortunately, repeated attempts to contact the family failed, and additional surveys on nearby public land in

2015 were not fruitful (J. Kapfer, unpublished data). Regardless, the Newville area is a popular travel destination for Illinois residents. It seems likely that the lizards observed were due to accidental or intentional translocation from farther south. Thus, we do not currently consider the Prairie Lizard a member of Wisconsin's herpetofauna.

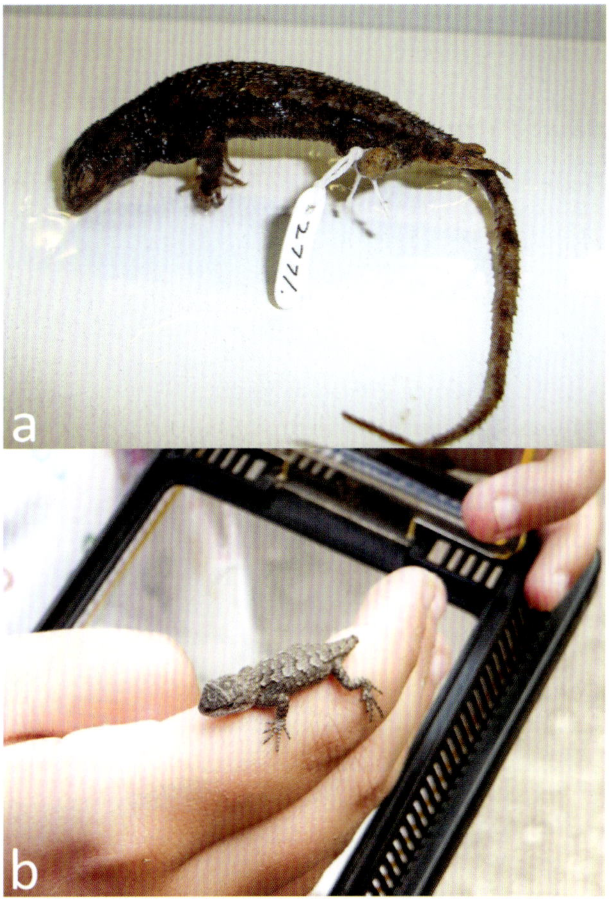

Figure 372. Prairie Lizards (*Sceloperus consobrinus*). (a) Adult collected in Kenosha County (MPM 2771; photo by J. Kapfer). (b) Juvenile recently collected in Rock County (photo by D. Bush).

CLASS REPTILIA—ORDER SQUAMATA—SNAKES

Kirtland's Snake (*Clonophis kirtlandii*)

Hoy (1883) reported that a single specimen was collected near the Fox River along the Wisconsin-Illinois border (likely Kenosha County). Higley (1889) indicated that this species was found in "Walworth County, probably throughout the southern counties; rare." Cahn (1929) reported finding one individual adjacent to the Oconomowoc River (Waukesha County) in 1926. He referred to it as "very rare." Unfortunately, none of these authors collected specimens that are currently known. Both Pope and Dickinson (1928) and Dickinson (1949) included it but cited the previously mentioned authors and did not add new data. Kirtland's Snakes are endemic to the Upper Midwest and found adjacent to Wisconsin (e.g., in eastern Illinois, Indiana, and Ohio and into the southern reaches of Michigan's Lower Peninsula; Ernst and Ernst 2003). They have been reported just south of the Wisconsin-Illinois border in Lake, DuPage, Cook, and Will Counties since 1980 (INHS 2014), although cataloged specimens only exist for Cook (INHS 15513, collected in 1985) and Will (INHS 21131, collected in 2007; INHS 2014). Anton (1999) did not find them during more recent surveys in Cook County but believed they were still present. Anton et al. (2014) also found a specimen collected in 1961 from nearby DeKalb County, Illinois, housed in the Michigan State University Museum herpetology collection (MSUM HE.14207). These small secretive snakes are often associated with wet prairies and grassy areas near wetlands (Phillips et al. 1999), habitats that exist in southern Wisconsin. Minton (2001) indicated that the largest known Indiana populations occured in urban landscapes. Given its apparent resilience to disturbance elsewhere and proximity of known populations to past anecdotal reports from Wisconsin, Kirtland's Snake could still be documented in the extreme southern parts of the state. However, surveys conducted in Kenosha County as part of a larger USFWS project on this species did not yield any individuals (Wilsmann and Sellers 1988). More recent surveys using a combination of coverboards, drift fences, and visual searches at state wildlife and state natural areas across several southern counties (southern Dane, Rock, and Walworth Counties) and locations within or near the Southern

Kettle Moraine State Forest (Walworth, Jefferson, and Waukesha Counties) conducted sporadically from 2012 to 2019 were also unsuccessful (J. Kapfer, unpublished data). In addition, many other professional biologists have surveyed for snakes in south-central and southeastern Wisconsin, with no sightings of Kirtland's Snakes. Although it could yet be found in Wisconsin, Kirtland's Snake is currently considered only a hypothetical inhabitant of the state.

Western Diamond-backed Rattlesnake (*Crotalus atrox*)

Numerous authors have discussed the presence of this south-central/southwestern US species in Wisconsin (Cochran 2008a, Pope 1928, 1930, 1931, Schorger 1967–68, Vogt 1981). Many claims of diamond-backed rattlesnakes in the state are surely misidentifications of harmless species or of Wisconsin's known native rattlesnakes (Timber Rattlesnakes [*Crotalus horridus*] and Massasaugas [*Sistrurus catenatus*]). Despite this, the Western Diamond-backed Rattlesnake has been reported in Wisconsin intermittently for decades, with several voucher specimens referenced. Only one of these specimens currently remains, an individual in the MPM collection reportedly taken near Viroqua (Vernon County) in 1928 (MPM 2293; Pope 1930). The story associated with MPM 2293 is somewhat convoluted. It was apparently obtained by W. R. Spellum, who had purchased it from a third party before presenting it to Pope (then a curator at MPM; see Some Prominent Figures in Wisconsin Herpetology chapter). This third-party individual claimed it had been killed just three hours prior in a quarry on Highway 56, west of Viroquoa. The truth of this can only be speculated upon. Spellum apparently created a mold of the specimen (Museum Mold No. 367, which no longer exists; Pope 1930) and prepared its skin, along with the existing preserved head and tail (Figure 373a).

Figure 373. Western Diamond-backed Rattlesnakes (*Crotalus atrox*; photos by J. Kapfer). (a) An individual purportedly collected in Vernon County (MPM 2293). (b) An accidentally translocated juvenile found in Ozaukee County.

Pope (1930) recounted an instance of numerous Western Diamond-backed Rattlesnakes reportedly killed by road construction workers on Hogback Hill (US Highway 61, south of Viroqua). The construction superintendent apparently signed an affidavit indicating he believed these snakes were different from the Timber Rattlesnakes he was supposedly familiar with (Pope 1930). Spellum also told Pope that he had seen the tails of the dead snakes and identified them as belonging to Western Diamond-backed Rattlesnakes. Unfortunately, additional inspection was not possible because the tails were submitted for monetary compensation under the Wisconsin rattlesnake bounty, which was active at that time. Pope also received a report from H. H. Smith (then curator of botany at the MPM), who claimed to have observed what he believed was a Western Diamond-backed Rattlesnake in Vernon County in 1922. This occurred less than 1.6 km (1 mi) from Hogback Hill near Monument Rock, south of Liberty Pole (Vernon County). Unfortunately, no specimens were taken.

A year later, Pope published further information on this topic (Pope 1931). He had obtained a secondhand account from L. C. Stuart (Museum of Zoology, University of Michigan) that Western Diamond-backed Rattlesnakes had escaped a circus near Pittsville (Wood County) in the mid-1920s. This is several counties removed from the reports in Vernon County. There was an associated rumor that a population of escapees had established near Babcock Creek, but no evidence exists to support this. Even the great rattlesnake biologist Laurence Klauber of the Zoological Society of San Diego, California, corresponded with Pope on this subject and indicated that escaped captives were much more plausible than a natural range extension for this species (Pope 1931). Klauber also stated that he had heard of "specimens escaping in this manner rather frequently, although it is of course true that seldom do a sufficient number get away to form a new colony" (Pope 1931). Schorger (1967–68) dismissed the observational accounts that Pope had collected as either Timber Rattlesnakes with aberrant patterns or escaped captive individuals. Vogt (1981) referenced the escaped circus snakes and suggested that "they survived in Vernon and Wood Counties for about a decade before succumbing to the cold or wanton slaughter." Pope never reported seeing this species himself in the wild in Wisconsin. Aside from the secondhand reports he compiled (one backed by a signed affidavit), the only historical evidence of this species in the state is one specimen apparently collected in Vernon County. This specimen consists of a severed head and a tail that were presented to Pope by Spellum, who had acquired them from an unnamed third party. It's difficult *not* to question the validity of where this specimen was purportedly collected.

Despite this, recent confirmed observations of Western Diamond-backed Rattlesnakes in Wisconsin exist. In 2006 a "rattlesnake" was encountered by a WE Energies employee, who found it on the grounds of the Port Washington facility (Ozaukee County). A WDNR employee was sent to investigate the claim. Inspection of this snake revealed that it was, indeed, a juvenile Western Diamond-backed Rattlesnake (Figure 373b) that had been accidentally shipped in a crate with equipment from the southern US. Further accidental translocations such as this are always possible but unlikely. Regardless, it is doubtful that translocated individuals would survive the harsh Wisconsin winters and become established in the state. As such, Western Diamond-backed Rattlesnakes are not currently considered part of Wisconsin's herpetofauna.

Plains Hog-nosed Snake (*Heterodon nasicus*)

The evidence to support Plains Hog-nosed Snakes in Wisconsin is a single specimen collected from La Crosse County in 1981 (MPM 23356; Figure 374). It is documented across the Mississippi River in nearby Minnesota counties such as Wabasha (JFBM 12954, among others; Moriarty and Hall 2014) and in northwestern Illinois (Jo Daviess County) as recently as 2012 (INHS 22974; INHS 2014). Therefore, it seems equally likely that the La Crosse specimen was either accidentally transported by humans or part of an isolated relict population on the Wisconsin side of the Mississippi River. This western species prefers sandy, open, and well-drained habitats (Phillips et al. 1999), which are present in southwestern Wisconsin. The proximity to confirmed locations in nearby Illinois and

Minnesota and the presence of suitable habitat in Wisconsin support the possibility that this species could be found in Wisconsin in the future. As such, the Plains Hog-nosed Snake is considered a hypothetical member of Wisconsin's herpetofauna.

Southern Hog-nosed Snake (*Heterodon simus*)

Hoy (1883) referred to this as the "Hog-nosed Viper" and indicated that it was "found in dry, sandy locations." Neither Pope and Dickinson (1928) nor Vogt (1981) included it. This species is restricted to the southeastern US, and there are no further records to support its presence in Wisconsin. Any observations by earlier authors were likely misidentifications of other species, probably the Eastern Hog-nosed Snake (*Heterodon platirhinos*). This species is not considered part of the state's herpetofauna.

Prairie Kingsnake (*Lampropeltis calligaster*)

Figure 374. A Plains Hog-nosed Snake (*Heterodon nasicus*) collected in La Crosse County (MPM 23356) (photos by J. Kapfer). (a) Dorsal view. (b) Ventral view.

Prairie Kingsnakes are a grassland species typically found in the south-central and central US (Minton 2001, Ernst and Ernst 2003). Pope and Dickinson (1928) originally included this snake as part of the state's herpetofauna because Stejneger and Barbour (1917) indicated that Wisconsin was within its range. Pope (1931) later described a living specimen given to him that was apparently collected in Milwaukee County by R. Ward on 10 July 1920 (MPM 1833). This specimen was "kept alive at the museum on exhibition until October 10, 1920, then killed and skinned. A cast skin of this specimen is also on hand. Length 51 inches. This is the first record we have of the species within the state although it had been anticipated." Dickinson (1949) also included the Prairie Kingsnake in his booklet nearly 20 years later. His justification was the previously mentioned Milwaukee County specimen (MPM 1833) and another specimen from Wood County (MPM 2346), both of which he apparently reconfirmed as Prairie Kingsnakes. Yet MPM 2346 was later reidentified as an Eastern Foxsnake (*Pantherophis vulpinus*), according to MPM records. Unfortunately, this specimen could not be found for further investigation during recent on-site reviews of the collection, so it is considered missing (J. Kapfer, personal observation). Museum records further report that MPM 1833 was later reidentified as a Gophersnake (*Pituophis catenifer*), although this specimen also could not be found and is presumed missing (J. Kapfer, personal observation).

The Michigan State University Museum collection once housed a skeletonized specimen apparently taken from Walworth County (MSUM HE SH.1202, now housed at Illinois State University [ISU 444]; Laura Abraczinska, Michigan State University, personal communication). It was purportedly found 3.2 km (2 mi) west of Delavan in 1962 by J. Alan Holman; however, there is some confusion regarding the collection locality of this specimen because cities named Delavan exist in Wisconsin and Illinois. The remarks in the now-defunct online HerpNet database entry read: "The original catalog entry read 'Illinois, Delavan, 2 miles W of Delavan.' Delavan is in Southern Wisconsin, not Illinois." Correspondence revealed that the Michigan State Museum collection houses other Prairie Kingsnakes taken from Delavan, Illinois (Tazewell County; L. Abraczinska, personal communication). Because Holman's specimen was skeletonized, additional assessment of its identity is difficult. Furthermore, the ambiguous location data associated with this specimen bring its validity as a Wisconsin individual into question.

Prairie Kingsnakes have been confirmed in Illinois as far north as Rock Island County (INHS 50816 and 50817, both collected in 1958; INHS 2014), although these records are old and several counties removed from the Wisconsin border. Given the general proximity of Rock Island County (Illinois) to Wisconsin, this species could have once followed the Mississippi River north into the extreme southern part of the state. Yet according to Vogt (1981), "No specimens have ever been verified; those reported were misidentified milk snakes *Lampropeltis triangulum*." Because there is no further evidence of this snake's presence in Wisconsin, we do not include it as part of the state's herpetofauna. It is worth noting that the "Speckled Kingsnake" (Pope and Dickinson 1928) or "Say's King Snake" (Dickinson 1949) has been included as part of Wisconsin's herpetofauna in the past. Based on recent taxonomic revisions and geographic proximity, previous authors were likely referring to what is now considered the Eastern Kingsnake (*Lampropeltis getula*; reviewed by Crother 2017). Inclusion of this species by Pope and Dickinson (1928) was based on a secondhand account by Garman (1892) that Hoy had collected it in Wisconsin. Hoy makes no reference to this species in his publication. In fact, Dickinson (1949) later seemed more doubtful of its presence in the state and did not include a distributional map. Vogt (1981) did not consider it part of Wisconsin's herpetofauna, and no new information has arisen to suggest otherwise.

Plain-bellied Watersnake (*Nerodia erythrogaster*) and Diamondback Watersnake (*N. rhombifer*)

Higley (1889) was the first to mention both species in Wisconsin but reported no voucher specimens. He stated of Plain-bellied Watersnakes, "I have seen two specimens of this species; one from near Beloit, and the other from Lake Geneva. Rare," which was apparently the basis for Clay (1936) and subsequently Gibbons and Dorcas (2004) to consider Plain-bellied Watersnakes present in Rock and Walworth Counties. Higley further reported that Diamondback Watersnakes were found in "unfrequented damp places" in Walworth County (again providing no voucher specimen information) but stated that they were also rare. Interestingly, both Pope and Dickinson (1928) and Dickinson (1949) included Diamondback Watersnakes as part of Wisconsin's herpetofauna, based on Higley's report, but not Plain-bellied Watersnakes. Vogt (1981) considered these species hypothetical, indicating that they could occur (or previously occurred) in Wisconsin. He also mentioned that Plain-bellied Watersnakes could be confused with Common Watersnakes (*Nerodia sipedon*), the latter of which are prevalent in southern Wisconsin.

The distributions of Plain-bellied and Diamondback Watersnakes are more concentrated in the southern US (reviewed by Ernst and Ernst 2003). However, both are documented in

Rock Island County, which is adjacent to the Mississippi River in northwestern Illinois and only several counties removed from the Wisconsin border (*Nerodia erythrogaster*, INHS 21694, collected in 2010, and *N. rhombifer*, INHS 18931, collected in 2003; INHS 2014).[‡‡] Isolated populations of Plain-bellied Watersnakes also exist in northern Indiana and the southern Lower Peninsula of Michigan (Minton 2001). Both species generally associate with similar aquatic habitats, as do Common Watersnakes, and these habitats are largely available in southern Wisconsin. Plain-bellied Watersnakes are confirmed in a greater number of locations that are closer to Wisconsin than Diamondback Watersnakes. As such, Plain-bellied Watersnakes seem more likely to occur in the state than Diamondback Watersnakes and could hypothetically enter along river corridors in extreme southwestern or southeastern regions that cross the border from adjacent states (e.g., the Mississippi, Rock, Fox, and Des Plaines Rivers). Regardless, there is no evidence to confirm that either species ever occurred in Wisconsin, while past reports can be explained as misidentified Common Watersnakes. Given the lack of corroborating specimens and new information, both species remain only hypothetical members of the state's herpetofauna.

Rough Greensnake (*Opheodrys aestivus*)

Pope and Dickinson (1928) first mentioned this species in Wisconsin based on "four specimens donated by the Wisconsin Natural History Society as having been collected in the state from 1891 to 1899 but unfortunately no specific locality was given." Dickinson (1949) also included it and said of the already mentioned specimens, "If all had the same date, the excuse of mislabeling might be used, but the entries are several years apart." Unfortunately, neither publication provided specimen numbers, while the accompanying photograph of a Rough Greensnake in Pope and Dickinson (1928) was not of a museum specimen but of a wild snake (presumably from outside of Wisconsin). The MPM once possessed specimens listed as Rough Greensnakes from Wisconsin, but they no longer exist. For example, museum records indicate that MPM 1666 was a Rough Greensnake collected in Sheboygan County but also that this specimen was "disposed." This species prefers lightly wooded or brushy borders adjacent to open fields and wetlands and is generally found to the south, currently confirmed no farther north than central Illinois and Indiana (Minton 2001). No definitive documentation of this species in Wisconsin currently exists, and it is not considered part of the state's herpetofauna.

Graham's Crayfish Snake (*Regina grahamii*)

Hoy (1883) stated that Graham's Crayfish Snake occurred in Wisconsin but gave no information on location. Higley (1889) also considered it present in the state and said that it was found "mostly in western countries [*sic*]; not rare." Pope and Dickinson (1928) and Dickinson (1949) included it based on the reports of Hoy and Higley, although the latter publication indicated that previous authors may have confused it with the Queensnake

---

[‡‡] Although Gibbons and Dorcas (2004) reported a *Nerodia rhombifer* specimen from extreme northwestern Illinois (UMMZ 4008; Jo Daviess County), this specimen could not receive additional scrutiny because it is currently lost (Greg Schneider, University of Michigan Museum of Zoology, personal communication). Garman (1892) mentioned a *Nerodia rhombifer* specimen from Cook County in northeastern Illinois that was apparently housed at Northwestern University, but no specimen number was provided.

(*Regina septemvittata*). This southern species occurs throughout Illinois (although it is still considered uncommon there; Phillips et al. 1999), including the northeastern counties near the Wisconsin border (Cook County, FMNH 245712 and 247523; reviewed by Anton 1999). In a 1941 letter to George Wagner (a past zoology professor at UW–Madison), Karl P. Schmidt (past curator, FMNH) reported the existence of a "considerable colony within the limits of Chicago" (copies of correspondence provided by Laura Monahan, collections manager, UWZM). Given the proximity to Wisconsin, unidentified populations of this species could conceivably exist in the southern reaches of the state. Cook County is particularly close to a recent substantial survey effort at sites in southern Rock and Walworth Counties that did not yield any specimens (J. Kapfer, unpublished data). Currently, Graham's Crayfish Snake is only a hypothetical member of the state's herpetofauna.

Smooth Earthsnake (*Virginia valeriae*)

A single specimen of this species, collected in 1947, is all that is known from Wisconsin (MPM 2648; Dickinson 1949, 1950; Figure 375). It was taken in the town of Shelby, south of the city of La Crosse (La Crosse County) along the Mississippi River. Smooth Earthsnakes are documented in Illinois, the farthest north being Knox County in the west-central part of the state in 2006 (INHS 21093; INHS 2014). A specimen in the US Museum of Natural History was purportedly collected from Cook County in northeastern Illinois (USNM 7303), but this record has been rejected (Anton 1999, C. Phillips, personal communication). Regardless, both of these locations are a substantial distance from the specimen collected in Wisconsin. Smooth Earthsnakes associate with steep rocky slopes in open woodlands or along woodland edges (Minton 2001), habitats that are certainly present in western Wisconsin. It is feasible that accidental translocation brought the Wisconsin specimen north along the Mississippi River (Vogt 1981). It also seems likely that the MPM specimen is actually the buff color morph of the Smooth Greensnake (*Opheodrys vernalis*). Regardless, a Rough Earthsnake (*Virginia striatula*) specimen exists in the collection at the US Museum of Natural History that was apparently collected from Milwaukee County and cataloged in 1958 (USNM 2453). The identification of this specimen was confirmed (Kenneth Tighe, US Museum of Natural History, personal communication). However, the museum ledger has handwritten notes over the original entry that include "??" after the location and an additional note that states "not Wisconsin at all (FNB)," the initials likely being those of Frank N. Blanchard (K. Tighe, personal communication). The true origin of this specimen is, therefore, not confirmed. These small secretive snakes are difficult to detect, and it is conceivable that unidentified relict populations exist in Wisconsin. The discovery of Western Wormsnakes (*Carphophis vermis*) in 1977 and Lined Snakes (*Tropidoclonion lineatum*) in 2011, both also small and secretive, gives further reason to hope this is possible. Regardless, neither earthsnake species is currently considered part of the state's herpetofauna.

Figure 375. A purported Smooth Earthsnake (*Virginia valeriae*) collected in La Crosse County (MPM 2648) (photo by J. Kapfer).

CLASS REPTILIA—ORDER TESTUDINES

Spotted Turtle (*Clemmys guttata*)

Higley (1883) was the first to report this species in Wisconsin. He recounted knowledge of two individuals observed in Walworth County and indicated receipt of reports from Beloit, Madison, Eau Claire, and "other places," but apparently no voucher specimens were collected. A single nesting female was purportedly observed by Carr (1890) near Lake Wingra in Madison, but again no corroborating evidence of the report exists. Cahn (1929) outlined isolated observations of Spotted Turtles from two locations in Waukesha County: the state fish hatchery ponds in Delafield, which are no longer in active use, and "Skidmore ponds," a series of waterbodies just north of Okauchee Lake. Pope and Dickinson (1928) added further secondhand reports from Waukesha County and from the "Moon Lake Wild-life Refuge" in Fond du Lac County, which was later renamed Mauthe Lake (Hayes 1988). Even much later publications listed Spotted Turtles as possible inhabitants of Wisconsin (Dickinson 1965, Knudsen 1957d). It is plausible that at least some of these authors misidentified juvenile or small adult Blanding's Turtles (*Emydoidea blandingii*; Vogt 1981), while others seem to have simply parroted past reports. However, all of the authors make separate mention of Blanding's Turtles, which they should have been able to distinguish from Spotted Turtles.

Regardless, no cataloged Spotted Turtle specimens or photographs currently exist from Wisconsin. This turtle is typically considered a more eastern species that largely associates with shallow still water, such as wetlands and ditches (Minton 2001). In 1950 it was vouchered across the Wisconsin-Illinois border in Cook County (INHS 6013), although the populations in this county are now likely extinct (Anton 1999, C. Phillips, personal communication). In 2013 it was vouchered in nearby Will County, Illinois (INHS 23707; INHS 2014), but currently only one population there is known (Anton 1999). It is known across much of northern Indiana and the southern reaches of the Lower Peninsula of Michigan (Minton 2001). Given the proximity of the northern Illinois observations to Wisconsin, this species could have been historically present in the far southern portion of the state, although its current presence there is improbable. Adler (1968) reported a nuchal bone belonging to this species from an archaeological site in Sauk County. This bone was drilled, evidence that Adler (1968) used to suggest that it was part of an ornament and carried to this location from elsewhere by humans. Neither Adler (1968) nor Vogt (1981) believed that this bone provided strong evidence to support the occurrence of Spotted Turtles in Wisconsin. The lack of recent observations and verified specimens supports that it is not currently part of Wisconsin's herpetofauna.

Yellow Mud Turtle (*Kinosternum flavescens*)

No known reports exist of Yellow Mud Turtles in Wisconsin. They are confirmed in a handful of counties from the northern third of Illinois (Henry, Whiteside, Lee, and Lake Counties; INHS 2014). It is remotely possible that Yellow Mud Turtles once entered the southern reaches of the state, given Wisconsin's proximity to these locations (particularly Lake County), although Phillips et al. (1999) speculated that the Lake County individuals were introduced. Regardless, if this species was ever present, it is unlikely a current member of the state's herpetofauna.

Alligator Snapping Turtle (*Macrochelys temminckii*)

Hoy (1883) stated that this "ferocious turtle is occasionally found in the Mississippi river as far north as the mouth of the Wisconsin [River]." This is also the exact entry for this species given by Higley (1889), who cited Hoy, but neither author provided supporting evidence. Pope and Dickinson (1928) included it based on these past reports, as did Dickinson (1965), who indicated that it was a "Mississippi River form" but did not include a distributional map. Later, Vogt (1981) mentioned that this southern US species is incredibly rare even in southern Illinois and unlikely to range as far north as Wisconsin. The Monte L. Bean Life Science Museum at Brigham Young University houses two specimens previously identified as Alligator Snapping Turtles taken from Wisconsin (BYUH 38809 and BYUH 37569). These were collected on 23 and 29 May 1964 from Three Mile Creek, several miles north of Janesville, by Richard D. Porter. Examination of photographs of the specimens revealed they were actually Snapping Turtles (*Chelydra serpentina*). In his review of turtles from archaeological sites in the Great Lakes region, Adler (1968) suggested that this species was found in the upper reaches of the Mississippi River in prehistoric times but only as far north as northern Illinois. Given that there are still no specimens from Wisconsin, we do not consider it a member of the state's herpetofauna.

Eastern Box Turtle (*Terrapene carolina*)

The first mention of this eastern woodland species in Wisconsin comes from Higley (1889). He considered it "very rare" and reported it from a smattering of counties, including Brown, La Crosse, Milwaukee, and Walworth, but apparently no specimens were collected. He also reported it from "Pine Lake," which could represent many locations, as there are numerous waterbodies named Pine Lake across Wisconsin and the town of Pine Lake in Oneida County. As Vogt (1981) stated about Higley's account, "Since he recognized the ornate box turtle *Terrapene ornata*, Blanding's turtle, and spotted turtles, it is hard to ascertain what he called *T. carolina*." Cahn (1929) also considered it present in Waukesha County based on an individual apparently captured along the south shore of Oconomowoc Lake in 1928, which was brought to him for identification (again, no voucher specimen or photographs are known). There are several other Eastern Box Turtle reports from locations throughout Wisconsin. For example, MPM records mention specimens purportedly collected in the state: MPM 1885 was taken in 1923 (museum records list the location only as "Wisconsin"), although the specimen is now "Presumed Missing," and MPM VZP401, a photo specimen taken in Winnebago County in 1992, although whether the collector considered it wild or an escaped pet is unreported. There is also a specimen in the Carnegie Museum of Natural History (Pittsburgh, Pennsylvania) purportedly taken from Washington County, possibly near West Bend (CM 137101). Correspondence revealed that this is a deformed hatchling or late-stage embryo whose identity was confirmed as an Eastern Box Turtle (Steve Rogers, Carnegie Museum, personal communication), but again associated context is unknown.

Pope and Dickinson (1928) reportedly verified an individual that was captured near Milwaukee and brought to MPM staff in 1923. It is unknown if the missing specimen mentioned above (MPM 1885) is the one referenced by Pope and Dickinson. Dickinson (1965) later referred to the presence of this species in the state as "problematical," although he included it in his booklet (without a range map). Vogt (1981) included a photograph of an individual from Point Beach State Forest (Manitowoc County) at a location where this species had been occasionally reported for a period of over 30 years. Although a naturalist at Point Beach State Forest hypothesized that a relict population existed there, Vogt did not find additional specimens on-site despite more than 20 field surveys of this location. Furthermore, no additional sightings from this location have been reported in the last 20 years (Guy Willman, WDNR, personal communication). Despite this, Eastern Box Turtles are still encountered on occasion in Wisconsin. For example, an individual was found in a horse pasture in rural Clark County during the summer of 2001 (J. Kapfer, personal observation). Although it was not collected, a photograph was taken (Figure 376).

Figure 376. An Eastern Box Turtle (*Terrapene carolina*) encountered in Clark County (photo by J. Kapfer).

Vogt (1981) indicated that this inhabitant of eastern deciduous hardwood forests could occur in Wisconsin, given that its preferred habitat once "stretched around Lake Michigan in preglacial times." In 1989 it was vouchered just over the Wisconsin-Illinois border in Lake County (INHS 10802) and more recently in 1999 in Cook County (INHS 15561; INHS 2014). It ranges well into the Lower Peninsula of Michigan (Holman 2012) but has a spotty distribution in northern Indiana (Minton 2001). No archaeological evidence of this species is known from Wisconsin. Adler (1968) reidentified archaeological remains of reported Eastern Box Turtles from two sites in Sauk County as Ornate Box Turtles (*Terrapene ornata*; Parmalee 1959, 1960). The Eastern Box Turtle has traditionally been a favorite in the pet trade (Gibbons et al. 2000, Telecky 2001), and the previously mentioned individuals observed in more recent times could have been escaped pets. Even Vogt (1981) reported having two Eastern Box Turtles that escaped captivity in Dane County (only one was recovered, albeit as roadkill). Thus, in lieu of confirmed reproducing populations in the state, the Eastern Box Turtle is not included as part of Wisconsin's current herpetofauna.

Pond Slider (*Trachemys scripta*)

Pond Sliders (specifically, the subspecies *Trachemys scripta elegans* [Red-eared Slider]) have been sporadically reported from various locations in Wisconsin over many years, including recently (Rori Paloski, WDNR, personal communication). Although isolated reports, specimens, or photographs of this species from the state exist, many are believed to be escaped pets or the result of human introductions. For example, Cochran et al. (1987) found an individual in Brown County, which they referred to as "undoubtedly an escaped or released pet." Unfortunately, the corroborating specimen (BBWS RE01) was deposited in a collection at the Bay Beach Wildlife Sanctuary in Brown County, and this collection is apparently lost (Lori A. Bankson, Bay Beach Wildlife Sanctuary, personal communication). A Dane County specimen exists in the UWZM (UWZM 20138) that was taken west of Madison by Richard Vogt in early spring 1964. This juvenile is believed to be an escaped pet that died during the winter and is the only specimen of this species that Vogt obtained from Wisconsin (Richard Vogt, Instituto Nacional de Pesquisas da Amazônia, personal communication). A verified specimen from northern Wisconsin (Bayfield County) exists in the collection at Brigham Young University (BYUH 40345; Jack Sites, Monte L. Bean Life Science Museum, Brigham Young University, personal communication). Although the collectors of the specimen did not mention it as a probable escaped pet, the location of this record appears far removed from the most obvious avenues of ingress for this species into the state (i.e., southern Wisconsin). As such, its collection at this location is likely the result of human transport.

Numerous more recent observations also exist. For example, a dead individual was found along the shore of a Waukesha County wetland in spring 2016 (MPM VZP837a–c). Although the cause of death was undetermined, the wetland is located along a well-used county road, and winterkill of a naive released pet is plausible. Furthermore, a road-killed individual was found near the UW–Whitewater campus in 2016, and a live specimen was captured in a wetland on campus in 2017 (both in Jefferson County) after the end of the spring semester, which were likely pets released by college students before returning home for the summer (J. Kapfer, personal observation). Additional anecdotal observations of this turtle exist for Bayfield, Buffalo, Dane, Milwaukee, and Pierce Counties (Casper and Anton 2013).

At the time of this writing, the only other recent published report of a Pond Slider in the state, supported by a voucher specimen or photograph, is that of Casper and Anton (2012). The authors captured a gravid female in Kenosha County on a tributary of the Des Plaines River (FMNH 281241). They indicated that this represented "the first Wisconsin record from a stream system within the natural range . . . and is considered a natural occurrence in an industrial corridor without public access." Although Pond Sliders are found in the same river system farther south in Illinois, it seems unlikely that this Kenosha County locality truly had limited public access. According to the GPS coordinates provided by Casper and Anton (2012), this turtle was captured from a river/wetland complex that is in relatively close proximity to roads (including Interstate 94) and parking areas (all within ca. 130–300 m [427–984 ft]). Furthermore, there is a suburban housing tract that surrounds a large anthropogenic pond adjacent to the river/wetland complex where the turtle was captured and within ca. 1,000 m (3,281 ft) of the record. All of these could have effectively acted as release points for an unwanted captive turtle. Whether natural or introduced, the gravid Kenosha County Pond Slider reported by Casper and Anton (2012) provides evidence of possible reproduction in Wisconsin, assuming the captured individual was not a released pet that was gravid at the time of introduction. No additional data or subsequent observations of Pond Sliders at this location are known.

There is some question about whether Pond Sliders could have existed historically in Wisconsin. Both Adler (1968) and Holman (1994) believed that this species naturally had a larger geographic distribution in the past that included Wisconsin and parts of Michigan. Yet only sparse archaeological evidence currently exists to support that Pond Sliders naturally occurred in Wisconsin during prehistoric times. Adler (1968) reported the remains of this turtle from the Durst Rockshelter in Sauk County, but neither Parmalee (1960) nor Theler (2000) made mention of it in their assessments of the same archaeological site. Holman (1994) believed that archaeological remains from a similar latitude in Michigan were a natural occurrence, which supports the possibility of their existence in Wisconsin. Yet turtle shells were often transported by prehistoric cultures due to their use as ornaments or vessels, and evidence that humans have or have not used fossilized turtle remains in this fashion is difficult to secure from disarticulated shell bones (Steven Kuehn, Illinois State Archaeological Survey, personal communication). It is of further note that a review of over

100 archaeological site reports and other literature from Wisconsin did not yield any additional mention of Pond Slider remains (S. Kuehn, personal communication).

Despite the lack of archaeological evidence that Pond Sliders once existed in Wisconsin, populations of this turtle are identified in several adjacent states. The existence of these populations could support past occurrence of Pond Sliders in Wisconsin, or at least their ability to naturally colonize the state from a nearby source. Holman (2012) reported evidence of established Pond Slider populations in the Lower Peninsula of Michigan. Recent genetic analyses of Michigan populations help paint a clearer picture but still leave ambiguity on the origin of these populations (Patrick Terry, Eastern Michigan University, personal communication). These initial analyses indicated that the origin of Pond Sliders sampled in Michigan could be traced to both geographically distant and proximal locations, but *proximal* in this context might still represent translocations from possibly native populations in Ohio or Indiana (P. Terry, personal communication). Pond Sliders have further been documented in northern Illinois (Winnebago, DeKalb, Kane, Du Page, Cook, and Lake Counties; INHS 2014), but parsing natural occurrences from introductions in that area is also difficult. For example, Anton (1999) found Pond Sliders in Cook County but reported that these were likely introduced populations. The species is known to occur in nearby northwestern Indiana, which Minton (2001) considered part of its natural range. Casper (2008c) suggested that climate change could result in the expansion of Illinois populations into Wisconsin, which is surely a reasonable scenario. Yet further evidence is necessary to determine if the individuals currently observed in Wisconsin are part of expanding natural populations or expanding introduced populations or are the result of translocations. Given these points, until additional data are collected to support otherwise, the Pond Slider is not currently considered a member of Wisconsin's herpetofauna.

ISOLATED PECULIAR SPECIMENS

A specimen labeled as a Greater Siren (*Siren lacertina*) exists in the UWZM collection (UWZM 9401) that was apparently taken in 1896 from Rock County, near the town of Milton (Figure 377). The notes associated with this specimen state that it was removed from a "pond tributary to Rock River" and obtained by "L. Kumlien; from H. V. Ogden Collection." No mention of this species exists in any publication on Wisconsin amphibians and reptiles (this specimen was brought to my attention by Jeffrey Lorch). During correspondence in the 1920s between George Wagner (UW–Madison) and Karl P. Schmidt (FMNH), Wagner referenced this specimen and stated: "I suspect that Kumlien's Siren was put over on him by some student" (Armand Esai, Field Museum of Natural History, personal communication). Greater Sirens are normally found in the southeastern US, and Wisconsin seems an unlikely location for a natural population. The Lesser Siren (*Siren intermedia*), on the other hand, is known from central Illinois (Fulton County, INHS 1050,

Figure 377. Specimen UWZM 9401 at the UW–Madison Zoology Museum. Listed as a "Greater Siren" from Rock County (photo by J. Kapfer).

Figure 378. A Texas Horned Lizard (*Phrynosoma cornutum*) collected in Milwaukee County (MPM 2243) (photo by J. Kapfer).

and Tazewell County, INHS 9008; INHS 2014). It was also confirmed in northwestern Indiana and the southern portion of the Lower Peninsula of Michigan (Minton 2001). This species prefers wetlands and other lentic environments with soft bottoms and ample aquatic vegetation (Minton 2001), which currently occur throughout southern Wisconsin. Both siren species are cryptic and difficult to find; thus, it is possible that small relict populations in Wisconsin could go undetected. Of the two, it seems more plausible that the Lesser Siren was once present in the state. However, no other observations of either species in Wisconsin exist, and neither is considered part of the state's current herpetofauna.

A variety of other species have been captured in Wisconsin and appropriately vouchered, which are likely escaped captives or the result of translocation. Some of these are North American species that are normally found in habitat types and climates that are vastly different from what are present in Wisconsin (and unlikely to survive in the state). For example, a Gopher Tortoise (*Gopherus polyphemus*) was collected from "Lincoln Park" in Milwaukee on 30 May 1929. Normally associated with the southeastern US, this individual was apparently deposited at the MPM (MPM 2333), but the specimen is currently "presumed missing." A Texas Horned Lizard (*Phrynosoma cornutum*), a species of the southwestern US, was collected in 1928 from the Shorewood area of Milwaukee (MPM 2243) with the accompanying remark "escaped pet?" (Figure 378). A specimen identified as a Chicken Turtle (*Deirochelys reticularia*) captured along the Mississippi River (Vernon County) in 1977 by Richard C. Vogt (determined by T. G. Anton) exists in the UWZM collection (UWZM 22256). The known range of this southern turtle extends up the Mississippi River no farther than the southern tip of Illinois. It is probable that this individual was a waif or perhaps translocated by humans.

MPM records indicate current or prior possession of specimens collected from Wisconsin of species that do not naturally occur in North America. For example, two Geckos (*Sphaerodactylus* spp., MPM 2570 and 2571) that are native to the Caribbean, Central America, and South America were collected in 1936 and 1937, respectively; they had arrived in Milwaukee in a shipment of bananas. Museum records further indicate prior possession of specimens now presumed missing or disposed, such as a Boa Constrictor (*Boa constrictor*) collected in Milwaukee in 1930 (MPM 2349, likely an escaped captive) and two Marginated Tortoises (*Testudo marginata*), native to Greece and Italy, that were collected in 1903 and 1911, respectively, from Lake Nagawicka (MPM 746, MPM 1587). It is telling that most of these exotic species were found near Milwaukee, which was an important port town and also contained a prominent zoological gardens and museum of natural history as far back as the late 1800s.

Appendix

Museum voucher specimen numbers for amphibian and reptile county records in Wisconsin as of March 2022. Notes are provided when our listed specimen number deviates from the published number, when specimens could not be found, and when additional information is needed. Scientific names follow Crother (2017), and museum acronyms follow Sabaj (2016). Vogt (1981), the precursor to this book, did not provide museum voucher numbers. For records in Vogt (1981), we used voucher numbers from other publications when available, and sought to obtain voucher numbers for the remaining 1,053 records. We attempted to confirm the existence and correct identification of all vouchers through direct observation of specimens, photographs of specimens, or confirmation from qualified collections managers. We were unable to locate museum vouchers for nearly one hundred of these observation records. These county records, as well as additional published records with missing or lost vouchers and for which no replacement was found, are shown in the individual species county record maps as "*Voucher Needed.*"

Scientific Name	County	Museum Voucher	Publication	Notes
Acris blanchardi	Adams	USNM 311739	Badje et al. (2015)	
Acris blanchardi	Buffalo	MPM VZP874	Vogt (1981)	Audio file
Acris blanchardi	Columbia	UMMZ 63534	Vogt (1981)	
Acris blanchardi	Crawford	USNM 311659–311678	Suzuki (1951), Vogt (1981)	
Acris blanchardi	Dane	UWZM 17597	Vogt (1981)	
Acris blanchardi	Dodge	UIMNH 1447–48	Suzuki (1951), Vogt (1981)	
Acris blanchardi	Fond Du Lac	MPM 4519	Vogt (1981)	
Acris blanchardi	Grant	MPM 21200	Vogt (1981)	
Acris blanchardi	Green	UWZM 20320	Vogt (1981)	
Acris blanchardi	Green Lake	UNSM ZM22449	Casper (2015)	University of Nebraska State Museum
Acris blanchardi	Iowa	UWZM 23988	Vogt (1981)	
Acris blanchardi	Jefferson	MPM 3467	Vogt (1981)	
Acris blanchardi	Juneau		Suzuki (1951), Vogt (1981)	Suzuki (1951) originally reported voucher as "MPM uncatalogued"
Acris blanchardi	Kenosha	CHAS HERP5462	Necker (1939b), Vogt (1981)	

Scientific Name	County	Museum Voucher	Publication	Notes
Acris blanchardi	La Crosse	MPM VZP904	Vogt (1981)	Audio file
Acris blanchardi	Lafayette	MPM 17391	Vogt (1981)	
Acris blanchardi	Manitowoc	MPM 2887	Suzuki (1951), Vogt (1981)	
Acris blanchardi	Marquette	UWSP 666	Vogt (1981)	
Acris blanchardi	Milwaukee	MPM 2881	Vogt (1981)	
Acris blanchardi	Monroe	UWZM 19039	Vogt (1981)	
Acris blanchardi	Pepin		Vogt (1981)	
Acris blanchardi	Portage	UWSP 2622	Vogt (1981)	
Acris blanchardi	Racine	MPM 1497	Vogt (1981)	
Acris blanchardi	Richland	UWSP 673	Vogt (1981)	
Acris blanchardi	Rock	UWZM 20310	Vogt (1981)	
Acris blanchardi	Sauk	MPM 2560	Vogt (1981)	
Acris blanchardi	Trempealeau	UWZM 19038	Casper (2015)	
Acris blanchardi	Vernon	MPM 2298	Vogt (1981)	
Acris blanchardi	Walworth	MPM 2559	Vogt (1981)	
Acris blanchardi	Washington	MPM 2883	Suzuki (1951), Vogt (1981)	
Acris blanchardi	Waukesha	MPM 3455	Vogt (1981)	
Acris blanchardi	Waupaca	UWSP 2711	Suzuki (1951), Vogt (1981)	Suzuki (1951) originally reported voucher as CNHM 14729, which could not be located
Acris blanchardi	Waushara	CA 19310	Vogt (1981)	
Acris blanchardi	Wood	UWZM 4051	Vogt (1981)	
Ambystoma laterale	Adams	MPM VZP545	Casper (1997d)	Photograph
Ambystoma laterale	Ashland	MPM 6676	Briggs (1974), Vogt (1981)	
Ambystoma laterale	Barron	MPM VZP338	Casper (1996)	Photograph
Ambystoma laterale	Bayfield	MPM 6677	Briggs (1974), Vogt (1981)	Briggs (1974) reported MPM 6675 and 6677 from Bayfield County, but recent investigations of collections did not uncover 6675
Ambystoma laterale	Brown	UWZM 22615	Cochran et al. (1987)	
Ambystoma laterale	Buffalo	JFBM Herp Photo 469	Casper (2015)	Photograph
Ambystoma laterale	Burnett	MPM 822	Casper (1996)	
Ambystoma laterale	Calumet	MPM VZP942	Burns (2020)	Photograph
Ambystoma laterale	Chippewa	MPM 33354	Vogt (1981)	
Ambystoma laterale	Clark		Vogt (1981)	Specimen FMNH 13036 was collected in Clark County, but museum records identify specimen only as part of the "*A. jefferson* complex"

Scientific Name	County	Museum Voucher	Publication	Notes
Ambystoma laterale	Columbia	UWZM 16667	Suzuki (1951), Vogt (1981)	Suzuki (1951) originally reported voucher as "UWZM uncatalogued" and also incorrectly referred to specimen as *A. jeffersonianum*
Ambystoma laterale	Crawford		Vogt (1981)	
Ambystoma laterale	Dane	MPM 23527	Casper (1996)	
Ambystoma laterale	Dodge	UMMZ 123712	Vogt (1981)	
Ambystoma laterale	Door	UWZM 4349	Suzuki (1951), Vogt (1981)	Suzuki (1951) reported voucher as UWZM 4348, which could not be found and incorrectly refers to the specimen as *A. jeffersonianum*
Ambystoma laterale	Douglas	JFBM 1516	Suzuki (1951), Vogt (1981)	Suzuki (1951) originally reported voucher as UWZM 6163, which could not be found
Ambystoma laterale	Dunn		Vogt (1981)	
Ambystoma laterale	Eau Claire	MPM VZP428	Casper (1996)	Photograph
Ambystoma laterale	Florence	UWZM 22630	Cochran and Hodgson (1986)	
Ambystoma laterale	Fond Du Lac	MPM 6666	Briggs (1974), Vogt (1981)	
Ambystoma laterale	Forest	UWZM 22786	Suzuki (1951), Vogt (1981)	Suzuki (1951) originally reported voucher as MPM 2558, which is currently assigned to a different species; he also incorrectly referred to specimen as *A. jeffersonianum*
Ambystoma laterale	Grant	MPM 4996	Vogt (1981)	
Ambystoma laterale	Green	INHS 2015l	Kapfer et al. (2015)	Photograph
Ambystoma laterale	Green Lake	UIMNH 1410	Suzuki (1951), Vogt (1981)	Suzuki (1951) incorrectly referred to specimen as *A. jeffersonianum*
Ambystoma laterale	Iowa	INHS 2015m	Kapfer et al. (2015)	Photograph
Ambystoma laterale	Iron	USNM 33144	Vogt (1981)	
Ambystoma laterale	Jackson	MPM 33144	Cochran and Cochran (2001a)	
Ambystoma laterale	Jefferson	INHS 2015v	Kapfer et al. (2015)	Photograph
Ambystoma laterale	Juneau	MPM VZP604	Vogt (1981)	Photograph
Ambystoma laterale	Kenosha	MPM 3316	Casper (1996)	
Ambystoma laterale	Kewaunee	MPM 2807	Suzuki (1951), Vogt (1981)	Suzuki (1951) incorrectly referred to specimen as *A. jeffersonianum*
Ambystoma laterale	La Crosse	MPM VZP503	Casper (1996)	Photograph
Ambystoma laterale	Langlade	UWZM 22629	Cochran and Hodgson (1986)	
Ambystoma laterale	Lincoln	MPM VZP793	Vogt (1981)	Photograph

Scientific Name	County	Museum Voucher	Publication	Notes
Ambystoma laterale	Manitowoc	MPM 6665	Briggs (1974), Vogt (1981)	
Ambystoma laterale	Marathon	MPM 2809	Suzuki (1951), Vogt (1981)	Suzuki (1951) incorrectly referred to specimen as *A. jeffersonianum*
Ambystoma laterale	Marinette	UWZM 22633	Cochran and Lyons (1986)	
Ambystoma laterale	Marquette	MPM 27951	Casper (1996)	
Ambystoma laterale	Menominee	MPM 27982	Casper (1997a)	
Ambystoma laterale	Milwaukee	MPM 3359	Vogt (1981)	
Ambystoma laterale	Monroe	MPM VZP302	Casper (1996)	Photograph
Ambystoma laterale	Oconto	UMMZ 85566	Suzuki (1951), Vogt (1981)	Suzuki (1951) incorrectly referred to specimen as *A. jeffersonianum*
Ambystoma laterale	Oneida	MPM 5171	Vogt (1981)	
Ambystoma laterale	Outagamie	UWGB 119–126	Casper (1996)	
Ambystoma laterale	Ozaukee	MPM 23368	Vogt (1981)	
Ambystoma laterale	Polk	MPM 6664	Briggs (1974), Vogt (1981)	
Ambystoma laterale	Portage	MPM 22143	Vogt (1981)	
Ambystoma laterale	Price	MPM 9089	Vogt (1981)	
Ambystoma laterale	Racine	MPM 31929	Vogt (1981)	
Ambystoma laterale	Richland	INHS 2008e	Muehlfeld et al. (2008)	Photograph
Ambystoma laterale	Rock	INHS 2015ad	Kapfer et al. (2015)	Photograph
Ambystoma laterale	Rusk	MPM VZP511	Casper (1996)	Photograph
Ambystoma laterale	Sauk	MPM 2461	Suzuki (1951), Vogt (1981)	Suzuki (1951) incorrectly referred to this as *A. jeffersonianum*
Ambystoma laterale	Sawyer	UMMZ 77131	Suzuki (1951)	Suzuki (1951) incorrectly referred to specimen as *A. jeffersonianum*
Ambystoma laterale	Shawano	MPM VZP308	Casper (1996)	Photograph
Ambystoma laterale	Sheboygan	MPM 16312	Vogt (1981)	
Ambystoma laterale	St. Croix	MPM VZP371	Casper (1996)	Photograph
Ambystoma laterale	Taylor	MPM VZP347	Casper (1996)	Photograph
Ambystoma laterale	Trempealeau	MPM 30299	Casper (2015)	
Ambystoma laterale	Vernon	UWZM 20403	Vogt (1981)	
Ambystoma laterale	Vilas	UWZM 22603	Suzuki (1951), Vogt (1981)	Suzuki (1951) originally reported UWZM 6673 as voucher, which is missing; he further incorrectly referred to specimens as *A. jeffersonianum*
Ambystoma laterale	Walworth	MPM VZP2	Casper (1996)	Photograph

Scientific Name	County	Museum Voucher	Publication	Notes
Ambystoma laterale	Washburn	MPM 6684	Briggs (1974), Vogt (1981)	
Ambystoma laterale	Washington	MPM 30308	Vogt (1981)	
Ambystoma laterale	Waukesha	MPM 11137	Vogt (1981)	
Ambystoma laterale	Waupaca	UWSP 2681	Vogt (1981)	
Ambystoma laterale	Waushara	UWZM 9688	Suzuki (1951), Vogt (1981)	Suzuki (1951) incorrectly referred to specimen as *A. jeffersonianum*
Ambystoma laterale	Winnebago	MPM 2140	Vogt (1981)	
Ambystoma laterale	Wood	MPM 2347	Suzuki (1951), Vogt (1981)	Suzuki (1951) incorrectly referred to specimen as *A. jeffersonianum*
Ambystoma maculatum	Ashland	MPM 13604	Vogt (1981)	
Ambystoma maculatum	Barron	MPM VZP482	Casper (1996)	Photograph
Ambystoma maculatum	Bayfield	UWSP 3540	Vogt (1981)	
Ambystoma maculatum	Calumet	INHS 2015d	Kapfer et al. (2015)	Photograph
Ambystoma maculatum	Chippewa	MPM VZP592	Casper (1997d)	Photograph
Ambystoma maculatum	Clark	FMNH 14698	Vogt (1981)	
Ambystoma maculatum	Dodge	INHS 2015g	Kapfer et al. (2015)	Photograph
Ambystoma maculatum	Door	UWZM 1930	Vogt (1981)	
Ambystoma maculatum	Douglas	UMMZ 243156	Casper (1996a), Casper (2015)	Casper (1996a) reported voucher as MPM 27967, which recent investigations of the collection did not uncover; Casper (2015) republished the record with the voucher as UMMZ 243156
Ambystoma maculatum	Florence	UWGB 232	Casper (1996)	
Ambystoma maculatum	Fond Du Lac	MPM 17546	Casper (1996)	
Ambystoma maculatum	Forest	MPM 3312	Vogt (1981)	
Ambystoma maculatum	Iron	UWSP 4270	Vogt (1981)	
Ambystoma maculatum	Jackson	MPM 33142	Cochran and Cochran (2001b)	
Ambystoma maculatum	Langlade	UWZM 16670	Vogt (1981)	
Ambystoma maculatum	Lincoln	UWGB 233	Casper (1996)	

Scientific Name	County	Museum Voucher	Publication	Notes
Ambystoma maculatum	Manitowoc		Suzuki (1951), Vogt (1981)	Suzuki (1951) reported voucher as MU 1316, part of a historical collection at Marquette University that was either donated or eliminated according to correspondence with institution staff; USNM 260904–260907 were originally identified as *A. laterale* from this location, then reidentified as *A. maculatum* by G. Casper, but later reverted back to *A. laterale* by the original collector (J. Jacobs, USNM, retired)
Ambystoma maculatum	Marathon	UWZM 22761	Vogt (1981)	
Ambystoma maculatum	Marinette	MPM VZP700	Casper (2015)	Photograph
Ambystoma maculatum	Milwaukee	MPM 2470	Casper (1996)	
Ambystoma maculatum	Oconto	UMMZ 85569	Suzuki (1951), Casper (1996)	Specific locality data for UMMZ 85569 not provided by Suzuki (1951) but Casper (1996) reported several additional specimens from this county (e.g., UWGB 90, 154)
Ambystoma maculatum	Oneida	MPM 5148	Vogt (1981)	
Ambystoma maculatum	Ozaukee	MPM 22699	Casper (1996)	
Ambystoma maculatum	Polk	JFBM 3963	Lang (1972a), Vogt (1981)	
Ambystoma maculatum	Portage	MPM 23436	Casper (1996)	
Ambystoma maculatum	Price	MPM 9090	Vogt (1981)	
Ambystoma maculatum	Racine	USNM 3938	Kapfer et al. (2015)	
Ambystoma maculatum	Rusk	UWZM 23998	Vogt (1981)	
Ambystoma maculatum	Sawyer	MPM VZP417	Casper (1996)	Photograph
Ambystoma maculatum	Shawano	UMMZ 1285–1288	Casper and Rutherford (2018a)	Photograph
Ambystoma maculatum	Sheboygan	MPM VZP652	Casper (1999b)	Photograph
Ambystoma maculatum	Taylor	UWSP 4289	Wild (2012)	
Ambystoma maculatum	Vilas	UWZM 22781	Vogt (1981)	

Scientific Name	County	Museum Voucher	Publication	Notes
Ambystoma maculatum	Washington	UMMZ 243163–4	Casper (2015)	
Ambystoma maculatum	Waukesha	MPM 22537	Casper (1996)	
Ambystoma maculatum	Winnebago	MPM 2139	Vogt (1981)	
Ambystoma maculatum	Wood	UWSP 278	Vogt (1981)	
Ambystoma tigrinum	Barron	MPM VZP479	Casper (1996)	Photograph
Ambystoma tigrinum	Bayfield	MPM VZP973	Jensen (2020)	Photograph
Ambystoma tigrinum	Burnett		Vogt (1981)	Specimen MPM 26487 was from Burnett County, but museum notes indicated it was discarded because the "specimen dried"
Ambystoma tigrinum	Calumet	UWGB 254	Casper (1996)	
Ambystoma tigrinum	Chippewa	MPM VZP595	Casper (1997d)	Photograph
Ambystoma tigrinum	Columbia	USNM 207351	Vogt (1981)	
Ambystoma tigrinum	Dane	MPM 8798	Vogt (1981)	
Ambystoma tigrinum	Dodge	UIMNH 271–273	Suzuki (1951), Vogt (1981)	
Ambystoma tigrinum	Dunn	MPM VZP486	Casper (1996)	Photograph
Ambystoma tigrinum	Eau Claire		Suzuki (1951), Vogt (1981)	Specimens MU 1619 and MPM 2818 were reportedly from the correct location, but currently presumed missing by museum staff
Ambystoma tigrinum	Fond Du Lac	MPM 22512	Vogt (1981)	
Ambystoma tigrinum	Green Lake	UWZM 1932	Vogt (1981)	
Ambystoma tigrinum	Jefferson	MPM 10974	Vogt (1981)	
Ambystoma tigrinum	Kenosha	CHAS HERP5463	Necker (1939b), Vogt (1981)	
Ambystoma tigrinum	Manitowoc	UWGB 356	Casper (1996)	
Ambystoma tigrinum	Marathon		Suzuki (1951), Vogt (1981)	Specimens MU 1611 and MPM 2817 were reportedly from the correct location, but currently presumed missing by museum staff
Ambystoma tigrinum	Marquette	MPM 30316	Casper (1999a)	
Ambystoma tigrinum	Milwaukee	MPM 2472	Vogt (1981)	
Ambystoma tigrinum	Ozaukee	MPM VZP52	Casper (1996)	Photograph
Ambystoma tigrinum	Pepin	MPM VZP507	Casper (1996)	Photograph
Ambystoma tigrinum	Pierce	JFBM 963	Suzuki (1951)	
Ambystoma tigrinum	Polk	MPM 7909	Vogt (1981)	
Ambystoma tigrinum	Portage	UWSP 4088	Vogt (1981)	
Ambystoma tigrinum	Racine	MPM 2816	Vogt (1981)	
Ambystoma tigrinum	Rock	UWZM 20233	Vogt (1981)	

Scientific Name	County	Museum Voucher	Publication	Notes
Ambystoma tigrinum	Sauk	UWZM 2413	Vogt (1981)	
Ambystoma tigrinum	Sheboygan	MPM VZP33	Vogt (1981)	Photograph
Ambystoma tigrinum	St. Croix	MPM 3074	Vogt (1981)	
Ambystoma tigrinum	Trempealeau	MPM VZP549	Casper (1997d)	Photograph
Ambystoma tigrinum	Walworth	FMNH 11036	Schmidt and Necker (1935), Vogt (1981)	
Ambystoma tigrinum	Washburn	MPM VZP520	Casper (1996)	Photograph
Ambystoma tigrinum	Washington	MPM 2815	Vogt (1981)	
Ambystoma tigrinum	Waukesha	MPM 33281	Vogt (1981)	
Ambystoma tigrinum	Waupaca	UWSP 1778	Vogt (1981)	
Ambystoma tigrinum	Waushara	CHAS HERP19298	Suzuki (1951), Vogt (1981)	
Ambystoma tigrinum	Winnebago	UWZM 20238	Vogt (1981)	
Ambystoma tigrinum	Wood		Vogt (1981)	
Anaxyrus americanus	Adams	UWZM 2436	Vogt (1981)	
Anaxyrus americanus	Ashland	MPM 2819	Suzuki (1951), Vogt (1981)	
Anaxyrus americanus	Barron	MPM 2820	Suzuki (1951), Vogt (1981)	
Anaxyrus americanus	Bayfield	MPM 33256	Vogt (1981)	
Anaxyrus americanus	Brown	UWZM 22611	Cochran et al. (1987)	
Anaxyrus americanus	Buffalo	MPM 1170	Vogt (1981)	
Anaxyrus americanus	Burnett	UWZM 2608	Vogt (1981)	
Anaxyrus americanus	Calumet	USNM 207376	Vogt (1981)	
Anaxyrus americanus	Chippewa	MPM 33353	Vogt (1981)	
Anaxyrus americanus	Clark	MPM 4084	Vogt (1981)	
Anaxyrus americanus	Columbia	MPM 2825	Vogt (1981)	
Anaxyrus americanus	Crawford	FMNH 284062	Vogt (1981)	
Anaxyrus americanus	Dane	CM 70283	Vogt (1981)	
Anaxyrus americanus	Dodge	UIMNH 1438	Suzuki (1951), Vogt (1981)	
Anaxyrus americanus	Door	UWZM 2517	Vogt (1981)	
Anaxyrus americanus	Douglas	MPM 924	Vogt (1981)	
Anaxyrus americanus	Dunn	MPM 2828	Vogt (1981)	
Anaxyrus americanus	Eau Claire	UWZM 2519	Vogt (1981)	
Anaxyrus americanus	Florence	UWSP 482	Vogt (1981)	
Anaxyrus americanus	Fond Du Lac	MPM 5070	Vogt (1981)	
Anaxyrus americanus	Forest	FMNH 277127	Vogt (1981)	
Anaxyrus americanus	Grant	MPM 11507	Vogt (1981)	
Anaxyrus americanus	Green	UMMZ 66728	Suzuki (1951), Vogt (1981)	

Scientific Name	County	Museum Voucher	Publication	Notes
Anaxyrus americanus	Green Lake	MPM 2830, 2831, 4091–4100	Suzuki (1951), Vogt (1981)	
Anaxyrus americanus	Iowa	MPM 8792	Vogt (1981)	
Anaxyrus americanus	Iron	MPM 2832	Vogt (1981)	
Anaxyrus americanus	Jackson	UWZM 2462	Vogt (1981)	
Anaxyrus americanus	Jefferson	MPM 2833, 4101–4126	Suzuki (1951), Vogt (1981)	
Anaxyrus americanus	Juneau	UWZM 20463	Vogt (1981)	
Anaxyrus americanus	Kenosha	MPM 26281	Vogt (1981)	
Anaxyrus americanus	Kewaunee	MPM 2834, 4127	Suzuki (1951), Vogt (1981)	
Anaxyrus americanus	La Crosse	UWZM 16156	Suzuki (1951), Vogt (1981)	Suzuki (1951) reported voucher as "MU uncatalogued"
Anaxyrus americanus	Lafayette	MPM VZP786	Vogt (1981)	Photograph
Anaxyrus americanus	Langlade	MPM 3252	Vogt (1981)	
Anaxyrus americanus	Lincoln	UMMZ 64830	Vogt (1981)	
Anaxyrus americanus	Manitowoc	MPM 2837, 2838, 4128–4136	Suzuki (1951), Vogt (1981)	
Anaxyrus americanus	Marathon	MPM 2839	Vogt (1981)	
Anaxyrus americanus	Marinette	MPM 3198	Suzuki (1951), Vogt (1981)	
Anaxyrus americanus	Marquette	MPM 2362	Vogt (1981)	
Anaxyrus americanus	Menominee	MPM 3243	Vogt (1981)	
Anaxyrus americanus	Milwaukee	MPM 2315	Vogt (1981)	
Anaxyrus americanus	Monroe	UWSP 1606	Vogt (1981)	
Anaxyrus americanus	Oconto	MPM 4151	Vogt (1981)	
Anaxyrus americanus	Oneida	MPM 5170	Vogt (1981)	
Anaxyrus americanus	Outagamie	MPM 2842	Suzuki (1951), Vogt (1981)	
Anaxyrus americanus	Ozaukee	MPM 2840, 2847	Suzuki (1951), Vogt (1981)	
Anaxyrus americanus	Pepin		Vogt (1981)	Museum notes indicated that UWZM 2515 was from Pepin County, but this specimen was not found during recent investigations of the collection
Anaxyrus americanus	Pierce	MPM 7916	Vogt (1981)	
Anaxyrus americanus	Polk	MPM 2855	Vogt (1981)	
Anaxyrus americanus	Portage	UWSP 92	Vogt (1981)	
Anaxyrus americanus	Price	MPM 18246	Vogt (1981)	
Anaxyrus americanus	Racine	MPM 32277	Vogt (1981)	
Anaxyrus americanus	Richland	MPM 2848	Vogt (1981)	
Anaxyrus americanus	Rock	UWW HERP0158	Vogt (1981)	

Scientific Name	County	Museum Voucher	Publication	Notes
Anaxyrus americanus	Rusk	MPM VZP419	Casper (1996)	Photograph
Anaxyrus americanus	Sauk	UWZM 2440	Vogt (1981)	
Anaxyrus americanus	Sawyer	MPM 11597	Vogt (1981)	
Anaxyrus americanus	Shawano	MPM 23442	Vogt (1981)	
Anaxyrus americanus	Sheboygan	MPM 23339	Vogt (1981)	
Anaxyrus americanus	St. Croix	UWSP 4024	Vogt (1981)	
Anaxyrus americanus	Taylor	UWZM 20923	Vogt (1981)	
Anaxyrus americanus	Trempealeau	MPM 2851	Suzuki (1951), Vogt (1981)	
Anaxyrus americanus	Vernon	MPM 9169	Vogt (1981)	
Anaxyrus americanus	Vilas	MPM 776	Vogt (1981)	
Anaxyrus americanus	Walworth	CHAS HERP5143	Necker (1939b), Vogt (1981)	
Anaxyrus americanus	Washburn	MPM 12538	Vogt (1981)	
Anaxyrus americanus	Washington	MPM 2856	Suzuki (1951), Vogt (1981)	
Anaxyrus americanus	Waukesha	MPM 30343	Vogt (1981)	
Anaxyrus americanus	Waupaca	MPM 2859, 4170	Suzuki (1951), Vogt (1981)	
Anaxyrus americanus	Waushara	CHAS HERP18478	Vogt (1981)	
Anaxyrus americanus	Winnebago	LSUMZ 13491	Casper (1996)	
Anaxyrus americanus	Wood	MPM 30522	Vogt (1981)	
Apalone mutica	Adams		Vogt (1981)	
Apalone mutica	Buffalo		Vogt (1981)	
Apalone mutica	Columbia	MPM 17501	Vogt (1981)	
Apalone mutica	Crawford	UWZM 2050	Vogt (1981)	
Apalone mutica	Dane	INHS 2008k	Muehlfeld and Kapfer (2009)	Photograph
Apalone mutica	Grant	MPM 6774	Vogt (1981)	
Apalone mutica	Iowa	MPM VZP598	Casper (1997d)	Photograph
Apalone mutica	La Crosse	MPM 33150	Casper (2015)	
Apalone mutica	Pepin	UWZM 2002	Vogt (1981)	
Apalone mutica	Sauk	MPM VZP248	Casper (1996)	Photograph
Apalone mutica	Trempealeau	MPM 6870	Vogt (1981)	
Apalone mutica	Vernon	MPM 2279	Vogt (1981)	
Apalone spinifera	Adams	MPM 6788	Vogt (1981)	
Apalone spinifera	Ashland	MPM VZP868	Moe and Badje (2017)	Photograph
Apalone spinifera	Barron	MPM VZP527	Casper (1996)	Photograph
Apalone spinifera	Brown	UWGB 320	Cochran (1991a), Cochran (1994)	

Scientific Name	County	Museum Voucher	Publication	Notes
Apalone spinifera	Buffalo	CM 88562	Vogt (1981)	
Apalone spinifera	Burnett	MPM 950	Vogt (1981)	
Apalone spinifera	Chippewa	MPM 6841	Vogt (1981)	
Apalone spinifera	Clark	MPM 6883	Vogt (1981)	
Apalone spinifera	Columbia	MPM 11139	Vogt (1981)	
Apalone spinifera	Crawford	MPM 2050	Vogt (1981)	
Apalone spinifera	Dane	MPM 30322	Vogt (1981)	
Apalone spinifera	Dodge	MPM VZP630	Casper (1999b)	Photograph
Apalone spinifera	Douglas	MPM VZP598	Casper (1997d)	Photograph
Apalone spinifera	Dunn	MPM 6845	Vogt (1981)	
Apalone spinifera	Eau Claire	FMNH 15971	Vogt (1981)	
Apalone spinifera	Fond Du Lac	UIMNH 50611	Vogt (1981)	
Apalone spinifera	Grant	MPM 6785	Vogt (1981)	
Apalone spinifera	Green	MPM VZP99	Casper (1996)	Photograph
Apalone spinifera	Green Lake		Vogt (1981)	
Apalone spinifera	Iowa	MPM 6779	Vogt (1981)	
Apalone spinifera	Jackson	MPM 6881	Vogt (1981)	
Apalone spinifera	Jefferson	MPM 2414	Dickinson (1950), Vogt (1981)	
Apalone spinifera	Juneau	MPM VZP424	Casper (1996)	Photograph
Apalone spinifera	Kenosha	MPM VZP699	Casper (2015)	Photograph
Apalone spinifera	La Crosse	INHS 2014i	Vogt (1981)	Photograph
Apalone spinifera	Lincoln	MPM VZP848a–b	Vogt (1981)	Photograph
Apalone spinifera	Marathon	MPM VZP340	Casper (1996)	Photograph
Apalone spinifera	Marquette	MPM 2591	Casper (1996)	
Apalone spinifera	Milwaukee	MPM 22542	Drought (1987a)	
Apalone spinifera	Monroe		Vogt (1981)	
Apalone spinifera	Oneida	MPM 796	Vogt (1981)	
Apalone spinifera	Ozaukee	INHS 2012h	Casper (2015)	Photograph
Apalone spinifera	Pepin	MPM 6850	Vogt (1981)	
Apalone spinifera	Pierce	MPM VZP501	Casper (1996)	Photograph
Apalone spinifera	Polk	MPM 6842	Vogt (1981)	
Apalone spinifera	Portage	MPM 6853	Vogt (1981)	
Apalone spinifera	Price	MPM VZP704	Casper (2015)	Photograph
Apalone spinifera	Racine	CHAS HERP19124	Vogt (1981)	
Apalone spinifera	Richland		Vogt (1981)	
Apalone spinifera	Rock	UWZM 6303	Vogt (1981)	
Apalone spinifera	Rusk	MPM VZP959a–b	Badje (2020b)	Photograph
Apalone spinifera	Sauk	MPM 8819	Vogt (1981)	

Scientific Name	County	Museum Voucher	Publication	Notes
Apalone spinifera	Sawyer	UWZM 22505	Cochran (1982d)	
Apalone spinifera	Sheboygan	CU 11332	Casper (2015)	Cornell University Museum of Vertebrates
Apalone spinifera	St. Croix	MPM 6844	Vogt (1981)	
Apalone spinifera	Trempealeau	MPM 6880	Vogt (1981)	
Apalone spinifera	Vernon	UWZM 23091	Vogt (1981)	
Apalone spinifera	Vilas	MPM VZP657	Casper (2001a)	Photograph
Apalone spinifera	Walworth	MPM 2417	Dickinson (1950), Vogt (1981)	
Apalone spinifera	Washburn	MPM VZP453	Casper (1996)	Photograph
Apalone spinifera	Washington	INHS 2010q	Casper (2015)	Photograph
Apalone spinifera	Waukesha	MPM 17522	Casper (1996)	
Apalone spinifera	Waupaca	UWSP 3803	Kapfer et al. (2015)	
Apalone spinifera	Winnebago	MPM VZP937	Badje (2020a)	Photograph
Apalone spinifera	Wood	MPM VZP1034a–c	Badje (2022b)	Photograph
Aspidoscelis sexlineata	Buffalo		Vogt (1981)	
Aspidoscelis sexlineata	Crawford	UWZM 12946	Vogt (1981)	
Aspidoscelis sexlineata	Dane	UWGB R4	Vogt (1981)	
Aspidoscelis sexlineata	Dunn	MPM VZP815	Vogt (1981)	Photograph
Aspidoscelis sexlineata	Eau Claire		Vogt (1981)	
Aspidoscelis sexlineata	Grant	MPM 27998	Vogt (1981)	
Aspidoscelis sexlineata	Iowa		Vogt (1981)	
Aspidoscelis sexlineata	Monroe	MPM VZP266	Casper (1996)	Photograph
Aspidoscelis sexlineata	Pepin	MPM VZP402	Vogt (1981)	Photograph
Aspidoscelis sexlineata	Pierce	MPM 1110	Vogt (1981)	
Aspidoscelis sexlineata	Richland	FMNH 18436	Vogt (1981)	
Aspidoscelis sexlineata	Sauk	MPM 5124	Vogt (1981)	
Aspidoscelis sexlineata	St. Croix	MPM VZP301	Casper (1996)	Photograph
Aspidoscelis sexlineata	Vernon	MPM 2028	Vogt (1981)	
Carphophis vermis	Grant	MPM 14754	Sajdak (1978), Vogt (1981)	
Chelydra serpentina	Adams	MPM 6833	Vogt (1981)	
Chelydra serpentina	Ashland	MPM VZP490	Casper (1996)	Photograph
Chelydra serpentina	Barron	MPM VZP827	Vogt (1981)	Photograph
Chelydra serpentina	Bayfield	MPM 4553	Vogt (1981)	
Chelydra serpentina	Brown	UWZM 22627	Cochran et al. (1987)	
Chelydra serpentina	Buffalo	MPM VZP299	Casper (1996)	Photograph
Chelydra serpentina	Burnett	UMMZ 72510	Dickinson (1950), Vogt (1981)	

Scientific Name	County	Museum Voucher	Publication	Notes
Chelydra serpentina	Calumet	MPM VZP847	Vogt (1981)	Photograph
Chelydra serpentina	Chippewa	MPM 7287	Vogt (1981)	
Chelydra serpentina	Clark	MPM 6840	Vogt (1981)	
Chelydra serpentina	Columbia	MPM 17429	Vogt (1981)	
Chelydra serpentina	Crawford	MPM 2081	Vogt (1981)	
Chelydra serpentina	Dane	MPM VZP563	Vogt (1981)	Photograph
Chelydra serpentina	Dodge	UIMNH 1658	Vogt (1981)	
Chelydra serpentina	Door	UWGB 393	Watermolen (1990b)	
Chelydra serpentina	Douglas	UWSP 3977	Vogt (1981)	
Chelydra serpentina	Dunn	MPM VZP185	Casper (1996)	Photograph
Chelydra serpentina	Eau Claire	MPM VZP269	Casper (1996)	Photograph
Chelydra serpentina	Florence	MPM VZP285	Casper (1996)	Photograph
Chelydra serpentina	Fond Du Lac	MPM 11004	Vogt (1981)	
Chelydra serpentina	Forest	MPM 3216	Dickinson (1972), Vogt (1981)	
Chelydra serpentina	Grant	FMNH 8892	Vogt (1981)	
Chelydra serpentina	Green	FMNH 283800	Walley (2000a)	Walley (2000a) reported voucher as HDW-NIU 1680, which was later donated to FMNH and given a new catalogue number (A. Resetar)
Chelydra serpentina	Green Lake	MPM VZP640	Casper (1999b)	Photograph
Chelydra serpentina	Iowa	MPM 6834	Vogt (1981)	
Chelydra serpentina	Iron	MPM VZP828	Vogt (1981)	Photograph
Chelydra serpentina	Jackson	UWSP 4145	Vogt (1981)	
Chelydra serpentina	Jefferson	MPM VZP575	Vogt (1981)	Photograph
Chelydra serpentina	Juneau	MPM VZP785	Vogt (1981)	Photograph
Chelydra serpentina	Kenosha	MPM VZP18	Casper (1996)	Photograph
Chelydra serpentina	Kewaunee	UWGB 318	Watermolen (1990a)	
Chelydra serpentina	La Crosse	MPM VZP232	Casper (1996)	Photograph
Chelydra serpentina	Lafayette	UMMZ 64794	Vogt (1981)	
Chelydra serpentina	Langlade	UWGB 373	Casper (1996)	
Chelydra serpentina	Lincoln		Vogt (1981)	Specimen UWZM 20760 is from Lincoln County, but consists of preserved eggs that are difficult to use for independent confirmation of species identification
Chelydra serpentina	Manitowoc	MPM VZP234	Cochran and Marks (1990)	Photograph
Chelydra serpentina	Marathon	UWZM 2010	Vogt (1981)	

Scientific Name	County	Museum Voucher	Publication	Notes
Chelydra serpentina	Marinette	MPM VZP432	Casper (1996)	Photograph
Chelydra serpentina	Marquette	MPM VZP796	Vogt (1981)	Photograph
Chelydra serpentina	Menominee	MPM VZP509	Casper (1996)	Photograph
Chelydra serpentina	Milwaukee	MPM 22663	Casper (1996)	
Chelydra serpentina	Monroe	UWZM 2271	Vogt (1981)	
Chelydra serpentina	Oconto	UWGB 127	Casper (1996)	
Chelydra serpentina	Oneida	INHS 2014a	Vogt (1981)	Photograph
Chelydra serpentina	Outagamie	MPM VZP450	Casper (1996)	Photograph
Chelydra serpentina	Ozaukee	MPM 22518	Casper (1996)	
Chelydra serpentina	Pepin	UWZM 2002	Vogt (1981)	
Chelydra serpentina	Pierce	MPM 7271	Vogt (1981)	
Chelydra serpentina	Polk	MPM 7270	Vogt (1981)	
Chelydra serpentina	Portage	UWSP 3694	Vogt (1981)	
Chelydra serpentina	Price	UWSP 3851	Vogt (1981)	
Chelydra serpentina	Racine	MPM 2418	Vogt (1981)	
Chelydra serpentina	Richland	MPM VZP804	Vogt (1981)	Photograph
Chelydra serpentina	Rock	MPM VZP839	Vogt (1981)	
Chelydra serpentina	Rusk		Vogt (1981)	Specimen UWZM 23952 is from Rusk County, but specimen consists of preserved eggs and difficult to use for independent confirmation of species identification
Chelydra serpentina	Sauk	MPM VZP397	Vogt (1981)	Photograph
Chelydra serpentina	Sawyer	UWSP 3827	Vogt (1981)	
Chelydra serpentina	Shawano	MPM VZP440	Casper (1996)	Photograph
Chelydra serpentina	Sheboygan	MPM VZP29	Casper (1996)	Photograph
Chelydra serpentina	St. Croix	MPM 7288	Vogt (1981)	
Chelydra serpentina	Taylor	MPM 23464	Casper (1996)	Casper (1996) reported voucher as REO 12325, part of the R. E. Olson collection; this collection was later donated to MPM and specimens given new catalogue numbers
Chelydra serpentina	Trempealeau	UWZM 22634	Cochran and Lyons (1986)	
Chelydra serpentina	Vernon	MPM 10982	Vogt (1981)	
Chelydra serpentina	Vilas	UWSP 3865	Vogt (1981)	
Chelydra serpentina	Walworth	FMNH 501	Schmidt and Necker (1935), Vogt (1981)	
Chelydra serpentina	Washburn	MPM 23728	Vogt (1981)	
Chelydra serpentina	Washington	MPM 22133	Vogt (1981)	

Scientific Name	County	Museum Voucher	Publication	Notes
Chelydra serpentina	Waukesha	MPM 2060	Vogt (1981)	
Chelydra serpentina	Waupaca	UWSP 663	Vogt (1981)	
Chelydra serpentina	Waushara	UWSP 2778	Casper (1996)	Casper (1996) reported voucher as UWSP 994, but this specimen could not be found
Chelydra serpentina	Winnebago	MPM VZP632	Casper (1999b)	Photograph
Chelydra serpentina	Wood	UMMZ 64764	Vogt (1981)	
Chrysemys picta	Adams	MPM 7052	Vogt (1981)	
Chrysemys picta	Ashland	MPM 33135	Vogt (1981)	
Chrysemys picta	Barron	MPM VZP363	Casper (1996)	Photograph
Chrysemys picta	Bayfield	MPM 10989	Vogt (1981)	
Chrysemys picta	Brown	UWGB R327	Cochran et al. (1987)	Authors originally reported voucher as UWZM 22617, but this specimen is missing according to recent investigations of the collection
Chrysemys picta	Buffalo	MPM VZP240	Casper (1996)	Photograph
Chrysemys picta	Burnett	MPM 7265	Vogt (1981)	
Chrysemys picta	Calumet	UWGB 1483	Sobota (1990a)	
Chrysemys picta	Chippewa	MPM 6742	Vogt (1981)	
Chrysemys picta	Clark	MPM 6755	Vogt (1981)	
Chrysemys picta	Columbia	MPM 6934	Vogt (1981)	
Chrysemys picta	Crawford	MPM 2065	Vogt (1981)	
Chrysemys picta	Dane	MPM 11709	Vogt (1981)	
Chrysemys picta	Dodge	MPM 11698	Vogt (1981)	
Chrysemys picta	Door	UWGB 320	Watermolen (1990c)	
Chrysemys picta	Douglas	MPM 7250	Vogt (1981)	
Chrysemys picta	Dunn	MPM VZP194	Casper (1996)	Photograph
Chrysemys picta	Eau Claire	UWZM 2037	Vogt (1981)	
Chrysemys picta	Florence	UWGB 309-R	Cochran (1988a)	
Chrysemys picta	Fond Du Lac	MPM 5136	Vogt (1981)	
Chrysemys picta	Forest	MPM 3218	Dickinson (1972), Vogt (1981)	
Chrysemys picta	Grant	MPM 5117	Vogt (1981)	
Chrysemys picta	Green	FMNH 270549	Casper (2015)	
Chrysemys picta	Green Lake	UMNH 11715	Vogt (1981)	Utah Museum of Natural History
Chrysemys picta	Iowa	MPM 7277	Vogt (1981)	
Chrysemys picta	Iron	MPM 781	Vogt (1981)	
Chrysemys picta	Jackson	UMMZ 64771	Vogt (1981)	
Chrysemys picta	Jefferson	MPM VZP574	Vogt (1981)	Photograph

Scientific Name	County	Museum Voucher	Publication	Notes
Chrysemys picta	Juneau	UWZM S23178	Vogt (1981)	Skeletonized specimen
Chrysemys picta	Kenosha	FMNH 229528	Necker (1939b)	
Chrysemys picta	Kewaunee	MPM 6753	Vogt (1981)	
Chrysemys picta	La Crosse	INHS 2014j	Vogt (1981)	Photograph
Chrysemys picta	Lafayette	MPM VZP329	Casper (1996)	Photograph
Chrysemys picta	Langlade	UWGB 374	Casper (1996)	
Chrysemys picta	Lincoln	MPM 7056	Vogt (1981)	
Chrysemys picta	Manitowoc	UWGB 319	Watermolen (1990c)	
Chrysemys picta	Marathon	MPM 23517	Casper (1996)	
Chrysemys picta	Marinette	MPM 3279	Dickinson (1972), Vogt (1981)	
Chrysemys picta	Marquette	MPM VZP190	Casper (1996)	Photograph
Chrysemys picta	Menominee	MPM VZP289	Casper (1996)	Photograph
Chrysemys picta	Milwaukee	MPM 22541	Drought (1987b)	
Chrysemys picta	Monroe	MPM 11054	Vogt (1981)	
Chrysemys picta	Oconto	UWZM 2163	Vogt (1981)	
Chrysemys picta	Oneida	MPM VZP348	Casper (1996)	Photograph
Chrysemys picta	Outagamie	MPM 26387	Casper (1996)	
Chrysemys picta	Ozaukee	MPM VZP21	Casper (1996)	Photograph
Chrysemys picta	Pepin		Vogt (1981)	Museum notes indicated that UWZM 2030 was from Pepin County, but this specimen was not found during recent investigation of the collection
Chrysemys picta	Pierce	MPM 7278	Vogt (1981)	
Chrysemys picta	Polk	MPM 7956	Vogt (1981)	
Chrysemys picta	Portage	MPM 7266	Vogt (1981)	
Chrysemys picta	Price	MPM VZP405	Casper (1996)	Photograph
Chrysemys picta	Racine	UWZM 9866	Vogt (1981)	
Chrysemys picta	Richland	MPM VZP382	Casper (1996)	Photograph
Chrysemys picta	Rock	MPM VZP315	Casper (1996)	
Chrysemys picta	Rusk	MPM 8818	Vogt (1981)	
Chrysemys picta	Sauk	MPM 5105	Vogt (1981)	
Chrysemys picta	Sawyer	MPM 8813	Vogt (1981)	
Chrysemys picta	Shawano	MPM 23441	Casper (1996)	
Chrysemys picta	Sheboygan	MPM VZP26	Casper (1996)	Photograph
Chrysemys picta	St. Croix	MPM 23362	Casper (1996)	
Chrysemys picta	Taylor	MPM VZP307	Casper (1996)	Photograph
Chrysemys picta	Trempealeau	UWZM 2028	Vogt (1981)	
Chrysemys picta	Vernon	MPM 2323	Vogt (1981)	

Scientific Name	County	Museum Voucher	Publication	Notes
Chrysemys picta	Vilas	USNM 207649	Vogt (1981)	
Chrysemys picta	Walworth	MPM 4988	Necker (1939b), Vogt (1981)	Necker (1939) reported various *C. picta* specimens from Walworth County in the collection at Northwestern University (e.g., NU 850–3, etc.); these specimens were transferred to the herpetological collection at the Chicago Academy of Sciences, but we could not determine the fate of the specific *C. picta* specimens he referenced
Chrysemys picta	Washburn	MPM 11000	Vogt (1981)	
Chrysemys picta	Washington	MPM 11952	Vogt (1981)	
Chrysemys picta	Waukesha	MPM 12147	Vogt (1981)	
Chrysemys picta	Waupaca	MPM 6885	Vogt (1981)	
Chrysemys picta	Waushara	MPM 11055	Casper (1996)	
Chrysemys picta	Winnebago	MPM VZP293	Casper (1996)	Photograph
Chrysemys picta	Wood	MPM 7254	Vogt (1981)	
Coluber constrictor	Adams	UWZM 2402	Vogt (1981)	
Coluber constrictor	Buffalo	MPM 1114	Vogt (1981)	
Coluber constrictor	Clark	FMNH 14546	Vogt (1981)	
Coluber constrictor	Columbia	MPM 23016	Vogt (1981)	
Coluber constrictor	Crawford	UWZM 22560	Cochran (1983a)	
Coluber constrictor	Dane	MPM 11840	Vogt (1981)	
Coluber constrictor	Eau Claire	MPM 23203	Casper (1996)	
Coluber constrictor	Grant	MPM 5007	Vogt (1981)	
Coluber constrictor	Iowa	MPM 11779	Vogt (1981)	
Coluber constrictor	Juneau	MPM 23514	Casper (1996)	
Coluber constrictor	La Crosse	JFBM 1831	Casper (1996)	
Coluber constrictor	Lafayette	UWZM 17362	Casper (2015)	
Coluber constrictor	Monroe	MPM 23331	Casper (1996)	
Coluber constrictor	Pepin	UWZM 2127	Vogt (1981)	
Coluber constrictor	Pierce	MPM 1111	Casper (1996)	
Coluber constrictor	Richland	UWZM S28147	Vogt (1981)	
Coluber constrictor	Rock	MPM VZP854a–c	Staffen and Paloski (2017a)	Photograph
Coluber constrictor	Sauk	MPM 23461	Dickinson (1950), Vogt (1981)	Dickinson (1950) reported voucher as MPM 2595, which was not found during recent investigations of collection
Coluber constrictor	St. Croix	MPM VZP303	Casper (1996)	Photograph
Coluber constrictor	Trempealeau	MPM VZP344	Casper (1996)	Photograph
Coluber constrictor	Vernon	MPM 2091	Casper (1996)	

Scientific Name	County	Museum Voucher	Publication	Notes
Coluber constrictor	Walworth	CHAS HERP 5977	Vogt (1981)	
Coluber constrictor	Waushara	FMNH 18435	Vogt (1981)	
Crotalus horridus	Buffalo	MPM VZP225	Vogt (1981)	Photograph
Crotalus horridus	Columbia		Casper (1996)	Casper (1996) reported voucher for Columbia County as MPM 2554 (a specimen that museum records identified as a dried "skin or hide"); recent investigations of museum collections revealed a tattered skin in the MPM basement that was collected in Vernon County by W. R. Spellum; the hide was associated with a tag labelled MPM 2292, with the number 2554 written in pencil beneath
Crotalus horridus	Crawford	MPM VZP435	Casper (1996)	Photograph
Crotalus horridus	Dane	MPM VZP165–66, VZP272	Casper (1996)	Photograph
Crotalus horridus	Grant	UWZM S22398	Vogt (1981)	Skeletonized specimen with associated rattle segments used for confirmation of species identification
Crotalus horridus	Iowa	MPM 23017	Vogt (1981)	
Crotalus horridus	La Crosse	MPM 23357	Casper (1996)	
Crotalus horridus	Pepin	MPM VZP223	Vogt (1981)	Photograph
Crotalus horridus	Pierce	MPM VZP219	Vogt (1981)	Photograph
Crotalus horridus	Richland	MPM VZP170	Casper (1996)	Photograph
Crotalus horridus	Sauk	UWZM 2304	Vogt (1981)	
Crotalus horridus	Trempealeau	MPM VZP229	Vogt (1981)	Photograph
Crotalus horridus	Vernon	MPM 26564	Casper (1996)	
Diadophis punctatus	Ashland	MPM 33255	Vogt (1981)	
Diadophis punctatus	Bayfield	MPM 13544	Vogt (1981)	
Diadophis punctatus	Brown	UWZM 22618	Cochran et al. (1987)	
Diadophis punctatus	Buffalo		Vogt (1981)	
Diadophis punctatus	Chippewa	UWZM 6501	Vogt (1981)	
Diadophis punctatus	Clark	AMNH 17083	Vogt (1981)	
Diadophis punctatus	Crawford	MPM 2358	Vogt (1981)	
Diadophis punctatus	Door	UWSP 3178	Vogt (1981)	
Diadophis punctatus	Douglas	MPM VZP536	Casper (1997d)	Photograph
Diadophis punctatus	Florence	MPM 11539	Vogt (1981)	
Diadophis punctatus	Forest	MPM 26450	Casper (1996)	
Diadophis punctatus	Grant	MPM 1566	Vogt (1981)	
Diadophis punctatus	Iron	MPM 2654	Casper (1996)	
Diadophis punctatus	Jackson	MPM VZP610	Casper and Burnham (2001a)	Photograph

Scientific Name	County	Museum Voucher	Publication	Notes
Diadophis punctatus	Langlade	MPM 2334	Casper (1996)	
Diadophis punctatus	Lincoln	BMR B180001	Casper (1996)	Casper (1996) reported voucher as BMR Z56R035, but the associated institution later gave this a new specimen number
Diadophis punctatus	Manitowoc	UNR 6343	Casper (2015)	University of Nevada, Museum of Biology, Reno
Diadophis punctatus	Marathon	MPM VZP 962a–b	Vogt (1981)	Photograph
Diadophis punctatus	Marinette	MPM 3059	Dickinson (1972), Vogt (1981)	
Diadophis punctatus	Milwaukee	MPM 103	Casper (1996)	
Diadophis punctatus	Monroe	INHS 2008h	Wilder et al. (2009)	Photograph
Diadophis punctatus	Oconto	CM 36867	Casper (2015)	
Diadophis punctatus	Oneida	MPM 27966	Vogt (1981)	
Diadophis punctatus	Pepin	MPM VZP840	Breitenbach and Wild (2016a)	Photograph
Diadophis punctatus	Portage	MPM VZP689	Vogt (1981)	Photograph
Diadophis punctatus	Price		Vogt (1981)	
Diadophis punctatus	Sauk	MPM 1853	Vogt (1981)	
Diadophis punctatus	Sawyer	MPM VZP418	Casper (1996)	Photograph
Diadophis punctatus	Shawano	MPM VZP281	Casper (1996)	Photograph
Diadophis punctatus	Sheboygan	UWZM 22754	Vogt (1981)	
Diadophis punctatus	Taylor	MPM VZP547	Casper (1997d)	Photograph
Diadophis punctatus	Vernon	INHS 2014z	Kapfer et al. (2015)	Photograph
Diadophis punctatus	Vilas	UWZM 19058	Vogt (1981)	
Diadophis punctatus	Washburn	MPM VZP333	Casper (1996)	Photograph
Diadophis punctatus	Waukesha		Vogt (1981)	
Emydoidea blandingii	Adams	MPM 23434	Casper (1996)	
Emydoidea blandingii	Ashland	MPM VZP909	Mertig et al. (2019)	Photograph
Emydoidea blandingii	Barron	MPM VZP590	Casper (1997d)	Photograph
Emydoidea blandingii	Bayfield	MPM VZP907		Photograph
Emydoidea blandingii	Brown		Cochran et al. (1987)	Authors reported voucher photographs BBWS J1108–20, J1201, J1204, but the associated collection no longer exists according to correspondence with Bay Beach Wildlife Sanctuary staff
Emydoidea blandingii	Buffalo	UWZM S20196	Vogt (1981)	Skeletonized specimen with shell and scutes for confirmation of species

Scientific Name	County	Museum Voucher	Publication	Notes
Emydoidea blandingii	Burnett	MPM VZP313	Vogt (1981)	Photograph
Emydoidea blandingii	Chippewa	MPM VZP191	Casper (1996)	Photograph
Emydoidea blandingii	Clark	MPM 23471	Casper (1996)	
Emydoidea blandingii	Columbia	MPM VZP627	Vogt (1981)	Photograph
Emydoidea blandingii	Crawford	MPM VZP638	Vogt (1981)	Photograph
Emydoidea blandingii	Dane	MPM 1680	Vogt (1981)	
Emydoidea blandingii	Dodge	MPM VZP913a–g	Bartholmai and Badje (2019)	Photograph
Emydoidea blandingii	Door	MPM VZP275	Casper (1996)	Photograph
Emydoidea blandingii	Douglas	MPM VZP429	Casper (1996)	Photograph
Emydoidea blandingii	Dunn	MPM VZP523	Casper (1996)	Photograph
Emydoidea blandingii	Eau Claire	MPM 23199	Casper (1996)	
Emydoidea blandingii	Florence	MPM VZP461	Casper (1996)	Photograph
Emydoidea blandingii	Fond Du Lac	MPM VZP602	Casper (1997d)	Photograph
Emydoidea blandingii	Grant	MPM VZP388	Vogt (1981)	Photograph
Emydoidea blandingii	Green	MPM VZP178	Casper (1996)	Photograph
Emydoidea blandingii	Green Lake	MPM VZP412	Vogt (1981)	Photograph
Emydoidea blandingii	Iowa	MPM 4555	Vogt (1981)	
Emydoidea blandingii	Jackson	UWZM 22635	Casper (1996)	
Emydoidea blandingii	Jefferson	MPM VZP339	Casper (1996)	Photograph
Emydoidea blandingii	Juneau	MPM VZP35	Casper (1996)	Photograph
Emydoidea blandingii	Kenosha	MPM VZP9	Casper (1996)	Photograph
Emydoidea blandingii	La Crosse	MPM VZP391	Vogt (1981)	Photograph
Emydoidea blandingii	Lafayette	MPM 24397	Casper (1996)	
Emydoidea blandingii	Marathon	MPM 23516	Casper (1996)	
Emydoidea blandingii	Marinette	MPM VZP335	Vogt (1981)	Photograph
Emydoidea blandingii	Marquette	MPM 2511	Dickinson (1950), Vogt (1981)	
Emydoidea blandingii	Menominee	MPM VZP276	Casper (1996)	Photograph
Emydoidea blandingii	Milwaukee	MPM VZP206	Casper (1996)	Photograph
Emydoidea blandingii	Monroe	MPM VZP140	Casper (1996)	Photograph
Emydoidea blandingii	Oconto	UWZM 22626	Casper (1996)	
Emydoidea blandingii	Oneida	MPM VZP491	Casper (1996)	Photograph
Emydoidea blandingii	Outagamie	MPM VZP204	Casper (1996)	Photograph
Emydoidea blandingii	Ozaukee	INHS 2007a	Casper et al. (2007)	Photograph
Emydoidea blandingii	Pepin	MPM VZP773a–b	Vogt (1981)	Photograph
Emydoidea blandingii	Pierce	MPM VZP869	Lundeen and Badje (2017)	Photograph
Emydoidea blandingii	Polk	MPM VZP287	Casper (1996)	Photograph
Emydoidea blandingii	Portage	FMNH 14714	Vogt (1981)	

Scientific Name	County	Museum Voucher	Publication	Notes
Emydoidea blandingii	Racine		Vogt (1981)	Museum notes indicated that MPM 32287 was from Racine County, but this specimen was not found during recent investigations of the collection
Emydoidea blandingii	Richland	MPM VZP829	Vogt (1981)	Photograph
Emydoidea blandingii	Rock	MPM VZP645	Casper (1999b)	Photograph
Emydoidea blandingii	Rusk	MPM 1995	Vogt (1981)	
Emydoidea blandingii	Sauk	MPM 5108	Vogt (1981)	
Emydoidea blandingii	Sawyer	MPM VZP288	Casper (1996)	Photograph
Emydoidea blandingii	Shawano	UWGB 307–R	Korb (1988a)	
Emydoidea blandingii	Sheboygan	MPM 23438	Casper (1996)	
Emydoidea blandingii	St. Croix	MPM VZP774a–b	Vogt (1981)	Photograph
Emydoidea blandingii	Taylor	MPM VZP938	Badje (2020c)	Photograph
Emydoidea blandingii	Trempealeau	MPM VZP146	Vogt (1981)	
Emydoidea blandingii	Vernon	MPM VZP283	Casper (1996)	Photograph
Emydoidea blandingii	Walworth	FMNH 21690	Necker (1939b), Vogt (1981)	
Emydoidea blandingii	Washburn	MPM VZP271	Vogt (1981)	Photograph
Emydoidea blandingii	Washington	MPM 17457	Casper (1996)	
Emydoidea blandingii	Waukesha	MPM 27987	Vogt (1981)	
Emydoidea blandingii	Waupaca	MPM VZP504	Vogt (1981)	Photograph
Emydoidea blandingii	Waushara	MPM VZP337	Casper (1996)	Photograph
Emydoidea blandingii	Winnebago	MPM VZP431	Casper (1996)	Photograph
Emydoidea blandingii	Wood	MPM 8807	Vogt (1981)	
Glyptemys insculpta	Adams		Vogt (1981)	
Glyptemys insculpta	Ashland	MPM 26361	Casper (1996)	
Glyptemys insculpta	Barron	MPM VZP577	Casper (1997d)	Photograph
Glyptemys insculpta	Bayfield	MPM 4556	Vogt (1981)	
Glyptemys insculpta	Brown	UWZM 22628	Cochran et al. (1987)	
Glyptemys insculpta	Buffalo	MPM 23198	Casper (1996)	Broken shell
Glyptemys insculpta	Burnett	MPM VZP425	Casper (1996)	Photograph
Glyptemys insculpta	Chippewa	MPM VZP247	Casper (1996)	Photograph
Glyptemys insculpta	Clark	MPM VZP916a–b	Badje (2019)	Photograph
Glyptemys insculpta	Columbia	CM 96230	Vogt (1981)	
Glyptemys insculpta	Crawford	MPM VZP336	Casper (1996)	Photograph
Glyptemys insculpta	Dane		Vogt (1981)	
Glyptemys insculpta	Douglas	MPM 27981	Vogt (1981)	Bag of dried fragments
Glyptemys insculpta	Dunn	MPM VZP912	Schuler and Badje (2019)	Photograph
Glyptemys insculpta	Eau Claire		Vogt (1981)	

Scientific Name	County	Museum Voucher	Publication	Notes
Glyptemys insculpta	Florence	UWGB 316–R	Cochran (1988b)	
Glyptemys insculpta	Forest	MPM VZP260	Dickinson (1950)	Dickinson (1950) listed voucher as MPM 5686, which could not be found during recent investigations of the collection
Glyptemys insculpta	Iron	MPM VZP811a	Vogt (1981)	Photograph
Glyptemys insculpta	Jackson	FMNH 19213	Vogt (1981)	
Glyptemys insculpta	La Crosse	INHS 2014o	Vogt (1981)	Photograph
Glyptemys insculpta	Langlade	MPM VZP914	Arrowood et al. (2019)	Photograph
Glyptemys insculpta	Lincoln	MPM VZP809a	Vogt (1981)	Photograph
Glyptemys insculpta	Marathon	MPM 26356	Vogt (1981)	
Glyptemys insculpta	Marinette	MPM VZP714	Vogt (1981)	Photograph
Glyptemys insculpta	Menominee	MPM 1856	Vogt (1981)	
Glyptemys insculpta	Monroe	MPM VZP37	Vogt (1981)	Photograph
Glyptemys insculpta	Oconto	MPM VZP629	Vogt (1981)	Photograph
Glyptemys insculpta	Oneida	MPM VZP42	Casper (1996)	Photograph
Glyptemys insculpta	Pierce	MPM VZP230	Casper (1996)	Photograph
Glyptemys insculpta	Polk	UWZM 2051	Vogt (1981)	
Glyptemys insculpta	Portage	MPM VZP611	Vogt (1981)	Photograph
Glyptemys insculpta	Price		Vogt (1981)	Museum notes indicated MPM VZP705 was a photographic specimen from Price County, but the photograph could not be found
Glyptemys insculpta	Richland		Vogt (1981)	
Glyptemys insculpta	Rock	UIMNH 2287	Vogt (1981)	
Glyptemys insculpta	Rusk	MPM VZP469	Casper (1996)	Photograph
Glyptemys insculpta	Sauk	UWWHERP 0272	Vogt (1981)	Also MPM VZP813a–b, VZP814
Glyptemys insculpta	Sawyer	MPM VZP552	Vogt (1981)	Photograph
Glyptemys insculpta	Shawano	MPM VZP43	Casper (1996)	Photograph
Glyptemys insculpta	St. Croix	MPM VZP512	Vogt (1981)	Photograph
Glyptemys insculpta	Taylor	MPM VZP325	Casper (1996)	Photograph
Glyptemys insculpta	Trempealeau	UWZM 17559	Vogt (1981)	
Glyptemys insculpta	Vernon	INHS 2015a	Johnston et al. (2015)	Photograph
Glyptemys insculpta	Vilas	MPM VZP370	Casper (1996)	Photograph
Glyptemys insculpta	Washburn	MPM VZP810a	Vogt (1981)	Photograph
Glyptemys insculpta	Waupaca	FMNH 14715	Vogt (1981)	
Glyptemys insculpta	Wood	UWZM 17018	Vogt (1981)	
Graptemys geographica	Adams	MPM VZP927–31	Watson et al. (2020)	Photographs
Graptemys geographica	Buffalo	UWZM 2139	Vogt (1981)	
Graptemys geographica	Burnett	UWSP 4141	Vogt (1981)	

Scientific Name	County	Museum Voucher	Publication	Notes
Graptemys geographica	Chippewa	MPM 6836	Vogt (1981)	Casper (2015) reported MPM 6836 as voucher for the Chippewa County record of *G. pseudogeographica*, but this specimen is a misidentified *G. geographica*
Graptemys geographica	Columbia	MPM 6922	Vogt (1981)	
Graptemys geographica	Crawford	USNM 288149	Casper (1997d)	
Graptemys geographica	Dane	UWZM 4906	Vogt (1981)	
Graptemys geographica	Dunn	MPM VZP1030a–c		Photograph
Graptemys geographica	Grant	MPM 1574	Vogt (1981)	
Graptemys geographica	Green	MPM VZP156	Casper (1996)	Photograph
Graptemys geographica	Green Lake	MPM VZP783	Vogt (1981)	Photograph
Graptemys geographica	Iowa	MPM 8817	Vogt (1981)	
Graptemys geographica	Jackson	MPM VZP864	Raimond (2017)	Photograph
Graptemys geographica	Juneau	MPM 25658	Casper (1996)	
Graptemys geographica	Kenosha	MPM 29588	Bolek (1997a), Casper (1999b)	
Graptemys geographica	La Crosse	MSB 83346	Vogt (1981)	
Graptemys geographica	Marathon	MPM VZP977, 978a–b, 979	Badje (2020d)	Photograph
Graptemys geographica	Marquette	MPM VZP186	Casper (1996)	Photograph
Graptemys geographica	Milwaukee	INHS 2012m	Casper (2015)	Photograph
Graptemys geographica	Outagamie	MPM VZP449	Casper (1996)	
Graptemys geographica	Pepin	MPM VZP974a–e	Badje and Staffen (2021)	Photograph
Graptemys geographica	Pierce	JFBM R490	Casper (2015)	
Graptemys geographica	Polk	CM 95123	Vogt (1981)	
Graptemys geographica	Portage	UWSP 3376	Vogt (1981)	
Graptemys geographica	Richland	MPM VZP381	Casper (1996)	Photograph
Graptemys geographica	Sauk	UWZM 2704	Vogt (1981)	
Graptemys geographica	Shawano	MPM 27978	Casper (1997b)	
Graptemys geographica	St. Croix	CM 95121	Vogt (1981)	
Graptemys geographica	Trempealeau	CM 107666	Vogt (1981)	
Graptemys geographica	Vernon	UWZM 21952	Vogt (1981), Vogt (1993)	
Graptemys geographica	Waukesha	MPM VZP867	Skryja (2017)	Photograph
Graptemys geographica	Waupaca	UWSP 4081	Vogt (1981)	
Graptemys geographica	Waushara	MPM VZP556	Casper (1997d)	Photograph
Graptemys geographica	Winnebago	MPM 2377	Dickinson (1950), Vogt (1981)	

Scientific Name	County	Museum Voucher	Publication	Notes
Graptemys geographica	Wood	MPM VZP1032–33	Badje (2021c)	Photograph
Graptemys ouachitensis	Adams	MPM VZP932, 933a–b	Badje et al. (2020)	Photograph
Graptemys ouachitensis	Buffalo	CM 28807	Vogt (1993)	
Graptemys ouachitensis	Columbia	CM 95110	Vogt (1981), Vogt (1993)	
Graptemys ouachitensis	Crawford	MPM VZP1022	Buchanan et al. (2022)	Photograph
Graptemys ouachitensis	Dane	INHS 2014e	Vogt (1981)	Photograph
Graptemys ouachitensis	Grant	CM 95100	Vogt (1981), Vogt (1993)	Vogt (1993) reported UWZM 21930 for this county, which could not be found during recent investigations of collection
Graptemys ouachitensis	Iowa	CM 95103	Vogt (1981), Vogt (1993)	
Graptemys ouachitensis	Jackson	MPM VZP934	Badje and Flory (2020)	Photograph
Graptemys ouachitensis	Juneau	MPM VZP925	Badje et al. (2020)	Photograph
Graptemys ouachitensis	La Crosse	INHS 2014k	Vogt (1981)	Photograph
Graptemys ouachitensis	Pepin	UWZM 2033	Vogt and Kapfer (2016)	Original museum notes incorrectly identified this specimen as *G. pseudogeographica*, while the label on the specimen jar incorrectly identified it as *G. kohni*.
Graptemys ouachitensis	Portage	CM 95113	Kapfer et al. (2015)	
Graptemys ouachitensis	Richland	INHS 14043	Vogt (1981)	
Graptemys ouachitensis	Sauk	MPM VZP800	Vogt (1981)	
Graptemys ouachitensis	Trempealeau	CM 107668	Vogt (1981), Vogt (1993)	
Graptemys ouachitensis	Vernon	UWZM 21212	Vogt (1981), Vogt (1993)	
Graptemys pseudogeographica	Buffalo	JFBM Herp Photo 472	Casper (2015)	Photograph
Graptemys pseudogeographica	Chippewa		Casper (2015)	Casper (2015) reported MPM 6836 as the voucher specimen for Chippewa County, but this specimen is a misidentified *G. geographica*
Graptemys pseudogeographica	Columbia	MPM 11951	Vogt (1981)	
Graptemys pseudogeographica	Crawford	USNM 288153	Casper (1997d)	
Graptemys pseudogeographica	Dane		Vogt (1981)	Museum records indicated that specimen UWZM 17600 was from Dane County, but this specimen could not be found during recent investigations of the collection

Scientific Name	County	Museum Voucher	Publication	Notes
Graptemys pseudogeographica	Eau Claire		Vogt (1981)	Casper (1996) reported MPM VZP208–11 for this county, but followed earlier convention that treated *G. pseudogeographica* and *G. ouachitensis* as subspecies of *G. pseudogeographica* on the range map and in the appendix (but referred to the specimen as *G. pseudogeographica* in text)
Graptemys pseudogeographica	Grant	CM 95117	Vogt (1981), Vogt (1993)	Vogt (1993) reported personal specimen numbers for this county that could not be linked to catalogued museum specimens.
Graptemys pseudogeographica	Iowa		Vogt (1981)	Vogt (1993) listed voucher as CM 95120, but museum notes indicated that the specimen in question is *G. ouachitensis*, not *G. pseudogeographica*; pictures of the specimen support this
Graptemys pseudogeographica	La Crosse	MPM VZP830	Vogt (1981)	Photograph
Graptemys pseudogeographica	Pepin		Vogt (1981)	Museum notes incorrectly identified UWZM 2033 as *G. pseudogeographica* from Pepin County but specimen was actually *G. ouachitensis* (R. Vogt)
Graptemys pseudogeographica	Pierce	CM 87555	Kapfer et al. (2015)	
Graptemys pseudogeographica	Polk	CM 95116	Vogt (1993)	
Graptemys pseudogeographica	Portage	CM 95112	Vogt (1981), Vogt (1993)	Vogt (1993) reported personal specimen numbers for this county that could not be linked to catalogued museum specimens.
Graptemys pseudogeographica	Richland		Vogt (1981)	Museum notes indicated that MPM VZP395 is from Richland County, but this photographic specimen only depicts a top-down view of the carapace so cannot be used to definitively identify as *G. pseudogeographica*
Graptemys pseudogeographica	Sauk		Vogt (1981)	
Graptemys pseudogeographica	St. Croix	UMMZ 72505	Dickinson (1950), Vogt (1981)	
Graptemys pseudogeographica	Trempealeau	CM 107661	Vogt (1981), Vogt (1993)	
Graptemys pseudogeographica	Vernon	UWZM 21198	Vogt (1981), Vogt (1993)	

Scientific Name	County	Museum Voucher	Publication	Notes
Hemidactylium scutatum	Adams	MPM VZP897a–b	Staffen and Paloski (2017b)	Photograph
Hemidactylium scutatum	Ashland	MPM 11451	Vogt (1981), Casper (1993a)	
Hemidactylium scutatum	Barron	UWWHERP 0266	Wild (2016a)	
Hemidactylium scutatum	Bayfield	JFBM 3958	Lang (1972b), Vogt (1981)	
Hemidactylium scutatum	Brown	UWGB 427–R	Casper (1993a)	
Hemidactylium scutatum	Burnett	MPM VZP1027a–b		Photograph
Hemidactylium scutatum	Chippewa	MPM VZP550	Casper (1997d)	Photograph
Hemidactylium scutatum	Clark	FMNH 14701	Vogt (1981), Casper (1993a)	
Hemidactylium scutatum	Columbia	CM 62536	Casper (2015)	
Hemidactylium scutatum	Crawford	UWZM 24018	Casper (2015)	
Hemidactylium scutatum	Dane	UWZM 17390	Vogt (1981), Casper (1993a)	
Hemidactylium scutatum	Door	UWGB 513–R	Price (2000), Price (2001)	
Hemidactylium scutatum	Douglas	MPM 27960	Casper (1996b), Casper (2015)	
Hemidactylium scutatum	Eau Claire	MPM 27939	Casper (1996)	
Hemidactylium scutatum	Fond Du Lac	MPM VZP779a	Vogt (1981)	Photograph
Hemidactylium scutatum	Forest	MPM 30220	Casper (1996)	Casper (1996) reported voucher as MPM 26485, which was not found during recent investigations of collection
Hemidactylium scutatum	Iowa	INHS 2015o	Kapfer et al. (2015)	Photograph
Hemidactylium scutatum	Iron	UWSP 3745	Wild (2007)	
Hemidactylium scutatum	Jackson	MPM 30246	Casper and Kirk (2001)	
Hemidactylium scutatum	Juneau		Rossman and Rossman (2001)	Authors reported voucher as MPM VZP694, a photograph that could not be found
Hemidactylium scutatum	Lincoln	UMMZ 243203	Casper (2015)	
Hemidactylium scutatum	Marathon	MPM VZP1037	LaFleur and Kutok (2022)	Photograph
Hemidactylium scutatum	Marinette	MPM VZP701	Casper (2015)	Photograph

Scientific Name	County	Museum Voucher	Publication	Notes
Hemidactylium scutatum	Monroe	UWWHERP 0261	Wild (2016a)	
Hemidactylium scutatum	Oconto	MPM 2805	Vogt (1981), Casper (1993a)	
Hemidactylium scutatum	Oneida	MPM VZP462	Casper (1996)	Photograph
Hemidactylium scutatum	Ozaukee	MPM VZP1036a–e	Cannizzaro and Schmanski (2022)	Photograph
Hemidactylium scutatum	Polk	UWZM 16901	Vogt (1981), Casper (1993a)	
Hemidactylium scutatum	Portage		Vogt (1981), Casper (1993a)	Casper (1993a) reported voucher as UWSP 940, but according to current museum records this catalogue number is associated with a different species
Hemidactylium scutatum	Price	INHS 2015ab	Kapfer et al. (2015)	Photograph
Hemidactylium scutatum	Richland	FMNH 14716	Vogt (1981), Casper (1993a)	
Hemidactylium scutatum	Rusk	UWSP 4300	Wild (2013a)	
Hemidactylium scutatum	Sauk	UWZM 13828	Vogt (1981), Casper (1993a),	
Hemidactylium scutatum	Sawyer	MPM 32903	Casper and Kirk (2001), Casper (2015)	
Hemidactylium scutatum	Shawano	UMMZ 1289–1290	Casper and Rutherford (2018b)	Photograph
Hemidactylium scutatum	Sheboygan		Vogt (1981), Casper (1993a)	Casper (1993a) reported voucher as SIUCH 3182 (Southern Illinois University), but repeated attempts to contact the institution regarding this specimen were ignored
Hemidactylium scutatum	Taylor	MPM 23204	Casper (1993a)	
Hemidactylium scutatum	Vernon	MPM 2329	Vogt (1981), Casper (1993a)	
Hemidactylium scutatum	Vilas	UWZM 17389	Casper (1993a)	
Hemidactylium scutatum	Walworth	CHAS HERP2560	Vogt (1981)	
Hemidactylium scutatum	Washburn	JFBM 3961	Lang (1972b), Vogt (1981)	
Hemidactylium scutatum	Wood	UMMZ i1284	Reid and Riddle (2017a)	Photograph
Heterodon platirhinos	Adams	MPM 6938	Vogt (1981)	

Scientific Name	County	Museum Voucher	Publication	Notes
Heterodon platirhinos	Ashland	MPM 30244	Casper and Ames (1998)	
Heterodon platirhinos	Bayfield	UWSP 3537	Vogt (1981)	
Heterodon platirhinos	Buffalo	INHS 2009o	Becker et al. (2009)	Photograph
Heterodon platirhinos	Burnett	MPM 959	Vogt (1981)	
Heterodon platirhinos	Clark	FMNH 14544	Vogt (1981)	
Heterodon platirhinos	Columbia	FMNH 15992	Vogt (1981)	
Heterodon platirhinos	Crawford	MPM 2136	Vogt (1981)	
Heterodon platirhinos	Dane	UWZM 1961	Vogt (1981)	
Heterodon platirhinos	Douglas	MPM 25653	Vogt (1981)	
Heterodon platirhinos	Dunn		Vogt (1981)	
Heterodon platirhinos	Eau Claire	UWZM 2251	Vogt (1981)	
Heterodon platirhinos	Florence	MPM 2606	Casper (1996)	
Heterodon platirhinos	Grant	MPM VZP615	Vogt (1981)	Photograph
Heterodon platirhinos	Green	UWZM 22788	Kapfer et al. (2015)	
Heterodon platirhinos	Green Lake		Vogt (1981)	
Heterodon platirhinos	Iowa	MPM 7874	Vogt (1981)	
Heterodon platirhinos	Jackson	MPM VZP98	Casper (1996)	Photograph
Heterodon platirhinos	Jefferson	INHS 2015x	Kapfer et al. (2015)	Photograph
Heterodon platirhinos	Juneau	CM 70414	Vogt (1981)	
Heterodon platirhinos	La Crosse	INHS 2015h	Kapfer et al. (2015)	Photograph
Heterodon platirhinos	Langlade	MPM VZP312	Casper (1996)	Photograph
Heterodon platirhinos	Marathon	UWSP 291	Vogt (1981)	
Heterodon platirhinos	Marinette	MPM 3075	Dickinson (1950), Vogt (1981)	Dickinson (1950) reported voucher as MPM 2606, which could not be found during recent investigations of the collection
Heterodon platirhinos	Marquette	MPM 23467	Casper (1996)	Casper (1996) reported voucher as REO 11668 (Robert E. Olson collection), which was later donated to MPM and then assigned an MPM catalogue number
Heterodon platirhinos	Menominee	MPM 4535	Vogt (1981)	
Heterodon platirhinos	Monroe	MPM 2519	Dickinson (1950)	
Heterodon platirhinos	Oconto	INHS VZ2014ab	Vogt (1981)	Photograph
Heterodon platirhinos	Pepin	MPM VZP976a–d	White et al. (2020a)	Photograph
Heterodon platirhinos	Pierce	MPM 1117	Vogt (1981)	
Heterodon platirhinos	Polk	MPM 23726	Casper (1996)	

Scientific Name	County	Museum Voucher	Publication	Notes
Heterodon platirhinos	Portage	MPM 2043	Vogt (1981)	
Heterodon platirhinos	Richland	MPM VZP542	Casper (1997d)	Photograph
Heterodon platirhinos	Rock	UWZM 9796	Vogt (1981)	
Heterodon platirhinos	Sauk	MPM 26357	Vogt (1981)	
Heterodon platirhinos	Shawano	MPM 23668	Casper (1996)	
Heterodon platirhinos	Trempealeau	MPM VZP345	Casper (1996)	Photograph
Heterodon platirhinos	Vernon	MPM 2034	Vogt (1981)	
Heterodon platirhinos	Walworth	MPM 1741	Vogt (1981)	
Heterodon platirhinos	Washburn	UWSP 3586	Vogt (1981)	
Heterodon platirhinos	Waukesha	MPM 22513	Vogt (1981)	
Heterodon platirhinos	Waupaca	UWGB 317–R	Korb (1988b)	
Heterodon platirhinos	Waushara	MPM VZP586	Vogt (1981)	Photograph
Heterodon platirhinos	Wood	UMMZ i1284	Reid and Riddle (2017b)	Photograph
Hyla chrysoscelis	Barron	MPM VZP998	Paloski (2022)	Audio file
Hyla chrysoscelis	Burnett	UWZM 20682	Vogt (1981)	Museum records indicated species identification confirmed via call
Hyla chrysoscelis	Chippewa		Casper (1997d)	Casper (1997d) reported voucher as "MPM audio tape (Cat. No. pending)," but correspondences with museum staff supports that audio files are missing (J. Colby)
Hyla chrysoscelis	Columbia	MPM 12731	Vogt (1981)	
Hyla chrysoscelis	Dane		Vogt (1981)	Museum records indicated that UWZM 20283 was from Dane County (identifiication confirmed via call), but this specimen was not found during recent investigations of the collection
Hyla chrysoscelis	Douglas	MPM 16596	Vogt (1981)	
Hyla chrysoscelis	Dunn		Vogt (1981)	
Hyla chrysoscelis	Eau Claire	MPM VZP967	Casper (1997d)	Audio file; Casper (1997d) reported voucher as "MPM audio tape (Cat. No. pending)," but correspondences with museum staff supports that this audio file is missing (J. Colby)
Hyla chrysoscelis	Grant		Vogt (1981)	
Hyla chrysoscelis	Green	MPM 12735	Vogt (1981)	Museum records indicated species identification confirmed via call
Hyla chrysoscelis	Green Lake	MPM VZP1031a–b	Badje (2022a)	Audio file
Hyla chrysoscelis	Iowa	UWZM 20715	Vogt (1981)	
Hyla chrysoscelis	Jackson	MPM 12736	Vogt (1981)	

Scientific Name	County	Museum Voucher	Publication	Notes
Hyla chrysoscelis	Jefferson	INHS 2014Y	Casper (2015)	Audio file; Casper (2015) reported voucher as MPM VZP658 (a photograph with "associated audio recording"), but MPM audio files are missing, according to correspondence with museum staff (J. Colby)
Hyla chrysoscelis	Juneau	CM 70373	Vogt (1981)	
Hyla chrysoscelis	Kenosha	MPM VZP616–18	Casper (1999b)	Photographs; Casper (1999b) references associated audio files on record that are missing according to museum staff (J. Colby)
Hyla chrysoscelis	Marinette	MPM 12703	Vogt (1981)	Museum records indicated species identification confirmed via call
Hyla chrysoscelis	Marquette	MPM 8731	Vogt (1981)	Museum records indicated species identification confirmed via call
Hyla chrysoscelis	Monroe		Vogt (1981)	
Hyla chrysoscelis	Oconto	MPM 11639	Vogt (1981)	
Hyla chrysoscelis	Pepin		Casper (1997d)	Casper (1997d) reported voucher as "MPM audio tape (Cat. No. pending)," but correspondences with museum staff supports that audio files are missing (J. Colby)
Hyla chrysoscelis	Racine	MPM VZP872	Badje (2017a)	Audio file
Hyla chrysoscelis	Rock	MPM 18210	Vogt (1981)	
Hyla chrysoscelis	Rusk		Casper (1997d)	Casper (1997d) reported voucher as "MPM audio tape (Cat. No. pending)," but correspondences with museum staff supports that audio files are missing (J. Colby)
Hyla chrysoscelis	St. Croix	UWZM 21177	Vogt (1981)	
Hyla chrysoscelis	Walworth	MPM 8331	Vogt (1981)	
Hyla chrysoscelis	Waukesha	MPM 29589	Bolek (1997b)	Bolek (1997b) also referenced vouchered slides with blood samples housed at MPM, but collections staff indicated that these slides are currently missing (J. Colby)
Hyla chrysoscelis	Waupaca	MPM 12686	Vogt (1981)	Museum records indicated species identification confirmed via call
Hyla chrysoscelis	Waushara	UWZM 20686	Vogt (1981)	Museum records indicated species identification confirmed via call
Hyla chrysoscelis	Wood	MPM VZP892	Vogt (1981)	Audio file
Hyla versicolor	Ashland	MPM 18154	Vogt (1981)	

Scientific Name	County	Museum Voucher	Publication	Notes
Hyla versicolor	Barron	MPM VZP1018	Casper (1996)	Audio file; Casper (1996) originally reported voucher as "MPM audio tape," but correspondences with museum staff supports that audio files are missing (J. Colby)
Hyla versicolor	Bayfield	MPM VZP1008	Vogt (1981)	Audio file
Hyla versicolor	Brown		Cochran et al. (1987)	Authors reported voucher as UWZM 22612, but museum records indicated this catalogue number was assigned to a *H. chrysoscelis* specimen
Hyla versicolor	Buffalo	MPM VZP875	Badje (2017b)	Audio file
Hyla versicolor	Burnett	MPM 949	Vogt (1981)	
Hyla versicolor	Chippewa	MPM VZP1019	Vogt (1981)	Audio file
Hyla versicolor	Clark	MPM VZP955	Vogt (1981)	Audio file
Hyla versicolor	Columbia	MPM 2879	Suzuki (1951), Vogt (1981)	
Hyla versicolor	Crawford	FMNH 284050	Walley and Scalleta (2000)	Walley and Scalleta (2000) reported voucher as HDW-NIU 1678, but this specimen was later donated to the FMNH and given a new number
Hyla versicolor	Dane	MPM VZP948	Paloski (2020)	Audio file
Hyla versicolor	Door	UMMZ HerpPhoto1261	Casper (2015)	Photograph with sonogram and audio file
Hyla versicolor	Douglas	MPM VZP1011	Vogt (1981)	Audio file
Hyla versicolor	Dunn	MPM VZP1020	Vogt (1981)	Audio file
Hyla versicolor	Eau Claire	MPM VZP968	Casper (1997d)	Audio file; Casper (1997d) reported voucher as "MPM audio tape (Cat. No. pending)," but correspondences with museum staff supports that this audio file is missing (J. Colby)
Hyla versicolor	Florence	UWGB 311–R	Cochran (1988e)	Author indicated that species identification confirmed via call
Hyla versicolor	Fond Du Lac		Vogt (1981)	
Hyla versicolor	Forest	MPM VZP969	Casper (1996)	Audio file; Casper (1996) reported several adult and larval whole specimen vouchers
Hyla versicolor	Grant		Vogt (1981)	Museum records indicated specimens MPM 1583 and 14753 were collected in Grant County, but the former was disposed of and the latter could not be found
Hyla versicolor	Green	UWZM 20665	Vogt (1981)	Museum records indicated species identification confirmed via call
Hyla versicolor	Green Lake	MPM VZP965	Badje (2021a)	Audio file

Scientific Name	County	Museum Voucher	Publication	Notes
Hyla versicolor	Iowa	UWZM 20710	Vogt (1981)	Museum records indicated species identification confirmed via call
Hyla versicolor	Iron	MPM 2878	Suzuki (1951), Vogt (1981)	
Hyla versicolor	Jackson	UWZM 20762	Vogt (1981)	
Hyla versicolor	Jefferson	INHS 2008g	Kapfer et al. (2008b)	Audio file
Hyla versicolor	Juneau	UWZM 20332	Vogt (1981)	Museum records indicated species identification confirmed via call
Hyla versicolor	Kewaunee	MPM 6902	Vogt (1981)	
Hyla versicolor	La Crosse	MPM VZP873	Vogt (1981)	Audio file
Hyla versicolor	Lafayette	MPM 11554	Vogt (1981)	
Hyla versicolor	Langlade	MPM VZP953	Badje (2020e)	Audio file
Hyla versicolor	Lincoln	MPM VZP952	Vogt (1981)	Audio file
Hyla versicolor	Manitowoc	MPM 2877	Suzuki (1951), Vogt (1981)	
Hyla versicolor	Marathon	MPM VZP947	Vogt (1981)	Audio file
Hyla versicolor	Marinette	MPM VZP958	Suzuki (1951), Vogt (1981)	Audio file; Suzuki (1951) reported voucher as MU 1015, part of a historical collection at Marquette University that was either donated or eliminated, according to correspondence with institution staff
Hyla versicolor	Marquette	MPM 18162	Vogt (1981)	
Hyla versicolor	Menominee	MPM VZP951	Badje (2020e)	Audio file
Hyla versicolor	Milwaukee	UWGB Photo 1498	Casper (2016a)	Photograph with accompanying audio file
Hyla versicolor	Monroe	MPM VZP1021	Vogt (1981)	Audio file
Hyla versicolor	Oconto	MPM VZP949–50	Suzuki (1951), Vogt (1981)	Audio file
Hyla versicolor	Oneida	MPM VZP1015	Vogt (1981)	Audio file
Hyla versicolor	Outagamie	UWZM 20771	Vogt (1981)	Museum records indicated species identification confirmed via call
Hyla versicolor	Ozaukee	MPM VZP534	Casper (1996)	Photograph; Museum records indicated species identification confirmed via call
Hyla versicolor	Pepin	MPM VZP880	Casper (1996)	Audio file; Casper (1996) originally reported voucher as "MPM audio tape," but these audio files are currently missing (J. Colby)
Hyla versicolor	Pierce	UWZM 20692	Vogt (1981)	
Hyla versicolor	Polk	MPM 2874	Vogt (1981)	Museum records indicated species identification confirmed via call
Hyla versicolor	Portage	MPM VZP946a–b	Badje (2020e)	Audio file

Scientific Name	County	Museum Voucher	Publication	Notes
Hyla versicolor	Price	MPM VZP960–61	Vogt (1981)	Audio file
Hyla versicolor	Richland	MPM 12705	Vogt (1981)	
Hyla versicolor	Rusk	MPM VZP957	Suzuki (1951), Vogt (1981)	Audio file; Suzuki (1951) reported voucher as UWZM "Uncatalogued"
Hyla versicolor	Sauk	MPM 12706	Vogt (1981)	Museum records indicated species identification confirmed via call
Hyla versicolor	Sawyer	MPM VZP956	Vogt (1981)	Audio file
Hyla versicolor	Shawano	UMMZ 1293	Casper and Rutherford (2018c)	Audio file and spectrogram image
Hyla versicolor	Sheboygan	MPM VZP101	Vogt (1981)	Photograph
Hyla versicolor	St. Croix	UWZM 20849	Vogt (1981)	
Hyla versicolor	Taylor	MPM VZP963–4	Suzuki (1951), Vogt (1981)	Audio file
Hyla versicolor	Trempealeau	MPM VZP877–8	Suzuki (1951), Vogt (1981)	Audio file; Suzuki (1951) reported voucher as MU 1225, part of a historical Marquette University collection that was either donated or eliminated according to correspondences with institution staff
Hyla versicolor	Vernon	MPM VZP971	Vogt (1981)	Audio file
Hyla versicolor	Vilas	MPM VZP895	Vogt (1981)	Audio file
Hyla versicolor	Washburn	MPM 2660	Suzuki (1951), Vogt (1981)	
Hyla versicolor	Washington	MPM VZP181	Casper (1996)	Photograph; Museum records indicated species identification confirmed via call
Hyla versicolor	Waukesha	MPM VZP881	Fritz (2017)	Audio file
Hyla versicolor	Waupaca	MPM VZP1016	Vogt (1981)	Audio file
Hyla versicolor	Waushara	UWZM 20673	Vogt (1981)	Museum records indicated species identification confirmed via call
Hyla versicolor	Winnebago	MPM VZP888	Badje (2017b)	Audio file
Hyla versicolor	Wood	UWSP 2581	Vogt (1981)	Museum records indicated species identification confirmed via call
Lampropeltis triangulum	Barron	JFBM 2575	Vogt (1981)	
Lampropeltis triangulum	Brown	UWGB R246	Cochran et al. (1987)	Cochran et al (1987) reported voucher as BBWS RE09, but this collection no longer exists according to recent correspondences with Bay Beach Wildlife Sanctuary staff
Lampropeltis triangulum	Buffalo	UWZM 13542	Vogt (1981)	

Scientific Name	County	Museum Voucher	Publication	Notes
Lampropeltis triangulum	Calumet	MPM VZP378	Casper (1996)	Photograph
Lampropeltis triangulum	Clark	MPM VZP1038	LaFleur (2022)	Photograph
Lampropeltis triangulum	Columbia	UWSP 1912	Vogt (1981)	
Lampropeltis triangulum	Crawford	MPM 26377	Vogt (1981)	
Lampropeltis triangulum	Dane	MPM VZP465	Vogt (1981)	Photograph
Lampropeltis triangulum	Dodge	MPM 11718	Vogt (1981)	
Lampropeltis triangulum	Door	MPM 1529	Vogt (1981)	
Lampropeltis triangulum	Dunn	MPM 23200	Casper (1996)	
Lampropeltis triangulum	Eau Claire	MPM 26323	Casper (1996)	
Lampropeltis triangulum	Fond Du Lac	MPM 5016	Vogt (1981)	
Lampropeltis triangulum	Grant	MPM VZP321	Casper (1996)	Photograph
Lampropeltis triangulum	Green	MPM 27954	Vogt (1981)	
Lampropeltis triangulum	Green Lake	UNSM ZM16472	Casper (2015)	University of Nebraska State Museum
Lampropeltis triangulum	Iowa	MPM 27997	Vogt (1981)	
Lampropeltis triangulum	Jefferson	MPM VZP690	Vogt (1981)	Photograph
Lampropeltis triangulum	Juneau	MPM VZP39	Casper (1996)	Photograph
Lampropeltis triangulum	Kenosha	CHAS HERP1392	Schmidt and Necker (1935), Vogt (1981)	
Lampropeltis triangulum	La Crosse	MPM VZP411	Casper (1996)	Photograph
Lampropeltis triangulum	Lafayette	MPM 28004	Casper (1997d)	
Lampropeltis triangulum	Manitowoc	UF 25058	Casper (1996)	
Lampropeltis triangulum	Marathon	MPM 23515	Vogt (1981)	
Lampropeltis triangulum	Marinette		Vogt (1981)	
Lampropeltis triangulum	Marquette	CHAS HERP18896	Vogt (1981)	

Scientific Name	County	Museum Voucher	Publication	Notes
Lampropeltis triangulum	Milwaukee	MPM VZP7	Casper (1996)	
Lampropeltis triangulum	Monroe	UWSP 1722	Vogt (1981)	
Lampropeltis triangulum	Oconto	UWGB 172	Casper (1996)	
Lampropeltis triangulum	Ozaukee	MPM 22517	Casper (1996)	
Lampropeltis triangulum	Pepin	MPM VZP841	Breitenbach and Wild (2016b)	Photograph
Lampropeltis triangulum	Pierce	MPM VZP358	Casper (1996)	Photograph
Lampropeltis triangulum	Polk	MPM 23468	Vogt (1981)	
Lampropeltis triangulum	Racine	MPM 2452	Dickinson (1950), Vogt (1981)	
Lampropeltis triangulum	Richland	USNM 99807	Casper (1997d)	
Lampropeltis triangulum	Rock	MPM 22502	Vogt (1981)	
Lampropeltis triangulum	Sauk	MPM 2108	Vogt (1981)	
Lampropeltis triangulum	Shawano	MPM 26292	Casper (1996)	
Lampropeltis triangulum	Sheboygan	MPM 22432	Vogt (1981)	
Lampropeltis triangulum	St. Croix	MPM VZP524	Casper (1996)	Photograph
Lampropeltis triangulum	Trempealeau	INHS 2014w	Kapfer et al. (2015)	Photograph
Lampropeltis triangulum	Vernon	MPM 2036	Vogt (1981)	
Lampropeltis triangulum	Walworth	CHAS HERP526	Dickinson (1950), Vogt (1981)	
Lampropeltis triangulum	Washington	MPM 2514	Vogt (1981)	
Lampropeltis triangulum	Waukesha	MPM 16301	Vogt (1981)	
Lampropeltis triangulum	Waushara	MPM VZP135	Casper (1996)	Photograph
Lampropeltis triangulum	Winnebago	MPM VZP641	Casper (1999b)	Photograph
Lampropeltis triangulum	Wood	UWSP 2737	Vogt (1981)	

Scientific Name	County	Museum Voucher	Publication	Notes
Lithobates catesbeianus	Ashland		Vogt (1981)	Museum records indicated that UWZM 20635 was from Ashland County, but this specimen was not found during recent investigations of collection
Lithobates catesbeianus	Barron	CHAS HERP19341	Suzuki (1951), Vogt (1981)	
Lithobates catesbeianus	Bayfield	UWZM 3528	Vogt (1981)	
Lithobates catesbeianus	Brown	UWGB 1488s	McFarland and Bishop (1990)	
Lithobates catesbeianus	Chippewa	MPM 3491	Casper (1996)	
Lithobates catesbeianus	Clark		Vogt (1981)	
Lithobates catesbeianus	Crawford	UWZM 2856	Vogt (1981)	
Lithobates catesbeianus	Dane	UWZM 16708	Vogt (1981)	
Lithobates catesbeianus	Dodge		Vogt (1981)	
Lithobates catesbeianus	Douglas	MPM VZP891	Kessler and Badje (2017)	Audio file
Lithobates catesbeianus	Dunn		Casper (1996)	Casper (1996) reported voucher as "MPM audio tape," but correspondences with museum staff supported that audio files are missing (J. Colby)
Lithobates catesbeianus	Eau Claire		Vogt (1981)	
Lithobates catesbeianus	Fond Du Lac	MPM VZP972	Vogt (1981)	Audio file
Lithobates catesbeianus	Grant	JFBM HerpPhoto470	Casper (2015)	Photograph
Lithobates catesbeianus	Green	MPM VZP882a–b	Paloski et al. (2018)	Photograph
Lithobates catesbeianus	Green Lake	UWGB Photo 1499	Casper and Schmidt (2016)	Photograph
Lithobates catesbeianus	Iron	UWZM 2911	Vogt (1981)	
Lithobates catesbeianus	Jefferson	MPM 30342	Vogt (1981)	
Lithobates catesbeianus	Juneau	MPM VZP871	Stevenson and Badje (2017)	Audio file
Lithobates catesbeianus	Kenosha	MPM 26282	Suzuki (1951), Vogt (1981)	Suzuki (1951) reported voucher as MU 1260, but this historical Marquette University collection was either donated or eliminated according to correspondences with institution staff
Lithobates catesbeianus	La Crosse	MPM VZP896	Casper (1996)	Casper (1996) reported voucher as MPM VZP553, which consisted of photographs that appear to be pictures of a newspaper article
Lithobates catesbeianus	Lafayette	MPM VZP331	Casper (1996)	Photograph
Lithobates catesbeianus	Lincoln	UWSP 572	Vogt (1981)	
Lithobates catesbeianus	Marathon	MPM 4032	Vogt (1981)	

Scientific Name	County	Museum Voucher	Publication	Notes
Lithobates catesbeianus	Marinette	MPM 33384	Suzuki (1951), Vogt (1981)	Suzuki (1951) reported MU 1272, but this historical Marquette University collection was either donated or eliminated according to correspondences with institution staff
Lithobates catesbeianus	Marquette	MPM VZP890	Badje (2017c)	Audio file
Lithobates catesbeianus	Milwaukee	MPM VZP162	Casper (1996)	Photograph
Lithobates catesbeianus	Oneida	UWZM 20364	Vogt (1981)	
Lithobates catesbeianus	Ozaukee	MPM VZP49	Casper (1996)	Photograph
Lithobates catesbeianus	Polk	JFBM 5242	Vogt (1981)	
Lithobates catesbeianus	Portage	MPM VZP883	Stevenson et al. (2017)	Audio file
Lithobates catesbeianus	Price		Vogt (1981)	
Lithobates catesbeianus	Racine	USNM 3514	Vogt (1981)	
Lithobates catesbeianus	Rock	UWZM 20203	Vogt (1981)	
Lithobates catesbeianus	Rusk	FMNH 16361	Vogt (1981)	
Lithobates catesbeianus	Sauk	UWZM 16210	Vogt (1981)	
Lithobates catesbeianus	Sawyer	UWZM 22504	Cochran (1982c)	
Lithobates catesbeianus	Shawano	UMMZ 1294	Casper (2018)	Audio file and spectrogram image
Lithobates catesbeianus	St. Croix		Casper (1996)	Casper (1996) reported voucher as "MPM audio tape," which is missing according to correspondences with museum staff (J. Colby)
Lithobates catesbeianus	Trempealeau		Casper (2015)	Voucher reported as CHAS HERP 19340, which the author stated was currently housed at FMNH; correspondence with the FMNH collections manager supported that no such specimen existed in their collection (A. Resetar), while correspondence with CHAS collections manager indicated that, although CHAS HERP 19340 existed in their collection, the catalogue number was assigned to a *L. clamitans* specimen (D. Roberts)
Lithobates catesbeianus	Vernon	MPM 2078	Vogt (1981)	
Lithobates catesbeianus	Vilas	MPM 11573	Vogt (1981)	
Lithobates catesbeianus	Walworth	MPM 4990	Vogt (1981)	
Lithobates catesbeianus	Washburn	MPM 11606	Vogt (1981)	
Lithobates catesbeianus	Washington	MPM 30270	Vogt (1981)	
Lithobates catesbeianus	Waukesha	MPM 5142	Vogt (1981)	
Lithobates catesbeianus	Waushara	UWSP 1918	Vogt (1981)	
Lithobates catesbeianus	Winnebago	MPM VZP889	Badje (2017c)	Audio file

Scientific Name	County	Museum Voucher	Publication	Notes
Lithobates catesbeianus	Wood	MPM VZP870	Stevenson and Badje (2017)	Audio file
Lithobates clamitans	Adams	UWZM 3840	Vogt (1981)	
Lithobates clamitans	Ashland	MPM 33118	Vogt (1981)	
Lithobates clamitans	Barron	MPM 2977	Suzuki (1951), Vogt (1981)	
Lithobates clamitans	Bayfield	MPM 33182	Vogt (1981)	
Lithobates clamitans	Brown	UWZM 20227	Vogt (1981)	
Lithobates clamitans	Buffalo	MPM 1171	Vogt (1981)	
Lithobates clamitans	Burnett	MPM 929	Vogt (1981)	
Lithobates clamitans	Calumet	MPM 2921	Suzuki (1951), Vogt (1981)	
Lithobates clamitans	Chippewa	MPM 33347	Vogt (1981)	
Lithobates clamitans	Clark	MPM 7889	Vogt (1981)	
Lithobates clamitans	Columbia	MPM 11126	Vogt (1981)	
Lithobates clamitans	Crawford	UWZM 2753	Vogt (1981)	
Lithobates clamitans	Dane	MPM 10799	Vogt (1981)	
Lithobates clamitans	Dodge	MPM 10797	Vogt (1981)	
Lithobates clamitans	Door	UWZM 21241	Vogt (1981)	
Lithobates clamitans	Douglas	MPM 30317	Vogt (1981)	
Lithobates clamitans	Dunn	UWZM 3329	Vogt (1981)	
Lithobates clamitans	Eau Claire	MPM 2984	Vogt (1981)	
Lithobates clamitans	Florence	UWSP 487	Vogt (1981)	
Lithobates clamitans	Fond Du Lac	MPM 3508	Vogt (1981)	
Lithobates clamitans	Forest	UWSP 3805	Dickinson (1972), Vogt (1981)	
Lithobates clamitans	Grant	MPM 17384	Vogt (1981)	
Lithobates clamitans	Green	MPM 11470	Vogt (1981)	
Lithobates clamitans	Green Lake	UWSP 4054	Vogt (1981)	
Lithobates clamitans	Iowa	MPM 16057	Vogt (1981)	
Lithobates clamitans	Iron	UWZM 9644	Vogt (1981)	
Lithobates clamitans	Jackson	MPM 13928	Vogt (1981)	
Lithobates clamitans	Jefferson	MPM 13209	Vogt (1981)	
Lithobates clamitans	Juneau	UWZM 22279	Vogt (1981)	
Lithobates clamitans	Kenosha	CHAS NWU1066	Necker (1939b), Vogt (1981)	
Lithobates clamitans	Kewaunee	MPM 2989	Suzuki (1951), Vogt (1981)	
Lithobates clamitans	La Crosse	JFBM 1923	Suzuki (1951), Vogt (1981)	Suzuki (1951) reported voucher as "MU uncatalogued"

Scientific Name	County	Museum Voucher	Publication	Notes
Lithobates clamitans	Lafayette	MPM 2901	Suzuki (1951), Vogt (1981)	
Lithobates clamitans	Langlade	MPM 14695	Vogt (1981)	
Lithobates clamitans	Lincoln	MPM 2993	Vogt (1981)	
Lithobates clamitans	Manitowoc	MPM 2994	Suzuki (1951), Vogt (1981)	
Lithobates clamitans	Marathon	MPM 3540	Vogt (1981)	
Lithobates clamitans	Marinette	MPM 2946	Suzuki (1951), Vogt (1981)	
Lithobates clamitans	Marquette	UWSP 4048	Vogt (1981)	
Lithobates clamitans	Menominee	MPM 3669	Vogt (1981)	
Lithobates clamitans	Milwaukee	MPM 3006	Vogt (1981)	
Lithobates clamitans	Monroe	MPM 7890	Vogt (1981)	
Lithobates clamitans	Oconto	MPM 11716	Vogt (1981)	
Lithobates clamitans	Oneida	MPM 12453	Vogt (1981)	
Lithobates clamitans	Outagamie	MPM 3008	Suzuki (1951), Vogt (1981)	
Lithobates clamitans	Ozaukee	MPM 32899	Vogt (1981)	
Lithobates clamitans	Pepin	UWSP 4070	Vogt (1981) [unclear], Casper (1996)	
Lithobates clamitans	Pierce	JFBM 5199	Vogt (1981)	
Lithobates clamitans	Polk	MPM 3565	Vogt (1981)	
Lithobates clamitans	Portage	UWSP 3563	Vogt (1981)	
Lithobates clamitans	Price	UMMZ 69588	Suzuki (1951), Vogt (1981)	
Lithobates clamitans	Racine	MPM 3569	Vogt (1981)	
Lithobates clamitans	Richland	MPM 7823	Vogt (1981)	
Lithobates clamitans	Rock	MPM 9042	Vogt (1981)	
Lithobates clamitans	Rusk	UMMZ 69592	Suzuki (1951), Vogt (1981)	
Lithobates clamitans	Sauk	MPM 11414	Vogt (1981)	
Lithobates clamitans	Sawyer	UWSP 3828	Vogt (1981)	
Lithobates clamitans	Shawano	MPM 3007	Vogt (1981)	
Lithobates clamitans	Sheboygan	UWZM 1977	Suzuki (1951), Vogt (1981)	
Lithobates clamitans	St. Croix	JFBM A–843	Suzuki (1951), Vogt (1981)	
Lithobates clamitans	Taylor	UMMZ 64823	Suzuki (1951), Vogt (1981)	
Lithobates clamitans	Trempealeau	MPM 3016	Vogt (1981)	
Lithobates clamitans	Vernon	MPM 2270	Suzuki (1951), Vogt (1981)	Suzuki (1951) reported voucher as "UWZM uncatalogued"

Scientific Name	County	Museum Voucher	Publication	Notes
Lithobates clamitans	Vilas	MPM 11467	Vogt (1981)	
Lithobates clamitans	Walworth	CHAS HERP5145	Necker (1939b), Vogt (1981)	
Lithobates clamitans	Washburn	MPM 11612	Vogt (1981)	
Lithobates clamitans	Washington	MPM 23019	Vogt (1981)	
Lithobates clamitans	Waukesha	MPM 3631	Vogt (1981)	
Lithobates clamitans	Waupaca	MPM 3021	Suzuki (1951), Vogt (1981)	
Lithobates clamitans	Waushara	CHAS HERP18888	Suzuki (1951), Vogt (1981)	Suzuki (1951) reported voucher as NU 462, but this specimen was later incorporated into the CHAS collection and given a new catalogue number
Lithobates clamitans	Winnebago	LSUMZ 11305	Casper (1996)	
Lithobates clamitans	Wood	MPM 2553	Vogt (1981)	
Lithobates palustris	Adams	USNM 312213	Kapfer et al. (2015)	
Lithobates palustris	Buffalo	MPM 16393	Vogt (1981)	
Lithobates palustris	Burnett		Vogt (1981)	
Lithobates palustris	Chippewa	MPM 2897	Suzuki (1951), Vogt (1981)	
Lithobates palustris	Clark	MPM 7883	Vogt (1981)	
Lithobates palustris	Columbia	UWZM 19172	Suzuki (1951), Vogt (1981)	Suzuki (1951) reported voucher as UWZM "uncatalogued"
Lithobates palustris	Crawford	FMNH 284059–61	Walley (2000b)	Walley (2000b) reported voucher as HDW-NIU 1682–3, but investigations into the fate of this specimen supports it is missing
Lithobates palustris	Dane	MPM 9034	Vogt (1981)	
Lithobates palustris	Douglas	MPM 940	Vogt (1981)	
Lithobates palustris	Dunn	USNM 131968	Vogt (1981)	
Lithobates palustris	Eau Claire	MPM 23201	Casper (1996)	
Lithobates palustris	Fond Du Lac	MPM 32883	Casper (2015)	
Lithobates palustris	Grant	MPM 22162	Vogt (1981)	
Lithobates palustris	Green	UWZM 18918	Vogt (1981)	
Lithobates palustris	Iowa	MPM 32203	Vogt (1981)	
Lithobates palustris	Jackson	UWZM 3501	Casper (2015)	
Lithobates palustris	Juneau	MPM VZP40	Casper (1996)	Photograph
Lithobates palustris	La Crosse		Vogt (1981)	Museum records indicated JFBM 2142 and 2143 were from La Crosse County, but museum staff confirmed these specimens are currently missing (B. Lowe)
Lithobates palustris	Lafayette	MPM 17381	Vogt (1981)	
Lithobates palustris	Langlade		Vogt (1981)	

Scientific Name	County	Museum Voucher	Publication	Notes
Lithobates palustris	Marinette	MPM 3726	Suzuki (1951), Vogt (1981)	
Lithobates palustris	Marquette	MPM VZP795a	Vogt (1981)	Photograph
Lithobates palustris	Monroe	MPM 7898	Vogt (1981)	
Lithobates palustris	Pierce	LSUMZ 47088	Casper (1996)	
Lithobates palustris	Portage	UWSP 1798	Vogt (1981)	
Lithobates palustris	Racine	USNM 167616	Vogt (1981)	
Lithobates palustris	Richland	MPM 11547	Casper (1996)	
Lithobates palustris	Rock	MPM 22159	Casper (1996)	
Lithobates palustris	Rusk		Vogt (1981)	
Lithobates palustris	Sauk	MPM VZP406	Vogt (1981)	Photograph
Lithobates palustris	Sheboygan	MSUM HE.2064	Vogt (1981)	
Lithobates palustris	Trempealeau	MPM 16619	Vogt (1981)	
Lithobates palustris	Vernon	MPM 2296	Vogt (1981)	
Lithobates palustris	Walworth	UWZM 9616	Necker (1939b), Vogt (1981)	Necker (1939b) reported voucher as FMNH 22414, which is a skeletonized specimen that was difficult to use for independent confirmation of species identification (A. Resetar)
Lithobates palustris	Washington	MPM 2302	Suzuki (1951), Vogt (1981)	
Lithobates palustris	Waukesha	MPM 1514	Vogt (1981)	
Lithobates palustris	Waupaca	MPM 2896	Suzuki (1951), Vogt (1981)	
Lithobates palustris	Waushara	CHAS HERP18886	Suzuki (1951), Vogt (1981)	Suzuki (1951) reported voucher as NU 460, but this specimen was transferred to CHAS and given a new catalogue number
Lithobates pipiens	Adams	UWZM 3816	Vogt (1981)	
Lithobates pipiens	Ashland	MPM 32954	Vogt (1981)	
Lithobates pipiens	Barron	MPM 2917	Suzuki (1951), Vogt (1981)	
Lithobates pipiens	Bayfield	USNM 39270	Pace (1974), Vogt (1981)	
Lithobates pipiens	Brown	MPM 2920	Suzuki (1951), Vogt (1981)	
Lithobates pipiens	Buffalo	MPM 8778	Vogt (1981)	
Lithobates pipiens	Burnett	MPM 11516	Vogt (1981)	
Lithobates pipiens	Calumet	MPM 2978	Suzuki (1951), Vogt (1981)	
Lithobates pipiens	Chippewa	AMNH 6842	Pace (1974), Vogt (1981)	
Lithobates pipiens	Clark	MPM 3791	Vogt (1981)	

Scientific Name	County	Museum Voucher	Publication	Notes
Lithobates pipiens	Columbia	MPM 11124	Vogt (1981)	
Lithobates pipiens	Crawford	MPM 3820	Vogt (1981)	
Lithobates pipiens	Dane	MPM 11697	Vogt (1981)	
Lithobates pipiens	Dodge	CHAS HERP19321	Suzuki (1951), Vogt (1981)	
Lithobates pipiens	Door	MPM 8774	Vogt (1981)	
Lithobates pipiens	Douglas	MPM 908	Vogt (1981)	
Lithobates pipiens	Dunn	MPM 2932	Vogt (1981)	
Lithobates pipiens	Eau Claire	MPM 3861	Vogt (1981)	
Lithobates pipiens	Florence	UWSP 493	Vogt (1981)	
Lithobates pipiens	Fond Du Lac	MPM 8770	Vogt (1981)	
Lithobates pipiens	Forest	MPM 2565	Suzuki (1951), Vogt (1981)	
Lithobates pipiens	Grant	MPM 4574	Vogt (1981)	
Lithobates pipiens	Green	UMMZ 69631	Suzuki (1951), Vogt (1981)	
Lithobates pipiens	Green Lake	MPM 2934	Suzuki (1951), Vogt (1981)	
Lithobates pipiens	Iowa	UWZM 22793	Vogt (1981)	
Lithobates pipiens	Iron	MPM 2937	Vogt (1981)	
Lithobates pipiens	Jackson	UWZM 2560	Vogt (1981)	
Lithobates pipiens	Jefferson	MPM 2939	Suzuki (1951), Vogt (1981)	Suzuki (1951) reported MU 1543–44, but this historical Marquette University collection was either donated or eliminated according to correspondences with institution staff
Lithobates pipiens	Juneau	UWZM 20467	Vogt (1981)	
Lithobates pipiens	Kenosha	CHAS HERP2634	Schmidt and Necker (1935), Vogt (1981)	
Lithobates pipiens	Kewaunee	MPM 2941	Suzuki (1951), Vogt (1981)	
Lithobates pipiens	La Crosse	JFBM 1957	Suzuki (1951), Vogt (1981)	Suzuki (1951) reported voucher as "MU uncatalogued"
Lithobates pipiens	Lafayette	UMMZ 69629	Suzuki (1951), Vogt (1981)	
Lithobates pipiens	Langlade	MPM 8670	Vogt (1981)	
Lithobates pipiens	Lincoln	MPM 8764	Vogt (1981)	
Lithobates pipiens	Manitowoc	MPM 2949	Suzuki (1951), Vogt (1981)	
Lithobates pipiens	Marathon	MPM 8695	Vogt (1981)	
Lithobates pipiens	Marinette	MPM 2894	Suzuki (1951), Vogt (1981)	
Lithobates pipiens	Marquette	MPM 8729	Vogt (1981)	

Scientific Name	County	Museum Voucher	Publication	Notes
Lithobates pipiens	Menominee	MPM 1906	Vogt (1981)	
Lithobates pipiens	Milwaukee	MPM 5143	Vogt (1981)	
Lithobates pipiens	Monroe	UWZM 2593	Vogt (1981)	
Lithobates pipiens	Oconto	MPM 8681	Vogt (1981)	
Lithobates pipiens	Oneida	MPM 8739	Vogt (1981)	
Lithobates pipiens	Outagamie	MPM 3934	Suzuki (1951), Vogt (1981)	
Lithobates pipiens	Ozaukee	MPM 8711	Vogt (1981)	
Lithobates pipiens	Pepin	UWZM 2600	Vogt (1981)	
Lithobates pipiens	Pierce	MPM 1191	Vogt (1981)	
Lithobates pipiens	Polk	MPM 11509	Vogt (1981)	
Lithobates pipiens	Portage	MPM 8701	Vogt (1981)	
Lithobates pipiens	Price	UWZM 2926	Vogt (1981)	
Lithobates pipiens	Racine	USNM 3421	Pace (1974), Vogt (1981)	
Lithobates pipiens	Richland	MPM 2964	Vogt (1981)	
Lithobates pipiens	Rock	MPM 22158	Vogt (1981)	
Lithobates pipiens	Rusk	MPM 8796	Vogt (1981)	
Lithobates pipiens	Sauk	MPM 7829	Vogt (1981)	
Lithobates pipiens	Sawyer	UWZM 3189	Vogt (1981)	
Lithobates pipiens	Shawano	MPM 8744	Vogt (1981)	
Lithobates pipiens	Sheboygan	MPM 23340	Vogt (1981)	
Lithobates pipiens	St. Croix	MPM 1190	Vogt (1981)	
Lithobates pipiens	Taylor	UMMZ 69630	Suzuki (1951), Vogt (1981)	
Lithobates pipiens	Trempealeau	MPM 8779	Vogt (1981)	
Lithobates pipiens	Vernon	MPM 1541	Vogt (1981)	
Lithobates pipiens	Vilas	MPM 11468	Vogt (1981)	
Lithobates pipiens	Walworth	CHAS HERP5148	Necker (1939b), Vogt (1981)	
Lithobates pipiens	Washburn	UWSP 3974	Vogt (1981)	
Lithobates pipiens	Washington	MPM 23020	Vogt (1981)	
Lithobates pipiens	Waukesha	MPM 3981	Vogt (1981)	
Lithobates pipiens	Waupaca	MPM 2972	Suzuki (1951), Vogt (1981)	
Lithobates pipiens	Waushara	CHAS HERP19318	Suzuki (1951), Vogt (1981)	Suzuki (1951) reported voucher as NU 893, which was transferred to CHAS and given a new catalogue number
Lithobates pipiens	Winnebago	AMNH 51294	Pace (1974), Vogt (1981)	
Lithobates pipiens	Wood	MPM 2974	Vogt (1981)	

Scientific Name	County	Museum Voucher	Publication	Notes
Lithobates septentrionalis	Ashland	UWSP 778	Casper (1996)	
Lithobates septentrionalis	Bayfield	MPM 11605	Vogt (1981)	
Lithobates septentrionalis	Burnett	FMNH 11075	Suzuki (1951), Vogt (1981)	
Lithobates septentrionalis	Chippewa	MPM VZP593	Casper (1997d)	Photograph
Lithobates septentrionalis	Douglas	MPM 16529	Vogt (1981)	
Lithobates septentrionalis	Florence		Vogt (1981)	
Lithobates septentrionalis	Forest	MPM 2566	Suzuki (1951), Vogt (1981)	
Lithobates septentrionalis	Iron	MPM 2303	Vogt (1981)	
Lithobates septentrionalis	Langlade	MPM 3232	Vogt (1981)	
Lithobates septentrionalis	Marathon	MPM 3474	Kapfer et al. (2015)	
Lithobates septentrionalis	Marinette	MPM 3221	Dickinson (1972), Vogt (1981)	
Lithobates septentrionalis	Menominee	MPM 3999	Vogt (1981)	
Lithobates septentrionalis	Oconto		Vogt (1981)	
Lithobates septentrionalis	Oneida	UWZM 20993	Vogt (1981)	
Lithobates septentrionalis	Polk	MPM VZP846	Collins and Badje (2016)	Audio file
Lithobates septentrionalis	Price	UNSM ZM19846	Vogt (1981)	University of Nebraska State Museum
Lithobates septentrionalis	Rusk	MPM VZP565	Casper (1997d)	Photograph
Lithobates septentrionalis	Sawyer		Vogt (1981)	
Lithobates septentrionalis	Taylor	UWZM 20275	Vogt (1981)	
Lithobates septentrionalis	Vilas	UWZM 20377	Vogt (1981)	
Lithobates septentrionalis	Washburn	MPM 8799	Vogt (1981)	
Lithobates sylvaticus	Adams	MPM 11443	Vogt (1981)	
Lithobates sylvaticus	Ashland	MPM 33192	Vogt (1981)	
Lithobates sylvaticus	Barron	MPM 13876	Casper (1996)	

Scientific Name	County	Museum Voucher	Publication	Notes
Lithobates sylvaticus	Bayfield	AMNH 77321	Schueler and Cook (1980), Vogt (1981)	
Lithobates sylvaticus	Brown	UWZM 22614	Cochran et al. (1987)	
Lithobates sylvaticus	Buffalo	MPM VZP246	Casper (1996)	Photograph
Lithobates sylvaticus	Burnett	MPM 945	Vogt (1981)	
Lithobates sylvaticus	Calumet	MPM VZP444	Casper (1996)	Photograph
Lithobates sylvaticus	Chippewa	MPM 33338	Vogt (1981)	
Lithobates sylvaticus	Clark	FMNH 8237	Vogt (1981)	
Lithobates sylvaticus	Columbia	USNM 312623	Vogt (1981)	
Lithobates sylvaticus	Crawford		Vogt (1981)	
Lithobates sylvaticus	Dane	MPM 27950	Wolter and Watermolen (1996)	
Lithobates sylvaticus	Dodge	MPM VZP502	Casper (1996)	Photograph
Lithobates sylvaticus	Door	UWSP 2603	Vogt (1981)	
Lithobates sylvaticus	Douglas	UWZM 3134	Vogt (1981)	
Lithobates sylvaticus	Dunn	MPM 14671	Vogt (1981)	
Lithobates sylvaticus	Eau Claire	MPM 23194	Casper (1996)	
Lithobates sylvaticus	Florence	UWSP 486	Vogt (1981)	
Lithobates sylvaticus	Fond Du Lac	MPM 5068	Vogt (1981)	
Lithobates sylvaticus	Forest	UWZM 1986	Vogt (1981)	
Lithobates sylvaticus	Grant		Vogt (1981)	Museum records indicated that MPM 4997 was collected in Grant County, but also that this specimen had been "on loan" to "D. Green" since 2004 and so was not available for further investigation
Lithobates sylvaticus	Green Lake	UWSP 4231	Neumeyer and Verjinski (2012)	
Lithobates sylvaticus	Iowa	MPM 11412	Vogt (1981)	
Lithobates sylvaticus	Iron	MPM 4005	Vogt (1981)	
Lithobates sylvaticus	Jackson	MPM 30302	Suzuki (1951), Vogt (1981)	Suzuki (1951) reported voucher as UWZM "uncatalogued"
Lithobates sylvaticus	Juneau	UWZM 22476	Vogt (1981)	
Lithobates sylvaticus	Kewaunee	MPM 23472	Watermolen (1991a)	
Lithobates sylvaticus	La Crosse	INHS 2014q	Casper (1996)	Photograph; Casper (1996) reported voucher as "MPM audio tape," but correspondences with museum staff indicated these audio files are currently missing (J. Colby)
Lithobates sylvaticus	Langlade	MPM 3270	Vogt (1981)	

Scientific Name	County	Museum Voucher	Publication	Notes
Lithobates sylvaticus	Lincoln	UMMZ 64808	Suzuki (1951), Vogt (1981)	
Lithobates sylvaticus	Manitowoc	MPM 2906	Suzuki (1951), Vogt (1981)	
Lithobates sylvaticus	Marathon	MPM 2904	Vogt (1981)	
Lithobates sylvaticus	Marinette	MPM 3258	Vogt (1981)	
Lithobates sylvaticus	Marquette	MPM 27952	Casper (1996)	
Lithobates sylvaticus	Menominee	MPM 3247	Vogt (1981)	
Lithobates sylvaticus	Milwaukee	MPM 1263	Vogt (1981)	
Lithobates sylvaticus	Monroe	MPM VZP407	Casper (1996)	Photograph
Lithobates sylvaticus	Oconto	UWGB 181	Casper (1996)	
Lithobates sylvaticus	Oneida	MPM 3220	Vogt (1981)	
Lithobates sylvaticus	Outagamie	UIMNH 1495	Suzuki (1951), Vogt (1981)	
Lithobates sylvaticus	Ozaukee	MPM 32900	Vogt (1981)	
Lithobates sylvaticus	Pepin	MPM VZP522	Casper (1996)	Photograph
Lithobates sylvaticus	Pierce	MPM VZP496	Casper (1996)	Photograph
Lithobates sylvaticus	Polk	UWZM 3275	Vogt (1981)	
Lithobates sylvaticus	Portage	UWSP 4093	Vogt (1981)	
Lithobates sylvaticus	Price	MPM 17988	Vogt (1981)	
Lithobates sylvaticus	Racine	USNM 3385	Kapfer et al. (2015)	Historical specimen, species many no longer be present in county
Lithobates sylvaticus	Richland	UWSP 676	Vogt (1981)	
Lithobates sylvaticus	Rock		Vogt (1981)	
Lithobates sylvaticus	Rusk	UMMZ 69619	Suzuki (1951), Casper (1997b)	Casper (1997b) reported voucher as "MPM audio tape (Cat. No. pending)," but audio files are missing, according to museum staff (J. Colby)
Lithobates sylvaticus	Sauk	UWZM 14007	Vogt (1981)	
Lithobates sylvaticus	Sawyer	UWSP 3846	Vogt (1981)	
Lithobates sylvaticus	Shawano	MPM 2908	Vogt (1981)	
Lithobates sylvaticus	Sheboygan	MPM 2413	Suzuki (1951), Vogt (1981)	
Lithobates sylvaticus	St. Croix		Vogt (1981)	
Lithobates sylvaticus	Taylor	UMMZ 69621	Suzuki (1951), Vogt (1981)	
Lithobates sylvaticus	Trempealeau	MPM VZP485	Casper (1996)	Photograph
Lithobates sylvaticus	Vernon	UWZM 24091	Vogt (1981)	
Lithobates sylvaticus	Vilas	MPM 2913	Vogt (1981)	

Scientific Name	County	Museum Voucher	Publication	Notes
Lithobates sylvaticus	Walworth	MPM VZP538	Casper (2015)	Photograph of egg masses; in the original museum records, this photo is assigned to *L. pipiens*, not *L. sylvaticus*
Lithobates sylvaticus	Washburn	UWSP 3972	Vogt (1981)	
Lithobates sylvaticus	Washington	MPM 23184	Casper (1996)	
Lithobates sylvaticus	Waukesha	MPM 25307	Reed and Lang (1992)	
Lithobates sylvaticus	Waupaca	MPM 4018	Vogt (1981)	
Lithobates sylvaticus	Waushara	MPM VZP139	Casper (1996)	Photograph
Lithobates sylvaticus	Winnebago	MPM VZP319	Casper (1996)	Photograph
Lithobates sylvaticus	Wood	MPM 2915	Vogt (1981)	
Necturus maculosus	Adams		Watermolen (2015b)	Reported observations without voucher specimen
Necturus maculosus	Ashland	MPM 31830	Cochran et al. (2002)	
Necturus maculosus	Barron	MPM VZP898	Vogt (1981)	Photograph
Necturus maculosus	Bayfield	MPM 13825	Vogt (1981)	
Necturus maculosus	Brown		Reed and Hodgson (1987)	Authors reported voucher as BBWS 626, but this collection no longer exists according to correspondence with Bay Beach Wildlife Sanctuary staff
Necturus maculosus	Burnett	MPM 820	Casper (1996)	
Necturus maculosus	Chippewa	MPM 2789	Suzuki (1951), Vogt (1981)	
Necturus maculosus	Clark	MPM 16394	Casper (1996)	
Necturus maculosus	Columbia	MPM VZP253	Cochran and Lyons (1990)	Photograph
Necturus maculosus	Crawford		Cochran and Lyons (1986), Watermolen (2015b)	Reported observations without voucher specimen
Necturus maculosus	Dane	UWZM 20243	Vogt (1981)	
Necturus maculosus	Dodge	UIMNH 1402	Vogt (1981)	
Necturus maculosus	Door	MPM 2781	Suzuki (1951), Vogt (1981)	
Necturus maculosus	Douglas		Vogt (1981)	
Necturus maculosus	Dunn	UWZM 22503	Cochran (1982b)	
Necturus maculosus	Eau Claire	MPM 30341	Casper (1999b)	
Necturus maculosus	Forest	UIMNH 18032	Vogt (1981)	
Necturus maculosus	Grant	UWZM 20234	Vogt (1981)	
Necturus maculosus	Green Lake	UWSP 471	Vogt (1981)	
Necturus maculosus	Iron	MPM VZP576	Casper (1997d)	Photograph

Scientific Name	County	Museum Voucher	Publication	Notes
Necturus maculosus	Jackson	MPM 16395	Vogt (1981)	
Necturus maculosus	Jefferson	MPM 2373	Vogt (1981)	
Necturus maculosus	Juneau	MPM VZP233	Suzuki (1951)	Photograph
Necturus maculosus	Kenosha	MPM 743	Vogt (1981)	
Necturus maculosus	Kewaunee	MPM 2790	Suzuki (1951), Vogt (1981)	
Necturus maculosus	La Crosse	MPM VZP531	Casper (1996)	
Necturus maculosus	Langlade		Watermolen (2015b)	Reported observations without voucher specimen
Necturus maculosus	Lincoln	MPM 2791	Vogt (1981)	
Necturus maculosus	Manitowoc	MPM 2785	Suzuki (1951), Vogt (1981)	
Necturus maculosus	Marathon	MPM VZP369	Casper (1996)	Photograph
Necturus maculosus	Marinette	UWGB 190	Casper (1996)	
Necturus maculosus	Marquette	MPM VZP642	Casper (1999b)	Photograph
Necturus maculosus	Milwaukee	MPM 814	Vogt (1981)	
Necturus maculosus	Oconto	MPM VZP899a–b–900a–c	Watermolen (2018b)	Photograph
Necturus maculosus	Oneida	MPM 12452	Casper (1996)	
Necturus maculosus	Outagamie	UWSP 2494	Casper (1996)	Casper (1996) reported as UWSP 1061, but review of museum records and discussion with collection manager supports this specimen is currently presumed missing (P. Zani)
Necturus maculosus	Ozaukee	MPM 29	Casper (1996)	
Necturus maculosus	Pierce		Vogt (1981)	
Necturus maculosus	Polk	MPM 2784	Vogt (1981)	
Necturus maculosus	Portage	UWSP 4107	Vogt (1981)	
Necturus maculosus	Price	INHS 10041	Casper (1996)	
Necturus maculosus	Racine	MPM 13878	Vogt (1981)	
Necturus maculosus	Richland	UWZM 22610	Cochran and Lyons (1986)	
Necturus maculosus	Rock	MPM 22167	Casper (1996)	
Necturus maculosus	Rusk	MPM VZP568	Casper (1997d)	Photograph
Necturus maculosus	Sauk	MPM 2120	Casper (1996)	
Necturus maculosus	Sawyer	MPM 17392	Vogt (1981)	
Necturus maculosus	Shawano	MPM 26489	Casper (1996)	
Necturus maculosus	Sheboygan		Vogt (1981)	
Necturus maculosus	St. Croix	MPM VZP304	Casper (1996)	Photograph
Necturus maculosus	Taylor	MPM 16624	Casper (1996)	
Necturus maculosus	Trempealeau	YPM 118	Casper (2015)	Yale Peabody Museum
Necturus maculosus	Vilas	UWSP 519	Vogt (1981)	

Scientific Name	County	Museum Voucher	Publication	Notes
Necturus maculosus	Walworth	USNM 10896	Vogt (1981)	
Necturus maculosus	Washburn	UMMZ 69624	Suzuki (1951)	
Necturus maculosus	Washington	UWGB 797R	Casper (2016b)	
Necturus maculosus	Waukesha	MPM 17545	Vogt (1981)	
Necturus maculosus	Waupaca	UWSP 232	Vogt (1981)	
Necturus maculosus	Waushara		Watermolen (2015b, 2018a)	Reported observations without voucher specimen
Necturus maculosus	Winnebago	MPM 2787	Vogt (1981)	
Necturus maculosus	Wood	MPM VZP1006	Cannizzaro et al. (2021)	Photograph
Nerodia sipedon	Adams	INHS 2008j	Kapfer et al. (2009b)	Photograph
Nerodia sipedon	Barron	MPM VZP553	Casper (1997d)	Photograph
Nerodia sipedon	Bayfield		Gibbons and Dorcas (2004)	Gibbons and Dorcas (2004) reported supporting documentation as a personal observation by G. Casper
Nerodia sipedon	Buffalo	MPM 1168	Vogt (1981), Gibbons and Dorcas (2004)	
Nerodia sipedon	Burnett	FMNH 21697	Vogt (1981), Gibbons and Dorcas (2004)	
Nerodia sipedon	Chippewa	UMMZ 64740	Gibbons and Dorcas (2004), Casper 2015	
Nerodia sipedon	Clark	FMNH 14514	Vogt (1981), Gibbons and Dorcas (2004)	
Nerodia sipedon	Columbia	FMNH 42165	Vogt (1981), Gibbons and Dorcas (2004)	
Nerodia sipedon	Crawford	MPM 1972	Vogt (1981), Gibbons and Dorcas (2004)	
Nerodia sipedon	Dane	CM 70294	Vogt (1981), Gibbons and Dorcas (2004)	
Nerodia sipedon	Dodge	UIMNH 1576	Vogt (1981), Gibbons and Dorcas (2004)	
Nerodia sipedon	Door	MPM 1044	Vogt (1981), Gibbons and Dorcas (2004)	
Nerodia sipedon	Douglas	MPM 956	Vogt (1981), Gibbons and Dorcas (2004)	

Scientific Name	County	Museum Voucher	Publication	Notes
Nerodia sipedon	Dunn	MPM VZP375	Casper (1996)	Photograph
Nerodia sipedon	Eau Claire	FMNH 14747	Vogt (1981), Gibbons and Dorcas (2004)	
Nerodia sipedon	Fond Du Lac	UWZM 20982	Vogt (1981), Gibbons and Dorcas (2004)	
Nerodia sipedon	Forest		Vogt (1981)	
Nerodia sipedon	Grant	FMNH 8830	Vogt (1981), Gibbons and Dorcas (2004)	
Nerodia sipedon	Green	UWZM 14721	Cochran (1983c)	Cochran (1983c) reported voucher as UWZM 22559, but this specimen could not be found during recent investigations of the collection
Nerodia sipedon	Green Lake	CHAS HERP 18893	Vogt (1981)	
Nerodia sipedon	Iowa	MPM 5122	Vogt (1981)	
Nerodia sipedon	Iron	MPM VZP492	Casper (1996)	Photograph
Nerodia sipedon	Jackson	FMNH 19211	Vogt (1981), Gibbons and Dorcas (2004)	
Nerodia sipedon	Jefferson	CM 75841	Vogt (1981), Gibbons and Dorcas (2004)	
Nerodia sipedon	Juneau	CM 70415	Vogt (1981), Gibbons and Dorcas (2004)	
Nerodia sipedon	Kenosha		Gibbons and Dorcas (2004)	Gibbons and Dorcas (2004) listed MPM 1079, but museum records indicated this specimen was disposed of
Nerodia sipedon	La Crosse	INHS 2014m–n	Vogt (1981)	Photograph
Nerodia sipedon	Lafayette	MPM VZP264	Casper (1996)	Photograph
Nerodia sipedon	Manitowoc		Gibbons and Dorcas (2004)	Gibbons and Dorcas (2004) reported supporting documentation as a personal observation by G. Casper
Nerodia sipedon	Marinette		Gibbons and Dorcas (2004)	Gibbons and Dorcas (2004) reported supporting documentation as a personal observation by G. Casper
Nerodia sipedon	Marquette	MPM VZP368	Casper (1996)	
Nerodia sipedon	Menominee	MPM 13517	Casper (1996)	
Nerodia sipedon	Milwaukee	MPM 22855	Drought (1987c)	
Nerodia sipedon	Monroe	MPM 30311	Vogt (1981), Gibbons and Dorcas (2004)	

Scientific Name	County	Museum Voucher	Publication	Notes
Nerodia sipedon	Oconto	UWZM 2406	Vogt (1981), Gibbons and Dorcas (2004)	
Nerodia sipedon	Oneida	UWZM 17379	Vogt (1981)	
Nerodia sipedon	Ozaukee	MPM VZP50	Casper (1996)	Photograph
Nerodia sipedon	Pepin	MPM 23459	Vogt (1981), Gibbons and Dorcas (2004)	
Nerodia sipedon	Pierce	MPM 1150	Casper (1996)	
Nerodia sipedon	Polk	JFBM R1294	Casper (2015)	
Nerodia sipedon	Portage	UWSP 1669	Vogt (1981), Gibbons and Dorcas (2004)	Gibbons and Dorcas (2004) reported UWSP 187, but this catalogue number is associated with a different species, according to collection records
Nerodia sipedon	Price	MPM VZP404	Casper (1996)	Photograph
Nerodia sipedon	Racine	MPM 26374	Vogt (1981), Gibbons and Dorcas (2004)	
Nerodia sipedon	Richland	CM 75839	Vogt (1981), Gibbons and Dorcas (2004)	
Nerodia sipedon	Rock	MPM 1481	Vogt (1981), Gibbons and Dorcas (2004)	
Nerodia sipedon	Rusk	FMNH 16364	Vogt (1981), Gibbons and Dorcas (2004)	
Nerodia sipedon	Sauk	MPM 7952	Vogt (1981)	
Nerodia sipedon	Sawyer		Vogt (1981)	
Nerodia sipedon	Shawano	MPM 26322	Casper (1996)	
Nerodia sipedon	Sheboygan	MPM 11711	Vogt (1981), Gibbons and Dorcas (2004)	
Nerodia sipedon	St. Croix	JFBM R1024	Casper (2015)	
Nerodia sipedon	Trempealeau	MPM 18288	Vogt (1981), Gibbons and Dorcas (2004)	
Nerodia sipedon	Vernon	CM 70520	Vogt (1981), Gibbons and Dorcas (2004)	
Nerodia sipedon	Vilas	FMNH 42396	Vogt (1981), Gibbons and Dorcas (2004)	
Nerodia sipedon	Walworth	CHAS HERP5078	Necker (1939b), Vogt (1981)	

Scientific Name	County	Museum Voucher	Publication	Notes
Nerodia sipedon	Washburn	UMMZ 68710	Vogt (1981), Gibbons and Dorcas (2004)	
Nerodia sipedon	Washington	MPM 12416	Vogt (1981), Gibbons and Dorcas (2004)	
Nerodia sipedon	Waukesha	MPM 7471	Vogt (1981), Gibbons and Dorcas (2004)	
Nerodia sipedon	Waupaca	MPM 27994	Casper (1997d)	
Nerodia sipedon	Waushara	MPM 4537	Casper (1996)	
Notophthalmus viridescens	Ashland	FMNH 35830	Suzuki (1951), Vogt (1981)	
Notophthalmus viridescens	Barron	MPM VZP484	Casper (1996)	Photograph
Notophthalmus viridescens	Bayfield	MPM 9108	Vogt (1981)	
Notophthalmus viridescens	Buffalo	MPM VZP231	Casper (1996)	Photograph
Notophthalmus viridescens	Burnett		Vogt (1981)	
Notophthalmus viridescens	Chippewa	MPM VZP591	Casper (1997d)	Photograph
Notophthalmus viridescens	Columbia	INHS 2015e	Kapfer et al. (2015)	Photograph
Notophthalmus viridescens	Crawford	MPM 31623	Casper (2001b)	
Notophthalmus viridescens	Dane	MPM VZP778a	Vogt (1981)	Photograph
Notophthalmus viridescens	Dodge	UIMNH 1407	Suzuki (1951), Vogt (1981)	Suzuki (1951) reported voucher as UIMNH 1404–06, but these catalogue numbers are currently assigned to according to correspondence with collection staff
Notophthalmus viridescens	Door	UWGB 10	Watermolen and Casper (1993)	
Notophthalmus viridescens	Eau Claire	MPM 23195	Casper (1996)	
Notophthalmus viridescens	Fond Du Lac	MPM 4995	Vogt (1981)	
Notophthalmus viridescens	Forest	MPM 30221	Vogt (1981)	
Notophthalmus viridescens	Iowa	UWZM 20983	Vogt (1981)	
Notophthalmus viridescens	Jefferson	MPM 2792	Suzuki (1951), Vogt (1981)	Suzuki (1951) reported voucher as "MU uncatalogued"

Scientific Name	County	Museum Voucher	Publication	Notes
Notophthalmus viridescens	Juneau	MPM VZP606	Vogt (1981)	Photograph
Notophthalmus viridescens	Kenosha		Suzuki (1951)	Suzuki (1951) reported MU 1023–24 as vouchers, but this historical Marquette University collection was either donated or eliminated according to correspondences with institution staff
Notophthalmus viridescens	La Crosse	MPM VZP460	Casper (2015)	Photograph; The catalogue number MPM VZP460 was first published by Casper (1996) as the voucher for Trempeleau County, which was later reported as the voucher for La Crosse County after reassessment of the location where this specimen was collected (Casper 2015)
Notophthalmus viridescens	Langlade	UWSP 2661	Vogt (1981)	
Notophthalmus viridescens	Lincoln	MSUM 6556	Casper (1996)	
Notophthalmus viridescens	Manitowoc	CM 70490	Vogt (1981)	
Notophthalmus viridescens	Marathon	MPM VZP844a–b	Schiltz and Lorch (2016)	Photograph
Notophthalmus viridescens	Marinette	MPM 2793	Suzuki (1951), Vogt (1981)	Suzuki (1951) reported the voucher as "MU uncatalogued"
Notophthalmus viridescens	Marquette	MPM 30362	Casper (2001b), Casper (2015)	
Notophthalmus viridescens	Menominee	MPM 30223	Vogt (1981)	
Notophthalmus viridescens	Milwaukee	MPM 1360	Vogt (1981)	
Notophthalmus viridescens	Oconto	UWGB 11	Casper (1996)	
Notophthalmus viridescens	Oneida	UMMZ HerpPhoto1259–60	Casper (2015)	Photograph
Notophthalmus viridescens	Ozaukee	MPM VZP55	Casper (1996)	Photograph
Notophthalmus viridescens	Polk	USNM 57293	Vogt (1981)	
Notophthalmus viridescens	Portage	UWSP 4189	Vogt (1981)	
Notophthalmus viridescens	Racine	USNM 277499	Vogt (1981)	
Notophthalmus viridescens	Richland	INHS 2015ac	Kapfer et al. (2015)	Photograph

Scientific Name	County	Museum Voucher	Publication	Notes
Notophthalmus viridescens	Rusk	MPM VZP879	Engelhardt and Engelhardt (2017)	Photograph
Notophthalmus viridescens	Sawyer	UWSP 4282	Wild (2013b)	
Notophthalmus viridescens	Shawano	UMMZ 1292	Rutherford and Casper (2018)	Photograph
Notophthalmus viridescens	Sheboygan	MPM 14702	Casper (1996)	
Notophthalmus viridescens	Trempealeau		Casper (1996)	Casper (1996) reported MPM VZP460 as the voucher for Trempealeau County, but after reexamination of the location where this specimen was collected, he determined it was more accurately from La Crosse County (Casper 2015)
Notophthalmus viridescens	Vernon	MPM 2330	Vogt (1981)	
Notophthalmus viridescens	Vilas	UWZM 17392	Vogt (1981)	
Notophthalmus viridescens	Walworth	FMNH 12365	Schmidt and Necker (1935), Vogt (1981)	
Notophthalmus viridescens	Washburn	MPM VZP354	Casper (1996)	Photograph
Notophthalmus viridescens	Washington	MPM 3332	Vogt (1981)	
Notophthalmus viridescens	Waukesha	MPM 22503	Suzuki (1951)	Suzuki (1951) reported voucher as MPM 2415, but this specimen could not be found during recent investigations of collection
Notophthalmus viridescens	Wood	UMMZ 64813–9	Casper (2015)	
Opheodrys vernalis	Adams	MPM 26279	Vogt (1981)	
Opheodrys vernalis	Ashland	MPM 33088	Casper (2015)	
Opheodrys vernalis	Bayfield	MPM 13548	Vogt (1981)	
Opheodrys vernalis	Brown	UWZM 22619	Cochran et al. (1987)	
Opheodrys vernalis	Burnett	MPM 27940	Vogt (1981)	
Opheodrys vernalis	Chippewa		Vogt (1981)	Museum records indicated FMNH 8936 is reportedly from Chippewa County, but this specimen consists of only a skull and identification of species could not be confirmed by museum staff (A. Resetar)
Opheodrys vernalis	Clark	FMNH 14575	Vogt (1981)	

Scientific Name	County	Museum Voucher	Publication	Notes
Opheodrys vernalis	Columbia	INHS 2014d	Vogt (1981)	Photograph
Opheodrys vernalis	Crawford	MPM VZP824–5	Raimond and Thomsen (2016)	Photograph
Opheodrys vernalis	Dane	INHS 2014u	Vogt (1981)	Photograph
Opheodrys vernalis	Dodge	MPM 645	Casper (1996)	
Opheodrys vernalis	Door	MPM VZP413	Casper (1996)	Photograph
Opheodrys vernalis	Douglas	MPM VZP317	Casper (1996)	Photograph; Casper (1996) reported voucher as UWS 508 (UW–Superior) but the institution did not respond to our repeated requests for information on this specimen; as such, we could not independently confirm its identity
Opheodrys vernalis	Eau Claire	MPM VZP455	Casper (1996)	Photograph
Opheodrys vernalis	Florence	MPM 7452	Vogt (1981)	
Opheodrys vernalis	Fond Du Lac	MPM 22128	Casper (1996)	
Opheodrys vernalis	Forest	MPM 2200	Dickinson (1972), Vogt (1981)	
Opheodrys vernalis	Green	MPM 27976	Casper (1996c)	
Opheodrys vernalis	Iron	MPM 2199	Casper (1996)	
Opheodrys vernalis	Jackson	MPM 23466	Vogt (1981)	
Opheodrys vernalis	Jefferson	MPM 22130	Vogt (1981)	
Opheodrys vernalis	Juneau	MPM 23190	Vogt (1981)	
Opheodrys vernalis	Kenosha	MPM VZP17	Vogt (1981)	Photograph
Opheodrys vernalis	La Crosse	MPM 23519	Casper (2015)	
Opheodrys vernalis	Langlade	MPM 5075	Vogt (1981)	
Opheodrys vernalis	Marathon	UWZM 20143	Vogt (1981)	
Opheodrys vernalis	Marinette	MPM 2775	Dickinson (1972), Vogt (1981)	
Opheodrys vernalis	Marquette	MPM 23465	Casper (1996)	Casper (1996) reported voucher as REO 11669, but this collection (the Robert E. Olson Collection) was donated to MPM and the specimen was given a new catalogue number
Opheodrys vernalis	Menominee	UWGB 798R	Casper and Rutherford (2016)	
Opheodrys vernalis	Milwaukee	MPM 648	Vogt (1981)	
Opheodrys vernalis	Monroe	MPM VZP38	Casper (1996)	Photograph
Opheodrys vernalis	Oconto	UWGB R50	Vogt (1981)	
Opheodrys vernalis	Oneida	MPM 2201	Vogt (1981)	
Opheodrys vernalis	Portage	UWSP 4178	Vogt (1981)	

Scientific Name	County	Museum Voucher	Publication	Notes
Opheodrys vernalis	Price	MPM 2582	Dickinson (1950), Vogt (1981)	
Opheodrys vernalis	Racine	MPM 23347	Vogt (1981)	
Opheodrys vernalis	Richland	MPM VZP392	Casper (1996)	Photograph
Opheodrys vernalis	Rusk	MPM VZP420	Casper (1996)	Photograph
Opheodrys vernalis	Sauk	MPM 2337	Casper (1996)	
Opheodrys vernalis	Sawyer	MPM 2449	Dickinson (1950)	
Opheodrys vernalis	Shawano	MPM 2204	Casper (1996)	
Opheodrys vernalis	St. Croix	MPM VZP943	Breitenbach et al. (2020)	Photograph
Opheodrys vernalis	Vernon	INHS 2014ae	Beck et al. (2015)	
Opheodrys vernalis	Vilas	MPM 1773	Vogt (1981)	
Opheodrys vernalis	Walworth	FMNH 19893	Schmidt and Necker (1935), Vogt (1981)	
Opheodrys vernalis	Washburn	JFBM 987	Dickinson (1950), Vogt (1981)	Dickinson (1950) reported voucher as MPM 2662, which more current museum records indicated was a *S. occipitomaculata*; although museum records indicated that MPM 2663 and 2668 were from Washburn County, these specimens are currently "presumed missing"
Opheodrys vernalis	Waukesha	MPM 22129	Vogt (1981)	
Opheodrys vernalis	Waupaca	MPM VZP380	Casper (1996)	Photograph
Opheodrys vernalis	Waushara	UWSP 3330	Vogt (1981)	
Opheodrys vernalis	Wood	MPM 30304	Vogt (1981)	
Ophisaurus attenuatus	Adams	UWZM 22802	Vogt (1981)	
Ophisaurus attenuatus	Columbia	MPM 27999	Vogt (1981)	
Ophisaurus attenuatus	Green Lake	MPM 2448	Dickinson (1950), Vogt (1981)	
Ophisaurus attenuatus	Jackson	UWZM 7304	Casper (1996)	
Ophisaurus attenuatus	Juneau	UWZM 24157	Vogt (1981)	
Ophisaurus attenuatus	La Crosse	MPM 23409	Vogt (1981)	
Ophisaurus attenuatus	Marquette	MPM 27953	Vogt (1981)	
Ophisaurus attenuatus	Monroe	MPM 23405	Casper (1996)	
Ophisaurus attenuatus	Sauk	MPM 2385	Dickinson (1950), Vogt (1981)	
Ophisaurus attenuatus	Waushara	MPM 4536	Vogt (1981)	
Pantherophis spiloides	Crawford	MPM 2061	Vogt (1981)	
Pantherophis spiloides	Grant	MPM VZP309	Casper (1996)	Photograph

Scientific Name	County	Museum Voucher	Publication	Notes
Pantherophis spiloides	Iowa	MPM 8805	Vogt (1981)	
Pantherophis spiloides	La Crosse	MPM 23355	Casper (1996)	
Pantherophis spiloides	Richland	MPM VZP648	Vogt (1981)	Photograph
Pantherophis spiloides	Sauk	MPM 2137	Vogt (1981)	
Pantherophis spiloides	Vernon	MPM 2049	Vogt (1981)	
Pantherophis vulpinus	Adams		Vogt (1981)	Museum records indicated that MPM 3099 was from Adams County, but the specimen is currently "presumed missing"
Pantherophis vulpinus	Ashland	INHS 2011j	Casper (2015)	Photograph
Pantherophis vulpinus	Barron	MPM VZP367	Casper (1996)	Photograph
Pantherophis vulpinus	Bayfield	MPM 16399	Casper (1996)	Casper (1996) reported voucher as "CNHM unnumbered specimen" (Cable Natural History Museum), but the institution did not respond to our repeated requests for information on specimens in their collection
Pantherophis vulpinus	Brown	FMNH 267513	Cochran et al. (1987)	Authors reported voucher as BBWS RE223 (live animal on display) and BBWS J1219 (photograph), but recent corresponenced with Bay Beach Wildlife Sanctuary staff indicated the specimens no longer exist
Pantherophis vulpinus	Buffalo	UWZM 2132	Vogt (1981)	
Pantherophis vulpinus	Burnett	MPM VZP386	Vogt (1981)	Photograph
Pantherophis vulpinus	Calumet	UWGB 1486	Sobota (1990b)	Photograph
Pantherophis vulpinus	Chippewa	MPM 33337	Vogt (1981)	
Pantherophis vulpinus	Clark	INHS 2014x	Vogt (1981)	Photograph
Pantherophis vulpinus	Columbia	MPM 23628	Vogt (1981)	
Pantherophis vulpinus	Crawford	MPM VZP1035a–b	LaFleur and Kapfer (2022)	Photograph; originally, Walley and Walley (2004a) reported this county record supported by voucher HDW-NIU 1937, a juvenile specimen transferred to the Field Museum and given the new catalogue number FMNH 284199; the specimen was later reidentified as a juvenile *P. spiloides* (A. Resetar), which invalidated the original county record
Pantherophis vulpinus	Dane	MPM 26181	Vogt (1981)	
Pantherophis vulpinus	Door	MPM 1043	Vogt (1981)	
Pantherophis vulpinus	Douglas	UWSP 3549	Vogt (1981)	
Pantherophis vulpinus	Dunn	USNM 574422	Vogt (1981)	
Pantherophis vulpinus	Eau Claire	MPM 23197	Casper (1996)	

Scientific Name	County	Museum Voucher	Publication	Notes
Pantherophis vulpinus	Florence	CM 75846	Vogt (1981)	
Pantherophis vulpinus	Forest	MPM 2336	Dickinson (1972), Vogt (1981)	Dickinson (1972) reported voucher as UWSP 208, but this catalogue number was assigned to a different species according to collections records
Pantherophis vulpinus	Grant	MPM VZP780	Vogt (1981)	Photograph
Pantherophis vulpinus	Green	MPM 26353	Vogt (1981)	
Pantherophis vulpinus	Green Lake	MPM 7947	Vogt (1981)	
Pantherophis vulpinus	Iowa	MPM 26368	Vogt (1981)	
Pantherophis vulpinus	Jackson	UMMZ 64760	Vogt (1981)	
Pantherophis vulpinus	Juneau	MPM 7873	Vogt (1981)	
Pantherophis vulpinus	Kenosha	INHS 2014ad	Casper (1996)	Photograph; Casper (1996) reported voucher as SIUCH R–2176, but we received no response from this institution to repeated requests for collection information
Pantherophis vulpinus	Kewaunee	MPM 6929	Vogt (1981)	
Pantherophis vulpinus	La Crosse	UWZM 20144	Vogt (1981)	
Pantherophis vulpinus	Lafayette	MPM 11780	Casper (1996)	
Pantherophis vulpinus	Langlade	CHAS HERP3852	Vogt (1981)	
Pantherophis vulpinus	Lincoln	MPM VZP921a–d	Millis and Badje (2019)	Photograph
Pantherophis vulpinus	Manitowoc	MPM 23448	Cochran (1990)	
Pantherophis vulpinus	Marathon	UWSP 4165	Vogt (1981)	
Pantherophis vulpinus	Marinette	UWGB 314–R	Cochran (1988c)	
Pantherophis vulpinus	Marquette	MPM 2458	Dickinson (1950)	
Pantherophis vulpinus	Menominee	MPM VZP278	Casper (1996)	Photograph
Pantherophis vulpinus	Milwaukee	LSUMZ 27725	Casper (1996)	
Pantherophis vulpinus	Monroe	MPM 23329	Dickinson (1950)	Dickinson (1950) reported voucher as MPM 2509, but museum records indicated this specimen was "presumed missing"
Pantherophis vulpinus	Oconto	MPM 2595	Vogt (1981)	
Pantherophis vulpinus	Oneida	MPM VZP585	Casper (1997d)	Photograph
Pantherophis vulpinus	Outagamie	UWGB 315	Cochran (1988c)	
Pantherophis vulpinus	Ozaukee	MPM 30350	Casper (1999b)	
Pantherophis vulpinus	Pepin		Vogt (1981)	
Pantherophis vulpinus	Pierce	MPM 23458	Vogt (1981)	
Pantherophis vulpinus	Polk	MPM VZP237	Casper (1996)	Photograph
Pantherophis vulpinus	Portage	UWSP 4192	Vogt (1981)	

Scientific Name	County	Museum Voucher	Publication	Notes
Pantherophis vulpinus	Price	MPM 30385	Casper and Burnham (2001b), Casper (2015)	
Pantherophis vulpinus	Racine	USNM 7269	Vogt (1981)	
Pantherophis vulpinus	Richland	MPM 6889	Vogt (1981)	
Pantherophis vulpinus	Rock	UWW HERP0043	Rutzen et al. (2013)	Rutzen et al. (2013) list reported voucher as a photographic specimen, but is a fluid specimen
Pantherophis vulpinus	Rusk	MPM 1993	Casper (1996)	
Pantherophis vulpinus	Sauk	MPM 26297	Vogt (1981)	
Pantherophis vulpinus	Sawyer	MPM 16400	Vogt (1981)	
Pantherophis vulpinus	Shawano	MPM 30267	Dickinson (1950), Vogt (1981)	
Pantherophis vulpinus	St. Croix	MPM 23361	Casper (1996)	
Pantherophis vulpinus	Taylor	MPM 26177	Casper (1996)	
Pantherophis vulpinus	Trempealeau	UWZM 17301	Vogt (1981)	
Pantherophis vulpinus	Vernon	MPM 8804	Vogt (1981)	
Pantherophis vulpinus	Vilas	UWSP 1925	Vogt (1981)	
Pantherophis vulpinus	Walworth	MPM VZP924a–b	Badje (2020f)	Photographs
Pantherophis vulpinus	Washburn	UWSP 4158	Vogt (1981)	
Pantherophis vulpinus	Washington	MPM 26488	Casper (1996)	
Pantherophis vulpinus	Waupaca	UWSP 3814	Vogt (1981)	
Pantherophis vulpinus	Waushara	MPM VZP582	Casper (1997d)	Photograph
Pantherophis vulpinus	Wood	MPM 2346	Vogt (1981)	
Pituophis catenifer	Burnett	MPM 27941	Vogt (1981)	
Pituophis catenifer	Columbia	INHS 2014aa	Vogt (1981)	Photograph
Pituophis catenifer	Crawford	MPM 2032	Vogt (1981)	
Pituophis catenifer	Dane	UWZM 13612	Vogt (1981)	
Pituophis catenifer	Dunn	MPM VZP561	Casper (1997d)	Photograph
Pituophis catenifer	Eau Claire	MPM VZP195	Casper (1996)	Photograph
Pituophis catenifer	Grant	MPM 30213	Vogt (1981)	
Pituophis catenifer	Iowa	UWZM 9975	Vogt (1981)	
Pituophis catenifer	Jackson	INHS 2014c	Vogt (1981)	Photograph
Pituophis catenifer	Jefferson		Vogt (1981)	
Pituophis catenifer	La Crosse	MPM 23358	Casper (1996)	
Pituophis catenifer	Lafayette	MPM VZP935a–c	Staffen and Staffen (2020)	Photographs
Pituophis catenifer	Monroe	INHS 2014r	Vogt (1981)	Photograph

Scientific Name	County	Museum Voucher	Publication	Notes
Pituophis catenifer	Pepin	INHS 2010i	Bauer et al. (2010)	Photograph
Pituophis catenifer	Pierce	JFBM 13235	Casper (1996)	
Pituophis catenifer	Polk	UWSP 4209	Wild (2011)	
Pituophis catenifer	Richland	MPM VZP597	Casper (1997d)	Photograph
Pituophis catenifer	Sauk	MPM 2019	Vogt (1981)	
Pituophis catenifer	Trempealeau	INHS 2008f	Kapfer et al. (2008d)	Photograph
Pituophis catenifer	Vernon	MPM 2092	Vogt (1981)	
Plestiodon fasciatus	Adams	MPM 23345	Casper (1996)	
Plestiodon fasciatus	Chippewa	UTADC 8161	Woghan and Ray (2014)	Photographic voucher of marginal quality and difficult to use for independent confirmation of species; despite this, the organism pictured resembles *P. septentrionalis* more so than *P. fasciatus*
Plestiodon fasciatus	Clark	UWZM 13539	Vogt (1981)	
Plestiodon fasciatus	Crawford	MPM VZP822	Walley and Rozelle (2004)	Photograph; Walley and Rozelle (2004) reported voucher as HDW-NIU 2002.10, which investigations with numerous institutions (including Northern Illinois University) supported was missing
Plestiodon fasciatus	Dane	UWZM 17589	Vogt (1981)	
Plestiodon fasciatus	Eau Claire	MPM VZP456	Casper (1996)	Photograph
Plestiodon fasciatus	Green Lake	MPM 22511	Casper (1996)	
Plestiodon fasciatus	Iowa		Vogt (1981)	
Plestiodon fasciatus	Jackson	MPM 6942	Vogt (1981)	
Plestiodon fasciatus	Juneau	MPM VZP607	Vogt (1981)	Photograph
Plestiodon fasciatus	Langlade		Vogt (1981)	
Plestiodon fasciatus	Marinette	MPM 26178	Vogt (1981)	Embryo collected by P. Cochran that is clearly a lizard, but difficult to identify to species; see further discussion in Cochran (1989)
Plestiodon fasciatus	Marquette	INHS 2015aa	Kapfer et al. (2015)	Photograph
Plestiodon fasciatus	Milwaukee	MPM 491	Graenicher (1911), Casper (1996)	
Plestiodon fasciatus	Monroe	MPM 30298	Casper and Burnham (2001c)	
Plestiodon fasciatus	Oconto	UWGB 38	Casper (1996)	
Plestiodon fasciatus	Polk	MPM 28005	Vogt (1981)	
Plestiodon fasciatus	Portage	UWSP 2578	Vogt (1981)	
Plestiodon fasciatus	Sauk		Vogt (1981)	
Plestiodon fasciatus	Shawano	MPM 2424	Dickinson (1950)	

Scientific Name	County	Museum Voucher	Publication	Notes
Plestiodon fasciatus	Washington		Graenicher (1911), Casper (1996)	Authors reported vouchers as MPM 489 & 490 collected in 1888, but the location information in museum records includes only "Cedar Lake" and there are numerous cedar lakes in Wisconsin
Plestiodon fasciatus	Waupaca	UWZM 14718	Kapfer et al. (2015)	
Plestiodon fasciatus	Waushara	UWZM 9908	Vogt (1981)	
Plestiodon fasciatus	Winnebago	UWZM 24251	Vogt (1981)	
Plestiodon fasciatus	Wood	UWZM 4080	Vogt (1981)	
Plestiodon septentrionalis	Barron	MPM VZP296	Casper (1996)	Photograph
Plestiodon septentrionalis	Bayfield	MPM 13559	Vogt (1981)	
Plestiodon septentrionalis	Buffalo	JFBM HerpPhoto473	Casper (2015)	Photograph
Plestiodon septentrionalis	Burnett	MPM VZP314	Vogt (1981)	Photograph
Plestiodon septentrionalis	Chippewa	MPM VZP518	Casper (1996)	Photograph
Plestiodon septentrionalis	Douglas	MPM 3035	Vogt (1981)	
Plestiodon septentrionalis	Dunn	MPM VZP850	Vogt (1981)	Photograph
Plestiodon septentrionalis	Eau Claire	MPM VZP352	Vogt (1981)	Photograph
Plestiodon septentrionalis	Pepin	MPM VZP975a–c	White et al. (2020b)	Photograph
Plestiodon septentrionalis	Polk	UWSP 3289	Vogt (1981)	
Plestiodon septentrionalis	Portage	UWSP 4003	Orlofske and Anderson (2005)	
Plestiodon septentrionalis	Sawyer	MPM VZP360	Cochran (1992)	Photograph
Plestiodon septentrionalis	St. Croix	MPM VZP263	Casper (1996)	Photograph
Plestiodon septentrionalis	Washburn	MPM 2618	Dickinson (1950), Vogt (1981)	
Plethodon cinereus	Ashland	MPM 13568	Vogt (1981)	
Plethodon cinereus	Barron	UWWHERP 0265	Wild (2016b)	
Plethodon cinereus	Bayfield	MPM 18049	Suzuki (1951), Vogt (1981)	Suzuki (1951) originally reported voucher as "UWZM uncatalogued"
Plethodon cinereus	Brown	UWZM 22616	Cochran et al. (1987)	

Scientific Name	County	Museum Voucher	Publication	Notes
Plethodon cinereus	Burnett	MPM 824	Casper (1996)	
Plethodon cinereus	Calumet	INHS 2015c	Kapfer et al. (2015)	Photograph
Plethodon cinereus	Chippewa	MPM VZP551	Casper (1997d)	Photograph
Plethodon cinereus	Clark	MPM VZP802	Vogt (1981)	Photograph
Plethodon cinereus	Dane	USNM 10903	Kapfer et al. (2015)	Historical specimen; may not still be present in county
Plethodon cinereus	Door	MPM 2593	Suzuki (1951), Vogt (1981)	
Plethodon cinereus	Douglas	UWZM 13836	Suzuki (1951), Vogt (1981)	Suzuki (1951) originally reported voucher as "UWZM uncatalogued"
Plethodon cinereus	Dunn	MPM VZP373	Casper (1996)	Photograph
Plethodon cinereus	Florence	MPM VZP342	Casper (1996)	Photograph
Plethodon cinereus	Forest	MPM 2569	Suzuki (1951), Vogt (1981)	
Plethodon cinereus	Iron	UWSP 4269	Vogt (1981)	
Plethodon cinereus	Jackson	MPM VZP621	Vogt (1981)	Photograph
Plethodon cinereus	Kewaunee	MPM 11714	Vogt (1981)	
Plethodon cinereus	Langlade		Vogt (1981)	MPM 3251 was collected in Langlade County, but museum records indicate it is currently "presumed missing"
Plethodon cinereus	Lincoln	BMR B325002	Casper (1996)	Casper (1996) originally reported the specimen number for this voucher as BMR Z56R033, but the institution later gave this specimen the new number listed here
Plethodon cinereus	Manitowoc	MPM 823	Suzuki (1951), Vogt (1981)	
Plethodon cinereus	Marathon	CHAS HERP 19299	Suzuki (1951), Vogt (1981)	Suzuki (1951) reported voucher as NU 874, but this specimen was later transferred to CHAS and given the new catalogue number listed here
Plethodon cinereus	Marinette	MPM 2667	Suzuki (1951), Vogt (1981)	
Plethodon cinereus	Menominee	UWZM 17592	Kapfer et al. (2015)	
Plethodon cinereus	Monroe	MPM VZP250	Casper (1996)	Photograph
Plethodon cinereus	Oconto	UMMZ 85567	Suzuki (1951), Vogt (1981)	
Plethodon cinereus	Oneida	MPM 3388	Vogt (1981)	
Plethodon cinereus	Pepin	MPM VZP849a–b	Wild (2016c)	Photographs
Plethodon cinereus	Pierce	MPM VZP500	Casper (1996)	Photograph

Scientific Name	County	Museum Voucher	Publication	Notes
Plethodon cinereus	Polk	JFBM 386	Suzuki (1951), Vogt (1981)	
Plethodon cinereus	Portage	MPM 22138	Vogt (1981)	
Plethodon cinereus	Price	MPM 9088	Suzuki (1951), Vogt (1981)	Suzuki (1951) reported voucher as USNM "uncatalogued"
Plethodon cinereus	Racine	USNM 3788	Kapfer et al. (2015)	Historical record; may not currently be present in county
Plethodon cinereus	Rusk	MPM 23407	Casper (1996)	
Plethodon cinereus	Sawyer	UWSP 4279	Vogt (1981)	
Plethodon cinereus	Shawano	MPM VZP134	Casper (1996)	Photograph
Plethodon cinereus	Sheboygan	MPM 14705	Vogt (1981)	
Plethodon cinereus	Taylor	MPM 23192	Casper (1996)	
Plethodon cinereus	Vilas	UWZM 20242	Suzuki (1951), Vogt (1981)	
Plethodon cinereus	Washburn	MPM VZP426	Casper (1996)	Photograph
Plethodon cinereus	Washington	INHS 2015ak	Kapfer et al. (2015)	Photograph
Plethodon cinereus	Waupaca	UWZM 20406	Vogt (1981)	
Plethodon cinereus	Wood	LSUMZ 32920	Vogt (1981)	
Pseudacris crucifer	Adams	MPM 11392	Vogt (1981)	
Pseudacris crucifer	Ashland	MPM 33215	Vogt (1981)	
Pseudacris crucifer	Barron		Casper (1996)	Casper (1996) reported voucher as "MPM audio tape," which is missing according to museum staff (J. Colby)
Pseudacris crucifer	Bayfield	MPM 16487	Vogt (1981)	
Pseudacris crucifer	Brown	UWZM 22640	Cochran and Knutsen (1987)	
Pseudacris crucifer	Buffalo	MPM 2868	Suzuki (1951), Vogt (1981)	
Pseudacris crucifer	Burnett	UWSP 4284	Vogt (1981)	
Pseudacris crucifer	Chippewa	MPM 33362	Casper (1996)	Casper (1996) reported voucher as "MPM audio tape," which is missing according to museum staff (J. Colby)
Pseudacris crucifer	Clark	FMNH 12654	Vogt (1981)	
Pseudacris crucifer	Columbia	CM 75760	Vogt (1981)	
Pseudacris crucifer	Crawford	MPM VZP488	Casper (1996)	Photograph
Pseudacris crucifer	Dane	MPM 2340	Vogt (1981)	
Pseudacris crucifer	Dodge	MPM 32932	Casper (2001c)	
Pseudacris crucifer	Door	UWSP 3188	Vogt (1981)	
Pseudacris crucifer	Douglas	MPM 16605	Vogt (1981)	
Pseudacris crucifer	Dunn	UWZM 20917	Vogt (1981)	

Scientific Name	County	Museum Voucher	Publication	Notes
Pseudacris crucifer	Eau Claire	MPM 18226	Casper (1996)	
Pseudacris crucifer	Florence	UWGB 310–R	Cochran (1988d)	
Pseudacris crucifer	Fond Du Lac	MPM 4980	Vogt (1981)	
Pseudacris crucifer	Forest	MPM 3269	Vogt (1981)	
Pseudacris crucifer	Grant	USNM 207452	Vogt (1981)	
Pseudacris crucifer	Green	USNM 207462	Vogt (1981)	
Pseudacris crucifer	Green Lake	UNSM ZM22864	Casper (2015)	University of Nebraska State Museum
Pseudacris crucifer	Iowa	MPM 11405	Vogt (1981)	
Pseudacris crucifer	Iron	UWZM 9540	Vogt (1981)	
Pseudacris crucifer	Jackson	AMNH 77320	Vogt (1981)	
Pseudacris crucifer	Jefferson	INHS 2015ap	Vogt (1981)	Audio/Visual file of calling males
Pseudacris crucifer	Juneau	UWZM 20918	Vogt (1981)	
Pseudacris crucifer	Kenosha		Hendricks (2000)	Hendricks (2000) reported voucher as MPM A1, which was likely an audio file that is now lost according to museum staff (J. Colby)
Pseudacris crucifer	Kewaunee	UWSP 3933	Saeland and Doperalski (2001), Saeland and Doperalski (2002)	
Pseudacris crucifer	La Crosse	INHS 2014p	Casper (1996)	Photograph; Casper (1996) reported voucher as "MPM audio tape," which is missing according to museum staff (J. Colby)
Pseudacris crucifer	Lafayette	USNM 207465	Vogt (1981)	
Pseudacris crucifer	Langlade	UWGB H187	Vogt (1981)	
Pseudacris crucifer	Lincoln	MPM VZP794	Vogt (1981)	Photograph
Pseudacris crucifer	Manitowoc	MPM 2871	Suzuki (1951), Vogt (1981)	
Pseudacris crucifer	Marathon	UWZM 16776	Vogt (1981)	
Pseudacris crucifer	Marinette	MPM 11474	Casper (1996)	Casper (1996) reported voucher as "MPM Audio Tape," but specimens MPM 11474–5 already existed (collected by R. Vogt in 1973)
Pseudacris crucifer	Marquette	UWSP 3670	Vogt (1981)	
Pseudacris crucifer	Milwaukee	MPM 54	Vogt (1981)	
Pseudacris crucifer	Monroe	MPM VZP799	Vogt (1981)	Photograph
Pseudacris crucifer	Oconto	UWZM 3818	Vogt (1981)	
Pseudacris crucifer	Oneida	MPM 5163	Vogt (1981)	
Pseudacris crucifer	Ozaukee	MPM 22538	Suzuki (1951), Vogt (1981)	

Scientific Name	County	Museum Voucher	Publication	Notes
Pseudacris crucifer	Pepin		Casper (1997d)	Casper (1997d) reported voucher as "MPM audio tape (Cat. No. pending)," which is missing according to museum staff (J. Colby)
Pseudacris crucifer	Pierce		Vogt (1981)	
Pseudacris crucifer	Polk	MPM 11510	Vogt (1981)	
Pseudacris crucifer	Portage	UWSP 4185	Vogt (1981)	
Pseudacris crucifer	Price	MPM 11438	Vogt (1981)	
Pseudacris crucifer	Racine		Casper (1999b)	Casper (1999b) reported voucher as "MPM Herpetology Audio Tape 4–16–97," which is missing according to museum staff (J. Colby); Museum records indicated that MPM 32285 was from Racine County, but specimen was not found during recent investigations of collection
Pseudacris crucifer	Richland	USNM 207472	Vogt (1981)	
Pseudacris crucifer	Rock	MPM VZP832	Casper (1997d)	Photograph; Casper (1997d) reported voucher as "MPM audio tape (Cat. No. pending)," but correspondences with museum staff supports that audio files are missing (J. Colby)
Pseudacris crucifer	Rusk	MPM VZP566	Casper (1997d)	Photograph
Pseudacris crucifer	Sauk	UWZM 3700	Vogt (1981)	
Pseudacris crucifer	Sawyer	UMMZ 71437	Suzuki (1951), Vogt (1981)	
Pseudacris crucifer	Shawano	UWZM 22631	Hodgson and Cochran (1986)	
Pseudacris crucifer	Sheboygan	MPM 27936	Casper (1996)	
Pseudacris crucifer	St. Croix		Vogt (1981)	
Pseudacris crucifer	Taylor	UWZM 20736	Vogt (1981)	
Pseudacris crucifer	Trempealeau	CHAS HERP19317	Suzuki (1951), Vogt (1981)	Suzuki (1951) reported voucher as NU 892, part of a collection that was transitioned to CHAS and given a new catalogue number
Pseudacris crucifer	Vernon	UWZM 21166	Suzuki (1951), Vogt (1981)	Suzuki (1951) reported voucher as "UWZM uncatalogued"
Pseudacris crucifer	Vilas	UWZM 6762	Vogt (1981)	
Pseudacris crucifer	Walworth	MPM 4994	Vogt (1981)	
Pseudacris crucifer	Washburn	MPM 12559	Vogt (1981)	
Pseudacris crucifer	Washington	MPM 2561	Suzuki (1951), Vogt (1981)	
Pseudacris crucifer	Waukesha	MPM 23532	Reed and Lang (1991)	

Scientific Name	County	Museum Voucher	Publication	Notes
Pseudacris crucifer	Waupaca	MPM 2869	Suzuki (1951), Vogt (1981)	
Pseudacris crucifer	Waushara	CHAS HERP19312	Suzuki (1951), Vogt (1981)	Suzuki (1951) reported voucher as NU 887–90, part of a collection that was trasitioned to CHAS
Pseudacris crucifer	Wood	MPM 16165	Vogt (1981)	
Pseudacris maculata	Adams	UWZM 3856	Suzuki (1951), Vogt (1981)	
Pseudacris maculata	Ashland	MPM 9171	Vogt (1981)	
Pseudacris maculata	Barron		Casper (1996)	Casper (1996) reported voucher as "MPM audio tape," which is missing according to museum staff (J. Colby)
Pseudacris maculata	Bayfield	MPM 9173	Vogt (1981)	
Pseudacris maculata	Brown	UWZM 22613	Cochran et al. (1987)	
Pseudacris maculata	Buffalo	USNM 311975	Vogt (1981)	
Pseudacris maculata	Burnett	FMNH 21694	Smith (1956), Vogt (1981)	
Pseudacris maculata	Calumet	USNM 207494	Vogt (1981)	
Pseudacris maculata	Chippewa	AMNH 5380	Casper (1996)	Casper (1996) reported voucher as "MPM audio tape," which is missing according to museum staff (J. Colby)
Pseudacris maculata	Clark	MPM 2863	Suzuki (1951), Vogt (1981)	
Pseudacris maculata	Columbia	CM 75761	Vogt (1981)	
Pseudacris maculata	Crawford	MPM 5009	Vogt (1981)	
Pseudacris maculata	Dane	UWZM 6505	Suzuki (1951), Vogt (1981)	
Pseudacris maculata	Dodge	UIMNH 1452	Suzuki (1951), Vogt (1981)	
Pseudacris maculata	Douglas	MPM 16588	Casper (1996)	
Pseudacris maculata	Dunn	JFBM 5268	Vogt (1981)	
Pseudacris maculata	Eau Claire	MPM 23193	Casper (1996)	
Pseudacris maculata	Fond Du Lac	MPM 4983	Vogt (1981)	
Pseudacris maculata	Forest		Casper (1996)	Casper (1996) reported voucher as "MPM audio tape," which is missing according to museum staff (J. Colby)
Pseudacris maculata	Grant	USNM 207514	Vogt (1981)	
Pseudacris maculata	Green	USNM 207662	Vogt (1981)	
Pseudacris maculata	Green Lake	UNSM ZM22884	Vogt (1981)	University of Nebraska State Museum
Pseudacris maculata	Iowa	MPM 11411	Vogt (1981)	

Scientific Name	County	Museum Voucher	Publication	Notes
Pseudacris maculata	Iron		Casper (1997d)	Casper (1997d) reported voucher as "MPM audio tape (Cat. No. pending)," which is missing according to museum staff (J. Colby)
Pseudacris maculata	Jackson	MPM VZP902	Badje (2018)	Audio file
Pseudacris maculata	Jefferson	INHS 2014v	Vogt (1981)	Audio/video file
Pseudacris maculata	Juneau	UWZM 20328	Vogt (1981)	
Pseudacris maculata	Kenosha	MPM 4517	Vogt (1981)	
Pseudacris maculata	La Crosse	MPM VZP903	Casper (1996)	Audio file; Casper (1996) reported voucher as "MPM audio tape," which is missing according to museum staff (J. Colby)
Pseudacris maculata	Lafayette	MPM 11471	Vogt (1981)	
Pseudacris maculata	Langlade		Casper (1996)	Casper (1996) reported voucher as "MPM audio tape," which is missing according to museum staff (J. Colby)
Pseudacris maculata	Lincoln	MPM VZP917	Vogt (1981)	Audio file
Pseudacris maculata	Manitowoc	MPM 2864	Suzuki (1951), Vogt (1981)	
Pseudacris maculata	Marathon	UWZM 16777	Vogt (1981)	
Pseudacris maculata	Marquette	UWSP 3663	Vogt (1981)	
Pseudacris maculata	Milwaukee	MPM 1260	Vogt (1981)	
Pseudacris maculata	Monroe	MPM VZP798	Vogt (1981)	Photograph
Pseudacris maculata	Oneida	UWZM 6760	Suzuki (1951), Casper (1997c)	Casper (1997c) reported as MPM "Audio Tape Collection (Cat. No. pending)," which is missing according to museum staff (J. Colby)
Pseudacris maculata	Outagamie		Vogt (1981)	
Pseudacris maculata	Ozaukee	MPM 22524	Vogt (1981)	
Pseudacris maculata	Pepin		Casper (1996)	Casper (1996) reported voucher as "MPM audio tape," which is missing according to museum staff (J. Colby)
Pseudacris maculata	Pierce	UWSP 4005	Vogt (1981)	
Pseudacris maculata	Polk	JFBM 12169	Vogt (1981)	
Pseudacris maculata	Portage	UWSP 91	Vogt (1981)	
Pseudacris maculata	Price	MPM 12561	Vogt (1981)	
Pseudacris maculata	Racine	MPM 32278	Vogt (1981)	
Pseudacris maculata	Richland	USNM 207549	Vogt (1981)	
Pseudacris maculata	Rock	UWZM 9514	Suzuki (1951), Vogt (1981)	

Scientific Name	County	Museum Voucher	Publication	Notes
Pseudacris maculata	Rusk		Casper (1997d)	Casper (1997d) reported voucher as "MPM audio tape (Cat. No. pending)," which is missing according to museum staff (J. Colby)
Pseudacris maculata	Sauk	USNM 207557	Vogt (1981)	
Pseudacris maculata	Sawyer	UWZM 22578	Casper (1996)	
Pseudacris maculata	Shawano		Casper (1996)	Casper (1996) reported voucher as "MPM audio tape," which is missing according to museum staff (J. Colby)
Pseudacris maculata	Sheboygan	MPM 25656	Casper (1996)	
Pseudacris maculata	St. Croix	TNHC 65814	Vogt (1981)	
Pseudacris maculata	Taylor	MPM VZP966	Badje (2021b)	Audio file
Pseudacris maculata	Trempealeau	MPM VZP826	Sheldon and Badje (2016)	Photograph
Pseudacris maculata	Vernon	MPM 11466	Vogt (1981)	
Pseudacris maculata	Walworth	MPM 32274	Vogt (1981)	
Pseudacris maculata	Washburn	UWZM 20421	Vogt (1981)	
Pseudacris maculata	Washington	MPM VZP182	Casper (1996)	Photograph
Pseudacris maculata	Waukesha	UWZM 9820	Suzuki (1951), Vogt (1981)	Suzuki (1951) reported voucher as UWZM 9521, which could not be found during recent investigations of the collection
Pseudacris maculata	Waupaca	MPM 2865	Suzuki (1951), Vogt (1981)	
Pseudacris maculata	Waushara	MPM 11911	Vogt (1981)	
Pseudacris maculata	Winnebago	MPM VZP517	Casper (1996)	Photograph
Pseudacris maculata	Wood	UWSP 1761	Vogt (1981)	
Regina septemvittata	Fond Du Lac		Gibbons and Dorcas (2004)	Gibbons and Dorcas (2004) reported supporting documentation as a personal observation by G. Casper
Regina septemvittata	Green	JFBM HerpPhoto471	Gibbons and Dorcas (2004), Casper (2015)	Photograph; Gibbons and Dorcas (2004) originally reported this record but with supporting documentation as a personal observation by G. Casper
Regina septemvittata	Jefferson		Vogt (1981), Gibbons and Dorcas (2004)	Museum records indicated that UWZM 9709 was from Jefferson County, but this specimen could not be found after recent investigations of the collection
Regina septemvittata	Milwaukee	MPM 99	Casper (1996)	
Regina septemvittata	Ozaukee	MPM 2225	Casper (1996)	
Regina septemvittata	Racine	MPM 4547	Vogt (1981), Gibbons and Dorcas (2004)	

Scientific Name	County	Museum Voucher	Publication	Notes
Regina septemvittata	Rock	INHS 2008i	Gibbons and Dorcas (2004), Kapfer et al. (2009c)	Photograph; Gibbons and Dorcas (2004) originally reported this record but with supporting documentation as a personal observation by G. Casper
Regina septemvittata	Sheboygan		Gibbons and Dorcas (2004)	Gibbons and Dorcas (2004) reported supporting documentation as a personal observation by G. Casper
Regina septemvittata	Walworth	CHAS HERP501	Schmidt and Necker (1935), Vogt (1981)	
Regina septemvittata	Washington	MPM VZP34	Casper (1996)	Photograph
Regina septemvittata	Waukesha	UWZM 9707	Vogt (1981)	
Sistrurus catenatus	Buffalo	CM 73913	Vogt (1981)	
Sistrurus catenatus	Clark	UWZM 22776	Vogt (1981)	
Sistrurus catenatus	Columbia	MPM 2209	Vogt (1981), Casper (1993b)	
Sistrurus catenatus	Dane	UWZM 17584	Casper (1993b)	
Sistrurus catenatus	Iowa	MPM 2115	Casper (1993b)	
Sistrurus catenatus	Jackson	MPM 23402	Vogt (1981), Casper (1993b)	
Sistrurus catenatus	Juneau	UWZM 17586	Vogt (1981), Casper (1993b)	
Sistrurus catenatus	La Crosse	MPM 2650	Dickinson (1950)	Dickinson (1950) reported voucher as MPM 2650, but associated location is only given at county level; other specimens, including MPM 23354 and UWZM 13011, exist that are accompanied with detailed location data
Sistrurus catenatus	Monroe	MPM VZP423	Casper (1996)	Photograph
Sistrurus catenatus	Pepin		Vogt (1981)	
Sistrurus catenatus	Portage	CM 70492	Casper (2015)	
Sistrurus catenatus	Racine	USNM 525	Vogt (1981)	
Sistrurus catenatus	Sauk	MPM 2425	Dickinson (1950)	
Sistrurus catenatus	Trempealeau		Vogt (1981), Casper (1993b)	Casper (1993b) reported voucher as "UWEC unnumbered"; note that a specimen with the catalogue number UWEC 1455 (part of the UW–Eau Claire collection) purportedly once existed, but the specimen is presumed missing according to correspondences with UWEC staff
Sistrurus catenatus	Walworth	FMNH 12698	Schmidt and Necker (1935)	

Scientific Name	County	Museum Voucher	Publication	Notes
Sistrurus catenatus	Wood	UWZM 23899	Vogt (1981)	
Sternotherus odoratus	Adams	MPM VZP923	Skawinski (2020)	Photograph
Sternotherus odoratus	Columbia		Casper (1997d)	Casper (1997d) reported voucher as MPM VZP600, but this is missing according to museum staff (J. Colby)
Sternotherus odoratus	Dane	UWZM 20307	Vogt (1981)	
Sternotherus odoratus	Fond Du Lac	MPM 17471	Casper (1996)	
Sternotherus odoratus	Grant	MPM VZP781	Vogt (1981)	Photograph
Sternotherus odoratus	Green	MPM VZP159	Casper (1996)	Photograph
Sternotherus odoratus	Iowa	MPM 6928	Vogt (1981)	
Sternotherus odoratus	Jefferson	MPM 26150	Casper (1996)	
Sternotherus odoratus	Kenosha	MPM VZP13	Casper (1996)	Photograph
Sternotherus odoratus	La Crosse	MPM 30227	Casper (1997e)	
Sternotherus odoratus	Manitowoc	INHS 2007d	Surendonk and Kapfer (2008)	Photograph
Sternotherus odoratus	Marquette	MPM 2372	Casper (1996)	
Sternotherus odoratus	Milwaukee	MPM 26363	Casper (1996)	
Sternotherus odoratus	Oconto	INHS 2007b	Ewert et al. (2007)	Photograph
Sternotherus odoratus	Outagamie	MPM VZP906	Ziesmer (2019)	Photograph
Sternotherus odoratus	Racine	CHAS HERP18749	Vogt (1981)	
Sternotherus odoratus	Richland	MPM VZP569	Casper (1997d)	Photograph
Sternotherus odoratus	Sauk	UWZM 24307	Casper (1996)	
Sternotherus odoratus	Shawano	MPM VZP836a	Brockman-Mederas (2016)	Photograph
Sternotherus odoratus	Trempealeau	MPM VZP323	Casper (1996)	Photograph
Sternotherus odoratus	Walworth	MPM 30344	Necker (1939b), Vogt (1981)	
Sternotherus odoratus	Washington	MPM 17459	Casper (1996)	
Sternotherus odoratus	Waukesha	UWZM 9820	Vogt (1981)	
Sternotherus odoratus	Waupaca	UWSP 3817	Vogt (1981)	
Sternotherus odoratus	Waushara	CHAS HERP10368	Vogt (1981)	
Storeria dekayi	Adams	INHS 2015b	Kapfer et al. (2015)	
Storeria dekayi	Brown		Cochran et al. (1987)	Authors reported voucher as BBWS RE08, which is part of a collection that no longer exists, according to correspondence with staff at the Bay Beach Wildlife Sanctuary
Storeria dekayi	Buffalo	CM 75883	Vogt (1981)	

Scientific Name	County	Museum Voucher	Publication	Notes
Storeria dekayi	Burnett		Vogt (1981)	Museum records indicated that UWZM 2093 and 2099 are from the appropriate location, but further investigation revealed both are misidentified *S. occipitomaculata*
Storeria dekayi	Calumet	UWGB Slide5	Casper (1996)	Photograph
Storeria dekayi	Chippewa	MPM VZP594	Casper (1997d)	Photograph
Storeria dekayi	Clark	FMNH 14593	Vogt (1981)	Museum records indicated that UWZM 2159 is from the appropriate location, but further investigation revealed it is a misidentified *S. occipitomaculata*
Storeria dekayi	Columbia	CM 33409	Vogt (1981)	
Storeria dekayi	Crawford	UWSP 481	Vogt (1981)	
Storeria dekayi	Dane	UMMZ 57862	Trapido (1944), Vogt (1981)	
Storeria dekayi	Dodge	UIMNH 1585	Vogt (1981)	
Storeria dekayi	Door	UWGB 25	Casper (1996)	
Storeria dekayi	Douglas		Vogt (1981)	
Storeria dekayi	Dunn	MPM VZP374	Casper (1996)	Photograph
Storeria dekayi	Fond Du Lac	MPM 6408	Casper (1996)	
Storeria dekayi	Grant	UMMZ 69642	Trapido (1944), Vogt (1981)	
Storeria dekayi	Green	INHS 2014s	Vogt (1981)	Photograph
Storeria dekayi	Green Lake	MPM 7934	Casper (2015)	
Storeria dekayi	Iowa	MPM 5099	Vogt (1981)	
Storeria dekayi	Jackson	MPM 33146	Cochran et al. (2001)	
Storeria dekayi	Jefferson	MPM VZP784	Vogt (1981)	Photograph
Storeria dekayi	Juneau	UWZM 24252	Vogt (1981)	
Storeria dekayi	Kenosha	MPM 26284	Schmidt and Necker (1935)	Schmidt and Necker (1935) reported voucher as CHAS HERP 1577, but correspondences with museum staff indicated this specimen was collected in Waukesha County (D. Roberts)
Storeria dekayi	La Crosse	UWZM 20164	Vogt (1981)	
Storeria dekayi	Lafayette	MPM 11548	Vogt (1981)	
Storeria dekayi	Manitowoc	MPM VZP349	Casper (1996)	Photograph
Storeria dekayi	Marathon	UWZM 20160	Vogt (1981)	
Storeria dekayi	Marinette	MPM 2773	Dickinson (1972), Vogt (1981)	
Storeria dekayi	Marquette	INHS 12812	Petzing et al. (1998)	

Scientific Name	County	Museum Voucher	Publication	Notes
Storeria dekayi	Milwaukee	MPM 2451	Vogt (1981)	
Storeria dekayi	Monroe	UWSP 2151	Vogt (1981)	
Storeria dekayi	Oconto	UWGB 48	Casper (1996)	
Storeria dekayi	Oneida		Vogt (1981)	Museum records indicated that while UWZM 1915 was taken from the appropriate location, further investigation revealed the specimen is a misidentified *S. occipitomaculata*
Storeria dekayi	Ozaukee	MPM 22515	Casper (1996)	
Storeria dekayi	Pepin	MPM VZP631	Casper (1999b)	Photograph
Storeria dekayi	Pierce	MPM VZP497	Casper (1996)	Photograph
Storeria dekayi	Polk	MPM VZP385	Vogt (1981)	Photograph
Storeria dekayi	Racine	USNM 1958	Trapido (1944), Vogt (1981)	
Storeria dekayi	Richland	MPM 23433	Vogt (1981)	
Storeria dekayi	Rock	INHS 2014t	Vogt (1981)	Photograph
Storeria dekayi	Sauk	MPM 2133	Vogt (1981)	
Storeria dekayi	Shawano	MPM VZP279	Casper (1996)	Photograph
Storeria dekayi	Sheboygan	MPM 1783	Vogt (1981)	
Storeria dekayi	St. Croix	JFBM R1019	Casper (2015)	
Storeria dekayi	Trempealeau		Vogt (1981)	
Storeria dekayi	Vernon	MPM 2089	Vogt (1981)	
Storeria dekayi	Walworth	MPM 26326	Vogt (1981)	
Storeria dekayi	Washburn		Casper (2015)	Casper (2015) reported voucher as UWZM 24065, but investigation of this specimen revealed it is a misidentified *S. occipitomaculata*
Storeria dekayi	Washington	MPM 23379	Vogt (1981)	
Storeria dekayi	Waukesha	MPM 2412	Vogt (1981)	
Storeria dekayi	Waushara		Vogt (1981)	
Storeria dekayi	Wood	UWSP 450	Vogt (1981)	
Storeria occipitomaculata	Adams	MPM 4530	Vogt (1981)	
Storeria occipitomaculata	Ashland	FMNH 18434	Trapido (1944), Vogt (1981)	
Storeria occipitomaculata	Barron	MPM VZP286	Casper (1996)	Photograph
Storeria occipitomaculata	Bayfield	UMMZ 74552	Trapido (1944), Vogt (1981)	
Storeria occipitomaculata	Brown	UWZM 22620	Cochran et al. (1987)	
Storeria occipitomaculata	Burnett	MPM 11514	Vogt (1981)	

Scientific Name	County	Museum Voucher	Publication	Notes
Storeria occipitomaculata	Calumet	MPM VZP445	Casper (1996)	Photograph
Storeria occipitomaculata	Chippewa	MPM 33348	Vogt (1981)	
Storeria occipitomaculata	Clark	FMNH 14596	Trapido (1944), Vogt (1981)	
Storeria occipitomaculata	Columbia	MSUM 6274	Casper (1996)	
Storeria occipitomaculata	Crawford		Vogt (1981)	
Storeria occipitomaculata	Dane	MPM 11538	Vogt (1981)	
Storeria occipitomaculata	Dodge	UIMNH 1593	Vogt (1981)	
Storeria occipitomaculata	Door	UWSP 3175	Vogt (1981)	
Storeria occipitomaculata	Douglas	MPM 25654	Vogt (1981)	
Storeria occipitomaculata	Dunn	MPM VZP376	Casper (1996)	Photograph
Storeria occipitomaculata	Eau Claire	MPM 23196	Casper (1996)	
Storeria occipitomaculata	Florence	UWSP 1735	Vogt (1981)	
Storeria occipitomaculata	Fond Du Lac	MPM 2388	Dickinson (1950)	
Storeria occipitomaculata	Forest	MPM 763	Dickinson (1972), Vogt (1981)	
Storeria occipitomaculata	Grant	MPM VZP782	Vogt (1981)	Photograph
Storeria occipitomaculata	Green	MPM VZP441	Casper (1996)	Photograph
Storeria occipitomaculata	Green Lake	MPM VZP639	Casper (1999b)	Photograph
Storeria occipitomaculata	Iowa	UWZM 17382	Vogt (1981)	
Storeria occipitomaculata	Iron	UMMZ 58589	Trapido (1944), Vogt (1981)	
Storeria occipitomaculata	Jackson	MPM 11561	Vogt (1981)	
Storeria occipitomaculata	Jefferson	MPM 26287	Casper (1996)	
Storeria occipitomaculata	Juneau	UWZM 13826	Vogt (1981)	
Storeria occipitomaculata	Kenosha	MPM VZP217	Casper (1996)	Photograph

Scientific Name	County	Museum Voucher	Publication	Notes
Storeria occipitomaculata	Kewaunee	UWGB 77	Casper (1996)	
Storeria occipitomaculata	Lafayette	INHS 2015z	Kapfer et al. (2015)	Photograph
Storeria occipitomaculata	Langlade	CHAS HERP3834	Trapido (1944), Vogt (1981)	
Storeria occipitomaculata	Lincoln	MPM 7953	Vogt (1981)	
Storeria occipitomaculata	Manitowoc	UWZM 19022	Vogt (1981)	
Storeria occipitomaculata	Marathon	UMMZ 64745	Trapido (1944), Vogt (1981)	
Storeria occipitomaculata	Marinette	MPM 33297	Vogt (1981)	
Storeria occipitomaculata	Marquette	MPM 31624	Casper (2001d)	
Storeria occipitomaculata	Menominee	MPM 1903	Casper (2015)	
Storeria occipitomaculata	Milwaukee	MPM 1672	Vogt (1981)	
Storeria occipitomaculata	Monroe	MPM 23404	Casper (1996)	
Storeria occipitomaculata	Oconto	FMNH 29370	Trapido (1944), Vogt (1981)	
Storeria occipitomaculata	Oneida	FMNH 26045	Trapido (1944), Vogt (1981)	
Storeria occipitomaculata	Outagamie	MPM VZP620	Casper (1999b)	Photograph
Storeria occipitomaculata	Ozaukee	MPM VZP75	Casper (1996)	Photograph
Storeria occipitomaculata	Pierce	MPM VZP499	Casper (1996)	Photograph
Storeria occipitomaculata	Polk	UWSP 4208	Casper (1996)	Casper (1996) reported voucher as MPM 23462, which recent investigation revealed has become dehydrated and difficult to identify to species
Storeria occipitomaculata	Portage	UWSP 201	Vogt (1981)	
Storeria occipitomaculata	Price	MPM 2508	Dickinson (1950), Vogt (1981)	
Storeria occipitomaculata	Racine	USNM 7281	Trapido (1944), Vogt (1981)	
Storeria occipitomaculata	Richland	MPM 6939	Vogt (1981)	
Storeria occipitomaculata	Rusk	MPM 26343	Casper (1996)	

Scientific Name	County	Museum Voucher	Publication	Notes
Storeria occipitomaculata	Sauk	USNM 10715	Trapido (1944), Vogt (1981)	
Storeria occipitomaculata	Sawyer	MPM 26365	Casper (1996)	
Storeria occipitomaculata	Shawano	MPM 2460	Dickinson (1950), Vogt (1981)	
Storeria occipitomaculata	Sheboygan	MPM 16334	Casper (1996)	
Storeria occipitomaculata	St. Croix	MPM VZP300	Casper (1996)	Photograph
Storeria occipitomaculata	Taylor	UMMZ 69625	Dickinson (1950), Vogt (1981)	
Storeria occipitomaculata	Vernon	JFBM 14450	Cochran et al. (2006)	
Storeria occipitomaculata	Vilas	USNM 55788	Trapido (1944), Vogt (1981)	
Storeria occipitomaculata	Walworth	MPM 32206	Vogt (1981)	
Storeria occipitomaculata	Washburn	UNSM ZM593	Dickinson (1950)	University of Nebraska State Museum; Dickinson (1950) reported voucher as MPM 2663, but recent investigations of the collection support this specimen is missing
Storeria occipitomaculata	Washington	MPM 23380	Casper (1996)	
Storeria occipitomaculata	Waukesha	MPM 22543	Vogt (1981)	
Storeria occipitomaculata	Waupaca	MPM 27989	Casper (1997d)	
Storeria occipitomaculata	Waushara	UWGB 394	Watermolen (1990d)	
Storeria occipitomaculata	Wood	MPM 30305	Vogt (1981)	
Terrapene ornata	Columbia	UWZM 2167	Vogt (1981)	
Terrapene ornata	Crawford		Vogt (1981)	
Terrapene ornata	Dane	UWZM 14724	Vogt (1981)	
Terrapene ornata	Grant	MPM VZP544	Vogt (1981)	Photograph
Terrapene ornata	Iowa	UWZM 17022	Vogt (1981)	
Terrapene ornata	Jackson	UWGB R451	Kapfer et al. (2015)	Historical specimen and species unlikely still present in county
Terrapene ornata	Rock	MPM VZP439	Casper (1996)	Photograph
Terrapene ornata	Sauk	UWZM 22003	Vogt (1981)	
Thamnophis butleri	Dodge		Sloss (2011)	Genetic evidence

Scientific Name	County	Museum Voucher	Publication	Notes
Thamnophis butleri	Fond Du Lac	MPM 11848	Vogt (1981)	
Thamnophis butleri	Jefferson	UWSP 4534	Kapfer et al. (2013c)	Tissue sample, with associated photographs UWSP 0162–0165
Thamnophis butleri	Juneau	UWSP 546	Casper (1996)	Casper (1996) reported voucher as UWSP 237, but this catalogue number is not currently listed in institution's collection records
Thamnophis butleri	Kenosha		Vogt (1981)	Currently known museum specimens may be hybrids
Thamnophis butleri	Milwaukee	MPM 30388	Vogt (1981)	
Thamnophis butleri	Ozaukee	MPM 23206	Casper (1996)	
Thamnophis butleri	Racine		Vogt (1981)	Currently known museum specimens may be hybrids
Thamnophis butleri	Sheboygan	UWZM 9906	Burghardt et al. (2006)	
Thamnophis butleri	Walworth		Vogt (1981)	
Thamnophis butleri	Washington		Vogt (1981)	
Thamnophis butleri	Waukesha	AMNH 76178	Rossman et al. (1981), Vogt (1981)	
Thamnophis proximus	Clark	FMNH 14580	Vogt (1981)	
Thamnophis proximus	Dane	UWZM 21232		Labels for both *T. proximus* and *T. saurita* in jar with UWZM 21232; determined as *T. proximus* by J. Lorch
Thamnophis proximus	Jefferson		Casper (2015)	Casper (2015) reported voucher as BYU 40420, but correspondences with BYU collection staff indicate the specimen is missing and cannot be further scrutinized (J. Sites)
Thamnophis proximus	Racine	USNM 776	Vogt (1981)	
Thamnophis proximus	Sauk		Vogt (1981)	Museum records indicated that CM 70518 is from Sauk County, but species identity cannot be confirmed
Thamnophis radix	Columbia	MSUM HE.6274	Vogt (1981)	
Thamnophis radix	Crawford		Walley and Walley (2004b)	Walley and Walley (2004b) reported voucher as HDW-NIU 1936, but correspondences with numerous institutions (including Northern Illinois University) supports this specimen is missing
Thamnophis radix	Dane	USNM 17423	Smith (1949), Vogt (1981)	
Thamnophis radix	Dodge	INHS 2014h	Vogt (1981)	Photograph
Thamnophis radix	Dunn	UWZM 2189	Vogt (1981)	
Thamnophis radix	Green	MPM 23601	Smith (1949), Vogt (1981)	

Scientific Name	County	Museum Voucher	Publication	Notes
Thamnophis radix	Green Lake	CHAS HERP6657	Smith (1949), Vogt (1981)	
Thamnophis radix	Iowa	UWZM 17287	Vogt (1981)	
Thamnophis radix	Jefferson	MPM VZP884a–d	Paloski et al. (2020)	Photographs of specimen that genetic data obtained from. No morph analyses conducted.
Thamnophis radix	Kenosha	CHAS HERP5453	Necker (1939b), Vogt (1981)	
Thamnophis radix	Lafayette	UWZM 17284	Casper (2015)	
Thamnophis radix	Milwaukee		Smith (1949), Vogt (1981)	Smith (1949) reported voucher as MPM 552 and 2518, which museum records indicated were redetermined as *T. buteri* (G. Casper)
Thamnophis radix	Racine	USNM 719	Smith (1949), Vogt (1981)	
Thamnophis radix	Rock	MPM 18313	Vogt (1981)	
Thamnophis radix	Sauk	UWZM 17277	Casper (2015)	
Thamnophis radix	Walworth	CHAS HERP5079	Necker (1939b), Vogt (1981)	
Thamnophis radix	Waukesha	MPM VZP852a–g	Paloski et al. (2017)	
Thamnophis saurita	Adams	USNM 307855	Casper (1997d)	
Thamnophis saurita	Columbia	UMMZ 1262–4	Casper (2015)	Photographs
Thamnophis saurita	Door	UMMZ 1257–8	Casper (2015)	Photographs
Thamnophis saurita	Marinette	MPM 2777	Rossman (1963b), Vogt (1981)	
Thamnophis saurita	Milwaukee	MPM 2605	Casper (1996)	
Thamnophis saurita	Sauk	UWZM 21231		Vogt (1981) reported *T. proximus* in Sauk County, but no specimen is known; perhaps that is based on a misidentification of this specimen
Thamnophis saurita	Sheboygan	MPM 18295	Vogt (1981)	
Thamnophis saurita	Washington	MPM 1007	Casper (1996)	
Thamnophis sirtalis	Adams	MPM 6941	Vogt (1981)	
Thamnophis sirtalis	Ashland	MPM 33073	Vogt (1981)	
Thamnophis sirtalis	Barron	MPM VZP364	Vogt (1981)	Photograph
Thamnophis sirtalis	Bayfield	MPM 11770	Vogt (1981)	
Thamnophis sirtalis	Brown	UWZM 22622	Cochran et al. (1987)	
Thamnophis sirtalis	Buffalo	MPM 1143	Vogt (1981)	
Thamnophis sirtalis	Burnett	MPM 26348	Vogt (1981)	
Thamnophis sirtalis	Calumet	UWGB 1485	Sobota (1990c)	Photograph
Thamnophis sirtalis	Chippewa	MPM 33367	Vogt (1981)	
Thamnophis sirtalis	Clark	UWZM 9971	Vogt (1981)	

Scientific Name	County	Museum Voucher	Publication	Notes
Thamnophis sirtalis	Columbia	MPM 11655	Vogt (1981)	
Thamnophis sirtalis	Crawford	MPM 5014	Vogt (1981)	
Thamnophis sirtalis	Dane	MPM 554	Vogt (1981)	
Thamnophis sirtalis	Dodge	UWZM 24068	Vogt (1981)	
Thamnophis sirtalis	Door	MPM 7951	Vogt (1981)	
Thamnophis sirtalis	Douglas	MPM 960	Vogt (1981)	
Thamnophis sirtalis	Dunn	UWZM 2188	Vogt (1981)	
Thamnophis sirtalis	Eau Claire	UWZM 2222	Vogt (1981)	
Thamnophis sirtalis	Florence	MPM 23406	Dickinson (1972), Vogt (1981)	Dickinson (1972) reported voucher as UWZM "unnumbered"
Thamnophis sirtalis	Fond Du Lac	MPM 32882	Vogt (1981)	
Thamnophis sirtalis	Forest	MPM 3214	Dickinson (1972), Vogt (1981)	
Thamnophis sirtalis	Grant	MPM 6886	Vogt (1981)	
Thamnophis sirtalis	Green	MPM VZP605	Vogt (1981)	Photograph
Thamnophis sirtalis	Green Lake	MPM 7935	Vogt (1981)	
Thamnophis sirtalis	Iowa	MPM 18309	Vogt (1981)	
Thamnophis sirtalis	Iron	MPM 11773	Vogt (1981)	
Thamnophis sirtalis	Jackson	UWZM 2284	Vogt (1981)	
Thamnophis sirtalis	Jefferson	MPM 18304	Vogt (1981)	
Thamnophis sirtalis	Juneau	UWZM 20894	Vogt (1981)	
Thamnophis sirtalis	Kenosha	MPM 11723	Vogt (1981)	
Thamnophis sirtalis	Kewaunee	MPM 6901	Vogt (1981)	
Thamnophis sirtalis	La Crosse	USNM 81990	Vogt (1981)	
Thamnophis sirtalis	Lafayette	MPM VZP251	Casper (1996)	Photograph
Thamnophis sirtalis	Langlade	UWZM 17295	Vogt (1981)	
Thamnophis sirtalis	Lincoln	MPM 7954	Vogt (1981)	
Thamnophis sirtalis	Manitowoc	MPM 6890	Vogt (1981)	
Thamnophis sirtalis	Marathon	MPM 6348	Vogt (1981)	
Thamnophis sirtalis	Marinette	MPM 2428	Dickinson (1972), Vogt (1981)	Dickinson (1972) reported voucher as MPM 3278, but recent investigations of the collection support the specimen is missing
Thamnophis sirtalis	Marquette	MPM 2530	Dickinson (1950), Vogt (1981)	
Thamnophis sirtalis	Menominee	UWGB 279	Casper (1996)	
Thamnophis sirtalis	Milwaukee	MPM 9755	Dickinson (1950), Vogt (1981)	Dickinson (1950) reported voucher as MPM 2600, 2605, but recent investigations of the collection support specimens are missing

Scientific Name	County	Museum Voucher	Publication	Notes
Thamnophis sirtalis	Monroe	UWSP 1720	Vogt (1981)	
Thamnophis sirtalis	Oconto	MPM 11637	Vogt (1981)	
Thamnophis sirtalis	Oneida	UWSP 1657	Vogt (1981)	
Thamnophis sirtalis	Outagamie	MPM VZP199	Casper (1996)	Photograph
Thamnophis sirtalis	Ozaukee	MPM 22535	Casper (1996)	
Thamnophis sirtalis	Pepin	UWZM 2344	Vogt (1981)	
Thamnophis sirtalis	Pierce	MPM 1145	Vogt (1981)	
Thamnophis sirtalis	Polk	MPM 984	Vogt (1981)	
Thamnophis sirtalis	Portage	UWSP 4187	Vogt (1981)	
Thamnophis sirtalis	Price	MPM 11778	Vogt (1981)	
Thamnophis sirtalis	Racine	MPM 32280	Vogt (1981)	
Thamnophis sirtalis	Richland	MPM 6891	Vogt (1981)	
Thamnophis sirtalis	Rock	UMMZ 46471	Vogt (1981)	
Thamnophis sirtalis	Rusk	MPM 26337	Casper (1996)	
Thamnophis sirtalis	Sauk	MPM 7955	Vogt (1981)	
Thamnophis sirtalis	Sawyer	UWSP 3847	Vogt (1981)	
Thamnophis sirtalis	Shawano	MPM 23445	Vogt (1981)	
Thamnophis sirtalis	Sheboygan	MPM 25655	Vogt (1981)	
Thamnophis sirtalis	St. Croix	MPM VZP48	Casper (1996)	Photograph
Thamnophis sirtalis	Taylor	MPM VZP801	Vogt (1981)	Photograph
Thamnophis sirtalis	Trempealeau	MPM VZP324	Vogt (1981)	Photograph
Thamnophis sirtalis	Vernon	MPM 2073	Vogt (1981)	
Thamnophis sirtalis	Vilas	UWSP 1886	Vogt (1981)	
Thamnophis sirtalis	Walworth	FMNH 29250	Necker (1939b), Vogt (1981)	
Thamnophis sirtalis	Washburn	UWZM 2190	Vogt (1981)	
Thamnophis sirtalis	Washington	MPM 1477	Vogt (1981)	
Thamnophis sirtalis	Waukesha	MPM 4578	Vogt (1981)	
Thamnophis sirtalis	Waupaca	MPM 27995	Vogt (1981)	
Thamnophis sirtalis	Waushara		Vogt (1981)	Museum records indicated that UWZM 9912 was from Waushara County, but recent investigations of the collection support this specimen is missing
Thamnophis sirtalis	Winnebago	MPM VZP519	Casper (1996)	Photograph
Thamnophis sirtalis	Wood	UWZM 17289	Vogt (1981)	
Tropidoclonion lineatum	Dane	UMMZ 1269–1273	Casper and Kleinhenz (2016)	Photographs
Tropidoclonion lineatum	Iowa	INHS 2011m	Raimond and Lorch (2012)	Photograph

Glossary

Except where identified by *, definitions were used with permission from H. B. Lillywhite, 2008, Dictionary of Herpetology, Krieger Publishing, Malabar, Florida, USA.

abundance: *n.* The number of individuals belonging to a population.

advertisement call: Reference to vocalization of male anurans used in courtship and territorial behavior.

aestivation: *n.* (*v.* estivate) A state of torpor characterized by reduced activity and metabolism associated with summer or periods of high temperatures or drought.

allele: *n.* A shorthand form of *allelomorph*, representing one of a series of possible alternate (heritable) forms of a given gene. Alleles differ in DNA sequences and affect the functioning of RNA and/or protein. This term is sometimes used interchangeably with *gene*.

allopolypoid: *adj.* A *polyploid* organism arising from the combination of chromosome sets that are genetically distinct (from two different species).

amplexus: *n.* Precopulation pairing; usually used in reference to sexual embrace of amphibians, with grip of males on female being usually either axillary (pectoral) or inguinal (pelvic) in position.

amplectant: *adj.* Term used to describe amphibians engaged in amplexus.

anal plate: The plate or scute which covers the vent of snakes.

anoxia: *n.* (*adj.* -ic) Lack of oxygen. Cf. *hypoxia.*

aposematic: *adj.* Reference to characters, especially coloration, that advertise an organism as dangerous or unpleasant to potential predators ("warning" characters).

aquatic: *adj.* Living in water.

***arthropod:** *n.* A member of the phylum Arthropoda, which includes many nonvertebrate animals, such as arachnids and insects.

atretic ova: Egg follicles that have regressed before ovulation.

auditory: *adj.* Pertaining to the sensory reception of sound.

autotomy: *n.* (*adj.* -ic, -ous) Self division or fracture, as in the ability to release a body part such as the tail breakage that occurs in many lizard species and snake species when the animal is seized by a predator (*caudal autotomy*). The definition has been proposed as descriptive of taxa having intravertebral breakage followed by tail regeneration (J. M. Savage and J. B. Slowinski, 1996, Biological Journal of the Linnean Society 57:129–194). Syn. *tail-dropping.* Cf. *pseudoautotomy, urotomy.*

beak: *n.* The horny covering of the jaws, also called *rhamphotheca* or *tomium* in turtles.

biennial: *adj.* Lasting for two years, or occurring on a cycle every two years.

bifurcated: *adj.* Divided into two branches or projections, as the tongue of snakes and some lizards.

biological diversity (biodiversity): *n.* Denoting the variety of living organisms (species diversity) and, connotatively, their genetic diversity, ecological diversity, etc. The total variety of life and its processes.

bottleneck: See *population bottleneck.*

bridge: *n.* The part of the shell connecting the carapace and plastron in turtles.

brood: *n., v.* 1. The offspring from a single birth or from a single clutch of eggs. 2. To incubate, cover, remain with, and perhaps care for a clutch of eggs. Cf. *litter.*

CTmax: See *critical thermal maximum.*

CTmin: See *critical thermal minimum.*

carapace: *n.* The domelike dorsal component of the shell of turtles, usually comprised of both epidermal and bony plates. Cf. *plastron.*

caruncle: *n.* A temporary keratinized protuberance on the apical snout of turtles, crocodilians, and tuataras, used to cut the eggshell at the time of hatching. This structure is not calcified and thus differs from the egg tooth of hatchling squamates.

caudal autotomy: See *autotomy.*

caudal pseudoautotomy: See *autotomy.*

caudocephalic wave: A horizontal undulating wave of the body that results in gentle body contacts between a female snake and a courting male that produces movement while lying beside or upon her. The undulation moves from posterior to anterior and is thought to stimulate mating. Used by J. A. Oliver (1955, The Natural History of North American Amphibians and Reptiles, p. 225). Cf. *cardiovascular facilitative movement.*

clade: *n.* 1. Any group of organisms defined by characters exclusive to all its members and that distinguish the group from all others. 2. In evolutionary biology, a taxon or other group consisting of a single species and its descendants, representing a distinct branch on a *cladogram* or *phylogenetic tree.* A *monophyletic group.* See *cladogram, phylogenetic tree.*

cladogram: *n.* A branching diagram constructed by cladistics methods. A cladogram represents a diagrammatic analysis of phylogenetic relationships among organisms that relates them according to suites of common characters. The construction attempts to represent the true evolutionary branchings of a lineage through time from its ancestral taxon, based on shared characters derived from a common ancestor. Cladograms show common ancestry but do not indicate the amount of evolutionary time separating taxa. Cf. *phylogram, phylogenetic tree.*

classification: *n.* 1. An arrangement of living organisms into hierarchical groups according to their similarities and evolved ancestry. 2. A node-based name based on such an arrangement. 3. The process of grouping organisms according to features they have in common, or on the basis of their ancestry, or both.

clinal variation: Reference to a geographic character gradient.

cline: *n.* (*adj.* -al) A gradient or change in a variable over distance, as in thermocline, topocline, etc. Common usage in ecology is with reference to a continuous or gradual change in a character among members of a population or species over a given geographical or ecological range.

cloaca (*pl.* -cae): *n.* The common chamber and passageway into which the reproductive, urinary, and digestive ducts or canals release their contents, opening to the exterior through the vent.

clutch, clutch size: *n.* The collective number of eggs ovulated and laid by a single female at one time.

coitus: *n.* Copulation or sexual union in vertebrates. Syn. *coition, copulation, sexual intercourse.*

cold-blooded: *adj.* An obsolete term sometimes used inappropriately to describe the nonhomeostatic condition of body temperature in amphibians and reptiles. This term should be replaced by *ectothermic* or *poikilothermic.* See *ectothermy.*

community: *n.* See *ecological community.*

conduction: *n.* Transfer of heat directly from one elementary particle to another, thus between objects that are in direct physical contact.

conductivity, thermal: The intrinsic ability of a substance to conduct heat.

congener, congeneric: *n., adj.* Belonging to the same genus.

connectivity: *n.* In ecology, the interrelationships between different components or compartments of an ecosystem.

conspecific: *adj., n.* Belonging to the same species. Sometimes used as a noun.

constrictor: *n.* Term applied to any snake that subdues prey by constriction.

copulation: *n.* (*v.* copulate) Coitus involving an intromittent organ.

critical thermal maximum: The mean of collective temperatures at which a species, when heated gradually, becomes incapacitated such that long-term survival is not possible at this or at higher temperatures. Typically, locomotor activity becomes disorganized, and the animal loses its ability to escape from conditions that will promptly lead to its death. Originally the *critical maximum,* used by R. B. Cowles and C. M. Bogert (1944, Bulletin of the American Museum of Natural

History 83:263–296) and modified by C. H. Lowe and V. J. Vance (1955, Science 122: 73–74) and V. L. Hutchinson (1961, Physiological Zoology 34:92–125; 1980, American Journal of Physiology 247:R367–R368). Cf. *critical thermal minimum.*

critical thermal minimum: The temperature at which a species, when cooled gradually, becomes incapacitated such that long-term survival is not possible at this or lower temperatures. The low temperature produces cold nacrosis and prevents locomotion. Origin of term is similar to *critical thermal maximum.*

cryo-: Prefix meaning "cold," or "pertaining to cold."

crypsis: *n.* (*adj.* -tic) Matching between the color and pattern of an organism and a random sample of the background it is viewed against, as perceived by another organism. Concealment. See J. A. Endler, 1978, Evolutionary Biology 11: 319–364. Syn. *camouflage, cryptic coloration.*

cryptic species: See *sibling species.*

dehydrate: *v.* (*n.* dehydration) The act of drying, or replacement of water in a specimen by a nonaqueous medium. To remove or lose water. Cf. *desiccate.*

density: *n.* The number of individuals or other item of observation per unit of area or volume.

deoxyribonucleic acid (DNA): The class of nucleic acids responsible for hereditary transmission and for the coding of amino acid sequences of proteins. The sequences of nucleotides adenine, guanine, cytosine, and thymine—contained in two polynucleotide chains forming a double helix—is the basis of the genetic code.

desiccate: *v.* (*n.* desiccation) Reference to total drying. Cf. *dehydrate.*

dextral: *adj.* Of or pertaining to the right side. Cf. *sinistral.*

diploid (2N) *adj.* (*n.* diploidy, diploid number): The condition of having two sets of homologous chromosomes, one derived from each parent, in somatic cells of most organisms. Having two copies of a gene per locus per genome (may or may not be the same *allele*).

direct development: Complete embryonic development within the egg of some amphibians, and thus a free-living larval stage is absent.

distribution: *n.* 1. The total geographic range in which a given species or group of organisms occurs. 2. The spatial pattern related to members of a population or group. Cf. *range.*

dispersal: *n.* Movements of individuals away from a specific location to another related to various factors such as territoriality, departure from natal or denning sites, resources, weather, or seasonal phenomena, etc. Cf. *migration.*

diurnal: *adj.* Active during daylight hours. Cf. *nocturnal.*

diversity: *n.* See *biological diversity* and *species diversity.*

DNA: See *deoxyribonucleic acid.*

dorsal: *adj., n.* Reference to the upper surfaces of the body; the back or spinal regions of vertebrates. Cf. *ventral.*

dorsolateral: *adj.* A position that is intermediate between dorsal and lateral.

dorsum: *n.* Reference to the upper surface of an animal.

ecdysis: *n.* The sloughing or shedding of the stratum corneum of the epidermis, related to periodic renewal of this outer generation layer by replacement inner generation. In snakes and in some other squamates and amphibians ecdysis is a synchronous, pan-body event, whereas in other reptiles and amphibians the skin may be sloughed in patches. Syn. *molting, shedding, skin shedding, sloughing, exfoliation.*

ecological community: An assemblage of interacting species (plants and animals) that occur together in space and time.

ecotone: *n.* A transition zone between two ecological conditions or community types.

ectotherm: *n.* (*adj.* -ic) An organism whose body temperature is dependent on heat exchange with the environment. Thus, regulation of comparatively high body temperature depends on external heat sources such as sunlight or warm substrates. (older, but not strictly equivalent terms, are *cold-blooded* and *poikilotherm*). Cf. *endotherm.*

ectothermy: *n.* The mode of temperature regulation in which body temperature depends primarily on absorption of heat energy from the environment.

electrophoresis: *n.* (*adj.* electrophoretic) A technique for separating charged molecules such as proteins by means of differential movement in a solution subjected to an electrical field. The medium consists of either a gel or a solution that is held by a porous support such as filter paper. The molecules are separated based on differences in electric charge and also by the size or geometry of the molecule.

electroreception: *n.* The ability to detect electric fields.

***emergent plants:** Plant species that are typically associated with aquatic habitats whose roots are submerged or in damp soil, but whose stems and leaves are often above water. Emergent plants grow in shallow water environments and are frequently found close to shore.

endotherm: *n.* (*adj.* -ic) An organism in which the body temperature is determined largely by internal heat production of metabolism. The term usually applies to birds and mammals, but some reptiles are facultatively endothermic. Examples are brooding female pythons, large sea turtles while swimming, and certain newborn or hatchling snakes while assimilating yolk for a brief period following birth or hatching.

endothermy: *n.* The mode of temperature regulation in which body temperature is determined largely by a balance of rates of metabolic heat production and its dissipation to the environment.

envenomation: *n.* The act or state of having venom injected into tissues, as in venomous snake bit.

ephemeral: *adj.* short-lived or transient.

erythrocyte: *n.* A mature red blood cell. These cells are formed elements that contain hemoglobin and circulate in the blood of all vertebrates (except for certain Antarctic fishes). Syn. *red blood cell.*

estivation: See *aestivation.*

extant: *adj.* Still in existence; currently living. Cf. *extinct.*

extinct: *adj.* Not in existence in a particular area; no longer living. Cf. *extant.*

extinction: *n.* Death or disappearance of a species or taxonomic group, regionally over a particular area or world-wide.

extirpation: *n.* Local extinction in contrast with global extinction of a species.

extrinsic: *adj.* Originating outside of the body or part on which it acts. From the environment. Cf. *intrinsic.*

***exurb:** *n.* (*adj.* exurban) A primarily residential district that may be associated with a city or town but is located beyond the boundaries of the city or town's suburbs. Exurban developments can occur in otherwise rural regions. Cf. *suburb.*

fitness: *n.* The ability of an organism to contribute its genes to the next generation, roughly the number of offspring contributed to the next generation weighted by the probability of surviving to reproduce. Absolute fitness is the absolute contribution to the next generation, or the per capita rate of a given genotype. Relative fitness usually is calculated as fitness in relation to a reference genotype, usually the most fit or average fit genotype. Relative fitness also is measured by the average contribution of other phenotypes.

***floating plants:** Plant species in aquatic habitats whose roots are either attached to the substrate underwater, or are suspended in the water column, but whose leaves float at the water's surface (such as water lilies).

***floodplain:** The area in the immediate vicinity of a waterbody, such as a river or stream, that holds water if the waterbody experiences flooding.

***fluid specimen:** A preserved biological specimen housed in a container with fluid preservative (such as ethanol). Museum collections of herpetofauna often contain fluid specimens (rather than, for example, taxidermy specimens). *Whole fluid specimens* are those with most of or all of the preserved organism's body intact.

food chain: A hierarchical series of organisms that feed one on another within a natural community. Energy is thereby transferred from one level to the next, and a food chain expresses these relations in linear form. Each stage or level is called a *trophic level.*

fossorial: *adj.* Reference to habits or structural adaptations related to burrowing, digging, or subterranean life beneath ground vegetation or substratum.

freezing tolerance: *n.* Reference to ability to withstand freezing without cellular damage. A number of species of amphibians have been shown to tolerate repeated freezing and thawing, related to complex biochemical responses to freezing that distributes "antifreeze" compounds and protects intracellular structures from damage.

friable: *adj.* Easily crumbled or broken apart.

gape limited: Reference to a predator that cannot ingest prey larger than a maximum size set by the functional size of its mouth.

gene flow: The exchange of genes between different populations attributable to emigration and immigration of individuals.

genetics: *n.* The study of heredity, concerned with the physical and chemical nature of genes, their products, control, replication, and consequences for development of organisms and evolutionary changes in populations.

***genial:** *n.* A term used to reference the chin region of an organism.

genome: *n.* 1. The total set of genes carried by a single gamete, which includes both nuclear and mitochondrial DNA. 2. The complete set of loci, or collection of genes in an organism. 3. The complete DNA sequence of an organism.

genotype: *n.* The genetic makeup of an individual, which, along with other factors such as environment, determines an organism's phenotype. 2. The set of genes an organism possesses at a locus or loci of interest. Cf. *phenotype.*

genus: (*pl.* genera) *n.* A category of taxonomy consisting of a grouping of closely related species and compromising a level of classification above species and below family. The genus comprises the first name in a scientific name (e.g., the species *Vipera berus* belongs to the genus *Vipera*).

geographic distribution: See *distribution* and *range.*

geographic range: See *distribution* and *range.*

granular gland: A major type of alveolar gland found in amphibian skin, often concentrated into macroscopic clusters and typically producing defensive, toxic, multiple, holocrine secretions including amines, alkaloids, and peptides. Syn. *poison gland.*

gravid: *adj.* The condition of a female being swollen from carrying eggs or accumulated embryos. Syn. *pregnant.*

***Great Lakes Region:** A region of North America that surrounds the Great Lakes of Superior, Michigan, Huron, Ontario, and Erie. This region includes the US states of Illinois, Indiana, Michigan, Minnesota, New York, Ohio, Pennsylvania, and Wisconsin, and the Canadian province of Ontario.

gular fold: A fold of skin running transversely across the throat immediately anterior to the insertion of the forelegs.

head-starting: The experimental practice of hatching eggs in captivity and rearing hatchlings to a size such that, when they are released into the wild, they are less vulnerable to predation and other dangers.

hedonic gland: Any gland which produces a secretion acting as a sexual stimulant.

hemipenis: (*pl.* -es) *n.* Either of paired copulatory organs lying in a cavity at the base of the tail in squamate reptiles.

herpetoculture: *n.* A term used colloquially for the practice of breeding amphibians and reptiles, usually for commercial sale. A person who engages such activity is called a *herpetoculturist.*

herpetofauna: *n.* A biogeographic term to designate all populations of amphibians and reptiles that inhabit a particular region.

herpetofaunal: See *herpetofauna.*

hibernaculum: *n.* (*pl.* -a) The winter retreat where an animal hibernates or overwinters.

histology: *n.* (*adj.* -ical) The study of cells and tissues and their compositional features, usually by microscopy in conjunction with thin slices of tissues using a microtome.

holotype: *n.* The single specimen that is designated as the type specimen and used for the formal description of a nominal species.

home range: The area in which an individual organism is typically active, including routine activity, foraging, and occasional visits to more peripheral points.

homeotherm: *n.* (*adj.* -ic) An animal that regulates its body temperature within a relatively narrow range, regardless of environmental conditions, by controlling metabolic heat production and heat loss. By convention this term has been used in reference to birds and mammals, but it may have application to non-avian reptiles in limited circumstances (e.g., brooding pythons, large sea turtles, possibly some dinosaurs). Cf. *endotherm, poikilotherm.*

hula display: A behavior performed by male newts (*Notophthalmus viridescens*) involving undulating motions of the lower body and tail during courtship. The behavior was described by P. A. Verrell (1982, Animal Behaviour 30:1224–1236) and resembles "vent shuffling" described in ambystomatid salamanders.

hydrology: *n.* Formal study of the properties, circulation, and distribution of water. Hydrology is concerned with the movement and distribution of water at or near the Earth's surface, whereas *meteorology* deals with water as it moves through the atmosphere.

***hydroperiod:** The period of time during which an area of land is wet or holds standing water.

***hylid:** A species of frog that is a member of the family Hylidae.

hypoxia: *n.* (*adj.* -ic) Reduced or inadequate oxygen levels (below normal). Cf. *anoxia.*

***immune response:** A physiological response, or suite of physiological responses, mounted by a living organism's immune system. The immune response can involve production of defense cells that combat (typically) foreign organisms that can cause pathogenesis (such as viruses and bacteria).

immune system: The organs, tissues, cells, and molecules responsible for immunity or protection from foreign substances.

immunology: *n.* The science of dealing with immunity, immunological reactions, and all of their related aspects.

insectivore: *n.* (*adj.* -ous) Feeding on insects; often inclusive of all arthropods.

insolation: *n.* Solar radiation reaching the Earth from the sun.

interorbital: *adj.* Reference to the dorsal region of the skull between the eyes.

interspecific: *adj.* Reference to comparisons between or among different species. Cf. *intraspecific.*

intra-: Prefix meaning "within."

intraguild predation: Reference to competitors that exploit common, limiting resources and also interact with one another as predator and prey. The phenomenon is common and can either reverse or reinforce the outcome of competitive interactions between species.

intraspecific: *adj.* Reference to comparisons within a single species (e.g., different populations). Cf. *interspecific.*

intrinsic: *adj.* Belonging entirely to a structure, individual or system originating within it. Cf. *extrinsic.*

intromission: *n.* Literally the introduction of one part or object into another, applied in biological literature to sexual union.

***invertebrate:** A term applied to any animal that lacks vertebrae, which are the bones that make up the spine (or backbone) found in vertebrate animals.

iridophore: *n.* A chromatophore characterized by the presence of the purine guanine. Crystals of guanine import a noniridescent bluish to silvery white color that varies depending on other surrounding pigments in the skin.

junior synonym: In taxonomy, any of two or more synonyms other than the earliest published. Cf. *synonym.*

karyology: *n.* (*adj.* -ical) The branch of cytology that entails study of nuclei, especially the structure of chromosomes.

karyotype: *n.* 1. The chromosome complement of a cell or individual. 2. Structural characteristics of chromosome set. 3. Reference to those individuals having an identical complement of chromosomes. 4. A *karyogram.*

keel: *n.* (*adj.* -ed). A prominent ridgelike process or raised line. Often used in reference to protruding midlines of scales, plates, or tail fins.

labial: *adj., n.* 1. Reference to the upper or lower lip. 2. As a noun, labial refers to any of the scales in rows that border the lips of reptiles.

lacustrine: *adj.* Reference to, or living in, lakes or ponds.

larva: (*pl.* -ae) *n.* An immature but free-living stage that is different in form from the adult and, unlike the embryo, is able to procure its own nourishment.

lateral line system: A collection of epidermal sense organs distributed over the head and along the body of aquatic amphibians (and fishes). Lateral line organs (*neuromasts*) function as mechanoreceptors or electroreceptors (*ampullary organs*) in some larval caecilians and aquatic salamanders. The lateral line system of amphibians differs from those of fishes by not having "pores" or channels. Thus, terminologies applied to amphibians should not include the term "pore."

lecithotrophy: *n.* (*adj.* -ic) A fetal nutritional pattern in which nutrients are supplied to the embryo primarily by yolk supplied to the egg prior to ovulation. Terminology is from J. P. Wourms (1981, American Zoologist 21:473–515).

lentic: *adj.* Pertaining to slow-moving or standing aquatic habitats, such as a pond. Cf. *lotic.*

litter: *n.* 1. Collective animals that are produced at a multiple birth. 2. Fallen plant material that is recent and only partially decomposed, forming a surface layer on some soils.

littoral, littoral zone: *adj., n.* Pertaining to the shore or intertidal; the shallow open-water region lakeward or seaward of the shoreline.

loreal: *n.* A scale, or reference to a region, between the *nasals* and *preoculars* on each side of the head. The scale is usually single in snakes but several in lizards, and it has taxonomic importance. It is typically present in colubrids, but always absent in elapids.

lotic: *adj.* Of, relating to, or living in actively moving water such as a stream. Cf. *lentic.*

***macrophyte:** A plant that is not microscopic, and thus can be seen by the naked eye (i.e., without the aid of a microscope).

MCP: See *minimum convex polygon.*

***mean:** The arithmetic average; computed by summing a series of numbers and dividing by the count of the series.

***median:** The middle number in a series of numbers that are sorted from smallest to largest. If there is an even number of observations in the series, the median represents the arithmetic average of the two middle numbers. See *mean.*

melanism: *n.* (*adj.* melanistic) Excessive pigmentation or blackening of the skin or other tissues, usually of genetic origin.

Mendelian (**Mendelian genetics** or **Mendelian inheritance**): Reference to the inheritance of genes according to the laws governing the transmission of chromosomes to successive generations. Named in honor of G. Mendel, the Austrian monk who studied inheritance of traits in garden peas during the period 1856–1866. Syn. *chromosomal inheritance, Mendelism.*

mental gland: A gland beneath the skin of the intermandibular area in certain male salamanders, usually distinguished by slight bulging and different coloration of the skin. The secretions of this gland are sexually stimulating to females.

mesic: *adj.* Reference to a habitat or a condition that is moderately moist. The term is used to describe organisms that occupy a moist habitat.

***mesopredator:** A smaller predatory species that typically feeds in mid-trophic levels (i.e., the middle of the food chain), such as raccoons. Note that this is in contrast to upper trophic level predators that feed at the top of the food chain, such as wolves.

metamorphosis: *n.* (*adj.* metamorphic) The marked structural transformation of an animal from one stage of its life history to another. Strictly, a change in the structure or shape, but used most often in reference to the transformation from larval to adult life stages in amphibians.

metapopulation: *n.* A group of several local populations that are linked by immigration and emigration; a group of conspecific populations coexisting in time but not in space. The original meaning was used in a stricter sense as a collection of populations that experience local extinction and recolonization (D. Levins, 1969, Bulletin of the Entomological Society of America 15:237–240).

microhabitat: *n.* The immediate surroundings in which an organism lives. Syn. *microenvironment.*

***Midwest:** *n.* A primarily central to north-central region of the US typically reported to include the states of Ohio, Indiana, Illinois, Michigan, Wisconsin, Minnesota, Iowa, Missouri, Kansas, and Nebraska.

midventral: *adj.* Reference to the portion of the ventral surface of a body or structure that lies along the midline.

migration: *n.* (*adj.* migratory) The directed movement of individuals, hence gene flow, between different regions or populations. Reference may also be to a behavior pattern that involves a regular or recurring movement from on location to another by a group of animals of a given species in response to changes in weather, season, food availability, or other environmental factors that stimulate the behavior.

***minimum convex polygon:** In ecology, this term is typically applied to the spatial arrangement of an organism's movement patterns, wherein the outer most movement locations are connected to enclose the remaining movement locations. Minimum convex polygons are frequently used to delineate an organism's home range (often reported as MCP). See also *home range*.

molecular biology: *n.* A branch of modern biology that attempts to understand biological phenomena in terms of molecular mechanisms. Biochemical and physical techniques are often used to investigate problems related to genetics.

***molecular data:** Data collected for molecular biology.

monophyletic group: A natural taxon composed of two or more species, including the known or hypothesized ancestral species and all of its descendants sharing a single common ancestor. Members of a monophyletic group are sister taxa, and a monophyletic group is also called a *clade*.

monophyly: *n.* (*adj.* -etic) A condition where a group of species have all been derived from a single common ancestor. Having one origin. Monophyletic groups are identified by shared, uniquely derived character states.

morphology: *n.* (*adj.* morphological) 1. The study of form and structure. A science dealing with the structures of organisms, including their developmental and evolutionary histories. 2. The form of an organism, especially considered as a whole.

mortality: *n.* Generally used in reference to death.

***mortality rate:** The number of individuals experiencing mortality per unit of time (e.g., per year), often reported as a percentage or proportion of the total population.

nape: *n.* The back of the neck.

naris: (*pl.* -es). *n.* The nostril or nasal opening, usually paired. Nares that open to the outside are termed *external nares* or *nostrils*, whereas those that open into the mouth or pharynx are termed *internal nares* or *choanae*.

nasolabial grooves: A furrow or groove running from the external *naris* to the edge of the upper lip in plethodontid salamanders. The groove permits quick drainage of the *nares* and, in some species, extends well onto the *cirrus* of the lip.

neonate: *n.* (*adj.* neonatal) A newly born individual, also used by many authors in reference to hatchlings. Neonates comprise an age class distinct from juveniles. D. J. Morafka et al. (2000, Herpetology Monographs 14:353–370) define neonate as the first post-paritive age class, which expresses the effects of pre-paritive investment, influences of pre-paritive environment (egg, nest, etc.), and in which transitory aptive features particularly enhance the success of partition and, or, first dispersal. The maximal extent of the neonatal stage is 10% of ontogeny of a young reptile prior to sexual maturity.

neotene: *n.* A neotenic organism.

neotenic, neotenous: *adj.* Reference to the condition of neoteny.

neoteny: *n.* A developmental process or state in which a larva grows and becomes sexually mature while retaining an otherwise larval body form or features. Metamorphosis and development of the adult body form is either delayed or absent. Cf. *paedomorphosis* (*-ic*).

nocturnal: *adj.* Active primarily or exclusively at night. Cf. *diurnal*.

ocellus: *n.* (*pl.* -i) A circular, eye-like spot consisting of concentric rings of contrasting color.

olfaction: *n.* (*adj.* -ory) The sense of smell, or sensory perception of molecules released at some distance from an animal. Chemoreception of the nasal and vomeronasal organs.

omnivore: *n.* (*adj.* -ous) An animal that feeds on both animal prey and plant materials.

ontogeny: *n.* (*adj.* -ous or -etic) References to complete development of an organism, from fertilized egg to adult. The life history of an individual, both embryonic and postnatal.

oophagous: *adj.* Feeding on eggs.

ophiophagy: *n.* (*adj.* -ous) The eating of snakes as prey items. Of or pertaining to a predator that eats snakes.

osmosis: *n.* The movement of pure solvent from an area of low solute concentration to an area of high solute concentration through a semipermeable membrane.

oviparity: *n.* (*adj.* oviparous) A mode of reproduction in which underdeveloped eggs are laid or spawned by the female, encased within membranes sometimes overlain by a shell. Embryonic development takes place outside of the female, and nutrition is *lecithotrophic*. Cf. *ovoviparity*, *viviparity*.

oviposition: *n.* (*v.* oviposit) The act of laying or depositing eggs.

ovoviviparity: *n.* (*adj.* ovoviviparous). A mode of reproduction in which fertilized eggs remain in the female reproductive tract, where they develop without substantial nourishment from the mother and there is not placenta. The young hatch or break through the egg membranes immediately before, during, or just after, the eggs are laid. Cf. *oviparity, viviparity*). See D. G. Blackburn (1994, Herpetological Journal 4:65–72) for opinions on terminology. Currently, ovoviviparity is a term that is no longer used by persons studying reptiles.

paedomorphosis: *n.* (*adj.* paedomorphic) An evolutionary change that results in truncated development in descendant species compared with ancestral species, and therefore adults retain immature or juvenile characteristics of ancestors. Paedomorphosis was defined originally by W. Garstang (1922, Zoological Journal of the Linnean Society 35:81–101) to mean "shaped like a child." Cf. *neoteny.*

palatine: *adj., n.* 1 Relating to or located in the palate (roof of the mouth). 2. Either of a pair of dermal bones in the roof of the mouth, situated immediately behind the vomers (or prevomers) and usually separated by contact with intervening vomer and pterygoid. This is a very distinctive element in snakes, bearing teeth and forming the anterior part of the pterygoid-palatine arch.

paraphyletic group: A natural taxon composed of two or more species, including the known or hypothesized shared common ancestral species, but not all of its descendants. See *paraphyly.*

paraphyly: *n.* (*adj.* -etic) A condition in which several species of a group are derived from a hypothetical common ancestor, but not all sister species are in the group. An incomplete clade.

paravertebral: *n.* Any scales in the series the run longitudinally on either side of the ventral scales of snakes. Cf. *vertebral.*

parturition: *n.* The process of giving birth by *viviparity*. Cf. *oviposition.*

pathogen: *n.* (*adj.* pathogenic) That which produces disease or toxic symptoms.

phenology: *n.* (*adj.* -ical) Reference to natural phenomena that recur periodically, e.g. migration, reproduction.

phenotype: *n.* The physical characteristics of an organism, which comprise the manifestation of genetic expression. Any property of an organism that can be measured. Cf. *genotype.*

phenotypic plasticity: Phenotypic variation due to environmental influence on the phenotype associated with a particular genotype, representing genotype-environment interaction.

philopatric: *adj.* Reference to organisms that tend to remain in the location where they were born.

photoperiod: *n.* The period of light in 24 hours.

phylogenetic tree: A diagram that depicts the hypothesized genealogical ties and sequence of historical ancestor-descendent relationships linking individual organisms, populations or taxa. When species are considered, they are represented by line segments, and points of branching correspond to speciation events, with a measure of relative or absolute time on one axis.

phylogeny: *n.* The relationships of groups of organisms according to their evolutionary history. Evolutionary history of a lineage or organism.

plasticity: *n.* Generally, the capacity for change in response to an external influence. See *phenotypic plasticity.*

plastron: (*pl.* -ra) *n.* The ventral turtle shell, consisting of paired bones overlain by a matching series of laminae or scutes. Cf. *carapace.*

-ploid: A suffix used in genetics and cytology to designate a specified multiple of the chromosome set of the nucleus of an organism.

ploidy: *n.* See *polyploidy.*

poikilotherm: *n.* (*adj.* -ic) An animal in which body temperature tends to fluctuate with the surrounding ambient temperature. Cf. *ectotherm, homeotherm.*

poison: *n.* A substance that is toxic when ingested or absorbed. Not to be confused with *venom.* A venomous animal actually delivers a toxic secretion, whereas a poisonous animal relies on passive means of harming another individual. Both poisons and venoms contain toxins and are mixtures of biologically active an inert substances. Cf. *venom.*

poisonous: *adj.* Reference to any living organism having secretions or substances that, by means of ingestion, inhalation, absorption, injection, or other application, are capable of causing physiological disturbance or structural damage. It is becoming commonplace to restrict this term to organisms that have a harmful effect when partially or completely eaten by another animal and to use the term *venomous* for organisms that introduce venom into the body of another animal by means of fangs (as in reptiles) or stingers.

polyandry: *n.* (*adj.* -ous) A reproductive system in which a female mates with more than one male during a breeding season. Cf. *polygamy, polygyny.*

polygyny: *n.* (*adj.* -ous) A mating system in which the male may mate with more than one female during a breeding season. Cf. *polyandry, polygamy.*

polygamous: *n.* (*adj.* -ous) A reproductive system in which animals have more than one mate during a breeding season (*polyandry* and/or *polygyny*).

polyploidy: *n.* (*adj.* polyploid) A condition wherein the number of chromosome sets is greater than two.

population: *n.* A local, geographically defined group of conspecific organisms sharing a common gene pool and exhibiting reproductive continuity from generation to generation. Cf. *metapopulation.*

population bottleneck: Reference to a sharp decrease in the size of a population with consequent reduction in size of the gene pool.

population density: See *density.*

population size: The number of individuals in a finite population, denoted by the letter *N*.

postocular: *n., adj.* 1. A scale or scales at the posterior boundary of the orbit. 2. Of or pertaining to the region immediately behind the eye. Syn. *postorbital.*

premaxilla, premaxillary: *n.* Either of the two dermal bones at the anterior end of the skull, located between the maxillary bones and often bearing teeth. These elements meet at the midline and form the anterior angle of the upper jaw in amphibians and reptiles.

preocular: *n., adj.* 1. Generally, any scale or scales bounding the anterior margin of the orbit. 2. Of or pertaining to the area immediately in front of the nasal region.

***radio-tagged:** An organism that is affixed with a radio transmitter that emits a signal allowing researchers to identify the location of the organism. See *radio telemetry, transmitters.*

***radio telemetry:** A technique commonly employed by ecologists, zoologists, and wildlife biologists to track the movement patterns, behavior, and habitat selection of organisms. Typically, it requires the use of transmitters that emit specific frequencies (attached to an organism) and receivers (that are held by researchers) which pick up transmitter frequencies so that researchers can repeatedly find and observe radio-tagged organisms. See *radio-tagged, transmitters.*

range: *n.* The limits of geographical distribution of a species or taxon.

ranid: *n.* A member of, or pertaining to, Ranidae.

refuge, refugium: (*pl.* refugia) *n.* An area in which prey may escape from, or avoid, a predator or deleterious effects of the environment.

riverine: *adj.* Living in, or pertaining to, rivers.

rostral: *n., adj.* 1. The scale at the tip of the snout, bordering the mouth and separating the labial rows. 2. Situated toward the rostrum (oral-nasal region), which may be anterior, dorsal, or ventral with respect to head structures, cephalad with respect to other regions of the body. Cf. *rostrum.*

rostrum: *n.* The snout, or a beak-shaped process. Cf. *rostral.*

scute: *n.* Any large, flat, horny scale such as a plate, lamina, or shield on the surface of a reptile.

semi-: Prefix meaning "half" or "partly."

senescence: *n.* Physiological deterioration associated with aging of individuals in post-reproductive ages.

serrate, serrated: *adj.* Notched or having a sawlike edge.

sibling species: Reference to species that are reproductively isolated but nearly identical in morphology. Syn. *cryptic species.*

sinistral: *adj.* Of or pertaining to the left side. This term is often used in relation to location of the spiracle of tadpoles. Cf. *dextral.*

species complex: A collective group of closely related species, usually having a recent common ancestor.

species diversity: In general, reference to the number of species. The term incorporates three concepts. *Species richness* refers to the total number of species. *Species evenness* refers to the relative abundance of each species. *Species dominance* refers to the most abundant species.

species evenness: See *species diversity.*

species richness: See *species diversity.*

spermatophore: *n.* A structure containing sperm, usually a gelatinous mass with a sperm cap on top, secreted by cloacal glands of a male salamander and deposited on a moist surface from which it is picked up by the cloacal lips of the female. Usually only the cap is taken into the cloaca where the spermatozoa are held within the spermatheca.

subcaudal scales: *n., adj.,* 1. Any scale on the ventral side of the tail, either in a single or divided (two) series in snakes. 2. Of or pertaining to the ventral surface of the tail. 3. An older term for the anal lamina on the plastron of turtles.

subnivean: *adj.* Below snow.

***suburb:** *n.* (*adj.* suburban): A primarily residential housing district associated with a city or town that is typically located on the immediate outskirts of a more urban district. Typically, suburban areas have lower human population densities than urban areas, but due to the larger area of impact associated with each home (i.e., larger home and yard size), they can still disturb a substantial area of land. Cf. *exurb.*

succession: *n.* An orderly process of development or change in community structure through time. In its general form, succession is a process of community or ecosystem change starting from an unoccupied or incompletely occupied site and terminating in a more or less stable condition (known as climax).

supercooling: *n.* Reference to cooling below the physical freezing (=melting) point of body fluids without formation of ice crystals.

survival rate: The percentage of individuals surviving from one development stage, year class, or life stage to the next stage or succeeding period.

survivorship: *n.* The proportion of individuals from a given cohort that survive at a given time or age.

sympatry: *n.* (*adj.* -ic) Living in the same geographic location, with reference usually to the overlap in geographic range of two closely related species, which remain otherwise distinct.

synapomorphy: *n.* (*pl.* -ies) (*adj.* -ic) A derived character state or characters (apomorphies) shared by two or more taxa.

synchronous breeding: Reference to a circumstance when all females arrive at the breeding site roughly at the same time.

syncrhonous calls: Reference to breeding calls in which male anurans respond to one another rapidly such that calls overlap and produce rhythmic bursts of sound.

synonym (syn): *n.* In taxonomy, one of two or more scientific names applied to the same taxon. A *senior synonym* is the earliest available name for a taxon, whereas a *junior synonym* is any available name other than the senior synonym.

systematics: *n.* The study of phylogeny and classification of organisms based on evolutionary relationships. The field also may be described as the study of biological diversity and its origins. Refers to systems of classification. See *taxonomy.*

tadpole: *n.* Term for the nonreproductive larval life stage of any anuran. The term does not refer to immature of other amphibians.

taxon: (*pl.* taxa) *n.* A general term for a group of organisms recognized as a unit of classification. A hierarchical category in a classification. Specifically, a group of organisms classified together on the basis of sharing a single common ancestor.

taxonomy: *n.* (*adj.* -ic) The study of the naming and classification of living organisms according to an established set of principles based on theory and techniques. Classically, the description and classification of organisms was based on morphology, but increasingly considers genetic variation and interrelationships. The term was coined in 1813 by A. P. de Candolle (a botanist) for the process of classification. See *systematics.*

telemetry: see *radio telemetry.*

temporal: *adj., n.* 1. Reference to the region of the skull behind the eye. 2. Used to denote the squamosal bone of the skull. 3. Reference to scale or scales behind the postoculars, below the parietal, and above the upper labials in snakes and lizards. The term has been used with other adjectives or prefixes to describe relative position of scales, which are often arranged vertically in two or more rows. 4. Of or relating to time.

territorial behavior: Any behavior associated with the establishment, possession or defense of an area that is defended against incursion by other members of the species. Such behaviors typically involve ritualized displays or combat. Syn. *territoriality.*

territoriality: *n.* 1. Defense of an area by an individual or group for its exclusive use. 2. An instinctive behavior related to the occupation and defense of a space referred to as a *territory*.

territory: *n.* (*adj.* -ial) An area of habitat that is defended against entry usually by other members of the same species.

***tetraploid:** An organism with four homologous sets of chromosomes in its cells or cell nuclei.

tetrapod: *n.* A four-limbed vertebrate animal, including members of the *Tetrapoda:* amphibians, reptiles, birds, and mammals.

Tetrapoda: *n.* A superclass of vertebrates having four limbs with carpals, tarsals, and digits. This taxon incorporates amphibians, reptiles, birds, and mammals.

thermal: *adj.* Relating to heat.

thermocline: *n.* A mid-layer of water in a lake or ocean characterized by rapidly decreasing temperature (with depth).

thermogenesis: *n.* Literally the production of heat, used in relation to heat production by metabolic means (e.g., muscle contractions of brooding pythons).

thermoregulation: *n.* Regulation of body temperature.

***transmitters:** *n.* In ecological research terms, transmitters are pieces of equipment affixed to individual animals that researchers want to relocate repeatedly. The transmitter emits a unique signal that is detected by a specialized receiver used by researchers to track the animal's position. See *radio-tagged, radio telemetry.*

trophic: *adj.* Reference to feeding, nutrition, or predator-prey interactions.

trophic level: The position of an organism or species within a food chain. See *food chain.*

turbidity: *n.* (*adj.* turbid) A quality of opaqueness in water due to suspended particulate matter or stirred-up sediment.

***Upper Midwest:** *n.* A region within the midwestern US that is comprised of the states North Dakota, South Dakota, Minnesota, Iowa, Wisconsin, and Michigan, but is sometimes considered to include only Minnesota, Wisconsin, and Michigan. See *Midwest.*

vagility: *n.* (*adj.* vagile) The ability to move from place to place.

venom: *n.* A toxic or potentially toxic substance normally secreted by an animal and injected by a sting or bite. Highly evolved snake venoms are a complex mixture of hydrolytic enzymes and systemic toxins. Cf. *poison.*

venomous: *adj.* Pertaining to an animal that possesses venom and envenomates other animals by use of a fang or sting.

vent: *n.* The external opening of the cloaca.

venter: *n.* The abdomen or complete undersurface of an animal. Syn. *ventrum.* See *ventral.*

ventral: *adj., n.* 1. Pertaining to the abdomen or venter; directed toward or situated on the belly surface, in contrast to *dorsal.* Cf. *dorsal.* 2. Any of the large scales or scutes on the belly surface of a snake, between the head and anal plate or (some authors) tail tip.

ventrolateral: *adj.* Of or pertaining to a position intermediate between the lower surface and lateral surface of a body or structure.

***vermivorous:** *adj.* An organism whose diet primarily consists of worms (such as earthworms).

vernal pool: A temporary pool that is formed during spring from meltwater or flood water.

vertebral: *adj., n.* 1. Generally, reference to the vertebral column or middorsal line of the body. 2. The term is commonly used to denote any of the scale rows located on the middorsal line of the body in snakes and lizards, and is sometimes used in reference to the centrals and neurals of the turtle carapace.

vertebrate: *n.* Any member of the subphylum Vertebrata, compromising all animals that have a vertebral column, including fishes, amphibians, reptiles, birds, and mammals.

vitellus: *n.* The yolk of the egg.

vitellogenesis: *n.* The production of yolk or "yolking" of an egg in the ovary. Deposition of fat reserves in the ovarian follicles.

viviparity: *n.* (*adj.* viviparous) The retention of embryos within the oviduct until development is complete, thus giving birth to developed young as opposed to laying eggs. In the past, viviparity has been defined with consideration of whether a placenta is involved or not. Most workers now prefer a broad definition of viviparity that includes numerous modes of live-bearing. Cf. *oviparity* and *ovoviviparity*.

vomerine teeth: Short, conical projections of the vomer, located near the internal nasal openings in the palate of amphibians.

voucher specimen: Any specimen that is identified by a recognized authority for the purposes of forming a reference collection. A voucher specimen may also be deposited in a museum to verify the species identification of animals used in a particular study, especially when the species is unusual or exotic.

***whole fluid specimen:** See *fluid specimen*.

xeric: *adj*. Reference to dry or arid conditions, lacking in moisture.

Literature Cited

Abel, B. 1992. Snapping Turtle attacks on Trumpeter Swan cygnets in Wisconsin. Passenger Pigeon 54:209–213.

[ABPC] American Biographical Publishing Company. 1877. United States biographical dictionary and portrait gallery of eminent and self-made men: Wisconsin volume. Volume 1. American Biographical Publishing Company, Chicago, Illinois, USA.

Acker, P. M., K. C. Kruse, and E. B. Krehbiel. 1986. Aging *Bufo americanus* by skeletochronology. Journal of Herpetology 20:570–574.

Adams, A. E. 1940. Sexual conditions in *Triturus viridescens*. III. The reproductive cycle of the adult aquatic form of both sexes. American Journal of Anatomy 66:235–275.

Adams, C. E., and K. J. Lindsey. 2010. Urban wildlife management. Second edition. CRC Press, New York, USA.

Adams, C. K., D. Saenz, and R. N. Conner. 2011. Palatability of twelve species of anuran larvae in eastern Texas. American Midland Naturalist 166:211–223.

Adams, M. J., C. A. Pearl, and R. B. Bury. 2003. Indirect facilitation of an anuran invasion by non-native fishes. Ecology Letters 6:343–351.

Adams, M. J., D. A. W. Miller, E. Muths, P. S. Corn, E. H. C. Grant, L. L. Bailey, G. M. Fellers, R. N. Fisher, W. J. Sadinski, H. Waddle, and S. C. Walls. 2013. Trends in amphibian occupancy in the United States. PLoS ONE 8:e64347.

Adams, S. R., and M. C. Decarvalho, Jr. 1984. Rates of heat exchange in the Ornate Box Turtle, *Terrapene ornata*. Comparative Biochemistry and Physiology 79A:359–361.

Ade, C. M., M. D. Boone, and H. J. Puglis. 2010. Effects of an insecticide and potential predators on Green Frogs and Northern Cricket Frogs. Journal of Herpetology 44:591–600.

Adler, K. K. 1968. Turtles from archeological sites in the Great Lakes region. Michigan Archaelogist 14:147–163.

Adler, K. K., and S. G. Tilley. 1960. A fish and a snail in the diet of *Natrix septemvittata* (Say). Journal of the Ohio Herpetological Society 2:28–29.

Agassiz, L. 1850. Lake Superior: its physical character, vegetation, and animals, compared with those of other and similar regions. Gould, Kendall and Lincoln, Boston, Massachusetts, USA.

Agassiz, L. 1857. Contributions to the natural history of the United States of America. Little, Brown and Company, Boston, Massachusetts, USA.

Akre, T. S. B. 2002. Growth, maturity, and reproduction of the Wood Turtle, *Clemmys insculpta* (LeConte 1830) in Virginia. Dissertation, George Mason University, Fairfax, Virginia, USA.

Akresh, M. E., D. I. King, B. C. Timm, and R. T. Brooks. 2017. Fuels management and habitat restoration activities benefit Eastern Hognose Snakes (*Heterodon platirhinos*) in a disturbance-dependent ecosystem. Journal of Herpetology 51:468–476.

Alberch, P., and E. A. Gale. 1985. A developmental analysis of an evolutionary trend: digital reduction in amphibians. Evolution 39:8–23.

Albert, D. A. 1995. Regional landscape ecosystems of Michigan, Minnesota, and Wisconsin: a working map and classification. US Forest Service General Technical Report NC-178, St. Paul, Minnesota, USA.

Albright, J. D. 2001. Microsatellite DNA markers, multiple paternity, and the inheritance of morphology and behavior in Butler's Garter Snake (*Thamnophis butleri*). Thesis, University of Tennessee, Knoxville, USA.

Aldrich, J. M. 1916. *Sarcophaga* and allies in North America. Thomas Say Foundation, Lafayette, Indiana, USA.

Aldridge, R. D., and D. Duvall. 2002. Evolution of the mating season in the pitvipers of North America. Herpetological Monographs 16:1–25.

Aldridge, R. D., and D. E. Metter. 1973. The reproductive cycle of the Western Worm Snake, *Carphophis vermis*, in Missouri. Copeia 1973:472–477.

Aldridge, R. D., and W. S. Brown. 1995. Male reproductive cycle, age at maturity, and cost of reproduction in the Timber Rattlesnake (*Crotalus horridus*). Journal of Herpetology 29:399–407.

Aldridge, R. D., A. P. Bufalino, and A. Reeves. 2005. Pheromone communication in the watersnake, *Nerodia sipedon*: a mechanistic difference between semi-aquatic and terrestrial species. American Midland Naturalist 154:412–422.

Aldridge, R. D., B. C. Jellen, M. C. Allender, M. J. Dreslik, D. B. Shepard, J. M. Cox, and C. A. Phillips. 2008. Reproductive biology of the Massasauga (*Sistrurus catenatus*) from south-central Illinois. Pages 403–412 *in* W. K. Hayes, K. R. Beaman, M. D. Cardwell, and S. P. Bush, editors. The biology of rattlesnakes. Loma Linda University Press, Loma Linda, California, USA.

Aldridge, R. D., S. R. Goldberg, S. S. Wisniewski, A. P. Bufalino, and C. B. Dillman. 2009. The reproductive cycle and estrus in the colubrid snakes of temperate North America. Contemporary Herpetology 2009:1–31.

Alexander, J. S., R. C. Wilson, and W. R. Green. 2012. A brief history and summary of the effects of river engineering and dams on the Mississippi River system and delta. US Geological Survey Report Circular 1375, Reston, Virginia, USA. Archived by WebCite at http://www.webcitation.org/73doWEi0G. Accessed 2 Nov 2018.

Alfaro, M. E. 2003. Sweeping and striking: a kinematic study of the trunk during prey capture in three thamnophiine snakes. Journal of Experimental Biology 206:2381–2392.

Alfaro, M. E., and S. J. Arnold. 2001. Molecular systematics and evolution of *Regina* and the Thamnophiine snakes. Molecular Phylogenetics and Evolution 21:408–423.

Alford, R. A. 2010. Declines and the global status of amphibians. Pages 13–45 *in* D. W. Sparling, G. Linder, C. A. Bishop, and S. K. Krest, editors. Ecotoxicology of Amphibians and Reptiles. Society for the Study of Amphibians and Reptiles, Pensacola, USA.

Alford, R. A., and S. J. Richards. 1999. Global amphibian declines: a problem in applied ecology. Annual Review of Ecology, Evolution, and Systematics 30:133–165.

Allan, J. D. 2004. Landscapes and riverscapes: the influence of land use on stream ecosystems. Annual Review of Ecology, Evolution, and Systematics 35:257–284.

Allan, J. D., and M. M. Castillo. 2007. Stream ecology: structure and function of running waters. Second edition. Springer Publishing, Dondrecht, Netherlands.

Allard, H. A. 1909. Notes on two common turtles of eastern United States. Science 30:453–454.

Allen, A. C. 1963. The Amphibia of Wayne County, Ohio. Journal of the Ohio Herpetological Society 4:23–30.

Allen, J. A. 1868. Catalogue of the reptiles and batrachians found in the vicinity of Springfield, Mass., with notices of all other species known to inhabit the state. Proceedings of the Boston Society of Natural History 12:171–204.

Allen, M. J. 1932. A survey of the amphibians and reptiles of Harrison County, Mississippi. American Museum Novitates 542:1–20.

Allender, M., C. Phillips, S. Baker, D. Wylie, A Narotsky, and M. Dreslik. 2016a. Hematology and emergence of Ophidiomyces in an Eastern Massasauga (*Sistrurus catenatus*) population. Journal of Wildlife Diseases 52:258–269.

Allender, M. C., D. B. Raudabaugh, F. H. Gleason, and A. N. Miller. 2015. The natural history, ecology, and epidemiology of *Ophidiomyces ophiodiicola* and its potential impact on free-ranging snake populations. Fungal Ecology 17:187–196.

Allender, M. C., E. T. Hileman, J. Moore, and S. Tetzlaff. 2016b. Detection of Ophidiomyces, the causative agent of Snake Fungal Disease, in the Eastern Massasauga (*Sistrurus catenatus*) in Michigan, USA, 2014. Journal of Wildlife Diseases 52:694–698.

Allender, M. C., M. Dreslik, S. Wylie, C. Phillips, D. B. Wylie, C. Maddox, M. A. Delaney, and M. J. Kinsel. 2011. *Chrysosporium* sp. infection in

Eastern Massasauga Rattlesnakes. Emerging Infectious Diseases 17:2383.

Allender, M. C., M. J. Dreslik, D. B. Wylie, S. J. Wylie, J. W. Scott, and C. A. Phillips. 2013. Ongoing health assessment and prevalence of *Chrysosporium* in the Eastern Massasauga (*Sistrurus catenatus catenatus*). Copeia 2013:97–102.

Allran, J. W., and W. H. Karasov. 2001. Effects of atrazine on embryos, larvae, and adults of anuran amphibians. Environmental Toxicology and Chemistry 20:769–775.

Allsteadt, J., A. H. Savitsky, C. E. Petersen, and D. N. Naik. 2006. Geographic variation in the morphology of *Crotalus horridus* (Serpentes: Viperidae). Herpetological Monographs 20:1–63.

Altig, R. 1970. A key to the tadpoles of the continental United States and Canada. Herpetologica 26:180–207.

Altig, R., and R. W. McDiarmid. 2015. Handbook of larval amphibians of the United States and Canada. Cornell University Press, Ithaca, New York, USA.

Amato, M. L., R. J. Brooks, and J. Z. Fu. 2008. A phylogeographic analysis of populations of the Wood Turtle (*Glyptemys insculpta*) throughout its range. Molecular Ecology 17:570–581.

Amburgey, S. M., D. A. W. Miller, E. H. Campbell Grant, T. A. G. Rittenhouse, M. F. Bernard, J. L. Richardson, M. C. Urban, W. Hughson, A. B. Brand, C. J. Davis, C. R. Hardin, P. W. C. Paton, C. J. Raithel, R. A. Relyea, A. F. Scott, D. K. Skelly, D. E. Skidds, C. K. Smith, and E. E. Werner. 2018. Range position and climate sensitivity: the structure of among-population demographic responses to climatic variation. Global Change Biology 24:439–454.

Amburgey, S., W. C. Funk, M. Murphy, and E. Muths. 2012. Effects of hydroperiod duration on survival, developmental rate, and size at metamorphosis in Boreal Chorus Frog tadpoles (*Pseudacris maculata*). Herpetologica 68:456–467.

Amin, O. M. 1985. Hosts and geographic distribution of *Acanthocephalus* (Ancanthocephala: Echinorhynchidae) from North American freshwater fishes, with a discussion of species relationships. Proceedings of the Helminthological Society of Washington 52:210–220.

Anderson, A. L., and W. D. Brown. 2009. Plasticity of hatching in Green Frogs (*Rana clamitans*) to both egg and tadpole predators. Herpetologica 65:207–213.

Anderson, B. J. 2005. The historical development of the tension zone concept in the Great Lakes region of North America. Michigan Botanist 44:127–138.

Anderson, C. D., and M. J. Rosenberg. 2011. Variation in association with anthropogenic habitat edges exhibited by the Timber Rattlesnake (*Crotalus horridus*) in St. Louis County, Missouri. Journal of Herpetology 45:50–55.

Anderson, C. S., H. L. Gibbs, and J. Chiucchi. 2010. Nineteen polymorphic microsatellite loci isolated from the Eastern Massasauga Rattlesnake, *Sistrurus c. catenatus*. Conservation Genetics Resources 2010:243–245.

Anderson, J. D. 1967. *Ambystoma maculatum* (Shaw) Spotted Salamander. Catalogue of American Amphibians and Reptiles 51.1–51.4.

Anderson, J. D., D. D. Hassinger, and G. H. Dalrymple. 1971. Natural mortality of eggs and larvae of *Ambystoma t. tigrinum*. Ecology 52:1107–1112.

Anderson, L. R., and J. A. Arruda. 2006. Land use and anuran biodiversity in southeast Kansas, USA. Amphibian and Reptile Conservation 4:46–59.

Anderson, N. J., and M. A. Rozmarynowski. 2011. *Nerodia sipedon* (Northern Watersnake). Response to mobbing by fish. Herpetological Review 42:442.

Anderson, P. 1965. The reptiles of Missouri. University of Missouri Press, Columbus, USA.

Anderson, R. C. 1983. The eastern prairie-forest transition: an overview. Proceedings of the North American Prairie Conference 8:86–92.

Anderson, R. K., D. K. Jansen, and R. L. Hine. 1981. An inventory method for Wisconsin frog and toads. Pages 48–55 *in* A. Elwell, K. Cram, and C. Johnson, editors. Proceedings of a symposium on ecology of reptiles and amphibians in Minnesota. Bald Eagle Outdoor Learning Center, 13–15 March 1981, Cass Lake, Minnesota, USA.

Anderson, R. V., M. L. Gutierrez, and M. A. Romano. 2002. Turtle habitat use in a reach of the Upper Mississippi River. Journal of Freshwater Ecology 17:171–177.

Andren, H. 1994. Effects of habitat fragmentation on birds and mammals in landscapes with different proportions of suitable habitat: a review. Oikos 71:355–366.

Andres, K. M., and R. M. Chambers. 2006. A test of philopatry by Common Musk Turtles. American Midland Naturalist 156:45–51.

Andrews, K. M., and J. W. Gibbons. 2005. How do highways influence snake movement? Behavioral responses to roads and vehicles. Copeia 2005:772–782.

Andrews, K. M., and J. W. Gibbons. 2008. Roads as catalysts of urbanization: snakes on roads face differential impacts due to inter- and intraspecific ecological attributes. Pages 145–153 *in* J. C. Mitchell, R. E. Jung Brown, and B. Bartholomew, editors. Urban Herpetology. Society for the Study of Amphibians and Reptiles, Herpetological Conservation 3, Salt Lake City, Utah, USA.

Andrews, K. M., J. W. Gibbons, and D. M. Jochimsen. 2008. Ecological effects of roads on amphibians and reptiles: a literature review. Pages 121–143 *in* J. Mitchell, R. Jung Brown, and B. Bartholomew, editors. Urban Herpetology. Society for the Study of Amphibians and Reptiles, Herpetological Conservation 3, Salt Lake City, Utah, USA.

Andrews, O. V. 1915. An ecological survey of Lake Butte des Morts bog, Oshkosh, Wisconsin. Bulletin of the Wisconsin Natural History Society 13:196–211.

Anholt, B. R., D. K. Skelly, and E. E. Werner. 1996. Factors modifying antipredator behavior in larval toads. Herpetologica 52:301–313.

Anholt, B. R., E. Werner, and D. K. Skelly. 2000. Effect of food predators on the activity of four larval ranid frogs. Ecology 81:3509–3521.

Annen, C. A., R. W. Tyser, and E. M. Kirsch. 2005. Effects of a selective herbicide, sethoxydim, on Reed Canarygrass. Ecological Restoration 23:99–102.

Anonymous Biography. 1970. William E. Dickinson. Lore 20:43–44.

Anonymous Obituary. 1892. Dr. P. R. Hoy dead. Racine Daily Journal. 8 December; page 47.

Anonymous Obituary. 1958. In Memoriam: Thomas Edward Burt Pope. Brown Alumni Monthly 58:92.

Anonymous Obituary. 1974. William E. Dickinson. Milwaukee Sentinel. 24 July; page 10.

Anthony, C. D., and R. A. Pfingsten. 2013. Eastern Red-backed Salamander. *Plethodon cinereus* (Green 1818). Pages 335–360 *in* R. A. Pfingsten, J. G. Davis, T. O. Matson, G. J. Lipps, Jr., D. Wynn, and B. J. Armitage, editors. Amphibians of Ohio. Ohio Biological Survey, Bulletin New Series Volume 17, Number 1, Columbus, Ohio, USA.

Anthonysamy, W. J. B., M. J. Dreslik, and C. A. Phillips. 2013. Disruptive influences of drought on the activity of a freshwater turtle. American Midland Naturalist 169:322–335.

Anthonysamy, W. J. B., M. J. Dreslik, D. Mauger, and C. A. Phillips. 2014a. A preliminary assessment of habitat partitioning in a freshwater turtle community at an isolated preserve. Copeia 2014:269–278.

Anthonysamy, W. J. B., M. J. Dreslik, M. R. Douglas, N. K. Marioni, and C. A. Phillips. 2014b. Reproductive ecology of an endangered turtle in a fragmented landscape. Copeia 2014:437–446.

Anton, T. G. 1993. Massasaugas in Lake County, Illinois: the Ryerson population. Pages 71–77 *in* B. Johnson and V. Menzies, editors. Proceedings of the International Symposium and Workshop on the Conservation of the Eastern Massasauga Rattlesnake *Sistrurus catenatus*. Metro Toronto Zoo, Ontario, Canada.

Anton, T. G. 1999. Current distribution and status of amphibians and reptiles in Cook County, Illinois. Transactions of the Illinois State Academy of Science 92:211–232.

Anton, T. G., T. F. Beauvais, and G. Casper. 2014. *Clonophis kirtlandii*. Geographic distribution. Herpetological Review 45:465.

Applegate, R. D. 1990. *Rana catesbeiana*, *Rana palustris* (Bullfrog, Pickerel Frog). Predation. Herpetological Review 21:90–91.

Aresco, M. J. 2005. Mitigation measures to reduce highway mortality of turtles and other herpetofauna at a north Florida lake. Journal of Wildlife Management 69:549–560.

Arias, E., N. P. Zaffaroni, and T. Zavanella. 1989. Teratological evaluation of the phenoxyacid herbicide MCPA in the regenerating forelimb of the adult newt. Ecotoxicology and Environmental Safety 17:30–37.

Arndt, R. G. 1980. A hibernating Eastern Hognose Snake, *Heterodon platirhinos*. Herpetological Review 11:30–32.

Arnold, S. J. 1976. Sexual behavior, sexual interference, and sexual defense in the salamanders *Ambystoma maculatum*, *Ambystoma tigrinum*, and *Plethodon jordani*. Zeitschrift fur Tierpsychologie 42:247–300.

Arnold, S. J. 1977. The evolution of courtship behavior in New World salamanders with some comments on Old World salamandrids. Pages 141–183 *in* D. H. Taylor and S. I. Guttman, editors. The reproductive biology of amphibians. Plenum Press, New York, USA.

Arnold, S. J., and A. F. Bennett. 1984. Behavioural variation in natural populations III: antipredator displays in the Garter Snake *Thamnophis radix*. Animal Behaviour 32:1108–1118.

Aronson, L. R. 1943. The sexual behavior of Anura. 4. Oviposition in the Mink Frog, *Rana septentrionalis* Baird. American Midland Naturalist 29:242–244.

Aronson, L. R. 1944. The sexual behavior of Anura 6. The mating pattern of *Bufo americanus*, *Bufo fowleri*, and *Bufo terrestris*. American Museum Novitates 1250:1–15.

Arrowood, C., A. Arrowood, and A. F. Badje. 2019. *Glyptemys insculpta* (Wood Turtle). Geographic distribution. Herpetological Review 50:324.

Arvisais, M., E. Lévesque, J.-C. Bourgeois, C. Daigle, D. Masse, and J. Jutras. 2004. Habitat selection by the Wood Turtle (*Clemmys insculpta*) at the northern limit of its range. Canadian Journal of Zoology 82:391–398.

Arvisais, M., J-C. Bourgeois, E. Lévesque, C. Digle, D. Masse, and J. Jutras. 2002. Home range and movements of a Wood Turtle (*Clemmys insculpta*) population at the northern limit of its range. Canadian Journal of Zoology 80:402–408.

Ash, A. N., and R. C. Bruce. 1994. Impacts of timber harvesting on salamanders. Conservation Biology 8:300–301.

Ashley, E. P., and J. T. Robinson. 1996. Road mortality of amphibians, reptiles and other wildlife on the Long Point Causeway, Lake Erie, Ontario. Canadian Field-Naturalist 110:403–412.

Ashley, E. P., A. Kosloski, and S. A. Petrie. 2007. Incidence of intentional vehicle-reptile collisions. Human Dimensions of Wildlife 12:137–143.

Ashton, K. G., and A. C. S. Knipps. 2011. Effects of fire history on amphibian and reptile assemblages in Rosemary scrub. Journal of Herpetology 45:497–503.

Ashton, R. E., Jr., A. L. Braswell, and S. I. Gutunan. 1980. Electrophoretic analysis of three species of *Necturus* (Amphibia: Proteidae), and the taxonomic status of *Necturus lewisi* (Brimley). Brimleyana 4:43–46.

Asplund, K. K. 1970. Metabolic scope and body temperatures of whiptail lizards (*Cnemidophorus*). Herpetologica 26:403–411.

Attig, J. W., M. Bricknell, E. C. Carson, L. Clayton, M. D. Johnson, D. M. Mickelson, and K. M. Syverson. 2011. Glaciation of Wisconsin. Fourth edition. Wisconsin Geological and Natural History Survey, Educational Series 36, Madison, USA.

Attum, O., C. D. Cutshall, K. Eberly, H. Day, and B. Tietjen. 2013. Is there really no place like home? Movement, site fidelity, and survival probability of translocated and resident turtles. Biodiversity Conservation 22:3185–3195.

Attum, O., Y. M. Lee, J. H. Roe, and B. A. Kingsbury. 2007. Upland-wetland linkages: relationship of upland and wetland characteristics with watersnake abundance. Journal of Zoology 271:134–139.

Attum, O., Y. M. Lee, J. H. Roe, and B. A. Kingsbury. 2008. Wetland complexes and upland-wetland linkages: landscape effects on the distribution of rare and common wetland reptiles. Journal of Zoology 275:245–251.

Austin, J. D., and K. R. Zamudio. 2008. Incongruence in the pattern and timing of intra-specific diversification in Bronze Frogs and Bullfrogs (Ranidae). Molecular Phylogenetics and Evolution 48:1041–1053.

Austin, J. D., J. A. Davila, S. C. Lougheed, and P. T. Boag. 2003a. Genetic evidence for female-biased dispersal in the Bullfrog, *Rana catesbeiana* (Ranidae). Molecular Ecology 12:3165–3172.

Austin, J. D., S. C. Lougheed, and P. T. Boag. 2004. Discordant temporal and geographic patterns in maternal lineages of eastern North American frogs, *Rana catesbeiana* (Ranidae) and *Pseudacris crucifer* (Hylidae). Molecular Phylogenetics and Evolution 32:799–816.

Austin, J. D., S. C. Lougheed, L. Niedrauer, A. A. Chek, and P. T. Boag. 2002. Cryptic lineages in a small frog; the post-glacial history of the Spring Peeper, *Pseudacris crucifer* (Anura: Hylidae). Molecular Phylogenetics and Evolution 25:316–329.

Austin, J. D., S. C. Lougheed, P. E. Moler, and P. T. Boag. 2003b. Phylogenetics, zoogeography, and the role of dispersal and vicariance in the evolution of the *Rana catesbeiana* (Anura: Ranidae) species group. Biological Journal of the Linnean Society 80:601–624.

Awan, A. R., and G. R. Smith. 2007. The effect of group size on the responses of Wood Frog tadpoles to fish. American Midland Naturalist 158:79–84.

Babbitt, K. J. 2005. The relative importance of wetland size and hydroperiod for amphibians in southern New Hampshire, USA. Wetlands Ecology and Management 13:269–279.

Babbitt, K. J., M. J. Baber, and T. L. Tarr. 2003. Patterns of larval amphibian distribution along a wetland hydroperiod gradient. Canadian Journal of Zoology 81:1539–1552.

Babbitt, L. H. 1932. Some remarks on Connecticut herpetology. Bulletin of the Boston Society of Natural History 63:23–28.

Babbitt, L. H. 1937. The amphibia of Connecticut. Connecticut State Geological and Natural History Survey Bulletin 57:1–50.

Babbitt, L. H., and C. H. Babbitt. 1951. A herpetological study in burned-over areas in Dade County, Florida. Copeia 1951:79.

Bachmann, K., and J. P. Bogart. 1975. Comparative cytochemical measurements in the diploid-tetraploid species pair of hylid frogs *Hyla chrysoscelis* and *H. versicolor*. Cytogenetic and Genome Research 15:186–194.

Bachmann, M. D. 1984. Defensive behavior of brooding female Red-backed Salamanders (*Plethodon cinereus*). Herpetologica 1984:436–443.

Bachmann, M. D., R. G. Carlton, J. M. Burkholder, and R. G. Wetzel. 1986. Symbiosis between salamander eggs and green algae: microelectrode measurements made inside eggs demonstrates effect of photosynthesis on oxygen concentration. Canadian Journal of Zoology 64:1586–1588.

Badje, A. F. 2017a. *Hyla chrysoscelis* (Cope's Gray Treefrog). Geographic distribution. Herpetological Review 48:584.

Badje, A. F. 2017b. *Hyla versicolor* (Gray Treefrog). Geographic distribution. Herpetological Review 48:806–807.

Badje, A. F. 2017c. *Lithobates catesbeianus* (American Bullfrog). Geographic distribution. Herpetological Review 48:808.

Badje, A. F. 2018. *Pseudacris maculata* (Boreal Chorus Frog). Geographic distribution. Herpetological Review 49:283.

Badje, A. F. 2019. *Glyptemys insculpta* (Wood Turtle). Geographic distribution. Herpetological Review 50:522–523.

Badje, A. F. 2020a. *Apalone spinifera* (Spiny Softshell). Geographic distribution. Herpetological Review 51:72.

Badje, A. F. 2020b. *Apalone spinifera* (Spiny Softshell). Geographic distribution. Herpetological Review 51:535.

Badje, A. F. 2020c. *Emydoidea blandingii* (Blanding's Turtle). Geographic distribution. Herpetological Review 51:73.

Badje, A. F. 2020d. *Graptemys geographica* (Northern Map Turtle). Geographic distribution. Herpetological Review 51:771.

Badje, A. F. 2020e. *Hyla versicolor* (Gray Treefrog). Geographic distribution. Herpetological Review 51:533.

Badje, A. F. 2020f. *Pantherophis vulpinus* (Eastern Foxsnake). Geographic distribution. Herpetological Review 51:80.

Badje, A. F. 2021a. *Hyla versicolor* (Gray Treefrog). Geographic distribution. Herpetological Review 52:77.

Badje, A. F. 2021b. *Pseudacris maculata* (Boreal Chorus Frog). Geographic distribution. Herpetological Review 52:78.

Badje, A. F. 2021c. *Graptemys geographica* (Northern Map Turtle). Geographic distribution. Herpetological Review. 52:793–794.

Badje, A. F. 2022a. *Hyla chrysocelis* (Cope's Gray Treefrog). Geographic distribution. Herpetological Review 53:71.

Badje, A. F. 2022b. *Apalone spinifera* (Spiny Softshell). Geographic distribution. Herpetological Review 53:75.

Badje, A. F., and A. I. Dyke. 2018. *Anaxyrus fowleri* (Fowler's Toad). Geographic distribution. Herpetological Review 49:706–707.

Badje, A. F., and J. A. Flory. 2020. *Graptemys ouachitensis* (Ouachita Map Turtle). Geographic distribution. Herpetological Review 51:73–74.

Badje, A. F., and R. A. Staffen. 2021. *Graptemys geographica* (Northern Map Turtle). Geographic distribution. Herpetological Review 52:79.

Badje, A. F., R. A. Paloski, J. Karow, G. Jensen, J. Robaidek, and J. Watson. 2020. *Graptemys ouachitensis* (Ouachita Map Turtle). Geographic distribution. Herpetological Review 51:73.

Badje, A. F., T. J. Brandt, T. L. Bergeson, R. A. Paloski, J. M. Kapfer, and G. W. Schuurman. 2016. Blanchard's Cricket Frog *Acris blanchardi* overwintering ecology in southwestern Wisconsin. Herpetological Conservation and Biology 11:101–111.

Badje, A. F., T. L. Bergeson, R. A. Paloski, and J. M. Kapfer. 2015. *Acris blanchardi* (Blanchard's Cricket Frog). Geographic distribution. Herpetological Review 46:375.

Badje, A. F., T. J. Brandt, T. L. Bergeson, R. A. Paloski, and J. M. Kapfer. 2021. Spring movement ecology of Blanchard's Cricket Frog (*Acris blanchardi*) in southwestern Wissconsin, USA. Herpetological Conservation and Biology 16:271–280.

Bahn, V., R. J. O'Connor, and W. B. Krohn. 2006. Effect of dispersal at range edges on the structure of species ranges. Oikos 115:89–96.

Bailey, R. L. 2010. Modeling habitat suitability and population demographics of the Eastern Massasauga Rattlesnake in managed lands in southwestern Michigan. Thesis, Michigan State University, East Lansing, Michigan, USA.

Bailey, R. L., H. Campa, III, K. M. Bissell, and T. M. Harrison. 2012. Resource selection by the Eastern Massasauga Rattlesnake on managed land in southwestern Michigan. Journal of Wildlife Management 76:414–421.

Bailey, R. L., H. Campa, III, T. M. Harrison, and K. Bissell. 2011. Survival of Eastern Massasauga Rattlesnakes (*Sistrurus catenatus catenatus*) in Michigan. Herpetologica 67:167–173.

Bailey, R. M. 1942. An intergeneric hybrid rattlesnake. American Midland Naturalist 76:376–285.

Bailey, R. M. 1948. Winter mortality in the snake *Storeria dekayi*. Copeia 1948:215.

Bailey, R. M. 1949. Temperature toleration of garter snakes in hibernation. Ecology 30:238–242.

Baird, S. F. 1850. Revision of the North American tailed-Batrachia, with descriptions of new genera and species. Journal of the Academy of Natural Sciences of Philadelphia, 2nd Series 1:281–294.

Baird, S. F. 1854. Descriptions of new genera and species of North American frogs. Proceedings of the Academy of Natural Sciences of Philadelphia 7:59–62.

Baird, S. F. 1858. Descriptions of new genera and species of North American lizards in the museum of the Smithsonian Institution. Proceedings of the National Academy of Sciences of Philadelphia 10:253–256.

Baird, S. F., and C. Girard. 1853. Catalogue of North American reptiles in the museum of the Smithsonian Institution, part 1, serpents. Smithsonian Miscellaneous Collections 2:1–172.

Bais, H. P., R. Vepachedu, S. Gilroy, R. M. Callaway, and J. M. Vivanco. 2003. Allelopathy and exotic plant invasion: from molecules and genes to species interactions. Science 301:1377–1380.

Baker, E. C. 1964. Ecological observations on the Queen Snake, *Natrix septemvittata* (Say), in Madison County, Kentucky. Thesis, Eastern Kentucky State College, Richmond, USA.

Baker, F. C. 1928. The fresh water Mollusca of Wisconsin, part 2: Pelecypoda. Bulletin of the University of Wisconsin 1527:1–495.

Baker, M. R. 1979. Seasonal population changes in *Rhabdias ranae* Walton, 1929 (Nematoda: Rhabdiasidae) in *Rana sylvatica* of Ontario. Canadian Journal of Zoology 57:179–183.

Baker, R. E., and J. C. Gillingham. 1983. An analysis of courtship behavior in Blanding's Turtle, *Emydoidea blandingii*. Herpetologica 39:166–173.

Baker, R. G., L. J. Maher, C. A. Chumbley, and K. L. Vanzant. 1992. Patterns of Holocene environmental-change in the midwestern United States. Quaternary Research 37:379–389.

Baker, S. J., E. K. Haynes, S. M. LaGrange, M. Britton, and M. C. Allender. 2018. First confirmed occurrence of *Ophidiomyces ophiodiicola* in Indiana, USA. Herpetological Review 49:470–473.

Baker, S. J., M. J. Dreslik, D. B Wylie, and C. A. Phillips. 2016. Sources of mortality in the endangered Eastern Massasauga (*Sistrurus catenatus*) in Illinois. Herpetological Conservation and Biology 11:335–343.

Bakkegard, K. A., and A. P. Pessier. 2010. *Batrachochytrium dendrobatidis* in adult *Notophthalmus viridescens* in north-central Alabama, USA. Herpetological Review 41:45–47.

Bakken, G. S., and A. R. Krochmal. 2007. The imaging properties and sensitivity of the facial pits of pitvipers as determined by optical and heat-transfer analysis. Journal of Experimental Biology 210:2801–2810.

Balas, C. J., N. H. Euliss, Jr., and D. M. Mushet. 2012. Influence of conservation programs on amphibians using seasonal wetlands in the prairie pothole region. Wetlands 32:333–345.

Baldauf, R. J. 1952. Climatic factors influencing the breeding migration of the Spotted Salamander, *Ambystoma maculatum* (Shaw). Copeia 1952:1778–181.

Baldwin, R. F., and A. J. K. Calhoun. 2002. *Ambystoma laterale* (Blue-spotted Salamander) and *Ambystoma maculatum* (Spotted Salamander). Predation. Herpetological Review 33:44–45.

Baldwin, R. F., and D. Vasconcelos. 2003. *Ambystoma maculatum* (Spotted Salamander) and *Rana sylvatica* (Wood Frog). Habitat. Herpetological Review 34:353–354.

Baldwin, R. F., A. J. K. Calhoun, and P. G. deMaynadier. 2006a. Conservation planning for amphibian species with complex habitat requirements: a case study using movements and habitat selection of the Wood Frog *Rana sylvatica*. Journal of Herpetology 40:443–454.

Baldwin, R. F., A. J. K. Calhoun, and P. G. deMaynadier. 2006b. The significance of hydroperiod and stand maturity for pool-breeding amphibians in forested landscapes. Canadian Journal of Zoology 84:1604–1615.

Baldwin, T. 2007. Ecology and morphological comparison between Rough Greensnakes (*Opheodrys a. aestivus*) and Eastern Smooth Greensnakes (*Opheodrys v. vernalis*) in West Virginia. Thesis, Marshall University, Huntington, West Virginia, USA.

Balent, K. L., and P. T. Andreadis. 1998. The mixed foraging strategy of juvenile Northern Water Snakes. 1998. Journal of Herpetology 32:575–579.

Ball, S. C. 1937. Amphibians of Gaspe County, Quebec. Copeia 1937:230.

Ballinger, R. E., J. D. Lynch, and G. R. Smith. 2010. Amphibians and reptiles of Nebraska. Rusty Lizard Press, Oro Valley, Arizona, USA.

Ballinger, R. E., J. W. Nietfeldt, and J. J. Krupa. 1979. An experimental analysis of the role of the tail in attaining high running speed in *Cnemidophorus sexlineatus* (Reptilia: Squamata: Lacertilia). Herpetologica 35:114–116.

Bancroft, G. T., J. S. Godley, D. T. Gross, N. N. Rojas, D. A. Sutphen, and R. W. McDiarmid. 1983. Large scale operations management test of use of the White Amur for control of problem aquatic plants. Report 1, baseline studies, volume V: the herpetofauna of Lake Conway, Florida. US Army Engineer Waterways Experiment Station, Environmental Laboratory Technical Report A-78-2, Vicksburg, Mississippi, USA. Archived by WebCite at http://www.webcitation.org/73dottw2l. Accessed 3 Nov 2018.

Banning, J. L., A. L. Weddle, G. W. Wahl, III, M. A. Simon, A. Lauer, R. L. Walters, and R. N. Harris. 2008. Antifungal skin bacteria, embryonic survival, and communal nesting in Four-toed Salamanders, *Hemidactylium scutatum*. Oecologia 156:423–429.

Banta, A. M. 1914. Sex recognition and the mating behavior of the Wood Frog, *Rana sylvatica*. Biological Bulletin 26:171–183.

Banta, B. H., and G. Carl. 1967. Death-Feigning behavior in the Eastern Gray Treefrog *Hyla versicolor versicolor*. Herpetologica 23:317–318.

Barach, J. P. 1951. The value of the skin secretions of the Spotted Salamander. Herpetologica 7:58.

Barbour, R. W. 1950. The reptiles of Big Black Mountain, Harlan County, Kentucky. Copeia 1950:100–107.

Barbour, R. W. 1971. Amphibians and reptiles of Kentucky. University Press of Kentucky, Lexington, USA.

Barden, A. 1942. Activity of the lizard, *Cnemidophorus sexlineatus*. Ecology 23:336–344.

Barko, V. A., and J. T. Briggler. 2006. Midland Smooth Softshell (*Apalone mutica*) and Spiny Softshell (*Apalone spinifera*) turtles in the Middle Mississippi River: habitat associations, population structure, and implications for conservation. Chelonian Conservation and Biology 5:225–231.

Barko, V. A., J. T. Briggler, and D. E. Ostendorf. 2004. Passive fishing techniques: a cause of turtle mortality in the Mississippi River. Journal of Wildlife Management 68:1145–1150.

Barrett, G. C., and M. R. Villarroul. 1994. *Storeria occipitomaculata occipitomaculata* (Northern Red-bellied Snake). Predation. Herpetological Review 25:29–30.

Barron, J. N., and G. M. Andraso. 2001. The influence of fall foraging success on follicle number in the Northern Water Snake, *Nerodia sipedon*. Journal of Herpetology 35:504–507.

Barrow, J. H., and J. J. Stockton. 1960. The influences of temperature on the host-parasite relationships of several species of snakes infected with *Entamoeba invadens*. Journal of Protozoology 7:377–383.

Barry, F. E., P. J. Weatherhead, and D. P. Philipp. 1992. Multiple paternity in a wild population of Northern Water Snakes, *Nerodia sipedon*. Behavioral Ecology and Sociobiology 30:193–199.

Barta, J. R., and S. S. Desser. 1984. Blood parasites of amphibians from Algonquin Park, Ontario. Journal of Wildlife Diseases 20:180–189.

Bartelt, P. E., and R. W. Klaver. 2017. Response of anurans to wetland restoration on a midwestern agricultural landscape. Journal of Herpetology 51:504–514.

Bartholmai, J., and A. F. Badje. 2019. *Emydoidea blandingii* (Blanding's Turtle). Geographic distribution. Herpetological Review 50:324.

Bartsch, A. F. 1972. Role of phosphorus in eutrophication. US Environmental Protection Agency, National Environmental Research Center Report EPA-R3-72-001, Corvallis, Oregon, USA.

Bartz, A. D., and R. A. Sajdak. 2004. *Crotalus horridus* (Timber Rattlesnake). Arboreality, courtship. Herpetological Review 35:61.

Basir, Y. J., and J. M. Conlon. 2003. Peptidomic analysis of the skin secretions of the Pickerel Frog *Rana palustris* identifies six novel families of structurally related peptides. Peptides 24:379–383.

Basir, Y. J., F. C. Knoop, J. Dulka, and J. M. Conlon. 2000. Multiple antimicrobial peptides and peptides related to bradykinin and neuromedin N isolated from skin secretions of the Pickerel Frog, *Rana palustris*. Biochimica et Biophysica Acta 1543:95–105.

Bastien, H., and R. Leclair, Jr. 1992. Aging Wood Frogs (*Rana sylvatica*) by skeletochronology. Journal of Herpetology 26:222–225.

Bateman, P., P. Fleming, and B. Rolek. 2014. Bite me: blue tails as a "risky-decoy" defense tactic for lizards. Current Zoology 60:333–337.

Bateman, P. W., and P. A. Fleming. 2009. To cut a long tail short: a review of lizard caudal autotomy studies carried out over the last 20 years. Journal of Zoology 277:1–14.

Bateson, Z. W., J. D. Krenz, and R. E. Sorenson. 2011. Multiple paternity in the Common Five-lined Skink (*Plestiodon fasciatus*). Journal of Herpetology 45:504–511.

Battin, J. 2004. When good animals love bad habitats: ecological traps and the conservation of animal populations. Conservation Biology 18:1482–1491.

Bauman, M. A., and D. E. Metter. 1975. Economics, feeding, and population structure of *Natrix s. sipedon* in a Goldfish hatchery. Progressive Fish-Culturist 37:197–201.

Bauman, M. A., and D. E. Metter. 1977. Reproductive cycle of the Northern Watersnake, *Natrix s. sipedon* (Reptilia, Serpentes, Colubridae). Journal of Herpetology 11:51–59.

Bauer, B., J. M. Kapfer, R. A. Staffen, and G. S. Casper. 2010. *Pituophis catenifer sayi* (Bullsnake). Geographic distribution. Herpetological Review 41:517–518.

Baur, G. 1890. Two new species of tortoises from the South. Science 16:262–263.

Bayless, L. E. 1969. Post-metamorphic growth of *Acris crepitans*. American Midland Naturalist 81:590–592.

Beachy, C. K. 1991. *Ambystoma maculatum* (Spotted Salamander). Predation. Herpetological Review 22:128.

Beamer, D. A., and M. J. Lannoo. 2005. *Plethodon glutinosus* (Green, 1818), Northern Slimy Salamander. Pages 627–632 *in* M. Lannoo, editor. Amphibian declines: the conservation status of United States species. University of California Press, Berkeley, USA.

Beane, J. C. 1990. *Rana palustris* (Pickerel Frog). Predation. Herpetological Review 21:59.

Beane, J. C., T. J. Thorp, and D. A. Jackan. 1998. *Heterodon simus* (Southern Hognose Snake). Diet. Herpetological Review 29:44–45.

Beasley, V. R., S. A. Faeh, B. Wikoff, C. Staehle, J. Eisold, D. Nichols, R. Cole, A. M. Schotthoefer, M. Greenwell, and L. E. Brown. 2005. Risk factors and declines in Northern Cricket Frogs (*Acris crepitans*). Pages 75–86 *in* M. Lannoo, editor. Amphibian declines: the conservation status of United States species. University of California Press, Berkeley, USA.

Beattie, A. M., M. R. Whiles, and P. W. Willink. 2017. Diets, population structure, and seasonal activity patterns of Mudpuppies (*Necturus maculosus*) in an urban, Great Lakes coastal habitat. Journal of Great Lakes Research 43:132–143.

Beaudry, F., P. G. deMaynadier, and M. L. Hunter, Jr. 2009. Seasonally dynamic habitat use by Spotted (*Clemmys guttata*) and Blanding's Turtles (*Emydoidea blandingii*) in Maine. Journal of Herpetology 43:636–645.

Beaudry, F., P. G. deMaynadier, and M. L. Hunter, Jr. 2010. Nesting movements and the use of anthropogenic nesting sites by Spotted Turtles (*Clemmys guttata*) and Blanding's Turtles (*Emydoidea blandingii*). Herpetological Conservation and Biology 5:1–8.

Beaupre, S. J., and L. E. Douglas. 2012. Responses of Timber Rattlesnakes to fire: lessons from two prescribed burns. Pages 192–204 *in* D. C. Day, M. C. Stambaugh, S. L. Clark, and C. J. Schweitzer, editors. Proceedings of the 4th Fire in Eastern Oak Forests Conference. US Forest Service, Northern Research Station General Technical Report NRS-P-102, Newtown Square, Pennsylvania, USA.

Beaupre, S. J., and F. Zaidan III. 2012. Digestive performance in the Timber Rattlesnake (*Crotalus horridus*) with reference to temperature dependence and bioenergetic cost of growth. Journal of Herpetology 46:637–642.

Beck, T., B. Johnston, A. F. Badje, and R. A. Staffen. 2015. *Opheodrys vernalis* (Smooth Greensnake). Geographic distribution. Herpetological Review 46:62–63.

Becker, C. G., C. R. Fonseca, C. F. B. Haddad, R. F. Batista, and P. I. Prado. 2007. Habitat split and the global decline of amphibians. Science 318:1775–1777.

Becker, C. G., D. Rodriguez, A. V. Longo, A. L. Talaba, and K. R. Zamudio. 2012. Disease risk in temperate amphibian populations is higher at closed-canopy sites. PLoS ONE 7:e48205.

Becker, J., P. Thorsell, L. Thorsell, and J. M. Kapfer. 2009. *Heterodon platirhinos* (Eastern Hog-nosed Snake). Geographic distribution. Herpetological Review 40:456.

Becker, M. N., and R. N. Harris. 2010. Cutaneous bacteria of the Redback Salamander prevent morbidity associated with a lethal disease. PLoS ONE 5:e10957.

Bee, M. A. 2007a. Selective phonotaxis by male Wood Frogs (*Rana sylvatica*) to the sound of a chorus. Behavioral Ecology and Sociobiology 61:955–966.

Bee, M. A. 2007b. Sound source segregation in Grey Treefrogs: spatial release from masking by the sound of a chorus. Animal Behaviour 74:549–558.

Bee, M. A. 2008. Finding a mate at a cocktail party: spatial release from masking improves acoustic mate recognition in Grey Treefrogs. Animal Behaviour 75:1781–1791.

Bee, M. A., and S. A. Perrill. 1996. Responses to conspecific advertisement calls in the Green Frog (*Rana clamitans*) and their role in male-male communication. Behaviour 133:283–301.

Bee, M. A., S. A. Perrill, and P. C. Owen. 2000. Male green frogs lower the pitch of acoustic signals in defense of territories: a possible dishonest signal of size? Behavioral Ecology 11:169–177.

Beebee, T. J. C. 2002. Amphibian phenology and climate change. Conservation Biology 16:1454.

Beebee, T. J. C. 2013. Effects of road mortality and mitigation measures on amphibian populations. Conservation Biology 27:657–668.

Behler, J. L., and C. M. Castellano. 2005. *Glyptemys insculpta* (North American Wood Turtle). Terrestrial movement. Herpetological Review 36:311.

Beiswenger, R. E. 1977. Diel patterns of aggregative behavior in tadpoles of *Bufo americanus*, in relation to light and temperature. Ecology 58:98–108.

Bekker, K. A. 2007. Comparative growth and demographics of two sympatric natricine snakes. Thesis, Bowling Green State University, Ohio, USA.

Belasen, A., E. Burkett, A. Injaian, K. Li, D. Allen, and I. Perfecto. 2013. Effect of sub-canopy on habitat selection in the Blue-spotted Salamander (*Ambystoma laterale-jeffersonianum* unisexual complex). Copeia 2013:254–261.

Belden, L. K. 2006. Impact of eutrophication on Wood Frog, *Rana sylvatica*, tadpoles infected with *Echinostoma trivolvis* cercariae. Canadian Journal of Zoology 84:1315–1321.

Belden, L. K., and J. M. Wojdak. 2011. The combined influence of trematode parasites and predatory salamanders on Wood Frog (*Rana sylvatica*) tadpoles. Oecologia 166:1077–1086.

Bell, S. L. M., T. B. Herman, and R. J. Wassersug. 2007. Ecology of *Thamnophis sauritus* (Eastern Ribbon Snake) at the northern limit of its range. Northeastern Naturalist 4:279–292.

Bellis, E. D. 1959. A study of movement of American Toads in a Minnesota bog. Copeia 1959:173–174.

Bellis, E. D. 1961. Growth of the Wood Frog, *Rana sylvatica*. Copeia 1961:74–77.

Bellis, E. D. 1962. The influence of humidity on Wood Frog activity. American Midland Naturalist 68:139–148.

Bellis, E. D. 1964. A summer Six-Lined Racerunner (*Cnemidophorus sexlineatus*) population in South Carolina. Herpetologica 20:9–16.

Bellis, E. D. 1965. Home range and movements of the Wood Frog in a northern bog. Ecology 46:90–98.

Bellis, E. D. 1968. Summer movement of Red-spotted Newts in a small pond. Journal of Herpetology 1:68–91.

Bellocq, M. I., K. Kloosterman, and S. M. Smith. 2000. The diet of coexisting species of amphibians in Canadian Jack Pine forests. Herpetological Journal 10:63–68.

Beltz, E. 2006. Scientific and common names of the reptiles and amphibians of North America—explained. Archived by WebCite at http://www.webcitation.org/73bhmRlc9. Accessed 1 Nov 2018.

Bennett, A. F., J. Q. Radford, and A. Haslem. 2006. Properties of land mosaics: implications for nature conservation in agricultural environments. Biological Conservation 133:250–264.

Bennett, A. M., M. Keevil, and J. D. Litzgus. 2010. Spatial ecology and population genetics of Northern Map Turtles (*Graptemys geographica*) in fragmented and continuous habitats in Canada. Chelonian Conservation and Biology 9:185–195.

Bennett, A. M., M. G. Keevil, and J. D. Litzgus. 2015. *Sternotherus odoratus* (Eastern Musk Turtle). Home range. Herpetological Review 46:245–246.

Bennett, D. H., J. W. Gibbons, and J. C. Franson. 1970. Terrestrial activity in aquatic turtles. Ecology 51:738–740.

Berg, S. W. 2014. Population structure, size, and the thermal ecology of Iowa Wood Turtles (*Glyptemys insculpta*): a comparison between suburban and rural populations. Thesis, University of Northern Iowa, Cedar Falls, USA.

Bergeson, M. T. 2001. Red-backed Salamanders (*Plethodon cinereus*) in hardwood forests in northeastern Wisconsin and the Upper Peninsula of Michigan. Thesis, University of Wisconsin, Madison, USA.

Bergeson, T., and R. Paloski. 2004. Blanchard's Cricket Frog (*Acris crepitans blanchardi*) surveys in southwestern Wisconsin: gaining knowledge to direct future conservation efforts. Wisconsin Department of Natural Resources, Bureau of Endangered Resources, Madison, USA.

Bergeson, T., and R. Paloski. 2005. Blanchard's Cricket Frog (*Acris crepitans blanchardi*) surveys in southwestern Wisconsin: gaining knowledge to direct future conservation efforts. Wisconsin Department of Natural Resources, Bureau of Endangered Resources, Madison, USA.

Bergeson, T., R. Paloski, A. Badje, and T. Brandt. 2013. Northern Cricket Frog (*Acris crepitans*) surveys in southwestern Wisconsin. Wisconsin Department of Natural Resources, Bureau of Natural Heritage Conservation Report 153, Madison, USA.

Bergquist, L. 2005. DNR compromises to protect garter snake, development. New policy eases costs, hassles for builders. Archived by WebCite at http://www.webcitation.org/73biIaFNf. Accessed 1 Nov 2018.

Bergquist, L. 2006a. Garter snake won't lose protection; committee rescinds vote to strip safeguards. Milwaukee Journal Sentinel. 29 November 2006; section News:B1.

Bergquist, L. 2006b. Snake may lose status: legislators want DNR to roll back protection for species. Milwaukee Journal Sentinel. 28 July 2006; section News:A1.

Berns, M. W., and L. D. Uhler. 1966. Blue frogs of the genus *Rana*. Herpetologica 22:181–183.

Bernstein, N. P., R. J. Richtsmeier, R. W. Black, and B. R. Montgomery. 2007. Home range and philopatry in the Ornate Box Turtle, *Terrapene ornata ornata*, in Iowa. American Midland Naturalist 157:162–174.

Berrill, M., D. Coulson, L. McGillivray, and B. Pauli. 1998. Toxicity of endosulfan to aquatic stages of anuran amphibians. Environmental Toxicology and Chemistry 17:1738–1744.

Berrill, M., S. Bertram, B. Pauli, D. Coulson, M. Kolohon, and D. Ostrander. 1995. Comparative sensitivity of amphibian tadpoles to single and pulsed exposures of the forest-use insecticide fenithrothion. Environmental Toxicology and Chemisty 14:1011–1018.

Bertram, S., and M. Berrill. 1997. Fluctuations in a northern population of Gray Treefrogs, *Hyla versicolor*. Pages 57–63 *in* D. M. Green, editor. Amphibians in decline: Canadian studies of a global problem. Society for the Study of Amphibians and Reptiles, Herpetological Conservation 1, St. Louis, Missouri, USA.

Berven, K. A. 1981. Mate choice in the Wood Frog, *Rana sylvatica*. Evolution 35:707–722.

Berven, K. A. 1982a. The genetic basis of altitudinal variation in the Wood Frog *Rana sylvatica*. I. An experimental analysis of life history traits. Evolution 36:962–983.

Berven, K. A. 1982b. The genetic basis of altitudinal variation in the Wood Frog *Rana sylvatica* II. An experiment analysis of larval development. Oecologia 52:360–369.

Berven, K. A. 1990. Factors affecting population fluctuations in larval and adult stages of the Wood Frog (*Rana sylvatica*). Ecology 71:1599–1608.

Berven, K. A. 1995. Population regulation in the Wood Frog, *Rana sylvatica*, from three diverse geographic localities. Australian Journal of Ecology 20:385–392.

Berven, K. A. 2009. Density dependence in the terrestrial stage of Wood Frogs: evidence from a 21-year population study. Copeia 2009:328–338.

Berven, K. A., and B. G. Chadra. 1988. The relationship among egg size, density and food level on larval development in the Wood Frog (*Rana sylvatica*). Oecologia 75:67–72.

Berven, K. A., and R. S. Boltz. 2001. Interactive effects of leech (*Desserobdella picta*) infection on Wood Frog (*Rana sylvatica*) tadpole fitness traits. Copeia 2001:907–915.

Berven, K. A., and T. A. Grudzien. 1990. Dispersal in the Wood Frog (*Rana sylvatica*): implications for genetic population structure. Evolution 44:2047–2056.

Berven, K. A., D. E. Gill, and S. J. Smith-Gill. 1979. Countergradient selection in the Green Frog, *Rana clamitans*. Evolution 33:609–623.

Bever, G. S., T. R. Lyson, D. J. Field, and B-A. S. Bhullar. 2015. Evolutionary origin of the turtle skull. Nature 525:239–242.

Bevier, C. R., and A. M. Gorman-Gelder. 2018. The effect of parasite infection on phonotactic response in the Mink Frog, *Lithobates septentrionalis*. Journal of Herpetology 51:34–39.

Bevier, C. R., D. C. Tierney, L. E. Henderson, and H. E. Reid. 2006. Chorus attendance and site fidelity in the Mink Frog, *Rana septentrionalis*: are males territorial? Journal of Herpetology 40:160–164.

Bevier, C. R., K. Larson, K. Reilly, and S. Tat. 2004. Vocal repertoire and calling activity of the Mink Frog, *Rana septentrionalis*. Amphibia-Reptilia 25:255–264.

Beyer, S. E., C. A. Phillips, and R. L. Schooley. 2015. Canopy cover and drought influence the landscape epidemiology of an amphibian chytrid fungus. Ecosphere 6:1–18.

Bi, K., J. P. Bogart, and J. Z. Fu. 2007. Intergenomic translocations in unisexual salamanders of the genus *Ambystoma* (Amphibia: Caudata). Cytogenetic and Genome Research 116:289–297.

Bi, K., J. P. Bogart, and J. Z. Fu. 2008. The prevalence of genome replacement in unisexual salamanders of the genus *Ambystoma* (Amphibia, Caudata) revealed by nuclear gene genealogy. BMC Evolutionary Biology 8:158.

Bickham, J. W., and J. L. Carr. 1983. Taxonomy and phylogeny of the higher categories of cryptodiran turtles based on a cladistics analysis of chromosomal data. Copeia 1983:918–932.

Bickham, J. W., T. Lamb, P. Minx, and J. C. Patton. 1996. Molecular systematics of the genus *Clemmys* and the intergeneric relationships of emydid turtles. Herpetologica 52:89–97.

Bider, R., and S. Matte. 1996. The atlas of amphibians and reptiles of Québec. St. Lawrence Valley Natural History Society and Ministere de l'Environnement et de la Faune, Quebec, Canada.

Bieser, N. D. 2008. Spatial ecology and survival of resident juvenile and headstarted Eastern Massasaugas (*Sistrurus catenatus catenatus*) in northern Michigan. Thesis, Purdue University, Fort Wayne, Indiana, USA.

Binckley, C. A., and W. J. Resetarits. 2003. Functional equivalence of non-lethal effects: generalized fish avoidance determines distribution of Gray Treefrog, *Hyla chrysoscelis*, larvae. Oikos 102:623–629.

Birchfield, G. L., and J. E. Deters. 2005. Movement paths of displaced Northern Green Frogs (*Rana clamitans melanota*). Southeastern Naturalist 4:63–76.

Bishop, C. A., and A. D. Gendron. 1998. Reptiles and amphibians: shy and sensitive vertebrates of the Great Lakes Basin and St. Lawrence River. Environmental Monitoring and Assessment 53:225–244.

Bishop, C. A., J. Struger, D. R. Barton, L. J. Shirose, L. Dunn, A. L. Lang, and D. Shepherd. 2000. Contamination and wildlife communities in stormwater detention ponds in Guelph and the Greater Toronto Area, Ontario, 1997 and 1998. Part I—Wildlife communities. Water Quality Research Journal of Canada 35:399–435.

Bishop, C. A., N. A. Mahony, J. Struger, P. Ng, and K. E. Pettit. 1999. Anuran development, density and diversity in relation to agricultural activity in the Holland River watershed, Ontario, Canada (1990–1992). Environmental Monitoring and Assessment 57:21–43.

Bishop, S. C. 1919. Notes on the habits and development of the Four-toed Salamander, *Hemidactylium scutatum* (Schlegel). New York State Museum Bulletin 219:251–282.

Bishop, S. C. 1926. Notes on the habits and development of the Mudpuppy, *Necturus maculosus* (Rafinesque). New York State Museum Bulletin 258:5–60.

Bishop, S. C. 1941a. Notes on salamanders, with descriptions of several new forms. University of Michigan, Occasional Papers of the Museum of Zoology 451:1–21.

Bishop, S. C. 1941b. The salamanders of New York. New York State Museum Bulletin 324:1–365.

Bishop, S. C. 1943. Handbook of salamanders. Comstock Publishing Company, Inc., Ithaca, New York, USA.

Bishop, S. C. 1948. Record *Necturus* from Wisconsin waters. Copeia 1948:66.

Bishop, S. C., and F. J. W. Schmidt. 1931. The painted turtles of the genus *Chrysemys*. Field Museum of Natural History Publication 293, Zoological Series Volume 18, Number 4, Chicago, Illinois, USA.

Bishop, S. C., and W. J. Schoonmacher. 1921. Turtle hunting in midwinter. Copeia 96:37–38.

Bissell, K. M. 2006. Modeling habitat ecology and population viability of the Eastern Massasauga Rattlesnake in southwestern lower Michigan. Thesis, Michigan State University, East Lansing, Michigan, USA.

Bittner, T. D., R. B. King, and J. M. Kerfin. 2002. Effects of body size and melanism on the thermal biology of Garter Snakes (*Thamnophis sirtalis*). Copeia 2002:477–482.

Black, J. H. 1976. The Lined Snake as food for birds. Bulletin of the Oklahoma Ornithological Society 9:17–18.

Black, M. R., and E. J. Judziewicz. 2009. Wildflowers of Wisconsin and the Great Lakes Region. Second edition. University of Wisconsin Press, Madison, USA.

Blackburn, D. G. 2000. Classification of the reproductive patterns of amniotes. Herpetological Monographs 14:371–377.

Blackburn, L. M., C. D. Hoefler, and P. Nanjappa. 2002. *Acris crepitans blanchardi* (Blanchard's Cricket Frog). Predation. Herpetological Review 33:299.

Blackwell, E. A., G. R. Cline, and K. R. Marion. 2004. Annual variation in population estimators for a southern population of *Ambystoma maculatum*. Herpetologica 60:304–311.

Blackwell, E. A., R. A. Angus, G. R. Cline, and K. R. Marion. 2003. Natural growth rates of *Ambystoma maculatum* in Alabama. Journal of Herpetology 37:608–612.

Blair, J., and R. J. Wassersug. 2000. Variation in the pattern of predator-induced damage to tadpole tails. Copeia 2000:390–401.

Blair, W. F. 1958. Call differences as an isolation mechanism in Florida species of hylid frogs. Journal of the Florida Academy of Sciences 21:32–48.

Blair, W. F. 1963. Evolutionary relationships of North American toads of the genus *Bufo*: a progress report. Evolution 17:1–16.

Blair, W. F. 1976. Some aspects of the biology of the Ornate Box Turtle, *Terrapene ornata*. Southwestern Naturalist 21:89–104.

Blais, B. 2015. Phylogeography of the Smooth Greensnake, *Opheodrys vernalis*. Thesis, Black Hills State University, Spearfish, South Dakota, USA.

Blais, B. 2016. *Anaxyrus americanus* (American Toad). Ectoparasites. Herpetological Review 47:435–436.

Blakemore, L. A. 1940. Barred Owl food habits in Glenwood Park, Minneapolis, Minnesota. Flicker 12:21–23.

Blanchard, F. N. 1923. The life history of the Four-toed Salamander. American Naturalist 57:262–268.

Blanchard, F. N. 1928. Topics from the life history and habits of the Red-backed Salamander in southern Michigan. American Naturalist 62:156–164.

Blanchard, F. N. 1930a. Further studies of the eggs and young of the Eastern Ring-neck Snake, *Diadophis punctatus edwardsii*. Bulletin of the Antivenin Institute of America 4:4–10.

Blanchard, F. N. 1930b. The stimulus to the breeding migration of the Spotted Salamander, *Ambystoma maculatum*. American Naturalist 62:156–164.

Blanchard, F. N. 1931. Secondary sex characteristics of certain snakes. Bulletin of the Antivenom Institute of America 4:95–104.

Blanchard, F. N. 1933a. Eggs and young of the Smooth Green Snake (*Liopeltis vernalis*) Harlan. Papers of the Michigan Academy of Science, Arts and Letters 17:493–514.

Blanchard, F. N. 1933b. Late autumn collections and hibernating situations of the salamander *Hemidactylium scutatum* (Schlegel) in southern Michigan. Copeia 1933:216.

Blanchard, F. N. 1933c. Spermatophores and the mating season of the salamander *Hemidactylium scutatum* (Schlegel). Copeia 1933:40.

Blanchard, F. N. 1934a. The relation of the female Four-toed Salamander to her nest. Copeia 1934:137–138.

Blanchard, F. N. 1934b. The spring migration of the Four-toed Salamander, *Hemidactylium scutatum* (Schlegel). Copeia 1934:50.

Blanchard, F. N. 1935. The sex ratio in the salamander *Hemidactylium scutatum* (Schlegel). Copeia 1935:103.

Blanchard, F. N. 1937a. Data on the natural history of the Red-bellied Snake, *Storeria occipitomaculata* (Storer), in northern Michigan. Copeia 1937:151–162.

Blanchard, F. N. 1937b. Eggs and natural nests of the Eastern Ring-neck Snake, *Diadophis punctatus edwardsii*. Papers of the Michigan Academy of Sciences, Arts and Letters 22:521–532.

Blanchard, F. N. 1942. The ring-neck snakes, genus *Diadophis*. Bulletin of the Chicago Academy of Science 7:1–144.

Blanchard, F. N., and F. C. Blanchard. 1931. Conservation groups and their characteristics in the salamander *Hemidactylium scutatum* (Schlegel). American Naturalist 65:149–164.

Blanchard, F. N., and F. C. Blanchard. 1942. Mating of the Garter Snake (*Thamnophis sirtalis sirtalis*). Papers of the Michigan Academy of Sciences, Arts and Letters 27:215–234.

Blanchard. F. N., M. R. Gilreath, and F. C. Blanchard. 1979. The Eastern Ring-neck Snake (*Diadophis punctatus edwardsii*) in northern Michigan (Reptilia Serpentes, Colubridae). Journal of Herpetology 13:377–402.

Blasus, R. E., J. Gerholdt, and M. Crawford. 2004. *Emydoidea blandingii* (Blanding's Turtle). Maximum size. Herpetological Review 35:54.

Blaustein, A. R., and P. J. Johnson. 2003. The complexity of deformed amphibians. Frontiers in Ecology and the Environment 1:87–93.

Blaustein, A. R., B. A. Han, R. A. Relyea, P. T. J. Johnson, J. C. Buck, S. S. Gervasi, and L. B. Kats. 2011. The complexity of amphibian population declines: understanding the role of cofactors in driving amphibian losses. Annals of the New York Academy of Sciences 1223:108–119.

Blaustein, A. R., D. B. Wake, and W. P. Sousa. 1994. Amphibian declines: judging stability, persistence, and susceptibility of populations to local and global extinctions. Conservation Biology 8:60–71.

Blaustein, A. R., J. M. Romansic, J. M. Kiesecker, and A. C. Hatch. 2003. Ultraviolet radiation, toxic chemicals and amphibian population declines. Diversity and Distributions 9:123–140.

Blaustein, A. R., L. K. Belden, D. H. Olson, D. M. Green, T. L. Root, and J. M. Kiesecker. 2001. Amphibian breeding and climate change. Conservation Biology 15:1804–1809.

Bleakney, J. S. 1963. Notes on the distribution and life histories of turtles in Nova Scotia. Canadian Field-Naturalist 77:67–76.

Bleakney, S. 1952. The amphibians and reptiles of Nova Scotia. Canadian Field-Naturalist 66:125–129.

Bleakney, S. 1958. Cannibalism in *Rana sylvatica* tadpoles a well known phenomenon. Herpetologica 14:34.

Blomquist, S. M., and M. L. Hunter, Jr. 2007. Externally attached radio-transmitters have limited effects on the antipredator behavior and vagility of *Rana pipiens* and *Rana sylvatica*. Journal of Herpetology 41:430–438.

Blomquist, S. M., and M. L. Hunter, Jr. 2009. A multi-scale assessment of habitat selection and movement patterns by Northern Leopard Frogs (*Lithobates* [*Rana*] *pipiens*) in a managed forest. Herpetological Conservation and Biology 4:142–160.

Blomquist, S. M., and M. L. Hunter, Jr. 2010. A multiscale assessment of amphibian habitat selection: Wood Frog response to timber harvesting. Ecoscience 17:251–264.

Blouin-Demers, G., and P. J. Weatherhead. 2001a. An experimental test of the link between foraging, habitat selection and thermoregulation in Black Rat Snakes *Elaphe obsoleta obsoleta*. Journal of Animal Ecology 70:1006–1013.

Blouin-Demers, G., and P. J. Weatherhead. 2001b. Habitat use by Black Rat Snakes (*Elaphe obsoleta obsoleta*) in fragmented forests. Ecology 82:2822–2896.

Blouin-Demers, G., and P. J. Weatherhead. 2001c. Thermal ecology of Black Rat Snakes (*Elaphe obsoleta*) in a thermally challenging environment. Ecology 82:3025–2043.

Blouin-Demers, G., and P. J. Weatherhead. 2002. Habitat-selection behavioural thermoregulation by Black Rat Snakes (*Elaphe obsoleta obsoleta*). Oikos 97:59–68.

Blouin-Demers, G., K. A. Prior, and P. J. Weatherhead. 2000. Patterns of variation in spring emergence by Black Rat Snakes (*Elaphe obsoleta obsoleta*). Herpetologica 56:175–188.

Blouin-Demers, G., K. A. Prior, and P. J. Weatherhead. 2002. Comparative demography of Black Rat Snakes (*Elaphe obsoleta*) in Ontario and Maryland. Journal of Zoology 256:1–10.

Blouin-Demers, G., L. P. G. Bjorgan, and P. J. Weatherhead. 2007. Changes in habitat use and movement patterns with body size in Black Ratsnakes (*Elaphe obsoleta*). Herpetologica 63:421–429.

Bluett, R. D., W. E. Louis, D. A. Newhouse, C. J. Handel, Jr., and J. H. Kube. 2013. Longitudinal structuring of turtle assemblages in an altered river in Central Illinois, USA: implications for conservation. Transactions of the Illinois State Academy of Science 106:47–53.

Boczulak, S. A., M. C. Vanderwel, and B. D. Hall. 2017. Survey of mercury in Boreal Chorus Frog (*Pseudacris maculata*) and Wood Frog (*Rana sylvatica*) tadpoles from wetland ponds in the Prairie Pothole Region of Canada. FACETS 2:315–329.

Bodie, J. R. 2001. Stream and riparian management for freshwater turtles. Journal of Environmental Management 62:443–455.

Bodie, J. R., and R. D. Semlitsch. 2001. Spatial and temporal use of floodplain habitats by lentic and lotic species of aquatic turtles. Oecologia 122:138–146.

Bodie, J. R., R. D. Semlitsch, and R. B. Renken. 2000. Diversity and structure of turtle assemblages: associations with wetland characters across a floodplain landscape. Ecography 23:444–456.

Bogart, J. P. 2019. Unisexual salamanders in the genus *Ambystoma*. Herpetologica 75:259–267.

Bogart, J. P., and A. O. Wasserman. 1972. Diploid-polyploid cryptic species pairs: a possible clue to evolution by polyploidization in anuran amphibians. Cytogenetic and Genome Research 11:7–24.

Bogart, J. P., and A. P. Jaslow. 1979. Distribution and call parameters of *Hyla chrysoscelis* and *Hyla versicolor* in Michigan. Royal Ontario Museum Life Sciences Contribution 117:13.

Bogart J. P., and M. W. Klemens. 2008. Additional distributional records of *Ambystoma laterale*, *A. jeffersonianum* (Amphibia: Caudata) and their unisexual kleptogens in northeastern North America. American Museum Novitates 3627:1–58.

Bogart, J. P., J. Bartoszek, D. W. A. Noble, and K. Bi. 2009. Sex in unisexual salamanders: discovery of a new sperm donor with ancient affinities. Heredity 103:483–493.

Bogart, J. P., K. Bi, J. Z. Fu, D. W. A. Noble, and J. Niedzwiecki. 2007. Unisexual salamanders (genus *Ambystoma*) present a new reproductive mode for eukaryotes. Genome 50:119–136.

Bogert, C. M., and V. D. Roth. 1966. Ritualistic combat of male Gopher Snakes (*Pituophis melanoleucus affinis*). American Museum Novitates 2245:1–28.

Bohnsack, K. K. 1951. Temperature data on the terrestrial hibernation of the Green Frog, *Rana clamitans*. Copeia 1951:236–239.

Bohnsack, K. K. 1952. Terrestrial hibernation of the Bullfrog, *Rana catesbeiana* Shaw. Copeia 1952:114.

Bolek, M. G. 1997a. *Graptemys geographica* (Common Map Turtle). Geographic distribution. Herpetological Review 28:94.

Bolek, M. G. 1997b. *Hyla chrysoscelis* (Cope's Gray Treefrog). Geographic distribution. Herpetological Review 28:93.

Bolek, M. G. 1997c. Seasonal occurrence of *Cosmocercoides dukae* and prey analysis in the Blue-Spotted Salamander, *Ambystoma laterale*, in southeastern Wisconsin. Journal of the Helminthological Society of Washington 64:292–295.

Bolek, M. G. 2001. *Chelydra serpentina* (Common Snapping Turtle) and *Emydoidea blandingii* (Blanding's Turtle). Parasites. Herpetological Review 32:37–38.

Bolek, M. G., and J. Janovy, Jr. 2004. Observations on myiasis by the calliphorids, *Bufolucilia silvarum* and *Bufolucilia elongate*, in Wood Frogs, *Rana sylvatica*, from southeastern Wisconsin. Journal of Parasitology 90:1169–1171.

Bolek, M. G., and J. Janovy, Jr. 2005. New host and distribution records for the amphibian leech *Desserobdella picta* (Rhynchobdellida: Glossiphoniidae) from Nebraska and Wisconsin. Journal of Freshwater Ecology 20:187–189.

Bolek, M. G., and J. Janovy, Jr. 2007. *Rana catesbeiana* (American bullfrog). Diet. Herpetological Review 38:325–326.

Bolek, M. G., and J. R. Coggins. 1998. Endoparasites of Cope's Gray Treefrog, *Hyla chrysoscelis*, and Western Chorus Frog, *Pseudacris t. triseriata*, from southeastern Wisconsin. Journal of the Helminthological Society of Washington 65:212–218.

Bolek, M. G., and J. R. Coggins. 2000. Seasonal occurrence and community structure of helminth parasites from the Eastern American Toad, *Bufo americanus americanus*, from southeastern Wisconsin, U.S.A. Comparative Parasitology 67:202–209.

Bolek, M. G., and J. R. Coggins. 2001. Seasonal occurrence and community structure of helminth parasites in Green Frogs, *Rana clamitans melanota*, from southeastern Wisconsin, USA. Comparative Parasitology 68:164–172.

Bolek, M. G., and J. R. Coggins. 2003. Helminth community structure of sympatric Eastern American Toad, *Bufo americanus americanus*, Northern Leopard Frog, *Rana pipiens*, and Blue-spotted Salamander, *Ambystoma laterale*, from southeastern Wisconsin. Journal of Parasitology 89:673–680.

Bolek, M. G., S. D. Snyder, and J. Janovy, Jr. 2009. Alternative life cycle strategies and colonization of young anurans by *Gorgoderina attenuata* in Nebraska. Journal of Parasitology 95:604–616.

Bolliger, J., and D. J. Mladenoff. 2005. Quantifying spatial classification uncertainties of the historical Wisconsin landscape (USA). Ecography 28:141–156.

Bolton, R. M., and R. J. Brooks. 2010. Impact of the seasonal invasion of *Phragmites australis* (Common Reed) on turtle reproductive success. Chelonian Conservation and Biology 92:238–243.

Bond, G. R., and K. K. Burkhart. 1997. Thrombocytopenia following Timber Rattlesnake envenomation. Annals of Emergency Medicine 30:40–44.

Bonin, J., J.-L. Desgranges, C. A. Bishop, J. Rodrigue, A. Gendron, and J. E. Elliott. 1995. Comparative study of contaminants in the Mudpuppy (Amphibia) and the Common Snapping Turtle (Reptilia), St. Lawrence River, Canada. Archives in Environmental Contamination and Toxicology 28:184–194.

Bonin, J., J.-L. Desgranges, J. Rodrigue, and A. Gendron. 1994. Potentiel d'ultilisation du necture tachete (*Necturus maculosus*) comme bioindicateur de la contamination du Flueve Saint-Laurent. Canadian Wildlife Service, Serie de Rapports Techniques 190, Montreal, Canada.

Bonin, J., J. L. DeGranges, J. Rodrigue, and M. Ouellet. 1997. Anuran species richness in agricultural landscapes of Québec: foreseeing long-term results of road call surveys. Pages 141–148 *in* D. M. Green, editor. Amphibians in decline: Canadian studies of a global problem. Society for the Study of Amphibians and Reptiles, Herpetological Conservation 1, St. Louis, Missouri, USA.

Bonnet, X., and G. Naulleau. 1996. Catchability in snakes: consequences for estimates of breeding frequency. Canadian Journal of Zoology 74:233–239.

Bonnet, X., D. Bradshaw, and R. Shine. 1998. Capital versus income breeding: an ectothermic perspective. Oikos 83:333–342.

Bonnet, X., G. Naulleau, and R. Shine. 1999. The dangers of leaving home: dispersal and mortality in snakes. Biological Conservation 89:39–50.

Boogaard, M. A., T. D. Bills, and D. A. Johnson. 2003. Acute toxicity of TFM and a TFM/niclosamide mixture to selected species of fish, including Lake Sturgeon (*Acipenser fulvescens*) and Mudpuppies (*Necturus maculosus*), in laboratory and field exposures. Journal of Great Lakes Research 29:529–541.

Booms, T. L. 1999. Vertebrates removed by mechanical weed harvesting in Lake Keesus, Wisconsin. Journal of Aquatic Plant Management 37:34–36.

Boone, M. D., and C. M. Bridges-Britton. 2006. Examining multiple sublethal contaminants on the Gray Treefrog (*Hyla Versicolor*): effects of an insecticide, herbicide, and fertilizer. Environmental Toxicology and Chemistry 25:3261–3265.

Boone, M. D., and S. M. James. 2003. Interactions of an insecticide, herbicide, and natural stressors in amphibian community mesocosms. Ecological Applications 13:829–841.

Boone, M. D., R. D. Semlitsch, and C. Mosby. 2008. Suitability of golf course ponds for amphibian metamorphosis when Bullfrogs are removed. Conservation Biology 22:172–179.

Boone, M. D., R. D. Semlitsch, E. E. Little, and M. C. Doyle. 2007. Multiple stressors in amphibian communities: effects of chemical contamination, Bullfrogs, and fish. Ecological Applications 17:291–301.

Boone, M. D., S. A. Hammond, N. Veldhoen, M. Youngquist, and C. C. Helbing. 2013. Specific time of exposure during tadpole development influences biological effects of the insecticide carbaryl in Green Frogs (*Lithobates clamitans*). Aquatic Toxicology 130:139–148.

Borkowski, R. 1997. Lead poisoning and intestinal perforations in a Snapping Turtle (*Chelydra serpentina*) due to fishing gear ingestion. Journal of Zoo and Wildlife Medicine 28:109–113.

Botkin, D. B., and E. A. Keller. 2003. Environmental Science: earth as a living planet. Fourth edition. John Wiley and Sons Publishing, Hoboken, New Jersey, USA.

Bouchard, J., A. T. Ford, F. E. Eigenbrod, and L. Fahrig. 2009. Behavioral responses of Northern Leopard Frogs (*Rana pipiens*) to roads and traffic: implications for population persistence. Ecology and Society 14:23.

Bouchard, J. L. 1951. The Platyhelminthes parasitizing some northern Maine amphibia. Transactions of the American Microscopical Society 70:245–250.

Bougie, T. A., N. W. Byer, C. N. Lapin, M. Z. Peery, J. E. Woodford, and J. N. Pauli. 2020. Wood Turtle (*Glyptemys insculpta*) nest protection reduces depredation and increases success, but annual variation influences its effectiveness. Canadian Journal of Zoology 98:715–724.

Boulenger, G. A. 1882. Catalogue of the Batrachia Salientia s. Ecaudata in the collection of the British Museum. Second Edition. London, England.

Bovbjerg, R. V. 1965. Experimental studies on the dispersal of the frog, *Rana pipiens*. Journal of the Iowa Academy of Science 72:412–418.

Bowen, K. D. 2004a. Geographic and size-related variation in the diet of insular populations of the Northern Watersnake. American Midland Naturalist 152:418–424.

Bowen, K. D. 2004b. *Tropidoclonion lineatum* (Lined Snake). Geographic distribution. Herpetological Review 35:413.

Bowen, K. D., and F. J. Janzen. 2008. Human recreation and the nesting ecology of a freshwater turtle (*Chrysemys picta*). Chelonian Conservation and Biology 7:95–100.

Bowen, K. D., and J. C. Gillingham. 2004. R9 species conservation assessment for Wood Turtle—*Glyptemys insculpta* (LeConte, 1830). US Forest Service, Eastern Region of the Forest Service Threatened and Endangered Species Program, Milwaukee, Wisconsin, USA.

Bowen, K. D., P. L. Colbert, and F. J. Janzen. 2004. Survival and recruitment in a human-impacted population of Ornate Box Turtles, *Terrapene ornata*, with recommendations for conservation and management. Journal of Herpetology 38:562–568.

Bowers, B. B., A. E. Bledsoe, and G. M. Burghardt. 1993. Responses to escalating predatory threat in garter and ribbon snakes (*Thamnophis*). Journal of Comparative Psychology 107:25–33.

Bowers, D. G., D. E. Andersen, and N. H. Euliss, Jr. 1998. Anurans as indicators of wetland condition in the Prairie Pothole Region of North Dakota: an environmental monitoring and assessment program pilot project. Pages 369–378 *in* M. J. Lannoo, editor. Status and distribution of midwestern amphibians. University of Iowa Press, Iowa City, USA.

Bowker, R. G., and O. W. Johnson. 1980. Thermoregulatory precision in three species of whiptail lizards (Lacertilia: Teiidae). Physiological Zoology 53:176–185.

Bowler, J. K. 1977. Longevity of reptiles and amphibians in North American collections as of 1 November, 1975. Society for the Study of Amphibians and Reptiles, Herpetological Circular 6, Lawrence, Kansas, USA.

Bowne, D. R., M. A. Bowers, and J. E. Hines. 2006. Connectivity in an agricultural landscape as reflected by interpond movements of a freshwater turtle. Conservation Biology 20:780–791.

Box, G. E. P., and R. Draper. 1987. Empirical model-building and response surfaces. John Wiley & Sons, New York, New York, USA.

Bozek, M. A., T. M. Burri, and R. V. Frie. 1999. Diets of Muskellunge in northern Wisconsin lakes. North American Journal of Fisheries Management 19:258–270.

Brackin, M. F. 1978. The relation of rank to physiological state in *Cnemidophorus sexlineatus* dominance hierarchies. Herpetologica 34:185–191.

Brackin, M. F. 1979. The seasonal reproductive, fat body, and adrenal cycles of male Six-lined Racerunners (*Cnemidophorus sexlineatus*) in central Oklahoma. Herpetologica 35:216–222.

Bradke D., B. Kiel, and J. A. Moore. 2013. Population demography of Eastern Massasauga Rattlesnakes in relation to habitat management. Unpublished Report to the Pierce Cedar Creek Institute, Hastings, Michigan, USA.

Bradke D., J. Bartman, N. Kudla and, J. A. Moore. 2015. Abundance and density estimates of Eastern Massasaugas at Pierce Cedar Creek Institute. Unpublished Report to the Pierce Cedar Creek Institute, Hastings, Michigan, USA.

Bradley, J., and J. B. Lichtenhan. 1989. Venomous snake bites in Kansas. Kansas Medicine May:137–140.

Bradley, N. L., A. C. Leopold, J. Ross, and W. Huffaker. 1999. Phenological changes reflect climate change in Wisconsin. Proceedings of the National Academy of Sciences of the United States of America 96:9701–9704.

Brady, J. F., G. S. LeMasters, R. K. Williams, J. H. Pittman, J. P. Daubert, M. W. Cheung, D. H. Skinner, J. Turner, M. A. Rowland, J. Lange, and S. M. Sobek. 1995. Immunosassay analysis and gas chromatography confirmation of atrazine residues in water samples from a field study conducted in the state of Wisconsin. Journal of Agriculture and Food Chemistry 43:268–274.

Brady, S. P. 2012. Road to evolution? Local adaptation to road adjacency in an amphibian (*Ambystoma maculatum*). Scientific Reports 2:235.

Bragg, A. N. 1943. Observations on the ecology and natural history of Anura XV. The hylids and microhylids in Oklahoma. Great Basin Naturalist 4:62–80.

Bragg, A. N. 1960. Is *Heterodon* venomous? Herpetologica 16:121–123.

Bragg, T. B., and A. A. S. Teuter. 1996. Prairie ecology—the mixed prairie. Pages 53–64 *in* F. B. Samson and F. L. Knopf, editors. Prairie conservation: preserving North America's most endangered ecosystem. Island Press, Washington, D.C., USA.

Branch, L. C., and R. Altig. 1981. Nocturnal stratification of three species of *Ambystoma* larvae. Copeia 1981:870–873.

Brand, A. B., and J. W. Snodgrass. 2010. Value of artificial habitats for amphibian reproduction in altered landscapes. Conservation Biology 24:295–301.

Brand, A. B., J. W. Snodgrass, M. T. Gallagher, R. E. Casey, R. and A. B. Van Meter. 2010. Lethal and sublethal effects of embryonic and larval exposure of *Hyla versicolor* to stormwater pond sediments. Archives of Environmental Contamination and Toxicology 58:325–331.

Brandley, M. C., A. Schmitz, and T. W. Reeder. 2005. Partitioned Bayesian analyses, partition choice, and the phylogenetic relationships of scincid lizards. Systematic Biology 54:373–390.

Brandon, R. A. 1961. A comparison of the larva of five northeastern species of *Ambystoma* (Amphibia, Caudata). Copeia 1961:377–383.

Brandon, R. A. 1964. An annotated and illustrated key to multistage larvae of Ohio salamanders. Ohio Journal of Science 60:252–258.

Brandon, R. A., and D. J. Bremer. 1966. Neotenic newts, *Notophthalmus viridescens louisianensis,* in southern Illinois. Herpetologica 22:213–217.

Brandon, R. A., G. M. Labanick, and J. E. Huheey. 1979. Learned avoidance of brown efts, *Notophthalmus viridescens louisianensis* (Amphibia, Urodela, Salamandridae), by chickens. Journal of Herpetology 13:171–176.

Branin, M. L. 1935. Courtship activities and extra-seasonal ovulation in the Four-toed Salamander, *Hemidactylium scutatum* (Schlegel). Copeia 1935:172–175.

Brannon, M. P. 2006. *Bufo americanus* (Eastern American Toad). Leucism. Herpetological Review 37:333–334.

Branson, B. A., and E. C. Baker. 1974. An ecological study of the Queen Snake, *Regina septemvittata* (Say) in Kentucky. Tulane Studies in Zoology and Botany 18:153–171.

Brattstrom, B. H. 1963. A preliminary review of the thermal requirements of amphibians. Ecology 44:238–255.

Brattstrom, B. H. 1965. Body temperatures of reptiles. American Midland Naturalist 73:376–422.

Braun, A. P., and Q. E. Phelps. 2016. Habitat use by five turtle species in the Middle Mississippi River. Chelonian Conservation and Biology 15:62–68.

Brazner, J. C., N. P. Danz, G. J. Niemi, R. R. Regal, A. S. Trebitz, R. W. Howe, J. M. Hanowski, L. B. Johnson, J. J. H. Ciborowski, C. A. Johnston, E. D. Reavie, V. J. Brady, and G. V. Sagro. 2007. Evaluation of geographic, geomorphic and human influences on Great Lakes wetland indicators: a multi-assemblage approach. Ecological Indicators 7:610–635.

Brecke, B., and J. J. Moriarty. 1989. *Emydoidea blandingii* (Blanding's Turtle). Longevity. Herpetological Review 20:53.

Breckenridge, W. J. 1943. The life history of the Black-banded Skink *Eumeces septentrionalis septentrionalis* (Baird). American Midland Naturalist 29:591–606.

Breckenridge, W. J. 1944. Reptiles and amphibians of Minnesota. University of Minnesota Press, Minneapolis, USA.

Breder, R. B. 1927. The courtship of the Spotted Salamander. Bulletin of the New York Zoological Society 30:51–56.

Breitenbach, G. L. 1982. The frequency of communal nesting and solitary brooding in the salamander, *Hemidactylium scutatum*. Journal of Herpetology 16:341–346.

Breitenbach, L., and E. R. Wild. 2016a. *Diadophis punctatus* (Ring-necked Snake). Geographic distribution. Herpetological Review 47:262.

Breitenbach, L., and E. R. Wild. 2016b. *Lampropeltis triangulum* (Eastern Milknake). Geographic distribution. Herpetological Review 47:263.

Breitenbach, L. R. 2011. Sampling methods and habitat selection by the Western Slender Glass Lizard (*Ophisaurus attenuatus*) at Fort McCoy, Wisconsin. Thesis, University of Wisconsin–Stevens Point, USA.

Breitenbach, L. R., R. A. Paloski, and E. R. Wild. 2020. *Opheodrys vernalis* (Smooth Greensnake). Geographic distribution. Herpetological Review 51:782.

Brenes, R., M. J. Gray, T. B. Waltzek, R. P. Wilkes, and D. L. Miller. 2014. Transmission of ranavirus between ectothermic vertebrate hosts. PLoS ONE 9:e92476.

Brennan, J. M., and L. Tischendorf. 2004. Eastern Massasauga Rattlesnake *Sistrurus catenatus* population viability in the Ojibway Prairie Complex. Environment Canada, Windsor, Ontario, Canada.

Brenner, F. J. 1969. The role of temperature and fat deposition in hibernation and reproduction in two species of frogs. Herpetologica 25:105–113.

Brenowitz, E. A., W. Wilczynski, and H. H. Zakon. 1984. Acoustic communication in Spring Peepers: environmental and behavioral aspects. Journal of Comparative Physiology 155A:585–592.

Brewster, C. M. 1985. Wood Turtle, *Clemmys insculpta*, research in northern Wisconsin. Bulletin of the Chicago Herpetological Society 20:13–20.

Brewster, K. N. 1984. Reptilian ecology on a Wisconsin sand prairie. Bulletin of the Chicago Herpetological Society 19:115–124.

Brewster, K. N., and C. M. Brewster. 1986. *Clemmys insculpta* (Wood Turtle). Ectoparasitism. Herpetological Review 17:48.

Brewster, K. N., and C. M. Brewster. 1991. Movement and microhabitat use by juvenile Wood Turtles introduced into a riparian habitat. Journal of Herpetology 25:379–382.

Bridges, C. M., and R. D. Semlitsch. 2000. Variation in pesticide tolerance of tadpoles among and within species of Ranidae and patterns of amphibian decline. Conservation Biology 14:1490–1499.

Bridges, C. M., and R. D. Semlitsch. 2005. Xenobiotics. Pages 89–92 *in* M. Lannoo, editor. Amphibian declines: the conservation and status of United States species. University of California Press, Berkeley, USA.

Bridgwater, D. D., and D. F. Penny. 1966. Predation by *Citellus tridecemlineatus* on other vertebrates. Journal of Mammalogy 47:345–346.

Briggler, J. T., and W. E. Moser. 2008. *Necturus maculosus* (Red River Mudpuppy). Host. Herpetological Review 39:205.

Briggs, C. J., V. T. Vredenburg, R. A. Knapp, and L. J. Rachowicz. 2005. Investigating the population-level effects of chytridiomycosis: an emerging infectious disease of amphibians. Ecology 86:3149–3159.

Briggs, J., and H. Young. 1976. Amphibians and reptiles of the Pigeon Lake Region. Wisconsin Academy of Sciences, Arts and Letters 64:277–280.

Briggs, J. L. 1972. Tremblay's Salamander (*Ambystoma tremblayi*). Geographic Distribution. Herpetological Review 4:23.

Briggs, J. L. 1974. Distribution records of *Ambystoma laterale* and *A. tremblayi* in Wisconsin. Herpetological Review 5:18–19.

Briggs, J. L. 1977. Tremblay's Salamander (*Ambystoma tremblayi*). Geographic Distribution. Herpetological Review 8:13.

Briggs, J. M., A. K. Knapp, and B. L. Brock. 2002. Expansion of woody plants in tallgrass prairie: a fifteen year study of fire and fire-grazing interactions. American Midland Naturalist 147:287–294.

Britton, D. 2001. COSEWIC update status report on the Northern Cricket Frog *Acris crepitans* in Canada. Committee on the Status of Endangered Wildlife in Canada, Ontario, Canada. Archived by WebCite at http://www.webcitation.org/73bkV7bnk. Accessed 1 Nov 2018.

Brock, T. D., and K. M. Brock. 2004. Oak savanna restoration: a case study. Proceedings of the North American Prairie Conference 19:178–183.

Brock, T. L., and C. D. Camp. 2018. *Diadophis punctatus edwardsii* (Northern Ring-necked Snake). Envenomation. Herpetological Review 49:340–341.

Brockman-Mederas, K. M. 2016. *Sternotherus odoratus* (Eastern Musk Turtle). Geographic distribution. Herpetological Review 47:423.

Brodie, E. D., Jr. 1968. Investigation on the skin toxin of the Red-spotted Newt, *Notophthalmus viridescens viridescens*. American Midland Naturalist 80:276–280.

Brodie, E. D., Jr. 1971. Two more toxic salamanders, *Ambystoma maculatum* and *Cryptobranchus alleganiensis*. Herpetological Review 3:8.

Brodie, E. D., Jr. 1977. Salamander antipredator postures. Copeia 1977:523–535.

Brodie, E. D., Jr. 1983. Antipredator adaptations of salamanders: evolution and convergence among terrestrial species. Pages 109–133 *in* N. S. Margaris, M. Arianoutsou-Faraggitaki, and R. J. Reiter, editors. Plant, animal, and microbial adaptations to terrestrial environments. Plenum Publishing Corporation, New York, New York, USA.

Brodie, E. D., Jr., and D. R. Formanowicz. 1981a. Antipredator behavior of the treefrog *Hyla versicolor* to the shrew *Blarina brevicauda*. Journal of Herpetology 15:235–236.

Brodie, E. D., Jr., and D. R. Formanowicz, Jr. 1981b. Larvae of the Predaceous Diving Beetle *Dytiscus verticalis* acquire an avoidance response to skin secretions of the newt *Notophthalmus viridescens*. Herpetologica 37:172–176.

Brodie, E. D., Jr., and D. R. Formanowicz, Jr. 1983. Prey size preference of predators: differential vulnerability of larval anurans. Herpetologica 39:67–75.

Brodie, E. D., Jr., and D. R. Formanowicz, Jr. 1987. Antipredator mechanisms of larval anurans: protection of palatable individuals. Herpetologica 43:369–373.

Brodie, E. D., Jr., and E. D. Brodie, III. 1980. Differential avoidance of mimetic salamanders by free-ranging birds. Science 208:181–182.

Brodie, E. D., Jr., and R. R. Howard. 1972. Behavioral mimicry in the defensive displays of the urodele amphibians *Nophthalmus viridescens* and *Pseudotriton ruber*. BioScience 22:666–667.

Brodie, E. D., Jr., D. R. Formanowicz, Jr., and E. D. Brodie, III. 1978. The development of noxiousness of *Bufo americanus* tadpoles to aquatic insect predators. Herpetologica 34:302–306.

Brodie, E. D., Jr., J. A. Johnson, and C. K. Dodd, Jr. 1974. Immobility as a defensive behavior in salamanders. Herpetologica 30:79–85.

Brodie E. D., Jr., R. A. Nussbaum, and R. M. Storm. 1969. An egg-laying aggregation of five species of Oregon reptiles. Herpetologica 25:223–227.

Brodie, E. D., Jr., R. T. Nowak, and W. R. Harvey. 1979. The effectiveness of antipredator secretions and behavior of selected salamanders against shrews. Copeia 1979:270–274.

Brodie, E. D., III, and P. K. Ducey. 1989. Allocation of reproductive investments in the Redbelly Snake *Storeria occipitomaculata*. American Midland Naturalist 122:51–58.

Brodman, R. 1993. The effect of acidity on the interactions of *Ambystoma* salamander larvae. Journal of Freshwater Ecology 8:209–214.

Brodman, R. 1995. Annual variation in breeding success of two syntopic species of *Ambystoma* salamanders. Journal of Herpetology 29:111–113.

Brodman, R. 1996. Effects of intraguild interactions on fitness and microhabitat use of larval *Ambystoma* salamanders. Copeia 1996:372–378.

Brodman, R. 1999. Food and space dependent effects during the interactions of two species of larval salamanders. Journal of Freshwater Ecology 14:431–437.

Brodman, R. 2002. A 12 year study monitoring two populations of mole salamanders in northeast Ohio. Ohio Journal of Science 102:123–127.

Brodman, R. 2003. Amphibians and reptiles from 23 counties of Indiana. Proceedings of the Indiana Academy of Science 112:43–54.

Brodman, R. 2004. Intraguild predation on congeners affects size, aggression and survival among *Ambystoma* salamander larvae. Journal of Herpetology 38:21–26.

Brodman, R. 2005a. *Ambystoma jeffersonianum* (Green, 1827), Jefferson Salamander. Pages 611–613 *in* M. Lannoo, editor. Amphibian declines: the conservation status of United States species. University of California Press, Berkeley, USA.

Brodman, R. 2005b. *Ambystoma laterale*, Blue-spotted Salamander. Pages 614–616 *in* M. Lannoo, editor. Amphibian declines: the conservation and status of United States species. University of California Press, Berkeley, USA.

Brodman, R. 2009. A 14-year study of amphibian populations and metacommunities. Herpetological Conservation and Biology 4:106–119.

Brodman, R. 2010. The importance of natural history, landscape factors, and management practices, in conserving pond-breeding salamander diversity. Herpetological Conservation and Biology 5:501–514.

Brodman, R. 2013. Spotted Salamander, *Ambystoma maculatum* Shaw. Pages 113–127 *in* R. A. Pfingsten, J. G. Davis, T. O. Matson, G. Lipps Jr., D. Wynn, and B. J. Armitage, editors. Amphibians of Ohio. Ohio Biological Survey, Bulletin New Series Volume 17, Number 1, Columbus, Ohio, USA.

Brodman, R., and H. D. Krouse. 2007. How Blue-spotted and Small-mouthed Salamander larvae coexist with their unisexual counterparts. Herpetologica 63:135–143.

Brodman, R., and J. M. Jaskula. 2002. Activity and microhabitat use during interactions among five species of pond-breeding salamander larvae. Herpetologica 58:346–354.

Brodman, R., and M. Kilmurry. 1998. Status of amphibians in northwestern Indiana. Pages 125–136 *in* M. Lannoo, editor. Status and conservation of midwestern amphibians. University of Iowa Press, Iowa City, USA.

Brodman, R., and R. Dorton. 2006. The effectiveness of pond-breeding salamanders as agents of larval mosquito control. Journal of Freshwater Ecology 21:467–474.

Brodman, R., and R. A. Pfingsten. 2010. *Ambystoma tigrinum* (Tiger Salamander). Predation. Herpetological Review 41:186.

Brodman, R., M. Parrish, H. Kraus, and S. Cortwright. 2006. Amphibian biodiversity recovery in a large-scale ecosystem restoration. Herpetological Conservation and Biology 1:101–108.

Brodman, R., S. Cortwright, and A. Resetar. 2002. Historical changes of reptiles and amphibians of northwest Indiana fish and wildlife properties. American Midland Naturalist 147:135–144.

Brooker, R. W., F. T. Maestre, R. M. Callaway, C. L. Lortie, L. A. Cavieres, G. Kunstler, P. Liancourt, K. Tielborger, J. M. J. Travis, F. Anthelme, C. Armas, L. Coll, E. Corcket, S. Delzon, E. Forey, Z. Kikvidze, J. Olofsson, F. Pugnaire, C. L. Quiroz, P. Saccone, K. Schiffers, M. Seifan, B. Touzard, and R. Michalet. 2008. Facilitation in plant communities: the past, the present, and the future. Journal of Ecology 96:18–34.

Brooks, D. R. 1979. New records for amphibian and reptile trematodes. Proceedings of the Helminthological Society of Washington 46:286–289.

Brooks, R. J., C. M. Shilton, G. P. Brown, and N. W. S. Quinn. 1992. Body size, age distribution, and reproduction in a northern population of Wood Turtles (*Clemmys insculpta*). Canadian Journal of Zoology 70:462–469.

Brooks, R. J., G. P. Brown, and D. A. Galbraith. 1991. Effects of sudden increase in natural mortality of adults on a population of the Common Snapping Turtle (*Chelydra serpentina*). Canadian Journal of Zoology 69:1214–1320.

Brooks, R. T., and T. D. Kyker-Snowman. 2008. Forest floor temperature and relative humidity following timber harvesting in southern New England, USA. Forest Ecology and Management 254:65–73.

Brophy, T. E. 1980. Food habits of sympatric larval *Ambystoma tigrinum* and *Notophthalmus viridescens*. Journal of Herpetology 14:1–6.

Browder, L. W., J. C. Underhill, and D. J. Merrell. 1966. Mid-dorsal stripe in the Wood Frog. Journal of Heredity 57:65–67.

Brown, A. D., K. Temple-Miller, W. M. Roosenburg, and M. M. White. 2012a. Mitochondrial DNA variation in the Ouachita Map Turtle. Copeia 2012:301–306.

Brown, A. L., and J. A. Litvaitis. 1995. Habitat features associated with predation of New England Cottontails: what scale is appropriate? Canadian Journal of Zoology 73:1005–1011.

Brown, C. J., B. Blossey, J. C. Maerz, and S. J. Joule. 2006. Invasive plant and experimental venue affect tadpole performance. Biological Invasions 8:327–338.

Brown, C. W., and C. H. Ernst 1986. A study of variation in Eastern Timber Rattlesnakes, *Crotalus horridus* Linnae (Serpentes: Viperidae). Brimleyana 12:57–74.

Brown, D. J., A. Duarte, I. Mali, M. C. Jones, and M. R. J. Forstner. 2014. Potential impacts of a high severity wildfire on abundance, movement, and diversity of herpetofauna in the Lost Pines ecoregion of Texas. Herpetological Conservation and Biology 9:192–205.

Brown, D. J., G. M. Street, R. W. Nairn, and M. R. J. Forstner. 2012b. A place to call home: amphibian use of created and restored wetlands. International Journal of Ecology 2012:Article ID 989872.

Brown, D. J., M. C. Jones, J. Bell, and M. R. J. Forstner. 2012c. Feral hog damage to endangered Houston Toad habitat in the Lost Pines of Texas. Texas Journal of Science 64:73–88.

Brown, D. J., M. D. Nelson, D. J. Rugg, R. R. Buech, and D. M. Donner. 2016. Spatial and temporal habitat-use patterns of Wood Turtles at the western edge of their distribution. Journal of Herpetology 50:347–356.

Brown, D. J., M. M. Cochrane, and R. A. Moen. 2017. Survey and analysis design for Wood Turtle population monitoring. Journal of Wildlife Management 81:868–877.

Brown, D. J., M. Schrage, D. Ryan, R. A. Moen, M. D. Nelson, and R. R. Buech. 2015. *Glyptemys insculpta* (Wood Turtle). Longevity in the wild. Herpetological Review 46:243–244.

Brown, D. J., V. R. Farallo, J. R. Dixon, J. T. Baccus, T. R. Simpson, and M. R. J. Forstner. 2011a. Freshwater turtle conservation in Texas: harvest effects and efficacy of the current management regime. Journal of Wildlife Management 75:486–494.

Brown, E. E. 1956. Nests and young of the Six-lined Racerunner, *Cnemidophorus sexlineatus* Linnaeus. Journal of the Elisha Mitchell Scientific Society 72:30–40.

Brown, E. E. 1979. Stray food records from New York and Michigan snakes. American Midland Naturalist 102:200–203.

Brown, E. E. 1992a. Notes on amphibians and reptiles of the western Piedmont of North Carolina. Journal of the Elisha Mitchell Scientific Society 108:38–54.

Brown, G. P., and P. J. Weatherhead. 1997. Effects of reproduction on survival and growth of female Northern Water Snakes, *Nerodia sipedon*. Canadian Journal of Zoology 75:424–432.

Brown, G. P., and P. J. Weatherhead. 1999a. Demography and sexual size dimorphism in Northern Water Snakes, *Nerodia sipedon*. Canadian Journal of Zoology 77:1358–1366.

Brown, G. P., and P. J. Weatherhead. 1999b. Female distribution affects mate searching and sexual selection in male Northern Water Snakes (*Nerodia sipedon*). Behavioral Ecology and Sociobiology 47:9–16.

Brown, G. P., and P. J. Weatherhead. 1999c. Growth and sexual size dimorphism in Northern Water Snakes (*Nerodia sipedon*). Copeia 1999:723–732.

Brown, G. P., and P. J. Weatherhead. 2000. Thermal ecology and sexual size dimorphism in Northern Water Snakes, *Nerodia sipedon*. Ecological Monographs 70:311–330.

Brown, G. P., and P. J. Weatherhead. 2004. Sexual abstinence and the cost of reproduction in adult male Water Snakes, *Nerodia sipedon*. Oikos 104:269–276.

Brown, G. P., and R. J. Brooks. 1993. Sexual and seasonal differences in activity in a northern population of Snapping Turtles, *Chelydra serpentina*. Herpetologica 49:311–318.

Brown, G. P., and R. J. Brooks. 1994. Characteristics of and fidelity to hibernacula in a northern population of Snapping Turtles, *Chelydra serpentina*. Copeia 1994:222–226.

Brown, G. P., C. A. Bishop, and R. J. Brooks. 1994. Growth rate, reproductive output, and temperature selection of Snapping Turtles in habitats of different productivities. Journal of Herpetology 28:405–410.

Brown, G. P., R. J. Brooks, and J. A Layfield. 1990. Radiotelemetry of body temperatures of free-ranging Snapping Turtles (*Chelydra serpentina*) during summer. Canadian Journal of Zoology 68:1659–1663.

Brown, G. W., J. W. Dorrough, and D. S. L. Ramsay. 2011b. Landscape and local influences on patterns of reptile occurrence in grazed temperature woodlands of southern Australia. Landscape and Urban Planning 103:277–288.

Brown, J. R., and S. Archer. 1999. Shrub invasion of grassland: recruitment is continuous and not regulated by the herbaceous biomass or density. Ecology 80:2385–2396.

Brown, J. S., and H. T. Boschung, Jr. 1954. *Rana palustris* in Alabama. Copeia 1954:226.

Brown, K. D., P. L. Colbert, and F. J. Janzen. 2004. Survival and recruitment in a human-impacted population of Ornate Box Turtles, *Terrapene ornata*, with recommendations for conservation and management. Journal of Herpetology 38:562–568.

Brown, L. E., and C. A. Phillips. 2012. Distribution, habitat, and zoogeography of the semifossorial Red-bellied Snake *Storeria occipitomaculata* (Storer) in Illinois. Illinois Natural History Survey Bulletin 39:297–322.

Brown, L. E., and J. R. Brown. 1972. Mating calls and distributional records of treefrogs of the *Hyla versicolor* complex in Illinois. Journal of Herpetology 6:233–234.

Brown, L. E., and J. R. Brown. 1975. Evidence of climbing ability by Western Fox Snakes (*Elaphe vulpina vulpina*). Bulletin of the Maryland Herpetological Society 11:179.

Brown, L. E., and K. A. Brown. 1995. Fox Snake winter activity in central Illinois. Herpetological Review 26:134–135.

Brown, W. S. 1982. Overwintering body temperatures of Timber Rattlesnakes (*Crotalus horridus*) in northeastern New York. Journal of Herpetology 16:145–150.

Brown, W. S. 1992b. Emergence, ingress, and seasonal captures at dens of Northern Timber Rattlesnakes, *Crotalus horridus*. Herpetologica 47:101–115.

Brown, W. S. 1993. Biology, status, and management of the Timber Rattlesnake (*Crotalus horridus*): a guide for conservation. Society for the Study of Amphibians and Reptiles, Herpetological Circular 22, Oxford, Ohio, USA.

Brown, W. S. 2016. Lifetime reproduction in a northern metapopulation of Timber Rattlesnakes (*Crotalus horridus*). Herpetologica 72:331–342.

Brown, W. S., and F. M. MacLean. 1983. Conspecific scent-trailing by newborn Timber Rattlesnakes, *Crotalus horridus*. Herpetologica 39:430–436.

Brown, W. S., and W. S. Parker. 1976. Movement ecology of *Coluber constrictor* near communal hibernacula. Copeia 1976:225–242.

Brown, W. S., and W. S. Parker. 1982. Niche dimensions and resource partitioning in a Great Basin Desert snake community. Herpetological Communities. US Fish and Wildlife Service, Wildlife Research Report 13:59–81.

Brown, W. S., M. Kery, and J. E. Hines. 2007. Survival of Timber Rattlesnakes (*Crotalus horridus*) estimated by capture-recapture models in relation to age, sex, color morph, time, and birthplace. Copeia 2007:656–671.

Brown, W. S., W. S. Parker, and J. A. Elder. 1974. Thermal and spatial relationships of two species of colubrid snakes during hibernation. Herpetologica 30:32–38.

Browne, C. L., and S. J. Hecnar. 2007. Species loss and shifting population structure of freshwater turtles despite habitat protection. Biological Conservation 138:421–429.

Browne, C. L., C. A. Paszowski, A. L. Foote, A. Moenting, and S. M. Boss. 2009. The relationship of amphibian abundance to habitat features across spatial scales in the Boreal Plains. Ecoscience 16:209–223.

Bruce, R. C. 2008. Intraguild interactions and population regulation in plethodontid salamanders. Herpetological Monographs 2008:31–53.

Brudvig, L. A., and H. Asbjornsen. 2009. The removal of woody encroachment restores biophysical gradients in midwestern oak savannas. Journal of Applied Ecology 46:231–240.

Bruggers, R. L., and W. B. Jackson. 1974. Eye-lens weight of the Bullfrog (*Rana catesbeiana*) related to larval development, transformation, and age of adults. Ohio Journal of Science 74:282–286.

Bruneau, M., and E. Magnin. 1980a. Croissance, nutrition, et reproduction des ouaouarons *Rana catesbeiana* Shaw (Amphibia Anura) des Laurentides au nord du Montreal. Canadian Journal of Zoology 58:175–183.

Bruneau, M., and E. Magnin. 1980b. Vie larvaire des ouaouarons *Rana catesbeiana* Shaw (Amphibia Anura) des Laurentides, Quebec. Canadian Journal of Zoology 58:169–174.

Brunner, J. L., K. E. Barnett, S. A McNulty, M. J. Rubbo, and M. B. Kolozsvary. 2011. Ranavirus infection in die-offs of vernal pool amphibians in New York, USA. Herpetological Review 42:76–79.

Brunson, C., and L. Comber. 2015. An introduction to R for spatial analysis & mapping. SAGE Publications Inc., Thousand Oaks, California, USA.

Brust, T. J. 2013. Dietary Preference of the Queensnake (*Regina septemvittata*). Thesis, Marshall University, West Virginia, USA.

Buchanan, J. L., E. C. Pauly-Hubbard, R. A. Paloski, and A. F. Badje. 2022. *Graptemys ouachitensis* (Ouachita Map Turtle). Geographic distribution. Herpetological Review 53:76.

Buchanan, S. W., B. C. Timm, R. P. Cook, R. Couse, and L. C. Hazard. 2016. Surface activity and body temperature of Eastern Hognose Snakes (*Heterodon platirhinos*) at Cape Cod National Seashore, MA, USA. Journal of Herpetology 50:17–25.

Buckhardt, E. M., G. A. Glowacki, and J. P. Gibbs. 2020. Environmental cues that trigger nesting by Blanding's Turtles (*Emydoidea blandingii*). Chelonian Conservation and Biology 19:67–71.

Buckle, J. 1971. A recent introduction of frogs to Newfoundland. Canadian Field-Naturalist 85:72–74.

Budischak, S. A., L. K. Belden, and W. A. Hopkins. 2008. Effects of malathion on embryonic development and latent susceptibility to trematode parasites in ranid tadpoles. Environmental Toxicology and Chemistry 27:2496–2500.Budischak, S. A., L. K. Belden, and W. A. Hopkins. 2009. Relative toxicity of malathion to trematode-infected and noninfected *Rana palustris* tadpoles. Archives of Environmental Contamination and Toxicology 56:123–128.Buech, R. R. 1995. The Wood Turtle: its life history, status, and relationship with forest management. Pages 118–127 *in* Proceedings of the 1995 NCASI central-lake states regional meeting. National Council of the Paper Industry for Air and Stream Improvement, Inc., Research Triangle Park, North Carolina, USA.

Buech, R. R., L. G. Hanson, and M. D. Nelson. 1997. Identification of Wood Turtle nesting areas for protection and management. Pages 383–391 *in* J. Van Abbema, editor. Proceedings: conservation, restoration, and management of tortoises and turtles—an international conference. Joint Publication of the New York Turtle and Tortoise Society and the WCS Turtle Recovery Program, New York, USA.

Buech, R. R., M. D. Nelson, L. G. Hanson, and M. A. Ewert. 2004. *Clemmys insculpta* (Wood Turtle). Hatchling failure. Herpetological Review 35:54.

Buhlmann, K. A., and C. P. Osborn. 2011. Use of an artificial nesting mound by Wood Turtles (*Glyptemys insculpta*): a tool for turtle conservation. Northeastern Naturalist 18:315–334.

Buhlmann, K. A., and J. W. Gibbons. 2001. Terrestrial habitat use by aquatic turtles from a seasonally fluctuating wetland: implications for wetland conservation boundaries. Chelonian Conservation and Biology 4:115–127.

Bull, E. L., J. W. Deal, and J. E. Hohmann. 2001. Avian and amphibian use of fenced and unfenced stock ponds in northeastern Oregon forests. US Forest Service, Pacific Northwest Research Station Research Paper PNW-RP-539, Portland, Oregon, USA.

Bull, J. J., and R. C. Vogt. 1979. Temperature dependent sex determination in turtles. Science 206:1186–1188.

Bull, J. J., and R. C. Vogt. 1981. Temperature sensitive periods of sex determination in emydid turtles. Journal of Experimental Zoology 218:435–440.

Bull, J. J., J. M. Legler, and R. C. Vogt. 1985. Non-temperature dependent sex determination in two suborders of turtles. Copeia 1985:784–786.

Bull, J. J., R. C. Vogt, and C. J. McCoy. 1982a. Sex determining temperature in turtles: a geographic comparison. Evolution 36:326–332.

Bull, J. J., R. C. Vogt, and M. G. Bulmer. 1982b. Heritability of sex ratio in turtles with environmental sex determination: a geographic comparison. Evolution 36:333–341.

Bullock, T. H., D. A. Bodznick, and R. G. Northcutt. 1983. The phylogenetic distribution of electroreception: evidence for convergent evolution of a primative vertebrate sense modality. Brain Research Reviews 6:25–46.

Bulté, G., and G. Blouin-Demers. 2008. Northern Map Turtles (*Graptemys geographica*) derive energy from the pelagic pathway through predation on zebra mussels (*Dreissena polymorpha*). Freshwater Biology 53:497–508.

Bulté, G., and G. Blouin-Demers. 2009. Does sexual bimaturation affect the cost of growth and the operational sex ratio in an extremely size-dimorphic reptile? Ecoscience 16:175–182.

Bulté, G., and G. Blouin-Demers. 2010. Estimating the energetic significance of basking behavior in a temperate-zone turtle. Ecoscience 17:387–393.

Bulté, G., M. A. Carrière, and G. Blouin-Demers. 2010. Impact of recreational power boating on two populations of Northern Map Turtles (*Graptemys geographica*). Aquatic Conservation: Marine and Freshwater Ecosystems 20:31–38.

Bulté, G., M. A. Gravel, and G. Blouin-Demers. 2008. Intersexual niche divergence in Northern Map Turtles (*Graptemys geographica*): the roles of diet and habitat. Canadian Journal of Zoology 86:1235–1243.

Bulté, G., R. R. Germain, C. M. O'Conner, and G. Blouin-Demers. 2013. Sexual dichromatism in the Northern Map Turtle, *Graptemys geographica*. Chelonian Conservation and Biology 12:187–192.

Burbrink, F. T. 2001. Systematics of the Eastern Ratsnake complex (*Elaphe obsoleta*). Herpetological Monographs 15:1–53.

Burbrink, F. T., and B. I. Crother. 2011. Evolution and taxonomy of snakes. Pages 19–53 *in* R. D. Aldridge and D. M. Sever, editors. Reproductive biology and phylogeny of snakes. CRC Press, Boca Raton, Florida, USA.

Burbrink, F. T., C. A. Phillips, and E. J. Heske. 1998. A riparian zone in southern Illinois as a potential dispersal corridor for reptiles and amphibians. Biological Conservation 86:107–115.

Burbrink, F. T., F. Fontanella, R. A. Pyron, T. J. Guiher, and C. Jimenez. 2008. Phylogeography across a continent: the evolutionary and demographic history of the North American Racer (Serpentes: Colubridae: *Coluber constrictor*). Molecular Phylogenetics and Evolution 47:274–288.

Burdick, S. L., and D. L. Swanson. 2010. Status, distribution and microhabitats of Blanchard's Cricket Frog *Acris blanchardi* in South Dakota. Herpetological Conservation and Biology 5:9–16.

Burger, A. C., B. W. Marlowe, E. P. Tristano, G. R. Smith, and J. E. Rettig. 2017. Effect of recreational trail traffic level on Eastern Red-backed Salamander (*Plethodon cinereus*) relative abundance. Journal of North American Herpetology 2017:1–4.

Burger, J., and R. T. Zappalorti. 1988. Habitat use in free-ranging Pine Snakes, *Pituophis melanoleucus*, in New Jersey pine barrens. Herpetologica 44:48–55.

Burger, J., and S. D. Garber. 1995. Risk assessment, life history strategies, and turtles: could declines be prevented or predicted? Journal of Toxicology and Environmental Health 46:483–500.

Burger, J., C. Jeitner, H. Jensen, M. Fitzgerald, S. Carlucci, S. Shukla, S. Burke, R. Ramos, and M. Gochfeld. 2004. Habitat use in basking Northern Water (*Nerodia sipedon*) and Eastern Garter (*Thamnophis sirtalis*) Snakes in urban New Jersey. Urban Ecosystems 7:17–27.

Burgett, A. A., C. D. Wright, G. R. smith, D. T. Fortune, and S. L. Johnson. 2007. Impact of ammonium nitrate on Wood Frog (*Rana sylvatica*) tadpoles: effects on survivorship and behavior. Herpetological Conservation and Biology 2:29–34.

Burggren, W. 1985. Gas exchange, metabolism, and "ventilation" in gelatinous frog egg masses. Physiological Zoology 58:503–518.

Burghardt, G. 1964. Effects of prey size and movement on the feeding behavior of the lizards *Anolis carolinensis* and *Eumeces fasciatus*. Copeia 1964:576–578.

Burghardt, G. M. 1966. Stimulus control of the prey attack response in naïve garter snakes. Psychonomic Science 4:37–38.

Burghardt, G. M. 1967. Chemical-cue preferences of inexperienced snakes: comparative aspects. Science 157:718–721.

Burghardt, G. M. 1968. Chemical preference studies on newborn snakes of three sympatric species of *Natrix*. Copeia 1968:732–737.

Burghardt, G. M. 1969. Comparative prey-attack studies in newborn snakes of the genus *Thamnophis*. Behaviour 33:77–114.

Burghardt, G. M. 1970. Intraspecific geographical variation in chemical food cue preferences of newborn Garter Snakes (*Thamnophis sirtalis*). Behaviour 36:246–257.

Burghardt, G. M. 1983. Aggregation and species discrimination in newborn snakes. Zeitschrift Fur Tierpsychologie 61:89–101.

Burghardt, G. M. 1993. The comparative imperative: genetics and ontogeny of chemoreceptive prey responses in natricine snakes. Brain, Behavior and Evolution 41:138–146.

Burghardt, G. M., and C. H. Pruitt. 1975. Role of the tongue and senses in feeding of naive and experienced garter snakes. Physiology and Behavior 14:185–194.

Burghardt, G. M., and M. A. Krause. 1999. Plasticity of foraging behavior in Garter Snakes (*Thamnophis sirtalis*) reared on different diets. Journal of Comparative Psychology 113:277–285.

Burghardt, G. M., J. B. Murphy, D. Chiszar, and M. Hutchins. 2009. Combating ophiophobia: origins, treatment, education and conservation tools. Pages 262–280 *in* S. J. Mullin and R. A. Seigel, editors. Snakes: ecology and conservation. Cornell University Press, Ithaca, New York, USA.

Burghardt, G. M., J. S. Placyk, Jr., G. S. Casper, R. L. Small, and K. Taylor. 2006. Genetic structure of Great Lakes region *Thamnophis butleri* and *Thamnophis radix* based on mtDNA sequence data: conservation implications for Wisconsin Butler's Gartersnake. Unpublished Report to the Wisconsin Department of Natural Resources, Madison, USA.

Burke, R. L. 1991. Relocations, repatriations, and translocations of amphibians and reptiles: taking a broader view. Herpetologica 47:350–357.

Burkett, R. D. 1969. An ecological study of the Cricket Frog, *Acris crepitans*, in northeastern Kansas. Dissertation, University of Kansas, Lawrence, USA.

Burkett, R. D. 1984. An ecological study of the Cricket Frog, *Acris crepitans*. Pages 89–103 *in* S. Fitch, R. A. Siegel, L. E. Hunt, J. L. Knight, L. Malaret, and N. L. Zuschlag, editors. Vertebrate ecology and systematics: a tribute to Henry S. Fitch. University of Kansas, Museum of Natural History Special Publication 10, Lawrence, USA.

Burkett, R. D. 1991. Ecological comparisons of some closely related species of amphibians in Land Between the Lakes and the Midsouth. Journal of the Tennessee Academy of Science 66:161–164.

Burkholder, G. 1998. *Hyla chrysoscelis* (Gray Treefrog). Hibernacula. Herpetological Review 29:231.

Burley, L. A., A. T. Moyer, and J. W. Petranka. 2006. Density of an intraguild predator mediates feeding group size, intraguild egg predation, and intra- and interspecific competition. Oecologia 148:641–649.

Burmeister, S., J. Konieczka, and W. Wilczynski. 1999. Agonistic encounters in a Cricket Frog (*Acris crepitans*) chorus: behavioral outcomes vary with local competition and within the breeding season. Ethology 105:335–347.

Burns, G. 2020. *Ambystoma laterale* (Blue Spotted Salamander). Geographic distribution. Herpetological Review 51:66–67.

Burri, T. M. 1997. Food habits of Muskellunge in Wisconsin. Thesis, University of Wisconsin, Stevens Point, USA.

Burt, C. E. 1928. Insect food of Kansas lizards with notes on feeding habits. Journal of the Kansas Entomological Society 1:50–68.

Burt, C. E. 1935. Further records of the ecology and distribution of amphibians and reptiles in the middle west. American Midland Naturalist 16:311–336.

Burtelow, A. E., P. J. Bohlen, and P. M. Groffman. 1998. Influence of exotic earthworm invasion on soil organic matter, microbial biomass and denitrification potential in forest soils of the northeastern United States. Applied Soil Ecology 9:97–202.

Burton, E. C., M. J. Gray, A. C. Schmutzer, and D. L. Miller. 2009. Differential responses of postmetamorphic amphibians to cattle grazing in wetlands. Journal of Wildlife Management 73:269–277.

Burton, T. M. 1977. Population estimates, feeding habits and nutrient and energy relationships of *Notophthalmus v. viridescens*, in Mirror Lake, New Hampshire. Copeia 1977:139–143.

Burton, T. M., and G. E. Likens. 1975. Salamander populations and biomass in the Hubbard Brook Experimental Forest, New Hampshire. Copeia 1975:541–546.

Bury, R. B. 1979. Population ecology of freshwater turtles. Pages 571–602 *in* M. Harless and H. Morlock, editors. Turtles: perspectives and research. John Wiley and Sons, New York, New York, USA.

Bury, R. B., and J. A. Whelan. 1984. Ecology and management of the Bullfrog. US Fish and Wildlife Service Resource Publication 155, Washington, D.C., USA.

Bury, R. B., and R. A. Luckenbach. 1976. Introduced amphibians and reptiles in California. Biological Conservation 10:1–14.

Bury, R. B., A. V. Nebeker, and M. J. Adams. 2000. Response of hatchling and yearling turtles to thermal gradients: comparison of *Chelydra serpentina* and *Trachemys scripta*. Journal of Thermal Biology 25:221–225.

Bush, S. P., V. H. Wu, and S. W. Corbett. 2000. Rattlesnake venom-induced thrombocytopenia response to antivenin (Crotalidae) polyvalent: a case series. Academic Emergency Medicine 7:181–185.

Butler, J. A. 1978. A physiological analysis of the Queen Snake, *Regina septemvittata* (Say). Dissertation, Ohio State University, Columbus, USA.

Butler, J. A., T. W. Hull, and R. Franz. 1995. Neonate aggregations and maternal attendance of young in the Eastern Diamondback Rattlesnake, *Crotalus adamanteus*. Copeia 1995:196–198.

Butterfield, B. P., M. J. Lannoo, and P. Nanjappa. 2005. *Pseudacris crucifer*. Pages 472–474 *in* M. Lannoo, editor. Amphibian declines: the conservation status of United States species. University of California Press, Berkeley, USA.

Butterfield, C. W. 1880. Wisconsin antiquities. Pages 19–20 *in* History of Dane County, Wisconsin. Western Historical Company, Chicago, Illinois, USA.

Byer, N. W., B. N. Reid, R. P. Thiel, and M. Z. Peery. 2020. Strong climate associations but no temporal trends in nesting phenology of Blanding's Turtles (*Emydoidea blandingii*). Herpetologica 76:396–402.

Cabe, P. R., R. B. Page, T. J. Hanlon, M. E. Aldrich, L. Connors, and D. M. Marsh. 2007. Fine-scale population differentiation and gene flow in a terrestrial salamander (*Plethodon cinereus*) living in a continuous habitat. Heredity 98:53–60.

Cáceres-Charneco, R. I., and T. S. Ransom. 2010. The influence of habitat provisioning: use of earthworm burrows by the terrestrial salamander, *Pelthodon cinereus*. Population Ecology 52:517–526.

Caetano, M. H., and R. Leclair, Jr. 1996. Growth and population structure of Red-spotted Newts (*Notophthalmus viridescens*) in permanent lakes of the Laurentian Shield, Quebec. Copeia 1996:866–874.

Cagle, F. R. 1937. Egg laying habits of the Slider Turtle (*Pseudemys troostii*), the Painted Turtle (*Chrysemys picta*) and the Musk Turtle (*Sternotherus odoratus*). Journal of the Tennessee Academy of Science 12:87–95.

Cagle, F. R. 1940. Eggs and natural nests of *Eumeces fasciatus*. American Midland Naturalist 23:227–233.

Cagle, F. R. 1942. Turtle populations in southern Illinois. Copeia 1942:155–162.

Cagle, F. R. 1944. Home range, homing behavior, and migration in turtles. University of Michigan, Miscellaneous Publications of the Museum of Zoology 61:1–34.

Cagle, F. R. 1945. Recovery from serious injury in the Painted Turtle. Copeia 1945:45.

Cagle, F. R. 1953a. The status of the turtle *Graptemys oculifera* (Baur). Zoologica 38:137–144.

Cagle, F. R. 1953b. Two new subspecies of *Graptemys pseudogeographica*. University of Michigan, Occasional Papers of the Museum of Zoology 546:1–17.

Cagle, F. R. 1954. Observations on the life history of the salamander, *Necturus louisianensis*. Copeia 1954:257–260.

Cagle, N. L. 2008. Snake species distributions and temperate grasslands: a case study from the American tallgrass prairie. Biological Conservation 141:744–755.

Cahn, A. R. 1926. A set of albino frog eggs. Copeia 1926:107–109.

Cahn, A. R. 1929. The herpetology of Waukesha County, Wisconsin. Copeia 1929:4–8.

Cahn, A. R. 1937. The turtles of Illinois. Illinois Biological Monographs 16:1–218.

Cahn, A. R., and W. Shumway. 1926. Color variations in larvae of *Necturus maculosus*. Copeia 1926:106–107.

Cahoe, R. D., and N. H. Troelstrup, Jr. 2004. Habitat use and population biology of the Northern Redbelly Snake at Oaklake Field Station, SD.

Proceedings of the South Dakota Academy of Science 83:137–147.

Caldwell, J. P. 1982. Disruptive selection: a tail color polymorphism in *Acris* tadpoles in response to differential predation. Canadian Journal of Zoology 60:2818–2827.

Caldwell, R. S. 1975. Observations on the winter activity of the Red-backed Salamander, *Plethodon cinereus*, in Indiana. Herpetologica 1975:21–22.

Calfee, R. D., C. M. Bridges, and E. E. Little. 2006. Sensitivity of two salamander (*Ambystoma*) species to ultraviolet radiation. Journal of Herpetology 40:35–42.

Calhoun, A. J. K., J. Arrigoni, R. P. Brooks, M. L. Hunter, and S. C. Richter. 2014. Creating successful vernal pools: a literature review and advice for practitioners. Wetlands 34:1027–1038.

Calhoun, A. J. K., N. A. Miller, and M. W. Klemens. 2005. Conserving pool-breeding amphibians in human-dominated landscapes through local implementation of best development practices. Wetlands Ecology and Management 13:291–304.

Camp, R. F. 1970. Rattlesnake hunt, Wisconsin style. Wisconsin Conservation Bulletin 35:21–23.

Campbell, K. R., and T. S. Campbell. 2000. Lizard contaminant data for ecological risk assessment. Reviews of Environmental Contamination and Toxicology 165:39–116.

Campbell, K. R., and T. S. Campbell. 2001. The accumulation and effects of environmental contaminants on snakes: a review. Environmental Monitoring and Assessment 70:253–301.

Campbell, K. R., T. S. Campbell, and J. Burger. 2005. Heavy metal concentrations in Northern Water Snakes (*Nerodia sipedon*) from East Fork Popular Creek and the Little River, East Tennessee, USA. Archives of Environmental Contamination and Toxicology 49:239–248.

Camper, J. D. 1986. *Ambystoma tigrinum tigrinum* (Eastern Tiger Salamander). Feeding. Herpetological Review 17:19.

Canfield, W. H. 1861. Outline sketches of Sauk County; including its history, from the first marks of man's hand to 1861, and its topography, both written and illustrated. Republic Office, Baraboo, Wisconsin, USA.

Cannatella, D. 2008. Anura. Version 11 January 2008. The Tree of Life Web Project. http://tolweb.org/Anura/16963. Accessed 9 Dec 2018.

Cannizzaro, J. S., R. LaFleur, and D. J. Watermolen. 2021. *Necturus maculosus* (Mudpuppy). Geographic distribution. Herpetological Review 52:571.

Cannizzaro IV, J. S., and M. Schmanski. 2022. *Hemidactylium scutatum* (Four-toed Salamander). Geographic Distribution. Herpetological Review 53:68.

Carfagno, G. L. F., and P. J. Weatherhead. 2006. Intraspecific and interspecific variation in use of forest-edge habitat by snakes. Canadian Journal of Zoology 84:1440–1452.

Carfagno, G. L. F., and P. J. Weatherhead. 2008. Energetics and space use: intraspecific and interspecific comparisons of movements and home ranges of two colubrid snakes. Journal of Animal Ecology 77:416–424.

Carfagno, G. L. F., and P. P. Fong. 2014. Growth inhibition of tadpoles exposed to sertraline in the presence of conspecifics. Journal of Herpetology 48:571–576.

Carfagno, G. L. F., E. J. Heske, and P. J. Weatherhead. 2006. Does mammalian prey abundance explain forest-edge use by snakes? Ecoscience 13:293–297.

Carfagno, G. L. F., J. M. Carithers, L. J. Mycoff, and R. M. Lehtinen. 2011. How the Cricket Frog lost its spot: the inducible defense hypothesis. Herpetologica 67:386–396.

Cargo, D. G. 1960. Predation of eggs of the Spotted Salamander, *Ambystoma maculatum* by the leech, *Macrobdella decora*. Chesapeake Science 1:119–120.

Carlisle, M., R. Sheffer, K. Morgan, D. Wahlstrom, and J. M. Kapfer. 2008. *Pantherophis vulpina vulpina* (Western Fox Snake). Mortality. Herpetological Review 39:98–99.

Carlson, A. M., and D. L. Gorchov. 2004. Effects of herbicide on the invasive biennial *Alliaria petiolata* (Garlic Mustard) and initial responses of native plants in a southwestern Ohio forest. Restoration Ecology 12:559–567.

Caroffino, D. C., T. M. Sutton, R. E. Elliott, and M. C. Donofrio. 2010. Predation on early life stages of Lake Sturgeon in the Peshtigo River, Wisconsin. Transactions of the American Fisheries Society 139:1846–1856.

Carpenter, C. C. 1952a. Comparative ecology of the Common Garter Snake (*Thamnophis s. sirtalis*), the Ribbon Snake (*Thamnophis s. sauritus*), and Butler's Garter Snake (*Thamnophis butleri*) in mixed populations. Ecological Monographs 22:235–258.

Carpenter, C. C. 1952b. Growth and maturity of the three species of *Thamnophis* in Michigan. Copeia 1952:237–243.

Carpenter, C. C. 1953. A study of hibernacula and hibernating associations of snakes and amphibians in Michigan. Ecology 34:74–80.

Carpenter C. C. 1955. Aposematic behavior in the salamander *Ambystoma tigrinum melanostictum*. Copeia 1955:311.

Carpenter, C. C. 1956. Body temperatures of three species of *Thamnophis*. Ecology 37:732–735.

Carpenter, C. C. 1959. A population of the Six-lined Racerunner (*Cnemidophorus sexlineatus*). Herpetologica 15:81–86.

Carpenter, C. C. 1962. Patterns of behavior of two Oklahoma lizards. American Midland Naturalist 67:132–151.

Carpenter, C. C. 1982. The Bullsnake as an excavator. Journal of Herpetology 16:394–401.

Carpenter, C. C., and D. E. Delzell. 1951. Road records as indicators of differential spring migrations of amphibians. Herpetologica 7:63–64.

Carpenter, H. L., and E. O., Morrison. 1973. Feeding behavior of the Bullfrog, *Rana catesbeiana*, in northcentral Texas. Bios 44:188–193.

Carr, A. F. 1952. Handbook of turtles: the turtles of the United States, Canada and Baja California. Cornell University Press, Ithaca, New York, USA.

Carr, C. F. 1890. A list of turtles observed in the vicinity of Madison, Wis. Wisconsin Naturalist 1:11–12.

Carr, L. W., and L. Fahrig. 2001. Effect of road traffic on two amphibian species of differing vagility. Conservation Biology 15:1071–1078.

Carreno, C. A., and R. N. Harris. 1998. Lack of nest defense behavior and attendance patterns in a joint nesting salamander, *Hemidactylium scutatum* (Caudata: Plethodontidae). Copeia 1998:183–189.

Carrière, M-A., and G. Blouin-Demers. 2010. Habitat selection at multiple spatial scales in Northern Map Turtles (*Graptemys geographica*). Canadian Journal of Zoology 88:846–854.

Carrière, M-A., N. Rollinson, A. N. Suley, and R. J. Brooks. 2008. Thermoregulation when the growing season is short: sex-biased basking patterns in a northern population of Painted Turtles (*Chrysemys picta*). Journal of Herpetology 42:206–209.

Carroll, D. M., and G. R. Ultsch. 2006. *Glyptemys insculpta* (Wood Turtle). Predation. Herpetological Review 37:215–216.

Carroll, D. M., and G. R. Ultsch. 2007. Emergence season and survival in the nest of hatchling turtles in southcentral New Hampshire. Northeastern Naturalist 14:307–310.

Carroll, T. E., and D. W. Ehrenfeld. 1978. Intermediate-range homing in the Wood Turtle, *Clemmys insculpta*. Copeia 1978:117–126.

Caruso, N. M., M. W. Sears, D. C. Adams, and K. R. Lips. 2014. Widespread rapid reductions in body size of adult salamanders in response to climate change. Global Change Biology 20:1751–1759.

Carver, J. 1781. Travels through the interior parts of North America in the years 1766, 1767, and 1768. Third edition. C. Dilly, H. Payne, and J. Phillips, London, England.

Cary Coyle, T. L., and W. H. Karasov. 2010. Chronic, dietary polybrominated diphenyl ether exposure affects survival, growth, and development of *Rana pipiens* tadpoles. Environmental Toxicology and Chemistry 29:133–141.

Casebere, L. A., and M. J. Lodato. 2011. The Four-toed Salamander (*Hemidactylium scutatum*) in Indiana: past and present. Proceedings of the Indiana Academy of Science 119:111–129.

Cash, M. N., and J. P. Bogart. 1978. Cytological differentiation of the diploid-tetraploid species pair of North American treefrogs (Amphibia, Anura, Hylidae). Journal of Herpetology 12:555–558.

Casper, G. S. 1983. Wisconsin rattlesnake bounty barely defeated. Minnesota Herpetological Society Newsletter 3:11–12.

Casper, G. S. 1989. Concern over publishing locality data. Bulletin of the Chicago Herpetological Society 24:178.

Casper, G. S. 1993a. A preliminary summary of available records for the Four-toed Salamander, Hemidactylium scutatum, in Wisconsin. Bulletin of the Chicago Herpetological Society 28:1–3.

Casper, G. S. 1993b. Massasauga Rattlesnake records collected by the Wisconsin Herpetological Atlas Project, 1986–1991. Pages 80–85 *in* B. Johnson and V. Menzies, editors. Proceedings of the International Symposium and Workshop on the

Conservation of the Eastern Massasauga Rattlesnake *Sistrurus catenatus catenatus*. Metro Toronto Zoo, Ontario, Canada.

Casper, G. S. 1993c. The reptiles and amphibians of the Baraboo Hills, Sauk County, Wisconsin: historical review, inventory and management recommendations. Unpublished Report to The Nature Conservancy, Wisconsin Chapter, Madison, USA.

Casper, G. S. 1996. Geographic distributions of the amphibians and reptiles of Wisconsin: an interim report of the Wisconsin Herpetological Atlas Project. Milwaukee Public Museum, Milwaukee, Wisconsin, USA.

Casper, G. S. 1996a. *Ambystoma maculatum* (Spotted Salamander). Geographic distribution. Herpetological Review 27:208.

Casper, G. S. 1996b. *Hemidactylium scutatum* (Four-toed Salamander). Geographic distribution. Herpetological Review 27:208.

Casper, G. S. 1996c. *Opheodrys vernalis* (Smooth Green Snake). Geographic distribution. Herpetological Review 27:214.

Casper, G. S. 1997a. *Ambystoma laterale* (Blue-spotted Salamander). Geographic distribution. Herpetological Review 28:47.

Casper, G. S. 1997b. *Graptemys geographica* (Common Map Turtle). Geographic distribution. Herpetological Review 28:49.

Casper, G. S. 1997c. *Pseudacris triseriata triseriata* (Western Chorus Frog). Geographic distribution. Herpetological Review 28:48–49.

Casper, G. S. 1997d. Recent amphibian and reptile distribution records for Wisconsin. Herpetological Review 28:214–216.

Casper, G. S. 1997e. *Sternotherus odoratus* (Common Musk Turtle). Geographic distribution. Herpetological Review 28:209.

Casper, G. S. 1998. Review of the status of Wisconsin amphibians. Pages 199–205 *in* M. J. Lannoo, editor. The status and conservation of Midwest amphibians. Iowa University Press, Iowa City, USA.

Casper, G. S. 1999a. *Ambystoma tigrinum* (Eastern Tiger Salamander). Geographic distribution. Herpetological Review 30:49.

Casper, G. S. 1999b. New distribution records from the Wisconsin Herpetology Atlas Project, 1998. Herpetological Review 30:181–182.

Casper, G. S. 2001a. *Apalone spinifera spinifera* (Eastern Spiny Softshell). Geographic distribution. Herpetological Review 32:191.

Casper, G. S. 2001b. *Notophthalmus viridescens louisianensis* (Central Newt). Geographic distribution. Herpetological Review 32:189.

Casper, G. S. 2001c. *Pseudacris crucifer crucifer* (Northern Spring Peeper). Geographic distribution. Herpetological Review 32:190.

Casper, G. S. 2001d. *Storeria occipitomaculata occipitomaculata* (Northern Red-bellied Snake). Geographic distribution. Herpetological Review 32:196.

Casper, G. S. 2002. A review of the amphibians and reptiles of the Lake Superior watershed. Technical Report provided to the Terrestrial Wildlife Community Committee, Lake Superior Lakewide Management Plan. Milwaukee Public Museum, Wisconsin, USA.

Casper, G. S. 2003. Analysis of amphibian and reptile distributions using presence-only data. Dissertation, University of Wisconsin, Milwaukee, USA.

Casper, G. S. 2005. *Rana septentrionalis*. Pages 581–584 *in* M. Lannoo, editor. Amphibian declines: the conservation status of United States species. University of California Press, Berkeley, USA.

Casper, G. S. 2007. On the herpetofaunal history of the University of Wisconsin–Milwaukee Field Station (Ozaukee County, Wisconsin). Bulletin of the Chicago Herpetological Society 42:21–30.

Casper, G. S. 2008a. An amphibian and reptile inventory of Isle Royale National Park. National Park Service, Natural Resource Technical Report NPS/GLKN/NRTR—2008/146, Fort Collins, Colorado, USA.

Casper, G. S. 2008b. An annotated checklist of amphibians and reptiles of Wisconsin. Bulletin of the Chicago Herpetological Society 43:73–79.

Casper, G. S. 2008c. Changes in amphibian and reptile communities. Pages 287–300 *in* D. M. Waller and T. P. Rooney, editors. The vanishing present: Wisconsin's changing lands, waters, and wildlife. University of Chicago Press, Chicago, Illinois, USA.

Casper, G. S. 2008d. Morphology of gartersnake populations in the Wisconsin zone of overlap between *Thamnophis butleri* and *Thamnophis radix*. Final Report to the Wisconsin Department of Natural Resources, Madison, USA.

Casper, G. S. 2015. New county distribution records for amphibians and reptiles in Wisconsin. Herpetological Review 46:582–586.

Casper, G. S. 2016a. *Hyla versicolor* (Gray Treefrog). Geographic distribution. Herpetological Review 47:419.

Casper, G. S. 2016b. *Necturus maculosus* (Mudpuppy). Geographic distribution. Herpetological Review 47:417.

Casper, G. S. 2018. *Lithobates catesbeianus* (American Bullfrog). Geographic distribution. Herpetological Review 49:282.

Casper, G. S., and J. Burnham. 2001a. *Diadophis punctatus* (Ringneck Snake). Geographic distribution. Herpetological Review 32:194.

Casper, G. S., and J. Burnham. 2001b. *Elaphe vulpina* (Western Fox Snake). Geographic distribution. Herpetological Review 32:195.

Casper, G. S., and J. Burnham. 2001c. *Eumeces fasciatus* (Five-lined Skink). Geographic distribution. Herpetological Review 32:193.

Casper, G. S., and K. Kirk. 2001. *Hemidactylium scutatum* (Four-toed Salamander). Geographic distribution. Herpetological Review 32:189.

Casper, G. S., and M. Schmidt. 2016. *Lithobates catesbeianus* (American Bullfrog). Geographic distribution. Herpetological Review 47:420.

Casper, G. S., and P. Kleinhenz. 2016. *Tropidoclonion lineatum* (Lined Snake). Geographic distribution. Herpetological Review 47:429.

Casper, G. S., and R. D. Rutherford. 2016. *Opheodrys vernalis* (Smooth Greensnake). Geographic distribution. Herpetological Review 47:428.

Casper, G. S., and R. D. Rutherford. 2018a. *Ambystoma maculatum* (Spotted Salamander). Geographic distribution. Herpetological Review 49:280.

Casper, G. S., and R. D. Rutherford. 2018b. *Hemidactylium scutatum* (Four-toed Salamander). Geographic distribution. Herpetological Review 49:280.

Casper, G. S., and R. D. Rutherford. 2018c. *Hyla versicolor* (Gray Treefrog). Geographic distribution. Herpetological Review 49:282.

Casper, G. S., and S. Ames. 1998. *Heterodon platirhinos* (Eastern Hognose Snake). Geographic distribution. Herpetological Review 29:53.

Casper, G. S., and T. G. Anton. 2012. *Trachemys scripta elegans* (Red-eared Slider). Geographic distribution. Herpetological Review 43:100.

Casper, G. S., and T. G. Anton. 2013. Current scientific and standard common names of Wisconsin amphibians and Reptiles. Wisconsin Department of Natural Resources, Bureau of Science Services PUB-SS-1121 2013, Madison, USA. Archived by WebCite at http://www.webcitation.org/73dq0zqoF. Accessed 3 Nov 2018.

Casper, G. S., J. B. LeClere, and J. C. Gillingham. 2015a. *Thamnophis sirtalis* (Common Gartersnake). Diet/scavenging. Herpetological Review 46:653–654.

Casper, G. S., J. M. Kapfer, and T. Muehlfeld. 2007. *Emydoidea blandingii* (Blanding's Turtle). Geographic distribution. Herpetological Review 38:99–100.

Casper, G. S., R. D. Rutherford, and T. G. Anton. 2015b. Baseline distribution records for amphibians and reptiles in the Upper Peninsula of Michigan. Herpetological Review 46:391–406.

Casper, G. S., W. A. Smith, and R. A. Staffen. 2011. Queensnakes on the edge. Final report to the Wisconsin Department of Natural Resources, Madison, USA.

Castellano, C. M., and J. L. Behler. 2003. *Glyptemys insculpta* (North American Wood Turtle). Diet. Herpetological Review 34:362.

Castellano, C. M., J. L. Behler, and G. R. Ultsch. 2008. Terrestrial movements of hatchling Wood Turtles (*Glyptemys insculpta*) in agricultural fields in New Jersey. Chelonian Conservation and Biology 7:113–118.

Caster, P. T., G. A. Heidt, and K. D. Stone. 1994. Faunal use of nest boxes in the Ouachita Mountains of central Arkansas. Southwestern Naturalist 39:380–382.

Cathro, A., and E. Lindquist. 2016. *Diadophis punctatus edwardsii* (Northern Ring-necked Snake). Diet. Herpetological Review 47:681.

Catling, P. M., and B. Freedman. 1980a. Food and feeding behavior of sympatric snakes at Amherstburg, Ontario. Canadian Field-Naturalist 94:28–33.

Catling, P. M., and B. Freedman. 1980b. Variation in distribution and abundance of four sympatric species of snakes at Amherstburg, Ontario. Canadian Field-Naturalist 94:19–27.

Catling, P. M., and W. Freedman. 1977. Melanistic Butler's Garter Snakes (*Thamnophis butleri*) at

Amherstburg, Ontario. Canadian Field-Naturalist 91:397–399.

Cavitt, J. F. 2000. Fire and a tallgrass prairie reptile community: effects on relative abundance and seasonal activity. Journal of Herpetology 34:12–20.

Ceballos, C. P., and L. A. Fitzgerald. 2004. The trade in native and exotic turtles in Texas. Wildlife Society Bulletin 32:881–891.

Cecil, S. G., and J. J. Just. 1979. Survival rate, population density and development of a naturally occurring anuran larvae (*Rana catesbeiana*). Copeia 1979:447–453.

Ceriaco, L. M. P. 2012. Human attitudes towards herpetofauna: the influence of folklore and negative values on the conservation of amphibians and reptiles in Portugal. Journal of Ethnobiology and Ethnomedicine 8:8.

Chadwick, C. S. 1944. Observations on the life cycle of the Common Newt in western North Carolina. American Midland Naturalist 32:491–494.

Chalmers, R. J., and C. S. Loftin. 2006. Wetland and microhabitat use by nesting Four-toed Salamanders in Maine. Journal of Herpetology 40:478–485.

Chambers, D. L. 2008. Logging road effects on breeding-site selection in *Notophthalmus viridescens* (Red-spotted Newt) and three ambystomatid salamanders in south-central Pennsylvania. Northeastern Naturalist 15:123–130.

Chance, K. B. 2002. A telemetric study of winter habitat selection by the American Bullfrog, *Rana catesbeiana*, in east-central Kansas. Thesis, Emporia State University, Kansas, USA.

Charland, M. B. 1995. Thermal consequences of reptilian viviparity: thermoregulation in gravid and nongravid garter snakes (*Thamnophis*). Journal of Herpetology 29:383–390.

Charland, M. B., and P. T. Gregory. 1995. Movements and habitat use in gravid and nongravid female garter snakes. (Colubridae: *Thamnophis*). Journal of Zoology 236:543–561.

Charlesworth, B. 1980. Evolution in age-structured populations. Cambridge University Press, New York, New York, USA.

Charney, N. D., J. J. Castorino, M. J. Dobro, and S. L. Steely. 2014. Embryo development inside female salamander (*Ambystoma jeffersonianum-laterale*) prior to egg laying. PLoS ONE 9:e91919.

Charnov, E. L. 1990. On evolution of age at maturity and adult lifespan. Journal of Evolutionary Biology 3:139–144.

Chase, J. M., and M. A. Leibold. 2003. Ecological niches: linking classical and contemporary approaches. University of Chicago Press, Chicago, Illinois, USA.

Chatfield, M. W. H., P. Moler, and C. L. Richards-Zawacki. 2012. The amphibian chytrid fungus, *Batrachochytrium dendrobatidis*, in fully aquatic salamanders from southeastern North America. PLoS ONE 7:e44821.

Chellman, I. C. 2011. Population demographics and genetic diversity of Mudpuppies (*Necturus maculosus*) in Vermont. Thesis, University of Vermont, Berlington, USA.

Chellman, I. C., D. L. Parrish, and T. M. Donovan. 2017. Estimating Mudpuppy (*Necturus maculosus*) abundance in the Lamoille River, Vermont, USA. Herpetological Conservation and Biology 12:422–434.

Chen, T-H., J. A. Gross, and W. H. Karasov. 2006. Sublethal effects of lead on Northern Leopard Frog (*Rana pipiens*) tadpoles. Environmental Toxicology and Chemistry 25:1383–1389.

Chen, T-H., J. A. Gross, and W. H. Karasov. 2007. Adverse effects of chronic copper exposure in larval Northern Leopard Frogs (*Rana pipiens*). Environmental Toxicology and Chemistry 26:1470–1475.

Chen, T-H., J. A. Gross, and W. H. Karasov. 2009. Chronic exposure to pentavalent arsenic of larval Leopard Frogs (*Rana pipiens*): bioaccumulation and reduced swimming performance. Ecotoxicology 18:587–593.

Chin, S. Y., J. D. Willson, D. A. Cristol, D. V. V. Drewett, and W. A. Hopkins. 2013a. Altered behavior of neonatal Northern Watersnakes (*Nerodia sipedon*) exposed to maternally transferred mercury. Environmental Pollution 176:144–150.

Chin, S. Y., J. D. Willson, D. A. Cristol, D. V. V. Drewett, and W. A. Hopkins. 2013b. High levels of maternally transferred mercury do not affect reproductive output or embryonic survival of Northern Watersnakes (*Nerodia sipedon*). Environmental Toxicology and Chemistry 32:619–626.

Chiszar, D., C. W. Radcliffe, and K. Scudder. 1980. Use of the vomeronasal system during predator episodes by Bullsnakes (*Pituophis melanoleucus*). Bulletin of the Psychonomic Society 15:35–36.

Chiucchi, J. E., and H. L. Gibbs. 2010. Similarity of contemporary and historical gene flow among highly fragmented populations of an endangered rattlesnake. Molecular Ecology 19:5345–5358.

Chivers, D. P., and R. S. Mirza. 2001. Importance of predator diet cues in responses of larval Wood Frogs to fish and invertebrate predators. Journal of Chemical Ecology 27:45–51.

Chouinard, A. J. 2012. Rapid onset of mate quality assessment via chemical signals in a woodland salamander (*Plethodon cinereus*). Behavioral Ecology and Sociobiology 66:765–775.

Christiansen, J. L. 1981. Population trends among Iowa's amphibians and reptiles. Proceedings of the Iowa Academy of Science 88:24–27.

Christein, D., and D. H. Taylor. 1978. Population dynamics in breeding aggregations of the American Toad, *Bufo americanus* (Amphibian, Anura, Bufonidae). Journal of Herpetology 12:17–24.

Christen, M. S., A. D. Gendron, P. Brousseau, L. Menard, D. J. Marcogliese, D. Cyer, S. Ruby, and M. Fournier. 2003. Effects of agricultural pesticides on the immune system of *Rana pipiens* and on its resistance to parasitic infection. Environmental Toxicology and Chemistry 22:1127–1133.

Christens, E., and J. R. Bider. 1987. Nesting activity and hatching success of the Painted Turtle (*Chrysemys picta marginata*) in southwestern Quebec. Herpetologica 43:55–65.

Christenson, T. A., M. E. Horton, B. C. Jackson, G. R. Smith, and J. E. Rettig. 2014. Effects of Cutrine-Plus algaecide and predators on Wood Frog (*Lithobates sylvaticus*) tadpole survival and growth. Environmental Science and Pollution Research 21:12472–12478.

Christiansen, J., and J. Fieselmann. 1993. Massasauga rattlesnake bites in Iowa. *Iowa Medicine* 5:187–191.

Christiansen, J. L. 1983. Temporal separation of sympatric hognose snake species (*Heterodon*) in Iowa. Proceedings of the Iowa Academy of Science 90:Article 6.

Christiansen, J. L., and R. R. Burken. 1979. Growth and maturity of the Snapping Turtle (*Chelydra serpentina*) in Iowa. Herpetologica 35:261–266.

Christman, S. P. 1982. *Storeria dekayi*. Catalogue of American Amphibians and Reptiles 306.1–306.4.

Christoffel, R., and R. Hay. 1995a. 1994 census of Blanchard's Cricket Frog (*Acris crepitans blanchardi*) in southwestern Wisconsin. Wisconsin Department of Natural Resources, Endangered Resources Report 110, Madison, USA.

Christoffel, R., and R. Hay. 1995b. The effectiveness of placing turtle mortality barriers and crossing signs along highways to reduce turtle mortality. Wisconsin Department of Natural Resources, Endangered Resources Report 107, Madison, USA.

Christoffel, R., R. Hay, and L. Ramirez. 2000. Snakes of Wisconsin. Wisconsin Department of Natural Resources, Bureau of Endangered Resources Pub-ER-100 00, Madison, USA.

Christoffel, R., R. Hay, and M. Monroe. 2002. Turtles & lizards of Wisconsin. Wisconsin Department of Natural Resources, Bureau of Endangered Resources PUB-ER-104, Madison, USA.

Christoffel, R., R. Hay, R. Paloski, and L. Ramirez. 2008. Snakes of Wisconsin. Second edition. Wisconsin Department of Natural Resources, Bureau of Endangered Resources PUB-ER-100 2008, Madison, USA.

Christoffel, R., R. Hay, R. Paloski, and M. Wolfgram. 2009. Amphibians of Wisconsin. Second edition. Wisconsin Department of Natural Resources, Bureau of Endangered Resources PUB-ER-105, Madison, USA.

Chu, T. 1936. A review of the status of the reptilian nematodes of the genus *Rhabdias* with a redescription of *Rhabdias fuscovenosa* var. *catanensis* (Rizzo, 1902) new rank. Journal of Parasitology 22:130–139.

Church, D. R. 2008. Role of current versus historical hydrology in amphibian species turnover within local pond communities. Copeia 2008:115–125.

Church, D. R., L. L. Bailey, H. M. Bailey, W. L. Kendall, and J. E. Hines. 2007. Iteroparity in the variable environment of the salamander *Ambystoma tigrinum*. Ecology 88:891–903.

Church, S. A., J. M. Kraus, J. C. Mitchell, D. R. Church, and D. R. Taylor. 2003. Evidence for multiple Pleistocene refugia in the postglacial expansion of the Eastern Tiger Salamander, *Ambystoma tigrinum tigrinum*. Evolution 57:372–383.

Churchill, T. A., and K. B. Storey. 1992. Freezing survival of the Garter Snake *Thamnophis sirtalis parietalis*. Canadian Journal of Zoology 70:99–105.

Churchill, T. A., and K. B. Storey. 1993. Dehydration tolerance in Wood Frogs: a new perspective on development of amphibian freeze tolerance. American Journal of Physiology-Regulatory, Integrative and Comparative Physiology 265:1324–1332.

Churchill, T. A., and K. B. Storey. 1996. Organ metabolism and cryoprotectant synthesis during freezing in Spring Peepers *Pseudacris crucifer*. Copeia 1996:517–524.

Cieslak, E. S. 1945. Relations between the reproductive cycle and the pituitary gland in the snake *Thamnophis*. Physiological Zoology 18:299–329.

[CITES] Convention on International Trade in Endangered Species of Wild Fauna and Flora. 2013. Homepage. http://www.cites.org. Accessed 1 Nov 2018.

City of Sioux Falls. 2002. Faunal survey for rare, threatened, or endangered species with special emphasis on the lined snake (*Tropidoclonion lineatum*): Sioux Falls east side corridor environmental assessment. City of Sioux Falls, Sioux Falls, South Dakota, USA.

Clark, A. M., P. E. Moler, E. E. Possardt, A. H. Savitzky, W. S. Brown, and B. W. Bowen. 2003. Phylogeography of the Timber Rattlesnake (*Crotalus horridus*) based on mtDNA sequences. Journal of Herpetology 37:145–154.

Clark, D. R., Jr. 1967. Experiments into selection of soil type, soil moisture level, and temperature by five species of small snakes. Transactions of the Kansas Academy of Science 70:490–496.

Clark, D. R., Jr. 1968. A proposal of specific status for the Western Worm Snake, *Carphophis amoenus vermis* (Kennicott). Herpetologica 24:104–112.

Clark, D. R., Jr. 1970. Ecological study of the Worm Snake *Carphophis vermis* (Kennicott). University of Kansas Publications, Museum of Natural History 19:85–194.

Clark, D. R., Jr. 1974. The Western Ribbon Snake (*Thamnophis proximus*): ecology of a Texas population. Herpetologica 30:372–379.

Clark, D. R., Jr. 1976. Ecological observations on a Texas population of Six-Lined Racerunners, *Cnemidophorus sexlineatus* (Reptilia, Lacertilia, Teiidae). Journal of Herpetology 10:133–138.

Clark, D. R., Jr. 1995. Dicofol and DDT residues in lizard carcasses and bird eggs from Texas, Florida, and California. Bulletin of Environmental Contamination and Toxicology 54:817–824.

Clark, D. R., Jr., and R. J. Hall. 1970. Function of the blue tail-coloration of the five-lined skink (*Eumeces fasciatus*). Herpetologica 26:271–274.

Clark, D. R., Jr., C. M. Bunck, and R. J. Hall. 1997. Female reproductive dynamics in a Maryland population of Ringneck Snakes (*Diadophis punctatus*). Journal of Herpetology 31:476–483.

Clark, K. L. 1986. Responses of *Ambystoma maculatum* populations in central Ontario to habitat acidity. Canadian Field-Naturalist 100:463–469.

Clark, K. L., and R. J. Hall. 1985. Effects of elevated hydrogen ion and aluminum concentrations on the survival of amphibian embryos and larvae. Canadian Journal of Zoology 63:116–123.

Clark, P. J., J. M. Reed, B. G. Tavernia, B. S. Windmiller, and J. V. Regosin. 2008a. Urbanization effects on Spotted Salamander and Wood Frog presence and abundance. Pages 67–75 *in* J. C. Mitchell, R. E. Jung Brown, and B. Bartholomew, editors. Urban Herpetology. Society for the Study of Amphibians and Reptiles, Herpetological Conservation 3, Salt Lake City, Utah, USA.

Clarke, R. 1955. Observations on *Eumeces s. septentrionalis* in Kansas. Herpetologica 11:161–164.

Clark, R. W. 2002. Diet of the Timber Rattlesnake, *Crotalus horridus*. Journal of Herpetology 36:494–499.

Clark, R. W. 2004a. Feeding experience modifies assessment of ambush sites by the Timber Rattlesnake, a sit-and-wait predator. Ethology 110:471–483.

Clark, R. W. 2004b. Timber Rattlesnakes (*Crotalus horridus*) use chemical cues to select ambush sites. Journal of Chemical Ecology 30:607–617.

Clark, R. W. 2006. Fixed videography to study predation behavior of an ambush foraging snake, *Crotalus horridus*. Copeia 2006:181–187.

Clark, R. W., M. N. Marchand, B. J. Clifford, R. Stechert, and S. Stephens. 2011. Decline of an isolated Timber Rattlesnake (*Crotalus horridus*) population: interactions between climate change, disease, and loss of genetic diversity. Biological Conservation 144:886–891.

Clark, R. W., W. S. Brown, R. Stechert, and K. R. Zamudio. 2008b. Integrating individual behaviour and landscape genetics: the population structure of Timber Rattlesnake hibernacula. Molecular Ecology 17:719–730.

Clark, R. W., W. S. Brown, R. Stechert, and K. R. Zamudio. 2010. Roads, interrupted dispersal, and genetic diversity in Timber Rattlesnakes. Conservation Biology 24:1059–1069.

Clarke, A. H. 1985. The tribe Alasmidontini (Unionidae: Anodontinae), Part II: *Lasmigona* and *Simpsonais*. Smithsonian Contributions to Zoology 399, Washington, D.C., USA.

Clarke, R. F. 1958. An ecological study of reptiles and amphibians in Osage County, Kansas. Emporia State Research Studies 7:1–52.

Clause, A. R., and E. A. Capaldi. 2006. Caudal autotomy and regeneration in lizards. Journal of Experimental Zoology, Part A. Comparative Experimental Biology 305:965–973.

Clausen, H. J. 1936. Observations on the Brown Snake *Storeria dekayi* (Holbrook), with especial reference to the habits and birth of young. Copeia 1936:98–102.

Clawson, R. G., B. G. Lockaby, and R. H. Jones. 1997. Amphibian responses to helicopter harvesting in forested floodplains of low order, blackwater streams. Forest Ecology and Management 90:225–235.

Clay, W. M. 1936. The taxonomy and phylogenetic relationships of the water snakes *Natrix erythrogaster* and *N. sipedon*. Dissertation, University of Michigan, Ann Arbor, USA.

Clayton, L., J. W. Attig, and D. M. Mickelson. 2001. Effects of late Pleistocene permafrost on the landscape of Wisconsin. Boreas 30:173–188.

Cleland, C. E. 1966. The prehistoric animal ecology and ethnozoology of the upper Great Lakes region. Museum of Anthropology, University of Michigan, Ann Arbor, USA.

Cleland, D. T., J. A. Freeouf, J. E. Keys, Jr., G. J. Nowacki, C. Carpenter, and W. H. McNab. 2007. Ecological subregions: sections and subsections of the conterminous United States. US Forest Service General Technical Report WO-76, Washington, D.C., USA.

Clements, F. E., and V. E. Shelford. 1939. *Bio-Ecology*. John Wiley and Sons, London, England.

Clesson, D., A. Bautista, D. D. Baleckaitus, and R. W. Krohmer. 2002. Reproductive biology of male Eastern Garter Snakes (*Thamnophis sirtalis sirtalis*) from a denning population in central Wisconsin. American Midland Naturalist 147:376–386.

Clevenger, A. P., and M. P. Huijser. 2011. Wildlife crossing structure handbook: design and evaluation in North America. US Department of Transportation, Federal Highway Administration, Central Federal Lands Highway Division Publication FHWA-CFL/TD-11-003, Lakewood, Colorado, USA. Archived by WebCite at http://www.webcitation.org/73dqGDGNV. Accessed 3 Nov 2018.

Clevenger, A. P., M. McIvor, D. McIvor, B. Chruszcz, and K. Gunson. 2001. Tiger Salamander, *Ambystoma tigrinum*, movements and mortality on the trans-Canada highway in southwestern Alberta. Canadian Field-Naturalist 115:199–204.

Cliburn, J. W. 1957. Leaping ability of the Glass Lizard. Herpetologica 13:24.

Cliburn, J. W. 1973. *Natrix septemvittata* in Tishomingo County, Mississippi. Journal of the Mississippi Academy of Sciences 19:159–160.

Cline, G. R. 2005. *Hyla versicolor* LeConte, 1825. Pages 458–461 *in* M. J. Lannoo, editor. Amphibian declines: the conservation status of United States species. University of California Press, Berkeley, USA.

Cobb, V. A., J. J. Green, T. Worrall, J. Pruett, and B. Glorioso. 2005. Initial den location behavior in a litter of neonate *Crotalus horridus* (Timber Rattlesnakes). Southeastern Naturalist 4:723–730.

Cochran, A. G., and P. A. Cochran. 2001b. *Ambystoma maculatum* (Spotted Salamander). Geographic distribution. Herpetological Review 32:267.

Cochran, A. G., P. A. Cochran, J. A. Cochran, and C. D. Knight. 2001. *Storeria dekayi* (Brown Snake). Geographic distribution. Herpetological Review 32:280.

Cochran, D. M., and C. J. Goin. 1970. The new field book of reptiles and amphibians. G.P. Putnam's Sons, New York, USA.

Cochran, J. A., P. A. Cochran, and G. S. Casper. 2006. *Storeria occipitomaculata* (Red-bellied Snake). Geographic distribution. Herpetological Review 37:243.

Cochran, J. D. 1982a. A brief note on an unusual breeding occurrence in the American Toad, *Bufo a. americanus*. Bulletin of the Maryland Herpetological Society 18:165–166.

Cochran, P. A. 1982b. *Necturus maculosus* (Mudpuppy). Geographic distribution. Herpetological Review 13:79.

Cochran, P. A. 1982c. *Rana catesbeiana* (Bullfrog). Geographic distribution. Herpetological Review 13:79–80.

Cochran, P. A. 1982d. *Trionyx spiniferus* (Spiny Softshell Turtle). Geographic distribution. Herpetological Review 13:80.

Cochran, P. A. 1983a. *Coluber constrictor* (Racer). Geographic distribution. Herpetological Review 14:123.

Cochran, P. A. 1983b. *Eumeces septentrionalis* (Northern Prairie Skink). Agonistic behavior. Herpetological Review 14:119.

Cochran, P. A. 1983c. *Nerodia sipedon* (Northern Watersnake). Geographic distribution. Herpetological Review 14:124.

Cochran, P. A. 1986a. Some observations on the herpetofauna of the University of Wisconsin–Madison campus. Bulletin of the Chicago Herpetological Society 22:34–35.

Cochran, P. A. 1986b. The herpetofauna of the Weaver Dunes, Wabasha County, Minnesota. Prairie Naturalist 18:143–150.

Cochran, P. A. 1987. *Graptemys geographica* (Map Turtle). Adult mortality. Herpetological Review 18:37.

Cochran, P. A. 1988a. *Chrysemys picta* (Painted Turtle). Geographic distribution. Herpetological Review 19:21.

Cochran, P. A. 1988b. *Clemmys insculpta* (Wood Turtle). Geographic distribution. Herpetological Review 19:21.

Cochran, P. A. 1988c. *Elaphe vulpina* (Fox Snake). Geographic distribution. Herpetological Review 19:20.

Cochran, P. A. 1988d. *Hyla crucifer* (Spring Peeper). Geographic distribution. Herpetological Review 19:16.

Cochran, P. A. 1988e. *Hyla versicolor* (Gray Treefrog). Geographic distribution. Herpetological Review 19:16.

Cochran, P. A. 1989. Notes on the phenology of the Chorus Frog, *Pseudacris triseriata*, in Dupage County, Illinois. Bulletin of the Chicago Herpetological Society 24:89–90.

Cochran, P. A. 1990. *Elaphe vulpina* (Western Fox Snake). Geographic distribution. Herpetological Review 21:41.

Cochran, P. A. 1991a. *Apalone spinifera* (Spiny Softshell). Geographic distribution. Herpetological Review 22:62.

Cochran, P. A. 1991b. Some dietary records for the Mudpuppy (*Necturus maculosus*), with comments on field techniques. Bulletin of the Chicago Herpetological Society 26:221–223.

Cochran, P. A. 1992. *Eumeces septentrionalis septentrionalis* (Northern Prairie Skink). Geographic distribution. Herpetological Review 23:123.

Cochran, P. A. 1994. *Apalone spinifera* (Spiny Softshell). Geographic distribution. Herpetological Review 25:75.

Cochran, P. A. 1999. Mudpuppies (*Necturus maculosus*) and barrier falls in Wisconsin. Bulletin of the Maryland Herpetological Society 35:4–13.

Cochran, P. A. 2008a. A Cottonmouth (*Agkistrodon piscivorus*) in Minnesota, and historical reports of other pit vipers unexpected in the Upper Midwest. Northeastern Naturalist 15:461–464.

Cochran, P. A. 2008b. An unusual microhabitat for an American Toad (*Anaxyrus americanus*). Bulletin of the Chicago Herpetological Society 43:62.

Cochran, P. A. 2008c. Foraging by Northern Prairie Skinks and Five-lined Skinks may include simultaneous search and prey handling. Journal of Kansas Herpetology 25:19–20.

Cochran, P. A. 2013a. Additional cases of snakes entrapped in garden netting: *Pantherophis vulpinus* and *Crotalus horridus* in southeastern Minnesota. Bulletin of the Chicago Herpetological Society 48:129.

Cochran, P. A. 2013b. *Storeria dekayi* (Dekay's Brownsnake). Seasonal activity. Herpetological Review 44:697–698.

Cochran, P. A., and A. G. Cochran. 2001a. *Ambystoma laterale* (Blue-spotted Salamander). Geographic distribution. Herpetological Review 32:188.

Cochran, P. A., and G. M. Borash. 2014a. *Necturus maculosus* (Mudpuppy). Diet. Herpetological Review 45:300–301.

Cochran, P. A., and G. M. Borash. 2014b. *Necturus maculosus* (Mudpuppy). Habitat. Herpetological Review 45:301–302.

Cochran, P. A., and J. Marks. 1990. *Chelydra serpentina* (Snapping Turtle). Geographic distribution. Herpetological Review 21:38.

Cochran, P. A., and J. A. Cochran. 2003. *Pseudacris crucifer* (Spring Peeper). Predation. Herpetological Review 34:360.

Cochran, P. A., and J. A. Cochran. 2007. *Bufo americanus* (American Toad). Predation. Herpetological Review 38:178.

Cochran, P. A., and J. D. Lyons. 1985. *Necturus maculosus* (Mudpuppy). Juvenile Ecology. Herpetological Review 16:53.

Cochran, P. A., and J. D. Lyons. 1986. New distributional records for Wisconsin amphibians and reptiles. Wisconsin Academy of Sciences, Arts and Letters 74:138–141.

Cochran, P. A., and J. D. Lyons. 1990. *Necturus maculosus* (Mudpuppy). Geographic distribution. Herpetological Review 21:37.

Cochran, P. A., and J. R. Hodgson. 1986. *Ambystoma laterale* (Blue-spotted Salamander). Geographic distribution. Herpetological Review 17:26.

Cochran, P. A., and K. J. Knutsen. 1987. *Hyla crucifer* (Spring Peeper). Geographic distribution. Herpetological Review 18:55.

Cochran, P. A., and S. J. Schmitt. 2014. *Crotalus horridus* (Timber Rattlesnake). Diel Activity. Herpetological Review 45:708–709.

Cochran, P. A., and S. J. Schmitt. 2016. *Aspidoscelis sexlineata* (Six-lined Racerunner). Temporal habitat use. Herpetological Review 47:132–133.

Cochran, P. A., B. D. Jones, R. Otto, and A. P. Kinziger. 2015. Observations on the diet of the Wood Turtle (*Glyptemys insculpta*) in northeastern Wisconsin and on its potential role as a seed disperser. Collinsorum 3:28–31.

Cochran, P. A., J. Lyons, and M. G. Bolek. 2002. *Necturus maculosus maculosus* (Common Mudpuppy). Herpetological Review 33:144.

Cochran, P. A., J. B. Hodgson, and R. M. Korb. 1987. New distributional records for reptiles and amphibians in Brown County, Wisconsin. Herpetological Review 18:21–23.

Cochrane, M. M., D. J. Brown, and R. A. Moen. 2019. GPS technology for semi-aquatic turtle research. Diversity 11:Article 34.

Cochrane, M. M., R. A. Moen, and D. J. Brown. 2015. *Glyptemys insculpta* (Wood Turtle). Nest predation. Herpetological Review 46:618.

Cochrane, T. S., and H. H. Iltis. 2000. Atlas of the Wisconsin prairie and savanna flora. Wisconsin Department of Natural Resources Technical Bulletin 191, Madison, USA.

Cochrane, T. S., K. Elliot, and C. Lipke. 2006. Prairie plants of the University of Wisconsin–Madison Arboretum. University of Wisconsin Press, Madison, USA.

Cockerell, T. D. A. 1937. Obituary: Sigmund Graenicher. Science 86:364–365.

Cocroft, R. B. 1994. A cladistic analysis of chorus frog phylogeny (Hylidae: *Pseudacris*). Herpetologica 50:420–437.

Coggins, J. R., and R. A. Sajdak. 1982. A survey of Helminth parasites in the salamanders and certain anurans from Wisconsin. Proceedings of the Helminthological Society of Washington 49:99–102.

Coleman, J. T. H., C. M. Adams, M. Kandel, and M. E. Richmond. 2012. Eating the invaders: the prevalence of Round Goby (*Apollonia melanostomus*) in the diet of Double-crested Cormorants on the Niagara River. Waterbirds 35:103–113.

Colley, M., S. C. Lougheed, K. Otterbein, and J. D. Litzgus. 2017. Mitigation reduces road mortality of a threatened rattlesnake. Wildlife Research 44:48–59.

Collins, B. M., and A. F. Badje. 2016. *Lithobates septentrionalis* (Mink Frog). Geographic distribution. Herpetological Review 47:623.

Collins, J. P., and J. E. Cheek. 1983. Effect of food and density on development of typical cannibalistic salamander larvae in *Ambystoma tigrinum nebulosum*. American Zoologist 23:77–84.

Collins, J. P., and J. R. Holomuzki. 1984. Intraspecific variation in diet within and between trophic morphs in larval Tiger Salamanders (*Ambystoma tigrinum nebulosum*). Canadian Journal of Zoology 62:168–174.

Collins, J. T. 1991. Viewpoint: a new taxonomic arrangement for some North American amphibians and reptiles. Herpetological Review 22:42–43.

Collins, J. T. 1993. Amphibians & reptiles in Kansas. Third edition, Revised. University Press of Kansas, Lawrence, USA.

Collins, J. T., S. L. Collins, and K. J. Irwin. 2006. A survey of the amphibians, turtles, and reptiles of the eastern portion of the Kiowa National Grassland of New Mexico and the Rita Blanca National Grassland of adjacent Oklahoma and Texas. Journal of Kansas Herpetology 18:10–20.

Collins, J. T., S. L. Collins, and T. W. Taggart. 2010. Amphibians, reptiles, and turtles in Kansas. Eagle Mountain Publishing, LC, Eagle Mountain, Utah, USA.

Collins, M. K. 2017. Searching for a salamander: distribution and habitat of the Common Mudpuppy (*Necturus maculosus*) in southeast Ohio using environmental DNA. Thesis, Ohio University, Athens, USA.

Collins, S. J., and L. Fahrig. 2017. Responses of anurans to composition and configuration of agricultural landscapes. Agriculture, Ecosystems and Environment 239:399–409.

Colteaux, B. C., and D. M. Johnson. 2017. Commercial harvest and export of Snapping Turtles (*Chelydra serpentina*) in the United States: trends and the efficacy of size limits at reducing harvest. Journal for Nature Conservation 35:13–19.

Compton, B. W. 1999. Ecology and conservation of the Wood Turtle (*Clemmys insculpta*) in Maine. Thesis, University of Maine, Orono, USA.

Compton, B. W., J. M. Rhymer, and M. McCollough. 2002. Habitat selection by Wood Turtles (*Clemmys insculpta*): an application of paired logistic regression. Ecology 83:833–843.

Compton, B. W., K. McGarigal, S. A. Cushman, and L. R. Gamble. 2007. A resistant-kernel model of connectivity for amphibians that breed in vernal pools. Conservation Biology 21:788–799.

Conant, R. 1938. The reptiles of Ohio. American Midland Naturalist 20:1–200.

Conant, R. 1940. A new subspecies of the Fox Snake, *Elaphe vulpina* Baird and Girard. Herpetologica 2:1–14.

Conant, R. 1951. The reptiles of Ohio. Second edition. University Notre Dame Press, Notre Dame, Indiana, USA.

Conant, R. 1958. A field guide to reptiles and amphibians of the United States and Canada east of the 100th meridian. Houghton Mifflin Company, Boston, Massachusetts, USA.

Conant, R. 1960. The Queensnake, *Natix septemvittata*, in the interior highlands of Arkansas and Missouri, with comments upon similar disjunct distributions. Academy of Natural Sciences of Philadelphia 112:25–40.

Conant, R., and J. T. Collins. 1991. A field guide to reptiles and amphibians, eastern and central North America. Third edition. Houghton Mifflin Company, New York, New York, USA.

Conant, R., and J. T. Collins. 1998. A field guide to reptiles and amphibians: eastern/central North America. Third edition (expanded). Houghton Mifflin Company, New York, New York, USA.

Conant, R., and R. G. Hudson. 1949. Longevity records for reptiles and amphibians in the Philadelphia Zoological Gardens. Herpetologica 5:1–8.

Conant, R., E. S. Thomas, and R. L. Rausch. 1945. The Plains Garter Snake, *Thamnophis radix*, in Ohio. Copeia 1945:61–68.

Congdon, J. D., and D. A. Keinath. 2006. Blanding's Turtles (*Emydoidea blandingii*): a technical conservation assessment. Report to the US Forest Service, Rocky Mountain Region, Lakewood, Colorado, USA.

Congdon, J. D., and R. C. van Loben Sels. 1991. Growth and body size in Blanding's Turtles (*Emydoidea blandingii*): relationships to reproduction. Canadian Journal of Zoology 69:239–245.

Congdon, J. D., and R. C. van Loben Sels. 1993. Relationship of reproductive traits and body size with attainment of sexual maturity and age in Blanding's Turtles (*Emydoidea blandingii*). Journal of Evolutionary Biology 6:547–557.

Congdon, J. D., A. E. Dunham, and R. C. van Loben Sels. 1993. Delayed sexual maturity and demographics of Blanding's Turtles (*Emydoidea blandingii*): implications for conservation and management of long-lived organisms. Conservation Biology 7:826–833.

Congdon, J. D., A. E. Dunham, and R. C. van Loben Sels. 1994. Demographics of Common Snapping Turtles (*Chelydra serpentina*): implications for conservation and management of long-lived organisms. American Zoologist 34:397–408.

Congdon, J. D., D. W. Tinkle, G. L. Breitenbach, and R. C. van Loben Sels. 1983. Nesting ecology and hatchling success in the turtle, *Emydoidea blandingi*. Herpetologica 39:417–429.

Congdon, J. D., G. L. Breitenback, R. C. van Loben Sels, and D. W. Tinkle. 1987. Reproduction and nesting ecology of Snapping Turtles (*Chelydra serpentina*) in southeastern Michigan. Herpetologica 43:39–54.

Congdon, J. D., J. L. Greene, and J. W. Gibbons. 1986. Biomass of freshwater turtles: a geographic comparison. American Midland Naturalist 115:165–173.

Congdon, J. D., M. Pappas, B. Brecke, and J. Capps. 2011a. Conservation implications of initial orientation of naïve hatchling Snapping Turtles (*Chelydra serpentina*) and Painted Turtles (*Chrysemys picta belli*) dispersing from experimental nests. Chelonian Conservation and Biology 10:42–53.

Congdon, J. D., O. M. Kinney, and R. D. Nagle. 2011b. Spatial ecology and core-area protection of Blanding's Turtle (*Emydoidea blandingii*). Canadian Journal of Zoology 89:1098–1106.

Congdon, J. D., R. D. Nagle, O. M. Kinney, M. Ostenoski, H. W. Avery, and R. C. van Loben Sels. 2000. Nesting ecology and embryo mortality: implications for hatching success and demography of Blanding's Turtles (*Emydoidea blandingii*). Chelonian Conservation and Biology 3:569–579.

Congdon, J. D., R. D. Nagle, O. M. Kinney, and R. C. van Loben Sels. 2001. Hypotheses of aging in a long-lived vertebrate, Blanding's Turtle (*Emydoidea blandingii*). Experimental Gerontology 36:813–827.

Congdon, J. D., R. D. Nagle, O. M. Kinney, R. C. van Loben Sels, T. Quinter, and D. W. Tinkle. 2003. Testing hypotheses of aging in long-lived Painted Turtles (*Chrysemys picta*). Experimental Gerontology 38:765–772.

Congdon, J. D., S. W. Gotte, and R. W. McDiarmid. 1992. Ontogenetic changes in habitat use by juvenile turtles, *Chelydra serpentina* and *Chrysemys picta*. Canadian Field-Naturalist 106:241–248.

Congdon, J. D., T. E. Graham, T. B. Herman, J. W. Lang, M. J. Pappas, and B. J. Brecke. 2008. *Emydoidea blandingii* (Holbrook 1838)—Blanding's Turtle. Chelonian Research Monographs 5:015.1–015.12.

Conger, P. S. W. 1921. The distribution of amphibians in Wisconsin. Thesis, University of Wisconsin, Madison, USA.

Conner, C. A., B. A. Douthitt, and T. J. Ryan. 2005. Descriptive ecology of a turtle assemblage in an urban landscape. American Midland Naturalist 153:426–435.

Connior, M. B., S. W. Chordas III. 2012. *Aspidoscelis sexlineata* (Six-lined Racerunner). Commensalism. Herpetological Review 43:644.

Conrad, J. L., J. C. Ast, S. Montanari, and M. A. Norell. 2011. A combined evidence phylogenetic analysis of Anguimorpha (Reptilia: Squamata). Cladistics 27:230–277.

Converse S. J., J. B. Iverson, and J. A. Savidge. 2002. Activity, reproduction and over-wintering behavior of Ornate Box Turtles (*Terrapene ornata ornata*) in the Nebraska sandhills. American Midland Naturalist 148:416–422.

Converse, S. J., J. B. Iverson, and J. A. Savidge. 2005. Demographics of an Ornate Box Turtle population experiencing minimal human-induced disturbances. Ecological Applications 15:2171–2179.

Cook, F. R. 1964. Communal egg laying in the Smooth Green Snake. Herpetologica 20:206.

Cook, F. R. 1984. Introduction to Canadian amphibians and reptiles. National Museums of Canada, Ottawa, Ontario.

Cook, K., S. Droege, and A. R. Kellogg. 2002. Stomach content analysis of 13 North American toad species. Bulletin of the Maryland Herpetological Society 38:75–85.

Cook, R. P. 2008. Potential and limitations of herpetofaunal restoration in an urban landscape. Pages 465–478 *in* J. C. Mitchell, R. E. Jung Brown, and B. Bartholomew, editors. Urban Herpetology. Society for the Study of Amphibians and Reptiles, Herpetological Conservation 3, Salt Lake City, Utah, USA.

Cooley, C. R., A. O. Floyd, A. Dolinger, and P. B. Tucker. 2003. Demography and diet of the Painted Turtle (*Chrysemys picta*) at high-elevation sites in southwestern Colorado. Southwestern Naturalist 48:47–53.

Cooper, J. E. 1956. A Maryland hibernation site for herptiles. Herpetologica 12:238.

Cooper, W. E., Jr., and L. J. Vitt. 1985. Blue tails and autotomy: enhancement of predation avoidance in juvenile skinks. Zeitschrift für Tierpsychologie 70:265–276.

Cooper, W. E., Jr., and S. Secor. 2007. Strong response to anuran chemical cues by an extreme dietary specialist, the Eastern Hog-nosed Snake (*Heterodon platirhinos*). Canadian Journal of Zoology 85:619–625.

Cooper, W. E., Jr., G. M. Burghardt, and W. S. Brown. 2000. Behavioural responses by hatchling Racers (*Coluber constrictor*) from two geographically distinct populations to chemical stimuli from potential prey and predators. Amphibia-Reptilia 21:103–115.

Cooper, W. E., Jr., O. Attum, and B. Kingsbury. 2008. Escape behaviors and flight initiation distance in the Common Water Snake *Nerodia sipedon*. Journal of Herpetology 42:493–500.

Cope, E. D. 1859. On the primary divisions of the Salamandridae, with descriptions of two new species. Proceedings of the Academy of Natural Sciences of Philadelphia 11:122–128.

Cope, E. D. 1865. Third contribution to the herpetology of tropical America. Proceedings of the Academy of Natural Sciences of Philadelphia 1865:185–198.

Cope, E. D. 1875. Checklist of North American Batrachia and Reptilia. Bulletin of the United States National Museum 1:1–104.

Cope, E. D. 1880. On the zoological position of Texas. Bulletin of the United States National Museum 17:1–51.

Cope, E. D. 1889. The Batrachia of North America. Bulletin of the US National Museum 34:1–525.

Cope, E. D. 1891. A critical review of the characters and variations of the snakes of North America. Proceedings of the United States National Museum 14:589–694.

Corin, C. W. 1976. The land vertebrates of the Huron Islands, Lake Superior. Jack-Pine Warbler 54:138–147.

Corn, P. S. 1986. Genetic and developmental studies of albino chorus frogs. Journal of Heredity 77:164–168.

Corn, P. S. 2005. Climate change and amphibians. Animal Biodiversity and Conservation 28:59–67.

Corn, P. S., and J. C. Fogleman. 1984. Extinction of montane populations of the Northern Leopard Frog (*Rana pipiens*) in Colorado. Journal of Herpetology 18:147–152.

Corn, P. S., and L. J. Livo. 1989. Leopard Frog and Wood Frog reproduction in Colorado and Wyoming. Northwestern Naturalist 70:1–9.

Corn, P. S., E. Muths, A. M. Kissel, and R. D. Scherer. 2011. Breeding chorus indices are weakly related to estimated abundance of Boreal Chorus Frogs. Copeia 2011:365–371.

Corser, J. D., and C. K. Dodd, Jr. 2004. Fluctuations in a metapopulation of nesting Four-toed Salamanders, *Hemidactylium scutatum*, in the Great Smoky Mountains National Park, USA, 1999–2003. Natural Areas Journal 24:135–140.

Cortwright, S. A. 1998. Ten- to eleven-year population trends of two pond-breeding amphibian species, Red-spotted Newts and Green Frogs. Pages 61–71 *in* M. J. Lannoo, editor. Status and Conservation of Midwestern Amphibians. University of Iowa Press, Iowa City, Iowa, USA.

Cortwright, S. A., and C. E. Nelson. 1990. An examination of multiple factors affecting community structure in an aquatic amphibian community. Oecologia 83:123–131.

Cory, B. L., and J. J. Manion. 1953. Predation on eggs of the Wood Frog, *Rana sylvatica*, by leeches. Copeia 1953:66.

Cosentino, B. J., R. L. Schooley, and C. A. Phillips. 2010. Wetland hydrology, area, and isolation influence occupancy and spatial turnover of the Painted Turtle, *Chrysemys picta*. Landscape Ecology 25:1589–1600.

[COSEWIC] Committee on the Status of Endangered Wildlife in Canada. 2004. COSEWIC status report on Northern Prairie Skink *Eumeces septentrionalis septentrionalis* in Canada. Committee on the Status of Endangered Wildlife in Canada, Ontario, Canada. Archived by WebCite at http://www.webcitation.org/73blAX8d3. Accessed 1 Nov 2018.

[COSEWIC] Committee on the Status of Endangered Wildlife in Canada. 2007. COSEWIC assessment and update status report on the Eastern Hog-nosed Snake *Heterodon platirhinos* in Canada. Committee on the Status of Endangered Wildlife in Canada, Ottawa, Canada. Archived by WebCite at http://www.webcitation.org/73blUTc21. Accessed 1 Nov 2018.

[COSEWIC] Committee on the Status of Endangered Wildlife in Canada. 2010a. COSEWIC assessment and update status report on the Butler's Gartersnake *Thamnophis butleri* in Canada. Committee on the Status of Endangered Wildlife in Canada, Ontario, Canada. Archived by WebCite at http://www.webcitation.org/73bm4O1dQ. Accessed 1 Nov 2018.

[COSEWIC] Committee on the Status of Endangered Wildlife in Canada. 2010b. COSEWIC assessment and status report on the Queensnake *Regina septemvittata* in Canada. Committee on the Status of Endangered Wildlife in Canada, Ontario, Canada. Archived by WebCite at http://www.webcitation.org/73bliY59B. Accessed 1 Nov 2018.

Costanzo, J. P. 1985. The bioenergetics of hibernation in the Eastern Garter Snake *Thamnophis sirtalis sirtalis*. Physiological Zoology 58:682–692.

Costanzo, J. P. 1986. Influences of hibernaculum microenvironment on the winter life history of the Garter Snake (*Thamnophis sirtalis*). Ohio Journal of Science 86:199–204.

Costanzo, J. P. 1988. Recovery from ice-entombment in garter snakes. Herpetological Review 19:76–77.

Costanzo, J. P. 1989a. A physiological basis for prolonged submergence in hibernating Garter Snakes *Thamnophis sirtalis*: evidence for an energy-sparing adaptation. Physiological Zoology 62:580–592.

Costanzo, J. P. 1989b. Conspecific scent trailing by Garter Snakes (*Thamnophis sirtalis*) during autumn: further evidence for use of pheromones in den location. Journal of Chemical Ecology 15:2531–2538.

Costanzo, J. P. 1989c. Effects of humidity, temperature, and submergence behavior on survivorship and energy use in hibernating Garter Snakes, *Thamnophis sirtalis*. Canadian Journal of Zoology 67:2486–2492.

Costanzo, J. P., and R. E. Lee, Jr. 1988. Natural freeze tolerance in a reptile. Cryo-Letters 9:380–385.

Costanzo, J. P., J. D. Litzgus, J. B. Iverson, and R. E. Lee, Jr. 2001. Cold-hardiness and evaporative water loss in hatchling turtles. Physiological and Biochemical Zoology 74:510–519.

Costanzo, J. P., M. F. Wright, and R. E. Lee, Jr. 1992. Freeze tolerance as an overwintering adaptation in Cope's Grey Treefrog (*Hyla chrysoscelis*). Copeia 1992:565–569.

Coupe, B. 2002. Pheromones, search patterns, and old haunts: how do male Timber Rattlesnakes (*Crotalus horridus*) locate mates? Pages 139–148 *in* G. W. Schuett, M. Hoggren, M. E. Douglas, and H. W. Greene, editors. Biology of the vipers. Eagle Mountain Publishing, LC, Eagle Mountain, Utah, USA.

Court, F. E. 2012. Pioneers of ecological restoration: the people and legacy of the University of Wisconsin Arboretum. University of Wisconsin Press, Madison, USA.

Craig, J. M., D. A. Mifsud, A. S. Briggs, J. Boase, and G. Kennedy. 2015. Mudpuppy (*Necturus maculosus maculosus*) spatial distribution, breeding water depth, and use of artificial spawning habitat in the Detroit River. Herpetological Conservation and Biology 10:926–934.

Crawford, J. A., C. A. Phillips, W. E. Peterman, I. E. MacAllister, N. A. Wesslund, A. R. Kuhns, and M. J. Dreslik. 2017. Chytrid infection dynamics in cricket frogs on military and public lands in the midwestern United States. Journal of Fish and Wildlife Management 8:344–352.

Crawford, K. M. 1991. The winter environment of Painted Turtles, *Chrysemys picta*: temperature, dissolved oxygen, and potential cues for emergence. Canadian Journal of Zoology 69:2493–2498.

Crawford, K. M. 1994. Patterns of energy substrate utilization in overwintering Painted Turtles, *Chrysemys picta*. Comparative Biochemistry and Physiology 109A:495–502.

Crawford, R. 1936. A study of the Massasauga as found in Trumbull County, Ohio. Thesis, University of Southern California, Los Angeles, USA.

Creer, D. A. 2005. Correlations between ontogenetic change in color pattern and antipredator behavior in the Racer, *Coluber constrictor*. Ethology 111:287–300.

Crespi, E. J., L. J. Rissler, N. M. Mathes, K. Engbrecht, S. I. Duncan, T. Seaborn, E. M. Hall, J. D. Peterson, and J. L. Brunner. 2015. Geophysiology of Wood Frogs: landscape patterns of prevalence of disease and circulating hormone concentrations across the eastern range. Integrative and Comparative Biology 55:602–617.

Criddle, S. 1937. Snakes from an ant hill. Copeia 1937:11–12.

Crookston, N. L., G. E. Rehfeldt, G. E. Dixon, and A. R. Weiskittel. 2010. Addressing climate change in the forest vegetation simulator to assess impacts on landscape forest dynamics. Forest Ecology and Management 260:1198–1211.

Cross, M. D., K. V. Root, C. J. Mehne, J. McGowan-Stinski, D. Pearsall, and J. C. Gillingham. 2015.

Multi-scale responses of Eastern Massasauga Rattlesnakes (*Sistrurus catenatus*) to prescribed fire. American Midland Naturalist 173:346–362.

Crother, B. I., committee chair. 2012. Scientific and standard English and French names of amphibians and reptiles of North America north of Mexico, with comments regarding confidence in our understanding. Seventh edition. Society for the Study of Amphibians and Reptiles, Herpetological Circular 39, Shoreview, Minnesota, USA.

Crother, B. I., committee chair. 2017. Scientific and standard English names of amphibians and reptiles of North America north of Mexico, with comments regarding confidence in our understanding. Eighth edition. Society for the Study of Amphibians and Reptiles, Herpetological Circular 43, Shoreview, Minnesota, USA.

Crother, B. I., M. E. White, J. M. Savage, M. E. Eckstut, M. R. Graham, and D. W. Gardner. 2011. A reevaluation of the status of the foxsnakes *Pantherophis gloydi* Conant and *P. vulpinus* Baird and Girard (Lepidosauria). ISRN Zoology 2011:Article ID 436049.

Crouch, W. B., and P. W. C. Paton. 2000. Using egg-mass counts to monitor Wood Frog populations. Wildlife Society Bulletin 28:895–901.

Crump, D., M. Berrill, D. Coulson, D. Lean, L. McGillivray, and A. Smith. 1999. Sensitivity of amphibian embryos, tadpoles, and larvae to enhanced UV-B radiation in natural pond conditions. Canadian Journal of Zoology 77:1956–1966.

Culley, D. D., Jr. 1973. Use of Bullfrogs in biological research. American Zoologist 13:85–90.

Culley, D. D., Jr. 1981. Have we turned the corner on Bullfrog culture? Aquaculture Magazine 7:20–24.

Cummins, H. 1920. The role of voice and coloration in spring migrations and sex recognition in frogs. Journal of Experimental Zoology 30:325–343.

Cummins, L., T. Craig, A. Sacerdote, K. Terio, and K. Gamble. 2012. Hypobiotic enteric rhabditiform nematodes in wild-caught Smooth Greensnakes (*Opheodrys vernalis*). Presentation to the Association of Reptile and Amphibian Veterinarians, Oakland, California, USA.

Cundall, D., and A. Pattishall. 2011. Foraging time investment in an urban population of Watersnakes (*Nerodia sipedon*). Journal of Herpetology 45:174–177.

Cunjak, R. A. 1986. Winter habitat of Northern Leopard Frogs, *Rana pipiens*, in a southern Ontario stream. Canadian Journal of Zoology 64:255–257.

Cunningham, D. S., and G. M. Burghardt. 1999. A comparative study of facial grooming after prey ingestion in colubrid snakes. Ethology 105:913–936.

Cunnington, D. C., and R. J. Brooks. 2000. Optimal egg size theory: does predation by fish affect egg size in *Ambystoma maculatum*? Journal of Herpetology 34:46–53.

Cunnington, G. M., and J. E. Cebek. 2005. Mating and nesting behavior of the Eastern Hognose Snake (*Heterodon platirhinos*) in the northern portion of its range. American Midland Naturalist 154:474–478.

Cupp, P. V., Jr. 2001. Chemical detection of predatory salamanders by the Four-toed Salamander, *Hemidactylium scutatum*. Abstract. American Zoologist 41:1420.

Cupp, P. V., Jr., J. P. Ritzert, D. Dourson, and J. Dourson. 2012. *Storeria occipitomaculata* (Red-bellied Snake). Behavior and reproduction. Herpetological Review 43:152.

Curnutt, J. L., S. L. Pimm, and B. A. Maurer. 1996. Population variability of sparrows in space and time. Oikos 76:131–144.

Currie, W., and E. D. Bellis. 1969. Home range and movements of the Bullfrog, *Rana catesbeiana* Shaw, in an Ontario pond. Copeia 1969:688–692.

Curtin, C. G. 1997. Biophysical analysis of the impact of shifting land use on Ornate Box Turtles, Wisconsin, USA. Pages 31–36 *in* J. Van Abbema, editor. Proceedings: conservation, restoration, and management of tortoises and turtles—an international conference. Joint Publication of the New York Turtle and Tortoise Society and the WCS Turtle Recovery Program, New York, New York, USA.

Curtin, C. G. 1998. Plasticity in Ornate Box Turtle thermal preference. Journal of Herpetology 32:298–301.

Curtis, J. T. 1959. The vegetation of Wisconsin: an ordination of plant communities. University of Wisconsin Press, Madison, USA.

Curtis, L. 1949. The snakes of Dallas County, Texas. Field and Laboratory 17:1–13.

Cushman, S. A. 2006. Effects of habitat loss and fragmentation on amphibians: a review and prospectus. Biological Conservation 128:231–240.

Cutright, N. J., B. R. Harriman, and R. W. Howe. 2006. Atlas of the breeding birds of Wisconsin. Wisconsin Society for Ornithology, Inc., Waukesha, USA.

Cutter, W. L. 1958. Food habits of the Swift Fox in northern Texas. Journal of Mammalogy 39:527–532.

Da Silva, H. R. 1997. Two character states new for hylines and the taxonomy of the genus *Pseudacris*. Journal of Herpetology 31:609–613.

Dahl, T. E. 1990. Wetland losses in the United States: 1780s to 1980s. Report to Congress. US Fish and Wildlife Service, Washington, D.C., USA. Archived by WebCite at http://www.webcitation.org/73bmIj9ge. Accessed 1 Nov 2018.

Daigle, C. 1997a. Distribution and abundance of the Chorus Frog, *Pseudacris triseriata*, in Quebec. Pages 73–77 *in* D. M. Green, editor. Amphibians in decline: Canadian studies of a global problem. Society for the Study of Amphibians and Reptiles, Herpetological Conservation 1, St. Louis, Missouri, USA.

Daigle, C. 1997b. Size and characteristics of a Wood Turtle, *Clemmys insculpta*, population in southern Québec. Canadian Field-Naturalist 111:440–444.

Daigle, C., and J. Jutras. 2005. Quantitative evidence of decline in a southern Québec Wood Turtle (*Glyptemys insculpta*) population. Journal of Herpetology 39:130–132.

Dale, J. M., B. Freedman, and J. Kerekes. 1985. Acidity and associated water chemistry of amphibian habitats in Nova Scotia. Canadian Journal of Zoology 63:97–105.

Dalrymple, G. H., and N. G. Reichenbach. 1981. Interactions between the Prairie Garter Snake (*Thamnophis radix*) and the Common Garter Snake (*T. sirtalis*) in Killdeer Plains, Wyandot County, Ohio. Ohio Biological Survey Biological Notes 15:244–250.

Dalrymple, G. H., T. M. Steiner, R. J. Nodell, and F. S. Bernardino, Jr. 1991. Seasonal activity of the snakes of Long Pine Key, Everglades National Park. Copeia 1991:294–302.

Dam, A. F. 2005. Landscape structure affects different eastern Ontario anuran species at different spatial scales. Thesis, Carleton University, Ontario, Canada.

Dananay, K. L., K. L. Krynak, T. L. Krynak, and M. F. Bernard. 2015. Legacy of road salt: apparent positive larval effects counteracted by negative postmetamorphic effects in Wood Frogs. Environmental Toxicology and Chemistry 34:2417–2424.

Dare, O. K., and M. R. Forbes. 2009. Patterns of trematode and nematode lungworm infections in Northern Leopard Frogs and Wood Frogs from Ontario, Canada. Journal of Helminthology 83:339–343.

Dart, R. C., and J. McNally. 2001. Efficacy, safety, and use of snake antivenoms in the United States. Annals of Emergency Medicine 37:181–188.

Darwin, C. R. 1859. On the origin of species by means of natural selection, or the preservation of favoured races in the struggle for life. John Murray, London, England.

Daszak, P., A. Strieby, A. A. Cunningham, J. E. Longcore, C. Brown., and D. Porter. 2004. Experimental evidence that the Bullfrog (*Rana catesbeiana*) is a potential carrier of chytridiomycosis, an emerging fungal disease of amphibians. Herpetological Journal 14:201–207.

Daszak, P., D. E. Scott, A. M. Kilpatrick, C. Faggioni, J. W. Gibbons, and D. Porter. 2005. Amphibian population declines at Savannah River site are linked to climate, not chytridiomycosis. Ecology 86:3232–3237.

Datta, G., and T. Tu. 1993. Toxicology and biochemistry of Colubridae venom. Journal of Toxicology and Toxin Review 12:63–89.

Daudin F. M. 1803. Histoire Naturelle générale et particuliére des Reptiles. F. Dufart, Paris 8:1–439.

Daufresne, M., K. Lengfellner, and U. Sommer. 2009. Global warming benefits the small in aquatic ecosystems. Proceedings of the National Academy of Sciences of the United States of America 106:12788–12793.

Davic, R. D., and H. H. Welsh, Jr. 2004. On the ecological role of salamanders. Annual Review of Ecology, Evolution, and Systematics 35:405–434.

David, R. S., and R. G. Jaeger. 1981. Prey location through chemical cues by a terrestrial salamander. Copeia 1981:435–440.

Davidson, E. W., M. Parris, J. P. Collins, J. E. Longcore, A. P. Pessier, and J. Brunner. 2003. Pathogenicity and transmission of chytridiomycosis in Tiger Salamanders (*Ambystoma tigrinum*). Copeia 2003:601–607.

Davis, A. K. 2007. Walking trails in a nature preserve alter terrestrial salamander distributions. Natural Areas Journal 27:385–289.

Davis, C. L., E. W. Teitsworth, and D. A. Miller. 2018. Combining data sources to understand drivers of Spotted Salamander (*Ambystoma maculatum*) population abundance. Journal of Herpetology 52:116–126.

Davis, D. D. 1936. Courtship and mating behavior in snakes. Field Museum of Natural History, Zoology Series 20:257–290.

Davis, M. 1981. Aspects of the social and spatial experience of Eastern Box Turtles, *Terrapene carolina carolina*. Dissertation. University of Tennessee, Knoxville, USA.

Dawson, E. T. 1953. Wildlife in Texas. Texas Game and Fish Commission Bulletin 28, Austin, Texas, USA.

de Buffrénil, V., A. Canoville, S. E. Evans, and M. Laurin. 2015. Histological study of karaurids, the oldest known (stem) Urodeles. Historical Biology 27:109–114.

de Queiroz, A. 1984. Effects of prey type on the prey-handing behavior of the Bullsnake, *Pituophis melanoleucus*. Journal of Herpetology 18:333–336.

de Quieroz, A., and R. Lawson. 1994. Phylogenetic relationships of the garter snakes based on DNA sequence and allozyme variation. Biological Journal of the Linnean Society 53:209–229.

de Queiroz, A., R. Lawson, and J. A. Lemos-Espinal. 2002. Phylogenetic relationships of North American garter snakes (*Thamnophis*) based on four mitochondrial genes: how much DNA sequence is enough? Molecular Phylogenetics and Evolution 22:315–329.

de Queiroz, K. 1998. The general lineage concept of species, species criteria, and the process of speciation: a conceptual unification and terminological recommendations. Pages 57–75 *in* D. J. Howard and S. H. Berlocher, editors. Endless forms: species and speciation. Oxford University Press, New York, New York, USA.

de Solla, S. R., and K. J. Fernie. 2004. Characterization of contaminants in Snapping Turtles (*Chelydra serpentina serpentina*) from Canadian Lake Erie areas of concern: St. Clair River, Detroit River, and Wheatley Harbour. Environmental Pollution 132:101–112.

de Solla, S. R., and P. A. Martin. 2007. Toxicity of nitrogenous fertilizers to eggs of Snapping Turtles (*Chelydra serpentina*) in field and laboratory exposures. Environmental Toxicology and Chemistry 26:1890–1895.

de Solla, S. R., K. J. Fernie, R. J. Letcher, S. G. Chu, K. G. Drouillard, and S. Shahmiri. 2007. Snapping Turtles (*Chelydra serpentina*) as bioindicators in Canadian areas of concern in the Great Lakes Basin. 1. Plybrominated diphenyl ethers, polychlorinated biphenyls, and organochlorine pesticies in eggs. Environmental Science and Technology 41:7252–7259.

Dearborn, N. 1932. Foods of some predatory furbearing animals in Michigan. University of Michigan School of Forestry and Conservation Bulletin 1:1–52.

DeBruyne, R. L., J. T. H. Coleman, J. R. Jackson, L. G. Rudstam, and A. J. VanDeValk. 2013. Analysis of prey selection by Double-crested Cormorants: a 15-year diet study in Oneida Lake, New York. Transactions of the American Fisheries Society 142:430–446.

DeFoliart, G. R. 1967. *Aedes canadensis* (Theobald) feeding on Blanding's Turtles. Journal of Medical Entomology 4:31.

DeGraaf, J. D., and D. G. Nein. 2010. Predation of Spotted Turtle (*Clemmys guttata*) hatchling by Green Frog (*Rana clamitans*). Northeastern Naturalist 17:667–670.

DeGraaf, R. M., and D. D. Rudis. 1983. Amphibians and reptiles of New England. Habitats and natural history. University of Massachusetts Press, Amherst, USA.

DeGregorio, B. A., J. V. Manning, N. Bieser, and B. A. Kingsbury. 2011a. The spatial ecology of the Eastern Massasauga (*Sistrurus c. catenatus*) in northern Michigan. Herpetologica 67:71–79.

DeGregorio, B. A., B. J. Putnam, and B. A. Kingsbury. 2011b. Which habitat selection method is most applicable to snakes? Case studies of the Eastern Massasauga (*Sistrurus catenatus*) and Eastern Fox Snake (*Pantherophis gloydi*). Herpetological Conservation and Biology 6:372–382.

DeGregorio, B. A., P. J. Weatherhead, and T. D. Tuberville. 2013. Time in captivity affects foraging behavior of ratsnakes: implications for translocation. Herpetological Conservation and Biology 8:581–590.

Deiss, W. A. 1985. The making of a naturalist: Spencer F. Baird, the early years. Archives of Natural History 1:141–148.

Delvinquier, B. L. J., and S. S. Desser. 1996. Opalinidae (Sacromastigophora) in North American amphibia. Genus *Opalina* Purkinje & Valentin, 1835. Systematic Parasitology 33:33–51.

Delzell, D. E. 1958. Spatial movement and growth of *Hyla crucifer*. Thesis, University of Michigan, Ann Arbor, USA.

Demastes, J. W., J. M. Eastman, J. S. East, and C. Spolsky. 2007. Phylogeography of the Blue-spotted Salamander, *Ambystoma laterale* (Caudata: Ambystomatidae). American Midland Naturalist 157:149–161.

deMaynadier, P. G., and M. L. Hunter, Jr. 1995. The relationship between forest management and amphibian ecology: a review of the North American literature. Environmental Review 3:230–261.

deMaynadier, P. G., and M. L. Hunter, Jr. 1998. Effects of silvicultural edges on the distribution and abundance of amphibians in Maine. Conservation Biology 12:340–352.

deMaynadier, P. G., and M. L. Hunter, Jr. . 1999. Forest canopy closure and juvenile emigration by pool-breeding amphibians in Maine. Journal of Wildlife Management 63:441–450.

Densmore, C. L., and D. E. Green. 2007. Diseases of amphibians. ILAR Journal 48:235–254.

Denton, R. D., K. L. Greenwald, and H. L. Gibbs. 2017. Locomotor endurance predicts differences in realized dispersal between sympatric sexual and unisexual salamanders. Functional Ecology 31:915–926.

Dernehl, P. H. 1902. Place-modes of *Acris gryllus* for Madison. Bulletin of the Wisconsin Natural History Society 2:75–83.

Desser, S. S., and J. R. Barta. 1984. An intraerythrocytic virus and rickettsia of frogs from Algonquin Park, Ontario. Canadian Journal of Zoology 62:1521–1524.

Deutschman, M. R., and J. J. Peterka. 1988. Secondary production of Tiger Salamanders (*Ambystoma tigrinum*) in three North Dakota prairie lakes. Canadian Journal of Fisheries and Aquatic Sciences 45:691–697.

DeVault, T. L., and A. R. Krochmal. 2002. Scavenging by snakes: an examination of the literature. Herpetologica 58:429–436.

Devillars, J., and J. M. Exbryant. 1992. Ecotoxicology of chemicals to amphibians. Gordon and Breach, Philadelphia, Pennsylvania, USA.

Devine, M. C. 1975. Copulatory plugs in snakes: enforced chastity. Science 187:844–845.

Devine, M. C. 1977. Copulatory plugs, restricted mating opportunities and reproductive competition among male garter snakes. Nature 267:345–346.

DeVito, J. 2003. Metamorphic synchrony and aggregation as antipredator responses in American Toads. Oikos 103:75–80.

Devitt, T., B. I. Crother, A. Meier, F. Burbrink, and J. Boundy. 2007. *Pituophis catenifer* (Bullsnake). Maximum length. Herpetological Review 38:209–210.

Diamond, J. 1989. Overview of recent extinctions. Pages 37–41 *in* D. Western, and M. Pearl, editors. Conservation for the Twenty-first Century. Wildlife Conservation International, New York, New York, USA.

Diamond, S. A., G. S. Peterson, J. E. Tietge, and G. T. Ankley. 2002. Assessment of the risk of solar ultraviolet radiation to amphibians. III. Predictions of impacts in selected northern midwestern wetlands. Environmental Science and Technology 36:2866–2874.

Diana, S. G., W. J. Resetarits, Jr., D. J. Schaeffer, K. B. Beckmen, and V. R. Beasley. 2000. Effects of atrazine on amphibian growth and survival in artificial aquatic communities. Environmental Toxicology and Chemistry 19:2961–2967.

Diaz, D. E. 2016. Northern snakes are more abundant in old fields than in forest. Thesis, University of Ottawa, Ontario, Canada.

Dickerson, M. C. 1906. The frog book: North American toads and frogs with a study of the habits and life histories of those of the northeastern United States. Doubleday, Page & Company, New York, New York, USA.

Dickerson, M. C. 1969. The frog book. Dover Publications, New York, New York, USA.

Dickinson, W. E. 1944. The mosquitoes of Wisconsin. Bulletin of the Public Museum of the City of Milwaukee 8:269–365.

Dickinson, W. E. 1946. The lizards and snakes of Wisconsin. Thesis, Marquette University, Milwaukee, Wisconsin, USA.

Dickinson, W. E. 1949. Field guide to the lizards and snakes of Wisconsin. Milwaukee Public Museum Press, Popular Science Handbook Series 2, Milwaukee, Wisconsin, USA.

Dickinson, W. E. 1950. Recent additions to the records of the distribution of the reptiles in Wisconsin. Transactions of the Wisconsin Academy of Sciences, Arts and Letters 40:71–77.

Dickinson, W. E. 1958. T. E. B. Pope 1879–1958. Lore 8:138.

Dickinson, W. E. 1960. Handbook of Wisconsin fishes. Milwaukee Public Museum Press, Popular Science Handbook Series 8, Milwaukee, Wisconsin, USA.

Dickinson, W. E. 1965. Handbook of amphibians and turtles of Wisconsin. Milwaukee Public Museum Press, Popular Science Handbook Series 10, Milwaukee, Wisconsin, USA.

Dickinson, W. E. 1972. The amphibians and reptiles of Forest, Florence, and Marinette counties with special reference to the Pine, Popple, and Pike watersheds. Transactions of the Wisconsin Academy of Sciences, Arts and Letters 60:303–308.

Dickinson, W. E. 1975. The amphibians and reptiles of Forest, Florence and Marinette Counties with special reference to the Pine, Popple and Pike Watersheds. Wisconsin Academy of Sciences, Arts and Letters 60:303–308.

Dickson, N. J. 2002. The natural history and possible extirpation of Blanchard's Cricket Frog, *Acris crepitans blanchardi*, in West Virginia. Thesis, Marshall University, Huntington, USA.

Dieter, C. D., L. A. Dixon, S. I. Ronningen, and T. Ronningen. 2014. Survey of turtles nesting on the Missouri River on the South Dakota-Nebraska border. Great Plains Research 24:111–118.

Dijak, W. D., and F. R. Thompson, III. 2000. Landscape and edge effects of the distribution of mammalian predators in Missouri. Journal of Wildlife Management 64:209–216.

Dileo, M. F., J. R. Row, and S. C. Lougheed. 2010. Discordant patterns of population structure for two co-distributed snake species across a fragmented Ontario landscape. Diversity and Distributions 16:571–581.

Diller, L. V., and R. L. Wallace. 1996. Comparative ecology of two snake species (*Crotalus viridis* and *Pituophis melanoleucus*) in southwestern Idaho. Herpetologica 52:343–360.

Dillon, R. M., S. P. Boyle, J. D. Litzgus, and D. Lesbarrères. 2020. Build it and some will use it: a test of road ecopassages for Eastern Gartersnakes. Journal of Herpetology 54:19–23.

DiMauro, D., and M. L. Hunter, Jr. 2002. Reproduction of amphibians in natural and anthropogenic temporary pools in managed forests. Forest Science 48:397–406.

Dinkelacker, S. A., J. P. Costanzo, J. B. Iverson, and R. E. Lee, Jr. 2004. Cold-hardiness and dehydration resistance of hatchling Blanding's Turtles (*Emydoidea blandingii*): implications for overwintering in a terrestrial habitat. Canadian Journal of Zoology 82:594–600.

Dinkelacker, S. A., J. P. Costanzo, J. B. Iverson, and R. E. Lee, Jr. 2005. Survival and physiological responses of hatchling Blanding's Turtles (*Emydoidea blandingii*) to submergence in normoxic and hypoxic water under simulated winter conditions. Physiological and Biochemical Zoology 78:356–363.

Dirig, R. 1978. A large diurnal breeding assemblage of American Toads, *Bufo americanus americanus*, in the Catskill Mountains. Pitch Pine Naturalist 4:1–2.

Distel, C. A., and M. D. Boone. 2009. Effects of aquatic exposure to the insecticide Carbaryl and density on aquatic and terrestrial growth and survival in American Toads. Environmental Toxicology and Chemistry 28:1963–1969.

Ditmars, R. L. 1907. The reptile book. Doubleday, Page & Co., New York, New York, USA.

Ditmars, R. L. 1912. Feeding habits of serpents. Zoologica 1:196.

Ditmars, R. L. 1936. The reptiles of North America. Doubleday, Garden City, New York, USA.

Ditmars, R. L. 1939. Thrills of a naturalist's quest. Macmillan Publishing Company, United Kingdom, London, England.

DiTomaso, J. M., M. L. Brooks, E. B. Allen, R. Minnich, P. M. Rice, and G. B. Kyser. 2006. Control of invasive weeds with prescribed burning. Weed Technology 20:535–548.

Dix, R. L., and J. E. Butler. 1954. The effects of fire on a dry, thin-soil prairie in Wisconsin. Journal of Rangeland Management 7:265–268.

do Amaral, J. P. S. 1999. Lip-curling in Redbelly Snakes (*Storeria occipitomaculata*): functional morphology and ecological significance. Journal of Zoology 248:289–293.

Dobbs, E. K., M. G. Brown, J. W. Snodgrass, D. R. Ownby, and W. Lutterschmidt. 2012. Salt toxicity to treefrogs (*Hyla chrysoscelis*) depends on depth. Herpetologica 68:22–30.

Dodd, C. K., Jr. 1977. Preliminary observations on the reactions of certain salamanders of the genus *Ambystoma* (Amphibia, Urodela, Ambystomatidae), to a small colubrid snake (Reptilia, Serpentes, Colubridae). Journal of Herpetology 11:222–223.

Dodd, C. K., Jr. 1993. Strategies for snake conservation. Pages 363–388 *in* R. A. Seigel, and J. T. Collins, editors. Snakes: ecology and behavior. McGraw-Hill, Inc., New York, USA.

Dodd, C. K., Jr. 2001. North American box turtles: a natural history. University of Oklahoma Press, Animal Natural History Series Volume 6, Norman, USA.

Dodd, C. K., Jr. 2004. The amphibians of Great Smoky Mountains National Park. University of Tennessee Press, Knoxville, USA.

Dodd, C. K., Jr. 2005. Population manipulations. Pages 265–270 *in* M. Lannoo, editor. Amphibian Declines: the conservation and status of United States species. University of California Press, Berkeley, USA.

Dodd, C. K., Jr. 2013. Frogs of the United States and Canada. Johns Hopkins University Press, Baltimore, Maryland, USA.

Dodd, C. K., Jr., and R. A. Seigel. 1991. Relocation, repatriation, and translocation of amphibians and reptiles: are they conservation strategies that work? Herpetologica 47:336–350.

Dodd, C. K., Jr., R. Franz, and S. A. Johnson. 1997. Shell injuries and anomalies in an insular population of Florida Box Turtles (*Terrapene carolina bauri*). Herpetological Natural History 5:66–72.

Dodd, C. K., Jr., W. J. Barichivich, and L. L. Smith. 2004. Effectiveness of a barrier wall and culverts in reducing wildlife mortality on a heavily traveled highway in Florida. Biological Conservation 118:619–631.

Dodd, C. K., Jr., W. J. Barichivich, S. A. Johnson, and J. S. Staiger. 2007. Changes in a northwestern Florida gulf coast herpetofaunal community over a 28-y period. American Midland Naturalist 158:29–48.

Dodson, S. I. 2008. Biodiversity in southern Wisconsin storm-water retention ponds: correlations with watershed cover and productivity. Lake and Reservoir Management 24:370–380.

Dodson, S. I., and R. A. Lillie. 2001. Zooplankton communities of restored depressional wetlands in Wisconsin. Wetlands 21:292–300.

Doherty, J. A., and H. C. Gerhardt. 1984. Evolutionary and neurobiological implications of selective phonotaxis in the Spring Peeper (*Hyla crucifer*). Animal Behaviour 32:875–881.

Dole, J. W. 1965. Spatial relations in natural populations of the Leopard Frog, *Rana pipiens* Schreber, in northern Michigan. American Midland Naturalist 74:464–478.

Dole, J. W. 1971. Dispersal of recently metamorphosed Leopard Frogs, *Rana pipiens*. Copeia 1971:221–228.

Donner, D. M., C. A. Ribic, A. J. Beck, D. Higgins, D. Eklund, and S. Reinecke. 2015. Woodland pond salamander abundance in relation to forest management and environmental conditions in northern Wisconsin. Journal of North American Herpetology 2015:34–42.

DonnerWright, D. M., M. A. Bozek, J. R. Probst, and E. M. Anderson. 1999. Responses of turtle assemblage to environmental gradients in the St. Croix River in Minnesota and Wisconsin, U.S.A. Canadian Journal of Zoology 77:989–1000.

Dorland, A., T. Rytwinski, and L. Fahrig. 2014. Do roads reduce Painted Turtle (*Chrysemys picta*) populations? PLoS ONE 9:e98414.

Doroff, A. M., and L. B. Keith. 1990. Demography and ecology of an Ornate Box Turtle (*Terrapene ornata*) population in south-central Wisconsin. Copeia 1990:387–399.

Doughty, P. 1994. Critical thermal minima of garter snakes (*Thamnophis*) depend on species and body size. Copeia 1994:537–540.

Douglas, M. E. 1979. Migration and sexual selection in *Ambystoma jeffersonianum*. Canadian Journal of Zoology 57:2303–2310.

Douglas, M. E., and B. L. Monroe. 1981. A comparative study of topographical orientation in *Ambystoma* (Amphibia: Caudata). Copeia 1981:460–463.

Douglass, N. J., and H. K. Reinert. 1982. The utilization of fallen logs as runways by small mammals. Proceedings of the Pennsylvania Academy of Science 56:162–164.

Dovčiak, M., P. A. Osborne, D. A. Patrick, and J. P. Gibbs. 2013. Conservation potential of prescribed fire for maintaining habitats and populations of an endangered rattlesnake, *Sistrurus c. catenatus*. Endangered Species Research 22:51–60.

Dowling, H. G. 1951. A practical solution of the nomenclature of the Common Gartersnake and Ribbonsnake. Copeia 1951:309–310.

Dowling, Z., T. Hartwig, E. Kiviat, and F. Keesing. 2010. Experimental management of nesting habitat for the Blanding's Turtle (*Emydoidea blandingii*). Ecological Restoration 28:154–159.

Downs, F. L. 1989a. *Ambystoma laterale* (Hallowell) Blue-spotted Salamander. Pages 102–107 *in* R. A. Pfingsten and F. L. Downs, editors. Salamanders of Ohio. Ohio State University, Bulletin of the Ohio Biological Survey New Series 7, Columbus, Ohio, USA.

Downs, F. L. 1989b. *Ambystoma maculatum* (Shaw), Spotted Salamander. Pages 108–125 *in* R. A. Pfingsten and F. L. Downs, editors. Salamanders of Ohio. Bulletin of the Ohio Biological Survey New Series 7, Columbus, USA.

Downs. F. L. 1989c. *Ambystoma tremblayi* Comeau Tremblay's Salamander. Pages 166–172 *in* R. A. Pfingsten and F. L. Downs, editors. Salamanders of Ohio. Ohio State University, Bulletin of the Ohio Biological Survey New Series 7, Columbus, Ohio, USA.

Drake, D. L., B. H. Ousterhout, J. R. Johnson, T. L. Anderson, W. E. Peterman, C. D. Shulse, D. J. Hocking, K. L. Lohraff, E. B. Harper, T. A. G. Rittenhouse, B. B. Rothermel, L. S. Eggert, and R. D. Semlitsch. 2015. Pond-breeding amphibian community composition in Missouri. American Midland Naturalist 174:180–187.

Dreslik, M. J. 2005. Ecology of the Eastern Massasauga (*Sistrurus catenatus catenatus*) from Carlyle Lake, Clinton County, Illinois. Dissertation, University of Illinois, Urbana-Champaign, Illinois, USA.

Dreslik, M. J., and C. A. Phillips. 2005. Turtle communities in the Upper Midwest, USA. Journal of Freshwater Ecology 20:149–164.

Dreslik, M. J., A. R. Kuhns, and C. A. Phillips. 2005. Structure and composition of a southern Illinois freshwater turtle assemblage. Northeastern Naturalist 12:173–186.

Dreslik, M. J., D. B. Shepard, B. C. Jellen, and C. A. Phillips. 2016a. Movement and home range of the Eastern Massasauga (*Sistrurus catenatus*) at its southern range limit. Pages 168–178 *in* M. J. Dreslik, W. K. Hayes, S. J. Beaupre, and S. P. Mackessy, editors. The Biology of Rattlesnakes II. ECO Herpetological Publishing and Distribution, Rodeo, New Mexico, USA.

Dreslik, M. J., S. J. Baker, D. B. Wylie, D. B. Shepard, and C. A. Phillips. 2016b. Temporal trends in population size and structure of the Eastern Massasauga (*Sistrurus catenatus*) in south-central Illinois. Pages 26–136 *in* M. J. Dreslik, W. K. Hayes, S. J. Beaupre, and S. P. Mackessy, editors. The Biology of Rattlesnakes II. ECO Herpetological Publishing and Distribution, Rodeo, New Mexico.

Driscoll, D. A. 2004. Extinction and outbreaks accompany fragmentation of a reptile community. Ecological Applications 14:220–240.

Driskell, E. A., D. L. Miller, S. L. Swist, and Z. S. Gyimesi. 2009. PCR detection of ranavirus in adult anurans from the Louisville Zoological Garden. Journal of Zoo and Wildlife Medicine 40:559–563.

Drought, J. F. 1987a. *Trionyx spinferus spiniferus* (Eastern Spiny Softshell). Geographic distribution. Herpetological Review 18:21.

Drought, J. F. 1987b. *Chrysemys picta marginata* (Midland Painted Turtle). Geographic distribution. Herpetological Review 18:21.

Drought, J. F. 1987c. *Nerodia sipedon sipedon* (Northern Water Snake). Geographic distribution. Herpetological Review 18:41.

Drummond, H. 1983. Aquatic foraging in garter snakes: a comparison of specialists and generalists. Behaviour 86:1–30.

DuBois, R. D. 1996. Recent observations of deformed anurans in Wisconsin. Presentation at the Third Annual Meeting of the North American Amphibian Monitoring Program, Patuxent Wildlife Research Center, Laurel, Maryland, USA.

Dubois, Y., G. Blouin-Demers, B. Shipley, and D. Thomas. 2009. Thermoregulation and habitat selection in Wood Turtles *Glyptemys insculpta*: chasing the sun slowly. Journal of Animal Ecology 78:1023–1032.

Dubrovsky, N. M., K. R. Burow, G. M. Clark, J. M. Gronberg, P. A. Hamilton, K. J. Hitt, D. K. Mueller, M. D. Munn, B. T. Nolan, L. J. Puckett, M. G. Rupert, T. M. Short, N. E. Spahr, L. A. Sprague, and W. G. Wilber. 2010. The quality of our nation's waters. Nutrients in the nation's streams and groundwater, 1992–2004. US Geological Survey Circular 1350, Reston, Virginia, USA.

Ducey, P. K., and E. D. Brodie, Jr. 1983. Salamanders respond selectively to contacts with snakes: survival advantage of alternative antipredator strategies. Copeia 1983:1036–1041.

Ducey, P. K., and J. Dulkiewicz. 1994. Ontogenetic variation in antipredator behavior of the newt *Notophthalmus viridescens*: comparisons of terrestrial adults and efts in field and laboratory studies. Journal of Herpetology 28:530–533.

Ducey, P. K., and J. Heuer. 1991. Effects of food availability on interspecific aggression in salamanders of the genus *Ambystoma*. Canadian Journal of Zoology 69:288–290.

Ducey, P. K., and P. Ritsema. 1988. Intraspecific aggression and responses to marked substrates in *Ambystoma maculatum* (Caudata: Ambystomatidae). Copeia 1988:1008–1013.

Ducey, P. K., K. Schramm, and N. Cambry. 1994. Interspecific aggression between the sympatric salamanders, *Ambystoma maculatum* and *Plethodon cinereus*. American Midland Naturalist 131:320–329.

Ducey, P. K., M. Messere, K. Lapoint, and S. Noce. 1999. Lumbricid prey and potential herpetofaunal predators of the invading terrestrial flatworm *Bipalium adventitium* (Turbellaria: Tricladida: Terricola). American Midland Naturalist 141:305–314.

Duellman, W. E. 1954. Observations on autumn movements of the salamander *Ambystoma tigrinum tigrinum* in southeastern Michigan. Copeia 1954:156–157.

Duellman, W. E., and L. Trueb. 1986. Biology of amphibians. McGraw-Hill Publishing Company, New York, New York, USA.

Duellman, W. E., and L. Trueb. 1994. Biology of amphibians. Second Edition. Johns Hopkins University Press, Baltimore, Maryland, USA.

Duellman, W. E., and R. G. Zweifel. 1962. A synopsis of the lizards of the *sexlineatus* group (genus *Cnemidophorus*). Bulletin of the American Museum of Natural History 123:155–210.

Duellman, W. E., A. B. Marion, and S. B. Hedges. 2016. Phylogenetics, classification, and biogeography of the treefrogs (Amphibia: Anura: Arboranae). Zootaxa 4104:1–109.

Duffus, A. L. J., B. D. Pauli, K. Wozney, C. R. Brunetti, and M. Berrill. 2008. Frog virus 3-like infections in aquatic amphibian communities. Journal of Wildlife Diseases 44:109–120.

Dukart, M. G. 1984. Some aspects of the ecology of the Plains Gartersnake *Thamnophis radix haydeni* (Kennicott) in southeastern Saskatchewan. Thesis, University of Regina, Saskachewan, Canada.

Duncan, R. B. 2005. *Ambystoma maculatum* (Spotted Salamander). Predation. Herpetological Review 36:294.

Dundee, H. A. 1947. Note on salamanders collected in Oklahoma. Copeia 1947:117–120.

Dundee, H. A., and M. C. Miller. 1968. Aggregative behavior and habitat conditioning by the Prairie Ringneck Snake, *Diadophis punctatus arnyi*. Tulane Studies in Zoology and Botany 15:41–58.

Dunlap, D. G., and J. E. Platz. 1981. Geographic variation of proteins and call in *Rana pipiens* from the northcentral United States. Copeia 4:876–879.

Dunn, E. R. 1935. The survival value of specific characteristics. Copeia 1935:85–98.

Dunson, W. A. 1960. Aquatic respiration in *Trionyx spinifer asper*. Herpetologica 16:277–283.

Durbian, F. E. 2006. Effects of mowing and summer burning on the Massasauga (*Sistrurus catenatus*). American Midland Naturalist 155:329–334.

Durbian, F. E., R. S. King, T. H. Crabill, H. Lambert-Doherty, and R. A. Seigel. 2008. Massasauga home range patterns in the Midwest. Journal of Wildlife Management 72:754–759.

Durham, L., and G. W. Bennett. 1963. Age, growth, and homing in the Bullfrog. Journal of Wildlife Management 27:107–123.

Durner, G. M., and J. E. Gates 1993. Spatial ecology of Black Rat Snakes on Remmington Farms,

Maryland. Journal of Wildlife Management 57:812–826.

DuShane, D. P., and C. Hutchinson. 1944. Differences in size and developmental rate between eastern and midwestern embryos of *Ambystoma maculatum*. Ecology 25:414–424.

Dusi, J. L. 1949. The natural occurrence of "redleg," *Pseudomonas hydrophila*, in a population of American Toads, *Bufo americanus*. Ohio Journal of Science 49:70–71.

Duvall, D., R. Herskowitz, and J. Trupiano-Duvall. 1980. Responses of Five-lined Skinks (*Eumeces fasciatus*) and Ground Skinks (*Scincella lateralis*) to conspecific and interspecific chemical cues. Journal of Herpetology 14:121–127.

Dyal, L. A. 2006. Novel courtship behaviors in three small eastern *Plethodon* species. Journal of Herpetology 40:55–65.

Dyer, J. M. 2006. Revisiting the deciduous forests of eastern North America. BioScience 56:341–352.

Dyer, W. G. 1994. Some helminths of the Racer, *Coluber constrictor*, and the Eastern Hognose Snake, *Heterodon platirhinos*, in southern Illinois. Transactions of the Illinois State Academy of Science 87:43–45.

Dymond, A. J. R., and F. E. J. Fry. 1932. Notes on the breeding habits of the Green Snake (*Liopeltis vernalis*). Copeia 1932:1–2.

Dyrkacz, S. 1975. The second recorded occurrence of an albino *Thamnophis radix radix* (Bard and Girard). Herpetological Review 6:10.

Dyrkacz, S. 1977. The natural history of the Eastern Milk Snake (Reptilia, Serpentes, Colubridae) in a disturbed environment. Journal of Herpetology 11:155–159.

Dyrkacz, S. 1981. Recent instances of albinism in North American amphibians and reptiles. Society for the Study of Amphibians and Reptiles, Herpetological Circular 11, Lawrence, Kansas, USA.

Eagle, A. C., E. M. Hay-Chmielewski, K. T. Cleveland, A. L. Derosier, M. E. Herbert, and R. A. Rustem. 2005. Michigan's wildlife action plan. Michigan Department of Natural Resources, Lansing, USA.

Earl, J. E., and H. H. Whiteman. 2015. Are commonly used fitness predictors accurate? A meta-analysis of amphibian size and age at metamorphosis. Copeia 103:297–309.

Earl, J. E., and M. J. Gray. 2014. Introduction of ranavirus to isolated Wood Frog populations could cause local extinction. EcoHealth 11:581–592.

Earl, J. E., and R. D. Semlitsch. 2015. Importance of forestry practices relative to microhabitat and microclimate changes for juvenile pond-breeding amphibians. Forest Ecology and Management 357:151–160.

Easterla, D. A. 1971. A breeding concentration of Four-toed Salamanders, *Hemidactylium scutatum*, in southeastern Missouri. Journal of Herpetology 5:194–195.

Eaton, B. R., W. M. Tonn, C. A. Paszkowski, A. J. Danylchuk, and S. M. Boss. 2005. Indirect effects of fish winterkills on amphibian populations in boreal lakes. Journal of Zoology 83:1532–1539.

Echaubard, P., K. Little, B. Pauli, and D. Lesbarreres. 2010. Context-dependent effects of ranaviral infection on Northern Leopard Frog life history traits. PLoS ONE 5:e13723.

Eckblad, J. W. 1986. The ecology of Pools 11–13 of the Upper Mississippi River: a community profile. US Fish and Wildlife Service, Biological Report 85(7.8), Slidell, Louisiana, USA.

Edblom, T. L. 2001. Exposure of Northern Leopard Frog (*Rana pipiens*) to a pollution gradient in the Green Bay and Fox River ecosystem, Wisconsin (USA): early development, sexual development, and impacts of ultraviolet-B radiation. Thesis, University of Wisconsin, Madison, USA.

Eddy, S. B. 1976. Population ecology of the Leopard Frog, *Rana pipiens pipiens* Schreber, at Delta Marsh, Manitoba. Thesis, University of Manitoba, Winnipeg, Canada.

Edge, C. B., B. D. Steinberg, R. J. Brooks, and J. D. Litzgus. 2010. Habitat selection by Blanding's Turtles (*Emydoidea blandingii*) in a relatively pristine landscape. EcoScience 17:90–99.

Edge, C. B., B. D. Steinberg, R. J. Brooks, and J. D. Litzgus. 2009. Temperature and site selection by Blanding's Turtles (*Emydoidea blandingii*) during hibernation near the species' northern range limit. Canadian Journal of Zoology 87:825–834.

Edginton, A. N., P. M. Sheridan, G. R. Stephenson, D. G. Thompson, and H. J. Boermans. 2004.

Comparative effects of pH and Vision herbicide on two life stages of four anuran amphibian species. Environmental Toxicology and Chemistry 23:815–822.

Edgren, R. A. 1942. A nesting rendezvous of the Musk Turtle. Chicago Naturalist 5:63.

Edgren, R. A. 1944. Notes on amphibians and reptiles in Wisconsin. American Midland Naturalist 32:495–498.

Edgren, R. A. 1949. Variation in the size of eggs of the turtles *Chelydra s. serpentina* (Linne) and *Sternotherus odoratus* (Latreille). Natural History Miscellanea 53:1.

Edgren, R. A. 1955a. Possible thermo-regulatory burrowing in the lizard *Cnemidophorus sexlineatus*. Natural History Miscellanea 141:1–2.

Edgren, R. A. 1955b. The natural history of the hog-nosed snakes, genus *Heterodon*: a review. Herpetologica 11:105–117.

Edgren, R. A. 1956. Egg size in the Musk Turtle, *Sternotherus odoratus* Latreille. Natural History Miscellanea 152:1–3.

Edgren, R. A. 1957. Melanism in Hog-nosed Snakes. Herpetologica 13:131–135.

Edgren, R. A., M. K. Edgren, and L. H. Tiffany. 1953. Some North American turtles and their epizoophytic algae. Ecology 34:733–740.

Edington, J. M. 1965. Some observations on stream temperature. Oikos 15:265–273.

Edmond, B. S., and L. M. Bowe. 2012. Predation of an adult Blanchard's Cricket Frog (*Acris blanchardi*) by a juvenile Northern Watersnake (*Nerodia sipedon*) in Missouri. Missouri Herpetological Association Newsletter 25:21.

Edmonds, J. H. 1998. Population ecology of the Stinkpot Turtle (*Sternotherus odoratus*) in Georgian Bay, Ontario. Thesis, University of Guelph, Canada.

Edmonds, J. H., and R. J. Brooks. 1996. Demography, sex ratio, and sexual size dimorphism in a northern population of Common Musk Turtles (*Sternotherus odoratus*). Canadian Journal of Zoology 74:918–925.

Edwards, J. R., K. L. Koster, and D. L. Swanson. 2000. Time course for cryoprotectant synthesis in the freeze-tolerant Chorus Frog, *Pseudacris triseriata*. Comparative Biochemistry and Physiology Part A: Physiology 125:367–375.

Egan, R. S., and P. W. C. Paton. 2004. Within-pond parameters affecting oviposition by Wood Frogs and Spotted Salamanders. Wetlands 24:1–13.

Egstad, H. M. 1933. In the alumni world. Wisconsin Alumni Magazine 34:184.

Ehrenfeld, J. G., and D. W. Ehrenfeld. 1973. Externally secreting glands of freshwater and sea turtles. Copeia 1973:305–314.

Ehrlich, D. 1979. Predation by Bullfrog tadpoles (*Rana catesbeiana*) on eggs and newly hatched larvae of the Plains Leopard Frog (*Rana blairi*). Bulletin of the Maryland Herpetological Society 15:25–26.

Eigenbrod, F., S. J. Hecnar, and L. Fahrig. 2008. The relative effects of road traffic and forest cover on anuran populations. Biological Conservation 141:35–46.

Eigenbrod, F., S. J. Hecnar, and L. Fahrig. 2011. Sub-optimal study design has major impacts on landscape-scale inference. Biological Conservation 144:298–305.

Eisenrich, K. M., S. M. Kelley, and C. L. Rowe. 2009. Latent mortality of juvenile Snapping Turtles from the upper Hudson River, New York, exposed maternally and via the diet to Polychlorinated Biphenyls (PCBs). Environmental Science and Technology 43:6052–6057.

Eisner, T., W. E. Conner, K. Hicks, K. R. Dodge, H. I. Rosenberg, T. H. Jones, M. Cohen, and J. Meinwald. 1977. Stink of the Stinkpot Turtle identified: α-phenylalkanoic acids. Science 196:1347–1349.

Elick, G. E., and J. A. Sealander. 1972. Comparative water loss related to habitat in small colubrid snakes. American Midland Naturalist 88:429–439.

Ellis, M. M., and J. Henderson. 1915. Amphibia and Reptilia of Colorado. Part 2. University of Colorado Studies 11:253–263.

Ellison, K. S., and C. A. Ribic. 2012. Nest defense: grassland bird responses to snakes. Pages 149–160 *in* C. A. Ribic, F. R. Thompson, III, and P. J. Pietz, editors. Video Surveillance of nesting birds. Studies in Avian Biology 43. University of California Press, Berkeley, USA.

Ellner, L. R., and W. H. Karasov. 1993. Latitudinal variation in the thermal biology of Ornate Box Turtles. Copeia 1993:447–455.

Emery, A. R., A. H. Berst, and K. Kodaira. 1972. Under-ice observations of wintering sites of Leopard Frogs. Copeia 1972:123–126.

Emery, S. M., and K. L. Gross. 2005. Effects of timing of prescribed fire on the demography of an invasive plant, Spotted Knapweed *Centaurea maculosa*. Journal of Applied Ecology 42:60–69.

Emlen, S. T. 1968. Territoriality in the Bullfrog, *Rana catesbeiana*. Copeia 1968:240–243.

Emlen, S. T. 1976. Lek organization and mating strategies in the Bullfrog. Behavioral Ecology and Sociobiology 1:283–313.

Engel, T. C. 1980. Effects of urban development on vertebrate wildlife populations in Schmeeckle Reserve, University of Wisconsin–Stevens Point. Thesis, University of Wisconsin, Stevens Point, USA.

Engelhardt, R. L., and J. W. Engelhardt. 2017. *Notophthalmus viridescens* (Eastern Newt). Geographic distribution. Herpetological Review 48:582.

Engeman, R. M., E. R. Martin, H. T. Smith, J. Woolard, C. K. Crady, B. Constantin, M. Stahl, and N. P. Groninger. 2006. Impact of predation of sea turtle nests when predator control was removed midway through the nesting season. Wildlife Research 33:187–193.

Enriquez, D. 2006. Water, snakes delay hospital; site may house threatened species. Milwaukee Journal Sentinel. 9 March 2006; section News:B1.

Epstein, E., W. Smith, J. Dobberpuhl, and A. Galvin. 1999a. Biotic inventory and analysis of the Brule River State Forest. Wisconsin Department of Natural Resources, Bureau of Endangered Resources PUB-ER-799 1999, Madison, USA. Archived by WebCite at http://www.webcitation.org/73bnCwY9d. Accessed 1 Nov 2018.

Epstein, E., W. Smith, J. Dobberpuhl, and A. Galvin. 1999b. Biotic inventory and analysis of the Northern Highland-American Legion State Forest. Wisconsin Department of Natural Resources, Bureau of Endangered Resources PUB-ER-093 99, Madison, USA. Archived by WebCite at http://www.webcitation.org/73bn1TlM4. Accessed 1 Nov 2018.

Epstein, E. J., E. J. Judziewicz, and E. A. Spencer. 2002. Wisconsin's natural communities. Wisconsin Department of Natural Resources, Bureau of Endangered Resources, Madison, Wisconsin, USA. Archived by WebCite at http://www.webcitation.org/73bmk0xSi. Accessed 1 Nov 2018.

Ernst, C. H. 1985. Midland Smooth Softshell *Trionyx muticus* Le Sueur. Pages 294–295 *in* H. H. Genoways and F. J. Brenner, editors. Species of concern in Pennsylvania. Carnegie Museum of Natural History Special Publication 11, Pittsburgh, Pennsylvania, USA.

Ernst, C. H. 1986a. Ecology of the turtle, *Sternotherus odoratus,* in southeastern Pennsylvania. Journal of Herpetology 20:341–352.

Ernst, C. H. 1986b. Environmental temperatures and activities in the Wood Turtle, *Clemmys insculpta*. Journal of Herpetology 20:222–229.

Ernst, C. H. 2001a. An overview of the North American turtle Genus *Clemmys* Ritgen, 1828. Chelonian Conservation and Biology 4:211–216.

Ernst, C. H. 2001b. Some ecological parameters of the Wood Turtle, *Clemmys insculpta,* in southeastern Pennsylvania. Chelonian Conservation and Biology 4:94–99.

Ernst, C. H. 2002. *Storeria occipitomaculata*. Catalogue of American Amphibians and Reptiles 759.1.

Ernst, C. H. 2003a. Ecological parameters of the Northern Brown Snake, *Storeria dekayi*. Herpetological Bulletin 86:10–19.

Ernst, C. H. 2003b. Natural history of the Queen Snake, *Regina septemvittata,* in southeastern Pennsylvania, USA. Herpetological Bulletin 85:2–11.

Ernst, C. H. 2008. Systematics, taxonomy, and geographic distribution of the Snapping Turtles, family Chelydridae. Pages 5–13 *in* A. C. Steyermark, M. S. Finkler, and R. J. Brooks, editors. Biology of the Snapping Turtle. Johns Hopkins University Press, Baltimore, Maryland, USA.

Ernst, C. H. 2012. *Storeria*. Catalogue of American Amphibians and Reptiles 900.1.

Ernst, C. H., and E. M. Ernst. 2003. Snakes of the United States and Canada. Smithsonian Books, Washington, D.C., USA.

Ernst, C. H., and E. M. Ernst. 2006. Synopsis of helminths endoparasitic in snakes of the United States and Canada. Society for the Study of Amphibians and Reptiles, Herpetological Circular 34, Shoreview, Minnesota, USA.

Ernst, C. H., and E. M. Ernst. 2011. Venomous reptiles of the United States, Canada, and northern Mexico. John Hopkins University Press, Baltimore, Maryland, USA.

Ernst, C. H., and J. E. Lovich. 2009. Turtles of the United States and Canada. Second edition. Johns Hopkins University Press, Baltimore, Maryland, USA.

Ernst, C. H., and R. W. Barbour. 1972. Turtles of the United States and Canada. University Press of Kentucky, Lexington, USA.

Ernst, C. H., A. F. Laemmerzahl, and T. R. Creque. 2006. A review of morphological and pattern variation in the Painted Turtle, *Chrysemys picta*, in Missouri, USA, with an alternate hypothesis of the origin of *Chrysemys picta marginata*. Herpetological Bulletin 95:6–15.

Ernst, C. H., J. E. Lovich, and R. W. Barbour. 1994. Turtles of the United States and Canada. Smithsonian Institution Press, Washington, D.C., USA.

Ernst, C. H., J. M. Orr, and T. R. Creque. 2003. *Carphophis vermis*. Catalogue of American Amphibians and Reptiles 775.1–775.4.

Ernst, C. H., S. C. Belfit, S. W. Sekscienski, and A. F. Laemmerzahl. 1997. The amphibians and reptiles of Ft. Belvoir and Northern Virginia. Bulletin of the Maryland Herpetological Society 33:1–62.

Ernst, C. H., T. P. Boucher, S. W. Sekscienski, and J. C. Wilgembusch. 1995. Fire ecology of the Florida Box Turtle (*Terrapene carolina bauri*). Herpetological Review 26:185–187.

Ernst, C. H., T. S. B. Akre, J. C. Wilgenbusch, T. P. Wilson, and K. Mills. 1999. Shell disease in turtles in the Rappahanock River, Virginia. Herpetological Review 30:214–215.

Erwin, W. J., and R. H. Stasiak. 1979. Vertebrate mortality during the burning of a reestablished prairie in Nebraska. American Midland Naturalist 101:247–249.

Esch, G. W., and J. C. Fernandez. 1994. Snail-trematode interactions and parasite community dynamics in aquatic systems: a review. American Midland Naturalist 131:209–237.

Eskew, E. A., S. J. Price, and M. E. Dorcas. 2010. Survival and recruitment of semi-aquatic turtles in an urbanized region. Urban Ecosystems 13:365–374.

Etheridge, K., and L. C. Wit. 1982. *Cnemidophorus sexlineatus* (Six-lined Racerunner). Cannibalism. Herpetological Review 13:19.

Etheridge, K., L. C. Wit, and J. C. Sellers. 1983. Hibernation in the lizard *Cnemidophorus sexlineatus* (Lacertilia: Teiidae). Copeia 1983:206–214.

Evermann, G. W., and H. W. Clark. 1916. The turtles and batrachians of the Lake Maxinkuckee region. Proceedings of the Indiana Academy of Science 1916:472–518.

Evrard, J. O. 2000. Birds and amphibians of selected pine barrens wetlands. Transactions of the Wisconsin Academy of Sciences, Arts and Letters 88:37–48.

Evrard, J. O., and M. E. Canfield. 2000. Blanding's Turtles in Crex Meadows Wildlife Area. Transactions of the Wisconsin Academy of Sciences, Arts and Letters 88:49–55.

Evrard, J. O., and S. R. Hoffman. 2000. Amphibians and reptiles captured in drift fences in northwest Wisconsin pine barrens. Journal of the Iowa Academy of Science 107:182–186.

Ewert, B., R. Korb, L. S. Meyers, R. M. Lange, and J. M. Kapfer. 2007. *Sternotherus odoratus* (Stinkpot). Geographic distribution. Herpetological Review 38:350.

Ewert, M. A. 1969. Seasonal movements of the toads *Bufo americanus* and *B. cognatus* in northwestern Minnesota. Dissertation, University of Minnesota, Minneapolis, USA.

Ewert, M. A. 1979. *Trionyx muticus muticus* (Midland Smooth Softshell). Herpetological Review 10:102.

Ewert, M. A. 2005. *Sternotherus odoratus* (Common Musk Turtle). Size and reproduction. Herpetological Review 36:314.

Ewert, M. A., and C. E. Nelson. 1991. Sex determination in turtles: diverse patterns and some possible adaptive values. Copeia 1991:50–69.

Eycleshymer, A. C. 1906. The habits of *Necturus maculosus*. American Naturalist 40:123–136.

Faccio, S. D. 2003. Postbreeding emigration and habitat use by Jefferson and Spotted Salamanders in Vermont. Journal of Herpetology 37:479–489.

Fadool, D. A., M. Wachiowiak, and J. H. Brann. 2001. Patch-clamp analysis of voltage-activated and chemically activated currents in the vomeronasal organ of *Sternotherus odoratus* (Stinkpot/Musk Turtle). Journal of Experimental Biology 204:4199–4212.

Fahrig, L., J. H. Pedlar, S. E. Pope, P. D. Taylor, and J. F. Wegner. 1995. Effect of road traffic on amphibian density. Biological Conservation 73:177–182.

Faisal, M. 2006. Aquatic animal health laboratory: preliminary laboratory report. Report to the

Michigan Department of Natural Resources, Lansing, Michigan, USA.

Faivovich, J., C. F. B. Haddad, P. C. A. Garcia, D. R. Frost, J. A. Campbell, and W. C. Wheeler. 2005. Systematic review of the frog family Hylidae, with special reference to Hylinae: phylogenetic analysis and taxonomic revision. Bulletin of the American Museum of Natural History 294:1–240.

Faragher, S. G., and R. G. Jaeger. 1997. Distributions of adult and juvenile Red-backed Salamanders: testing new hypotheses regarding territoriality. Copeia 1997:410–414.

Farrell, R. F., and T. E. Graham. 1991. Ecological notes on the turtle *Clemmys insculpta* in northwestern New Jersey. Journal of Herpetology 25:1–9.

Fassbender, R., and D. J. Watermolen. 2002. Bird predation on the Mudpuppy (*Necturus maculosus maculosus*). Bulletin of the Chicago Herpetological Society 37:137–138.

Faust, L., J. Szymanski, and M. Redmer. 2011. Range wide extinction risk modeling for the Eastern Massasauga Rattlesnake (*Sistrurus catenatus catenatus*). Alexander Center for Applied Population Biology, Lincoln Park Zoo, Chicago, Illinois, USA.

Feaver, P. E. 1977. The demography of a Michigan population of *Natrix sipedon* with discussions of ophidian growth and reproduction. Dissertation, University of Michigan, Ann Arbor, USA.

Federle, W., W. J. P. Barnes, W. Baumgartner, P. Drechsler, and J. M. Smith. 2006. Wet but not slippery: boundary friction in tree frog adhesive toe pads. Journal of the Royal Society Interface 3:689–697.

Feldman, C. R., and G. S. Spicer. 2007. Comparative phylogeography of woodland reptiles in California. Molecular Ecology 15:2201–2222.

Feldman, C. R., and J. F. Parham. 2002. Molecular phylogenetics of emydine turtles: taxonomic revision of the evolution of shell kinesis. Molecular Phylogenetics and Evolution 22:388–398.

Fellers, G. M. 1979a. Aggression, territoriality, and mating behaviour in North American treefrogs. Animal Behavior 27:107–119.

Fellers, G. M. 1979b. Mate selection in the Gray Treefrog, *Hyla versicolor*. Copeia 1979:286–290.

Fenoilio, D. B., G. O. Graening, and J. F. Stout. 2005. Seasonal movement patterns of Pickerel Frogs (*Rana palustris*) in an Ozark cave and trophic implications supported by stable isotope evidence. Southwestern Naturalist 50:385–389.

Ferguson, D. E., J. P. McKeown, O. S. Bosarge, and H. F. Landreth. 1968. Sun-compass orientation of Bullfrogs. Copeia 1968:230–235.

Ferrara, C. R., R. C. Vogt, M. R. Harfush, R. S. Sousa-Lima, E. Albavera, and A. Tavera. 2014. First evidence of Leatherback Turtle (*Dermochelys coriacea*) embryos and hatchlings emitting sounds. Chelonian Conservation and Biology 13:110–114.

Feuer, R. C. 1971. Ecological factors in success and dispersal of the Snapping Turtle *Chelydra serpentina* (Linnaeus). Bulletin of the Philadelphia Herpetological Society 19:3–14.

Ficetola, G. F., C. Coic, M. Detaint, M. Berroneau, O. Lorvelec, and C. Miaud. 2007. Pattern and distribution of the American Bullfrog *Rana catesbeiana* in Europe. Biological Invasions 9:767–772.

Figiel, C. R., Jr., and R. D. Semlitsch. 1990. Population variation in survival and metamorphosis of larval salamanders (*Ambystoma maculatum*). Copeia 1990:818–826.

Figueroa, A., A. D. McKelvy, L. L. Grismer, C. D. Bell, and S. P. Lailvaux. 2016. A species-level phylogeny of extant snakes with description of a new colubrid subfamily and genus. PLoS ONE 11:e0161070.

Figueras, M., and R. L. Burke. 2017. *Glyptemys insculpta* (Wood Turtle). Long-term sperm storage. Herpetological Review 48:175.

Findlay, C. S., L. Lenton, and L. Zheng. 2001. Land-use correlates of anuran community richness and composition in southeastern Ontario wetlands. Ecoscience 8:336–343.

Finkler, M. S., and D. L. Claussen. 1999. Influence of temperature, body size, and inter-individual variation on forced and voluntary swimming and crawling speeds in *Nerodia sipedon* and *Regina septemvittata*. Journal of Herpetology 33:62–71.

Finkler, M. S., and J. J. Kolbe. 2008. Physiology and ecology of hatchling Snapping Turtles. Pages 158–167 *in* A. C. Steyermark, M. S. Finkler, and R. J. Brooks, editors. Biology of the Snapping Turtle. John Hopkins University Press, Baltimore, Maryland, USA.

Finkler, M. S., C. J. Hayes, and L. Rifai. 2014. Sexual dimorphisms in metabolism, organ mass, and reproductive energetics in pre-breeding

American Toads (*Anaxyrus americanus*). Copeia 2014:447–453.

Finley, R. W. 1976. Original vegetation cover of Wisconsin, compiled from US General Land Office notes. Map (1:500,000). US Forest Service, North Central Forest Experiment Station, St. Paul, Minnesota, USA.

Finneran, L. C. 1949. A sexual aggregation of the garter snake *Thamnophis butleri* (Cope). Copeia 1949:141–144.

Fischer, A., P. Marshall, and A. Camp. 2013. Disturbances in deciduous temperate forest ecosystems of the northern hemisphere: their effects on both recent and future forest development. Biodiversity Conservation 22:1863–1893.

Fischer, J., and D. B. Lindenmayer. 2007. Landscape modification and habitat fragmentation: a synthesis. Global Ecology and Biogeography 16:265–280.

Fish & Wildlife Compensation Program. 2010. Turtle mortalities: the deadly reach of invasive plants. 2010:Spring/Summer Newsletter.

Fishbeck, D. W., and J. C. Underhill. 1971. Distribution of stripe polymorphisim in Wood Frogs, *Rana sylvatica* LeConte, from Minnesota. Copeia 1971:253–259.

Fisher, M. C., and T. W. J. Garner. 2007. The relationship between the emergence of *Batrachochytrium dendrobatidis*, the international trade in amphibians and introduced amphibian species. Fungal Biology Reviews 21:2–9.

Fisher, M. C., T. W. J. Garner, and S. F. Walker. 2009. Global emergence of *Batrachochytrium dendrobatidis* and amphibian chytridiomycosis in space, time, and host. Annual Review of Microbiology 63:291–310.

Fitch, H. S. 1954. Life history and ecology of the Five-lined Skink, *Eumeces fasciatus*. University of Kansas Publications, Museum of Natural History 8:1–156.

Fitch, H. S. 1956. Temperature responses of free-living amphibians and reptiles of northeastern Kansas. University of Kansas Publications, Museum of Natural History 8:417–476.

Fitch, H. S. 1958a. Home ranges, territories, and seasonal movements of vertebrates of the Natural History Reservation. University of Kansas Publications, Museum of Natural History 11:63–326.

Fitch, H. S. 1958b. Natural history of the Six-Lined Racerunner (*Cnemidophorus sexlineatus*). University of Kansas Publications, Museum of Natural History 11:11–62.

Fitch, H. S. 1963a. Natural history of the Black Rat Snake (*Elaphe o. obsoleta*) in Kansas. Copeia 1963:649–658.

Fitch, H. S. 1963b. Natural history of the Racer, *Coluber constrictor*. University of Kansas Publications, Museum of Natural History 15:351–468.

Fitch, H. S. 1965. An ecological study of the Garter Snake *Thamnophis sirtalis*. University of Kansas Publications, Museum of Natural History 15:493–564.

Fitch, H. S. 1970. Reproductive cycles in lizards and snakes. University of Kansas, Museum of Natural History Miscellaneous Publications 52:1–247.

Fitch, H. S. 1974. Observations on the food and nesting of the Broad-winged Hawk (*Buteo platypterus*) in northeastern Kansas. Condor 76:331–333.

Fitch, H. S. 1975. A demographic study of the Ringneck Snake (*Diadophis punctatus*) in Kansas. University of Kansas, Museum of Natural History Miscellaneous Publications 62:1–53.

Fitch, H. S. 1980. *Thamnophis sirtalis*. Common Garter Snake. Catalogue of American Amphibians and Reptiles 270.1–270.3.

Fitch, H. S. 1982. Resources of a snake community in prairie-woodland habitat of northeastern Kansas. Pages 83–97 *in* N. J. Scott, Jr., editor. Herpetological communities. US Fish and Wildlife Service Research Report 13, Washington, D.C., USA.

Fitch, H. S. 1985. Variation in clutch and litter size in new world reptiles. University of Kansas, Museum of Natural History Miscellaneous Publications 76:1–76.

Fitch, H. S. 1989. A field study of the Slender Glass Lizard, *Ophisaurus attenuatus*, in northeastern Kansas. University of Kansas, Occasional Papers of the Museum of Natural History 125:1–50.

Fitch, H. S. 1993. Relative abundance of snakes in Kansas. Transactions of the Kansas Academy of Science 96:213–224.

Fitch, H. S. 1999. A Kansas snake community: composition and changes over 50 years. Krieger Publishing Company, Malabar, Florida, USA.

Fitch, H. S. 2003a. A comparative study of loss and regeneration of lizard tails. Journal of Herpetology 37:395–399.

Fitch, H. S. 2003b. Reproduction in snakes of the Fitch Natural History Reservation in northeastern Kansas. Journal of Kansas Herpetology 6:21–24.

Fitch, H. S. 2006a. Collapse of a fauna: reptiles and turtles of the University of Kansas Natural History Reservation. Journal of Kansas Herpetology 17:10–13.

Fitch, H. S. 2006b. Ecological succession on a natural area in northeastern Kansas from 1948 to 2006. Herpetological Conservation and Biology 1:1–5.

Fitch, H. S. 2006c. Gopher Snakes, Bullsnakes and Pine Snakes. Journal of Kansas Herpetology 17:16–17.

Fitch, H. S., and A. F. Echelle. 2006. Abundance and biomass of twelve species of snakes native to northeastern Kansas. Herpetological Review 37:161–165.

Fitch, H. S., and H. W. Shirer. 1971. A radiotelemetric study of spatial relationships in some common snakes. Copeia 1971:118–128.

Fitch, H. S., and M. V. Plummer. 1975. A preliminary ecological study of the Soft-shelled Turtle *Trionyx muticus* in the Kansas River. Israel Journal of Zoology 24:28–42.

Fitch, H. S., and P. L. von Achen. 1977. Spatial relationships and seasonality in the skinks *Eumeces fasciatus* and *Scincella laterale* in northeastern Kansas. Herpetologica 33:303–313.

Fitch, H. S., and R. O. Bare. 1978. A field study of the Red-tailed Hawk in eastern Kansas. Transactions of the Kansas Academy of Science 81:1–13.

Fitch, H. S., and R. R. Fleet. 1970. Natural history of the Milk Snake (*Lampropeltis triangulum*) in northeastern Kansas. Herpetologica 26:387–396.

Fitch, K. L. 1959. Observations on the nesting habits of the Mudpuppy, *Necturus maculosus* Rafinesque. Copeia 1959:339–340.

FitzGerald, G. J., and J. R. Bider. 1974a. Influence of moon phase and weather factors on locomotory activity in *Bufo americanus*. Oikos 25:338–340.

FitzGerald, G. J., and J. R. Bider. 1974b. Seasonal activity of the toad *Bufo americanus* in southern Quebec as revealed by a sand-transect technique. Canadian Journal of Zoology 52:1–5.

Fitzgerald, K. T., H. M. Smith, and L. J. Guillette. 1981. Nomenclature of the diploid species of the diploid-tetraploid *Hyla versicolor* complex. Journal of Herpetology 15:356–360.

Fitzinger, L. J. F. L. 1843. Systema reptilium. Fasciculus primus. Braumüller et Seidel, Wien, Vienna, Austria.

Fitzpatrick, B. M., J. S. Placyk, Jr., M. L. Niemiller, G. S. Casper, and G. M. Burghardt. 2008. Distinctiveness in the face of gene flow: hybridization between specialist and generalist gartersnakes. Molecular Ecology 17:4107–4117.

Fitzpatrick, B. M., K. Shook, and R. Izally. 2009. Frequency-dependent selection by wild birds promotes polymorphism in model salamanders. BMC Ecology 9:12.

Flageole, S., and R. Le Clair, Jr. 1992. Etude demographique d'une population de salamanders (*Ambystoma maculatum*) a l'aide de la methode squelatto-chronologique. Canadian Journal of Zoology 70:740–749.

Flaherty, N. C., and J. R. Bider. 1984. Physical structures and the social factor as determinants of habitat use by *Graptemys geographica* in southwestern Quebec. American Midland Naturalist 111:259–266.

Fleming, M. M., L. B. Mills, J. K. Russell, G. R. Smith, and J. E. Rettig. 2011. Effects of trails on Eastern Red-backed Salamanders (*Plethodon cinereus* Green). Herpetology Notes 4:229–232.

Flickinger, E. L., K. A. King, W. F. Stout, and M. M. Mohn. 1980. Wildlife hazards from Furadan 3G applications to rice in Texas. Journal of Wildlife Management 44:190–197.

Flores, G. 1978. The reproductive ecology of the Northern Spring Peeper, *Hyla crucifer crucifer*, in southern Connecticut. Bulletin of the New York Herpetological Society 14:22–25.

Florsheim, J. L., J. F. Mount, and A. Chin. 2008. Bank erosion as a desirable attribute of rivers. BioScience 58:519–529.

Fogell, D. D., T. J. Leonard, and J. D. Fawcett. 2002. *Crotalus horridus horridus* (Timber Rattlesnake). Habitat. Herpetological Review 33:211–212.

Fong, P. P., L. B. Thompson, G. L. F. Carfagno, and A. J. Sitton. 2016. Long-term exposure to gold nanoparticles accelerates larval metamorphosis without affecting mass in Wood Frogs (*Lithobates sylvaticus*) at environmentally relevant concentrations. Environmental Toxicology and Chemistry 35:2304–2310.

Fontanella, F. M., C. R. Feldman, M. E. Siddall, and F. T. Burbrink. 2008. Phylogeography of *Diadophis punctatus*: extensive lineage diversity and repeated patterns of historical demography in a trans-continental snake. Molecular Phylogenetics and Evolution 46:1049–1070.

Force, E. R. 1925. Notes on reptiles and amphibians of Okmulgee County, Oklahoma. Copeia 1925:25–27.

Force, E. R. 1930. The amphibians and reptiles of Tulsa County, Oklahoma, and vicinity. Copeia 1930:25–39.

Force, E. R. 1931. Habits and birth of young of the lined snake, *Tropidoclonion lineatum* (Hallowell). Copeia 1931:51–53.

Ford, D. K., and D. Moll. 2004. Sexual and seasonal variation in foraging patterns in the Stinkpot, *Sternotherus odoratus*, in southwestern Missouri. Journal of Herpetology 38:296–301.

Ford, N. B. 1981. Seasonality of pheromone trailing behavior in two species of garter snake, *Thamnophis* (Colubridae). Southwestern Naturalist 26:385–388.

Ford, N. B. 1982a. Courtship behavior of the Queen Snake, *Regina septemvittata*. Herpetological Review 13:72.

Ford, N. B. 1982b. Species specificity of sex pheromone trails of sympatric and allopatric garter snakes (*Thamnophis*). Copeia 1982:10–13.

Ford, N. B. 1986. The role of pheromone trails in the sociobiology of snakes. Pages 261–278 *in* D. Duvall, D. Müller-Schwarze, and R. M. Silverstein, editors. Chemical signals in vertebrates 4: ecology, evolution, and comparative biology. Springer, New York, New York, USA.

Ford, N. B., and C. W. Schofield. 1984. Species specificity of sex pheromone trails in the Plains Gartersnake, *Thamnophis radix*. Herpetologica 40:51–55.

Ford, N. B., and D. W. Killebrew. 1983. Reproductive tactics and female body size in Butler's Garter Snake, *Thamnophis butleri*. Journal of Herpetology 17:271–275.

Ford, N. B., and P. M. Hampton. 2008. Ontogenetic and sexual differences in diet in an actively foraging snake, *Thamnophis proximus*. Canadian Journal of Zoology 87:254–261.

Ford, N. B., V. A. Cobb, and J. Stout. 1991. Species diversity and seasonal abundance of snakes in a mixed pine-hardwood forest of eastern Texas. Southwestern Naturalist 36:171–177.

Ford, N. B., V. A. Cobb, and W. W. Lamar. 1990. Reproductive data on snakes from northeastern Texas. Texas Journal of Science 42:355–368.

Foreman, D. 2004. Rewilding North America: a vision for conservation in the 21st century. Island Press, Washington, USA.

Forester, D. C., and D. V. Lykens. 1986. Significance of satellite males in a population of Spring Peepers (*Hyla crucifer*). Copeia 1986:719–724.

Forester, D. C., and D. V. Lykens. 1988. The ability of Wood Frog eggs to withstand prolonged terrestrial stranding: an empirical study. Canadian Journal of Zoology 66:1733–1735.

Forester, D. C., and D. V. Lykens. 1991. Age structure in a population of Red-spotted Newts from the Allegheny Plateau of Maryland. Journal of Herpetology 25:373–376.

Forester, D. C., and K. J. Thompson. 1998. Gauntlet behaviour as a male sexual tactic in the American Toad (Amphibia: Bufonidae). Behaviour 135:99–119.

Forester, D. C., and R. Czarnowsky. 1985. Sexual selection in the Spring Peeper, *Hyla crucifer* (Amphibia, Anura): role of the advertisement call. Behaviour 92:112–127.

Forester, D. C., J. W. Snodgrass, K. Marsalek, and Z. Lanham. 2006. Post-breeding dispersal and summer home range of female American Toads (*Bufo americanus*). Northeastern Naturalist 13:59–72.

Forman, R. T. T., and L. E. Alexander. 1998. Roads and their major ecological effects. Annual Review of Ecology and Systematics 29:207–231.

Forman, R. T. T., D. Sperling, J. A. Bissonette, A. P. Clevenger, C. D. Cutshall, V. H. Dale, L. Fahrig, R. France, C. R. Goldman, K. Heanue, J. A. Jones, F. J. Swanson, T. Turrentine, and T. C. Winter. 2003. Road ecology: science and solutions. Island Press, Washington, D.C., USA.

Formanowicz, D. R., Jr. 1979. Palatability and antipredator behavior of selected *Rana* to the shrew *Blarina*. American Midland Naturalist 101:456–458.

Formanowicz, D. R., Jr. 1986. Anuran tadpole/aquatic insect predator-prey interactions: tadpole size and predator capture success. Herpetologica 42:367–373.

Formanowicz, D. R., Jr., and E. D. Brodie, Jr. 1979. Palatability and antipredator behavior of selected *Rana* to the shrew *Blarina*. American Midland Naturalist 101:456–458.

Formanowicz, D. R., Jr., and E. D. Brodie, Jr. 1982. Relative palatabilities of members of a larval amphibian community. Copeia 1982:91–97.

Forrester, J. A., C. P. Holstege, and J. D. Forrester. 2012. Fatalities from venomous and nonvenomous animals in the United States (1999–2007). Wilderness and Environmental Medicine 23:146–152.

Forson, D. D., and A. Storfer. 2006. Atrazine increases ranavirus susceptibility in the Tiger Salamander, *Ambystoma tigrinum*. Ecological Applications 16:2325–2332.

Forzan, M. J., and J. Wood. 2013. Low detection of ranavirus DNA in wild postmetamorphic Green Frogs, *Rana* (*Lithobates*) *clamitans*, despite previous or concurrent tadpole mortality. Journal of Wildlife Diseases 49:879–886.

Fouquette, M. J., Jr. 1954. Food competition among four sympatric species of garter snakes, genus *Thamnophis*. Texas Journal of Science 6:172–189.

Fowler, H. W. 1918. An albino spring frog in winter. Copeia 61:84.

Fowler, J. A. 1966. A communal nesting site for the Smooth Green Snake in Michigan. Herpetologica 22:12–13.

Fox, J. J. 1985. Ecology and management of the Bullsnake, *Pituophis melanoleucus sayi*, in the Nebraska sandhills. Interim Report to the Crescent Lake National Wildlife Refuge, Ellsworth, Nebraska, USA.

Fox, J. J. 1986. Ecology and management of the Bullsnake, *Pituophis melanoleucus sayi*, in the Nebraska sandhills. Final Report to the Crescent Lake National Wildlife Refuge, Ellsworth, Nebraska, USA.

Fox, S. R. 1981. The American conservation movement: John Muir and his legacy. University of Wisconsin Press, Madison, USA.

Fraker, M. A. 1970. Home range and homing in the Watersnake, *Natrix sipedon sipedon*. Copeia 1970:665–673.

Fraker, M. E. 2009. Predation risk by Green Frog (*Rana clamitans*) tadpoles through chemical cues produced by multiple prey. Behavioral Ecology and Sociobiology 63:1397–1402.

Fraker, M. E. 2010. Risk assessment and anti-predator behavior of Wood Frog (*Rana sylvatica*) tadpoles: a comparison with Green Frog (*Rana clamitans*) tadpoles. Journal of Herpetology 44:390–398.

Franklin, L. A. 2010. Endangered species spawns bridge. Northwest Indiana Times. 17 August 2010. Archived by WebCite at http://www.webcitation.org/73bnXebK2. Accessed 1 Nov 2018.

Franz, R. 1971. Notes on the distribution and ecology of the herpetofauna of northwestern Montana. Bulletin of the Maryland Herpetological Society 7:1–10.

Fraser, D. F. 1980. On the environmental control of oocyte maturation in a plethodontid salamander. Oecologia 46:302–307.

Fratt, T. W. 1991. Trophic ecology of Burbot and Lake Trout in Green Bay and Lake Michigan with information for Burbot on fecundity, gonadosomatic index, Sea Lamprey marks, and intestinal acanthocephalans. Thesis, University of Wisconsin, Stevens Point, USA.

Freda, J. 1983. Diet of larval *Ambystoma maculatum* in New Jersey. Journal of Herpetology 17:1177–179.

Freda, J. 1986. The influence of acidic pond water on amphibians: a review. Water, Air, and Soil Pollution 30:439–450.

Freda, J., and D. G. McDonald. 1993. Toxicity of amphibian breeding ponds in the Sudbury region. Canadian Journal of Fisheries and Aquatic Sciences 50:1497–1503.

Freda, J., and D. H. Taylor. 1992. Behavioral response of amphibian larvae to acidic water. Journal of Herpetology 26:429–433.

Freedberg, S., T. J. Greives, M. A. Ewert, G. E. Demas, N. Beecher, and C. E. Nelson. 2008. Incubation environment affects immune system development in a turtle with environmental sex determination. Journal of Herpetology 42:536–541.

Freedman, B., and P. M. Catling. 1978. Population size and structure of four sympatric species of snakes at Amherstburg, Ontario. Canadian Field-Naturalist 92:167–173.

Freedman, B., and P. M. Catling. 1979. Movements of sympatric species of snakes at Amherstburg, Ontario. Canadian Field-Naturalist 93:399–404.

Frese, P. W. 2003. *Eumeces septentrionalis septentrionalis* (Northern Prairie Skink). Nesting behavior. Herpetological Review 34:143.

Friedel, P., B. A. Young, and J. L. van Hemmen. 2008. Auditory localization of ground-borne vibrations in snakes. Physical Review Letters 100:048701.

Friederici, P. 2008. How rattlesnakes rattle. National Wildlife Federation, Merrifield, Virginia, USA. Archived by WebCite at http://www.webcitation.org/73bnryMEo. Accessed 1 Nov 2018.

Frisbie, M. P., J. P. Costanzo, and R. E. Lee, Jr. 2000. Physiological and ecological aspects of low-temperature tolerance in embryos of the Wood Frog, *Rana sylvatica*. Canadian Journal of Zoology 78:1032–1041.

Fritts, T. 1968. Intrabrood variation in *Opheodrys vernalis* (Harlan). Herpetologica 24:79–82.

Fritz, D. J. 2017. *Hyla versicolor* (Gray Treefrog). Geographic distribution. Herpetological Review 48:584–585.

Fritz, U., C. Schmidt, and C. H. Ernst. 2011. Competing generic concepts for Blanding's, Pacific, and European Pond Turtles (*Emydoidea*, *Actinemys*, and *Emys*)—which is best? Zootaxa 2791:41–53.

Froese, A. D. 1978. Habitat preferences of the Common Snapping Turtle, *Chelydra s. serpentina* (Reptilia, Testudines, Chelydridae). Journal of Herpetology 12:53–58.

Froese, A. D., and G. M. Burghardt. 1974. Food competition in captive juvenile Snapping Turtles, *Chelydra serpentina*. Animal Behavior 22:735–740.

Froese, A. D., and G. M. Burghardt. 1975. A dense natural population of the Common Snapping Turtle (*Chelydra s. serpentina*). Herpetologica 31:204–208.

Froom, B. 1972. Snakes of Canada. McClelland and Stewart Publishing, Toronto, Ontario, Canada.

Froom, B. 1982. Amphibians of Canada. McClelland and Stewart Ltd., Toronto, Ontario, Canada.

Frost, D. 1983. Past occurrence of *Acris crepitans* (Hylidae) in Arizona. Southwest Naturalist 28:105.

Frost, D. R., and D. M. Hillis. 1990. Species in concept and practice: herpetological applications. Herpetologica 46:87–104.

Frost, D. R., R. W. McDiarmid, and J. R. Mendelson, III. 2009. Response to the point of view of Gregory B. Pauly, David M. Hillis, and David C. Cannatella, by the anuran subcommittee of the SSAR/HL/ASIH scientific and standard English names list. Herpetologica 65:136–153.

Frost, D. R., T. Grant, J. Faivovich, R. H. Bain, A. Haas, C. F. B. Haddad, R. O. De Sa, A. Channing, M. Wilkinson, S. C. Donnellan, C. J. Raxworthy, J. A. Campbell, B. L. Blotto, P. Moler, R. C. Drewes, R. A. Nussbaum, J. D. Lynch, D. M. Green, and W. C. Wheeler. 2006. The amphibian tree of life. Bulletin of the American Museum of Natural History 297:1–371.

Froyd, C. A., and K. J. Willis. 2008. Emerging issues in biodiversity and conservation management: the need for a palaeoecological perspective. Quarternary Science Reviews 27:1723–1732.

Fry, J., G. Xian, S. Jin, J. Dewitz, C. Homer, L. Yang, C. Barnes, N. Herold, and J. Wickham. 2011. Completion of the 2006 National Land Cover Database for the conterminous United States. PE&RS 77:858–864.

Frye, A., K. Hardy, A. Hedrick, and J. B. Iverson. 2017. Factors affecting nesting times in the Painted Turtle *Chrysemys picta* in Nebraska. Chelonian Conservation and Biology 16:44–51.

Fuerst, G. S., and C. C. Austin. 2004. Population genetic structure of the Prairie Skink (*Eumeces septentrionalis*): nested clade analysis of post Pleistocene populations. Journal of Herpetology 38:257–268.

Fuhlendorf, S. D., and D. M. Engle. 2001. Restoring heterogeneity on rangelands: ecosystem management based on evolutionary grazing patterns. BioScience 51:625–632.

Fulk, F. D., and J. O. Whitaker, Jr. 1969. The food of *Rana catesbeiana* in three habitats in Owen County, Indiana. Indiana Academy of Science 78:491–496.

Funk, R. S., and J. K. Tucker. 1978. Variation in a large brood of lined snakes, *Tropidoclonion lineatum* (Reptilia, Serpentes, Colubridae). Journal of Herpetology 12:115–117.

Funk, W. C., T. S. Garcia, G. A. Cortina, and R. H. Hill. 2011. Population genetics of introduced

Bullfrogs, *Rana* (*Lithobates*) *catesbeianus*, in the Willamette Valley, Oregon, USA. Biological Invasions 13:651–658.

Fuselier, L., and D. Edds. 1994. Habitat partitioning among three sympatric species of map turtles, genus *Graptemys*. Journal of Herpetology 28:154–158.

Gabor, C. R. 1995. A correlational test of Mathis' hypothesis that bigger salamanders have better territories. Copeia 1995:729–735.

Gabor, C. R., and C. C. Nice. 2004. Genetic variation among populations of Eastern Newts, *Notophthalmus viridescens*: a preliminary analysis based on allozymes. Herpetologica 60:373–386.

Gabor, C. R., and R. G. Jaeger. 1995. Resource quality affects the agonistic behaviour of territorial salamanders. Animal Behavior 49:71–79.

Gage, S. H. 1891. The life history of the Vermilion-spotted Newt (*Diemyctylus viridescens* Raf.). American Naturalist 25:1084–1103.

Gagné, S. A., and L. Fahrig. 2007. Effect of landscape context on anuran communities in breeding ponds in the National Capital Region, Canada. Landscape Ecology 22:205–215.

Gagné, S. A., and L. Fahrig. 2010. Effects of time since urbanization on anuran community composition in remnant urban ponds. Environmental Conservation 37:128–135.

Gahl, M. K. 2007. Spatial and temporal patterns of amphibian disease in Acadia National Park wetlands. Dissertation, University of Maine, Orono, USA.

Gahl, M. K., J. E. Longcore, and J. E. Houlahan. 2012. Varying responses of northeastern North American amphibians to the Chytrid pathogen *Batrachochytrium dendrobatidis*. Conservation Biology 26:135–141.

Gaietto, K. M., S. L. Rumschlag, and M. D. Boone. 2014. Effects of pesticide exposure and the amphibian chytrid fungus on Gray Treefrog (*Hyla chrysoscelis*) metamorphosis. Environmental Toxicology and Chemistry 33:2358–2362.

Gailbraith, D. A., and R. J. Brooks. 1987. Survivorship of adult females in a northern population of Common Snapping Turtles, *Chelydra serpentina*. Canadian Journal of Zoology 65:1581–1586.

Galarneau, S., and J. Masterson. 2001. Water resources of the Sheboygan River basin: supplement to the state of the Sheboygan River basin. Wisconsin Department of Natural Resources PUB-WR-669–01, Madison, USA. Archived by WebCite at http://www.webcitation.org/73dreQMd9. Accessed 3 Nov 2018.

Galbraith, D. A., B. N. White, R. J. Brooks, and P. T. Boag. 1993. Multiple paternity in clutches of Snapping Turtles (*Chelydra serpentina*) detected using DNA fingerprints. Canadian Journal of Zoology 71:318–324.

Galbraith, D. A., C. A. Bishop, R. J. Brooks, W. L. Simser, and K. P. Lampman. 1988. Factors affecting the density of populations of Common Snapping Turtles (*Chelydra serpentina serpentina*). Canadian Journal of Zoology 66:1233–1240.

Galbraith, D. A., R. J. Brooks, and G. P. Brown. 1997. Can management intervention achieve sustainable exploitation of turtles? Pages 186–194 *in* J. Van Abbema, editor. Proceedings of conservation, restoration, and management of tortoises and turtles: an international conference. Joint publication of the New York Turtle and Tortoise Society and WCS Turtle Recovery Program, New York, New York, USA.

Galbraith D. A., R. J. Brooks, and M. E. Obbard. 1989. The influence of growth rate on age and body size at maturity in female Snapping Turtles (*Chelydra serpentina*). Copeia 1989:896–904.

Gallant, A. L., R. W. Klaver, G. S. Casper, and M. J. Lannoo. 2007. Global rates of habitat loss and implications for amphibian conservation. Copeia 2007:967–979.

Gallie, J. A., R. L. Mumme, and S. A. Wissinger. 2001. Experience has no effect on the development of chemosensory recognition of predators by tadpoles of the American Toad, *Bufo americanus*. Herpetologica 57:376–383.

Gamble, L. R., K. McGarigal, C. L. Jenkins, and B. C. Timm. 2006. Limitations of regulated "buffer zones" for the conservation of Marbled Salamanders. Wetlands 26:298–306.

Gamble, T., and A. M. Simons. 2004. Comparison of harvested and nonharvested Painted Turtle populations. Wildlife Society Bulletin 32:1269–1277.

Gamble, T., P. B. Berendzen, H. B. Shaffer, D. E. Starkey, and A. M. Simons. 2008. Species limits and phylogeography of North American cricket frogs (*Acris*: Hylidae). Molecular Phylogenetics and Evolution 48:112–125.

Gans, C., and J. P. Gasc. 1990. Tests on the locomotion of the elongate and limbless reptile *Ophisaurus apodus* (Sauria: Anguidae). Journal of Zoology 220:517–536.

Garber, S. D., and J. Burger. 1995. A 20-yr study documenting the relationship between turtle decline and human recreation. Ecological Applications 5:1151–1162.

Gardiner, D. M., and D. M. Hoppe. 1999. Environmentally induced limb malformations in Mink Frogs (*Rana septentrionalis*). Journal of Experimental Zoology 284:207–216.

Gardiner, L. E., C. M. Somers, J. A. Martino, D. L. Parker, and R. G. Poulin. 2013. Balancing the dumbbell: summer habitats need protection in addition to winter dens for northern snake communities. Journal of Wildlife Management 77:975–982.

Gardner, S. C., and E. Oberdoster. 2005. Toxicology of Reptiles. CRC Press, Boca Raton, Florida, USA.

Gardner, T. A., J. Barlow, and C. A. Peres. 2007. Paradox, presumption in conservation biology: the importance of habitat change for amphibians and reptiles. Biological Conservation 138:166–179.

Gardner-Santana, L. C., and S. J. Beaupre. 2009. Timber Rattlesnakes (*Crotalus horridus*) exhibit elevated and less variable body temperatures during pregnancy. Copeia 2009:363–368.

Garman, H. 1892. A synopsis of the reptiles and amphibians of Illinois. Bulletin of the Illinois State Laboratory of Natural History. Volume 3. Urbana, Illinois, USA.

Garnier, J. H. 1883. The Mink or Hoosier Frog. American Naturalist 17:945–954.

Gartside, D. F., J. S. Rogers, and H. C. Dessauer. 1977. Speciation with little genic and morphologic differentiation in the ribbon snakes *Thamnophis proximus* and *T. sauritus* (Colubridae). Copeia 1977:697–705.

Gartska, W. R., B. Camazine, and D. Crews. 1982. Interactions of behavior and physiology during the annuaa reproductive cycle of the Red-sided Garter Snake (*Thamnophis sirtalis parietalis*). Herpetologica 38:104–123.

Gassen, J. A., E. C. Lawson, and F. W. Madison. 1987. Soils of St. Croix County and their ability to attenuate contaminants. Wisconsin Geological and Natural History Survey, Madison, Wisconsin, USA.

Gates, J. E., and E. L. Thompson. 1982. Small pool habitat selection by Red-spotted Newts in western Maryland. Journal of Herpetology 16:7–15.

Gatten, R. E., Jr. 2008. Energetics of the Snapping Turtle. Pages 80–90 *in* A. C. Steyermark, M. S. Finkler, and R. J. Brooks, editors. Biology of the Snapping Turtle. John Hopkins University Press, Baltimore, Maryland, USA.

Gatz, A. J., Jr. 1971. Critical thermal maxima of *Ambystoma maculatum* (Shaw) and *Ambystoma jeffersonianum* (Green) in relation to timing of breeding. Herpetologica 27:157–160.

Gatz, A. J., Jr. 1973. Algal entry into the eggs of *Ambystoma maculatum*. Journal of Herpetology 7:137–138.

Gatz, A. J., Jr. 1981a. Non-random mating by size in American Toads, *Bufo americanus*. Animal Behaviour 29:1004–1012.

Gatz, A. J., Jr. 1981b. Size selective mating in *Hyla versicolor* and *Hyla crucifer*. Journal of Herpetology 15:114–116.

Gaul, R. W. 2008. Ecological observations of the Northern Brownsnake (*Storeria dekayi*) in an urban environment in North Carolina, USA. Pages 361–363 *in* J. C. Mitchell, R. E. Jung Brown, and B. Bartholomew, editors. Urban Herpetology. Society for the Study of Amphibians and Reptiles, Herpetological Conservation 3, Salt Lake City, Utah, USA.

Gayou, D. C. 1984. Effects of temperature on the mating call of *Hyla versicolor*. Copeia 1984:733–738.

Gehlbach, F. R. 1965. *Ambystoma tigrinum*. Catalogue of American Amphibians and Reptiles 52.1–52.4.

Gehlbach, F. R. 1970. Death-feigning and erratic behavior in leptotyphlopid, colubrid and elapid snakes. Herpetologica 26:24–34.

Gehrt, S. D., G. F. Hubert, Jr., and J. A. Ellis. 2002. Long-term population trends of Raccoons in Illinois. Wildlife Society Bulletin 30:457–463.

Geller, G. A. 2012a. Notes on the nest predation of *Graptemys* at two Wisconsin sites using trail camera monitoring. Chelonian Conservation and Biology 11:197–205.

Geller, G. A. 2012b. Notes on the nesting ecology of Ouachita Map Turtles (*Graptemys ouachitensis*) at two Wisconsin sites using trail camera monitoring. Chelonian Conservation and Biology 11:206–213.

Geller, G. A. 2012c. Reducing predation of freshwater turtle nests with a simple electric fence. Herpetological Review 43:398–403.

Geller, G. A. 2014. *Graptemys ouachitensis* (Ouachita Map Turtle). Interaction with predator. Herpetological Review 45:484–485.

Geller, G. A. 2020. *Graptemys ouachitensis* (Ouachita Map Turtle). Predation. Herpetological Review 51:314–315.

Geller, G. A., G. S. Casper, and B. J. Halstead. 2020a. Dispersal of hatchling Ouachita Map Turtles (*Graptemys ouachitensis*) from natural nests on the lower Wisconsin River, Wisconsin. Chelonian Conservation and Biology 19:236–245.

Geller, G. A., G. S. Casper, and B. J. Halstead. 2020b. Hatchling emergence ecology of Ouachita Map Turtles (*Graptemys ouachitensis*) on the lower Wisconsin River, Wisconsin. Chelonian Conservation and Biology 19:217–235.

Geller, G. A., M. J. Pappas, and R. C. Vogt. 2020c. *Graptemys ouachitensis* (Ouachita Map Turtle). Clutch size. Herpetological Review 51:314.

Geluso, K., and M. J. Harner. 2013. Reexamination of herpetofauna on Mormon Island, Hall County, Nebraska, with notes on natural history. Transactions of the Nebraska Academy of Sciences 33:7–20.

George, C. J., C. W. Boylen, and R. B. Sheldon. 1977. The presence of the Red-spotted Newt, *Notophthalmus viridescens* Rafinesque (Amphibia, Urodela, Salamandridae) in waters exceeding 12 meters in Lake George, New York. Journal of Herpetology 11:87–90.

Gerald, G. W. 2008. Feign versus flight: influences of temperature, body size and locomotor abilities on death feigning in neonate snakes. Animal Behaviour 75:647–654.

Gerald, G. W., M. A. Bailey, and J. N. Holmes. 2006. Habitat utilization of *Pituophis melanoleucus melanoleucus* (Northern Pinesnakes) on Arnold Air Force Base in middle Tennessee. Southeastern Naturalist 5:253–264.

Gergits, W. F., and R. G. Jaeger. 1990a. Field observations of the behavior of the Red-backed Salamander (*Plethodon cinereus*): courtship and agonistic interactions. Journal of Herpetology 24:93–95.

Gergits, W. F., and R. G. Jaeger. 1990b. Site attachment by the Red-backed Salamander, *Plethodon cinereus*. Journal of Herpetology 24:91–93.

Gerhardt, H. C. 1974. Mating call differences between eastern and western populations of the treefrog *Hyla chrysoscelis*. Copeia 534–536.

Gerhardt, H. C. 1978. Temperature coupling in the vocal communication system of the Gray Tree Frog, *Hyla versicolor*. Science 199:992–994.

Gerhardt, H. C. 2005. Advertisement-call preferences in diploid-tetraploid treefrogs (*Hyla chrysoscelis* and *Hyla versicolor*): implications for mate choice and the evolution of communication systems. Evolution 59:395–408.

Gerhardt, H. C., B. Diekamp, and M. Ptacek. 1989. Inter-male spacing in choruses of the Spring Peeper, *Pseudacris* (*Hyla*) *crucifer*. Animal Behaviour 38:1012–1024.

Gerhardt, H. C., M. B. Ptacek, L. Barnett, and K. G. Torke. 1994. Hybridization in the diploid-tetraploid treefrogs *Hyla chrysoscelis* and *Hyla versicolor*. Copeia 1994:51–59.

Gerhardt, H. C., M. L. Dyson, and S. D. Tanner. 1996. Dynamic properties of the advertisement calls of Gray Tree Frogs: patterns of variability and female choice. Animal Ecology 7:7–18.

Gerhardt, H. C., S. D. Tanner, C. M. Corrigan, and H. C. Walton. 2000. Female preference functions based on call duration in the Gray Tree Frog (*Hyla versicolor*). Behavioral Ecology 11:663–669.

Germaine, S., and D. Hays. 2007. Distribution and post-breeding environmental relationships of Northern Leopard Frogs (*Rana pipiens*) in Grant County, Washington. Interim Report to the Washington Department of Fish and Wildlife, Olympia, USA.

Germaine, S. S., and D. W. Hays. 2009. Distribution and postbreeding environmental relationships of Northern Leopard Frogs (*Rana* [*Lithobates*] *pipiens*) in Washington. Western North American Naturalist 69:537–547.

Germano, D. J., and P. J. Bishop. 2009. Suitability of amphibians and reptiles for translocation. Conservation Biology 23:7–15.

Germano, D. J., and R. B. Bury. 1998. Age determination in turtles: evidence for annual deposition of scute rings. Chelonian Conservation and Biology 3:123–132.

Germano, D. J., R. B. Bury, and M. Jennings. 2000. Growth and population structure of *Emydoidea blandingii* from western Nebraska. Chelonian Conservation and Biology 3:618–625.

Gertzog, B. J., L. J. Kaplan, D. Nichols, G. R. Smith, and J. E. Rettig. 2011. Avoidance of three herbicide formulations by Eastern Red-backed Salamanders (*Plethodon cinereus*). Herpetological Conservation and Biology 6:237–241.

Getz, L. L. 1958. The winter activities of *Rana clamitans* tadpoles. Copeia 1958:219.

Gibbons, J. W. 1988. The management of amphibians, reptiles, and small mammals in North America: the need for an environmental attitude adjustment. Pages 4–10 *in* R. Szaro, K. Severson, and D. Patton, editors. Management of amphibians, reptiles, and small mammals in North America. US Forest Service General Technical Report RM-166, Flagstaff, Arizona, USA.

Gibbons, J. W. 2003. Terrestrial habitat: a vital component for herpetofauna of isolated wetlands. Wetlands 23:630–635.

Gibbons, J. W., and D. H. Bennett. 1974. Determination of anuran terrestrial activity patterns by a drift fence. Copeia 1974:236–243.

Gibbons, J. W., and J. H. Nelson. 1978. The evolutionary significance of delayed emergence from the nest by hatchling turtles. Evolution 32:297–303.

Gibbons, J. W., and M. E. Dorcas. 2004. North American watersnakes: a natural history. University of Oklahoma Press, Norman, USA.

Gibbons, J. W., and R. D. Semlitsch. 1987. Activity patterns. Pages 396–421 *in* R. A. Seigel, J. T. Collins, and S. S. Novak, editors. Snakes: ecology and evolutionary biology. Macmillan Publishing Company, New York, New York, USA.

Gibbons, J. W., and S. Nelson, Jr. 1968. Observations on the Mudpuppy, *Necturus maculosus*, in a Michigan Lake. American Midland Naturalist 80:562–564.

Gibbons, J. W., D. E. Scott, T. J. Ryan, K. A. Buhlmann, T. D. Tuberville, B. S. Metts, J. L. Greene, T. Mills, Y. Leiden, S. Poppy, and C. T. Winne. 2000. The global decline of reptiles, Déjà Vu amphibians. Bioscience 50:653–666.

Gibbons, J. W., J. L. Greene, and J. D. Congdon. 1983. Drought-related responses of aquatic turtle populations. Journal of Herpetology 17:242–246.

Gibbs, E. L., G. W. Nace, and M. B. Emmons. 1971. The live frog is almost dead. BioScience 21:1027–1034.

Gibbs, H. L., and J. E. Chiucchi. 2011. Deconstructing a complex molecular phenotype: population-level variation in individual venom proteins in Eastern Massasauga Rattlesnakes (*Sistrurus c. catenatus*). Journal of Molecular Evolution 72:383–397.

Gibbs, H. L., and R. D. Denton. 2016. Cryptic sex? Estimates of genome exchange in unisexual mole salamanders (*Ambystoma* sp.). Molecular Ecology 25:2805–2815.

Gibbs, H. L., and S. P. Mackessy. 2009. Functional basis of a molecular adaptation: prey-specific toxic effects of venom from *Sistrurus* rattlesnakes. Toxicon 53:672–679.

Gibbs, H. L., S. J. Corey, G. Blouin-Demers, K. A. Prior, and P. J. Weatherhead. 2006. Hybridization between mtDNA-defined phylogeographic lineages of Black Ratsnakes (*Pantherophis* spp.). Molecular Ecology 15:3755–3767.

Gibbs, J. P. 1998a. Amphibian movements in response to forest edges, roads, and streambeds in southern New England. Journal of Wildlife Management 62:584–589.

Gibbs, J. P. 1998b. Distribution of woodland amphibians along a forest fragmentation gradient. Landscape Ecology 13:263–268.

Gibbs, J. P. 1998c. Genetic structure of Redback Salamander *Plethodon cinereus* populations in continuous and fragmented forests. Biological Conservation 86:77–81.

Gibbs, J. P., and A. R. Breisch. 2001. Climate warming and calling phenology of frogs near Ithaca, New York, 1900–1999. Conservation Biology 15:1175–1178.

Gibbs, J. P., and D. A. Steen. 2008. Trends in sex ratios of turtles in the United States: implications of road mortality. Conservation Biology 19:552–556.

Gibbs, J. P., and W. G. Shriver. 2002. Estimating the effects of road mortality on turtle populations. Conservation Biology 16:1647–1652.

Gibbs, J. P., and W. G. Shriver. 2005. Can road mortality limit populations of pool-breeding amphibians? Wetlands Ecology and Management 13:281–289.

Gibbs, J. P., A. R. Breisch, P. K. Ducey, G. Johnson, J. L. Behler, and R. C. Bothner. 2007. The amphibians and reptiles of New York State: identification, natural history, and conservation. Oxford University Press, New York, New York, USA.

Gibbs, J. P., K. K. Whiteleather, and F. W. Schueler. 2005. Changes in frog and toad populations over 30 years in New York state. Ecological Applications 15:1148–1157.

Gibson, J. D., and D. A. Merkle. 2004. Road mortality of snakes in central Virginia. Banisteria 24:8–14.

Gibson, J. D., and D. A. Merkle. 2005. *Ambystoma maculatum* (Spotted Salamander). Reproduction. Herpetological Review 36:294.

Gibson, J. D., and J. C. Mitchell. 2006. Opportunistic scavenging by Eastern American Toad (*Bufo americanus americanus*) tadpoles on a dead Red-spotted Newt (*Notophthalmus viridescens viridescens*). Banisteria 28:59–60.

Gibson, S. E., Z. Walker, and B. A. Kingsbury. 2008. Microhabitat preferences of the Timber Rattlesnake (*Crotalus horridus*) in the hardwood forests of Indiana. Pages 275–286 *in* W. K. Hayes, K. R. Beaman, M. D. Cardwell, and S. P. Bush, editors. The biology of rattlesnakes. Loma Linda University Press, Loma Linda, California, USA.

Giery, S. T., and R. S. Ostfeld. 2007. The role of lizards in the ecology of lyme disease in two endemic zones of the northeastern United States. Journal of Parasitology 93:511–517.

Gilbert, E. A., S. L. Payne, and M. K. Vickaryous. 2013. The anatomy and histology of caudal autotomy and regeneration in lizards. Physiological and Biochemical Zoology 86:631–644.

Gilbert, P. W. 1941. Eggs and nests of *Hemidactylium scutatum* in the Ithaca region. Copeia 1941:47.

Gilbert, P. W. 1942. Observations on the eggs of *Ambystoma maculatum* with special reference to the green algae found within the eggs' envelopes. Ecology 25:215–227.

Gilbert, P. W. 1944. The alga-egg relationship in *Ambystoma maculatum*. Ecology 25:366–369.

Gilderhus, A., and B. G. H. Johnson. 1980. Effects of Sea Lamprey (*Petromyzon marinus*) control in the Great Lakes on aquatic plants, invertebrates, and amphibians. Canadian Journal of Fisheries and Aquatic Sciences 37:1895–1905.

Giles, L. 1940. Food habits of the Raccoon in eastern Iowa. Journal of Wildlife Management 4:375–382.

Gilhen, J. 1974. Distribution, natural history, and morphology of the Blue-spotted Salamanders, *Ambystoma laterale* and *A. tremblayi* in Nova Scotia. Nova Scotia Museum Curatorial Report Publication 22, Halifax, Canada.

Gilhen, J. 1984. Amphibians and reptiles of Nova Scotia. Nova Scotia Museum, Halifax, Canada.

Gilhen, J., G. Caswell, C. Drake, M. MacDonald, and H. McKinnon-Ramshaw. 2012. A partial albino hatchling Northern Ring-necked Snake, *Diadophis punctatus ewardsii*, from Big Tancook Island, Mahone Bay, Lunenburg County, Nova Scotia, Canada. Canadian Field-Naturalist 126:336–339.

Gill, D. E. 1978a. Effective population size and interdemic migration rates in a metapopulation of the Red-spotted Newt, *Notophthalmus viridescens* (Rafinesque). Evolution 32:839–849.

Gill, D. E. 1978b. The metapopulation ecology of the Red-spotted Newt, *Notophthalmus viridescens* (Rafinesque). Ecological Monographs 48:145–166.

Gill, D. E. 1979. Density dependence and homing behavior in adult Red-spotted Newts *Notophthalmus viridescens* (Rafinesque). Ecology 60:800–813.

Gill, D. E. 1985. Interpreting breeding patterns from census data: a solution to the Husting dilemma. Ecology 66:344–354.

Gill, D. E., K. A. Berven, and B. A. Mock. 1983. The environmental component of evolutionary biology. Pages 1–36 *in* C. E. King and P. S. Dawson, editors. Population biology—retrospect and prospect. Columbia University Press, New York, New York, USA.

Gillette J. R., and M. G. Peterson. 2001. The benefits of transparency: candling as a simple method for determining sex in Red-backed Salamanders (*Plethodon cinereus*). Herpetological Review 32:233–235.

Gillette, J. R., R. G. Jaeger, and M. G. Peterson. 2000. Social monogamy in a territorial salamander. Animal Behavior 59:1241–1250.

Gillingham, J. C. 1974. Reproductive behavior of the Western Fox Snake, *Elaphe v. vulpina*. Herpetologica 30:309–313.

Gillingham, J. C. 1980. Communication and combat behavior of the Black Rat Snake (*Elaphe obsoleta*). Herpetologica 36:120–127.

Gillingham, J. C., C. C. Carpenter, B. J. Brecke, and J. B. Murphy. 1977. Courtship and copulatory behavior of the Mexican Milk-snake, *Lampropeltis triangulum sinaloae*. Southwestern Naturalist 22:187–194.

Gillingham, J. C., J. Rowe, and M. A. Weins. 1990. Chemosensory orientation and earthworm location by foraging Eastern Garter Snakes, *Thamnophis s. sirtalis*. Pages 522–532 *in* D. W. MacDonald, D. Muller-Schwarze, and S. E. Natynczuk, editors. Chemical signals in vertebrates 5. Oxford University Press, New York, New York, USA.

Gist, D. H., and J. D. Congdon. 1998. Oviductal sperm storage as a reproductive tactic of turtles. Journal of Experimental Zoology 282:526–534.

Given, M. F. 2005. Vocalizations and reproductive behavior of male Pickerel Frogs, *Rana palustris*. Journal of Herpetology 39:223–233.

Given, M. F. 2009. Does physical or acoustical disturbance cause male Pickerel Frogs, *Rana palustris*, to vocalize underwater? Amphibia-Reptilia 29:177–184.

Glass, T. G., Jr. 1991. Management of poisonous snakebite. Glass Publishing Company, San Antonio, Texas, USA.

Gleeson, T. T., and A. F. Bennett. 1982. Acid-base imbalances in lizards during activity and recovery. Journal of Experimental Biology 98:439–453.

Glenn, J. L., and R. C. Straight. 1982. The rattlesnakes and their venom yield and lethal toxicity. Pages 3–119 *in* A. T. Tu, editor. Rattlesnake venoms: their actions and treatment. Marcel Dekker, New York, New York, USA.

Glennemeier, K. A., and L. J. Begnoche. 2002. Impact of organochlorine contamination on amphibian populations in southwestern Michigan. Journal of Herpetology 36:233–244.

Glenney, G. W., J. T. Julian, and W. M. Quartz. 2010. Preliminary amphibian health survey in the Delaware Water Gap National Recreation Area. Journal of Aquatic Animal Health 22:102–114.

Glista, D. J., T. L. DeVault, and J. A. DeWoody. 2008. Vertebrate road mortality predominantly impacts amphibians. Herpetological Conservation and Biology 3:77–87.

Glowacki, G., and A. R. Kuhns. 2010. Recovery of the Blanding's Turtle (*Emydoidea blandingii*) at Spring Bluff Nature Preserve, Lake County Forest Preserves. Final Report to the Illinois Department Natural Resources, Springfield, USA. Archived by WebCite at http://www.webcitation.org/73cA9A6n5. Accessed 1 Nov 2018.

Godley, J. S., R. W. McDiarmid, and N. N. Rojas. 1984. Estimating prey size and number in crayfish-eating snakes, genus *Regina*. Herpetologica 40:82–88.

Godwin, G. J., and S. M. Roble. 1983. Mating success in male treefrogs, *Hyla chrysoscelis* (Anura: Hylidae). Herpetologica 39:141–146.

Gold, B. S., R. C. Dart, and R. A. Barish. 2002. Bites of venomous snakes. New England Journal of Medicine 347:347–356.

Goldsmith, W. M. 1944. Notes on the egg laying habitats of the soft-shell turtles. Proceedings of the Iowa Academy of Science 51:447–449.

Golet, W. J., and T. A. Haines. 2000. Snapping Turtles (*Chelydra serpentina*) as monitors for mercury contamination of aquatic environments. Environmental Monitoring and Assessment 71:211–220.

Gomez-Mestre, I., J. C. Touchon, and K. M. Warkentin. 2006. Amphibian embryo and parental defenses and a larval predator reduce egg mortality from water mold. Ecology 87:2570–2581.

Gonzalez-Abraham, C. E., V. C. Radeloff, T. J. Hawbaker, R. B. Hammer, S. I. Stewart, and M. K. Clayton. 2007. Patterns of houses and habitat loss from 1937 to 1999 in northern Wisconsin, USA. Ecological Applications 17:2011–2023.

Goodburn, J. M., and C. G. Lorimer. 1998. Cavity trees and coarse woody debris in old-growth and managed northern hardwood forests in Wisconsin and Michigan. Canadian Journal of Forest Research 28:427–438.

Goode, M. J., D. E. Swann, and C. R. Schwalbe. 2004. Effects of destructive collecting practices on reptiles: a field experiment. Journal of Wildlife Management 68:429–434.

Goode, M. J., W. C. Horrace, M. J. Sredl, and J. M. Howland. 2005. Habitat destruction by collectors associated with decreased abundance of rock-dwelling lizards. Biological Conservation 125:47–54.

Goodman, R. M. 2006. Effects of tail loss on growth and spring speed of juvenile *Eumeces fasciatus* (Scincidae). Journal of Herpetology 40:99–102.

Goodman, R. M., and Y. T. Ararso. 2012. Survey of ranavirus and the fungus *Batrachochytrium dendrobatidis* in frogs of central Virginia, USA. Herpetological Review 43:78–80.

Gopurenko, D., R. N. Williams, C. R. McCormick, and J. A. DeWoody. 2006. Insights into the mating habits of the Tiger Salamander (*Ambystoma tigrinum tigrinum*) as revealed by genetic parentage analyses. Molecular Ecology 15:1917–1928.

Gordon, D. M., and F. R. Cook. 1980. An aggregation of gravid snakes in the Quebec Laurentians. Canadian Field-Naturalist 94:456–457.

Gordon, D. M., and R. D. MacCulloch. 1980. An investigation of the ecology of the Map Turtle, *Graptemys geographica* (Le Sueur), in the northern part of its range. Canadian Journal of Zoology 58:2210–2219.

Gordon, N. M. 2004. The effect of supplemental feeding on the territorial behavior of the Green Frog (*Rana clamitans*). Amphibia-Reptilia 25:55–62.

Gordon, R. E. 1968. Terrestrial activity of the Spotted Salamander, *Ambystoma maculatum*. Copeia 1968:879–880.

Gorman, W. L. 1986. Patterns of color polymorphism in the Cricket Frog, *Acris crepitans*, in Kansas. Copeia 1986:995–999.

Gosner, K. L. 1942. Lip curling of the Red-bellied Snake. Copeia 1942:181–183.

Gosner, K. L. 1959. Systematic variations in tadpole teeth with notes on food. Herpetologica 15:203–210.

Gosner, K. L. 1960. A simplified table for staging anuran embryos and larvae with notes on identification. Herpetologica 16:183–190.

Gosner, K. L., and D. A. Rossman. 1960. Eggs and larval development of the treefrogs *Hyla crucifer* and *Hyla ocularis*. Herpetologica 16:225–232.

Gosner, K. L., and I. H. Black. 1957a. Larval development in New Jersey Hylidae. Copeia 1957:31–36.

Gosner, K. L., and I. H. Black. 1957b. The effects of acidity on the development of New Jersey frogs. Ecology 38:256–262.

Gossen, M. L., and P. A. Cochran. 2013. *Aspidoscelis sexlineata* (Six-lined Racerunner). Artificial Habitat. Herpetological Review 44:668.

Goulet, C., J. A. Litvaitis, and M. N. Marchand. 2015. Habitat associations of the Eastern Hognose Snake at the northern edge of its geographic distribution: should remnant populations guide restoration? Northeastern Naturalist 22:530–540.

Government of Canada. 2018. Species at Risk Act (S. C. 2002, c. 29). Archived by WebCite at http://www.webcitation.org/73cAlgSVY. Accessed 1 Nov 2018.

Graenicher, S. 1911. Some records of Wisconsin lizards. Bulletin of the Wisconsin Natural History Society 9:78–81.

Graham, E. R., S. A. Fay, A. Davey, and R. W. Sanders. 2013. Intracapsular algae provide fixed carbon to developing embryos of the salamander *Ambystoma maculatum*. Journal of Experimental Biology 216:452–459.

Graham, T. E., and J. E. Forsberg. 1991. Aquatic oxygen uptake by naturally wintering Wood Turtles *Clemmys insculpta*. Copeia 1991:836–838.

Graham, T. E., and T. S. Doyle. 1977. Growth and population characteristics of Blanding's Turtle, *Emydoidea blandingii*, in Massachusetts. Herpetologica 33:410–414.

Graham, T. E., and T. S. Doyle. 1979. Dimorphism, courtship, eggs, and hatchlings of the Blanding's Turtle, *Emydoidea blandingii* (Reptilia, Testudines, Emydidae) in Massachusetts. Journal of Herpetology 13:125–127.

Graham, T. E., C. B. Graham, C. E. Crocker, and G. R. Ultsch. 2000. Dispersal from and fidelity to a hibernaculum in a northern Vermont population of Common Map Turtles, *Graptemys geographica*. Canadian Field-Naturalist 114:405–408.

Graham, T. E., J. E. Forsberg, and J. J. Albright. 1987. Updated distribution of Blanding's Turtle, *Emydoidea blandingii*, in Maine. Bulletin of the Maryland Herpetological Society 23:119–121.

Grange, W. B. 1948. Wisconsin grouse problems. Wisconsin Conservation Department PUB-328, A-1948, Madison, USA.

Grant, J. B. 2001. *Rana palustris* (Pickerel Frog). Production of odor. Herpetological Review 32:183.

Grant, K. P., and L. E. Licht. 1995. Effects of ultraviolet radiation on life-history stages of anurans from Ontario, Canada. Canadian Journal of Zoology 73:2292–2301.

Grant, W. C., Jr. 1955. Territorialism in two species of salamanders. Science 121:137–138.

Gravel, M., M. J. Mazerolle, and M. A. Villard. 2012. Interactive effects of roads and weather on juvenile amphibian movements. Amphibia-Reptilia 33:113–127.

Graves, B. M., and M. Halpern. 1988. Neonate Plains Gartersnakes (*Thamnophis radix*) are attracted to conspecific skin extracts. Journal of Comparative Psychology 102:251–253.

Gray, B. S. 2008. A note on site fidelity for ecdysis in the Northern Brown Snake, *Storeria dekayi dekayi*. Bulletin of the Chicago Herpetological Society 43:39–41.

Gray, B. S. 2013. *Storeria dekayi* (Dekay's Brownsnake). Diet. Herpetological Review 44:527.

Gray, B. S. 2014a. Natural history of Dekay's Brownsnake, *Storeria dekayi* (Holbrook, 1836), at a site in northwestern Pennsylvania. Journal of North American Herpetology 2014:28–39.

Gray, B. S. 2014b. *Storeria dekayi* (Dekay's Brownsnake). Cold tolerance. Herpetological Review 45:148.

Gray, B. S. 2015. A study of the defensive behaviors of free-ranging Dekay's Brownsnakes, *Storeria dekayi* (Holbrook, 1936). Journal of North American Herpetology 2015:43–52.

Gray, B. S., and M. Lethaby. 2008a. *Storeria occipitomaculata occipitomaculata* (Northern Red-bellied Snake). Maximum size. Herpetological Review 39:101.

Gray, B. S., and M. Lethaby. 2008b. The amphibians and reptiles of Erie County, Pennsylvania. Bulletin of the Maryland Herpetological Society 44:49–69.

Gray, B. S., M. Lethaby, N. McAllister, and S. Bloomstine. 2013. *Storeria occipitomaculata* (Red-bellied Snake). Diet. Herpetological Review 44:335.

Gray, J. E. 1831. Synopsis reptilium; or short descriptions of the species of reptiles. Part 1: cataphracta, tortoises, crocodiles, and enaliosaurians. Treuttel, Wurtz, and Co., London, England.

Gray, M. J., D. L. Miller, and J. T. Hoverman. 2009. Ecology and pathology of amphibian ranaviruses. Diseases of Aquatic Organisms 87:243–266.

Gray, R. H. 1971. Fall activity and overwintering of the Cricket Frog (*Acris crepitans*) in central Illinois. Copeia 1971:748–750.

Gray, R. H. 1983. Seasonal, annual and geographic variation in color morph frequencies of the Cricket Frog, *Acris crepitans*, in Illinois. Copeia 1983:300–311.

Gray, R. H. 2000a. Historical occurrence of malformations in the Cricket Frog, *Acris crepitans*, in Illinois. Transactions of the Illinois State Academy of Science 93:279–284.

Gray, R. H. 2000b. Morphological abnormalities in Illinois Cricket Frogs, *Acris crepitans*, 1968–1971. Journal of the Iowa Academy of Science 107:92–95.

Gray, R. H., and L. E. Brown. 2005. Decline of Northern Cricket Frogs (*Acris crepitans*). Pages 47–54 *in* M. Lannoo, editor. Amphibian declines: the conservation status of United States species. University of California Press, Berkeley, USA.

Gray, R. H., L. E. Brown, and L. Blackburn. 2005. *Acris crepitans*. Pages 441–443 *in* M. Lannoo, editor. Amphibian declines: the conservation status of United States species. University of California Press, Berkeley, USA.

Grayson, K. L., and H. M. Wilbur. 2009. Sex and context-dependent migration in a pond-breeding amphibian. Ecology 90:306–312.

Greaves, W. F., and J. D. Litzgus. 2007. Overwintering ecology of Wood Turtles (*Glyptemys insculpta*) at the species' northern range limit. Journal of Herpetology 41:32–40.

Green, D. E., K. A. Converse, and A. K. Schrader. 2002. Epizootiology of sixty-four amphibian morbidity and mortality events in the USA, 1996–2001. Annals of the New York Academy of Sciences 969:323–339.

Green, D. M. 1980. Size differences in adhesive toe-pad cells of treefrogs of the diploid-polyploid *Hyla versicolor* complex. Journal of Herpetology 14:15–19.

Green, D. M. 1981. Adhesion and the toe-pads of treefrogs. Copeia 790–796.

Green, D. M. 2005. *Bufo americanus* (Holbrook, 1836)—American Toad. Pages 386–390 *in* M. Lannoo, editor. Amphibian declines: the conservation status of United States species. University of California Press, Berkeley, USA.

Green, D. M., and C. Pustowka. 1997. Correlated morphological and allozyme variation in the hybridizing toads *Bufo americanus* and *Bufo hemiophrys*. Herpetologica 53:218–228.

Green, D. M., and J. Carson. 1988. The adhesion of treefrog toe-pads to glass: cryogenic examination of a capillary adhesion system. Journal of Natural History 22:131–135.

Green, D. M., L. A. Weir, G. S. Casper, and M. J. Lannoo. 2013. North American amphibians: distribution and diversity. University of California Press, Berkeley, USA.

Green, N. B., and T. K. Pauley. 1987. Amphibians and reptiles in West Virginia. University of Pittsburgh Press, Pittsburgh, Pennsylvania, USA.

Greenberg, C. H., and T. A. Waldrop. 2008. Short-term response of reptiles and amphibians to prescribed fire and mechanical fuel reduction in a southern Appalachian upland hardwood forest. Forest Ecology and Management 255:2883–2893.

Greenberg, J. 2008. Of prairie, woods, and water: two centuries of Chicago nature writing. University of Chicago Press, Chicago, Illinois, USA.

Greene, H. W., P. G. May, D. L. Hardy, Sr., J. M. Sciturro, and T. M. Farrell. 2002. Parental behavior by vipers. Pages 179–206 *in* G. W. Schuett, M. Hoggren, M. E. Douglas, and H. W. Greene, editors. Biology of the vipers. Eagle Mountain Publishing, LC, Eagle Mountain, Utah, USA.

Greenspan, S. E., A. J. K. Calhoun, J. E. Longcore, and M. G. Levy. 2012. Transmission of *Batrachochytrium dendrobatidis* to Wood Frogs (*Lithobates sylvaticus*) via a Bullfrog (*L. catesbeianus*) vector. Journal of Wildlife Diseases 48:575–582.

Greenwald, K. R., and H. L. Gibbs. 2012. A single nucleotide polymorphism assay for the identification of unisexual *Ambystoma* salamanders. Molecular Ecology Resources 12:354–362.

Greenwald, K. R., R. D. Denton, and H. L. Gibbs. 2016. Niche partitioning between sexual and unisexual *Ambystoma* salamanders. Ecosphere 7:1–17.

Greenwell, M., V. Beasley, and L. E. Brown. 1996. Cricket Frog research: the mysterious decline of the Cricket Frog. Aquaticus 26:48–54.

Greer, A. L., M. Berrill, and P. J. Wilson. 2005. Five amphibian mortality events associated with ranavirus infection in south central Ontario, Canada. Diseases of Aquatic Organisms 67:9–14.

Gregory, P. T. 1975. Aggregations of gravid snakes in Manitoba, Canada. Copeia 1975:185–186.

Gregory, P. T. 1977a. Life history observations of three species of snakes in Manitoba. Canadian Field-Naturalist 91:19–27.

Gregory, P. T. 1977b. Life-history parameters of the Red-sided Garter Snake (*Thamnophis sirtalis parietalis*) in an extreme environment, the Interlake region of Manitoba. National Museum of Canada, Publications in Zoology 13, Ottawa.

Gregory, P. T. 1982. Reptilian hibernation. Pages 53–154 *in* C. Gans and F. H. Pough, editors. Biology of the Reptilia. Volume 13. Academic Press, London, England.

Gregory, P. T. 1984. Communal denning in snakes. Pages 57–75 *in* R. A. Seigel, L. E. Hunt, J. L. Knight, L. Malaret, and N. L. Zuschlag, editors. Vertebrate ecology and systematics: a tribute to Henry S. Fitch. University of Kansas, Museum of Natural History Special Publication 10, Lawrence, USA.

Gregory, P. T. 2004. Analysis of patterns of aggregation under cover objects in an assemblage of six species of snakes. Herpetologica 60:178–186.

Gregory, P. T. 2009. Northern lights and seasonal sex: the reproductive ecology of cool-climate snakes. Herpetologica 65:1–13.

Gregory, P. T. 2015. Natural history notes on a northern population of the Milksnake (*Lampropeltis triangulum*), with evidence for ontogenetic shifts in cover use and diet. Herpetological Review 46:506–509.

Gregory, P. T., and K. W. Larsen. 1996. Are there any meaningful correlates of geographic life-history variation in the Garter Snake, *Thamnophis sirtalis*? Copeia 1996:183–189.

Grgurovic, M., and P. R. Sievert. 2005. Movement patterns of Blanding's Turtles (*Emydoidea blandingii*) in the suburban landscape of eastern Massachusetts. Urban Ecosystems 8:203–213.

Griesbaum, J. 2012. Effects of grassland management on snake species diversity and habitat use, home range and movements of Western Fox Snakes. Thesis, University of Illinois, Urbana-Champaign, USA.

Griffis-Kyle, K. L. 2005. Ontogenic delays in effects of nitrite exposure on Tiger Salamanders (*Ambystoma tigrinum tigrinum*) and Wood Frogs (*Rana sylvatica*). Environmental Toxicology and Chemistry 24:1523–1527.

Griffis-Kyle, K. L., and M. E. Ritchie. 2007. Amphibian survival, growth and development in response to mineral nitrogen exposure and predator cues in the field: an experimental approach. Oecologia 152:633–642.

Gritters, S. A., and L. M. Mauldin. 1994. Four years of turtle collections on navigation pool 13 of the Upper Mississippi River. Environmental Management Technical Center, Long Term Resource Monitoring Special Report, Onalaska, Wisconsin, USA.

Grizzell, R. A., Jr. 1949. Hibernation sites of three snakes and a salamander. Copeia 1949:231–232.

Grobman, A. B. 1941. A contribution to the knowledge of variation in *Opheodrys vernalis* (Harlan), with the description of a new subspecies. University of Michigan, Miscellaneous of the Museum of Zoology 50, Ann Arbor, USA.

Grobman, A. B. 1990. The effect of soil temperatures on emergence from hibernation of *Terrapene carolina* and *T. ornata*. American Midland Naturalist 124:366–371.

Grobman, A. B. 1992. Metamerism in the snake *Opheodrys vernalis*. Journal of Herpetology 26:175–186.

Groff, L. A., A. J. K. Calhoun, and C. S. Loftin. 2016. Hibernal habitat selection by Wood Frogs (*Lithobates sylvaticus*) in a northern New England montane landscape. Journal of Herpetology 50:559–569.

Grogan, W. L. 1974. Effects of accidental envenomation from the saliva of the Eastern Hognose Snake, *Heterodon platyrhinos*. Herpetologica 30:248–249.

Groner, M. L., and R. A. Relyea. 2010. *Batrachochytrium dendrobatidis* is present in northwest Pennsylvania, USA, with high prevalence in *Notophthalmus viridescens*. Herpetological Review 41:462–465.

Gross, I. P. 2013. Taking the road most travelled: understanding patterns of snake (Colubridae; *Storeria*) movement and road mortality in a state park. Senior Honors Thesis, Eastern Illinois University, Charleston, USA.

Gross, J. A., P. T. J. Johnson, L. K. Prahl, and W. H. Karasov. 2009. Critical period of sensitivity for effects of cadmium on frog growth and development. Environmental Toxicology and Chemistry 28:1227–1232.

Gross, J. A., T-H. Chen, and W. H. Karasov. 2007. Lethal and sublethal effects of chronic cadmium exposure on Northern Leopard Frog (*Rana pipiens*) tadpoles. Environmental Toxicology and Chemistry 26:1192–1197.

Grosse, A. M., S. C. Sterrett, and J. C. Maerz. 2010. Effects of turbidity on the foraging success of the Eastern Painted Turtle. Copeia 2010:463–467.

Grossmann, E. B., and D. J. Mladenoff. 2008. Farms, fires, and forestry: disturbance legacies in the soils of the northwest Wisconsin (USA) sand plain. Forest Ecology and Management 256:827–836.

Grover, M. C. 1998. Influence of cover and moisture on abundances of the terrestrial salamanders *Plethodon cinereus* and *Plethodon glutinosus*. Journal of Herpetology 32:489–497.

Grover, M. C., and H. M. Wilbur. 2002. Ecology of ecotones: interactions between salamanders on a complex environmental gradient. Ecology 83:2112–2123.

Groves, J. D. 1980. Mass predation of a population of the American Toad, *Bufo americanus*. American Midland Naturalist 103:202–203.

Groves, J. D. 1982. Egg-eating behavior of brooding Five-lined Skinks, *Eumeces fasciatus*. Copeia 1982:969–971.

Grubb, J. C. 1972. Differential predation by *Gambusia affinis* on the eggs of seven species of anuran amphibians. American Midland Naturalist 88:103–108.

Grudzien, T. A., and P. J. Owens. 1991. Genic similarity in the gray and brown color morphs of the snake *Storeria occipitomaculata*. Journal of Herpetology 25:90–92.

Guerry, A. D., and M. L. Hunter, Jr. 2002. Amphibian distributions in a landscape of forests and agriculture: an examination of landscape composition and configuration. Conservation Biology 16:745–754.

Gundale, M. J., W. M. Jolly, and T. H. Deluca. 2005. Susceptibility of a northern hardwood forest to exotic earthworm invasion. Conservation Biology 19:1075–1083.

Guo, P., Q. Liu, X. Xu, K. Jiang, M. Hou, L. Ding, R. A. Pyron, and F. T. Burbrink. 2012. Out of Asia: natricine snakes support the Cenozoic Beringian dispersal hypothesis. Molecular Phylogenetics and Evolution 63:825–833.

Gustafson, K. D., R. A. Newman, and V. V. Tkach. 2013. Effects of host species and life stage on the helminth communities of sympatric Northern Leopard Frogs (*Lithobates pipiens*) and Wood Frogs (*Lithobates sylvaticus*) in the Sheyenne National Grasslands, North Dakota. Journal of Parasitology 99:587–594.

Guthrie, J. E. 1927. Rattlesnake eggs in Iowa. Copeia 162:12–14.

Guthrie, J. E. 1932. Snakes versus birds: birds versus snakes. Wilson Bulletin June:88–113.

Guttman, S. I., L. A. Weigt, P. E. Moler, R. E. Ashton, Jr., B. W. Mansell, and J. Peavy. 1990. An electrophoretic analysis of *Necturus* from the southeastern United States. Journal of Herpetology 24:163–175.

Gutzke, W. H. N., and G. C. Packard. 1987. The influence of temperature on eggs and hatchlings of Blanding's Turtles *Emydoidea blandingii*. Journal of Herpetology 21:161–163.

Hacker, V. A. 1957. Biology and management of Lake Trout in Green Lake, Wisconsin. Transactions of the American Fisheries Society 86:71–83.

Hager, H. A. 1998. Area-sensitivity of reptiles and amphibians: are there indicator species for habitat fragmentation? Ecoscience 5:139–147.

Hailman, J. P., and R. G. Jaeger. 1974. Phototactic responses to spectrally dominant stimuli and use of colour vision by adult anuran amphibians: a comparative survey. Animal Behaviour 22:757–795.

Hale, C. 2007. Earthworms of the Great Lakes. Kollath+Stensaas Publishing, Duluth, Minnesota, USA.

Hall, C., and B. Carlson. 2004. Forest management guidelines for the protection of Four-toed and Spotted Salamander populations. Minnesota Department of Natural Resources, St. Paul, USA. Archived by WebCite at http://www.webcitation.org/73cB57eq3. Accessed 1 Nov 2018.

Hall, C. D., and F. J. Cuthbert. 2000. Impacts of a controlled wetland drawdown on Blanding's Turtles in Minnesota. Chelonian Conservation and Biology 3:643–649.

Hall, E. M., S. P. Brady, N. M. Mattheus, R. L. Earley, M. Diamond, and E. J. Crespi. 2017. Physiological consequences of exposure to salinized roadside ponds on Wood Frog larvae and adults. Biological Conservation 209:98–106.

Hallmen, M. 2002. Beobachtungen zur vererbung des albinismus bei der Prarie-Strumpfbandnatter *Thamnophis radix* (Baird and Girard, 1853). The Garter Snake 2/02:20–25.

Hallock, L. A. 1991. Habitat utilization, diet and behavior of the Eastern Massasauga (*Sistrurus catenatus catenatus*) in southern Michigan. Thesis, Michigan State University, Lansing, USA.

Halloy, M., and G. M. Burghardt. 1990. Ontogeny of fish capture and ingestion in four species of garter snakes (*Thamnophis*). Behaviour 112:299–318.

Halstead, N. T. 2007. Long-term effects of prescribed fire on reptile and amphibian communities in Florida sandhill. Thesis, University of South Florida, Tampa, USA.

Halstead, N. T., H. R. Mushinsky, and E. D. McCoy. 2008. Sympatric *Masticophis flagellum* and *Coluber constrictor* select vertebrate prey at different levels of taxonomy. Copeia 2008:897–908.

Hallmen, M., and J. Chlebowy. 2001. Seltene farbformen der Prarie-Strumpfbandnatter *Thamnophis radix* gezuchtet. Elaphe 9:18–20.

Hallowell, E. 1856. Notice of a collection of reptiles from Kansas and Nebraska, presented to the Academy of Natural Sciences, by Dr. Hammond, U.S.A. Proceedings of the Academy of Natural Sciences of Philadelphia 8:238–253.

Halvorsen, H. J., and R. K. Anderson. 1980. Evaluation of grassland management for wildlife in central Wisconsin. Proceedings of the North America Prairie Conference 7:267–279.

Hamel, J. 2009. *Anaxyrus americanus* (American Toad). Egg cannibalism. Herpetological Review 40:67–68.

Hamer, A. J., R. van der Ree, M. J. Mahony, and T. Langton. 2014. Usage rates of an under-road tunnel by three Australian frog species: implications for road mitigation. Animal Conservation 17:379–387.

Hamernick, M. G. 1991. Home ranges and habitat selection of Blanding's Turtles (*Emydoidea blandingii*) at the Weaver Dunes, Minnesota. Report to the Minnesota Department of Natural Resources, St. Paul, USA.

Hamerstrom, F. 1983. Birds of prey of Wisconsin. Wisconsin Department of Natural Resources Report PUB-1–8100(04), Madison, USA.

Hamerstrom, F., and O. Mattson. 1939. Food of central Wisconsin horned owls. American Midland Naturalist 22:700–702.

Hamilton, B. T., R. Hart, and J. W. Sites, Jr. 2012. Feeding ecology of the Milksnake (*Lampropeltis triangulum*, Colubridae) in the western United States. Journal of Herpetology 46:515–522.

Hamilton, C. M. 2009. Home range size and habitat selection of Timber Rattlesnakes (*Crotalus hor-*

ridus) in southwestern Wisconsin. Thesis, University of Wisconsin, Stevens Points, USA.

Hamilton, W. J., Jr. 1930. Notes on the food of the American Toad. Copeia 1930:45.

Hamilton, W. J., Jr. 1932. The food and feeding habits of some eastern salamanders. Copeia 1932:83–86.

Hamilton, W. J., Jr. 1934. The rate of growth of the toad (*Bufo americanus americanus* Holbrook) under natural conditions. Copeia 1934:88–90.

Hamilton, W. J., Jr. 1940a. Observations on the reproductive behavior of the Snapping Turtle. Copeia 1940:124–126.

Hamilton, W. J., Jr. 1940b. The feeding habit of larval newts with reference to availability and predilection of food items. Ecology 21:351–356.

Hamilton, W. J., Jr. 1947. Hibernation of the Lined Snake. Copeia 1947:209–210.

Hamilton, W. J., Jr. 1948. The food and feeding behavior of the Green Frog, *Rana clamitans* Latreille, in New York State. Copeia 1948:203–207.

Hamilton, W. J., Jr. 1951. The food of the Opossum in New York. Journal of Wildlife Management 15:258–264.

Hamilton, W. J., Jr., and J. A. Pollack. 1956. The food of some colubrid snakes from Fort Benning, Georgia. Ecology 37:519–526.

Hamilton, W. J., Jr., and J. A. Pollack. 1961. The food of some lizards from Fort Benning, Georgia. Herpetologica 17:99–106.

Hammer, D. A. 1969. Parameters of a marsh Snapping Turtle population, LaCreek Refuge, South Dakota. Journal of Wildlife Management 33:995–1005.

Hammerson, G. 1981. An ecographic analysis of the herpetofauna of Colorado. Dissertation, University of Colorado, Boulder, USA.

Hammerson, G. A. 1982. Bullfrog eliminating Leopard Frogs in Colorado? Herpetological Review 13:115–116.

Hammerson, G. A., 1999. Amphibians and reptiles in Colorado. Second edition. University Press of Colorado, Niwot, USA.

Hammerson, G. A. 2007. *Thamnophis sauritus*. The IUCN Red List of Threatened Species 2007: e.T63991A12727431. www.iucnredlist.org. Accessed 1 Nov 2018.

Hammerson, G. A., and L. J. Livo. 1999. Conservation status of the Northern Cricket Frog (*Acris crepitans*) in Colorado and adjacent areas at the northwestern extent of the range. Herpetological Review 30:78–80.

Hampton, P. M. 2008. Prey items of the Western Ribbon Snake, *Thamnophis proximus*. Southwestern Naturalist 53:115–118.

Hampton, P. M. 2011. Feeding performance in the Western Ribbon Snake (*Thamnophis proximus*): ontogeny and the effects of prey type and size. Canadian Journal of Zoology 89:945–950.

Hankin, F. M., M. D. Smith, J. A. Penner, and D. S. Louis. 1987. Eastern Massasauga rattlesnake bites. Journal of Pediatric Orthopedics 7:201–205.

Hanlon, S. M., and M. J. Parris. 2014. The interactive effects of chytrid fungus, pesticides, and exposure timing on Gray Treefrog (*Hyla versicolor*) larvae. Environmental Toxicology and Chemistry 33:216–222.

Hanna, G., and W. J. Barnes. 1991. Adhesion and detachment of the toe pads of tree frogs. Journal of Experimental Biology 155:103–125.

Hansen, A. J., R. L. Knight, J. M. Marzluff, S. Powell, K. Brown, P. H. Gude, and K. Jones. 2005. Effects of exurban development on biodiversity: patterns, mechanisms, and research needs. Ecological Applications 15:1893–1905.

Harding, J. H. 1985. *Clemmys insculpta* (Wood Turtle). Predation-mutilation. Herpetological Review 16:30.

Harding, J. H. 1997. Amphibians and reptiles of the Great Lakes region. University of Michigan Press, Ann Arbor, USA.

Harding, J. H., and J. A. Holman. 1992. Michigan frogs, toads, and salamanders: a field guide and pocket reference. Michigan State University Museum, Lansing, USA.

Harding, J. H., and S. K. Davis. 1999. *Clemmys insculpta* (Wood Turtle) and *Emydoidea blandingii* (Blanding's Turtle). Hybridization. Herpetological Review 30:225–226.

Harding, J. H., and T. J. Bloomer. 1979. The Wood Turtle, *Clemmys insculpta* . . . a natural history. Bulletin of the New York Herpetological Society 15:9–26.

Hardy, D. F. 1962. Ecology and behavior of the Six-Lined Racerunner, *Cnemidophorus sexlineatus*. University of Kansas Science Bulletin 45:1–73.

Hardy, J. D., Jr. 1952. A concentration of juvenile Spotted Salamanders *Ambystoma maculatum* (Shaw). Copeia 1952:181.

Hardy, J. D., Jr., and R. J. Borroughs. 1986. Systematic status of the Spring Peeper, *Hyla crucifer* (Amphibia: Hylidae). Bulletin of the Maryland Herpetological Society 22:68–89.

Hardy, L. M., and L. R. Raymond. 1991. Observations on the activity of the Pickerel Frog, *Rana palustris* (Anura: Ranidae), in northern Louisiana. Journal of Herpetology 25:220–222.

Hardy, L. M., and M. C. Lucas. 1991. A crystalline protein is responsible for dimorphic egg jellies in the Spotted Salamander, *Ambystoma maculatum* (Shaw) (Caudata: Ambystomatidae). Comparative Biochemistry and Physiology A 100:653–660.

Harlan, R. 1827. Genera of North American Reptilia and a synopsis of the species. Journal of the Academy of Natural Sciences of Philadelphia 5:317–372.

Harlan, R. 1828. Description of a new species of *Salamandra*. Journal of the Academy of Natural Sciences of Philadelphia 6:101.

Harner, M. J., and K. Geluso. 2011. Response of herpetofauna to grazing and fire in wet, tallgrass prairies along the Platte River. Final Report to the Nebraska Game and Parks Commission, Lincoln, USA.

Harper, E. B., T. A. G. Rittenhouse, and R. D. Semlitsch. 2008. Demographic consequences of terrestrial habitat loss for pool-breeding amphibians: predicting extinction risks associated with inadequate size of buffer zones. Conservation Biology 22:1205–1215.

Harper, F. 1939. A southern subspecies of the Spring Peeper (*Hyla crucifer*). Notulae Naturae 27:1–4.

Harper, F. 1947. A new cricket frog (*Acris*) from the middle western states. Proceedings of the Biological Society of Washington 60:39–40.

Harper, F. 1963. Amphibians and reptiles of Keewatin and northern Manitoba. Proceedings of the Biological Society of Washington 76:159–168.

Harrah, E. C. 1922. North American monostomes. Primarily from fresh-water hosts. Illinois Biological Monographs 7:1–107.

Harriot, S. C. 1940. Chipmunk eating a Red-bellied Snake. Journal of Mammalogy 21:92.

Harris, J. P., Jr. 1959. The natural history of *Necturus*: III. Food and feeding. Field and Laboratory 27:105–111.

Harris, M. L., C. A. Bishop, J. Struger, M. R. van den Huevel, G. J. Van Der Kraak, D. G. Dixon, B. Ripley, and J. P. Bogart. 1998. The functional integrity of Northern Leopard Frog (*Rana pipiens*) and Green Frog (*Rana clamitans*) populations in orchard wetlands. I. Genetics, physiology, and biochemistry of breeding adults and young-of-the-year. Environmental Toxicology and Chemistry 17:1338–1350.

Harris, P. M. 1995. Are autecologically similar species also functionally similar? A test in pond communities. Ecology 76:544–552.

Harris, R. N. 1980. The consequences of within-year timing of breeding in *Ambystoma maculatum*. Copeia 1980:719–722.

Harris, R. N. 1987. Density-dependent paedomorphosis in the salamander *Notophthalmus viridescens dorsalis*. Ecology 68:705–712.

Harris, R. N. 2005. *Hemidactylium scutatum*. Pages 780–781 *in* M. J. Lannoo, editor. Amphibian declines: the conservation status of United States species. University of California Press, Berkeley, USA.

Harris, R. N., and D. E. Gill. 1980. Communal nesting, brooding behavior, and embryonic survival of the Four-toed Salamander *Hemidactylium scutatum*. Herpetologica 36:141–144.

Harris, R. N., and P. M. Ludwig. 2004. Resource level and reproductive frequency in female Four-toed Salamanders, *Hemidactylium scutatum*. Ecology 85:1585–1590.

Harris, R. N., R. A. Alford, and H. M. Wilbur. 1988. Density and phenology of *Notophthalmus viridescens dorsalis* in a natural pond. Herpetologica 44:234–242.

Harris, R. N., T. J. Vess, J. I. Hammond, and C. J. Lindermuth. 2003. Context-dependent kin discrimination in larval Four-toed Salamanders *Hemidactylium scutatum* (Caudata: Plethodontidae). Herpetologica 59:164–177.

Harris, R. N., T. Y. James, A. Lauer, M. A. Simon, and A. Patel. 2006. Amphibian pathogen *Batrachochytrium dendrobatidis* is inhibited by the cutaneous bacteria of amphibian species. EcoHealth 3:53–56.

Harris, R. N., W. W. Hames, I. T. Knight, C. A. Carreno, and T. J. Vess. 1995. An experimental analysis of joint nesting in the salamander *Hemidactylium scutatum* (Caudata: Plethodontidae): the effects of population density. Animal Behaviour 50:1309–1216.

Hart, D. R. 1979. Niche relationships of *Thamnophis radix haydeni* and *Thamnophis sirtalis parietalis* in the interlake district of Manitoba. Tulane Studies in Zoology and Botany 21:125–140.

Hartman, F. A. 1906. Food habits of Kansas lizards and batrachians. Transactions of the Kansas Academy of Science 20:225–229.

Hartson, R. B., S. A. Orlofske, V. E. Melin, R. T. Dillon Jr., and P. T. Johnson. 2011. Land use and wetland spatial position jointly determine amphibian parasite communities. EcoHealth 8:485–500.

Hartup, B. K. 1996. Rehabilitation of native reptiles and amphibians in DuPage County, Illinois. Journal of Wildlife Diseases 32:109–112.

Hartwig, T. S., and E. Kiviat. 2007. Microhabitat association of Blanding's Turtles in natural and constructed wetlands in southeastern New York. Journal of Wildlife Management 71:576–582.

Hartzell, S. M. 2016. *Chelydra serpentina* (Snapping Turtle). Foraging behavior. Herpetological Review 47:444.

Harvey, D. S., and P. J. Weatherhead. 2006a. A test of the hierarchical model of habitat selection using Eastern Massasauga Rattlesnakes (*Sistrurus c. catenatus*). Biological Conservation 130:206–216.

Harvey, D. S., and P. J. Weatherhead. 2006b. Hibernation site selection by Eastern Massasauga Rattlesnakes (*Sistrurus c. catenatus*) near their northern range limit. Journal of Herpetology 40:66–73.

Harvey, D. S., A. M. Lentini, K. Cedar, and P. J. Weatherhead. 2013. Moving Massasaugas: insight into rattlesnake relocation using *Sistrurus c. catenatus*. Herpetological Conservation and Biology 9:67–75.

Harwood, P. D. 1945. The behavior and growth of young Dekay's Snakes (*Storeria dekayi dekayi*) in captivity. American Midland Naturalist 34:523–525.

Hastings, J. 2009. Nongame wildlife habitat guide: complementary opportunities for stream restoration projects. Trout Unlimited, Westby, Wisconsin, USA. Archived by WebCite at http://www.webcitation.org/73cFTbbgs. Accessed 1 Nov 2018.

Hatch, K. 1997. Translocation, reintroduction and habitat use of Ornate Box Turtles. Report to the Zoological Society of Milwaukee County, Milwaukee, Wisconsin, USA.

Hatch, K. A., and J. D. Stein. 1994. *Cnemidophorus sexlineatus* (Six-lined Racerunner). Oophagy. Herpetological Review 25:65.

Hausfater, G., H. C. Gerhardt, and G. M. Klump. 1990. Parasites and mate choice in Gray Treefrogs, *Hyla versicolor*. American Zoologist 30:299–311.

Haxton, T. 2000a. A radiotelemetric study of activity and movements of the Blanding's Turtle (*Emydoidea blandingii*) in northeastern Illinois. Canadian Field-Naturalist 114:106–110.

Haxton, T. 2000b. Road mortality of Snapping Turtles, *Chelydra serpentina*, in central Ontario during their nesting period. Canadian Field-Naturalist 100:106–110.

Hay, O. P. 1892. The batrachians and reptiles of the State of Indiana. Annual Report of the Indiana Department of Geology and Natural Resources 17:412–602.

Hay, R. 1998. Blanchard's Cricket Frogs in Wisconsin: a status report. Pages 79–82 *in* M. Lannoo, editor. Status and conservation of midwestern amphibians. University of Iowa Press, Iowa City, USA.

Hayes, P. G. 1988. Scoring a glacial gift. The Milwaukee Journal. 21 August 1988; pages 10–42.

Hayes, T. B. 2004. There is no denying this: defusing the confusion about atrazine. BioScience 54:1138–1149.

Hayes, T. B., A. Collins, M. Lee, M. Mendoza, N. Noriega, A. A. Stuart, and A. Vonk. 2002. Hermaphroditic, demasculinzed frogs after exposure to the herbicide atrazine at low ecological relevant doses. Proceedings of the National Academy of Sciences of the United States of America 99:5476–5480.

Hayes, T. B., K. Haston, M. Tsui, A. Hoang, C. Haeffele, and A. Vonk. 2003. Atrazine-induced hermaphroditism at 0.1 ppb in American Leopard Frogs (*Rana pipiens*): laboratory and field evidence. Environmental Health Perspectives 111:568–575.

Hayes, T. B., V. Khoury, A. Narayan, M. Nazir, A. Park, T. Brown, L. Adame, E. Chan, D. Buchholz,

T. Stueve, and S. Gallipeau. 2010. Atrazine induces complete feminization and chemical castration in mail African Clawed Frogs (*Xenous laevis*). Proceedings of the National Academy of Sciences of the United States of America 107:4612–4617.

Hays, S. M. 2010. Center pivot irrigation revolutionizes agriculture. The Fence Post. 25 May 2010. Archived by WebCite at http://www.webcitation.org/73cFjnnqq. Accessed 1 Nov 2018.

Healy, W. R. 1970. Reduction in neoteny in Massachusetts populations of *Notophthalmus viridescens*. Copeia 1970:578–581.

Healy, W. R. 1974a. Population consequences of alternative life histories in *Notophthalmus v. viridescens*. Copeia 1974:221–229.

Healy, W. R. 1974b. Sex ratio variation in samples of adult *Notophthalmus viridescens*. American Midland Naturalist 92:492–495.

Healy, W. R. 1975. Terrestrial activity and home range in efts of *Notophthalmus viridescens*. American Midland Naturalist 93:131–138.

Heath, A. G. 1975. Behavioral thermoregulation in high altitude Tiger Salamanders, *Ambystoma tigrinum*. Herpetologica 31:84–93.

Heath, D. R., D. A. Saugey, and G. A. Heidt. 1986. Abandoned mine fauna of the Ouachita Mountains, Arkansas: vertebrate taxa. Proceedings of the Arkansas Academy of Sciences 40:33–36.

Heatwole, H. 1960. Burrowing ability and behavioral responses to desiccation of the salamander *Plethodon cinereus*. Ecology 41:661–668.

Heatwole, H. 1961. Habitat selection and activity of the Wood Frog, *Rana sylvatica* Le Conte. American Midland Naturalist 66:301–313.

Heatwole, H. 1962. Environmental factors influencing local distribution and activity of the salamander, *Plethodon cinereus*. Ecology 43:460–472.

Hecht, M. K. 1958. A synopsis of the Mud Puppies of eastern North America. Proceedings of the Staten Island Institute of Arts and Science 21:5–38.

Hecnar, S. J. 1992. Nest distribution, site selection, and brooding in the Five-lined Skink (*Eumeces fasciatus*). Canadian Journal of Zoology 72:1510–1516.

Hecnar, S. J. 1995. Acute and chronic toxicity of ammonium nitrate fertilizer to amphibians from southern Ontario. Environmental Toxicology and Chemistry 14:2131–2137.

Hecnar, S. J. 2004. Great Lakes wetlands as amphibian habitats: a review. Aquatic Ecosystem Health & Management 7:289–303.

Hecnar, S. J., and D. R. Hecnar. 2011. Microhabitat selection of woody debris by Dekay's Brownsnake (*Storeria dekayi*) in a dune habitat in Ontario, Canada. Journal of Herpetology 45:478–483.

Hecnar, S. J., and R. T. M'Closkey. 1996. Regional dynamics and the status of amphibians. Ecology 77:2091–2097.

Hecnar, S. J., and R. T. M'Closkey. 1997a. Spatial scale and determination of species status of the Green Frog. Conservation Biology 11:670–682.

Hecnar, S. J., and R. T. M'Closkey. 1997b. The effects of predatory fish on amphibian species richness and distribution. Biological Conservation 79:123–131.

Hecnar, S. J., and R. T. M'Closkey. 1998. Effects of human disturbance on Five-lined Skink, *Eumeces fasciatus*, abundance and distribution. Biological Conservation 85:213–222.

Hecnar, S. J., G. S. Casper, R. W. Russell, D. R. Hecnar, and J. N. Robinson. 2002. Nested species assemblages of amphibians and reptiles on islands in the Laurentian Great Lakes. Journal of Biogeography 29:475–489.

Hedeen, S. E. 1967. Feeding behavior of the Great Blue Heron in Itasca State Park, Minnesota. Loon 39:116–120.

Hedeen, S. E. 1971a. Body temperatures of the Mink Frog, *Rana septentrionalis* Baird. Journal of Herpetology 5:11–212.

Hedeen, S. E. 1971b. Growth of the tadpoles of the Mink Frog, *Rana septentrionalis*. Herpetologica 27:160–165.

Hedeen, S. E. 1972a. Escape behavior and causes of death of the Mink Frog, *Rana septentrionalis*. Herpetologica 28:261–262.

Hedeen, S. E. 1972b. Food and feeding behavior of the Mink Frog, *Rana septentrionalis* Baird, in Minnesota. American Midland Naturalist 88:291–300.

Hedeen, S. E. 1972c. Postmetamorphic growth and reproduction of the Mink Frog, *Rana septentrionalis* Baird. Copeia 1972:169–175.

Hedeen, S. E. 1986. The southern geographic limit of the Mink Frog, *Rana septentrionalis*. Copeia 1986:239–244.

Hedges, S. B. 1986. An electrophorectic analysis of holarctic hylid frog evolution. Systematic Zoology 35:1–21.

Hedges, S. B., and N. Vidal. 2009. Lizards, snakes, and amphisbaenians (Squamata). Pages 383–389 *in* B. Hedges, and S. Kumar, editors. The Timetree of Life. Oxford University Press, Oxford, England, UK.

Heinen, J. T. 1993. Substrate choice and predation risk in newly metamorphosed American Toads *Bufo americanus*: an experimental analysis. American Midland Naturalist 130:184–192.

Heinen, J. T. 1994a. Antipredator behavior of newly metamorphosed American Toads (*Bufo a. americanus*) and mechanisms of hunting by Eastern Garter Snakes (*Thamnophis s. sirtalis*). Herpetologica 50:137–145.

Heinen, J. T. 1994b. The significance of color change in newly metamorphosed American Toads (*Bufo a. americanus*). Journal of Herpetology 28:87–93.

Heinen, J. T. 2014. *Chelydra serpentina* (Common Snapping Turtle). Nesting aggregation. Herpetological Review 45:685–686.

Heinen, J. T., and G. Hammond. 1997. Antipredator behaviors of newly metamorphosed Green Frogs (*Rana clamitans*) and Leopard Frogs (*R. pipiens*) in encounters with Eastern Garter Snakes (*Thamnophis s. sirtalis*). American Midland Naturalist 137:136–144.

Heinen, J. T., and J. A. Abdella. 2005. On the advantages of putative cannibalism in American Toad tadpoles (*Bufo a. americanus*): is it active or passive and why? American Midland Naturalist 153:338–347.

Heintzelman, D. S. 1964. Spring and summer Sparrow Hawk food habits. Wilson Bulletin 76:323–330.

Helgen, J., R. G. McKinnell, and M. C. Gernes. 1998. Investigation of malformed Northern Leopard Frogs in Minnesota. Pages 288–297 *in* M. J. Lannoo, editor. Status and conservation of midwestern amphibians. University of Iowa Press, Iowa City, USA.

Hemesath, L. M. 1998. Iowa's frog and toad survey, 1991–1994. Pages 206–216 *in* M. Lannoo, editor. Status and conservation of midwestern amphibians. University of Iowa Press, Iowa City, USA.

Henderson, R. W. 1970. Feeding behavior, digestion, and water requirements of *Diadophis punctatus arnyi* Kennicott. Herpetologica 26:520–526.

Henderson, R. W. 2015. Natural history of neotropical treeboas (genus *Corallus*). Edition Chimaira, Frankfurt, Germany.

Henderson, R. W., and R. Powell. 2009. Natural history of West Indian reptiles and amphibians. University Press of Florida, Gainesville, USA.

Henderson, R. W., M. H. Binder, R. A. Sadjak, and J. A. Buday. 1980. Aggregating behavior and exploitation of subterranean habitat by gravid Eastern Milksnakes (*Lampropeltis t. triangulum*). Milwaukee Public Museum Contributions in Biology and Geology 32:1–9.

Hendricks, III, R. 2000. *Pseudacris crucifer* (Spring Peeper). Geographic distribution. Herpetological Review 31:251.

Hennepin, L. 1698. A new discovery of a vast country in America. Coles Publication Limited, Toronto, Canada.

Henning, B. M., and A. J. Remsburg. 2009. Lakeshore vegetation effects on avian and anuran populations. American Midland Naturalist 161:123–133.

Hensley, M. M. 1959. Albinism in North American amphibians and reptiles. Michigan State University, Publications of the Museum Biological Series 1:133–159.

Heppell, S. S. 1998. Application of life-history theory and population model analysis to turtle conservation. Copeia 1998:367–375.

Heppell, S. S., L. B. Crowder, and D. T. Crouse. 1996. Models to evaluate headstarting as a management tool for long-lived turtles. Ecological Applications 6:556–565.

Herman, R. L. 1984. Ichthyophonus-like infection in newts (*Notophthalmus viridescens* Rafinesque). Journal of Wildlife Diseases 20:55–56.

Herman, T. A., and J. L. Bouzat. 2016. Range-wide phylogeography of the Four-toed Salamander: out of Appalachia and into the glacial aftermath. Journal of Biogeography 43:666–678.

Herreid, C. F., and S. Kinney. 1966. Survival of Alaskan Wood Frog (*Rana sylvatica*) larvae. Ecology 47:1039–1041.

Herreid, C. F., and S. Kinney. 1967. Temperature and development of the Wood Frog, *Rana sylvatica*, in Alaska. Ecology 48:579–590.

Herrel, A., and A. van der Meijden. 2014. An analysis of the live reptile and amphibian trade in the USA compared to the global trade in endangered species. Herpetological Journal 24:103–110.

Herrick, S. Z., K. D. Wells, T. E. Farkas, and E. T. Schultz. 2018. Noisy neighbors: acoustic interference and vocal interactions between two syntopic species of ranid frogs, *Rana clamitans* and *Rana catesbeiana*. Journal of Herpetology 52:176–184.

Herrmann, H. L., K. J. Babbitt, M. J. Baber, and R. G. Congalton. 2005. Effects of landscape characteristics on amphibian distribution in a forest-dominated landscape. Biological Conservation 123:139–149.

Herwig, B. R., L. W. Schroeder, K. D. Zimmer, M. A. Hanson, D. F. Stamples, R. G. Wright, and J. A. Younk. 2013. Fish influences on amphibian presence and abundance in prairie and parkland landscapes of Minnesota, USA. Journal of Herpetology 47:489–497.

Herzog, H. A., Jr., and G. M. Burghardt. 1974. Prey movement and predatory behavior of juvenile Western Yellow-bellied Racers, *Coluber constrictor mormon*. Herpetologica 30:285–289.

Herzog, H. A., Jr., and G. M. Burghardt. 1986. Development of antipredator responses in snakes: I. Defensive and open-field behaviors in newborns and adults of three species of garter snakes (*Thamnophis melanogaster, T. sirtalis, T. butleri*). Journal of Comparative Psychology 100:372–379.

Herzog, H. A., Jr., B. B. Bowers, and G. M. Burghardt. 1989a. Development of antipredator responses in snakes: IV. Interspecific and intraspecific differences in habituation of defensive behavior. Developmental Psychobiology 22:489–508.

Herzog, H. A., Jr., B. B. Bowers, and G. M. Burghardt. 1989b. Stimulus control of antipredator behavior in newborn and juvenile garter snakes (*Thamnophis*). Journal of Comparative Psychology 103:233–242.

Herzog, H. A., Jr., B. B. Bowers, and G. M. Burghardt. 1992. Development of antipredator responses in snakes: V. Species differences in ontogenetic trajectories. Developmental Psychobiology 25:199–211.

Heske, E. J., and W. L. Keller. 1999. Habitat use and hibernacula of three species of snakes at the Middle Fork Fish and Wildlife Area. Illinois Natural History Survey, Center for Wildlife Ecology.

Hess, Z. J., and R. N. Harris. 2000. Eggs of *Hemidactylium scutatum* (Caudata: Plethodontidae) are unpalatable to insect predators. Copeia 2000:597–600.

Heyer, W. R., R. W. McDiarmid, and D. L. Weigmann. 1975. Tadpoles, predation and pond habitats in the tropics. Biotropica 7:100–111.

Hickerson, C. M., C. D. Anthony, and R. M. Walton. 2017. Eastern Red-backed Salamanders regulate top-down effects in a temperate forest-floor community. Herpetologica 73:180–189.

Higginbotham, A. C. 1939. Studies on amphibian activity. I. Preliminary report on the rhythmic activity of *Bufo americanus americanus* Holbrook, and *Bufo fowleri* Hinckley. Ecology 20:58–70.

Highton, R. 1975. Geographic variation in genetic dominance of the color morphs of the Red-backed Salamander, *Plethodon cinereus*. Genetics 80:363–374.

Highton, R. 1995. Speciation in eastern North American salamanders of the genus *Plethodon*. Annual Review of Ecology and Systematics 26:579–600.

Highton, R. 1999. Geographic protein variation and speciation in the salamanders of the *Plethodon cinereus* group with the description of two new species. Herpetologica 55:43–90.

Highton, R. 2005. Declines of eastern North American woodland salamanders (*Plethodon*). Pages 34–46 *in* M. J. Lannoo, editor. Amphibian declines: the conservation status of United States species. University of California Press, Berkeley, USA.

Higley, E., A. R. Tompsett, J. P. Giesy, M. Hecker, and S. Wiseman. 2013. Effects of triphenyltin on growth and development of the Wood Frog (*Lithobates sylvaticus*). Aquatic Toxicology 144–145:155–161.

Higley, W. K. 1889. Reptilia and Batrachia of Wisconsin. Transactions of the Wisconsin Academy of Sciences, Arts and Letters 7:155–176.

Hileman, E. T. 2010. Abundance and survivorship of Butler's Gartersnake (*Thamnophis butleri*) in Wisconsin. Thesis, University of Wisconsin, Milwaukee, USA.

Hileman, E. T. 2014. Habitat protection and restoration in Milwaukee River Estuary Area of Concern. Report to the Urban Ecology Center—Riverside Park, Milwaukee, Wisconsin, USA. Archived by WebCite at http://www.webcitation.org/73cGHgjuZ. Accessed 1 Nov 2018.

Hileman, E. T., D. R. Bradke, D. M. Delaney, and R. B. King. 2015a. Protection by association: implications of scent trailing in neonate Eastern Massasaugas (*Sistrurus catenatus*). Herpetological Conservation Biology 10:654–660.

Hileman, E. T., J. M. Kapfer, T. C. Muehlfeld, and J. H. Govanni. 2015b. Recouping lost information when mark-recapture data are pooled: a case study of Milksnakes (*Lampropeltis triangulum*) in the upper midwestern United States. Journal of Herpetology 49:428–436.

Hileman, E. T., R. B. King, J. M. Adamski, T. G. Anton, R. L. Bailey, S. J. Baker, N. D. Bieser, T. A. Bell Jr., K. M. Bissell, D. R. Bradke, H. Campa III, G. S. Casper, K. Cedar, M. D. Cross, B. A. DeGregorio, M. J. Dreslik, L. J. Faust, D. S. Harvey, R. W. Hay, B. C. Jellen, B. D. Johnson, G. Johnson, B. D. Kie, B. A. Kingsbury, M. J. Kowalski, Y. M. Lee, A. M. Lentini, J. C. Marshal, D. Mauger, J. A. Moore, R. A. Paloski, C. A. Phillips, P. D. Pratt, T. Preney, K. A. Prior, A. Promaine, M. Redmer, H. K. Reinert, J. D. Rouse, K. T. Shoemaker, S. Sutton, T. J. VanDeWalle, P. J. Weatherhead, D. Wynn, and A.Yagi. 2017. Climatic and geographic predictors of life history variation in Eastern Massasauga (*Sistrurus catenatus*): a rangewide synthesis. PLoS ONE 12:e0172011.

Hillis, D. M. 1977. Sex ratio, mortality rate, and breeding stimulus in a Maryland population of *Ambystoma maculatum*. Bulletin of the Maryland Herpetological Society 13:84–91.

Hillis, D. M. 1988. Systematics of the *Rana pipiens* complex: puzzle and paradigm. Annual Review of Ecology and Systematics 19:39–63.

Hillis, D. M., and T. P. Wilcox. 2005. Phylogeny of the New World true frogs (*Rana*). Molecular Phylogenetics and Evolution 34:299–314.

Hilty, J. A., W. Z. Lidicker, Jr., and A. M. Merenlender. 2006. Corridor ecology: the science and practice of linking landscapes for biodiversity conservation. Island Press, Washington, D.C., USA.

Himes, J. C. 2004. The non-fish, vertebrate diet of sympatric populations of the Cottonmouth (*Agkistrodon piscivorus*) and Northern Watersnake (*Nerodia sipedon*). Herpetological Review 35:123–128.

Hine, R. L., B. L. Les, and B. F. Hellmich. 1981. Leopard Frog populations and mortality in Wisconsin, 1974–76. Wisconsin Department of Natural Resources Technical Bulletin 122, Madison, USA.

Hine, R. L., B. L. Les, B. F. Hellmich, and R. C. Vogt. 1975. Preliminary report on Leopard Frog (*Rana pipiens*) populations in Wisconsin. Wisconsin Department of Natural Resources Research Report 81, Madison, USA.

Hinshaw, S. H., and B. K. Sullivan. 1990. Predation on *Hyla versicolor* and *Pseudacris crucifer* during reproduction. Journal of Herpetology 24:196–197.

Hisaw, F. L., and H. K. Gloyd. 1926. The Bull Snake as a natural enemy of injurious rodents. Journal of Mammalogy 7:200–205.

Hobbs, H. H., III. 1993. Trophic relationships of North American freshwater crayfishes and shrimps. Milwaukee Public Museum Contributions in Biology and Geology 85, Milwaukee, Wisconsin, USA.

Höbel, G., and R. C. Kolodziej. 2013. Wood Frogs (*Lithobates sylvaticus*) use of water surface waves in their reproductive behaviour. Behaviour 150:471–483.

Höbel, G., and T. Barta. 2014. Adaptive plasticity in calling site selection in Gray Treefrogs (*Hyla versicolor*). Behaviour 151:741–754.

Hobert, J. P., C. E. Montgomery, and S. P. Mackessy. 2004. Natural history of the Massasauga, *Sistrurus catenatus edwardsii*, in southeastern Colorado. Southwestern Naturalist 49:321–326.

Hoddenbach, G. A. 1966. Reproduction in western Texas *Cnemidophorus sexlineatus* (Sauria: Teiidae). Copeia 1966:110–113.

Hodgson, J. R., and E. M. Hansen. 2005. Terrestrial prey items in the diet of Largemouth Bass *Micropterus salmoides*, in a small north temperate lake. Journal of Freshwater Ecology 20:793–794.

Hodgson, J. R., and P. A. Cochran. 1986. *Hyla crucifer* (Spring Peeper). Geographic distribution. Herpetological Review 17:26.

Hoff, J. G., and S. A. Moss. 1974. A distress call in the Bullfrog *Rana catesbeiana*. Copeia 1974:533–534.

Hoff, V. S. K., A. R. Blaustein, R. W. McDiarmid, and R. Altig. 1999. Behavior: interactions and their consequences. Pages 215–239 *in* R. W. McDiarmid and R. Altig, editors. Tadpoles: the biology of anuran larvae. University of Chicago Press, Chicago, Illinois, USA.

Hoffman, E. A., and M. S. Blouin. 2004. Evolutionary history of the Northern Leopard Frog: reconstruction of phylogeny, phylogeography, and historical changes in population demography from mitochondrial DNA. Evolution 58:145–149.

Hoffman, N. 2003. Frog fence along Vermont Route 2 in Sandbar Wildlife Management Area: collaboration between Vermont Agency of Transportation and Vermont Agency of Natural Resources. Pages 431–432 *in* C. L. Irwin, P. Garrett, and K. P. McDermott, editors. Proceedings of the 2003 International Conference on Ecology and Transportation. Center for Transportation and the Environment, North Carolina State University, Raleigh, USA.

Hoffman, R. 2002. Wisconsin's natural communities: how to recognize them, where to find them. University of Wisconsin Press, Madison, USA.

Hogg, D. M. 1975. The Snapping Turtles of Wye Marsh. Ontario Fish and Wildlife Review 14:15–20.

Hokit, D. G., B. M. Stith, and L. C. Branch. 1999. Effects of landscape structure in Florida scrub: a population perspective. Ecological Applications 9:124–134.

Holbrook, J. E. 1838. North American herpetology. Volume 3. J. Dobson, Philadelphia, Pennsylvania, USA.

Holland, M. P. 2009. Echinostome metacercariae cyst elimination in *Rana clamitans* (Green Frog) tadpoles is age-dependent. Journal of Parasitology 95:281–285.

Holland, M. P., D. K. Skelly, M. Kashgarian, S. R. Bolden, L. M. Harrison, and M. Capello. 2007. Echinostome infection in Green Frogs (*Rana clamitans*) is stage and age dependent. Journal of Zoology 271:455–462.

Holloway, A. K., D. C. Cannatella, H. C. Gerhardt, and D. M. Hillis. 2006. Polyploids with different origins and ancestors form a single sexual polyploid species. American Naturalist 167:E88–E101.

Holman, J. A. 1971. *Ophisaurus attenuatus*. Catalogue of American amphibians and reptiles 111.1–111.3.

Holman, J. A. 1976. Paleoclimatic implications of "ecologically incompatible" herpetological species (late Pleistocene: southeastern United States). Herpetologica 32:290–295.

Holman, J. A. 1992a. Late quaternary herpetofauna of the central Great Lakes region, U.S.A.: zoogeographical and paleoecological implications. Quaternary Science Reviews 2:345–351.

Holman, J. A. 1992b. Patterns of herpetological reoccupation of post-glacial Michigan: amphibians and reptiles come home. Michigan Academician 24:453–466.

Holman, J. A. 1994. Status of the Red-Eared Slider Turtle *Trachemys scripta elegans* (Wied) in Michigan: a preliminary report. Michigan Academician 26:471–477.

Holman, J. A. 1995. Pleistocene amphibians and reptiles in North America. Oxford University Press, New York, New York, USA.

Holman, J. A. 2000. Fossil snakes of North America: origin, evolution, distribution, paleoecology. Indiana University Press, Bloomington, USA.

Holman, J. A. 2001. In quest of Great Lakes ice age vertebrates. Michigan State University Press, East Lansing, USA.

Holman, J. A. 2003. Fossil frogs and toads of North America. Indiana University Press, Bloomington, USA.

Holman, J. A. 2012. The amphibians and reptiles of Michigan: a quarternary and recent faunal adventure. Wayne State University Press, Detroit, Michigan, USA.

Holman, J. A., and R. L. Richards. 1981. Late Pleistocene occurrence in southern Indiana of the Smooth Green Snake, *Opheodrys vernalis*. Journal of Herpetology 15:123–125.

Holman, J. A., and U. Fritz. 2001. A new emydine species from the medial Miocene (Barstovian) of Nebraska, USA with a new generic arrangement for the species of *Clemmys* sensu McDowell (1964) (Reptilia: Testudines: Emydidae). Zoologische Abhandlungen Staatliches Museum für Tierkunde Dresden 51:331–354.

Holomuzki, J. R. 1986. Predator avoidance and diel patterns of microhabitat use by larval Tiger Salamanders. Ecology 67:737–748.

Holomuzki, J. R. 1995. Oviposition sites and fish-deterrent mechanisms of two stream anurans. Copeia 1995:607–613.

Holomuzki, J. R. 1997. Habitat-specific life-histories and foraging by stream-dwelling American Toads. Herpetologica 53:445–453.

Holomuzki, J. R. 1998. Grazing effects by Green Frog tadpoles (*Rana clamitans*) in a woodland pond. Journal of Freshwater Ecology 13:1–8.

Holomuzki, J. R., and J. P. Collins. 1983. Diel movement of larvae of the Tiger Salamander, *Ambystoma tigrinum nebulosum*. Journal of Herpetology 17:276–278.

Holomuzki J. R., J. P. Collins, and P. E. Brunkow. 1994. Trophic control of fishless ponds by Tiger Salamander larvae. Oikos 71:55–64.

Holtzman, D. A. 1993. The ontogeny of nasal chemical senses in garter snakes. Brain, Behavior and Evolution 41:163–170.

Holtzman, D. A., and M. Halpern. 1990. Embryonic and neonatal development of the vomeronasal and olfactory systems in garter snakes (*Thamnophis* spp.). Journal of Morphology 203:123–140.

Holycross, A. T., and S. P. Mackessy. 2002. Variation in the diet of *Sistrurus catenatus* (Massasauga), with emphasis on *Sistrurus catenatus edwardsii* (Desert Massasauga). Journal of Herpetology 36:454–464.

Holzer, K. A., and S. P. Lawler. 2015. Introduced Reed Canary Grass attracts and supports a common native amphibian. Journal of Wildlife Management 79:1081–1090.

Holzwart, J., and K. D. Hall. 1984. The absence of glycerol in the hibernating American Toad (*Bufo americanus*). Bios 55:31–36.

Homan, R. N., B. S. Windmiller, and J. M. Reed. 2004. Critical thresholds associated with habitat loss for two vernal pool-breeding amphibians. Ecological Applications 14:1547–1553.

Homan, R. N., B. S. Windmiller, and J. M. Reed. 2007. Comparative life histories of two sympatric *Ambystoma* species at a breeding pond in Massachusetts. Journal of Herpetology 41:401–409.

Homan, R. N., J. V. Regosin, D. M. Rodrigues, J. M. Reed, B. S. Windmiller, and L. M. Romero. 2003. Impacts of varying habitat quality on the physiological stress of Spotted Salamanders (*Ambystoma maculatum*). Animal Conservation 6:11–18.

Homyack, J. A., and C. A. Haas. 2009. Long-term effects of experimental forest harvesting on abundance and reproductive demography of terrestrial salamanders. Biological Conservation 142:110–121.

Homyack, J. D., and W. M. Giuliano. 2002. Effect of streambank fencing on herpetofauna in pasture stream zones. Wildlife Society Bulletin 30:361–369.

Hoopes, I. 1938. Do you know the Mink Frog? New England Naturalist 1:4–6.

Hoover, J. J., and J. Killgore. 2002. Small floodplain pools as habitat for fishes and amphibians: methods for evaluation. US Army Engineer Research and Development Center, Technical Notes Collection ERDC TN–EMRRP–EM–03, Vicksburg, Mississippi, USA.

Hopey, M. E., and J. W. Petranka. 1994. Restriction of Wood Frogs to fish-free habitats: how important is adult choice? Copeia 1994:1023–1025.

Hoppe, D. M. 2005. Malformed frogs in Minnesota: history and interspecific differences. Pages 103–108 *in* M. Lannoo, editor. Amphibian Declines: the conservation and status of United States species. University of California Press, Berkeley, USA.

Horne, M. T., and W. A. Dunson. 1995. Effects of low pH, metals, and water hardness on larval amphibians. Archives of Environmental Contamination and Toxicology 29:500–505.

Houlahan, J. E., and C. S. Findlay. 2003. The effects of adjacent land use on wetland amphibian species richness and community composition. Canadian Journal of Fisheries and Aquatic Sciences 60:1078–1094.

Hoverman, J. T., M. J. Gray, and D. L. Miller. 2010. Anuran susceptibilities to ranaviruses: role of species identity, exposure route, and a novel virus isolate. Diseases of Aquatic Organisms 89:97–107.

Hoverman, J. T., M. J. Gray, D. L. Miller, and N. A. Haislip. 2012. Widespread occurrence of ranavirus in pond-breeding amphibian populations. EcoHealth 9:36–48.

Hoverman, J. T., M. J. Gray, N. A. Haislip, and D. L. Miller. 2011. Phylogeny, life history, and ecology contribute to differences in amphibian susceptibility to ranaviruses. EcoHealth 8:301–319.

Howard, A. D. 1915. Some exceptional cases of breeding among the Unionidae. Nautilus 29:4–11.

Howard, A. D. 1951. A river mussel parasitic on a salamander. Chicago Academy of Sciences Natural History Miscellanea 77, Chicago, Illinois, USA.

Howard, R. D. 1978a. The evolution of mating strategies in Bullfrogs, *Rana catesbeiana*. Evolution 32:850–871.

Howard, R. D. 1978b. The influence of male-defended oviposition sites on early embryo mortality in Bullfrogs, *Rana catesbeiana*. Ecology 32:789–798.

Howard, R. D. 1980. Mating behaviour and mating success in Wood Frogs, *Rana sylvatica*. Animal Behaviour 28:705–716.

Howard, R. D. 1981. Sexual dimorphism in Bullfrogs. Ecology 62:303–310.

Howard, R. D. 1988a. Reproductive success in two species of anurans. Pages 99–133 *in* T. H. Clutton-Brock, editor. Reproductive success: studies of individual variation in contrasting breeding systems. University of Chicago Press, Chicago, Illinois, USA.

Howard, R. D. 1988b. Sexual selection on male body size and mating behaviour in American Toads, *Bufo americanus*. Animal Behaviour 36:1796–1808.

Howard, R. D., and A. G. Kluge. 1985. Proximate mechanisms of sexual selection in Wood Frogs. Evolution 39:260–277.

Howard, R. D., and J. R. Young. 1998. Individual variation in male vocal traits and female mating preferences in *Bufo americanus*. Animal Behaviour 55:1165–1179.

Howard, R. R. 1971. Avoidance learning in Spotted Salamanders, *Ambystoma maculatum*, by domestic chickens. American Zoologist 11:637.

Howe, C. M., M. Berrill, B. D. Pauli, C. C. Helbing, K. Werry, and N. Veldoen. 2004. Toxicity of glyphosate-based pesticides to four North American frog species. Environmental Toxicology and Chemistry 23:1928–1938.

Howe, G. E., R. Gillis, and R. C. Mowbray. 1998. Effect of chemical synergy and larval stage on the toxicity of atrazine and alachlor to amphibian larvae. Environmental Toxicology and Chemistry 17:519–525.

Howe, H. F. 1994. Managing species diversity in tallgrass prairie: assumptions and implications. Conservation Biology 8:691–704.

Howes, B. J., B. Lindsay, and S. C. Lougheed. 2006. Range-wide phylogeography of a temperate lizard, the Five-lined Skink (*Eumeces fasciatus*). Molecular Phylogenetics and Evolution 40:183–194.

Howes, B. J., and S. C. Lougheed. 2004. The importance of cover rock in northern populations of the Five-lined Skink (*Eumeces fasciatus*). Herpetologica 60:287–294.

Hoy, P. R. 1874. Some of the peculiarities of the fauna near Racine. Wisconsin Academy of Sciences, Arts and Letters 2:120–122.

Hoy, P. R. 1882. Water Puppy (*Menobranchus lateralis* Say). Transactions of the Wisconsin Academy of Sciences, Arts and Letters 5:248–250.

Hoy, P. R. 1883. Catalogue of the cold-blooded vertebrates of Wisconsin. Geologic Survey of Wisconsin, 1873–1879 1:422–475.

Hsiang, A. Y., D. J. Field, T. H. Webster, A. D. B. Behlke, M. B. Davis, R. A. Racicot, and J. A. Gauthier. 2015. The origin of snakes: revealing the ecology, behavior, and evolutionary history of early snakes using genomics, phenomics, and the fossil record. BMC Evolutionary Biology 15:Article 87.

Huang, Y., D. J. Hoffman, and W. H. Karasov. 2007. Oxidative stress induced in PCB 126-exposed Northern Leopard Frogs, *Rana pipiens*. Journal of Toxicology and Environmental Health, Part A 70:676–681.

Hubbs, C. L., and T. E. B. Pope. 1937. The spread of the Sea Lamprey through the Great Lakes. Transactions of the American Fisheries Society 66:172–176.

Huey, R. B., and R. D. Stevenson. 1979. Integrating thermal physiology and ecology of ectotherms: a discussion of approaches. American Zoologist 19:357–366.

Hughes, D. F., A. K. Tegeler, and W. E. Meshaka, Jr. 2016. Differential use of ponds and movements by two species of aquatic turtles (*Chrysemys picta marginata* and *Chelydra serpentina serpentina*) and their role in colonization. Herpetological Conservation and Biology 11:214–231.

Hughes, G. N., W. F. Greaves, and J. D. Litzgus. 2009. Nest-site selection by Wood Turtles (*Glyptemys insculpta*) in a thermally limited environment. Northeastern Naturalist 16:321–338.

Hughes, M. S. 2008. Analysis of the parasite community of Adams Lake mudpuppies. Presentation at the 9th Annual Undergraduate Research Symposium, University of Wisconsin, Stevens Point, USA.

Hughes, R. C., J. W. Higginbotham, and J. W. Clary. 1942. The trematodes of reptiles, part I, systematic section. American Midland Naturalist 27:109–134.

Huheey, J. E., and A. Stupka. 1967. Amphibians and reptiles of the Great Smoky Mountains. Herpetologica 21:148–155.

Huijser, M. P., and J. S. Begley. 2015. Wildlife mitigation plan State HWY 49, Horicon Marsh, Wisconsin, USA. US Fish and Wildlife Service, Region 3 Report 4W4838–08, Bloomington, Minnesota, USA.

Hulse, A. C., and E. J. Routman. 1982. Leech (*Placobdella parasitica*) infestations on the Wood Turtle, *Clemmys insculpta*. Herpetological Review 13:116–117.

Humphrey, R. R. 1928. Ovulation in the Four-toed Salamander, *Hemidactylium scutatum*, and the external features of cleavage and gastrulation. Biological Bulletin 54:307–323.

Humphries, H. C., and P. S. Bourgeron. 2001. Methods for determining historical range of variability. Pages 273–291 *in* M. Jensen, and P. Bourgeron, editors. A guidebook for integrated ecological assessments. Springer-Verlag, New York, USA.

Hunsinger, T. W., and M. J. Lannoo. 2005. *Notophthalmus viridescens* (Rafinesque, 1820). Eastern Newt. Pages 889–894 *in* M. J. Lannoo, editor. Amphibian declines: the conservation status of United States species. University of California Press, Berkeley, USA.

Hupy, C. M., and C. H. Yansa. 2009. The last 17,000 years of vegetation history. Pages 91–105 *in* R. Schaetzl, J. Darden, and D. Brandt, editors. Michigan Geography and Geology. Pearson Custom Publishers, Boston, Massachusetts, USA.

Hurlbert, S. H. 1969. The breeding migrations and interhabitat wandering of the Vermilion-spotted Newt, *Notophthalmus viridescens* (Rafinesque). Ecological Monographs 39:465–488.

Hurter, J. 1911. Herpetology of Missouri. Transactions of the Academy of Science of St. Louis 20:59–274.

Hutchinson, A., and M. L. Hunter. 1992. The status of the Timber Rattlesnake in Maine. Page 6 *in* T. F. Tyning, editor. Conservation of the Timber Rattlesnake in the northeast. Massachusetts Audubon Society, Lincoln, USA.

Hutchison, V. H., and J. P. Ritchart. 1989. Annual cycle of thermal tolerance in the salamander, *Necturus maculosus*. Journal of Herpetology 23:73–76.

Hutchison, V. H., and S. D. Rowlan. 1975. Thermal acclimation and tolerance in the Mudpuppy, *Necturus maculosus*. Journal of Herpetology 9:367–368.

Hutchinson, V. H., A. Vinegar, and R. J. Kosh. 1966. Critical thermal maxima in turtles. Herpetologica 22:32–41.

Hutt, F. B. 1945. Complete albinism in the Mud Puppy, *Necturus*. Journal of Heredity 36:145–147.

Hyde, T., K. Kearns, R. Paloski, and W. Smith. 2012. Preliminary environmental assessment on proposed rule revision to update Wisconsin's endangered and threatened species list [ER-27–11]. Wisconsin Department of Natural Resources, Madison, USA. Archived by WebCite at http://www.webcitation.org/73cGYJi6V. Accessed 1 Nov 2018.

Hyde, T., R. Paloski, R. Hay, and P. Miller, editors. 2007. Butler's Gartersnake (*Thamnophis butleri*) in Wisconsin: population and habitat viability assessment—the stakeholder workshop report. IUCN/SSC Conservation Breeding Specialist Group, Apple Valley, Minnesota, USA.

Hyman, O. J. 2012. The ecology of chytridiomycosis in Boreal Chorus Frogs (*Pseudacris maculata*). Dissertation, Arizona State University, Tempe, USA.

Ihlow, F., J. Dambach, J. O. Engler, M. Flecks, T. Hartmann, S. Nekum, H. Rajaei, and D. Rodder. 2012. On the brink of extinction? How climate change may affect global chelonian species richness and distribution. Global Change Biology 18:1520–1530.

Imlay, T. L., J. Saroli, T. B. Herman, and S. W. Mockford. 2015. Movements of the Eastern Ribbonsnake (*Thamnophis sauritus*) in Nova Scotia. Canadian Field-Naturalist 129:379–385.

Indiana Legislature. 2011. 312 IAC 9-5-4 Endangered species of reptiles and amphibians. Indiana Administrative Code.

Ingle, D. J., and K. vS. Hoff. 1990. Visually elicited evasive behavior in frogs. BioScience 40:284–291.

Ingram, W. M., and E. C. Raney. 1943. Additional studies on the movement of tagged Bullfrogs, *Rana catesbeiana* Shaw. American Midland Naturalist 29:239–241.

[INHS] Illinois Natural History Survey. 2014. Amphibians and reptiles: Illinois species. Archived by WebCite at http://www.webcitation.org/73cH4m1eX. Accessed 1 Nov 2018.

[IPCC] Intergovernmental Panel on Climate Change. 2013. Climate change 2013: the physical science basis. Contribution of working group I to the fifth assessment report of the Intergovernmental Panel on Climate Change. T. F. Stocker, D. Qin, G.-K. Plattner, M. Tignor, S. K. Allen, J. Boschung, A. Nauels, Y. Xia, V. Bex, and P. M. Midgley, editors. Cambridge University Press, New York, New York, USA.

Ireland, D. H., A. J. Wirsing, and D. L. Murray. 2007. Phenotypically plastic responses of Green Frog embryos to conflicting predation risk. Oecologia 152:162–168.

Ireland, P. H. 1989. Larval survivorship in two populations of *Ambystoma maculatum*. Journal of Herpetology 23:209–215.

Ireland, P. H. 1991. Separate effects of acid-derived anions and cations on growth of larval salamanders of *Ambystoma maculatum*. Copeia 1991:132–137.

Irschick, D. J., and H. B. Shaffer. 1997. The polytypic species revisited: morphological differentiation among Tiger Salamanders (*Ambystoma tigrinum*) (Amphibia: Caudata). Herpetologica 53:30–49.

Irwin, J. T. 2005. Overwintering in Northern Cricket Frogs (*Acris crepitans*). Pages 55–58 *in* M. J. Lannoo, editor. Amphibian declines: the conservation status of United States species. University of California Press, Berkeley, USA.

Irwin, J. T., and R. E. Lee, Jr. 2003. Geographic variation in energy storage and physiological responses to freezing in the Gray Treefrogs *Hyla versicolor* and *H. chrysoscelis*. Journal of Experimental Biology 206:2859–2867.

Irwin, J. T., J. P. Costanzo, and R. E. Lee, Jr. 1999. Terrestrial hibernation in the Northern Cricket Frog, *Acris crepitans*. Canadian Journal of Zoology 77:1240–1246.

Irwin, J. T., J. P. Costanzo, and R. E. Lee, Jr. 2003. Postfreeze reduction of locomotor endurance in the freeze-tolerant Wood Frog, *Rana sylvatica*. Physiological and Biochemical Zoology 76:331–338.

Irwin, K. J., J. T. Briggler, and T. L. Crabill. 2014. *Cryptobranchus alleganiensis bishopi* (Ozark Hellbender). Diet. Herpetological Review 45:472.

[IUCN] International Union for Conservation of Nature. 2020. The IUCN Red List of Threatened Species. www.iucnredlist.org. Accessed 11 Nov 2020.

Iverson, J. B. 1977. Geographic variation in the Musk Turtle, *Sternotherus minor*. Copeia 1977:502–517.

Iverson, J. B. 1978. Reproductive notes on Florida snakes. Florida Scientist 4:201–207.

Iverson, J. B. 1988. Growth in the Common Map Turtle, *Graptemys geographica*. Transactions of the Kansas Academy of Science 91:153–157.

Iverson, J. B. 1991. Patterns of survivorship in turtles (order Testudines). Canadian Journal of Zoology 69:385–391.

Iverson, J. B. 1998. Molecules, morphology and mud turtle phylogenetics (Family Kinosternidae). Chelonian Conservation and Biology 3:113–117.

Iverson, J. B. 2014. *Pituophis catenifer sayi* (Bullsnake). Arboreality. Herpetological Review 45:342.

Iverson, J. B., and G. R. Smith. 2010. Population size and sex ratio of Snapping Turtles on the Crescent Lake National Wildlife Refuge. IRCF Reptiles & Amphibians 17:195–200.

Iverson, J. B., and M. A. Ewert. 1991. Physical characteristics of reptilian eggs and a comparison with avian eggs. Pages 87–100 *in* D. C. Deeming and M. W. J. Ferguson, editors. Egg incubation: its effect on embryonic development in birds and reptiles. Cambridge University Press, New York, New York, USA.

Iverson, J. B., and W. E. Meshaka, Jr. 2006. *Sternotherus odoratus*—Common Musk Turtle or Stinkpot. Pages 207–223 *in* P. A. Meylan, editor. Biology and conservation of Florida turtles. Chelonian Research Foundation, Chelonian Research Monograph 3, Lunenburg, Massachusetts, USA.

Iverson, J. B., G. R. Smith, and J. Rettig. 2019. Body size, growth, and longevity in Northern Map Turtles (*Graptemys geographica*) in Indiana. Journal of Herpetology 53:297–301.

Iverson, J. B., C. A. Young, T. S. B. Akre, and C. M. Griffiths. 2012. Reproduction by female Bullsnakes (*Pituophis catenifer sayi*) in the Nebraska sandhills. Southwestern Naturalist 57:58–73.

Iverson, J. B., H. Higgins, A. Sirulnik, and C. Griffiths. 1997. Local and geographic variation in the reproductive biology of the Snapping Turtle (*Chelydra serpentina*). Herpetologica 53:96–117.

Iverson, L. R., A. M. Prasad, S. N. Matthews, and M. Peters. 2008. Estimating potential habitat for 134 eastern US tree species under six climate scenarios. Forest Ecology and Management 254:390–406.

Iverson, L. R., M. P. Peters, S. Matthews, and A. Prasad. 2013. An overview of some concepts, potentials, issues, and realities of assisted migration for climate change adaptation in forests. Proceedings of the 60th Annual Western International Forest Disease Work Conference. Tahoe City, California, USA.

Jackrel, S. L., and H. K. Reinert. 2011. Behavioral responses of a dietary specialist, the Queen Snake (*Regina septemvittata*), to potential chemoattractants released by its prey. Journal of Herpetology 45:272–276.

Jackson, D. C., J. Allen, and P. K. Strupp. 1976. The contribution of non-pulmonary surfaces to CO_2 loss in 6 species of turtles at 20°C. Comparative Biochemistry and Physiology 55A:243–246.

Jackson, H. H. T. 1961. Mammals of Wisconsin. University of Wisconsin Press, Madison, USA.

Jackson, J. F., and J. A. Pounds. 1980. Of differential head-body pigmentation in snakes. Journal of Herpetology 14:307–311.

Jackson, N. D., and L. Fahrig. 2011. Relative effects of road mortality and decreased connectivity on population genetic diversity. Biological Conservation 144:3143–3148.

Jackson, S. D. 1996. Underpass systems for amphibians. Pages 224–227 *in* G. L. Evink, P. Garrett, D. Zeigler, and J. Barry, editors. Trends in addressing transportation related wildlife mortality: proceedings of the transportation related wildlife mortality seminar. Florida Department of Transportation Publication FL-ER-58.96, Tallahassee, USA.

Jackson, S. D., and T. F. Tyning. 1989. Effectiveness of drift fences and tunnels for moving Spotted Salamanders *Ambystoma maculatum* under roads. Pages 93–99 *in* T. E. S. Lanston, editor. Amphibians and roads. ACO (Ahlmann Company) Polymer Products, Shofford, United Kingdom.

Jackson, S. D., D. J. Smith, and K. E. Gunson. 2015. Mitigating road effects on small animals. Pages 177–207 *in* K. M. Andrews, P. Nanjappa, and S. P. D. Riley, editors. Roads and ecological infrastructure: concepts and applications for small animals. John Hopkins University Press, Baltimore, Maryland, USA.

Jackson-Smith, D. B., and B. Barham. 2000. The changing face of Wisconsin dairy farms: a summary of PATS research on structural change in the 1990s. University of Wisconsin, PATS Research Report 7, Madison, USA.

Jacobs, D. L. 1950. Status of gartersnakes in a forest-prairie ecotone in southern Minnesota. Copeia 1950:233–234.

Jacobs, L., and J. E. Houlahan. 2011. Adjacent land-use affects amphibian community composition and species richness in managed forests in New Brunswick, Canada. Canadian Journal of Forest Research 41:1687–1697.

Jacobs, P. M., J. A. Mason, and P. R. Hanson. 2011. Mississippi Valley regional source of loess on the southern Green Bay Lobe land surface, Wisconsin. Quaternary Research 75:574–583.

Jacobson, E., and W. G. Whitford. 1970. The effect of acclimation on physiological responses to temperature in snakes *Thamnophis proximus* and *Natrix rhombifera*. Comparative Biochemistry and Physiology 35A:439–449.

Jaeger, C. P., R. B. King, and M. R. Duvall. 2014. Initial characterization of major histocompatibility complex (MHC) class IIB exon 2 in and endangered rattlesnake, the Eastern Massasauga (*Sistrurus catenatus*). Journal of North American Herpetology 2014:98–104.

Jaeger, R. G. 1978. Plant climbing by salamanders: periodic availability of plant-dwelling prey. Copeia 1978:686–691.

Jaeger, R. G. 1980a. Fluctuations in prey availability and food limitation for a terrestrial salamander. Oecologia 44:335–341.

Jaeger, R. G. 1980b. Microhabitats of a terrestrial forest salamander. Copeia 1980:265–268.

Jaeger, R. G. 1984. Agonistic behavior of the Red-backed Salamander. Copeia 1984:309–314.

Jaeger, R. G., and D. C. Forester. 1993. Social behavior of plethodontid salamanders. Herpetologica 1993:163–175.

Jaeger, R. G., and D. E. Barnard. 1981. Foraging tactics of a terrestrial salamander: choice of diet in structurally simple environments. American Naturalist 117:639–664.

Jaeger, R. G., and J. K. Schwarz. 1991. Gradational threat postures by the Red-backed Salamander. Journal of Herpetology 25:112–114.

Jaeger, R. G., and J. P. Hailman. 1973. Effects of intensity on the phototactic responses of adult anuran amphibians: a comparative survey. Ethology 33:352–407.

Jaeger, R. G., C. R. Gabor, and H. M. Wilbur. 1998. An assemblage of salamanders in the southern Appalachian Mountains: competitive and predatory behavior. Behaviour 135:795–821.

Jaeger, R. G., J. A. Wicknick, M. R. Griffis, and C. D. Anthony. 1995. Socioecology of a terrestrial salamander: juveniles enter adult territories during stressful foraging periods. Ecology 76:533–543.

Jaeger, R. G., J. R. Gillette, and R. C. Cooper. 2002. Sexual coercion in a territorial salamander: males punish socially polyandrous female partners. Animal Behaviour 5:871–877.

Jaffe, A. L., G. J. Slater, and M. E. Alfaro. 2011. The evolution of island gigantism and body size variation in tortoises and turtles. Biology Letters 7:558–561.

James, S. W. 1995. Systematics, biogeography, and ecology of nearctic earthworms from eastern, central, southern, and southwestern United States. Pages 29–51 *in* P. F. Hendrix, editor. Earthworm ecology and biogeography. CRC Press, Boca Raton, Florida, USA.

Jameson, E. W., Jr. 1944. Food of the Red-backed Salamander. Copeia 1944:145–147.

Jameson, E. W., Jr. 1947. The food of the Western Cricket Frog. Copeia 3:212.

Jancovich, J. K., E. W. Davidson, A. Seiler, B. L. Jacobs, and J. P. Collins. 2001. Transmission of the *Ambystoma tigrinum* virus to alternative hosts. Diseases of Aquatic Organisms 46:159–163.

Jancovich, J. K., E. W. Davidson, J. F. Morado, B. L. Jacobs, and J. P. Collins. 1997. Isolation of a lethal virus from the endangered Tiger Salamander *Ambystoma tigrinum stebbinsi*. Diseases of Aquatic Organisms 31:161–167.

Jancovich, J. K., E. W. Davidson, N. Parameswaran, J. Mao, V. G. Chinchar, J. P. Collins, B. L. Jacobs, and A. Storfer. 2005. Evidence for emergence of an amphibian iridoviral disease because of human-enhanced spread. Molecular Ecology 14:213–225.

Janes, R. G. 1939. Studies on amphibian digestive system, IV. The effect of diet on the small intestine of *Rana sylvatica*. Copeia 1939:134–140.

Janowiak, M. K., L. R. Iverson, D. J. Mladenoff, E. Peters, K. R. Wythers, W. Xi, L. A. Brandt, P. R. Butler, S. D. Handler, P. D. Shannon, C. Swanston, L. R. Parker, A. J. Amman, B. Bogaczyk, C. Handler, E. Lesch, P. B. Reich, S. Matthews, M. Peters, A. Prasad, S. Khanal, F. Liu, T. Bal, D. Bronson, A. Burton, J. Ferris, J. Fosgitt, S. Hagan, E. Johnston, E. Kane, C. Matula, R. O'Connor, D. Higgins, M. St. Pierre, J. Daley, M. Davenport, M. R. Emery, D. Fehringer, C. L. Hoving, G. Johnson, D. Neitzel, M. Notaro, A. Rissman, C. Rittenhouse, and R. Ziel. 2014. Forest ecosystem vulnerability assessment and synthesis for northern Wisconsin and western Upper Michigan: a report from the Northwoods Climate Change Response Framework Project. US Forest Service, Northern Research Station General Technical Report NRS-136, Newtown Square, Pennsylvania, USA.

Jaskula, J. M., and R. Brodman. 2000. Density-dependent effects on microhabitat selection and activity of two species of larval *Ambystoma* salamanders. Journal of the Iowa Academy of Sciences 107:146–150.

Jaslow, A. P., and R. C. Vogt. 1977. Identification and distribution of *Hyla versicolor* and *Hyla chrysoscelis* in Wisconsin. Herpetologica 33:201–205.

Jass, J. P. 2004. Distributions of gastropods in Wisconsin. Milwaukee Public Museum Contributions in Biology and Geology 99:1–28.

Jayne, B. C., and A. F. Bennett. 1990. Selection on locomotor performance capacity in a natural population of Garter Snakes. Evolution 44:1204–1229.

Jefferson, D. M., K. A. Hobson, B. S. Demuth, M. C. O. Ferrari, and D. P. Chivers. 2014. Frugal cannibals: how consuming conspecific tissues can provide conditional benefits to Wood Frog tadpoles (*Lithobates sylvaticus*). Naturwissenschaften 101:291–303.

Jellen, B. C., and M. J. Kowalski. 2007. Movement and growth of neonate Eastern Massasaugas (*Sistrurus catenatus*). Copeia 2007:994–1000.

Jellen, B. C., and R. D. Aldridge. 2010. Paternity patterns. Pages 619–644 *in* R. D. Aldridge and D. M. Sever, editors. Reproductive biology and phylogeny of snakes. CRC Press, Boca Raton, Florida, USA.

Jellen, B. C., D. B. Shepard, M. J. Dreslik, and C. A. Phillips. 2007. Male movement and body size affect mate acquisition in the Eastern Massasauga (*Sistrurus catenatus*). Journal of Herpetology 41:451–457.

Jenkins, D. W. 1948. Trombiculid mites affecting man. I. Bionomics with reference to epidemiology in the United States. American Journal of Hygeine 48:22–35.

Jenkins, J. L., and D. L. Swanson. 2005. Liver glycogen, glucose mobilization and freezing survival in Chorus Frogs, *Pseudacris triseriata*. Journal of Thermal Biology 30:485–494.

Jennrich, R. I., and F. B. Turner. 1969. Measurement of non-circular home range. Journal of Theoretical Biology 22:227–237.

Jensen, E. L., P. Govindarajulu, and M. A. Russello. 2015. Genetic assessment of taxonomic uncertainty in Painted Turtles. Journal of Herpetology 49:314–324.

Jensen, J. B., C. D. Camp, W. Gibbons, and M. J. Elliott. 2008. Amphibians and reptiles of Georgia. University of Georgia Press, Athens, USA.

Jensen, J. B. 2020. *Ambystoma tigrinum* (Eastern Tiger Salamander). Geographic distribution. Herpetological Review 51:767.

Jenssen, T. A. 1967. Food habits of the Green Frog, *Rana clamitans*, before and during metamorphosis. Copeia 1967:214–218.

Jenssen, T. A., and W. D. Klimstra. 1966. Food habits of the Green Frog, *Rana clamitans*, in southern Illinois. American Midland Naturalist 76:169–182.

Jergenson, A. M., D. A. W. Miller, L. A. Neuman-Lee, D. A. Warner, and F. J. Janzen. 2014. Swimming against the tide: resilience of a riverine turtle to recurrent extreme environmental events. Biology Letters 10:20130782.

Jofré, M. B., and W. H. Karasov. 1999. Direct effects of ammonia on three species of North American anuran amphibians. Environmental Toxicology and Chemistry 18:1806–1812.

Jofré, M. B., and W. H. Karasov. 2008. Effect of mono-ortho and di-ortho substituted polychlorinated byphenyl (PCB) congeners on Leopard Frog survival and sexual development. Chemosphere 70:1609–1619.

John-Alder, H. B., P. J. Morin, and S. P. Lawlwer. 1988. Thermal physiology, phenology, and distribution of treefrogs. American Naturalist 132:506–520.

Johnson, B. 1989. Familiar amphibians and reptiles of Ontario. Natural Heritage/Natural History Inc., Toronto, Ontario, Canada.

Johnson, B. D., J. P. Gibbs, K. T. Shoemaker, and J. B. Cohen. 2016a. Demography of a small and isolated population of Eastern Massasauga Rattlesnakes (*Sistrurus catenatus*) threatened by vegetative succession. Journal of Herpetology 50:534–540.

Johnson, B. D., J. P. Gibbs, T. A. Bell, Jr., and K. T. Shoemaker. 2016b. Manipulation of basking sites for endangered Eastern Massasauga Rattlesnakes. Journal of Wildlife Management 80:803–811.

Johnson, B. K., and J. L. Christiansen. 1976. The food and food habits of Blanchard's Cricket Frog, *Acris crepitans blanchardi* (Amphibia, Anura, Hylidae), in Iowa. Journal of Herpetology 10:63–74.

Johnson, C. 1959. Genetic incompatibility in the call races of *Hyla versicolor* Le Conte in Texas. Copeia 1959:327–335.

Johnson, C. 1963. Additional evidence of sterility between the call-types in the *Hyla versicolor* complex. Copeia 1963:139–143.

Johnson, G. 1995. Spatial ecology, habitat preference, and habitat management of the Eastern Massasauga, *Sistrurus c. catenatus* in a New York weakly minerotrophic peatland. Dissertation, State University of New York, Syracuse, USA.

Johnson, G. 2000a. Spatial ecology of the Eastern Massasauga (*Sistrurus c. catenatus*) in a New York peatland. Journal of Herpetology 34:186–192.

Johnson, G., and D. J. Leopold. 1998. Habitat management for the Eastern Massasauga in a central New York peatland. Journal of Wildlife Management 62:84–97.

Johnson, G., C. Parent, B. A. Kingsbury, R. A. Seigel, R. S. King, and J. Szymanski. 2000. The Eastern Massasauga Rattlesnake: a handbook for land managers. US Fish and Wildlife Service, Minneapolis, Minnesota, USA.

Johnson, J. R. 2005. Multi-scale investigations of Gray Treefrog movements: patterns of migration, dispersal, and gene flow. Dissertation, University of Missouri, Columbia, USA.

Johnson, J. R., and R. D. Semlitsch. 2003. Defining core habitat of local populations of the Gray

Treefrog (*Hyla versicolor*) based on choice of oviposition site. Oecologia 137:205–210.

Johnson, J. R., J. H. Knouft, and R. D. Semlitsch. 2007a. Sex and seasonal differences in the spatial terrestrial distribution of Gray Treefrog (*Hyla versicolor*) populations. Biological Conservation 140:250–258.

Johnson, J. R., R. D. Mahan, and R. D. Semlitsch. 2008. Seasonal terrestrial microhabitat use by Gray Treefrogs (*Hyla versicolor*) in Missouri oak-hickory forests. Herpetologica 64:259–269.

Johnson, L. M. 1991a. Growth and development of larval Northern Cricket Frogs (*Acris crepitans*) in relation to phytoplankton abundance. Freshwater Biology 25:51–59.

Johnson, M. L., and R. Speare. 2005. Possible modes of dissemination of the amphibian chytrid *Batrachochytrium dendrobatidis* in the environment. Diseases of Aquatic Organisms 65:181–186.

Johnson, P. T. J., and D. R. Sutherland. 2003. Amphibian deformities and *Ribeiroia* infection: an emerging helminthiasis. Trends in Parasitology 19:332–336.

Johnson, P. T. J., and J. M. Chase. 2004. Parasites in the food web: linking amphibian malformations and aquatic eutrophication. Ecology Letters 7:521–526.

Johnson, P. T. J., and R. B. Hartson. 2009. All hosts are not equal: explaining differential patterns of malformations in an amphibian community. Journal of Animal Ecology 78:191–201.

Johnson, P. T. J., D. R. Sutherland, J. M. Kinsella, and K. B. Lunde. 2004. Review of the trematode genus *Ribeiroia* (Psilostomidae): ecology, life history and pathogenesis with special emphasis on the amphibian malformation problem. Advances in Parasitology 57:191–253.

Johnson, P. T. J., J. M. Chase, K. L. Dosch, R. B. Hartson, J. A. Gross, D. J. Larson, D. R. Sutherland, and S. R. Carpenter. 2007b. Aquatic eutrophication promotes pathogenic infection in amphibians. Proceedings of the National Academy of Sciences of the United States of America 104:15781–15786.

Johnson, P. T. J., V. J. McKenzie, A. C. Peterson, J. L. Kerby, J. Brown, A. R. Blaustein, and T. Jackson. 2011. Regional decline of an iconic amphibian associated with elevation, land-use change, and invasive species. Conservation Biology 25:556–566.

Johnson, T. R. 1987. The amphibians and reptiles of Missouri. Missouri Department of Conservation, Jefferson City, USA.

Johnson, T. R. 1991b. Western Smooth Green Snake (*Opheodrys vernalis blanchardi* Grobman): translocation to northwest Missouri. Final Report to the Missouri Department of Conservation, Natural History Division, Jefferson City, USA.

Johnson, T. R. 1998. Missouri toad and frog calling survey: the first year. Pages 357–359 *in* M. J. Lannoo, editor. Status and Conservation of Midwestern Amphibians. University of Iowa Press, Iowa City, USA.

Johnson, T. R. 2000b. The amphibians and reptiles of Missouri. Second edition. Missouri Department of Conservation, Jefferson City, USA.

Johnston, B., A. F. Badje, and R. A. Staffen. 2015. *Glyptemys insculpta* (Wood Turtle). Geographic distribution. Herpetological Review 46:564.

Jones, B., S. F. Fox, D. M. Leslie, Jr., D. M. Engle, and R. L. Lochmiller. 2000. Herpetofaunal responses to brush management with herbicide and fire. Journal of Range Management 53:154–158.

Jones, B. D., and P. A. Cochran. 2014. *Chrysemys picta* (Painted Turtle) and *Chelydra serpentina* (Snapping Turtle): interspecific interactions. Herpetological Review 45:311–312.

Jones, K. S. 2014. *Carphophis amoenus* (Eastern Wormsnake). Association with ants. Herpetological Review 45:705–706.

Jones, M. T. 2009. Spatial ecology, population structure, and conservation of the Wood Turtle, *Glyptemys insculpta*, in central New England. Dissertation, University of Massachusetts, Amherst, USA.

Jones, M. T., and B. W. Compton. 2010. *Glyptemys insculpta* (Wood Turtle). Maximum adult size. Herpetological Review 41:71.

Jones, M. T., and P. R. Sievert. 2009a. Effects of stochastic flood disturbance on adult Wood Turtles, *Glyptemys insculpta*, in Massachusetts. Canadian Field-Naturalist 123:313–322.

Jones, M. T., and P. R. Sievert. 2009b. *Glyptemys insculpta* (Wood Turtle). Diet. Herpetological Review 40:433–434.

Jones, M. T., and P. R. Sievert. 2012. Elevated mortality of hatchling Blanding's Turtles (*Emydoidea blandingii*) in residential settings. Herpetological Conservation and Biology 7:89–94.

Jones, P. C., R. B. King, and S. Sutton. 2017 Demographic analysis of imperiled Eastern Massasaugas (*Sistrurus catenatus catenatus*). Journal of Herpetology 51:383–387.

Joppa, L. N., and S. A. Temple. 2005. Use of upland habitat by the Butler's Gartersnake (*Thamnophis butleri*). Chicago Herpetological Bulletin 40:221–227.

Jordan, E. O. 1893. The habits and development of the newt (*Diemyctylus viridescens*). Journal of Morphology 8:269–366.

Jordan, O. R. 1967. The occurrence of *Thamnophis sirtalis* and *T. radix* in the prairie-forest ecotone west of Itasca State Park, Minnesota. Herpetologica 23:303–308.

Jordan, R., Jr. 1970. Death-feigning in a captive Red-bellied Snake, *Storeria occipitomaculata* (Storer). Herpetologica 26:466–468.

Joyal, L. A., M. McCollough, and M. L. Hunter, Jr. 2001. Landscape ecology approaches to wetland species conservation: a case study of two turtle species in southern Maine. Conservation Biology 15:1755–1762.

Judd, W. W. 1954. Observations on the food of the Little Brownsnake, *Storeria dekayi*, at London, Ontario. Copeia 1954:62–64.

Judd, W. W. 1960. Observations on the habitat, food, reproductive state and intestinal parasites of the Smooth Green Snake at London, Ontario. Canadian Field-Naturalist 74:100–106.

Judziewicz, E. J., R. W. Freckmann, M. R. Black, and L. G. Clark. 2014. Field guide to Wisconsin grasses. University of Wisconsin Press, Madison, USA.

Jung, R. E. 1993. Blanchard's Cricket Frogs (*Acris crepitans blanchardi*) in southwest Wisconsin. Transactions of the Wisconsin Academy of Sciences, Arts and Letters 81:79–87.

Kalb, H. J., and G. R. Zug. 1990. Age estimates for a population of American Toads, *Bufo americanus* (Salientia: Bufonidae), in northern Virginia. Brimleyana 16:79–86.

Kalff, J. 2002. Limnology: inland water ecosystems. Prentice Hall. Upper Saddle River, New Jersey, USA.

Kapfer, J. M. 2009. Body size and sexual size dimorphism of Bullsnakes (*Pituophis catenifer sayi*) in Wisconsin, USA. Herpetological Conservation and Biology 4:353–357.

Kapfer, J. M. 2010. A survey of gravid snakes at several sites in southern Wisconsin. IRCF Reptiles and Amphibians 17:22–25.

Kapfer, J. M., and C. Sloane. 2011. *Carphophis amoenus* (Eastern Wormsnake). Defensive behavior. Herpetological Review 42:437.

Kapfer, J. M., and C. T. Bennell. 2005. *Pituophis catenifer sayi* (Bullsnake). Predation. Herpetological Review 36:326.

Kapfer, J. M., and M. J. Pauers. 2006. Activity season, habitat use, and environmental temperature selection of the Prairie Racerunner (*Aspidoscelis sexlineata viridis*) at the northern limit of its range. Herpetological Review 37:420–423.

Kapfer, J. M., and R. A. Paloski. 2011. On the threat to snakes of mesh deployed for erosion control and wildlife exclusion. Herpetological Conservation and Biology 6:1–9.

Kapfer, J. M., and T. C. Muehlfeld. 2006. *Rana sylvatica* (Wood Frog) and *Ambystoma maculatum* (Spotted Salamander). Behavior. Herpetological Review 37:210–211.

Kapfer, J. M., B. L. Sloss, G. W. Schuurman, R. A. Paloski, and J. M. Lorch. 2013a. Evidence of hybridization between Common Gartersnakes (*Thamnophis sirtalis*) and Butler's Gartersnakes (*Thamnophis butleri*) in Wisconsin, USA. Journal of Herpetology 47:400–405.

Kapfer, J. M., C. W. Pekar, D. M. Reineke, J. R. Coggins and R. Hay. 2010. Modeling the relationship between habitat preferences and home-range size: a case study on a large mobile colubrid snake from North America. American Journal of Zoology 282:13–20.

Kapfer, J. M., G. S. Casper, and J. J. Kapfer. 2008b. *Hyla versicolor* (Gray Treefrog). Geographic distribution. Herpetological Review 39:478.

Kapfer, J. M., J. M. Lorch, E. R. Wild, D. J. Brown, L. Mitchem, N. Rudolph, K. Rutzen, and R. C. Vogt. 2015. Distributional records for amphibians and reptiles from Wisconsin, USA. Herpetological Review 46:587–590.

Kapfer, J. M., J. R. Coggins, and R. Hay. 2008a. Estimates of population size, measurements of sex ratios and reported mortality rates for Bullsnakes (*Pituophis catenifer sayi*) at a site in the upper midwestern United States. Journal of Herpetology 42:265–269.

Kapfer, J. M., J. R. Coggins, and R. Hay. 2008e. Spatial ecology and habitat selection of Bullsnakes (*Pituophis catenifer sayi*) at the northern periphery of their geographic range. Copeia 2008:815–826.

Kapfer, J. M., K. Doehler, and R. Hay. 2013b. The influence of habitat type and the presence of an invasive wetland plant (*Phalaris arundinacea*) on capture rates of sympatric rare and common gartersnake species (*Thamnophis butleri* and *Thamnophis sirtalis*). Journal of Herpetology 47:126–130.

Kapfer, J. M., K. Katovich, G. W. Schuurman, R. A. Paloski, and B. L. Sloss. 2013c. *Thamnophis butleri* (Butler's Gartersnake). Geographic distribution. Herpetological Review 44:276.

Kapfer, J. M., M. J. Pauers, D. M. Reineke, J. R. Coggins, and R. Hay. 2008c. Environmental, behavioral, and habitat variables influencing body temperature in radio-tagged Bullsnakes, *Pituophis catenifer sayi*. Journal of Thermal Biology 33:174–179.

Kapfer, J. M., M. J. Pauers, J. R. Coggins and R. Hay. 2009a. Microhabitat selection of Bullsnakes (*Pituophis catenifer sayi*) at a site in the upper midwestern United States. Herpetological Review 40:148–151.

Kapfer, J. M., R. Hay, and M. Wanatka. 2008d. *Pituophis catenifer sayi* (Bullsnake). Geographic distribution. Herpetological Review 39:486.

Kapfer, J. M., T. Muehlfeld, and S. Muehlfeld. 2009b. *Nerodia sipedon sipedon* (Northern Watersnake). Geographic distribution. Herpetological Review 40:238–239.

Kapfer, J. M., W. P. Mueller, M. Porzky, and G. S. Casper. 2009c. *Regina septemvittata* (Queensnake). Geographic distribution. Herpetological Review 40:116.

Kapus, E. J. 1964. Anatomical evidence for *Heterodon* being poisonous. Herpetologica 20:137.

Karasov, W. H., R. E. Jung, S. Vanden Langenberg, and T. L. E. Bergeson. 2005. Field exposure of frog embryos and tadpoles along a pollution gradient in the Fox River and Green Bay ecosystem in Wisconsin, USA. Environmental Toxicology and Chemistry 24:942–953.

Kardong, K. V. 1979. "Protovipers" and the evolution of snake fangs. Evolution 33:433–443.

Karns, D. R. 1992. Effects of acidic bog habitats on amphibian reproduction in a northern Minnesota peatland. Journal of Herpetology 26:401–412.

Karraker, N. E. 2007. Are embryonic and larval Green Frogs (*Rana clamitans*) insensitive to road deicing salt? Herpetological Conservation and Biology 2:35–41.

Karraker, N. E. 2008. Impacts of road deicing salts on amphibians and their habitat. Pages 211–223 *in* J. Mitchell, R. Jung, and B. Bartholomew, editors. Urban Herpetology. Society for the Study of Amphibians and Reptiles, Herpetological Conservation 3, Salt Lake City, Utah, USA.

Karraker, N. E., and J. P. Gibbs. 2009. Amphibian production in forested landscapes in relation to wetland hydroperiod: a case study of vernal pools and beaver ponds. Biological Conservation 142:2293–2302.

Karraker, N. E., and J. P. Gibbs. 2011. Contrasting road effect signals in reproduction of long-versus short-lived amphibians. Hydrobiologia 664:213–218.

Karraker, N. E., J. P. Gibbs, and J. R. Vonesh. 2008. Impacts of road deicing salt on pond-breeding amphibians. Ecological Applications 18:724–734.

Karstad, L. 1967. Fatal poisoning of a Fox Snake (*Elaphe vulpina*) by feeding a toad (*Bufo americanus*). Bulletin of the Wildlife Disease Association 3:73–74.

Kassulke, N., and K. Esposito. 2005. The forest of tomorrow. Wisconsin Natural Resources Magazine PUB-FR-337–2005, Madison, USA. Archived by WebCite at http://www.webcitation.org/73cJEq25a. Accessed 1 Nov 2018.

Kastle, M., J. Kapfer, A. R. Kuhns, W. Graser, G. Glowacki, A. Ibach, L. Mitchem, J. Mozuch, N. Rudolph, K. Rutzen, and R. King. 2021. Blanding's Turtle hatchling survival and movements following natural vs. artificial incubation. Journal of Herpetology 55:167–173.

Kats, L. B. 1988. The detection of certain predators via olfaction by Small-mouthed Salamander larvae (*Ambystoma texanum*). Behavioral and Neural Biology 50:126–131.

Kats, L. B., and R. G. Van Dragt. 1986. Background color-matching in the Spring Peeper, *Hyla crucifer*. Copeia 1986:109–115.

Kats, L. B., and R. P. Ferrer. 2003. Alien predators and amphibian declines: review of two decades of science and the transition to conservation. Diversity and Distributions 9:99–110.

Kats, L. B., J. W. Petranka, and A. Sih. 1988. Antipredator defenses and the persistence of amphibian larvae with fishes. Ecology 69:1865–1870.

Kaufmann, G. S., H. T. Smith, R. M. Engeman, W. E. Meshaka, and E. M. Cowan. 2007. *Ophisaurus ventralis* (Eastern Glass Lizard). Fire-induced mortality. Herpetological Review 38:460–461.

Kaufmann, J. H. 1986. Stomping for earthworms by Wood Turtles, *Clemmys insculpta*: a newly discovered foraging technique. Copeia 1986:1001–1004.

Kaufmann, J. H. 1992a. Habitat use by Wood Turtles in central Pennsylvania. Journal of Herpetology 26:315–327.

Kaufmann, J. H. 1992b. The social behavior of Wood Turtles, *Clemmys insculpta*, in central Pennsylvania. Herpetological Monographs 6:1–25.

Kaufmann, J. H. 1995. Home ranges and movements of Wood Turtles, *Clemmys insculpta*, in central Pennsylvania. Copeia 1995:22–27.

Kaza, N., and T. K. BenDor. 2013. The land value impacts of wetland restoration. Journal of Environmental Management 127:289–299.

Keegan, L. 1944. An albino Fox Snake from Iowa. Copeia 1944:59.

Keen, W. H. 1984. Influence of moisture on the activity of a plethodontid salamander. Copeia 1984:684–688.

Keenlyne, K. D. 1968. The Massasauga (*Sistrurus catenatus catenatus*) in Buffalo County, Wisconsin. Thesis, University of Minnesota, Minneapolis, USA.

Keenlyne, K. D. 1972. Sexual differences in feeding habitats of *Crotalus horridus*. Journal of Herpetology 6:234–237.

Keenlyne, K. D. 1978. Reproductive cycles in two species of rattlesnakes. American Midland Naturalist 100:368–375.

Keenlyne, K. D., and J. R. Beer. 1973a. Food habits of *Sistrurus catenatus catenatus*. Journal of Herpetology 7:382–384.

Keenlyne, K. D., and J. R. Beer. 1973b. Note on the size of *Sistrurus catenatus catenatus* at birth. Journal of Herpetology 7:381–382.

Keller, W. L., and E. J. Heske. 2000. Habitat use by three species of snakes at the Middle Fork Fish and Wildlife Area, Illinois. Journal of Herpetology 34:558–564.

Kempinger, J. J. 1988. Spawning and early life history of Lake Sturgeon in the Lake Winnebago system, Wisconsin. Pages 110–122 *in* R. D. Hoyt, editor. 11th Annual Larval Fish Conference, Symposium 5. American Fisheries Society, Bethesda, Maryland, USA.

Kennedy, J. P. 1964. Natural history notes on some snakes in eastern Texas. Texas Journal of Science 16:210–215.

Kennicott, R. 1859. Notes on *Coluber calligaster* of Say, and a description of new species of serpents in the collection of the North Western University of Evanston, Illinois. Proceedings of the Academy of Natural Science of Philadelphia 11:98–100.

Kenny, J. S. 1969. Feeding mechanisms in anuran larvae. Journal of Zoology 157:225–246.

Kenow, K. P., J. M. Kapfer, and C. E. Korschgen. 2009. Predation of radio-marked Mallard (*Anas platyrhynchos*) ducklings by Eastern Snapping Turtles (*Chelydra serpentina serpentina*) and Western Fox Snakes (*Pantherophis vulpinus*) on the Upper Mississippi River. Journal of Herpetology 43:154–158.

Kent, D. M. 1983. Wildlife response to secondarily treated sewage application on the Drummond Bog. Thesis, University of Wisconsin, Stevens Point, USA.

Keogh, J. S., and F. P. DeSerto. 1994. Temperature dependent defensive behavior in three species of North American colubrid snakes. Journal of Herpetology 28:258–261.

Kerby, J. L., A. J. Hart, and A. Storfer. 2011. Combined effects of virus, pesticide, and predator cue on the larval Tiger Salamander (*Ambystoma tigrinum*). Ecohealth 8:46–54.

Kern, M. M., A. A. Nassar, J. C. Guzy, and M. E. Dorcas. 2013. Oviposition site selection by Spotted Salamanders (*Ambystoma maculatum*) in an isolated wetland. Journal of Herpetology 47:445–449.

Kessler, G. P., and A. F. Badje. 2017. *Lithobates catesbeianus* (American Bullfrog). Geographic distribution. Herpetological Review 48:807–808.

Keyler, D. E. 2005. Venomous snakebites: Minnesota and upper Mississippi River Valley 1982–2002. Minnesota Herpetological Society Occasional Paper 7:1–28.

Keyler, D. E. 2008. Timber Rattlesnake (*Crotalus horridus*) envenomations in the Upper Mississippi River Valley. Pages 569–580 *in* W. K. Hayes, K. R. Beaman, M. D. Cardwell, and S. P. Bush, editors. The Biology of rattlesnakes. Loma Linda University Press, Loma Linda, California, USA.

Keyler, D. E., and B. L. Oldfield. 1992. Status of the Timber Rattlesnake (*Crotalus horridus*) in Minnesota. Page 7 *in* T. F. Tyning, editor. Conservation of the Timber Rattlesnake in the northeast. Massachusetts Audubon Society, Lincoln, USA.

Keyler, D. E., and M. E. Peterson. 2011. Crotalids (pit vipers). Pages 382–392 *in* H. Osweller and B. Lee, editors. Blackwell's five-minute veterinary consult clinical companion: small animal toxicology. Wiley-Blackwell, Ames, Iowa, USA.

Kiesecker, J. 1996. pH-mediated predator-prey interactions between *Ambystoma tigrinum* and *Pseudacris triseriata*. Ecological Applications 6:1325–1331.

Kiesecker, J. M. 2002. Synergism between trematode infection and pesticide exposure: a link to amphibian limb deformities in nature? Proceedings of the National Academy of Sciences 99:9900–9904.

Kiesecker, J. M., and D. K. Skelly. 2000. Choice of oviposition site by Gray Treefrogs: the role of potential parasitic infection. Ecology 81:2939–2943.

Kiesow, A. M. 2006. Field guide to amphibians and reptiles of South Dakota. South Dakota Department of Game, Fish and Parks, Pierre, USA.

Killebrew, F. C. 1975. Mitotic chromosomes of turtles, part 3: the Kinosternidae. Herpetologica 31:398–403.

King, R., C. Berg, and B. Hay. 2004. A repatriation study of the Eastern Massasauga (*Sistrurus catenatus catenatus*) in Wisconsin. Herpetologica 60:429–437.

King, R. B. 1986. Population ecology of the Lake Erie Water Snake, *Nerodia sipedon insularum*. Copeia 1986:757–772.

King, R. B. 1988a. Biogeography of reptiles on islands in Lake Erie. Pages 123–133 *in* J. F. Downhower, editor. The biogeography of the island region of western Lake Erie. Ohio State University Press, Columbus, USA.

King, R. B. 1988b. Polymorphic populations of the Garter Snake *Thamnophis sirtalis* near Lake Erie. Herpetologica 44:451–458.

King, R. B. 1993a. Determinants of offspring number and size in the Brown Snake, *Storeria dekayi*. Journal of Herpetology 27:175–185.

King, R. B. 1993b. Microgeographic, historical, and size-correlated variation in water snake diet composition. Journal of Herpetology 27:90–94.

King, R. B. 1997a. The effects of ecdysis on feeding frequency and behavior of the Common Garter Snake (*Thamnophis sirtalis*). Journal of Herpetology 31:310–312.

King, R. B. 1997b. Variation in Brown Snake (*Storeria dekayi*) morphology and scalation: sex, family, and microgeographic differences. Journal of Herpetology 31:335–346.

King, R. B. 2002a. Family, sex and testosterone effects on garter snake behavior. Animal Behaviour 64:345–359.

King, R. B. 2002b. Predicted and observed maximum prey size—snake size allometry. Functional Ecology 16:766–772.

King. R. B. 2003. Mendelian inheritance of melanism in the Garter Snake *Thamnophis sirtalis*. Herpetologica 59:484–489.

King, R. B., and B. King. 1991. Sexual differences in color and color change in Wood Frogs. Canadian Journal of Zoology 69:1963–1968.

King, R. B., and E. T. Hileman. 2012. Assessing management needs to enhance recovery for the Eastern Massasauga. Interim Report to the US Fish and Wildlife Service, DeKalb, Illinois, USA.

King, R. B., and E. T. Hileman. 2013. Assessing management needs to enhance recovery for the Eastern Massasauga. Interim Report to the US Fish and Wildlife Service, DeKalb, Illinois, USA.

King, R. B., and K. M. Stanford. 2006. Headstarting as a management tool: a case study of the Plains Gartersnake. Herpetologica 62:282–292.

King, R. B., and M. L. Niiro. 2013. Predicting climate-change induced distributional shifts in Great Lakes Region reptiles. Final Report to the US Fish and Wildlife Service, DeKalb, Illinois, USA.

King, R. B., and R. Lawson. 2001. Patterns of population subdivision and gene flow in three sympatric natricine snakes. Copeia 2001:602–614.

King, R. B., and R. M. Bowden. 2011. Seasonal, condition-dependent, and individual variation in testosterone in a natricine snake. Journal of Herpetology 47:179–183.

King, R. B., and J. P. Vanek. 2020. Responses of grassland snakes to tallgrass prairie restoration. Restoration Ecology 28:573–582.

King, R. B., J. M. Ray, and K. M. Stanford. 2006. Gorging on gobies: beneficial effects of alien prey on a threatened vertebrate. Canadian Journal of Zoology 84:108–115.

King, R. B., K. M. Stanford, P. C. Jones, and K. Bekker. 2016. Size matters: individual variation in ectotherm growth and asymptotic size. PLoS ONE 11:e0146299.

King, R. B., M. Dreslik, L. Fuast, G. Glowacki, D. Ludwig, A. Manlowski, M. Redmer, and B. Zeigler. 2013. Illinois conservation assessment for the Blanding's Turtle (*Emydoidea blandingii*). Report to the Illinois Endangered Species Protection Board, DeKalb, Illinois, USA. Archived by WebCite at http://www.webcitation.org/73cJTnBqi. Accessed 1 Nov 2018.

King, R. B., T. D. Bittner, A. Querel-Regil, and J. H. Cline. 1999. Sexual dimorphism in neonate and adult snakes. Journal of Zoology 247:19–28.

King, R. B., W. B. Milstead, H. L. Gibbs, M. R. Prosser, G. M. Burghardt, and G. F. McCracken. 2001. Application of microsatellite DNA markers to discriminate between maternal and genetic effects on scalation and behavior in multiply sired garter snake litters. Canadian Journal of Zoology 79:121–128.

King, R. S. 1999. Habitat use and movement patterns of the Eastern Massasauga in Wisconsin. Page 80 *in* B. Johnson and V. Menzies, editors. Proceedings of the International Symposium and Workshop on the Conservation of the Eastern Massasauga Rattlesnake *Sistrurus catenatus catenatus*. Metro Toronto Zoo, Ontario, Canada.

King, S. K. 2012. Four-toed Salamander (*Hemidactylium scutatum*) nest site characteristics in natural and constructed wetlands in eastern Kentucky. Thesis, Eastern Kentucky University, Richmond, USA.

Kingsbury, B. A., and J. Gibson, editors. 2012. Habitat management guidelines for amphibians and reptiles of the midwestern United States. Second edition. Partners in Amphibian and Reptile Conservation Technical Publication HMG-1.

Kinkead, K. E., and D. L. Otis. 2007. Estimating superpopulation size and annual probability of breeding for pond-breeding salamanders. Herpetologica 63:151–162.

Kinney, O. M., R. D. Nagle, and J. D. Congdon. 1998. Water transport by nesting Painted Turtles (*Chrysemys picta marginata*) in Michigan. Chelonian Conservation and Biology 3:71–76.

Kirby, L. 2005. A comparative study of behavior in neonate gartersnakes, *Thamnophis butleri* and *T. radix* (Colubridae), in an area of potential hybridization. Thesis, University of Tennessee, Knoxville, USA.

Kirkland, G. L., Jr., H. W. Snoddy, and T. L. Amsler. 1996. Impact of fire on small mammals and amphibians in a central Appalachian deciduous forest. American Midland Naturalist 135:253–260.

Kissner, K. J., and P. J. Weatherhead. 2005. Phenotypic effects on survival of neonatal Northern Watersnakes *Nerodia sipedon*. Journal of Animal Ecology 74:259–265.

Kissner, K. J., P. J. Weatherhead, and H. L. Gibbs. 2005. Experimental assessment of ecological and phenotypic factors affecting male mating success and polyandry in Northern Watersnakes, *Nerodia sipedon*. Behavioral Ecology and Sociobiology 59:207–214.

Kitchell, J., A. Badje, and T. Bergeson. 2013. Frog and toad survey, 2013. Wisconsin Department of Natural Resources, Wisconsin Wildlife Surveys, Madison, USA.

Kitchell, J., A. Badje, and T. Bergeson. 2014. Frog and toad survey, 2014. Wisconsin Department of Natural Resources, Wisconsin Wildlife Surveys, Madison, USA.

Kitchens, C. S. 1996. Treatment of pit viper envenomation. Journal of the Florida Medical Association 83:174–177.

Kiviat, E. 1978. Vertebrate use of Muskrat lodges and burrows. Estuaries 1:196–200.

Kiviat, E., G. Stevens, R. Brauman, S. Hoeger, P. J. Petokas, and G. G. Hollands. 2000. Restoration of wetland and upland habitat for the Blanding's Turtle, *Emydoidea blandingii*. Chelonian Conservation and Biology 3:650–657.

Kjoss, V. A., and J. A. Litvaitis. 2001. Community structure of snakes in a human-dominated landscape. Biological Conservation 98:285–292.

Klauber, L. M. 1948. Some misapplications of the Linnaean names applied to American snakes. Copeia 1948:1–14.

Kleeberger, S. R., and E. E. Werner. 1983. Post-breeding migration and summer movement of *Ambystoma maculatum*. Journal of Herpetology 17:176–177.

Kleeberger, S. R., and J. K. Werner. 1982. Home range and homing behavior of *Plethodon cinereus* in northern Michigan. Copeia 1982:409–415.

Klemens, M. W. 1993. Amphibians and reptiles of Connecticut and adjacent regions. State Geological and Natural History Survey of Connecticut Bulletin 112, Hartford, Connecticut, USA.

Klemish, J. L., B. L. Johnson, S. R. Siddons, and E. R. Wild. 2012. Occurrence of *Batrachochytrium dendrobatidis* among populations of *Lithobates clamitans* and *L. pipi*ens in Wisconsin, U.S.A. Herpetological Review 43:282–288.

Klimstra, W. D. 1959. Foods of the Racer, *Coluber constrictor*, in southern Illinois. Copeia 1959:210–214.

Kline, J. 1998. Monitoring amphibians in created and restored wetlands. Pages 360–368 *in* M. J. Lannoo, editor. Status and conservation of midwestern amphibians. University of Iowa Press, Iowa City, USA.

Klug, P. E., J. Fill, and K. A. With. 2011a. Spatial ecology of Eastern Yellow-Bellied Racer (*Coluber constrictor flaviventris*) and Great Plains Rat Snake (*Pantherophis emoryi*) in a contiguous tallgrass-prairie landscape. Herpetologica 67:428–439.

Klug, P. E., L. L. Wolfenbarger, and J. P. McCarty. 2010a. Snakes are important nest predators of Dickcissels in an agricultural landscape. Wilson Journal of Ornithology 122:799–803.

Klug, P. E., S. L. Jackrel, and K. A. With. 2010b. Linking snake habitat use to nest predation risk in grassland birds: the dangers of shrub cover. Oecologia 162:803–813.

Klug, P. E., S. M. Wisely, and K. A. With. 2011b. Population genetic structure and landscape connectivity of the Eastern Yellowbelly Racer (*Coluber constrictor flaviventris*) in the contiguous tallgrass prairie of northeastern Kansas, USA. Landscape Ecology 26:281–294.

Knight, A. J. 2008. "Bats, snakes and spiders, oh my!" How aesthetic and negativistic attitudes, and other concepts predict support for species protection. Journal of Environmental Psychology 28:94–103.

Knoot, T. G., and L. B. Best. 2011. A multiscale approach to understanding snake use of conservation buffer strips in an agricultural landscape. Herpetological Conservation and Biology 6:191–201.

Knowles, T. W., and P. D. Weigl. 1990. Thermal dependence of anuran burst locomotor performance. Copeia 1990:796–802.

Knudsen, G. J. 1954a. Fox Snake (*Elaphe vulpina*). Wisconsin Conservation Bulletin 19(12):35–36.

Knudsen, G. J. 1954b. Massasauga (*Sistrurus catenatus catenatus*). Wisconsin Conservation Bulletin 19(10):35–36.

Knudsen, G. J. 1954c. Northern Banded Watersnake (*Natrix sipedon sipedon*). Wisconsin Conservation Bulletin 19(11):37–38.

Knudsen, G. J. 1955a. Bull Snake (*Pituophis sayi sayi*). Wisconsin Conservation Bulletin 20(1):36–37.

Knudsen, G. J. 1955b. Common Garter Snake (*Thamnophis s. sirtalis*). Wisconsin Conservation Bulletin 20(5):35–36.

Knudsen, G. J. 1955c. Eastern Blue Racer (*Coluber constrictor flaviventris*). Wisconsin Conservation Bulletin 20(3):36–37.

Knudsen, G. J. 1955d. Eastern Ringneck Snake (*Diadophis punctatus edwardsii*). Wisconsin Conservation Bulletin 20(10):32–33.

Knudsen, G. J. 1955e. Green Snake (*Liopeltis vernalis*). Wisconsin Conservation Bulletin 20(9):36–37.

Knudsen, G. J. 1955f. Milk Snake (*Lampropeltis Triangulum triangulum*). Wisconsin Conservation Bulletin 20(2):35–36.

Knudsen, G. J. 1955g. Red-bellied Snake (*Storeria occipito-maculata*). Wisconsin Conservation Bulletin 20(12):29–30.

Knudsen, G. J. 1956a. Glass "snake" (*Ophisaurus ventralis*). Wisconsin Conservation Bulletin 21(2):36–37.

Knudsen, G. J. 1956b. Six-lined Lizard (*Cnemidophorus sexlineatus*). Wisconsin Conservation Bulletin 21(3):38–39.

Knudsen, G. J. 1956c. Snapping Turtle (*Chelydra serpentina*). Wisconsin Conservation Bulletin 21(5):36–37.

Knudsen, G. J. 1956d. Western Painted Terrapin (*Chrysemys bellii marginata*). Wisconsin Conservation Bulletin 21(12):30–31.

Knudsen, G. J. 1956e. Western Box Turtle (*Terrapene ornata*). Wisconsin Conservation Bulletin 21(10):36–37.

Knudsen, G. J. 1957a. American Toad (*Bufo d. americanus*). Wisconsin Conservation Bulletin 22(10):36–37.

Knudsen, G. J. 1957b. Blanding's Turtle (*Emys blandingii*). Wisconsin Conservation Bulletin 22(3):36–37.

Knudsen, G. J. 1957c. Brown Soft-shelled Turtle (*Amyda mutica*). Wisconsin Conservation Bulletin 22(1):35–36.

Knudsen, G. J. 1957d. The Spotted Turtle (*Clemmys guttata*). Wisconsin Conservation Bulletin 22(9): 35–36.

Knudsen, G. J. 1957e. Wood Terrapin (*Clemmys insculpta*). Wisconsin Conservation Bulletin 22(2):36–37.

Knudsen, G. J. 1958a. American Bullfrog. Wisconsin Conservation Bulletin 23(3):36–37.

Knudsen, G. J. 1958b. Eastern Red-backed Salamander (*Plethodon c. cinereus*). Wisconsin Conservation Bulletin 23(12):38.

Knudsen, G. J. 1958c. Leopard Frog (*Rana p. pipiens*). Wisconsin Conservation Bulletin 23(5):38–39.

Knudsen, G. J. 1958d. Spring Peeper (*Hyla c. crucifer*). Wisconsin Conservation Bulletin 23(4):35–37.

Knudsen, G. J. 1958e. Striped Tree Frog (*Pseudacris nigrita triseriata*). Wisconsin Conservation Bulletin 23(9):36–37.

Knudsen, G. J. 1958f. Mink Frog (*Rana septentrionalis*). Wisconsin Conservation Bulletin 23(2):36–37.

Knudsen, G. J. 1958g. Western Cricket Frog (*Acris gryllus crepitans*). Wisconsin Conservation Bulletin 23(10):35–36.

Knudsen, G. J. 1962. Relationship of beaver to forests, trout and wildlife in Wisconsin. Wisconsin Conservation Department Technical Bulletin 24, Madison, USA.

Knudsen, G. J. 1963. History of beaver in Wisconsin. Wisconsin Conservation Department Miscellaneous Research Report 7, Madison, USA.

Knudsen, G. J. 1985. The Bullfrog, alias Mr. Jug-O'-Rum. Wisconsin Outdoors and Conservation News. 20 February 1985; pages 1–3.

Knudsen, G. J., and J. B. Hale. 1970. Food habits of Opossums in southern Wisconsin. Wisconsin Department of Natural Resources Miscellaneous Research Report 61, Madison, USA.

Knutson, M. G., J. H. Herner-Thogmartin, W. E. Thogmartin, J. M. Kapfer, and J. C. Nelson. 2018. Habitat selection, movement patterns, and hazards encountered by Northern Leopard Frogs (*Lithobates pipiens*) in an agricultural landscape. Herpetological Conservation and Biology 13:113–130.

Knutson, M. G., J. R. Sauer, D. A. Olsen, M. J. Mossman, L. M. Hemesath, and M. J. Lannoo. 1999. Effects of landscape composition and wetland fragmentation on frog and toad abundance and species richness in Iowa and Wisconsin, U.S.A. Conservation Biology 13:1437–1446.

Knutson, M. G., J. R. Sauer, D. A. Olsen, M. J. Mossman, L. M. Hemesath, and M. J. Lannoo. 2000. Landscape associations of frog and toad species in Iowa and Wisconsin, U.S.A. Journal of the Iowa Academy of Science 107:134–145.

Knutson, M. G., W. B. Richardson, D. M. Reineke, B. R. Gray, J. R. Parmelee, and S. E. Weick. 2004. Agricultural ponds support amphibian populations. Ecological Applications 14:669–684.

Koch, C. 2009. The year of the hognose. Bulletin of the Chicago Herpetological Society 44:17–20.

Kofron, C. P. 1979. Female reproductive biology of the Brown Snake, *Storeria dekayi*, in Louisiana. Copeia 1979:463–466.

Kofron, C. P., and A. A. Schreiber. 1985. Ecology of two endangered aquatic turtles in Missouri: *Kinosternon flavescens* and *Emydoidea blandingii*. Journal of Herpetology 19:27–40.

Kohn, N. R., R. G. Jaeger, and J. Franchebois. 2005. Effects of intruder number and sex on territorial behavior of female Red-backed Salamanders (*Plethodon cinereus*: Plethodontidae). Journal of Herpetology 39:645–648.

Kolbe, J. J., and F. J. Janzen. 2002a. Impact of nest-site selection on nest success and nest temperature in natural and disturbed habitats. Ecology 83:269–281.

Kolbe, J. J., and F. J. Janzen. 2002b. Spatial and temporal dynamics of turtle nest predation: edge effects. Oikos 99:538–544.

Kolozsvary, M. B., and R. K. Swihart. 1999. Habitat fragmentation and the distribution of amphibians: patch and landscape correlates in farmland. Canadian Journal of Zoology 77:1288–1299.

Kondolf, G. M. 1997. Hungry water: effects of dams and gravel mining on river channels. Environmental Management 21:533–551.

Koprivnikar, J., J. C. Redfern, and H. L. Mazier. 2014. Variation in anti-parasite behaviour and infection among larval amphibian species. Oecologia 174:1179–1185.

Koprivnikar, J., J. C. Redfern, and R. L. Baker. 2006a. On the efficacy of anti-parasite behaviour: a case study of tadpole susceptibility to cercariae of *Echinostoma trivolvis*. Canadian Journal of Zoology 84:1623–1629.

Koprivnikar, J., M. R. Forbes, and R. L. Baker. 2007. Contaminant effects on host-parasite interactions: atrazine, frogs, and trematodes. Environmental Toxicology and Chemistry 26:2166–2170.

Koprivnikar, J., R. L. Baker, and M. R. Forbes. 2006b. Environmental factors influencing trematode prevalence in Grey Tree Frog (*Hyla versicolor*) tadpoles in southern Ontario. Journal of Parasitology 92:997–1001.

Korb, R. M. 1988a. *Emydoidea blandingii* (Blanding's Turtle). Geographic distribution. Herpetological Review 19:21.

Korb, R. M. 1988b. *Heterodon platyrhinos* (Eastern Hognose Snake). Geographic distribution. Herpetological Review 19:20.

Kornilev, Y. V., S. J. Price, E. P. Hill, and M. E. Dorcas. 2006. *Ambystoma maculatum* (Spotted Salamander). Mortality. Herpetological Review 37:196.

Kost, M. A., and D. de Steven. 2000. Plant community responses to prescribed burning in Wisconsin sedge meadows. Natural Areas Journal 20:36–45.

Kozak, K. H. 2017. What drives variation in plethodontid salamander species richness over space and time? Herpetologica 73:220–228.

Kramek, W. C. 1972. Food of the frog *Rana septentrionalis* in New York. Copeia 1972:390–392.

Kramek, W. C., and M. M. Stewart. 1980. Ontogenetic and sexual differences in the pattern of *Rana septentrionalis*. Journal of Herpetology 14:369–375.

Kramer, D. C. 1973. Movements of Western Chorus Frogs *Pseudacris triseriata triseriata* tagged with Co^{60}. Journal of Herpetology 7:231–235.

Kramer, D. C. 1974. Home range of the Western Chorus Frog *Pseudacris triseriata triseriata*. Journal of Herpetology 8:245–246.

Krause Danielsen, A., P. Rutherford, and N. Koper. 2014. The importance of vegetation structure and artificial cover for Prairie Skinks (*Plestiodon septentrionalis*) on exurban land. Journal of Herpetology 48:67–73.

Krause, M. A., G. M. Burghardt, and J. C. Gillingham. 2003. Body size plasticity and local variation of relative head and body size sexual dimorphism in Garter Snakes (*Thamnophis sirtalis*). Journal of Zoology 261:399–407.

Krohmer, R. W. 1989. Body temperature relationships in the Lined Snake, *Tropidoclonion lineatum*. Comparative Biochemistry and Physiology Part A: Physiology 92:541–543.

Krohmer, R. W., and R. D. Aldridge. 1985a. Female reproductive cycle of the Lined Snake (*Tropidoclonion lineatum*). Herpetologica 41:39–44.

Krohmer, R. W., and R. D. Aldridge. 1985b. Male reproductive cycle of the Lined Snake (*Tropidoclonion lineatum*). Herpetologica 41:33–38.

Kruse, K. C. 1981. Mating success, fertilization potential, and male body size in the American Toad (*Bufo americanus*). Herpetologica 37:228–233.

Kruse, K. C. 1982. Male-male competition in the American Toad, *Bufo americanus*. Biologist 64:78–82.

Kruse, K. C. 1983. Optimal foraging by Predaceous Diving Beetle larvae on toad tadpoles. Oecologia 58:383–388.

Kruse, K. C., and M. Mounce. 1982. The effects of multiple matings on fertilization capability in male American Toads (*Bufo americanus*). Journal of Herpetology 16:410–412.

Krynak, K. L., D. J. Burke, and M. F. Benard. 2017. Rodeo herbicide negatively affects Blanchard's Cricket Frogs (*Acris blanchardi*) survival and alters the skin-associated bacterial community. Journal of Herpetology 51:402–410.

Kubatko, L. S., H. L. Gibbs, and E. W. Bloomquist. 2011. Inferring species-level phylogenies and taxonomic distinctiveness using multilocus data in *Sistrurus* rattlesnakes. Systematic Biology 60:393–409.

Kubie, J. L., and M. Halpern. 1979. Chemical senses involved in garter snake prey trailing. Journal of Comparative and Physiological Psychology 93:648–667.

Kumar, D. T., S. S. Kumar, and M. R. Prasad. 2014. Current status and possible causes of reptile's decline. International Research Journal of Environmental Sciences 3:75–79.

Kunkel, D. B., S. C. Curry, M. V. Vance, and P. J. Ryan. 1984. Reptile envenomations. Journal of Toxicology: Clinical Toxicology 21:503–526.

Kuo, C., and F. J. Janzen. 2004. Genetic effects of a persistent bottleneck on a natural population of Ornate Box Turtles (*Terrapene ornata*). Conservation Genetics 5:425–437.

Kurta, A. 1995. Mammals of the Great Lakes Region. University of Michigan Press, Ann Arbor, USA.

Kutka, F. J. 1994. Low pH effects on swimming activity of *Ambystoma* salamander larvae. Environmental Toxicology and Chemistry 13:1821–1824.

Kutka, F. J. 1998. Habitat acidification hazard to amphibians breeding in woodland ponds in northern Wisconsin. Thesis, Iowa State University, Ames, USA.

Kutka, F. J., and M. D. Bachmann. 1990. Acid sensitivity and water chemistry correlates of amphibian breeding ponds in northern Wisconsin, USA. Hydrobiologia 208:153–160.

Kuzmin, Y., V. V. Tkach, and S. D. Snyder. 2001. *Rhabdias ambystomae* sp. n. (Nematoda: Rhabdiasidae) from the North American Spotted Salamander *Ambystoma maculatum* (Amphibia: Ambystomatidae). Comparative Parasitology 68:228–235.

Kyriakopoulou-Sklavounou, P., and N. Loumbourdis. 1990. Annual ovarian cycle in the frog, *Rana ridibunda*, in northern Greece. Journal of Herpetology 24:185–191.

Labanick, G. M. 1976. Prey availability, consumption and selection in the Cricket Frog, *Acris crepitans* (Amphibia, Anura, Hylida). Journal of Herpetology 10:293–298.

Lacey, J. R., C. B. Marlow, and J. R. Lane. 1989. Influence of Spotted Knapweed (*Centaurea maculosa*) on surface runoff and sediment yield. Weed Technology 4:627–631.

Lachner, E. A. 1942. An aggregation of snakes and salamanders during hibernation. Copeia 1942:262–263.

Lacki, M. J., J. W. Hummer, and H. J. Webster. 1992. Mine-drainage treatment wetland as habitat for herpetofaunal wildlife. Environmental Management 16:513–520.

LaFleur, R. M. 2022. *Lampropeltis triangulum* (Eastern Milksnake). Geographic distribution. Herpetological Review 53:79.

LaFleur, R. M., and J. M. Kapfer. 2022. *Pantherophis vulpinus* (Eastern Foxsnake). Geographic distribution. Herpetological Review 53:81.

LaFleur, R. M., and N. J. Kutok. 2022. *Hemidactylium scutatum* (Four-toed Salamander). Geographic distribution. Herpetological Review 53:68.

Lagler, K. F. 1943. Food habits and economic relations of the turtles of Michigan with special reference to fish management. American Midland Naturalist 29:257–312.

Lagler, K. F., and B. T. Ostenson. 1942. Early spring food of the Otter in Michigan. Journal of Wildlife Management 6:244–254.

Lagler, K. F., and J. C. Salyer, II. 1947. Food and habits of the Common Watersnake, *Natrix s. sipedon*, in Michigan. Papers of the Michigan Academy of Science, Arts and Letters 31:169–180.

Lagler, K. F., and K. E. Goellner. 1941. Notes on *Necturus maculosus* (Rafinesque), from Evans Lake, Michigan. Copeia 1941:96–98.

Lagory, K. E., L. J. Walston, C. Goulet, R. A. Van Lonkhuyzen, S. Najjar, and C. Andrews. 2009. An examination of scale-dependent resource use by Eastern Hognose Snakes in southcentral New Hampshire. Journal of Wildlife Management 73:1387–1393.

LaGro, J. A., Jr. 1996. Designing without nature: unsewered residential development in rural Wisconsin. Landscape and Urban Planning 35:1–9.

Lamb, T., C. Lydeard, R. B. Walkder, and J. W. Gibbons. 1994. Molecular systematics of map turtles (*Graptemys*): a comparison of mitochondrial restriction site versus sequence data. Systematic Biology 43:543–559.

Lambert, M. R. 2015. Clover root exudate produces male-biased sex ratios and accelerates male metamorphic timing in Wood Frogs. Royal Society Open Science 2:150433.

Lambert, M. R., and G. S. J. Giller. 2015. *Chelydra serpentina* (Snapping Turtle), *Chrysemys picta picta* (Eastern Painted Turtle), *Sternotherus odoratus* (Eastern Musk Turtle), *Trachemys scripta elegans* (Red-eared Slider). Urban Population Density. Herpetological Review 46:242–243.

Lambert, M. R., A. B. Stoler, M. S. Smylie, R. A. Relyea, and D. K. Skelly. 2017a. Interactive effects of road salt and leaf litter on Wood Frog sex ratios and sexual size dimorphism. Canadian Journal of Fisheries and Aquatic Sciences 74:141–146.

Lambert, M. R., B. E. Carlson, M. S. Smylie, and L. Swierk. 2017b. Ontogeny of sexual dichromatism in the explosively breeding Wood Frog. Herpetological Conservation and Biology 12:447–456.

Lambert, M. R., G. S. J. Giller, D. K. Skelly, and R. G. Bribiescas. 2016. Septic systems, but not sanitary sewer lines, are associated with elevated estradiol in male frog metamorphs from suburban ponds. General and Comparative Endocrinology 232:109–114.

Lambert, M. R., G. S. J. Giller, L. B. Barber, K. C. Fitgerald, and D. K. Skelly. 2015. Suburbanization, estrogen contamination, and sex ratio in wild amphibian populations. Proceedings of the National Academy of Sciences of the United States of America 112:11881–11886.

Lambert, M. R., S. N. Nielsen, A. N. Wright, R. C. Thomson, and H. B. Shaffer. 2013. Habitat features determine the basking distribution of introduced Red-eared Sliders and native Western Pond Turtles. Chelonian Conservation and Biology 12:192–199.

Lamore, D. H. 1963. Prey of a Sparrow Hawk family when raising young. Wilson Bulletin 75:461.

Lamoureux, V. S., and D. M. Madison. 1999. Overwintering habitats of radio-implanted Green Frogs, *Rana clamitans*. Journal of Herpetology 33:430–435.

Lamoureux, V. S., J. C. Maerz, and D. M. Madison. 2002. Premigratory autumn foraging forays in the Green Frog, *Rana clamitans*. Journal of Herpetology 36:245–254.

Lampe, R. P. 1982. Food habits of Badgers in east central Minnesota. Journal of Wildlife Management 46:790–795.

Lancaster, D. L. 1994. The role of substrate odorants in the predator-prey relationship of *Diadophis punctatus* and *Plethodon cinereus*: failure to detect potential predators and the detection of potential prey. Dissertation, University of Southwestern Louisiana, Lafayette, USA.

Lancaster, D. L., and S. E. Wise. 1996. Differential response by the Ringneck Snake, *Diadophis punctatus*, to odors of tail-autotomizing prey. Herpetologica 52:98–108.

Landreth, H. F. 1972. Physiological responses of *Elaphe obsoleta* and *Pituophis melanoleucus* to lowered ambient temperatures. Herpetologica 28:376–380.

Lang, C., and R. G. Jaeger. 2000. Defense of territories by male-female pairs in the Red-Backed Salamander (*Plethodon cinereus*). Copeia 2000:169–177.

Lang, J. W. 1969. Hibernation and movements of *Storeria occipitomaculata* in northern Minnesota. Journal of Herpetology 3:196–197.

Lang, J. W. 1971. Overwintering of three species of snakes in Northwestern Minnesota. Thesis, University of North Dakota, Grand Forks, USA.

Lang, J. W. 1972a. *Ambystoma maculatum*. Geographic distribution 4:170.

Lang, J. W. 1972b. *Hemidactylium scutatum*. Geographic distribution 4:170.

Lang, J. W. 1982. Distribution and abundance of the Five-lined Skink (*Eumeces fasciatus*) in Minnesota. Final Report to Minnesota Department of Natural Resources, St. Paul, USA.

Lang, J. W. 2000. Blanding's Turtles, roads, & culverts at Weaver Dunes. Final Report to The Nature Conservancy and Minnesota Department of Natural Resources, St. Paul, USA.

Lange, K. I. 2004. A naturalist's journey. New Past Press, Inc., Friendship, Wisconsin, USA.

Langen, T. A., K. M. Andrews, S. P. Brady, N. E. Karraker, and D. J. Smith. 2015. Road effects on habitat quality for small animals. Pages 57–93 *in* K. M. Andrews, P. Nanjappa, and S. P. D. Riley, editors. Roads and ecological infrastructure: concepts and applications for small animals. John Hopkins University Press, Baltimore, Maryland, USA.

Langford, G. J., J. A. Borden, and D. H. Nelson. 2011. Ecology of the Eastern Ribbonsnake (*Thamnophis sauritus*) in southern Alabama with evidence of seasonal multiple broods. Herpetological Conservation and Biology 6:400–409.

Langhans, M., B. Peterson, A. Walker, G. R. Smith, and J. E. Rettig. 2009. Effects of salinity on survivorship on Wood Frog (*Rana sylvatica*) tadpoles. Journal of Freshwater Ecology 24:335–337.

Langley, R. L. 2009. Deaths from reptile bites in the United States, 1997–2004. Clinical Toxicology 47:44–47.

Langley, W. M., H. W. Lipps, and J. F. Theis. 1989. Reponses of Kansas motorists to snake models on a rural highway. Transactions of the Kansas Academy of Science 92:43–48.

Langlois, T. H. 1964. Amphibians and reptiles of the Erie Islands. Ohio Journal of Science 64:11–25.

Lannoo, M. J. 1987. Neuromast topography in urodele amphibians. Journal of Morphology 191:247–263.

Lannoo, M. J. 1996. Okoboji wetlands: a lesson in natural history. University of Iowa Press, Iowa City, USA.

Lannoo, M. J. 1997. A fish fry: the role of exotic species and aquacultural practices in producing amphibian declines in the Upper Midwest. Pages 25–27 *in* J. J. Moriarty and D. Jones, editors. Minnesota's amphibians and reptiles: their conservation and status. Serpent's Tale Press, Excelsior, Minnesota, USA.

Lannoo, M. J. 1998. Amphibian conservation and wetland management in the Upper Midwest: a catch-22 for the Cricket Frog. Status and conservation of midwestern amphibians. University of Iowa Press, Iowa City, USA.

Lannoo, M. J. 2008. Malformed frogs: the collapse of aquatic ecosystems. University of California Press, Berkeley, USA.

Lannoo, M. J., and C. A. Phillips. 2005. *Ambystoma tigrinum*, Tiger Salamander. Pages 636–639 *in* M. Lannoo, editor. Amphibian declines: the conservation and status of United States species. University of California Press, Berkeley, USA.

Lannoo, M. J., and M. D. Bachmann. 1984. Aspects of cannibalistic morphs in a population of *Ambystoma t. tigrinum* larvae. American Midland Naturalist 112:103–109.

Lannoo, M. J., C. Petersen, R. E. Lovich, P. Nanjappa, C. Phillips, J. C. Mitchell, and I. Macallister. 2011. Do frogs get their kicks on Route 66? Continental U.S. transect reveals spatial and temporal patterns of *Batrachochytrium dendrobatidis* infection. PLoS ONE 6:e22211.

Lannoo, M. J., D. R. Sutherland, P. Jones, D. Rosenberry, R. W. Klaver, D. M. Hoppe, P. T. J. Johnson, K. B. Lunde, C. Facemire, and J. M. Kapfer. 2003. Multiple causes for the malformed frog phenomenon. Pages 233–262 *in* G. Linder, E. Little, S. Krest, and D. Sparling, editors. Multiple stressors in relation to declining amphibian populations. American Society for Testing and Materials ASTM STP 1443, Conshohocken, Pennsylvania, USA.

Lannoo, M. J., J. A. Holman, G. S. Casper, and E. Johnson. 1998. Mummification following winterkill of adult Green Frogs (Ranidae: *Rana clamitans*). Herpetological Review 29:82–84.

Lannoo, M. J., K. Lang, T. Waltz and G. S. Phillips. 1994. An altered amphibian assemblage: Dickinson County, Iowa, 70 years after Frank Blanchard's survey. American Midland Naturalist 131:311–319.

Lapham, I. A. 1853. Fauna and flora of Wisconsin. Volume 2. Transactions of the Wisconsin State Agricultural Society, Madison, USA.

Lapin, C. N., J. Tamplin, M. M. Cochrane, J. E. Woodford, D. J. Brown, and R. A. Moen. 2019. A regional analysis of *Glyptemys insculpta* (Wood Turtle) survival in the Upper Midwest of the USA. Herpetological Conservation and Biology 14:668–679.

Laposata, M. M., and W. A. Dunson. 1998. Effects of boron and nitrate on hatching success of amphibian eggs. Archives of Environmental Contamination and Toxicology 35:615–619.

Larkin, J. 2011. Microhabitat preferences of the Prairie Skink (*Plestiodon septentrionalis*) in southwestern Manitoba. Thesis, University of Manitoba, Winnipeg, Canada.

Larsen, K. W. 1987. Movements and behavior of migratory Garter Snakes, *Thamnophis sirtalis*. Canadian Journal of Zoology 65:2241–2247.

Larsen, K. W., and P. T. Gregory. 1989. Population size and survivorship of the Common Garter Snake, *Thamnophis sirtalis*, near the northern limit of its distribution. Holarctic Ecology 12:81–86.

Larson, A., D. Wake, and T. Devitt. 2006. Caudata. Salamanders. Version 05 September 2006. The Tree of Life Web Project. http://tolweb.org/Caudata. Accessed 9 Dec 2018.

Larson, D. 2014. Grassland fire and cattle grazing regulate reptile and amphibian assembly among patches. Environmental Management 54:1434–1444.

Larson, D. J., L. Middle, H. Vu, W. Zhang, A. S. Serianni, J. Duman, and B. M. Barnes. 2014. Wood Frog adaptations to overwintering in Alaska: new limits to freezing tolerance. Journal of Experimental Biology 217:2193–2200.

Larson, D. L., S. McDonald, S. J. Hamilton, A. J. Fivizzani, and W. E. Newton. 1998. Effects of the herbicide atrazine on *Ambystoma tigrinum* metamorphosis: duration, larval growth and hormonal response. Physiological and Biochemical Zoology 71:671–679.

Lathrop, R. C., S. B. Nehls, C. L. Brynildson, and K. R. Plass. 1992. The fishery of the Yahara lakes. Wisconsin Department of Natural Resources Technical Bulletin 181, Madison, USA.

Latreille, P. A. 1801. XXII Genre. Hétérodon, Heterodon. Pages 32–37 *in* C. S. Sonnini and P. A. Latreille, editors. Histoire naturelle des reptiles avec figures dessinées d'après nature. Volume 4. Chez Deterville, Paris, France.

Laughlin, M. M., E. R. Olson, and J. G. Martin. 2017. Arboreal camera trapping expands *Hyla versicolor* complex (Hylidae) canopy use to new heights. Ecology 98:2221–2223.

Lawing, A. M., and P. D. Polly. 2011. Pleistocene climate, phylogeny, and climate envelope models: an integrative approach to better understand species' response to climate change. PLoS ONE 6:e28554.

Lawler, J. J., A. S. Ruesch, J. D. Olden, and B. H. McRae. 2013. Projected climate-driven faunal movement routes. Ecology Letters 16:1014–1022.

Lawler, J. J., S. L. Shafer, and A. R. Blaustein. 2010. Projected climate impacts for the amphibians of the western hemisphere. Conservation Biology 24:38–50.

Lawler, S. P. 1989. Behavioural responses to predators and predation risk in four species of larval anurans. Animal Behaviour 38:1039–1047.

Lawson, P. A., and D. M. Secoy. 1991. The use of solar cues as migratory orientation guides by the Plains Gartersnake, *Thamnophis radix*. Canadian Journal of Zoology 69:2700–2702.

Lawson, R. 1987. Molecular studies of thamnophiine snakes: the phylogeny of the genus *Nerodia*. Journal of Herpetology 21:140–157.

Layne, J. R., Jr., and J. Kefauver. 1997. Freeze tolerance and postfreeze recovery in the frog *Pseudacris crucifer*. Copeia 1997:260–264.

Layne, J. R., Jr., and M. C. First. 1991. Resumption of physiological functions in the Wood Frog (*Rana sylvatica*) after freezing. American Journal of Physiology—Regulatory, Integrative and Comparative Physiology 261:134–137.

Layne J. R., Jr., and N. B. Ford. 1984. Flight distance of the Queen Snake, *Regina septemvittata*. Journal of Herpetology 18:496–498.

Layne, J. R., Jr., and R. E. Lee. 1989. Seasonal variation in freeze tolerance and ice content of the tree frog *Hyla versicolor*. Journal of Experimental Zoology 249:133–137.

Layne, J. R., Jr., R. E. Lee, Jr., and T. L. Heil. 1989. Freezing-induced changes in the heart rate of Wood Frogs (*Rana sylvatica*). American Journal of Physiology—Regulatory, Integrative and Comparative Physiology 257:1046–1049.

Lazell, J. 1993. *Heterodon platirhinos* (Eastern Hognose Snake). Melanism heredity. Herpetological Review 24:35.

Lazure, L., P. Paré, L. Bouthillier, and P. Galois. 2019. Nesting biology and conservation of a northern population of Spiny Softshell Turtles (*Apalone spinifera*). Herpetological Conservation and Biology 14:659–667.

Leach, C. 2012. Influences of temperature and precipitation on Wood Frog (*Rana sylvatica*) breeding phenology with predictions of climate change impacts. McNair Scholars Journal of the University of Wisconsin–Superior 2012:1–12.

Leach, M. K., and T. J. Givnish. 1999. Gradients in the composition, structure, and diversity of remnant oak savannas in southern Wisconsin. Ecological Monographs 69:353–374.

Leaché, A. D., and T. W. Reeder. 2002. Molecular systematics of the Eastern Fence Lizard (*Sceloporus undulatus*): a comparison of parsimony, likelihood, and Bayesian approaches. Systematic Biology 51:44–68.

Leclair, M. H., M. Levasseur, and R. Leclair, Jr. 2006. Life-history traits of *Plethodon cinereus* in the northern parts of its range: variations in population structure, age, and growth Herpetologica 62:266–282.

Leclair, M. H., M. Levsseur, and R. Leclair, Jr. 2008. Activity and reproductive cycles in northern populations of the Red-backed Salamander, *Plethodon cinereus*. Journal of Herpetology 42:31–38.

Leclair, R., Jr., and G. Laurin. 1996. Growth and body size in populations of Mink Frog *Rana septentrionalis* from two latitudes. Ecography 19:296–304.

Leclair, R., Jr., and J. Castanet. 1987. A skeletochronological assessment of age and growth in the frog *Rana pipiens* Schreber (Amphibia, Anura) from southwestern Quebec. Copeia 1987:361–369.

LeClair, R., Jr., and J. P. Bourassa. 1981. Observation et analyse de la larves de Diptera chirnomides, dans la region de Trois-Rivieres (Quebec). Canadian Journal of Zoology 59:1339–1343.

LeClere, J. B. 2013. A field guide to the amphibians and reptiles of Iowa. ECO Herpetological Publishing and Distribution, Rodeo, New Mexico, USA.

LeClere, J. B., E. P. Hoaglund, J. Scharosch, C. E. Smith, and T. Gamble. 2012. Two naturally occurring intergeneric hybrid snakes (*Pituophis catenifer sayi* x *Pantherophis vulpinus*; Lampropeltini, Squamata) from the midwestern United States. Journal of Herpetology 46:257–262.

LeConte, J. 1825. Remarks on the American species of the genera *Hyla* and *Rana*. Annals of the Lyceum of Natural History of New York 1:278–282.

LeDee, O. E., and C. A. Ribic. 2015. Determining climate change management priorities: a case study from Wisconsin. Journal of Conservation Planning 11:1–12.

LeDee, O. E., S. Hagell, K. Martin, D. MacFarland, M. Meyer, A. Paulios, C. A. Ribic, D. Sample, and T. Van Deelen. 2013. Climate change impacts on Wisconsin's wildlife: a preliminary assessment. Wisconsin Department of Natural Resources Technical Bulletin 197, Madison, Wisconsin, USA. Archived by WebCite at http://www.webcitation.org/73fzF49fj. Accessed 4 Nov 2018.

Ledford, D. P. 2011. Spatial ecology of the Spotted Salamander, *Ambystoma maculatum*, during the nonbreeding season. Thesis, East Tennessee State University, Johnson City, USA.

Lee, D. S., and L. R. Franz. 1973. Red leg in a subterranean population of Pickerel Frogs. Bulletin of the Maryland Herpetological Society 9:111.

Lee, R. E., Jr., J. P. Costanzo, E. C. Davidson, and J. R. Layne, Jr. 1992. Dynamics of body water during freezing and thawing in a freeze-tolerant frog (*Rana sylvatica*). Journal of Thermal Biology 17:263–266.

Lee, Y. M. 2004. Monitoring of Eastern Fox Snakes (*Elaphe vulpina gloydi*) in response to habitat restoration at Sterling State Park in Michigan. Michigan Department of Natural Resources, Parks and Recreation Bureau Report 2004–20, Lansing, USA.

Lee-Yaw, J. A., J. T. Irwin, and D. M. Green. 2008. Postglacial range expansion from northern refugia by the Wood Frog, *Rana sylvatica*. Molecular Ecology 17:867–884.

Lefebvre, J., T. S. Avery, and T. B. Herman. 2011. Size dimorphism and growth rates in distinct populations of Blanding's Turtles (*Emydoidea blandingii*) in Nova Scotia in relation to Environment. Herpetological Conservation and Biology 6:465–472.

Legler, J. M. 1960. Natural history of the Ornate Box Turtle *Terrapene ornata ornata* Agassiz. University of Kansas Publications Museum of Natural History 11:527–669.

Legler, J. M., and R. C. Vogt. 2013. The turtles of Mexico: land and freshwater forms. University of California Press, Berkeley, USA.

Lehtinen, R. M. 2002. A historical study of the distribution of Blanchard's Cricket Frog (*Acris crepitans*) in southeastern Michigan. Herpetological Review 33:194–197.

Lehtinen, R. M., and A. A. Skinner. 2006. The enigmatic decline of Blanchard's Cricket Frog (*Acris crepitans blanchardi*): a test of the habitat acidification hypothesis. Copeia 2006:159–167.

Lehtinen, R. M., and M. C. MacDonald. 2011. Live fast, die young? A six-year field study of longevity and survivorship in Blanchard's Cricket Frog (*Acris crepitans blanchardi*). Herpetological Review 42:504–507.

Lehtinen, R. M., and S. M. Galatowitsch. 2001. Colonization of restored wetlands by amphibians in Minnesota. American Midland Naturalist 145:388–396.

Lehtinen, R. M., S. M. Galatowitsch, and J. R. Tester. 1999. Consequences of habitat loss and fragmentation for wetland amphibian assemblages. Wetlands 19:1–12.

Lemmon, A. R., and E. M. Lemmon. 2008. A likelihood framework for estimating phylogeographic structure on a continuous landscape. Systematic Biology 57:544–561.

Lenhart, C. F., J. R. Naber, and J. L. Nieber. 2013. Impacts of hydrologic change on sandbar nesting availability for riverine turtles in eastern Minnesota, USA. Water 5:1243–1261.

Lenker, M. A., A. E. Savage, C. G. Becker, D. Rodriguez, and K. R. Zamudio. 2014. *Batrachochytrium dendrobatidis* infection dynamics vary seasonally in upstate New York, USA. Diseases of Aquatic Organisms 111:51–60.

Lentini A. M., and J. Earnhardt. 2013. Eastern Massasauga Rattlesnake (*Sistrurus catenatus catenatus*) care manual. Association of Zoos and Aquariums, Silver Spring, Maryland, USA. Archived by WebCite at http://www.webcitation.org/73fzPab5A. Accessed 4 Nov 2018.

Leopold, A. 1933. Game Management. Charles Scribner's Sons, New York, New York, USA.

Leopold, A. 1972. Round River. Oxford University Press, Cary, North Carolina, USA.

Leopold, A. 1887–1948. Aldo Leopold papers. University of Wisconsin, Archival Resources in Wisconsin, Madison, USA.

Leopold, A. 1936. Franklin J. Schmidt. Wilson Bulletin 48:181–186.

Lesch, M. E., and J. D. Fawcett. 1978. Preliminary observations on the life history of the Western Plains Garter Snake, *Thamnophis radix haydeni*, in Douglas County, Nebraska. Proceedings of the Nebraska Academy of Sciences 88:17.

Lesica, P., and F. W. Allendorf. 1995. When are peripheral populations valuable for conservation? Conservation Biology 9:753–760.

Letcher, R. J., Z. Lu, S. R. de Solla, C. D. Sandau, and K. J. Fernie. 2015. Snapping Turtles (*Chelydra serpentina*) from Canadian Areas of Concern across the southern Laurentian Great Lakes: chlorinated and brominated hydrocarbon contaminants and metabolites in relation to circulating concentrations of thyroxine and vitamin A. Environmental Research 143:266–278.

Lett, M. S., and A. K. Knapp. 2005. Woody plant encroachment and removal in mesic grassland: production and composition responses of herbaceous vegetation. American Midland Naturalist 153:217–231.

Lettow, M. C., L. A. Brudvig, C. A. Bahlai, and D. A. Landis. 2014. Oak savanna management strategies and their differential success on vegetative structure, understory light, and flowing forbs. Forest Ecology and Management 329:89–98.

Levell, J. P. 2000. Commercial exploitation of Blanding's Turtle, *Emydoidea blandingii*, and the Wood Turtle, *Clemmys insculpta*, for the live animal trade. Chelonian Conservation and Biology 3:665–674.

Lewis, C. H., S. F. Molloy, R. M. Chambers, and J. Davenport. 2007. Response of Common Musk Turtles (*Sternotherus odoratus*) to intraspecific chemical cues. Journal of Herpetology 41:349–353.

Lewis, E. L., and J. B. Iverson. 2018. *Terrapene ornata* (Ornate Box Turtle). Longevity. Herpetological Review 49:530.

Lewis, W. M. 1962. Stomach contents of Bullfrogs (*Rana catesbeiana*) taken from a minnow hatchery. Transactions of the Illinois State Academy of Science 55:80–83.

Ley, D. 2004. Knudsen, George J. Interview #676. University of Wisconsin, Oral History Project Madison Archives, Madison, USA.

Li, Y., J. M. Cohen, and J. R. Rohr. 2013. Review and synthesis of the effects of climate change on amphibians. Integrative Zoology 8:145–161.

Licht, L. E. 1969. Palatability of *Rana* and *Hyla* eggs. American Midland Naturalist 82:296–298.

Licht, L. E. 1976. Sexual selection in toads (*Bufo americanus*). Canadian Journal of Zoology 54:1277–1284.

Licht, L. E. 2003. Shedding light on ultraviolet radiation and amphibian embryos. BioScience 53:551–561.

Licht L. E., and J. P. Bogart. 1987. Ploidy and developmental rate in a salamander hybrid complex (genus *Ambystoma*). Evolution 41:918–920.

Licht, L. E., and K. P. Grant. 1997. The effects of ultraviolet radiation on the biology of amphibians. American Zoologist 37:137–145.

Lieberman, S. 1994. Can CITES save the box turtle? Species Technical Bulletin 19:16–17.

Liebgold, E. B. 2014. The influence of social environment: behavior of unrelated adults affects future juvenile behaviors. Ethology 120:388–399.

Liebgold, E. B., and C. Dibble. 2011. Better the devil you know: familiarity affects foraging activity of adult and juvenile Red-backed Salamanders, *Plethodon cinereus*. Animal Behavior 82:1059–1066.

Liebgold, E. B., and P. R. Cabe. 2008. Familiarity with adults, but not relatedness, affects the growth of juvenile Red-backed Salamanders (*Plethodon cinereus*). Behavioral Ecology and Sociobiology 63:277–284.

Liebgold, E. B., and R. G. Jaeger. 2007. Juvenile movements and potential inter-age class associations of Red-backed Salamanders. Herpetologica 63:51–55.

Liebgold, E. B., E. D. Brodie, III, and P. R. Cabe. 2011. Female philopatry and male-biased dispersal in a direct-developing salamander, *Plethodon cinereus*. Molecular Ecology 20:249–257.

Lillywhite, H. B. 1970. Behavioral temperature regulation in the Bullfrog, *Rana catesbeiana*. Copeia 1970:158–168.

Lillywhite, H. B. 1985. Trailing movements and sexual behavior in *Coluber constrictor*. Journal of Herpetology 19:306–308.

Lillywhite, H. B. 1987. Temperature, energetics and physiological ecology. Pages 422–477 *in* R. A. Seigel, J. T. Collins, and S. S. Novak, editors. Snakes: ecology and evolutionary biology. Macmillan Publishing Company, New York, New York, USA.

Linck, M. H. 2000. Reduction in road mortality in a Northern Leopard Frog population. Journal of the Iowa Academy of Science 107:209–211.

Lindeman, P. V. 1991. Survivorship of overwintering hatchling Painted Turtles (*Chrysemys picta*) in northern Idaho. Canadian Field-Naturalist 105:263–266.

Lindeman, P. V. 1992. Nest-site fidelity among Painted Turtles (*Chrysemys picta*) in northern Idaho. Northwestern Naturalist 73:27–30.

Lindeman, P. V. 1999a. Growth curves for *Graptemys*, with a comparison to other emydid turtles. American Midland Naturalist 142:141–151.

Lindeman, P. V. 1999b. Surveys of basking map turtles *Graptemys* spp. in three river drainages and the importance of deadwood abundance. Biological Conservation 88:33–42.

Lindeman, P. V. 2000. Resource use of five sympatric turtle species: effects of competition, phylogeny, and morphology. Canadian Journal of Zoology 78:992–1008.

Lindeman, P. V. 2001. A contrast in the basking habits of the sympatric Trionychid turtles *Apalone mutica* and *A. spinifera*. Herpetological Natural History 8:87–89.

Lindeman, P. V. 2013. The map turtle and sawback atlas. University of Oklahoma Press, Norman, USA.

Lindeman, P. V. 2017. *Graptemys geographica* (Northern Map Turtle). Epibiont. Herpetological Review 48:177–178.

Ling, R. W., J. P. VanAmberg, and J. K. Werner. 1986. Pond acidity and its relationship to larval development of *Ambystoma maculatum* and *Rana sylvatica* in upper Michigan. Journal of Herpetology 20:230–236.

Lingenfelter, A. R., K. Geluso, M. P. Nenneman, B. C. Peterson, and J. L. Kerby. 2014. Distribution, diet, and prevalence of amphibian chytrid fungus in non-native American Bullfrogs (*Lithobates catesbeianus*) at the Valentine National Wildlife Refuge, Nebraska, USA. Journal of North American Herpetology 2014:81–86.

Linnaeus, C. 1758. Systema naturae per regna tria naturae: secundum classes, ordines, genera, species, cum characteribus, differentiis, synonymis, locis. Tenth edition. Volume 1. Laurentius Salvius, Stockholm, Sweden.

Linnaeus, C. 1766. Systema Naturae. Twelfth edition. Homiae, Salvii. Stockholm, Sweden.

Linsdale, J. M. 1927. Amphibians and reptiles of Doniphan County, Kansas. Copeia 1927:75–81.

Little, M. J. 2002. *Necturus maculosus* (Mudpuppy). Predation. Herpetological Review 33:122.

Littlejohn, M. J. 1958. Mating behavior in the treefrog, *Hyla versicolor*. Copeia 1958:222–223.

Liu, F., D. J. Mladenoff, N. S. Keuler, and L. S. Moore. 2011. Broadscale variability in tree data of the historical public land survey and its consequences for ecological studies. Ecological Monographs 81:259–275.

Livezey, R. L. 1950. The eggs of *Acris gryllus crepitans*, Baird. Herpetologica 6:139–140.

Livezey, R. L., and A. H. Wright. 1947. A synoptic key to the salientian eggs of the United States. American Midland Naturalist 37:179–222.

Lockhart, O. M. 2000. *Necturus maculosus* (Mudpuppy). Egg predation. Herpetological Review 31:98.

Loda, J. L., and D. L. Otis. 2009. Low prevalence of amphibian chytrid fungus (*Batrachochytrium dendrobatidis*) in Northern Leopard Frog (*Rana pipiens*) populations of north-central Iowa. Herpetological Review 40:428–431.

Loftin, C. S., A. J. K. Calhoun, S. J. Nelson, A. A. Elskus, and K. Simon. 2012. Mercury bioaccumulation in Wood Frogs developing in seasonal pools. Northeastern Naturalist 19:579–600.

Logier, E. B. S. 1925. Notes on the herpetology of Point Pelee, Ontario. Canadian Field-Naturalist 29:91–95.

Logier, E. B. S. 1931. The amphibians and reptiles of Long Point. A faunal investigation of Long Point and vicinity, Norfolk County, Ontario. Transactions of the Royal Canadian Institute 18:299–236.

Logier, E. B. S. 1939. The Reptiles of Ontario. University of Toronto Press, Ontario, Canada.

Loncke, D. J., and M. E. Obbard. 1977. Tag success, dimensions, clutch size and nesting site fidelity for the Snapping Turtles, *Chelydra serpentina* (Reptilia, Testudines, Chelydridae), in Algonquin Park, Ontario, Canada. Journal of Herpetology 11:243–244.

Long, C. A. 1982. Rare gigantic toads, *Bufo americanus*, from Lake Michigan isles. University of Wisconsin, Reports of the Museum of Natural History 18:15–19.

Long, C. A. 2008. Wild mammals of Wisconsin. University of Wisconsin–Stevens Point Museum of Natural History Publication 56. Pensoft Publishers, Sofia, Bulgaria.

Long, C. A., and C. A. Long. 1976. Some amphibians and reptiles collected on islands in Green Bay, Lake Michigan. Jack-Pine Warbler 54:54–58.

Long, C. A., J. E. Long, and C. A. Long. 1978. Some amphibians and reptiles from Chambers Island, Green Bay, Lake Michigan. Jack-Pine Warbler 56:46–47.

Long, C. A., J. E. Long, and C. A. Long. 1987. A listing of known amphibians and reptiles from Washington Island, Lake Michigan, with new records of the Milk Snake and Red-bellied Snake. Jack-Pine Warbler 65:39.

Long, C. J., M. J. Power, and B. McDonald. 2011. Millennial-scale fire and vegetation history from a mesic hardwood forest of southeastern Wisconsin, USA. Journal of Quaternary Science 26:318–325.

Longcore, J. R., J. E. Longcore, A. P. Pessier, and W. A. Halteman. 2007. Chytridiomycosis widespread in anurans of northeastern United States. Journal of Wildlife Management 71:435–444.

Lorch, J. M., J. Lankton, K. Werner, E. A. Falendysz, K. McCurley, and D. S. Blehert. 2015. Experimental infection of snakes with *Ophidiomyces ophiodiicola* causes pathological changes that typify snake fungal disease. mBio 6:e01534–15.

Lorch, J. M., S. Knowles, J. S. Lankton, K. Michell, J. L. Edwards, J. M. Kapfer, R. A. Staffen, E. R. Wild, K. Z. Schmidt, A. E. Ballmann, D. Blodgett, T. M. Farell, B. M. Glorioso, L. A. Last, S. J. Price, K. L. Schuler, C. E. Smith, J. F. X. Wellehan, Jr., and D. S. Blehert. 2016. Snake fungal disease: an emerging threat to wild snakes. Philosophical Transactions of the Royal Society B: Biological Sciences 371:20150457.

Loss, S. R., T. Will, and P. P. Marra. 2013. The impact of free-ranging domestic cats on wildlife of the United States. Nature Communications 4:Article 1396.

Loudon, A. H., D. C. Woodhams, L. W. Parfrey, H. Archer, R. Knight, V. McKenzie, and R. N. Harris. 2014. Microbial community dynamics and effect of environmental microbial reservoirs on Red-backed Salamanders (*Plethodon cinereus*). ISME Journal 8:830–840.

Lovich, J. E., C. H. Ernst, and J. F. McBreen. 1990. Growth, maturity, and sexual dimorphism in the Wood Turtle, *Clemmys insculpta*. Canadian Journal of Zoology 68:672–677.

Lovich, J. E., C. H. Ernst, E. M. Ernst, and J. L. Riley. 2014. A 21-year study of seasonal and interspecific variation of hatchling emergence in a Nearctic freshwater turtle community: to overwinter or not to overwinter? Herpetological Monographs 28:93–109.

Lowcock L. A., H. Griffith, and W. Murphy. 1992. Size in relation to sex, hybridity, ploidy, and breeding dynamics in central Ontario populations of the *Ambystoma laterale-jeffersonianum* complex. Journal of Herpetology 26:46–53.

Lowcock, L. A., L. E. Licht, and J. P. Bogart. 1987. Nomenclature in hybrid complexes of *Ambystoma* (Urodela: Ambystomatidae): no case for the erection of hybrid "species." Systematic Zoology 36:328–336.

Lowcock, L. A., T. F. Sharbel, J. Bonin, M. Oullet, J. Rodrigue, and J-L. DesGranges. 1997. Flow cytometric assay for in vivo genotoxic effects of pesticides in Green Frogs (*Rana clamitans*). Aquatic Toxicology 38:241–255.

Lowe, S., M. Browne, S. Boudjelas, and M. De Poorter. 2000. 100 of the world's worst invasive alien species: a selection from the Global Invasive Species Database. University of Aukland, Invasive Species Specialist Group, Aukland, New Zealand. Archived by WebCite at http://www.webcitation.org/73cKKFq6d. Accessed 1 Nov 2018.

Ludwig, D. R., M. Redmer, R. Domazlicky, S. Kobal and B. Conklin. 1992. Current status of amphibians and reptiles in DuPage County, Illinois. Transactions of the Illinois State Academy of Sciences 85:187–199.

Lueth, F. X. 1941. Effects of temperature on snakes. Copeia 1941:125–132.

Lundeen, D., and A. F. Badje. 2017. *Emydoidea blandingii* (Blanding's Turtle). Geographic distribution. Herpetological Review 48:386.

Lurgi, M., B. C. Lopéz, and J. M. Montoya. 2012. Novel communities from climate change. Philosophical Transactions of the Royal Society B: Biological Sciences 367:2913–2922.

Lurie, N. O. 1983. A special style: the Milwaukee Public Museum, 1882–1982. Milwaukee Public Museum Press, Milwaukee, Wisconsin, USA.

Lyderson, K. 2006. Politics put status of snake at risk. The Washington Post. 30 September 2006; section A.2.

Lykens, D. V., and D. C. Forester. 1987. Age structure in the Spring Peeper: do males advertise longevity? Herpetologica 43:216–223.

Lyman-Henley, L. P., and G. M. Burghardt. 1995. Diet, litter, and sex effects on chemical prey preference, growth, and site selection in two sympatric species of *Thamnophis*. Herpetological Monographs 9:140–160.

Lynch, J. D. 1966. Communal egg laying in the Pilot Blacksnake, *Elaphe obsoleta obsoleta*. Herpetologica 22:305.

Lynch, J. D. 1985. Annotated checklist of the amphibians and reptiles of Nebraska. Transactions of the Nebraska Academy of Sciences 13:33–57.

Lynch, W. 1978. Death-feigning in the Eastern Yellow-bellied Racer. Blue Jay 36:92–93.

Lyon, W. I. 1927. The Fox Snake feeds on eggs of the Common Tern. Wilson Bulletin 39:186.

Lytwyn, J. 2010. Remote sensing and GIS investigation of glacial features in the region of Devil's Lake State Park, south-central Wisconsin, USA. Geomorphology 123:46–60.

MacArthur, D. L., and J. W. T. Dandy. 1982. Physiological aspects of overwintering in the Boreal Chorus Frog (*Pseudacris triseriata maculata*). Comparative Biochemistry and Physiology Part A: Physiology 72:137–141.

MacArthur, R. H., and E. O. Wilson. 1967. The theory of island biogeography. Princeton University Press, New Jersey, USA.

Macartney, J. M., P. T. Gregory, and K. W. Larsen. 1988. A tabular survey of data on movements and home ranges of snakes. Journal of Herpetology 22:61–73.

MacCulloch, R. D. 2002. The ROM field guide to amphibians and reptiles of Ontario. Royal Ontario Museum, Toronto, Canada.

MacCulloch, R. D., and D. M. Secoy. 1983. Movement in a river population of *Chrysemys picta belli* in southern Saskatchewan. Journal of Herpetology 17:283–285.

MacCulloch, R. D., and W. F. Weller. 1988. Some aspects of reproduction in a Lake Erie population of Blanding's Turtle, *Emydoidea blandingii*. Canadian Journal of Zoology 66:2317–2319.

MacDonald, N. W., L. M. Martin, C. K. Kapolka, T. F. Botting, and T. E. Brown. 2013. Hand pulling following mowing and herbicide treatments increases control of Spotted Knapweed (*Centaurea stoebe*). Invasive Plant Science and Management 6:470–479.

Macey, J. R., J. A. Schulte, II, A. Larson, B. S. Tuniyev, N. Orlov, and T. J. Papenfuss. 1999. Molecular phylogenetics, tRNA evolution, and historical biogeography in anguid lizards and related taxonomic families. Molecular Phylogenetics and Evolution 12:250–272.

MacGowan, B., B. Kingsbury, and R. Williams. 2005. Turtles of Indiana. Purdue University Cooperative Extension Service Publication FNR-243, West Lafayette, Indiana, USA.

Mack, E. W. 2015. Delayed neonatal feeding and growth in the Lake Erie Watersnake, *Nerodia sipedon insularum*. Thesis, Northern Illinois University, Dekalb, USA.

Mack, E. W., J. L. Beck, K. M. Stanford, and R. B. King. 2017. Maternal investment and delayed feeding in neonatal Lake Erie Watersnakes: a life history strategy. Journal of Zoology 301:150–156.

Mackenzie, C. A., M. Berrill, C. Metcalfe, and B. D. Pauli. 2003. Gonadal differentiation in frogs exposed to estrogenic and antiestrogenic compounds. Environmental Toxicology and Chemistry 22:2466–2475.

MacKinnon, C. A., A. Lawson., E. D. Stevens, and R. J. Brooks. 2006. Body temperature fluctuations in free-ranging Eastern Foxsnakes (*Elaphe gloydi*) during cold-water swimming. Canadian Journal of Zoology 84:9–19.

MacNamara, M. C. 1977. Food habits of terrestrial adult migrants and immature red efts of the Red-spotted Newt *Notophthalmus viridescens*. Herpetologica 33:127–132.

MacNeil, J., B. J. MacGowan, A. Currylow, and R. N. Williams. 2013. Forest management for reptiles and amphibians: a technical guide for the Midwest. Purdue University Cooperative Extension Service Publication FNR-480-W, West Lafayette, Indiana, USA. Archived by WebCite at http://www.webcitation.org/73g00UJUo. Accessed 4 Nov 2018.

MacNeil, J. E., and R. N. Williams. 2014. Effects of timber harvests and silvicultural edges on terrestrial salamanders. PLoS ONE 9:e114683.

MacNeill, A. L., E. W. Uhl, H. Kolenda-Roberts, and E. Jacobson. 2002. Mortality in a Wood Turtle (*Clemmys insculpta*) collection. Veterinary Clinical Pathology 31:133–136.

Maddalena, M. 2018. Multi-scale patterns of Eastern Milksnake (*Lampropeltis triangulum*) habitat selection and behavioural responses to habitat fragmentation. Thesis, University of Waterloo, Ontario, Canada.

Madison, D. M. 1997. The emigration of radio-implanted Spotted Salamanders, *Ambystoma maculatum*. Journal of Herpetology 31:542–551.

Madison, D. M., and L. Farrand, III. 1998. Habitat use during breeding and emigration in radio-implanted Tiger Salamanders, *Ambystoma tigrinum*. Copeia 1998:402–410.

Maerz, J. C., B. Blossey, and V. Nuzzo. 2005a. Green Frogs show reduced foraging success in habitats invaded by Japanese Knotweed. Biodiversity and Conservation 14:2901–2911.

Maerz, J. C., C. J. Brown, C. T. Chapin, and B. Blossey. 2005b. Can secondary compounds of an invasive plant affect larval amphibians? Functional Ecology 19:970–975.

Maerz, J. C., J. S. Cohen, and B. Blossey. 2010. Does detritus quality predict the effect of native and non-native plants on the performance of larval amphibians? Freshwater Biology 55:1694–1704.

Maerz, J. C., V. A. Nuzzo, and B. Blossey. 2009. Declines in woodland salamander abundance associated with non-native earthworm and plant invasions. Conservation Biology 23:975–981.

Maglia, A. M. 1996. Ontogeny and feeding ecology of the Red-Backed Salamander *Plethodon cinereus*. Copeia 1996:576–586.

Magnuson, J. J., K. E. Webster, R. A. Assel, C. J. Bowser, P. J. Dillon, J. G. Eaton, H. E. Evans, E. J. Fee, R. I. Hall, L. R. Mortsch, D. W. Schindler, and F. H. Quinn. 1997. Potential effects of climate changes on aquatic systems: Laurentian Great Lakes and Precambrian Shield Region. Hydrological Processes 11:825–871.

Mahan, R. D., and J. R. Johnson. 2007. Diet of the Gray Treefrog (*Hyla versicolor*) in relation to foraging site location. Journal of Herpetology 41:16–23.

Mahmoud, I. Y. 1967. Courtship behavior and sexual maturity in four species of kinosternid turtles. Copeia 1967:314–319.

Mahmoud, I. Y. 1968a. Feeding behavior in kinosternid turtles. Herpetologica 24:300–305.

Mahmoud, I. Y. 1968b. Nesting behavior in the Western Painted Turtle, *Chrysemys picta belli*. Herpetologica 24:158–162.

Mahmoud, I. Y. 1969. Comparative ecology of the kinosternid turtles of Oklahoma. Southwestern Naturalist 14:31–66.

Mahmoud, I. Y., and P. Licht. 1997. Seasonal changes in gonadal activity and the effects of stress on reproductive hormones in the Common Snapping Turtle, *Chelydra serpentina*. General and Comparative Endocrinology 107:359–372.

Mahmoud, I. Y., A. E. Colas, M. J. Woller, and R. V. Cyrus. 1986. Cytoplasmic progesterone receptors in uterine tissues of the Snapping Turtle (*Chelydra serpentina*). Journal of Endocrinology 109:385–392.

Mahmoud, I. Y., J. R. Paulson, M. Dudley, J. S. Patzlaff, and A. Y. A. Al-Kindi. 2004. Secretory proteins in the reproductive tract of the Snapping Turtle, *Chelydra serpentina*. Comparative Biochemistry and Physiology, Part A 139:487–494.

Maiorana, V. C. 1976. Size and environmental predictability for salamanders. Evolution 30:599–613.

Maisonneuve, C., and S. Rioux. 2001. Importance of riparian habitats for small mammal and herpetofaunal communities in agricultural landscapes of southern Quebec. Agriculture, Ecosystems and Environment 83:165–175.

Maizel, M., R. D. White, S. Gage, L. Osborne, R. Root, S. Stitt, and G. Muehlbach. 1998. Historical interrelationships between population settlement and farmland in the conterminous United States, 1790 to 1992. Pages 5–12 *in* T. D. Sisk, editor. Perspectives on the land use history of North America: a context for understanding our changing environment. US Geological Survey, Biological Resources Division Biological Science Report USGS/BRD/BSR-1998–0003, Flagstaff, Arizona, USA.

Mali, I., M. W. Vandewege, S. K. Davis, and M. R. J. Forstner. 2014. Magnitude of the freshwater turtle exports from the US: long term trends and early effects of newly implemented harvest management regimes. PLoS ONE 9:e86478.

Mallory, M. L., and R. Lariviere. 1998. Wood Duck, *Aix sponsa*, eats Mink Frogs, *Rana septentrionalis*. Canadian Field-Naturalist 112:714–715.

Manion, J. J. 1952. Comparative ecological studies on the amphibians of Cass County, Michigan, and vicinity. Dissertation, University of Notre Dame, South Bend, Indiana, USA.

Manning, G. J., C. J. Cole, H. C. Dessauer, and J. M. Walker. 2005. Hybridization between parthenogenetic lizards (*Aspidoscelis neomexicana*) and gonochoristic lizards (*Aspidoscelis sexlineata viridis*) in New Mexico: ecological, morphological, cytological, and molecular context. American Museum Novitates 3492:1–56.

Manning, G. J., J. M. Walker, and S. R. Goldberg. 2009. *Aspidoscelis tesselata* (Checkered Whiptail) x *Aspidoscelis sexlineata viridis* (Prairie Racerunner). Reproductive potential. Herpetological Review 40:340.

Maple, W. T., and L. P. Orr. 1968. Overwintering adaptations of *Sistrurus c. catenatus* in northeastern Ohio. Journal of Herpetology 2:179–180.

Marchand, M. N., and J. A. Litvaitis. 2004a. Effects of landscape composition, habitat features, and nest distribution on predation rates of simulated turtle nests. Biological Conservation 117:243–251.

Marchand, M. N., and J. A. Litvaitis. 2004b. Effects of habitat features and landscape composition on the population structure of a common aquatic turtle in a region undergoing rapid development. Conservation Biology 18:758–767.

Marchand, M. N., J. A. Litvaitis, T. J. Maier, and R. M. DeGraaf. 2002. Use of artificial nests to investigate predation of freshwater turtle nests. Wildlife Society Bulletin 30:1092–1098.

Marchisin, A., and J. D. Anderson. 1978. Strategies employed by frogs and toads (Amphibia, Anura) to avoid predation by snakes (Reptilia, Serpentes). Journal of Herpetology 12:151–155.

Marco, A., C. Quilchano, and A. R. Blaustein. 1999. Sensitivity to nitrate and nitrite in pond-breeding amphibians from the Pacific Northwest, USA. Environmental Toxicology and Chemistry 18:2836–2839.

Marcogliese, D. J., K. C. King, H. M. Salo, M. Fournier, P. Brousseau, P. Spear, L. Champoux, J. D. McLaughlin, and M. Boily. 2009. Combined effects of agricultural activity and parasites on biomarkers in the Bullfrog, *Rana catesbeiana*. Aquatic Toxicology 91:126–134.

Maret, T. J., J. D. Snyder, and J. P. Collins. 2006. Altered drying regime controls distribution of endangered salamanders and introduced predators. Biological Conservation 127:129–138.

Marino, J. A., Jr., and E. E. Werner. 2013. Synergistic effects of predators and trematode parasites on larval Green Frog (*Rana clamitans*) survival. Ecology 94:2697–2708.

Marion, Z. H., and M. E. Hay. 2011. Chemical defense of the Eastern Newt (*Notophthalmus viridescens*): variation in efficiency against different consumers and in different habitats. PLoS ONE 6:e27581.

Markle, C. E., S. D. Gillingwater, R. Levick, and P. Chow-Fraser. 2017. The true cost of partial fencing: evaluating strategies to reduce reptile road mortality. Wildlife Society Bulletin 41:342–250.

Marsh, D. M., and N. G. Beckman. 2004. Effects of forest roads on the abundance and activity of terrestrial salamanders. Ecological Applications 14:1882–1891.

Marsh, D. M., B. J. Cosentino, K. S. Jones, J. J. Apodaca, K. H. Beard, J. M. Bell, C. Bozarth, D. Carper, J. F. Charbonnier, A. Dantas, E. A. Forys, M. Foster, J. General, K. S. Genet, M. Hanneken, K. R. Hess, S. A. Hill, F. Iqbal, N. E. Karraker, E. S. Kilpatrick, T. A. Langren, J. Langford, K. Lauer, A. J. McCarthy, J. Neale, S. Patel, A. Patton, C. Southwick, N. Stearrett, N. Steijn, M. Tasleem, J. M. Taylor, and J. R. Vonesh. 2017. Effects of roads and land use on frog distributions across spatial scales and regions in the eastern and central United States. Diversity and Distributions 23:158–170.

Marsh, D. M., G. S. Milam, N. A. Gorham, and N. G. Beckman. 2005. Forest roads are partial barriers to salamander movement. Conservation Biology 19:2004–2008.

Marsh, D. M., K. A. Thakur, K. C. Bulka, and L. B. Clarke. 2004. Dispersal and colonization through open fields by a terrestrial, woodland salamander. Ecology 85:3396–3405.

Marsh, D. M., R. B. Page, T. J. Hanlon, R. Corritone, E. C. Little, D. E. Seifert, and P. R. Cabe. 2008. Effects of roads on patterns of genetic differentiation in Red-backed Salamanders, *Plethodon cinereus*. Conservation Genetics 9:603–613.

Marshall, J. C., Jr., J. V. Manning, and B. A. Kingsbury. 2006. Movement and macrohabitat selection of the Eastern Massasauga in a fen habitat. Herpetologica 62:141–150.

Marshall, W. H., and M. F. Buell. 1955. A study of the occurrence of amphibians in relation to a bog succession, Itasca State Park, Minnesota. Ecology 36:381–387.

Martel, A., A. S-v. der Sluijs, M. Blooi, W. Bert, R. Ducatelle, M. C. Fisher, A. Woeltjes, W. Bosman, K. Chiers, F. Bossuyt, and F. Pasmans. 2013. *Batrachochytrium salamandrivorans* sp. nov. causes lethal chytridiomycosis in amphibians. Proceedings of the National Academy of Sciences of the United States of America 110:15325–15329.

Martel, A., M. Blooi, C. Adriaensen, P. Van Rooij, W. Beukema, M. C. Fisher, R. A. Farrer, B. R. Schmidt, U. Tobler, K. Goka, K. R. Lips, C. Muletz, K. R. Zamudio, J. Bosch, S. Lötters, E. Wombwell, T. W. J. Garner, A. A. Cunningham, A. Spitzen-van der Sluijs, S. Salvidio, R. Ducatelle, K. Nishikawa, T. T. Nguyen, J. E. Kolby, I. Van Bocxlaer, F. Bossuyt, and F. Pasmans. 2014. Recent introduction of a chytrid fungus endangers western Palearctic salamanders. Science 346:630–631.

Martin, J. B., N. B. Witherspoon, and M. H. A. Keenleyside. 1974. Analysis of feeding behavior in the newt *Notophthalmus viridescens*. Canadian Journal of Zoology 52:277–281.

Martin, K. 1979. Common Garter Snake predation on Robin nestlings. Canadian Field-Naturalist 93:70–71.

Martin L. J., and B. Blossey. 2013. Intraspecific variation overrides origin effects in impacts of litter-derived secondary compounds on larval amphibians. Oecologia 173:449–459.

Martin, T. B., N. P. Bernstein, R. D. Birkhead, J. F. Koukl, S. M. Mussman, and J. S. Placyk, Jr. 2013. Sequence-based molecular phylogenetics and phylogeography of the American box turtles (*Terrapene* spp.) with support from DNA barcoding. Molecular Phylogenetics and Evolution 68:119–134.

Martin, W. H. 2002. Life history constraints on the Timber Rattlesnake (*Crotalus horridus*) at its climatic limits. Pages 285–306 *in* G. W. Schuett, M. Höggren, M. E. Douglas, and H. W. Greene, editors. Biology of the vipers. Eagle Mountain Publishing, LC, Eagle Mountain, Utah, USA.

Martin, W. H., W. H. Smith, S. H. Harwig, R. O. Magram, and R. Stechert. 1990. Distribution and status of the Timber Rattlesnake (*Crotalus horridus*) in Pennsylvania. Unpublished Report to the Carnegie Museum of Natural History and Pennsylvania Fish Commission, Pittsburgh, USA.

Martof, B. 1952. Early transformation of the Green Frog, *Rana clamitans* Latreille. Copeia 1952:115–116.

Martof, B. 1954. Variation in a large litter of Gartersnakes, *Thamnophis sirtalis sirtalis*. Copeia 1954:100–105.

Martof, B. S. 1953a. Home range and movements of the Green Frog, *Rana clamitans*. Ecology 34:529–543.

Martof, B. S. 1953b. Territoriality in the Green Frog, *Rana clamitans*. Ecology 34:165–174.

Martof, B. S. 1956a. Factors influencing the size and composition of populations of *Rana clamitans*. American Midland Naturalist 56:224–245.

Martof, B. S. 1956b. Growth and development of the Green Frog, *Rana clamitans*, under natural conditions. American Midland Naturalist 55:101–117.

Martof, B. S. 1970. *Rana sylvatica*. Catalogue of American Amphibians and Reptiles 86.1–86.4.

Martof, B. S., and R. L. Humphries. 1959. Geographic Variation in the Wood Frog *Rana sylvatica*. American Midland Naturalist 61:350–389.

Martof, B. S., W. M. Palmer, J. R. Bailey, J. R. Harrisson, III, and J. Dermid. 1980. Amphibians and reptiles of the Carolinas and Virginia. University of North Carolina Press, Chapel Hill, USA.

Maryland Natural Heritage Program. 2010. Rare, threatened and endangered animals of Maryland. Maryland Department of Natural Resources, Wildlife and Heritage Service, Annapolis, USA.

Mascaro, J., and S. A. Schnitzer. 2007. *Rhamnus cathartica* L. (Common Buckthorn) as an ecosystem dominant in southern Wisconsin forests. Northeastern Naturalist 14:387–402.

Mason, R. T., and D. Crews. 1985. Female mimicry in garter snakes. Nature 316:59–60.

Massey, A. 1990. Notes on the reproductive ecology of Red-spotted Newts (*Notophthalmus viridescens*). Journal of Herpetology 24:106–107.

Mathews, R. C., and A. C. Echternacht. 1984. Herpetofauna of the spruce-fir ecosystem in the southern Appalachian Mountain regions, with emphasis on the Great Smoky Mountains National Park. Pages 155–167 *in* P. S. White, editor. The southern Appalachian spruce-fir ecosystem: its biology and threats. National Park Service, Research/Resources Management Report SER-71, Atlanta, Georgia, USA.

Mathiak, H. A. 1966. Muskrat population studies at Horicon Marsh, Wisconsin. Wisconsin Conservation Department Technical Bulletin 36, Madison, USA.

Mathiak, H. A. 1979. A river survey of the unionid mussels of Wisconsin, 1973–1977. Sand Shell Press, Horicon, Wisconsin, USA.

Mathis, A. 1990. Territoriality in a terrestrial salamander: the influence of resource quality and body size. Behavior 112:162–175.

Mathis, A. 1991. Territories of male and female terrestrial salamanders: costs, benefits, and intersexual spatial associations. Oecologia 86:433–440.

Mathis, A. 2003. Use of chemical cues in detection of conspecific predators and prey by newts, *Notophthalmus viridescens.* Chemoecology 13:193–197.

Mathis, A., and F. Vincent. 2000. Differential use of visual and chemical cues in predator recognition and threat-sensitive predator-avoidance responses by larval newts (*Notophthalmus viridescens*). Canadian Journal of Zoology 78:1646–1652.

Matson, T. O. 1990a. A morphometric comparison of Gray Treefrogs, *Hyla chrysoscelis* and *H. versicolor,* from Ohio. Ohio Journal of Science 90:98–101.

Matson, T. O. 1990b. Erythrocyte size as a taxonomic character in the identification of Ohio *Hyla chrysoscelis* and *H. versicolor*. Herpetologica 46:457–462.

Matson, T. O. 1990c. Estimation of numbers for a riverine *Necturus* population before and after TFM lampricide exposure. Kirtlandia 43:33–45.

Matson, T. O. 1998. Evidence for home ranges in Mudpuppies and implications for impacts due to episodic applications of the lampricide TFM. Pages 278–287 *in* M. J. Lannoo, editor. Status and conservation of midwestern amphibians. University of Iowa Press, Iowa City, USA.

Matson, T. O. 2005. *Necturus maculosus* (Rafinesque, 1818) Mudpuppy. Pages 870–871 *in* M. J. Lannoo, editor. Amphibian declines: the conservation status of United States species. University of California Press, Berkeley, USA.

Matthews, C. E., C. E. Moorman, C. H. Greenberg, and T. A. Waldrop. 2010. Response of reptiles and amphibians to repeated fuel reduction treatments. Journal of Wildlife Management 74:1301–1310.

Matthews, T. C. 1971. Genetic changes in a population of Boreal Chorus Frogs (*Pseudacris triseriata*) polymorphic for color. American Midland Naturalist 85:208–221.

Maxson, L. R., P. E. Moler, and B. W. Mansell. 1988. Albumin evolution in salamanders of the genus *Necturus.* Journal of Herpetology 22:231–235.

May, D. D. 1952. Occurrence of an albino garter snake, *Thamnophis radix,* in Cook County, Illinois. Chicago Academy of Sciences Bulletin, Natural History Miscellanea 111:1–3.

Mayden, R. L. 1997. A hierarchy of species concepts: the denouement in the saga of the species problem. Pages 381–424 *in* M. F. Claridge, H. A. Dawah, and M. R. Wilson, editors. Species: the units of biodiversity. Chapman and Hall Ltd., London, England.

Maynard, E. A. 1934. The aquatic migration of the toad, *Bufo americanus.* Copeia 1934:174–177.

Mazerolle, M. J. 2004. Amphibian road mortality in response to nightly variations in traffic intensity. Herpetologica 60:45–53.

Mazerolle, M. J., and A. Desrochers. 2005. Landscape resistance to frog movements. Canadian Journal of Zoology 83:455–464.

Mazerolle, M. J., and M. Villard. 1999. Patch characteristics and landscape context as predictors of species presence and abundance: a review. Ecoscience 6:117–124.

Mazerolle, M. J., M. Huot, and M. Gravel. 2005. Behavior of amphibians on the road in response to car traffic. Herpetologica 61:380–388.

McAlister, W. H. 1963a. A post-breeding concentration of the Spring Peeper. Herpetologica 19:293.

McAlister, W. H. 1963b. Evidence of mild toxicity in the saliva of the Hognose Snake (*Heterodon*). Herpetologica 19:132–137.

McAllister, A. J. 1995. Wetland use by the Black Rat Snake, *Elaphe obsoleta*, in eastern Ontario. Canadian Field-Naturalist 109:449–451.

McAllister, C. A., S. E. Trauth, and C. R. Bursey. 1995. Parasites of the Pickerel Frog, *Rana palustris* (Anura: Ranidae), from the southern part of its range. Southwestern Naturalist 40:111–116.

McAlpine, D. F. 1997. Helminth communities in Bullfrogs (*Rana catesbieana*), Green Frogs (*Rana clamitans*), and Leopard Frogs (*Rana pipiens*) from New Brunswick, Canada. Canadian Journal of Zoology 75:1883–1890.

McAlpine, D. F., and D. B. Burt. 1998. Helminths of Bullfrogs, *Rana catesbeiana*, Green Frogs, *R. clamitans*, and Leopard Frogs, *R. pipiens* in New Brunswick. Canadian Field-Naturalist 112:50–68.

McAlpine, D. F., and T. G. Dilworth. 1989. Microhabitat and prey size among three species of *Rana* (Anura: Ranidae) sympatric in eastern Canada. Canadian Journal of Zoology 67:2244–2252.

McAlpine, D. F., N. M. Burgess, and D. G. Busby. 1998. Densities of Mink Frogs, *Rana septentrionalis*, in New Brunswick forest ponds sprayed with the insecticide fenitrothion. Bulletin of Environmental Contamination and Toxicology 60:30–36.

McAtee, W. L. 1921. Homing and other habits of the Bullfrog. Copeia 96:39–40.

McBride, M. P., B. Kimberlee, D. V. M. Wojick, T. A. Georoff, J. Kimbro, M. M. Garner, X. Wang, A. L. Childress, and J. F. X. Wellehan, Jr. 2015. *Ophidiomyces ophiodiicola* dermatitis in eight free-ranging Timber Rattlesnakes (*Crotalus horridus*) from Massachusetts. Journal of Zoo and Wildlife Medicine 46:86–94.

McCallum, M. L. 2010. Future climate change spells catastrophe for Blanchard's Cricket Frog, *Acris blanchardi* (Amphibia: Anura: Hylidae). Acta Herpetologica 5:119–130.

McCallum, M. L., and S. E. Trauth. 2003a. A forty-three year museum study of Northern Cricket Frog (*Acris crepitans*) abnormalities in Arkansas: upward trends and distributions. Journal of Wildlife Diseases 39:522–528.

McCallum, M. L., and S. E. Trauth. 2003b. *Acris crepitans* (Northern Cricket Frog). Communal hibernaculum. Herpetological Review 34:228.

McCallum, M. L., and S. E. Trauth. 2004. Blanchard's Cricket Frog in Nebraska and South Dakota. Prairie Naturalist 36:129–135.

McCallum, M. L., and S. E. Trauth. 2006. An evaluation of the subspecies *Acris crepitans blanchardi* (Anura, Hylidae). Zootaxa 1104:1–21.

McCallum, M. L., and S. E. Trauth. 2021. Complex associations of environmental factors may explain Blanchard's Cricket Frog, *Acris blanchardi* declines and drive population recovery. Journal of North American Herpetology 2021:26–49.

McCallum, M. L., C. Brooks, R. Mason, and S. E. Trauth. 2011. Growth, reproduction, and life span in Blanchard's Cricket Frog (*Acris blanchardi*) with notes on the growth of the Northern Cricket Frog (*Acris crepitans*). Herpetology Notes 4:25–35.

McCallum, M. L., S. E. Trauth, and R. G. Neal. 2006. Tail-coiling in ringneck snakes: flash display or decoy? Herpetological Natural History 10:91–94.

McClelland, B. E., and W. Wilczynski. 1989. Release call characteristics of male and female *Rana pipiens*. Copeia 1989:1045–1049.

McCollum, S. A. 1994. Ecological consequences of predator-induced polyphenism in larval hylid frogs. Dissertation, Duke University, Durham, North Carolina, USA.

McCollum, S. A., and J. Van Buskirk. 1996. Costs and benefits of a predator-induced polyphenism in the Gray Treefrog *Hyla chrysoscelis*. Evolution 50:583–593.

McConkey, E. H. 1954. A systematic study of the North American lizards of the genus *Ophisaurus*. American Midland Naturalist 51:133–171.

McCoy, C. J., and R. C. Vogt. 1990. *Graptemys geographica*. Catalogue of American Amphibians and Reptiles 484.1–484.4.

McCoy, K. A., and R. N. Harris. 2003. Integrating developmental stability analysis and current amphibian monitoring techniques: an experimental evaluation with the salamander *Ambystoma maculatum*. Herpetologica 59:22–36.

McCrary, J. E., and M. G. Heagler. 1997. The use of a simultaneous multiple species acute toxicity test to compare the relative sensitivities of aquatic organisms to mercury. Journal of Environmental Science and Health 32:73–81.

McCrudden, C. M., M. Zhou, T. Chen, M. O'Rourke, B. Walker, D. Hirst, and C. Shaw. 2007. The complex array of bradykinin-related peptides (BRPs) in the peptidome of Pickerel Frog (*Rana palustris*) skin secretion is the product of transcriptional economy. Peptides 28:1275–1281.

McCumber, E., and R. Hay 2002. Eastern Massasauga Rattlesnake status survey and telemetry study, lower Chippewa River, Buffalo Co. WI, 2001 interim report. Interim Report to the US Fish and Wildlife Service.

McCumber, E., and R. Hay 2003. Eastern Massasauga Rattlesnake status survey and telemetry study, lower Chippewa River, Buffalo Co. WI, 2002 interim report. Interim Report to the US Fish and Wildlife Service.

McDaniel, T. V., P. A. Martin, G. C. Barrett, K. Hughes, A. D. Gendron, L. Shirose, and C. A. Bishop. 2009. Relative abundance, age structure, and body size in Mudpuppy populations in southwestern Ontario. Journal of Great Lakes Research 35:182–189.

McDaniel, T. V., P. A. Martin, N. Ross, S. Brown, S. Lesage, and B. D. Pauli. 2004. Effects of chlorinated solvents on four species of North American amphibians. Archives of Environmental Contamination and Toxicology 47:101–109.

McDowell, S. B. 1964. Partition of the genus *Clemmys* and related problems in the taxonomy of the aquatic Testudinidae. Proceedings of the Zoological Society of London 143:239–278.

McFarland, C. G., and W. G. Reeder. 1974. Cleaning symbiosis involving Galapagos Tortoises and two species of Darwin's Finches. Zeitschrift fur Tierpsychologie 34:464–483.

McFarland, M., and R. Bishop. 1990. *Rana catesbeiana* (Bullfrog). Geographic distribution. Herpetological Review 21:22.

McGuire, J. M., K. T. Scribner, and J. D. Congdon. 2013. Spatial aspects of movements, mating patterns, and nest distributions influence gene flow among population subunits of Blanding's Turtles (*Emydoidea blandingii*). Conservation Genetics 14:1029–1042.

McKenney, T. L. 1827. Sketches of a tour of the lakes, of the character and costumes of the Chippeway Indians, and of incidents connected with the Treaty of Fond du Lac. Fielding Lucus, Jr., Baltimore, Maryland, USA.

McKenny, H. C., W. S. Keeton, and T. M. Donovan. 2006. Effects of structural complexity enhancement on Eastern Red-backed Salamander (*Plethodon cinereus*) populations in northern hardwood forests. Forest Ecology and Management 230:186–196.

McKinnell, R. G., and D. C. Dapkus. 1973. The distribution of burnsi and kandiyohi frogs in Minnesota and contiguous states. American Zoologist 13:81–84.

McKinstry, D. M. 1978. Evidence of toxic saliva in some colubrid snakes of the United States. Toxicon 16:523–534.

McLeod, D. S. 2005. Nebraska's declining amphibians. Pages 292–294 *in* M. Lannoo, editor. Amphibian declines: the conservation status of United States species. University of California Press, Berkeley, USA.

McLeod, R. F., and J. E. Gates. 1998. Response of herpetofaunal communities to forest cutting and burning at Chesapeake Farms, Maryland. American Midland Naturalist 139:164–177.

McMaster, N. L., and T. B. Herman. 2000. Occurrence, habitat selection, and movement patterns of juvenile Blanding's Turtles (*Emydoidea blandingii*) in Kejimkujik National Park, Nova Scotia. Chelonian Conservation and Biology 3:602–610.

McNab, W. H., D. T. Cleland, J. A. Freeouf, J. E. Keys, Jr., G. J. Nowacki, and C. A. Carpenter. 2007. Description of "ecological subregions: sections of the conterminous United States." US Forest Service General Technical Report WO-76B, Washington, D.C., USA.

McNeil, J. A., T. B. Herman, and K. L. Standing. 2000. Movement of hatchling Blanding's Turtles (*Emydoidea blandingii*) in Nova Scotia in response to proximity to open water: a manipulative experiment. Chelonian Conservation and Biology 3:611–617.

McVay, J. D. 2013. Molecular phylogenetics of the North American snake tribe Thamnophiini. Thesis, Louisiana State University, Baton Rouge, USA.

McVay, J. D., and B. Carstens. 2013. Testing monophyly without well-supported gene trees: evidence from multi-locus nuclear data conflicts with existing taxonomy in the snake tribe Thamnophiini. Molecular Phylogenetics and Evolution 68:425–431.

McVay, J. D., O. Flores-Villela, and B. Carstens. 2015. Diversification of North American natricine snakes. Biological Journal of the Linnean Society 116:1–12.

Mecham, J. S. 1954. Geographic variation in the Green Frog, *Rana clamitans* Latreille. Texas Journal of Science 6:1–24.

Mecham, J. S. 1960. Natural hybridization between the tree frogs *Hyla versicolor* and *Hyla avivoca*. Journal of the Elisha Mitchell Scientific Society 76:64–67.

Mecham, J. S. 1967. *Notophthalmus viridescens*. Catalogue of American Amphibians and Reptiles 53.1–53.4.

Medsger, O. P. 1922. Milk Snake and Red-bellied Snake. Copeia 1920:79–80.

Meeks, D. E., and J. W. Nagel. 1973. Reproduction and development of the Wood Frog, *Rana sylvatica*, in eastern Tennessee. Herpetologica 29:188–191.

Meeks, R. L., and G. R. Ultsch. 1990. Overwintering behavior of Snapping Turtles. Copeia 3:880–884.

Meine, C. 2010. Aldo Leopold: his life and work. University of Wisconsin Press, Madison, USA.

Menasco, D. J. 1975. Geographic distribution and variation of the *Hyla versicolor* complex in Wisconsin. Thesis, University of Wisconsin, Whitewater, USA.

Mendonça, M. T. 1987. Photothermal effects on the ovarian cycle of the Musk Turtle, *Sternotherus odoratus*. Herpetologica 43:82–90.

Mendonça, M. T., and P. Licht. 1986. Photothermal effects on the testicular cycle in the Musk Turtle, *Sternotherus odoratus*. Journal of Experimental Zoology 239:117–130.

Menzel, B. W., and K. E. Goellner. 1976. Occurrence of the Blue-spotted Salamander, *Ambystoma laterale*, in Iowa. Proceedings of the Iowa Academy of Science 82:182–186.

Merrell, D. J. 1968. A comparison of the estimated size and the "effective size" of breeding populations of the Leopard Frog, *Rana pipiens*. Evolution 22:274–283.

Merrell, D. J. 1977. Life history of the Leopard Frog, *Rana pipiens*, in Minnesota. Bell Museum of Natural History Occasional Papers 15:1–23.

Merrill, L. 2014. *Chelydra serpentina* (Snapping Turtle). Ritualized aggressive behavior. Herpetological Review 45:686.

Merritt, D. A., Jr. 1980. The Wood Turtle, *Clemmys insculpta* (LeConte): natural history, behavior, and food habits. Bulletin of the Chicago Herpetological Society 15:6–9.

Merrow, J. S., and T. Aubertin. 2005. *Crotalus horridus* (Timber Rattlesnake). Reproduction. Herpetological Review 36:192.

Mertig, G., C. Nelson, R. J. Magana, and R. A. Staffen. 2019. *Emydoidea blandingii* (Blanding's Turtle). Geographic distribution. Herpetological Review 50:99–100.

Meshaka, W. E., Jr. 1994. Clutch parameters of *Storeria dekayi* Holbrook (Serpentes: Colubridae) from southcentral Florida. Brimleyana 21:73–76.

Meshaka, W. E., Jr., and A. N. Klippel. 2011. Seasonal activity, reproduction, and growth of the Northern Red-bellied Snake, *Storeria occipitomaculata* (Storer, 1839), from Pennsylvania. Journal of Kansas Herpetology 37:16–20.

Meshaka, W. E., Jr., and P. R. Delis. 2014. Snake assemblage structures and seasonal activity patterns on a military base in south-central Pennsylvania: land management implications for snake conservation. Journal of North American Herpetology 2014:7–20.

Meshaka, W. E., Jr., N. Edwards, and P. R. Delis. 2012. Seasonal activity, reproductive cycles and growth of the Pickerel Frog *Lithobates palustris* (LeConte, 1825), from Pennsylvania. Herpetological Bulletin 119:1–8.

Meshaka, W. E., Jr., S. D. Marshall, and T. J Guiher. 2008. Seasonal activity and reproductive characteristics of an oldfield-grassland snake assemblage: implications for land management. Herpetological Bulletin 105:35–40.

Meshaka, W. E., Jr., S. D. Marshall, T. J. Guiher, and L. Zemba. 2009. Grassland snake assemblages in central and western Pennsylvania and northeastern Ohio, USA. Herpetological Bulletin 110:8–19.

Mesner, P. W., I. Y. Mahmoud, and R. V. Cyrus. 1993. Seasonal testosterone levels in Leydig and Sertoli cells of the Snapping Turtle (*Chelydra serpentina*) in natural populations. Journal of Experimental Zoology 266:266–276.

Messeling, E. 1953. Rattlesnakes in southwestern Wisconsin. Wisconsin Conservation Bulletin 18:21–23.

Metcalf, E., and A. L. Metcalf. 1970. Observations on Ornate Box Turtles (*Terrapene ornata ornata,* Agassiz). Transactions of the Kansas Academy of Science 73:96–117.

Metcalf, E. L., and A. L. Metcalf. 1979. Mortality in hibernating Ornate Box Turtles, *Terrapene ornata.* Herpetologica 35:93–96.

Metcalf, M. M. 1923. The opalinid ciliate infusorians. Bulletin of the United States National Museum 120:1–484.

Metts, B. S., W. A. Hopkins, and J. P. Nestor. 2005. Interaction of an insecticide with larval density in pond-breeding salamanders (*Ambystoma*). Freshwater Biology 50:685–696.

Meyer, D. T., E. J. Henderson, and Z. J. Loughman. 2019. Prey selection in the Queensnake, *Regina septemvittata* in streams of the northern panhandle of West Virginia, USA. Proceedings of the West Virginia Academy of Science 91: 94th Annual Meeting Abstracts.

Meyer, S. L. 1938. *Haptophrya virginiensis* nov. sp., a protozoan parasite of the Pickerel Frog, *Rana palustris* Le Conte. Anatomical Record 72:54–55.

Meyer, S. L. 1939. Description of *Haptophrya virginiensis,* a protozoan parasite of the Pickerel Frog, *Rana palustris* Le Conte. Journal of Parasitology 25:141–144.

Meylan, P. A. 1996. Book review: turtles of the United States and Canada (by C. H. Ernst, J. E. Lovich, and R. W. Barbour). Herpetological Review 27:41–42.

Meylan, P. A. 2001. Testudines. Turtles, tortoises and terrapins. Version 31 May 2001. The Tree of Life Web Project. http://tolweb.org/Testudines. Accessed 4 Dec 2018.

Michell, K., and R. G. Michell. 2015. Use of radio-telemetry and recapture to determine the success of head-started Wood Turtles (*Glyptemys insculpta*) in New York. Herpetological Conservation and Biology 10:525–534.

Middleton, J., and J. Y. Chu. 2004. Population viability analysis (PVA) of the Eastern Massasauga Rattlesnake, *Sistrurus catenatus catenatus,* in Georgian Bay Islands National Park and elsewhere in Canada. Parks Canada, Eastern Massasauga Rattlesnake Species Recovery Team, Gatineau, Quebec, Canada.

Mierzwa, K. S. 1985. Robert Kennicott's letters to Spencer F. Baird, 1853–1865. Bulletin of the Chicago Herpetological Society 20:61–81.

Mierzwa, K. S. 1998. Status of northeastern Illinois amphibians. Pages 115–124 *in* M. J. Lannoo, editor. Status and conservation of midwestern amphibians. University of Iowa Press, Iowa City, USA.

Mifsud, D. A., and J. C. Thomas. 2013. Persistence of herpetofauna in the urbanized Rouge River ecosystem. Open Journal of Ecology 3:234–241.

Mikaelian, I., M. Ouellet, B. Pauli, J. Rodrigue, J. C. Harshbarger, and D. M. Green. 2000. *Ichthyophonus*-like infection in wild amphibians from Québec, Canada. Disease of Aquatic Organisms 40:195–201.

Millar, C. S., and G. Blouin-Demers. 2011. Spatial ecology and seasonal activity of Blanding's Turtles (*Emydoidea blandingii*) in Ontario, Canada. Journal of Herpetology 45:370–378.

Millar, C. S., and G. Blouin-Demers. 2012. Habitat suitability modelling for species at risk is sensitive to algorithm and scale: a case study of Blanding's Turtle, *Emydoidea blandingii,* in Ontario, Canada. Journal for Nature Conservation 20:18–29.

Miller, D., M. Gray, and A. Storfer. 2011. Ecopathology of ranaviruses infecting amphibians. Viruses 3:2351–2373.

Miller, N. 1909a. The American Toad (*Bufo lentiginosus americanus,* LeConte). A study in dynamic biology. American Naturalist 43:641–688.

Miller, N. 1909b. The American Toad (*Bufo americanus,* LeConte). II. A study in dynamic biology. American Naturalist 43:730–745.

Miller, R., and G. Grall. 1978. Reproductive data on *Lampropeltis triangulum temporalis* from Maryland. Bulletin of the Maryland Herpetological Society 14:36–38.

Millis, B. S., and A. F. Badje. 2019. *Pantherophis vulpinus* (Eastern Foxsnake). Geographic distribution. Herpetological Review 50:530–531.

Milner, J. W. 1874. Report of the fisheries of the Great Lakes; the results of inquiries prosecuted in 1871 and 1872. Appendix A. Report to US Commissioner of Fish and Fisheries, Washington, D.C., USA.

Minnesota Department of Natural Resources. 2014. Lined Snake. Archived by WebCite at http://www.webcitation.org/73cKkh4g7. Accessed 1 Nov 2018.

Minnesota Department of Natural Resources. 2016. Minnesota's wildlife action plan 2015–2025. Minnesota Department of Natural Resources, Division of Ecological and Water Resources, St. Paul, USA.

Minton, S. A., Jr. 1968. The fate of amphibians and reptiles in a suburban area. Journal of Herpetology 2:113–116.

Minton, S. A., Jr. 1972. Amphibians and reptiles of Indiana. Indiana Academy of Sciences Monograph 3:1–346.

Minton, S. A., Jr. 1998. Observations on Indiana amphibian populations: a forty-five-year overview. Pages 217–220 *in* M. Lannoo, editor. Status and conservation of midwestern amphibians. University of Iowa Press, Iowa City, USA.

Minton, S. A., Jr. 2001. Amphibians and reptiles of Indiana. Second edition. Indiana Academy of Science, Indianapolis, USA.

Minton, S. A., Jr., J. C. List, and M. J. Lodato. 1982. Recent records and status of amphibians and reptiles in Indiana. Proceedings of the Indiana Academy of Science 92:489–498.

Mirk, W. 1997. An introduction to the tallgrass prairie of the Upper Midwest: its history, ecology, preservation, and reconstruction. The Prairie Enthusiasts, Boscobel, Wisconsin, USA.

Mitchell, J., and W. Gibbons. 2010. Salamanders of the southeast. University of Georgia Press, Athens, USA.

Mitchell, J. C. 1985. Female reproductive cycle and life history attributes in a Virginia population of Stinkpot Turtles, *Sternotherus odoratus*. Copeia 1985:941–949.

Mitchell, J. C. 1988. Population ecology and life histories of the freshwater turtles *Chrysemys picta* and *Sternotherus odoratus* in an urban lake. Herpetological Monographs 2:40–61.

Mitchell, J. C. 1994. The reptiles of Virginia. Smithsonian Institution Press, Washington, D.C., USA.

Mitchell, J. C., and J. D. Groves. 1993. Intraspecific oophagy in reptiles. Herpetological Review 24:126–130.

Mitchell, J. C., and R. A. Beck. 1992. Free-ranging domestic cat predation on native vertebrates in rural and urban Virginia. Virginia Journal of Science 43:197–207.

Mitchell, J. C., and R. F. Mueller. 1996. *Clemmys insculpta* (Wood Turtle). Reproductive behavior. Herpetological Review 27:197.

Mitchell, J. C., and W. S. Woolcott. 1985. Observations of the microdistribution diet and predator-prey size relationships in the salamander *Plethodon cinereus* from the Virginia Piedmont. Virginia Journal of Science 36:281–288.

Mitchell, J. C., R. E. Jung, and B. Bartholomew, editors. 2008. Urban Herpetology. Society for the Study of Amphibians and Reptiles, Herpetological Conservation 3, Salt Lake City, Utah, USA.

Mockford, S. W., T. B. Herman, M. Snyder, and J. M. Wright. 2007. Conservation genetics of Blanding's Turtle and its application in the identification of evolutionarily significant units. Conservation Genetics 8:209–219.

Moe, K., and A. F. Badje. 2017. *Apalone spinifera* (Spiny Softshell). Geographic distribution. Herpetological Review 48:386.

Moldowan, P. D., M. G. Keevil, P. B. Mills, R. J. Brooks, and J. D. Litzgus. 2015. Diet and feeding behavior of Snapping Turtles (*Chelydra serpentina*) and Midland Painted Turtles (*Chrysemys picta marginata*) in Algonquin Provincial Park, Ontario. Canadian Field-Naturalist 129:403–408.

Moll, D. 1977. Ecological investigations of turtles in a polluted ecosystem: the central Illinois River and adjacent flood plain lakes. Dissertation, Illinois State University, Normal, USA.

Moll, D. L. 1980. Dirty river turtles. Natural History 89:42–49.

Moll, E. O. 1973. Latitudinal and intersubspecific variation in reproduction of the Painted Turtle, *Chrysemys picta*. Herpetologica 29:307–318.

Moll, E. O., and D. Moll. 2000. Conservation of river turtles. Pages 126–155 *in* M. W. Klemens, editor. Turtle conservation. Smithsonian Institution Press, Washington, D.C., USA.

Montieth, K. E., and P. W. C. Paton. 2006. Emigration behavior of Spotted Salamanders on golf courses in southern Rhode Island. Journal of Herpetology 40:195–205.

Mooi, R. D., J. P. Wiens, and G. S. Casper. 2011. Extreme color variation within populations of the Common Gartersnake, *Thamnophis sirtalis*, in central North America, with implications for subspecies status. Copeia 2011:187–200.

Moore, A. L., C. E. Williams, T. H. Martin, and W. J. Moriarty. 2001. Influence of season, geomorphic surface, and cover item on capture, size, and weight of *Desmognathus ochrophaeus* and *Plethodon cinereus* in Allegheny Plateau riparian forests. American Midland Naturalist 145:39–45.

Moore, B. 1993. 1992 Ornate Box Turtle project: results and recommendations. Wisconsin Endangered Resources Report 96, Madison, USA.

Moore, H. F., and T. E. B. Pope. 1910. Oyster culture experiments and investigations in Louisiana. Department of Commerce and Labor, Bureau of Fisheries Document 731, Washington, D.C., USA.

Moore, J-D., and M. Ouellet. 2014. A review of colour phenotypes of the Eastern Red-backed Salamander, *Plethodon cinereus*, in North America. Canadian Field-Naturalist 128:250–259.

Moore, J. A. 1939. Temperature tolerance and rates of development in the eggs of Amphibia. Ecology 20:459–478.

Moore, J. A. 1940. Adaptive differences in the egg membranes of frogs. American Midland Naturalist 74:89–93.

Moore, J. A. 1941. Developmental rate of hybrid frogs. Journal of Experimental Zoology 86:405–422.

Moore, J. A. 1942. Embryonic temperature tolerance and rate of development in *Rana catesbeiana*. Biological Bulletin 83:375–388.

Moore, J. A. 1946. Hybridization between *Rana palustris* and different geographical forms of *Rana pipiens*. Proceedings of the National Academy of Sciences of the United States of America 32:209–212.

Moore, J. A. 1952. An analytical study of the geographic distribution of *Rana septentrionalis*. American Naturalist 86:5–22.

Moore, J. A. 1955. Abnormal combinations of nuclear and cytoplasmic systems in frogs and toads. Advances in Genetics 7:139–182.

Moore, J. A., and J. C. Gillingham. 2006. Spatial ecology and multi-scale habitat selection by a threatened rattlesnake: the Eastern Massasauga (*Sistrurus catenatus catenatus*). Copeia 2006:742–751.

Moore, J. E., and E. H. Strickland. 1954. Notes on the food of three species of Alberta amphibians. American Midland Naturalist 52:221–224.

Moore, J. E., and E. H. Strickland. 1955. Further notes on the food of Alberta amphibians. American Midland Naturalist 54:253–256.

Moore, M. K., and V. R. Townsend, Jr. 1998. The interaction of temperature, dissolved oxygen and predation pressure in an aquatic predator-prey system. Oikos 81:329–336.

Moosman, D. L., and P. R. Moosman, Jr. 2005. *Bufo americanus* (American Toad) and *Rana catesbeiana* (Bullfrog). Microhabitat. Herpetological Review 36:298.

Morgan, A. H., and M. C. Grierson. 1932. Winter habits and yearly food consumption of adult Spotted Newts, *Triturus viridescens*. Ecology 13:54–62.

Morgan, S. K., M. W. Pugh, M. M. Gangloff, and L. Siefferman. 2014. The spots of the Spotted Salamander are sexually dimorphic. Copeia 2014:251–256.

Moriarty E., and M. J. Lannoo. 2005. *Pseudacris triseriata* complex (including *feriarum*, *kalmi*, *triseriata*, and *maculata*) striped (upland, New Jersey, western, boreal) chorus frog. Pages 485–488 *in* M. Lannoo, editor. Amphibian declines: the conservation status of United States species. University of California Press, Berkeley, California, USA.

Moriarty, E. C., and D. C. Cannatella. 2004. Phylogenetic relationships of the North American chorus frogs (*Pseudacris*: Hylidae). Molecular Phylogenetics and Evolution 30:409–420.

Moriarty, J. J. 1998. Status of amphibians in Minnesota. Pages 166–168 *in* M. Lannoo, editor. Status and conservation of midwestern amphibians. University of Iowa Press, Iowa City, USA.

Moriarty, J. J., and C. D. Hall. 2014. Amphibians and reptiles in Minnesota. University of Minnesota Press, Minneapolis, USA.

Moriarty, J. J., and M. Linck. 1997. Reintroduction of Bullsnakes into a recreated prairie. Pages 43–52 *in* J. J. Moriarty and D. Jones, editors. Minnesota's amphibians and reptiles, their conservation and status: proceedings of a symposium. Special Publication of the Minnesota Herpetological Society, Lanesboro, USA.

Moriarty Lemmon, E., A. R. Lemmon, and D. C. Cannatella. 2007a. Geological and climatic forces driving speciation in the continentally distributed trilling chorus frogs (*Pseudacris*). Evolution 61:2086–2103.

Moriarty Lemmon, E., A. R. Lemmon, J. T. Collins, and D. C. Cannatella. 2008. A new North American chorus frog species (Amphibia: Hylidae: *Pseudacris*) from the south-central United States. Zootaxa 1675:1–30.

Moriarty Lemmon, E., A. R. Lemmon, J. T. Collins, J. A. Lee-Yaw, and D. C. Cannatella. 2007b. Phylogeny-based delimitation of species boundaries and contact zones in the trilling chorus frogs (*Pseudacris*). Molecular Phylogenetics and Evolution 44:1068–1082.

Morin, P. J. 1983a. Competitive and predatory interactions in natural and experimental populations of *Notophthalmus viridescens dorsalis* and *Ambystoma tigrinum*. Copeia 1983:628–639.

Morin, P. J. 1983b. Predation, competition, and the composition of larval anuran guilds. Ecological Monographs 53:119–138.

Morin, P. J. 1986. Interactions between intraspecific competition and predation in an amphibian predator-prey system. Ecology 67:713–720.

Morin, P. J. 1987. Predation, breeding asynchrony, and the outcome of competition among treefrog tadpoles. Ecology 68:675–683.

Morin, P. J., H. M. Wilbur, and R. N. Harris. 1983. Salamander predation and the structure of experimental communities: responses of *Notophthalmus* and microcrustacea. Ecology 64:1430–1436.

Morris, K., and T. J. Maret. 2007. Effects of timber management on pond-breeding salamanders. Journal of Wildlife Management 71:1034–1041.

Morris, M. A. 1974. Observations on a large litter of the snake *Storeria dekayi*. Transactions of the Illinois State Academy of Science 67:359–360.

Morris, M. A. 1975. Variation in a population of *Thamnophis radix*. Bulletin of the Chicago Herpetological Society 10:14–16.

Morris, M. A. 1976. A case of melanism in *Thamnophis radix*. Transactions of the Illinois Academy of Sciences 69:119.

Morris, M. A., and S. M. Meyer. 1980. Longevity records for captive *Bufo americanus* and *Uta stansburiana*. Bulletin of the Chicago Herpetological Society 15:79–80.

Morris, M. R., and S. L. Yoon. 1989. A mechanism for female choice of large males in the treefrog *Hyla chrysoscelis*. Behavioral Ecology and Sociobiology 25:65–71.

Morrison, M. L., B. G. Marcot, and R. W. Mannan. 2006. Wildlife-habitat relationships: concepts and applications. Third edition. Island Press, Washington, D.C., USA.

Mosimann, J. E., and J. R. Bider. 1960. Variation, sexual dimorphism, and maturity in a Quebec population of the Common Snapping Turtle, *Chelydra serpentina*. Canadian Journal of Zoology 38:19–38.

Moss, S. T., G. Bogden, R. C. Dart, S. P. Norek, S. R. Williams, and R. F. Clark. 1997. Association of rattlesnake bite location with severity of clinical manifestations. Annals of Emergency Medicine 30:58–61.

Mossman, M. J., and R. L. Hine. 1984. The Wisconsin frog and toad survey: establishing a long-term monitoring program. Wisconsin Department of Natural Resources Endangered Resources Report 9, Madison, USA.

Mossman, M. J., and R. L. Hine. 1985. Wisconsin's frog and toad survey, 1984. Wisconsin Department of Natural Resources Endangered Resources Report 16, Madison, USA.

Mossman, M. J., L. M. Hartman, R. Hay, J. R. Sauer, and B. J. Dhuey. 1998. Monitoring long-term trends in Wisconsin frog and toad populations. Pages 169–198 *in* M. J. Lannoo, editor. Status and Conservation of Midwestern Amphibians. University of Iowa Press, Iowa City, USA.

Most, M. G. 2013. Activity patterns and spatial resource selection of the Eastern Garter Snake (*Thamnophis sirtalis sirtalis*). Thesis, Loyola University, Chicago, USA.

Mothes, C. C., H. J. Howell, and C. A. Searcy. 2020. Habitat suitability models for the imperiled Wood Turtle (*Glyptemys insculpta*) raise concerns for the species' persistence under future climate change. Global Ecology and Conservation 24:e01247.

Mount, R. H. 1975. The reptiles and amphibians of Alabama. Auburn University Agricultural Experiment Station, Auburn, Alabama, USA.

[MPM] Milwaukee Public Museum. 1925. Yearbook of the Milwaukee Public Museum, Volumes 5–7. Stevens Printing Company, Inc., Milwaukee, WI, USA.

[MPM] Milwaukee Public Museum. 2017. Invertebrate zoology collections overview. Archived by WebCite at http://www.webcitation.org/73cLtIaw3. Accessed 1 Nov 2018.

Muehlfeld, T., and J. M. Kapfer. 2009. *Apalone mutica mutica* (Midland Smooth Softshell). Geographic distribution. Herpetological Review 40:235.

Muehlfeld, T., M. Muehlfeld, and J. M. Kapfer. 2008. *Ambystoma laterale* (Blue-spotted Salamander). Geographic distribution. Herpetological Review 39:476.

Mueller, D. K., and D. R. Helsel. 1999. Nutrients in the nation's waters: too much of a good thing? US Geological Survey Circular 1136, Denver, Colorado, USA. Archived by WebCite at http://www.webcitation.org/73cM8kld0. Accessed 1 Nov 2018.

Mueller, G. A., J. Carpenter, and D. Thornbrugh. 2006. Bullfrog tadpole (*Rana catesbeiana*) and Red Swamp Crayfish (*Procambarus clarkii*) predation on early life stages of endangered Razorback Sucker (*Xyrauchen texanus*). Southwestern Naturalist 51:258–261.

Mueller, N. S. 1980. Mallards capture and eat American Toads. Wilson Bulletin 92:523–524.

Mui, A. B., C. B. Edge, J. E. Paterson, B. Caverhill, B. Johnson, J. D. Litzgus, and Y. He. 2016. Nesting sites in agricultural landscapes may reduce the reproductive success of populations of Blanding's Turtles (*Emydoidea blandingii*). Canadian Journal of Zoology 94:61–67.

Muir, J. 1901. Our national parks. Houghton Mifflin Company, Boston, Massachusetts, USA.

Muir, J. 1913. The story of my boyhood and youth. Houghton Mifflin Company, Boston, Massachusetts, USA.

Muir, J. H. 1984. Commercial exploitation of *Graptemys* and *Sternotherus* turtles. Bulletin of the Chicago Herpetological Society 19:98–100.

Mulcare, D. J. 1966. The problem of toxicity in *Rana palustris*. Indiana Academy of Science 75:319–324.

Muletz, C., N. M. Caruso, R. C. Fleischer, R. W. McDiarmid, and K. R. Lips. 2014. Unexpected rarity of the pathogen *Batrachochytrium dendrobatidis* in Appalachian *Plethodon* salamanders: 1957–2011. PLoS ONE 9:e103728.

Muller, J. F. 1921. Notes on the habits of the Soft-shell Turtle *Amyda mutica*. American Midland Naturalist 7:180–184.

Mullhouse, J. M., and S. M. Galatowitsch. 2003. Revegetation of prairie pothole wetlands in the mid-continental US: twelve years post-reflooding. Plant Ecology 169:143–159.

Mullin, S. J. 1996. Adaptations facilitating facultative oophagy in the Gray Rat Snake, *Elaphe obsoleta spiloides*. Amphibia-Reptilia 17:387–393.

Mullin, S. J., and R. J. Cooper. 2000. The foraging ecology of the Gray Rat Snake (*Elaphe obsoleta spiloides*). II. Influence of habitat structural complexity when searching for arboreal avian prey. Amphibia-Reptilia 21:211–222.

Mullin, S. J., J. B. Towey, and R. E. Szafoni. 2004. Using Rotenone to enhance native amphibian breeding habitat in ponds (Illinois). Ecological Restoration 22:305–306.

Muñoz, D. J., K. M. Hesed, E. H. C. Grant, and D. A. W. Miller. 2016. Evaluating within-population variability in behavior and demography for the adaptive potential of a dispersal-limited species to climate change. Ecology and Evolution 6:8740–8755.

Munz, P. A. 1920. A study of the food habits of the Ithacan species of Anura during transformation. Journal of Entomology and Zoology 12:33–56.

Murphy, F. A., K. Tucker, and D. A. Fadool. 2001. Sexual dimorphism and developmental expression of signal-transduction machinery in the vomeronasal organ. Journal of Comparative Neurology 432:61–74.

Murphy, J. C., and R. M. Curry. 2000a. A case of parthenogenesis in the Plains Gartersnake, *Thamnophis radix*. Bulletin of the Chicago Herpetological Society 35:17–19.

Murphy, J. C., and R. M. Curry. 2000b. Erratum: a case of parthenogenesis in the Plains Gartersnake, *Thamnophis radix*. Bulletin of the Chicago Herpetological Society 35:49.

Murphy, T. D. 1961. Predation on eggs of the salamander, *Ambystoma maculatum*, by caddisfly larvae. Copeia 1961:495–496.

Murphy, T. D. 1963. Amphibian populations and movements at a small semi-permanent pond in Orange County, North Carolina. Dissertation, Duke University, Durham, North Carolina, USA.

Murphy, T. D. 1965. High incidence of two parasitic infestations and morphological abnormalities in a population of the frog, *Rana palustris* Le Conte. American Midland Naturalist 74:233–239.

Murphy, T. D. 1968. *Hyla versicolor* uses bird house for shelter. Herpetologica 24:78–79.

Mushinsky, H. R. 1979. Mating behavior of the Common Water Snake, *Nerodia sipedon sipedon*, in eastern Pennsylvania (Reptilia, Serpentes, Colubridae). Journal of Herpetology 13:127–129.

Mushinsky, H. R. 1985. Fire and the Florida sandhill herpetofaunal community: with special attention to responses of *Cnemidophorus sexlineatus*. Herpetologica 41:333–342.

Mushinsky, H. R. 1992. Natural history and abundance of southeastern Five-lined Skinks, *Eumeces inexpectatus*, on a periodically burned sandhill in Florida. Herpetologica 48:307–312.

Muths, E., R. D. Scherer, S. M. Amburgey, T. Mathews, A. W. Spencer, and P. S. Corn. 2016. First estimates of the probability of survival in a small-bodied, high-elevation frog (Boreal Chorus Frog, *Pseudacris maculata*), or how historical data can be useful. 2016. Canadian Journal of Zoology 94:599–606.

Muths, E., T. L. Johnson, and P. S. Corn. 2001. Experimental repatriation of Boreal Toad (*Bufo boreas*) eggs, metamorphs, and adults in Rocky Mountain National Park. Southwestern Naturalist 46:106–113.

Muzzall, P. M. 1991. Helminth infracommunities of the newt, *Notophthalmus viridescens*, from Turkey Marsh, Michigan. Journal of Parasitology 77:87–91.

Muzzall, P. M. 2005. Parasites of amphibians and reptiles from Michigan: a review of the literature 1916–2003. Michigan Department of Natural Resources, Fisheries Research Report 2077, Ann Arbor, USA. Archived by WebCite at http://www.webcitation.org/73g0rJceP. Accessed 4 Nov 2018.

Muzzall, P. M., and C. R. Peebles. 1991. Helminths of the Wood Frog, *Rana sylvatica*, and Spring Peeper, *Pseudacris c. crucifer*, from southern Michigan. Journal of the Helminthological Society of Washington 58:263–265.

Muzzall, P. M., M. G. Gillilland, III, C. S. Summer, and C. J. Mehne. 2001. Helminth communities of Green Frogs *Rana clamitans* Latreille, from southwestern Michigan. Journal of Parasitology 87:962–968.

Myers, C. W. 1965. Biology of the Ringneck Snake, *Diadophis punctatus*, in Florida. Bulletin of the Florida State Museum of Biological Science 10:43–90.

Myers, E. M., and K. R. Zamudio. 2004. Multiple paternity in an aggregate breeding amphibian: the effect of reproductive skew on estimates of male reproductive success. Molecular Ecology 13:1951–1963.

Nagel, J. W. 1977. Life history of the Red-backed Salamander, *Plethodon cinereus*, in northeastern Tennessee. Herpetological 33:13–18.

Nagle, R. D., and J. D. Congdon. 2016. Reproductive ecology of *Graptemys geographica* of the Juniata River in central Pennsylvania, with recommendations for conservation. Herpetological Conservation and Biology 11:232–243.

Nagle, R. D., C. L. Lutz, and A. I. Pyle. 2004. Overwintering in the nest by hatchling Map Turtles (*Graptemys geographica*). Canadian Journal of Zoology 82:1211–1218.

Nagy, Z. T., R. Lawson, U. Joger, and M. Wink. 2004. Molecular systematics of racers, whipsnakes and relatives (Reptilia: Colubridae) using mitochondrial and nuclear markers. Journal of Zoological Systematics and Evolutionary Research 42:223–233.

Najarian, H. H. 1955. Trematodes parasitic in the Salientia in the vicinity of Ann Arbor, Michigan. American Midland Naturalist 53:195–197.

NatureServe. 2020. NatureServe Explorer: an online encyclopedia of life. NatureServe, Arlington, Virginia. http://explorer.natureserve.org. Accessed 11 Nov 2020.

Navarro-Martín, L., C. Lanctôt, P. Jackman, B. J. Park, K. Doe, B. D. Pauli, and V. L. Trudeau. 2014. Effects of glyphosate-based herbicides on survival, development, growth and sex ratios of Wood Frog (*Lithobates sylvaticus*) tadpoles. I: chronic laboratory exposures to VisionMax. Aquatic Toxicology 154:278–290.

Necker, W. L. 1939a. Buff green-snakes. Chicago Naturalist 2:27.

Necker, W. L. 1939b. Records of the amphibians and reptiles of the Chicago region. Bulletin of the Chicago Academy of Sciences 6:1–10.

Nedland, T. S., A. Wolf, and T. Reed. 2007. A reexamination of restored wetlands in Manitowoc County, Wisconsin. Wetlands 27:999–1015.

Neill, W. T. 1948. Hibernation of amphibians and reptiles in Richmond County, Georgia. Herpetologica 4:107–114.

Neill, W. T. 1955. Posture of chilled newts (*Diemyctylus viridescens louisianensis*). Copeia 1955:61.

Neill, W. T. 1963a. A new subspecies of the Queen Snake, *Natrix septemvittata*, from southern Alabama. Herpetologica 19:1–9.

Neill, W. T. 1963b. *Hemidactylium*. Catalogue of American Amphibians and Reptiles 1.1.

Neill, W. T. 1963c. *Hemidactylium scutatum*. Catalogue of American Amphibians and Reptiles 2.1–2.2.

Nelson, T. C. 1915. *Rana palustris* in Wisconsin. Copeia 19:13–14.

Nelson, W. F. 1963. Natural history of the Northern Prairie Skink, *Eumeces septentrionalis septentrionalis* (Baird). Dissertation, University of Minnesota, Minneapolis, USA.

Nelson, W. F. 1969. Notes on parturition and brood sizes in *Storeria occipitomaculata*. Journal of the Tennessee Academy of Science 44:20–21.

Neumann-Lee, L. A., and F. J. Janzen. 2011. Atrazine exposure impacts behavior and survivorship of neonatal turtles. Herpetologica 67:23–31.

Neumeyer, C. H., and S. C. Verjinski. 2012. *Lithobates sylvaticus* (Wood Frog). Geographic distribution. Herpetological Review 43:441.

Neves, R. J., A. E. Bogan, J. D. Williams, S. A. Ahlstedt, and P. W. Hartfield. 1997. Status of aquatic mollusks in the southeastern United States: a downward spiral of diversity. Pages 43–86 *in* G. W. Benz, editor. Aquatic fauna in peril: the southeastern perspective. Southeast Aquatic Research Institute Special Publication 1, Decatur, Georgia, USA.

Nevo, E. 1973. Adaptive color polymorphism in Cricket Frogs. Evolution 27:353–367.

Newcomb Homan, R., J. M. Reed, and B. S. Windmiller. 2003. Analysis of Spotted Salamander (*Ambystoma maculatum*) growth rates based on bone growth rings. Journal of Herpetology 37:617–621.

Newcomber, R. T., D. H. Taylor, and S. I. Guttman. 1974. Celestial orientation in two species of water snakes (*Natrix sipedon* and *Regina septemvittata*). Herpetologica 30:194–200.

Ng, M. V., and H. M. Wilbur. 1995. The cost of brooding in *Plethodon cinereus*. Herpetologica 51:1–8.

Niederberger, A. J., and M. E. Seidel. 1999. Ecology and status of a Wood Turtle (*Clemmys insculpta*) population in West Virginia. Chelonian Conservation and Biology 3:414–418.

Nigrelli, R. F. 1954. Some longevity records of vertebrates. Transactions of the New York Academy of Sciences, Series II 16:296–299.

[NOAA] National Oceanic and Atmospheric Administration. 2015. Fisheries historical highlights: 1940s. Archived by WebCite at http://www.webcitation.org/73cMWafbD. Accessed 1 Nov 2018.

Noble, G. K. 1926. The Long Island Newt: a contribution to the life history of *Triturus viridescens*. American Museum Novitates 228:1–11.

Noble, G. K. 1937. The sense organs involved in the courtship of *Storeria*, *Thamnophis* and other snakes. Bulletin of the American Museum of Natural History 73:673–725.

Noble, G. K., and E. J. Farris. 1929. The method of sex recognition in the Wood Frog, *Rana sylvatica* LeConte. American Museum of Natural History 363:1–17.

Noble, G. K., and E. R. Mason. 1933. Experiments on the brooding habits of the lizards *Eumeces* and *Ophisaurus*. American Museum Novitates 619:1–29.

Noble, G. K., and H. J. Clausen. 1936. The aggregation behavior of *Storeria dekayi* and other snakes with especial reference to the sense organs involved. Ecological Monographs 6:269–316.

Noble, G. K., and H. T. Bradley. 1933. The mating behavior of lizards; its bearing on the theory of sexual selection. Annals of the New York Academy of Sciences 35:25–100.

Noble, G. K., and L. R. Aronson. 1942. The sexual behavior of Anura. 1. The normal mating pattern of *Rana pipiens*. Bulletin of the American Museum of Natural History 80:127–142.

Noland, R., and G. R. Ultsch. 1981. The roles of temperature and dissolved oxygen in microhabitat selection by the tadpoles of a frog (*Rana pipiens*) and a toad (*Bufo terrestris*). Copeia 1981:645–652.

Northcutt, R. G., K. Brändle, and B. Fritzsch. 1995. Electroreceptors and mechanosensory lateral line organs arise from single placodes in Axolotls. Developmental Biology 168:358–373.

Nosil, P. 2008. Speciation with gene flow could be common. Molecular Ecology 17:2103–2106.

Notaro, M., J. W. Williams, and D. Lorenz. 2012. Vegetation and land carbon projections for Wisconsin, USA, in the 21st century. Climate Research 54:149–165.

Nussbaum, S. E., B. H. Ousterhout, and R. D. Semlitsch. 2016. Agonistic behavior and resource defense among sympatric juvenile pond-breeding salamanders. Journal of Herpetology 50:388–393.

Nyman, S. 1986. Mass mortality in larval *Rana sylvatica* attributable to the bacterium, *Aeromonas hydrophila*. Journal of Herpetology 20:196–201.

Nyman, S. 1991. Ecological aspects of syntopic larvae of *Ambystoma maculatum* and the *A.laterale-jeffersonianum* complex in two New Jersey ponds. Journal of Herpetology 25:505–509.

O'Brien, T. K. 2014. The importance of pre- and postnatal thermal conditions in determining growth trajectories in three viviparous grassland snakes. Thesis, Northern Illinois University, DeKalb, USA.

O'Bryan, C. J., M. J. Gray, and C. S. Brooks. 2012. Further presence of ranavirus infection in amphibian populations of Tennessee. Herpetological Review 43:293–295.

O'Connor, J. H., and T. A. G. Rittenhouse. 2016. Snow cover and late fall movement influence Wood Frog survival during an unusually cold winter. Oecologia 181:635–644.

O'Connor, K. M., T. A. G. Rittenhouse, and J. L. Brunner. 2016. Ranavirus common in Wood Frog (*Lithobates sylvaticus*) tadpoles throughout Connecticut. Herpetological Review 47:394–397.

O'Donnell, R. P., K. Staniland, and R. T. Mason. 2007. Experimental evidence that oral secretions of Northwestern Ring-necked Snakes (*Diadophis punctatus occidentalis*) are toxic to their prey. Toxicon 50:810–815.

O'Donnell, R. P., R. Shine, and R. T. Mason. 2004. Seasonal anorexia in the male Red-sided Garter Snake, *Thamnophis sirtalis parietalis*. Behavioral and Ecological Sociobiology 56:413–419.

O'Laughlin, B. E., and R. N. Harris. 2000. Models of metamorphic timing: an experimental evaluation with the pond-dwelling salamander *Hemidactylium scutatum* (Caudata: Plethodontidae). Oecologia 124:343–350.

O'Neill, E. M., R. Schwartz, T. T. Bullock, J. S. Williams, H. B. Shaffer, X. Aguilar-Miguel, G. Parra-Olea, and D. W. Weisrock. 2012. Parallel tagged amplicon sequencing reveals major lineages and phylogenetic structure in the North American Tiger Salamander (*Ambystoma tigrinum*) species complex. Molecular Ecology 22:111–129.

Obbard, M. E. 1983. Population ecology of the Common Snapping Turtle, *Chelydra serpentina*, in northcentral Ontario. Dissertation, University of Guelph, Ontario, Canada.

Obbard, M. E., and R. J. Brooks 1979. Factors affecting basking in a northern population of the Common Snapping Turtle, *Chelydra serpentina*. Canadian Journal of Zoology 57:435–440.

Obbard, M. E., and R. J. Brooks. 1980. Nesting migrations of the Snapping Turtle (*Chelydra serpentina*). Herpetologica 36:158–162.

Obbard, M. E., and R. J. Brooks. 1981. A radio-telementry and mark-recapture study of activity in the Common Snapping Turtle, *Chelydra serpentina*. Copeia 1981:630–637.

Odlaug, T. O. 1954. Parasites of some Ohio Amphibia. Ohio Journal of Science 54:126–128.

Oehler, J. P., and J. A. Litvaitis. 1996. The role of spatial scale in understanding responses of medium-sized carnivores to forest fragmentation. Canadian Journal of Zoology 74:2070–2079.

Oldfield, B., and J. J. Moriarty. 1994. Amphibians and reptiles native to Minnesota. University of Minnesota Press, Minneapolis, USA.

Oldfield, B. L., and D. E. Keyler. 1989. Survey of Timber Rattlesnake (*Crotalus horridus*) distribution along the Mississippi River in western Wisconsin. Transactions of the Wisconsin Academy of Sciences, Arts and Letters 77:27–34.

Oldham, M. J. 1992. Declines in Blanchard's Cricket Frog in Ontario. Pages 30–31 *in* C. A. Bishop and K. E. Pettit, editors. Declines in Canadian amphibian populations: designing a national monitoring strategy. Canadian Wildlife Service Occasional Paper 76, Ottawa, Ontario, Canada.

Oldham, M. J., and C. A. Campbell. 1990. Committee on the Status of Endangered Wildlife in Canada (COSEWIC) status report on Northern Cricket Frog *Acris crepitans* in Canada. Committee on the Status of Endangered Wildlife in Canada, Ottawa, Ontario, Canada.

Oldham, R. S. 1966. Spring movements in the American Toad, *Bufo americanus*. Canadian Journal of Zoology 44:63–100.

Oldham, R. S. 1967. Orienting mechanisms of the Green Frog, *Rana clamitans*. Ecology 48:477–491.

Oldham, R. S. 1969. Initiation of breeding behavior in the American Toad, *Bufo americanus*. Canadian Journal of Zoology 47:1083–1086.

Olin, N. 1930. Reminiscences of Milwaukee in 1835–1836. Wisconsin Magazine of History 13:201–223.

Olmo, E. 1984. Genomic composition of reptiles: evolutionary perspectives. Journal of Herpetology 18:20–32.

Olson, E. R. 2010. *Rana pipiens* (Northern Leopard Frog). Winter activity. Herpetological Review 41:206.

Olson, R. E. 1984. *Hyla chrysoscelis* (Cope's Gray Treefrog). Geographic distribution. Herpetological Review 15:76.

Olson, R. E. 1995. Important herpetological notes for northern Illinois and southern Wisconsin. Bulletin of the Maryland Herpetological Society 31:96.

Olson, Z. H., B. J. MacGowan, M. T. Hamilton, A. F. T. Currylow, and R. N. Williams. 2015. Survival of Timber Rattlesnakes (*Crotalus horridus*): investigating individual, environmental, and ecological effects. Herpetologica 71:274–279.

Oplinger, C. S. 1967. Food habits and feeding activity of recently transformed and adult *Hyla crucifer crucifer* Wied. Herpetologica 23:209–217.

Orlofske, S. A., and C. G. Anderson. 2005. *Eumeces septentrionalis septentrionalis*. Geographic distribution. Herpetological Review 36:467.

Orlofske, S. A., and W. A. Hopkins. 2009. Energetics of metamorphic climax in the Pickerel Frog (*Lithobates palustris*). Comparative Biochemistry and Physiology. Part A, Molecular & Integrative Physiology 154:191–196.

Orlofske, S. A., L. K. Beldon, and W. A. Hopkins. 2009. Moderate *Echinostoma trivolvis* infection has no effects on physiology and fitness-related traits of larval Pickerel Frogs (*Rana palustris*). Journal of Parasitology 95:787–792.

Orlofske, S. A., L. K. Belden, and W. A. Hopkins. 2013. Larval Wood Frog (*Rana [=Lithobates] sylvatica*) development and physiology following infection with the trematode parasite, *Echinostoma trivolvis*. Comparative Bioschemistry and Physiology. Part A, Molecular & Integrative Physiology 164:529–536.

Orlofske, S. A., L. K. Belden, and W. A. Hopkins. 2017. Effects of *Echinostoma trivolvis* metacercariae infection during development and metamorphosis of the Wood Frog (*Lithobates sylvaticus*). Comparative Biochemistry and Physiology. Part A, Molecular & Integrative Physiology 203:40–48.

Ortiz-Santaliestra, M. E., T. A. G. Rittenhouse, T. L. Cary, and W. H. Karasov. 2013. Interspecific and postmetamorphic variation in susceptibility of three North American anurans to *Batrachochytrium dendrobatidis*. Journal of Herpetology 47:286–292.

Osbourn, M. S., G. M. Connette, and R. D. Semlitsch 2014. Effects of fine-scale forest habitat quality on movement and settling decisions in juvenile pond-breeding salamanders. Ecological Applications 24:1719–1729.

Oseen, K. L., and R. J. Wassersug. 2002. Environmental factors influencing calling in sympatric anurans. Oecologia 133:616–625.

Otten, J. G., and T. J. VanDeWalle. 2014. *Apalone spinifera* (Spiny Softshell), *Chelydra serpentina* (Snapping Turtle), and *Sternotherus odoratus* (Eastern Musk Turtle). Homing behavior. Herpetological Review 45:309–311.

Ouellette, M., and J. A. Cardille. 2011. The complex linear home range estimator: representing the home range of river turtles moving in multiple channels. Chelonian Conservation and Biology 10:259–265.

Ouellet, M., C. Fortin, and M-J. Grimard. 2009. Distribution and habitat use of the Boreal Chorus Frog (*Pseudacris maculata*) at its extreme northeastern range limit. Herpetological Conservation and Biology 4:277–284.

Ouellet, M., I. Mikaelian, B. D. Pauli, J. Rodrigue, and D. M. Green. 2005. Historical evidence of widespread chytrid infection in North American amphibian populations. Conservation Biology 19:1431–1440.

Ouellet, M., J. Bonin, J. Rodrigue, J-L. DesGranges, and S. Lair. 1997. Hindlimb deformities (ectromelia, ectrodactyly) in free-living anurans from agricultural habitats. Journal of Wildlife Diseases 33:95–104.

Ousterhout, B. H., and E. B. Liebgold. 2010. Dispersal versus site tenacity of adult and juvenile

Red-backed Salamanders (*Plethodon cinereus*). Herpetologica 66:269–275.

Ousterhout, B. H., T. L. Anderson, D. L. Drake, W. E. Peterman, and R. D. Semlitsch. 2015. Habitat traits and species interactions differentially affect abundance and body size in pond-breeding amphibians. Journal of Animal Ecology 84:914–924.

Overton, F. 1914. Long Island Fauna and Flora-III: the Frogs and Toads. Museum of the Brooklyn Institute of Arts and Sciences 2:21–40.

Owen, P. C., and M. Austin. 2013. Green Frog, *Lithobates clamitans* (Latreille 1801). Pages 601–644 *in* R. A. Pfingsten, J. G. Davis, T. O. Matson, G. Lipps, Jr., D. Wynn, and B. J. Armitage, editors. Amphibians of Ohio. Ohio Biological Survey, Bulletin New Series Volume 17, Number 1, Columbus, USA.

Owen, P. C., and N. M. Gordon. 2005. The effect of perceived intruder proximity and resident body size on the aggressive responses of male Green Frogs, *Rana clamitans* (Anura: Ranidae). Behavioral Ecology and Sociobiology 58:446–455.

Pace, A. E. 1974. Systematic and biological studies of the leopard frogs (*Rana pipiens* complex) of the United States. University of Michigan, Miscellaneous Publication of the Museum of Zoology No 148, Ann Arbor, USA.

Packard, G. C., M. J. Packard, and J. W. Lang. 2000. Why hatchling Blanding's Turtles don't overwinter inside their nest. Herpetologica 56:367–374.

Packard, G. C., M. J. Packard, and W. H. N. Gutzke. 1985. Influence of hydration of the environment on eggs and embryos of the terrestrial turtle *Terrapene ornata*. Physiological Zoology 58:564–575.

Padgett, D. J., J. J. Carboni, and D. J. Schepis. 2010. The dietary composition of *Chrysemys picta picta* (Eastern Painted Turtles) with special reference to the seeds of aquatic macrophytes. Northeastern Naturalist 17:305–312.

Paisley, R. N., J. F. Wetzel, J. S. Nelson, C. Stetzer, M. G. Hammer, and B. P. Anderson. 2009. Survival and spatial ecology of the Snapping Turtle, *Chelydra serpentina*, on the upper Mississippi River. Canadian Field-Naturalist 123:329–337.

Palis, J. G. 2007. If you build it, they will come: herpetofaunal colonization of constructed wetlands and adjacent terrestrial habitat in the Cache River Drainage of southern Illinois. Transactions of the Illinois State Academy of Science 100:177–189.

Pallas, D. C. 1960. Observations on a nesting of the Wood Turtle, *Clemmys insculpta*. Copeia 1960:155–156.

Palmer, W. M., and A. L. Braswell. 1995. Reptiles of North Carolina. University of North Carolina Press, Chapel Hill, USA.

Paloski, R. A. 2006. Relationship between lakeshore development and anuran populations in Portage County, WI. Thesis, University of Wisconsin, Stevens Point, USA.

Paloski, R. A. 2008. Case study 1: relationship between lakeshore development and frog populations in central Wisconsin. Pages 77–83 *in* J. C. Mitchell, R. E. Jung Brown, and B. Bartholomew, editors. Urban Herpetology. Society for the Study of Amphibians and Reptiles, Herpetological Conservation 3, Salt Lake City, Utah, USA.

Paloski, R. A. 2020. *Hyla versicolor* (Gray Treefrog). Geographic distribution. Herpetological Review 51:769.

Paloski, R. A. 2022. *Hyla chrysoscelis* (Cope's Gray Treefrog). Geographic distribution. Herpetological Review 53:71.

Paloski, R. A., A. F. Badje, R. A. Staffen, and T. L. E. Bergeson. 2018. *Lithobates catesbeianus* (American Bullfrog). Geographic distribution. Herpetological Review 49:70.

Paloski, R. A., E. T. Hileman, G. C. Mayer, G. W. Schurrman, and J. M. Kapfer. 2017. *Thamnophis radix* (Plains Gartersnake). Geographic distribution. Herpetological Review 48:591–592.

Paloski, R. A., G. W. Schuurman, and J. M. Kapfer. 2020. *Thamnophis radix* (Plains Gartersnake). Geographic distribution. Herpetological Review 51:82.

Paloski, R. A., T. L. E. Bergeson, M. Mossman, and R. Hay, editors. 2014. Wisconsin frog and toad survey: phenological survey manual. Wisconsin Department of Natural Resources, Bureau of Natural Heritage Conservation PUB-NH-649, Madison, USA. Archived by WebCite at http://www.webcitation.org/73g1BmetZ. Accessed 4 Nov 2018.

Panik, H. R., and S. Barrett. 1994. Distribution of amphibians and reptiles along the Truckee River system. Northwest Science 68:197–204.

Panyushkina, I. P., and S. W. Leavitt. 2013. Ancient boreal forests under the environmental instability of the glacial to postglacial transition in the Great

Lakes region (14 000–11 000 years BP). Canadian Journal of Forest Research 43:1032–1039.

Pappas, M. J., and B. J. Brecke. 1992. Habitat selection of juvenile Blanding's Turtles, *Emydoidea blandingii*. Journal of Herpetology 26:233–234.

Pappas, M. J., B. J. Brecke, and J. D. Congdon. 2000. The Blanding's Turtles (*Emydoidea blandingii*) of Weaver Dunes, Minnesota. Chelonian Conservation and Biology 3:557–568.

Pappas, M. J., J. D. Congdon, B. J. Brecke, and J. D. Capps. 2009. Orientation and dispersal of hatchling Blanding's Turtles (*Emydoidea blandingii*) from experimental nests. Canadian Journal of Zoology 87:755–766.

Parent, C., and P. J. Weatherhead. 2000. Behavioral and life history responses of Eastern Massasauga Rattlesnakes (*Sistrurus catenatus catenatus*) to human disturbance. Oecologia 125:170–178.

Parker, I. M., S. K. Mertens, and D. W. Schemske. 1993. Distribution of seven native and two exotic plants in a tallgrass prairie in southeastern Wisconsin: the importance of human disturbance. American Midland Naturalist 130:43–55.

Parker, W. S., and M. V. Plummer. 1987. Population ecology. Pages 253–301 *in* R. A. Seigel, J. T. Collins, and S. S. Novak, editors. Snakes: ecology and evolutionary biology. Macmillan Publishing Company, New York, New York, USA.

Parker, W. S., and W. S. Brown. 1980. Comparative ecology of two colubrid snakes, *Masticophis t. taeniatus* and *Pituophis melanoleucus deserticola*, in northern Utah. Milwaukee Public Museum Publications in Biology and Geology 7:1–104.

Parmalee, P. W. 1959. Animal remains from the Raddatz Rockshelter, Sk5, Wisconsin. Wisconsin Archeologist 40:83–90.

Parmalee, P. W. 1960. Animal remains from the Durst Rockshelter, Sauk County, Wisconsin. Wisconsin Archeologist 41:11–17.

Parmalee, P. W. 1963. Vertebrate remains from the Bell Site, Winnebago County, Wisconsin. Wisconsin Archeologist 44:58–69.

Parmalee, P. W., and W. E. Klippel. 1981. Remains of the Wood Turtle *Clemmys insculpta* (Le Conte) from a late Pleistocene deposit in middle Tennessee. American Midland Naturalist 105:413–416.

Parmelee, J. R. 1990. Lack of burrowing ability in the Blue-spotted Salamander, *Ambystoma laterale*. Bulletin of the Chicago Herpetological Society 25:81–83.

Parmelee, J. R. 1993. Microhabitat segregation and spatial relationships among four species of mole salamander (genus *Ambystoma*). University of Kansas, Occasional Papers of the Museum of Natural History 160, Lawrence, USA.

Parmelee, J. R., M. G. Knutson, and J. E. Lyon. 2002. A field guide to amphibian larvae and eggs of Minnesota, Wisconsin, and Iowa. US Geological Survey, Biological Resources Division Information and Technology Report USGS/BRD/ITR-2002–0004, Washington, D.C., USA. Archived by WebCite at http://www.webcitation.org/73g1IYNZw. Accessed 4 Nov 2018.

Parren, S. G. 2013. A twenty-five year study of the Wood Turtle (*Glyptemys insculpta*) in Vermont: movements, behavior, injuries, and death. Herpetological Conservation and Biology 8:176–190.

Parren, S. G., and M. A. Rice. 2004. Terrestrial overwintering of hatchling turtles in Vermont nests. Northeastern Naturalist 11:229–233.

Parris, M. J., and D. R. Baud. 2004. Interactive effects of a heavy metal and chytridiomycosis on Gray Treefrog larvae (*Hyla chrysoscelis*). Copeia 2004:344–350.

Parrish, H. M. 1965a. Frequency of snakebites in Wisconsin. Wisconsin Medical Journal 64:216–220.

Parrish, H. M. 1965b. Rarity of snakebites in Minnesota. Minnesota Medicine 48:1071–1076.

Pastor, J., and D. J. Mladenoff. 1992. The southern boreal-northern hardwood forest border. Pages 216–240 *in* H. H. Shugart, R. Leemans, and G. B. Bonan, editors. A systems analysis of the global boreal forest. Cambridge University Press, New York, New York, USA.

Paterson, J. E., B. D. Steinberg, and J. D. Litzgus. 2012. Revealing a cryptic life-history stage: differences in habitat selection and survivorship between hatchlings of two turtle species at risk (*Glyptemys insculpta* and *Emydoidea blandingii*). Wildlife Research 39:408–418.

Paterson, J. E., B. D. Steinberg, and J. D. Litzgus. 2013. Not just any old pile of dirt: evaluating the use of artificial nesting mounds as conservation tools for freshwater turtles. Oryx 47:607–615.

Paton, P. W., and W. B. Crouch, III. 2002. Using the phenology of pond-breeding amphibians to develop conservation strategies. Conservation Biology 16:194–204.

Paton, P. W., S. Stevens, and L. Longo. 2000. Seasonal phenology of amphibian breeding and recruitment at a pond in Rhode Island. Northeastern Naturalist 7:255–269.

Patrick, D. A., and J. P. Gibbs. 2009. Snake occurrences in grassland associated with road versus forest edges. Journal of Herpetology 43:716–720.

Patrick, D. A., and J. P. Gibbs. 2010. Population structure and movements of freshwater turtles across a road-density gradient. Landscape Ecology 25:791–801.

Patrick, D. A., C. M. Schalk, J. P. Gibbs, and H. W. Wolz. 2010. Effective culvert placement and design to facilitate passage of amphibians across roads. Journal of Herpetology 44:618–626.

Patrick, D. A., E. B. Harper, V. D. Popescu, S. Bozic, A. Byrne, J. Daub, A. Lecheminant, and J. Pierce. 2012. The ecology of the Mink Frog, *Lithobates septentrionalis,* in the Adirondack Park New York, with notes on conducting experimental research. Herpetological Review 43:396–398.

Patrick, D. A., M. L. Hunter, Jr., and A. J. K. Calhoun. 2006. Effects of experimental forestry treatments on a Maine amphibian community. Forest Ecology and Management 234:323–332.

Pattishall, A., and D. Cundall. 2008. Spatial biology of Northern Watersnakes (*Nerodia sipedon*) living along an urban stream. Copeia 2008:752–762.

Pattishall, A., and D. Cundall. 2009. Habitat use by synurbic watersnakes (*Nerodia sipedon*). Herpetologica 65:183–198.

Pauley, L. R., J. E. Earl, and R. D. Semlitsch. 2015. Ecological effects and human use of commercial mosquito insecticides in aquatic communities. Journal of Herpetology 49:28–35.

Pauley, T. K., and M. B. Watson. 2005. *Eurycea cirrigera* (Green, 1830), Southern Two-lined Salamander. Pages 740–743 *in* M. Lannoo, editor. Amphibian declines: the conservation status of United States species. University of California Press, Berkeley, USA.

Pauley, T. K., and M. J. Lannoo. 2005. *Rana clamitans* Latreille, 1801. Green Frog. Pages 549–552 *in* M. Lannoo, editor. Amphibian Declines: the conservation status of United States species. University of California Press, Berkeley, USA.

Pauli, B. D., D. R. Coulson, and M. Berrill. 1999. Sensitivity of amphibian embryos and tadpoles to Mimic 240 LC insecticide following single or double exposure. Environmental Toxicology and Chemistry 18:2538–2544.

Paulissen, M. A. 1987a. Diet of adult and juvenile Six-Lined Racerunners, *Cnemidophorus sexlineatus* (Sauria: Teiidae). Southwestern Naturalist 32:395–397.

Paulissen, M. A. 1987b. Optimal foraging and intraspecific diet differences in the lizard *Cnemidophorus sexlineatus.* Oecologia 71:439–446.

Paulissen, M. A. 1988a. Ontogenetic and seasonal comparisons of daily activity patterns of the Six-lined Racerunner, *Cnemidophorus sexlineatus.* American Midland Naturalist 120:355–361.

Paulissen, M. A. 1988b. Ontogenetic and seasonal shifts in microhabitat use by the lizard *Cnemidophorus sexlineatus.* Copeia 1988:1021–1029.

Paulissen, M. A. 1988c. Ontogenetic comparison of body temperature selection and thermal tolerance of *Cnemidophorus sexlineatus.* Journal of Herpetology 22:473–476.

Paulissen, M. A., and B. C. Harvey. 1985. *Cnemidophorus sexlineatus viridis* (Prairie-lined Racerunner). Arboreality. Herpetological Review 16:27.

Paull, R. K., and R. A. Paull. 1977. Geology of Wisconsin and upper Michigan: including parts of adjacent states. Kendall Hunt Publishing Company, Dubuque, Iowa, USA.

Pauly, G. B., D. M. Hillis, and D. C. Cannatella. 2009. Taxonomic freedom and the role of official lists of species names. Herpetologica 64:115–128.

Peacor, S. D., and C. A. Pfister. 2006. Experimental and model analyses of the effects of competition on individual size variation in Wood Frog (*Rana sylvatica*) tadpoles. Journal of Animal Ecology 75:990–999.

Peacor, S. D., and E. E. Werner. 2000. Induced changes in consumer foraging behavior. Ecology 81:1998–2010.

Pearman, P. B. 1993. Effects of habitat size on tadpole populations. Ecology 74:1982–1991.

Pearse, A. S. 1921. Habits of the Mud-puppy *Necturus,* an enemy of food fishes. Department of Commerce, Bureau of Fisheries Economic Circular 49, Washington, D.C., USA.

Pearse, A. S. 1923. The abundance and migration of turtles. Ecology 4:24–28.

Pearson, O. P. 1942. On the cause and nature of a poisonous action produced by the bite of a shrew (*Blarina brevicauda*). Journal of Mammalogy 23:159–166.

Pechmann, J. H. K., and H. M. Wilbur. 1994. Putting declining amphibian populations in perspective: natural fluctuations and human impacts. Herpetologica 50:65–84.

Pechmann, J. H. K., and R. D. Semlitsch. 1986. Diel activity patterns in the breeding migrations of winter-breeding anurans. Canadian Journal of Zoology 64:1116–1120.

Peckham, R. S., and C. F. Dineen. 1954. Spring migrations of salamanders. Proceedings of the Indiana Academy of Science 62:278–280.

Pedler, J. H., L. Fahrig, and H. G. Merriam. 1997. Raccoon habitat use at two spatial scales. Journal of Wildlife Management 61:102–112.

Peet-Paré, C. A. 2010. Nest-site selection in Eastern Hog-nosed Snakes (*Heterodon platirhinos*). Canadian Journal of Zoology 90:1215–1220.

Peet-Paré, C. A., and G. Blouin-Demers. 2012. Female Eastern Hog-nosed Snakes (*Heterodon platirhinos*) choose nest sites that produce offspring with phenotypes likely to improve fitness. Canadian Journal of Zoology 90:1215–1220.

Peles, J. D. 2013. Effects of chronic aluminum and copper exposure on growth and development of Wood Frog (*Rana sylvatica*) larvae. Aquatic Toxicology 140–141:242–248.

Penn, G. H. 1950. Utilization of crawfishes by cold-blooded vertebrates in the eastern United States. American Midland Naturalist 44:643–658.

Pentecost, E. D., and R. C. Vogt. 1976. Environmental status of the Lake Michigan region. Volume 16. Amphibians and reptiles of the Lake Michigan drainage basin. US Energy Research and Development Administration ANL/ES-40, Argonne, Illinois, USA.

Perdzock, A. 2015. Best management practices for boats, gear, and equipment decontamination. Wisconsin Department of Natural Resources, Bureau of Water Quality, Madison, USA. Archived by WebCite at http://www.webcitation.org/73cMtTZKQ. Accessed 1 Nov 2018.

Perez, A., M. J. Mazerolle, and J. Brisson. 2013. Effects of exotic Common Reed (*Phragmites australis*) on Wood Frog (*Lithobates sylvaticus*) tadpole development and food availability. Journal of Freshwater Ecology 28:165–177.

Pérez-Ramos, E., A. N-M. de Oca, J. A. Vargas-Contreras, J. E. Cordes, M. A. Paulissen, and J. M. Walker. 2010. *Aspidoscelis sexlineata* (Sauria: Teiidae) in Mexico: distribution, habitat, morphology, and taxonomy. Southwestern Naturalist 55:419–425.

Perkins, D. W., and M. L. Hunter, Jr. 2006. Effects of riparian timber management on amphibians in Maine. Journal of Wildlife Management 70:657–670.

Perkins, M. J., and B. D. Palmer. 1996. Histology and functional morphology of the oviduct of an oviparous snake, *Diadophis punctatus*. Journal of Morphology 227:67–79.

Perrill, S. A., and M. Magier. 1988. Male mating behavior in *Acris crepitans*. Copeia 1988:245–248.

Perry, C. H., V. A. Everson, B. J. Butler, S. J. Crocker, S. E. Dahir, A. L. Diss-Torrance, G. M. Domke, D. D. Gormanson, M. A. Hatfield, S. K. Herrick, S. S. Hubbard, T. R. Mace, P. D. Miles, M. D. Nelson, R. B. Rideout, L. T. Saunders, K. M. Stueve, B. T. Wilson, and C. W. Woodall. 2012. Wisconsin's forests 2009. US Forest Service, Northern Research Station Resource Bulletin NRS-67, Newtown Square, Pennsylvania, USA.

Perry, G., B. W. Buchanan, R. N. Fisher, M. Salmon, and S. E. Wise. 2008. Effects of artificial night lighting on amphibians and reptiles in urban environments. Pages 239–256 *in* J. Mitchell, R. Jung Brown, and B. Bartholomew, editors. Urban Herpetology. Society for the Study of Amphibians and Reptiles, Herpetological Conservation 3, Salt Lake City, Utah, USA.

Peterman, W. E., T. A. G. Rittenhouse, J. E. Earl, and R. D. Semlitsch. 2013. Demographic network and multi-season occupancy modeling of *Rana sylvatica* reveal spatial and temporal patterns of population connectivity and persistence. Landscape Ecology 28:1601–1613.

Peterman, W. E., T. L. Anderson, D. L. Drake, B. H. Ousterhout, and R. D. Semlitsch. 2014. Maximizing pond biodiversity across the landscape: a case study of larval ambystomatid salamanders. Animal Conservation 17:275–285.

Peterson, D. W., and P. B. Reich. 2001. Prescribed fire in oak savanna: fire frequency effects on stand structure and dynamics. Ecological Applications 11:914–927.

Petokas, P. J. 1981. A. S. Pearse's density estimates for turtles; the correction of a long-standing error. Bulletin of the Maryland Herpetological Society 17:68–70.

Petranka, J. W. 1989a. Density-dependent growth and survival of larval *Ambystoma*: evidence from whole-pond manipulations. Ecology 70:1752–1767.

Petranka, J. W. 1989b. Response of toad tadpoles to conflicting chemical stimuli: predator avoidance versus "optimal" foraging. Herpetologica 45:283–292.

Petranka, J. W. 1998. Salamanders of the United States and Canada. Smithsonian Press, Washington, D.C., USA.

Petranka, J. W., and C. A. Kennedy. 1999. Pond tadpoles with generalized morphology: is it time to reconsider their functional roles in aquatic communities? Oecologia 120:621–631.

Petranka, J. W., and C. T. Holbrook. 2006. Wetland restoration for amphibians: should local sites be designed to support metapopulations or patchy populations? Restoration Ecology 14:404–411.

Petranka, J. W., and D. A. G. Thomas. 1995. Explosive breeding reduces egg and tadpole cannibalism in the Wood Frog, *Rana sylvatica*. Animal Behaviour 50:731–739.

Petranka, J. W., and K. Fakhoury. 1991. Evidence of a chemically mediated avoidance response of ovipositing insects to Blue-gills and Green Frog tadpoles. Copeia 1991:234–239.

Petranka, J. W., and L. Hayes. 1998. Chemically mediated avoidance of a predatory odonate (*Anax junius*) by American Toad (*Bufo americanus*) and Wood Frog (*Rana sylvatica*) tadpoles. Behavioral Ecology and Sociobiology 42:263–271.

Petranka, J. W., and S. S. Murray. 2001. Effectiveness of removal sampling for determining salamander density and biomass: a case study in an Appalachian streamside community. Journal of Herpetology 35:36–44.

Petranka, J. W., A. W. Rushlow, and M. E. Hopey. 1998. Predation by tadpoles of *Rana sylvatica* on embryos of *Ambystoma maculatum*: implications of ecological role reversals by *Rana* (predator) and *Ambystoma* (prey). Herpetologica 54:1–13.

Petranka, J. W., C. A. Kennedy, and S. S. Murray. 2003. Responses of amphibians to restoration of a southern Appalachian wetland: a long-term analysis of community dynamics. Wetlands 23:1030–1042.

Petranka, J. W., C. K. Smith, and A. F. Scott. 2004. Identifying the minimal demographic unit for monitoring pond-breeding amphibians. Ecological Applications 14:1065–1078.

Petranka, J. W., E. M. Harp, C. T. Holbrook, and J. A. Hamel. 2007. Long-term persistence of amphibian populations in a restored wetland complex. Biological Conservation 138:371–380.

Petranka, J. W., M. E. Eldridge, and K. E. Haley. 1993. Effects of timber harvesting on southern Appalachian salamanders. Conservation Biology 7:363–370.

Petranka, J. W., M. E. Hopey, B. T. Jennings, S. D. Baird, and S. J. Boone. 1994. Breeding habitat segregation of Wood Frogs and American Toads: the role of interspecific tadpole predation and adult choice. Copeia 1994:691–697.

Pettit, K. E., C. A. Bishop, and R. J. Brooks. 1995. Home range and movements of the Common Snapping Turtle, *Chelydra serpentina*, in a coastal wetland of Hamilton Harbour, Lake Ontario, Canada. Canadian Field-Naturalist 109:192–200.

Pettus, D., and G. M. Angleton. 1967. Comparative reproductive biology of montane and piedmont chorus frogs. Evolution 21:500–507.

Petzing, J. E., M. H. Sabaj, and R. Doerscheln. 1998. *Storeria dekayi* (Brown Snake). Geographic distribution. Herpetological Review 29:55.

Peyton, M. M. 1989. *Storeria occipitomaculata*. Geographic distribution. Herpetological Review 20:13.

Pfingsten, R. A. 1989. *Plethodon cinereus* (Green), Redback Salamander. Pages 229–242 *in* R. A. Pfingsten and F. L. Downs, editors. Salamanders of Ohio. Ohio Biological Survey, Bulletin New Series Volume 7, Columbus, Ohio, USA.

Pfingsten, R. A. 1998. Distribution of Ohio amphibians. Pages 221–255 *in* M. J. Lannoo, editor. Status and conservation of midwestern amphibians. University of Iowa Press, Iowa City, USA.

Pfingsten, R. A., and A. M. White. 1989. *Necturus maculosus* (Rafinesque), Mudpuppy. Pages 72–78 *in* R. A. Pfingsten and F. L. Downs, editors. Salamanders of Ohio. Ohio Biological Survey, Bulletin New Series Volume 7, Columbus, USA.

Phillips, C. A. 1994. Geographic distribution of mitochoindrial DNA variants and the historical biogeography of the Spotted Salamander, *Ambystoma maculatum*. Evolution 48:597–607.

Phillips, C. A., and J. Mui. 2005. Unisexual members of the *Ambystoma jeffersonianum* complex. Pages 640–642 *in* M. Lannoo, editor. Amphibian declines: the conservation status of United States species. University of California Press, Berkeley, USA.

Phillips, C. A., and O. J. Sexton. 1989. Orientation and sexual differences during breeding migrations of the Spotted Salamander, *Ambystoma maculatum*. Copeia 1989:17–22.

Phillips, C. A., N. A. Wesslund, and I. E. MacAllister. 2014. Occurrence of the chytrid fungus *Batrachochytrium dendrobatidis* in amphibians in Illinois, USA. Herpetological Review 45:238–240.

Phillips, C. A., R. A. Brandon, and E. O. Moll. 1999. Field guide to amphibians and reptiles of Illinois. Illinois Natural History Survey Manual 8, Champaign, USA.

Phillips, P. W. 1961. *Sternothaerus odoratus* (Latreille), Stinkpot or Musk Turtle. Pages 122–124 *in* P. A. Smith, editor. The amphibians and reptiles of Illinois. Illinois Natural History Survey Bulletin 28, Urbana, Illinois, USA.

Pianka, E. P., and L. J. Vitt. 2003. Lizards: windows to the evolution of diversity. University of California Press, Oakland, California, USA.

Picard, G., M. Carrière, and G. Blouin-Demers. 2011. Common Musk Turtles (*Sternotherus odoratus*) select habitats of high thermal quality at the northern extreme of their range. Amphibia-Reptilia 32:83–92.

Pieau, C. 1974. Sur la différenciation sexuelle des embryons d'*Emys orbicularis* L. (Chelonien) issus d'oeufs incubus dans le sol an cours de l'été. 1973. Bulletin De La Societe Zoologique De France 99:363–376.

Piepgras, S. J., and J. W. Lang. 2000. Spatial ecology of Blanding's Turtle in central Minnesota. Chelonian Conservation and Biology 3:589–601.

Pierce, A. M., and P. B. Reich. 2010. The effects of Eastern Red Cedar (*Juniperus virginiana*) invasion and removal on a dry bluff prairie ecosystem. Biological Invasions 12:241–252.

Pierce, P. E., and D. A. Ross. 1989. Bald Eagles prey on snakes. Passenger Pigeon 51:155–156.

Pilgrim, M. A., T. M. Farrell, P. G. May, M. R. Vollman, and R. A. Seigel. 2011. Secondary sex ratios in six snake species. Copeia 2011:553–558.

Pilliod, D. S., R. B. Bury, E. J. Hyde, C. A. Pearl, and P. S. Corn. 2003. Fire and amphibians in North America. Forest Ecology and Management 178:163–181.

Pillsbury, F. C., and J. R. Miller. 2008. Habitat and landscape characteristics underlying anuran community structure along an urban-rural gradient. Ecological Applications 18:1107–1118.

Pimentel, D., R. Zuniga, and D. Morrison. 2005. Update on the environmental and economic costs associated with alien-invasive species in the United States. Ecological Economics 52:273–288.

Pinder, A. W., and S. C. Friet. 1994. Oxygen transport in egg masses of the amphibians *Rana sylvatica* and *Ambystoma maculatum*: convection, diffusion and oxygen production by algae. Journal of Experimental Biology 197:17–30.

Pinder, M. J. 1996. *Regina septemvittata* (Queen Snake). Catesbiana 16:9.

Pinder, R. 2010. *Rana clamitans* (Green Frog). Coloration. Herpetological Review 41:66.

Pinou, T., C. A. Hass, and L. R. Maxson. 1995. Geographic variation of serum albumin in the monotypic snake genus *Diadophis* (Colubridae: Xenodontinae). Journal of Herpetology 29:105–110.

Pintar, M. R., and W. J. Resetarits. 2017a. Larval development varies across pond age and larval density in Cope's Gray Treefrogs, *Hyla chrysoscelis*. Herpetologica 73:291–296.

Pintar, M. R., and W. J. Resetarits. 2017b. Out with the old, in with the new: oviposition preference matches larval success in Cope's Gray Treefrog, *Hyla chrysoscelis*. Journal of Herpetology 51:186–189.

Pinto, B. J., J. L. Banks, T. Gamble, and D. Becker. 2018. *Pituophis catenifer sayi* (Bullsnake) and *Pantherophis vulpinus* (Western Foxsnake). Intergeneric hybridization. Herpetological Review 49:350–351.

Pisani, G. R. 2009a. Use of an active ant nest as a hibernaculum by small snake species. Transactions of the Kansas Academy of Science 112:113–118.

Pisani, G. R. 2009b. *Virginia valeriae* and *Storeria dekayi* in a northeast Kansas grassland community: ecology and conservation implications. Journal of Kansas Herpetology 32:20–36.

Pisani, G. R., and G. Pittman. 2009. Early activity of *Storeria dekayi* in Jefferson County, Kansas. Journal of Kansas Herpetology 29:10–11.

Pisani, G. R., J. T. Collins, and S. R. Edwards. 1972. A re-evaluation of the subspecies of *Crotalus horridus*. Transactions of the Kansas Academy of Science 75:255–263.

Pitkin, R. B., and S. G. Tilley. 1982. An unusual congregation of adult *Notophthalmus viridescens*. Copeia 1982:185–186.

Pitt, A. L., and M. A. Nickerson. 2012. Reassessment of the turtle community in the North Fork of the White River, Ozark County, Missouri. Copeia 2012:367–374.

Pitt, A. L., and M. A. Nickerson. 2013. Potential recovery of a declined turtle population diminished by a community shift towards more generalist species. Amphibia-Reptilia 34:192–200.

Pitt, A. L., and M. A. Nickerson. 2014. Description and comparison of turtle assemblages and populations located within a spring-fed river. Herpetological Conservation and Biology 9:475–483.

Pitt, A. L., J. J. Tavano, R. F. Baldwin, and T. A. Waldrop. 2013. Effects of fuel reduction treatments on movement and habitat use of American Toads in a southern Appalachian hardwood forest. Forest Ecology and Management 310:289–299.

Pitt, W. C. 2001. Density of Prairie Skinks (*Eumeces septentrionalis*) in old-field habitats. American Midland Naturalist 146:86–93.

Pittman, S. E., A. L. Jendrek, S. J. Price, and M. E. Dorcas. 2008. Habitat selection and site fidelity of Cope's Gray Treefrog (*Hyla chrysoscelis*) at the aquatic-terrestrial ecotone. Journal of Herpetology 42:378–385.

Placyk, J. S., Jr., and B. M. Graves. 2001. Foraging behavior of the Red-backed Salamander (*Plethodon cinereus*) under various lighting conditions. Journal of Herpetology 35:524–529.

Placyk, J. S., Jr., B. M. Fitzpatrick, G. S. Casper, R. L. Small, R. G. Reynolds, D. W. Noble, R. J. Brooks, and G. M. Burghardt. 2012. Hybridization between two gartersnake species (*Thamnophis*) of conservation concern: a threat or an important natural interaction? Conservation Genetics 13:649–663.

Placyk, J. S., Jr., G. M. Burghardt, R. L. Small, R. B. King, G. S. Casper, and J. W. Robinson. 2007. Post-glacial recolonization of the Great Lakes region by the Common Gartersnake (*Thamnophis sirtalis*) inferred from mtDNA sequences. Molecular Phylogenetics and Evolution 43:452–467.

Platt, D. R. 1969. Natural history of the hognose snakes *Heterodon platyrhinos* and *Heterodon nasicus*. University of Kansas Publications, Museum of Natural History 18:253–420.

Platt, D. R. 1984. Growth of Bullsnakes (*Pituophis melanoleucus sayi*) on a sand prairie in south central Kansas. University of Kansas Publications, Special Publications of the Museum of Natural History 10:41–55.

Platt, D. R. 1985. History and spelling of the name *Heterodon platyrhinos*. Journal of Herpetology 19:417–418.

Platt, S. G., K. R. Russell, W. E. Snyder, L. W. Fontenot, and S. Miller. 1999. Distribution and conservation of selected amphibians and reptiles in the Piedmont of South Carolina. Journal of the Elisha Mitchell Scientific Society 115:8–19.

Platz, J. E. 1989. Speciation within the Chorus Frog *Pseudacris triseriata*: morphometric and mating call analyses of the boreal and western subspecies. Copeia 1989:704–712.

Pleyte, T. A. 1975. The Slender Glass Lizard (*Ophisaurus attenuatus*) in Waushara Co., Wisconsin. Thesis, University of Wisconsin, Milwaukee, USA.

Plummer, M. V. 1976. Some aspects of nesting success in the turtle, *Trionyx muticus*. Herpetologica 32:353–359.

Plummer, M. V. 1977a. Activity, habitat and population structure in the turtle, *Trionyx muticus*. Copeia 1977:431–440.

Plummer, M. V. 1977b. Notes on the courtship and mating behavior of the Soft-shell Turtle, *Trionyx muticus* (Reptilia: Testudines: Trionychidae). Journal of Herpetology 11:90–92.

Plummer, M. V. 1977c. Reproduction and growth in the turtle *Trionyx muticus*. Copeia 1977:440–447.

Plummer, M. V. 2007. Nest emergence of Smooth Softshell Turtle (*Apalone mutica*) hatchlings. Herpetological Conservation and Biology 2:61–64.

Plummer, M. V., and D. B. Farrar. 1981. Sexual dietary differences in a population of *Trionyx muticus*. Journal of Herpetology 15:175–179.

Plummer, M. V., and H. W. Shirer. 1975. Movement patterns in a river population of the Softshell Turtle, *Trionyx muticus*. University of Kansas, Occasional Papers of the Museum of Natural History 43:1–26.

Plummer, M. V., and J. W. Goy. 1997. *Sternotherus odoratus* (Common Musk Turtle). Mortality. Herpetological Review 28:88.

Plummer, M. V., and N. E. Mills. 1996. Observations on trailing and mating behaviors in Hognose Snakes (*Heterodon platirhinos*). Journal of Herpetology 30:80–82.

Plummer, M. V., and N. E. Mills. 2000. Spatial ecology and survivorship of resident and translocated Hognose Snakes (*Heterodon platyrhinos*). Journal of Herpetology 34:565–575.

Plummer, M. V., and N. E. Mills. 2006. *Heterodon platirhinos* (Eastern Hog-nosed Snake). Road crossing behavior. Herpetological Review 37:352.

Plummer, M. V., and N. E. Mills. 2010. Body temperature variation in free-ranging Hognose Snakes (*Heterodon platirhinos*). Journal of Herpetology 44:471–474.

Plummer, M. V., T. N. Lee, and N. E. Mills. 2008. Effect of a sand substrate on the growth and condition of *Apalone mutica* hatchlings. Journal of Herpetology 42:550–554.

Pluto, T. G., and E. D. Bellis. 1988. Seasonal and annual movements of riverine Map Turtles, *Graptemys geographica*. Journal of Herpetology 22:152–158.

Pluto, T. G., and H. Rothenbacher. 1976. *Eimeria juniataensis* sp. n. (Protozoa: Eimeriidae) from the Map Turtle, *Graptemys geographica*, in Pennsylvania. Journal of Parasitology 62:207–208.

Pomara, L. Y., O. E. LeDee, K. J. Martin, and B. Zuckerberg. 2014. Demographic consequences of climate change and land cover help explain a history of extirpations and range contraction in a declining snake species. Global Change Biology 20:2087–2099.

Pope, C. H. 1944. Amphibians and reptiles of the Chicago area. Chicago Natural History Museum Press, Chicago, Illinois, USA.

Pope, C. H. 1947. Amphibians and reptiles of the Chicago area. Chicago Natural History Museum Press, Chicago, Illinois, USA.

Pope, C. H. 1964. Amphibians and reptiles of the Chicago Area. Chicago Natural History Museum Press, Chicago, Illinois, USA.

Pope, S. E., L. Fahrig, and H. G. Merriam. 2000. Landscape complementation and metapopulation effects on Leopard Frog populations. Ecology 81:2498–2508.

Pope, T. E. B. 1923. Hunting Wisconsin rattlesnakes. Milwaukee Public Museum Yearbook 3:13–26.

Pope, T. E. B. 1926. The Massasauga in Wisconsin. Milwaukee Public Museum Yearbook 6:171–181.

Pope, T. E. B. 1928. The Diamond-back Rattlesnake in Wisconsin. Milwaukee Public Museum Yearbook 8:167–177.

Pope, T. E. B. 1930. Wisconsin herpetological notes. Transactions of the Wisconsin Academy of Sciences, Arts and Letters 25:273–284.

Pope, T. E. B. 1931. Wisconsin herpetological notes. Wisconsin Academy of Sciences, Arts and Letters 26:321–329.

Pope, T. E. B., and W. E. Dickinson. 1928. The amphibians and reptiles of Wisconsin. Bulletin of the Public Museum of the City of Milwaukee 8:1–138.

Popescu, V. D., and J. P. Gibbs. 2009. Interactions between climate, beaver activity, and pond occupancy by the cold-adapted Mink Frog in New York State, USA. Biological Conservation 142:2059–2068.

Popescu, V. D., and M. L. Hunter, Jr. 2011. Clear-cutting affects habitat connectivity for a forest amphibian by decreasing permeability to juvenile movements. Ecological Applications 21:1283–1295.

Porej, D., and T. E. Hetherington. 2005. Designing wetlands for amphibians: the importance of predatory fish and shallow littoral zones in structuring of amphibian communities. Wetlands Ecology and Management 13:445–455.

Porej, D., M. Micacchion, and T. E. Hetherington. 2004. Core terrestrial habitat for conservation of local populations of salamanders and Wood Frogs in agricultural landscapes. Biological Conservation 120:399–409.

Poser, J. 1932. Classes. Wisconsin Badger 47:53.

Pough, F. H. 1974. Natural daily temperature acclimation of eastern red efts, *Notophthalmus v. viridescens*

(Rafinesque) (Amphibia: Caudata). Comparative Biochemistry and Physiology 47:71–78.

Pough, F. H. 1976. Acid precipitation and embryonic mortality of Spotted Salamanders. Science 192:68–70.

Pough, F. H., and R. E. Wilson. 1970. Natural daily temperature stress, dehydration, and acclimation in juvenile *Ambystoma maculatum* (Shaw) (Amphibia: Caudata). Physiological Zoology 43:194–205.

Pough, F. H., and S. Kamel. 1984. Post-metamorphic change in activity metabolism of anurans in relation to life history. Oecologia 65:138–144.

Pough, F. H., E. M. Smith, D. H. Rhodes, and A. Collazo. 1987. The abundance of salamanders in forest stands with different histories of disturbance. Forest Ecology and Management 20:1–9.

Powell, G. L., and A. P. Russell. 2007. Life history implications for conservation and monitoring of lizards in Canada. Pages 23–40 *in* C. N. L. Seburn and C. A. Bishop, editors. Ecology, conservation, and status of reptiles in Canada. Society for the Study of Amphibians and Reptiles, Herpetological Conservation 2, Salt Lake City, Utah, USA.

Powell, R., J. T. Collins, and E. D. Hooper, Jr. 1998. A key to amphibians and reptiles of the continental United States and Canada. University of Kansas Press, Lawrence, USA.

Powell, R., R. Conant, and J. T. Collins. 2016. A field guide to reptiles and amphibians of eastern and central North America. Fourth edition. Houghton Mifflin Harcourt Publishing Co., New York, New York, USA.

Prange, S., and S. D. Gehrt. 2004. Changes in mesopredator-community structure in response to urbanization. Canadian Journal of Zoology 82:1804–1817.

Prange, S., S. D. Gehrt, and E. P. Wiggers. 2003. Demographic factors contributing to high Raccoon densities in urban landscapes. Journal of Wildlife Management 67:324–333.

Predick, K. I., and M. G. Turner. 2008. Landscape configuration and flood frequency influence invasive shrubs in floodplain forests of the Wisconsin River (USA). Journal of Ecology 96:91–102.

Price, B., A. S. Kutt, and C. A. McAlpine. 2010. The importance of fine-scale savanna heterogeneity for reptiles and small mammals. Biological Conservation 143:2504–2513.

Price. S. J. 2000. *Hemidactylium scutatum* (Four-toed Salamander). Geographic distribution. Herpetological Review 31:249.

Price. S. J. 2001. *Hemidactylium scutatum* (Four-toed Salamander). Geographic distribution. Herpetological Review 32:53.

Price, S. J., D. R. Marks, R. W. Howe, J. M. Hanowski, and G. J. Niemi. 2004. The importance of spatial scale for conservation and assessment of anuran populations in coastal wetlands of the western Great Lakes, USA. Landscape Ecology 20:441–454.

Price, S. J., J. Guzy, L. Witczak, and M. E. Dorcas. 2013. Do ponds on golf courses provide suitable habitat for wetland-dependent animals in suburban areas? An assessment of turtle abundances. Journal of Herpetology 47:243–250.

Price, S. J., R. W. Howe, J. Hanowski, R. R. Regal, G. J. Niemi, and C. R. Smith. 2007. Are anurans of Great Lakes coastal wetlands reliable indicators of ecological condition? Journal of Great Lakes Research 33:211–223.

Prior, K. A., and C. M. Shilton. 1996. Post-hibernation mortality in Black Rat Snakes, *Elaphe o. obsoleta*. Journal of Herpetology 30:275–278.

Prior, K. A., and P. J. Weatherhead. 1994. Response of free-ranging Eastern Massasauga Rattlesnakes to human disturbance. Journal of Herpetology 28:255–257.

Prior, K. A., and P. J. Weatherhead. 1996. Habitat features of Black Rat Snake hibernacula in Ontario. Journal of Herpetology 30:211–218.

Pritchard, P. C. H. 1979. Encyclopedia of turtles. T. F. H. Publications, Inc., Neptune, New Jersey, USA.

Pritchard, P. C. H. 2003. Akinesis and plastral scute homologies in *Sternotherus* (Testudines: Kinosternidae). Chelonian Conservation and Biology 4:671–674.

Pritchard, P. C. H. 2008. Evolution and structure of the turtle shell. Pages 45–84 *in* J. Wyneken, M. H. Godfrey, and V. Bels, editors. Biology of turtles: from structures to strategies of life. CRC Press, Boca Raton, Florida, USA.

Prober, S. M., I. D. Lunt, and K. R. Thiele. 2002. Determining reference conditions for management and restoration of temperate grassy woodlands: relationships among trees, topsoils, and understorey flora in little-grazed remnants. Australian Journal of Botany 50:687–697.

Prosser, M. R., P. J. Weatherhead, H. L. Gibbs, and G. P. Brown. 2002. Genetic analysis of the mating system and opportunity for sexual selection in Northern Water Snakes (*Nerodia sipedon*). Behavioral Ecology 13:800–807.

Proulx, C. L., G. Fortin, and G. Blouin-Demers. 2014. Blanding's Turtle (*Emydoidea blandingii*) avoid crossing unpaved and paved roads. Journal of Herpetology 48:267–271.

Pryor, S. C., D. Scavia, C. Downer, M. Gaden, L. Iverson, R. Nordstrom, J. Patz, and G. P. Robertson. 2014. Midwest. Pages 418–440 *in* J. M. Melillo, T. C. Richmond, and G. W. Yohe, editors. Climate change impacts in the United States. US Global Change Research Program, US National Climate Assessment, Washington, D.C., USA.

Ptacek, M. B. 1992. Calling sites used by male Gray Treefrogs, *Hyla versicolor* and *Hyla chrysoscelis*, in sympatry and allopatry in Missouri. Herpetologica 48:373–382.

Ptacek, M. B. 1996. Interspecific similarity in life-history traits in sympatric populations of Gray Treefrogs *Hyla chrysoscelis* and *Hyla versicolor*. Herpetologica 52:323–332.

Ptacek, M. S. 1984. Reproductive ecology and habitat analysis of the Northern Spring Peeper (*Hyla c. crucifer*) in southeastern Kansas. Thesis, Emporia State University, Emporia, USA.

Puckette, B. G. 1962. Ophiophagy in *Hyla versicolor*. Herpetologica 18:143.

Pulis, E. E., V. V. Tkach, and R. A. Newman. 2011. Helminth parasites of the Wood Frog, *Lithobates sylvaticus*, in prairie pothole wetlands of the northern Great Plains. Wetlands 31:675–685.

Pultz, S. 2001. Incentives for conservation on private lands. Endangered Species Bulletin 2001:4–5.

Punzo, F. 1975. Studies on the feeding behavior, diet, nesting habits and temperature relationships of *Chelydra serpentina osceola* (Chelonia: Chelydridae). Journal of Herpetology 9:207–210.

Punzo, F. 1990. Feeding ecology of the Six-lined Racerunner (*Cnemidophorus sexlineatus*) in southern Florida. Herpetological Review 21:33–35.

Purgue, A. P. 1997. Tympanic sound radiation in the Bullfrog *Rana catesbeiana*. Journal of Comparative Physiology 181A:438–445.

Purrenhage, J. L., and M. D. Boone. 2005. Effects of habitat structure and competition on larval amphibian development. Integrative and Comparative Biology 45:1180.

Purrenhage, J. L., and M. D. Boone. 2009. Amphibian community response to variation in habitat structure and competitor density. Herpetologica 65:14–30.

Pyburn, W. F. 1958. Size and movements of a local population of Cricket Frogs (*Acris crepitans*). Texas Journal of Science 10:325–342.

Pyburn, W. F. 1960. Hybridization between *Hyla versicolor* and *H. femoralis*. Copeia 1:55–56.

Pyron, R. A., and J. J. Wiens. 2013. Large-scale phylogenetic analyses reveal the causes of high tropical amphibian diversity. Proceedings of the Royal Society of London B: Biological Sciences 280:20131622.

Pyron, R. A., F. T. Burbrink, and J. J. Wiens. 2013. A phylogeny and revised classification of Squamata, including 4161 species of lizards and snakes. BMC Evolutionary Biology 13:93.

Pyron, R. A., F. T. Burbrink, G. R. Colli, A. N. M. de Oca, L. J. Vitt, C. A. Kuczynski, and J. J. Wiens. 2011. The phylogeny of advanced snakes (Colubroidea), with discovery of a new subfamily and comparison of support methods for likelihood trees. Molecular Phylogenetics and Evolution 58:329–342.

Quaife, M. M. 1921. Increase Allen Lapham, father of forest conservation. Wisconsin Magazine of History 5:104–108.

Quammen, J. K., and R. D. Durtsche. 2003. Diet comparisons in three tadpole species, *Rana sylvatica*, *Bufo americanus*, and *Pseudacris crucifer*, in a northern temperate climate. Norse Scientist 1:45–50.

Quaranta, A., V. Bellantuono, G. Cassano, and C. Lippe. 2009. Why amphibians are more sensitive than mammals to xenobiotics. PLoS ONE 4:e7699.

Quinn, H. R., H. Quinn, and A. Higa. 2014. Notes on reproduction and growth of South Dakota Ornate Box Turtles (*Terrapene ornata*). Chelonian Conservation and Biology 13:65–71.

Quinn, N. W. S., and D. P. Tate. 1991. Seasonal movements and habitat of Wood Turtles (*Clemmys insculpta*) in Algonquin Park, Canada. Journal of Herpetology 25:217–220.

Quirt, K. C., G. Blouin-Demers, B. J. Howes, and S. C. Lougheed. 2006. Microhabitat selection of Five-lined Skinks in northern peripheral populations. Journal of Herpetology 40:335–342.

Radaj, R. H. 1981a. *Opheodrys v. vernalis* (Smooth Green Snake). Reproduction. Herpetological Review 12:80.

Radaj, R. H. 1981b. *Rana clamitans* (Green Frog). Coloration. Herpetological Review 12:79.

Radeloff, V. C., D. J. Mladenoff, H. S. He, and M. S. Boyce. 1999. Forest landscape change in the northwestern Wisconsin pine barrens from pre-European settlement to the present. Canadian Journal of Forest Research 29:1649–1659.

Radeloff, V. C., D. J. Mladenoff, K. L. Manies, and M. S. Boyce. 1998. Analyzing forest landscape restoration potential: pre-settlement and current distribution of oak in the northwest Wisconsin pine barrens. Transactions of the Wisconsin Academy of Sciences, Arts and Letters 86:189–205.

Radeloff, V. C., R. B. Hammer, P. R. Voss, A. E. Hagen, D. R. Field, and D. J. Mladenoff. 2001. Human demographic trends and landscape level forest management in the northwest Wisconsin pine barrens. Forest Sciences 47:229–261.

Raffel, T. R., J. R. Dillard, and P. J. Hudson. 2006. Field evidence for leech-borne transmission of amphibian *Ichthyophonus* sp. Journal of Parasitology 92:1256–1264.

Raffel, T. R., J. T. Hoverman, N. T. Halstead, P. J. Michel, and J. R. Rohr. 2010a. Parasitism in a community context: trait-mediated interactions with competition and predation. Ecology 91:1900–1907.

Raffel, T. R., P. J. Michel, E. W. Sites, and J. R. Rohr. 2010b. What drives chytrid infections in newt populations? Associations with substrate, temperature, and shade. EcoHealth 7:526–536.

Raffel, T. R., T. Bommarito, D. S. Barry, S. M. Witiak, and L. A. Schackelton. 2008. Widespread infection of the Eastern Red-spotted Newt (*Notophthalmus viridescens*) by a new species of *Amphibiocystidium*, a genus of fungus-like mesomycetozoan parasites not previously reported in North America. Parasitology 135:203–215.

Rafinesque, C. S. 1818. Discoveries in natural history, made during a journey through the western region of the United States. American Monthly Magazine and Critical Review 3:354–355.

Rafinesque, C. S. 1820. Annual synopsis of new genera and species of animals and plants discovered in North America. Annals of Nature 1:1–20.

Raimond, C. 2017. *Graptemys geographica* (Northern Map Turtle). Geographic distribution. Herpetological Review 48:386.

Raimond, C., and E. Thomsen. 2016. *Ophyodrys vernalis* (Smooth Green Snake). Geographic distribution. Herpetological Review 47:631.

Raimond, C., and J. M. Lorch. 2012. *Tropidoclonion lineatum* (Lined Snake). Geographic distribution. Herpetological Review 43:108.

Raimondo, S. M., C. L. Rowe, and J. D. Congdon. 1998. Exposure to coal ash impacts swimming performance and predator avoidance in larval Bullfrogs (*Rana catesbeiana*). Journal of Herpetology 32:289–292.

Raithel, C. J. 1992. The extirpation of the Timber Rattlesnake from Rhode Island. Pages 5–6 *in* T. F. Tyning, editor. Conservation of the Timber Rattlesnake in the northeast. Massachusetts Audubon Society, Lincoln, USA.

Ralin, D. B. 1976. Comparative hybridization of a diploid-tetraploid cryptic species pair of treefrogs. Copeia 1976:191–196.

Ralin, D. B. 1977. Evolutionary aspects of mating call variation in a diploid-tetraploid species complex of treefrogs (Anura). Evolution 31:721–736.

Ralin, D. B., and J. S. Rogers. 1972. Aspects of tolerance to desiccation in *Acris crepitans* and *Pseudacris streckeri*. Copeia 1972:519–525.

Ramminger, A. 1992. Habitat evaluation of Ornate Box Turtles in Wisconsin. Wisconsin Department of Natural Resources, Madison, USA.

Ramsey, L. W. 1946. Captive specimens of *Tropidoclonion lineatum*. Herpetologica 3:112.

Ramsey, L. W. 1947. Feeding behavior of *Tropidoclonion lineatum*. Herpetologica 4:15–18.

Ramsey, L. W. 1953. The Lined Snake, *Tropidoclonion lineatum* (Hallowell). Herpetologica 9:7–24.

Raney, E. C. 1940. Summer movements of the Bullfrog, *Rana catesbeiana* Shaw, as determined by the jaw-tag method. American Midland Naturalist 23:733–745.

Raney, E. C., and R. A. Josephson. 1954. Record of combat in the Snapping Turtle, *Chelydra serpentina*. Copeia 3:228.

Raney, E. C., and R. M. Roecker. 1947. Food and growth of two species of watersnakes from western New York. Copeia 1947:171–174.

Raney, E. C., and W. M. Ingram. 1941. Growth of tagged frogs (*Rana catesbeiana* Shaw and *Rana clamitans* Daudin) under natural conditions. American Midland Naturalist 26:201–206.

Rankin, J. S., Jr. 1945. An ecological study of the helminth parasites of amphibians and reptiles of western Massachusetts and vicinity. Journal of Parasitology 31:142–150.

Ransom, R. E., and M. V. Plummer. 1999. *Cnemidophorus sexlineatus*. Activity. Herpetological Review 30:96–97.

Ransom, T. S. 2011. Earthworms, as ecosystem engineers, influence multiple aspects of a salamander's ecology. Oecologia 165:745–754.

Ransom, T. S., and R. G. Jaeger. 2006. An assemblage of salamanders in the southern Appalachian Mountains revisited: competitive and predatory behavior? Behaviour 143:1357–1382.

Rao, R. B., M. B. Palmer, and M. Touger. 1998. Thrombocytopenia after rattlesnake envenomation [letter]. Annals of Emergency Medicine 31:139–140.

Ratner, N., and R. Anderson. 1978. Population, nesting, movement, habitat, and thermoregulation studies of the Wood Turtles in west central Wisconsin. University of Wisconsin Resource Report, Stevens Point, USA.

Rau, M. E., and D. M. Gordon. 1980. Host specificity among the helminth parasites for four species of snakes. Canadian Journal of Zoology 58:929–930.

Rausch, R. 1946. New host records for *Microphallus ovatus* Osborn, 1919. Journal of Parasitology 32:93–94.

Rausch, R. 1947. Observations on some helminths parasitic in Ohio turtles. American Midland Naturalist 38:434–442.

Ravesi, M. J., S. J. Tetzlaff, and B. A. Kingsbury. 2015. *Sistrurus catenatus* (Massasauga). Overwintering site. Herpetological Review 46:454.

Ravesi, M. J., S. J. Tetzlaff, M. C. Allender, and B. A. Kingsbury. 2016. Detection of snake fungal disease from a *Lampropeltis triangulum* (Eastern Milksnake) in northern Michigan. Northeastern Naturalist Notes 23:N18–N23.

Rawling, J. E., III, P. R. Hanson, A. R. Young, and J. W. Attig. 2008. Late Pleistocene dune construction in the central sand plain of Wisconsin, USA. Geomorphology 100:494–505.

Ray J. W., R. B. King, M. B. Duvall, J. W. Robinson, C. P. Jaeger, M. J. Dreslik, B. J. Swanson, and D. Mulkerin. 2013. Genetic analysis and captive breeding program design for the Eastern Massasauga *Sistrurus catenatus catenatus*. Journal of Fish and Wildlife Management 1013:104–113.

Raymond, L. R., and L. M. Hardy. 1991. Effects of a clearcut on a population of the Mole Salamander, *Ambystoma talpoideum*, in an adjacent unaltered forest. Journal of Herpetology 25:509–512.

Reading, C. J., L. M. Luiselli, G. C. Akani, X. Bonnet, G. Amori, J. M. Ballouard, E. Filippi, G. Naulleau, D. Pearson, and L. Rugiero. 2010. Are snake populations in widespread decline? Biology Letters 6:777–780.

Ream, C., and R. Ream. 1966. The influence of sampling methods on the estimation of population structure in Painted Turtles. American Midland Naturalist 75:325–338.

Ream, C. H. 1967. Some aspects of the ecology of Painted Turtles, Lake Mendota, Wisconsin. Dissertation, University of Wisconsin, Madison, USA.

Redder, A. J., B. E. Smith, and D. A. Keinath. 2006. Smooth Green Snake (*Opheodrys vernalis*): a technical conservation assessment. US Forest Service, Rocky Mountain Region, Laramie, Wyoming, USA.

Redmer, M. 1987. Cannibalism in captive *Hyla crucifer*. Bulletin of the Chicago Herpetological Society 22:75.

Redmer, M. 1998. Status and distribution of two uncommon frogs, Pickerel Frogs and Wood Frogs, in Illinois. Pages 83–90 *in* M. J. Lannoo, editor. Status and conservation of midwestern amphibians. University of Iowa Press, Iowa City, USA.

Redmer, M. 2005. *Rana palustris* (LeConte, 1825), Pickerel Frog. Pages 568–570 *in* M. J. Lannoo, editor. Amphibian declines: the conservation status of United States species. University of California Press, Berkeley, USA.

Redmer, M., and J. P. Zaworski. 1987. Notes on two red Plains Garter Snakes, *Thamnophis radix radix*, from Illinois. Bulletin of the Chicago Herpetological Society 22:179.

Redmer, M., and K. S. Mierzwa. 1994. A review of the distribution and zoogeography of the Pickerel Frog, *Rana palustris*, in northern Illinois. Bulletin of the Chicago Herpetological Society 29:21–30.

Reed, D. M., and R. E. Lang. 1991. *Pseudacris crucifer crucifer* (Northern Spring Peeper). Geographic distribution. Herpetological Review 22:102.

Reed, D. M., and R. E. Lang. 1992. *Rana sylvatica* (Wood Frog). Geographic distribution. Herpetological Review 23:25.

Reed, M., and J. R. Hodgson. 1987. *Necturus maculosus maculosus* (Mudpuppy). Geographic distribution. Herpetological Review 18:39.

Reeder, A. L., M. O. Ruiz, A. Pessier, L. E. Brown, J. M. Levengood, C. A. Phillips, M. B. Wheeler, R. E. Warner and V. R. Beasley. 2005. Intersexuality and the Cricket Frog decline: historic and geographic trends. Environmental Health Perspectives 113:261–265.

Reeder, T. W., C. J. Cole, and H. C. Dessauer. 2002. Phylogenetic relationships of whiptail lizards of the genus *Cnemidophorus* (Squamata: Teiidae): a test of monophyly, reevaluation of karyotypic evolution, and review of hybrid origins. American Museum Novitates 3365:1–61.

Reehl, M., J. Thompson, and J. K. Tucker. 2006. A three year survey of aquatic turtles in a riverside pond. Transactions of the Illinois State Academy of Science 99:145–152.

Reese, S. A., C. E. Crocker, M. E. Carwile, D. C. Jackson, and G. R. Ultsch. 2001. The physiology of hibernation in Common Map Turtles (*Graptemys geographica*). Comparative Biochemistry and Physiology Part A: Molecular & Integrative Physiology 130:331–340.

Reese, S. A., D. C. Jackson, and G. R. Ultsch. 2003. Hibernation in freshwater turtles: Softshell Turtles (*Apalone spinifera*) are the most intolerant of anoxia among North American species. Journal of Comparative Physiology 173:263–268.

Reese, S. A., E. R. Stewart, C. E. Crocker, D. C. Jackson, and G. R. Ultsch. 2004. Geographic variation of the physiological response to overwintering in the Painted Turtle (*Chrysemys picta*). Physiological and Biochemical Zoology 77:619–630.

Refsnider, J. M. 2009. High frequency of multiple paternity in Blanding's Turtle (*Emys blandingii*). Journal of Herpetology 43:74–81.

Refsnider, J. M., and M. H. Linck. 2012. Habitat use and movement patterns of Blanding's Turtles (*Emydoidea blandingii*) in Minnesota, USA: a landscape approach to species conservation. Herpetological Conservation and Biology 7:185–195.

Refsnider, J. M., J. Strickland, and F. J. Janzen. 2012. Home range and site fidelity of imperiled Ornate Box Turtles (*Terrapene ornata*) in northwestern Illinois. Chelonian Conservation and Biology 11:78–83.

Regester, K. J., H. C. Eisenman, J. L. Stragand, and C. S. Brooks. 2016. First reported detection of the amphibian pathogens *Batrachochytrium dendrobatidis* and *Ranavirus* in the common Mudpuppy (*Necturus m. maculosus*). Herpetological Review 47:397–400.

Regester, K. J., K. R. Lipps, and M. R. Whiles. 2006. Energy flow and subsidies associated with complex life cycle of ambystomatid salamanders in ponds and adjacent forest in southern Illinois. Oecologia 147:303–314.

Regosin, J. V., B. S. Windmiller, R. N. Homan, and J. M. Reed. 2003. Terrestrial habitat use and winter densities of the Wood Frog (*Rana sylvatica*). Journal of Herpetology 37:390–394.

Regosin, J. V., B. S. Windmiller, R. N. Homan, and J. M. Reed. 2005. Variation in terrestrial habitat use by four pool-breeding amphibian species. Journal of Wildlife Management 69:1481–1493.

Reichenbach, N. G., and G. H. Dalrymple. 1986. Energy use, life-histories, and the evaluation of potential competition in two species of garter snake. Journal of Herpetology 20:133–153.

Reid, B. N., and M. Z. Peery. 2014. Land use patterns skew sex ratios, decrease genetic diversity and trump the effects of recent climate change in an endangered turtle. Diversity and Distributions 20:1425–1437.

Reid, B. N., and P. C. Riddle. 2017a. *Hemidactylium scutatum* (Four-toed Salamander). Geographic distribution. Herpetological Review 48:805.

Reid, B. N., and P. C. Riddle. 2017b. *Heterodon platirhinos* (Eastern Hog-nosed Snake). Geographic distribution. Herpetological Review 48:814.

Reid B. N., R. Thiel, and M. Z. Perry. 2016. Population dynamics of endangered Blanding's Turtles in a restored area. Journal of Wildlife Management 80:553–562.

Reid, G. K. 1961. Ecology of inland waters and estuaries. Reinhold Book Corporation, New York, New York, USA.

Reigle, N. J. 1967. The occurrence of *Necturus* in the deeper waters of Green Bay. Herpetologica 23:232–233.

Reilly, S. M., G. V. Lauder, and J. P. Collins. 1992. Performance consequences of a trophic polymorphism: feeding behavior in typical and cannibal phenotypes of *Ambystoma tigrinum*. Copeia 1992:672–679.

Reinartz, J. A. 1997. Controlling Glossy Buckthorn (*Rhamnus frangula* L.) with winter herbicide treatments of cut stumps. Natural Areas Journal 17:38–41.

Reinert, H. K. 1981. Reproduction by the Massasauga (*Sistrurus catenatus catenatus*). American Midland Naturalist 105:393–395.

Reinert, H. K. 1984a. Habitat separation between sympatric snake populations. Ecology 65:478–486.

Reinert, H. K. 1984b. Habitat variation within sympatric snake populations. Ecology 65:1673–1682.

Reinert, H. K. 1991. Translocation as a conservation strategy for amphibians and reptiles: some comments, concerns, and observations. Herpetologica 7:337–363.

Reinert, H. K. 2005. A telemetric study of the survivorship, behavior, and spatial ecology of neonatal Timber Rattlesnakes, *Crotalus horridus*. Page 45 *in* Program and Abstracts of the Biology of the Rattlesnakes Symposium. Loma Linda University, Loma Linda, California.

Reinert, H. K., and R. R. Rupert. 1999. Impacts of translocation on behavior and survival of Timber Rattlesnakes, *Crotalus horridus*. Journal of Herpetology 33:45–61.

Reinert, H. K., and W. R. Kodrich. 1982. Movements and habitat utilization by the Massasauga, *Sistrurus catenatus catenatus*. Journal of Herpetology 16:162–171.

Reinert, H. K., D. Cundall, and L. M. Bushar. 1984. Foraging behavior of the Timber Rattlesnake, *Crotalus horridus*. Copeia 1984:976–981.

Relyea, R. A. 2000a. Morphological plasticity in four larval anurans distributed along an environmental gradient. Copeia 2000:178–190.

Relyea, R. A. 2000b. Trait-mediated indirect effects in larval anurans: revising competition with the threat of predation. Ecology 81:2278–2289.

Relyea, R. A. 2001a. Morphological and behavioral plasticity of larval anurans in response to different predators. Ecology 82:523–540.

Relyea, R. A. 2001b. The lasting effects of adaptive plasticity: predator-induced tadpoles become long-legged frogs. Ecology 82:1947–1955.

Relyea, R. A. 2001c. The relationship between predation risk and antipredator responses larval anurans. Ecology 82:541–554.

Relyea, R. A. 2005a. The impact of insecticides and herbicides on the biodiversity and productivity of aquatic communities. Ecological Applications 15:618–627.

Relyea, R. A. 2005b. The lethal impact of Roundup on aquatic and terrestrial amphibians. Ecological Applications 15:1118–1124.

Relyea, R. A. 2005c. The lethal impacts of Roundup and predatory stress on six species of North American tadpoles. Archives of Environmental Contamination and toxicology 48:351–357.

Relyea, R. A. 2012. New effects of Roundup on amphibians: predators reduce herbicide mortality; herbicides induce antipredator morphology. Ecological Applications 22:634–647.

Relyea, R. A., and D. K. Jones. 2009. The toxicity of Roundup Original Max to 13 species of larval amphibians. Environmental Toxicology and Chemistry 28:2004–2008.

Relyea, R. A., and E. E. Werner. 1999. Quantifying the relation between predator-induced behavior and growth performance in larval anurans. Ecology 80:2117–2124.

Relyea, R. A., and J. R. Auld. 2004. Having the guts to compete: how intestinal plasticity explains costs of inducible defences. Ecology Letters 7:869–875.

Relyea, R. A., and K. L. Yurewicz. 2002. Predicting community outcomes from pairwise interactions: integrating density- and trait-mediated effects. Oecologia 131:569–579.

Relyea, R. A., and N. Mills. 2001. Predator-induced stress makes the pesticide carbaryl more deadly to Gray Treefrog tadpoles (*Hyla versicolor*). Proceedings of the National Academy of Sciences of the United States of America 98:2491–2496.

Relyea, R. A., N. M. Schoeppner, and J. T. Hoverman. 2005. Pesticides and amphibians: the importance of community context. Ecological Applications 15:1125–1134.

Remsberg, A. J., T. L. Lewis, P. W. Huber, and K. A. Asmus. 2006. Home ranges of Wood Turtles (*Glyptemys insculpta*) in northern Michigan. Chelonian Conservation and Biology 5:42–47.

Renfrew, R. B., and C. A. Ribic. 2003. Grassland passerine nest predators near pasture edges identified on videotape. Auk 120:371–383.

Renken, R. B., W. K. Gram, D. K. Fantz, S. C. Richter, T. J. Miller, K. B. Ricke, B. Russell, and X. Wang. 2004. Effects of forest management on amphibians and reptiles in Missouri Ozark forests. Conservation Biology 18:174–188.

Resetarits, W. J. 1998. Differential vulnerability of *Hyla chrysoscelis* eggs and hatchlings to larval insect predators. Journal of Herpetology 32:440–443.

Resetarits, W. J., and H. M. Wilbur. 1989. Choice of oviposition site by *Hyla chrysoscelis*: role of predators and competitors. Ecology 70:220–228.

Reynolds, J. W. 2012. The status of terrestrial earthworms (Oligochaeta) surveys in North America and some Caribbean countries. Megadrilogica 15:227–249.

Reynolds, J. W., and M. J. Wetzel. 2012. Terrestrial Oligochaeta (Annelida: Clitellata) in North America, including Mexico, Puerto Rico, Hawaii, and Bermuda. III. Megadrilogica 15:191–211.

Reynolds, M., C. Blakeslee, T. Paciorek, and S. P. McRobert. 2011. The effects of population density on survival and metamorphosis in American Toad (*Bufo americanus*) tadpoles. Russian Journal of Herpetology 18:241–246.

Rhemtulla, J. M., D. J. Mladenoff, and M. K. Clayton. 2007. Regional land-cover conversion in the US upper Midwest: magnitude of change and limited recovery (1850–1935–1993). Landscape Ecology 22:57–75.

Rhemtulla, J. M., D. J. Mladenoff, and M. K. Clayton. 2009. Legacies of historical land use on regional forest composition and structure in Wisconsin, USA (mid-1800s-1930s-2000s). Ecological Applications 19:1061–1078.

Rhodin, A. G. J., J. B. Iverson, R. Bour, U. Fritz, A. Georges, H. B. Shaffer, and P. P. van Dijk. 2017. Turtles of the world: annotated checklist and atlas of taxonomy, synonymy, distribution, and conservation status. Eighth edition. Chelonian Research Monographs 7:1–292.

Ribeiro, M. C., J. P. Metzger, A. C. Martensen, F. J. Ponzoni, and M. M. Hirota. 2009. The Brazilian Atlantic Forest: how much is left, and how is the remaining forest distributed? Implications for conservation. Biological Conservation 142:1141–1153.

Richards, C., L. B. Johnson, and G. E. Host. 1996. Landscape-scale influences on stream habitats and biota. Canadian Journal of Fisheries and Aquatic Sciences 53:295–311.

Richards, C. M., and G. W. Nace. 1983. Dark pigment variants in anurans: classification, new descriptions, color changes, and inheritance. Copeia 1983:979–990.

Richards-Dimitrie, T. M., S. E. Gresens, S. A. Smith, and R. A. Seigel. 2013. Diet of Northern Map Turtles (*Graptemys geographica*): sexual differences and potential impacts of an altered river system. Copeia 2013:477–484.

Richards-Hrdlicka, K. L., J. L. Richardson, and L. Mohabir. 2013. First survey for the amphibian chytrid fungus *Batrachochytrium dendrobatidis* in Connecticut (USA) finds widespread prevalence. Diseases of Aquatic Organisms 102:169–180.

Richardson, D. L. 2001a. The relative roles of adaptation and phylogeny in determination of larval traits in diversifying amphibian lineages. American Naturalist 157:282–299.

Richardson, D. M., P. Pysek, M. Rejmanek, M. G. Barbour, F. D. Panetta, and C. J. West. 2000. Naturalization and invasion of alien plants: concepts and definitions. Diversity and Distributions 6:93–107.

Richardson, J. L. 2012. Divergent landscape effects on population connectivity in two co-occurring amphibian species. Molecular Ecology 18:4437–4451.

Richardson, J. L., and M. M. Brinson. 2000. Wetland soils and the hydrogeomorphic classification of wetlands. Pages 209–227 *in* J. L. Richardson, and M. J. Vepraskas, editors. Wetland soils: genesis, hydrology, landscapes, and classification. Lewis Publishers, Boca Raton, Florida, USA.

Richardson, J. M. L. 2001b. A comparative study of activity levels in larval anurans and response to the presence of different predators. Behavioral Ecology 12:51–58.

Richmond, L. S., and S. C. Trombulak. 2009. Distribution of Red-backed Salamander (*Plethodon cinereus*) with respect to cover-object characteristics in the Green Mountains of Vermont. Northeastern Naturalist 16:13–26.

Richtsmeier, R. J., N. P. Bernstein, J. W. Demastes, and R. W. Black. 2008. Migration, gene flow, and genetic diversity within and among Iowa populations of Ornate Box Turtles (*Terrapene ornata ornata*). Chelonian Conservation and Biology 7:3–11.

Rickard, A. 2006a. *Acris crepitans blanchardi* (Blanchard's Cricket Frog). Predation. Herpetological Review 37:439.

Rickard, A. 2006b. Assessment of translocations of Blanchard's Cricket Frog (*Acris crepitans blanchardi*) in southeast Michigan. Endangered Species UPDATE 23:3–14.

Riedel, B. L., K. R. Russell, and W. M. Ford. 2012. Physical condition, sex, and age-class of Eastern Red-backed Salamanders (*Plethodon cinereus*) in forested and open habitats of West Virginia, USA. International Journal of Zoology 2012:1–8.

Riedle, J. D. 2012. Demography of an urban population of Ring-necked Snakes (*Diadophis punctatus*) in Missouri. Herpetological Conservation and Biology 9:278–284.

Riedle, J. D., P. A. Shipman, and S. F. Fox. 2009. Habitat associations of aquatic turtle communities in eastern Oklahoma. Proceedings of the Oklahoma Academy of Sciences 89:11–22.

Riley, C. M., R. V. Anderson, and S. E. Jenkins. 2005. Nesting habitat of the Softshell Turtle, *Trionyx mutica*, below Lock and Dam 19, Upper Mississippi River. Journal of Freshwater Ecology 20:513–517.

Riley, E. E., and M. R. Weil. 1986. The effects of thiosemicarbazide on development in the Wood Frog, *Rana sylvatica*. I. Concentration effects. Ecotoxicology and Environmental Safety 12:154–160.

Riley, J. L., P. D. Moldowan, and J. D. Litzgus, 2015. *Sistrurus catenatus* (Eastern Massasauga) and *Nerodia sipedon* (Northern Watersnake). Diet and foreign object. Herpetological Review 46:107.

Rimer, R. L., and J. T. Briggler. 2010. Occurrence of amphibian chytrid fungus (*Batrachochytrium dendrobatidis*) in Ozark caves, Missouri, USA. Herpetological Review 41:175–177.

Rinella, M. J., B. D. Maxwell, P. K. Fay, T. Weaver, and R. L. Sheley. 2009. Control effort exacerbates invasive-species problem. Ecological Applications 19:155–162.

Ríos-Saldaña, C. A., M. Delibes-Mateos, and C. C. Ferreira. 2018. Are fieldwork studies being regulated to second place in conservation science? Global Ecology and Conservation 14:e00389.

Ripple, W. J., and R. L. Beschta. 2011. Trophic cascades in Yellowstone: the first 15 years after wolf reintroduction. Biological Conservation 145:205–213.

Risley, P. L. 1933. Observations on the natural history of the Common Musk Turtle, *Sternotherus odoratus* (Latreille). Papers of the Michigan Academy of Science, Arts and Letters 17:685–711.

Ritke, M. E., and C. A. Lessman. 1994. Longitudinal study of ovarian dynamics in female Gray Treefrogs (*Hyla chrysoscelis*). Copeia 1994:1014–1022.

Ritke, M. E., and J. G. Babb. 1991. Behavior of the Gray Treefrog (*Hyla chrysoscelis*) during the nonbreeding season. Herpetological Review 22:5–8.

Ritke, M. E., and R. D. Semlitsch. 1991. Mating behavior and determinants of male mating success in the Gray Treefrog, *Hyla chrysoscelis*. Canadian Journal of Zoology 69:246–250.

Ritke, M. E., J. G. Babb, and M. K. Ritke. 1990. Life history of the Gray Treefrog (*Hyla chrysoscelis*) in western Tennessee. Journal of Herpetology 24:135–141.

Ritke, M. E., J. G. Babb, and M. K. Ritke. 1992. Temporal patterns of reproductive activity in the Gray Treefrog (*Hyla chrysoscelis*). Journal of Herpetology 26:107.

Rittenhouse, T. A. G. 2011. Anuran larval habitat quality when Reed Canary Grass is present in wetlands. Journal of Herpetology 45:491–496.

Rittenhouse, T. A. G., and R. D. Semlitsch. 2006. Grasslands as movement barriers for a forest-associated salamander: migration behavior of adult and juvenile salamanders at a distant habitat edge. Biological Conservation 131:14–22.

Rittenhouse, T. A. G., and R. D. Semlitsch. 2007. Postbreeding habitat use of Wood Frogs in a

Missouri oak-hickory forest. Journal of Herpetology 41:645–653.

Rittenhouse, T. A. G., and R. D. Semlitsch. 2009. Behavioral response of migrating Wood Frogs to experimental timber harvest surrounding wetlands. Canadian Journal of Zoology 87:618–625.

Rittenhouse, T. A. G., E. B. Harper, L. R. Rehard, and R. D. Semlitsch. 2008. The role of microhabitats in the desiccation and survival of anurans in recently harvested oak-hickory forest. Copeia 2008:807–814.

Rittenhouse, T. A. G., M. C. Doyle, C. R. Mank, B. B. Rothermel, and R. D. Semlitsch. 2004. Substrate cues influence habitat selection by Spotted Salamanders. Journal of Wildlife Management 68:1151–1158.

Rittenhouse, T. A. G., R. D. Semlitsch, and F. R. Thompson, III. 2009. Survival costs associated with Wood Frog breeding migrations: effects of timber harvest and drought. Ecology 90:1620–1630.

Rittman, S. E., E. Muths, and D. E. Green. 2003. *Pseudacris triseriata* (Western Chorus Frog) and *Rana sylvatica* (Wood Frog). Chytridiomycosis. Herpetological Review 34:53.

Rizkalla, C. E., and R. K. Swihart. 2006. Community structure and differential responses of aquatic turtles to agriculturally induced habitat fragmentation. Landscape Ecology 21:1361–1375.

Robbins, C. S., D. Bystrak, and P. Geissler. 1986. The breeding bird survey: its first fifteen years, 1965–1979. US Fish and Wildlife Service Resource Publication 157, Washington, D.C., USA.

Roberts, A. M., and E. B. Liebgold. 2008. The effects of perceived mortality risk on habitat selection in a terrestrial salamander. Behavioral Ecology 19:621–626.

Roberts, H. P., M. T. Jones, L. L. Willey, T. S. B. Akre, P. R. Sievert, P. deMaynadier, K. D. Gipe, G. Johnson, J. Kloepfer, M. Marchand, J. Megyesy, S. Parren, E. Thompson, C. Urban, D. Yorks, B. Zarate, L. Erb, A. M. Ross, J. Dragon, L. Johnson, E. Lassiter, and E. Lassiter. 2021. Large-scale collaboration reveals landscape-level effects of land-use on turtle demography. Global Ecology and Conservation 30:e01759.

Roberts, W., and V. Lewin. 1979. Habitat utilization and population densities of the amphibians of northeastern Alberta. Canadian Field-Naturalist 93:144–154.

Robertson, I. C., and P. J. Weatherhead. 1992. The role of temperature in microhabitat selection by Northern Water Snakes (*Nerodia sipedon*). Canadian Journal of Zoology 70:417–422.

Robertson, J. 2010. Mimicry: experimental design and scientific logic. Pages 51–62 *in* S. Braude and B. S. Low, editors. An introduction to methods and models in ecology, evolution, and conservation biology. Princeton University Press, Princeton, New Jersey, USA.

Robillard, A., and B. Johnson. 2015. Edge effects on Eastern Massasauga Rattlesnakes basking in managed habitat. Northeastern Naturalist 22:200–208.

Robinson, C., and J. R. Bider. 1988. Nesting synchrony: a strategy to decrease predation of Snapping Turtle (*Chelydra serpentina*) nests. Journal of Herpetology 22:470–473.

Robinson, J. 2006. Fish and wildlife deaths due to botulism Type E. Pages 15–19 *in* Lake Erie lakewide management plan. US Environmental Protection Agency, Chicago, Illinois, USA.

Robinson, S. K., and D. S. Wilcove. 1994. Forest fragmentation in the temperate zone and its effects on migratory songbirds. Bird Conservation International 4:233–249.

Roble, S. M. 1979. Dispersal movements and plant associations of juvenile Gray Treefrogs, *Hyla versicolor* Le Conte. Transactions of the Kansas Academy of Science 82:235–245.

Roble, S. M. 1985. Observations on satellite males in *Hyla chrysoscelis*, *Hyla picta*, and *Pseudacris triseriata*. Journal of Herpetology 19:432–436.

Robson, L. E. 2011. The spatial ecology of Eastern Hognose Snakes (*Heterodon platirhinos*): habitat selection, home range size, and the effect of roads on movement patterns. Thesis, University of Ottowa, Ottawa, Ontario, Canada.

Robson, L. E., and G. Blouin-Demers. 2013. Eastern Hognose Snakes (*Heterodon platirhinos*) avoid crossing paved roads, but not unpaved roads. Copeia 3:507–511.

Rocek, Z. 2000. Mesozoic anurans. Pages 1295–1331 *in* H. Heatwole and R. L. Carroll. Amphibian biology. Volume 4. Paleontology: the evolutionary history of amphibians. Surrey Beatty & Sons Pty Ltd, Chipping Norton, England.

Rodriguez, E. M., T. Gamble, M. V. Hirt, and S. Cotner. 2009. Presence of *Batrachochytrium dendronatidis* at the headwaters of the Mississippi River, Itasca State Park, Minnesota, USA. Herpetological Review 40:48–50.

Rodríguez-Robles, J. A. 2002. Feeding ecology of North American Gopher Snakes (*Pituophis catenifer*, Colubridae). Biological Journal of the Linnean Society 77:165–183.

Rodríguez-Robles, J. A., and J. M. De Jesús-Escobar. 1999. Molecular systematics of new world lampropeltine snakes (Colubridae): implications for biogeography and evolution of food habits. Biological Journal of the Linnean Society 68:355–385.

Roe, A. W., and K. L. Grayson. 2008. Terrestrial movements and habitat use of juvenile and emigrating adult Eastern Red-spotted Newts, *Notophthalmus viridescens.* Journal of Herpetology 42:22–30.

Roe, J. H., B. A. Kingsbury, and N. R. Herbert. 2003. Wetland and upland use patterns in semi-aquatic snakes: implications for wetland conservation. Wetlands 23:1003–1014.

Roe, J. H., B. A. Kingsbury, N. R. Herbert. 2004. Comparative water snake ecology: conservation of mobile animals that use temporally dynamic resources. Biological Conservation 118:79–89.

Roe, J. H., J. Gibson, and B. A. Kingsbury. 2006. Beyond the wetland border: estimating the impact of roads for two species of water snakes. Biological Conservation 130:161–168.

Roe, J. H., M. R. Frank, and B. A. Kingsbury. 2015. Experimental evaluation of captive-rearing practices to improve success of snake reintroductions. Herpetological Conservation and Biology 10:711–722.

Rogers, L. L., S. A. Mansfield, K. Hornby, S. Hornby, T. D. Debruyn, M. Mize, R. W. Clark, and G. M. Burghardt. 2014. Black Bear reactions to venomous and non-venomous snakes in eastern North America. Ethology 120:641–651.

Rohr, J. R., A. M. Schotthoefer, T. R. Raffel, H. J. Carrick, N. Halstead, J. T. Hoverman, C. M. Johnson, L. B. Johnson, C. Lieske, M. D. Piwoni, P. K. Schoff, and V. R. Beasley. 2008a. Agrochemicals increase trematode infections in a declining amphibian species. Nature 455:1235–1239.

Rohr, J. R., T. R. Raffel, and C. A. Hall. 2010. Developmental variation in resistance and tolerance in a multi-host-parasite system. Functional Ecology 24:1110–1121.

Rohr, J. R., T. R. Raffel, S. K. Sessions, and P. J. Hudson. 2008b. Understanding the net effects of pesticides on amphibian trematode infections. Ecological Applications 18:1743–1753.

Rojas, S., K. Richards, J. K. Jancovich, and E. W. Davidson. 2005. Influences of temperature on ranavirus infection in larval salamanders *Ambystoma tigrinum.* Diseases of Aquatic Organisms 63:95–100.

Rollins-Smith, L. A., and J. M. Conlon. 2005. Antimicrobial peptide defenses against chytridiomycosis, an emerging infectious disease of amphibian populations. Developmental and Comparative Immunology 29:589–598.

Rollins-Smith, L. A., C. Carey, J. Longcore, J. K. Doersam, A. Boutte, J. E. Bruzgal, and J. M. Conlon. 2002. Activity of antimicrobial skin peptides from ranid frogs against *Batrachochytrium dendrobatidis*, the chytrid fungus associated with global amphibian declines. Developmental and Comparative Immunology 26:471–479.

Rood, S. B., G. M. Samuelson, J. H. Braatne, C. R. Gourley, F. M. Hughes, and J. M. Mahoney. 2005. Managing river flows to restore floodplain forests. Frontiers in Ecology and the Environment 3:193–201.

Root, R. W. 1949. Aquatic respiration in the Musk Turtle. Physiological Zoology 22:172–178.

Rorabaugh, J. C. 2005. *Rana pipiens* Schreber, 1782. Pages 570–577 *in* M. J. Lannoo, editor. Amphibian declines: conservation status of United States species. University of California Press, Berkeley, USA.

Rose, B. E. 1922. Reptiles of Wisconsin. Thesis, University of Wisconsin, Madison, USA.

Rose, F. L. 1986. Carapace regeneration in *Terrapene* (Chelonia: Testudinidae). Southwestern Naturalist 31:131–134.

Rosen, M., and R. E. Lemon. 1974. The vocal behavior of Spring Peepers, *Hyla crucifer*. Copeia 1974:940–950.

Rosen, P. C. 1991a. Comparative ecology and life history of the Racer (*Coluber constrictor*) in Michigan. Copeia 1991:897–909.

Rosen, P. C. 1991b. Comparative field study of thermal preferenda in garter snakes (*Thamnophis*). Journal of Herpetology 25:301–312.

Rosenberg, D. K., B. R. Noon., and E. C. Meslow. 1997. Biological corridors: form, function, and efficacy. BioScience 47:677–687.

Rosenfield, R. N., M. W. Gratson, and L. B. Carson. 1984. Food brought by Broad-Winged Hawks to a Wisconsin nest. Journal of Field Ornithology 55:246–247.

Ross, D. A. 1988. *Chrysemys picta* (Painted Turtle). Predation. Herpetological Review 19:85–87.

Ross, D. A. 1989a. Amphibians and reptiles in the diets of North American raptors. Wisconsin Department of Natural Resources Endangered Resources Report 59, Madison, USA.

Ross, D. A. 1989b. Population ecology of Painted and Blanding's Turtles (*Chrysemys picta* and *Emydoidea blandingii*) in central Wisconsin. Wisconsin Academy of Sciences, Arts and Letters 77:77–84.

Ross, D. A., K. N. Brewster, R. K. Anderson, N. Ratner, and C. M. Brewster. 1991. Aspects of the ecology of Wood Turtles, *Clemmys insculpta*, in Wisconsin. Canadian Field-Naturalist 105:363–367.

Ross, D. A., and R. K. Anderson. 1990. Habitat use, movements, and nesting of *Emydoidea blandingi* in central Wisconsin. Journal of Herpetology 24:6–12.

Ross, P., Jr., and D. Crews. 1978. Stimuli influencing mating behavior in the Garter Snake, *Thamnophis radix*. Behavioral Ecology and Sociobiology 4:133–142.

Rossi, J. V., and R. Rossi. 1995. Snakes of the United States and Canada. Volume 2. Western area. Krieger Publishing Co., Malabra, Florida, USA.

Rossman, D. A. 1962. *Thamnophis proximus* (Say), a valid species of garter snake. Copeia 1962:741–748.

Rossman, D. A. 1963a. Relationships and taxonomic status of the North American natricine snake genera *Liodytes*, *Regina*, and *Clonophis*. Louisiana State University, Occasional Papers of the Museum of Zoology 29:1–29.

Rossman, D. A. 1963b. The colubrid snake genus *Thamnophis*: a revision of the *sauritus* group. Bulletin of the Florida State Museum of Biological Sciences 7:99–178.

Rossman, D. A. 1970a. *Thamnophis proximus*. Catalogue of American amphibians and reptiles 98.1–98.3.

Rossman, D. A. 1970b. *Thamnophis sauritus*. Catalogue of American amphibians and reptiles 99.1–99.2.

Rossman, D. A. 1973. Evidence of the conspecificity of *Carphophis amoenus* (Say) and *Carphophis vermis* (Kennicott). Journal of Herpetology 7:140–141.

Rossman, D. A., and P. A. Myer. 1990. Behavioral and morphological adaptations for snail extraction in the North American brown snakes (genus *Storeria*). Journal of Herpetology 24:434–438.

Rossman, D. A., N. B. Ford, and R. A. Seigel. 1996. The garter snakes: evolution and ecology. University of Oklahoma Press, Norman, USA.

Rossman, N. J., D. A. Rossman, and N. K. Keith. 1981. Comparative visceral topography of the New World snake tribe Thamnophiini (Colubridae, Natricinae). Tulane Studies in Zoology and Botany 23:123–164.

Rossman, S. C., and D. A. Rossman. 2001. *Hemidactylium scutatum* (Four-toed Salamander). Geographic distribution. Herpetological Review 32:53–54.

Roster, N. O., C. L. Clark, and J. C. Gillingham. 1995. Prey-catching behavior in frogs and toads using video-simulated prey. Copeia 1995:496–498.

Roth, T. C., II, and B. D. Greene. 2006. Movement patterns and home range use of the Northern Watersnake (*Nerodia sipedon*). Copeia 2006:544–551.

Rothermel, B. B. 2004. Migratory success of juveniles: a potential constraint on connectivity for pond-breeding amphibians. Ecological Applications 14:1535–1546.

Rothermel, B. B., and R. D. Semlitsch. 2002. An experimental investigation of landscape resistance of forest versus old-field habitats to emigrating juvenile amphibians. Conservation Biology 16:1324–1332.

Rothermel, B. B., E. R. Travis, D. L. Miller, R. L. Hill, J. L. McGuire, and M. J. Yabsley. 2013. High occupancy of stream salamanders despite high *Ranavirus* prevalence in a southern Appalachians watershed. EcoHealth 10:184–189.

Rothermel, B. B., S. C. Walls, J. C. Mitchell, C. K. Dodd, Jr., L. K. Irwin, D. E. Green, V. M. Vazquez, J. W. Petranka, and D. J. Stevenson. 2008. Widespread occurrence of the amphibian chytrid fungus *Batrachochytrium dendrobatidis* in the southeastern USA. Diseases of Aquatic Organisms 82:3–18.

Rouse, J. D., R. J. Willson, R. Black and R. J. Brooks. 2011. Movement and spatial dispersion of *Sistrurus catenatus* and *Heterodon platirhinos*: implications for interactions with roads. Copeia 2011:443–456.

Routman, E. 1993. Population structure and genetic diversity of metamorphic and paedomorphic populations of the Tiger Salamander, *Ambystoma tigrinum*. Journal of Evolutionary Biology 6:329–357.

Row, J. R., and B. Blouin-Demers. 2006a. Thermal quality influences effectiveness of thermoregulation, habitat use, and behavior in Milk Snakes. Oecologia 148:1–11.

Row, J. R., and G. Blouin-Demers. 2006b. Thermal quality influences habitat selection at multiple spatial scales in Milksnakes. Ecoscience 13:443–450.

Row, J. R., and S. Lougheed. 2006. Demography and landscape genetics of Eastern Foxsnakes (*Elaphe gloydi*). Report to the Endangered Species Recovery Fund, Toronto, Ontario, Canada.

Row, J. R., G. Blouin-Demers, and P. J. Weatherhead. 2007. Demographic effects of road mortality in Black Ratsnakes (*Elaphe obsoleta*). Biological Conservation 137:117–124.

Row, J. R., G. Blouin-Demers, and S. C. Lougheed. 2010. Habitat distribution influences dispersal and fine-scale genetic population structure of Eastern Foxsnakes (*Mintonius gloydi*) across a fragmented landscape. Molecular Ecology 19:5157–5171.

Row, J. R., G. Blouin-Demers, and S. C. Lougheed. 2012. Movements and habitat use of Eastern Foxsnakes (*Pantherophis gloydi*) in two areas varying in size and fragmentation. Journal of Herpetology 46:94–99.

Row, J. R., R. J. Brooks, C. A. MacKinnon, A. Lawson, B. I. Crother, M. White, and S. C. Lougheed. 2011. Approximate Bayesian computation reveals the factors that influence genetic diversity and population structure of foxsnakes. Journal of Evolutionary Biology 24:2364–2377.

Rowe, C. L., and W. A. Dunson. 1993. Relationships among biotic parameters and breeding effort by three amphibians in temporary wetlands of central Pennsylvania. Wetlands 13:237–246.

Rowe, C. L., and W. A. Dunson. 1995. Impacts of hydroperiod on growth and survival of larval amphibians in temporary ponds of central Pennsylvania. Oecologia 102:397–403.

Rowe, C. L., A. Heyes, and J. Hilton. 2011. Differential patterns of accumulation and depuration of dietary selenium and vanadium during metamorphosis in the Gray Treefrog (*Hyla versicolor*). Archives of Environmental Contamination and Toxicology 60:336–342.

Rowe, C. L., W. A. Hopkins, and V. R. Coffman. 2001. Failed Recruitment of Southern Toads (*Bufo terrestris*) in a trace element-contaminated breeding habitat: direct and indirect effects that may lead to a local population sink. Archives of Environmental Contamination and Toxicology 40:399–405.

Rowe, C. L., W. J. Sadinski, and W. A. Dunson. 1994. Predation on larval and embryonic amphibians by acid-tolerant caddisfly larvae (*Ptilostomis postica*). Journal of Herpetology 28:357–364.

Rowe, J. W. 1987. Seasonal and daily activity in a population of Blanding's Turtle (*Emydoidea blandingii*) in Northern Illinois. Thesis, Eastern Illinois University, Charleston, USA.

Rowe, J. W. 1992a. Dietary habits of the Blanding's Turtle (*Emydoidea blandingii*) in northern Illinois. Journal of Herpetology 26:111–114.

Rowe, J. W. 1992b. Observations of body size, growth, and reproduction in Blanding's Turtles (*Emydoidea blandingii*) from western Nebraska. Canadian Journal of Zoology 70:1690–1695.

Rowe, J. W. 2003. Activity and movements of Midland Painted Turtles (*Chrysemys picta marginata*) living in a small marsh system on Beaver Island, Michigan. Journal of Herpetology 37:342–353.

Rowe, J. W., and E. O. Moll. 1991. A radiotelemetric study of activity and movements of the Blanding's Turtle (*Emydoidea blandingii*) in northeastern Illinois. Herpetologica 25:178–185.

Rowe, J. W., and S. F. Dalgarn. 2010. Home range size and daily movements of Midland Painted Turtles (*Chrysemys picta marginata*) in relation to body size, sex, and weather patterns. Herpetological Conservation and Biology 5:461–473.

Rowe, J. W., G. C. Lehr, P. M. McCarthy, and P. M. Converse. 2009. Activity, movements, and activity area size in Stinkpot Turtles (*Sternotherus odoratus*) in a southwestern Michigan Lake. American Midland Naturalist 162:266–275.

Rowell, J. C. 2012. The snakes of Ontario: natural history, distribution, and status. Art Bookbindery, Winnipeg, Canada.

Ruane, S., R. W. Bryson, Jr., R. A. Pyron, and F. T. Burbrink. 2013. Coalescent species delimitation in milksnakes (genus *Lampropeltis*) and impacts on phylogenetic comparative analyses. Systemic Biology 63:231–250.

Ruane, S., S. A. Dinkelacker, and J. B. Iverson. 2008. Demographic and reproductive traits of Blanding's Turtles, *Emydoidea blandingii*, at the western edge of the species' range. Copeia 2008:771–779.

Rubbo, M. J., and J. M. Kiesecker. 2005. Amphibian breeding distribution in an urbanized landscape. Conservation Biology 19:504–511.

Rubbo, M. J., J. L. Lanterman, R. C. Falco, and T. J. Daniels. 2011. The influence of amphibians on mosquitoes in seasonal pools: can wetlands protection help to minimize disease risk? Wetlands 31:799–804.

Rubbo, M. J., V. R. Townsend, Jr., S. D. Smyers, and R. G. Jaeger. 2003. An experimental assessment of invertebrate/vertebrate predation: the interaction between Wolf Spiders (*Gladicosa pulchra*) and terrestrial salamanders (*Ambystoma maculatum*). Journal of Zoology 261:1–5.

Rubin, C. S., R. E. Wagner, J. L. Bouzat, and K. N. Paige. 2001a. Population genetic structure of Blanding's Turtles (*Emydoidea blandingii*) in an urban landscape. Biological Conservation 99:323–330.

Rubin, C. S., R. Warner, and D. R. Ludwig. 2001b. Habitat use and movements of radiotagged Blanding's Turtles (*Emydoidea blandingii*) in a suburban landscape. Chelonian Conservation and Biology 4:146–141.

Rudd, H., J. Vala, and V. Schaefer. 2002. Importance of backyard habitat in a comprehensive biodiversity conservation strategy: a connectivity analysis of urban green spaces. Restoration Ecology 10:368–275.

Rudolph, D. C., and S. J. Burgdorf. 1997. Timber Rattlesnakes and Louisiana Pine Snakes of the west Gulf Coastal plain: hypotheses of decline. Texas Journal of Science 49:111–122.

Rudolph, R., G. E. Neal, J. S. Williams, and A. P. McMahan. 1995. Snakebite treatment at a southwestern regional referral center. Annals of Surgery 61:767–772.

Ruffner, B. M. 1933. Practical frog raising. Southern Frog Farms, Jennings, Louisiana, USA.

Rugh, R. 1941. Experimental studies on the reproductive physiology of the male Spring Peeper, *Hyla crucifer*. Proceedings of the American Philosophical Society 84:617–632.

Rumschlag, S. L., and M. D. Boone 2015. How time of exposure to the amphibian chytrid fungus affects *Hyla chrysoscelis* in the presence of an insecticide. Herpetologica 71:169–176.

Russell, F. E. 1983. Snake venom poisoning. Second edition. Scholium International, Great Neck, New York, USA.

Russell, F. E. 1990. When a snake strikes. Emergency Medicine June:21–43.

Russell, F. E. 1991. Snake venom poisoning. Veterinary and Human Toxicology 33:584–586.

Russell, K. R., and H. G. Hanlin. 1999. Aspects of the ecology of Worm Snakes (*Carphophis amoenus*) associated with small isolated wetlands in South Carolina. Journal of Herpetology 33:339–344.

Russell, K. R., D. H. Van Lear, and D. C. Guynn, Jr. 1999. Prescribed fire effects on herpetofauna: review and management implications. Wildlife Society Bulletin 27:374–384.

Russell, R. W., G. J. Lipps, Jr., S. J. Hecnar, and G. D. Haffner. 2002. Persistent organic pollutants in Blanchard's Cricket Frogs (*Acris crepitans blanchardi*) from Ohio. Ohio Journal of Science 102:119–122.

Rustigian, H. L., M. V. Santelmann, and N. H. Schumaker. 2003. Assessing the potential impacts of alternative landscape designs on amphibian population dynamic. Landscape Ecology 18:65–81.

Ruth, B. C., W. A. Dunson, C. L. Rowe, and S. B. Blair. 1993. A molecular and functional evaluation of the egg mass color polymorphism of the Spotted Salamander, *Ambystoma maculatum*. Journal of Herpetology 27:306–314.

Rutherford, J. L. 2012. *Glyptemys insculpta* (Wood Turtle). Aggressive behavior. Herpetological Review 43:326–327.

Rutherford, J. L., G. S. Casper, and B. Graves. 2016. Factors affecting predation on Wood Turtle (*Glyptemys insculpta*) nests in the Upper Peninsula of Michigan. Chelonian Conservation and Biology 15:181–186.

Rutherford, R. D., and G. S. Casper. 2018. *Notophthalmus viridescens louisianensis* (Central Newt). Geographic distribution. Herpetological Review 49:281.

Ruthig, G. R., L. J. Bode, A. Cvetkovic, and P. Shrestha. 2020. Dead Mink Frogs (*Lithobates septentrionalis*) found in northern Minnesota were infected with both *Batrachochytrium dendrobatidis* and Ranavirus. Herpetological Review 51:744–746.

Ruthven, A. G. 1906. The cold-blooded vertebrates of the Porcupine Mountains and adjacent Isle Royale, Michigan. Pages 107–112 *in* A. C. Lane, editor. An ecological survey in northern Michigan. University of Michigan, Lansing, USA.

Ruthven, A. G. 1908. Variations and genetic relationships of the garter-snakes. United States National Museum Bulletin 61:1–201.

Ruthven, A. G. 1911. A biological survey of the sand-dune region in the south shore of Saginaw Bay, Michigan. Michigan Geological and Biological Survey 4:1–347.

Ruthven, A. G. 1912. On the breeding habits of Butler's Garter-snake. Biological Bulletin 24:18–20.

Ruthven, A. G., C. Thompson, and H. T. Gaige. 1928. The herpetology of Michigan. University of Michigan, University Museums Michigan Handbook Series 3, Ann Arbor, USA.

Ruthven, D. C., III, R. T. Kazmaier, and M. W. Janis. 2008. Short-term response of herpetofauna to various burning regimes in the south Texas plains. Southwestern Naturalist 53:480–487.

Rutzen, K., L. Mitchem, N. Rudolph, K. Katovich, and J. M. Kapfer. 2015. *Emydoidea blandingii* (Blanding's Turtle). Juvenile diet. Herpetological Review 46:79–80.

Rutzen, K. M., P. K. Pieper, and J. M. Kapfer. 2013. *Pantherophis vulpinus* (Eastern Foxsnake). Geographic distribution. Herpetological Review 44:629.

Ryan, K. J., and A. J. K. Calhoun. 2014. Postbreeding habitat use of the rare, pure-diploid Blue-spotted Salamander (*Ambystoma laterale*). Journal of Herpetology 48:556–566.

Ryan, K. J., J. D. Zydlewski, and A. J. K. Calhoun. 2014a. Using passive integrated transponder (PIT) systems for terrestrial detection of Blue-spotted Salamanders (*Ambystoma laterale*) in situ. Herpetological Conservation and Biology 9:97–105.

Ryan, M. J. 1980. The reproductive behavior of the Bullfrog (*Rana catesbeiana*). Copeia 1980:108–144.

Ryan, R. A. 1953. Growth rates of some ranids under natural conditions. Copeia 1953:73–80.

Ryan, T. J., C. A. Conner, B. A. Douthitt, S. C. Sterrett, and C. M. Salsbury. 2008. Movement and habitat use of two aquatic turtles (*Graptemys geographica* and *Trachemys scripta*) in an urban landscape. Urban Ecosystems 11:213–225.

Ryan, T. J., W. E. Peterman, J. D. Stephens, and S. C. Sterrett. 2014b. Movement and habitat use of the Snapping Turtle in an urban landscape. Urban Ecosystems 17:613–623.

Ryberg, W. A., J. A. Harvey, A. Blick, T. J. Hibbitts, and G. Voelker. 2015. Genetic structure is inconsistent with subspecies designation in the Western Massasauga *Sistrurus tergeminus*. Journal of Fish and Wildlife Management 6:350–359.

Rytwinski, T., and L. Fahrig. 2012. Do species life history traits explain population responses to roads? A meta-analysis. Biological Conservation 147:87–98.

Sabaj, M. H. 2016. Standard symbolic codes for institutional resource collections in herpetology and ichthyology: an online reference. Version 6.5 (16 August 2016). American Society of Ichthyologists and Herpetologists, Washington, D.C., USA. Archived by WebCite at http://www.webcitation.org/73cNc9Fe6. Accessed 1 Nov 2018.

Saber, P. A., and W. A. Dunson. 1978. Toxicity of bog water to embryonic and larval amphibians. Journal of Experimental Zoology 204:33–42.

Saccucci M. J., R. D. Denton, M. L. Holding, and H. L. Gibbs. 2016. Polyploid unisexual salamanders have higher tissue regeneration rates than diploid sexual relatives. Journal of Zoology 300:77–81.

Sacerdote, A. B., and R. B. King. 2009. Dissolved oxygen requirements for hatching success of two ambystomatid salamanders in restored emphemeral ponds. Wetlands 29:1202–1213.

Sacerdote, A. B., and R. B. King. 2014. Direct effects of an invasive European Buckthorn metabolite on embryo survival and development in *Xenopus laevis* and *Pseudacris triseriata*. Journal of Herpetology 48:51–58.

Sacerdote, A. B., G. Glowacki, and T. Schmidtz. 2012. Communal nesting in the Smooth Greensnake (*Opheodrys vernalis*). Herpetological Review 43:661.

Sacerdote-Velat, A., J. M. Earnhardt, D. Mulkerin, D. Boehm, and G. Glowacki. 2014. Evaluation of headstarting and release techniques for population augmentation and reintroduction of the Smooth Green Snake. Animal Conservation 17(S1):65–73.

Sadinski, W., M. Roth, S. Treleven, J. Theyerl, and P. Dummer. 2010. Detection of the chytrid fungus, *Batrachochytrium dendrobatidis,* on recently metamorphosed amphibians in the north-central United States. Herpetological Review 41:170–175.

Sadinski, W., M. Roth, T. Hayes, P. Jones, and A. Galland. 2014. Indicators of the statuses of amphibian populations and their potential for exposure to atrazine in four midwestern U.S. conservation areas. PLoS ONE 9:e107018.

Sadinski, W. J., and W. A. Dunson. 1992. A multilevel study of effects of low pH on amphibians of temporary ponds. Journal of Herpetology 26:413–422.

Saeland, M. J., and M. W. Doperalski. 2001. *Pseudacris crucifer* (Spring Peeper). Geographic distribution. Herpetological Review 32:272.

Saeland, M. J., and M. W. Doperalski. 2002. *Pseudacris crucifer* (Spring Peeper). Geographic distribution. Herpetological Review 33:63.

Saenz, D., C. S. Collins, and R. N. Conner. 1999. A bark-shaving technique to deter rat snakes from climbing red-cockaded woodpecker cavity trees. Wildlife Society Bulletin 27:1069–1073.

Sagarin, R. D., S. D. Gaines, and B. Gaylord. 2006. Moving beyond assumptions to understand abundance distributions across the ranges of species. Trends in Ecology & Evolution 21:524–530.

Sagor, E. S., M. Ouellet, E. Barten, and D. M. Green. 1998. Skeletochronology and geographic variation in age structure in the Wood Frog, *Rana sylvatica.* Journal of Herpetology 32:469–474.

Sajdak, R., and C. Berg. 2005. Rattlesnakes on the bluffs: Wisconsin Timber Rattlesnakes. Iguana 12:90–97.

Sajdak, R. A. 1978. *Carphophis amoenus vermis* (Western Worm Snake). Geographic distribution. Herpetological Review 9:62.

Sajdak, R. A. 1982. Seasonal activity patterns, habitat selection, and population structure of the Mudpuppy, *Necturus maculosus,* in a Wisconsin stream. Thesis, University of Wisconsin, Milwaukee, USA.

Sajdak, R. A. 1999. Wisconsin Timber Rattlesnake study, Rush Creek 1999 annual report. Wisconsin Department of Natural Resources, Bureau of Endangered Resources, Madison, USA.

Sajdak, R. A. 2000. Wisconsin Timber Rattlesnake study, Rush Creek 2000 annual report. Wisconsin Department of Natural Resources, Bureau of Endangered Resources, Madison, USA.

Sajdak, R. A. 2010. Hunters in the trees: a natural history of arboreal snakes. Krieger Publishing, Malabar, Florida, USA.

Sajdak, R. A., and A. Bartz. 2001. Wisconsin Timber Rattlesnake study, Rush Creek 2001 annual report. Wisconsin Department of Natural Resources, Bureau of Endangered Resources, Madison, USA.

Sajdak, R. A., and A. Bartz. 2004. *Crotalus horridus* (Timber Rattlesnake). Arboreality, diet. Herpetological Review 35:60–61.

Sajdak, R. A., and J. A. Buday. 1979. Tremblay's Salamander (*Ambystoma tremblayi*). Geographic Distribution. Herpetological Review 10:59.

Sajdak, R. A., and L. Sajdak. 1999. *Thamnophis sirtalis sirtalis* (Eastern Garter Snake). Carrion feeding. Herpetological Review 30:229.

Sajwaj, T. D., and J. W. Lang. 2000. Thermal ecology of Blanding's Turtle in central Minnesota. Chelonian Conservation and Biology 3:626–636.

Salthe, S. N., and N. O. Kaplan. 1966. Immunology and rates of enzyme evolution in the Amphibia in relation to the origins of certain taxa. Evolution 20:603–616.

Sams, E., and M. D. Boone. 2010. Interactions between recently metamorphosed Green Frogs and American Toads under laboratory conditions. American Midland Naturalist 163:269–279.

Samson, F., and F. Knopf. 1994. Prairie conservation in North America. BioScience 44:418–421.

Sandinski, W., M. Roth, S. Treleven, J. Theyerl, and P. Dummer. 2010. Detection of chytrid fungus, *Batrachochytrium dendrobatidis,* on recently metamorphosed amphibians in the north-central United States. Herpetological Review 41:170–175.

Sanzo, D., and S. J. Hecnar. 2006. Effects of road deicing salt (NaCl) on larval Wood Frogs (*Rana sylvatica*). Environmental Pollution 140:247–256.

Sargent, L. G. 2000. Frog and toad population monitoring in Michigan. Journal of the Iowa Academy of Science 107:195–199.

Saumure, R. A., and J. R. Bider. 1998. Impact of agricultural development on a population of Wood Turtles (*Clemmys insculpta*) in southern Québec, Canada. Chelonian Conservation and Biology 3:37–45.

Saumure, R. A., T. B. Herman, and R. D. Titman. 2007. Effects of haying and agricultural practices on a declining species: the North American Wood Turtle, *Glyptemys insculpta*. Biological Conservation 135:581–591.

Savage, W. K., and K. R. Zamudio. 2005. *Ambystoma maculatum* (Shaw, 1802), Spotted Salamander. Pages 621–627 *in* M. J. Lannoo editor. Amphibian declines: the conservation status of United States species. University of California Press, Berkeley, USA.

Savage, W. K., F. W. Quimby, and A. P. DeCaprio. 2002. Lethal and sublethal effects of polychlorinated biphenyls on *Rana sylvatica* tadpoles. Environmental Toxicology and Chemistry 21:168–174.

Saviola, A. J., V. J. McKenzie, and D. Chiszar. 2012. Chemosensory responses to chemical and visual stimuli in five species of colubrid snakes. Acta Herpetologica 7:91–103.

Say, T. 1823. Pages 185–187 *in* E. James, editor. An account of an expedition from Pittsburgh to the Rocky Mountains, performed in the years 1819 and '20: by order of the Hon. Volume 1. H. C. Carey and I. Lea, Philadelphia, Pennsylvania, USA.

Sayler, A. 1966. The reproductive ecology of the Red-backed Salamander, *Plethodon cinereus*, in Maryland. Copeia 1966:183–193.

Schaaf, R. T., Jr., and J. S. Garton. 1970. Raccoon predation on the American Toad, *Bufo americanus*. Herpetologica 26:334–335.

Schaaf, R. T., Jr., and P. W. Smith. 1970. Geographic variation in the Pickerel Frog. Herpetologica 26:240–254.

Schaetzl, R. J., and J. W. Attig. 2013. The loess cover of northeastern Wisconsin. Quaternary Research 79:199–214.

Schalk, G., M. R. Forbes, and P. J. Weatherhead. 2002. Developmental plasticity and growth rates of Green Frog (*Rana clamitans*) embryos and tadpoles in relation to a leech (*Macrobdella decora*) predator. Copeia 2002:445–449.

Scheffers, B. R., and C. A. Paszkowski. 2016. Large body size for metamorphic Wood Frogs in urban stormwater wetlands. Urban Ecosystems 19:347–359.

Scheller, R. M., and D. J. Mladenoff. 2005. A spatially interactive simulation of climate change, harvesting, wind, and tree species migration and projected changes to forest composition and biomass in northern Wisconsin, USA. Global Change Biology 11:307–321.

Scherer, R. D., E. Muths, and B. R. Noon. 2012. The importance of local and landscape-scale processes to the occupancy of wetlands by pond-breeding amphibians. Population Ecology 54:487–498.

Schieltz, J. M., L. M. B. Haywood, and D. M. Marsh. 2010. Effects of cover object spacing on the socioecology of the Red-backed Salamander, *Plethodon cinereus*. Herpetologica 66:276–282.

Schiesari, L., E. E. Werner, and G. W. Kling. 2009. Carnivory and resource-based niche differentiation in anuran larvae: implications for food web and experimental ecology. Freshwater Biology 54:572–586.

Schiltz, N. G., and J. M. Lorch. 2016. *Notophthalmus viridescens* (Eastern Newt). Geographic distribution. Herpetological Review 47:621.

Schlaepfer, M. A., C. Hoover, and C. K. Dodd, Jr. 2005a. Challenges in evaluating the impact of the trade in amphibians and reptiles on wild populations. BioScience 55:256–264.

Schlaepfer, M. A., P. W. Sherman, B. Blossey, and M. C. Runge. 2005b. Introduced species as evolutionary traps. Ecology Letters 8:241–246.

Schlauch, F. C. 1975. Agonistic behavior in a suburban Long Island population of the Smooth Green Snake, *Opheodrys vernalis*. Engelhardtia 6:25–26.

Schmaier, A. H., and R. W. Coleman. 1980. Crotalocytin: characterization of the Timber Rattlesnake platelet activating protein. Blood 56:1020–1028.

Schmid, W. D. 1965. Some aspects of the water economies of nine species of amphibians. Ecology 46:261–269.

Schmid, W. D. 1982. Survival of frogs in low temperature. Science 215:697–698.

Schmidt, F. J. W. 1926. List of the amphibians and reptiles of Worden Township, Clark County, Wisconsin. Copeia 154:131–132.

Schmidt, K. P. 1930. The salamanders of the Chicago area. Chicago Field Museum of Natural History 12:1–16.

Schmidt, K. P., and W. L. Necker. 1935. Amphibians and reptiles of the Chicago region. Bulletin of the Chicago Academy of Sciences 5:57–77.

Schmidt, R. S. 1990. Releasing (unclasping) in male American Toads: a neural substrate in the lateral subtoral tegmentum. Brain, Behavior and Evolution 36:307–314.

Schneider, A. C., K. D. Frahm, B. J. Swanson, J. J. Moriarty, J. T. McGuire, and T. L. Lewis. 2019. Annual and diel activity of Spiny Softshell Turtles (*Apalone spinifera*) and Snapping Turtles (*Chelydra serpentina*) in an urban Minnesota lake. Herpetological Review 50:709–712.

Schorger, A. W. 1944. Philo Romayne Hoy: Wisconsin's greatest pioneer zoologist. Passenger Pigeon 6:55–59.

Schorger, A. W. 1967–68. Rattlesnakes in early Wisconsin. Transactions of the Wisconsin Academy of Sciences, Arts and Letters 56:29–48.

Schotthoefer, A. M., M. G. Bolek, R. C. Cole, and V. R. Beasley. 2009. Parasites of the Mink Frog (*Rana septemtrionalis*) in Minnesota. Comparative Parasitology 76:240–246.

Schreber, H. 1782. Beytrag zur Naturgeschichte der Frösche. Der Naturforscher. Halle 18:182–193.

Schroder, R. C. 1950. Hibernation of Blue Racers and Bull Snakes in western Illinois. Chicago Academy of Sciences, Natural History Miscellanea 75:1–2.

Schroeder, E. E. 1966. Age determination and population structure in Missouri Bullfrogs. Dissertation. University of Missouri, Columbia, USA.

Schroeder, E. E. 1976. Dispersal and movement of newly transformed Green Frogs, *Rana clamitans*. American Midland Naturalist 95:471–474.

Schroeder, E. E., and T. S. Baskett. 1968. Age estimation, growth rates, and population structure in Missouri Bullfrogs. Copeia 1968:583–592.

Schueler, F. W. 1973. Frogs of the Ontario coast of Hudson Bay and James Bay. Canadian Field-Naturalist 87:409–418.

Schueler, F. W. 1975. Geographic variation in the size of *Rana septentrionalis* in Quebec, Ontario, and Manitoba. Journal of Herpetology 9:177–185.

Schueler, F. W., and F. R. Cook. 1980. Distribution of middorsal stripe dimorphism in the Wood Frog, *Rana sylvatica*, in eastern North America. Canadian Journal of Zoology 58:1643–1651.

Schuett, G. W., and R. E. Gatten, Jr. 1980. Thermal preferences in Snapping Turtles (*Chelydra serpentina*). Copeia 1980:149–152.

Schuler, A., and A. F. Badje. 2019. *Glyptemys insculpta* (Wood Turtle). Geographic distribution. Herpetological Review 50:324.

Schuler, M., and R. P. Thiel. 2008. Annual vs. multiple-year home range sizes of individual Blanding's Turtles, *Emydoidea blandingii*, in central Wisconsin. Canadian Field-Naturalist 122:61–64.

Schulte, L. A., D. J. Mladenoff, and E. V. Nordheim. 2002. Quantitative classification of a historic northern Wisconsin (U.S.A.) landscape: mapping forests at regional scales. Canadian Journal of Forest Research 32:1616–1638.

Schultze, S. 2005. MMSD, contractor face $1.65 million in fines; attorney general accuses them of environmental violations. Milwaukee Journal Sentinel. 23 February 2005; section News:03.

Schulz, R. A. 1976. Animal ecology of Bear Island, Wisconsin, Apostle Islands National Lakeshore. Thesis, Michigan Technological University, Houghton, USA.

Schuster, Z. T., K. W. Potter, and D. S. Liebl. 2012. Assessing the effects of climate change in precipitation and flood damage in Wisconsin. Journal of Hydrologic Engineering 17:888–894.

Schwartz, A., and W. E. Duellman. 1952. The taxonomic status of the newts, *Diemictylus viridescens*, of peninsular Florida. Bulletin of the Chicago Academy of Sciences 9:217–227.

Schwartz, J. J., B. W. Buchanan, H. C. Gerhardt, J. J. Schwartz, H. C. Gerhardt, and B. W. Buchanan. 2001. Female mate choice in the Gray Treefrog (*Hyla versicolor*) in three experimental environments. Behavioral Ecology and Sociobiology 49:443–455.

Schwartz, M. W. 1989. Predicting tree frequencies from pollen frequency: an attempt to validate the R value method. New Phytologist 112:129–143.

Schwarz, A. 1957. Chorus Frogs (*Pseudacris nigrita* LeConte) in South Carolina. American Museum Novitates 1838:1–12.

Schwarzkopf, L., and R. J. Brooks. 1985. Application of operative environmental temperature to analysis of basking behavior in *Chrysemys picta*. Herpetologica 41:206–212.

Scott, D. E. 2005. *Ambystoma opacum* (Gravenhorst, 1807), Marbled Salamander. Pages 627–632 *in*

M. Lannoo, editor. Amphibian declines: the conservation status of United States species. University of California Press, Berkeley, USA.

Scott, T. G., and R. B. Sheldahl. 1937. Black-banded Skink in Iowa. Copeia 1937:192.

Seale, D. B. 1980. Influence of Amphibian larvae on primary production, nutrient flux, and competition in a pond ecosystem. Ecology 61:1531–1550.

Seale, D. B. 1982. Physical factors influencing oviposition by the Wood Frog, *Rana sylvatica*, in Pennsylvania. Copeia 1982:627–635.

Sealy, J. 2002. Ecology and behavior of the Timber Rattlesnake (*Crotalus horridus*) in the upper Piedmont of North Carolina: identified threats and conservation recommendations. Pages 561–578 *in* G. W. Schuett, M. Hoggren, M. E. Douglas, and H. W. Greene, editors. Biology of the vipers. Eagle Mountain Publishing, LC, Eagle Mountain, Utah, USA.

Searle, C. L., S. S. Gervasi, J. Hua, J. I. Hammond, R. A. Relyea, D. H. Olson, and A. R. Blaustein. 2011. Differential host susceptibility to *Batrachochytrium dendrobatidis*, an emerging amphibian pathogen. Conservation Biology 25:965–974.

Seburn, C. N. L. 1993. Spatial distribution and microhabitat use in the Five-lined Skink (*Eumeces fasciatus*). Canadian Journal of Zoology 71:445–450.

Seburn, C. N. L., D. C. Seburn, and C. A. Paszkowski. 1997. Northern Leopard Frog (*Rana pipiens*) dispersal in relation to habitat. Pages 64–72 *in* D. M. Green, editor. Amphibians in decline: Canadian studies of a global problem. Society for the Study of Amphibians and Reptiles, Herpetological Conservation 1, St. Louis, Missouri, USA.

Seburn, D. 2009. Recovery strategy for the Eastern Hog-nosed Snake (*Heterodon platirhinos*) in Canada. Parks Canada, Species at Risk Act Recovery Strategy Series, Ottawa, Ontario, Canada.

Seburn, D. C., C. N. L. Seburn, and W. F. Weller. 2008. A localized decline in the Western Chorus Frog, *Pseudacris triseriata*, in eastern Ontario. Canadian Field-Naturalist 122:158–161.

Seibert, H. C. 1950. Population density of snakes in an area near Chicago. Copeia 1950:229–230.

Seibert, H. C., and C. W. Hagen, Jr. 1947. Studies on a population of snakes in Illinois. Copeia 1947:6–22.

Seibert, H. C., and R. A. Brandon. 1960. The salamanders of southeastern Ohio. Ohio Journal of Science 60:291–303.

Seidel, M. E., and R. V. Lucchino. 1981. Allozymic and morphological variation among the Musk Turtles *Sternotherus carinatus*, *S. depressus*, and *S. minor* (Kinosternidae). Copeia 1981:119–128.

Seidel, M. E., S. L. Reynolds, and R. V. Lucchino. 1981. Phylogenetic relationships among musk turtles (genus *Sternotherus*) and genic variation in *Sternotherus odoratus*. Herpetologica 37:161–165.

Seigel, R. A. 1983. Natural survival of eggs and tadpoles of the Wood Frog, *Rana sylvatica*. Copeia 1983:1096–1098.

Seigel, R. A. 1986. Ecology and conservation of an endangered rattlesnake, *Sistrurus catenatus*, in Missouri, USA. Biological Conservation 35:333–346.

Seigel, R. A., and C. A. Sheil. 1999. Population viability analysis: applications for the conservation of massasaugas. Pages 17–22 *in* B. Johnson and M. Wright, editors. Second international symposium and workshop on the conservation of the Eastern Massasauga Rattlesnake, *Sistrurus catenatus catenatus*: population and habitat management issues in urban, bog, prairie and forested ecosystems. Toronto Zoo, Scarborough, Ontario, Canada.

Seigel, R. A., and C. K. Dodd, Jr. 2002. Translocations of amphibians: proven management method or experimental technique? Conservation Biology 16:552–554.

Seigel, R. A., and H. S. Fitch. 1985. Annual variation in reproduction in snakes in a fluctuating environment. Journal of Animal Ecology 54:497–505.

Seigel, R. A., and M. A. Pilgrim. 2002. Long-term changes in movement patterns of Massasaugas (*Sistrurus catenatus*). Biology of the Vipers 2002:405–412.

Sekowska, J. M., M. C. Allender, and C. A. Phillips. 2014. Prevalence and spatial distribution of Frog Virus 3 in free-ranging populations of Northern Cricket Frogs (*Acris crepitans*), American Bullfrogs (*Lithobates catesbeianus*), and Northern Leopard Frogs (*Lithobates pipiens*) in western Illinois and Wisconsin, USA. Herpetological Review 45:245–247.

Semlitsch, R. D. 1983a. Burrowing ability and behavior of salamanders of the genus *Ambystoma*. Canadian Journal of Zoology 61:616–620.

Semlitsch, R. D. 1983b. Structure and dynamics of two breeding populations of the Eastern Tiger Salamander, *Ambystoma tigrinum*. Copeia 1983:608–616.

Semlitsch, R. D. 1983c. Terrestrial movements of an Eastern Tiger Salamander, *Ambystoma tigrinum*. Herpetological Review 14:112–113.

Semlitsch, R. D. 1988. Allopatric distribution of two salamanders: effects of fish predation and competitive interactions. Copeia 1988: 290–298.

Semlitsch, R. D. 1990. Effects of body size, sibship, and tail injury on the susceptibility of tadpoles to dragonfly predation. Canadian Journal of Zoology 68:1027–1030.

Semlitsch, R. D. 1998. Biological delineation of terrestrial buffer zones for pond-breeding salamanders. Conservation Biology 12:1113–1119.

Semlitsch, R. D., and D. L. Drake. 2015. Structure and dynamics of *Lithobates sylvaticus* (Wood Frog) at the periphery of its range in Missouri. Southeastern Naturalist 14:329–341.

Semlitsch, R. D., and G. B. Moran. 1984. Ecology of the Red-bellied Snake (*Storeria occipitomaculata*) using mesic habitats in South Carolina. American Midland Naturalist 111:33–40.

Semlitsch, R. D., and J. R. Bodie. 1998. Are small, isolated wetlands expendable? Conservation Biology 12:1129–1133.

Semlitsch, R. D., and J. R. Bodie. 2003. Biological criteria for buffer zones around wetlands and riparian habitats for amphibians and reptiles. Conservation Biology 17:1219–1228.

Semlitsch, R. D., B. D. Todd, S. M. Blomquist, A. J. K. Calhoun, J. W. Gibbons, J. P. Gibbs, G. J. Graeter, E. B. Harper, D. J. Hocking, M. L. Hunter, Jr., D. A. Patrick, T. A. G. Rittenhouse, and B. B. Rothermal. 2009. Effects of timber harvest on amphibian populations: understanding mechanisms from forest experiments. BioScience 59:853–862.

Semlitsch, R. D., C. A. Conner, D. J. Hocking, T. A. G. Rittenhouse, and E. B. Harper. 2008. Effects of timber harvesting on pond-breeding amphibian persistence: testing the evacuation hypothesis. Ecological Applications 18:283–289.

Semlitsch, R. D., C. M. Bridges, and A. M. Welch. 2000. Genetic variation and a fitness tradeoff in the tolerance of Gray Treefrog (*Hyla versicolor*) tadpoles to the insecticide carbaryl. Oecologia 125:179–185.

Semlitsch, R. D., D. E. Scott, J. H. K. Pechmann, and J. W. Gibbons. 1996. Structure and dynamics of an amphibian community: evidence from a 16-year study of a natural pond. Pages 217–248 *in* M. L. Cody and J. A. Smallwood, editors. Long-term studies of vertebrate communities. Academic Press, San Diego, California, USA.

Sessions, S. K. 1980. Evidence for a highly differentiated sex chromosome heteromorphism in the salamander *Necturus maculosus* (Rafinesque). Chromosoma 77:157–168.

Sessions, S. K., and J. E. Wiley. 1985. Chromosome evolution in salamanders of the genus *Necturus*. Brimleyana 10:37–52.

Sessions, S. K., R. A. Franssen, and V. L. Horner. 1999. Morphological clues from multilegged frogs: are retinoids to blame? Science 281:800–802.

Setser, K., and J. F. Cavitt. 2003. Effects of burning on snakes in Kansas, USA, tallgrass prairie. Natural Areas Journal 23:315–319.

Sever, D. M. 1997. Sperm storage in the spermatheca of the Red-backed Salamander, *Plethodon cinereus* (Amphibia: Plethodontidae). Journal of Morphology 234:131–146.

Sever, D. M. 2005. *Eurycea bislineata* (Green, 1818), Northern Two-lined Salamander. Pages 735–738 *in* M. Lannoo, editor. Amphibian declines: the conservation status of United States species. University of California Press, Berkeley, USA.

Sever, D. M. 2006. The "false breeding season" of the Eastern Newt, *Notophthalmus viridescens*. Chicago Herpetological Society Bulletin 41:149–153.

Sever, D. M., and C. F. Dineen. 1978. Reproductive ecology of the Tiger Salamander, *Ambystoma tigrinum* in northern Indiana. Proceedings of the Indiana Academy of Sciences 87:189–203.

Sexton, O. J. 1959. Spatial and temporal movements of a population of the Painted Turtle, *Chrysemys picta marginata* (Agassiz). Ecological Monographs 29:113–140.

Sexton, O. J. 1979. Remarks on defensive behavior of hognose snakes, *Heterodon*. Herpetological Review 10:86–87.

Sexton, O. J. 1995. Miscellaneous comments on the natural history of Blanding's Turtles (*Emydoidea blandingii*). Transactions of the Missouri Academy of Science 29:1–13.

Sexton, O. J., and C. Phillips. 1986. A qualitative study of fish-amphibian interaction in 3 Missouri ponds. Transactions of the Missouri Academy of Science 20:25–35.

Sexton, O. J., and L. Claypool. 1978. Nest sites of a northern population of an oviparous snake, *Opheodrys vernalis* (Serpentes, Colubridae). Journal of Natural History 1978:365–370.

Sexton, O. J., C. A. Phillips, T. J. Bergman, E. B. Wattenberg, and R. E. Preston. 1998. Abandon not hope: status of repatriated populations of Spotted Salamanders and Wood Frogs at the Tyson Research Center, St. Louis County, Missouri. Pages 340–344 *in* M. J. Lannoo, editor. Status and conservation of midwestern amphibians. University of Iowa Press, Iowa City, USA.

Shaffer, B., D. Starkey, and M. Fujita. 2008. Molecular insights into the systematics of the snapping turtles (Chelydridae). Pages 44–49 *in* A. C. Steyermark, M. S. Finkler, and R. J. Brooks, editors. Biology of the Snapping Turtle. John Hopkins University Press, Baltimore, Maryland, USA.

Shaffer, H. B., and M. L. McKnight. 1996. The polytypic species revisited: genetic differentiation and molecular phylogenetics of the Tiger Salamander *Ambystoma tigrinum* (Amphibia: Caudata) complex. Evolution 50:417–433.

Shaffer, S. A., J. T. Briggler, R. A. Gitzen, and J. J. Millspaugh. 2017. Abundance and harvest proportion of river turtles in Missouri. Journal of Freshwater Ecology 32:541–555.

Shaw, C. E. 1951. Male combat in American colubrid snakes with remarks on combat in other colubrid and elapid snakes. Herpetologica 7:149–168.

Shearin, A. F., A. J. K. Calhoun, and C. S. Loftin. 2012. Evaluation of listener-based anuran surveys with automated audio recording devices. Wetlands 32:737–751.

Sheeley, D. 2007. Retzer, snakes to benefit from land grant; threatened species gets a little room. Milwaukee Journal Sentinel. 14 January 2007; section Waukesha County.

Sheldon, A. B., and A. F. Badje. 2016. *Pseudacris maculata* (Boreal Chorus Frog). Geographic distribution. Herpetological Review 47:624.

Shelley, L. O. 1938. Milk Snakes vs. birds. Auk 55:548.

Shepard, D. B., A. R. Kuhns, M. J. Dreslik, and C. A. Phillips. 2008a. Roads as barriers to animal movement in fragmented landscapes. Animal Conservation 11:288–296.

Shepard, D. B., C. A. Phillips, M. J. Dreslik, and B. C. Jellen. 2004. Prey preference and diet of neonate Eastern Massasaugas (*Sistrurus c. catenatus*). American Midland Naturalist 152:360–368.

Shepard, D. B., M. J. Dreslik, B. C. Jellen, and C. A. Phillips. 2008b. Reptile road mortality around an oasis in the Illinois corn desert with emphasis on the endangered Eastern Massasauga. Copeia 2008:350–359.

Shepard, D. B., M. J. Dreslik, C. A. Phillips, and B. C. Jellen. 2003. *Sistrurus catenatus catenatus* (Eastern Massasauga): male-male aggression. Herpetological Review 34:155–156.

Sherman, E. 2009. Fluctuating asymmetry in *Ichthyophonus*-sp. infected newts, *Notophthalmus viridescens*, from Vermont. Applied Herpetology 6:369–378.

Sherrod, S. K. 1978. Diets of North American Falconiformes. Raptor Research 12:49–121.

Shew, J. J., B. D. Greene, and F. E. Durbian. 2012. Spatial ecology and habitat use of the Western Foxsnake (*Pantherophis vulpinus*) on Squaw Creek National Wildlife Refuge (Missouri). Journal of Herpetology 46:539–548.

Shewchuk, C. H. 1996. The natural history of reproduction and movement patterns in the Gopher Snake (*Pituophis melanoleucus*) in southern British Columbia. Thesis, University of Victoria, British Columbia, Canada.

Shields, F. D., Jr., S. S. Knight, and C. M. Cooper. 1998. Rehabilitation of aquatic habitats in warmwater streams damaged by channel incision in Mississippi. Hydrobiologia 382:63–86.

Shine, R. 1991. Intersexual dietary divergence and the evolution of sexual dimorphism in snakes. American Naturalist 138:103–122.

Shine, R. 1978. Sexual size dimorphism and male combat in snakes. Oecologia 33:269–277.

Shine, R. 1994. Sexual size dimorphism in snakes revisited. Copeia 1994:326–346.

Shine, R. 1995. A new hypothesis for the evolution of viviparity in reptiles. American Naturalist 145:809–823.

Shine, R. 2003. Reproductive strategies in snakes. Proceedings of the Royal Society of London, B: Biological Sciences 270:995–1004.

Shine, R., and R. T. Mason. 2001. Serpentine cross-dressers. Natural History 110:56–61.

Shine, R., and R. T. Mason. 2004. Patterns of mortality in a cold-climate population of Garter Snakes (*Thamnophis sirtalis parietalis*). Biological Conservation 120:201–210.

Shine, R., B. Phillips, H. Waye, M. LeMaster, and R. T. Mason. 2001a. Animal behavior: benefits of female mimicry in snakes. Nature 414:267.

Shine, R., M. LeMaster, M. Wall, T. Langkilde, and R. Mason. 2004. Why did the snake cross the road? Effects of roads on movement and location of mates by Garter Snakes (*Thamnophis sirtalis parietalis*). Ecology and Society 9:9.

Shine, R., M. J. Elphick, P. S. Harlow, I. T. Moore, M. P. LeMaster, and R. T. Mason. 2001b. Movements, mating and dispersal of Red-sided Gartersnakes from a communal den in Manitoba. Copeia 2001:82–91.

Shine, R., M. M. Olsson, M. P. Lemaster, I. T. Moore, and R. T. Mason. 2000a. Effects of sex, body size, temperature, and location on the antipredator tactics of free-ranging Gartersnakes (*Thamnophis sirtalis*, Colubridae). Behavioral Ecology 11:239–245.

Shine, R., P. S. Harlow, M. J. Elphick, M. M. Olsson, and R. T. Mason. 2000b. Conflicts between courtship and thermoregulation: the thermal ecology of amorous male Garter Snakes (*Thamnophis sirtalis parietalis*, Colubridae). Physiological and Biochemical Zoology 73:508–516.

Shine, R., T. Langkilde, and R. T. Mason. 2012. Facultative pheromonal mimicry in snakes: "she-males" attract courtship only when it is useful. Behavioral and Ecological Sociobiology 66:691–695.

Shirose, L. J., and R. J. Brooks. 1995. Age structure, mortality, and longevity in syntopic populations of three species of ranid frogs in central Ontario. Canadian Journal of Zoology 73:1878–1886.

Shirose, L. J., and R. J. Brooks. 1995. Growth rate and age at maturity in syntopic populations of *Rana clamitans* and *Rana septentrionalis* in central Ontario. Canadian Journal of Zoology 73:1468–1473.

Shirose, L. J., and R. J. Brooks. 1997. Fluctuations in abundance and age structure in three species of frogs (Anura: Ranida) in Algonquin Park, Canada, from 1985 to 1993. Pages 16–26 *in* D. M. Green, editor. Amphibians in decline: Canadian studies of a global problem. Society for the Study of Amphibians and Reptiles, Herpetological Conservation 1, St. Louis, Missouri, USA.

Shively, S. 2014. Freshwater turtle species. Predation. Herpetological Review 45:313–314.

Shoemaker, K. T., G. Johnson, and K. A. Prior. 2009. Habitat manipulation as a viable conservation strategy. Pages 221–243 *in* S. J. Mullin and R. A. Seigel, editors. Snakes: ecology and conservation. Comstock Publishing Associates, Ithaca, New York, USA.

Shonfield, J., W. King, and W. R. Koski. 2019. Habitat use and movement patterns of Butler's Gartersnake (*Thamnophis butleri*) in southewestern Ontario, Canada. Herpetological Conservation and Biology 14:680–690.

Shoop, C. R. 1965a. Aspects of reproduction in Louisiana *Necturus* populations. American Midland Naturalist 74:357–367.

Shoop, C. R. 1965b. Orientation of *Ambystoma maculatum*: movements to and from breeding ponds. Science 149:558–559.

Shoop, C. R. 1968. Migratory orientation of *Ambystoma maculatum* movements near breeding ponds and displacements of migrating individuals. Biological Bulletin 135:230–238.

Shoop, C. R. 1974. Yearly variation in larval survival of *Ambystoma maculatum*. Ecology 55:440–444.

Shulse, C. D., R. D. Semlitsch, K. M. Trauth, and A. D. Williams. 2010. Influence of design and landscape placement parameters on amphibian abundance in constructed wetlands. Wetlands 30:915–928.

Shulse, C. D., R. D. Semlitsch, K. M. Trauth, and J. E. Gardner. 2012. Testing wetland features to increase amphibian reproductive success and species richness for mitigation and restoration. Ecological Applications 22:1675–1688.

Simões, T. R., M. W. Caldwell, M. Tałanda, M. Bernardi, A. Palci, O. Vernygora, F. Bernardini, L. Mancini, and R. L. Nydam. 2018. The origin of squamates revealed by a middle Triassic lizard from the Italian Alps. Nature 557:706–709.

Simon, J. A., J. W. Snodgrass, R. E. Casey, and D. W. Sparling. 2009. Spatial correlates of amphibian use of constructed wetlands in an urban landscape. Landscape Ecology 24:361–373.

Sinclair, B. J., J. R. Stinziano, C. M. Willians, H. A. MacMillan, K. E. Marshall, and K. B. Storey. 2013. Real-time measurement of metabolic rate during freezing and thawing of the Wood Frog, *Rana sylvatica*: implications for overwinter energy use. Journal of Experimental Biology 216:292–302.

Sing, K., T. Erickson, S. Aks Do, H. Rothenberg, and J. Lipscomb. 1994. Eastern Massasauga Rattlesnake envenomations in an urban wilderness. Journal of Wilderness Medicine 5:77–87.

Sites, J. W., Jr., J. W. Bickham, M. W. Haiduk, and J. B. Iverson. 1979. Banded karyotypes of six taxa of kinosternid turtles. Copeia 1979:692–698.

Sites, J. W., Jr., M. Morando, R. Highton, F. Huber, and R. E. Jung. 2004. Phylogenetic relationships of the endangered Shenandoah Salamander (*Plethodon Shenandoah*) and other salamanders of the *Plethodon cinereus* group (Caudata: Plethodontidae). Journal of Herpetology 38:96–105.

Sites, J. W., Jr., T. W. Reeder, J. J. Wiens. 2011. Phylogenetic insights on evolutionary novelties in lizards and snakes: sex, birth, bodies, niches, and venom. Annual Review of Ecology Evolution and Systematics 42:227–244.

Skawinski, P. M. 2020. *Sternotherus odoratus* (Eastern Musk Turtle). Geographic distribution. Herpetological Review 51:75.

Skelly, D. K. 1992. Field evidence for a cost of behavioral antipredator response in a larval amphibian. Ecology 73:704–708.

Skelly, D. K. 1994. Activity level and susceptibility of anuran larvae to predation. Animal Behaviour 47:465–468.

Skelly, D. K. 1995. Competition and the distribution of Spring Peeper larvae. Oecologia 103:203–207.

Skelly, D. K. 1996. Pond drying, predators, and the distribution of *Pseudacris* tadpoles. Copeia 1996:599–605.

Skelly, D. K. 2004. Microgeographic countergradient variation in the Wood Frog, *Rana sylvatica*. Evolution 58:160–165.

Skelly, D. K., and E. E. Werner. 1990. Behavioral and life-historical responses of larval American Toads to an odonate predator. Ecology 71:2313–2322.

Skelly, D. K., E. E. Werner, and S. A. Cortwright. 1999. Long-term distributional dynamics of a Michigan amphibian assemblage. Ecology 80:2326–2337.

Skelly, D. K., L. K. Freidenburg, and J. M. Kiesecker. 2002. Forest canopy and the performance of larval amphibians. Ecology 83:983–992.

Skelly, D. K., S. R. Bolden, and K. B. Dion. 2010. Intersex frogs concentrated in suburban and urban landscapes. EcoHealth 7:374–379.

Skerratt, L. F., L. Berger, R. Speare, S. Cashins, K. R. McDonald, A. D. Phillott, H. B. Hines, and N. Kenyon. 2007. Spread of chytridiomycosis has caused the rapid global decline and extinction of frogs. EcoHealth 4:125–134.

Skidds, D. E., and F. C. Golet. 2005. Estimating hydroperiod suitability for breeding amphibians in southern Rhode Island seasonal forest ponds. Wetland Ecology and Management 13:349–366.

Skidds, D. E., F. C. Golet, P. W. C. Paton, and J. C. Mitchell. 2007. Habitat correlates of reproductive effort in Wood Frogs and Spotted Salamanders in an urbanizing watershed. Journal of Herpetology 41:439–450.

Skryja, D. 2017. *Graptemys geographica* (Northern Map Turtle). Geographic distribution. Herpetological Review 48:386.

Sleeman, J. 2013. Snake fungal disease in the United States. US Geological Survey, National Wildlife Health Center Wildlife Health Bulletin 2013–02, Madison, Wisconsin, USA.

Sloss, B. L. 2011. Genetic identity of Wisconsin gartersnakes (*Thamnophis* spp.) using microsatellite genetic markers. Wisconsin Department of Natural Resources Research Report 192, Madison, USA. Archived by WebCite at http://www.webcitation.org/73cO82uZM. Accessed 1 Nov 2018.

Smallwood, W. M. 1928. Notes on the food of some Onondaga Urodela. Copeia 1928:89–98.

Smiley, P. C., Jr., S. S. Knight, F. D. Shields, Jr., and C. M. Cooper. 2009. Influence of gully erosion control on amphibian and reptile communities within riparian zones of channelized streams. Ecohydrology 2:303–312.

Smith, A. G. 1949. The subspecies of the Plains Garter Snake, *Thamnophis radix*. Bulletin of the Chicago Academy of Sciences 8:285–300.

Smith, A. K. 1975. Incidence of tail coiling in a population of Ringneck Snakes (*Diadophis punctatus*).

Transactions of the Kansas Academy of Science 77:237–238.

Smith, A. L., C. M. Bull, and D. A. Driscoll. 2013. Successional specialization in a reptile community cautions against widespread planned burning and complete fire suppression. Journal of Applied Ecology 50:1178–1186.

Smith, B., and N. Stephens. 2003. Conservation assessment for the Redbelly Snake in the Black Hills National Forest, South Dakota and Wyoming. Report to the US Forest Service, Rocky Mountain Region and Black Hills National Forest, Custer, South Dakota, USA.

Smith, B. E. 2003. Conservation assessment for the Northern Leopard Frog in the Black Hills National Forest, South Dakota and Wyoming. Report to the US Forest Service, Rocky Mountain Region and Black Hills National Forest, Custer, South Dakota, USA.

Smith, B. G. 1911. The nests and larvae of *Necturus*. Biological Bulletin 20:191–200.

Smith, C. E., and M. B. Zimmer. 2013. *Thamnophis radix* (Plains Gartersnake). Reproduction / litter size. Herpetological Review 44:454.

Smith, C. E., and G. S. Casper. 2015. *Chelydra serpentina* (Snapping Turtle). Hunting behavior. Herpetological Review 46:241–242.

Smith, C. J., and R. A. Pfingsten. 1989. *Notophthalmus viridescens* (Rafinesque). Eastern Newt. Pages 79–87 *in* R. A. Pfingsten and F. L. Dawns, editors. Salamanders of Ohio. Ohio Biological Survey, Bulletin New Series Volume 7, Columbus, USA.

Smith, C. S. 2009. Hibernation of the Eastern Massasauga Rattlesnake (*Sistrurus catenatus catenatus*) in northern Michigan. Thesis, Purdue University, Lafayette, Indiana, USA.

Smith, D. C. 1983. Factors controlling tadpole populations of the Chorus Frog (*Pseudacris triseriata*) on Isle Royale, Michigan. Ecology 64:501–510.

Smith, D. C. 1987. Adult recruitment in Chorus Frogs: effects of size and date at metamorphosis. Ecology 68:344–350.

Smith, D. C., and J. Van Buskirk. 1995. Phenotypic design, plasticity, and ecological performance in two tadpole species. American Naturalist 145:211–233.

Smith, D. H. V., B. Jones, L. Randall, and D. R. C. Prescott. 2014. Difference in detection and occupancy between two anurans: the importance of species-specific monitoring. Herpetological Conservation and Biology 9:267–277.

Smith, G. R. 2002. *Pseudacris crucifer* (Spring Peeper). Predation. Herpetological Review 33:48.

Smith, G. R., and A. R. Awan. 2009. The roles of predator identity and group size in the antipredator responses of American Toad (*Bufo americanus*) and Bullfrog (*Rana catesbeiana*) tadpoles. Behaviour 146:225–243.

Smith, G. R., and J. B. Iverson. 1993. Reactions to odor trails in Bullsnakes. Journal of Herpetology 27:333–335.

Smith, G. R., and J. B. Iverson. 2002. Sex ratio of Common Musk Turtles (*Sternotherus odoratus*) in a north-central Indiana lake: a long-term study. American Midland Naturalist 148:185–189.

Smith, G. R., and J. B. Iverson. 2004. Diel activity patterns of the turtle assemblage of a northern Indiana lake. American Midland Naturalist 152:156–164.

Smith, G. R., and J. E. Rettig. 1998. Observations on egg masses of the American Toad (*Bufo americanus*). Herpetological Natural History 6:61–64.

Smith, G. R., and J. J. Schulte. 2008. Conservation management implications of substrate choice in recently metamorphosed American Toads (*Bufo americanus*). Applied Herpetology 5:87–92.

Smith, G. R., A. Todd, J. E. Rettig, and F. Nelson. 2003a. Microhabitat selection by Northern Cricket Frogs (*Acris crepitans*) along a west-central Missouri creek: field and experimental observations. Journal of Herpetology 37:383–385.

Smith, G. R., A. J. Terlecky, C. B. Dayer, A. Boyd, M. E. Ogle, and C. J. Dibble. 2011. Effects of Mosquitofish and ammonium nitrate on activity of Green Frog (*Lithobates clamitans*) tadpoles: a mesocosm experiment. Journal of Freshwater Ecology 26:59–63.

Smith, G. R., H. A. Dingfelder, and D. A. Vaala. 2003b. Distribution and abundance of amphibian larvae within two temporary ponds in central Ohio, USA. Journal of Freshwater Biology 18:491–496.

Smith, G. R., H. A. Dingfelder, and D. A. Vaala. 2004. Asymmetric competition between *Rana clamitans* and *Hyla versicolor* tadpoles. Oikos 105:626–632.

Smith, G. R., J. B. Iverson, and J. E. Rettig. 2006. Changes in a turtle community from a northern

Indiana lake: a long-term study. Journal of Herpetology 40:180–185.

Smith, G. R., J. E. Retting, G. G. Mittelbach, J. L. Valiulis, and S. R. Schaack. 1999. The effects of fish on assemblages of amphibians in ponds: a field experiment. Freshwater Biology 41:829–837.

Smith, H. M. 1934. The amphibians of Kansas. American Midland Naturalist 15:377–528.

Smith, H. M. 1946. Handbook of lizards: lizards of the United States and Canada. Cornell University Press, Ithaca, New York, USA.

Smith, H. M. 1950. Handbook of amphibians and reptiles of Kansas. University of Kansas, Museum of Natural History Miscellaneous Publication 2, Lawrence, USA.

Smith, H. M. 1958. Total regeneration of the carapace in the box turtle. Turtox News 36:234–237.

Smith, H. M. 1963. The identity of the Black Hills population of *Storeria occipitomaculata*, the Red-bellied Snake. Herpetologica 19:17–21.

Smith, H. M. 1964. The taxonomic status of the Black Hills population of Smooth Greensnakes. Herpetologica 19:256–261.

Smith, H. M. 1965. Two new colubrid snakes from the United States and Mexico. Journal of the Ohio Herpetological Society 5:1–4.

Smith, P. W. 1956. The status, correct name, and geographic range of the Boreal Chorus Frog. Proceedings of the Biological Society of Washington 69:169–176.

Smith, P. W. 1957. An analysis of post-Wisconsin biogeography of the Prairie Peninsula region based on distributional phenomena among terrestrial vertebrate populations. Ecology 38:205–218.

Smith, P. W. 1961. The amphibians and reptiles of Illinois. Illinois Natural History Survey Bulletin 28:1–298.

Smith, P. W., and D. M. Smith. 1952. The relationship of the chorus frogs, *Pseudacris nigrita feriarum* and *Pseudacris n. triseriata*. American Midland Naturalist 48:165–180.

Smith, P. W., and S. A. Minton, Jr. 1957. A distributional summary of the herpetofauna of Indiana and Illinois. American Midland Naturalist 58:341–351.

Smith, W. H. 1882. Report on the reptiles and amphibians of Ohio. Pages 629–734 *in* J. S. Newberry, E. Orton, E. B. Andrews, T. G. Wormley, and F. B. Meek, editors. Report of the Geological Survey of Ohio. Volume 4. Zoology and Botany. Part I. Zoology. Nevins & Myers, Columbus, Ohio, USA.

Smith-Gill, S. J., and K. A. Berven. 1979. Predicting amphibian metamorphosis. American Naturalist 113:563–585.

Snider, A. T., and J. K. Bowler 1992. Longevity of reptiles and amphibians in North American Collections. Second edition. Society for the Study of Amphibians and Reptiles, Herpetological Circular 21, St. Louis, Missouri, USA.

Snodgrass, J. W., M. J. Komoroski, A. L. Bryan, Jr., and J. Burger. 2000. Relationships among isolated wetland size, hydroperiod, and amphibian species richness: implications for wetland regulations. Conservation Biology 14:414–419.

Snodgrass, J. W., R. E. Casey, D. Joseph, and J. A. Simon. 2008. Microcosm investigations of stormwater pond sediment toxicity to embryonic and larval amphibians: variation in sensitivity among species. Environmental Pollution 154:291–297.

Sobota, M. 1990a. *Chrysemys picta* (Painted Turtle). Geographic distribution. Herpetological Review 21:24.

Sobota, M. 1990b. *Elaphe vulpina* (Fox Snake). Geographic distribution. Herpetological Review 21:23.

Sobota, M. 1990c. *Thamnophis sirtalis* (Common Garter Snake). Geographic distribution. Herpetological Review 21:24.

Solis, K. 1998. Update: new results indicate flowering garlic mustard should be bagged and destroyed (Wisconsin). Restoration and Management Notes 16:223–224.

Somma, L. A. 1985. Notes on maternal behavior and post-brooding aggression in the Prairie Skink (*Eumeces septentrionalis*). Nebraska Herpetological Newsletter 6:9–12.

Somma, L. A. 1987a. Maternal care of the neonates in the Prairie Skink, *Eumeces septentrionalis*. Great Basin Naturalist 47:536–537.

Somma, L. A. 1987b. Reproduction of the Prairie Skink, *Eumeces septentrionalis*, in Nebraska. Great Basin Naturalist 47:373–374.

Somma, L. A. 1989. Influence of substrate water content on neonate size in the Prairie Skink, *Eumeces septentrionalis*. Great Basin Naturalist 49:198–200.

Somma, L. A. 1990. Observations on the nesting ecology of the Prairie Skink (*Eumeces septentrionalis*) in Nebraska. Bulletin of the Chicago Herpetological Society 25:77–80.

Somma, L. A. 1991. *Eumeces septentrionalis.* Piscivory. Herpetological Review 22:58–59.

Somma, L. A., and J. D. Fawcett. 1989. Brooding behavior of the Prairie Skink, *Eumeces septentrionalis*, and its relationship to the hydric environment of the nest. Zoological Journal of the Linnean Society 95:245–256.

Sonntag, E. 2006. In response to: assessment of translocations of Blanchard's Cricket Frog (*Acris crepitans blanchardi*) in southeast Michigan. Endangered Species UPDATE 23:15–24.

Souder, W. 1999. A Plague of frogs: the horrifying true story. Hyperion Press, New York, New York, USA.

South Dakota Game, Fish, and Parks. 2018. Threatened and endangered species. Archived by WebCite at http://www.webcitation.org/73dajh7Ke. Accessed 2 Nov 2018.

Sower, S. A., K. L. Reed, and K. J. Babbitt. 2000. Limb malformations and abnormal sex hormone concentrations in frogs. Environmental Health Perspectives 108:1085–1090.

Sparling, D. W., G. Linder, C. A. Bishop, and S. Krest, editors. 2010. Ecotoxicology of amphibians and reptiles. Second Edition. CRC Press, Boca Raton, Florida, USA.

Spear, P. A., M. Boily, I. Giroux, C. DeBlois, M. H. Leclair, M. Levasseur, and R. Leclair. 2009. Study design, water quality, morphometrics and age of the Bullfrog, *Rana catesbeiana*, in sub-watersheds of the Yamaska River drainage basin, Quebec, Canada. Aquatic Toxicology 91:110–117.

Speare, R., and L. Berger. 2013. Global distribution of chytridiomycosis in amphibians. James Cook University, Townsville, Australia.

Spencer, R. J., J. U. Van Dyke, and M. B. Thompson. 2017. Critically evaluating best management practices for preventing freshwater turtle extinctions. Conservation Biology 31:1340–1349.

Spencer, W. A. 1964. Movement in a population of *Pseudacris triseriata*. Journal of the Colorado-Wyoming Academy of Science 5:44.

Spencer, W. A. 1971. Boreal Chorus Frogs (*Pseudacris triseriata*) breeding in the alpine in southwestern Colorado. Arctic and Alpine Research 3:353.

Sperry, J. H., M. P. Ward, and P. J. Weatherhead. 2013. Effects of temperature, moon phase, and prey on nocturnal activity in ratsnakes: an automated telemetry study. Journal of Herpetology 47:105–111.

Spetich, M. A., S. R. Shifley, and G. R. Parker. 1999. Regional distribution and dynamics of coarse woody debris in midwestern old-growth forests. Forest Science 45:302–313.

Spinks, P. Q., G. B. Pauly, J. J. Crayon, and H. B. Shaffer. 2003. Survival of the western pond turtle (*Emys marmorata*) in an urban California environment. Biological Conservation 113:257–267.

Spohr, S. M. 2005. *Ambystoma maculatum* (Spotted Salamander). Predation. Herpetological Review 36:294–295.

Spotila, J. R., and B. A. Bell. 2008. Thermal ecology and feeding of the Snapping Turtle, *Chelydra serpentina*. Pages 71–79 *in* A. C. Steyermark, M. S. Finkler, and R. J. Brooks, editors. Biology of the Snapping Turtle. John Hopkins University Press, Baltimore, Maryland, USA.

Spradling, T. A., J. W. Tamplin, S. S. Dow, and K. J. Meyer. 2010. Conservation genetics of a peripherally isolated population of the Wood Turtle (*Glyptemys insculpta*) in Iowa. Conservation Genetics 11:1667–1677.

Sprent, J. F. A. 1988. Ascaridoid nematodes of amphibians and reptiles: *Ophidascaris* Baylis, 1920. Systematic Parasitology 11:165–213.

Staffen, A. D., and R. A. Staffen. 2020. *Pituophis catenifer* (Gophersnake). Geographic distribution. Herpetological Review 51:81.

Staffen, R. A., and R. A. Paloski. 2017a. *Coluber constrictor* (North American Racer). Geographic distribution. Herpetological Review 48:127.

Staffen, R. A., and R. A. Paloski. 2017b. *Hemidactylium scutatum* (Four-toed Salamander). Geographic distribution. Herpetological Review 48:805.

Stake, M. M., F. R. Thompson, III, J. Faaborg, and D. E. Burhans. 2005. Patterns of snake predation at songbird nests in Missouri and Texas. Journal of Herpetology 39:215–222.

Standing, K. L., T. B. Herman, and I. P. Morrison. 2000a. Predation of neonate Blanding's Turtles (*Emydoidea blandingii*) by Short-tailed Shrews (*Blarina brevicauda*). Chelonian Conservation and Biology 3:658–660.

Standing, K. L., T. B. Herman, M. Shallow, T. Power, and I. P. Morrison. 2000b. Results of the nest protection program for Blanding's Turtles in Kejimkujik National Park, Canada: 1987–1997. Chelonian Conservation and Biology 3:637–642.

Standish, I., E. Leis, N. Schmitz, J. Credico, S. Erickson, J. Bailey, J. Kerby, K. Phillips, and T. Lewis. 2018. Optimizing, validating, and field testing a multiplex qPCR for the detection of amphibian pathogens. Diseases of Aquatic Organisms 129:1–13.

Stanford, K. M. 2002. Demography and life history of an urban population of Plains Garter Snakes, *Thamnophis radix*. Thesis, Northern Illinois University, DeKalb, USA.

Stanford, K. M., and R. B. King. 2004. Growth, survival, and reproduction in a northern Illinois population of the Plains Garter Snake, *Thamnophis radix*. Copeia 2004:465–478.

Stanford, K. M., R. B. King, and D. Wynn. 2010. Summer and winter spatial habitat use by the Lake Erie Watersnake. Journal of Fish and Wildlife Management 1:122–130.

Stangel, P. W. 1983. Least Sandpiper predation on *Bufo americanus* and *Ambystoma maculatum* larvae. Herpetological Review 14:112.

Stangel, P. W. 1988. Premetamorphic survival of the salamander *Ambystoma maculatum*, in eastern Massachusetts. Journal of Herpetology 22:345–347.

Starkey, D. E., H. B. Shaffer, R. L. Burke, M. R. J. Forstner, J. B. Iverson, F. J. Janzen, A. G. J. Rhodin, and G. R. Ultsch. 2003. Molecular systematics, phylogeography, and the effects of Pleistocene glaciation in the Painted Turtle (*Chrysemys picta*) complex. Evolution 57:119–128.

Stedman, A. L., C. P. Jaeger, E. T. Hileman, B. C. Jellen, B. J. Swanson, and R. B. King. 2016. Multiple paternity in three wild populations of Eastern Massasauga (*Sistrurus catenatus*). Herpetological Conservation Biology 11:160–167.

Steen, D. 2011. Aging rattlesnakes: don't bother counting their rattles. Living alongside wildlife. Archived by WebCite at http://www.webcitation.org/73daudFVQ. Accessed 2 Nov 2018.

Steen, D. A., and J. P. Gibbs. 2004. Effects of roads on the structure of freshwater turtle populations. Conservation Biology 18:1143–1148.

Steen, D. A., C. J. W. McClure, J. C. Brock, D. C. Rudolph, J. B. Pierce, J. R. Lee, W. J. Humphries, B. B. Gregory, W. B. Sutton, L. L. Smith, D. L. Baxley, D. J. Stevenson, and C. Guyer. 2012a. Landscape-level influences of terrestrial snake occupancy within the southeastern United States. Ecological Applications 22:1084–1097.

Steen, D. A., J. P. Gibbs, K. A. Buhlmann, J. L. Carr, B. W. Compton, J. D. Congdon, J. S. Doody, J. C. Godwin, K. L. Holcomb, D. R. Jackson, F. J. Janzen, G. Johnson, M. T. Jones, J. T. Lamer, T. A. Langen, M. V. Plummer, J. W. Rowe, R. A. Saumure, J. K. Tucker, and D. S. Wilson. 2012b. Terrestrial habitat requirements of nesting freshwater turtles. Biological Conservation 150:121–128.

Steen, D. A., L. L. Smith, G. Morris, L. M. Conner, A. R. Litt, S. Pokswinski, and C. Guyer. 2013. Response of Six-lined Racerunner (*Aspidoscelis sexlineata*) to habitat restoration in fire-suppressed Longleaf Pine (*Pinus palustris*) sandhills. Restoration Ecology 21:457–463.

Steen, D. A., M. J. Aresco, S. G. Beilke, B. W. Compton, E. P. Condon, C. K. Dodd, Jr., H. Forrester, J. W. Gibbons, J. L. Greene, G. Johnson, J. A. Langen, M. J. Oldham, D. N. Oxier, R. A. Saumure, F. W. Schueler, J. M. Sleeman, L. L. Smith, J. K. Tucker, and J. P. Gibbs. 2006. Relative vulnerability of female turtles to road mortality. Animal Conservation 9:269–273.

Steen, D. A., P. A. Osborne, M. Dovciak, D. A. Patrick, and J. P. Gibbs. 2015. A preliminary investigation into the short-term effects of a prescribed fire on habitat quality for a snake assemblage. Herpetological Conservation and Biology 10:263–272.

Stein, J. 1995. A field study of the Six-lined Racerunner, *Cnemidophorus sexlineatus*, in southwestern Wisconsin. Thesis, University of Wisconsin, Madison, USA.

Stein, K. F. 1938. Migration of *Triturus viridescens*. Copeia 1938:86–88.

Stein, R. J. 1985. *Rana sylvatica* (Wood Frog). Predation. Herpetological Review 16:109–111.

Steiner, S. L., and R. M. Lehtinen. 2008. Occurrence of the amphibian pathogen *Batrachochytrium dendrobatidis* in Blanchard's Cricket Frog (*Acris crepitans blanchardi*) in the US Midwest. Herpetological Review 39:193–196.

Steinwascher, K. 1979. Competitive interactions among tadpoles: responses to resource level. Ecology 60:1172–1183.

Steinwascher, K., and J. Travis. 1983. Influence of food quality and quantity on early larval growth of two anurans. Copeia 1:238–242.

Stejneger, L., and T. Barbour. 1917. A check list of North American amphibians and reptiles. Harvard University Press, Cambridge, Massachusetts, USA.

Stejneger, L., and T. Barbour. 1943. A check list of North American amphibians and reptiles. Fifth edition. Harvard University Press, Cambridge, Massachusetts, USA.

Stenhouse, S. L. 1985. Interdemic variation in predation on salamander larvae. Ecology 66:1706–1717.

Stenhouse, S. L. 1987. Embryo mortality and recruitment of juveniles of *Ambystoma maculatum* and *Ambystoma opacum* in North Carolina. Herpetologica 43:496–501.

Stenhouse, S. L., N. G. Hairston, and A. E. Cobey. 1983. Predation and competition in *Ambystoma* larvae: field and laboratory experiments. Journal of Herpetology 17:210–220.

Sterrett, S. C., L. L. Smith, S. W. Golladay, S. H. Schweitzer, and J. C. Maerz. 2010. The conservation implications of riparian land use on river turtles. Animal Conservation 14:38–46.

Stevens, C. E., A. W. Diamond, and T. S. Gabor. 2002. Anuran call surveys on small wetlands in Prince Edward Island, Canada restored by dredging of sediments. Wetlands 22:90–99.

Stevenson, A. T., and A. F. Badje. 2017. *Lithobates catesbeianus* (American Bullfrog). Herpetological Review 48:384.

Stevenson, A. T., S. Lee, A. Greuel, and A. F. Badje. 2017. *Lithobates catesbeianus* (American Bullfrog). Geographic distribution. Herpetological Review 48:585.

Stevenson, D. J., and K. J. Dyer. 2002. *Masticophis flagellum flagellum* (Eastern Coachwhip). Diet. Herpetological Review 33:57.

Stewart, E. R., S. A. Reese, and G. R. Ultsch. 2004. The physiology of hibernation in Canadian Leopard Frogs (*Rana pipiens*) and Bullfrogs (*Rana catesbeiana*). Physiological and Biochemical Zoology: Ecological and Evolutionary Approaches 77:65–73.

Stewart, G. R. 1965. Thermal ecology of the garter snakes *Thamnophis sirtalis concinnus* (Hallowell) and *Thamnophis ordinoides* (Baird and Girard). Herpetologica 21:81–102.

Stewart, M. M. 1956. The separate effects of food and temperature differences on development of Marbled Salamander larvae. Journal of the Elisha Mitchell Scientific Society 72:47–56.

Stewart, M. M., and P. S. Sandison. 1972. Comparative food habits of sympatric Mink Frogs, Bullfrogs, and Green Frogs. Journal of Herpetology 6:241–244.

Stickel, L. F., W. H. Stickel, and F. C. Schmid. 1980. Ecology of a Maryland population of Black Rat Snakes (*Elaphe o. obsoleta*). American Midland Naturalist 103:1–14.

Stiles, R. M., C. H. La Rue, M. J. Hawkins, W. A. Mitchell, and M. J. Lannoo. 2016. Amphibian response to a large-scale habitat restoration in the Prairie Pothole region. Journal of North American Herpetology 2016:70–79.

Stille, W. 1954a. Observations on the reproduction and distribution of the Green Snake *Opheodrys vernalis* (Harlan). Chicago Academy of Sciences Natural History Miscellanea 127:1–11.

Stille, W. T. 1947. The aquatic habits of *Cnemidophorus sexlineatus*. Copeia 1947:143.

Stille, W. T. 1954b. Eggs of the salamander *Ambystoma jeffersonianum* in the Chicago area. Copeia 1954:300.

Stinner, J., N. Zarlinga, and S. Orcutt. 1994. Overwintering behavior of adult Bullfrogs, *Rana catesbeiana*, in northeastern Ohio. Ohio Journal of Science 94:8–13.

Stinson, K., S. Kaufman, L. Durbin, and F. Lowenstein. 2007. Impacts of Garlic Mustard invasion on a forest understory community. Northeastern Naturalist 14:73–88.

Stoddard, J. L., D. P. Larsen, C. P. Hawkins, R. K. Johnson, and R. H. Norris. 2006. Setting expectations for the ecological condition of streams: the concept of the reference condition. Ecological Applications 16:1267–1276.

Stohlgren, T. J., D. Barnett, C. Flather, J. Kartesz, and B. Peterjohn. 2005. Plant species invasions along the latitudinal gradient in the United States. Ecology 86:2298–2309.

Stokes, G. D., and W. A. Dunson. 1982. Permeability and channel structure of reptilian skin. American Journal of Physiology 242:F681–F689.

Stone, P. A., and J. B. Iverson. 1999. Cutaneous surface area in freshwater turtles. Chelonian Conservation and Biology 3:512–515.

Stone, P. A., J. B. Hauge, A. F. Scott, G. Guyer, and J. L. Dobie. 1993. Temporal changes in two turtle assemblages. Journal of Herpetology 27:13–27.

Stone, P. A., J. L. Dobie, and R. P. Henry. 1992. The effect of aquatic O_2 levels on diving and ventilator behavior in Soft-shelled (*Trionyx spiniferus*), Stinkpot (*Sternotherus odoratus*), and Mud Turtles (*Kinosternon subrubrum*). Physiological Zoology 65:331–345.

Storey, J. M., and K. B. Storey. 1985. Adaptations of metabolism for freeze tolerance in the Gray Tree Frog. Canadian Journal of Zoology 63:49–54.

Storey, K. B. 1987. Glycolysis and the regulation of cryoprotectant synthesis in liver of the freeze tolerant Wood Frog. Journal of Comparative Physiology B 157:373–380.

Storey, K. B. 1990. Life in a frozen state: adaptive strategies for natural freeze tolerance in amphibians and reptiles. American Journal of Physiology 258:559–568.

Storey, K. B. 1997 Organic solutes in freezing tolerance. Comparative Biochemistry and Physiology Part A: Physiology 117:319–326.

Storey, K. B., and J. M. Storey. 1984. Biochemical adaptation for freezing tolerance in the Wood Frog, *Rana sylvatica*. Journal of Comparative Physiology B 155:29–36.

Storey, K. B., and J. M. Storey. 1986. Freeze tolerance and intolerance as strategies of winter survival in terrestrially hibernating amphibians. Comparative Biochemistry and Physiology Part A: Physiology 83:613–617.

Storey, K. B., and J. M. Storey. 1987. Persistence of freeze tolerance in terrestrially hibernating frogs after spring emergence. Copeia 1987:720–726.

Storez, R. A. 1969. Observations on the courtship of *Ambystoma laterale*. Journal of Herpetology 3:87–96.

Storrs, S. I., and J. M. Kiesecker. 2004. Survivorship patterns of larval amphibians exposed to low concentrations of atrazine. Environmental Health Perspectives 112:1054–1057.

Storrs Mendéz, S. I., D. E. Tillitt, T. A. G. Rittenhouse, and R. D. Semlitsch. 2009. Behavioral response and kinetics of terrestrial atrazine exposure in American Toads (*Bufo americanus*). Archives of Environmental Contamination and Toxicology 57:590–597.

Stout, B. M., III, K. K. Stout, and C. W. Stihler. 1992. Predation by the Caddisfly *Banksiola dossuaria* on egg masses of the Spotted Salamander *Ambystoma maculatum*. American Midland Naturalist 127:368–372.

Strahler, A. N. 1952. Hypsometric (area-altitude) analysis of erosional topology. Geological Society of America Bulletin 63:1117–1142.

Strain, G. F., J. T. Anderson, E. D. Michael, and P. J. Turk. 2012. Hibernacula use and hibernation phenology in the Common Snapping Turtle (*Chelydra serpentina*) in Canaan Valley, West Virginia. Journal of Herpetology 46:269–274.

Strauss, A., and K. G. Smith. 2013. Why does amphibian chytrid (*Batrachochytrium dendrobatidis*) not occur everywhere? An exploratory study in Missouri Ponds. PLoS ONE 8:e76035.

Strickland, J. T., and F. J. Janzen. 2010. Impacts of anthropogenic structures on predation of Painted Turtle (*Chrysemys picta*) nests. Chelonian Conservation and Biology 9:131–135.

Strickler-Shaw, S., and D. H. Taylor. 1990. Sublethal exposure of lead inhibits acquisition and retention of discriminate avoidance learning in Green Frog (*Rana clamitans*) tadpoles. Environmental Toxicology and Chemistry 9:47–52.

Stuart, J. N., and C. W. Painter. 1993. *Rana catesbeiana* (Bullfrog). Cannibalism. Herpetological Review 24:103.

Stuart, S. N., J. S. Chanson, N. A. Cox, B. E. Young, A. S. L. Rodrigues, D. L. Fischman, and R. W. Waller. 2004. Status and trends of amphibian declines and extinctions worldwide. Science 306:1783–1786.

Stull, O. G. 1940. Variations and relationships in the snakes of the genus *Pituophis*. Bulletin of the United States National Museum 175:1–225.

Stunkard, H. W. 1917. Studies on North American Polystomidae, Aspidogastridae and Paramphistomidae. Dissertation, University of Illinois, Urbana-Champaign, USA.

Sugalski, M. T., and D. L. Claussen. 1997. Preference for soil moisture, soil pH, and light intensity by the salamander, *Plethodon cinereus*. Journal of Herpetology 31:245–250.

Suhre, D. O. 2010. Dispersal and demography of the American Bullfrog (*Rana catesbeiana*) in a semi-arid grassland. Thesis, University of Arizona, Tucson, USA.

Sullivan, A. M., J. C. Maerz, and D. M. Madison. 2002. Anti-predator response of Red-backed Salamanders (*Plethodon cinereus*) to chemical cues from Garter Snakes (*Thamnophis sirtalis*): laboratory and field experiments. Behavioral Ecology and Sociobiology 51:227–233.

Sullivan, B. K. 1992. Sexual selection and calling behavior in the American Toad (*Bufo americanus*). Copeia 1992:1–7.

Sullivan, B. K., and S. H. Hinshaw. 1992. Female choice and selection on male calling behaviour in the Grey Treefrog *Hyla versicolor*. Animal Behaviour 44:733–744.

Surdick, J. A., Jr. 2005. Amphibian and avian species composition of forested depressional wetlands and circumjacent habitat: the influence of land use type and intensity. Dissertation, University of Florida, Gainesville, USA.

Surendonk, S., and J. M. Kapfer. 2008. *Sternotherus odoratus* (Eastern Musk Turtle). Geographic distribution. Herpetological Review 39:108.

Surface, H. A. 1906. The serpents of Pennsylvania. Pennsylvania Department of Agriculture, Division of Zoology Monthly Bulletin 4:114–208.

Surface, H. A. 1908. First report on the economic features of the turtles of Pennsylvania. Pennsylvania Department of Agriculture, Division of Zoology Monthly Bulletin 6:105–196.

Sutherland, D. 2005. Parasites of North American frogs. Pages 109–123 *in* M. Lannoo, editor. Amphibian declines: the conservation status of United States species. University of California Press, Berkeley, USA.

Sutherland, W. J., R. Aveling, T. M. Brooks, M. Clout, L. V. Dicks, L. Fellman, E. Fleishman, D. W. Gibbons, B. Keim, F. Lickorish, K. A. Monk, D. Mortimer, L. S. Peck, J. Pretty, J. Rockstrom, J. P. Rodriguez, R. K. Smith, M. D. Spalding, F. H. Tonneijck, and A. R. Watkinson. 2014. A horizon scan of global conservation issues for 2014. Trends in Ecology and Evolution 29:15–22.

Sutton, W. B. 2004. The ecology and natural history of the Northern Leopard Frog, *Rana pipiens* Schreber, in West Virginia. Thesis, Marshall University, Huntington, West Virginia, USA.

Suzuki, H. K. 1951. Recent additions to the records of the distribution of the amphibians in Wisconsin. Wisconsin Academy of Sciences, Arts and Letters 40:215–234.

Swannack, T. M., and M. R. J. Forstner. 2007. Possible cause for the sex-ratio disparity of the endangered Houston Toad (*Bufo houstonensis*). Southwestern Naturalist 52:386–392.

Swanson, D. L., and S. L. Burdick. 2010. Overwintering physiology and hibernacula microclimates of Blanchard's Cricket Frogs at their northwestern range boundary. Copeia 2010:247–253.

Swanson, D. L., B. M. Graves, and K. L. Koster. 1996. Freezing tolerance/intolerance and cryoprotectant synthesis in terrestrially overwintering anurans in the Great Plains, USA. Journal of Comparative Physiology B: Biochemical, Systems, and Environmental Physiology 116:110–119.

Swanson, E. M., S. M. Tekmen, and M. A. Bee. 2007. Do female frogs exploit inadvertent social information to locate breeding aggregations? Canadian Journal of Zoology 85:921–932.

Swanson, P. L. 1952. The reptiles of Venago County, Pennsylvania. American Midland Naturalist 47:161–182.

Swanston, C., M. Janowiak, L. Iverson, L. Parker, D. Mladenoff, L. Brandt, P. Butler, M. St. Pierre, A. Prasad, S. Matthews, M. Peters, D. Higgins, and A. Dorland. 2011. Ecosystem vulnerability assessment and synthesis: a report from the Climate Change Response Framework Project in northern Wisconsin. US Forest Service General Technical Report NRS-82, Newtown Square, Pennsylvania, USA.

Swihart, R. K., J. J. Lusk, J. E. Duchamp, C. E. Rizkalla, and J. E. Moore. 2006. The roles of landscape context, niche breadth, and range boundaries in predicting species responses to habitat alteration. Diversity and Distributions 12:277–287.

Swihart, R. K., T. M. Gehring, M. B. Kolozsvary, and T. E. Nupp. 2003. Responses of "resistant" vertebrates to habitat loss and fragmentation: the importance on niche breadth and range boundaries. Diversity and Distributions 9:1–18.

Syverson, K. M., and P. M. Colgan. 2004. The quaternary of Wisconsin: a review of stratigraphy and glaciations history. Pages 295–311 *in* J. Ehlers and P. L. Gibbard, editors. Quaternary glaciations—extent and chronology, part II: North America. Elsevier B.V., Amsterdam, Netherlands.

Szczygielski, T., and T. Sulej. 2016. Revision of the Triassic European turtles *Proterochersis* and *Murrhardtia* (Reptilia, Testudinata, Proterochersidae), with the description of new taxa from Poland and Germany. Zoological Journal of the Linnean Society 177:395–427.

Szuroczki, D., and J. M. L. Richardson. 2009. The role of trematode parasites in larval amphibian communities: an aquatic ecologist's guide to the major players. Oecologia 161:371–385.

Szuroczki, D., and J. M. L. Richardson. 2011. Palatability of the larvae of three species of *Lithobates*. Herpetologica 67:213–221.

Szuroczki, D., and J. M. L. Richardson. 2012. The behavioral response of larval amphibians (Ranidae) to threats from predators and parasites. PLoS ONE:e49592.

Szymanski, J. 1998. Status assessment for the Eastern Massasauga (*Sistrurus c. catenatus*). US Fish and Wildlife Service, Fort Snelling, Minnesota, USA. Archived by WebCite at http://www.webcitation.org/73db9XzXC. Accessed 2 Nov 2018.

Tabachnick, W. J. 1977. Geographic variation of five biochemical polymorphisms in *Notophthalmus viridescens*. Journal of Heredity 68:117–122.

Taggart, T. W. 1992. Observations on Kansas amphibians and reptiles. Kansas Herpetological Society Newsletter 88:13–15.

Taigen, T. L., and F. H. Pough. 1981. Activity metabolism of the toad (*Bufo americanus*): ecological consequences and ontogenetic change. Journal of Comparative Physiology B: Biochemical, Systems, and Environmental Physiology 144:247–252.

Taigen, T. L., and F. H. Pough. 1985. Metabolic correlates of anuran behavior. American Zoologist 25:987–997.

Takahashi, M. K., and M. J. Parris. 2008. Life cycle polyphenism as a factor affecting ecological divergence within *Notophthalmus viridescens*. Oecologia 158:23–34.

Talentino, K., and E. Landre. 1991. Comparative development of two species of sympatric *Ambystoma* salamanders. Journal of Freshwater Ecology 6:395–401.

Talley, B. L., C. R. Muletz, V. T. Vredenburg, R. C. Fleischer, and K. R. Lips. 2015. A century of *Batrachochytrium dendrobatidis* in Illinois amphibians (1888–1989). Biological Conservation 182:254–261.

Tarr, T. L., and K. J. Babbitt. 2002. Effects of habitat complexity and predator identity on predation of *Rana clamitans* larvae. Amphibia-Reptilia 23:13–20.

Tate, L. S. 1994. Habitat fragmentation and the distribution of the Spring Peeper (*Hyla crucifer*) in south-central Wisconsin. Thesis, University of Wisconsin, Madison, USA.

Tattersall, G. J., and G. R. Ultsch. 2008. Physiological ecology of aquatic overwintering in ranid frogs. Biological Reviews 83:119–140.

Tattersall, G. J., and P. A Wright. 1996. The effects of ambient pH on nitrogen excretion in early life stages of the American Toad (*Bufo americanus*). Comparative Biochemistry and Physiology A 113:369–374.

Taub, A. M. 1967. Comparative histological studies on Duvernoy's gland of colubrid snakes. Bulletin of American Museum of Natural History 138:1–50.

Taylor, B., D. Skelly, L. K. Demarchis, M. D. Slade, D. Galusha, and P. M. Rabinowitz. 2005. Proximity to pollution sources and risk of amphibian limb malformation. Environmental Health Perspectives 113:1497–1501.

Taylor, C. N., K. L. Oseen, and R. J. Wassersug. 2004. On the behavioural response of *Rana* and *Bufo* tadpoles to echinostomatoid cercariae: implications to synergistic factors influencing trematode infections in anurans. Canadian Journal of Zoology 82:701–706.

Taylor, D. H., C. W. Steele, and S. Strickler-Shaw. 1990. Responses of Green Frog (*Rana clamitans*) tadpoles to lead-polluted water. Environmental Toxicology and Chemistry 9:87–93.

Taylor, H. J. 1937. Ludwig Kumlien. Wilson Bulletin 49:85–90.

Taylor, R. B. 1999. Seasonal diets and food habits of feral swine. Pages 58–66 *in* Proceedings of the First National Feral Swine Conference. Fort Worth, Texas, USA.

Taylor, R. B., and E. C. Hellgren. 1997. Diet of feral hogs in the western south Texas plains. Southwestern Naturalist 42:33–39.

Tekiela, S. 2004. Reptiles and amphibians of Wisconsin. Adventure Publications, Cambridge, Minnesota, USA.

Telecky, T. M. 2001. United States import and export of live turtles and tortoises. Turtle and Tortoise Newsletter 4:8–13.

Telford, S. R., Jr. 2010. Three new *Hepatozoon* species (Apicomplexa: Hepatozoidae) infecting the Florida Kingsnake, *Lampropeltis getula floridana*. Journal of Parasitology 96:162–169.

Temminck, C. J., and H. Schlegel. 1838. Fauna Japonica, sive, descriptio animalium, quae in itinere per Japonianum, jussu et auspiciis, superiorum, qui summum in India Batava imperium tenent, suscepto, annis 1823–1830 colleget, notis observationibus et adumbrationibus illustratis. Volume three. Chelonia, Ophidia, Sauria, Batrachia. J. G. Lalau, Leiden, Holland.

Temple, S. A. 1987. Predation on turtle nests increases near ecological edges. Copeia 1987:250–252.

Templeton, A. R. 1996. Translocation in conservation. Pages 315–325 *in* R. C. Szaro and D. W. Johnston, editors. Biodiversity in managed landscapes: theory and practice. Oxford University Press, New York, New York, USA.

Tennant, A., and R. D. Bartlett. 2000. Snakes of North America, eastern and central regions. Gulf Publishing Company, Houston, Texas, USA.

Tennessen, J. A., and K. R. Zamudio. 2003. Early male precedence, multiple paternity, and sperm storage in an amphibian aggregate breeder. Molecular Ecology 12:1567–1576.

Test, F. H. 1958. Butler's Garter Snake eats amphibian. Copeia 1958:151–152.

Test, F. H., and B. A. Bingham. 1948. Census of a population of the Red-backed Salamander (*Plethodon cinereus*). American Midland Naturalist 39:362–372.

Test, F. H., and H. Heatwold. 1962. Nesting sites of the Red-backed Salamander, *Plethodon cinereus*, in Michigan. Copeia 1962:206–207.

Test, F. H., and R. G. McCann. 1976. Foraging behavior of *Bufo americanus* tadpoles in response to high densities of micro-organisms. Copeia 1976:576–578.

Tetzlaff, S. J., M. Allender, M. Ravesi, J. Smith, and B. Kingsbury. 2015. First report of snake fungal disease from Michigan, USA involving Massasaugas, *Sistrurus catenatus* (Rafinesque 1818). Herpetology Notes 8:31–33.

Tetzlaff, S. J., M. J. Ravesi, M. C. Allender, E. T. Carter, B. A. Degregorio, J. M. Josimovich, and B. A. Kingsbury. 2017. Snake fungal disease affects behavior of free-ranging Massasauga Rattlesnakes (*Sistrurus catenatus*). Herpetological Conservation and Biology 12:624–634.

Tews, J., U. Brose, V. Grimm, K. Tielbörger, M. C. Wichmann, M. Schwager, and F. Jeltsch. 2004. Animal species diversity driven by habitat heterogeneity/diversity: the importance of keystone structures. Journal of Biogeography 31:79–92.

Theler, J. L. 2000. Animal remains from Native American archaeological sites in western Wisconsin. Transactions of the Wisconsin Academy of Sciences, Arts and Letters 88:121–142.

Thiel, R. P., and T. Wilder. 2010. Over-wintering characteristics of west-central Wisconsin Blanding's Turtles, *Emydoidea blandingii.* Canadian Field-Naturalist 124:134–138.

Thiemann, G. W., and R. J. Wassersug. 2000. Patterns and consequences of behavioural reponses to predators and parasites in *Rana* tadpoles. Biological Journal of the Linnean Society 71:513–528.

Thomas, C. D. 2010. Climate, climate change and range boundaries. Diversity and Distributions 16:488–495.

Thomas, E. S., and M. B. Trautman. 1937. Segregated hibernaculum of *Sternotherus odoratus* (Latreille). Copeia 1937:231.

Thomas, J. T., and G. Höbel. 2014. *Hyla versicolor* (Eastern Gray Treefrog). Lacrimal spot. Herpetological Review 45:112.

Thomas, M. A. 2014. Aspects of the trophic ecology of an invertivorous snake community. Thesis, Eastern Illinois University, Charleston, USA.

Thomas, M. A., K. M. Standford, and K. Bekker. 2007. *Necturus maculosus* (Mudpuppy). Body size and diet. Herpetological Review 38:433–434.

Thompson, I. D., J. A. Baker, and M. Ter-Mikaelian. 2003. A review of the long-term effects of post-harvest silviculture on vertebrate wildlife, and predictive models, with an emphasis on boreal forests in Ontario, Canada. Forest Ecology and Management 177:441–469.

Thompson, L. B., G. L. F. Carfagno, K. Andresen, A. J. Sitton, T. Bury, L. L. Lee, K. T. Lerner, and P. P. Fong. 2017a. Differential uptake of gold nanoparticles by 2 species of tadpoles, the Wood Frog (*Lithobates sylvaticus*) and the Bullfrog (*Lithobates catesbeianus*). Environmental Toxicology and Chemistry 36:3351–3358.

Thompson, M. M., B. H. Coe, J. D. Congdon, D. F. Stauffer, and W. A. Hopkins. 2017b. Nesting ecology and habitat use of *Chelydra serpentina* in an area modified by agriculture and industrial activity. Herpetological Conservation and Biology 12:292–306.

Thorson, T. B. 1955. The relationship of water economy to terrestrialism in amphibians. Ecology 36:100–116.

Threlfall, M., L. Severson, and D. Samuelsen. 1973. Plant and animal survey in Cherokee Marsh 1973: a preliminary report. The Friends of Cherokee Marsh, Madison, Wisconsin, USA. Archived by WebCite at http://www.webcitation.org/73dbRDskT. Accessed 2 Nov 2018.

Thurow, G. 1997. Observations on *Hemidactylium scutatum* habitat and distribution. Bulletin of the Chicago Herpetological Society 32:1–6.

Tiebout, H. M., III. 1983. *Nerodia sipedon sipedon* (Northern Water Snake). Reproduction. Herpetological Review 14:121.

Tiebout, H. M., III, and J. R. Cary. 1987. Dynamic spatial ecology of the Water Snake, *Nerodia sipedon*. Copeia 1987:1–18.

Tietge, J. E., S. A. Diamond, G. T. Ankley, D. L. DeFoe, G. W. Holcombe, K. M. Jensen, S. J. Degitz, G. E. Elonen, and E. Hammer. 2001. Ambient solar UV radiation causes mortality in larvae of three species of *Rana* under controlled exposure conditions. Photochemistry and Photobiology 74:261–268.

Tilley, S. G., B. G. Lundrigan, and L. P. Brower. 1982. Erythrism and mimicry in the salamander *Plethodon cinereus*. Herpetologica 38:409–417.

Timm, B. C., K. McGarigal, and L. R. Gamble. 2007a. Emigration timing of juvenile pond-breeding amphibians in western Massachusetts. Journal of Herpetology 41:243–250.

Timm, B. C., K. McGarigal, and B. W. Compton. 2007b. Timing of large movement events of pond-breeding amphibians in western Massachusetts. Biological Conservation 136:442–454.

Tingley, R., D. G. McCurdy, M. D. Pulsifer, and T. B. Herman. 2009. Spatio-temporal differences in the use of agricultural fields by male and female Wood Turtles (*Glyptemys insculpta*) inhabiting an agri-forest mosaic. Herpetological Conservation and Biology 4:185–190.

Tingley, R., T. B. Herman, M. D. Pulsifer, D. G. McCurdy, and J. P. Stephens. 2010. Intra-specific niche partitioning obscures the importance of fine-scale habitat data in species distribution models. Biodiversity Conservation 19:2455–2467.

Tinkle, D. W. 1957. Ecology, maturation and reproduction of *Thamnophis sauritus proximus*. Ecology 38:69–76.

Tinkle, D. W. 1958. The systematics and ecology of the *Sternotherus carinatus* complex (Testudinae, Chelydridae). Tulane Studies in Zoology 6:3–56.

Tinkle, D. W. 1961. Geographic variation in reproductive size, sex ratio and maturity of *Sternotherus odoratus* (Testidinata: Chelydridae). Ecology 42:68–76.

Tinkle, D. W., and R. G. Webb. 1955. A new species of *Sternotherus* with a discussion of the *Sternotherus carinatus* complex (Chelonia, Kinosternidae). Tulane Studies in Zoology 3:53–67.

Tinklepaugh, O. L. 1932. Maze learning of a turtle. Journal of Comparative Psychology 13:201–206.

Tkach, V. V., Y. Kuzmin, and E. E. Pulis. 2006. A new species of *Rhabdias* from the lungs of the Wood Frog, *Rana sylvatica*, in North America: the last sibling of *Rhabdias ranae*? Journal of Parasitology 92:631–636.

Todd, B. D., and C. T. Winne. 2006. Ontogenetic and interspecific variation in timing of movement and responses to climatic factors during migrations by pond-breeding amphibians. Canadian Journal of Zoology 84:715–722.

Todd, B. D., and K. M. Andrews. 2008. Response of a reptile guild to forest harvesting. Conservation Biology 22:753–761.

Todd, B. D., D. E. Scott, J. H. K. Pechmann, and J. W. Gibbons. 2011. Climate change correlates with rapid delays and advancements in reproductive timing in an amphibian community. Proceedings of the Royal Society B: Biological Sciences 278:2191–2197.

Todd, B. D., S. M. Blomquist, E. B. Harper, and M. S. Osbourn. 2014. Effects of timber harvesting on terrestrial survival of pond-breeding amphibians. Forest Ecology and Management 313:123–131.

Todd, B. D., T. M. Luhring, B. B. Rothermal, and J. W. Gibbons. 2009. Effects of forest removal on amphibian migrations: implications for habitat and landscape connectivity. Journal of Applied Ecology 46:554–561.

Todd, J., and R. Wassersug. 2010. Caudal pseudoautotomy in the Eastern Ribbon Snake *Thamnophis sauritus*. Amphibia-Reptilia 31:213–215.

Toll, S. J., R. G. Jaeger, and J. R. Gillette. 2000. Socioecology of a terrestrial salamander: females and males as territorial residents and invaders. Copeia 2000:276–281.

Tompsett, A. R., S. Wiseman, E. Higley, J. P. Giesy, and M. Hecker. 2013. Effects of exposure to 17α-ethynylestradiol during larval development on growth, sexual differentiation, and abundances of transcripts in the liver of the Wood Frog (*Lithobates sylvaticus*). Aquatic Toxicology 126:42–51.

Tordoff, W., III. 1980. Selective predation of Gray Jays, *Perisoreus canadensis*, upon Boreal Chorus Frogs, *Pseudacris triseriata*. Evolution 34:1004–1008.

Tordoff, W., III, and D. Pettus. 1977. Temporal stability of phenotypic frequencies in *Pseudacris triseriata* (Amphibia, Anura, Hylidae). Journal of Herpetology 11:161–168.

Tornick, J. K. 2010. Factors affecting aggression during nest guarding in the Eastern Red-backed Salamander (*Plethodon cinereus*). Herpetologica 66:385–392.

Tran, S. L., D. L. Moorhead, and K. C. McKenna. 2007. Habitat selection by native turtles in a Lake Erie wetland, USA. American Midland Naturalist 158:16–28.

Trapido, H. 1940. Mating time and sperm viability in *Storeria*. Copeia 1940:107–109.

Trapido, H. 1944. The snakes of the genus *Storeria*. American Midland Naturalist 31:1–84.

Trauth, S. E. 1977. Winter collection of *Cnemidophorus sexlineatus* eggs from Arkansas. Herpetological Review 8:33.

Trauth, S. E. 1980. Geographic variation and systematics of the lizard *Cnemidophorus sexlineatus* (Linnaeus) in the United States. Dissertation, Auburn University, Alabama, USA.

Trauth, S. E. 1982. *Ambystoma maculatum* (Ambystomatidae) in the diet of *Heterodon platyrhinos* (Colubridae) from northern Arkansas. Southwest Naturalist 27:230.

Trauth, S. E. 1983. Nesting habitat and reproductive characteristics of the lizard *Cnemidophorus sexlineatus* (Lacertilia: Teiidae). American Midland Naturalist 109:289–299.

Trauth, S. E. 1984. Seasonal incidence and reproduction in the Western Slender Glass Lizard (*Ophisaurus attenuatus attenuatus*), in Arkansas. Southwestern Naturalist 29:271–275.

Trauth, S. E. 1991. Distribution, scutellation, and reproduction in the Queen Snake, *Regina septemvittata* (Serpentes: Colubridae), from Arkansas. Proceedings of the Arkansas Academy of Science 45:103–106.

Trauth, S. E., and B. G. Cochran. 1991. *Hemidactylium scutatum* (Four-toed Salamander). Predation. Herpetological Review 22:55.

Trauth, S. E., and C. T. McAllister. 1996. *Cnemidophorus sexlineatus*. Catalogue of American Amphibians and Reptiles 628.1–628.12.

Trauth, S. E., and R. G. Neal. 2004. Geographic range expansion and feeding response by the leech *Macrobdella diplotertia* (Annelida: Hirudinea) to Wood Frog and Spotted Salamander egg masses. Journal of the Arkansas Academy of Science 58:139–141.

Trauth, S. E., H. W. Robinson, and M. V. Plummer. 2004. The amphibians and reptiles of Arkansas. University of Arkansas Press, Fayetteville, USA.

Trauth, S. E., J. M. Walker., and J. E. Cordes. 2013. *Aspidoscelis sexlineata viridis* (Prairie Racerunner). Burrow site fidelity. Herpetological Review 44:668–669.

Trauth, S. E., J. W. Wildhide, L. C. Hunt, A. Holt, T. L. Klotz, and S. A. Woolbright. 1996. *Cnemidophorus sexlineatus* (Six-lined Racerunner). Aquatic behavior. Herpetological Review 27:20–21.

Trauth, S. E., M. L. McCallum, and M. E. Cartwright. 2000. Breeding mortality in the Wood Frog, *Rana sylvatica* (Anura: Ranidae), from northcentral Arkansas. Journal of the Arkansas Academy of Science 54:154–156.

Trauth, S. E., R. L. Cox, Jr., J. D. Wilhide, and H. J. Worley. 1995. Egg mass characteristics of terrestrial morphs of the Mole Salamander, *Ambystoma talpoideum* (Caudata: Ambystomatidae), from northern Arkansas and clutch comparisons with other *Ambystoma* species. Proceedings of the Arkansas Academy of Science 49:193–196.

Trenham, P. C., W. D. Koenig, M. J. Mossman, S. L. Stark, and L. A. Jagger. 2003. Regional dynamics of wetland-breeding frogs and toads: turnover and synchrony. Ecological Applications 13:1522–1532.

Triplehorn, C. A. 1949. A large specimen and a high embryo count for the Queen Snake. Copeia 1949:76.

Trombulak, S. C., and C. A. Frissell. 2000. Review of ecological effects of roads on terrestrial and aquatic communities. Conservation Biology 14:18–30.

Tronzo, R. D. 1993. *Clemmys insculpta* (Wood Turtle). Reproduction. Herpetological Review 24:149.

Tscharntke, T., A. M. Klein, A. Kruess, I. Steffen-Dewenter, and C. Thies. 2005. Landscape perspectives on agricultural intensification and biodiversity—ecosystem service management. Ecology Letters 8:857–874.

Tschudi, J. J., von. 1838. Classification der Batrachier mit Berücksichtigung der fossilen Thiere dieser Abtheilung der Reptilien. Petitpierre, Neuchâtel, Switzerland.

Tucker, C. R., J. T. Strickland, B. S. Edmond, D. K. Delaney, and D. B. Ligon. 2015. Activity patterns of Ornate Box Turtles (*Terrapene ornata*) in Northwestern Illinois. Copeia 103:502–511.

Tucker, J. K. 1995. Notes on road-killed snakes and their implications on habitat modification due to summer flooding on the Mississippi River in west central Illinois. Transactions of the Illinois State Academy of Science 88:61–71.

Tucker, J. K. 1999. Reproductive output of *Terrapene carolina*, *Chrysemys picta*, and *Sternotherus odoratus* from west-central Illinois. Bulletin of the Maryland Herpetological Society 35:61–75.

Tucker, J. K. 2000. Illinois snakes: young-of-year morphology and food habits. Herpetological Review 31:106–107.

Tucker, J. K., and J. T. Lamer. 2004. Another challenge in Snapping Turtle (*Chelydra serpentina*) conservation. Turtle and Tortoise Newsletter 8:10–11.

Tucker, J. K., G. L. Paukstis, and F. J. Janzen. 2001. Interspecific variation in egg and hatchling size in four sympatric chelonians: influence of the hydric environment of incubation. Bulletin of the Maryland Herpetological Society 37:83–100.

Tucker, J. K., S. A. Gritters, and R. A. Hrabik. 2008. Turtle communities in the Upper Mississippi River System, 1992–1995. Illinois Natural History Survey Technical Report 2008 (30), Champaign, USA.

Turner, F. B. 1977. The dynamics of populations of squamates, crocodilians, and rhynchocephalians. Pages 157–264 *in* C. Gans and D. W. Tinkle, editors. Biology of the Reptilia. McGraw-Hill, New York, New York, USA.

Turtle, S. 2000. Embryonic survivorship of the Spotted Salamander (*Ambystoma maculatum*) in roadside and woodland vernal pools in southeastern New Hampshire. Journal of Herpetology 34:60–67.

Tuttle, K. N., and P. T. Gregory. 2009. Food habits of the Plains Garter Snake (*Thamnophis radix*) at the northern limit of its range. Journal of Herpetology 43:65–73.

Tuttle, K. N., and P. T. Gregory. 2014. Reproduction of the Plains Garter Snake, *Thamnophis radix*, near its northern range limit: more evidence for a "fast" life history. Copeia 2014:130–135.

Tuttle, S. E., and D. M. Carroll. 2005. Movements and behavior of hatchling Wood Turtles (*Glyptemys insculpta*). Northeastern Naturalist 12:331–348.

Tyning, T. F. 1990. A guide to amphibians and reptiles. Little, Brown & Company, Boston, Massachusetts, USA.

Tyser, R. W., and C. H. Key. 1988. Spotted Knapweed in natural area fescue grasslands: an ecological assessment. Northwest Science 62:151–159.

Uhler, F. M., C. Cottam, and T. E. Clarke. 1939. Food of snakes of the George Washington National Forest, Virginia. Transactions of the North American Wildlife Conference 4:605–622.

Ultsch, G R. 2006. The ecology of overwintering among turtles: where turtles overwinter and its consequences. Biological Reviews 81:339–367.

Ultsch, G. R., and S. A. Reese. 2008. Ecology and physiology of overwintering. Pages 91–99 *in*

A. C. Steyermark, M. S. Finkler, and R. J. Brooks, editors. Biology of the Snapping Turtle. John Hopkins University Press, Baltimore, Maryland, USA.

Ultsch, G. R., M. Draud., and B. Wicklow. 2007. Post-emergence movements and overwintering of Snapping Turtle, *Chelydra serpentina*, hatchlings in New York and New Hampshire. Canadian Field-Naturalist 12:178–181.

Ultsch, G. R., T. E. Graham, and C. E. Crocker. 2000. Aggregation of overwintering Leopard Frogs, *Rana pipiens*, and Common Map Turtles, *Graptemys geographica*, in northern Vermont. Canadian Field-Naturalist 114:314–315.

Underwood, G. 1967. A comprehensive approach to the classification of higher snakes. Herpetologica 23:161–168.

University of Wisconsin–Madison Zoological Museum. 2015. Brief sketch of the history of the University of Wisconsin Zoological Museum (UWZM). Archived by WebCite at http://www.webcitation.org/73dbkj08f. Accessed 2 Nov 2018.

Upton, D., R. Black, F. Burrows, K. Cedar, K. Frohlich, R. Gray, A. Lentini, P. Pratt, K. Prior, and A. Yagi. 2002. The Eastern Massasauga Rattlesnake stewardship guide: a resource and field guide for living with rattlesnakes in Ontario. Toronto Zoo, Toronto, Ontario, Canada.

US Environmental Protection Agency. 2018. Polychlorinated biphenyls (BCPs). Archived by WebCite at http://www.webcitation.org/73dbxlBks. Accessed 2 Nov 2018.

[USFS] US Forest Service. 2014. Forest management: silviculture. https://tinyurl.com/ya36a84e. Accessed 1 Nov 2018.

[USFS] US Forest Service. 2015. Forest management 101: a handbook to forest management in the North Central Region. https://tinyurl.com/y7h9u9e5. Accessed 1 Nov 2018.

[USFWS] US Fish and Wildlife Service. 1989. Red Wolf recovery plan. US Fish and Wildlife Service, Atlanta, Georgia, USA.

[USFWS] US Fish and Wildlife Service. 1995. Changes in the list of species in appendices to the Convention on International Trade in Endangered Species of Wild Fauna and Flora. Federal Register 60:665–670.

[USFWS] US Fish and Wildlife Service. 1996. California Condor recovery plan. Third Revision. US Fish and Wildlife Service, Portland, Oregon, USA.

[USFWS] US Fish and Wildlife Service. 2015. Environmental conservation online system. Eastern Massasauga (=rattlesnake) (*Sistrurus catenatus*). Archived by WebCite at http://www.webcitation.org/73dcmqWmG. Accessed 2 Nov 2018.

[USFWS] US Fish and Wildlife Service. 2018. Endangered species. https://www.fws.gov/endangered/. Accessed 2 Mar 2019.

[USGS] US Geological Survey. 2016. Earth's water: rivers and streams. US Geological Survey, The USGS Water Science School. Archived by WebCite at http://www.webcitation.org/73dcvyT3I. Accessed 2 Nov 2018.

Utiger, U., B. Schätti, and N. Helfenberger. 2005. The oriental colubrine genus *Coelognathus* Fitzinger, 1843 and classification of Old and New World racers and ratsnakes (Reptilia, Squamata, Colubridae, Colubrinae). Russian Journal of Herpetology 12:39–60.

Utiger, U., N. Helfenberger, B. Schätti, C. Schmidt, M. Ruf, and V. Ziswiler. 2002. Molecular systematics and phylogeny of Old World and New World ratsnakes, *Elaphe* Auct., and related genera (Reptilia, Squamata, Colubridae). Russian Journal of Herpetology 9:105–124.

Uzzell, T. M., Jr. 1964. Relations of the diploid and triploid species of the *Ambystoma jeffersonianum* complex (Amphibia, Caudata). Copeia 1964:257–300.

Uzzell, T. M., Jr. 1969. Notes on spermatophore production by salamanders of the *Ambystoma jeffersonianum* complex. Copeia 1969:601–612.

Van Bocxlaer, I., S. P. Loader, K. Roelants, S. D. Biju, M. Menegon, and F. Bossuyt. 2010. Gradual adaptation toward a range-expansion phenotype initiated the global radiation of toads. Science 327:679–682.

Van Buskirk, J., and D. C. Smith. 1991. Density-dependent population regulation in a salamander. Ecology 72:1747–1756.

Van Buskirk, J., S. A. McCollum, and E. E. Werner. 1997. Natural selection for environmentally induced phenotypes in tadpoles. Evolution 51:1983–1992.

van der Schalie, H., and A. van der Schalie. 1950. The mussels of the Mississippi River. American Midland Naturalist 44:448–466.

van Dijk, P. P., and J. Harding. 2013. *Glyptemys insculpta*. The IUCN Red List of Threatened Species 2011: e.T4965A97416259. www.iucnredlist.org. Accessed 1 Nov 2018.

Van Gorp, C. D. 2001. Changes in the distribution of *Acris crepitans blanchardi*, with studies of nesting microhabitat and the possible impact of ultraviolet-B radiation. Thesis, Drake University, Des Moines, Iowa, USA.

VanDeWalle, T. J. 1993. The impact of stream modification on the distribution of riverine turtles in Iowa. Thesis, Drake University, Des Moines, Iowa, USA.

VanDeWalle, T. J. 2005. *Sistrurus catenatus catenatus* (Eastern Massasauga). Male-male combat. Herpetological Review 36:196–197.

VanDeWalle, T. J., and J. L. Christiansen. 1996. A relationship between river modification and species richness of freshwater turtles in Iowa. Journal of the Iowa Academy of Sciences 103:1–8.

Vandre, G. W. 1975. Effects of channel redredging on wildlife and wildlife habitat of Buena Vista Marsh. Thesis, University of Wisconsin, Stevens Point, USA.

Vanek, J. P., and D. K. Wasko. 2017. Spatial ecology of the Eastern Hog-nosed Snake (*Heterodon platyrhinos*) at the northeastern limit of its range. Herpetological Conservation and Biology 12:109–118.

Vannote, R. L., G. W. Minshall, K. W. Cummins, J. R. Sedell, and C. E. Cushing. 1980. The river continuum concept. Canadian Journal of Fisheries and Aquatic Sciences 37:130–137.

Vasconcelos, D., and A. J. K. Calhoun. 2004. Movement patterns of adult and juvenile *Rana sylvatica* (LeConte) and *Ambystoma maculatum* (Shaw) in three restored seasonal pools in Maine. Journal of Herpetology 38:551–561.

Vasconcelos, D., and A. J. K. Calhoun. 2006. Monitoring created seasonal pools for functional success: a six year case study of amphibian responses, Sears Island, Maine, USA. Wetlands 26:992–1003.

Vaughan, T. A. 1961. Vertebrates inhabiting Pocket Gopher burrows in Colorado. Journal of Mammalogy 42:171–174.

Venesky, M. D., and C. D. Anthony. 2007. Antipredator adaptations and predator avoidance by two color morphs of the Eastern Red-backed Salamander *Plethodon cinereus*. Herpetologica 63:450–458.

Vermont Fish and Wildlife. 2015. Vermont's wildlife action plan: draft 10/1/2015. Vermont Fish and Wildlife, Montpelier, USA. Archived by WebCite at http://www.webcitation.org/73g3eZdsF. Accessed 4 Nov 2018.

Vernberg, F. J. 1953. Hibernation studies of two species of salamanders, *Plethodon cinereus cinereus* and *Eurycea bislineata bislineata*. Ecology 34:55–62.

Verrell, P. A. 1982a. Male newts prefer large females as mates. Animal Behaviour 30:1254–1255.

Verrell, P. A. 1982b. The sexual behavior of the Red-spotted Newt, *Notophthalmus viridescens* (Amphibia: Urodela: Salamandridae). Animal Behaviour 30:1224–1236.

Verrell, P. A. 1983. The influence of the ambient sex ratio and intermale competition on the sexual behavior of the Red-spotted Newt, *Notophthalmus viridescens* (Amphibia: Urodela: Salamandridae). Behavioral Ecology and Sociobiology 13:307–313.

Verrell, P. A. 1985. Male mate choice for large, fecund females in the Red-spotted Newt, *Notophthalmus viridescens*: how is size assessed? Herpetologica 41:382–386.

Verrell, P. A. 1990. A note on the courtship of the Broken-striped Newt, *Notophthalmus viridescens dorsalis* (Harlan). Journal of Herpetology 24:215–217.

Vervust, B., J. Brecko, and A. Herrell. 2011. Temperature effects on snapping performance in the Common Snapper *Chelydra serpentina* (Reptilia, Testudines). Journal of Experimental Zoology Part A: Ecological Genetics and Physiology 315:41–47.

Vieites, D. R., S. N. Román, M. H. Wake, and D. B. Wake. 2011. A multigenic perspective on phylogenetic relationships in the largest family of salamanders, the Plethodontidae. Molecular Phylogenetics and Evolution 59:623–635.

Villamizar-Gomez, A., M. R. J. Forstner, T. Suriyamongkol, K. N. Forks, W. E. Grant, H. Wang, and I. Mali. 2016. Prevalence of *Batrachochytrium dendrobatidis* in two sympatric treefrog species, *Hyla cinerea* and *Hyla versicolor*. Herpetological Review 47:601–605.

Vitale, K. O. 2013. Habitat use and seasonal movement patterns of Four-toed Salamanders (*Hemidactylium scutatum*) in Massachusetts. Thesis, University of Massachusetts, Amherst, USA.

Vitt, L. J. 1987. Communities. Pages 335–365 *in* R. A. Seigel, J. T. Collins, and S. S. Novak, editors. Snakes: ecology and evolutionary biology. Macmillan Publishing Company, New York, New York, USA.

Vitt, L. J., and W. E. Cooper, Jr. 1986. Tail loss, tail color, and predator escape in *Eumeces* (Lacertilia: Scincidae): age-specific differences in cost and benefits. Canadian Journal of Zoology 64:583–592.

Vitt, L. J., and W. E. Cooper, Jr. 1989. Maternal care in skinks (*Eumeces*). Journal of Herpetology 23:29–34.

Vogt, R. C. 1978. Systematics and ecology of the False Map Turtle complex *Graptemys pseudogeographica*. Dissertation, University of Wisconsin, Madison, USA.

Vogt, R. C. 1979. Cleaning/feeding symbiosis between Grackles (*Quiscalus*:Icteridae) and map turtles (*Graptemys*: Emydidae). Auk 96:608–609.

Vogt, R. C. 1980. Natural history of the map turtles *Graptemys pseudogeographica* and *Graptemys ouachitensis* in Wisconsin. Tulane Studies in Zoology and Botany 22:17–48.

Vogt, R. C. 1981a. Food partitioning in three sympatric species of map turtles, genus *Graptemys* (Testudinata, Emydidae). American Midland Naturalist 105:102–111.

Vogt, R. C. 1981. Natural history of amphibians and reptiles in Wisconsin. Milwaukee Public Museum Press, Milwaukee, Wisconsin, USA.

Vogt, R. C. 1981b. Turtle egg (*Graptemys*: Emydidae) infestation by fly larvae. Copeia 1981:457–459.

Vogt, R. C. 1993. Systematics of the False Map Turtle (*Graptemys pseudogeographica* complex: Reptilia, Testudines, Emydidae). Annals of Carnegie Museum 62:1–46.

Vogt, R. C. 1994. Temperature controlled sex determination as a tool for turtle conservation. Chelonian Conservation and Biology 1:159–162.

Vogt, R. C. 1995. *Graptemys pseudogeographica*. Catalogue of American Amphibians and Reptiles 604.1–604.6.

Vogt, R. C., and C. J. McCoy. 1980. Status of the emydine turtle genera *Chrysemys* and *Pseudemys*. Annals of Carnegie Museum 49:93–102.

Vogt, R. C., and J. J. Bull. 1982. Temperature controlled sex-determination in turtles: ecological and behavioral aspects. Herpetologica 38:156–164.

Vogt, R. C., and J. J. Bull. 1984. Ecology of hatchling sex ratio in map turtles. Ecology 65:582–587.

Vogt, R. C., and J. M. Kapfer. 2016. *Graptemys ouachitensis* (Ouachita Map Turtle). Geographic distribution. Herpetological Review 47:422.

Vogt, R. C., and R. L. Hine. 1977. Amphibian and reptile indices. Pages 103–158 *in* M. Bruner, R. Hine, J. R. March, M. A. Martin, G. R. Miller, D. H. Rusch, F. Stearns, N. G. Tilghman, and R. Vogt, editors. Development and evaluation of techniques for assessment of impact of highway development on wildlife populations and habitats in the Upper Midwest. Report to the US Fish and Wildlife Service, Madison, Wisconsin, USA.

Vogt, R. C., J. J. Bull, C. J. McCoy, and T. W. Houseal. 1982. Incubation temperature influences sex determination in kinosternid turtles. Copeia 1982:480–482.

von Achen, P. H., and J. L. Rakestraw. 1984. The role of chemoreception in the prey selection of neonate reptiles. Pages 163–172 *in* R. A. Seigel, L. E. Hunt, J. L. Knight, L. Malaret, and N. L. Zuschlag, editors. Vertebrate ecology and systematics: a tribute to Henry S. Fitch. University of Kansas Museum of Natural History Special Publication 10, Lawrence, USA.

Voss, S. R. 1993. Effect of temperature on body size, developmental stage, and timing of hatching in *Ambystoma maculatum*. Journal of Herpetology 27:329–333.

Vraniak, D. A., C. N. Lapin, and T. Bougie. 2017. *Glyptemys insculpta* (Wood Turtle). Nest predation. Herpetological Review 48:424–425.

Wacasey, J. W. 1961. An ecological study of two sympatric species of salamanders, *Ambystoma maculatum* and *Ambystoma jeffersonianum*, in southern Michigan. Dissertation, Michigan State University, East Lansing, USA.

Wacha, R. S., and J. L. Christiansen. 1976. Coccidian parasites from Iowa turtles: systematics and prevalence. Journal of Protozoology 23:57–63.

Wagner, G. 1922. On the present status of *Ophisaurus* in Wisconsin. Copeia 113:90–91.

Wahl, G. W., III, R. N. Harris, and T. Nelms. 2008. Nest site selection and embryonic survival in

Four-toed Salamanders, *Hemidactylium scutatum* (Caudata: Plethodontidae). Herpetologica 64:12–19.

Wake, D. B. 2012a. Facing extinction in real time. Science 335:1052–1053.

Wake, D. B. 2012b. Taxonomy of the salamanders of the family Plethodontidae (Amphibia: Caudata). Zootaxa 3484:75–82.

Wake, D. B., and V. T. Vredenburg. 2008. Are we in the midst of the sixth mass extinction? A view from the world of amphibians. Proceedings of the National Academy of Sciences of the United States of America 105:11466–11473.

Wakely, J. F., G. J. Fuhrman, F. A. Fuhrman, H. G. Fischer, and H. S. Mosher. 1966. The occurrence of tetrodotoxin (tarichatoxin) in Amphibia and the distribution of the toxin in the organs of newts (*Taricha*). Toxicon 3:195–203.

Walde, A. D., and R. A. Saumure. 2008. *Glyptemys insculpta* (Wood Turtle). Maximum clutch size. Herpetological Review 39:82.

Walde, A. D., J. R. Bider, C. Daigle, D. Masse, J. C. Bourgeois, J. Jutras, and R. D. Titman. 2003. Ecological aspects of a Wood Turtle, *Glyptemys insculpta*, population at the northern limit of its range in Québec. Canadian Field-Naturalist 117:377–388.

Walde, A. D., J. R. Bider, D. Masse, R. A. Saumure, and R. D. Titman. 2007. Nesting ecology and hatching success of the Wood Turtle, *Glyptemys insculpta*, in Québec. Herpetological Conservation and Biology 2:49–60.

Walde, A. D., P. J. Baker, and R. A. Saumure. 2008. *Glyptemys insculpta* (Wood Turtle). Hatchling behavior. Herpetological Review 39:81–82.

Waldick, R. C., B. Freedman, and R. J. Wassersug. 1999. The consequences for amphibians of the conversion of natural, mixed-species forests to conifer plantations in southern New Brunswick. Canadian Field-Naturalist 113:408–418.

Waldman, B. 1982. Adaptive significance of communal oviposition in Wood Frogs (*Rana sylvatica*). Behavioral Ecology and Sociobiology 10:169–174.

Waldman, B., and M. J. Ryan. 1983. Thermal advantages of communal egg mass deposition in Wood Frogs (*Rana sylvatica*). Journal of Herpetology 17:70–72.

Walker, C. F. 1946. The amphibians of Ohio. Part 1. Frogs and toads. Ohio State Museum of Science Bulletin 1:1–109.

Walker, J. M., J. E. Cordes, S. E. Trauth, and G. J. Manning. 2013. *Aspidoscelis neomexicana* (New Mexico Whiptail) x *Aspidoscelis sexlineata viridis* (Prairie Racerunner). Duration of hybridization. Herpetological Review 44:505–506.

Walker, R. F., and W. G. Whitford. 1970. Soil water absorption capabilities in selected species of anurans. Herpetologica 26:411–418.

Waller, D. M., and T. P. Rooney. 2008. The vanishing present: Wisconsin's changing lands, waters, and wildlife. University of Chicago Press, Chicago, Illinois, USA.

Walley, H. D. 1962. The Northern Prairie Skink, *Eumeces septentrionalis septentrionalis*, in Wisconsin and Illinois. Herpetologica 18:281–283.

Walley, H. D. 2000a. *Chelydra serpentina* (Common Snapping Turtle). Geographic distribution. Herpetological Review 31:184.

Walley, H. D. 2000b. *Rana palustris* (Pickerel Frog). Geographic distribution. Herpetological Review 31:252.

Walley, H. D., and B. L. Rozelle. 2004. *Eumeces fasciatus* (Five-lined Skink). Geographic distribution. Herpetological Review 35:408.

Walley, H. D., and J. Scalleta. 2000. *Hyla versicolor* (Eastern Gray Treefrog). Geographic distribution. Herpetological Review 31:182.

Walley, H. D., and J. R. Walley. 2004a. *Elaphe vulpina* (Western Fox Snake). Geographic distribution. Journal of Kansas Herpetology 11:14.

Walley, H. D., and J. R. Walley. 2004b. *Thamnophis radix* (Plains Gartersnake). Geographic distribution. Journal of Kansas Herpetology 11:14.

Walls, G. L. 1934. The reptilian retina: a new concept of visual-cell evolution. American Journal of Ophthalmology 17:892–915.

Walls, S. C. 1990. Interference competition in postmetamorphic salamanders: interspecific differences in aggression by coexisting species. Ecology 71:307–314.

Walls, S. C., C. A. Mathis, R. G. Jaeger, and W. F. Gergits. 1989. Male salamanders with high quality diets have faeces attractive to females. Animal Behavior 38:546–548.

Walpole, A. A., J. Bowman, D. C. Tozer, and D. S. Badzinski. 2012. Community-level response to climate change: shifts in anuran calling phenology. Herpetological Conservation and Biology 7:249–257.

Walston, L. J., and S. J. Mullin. 2007a. Population responses of Wood Frog (*Rana sylvatica*) tadpoles to overwintered Bullfrog (*Rana catesbeiana*) tadpoles. Journal of Herpetology 41:24–31.

Walston, L. J., and S. J. Mullin. 2007b. Responses of a pond-breeding amphibian community to experimental removal of predatory fish. American Midland Naturalist 157:63–73.

Walter, F. G., U. Stolz, F. Shirazi, and J. McNally. 2009. Epidemiology of severe and fatal rattlesnake bites published in the American Association of Poison Control Centers' annual reports. Clinical Toxicology 47:663–669.

Walters, B. 1975. Studies of interspecific predation within an amphibian community. Journal of Herpetology 9:267–279.

Walton, A. C. 1929. Studies on some nematodes of North American frogs. I. Journal of Parasitology 15:227–240.

Walton, B. M. 2013. Top-down regulation of litter invertebrates by a terrestrial salamander. Herpetologica 69:127–146.

Warburton, F. E. 1951. Intimidation display of Red-bellied Snake. Canadian Field-Naturalist 65:79.

Ward, D., and O. J. Sexton. 1981. Anti-predator role of salamander egg membranes. Copeia 1981:724–726.

Ward, J. V., K. Tockner, and F. Schiemer. 1999. Biodiversity of floodplain river ecosystems: ecotones and connectivity. Regulated Rivers: Research and Management 15:125–139.

Warkentin, K. M. 1992a. Effects of temperature and illumination on feeding rates of Green Frog tadpoles (*Rana clamitans*). Copeia 1992:725–730.

Warkentin, K. M. 1992b. Microhabitat use and feeding variation in Green Frog tadpoles (*Rana clamitans*). Copeia 1992:731–740.

Warner, D. A. 2000. Ecological observations on the Six-lined Racerunner (*Cnemidophorus sexlineatus*) in northwestern Illinois. Transactions of the Illinois State Academy of Science 93:239–248.

Wasserman, A. O. 1970. Polyploidy in the common tree toad *Hyla versicolor* Le Conte. Science 167:385–386.

Wassersug, R. J., and E. A. Seibert 1975. Behavioral responses of amphibian larvae to variation in dissolved oxygen. Copeia 1975:86–103.

Wassersug, R. J., and K. Hoff. 1979. A comparative study of the buccal pumping mechanism of tadpoles. Biological Journal of the Linnean Society 12:225–229.

Watermolen, D. J. 1990a. *Chelydra serpentina* (Snapping Turtle). Geographic distribution. Herpetological Review 21:24.

Watermolen, D. J. 1990b. *Chelydra serpentina* (Snapping Turtle). Geographic distribution. Herpetological Review 21:38.

Watermolen, D. J. 1990c. *Chrysemys picta* (Painted Turtle). Geographic distribution. Herpetological Review 21:26.

Watermolen, D. J. 1990d. *Storeria occipitomaculata* (Red-belly Snake). Geographic distribution. Herpetological Review 21:42.

Watermolen, D. J. 1991a. *Rana sylvatica* (Wood Frog). Geographic distribution. Herpetological Review 22:24.

Watermolen, D. J. 1991b. *Storeria occipitomaculata occipitomaculata*. Defense behavior. Herpetological Review 22:61.

Watermolen, D. J. 1992a. *Elaphe vulpina* (Fox Snake). Predation. Herpetological Review 23:120.

Watermolen, D. J. 1992b. Wisconsin herpetology: a bibliographical update with taxonomic, subject and geographical indices. Wisconsin Department of Natural Resources, Bureau of Endangered Resources Report 87, Madison, USA.

Watermolen, D. J. 1995. A key to the eggs of Wisconsin's amphibians. Wisconsin Department of Natural Resources Research Report 165, Madison, USA.

Watermolen, D. J. 1996. Osprey (*Pandion haliaetus*) preys on Mudpuppy (*Necturus maculosus maculosus*) in Menominee County. Michigan Birds and Natural History 3:157–158.

Watermolen, D. J. 2004. Softshell turtles (*Apalone* spp.) as bald eagle prey. Bulletin of the Chicago Herpetological Society 39:69–70.

Watermolen, D. J. 2013. Wisconsin herpetology, 2001–2010: a bibliography with taxonomic, geographic, and subject indices. Wisconsin Department of Natural Resources, Bureau of Science Services Miscellaneous Publication PUB-SS-1127,

Madison, USA. Archived by WebCite at http://www.webcitation.org/73g3vuOsJ. Accessed 4 Nov 2018.

Watermolen, D. J. 2014. Parasites and disease-causing organisms reported from Wisconsin amphibians and reptiles. Wisconsin Department of Natural Resources Miscellaneous Publication PUB-SS-1145 2014, Madison, USA. Archived by WebCite at http://www.webcitation.org/73g43v6gg. Accessed 4 Nov 2018.

Watermolen, D. J. 2015a. In memoriam: Philip A. Cochran. Bulletin of the Chicago Herpetological Society 50:112–115.

Watermolen, D. J. 2015b. Observational records of Mudpuppies (*Necturus maculosus*) from some Wisconsin counties. Bulletin of the Chicago Herpetological Society 50:125–126.

Watermolen, D. J. 2018a. Additional Mudpuppy (*Necturus maculosus*) records from fisheries assessment surveys conducted in Wisconsin. Bulletin of the Chicago Herpetological Society 53:57–60.

Watermolen, D. J. 2018b. *Necturus maculosus* (Mudpuppy). Geographic distribution. Herpetological Review 49:705.

Watermolen, D. J., and G. S. Casper. 1993. *Notophthalmus viridescens louisianensis* (Central Newt). Geographic distribution. Herpetological Review 24:107.

Watermolen, D. J., and H. Gilbertson. 1996. Keys for the identification of Wisconsin's larval amphibians. Wisconsin Department of Natural Resources, Bureau of Endangered Resources Endangered Species Report 109, Madison, USA. Archived by WebCite at http://www.webcitation.org/73g49QG3t. Accessed 4 Nov 2018.

Waters, R. 1993. Season prey preference by the Smooth Green Snake. Thesis, Central Michigan University, Mount Pleasant, USA.

Watkins-Colwell, G. J. 2002. A partial albino Northern Red-bellied Snake from Vermont. Northeastern Naturalist 9:221–224.

Watson, C. M. 1995. Habitat use and movement patterns of the Eastern Fox Snake (*Elaphe vulpina gloydi*) at Point Pelee National Park, Ontario. Thesis, University of Windsor, Windsor, Ontario, Canada.

Watson, C. M., C. E. Roelke, P. N. Pasichnyk, and C. L. Cox. 2012. The fitness consequences of the autotomous blue tail in lizards: an empirical test of predator response using clay models. Zoology 115:339–344.

Watson, J., R. A. Paloski, J. Karow, G. Jensen, J. Robaidek, and A. F. Badje. 2020. *Graptemys geographica* (Northern Map Turtle). Geographic distribution. Herpetological Review 51:73.

Wayne, H. L. 2001.Teflon tubing as radio transmitter belt material for Northern Leopard Frogs (*Rana pipiens*). Herpetological Review 32:88–89.

[WDATCP] Wisconsin Department of Agriculture, Trade and Consumer Protection. 2015. Water quality: atrazine. Archived by WebCite at http://www.webcitation.org/73ddYegvm. Accessed 2 Nov 2018.

[WDNR] Wisconsin Department of Natural Resources. 1977. Tree planting in Wisconsin. Wisconsin Department of Natural Resources Pub. 1–2400 (77), Madison, USA.

[WDNR] Wisconsin Department of Natural Resources. 2004. Wisconsin amphibian and reptile Regulations. Wisconsin Department of Natural Resources, Bureau of Natural Heritage Conservation Publication ER-102, Madison, USA.

[WDNR] Wisconsin Department of Natural Resources. 2005a. Biotic inventory and analysis of the Black River State Forest / Meadow Valley Landscape. Wisconsin Department of Natural Resources Publication ER-805, Madison, USA.

[WDNR] Wisconsin Department of Natural Resources. 2005b. Wisconsin's strategy for wildlife species of greatest conservation need. Wisconsin Department of Natural Resources, Madison, USA.

[WDNR] Wisconsin Department of Natural Resources. 2010. Wisconsin Blanchard's Cricket Frog species guidance. Wisconsin Department of Natural Resources, Bureau of Endangered Resources Publication ER-666, Madison, USA.

[WDNR] Wisconsin Department of Natural Resources. 2011a. Grassland and savanna protocols: protocols for incidental take permit and authorization—Blanding's Turtles (*Emydoidea blandingii*). Wisconsin Department of Natural Resources, Madison, USA.

[WDNR] Wisconsin Department of Natural Resources. 2011b. Species status assessment worksheet—Blanding's Turtles (*Emydoidea *blandingii*). Wisconsin Department of Natural Resources, Madison, USA.

[WDNR] Wisconsin Department of Natural Resources. 2012. Ecological landscapes of Wisconsin. Wisconsin Department of Natural Resources Handbook 1805.1, Madison, USA.

[WDNR] Wisconsin Department of Natural Resources. 2013a. Report on commercial fishing on the Mississippi River from 1953 to 2013. Wisconsin Department of Natural Resources, Madison, USA.

[WDNR] Wisconsin Department of Natural Resource. 2013b. Wisconsin Eastern Ribbonsnake species guidance. Wisconsin Department of Natural Resources, Bureau of Natural Heritage Conservation PUB-ER-710, Madison, USA.

[WDNR] Wisconsin Department of Natural Resources. 2014a. Revisions to endangered and threatened list s. 29.604, stats. Wisconsin Department of Natural Resources, Madison, USA.

[WDNR] Wisconsin Department of Natural Resources. 2014b. Wisconsin Butler's Gartersnake species guidance. Wisconsin Department of Natural Resources, Bureau of Natural Heritage Conservation PUB-ER-665, Madison, USA.

[WDNR] Wisconsin Department of Natural Resources. 2015a. Guide to Wisconsin spearing, netting, and bait harvest regulations, 2014–2015. Wisconsin Department of Natural Resources PUB-FH-204 2014, Madison, USA. Archived by WebCite at http://www.webcitation.org/73g4xXBHH. Accessed 4 Nov 2018.

[WDNR] Wisconsin Department of Natural Resources. 2015b. The ecological landscapes of Wisconsin: an assessment of ecological resources and a guide to planning sustainable management. Wisconsin Department of Natural Resources PUB-SS-1131 2015, Madison, USA.

[WDNR] Wisconsin Department of Natural Resources. 2015c. Wisconsin endangered and threatened species laws and list. Wisconsin Department of Natural Resources PUB-ER-001 2004 REV June 2015, Madison, USA. Archived by WebCite at http://www.webcitation.org/73dgagbbY. Accessed 2 Nov 2018.

[WDNR] Wisconsin Department of Natural Resources. 2015d. Wisconsin Wood Turtle species guidance. Wisconsin Department of Natural Resources, Bureau of Natural Heritage Conservation PUB-ER-684, Madison, USA.

[WDNR] Wisconsin Department of Natural Resources. 2016a. Wisconsin natural heritage working list. https://dnr.wisconsin.gov/topic/NHI/WList. Accessed 11 Nov 2020.

[WDNR] Wisconsin Department of Natural Resources. 2016b. Wisconsin's natural heritage inventory (NHI). https://dnr.wi.gov/topic/nhi/. Accessed 2 Nov 2018.

[WDNR] Wisconsin Department of Natural Resources. 2017a. Broad incidental take permit/authorization for grassland and savanna management. Archived by WebCite at http://www.webcitation.org/73dgxD9UX. Accessed 2 Nov 2018.

[WDNR] Wisconsin Department of Natural Resources. 2017b. Slender Glass Lizard (*Ophisaurus attenuatus*) species guidance. Wisconsin Department of Natural Resources, Bureau of Natural Heritage Conservation PUB-ER-682, Madison, USA. Archived by WebCite at http://www.webcitation.org/73dfq0Xkf. Accessed 2 Nov 2018.

[WDNR] Wisconsin Department of Natural Resources. 2018a. Feral pig hunting. Archived by WebCite at http://www.webcitation.org/73dj4UI54. Accessed 2 Nov 2018.

[WDNR] Wisconsin Department of Natural Resources. 2018b. Four-toed Salamander (*Hemidactylium scutatum*). Archived by WebCite at http://www.webcitation.org/73djM3°cr. Accessed 2 Nov 2018.

[WDNR] Wisconsin Department of Natural Resources. 2018c. Grassland and savanna protocol for incidental take permit and authorization: Ornate Box Turtle (*Terrapene ornata ornata*). Wisconsin Department of Natural Resources, Madison, USA. Archived by WebCite at http://www.webcitation.org/73dlXtJNB. Accessed 2 Nov 2018.

[WDNR] Wisconsin Department of Natural Resources. 2018d. Mudpuppy (*Necturus maculosus*). Archived by WebCite at http://www.webcitation.org/73dk7CA2r. Accessed 2 Nov 2018.

[WDNR] Wisconsin Department of Natural Resources. 2018e. Wisconsin's natural communities. Archived by WebCite at http://www.webcitation.org/73dkxiicp. Accessed 2 Nov 2018.

Weatherhead, P. J., and D. J. Hoysak. 1989. Spatial and activity patterns of Black Rat Snakes (*Elaphe obsoleta*) from radiotelemetry and recapture data. Canadian Journal of Zoology 67:463–468.

Weatherhead, P. J., and I. C. Robertson. 1990. Homing to food by Black Rat Snakes (*Elaphe obsoleta*). Copeia 1990:1164–1165.

Weatherhead, P. J., and K. A. Prior. 1992. Preliminary observations of habitat use and movements of the Eastern Massasauga Rattlesnake (*Sistrurus c. catenatus*). Journal of Herpetology 26:447–452.

Weatherhead, P. J., and M. B. Charland. 1985. Habitat selection in an Ontario population of the snake, *Elaphe obsoleta*. Journal of Herpetology 19:12–19.

Weatherhead, P. J., F. E. Barry, G. P. Brown, and M. R. L. Forbes. 1995. Sex ratios, mating behavior and sexual size dimorphism of the Northern Water Snake, *Nerodia sipedon*. Behavioral Ecology and Sociobiology 36:301–311.

Weatherhead, P. J., G. Blouin-Demers, and K. M. Cavey. 2003. Seasonal and prey-size dietary patterns of Black Ratsnakes (*Elaphe obsoleta obsoleta*). American Midland Naturalist 150:275–281.

Weatherhead, P. J., G. P. Brown, M. R. Prosser, and K. J. Kissner. 1998. Variation in offspring sex ratios in the Northern Water Snake (*Nerodia sipedon*). Canadian Journal of Zoology 76:2200–2206.

Weatherhead, P. J., G. P. Brown, M. R. Prosser, and K. J. Kissner. 1999. Factors affecting neonate size variation in Northern Water Snakes, *Nerodia sipedon*. Journal of Herpetology 33:577–589.

Weatherhead, P. J., J. M. Knox, D. S. Harvey, D. Wynn, J. Chiucchi, and H. Gibbs. 2009. Diet of *Sistrurus catenatus* in Ontario and Ohio: effects of body size and habitat. Journal of Herpetology 43:693–697.

Weaver, D. A., D. R. Stroup, B. A. Slafka, J. Kuzeniewski, and C. Hughes. 1991. Case review—Timber Rattlesnake bite to the hand with secondary coagulopathy and serum sickness. Journal of Emergency Nursing 4:193–196.

Weaver, R. E. 2010. Chemosensory and behavioral ecology of the dipsadid snakes: *Contia taenuis*, *Diadophis punctatus*, and *Hypsiglena chlorophaea*. Dissertation, Washington State University, Pullman, USA.

Weaver, W. G., Jr. 1965. The cranial anatomy of the hognose snakes (*Heterodon*). Bulletin of the Florida State Museum, Biological Sciences 9:275–304.

Webb, J. K., R. Shine, and R. M. Pringle. 2005. Canopy removal restores habitat quality for an endangered snake in a fire suppressed landscape. Copeia 2005:894–900.

Webb, R. G. 1962. North American recent soft-shelled turtles (family Trionychidae). University of Kansas Publications, Museum of Natural History 13:429–611.

Webb, R. G. 1990. *Trionyx*. Catalogue of American Amphibians and Reptiles 487:1–7.

Webb, S. L. 1987. Beech range extension and vegetation history: pollen stratigraphy of two Wisconsin lakes. Ecology 68:1993–2005.

Webb, T., III. 1974. A vegetational history from northern Wisconsin: evidence from modern and fossil pollen. American Midland Naturalist 92:12–34.

Webb, T., III, E. J. Cushing, and H. E. Wright, Jr. 1983. Holocene changes in the vegetation of the Midwest. Pages 142–165 *in* H. E. Wright, Jr., editor. Late Quarternary environments of the United States. Volume 2. The Holocene. University of Minnesota Press, Minneapolis, USA.

Weber, A. S., and J. B. Layzer. 2011. A comparison of turtle sampling methods in a small lake in Standing Stone State Park, Overton County, Tennessee. Journal of the Tennessee Academy of Science 86:45–52.

Weed, A. C. 1922. Reptile notes. Copeia 1922:84–87.

Weil, M. R., and R. D. Aldridge. 1981. Seasonal androgenesis in the male Water Snake, *Nerodia sipedon*. General and Comparative Endocrinology 44:44–53.

Weinstein, S. A., and D. E. Keyler. 2009. Local envenoming by the Western Hognose Snake (*Heterodon nasicus*): a case report and review of medically significant *Heterodon* bites. Toxicon 54:354–360.

Weinstein, S. A., and K. V. Kardong. 1994. Properties of the Duvernoy's secretions from opisthoglyphous and aglyphous colubrid snakes. Toxicon 32:1161–1185.

Weinstein, S. A., C. F. DeWitt, and L. A. Smith. 1992. Variability of venom-neutralizing properties of serum from snakes of the colubrid genus *Lampropeltis*. Journal of Herpetology 26:452–461.

Weir, L., I. J. Fiske, and J. A. Royle. 2009. Trends in anuran occupancy from northeastern states of the North American Amphibian Monitoring Program. Herpetological Conservation and Biology 4:389–402.

Weir, L. A., and M. J. Mossman. 2005. North American amphibian monitoring program (NAAMP). Pages 307–313 *in* M. J. Lannoo, editor. Amphibian declines: conservation status of United States species. University of California Press, Berkeley, USA.

Welch, A. M., R. D. Semlitsch, and H. C. Gerhardt. 1998. Call duration as an indicator of genetic quality in male Gray Treefrogs. Science 392:1928–1930.

Welch, R. J. 1987. Food habits of the Red-shouldered Hawk in Wisconsin. Passenger Pigeon 49:81–91.

Weldon, P. J., B. J. Demeter, and R. Rosscoe. 1993. A survey of shed skin-eating (dermatophagy) in amphibians and reptiles. Journal of Herpetology 27:219–228.

Weller, W. F., S. J. Hecnar, D. R. Hecnar, G. S. Casper, and F. N. Dawson. 2010. Quantitative assessment of intergradation between two subspecies of Painted Turtles, *Chrysemys picta belli* and *C. p. marginata*, in the Algoma District of west central Ontario, Canada. Herpetological Conservation and Biology 5:166–173.

Wells, C. S., and R. N. Harris. 2001. Activity level, growth rate, and vulnerability of predation in two species of salamanders. Herpetologica 57:116–127.

Wells, K. D. 1976. Multiple egg clutches in the Green Frog (*Rana clamitans*). Herpetologica 32:85–87.

Wells, K. D. 1977a. Territoriality and male mating success in the Green Frog (*Rana clamitans*). Ecology 58:750–762.

Wells, K. D. 1977b. The social behaviour of anuran amphibians. Animal Behaviour 25:666–693.

Wells, K. D. 1978. Territoriality in the Green Frog (*Rana clamitans*): vocalizations and aggressive behavior. Animal Behaviour 26:1051–1063.

Wells, K. D. 2007. The ecology and behavior of amphibians. University of Chicago Press, Chicago, Illinois, USA.

Wells, K. D., and T. L. Taigen. 1984. Reproductive behavior and aerobic capacities of male American Toads (*Bufo americanus*): is behavior constrained by physiology? Herpetologica 40:292–298.

Welsh, H. H., and S. Droege. 2001. A case for using plethodontid salamanders for monitoring biodiversity and ecosystem integrity of North American forests. Conservation Biology 15:558–569.

Wendelken, P. W. 1978. On prey-specific hunting behavior in the Western Ribbon Snake, *Thamnophis proximus* (Reptilia, Serpentes, Colubridae). Journal of Herpetology 12:577–578.

Werner, E. E. 1994. Ontogenetic scaling of competitive relations: size-dependent effects and responses in two anuran larvae. Ecology 75:197–213.

Werner, E. E., and K. S. Glennemeier. 1999. Influence of forest canopy cover on the breeding pond distributions of several amphibian species. Copeia 1999:1–12.

Werner, E. E., and M. A. McPeek. 1994. Direct and indirect effects of predators on two anuran species along an environmental gradient. Ecology 75:1368–1382.

Werner, E. E., G. A. Wellborn, and M. A. McPeek. 1995. Diet composition in postmetamorphic Bullfrogs and Green Frogs: implications for interspecific predation and competition. Journal of Herpetology 29:600–607.

Werner, E. E., K. L. Yurewicz, D. K. Skelly, and R. A. Relyea. 2007. Turnover in an amphibian metacommunity: the role of local and regional factors. Oikos 116:1713–1725.

Werner, E. E., R. A. Relyea, K. L. Yurewicz, D. K. Skelly, and C. J. Davis. 2009. Comparative landscape dynamics of two anuran species: climate-driven interaction of local and regional processes. Ecological Monographs 79:503–521.

Werner, J. K. 1971. Notes on the reproductive cycle of *Plethodon cinereus* in Michigan. Copeia 1971:161–162.

Werner, J. K. 1975. Vertebrate animal populations of the McCormick Forest. US Forest Service, North Central Forest Experiment Station Research Paper NC-118, St. Paul, Minnesota, USA.

Werner, J. K., and M. B. McCune. 1979. Seasonal changes in anuran populations in a Northern Michigan pond. Journal of Herpetology 13:101–104.

Werth, R. J. 1972. Lizard ecology: evidence of competition. Transactions of the Kansas Academy of Science 75:283–300.

West, R. M., and J. E. Dallman. 1980. Late Pleistocene and Holocene vertebrate fossil record of Wisconsin. Geoscience Wisconsin 4:25–45.

Westerman, A. G., A. J. Wigginton, D. J. Price, G. Linder, and W. J. Birge. 2003. Integrating amphibians into ecological risk assessment strategies. Pages 283–313 *in* G. Linder, S. K. Krest, and

D. W. Sparling, editors. Amphibian decline: an integrated analysis of multiple stressor effects. Society of Environmental Toxicology and Chemistry, Pensacola, Florida, USA.

Western Historical Company. 1880. The history of Dane County, Wisconsin. Culver, Page, and Hoyne Printers, Chicago, Illinois, USA.

Western Historical Company. 1882. The history of Walworth County, Wisconsin. Culver, Page, and Hoyne Printers, Chicago, Illinois, USA.

Wetzel, R. G. 2001. Limnology: lake and river ecosystems. Third edition. Academic Press, San Diego, California, USA.

Weyrauch, S. L., and T. C. Grubb, Jr. 2004. Patch and landscape characteristics associated with the distribution of woodland amphibians in an agricultural fragmented landscape: an information-theoretic approach. Biological Conservation 115:443–450.

Wheeler, W. E. 1984. Duck egg predation by Fox Snakes in Wisconsin. Wildlife Society Bulletin 12:77–78.

Wheelwright, N. T., M. J. Gray, R. D. Hill, and D. L. Miller. 2014. Sudden mass die-off of a large population of Wood Frog (*Lithobates sylvaticus*) tadpoles in Maine, USA, likely due to ranavirus. Herpetological Review 45:240–242.

Whitaker, J. O., Jr. 1961. Habitat and food of mouse-trapped young *Rana pipiens* and *Rana clamitans*. Herpetologica 17:173–179.

Whitaker, J. O., Jr. 1971. A study of the Western Chorus Frog, *Pseudacris triseriata*, in Vigo County, Indiana. Journal of Herpetology 5:127–150.

White, D., P. G. Minotti, M. J. Barczak, J. C. Sifneos, K. E. Freemark, M. V. Santelmann, C. F. Steinitz, A. R. Kiester, and E. M. Preston. 1997. Assessing risks to biodiversity from future landscape change. Conservation Biology 11:349–360.

White, J. P., H. M. Kaarakka, R. A. Staffen, J. E. Krumrie, M. G. Geary, M. Heerema, and A. F. Badje. 2020a. *Heterodon platihrinos* (Eastern Hog-nosed Snake). Geographic distribution. Herpetological Review 51:781.

White, J. P., H. M. Kaarakka, R. A. Staffen, J. E. Krumrie, M. G. Geary, M. Heerema, and A. F. Badje. 2020b. *Plestiodon septentrionalis* (Prairie Skink). Geographic distribution. Herpetological Review 51:777–778.

White, J. B., and G. G. Murphy. 1973. The reproductive cycle and sexual dimorphism of the Common Snapping Turtle, *Chelydra serpentina serpentina*. Herpetologica 29:240–246.

White, M. A., and D. J. Mladenoff. 1994. Old-growth forest landscape transitions from pre-European settlement to present. Landscape Ecology 9:191–205.

Whitford, W. G., and A. Vinegar. 1966. Homing, survivorship and over-wintering larvae in Spotted Salamanders, *Ambystoma maculatum*. Copeia 1966:515–519.

Whiting, A. 2004. Population ecology of the Western Chorus Frog, *Pseudacris triseriata*. Thesis, McGill University, Montreal, Quebec, Canada.

Whiting, M. J., J. C. Godwin, and M. K. Coldren. 1991. *Cnemidophorus sexlineatus* (Six-lined Racerunner) and *Cophosaurus texanus* (Texas Earless Lizard). Spider predation. Herpetological Review 22:58.

Whitmore, S. S., S. Losee, L. Meyer, and T. A. Spradling. 2013. Conservation genetics of the Central Newt (*Notophthalmus viridescens*) in Iowa: the importance of a biogeographic framework. Conservation Genetics 14:771–781.

Whittier, J. M., R. T. Mason, and D. Crews. 1985. Mating in the Red-sided Garter Snake, *Thamnophis sirtalis parietalis*: differential effects on male and female sexual behavior. Behavioral Ecology and Sociobiology 16:257–261.

[WIATRI] Wisconsin Aquatic and Terrestrial Resources Inventory. Wisconsin frog and toad survey. 2016. https://tinyurl.com/y73f873g. Accessed 2 Nov 2018.

[WICCI] Wisconsin Initiative on Climate Change Impacts. 2011. Wisconsin's changing climate: impacts and adaptation. Nelson Institute for Environmental Studies and Wisconsin Department of Natural Resources, Madison, USA.

Wied-Neuwied, M. A. D. 1838. Reise in das innere Nord-America in den jahren 1832 bis 1834. Volume 1. J. Hoelscher, Coblenz, Germany.

Wiener, J. G., and M. B. Sandheinrich. 2010. Contaminants in the upper Mississippi River: historic trends, responses to regulatory controls, and emerging concerns. Hydrobiologia 640:49–70.

Wiens, J. J., C. A. Kuczynski, and P. R. Stephens. 2010. Discordant mitochondrial and nuclear gene phylogenies in emydid turtles: implications for speciation and conservation. Biological Journal of the Linnean Society 99:445–461.

Wiewandt, T. A. 1969. Vocalization, aggressive behavior, and territoriality of the Bullfrog, *Rana catesbeiana*. Copeia 1969:276–285.

Wilbur, H. M. 1971. The ecological relationship of the salamander *Ambystoma laterale* to its all-female, gynogenetic associate. Evolution 25:168–179.

Wilbur, H. M. 1972. Competition, predation, and the structure of the *Ambystoma-Rana sylvatica* community. Ecology 53:3–21.

Wilbur, H. M. 1976. Density-dependent aspects of metamorphosis in *Ambystoma* and *Rana sylvatica*. Ecology 57:1289–1296.

Wilbur, H. M. 1977a. Density-dependent aspects of growth and metamorphosis in *Bufo americanus*. Ecology 58:196–200.

Wilbur, H. M. 1977b. Interactions of food level and population density in *Rana sylvatica*. Ecology 58:206–209.

Wilbur, H. M. 1977c. Propagule size, number, and dispersion pattern in *Ambystoma* and *Asclepias*. American Naturalist 111:43–68.

Wilbur, H. M. 1987. Regulation of structure in complex systems: experimental temporary pond communities. Ecology 68:1437–1452.

Wilbur, H. M. 1997. Experimental ecology of food webs: complex systems in temporary ponds. Ecology 78:2279–2302.

Wilbur, H. M., and J. E. Fauth. 1990. Experimental aquatic food webs: interactions between two predators and two prey. American Naturalist 135:176–204.

Wilbur, H. M., and J. P. Collins. 1973. Ecological aspects of amphibian metamorphosis: nonnormal distributions of competitive ability reflect selection for facultative metamorphosis. Science 182:1305–1314.

Wilbur, H. M., and R. A. Alford. 1985. Priority effects in experimental pond communities: responses of *Hyla* to *Bufo* and *Rana*. Ecology 66:1106–1114.

Wilcove, D. S. 1985. Nest predation in forest tracts and the decline of migratory songbirds. Ecology 66:1211–1214.

Wilcove, D. S., D. Rothstein, J. Dubow, A. Phillips, and E. Loses. 1998. Quantifying threats to imperiled species in the United States. BioScience 48:607–615.

Wild, E. R. 2002. Appendix H: herpetological survey of the Wolf River Geographic Management Unit with emphasis on rare species. Pages H-1–H-14 *in* E. Epstein, W. Smith, C. Anderson, E. Spencer, J. Lyons, and D. Feldkirchner, editors. Wolf River basin biotic inventory and analysis. Wisconsin Department of Natural Resources, Bureau of Endangered Resources PUBL ER-802 2002, Madison, USA.

Wild, E. R. 2007. *Hemidactylium scutatum* (Four-toed Salamander). Geographic distribution. Herpetological Review 38:472.

Wild, E. R. 2011. *Pituophis catenifer* (Gopher Snake). Geographic distribution. Herpetological Review 42:574.

Wild, E. R. 2012. *Ambystoma maculatum* (Spotted Salamander). Geographic distribution. Herpetological Review 43:611.

Wild, E. R. 2013a. *Hemidactylium scutatum* (Four-toed Salamander). Geographic distribution. Herpetological Review 44:269.

Wild, E. R. 2013b. *Notophthalmus viridescens* (Eastern Newt). Geographic distribution. Herpetological Review 44:269.

Wild, E. R. 2016a. *Hemidactylium scutatum* (Four-toed Salamander). Geographic distribution. Herpetological Review 47:245.

Wild, E. R. 2016b. *Plethodon cinereus* (Eastern Red-backed Salamander). Geographic distribution. Herpetological Review 47:246.

Wild, E. R. 2016c. *Plethodon cinereus* (Eastern Red-backed Salamander). Geographic distribution. Herpetological Review 47:621.

Wilder, T. T., M. Heeter, L. Breitenbach, G. S. Casper, and J. M. Kapfer. 2009. *Diadophis punctatus edwardsii* (Northern Ring-necked Snake). Geographic distribution. Herpetological Review 40:113–114.

Wildlife Health Information Sharing Partnership (WHISPers). 2017. On-line database query. http://www.nwhc.usgs.gov/whispers/. Accessed 3 March 2017.

Wiley, E. O., and B. S. Lieberman. 2011. Phylogenetics: theory and practice of phylogenetic systematics. Second edition. John Wiley and Sons, Inc. Hoboken, New Jersey, USA.

Wilgers, D. J., and E. A. Horne. 2006. Effects of different burn regimes on tallgrass prairie herpetofaunal species diversity and community composition in the Flint Hills, Kansas. Journal of Herpetology 40:73–84.

Williams, A. H. 1996. Conservation of the plant *Onosmodium molle* A. Michaux (Boraginaceae) and the beetle *Longitarsus subrufus* LeConte (Chrysomelidae) in Wisconsin. Thesis, University of Wisconsin, Madison, USA.

Williams, B. K., T. A. G. Rittenhouse, and R. D. Semlitsch. 2008. Leaf litter input mediates tadpole performance across forest canopy treatments. Oecologia 155:377–384.

Williams, D. D., and S. J. Taft. 1980. Helminths of anurans from NW Wisconsin. Proceedings of the Helminthological Society of Washington 47:278.

Williams, E. C. 1947. The terrestrial form of the newt, *Triturus viridescens*, in the Chicago region. Chicago Academy of Sciences, Natural History Miscellanea 5, Chicago, Illinois, USA.

Williams, J. E. 1952. Homing behavior of the Painted Turtle and Musk Turtle in a lake. Copeia 1952:76–82.

Williams, J. W., A. Ordonez, M. Notaro, S. Veloz, and D. Vimont. 2012a. Climate analogs, climate velocity, and potential shifts in vegetation structure and biomass for Wisconsin under 21st-century climate-change scenarios. Final Report to the Environmental and Economic Research and Development Program of Wisconsin's Focus on Energy and the Bryson Climate, People, and Environment Program, University of Wisconsin, Madison, USA.

Williams, J. W., B. Shuman, and P. J. Bartlein. 2009a. Rapid responses of the prairie-forest ecotone to early Holocene aridity in mid-continental North America. Global and Planetary Change 66:195–207.

Williams, K. E., K. E. Hodges, and C. A. Bishop. 2012b. Small reserves around hibernation sites may not adequately protect mobile snakes: the example of Great Basin Gophersnakes (*Pituophis catenifer deserticola*) in British Columbia. Canadian Journal of Zoology 90:304–312.

Williams, K. E., K. E. Hodges, and C. A. Bishop. 2014. Phenology and demography of Great Basin Gophersnakes (*Pituophis catenifer deserticola*) at the northern edge of their range. Herpetological Conservation and Biology 9:246–256.

Williams, K. E., K. E. Hodges, and C. A. Bishop. 2015. Hibernation and oviposition sites of Great Basin Gophersnakes (*Pituophis catenifer deserticola*) near their northern range limit. Journal of Herpetology 49:207–216.

Williams, P. K. 1973. Seasonal movements and population dynamics of four sympatric mole salamanders, genus *Ambystoma*. Dissertation, Indiana University, Bloomington, USA.

Williams, R. N., D. Gopurenko, K. R. Kemp, B. Williams, and J. A. DeWoody. 2009b. Breeding chronology, sexual dimorphism, and genetic diversity of congeneric ambystomatid salamanders. Journal of Herpetology 43:438–449.

Williams, T. A., and J. L. Christiansen. 1981. The niches of two sympatric softshell turtles, *Trionyx muticus* and *Trionyx spiniferus*, in Iowa. Journal of Herpetology 15:303–308.

Willis, L. S., T. Threlkeld, and C. C. Carpenter. 1982. Tail loss patterns in *Thamnophis* (Reptilia: Colubridae) and the probable fate of injured individuals. Copeia 1:98–101.

Willis, Y. L., D. L. Moyle, and T. S. Baskett. 1956. Emergence, breeding, hibernation, movements, and transformation of the Bullfrog, *Rana catesbeiana*, in Missouri. Copeia 1956:30–41.

Willson, J. D., and M. E. Dorcas. 2004. Aspects of the ecology of small fossorial snakes in the western Piedmont of North Carolina. Southeastern Naturalist 3:1–12.

Willson, R. J., and R. J. Brooks. 2006. Thermal biology of reproduction in female Eastern Foxsnakes (*Elaphe gloydi*). Journal of Herpetology 40:285–289.

Wilson, L. D. 1978. *Coluber constrictor*. Catalogue of American Amphibians and Reptiles 218.1–218.4.

Wilson, R. 1985. Yellow Warbler nestling predation by Eastern Fox Snake. Ontario Birds 3:73–75.

Wilson, R. E. 1976. An ecological study of *Ambystoma maculatum* and *Ambystoma jeffersonianum*. Dissertation, Cornell University, Ithaca, New York, USA.

Windmiller, B., R. N. Homan, J. V. Regosin, L. A. Willitts, D. L. Wells, and J. M. Reed. 2008. Breeding amphibian population declines following loss of upland forest habitat around vernal pools in Massachusetts, USA. Pages 41–51 *in* J. C. Mitchell, R. E. Jung Brown, and B. Bartholomew, editors. Urban Herpetology. Society for the Study of

Amphibians and Reptiles, Herpetological Conservation 3, Salt Lake City, Utah, USA.

Windmiller, B. S. 1996. The pond, the forest, and the city: Spotted Salamander ecology and conservation in a human-dominated landscape. Dissertation, Tufts University, Medford, Massachusetts, USA.

Wirsing, A. J., J. R. Phillips, M. E. Obbard, and D. L. Murray. 2012. Incidental nest predation in freshwater turtles: inter- and intraspecific differences in vulnerability are explained by relative crypsis. Oecologia 168:977–988.

Wisconsin Department of Transportation. 2013. Motor vehicle-deer crashes 2013. Archived by WebCite at http://www.webcitation.org/73dnNGqEy. Accessed 2 Nov 2018.

Wisconsin Legislature. 2010. Administrative Rules Chapter NR 16.11: captive wildlife. Archived by WebCite at http://www.webcitation.org/73dndP2u0. Accessed 2 Nov 2018.

Wisconsin Reed Canary Grass Management Working Group. 2009. Reed Canary Grass (*Phalaris arundinacea*) management guide: recommendation for landowners and restoration professionals. Wisconsin Department of Natural Resources Publication PUB–FR–428, Madison, USA. Archived by WebCite at http://www.webcitation.org/73g5KOydr. Accessed 4 Nov 2018.

Wise, R. S., S. L. Rumschlag, and M. D. Boone. 2014. Effects of amphibian chytrid fungus exposure on American Toads in the presence of an insecticide. Environmental Toxicology and Chemistry 33:2541–2544.

Wissinger, S. A., H. H. Whiteman, G. B. Sparks, G. L. Rouse and W. S. Brown. 1999. Foraging trade-offs along a predator-permanence gradient in subalpine wetlands. Ecology 80:2102–2116.

Wistrand, H. E. 1972. Predation on a snake by *Spermophilus tridecemlineatus*. American Midland Naturalist 88:511–512.

Witz, B. W. 1996. The functional response of *Cnemidophorus sexlineatus*: laboratory versus field measurements. Journal of Herpetology 30:498–506.

Wogahn, T. J., and J. M. Ray. 2014. *Plestiodon fasciatus* (Common Five-lined Skink). Geographic distribution. Herpetological Review 45:463.

Wolter, P. B., and D. J. Watermolen. 1996. *Rana sylvatica* (Wood Frog). Geographic distribution. Herpetological Review 27:30.

Wolterstorff, W. 1914. Über *Diemyctylus viridescens* Raf., *subsp. louisianensis n. subsp*. Abhandlungen und Berichte aus dem Museum für Natur- und Heimatkunde und dem Naturwissenschatflichen Verein in Madgeburg 2:383–392.

Woltz, H. W., G. P. Gibbs, and P. K. Ducey. 2008. Road crossing structures for amphibians and reptiles: informing design through behavioral analysis. Biological Conservation 141:2745–2750.

Woo, P. T. K. 1969. Trypanosomes in amphibians and reptiles in southern Ontario. Canadian Journal of Zoology 47:1139–1151.

Wood, J. T. 1944. Fall aggregation of the Queen Snake. Copiea 1944:253.

Wood, J. T. 1949. Observations on *Natrix septemvittata* (Say) in southwestern Ohio. American Midland Naturalist 42:744–750.

Wood, J. T. 1953. Observations on the complements of ova and nesting of the Four-toed Salamander in Virginia. American Naturalist 87:77–86.

Wood, J. T. 1955. The nesting of the Four-toed Salamander, *Hemidactylium scutatum* (Schlegel), in Virginia. American Midland Naturalist 53:381–389.

Wood, J. T., and O. K. Goodwin. 1954. Observations on the abundance, food, and feeding behaviour of the newt *Notophthalmus viridescens* in Virginia. Journal of the Elisha Mitchell Scientific Society 70:27–30.

Wood, J. T., and W. E. Duellman. 1950. Size and scutellation in *Natrix septemvittata* (Say) in southwestern Ohio. American Midland Naturalist 43:173–178.

Wood, W. F., J. M. Parker, and P. J. Weldon. 1995. Volatile components in scent gland secretions of garter snakes (*Thamnophis* spp.). Journal of Chemical Ecology 21:213–219.

Woodford, J. E., and M. W. Meyer. 2003. Impact of lakeshore development on Green Frog abundance. Biological Conservation 110:277–284.

Woodhams, D. C., A. D. Hyatt, D. G. Boyle, and L. A. Rollins-Smith. 2008. The Northern Leopard Frog *Rana pipiens* in a widespread reservoir species harboring *Batrachochytrium dendrobatidis* in North America. Herpetological Review 39:66–68.

Woodhead, A. E. 1928. *Haptophrya michiganensis* sp. Nov. A protozoan parasite of the Four-toed Salamander. Journal of Parasitology 14:177–182.

Woodhead, A. E., and F. Kruidenier. 1936. The probable method of infection of the Four-toed Salamander with the protozoon, *Haptophrya michiganensis*. Journal of Parasitology 22:107–108.

Woods, G. T. 1945. Rate of travel of the Wood Turtle. Copeia 1945:49.

Woolbright, L. L., and C. P. Martin. 2014. Seasonal migration by Red-backed Salamanders, *Plethodon cinereus*. Journal of Herpetology 48:546–551.

Worthington, R. D. 1968. Observations on the relative size of three species of salamander larvae in a Maryland pond. Herpetologica 24:242–246.

Worthington, R. D. 1969. Additional observations on sympatric species of salamander larvae in a Maryland pond. Herpetologica 25:227–229.

Worthington, R. D. 1973. Remarks on the distribution of the Smooth Green Snake, *Opheodrys vernalis blanchardi* Grobman in Texas. Southwestern Naturalist 18:344–346.

Wright, A. H. 1914. North American Anura: life-histories of the Anura of Ithaca, New York. Carnegie Institution of Washington Publication 197, Washington, D.C., USA.

Wright, A. H. 1918. Notes on Muhlenberg's Turtle. Copeia 52:7.

Wright, A. H. 1920. Frogs: their natural history and utilization. Appendix VI to the Report of the U.S. Commissioner of Fisheries for 1919. Government Printing Office, Bureau of Fisheries Document 888, Washington, D.C., USA.

Wright, A. H. 1929. Synopsis and description of North American tadpoles. Proceedings of the United States National Museum 74:1–70.

Wright, A. H. 1932. Life-histories of the frogs of Okefinokee Swamp, Georgia. Macmillan Publishing Company, New York, New York, USA.

Wright, A. H., and A. A. Allen. 1909. The early breeding habits of *Ambystoma punctatum*. American Naturalist 43:687–692.

Wright, A. H., and A. A. Wright. 1949. Handbook of frogs and toads of the United States and Canada. Third edition. Comstock Publishing Associates, Ithaca, New York, USA.

Wright, A. H., and A. A. Wright. 1957. Handbook of snakes of the United States and Canada. Volume 1. Comstock Publishing Associates, Ithaca, New York, USA.

Wright, A. H., and S. C. Bishop. 1915. A biological reconnaissance of the Okefinokee Swamp in Georgia: the reptiles. Proceedings of the Academy of Natural Sciences of Philadelphia 67:108–192.

Wright, H. A. 1974. Range burning. Journal of Range Management 27:5–11.

Wright, H. E., Jr. 1992. Patterns of Holocene climatic change in the midwestern United States. Quaternary Research 38:129–134.

Wright, J. D. 2008. Bullsnake, *Pituophis catenifer sayi*, nesting biology in Alberta. Canadian Field Naturalist 122:138–141.

Wuerthner, V. P., J. Hua, and J. T. Hoverman. 2017. The benefits of coinfection: trematodes alter disease outcomes associated with virus infection. Journal of Animal Ecology 86:921–931.

Wusterbarth, T. L., R. B. King, M. R. Duvall, W. S. Grayburn, and G. M. Burghardt. 2010. Phylogenetically widespread multiple paternity in New World Natricine Snakes. Herpetological Conservation and Biology 5:86–93.

Wyman, R. L., and D. S. Hawksley-Lescault. 1987. Soil acidity affects distribution, behavior, and physiology of the salamander, *Plethodon cinereus*. Ecology 68:1819–1827.

Xie, G. Y., D. H. Olson, and A. R. Blaustein. 2016. Projecting the global distribution of the emerging amphibian fungal pathogen, *Batrachochytrium dendrobatidis*, based on IPCC climate futures. PLoS ONE 11:e0160746.

Xuereb, A., J. R. Row, R. J. Brooks, C. MacKinnon, and S. C. Lougheed. 2012. Relation between parasitism, stress, and fitness correlates of the Eastern Foxsnake (*Pantherophis gloydi*) in Ontario. Journal of Herpetology 46:555–561.

Xuereb, A. T. J., J. D. Rouse, G. Cunnington and S. C. Lougheed. 2015. Population genetic structure at the northern range limit of the threatened Eastern Hog-nosed Snake (*Heterodon platirhinos*). Conservation Genetics 16:1265–1276.

Yahner, R. H., W. C. Bramble, and W. R. Byrnes. 2001. Effect of vegetation maintenance of an electric transmission right-of-way on reptile and amphibian populations. Journal of Arboriculture 27:24–29.

Yarrow, H. C. 1882. Checklist of North American Reptilia and Batrachia, with catalogue of specimens in the U.S. National Museum. Bulletin of the United States National Museum 24:1–249.

Yntema, C. L. 1970. Observations on females and eggs of the Common Snapping Turtle, *Chelydra serpentina*. American Midland Naturalist 84:68–76.

Yntema, C. L. 1976. Effects of incubation temperature on sexual differentiation in the turtle, *Chelydra serpentina*. Journal of Morphology 150:453–461.

Yntema, C. L. 1979. Temperature levels and periods of sex determination during incubation of eggs of *Chelydra serpentina*. Journal of Morphology 159:17–28.

Yoder, H. R., and J. R. Coggins. 1996. Helminth communities in the Northern Spring Peeper, *Pseudacris c. crucifer* Weid, and the Wood Frog, *Rana sylvatica* Le Conte, from southeastern Wisconsin. Journal of the Helminthological Society of Washington 63:211–214.

Yoder, H. R., and J. R. Coggins. 2007. Helminth communities in five species of sympatric amphibians from three adjacent ephemeral ponds in southeastern Wisconsin. Journal of Parasitology 93:755–760.

Yoder, H. R., J. R. Coggins, and J. C. Reinbold. 2001. Helminth parasites of the Green Frog (*Rana clamitans*) from southeastern Wisconsin, USA. Comparative Parasitology 68:269–272.

Young, A. G., and G. M. Clarke, editors. 2000. Genetics, demography and viability of fragmented populations. Cambridge University Press, New York, New York, USA.

Young, R. A. 1992. Effects of Duvernoy's gland secretions from the Eastern Hognose Snake, *Heterodon platirhinos*, on smooth muscle and neuromuscular junction. Toxicon 30:775–779.

Youngquist, M. B., and M. D. Boone. 2014. Movement of amphibians through agricultural landscapes: the role of habitat on edge permeability. Biological Conservation 175:148–155.

Youngquist, M. B., and M. I. Sitvarin. 2015. Predicting predatory outcomes in the context of carryover effects: interactions between juvenile frogs and spider predators. Ethology 121:601–608.

Youngquist, M. B., K. Downard, and M. D. Boone. 2015. Competitive interactions between Cricket Frogs (*Acris blanchardi*) and other anurans. Herpetologica 71:260–267.

Youngstrom, K. A., and H. M. Smith. 1936. Description of the larvae of *Pseudacris triseriata* and *Bufo woodhousii woodhousii* (Anura). American Midland Naturalist 17:629–633.

Yu, Z. C. 2000. Ecosystem response to lateglacial and early Holocene climate oscillations in the Great Lakes region of North America. Quaternary Science Reviews 19:1723–1747.

Yuan, Z-Y., W-W. Zhou, X. Chen, N. A. Poyarkov, Jr., H-M. Chen, N-H. Jang-Liaw, W-H. Chou, N. J. Matzke, K. Iizuka, M-S. Min, S. L. Kuzmin, Y-P. Zhang, D. C. Cannatella, D. M. Hillis, and J. Che. 2016. Spatiotemporal diversification of the true frogs (genus *Rana*): a historical framework for a widely studied group of model organisms. Systematic Biology 65:824–842.

Yurewicz, K. L. 2004. A growth/mortality trade-off in larval salamanders and the coexistence of intraguild predators and prey. Oecologia 138:102–111.

Yurkonis, K. A., and S. J. Meiners. 2006. Drought impacts and recovery are driven by local variation in species turnover. Plant Ecology 184:325–336.

Zadnik, A. K., J. T. Anderson, P. B. Wood, and K. Bledsoe. 2009. Wildlife use of back channels associated with islands on the Ohio River. Wetlands 29:543–551.

Zaga, A., E. E. Little, C. E. Rabeni, and M. R. Ellersieck. 1998. Photoenhanced toxicity of a carbamate insecticide to early life stage anuran amphibians. Environmental Toxicology and Chemistry 17:2543–2553.

Zamudio. K. R., and W. K. Savage. 2003. Historical isolation, range expansion, and secondary contact of two highly divergent mitochondrial lineages in Spotted Salamanders (*Ambystoma maculatum*). Evolution 57:1631–1652.

Zappalorti, R. T., and J. C. Mitchell. 2008. Snake use of urban habitats in the New Jersey pine barrens. Pages 355–360 *in* J. C. Mitchell, R. E. Jung Brown, and B. Bartholomew, editors. Urban Herpetology. Society for the Study of Amphibians and Reptiles, Herpetological Conservation 3, Salt Lake City, Utah, USA.

Zaremba, T. 1978. A Fox Snake hibernaculum. Bulletin of the Chicago Herpetological Society 13:87–90.

Zavaleta, E. S., R. J. Hobbs, and H. A. Mooney. 2001. Viewing invasive species removal in a whole-ecosystem context. Trends in Ecology and Evolution 16:454–459.

Zellmer, A. J., C. L. Richards, and L. M. Martens. 2008. Low prevalence of *Batrachochytrium dendrobatidis* across *Rana sylvatica* populations in southeastern Michigan, USA. Herpetological Review 39:196–199.

Zeuthen, E. 1953. Oxygen uptake as related to body size in organisms. Quarterly Review of Biology 28:1–12.

Zhao, T., and M. D. Schwartz. 2003. Examining the onset of spring in Wisconsin. Climate Research 24:59–70.

Zheng, Y., and J. J. Wiens. 2016. Combining phylogenomic and supermatrix approaches, and a time-calibrated phylogeny for squamate reptiles (lizards and snakes) based on 52 genes and 4,162 species. Molecular Phylogenetics and Evolution 94:537–547.

Zhour, Z., and Z. Jiang. 2008. Characteristics and risk assessment of international trade in tortoises and freshwater turtles in China. Chelonian Conservation and Biology 7:28–36.

Ziemba, J. L., A. C. Cameron, K. L. Peterson, C. A. M. Hickerson, and C. D. Anthony. 2015. Invasive Asian earthworms of the genus *Amynthas* alter microhabitat use by terrestrial salamanders. Canadian Journal of Zoology 93:805–811.

Ziesmer, J. H. 2019. *Sternotherus odoratus* (Common Musk Turtle). Geographic distribution. Herpetological Review 50:101.

Zimmer-Shaffer, S. A., J. T. Briggler, and J. J. Millspaugh. 2014. Modeling the effects of commercial harvest on population growth of river turtles. Chelonian Conservation and Biology 13:227–236.

Zimmerman, J. H. 1950. Outdoor calendar: a look at the approaching season. Passenger Pigeon 12:147–152.

Zimmerman, J. H. 1951. Outdoor calendar. Passenger Pigeon 13:9–14.

Zimmerman, S. L., J. Frisbie, D. L. Goldstein, J. West, K. Rivera, and C. M. Krane. 2007. Excretion and conservation of glycerol, and expression of aquaporins and glyceroporins, during cold acclimation in Cope's Gray Tree Frog *Hyla chrysoscelis*. American Journal of Physiology-Regulatory, Integrative and Comparative Physiology 292:544–555.

Zippel, K., and C. Tabaka. 2008. Amphibian chytridiomycosis in captive *Acris crepitans blanchardi* (Blanchard's Cricket Frog) collected from Ohio, Missouri, and Michigan, USA. Herpetological Review 39:192–193.

Zippel, K., K. Johnson, R. Gagliardo, R. Gibson, M. McFadden, R. Browne, C. Martinez, and E. Townsend. 2011. The Amphibian Ark: a global community for ex situ conservation of amphibians. Herpetological Conservation and Biology 6:340–352.

Zug, G. R. 1994. North American box turtles: a continuing concern. Virginia Herpetological Society Newsletter 4:1–2.

Contributor Biographies

Noah J. Anderson earned his doctorate from Idaho State University in 2009 in Biology Education. His eclectic research interests span the ecological physiology of reptiles and amphibians to understanding what draft animal power can teach us about living sustainably on the land. He was an Associate Professor at University of Wisconsin–Platteville Baraboo Sauk County and is currently a member of the Biology Department at Winona State University.

Andrew F. Badje is a Conservation Biologist with the Wisconsin Department of Natural Resource's Bureau of Natural Heritage Conservation. Andrew received his B.S. in Biological Aspects of Conservation and Geographic Information Systems Certificate at the University of Wisconsin–Madison. His efforts are dedicated to the inventory, monitoring, and management of Wisconsin's rare and common amphibians and reptiles. Additional conservation efforts Andrew contributes to are white-nose syndrome in North American bats and citizen-based monitoring.

Craig S. Berg served the Milwaukee County Zoo (MCZ) as the Aquarium/Reptile Curator from 1996 to 2017. He is currently retired. During his tenure, the MCZ became involved in conservation projects at both local (Timber Rattlesnake, Eastern Massasauga, Ornate Box Turtle) and international (Caribbean herpetofauna and icthyofauna). He was a founding member of the Association of Zoos and Aquariums' Eastern Massasauga Rattlesnake Species Survival Plan. He was also a participant in the Amphibian Ark Conservation Needs Assessment workshop of Caribbean amphibians held in the Dominican Republic in 2011. He has collaborated on projects with the Wisconsin Department of Natural Resources, US Fish and Wildlife Service, Milwaukee Public Museum, Wisconsin Lutheran College, and St. George's University (Grenada, West Indies). He is a coauthor on a number of publications related to the previously mentioned conservation activities.

Tara L. E. Bergeson is a Conservation Biologist and Invasive Species Team Leader at the Wisconsin Department of Natural Resources. She received a B.S. in Natural Resources and a M.S. in Wildlife Ecology from the University of Wisconsin–Madison, where she studied the impacts of contaminants on amphibians. Tara's herpetofaunal interests include amphibian ecology and conservation, climate change impacts and adaptation strategies, conservation policy, and effectiveness measures for conservation actions.

Danielle R. Bradke earned her M.S. in biology from Grand Valley State University (Michigan) and is presently a Ph.D. candidate at the University of Georgia. Prior to beginning her Ph.D., Danielle worked as a biologist at the US Geological Survey Brown Treesnake Laboratory in Guam. Her research interests include using population models and genetic techniques to better understand population processes and improve monitoring methods for herpetofauna.

Robert D. Brodman earned a Ph.D. in ecology from Kent State University. His research interests are on the ecology and conservation of amphibians and reptiles. Robert was a Professor of Biology and Environmental Science at Saint Joseph's College in Indiana, and he has also been a Midwest Regional Co-chair and advisory board member as well as a Joint National Steering Committee member for Partners in Amphibian and Reptile Conservation. He is an Associate Editor and Steering Committee member for Herpetological Conservation & Biology and has also served as Indiana State Coordinator for the Declining Amphibian Populations Task Force, and as chair and member of the Technical Advisory Committee on herpetofauna for the Indiana Department of Natural Resources.

Jeff Canfield graduated from St. Olaf College (Minnesota) with a major in Biology and concentration in Statistics. He started as an intern with the Minnesota Pollution Control Agency (MPCA) his senior year of college and subsequently worked with the MPCA for four years on the malformed frog project. His passion for quality control through his work at the MPCA led him to a career in software development where he currently continues working as a manager of software development teams.

Rebecca A. Christoffel earned her B.S. and M.S. degrees in Wildlife Ecology from the University of Wisconsin–Madison, and her Ph.D. in Fisheries and Wildlife from Michigan State University. She established Christoffel Conservation in 2015 to help take the mystery out of mysterious wildlife such as turtles and snakes, and to assist landowners in managing their properties with amphibians and reptiles in mind. Rebecca founded Snake Conservation Society in 2012, with the aim of changing the image of snakes in the public's imagination. She is also the Co-Director of Turtles for Tomorrow, which is dedicated to the conservation and management of Wisconsin's rare turtles.

Deahn M. Donner's enthusiasm for the outdoors began at an early age while hiking, fishing, hunting, and camping along the lower Wisconsin River with her family. She earned her Ph.D. from the University of Wisconsin–Madison. Current research interests include examining how forest management and restoration activities affect large-scale wildlife movement, habitat use, and conservation for multiple species groups. Of special interest is understanding the impacts of removing harvest residues (woody biomass) from native forests on biodiversity, especially reptiles and amphibians, and other ecological services the forests provide.

Samantha Foster is a Conservation Biologist and Educator. While pursuing her biology degree from the University of Wisconsin–Whitewater and teacher education certifications from Concordia University, as a returning adult college student, Samantha was fortunate to participate in telemetry studies on Blanding's Turtles, and demographic surveys for various snakes and turtles, including the endangered Ornate Box Turtle. She has also been involved in headstarting projects focused on the protected Wood Turtle in Wisconsin. Samantha continues to take part in volunteer work with the Wisconsin Department of Natural Resources, while remaining involved in numerous additional conservation and habitat restoration projects.

Tony Gamble studies evolution in lizards and snakes, with an emphasis on geckos. He uses large genetic datasets to investigate reptile sex determination.

James H. Harding received his B.S. and M.S. degrees from the Department of Fisheries and Wildlife at Michigan State University. He has served as instructor for the undergraduate herpetology course at MSU since 1997. Jim has formally studied a population of Wood Turtles in Michigan since 1969 and has also studied additional turtle species and surveyed herpetofaunal populations in the state. He has been actively involved in herpetological conservation and education for many decades and has authored or coauthored several books on Michigan and Great Lakes amphibians and reptiles.

Billie C. Harrison works for the Milwaukee Public Museum and is pursuing her M.S. with an emphasis in herpetology and arachnology. She currently serves as the Outreach Coordinator for the Eastern Massasauga Species Survival Plan, working to advocate for the conservation of this federally listed rattlesnake. She has worked for the Racine and Milwaukee County Zoos, specializing in herpetofauna, invertebrates, and fish. Billie has been fortunate to participate in herpetofaunal research and conservation in Wisconsin and will continue to spend her summer evenings flipping cover boards forever seeking snakes.

Robert W. Hay graduated from the University of Wisconsin–Stevens Point with a B.S. in Natural Resource Management and is a self-taught herpetologist. He was employed by the Wisconsin Department of Natural Resources as a herpetologist for nearly 19 years. After retiring, Bob did contract work for the Wisconsin Department of Natural Resources before starting a non-profit called Turtles For Tomorrow in 2012, and he currently spends much of each field season creating and managing nesting sites for Wood Turtles. Bob is a volunteer steward for a state natural area in Dane County, WI where he and his team work year-round to manage an Oak Barrens.

Eric T. Hileman is an Assistant Research Professor in the Department of Wildlife, Fisheries and Aquaculture at Mississippi State University and serves as the Quantitative Ecologist for the Eastern Massasauga Rattlesnake Species Survival Plan. Prior to his current appointment, he was a Mendenhall Research Fellow at the USGS Fort Collins Science Center and worked in the Brown Treesnake Laboratory in Yigo, Guam. He earned his doctorate from Northern Illinois University and holds B.S. and M.S. degrees from the University of Wisconsin–Parkside and the University of Wisconsin–Milwaukee, respectively. His research centers on population ecology and factors that regulate vital rates, population dynamics, and the distribution of species.

Collin P. Jaeger holds a Ph.D. from Northern Illinois University (2014), M.S. from Middle Tennessee State University (2008), and B.S. from Western Illinois University (2006). He is broadly interested in ecology and evolutionary biology, specifically with regard to snakes and turtles. In addition, he enjoys teaching and working with colleagues in the field of biology education research. When he's not working (or coding in R or Python), Collin likes to spend time outdoors with his wife and their three children.

Tom R. Johnson grew up on a farm in Portage County, Wisconsin. He joined the US Navy after high school and used the GI Bill to attend the University of Wisconsin–Stevens Point, where he graduated in 1970 with a degree in Biology. Tom worked in three midwestern zoos and, while working at the St. Louis Zoo, wrote *The Amphibians of Missouri* (1977, University of Kansas Press). In 1977 he became state herpetologist with the Missouri Department of Conservation, and published his second book, *The Amphibians and Reptiles of Missouri*, in 1987. He retired in 2000 and currently resides in Jefferson City, Missouri, USA.

Emmet J. Judziewicz earned a Ph.D. in plant taxonomy at the University of Wisconsin–Madison in 1987. He worked for the Wisconsin Department of Natural Resources and other state agencies intermittently from 1982 to 2001, doing rare plant and community surveys throughout the state. He later taught at the University of Wisconsin–Stevens Point from 2001 to 2015, where he is currently Emeritus Professor of Biology.

Heather M. Kaarakka obtained her B.S. in Conservation Biology from University of Wisconsin–Madison in 2009 and started work for the Wisconsin Department of Natural Resources that summer. Her first position for the department involved surveys for Butler's Gartersnakes and other herpetofauna. After a summer chasing snakes, Heather moved into the Wisconsin Bat Program, where she currently coordinates citizen-based bat monitoring projects and summer fieldwork for bats in the state, but continues to dabble in herpetofaunal surveys when time allows. Heather earned her M.S. from the University of Wisconsin–Madison in Wildlife Ecology in 2019, researching the summer habitats used by two rare bat species in Wisconsin.

Daniel E. Keyler's interest in herpetology and the toxicology of venoms was spurred early on by the late Sherman A. Minton Jr., M.D., a friend and mentor until his passing. He earned degrees from Purdue University and the University of Minnesota. He retired from Hennepin County Medical Center, Minneapolis, Minnesota in 2009. Dan is the past President of the North American Society of Toxinology, a Fellow of the American Academy of Clinical Toxicology, and an Adjunct Professor of Experimental and Clinical Pharmacology, University of Minnesota. He has published scientific literature and book chapters involving immunotherapeutics, animal toxins, and venomous snakebite. He has coauthored two books, *Venomous Bites from Non-Venomous Snakes: A Critical Analysis of Risk Management of Colubrid Snake Bites*, and *Venomous Snakebite in the Western United States*. He was also the envenomation Section Chair with the American Academy of Clinical Toxicology from 2002 to 2007, and was an author and member of the Medical Advisory Committee to the Online Antivenom Index.

Richard B. King is a Distinguished Research Professor at Northern Illinois University, DeKalb, Illinois. His research interests center on ecological and evolutionary processes at local and regional scales and the conservation biology of midwestern amphibians and reptiles.

Richard S. King is a member of the Partners in Amphibian and Reptile Conservation, the Wildlife Society, and the Society for Ecological Restoration. He has conducted research on some of Wisconsin's rarest animals including the Eastern Massasauga, Cerulean Warbler, Karner Blue Butterfly, and Whooping Crane. Rich and his wife Lori own property on the Mississippi River in the heart of the Driftless Area. They manage the property for a variety of wildlife and habitats including turtles, peregrine falcons, and colluvium bluff prairie.

Jaimie L. Klemish specializes in herpetology and aquatic ecotoxicology. She is currently a postdoctoral fellow at the University of Lethbridge studying the effects of environmental contaminants, such as metals and agrichemicals, on aquatic organisms.

Michael J. Lannoo is Professor of Anatomy and Cell Biology at Indiana University and an affiliate of the Field Museum of Natural History in Chicago, the Illinois Natural History Survey at the University of Illinois, and Purdue University. He has considerable tropical and polar field experience in addition to his primary research emphasis on temperate systems. In 2001 Lannoo received the Parker/Gentry Award for Excellence and Innovation in Conservation Biology through The Field Museum. This award honors "an outstanding individual, team or organization whose efforts are distinctive and courageous and have had a significant impact on preserving the world's natural heritage, and whose actions and approaches can serve as a model to others."

Carly N. Lapin is a Regional Ecologist in beautiful north-central Wisconsin for the Wisconsin Department of Natural Resources. She is based in Rhinelander and manages a number of State Natural Areas and monitors Wood Turtles on several local streams.

Jeffrey M. Lorch has had a lifelong interest in herpetofauna, conducting field survey work and contributing to several projects related to herpetofaunal conservation. He holds a Ph.D. in molecular and environmental toxicology and has dual B.S. degrees in microbiology and wildlife ecology from the University of Wisconsin–Madison. Jeffrey currently works at the US Geological Survey National Wildlife Health Center in Madison, where he investigates and studies emerging wildlife diseases, with a special interest in diseases that affect herpetofaunal populations.

Elizabeth Mack McKeag has a B.S. in Wildlife and Fisheries Management from Texas Tech University and a M.S. in Biology from Northern Illinois University. For her master's thesis, McKeag studied the life history of the Lake Erie Watersnake, a subspecies of the Common Watersnake. After graduate school, she worked for the US Fish and Wildlife Service in Wyoming for three years as the lead biologist for the recovery of the critically endangered Wyoming Toad. McKeag now lives in Idaho and works for the Idaho Wildlife Federation.

Gabriel (Gabe) Miller received a B.S. in Wildlife Management from the University of Wisconsin–Stevens Point and a M.S. in Wildlife Ecology and Conservation from the University of Florida. He has worked in the natural resources field since 1998 having served state agencies, including the Minnesota Department of Natural Resources and Florida Fish and Wildlife Conservation Commission, and private organizations including J.W. Jones Ecological Research Center in Georgia. Much of his early career has had an emphasis on reptile and amphibian research. Since 2009 he has served the Prairie Island Indian Community's Department of Land and Environment as wildlife biologist and conservationist, and currently is the Environmental Program Manager working on a wide variety of environmental topics.

Michael J. Mossman is an ecologist, conservation biologist, and historian who has worked on Wisconsin's herpetofauna, small mammals, and birds for over 40 years. As a Wisconsin Department of Natural Resources biologist in the early 1980s he initiated the Wisconsin Herp Working Group and, with Ruth Hine, the Wisconsin Frog and Toad Survey. Since retiring in 2016, he works on conservation in southwestern Wisconsin, especially the Baraboo Hills, former Badger Army Ammunition Plant, and the Lower Wisconsin Riverway. He currently lives in Sauk County, where he plays music and manages the wetlands, bluff prairie, and woodland of his retired farm with wife Lisa Hartman and son Angus.

Patrick C. Owen earned his Ph.D. from the University of Connecticut in ecology. His research includes studying aggressive behavior in frogs and examining interactions between salamanders and invasive earthworms.

R. Neal Paisley, retired Wisconsin Department of Natural Resources wildlife manager, began his career in 1983 with the US Fish and Wildlife Service conducting waterfowl management activities on the Upper Mississippi River. Subsequent professional work included prairie waterfowl nesting research and management in North Dakota and Montana, wild turkey restoration and research in southwestern Wisconsin, and snapping turtle research on the Upper Mississippi River. He culminated his career as supervisor of the Sandhill/Meadow Valley Work Unit in central Wisconsin from 2001 to 2017.

Rori A. Paloski earned her B.S. in Biology from the University of Wisconsin–Eau Claire and her M.S. in Wildlife Ecology from the University of Wisconsin–Stevens Point, where her graduate research focused on the impacts of lakeshore development on frog populations in central Wisconsin. She has worked for the Wisconsin Department of Natural Resources as a Conservation Biologist, specializing in herpetology and endangered species regulation, since 2004. This position has involved work on a wide variety of amazing herpetofaunal projects, including Blanchard's Cricket Frog calling surveys and habitat assessments, Eastern Massasauga monitoring and research, Ornate Box Turtle surveys, gartersnake genetics research, citizen-based herpetofaunal monitoring, and biotic inventory herpetofaunal surveys on state properties throughout Wisconsin.

John D. Peterson is an Associate Professor of Biology at the University of Wisconsin–Platteville. He teaches courses in herpetology, anatomy, and physiology. He has published on the parasites of newts, anatomical and physiological impacts of coal waste on amphibian development and metamorphosis, and physiological impacts of the amphibian chytrid fungus on frogs. He currently conducts research with an army of undergraduates on frog acoustic communication, tadpole body morphology and oral abnormalities, river turtle natural history, and snake fungal disease ecology. He co-advises an animal care and outreach program working with amphibians, reptiles, and invertebrates, called the Animal House. He volunteers on two Wisconsin Frog and Toad Survey routes and collaborates on conservation work with the Wisconsin Department of Natural Resources, Wisconsin Natural Resource Conservation Service, Trout Unlimited, and private land owners.

Steven J. Price is an Associate Professor in the Department of Forestry and Natural Resources at the University of Kentucky. He received a B.S. (2000) and M.S. (2003) in Environmental Science from the University of Wisconsin–Green Bay, and a Ph.D. in Biology from Wake Forest University (2011). Additionally, Steven was employed as research coordinator in the Biology Department at Davidson College (North Carolina) from 2004 to 2012. His research focuses on the ecology, conservation biology, and management of aquatic and semi-aquatic animals.

Corey Raimond has worked as an ecologist for several universities and the Wisconsin Department of Natural Resources. He was the first person to document Lined Snakes in Wisconsin. He currently teaches biology at Sparta High School in western Wisconsin.

Christine A. Ribic is a wildlife biologist who investigates the impact of the changing climate on terrestrial vertebrates in Wisconsin.

Tracy A. G. Rittenhouse is a wildlife ecologist who studies where animals live and how animals travel through habitats. While a graduate student, she used radio telemetry to describe the terrestrial, nonbreeding habitat of Spotted Salamanders and Wood Frogs in both continuous forest and in response to timber harvest. Her current research is motivated by this core question: *Why do some wildlife populations increase while others decrease in human-dominated landscapes?* While addressing this topic of global relevance, she remains focused on the uniqueness of local places and the research needs of people charged with managing local populations and habitats.

Allison B. Sacerdote-Velat is the Curator of Herpetology at Chicago Academy of Sciences. Her research focuses on applied conservation, population ecology, and restoration management of Smooth Greensnakes, other grassland snakes, and forest amphibians. Allison coordinates the Calling Frog Survey citizen science program and serves on the steering committee of Midwest Partners in Amphibian and Reptile Conservation.

Richard A. Sajdak retired from the Milwaukee County Zoo as Aquarium/Reptile Curator. He earned a B.A. in Conservation and M.S. in Zoology from the University of Wisconsin–Milwaukee. His research interests include Mudpuppy ecology, radiolocation studies of Timber Rattlesnakes, and ecology of Grenada Bank Treeboas.

Gregor W. Schuurman is an ecologist with the National Park Service Climate Change Response Program, where he works with national parks and partners to understand and adapt to a wide range of climate change impacts and vulnerabilities. Before that, he worked for the Wisconsin Department of Natural Resources, and focused on research, monitoring, and management guidance for a diversity of rare species, including amphibians and reptiles.

Spencer R. Siddons began his career in herpetology while pursuing a B.S. degree at the University of Wisconsin–Stevens Point, where he contributed to a field study on chytrid fungus in Wisconsin frog species. This experience propelled Spencer to study amphibian disease ecology at the University of South Dakota for his M.S. degree, and then at Purdue University where he is currently a Ph.D. Candidate. Spencer's future career goals include teaching and conducting research at a university, with a continued focus on herpetology and disease ecology.

Richard A. Staffen has been a Conservation Biologist for the Wisconsin Department of Natural Resource's Natural Heritage Conservation program since 2002. He helps coordinate and performs inventory and monitoring projects across the state focused on rare vertebrate animals. Richard has led efforts to better understand the extent of snake fungal disease in Wisconsin, assisted in the recovery of the Eastern Massasauga rattlesnake, and performed surveys of numerous other rare reptiles in the state.

Rochelle M. Stiles currently works for San Francisco Zoological Society, head-starting local, threatened herpetofauna for release and researching their behavior in order to improve reintroduction techniques. In 2016, she completed her Ph.D. at Indiana State University, where she assessed the amphibian response to wetlands in restored and reclaimed grasslands, and implications for reintroductions.

Paul M. Skawinski is an Outreach Specialist and Coordinator of the Wisconsin Citizen Lake Monitoring network with the University of Wisconsin Extension Lakes Program at the University of Wisconsin–Stevens Point. Paul also teaches Aquatic Botany and is the author of *Aquatic Plants of the Upper Midwest: A Photographic Field Guide to Our Underwater Forests.*

Jonathan W. Swan specializes in herpetology and limnology. He currently resides in the Catskill Mountains of New York.

Phoebe Vanselow graduated from Macalester College (Minnesota) with a Biology and Environmental Studies double major. She worked two seasons on the Minnesota Pollution Control Agency's malformed frogs project before escaping to Alaska, where she has worked for Glacier Bay National Park as a coastwalker, humpback whale biotech, and now IT Specialist.

Sara E. Viernum is a herpetologist and wildlife biologist with Stantec in Portland, Oregon. She has worked with various amphibian, reptile, bird, mammal, and invertebrate species in numerous ecosystems including desert, montane, and northwestern rainforest. When not chasing snakes and frogs in the field, she can be found either hiking, cross-stitching salamanders, traveling the world with her partner, or relaxing at home with their adorable milksnake.

Richard C. Vogt (1949–2021) received his Ph.D. from the University of Wisconsin–Madison in 1978. His doctoral research focused on the systematics and ecology of the *Graptemys pseudogeographica* complex in the Mississippi River Basin. He conducted postdoctoral research with the Section of Amphibians and Reptiles at the Carnegie Museum of Natural History from 1978 to 1980. Shortly thereafter he published *Natural History of Reptiles and Amphibians in Wisconsin* (1981). From 1980 to 2000, Vogt was a Herpetologist at Estacion de Biologia Tropical Los Tuxtlas, Universidad Nacional Autonoma de Mexico. Since 2000, he was the curator of reptiles and amphibians at the zoology collection and a research professor at the Brazilian Institute for Amazon Research in Manaus, Amazonas, Brazil. He was also Director of the Center for Amazon Turtle Conservation. He authored or coauthored numerous scientific works, including the book *Turtles of Mexico, Freshwater and Land Forms* with John M. Legler (2013). He passed away in January of 2021 at the age of 71.

Dreux J. Watermolen is an ecologist and chief of the Analysis Services section at the Wisconsin Department of Natural Resources, where he has worked in various programs since 1986. He earned his Bachelor's degree at St. Norbert College and pursued graduate studies at the University of Wisconsin–Green Bay. He holds an honorary Research Affiliate position with the Milwaukee Public Museum and is an Associate Academic Research Fellow at the University of Wisconsin. His research interests focus on the zoogeography, life history, and taxonomy of Wisconsin's rich biological diversity.

Erik R. Wild—biologist, artist, educator—is a fellow of the Biodiversity Institute of the Museum of Natural History at the University of Kansas, an Honorary Associate of the Department of Biology at the University of Wisconsin–Stevens Point, and a Fulbright Scholar. He is originally from River Falls, Wisconsin, but now resides in Belo Horizonte, Brazil.

Index

Joshua M. Kapfer has lived most of his life in southern Wisconsin. He was born in Stoughton, where his interest in wildlife was fueled by catching amphibians and reptiles encountered near his childhood home. He earned his B.S. and M.S. degrees from the University of Wisconsin–La Crosse (1999, 2002) in Biology and Aquatic Science. His master's research focused on the survival of amphibian larvae in agricultural ponds. Josh then went on to earn a Ph.D. from the University of Wisconsin–Milwaukee (2007) in Ecology and Evolution. His doctoral research investigated numerous aspects of Gophersnake ecology and conservation in Sauk County. After earning his doctoral degree, Josh worked for the Wisconsin Department of Natural Resources, NRC Environmental Consulting (now Stantec), and Elon University in North Carolina. Since 2011, Josh has been with the Department of Biological Sciences at the University of Wisconsin–Whitewater as a professor and wildlife biologist, where he teaches various lecture and field-based ecology courses and conducts research with undergraduate students. He is broadly interested in vertebrate ecology and conservation, with publications centered on amphibians, reptiles, fish, and mammals (although he will admit to a particular fondness for Wisconsin's reptiles). He currently lives in Walworth County with his family.

Donald J. Brown received his B.S. in Fisheries & Wildlife from the University of Minnesota-Twin Cities in 2007, M.S. in Wildlife Ecology from Texas State University in 2008, and Ph.D. in Aquatic Resources from Texas State University in 2013. His graduate research in Texas focused on freshwater turtle conservation issues, amphibian and reptile responses to prescribed fire and high severity wildfires, and improving monitoring and conservation of the federally endangered Houston Toad. After graduate school, Donald moved to Wisconsin for a postdoctoral research position with University of Wisconsin–Madison and the US Forest Service Northern Research Station in Rhinelander, Wisconsin. His postdoctoral work focused on assessing population viability of the Kirtland's Warbler under potential future environmental and management changes, improving monitoring and conservation of Wood Turtles in Minnesota, and beginning this book project. Donald is currently a Research Assistant Professor in a joint position with West Virginia University and the US Forest Service Northern Research Station. His current research projects focus on salamander and freshwater turtle conservation issues in Central Appalachia and the Upper Midwest.

Douglas
Bayfield
Iron
Ashland
Vilas
Burnett
Washburn
Sawyer
Florence
Price
Oneida
Forest
Polk
Barron
Rusk
Marinette
Door
Lincoln
Langlade
Taylor
Saint Croix
Chippewa
Dunn
Menominee
Oconto
Door
Marathon
Door
Shawano
Pierce
Eau Claire
Clark
Pepin
Portage
Wood
Buffalo
Trempealeau
Waupaca
Outagamie
Brown
Kewaunee
Jackson
Waushara
Winnebago
Calumet
Manitowoc
La Crosse
Monroe
Juneau
Adams
Marquette
Green Lake
Fond du Lac
Sheboygan
Vernon
Richland
Sauk
Columbia
Dodge
Washington
Ozaukee
Crawford
Dane
Iowa
Jefferson
Waukesha
Milwaukee
Grant
Racine
Lafayette
Green
Rock
Walworth
Kenosha